22. $$\int \frac{dx}{x\sqrt{ax+b}} = \frac{2}{\sqrt{-b}} \tan^{-1} \sqrt{\frac{ax+b}{-b}} \qquad b < 0$$

23. $$\int \frac{dx}{x^2\sqrt{ax+b}} = -\frac{\sqrt{ax+b}}{bx} - \frac{a}{2b} \int \frac{dx}{x\sqrt{ax+b}}$$

24. $$\int \sqrt{\frac{cx+d}{ax+b}}\, dx = \frac{\sqrt{ax+b}\sqrt{cx+d}}{a} + \frac{ad-bc}{2a} \int \frac{dx}{\sqrt{ax+b}\sqrt{cx+d}}$$

Expressions Containing $ax^2 + c$, $x^2 \pm p^2$, and $p^2 - x^2$, $p > 0$

25. $$\int \frac{dx}{p^2 - x^2} = \frac{1}{2p} \ln \left| \frac{p+x}{p-x} \right|$$

26. $$\int \frac{dx}{ax^2 + c} = \frac{1}{\sqrt{ac}} \tan^{-1} \left(x\sqrt{\frac{a}{c}} \right) \qquad a, c > 0$$

27. $$\int \frac{dx}{ax^2 + c} = \begin{cases} \dfrac{1}{2\sqrt{-ac}} \ln \left| \dfrac{x\sqrt{a} - \sqrt{-c}}{x\sqrt{a} + \sqrt{-c}} \right| & a > 0,\ c < 0 \\ \dfrac{1}{2\sqrt{-ac}} \ln \left| \dfrac{\sqrt{c} + x\sqrt{-a}}{\sqrt{c} - x\sqrt{-a}} \right| & a < 0,\ c > 0 \end{cases}$$

28. $$\int \frac{dx}{(ax^2 + c)^n} = \frac{1}{2(n-1)c} \frac{x}{(ax^2+c)^{n-1}} + \frac{2n-3}{2(n-1)c} \int \frac{dx}{(ax^2+c)^{n-1}} \qquad n > 1$$

29. $$\int x(ax^2 + c)^n\, dx = \frac{1}{2a} \frac{(ax^2+c)^{n+1}}{n+1} \qquad n \neq -1$$

30. $$\int \frac{x}{ax^2 + c}\, dx = \frac{1}{2a} \ln |ax^2 + c|$$

31. $$\int \sqrt{x^2 \pm p^2}\, dx = \tfrac{1}{2}[x\sqrt{x^2 \pm p^2} \pm p^2 \ln |x + \sqrt{x^2 \pm p^2}|]$$

32. $$\int \sqrt{p^2 - x^2}\, dx = \frac{1}{2}\left(x\sqrt{p^2 - x^2} + p^2 \sin^{-1} \frac{x}{p} \right)$$

33. $$\int \frac{dx}{\sqrt{x^2 \pm p^2}} = \ln |x + \sqrt{x^2 \pm p^2}|$$

34. $$\int (p^2 - x^2)^{3/2}\, dx = \frac{x}{4}(p^2 - x^2)^{3/2} + \frac{3p^2x}{8}\sqrt{p^2 - x^2} + \frac{3p^4}{8} \sin^{-1} \frac{x}{p}$$

Expressions Containing $ax^2 + bx + c$

35. $$\int \frac{dx}{ax^2 + bx + c} = \frac{1}{\sqrt{b^2 - 4ac}} \ln \left| \frac{2ax + b - \sqrt{b^2 - 4ac}}{2ax + b + \sqrt{b^2 - 4ac}} \right| \qquad b^2 > 4ac$$

36. $$\int \frac{dx}{ax^2 + bx + c} = \frac{2}{\sqrt{4ac - b^2}} \tan^{-1} \frac{2ax + b}{\sqrt{4ac - b^2}} \qquad b^2 < 4ac$$

37. $$\int \frac{dx}{ax^2 + bx + c} = -\frac{2}{2ax + b} \qquad b^2 = 4ac$$

38. $$\int \frac{dx}{(ax^2 + bx + c)^{n+1}} = \frac{2ax + b}{n(4ac - b^2)(ax^2 + bx + c)^n} + \frac{2(2n-1)a}{n(4ac - b^2)} \int \frac{dx}{(ax^2 + bx + c)^n}$$

39. $$\int \frac{x\, dx}{ax^2 + bx + c} = \frac{1}{2a} \ln |ax^2 + bx + c| - \frac{b}{2a} \int \frac{dx}{ax^2 + bx + c}$$

40. $\displaystyle\int \frac{dx}{\sqrt{ax^2+bx+c}} = \frac{1}{\sqrt{a}} \ln |2ax + b + 2\sqrt{a}\sqrt{ax^2+bx+c}| \qquad a > 0$

41. $\displaystyle\int \frac{dx}{\sqrt{ax^2+bx+c}} = \frac{1}{\sqrt{-a}} \sin^{-1} \frac{-2ax-b}{\sqrt{b^2-4ac}} \qquad a < 0$

42. $\displaystyle\int \frac{x\,dx}{\sqrt{ax^2+bx+c}} = \frac{\sqrt{ax^2+bx+c}}{a} - \frac{b}{2a}\int \frac{dx}{\sqrt{ax^2+bx+c}}$

43. $\displaystyle\int \sqrt{ax^2+bx+c}\,dx = \frac{2ax+b}{4a}\sqrt{ax^2+bx+c} + \frac{4ac-b^2}{8a}\int \frac{dx}{\sqrt{ax^2+bx+c}}$

Expressions Containing Powers of Trigonometric Functions

44. $\displaystyle\int \sin^2 ax\,dx = \frac{x}{2} - \frac{\sin 2ax}{4a}$

45. $\displaystyle\int \sin^3 ax\,dx = -\frac{1}{a}\cos ax + \frac{1}{3a}\cos^3 ax$

46. $\displaystyle\int \sin^n ax\,dx = -\frac{\sin^{n-1} ax \cos ax}{na} + \frac{n-1}{n}\int \sin^{n-2} ax\,dx \qquad n \text{ positive integer}$

47. $\displaystyle\int \frac{dx}{1 \pm \sin ax} = \mp\frac{1}{a}\tan\left(\frac{\pi}{4} \mp \frac{ax}{2}\right)$

48. $\displaystyle\int \cos^2 ax\,dx = \frac{x}{2} + \frac{\sin 2ax}{4a}$

49. $\displaystyle\int \cos^3 ax\,dx = \frac{1}{a}\sin ax - \frac{1}{3a}\sin^3 ax$

50. $\displaystyle\int \cos^n ax\,dx = \frac{\cos^{n-1} ax \sin ax}{na} + \frac{n-1}{n}\int \cos^{n-2} ax\,dx$

51. $\displaystyle\int \tan^2 ax\,dx = \frac{1}{a}\tan ax - x$

52. $\displaystyle\int \tan^3 ax\,dx = \frac{1}{2a}\tan^2 ax + \frac{1}{a}\ln|\cos ax|$

53. $\displaystyle\int \tan^n ax\,dx = \frac{\tan^{n-1} ax}{a(n-1)} - \int \tan^{n-2} ax\,dx \qquad n \neq 1$

54. $\displaystyle\int \sec^2 ax\,dx = \frac{1}{a}\tan ax$

55. $\displaystyle\int \sec^3 ax\,dx = \frac{1}{2a}\sec ax \tan ax + \frac{1}{2a}\ln|\sec ax + \tan ax|$

56. $\displaystyle\int \sec^n ax\,dx = \frac{\sec^{n-2} ax \tan ax}{a(n-1)} + \frac{n-2}{n-1}\int \sec^{n-2} ax\,dx \qquad n \neq 1$

Expressions Containing Algebraic and Trigonometric Functions

57. $\displaystyle\int x \sin ax\,dx = \frac{1}{a^2}\sin ax - \frac{1}{a}x\cos ax$

58. $\displaystyle\int x \cos ax\,dx = \frac{1}{a^2}\cos ax + \frac{1}{a}x\sin ax$

59. $\displaystyle\int x^n \sin ax\,dx = -\frac{1}{a}x^n\cos ax + \frac{n}{a}\int x^{n-1}\cos ax\,dx \qquad n \text{ positive}$

CALCULUS
AND ANALYTIC GEOMETRY

SCHAUM'S SOLVED PROBLEMS BOOKS

CALCULUS

AND ANALYTIC GEOMETRY

FIFTH EDITION

SHERMAN K. STEIN
University of California, Davis

ANTHONY BARCELLOS
American River College

McGRAW-HILL, INC.

*New York St. Louis San Francisco Auckland Bogotá
Caracas Lisbon London Madrid Mexico Milan Montreal
New Delhi Paris San Juan Singapore Sydney Tokyo Toronto*

CALCULUS AND ANALYTIC GEOMETRY

6 7 8 9 10 VH VH 08 07 06 05 04 03 02

ISBN 0-07-061175-0

This book was set in Times Roman
by York Graphic Services, Inc.
The editors were Richard Wallis and James W. Bradley;
the production supervisor was Annette Mayeski.
The cover was designed by Joan E. O'Connor.
Cover photograph by Hiroshi Yagi/Superstock, Inc.
New drawings were done by Fine Line Illustrations, Inc.
Von Hoffmann Press, Inc., was printer and binder.

Library of Congress Cataloging-in-Publication Data

Stein, Sherman K.
Calculus and analytic geometry.—5th ed. / Sherman K. Stein,
Anthony Barcellos.
p. cm.
Includes index.
ISBN 0-07-061175-0
1. Calculus. 2. Geometry, Analytic. I. Barcellos, Anthony.
II. Title.
QA303.S857 1992
515'. 16—dc20 91-11421

ABOUT THE AUTHORS

Sherman K. Stein is Professor of Mathematics at the University of California, Davis, where he has been on the faculty since 1953. He has degrees in mathematics from the California Institute of Technology and Columbia University. In 1975 he received an honorary doctor of letters degree from Marietta College in recognition of his textbooks. Stein was chosen as one of the Distinguished Teachers on the UC Davis campus in 1974. He was selected in 1975 for the Mathematical Association of America's Lester R. Ford Award for expository writing.

In addition to his calculus book, Stein has written a series of high-school algebra and geometry books in collaboration with G. D. Chakerian and C. D. Crabill, as well as *Mathematics: The Man-made Universe,* a text for "mathematical ideas" courses. Stein's principal research field is abstract algebra, with a special emphasis on tiling and packing problems.

Stein has lectured and written on the teaching of mathematics. His talk, "Gresham's Law: Algorithm Drives Out Thought," delivered at the 1987 meeting of the American Mathematical Society in San Antonio, was published in *For the Learning of Mathematics,* 1987, and in the *Journal of Mathematical Behavior,* 1988.

Anthony Barcellos has been a member of the mathematics faculty of American River College in Sacramento, California, since 1987. He holds degrees in mathematics from Porterville College, the California Institute of Technology, and California State University, Fresno. Barcellos began his student teaching at UC Davis under the supervision of Alan H. Schoenfeld in 1975 and later became Sherman Stein's teaching assistant. While working at Davis, he received both the departmental and the campus citations for distinguished teaching by a teaching assistant. His textbook experience began with the *Student's Solutions Manual* to the second edition of Stein's calculus book in 1977; he has participated in each subsequent edition.

Barcellos worked as a science journalist for the *Albuquerque Journal* in 1978 under the sponsorship of the American Association for the Advancement of Science's Mass Media Fellowship Program. In 1979 he joined the staff of the California State Senate as a Senate Fellow and worked thereafter for the Commission on State Finance, California's principal econometric forecasting agency.

Barcellos has served on the board of editors of the Mathematical Association of America's *College Mathematics Journal.* He received both the George Pólya Award (1985) and Merten M. Hasse Prize (1987) for expository writing for an article on Mandelbrot's fractal geometry. His interviews of Benoit Mandelbrot, Martin Gardner, and Stanislaw Ulam were published in *Mathematical People* (Birkhäuser Boston). From 1985 to 1990, Barcellos was editor of *Sacra Blue,* the monthly magazine of the Sacramento Personal Computer Users Group.

To
Joshua,
Rebecca,
and
Susanna

To
My
Parents

The great body of physics, a great deal of the essential fact of financial science, and endless social and political problems are only accessible and only thinkable to those who have had a sound training in mathematical analysis, and the time may not be very remote when it will be understood that for complete initiation as an efficient citizen of one of the new great complex worldwide states that are now developing, it is as necessary to be able to compute, to think in averages and maxima and minima, as it is now to be able to read and write.

H. G. Wells
Mankind in the Making, p. 192, Scribner's, New York, 1904.

. . . at about the age of sixteen, I was offered a choice which, in retrospect, I can see that I was not mature enough, at the time, to make wisely. The choice was between starting on the calculus and, alternatively, giving up mathematics altogether and spending the time saved from it on reading Latin and Greek literature more widely. I chose to give up mathematics, and I have lived to regret this keenly after it has become too late to repair my mistake. The calculus, even a taste of it, would have given me an important and illuminating additional outlook on the Universe, whereas, by the time at which the choice was presented to me, I had already got far enough in Latin and Greek to have been able to go farther with them unaided. So the choice that I made was the wrong one, yet it was natural that I should choose as I did. I was not good at mathematics; I did not like the stuff. . . . Looking back, I feel sure that I ought not to have been offered the choice; the rudiments, at least, of the calculus ought to have been compulsory for me. One ought, after all, to be initiated into the life of the world in which one is going to have to live. I was going to have to live in the Western World . . . and the calculus, like the full-rigged sailing ship, is . . . one of the characteristic expressions of the modern Western genius.

Arnold Toynbee
Experiences, pp. 12–13, Oxford University Press, London, 1969.

CONTENTS

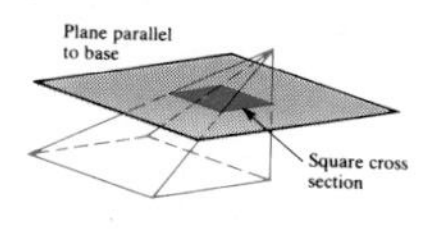
Plane parallel to base
Square cross section

TO THE INSTRUCTOR

We live in a fascinating time in which to teach calculus. Today, the way calculus is taught has become the focus of much discussion if not of a real revolution. Instructors have been urged to use calculators and computers, add meaningful applications for motivation, encourage students to express themselves in writing, and—above all—emphasize conceptual learning.

While the calculus debate encourages authors to depart from tradition, it also tempts them to follow what might be a passing fancy, or to hear only the most strident voices. In this fifth edition, we have taken advantage of the increased freedom but resisted the lure of change for its own sake. We have adhered to one guiding principle: *Make concepts as clear to the student as we can.* That is a theme common to all the reformers, and we have tried to implement it in every area of the text: organization, examples, proofs, applications, exercises, and illustrations.

Student-Oriented Text. This book has been designed to be read and used. Students entering a calculus course differ widely in background, interest, and study habits. We have kept this in mind and have written this text to make the concepts accessible to a broad spectrum of students. To help achieve our goals we brought students into the editorial process even more than before. We surveyed over one hundred past and present student users of the fourth edition, soliciting their comments on that edition and on drafts of this one. Thus we have had not only the suggestions of professors, which publishers usually gather and colleagues happily volunteer, but the advice of those who will buy the book and read it. (Most students *do* read the text, not just its examples and exercises.)

Sometimes the students and the professors agreed; often they did not. The professor said, ''Make the book shorter.'' The student said, ''Add more explanations, more motivation. Tell us what we will do and why. Then tell us the core of what we have done.'' The professor said, ''Get rid of wide margins; save trees.'' The student said, ''We need the margins. We want more white space. Then it's easier to read and we have space to write in.'' The professor said, ''String the steps of a calculation on a line; save space.'' The student said, ''Display them below each other so they're easier to follow.'' The professor said, ''Cut the summaries. They're redundant.'' The student said, ''The summaries put it all together.'' In these cases, *the students prevailed.*

Also in response to student suggestions, we have added many illustrations ("for those of us who think geometrically, not linearly"), expanded chapter and section introductions, and included many new summaries for most of the sections. An example of a newly written chapter introduction is the opening section of Chapter 10, which previews the two chapters on series and shows why series are important.

The chapter summaries in this book are the most extensive of any calculus text. Although instructors may use them only as another source of exercises, students find these overviews very helpful. Taken together with the concise section summaries, they comprise a built-in study guide.

In order to enhance readability, we did not clutter the exposition with too many digressions. The students have enough to read as it is. Some of the illustrative material and applications are set off in boxes, where they make their point without interrupting the exposition. (See, for instance, the conversation about half-life in Section 6.7, the relation of k^n to chain reactions in Section 10.2, and the role of complex numbers in alternating currents in Section 11.6.)

In the sense that students can read this text, it might be termed "easy." However, because of its ample supply of both routine and challenging exercises, the instructor can choose the level of the course. (Of course, that level is also influenced by other factors, such as the preparation of the student, the speed at which material is covered, and whether certain topics—such as ϵ, δ—are included. These options are discussed in the *Instructor's Manual*.)

Organization. The text remains fairly standard, yet it permits a couple of variations by the instructor. As before, each section focuses on one main idea and generally corresponds to one lecture. Except for the section on natural growth (6.7), Chapter 6 can be covered before Chapter 5. That permits a full treatment of differential calculus that includes *all* the elementary functions quite early. The small portion of Chapter 6 that uses Chapter 5 is clearly indicated. Also, Chapter 8, Applications of the Definite Integral, can precede Chapter 7, Computing Antiderivatives. Exercises in Chapter 8 that use material from Chapter 7 are so indicated.

There are several changes in organization in this fifth edition. The opening chapter is now a survey of calculus, introducing informally the derivative, the definite integral, and series. It may be covered in one to three lectures or merely suggested as reading at the start of the term. Methods of estimating an integral, including the trapezoidal method and Simpson's method, now come much earlier (5.4), preceding the fundamental theorem (5.7). In this way, the student is encouraged to view the definite integral as a function of its right-hand endpoint and to conjecture what its derivative is. To introduce the fundamental theorem, we added a new section (5.6), "Background for the Fundamental Theorem of Calculus." In it, the student has a chance to become familiar with the idea that "area is a function of x" and to carry out experiments with a calculator that suggest the fundamental theorem.

We have added several other new sections. In Section 4.10 we use an analogy with racing cars to show why the second derivative measures the error in using a differential or linear approximation. This is a prelude to the new Section 11.3, which shows why the error in Taylor series is controlled by a higher derivative. Section 12.2, on projections, is also new. The notion of a projection is so important in theory and applications that we have now collected the properties of the various projections in one place, permitting a unified treatment of a topic that recurs throughout subsequent sections.

We have limited the coverage of differential equations in the main body of the text to a brief treatment in Section 6.7 of separable differential equations (needed for studying natural growth); material on linear differential equations with constant coefficients now appears as Appendix M. This is in keeping with prevailing practice, where differential equations constitutes a full sophomore course in its own right. While the same could be said of vector analysis, the integral theorems of vector analysis fit so naturally into multivariate calculus that it is easy to include them in an introductory text. We feel it is better to present vector analysis well, with adequate motivation, than to attempt to cover both vector analysis and differential equations in the perfunctory way that space and time limitations would require. Students would gain little from abbreviated accounts of either. On many campuses the vector analysis course is incorporated in the third semester or fourth quarter of the introductory calculus sequence; this text supports that natural progression.

We have split infinite series into two chapters (10 and 11), corresponding roughly to series with constant terms and power series. We also split into two sections the vector treatment of lines and planes (Sections 12.4 and 12.7) and the chain rule for partial derivatives (Sections 14.6 and 14.11), with Section 14.11 devoted to the case where a variable is both intermediate and terminal. We have emphasized the ∂ notation for partial derivatives in Chapter 14, in keeping with the notation favored by those who use them in applications (and prefer to reserve subscripts for vector components). The subscript notation is retained, however, as an alternative for those who prefer it.

The former chapter on Green's theorem, the divergence theorem, and Stokes' theorem is also now split in two. Chapter 16 covers Green's theorem and concerns mostly vector fields in the plane, where the diagrams, arguments, and calculations are much easier. The student who has mastered this chapter—in particular the relation of Green's theorem and Stokes' theorem in the plane to fluid flow, work, subtended angle, central fields, and conservative fields—is well prepared for vector analysis over surfaces and solids, which constitutes Chapter 17.

The discussion of vector analysis has been extensively revised, mainly at the insistence of colleagues in physics and engineering who wish to emphasize concepts over routine calculations. There are more sections than before (11 versus 7 in the previous edition), but they are now shorter and more focused. The only new concept is "solid angle."

The mention of solid angles reminds us of other topics needed in applications and assumed there, although they are not always covered in a calculus course. Complex numbers is such a topic. Few students have seen them in high school, and, if they did, they seldom acquired any feel for the geometry. That is why we devote a full section (11.6) to them. The following section presents the relation $e^{i\theta} = \cos\,\theta + i\,\sin\,\theta$, which deserves emphasis in freshman calculus for several reasons: the most immediate is its application in sophomore physics and engineering courses.

Proofs. Proofs are as much a part of the course as are examples and exercises. They are not ornaments. They are not included to assure the student that a statement is true. (After all, students seldom express the fear that a theorem that has been around for at least a century might be false.) A proof should be included in an introductory text not to show that something is true, but to show *why* it is true, to reinforce definitions and principles, and to tie the concepts together. Proofs should not appear to the student to be pulled out of a hat, however much the trick may appeal to the instructor. For instance, we have shifted from the usual ''neat'' proof that absolute convergence implies convergence [''$a_n = (a_n + |a_n|) - |a_n|$''] to a more intuitive, though slightly longer, proof. (See Section 10.7.) Instead of treating $\lim_{x \to 0} (1 - \cos x)/x = 0$ as a sleight-of-hand consequence of $\lim_{x \to 0} (\sin x)/x = 1$, we examine it as a limit in its own right and show the student how similar situations can lead to very different outcomes. (See Section 2.7.)

For the same reasons, we obtain the formula for the dot product in terms of components in a more instructive way than by the law of cosines. The proof depends on the simplest property of a projection. Section 12.2 treats the projection of a line segment on a line or plane, a vector on a vector or plane, and a flat surface on a plane.

At the Tulane conference on ''Lean and Lively Calculus'' in 1986 we heard the engineers say, ''Teach the concepts. We'll take care of the applications.'' Steve Whitaker, in the engineering department at Davis, advised us, ''Emphasize proofs, because the ideas that go into the proofs are often the ideas that go into the applications.'' Oddly, mathematicians suggest that we emphasize applications, and the applied people suggest that we emphasize concepts. We have tried to strike a reasonable balance that gives the instructor flexibility to move in either direction.

Exercises. We have included more routine exercises for students who need additional drill to hone their skills. But we have also added exercises that require students to provide written explanations, discussions, and conclusions. These are marked with a pencil: ✎. Just as there are both routine and challenging computational problems, there are routine and challenging writing exercises.

Some of the exercises simply ask the student to draw a diagram. These are indicated with a ✎. The ability to sketch a quick and accurate working diagram will serve the student well in calculus and later studies, so we have given many practical drawing tips at appropriate points in the text. Often students cannot solve a problem because their drawings are too sloppy or too small. The drawing exercises are intended to develop skill in sketching useful diagrams. (See, for instance, Section 8.2.) In addition, there are more exercises that exploit the calculator or computer. These are marked with a ▦ or ■.

''Exploration'' problems are marked with a compass: ✱. These exercises usually fit the so-called ''tri-ex'' format: ''experiment, extract, explain.'' The student is encouraged to *experiment* with various cases, *extract* a general principle, and then *explain* why the principle is valid. The student is given more latitude in tri-ex problems than in typical ''guided proof'' problems (where the prescribed steps inexorably lead the student to the desired conclusion). While we include guided proof problems where they seem useful, the tri-ex problems are intended to develop more self reliance. (Their solutions are in the *Instructor's Manual,* not the *Student's Solutions Manual.*)

As before, a serious effort has been made to rank the exercises. In general, each exercise set begins with paired exercises that students can use for drill and practice. A horizontal color bar marks the end of these exercises. Exercises appearing after the bar occur singly rather than in matched pairs. They may offer an additional perspective, complete a proof, involve longer calculations, illustrate an application, be more theoretical, invite an explanation, and so on. Thus they are not necessarily harder than exercises before the bar, but simply serve a different purpose.

The difficulty level of the exercises tends to rise throughout each exercise set, but sometimes the effort to group similar problems must interrupt such a gradation. Also, we do not think it is a good idea to tell students that certain exercises are hard while others are easy. Students are then discouraged when ''easy'' exercises turn out to be difficult for them and may give up too easily on exercises that are described as ''hard.'' To assist the instructor in assigning homework problems, the *Instructor's Manual* contains notes on the exercises.

Illustrations and Design. We have improved most figures in the text and added many new ones. We have made a much greater use of color, even in diagrams of some plane figures, where a third or fourth color can highlight a feature. For instance, we use colors to distinguish a curve from a tangent and the tangent from its approximating secants. Many diagrams of lines, vectors, and surfaces in space benefit from the use of several colors. In addition, we have expanded the use of computer graphics, without succumbing to the temptation of adding computer-created function graphs merely to demonstrate the power of modern software. Unnecessarily complex art may easily discourage students from developing their own ability to draw neat, useful diagrams.

Computers and Calculators. These execute algorithms and sketch graphs far more rapidly and accurately than humans can with pencil and paper. We have included exercises that exploit these tools and also describe some of the software available. However, we do not make the text dependent on them. First of all, the software is changing rapidly. Second, the hardware and its accessibility vary from campus to campus. Third, incorporating a computer on a daily basis demands extra labor and careful coordination. Instructors of large classes already have their hands full simply giving their lectures, conducting office hours, and reading papers or coordinating readers and teaching assistants. So, in many cases, we cite some of the related software and suggest exercises, but we leave it to the instructor to decide how to exploit the calculator and computer. For further information, see ''Computers and Calculators'' on page xxxiii.

Supplements. Calculus texts today are accompanied by ancillaries designed to assist the student and instructor. The *Instructor's Manual* and *Student's Solutions Manual* are principal among these. We have all seen examples where answers in manuals disagree with answers in the text, and where solutions in manuals use techniques not available to the students. We have worked to eliminate such discrepancies.

The *Instructor's Manual* contains our own pedagogical notes to the instructor on alternative approaches to the material in the text, plus solution sketches for even-numbered exercises (and for exploration problems omitted from the *Student's Solutions Manual*).

Also available to instructors adopting the text are a test-item file (in both printed and computerized formats) and an extensive system of color overhead transparencies for use in the classroom. Software packages for numerical and graphical displays are available to adopters by contacting the local McGraw-Hill sales representative.

Related McGraw-Hill titles that will be of interest to users of this book are *Introduction to Math CAD for Scientists and Engineers* by Sol Wieder (1992). *Discovering Calculus with the HP-28 and HP-48* by Robert T. Smith and Roland B. Minton (1992), and *Explorations in Calculus with a Computer Algebra System* by Donald B. Small and John M. Hosack (1991).

Acknowledgments. In conclusion, we wish to acknowledge the assistance of the many people who have helped us prepare this edition. Rodney Cole of the physics department and Stephen Whitaker of the engineering department at the University of California, Davis, made many suggestions, as did Henry Alder, Carlos Borges, Dean Hickerson, Lawrence Marx, and Howard Weiner of the mathematics department. Phil R. Smith of the mathematics department at American River College and Sándor Szabó at the University of the Pacific provided helpful comments.

Other reviewers included Elizabeth B. Appelbaum, Longview Community College; Julia Brown, Atlantic Community College; Doug Child, Rollins College; Peter R. Christopher, Worcester Polytechnic Institute; Charles C. Clever, South Dakota State University; Michael R. Colvin, California Polytechnic State University; David B. Cooke, Hastings College; Carol G. Crawford, United States Naval Academy; Daniel S. Drucker, Wayne State University; Betty Hawkins, Shoreline Community College; Russell Hendel, Dowling College; Rahim G. Karimpour, Southern Illinois University; Frank J. Kelly, Jr., University of New Mexico; Michael G. Kerckhove, University of Richmond; Harvey B. Keynes, University of Minnesota; Jimmie D. Lawson, Louisiana State University; Jonathan W. Lewin, Kennesaw State College; Daniel P. Maki, Indiana University; Andrew S. Miller, Berkshire Community College; Sandra A. Monteferrante, Dowling College; John W. Petro, Western Michigan University; Ralph A. Raimi, University of Rochester; Thomas W. Rishel, Cornell University; Walter A. Rosenkrantz, University of Massachusetts; John Schneider, Hastings College; Dorothy Schwellenbach, Hartnell College; Tatiana Shubin, San Jose State University; Joseph F. Stokes, Western Kentucky University; Ray Talavera, Laney College; and James T. Vance, Jr., Wright State University.

At least as important were the comments of students, collected in an anonymous survey.

Assisting with the preparation of the manuscript were undergraduates Michèle Brown and Keith Sollers, and graduates Mallory Austin and Masato Kimura. Patrick Reardon and Philippe Bérard helped to prepare the art manuscript. Duane Kouba, Ali Dad-del, Aaron Klebanoff, Don Johnson, Linda Namikas, Kathy Smith, Tom Schiller, Denise Quigley, and Pat Rhodes checked the problems. Richard and Judith Kinter, John Lynskey, Gordon Nelder-Adams, and Phil R. Smith helped to type the manuscript.

We are indebted to the McGraw-Hill staff. Former mathematics editor, Robert Weinstein, who believes, rightly so, that publishers are teachers—not just producers of books—encouraged us to undertake this substantial revision, and his successor, Richard Wallis, has maintained his high standards.

Sherman K. Stein
Anthony Barcellos

SAMPLE COURSE OUTLINES

The outlines sketch three courses:

Full This includes a review of precalculus material. The pace permits a thorough treatment, including the assignment of conceptual, exploratory, and writing exercises.

Core Here we assume that the students' precalculus background is sound (functions, graphs, trigonometry, and conics) and that the instructor wishes to emphasize only the most essential concepts and de-emphasize computations, which are consigned to calculators or a later numerical analysis course.

Medium This is one of the many possibilities that lie between the two extremes. The syllabi will suggest others, such as the Core at the pace of the Full.

Chapter	*Full:Lectures*	*Medium:Lectures*	*Core:Lectures*
1	3	1	0 (to be read)
2	13 (2 on 2.4, 2.6, 2.8)	9 (2.9 & 2.10 together)	5 (omit 2.2, 2.5, 2.6, 2.9, 2.10)
3	8 (2 on 3.2, 3.4)	6	5 (3.1 & 3.2 together)
4	13 (2 on 4.1, 4.7, 4.9)	9 (omit 4.4)	7 (omit 4.4, 4.6, 4.10)
5	8 (2 on 5.1)	7	5 (omit 5.4, 5.6)
6	12 (2 on 6.3, 6.7, 6.8)	9	4 (6.1 & 6.2 together, omit 6.4, 6.7, 6.8, 6.9)
7	9 (2 on 7.2, 7.7)	7	6 (omit 7.7)
8	9 (2 on 8.8)	8	4 (8.2 to be read, omit 8.5, 8.6, 8.7)
9	7	5 (omit 9.5, 9.7)	4 (omit 9.5, 9.6, 9.7)
10	8 (2 on 10.7)	8 (2 on 10.7)	7
11	8 (2 on 11.6)	7	4 (omit 11.3, 11.6, 11.7)
12	10 (2 on 12.2, 12.3, 12.4)	8 (2 on 12.4)	5 (omit 12.5, 12.7)
13	6 (2 on 13.5)	5	2 (omit 13.3, 13.4, 13.5)
14	14 (2 on 14.6, 14.7, 14.9)	13 (2 on 14.7, 14.9)	7 (14.2, 14.4 to be read, omit 14.10, 14.11)
15	7	7	4 (15.1 & 15.2 together, omit 15.3, 15.6, 15.7
16	10 (2 on 16.2, 16.4, 16.5, 16.6)	9 (2 on 16.2, 16.4, 16.5)	6
17	6 (2 on 17.1, 17.2)	5 (2 on 17.2)	4
Total	151 (3 semesters, 4/week)	123 (4 quarters, 3/week)	79 (1 year, 3/week)

Appendices	*Full:Lectures*	*Medium:Lectures*	*Core:Lectures*
A	1	1	0
B	2	1	0
C	2	1	0
D	1	1	0
E	1	0 (to be read)	0
F	1	0 (to be read)	0
G	4	3 (omit G.4)	0
H	2	1	0
I	2	0	0
J	2	1	0
K	2	1	0
L	3	2	0
M	2	1	0
Total	25	13	0

TO THE STUDENT

Welcome to calculus, one of the most powerful and important mathematical tools ever developed. You will learn concepts and techniques that go far beyond the mathematics you presently know. Right now you can find the area of a polygon or circle. With calculus you can compute the areas of much more general regions or even the volumes of many solid objects (not just boxes). With calculus you can discover when some varying quantity reaches its maximum value. Many applications make direct use of these maximization techniques, and you'll meet several in examples and exercises.

Calculus has the reputation of being difficult. Don't panic. The reputation is not deserved. The trouble some students have lies not in the subject but in their preparation or study habits. If you have done well in your precalculus mathematics you should do well in calculus, though you may need to change the way you approach mathematics.

You may look at this book, lift it, and moan, "Oh, there's so much to learn"; but the book is shorter than it appears. The core of the book—leaving out introductions, summaries, exercises, and historical asides—is only about 400 pages, and you will have at least a year to study it. Besides, by the time you complete Chapter 5 you will have met the core of calculus. It is in the early chapters that you will find the three central ideas of limit, derivative, and integral. Master those chapters and you will find practically everything that follows is just a variation on some definition or theorem you learned there.

New Learning Skills. What does it mean to "master" a mathematical concept? It does not mean that you have memorized some formulas and can then simply plug them into the exercises. Mastery means that you can apply the concept and that you understand why it is true. From kindergarten through high school the emphasis was on computation. You learned how to add, multiply, subtract, and divide decimals and fractions; in algebra, how to simplify algebraic expressions and solve equations; and in trigonometry, the definitions of the trigonometric functions and their applications. Your homework and exams probably concentrated on computations.

In such classes a teacher might have done an example on the board and then asked you to work many exercises just like it. You could probably do them without looking at your text outside of class. In short, the emphasis was on accuracy and speed in executing *algorithms,* which is just a word to describe mechanical procedures that require accuracy but little thought. You probably feel, therefore, that "mastering mathematics" means being good at carrying out routine computations.

New Study Habits. You may be dismayed, having done well in mathematics, to learn that the study habits appropriate for practicing computations are inadequate for mastering calculus. Of course, you will meet algorithms while studying calculus, and you may learn to apply these algorithms in situations that duplicate examples you have memorized. But computer software can execute these algorithms even better than you can. If algorithms were all there is to calculus, we could do without people. (See "Computers and Calculators," page xxxiii, and Section 7.1 for more information about computer calculus.)

A New York subway poster shows the circuitry of a calculator and asks an intimidating question:

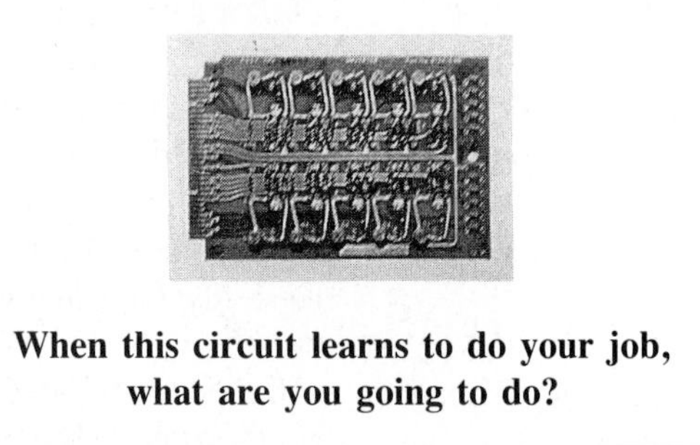

When this circuit learns to do your job, what are you going to do?

Does this question suggest that humans will soon be obsolete? Not at all. A computer is simply a rapid idiot; it has an IQ of zero. Human beings can conceptualize, analyze, and make choices. A computer has no more judgment than a vending machine. This brings us to the second aspect of mathematics, one that is particularly important to calculus.

By studying the "why" behind the algorithms, you will be developing your ability to analyze problems. You will see many examples of closely reasoned arguments. (Daily life offers very few.) Some of the exercises will ask you to express your own reasoning in words, a skill that can serve you well whether you become an engineer, a physician, an economist, a lawyer, or, for that matter, the head of a hotel chain. As Conrad Hilton, who founded the Hilton hotels, wrote in *Be My Guest,*

> I'm not out to convince anyone that calculus or even algebra and geometry are necessities in the hotel business. But I will argue long and loud that they are not useless ornaments pinned onto an average man's education. For me, the ability to formulate quickly, to reduce any problem to its simplest, clearest form, has been exceedingly useful. . . . I found mathematics the best possible exercise for developing the mental muscles necessary to this process.

Keep an open mind. View your study as an investment in the future. You may use some of the specific facts, or not. You may, like Hilton, use only the *style* of thinking developed by studying mathematics. As Tony Wexler, a professor of mechanical engineering at the University of Delaware, advises,

> Don't keep asking, "What's the use of this to me?" You can't predict which math will be of practical use after you graduate. In my eleven years in the "real world" I've worked on the flow of jet fuel, solar energy, computers, statistical software, air pollution, and modeling the physiological functions of the kidney. In this last application I needed to compute the partial derivatives of functions defined implicitly [as in Exercise 11 of Sec. 14.11]. When I was a student I never would have guessed anything this far out would someday come in handy.

Steve Whitaker, a University of California engineering professor, urges both instructor and student to "emphasize proofs, because the ideas that go into the proofs are often the ideas that go into the applications."

Good Study Techniques. Students develop various effective ways to study calculus. Some read a section or a chapter in advance. Even though they don't try to grasp every detail, they get a feel for the main ideas. In this way, they are better prepared to understand the lectures.

Some keep a separate notebook of the definitions and theorems. It is surprising how few pages are needed to record the essentials of calculus. This notebook helps students maintain a larger view of the course. It is especially useful for reviewing.

Some students form study groups of two, three, or four members in which they can talk about the homework or review for exams. Having to express your thoughts or questions out loud can often clarify matters. Working on homework together is a good idea, as long as you don't depend on one person to do all the work, but work on each problem yourself. In high school you could earn an A or B without doing any "homework" at home. But in college you can expect at least two hours of homework for each hour in class. Find a quiet place where you can concentrate, so these hours will be fruitful.

When you do your homework, follow these steps in the order shown:

1 Read the text and your lecture notes.
2 Study the examples, first trying to work them without looking at their solutions.
3 Begin the homework problems.

We emphasize this order because it is the very reverse of the order that many students follow.

In high school, the homework exercises were almost always identical to the examples done in the text or by the teacher. It was seldom necessary to read the text. A student would start with step 3; then, if stuck, go to step 2; and if that did not do the trick, the student would reluctantly go to step 1, finally reading the text.

The first step, "Read the text," requires an explanation. Don't read it as you would read a newspaper or magazine. Instead, make sure you have paper and pencil ready. When you read the statement of a theorem, try to prove it before looking at the proof. Even if you don't succeed, you will be in a better position to understand the proof when you do read it. After you read it, see if you can write the proof without looking at it again.

Proofs are important, but not because they prove that a theorem is true—after all, some of the theorems have been around since 1665 and no one is questioning their validity. Proofs in elementary calculus are important for two reasons: First, they give you a chance to review basic principles; second, they reduce the strain on the memory. Instead of memorizing a lot of seemingly unrelated facts, you remember a few basic ideas from which the facts can be derived. If you forget a formula, you can quickly derive it from scratch. This skill is useful not just during exams but long after the course is over. That is another reason why you should pay close attention to the proofs and to the definitions on which they rest.

The second step, "Study the examples," should be carried out with the same care as the first step. Try to work them before reading their solutions.

Only after these two steps are completed (and the definitions and theorems now in your grasp and perhaps in your notebook) should you start the third step, "Begin the homework." After all, the homework is supposed to reinforce the ideas; if you have not studied the ideas, there is nothing to reinforce and the homework will not do you much good.

Perhaps the relation of homework to the material is distorted because its score usually affects the grade in the course. You concentrate on getting as high a score as possible rather than on learning the principles. Since mastery of the ideas generally leads to better homework and exam scores, you are not jeopardizing your grade by giving understanding the priority it deserves.

Similarly, students often allow exams to interfere with genuine study. They do countless problems with the goal of passing tomorrow's exam, rather than the goal of understanding what is going on. Just before the final examination, students besiege instructors with requests for practice exams, as though the final is an alien beast from outer space, rather than a set of questions based on the material presented in class. Don't fall into this trap. If you master the ideas, you will do well on the exam.

If you highlight with a felt pen, don't wait until the night before a test to think about what you marked—and don't overmark. *Mastering the material as you go along will deepen your understanding and reduce stress.* Understanding cannot be achieved through cramming.

Keep It in Perspective. You could read Chapter 1, a survey of calculus, even before the course starts. Return to it from time to time, as you read the later chapters, to help maintain your perspective. Try to fit the new material into the framework provided by Chapter 1's overview. When you reach Chapter 5, the key chapter in the entire text, be sure to read *The long road to calculus,* which sketches the history of the creation of this powerful tool. The chapter summaries will help you develop a larger view than the individual sections can offer.

It is natural that a calculus text begin with calculus itself, instead of devoting lots of time to precalculus material. In case you need to refresh your memory, however, or have missed some topics along the way, several of the appendices treat precalculus material. For instance, if you are not familiar with the various notations for intervals on the x axis, you can find them in Appendix A. Some algebra is reviewed in Appendix C. Trigonometry, however, has a section in Chapter 2, and logarithms are discussed in Chapter 6. You might read those sections before they are needed.

Answers to almost all the odd-numbered exercises are in the back of the book; complete solutions are in the *Student's Solutions Manual* by Anthony Barcellos.

Exercises come in many flavors and levels of difficulty. Some have been marked with special icons to help you understand what is expected of you. There are "essay" exercises that require answers written out in complete sentences. These are marked with a pencil: ✎. Your answers might be as short as a couple of sentences or as long as several paragraphs. "Exploration" problems are marked with a compass: ✷. These exercises fit the so-called "tri-ex" format: "experiment, extract, explain." You are expected to *experiment* with various cases, *extract* a general principle, and then *explain* why the principle is valid. Exercises that exploit the calculator or computer are marked with a 🖩 or a ◘, respectively.

Some of the problems ask you to draw a diagram. These are indicated with a 📏. Drawing is a much neglected skill. The ability to sketch a quick and accurate working diagram will serve you well in calculus and later studies, so we have given many practical drawing tips in the text. At the very least, use a ruler to draw lines. A plastic template or compass will also come in handy. For some drawings you may want to use more than one color or, if you stick to one color, draw some curves solid, some with dashes, and others with dots.

In general, each exercise set begins with paired exercises that you can use for drill and practice. They help you check your basic understanding of the definitions, algorithms, and concepts in the section. A horizontal bar marks the end of these paired exercises. Problems appearing after the bar may still be fairly routine, but occur singly rather than in matched pairs. Some suggest different views of the ideas, some may invite exploration, and some may require putting the ideas of several sections together. The difficulty level tends to rise throughout each exercise set, but this is merely a guideline. We do not think it is a good idea to suggest that certain exercises are hard while others are easy. That judgment varies from person to person.

Before you begin your study, read the remarks of Arnold Toynbee, the historian, and of H. G. Wells, the social philosopher, quoted on page xi.

We have made a special effort to root out mistakes, which are a nuisance to instructor and student and an embarrassment to the authors. Nevertheless, some mistakes may have slipped by, and we encourage you to tell us of any you may find.

Calculus is exciting, powerful, and beautiful. This book was written in the hope that we could share with you some of that excitement, power, and beauty. Enjoy your study of calculus, which is one of the grandest achievements of the human mind.

Sherman K. Stein
Anthony Barcellos

COMPUTERS AND CALCULATORS

Personal computers and programmable calculators continue to grow in scope and power. They can now perform many operations that once had to be done by hand, even including differentiation and some kinds of formal integration.

One must recognize the utility of these tools and make preparations for their use. However, hardware and software evolve rapidly, while an edition of a calculus text has a typical lifetime of four to six years. Furthermore, particularly in large classes, many students have limited access to computers. A text needs, therefore, to be flexible in its approach to ''machine-based'' calculus.

We have designated certain exercises for computer work without stipulating what system to use. (The ■ is used to mark such problems.) These are typically numerical in nature and call for such things as Simpson's rule, difference quotients, or partial sums of power series. Some are explorations (denoted ✷) that begin with a gathering of data with the aid of a calculator. Exercises marked with a ▦ require a scientific calculator to produce numerical answers to a specified number of decimal places. Programmable calculators can, in most cases, solve computer problems as well. Many exercises that are not flagged as computer problems can nevertheless be used with systems that permit more advanced symbolic manipulation. We leave it to the instructor to decide how extensively the computer is used in teaching calculus.

The following list of selected hardware and software should serve as a starting point for the instructor who wants to take advantage of the computer exercises in the text. Keep in mind that all items in the list can be expected to change yearly, adding more capability with each revision. Check with the manufacturer or publisher before deciding what to use.

Programmable Calculators

TI-81
Texas Instruments, Inc.
13500 North Central Parkway
Dallas, TX 75243
(214) 995-2011
Plots up to four functions on a set of axes. Programmable for various numerical methods.

HP-28s, HP-48s, and HP-48sx
Hewlett-Packard Company
Portable Computer Division
1000 N.E. Circle Blvd.
Corvallis, OR 97330
(800) 367-4772
Performs symbolic differentiation and some integration. Plots graphs. Programmable for various numerical methods.

Personal Computer Software

Maple
Brooks-Cole Publishing Company
Pacific Grove, CA
(408) 373-0728
Symbolic mathematics package for Macintosh computers. Includes calculus functions and two-dimensional graphs. Some programmability. Can use a mathematics coprocessor on computers so equipped.

MathCAD
MathSoft, Inc.
One Kendall Square
Cambridge, MA 02139
(617) 577-1017
Symbolic mathematics package for Macintosh and IBM personal computers and compatibles. Includes calculus functions and graphs in two or three dimensions. Can produce documents with mixed text and graphics. Can use a mathematics coprocessor on computers so equipped.

Derive
Soft Warehouse, Inc.
3615 Harding Avenue, Suite 505
Honolulu, HI 96816-3735
(808) 734-5801
Symbolic mathematics package for IBM personal computers and compatibles. Includes calculus operations and graphics in two or three dimensions. Some programmability.

PC MACSYMA
Symbolics Inc.
Burlington, MA
(617) 221-1250
Symbolic mathematics package for 386-based IBM personal computers and compatibles. Includes calculus operations and graphics in two or three dimensions. Full programming language.

Mathematica
Wolfram Research, Inc.
P.O. Box 6059
Champaign, IL 61826
(217) 398-0700

Symbolic mathematics package for Macintosh and 386-based IBM personal computers or compatibles. Includes calculus operations and graphics in two or three dimensions, including animation. Can produce documents with mixed text and graphics. Full programming language. Can use mathematics coprocessors on systems so equipped.

Theorist
Prescience Corporation
814 Castro Street, #101
San Francisco, CA 94114
(415) 282-5864
Symbolic math and graphics package for Macintosh personal computers. Many of the same features as Mathematica.

The Math Utilities
Bridge Software
P.O. Box 118
New Town Branch
Boston, MA 02258
(617) 527-1585
Graphics package for IBM-compatible personal computers. Includes CURVES for two-dimensional plots and SURFS for three-dimensional graphs.

CoPlot
CoHort Software
P.O. Box 1149
Berkeley, CA 94701
(415) 524-9878
A scientific graphics package for IBM-compatible computers that can generate various rectangular and polar plots, as well as several other graph types, including three-dimensional. Several graphs can be overlaid on a single axis system.

Axum
TriMetrix, Inc.
444 N.E. Ravenna Blvd., #210
Seattle, WA 98115
(206) 527-1801
A scientific graphics package for IBM-compatible personal computers. Many graphics formats including two-dimensional and three-dimensional plots.

CALCULUS
AND ANALYTIC GEOMETRY

1

AN OVERVIEW OF CALCULUS

There are two main concepts in calculus: the derivative and the integral. Underlying both is the theme of limits. This chapter introduces these ideas informally, tells where they appear in the text, and offers a glimpse into their history. You may wish to turn back to these pages from time to time to maintain a broad perspective, which you may otherwise easily lose in the day-by-day details of definitions, theorems, and applications.

1.1 THE DERIVATIVE

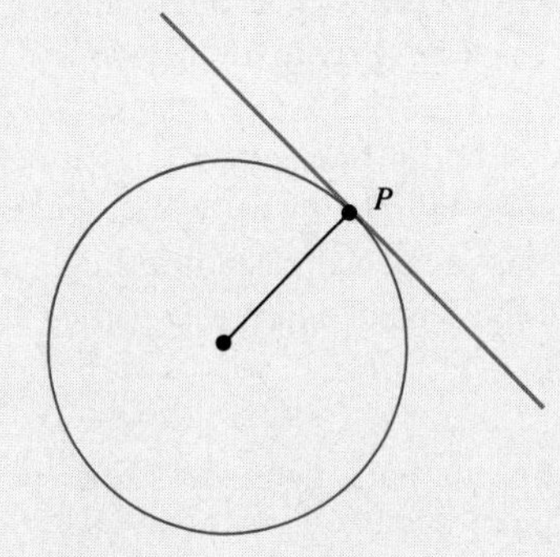

Figure 1

The tangent line to a circle at a given point P can be found as follows. First draw the radius from the center of the circle to P and then construct the line through P perpendicular to that radius. That line is tangent to the circle. (See Fig. 1.)

But how would we construct the tangent line at a point P on a curve that is not a circle? For instance, how would we find the tangent line at the point P on the curve in Fig. 2, which is described by the equation $y = x^2$?

The ancient Greeks solved this problem. They also worked with tangents to ellipses, hyberbolas, and certain spirals. In the first half of the seventeenth century mathematicians struggled to find ways to construct tangents to other curves, such as the graph of a polynomial.

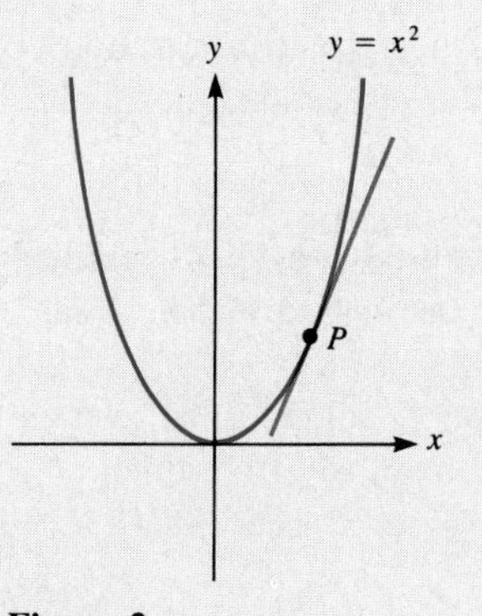

Figure 2

You may wonder, ''Why was anyone interested in finding a tangent to a curve?'' After all, there was no urgent practical need for solving the problem. The mathematicians of the time, as many mathematicians before and since, were driven by curiosity, by the desire to unravel a mystery. Little did they expect that the tools that they fashioned would be exploited as early as the eighteenth century, in the study of motion, fluid flow, heat, and astronomy.

The mathematician Pierre Fermat (pronounced *Fair-MAH*) published a general method for finding the tangent to a curve in 1637 in his book *Disquirendam Maximam et Minimam* (A Method for Finding Maxima and Minima). This method was closely connected with finding the steepness or ''slope'' of a curve at any point on that curve. Once we know the slope, we can easily draw the tangent line. For instance, if the slope of the tangent line at a certain point P is

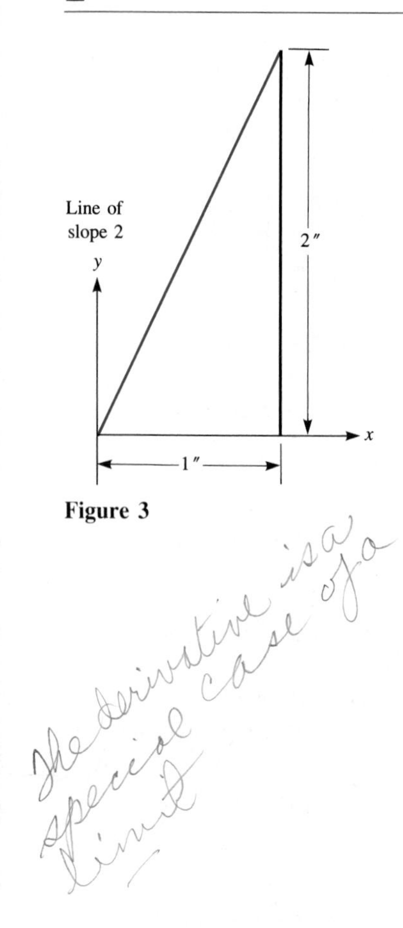

Figure 3

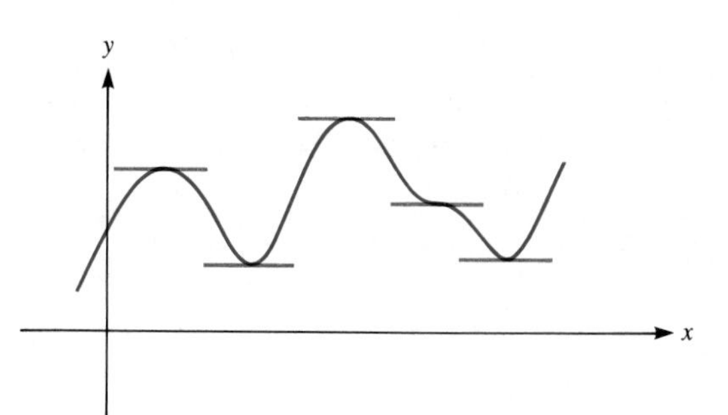

Figure 4

2, we use a ruler and lay off a triangle whose horizontal side is 1 inch long and whose vertical side is 2 inches long, as in Fig. 3.

As the title of Fermat's work may indicate, information about tangent lines can help in finding high and low points on a curve. A glance at Fig. 4 suggests that at such points the tangent is horizontal. Thus the search for high and low points on a curve can be simplified: first find where the tangent is horizontal. Chapter 4 develops and applies this technique.

Finding tangent lines is only one of the many applications of the *derivative*, one of the two central concepts of calculus. For instance, the derivative also serves to measure how rapidly a quantity changes. To be specific, consider an object that falls under the influence of gravity. As it falls, its velocity increases. How can we find that varying velocity if we know the distance the object falls during the first t seconds for any positive number t? The derivative provides a formula for that velocity.

Chapter 3 introduces the derivative, and Chaps. 4 and 6 present some of its main applications. Chapter 2, on limits, provides the foundation on which the derivative rests. It turns out that the derivative is just a special case of a limit.

The first example illustrates the notion of a limit; the second explores a particular limit that arises in the study of the derivative.

EXAMPLE 1 Examine what happens to $\sqrt{x^2+3x}-x$ as x gets very large.

x	$\sqrt{x^2+3x}$	$\sqrt{x^2+3x}-x$
10	11.402	1.402
20	21.448	1.448
50	51.478	1.478
100	101.489	1.489
1000	1001.499	1.499

SOLUTION Make a table to three decimal places with the aid of a calculator. For instance, for $x = 10$, we find that

$$\sqrt{x^2+3x}-x=\sqrt{10^2+3\cdot 10}-10=\sqrt{130}-10\approx 11.402-10=1.402.$$

It appears that, as x gets larger $\sqrt{x^2+3x}-x$ gets nearer and nearer to 1.5. Though we are not sure, this seems a likely guess. We suspect that the "limit" of $\sqrt{x^2+3x}-x$ as x gets arbitrarily large is 1.5. ■

EXAMPLE 2 An object moves t^3 feet during the first t minutes of its journey. Estimate how fast it is moving after 2 minutes.

SOLUTION To estimate the "instantaneous" velocity, compute the average velocity of the object during a very small interval of time, say from time $t = 2$ minutes to time $t = 2.1$ minutes. (See Fig. 5.)

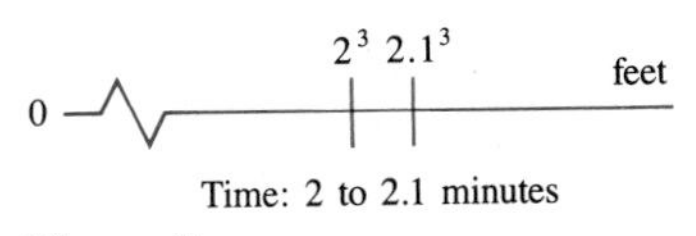

Figure 5

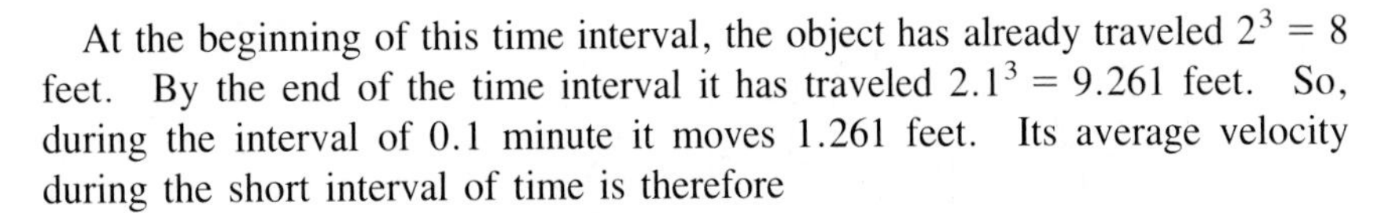

At the beginning of this time interval, the object has already traveled $2^3 = 8$ feet. By the end of the time interval it has traveled $2.1^3 = 9.261$ feet. So, during the interval of 0.1 minute it moves 1.261 feet. Its average velocity during the short interval of time is therefore

$$\text{Average velocity} = \frac{\text{Distance}}{\text{Time}} = \frac{1.261}{0.1} = 12.61 \text{ feet per minute.}$$

This provides an estimate of the velocity at time $t = 2$ minutes.

To get better estimates, use smaller intervals of time.

For the time interval from $t = 2$ to $t = 2.01$ minutes, the average velocity is

$$\frac{2.01^3 - 2^3}{0.01} = \frac{8.120601 - 8}{0.01} = \frac{0.120601}{0.01} = 12.0601 \text{ feet per minute.}$$

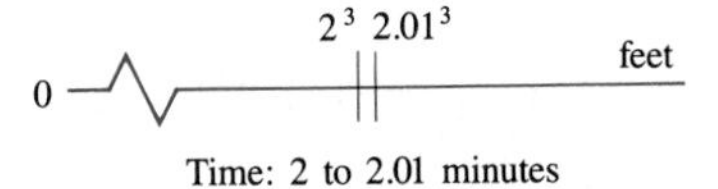

Figure 6

(See Fig. 6.)

But we could consider even shorter intervals of time. Moreover, we could consider short intervals of time that *end* at $t = 2$ rather than start at $t = 2$. For instance, the average velocity from $t = 1.99$ to $t = 2$ minutes is

$$\frac{2^3 - 1.99^3}{0.01} = \frac{8 - 7.880599}{0.01} = \frac{0.119401}{0.01} = 11.9401 \text{ feet per minute.}$$

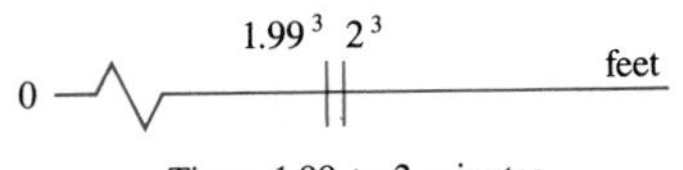

Figure 7

(See Fig. 7.)

The estimates 12.61, 12.0601, and 11.9401 feet per minute are just approximations of the velocity at time $t = 2$ minutes.

What we really want to find is what happens to the quotient

$$\frac{t^3 - 8}{t - 2}$$

as t gets nearer and nearer the number 2. Just as in Example 1, we are studying a "limit," in this case the limit of $(t^3 - 8)/(t - 2)$ as t approaches 2. The data suggest that this limit may be 12.

This particular limit is called "the derivative of t^3 at $t = 2$." ■

EXERCISES FOR SEC. 1.1: THE DERIVATIVE

A calculator would come in handy in these exercises.

1 (*a*) Complete this table.

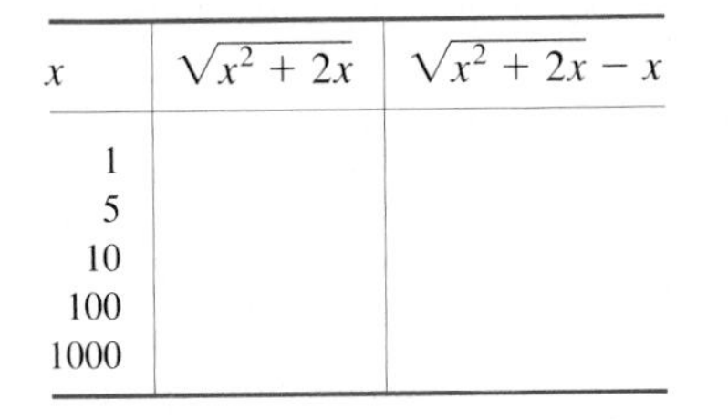

x	$\sqrt{x^2 + 2x}$	$\sqrt{x^2 + 2x} - x$
1		
5		
10		
100		
1000		

(*b*) On the basis of (*a*), what number do you think $\sqrt{x^2 + 2x} - x$ approaches as x gets very large?

2 (*a*) Complete this table.

x	$\sqrt[3]{x^3 + x}$	$\sqrt[3]{x^3 + x} - x$
1		
10		
100		
1000		

(*b*) On the basis of (*a*), what number do you think $\sqrt[3]{x^3 + x} - x$ approaches as x gets very large?

3 (*a*) Complete this table.

x	$x^3 - 1$	$x - 1$	$(x^3 - 1)/(x - 1)$
0.5			
0.9			
0.99			
0.999			

(*b*) On the basis of (*a*), what number do you think $(x^3 - 1)/(x - 1)$ approaches as x gets nearer and nearer to 1?

4 (*a*) Complete this table.

x	$x^3 - 1$	$x^2 - 1$	$(x^3 - 1)/(x^2 - 1)$
2			
1.5			
1.1			
1.01			
1.001			
0.9			
0.99			

(*b*) On the basis of (*a*), what number do you think $(x^3 - 1)/(x^2 - 1)$ approaches as x gets nearer and nearer to 1?

5 An object travels t^3 feet in the first t minutes of its journey. Estimate its speed when $t = 1$ as follows.

(*a*) How far does it move during the time interval from $t = 1$ to $t = 1.01$ minutes?

(*b*) What is its average speed during the time interval given in (*a*)?

(*c*) What is its average speed during the time interval from $t = 1$ to $t = 1.001$ minutes?

(*d*) What is its average speed during the time interval from $t = 0.999$ minute to $t = 1$ minute?

(*e*) On the basis of (*b*), (*c*), and (*d*), what do you think its speed at time $t = 1$ minute is?

6 An object falls $16t^2$ feet in t seconds.

(*a*) How far does it fall during the interval of time from $t = 2$ seconds to $t = 2.01$ seconds?

(*b*) What is its average speed during the time interval from $t = 2$ seconds to $t = 2.01$ seconds?

(*c*) What is its average speed during the time interval from $t = 2$ seconds to $t = 2.001$ seconds?

(*d*) What is its average speed during the time interval from $t = 1.999$ seconds to $t = 2$ seconds?

(*e*) On the basis of (*b*), (*c*), and (*d*), what do you think its speed at time $t = 2$ seconds is?

7 (*a*) Complete this table.

x	2^x	$2^x - 1$	$\frac{2^x - 1}{x}$
1			
0.5			
0.1			
0.01			
0.001			
−0.001			

(*b*) On the basis of (*a*), what do you think happens to $(2^x - 1)/x$ as x gets closer and closer to 0?

8 In this exercise angles are measured in degrees. For instance $\sin 30° = 0.5$.

(*a*) Complete this table.

x	$\sin x$	$\frac{\sin x}{x}$
30°		
10°		
5°		
1°		
0.1°		

(*b*) On the basis of (*a*), what do you think happens to $(\sin x)/x$ as x gets closer and closer to 0? (Express your guess to five decimal places.)

9 In this exercise angles are measured in radians. For instance, $\sin(\pi/2) = 1$ (so $\sin 1.57$ is close to 1).

(*a*) Evaluate $(\sin x)/x$ for $x = 1$, 0.1, and 0.01.

(*b*) What do you think happens to $(\sin x)/x$ as x gets nearer and nearer to 0? (Chap. 2 treats this limit.)

10 In this exercise angles are measured in radians.

(*a*) Evaluate $(1 - \cos x)/x^2$ for $x = 1$, 0.1, and 0.01.

(*b*) What do you think happens to $(1 - \cos x)/x^2$ as x gets nearer and nearer to 0?

11 An object travels $3^t - 1$ meters during the first t seconds of motion. For instance, it travels 8 meters in the first 2 seconds.

(*a*) What is its average velocity during the time interval from $t = 2$ to $t = 2.1$ seconds? From $t = 2$ to $t = 2.01$ seconds?

(*b*) Using a smaller interval of time, make another estimate of its speed when $t = 2$ seconds.

Note: In Sec. 6.5 it will be shown that this limit can be expressed in terms of logarithms.

1.2 THE INTEGRAL

The derivative provides information at a point or at a particular instant, so-called *local* information. The *integral*, the other major concept of calculus, does just the opposite. It is a tool for obtaining the numerical value of some overall quantity from local information. For instance, we may know that the length of the vertical line segment in Fig. 1 is x^2 for each x and want to find the total shaded area in Fig. 2. What makes the problem both difficult and interesting is that the length x^2 is not constant but varies with x. This should be contrasted with the problem of finding the area of a rectangle, which is simple to calculate because the parallel line segments all have the same length, as shown in Fig. 3. With the aid of the integral, developed in Chap. 5, we will be able to find the shaded area in Fig. 2 almost as easily as we calculate the area of the rectangle in Fig. 3.

The integral can be used to find area.

The integral is also used in the theory of motion. Assume that we know the speed of a moving object at any time and wish to calculate the total distance the object travels during a certain interval of time. If the speed is constant, the problem is easy: The distance is then obtained by multiplying the fixed speed by the length of time. But if the speed is not constant, if it changes from moment to moment, the problem is not so simple. However, with the aid of the integral, we will be able to find the total distance if we have a formula for the speed at each instant.

The integral can be used to find the distance an object moves if we know its speed at any time.

The derivative takes us from the global to the local, for instance, from total distance to the speed at any time. The integral takes us from the local to the global, for instance, from the speed at any time to total distance. This close connection is exploited in Chap. 5 to show that derivatives can be used to evaluate integrals, a result known as the Fundamental Theorem of Calculus. This intimate connection between the derivative and the integral was observed by Isaac Newton in 1669 and, independently, by Gottfried Leibniz (pronounced

The fundamental theorem of calculus links the derivative and the integral.

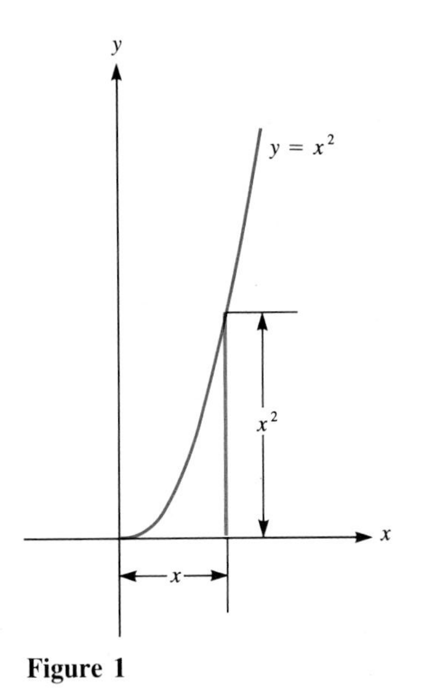

Figure 1

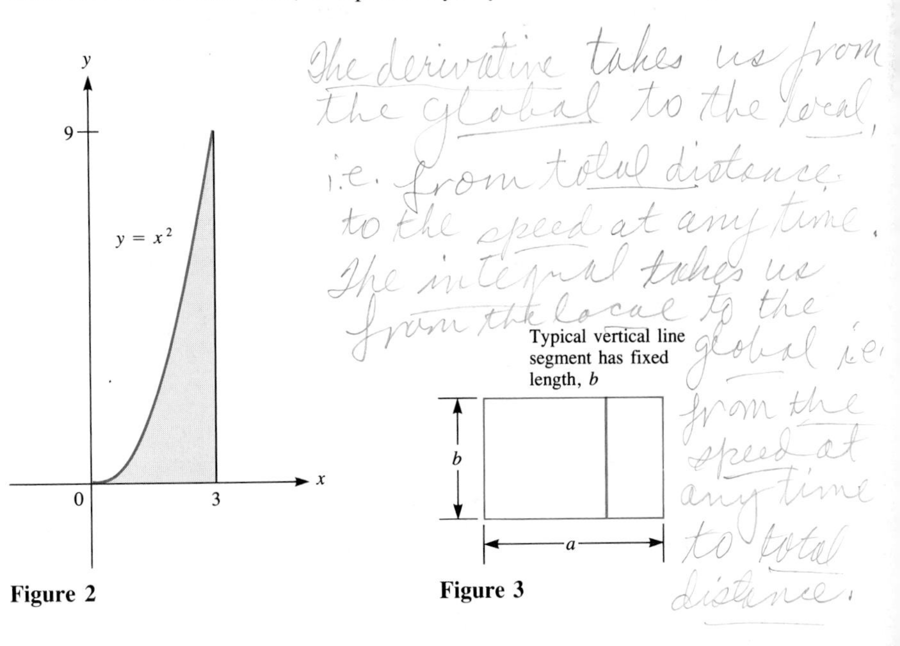

Figure 2

Figure 3

"LIBE-nits") in 1673 during their development of calculus into a structure with techniques applicable to a broad class of problems.

Before the invention of calculus the area under each curve was found by some special technique. For instance, Fermat in 1636 and Cavalieri in 1639 managed to find the area under a curve of the form $y = x^n$ and James Gregory in 1668 found the area under the curve $y = 1/(\cos x)$. Calculus develops a uniform approach for computing such areas with the aid of integrals.

In Chap. 7 techniques for computing some integrals are presented. In Chaps. 8 and 9 the integral is applied in geometric problems, such as calculating the volume and surface area of a sphere and the length of a curve, as well as in other disciplines, such as physics, economics, and biology.

To give some idea of the integral—and why it, like the derivative, involves limits—let us try to estimate the area of the shaded region in Fig. 2.

Since we can compute the area of a rectangle precisely, let us estimate the area in Fig. 2 by approximating it with a staircase of narrow rectangles, as in Figs. 4 and 5. The rectangles in Fig. 4 *underestimate* the area under the curve; the rectangles in Fig. 5 *overestimate* the area. In Examples 1 and 2 the computations for specific choices of rectangles are carried out in detail.

EXAMPLE 1 Figure 6 shows three rectangles, all of width 1. Find their total area.

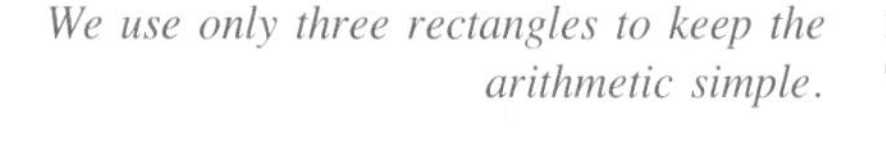
We use only three rectangles to keep the arithmetic simple.

SOLUTION Each rectangle has width 1. The heights of the rectangles can be found by using the formula for the curve, $y = x^2$. The height of the lowest rectangle is therefore 1^2, the next has height 2^2, and the tallest has height 3^2. Their total area is

$$\underbrace{1^2}_{\text{height}} \cdot \underbrace{1}_{\text{width}} + \underbrace{2^2}_{\text{height}} \cdot \underbrace{1}_{\text{width}} + \underbrace{3^2}_{\text{height}} \cdot \underbrace{1}_{\text{width}} = 1 + 4 + 9 = 14.$$

The area under the curve is therefore less than 14. ■

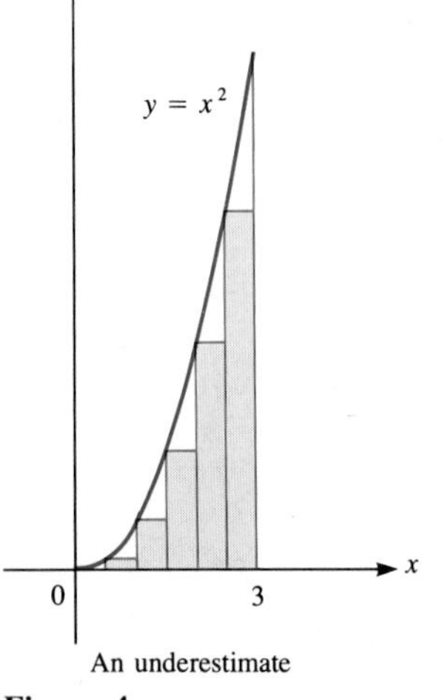

An underestimate

Figure 4

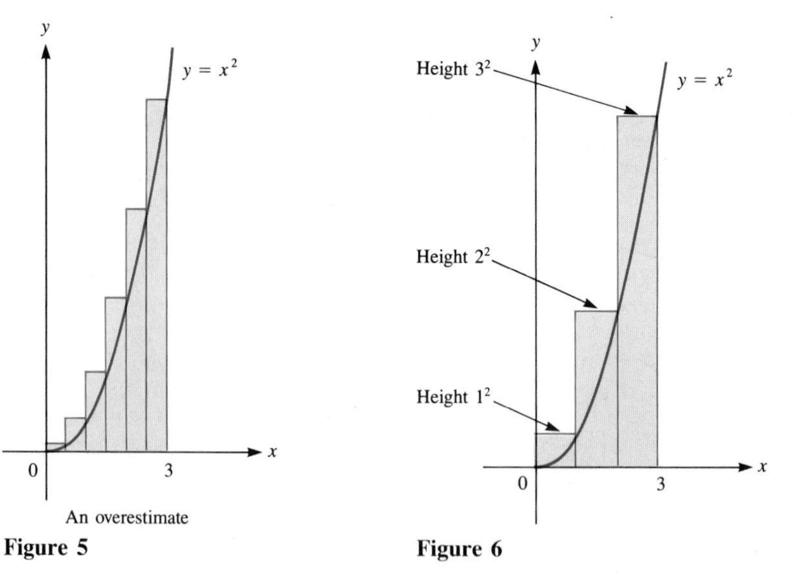

An overestimate

Figure 5

Figure 6

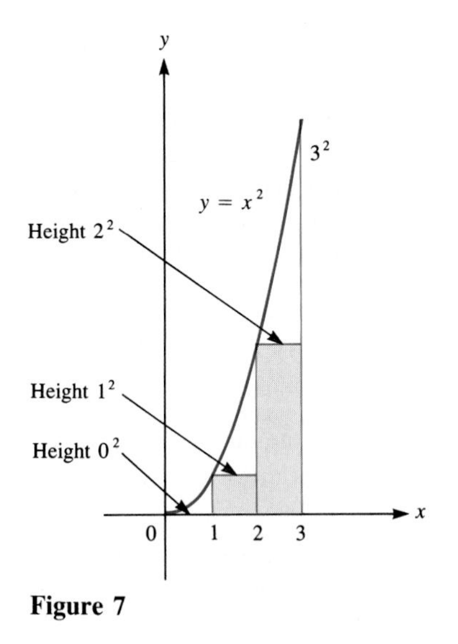

Figure 7

EXAMPLE 2 Figure 7 shows three rectangles, all of width 1. Find their total area.

SOLUTION (The first rectangle, of height 0, is just a line segment.) In this case the sum of the areas of the rectangles is

$$\underbrace{0^2}_{\text{height}} \cdot \underbrace{1}_{\text{width}} + \underbrace{1^2}_{\text{height}} \cdot \underbrace{1}_{\text{width}} + \underbrace{2^2}_{\text{height}} \cdot \underbrace{1}_{\text{width}} = 0 + 1 + 4 = 5.$$

The area under the curve is therefore larger than 5. ■

The computations in Examples 1 and 2 show that the area under the curve is somewhere between 5 and 14, which is quite a wide margin. *To get closer bounds, use narrower rectangles.* Figures 8 and 9 each show 10 rectangles, all of width $\frac{3}{10} = 0.3$. The total area of the rectangles in Fig. 8 is 10.395, and the total area of those in Fig. 9 is 7.695. So the area under the curve is trapped between the numbers 10.395 and 7.695, which represents a much smaller margin of error than 5 and 14, which were found in Examples 1 and 2.

It seems likely that the more rectangles we use and the narrower they are, the better their total area will approximate the area under the curve. This suggests that, to find the area under the curve, we should discover what happens to the total area of the approximating staircase of rectangles as the widths of the rectangles are chosen smaller and smaller. Just as in the preceding section we again are confronting the notion of a limit. The limit of the total area of the staircase of rectangles as their widths approach 0 will give us the area under the curve. This limit is called "the integral of x^2 from 0 to 3."

Chapter 5 shows how to compute such a limit.

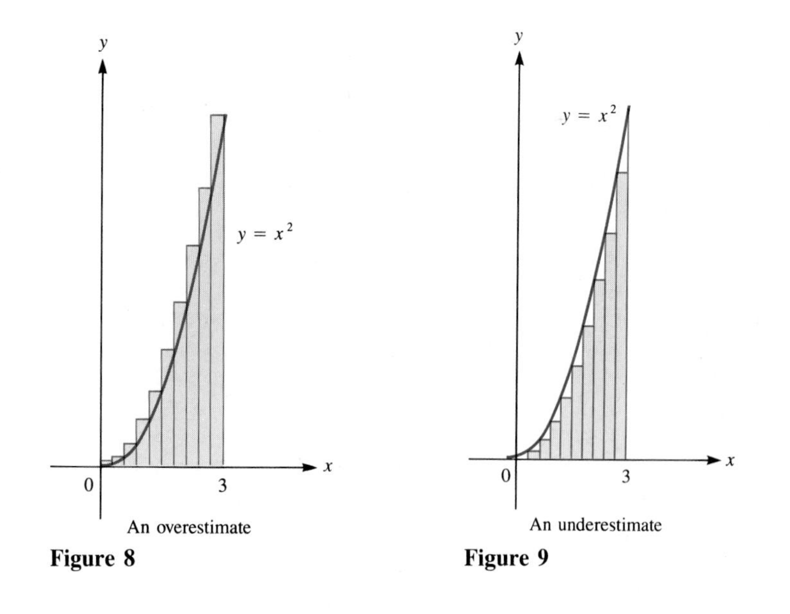

Figure 8

Figure 9

EXERCISES FOR SEC. 1.2: THE INTEGRAL

1 (*a*) Draw the four rectangles of width $\frac{3}{4}$ which, like the rectangles of Example 1, *overestimate* the area under $y = x^2$ from 0 to 3.

(*b*) Complete this table. The rectangles are labeled from left to right.

Rectangle	*Height*	*Width*	*Area*
First			
Second			
Third			
Fourth			

(*c*) Find the total area of the rectangles in (*b*).

2 (*a*) Draw the four rectangles of width $\frac{3}{4}$ which, like the rectangles in Example 2, *underestimate* the area under $y = x^2$ from 0 to 3.

(*b*) Find their total area. (The first one has area 0.)

3 Like Exercise 1, with five rectangles of width $\frac{3}{5}$ instead of four.

4 Like Exercise 2, with five rectangles of width $\frac{3}{5}$ instead of four.

5 Using 10 rectangles all of width $\frac{3}{10}$, verify that the area under the curve $y = x^2$ from 0 to 3 is (*a*) less than 10.395 and (*b*) more than 7.695.

6 Consider the area under the curve $y = x^2$ from 1 to 2.

(*a*) Using five rectangles of width $\frac{1}{5}$, find an estimate of the area that is too large.

(*b*) Again using five rectangles of width $\frac{1}{5}$, find an estimate of the area that is too small.

7 Consider the area of the region under the curve $y = x^3$ from $x = 0$ to $x = 1$.

(*a*) Compute the total area of the four rectangles in Fig. 10. Each has width $\frac{1}{4}$.

(*b*) Using eight rectangles of width $\frac{1}{8}$ that lie *below* the curve $y = x^3$, show that the area under $y = x^3$ from 0 to 1 is greater than $\frac{49}{256} = 0.19140625$.

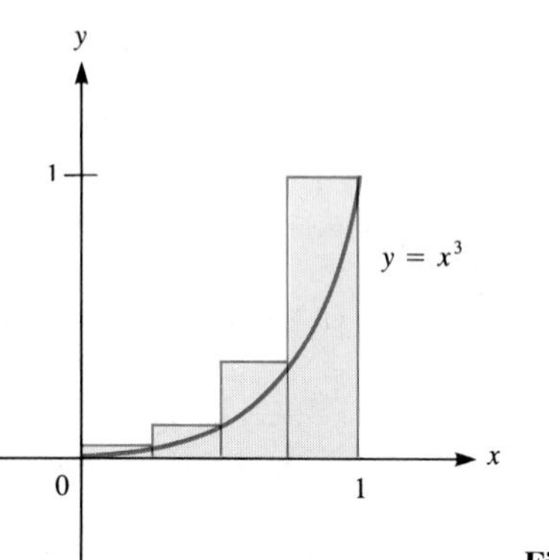

Figure 10

8 Estimate the area of the region under the curve $y = 1/x$ from $x = 1$ to $x = 2$ as follows:

(*a*) Use 5 rectangles of width $\frac{1}{5}$ that lie below the curve.

(*b*) Like (*a*), but use 10 rectangles instead of 5.

(*c*) Like (*b*), but use 10 rectangles whose top edges lie *above* the curve.

(*d*) What do (*b*) and (*c*) imply about the area under the curve?

9 (*a*) Estimate the area of the region under the curve $y = 1/x$ from $x = 3$ to $x = 6$ using 10 rectangles of equal width that lie below the curve.

(*b*) Compare the result of (*a*) with Exercise 8(*b*) and comment.

10 Estimate the shaded area under the curve $y = 1/(x^3 + 1)$, using the rectangles shown in (*a*) Fig. 11 and (*b*) Fig. 12. The rectangles have equal widths.

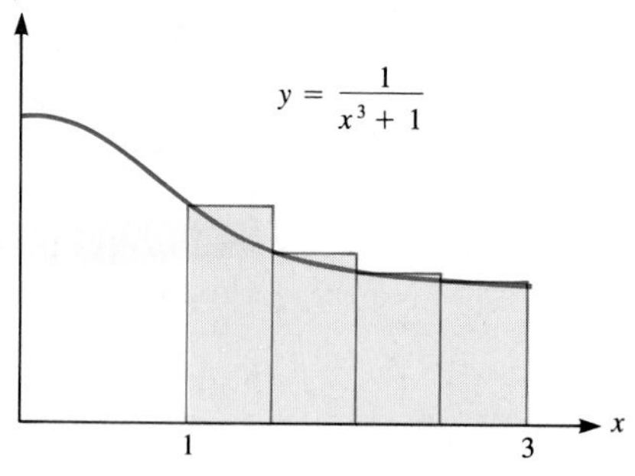

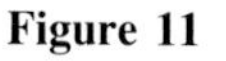

Figure 11

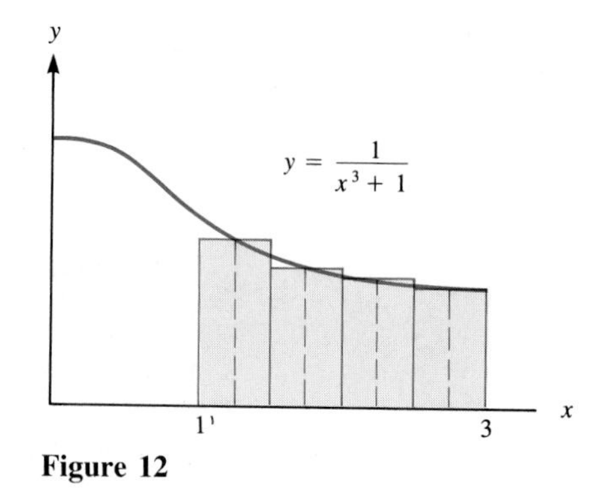

Figure 12

11 Estimate the area of the shaded region under the curve $y = \frac{1}{3}\sqrt[3]{x^6 + 1}$ shown in Fig. 13 using the 10 rectangles of equal widths. Their heights are determined by the values of $\frac{1}{3}\sqrt[3]{x^6 + 1}$ at the midpoints of their bases.

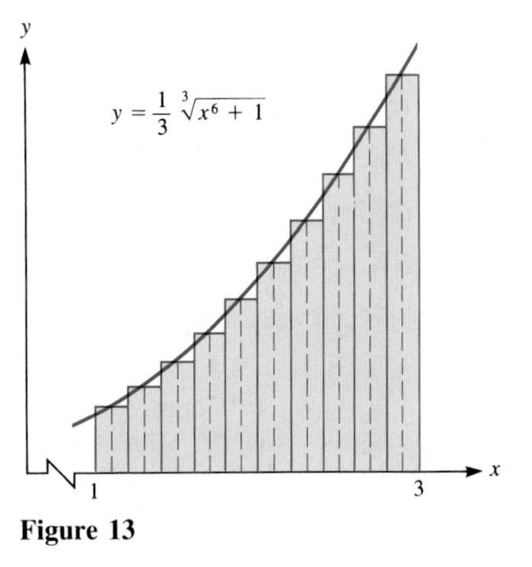

Figure 13

1.3 SURVEY OF THE TEXT

The theory and application of the derivative and integral occupy Chaps. 3 to 9. Chapter 2 develops the limit concept, which is the foundation for the definitions of the derivative in Chap. 3 and the definite integral in Chap. 5. Chapters 3 to 5 are the core of calculus.

Chapters 10 and 11 on series have a different flavor, though they make significant use of the derivative and integral. It turns out that certain quantities, though not themselves polynomials, can be very closely approximated by polynomials. In particular, if we consider polynomials of "infinite degree" (that is, the terms go on forever), we can get exact equality instead of just approximate equality. For instance,

$$\sin x = x - \frac{x^3}{1 \cdot 2 \cdot 3} + \frac{x^5}{1 \cdot 2 \cdot 3 \cdot 4 \cdot 5} - \frac{x^7}{1 \cdot 2 \cdot 3 \cdot 4 \cdot 5 \cdot 6 \cdot 7} + \cdots.$$

(The angle is in radians, not degrees.) This formula enables us to evaluate $\sin x$ to as many decimal places as we please, using the four basic operations of arithmetic: addition, subtraction, multiplication, and division. Calculators and computers, as well as we humans, are programmed to execute these operations. So Chaps. 10 and 11 examine what happens when you add more and more numbers that are getting smaller and smaller. Do the sums get arbitrarily large? Do they approach some specific number? Both outcomes are possible, and we will learn to distinguish between them.

Chapters 14 and 15 extend the notions of the derivative and the integral to higher dimensions.

Most of the ideas and techniques in the first 11 chapters were known by the middle of the eighteenth century, the contributions of Newton and Leibniz and their successors, such as the Bernoulli brothers, Brook Taylor, and Leonhard Euler.

Chapter 12 introduces vectors, which represent both a magnitude and a direction. This chapter illustrates how vectors can be used in dealing with lines and planes in space.

Chapters 13, 16, and 17, further generalizing the derivative and integral, introduce vector analysis, which was put into its present form at the turn of the century. Vector analysis is the language of electricity, magnetism, gravitation, and fluid flow.

Some of the appendixes review precalculus topics, such as slope, absolute value, interval notation, exponents, induction, and conics. (Trigonometry is reviewed in Chap. 2 and logarithms in Chap. 6.) Some of the other appendixes are intended as references for theorems that are not usually covered in lower division calculus but are often applied. Appendix F points out the difference between "A implies B" and "B implies A," a distinction that is critical in several chapters. Read it early in the course and from time to time as the course progresses.

The reader who wishes to learn the history of calculus, which goes back to before Archimedes (287–212 B.C.), may wish to read *The long road to calculus* (pp. 310–311) and consult the references at the end of this section.

Since calculus may be the last mathematics some students meet in their studies, it should be emphasized that it is by no means all of mathematics. Algebra, combinatorics, logic, computations and algorithms, topology, analysis, and geometry are a few of the other branches of mathematics that are rich in applications. New mathematical discoveries are still being made throughout the world at the rate of over 150 new theorems or applications per day (including weekends). Since each question answered raises more questions, our ignorance grows at least as fast as our knowledge. For instance, it is known that there is no end to primes of the form $4n + 1$, where n is an integer (5, 13, 17, and 29 are four such primes). It is not known whether there is an end to primes of the form $n^2 + 1$. There remain deep questions about calculus, too complicated to state here, that have resisted the assaults of countless mathematicians.

Some research is motivated by applications in such areas as physics, biology, economics, computer science, and logic, but much of it is motivated by sheer curiosity. Surprisingly, many of the results obtained to satisfy curiosity turn out, years or decades after their discovery, to be exactly what is needed in some applied area.

In spite of the turmoil of this century, even the upheaval of two world wars and innumerable lesser conflicts, we still live in a golden age of mathematics that began some three or four centuries ago.

EXERCISES FOR SEC. 1.3: SURVEY OF THE TEXT

1 An angle of 1 radian is about 57° ($180/\pi$, to be exact). To estimate the sine of this angle, we may use the formula quoted at the beginning of this section, which gives

$$\sin x = x - \frac{x^3}{1\cdot 2\cdot 3} + \frac{x^5}{1\cdot 2\cdot 3\cdot 4\cdot 5} - \frac{x^7}{1\cdot 2\cdot 3\cdot 4\cdot 5\cdot 6\cdot 7} + \frac{x^9}{1\cdot 2\cdot 3\cdot 4\cdot 5\cdot 6\cdot 7\cdot 8\cdot 9} + \cdots.$$

What decimal estimate do you get when you use (*a*) two summands? (*b*) three summands? (*c*) four summands? (*d*) five summands? (*e*) What does your calculator give as the value of sin 1?

2 In Chap. 11 it is also shown that for $-1 \le x \le 1$, when angle is measured in radians, the angle whose tangent is x equals

$$x - \frac{x^3}{3} + \frac{x^5}{5} - \frac{x^7}{7} + \cdots.$$

(*a*) Deduce that

$$\frac{\pi}{4} = 1 - \frac{1}{3} + \frac{1}{5} - \frac{1}{7} + \cdots.$$

This means that as you alternately keep adding or subtracting the reciprocals of the odd integers the sum gets nearer and nearer to $\pi/4$.

(*b*) Use (*a*) to estimate π.

(*c*) What equation results if you replace x by $1/\sqrt{3}$ instead of by 1?

(*d*) π is the ratio of the circumference of a circle to its diameter. Draw some circles with a compass and use a tape measure to find the circumference and diameter of each. Is this a good way to find the value of π? Discuss the advantages or disadvantages of this approach compared to the one in (*b*).

(*e*) Using

$$x - \frac{x^3}{3} + \frac{x^5}{5} - \frac{x^7}{7} + \cdots$$

with $x = \frac{1}{2}$, estimate the angle whose tangent is $\frac{1}{2}$. What does your calculator give for the value of the angle whose tangent is $\frac{1}{2}$?

REFERENCES

History of Calculus

E. T. Bell, *Men of Mathematics,* Simon and Schuster, New York, 1937. (In particular, chap. 6 on Newton and chap. 7 on Leibniz.)

Carl B. Boyer, *The History of Calculus and Its Conceptual Development,* Dover, New York, 1959.

Carl B. Boyer, *A History of Mathematics,* Wiley, New York, 1968.

M. J. Crowe, *A History of Vector Analysis,* Notre Dame, 1967. (In particular, chap. 5.)

C. H. Edwards, Jr., *The Historical Development of the Calculus,* Springer-Verlag, New York, 1979. (In particular, chap. 8, "The Calculus According to Newton," and chap. 9, "The Calculus According to Leibniz.")

Morris Kline, *Mathematical Thought from Ancient to Modern Times,* Oxford, New York, 1972. (In particular, chap. 17.)

Introductions to Other Areas of Mathematics

Donald J. Albers and G. L. Alexanderson, eds., *Mathematical People,* Birkhäuser Boston, Cambridge, 1984.

Richard Courant and Herbert Robbins, *What Is Mathematics?,* Oxford, New York, 1960.

Philip J. Davis and Reuben Hersh, *The Mathematical Experience,* Birkhäuser Boston, Cambridge, 1981.

Ross Honsberger, ed., *Mathematical Plums,* Dolciani Mathematical Expositions, Number 4, Mathematical Association of America, 1979.

Morris Kline, *Mathematics in the Modern World, Readings from Scientific American,* W. H. Freeman, New York, 1968.

James R. Newman, *The World of Mathematics,* Microsoft Press, Redmond, Washington, 1989.

Lynn Arthur Steen, *Mathematics Today, Twelve Informal Essays,* Springer-Verlag, New York, 1979.

Sherman K. Stein, *Mathematics, the Man-made Universe,* W. H. Freeman, New York, 1975.

2

FUNCTIONS, LIMITS, AND CONTINUITY

Calculus is the study of change, of how rapidly or slowly one quantity changes as another varies. If you drop a rock, its height varies and you may want to know how rapidly that height is changing. The height depends on time; so does the speed. Height and speed are functions of time.

Section 2.1 develops the notion of a function, and Sec. 2.2 shows how functions can be built up from simpler functions.

Sections 2.3 and 2.4 concern limits of functions, that is, how their values behave as a variable approaches a specific number or, perhaps, gets arbitrarily large. Section 2.5 presents techniques for sketching the graph of a function, including the use of limits.

In Sec. 2.6 we pause to review the essentials of trigonometry, in particular, the functions $\sin \theta$ and $\cos \theta$ and their graphs. Then, in Sec. 2.7, we examine what happens to the quotient $(\sin \theta)/\theta$ when θ is small. This particular limit is the basis for the calculus of trigonometric functions, developed in Chap. 3.

Section 2.8 defines the continuous functions. Actually, just about all functions met in applications are continuous. We highlight the notion of continuity because it is this property that is needed in several important results in later chapters.

The chapter concludes with Secs. 2.9 and 2.10, which present a more precise definition of limit than those given in Secs. 2.3 and 2.4.

With Chap. 2 as a foundation, we will then be ready to introduce the derivative of a function in Chap. 3.

2.1 FUNCTIONS

The area A of a square depends on the length of its side x and is given by the formula $A = x^2$. See Fig. 1.

Similarly, the distance s (in feet) that a freely falling object drops in the first t seconds is described by the formula $s = 16t^2$. Each choice of t determines a unique value for s. For instance, when $t = 3$ seconds, $s = 16 \cdot 3^2 = 144$ feet.

Both these formulas illustrate the mathematical notion of a function.

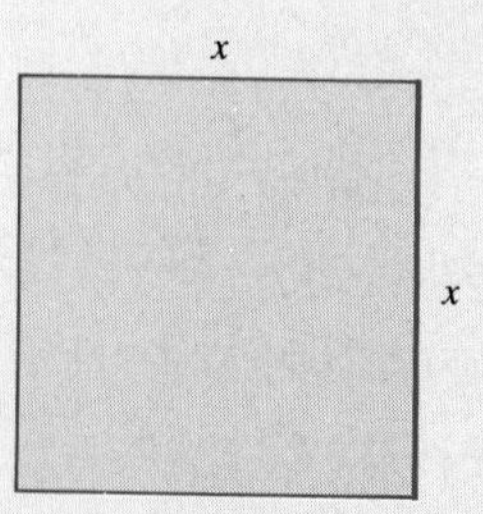

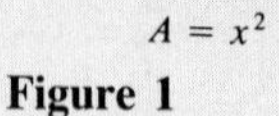
Figure 1

> **Definition** Let X and Y be sets. A **function** from X to Y is a rule (or method) for assigning one (and only one) element in Y to each element in X.

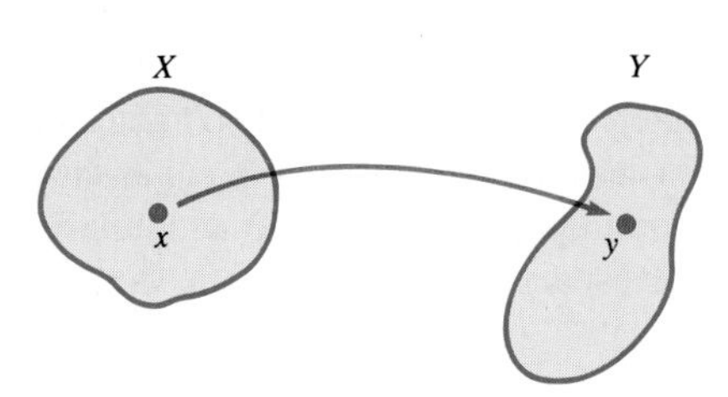

Figure 2

The notion of a function is illustrated in Fig. 2, where the element y in Y is assigned to the element x in X. Usually X and Y will be sets of numbers.

A function is often denoted by the symbol f. The element that the function assigns to the element x is denoted $f(x)$ (read "f of x"). In practice, though, almost everyone speaks interchangeably of the function f or the function $f(x)$.

If $f(x) = y$, x is called the **input** or **argument** and y is called the **output** or **value** of the function at x. Also, x is called the **independent variable** and y the **dependent variable**.

A function may be given by a formula, as is the function $A = x^2$. Since A depends on x, we say that "A is a function of x." Since A depends on only one number, x, it is called a function of a single variable. The function $s = 16t^2$ is also a function of a single variable, since s depends only on t. Throughout the first 11 chapters we confine our attention to functions of a single variable. The area A of a rectangle depends on the length l and the width w of the rectangle. It is a function of *two* variables, $A = lw$. Such functions are treated in Chap. 14.

EXAMPLE 1 Let $f(x) = x^2$ for each real number x. Compute (*a*) $f(3)$, (*b*) $f(2)$, and (*c*) $f(-2)$.

SOLUTION

(*a*) $f(3) = 3^2 = 9$.

(*b*) $f(2) = 2^2 = 4$.

(*c*) $f(-2) = (-2)^2 = 4$. ■

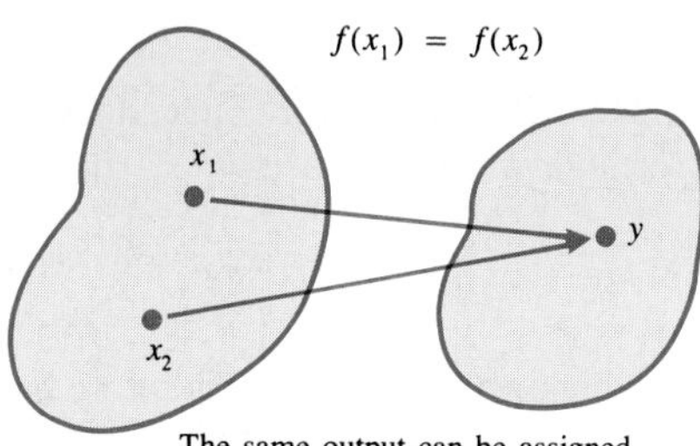

The same output can be assigned to different inputs.

Figure 3

From (*b*) and (*c*) in the example we see that two different values of x can have the same result for $f(x)$. The definition of a function said that each element of X is assigned one element of Y; it did not say that elements of Y cannot be used more than once. See Fig. 3.

Functions on a Calculator

Calculators have keys for several functions. For instance, the x^2-key in the calculator shown in Fig. 4 controls the "squaring function." When you enter

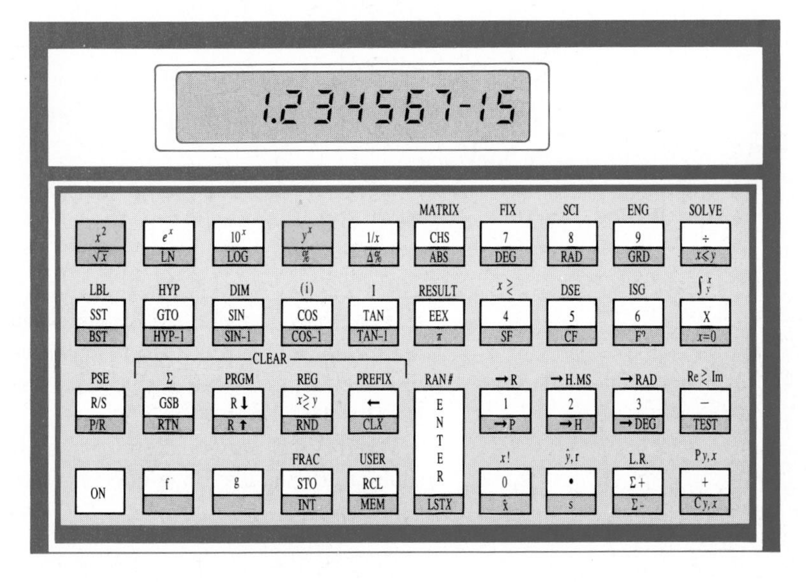

Figure 4

the input 7, for instance, and press the x^2-key, the display shows the output, 49. The y^x-key represents a function of two variables. You must enter two numbers, y and x, and the calculator displays y^x. For instance, if you enter 2 and 3, then press the y^x-key, the display will be 8. You might check which function keys your own calculator has and whether they correspond to functions of one variable or two variables.

Domain and Range

If f is a function from X to Y, we will often be concerned with the set of inputs (X) and the set of possible outputs. For this reason we give names to the two sets.

Definition **Domain** and **range**. Let X and Y be sets and let f be a function from X to Y. The set X is called the *domain* of the function. The set of all outputs of the function is called the *range* of the function.

The range is not necessarily all of Y.

When the function is given by a formula, the domain is usually understood to consist of all the numbers for which the formula is defined.

EXAMPLE 2 Find the domain and range of the function $f(x) = 1/x^2$.

SOLUTION The expression $1/x^2$ is defined whenever x is not 0. Therefore the domain of the function consists of all numbers except 0.

Domain

To find the range of the function, we must describe the set of all possible outputs. A number b is an output if we can find a number x such that $b = f(x)$, that is,

$$b = \frac{1}{x^2}.$$

In other words, b is an output if we can solve the equation $b = 1/x^2$ for x. If $b = 1/x^2$, we have, by algebra,

$$x^2 b = 1$$

$$x^2 = \frac{1}{b}$$

$$x = \pm\sqrt{\frac{1}{b}}.$$

If b is negative, there is no (real) solution, since the negative number $1/b$ does not have a square root. If b is positive, it is in the range of the function. For instance, with $b = 9$, $x = \pm\sqrt{\frac{1}{9}} = \pm\frac{1}{3}$. (As a check, $f(\frac{1}{3}) = 1/(\frac{1}{3})^2 = 9$ and $f(-\frac{1}{3}) = 1/(-\frac{1}{3})^2 = 9$.

When $b = 0$, the equation $b = 1/x^2$ becomes $0 = 1/x^2$. This equation has no solution. Thus the range of $f(x) = 1/x^2$ consists of precisely the positive numbers. ■

Range

Example 2 illustrates how to find the domain and range of a function given by a formula $y = f(x)$.

> The domain consists of all x such that $f(x)$ is defined. So exclude those x for which the formula makes no sense (requiring, for instance, division by 0 or the square root of a negative number). The range consists of all numbers b such that the equation $b = f(x)$ has at least one solution.

Domain on a calculator

When using a calculator, you must pay attention to the domain corresponding to a function key. If you enter a negative number as x and press the $\sqrt{x}$-key to calculate the square root of x, your calculator will probably not be happy. It might display an E for "error" or start flashing on and off, the standard signal for distress. Your error was entering a number not in the domain of the square root function. (Try it. What does your calculator do? Some advanced calculators go into "complex number" mode to handle square roots of negative numbers.) You can also get into trouble if you enter 0 and press the $1/x$-key. (Try it. *No* calculator, however advanced, can permit division by zero.) The domain of the function $1/x$ consists of all numbers except 0.

EXAMPLE 3 Find the domain and range of the function given by the formula

$$f(x) = x^2.$$

The notation for intervals is in Appendix A.

SOLUTION First, let's find the domain. Since x^2 is defined for every real number, the domain of the squaring function x^2 consists of all real numbers, that is, the interval $(-\infty, \infty)$.

The range of f consists of all real numbers that are of the form x^2 for some real number x. The number 9 is in the range since $9 = 3^2$; the number 5 is in the range since $5 = (\sqrt{5})^2$. Zero is also in the range since $0 = 0^2$. In fact, every nonnegative real number b is in the range since the equation $b = x^2$ can be solved: $x = \sqrt{b}$ or $-\sqrt{b}$.

However, no negative number is in the range, since no negative number is the square of a real number [-9 is not 3^2 and it is not $(-3)^2$ either].

Thus the range consists precisely of the nonnegative real numbers, that is, the interval $[0, \infty)$. ■

EXAMPLE 4 Find the domain and range of $f(x) = 2 + \sqrt{x - 1}$.

SOLUTION For $2 + \sqrt{x-1}$ to be meaningful, the square root of $x - 1$ must make sense; that is, you must be able to take the square root of $x - 1$. Negative numbers do not have (real) square roots; 0 and positive numbers do have square roots. Thus, the domain consists of all numbers x such that

$$x - 1 \geq 0,$$

or, equivalently,

$$x \geq 1.$$

That is, the domain is the interval $[1, \infty)$.

As x varies from 1 to larger numbers, $f(x)$ increases from $f(1) = 2 + \sqrt{1 - 1}$

to arbitrarily large values. Thus the range of f consists of all numbers greater than or equal to 2, that is, the interval $[2, \infty)$. ■

EXAMPLE 5 Find the domain and range of

$$f(x) = \frac{1}{\sqrt{x^2 - 1}}.$$

SOLUTION The denominator is 0 when x is 1 or -1, so the domain of f does not contain 1 or -1. In order for $\sqrt{x^2 - 1}$ to be meaningful, we must have

$$x^2 - 1 \geq 0$$

or, equivalently,

$$x^2 \geq 1.$$

So, if $x^2 < 1$, x is *not* in the domain.

All told, the set of prohibited x is the closed interval $[-1, 1]$. The domain consists of all numbers except $[-1, 1]$. In other words, the domain consists of the two intervals $(-\infty, -1)$ and $(1, \infty)$.

What is the range of f? It is all outputs of the form $1/\sqrt{x^2 - 1}$ as x runs through the domain. The denominator $\sqrt{x^2 - 1}$ is always positive. It can be very small or very large. Thus $1/\sqrt{x^2 - 1}$ can be arbitrarily large or as near 0 as we please. The range therefore consists of all positive numbers; it is the interval $(0, \infty)$. ■

Graph of a Function

In case both the inputs and outputs of a function are numbers, we can draw a picture of the function, called its **graph**.

> **Definition** *Graph of a function.* Let f be a function whose inputs and outputs are numbers. The **graph** of f consists of those points (x, y) such that $y = f(x)$.

The next example shows how to use a table of inputs and outputs to graph a function.

EXAMPLE 6 Graph the function $f(x) = 1/(1 + x^2)$.

SOLUTION Since $1 + x^2$ is never 0, the domain of the function consists of the entire x axis. Pick a few convenient inputs and calculate the corresponding outputs, as shown in this table:

x	0	1	2	3	-1	-2	-3
$f(x) = 1/(1 + x^2)$	1	$\frac{1}{2}$	$\frac{1}{5}$	$\frac{1}{10}$	$\frac{1}{2}$	$\frac{1}{5}$	$\frac{1}{10}$

For any x, $x^2 \geq 0$, so $1 + x^2 \geq 1$ and $1/(1 + x^2) \leq 1$.

Plotting the seven calculated points suggests the general shape of the graph, shown in Fig. 5. (The dots record the seven points.) When x or $-x$ is large,

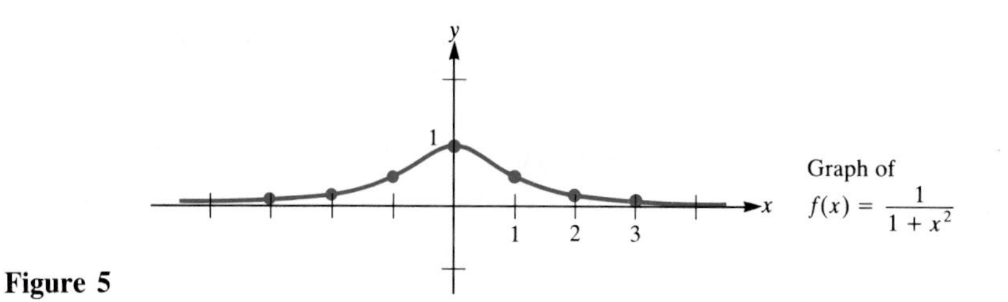

Graph of $f(x) = \frac{1}{1 + x^2}$

Figure 5

$y = 1/(1 + x^2)$ is small. This means that far to the right or left the graph approaches the x axis.

See Appendix B for more on graphs.

Note that the range of the function consists of all positive numbers less than or equal to 1, that is, the interval $(0, 1]$. ■

Nation	*Population in 1990 (in millions)*
China	1120
India	832
Soviet Union	291
United States	250
Indonesia	178
Brazil	150
Japan	123
Bangladesh	115
Pakistan	113
Nigeria	113
Mexico	89
Germany	80
Vietnam	65
Phillipines	61
Italy	57
Turkey	56
United Kingdom	56
Thailand	56
France	55

Ways of Representing a Function

In daily life we are surrounded by functions, but they are seldom given by a mathematical formula. Sometimes a function is given by a table that lists the inputs and their corresponding outputs. For example, the *World Almanac* lists the population of 103 nations. In this case the input x is a nation and the output $f(x)$ is the number giving the size of its population. Part of the data is shown in the table in the margin in descending size. For instance, $f(\text{China}) = 1120$ (millions).

A function can also be described pictorially, by a graph that summarizes numerical data, as in Fig. 6. The horizontal line gives a scale for inputs, the vertical for outputs.

The "accidents" curve records the variation in the number of fatal accidents as a function of the time of day. At time t, the output $f(t)$ is the percent of accidents that occur within half an hour of t. The curve labeled "traffic" records the percent of the day's total traffic within half an hour of time t. Figure 6 is based on S. K. Stein, "Risk Factors of Sober and Drunk Drivers by Time of Day," *Alcohol, Drugs, and Driving,* vol. 5, no. 3, 1989.

Not every curve is the graph of a function. For instance, the curve in Fig. 7 is not the graph of a function. The reason is that a function assigns to a given input a *single* number as the output.

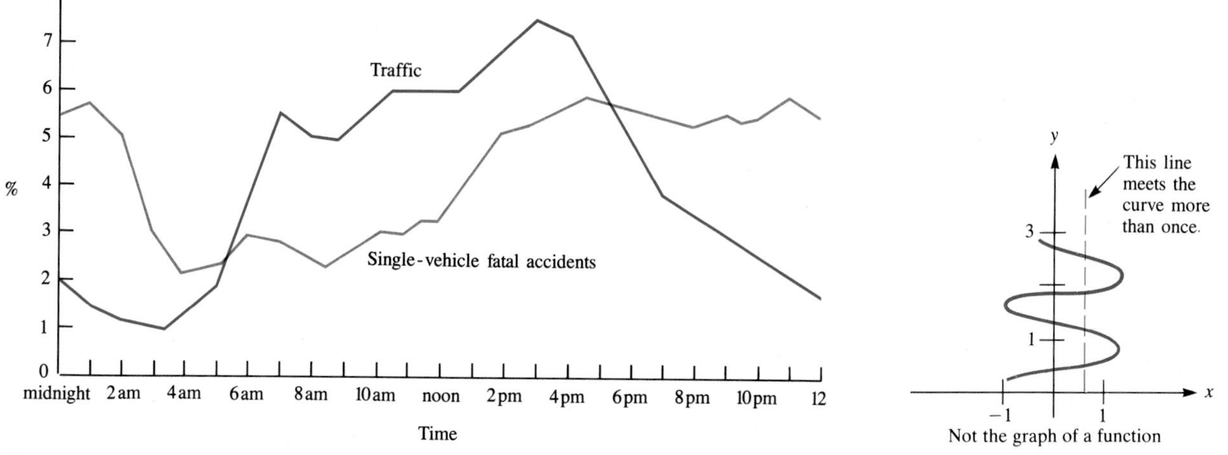

Figure 6

Not the graph of a function

Figure 7

A line parallel to the y axis therefore meets the graph of a function in at most one point. This observation provides a visual test for deciding whether a curve in the xy plane is the graph of a function $y = f(x)$. If some line parallel to the y axis meets the curve more than once, then the curve is *not* the graph of a function. Otherwise it is. No vertical line meets the curve in Fig. 8 more than once, so it is the graph of a function.

The "vertical line" test for a function

If you look at Fig. 8 and want to find $f(1)$, you can say that it is about 2.1. But if you look at Fig. 7 to find the output when the input is 1, you find only confusion: Is the output equal to 0.8, 1.3, 2.1, or 2.6? Such ambiguity prevents the curve in Fig. 7 from being the graph of a function.

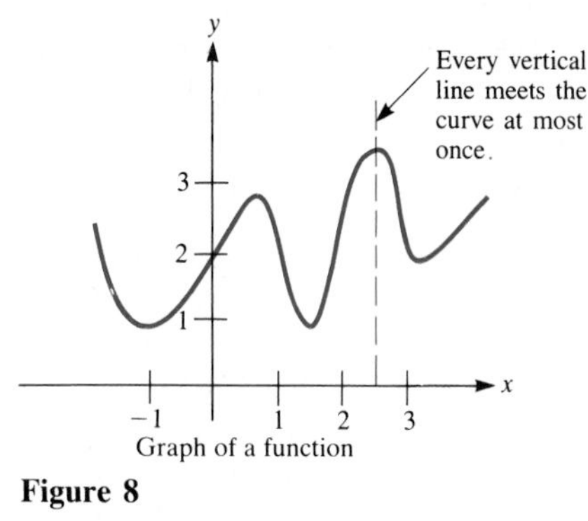

Figure 8

On several occasions it will be illuminating to picture a function f as a projection from a slide to a screen. Both the slide and screen are lines. The function, through some ingenious lens, projects the point on the slide given by the number x to the point on the screen given by the number $f(x)$, as shown in Fig. 9.

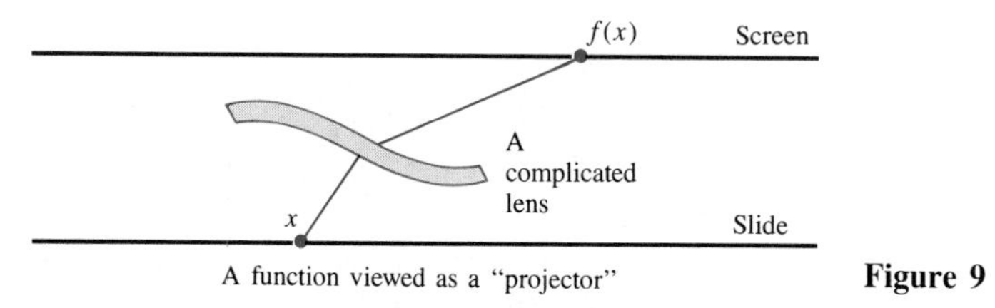

A function viewed as a "projector"

Figure 9

Practice with Functions

We shall often need to examine how the output of a function changes as we change the input. To prepare for the algebraic manipulations, we pause to illustrate some of the techniques.

EXAMPLE 7 Let f be the squaring function $f(x) = x^2$. Compute (*a*) $f(2 + 3)$ and (*b*) $f(2 + h)$.

SOLUTION

(*a*) For any number x, $f(x)$ is the square of that number. Thus

$$f(2 + 3) = (2 + 3)^2 = 5^2 = 25.$$

(*b*) Similarly, $$f(2 + h) = (2 + h)^2 = 4 + 4h + h^2. \quad ■$$

Warning about the value of a function when the input is a sum

Warning: A common error is to assume that $f(2 + 3)$ is somehow equal to $f(2) + f(3)$. For most functions there is no relation between the two numbers. In the case of the function x^2, $f(2) + f(3) = 2^2 + 3^2 = 4 + 9 = 13$, but $f(2 + 3) = f(5) = 25$.

The difference also shows up when you use a calculator. For example, to compute $f(2 + 3)$ on a Hewlett-Packard calculator, where f is given by the x^2-key you may press

$$2, \text{Enter}, 3, +, x^2,$$

but to compute $f(2) + f(3)$, you would press

$$2, x^2, 3, x^2, +.$$

The sequence of commands is very different in the two cases.

Advice: If you are careful and methodical, you do not need this warning. Just do what the function tells you. Each symbol in something like $f(2 + 3)$ matters, including the parentheses "(" and ")". Perhaps read it aloud so you know it is $f(2 + 3)$: "f of the quantity 2 plus 3"—rather than something else, such as $f(2) + 3$, which is read aloud as "f of 2 (*pause*) plus 3." If you have done any programming, you know the importance of every comma, semicolon, and so on. Treat formulas with just as much respect.

EXAMPLE 8 Let f be the cubing function $f(x) = x^3$. Evaluate the difference $f(2 + 0.1) - f(2)$.

SOLUTION
$$\begin{aligned} f(2 + 0.1) - f(2) &= f(2.1) - f(2) \\ &= (2.1)^3 - 2^3 \\ &= 9.261 - 8 \\ &= 1.261. \quad \blacksquare \end{aligned}$$

In Examples 7 and 8 the inputs are specific numbers. Often the inputs will be indicated by an algebraic expression instead. The next example shows how to deal with such inputs.

EXAMPLE 9 Let f be the squaring function $f(x) = x^2$. Simplify the expression $f(a + b) - f(a) - f(b)$ as far as possible.

SOLUTION
$$\begin{aligned} f(a + b) - f(a) - f(b) &= (a + b)^2 - a^2 - b^2 \\ &= a^2 + 2ab + b^2 - a^2 - b^2 \\ &= 2ab. \quad \blacksquare \end{aligned}$$

Functions from Geometry

In calculus, functions often originate in geometric or physical problems. The next example is typical.

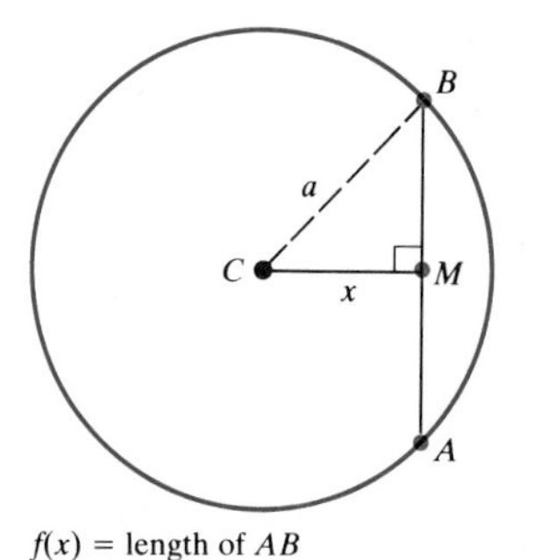

$f(x)$ = length of AB

Figure 10

EXAMPLE 10 Consider a circle of radius a, as shown in Fig. 10. Let $f(x)$ be the length of a chord AB of this circle at a distance x from the center of the circle. Find a formula for $f(x)$.

SOLUTION We are trying to find how the length $\overline{AB}$ varies as x varies. That is, we are looking for a *formula* for $\overline{AB}$, the length of AB, in terms of x.

Before searching for the formula, it is a good idea to calculate $f(x)$ for some easy inputs. These values can serve as a check on the formula we work out. In this case $f(0)$ and $f(a)$ can be read at a glance from Fig. 10: $f(0) = 2a$ and $f(a) = 0$. (Why?) Now let us find $f(x)$ for all x in $[0, a]$.

Let M be the midpoint of the chord AB and let C be the center of the circle. Observe that $\overline{CM} = x$ and $\overline{CB} = a$. By the pythagorean theorem, $\overline{BM} = \sqrt{a^2 - x^2}$. Hence $\overline{AB} = 2\sqrt{a^2 - x^2}$. Thus

$$f(x) = 2\sqrt{a^2 - x^2}.$$

As a check, note that $f(0) = 2\sqrt{a^2 - 0^2} = 2a$, which is correct. Similarly, $f(a) = 2\sqrt{a^2 - a^2} = 0$, which is also correct. The domain of the function f in our geometric setting is the interval $[0, a]$. (Although $\sqrt{a^2 - x^2}$ is defined for some negative inputs, that is irrelevant in this case.) ■

Section Summary

We discussed the notion of *function*, including *input, output, domain,* and *range*. If both the domain and range of f consist of real numbers, then f can be graphed on an xy coordinate system. We pointed out that a function may be given by a formula, a table, a picture, or geometrically.

EXERCISES FOR SEC. 2.1: FUNCTIONS

In Exercises 1 to 16 graph the functions.

1 $f(x) = 3x$
2 $f(x) = x/2$
3 $f(x) = -2x + 3$
4 $f(x) = 4x - 5$
5 $f(x) = 3x^2 + 1$
6 $f(x) = -2x^2 + 1$
7 $f(x) = \frac{1}{2}x^2 - 4$
8 $f(x) = \frac{1}{3}x^2 + 5$
9 $f(x) = x^2 - x$
10 $f(x) = 2x^2 + x$
11 $f(x) = x^2 + x + 1$
12 $f(x) = x^2 - x + 2$
13 $f(x) = 1/(2 + x^2)$
14 $f(x) = 1/(2x^2 + 1)$
15 $f(x) = x^2/(1 + 2x^2)$
16 $f(x) = 2x/(3 + x^2)$

In Exercises 17 to 28 give the domain and range of each of the functions.

17 $f(x) = \sqrt{x}$
18 $f(x) = \sqrt{2x}$
19 $f(x) = \sqrt{x + 1}$
20 $f(x) = \sqrt{x - 1}$
21 $f(x) = \sqrt{4 - x^2}$
22 $f(x) = \sqrt{x^2 - 4}$
23 $f(x) = 1/x$
24 $f(x) = 2/x$
25 $f(x) = 1/(x + 1)$
26 $f(x) = 1/(x - 2)$
27 $f(x) = 1/(1 - x^2)$
28 $f(x) = 1/(2 + x^2)$

In each of Exercises 29 to 32 compute as decimals the outputs of the given functions for the given inputs.

29 $f(x) = x + 1$ (*a*) -1 (*b*) 3 (*c*) 1.25 (*d*) 0

30 $f(x) = 1/(1 + x)$ (*a*) -3 (*b*) 3 (*c*) 9 (*d*) 99

31 $f(x) = x^3$ (*a*) $1 + 2$ (*b*) $4 - 1$

32 $f(x) = 1/x^2$ (*a*) $5 - 3$ (*b*) $4 - 6$

33 Let $f(x) = x^2$. Evaluate

$$\frac{f(3 + h) - f(3)}{h}$$

to four decimal places for h equal to (*a*) 1, (*b*) 0.01, (*c*) -0.01, and (*d*) 0.0001. (*e*) What do you think happens to $[f(3 + h) - f(3)]/h$ as h gets smaller and smaller?

34 Let $f(x) = x^3$. Evaluate

$$\frac{f(2 + h) - f(2)}{h}$$

to four decimal places for h equal to (*a*) 1, (*b*) 0.05, (*c*) 0.0001, and (*d*) -0.001. (*e*) What do you think happens to $[f(2 + h) - f(2)]/h$ as h gets smaller and smaller?

In Exercises 35 to 38 evaluate and simplify the given expressions for the given functions. (Assume that no denominator is 0.)

35 $f(x) = x^3$; $f(a + 1) - f(a)$

36 $f(x) = 1/x$; $f(a + h) - f(a)$

37 $f(x) = \dfrac{1}{x^2}$; $\dfrac{f(d) - f(c)}{d - c}$

38 $f(x) = \dfrac{1}{2x + 1}$; $\dfrac{f(x + h) - f(x)}{h}$

In Exercises 39 and 40 which curves are the graphs of functions and which are not?

39

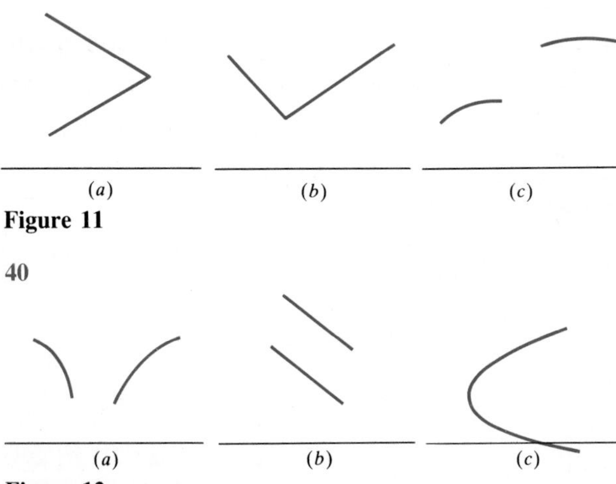

Figure 11

40

Figure 12

In Exercises 41 to 50 state the formula for the function f and give the domain of the function.

41 $f(x)$ is the perimeter of a circle of radius x.

42 $f(x)$ is the area of a circle of radius x.

43 $f(x)$ is the perimeter of a square of side x.
44 $f(x)$ is the volume of a cube of side x.
45 f is the total surface area of a cube of side x.
46 $f(x)$ is the length of the hypotenuse of the right triangle whose legs have lengths 3 and x.
47 $f(x)$ is the length of the side AB in the triangle in Fig. 13.

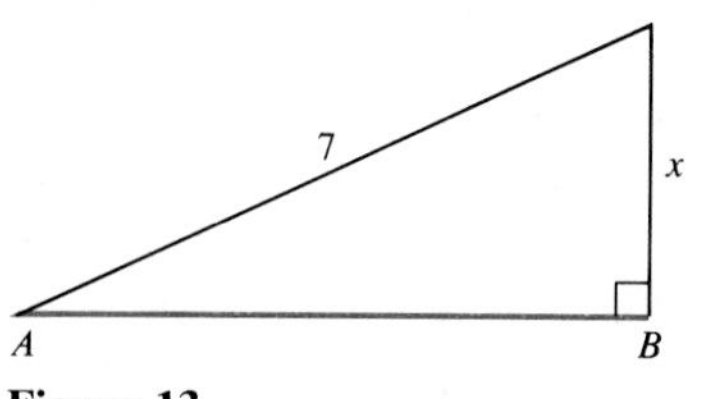

Figure 13

48 For $0 \le x \le 4$, $f(x)$ is the length of the path from A to B to C in Fig. 14.

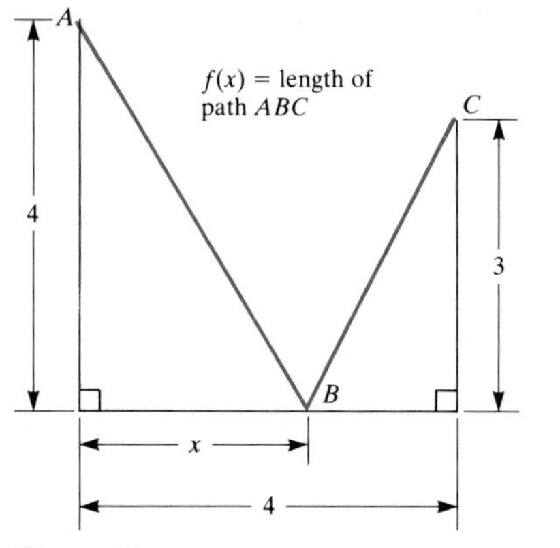

Figure 14

49 For $0 \le x \le 10$, $f(x)$ is the perimeter of the rectangle $ABCD$, one side of which has length x, inscribed in the circle of radius 5 shown in Fig. 15.

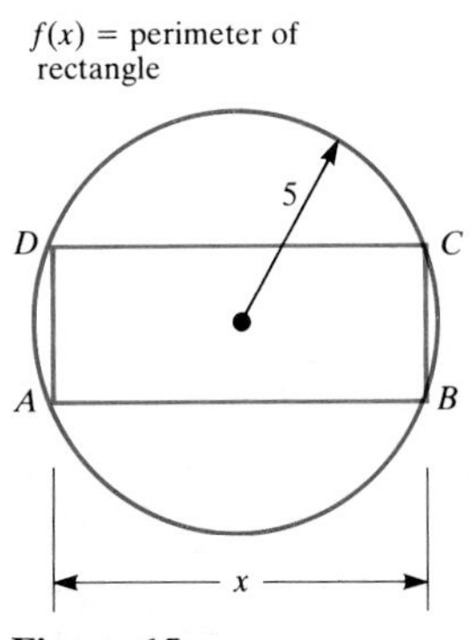

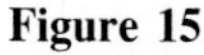
Figure 15

50 (See Fig. 16.) A person at point A in a lake is going to swim to the shore ST and then walk to point B. She swims at 1.5 miles per hour and walks at 4 miles per hour. If she reaches the shore at point P, x miles from S, let $f(x)$ denote the time for her combined swim and walk. Obtain an algebraic formula for $f(x)$.

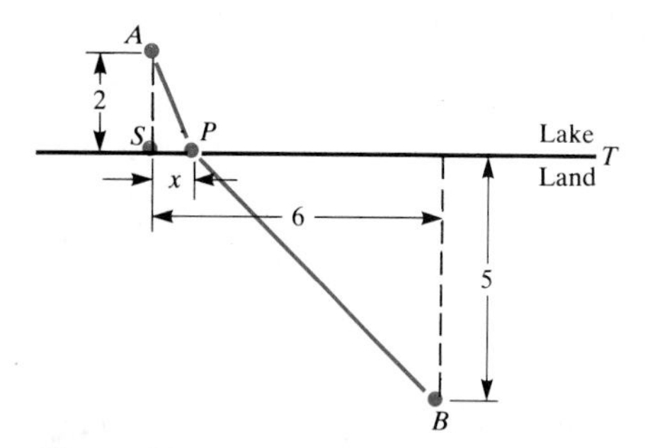

Figure 16

51 Find the formula for the function $g(x)$ that equals the length of the path APB in Fig. 16.

52 Let $f(x)$ be the length of the segment AB in Fig. 17.
(*a*) What are $f(0)$ and $f(a)$?
(*b*) What is $f(a/2)$?
(*c*) Find the formula for $f(x)$ and explain your solution.

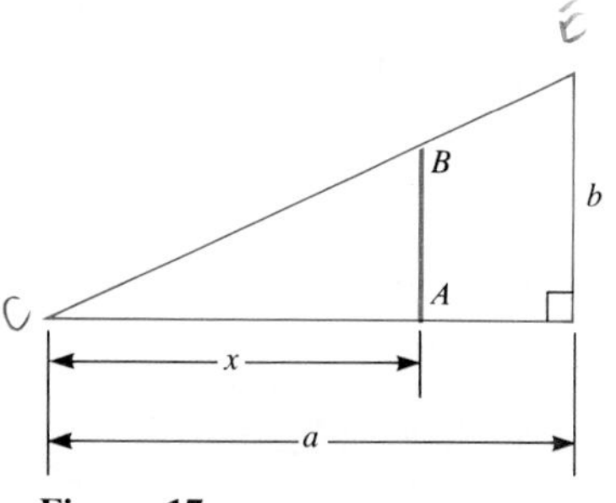

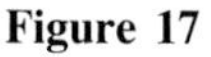
Figure 17

53 Let $f(x)$ be the area of the right circular cone cross section in Fig. 18.
(*a*) What are $f(0)$ and $f(h)$?
(*b*) Find a formula for $f(x)$ and explain your solution.

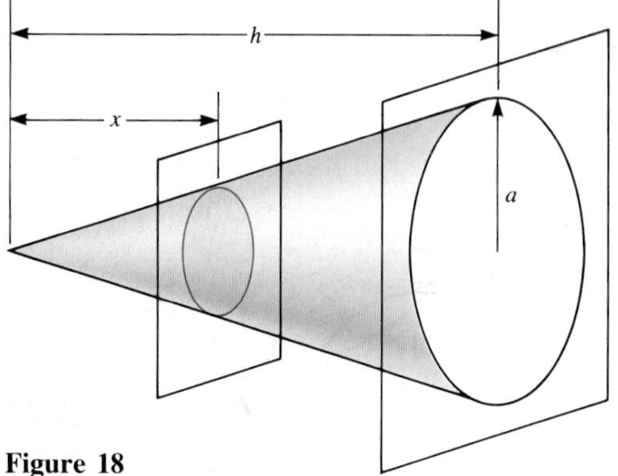

Figure 18

Exercises 54 and 55 refer to Fig. 6 in this section. Let $f(t)$ be the traffic function and $g(t)$ be the accident function.

54 (*a*) At what time is $f(t)$ at a maximum? a minimum?
(*b*) At what time is $g(t)$ at a maximum? a minimum?

55 (*a*) What expression defined in terms of f and g measures the danger or risk of driving as a function of time of day?
(*b*) On the basis of Fig. 6, estimate what is the most dangerous time to drive and what is the safest. How many times as risky is the first in comparison with the second?

56 Graph $f(x) = x(x + 1)(x - 1)$.
(*a*) For which values is $f(x) = 0$?
(*b*) Where does the graph cross the x axis?
(*c*) Where does the graph cross the y axis?

57 A complicated lens projects a linear slide to a linear screen as shown in Fig. 19, which indicates the paths of four of the light rays. Let $f(x)$ be the image on the screen of x on the slide.
(*a*) What are $f(0)$, $f(1)$, $f(2)$, and $f(3)$?
(*b*) Fill in this table:

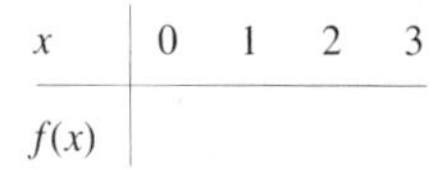

x	0	1	2	3
$f(x)$				

(*c*) Plot the four points in (*b*).
(*d*) Which is larger, $f(3) - f(2)$ or $f(2) - f(1)$?

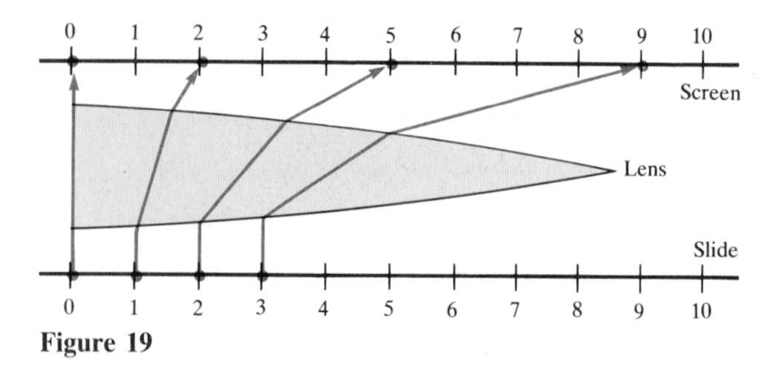

Figure 19

58 For which of the following functions is $f(a + b)$ equal to $f(a) + f(b)$ for all positive numbers a and b?
(*a*) $f(x) = x^2$
(*b*) $f(x) = 3x$
(*c*) $f(x) = -4x$
(*d*) $f(x) = \sqrt{x}$
(*e*) $f(x) = 2x + 1$

59 For which of the following functions is $f(ab)$ equal to $f(a)f(b)$ for all positive numbers a and b?
(*a*) $f(x) = 3x$
(*b*) $f(x) = x^3$
(*c*) $f(x) = 1/x$
(*d*) $f(x) = \sqrt{x}$
(*e*) $f(x) = x + 1$

In Exercises 60 to 62 give three examples of numerical functions f that meet the given condition.

60 $f(-x) = 1/f(x)$ for all numbers x

61 $f(x + 1) = 2f(x)$ for all numbers x

62 $f(xy) = f(x)f(y)$ for all numbers x and y

✱ **63** Find all the functions with a domain equal to the positive integers, 1, 2, 3, . . . , and with real numbers as outputs such that $f(x + y) = f(x) + f(y)$ for all positive integers x and y.
(*a*) Give at least three examples of such functions.
(*b*) Make a conjecture.
(*c*) Prove it.

64 The following table records two functions of time t: the number of sober drivers during the hour beginning at the indicated time and also a scaled quantity representing the number of single-vehicle fatal accidents involving sober drivers.

Time	*Traffic*	*Accidents*
Midnight	1343	76
1	936	90
2	770	66
3	800	58
4	957	65
5	1674	103
6	3492	103
7	5335	109
8	4850	114
9	4753	131
10	5194	130
11	5586	124
Noon	5723	119
1	5664	140
2	5856	152
3	6460	131
4	7030	153
5	6745	118
6	5452	112
7	4464	83
8	3420	101
9	2904	86
10	2376	90
11	1826	74

Source: Derived from S. K. Stein, "Risk Factors of Sober and Drunk Drivers by Time of Day," *Alcohol, Drugs, and Driving,* vol. 5, no. 3, 1989.

(*a*) When are there the most fatal accidents?
(*b*) When are the most sober drivers on the road?
(*c*) Determine during which hour sober drivers are at the greatest risk of a single-car fatal accident. (*Hint:* "Risk" should be in terms of accidents per driver.)
(*d*) Determine during which hour sober drivers are at the least risk of a single-car fatal accident.
(*e*) How many times as dangerous is the most risky hour relative to the least risky hour?
(*f*) Graph "risk" as a function of time.

2.2 COMPOSITE FUNCTIONS

Many functions are built up by applying one function to the output of another function. For instance, the function

$$y = (1 + x^2)^{100}$$

is built up by raising $1 + x^2$ to the hundredth power. That is,

$$y = u^{100}, \qquad \text{where } u = 1 + x^2.$$

This section develops this concept, called *composition of functions,* which plays a key role in Chap. 3.

Composite Functions

A few important functions are built into most calculators. For instance, the *squaring function,* with its key usually labeled "x^2," and the *reciprocal function,* with a key labeled "$1/x$." As you know, to obtain 3^2 you can press the key labeled "3" followed by the x^2-key. Similarly, $\frac{1}{3}$ is displayed when you press "3" followed by the $1/x$-key. Scientific calculators also have trigonometric, exponential, and logarithmic functions.

You can create a composite function by pressing more than one function key. For example, the function $1/x^2$ is built up in steps; see the table.

Step	*Result*
1 Take an input	x
2 Square it	x^2
3 Take its reciprocal	$1/x^2$

In short, you first apply the squaring function and then apply the reciprocal function to the initial result.

For instance, if the input x is 5 we obtain

$$5 \xrightarrow[\text{square}]{} 25 \xrightarrow[\text{reciprocal}]{} \tfrac{1}{25} = 0.04.$$

Of course, if you start with the input $x = 0$, you will run into division by 0. (The squaring function works, but the reciprocal function cannot.) Try this case on your calculator. What does it do?

This brings us to the definition of a composite function.

Definition *Composition of functions.* Let X, Y, and Z be sets (perhaps the same set). Let g be a function from X to Y and let f be a function from Y to Z. Then the function that assigns to each element x in X the element

$$f(g(x))$$

in Z is called the **composition** of f and g. It is denoted $f \circ g$. ($f \circ g$ is read as "f circle g" or as "f composed with g.")

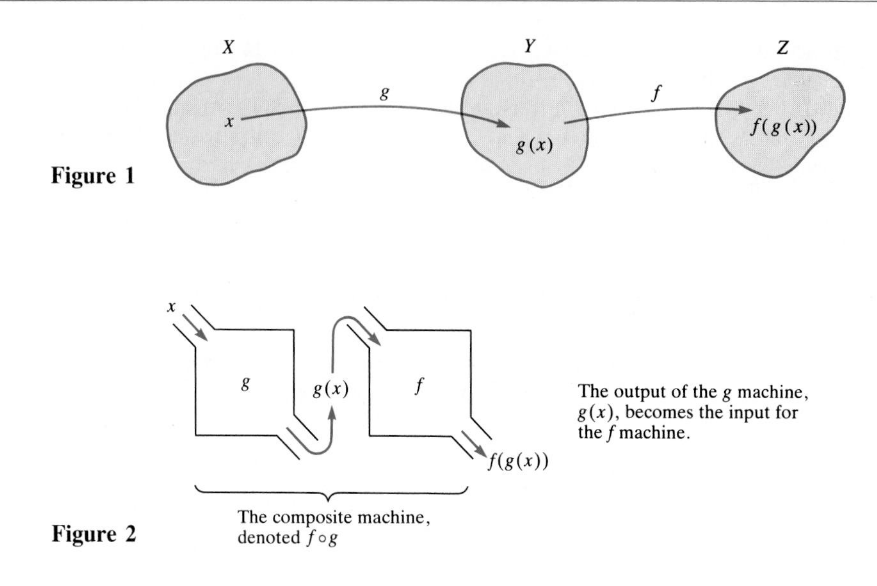

Figure 1

Figure 2

Figure 1 depicts the notion of a composite function.

If $g(x) = u$ and $f(u) = y$, then $(f \circ g)(x) = y$. In practical terms, the definition says: ''To compute $f \circ g$, first apply g and then apply f to the result.''

Thinking of functions as input-output machines, we may consider $f \circ g$ as the machine built by hooking the machine for f onto the machine for g, as shown in Fig. 2.

It is also instructive to think of composite functions in terms of projections from slides to screens. If g is interpreted as some complicated projection from a slide to a screen and f as a projection from that screen to a second screen, then $f \circ g$ is the projection from the slide to the second screen, as in Fig. 3.

Before looking at the solution of Example 1, carry out the necessary computations on your calculator.

EXAMPLE 1 Let f be the squaring function and g the ''add 3'' function, that is, $f(x) = x^2$ and $g(x) = x + 3$. Find (*a*) $f(g(5))$ and (*b*) $g(f(5))$.

SOLUTION

(*a*) $$5 \xrightarrow[\text{add 3}]{} 8 \xrightarrow[\text{square}]{} 64$$

(*b*) $$5 \xrightarrow[\text{square}]{} 25 \xrightarrow[\text{add 3}]{} 28 \quad \blacksquare$$

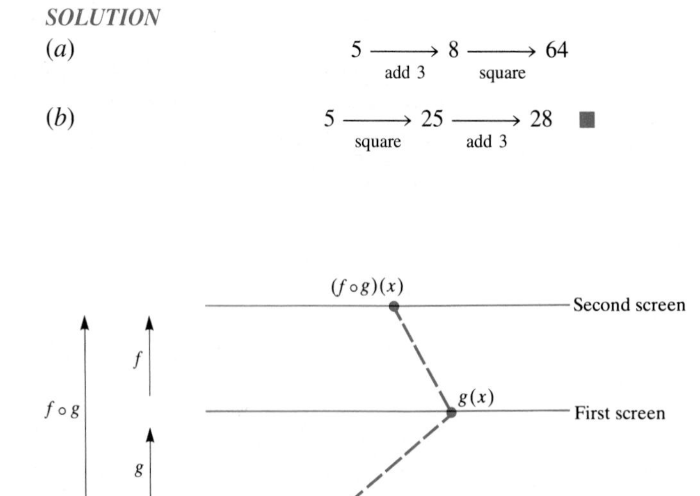

Figure 3

In Example 1, $f(x) = x^2$ and $g(x) = x + 3$. We have

$$f(g(x)) = f(x + 3) = (x + 3)^2$$

and

$$g(f(x)) = g(x^2) = x^2 + 3.$$

Both $(x + 3)^2$ and $x^2 + 3$ are obtained from f and g by composition. Clearly, the order in which we apply f and g is important.

EXAMPLE 2 Show how the function $\sqrt{1 + x^2}$ is built up by the composition of functions.

SOLUTION The function $\sqrt{1 + x^2}$ is obtained by applying the square-root function to the function $1 + x^2$. To be specific, let

$$g(x) = 1 + x^2 \quad \text{and} \quad f(u) = \sqrt{u}.$$

Then

$$f(g(x)) = f(1 + x^2)$$
$$= \sqrt{1 + x^2}. \quad \blacksquare$$

The next example shows that a function can be the composition of more than two functions.

EXAMPLE 3 Express $1/\sqrt{1 + x^2}$ as a composition of three functions.

SOLUTION Call the input x.

First, we compute the output $1 + x^2$.
Second, we take the square root of that output, getting $\sqrt{1 + x^2}$.
Third, we take the reciprocal of that result, getting $1/\sqrt{1 + x^2}$. In short, we form

$$u = 1 + x^2, \quad \text{then } v = \sqrt{u}, \quad \text{then } y = \frac{1}{v}.$$

We first apply the function $1 + x^2$, then the square-root function, then the reciprocal function. $\blacksquare$

The next example involves a rearrangement of the functions in Example 3.

EXAMPLE 4 What function do you get if you carry out the following steps in the indicated order?

Input x.
First, take the reciprocal.
Second, square and add 1.
Third, take the square root.

SOLUTION We use a diagram to work through the above steps:

$$x \xrightarrow[\text{reciprocal}]{} \frac{1}{x} \xrightarrow[\text{square and add 1}]{} \left(\frac{1}{x}\right)^2 + 1 \xrightarrow[\text{square root}]{} \sqrt{\left(\frac{1}{x}\right)^2 + 1}.$$

As we see, the resulting function is $\sqrt{(1/x)^2 + 1}$. ■

EXAMPLE 5 Using a calculator, compute (*a*) $\sqrt{1 + 5^2}$, (*b*) $(1 + \sqrt{5})^2$.

SOLUTION
(*a*) $\sqrt{1 + 5^2} = \sqrt{1 + 25} = \sqrt{26} \approx 5.099$.
(*b*) $(1 + \sqrt{5})^2 \approx (1 + 2.2361)^2 \approx 10.47$. ■

EXAMPLE 6 Let f be the cubing function, $f(x) = x^3$, and g the cube root function, $g(x) = \sqrt[3]{x}$. Compute $(f \circ g)(x)$ and $(g \circ f)(x)$.

SOLUTION

$$(f \circ g)(x) = f(g(x)) = f(\sqrt[3]{x}) = (\sqrt[3]{x})^3 = x.$$

$$(g \circ f)(x) = g(f(x)) = g(x^3) = \sqrt[3]{x^3} = x. \quad ■$$

In Example 6, f and g undo or reverse the effect of each other. For instance, $f(2) = 2^3 = 8$ and $g(8) = \sqrt[3]{8} = 2$. Two functions related this way are said to be "inverses" of each other. The concept of inverse functions will be developed in Sec. 6.4, for use in computing derivatives in later sections.

Traveler's Advisory We must be especially careful when composing functions when one of the functions is a trigonometric function. For instance, what is meant by "sin x^3"? Is it sin (x^3) or $(\sin x)^3$? Do we first cube x, then take the sine, or the other way around? There is general agreement that $\sin x^3$ stands for $\sin (x^3)$; you cube first, then take the sine.

Spoken aloud, $\sin x^3$ is usually "sine of x cubed," which is ambiguous. We can either insert a brief pause—"sine of (pause) x cubed"—to emphasize that x is cubed rather than sine x, or rephrase it as "sine of the cube of x." A reasonable, but yet more cumbersome alternative is "sine of the quantity x cubed."

On the other hand, $(\sin x)^3$, which is by convention usually written as $\sin^3 x$, is spoken aloud as "the cube of sin x" or "sine cubed of x."

Similar warnings apply to other trigonometric functions and logarithmic functions.

Section Summary

This section showed how we can build up functions by applying one function after another. Some function g takes an input x and produces an output u. Another function f then uses u as an input and produces an output y. Thus y is a function of x, which we may write as $y = f(g(x))$ or $y = (f \circ g)(x)$. The order in which we apply the functions usually affects the results.

EXERCISES FOR SEC. 2.2: COMPOSITE FUNCTIONS

In Exercises 1 to 6 state which keys you would press on your calculator to evaluate the given composite functions. List the keys in correct order. (There may be more than one correct answer, and it depends on your calculator.)

1 $\sin x^2$ at $x = 3$

2 $(\sin x)^2$ at $x = 3$

3 $\sqrt{1 + x^3}$ at $x = 4$

4 $(\sqrt{1 + x})^3$ at $x = 4$

5 $\sqrt[3]{\cos x^2}$ at $x = 2$

6 $(\cos \sqrt[3]{x})^2$ at $x = 2$

In Exercises 7 to 18 express y as a function of x.

7 $y = u^2$, $u = 1 + x$
8 $y = u^2$, $u = 2x + 1$
9 $y = 1/u$, $u = x^3$
10 $y = 1/u$, $u = 2x^2 - 3$
11 $y = u^2$, $u = x^3$
12 $y = u^2$, $u = \sqrt{x}$
13 $y = \sqrt{u}$, $u = \cos x$
14 $y = \sin u$, $u = \sqrt{x}$
15 $y = u^3$, $u = 1 + v$, $v = \sqrt{x}$
16 $y = u^2$, $u = v^2 - 1$, $v = 2x$
17 $y = \cos u$, $u = 1 + v^2$, $v = \tan x$
18 $y = u^3$, $u = \sin v$, $v = x + x^2$

In Exercises 19 to 26 show that the function is a composition of other functions by introducing the necessary symbols, such as u and v, and describing the functions with equations.

19 $(x^3 + x^2 - 2)^{50}$
20 $(\sqrt{x} + 1)^{10}$
21 $\sqrt{x + 3}$
22 $\sqrt[3]{1 + x^2}$
23 $\sin 2x$
24 $\sin^2 x$
25 $\cos^3 2x$
26 $\sqrt[3]{(1 + 2x)^{50}}$
27 These tables show some of the values of functions f and g:

x	1	2	3	4	5
$f(x)$	6	8	9	7	10

x	6	7	8	9	10
$g(x)$	4	3	2	5	1

(a) Find $f(g(7))$.
(b) Find $g(f(3))$.

28 Figure 4 shows the graphs of functions f and g.
(a) Estimate $f(g(0.6))$.
(b) Estimate $g(f(0.3))$.
(c) Estimate $f(f(0.5))$.

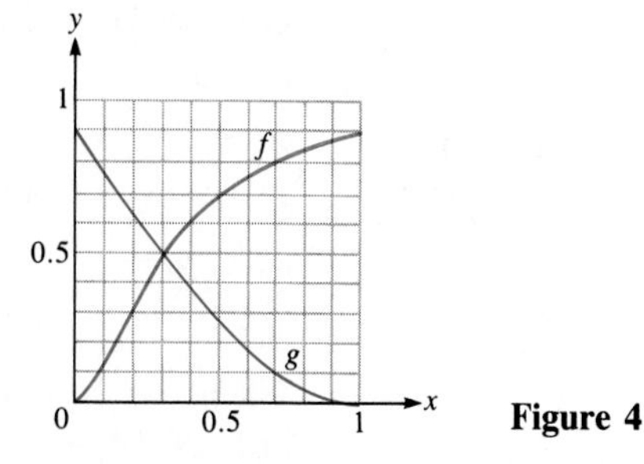

Figure 4

29 Let $f(x) = 2x^2 - 1$ and $g(x) = 4x^3 - 3x$. Show that $(f \circ g)(x) = (g \circ f)(x)$. [Rare indeed are pairs of polynomials that commute with each other under composition, as you may convince yourself by trying to find more. Of course, any two powers, such as x^3 and x^4, commute. (The composition of x^3 and x^4 in any order is x^{12}, as may be checked.)]

30 Let $f(x) = 1/(1 - x)$. What is the domain of f? of $f \circ f$? of $f \circ f \circ f$? Show that $(f \circ f \circ f)(x) = x$ for all x in the domain of $f \circ f \circ f$.

31 Let $g(x) = x^2$. Find all first-degree polynomials $f(x) = ax + b$, $a \neq 0$, such that $f \circ g = g \circ f$, that is, $f(g(x)) = g(f(x))$.

32 Let $g(x) = x^2$. Find all second-degree polynomials $f(x) = ax^2 + bx + c$, $a \neq 0$, such that $f \circ g = g \circ f$, that is, $f(g(x)) = g(f(x))$.

33 Let $f(x) = 2x + 3$. How many functions are there of the form $g(x) = ax + b$, a and b constants, such that $f \circ g = g \circ f$?

34 Let $f(x) = 2x + 3$. How many functions are there of the form $g(x) = ax^2 + bx + c$, a, b, and c constants, such that $f \circ g = g \circ f$?

35 Let $f(x) = x^5$. Is there a function $g(x)$ such that $(f \circ g)(x) = x$ for all numbers x? If so, how many such functions are there?

36 Let $f(x) = x^4$. Is there a function $g(x)$ such that $(f \circ g)(x) = x$ for all numbers x? If so, how many such functions are there?

2.3 THE LIMIT OF A FUNCTION

Three examples will introduce the notion of the limit of a function. After them, we will define the concept of a limit. This concept is the basis of the definition of the derivative in Chap. 3 and of the definite integral in Chap. 5.

Three Examples

EXAMPLE 1 Let $f(x) = 2x^2 + 1$. What happens to $f(x)$ as x is chosen closer and closer to 3?

SOLUTION Let us make a table of the values of $f(x)$ for some choices of x near 3:

x	3.1	3.01	3.001	2.999	2.99	2.9
$f(x)$	20.22	19.1202	19.012002	18.988002	18.8802	17.82

When x is close to 3, $2x^2 + 1$ is close to $2 \cdot 3^2 + 1 = 19$. We say that "the limit of $2x^2 + 1$ as x approaches 3 is 19" and write

The "limit" notation

$$\lim_{x \to 3} (2x^2 + 1) = 2 \cdot 3^2 + 1 = 19.$$

The arrow $\to$ stands for "approaches," so we could also write

$$\text{as } x \to 3, \ 2x^2 + 1 \to 19. \quad \blacksquare$$

Example 1 presented no obstacle. The next example offers a slight challenge.

EXAMPLE 2 Let $f(x) = (x^3 - 1)/(x^2 - 1)$. Note that this function is not defined when $x = 1$, for when $x = 1$, both numerator and denominator are 0. But we have every right to ask: How does $f(x)$ behave when x is *near* 1 but is *not* 1 itself?

SOLUTION Unlike Example 1, we can't just substitute $x = 1$ into the formula. To examine how $f(x)$ behaves, we first make a brief table of values of $f(x)$, to four decimal places, for x near 1. Choose some x larger than 1 and some x smaller than 1. For instance,

$$f(1.01) = \frac{1.01^3 - 1}{1.01^2 - 1} = \frac{1.030301 - 1}{1.0201 - 1} = \frac{0.030301}{0.0201} \approx 1.5075.$$

x	1.1	1.01	0.9	0.99
$\frac{x^3 - 1}{x^2 - 1}$	1.5762	1.5075	1.4263	1.4925

[If you have a calculator handy, evaluate $(x^3 - 1)/(x^2 - 1)$ at 1.001 and 0.999 as well.]

Two influences operate on $\frac{x^3 - 1}{x^2 - 1}$.

There are two influences acting on the fraction $(x^3 - 1)/(x^2 - 1)$ when x is near 1. *On the one hand, the numerator* $x^3 - 1$ *approaches* 0; *thus there is an influence pushing the fraction toward* 0. *On the other hand, the denominator* $x^2 - 1$ *also approaches* 0; *division by a small number tends to make a fraction large.* How do these two opposing influences balance out?

The algebraic identities

$$x^3 - 1 = (x^2 + x + 1)(x - 1)$$

and

$$x^2 - 1 = (x + 1)(x - 1)$$

enable us to answer the question.

Rewrite the quotient $(x^3 - 1)/(x^2 - 1)$ as follows: When $x \neq 1$, we have

$$\frac{x^3 - 1}{x^2 - 1} = \frac{(x^2 + x + 1)(x - 1)}{(x + 1)(x - 1)} = \frac{x^2 + x + 1}{x + 1},$$

so the behavior of $(x^3 - 1)/(x^2 - 1)$ for x near 1, but not equal to 1, is the same as the behavior of $(x^2 + x + 1)/(x + 1)$ for x near 1, but not equal to 1. Thus

$$\lim_{x \to 1} \frac{x^3 - 1}{x^2 - 1} = \lim_{x \to 1} \frac{x^2 + x + 1}{x + 1}.$$

Now, $(x^2 + x + 1)/(x + 1)$ behaves as nicely for x near 1 as $2x^2 + 1$ in Example 1 behaves for x near 3. As x approaches 1, $x^2 + x + 1$ approaches $1^2 + 1 + 1 = 3$ and $x + 1$ approaches $1 + 1 = 2$. Thus

$$\lim_{x \to 1} \frac{x^2 + x + 1}{x + 1} = \frac{3}{2},$$

from which it follows that

$$\lim_{x \to 1} \frac{x^3 - 1}{x^2 - 1} = \frac{3}{2}.$$

Note that $\frac{3}{2} = 1.5$, which is closely approximated by $f(1.01)$ and $f(0.99)$. ■

EXAMPLE 3 Let $f(x) = x/|x|$. Graph f and examine how $f(x)$ behaves as $x \to 0$.

SOLUTION First of all, note that $f(x)$ is *not* defined when $x = 0$, since division by 0 makes no sense.

Next, make a table of $f(x)$ for a few inputs x:

x	2	1	0.1	−0.1	−1	−2
$f(x) = x/\|x\|$	1	1	1	−1	−1	−1

The "hollow dot" notation for a missing point.

When x is positive, $f(x) = 1$. When x is negative, $f(x) = -1$. [See Fig. 1. The hollow circles at (0, 1) and at (0, −1) indicate that these points are not on the graph.]

Now we can describe how $f(x)$ behaves as $x \to 0$.

If x approaches 0 from the right, $f(x)$ is always 1. If x approaches 0 from the left, $f(x)$ is always −1. Since $f(x)$ does not approach some fixed number as $x \to 0$, $\lim_{x \to 0} x/|x|$ does *not* exist. ■

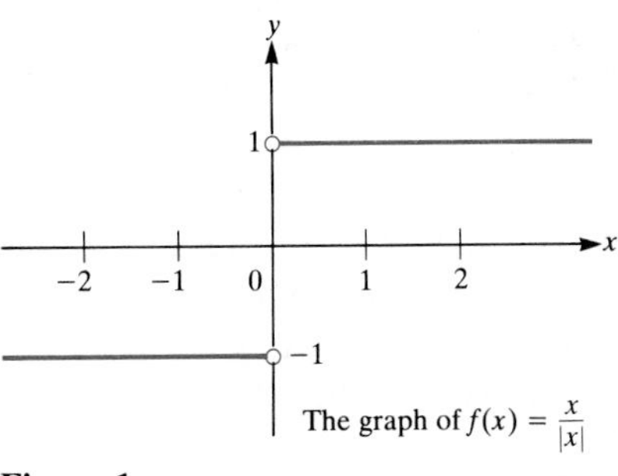

The graph of $f(x) = \frac{x}{|x|}$

Figure 1

Whether a function f has a limit at a has nothing to do with $f(a)$ itself. In fact, a might not even be in the domain of f. See, for instance, Examples 2 and 3. In Example 1, a (=3) happened to be in the domain of f, but that fact did not influence the reasoning. It is only the behavior of $f(x)$ for x *near* a that concerns us.

These three examples provide a background for describing the limit concept which will be used throughout the text.

Consider a function f and a number a which may or may not be in the domain of f. In order to discuss the behavior of $f(x)$ for x near a, we must know that the domain of f contains numbers arbitrarily close to a. Note how this assumption is built into each of the following definitions.

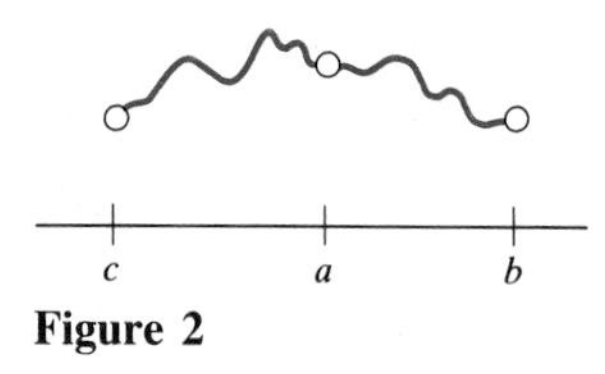

Figure 2

Definition *Limit of f(x) at a.* Let f be a function and a some fixed number. Assume that the domain of f contains open intervals (c, a) and (a, b), as shown in Fig. 2. If there is a number L such that as x approaches a, either from the right or from the left, $f(x)$ approaches L, then L is called the **limit** of $f(x)$ as x approaches a. This is written

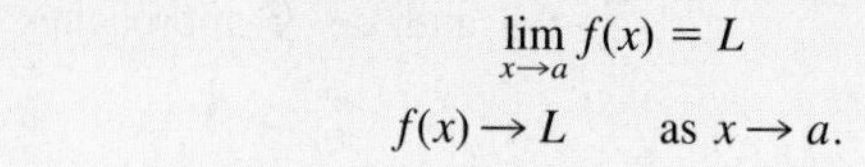

$$\lim_{x\to a} f(x) = L$$

or

$$f(x) \to L \qquad \text{as } x \to a.$$

In Example 1 we found $\lim_{x\to 3}(2x^2 + 1) = 19$, which illustrates this definition for $a = 3$ and $f(x) = 2x^2 + 1$. The fact that 3 happens to be in the domain of f is irrelevant. Example 2 showed that

$$\lim_{x\to 1} \frac{x^3 - 1}{x^2 - 1} = \frac{3}{2}$$

which illustrates this definition for $a = 1$ and $f(x) = (x^3 - 1)/(x^2 - 1)$. The fact that $f(x)$ is not defined for $x = 1$ did not affect the reasoning.

The next example illustrates a limit that will be discussed further in Chap. 6.

EXAMPLE 4 Examine $\lim_{x\to 0} \frac{2^x - 1}{x}$.

SOLUTION When x is near 0, 2^x is near 1. So the numerator is near 0. The denominator is near 0 when x is small. So it is not clear at a glance how the quotient $(2^x - 1)/x$ behaves when x is near 0.

In this case $2^x - 1$ is not a polynomial; no algebraic identity will magically give us the answer. So we resort to a calculator. The following table records some calculations for a few sample x values near 0:

x	0.5	0.1	0.05	0.01	−0.01	−0.05	−0.1	−0.5
$2^x - 1$	0.4142	0.0718	0.0353	0.006956	−0.006908	−0.003406	−0.06697	−0.2929
$\frac{2^x - 1}{x}$	0.8284	0.7177	0.7053	0.6956	0.6908	0.6813	0.6697	0.5858

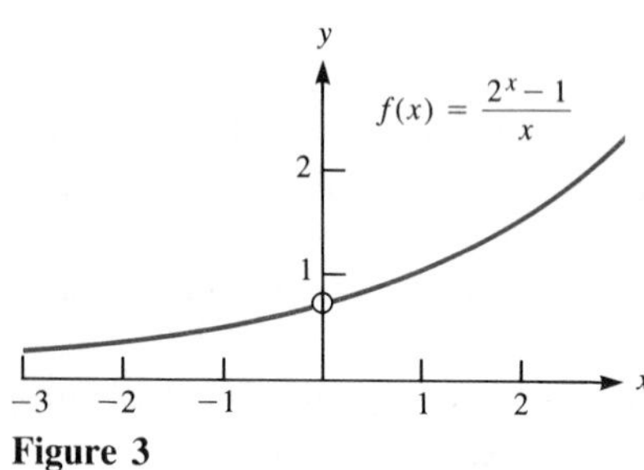

Figure 3

The data suggest that as x approaches 0 from the right, $(2^x - 1)/x$ decreases and approaches a number near 0.69; also, as x approaches 0 from the left, it seems that $(2^x - 1)/x$ increases and approaches, perhaps, the same number. If so, what is that number? It cannot be some famous fraction, since $\frac{2}{3}$ is the only famous fraction nearby—and it is too small.

Figure 3 shows the graph of $f(x) = (2^x - 1)/x$ for x near 0. This limit will be determined in Chap. 6. ■

The Limit of a Quotient

Example 2 concerned $\lim_{x\to 1} (x^3 - 1)/(x^2 - 1)$ and Example 4 treated $\lim_{x\to 0} (2^x - 1)/x$. In each case the numerator and denominator both approach 0. In the first case the quotient approaches $\frac{3}{2}$, and in the second case the quotient approaches a number whose decimal representation begins with 0.693. This raises the question: If a and b are both near 0, what can we say about the quotient a/b?

For instance, choose $a = 0.0001$. If $b = 0.0000001$, the quotient is

$$\frac{a}{b} = \frac{0.0001}{0.0000001} = 1000,$$

a *large* number.

If $a = 0.0001$ and $b = 0.001$, the quotient is

$$\frac{a}{b} = \frac{0.0001}{0.001} = 0.1,$$

a *small* number.

Finally, if $a = 0.0001$ and $b = 0.0001$, the quotient is

$$\frac{a}{b} = \frac{0.0001}{0.0001} = 1.$$

The quotient of two small numbers can be big or little.

As these computations show, *the quotient of two numbers near 0 can be large, small,* or *"medium."* Just knowing that two numbers are near 0 tells us nothing about their quotient. *This is important.* It tells us that if $\lim_{x\to 0} f(x) = 0$ and $\lim_{x\to 0} g(x) = 0$, we must usually do some work to determine the value of $\lim_{x\to 0} f(x)/g(x)$.

However, if $f(x) \to 3$ and $g(x) \to 4$, say, as $x \to a$, then we do not have to do any work to find $\lim_{x\to a} f(x)/g(x)$. It is $\frac{3}{4}$.

Beware the calculator.

The calculator was an indispensable tool in our exploration of the behavior of $(2^x - 1)/x$ in the previous example. However, a word of caution is in order. Beyond a certain point, calculators cannot be trusted. You might think that we could refine our estimate of the limit of $(2^x - 1)/x$ at 0 by continuing to take values of x ever closer to 0. If we attempt to put this into practice, we find instead that we get wildly varying results. At some point the errors begin to dominate the computations.

One step in the calculation involves subtraction, $2^x - 1$. As x draws closer to 0, 2^x becomes very close to 1. There is, of course, some round-off error in the computation of 2^x. The closer 2^x is to 1, the more the error will tend to dominate their difference. For example, one calculator gives 1.00000007 for the value of $2^{0.00000001}$. Upon subtracting 1 and dividing by 0.00000001, we obtain 0.7. Since the actual limit is about 0.693, we see from the table that the input 0.01 gives a better estimate than the input 0.00000001. Beware of this effect when using a calculator to estimate a limit.

One-Sided Limits

In Example 4, $f(x) = (2^x - 1)/x$ seems to approach a specific number, whether x approaches 0 from the right or the left. In Example 3, $f(x) = x/|x|$ behaves differently, since $x/|x| \to 1$ as x approaches 0 from the right, but $x/|x| \to -1$ as x approaches 0 from the left. This introduces the notion of one-sided limits.

Right-hand limit

Definition *Right-hand limit of $f(x)$ at a.* Let f be a function and a some fixed number. Assume that the domain of f contains an open interval (a, b). If, as x approaches a from the right, $f(x)$ approaches a specific number L, then L is called the **right-hand limit** of $f(x)$ as x approaches a. This is written

$$\lim_{x \to a^+} f(x) = L$$

or

$$\text{as } x \to a^+, \qquad f(x) \to L.$$

The assertion that

$$\lim_{x \to a^+} f(x) = L$$

is read "the limit of f of x as x approaches a from the right is L" or "as x approaches a from the right, $f(x)$ approaches L."

Left-hand limit

The left-hand limit is defined similarly. The only differences are that the domain of f must contain an open interval of the form (c, a) and $f(x)$ is examined as x approaches a from the left. The notations for the left-hand limit are

$$\lim_{x \to a^-} f(x) = L$$

or

$$\text{as } x \to a^-, \qquad f(x) \to L.$$

As Example 3 shows,

$$\lim_{x \to 0^+} \frac{x}{|x|} = 1 \qquad \text{and} \qquad \lim_{x \to 0^-} \frac{x}{|x|} = -1.$$

We could also write, for instance,

$$\text{as } x \to 0^+, \qquad \frac{x}{|x|} \to 1.$$

Note that if both the right-hand and the left-hand limits of f exist at a and are equal, then $\lim_{x \to a} f(x)$ exists. But if the right-hand and left-hand limits are not equal, then $\lim_{x \to a} f(x)$ does not exist. For instance, $\lim_{x \to 0} x/|x|$ does not exist.

The next example reviews the three limit concepts.

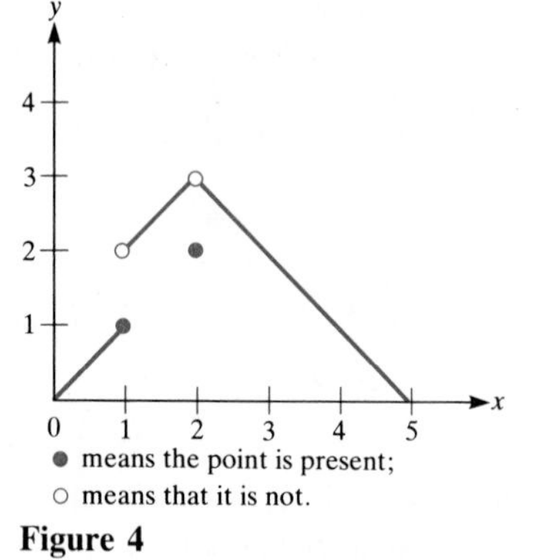

Figure 4

EXAMPLE 5 Figure 4 shows the graph of a function f whose domain is the closed interval $[0, 5]$.

(*a*) Does $\lim_{x \to 1} f(x)$ exist?
(*b*) Does $\lim_{x \to 2} f(x)$ exist?
(*c*) Does $\lim_{x \to 3} f(x)$ exist?

SOLUTION

(*a*) Inspection of the graph shows that

$$\lim_{x \to 1^-} f(x) = 1 \qquad \text{and} \qquad \lim_{x \to 1^+} f(x) = 2.$$

Although the two one-sided limits exist, they are not equal. Thus $\lim_{x\to 1} f(x)$ does not exist. In short, "f does not have a limit as $x \to 1$."

(*b*) Inspection of the graph shows that

$$\lim_{x\to 2^-} f(x) = 3 \qquad \text{and} \qquad \lim_{x\to 2^+} f(x) = 3.$$

Thus $\lim_{x\to 2} f(x)$ exists and is 3. Incidentally, the solid dot at (2, 2) shows that $f(2) = 2$. This information, however, plays no role in our examination of the limit of $f(x)$ as $x \to 2$.

(*c*) Inspection shows that

$$\lim_{x\to 3^-} f(x) = 2 \qquad \text{and} \qquad \lim_{x\to 3^+} f(x) = 2.$$

Thus $\lim_{x\to 3} f(x)$ exists and is 2. Incidentally, the fact that $f(3)$ is equal to 2 is irrelevant in determining $\lim f(x)$. ■

A function as wild as the one described in Example 5 will not be of major concern in calculus. However, it does serve to clarify the definitions of right-hand limit and left-hand limit, just as the notion of sickness illuminates our understanding of health.

By contrast, the tamest functions are the "constant" functions. A **constant function** assigns the same output to all inputs. If that fixed output is, say, L, then $f(x) = L$ for all x. The graph of this function is a line parallel to the x axis, as in Fig. 5. We have

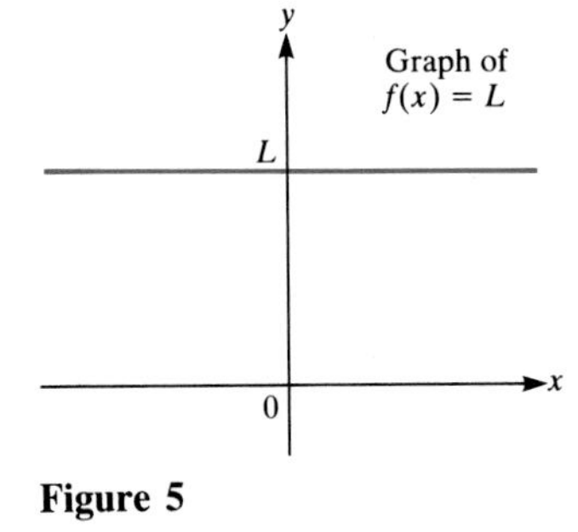

Figure 5

$$\lim_{x\to a} f(x) = L,$$

where a is any real number.

It may seem strange to say that "the limit of L is L," but in practice this offers no difficulty. For instance,

$$\lim_{x\to 5} \frac{1 + x^2}{1 + x^2} = \lim_{x\to 5} 1 = 1$$

and

$$\lim_{x\to 3} 1^x = \lim_{x\to 3} 1 = 1.$$

Section Summary

This section introduced the notion of the limit of a function $f(x)$ as x approaches a number a. The limit, if it exists, is denoted $\lim_{x\to a} f(x)$. Its definition in no way involves $f(a)$; in fact, a may not even be in the domain of f. To see whether $\lim_{x\to a} f(x)$ exists you "put your thumb over a" and see how $f(x)$ behaves nearby, for x close to a. For most common functions the limit exists and, if $f(a)$ is defined, equals $f(a)$.

Sometimes a limit is not easily determined at a glance. For instance, it may involve the quotient of two numbers, both of which approach 0.

We also introduced the right-hand limit $\lim_{x\to a^+} f(x)$ and the left-hand limit $\lim_{x\to a^-} f(x)$.

EXERCISES FOR SEC. 2.3: THE LIMIT OF A FUNCTION

In Exercises 1 to 14 find the limits, all of which exist. Use intuition and, if needed, algebra.

1 $\lim_{x\to 5} (x + 7)$ **2** $\lim_{x\to 1} (4x - 2)$

3 $\lim_{x\to 2} \dfrac{x^2 - 4}{x - 2}$ **4** $\lim_{x\to 3} \dfrac{x^2 - 9}{x - 3}$

5 $\lim_{x\to 1} \dfrac{x^4 - 1}{x^3 - 1}$ **6** $\lim_{x\to 1} \dfrac{x^6 - 1}{x^3 - 1}$

7 $\lim_{x\to 3} \dfrac{1}{x + 2}$ **8** $\lim_{x\to 5} \dfrac{3x + 5}{4x}$

9 $\lim_{x\to 3} 25$ **10** $\lim_{x\to 3} \pi^2$

11 $\lim_{x\to 0^+} \sqrt{x}$ **12** $\lim_{x\to 1^+} \sqrt{4x - 4}$

13 $\lim_{x\to 1^+} \dfrac{x - 1}{|x - 1|}$ **14** $\lim_{x\to 1^-} \dfrac{x - 1}{|x - 1|}$

In Exercises 15 to 20 decide whether the limits exist and, if they do, evaluate them.

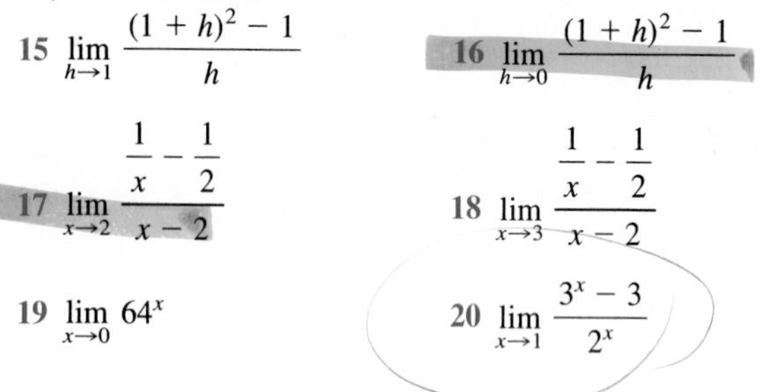

15 $\lim_{h\to 1} \dfrac{(1 + h)^2 - 1}{h}$ **16** $\lim_{h\to 0} \dfrac{(1 + h)^2 - 1}{h}$

17 $\lim_{x\to 2} \dfrac{\dfrac{1}{x} - \dfrac{1}{2}}{x - 2}$ **18** $\lim_{x\to 3} \dfrac{\dfrac{1}{x} - \dfrac{1}{2}}{x - 2}$

19 $\lim_{x\to 0} 64^x$ **20** $\lim_{x\to 1} \dfrac{3^x - 3}{2^x}$

In each of Exercises 21 and 22 there is a graph of a function. Decide which of the given limits exist, and evaluate those which do.

21 (See Fig. 6.) (*a*) $\lim_{x\to 0^+} f(x)$ (*b*) $\lim_{x\to 1} f(x)$ (*c*) $\lim_{x\to 2^-} f(x)$ (*d*) $\lim_{x\to 2^+} f(x)$

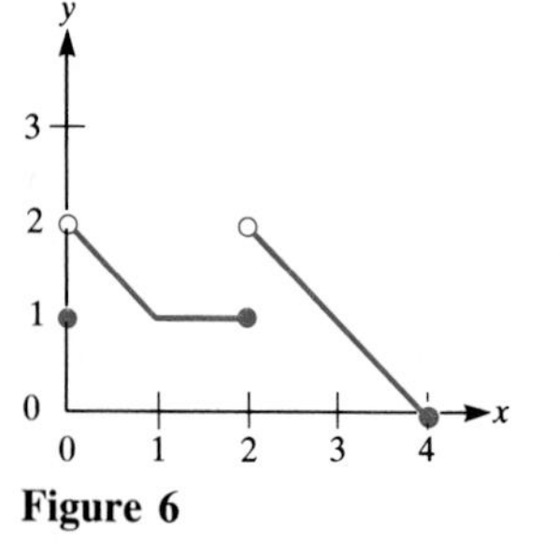

Figure 6

22 (See Fig. 7.) (*a*) $\lim_{x\to 1} f(x)$ (*b*) $\lim_{x\to 2} f(x)$ (*c*) $\lim_{x\to 3} f(x)$ (*d*) $\lim_{x\to 4^-} f(x)$

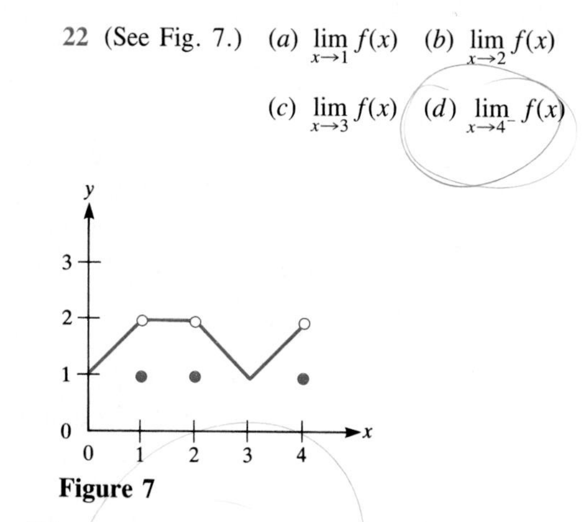

Figure 7

23 Examine $(3^x - 1)/x$ for x near 0. Do you think $\lim_{x\to 0} (3^x - 1)/x$ exists?

24 Examine $(3^x - 2^x)/x$ for x near 0. Do you think $\lim_{x\to 0} (3^x - 2^x)/x$ exists?

25 Let $f(x) = (3^x - 1)/(2^x - 1)$.

(*a*) Fill in this table:

x	1	0.1	0.01	0.001	−1	−0.01	−0.001
$f(x)$							

(*b*) On the basis of (*a*) do you think $\lim_{x\to 0} f(x)$ exists? If so, estimate this limit.

26 Let $f(x) = (1 + x)^{1/x}$ for $x > -1$, $x \neq 0$.

(*a*) Compute $f(x)$ for $x = 1$, 0.1, 0.01, and 0.001.

(*b*) Compute $f(x)$ for $x = -0.1$, -0.01, -0.001.

(*c*) Do you think $\lim_{x\to 0} (1 + x)^{1/x}$ exists? If so, what do you think it is?

---

27 Figure 8 shows a graph of a function that goes up and down infinitely often between the dashed lines, both to the right and to the left of 3. Does $\lim_{x\to 3} f(x)$ exist? If so, what is it?

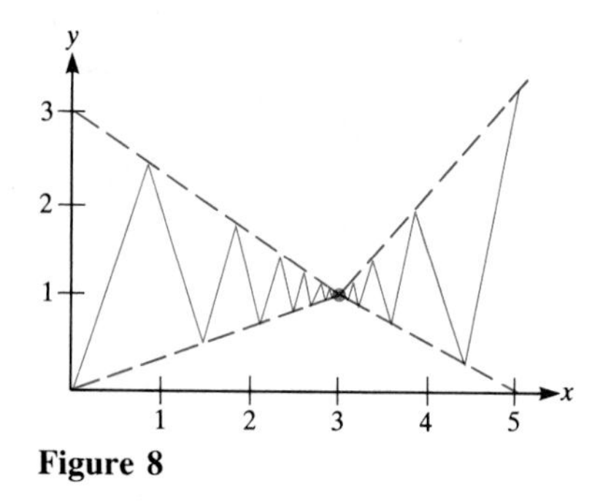

Figure 8

28 Define a certain function f as follows:

$$f(x) = \begin{cases} 1 & \text{if } x \text{ is an integer,} \\ 0 & \text{if } x \text{ is not an integer.} \end{cases}$$

(*a*) Graph f.
(*b*) Does $\lim_{x\to 3} f(x)$ exist?
(*c*) Does $\lim_{x\to 3.5} f(x)$ exist?
(*d*) For which numbers a does $\lim_{x\to a} f(x)$ exist?

29 (*a*) Graph the function f given by the formula

$$f(x) = |x| - x.$$

(*b*) For which numbers a does $\lim_{x\to a} f(x)$ exist?

30 Define f as follows:

$$f(x) = \begin{cases} x & \text{if } x \text{ is rational,} \\ -x & \text{if } x \text{ is not rational.} \end{cases}$$

(*a*) What does the graph of f look like? (A dotted curve may be used to indicate that points are missing.)
(*b*) Does $\lim_{x\to 1} f(x)$ exist?
(*c*) Does $\lim_{x\to \sqrt{2}} f(x)$ exist?
(*d*) Does $\lim_{x\to 0} f(x)$ exist?
(*e*) For which numbers a does $\lim_{x\to a} f(x)$ exist?

31 Define $f(x) = \begin{cases} x^2 & \text{if } x \text{ is rational,} \\ x^3 & \text{if } x \text{ is irrational.} \end{cases}$

(*a*) What does the graph of f look like? [See the advice for Exercise 30(*a*).]
(*b*) Does $\lim_{x\to 2} f(x)$ exist?
(*c*) Does $\lim_{x\to 1} f(x)$ exist?
(*d*) Does $\lim_{x\to 0} f(x)$ exist?
(*e*) For which numbers a does $\lim_{x\to a} f(x)$ exist?

32 Let $f(x) = x^x$ for $x > 0$.

(*a*) Fill in this table:

x	1.0	0.5	0.4	0.3	0.2	0.1	0.01
x^x							

(*b*) What do you think is the smallest value of x^x for x in (0, 1)?
(*c*) Do you think that $\lim_{x\to 0^+} x^x$ exists? If so, what do you think it is?

* **33** (*a*) Estimate $(4^x - 1)/x$ for x very near 0, both positive and negative.
(*b*) Estimate $\lim_{x\to 0} (4^x - 1)/x$ to three decimal places.
(*c*) Compare your answer to (*b*) with the result in Example 4 concerning $\lim_{x\to 0} (2^x - 1)/x$.
(*d*) What relation do you think there is between $\lim_{x\to 0} (4^x - 1)/x$ and $\lim_{x\to 0} (2^x - 1)/x$?
(*e*) Explain, in complete sentences, why your conjecture in (*d*) is correct (assuming both limits exist).

* **34** Let $f(x) = 2^x$.

(*a*) Estimate $\displaystyle\lim_{x\to a} \frac{f(x) - f(a)}{x - a}$ for $a = 0, 1, 2,$ and 3.
(*b*) Based on (*a*), what do you think the limit is when $a = 4$?
(*c*) Based on (*a*), make a conjecture about the limit for any number a.
(*d*) Assuming that the limit in (*a*) exists, explain, in complete sentences, why your conjecture in (*c*) is correct.

* **35** For a positive integer n let $f(n)$ be the sum of the reciprocals of all the integers from n to $2n$:

$$f(n) = \frac{1}{n} + \frac{1}{n+1} + \cdots + \frac{1}{2n}.$$

(*a*) Compute $f(n)$ for at least $n = 1, 2, 3, \ldots, 10$.
(*b*) What happens to $f(n)$ as n increases?
(*c*) Make a conjecture about $\lim_{n\to\infty} f(n)$.
(*d*) Explain, in complete sentences, why you feel that your conjecture in (*c*) is true.

2.4 COMPUTATIONS OF LIMITS

For convenience certain frequently used properties of limits should be put on the record.

Properties of limits

Theorem Let f and g be two functions and assume that

$$\lim_{x \to a} f(x) \quad \text{and} \quad \lim_{x \to a} g(x)$$

both exist. Then

1 $\lim_{x \to a} (f(x) + g(x)) = \lim_{x \to a} f(x) + \lim_{x \to a} g(x)$.

2 $\lim_{x \to a} (f(x) - g(x)) = \lim_{x \to a} f(x) - \lim_{x \to a} g(x)$.

3 $\lim_{x \to a} kf(x) = k \lim_{x \to a} f(x)$ for any constant k.

4 $\lim_{x \to a} f(x)g(x) = \lim_{x \to a} f(x) \lim_{x \to a} g(x)$.

5 $\lim_{x \to a} \frac{f(x)}{g(x)} = \frac{\lim_{x \to a} f(x)}{\lim_{x \to a} g(x)}$ if $\lim_{x \to a} g(x) \neq 0$.

6 $\lim_{x \to a} f(x)^{g(x)} = \left(\lim_{x \to a} f(x)\right)^{\lim_{x \to a} g(x)}$ if $\lim_{x \to a} f(x) > 0$. ■

These properties have been tacitly assumed in Sec. 2.3. Properties 1 to 5 are treated in Appendix I, which employs the precise definitions of limits given in Secs. 2.9 and 2.10. Property 6 depends on results in Appendix H.

Property 1, for instance, asserts that if $\lim_{x \to a} f(x)$ and $\lim_{x \to a} g(x)$ both exist, then $\lim_{x \to a} (f(x) + g(x))$ exists and equals the sum of the two given limits. This property extends to any finite sum of functions: For example, if $\lim_{x \to a} f(x)$, $\lim_{x \to a} g(x)$, and $\lim_{x \to a} h(x)$ exist, then

$$\lim_{x \to a} (f(x) + g(x) + h(x)) \quad \text{exists,}$$

and

$$\lim_{x \to a} (f(x) + g(x) + h(x)) = \lim_{x \to a} f(x) + \lim_{x \to a} g(x) + \lim_{x \to a} h(x).$$

Similarly, property 4 extends to the product of any finite number of functions.

EXAMPLE 1 Suppose that $\lim_{x \to 3} f(x) = 4$ and $\lim_{x \to 3} g(x) = 5$; discuss $\lim_{x \to 3} f(x)/g(x)$.

SOLUTION By property 5, $\lim_{x \to 3} f(x)/g(x)$ exists and

$$\lim_{x \to 3} \frac{f(x)}{g(x)} = \frac{\lim_{x \to 3} f(x)}{\lim_{x \to 3} g(x)} = \frac{4}{5}.$$

No further information is needed to determine the limit of $f(x)/g(x)$ as $x \to 3$. If we know only that $\lim_{x \to 3} f(x) = 4$ and $\lim_{x \to 3} g(x) = 5$, we know how the quotient $f(x)/g(x)$ behaves as $x \to 3$. ■

EXAMPLE 2 Suppose that $\lim_{x\to 3} f(x) = 0$ and $\lim_{x\to 3} g(x) = 0$; discuss $\lim_{x\to 3} f(x)/g(x)$.

SOLUTION In contrast to Example 1, in this case property 5 gives no information, since $\lim_{x\to 3} g(x) = 0$. *It is necessary to have more information about f and g.* We will give an example of functions f and g, where $f(x) \to 0$ and $g(x) \to 0$ as $x \to 3$ and $f(x)/g(x) \to 6$. Then we will give an example where $f(x) \to 0$ and $g(x) \to 0$ as $x \to 3$ and $f(x)/g(x) \to 0$.

For instance, if

$$f(x) = x^2 - 9 \qquad \text{and} \qquad g(x) = x - 3,$$

then

$$\lim_{x\to 3} f(x) = 0 \qquad \text{and} \qquad \lim_{x\to 3} g(x) = 0$$

and the limit of the quotient is

$$\lim_{x\to 3} \frac{x^2 - 9}{x - 3} = \lim_{x\to 3} \frac{(x + 3)(x - 3)}{x - 3}$$

$$= \lim_{x\to 3} (x + 3) = 6.$$

Loosely put, "when x is near 3, $x^2 - 9$ is about 6 times as large as $x - 3$."

A *different* choice of f and g could produce a *different* limit for the quotient $f(x)/g(x)$. To be specific, let

$$f(x) = (x - 3)^2 \qquad \text{and} \qquad g(x) = x - 3.$$

Then

$$\lim_{x\to 3} f(x) = 0 \qquad \text{and} \qquad \lim_{x\to 3} g(x) = 0,$$

and the limit of the quotient is

$$\lim_{x\to 3} \frac{(x - 3)^2}{x - 3} = \lim_{x\to 3} (x - 3)$$

$$= 0.$$

In this case we could say "$(x - 3)^2$ approaches 0 much faster than does $x - 3$, when $x \to 3$."

See also Examples 2 and 4 of the preceding section.

In short, the information that $\lim_{x\to 3} f(x) = 0$ and $\lim_{x\to 3} g(x) = 0$ is not enough to tell us how $f(x)/g(x)$ behaves as $x \to 3$. If we know only that $f(x) \to 0$ and $g(x) \to 0$ as $x \to a$, we do not know how $f(x)/g(x)$ behaves. ■

Determinate and Indeterminate Limits

In Example 1, knowing only that $\lim_{x\to 3} f(x) = 4$ and $\lim_{x\to 3} g(x) = 5$ is enough information to determine $\lim_{x\to 3} f(x)/g(x)$. We don't need to know anything more about the two functions. More generally, if

$$\lim_{x\to a} f(x) = A$$

and

$$\lim_{x\to a} g(x) = B \qquad B \neq 0$$

then, as property 5 of limits says,

$$\lim_{x\to a} \frac{f(x)}{g(x)} = \frac{A}{B}.$$

The situation is called a **determinate limit**, because the limit can be determined without further information.

However, as Example 2 shows, if

$$\lim_{x\to a} f(x) = 0$$

and

$$\lim_{x\to a} g(x) = 0,$$

we do *not* have enough information to determine

$$\lim_{x\to a} \frac{f(x)}{g(x)}.$$

For this reason, it is called an **indeterminate limit** of the form "zero over zero."

Warning! Some people, as they begin the study of limits, assume that if two quantities are both approaching 0, their quotient approaches 1. Example 2 shows that this is definitely not true. It takes some work on our part to find how their quotient actually behaves.

Limits as $x \to \infty$

Sometimes it is useful to know how $f(x)$ behaves when x is a very large positive number (or a negative number of large absolute value). Example 3 serves as an illustration and introduces a variation on the theme of limits.

EXAMPLE 3 Determine how $f(x) = 1/x$ behaves for (a) large positive inputs and (b) negative inputs of large absolute value.

SOLUTION

x	10	100	1000
$1/x$	0.1	0.01	0.001

(a) First make a table of values as shown in the margin. As x gets arbitrarily large, $1/x$ approaches 0.

(b) This is similar to (a). For instance,

$$f(-1000) = -0.001.$$

As negative numbers x are chosen of arbitrarily large absolute value, $1/x$ approaches 0. ■

The notation $\lim_{x\to\infty} f(x) = L$

Rather than writing "as x gets arbitrarily large through positive values, $f(x)$ approaches the number L," it is customary to use the shorthand

$$\lim_{x\to\infty} f(x) = L.$$

Since ∞ *is not a number, the case* $x \to \infty$ *is distinct from the case* $x \to a$.

This is read "as x approaches infinity, $f(x)$ approaches L," or "the limit of $f(x)$ as x approaches infinity is L." For instance,

$$\lim_{x\to\infty} \frac{1}{x} = 0.$$

More generally, for any fixed positive exponent a,

$$\lim_{x\to\infty} \frac{1}{x^a} = 0.$$

The notation $\lim_{x \to -\infty} f(x) = L$

Similarly, the assertion that "as negative numbers x are chosen of arbitrarily large absolute value, $f(x)$ approaches the number L" is abbreviated to

$$\lim_{x \to -\infty} f(x) = L.$$

For instance, $$\lim_{x \to -\infty} \frac{1}{x} = 0.$$

The six properties of limits stated at the beginning of the section hold when "$x \to a$" is replaced by "$x \to \infty$" or by "$x \to -\infty$."

It could happen that as $x \to \infty$, a function $f(x)$ becomes and remains arbitrarily large and positive. For instance, as $x \to \infty$, x^3 gets arbitrarily large. The shorthand for this is

The notation $\lim_{x \to \infty} f(x) = \infty$

$$\lim_{x \to \infty} f(x) = \infty.$$

For instance, $$\lim_{x \to \infty} x^3 = \infty.$$

It is important, when reading the shorthand

$$\lim_{x \to \infty} f(x) = \infty,$$

to keep in mind that "∞" is not a number. *The limit does not exist.* Properties 1 to 6 cannot, in general, be applied in such cases.

Other notations, such as $\lim_{x \to \infty} f(x) = -\infty$ or $\lim_{x \to -\infty} f(x) = \infty$ are defined similarly. For instance,

$$\lim_{x \to -\infty} x^3 = -\infty.$$

It can be shown that if, as $x \to \infty$, $f(x) \to \infty$ and $g(x) \to L > 0$, then $\lim_{x \to \infty} f(x)g(x) = \infty$. This fact is used in the next example.

EXAMPLE 4 Discuss the behavior of $f(x) = 2x^3 - 11x^2 + 12x$ when x is large.

SOLUTION First consider x positive and large. For instance, when $x = 10{,}000$, we obtain

$$\begin{aligned} f(10{,}000) &= 2(10{,}000^3) - 11(10{,}000^2) + 12(10{,}000) \\ &= 2{,}000{,}000{,}000{,}000 - 1{,}100{,}000{,}000 + 120{,}000 \qquad (1) \end{aligned}$$

The three numbers in (1) are "large," but 2,000,000,000,000 is by far the largest. It is 2 trillion (on the order of our national debt), while 1,100,000,000 is just over a billion (on the order of what is spent on defense each day). The numbers 1,100,000,000 and 120,000 scarcely influence the size of $f(10{,}000)$. The important term is the first, $2(10{,}000)^3$.

When x is large, the three terms $2x^3$, $-11x^2$, and $12x$ all become of large absolute value. To see how the function $2x^3 - 11x^2 + 12x$ behaves for large positive x, factor out x^3:

This factoring shows the importance of the highest power.

$$2x^3 - 11x^2 + 12x = x^3\left(2 - \frac{11}{x} + \frac{12}{x^2}\right). \qquad (2)$$

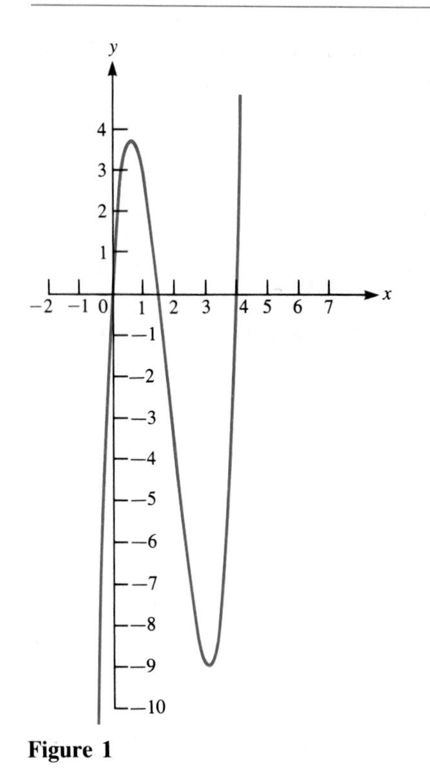

Figure 1

Now, since $11/x$ and $12/x^2 \to 0$ as $x \to \infty$,

$$\lim_{x\to\infty} \left(2 - \frac{11}{x} + \frac{12}{x^2}\right) = 2.$$

Moreover, as $x \to \infty$, $x^3 \to \infty$. Thus

$$\lim_{x\to\infty} x^3 \left(2 - \frac{11}{x} + \frac{12}{x^2}\right) = \infty;$$

hence

$$\lim_{x\to\infty} (2x^3 - 11x^2 + 12x) = \infty.$$

Now consider x negative and of large absolute value. The argument is similar. Use Eq. (2), and notice that $\lim_{x\to-\infty} x^3 = -\infty$ and

$$\lim_{x\to-\infty} \left(2 - \frac{11}{x} + \frac{12}{x^2}\right) = 2.$$

It follows that

$$\lim_{x\to-\infty} (2x^3 - 11x^2 + 12x) = -\infty.$$

So when $|x|$ is large, the behavior of $2x^3 - 11x^2 + 12x$ is determined by the behavior of $2x^3$.

The graph of $f(x) = 2x^3 - 11x^2 + 12x$ shows what is happening for $|x|$ large. (See Fig. 1.) Since $\lim_{x\to\infty} (2x^3 - 11x^2 + 12x) = \infty$, the graph rises arbitrarily high as $x \to \infty$. Since $\lim_{x\to-\infty} (2x^3 - 11x^2 + 12x) = -\infty$, the graph goes arbitrarily far down as $x \to -\infty$. ■

General form of a polynomial

Example 4 generalizes to any polynomial. A **polynomial** is a function of the form $a_n x^n + a_{n-1}x^{n-1} + \cdots + a_0$, where $a_0, a_1, \ldots, a_n$ are fixed real numbers and n is a nonnegative integer. If a_n is not 0, n is the **degree** of the polynomial. The numbers $a_0, a_1, \ldots, a_n$ are the **coefficients**.

Note: The expression $a_n x^n + a_{n-1}x^{n-1} + \cdots + a_0$ may be read aloud as "a sub n x to the n plus a sub n minus 1 x to the n minus 1 plus dot dot dot plus a sub zero."

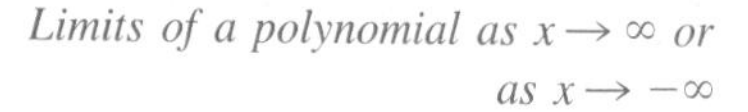

Limits of a polynomial as $x \to \infty$ or as $x \to -\infty$

Let $f(x) = a_n x^n + a_{n-1}x^{n-1} + \cdots + a_0$ be a polynomial of degree at least 1 and with the lead coefficient a_n *positive*. Then

$$\lim_{x\to\infty} f(x) = \infty.$$

If the degree of f is even, then

$$\lim_{x\to-\infty} f(x) = \infty.$$

But if the degree of f is odd, then

$$\lim_{x\to-\infty} f(x) = -\infty.$$

EXAMPLE 5 Determine how $f(x) = (x^3 + 6x^2 + 10x + 2)/(2x^3 + x^2 + 5)$ behaves for arbitrarily large positive numbers x.

A contest between a large numerator and a large denominator

SOLUTION As x gets large, the numerator $x^3 + 6x^2 + 10x + 2$ grows large, influencing the quotient to become large. On the other hand, the denominator also grows large, influencing the quotient to become small. An algebraic device will help reveal what happens to the quotient. We have

$$f(x) = \frac{x^3 + 6x^2 + 10x + 2}{2x^3 + x^2 + 5} = \frac{x^3\left(1 + \frac{6}{x} + \frac{10}{x^2} + \frac{2}{x^3}\right)}{x^3\left(2 + \frac{1}{x} + \frac{5}{x^3}\right)}$$

$$= \frac{1 + \frac{6}{x} + \frac{10}{x^2} + \frac{2}{x^3}}{2 + \frac{1}{x} + \frac{5}{x^3}} \qquad \text{for } x \neq 0.$$

Now we can see what happens to $f(x)$ when x is large.

As x increases, $6/x \to 0$, $10/x^2 \to 0$, $2/x^3 \to 0$, $1/x \to 0$, and $5/x^3 \to 0$. Thus

$$\lim_{x\to\infty} f(x) = \lim_{x\to\infty} \frac{x^3 + 6x^2 + 10x + 2}{2x^3 + x^2 + 5}$$

$$= \lim_{x\to\infty} \frac{1 + \frac{6}{x} + \frac{10}{x^2} + \frac{2}{x^3}}{2 + \frac{1}{x} + \frac{5}{x^3}}$$

$$= \frac{1 + 0 + 0 + 0}{2 + 0 + 0}$$

$$= \frac{1}{2}.$$

So, as x gets arbitrarily large through positive values, the quotient approaches $\frac{1}{2}$. In short,

$$\lim_{x\to\infty} \frac{x^3 + 6x^2 + 10x + 2}{2x^3 + x^2 + 5} = \frac{1}{2}. \quad ■$$

The technique used in Example 5 applies to any function that can be written as the quotient of two polynomials. Such a function is called a **rational function**.

How to find the limit of a rational function as $x \to \infty$ or as $x \to -\infty$.

Let $f(x)$ be a polynomial and let ax^n be its term of highest degree. Let $g(x)$ be another polynomial and let bx^m be its term of highest degree. Then

$$\lim_{x\to\infty} \frac{f(x)}{g(x)} = \lim_{x\to\infty} \frac{ax^n}{bx^m} \quad \text{and} \quad \lim_{x\to-\infty} \frac{f(x)}{g(x)} = \lim_{x\to-\infty} \frac{ax^n}{bx^m}.$$

(The proofs of these facts are similar to the argument used in Example 5.) In short, when working with the limit of a quotient of two polynomials as $x \to \infty$ or as $x \to -\infty$, disregard all terms except the one of highest degree in each of the polynomials. The next example illustrates this technique, which makes it quite easy to find limits of rational functions as $x \to \infty$ and $x \to -\infty$.

EXAMPLE 6 Examine the following limits:

(a) $\lim_{x\to\infty} \dfrac{3x^4 + 5x^2}{-x^4 + 10x + 5}$ (b) $\lim_{x\to\infty} \dfrac{x^3 - 16x}{5x^4 + x^3 - 5x}$

(c) $\lim_{x\to-\infty} \dfrac{x^4 + x}{6x^3 - x^2}$

SOLUTION By the preceding observations,

(a) $\lim_{x\to\infty} \dfrac{3x^4 + 5x^2}{-x^4 + 10x + 5} = \lim_{x\to\infty} \dfrac{3x^4}{-x^4} = \lim_{x\to\infty} (-3) = -3.$

(b) $\lim_{x\to\infty} \dfrac{x^3 - 16x}{5x^4 + x^3 - 5x} = \lim_{x\to\infty} \dfrac{x^3}{5x^4} = \lim_{x\to\infty} \dfrac{1}{5x} = 0.$

(c) $\lim_{x\to-\infty} \dfrac{x^4 + x}{6x^3 - x^2} = \lim_{x\to-\infty} \dfrac{x^4}{6x^3} = \lim_{x\to-\infty} \dfrac{x}{6} = -\infty.$ ■

The technique of factoring out the highest power of x applies more generally than just to polynomials, as the next example illustrates.

EXAMPLE 7 Examine (a) $\lim_{x\to\infty} (\sqrt{3x^2 + x}/x)$ and (b) $\lim_{x\to-\infty} (\sqrt{3x^2 + x}/x)$.

SOLUTION Before beginning the solution, note that if x is positive, $\sqrt{x^2} = x$, but if x is negative, $\sqrt{x^2} = -x$.

Recall that $\sqrt{a^2} = |a|$.

(a)
$$\lim_{x\to\infty} \frac{\sqrt{3x^2 + x}}{x} = \lim_{x\to\infty} \frac{\sqrt{x^2\left(3 + \frac{1}{x}\right)}}{x}$$
$$= \lim_{x\to\infty} \frac{x\sqrt{3 + 1/x}}{x}$$
$$= \lim_{x\to\infty} \sqrt{3 + 1/x} = \sqrt{3}.$$

(b)
$$\lim_{x\to-\infty} \frac{\sqrt{3x^2 + x}}{x} = \lim_{x\to-\infty} \frac{\sqrt{x^2\left(3 + \frac{1}{x}\right)}}{x}$$
$$= \lim_{x\to-\infty} \frac{-x\sqrt{3 + 1/x}}{x}$$
$$= \lim_{x\to-\infty} -\sqrt{3 + 1/x} = -\sqrt{3}.$$ ■

Infinite Limits at *a*

The next example concerns a case in which $f(x)$ becomes arbitrarily large as x approaches a fixed real number.

EXAMPLE 8 How does $f(x) = 1/x$ behave when x is near 0?

SOLUTION The reciprocal of a small number x has a large absolute value. For instance, when $x = 0.01$, $1/x = 100$; when $x = -0.01$, $1/x = -100$. Thus, as x approaches 0 from the right, $1/x$, which is positive, becomes arbitrarily large. The notation for this is

$$\lim_{x\to 0^+} \frac{1}{x} = \infty.$$

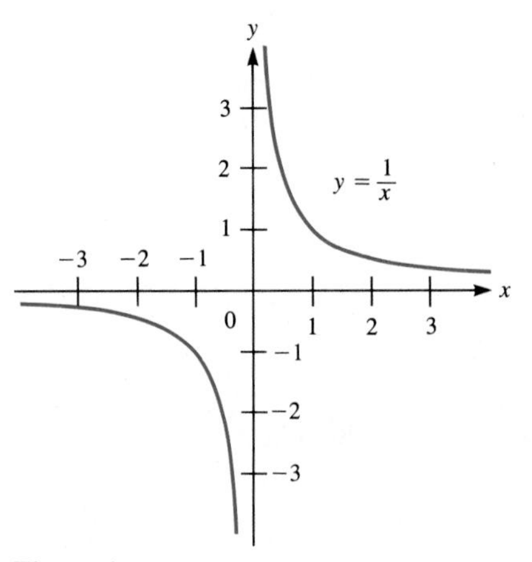

Figure 2

The notation "$\lim_{x \to 0} \frac{1}{x^2} = \infty$" is useful, though the limit does not exist since ∞ is not a number.

As x approaches 0 from the left, $1/x$, which is negative, has arbitrarily large absolute values. The notation for this is

$$\lim_{x \to 0^-} \frac{1}{x} = -\infty.$$

Figure 2, shows the graph of $y = 1/x$. (See also Example 3.) Note, that as $x \to 0$ from the right, $f(x) \to +\infty$, but as $x \to 0$ from the left, $f(x) \to -\infty$. ■

The behavior of $1/x$, described in Example 8, is quite different from that of $1/x^2$. Since x^2 is positive whether x is positive or negative, and since $1/x^2$ is large when x is near 0, we have

$$\lim_{x \to 0^+} \frac{1}{x^2} = \infty \quad \text{and} \quad \lim_{x \to 0^-} \frac{1}{x^2} = \infty.$$

In this case we may write

$$\lim_{x \to 0} \frac{1}{x^2} = \infty,$$

meaning that "as $x \to 0$, both from the right and from the left, $1/x^2$ becomes arbitrarily large through positive values." We can also write

$$\lim_{x \to 0} \frac{1}{|x|} = \infty,$$

but there is no corresponding statement for $\lim_{x \to 0} 1/x$.

The many different types of limits all have the same flavor. Rather than spell each out in detail, we list some typical cases.

Notation	*In Words*	*Concept*	*Example*
$\lim_{x \to a} f(x) = L$	As x approaches a, $f(x)$ approaches L.	$f(x)$ is defined in some open intervals (c, a) and (a, b) and, as x approaches a from the right or from the left, $f(x)$ approaches L.	$\lim_{x \to 3} (2x + 1) = 7$
$\lim_{x \to \infty} f(x) = L$	As x approaches positive infinity, $f(x)$ approaches L.	$f(x)$ is defined for all x beyond some number and, as x gets large through positive values, $f(x)$ approaches L.	$\lim_{x \to \infty} \frac{1}{x} = 0$
$\lim_{x \to -\infty} f(x) = L$	As x approaches negative infinity, $f(x)$ approaches L.	$f(x)$ is defined for all x to the left of some number and, as the negative number x takes on large absolute values, $f(x)$ approaches L.	$\lim_{x \to -\infty} \frac{2x + 1}{2} = 2$
$\lim_{x \to \infty} f(x) = \infty$	As x approaches infinity, $f(x)$ approaches positive infinity.	$f(x)$ is defined for all x beyond some number and, as x gets large through positive values, $f(x)$ becomes and remains arbitrarily large and positive.	$\lim_{x \to \infty} x^3 = \infty$
$\lim_{x \to a^+} f(x) = \infty$	As x approaches a from the right, $f(x)$ approaches positive infinity.	$f(x)$ is defined in some open interval (a, b), and, as x approaches a from the right, $f(x)$ becomes and remains arbitrarily large and positive.	$\lim_{x \to 0^+} \frac{1}{x} = \infty$
$\lim_{x \to a^+} f(x) = -\infty$	As x approaches a from the right, $f(x)$ approaches negative infinity.	$f(x)$ is defined in some open interval (a, b), and, as x approaches a from the right, $f(x)$ becomes negative and $\lvert f(x) \rvert$ becomes and remains arbitrarily large.	$\lim_{x \to 1^+} \frac{1}{1 - x} = -\infty$
$\lim_{x \to a} f(x) = \infty$	As x approaches a, $f(x)$ approaches positive infinity.	$f(x)$ is defined for some open intervals (c, a) and (a, b), and, as x approaches a from either side, $f(x)$ becomes and remains arbitrarily large and positive.	$\lim_{x \to 0} \frac{1}{x^2} = \infty$

Other notations, such as

$$\lim_{x\to a} f(x) = -\infty \qquad \lim_{x\to a^-} f(x) = \infty \qquad \text{and} \qquad \lim_{x\to\infty} f(x) = -\infty$$

are defined similarly.

EXERCISES FOR SEC. 2.4: COMPUTATIONS OF LIMITS

In Exercises 1 to 32 examine the given limits and compute those that exist. Indicate limits that are infinite (either by "$-\infty$" or "∞").

1 $\lim_{x\to 2} (3x^2 + 2)$

2 $\lim_{x\to 4} (2x^2 - 5)$

3 $\lim_{x\to 2} \dfrac{3x^2 + 1}{x + 3}$

4 $\lim_{x\to 3} \dfrac{7x^2 - 10}{x - 1}$

5 $\lim_{x\to 1} [(4x^2 + x)(x + 3)]$

6 $\lim_{x\to 5} [(x^2 - x)(2x - 7)]$

7 $\lim_{x\to\infty} (7x + 2)$

8 $\lim_{x\to\infty} (5x - 9)$

9 $\lim_{x\to\infty} (4x^2 - x + 3)$

10 $\lim_{x\to\infty} (3x^2 - 7x + 2)$

11 $\lim_{x\to\infty} (x^5 - 100x^4)$

12 $\lim_{x\to\infty} (-4x^5 + 35x^2)$

13 $\lim_{x\to-\infty} (6x^5 + 21x^3)$

14 $\lim_{x\to-\infty} (19x^6 + 5x)$

15 $\lim_{x\to-\infty} (-x^3)$

16 $\lim_{x\to-\infty} (-x^4)$

17 $\lim_{x\to\infty} \dfrac{6x^3 - x}{2x^{10} + 5x + 8}$

18 $\lim_{x\to\infty} \dfrac{100x^9 + 22}{x^{10} + 21}$

19 $\lim_{x\to\infty} \dfrac{x^4 + 1066x^2 - 1492}{2x^4 - 2001}$

20 $\lim_{x\to\infty} \dfrac{6x^3 - x^2 + 5}{3x^3 - 100x + 1}$

21 $\lim_{x\to\infty} \dfrac{x^3 + 1}{x^4 + 2}$

22 $\lim_{x\to-\infty} \dfrac{5x^3 + 2x}{x^2 + x + 7}$

23 $\lim_{x\to 0^+} \dfrac{1}{x^3}$

24 $\lim_{x\to 0^-} \dfrac{1}{x^3}$

25 $\lim_{x\to 0^+} \dfrac{1}{x^4}$

26 $\lim_{x\to 0^-} \dfrac{1}{x^4}$

27 (*a*) $\lim_{x\to 1^+} \dfrac{1}{x - 1}$ (*b*) $\lim_{x\to 1^-} \dfrac{1}{x - 1}$ (*c*) $\lim_{x\to 1} \dfrac{1}{x - 1}$

28 (*a*) $\lim_{x\to -1^+} \dfrac{1}{(x + 1)^2}$ (*b*) $\lim_{x\to -1^-} \dfrac{1}{(x + 1)^2}$ (*c*) $\lim_{x\to -1} \dfrac{1}{(x + 1)^2}$

29 $\lim_{x\to\infty} \dfrac{\sqrt{4x^2 + 2x + 1}}{3x}$

30 $\lim_{x\to-\infty} \dfrac{\sqrt{9x^2 + x + 3}}{6x}$

31 $\lim_{x\to\infty} \dfrac{\sqrt{4x^2 + x}}{\sqrt{9x^2 - 3x}}$

32 $\lim_{x\to-\infty} \dfrac{\sqrt{x^2 + 3x + 1}}{\sqrt{16x^2 + x + 2}}$

33 (*a*) "I am thinking of two numbers that are very near 0. What, if anything, can you say about their product?" Explain.
(*b*) "I am thinking of two numbers that are very near 0. What, if anything, can you say about their quotient?" Explain.

34 (*a*) "I am thinking of two very large positive numbers. What, if anything, can you say about their product?" Explain.
(*b*) "I am thinking of two very large positive numbers. What, if anything, can you say about their quotient?" Explain.
(*c*) "I am thinking of two very large positive numbers. What, if anything, can you say about their difference?" Explain.

35 Examine $\lim_{x\to\infty} x/2^x$.

36 Examine $\lim_{x\to\infty} 2^x/4^x$, $\lim_{x\to-\infty} 2^x/4^x$, and $\lim_{x\to 0} 2^x/4^x$.

37 (*a*) Determine $\lim_{x\to 0} \dfrac{2x^3 + x^2 + x}{3x^3 - x^2 + 2x}$.
(*b*) Check your answer by calculating the quotient when $x = 0.01$.

---

38 Two citizens are arguing about

$$\lim_{x\to\infty} \left(\frac{3x^2 + 2x}{x + 5} - 3x\right).$$

The first claims, "For large x, $2x$ is small in comparison to $3x^2$, and 5 is small in comparison to x. So the quotient $(3x^2 + 2x)/(x + 5)$ behaves like $3x^2/x = 3x$. Hence the limit in question is 0." Her companion replies, "Nonsense. After all,

$$\frac{3x^2 + 2x}{x + 5} = \frac{3x + 2}{1 + (5/x)}$$

which clearly behaves like $3x + 2$ for large x. Thus the limit in question is 2, not 0." Settle the argument.

39 Three citizens are arguing about limits in a case where $\lim_{x\to\infty} f(x) = 0$ and $\lim_{x\to\infty} g(x) = \infty$. The first citizen claims that, "$\lim_{x\to\infty} f(x)g(x) = 0$, since $f(x)$ is going toward 0." The second citizen cries out, "Rubbish! Since $g(x)$ gets large, it will turn out that $\lim_{x\to\infty} f(x)g(x) = \infty$." The third citizen interrupts with, "You're both wrong. The two influences will balance out and you will see that $\lim_{x\to\infty} f(x)g(x) = 1$." Settle the argument.

40 Two citizens are arguing about limits in a case where $f(x) \geq 1$ for $x > 0$, $\lim_{x\to 0^+} f(x) = 1$, and $\lim_{x\to 0} g(x) = \infty$. What can be said about $\lim_{x\to 0^+} f(x)^{g(x)}$? The first citizen says, "That's easy. Multiply a bunch of 1's and you get 1. So the limit will be 1." "Nonsense," says the other citizen, "since $f(x)$ may be bigger than 1 and you are multiplying it lots and lots of times, you will get a really large number. There's no doubt in my mind: $\lim_{x\to 0^+} f(x)^{g(x)} = \infty$." Settle the argument.

In Exercises 41 to 43 information is given about functions f and g. In each case decide whether the limit asked for can be determined on the basis of that information. If it can, give its value. If it cannot, show by specific choices of f and g that it cannot.

41 Given that $\lim_{x\to\infty} f(x) = 0$ and $\lim_{x\to\infty} g(x) = 1$, discuss

(*a*) $\lim_{x\to\infty} (f(x) + g(x))$ (*b*) $\lim_{x\to\infty} (f(x)/g(x))$

(*c*) $\lim_{x\to\infty} f(x)g(x)$ (*d*) $\lim_{x\to\infty} (g(x)/f(x))$

(*e*) $\lim_{x\to\infty} g(x)/|f(x)|$

42 Given that $\lim_{x\to\infty} f(x) = \infty$ and $\lim_{x\to\infty} g(x) = \infty$, discuss

(*a*) $\lim_{x\to\infty} (f(x) + g(x))$ (*b*) $\lim_{x\to\infty} (f(x) - g(x))$

(*c*) $\lim_{x\to\infty} f(x)g(x)$ (*d*) $\lim_{x\to\infty} (g(x)/f(x))$

43 Given that $\lim_{x\to\infty} f(x) = 1$ and $\lim_{x\to\infty} g(x) = \infty$, discuss

(*a*) $\lim_{x\to\infty} (f(x)/g(x))$ (*b*) $\lim_{x\to\infty} f(x)g(x)$

(*c*) $\lim_{x\to\infty} (f(x) - 1)g(x)$

44 Let $P(x)$ be a polynomial of degree n, with lead term ax^n, $a > 0$, and let $Q(x)$ be a polynomial of degree m, with lead term bx^m, $b > 0$. Examine $\lim_{x\to\infty} P(x)/Q(x)$ if (*a*) $m = n$, (*b*) $m < n$, (*c*) $m > n$.

45 A function f is defined as follows: $f(x) = 2$ if x is an integer and $f(x) = 3$ if x is not an integer. (*a*) Graph f. (*b*) Discuss $\lim_{x\to\infty} f(x)$. (*c*) Discuss $\lim_{x\to 2} f(x)$.

* **46** Examine $\lim_{x\to\infty} \sqrt{x^2 + 100} - x$, as follows.

(*a*) Calculate $\sqrt{x^2 + 100} - x$ for several large values of x.

(*b*) Make a conjecture.

(*c*) Show that your conjecture is true. (The algebraic identity $a - b = \dfrac{(a - b)(a + b)}{a + b}$ may help.)

* **47** Like the previous exercise, for $\lim_{x\to\infty} \sqrt{x^2 + 20x} - x$.

2.5 SOME TOOLS FOR GRAPHING

One way to graph a function $f(x)$ is to compute $f(x)$ at several inputs x, plot the points $(x, f(x))$ that you get, and draw a curve through them. This procedure may be tedious and, if you happen to choose inputs that give little information, may result in an inaccurate graph.

Another way is to use a calculator that has a graphing routine built in. However, only a portion of the graph is displayed and, if you have no idea what to expect, you may have asked it to compute a part of the graph that is misleading or of little interest. At points with large function values, the graph may be distorted by the calculator's choice of scale.

So it pays to be able to get some idea of the general shape of a graph quickly, without having to compute lots of values. This section describes some shortcuts.

Symmetry of Odd and Even Functions

Some functions have the property that when you replace x by $-x$, you don't change the value of the function. For instance, the function $f(x) = x^2$ has this property since

$$f(-x) = (-x)^2 = x^2 = f(x).$$

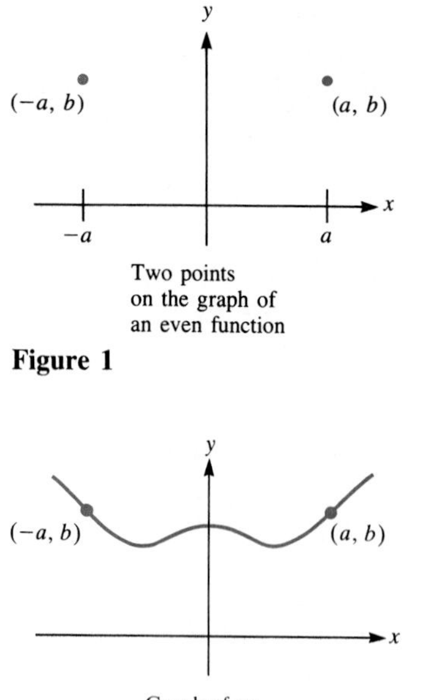

Two points on the graph of an even function

Figure 1

Graph of an even function

Figure 2

So does the function $f(x) = x^n$ for any *even* integer n. There are fancier functions, such as $3x^4 - 5x^2 + 6$ and $\cos x$, that also have this property.

> **Definition** *Even function.* A function f such that $f(-x) = f(x)$ is called an **even function.**

For an even function f, if $f(a) = b$, then $f(-a) = b$ also. In other words, if the point (a, b) is on the graph of f, so is the point $(-a, b)$, as indicated in Fig. 1.

This means that the graph of f is symmetric with respect to the y axis, as shown in Fig. 2. So if you notice that a function is even, you can save yourself some work in finding its graph. First graph it for positive x and then get the part for negative x free of charge by reflection across the y axis. If you wanted to graph $y = x^4/(1 - x^2)$, for example, first stick to $x > 0$, then reflect the result.

> **Definition** *Odd function.* A function f such that $f(-x) = -f(x)$ is called an **odd function.**

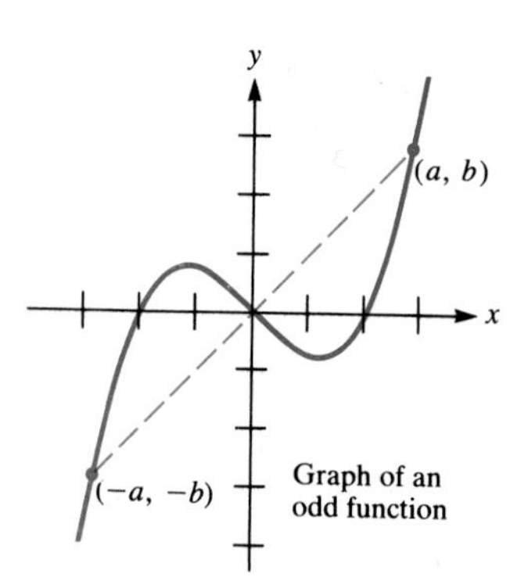

Graph of an odd function

Figure 3

The function $f(x) = x^3$ is odd since

$$f(-x) = (-x)^3 = -(x^3) = -f(x).$$

For any odd integer n, $f(x) = x^n$ is an odd function. The sine function is also odd, since $\sin(-x) = -\sin x$.

If the point (a, b) is on the graph of an odd function, so is the point $(-a, -b)$, since

$$f(-a) = -f(a) = -b.$$

(See Fig. 3.) Note that the origin $(0, 0)$ is the midpoint of (a, b) and $(-a, -b)$. The graph is said to be "symmetric with respect to the origin." If you work out the graph of an odd function for positive x, you can obtain the graph for negative x by reflecting it point by point through the origin. For example, if you graph $y = x^3$ for $x \geq 0$, as in Fig. 4, you can complete the graph by reflection in the origin, as indicated by the dashed lines.

Most functions are neither even nor odd. For instance, $x^3 + x^4$ is neither even nor odd since $(-x)^3 + (-x)^4 = -x^3 + x^4$, which is neither $x^3 + x^4$ nor $-(x^3 + x^4)$.

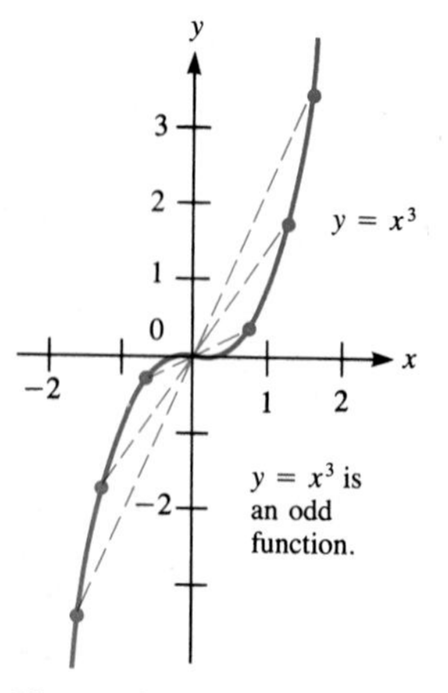

$y = x^3$ is an odd function.

Figure 4

Intercepts

The x coordinates of the points where the graph of a function meets the x axis are the x **intercepts** of the function. The y coordinates of the points where a graph meets the y axis are the y **intercepts** of the function.

EXAMPLE 1 Find the intercepts of the graph of $y = x^2 - 4x - 5$.

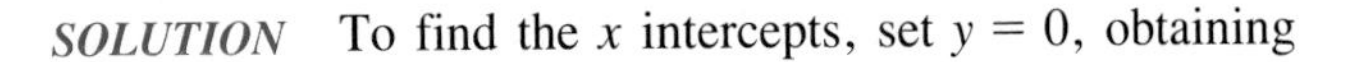

SOLUTION To find the x intercepts, set $y = 0$, obtaining

$$0 = x^2 - 4x - 5.$$

Factoring also works.

By the quadratic formula,

$$x = \frac{4 \pm \sqrt{16 + 20}}{2} = \frac{4 \pm \sqrt{36}}{2} = \frac{4 \pm 6}{2} = 5 \text{ and } -1.$$

Graph of $y = x^2 - 4x - 5$

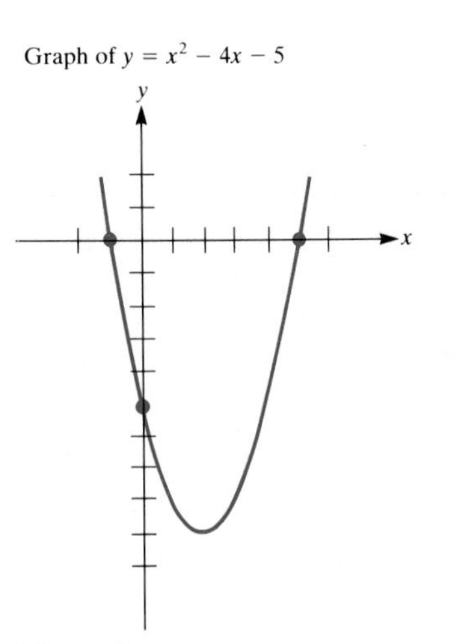

Figure 5

So the graph meets the x axis at $x = 5$ and at $x = -1$.

To find y intercepts, set $x = 0$, obtaining

$$y = 0^2 - 4 \cdot 0 - 5 = -5.$$

There is only one y intercept, namely, -5.

The intercepts in this case give us three points on the graph. Tabulating a few more points gives the parabola in Fig. 5, where the intercepts are shown as well. ■

If $f(x)$ is not defined when $x = 0$, there is no y intercept. If $f(x)$ is defined when $x = 0$, then it's easy to get the y intercept; just evaluate $f(0)$. While there is at most one y intercept, there may be many x intercepts. To find them, solve the equation $f(x) = 0$. In short,

> To find the y intercept, compute $f(0)$.
>
> To find the x intercepts, solve the equation $f(x) = 0$.

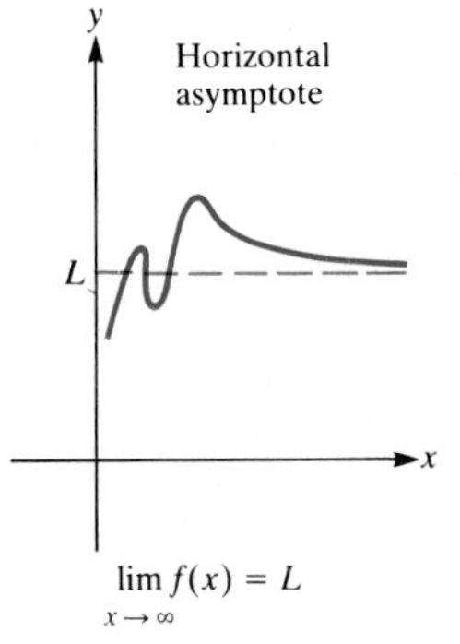

$\lim_{x \to \infty} f(x) = L$

Figure 6

Asymptotes

Horizontal asymptotes

If $\lim_{x\to\infty} f(x) = L$, where L is a real number, the graph of $y = f(x)$ gets arbitrarily close to the horizontal line $y = L$ as x increases. The line $y = L$ is called a **horizontal asymptote** of the graph of f. An asymptote is defined similarly if $f(x) \to L$ as $x \to -\infty$. (See Fig. 6.)

If a graph has an asymptote, we can draw it and use it as a guide in drawing the graph.

Vertical asymptote

If $f(x) \to \infty$ as $x \to a$, then the graph resembles the vertical line $x = a$ for x near a. The line $x = a$ is called a **vertical asymptote**. A similar definition holds if

$$\lim_{x\to a} f(x) = -\infty, \quad \lim_{x\to a^+} f(x) = \infty \text{ or } -\infty \qquad \text{or} \qquad \lim_{x\to a^-} f(x) = \infty \text{ or } -\infty.$$

Figures 7 to 10 illustrate these situations.

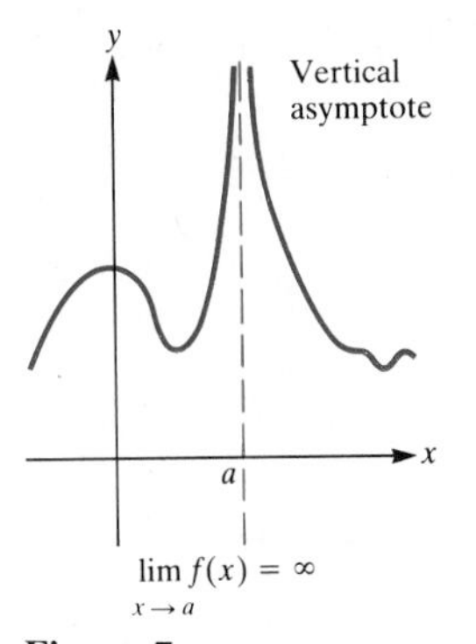

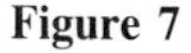

Figure 7

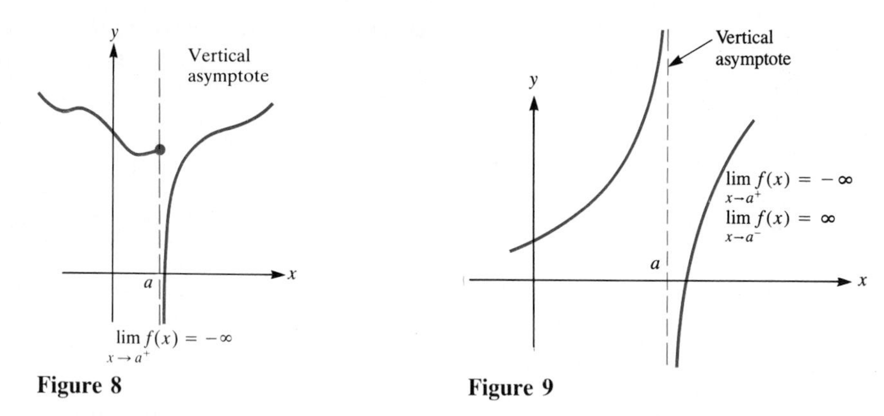

Figure 8

Figure 9

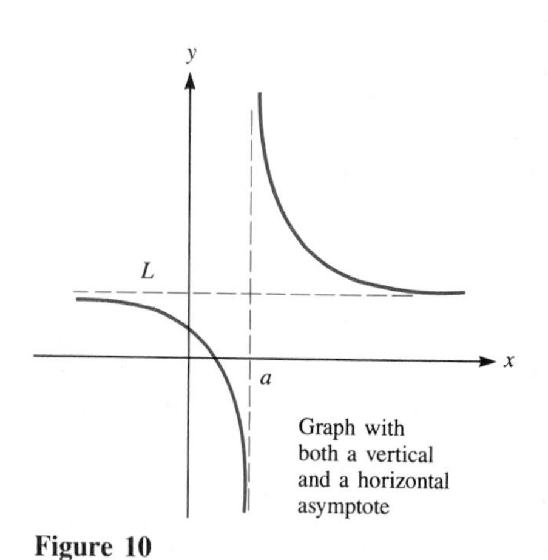

Figure 10

The next example illustrates the use of intercepts and asymptotes. Example 3 illustrates the use of symmetry as well.

EXAMPLE 2 Graph $f(x) = 1/(x-1)^2$.

SOLUTION To see if there is any *symmetry*, check whether $f(-x)$ is $f(x)$ or $-f(x)$. We have

$$f(-x) = \frac{1}{(-x-1)^2} = \frac{1}{(x+1)^2}.$$

Since $1/(x+1)^2$ is neither $1/(x-1)^2$ nor $-1/(x-1)^2$, the function $f(x)$ is neither even nor odd. Therefore the graph is *not* symmetric with respect to the y axis or with respect to the origin.

To determine the *y intercept* compute $f(0) = 1/(0-1)^2 = 1$. The y intercept is 1. To find any x intercepts, solve the equation $f(x) = 0$, that is,

$$\frac{1}{(x-1)^2} = 0.$$

Since no number has a reciprocal equal to 0, there are no x intercepts.

To search for a *horizontal asymptote* examine $\lim_{x\to\infty} 1/(x-1)^2$ and $\lim_{x\to-\infty} 1/(x-1)^2$: $\lim_{x\to\infty} 1/(x-1)^2 = 0$ and $\lim_{x\to-\infty} 1/(x-1)^2 = 0$. The line $y = 0$, that is, the x axis, is an asymptote both to the right and the left. Since $1/(x-1)^2$ is positive, the graph lies above the asymptote.

To discover any *vertical asymptotes*, find when the function $1/(x-1)^2$ "blows up"—that is, becomes arbitrarily large. This happens when the denominator $(x-1)^2$ becomes small. Setting the denominator equal to 0, $(x-1)^2 = 0$, we find that $x = 1$. The function is not defined at $x = 1$. The line $x = 1$ is a vertical asymptote.

To determine the shape of the graph near the line $x = 1$, we examine $\lim_{x\to1^+} 1/(x-1)^2$ and $\lim_{x\to1^-} 1/(x-1)^2$. Since the square of a nonzero number is always positive, we see that $\lim_{x\to1^+} 1/(x-1)^2 = +\infty$ and $\lim_{x\to1^-} 1/(x-1)^2 = +\infty$. All this information is displayed in Fig. 11. ■

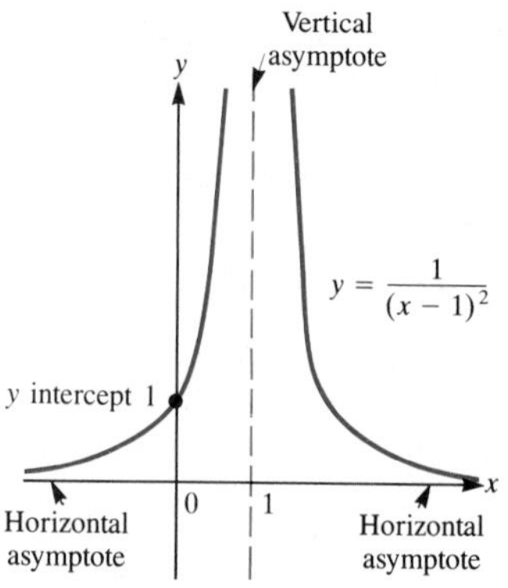

Vertical asymptote

$y = \frac{1}{(x-1)^2}$

y intercept 1

Horizontal asymptote

Horizontal asymptote

Figure 11

EXAMPLE 3 Graph $y = \dfrac{x^3 - x}{x^2 - 4}$.

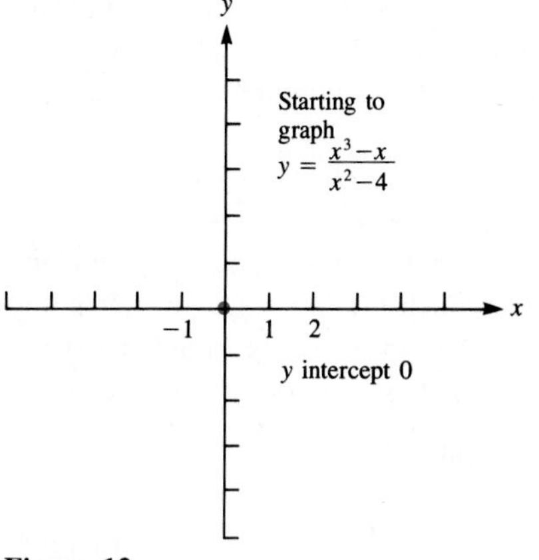

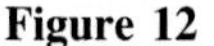
Figure 12

SOLUTION In this case $f(x) = (x^3 - x)/(x^2 - 4)$. To check for symmetry, see whether $f(-x)$ happens to equal $f(x)$ or $-f(x)$. We have

$$f(-x) = \frac{(-x)^3 - (-x)}{(-x)^2 - 4} = \frac{-x^3 + x}{x^2 - 4} = -f(x).$$

So f is an odd function and its graph is symmetric with respect to the origin. (We can graph $f(x)$ for $x > 0$ and then reflect that graph through the origin.)

How about intercepts? Let us check for a y intercept since that is easy to do; just evaluate $f(0)$:

$$f(0) = \frac{0^3 - 0}{0^2 - 4} = \frac{0}{-4} = 0.$$

So the y intercept is 0; the graph passes through the origin. (See Fig. 12.)

To find any x intercepts we must solve the equation

$$\frac{x^3 - x}{x^2 - 4} = 0.$$

The numerator must be 0:

$$x^3 - x = 0. \tag{1}$$

Since $x^3 - x = x(x^2 - 1) = x(x + 1)(x - 1)$, Eq. (1) is equivalent to

$$x(x + 1)(x - 1) = 0. \tag{2}$$

The product of several numbers is 0 only when at least one of them is 0, so (2) reduces to the three equations

$$x = 0 \qquad \text{or} \qquad x + 1 = 0 \qquad \text{or} \qquad x - 1 = 0.$$

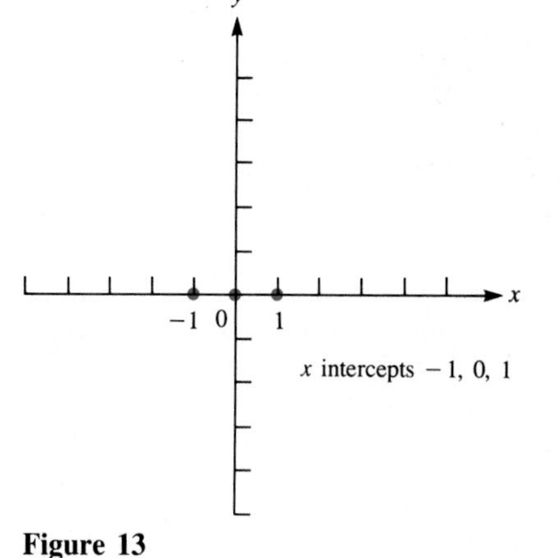

Figure 13

So there are three x intercepts: $x = 0$, -1, and 1. (See Fig. 13.)

Finally, check for asymptotes. To see whether there is a horizontal asymptote we examine

$$\lim_{x \to \infty} \frac{x^3 - x}{x^2 - 4} \qquad \text{and} \qquad \lim_{x \to -\infty} \frac{x^3 - x}{x^2 - 4}.$$

Since $(x^3 - x)/(x^2 - 4)$ is a rational function, the limit is determined by the limit of the quotient of their terms of highest degree. Thus

$$\lim_{x \to \infty} \frac{x^3 - x}{x^2 - 4} = \lim_{x \to \infty} \frac{x^3}{x^2} = \lim_{x \to \infty} x = \infty$$

and

$$\lim_{x \to -\infty} \frac{x^3 - x}{x^2 - 4} = \lim_{x \to -\infty} \frac{x^3}{x^2} = \lim_{x \to -\infty} x = -\infty.$$

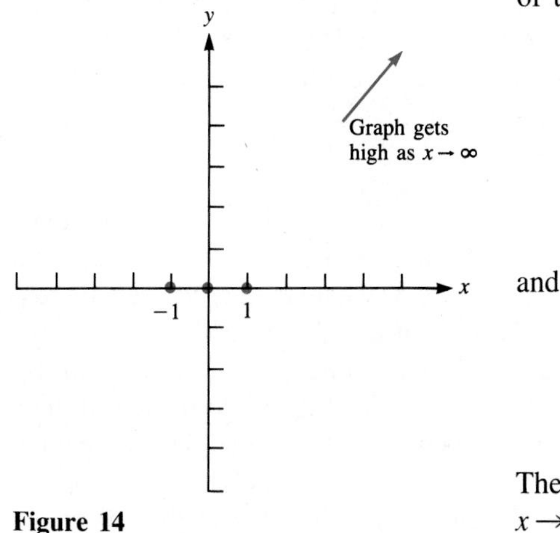

Figure 14

There are no horizontal asymptotes. But we do now know that f gets large as $x \to \infty$, as shown in Fig. 14.

Are there any vertical asymptotes? Such an asymptote occurs at a number a where the function ''blows up''—that is, at a number a near which $f(x)$ becomes arbitrarily large. In this case, this happens when the denominator of the quotient $(x^3 - x)/(x^2 - 4)$ is 0; that is, when

$$x^2 - 4 = 0.$$

The solutions of this equation are 2 and -2. How does $f(x)$ behave for x near 2? (We don't need to consider -2, since we will get the graph for negative x by reflecting the graph for positive x through the origin.) We know that $f(x)$ is large, but is it positive or negative? To see what's going on, we must examine

$$\lim_{x\to 2^+} f(x) \qquad \text{and} \qquad \lim_{x\to 2^-} f(x).$$

First let x approach 2 from the right. Pick a number x that is a little larger than 2, such as 2.01. For such a number, $x^3 - x$ is close to 6 and $x^2 - 4$ is a small positive number, since x^2 is larger than 4 (think of 2.01^2 and 4). So we can think of

$$\frac{x^3 - x}{x^2 - 4}$$

as

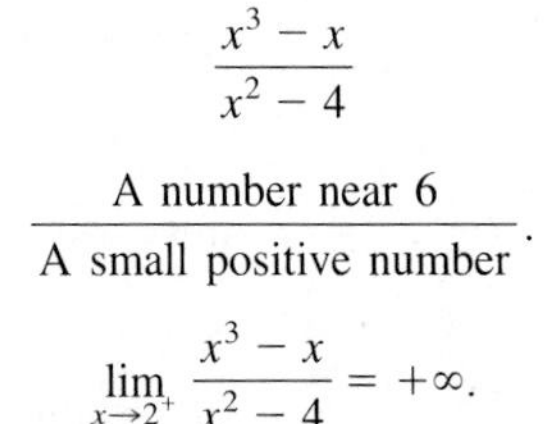

$$\frac{\text{A number near 6}}{\text{A small positive number}}.$$

Thus

$$\lim_{x\to 2^+} \frac{x^3 - x}{x^2 - 4} = +\infty.$$

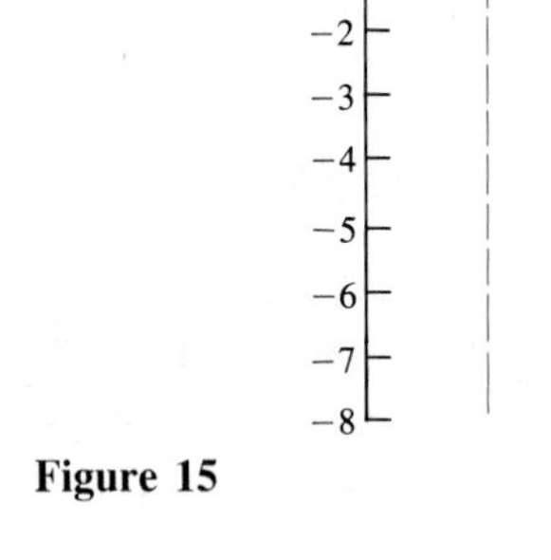

Figure 15

(See Fig. 15.)

Now let x approach 2 from the left. What if x is near 2 but smaller than 2? (Think of 1.99 for instance.) Then $x^3 - x$ is again near 6, but $x^2 - 4$ is now a small *negative* number. So

$$\lim_{x\to 2^-} \frac{x^3 - x}{x^2 - 4} = -\infty.$$

We can sketch a little more of the graph, adding a part near the line $x = 2$, but very low. (See Fig. 16.)

At this point we can guess the graph's shape, except for the part where x is in $[0, 1]$. So let us compute $f(x)$ for some choice of x in $[0, 1]$, say $x = \frac{1}{2}$. We have

$$f(\tfrac{1}{2}) = \frac{(\frac{1}{2})^3 - \frac{1}{2}}{(\frac{1}{2})^2 - 4} = \frac{\frac{1}{8} - \frac{1}{2}}{\frac{1}{4} - 4} = \frac{-\frac{3}{8}}{-\frac{15}{4}} = \frac{3}{8} \cdot \frac{4}{15} = \frac{1}{10}.$$

The point $(\frac{1}{2}, \frac{1}{10})$ is on the graph. Since $\frac{1}{10}$ is positive, the curve lies *above* the x axis for x in $(0, 1)$. How high does it go in that interval? We will not have the machinery to answer that question until Chap. 4 (where the derivative will help us out). Figure 17 records what we have learned. (We consider only $x \geq 0$.) We join the pieces, keeping in mind that the curve does not meet the positive x axis at any point other than $x = 1$. (See Fig. 18.) The part to the left of the y axis is then drawn by using the symmetry with respect to the origin, as shown in Fig. 19.

How low the curve goes for $x > 2$ can be determined with the aid of the derivative. (You will be asked to do this in Exercise 53 of Sec 4.2.) ■

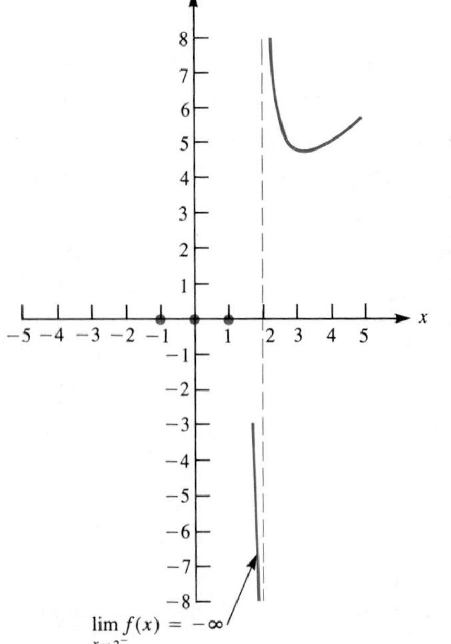

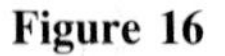

Figure 16

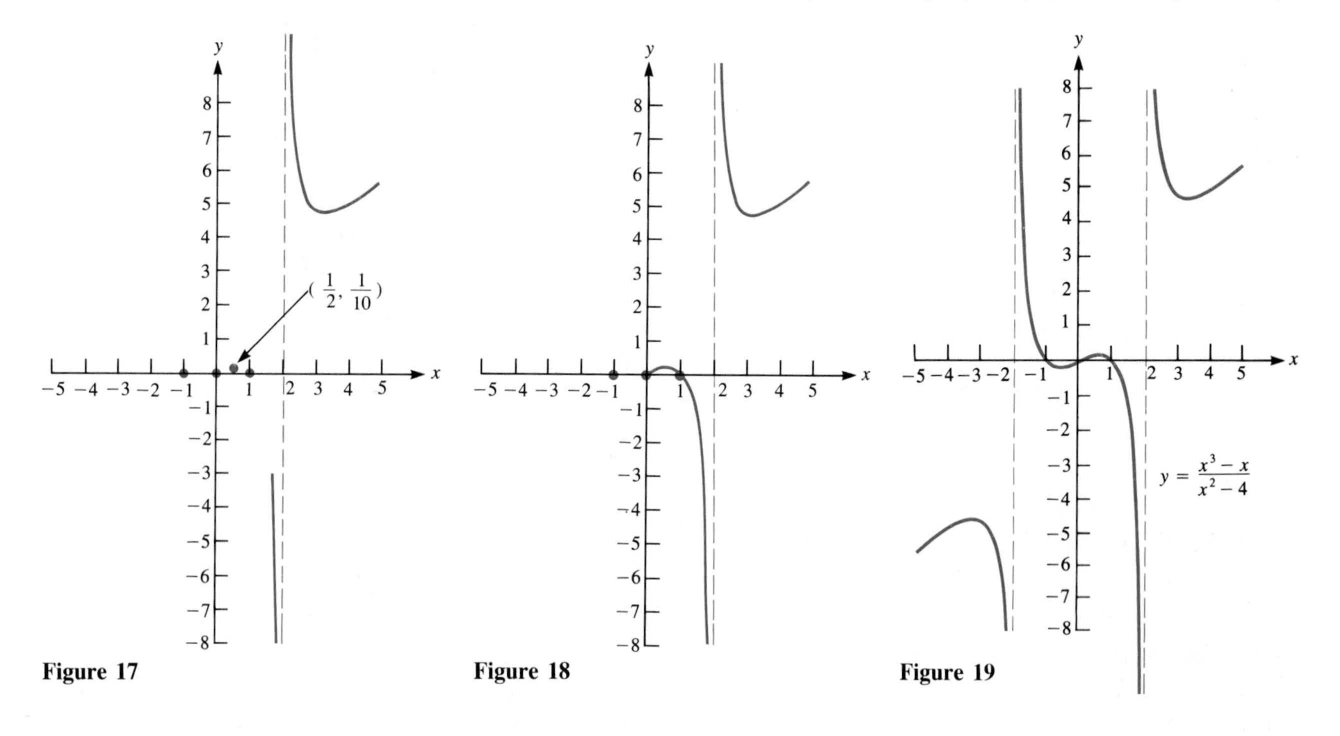

Figure 17

Figure 18

Figure 19

Observe that in this example we did not make an extensive table of values and plot points. Even if we had, we might not have noticed the vertical asymptotes, for the inputs we choose might not include −2 or 2. As we have noted, finding where "peaks" and "valleys" occur will require the derivative, the topic of Chap. 4.

Section Summary

This section presents three tools for making a quick sketch of the graph of $y = f(x)$.

1 *Check for symmetry.* Is $f(-x)$ equal to $f(x)$ or $-f(x)$?

2 *Check for intercepts.* Find $f(0)$ to get the y intercept. Solve $f(x) = 0$ to get the x intercepts.

3 *Check for asymptotes.* If $\lim_{x \to \infty} f(x) = L$ or $\lim_{x \to -\infty} f(x) = L$ (where L is some real number), then the line $y = L$ is a horizontal asymptote. If $\lim_{x \to a} f(x) = +\infty$ or $-\infty$, then the line $x = a$ is a vertical asymptote. This is also the case whenever $\lim_{x \to a^-} f(x)$ or $\lim_{x \to a^+} f(x)$ is $+\infty$ or $-\infty$.

EXERCISES FOR SEC. 2.5: SOME TOOLS FOR GRAPHING

1 Show that these are even functions.
(*a*) $x^2 + 2$
(*b*) $\sqrt{x^4 - 1}$
(*c*) $1/x^2$

2 Show that these are even functions.
(*a*) $\sqrt{1 - x^2}$
(*b*) $5x^4 - x^2$
(*c*) $7/x^6$

3 Show that these are odd functions.
(*a*) $x^3 + x$
(*b*) $x + 1/x$
(*c*) $\sqrt[3]{x}$

4 Show that these are odd functions.
(*a*) $2x - x^3$
(*b*) $\dfrac{x^3}{1 + x^2}$
(*c*) $\sqrt[5]{x}$

5 Show that these functions are neither odd nor even.
(*a*) $3 + x$
(*b*) $(x + 2)^2$
(*c*) $\dfrac{x}{x + 1}$

6 Show that these functions are neither odd nor even.
(*a*) $2x - 1$
(*b*) $x^3 + x^2$
(*c*) $x^2 + 1/x$

7 Label each function as even, odd, or neither.
(*a*) $x + x^3 + 5x^4$
(*b*) $7x^4 - 5x^2$
(*c*) $\sqrt[3]{x^2 + 1}$

8 Label each function as even, odd, or neither.
(*a*) $\dfrac{1 + x}{1 - x}$
(*b*) $\dfrac{4x^2 - 3x^4}{x^3}$
(*c*) $x\sqrt[3]{x}$

In Exercises 9 to 14 find the x and y intercepts, if any.

9 $y = 2x + 3$ **10** $y = 3x - 7$
11 $y = x^2 + 3x + 2$ **12** $y = 2x^2 + 5x + 3$
13 $y = 2x^2 + 1$ **14** $y = x^2 + x + 1$

In Exercises 15 to 20 find all the horizontal and vertical asymptotes.

15 $y = \dfrac{x + 2}{x - 2}$ **16** $y = \dfrac{x - 2}{x^2 - 9}$

17 $y = \dfrac{x}{x^2 + 1}$ **18** $y = \dfrac{2x + 3}{x^2 + 4}$

19 $y = \dfrac{x^2 + 1}{x^2 - 3}$ **20** $y = \dfrac{x}{x^2 + 2x + 1}$

In Exercises 21 to 28 graph the function.

21 $y = \dfrac{1}{x - 2}$ **22** $y = \dfrac{1}{x + 3}$

23 $y = \dfrac{1}{x^2 - 1}$ **24** $y = \dfrac{x}{x^2 - 2}$

25 $y = \dfrac{x^2}{1 + x^2}$ **26** $y = x^3 + x^{-1}$

27 $y = \dfrac{1}{x(x - 1)(x + 2)}$ **28** $y = \dfrac{x + 2}{x^3 + x^2}$

Exercises 29 to 35 concern even and odd functions.

29 If two functions are odd, what can you say about
(*a*) their sum? (*b*) their product? (*c*) their quotient?

30 If two functions are even, what can you say about
(*a*) their sum? (*b*) their product? (*c*) their quotient?

31 If f is odd and g is even, what can you say about
(*a*) $f + g$? (*b*) fg? (*c*) f/g?

32 What, if anything, can you say about $f(0)$ if
(*a*) f is an even function?
(*b*) f is an odd function?

33 Which polynomials are odd? Explain.

34 Which polynomials are even? Explain.

35 Is there a function that is both odd and even? Explain.

***36** Let $P(x)$ be a polynomial of degree m and $Q(x)$ a polynomial of degree n. For which m and n does the graph of $y = P(x)/Q(x)$ have a horizontal asymptote?

Exercises 37 to 40 concern tilted asymptotes. Let $A(x)$ and $B(x)$ be polynomials such that the degree of $A(x)$ is equal to 1 plus the degree of $B(x)$. Then when you divide $B(x)$ into $A(x)$, you get a quotient $Q(x)$, which is a polynomial of degree 1, and a remainder $R(x)$, which is a polynomial of degree less than the degree of $B(x)$.

For example, if $A(x) = x^2 + 3x + 4$ and $B(x) = 2x + 2$,

$$\begin{array}{r} \tfrac{1}{2}x + 1 \quad \leftarrow Q(x) \\ 2x + 2\,\overline{)\,x^2 + 3x + 4} \\ \underline{x^2 + x} \phantom{{}+4} \\ 2x + 4 \\ \underline{2x + 2} \\ 2 \quad \leftarrow R(x) \end{array}$$

Thus

$$x^2 + 3x + 4 = (\tfrac{1}{2}x + 1)(2x + 2) + 2.$$

This tells us that

$$\frac{x^2 + 3x + 4}{2x + 2} = \tfrac{1}{2}x + 1 + \frac{2}{2x + 2}.$$

When x is large, $2/(2x + 2) \to 0$. Thus the graph of $y = (x^2 + 3x + 4)/(2x + 2)$ is asymptotic to the line $y = \tfrac{1}{2}x + 1$. See Fig. 20.

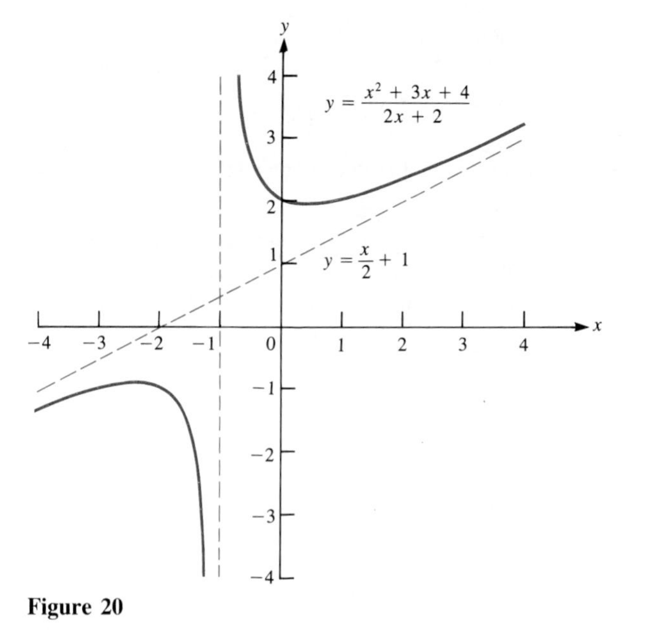

Figure 20

Whenever the degree of $A(x)$ exceeds the degree of $B(x)$ by exactly 1, the graph of $y = A(x)/B(x)$ has a tilted asymptote. You find it as we did in the example, by dividing $B(x)$ into $A(x)$, obtaining a quotient $Q(x)$ and a remainder $R(x)$. Then

$$\frac{A(x)}{B(x)} = Q(x) + \frac{R(x)}{B(x)}.$$

The asymptote is $y = Q(x)$. In each exercise graph the function, showing all asymptotes.

37 $y = \dfrac{x^2}{x - 1}$

38 $y = \dfrac{x^3}{x^2 - 1}$

39 $y = \dfrac{x^2 - 4}{x + 4}$

40 $y = \dfrac{x^2 + x + 1}{x - 2}$

41 Assume that you already have drawn the graph of a function $y = f(x)$. How would you obtain the graph of $y = g(x)$ from that graph if

(*a*) $g(x) = f(x) + 2$?
(*b*) $g(x) = f(x) - 2$?
(*c*) $g(x) = f(x - 2)$?
(*d*) $g(x) = f(x + 2)$?
(*e*) $g(x) = 2f(x)$?
(*f*) $g(x) = 3f(x - 2)$?

* **42** Is there a function f defined for all x such that $f(-x) = 1/f(x)$? If so, how many? If not, explain why there are no such functions.

* **43** Is there a function f defined for all x such that $f(-x) = 2f(x)$? If so, how many? If not, explain why there are no such functions.

44 Is there a constant k such that the function

$$f(x) = \frac{1}{3^x - 1} + k$$

is odd? even?

2.6 A REVIEW OF TRIGONOMETRY

In the next section we will examine an important limit involving a trigonometric function. For this reason we pause to review radian measure and the six trigonometric functions.

Radian Measure

In daily life angles are measured in degrees, with 360° measuring the full circular angle. The number 360 was chosen by the Babylonian astronomers, perhaps because there are close to 360 days in a year and 360 has many divisors. This somewhat arbitrary measure would complicate the calculus of trigonometric functions. In calculus a much more natural system is used, called **radian measure**. It is defined as follows.

Let ABC be an angle, as shown in Fig. 1. To measure its size draw a circle

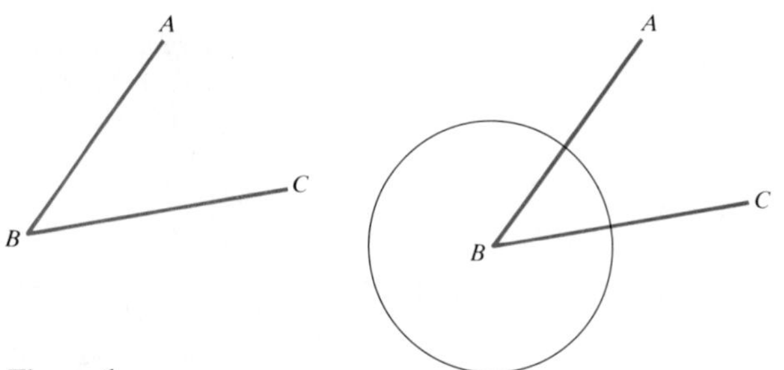

Figure 1

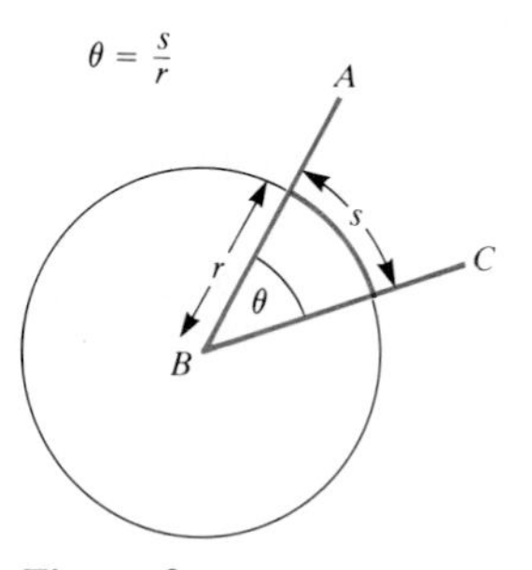

Figure 2

with center at B. If the circle has radius r and the angle intercepts an arc of length s, then the quotient s/r shall be the measure of the angle, and we say that the angle has a measure of s/r *radians*. It is frequently convenient to denote the measure of an angle by θ, and then write $\theta = s/r$. (See Fig. 2.) The radian measure of an angle does not depend on the size of the circle. Since both s and r measure lengths, their ratio θ is dimensionless.

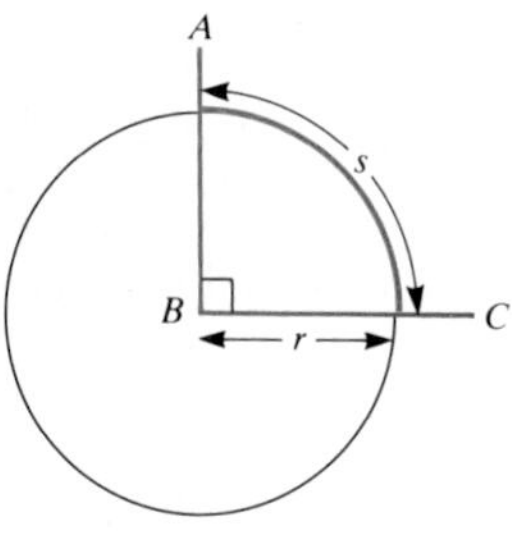

Figure 3

EXAMPLE 1 Find the radian measure of the right angle ABC in Fig. 3.

SOLUTION Draw a circle of radius r centered at B and compute the quotient s/r. The circumference of the circle of radius r is $2\pi r$. The right angle intercepts a quarter of the circumference. Thus

$$s = \frac{1}{4}(2\pi r) = \frac{\pi r}{2}.$$

It follows that

$$\theta = \frac{s}{r} = \frac{\pi r/2}{r} = \frac{\pi}{2}.$$

So a right angle has the measure $\pi/2$ radians (about 1.57 radians). ■

Translation Between Degrees and Radians

We grow up thinking of a right angle as a 90° angle. As Example 1 shows, a right angle is also $\pi/2$ radians. With practice, we can get used to thinking of a right angle as $\pi/2$ radians.

The straight angle, which has 180°, consists of two right angles and therefore has a measure of π radians. This fact is helpful in translating from degrees to radians and from radians to degrees: Simply use the proportion

$$\frac{\text{Degrees}}{180} = \frac{\text{Radians}}{\pi}. \tag{1}$$

EXAMPLE 2 What is the measure in radians of the 30° angle?

SOLUTION The proportion (1) in this case becomes

$$\frac{30}{180} = \frac{\text{Radians}}{\pi},$$

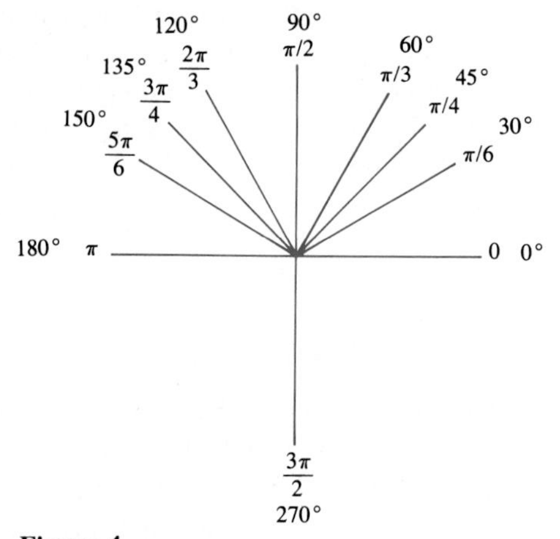

Figure 4

from which it follows that

$$\text{Radians} = \pi\frac{30}{180} = \frac{\pi}{6}. \quad ■$$

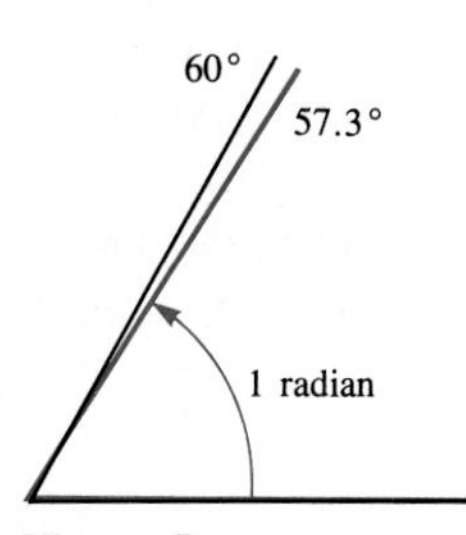

Figure 5

Computations like those in Example 2 can find the radian measure of any angle given in degrees. Figure 4 shows the radian measure of the more famous angles.

Equation (1) also can help us go from radians to degrees, as Example 3 illustrates.

EXAMPLE 3 How many degrees are there in an angle of 1 radian?

SOLUTION In this case, the proportion is

$$\frac{\text{Degrees}}{180} = \frac{1}{\pi};$$

hence

$$\text{Degrees} = \frac{180}{\pi} \approx \frac{180}{3.14} \approx 57.3°.$$

So an angle of 1 radian is about 57.3°, a little less than 60°. See Fig. 5. ■

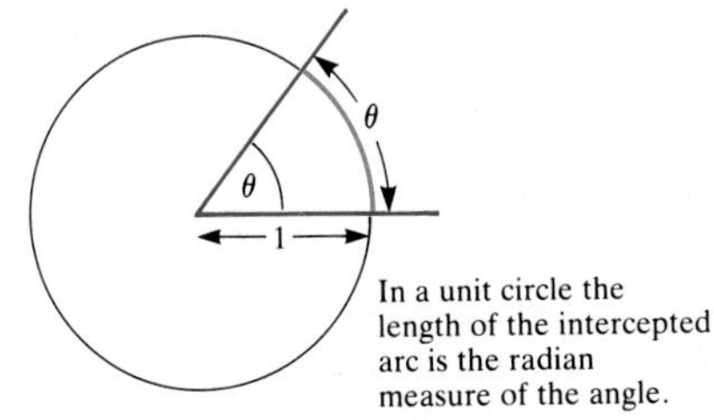

In a unit circle the length of the intercepted arc is the radian measure of the angle.

Figure 6

In the case of the **unit circle**, the circle whose radius is 1, the formula $\theta = s/r$ becomes $\theta = s/1 = s$. In that case, the length of arc intercepted equals the measure of the angle in radians, as shown in Fig. 6.

Angles Larger Than 2π

So far, an angle has been associated with each number θ in the interval $[0, 2\pi]$. We now will consider θ larger than 2π. For convenience, a unit circle will be used. One arm of the angle will be placed along the positive x axis. To draw the second arm of the angle, go around the circle in a counterclockwise direction a distance θ. For instance, if $\theta = 5\pi/2$, it is necessary to travel clear around the

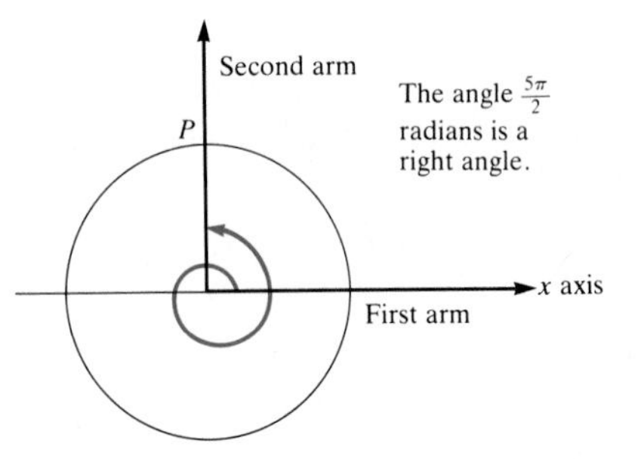

Figure 7

circle once and then reach the point P above the center of the circle. In this case we obtain the right angle, also described by $\pi/2$ radians and shown in Fig. 7.

Every time we travel around the unit circle, we increase the measure of an angle by 2π radians. Thus the right angle of $\pi/2$ radians has an endless supply of descriptions:

$$\frac{\pi}{2}, \quad \frac{\pi}{2} + 2\pi = \frac{5\pi}{2}, \quad \frac{\pi}{2} + 4\pi = \frac{9\pi}{2}, \quad \ldots .$$

Negative Angles

Negative θ

To associate angles with the negative number θ, go *clockwise* around the unit circle through an angle $|\theta|$. For instance, to draw the angle $-\pi/2$, start at the point (1, 0) and move along the unit circle clockwise through a right angle until reaching the point P directly below the center of the circle. Note in Fig. 8 that an angle of $-\pi/2$ radians is coterminal with an angle of $3\pi/2$ radians.

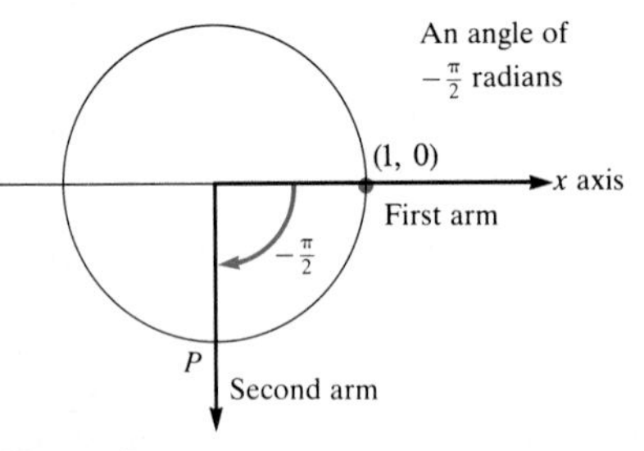

Figure 8

The Sine and Cosine Functions

The two fundamental functions of trigonometry, sine and cosine, can now be defined.

> **Definition** *The sine and cosine functions.* For each number θ, the sine and cosine of θ are defined as follows. Draw the angle of θ radians whose first arm is the positive x axis and whose vertex is at (0, 0). The second arm meets the unit circle whose center is at (0, 0) in a point P. The x coordinate of P is called the cosine of θ and is denoted $\cos\theta$. The y coordinate of P is called the sine of θ and is denoted $\sin\theta$. (See Fig. 9.)

This diagram is the most important in this section. It is the basis of trigonometry.

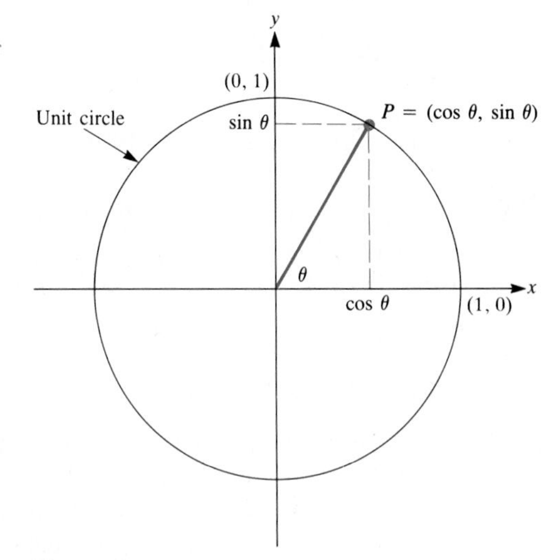

Figure 9

EXAMPLE 4 Find cos ($\pi/2$) and sin ($\pi/2$).

SOLUTION If $\theta = \pi/2$, then the angle is a right angle, and $P = (0, 1)$. Hence

$$\cos\frac{\pi}{2} = 0 \qquad \text{and} \qquad \sin\frac{\pi}{2} = 1. \qquad \blacksquare$$

EXAMPLE 5 Find cos ($-\pi$) and sin ($-\pi$).

SOLUTION If $\theta = -\pi$, P is the point $(-1, 0)$. Hence

$$\cos(-\pi) = -1 \qquad \text{and} \qquad \sin(-\pi) = 0. \qquad \blacksquare$$

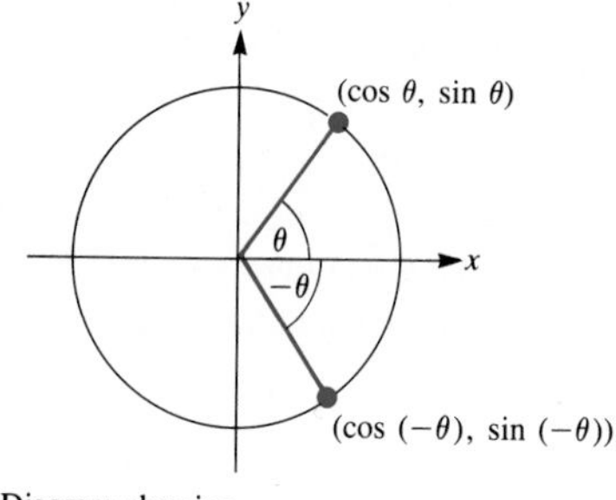

Diagram showing that $\cos(-\theta) = \cos\theta$ and $\sin(-\theta) = -\sin\theta$.

Figure 10

Some Properties of cos θ and sin θ

The trigonometric functions satisfy various identities. First of all, since a change of 2π in θ leads to the same point P on the circle,

$$\cos(\theta + 2\pi) = \cos\theta$$

and

$$\sin(\theta + 2\pi) = \sin\theta.$$

One says that the cosine and sine functions have period 2π. Second, inspection of the unit circle in Fig. 10 shows that

$$\cos(-\theta) = \cos\theta \qquad \text{and} \qquad \sin(-\theta) = -\sin\theta.$$

Hence $\cos\theta$ is an even function and $\sin\theta$ is an odd function. The numbers $\cos\theta$ and $\sin\theta$ are related by the equation

$$\cos^2\theta + \sin^2\theta = 1.$$

[$\cos^2\theta$ is short for $(\cos\theta)^2$.] To establish this, apply the pythagorean theorem to the right triangle OAP shown in Fig. 11.

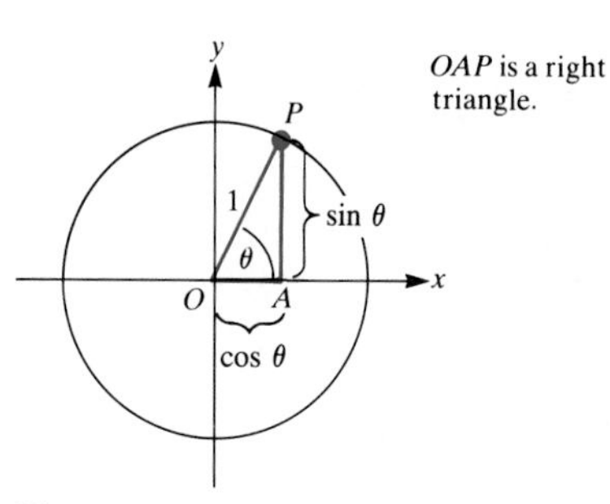

OAP is a right triangle.

Figure 11

With the aid of this relation between cos θ and sin θ the next two examples determine cos ($\pi/4$), sin ($\pi/4$), cos ($\pi/3$), and sin ($\pi/3$).

EXAMPLE 6 Find cos ($\pi/4$) and sin ($\pi/4$).

SOLUTION When the angle is $\pi/4$ (45°), a quick sketch shows that the cosine equals the sine:

$$\cos\frac{\pi}{4} = \sin\frac{\pi}{4}.$$

Thus

$$\cos^2\frac{\pi}{4} + \cos^2\frac{\pi}{4} = 1,$$

from which it follows that

$$\cos^2\frac{\pi}{4} = \frac{1}{2}.$$

Since cos ($\pi/4$) is positive,

$$\cos\frac{\pi}{4} = \sqrt{\frac{1}{2}} = \frac{\sqrt{2}}{2} \approx 0.707,$$

and

$$\sin\frac{\pi}{4} = \frac{\sqrt{2}}{2}. \qquad \blacksquare$$

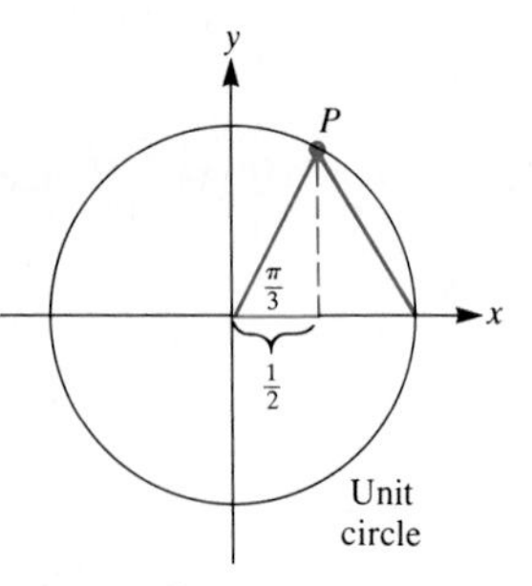

Figure 12

EXAMPLE 7 Find cos ($\pi/3$) and sin ($\pi/3$).

SOLUTION The angle $\pi/3$ (60°) is the angle in an equilateral triangle. Place such a triangle in the unit circle as shown in Fig. 12. Inspection of the figure shows that

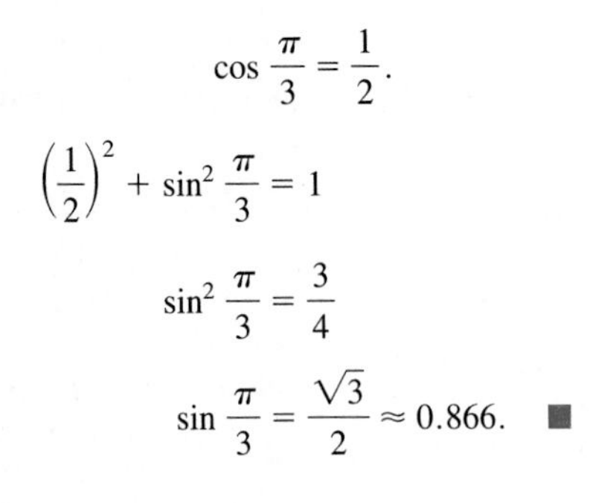

$$\cos\frac{\pi}{3} = \frac{1}{2}.$$

Then,
$$\left(\frac{1}{2}\right)^2 + \sin^2\frac{\pi}{3} = 1$$
$$\sin^2\frac{\pi}{3} = \frac{3}{4}$$
$$\sin\frac{\pi}{3} = \frac{\sqrt{3}}{2} \approx 0.866. \quad ■$$

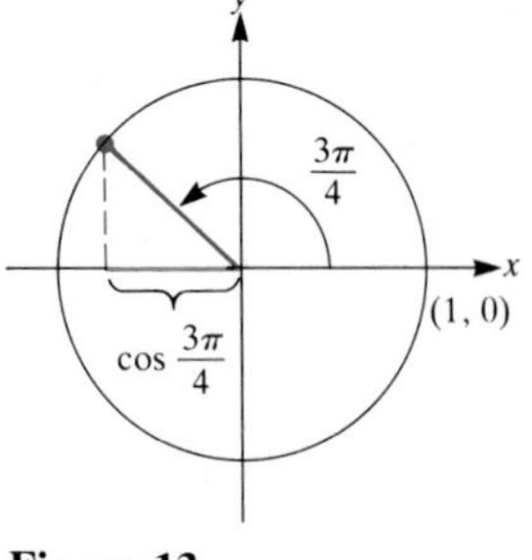

Figure 13

Once cos ($\pi/4$) is known, the cosines of multiples of $\pi/4$ can be found by sketching the unit circle. For instance, to find cos ($3\pi/4$), draw an angle of $3\pi/4$ radians, as in Fig. 13. It is clear that cos ($3\pi/4$) is negative and that $|\cos(3\pi/4)| = \sqrt{2}/2$. Hence

$$\cos\frac{3\pi}{4} = -\frac{\sqrt{2}}{2}.$$

A similar method can be used to compute the cosines of multiples of $\pi/6$. With the aid of such computations, the following table for the cosine function can be obtained:

θ	0	$\frac{\pi}{6}$	$\frac{\pi}{4}$	$\frac{\pi}{3}$	$\frac{\pi}{2}$	$\frac{2\pi}{3}$	$\frac{3\pi}{4}$	$\frac{5\pi}{6}$	π	$\frac{7\pi}{6}$	$\frac{4\pi}{3}$	$\frac{3\pi}{2}$	2π
$\cos\theta$	1	$\frac{\sqrt{3}}{2}$	$\frac{\sqrt{2}}{2}$	$\frac{1}{2}$	0	$\frac{-1}{2}$	$\frac{-\sqrt{2}}{2}$	$\frac{-\sqrt{3}}{2}$	-1	$\frac{-\sqrt{3}}{2}$	$\frac{-1}{2}$	0	1

This table provides enough information to graph the cosine function.

There is no need to compute $\cos\theta$ for θ outside the interval $[0, 2\pi]$. When you increase θ by 2π you get the same point on the unit circle and therefore the same value of the cosine. That is the meaning of the equation $\cos(\theta + 2\pi) = \cos\theta$. No polynomial of degree 1 or more is periodic, but trigonometric functions are periodic: increasing the input by a fixed amount does not change the output.

Graphs of cos θ and sin θ

The graph of $y = \cos\theta$ consists of the portion from 0 to 2π endlessly repeated. The graph of $y = \sin\theta$ is sketched in a similar manner. See Figs. 14 and 15. Comparison of these two graphs suggests that $\sin\theta = \cos(\theta - \pi/2)$.

Several important values of $\cos\theta$ and $\sin\theta$ are given in Fig. 16, which shows them as coordinates of points on the unit circle.

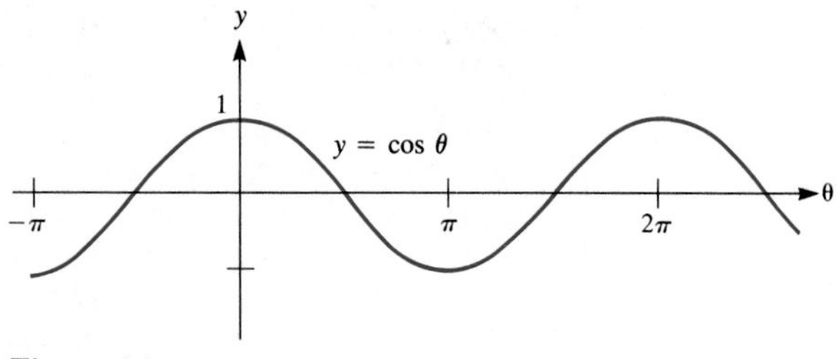

Figure 14

The graph of the sine function is that of the cosine function shifted π/2 to the right.

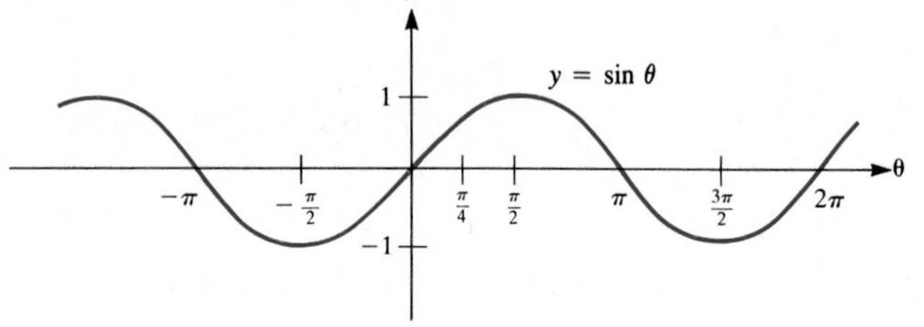

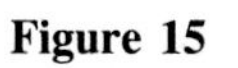

Figure 15

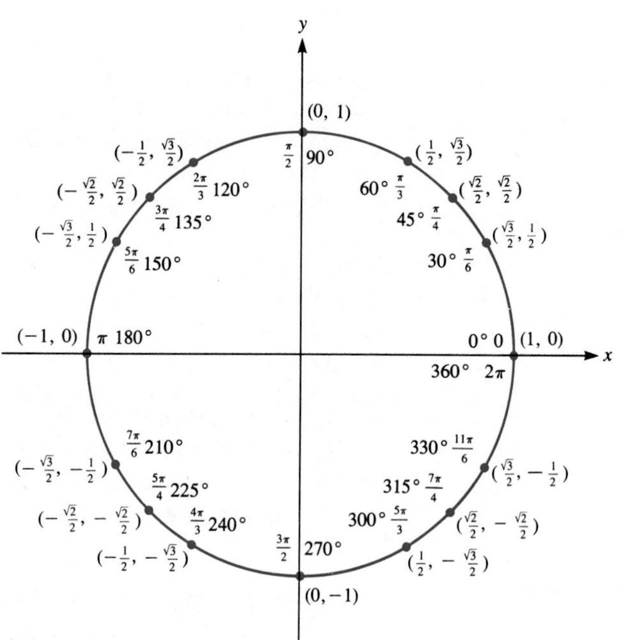

Figure 16

RICHARD FEYNMAN AND TRIGONOMETRY

"Another thing I did in high school was to invent problems and theorems . . . . I noticed some connection—perhaps it was $\sin^2 + \cos^2 = 1$—that reminded me of trigonometry. Now, a few years earlier, perhaps when I was eleven or twelve, I had read a book on trigonometry that I had checked out from the library, but the book was by now long gone. I remembered only that trigonometry had something to do with relations between sines and cosines. So I began to work out all the relations by drawing triangles, and each one I proved by myself. I also calculated the sine, cosine, and tangent of every five degrees, starting with the sine of five degrees as given, by addition and half-angle formulas that I had worked out . . . . While I was doing all this trigonometry, I didn't like the symbols for sine, cosine, tangent, and so on. To me "sin *f*" looked like *s* times *i* times *n* times *f*! So I invented another symbol, like a square root sign, that was a sigma with a long arm sticking out of it, and I put the *f* underneath."

From *"Surely You're Joking, Mr. Feynman!"*, by Richard P. Feynman (Nobel laureate in physics), Bantam Books, New York, 1986.

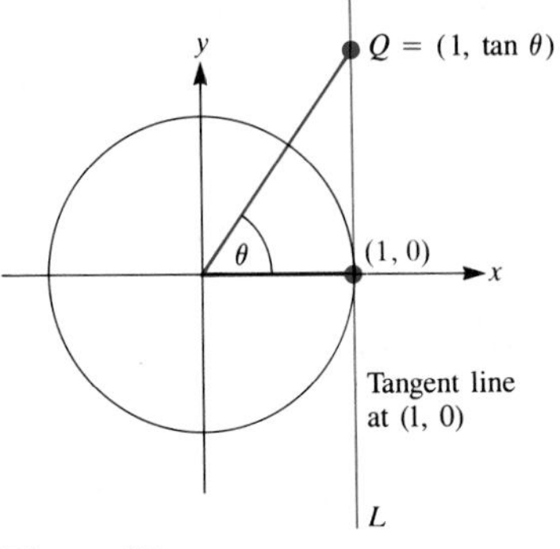

Figure 17

The Tangent Function

The trigonometric function next in importance to the cosine and sine is the tangent function.

Definition *The tangent function.* For each number θ that does not differ from $\pi/2$ by a multiple of π, the tangent of θ, denoted $\tan\theta$, is defined as follows: Draw the angle of θ radians whose first arm is the positive x axis and whose vertex is $(0, 0)$. Let L be the line through $(1, 0)$ parallel to the y axis. The line on the second arm of the angle meets the line L at a point Q. The y coordinate of Q is called $\mathbf{tan}\ \theta$. (See Fig. 17.)

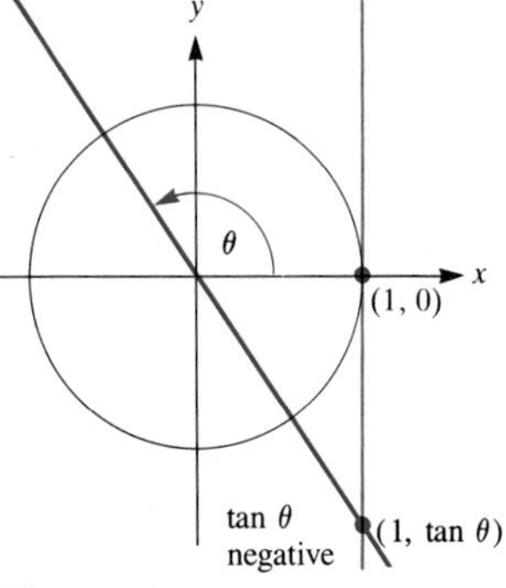

Figure 18

Note from Fig. 17 that for θ near $\pi/2$ but less than $\pi/2$, $\tan\theta$ becomes very large. While $\cos\theta$ and $\sin\theta$ never exceed 1, $\tan\theta$ takes arbitrarily large values. Note that for $\pi/2 < \theta < \pi$ (a second-quadrant angle), $\tan\theta$ is negative, as shown in Fig. 18. The behavior of $\tan\theta$ for θ near $\pi/2$ is described by these two limits:

$$\lim_{\theta\to(\pi/2)^-} \tan\theta = \infty$$

and

$$\lim_{\theta\to(\pi/2)^+} \tan\theta = -\infty.$$

It follows from the definition of the tangent function that

$$\tan(\theta + \pi) = \tan\theta.$$

While cosine and sine have period 2π, the tangent function has period π. (See Fig. 19.)

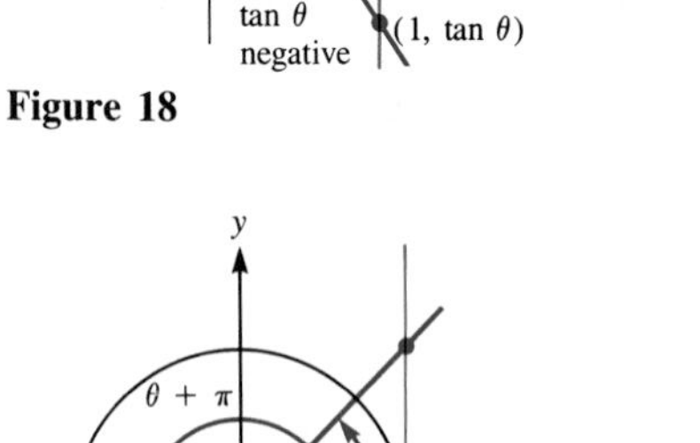

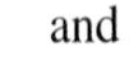

Figure 19

The functions $\cos\theta$, $\sin\theta$, and $\tan\theta$ are available on many calculators.

These functions are related by the equation

This equation is often used to define tan θ.

$$\boxed{\tan\theta = \frac{\sin\theta}{\cos\theta}}. \tag{2}$$

This can easily be deduced from inspection of Fig. 20. By the similarity of the two triangles in the figure,

$$\frac{\tan\theta}{\sin\theta} = \frac{1}{\cos\theta},$$

from which it follows that $\tan\theta = (\sin\theta)/(\cos\theta)$.

Equation (2) holds for all θ such that $\cos\theta$ is not 0.

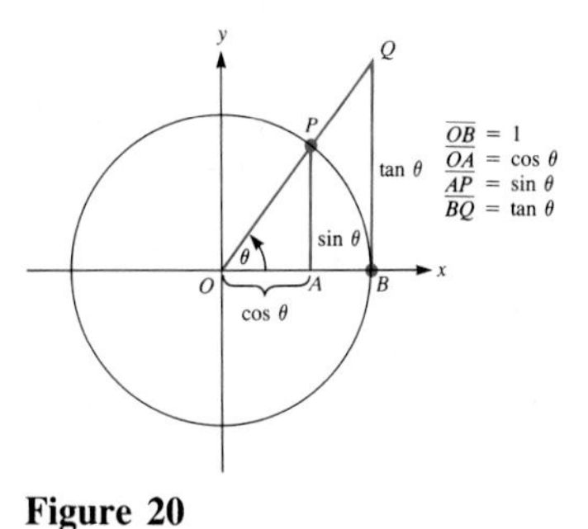

Figure 20

Graph of tan θ

Because $\tan(\theta + \pi) = \tan\theta$, we will graph $y = \tan\theta$ for θ in $[0, \pi]$. The rest of the graph consists of copies of this graph.

First compute $\tan\theta$ at a few famous angles. A glance at Fig. 20 shows that $\tan\theta = 0$. Figure 21 shows that $\tan \pi/4 = 1$, since triangle OAB is an isosceles triangle.

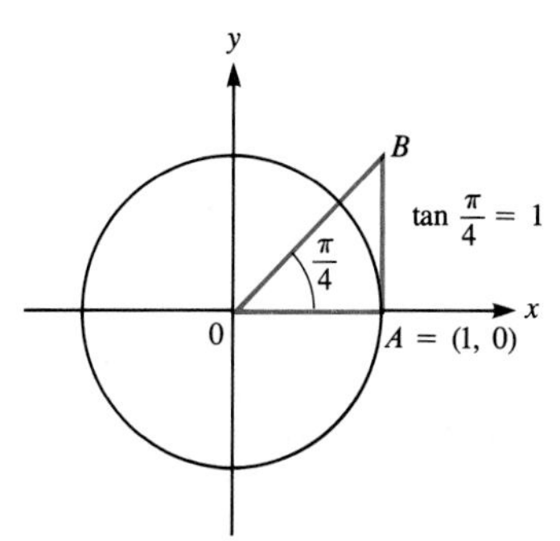

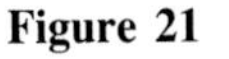

Figure 21

Equation (2) enables us to find a few other values of $\tan\theta$. For instance,

$$\tan\frac{\pi}{6} = \frac{\sin\frac{\pi}{6}}{\cos\frac{\pi}{6}} = \frac{\frac{1}{2}}{\frac{\sqrt{3}}{2}} = \frac{1}{\sqrt{3}} = \frac{\sqrt{3}}{3}$$

and

$$\tan\frac{\pi}{3} = \frac{\sin\frac{\pi}{3}}{\cos\frac{\pi}{3}} = \frac{\frac{\sqrt{3}}{2}}{\frac{1}{2}} = \sqrt{3}.$$

The rest of the following table is filled out in the same way:

θ	0	$\frac{\pi}{6}$	$\frac{\pi}{4}$	$\frac{\pi}{3}$	$\frac{\pi}{2}$	$\frac{2\pi}{3}$	$\frac{3\pi}{4}$	$\frac{5\pi}{6}$
$\tan\theta$	0	$\frac{\sqrt{3}}{3}$	1	$\sqrt{3}$	—	$-\sqrt{3}$	-1	$-\frac{\sqrt{3}}{3}$

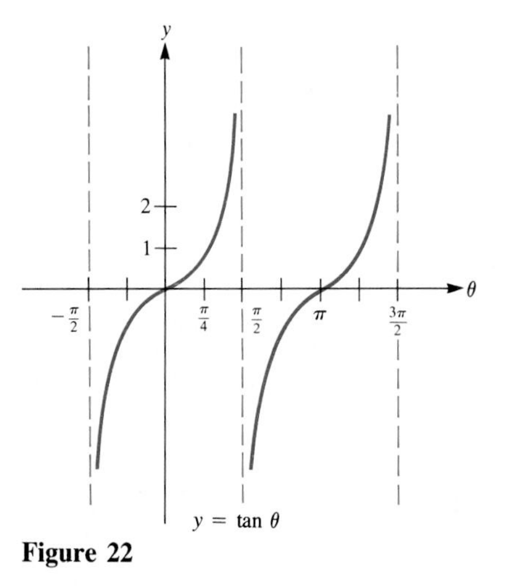

Figure 22

The graph of $y = \tan\theta$ is indicated in Fig. 22. Note that $\pm\pi/2, \pm 3\pi/2, \pm 5\pi/2, \ldots$ are not in the domain; there are vertical asymptotes at these numbers.

The Secant, Cosecant, and Cotangent Functions

Three other trigonometric functions will be needed in calculus. They are the reciprocals of the cosine, sine, and tangent functions, and are called, respec-

Since $|\cos\theta| \le 1$, $|\sec\theta| \ge 1$. *When* $\cos\theta = 1$, *then* $\sec\theta = 1$. *When* $\cos\theta$ *is near* 0, $|\sec\theta|$ *is large.*

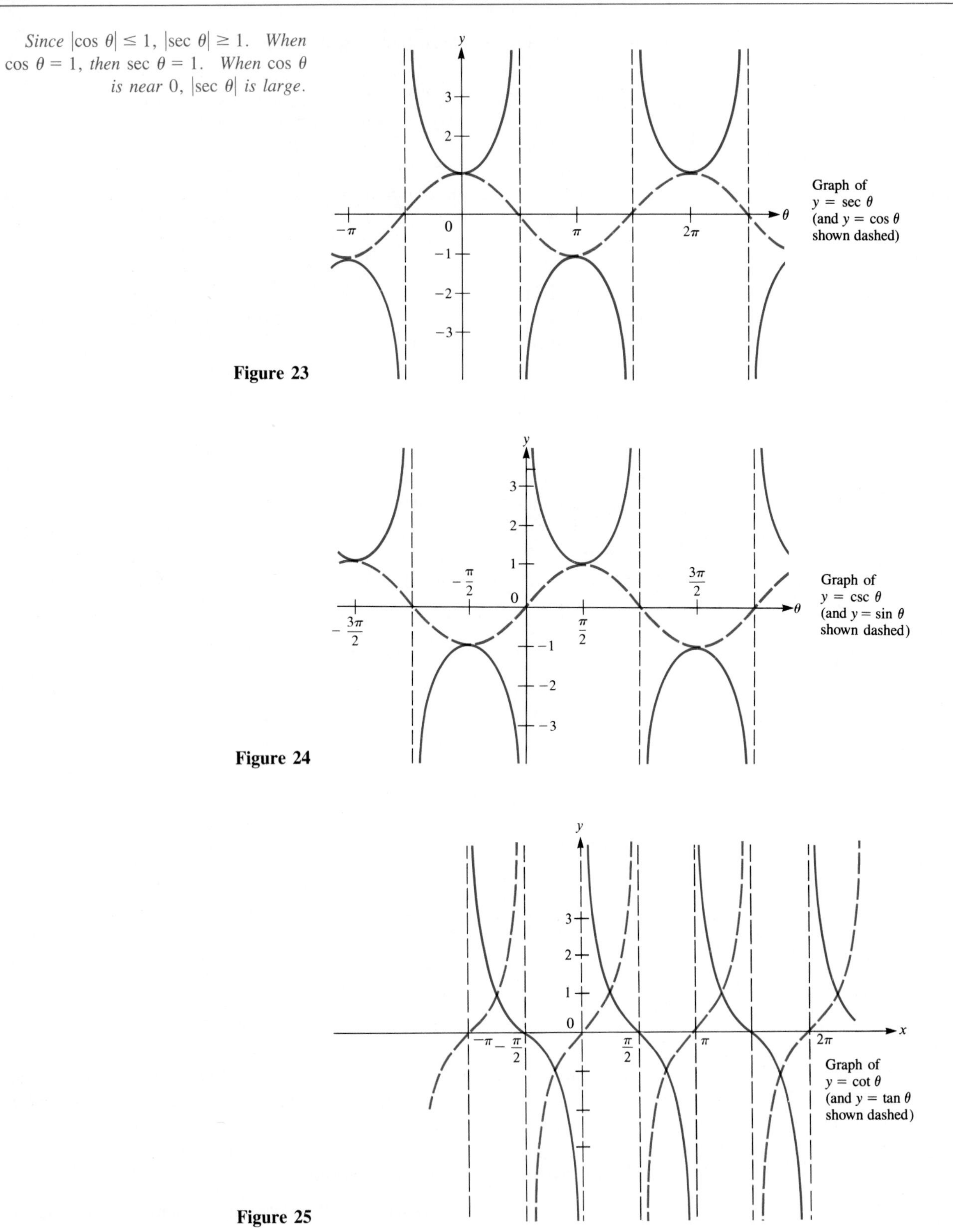

Figure 23

Figure 24

Figure 25

tively, **secant**, **cosecant**, and **cotangent**:

$$\sec\theta = \frac{1}{\cos\theta}, \qquad \csc\theta = \frac{1}{\sin\theta}, \qquad \cot\theta = \frac{1}{\tan\theta}.$$

Scientific calculators have keys for the functions $\cos\theta$, $\sin\theta$, and $\tan\theta$. To get $\sec\theta$, for instance, you first find $\cos\theta$, then press the $1/x$-key.

Consider the function $\sec\theta$. First of all, it is not defined when $\cos\theta = 0$. For instance, $\pi/2$ is not in the domain of $\sec\theta$. However, for θ near $\pi/2$, $\sec\theta$ is large. Second, since $|\cos\theta| \leq 1$ for all θ, $|\sec\theta| \geq 1$ for all θ for which $\sec\theta$ is defined. Third, $\sec\theta$ is positive when $\cos\theta$ is positive and negative when $\cos\theta$ is negative. Fourth, $\sec\theta$ has period 2π since $\cos\theta$ does. Figure 23 shows the graph of $y = \sec\theta$ and, for comparison, the graph of $\cos\theta$, which is shown dashed.

The graph of $y = \csc\theta$ bears a similar relationship to the graph of $y = \sin\theta$. It is simply the graph of $\sec\theta$ moved $\pi/2$ units to the right. (See Fig. 24.)

The graph of $y = \cot\theta$, along with that of $\tan\theta$, is shown in Fig. 25. Note that when $\tan\theta$ is large, $\cot\theta$ is near 0; when $\tan\theta$ is near 0, $\cot\theta$ is large.

Identities

Four important identities relate the cosine and sine of the sum and difference of two angles to the values of the cosine and sine of the angles:

$$\cos(A+B) = \cos A\cos B - \sin A\sin B, \qquad \cos(A-B) = \cos A\cos B + \sin A\sin B,$$

$$\sin(A+B) = \sin A\cos B + \cos A\sin B, \qquad \sin(A-B) = \sin A\cos B - \cos A\sin B.$$

(These identities are obtained in Exercises 43 to 48.) From these follow the "double angle" and "half angle" identities:

$$\cos 2\theta = \cos^2\theta - \sin^2\theta, \qquad \sin 2\theta = 2\sin\theta\cos\theta,$$

$$\cos 2\theta = 2\cos^2\theta - 1, \qquad \sin^2\theta = \frac{1-\cos 2\theta}{2},$$

$$\cos 2\theta = 1 - 2\sin^2\theta, \qquad \cos^2\theta = \frac{1+\cos 2\theta}{2}.$$

The following identities involving the tangent function are of use:

$$\tan(A+B) = \frac{\tan A + \tan B}{1 - \tan A\tan B}, \qquad \tan 2\theta = \frac{2\tan\theta}{1-\tan^2\theta},$$

$$\tan(A-B) = \frac{\tan A - \tan B}{1 + \tan A\tan B}, \qquad \tan\frac{\theta}{2} = \frac{\sin\theta}{1+\cos\theta}.$$

Finally, there is an identity that shows how to find the length of the side opposite an angle in a triangle if you know the angle and lengths of the sides next to the angle. If the lengths of two sides a and b of a triangle and the angle θ between these sides are known, the length of the third side c is determined. The formula for finding c, the **law of cosines**, is

$$c^2 = a^2 + b^2 - 2ab\cos\theta.$$

(When $\theta = \pi/2$, the law of cosines is the pythagorean theorem.) A proof is outlined in Exercise 53.

Section Summary

This section treated radian measure and the six basic trigonometric functions and their graphs. Radian measure θ is defined in terms of arclength s intercepted on a circle of radius r: $\theta = s/r$. Hence $s = r\theta$, a formula that we will use often. (Given the angle in radians and the radius, we can easily find the length of the intercepted arc.) The six functions are all defined in terms of a unit circle.

EXERCISES FOR SEC. 2.6: A REVIEW OF TRIGONOMETRY

1 What are the radian measures of the following angles?
(*a*) 90° (*b*) 30°
(*c*) 120° (*d*) 360°

2 What are the radian measures of the following angles?
(*a*) 180° (*b*) 60°
(*c*) 45° (*d*) 0°

3 How many degrees are there in an angle whose radian measure is (*a*) $3\pi/4$, (*b*) $\pi/3$, (*c*) $2\pi/3$, (*d*) 4π?

4 How many degrees are there in an angle whose radian measure is (*a*) $\pi/4$, (*b*) $\pi/6$, (*c*) 2π, (*d*) π?

5 An angle intercepts an arc of 5 inches in a circle of radius 3 inches.
(*a*) What is the measure of the angle in radians?
(*b*) What is the measure of the angle in degrees?

6 How long an arc of a circle of radius 3 inches is intercepted by an angle of 0.5 radian?

7 (*a*) Express an angle of 50° in radians.
(*b*) Express an angle of 2 radians in degrees.

8 (*a*) Express an angle of 3 radians in degrees.
(*b*) Express an angle of 1 degree in radians.

9 Find θ in Fig. 26.

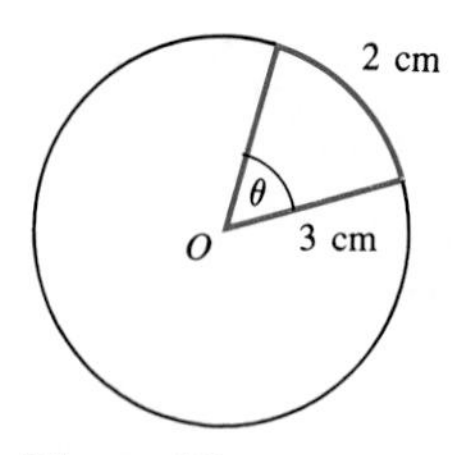

Figure 26

10 How would you draw an angle of 2 radians (*a*) with a protractor that shows angles in degrees? (*b*) with a string?

11 Draw an angle of (*a*) $13\pi/6$ radians, (*b*) -3π radians.

12 Draw an angle of (*a*) 6π radians, (*b*) $-\pi/3$ radians.

13 In Example 7 it was shown that $\cos(\pi/3) = \frac{1}{2}$. Using the identity $\sin\theta = \cos(\pi/2 - \theta)$, deduce that $\sin(\pi/6) = \frac{1}{2}$.

14 In Example 7 it was shown that $\sin(\pi/3) = \sqrt{3}/2$. Using the identity $\cos\theta = \sin(\pi/2 - \theta)$, deduce that $\cos(\pi/6) = \sqrt{3}/2$.

15 Fill in this table by making a quick sketch of the unit circle:

θ	0	$\frac{\pi}{6}$	$\frac{\pi}{4}$	$\frac{\pi}{3}$	$\frac{\pi}{2}$	π	$\frac{3\pi}{2}$	2π
$\sin\theta$								

16 By sketching the angles in a unit circle and using the information that $\cos(\pi/6) = \sqrt{3}/2$ and $\sin(\pi/6) = \frac{1}{2}$, find
(*a*) $\cos(-\pi/6)$ (*b*) $\sin(-\pi/6)$
(*c*) $\cos(5\pi/6)$ (*d*) $\sin(5\pi/6)$
(*e*) $\cos(13\pi/6)$ (*f*) $\sin(13\pi/6)$

17 Check the identity for $\cos(A + B)$ when $A = \pi/6$ and $B = \pi/3$.

18 Check the identity for $\sin(A + B)$ when $A = \pi/4$ and $B = \pi/4$.

19 Find $\sin\frac{5\pi}{12}$. *Hint:* $\frac{5\pi}{12} = \frac{\pi}{4} + \frac{\pi}{6}$.

20 Find $\sin\frac{\pi}{12}$. *Hint:* $\frac{\pi}{12} = \frac{\pi}{4} - \frac{\pi}{6}$.

21 Deduce the identity $\cos 2\theta = \cos^2\theta - \sin^2\theta$ from the identity for $\cos(A + B)$.

22 Deduce the identity $\sin 2\theta = 2\sin\theta\cos\theta$ from the identity for $\sin(A + B)$.

23 Deduce the identity $\cos 2\theta = 2\cos^2\theta - 1$ from Exercise 21.

24 Deduce the identity $\cos 2\theta = 1 - 2\sin^2\theta$ from Exercise 21.

25 Deduce from Exercise 23 that $\cos^2\theta = (1 + \cos 2\theta)/2$.

26 Deduce from Exercise 24 that $\sin^2\theta = (1 - \cos 2\theta)/2$.

27 Use a sketch of the angle in a unit circle to determine the sign (+ or −) of the function in each case:
(*a*) $\sin\theta$, $\pi < \theta < 2\pi$ (*b*) $\tan\theta$, $\pi < \theta < 3\pi/2$
(*c*) $\cos\theta$, $-\pi/2 < \theta < \pi/2$
(*d*) $\tan\theta$, $\pi/2 < \theta < \pi$

28 Give the sign (+ or −) of $\cos\theta$, $\sin\theta$, and $\tan\theta$ for
(*a*) $0 < \theta < \pi/2$, (*b*) $-\pi/2 < \theta < 0$.

29 (*a*) Using the definition of $\tan\theta$ and a sketch, show that $\tan(\pi/4) = 1$.
(*b*) Using the identity $\tan\theta = (\sin\theta)/\cos\theta$, show that $\tan(\pi/4) = 1$.

30 Find (*a*) tan ($\pi/6$), (*b*) tan ($\pi/3$).

Exercises 31 to 34 refer to angle of inclination, which is reviewed on page 559.

31 Find the slope of a line whose angle of inclination is
(*a*) 10° (*b*) 70° (*c*) 110°
(*d*) 135° (*e*) 0°

32 Find the angle of inclination, both in degrees and in radians, of a line whose slope is
(*a*) -1 (*b*) 1 (*c*) 2 (*d*) 3 (*e*) -3

33 Find the angle of inclination in degrees and in radians of a line whose slope is
(*a*) $\sqrt{3}$ (*b*) $1/\sqrt{3}$

34 Find the angle of inclination in degrees and in radians of a line whose slope is
(*a*) $\frac{1}{10}$ (*b*) 10

35 Consider an acute angle θ in any right triangle, as in Fig. 27. With the aid of the similar triangles in Fig. 28, show that (*a*) $\cos\theta = a/c$, (*b*) $\sin\theta = b/c$, and (*c*) $\tan\theta = b/a$. These formulas for cosine, sine, and tangent are sometimes taken as their definitions for θ in the first quadrant.

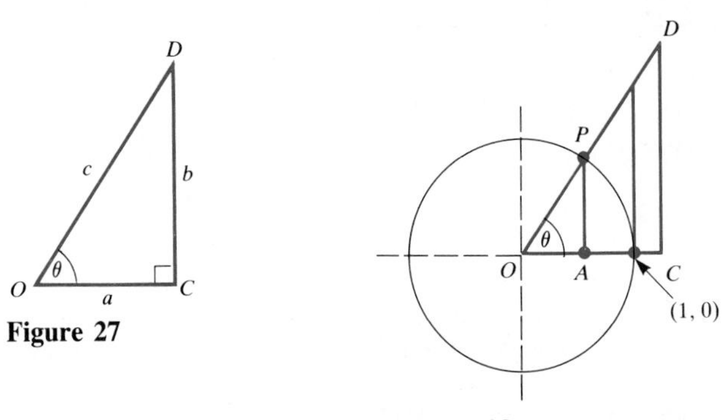

Figure 27

Figure 28

36 (See Exercise 35, Fig. 27.)
(*a*) Express b in terms of θ and c.
(*b*) Express a in terms of θ and c.
(*c*) Express b in terms of θ and a.

37 In Fig. 29 the two acute angles in a right triangle are labeled α and β.
(*a*) Express $\cos\alpha$ in terms of the lengths of the sides.
(*b*) Express $\sin\beta$ in terms of the lengths of the sides.
(*c*) Express $\tan\alpha$ in terms of the lengths of the sides.

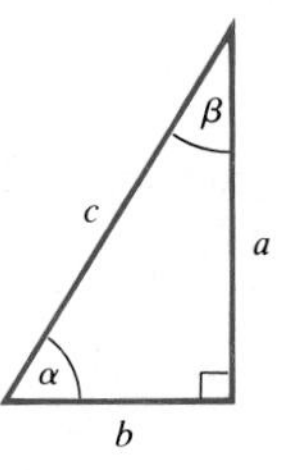

Figure 29

38 (See Exercise 35.) Use the triangle *OCD* in Fig. 30 to obtain cos ($\pi/3$), sin ($\pi/3$), and tan ($\pi/3$). (*ODE* is equilateral.) A quick sketch of *OCD* will remind you of the values of cos ($\pi/3$), sin ($\pi/3$), and tan ($\pi/3$) when you need them.

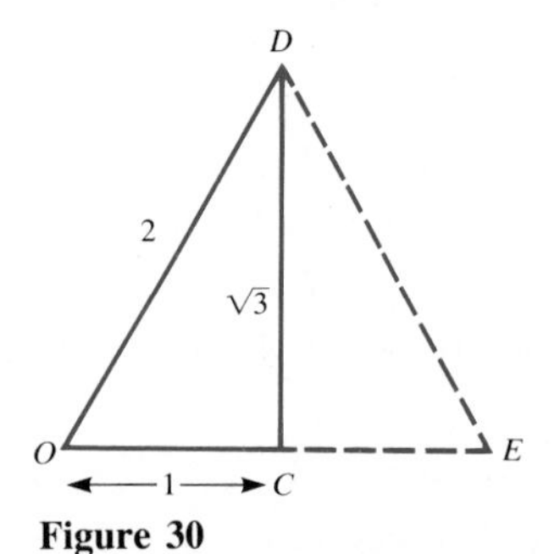

Figure 30

39 Use the triangle *OCD* in Exercise 38 to compute cos ($\pi/6$), sin ($\pi/6$), and tan ($\pi/6$). *Hint:* See Exercise 35.

40 (See Exercise 35.) Solve for the length x in each triangle in Fig. 31.

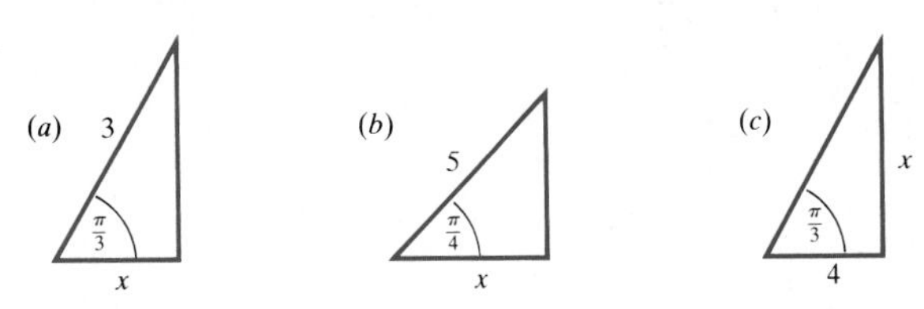

Figure 31

41 (*a*) Evaluate $\sec\theta$ for $\theta = 0, \pi/6, \pi/4, \pi/3$.
(*b*) Plot the four points given by (*a*) and graph $\sec\theta$.

42 (*a*) Evaluate $\csc\theta$ for $\theta = \pi/6, \pi/4, \pi/3, \pi/2$.
(*b*) Plot the four points given by (*a*) and graph $\csc\theta$.

---

43 This exercise outlines a proof that $\cos(A + B) = \cos A \cos B - \sin A \sin B$. Figure 32 shows two diagrams involving the unit circle.
(*a*) Show that the line segments *PQ* and *RS* in Fig. 32 have the same length.

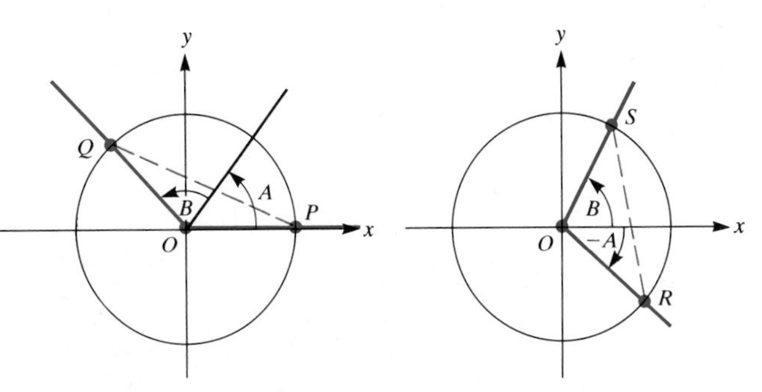

Figure 32

(*b*) Show that

$$Q = (\cos(A + B), \sin(A + B)),$$
$$S = (\cos B, \sin B),$$
$$R = (\cos A, -\sin A).$$

(*c*) Using the coordinates in (*b*), show that

$$\overline{PQ}^2 = 2 - 2\cos(A + B).$$

(*d*) Using the coordinates in (*b*), show that

$$\overline{RS}^2 = 2 - 2\cos A \cos B + 2\sin A \sin B.$$

(*e*) Deduce the identity for $\cos(A + B)$.

44 Replacing B by $-B$ in the identity for $\cos(A + B)$, obtain the identity

$$\cos(A - B) = \cos A \cos B + \sin A \sin B.$$

45 Using the identity for $\cos(A - B)$, show that

$$\sin\theta = \cos\left(\frac{\pi}{2} - \theta\right).$$

46 Using the identity $\sin\theta = \cos(\pi/2 - \theta)$ from Exercise 45, show that $\cos\theta = \sin(\pi/2 - \theta)$.

47 With the aid of the identities in Exercises 44 to 46, show that $\sin(A + B) = \sin A \cos B + \cos A \sin B$, as follows:
(*a*) Show that $\sin(A + B) = \cos[\pi/2 - (A + B)] = \cos[(\pi/2 - A) - B]$.
(*b*) Apply the identity in Exercise 44 to $\cos(x - y)$ with $x = \pi/2 - A$ and $y = B$.
(*c*) Use the identities in Exercises 45 and 46 to remove $\pi/2$ from the identity you obtain in (*b*).

48 Use the identity for $\sin(A + B)$ obtained in Exercise 47 to show that $\sin(A - B) = \sin A \cos B - \cos A \sin B$.

49 (*a*) From the identity $\cos 2\theta = 2\cos^2\theta - 1$ deduce that $\cos\theta = \pm\sqrt{(1 + \cos 2\theta)/2}$.
(*b*) Use the identity in (*a*) to find $\cos(\pi/4)$.
(*c*) Use the identity in (*a*) to find $\cos(3\pi/4)$.

50 (*a*) From the identity $\cos 2\theta = 1 - 2\sin^2\theta$ deduce that $\sin\theta = \pm\sqrt{(1 - \cos 2\theta)/2}$.
(*b*) Use the identity in (*a*) to find $\sin(\pi/4)$.
(*c*) Use the identity in (*a*) to find $\sin(-\pi/4)$.

51 Using the identities for $\cos(A - B)$ and $\sin(A - B)$, prove that

$$\tan(A - B) = \frac{\tan A - \tan B}{1 + \tan A \tan B}.$$

52 Show that
(*a*) $\sec^2\theta = 1 + \tan^2\theta$,
(*b*) $\csc^2\theta = 1 + \cot^2\theta$.

53 This exercise outlines a proof of the law of cosines in the case $0 < \theta < \pi/2$. Consider Fig. 33.
(*a*) Show that $\overline{CD} = a\cos\theta$ and $\overline{AD} = b - a\cos\theta$.
(*b*) Show that

$$a^2 - a^2\cos^2\theta = h^2 = c^2 - (b - a\cos\theta)^2.$$

(*c*) From (*b*) deduce the law of cosines.

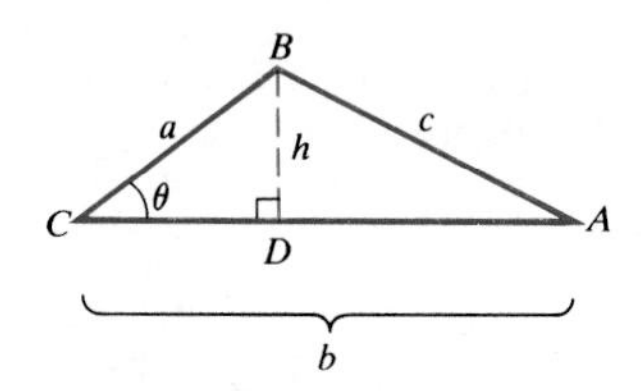

Figure 33

54 Imagine that the only trigonometric function on your calculator is the cosine function. How would you use your calculator to compute the other five trigonometric functions?

55 (*a*) Examine $(\sin\theta)/\theta$ for small values of θ, both positive and negative. (Be sure to set your calculator for radians.)
(*b*) Does $\lim_{\theta\to 0}(\sin\theta)/\theta$ seem to exist?

56 Which of the six basic trigonometric functions are even? odd? neither?

2.7 THE LIMIT OF $(\sin\theta)/\theta$ AS θ APPROACHES 0

In the next chapter we will need to know

$$\lim_{\theta\to 0}\frac{\sin\theta}{\theta} \quad\text{and}\quad \lim_{\theta\to 0}\frac{1 - \cos\theta}{\theta}.$$

This section determines these two limits. Both are indeterminate since in each case the numerator and denominator approach 0 as θ approaches 0.

Exploring $\lim_{\theta\to 0}(\sin\theta)/\theta$

A few computations on a calculator will suggest what happens to $(\sin\theta)/\theta$ as

$\theta \to 0$. (If you carry out some computations, be sure to set angles in radians, not degrees.) The table in the margin shows sin θ and (sin θ)/θ to five significant figures.

θ	1	0.1	0.01
$\sin\theta$	0.84147	0.099833	0.0099998
$\dfrac{\sin\theta}{\theta}$	0.84147	0.99833	0.99998

Note that

$$\frac{\sin(-\theta)}{-\theta} = \frac{-\sin\theta}{-\theta} = \frac{\sin\theta}{\theta}.$$

Thus the function (sin θ)/θ is an even function, and (sin θ)/θ behaves the same for negative θ as it does for positive θ. Judging, then, by the table, we expect that

$$\lim_{\theta\to 0}\frac{\sin\theta}{\theta} = 1.$$

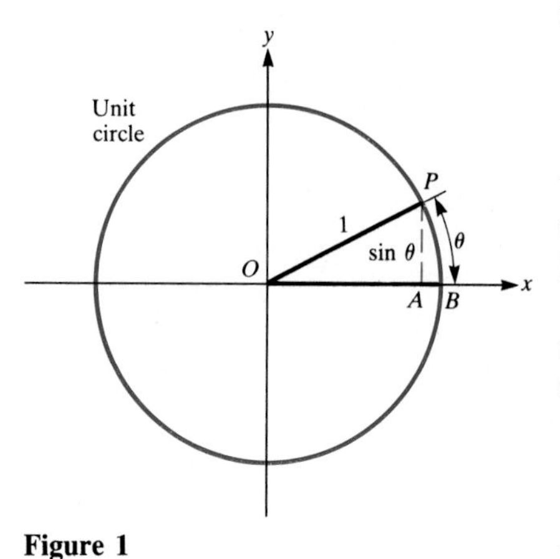

Figure 1

You might try $\theta = 0.001$ and 0.0001 to see if your data agree with this conjecture. Let us also look at (sin θ)/θ geometrically, using the definition of sin θ. Draw the angle θ and a unit circle, as in Fig. 1.

Since we measure angles in radians, and the circle has radius 1, the length of the arc PB is θ. By the definition of sin θ, the length of line segment PA is sin θ. Thus

$$\frac{\sin\theta}{\theta} = \frac{\text{Length of side } PA}{\text{Length of arc } PB}.$$

When θ is small, so are the lengths of PA and PB. However, for small θ, PA looks so much like the arc PB that it seems likely that the quotient of their lengths is near 1. This suggests once again that

$$\lim_{\theta\to 0}\frac{\sin\theta}{\theta} = 1.$$

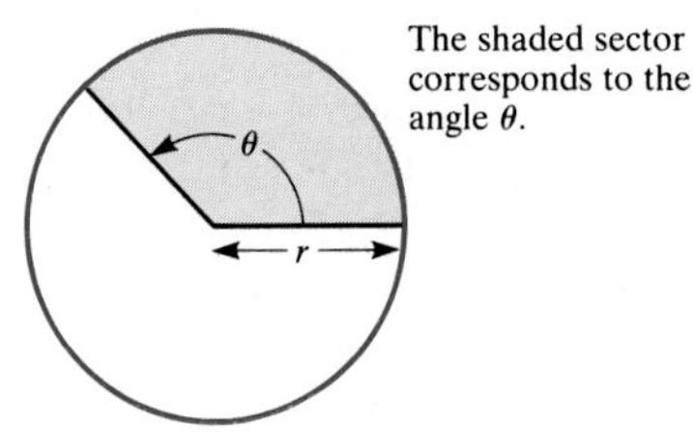

Figure 2

Determining $\lim_{\theta\to 0}$ (sin θ)/θ

The proof that $\lim_{\theta\to 0}$ (sin θ)/$\theta = 1$ depends on a comparison of three areas. For this reason it will be necessary to develop a formula for the area of a sector of a circle subtended by an angle of θ radians, as in Fig. 2.

When the angle is 2π, the sector is the entire circle of radius r; hence it has an area of πr^2. Since the area of a sector is proportional to θ, it follows that

$$\frac{\text{Area of sector}}{\pi r^2} = \frac{\theta}{2\pi}.$$

From this equation it follows that

Formula for area of a sector

$$\boxed{\text{Area of sector} = \frac{\theta}{2\pi}\pi r^2 = \frac{\theta r^2}{2}.}$$

Advisory This formula will be used in later chapters. It is safer to memorize the proportion that led to it rather than the formula itself. It is easy to forget the denominator 2 and also that the number π does not appear.

We deal with $(\sin \theta)/\theta$ by introducing two functions, $g(\theta)$ and $h(\theta)$, such that

$$g(\theta) \leq \frac{\sin \theta}{\theta} \leq h(\theta).$$

Moreover, these functions will be simpler than $(\sin \theta)/\theta$. We will see easily that

$$\lim_{\theta \to 0} g(\theta) = 1 \quad \text{and} \quad \lim_{\theta \to 0} h(\theta) = 1.$$

Then $(\sin \theta)/\theta$, being squeezed between two functions that approach 1, must also approach 1. For later use we state this as a general principle.

THE SQUEEZE PRINCIPLE

If $g(x) \leq f(x) \leq h(x)$ and

$$\lim_{x \to a} g(x) = L = \lim_{x \to a} h(x),$$

then

$$\lim_{x \to a} f(x) = L.$$

Figure 3 illustrates the spirit of the squeeze principle. A cow shies away from two herding dogs, staying always between them. If both dogs maintain their course toward the gate labeled L, the cow is forced to pass through the gate as well.

Theorem 1 Let $\sin \theta$ denote the sine of an angle of θ radians. Then

$$\lim_{\theta \to 0} \frac{\sin \theta}{\theta} = 1.$$

Proof It will be enough to consider only $\theta > 0$, since we previously established that $(\sin \theta)/\theta$ is an even function.

Moreover, it will be convenient to restrict θ to be less than $\pi/2$ so that $\sin \theta$ and $\cos \theta$ are not negative.

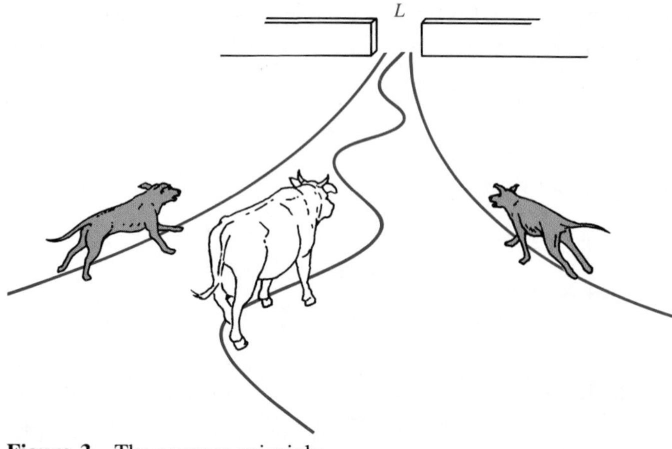

Figure 3 The squeeze principle

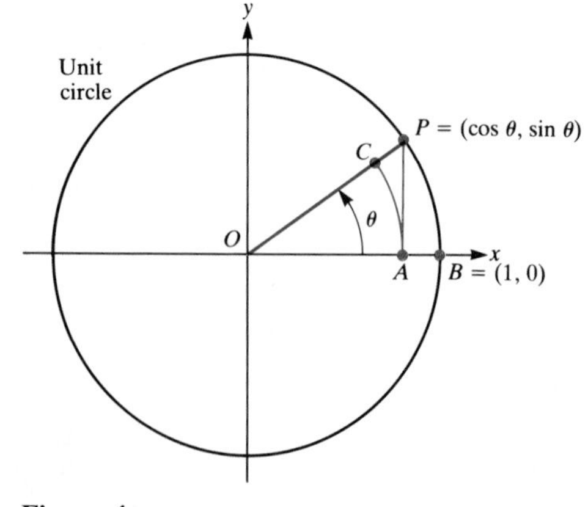

Figure 4

We shall compare the areas of the three regions in Fig. 4, two sectors and a triangle. One sector, OAC, has angle θ and radius equal to $\cos\theta$. Sector OBP has angle θ and radius 1. Triangle OAP has base $\cos\theta$ and altitude $\sin\theta$.

As inspection of Fig. 4 shows,

$$\text{Area of sector } OAC < \text{Area of triangle } OAP < \text{Area of sector } OBP. \tag{1}$$

Now, by the formula for the area of a sector,

$$\text{Area of sector } OAC = \frac{\theta(\cos\theta)^2}{2},$$

and

$$\text{Area of sector } OBP = \frac{\theta\cdot 1^2}{2} = \frac{\theta}{2}.$$

Also,

$$\text{Area of triangle } OAP = \frac{1}{2}\cdot\text{base}\cdot\text{altitude} = \frac{\cos\theta\cdot\sin\theta}{2}.$$

So inequalities (1) take the form

$$\frac{\theta\cos^2\theta}{2} < \frac{\cos\theta\sin\theta}{2} < \frac{\theta}{2}. \tag{2}$$

Next we manipulate these inequalities to make the middle term (sin θ)/θ. Multiplying by the positive number 2 and dividing by the positive number $\theta\cos\theta$ yields the inequalities

$$\cos\theta < \frac{\sin\theta}{\theta} < \frac{1}{\cos\theta}.$$

A glance at Fig. 4 shows that

$$\lim_{\theta\to 0}\cos\theta = 1,$$

and hence

$$\lim_{\theta\to 0}\frac{1}{\cos\theta} = 1.$$

Thus, as $\theta\to 0$, both $\cos\theta$ and $1/(\cos\theta)$ approach 1. Hence (sin θ)/θ, squeezed between $\cos\theta$ and $1/(\cos\theta)$, must also approach 1. Thus

$$\lim_{\theta\to 0}\frac{\sin\theta}{\theta} = 1,$$

as had been anticipated. ■

For a limit that may surprise you, see Exercise 20.

Determining $\lim_{\theta\to 0}(1-\cos\theta)/\theta$

We shall also need the limit

$$\lim_{\theta\to 0}\frac{1-\cos\theta}{\theta}$$

in the next chapter. It is not obvious what this limit is, if indeed it exists. As $\theta\to 0$, the numerator $1-\cos\theta$ approaches 0; so does the denominator. The numerator influences the quotient to become small, while the denominator influences the quotient to become large.

As usual, we make a table to see what the limit might be:

θ	0.5	0.1	0.01	−0.01	−0.1	−0.5
$\cos\theta$	0.87758	0.99500	0.99995	0.99995	0.99500	0.87758
$1 - \cos\theta$	0.12242	0.00500	0.00005	0.00005	0.00500	0.12242
$\dfrac{1-\cos\theta}{\theta}$	0.24483	0.04996	0.00500	−0.00500	−0.04996	−0.24483

It appears that

$$\lim_{\theta\to 0} \frac{1-\cos\theta}{\theta} = 0.$$

To see whether this is correct, let us go back to the definition of $\cos\theta$. Figure 5 shows a unit circle. Note that the length of AB is $1 - \cos\theta$, while the length of arc PB is θ. When θ is small, the lengths of AB and arc PB are both small. However, it does seem that AB is much smaller than arc PB, so that the ratio

$$\frac{1-\cos\theta}{\theta} = \frac{\text{Length of } AB}{\text{Length of arc } PB}$$

might be small.

We will use the geometry of Fig. 5 to show that $\lim_{\theta\to 0}(1-\cos\theta)/\theta$ is in fact 0, as suspected. We need not consider negative angles since

$$\frac{1-\cos(-\theta)}{-\theta} = \frac{1-\cos\theta}{-\theta} = -\frac{1-\cos\theta}{\theta}.$$

Thus if $(1-\cos\theta)/\theta$ approaches 0 as $\theta\to 0^+$, it approaches 0 as $\theta\to 0^-$.

In the argument we will need the following fact from basic geometry. Let O

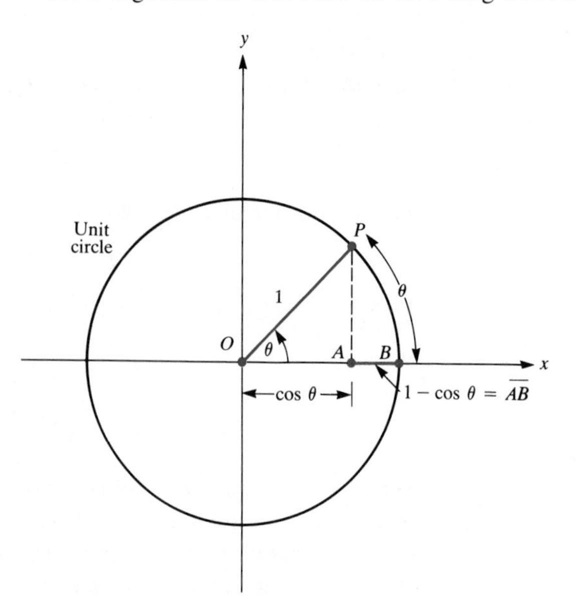

Figure 5

be the center of a circle; let P and B be points on the circle as shown in Fig. 6. Since triangle BOP is isosceles, angles OPB and OBP are equal. Thus angle OBP is $(\pi - \theta)/2$. (Recall that the sum of the angles of a triangle is π.)

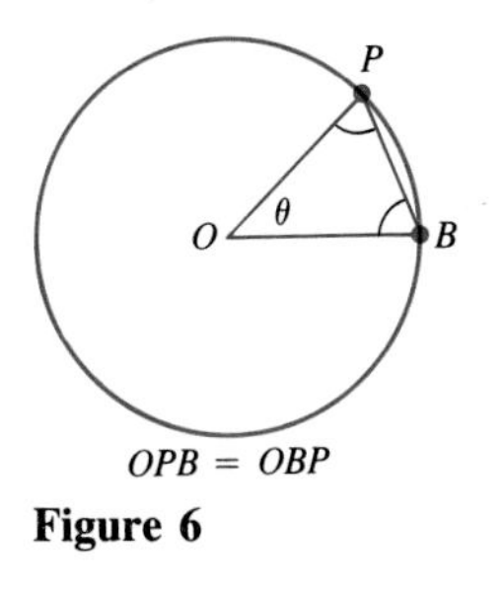

$OPB = OBP$

Figure 6

> **Theorem 2** Let $\cos\theta$ denote the cosine of an angle of θ radians. Then
> $$\lim_{\theta\to 0}\frac{1-\cos\theta}{\theta} = 0.$$

Proof As mentioned, we may restrict our attention to positive θ. Draw the line PB in Fig. 5, obtaining Fig. 7. Note that angle ABP is $(\pi - \theta)/2$.

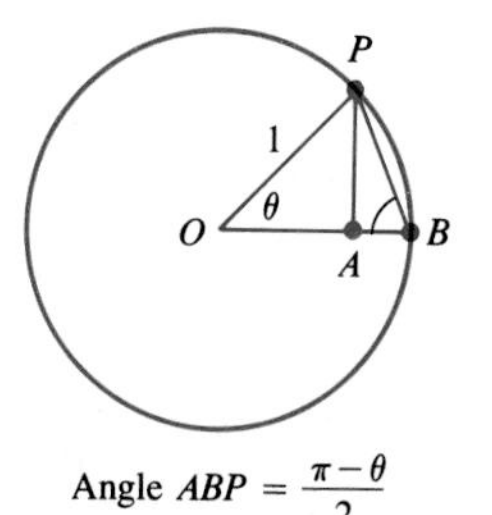

Angle $ABP = \frac{\pi-\theta}{2}$

Figure 7

Since the straight line is the shortest distance between two points, line segment PB is shorter than arc PB. Thus

$$\frac{1-\cos\theta}{\theta} = \frac{\text{Length of } AB}{\text{Length of arc } PB} < \frac{\text{Length of } AB}{\text{Length of } PB}.$$

Using right triangle PAB, we see that

$$\frac{\text{Length of } AB}{\text{Length of } PB} = \cos\left(\frac{\pi-\theta}{2}\right).$$

We therefore have

$$0 < \frac{1-\cos\theta}{\theta} < \cos\left(\frac{\pi-\theta}{2}\right). \tag{3}$$

But $$\lim_{\theta\to 0^+} 0 = 0 \quad\text{and}\quad \lim_{\theta\to 0^+}\cos\left(\frac{\pi-\theta}{2}\right) = \cos\frac{\pi}{2} = 0,$$

so we can apply the squeeze principle to (3), obtaining

$$\lim_{\theta\to 0}\frac{1-\cos\theta}{\theta} = 0,$$

Here is a case where the numerator approaches 0 much more rapidly than the denominator.

as claimed. ■

Theorem 2 tells us that $1 - \cos\theta$ approaches 0 much more rapidly than does θ—so rapidly that the ratio $(1 - \cos\theta)/\theta$ approaches 0 as $\theta \to 0$.

The Meaning of $\lim_{\theta\to 0}(\sin\theta)/\theta = 1$

When θ is small, the quotient $(\sin\theta)/\theta$ is nearly 1. This means that $\sin\theta$ is very close to θ. For instance, when $\theta = 0.1$, then $\sin\theta = 0.099833$ to five significant decimal places. The difference

$$\begin{aligned}\theta - \sin\theta &= 0.1 - 0.099833\\ &= 0.0000167\end{aligned}$$

is small, even when compared to $\theta = 0.1$. In other words, *θ is a good approximation to sin θ when θ is small* (provided that θ is in radians). If you want to estimate sin 0.08, just use 0.08 itself.

If you want to estimate the sine of a small angle given in degrees, *you must first express the angle in radians.*

EXAMPLE 1 Estimate $\sin 6°$.

SOLUTION First express the angle in radians:

$$\frac{6}{180} = \frac{\text{Radians}}{\pi},$$

hence

$$\text{Radians} = \frac{6\pi}{180} = \frac{\pi}{30}.$$

Thus

$$\sin 6° = \sin \frac{\pi}{30} \approx \frac{\pi}{30}.$$

Using 3.14 as an estimate of π, we have $\pi/30 \approx 3.14/30 \approx 0.105$. If you look up $\sin 6°$ on a calculator, you will get 0.1045 (to four decimal places). ■

Other Trigonometric Limits

With the aid of $\lim_{\theta\to 0} (\sin \theta)/\theta = 1$ we can determine other trigonometric limits, as the next example illustrates.

EXAMPLE 2 Find $\lim_{\theta\to 0} (\tan \theta)/\theta$.

SOLUTION Both $\tan \theta$ and θ approach 0 as $\theta \to 0$. Therefore the limit of their quotient, $(\tan \theta)/\theta$, is indeterminate. To find this limit we relate it to a limit we know, namely $\lim_{\theta\to 0} (\sin \theta)/\theta$. A little algebra does the trick:

$$\begin{aligned}
\lim_{\theta\to 0} \frac{\tan \theta}{\theta} &= \lim_{\theta\to 0} \frac{(\sin \theta)/(\cos \theta)}{\theta} \\
&= \lim_{\theta\to 0} \frac{\sin \theta}{\theta} \cdot \frac{1}{\cos \theta} \\
&= \lim_{\theta\to 0} \frac{\sin \theta}{\theta} \lim_{\theta\to 0} \frac{1}{\cos \theta} \\
&= 1 \cdot \frac{1}{1} \\
&= 1.
\end{aligned}$$

Thus

$$\lim_{\theta\to 0} \frac{\tan \theta}{\theta} = 1. \quad ■$$

Example 2 tells us that $\tan \theta$, just like $\sin \theta$, is well approximated by θ when θ is small (angle in radians).

EXAMPLE 3 Find $\lim_{x\to 0} (\sin 5x)/(5x)$.

SOLUTION Observe that as $x \to 0$, $5x \to 0$. Let $\theta = 5x$. Thus

$$\lim_{x\to 0} \frac{\sin 5x}{5x} = \lim_{\theta\to 0} \frac{\sin \theta}{\theta} = 1. \quad ■$$

The observation in Example 3 shows that $\lim_{x\to 0} (\sin ax)/(ax) = 1$ for any constant $a \neq 0$. Similarly, $\lim_{x\to 0} (\sin x^2)/x^2 = 1$. In other words, the quotient sin (*anything*)/*anything* approaches 1, provided that *anything* approaches 0.

EXAMPLE 4 Find $\lim_{x\to 0} (\sin 5x)/(2x)$.

SOLUTION A little algebra permits us to exploit the result found in Example 3:

$$\lim_{x\to 0} \frac{\sin 5x}{2x} = \lim_{x\to 0} \frac{\sin 5x}{2x} \cdot \frac{5x}{5x}$$

$$= \lim_{x\to 0} \frac{\sin 5x}{5x} \cdot \frac{5x}{2x}$$

$$= 1 \cdot \frac{5}{2} = \frac{5}{2}. \quad \blacksquare$$

EXAMPLE 5 Find $\lim_{x\to 0} (\sin 3x)/(\sin 2x)$.

SOLUTION First rewrite the quotient as follows:

$$\frac{\sin 3x}{\sin 2x} = \frac{\sin 3x}{3x} \cdot \frac{2x}{\sin 2x} \cdot \frac{1}{2x} \cdot 3x$$

$$= \frac{3}{2} \cdot \frac{\sin 3x}{3x} \cdot \frac{2x}{\sin 2x}$$

$$= \frac{3}{2} \cdot \frac{\sin 3x}{3x} \cdot \frac{1}{\left(\dfrac{\sin 2x}{2x}\right)}.$$

Thus

$$\lim_{x\to 0} \frac{\sin 3x}{\sin 2x} = \lim_{x\to 0} \frac{3}{2} \cdot \frac{\sin 3x}{3x} \cdot \frac{1}{\left(\dfrac{\sin 2x}{2x}\right)}$$

$$= \frac{3}{2} \cdot 1 \cdot \frac{1}{1}$$

$$= \frac{3}{2}. \quad \blacksquare$$

Section Summary

We determined two limits needed in Chap. 3, $\lim_{\theta\to 0} (\sin \theta)/\theta = 1$ and $\lim_{\theta\to 0} (1 - \cos \theta)/\theta = 0$. The first limit tells us that when θ is small, θ is a good estimate of $\sin \theta$ (angle in radians). It is also a good estimate of $\tan \theta$ for small θ. To estimate the sine when the angle is in degrees, first express the angle in radians.

EXERCISES FOR SEC. 2.7: THE LIMIT OF (sin θ)/θ AS θ APPROACHES 0

1 What is the area of a sector of a circle of (*a*) radius 3 and angle $\pi/2$? (*b*) radius 1 and angle θ? (*c*) radius 2 and angle θ?

2 What is the area of the sector of a circle of radius 6 inches subtended by an angle of (*a*) $\pi/4$ radians? (*b*) 3 radians? (*c*) 45°?

In Exercises 3 to 10 examine the limits.

3 $\lim_{x\to 0} \dfrac{\sin x}{2x}$ **4** $\lim_{x\to 0} \dfrac{\sin 2x}{x}$

5 $\lim_{x\to 0} \dfrac{\sin 3x}{5x}$ **6** $\lim_{x\to 0} \dfrac{2x}{\sin 3x}$

7 $\lim_{\theta\to 0} \dfrac{\sin^2 \theta}{\theta}$ **8** $\lim_{h\to 0} \dfrac{\sin h^2}{h^2}$

9 $\lim_{\theta\to 0} \dfrac{\tan^2 \theta}{\theta}$ **10** $\lim_{\theta\to 0} \theta \cot \theta$

In Exercises 11 to 14 estimate the given quantity and compare it to its value given by a calculator. Angles are in radians unless otherwise indicated. Use $\pi \approx 3.14$.

11 sin 0.05 **12** tan 0.04

13 sin 7° **14** tan 9°

---

15 Here is another proof that $\lim_{\theta\to 0} (1 - \cos \theta)/\theta = 0$. It is assumed that $\lim_{\theta\to 0} (\sin \theta)/\theta = 1$.

(*a*) Show that

$$\frac{1 - \cos \theta}{\theta} = \frac{\sin^2 \theta}{\theta(1 + \cos \theta)}$$

by multiplying $(1 - \cos \theta)/\theta$ by $(1 + \cos \theta)/(1 + \cos \theta)$.

(*b*) Use (*a*) to show that $\lim_{\theta\to 0} (1 - \cos \theta)/\theta = 0$.

16 (See Exercise 15.) Show that

$$\lim_{\theta\to 0} \frac{1 - \cos \theta}{\theta^2} = \frac{1}{2}.$$

17 Using Exercise 16, explain why $1 - \theta^2/2$ is a good estimate of $\cos \theta$ when θ is a small angle expressed in radians. Include some comparisons of this estimate with the value of $\cos \theta$ on a calculator.

18 Examine the behavior of $(\theta - \sin \theta)/\theta^3$ for θ near 0.

19 Examine the behavior of the quotient $(\cos \theta - 1 + \theta^2/2)/\theta^4$ for θ near 0.

20 (A test of intuition) An intuitive argument suggested that $\lim_{\theta\to 0} (\sin \theta)/\theta = 1$, which turned out to be correct. Try your intuition on another limit associated with the unit circle shown in Fig. 8.

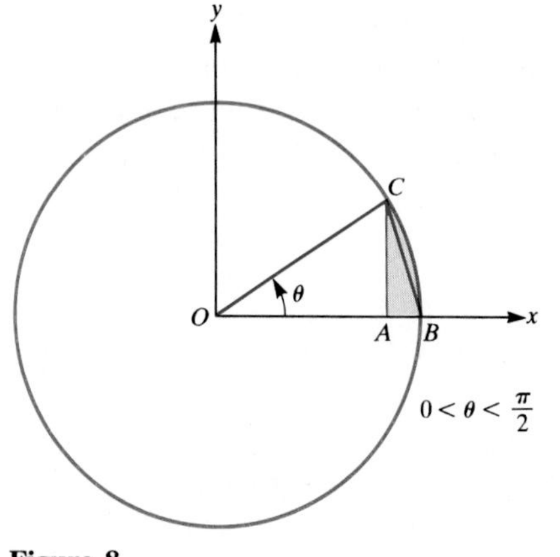

Figure 8

(*a*) What do you think happens to the quotient

$$\frac{\text{Area of triangle } ABC}{\text{Area of shaded region}}$$

as $\theta \to 0$? More precisely, what does your intuition suggest is the limit of that quotient as $\theta \to 0$?

(*b*) Estimate the limit in (*a*) using $\theta = 0.01$. *Note:* The limit is determined in Chap. 6. This question arose during some research in geometry. I guessed wrong, as has everyone I have asked.

21 (*a*) What is the domain of the function $f(x) = (\sin x)/x$?

(*b*) Show that f is an even function, that is, $f(-x) = f(x)$.

(*c*) Find $\lim_{x\to\infty} f(x)$. *Hint:* For all x, $|\sin x| \le 1$.

(*d*) For which x is $f(x) = 0$?

(*e*) Using a calculator, fill in the following table to two decimal places:

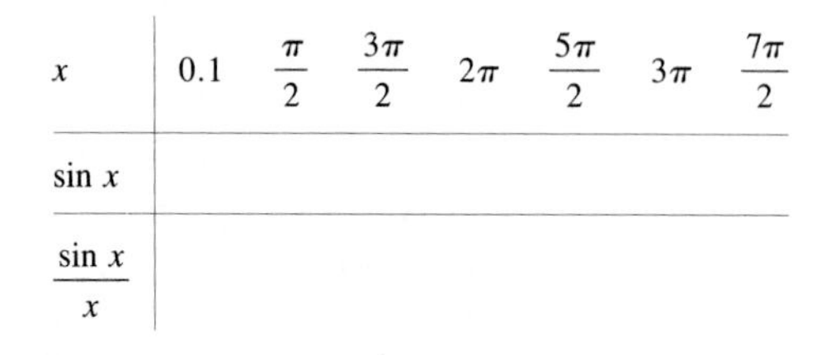

x	0.1	$\frac{\pi}{2}$	$\frac{3\pi}{2}$	2π	$\frac{5\pi}{2}$	3π	$\frac{7\pi}{2}$
$\sin x$							
$\frac{\sin x}{x}$							

(*f*) Graph f for $x > 0$.

(*g*) Graph f for $x < 0$.

(*h*) What is $\lim_{x\to 0} f(x)$?

22 (*a*) What is the domain of $g(x) = (1 - \cos x)/x$?

(*b*) Show that g is an odd function, that is, $g(-x) = -g(x)$.

(*c*) Find $\lim_{x\to\infty} g(x)$. *Hint:* $0 \le 1 - \cos x \le 2$ for all x.

(*d*) For which x is $g(x) = 0$?

(*e*) Using a calculator, fill in the following table to two decimal places:

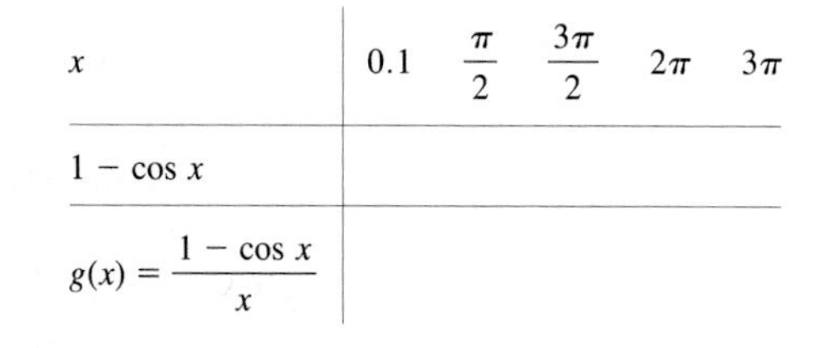

x	0.1	$\frac{\pi}{2}$	$\frac{3\pi}{2}$	2π	3π
$1 - \cos x$					
$g(x) = \frac{1 - \cos x}{x}$					

(*f*) Graph g for $x > 0$.

(*g*) Graph g for $x < 0$.

(*h*) What is $\lim_{x\to 0} g(x)$?

Exercises 23 to 27 concern the function $f(x) = \sin(1/x)$ and its graph. Use the same scale on both axes, with the distance from 0 to 1 at least 10 centimeters.

23 (*a*) Show that $f(1/(n\pi)) = 0$ for any nonzero integer n.

(*b*) Plot the points on the graph of f with x coordinate $1/\pi$, $1/(2\pi)$, $1/(3\pi)$, $1/(4\pi)$, and $1/(5\pi)$.

24 (*a*) Show that $f(1/[(2n + \frac{3}{2})\pi]) = -1$ for any integer n.

(*b*) Plot the points on the graph of f with $x = 1/(\frac{3}{2}\pi)$, $1/(\frac{7}{2}\pi)$, and $1/(\frac{11}{2}\pi)$.

25 (*a*) Show that, for any integer n, $f(1/[(2n + \frac{1}{2})\pi]) = 1$.
(*b*) Plot the points on the graph of f with $x = 1/(\pi/2)$, $1/(5\pi/2)$, and $1/(9\pi/2)$.

26 What is $\lim_{x\to\infty} \sin(1/x)$?

27 (*a*) With the aid of Exercises 23 to 26 graph f for $x > 0$.
(*b*) Does $\lim_{x\to 0^+} f(x)$ exist?

28 This exercise concerns the graph of $f(x) = x \sin x$.
(*a*) For which x is $f(x) = 0$?
(*b*) For which x is $f(x) = x$?
(*c*) For which x is $f(x) = -x$?
(*d*) For $x \geq 0$, plot the points given in (*a*), (*b*), and (*c*). (There are an infinite number of them; just sketch a few.)
(*e*) With the aid of (*d*) graph f for $x \geq 0$.
(*f*) Does $\lim_{x\to 0^+} f(x)$ exist?
(*g*) Does $\lim_{x\to\infty} f(x)$ exist?

2.8 CONTINUOUS FUNCTIONS

This section introduces the notion of a **continuous function**. Almost all functions met in practice are continuous. We begin with an informal, intuitive description and then give a more useful working definition.

An Informal Introduction to Continuous Functions

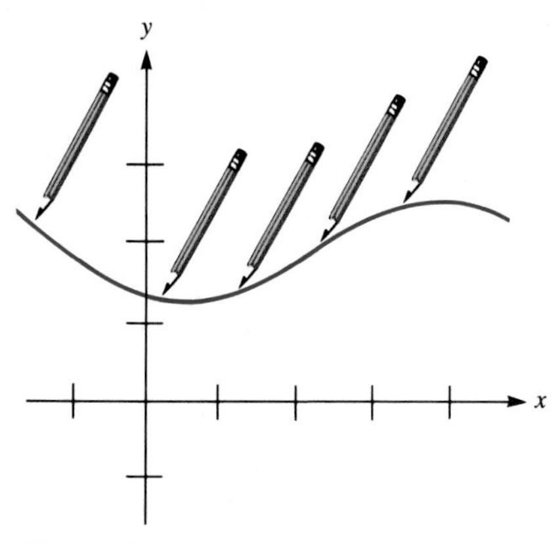

Figure 1

When we draw the graph of a function defined on some interval, we usually do not have to lift the pencil off the paper. Figure 1 shows this typical situation. The graph comes in one piece, like a length of wire. There are no gaps or jumps, although there may be sharp corners.

A function is said to be *continuous* if, when considered on any interval in its domain, its graph has no jumps—it can be traced without lifting pencil off paper. (The domain of the function may consist of several intervals.)

According to this definition any polynomial is continuous. So is each of the basic trigonometric functions, including $y = \tan x$, whose graph is shown in Fig. 22 of Sec. 2.6. You may be tempted to say "But tan x blows up at $\pi/2$ and I have to lift my pencil off the paper to draw the graph." However, $\pi/2$ is not in the domain of the tangent function. *On every interval in its domain, tan x behaves quite decently:* on such an interval we can sketch its graph without lifting pencil from paper. That is why tan x is continuous. The function $1/x$ is also continuous, since it "explodes" only at a number not in its domain, namely at 0.

The function whose graph is shown in Fig. 2 is not continuous. It is defined throughout the interval $[-2, 3]$, but to draw its graph you must lift the pencil from the paper when $x = 1$. Unlike the cases involving tan x or $1/x$, the problem lies in the domain of the function, which prevents it from being continuous. However, when you consider the function *only for x in* $[1, 3]$, *then it is continuous*—at the cost of throwing away part of its domain.

By the way, if you want a formula for the function given graphically in Fig. 2, here it is:

$$f(x) = \begin{cases} x + 1 & \text{for } x \text{ in } [-2, 1) \\ x & \text{for } x \text{ in } [1, 2) \\ -x + 4 & \text{for } x \text{ in } [2, 3]. \end{cases}$$

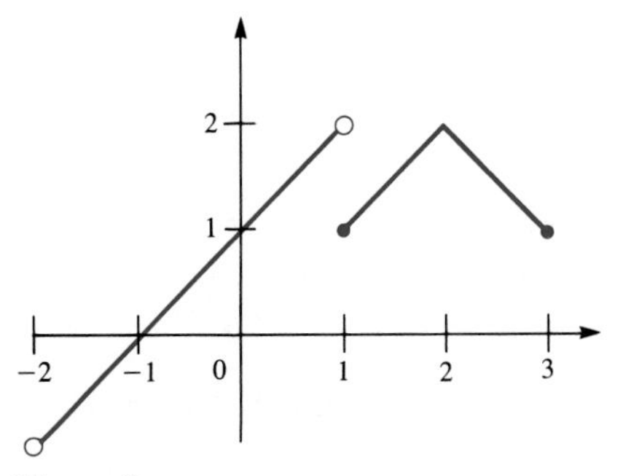

Figure 2

It is pieced together from three different continuous functions. Functions with "piecewise" definitions are often dismissed as unnatural or "artificial." However, such functions play a prominent role in applications. In real life scientists may need to splice together several different functions to fit the graph of actual research data. In Sec. 5.4 we will use functions defined piecewise to approximate the area under a curve.

The Definition of Continuity

Our informal "moving pencil" notion of a continuous function requires drawing a graph of the function. Our working definition does not require such a graph. Moreover, it easily generalizes to functions of more than one variable.

To get the feeling of the second definition, imagine that you had the information shown in the table in the margin about some function f. What would you expect the output $f(1)$ to be?

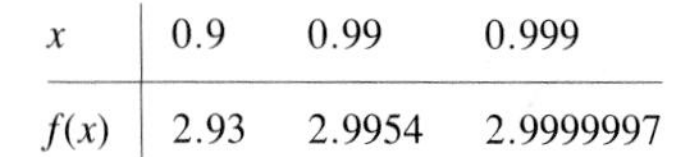

x	0.9	0.99	0.999
$f(x)$	2.93	2.9954	2.9999997

It would be quite a shock to be told that $f(1)$ is, say, 625. A reasonable function should present no such surprise. The expectation is that $f(1) = 3$. More generally, we expect the output of a function at the input a to be closely connected with the outputs of the functions at inputs that are near a. The functions of interest in calculus usually behave that way. With this in mind, we define the notion of *continuity at a number a*.

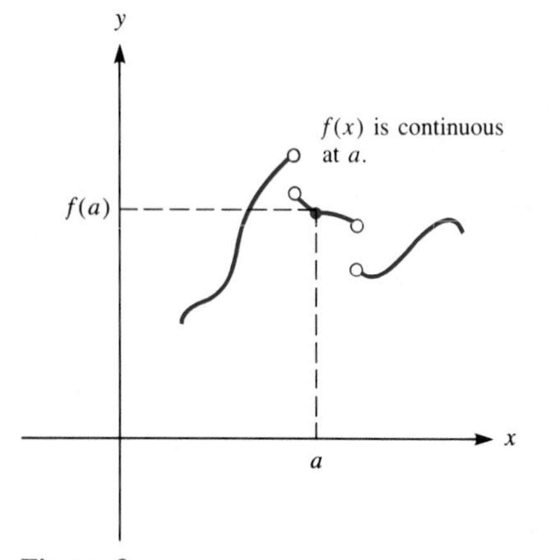

Figure 3

> **Definition** *Continuity at a number a.* Assume that $f(x)$ is defined in some open interval that contains the number a. Then the function f is *continuous at a* if $\lim_{x \to a} f(x) = f(a)$. This means that
>
> **1** $f(a)$ is defined (that is, a is in the domain of f).
> **2** $\lim_{x \to a} f(x)$ exists.
> **3** $\lim_{x \to a} f(x) = f(a)$.

As Fig. 3 shows, whether a function is continuous at a depends only on its behavior for inputs near a. Being continuous at a is a local matter, involving perhaps very tiny intervals about a.

To check whether a function f is continuous at a number a, we ask three questions:

1 Is a in the domain of f?
2 Does $\lim_{x \to a} f(x)$ exist?
3 Does $f(a)$ equal $\lim_{x \to a} f(x)$?

If the answer is "yes" to each of these questions, we say that f is continuous at a.

If a is in the domain of f and the answer to question 2 or 3 is "no," then f is said to be *discontinuous at a*. If a is not in the domain of f, we do not define either continuity or discontinuity there.

We are now ready to define a continuous function.

> **Definition** *Continuous function.* Let f be a function whose domain is the x axis or is made up of open intervals. Then f is a *continuous function* if it is continuous at each number a in its domain.

EXAMPLE 1 Use the definition of continuity to decide whether $f(x) = 1/x$ is continuous.

SOLUTION Let a be in the domain of $f(x)$. In other words, a is not 0. Since

$$\lim_{x \to a} \frac{1}{x} = \frac{1}{a},$$

the answer to question 2 is "yes."

Since

$$f(a) = \frac{1}{a},$$

the answer to question 3 is "yes." Thus $1/x$ is continuous at every number in its domain. Hence it is a continuous function.

Note that the conclusion agrees with our "moving pencil" picture of continuity. ■

Not every important function is continuous. For instance, let $f(x)$ be the greatest integer that is less than or equal to x. We have $f(1.8) = 1$, $f(1.9) = 1$, $f(2) = 2$, and $f(2.3) = 2$. This function is often used in number theory and computer science, where it is denoted $[x]$ or $\lfloor x \rfloor$ and called the *floor* of x. The next example examines where the floor function fails to be continuous.

Most people use the floor function every time they answer the question, "How old are you?"

EXAMPLE 2 Let $f(x) = [x]$. Graph f and find where it is continuous. Is f a continuous function?

SOLUTION We begin with the following table:

x	0	0.5	0.8	1	1.1	1.99	2	2.01
$[x]$	0	0	0	1	1	1	2	2

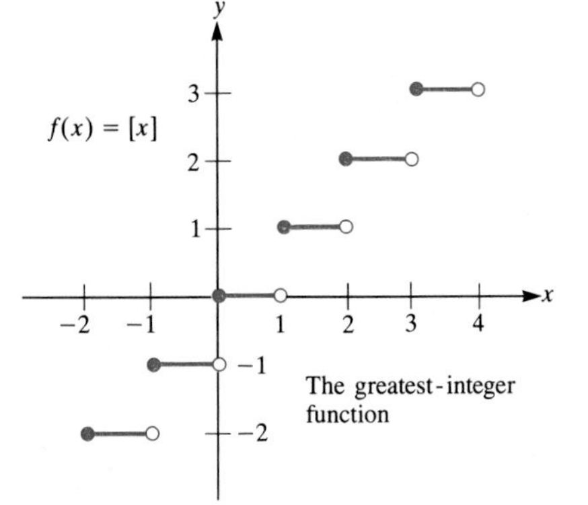

Figure 4

For $0 \le x < 1$, $[x] = 0$. But at the input $x = 1$ the output jumps to 1, since $[1] = 1$. For $1 \le x < 2$, $[x]$ remains at 1. Then at 2 it jumps to 2. More generally, $[x]$ has a jump at every integer, as shown in Fig. 4.

Let us show that $f(x)$ is not continuous at $a = 2$ by seeing which of the three conditions in the definition are not satisfied.

First of all, condition 1 is satisfied since 2 does lie in the domain of the function; indeed, $f(2) = 2$.

Does condition 2 hold? That is, does $\lim_{x \to 2} f(x)$ exist? We see that

$$\lim_{x \to 2^+} f(x) = 2 \qquad \text{and} \qquad \lim_{x \to 2^-} f(x) = 1.$$

But the right-hand and left-hand limits are not equal, so $\lim_{x \to 2} f(x)$ does not exist.

Already we know that the function is not continuous at $a = 2$. There is no point in checking the third condition, $\lim_{x \to 2} f(x) = f(2)$, since the limit does not exist. Thus $[x]$ is not a continuous function since it is not continuous at every number in its domain. It fails to be continuous at $x = a$, whenever a is an integer.

Is f continuous at a if a is not an integer? Let us take the case $a = 1.5$, for instance.

Condition 1 is satisfied, since $f(1.5)$ is defined. (It equals 1.)

Condition 2 is satisfied, since $\lim_{x\to 1.5} f(x) = 1$.

Condition 3 is satisfied, since $\lim_{x\to 1.5} f(x)$ does equal $f(1.5)$. (Both equal 1.)

The function $[x]$ is continuous at 1.5. Similarly, it is continuous at every number a that is not an integer.

Note that $[x]$ is continuous on any interval that does not include an integer. For instance, if we consider the function only on the interval (1.1, 1.9), it is continuous there. ■

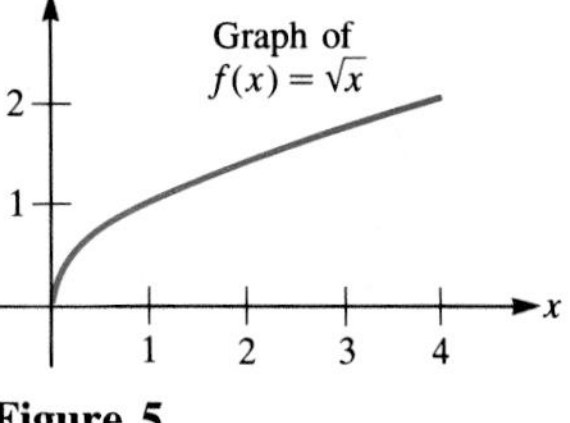

Figure 5

Continuity at an Endpoint

The function $f(x) = \sqrt{x}$ is graphed in Fig. 5 and $f(x) = \sqrt{1 - x^2}$ is graphed in Fig. 6. We would like to consider both of these functions continuous. However, there is a slight technical problem. The number 0 is in the domain of $\sqrt{x}$, but there is no open interval around 0 that lies completely in the domain, as our definition of continuity requires. We introduce a definition to cover such a case. We would not be interested in numbers x to the left of 0, since $f(x)$ is not defined for such x. Hence we use a one-sided definition of continuity.

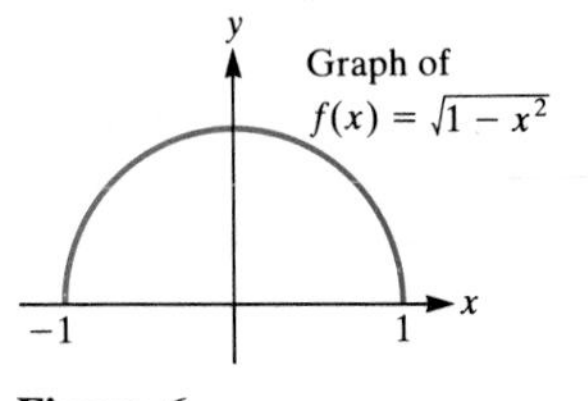

Figure 6

> **Definition** *Continuity from the right at a number a.* Assume that $f(x)$ is defined in some closed interval $[a, b]$. Then the function f is *continuous at a from the right* if
>
> **1** $f(a)$ is defined.
> **2** $\lim_{x\to a^+} f(x)$ exists.
> **3** $\lim_{x\to a^+} f(x) = f(a)$.

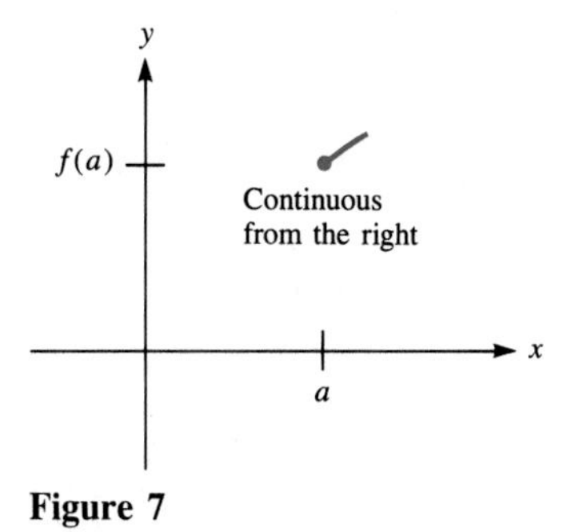

Figure 7

Figure 7 illustrates this definition.

This definition also takes care of the continuity of $\sqrt{1 - x^2}$ at -1 in Fig. 6. The next definition attends to the similar problem at $a = 1$.

> **Definition** *Continuity from the left at a number a.* Assume that $f(x)$ is defined at a and in some closed interval $[b, a]$. Then the function f is *continuous at a from the left* if
>
> **1** $f(a)$ is defined.
> **2** $\lim_{x\to a^-} f(x)$ exists.
> **3** $\lim_{x\to a^-} f(x) = f(a)$.

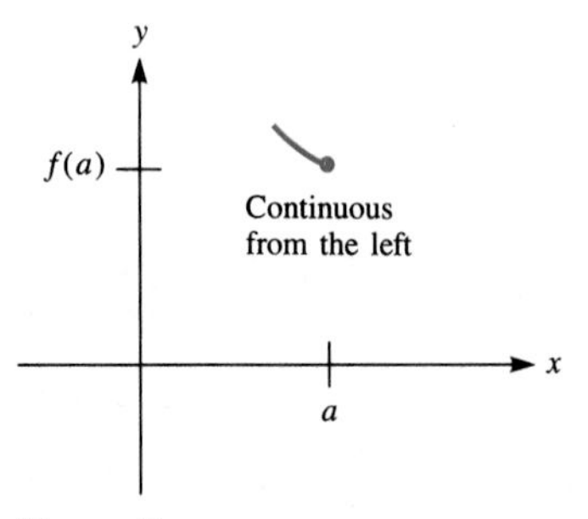

Figure 8

Figure 8 illustrates this definition.

With these two extra definitions to cover some special numbers in the domain, we can therefore extend the definition of continuous function to include those functions whose domains may contain some intervals that include their endpoints. We say, for instance, that $\sqrt{1 - x^2}$ is continuous because it is continuous at any number in $(-1, 1)$, is continuous from the right at -1, and continuous from the left at 1.

These special considerations are minor matters that will little concern us in the future. The key point is that $\sqrt{1 - x^2}$ and $\sqrt{x}$ are continuous functions. So are practically all the functions studied in calculus.

The following example reviews the notion of continuity.

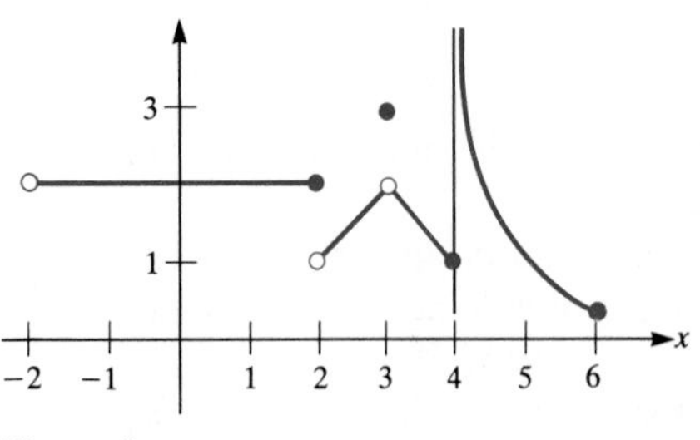

Figure 9

EXAMPLE 3 Figure 9 is the graph of a certain function $f(x)$ whose domain is the interval $(-2, 6]$. Discuss the continuity of $f(x)$ at (*a*) 6, (*b*) 4, (*c*) 3, (*d*) 2, (*e*) 1, and (*f*) -2.

SOLUTION

(*a*) Since $\lim_{x\to 6^-} f(x)$ exists and equals $f(6)$, $f(x)$ is continuous from the left at 6.

(*b*) Since $\lim_{x\to 4} f(x)$ does not exist, $f(x)$ is not continuous at 4.

(*c*) Inspection of the graph shows that $\lim_{x\to 3} f(x) = 2$. However, condition 3 for continuity is not satisfied, since $f(3) = 3$, which is *not* equal to $\lim_{x\to 3} f(x)$. Thus $f(x)$ is not continuous at 3.

(*d*) Though $\lim_{x\to 2^+} f(x)$ and $\lim_{x\to 2^-} f(x)$ both exist, they are not equal. (The first equals 1; the second is 2.) Thus $\lim_{x\to 2} f(x)$ does not exist. Since condition 2 for continuity is not satisfied, $f(x)$ is not continuous at 2.

(*e*) All three conditions are satisfied at 1: $\lim_{x\to 1} f(x)$ exists (it equals 2) and equals $f(1)$. $f(x)$ is continuous at 1.

(*f*) Since -2 is not in the domain of the function, we do not even speak of "continuity" or "discontinuity" at this number. ■

As Example 3 shows, a function can fail to be continuous at a given number a of its domain for either of two reasons: First, $\lim_{x\to a} f(x)$ might not exist. Second, when $\lim_{x\to a} f(x)$ does exist, $f(a)$ might not equal that limit.

Two Important Properties of Continuous Functions

Continuous functions have two properties of particular importance in calculus: the "maximum-value" property and the "intermediate-value" property. Both are quite plausible, and a glance at the graph of a "typical" continuous function may persuade us that they are true of all continuous functions. No proofs will be offered: they depend on the precise definitions of limits given in Secs. 2.9 and 2.10 and are part of an advanced calculus course.

The first theorem asserts that a function that is continuous throughout the closed interval $[a, b]$ takes on a largest value somewhere in the interval.

Maximum-Value Theorem Let f be continuous throughout the closed interval $[a, b]$. Then there is at least one number in $[a, b]$ at which f takes on a maximum value. That is, for some number c in $[a, b]$, $f(c) \geq f(x)$ for all x in $[a, b]$.

To persuade yourself that this theorem is plausible, imagine sketching the graph of a continuous function. As your pencil moves along the graph from

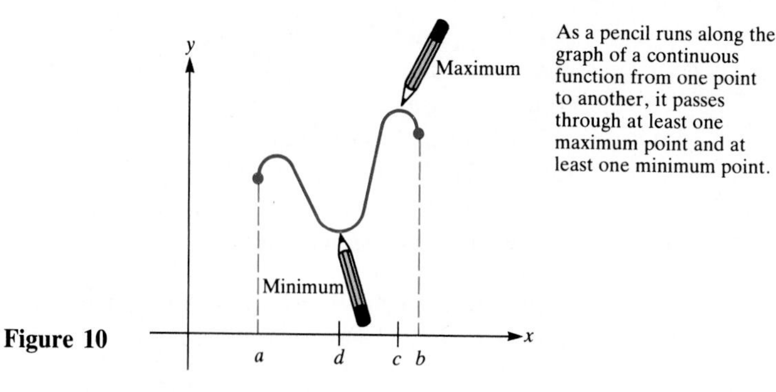

As a pencil runs along the graph of a continuous function from one point to another, it passes through at least one maximum point and at least one minimum point.

Figure 10

some point on the graph to some other point on the graph, it passes through a highest point. (See Fig. 10.)

The maximum-value theorem guarantees that a maximum value exists, but it does *not tell how* to find it. The problem of finding the maximum value is discussed in Chap. 4.

There is also a minimum-value theorem.

> **Minimum-Value Theorem** Let f be continuous throughout the closed interval $[a, b]$. Then there is at least one number in $[a, b]$ at which f takes on a minimum value. That is, for some number d in $[a, b]$, $f(d) \leq f(x)$ for all x in $[a, b]$.

See Fig. 10 for an illustration of this theorem.

These two theorems, taken together, are called the *extreme-value theorem.*

EXAMPLE 4 Let $f(x) = \cos x$ and $[a, b] = [0, 3\pi]$. Find all numbers in $[0, 3\pi]$ at which f takes on a maximum value. Also find all numbers in $[0, 3\pi]$ at which f takes on a minimum value.

SOLUTION Figure 11 is a graph of $\cos x$ for x in $[0, 3\pi]$. Inspection of the graph shows that the maximum value of $\cos x$ for $0 \leq x \leq 3\pi$ is 1, and it is attained when $x = 0$ and when $x = 2\pi$. The minimum value is -1, which is attained when $x = \pi$ and when $x = 3\pi$. ■

If you remove one of the two assumptions in the extreme-value theorem ("f is continuous" and "the domain is a closed interval"), the conclusion need not hold.

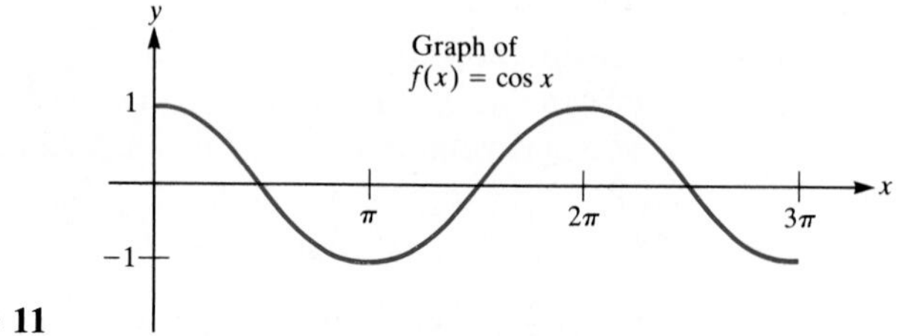

Figure 11

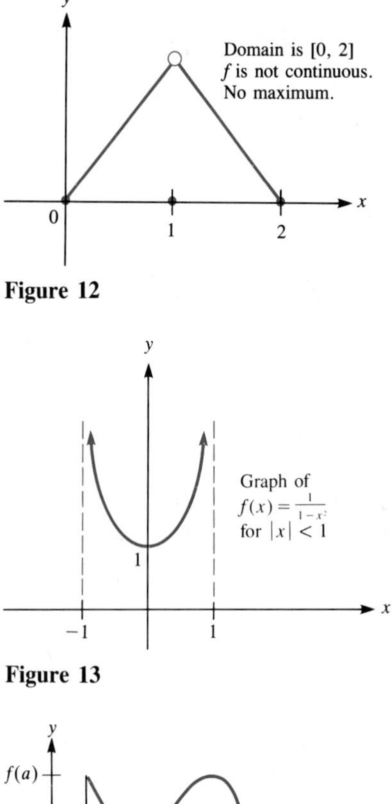

Figure 12

Figure 13

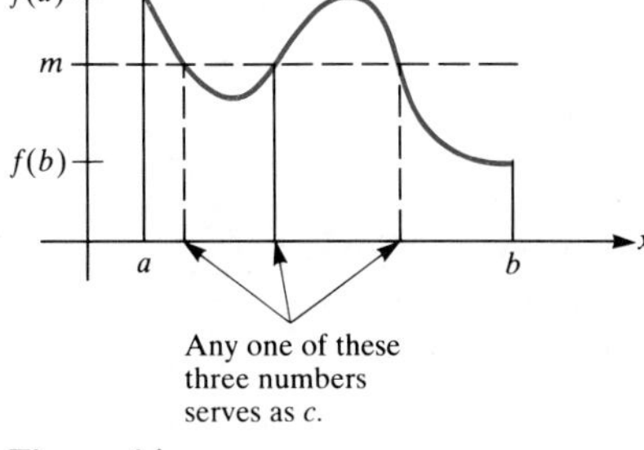

Figure 14

Figure 12 shows the graph of a function that is *not* continuous, is defined on a closed interval, but has no maximum value. On the other hand $f(x) = 1/(1 - x^2)$ is continuous on the domain $(-1, 1)$. It has no maximum value, as a glance at Fig. 13 shows. This does not violate the maximum-value theorem, since the domain $(-1, 1)$ is not a closed interval.

Intermediate-Value Theorem

Imagine graphing a continuous function f defined on the closed interval $[a, b]$. As your pencil moves from the point $(a, f(a))$ to the point $(b, f(b))$ the y coordinate of the pencil point goes through all values between $f(a)$ and $f(b)$. (Similarly, if you hike all day, starting at an altitude of 5,000 feet and ending at 11,000 feet, you must have been, say, at 7,000 feet at least once during the day.)

The next theorem states this in mathematical terms, not in terms of the pencil (or a hike). It says that a function which is continuous throughout an interval takes on all values between any two of its values.

> **Intermediate-Value Theorem** Let f be continuous throughout the closed interval $[a, b]$. Let m be any number between $f(a)$ and $f(b)$. [That is, $f(a) \le m \le f(b)$ if $f(a) \le f(b)$, or $f(a) \ge m \ge f(b)$ if $f(a) \ge f(b)$.] Then there is at least one c in $[a, b]$ such that $f(c) = m$. In other words, there is an input c in $[a, b]$ such that the corresponding output $f(c)$ is equal to the prescribed value m.

In ordinary English, the intermediate-value theorem reads: A continuous function defined on $[a, b]$ takes on all values between $f(a)$ and $f(b)$. Pictorially, it asserts that a horizontal line of height m must meet the graph of f at least once if m is between $f(a)$ and $f(b)$, as shown in Fig. 14.

Even though the theorem guarantees the existence of a certain number c, it does *not tell how* to find it. To find it, we must solve an equation, namely, $f(x) = m$.

EXAMPLE 5 Use the intermediate-value theorem to show that the equation $2x^3 + x^2 - x + 1 = 5$ has a solution in the interval $[1, 2]$.

SOLUTION Let $P(x) = 2x^3 + x^2 - x + 1$. Then

$$P(1) = 2 \cdot 1^3 + 1^2 - 1 + 1 = 3$$

and

$$P(2) = 2 \cdot 2^3 + 2^2 - 2 + 1 = 19.$$

The intermediate-value theorem can help guarantee a solution

Since P is continuous and $m = 5$ is between $P(1) = 3$ and $P(2) = 19$, we may apply the intermediate-value theorem to P in the case $a = 1$, $b = 2$, and $m = 5$. Thus there is at least one number c between 1 and 2 such that $P(c) = 5$. This completes the answer.

[To get a more accurate estimate for a number c such that $P(c) = 5$, find a shorter interval for which the intermediate-value theorem can be applied. For instance, $P(1.2) = 4.696$ and $P(1.3) = 5.784$. By the intermediate-value theorem, there is a number c in $[1.2, 1.3]$ such that $P(c) = 5$.] ■

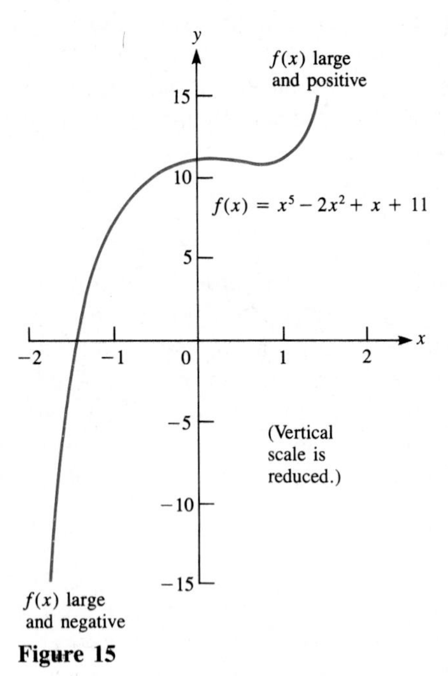

Figure 15

EXAMPLE 6 Show that the equation $x^5 - 2x^2 + x + 11 = 0$ has at least one real root; that is, the function $x^5 - 2x^2 + x + 11$ has an x intercept.

SOLUTION Let $f(x) = x^5 - 2x^2 + x + 11$. We wish to show that there is a number c such that $f(c) = 0$. In order to use the intermediate-value theorem, we need an interval $[a, b]$ for which 0 is between $f(a)$ and $f(b)$. Then we could apply that theorem, using $m = 0$.

We show that there are such numbers a and b, as follows. For x large and positive the polynomial $f(x) = x^5 - 2x^2 + x + 11$ is positive [since $\lim_{x\to\infty} f(x) = \infty$]. Thus there is a number b such that $f(b) > 0$. Similarly, for x negative and of large absolute value, $f(x)$ is negative [since $\lim_{x\to-\infty} f(x) = -\infty$]. Select a number a such that $f(a) < 0$. (See Fig. 15.)

The number 0 is between $f(a)$ and $f(b)$. Since f is continuous on the interval $[a, b]$, there is a number c in $[a, b]$ such that $f(c) = 0$. This number c is a solution to the equation $x^5 - 2x^2 + x + 11 = 0$. ■

Note that the argument in Example 6 applies to any polynomial of *odd* degree. Any polynomial of odd degree has a real root. The argument does not hold for polynomials of even degree, since the equations $x^2 + 1 = 0$, $x^4 + 1 = 0$, $x^6 + 1 = 0$, and so on have no real solutions.

EXAMPLE 7 Use the intermediate-value theorem to show that there is a number c such that $4 - c = 2^c$.

SOLUTION We wish to show that there is a number c where the function $4 - x$ has the same value as the function 2^x. Now, $4 - x = 2^x$ is equivalent to

$$4 - x - 2^x = 0.$$

The problem reduces to showing that the function $f(x) = 4 - x - 2^x$ has the value 0 for some input c.

We will proceed as we did in the previous example. We use the intermediate-value theorem for a suitable interval $[a, b]$. That is, we want to find numbers a and b such that $f(a)$ and $f(b)$ have opposite signs, one being positive, the other negative. Then $m = 0$ will lie between them.

To conduct the search for a and b, make a table:

x	0	1	2
$4 - x - 2^x$	3	1	−2

We see that $f(1) = 1$ is positive and $f(2) = -2$ is negative. Thus

$$f(1) \geq 0 \geq f(2).$$

Since $m = 0$ lies between $f(1)$ and $f(2)$, and f is continuous on $[1, 2]$, the intermediate-value theorem asserts that there is a number c, between 1 and 2, such that $f(c) = 0$. It follows that $4 - c = 2^c$. ■

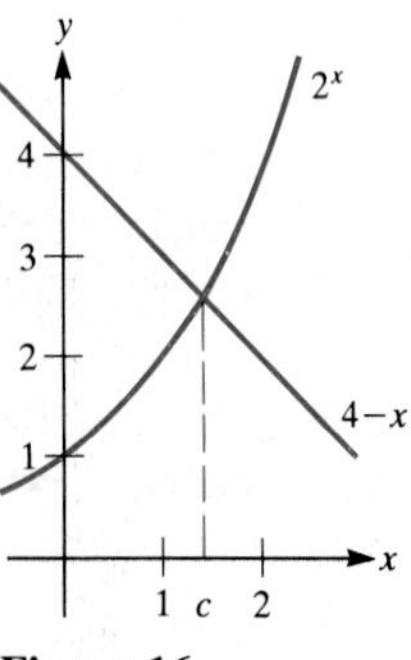

Figure 16

In Example 7 the intermediate-value theorem does not tell what c is. The graphs of $4 - x$ and 2^x in Fig. 16 suggest that c is unique and about 1.4. Further arithmetic, done on a calculator, shows that $c \approx 1.39$.

From Examples 6 and 7 we make two observations:

1 If a continuous function defined on an interval is positive somewhere in the interval and negative somewhere in the interval, then it must be 0 at some number in that interval.

2 To show that two functions are equal at some number in an interval, show that their difference is 0 at some number in the interval.

Section Summary

This section opened with an informal view of continuous functions, expressed in terms of a moving pencil. It then gave the definition that we will use throughout the text. This definition is phrased in terms of limits. It concluded with the statement of two important properties of continuous functions considered on closed intervals: the extreme-value theorem and the intermediate-value theorem.

EXERCISES FOR SEC. 2.8: CONTINUOUS FUNCTIONS

In each of Exercises 1 to 10 state which of the three conditions for continuity are satisfied at the given number. Decide whether the function is continuous at that number.

1 $a = \frac{1}{2}$

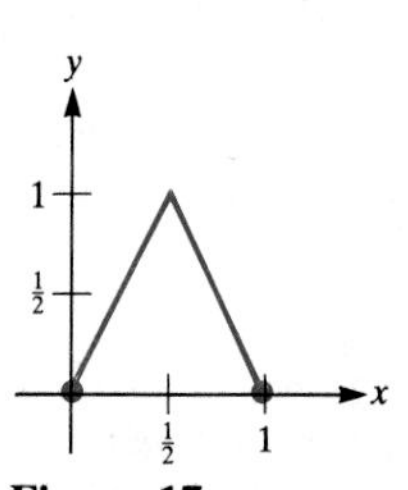

Figure 17

2 $a = \frac{1}{2}$

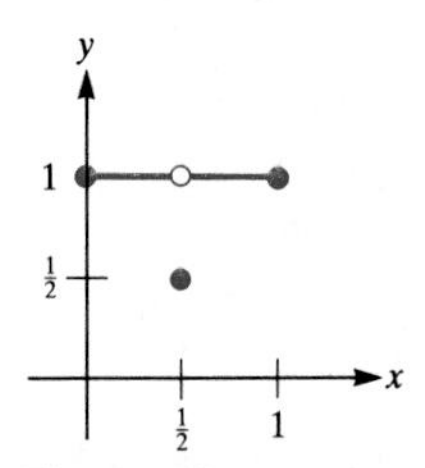

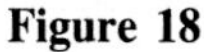

Figure 18

3 $a = \frac{1}{2}$

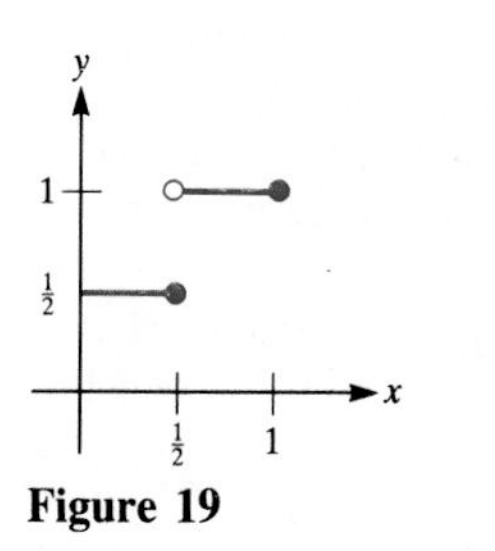

Figure 19

4 $a = 1$

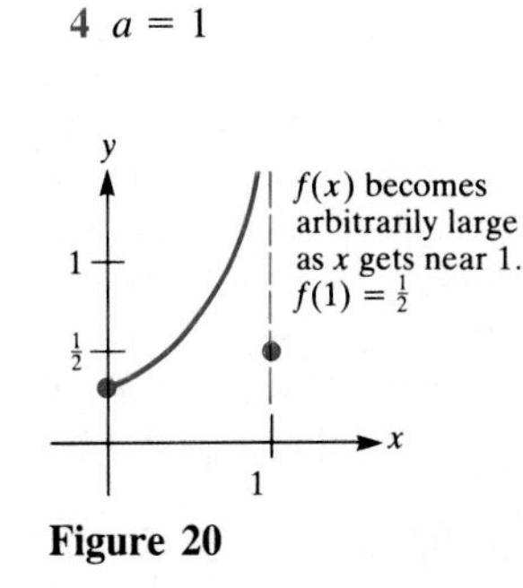

Figure 20

5 $a = 0$

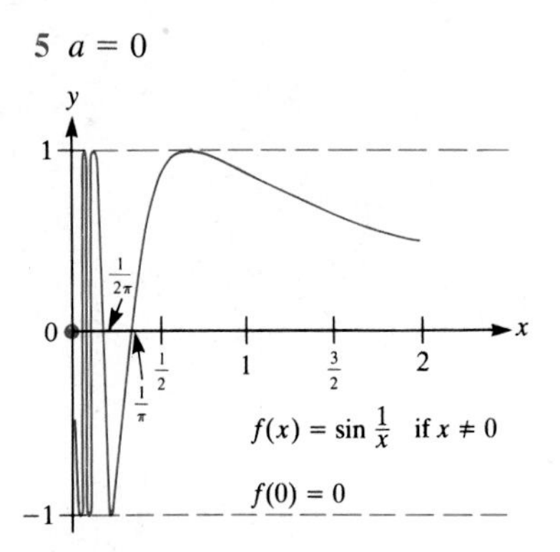

The graph oscillates infinitely often between the lines $y = 1$ and $y = -1$.

Figure 21

6 $a = 0$

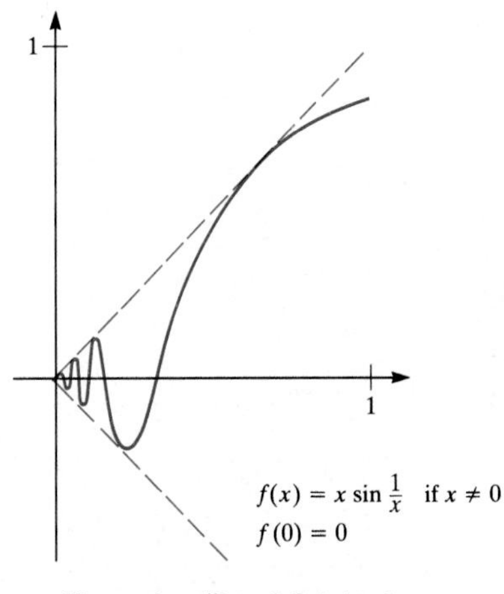

The graph oscillates infinitely often between the dashed lines.

Figure 22

7 $a = \frac{1}{4}$

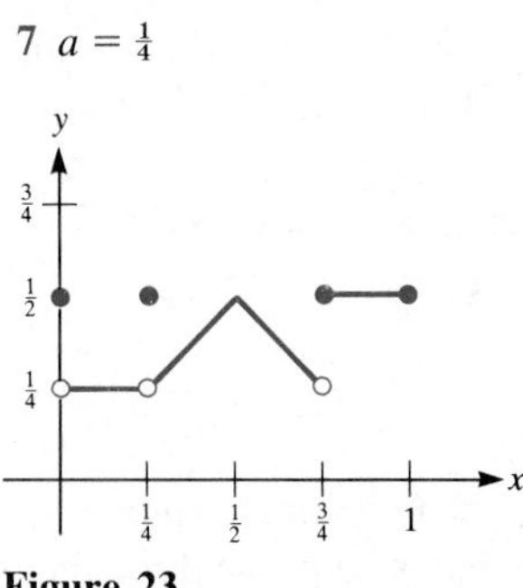

Figure 23

8 $a = \frac{1}{2}$

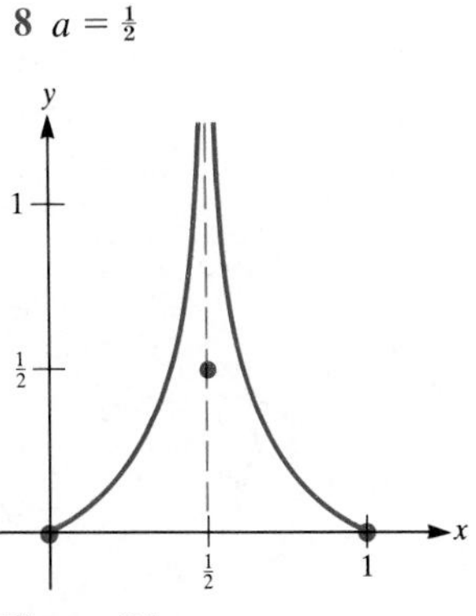

Figure 24

9 $a = 0$

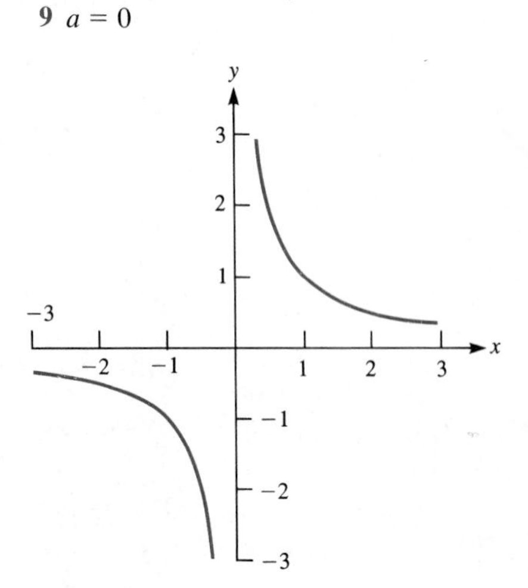

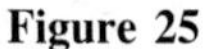

Figure 25

10 $a = 0$

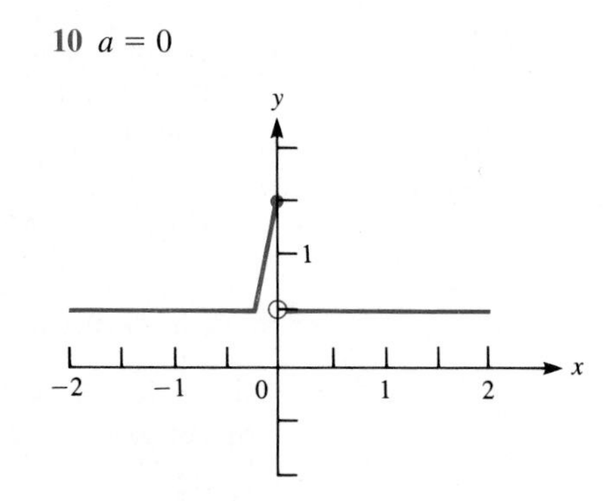

Figure 26

11 Let $f(x)$ equal the least integer that is greater than or equal to x. For instance, $f(3) = 3$, $f(3.4) = 4$, $f(3.9) = 4$. (This function is sometimes denoted $\lceil x \rceil$ and called the "ceiling" of x.)

(*a*) Graph f.
(*b*) Does $\lim_{x \to 4^-} f(x)$ exist? If so, what is it?
(*c*) Does $\lim_{x \to 4^+} f(x)$ exist? If so, what is it?
(*d*) Does $\lim_{x \to 4} f(x)$ exist? If so, what is it?
(*e*) Is f continuous at 4?
(*f*) Where is f continuous?
(*g*) Where is f not continuous?

12 Let f be the "nearest integer, with rounding down" function. That is,

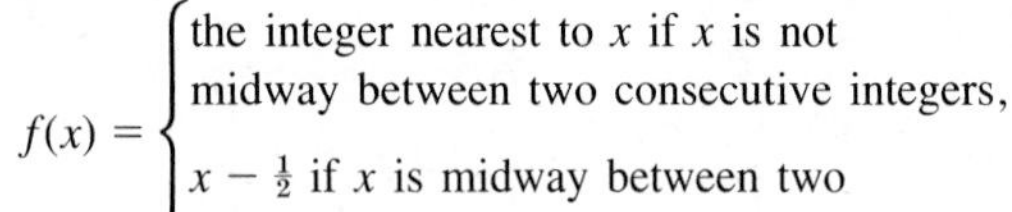

$$f(x) = \begin{cases} \text{the integer nearest to } x \text{ if } x \text{ is not midway between two consecutive integers,} \\ x - \frac{1}{2} \text{ if } x \text{ is midway between two consecutive integers.} \end{cases}$$

For instance, $f(1.4) = 1$, $f(1.5) = 1$, and $f(1.6) = 2$.

(*a*) Graph f.
(*b*) Does $\lim_{x \to 3.5^-} f(x)$ exist? If so, evaluate it.
(*c*) Does $\lim_{x \to 3.5^+} f(x)$ exist? If so, evaluate it.
(*d*) Does $\lim_{x \to 3.5} f(x)$ exist? If so, evaluate it.
(*e*) Is f continuous at 3.5?
(*f*) Where is f continuous?
(*g*) Where is f not continuous?

13 Let $f(x) = (1 - \cos x)/x$ for $x \neq 0$. Is it possible to define $f(0)$ in such a way that f is continuous throughout the x axis?

14 Let $f(x) = (x^3 - 1)/(x - 1)$ if $x \neq 1$. Is it possible to define $f(1)$ in such a way that f is continuous throughout the x axis?

15 Let $f(x) = x/|x|$ if x is not 0. Is it possible to define $f(0)$ in such a way that f is continuous throughout the x axis?

16 Let $f(x) = (x - 1)/|x - 1|$ if x is not 1. Is it possible to define $f(1)$ in such a way that f is continuous throughout the x axis?

17 For each of the given intervals, find the maximum value of $\sin x$ over that interval and find the value of x at which it occurs.

(*a*) $[0, \pi/4]$ (*b*) $[0, \pi]$

18 For each of the given intervals, find the maximum value of $\cos x$ over that interval and find the value of x at which it occurs.
(*a*) $[0, \pi/2]$ (*b*) $[0, 2\pi]$

19 Does the function $(x^3 + x^4)/(1 + 5x^2 + x^6)$ have (*a*) a maximum value for x in $[1, 4]$? (*b*) a minimum value for x in $[1, 4]$?

20 Does the function $2^x - x^3 + x^5$ have (*a*) a maximum value for x in $[-3, 10]$? (*b*) a minimum value for x in $[-3, 10]$?

21 Does the function x^3 have a maximum value for x in (*a*) $[2, 4]$? (*b*) $[-3, 5]$? (*c*) $(1, 6)$? If so, where?

22 Does the function x^4 have a minimum value for x in (*a*) $[-5, 6]$? (*b*) $(-2, 4)$? (*c*) $(3, 7)$? (*d*) $(-4, 4)$? If so, where?

23 Does the function $2 - x^2$ have (*a*) a maximum value for x in $(-1, 1)$? (*b*) a minimum value for x in $(-1, 1)$? If so, where?

24 Does the function $2 + x^2$ have (*a*) a maximum value for x in $(-1, 1)$? (*b*) a minimum value for x in $(-1, 1)$? If so, where?

25 Show that the equation $x^5 + 3x^4 + x - 2 = 0$ has at least one solution in the interval $[0, 1]$.

26 Show that the equation $x^5 - 2x^3 + x^2 - 3x + 1 = 0$ has at least one solution in the interval $[1, 2]$.

In Exercises 27 to 31 verify the intermediate-value theorem for the specified function f, the interval $[a, b]$, and the indicated value m. Find all c's in each case.

27 Function $3x + 5$; interval $[1, 2]$; $m = 10$.

28 Function $x^2 - 2x$; interval $[-1, 4]$; $m = 5$.

29 Function $\sin x$; interval $[\pi/2, 11\pi/2]$; $m = -1$.

30 Function $\cos x$; interval $[0, 5\pi]$; $m = \sqrt{3}/2$.

31 Function $x^3 - x$; interval $[-2, 2]$; $m = 0$.

32 Show that the equation $2^x - 3x = 0$ has a solution in the interval $[0, 1]$.

33 Does the equation $x + \sin x = 1$ have a solution?

34 Does the equation $x^3 = 2^x$ have a solution?

35 Use the intermediate-value theorem to show that the equation $3x^3 + 11x^2 - 5x = 2$ has a solution.

36 Let $f(x) = 1/x$, $a = -1$, $b = 1$, $m = 0$. Note that $f(a) \le 0 \le f(b)$. Is there at least one c in $[a, b]$ such that $f(c) = 0$? If so, find c; if not, does this imply that the intermediate-value theorem is sometimes false?

---

37 Let $f(x) = x + |x|$.
(*a*) Graph f.
(*b*) Is f continuous at 0?

38 Let $f(x) = 2^{1/x}$ for $x \neq 0$.
(*a*) Find $\lim_{x\to\infty} f(x)$.
(*b*) Find $\lim_{x\to-\infty} f(x)$.
(*c*) Does $\lim_{x\to0^+} f(x)$ exist?
(*d*) Does $\lim_{x\to0^-} f(x)$ exist?
(*e*) Graph f, incorporating the information in parts (*a*) to (*d*).
(*f*) Is it possible to define $f(0)$ in such a way that f is continuous throughout the x axis?

39 Let $P(x) = a_nx^n + a_{n-1}x^{n-1} + \cdots + a_0$ be a polynomial of odd degree n and with positive lead coefficient a_n. Show that there is at least one real number r such that $P(r) = 0$.

40 (This continues Exercise 39.) The *factor theorem* from algebra asserts that the number r is a root of the polynomial $P(x)$ if and only if $x - r$ is a factor of $P(x)$. For instance, 2 is a root of the polynomial $x^2 - 3x + 2$ and $x - 2$ is a factor of the polynomial: $x^2 - 3x + 2 = (x - 2)(x - 1)$. This is reviewed in Appendix C.
(*a*) Use the factor theorem and Exercise 39 to show that every polynomial of odd degree has a factor of degree 1.
(*b*) Show that none of the polynomials $x^2 + 1$, $x^4 + 1$, or $x^{100} + 1$ has a first-degree factor.
(*c*) Check $x^4 + 1 = (x^2 + \sqrt{2}x + 1)(x^2 - \sqrt{2}x + 1)$. (It can be shown using complex numbers that every polynomial is the product of polynomials of degrees at most 2.)

Exercises 41 to 47 concern convex sets and show how the intermediate-value theorem can give geometric information. A set in the plane bounded by a curve is *convex* if for any two points P and Q in the set the line segment joining them also lies in the set. (See Fig. 27.) The boundary of a convex set we will call a *convex curve*. (These definitions generalize to a solid and its boundary surface.)

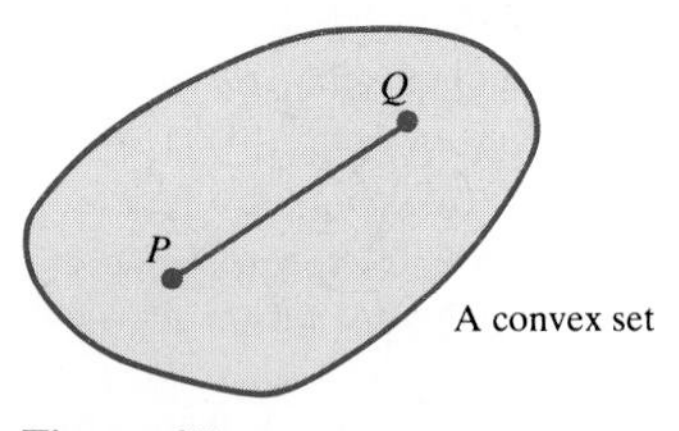

Figure 27

Disks, triangles, and parallelograms are convex sets. The quadrilateral shown in Fig. 28 is not convex. Convex sets will be referred to in the following exercises and occasionally in the exercises of later chapters.

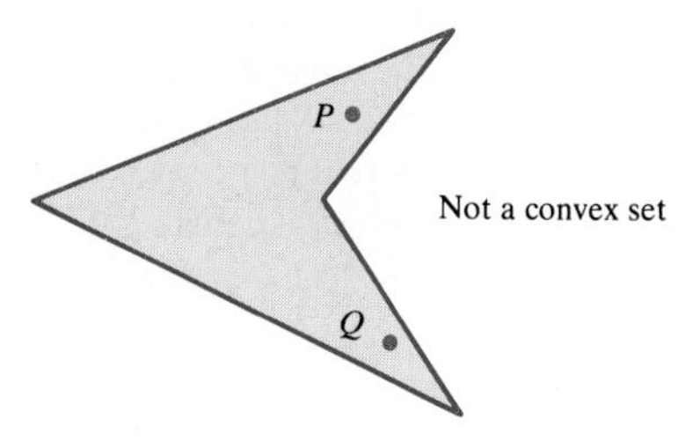

Figure 28

In Exercises 41 to 47 you will need to define various functions geometrically. *You may assume that they are continuous.*

41 Let L be a line in the plane and let K be a convex set. Show that there is a line parallel to L that cuts K into two pieces of equal area.

Follow these steps.

(*a*) Introduce an x axis perpendicular to L with its origin on L. Each line parallel to L and meeting K crosses the x axis at a number x. Label the line L_x. Let a be the smallest and b the largest of these numbers. (See Fig. 29.)

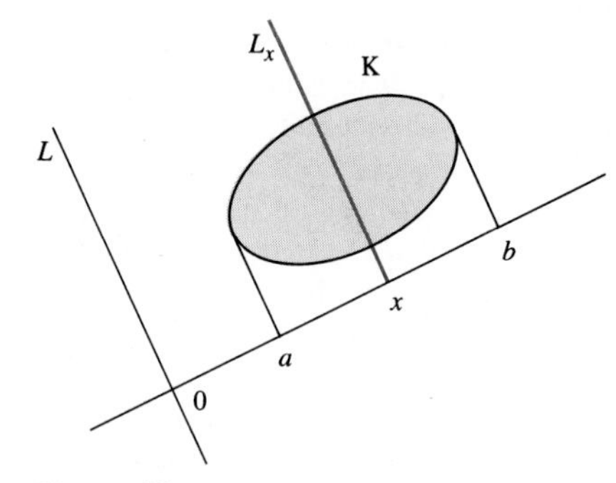

Figure 29

Let $A(x)$ be the area of K situated to the left of the line corresponding to x. Let the area of K be A. What is $A(a)$? $A(b)$?

(*b*) Use the intermediate-value theorem to show that there is an x in $[a, b]$ such that $A(x) = A/2$.

(*c*) Why does (*b*) show that there is a line parallel to L that cuts K into two pieces of equal area?

42 Solve the preceding exercise by applying the intermediate-value theorem to the function $f(x) = A(x) - B(x)$, where $B(x)$ is the area to the right of L_x.

43 Let P be a point in the plane and let K be a convex set. Is there a line through P that cuts K into two pieces of equal area?

44 Let K_1 and K_2 be two convex sets in the plane. Is there a line that simultaneously cuts K_1 into two pieces of equal area and cuts K_2 into two pieces of equal area? (This is known as the "two pancakes" question.)

45 Let K be a convex set in the plane. Show that there is a line that simultaneously cuts K into two pieces of equal area and cuts the boundary of K into two pieces of equal length.

46 Let K be a convex set. Show that there are two perpendicular lines that cut K into four pieces of equal area. (It is not known whether it is always possible to find two perpendicular lines that divide K into four pieces whose areas are $\frac{1}{8}$, $\frac{1}{8}$, $\frac{3}{8}$, and $\frac{3}{8}$ of the area of K, with the parts of equal area sharing an edge, as in Fig. 30.)

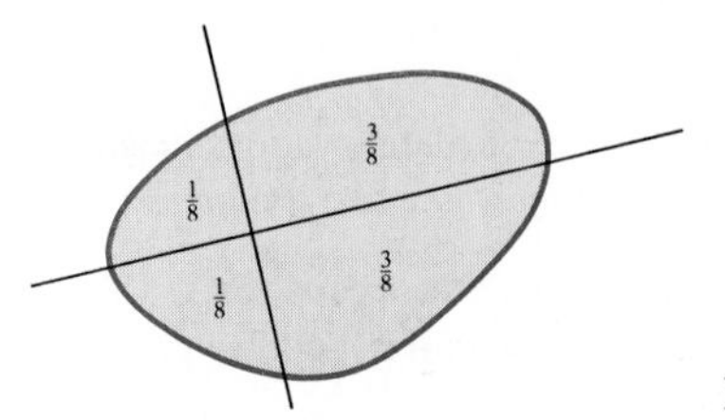

Figure 30

47 Let K be a convex set whose boundary contains no line segments. A polygon is said to circumscribe K if each edge of the polygon is tangent to the boundary of K.

(*a*) Is there necessarily a circumscribing equilateral triangle? If so, how many?

(*b*) Is there necessarily a circumscribing rectangle? If so, how many?

(*c*) Is there necessarily a circumscribing square?

48 Let f be a continuous function whose domain is the x axis and which has the property that

$$f(x + y) = f(x) + f(y)$$

for all numbers x and y. This exercise shows that f must be of the form $f(x) = cx$ for some constant c. [The function cx does satisfy the equation, $f(x + y) = f(x) + f(y)$, since $c(x + y) = cx + cy$.]

(*a*) Let $f(1) = c$. Show that $f(2) = 2c$.

(*b*) Show that $f(0) = 0$.

(*c*) Show that $f(-1) = -c$.

(*d*) Show that for any positive integer n, $f(n) = cn$.

(*e*) Show that for any negative integer n, $f(n) = cn$.

(*f*) Show that $f(\frac{1}{2}) = c/2$.

(*g*) Show that for any nonzero integer n, $f(1/n) = c/n$.

(*h*) Show that for any integer m and positive integer n, $f(m/n) = c(m/n)$.

(*i*) Show that for any irrational number x, $f(x) = cx$. (This is where the continuity of f enters.) Parts (*h*) and (*i*) together complete the solution.

49 *The reason 0^0 is not defined.* It might be hoped that if the positive number b and the number x are both close to 0, then b^x might be close to some fixed number. If that were so, it would suggest a definition of 0^0. Experiment with various choices of b and x near 0 and on the basis of your data write a paragraph on the theme, "Why 0^0 is not defined."

50 (*a*) Let f be a continuous function with the entire x axis for its domain. Is there necessarily a number x such that $f(x) = x$?

(*b*) Let f be a continuous function with domain $[0, 1]$ such that $f(0) = 1$ and $f(1) = 0$. Is there necessarily a number x such that $f(x) = x$?

2.9 PRECISE DEFINITIONS OF "$\lim_{x\to\infty} f(x) = \infty$" AND "$\lim_{x\to\infty} f(x) = L$"

One day I drew on the board the graph shown in Fig. 1. It is the graph of $x/2 + \sin x$. Then I asked my class whether they thought that

$$\lim_{x\to\infty} f(x) = \infty.$$

A third of the class voted "No" because "it keeps going up and down." A third voted "Yes" because "the function tends to get very large as x increases." A third didn't vote. Such a variety of views on such a fundamental concept suggests that we need a more precise definition of a limit than the ones developed in Secs. 2.3 and 2.4.

In the definitions of the limits considered in Secs. 2.3 and 2.4 appear such phrases as "x approaches a," "$f(x)$ approaches a specific number," "as x gets larger," and "$f(x)$ becomes and remains arbitrarily large." Such phrases, although appealing to the intuition and conveying the sense of a limit, are not precise. The definitions seem to suggest moving objects and call to mind the motion of a pencil point as it traces out the graph of a function.

This informal approach was adequate during the early development of calculus, from Leibniz and Newton in the seventeenth century through the Bernoullis, Euler, and Gauss in the eighteenth and early nineteenth centuries. But by the mid-nineteenth century, mathematicians, facing more complicated functions and more difficult theorems, no longer could depend solely on intuition. They realized that glancing at a graph was no longer adequate to understand the behavior of functions—especially if theorems covering a broad class of functions were needed.

It was Weierstrass who developed, in the period 1841–1856, a way to define limits without any hint of motion or of pencils tracing out graphs. His approach, on which he lectured after joining the faculty at the University of Berlin in 1859, has since been followed by pure and applied mathematicians throughout the world. Even an undergraduate advanced calculus course depends on Weierstrass's approach.

In this section we examine how Weierstrass would define the concepts:

$$\lim_{x\to\infty} f(x) = \infty \quad \text{and} \quad \lim_{x\to\infty} f(x) = L.$$

In the next section we consider "$\lim_{x\to a} f(x) = L$."

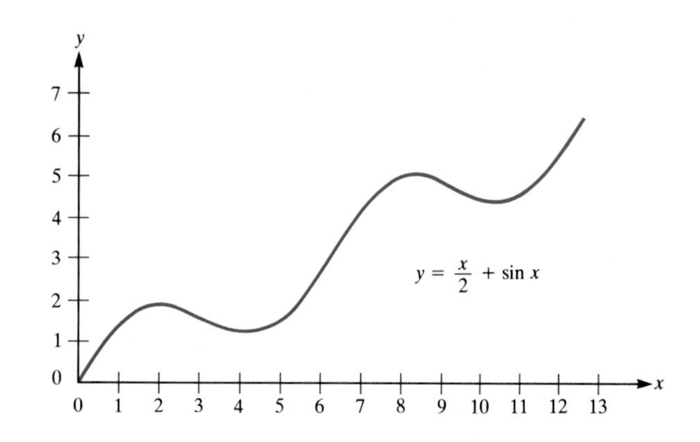

Figure 1

The Precise Definition of $\lim_{x\to\infty} f(x) = \infty$

Recall the definition of "$\lim_{x\to\infty} f(x) = \infty$" given in the table in Sec. 2.4.

Informal

Informal definition of $\lim_{x\to\infty} f(x) = \infty$:

$f(x)$ is defined for all x beyond some number and, as x gets large through positive values, $f(x)$ becomes and remains arbitrarily large and positive.

To take us part way to the precise definition, let us reword the informal definition, paraphrasing it in the following definition, which is still informal.

Reworded

Reworded informal definition of $\lim_{x\to\infty} f(x) = \infty$ [assume that $f(x)$ is defined for all x greater than some number c]:

If x is sufficiently large and positive, then $f(x)$ is necessarily large and positive.

The precise definition parallels the reworded definition.

Precise

> *Precise definition of* $\lim_{x\to\infty} f(x) = \infty$ [assume that $f(x)$ is defined for all x greater than some number c]:
>
> For each number E there is a number D such that for all $x > D$ it is true that
>
> $$f(x) > E.$$

The "challenge and reply" approach to limits

Think of the number E as a challenge and D as the reply. The *larger* E is, the *larger* D must usually be. Only if a number D (which depends on E) can be found for *every* number E can we make the claim that "$\lim_{x\to\infty} f(x) = \infty$." In other words, D could be expressed as a function of E.

To picture the idea behind the precise definition, consider the graph in Fig. 2 of a function f for which $\lim_{x\to\infty} f(x) = \infty$. For each possible choice of a horizontal line, say, at height E, if you are far enough to the right on the graph of f, you stay above that line. That is, there is a number D such that if $x > D$, then $f(x) > E$, as illustrated in Fig. 3.

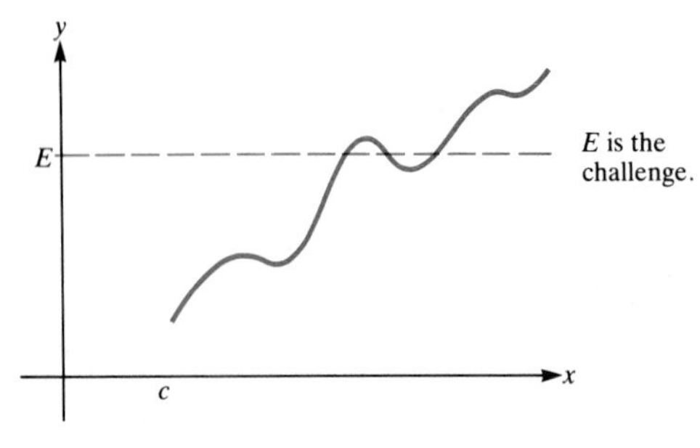

Figure 2

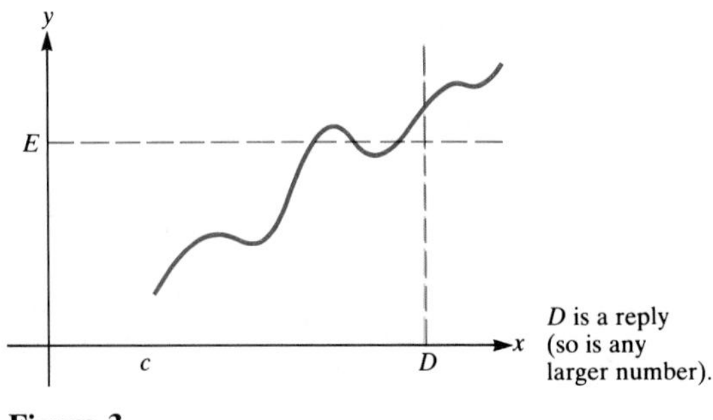

Figure 3

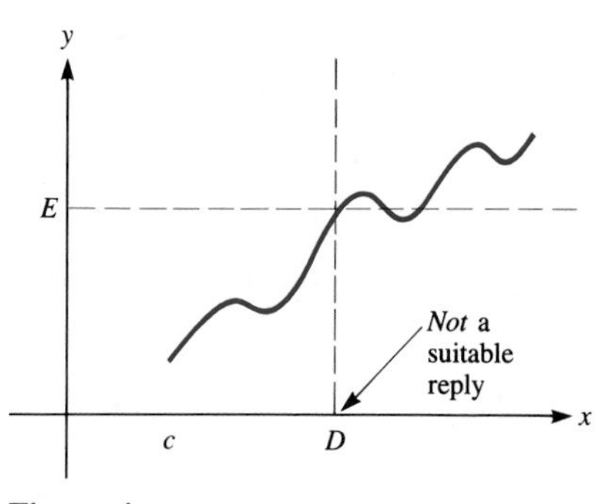

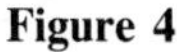
Figure 4

The number D in Fig. 4 is *not* a suitable reply. It is too small since there are values of $x > D$ such that $f(x) \leq E$.

Examples 1 and 2 illustrate how the precise definition is used.

EXAMPLE 1 Using the precise definition, show that $\lim_{x\to\infty} 2x = \infty$.

SOLUTION Let E be any number. We must show that there is a number D such that whenever $x > D$, it follows that $2x > E$. (For example, if $E = 100$, then $D = 50$ would do. It is indeed the case that if $x > 50$, then $2x > 100$.) The number D will depend on E.

Now, the inequality $2x > E$ is equivalent to

$$x > \frac{E}{2}.$$

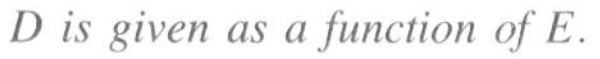
D is given as a function of E.

In other words, if $x > E/2$, then $2x > E$. So $D = E/2$ suffices. That is, for $x > D\ (= E/2)$, $2x > E$. We conclude immediately that

$$\lim_{x\to\infty} 2x = \infty.$$

The reply D is shown in Fig. 5. Any number larger than this particular D would also be a suitable reply. ■

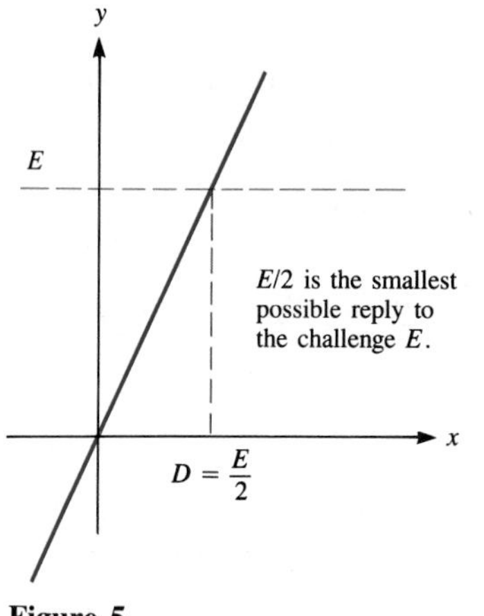

Figure 5

In Example 1 a formula was provided for a suitable D in terms of E, namely, $D = E/2$. For instance, when $E = 1000$, $D = 500$ suffices. In fact, any larger value of D also is suitable. If $x > 600$, it is still the case that $2x > 1000$ (since then $2x > 1200$). If one value of D is a satisfactory response to a given challenge E, then any larger value of D also is a satisfactory response.

Now that we have a precise definition of $\lim_{x\to\infty} f(x) = \infty$, we can settle the question, "Is $\lim_{x\to\infty} (x/2 + \sin x) = \infty$?"

EXAMPLE 2 Using the precise definition, show that $\lim_{x\to\infty} (x/2 + \sin x) = \infty$.

SOLUTION Let E be any number. We must exhibit a number D, depending on E, such that $x > D$ forces

$$\frac{x}{2} + \sin x > E. \tag{1}$$

Now, $\sin x \geq -1$ for all x. So, if we can force

$$\frac{x}{2} > E + 1, \tag{2}$$

then it will follow that

$$\frac{x}{2} + \sin x > E.$$

Inequality (2) is equivalent to

$$x > 2(E + 1).$$

D is given as a function of E.

Thus $D = 2(E + 1)$ will suffice. That is,

$$\text{If } x > 2(E + 1), \qquad \text{then } \frac{x}{2} + \sin x > E.$$

To verify this assertion, we must check that $D = 2(E + 1)$ is a satisfactory reply to E. Assume that $x > 2(E + 1)$. Then

$$\frac{x}{2} > E + 1$$

and $$\sin x \geq -1.$$

If $a > b$ and $c \geq d$, then $a + c > b + d$.

Adding these last two inequalities gives

$$\frac{x}{2} + \sin x > (E + 1) - 1,$$

or simply, $$\frac{x}{2} + \sin x > E,$$

which is inequality (1).

Thus we are permitted to assert that

$$\lim_{x \to \infty} \left(\frac{x}{2} + \sin x \right) = \infty.$$

As x increases, the function does tend to *become* and *remain* large, despite the small dips downward. ■

The Precise Definition of $\lim_{x \to \infty} f(x) = L$

Next, recall the definition of "$\lim_{x \to \infty} f(x) = L$" given in the table of Sec 2.4. L refers to a fixed number.

Informal

Informal definition of $\lim_{x \to \infty} f(x) = L$ [assume that $f(x)$ is defined for all x beyond some number c]:

As x gets large through positive values, $f(x)$ approaches L.

Again we reword this definition before offering the precise definition.

Reworded

Reworded informal definition of $\lim_{x \to \infty} f(x) = L$ [assume that there is a number c such that $f(x)$ is defined for all $x > c$]:

If x is sufficiently large and positive, then $f(x)$ is necessarily near L.

Again, the precise definition parallels the reworded informal definition. In order to make precise the phrase "$f(x)$ is necessarily near L," we shall use the absolute value of $f(x) - L$ to measure the distance from $f(x)$ to L. The following definition says that "if x is large enough, then $|f(x) - L|$ is as small as we please."

Precise

Precise definition of $\lim_{x \to \infty} f(x) = L$ [assume that $f(x)$ is defined for all x beyond some number c]:

For each positive number ϵ, there is a number D such that for all $x > D$ it is true that

$$|f(x) - L| < \epsilon.$$

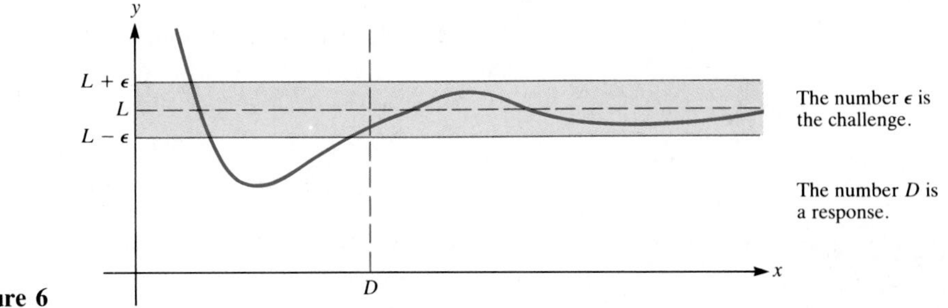

Figure 6

Draw two lines parallel to the x axis, one of height $L + \epsilon$ and one of height $L - \epsilon$. They are the two edges of an endless band of width 2ϵ. Assume that for each positive ϵ, a number D can be found, depending on ϵ, such that the part of the graph to the right of $x = D$ lies within the band. Then we say that "as x approaches ∞, $f(x)$ approaches L" and write

$$\lim_{x\to\infty} f(x) = L.$$

ϵ (epsilon) is the Greek letter corresponding to the English letter e.

The positive number ϵ is the challenge, and D is a response. The smaller ϵ is, the narrower the band, the larger D usually must be chosen. The geometric meaning of the precise definition of $\lim_{x\to\infty} f(x) = L$ is shown in Fig. 6.

EXAMPLE 3 Use the precise definition of "$\lim_{x\to\infty} f(x) = L$" to show that

$$\lim_{x\to\infty} \left(1 + \frac{1}{x}\right) = 1.$$

SOLUTION Here $f(x) = 1 + 1/x$, which is defined for all $x > 0$. The number L is 1. We must show that for each positive number ϵ, however small, there is a number D such that, for all $x > D$,

$$\left|\left(1 + \frac{1}{x}\right) - 1\right| < \epsilon. \tag{3}$$

Inequality (3) reduces to

$$\left|\frac{1}{x}\right| < \epsilon.$$

Since we shall consider only $x > 0$, this inequality is equivalent to

$$\frac{1}{x} < \epsilon. \tag{4}$$

Multiplying inequality (4) by the positive number x yields the equivalent inequality

$$1 < \epsilon x. \tag{5}$$

Division of inequality (5) by the positive number ϵ yields

$$\frac{1}{\epsilon} < x \qquad \text{or} \qquad x > \frac{1}{\epsilon}.$$

D is given as a function of ϵ.

These steps are reversible. This shows that $D = 1/\epsilon$ is a suitable reply to the challenge ϵ. If $x > 1/\epsilon$, then

$$\left|\left(1 + \frac{1}{x}\right) - 1\right| < \epsilon.$$

Figure 7

According to the precise definition of "$\lim_{x\to\infty} f(x) = L$," we may conclude that

$$\lim_{x\to\infty}\left(1+\frac{1}{x}\right) = 1. \quad ■$$

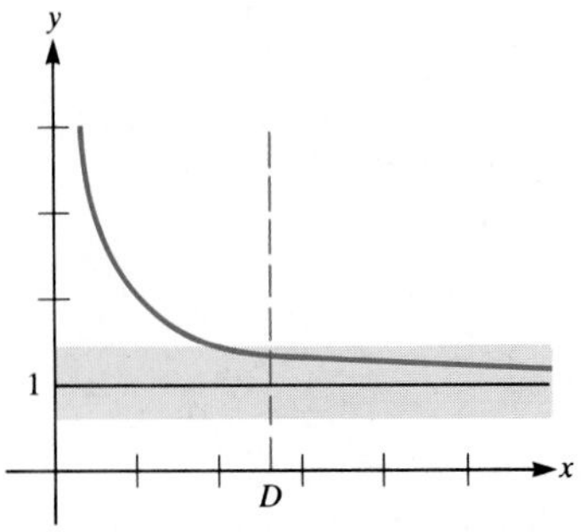

The graph of $f(x) = 1 + 1/x$, shown in Fig. 7, reinforces the argument. It seems plausible that no matter how narrow a band someone may place around the line $y = 1$, it will always be possible to find a number D such that the part of the graph to the right of $x = D$ stays within that band. In Fig. 7 the typical band is shown shaded.

The precise definitions also can be used to show that some claim about an alleged limit is false. The next example illustrates how this is done.

EXAMPLE 4 Show that the claim that $\lim_{x\to\infty} \sin x = 0$ is false.

SOLUTION To show that the claim is *false,* we must exhibit a number $\epsilon > 0$ for which no response D can be found. That is, we must exhibit a number $\epsilon > 0$ such that no D exists for which $|\sin x - 0| < \epsilon$ for all $x > D$.

Recall that $\sin(\pi/2) = 1$ and that $\sin x = 1$ whenever $x = \pi/2 + 2n\pi$ for any integer n. This means that there are arbitrarily large values of x for which $\sin x = 1$. This suggests how to exhibit an $\epsilon > 0$ for which no response D can be found. Simply pick ϵ to be some positive number less than or equal to 1. For instance, $\epsilon = 0.7$ will do.

For any number D there is always a number $x^* > D$ such that we have $\sin x^* = 1$. This means that $|\sin x^* - 0| = 1 > 0.7$. Hence no response can be found for $\epsilon = 0.7$. Thus the claim that $\lim_{x\to\infty} \sin x = 0$ is false. ■

EXERCISES FOR SEC. 2.9: PRECISE DEFINITIONS OF $\lim_{x\to\infty} f(x) = \infty$ AND $\lim_{x\to\infty} f(x) = L$

1 Let $f(x) = 3x$.
  (*a*) Find a number D such that, for $x > D$, it follows that $f(x) > 600$.
  (*b*) Find another number D such that, for $x > D$, it follows that $f(x) > 600$.
  (*c*) What is the smallest number D such that, for $x > D$, it follows that $f(x) > 600$?

2 Let $f(x) = 4x$.
  (*a*) Find a number D such that, for $x > D$, it follows that $f(x) > 1000$.
  (*b*) Find another number D such that, for $x > D$, it follows that $f(x) > 1000$.
  (*c*) What is the smallest number D such that, for $x > D$, it follows that $f(x) > 1000$?

3 Let $f(x) = 5x$. Find a number D such that, for all $x > D$, (*a*) $f(x) > 2000$, (*b*) $f(x) > 10{,}000$.

4 Let $f(x) = 6x$. Find a number D such that, for all $x > D$, (*a*) $f(x) > 1200$, (*b*) $f(x) > 1800$.

In Exercises 5 to 12 use the precise definition of the assertion "$\lim_{x\to\infty} f(x) = \infty$" to establish the given limits.

5 $\lim_{x\to\infty} 3x = \infty$

6 $\lim_{x\to\infty} 4x = \infty$

7 $\lim_{x\to\infty} (x + 5) = \infty$

8 $\lim_{x\to\infty} (x - 600) = \infty$

9 $\lim_{x\to\infty} (2x + 4) = \infty$

10 $\lim_{x\to\infty} (3x - 1200) = \infty$

11 $\lim_{x\to\infty} (4x + 100\cos x) = \infty$

12 $\lim_{x\to\infty} (2x - 300\cos x) = \infty$

13 Let $f(x) = x^2$.
  (*a*) Find a number D such that, for $x > D$, $f(x) > 100$.
  (*b*) Let E be any nonnegative number. Find a number D such that, for $x > D$, it follows that $f(x) > E$.
  (*c*) Let E be any negative number. Find a number D such that, for $x > D$, it follows that $f(x) > E$.

(*d*) Using the precise definition of "$\lim_{x\to\infty} f(x) = \infty$," show that $\lim_{x\to\infty} x^2 = \infty$.

14 Using the precise definition of "$\lim_{x\to\infty} f(x) = \infty$," show that $\lim_{x\to\infty} x^3 = \infty$.

Exercises 15 to 22 concern the precise definition of "$\lim_{x\to\infty} f(x) = L$."

15 Let $f(x) = 3 + 1/x$ if $x \neq 0$.

(*a*) Find a number D such that, for $x > D$, it follows that $|f(x) - 3| < \frac{1}{10}$.

(*b*) Find another number D such that, for $x > D$, it follows that $|f(x) - 3| < \frac{1}{10}$.

(*c*) What is the smallest number D such that, for $x > D$, it follows that $|f(x) - 3| < \frac{1}{10}$?

(*d*) Using the precise definition of "$\lim_{x\to\infty} f(x) = L$," show that $\lim_{x\to\infty} (3 + 1/x) = 3$.

16 Let $f(x) = 2/x$ if $x \neq 0$.

(*a*) Find a number D such that, for $x > D$, it follows that $|f(x) - 0| < \frac{1}{100}$.

(*b*) Find another number D such that, for $x > D$, it follows that $|f(x) - 0| < \frac{1}{100}$.

(*c*) What is the smallest number D such that, for $x > D$, it follows that $|f(x) - 0| < \frac{1}{100}$?

(*d*) Using the precise definition of "$\lim_{x\to\infty} f(x) = L$," show that $\lim_{x\to\infty} 2/x = 0$.

In Exercises 17 to 22 use the precise definition of "$\lim_{x\to\infty} f(x) = L$" to establish the given limits.

17 $\lim_{x\to\infty} (\sin x)/x = 0$ (*Hint:* $|\sin x| \leq 1$ for all x.)

18 $\lim_{x\to\infty} (x + \cos x)/x = 1$

19 $\lim_{x\to\infty} 4/x^2 = 0$

20 $\lim_{x\to\infty} (2x + 3)/x = 2$

21 $\lim_{x\to\infty} 1/(x - 100) = 0$

22 $\lim_{x\to\infty} (2x + 10)/(3x - 5) = \frac{2}{3}$

23 Using the precise definition of "$\lim_{x\to\infty} f(x) = \infty$," show that the claim that $\lim_{x\to\infty} x/(x + 1) = \infty$ is false.

24 Using the precise definition of "$\lim_{x\to\infty} f(x) = L$," show that the claim that $\lim_{x\to\infty} \sin x = \frac{1}{2}$ is false.

25 Using the precise definition of "$\lim_{x\to\infty} f(x) = L$," show that the claim that $\lim_{x\to\infty} 3x = 6$ is false.

26 Using the precise definition of "$\lim_{x\to\infty} f(x) = L$," show that for every number L the assertion "$\lim_{x\to\infty} 2x = L$" is false.

In Exercises 27 to 30 develop precise definitions of the given limits. Phrase your definitions in terms of a challenge number E or ϵ and a reply D. Show the geometric meaning of your definition on a graph.

27 $\lim_{x\to\infty} f(x) = -\infty$

28 $\lim_{x\to-\infty} f(x) = \infty$

29 $\lim_{x\to-\infty} f(x) = -\infty$

30 $\lim_{x\to-\infty} f(x) = L$

31 Let $f(x) = 5$ for all x. (*a*) Using the precise definition of "$\lim_{x\to\infty} f(x) = L$," show that $\lim_{x\to\infty} f(x) = 5$. (*b*) Using the precise definition of "$\lim_{x\to-\infty} f(x) = L$," show that $\lim_{x\to-\infty} f(x) = 5$. (See Exercise 30.)

32 Is this argument correct? "I will prove that $\lim_{x\to\infty} (2x + \cos x) = \infty$. Let E be given. I want

$$2x + \cos x > E,$$

or

$$2x > E - \cos x,$$

so

$$x > \frac{E - \cos x}{2}.$$

Thus, if $D = \dfrac{E - \cos x}{2}$, then $2x + \cos x > E$."

2.10 PRECISE DEFINITION OF "$\lim_{x\to a} f(x) = L$"

Recall the informal definition given in Sec. 2.3.

Informal

Informal definition of $\lim_{x\to a} f(x) = L$:

Let f be a function and a some fixed number. Assume that the domain of f contains open intervals (c, a) and (a, b) for some number $c < a$ and some number $b > a$.

If, as x approaches a, either from the left or from the right, $f(x)$ approaches a specific number L, then L is called the **limit** of $f(x)$ as x approaches a. This is written

$$\lim_{x\to a} f(x) = L.$$

Keep in mind that a need not be in the domain of f. Even if a happens to be in the domain of f, the value $f(a)$ plays no role in determining whether $\lim_{x \to a} f(x) = L$.

Reworded

Reworded informal definition of $\lim_{x \to a} f(x) = L$ [assume that $f(x)$ is defined for all x in some intervals (c, a) and (a, b)]:
If x is sufficiently close to a but not equal to a, then $f(x)$ is necessarily near L.

The "ϵ, δ" definition of "$\lim_{x \to a} f(x) = L$"

The precise definition parallels the reworded informal definition. The letter δ that appears in it is the lower case Greek "delta," equivalent to the English letter d.

> *Precise definition of* $\lim_{x \to a} f(x) = L$ [assume that $f(x)$ is defined in some intervals (c, a) and (a, b)]:
>
> For each positive number ϵ there is a positive number δ such that for all x that satisfy the inequality
>
> $$0 < |x - a| < \delta$$
>
> it is true that $$|f(x) - L| < \epsilon.$$

The meaning of $0 < |x - a| < \delta$

The inequality $0 < |x - a|$ that appears in the definition is just a fancy way of saying "x is not a." The inequality $|x - a| < \delta$ asserts that x is within a distance δ of a. The two inequalities may be combined as the single statement $0 < |x - a| < \delta$, which describes the open interval $(a - \delta, a + \delta)$ from which a is deleted. This deletion is made since the value $f(a)$ plays no role in the definition of $\lim_{x \to a} f(x)$.

Once again ϵ is the challenge. The response is δ. Usually, the smaller ϵ is, the *smaller* δ will have to be.

The geometric significance of the precise definition of "$\lim_{x \to a} f(x) = L$" is shown in Fig. 1. The narrow horizontal band of width 2ϵ is again the challenge. The response is a sufficiently narrow vertical band, of width 2δ, such that the part of the graph within that vertical band (except perhaps at $x = a$) also lies in the challenging horizontal band of width 2ϵ. In Fig. 2 the vertical band shown is not narrow enough to meet the challenge of the horizontal band shown. But the vertical band shown in Fig. 3 is sufficiently narrow.

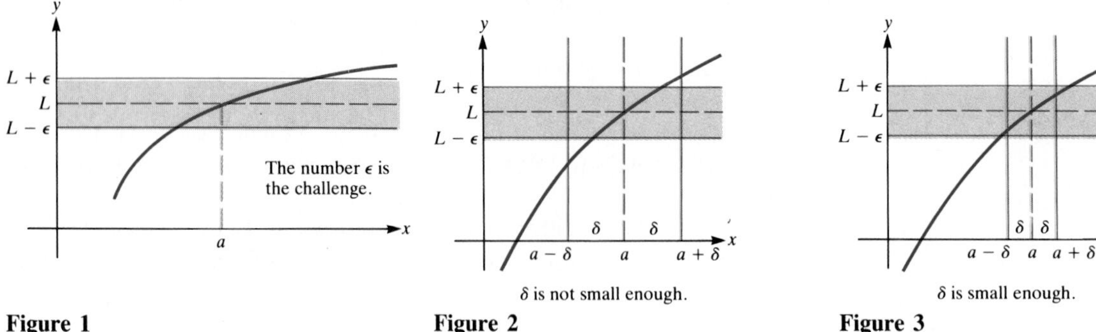

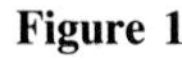

Figure 1

δ is not small enough.

Figure 2

δ is small enough.

Figure 3

Assume that for each positive number ϵ it is possible to find a positive number δ such that the parts of the graph between $x = a - \delta$ and $x = a$ and between $x = a$ and $x = a + \delta$ lie within the given horizontal band. Then we say that "as x approaches a, $f(x)$ approaches L." The narrower the horizontal band around the line $y = L$, the smaller δ usually must be.

EXAMPLE 1 Use the precise definition of "$\lim_{x \to a} f(x) = L$" to show that $\lim_{x \to 0} x^2 = 0$.

SOLUTION In this case $a = 0$ and $L = 0$. Let ϵ be a positive number. We wish to find a positive number δ such that for $0 < |x - 0| < \delta$ it follows that $|x^2 - 0| < \epsilon$.

Since $|x|^2 = |x^2|$, we are asking, "for which x is $|x|^2 < \epsilon$?" This inequality is satisfied when

$$|x| < \sqrt{\epsilon}.$$

In other words, when $|x| < \sqrt{\epsilon}$, it follows that $|x^2 - 0| < \epsilon$. Thus $\delta = \sqrt{\epsilon}$ suffices.

(For instance, when $\epsilon = 1$, $\delta = \sqrt{1} = 1$ is a suitable response. When $\epsilon = 0.01$, $\delta = 0.1$ suffices.) ■

EXAMPLE 2 Use the precise definition of "$\lim_{x \to a} f(x) = L$" to show that $\lim_{x \to 2} (3x + 5) = 11$.

SOLUTION Here $a = 2$ and $L = 11$. Let ϵ be a positive number. We wish to find a number $\delta > 0$ such that for $0 < |x - 2| < \delta$ we have $|(3x + 5) - 11| < \epsilon$.

So let us find out for which x it is true that $|(3x + 5) - 11| < \epsilon$. This inequality is equivalent to

$$|3x - 6| < \epsilon$$

or

$$3|x - 2| < \epsilon$$

or

$$|x - 2| < \frac{\epsilon}{3}.$$

Thus $\delta = \epsilon/3$ is an adequate response. If $0 < |x - 2| < \epsilon/3$, then $|(3x + 5) - 11| < \epsilon$. ■

The algebra of finding a response δ can be much more involved for other functions, such as $f(x) = ax^2 + bx + c$.

EXERCISES FOR SEC. 2.10: PRECISE DEFINITION OF "$\lim_{x\to a} f(x) = L$"

In Exercises 1 to 4 use the precise definition of "$\lim_{x\to a} f(x) = L$" to justify the statements.

1 $\lim_{x\to 2} 3x = 6$

2 $\lim_{x\to 3} (4x - 1) = 11$

3 $\lim_{x\to 1} (x + 2) = 3$

4 $\lim_{x\to 5} (2x - 3) = 7$

In Exercises 5 and 6 find a number δ such that the point $(x, f(x))$ lies in the shaded band for all x in the interval $(a - \delta, a + \delta)$.

5

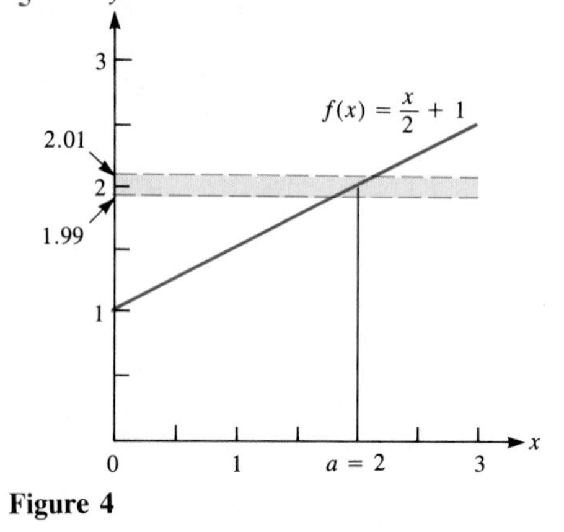

Figure 4

6

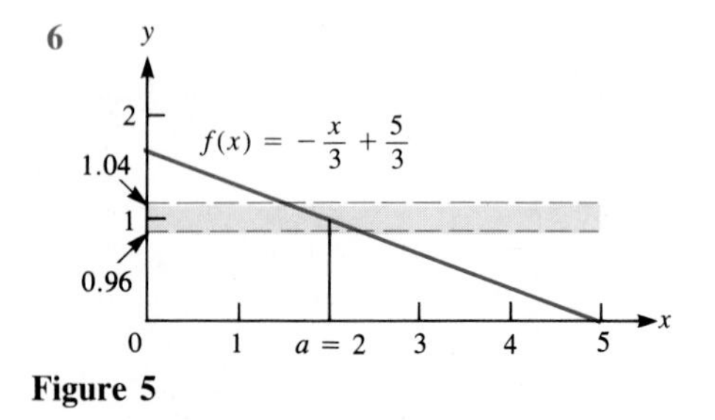

Figure 5

In Exercises 7 to 10 use the precise definition of "$\lim_{x\to a} f(x) = L$" to justify the statements.

7 $\lim_{x\to 0} \frac{x^2}{4} = 0$

8 $\lim_{x\to 0} 4x^2 = 0$

9 $\lim_{x\to 1} (3x + 5) = 8$

10 $\lim_{x\to 1} \frac{5x + 3}{4} = 2$

11 Give an example of a number $\delta > 0$ such that $|x^2 - 4| < 1$ if $0 < |x - 2| < \delta$.

12 Give an example of a number $\delta > 0$ such that $|x^2 + x - 2| < 0.5$ if $0 < |x - 1| < \delta$.

In Exercises 13 and 14 find a number δ such that the point $(x, f(x))$ lies in the shaded band for all x in the interval $(a - \delta, a + \delta)$.

13

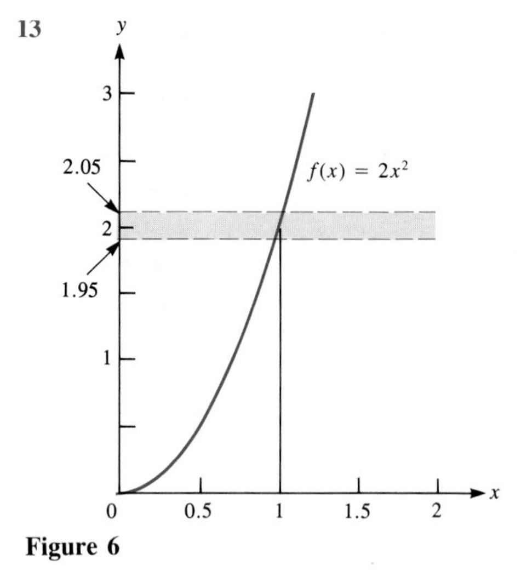

Figure 6

14

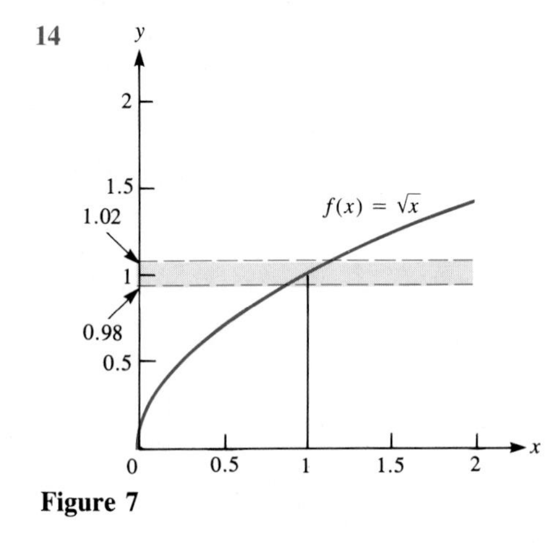

Figure 7

---

15 (*a*) Show that, if $0 < \delta < 1$ and $|x - 3| < \delta$, then $|x^2 - 9| < 7\delta$.
(*b*) From (*a*) deduce that $\lim_{x\to 3} x^2 = 9$.

16 (*a*) Show that, if $0 < \delta < 1$ and $|x - 4| < \delta$, then

$$|\sqrt{x} - 2| < \frac{\delta}{\sqrt{3} + 2}.$$

(*Hint:* Rationalize $\sqrt{x} - 2$.)
(*b*) From (*a*) deduce that $\lim_{x\to 4} \sqrt{x} = 2$.

17 (*a*) Show that, if $0 < \delta < 1$ and $|x - 3| < \delta$, then $|x^2 + 5x - 24| < 12\delta$. (*Hint:* Factor $x^2 + 5x - 24$.)
(*b*) From (*a*) deduce that $\lim_{x\to 3} (x^2 + 5x) = 24$.

18 (*a*) Show that, if $0 < \delta < 1$ and $|x - 2| < \delta$, then

$$\left|\frac{1}{x} - \frac{1}{2}\right| < \frac{\delta}{2}.$$

(*b*) From (*a*) deduce that $\lim_{x\to 2} 1/x = \frac{1}{2}$.

In Exercises 19 to 24 develop precise definitions of the given limits. Phrase your definitions in terms of a challenge, E or ϵ, and a response, δ.

19 $\lim_{x\to a^+} f(x) = L$

20 $\lim_{x\to a^-} f(x) = L$

21 $\lim_{x\to a} f(x) = \infty$

22 $\lim_{x\to a} f(x) = -\infty$

23 $\lim_{x\to a^+} f(x) = \infty$

24 $\lim_{x\to a^-} f(x) = \infty$

25 Let $f(x) = 9x^2$.

(*a*) Find $\delta > 0$ such that, for $0 < |x - 0| < \delta$ it follows that $|9x^2 - 0| < \frac{1}{100}$.

(*b*) Let ϵ be any positive number. Find a positive number δ such that, for $0 < |x - 0| < \delta$ we have $|9x^2 - 0| < \epsilon$.

(*c*) Show that $\lim_{x\to 0} 9x^2 = 0$.

26 Let $f(x) = x^3$.

(*a*) Find $\delta > 0$ such that for $0 < |x - 0| < \delta$ it follows that $|x^3 - 0| < \frac{1}{1000}$.

(*b*) Show that $\lim_{x\to 0} x^3 = 0$.

27 Show that the assertion "$\lim_{x\to 2} 3x = 5$" is false. To do this, it is necessary to exhibit a positive number ϵ such that there is no response number $\delta > 0$. (*Hint:* Draw a picture.)

28 Show that the assertion "$\lim_{x\to 2} x^2 = 3$" is false.

2.S SUMMARY

This chapter discussed functions and their limits. A function is any method for assigning to each element in some set X a unique element in some set Y. If both the inputs and the outputs are real numbers, the function is called a real function of a real variable.

The graph of a function f consists of all points of the form $(x, f(x))$.

Calculus is concerned mainly with functions that describe how one quantity depends on another.

Two central ideas of this chapter are limits and continuous functions. Limits were used to find asymptotes, as an aid in graphing. Two limits needed in the next chapter were determined:

$$\lim_{\theta\to 0} \frac{\sin\theta}{\theta} = 1 \quad \text{and} \quad \lim_{\theta\to 0} \frac{1 - \cos\theta}{\theta} = 0.$$

The definition of a continuous function is phrased in terms of behavior of the function at and near each number in an interval. Although the definition of continuity depends on the definition of limit, the easiest way to think of a function that is continuous *throughout an interval* is that its graph is a curve that can be drawn without lifting pencil from paper.

Two properties of continuous functions, known as the extreme-value property and the intermediate-value property, were discussed. They will be used often in later chapters.

The final two sections concerned precise definitions of limits.

Vocabulary and Symbols

function
domain
range
input (argument)
output (value)
graph
composite function
even function
odd function
symmetry
limit, $\lim_{x\to a} f(x)$, $\lim_{x\to\infty} f(x)$, etc.
right-hand limit, $\lim_{x\to a^+} f(x)$
left-hand limit, $\lim_{x\to a^-} f(x)$
constant function
polynomial function
rational function
asymptote (vertical, horizontal)
greatest integer less than or equal to x, $[x]$, or $\lfloor x\rfloor$
continuous at a
continuous on an interval
continuous
maximum value, minimum value, extreme value, extremum
maximum-value theorem
intermediate-value theorem

Key Facts

A curve in the xy plane which meets each vertical line at most once is the graph of a function.

A function can be viewed many ways: as a table of values, as a graph, as an input-output machine, as a projection from one line to another.

The graph of an even function is symmetric with respect to the y axis; the graph of an odd function is symmetric with respect to the origin.

For the definitions of the various limits, such as

$$\lim_{x\to a} f(x) = L, \qquad \lim_{x\to\infty} f(x) = L, \qquad \text{and} \qquad \lim_{x\to-\infty} f(x) = L,$$

see Secs. 2.3 and 2.4. (Precise definitions are given in Secs. 2.9 and 2.10, which are not covered in this summary.)

PROPERTIES OF LIMITS

If $\lim_{x\to a} f(x)$ and $\lim_{x\to a} g(x)$ both exist, then

$$\lim_{x\to a} (f(x) + g(x)) = \lim_{x\to a} f(x) + \lim_{x\to a} g(x)$$

$$\lim_{x\to a} (f(x) - g(x)) = \lim_{x\to a} f(x) - \lim_{x\to a} g(x)$$

$$\lim_{x\to a} f(x)g(x) = \lim_{x\to a} f(x) \lim_{x\to a} g(x)$$

$$\lim_{x\to a} f(x)/g(x) = \lim_{x\to a} f(x)/\lim_{x\to a} g(x), \qquad \text{if } \lim_{x\to a} g(x) \neq 0$$

$$\lim_{x\to a} f(x)^{g(x)} = \left(\lim_{x\to a} f(x)\right)^{\lim_{x\to a} g(x)}, \qquad \text{if } \lim_{x\to a} f(x) > 0.$$

Limits of Rational Functions

$$\lim_{x\to\infty} \frac{ax^n + \cdots}{bx^m + \cdots} = \lim_{x\to\infty} \frac{ax^n}{bx^m}$$

(The degree of the numerator is n and the degree of the denominator is m.)

Consequently,

$$\lim_{x\to\infty} \frac{ax^n + \cdots}{bx^m + \cdots} = \begin{cases} a/b & \text{if } m = n \\ 0 & \text{if } n < m \\ \infty \text{ or } -\infty & \text{if } n > m \text{ (depending on the signs of } a \text{ and } b). \end{cases}$$

Similar assertions hold for $x \to -\infty$.

Two Trigonometric Limits

$$\lim_{\theta\to 0} \frac{\sin\theta}{\theta} = 1$$

$$\lim_{\theta\to 0} \frac{1-\cos\theta}{\theta} = 0$$

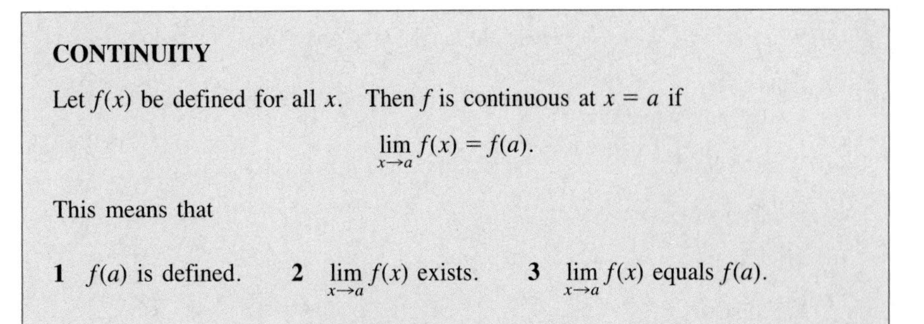

CONTINUITY

Let $f(x)$ be defined for all x. Then f is continuous at $x = a$ if

$$\lim_{x \to a} f(x) = f(a).$$

This means that

1 $f(a)$ is defined. **2** $\lim_{x \to a} f(x)$ exists. **3** $\lim_{x \to a} f(x)$ equals $f(a)$.

A similar definition holds if f, though not defined for all x, is defined at least on some open interval that includes the number a. For the definitions of "continuous from the right" and "continuous from the left" see Sec. 2.8.

A function that is continuous at each point of an open interval is said to be continuous on that interval. Similar definitions cover functions whose domains are closed intervals. In general, "continuous function" means a function that is continuous on its domain.

MAXIMUM-VALUE THEOREM

Let f be continuous throughout the closed interval $[a, b]$. Then there is at least one number in $[a, b]$ at which f takes on a maximum value. That is, for some number c in $[a, b]$, $f(c) \geq f(x)$, for all x in $[a, b]$.

A corresponding minimum-value theorem also holds.

INTERMEDIATE-VALUE THEOREM

Let f be continuous throughout the closed interval $[a, b]$. Let m be any number between $f(a)$ and $f(b)$. Then there is at least one number c in $[a, b]$ such that $f(c) = m$.

In particular, if f is continuous throughout $[a, b]$ and if one of $f(a)$ and $f(b)$ is negative and the other is positive, then there is a number c in $[a, b]$ such that $f(c) = 0$.

With the aid of the preceding fact, it was shown that a polynomial of odd degree has at least one real root. In other words, the graph of a polynomial of odd degree always crosses the x axis at least once.

GUIDE QUIZ ON CHAP. 2: FUNCTIONS, LIMITS, AND CONTINUITY

1 Give an example of functions $f(x)$ and $g(x)$ and a number a such that $\lim_{x \to a} f(x) = 0$, $\lim_{x \to a} g(x) = 0$, and
(*a*) $\lim_{x \to a} f(x)/g(x) = 0$,
(*b*) $\lim_{x \to a} f(x)/g(x) = 1$,
(*c*) $\lim_{x \to a} f(x)/g(x) = \infty$.

2 Graph $f(x) = \dfrac{x^2 - 4}{x^2 - 1}$, discussing symmetry, intercepts, and asymptotes.

3 Explain why a polynomial of odd degree has at least one real root.

4 (*a*) What is $\lim_{h\to 0} \frac{\sin h}{h}$ (where the angle h is given in radians)?
(*b*) Use (*a*) to estimate $\sin 5°$.

5 (*a*) What is $\lim_{h\to 0} \frac{1 - \cos h}{h}$ (where the angle h is given in radians)?
(*b*) What is $\lim_{h\to 0} \frac{1 - \cos h}{h^2}$ (where the angle h is given in radians)?
[*Suggestion:* Multiply by $(1 + \cos h)/(1 + \cos h)$.]
(*c*) Use (*b*) to estimate $\cos 0.2$.
(*d*) Use (*b*) to estimate $\cos 20°$.
(*e*) Compare your answer in (*d*) to that given by a calculator.

6 What is meant by an "indeterminate limit" and a "determinate limit"? Include examples in your discussion.

7 Let $f(x) = \sqrt{x^2 + 4x} - x$.
(*a*) Compute $f(x)$ for some large values of x.
(*b*) Determine $\lim_{x\to\infty} f(x)$.
(*c*) What is the domain of f?
(*d*) Graph $y = f(x)$.

8 Simplify the expression

$$\frac{f(x + h) - f(x)}{h}$$

for $h \neq 0$ as far as possible if (*a*) $f(x) = x^3$ for all x. (*b*) $f(x) = 1/x$ for all $x \neq 0$. Assume that $x + h \neq 0$.

9 Let $f(x)$ be defined for all x in $[0, 2]$.
(*a*) What is meant by "$f(x)$ is continuous at $a = 1/2$"?
(*b*) What is meant by "$f(x)$ is continuous at $a = 0$"?
(*c*) What is meant by "$f(x)$ is continuous at $a = 2$"?

10 Figure 1 is the graph of a function f whose domain is $[1, 5]$.
(*a*) Does $\lim_{x\to 2} f(x)$ exist?
(*b*) Is f continuous at 2?
(*c*) Does $\lim_{x\to 3} f(x)$ exist?
(*d*) Does $\lim_{x\to 5} f(x)$ exist?
(*e*) Is f continuous on $[1, 5]$?
(*f*) Is f continuous on $(3, 5)$?

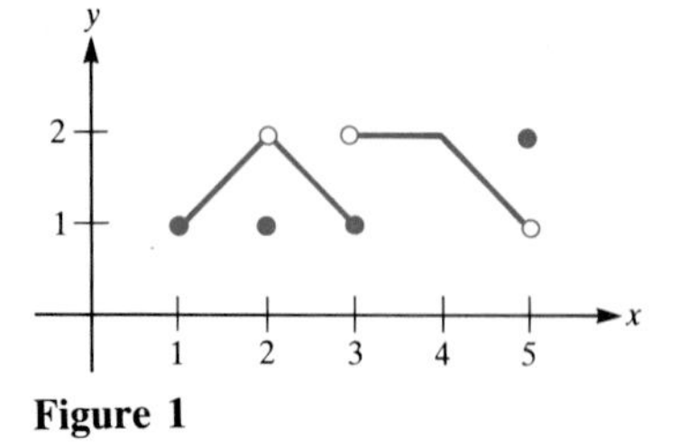

Figure 1

11 Examine the following limits:
(*a*) $\lim_{x\to 1} (x^2 + 5x)$ (*b*) $\lim_{x\to\infty} \frac{3x^4 - 100x + 3}{5x^4 + 7x - 1}$
(*c*) $\lim_{x\to 0} \frac{3x^4 - 100x + 3}{5x^4 + 7x - 1}$ (*d*) $\lim_{x\to -\infty} \frac{500x^3 - x^2 - 5}{x^4 + x}$
(*e*) $\lim_{x\to 0} \frac{\sin 3x}{6x}$ (*f*) $\lim_{x\to -\infty} \frac{-6x^5 + 4x}{x^2 + x + 5}$
(*g*) $\lim_{x\to\infty} 2^{-x}$ (*h*) $\lim_{x\to 0} \frac{x^3 + 8}{x + 2}$
(*i*) $\lim_{x\to -2} \frac{x^3 + 8}{x + 2}$ (*j*) $\lim_{x\to 0} \sin \frac{1}{x}$
(*k*) $\lim_{x\to\infty} \sin x$ (*l*) $\lim_{x\to\infty} \frac{1 + 3 \cos x}{x^2}$
(*m*) $\lim_{x\to\infty} (\sqrt{4x^2 + 5x} - \sqrt{4x^2 + x})$
(*n*) $\lim_{x\to 16} \frac{\sqrt{x} - 4}{x - 16}$

12 If $\lim_{x\to a} f(x) = 3$ and $\lim_{x\to a} g(x) = 4$, what, if anything, can be said about
(*a*) $\lim_{x\to a} f(x)g(x)$? (*b*) $\lim_{x\to a} f(x)/g(x)$?
(*c*) $\lim_{x\to a} [f(x) + g(x)]$? (*d*) $\lim_{x\to a} [f(x) - 3]/[g(x) - 4]$?
(*e*) $\lim_{x\to a} [f(x) - 3]^{g(x)}$?

13 If $\lim_{x\to a} f(x) = 0$ and $\lim_{x\to a} g(x) = \infty$, what, if anything, can be said about
(*a*) $\lim_{x\to a} [f(x) + g(x)]$? (*b*) $\lim_{x\to a} f(x)/g(x)$?
(*c*) $\lim_{x\to a} f(x)^{g(x)}$? (*d*) $\lim_{x\to a} [2 + f(x)]^{g(x)}$?
(*e*) $\lim_{x\to a} f(x)g(x)$?

14 (*a*) State the assumptions in the maximum-value theorem.
(*b*) State the conclusion of the maximum-value theorem.

15 (*a*) State the assumptions in the intermediate-value theorem.
(*b*) State the conclusion of the intermediate-value theorem.

16 Four squares, each of side x, are removed from the corners of a rectangular piece of metal, as shown in Fig. 2. The dimensions of the rectangle are 10 inches by 15 inches.

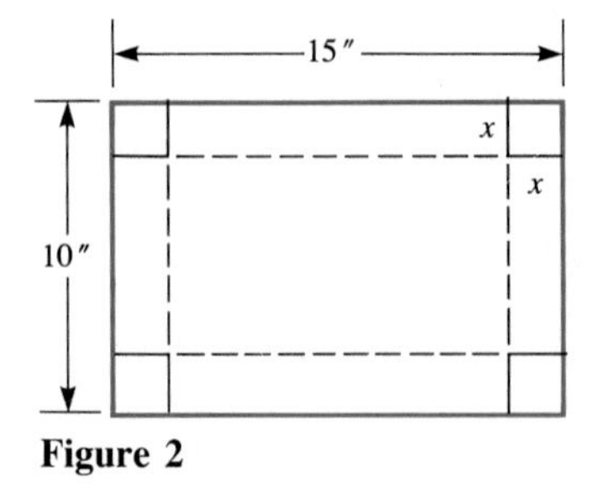

Figure 2

A box of height x is formed by folding up the four flaps created by the removal of the corner squares. The box has no top. Let $V(x)$ be the volume of the box.
(a) Find a formula for $V(x)$ in terms of x.
(b) What is the domain of the function V?

17 Show that $f(x) = \cos^2 \sqrt{x}$ is the composition of three functions found in scientific calculators.

18 Let $P(\theta)$ be the perimeter of the shaded sector in Fig. 3. Find a formula for $P(\theta)$.

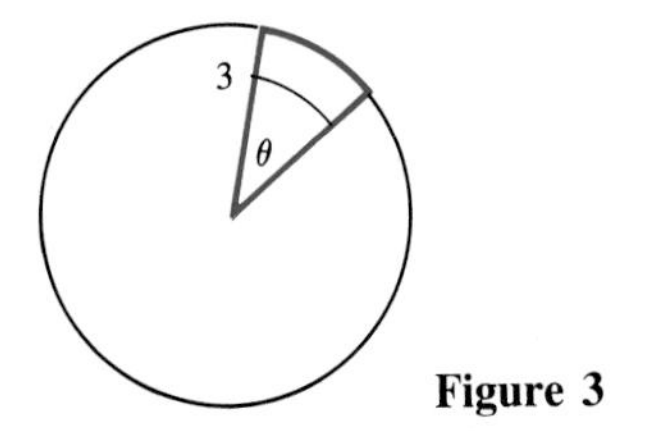

Figure 3

REVIEW EXERCISES FOR CHAP. 2: FUNCTIONS, LIMITS, AND CONTINUITY

1 Figure 4 shows a sector of a circle of radius 5. The central angle of the sector is $\pi/3$.

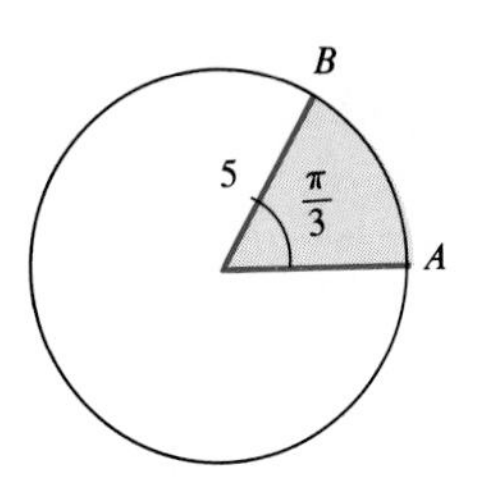

Figure 4

(a) What is the area of the sector?
(b) What is the length of arc AB?

2 Let $f(x) = 2x^2 - 1/x$. Express the fraction

$$\frac{f(x+h) - f(x)}{h}$$

as simply as possible. (Assume that none of h, x, and $x + h$ are 0.)

In each of Exercises 3 to 8 give the domain and range of the function.

3 $x^{1/5}$ **4** $x^{1/6}$
5 $\cos x$ **6** $\tan x$
7 $1/\sqrt{x+1}$ **8** $\sqrt{4 - x^2}$

In Exercises 9 and 10 evaluate and express as decimals.

9 $f(2 + 0.1) - f(2)$ if $f(x) = x^2$
10 $f(2 - 0.1) - f(2)$ if $f(x) = x^2 + 3$

In Exercises 11 to 14 evaluate and simplify using algebra.

11 $\dfrac{f(a+h) - f(a)}{h}$ if $f(x) = 1/(x+1)$

12 $\dfrac{f(2+h) - f(2)}{h}$ if $f(x) = 3 + 2x + x^2$

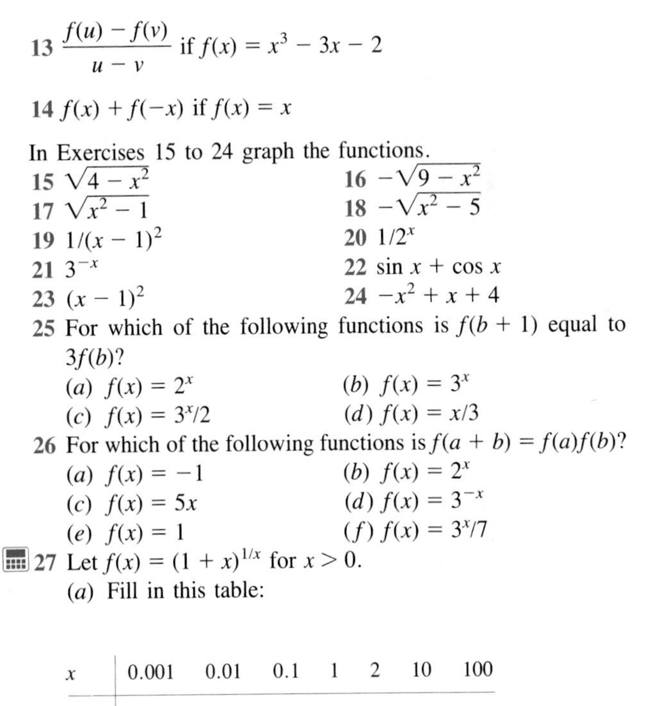

13 $\dfrac{f(u) - f(v)}{u - v}$ if $f(x) = x^3 - 3x - 2$

14 $f(x) + f(-x)$ if $f(x) = x$

In Exercises 15 to 24 graph the functions.

15 $\sqrt{4 - x^2}$ **16** $-\sqrt{9 - x^2}$
17 $\sqrt{x^2 - 1}$ **18** $-\sqrt{x^2 - 5}$
19 $1/(x-1)^2$ **20** $1/2^x$
21 3^{-x} **22** $\sin x + \cos x$
23 $(x-1)^2$ **24** $-x^2 + x + 4$

25 For which of the following functions is $f(b+1)$ equal to $3f(b)$?
(a) $f(x) = 2^x$ (b) $f(x) = 3^x$
(c) $f(x) = 3^x/2$ (d) $f(x) = x/3$

26 For which of the following functions is $f(a+b) = f(a)f(b)$?
(a) $f(x) = -1$ (b) $f(x) = 2^x$
(c) $f(x) = 5x$ (d) $f(x) = 3^{-x}$
(e) $f(x) = 1$ (f) $f(x) = 3^x/7$

27 Let $f(x) = (1 + x)^{1/x}$ for $x > 0$.
(a) Fill in this table:

x	0.001	0.01	0.1	1	2	10	100
$f(x)$							

(b) With the aid of (a) graph f.

28 Graph $f(x) = x/2^x$.

29 It is known that the perimeter of a rectangle is 100 inches. If one side has length x inches, denote the area of that rectangle as $f(x)$ square inches.
(a) Find a formula for $f(x)$.
(b) What is the domain of the function in the given context?

30 A rectangular box has sides of lengths x, y, and z inches. Obtain a formula for the function that describes (a) the total surface area of the box, (b) the total volume of the box, and (c) the total length of the edges of the box.

31 Is $\sin x$ ever equal to $3 \cos x$? Explain.

32 Let $f(x) = x^x$ for $x > 0$.

(*a*) Evaluate the function for the inputs 1, 0.5, 0.4, 0.3, 0.1, and 0.001.

(*b*) With the aid of (*a*) graph f.

In Exercises 33 to 68 examine the limits. Evaluate those which exist. Determine those which do not exist and, among these, the ones that are infinite. In Exercises 47 and 48, $[x]$ denotes the "greatest integer" function.

33 $\lim_{x\to 1} \dfrac{x^3+1}{x^2+1}$ **34** $\lim_{x\to 1} \dfrac{x^3-1}{x^2-1}$

35 $\lim_{x\to 2} \dfrac{x^4-16}{x^3-8}$ **36** $\lim_{x\to 0} \dfrac{x^4-16}{x^3-8}$

37 $\lim_{x\to\infty} \dfrac{x^7-x^2+1}{2x^7+x^3+300}$ **38** $\lim_{x\to-\infty} \dfrac{x^9+6x+3}{x^{10}-x-1}$

39 $\lim_{x\to-\infty} \dfrac{x^3+1}{x^2+1}$ **40** $\lim_{x\to-\infty} \dfrac{x^4+x^2+1}{3x^2+4}$

41 $\lim_{x\to 4} \dfrac{\sqrt{x}-2}{x-4}$ **42** $\lim_{x\to 81} \dfrac{x-81}{\sqrt{x}-9}$

43 $\lim_{x\to\infty} (\sqrt{x^2+2x+3} - \sqrt{x^2-2x+3})$

44 $\lim_{x\to\infty} (\sqrt{2x^2} - \sqrt{2x^2-6x})$

45 $\lim_{x\to 4^+} \dfrac{1}{x-4}$ **46** $\lim_{x\to 4^-} \dfrac{1}{x-4}$

47 $\lim_{x\to 3^-} [2x]$ **48** $\lim_{x\to 3^+} [2x]$

49 $\lim_{x\to 0^+} 2^{1/x}$ **50** $\lim_{x\to 0^-} 2^{1/x}$

51 $\lim_{x\to\infty} 2^{1/x}$ **52** $\lim_{x\to-\infty} 2^{1/x}$

53 $\lim_{x\to\infty} \dfrac{(x+1)(x+2)}{(x+3)(x+4)}$ **54** $\lim_{x\to-\infty} \dfrac{(x+1)^{100}}{(2x+50)^{100}}$

55 $\lim_{x\to\pi/2} \dfrac{\cos x}{1+\sin x}$ **56** $\lim_{x\to\pi/2} \dfrac{\cos x}{1-\sin x}$

57 $\lim_{x\to 0} \dfrac{\sin x}{3x}$ **58** $\lim_{x\to\infty} \dfrac{\sin x}{3x}$

59 $\lim_{x\to\pi/2^+} \cos x$ **60** $\lim_{x\to\pi/2^+} \sec x$

61 $\lim_{x\to 0^-} \sin x$ **62** $\lim_{x\to 0^-} \csc x$

63 $\lim_{x\to\infty} \sin \dfrac{1}{x}$ **64** $\lim_{x\to\infty} x \sin \dfrac{1}{x}$

65 $\lim_{x\to\pi/4} x^2 \cos x$ **66** $\lim_{x\to\infty} x^2 \cos x$

67 $\lim_{\theta\to\infty} (\cos^2\theta + \sin^2\theta)$ **68** $\lim_{\theta\to\infty} (\cos^2\theta - \sin^2\theta)$

In Exercises 69 to 74 exhibit specific functions f and g that meet all three conditions. (The answers are not unique.)

69 $\lim_{x\to 0} f(x) = 0$, $\lim_{x\to 0} g(x) = 0$, and $\lim_{x\to 0} f(x)/g(x) = 5$

70 $\lim_{x\to\infty} f(x) = 0$, $\lim_{x\to\infty} g(x) = \infty$, and $\lim_{x\to\infty} f(x)g(x) = 20$

71 $\lim_{x\to\infty} f(x) = 0$, $\lim_{x\to\infty} g(x) = \infty$, and $\lim_{x\to\infty} f(x)g(x) = \infty$

72 $\lim_{x\to\infty} f(x) = \infty$, $\lim_{x\to\infty} g(x) = \infty$, and $\lim_{x\to\infty} [f(x) - g(x)] = 3$

73 $\lim_{x\to\infty} f(x) = \infty$, $\lim_{x\to\infty} g(x) = \infty$, and $\lim_{x\to\infty} [f(x) - g(x)] = \infty$

74 $\lim_{x\to\infty} f(x) = \infty$, $\lim_{x\to\infty} g(x) = \infty$, and $\lim_{x\to\infty} f(x)/g(x) = \infty$

75 Does $x + \sin x$ have a maximum value for x in

(*a*) $[0, 100]$? (*b*) $[0, \infty)$?

76 Does $x^3 + x + 1$ have a minimum value for x in

(*a*) $[-100, 5]$? (*b*) $(-\infty, 5]$?

77 Does $1/(1 + x^2)$ have (*a*) a maximum value for x in $(-1, 1)$? (*b*) a minimum value for x in $(-1, 1)$?

78 Does $1/x^3$ have a maximum value for x in

(*a*) $[2, 100]$? (*b*) $[2, \infty)$?

A minimum value for x in

(*c*) $[2, 100]$? (*d*) $[2, \infty)$?

79 Show that the equation $x^5 = 2^x$ has a solution (*a*) less than 2, and (*b*) greater than 2.

80 Show that the equation $x^3 - 2x^2 - 3x + 1 = 0$ has a solution (*a*) less than 0, (*b*) in $[0, 2]$, and (*c*) larger than 2.

81 (*a*) Does $(\sin x - \cos x)^2$ have a maximum value? If so, find it.

(*b*) Does $(\sin x - \cos x)^2$ have a minimum value? If so, find it.

82 Assume that $\lim_{x\to 3} f(x) = 0$ and $\lim_{x\to 3} g(x) = 0$. What, if anything, can be said about

(*a*) $\lim_{x\to 3} [f(x) - g(x)]$? (*b*) $\lim_{x\to 3} \sin f(x)$?

(*c*) $\lim_{x\to 3} \cos f(x)$? (*d*) $\lim_{x\to 3} f(x)g(x)$?

(*e*) $\lim_{x\to 3} [f(x)]^3/g(x)$?

In each of Exercises 83 to 85 verify the intermediate-value theorem for the indicated function, closed interval, and value.

83 2^x, $[1, 8]$, $m = 4$ **84** $\sin x$, $[0, 9\pi/2]$, $m = \frac{1}{2}$

85 $\tan x$, $[0, \pi/3]$, $m = 1$ **86** Find $\lim_{x\to 0} \dfrac{\tan x - \sin x}{x^2}$.

87 Let $f(x) = \sin(1/x)$ if $x \neq 0$. Is it possible to define $f(0)$ in such a way that f is continuous throughout the x axis?

88 Let $f(x) = x \sin(1/x)$ if $x \neq 0$. Is it possible to define $f(0)$ in such a way that f is continuous throughout the x axis?

89 Let f be a continuous function such that $f(x)$ is in $[0, 1]$ when x is in $[0, 1]$. Prove that there is at least one number c in $[0, 1]$ such that $f(c) = c$. [*Hint:* Consider the function g given by $g(x) = f(x) - x$.]

90 Examine (*a*) $\lim_{x \to 0^+} (4^x + 3^x)^{1/x}$, (*b*) $\lim_{x \to 0^-} (4^x + 3^x)^{1/x}$, and (*c*) $\lim_{x \to \infty} (4^x + 3^x)^{1/x}$.

* **91** Let K be a bounded convex set and P any point inside K. (*a*) Is there always a chord through P that is divided by P into two pieces, one of which is twice as long as the other? (*b*) Is there always a chord through P that is divided by P into pieces of equal length?

3

THE DERIVATIVE

Section 3.1 presents four problems, all of which reduce to the same idea. This idea, the ''derivative of a function,'' is developed in Sec. 3.2. Section 3.3 examines the relation of derivatives to continuity and introduces the notion of an antiderivative, which will be needed beginning in Chap. 5. Section 3.4 develops some formulas for computing derivatives. The derivatives of the six trigonometric functions ($\sin x$, $\cos x$, . . .) are determined in Sec. 3.5. Section 3.6 obtains the most important tool for finding derivatives, the ''chain rule.'' Chapters 4 and 6 will illustrate some of the applications of the derivative, with the derivatives of the exponential and logarithmic functions being found in Chap. 6. (We delay them for two reasons: to keep Chap. 3 of reasonable length and to give you an earlier opportunity to see how the derivative is used.)

3.1 FOUR PROBLEMS WITH ONE THEME

This section discusses four problems which at first glance may seem unrelated. The first one concerns the slope of a tangent line to a curve. The second involves velocity. The final two concern magnification and density. A little arithmetic will quickly show that they are all just different versions of one mathematical idea.

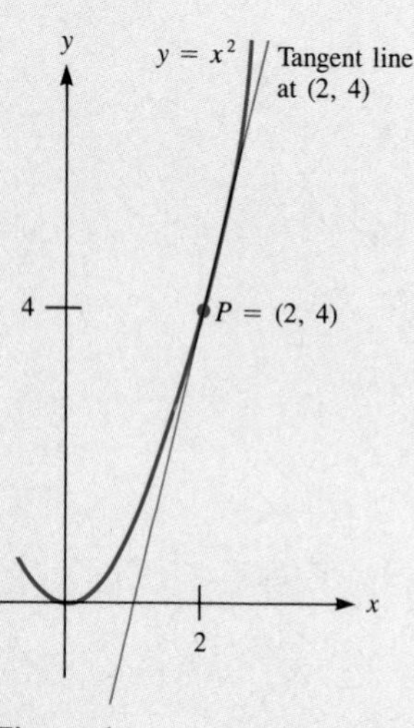

Figure 1

Slope

Our first problem is important because it is related to finding the straight line that most closely resembles a given graph near a given point on the graph.

PROBLEM 1 *Slope*. What is the slope of the tangent line to the graph of $y = x^2$ at the point $P = (2, 4)$, as shown in Fig. 1?

For the present, the **tangent line** to a curve at a point P on the curve shall mean the line through P that has the ''same direction'' as the curve at P. (Look again at Fig. 1.) This will be made precise in the next section.

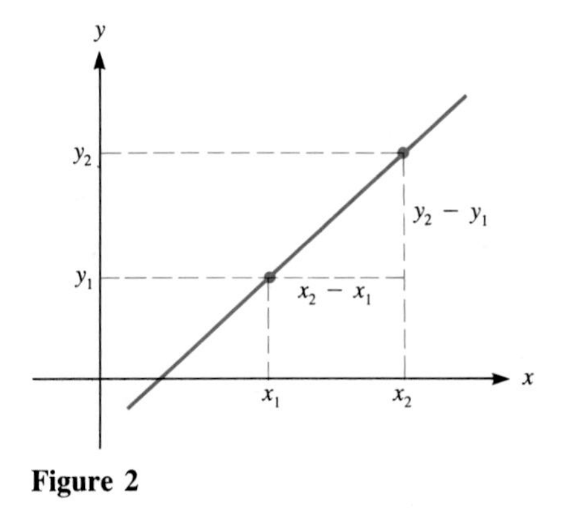

Figure 2

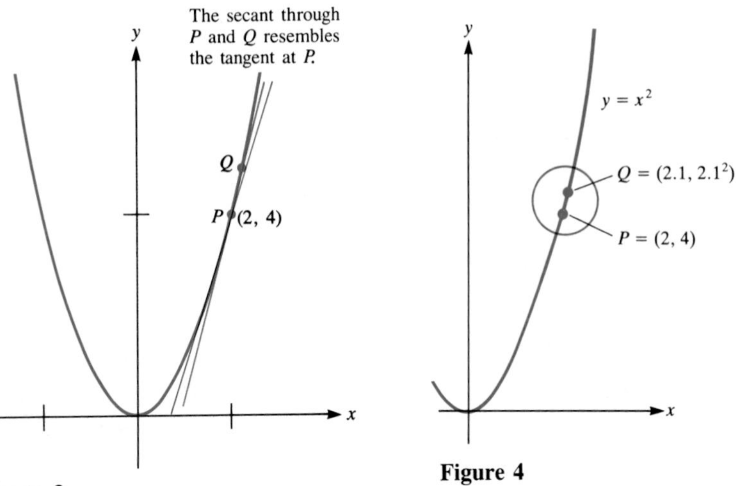

Figure 3

Figure 4

SOLUTION If we know two points (x_1, y_1) and (x_2, y_2) on a straight line, we can compute the slope of that line. The slope is ''change in y divided by change in x''; that is,

$$\text{Slope of line} = \frac{y_2 - y_1}{x_2 - x_1}.$$

See Fig. 2.

(Appendix B has a further discussion of the slope of a line.)

However, we know only one point on the tangent line at (2, 4), namely, just the point (2, 4) itself. To get around this difficulty we will choose a point Q on the parabola $y = x^2$ near P and compute the slope of the line through P and Q. Such a line is called a *secant*. As Fig. 3 suggests, such a secant line resembles the tangent line at (2, 4).

For instance, choose $Q = (2.1, 2.1^2)$ and compute the slope of the line through P and Q as shown in Fig. 4.

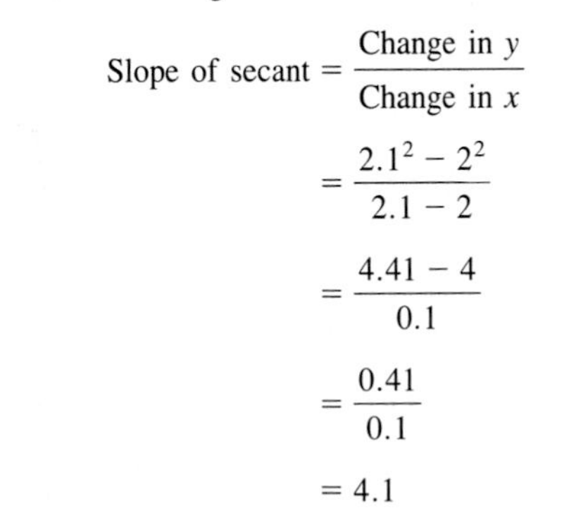

$$\begin{aligned}\text{Slope of secant} &= \frac{\text{Change in } y}{\text{Change in } x}\\ &= \frac{2.1^2 - 2^2}{2.1 - 2}\\ &= \frac{4.41 - 4}{0.1}\\ &= \frac{0.41}{0.1}\\ &= 4.1\end{aligned}$$

Thus an estimate of the slope of the tangent line is 4.1. Note that in making this estimate there was no need to draw the curve.

We can also choose the point Q on the parabola to be the left of $P = (2, 4)$. For instance, choose $Q = (1.9, 1.9^2)$. (See Fig. 5.) Then

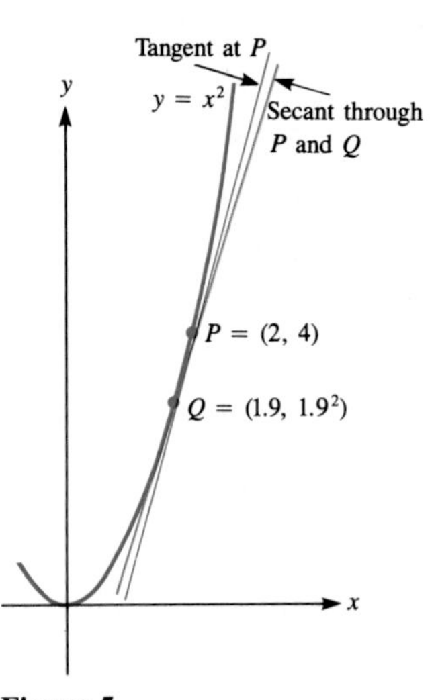

Figure 5

$$\text{Slope of secant} = \frac{\text{Change in } y}{\text{Change in } x}$$
$$= \frac{1.9^2 - 2^2}{1.9 - 2}$$
$$= \frac{3.61 - 4}{-0.1}$$
$$= \frac{-0.39}{-0.1}$$
$$= 3.9.$$

To obtain a better estimate, we could repeat the process using, for instance, the line through $P = (2, 4)$ and $Q = (2.01, 2.01^2)$. Rather than do this, it is simpler to consider a *typical point* Q. That is, consider the line through $P = (2, 4)$ and $Q = (2 + h, (2 + h)^2)$ when h is small, either positive or negative. (See Figs. 6 and 7.) This line has slope

$$\frac{(2 + h)^2 - 2^2}{(2 + h) - 2} = \frac{(2 + h)^2 - 2^2}{h}.$$

h could be > 0 or < 0.

To find out what happens to this quotient as h gets closer to 0, apply the techniques of limits. We have

$$\lim_{h \to 0} \frac{(2 + h)^2 - 2^2}{h} = \lim_{h \to 0} \frac{4 + 4h + h^2 - 4}{h} = \lim_{h \to 0} (4 + h) = 4.$$

Observe the use of limits.

Thus the tangent line at (2, 4) has slope 4.

Figure 8 indicates the idea of the solution by showing how secant lines approximate the tangent line. It suggests a blowup of a small part of the curve $y = x^2$.

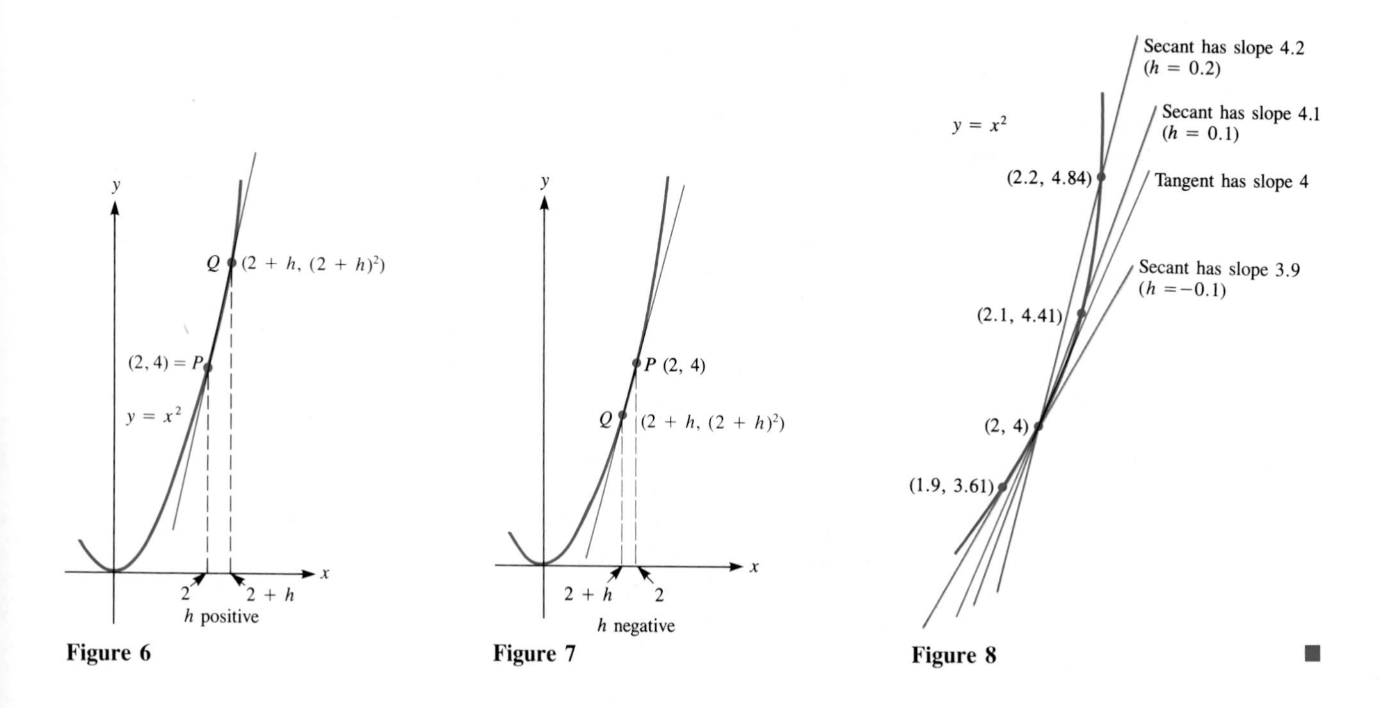

Figure 6 **Figure 7** **Figure 8** ■

Velocity

If an airplane or automobile is moving at a constant velocity, we know that "distance traveled equals velocity times time." Thus

$$\text{Velocity} = \frac{\text{Distance traveled}}{\text{Time}}.$$

If its velocity is not constant, we still may speak of its "average velocity," which is defined as

$$\text{Average velocity} = \frac{\text{Distance traveled}}{\text{Time}}.$$

For instance, if you drive from San Francisco to Los Angeles, a distance of 400 miles, in 8 hours, your average velocity is 400/8 = 50 miles per hour.

Suppose that at time t_1 you have traveled a distance D_1, while at time t_2 you have traveled a distance D_2, where $t_2 > t_1$. Then during the time interval $[t_1, t_2]$ the distance traveled is $D_2 - D_1$. Thus the average velocity during the time interval $[t_1, t_2]$, which has duration $t_2 - t_1$, is

$$\text{Average velocity} = \frac{D_2 - D_1}{t_2 - t_1}.$$

We see that the computations for average velocity are the same as those for finding the slope of a line.

The next problem shows how to find the velocity at any instant of time for an object whose velocity is not constant.

PROBLEM 2 *Velocity.* A rock initially at rest falls $16t^2$ feet in t seconds. What is its velocity after 2 seconds?

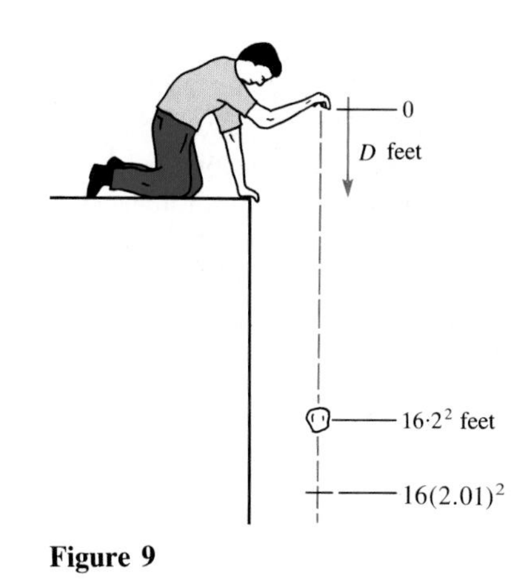

Figure 9

SOLUTION If the rock moves a distance of D feet in an interval of time of t seconds, we know what is meant by its average velocity during the time, namely, the quotient D/t feet per second. We will use this idea to deal with the much more abstract idea of "velocity at a given time," the so-called instantaneous velocity. In the next section the notion of velocity at a given instant will be made precise.

For practice, make an estimate by finding the average velocity of the rock during a short time interval, say from 2 to 2.01 seconds. At the start of this interval the rock has fallen $16(2^2) = 64$ feet. By the end it has fallen $16(2.01^2) = 16(4.0401) = 64.6416$ feet. So during an interval of 0.01 second it fell 0.6416 feet. Its average velocity during this time interval is

$$\frac{\text{Distance}}{\text{Time}} = \frac{0.6416}{0.01} = 64.16 \text{ feet per second},$$

an estimate of the velocity at time $t = 2$ seconds. (See Fig. 9.)

Although we will keep $h > 0$, estimates could just as well be made with $h < 0$.

Rather than make another estimate with the aid of a still shorter interval of time, let us consider the typical time interval from 2 to $2 + h$ seconds, $h > 0$. During this short time of h seconds the rock travels $16(2 + h)^2 - 16 \cdot 2^2 = 16[(2 + h)^2 - 2^2]$ feet, as shown in Fig. 10. The average velocity of the rock during this period is

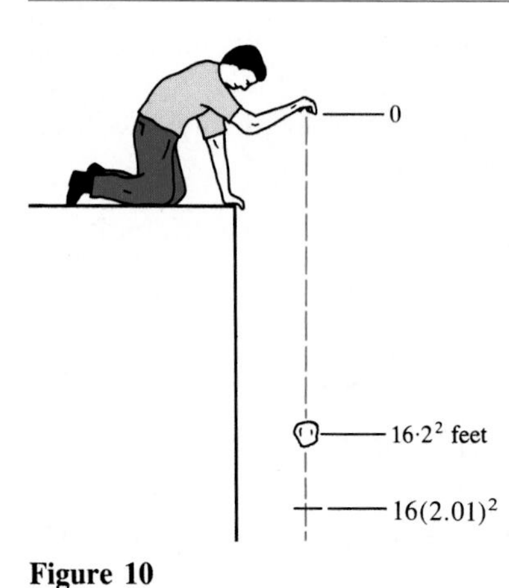

Figure 10

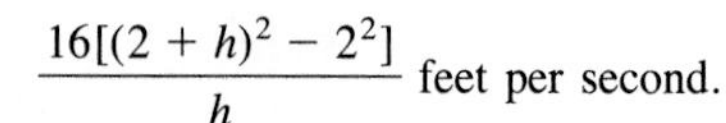

$$\frac{16[(2+h)^2 - 2^2]}{h} \text{ feet per second.}$$

When h is close to 0, what happens to this average velocity? It approaches

$$\lim_{h\to 0} \frac{16[(2+h)^2 - 2^2]}{h} = 16 \lim_{h\to 0} \frac{(2+h)^2 - 2^2}{h}$$

$$= 16 \lim_{h\to 0} (4 + h) = 16 \cdot 4 = 64 \text{ feet per second.}$$

We say that the velocity at time $t = 2$ is 64 feet per second. ■

Even though Problems 1 and 2 seem unrelated at first, their solutions turn out to be practically identical: The slope in Problem 1 is approximated by the quotient

$$\frac{(2+h)^2 - 2^2}{h}$$

and the velocity in Problem 2 is approximated by the quotient

$$\frac{16(2+h)^2 - 16 \cdot 2^2}{h} = 16\,\frac{(2+h)^2 - 2^2}{h}.$$

The only difference between the solutions is that the second quotient has an extra factor of 16. This may not be too surprising, since the functions involved, x^2 and $16t^2$, differ by a factor of 16.

Magnification

The third problem concerns magnification, a concept that occurs in everyday life. For instance, photographs can be blown up or reduced in size, with each part magnified or shrunk by the same factor. However, magnification may vary from point to point, as with a curved mirror. If the projection of an interval of length L is an interval of length L^*, we say that the interval on the photograph is magnified by the factor L^*/L. (See Fig. 11.)

As with slope and velocity, we treat "magnification at a point" intuitively. The next section will make it precise.

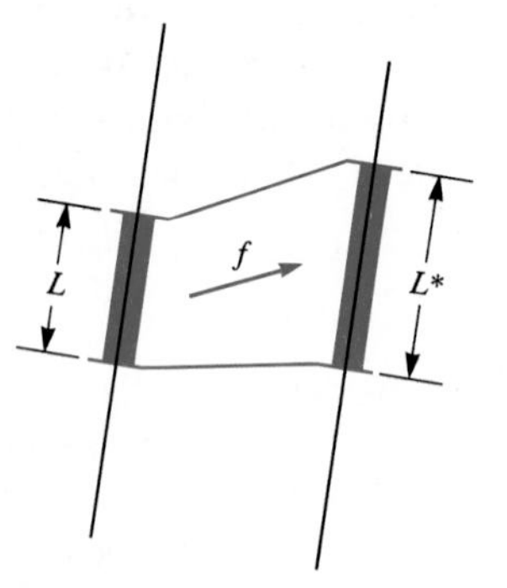

Figure 11

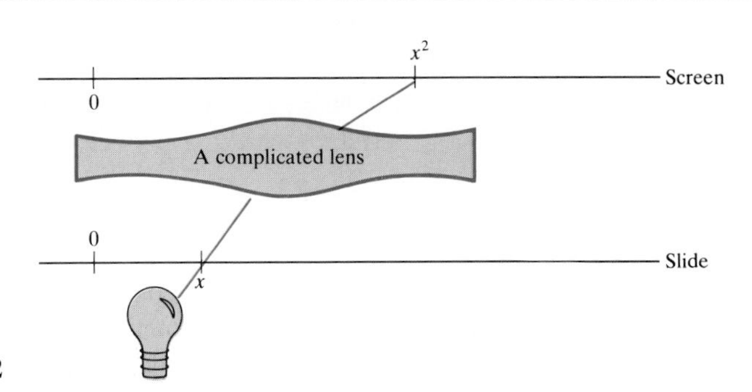

Figure 12

PROBLEM 3 *Magnification.* A light, two lines (a slide and a screen), and a complicated lens are placed as in Fig. 12. This arrangement projects the point on the bottom line, whose coordinate is x, to the point on the top line, whose coordinate is x^2. For example, 2 is projected onto 4, and 3 onto 9. The projection of the interval [2, 3] is the interval [4, 9]. Now [2, 3] has length $3 - 2 = 1$, while [4, 9] has length $9 - 4 = 5$, so the projection is five times as long. The **magnification** of the interval [2, 3] is said to be 5. (See Fig. 13.) The projection of the interval [0, 1/3] is [0, 1/9], which is only one-third as long. In this case, the interval is magnified by a factor of 1/3. For large x, the lens increases length to a great extent; for x near 0, the lens markedly reduces. What is the "magnification at $x = 2$"?

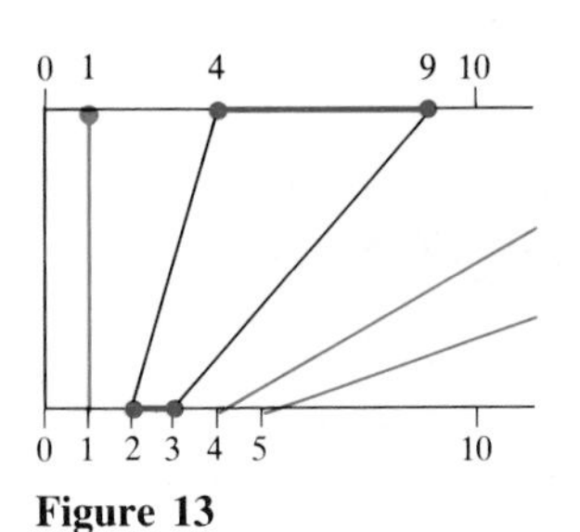

Figure 13

SOLUTION To estimate the magnification at 2 on the slide, examine the projection of a small interval in the vicinity of 2. Let us see what the image of [2, 2.1] is on the screen. Since the image of 2 is 2^2 and the image of 2.1 is 2.1^2, the image of the interval [2, 2.1] of length 0.1 is the interval $[2^2, 2.1^2]$ of length

$$2.1^2 - 2^2 = 0.41.$$

(See Fig. 14.)

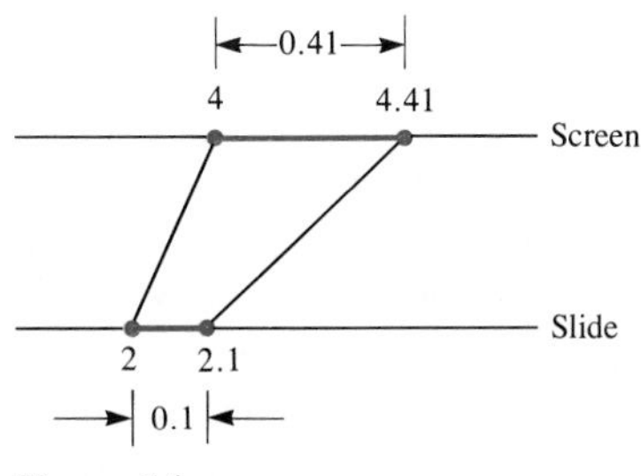

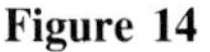

Figure 14

The magnifying factor over the interval [2, 2.1] is

$$\frac{\text{Length of image on screen}}{\text{Length of interval on slide}} = \frac{0.41}{0.1} = 4.1.$$

This number, 4.1, is an estimate of the magnification at $x = 2$.

You have probably guessed the next step. Rather than go on and consider the magnification of another specific interval, such as the interval [2, 2.01], we go directly to a typical interval with 2 as its left end.

For convenience we use $h > 0$. Estimates could also be made with $h < 0$.

The image of the interval $[2, 2 + h]$, where h is greater than 0, is the interval $[2^2, (2 + h)^2]$. Since $[2, 2 + h]$ has length h and $[2^2, (2 + h)^2]$ has length $(2 + h)^2 - 2^2$, the magnification of the interval $[2, 2 + h]$ is

$$\frac{(2 + h)^2 - 2^2}{h} = 4 + h, \qquad h > 0.$$

As already observed, when h approaches 0, this quotient approaches 4. Thus the magnification at 2 is 4. In terms of limits,

$$\text{Magnification at } 2 = \lim_{h \to 0} \frac{(2 + h)^2 - 2^2}{h} = 4. \quad \blacksquare$$

Density

The next problem is concerned with density, which is a measure of the heaviness of a material. Density is defined as mass divided by volume:

$$\text{Density} = \frac{\text{Total mass}}{\text{Total volume}}.$$

Water has a density of 1 gram per cubic centimeter, while the density of lead is 11.3 grams per cubic centimeter. The density of an object may vary from point to point. For instance, the density of matter near the center of the earth is much greater than that near the surface. The idea of density provides a concrete analog of several mathematical ideas and will be referred to frequently in later chapters.

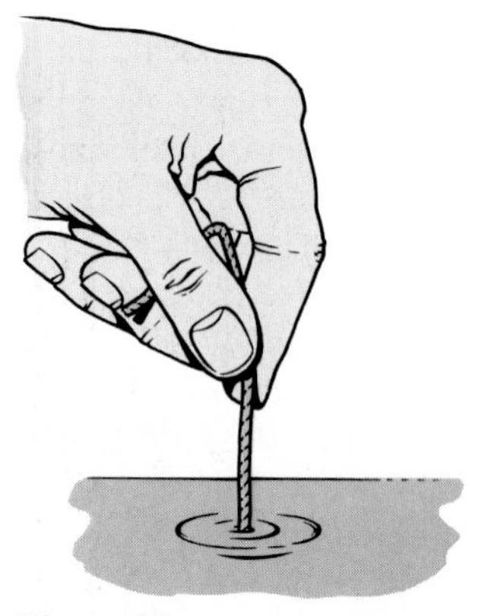

Figure 15

The next problem concerns a string of varying density. This density will be considered in terms of grams per linear centimeter rather than grams per cubic centimeter. The matter is imagined as a continuous distribution, not composed of isolated molecules.

Later chapters also will be concerned with density of matter distributed along a curve, in a flat object, and in a solid. The notion of "density at a point" will be made precise in the next section. For the moment, we deal with it intuitively.

Imagine that you dip a piece of string into a pot of water and then hang it up to dry, as in Fig. 15. The water in the string tends to move down in the string, so that the lower part is denser (wetter) than the upper part. You may have noticed that this happens when you hang a wet towel on a clothesline. In the next problem you may think of this string as being removed from the line and laid horizontally on a table, as in Fig. 16.

A nonhomogeneous string varies in density from point to point.

PROBLEM 4 *Density*. The mass of the left-hand x centimeters of a nonhomogeneous string 10 centimeters long is x^2 grams, as shown in Fig. 16.

For instance, the mass in the interval $[0, 5]$ has a mass of $5^2 = 25$ grams. The mass in the interval $[0, 6]$ is $6^2 = 36$ grams. Consequently the mass in the interval $[5, 6]$ is $36 - 25 = 11$ grams. Similarly, the mass in the interval $[6, 7]$ is $7^2 - 6^2 = 49 - 36 = 13$ grams. What is the density, in grams per centimeter, of the material at $x = 2$?

SOLUTION To estimate the density of the string 2 centimeters from its left end, examine the mass of the material in the short interval $[2, 2.1]$. (See Fig. 17.)

The material in the interval $[2, 2.1]$ has a mass of $2.1^2 - 2^2$ grams, which equals $4.41 - 4 = 0.41$ gram. Thus the average density for this interval is $0.41/0.1 = 4.1$ grams per centimeter.

Rather than make another estimate, consider the density in the typical small interval $[2, 2 + h]$. The mass in this interval is

$$(2 + h)^2 - 2^2 \text{ grams.}$$

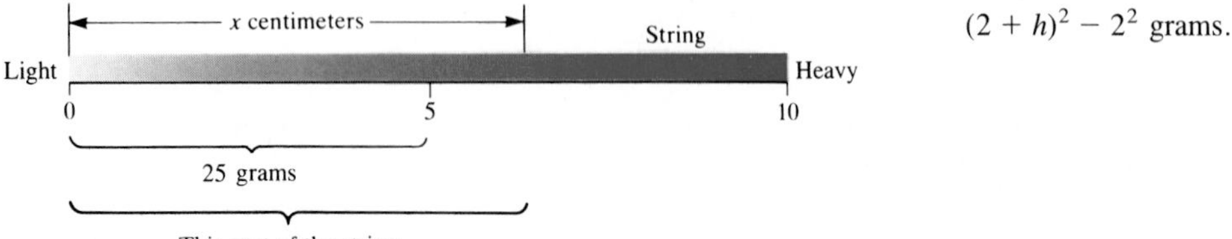

Figure 16

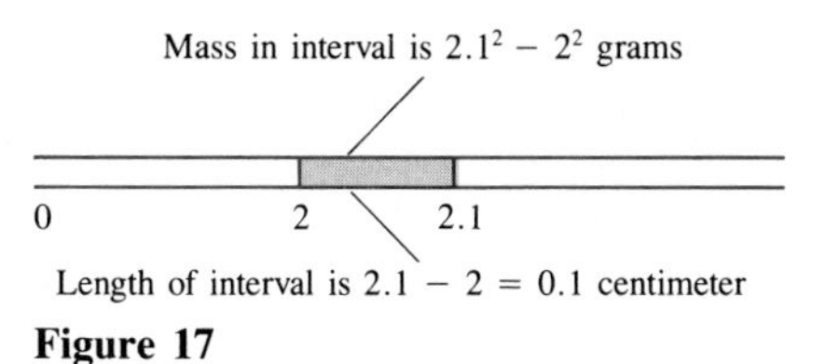

Figure 17

The interval has length h centimeters. Thus the density of matter in this interval is

$$\frac{(2+h)^2 - 2^2}{h} = 4 + h \text{ grams per centimeter.}$$

As h approaches 0, this quotient approaches 4, and we say that the density 2 centimeters from the left end of the string is 4 grams per centimeter.

In terms of limits,

$$\text{Density at } 2 = \lim_{h \to 0} \frac{(2+h)^2 - 2^2}{h} = 4. \quad ■$$

Section Summary

From a mathematical point of view, the problems of finding the slope of the tangent line, the velocity of the rock, the magnification of the lens, and the density along the string are the same. Each leads to the same type of quotient as an estimate. In each case, the behavior of this quotient is studied as h approaches 0. In each case, the answer is a limit.

The underlying mathematical theme is explored in the next section, which introduces the derivative and will enable us to give precise definitions of "slope," "velocity," "magnification," and "density" when these vary from point to point.

EXERCISES FOR SEC. 3.1: FOUR PROBLEMS WITH ONE THEME

Exercises 1 to 4 review the concept of slope of a line. (See Appendix B.)

1 What angle does a line make with the x axis if its slope is (*a*) 1? (*b*) 2? [For (*b*) you will need a calculator.]

2 What angle does a line make with the x axis if its slope is (*a*) -1? (*b*) -2? [Remember that the angle is in the second quadrant. For (*b*) you will need a calculator.]

3 Draw x and y axes and a line that is neither horizontal nor vertical. Using a ruler (preferably in centimeters), estimate the slope of the line you drew.

4 Draw the line through (1, 2) that has (*a*) slope $\frac{3}{2}$. (*b*) slope $-\frac{3}{2}$.

Exercises 5 to 12 concern slope. In each case use the technique of Problem 1 to find the slope of the tangent line to the curve at the point.

5 $y = x^2$ at the point $(3, 3^2) = (3, 9)$

6 $y = x^2$ at the point $(\frac{1}{2}, (\frac{1}{2})^2) = (\frac{1}{2}, \frac{1}{4})$

7 $y = x^2$ at the point $(-2, (-2)^2) = (-2, 4)$

8 $y = x^2$ at the point $(1, 1^2) = (1, 1)$

9 $y = x^3$ at $(2, 2^3) = (2, 8)$ [*Hint:* Recall that $(a+b)^3 = a^3 + 3a^2b + 3ab^2 + b^3$.]

10 $y = x^3$ at $(1, 1^3) = (1, 1)$

11 (*a*) $y = x^2$ at (0, 0)
(*b*) Sketch the graph of $y = x^2$ and the tangent line at (0, 0).

12 (*a*) $y = x^3$ at (0, 0)
(*b*) Sketch the graph of $y = x^3$ and the tangent line at (0, 0). [Be especially careful when sketching the graph near (0, 0).]

In Exercises 13 to 16 use the method of Problem 2 to find the velocity of the rock after

13 3 seconds
14 $\frac{1}{2}$ second
15 1 second
16 $\frac{1}{4}$ second

17 A certain object travels t^3 feet in the first t seconds.
(*a*) How far does it travel during the time interval from 2 to 2.1 seconds?
(*b*) What is its average velocity during that time interval?
(*c*) Let h be any positive number. Find the average velocity of the object from time 2 to time $2 + h$ seconds.
(*d*) Find the velocity of the object at time 2 seconds by letting h approach 0 in part (*c*).

18 A certain object travels t^3 feet in the first t seconds.
(*a*) Find its average velocity during the time interval from 3 to 3.01 seconds.
(*b*) Find its average velocity during the time interval from 3 to $3 + h$ seconds, $h > 0$.
(*c*) By letting h approach 0 in part (*b*), find the velocity of the object at time 3 seconds.

In the slope problem the nearby point Q was always pictured as being to the right of P. The point Q could just as well have been chosen to the left of P. Exercises 19 and 20 illustrate this case.

19 Consider the parabola $y = x^2$.
- (*a*) Find the slope of the line through $P = (2, 4)$ and $Q = (1.99, 1.99^2)$.
- (*b*) Find the slope of the line through $P = (2, 4)$ and $Q = (2 + h, (2 + h)^2)$, where $h < 0$.
- (*c*) Show that as h approaches 0, the slope in (*c*) approaches 4.

20 Consider the curve $y = x^3$.
- (*a*) Find the slope of the line through $P = (2, 2^3)$ and $Q = (1.9, 1.9^3)$.
- (*b*) Find the slope of the line through $P = (2, 2^3)$ and $Q = (2 + h, (2 + h)^3)$, where $h < 0$.
- (*c*) Show that as h approaches 0, the slope in (*b*) approaches 12.

21 (*a*) Find the slope of the tangent line to $y = x^2$ at $(4, 16)$.
- (*b*) Use it to draw the tangent line to the curve at $(4, 16)$.

22 (*a*) Find the slope of the tangent line to $y = x^2$ at $(-1, 1)$.
- (*b*) Use it to draw the tangent line to the curve at $(-1, 1)$.

Exercises 23 to 26 concern magnification.

23 By what factor does the lens in Problem 3 magnify the interval (*a*) $[1, 1.1]$? (*b*) $[1, 1.01]$? (*c*) $[1, 1.001]$? (*d*) Find the magnification at 1.

24 By what factor does the lens in Problem 3 magnify the interval (*a*) $[3, 3.1]$? (*b*) $[3, 3.01]$? (*c*) $[3, 3.001]$? (*d*) Find the magnification at 3.

25 By what factor does the lens in Problem 3 magnify the interval (*a*) $[0.49, 0.5]$? (*b*) $[0.499, 0.5]$? (*c*) Find the magnification at 0.5 by examining the magnification of intervals of the form $[0.5 + h, 0.5]$, where $h < 0$.

26 By what factor does the lens in Problem 3 magnify the interval (*a*) $[1.49, 1.5]$? (*b*) $[1.499, 1.5]$? (*c*) Find the magnification at 1.5 by examining the typical interval of the form $[1.5 + h, 1.5]$, where $h < 0$.

Exercises 27 and 28 concern density.

27 The left x centimeters of a string have a mass of x^2 grams.
- (*a*) What is the mass in the interval $[3, 3.01]$?
- (*b*) Using the interval $[3, 3.01]$, estimate the density at 3.
- (*c*) Using the interval $[2.99, 3]$, estimate the density at 3.
- (*d*) By considering intervals of the form $[3, 3 + h]$, $h > 0$, find the density at the point 3 centimeters from the left end.
- (*e*) By considering intervals of the form $[3 + h, 3]$, $h < 0$, find the density at the point 3 centimeters from the left end.

28 The left x centimeters of a string have a mass of x^2 grams.
- (*a*) What is the mass in the interval $[2, 2.01]$?
- (*b*) Using the interval $[2, 2.01]$, estimate the density at 2.
- (*c*) Using the interval $[1.99, 2]$, estimate the density at 2.
- (*d*) By considering intervals of the form $[2, 2 + h]$, $h > 0$, find the density at the point 2 centimeters from the left end.
- (*e*) By considering intervals of the form $[2 + h, 2]$, $h < 0$, find the density at the point 2 centimeters from the left end.

---

29 (*a*) What is the slope of the curve $y = x^3$ at $(1, 1)$? (Find it by the method of this section.)
- (*b*) What does the graph of $y = x^3$ look like near $(1, 1)$?

30 (*a*) Find the slope of the curve $y = x^3$ at $(-1, -1)$.
- (*b*) What does the graph of $y = x^3$ look like near $(-1, -1)$?

The next two exercises show that the idea common to the four problems in this section also appears in biology and economics.

31 A certain bacterial culture has a mass of t^2 grams after t minutes of growth.
- (*a*) How much does it grow during the time interval $[2, 2.01]$?
- (*b*) What is its rate of growth during the time interval $[2, 2.01]$?
- (*c*) What is its rate of growth when $t = 2$?

32 A thriving business has a profit of t^2 million dollars in its first t years. Thus from time $t = 3$ to time $t = 3.5$ (the first half of its fourth year) it has a profit of $(3.5)^2 - 3^2$ million dollars, giving an annual rate of

$$\frac{(3.5)^2 - 3^2}{0.5} = 6.5 \text{ million dollars per year.}$$

- (*a*) What is its annual rate of profit during the time interval $[3, 3.1]$?
- (*b*) What is its annual rate of profit during the time interval $[3, 3.01]$?
- (*c*) What is its annual rate of profit after 3 years?

33 (*a*) Graph the curve $y = 2x^2 + x$.
- (*b*) By eye, draw the tangent line to the curve at the point $(1, 3)$. Using a ruler, estimate the slope of the tangent line.
- (*c*) Sketch the line that passes through the point $(1, 3)$ and the point $(1 + h, 2(1 + h)^2 + (1 + h))$.
- (*d*) Find the slope of the line in (*c*).
- (*e*) Letting h get closer and closer to 0, find the slope of the tangent line at $(1, 3)$. How close was your estimate in (*b*)?

34 An object travels $2t^2 + t$ feet in t seconds.
- (*a*) Find its average velocity during the interval of time $[1, 1 + h]$, where h is greater than 0.
- (*b*) Letting h get closer and closer to 0, find the velocity at time 1.

35 Find the slope of the tangent line to the curve $y = x^2$ of Problem 1 at the typical point $P = (x, x^2)$. To do this, consider the slope of the line through P and the nearby point $Q = (x + h, (x + h)^2)$ and let h approach 0.

36 Find the velocity of the falling rock of Problem 2 at any time t. To do this, consider the average velocity during the time interval $[t, t + h]$ and then let h approach 0.

37 Find the magnification of the lens in Problem 3 at the typical point x by considering the magnification of the short interval $[x, x + h]$, where $h > 0$, and then let h approach 0.

38 Find the density of the string in Problem 4 at a typical point x centimeters from the left end. To do this, consider the mass in a short interval $[x, x + h]$, where $h > 0$, and let h approach 0.

39 (*a*) Sketch the curve $y = x^3 - x^2$.
(*b*) Using the method of the nearby point, find the slope of the tangent line to the curve at the point $(x, x^3 - x^2)$.
(*c*) Find all points on the curve where the tangent line is horizontal.
(*d*) Find all points where the tangent line has slope 1.

40 Answer the same questions as in Exercise 39 for the curve $y = x^3 - x$.

41 Does the tangent line to the curve $y = x^2$ at the point (1, 1) pass through the point (6, 12)?

42 An astronaut is traveling from left to right along the curve $y = x^2$. When she shuts off the engine, she will fly off along the line tangent to the curve at the point where she is at that moment. At what point should she shut off the engine in order to reach the point
(*a*) (4, 9)? (*b*) (4, −9)?

43 See Exercise 42. Where can an astronaut who is traveling from left to right along $y = x^3 - x$ shut off the engine and pass through the point (2, 2)?

44 With the aid of a calculator, estimate the slope of $y = 2^x$ at $x = 1$, using the intervals
(*a*) [1, 1.1]
(*b*) [1, 1.01]
(*c*) [0.9, 1]
(*d*) [0.99, 1].

45 With the aid of a calculator, estimate the slope of $y = (x + 1)/(x + 2)$ at $x = 2$, using the intervals
(*a*) [2, 2.1]
(*b*) [2, 2.01]
(*c*) [2, 2.001]
(*d*) [1.999, 2].

46 (*a*) Graph the curve $y = 2^x$ as well as you can for $-2 \leq x \leq 3$.
(*b*) Using a straight edge, draw as well as you can a tangent to the curve at (2, 4). Estimate the slope of this tangent by using a ruler to draw a "rise-and-run" triangle.
(*c*) Using a secant through (2, 4) and $(x, 2^x)$, for x near 2, estimate the slope of the tangent to the curve at (2, 4). (Choose particular values of x and use your calculator to create a table of your results.)

3.2 THE DERIVATIVE

The solution of the slope problem in Sec. 3.1 (as well as those of the magnification problem and the density problem) led to the limit

$$\lim_{h \to 0} \frac{(2 + h)^2 - 2^2}{h}.$$

The velocity problem involved a similar limit,

$$\lim_{h \to 0} \frac{16(2 + h)^2 - 16 \cdot 2^2}{h}.$$

These limits arose from the particular formulas x^2 and $16t^2$ that had been picked. In each case we formed a **difference quotient**,

$$\frac{\text{Difference in outputs}}{\text{Difference in inputs}},$$

and examined its limit as the change in the inputs was made smaller and smaller.

The whole procedure can be carried out for functions other than x^2 and $16t^2$ and at numbers other than 2.

The four problems in Sec. 3.1 had one theme in common.

The underlying common theme of the four problems in Sec. 3.1 is the important mathematical concept, the **derivative** of a numerical function, which will now be defined.

In the following definition x is fixed and h approaches 0.

The f' notation

Definition *The derivative of a function at the number x.* Let f be a function that is defined at least in some open interval that contains the number x. If

$$\lim_{h \to 0} \frac{f(x+h) - f(x)}{h}$$

exists, it is called the **derivative of f at x** and is denoted $f'(x)$. The function f is said to be **differentiable** at x.

The numerator, $f(x + h) - f(x)$, is the change, or difference, in the outputs; the denominator, h, is the change in the inputs. (See Fig. 1.) Keep in mind that $x + h$ can be either to the right or left of x. Similarly, $f(x + h)$ can be either larger or smaller than $f(x)$.

A few examples will illustrate the concept of the derivative.

EXAMPLE 1 Find the derivative of the squaring function at the number 2.

SOLUTION In this case, $f(x) = x^2$ for any input x. By definition, the derivative of this function at 2 is

$$\lim_{h \to 0} \frac{f(2+h) - f(2)}{h} = \lim_{h \to 0} \frac{(2+h)^2 - 2^2}{h} = \lim_{h \to 0} (4 + h) = 4.$$

We say that "the derivative of the function x^2 at 2 is 4." ■

The next example determines the derivative of the squaring function at any input, not just at 2.

EXAMPLE 2 Find the derivative of the function x^2 at any number x.

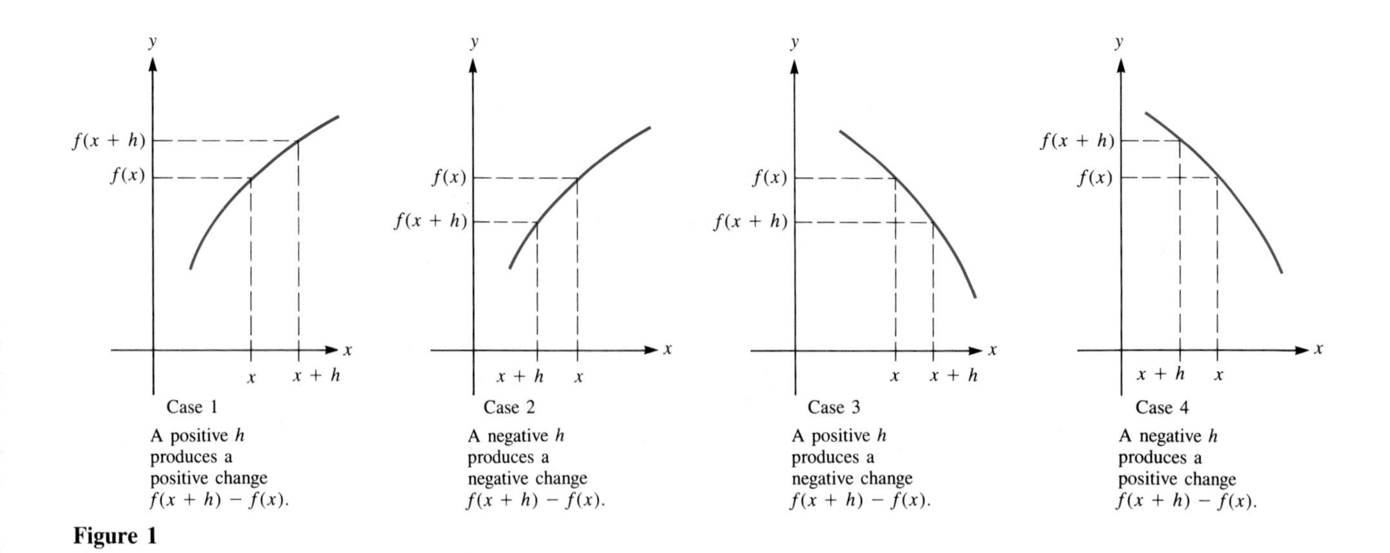

Figure 1

SOLUTION By definition of the derivative,

$$\begin{aligned} f'(x) &= \lim_{h\to 0} \frac{(x+h)^2 - x^2}{h} \\ &= \lim_{h\to 0} \frac{x^2 + 2hx + h^2 - x^2}{h} \\ &= \lim_{h\to 0} \frac{2hx + h^2}{h} \\ &= \lim_{h\to 0} (2x + h) \\ &= 2x. \end{aligned}$$

The derivative of the squaring function at x is $2x$. ■

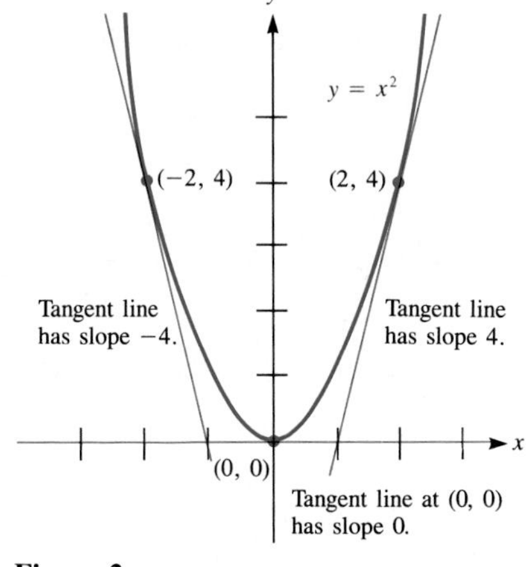

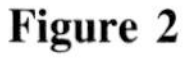
Figure 2

The fact that the derivative of the function x^2 is the function $2x$ is recorded in the notation

$$(x^2)' = 2x.$$

Read this aloud as "The derivative of the function x^2 is the function $2x$."

The result in Example 2 can be interpreted in terms of each of the four problems in Sec. 3.1. For example, we now know from Example 2 that the slope of the tangent line to the parabola $y = x^2$ at the point (x, x^2) is $2x$. In particular, the slope of the tangent line at $(2, 2^2)$ is $2 \cdot 2 = 4$, a result found in Sec. 3.1. Also, according to the formula for the derivative, $(x^2)' = 2x$, the slope of the tangent line to $y = x^2$ at $(-2, (-2)^2)$ is $2 \cdot (-2) = -4$ and at $(0, 0)$ is $2 \cdot 0 = 0$. A glance at Fig. 2 shows that these are reasonable results. The derivative of x^2 is a function. It assigns to the number x the slope of the tangent line to the parabola $y = x^2$ at the point (x, x^2).

The next two examples illustrate the idea of the derivative with functions other than x^2.

EXAMPLE 3 Find $f'(x)$ if $f(x) = x^3$.

Keep in mind that x is fixed and that $h \to 0$.

SOLUTION In this case, $f(x + h) = (x + h)^3$ and $f(x) = x^3$. Therefore the derivative of the function at x is

$$\begin{aligned} f'(x) = \lim_{h\to 0} \frac{(x+h)^3 - x^3}{h} &= \lim_{h\to 0} \frac{x^3 + 3x^2h + 3xh^2 + h^3 - x^3}{h} \\ &= \lim_{h\to 0} \frac{3x^2h + 3xh^2 + h^3}{h} \\ &= \lim_{h\to 0} (3x^2 + 3xh + h^2) \\ &= 3x^2. \end{aligned}$$

The derivative of x^3 at x is $3x^2$. ■

In view of Example 3, we may say that "the derivative of the function x^3 is the function $3x^2$" and write $(x^3)' = 3x^2$.

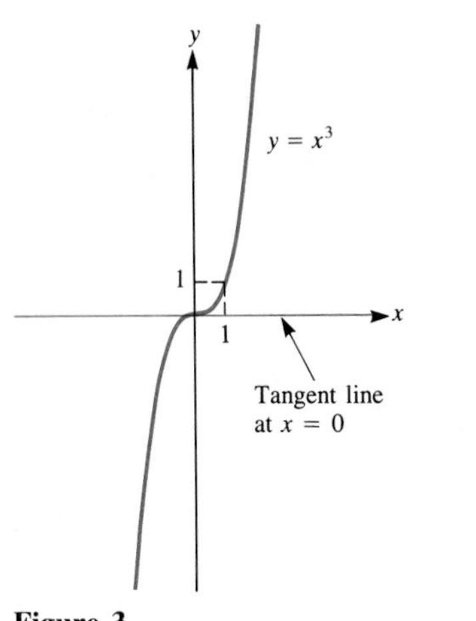

Figure 3

Example 3 tells us, for instance, that the slope of the graph of $y = x^3$ is never negative (since x^2 is never negative). Moreover, at $x = 0$ the slope is $3 \cdot 0^2 = 0$. Thus the tangent line to the curve $y = x^3$ at the origin is horizontal, as is shown in Fig. 3.

It may seem strange that a tangent line can *cross the curve,* as this tangent line at $x = 0$ does. However, the basic property of a tangent line is that it indicates the direction of a curve at a point. In high school geometry, where only tangent lines to circles are considered, the tangent line never crosses the curve.

Example 3 can also be interpreted in terms of the velocity of a moving object. If an object moves x^3 feet in the first x seconds, its velocity after x seconds is $3x^2$ feet per second.

EXAMPLE 4 Find the derivative of the square root function $f(x) = \sqrt{x}$.

SOLUTION Since the domain of $\sqrt{x}$ contains no negative numbers, assume that $x \geq 0$. If $x > 0$, then, by definition of the derivative,

When finding a derivative, be sure to write "lim" at each step.

$$f'(x) = \lim_{h \to 0} \frac{\sqrt{x+h} - \sqrt{x}}{h}$$

$$= \lim_{h \to 0} \left(\frac{\sqrt{x+h} - \sqrt{x}}{h}\right)\left(\frac{\sqrt{x+h} + \sqrt{x}}{\sqrt{x+h} + \sqrt{x}}\right) \qquad \text{rationalize numerator}$$

$$= \lim_{h \to 0} \frac{x + h - x}{h(\sqrt{x+h} + \sqrt{x})} \qquad (a-b)(a+b) = a^2 - b^2$$

$$= \lim_{h \to 0} \frac{1}{\sqrt{x+h} + \sqrt{x}}$$

$$= \frac{1}{2\sqrt{x}}.$$

Therefore we can write

$$(\sqrt{x})' = \frac{1}{2\sqrt{x}},$$

provided that $x > 0$.

If $x = 0$, the limit we are considering is

$$\lim_{h \to 0} \frac{1}{\sqrt{0+h} + \sqrt{0}},$$

which does not exist. We say that the derivative of $\sqrt{x}$ does not exist at 0. ■

According to Example 4, $(\sqrt{x})' = 1/(2\sqrt{x})$. Is this result reasonable? It says that when x is large, the slope of the tangent line at $(x, \sqrt{x})$ is near 0 [since $1/(2\sqrt{x})$ is near 0]. Let us draw the graph and see. First we make the table shown in the margin. With the aid of these six points, the graph can be sketched. (See Fig. 4.) For points far to the right on the graph, the tangent line is indeed almost horizontal, as the formula $1/(2\sqrt{x})$ suggests. When x is near 0, the derivative $1/(2\sqrt{x})$ is large. The graph gets steeper and steeper near $x = 0$.

x	0	1	4	9	16	25
$\sqrt{x}$	0	1	2	3	4	5

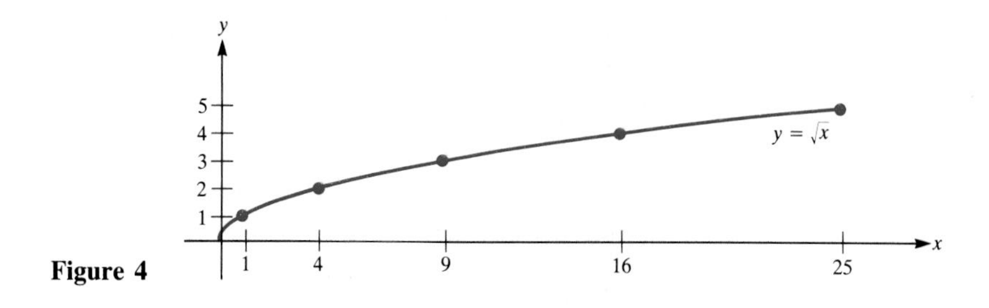

Figure 4

The Derivative of x^n

In Examples 2 and 3 it was shown that

$$(x^2)' = 2x \qquad \text{and} \qquad (x^3)' = 3x^2.$$

The next theorem obtains the derivative of x^n, where n is any positive integer. Its proof makes use of the form of $(x + h)^n$ when it is expanded. For example,

$$(x + h)^2 = x^2 + 2xh + h^2,$$

$$(x + h)^3 = x^3 + 3x^2h + 3xh^2 + h^3,$$

and

$$(x + h)^4 = x^4 + 4x^3h + 6x^2h^2 + 4xh^3 + h^4.$$

You can obtain these formulas either by multiplying out $(x + h)^2$, $(x + h)^3$, and $(x + h)^4$ or by using the binomial formula from Appendix C. All that we will need is that when n is a positive integer,

$$(x + h)^n = x^n + nx^{n-1}h + \text{terms that involve } h^2.$$

Theorem 1 For each positive integer n,

$$(x^n)' = nx^{n-1}.$$

Proof We must find what happens to the quotient

$$\frac{(x + h)^n - x^n}{h}$$

when $h \to 0$. We know that

$$\frac{(x + h)^n - x^n}{h} = \frac{x^n + nx^{n-1}h + (\text{terms that involve } h^2) - x^n}{h}$$

$$= nx^{n-1} + \text{terms involving } h.$$

Hence

$$\lim_{h\to 0} \frac{(x + h)^n - x^n}{h} = \lim_{h\to 0} (nx^{n-1} + \text{terms involving } h)$$

$$= nx^{n-1} + 0$$

$$= nx^{n-1}.$$

Consequently, the derivative of x^n is nx^{n-1}. ■

Direct application of this theorem yields, for instance,

The derivative of x^4 is $4x^{4-1} = 4x^3$.
The derivative of x^3 is $3x^{3-1} = 3x^2$.
The derivative of x^2 is $2x^{2-1} = 2x$.
The derivative of x^1 is $1x^0 = 1$ (in agreement with the fact that the line given by the formula $y = x$ has slope 1).

The Derivative of x^a for Any a

Does Theorem 1 hold for a negative exponent, such as -1? To see, let's find $(x^{-1})'$ by the definition of the derivative. In this case

$$f(x) = \frac{1}{x},$$

so we must examine

$$\lim_{h\to 0} \frac{f(x+h) - f(x)}{h} = \lim_{h\to 0} \frac{\dfrac{1}{x+h} - \dfrac{1}{x}}{h}.$$

Well,

$$\begin{aligned}
\frac{f(x+h) - f(x)}{h} &= \frac{\dfrac{1}{x+h} - \dfrac{1}{x}}{h} \\
&= \frac{\dfrac{x-(x+h)}{x(x+h)}}{h} \\
&= \frac{x-(x+h)}{hx(x+h)} && \frac{a/b}{c} = \frac{a}{bc} \\
&= \frac{-h}{hx(x+h)} \\
&= \frac{-1}{x(x+h)}. && \text{canceling } h
\end{aligned}$$

Since $x + h \to x$ as $h \to 0$,

$$\begin{aligned}
\lim_{h\to 0} \frac{f(x+h) - f(x)}{h} &= \lim_{h\to 0} \frac{-1}{x(x+h)} \\
&= -\frac{1}{x^2}.
\end{aligned}$$

Thus $(x^{-1})' = -1/x^2$.

Does this result agree with formula $(x^n)' = nx^{n-1}$ when $n = -1$? With $n = -1$ we have

$$nx^{n-1} = (-1)x^{-1-1} = -x^{-2} = -\frac{1}{x^2}.$$

So the formula $(x^n)' = nx^{n-1}$ is correct when $n = -1$.

Does Theorem 1 hold when the fixed exponent is a fraction? For instance, is

$$(x^{1/2})' = \tfrac{1}{2}x^{(1/2)-1}? \qquad (1)$$

Let us see. First of all, Eq. (1) can be written as

$$(x^{1/2})' = \frac{1}{2} \cdot \frac{1}{x^{1/2}},$$

or, with the square root sign, as

$$(\sqrt{x})' = \frac{1}{2} \cdot \frac{1}{\sqrt{x}}. \qquad (2)$$

Example 4 shows that (1) is correct.

The formula $(x^a)' = ax^{a-1}$ holds for all fixed exponents a, as will be shown in Chap. 6. We state this as a theorem, which will be used in Chaps. 3, 4, and 5.

Theorem 2 *The power rule.* For any fixed exponent a,

$$(x^a)' = ax^{a-1}$$

throughout any interval where x^a and x^{a-1} are both defined.

Theorem 2 tells us that to find the derivative of x^a, for fixed exponent a, lower the exponent by 1 and multiply by the original exponent.

EXAMPLE 5 Use Theorem 2 to differentiate $\sqrt[3]{x}$.

SOLUTION Write $\sqrt[3]{x}$ as $x^{1/3}$. By Theorem 2,

$$\begin{aligned}(x^{1/3})' &= \tfrac{1}{3}x^{(1/3)-1}\\ &= \tfrac{1}{3}x^{-2/3}\\ &= \frac{1}{3}\,\frac{1}{x^{2/3}}.\end{aligned}$$

If you want to express the answer with the radical sign, you could write $x^{2/3}$ as $\sqrt[3]{x^2}$ (or as $(\sqrt[3]{x})^2$). Then,

$$(\sqrt[3]{x})' = \frac{1}{3\sqrt[3]{x^2}}. \quad ■$$

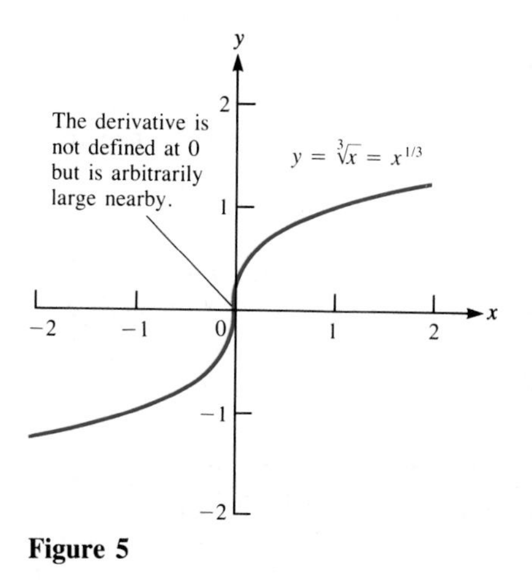

Figure 5

Remark The domain of the function $\sqrt[3]{x}$ is the entire x axis. However, the domain of its derivative does not contain the number 0. When x is near 0, the derivative is a large positive number. At 0 itself the derivative is not defined. A glance at Fig. 5 suggests that we should define the tangent line at (0, 0) to be vertical, that is, the y axis.

Remark A minor detail: According to Theorem 2, $(x^{5/2})' = \tfrac{5}{2}x^{3/2}$. This is valid even when $x = 0$. Now $f(x) = x^{5/2}$ is not defined when x is negative (since you can't take the square root of a negative number), so it is not defined in an open interval around 0. Strictly speaking, the definition of the derivative does not hold for this function at $x = 0$. (Recall that in the definition of the derivative,

$$\lim_{h \to 0} \frac{f(x + h) - f(x)}{h},$$

h is permitted to be positive or negative.) The expression $(0 + h)^{5/2}$ is meaningless when h is negative. In such a case, however, we just define the derivative using the right-hand limit

$$\lim_{h \to 0^+} \frac{f(x + h) - f(x)}{h}.$$

A similar adjustment is made when a left-hand limit is needed.

Four Applications of the Derivative

Now that we have the concept of the derivative, we are in a position to define **tangent line**, **velocity**, **magnification**, and **density**, terms used only intuitively until now. These definitions are suggested by the similarity of the computations made in the four problems in Sec. 3.1. *(In all five definitions it is assumed that the derivative exists.)*

Slope

Definition *Slope of a curve.* The **slope** of the graph of the function f at $(x, f(x))$ is the derivative of f at x.

Tangent

Definition *Tangent line to a curve.* The **tangent line** to the graph of the function f at the point $P = (x, f(x))$ is the line through P that has a slope equal to the derivative of f at x.

Velocity, speed

Definition *Velocity and speed of a particle moving on a line.* The **velocity** at time t of an object whose coordinate on an axis at time t is given by $f(t)$ is the derivative of f at time t. The **speed** of the particle is the absolute value of the velocity. (Thus velocity tells a direction—up or down, right or left—but speed does not.)

Note the distinction between velocity and speed. Velocity can be negative; speed is either positive or 0.

Magnification

Definition *Magnification of a linear projector.* The **magnification** at x of a lens that projects the point x of one line onto the point $f(x)$ of another line is the derivative of f at x.

Density

Definition *Density of material.* The **density** at x of material distributed along a line in such a way that the left-hand x centimeters have a mass of $f(x)$ grams is equal to the derivative of f at x.

Slope, velocity, magnification, and density are just interpretations, or applications, of the derivative. But biology, economics, chemistry, engineering, physics, computer models, and management use the derivative both to describe the concept "the rate at which some quantity is changing" and as a device for calculating that rate of change. The derivative itself is a purely mathematical concept; it is a special limit formed in a certain way from a function:

$$\lim_{h \to 0} \frac{f(x+h) - f(x)}{h}.$$

Section Summary

In this section we defined the fundamental concept of differential calculus, the derivative of a function. The derivative of the function f is another function f', defined as

$$f'(x) = \lim_{h \to 0} \frac{f(x+h) - f(x)}{h},$$

if that limit exists. We showed that for positive integer n, the derivative of x^n is nx^{n-1}. We also stated the more general result that $(x^a)' = ax^{a-1}$ for any fixed exponent a.

The concepts of slope of a curve, tangent to a curve, velocity, density, and magnification were then made precise with the aid of the derivative.

EXERCISES FOR SEC. 3.2: THE DERIVATIVE

In Exercises 1 to 16 use the definition of the derivative to find the derivatives of the given functions.

1 $2x$
2 $5x$
3 $4x + 4$
4 $3x - 1$
5 $5x^2$
6 $-x^2$
7 $x^2 + 2x$
8 $3x^2 - x$
9 $7\sqrt{x}$
10 $\frac{1}{2}\sqrt{x}$
11 $x^2 + 3\sqrt{x}$
12 $7x^2 - \sqrt{x}$
13 $x^3 + 3x$
14 $5x - x^3$
15 $x^2 + \dfrac{1}{x}$
16 $\dfrac{1}{x^2} + x$

In Exercises 17 to 28, use Theorem 2 to find the derivatives of the given functions at the given numbers.

17 x^4 at -1
18 x^4 at $\frac{1}{2}$
19 x^5 at a
20 x^5 at $\sqrt{2}$
21 $\sqrt[3]{t}$ at -8
22 $\sqrt[3]{t}$ at -1
23 x^π at 2
24 $x^{\sqrt{2}}$ at 3
25 $x^{2/3}$ at 8
26 $\sqrt[3]{x^2}$ at -27
27 $\sqrt[4]{x^5}$ at 16
28 $\sqrt[5]{x^4}$ at $\frac{1}{32}$
29 If $f(x) = 1/x^3$, find $f'(2)$, using Theorem 2.
30 If $f(x) = 1/\sqrt[3]{x}$, find $f'(8)$, using Theorem 2.

In Exercises 31 and 32 give the answers to three decimal places.

31 Find $f'(2)$ if $f(x) = x^{2.7}$.
32 Find $f'(3)$ if $f(x) = x^{\sqrt{2}}$.

33 Let $f(x) = x^4$. (*a*) What is the slope of the line joining (1, 1) to $(1.1, 1.1^4)$? (*b*) What is the slope of the tangent line to the curve at the point (1, 1)?

34 An object travels t^4 feet in the first t seconds.
(*a*) What is its average velocity from time $t = 2$ to time $t = 2.01$?
(*b*) What is its average velocity from time $t = 1.99$ to time $t = 2$?
(*c*) What is its velocity at time $t = 2$?

35 A lens projects x on the slide to x^4 on the screen.
(*a*) How much does it magnify the interval [1, 1.01]?
(*b*) What is its magnification at $x = 1$?

36 The left x centimeters of a string have a mass of x^3 grams.
(*a*) What is the average density of the interval [2, 2.01]?
(*b*) What is the average density of the interval [1.99, 2.01]?
(*c*) What is the density at $x = 2$?

37 (*a*) Show that the tangent line to the curve $y = x^3$ at (1, 1) passes through (2, 4).
(*b*) Use (*a*) to draw the tangent to the curve $y = x^3$ at (1, 1). (It is not necessary to draw the curve.)

38 Figure 6 is the graph of a function $y = f(x)$. Trace the curve on your paper and use a ruler to draw tangent lines. Estimate $f'(x)$ for (*a*) $x = 1$ (*b*) $x = 4$ (*c*) $x = 5$ (*d*) $x = 6$ (*e*) Sketch a graph of $y = f'(x)$.

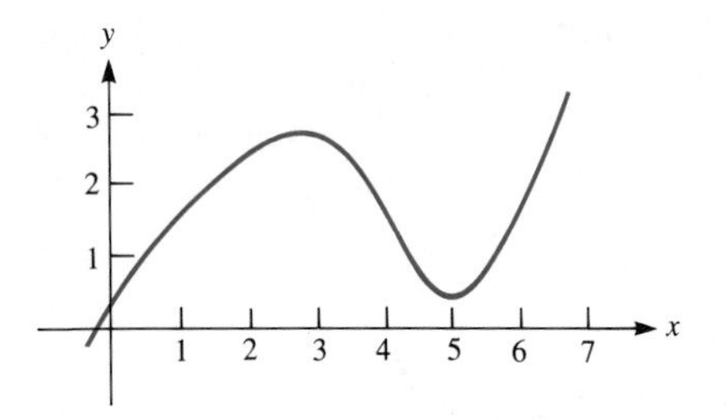

Figure 6

39 (See Exercise 38.) Estimate the derivative of the function whose graph is Fig. 7, at (*a*) 1 (*b*) 2 (*c*) 3 (*d*) 5 (*e*) 7 (*f*) 8 (*g*) Sketch a graph of $y = f'(x)$.

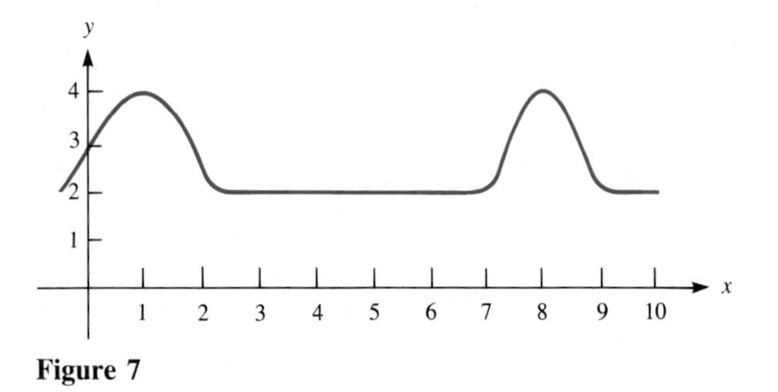

Figure 7

---

40 (*a*) Using the definition of the derivative, find the derivative of $x + x^2$.
(*b*) Graph $y = x + x^2$.
(*c*) For which x is the derivative in (*a*) equal to 0?
(*d*) Using (*c*), find at which point on the graph of $y = x + x^2$ the slope is 0.
(*e*) In view of (*d*), what do you think is the smallest possible value of $x + x^2$?

41 Say that you know the derivative of 3^x at 0 is approximately 1.0986. Show that you would then have an estimate of its derivative at any value of x. (*Hint:* What *must* be your first step?)

42 (*a*) What are the domains of $x^{3/5}$, $x^{3/4}$, and $x^{4/3}$? (Don't use a calculator. It might mislead you.) Explain with numerical illustrations.
(*b*) Let m and n be integers whose greatest common divisor is 1; that is, m/n is an irreducible fraction. Describe the domain of $x^{m/n}$. Justify your description.

43 (The trouble with x^a when x is negative and a is *not* rational) A number that is not the quotient of two integers is called *irrational*. The Greeks discovered around the year 450 B.C. that $\sqrt{2}$ is irrational. In the nineteenth century it was proved that π is irrational. Say we wanted to define $(-8)^{\sqrt{2}}$. We might pick a rational number r near $\sqrt{2}$ and compute $(-8)^r$. Then we would pick another rational number s nearer to $\sqrt{2}$, and compute $(-8)^s$. In this way we would try to find

$$\lim_{x \to \sqrt{2}} (-8)^x \qquad (x \text{ rational})$$

and use the limit as the definition of $(-8)^{\sqrt{2}}$.
(*a*) $\frac{7}{5}$ is near $\sqrt{2}$. What is $(-8)^{7/5}$? (*Hint:* How is it related to $8^{7/5}$?)
(*b*) 10/7 is also near $\sqrt{2}$. What is $(-8)^{10/7}$?
(*c*) Write an explanation, with more numerical examples, showing why $(-8)^{\sqrt{2}}$ is not defined and why the domain of x^a, when a is irrational, contains no negative numbers.

44 We asserted that for a positive integer n, $(x + h)^n = x^n + nx^{n-1}h + Q$, where the terms in Q all have h^2 as a factor. Use mathematical induction to prove this assertion. (Mathematical induction is discussed in Appendix E.)

3.3 THE DERIVATIVE AND CONTINUITY

After presenting other notations for the difference quotient

$$\frac{f(x + h) - f(x)}{h}$$

and the derivative $f'(x)$, this section shows the relation between "having a derivative" and "being continuous."

Same Ideas, New Notation

Δ is the capital Greek letter corresponding to the English D.

It is common to give the difference or change h the name Δx ("delta x"). The difference quotient then takes the form

$$\frac{f(x + \Delta x) - f(x)}{\Delta x},$$

Δx is not the product of Δ and x.

that is,

$$\frac{f(x+\Delta x)-f(x)}{\Delta x}=\frac{f(x+h)-f(x)}{h}=\frac{\text{Difference in outputs}}{\text{Difference in inputs}}.$$

See Fig. 1.

The derivative of f at x is then defined as

$$f'(x)=\lim_{\Delta x\to 0}\frac{f(x+\Delta x)-f(x)}{\Delta x}.$$

Figure 1

The difference in outputs is often named Δf ("delta f"),

$$\Delta f = f(x+\Delta x)-f(x),$$

and so

$$f(x+\Delta x)=f(x)+\Delta f.$$

This equation says that "the value of the function at $x+\Delta x$, namely $f(x+\Delta x)$, is equal to the value of the function at x plus the change in the function."

With Δx denoting the change in the inputs and Δf denoting the change in the outputs, we have

$$f'(x)=\lim_{\Delta x\to 0}\frac{\Delta f}{\Delta x}.$$

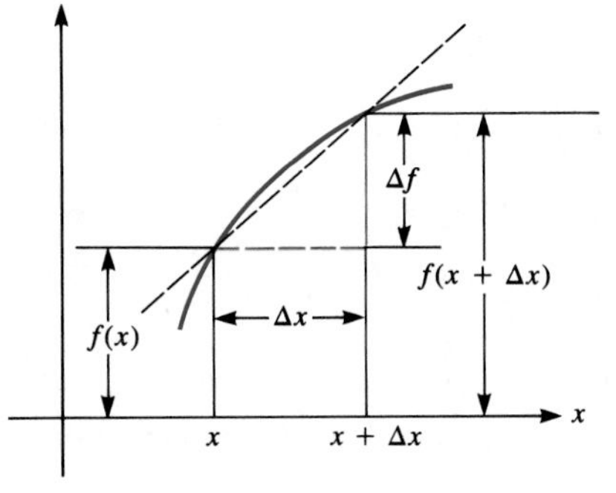

Figure 2

Figure 2 illustrates the Δ notation for the difference quotient.

Often the change in output, $f(x+\Delta x)-f(x)$, is denoted Δy, as in Fig. 3; then

$$f'(x)=\lim_{\Delta x\to 0}\frac{\Delta y}{\Delta x}.$$

EXAMPLE 1 Find $(x^2)'$ using the Δ notation.

SOLUTION By the definition of the derivative, the derivative of the squaring function at x is

$$\lim_{\Delta x\to 0}\frac{(x+\Delta x)^2-x^2}{\Delta x}.$$

Since $$(x+\Delta x)^2 = x^2+2x\cdot\Delta x+(\Delta x)^2,$$

the limit equals $$\lim_{\Delta x\to 0}\frac{x^2+2x\cdot\Delta x+(\Delta x)^2-x^2}{\Delta x}.$$

Then we have

$$\begin{aligned}(x^2)' &= \lim_{\Delta x\to 0}\frac{2x\,\Delta x+(\Delta x)^2}{\Delta x} && \text{algebra}\\ &= \lim_{\Delta x\to 0}\frac{\Delta x(2x+\Delta x)}{\Delta x} && \text{factoring}\\ &= \lim_{\Delta x\to 0}(2x+\Delta x) && \text{canceling } \Delta x\neq 0\\ &= 2x.\end{aligned}$$

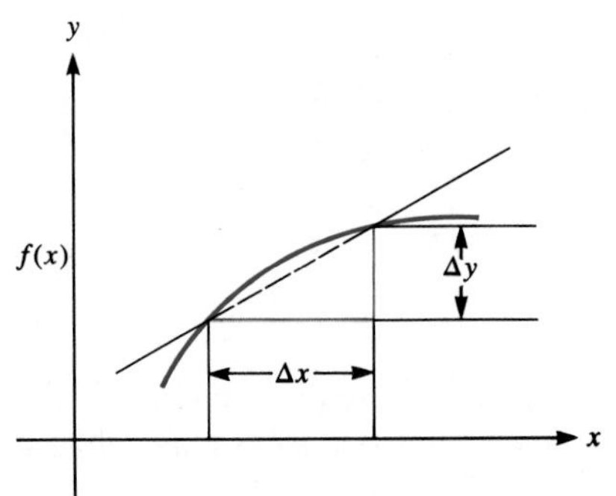

Figure 3

So the derivative of x^2 is $2x$, in agreement with the result in Example 2 of the preceding section. The only difference is that h has been replaced by Δx. ■

We have discussed various notations for the "change in inputs" and "change in outputs." Now we present some of the notations for the derivative.

Notations for the Derivative

The derivative $f'(x)$ is also commonly denoted

$$\frac{df}{dx} \quad \text{or} \quad D(f).$$

If $f(x)$ is denoted y, the derivative is also denoted

$\frac{dy}{dx}$ is the notation used by Leibniz.

$$\frac{dy}{dx} \quad \text{or} \quad D(y).$$

For instance, if $y = x^3$, we write

$$\frac{d(x^3)}{dx} = 3x^2, \qquad D(x^3) = 3x^2, \qquad \text{or} \qquad \frac{dy}{dx} = 3x^2.$$

If the formula for the function is long, it is customary to write

$$\frac{d}{dx}(f(x)) \quad \text{instead of} \quad \frac{d(f(x))}{dx}.$$

For instance, the notation

$$\frac{d}{dx}(x^3 - x^2 + 6x + \sqrt{x})$$

is less awkward than

$$\frac{d(x^3 - x^2 + 6x + \sqrt{x})}{dx}.$$

If the variable is denoted by some letter other than x, such as t (as in functions of time), we write, for instance,

$$\frac{d(t^3)}{dt} = 3t^2, \qquad \frac{d}{dt}(t^3) = 3t^2, \qquad \text{and} \qquad D(t^3) = 3t^2.$$

dy and dx are without individual meaning at this point.

Keep in mind that in the notations df/dx and dy/dx, the symbols df, dy, and dx have no meaning by themselves. The symbol dy/dx should be thought of as a single entity, just like the numeral 8, which we do not think of as formed of two 0's. Only later, in Sec. 4.9, will we give individual meaning to dy and dx.

The dot notation

In the study of motion, Newton's dot notation is often used. If x is a function of time t, then $\dot{x}$ denotes the derivative dx/dt.

Continuity and the Derivative

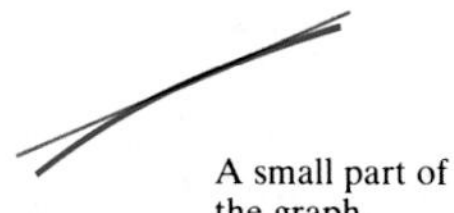

A small part of the graph resembles a line.

Figure 4

If f is differentiable at each number x in its domain, it is said to be *differentiable*.

A very small piece of the graph of a differentiable function looks almost like a straight line, namely, a tangent line, as shown in Fig. 4.

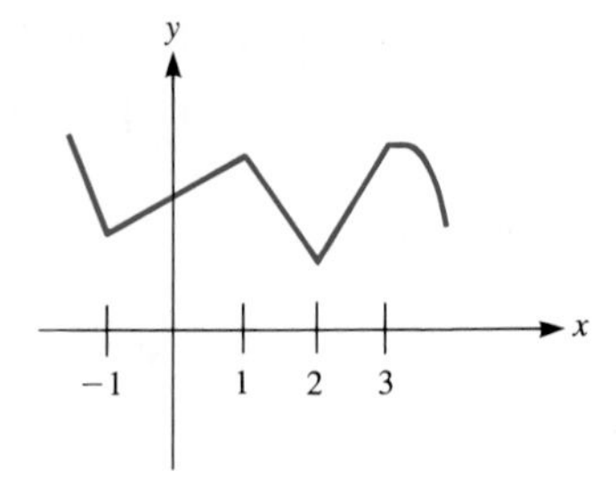

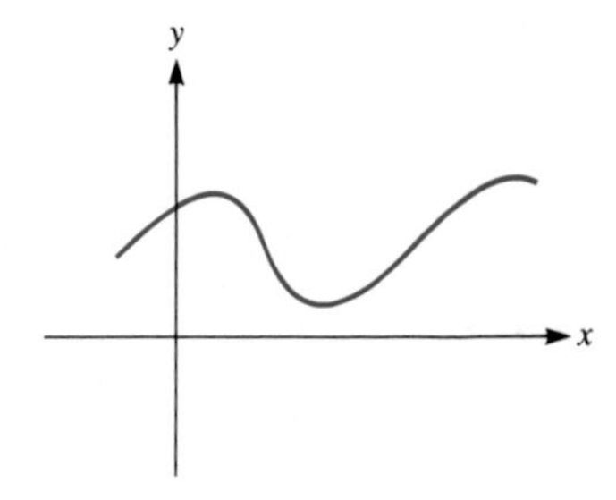

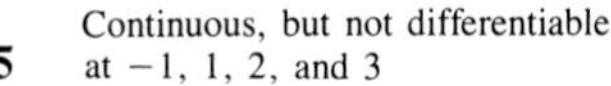

Figure 5 Continuous, but not differentiable at −1, 1, 2, and 3 | Differentiable (and continuous)

The graph of a continuous function *can* have sharp corners. (Look at the graph of $|x|$ in Fig. 7, page 126.) The graph of a differentiable function *does not* have sharp corners. It is smooth. Indeed, differential calculus can be thought of as the study of curves that "locally" resemble straight lines. Figure 5 contrasts a continuous function and a differentiable function.

It is reasonable to expect that a differentiable function is necessarily continuous. After all, locally, the graph of a differentiable function looks like a line—and a line doesn't have gaps.

To check that our expectation is correct, we must go back to the definitions of the derivative and of continuity.

Let a be an input where the function f is differentiable. That means

$$\lim_{\Delta x \to 0} \frac{\Delta f}{\Delta x}$$

exists, where $\Delta f = f(a + \Delta x) - f(a)$. To show f is continuous at a we have to show three things: (1) $f(a)$ is defined, (2) $\lim_{x \to a} f(x)$ exists, and (3) this limit equals $f(a)$.

We already know that $f(a)$ is defined. To show that $\lim_{x \to a} f(x)$ exists and equals $f(a)$, we show that

$$\lim_{x \to a} [f(x) - f(a)] = 0. \tag{1}$$

For if (1) holds, we could write

$$\begin{aligned} \lim_{x \to a} f(x) &= \lim_{x \to a} ([f(x) - f(a)] + f(a)) \\ &= \lim_{x \to a} [f(x) - f(a)] + \lim_{x \to a} f(a) \\ &= 0 + f(a) \\ &= f(a). \end{aligned}$$

Then all the conditions for continuity would hold. So, all that remains is to establish (1). This is done in the proof of Theorem 1.

Theorem 1 If f is differentiable at a, then it is continuous at a.

Proof By the preceding remarks, it suffices to prove that

$$\lim_{x \to a} [f(x) - f(a)] = 0,$$

or, in the Δ notation,

$$\lim_{\Delta x \to 0} \Delta f = 0.$$

We know that

$$\lim_{\Delta x \to 0} \frac{\Delta f}{\Delta x} = f'(a). \tag{2}$$

To make use of (2), we write

$$\Delta f = \frac{\Delta f}{\Delta x} \Delta x.$$

Thus

$$\begin{aligned} \lim_{\Delta x \to 0} \Delta f &= \lim_{\Delta x \to 0} \frac{\Delta f}{\Delta x} \Delta x \\ &= \lim_{\Delta x \to 0} \frac{\Delta f}{\Delta x} \cdot \lim_{\Delta x \to 0} \Delta x \\ &= f'(a) \cdot 0 \\ &= 0. \end{aligned}$$

This proves that f is continuous at a. ■

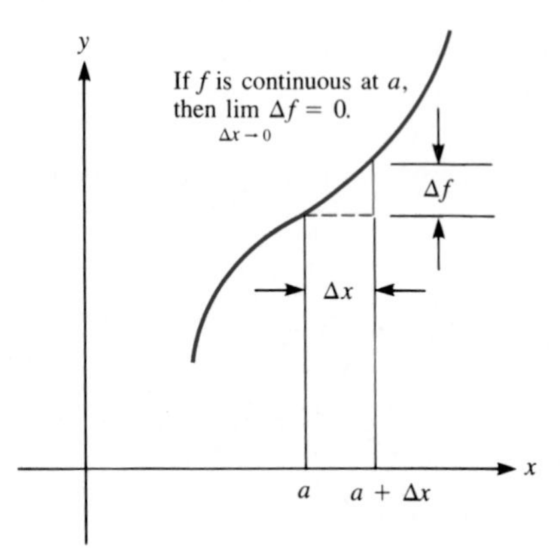

Figure 6

Remark The idea of the proof is this. If $\Delta f/\Delta x$ is approximately $f'(a)$, then Δf is approximately $f'(a)$ times Δx. Thus, if Δx is small, Δf must also be small. The trick of dividing and multiplying by Δx justifies the claim "Δf must also be small."

Remark Note that if $f(a)$ is defined and if f is continuous at a, then Δf approaches 0 as $\Delta x \to 0$. Moreover, if $\Delta f \to 0$ as $\Delta x \to 0$, then f is continuous at a. (See Fig. 6.)

The converse of Theorem 1 is not true. The function $|x|$ is continuous at 0, but, as the graph in Fig. 7 suggests, it is not differentiable at 0. This is demonstrated in Example 2.

EXAMPLE 2 Show that the absolute-value function $f(x) = |x|$ is not differentiable at 0.

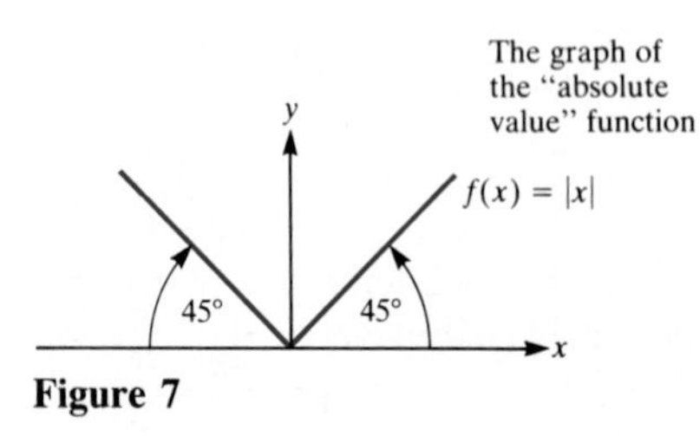

Figure 7

SOLUTION Since $\lim_{x \to 0} |x| = 0 = f(0)$, the function $f(x) = |x|$ is continuous at $x = 0$. To show that f is not differentiable at $x = 0$, we must show that

$$\lim_{\Delta x \to 0} \frac{f(0 + \Delta x) - f(0)}{\Delta x}$$

does not exist. Since f is the absolute-value function, we must show that

$$\lim_{\Delta x \to 0} \frac{|0 + \Delta x| - |0|}{\Delta x}$$

does not exist. Since $|0| = 0$ and $|0 + \Delta x| = |\Delta x|$, we must examine

$$\lim_{\Delta x \to 0} \frac{|\Delta x|}{\Delta x}.$$

For this limit to exist both the right-hand limit, $\lim_{\Delta x \to 0^+} |\Delta x|/\Delta x$, and the left-hand limit, $\lim_{\Delta x \to 0^-} |\Delta x|/\Delta x$, must exist, and these two limits must be equal to each other.

As $\Delta x \to 0$ from the right, Δx is positive. Thus

$$\lim_{\Delta x \to 0^+} \frac{|\Delta x|}{\Delta x} = \lim_{\Delta x \to 0^+} \frac{\Delta x}{\Delta x} = \lim_{\Delta x \to 0^+} 1 = 1.$$

On the other hand, as $\Delta x \to 0$ from the left, Δx is negative, and $|\Delta x| = -\Delta x$. We therefore have

$$\lim_{\Delta x \to 0} \frac{|\Delta x|}{\Delta x} = \lim_{\Delta x \to 0} \frac{-\Delta x}{\Delta x} = \lim_{\Delta x \to 0} (-1) = -1.$$

Since $\lim_{\Delta x \to 0^+} |\Delta x|/\Delta x = 1$ and $\lim_{\Delta x \to 0^-} |\Delta x|/\Delta x = -1$, these two limits are not equal. Hence

$$\lim_{\Delta x \to 0} \frac{f(0 + \Delta x) - f(0)}{\Delta x}$$

does not exist. That tells us that the absolute-value function is not differentiable at 0. ■

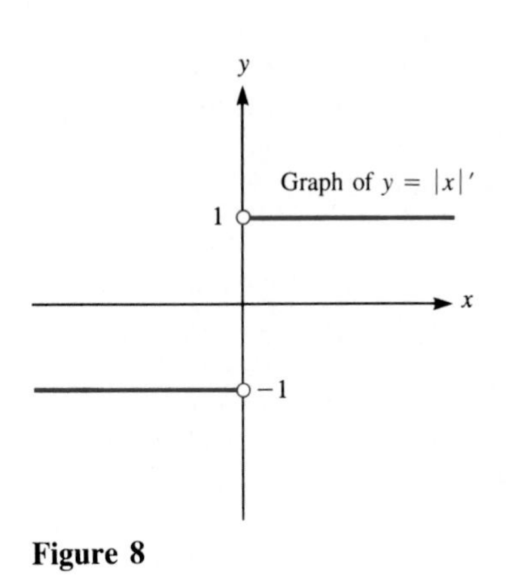

Figure 8

However, the absolute-value function, $f(x) = |x|$, is differentiable everywhere else. Its graph is made up of two rays, one with slope 1 and one with slope -1. So $f'(x) = 1$ for positive x and $f'(x) = -1$ for negative x. For good measure we graph $f'(x)$, the derivative of the absolute-value function, in Fig. 8. (It records the slope of the graph of $|x|$ in Fig. 7.)

Antiderivatives

As early as Chaps. 4 and 5 we will need to "reverse" the process of differentiation, when we study motion and area. Given a function f, we will seek a function F whose derivative is f. We will call F an *antiderivative* of f.

For instance, since $(x^3)' = 3x^2$, we say that "x^3 is an antiderivative of $3x^2$." As another example, since $(1/x)' = -1/x^2$, "$1/x$ is an antiderivative of $-1/x^2$." The next section will show how to find antiderivatives of polynomials. Chapter 7 presents techniques for finding antiderivatives for many other functions.

Section Summary

This section presented various notations for the quotient

$$\frac{f(x+h)-f(x)}{h},$$

such as

$$\frac{f(x+\Delta x)-f(x)}{\Delta x}, \quad \frac{\Delta f}{\Delta x}, \quad \text{and} \quad \frac{\Delta y}{\Delta x}.$$

It also introduced other notations for the derivatives, such as $D(f)$, df/dx, dy/dx, and $\dot{x}$.

BROWNIAN MOTION

Differentiable functions have smooth graphs and are easy to work with, so they serve as convenient models for many natural processes, like the motion of falling bodies and the density of solid objects. Nevertheless, nondifferentiable functions may also occur in nature. A function fails to have a derivative wherever its graph is so jagged that there is no clear direction. Such angular motion occurs at the molecular level of fluids, where it is known as *Brownian motion*. The mathematician Norbert Wiener studied Brownian motion and made the following observations, excerpted from his autobiography *I am a Mathematician* (The M.I.T. Press, 1956):

> To understand the Brownian motion, let us imagine a pushball in a field in which a crowd is milling around. Various people in the crowd will run into the pushball and will move it about. Some will push in one direction and some in another, and the balance of pushes is likely to be tolerably even. Nevertheless, notwithstanding these balanced pushes, the fact remains that they are pushes by individual people and that their balance will be only approximate. Thus, in the course of time, the ball will wander about the field like [a] drunken man . . . and have a certain irregular motion in which what happens in the future will have very little to do with what has happened in the past.
>
> Now consider the molecules of a fluid, whether gas or liquid. These molecules will not be at rest but will have a random irregular motion like that of the people in the crowd. This motion will become more active as the temperature goes up. Let us suppose that we have introduced into this fluid a small sphere which can be pushed about by the molecules in much the way that the pushball is agitated by the crowd. If this sphere is extremely small we cannot see it, and if it is extremely large and suspended in fluid, the collisions of the particles of the fluid with the sphere will average out sufficiently well so that no motion is observable. There is an intermediate range in which the sphere is large enough to be visible under the microscope in a constant irregular motion. This agitation, which indicates the irregular movement of the molecules, is known as the Brownian motion.
>
> [T]he French physicist Perrin in his book *Les Atomes* . . . said in effect that the very irregular curves followed by particles in the Brownian motion led one to think of the supposed continuous nondifferentiable curves of the mathematicians. He called the motion continuous because the particles never jump over a gap and nondifferentiable because at no time do they seem to have a well-defined direction of movement.

It defined a *differentiable function*, and showed that such a function is necessarily continuous. The converse is not true: For instance, the absolute-value function is continuous, but it is *not* differentiable at 0.

If f is continuous at a, then $\Delta f = f(a + \Delta x) - f(a)$ approaches 0 as $\Delta x \to 0$ (that is, $\lim_{\Delta x \to 0} \Delta f = 0$).

The section concluded with the definition of an antiderivative of a function.

EXERCISES FOR SEC. 3.3: THE DERIVATIVE AND CONTINUITY; ANTIDERIVATIVES

In Exercises 1 to 10 use the Δ notation to find the given derivatives.

1 $\dfrac{d(x^3)}{dx}$ **2** $\dfrac{d(x^4)}{dx}$

3 $\dfrac{d(\sqrt{x})}{dx}$ **4** $\dfrac{d(3\sqrt{x})}{dx}$

5 $\dfrac{d(5x^2)}{dx}$ **6** $\dfrac{d}{dx}(5x^2 + 3x + 2)$

7 $D\left(\dfrac{3}{x}\right)$ **8** $D\left(\dfrac{5}{x^2}\right)$

9 $\dfrac{d}{dx}\left(\dfrac{3}{x} - 4x + 2\right)$ **10** $\dfrac{d}{dx}(x^3 - 5x + 1992)$

11 Let $f(x) = x^2$. Find Δf if (*a*) $x = 1$ and $\Delta x = 0.1$, (*b*) $x = 3$ and $\Delta x = -0.1$.

12 Let $f(x) = 1/x$. Find Δf if (*a*) $x = 2$ and $\Delta x = \frac{1}{5}$, (*b*) $x = 2$ and $\Delta x = -\frac{1}{8}$.

13 For a certain function $y = f(x)$, $f(2) = 3$ and $f(2.1) = 3.05$.
(*a*) If $x = 2$ and $\Delta x = 0.1$, find Δy.
(*b*) Draw the two points $(x, f(x))$ and $(x + \Delta x, f(x + \Delta x))$ and show Δx and Δy on the diagram.

14 For a certain function, $y = f(x)$, $f(3) = 4$ and $f(2.8) = 4.3$.
(*a*) If $x = 3$ and $\Delta x = -0.2$, find Δy.
(*b*) Draw the two points $(x, f(x))$ and $(x + \Delta x, f(x + \Delta x))$ and show Δx and Δy on the diagram.

In Exercises 15 to 18 answer these questions for the hypothetical functions given by the graphs.
(*a*) For which numbers a does $\lim_{x \to a} f(x)$ exist, but f is not continuous at a? Explain.
(*b*) For which numbers a is f continuous at a, but not differentiable at a? Explain.

15

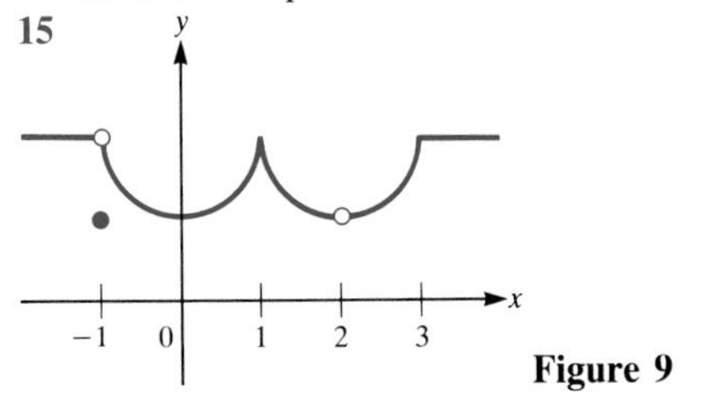

Figure 9

16

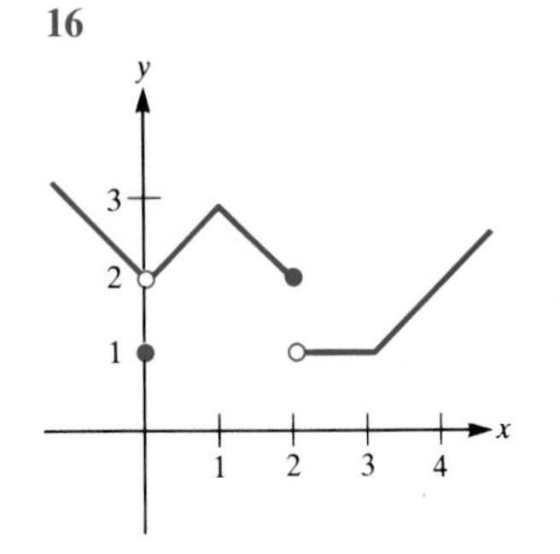

Figure 10

17

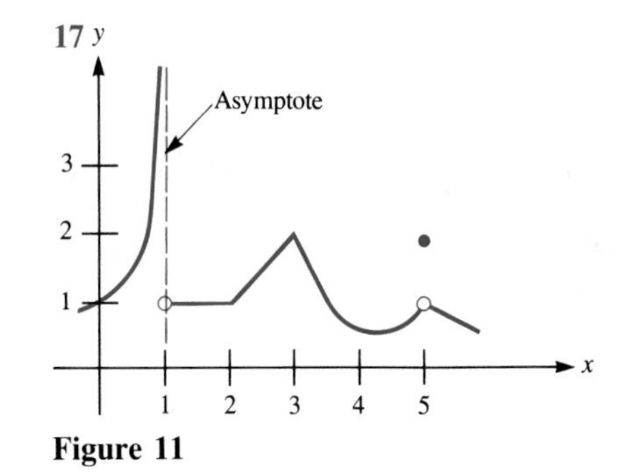

Figure 11

18

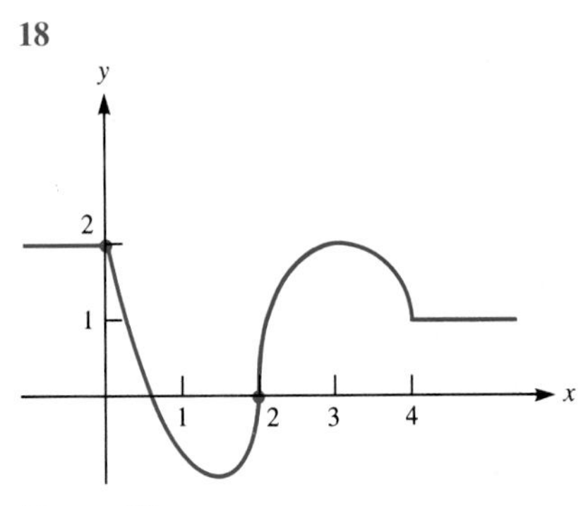

Figure 12

$c' = 0$ no matter how "fancy" the constant c may be.

for all x. Since f is constant, $f'(x) = 0$. Thus $(\pi^3)' = 0$. ■

Functions can be built up from simpler functions by addition, subtraction, multiplication, and division, as the function in the opening paragraph illustrates. In order to develop the differentiation formulas that will make our life simple, we need the following definitions.

Definition *Sum, difference, product, and quotient of functions.* Let f and g be two functions. The functions $f + g$, $f - g$, fg, and f/g are defined as follows:

$(f + g)(x) = f(x) + g(x)$ for x in the domains of both f and g

$(f - g)(x) = f(x) - g(x)$ for x in the domains of both f and g

$(fg)(x) = f(x)g(x)$ for x in the domains of both f and g

$\left(\frac{f}{g}\right)(x) = \frac{f(x)}{g(x)}.$ for x in the domains of both f and g, $g(x) \neq 0$

Derivatives of $f + g$ and $f - g$

The next theorem asserts that if the functions f and g have derivatives at a certain number x, then so does their sum $f + g$, and

$$\frac{d}{dx}(f + g) = \frac{df}{dx} + \frac{dg}{dx}.$$

In other words, "the derivative of the sum is the sum of the derivatives." A similar formula holds for the derivative of $f - g$.

Theorem 2 If f and g are differentiable functions, then so is $f + g$. Its derivative is given by the formula

$$\boxed{(f + g)' = f' + g'.}$$

Similarly,

$$\boxed{(f - g)' = f' - g'.}$$

Proof To prove this theorem we must go back to the definition of the derivative.

To begin, we give the function $f + g$ the name u, that is, $u(x) = f(x) + g(x)$. We have to examine

$$\lim_{\Delta x \to 0} \frac{u(x + \Delta x) - u(x)}{\Delta x}$$

or, equivalently,

$$\lim_{\Delta x \to 0} \frac{\Delta u}{\Delta x}. \tag{1}$$

In order to evaluate (1), we will express Δu in terms of Δf and Δg. Here are the details:

$$\begin{aligned}
\Delta u &= u(x + \Delta x) - u(x) \\
&= [f(x + \Delta x) + g(x + \Delta x)] - [f(x) + g(x)] && \text{definition of } u \\
&= [f(x + \Delta x) - f(x)] + [g(x + \Delta x) - g(x)] && \text{algebra} \\
&= \Delta f + \Delta g. && \text{definition of } \Delta f \text{ and } \Delta g
\end{aligned}$$

The change in u is the change in f plus the change in g.

All told, $\Delta u = \Delta f + \Delta g$.

The hard work is over. We can now evaluate (1):

$$\begin{aligned}
\lim_{\Delta x \to 0} \frac{\Delta u}{\Delta x} &= \lim_{\Delta x \to 0} \frac{\Delta f + \Delta g}{\Delta x} \\
&= \lim_{\Delta x \to 0} \frac{\Delta f}{\Delta x} + \lim_{\Delta x \to 0} \frac{\Delta g}{\Delta x} \\
&= f'(x) + g'(x).
\end{aligned}$$

Thus $u = f + g$ is differentiable and

$$u'(x) = f'(x) + g'(x).$$

A similar argument applies to $f - g$. ■

Theorem 2 extends to any finite number of differentiable functions. For example,

$$(f + g + h)' = f' + g' + h'$$

and

$$(f - g + h)' = f' - g' + h'.$$

EXAMPLE 2 Using Theorem 2, differentiate $x^2 + x^3$.

SOLUTION

$$\begin{aligned}
\frac{d}{dx}(x^2 + x^3) &= \frac{d}{dx}(x^2) + \frac{d}{dx}(x^3) && \text{Theorem 2} \\
&= 2x + 3x^2 && D(x^n) = nx^{n-1}.
\end{aligned}$$

■

EXAMPLE 3 Differentiate $x^4 - \sqrt{x} + 5$.

SOLUTION

$$\begin{aligned}
\frac{d}{dx}(x^4 - \sqrt{x} + 5) &= \frac{d}{dx}(x^4) - \frac{d}{dx}(\sqrt{x}) + \frac{d}{dx}(5) && \text{Theorem 2} \\
&= 4x^3 - \frac{1}{2\sqrt{x}} + 0 = 4x^3 - \frac{1}{2\sqrt{x}}.
\end{aligned}$$

■

The Derivative of fg

The following theorem concerning the derivative of the product of two functions may be surprising, for it turns out that the derivative of the product is *not* the product of the derivatives. The formula is more complicated than that for the derivative of the sum. (It asserts that the derivative of the product is "the first function times the derivative of the second plus the second function times the derivative of the first.")

Theorem 3 If f and g are differentiable functions, then so is fg. Its derivative is given by the formula

The product rule

$$\boxed{(fg)' = fg' + gf'.}$$

Proof The proof is similar to that for Theorem 2. This time we give the *product* fg the name u. Then we express Δu in terms of Δf and Δg. Finally, we determine $u'(x)$ by examining $\lim_{\Delta x \to 0} \Delta u/\Delta x$.

Call the function fg simply u. That is,

$$u(x) = f(x)g(x).$$

Then
$$u(x + \Delta x) = f(x + \Delta x)g(x + \Delta x).$$

Rather than subtract directly, first write

$$f(x + \Delta x) = f(x) + \Delta f \quad \text{and} \quad g(x + \Delta x) = g(x) + \Delta g.$$

Then
$$\begin{aligned} u(x + \Delta x) &= [f(x) + \Delta f][g(x) + \Delta g] \\ &= f(x)g(x) + f(x)\,\Delta g + g(x)\,\Delta f + \Delta f\,\Delta g. \end{aligned}$$

Hence
$$\begin{aligned} \Delta u &= u(x + \Delta x) - u(x) \\ &= f(x)g(x) + f(x)\,\Delta g + g(x)\,\Delta f + \Delta f\,\Delta g - f(x)g(x) \\ &= f(x)\,\Delta g + g(x)\,\Delta f + \Delta f\,\Delta g \end{aligned}$$

and
$$\frac{\Delta u}{\Delta x} = f(x)\,\frac{\Delta g}{\Delta x} + g(x)\,\frac{\Delta f}{\Delta x} + \Delta f\,\frac{\Delta g}{\Delta x}.$$

As $\Delta x \to 0$, $\Delta g/\Delta x \to g'(x)$ and $\Delta f/\Delta x \to f'(x)$. Furthermore, because f is differentiable (hence continuous), $\Delta f \to 0$ as remarked in the preceding section. It follows that

$$\lim_{\Delta x \to 0} \frac{\Delta u}{\Delta x} = f(x)g'(x) + g(x)f'(x) + 0 \cdot g'(x).$$

The formula for $(fg)'$ was discovered by Leibniz in 1676. (His first guess was wrong.)

Therefore, u is differentiable and

$$u' = fg' + gf'. \quad \blacksquare$$

Remark Figure 2 provides a picture to illustrate Theorem 3 and its proof. With f, Δf, g, and Δg taken to be positive, the inner rectangle has area $u = fg$ and the whole rectangle has area $u + \Delta u = (f + \Delta f)(g + \Delta g)$. The shaded region whose area is Δu is made up of rectangles of areas $f\Delta g$, $g\Delta f$, and $\Delta f\Delta g$. The little corner rectangle, of area $\Delta f\Delta g$, is negligible in comparison with the other two rectangles. Thus $\Delta u \approx f\Delta g + g\Delta f$, which suggests the form of the formula for the derivative of the product.

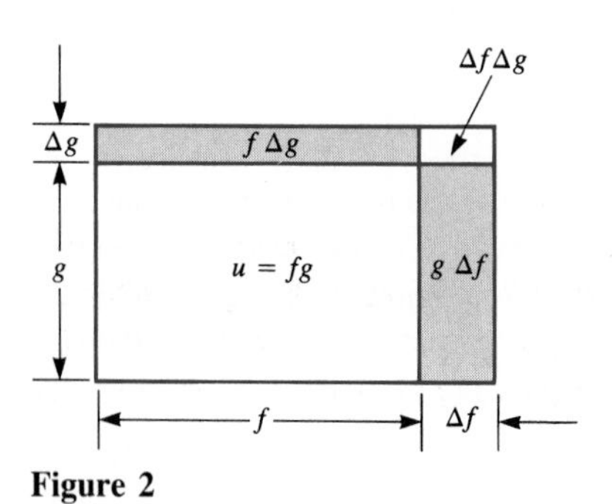

Figure 2

EXAMPLE 4 Find $\dfrac{d}{dx}\,[(x^2 + x^3)(x^4 - \sqrt{x} + 5)]$.

SOLUTION (Note that in Examples 2 and 3 the derivatives of both factors,

$x^2 + x^3$ and $x^4 - \sqrt{x} + 5$, were found.)

By Theorem 3,

$$\frac{d}{dx}[(x^2 + x^3)(x^4 - \sqrt{x} + 5)]$$

$$= (x^2 + x^3)\frac{d}{dx}(x^4 - \sqrt{x} + 5) + (x^4 - \sqrt{x} + 5)\frac{d}{dx}(x^2 + x^3)$$

$$= (x^2 + x^3)\left(4x^3 - \frac{1}{2\sqrt{x}}\right) + (x^4 - \sqrt{x} + 5)(2x + 3x^2). \quad \blacksquare$$

A special case of the formula $(fg)' = fg' + gf'$ occurs so frequently that it is singled out in Theorem 4.

Derivative of Constant Times f

Theorem 4 If c is a constant function and f is a differentiable function, then cf is differentiable and its derivative is given by the formula

The derivative of cf

$$\boxed{(cf)' = cf'.}$$

Proof We are dealing with a product of two functions, c and f. Therefore we may use Theorem 3, which tells us how to differentiate a product of any two differentiable functions. We have

$$\begin{aligned} (cf)' &= cf' + fc' && \text{derivative of a product} \\ &= cf' + f \cdot 0 && \text{derivative of constant is 0} \\ &= cf'. \quad \blacksquare \end{aligned}$$

In other notations for the derivative, Theorem 4 is expressed as

$$\frac{d(cf)}{dx} = c\frac{df}{dx} \quad \text{and} \quad D(cf) = cD(f).$$

A constant factor can go past the derivative symbol.

Theorem 4 asserts that "it is legal to move a constant factor outside the derivative symbol."

EXAMPLE 5 Find $D(6x^3)$.

SOLUTION

$$\begin{aligned} D(6x^3) &= 6D(x^3) && \text{6 is constant} \\ &= 6 \cdot 3x^2 && D(x^n) = nx^{n-1} \\ &= 18x^2. \quad \blacksquare \end{aligned}$$

Many students, when they differentiate $6x^3$, say to themselves, "This is the product of 6 and x^3, so I'll use the formula for differentiating a product."

$$\begin{aligned} (6x^3)' &= 6(x^3)' + 6'(x^3) \\ &= 6 \cdot 3x^2 + 0 \cdot x^3 \\ &= 18x^2. \end{aligned}$$

The result is, of course, correct. However, the student could have obtained it much more easily by using the formula $(cf)' = cf'$. The calculations would then be simply

$$(6x^3)' = 6(x^3)' = 6 \cdot 3x^2 = 18x^2.$$

In fact, the student, with a little practice, might simply write down

$$(6x^3)' = 18x^2.$$

EXAMPLE 6 Find $D(x^5/11)$.

SOLUTION

$$D\left(\frac{x^5}{11}\right) = D(\tfrac{1}{11}x^5) = \tfrac{1}{11}D(x^5)$$
$$= \tfrac{1}{11}(5x^4) = \tfrac{5}{11}x^4. \quad \blacksquare$$

Example 6 generalizes to the fact that for a nonzero c,

$\dfrac{d}{dx}\left(\dfrac{f}{c}\right) = \dfrac{1}{c}\dfrac{df}{dx}$

$$\boxed{\left(\frac{f}{c}\right)' = \frac{f'}{c}.}$$

The formula for the derivative of the product extends to the product of several differentiable functions. For instance,

$$\boxed{(fgh)' = f'gh + fg'h + fgh'.}$$

In each summand only one derivative appears. (See Exercise 55.) The next example illustrates the use of this formula.

EXAMPLE 7 Differentiate $\sqrt{x}\,(x^2 - 2)(2x^3 + 1)$.

SOLUTION By the preceding remark,

$$[\sqrt{x}(x^2 - 2)(2x^3 + 1)]'$$
$$= (\sqrt{x})'(x^2 - 2)(2x^3 + 1) + \sqrt{x}(x^2 - 2)'(2x^3 + 1) + \sqrt{x}(x^2 - 2)(2x^3 + 1)'$$
$$= \frac{1}{2\sqrt{x}}(x^2 - 2)(2x^3 + 1) + \sqrt{x}(2x)(2x^3 + 1) + \sqrt{x}(x^2 - 2)(6x^2). \quad \blacksquare$$

Any polynomial can be differentiated by the methods already developed, as Example 8 illustrates.

EXAMPLE 8 Differentiate $6x^8 - x^3 + 5x^2 + \pi^3$.

Differentiate a polynomial "term by term."

SOLUTION

$$(6x^8 - x^3 + 5x^2 + \pi^3)' = (6x^8)' - (x^3)' + (5x^2)' + (\pi^3)'$$
$$= 6 \cdot 8x^7 - 3x^2 + 5 \cdot 2x + 0$$
$$= 48x^7 - 3x^2 + 10x. \quad \blacksquare$$

Derivative of f/g

Next we show that if the functions f and g are differentiable at a number x, and if $g(x) \neq 0$, then f/g is differentiable at x. The formula for $(f/g)'$ is a bit messy; a suggestion for remembering it is given after the proof.

Theorem 5 If f and g are differentiable functions, then so is f/g and

The quotient rule

$$\boxed{\left(\frac{f}{g}\right)' = \frac{gf' - fg'}{g^2}} \qquad \text{where } g(x) \text{ is not } 0.$$

Proof The argument is similar to the proofs concerning $(f + g)'$ and $(fg)'$.
We give the function f/g the name u. Thus

$$u(x) = \frac{f(x)}{g(x)}$$

and

$$u(x + \Delta x) = \frac{f(x + \Delta x)}{g(x + \Delta x)}.$$

[Since we consider only values of x such that $g(x) \neq 0$ and g is continuous, $g(x + \Delta x) \neq 0$ whenever Δx is sufficiently small.]

Then we express Δu in terms of Δf and Δg. Finally we determine

$$u'(x) = \lim_{\Delta x \to 0} \frac{\Delta u}{\Delta x}.$$

Before computing $\lim_{\Delta x \to 0} (\Delta u/\Delta x)$, express the numerator Δu as simply as possible in terms of $f(x)$, Δf, $g(x)$, and Δg:

$$\Delta u = u(x + \Delta x) - u(x) \qquad \text{by definition of } \Delta u$$

$$= \frac{f(x + \Delta x)}{g(x + \Delta x)} - \frac{f(x)}{g(x)} \qquad \text{by definition of the function } u$$

$$= \frac{f(x) + \Delta f}{g(x) + \Delta g} - \frac{f(x)}{g(x)} \qquad \text{since } f(x + \Delta x) = f(x) + \Delta f \text{ and } g(x + \Delta x) = g(x) + \Delta g$$

$$= \frac{g(x)[f(x) + \Delta f] - f(x)[g(x) + \Delta g]}{[g(x) + \Delta g]g(x)} \qquad \text{placing over common denominator}$$

$$= \frac{g(x)f(x) + g(x)\,\Delta f - f(x)g(x) - f(x)\,\Delta g}{[g(x) + \Delta g]g(x)} \qquad \text{multiplying out}$$

$$= \frac{g(x)\,\Delta f - f(x)\,\Delta g}{[g(x) + \Delta g]g(x)}. \qquad \text{simplifying}$$

After this simplification, $u'(x)$ can be found as follows:

$$u'(x) = \lim_{\Delta x \to 0} \frac{\Delta u}{\Delta x} = \lim_{\Delta x \to 0} \frac{\left[\dfrac{g(x)\,\Delta f - f(x)\,\Delta g}{[g(x) + \Delta g]g(x)}\right]}{\Delta x}$$

$$= \lim_{\Delta x \to 0} \frac{g(x)\,\Delta f - f(x)\,\Delta g}{[g(x) + \Delta g]g(x)\,\Delta x} \qquad \text{algebra: } \frac{\left(\frac{a}{b}\right)}{c} = \frac{a}{bc}$$

$$= \lim_{\Delta x \to 0} \frac{g(x)\dfrac{\Delta f}{\Delta x} - f(x)\dfrac{\Delta g}{\Delta x}}{[g(x) + \Delta g]g(x)} \qquad \text{algebra: } \frac{ab - cd}{ef} = \frac{a\dfrac{b}{f} - c\dfrac{d}{f}}{e}$$

$$= \frac{g(x)f'(x) - f(x)g'(x)}{g(x)g(x)}. \qquad \text{since } \frac{\Delta f}{\Delta x} \to f'(x),\ \frac{\Delta g}{\Delta x} \to g'(x), \text{ and } \Delta g \to 0 \text{ as } \Delta x \to 0$$

This establishes the formula for $(f/g)'$. ■

A memory device for $(f/g)'$: When using the formula for $(f/g)'$, first write down the parts where g^2 and g appear:

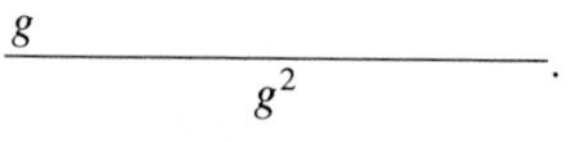

$$\frac{g \qquad\qquad\qquad\qquad}{g^2}.$$

In this way you will get the denominator correct and have a good start on the numerator. You may then go on to complete the numerator, *remembering that it has a minus sign:*

$$\left(\frac{f}{g}\right)' = \frac{gf' - fg'}{g^2}.$$

EXAMPLE 9 Compute $[x^2/(x^3 + 1)]'$, showing each step in detail.

SOLUTION

Step 1 $\left(\dfrac{x^2}{x^3 + 1}\right)' = \dfrac{(x^3 + 1)\cdots}{(x^3 + 1)^2}$ — write denominator and start numerator

Step 2 $= \dfrac{(x^3 + 1)(x^2)' - (x^2)(x^3 + 1)'}{(x^3 + 1)^2}$ — complete numerator, remembering minus sign

Step 3 $= \dfrac{(x^3 + 1)(2x) - x^2(3x^2)}{(x^3 + 1)^2}$ — compute derivatives

Step 4 $= \dfrac{2x^4 + 2x - 3x^4}{(x^3 + 1)^2}$ — algebra

Step 5 $= \dfrac{2x - x^4}{(x^3 + 1)^2}.$ — algebra: collecting ■

As Example 9 illustrates, the techniques for differentiating polynomials and quotients suffice to differentiate any rational function, that is, any quotient of polynomials.

The next example uses the formulas for the derivatives of the product and the quotient.

EXAMPLE 10 Differentiate $x\sqrt{x + 1}/(x^2 + 1)$.

SOLUTION

$$\left[\frac{x\sqrt{x+1}}{x^2+1}\right]' = \frac{(x^2+1)[x\sqrt{x+1}]' - [x\sqrt{x+1}](x^2+1)'}{(x^2+1)^2} \qquad D\left(\frac{f}{g}\right)$$

$$= \frac{(x^2+1)[x(\sqrt{x+1})' + (\sqrt{x+1})(x)'] - [x\sqrt{x+1}](2x)}{} \qquad D(fg)$$

$$= \frac{(x^2+1)\left[x\dfrac{1}{2\sqrt{x+1}} + (\sqrt{x+1})(1)\right] - 2x^2\sqrt{x+1}}{(x^2+1)^2}$$

$$= \frac{(x^2+1)[x/2 + (x+1)] - 2x^2(x+1)}{(x^2+1)^{5/2}}$$

$$= \frac{(x^2+1)(3x/2+1) - 2x^2(x+1)}{(x^2+1)^{5/2}}$$

$$= \frac{-x^3 - 2x^2 + 3x + 2}{2(x^2+1)^{5/2}}. \quad \blacksquare$$

The following theorem is just a special case of the formula for $(f/g)'$. Since it is needed often, it is worth memorizing.

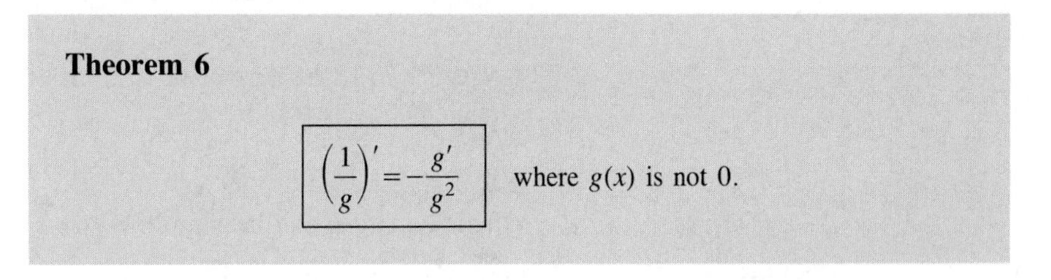

Theorem 6

The derivative of $1/g$

$$\left(\frac{1}{g}\right)' = -\frac{g'}{g^2}$$ where $g(x)$ is not 0.

Proof

$$\left(\frac{1}{g}\right)' = \frac{g\cdot(1)' - 1\cdot g'}{g^2} \qquad \text{derivative of a quotient}$$

$$= \frac{g\cdot 0 - g'}{g^2} \qquad \text{derivative of a constant}$$

$$= -\frac{g'}{g^2}. \quad \blacksquare$$

EXAMPLE 11 Find $\left(\dfrac{1}{2x^3+x+5}\right)'$.

SOLUTION By the formula for $(1/g)'$,

$$\left(\frac{1}{2x^3+x+5}\right)' = \frac{-(2x^3+x+5)'}{(2x^3+x+5)^2}$$

$$= \frac{-(6x^2+1)}{(2x^3+x+5)^2}. \quad \blacksquare$$

You could work Example 11 with the aid of the formula for the derivative of a quotient, f/g with $f = 1$, but the calculations would be longer. One of the goals of calculus is to make our lives easier, not harder.

In the next example we obtain an equation for a tangent line. (Appendix B describes various ways of obtaining an equation of a line.) We will need the "point-slope" case.

EXAMPLE 12 Find an equation of the line tangent to the curve $y = x^3/3 + x^2 - 2\sqrt{x}$ at the point $(1, -\frac{2}{3})$.

SOLUTION We already know a point on the tangent line, namely, $(1, -\frac{2}{3})$. If we find the slope of the tangent line, we will be able to use the point-slope formula to write down its equation. If the slope is m the equation is

$$y - (-\tfrac{2}{3}) = m(x - 1).$$

All that remains is to find the slope. That is where the derivative comes in handy.

If

$$y = \frac{x^3}{3} + x^2 - 2\sqrt{x},$$

then

$$\frac{dy}{dx} = \frac{3x^2}{3} + 2x - 2\frac{1}{2\sqrt{x}} = x^2 + 2x - \frac{1}{\sqrt{x}}.$$

When $x = 1$, the derivative is 2. By the point-slope formula, an equation of the tangent line is

$$y - (-\tfrac{2}{3}) = 2(x - 1).$$

This equation can be simplified to $y = 2x - \frac{8}{3}$. ■

An Antiderivative of x^a

Now that we have more formulas for differentiation we can find more antiderivatives, as the next example illustrates.

EXAMPLE 13 Find an antiderivative of x^6.

SOLUTION When we differentiate x^a we get ax^{a-1}. The exponent in the derivative, $a - 1$, is one less than the original exponent, a. So we expect an antiderivative of x^6 to involve x^7.

Now $(x^7)' = 7x^6$. This means that x^7 is an antiderivative of $7x^6$, not of x^6. We must get rid of that coefficient 7 in front of x^6. To accomplish this, divide x^7 by 7. We then have

$$\left(\frac{x^7}{7}\right)' = \frac{7x^6}{7} \qquad \left(\frac{f}{c}\right)' = \frac{f'}{c}$$

$$= x^6. \qquad \text{the 7's cancel}$$

We can state that $x^7/7$ is an antiderivative of x^6.

However, $x^7/7$ is not the only antiderivative of x^6. For instance,

$$\left(\frac{x^7}{7} + 1991\right)' = \frac{7x^6}{7} + 0 = x^6.$$

We can add any constant to $x^7/7$ and get another antiderivative of x^6. ■

The same reasoning as in Example 13 shows that

$$\frac{x^{a+1}}{a+1} + C \text{ is an antiderivative of } x^a \ (a \neq -1) \text{ for any constant } C.$$

In Chap. 6 we will find an antiderivative for x^a when $a = -1$, that is, an antiderivative for $1/x$.

Section Summary

$$(f+g)' = f' + g' \qquad (f-g)' = f' - g'$$

$$(fg)' = fg' + gf' \qquad \left(\frac{f}{g}\right)' = \frac{gf' - fg'}{g^2} \qquad \left(\frac{1}{g}\right)' = \frac{-g'}{g^2}$$

$$c' = 0 \qquad (cf)' = cf' \qquad \left(\frac{f}{c}\right)' = \frac{f'}{c} \qquad (c \text{ denotes a constant function})$$

The derivative of a polynomial is the sum of the derivatives of its terms.

$$\frac{x^{a+1}}{a+1} + C \text{ is an antiderivative of } x^a \text{ if } a \neq -1.$$

EXERCISES FOR SEC. 3.4: THE DERIVATIVES OF THE SUM, DIFFERENCE, PRODUCT, AND QUOTIENT

In Exercises 1 to 36 differentiate with the aid of formulas, *not* by using the definition of the derivative.

1 $x^2 + 5x$

2 $2x^2 - x$

3 $x^3 + 5x^2 + 2$

4 $2x^3 - x^2 + x$

5 $x^5 - 2x^2 + 3$

6 $x^6 + 5x^2 + 2$

7 $3x^4 - 6\sqrt{x}$

8 $2x^3 + 3\sqrt{x}$

9 $\dfrac{3}{x} + \sqrt[3]{x}$

10 $\dfrac{5}{x^2} + \sqrt[4]{x}$

11 $(2x + 1)(x - 4)$

12 $(3x - 4)(5x + 1)$

13 $(x^2 + x)(2x + 1)$

14 $(2x^2 - 1)(x^2 - 3)$

15 $(x^3 - 2x)(2x^5 + 5)$

16 $(5x^5 + x)(2x + 1)$

17 $(4 + 5\sqrt{x})(3x - x^2)$

18 $(x + 3\sqrt{x})(2x + 1)$

19 $\dfrac{2x + \sqrt[3]{x^2}}{4}$

20 $\dfrac{7x - \sqrt{x^3}}{6}$

21 $\dfrac{3 + x}{4 + x}$

22 $\dfrac{1 + x}{2 - x}$

23 $\dfrac{x^2 + x}{x + 3}$

24 $\dfrac{2x^2 - 7}{3x + 1}$

25 $\dfrac{t^2 - 3t + 1}{t^2 + 1}$

26 $\dfrac{s^2 - s}{5s^2 + s + 2}$

27 $\dfrac{6 + \sqrt{x}}{2x + 3}$

28 $\dfrac{2w + \sqrt{w^3}}{3w}$

29 $\dfrac{1}{x} + \dfrac{1}{x^2}$

30 $\dfrac{1}{x + \sqrt{x}}$

31 $\dfrac{1}{x^3 + 2x + 1}$

32 $\dfrac{1}{2x^2 - x + 5}$

33 $\dfrac{(2x^2 - 3)(x^2 - 1)}{5x + 7}$

34 $\dfrac{(2x + 9)(3x^2 - x)}{x^2}$

35 $\dfrac{1 + (1/x)}{1 - (1/x)}$

36 $\left(x + \dfrac{2}{x}\right)(x^2 + 6x + 1)$

In each of Exercises 37 to 40 find an equation of the tangent to the given curve at the given point.

37 $y = x^3 - x^2 + 2x$, at $(1, 2)$

38 $y = 1/(2x + 1)$, at $(2, 1/5)$

39 $y = \sqrt{x}(x^2 + 2)$, at $(4, 36)$
40 $y = (x + 1)/(x + 2)$, at $(-1, 0)$

In Exercises 41 and 42 the distance an object travels in the first t seconds is given by the formula. Find the velocity at the given time t.
41 $2t^4 + t^3 + 2t$, $t = 1$
42 $5\sqrt{t}$, $t = 9$

In each of Exercises 43 and 44 a lens projects x on the linear slide to a point on the linear screen given by the formula. Find the magnification at the given point x.
43 $\sqrt[3]{x}$, $x = 8$ **44** $4x^2 + x + 2$, $x = 2$

In each of Exercises 45 and 46 the mass of the left x centimeters of a string is given by the indicated formula. Find the density at the point x.
45 $x\sqrt[3]{x}$, $x = 8$ **46** $x/(x + 1)$, $x = 2$

In Exercises 47 and 48 give at least two antiderivatives for each of the given functions. Check your answers by differentiating.
47 (*a*) x^2
(*b*) x^3
(*c*) $1/x^2$
(*d*) $3x^2 + x$
48 (*a*) x^4
(*b*) $\sqrt{x}$
(*c*) $1/x^3$
(*d*) $3x + \dfrac{5}{x^2}$

---

49 This exercise obtains the formula $D(x^n) = nx^{n-1}$ for n a positive integer with the aid of the formula for $(fg)'$.
(*a*) Use the definition of the derivative to show that $x' = 1$.
(*b*) With the aid of (*a*) and the formula for the derivative of a product of two functions, obtain $(x^2)'$. (Write x^2 as $x \cdot x$.)
(*c*) Knowing x' and $(x^2)'$, use the same method to find $(x^3)'$ and then $(x^4)'$.
(*d*) If you are familiar with mathematical induction, obtain the formula $(x^n)' = nx^{n-1}$ for all positive integers. (See Appendix E.)
50 This exercise obtains the formula $D(x^n) = nx^{n-1}$, when n is a negative integer, with the aid of the formula for $(1/f)'$. Assume that we know that $D(x^n) = nx^{n-1}$ when n is a positive integer.
(*a*) Find $D(x^{-4})$ using the formula for $(1/f)'$ and $(x^n)'$, n a positive integer.
(*b*) Similarly, find $D(x^{-m})$, for any positive integer m.
51 Find $D(\sqrt[3]{x}\sqrt{x})$ in two ways:
(*a*) by the formula for the derivative of a product.
(*b*) by first writing $\sqrt[3]{x}\sqrt{x}$ in the form x^a.
52 Assume that $D(1/x) = -1/x^2$. Using the formula for the derivative of a product and mathematical induction, show that $D(1/x^n) = -n/x^{n+1}$ for all positive integers n.
53 Show that if f, g, and h are differentiable functions, then $(f + g + h)' = f' + g' + h'$. *Hint:* Use Theorem 2 twice after writing $f + g + h$ as $(f + g) + h$.
54 Using the definition of the derivative, prove that $(f - g)' = f' - g'$.
55 Show that if f, g, and h are differentiable functions, then $(fgh)' = f'gh + fg'h + fgh'$. *Hint:* First write fgh as $(fg)h$. Then use the formula for the derivative of the product of two functions.
56 (Economics: elasticity) Economists try to forecast the impact a change in price will have on demand. For instance, if the price of a gallon of gasoline is raised 10 cents, either by the oil company or by the government (in the form of a tax), how much petroleum will be conserved? To deal with such problems economists use the concept of the *elasticity of demand,* which will now be defined. Let $y = f(x)$ be the demand for a product as a function of the price x, that is, the amount that will be bought at the price x in a given time. Assume that y is a differentiable function of x.
(*a*) Is the derivative y' in general positive or is it negative?
(*b*) What are the dimensions of y' if y is measured in gallons and x is measured in cents?
(*c*) Why is the ratio $(\Delta y/y)/(\Delta x/x)$ called a "dimensionless quantity"? (*Note:* If gasoline is measured in liters instead of gallons, this ratio does not change.)
(*d*) The **elasticity of demand** at the price x is defined as

$$\epsilon = \lim_{\Delta x \to 0} \frac{\Delta y/y}{\Delta x/x}.$$

Show that $$\epsilon = \left(\frac{x}{y}\right)y'.$$

(*e*) Estimate ϵ if a 1 percent increase in the price causes a 2 percent decrease in demand.
(*f*) Estimate ϵ if a 2 percent increase in price causes a 1 percent decrease in demand.
(*g*) If $|\epsilon| > 1$, the demand is called **elastic**; if $|\epsilon| < 1$, it is called **inelastic**. Why?
(*h*) Show that $y = x^{-3}$ has a constant elasticity (that is, the elasticity is independent of x).
57 Figure 3 shows the typical tangent line to the curve $y = 1/x$ at a point P. Show that the area of the triangle OAB is constant, independent of the choice of point P.

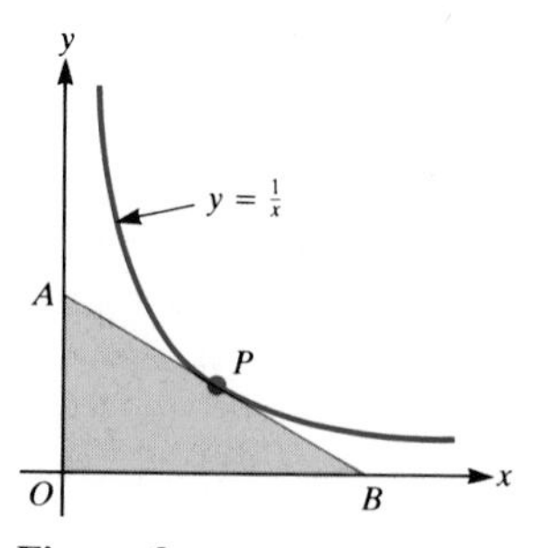

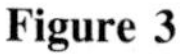
Figure 3

58 Is there a line that is simultaneously tangent to the curves $y = x^2$ and $y = -x^2 + 2x - 2$? (First sketch the curves.) If so, find all such lines.

59 (*a*) Find an equation of the tangent line to the curve $y = 2x^2 - 3x + 1$ at $(1, 0)$.
(*b*) Find an equation of the line through $(1, 0)$ perpendicular to the tangent line in (*a*). (Section B.2 discusses the slope of a line perpendicular to a given line.)
(*c*) At what points does the line in (*b*) intersect the curve?

60 (*a*) Find an equation of the tangent line to the curve $y = x^3$ at the point (a, a^3).
(*b*) Does the tangent line in (*a*) always meet the curve at another point than (a, a^3)? If not, when does it?

61 There are four points on the curve $y = x^4 - 8x^2$ where the associated tangent line passes through $(-\frac{11}{3}, 49)$. Find the x coordinates of these points. (*Hint:* Two are integers. See Appendix C for advice in searching for rational roots of a polynomial.)

3.5 THE DERIVATIVES OF THE TRIGONOMETRIC FUNCTIONS

In this section we will find the derivatives of the six trigonometric functions—$\sin x$, $\cos x$, $\tan x$, $\sec x$, $\csc x$, and $\cot x$.

In order to find $\frac{d}{dx}(\sin x)$ and $\frac{d}{dx}(\cos x)$, it will be necessary to make use of the limits

$$\lim_{\Delta x \to 0} \frac{\sin \Delta x}{\Delta x} = 1 \qquad \text{and} \qquad \lim_{\Delta x \to 0} \frac{1 - \cos \Delta x}{\Delta x} = 0,$$

which were found in Sec. 2.7.

The Derivative of sin x

The derivatives of $\sin x$ *and* $\cos x$

Theorem 1 The derivative of the sine function is the cosine function; symbolically,

$$(\sin x)' = \cos x \qquad \text{or} \qquad \frac{d}{dx}(\sin x) = \cos x.$$

Proof The derivative at x of a function f is defined as

$$\lim_{\Delta x \to 0} \frac{f(x + \Delta x) - f(x)}{\Delta x}.$$

In this case, f is the function "sine," and the limit under consideration is

$$\lim_{\Delta x \to 0} \frac{\sin(x + \Delta x) - \sin x}{\Delta x}.$$

Keep in mind that x is fixed while $\Delta x \to 0$. As $\Delta x \to 0$, the numerator approaches

$$\sin x - \sin x = 0,$$

When finding a derivative, we always run into "zero-over-zero."

while the denominator Δx also approaches 0. Since the expression 0/0 is meaningless, it is necessary to change the form of the quotient

$$\frac{\sin(x + \Delta x) - \sin x}{\Delta x}$$

before letting Δx approach 0.

Let us use the trigonometric identity

$$\sin(A+B) = \sin A \cos B + \cos A \sin B$$

in the case $A = x$ and $B = \Delta x$, obtaining

$$\sin(x+\Delta x) = \sin x \cos \Delta x + \cos x \sin \Delta x.$$

Then the numerator, $\sin(x+\Delta x) - \sin x$, takes the form

$$\begin{aligned}\sin(x+\Delta x) - \sin x &= (\sin x \cos \Delta x + \cos x \sin \Delta x) - \sin x \\ &= \sin x(\cos \Delta x - 1) + \cos x \sin \Delta x \\ &= -\sin x(1 - \cos \Delta x) + \cos x \sin \Delta x.\end{aligned}$$

Therefore,

$$\begin{aligned}\lim_{\Delta x \to 0} \frac{\sin(x+\Delta x) - \sin x}{\Delta x} &= \lim_{\Delta x \to 0} \frac{-\sin x(1-\cos \Delta x) + \cos x \sin \Delta x}{\Delta x} \\ &= \lim_{\Delta x \to 0} \left(-\sin x \frac{1-\cos \Delta x}{\Delta x} + \cos x \frac{\sin \Delta x}{\Delta x}\right) \\ &= (-\sin x)(0) + (\cos x)(1) \\ &= \cos x.\end{aligned}$$

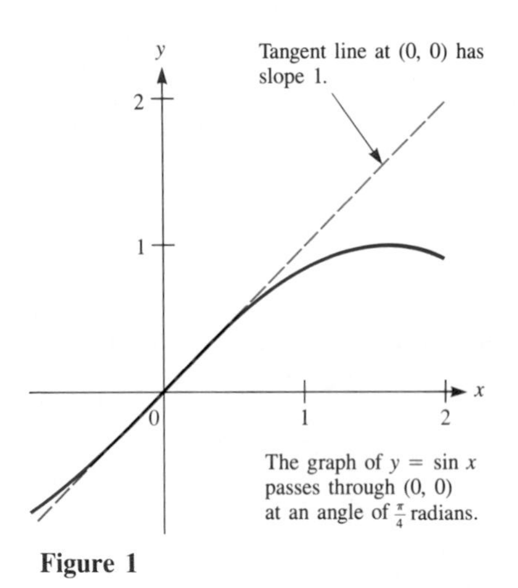

Figure 1

In short, the derivative of the sine function is the cosine function. This concludes the proof. ■

The formula obtained in Theorem 1 provides interesting information about the graph of $y = \sin x$. Since

$$\frac{d}{dx}(\sin x) = \cos x,$$

the derivative of the sine function when $x = 0$ is $\cos 0$, which is 1. This implies that the slope of the curve $y = \sin x$, when $x = 0$, is 1. Consequently, the graph of $y = \sin x$ passes through the origin at an angle of $\pi/4$ radians (45°). See Fig. 1.

The Derivative of cos x

Theorem 2 The derivative of the cosine function is the negative of the sine function; symbolically,

$$(\cos x)' = -\sin x.$$

We omit the proof, which is similar to that of Theorem 1. It makes use of the trigonometric identity

$$\cos(A+B) = \cos A \cos B - \sin A \sin B,$$

and is left as Exercise 43.

Recall that

$$\sec x = \frac{1}{\cos x}, \qquad \tan x = \frac{\sin x}{\cos x}, \qquad \csc x = \frac{1}{\sin x}, \qquad \cot x = \frac{\cos x}{\sin x}.$$

In the next two theorems the derivatives of these four functions are obtained.

The Derivatives of sec x and csc x

Theorem 3

$$(\sec x)' = \sec x \tan x \qquad \text{and} \qquad (\csc x)' = -\csc x \cot x.$$

Proof

$$(\sec x)' = \left(\frac{1}{\cos x}\right)' = \frac{-(\cos x)'}{\cos^2 x} \qquad \text{using } \left(\frac{1}{g}\right)' = \frac{-g'}{g^2}$$

$$= \frac{-(-\sin x)}{\cos^2 x}$$

$$= \frac{\sin x}{\cos^2 x}$$

$$= \frac{1}{\cos x}\,\frac{\sin x}{\cos x}$$

$$= \sec x \tan x.$$

Thus $(\sec x)' = \sec x \tan x$. The derivative of $\csc x$ is obtained similarly. ■

The Derivatives of tan x and cot x

Theorem 4

$$(\tan x)' = \sec^2 x \qquad \text{and} \qquad (\cot x)' = -\csc^2 x.$$

Proof

$$(\tan x)' = \left(\frac{\sin x}{\cos x}\right)'$$

$$= \frac{\cos x\,(\sin x)' - \sin x\,(\cos x)'}{\cos^2 x}$$

$$= \frac{\cos x \cos x - \sin x\,(-\sin x)}{\cos^2 x}$$

$$= \frac{\cos^2 x + \sin^2 x}{\cos^2 x} = \frac{1}{\cos^2 x} = \sec^2 x.$$

The derivative of $\cot x$ is obtained similarly. ■

This table summarizes the six formulas. Note that the derivatives of the cosine, cosecant, and cotangent have minus signs. (The derivatives of the co-functions have a $-$.)

The derivatives of the cofunctions (cosine, cotangent, cosecant) *have minus signs.*

TRIGONOMETRIC DERIVATIVES

$(\sin x)' = \cos x$	$(\cos x)' = -\sin x$
$(\tan x)' = \sec^2 x$	$(\cot x)' = -\csc^2 x$
$(\sec x)' = \sec x \tan x$	$(\csc x)' = -\csc x \cot x$

EXAMPLE 1 Differentiate $x - \sin x \cos x$.

SOLUTION

$$\begin{aligned}(x - \sin x \cos x)' &= (x)' - (\sin x \cos x)' \\ &= 1 - [\sin x\,(\cos x)' + \cos x\,(\sin x)'] \\ &= 1 - [\sin x\,(-\sin x) + \cos x\,(\cos x)] \\ &= 1 + \sin^2 x - \cos^2 x.\end{aligned}$$

Since $1 - \cos^2 x = \sin^2 x$, the last expression can be simplified to $\sin^2 x + \sin^2 x$, or simply $2 \sin^2 x$. ■

The next section will present a shortcut for differentiating $\sin 2x$ and will show why the derivative is *not* $\cos 2x$. The next example uses a trigonometric identity to find $(\sin 2x)'$.

EXAMPLE 2 Differentiate $\sin 2x$.

SOLUTION $\sin 2x = 2 \sin x \cos x.$ trigonometric identity

Thus

$$\begin{aligned}(\sin 2x)' &= (2 \sin x \cos x)' \\ &= 2\,(\sin x \cos x)' \\ &= 2\,[\sin x\,(\cos x)' + \cos x\,(\sin x)'] \\ &= 2\,[\sin x\,(-\sin x) + \cos x\,(\cos x)] \\ &= 2\,(\cos^2 x - \sin^2 x) = 2 \cos 2x.\end{aligned}$$

In short, the derivative of $\sin 2x$ is $2 \cos 2x$. ■

Warning: The derivative of $\sin 2x$ is $2 \cos 2x$, not $\cos 2x$.

EXAMPLE 3 Find $(x^3 \sec x)'$.

SOLUTION

$$\begin{aligned}(x^3 \sec x)' &= x^3\,(\sec x)' + \sec x\,(x^3)' \\ &= x^3 \sec x \tan x + \sec x\,(3x^2),\end{aligned}$$

which is usually written $x^3 \sec x \tan x + 3x^2 \sec x$ for clarity. ■

EXAMPLE 4 Find $\frac{d}{dx}\left(\frac{\csc x}{\sqrt{x}}\right)$.

SOLUTION

$$\frac{d}{dx}\left(\frac{\csc x}{\sqrt{x}}\right) = \frac{\sqrt{x}\left[\frac{d}{dx}(\csc x)\right] - \csc x\left[\frac{d}{dx}(\sqrt{x})\right]}{(\sqrt{x})^2}$$

$$= \frac{\sqrt{x}(-\csc x \cot x) - \csc x \frac{1}{2\sqrt{x}}}{x}.$$

Since $\frac{\sqrt{x}}{x} = \frac{1}{\sqrt{x}}$ and $x\sqrt{x} = x^{3/2}$, this can be simplified a little to

$$\frac{-\csc x \cot x}{\sqrt{x}} - \frac{\csc x}{2x^{3/2}}. \quad \blacksquare$$

Why Radian Measure Is Used in Calculus

Throughout this section angles are measured in radians, as is customary in calculus. If we measured angles in degrees instead, the formulas for the derivatives of the trigonometric functions would be more complicated. Each formula would have an extra factor, $\pi/180$, as we will now show.

In Sec. 2.7 it was shown that when angles are measured in radians,

$$\lim_{\theta \to 0} \frac{\sin \theta}{\theta} = 1.$$

When angles are measured in degrees, this limit is not 1. Let Sin θ denote the sine of an angle of θ degrees. The following table suggests that the limit is much smaller (angles measured in degrees; data to four significant figures):

θ	10	5	1	0.1 (degrees)
Sin θ	0.1736	0.08716	0.01745	0.001745
$\frac{\text{Sin } \theta}{\theta}$	0.01736	0.01743	0.01745	0.01745

The data suggest that $\lim_{\theta \to 0}$ (Sin θ)/θ is about 0.01745. As you will see when doing Exercise 47, the limit is precisely $\pi/180$. If you then go through this section and carry through the steps that determined the derivatives of the trigonometric functions, you will see that an extra factor of $\pi/180$ must be put into each formula. Clearly, the use of radians as the measure of angles gives the simplest formulas for trigonometric derivatives.

Section Summary

Using the fact that $\lim_{x \to 0} (\sin x)/x = 1$ and $\lim_{x \to 0} (1 - \cos x)/x = 0$, we obtained the derivatives of $\sin x$ and $\cos x$. Knowing these derivatives, we then computed the derivatives of the four other basic trigonometric functions.

EXERCISES FOR SEC. 3.5: THE DERIVATIVES OF THE TRIGONOMETRIC FUNCTIONS

In Exercises 1 to 18 differentiate the functions.

1 $5 \sin x$

2 $7 \cos x$

3 $2 \tan x$

4 $-6 \cot x$

5 $3 \sec x$

6 $5 \csc x$

7 $x^2 \sin x$

8 $x^3 \cos x$

9 $\dfrac{1 + \sin x}{\cos x}$

10 $\dfrac{1 - \sin x}{\cos x}$

11 $\dfrac{1 + 3 \sec x}{\tan x}$

12 $x^3 \sec x$

13 $\dfrac{\csc x}{\sqrt[3]{x}}$

14 $3 \csc x + 2 \tan x$

15 $\sin x \tan x$

16 $x^2 \cos x \cot x$

17 $\dfrac{\cot x}{1 + x^2}$

18 $\dfrac{x}{1 + \sec x}$

In each of Exercises 19 to 26 differentiate the function and simplify your answer.

19 $\sin x - x \cos x$

20 $\cos x + x \sin x$

21 $2x \sin x + 2 \cos x - x^2 \cos x$

22 $3x^2 \sin x - 6 \sin x - x^3 \cos x + 6x \cos x$

23 $\tan x - x$

24 $-\cot x - x$

25 $2x \cos x - 2 \sin x + x^2 \sin x$

26 $(3x^2 - 6) \cos x + (x^2 - 6) \sin x$

In Exercises 27 to 30 evaluate the given quantities. In some cases a calculator is needed; give such answers to three decimal places.

27 $D(\sin x)$ when x is (*a*) $\pi/6$, (*b*) 1.2, (*c*) $3\pi/4$.

28 $D(\cos x)$ when x is (*a*) $\pi/4$, (*b*) $-2\pi/3$, (*c*) 2.

29 $D(\tan x)$ when x is (*a*) $\pi/4$, (*b*) $\pi/6$, (*c*) 3.

30 $D(\sec x)$ when x is (*a*) $\pi/4$, (*b*) $3\pi/4$, (*c*) 0.5.

31 Find $D(\csc \theta)$, using Theorems 1 and 2.

32 Find $D(\cot \theta)$, using Theorems 1 and 2.

33 A mass bobbing up and down on the end of a spring has height $y = 3 \sin t$ centimeters at time t seconds. (*a*) How high does it go? (*b*) How low? (*c*) What is its velocity when $t = 0$ and when $t = \pi$? (*d*) What is its speed when $t = 0$ and when $t = \pi$?

34 The height of the ocean surface above (or below) mean sea level is, say, $y = 2 \sin t$ feet at t hours.
(*a*) Find the rate at which the tide is rising or falling at time t.
(*b*) Is the surface rising more rapidly at low tide or when it is at mean sea level?

35 At what angle does the graph of $y = \tan x$ cross the x axis?

36 (*a*) Find the slope of the curve $y = \tan x$ when $x = \pi/4$.
(*b*) Using (*a*), estimate the angle that the tangent line to the curve $y = \tan x$ at $(\pi/4, 1)$ makes with the x axis.

37 What is the angle of inclination of the tangent line to the curve $y = \sin x$ at the point $(\pi, 0)$?

38 What is the angle of inclination of the tangent line to the curve $y = \cos x$ at the point $(\pi/3, 1/2)$?

In Exercises 39 to 42 give an antiderivative for the given functions and check your answers by differentiation.

39 (*a*) $3 \sin x$ (*b*) $4 \cos x$

40 (*a*) $5 \sec^2 x$ (*b*) $6 \csc^2 x$

41 (*a*) $2 \sec x \tan x$ (*b*) $7 \csc x \cot x$

42 (*a*) $5 \sin x - 6 \cos x$ (*b*) $\dfrac{3}{\cos^2 x}$

---

43 Using the identity for $\cos (A + B)$, show that the derivative of $\cos x$ is $-\sin x$.

44 Using the definition of the derivative and the identity for $\sin (A + B)$, show that $(\sin 7x)' = 7 \cos 7x$.

45 Using the definition of the derivative and the identity for $\cos (A + B)$, show that $(\cos 11x)' = -11 \sin 11x$.

46 (*a*) Sketch the curve $y = \sin x$ for x in $[0, \pi]$ and the line L through (1, 1) and (4, 2).
(*b*) Use your graph from (*a*) to estimate the coordinates of the point P on the curve such that the tangent at P is parallel to L.
(*c*) Using the derivative and a calculator, estimate the coordinates of P.

47 Let $\text{Sin } \theta$ denote the sine of an angle of θ degrees. Since an angle of θ degrees is the same as an angle of $\pi\theta/180$ radians, $\text{Sin } \theta = \sin (\pi\theta/180)$, where $\sin x$ denotes the sine of an angle of x radians. Deduce that $(\text{Sin } \theta)' = (\pi/180) \text{ Cos } \theta$, where $\text{Cos } \theta$ denotes the cosine of an angle of θ degrees. *Hint:* First find $\lim_{\theta \to 0} (\text{Sin } \theta)/\theta$.

3.6 THE DERIVATIVE OF A COMPOSITE FUNCTION

This section presents the most important technique for finding the derivative of a function. It turns out that we can easily compute the derivative of a composite function if we know the derivatives of the functions from which it is composed. (Section 2.2 discusses composite functions.)

Differentiating a Composite Function

What is the derivative of $(1 + x^2)^{100}$? You might be tempted to guess that the answer would be $100(1 + x^2)^{99}$. *This cannot be right.* After all, when you multiply out $(1 + x^2)^{100}$ you get a polynomial of degree 200, so its derivative is a polynomial of degree 199. But when you multiply out $100(1 + x^2)^{99}$ you get a polynomial of degree 198. Something is wrong.

To see what the mistake is, we must learn how to differentiate a composite function. Note that $(1 + x^2)^{100}$ is a composite function, for it is of the form $y = u^{100}$, where $u = 1 + x^2$.

If f and g are differentiable functions, is the composite function $f \circ g$ also differentiable? If so, what is its derivative? More concretely: If $y = f(u)$ and $u = g(x)$, then y is a function of x. How can we find dy/dx?

Take a very simple case. The function $y = 6x$ can be built up by composing the functions $y = 3u$ and $u = 2x$, since $6x = 3(2x)$. In this case,

$$\frac{dy}{du} = 3, \qquad \frac{du}{dx} = 2, \qquad \text{and} \qquad \frac{dy}{dx} = 6.$$

So dy/dx is the product of the derivatives dy/du and du/dx. This observation suggests the all-important **chain rule**, which will be proved at the end of this section after several examples show how it is used.

THE CHAIN RULE (INFORMAL STATEMENT)

If y is a differentiable function of u and u is a differentiable function of x, then y is a differentiable function of x and

$$\frac{dy}{dx} = \frac{dy}{du} \cdot \frac{du}{dx}.$$

An easily remembered form of the chain rule

The equation

$$\frac{dy}{dx} = \frac{dy}{du} \cdot \frac{du}{dx}$$

is read "derivative of y with respect to x equals derivative of y with respect to u times derivative of u with respect to x."

As we have already remarked, the notation "dy/dx" is not a fraction, but rather a *notation* for the derivative of y with respect to x. The chain rule is a statement about derivatives, not about fractions, and we should not think of the "du" as "canceling out."

Example 1 shows how to find the derivative of $(1 + x^2)^{100}$ without multiplying it out.

EXAMPLE 1 Find $D((1 + x^2)^{100})$.

SOLUTION Here $y = (1 + x^2)^{100}$. So

$$y = u^{100} \qquad \text{where } u = 1 + x^2.$$

According to the chain rule

$$\begin{aligned}\frac{dy}{dx} &= \frac{d}{du}(u^{100}) \cdot \frac{d}{dx}(1 + x^2) \\ &= 100u^{99} \cdot 2x \\ &= 100(1 + x^2)^{99} \cdot 2x \\ &= 200x(1 + x^2)^{99}.\end{aligned}$$

As mentioned at the beginning of this section, the answer is *not* $100(1 + x^2)^{99}$. There is an extra factor of $2x$ that comes from differentiating $1 + x^2$. ■

The chain rule extends to a function built up as the composition of three or more functions. For instance, if

$$y = f(u), \qquad u = g(v), \qquad \text{and} \qquad v = h(x),$$

then y is a function of x and it can be shown that

An extended form of the chain rule

$$\frac{dy}{dx} = \frac{dy}{du}\frac{du}{dv}\frac{dv}{dx}.$$

The next example applies this fact.

EXAMPLE 2 Differentiate $\sqrt{(1 + x^2)^5}$.

SOLUTION $y = \sqrt{(1 + x^2)^5}$ can be expressed as

$$y = \sqrt{u}, \qquad u = v^5, \qquad \text{and} \qquad v = 1 + x^2.$$

Then

$$\frac{dy}{dx} = \frac{dy}{du}\frac{du}{dv}\frac{dv}{dx}$$

or

$$\begin{aligned}\frac{d}{dx}(\sqrt{(1 + x^2)^5}) &= \frac{d}{du}(\sqrt{u})\,\frac{d}{dv}(v^5)\,\frac{d}{dx}(1 + x^2) \\ &= \frac{1}{2\sqrt{u}} \cdot 5v^4 \cdot 2x \\ &= \frac{5v^4x}{\sqrt{u}} \\ &= \frac{5(1 + x^2)^4x}{\sqrt{v^5}} \\ &= \frac{5(1 + x^2)^4x}{\sqrt{(1 + x^2)^5}} \\ &= 5x(1 + x^2)^{3/2}.\end{aligned}$$ ■

EXAMPLE 3 Differentiate $\sin^2 3x$.

SOLUTION $y = \sin^2 3x$ can be written as

$$y = u^2, \qquad u = \sin v, \qquad \text{and} \qquad v = 3x.$$

Thus

$$\begin{aligned}\frac{d}{dx}(\sin^2 3x) &= \frac{d}{du}(u^2)\,\frac{d}{dv}(\sin v)\,\frac{d}{dx}(3x) \\ &= (2u)(\cos v)(3) \\ &= 6u \cos v \\ &= 6 \sin v \cos v \\ &= 6 \sin 3x \cos 3x.\end{aligned}$$ ■

EXAMPLE 4 Compute $\frac{d}{dx}(x^2 \sin^5 2x)$.

SOLUTION First of all, by the formula for the derivative of the product,

$$\frac{d}{dx}(x^2 \sin^5 2x) = x^2 \frac{d}{dx}(\sin^5 2x) + \sin^5 2x \frac{d}{dx}(x^2)$$

$$= x^2 \frac{d}{dx}(\sin^5 2x) + \sin^5 2x\,(2x).$$

The chain rule is needed for computing

$$\frac{d}{dx}(\sin^5 2x).$$

To differentiate $\sin^5 2x$ we use the chain rule twice. Let $y = \sin^5 2x$. Then

$$y = u^5 \qquad \text{where } u = \sin v \text{ and } v = 2x.$$

Thus

$$\frac{dy}{dx} = \frac{dy}{du}\frac{du}{dv}\frac{dv}{dx}$$
$$= 5u^4(\cos v)(2)$$
$$= 10u^4 \cos v$$
$$= 10(\sin v)^4 \cos v$$
$$= 10 \sin^4 2x \cos 2x.$$

Thus
$$\frac{d}{dx}(x^2 \sin^5 2x) = x^2(10 \sin^4 2x \cos 2x) + (\sin^5 2x)(2x)$$
$$= 10x^2 \sin^4 2x \cos 2x + 2x \sin^5 2x. \quad ■$$

As these examples suggest, the chain rule is one of the most frequently used tools in the computation of derivatives.

The table in the margin records a few special cases of the chain rule. They are used so often that they are worth memorizing. In each case u is a differentiable function of x.

y	$\frac{dy}{dx}$
u^n	$nu^{n-1}\frac{du}{dx}$
$\sin u$	$\cos u \frac{du}{dx}$
$\cos u$	$-\sin u \frac{du}{dx}$

For instance,

$$D[(1 + x^2)^{100}] = 100(1 + x^2)^{99}(2x)$$

$$D(\sin \sqrt{x}) = \left(\cos \sqrt{x}\right)\left(\frac{1}{2\sqrt{x}}\right)$$

$$D(\cos x^3) = -\sin x^3\,(3x^2).$$

Proof of the Chain Rule

We wish to prove that if $y = f(u)$ and $u = g(x)$, then

$$\frac{dy}{dx} = \frac{dy}{du} \cdot \frac{du}{dx}.$$

Because derivatives are defined as limits of certain quotients, we must show that

$$\lim_{\Delta x \to 0} \frac{\Delta y}{\Delta x} = \lim_{\Delta u \to 0} \frac{\Delta y}{\Delta u} \cdot \lim_{\Delta x \to 0} \frac{\Delta u}{\Delta x}. \tag{1}$$

Knowing that the two limits on the right side of (1) exist, we may be tempted to use the "limit of a product" rule and write

$$\frac{dy}{du}\cdot\frac{du}{dx} = \lim_{\Delta u\to 0}\frac{\Delta y}{\Delta u}\cdot\lim_{\Delta x\to 0}\frac{\Delta u}{\Delta x}$$

$$= \lim_{\Delta x\to 0}\frac{\Delta y}{\Delta u}\cdot\frac{\Delta u}{\Delta x} \qquad \Delta x\to 0 \text{ forces } \Delta u\to 0.$$

$$= \lim_{\Delta x\to 0}\frac{\Delta y}{\Delta x} \qquad \text{Cancel } \Delta u.$$

$$= \frac{dy}{dx}.$$

The result is correct, but there is a flaw in the logic.

Look carefully at the expression

$$\frac{\Delta y}{\Delta u}\cdot\frac{\Delta u}{\Delta x}. \tag{2}$$

Here Δx denotes "change in the input of x." This change induces a change in the output of $u = g(x)$, which we call Δu. Even though Δx is *not* 0, Δu *may* be 0. That means that the expression (2) may involve *division by* 0, which is a meaningless operation.

There are functions $u = g(x)$ for which a nonzero Δx induces $\Delta u = 0$. For instance, $u = c$, where c is a constant. Then *every* choice of Δx induces $\Delta u = 0$.

If $u = g(x)$ is a differentiable function with the property that as Δx approaches 0, you keep running into $\Delta u = 0$, then $\lim_{\Delta x\to 0}\Delta u/\Delta x$ must be 0. That is, $du/dx = 0$. If du/dx is not 0, then for Δx sufficiently small, Δu is not 0. In this case we avoid division by 0 and our earlier argument for the chain rule provides a valid proof. It runs as follows.

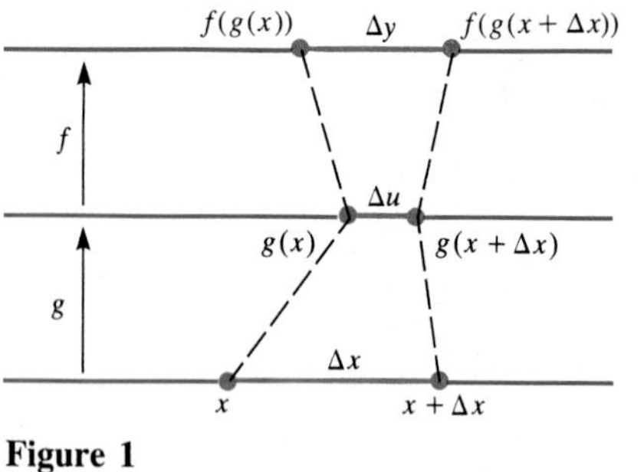

Figure 1

Proof of Chain Rule [Assuming $g'(x) \neq 0$.] (See Fig. 1.)

$$\frac{dy}{dx} = \lim_{\Delta x\to 0}\frac{\Delta y}{\Delta x}$$

$$= \lim_{\Delta x\to 0}\frac{\Delta y}{\Delta u}\cdot\frac{\Delta u}{\Delta x} \qquad \Delta u \neq 0 \text{ if } \Delta x \text{ is small enough}$$

$$= \lim_{\Delta x\to 0}\frac{\Delta y}{\Delta u}\cdot\lim_{\Delta x\to 0}\frac{\Delta u}{\Delta x} \qquad \text{since both limits exist}$$

$$= \lim_{\Delta u\to 0}\frac{\Delta y}{\Delta u}\cdot\lim_{\Delta x\to 0}\frac{\Delta u}{\Delta x} \qquad \text{since } \Delta u\to 0 \text{ as } \Delta x\to 0$$

$$= \frac{dy}{du}\cdot\frac{du}{dx}. \qquad \text{definition of derivative}$$

The special case where $g'(x) = 0$ is outlined in Exercise 55. ■

DIFFERENTIATION BY MACHINE

We calculate derivatives by a small number of rules obtained in this chapter and in Chap. 6. These rules may be programmed into computers or calculators, which can then do our differentiation for us. Such programs do "symbolic" rather than "numerical" mathematics. One of the most powerful symbolic programs is Mathematica, available from Wolfram Research, Inc., for the Apple Macintosh, 386-based IBM PS/2 models, and many other computer systems. In Mathematica, for example, the command $D[f, x]$ differentiates the function f with respect to x. Hence the input `D[x^2 + 3x, x]` produces the output `2x + 3`.

Symbolic differentiation is also available on hand-held calculators like the Hewlett-Packard HP-48sx. For more information on symbolic computer programs and calculators see the list on pp. xxvii to xxix.

Section Summary

We developed a formula called the chain rule for differentiating composite functions. The chain rule is the most commonly used rule for differentiating functions. When first working with the chain rule, write down every step, showing the various functions with the aid of the letters u, v, and so on. However, with practice, it will not be necessary to record every detail.

EXERCISES FOR SEC. 3.6: THE DERIVATIVE OF A COMPOSITE FUNCTION

In Exercises 1 to 40 differentiate the functions.

1 $(1 + 2x)^{100}$ **2** $(2 - 7x)^{20}$
3 $(2x^3 - 1)^{40}$ **4** $(x^4 - 6)^{100}$
5 $3 \sin \sqrt{x}$ **6** $5 \cos 2x$
7 $\sin^3 x$ **8** $\sin^3 2x$
9 $\cos^4 3x$ **10** $\cos^3 4x$
11 $\tan^2 3x$ **12** $\sec^5 2x$
13 $\sqrt{\cot x}$ **14** $\sqrt{\csc x}$
15 $\sqrt{x^3 + x + 2}$ **16** $\sqrt[3]{x^3 + 8}$
17 $\sin (3x + 2)^5$ **18** $\sin^5 (3x + 2)$
19 $\dfrac{x^3 \tan x}{1 + x^2}$ **20** $\sqrt{1 + x^2} \sec^3 5x$
21 $(2x + 1)^5(3x + 1)^7$ **22** $x^5(x^2 + 2)^3 \cos^3 5x$
23 $x^2 \sin^5 3x$ **24** $x^3 \cos^2 3x \sin^2 2x$
25 $\dfrac{1}{(2x + 3)^5}$ **26** $\dfrac{1}{(3x - 5)^7}$
27 $\dfrac{x^2}{(x^2 + 1)^3}$ **28** $\dfrac{(x^3 - 1)^5 \cot^3 x}{x}$
29 $\dfrac{\sqrt{1 - x^2} \sec^2 x}{1 + x}$ **30** $\left(\dfrac{1 + 2x}{1 + 3x}\right)^4$
31 $\left(\dfrac{x^2 + 3x + 5}{2x - 1}\right)^5$ **32** $\tan^2 \sqrt{x}$
33 $(2x + 1)^3 \cot^4 x^2$ **34** $\dfrac{1}{\sqrt{1 - x^2}}$
35 $\dfrac{x}{\sqrt{1 - x^2}}$ **36** $\dfrac{(2x + 1)^3}{(3x + 1)^4}$
37 $\sqrt{5(3x - 2)^4 + 1}$ **38** $\sqrt[3]{(2x + 1)^2 + x}$
39 $\cos 3x \sin 4x$ **40** $\sec 2x \tan 5x$

In Exercises 41 to 44 differentiate the given functions and simplify your answers as far as possible.

41 $\dfrac{2(9x - 2)}{135}\sqrt{(3x + 1)^3}$
42 $-\frac{1}{3} \cos 3x + \frac{1}{9} \cos^3 3x$
43 $\dfrac{3x}{8} - \dfrac{3 \sin 5x \cos 5x}{40} - \dfrac{\sin^3 5x \cos 5x}{20}$
44 $\frac{1}{20}(5 + 2x)^5 - \frac{5}{16}(5 + 2x)^4$

In Exercises 45 to 48 give an antiderivative of each of the given functions. Check by differentiating your answers. (*Hint:* Keep the chain rule in mind as you make your initial guess. You may have to adjust your first guess to fit the situation.)

45 $8x(1 + x^2)^3$ **46** $3 \sin^2 x \cos x$
47 $3x^2(1 + x^3)^4$ **48** $\cos^3 x \sin x$

49 It is known of two functions f and g that $f(3) = 2, f'(3) = 4$, $g(5) = 3$, and $g'(5) = 7$. At what input x is it possible to compute $(f \circ g)'$? What does it equal?

50 Let f and g be differentiable functions. Assume that $g(1) = 2$, $g(2) = 6$, $g'(1) = 3$, $g'(2) = 6$, $f(2) = 4$, $f(3) = 4$, $f'(2) = 5$, and $f'(3) = 7$.

(*a*) Draw a two-projection diagram, such as Fig. 2, and show the values of f and g on it.

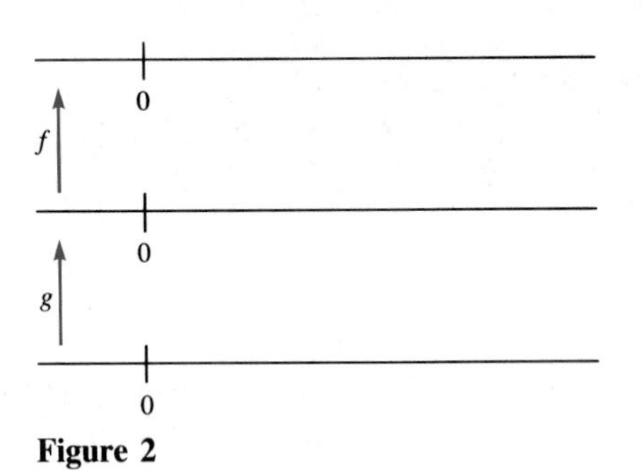

Figure 2

(*b*) Let $h(x) = f(g(x))$. At which input x can $h'(x)$ be computed? What is $h'(x)$ at that input?

51 Let f and g be differentiable functions. Define $h(x)$ to be $f(g(x))$. Assume that $g(2) = 3$, $g(2.03) = 3.04$, $f(3) = 5$, and $f(3.04) = 4.96$.

(*a*) Using a diagram like Fig. 2 in Exercise 50, enter the given data on the lines.

(*b*) Use the data to estimate $h'(2)$.

52 Let g be a differentiable function such that its derivative is $1/(x^3 + 1)$. Let $h(x) = g(x^2)$. Find $h'(x)$.

53 The temperature in a certain wire is not the same at all points. Assume that x centimeters from the left end the temperature is x^2 degrees Celsius. A bug crawls along the wire at the rate of 2 centimeters per minute. The bug notices that the temperature increases as it crawls. At what rate is the temperature increasing (degrees per minute) when the bug is 3 centimeters from the cool end? (See Fig. 3.)

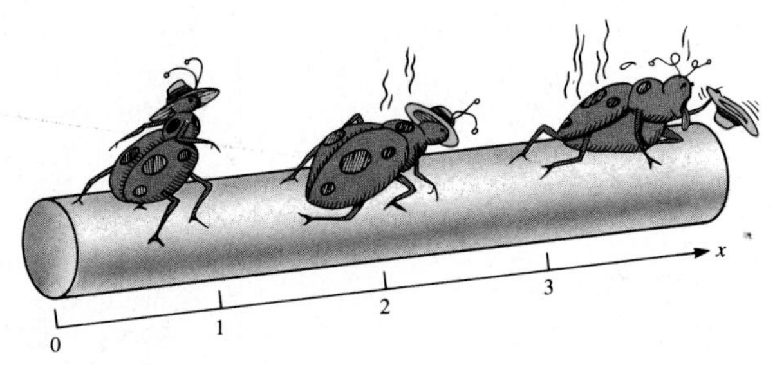

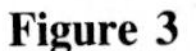

Figure 3

54 Air is blown into a spherical balloon at the rate of one cubic foot per second. At what rate is the radius of the balloon increasing when the radius is 3 feet? (The volume of a ball of radius r is $4\pi r^3/3$.)

55 This exercise outlines the proof of the chain rule when $g'(x) = 0$.

(*a*) Show that in this case it is sufficient to prove that

$$\lim_{\Delta x \to 0} \frac{\Delta y}{\Delta x} = 0.$$

(*b*) For some values of Δx, we have $\Delta u \neq 0$; for others, we have $\Delta u = 0$. Show that as $\Delta x \to 0$ through values of the first type, then $\Delta y/\Delta x \to 0$. [*Hint:* Write $\Delta y/\Delta x = (\Delta y/\Delta u)(\Delta u/\Delta x)$.]

(*c*) Show that, when Δx is of the second type, Δy is 0; hence

$$\frac{\Delta y}{\Delta x} = 0.$$

Thus, as $\Delta x \to 0$ through values of the second type, the quotient $\Delta y/\Delta x \to 0$.

(*d*) Combine (*b*) and (*c*) to show that as $\Delta x \to 0$, $\Delta y/\Delta x \to 0$.

3.S SUMMARY

This chapter defined the derivative of a function as

$$\lim_{h \to 0} \frac{f(x+h) - f(x)}{h} \qquad \text{or} \qquad \lim_{\Delta x \to 0} \frac{f(x+\Delta x) - f(x)}{\Delta x}$$

if this limit exists. Informally, the derivative is

$$\text{The limit of } \frac{\text{difference in outputs}}{\text{difference in inputs}},$$

as the difference in inputs approaches 0.

The derivative measures how quickly the value of a function changes. If a slight change in the input causes a relatively large change in the output, the derivative will be large. Most of the chapter was spent developing techniques for computing derivatives without having to return to the definition of the derivative and calculate a (perhaps horrendous) limit each time you want to differentiate some function. These labor-saving methods are listed later in this summary.

The derivative was motivated by slope, velocity, magnification, and density—concepts drawn from geometry or the physical world. The derivative has many more applications, and we will cite four more. Whenever the rate at which some quantity changes is studied, a derivative will surely enter the picture.

Biology

Let $P(t)$ be a differentiable function that estimates the size of a population at time t. Then the derivative $P'(t)$ tells how fast the population is increasing [if $P'(t) > 0$] or decreasing [if $P'(t) < 0$] at time t.

Physiology

Let $Q(t)$ be the amount of blood, in cubic centimeters, that flows through an artery during the first t seconds of an observation. Then the derivative $Q'(t)$ is the rate, in cubic centimeters per second, at which blood flows through the artery at time t.

Economics

Let $C(x)$ be the cost in dollars of producing x refrigerators. [In reality x is an integer; in economic theory and practice it is convenient to assume that $C(x)$ is defined for all real numbers in some interval and that $C(x)$ is a differentiable function.] The derivative $C'(x)$ is called the **marginal cost**. This marginal cost, as we will now show, is roughly the cost of producing the $(x+1)$st refrigerator. The actual cost of producing refrigerator number $x+1$ is the cost of producing the first $x+1$ refrigerators less the cost of producing the first x refrigerators. So the cost of producing the $(x+1)$st refrigerator is $C(x+1) - C(x)$, which equals

$$\frac{C(x+1) - C(x)}{1},$$

which, by the definition of the derivative, is an approximation of $C'(x)$. Or, looked at the opposite way, $C'(x)$ is an approximation of the ratio $[C(x+1) - C(x)]/1$, the cost of the $(x+1)$st refrigerator.

Similarly, if $R(x)$ is the total revenue received for x refrigerators, then the derivative $R'(x)$ is called the **marginal revenue**, which can be thought of as the extra revenue obtained by selling the $(x+1)$st refrigerator.

Energy

Let $Q(t)$ be the total amount of crude oil in the earth at time t, measured in barrels. (One barrel holds 42 gallons.) The derivative $Q'(t)$ tells how fast $Q(t)$ changes. If no new reserves are being formed, then $Q'(t)$ is negative, approximately $-50{,}000{,}000$ barrels per day. Estimates of $Q(t)$ for $t = 1980$ vary but are on the order of $2 \cdot 10^{12}$ barrels (two trillion barrels). If $Q'(t)$ remains constant, all known and conjectured reserves would be used up in about a century.

Predictions of the rate at which petroleum—or any other natural resource—will be used depend on estimates of derivatives.

The following table is worth careful study. The bottom row describes the derivative, which is the underlying mathematical concept, free of any particular interpretation. Each of the other lines describes one of its many *applications* or *interpretations*.

If we interpret x as . . .	*and f(x) as . . .*	*then* $\frac{f(x + \Delta x) - f(x)}{\Delta x}$ *is . . .*	*and, as* Δx *approaches* 0, *the quotient approaches . . .*
The abscissa of a point in the plane	The ordinate of that point	The slope of a certain line	The slope of a tangent line
Time	The location of a particle moving on a line	An average velocity over a time interval	The velocity at time x
A point on a linear slide	Its projection on a linear screen	An average magnification	The magnification at x
A location on a non-uniform string	The mass from 0 to x	An average density	The density at x
Time	Mass of a bacterial culture at time x	An average growth rate over a time interval	The growth rate at time x
Time	Total profit up to time x	An average rate of profit over a time interval	The rate of profit at time x
Just a number: the input	A number depending on x: the output	A quotient: the change in the output divided by the change in the input	The derivative evaluated at x (the rate of change of the function with respect to x)

Vocabulary and Symbols

difference quotient $\frac{f(x + \Delta x) - f(x)}{\Delta x}$

derivative f', $\frac{d}{dx}(f)$, $\frac{df}{dx}$, $\frac{dy}{dx}$, $D(f)$, and the dot notation $\dot{x} = \frac{dx}{dt}$

differentiable
velocity and speed
magnification
density
tangent to a curve
antiderivative
chain rule
Δx (change in input)
Δf (change in output)
slope of a curve

Key Facts

The derivative is defined as a limit:

$$\lim_{\Delta x \to 0} \frac{f(x + \Delta x) - f(x)}{\Delta x}$$

if the limit exists.

If $f'(x)$ exists at a particular number x, then f is said to be differentiable at that number. A function that is differentiable throughout an interval is said to be differentiable on that interval. Most functions met in applications are differentiable throughout their domains with perhaps the exception of a few isolated points. For instance, $\sqrt{x}$ is differentiable throughout $(0, \infty)$ but not at 0.

Wherever a function is differentiable it is necessarily continuous. However, a function can be continuous at a number yet not be differentiable there. For example $|x|$ is continuous at 0 but not differentiable there.

The computational formulas and techniques obtained in this chapter are recorded in the following two tables.

FORMULAS FOR DERIVATIVES

f	f'	*Remark*
c	0	Constant function.
x	1	
x^a	ax^{a-1}	a is a constant.
$\sqrt{x}$	$\frac{1}{2\sqrt{x}}$	Although a special case of x^a, worth memorizing.
$\frac{1}{x}$	$\frac{-1}{x^2}$	Although a special case of x^a, worth memorizing.
$a_nx^n + \cdots + a_1x + a_0$	$na_nx^{n-1} + \cdots + a_1$	Differentiate polynomials term by term.
$\sin x$	$\cos x$	
$\cos x$	$-\sin x$	Remember that the
$\tan x$	$\sec^2 x$	derivatives of the "co"
$\cot x$	$-\csc^2 x$	functions have the
$\sec x$	$\sec x \tan x$	minus sign.
$\csc x$	$-\csc x \cot x$	

Combining these formulas with the chain rule shows, for instance, that

$$\frac{d}{dx}((u(x))^a) = a(u(x))^{a-1}u'(x)$$

and

$$\frac{d}{dx}(\sin u(x)) = (\cos u(x))u'(x),$$

where $u(x)$ is a differentiable function of x.

TECHNIQUES OF DIFFERENTIATION

$$(cf)' = cf' \qquad \left(\frac{f}{c}\right)' = \frac{f'}{c} \qquad (c \text{ constant})$$

$$(f+g)' = f' + g' \qquad (f-g)' = f' - g'$$

$$(fg)' = fg' + gf' \qquad \left(\frac{f}{g}\right)' = \frac{gf' - fg'}{g^2} \qquad \left(\frac{1}{g}\right)' = -\frac{g'}{g^2}$$

And, most important, most often used, the chain rule: If $y = f(u)$ and $u = g(x)$, then

$$\frac{dy}{dx} = \frac{dy}{du}\frac{du}{dx}.$$

Any function F whose derivative is f is an "antiderivative" of f. For instance,

$$F(x) = \frac{x^{a+1}}{a+1} + C \qquad (a \neq -1)$$

is an antiderivative of x^a for any constant C.

GUIDE QUIZ ON CHAP. 3: THE DERIVATIVE

1 Define the derivative.

2 Use the definition of the derivative to compute

(*a*) $\frac{d}{dx}(5x^3 - 2x + 2)$ (*b*) $\frac{d}{dx}\left(\frac{5}{3x+2} + 6x\right)$

(*c*) $\frac{d}{dx}(3 \sin 2x)$ (*d*) $\frac{d}{dx}(x^{-2})$

3 Using formulas developed in the chapter, differentiate

(*a*) $5\sqrt{x}$ (*b*) $x^2\sqrt{3 - 2x^2}$ (*c*) $\cos 5x$

(*d*) $(1 + x^2)^{3/4}$ (*e*) $\sqrt[3]{\tan 6x}$ (*f*) $x^3 \sin 5x$

(*g*) $\frac{1}{\sqrt{2x+1}}$ (*h*) $(2x^5 - x^3)^{-4}$ (*i*) $\sqrt[3]{x^3 - 3}$

(*j*) $\frac{2x^3+2}{3x+1}$ (*k*) $\frac{1}{5x^2+1}$ (*l*) $\frac{1}{(3x+2)^{10}}$

(*m*) $(1 + 2x)^5 x^3 \sec 3x$ (*n*) $\csc \sqrt{x}$

(*o*) $(1 + 3 \cot 4x)^{-2}$

4 On a sketch of the graph of a typical function f,

(*a*) show the line whose slope is $[f(x + h) - f(x)]/h$;

(*b*) show the tangent line at the point $(x, f(x))$.

5 (*a*) Sketch the graph of $y = 3x^2 + 5x + 6$.

(*b*) By inspection of your graph, estimate the x coordinate of the point where the tangent line is horizontal.

(*c*) Using the derivative, solve (*b*) precisely.

6 (*a*) Without sketching the graph of $y = x^4$, draw the line that is tangent to it at the point $(\frac{1}{2}, \frac{1}{16})$.

(*b*) Find an equation for this line.

7 (*a*) Graph $y = x^3 - 12x$ with the aid of this table.

x	-2	-1	0	1	2	3
$x^3 - 12x$						

(*b*) Evaluate $(x^3 - 12x)'$.

(*c*) Find all x such that the derivative in (*b*) has the value 0.

(*d*) At what points on the graph in (*a*) is the tangent line horizontal? Specify both the x and y coordinates.

8 Let f be a differentiable function and let Δx be a positive number. Interpret $f(x + \Delta x) - f(x)$, Δx and their quotient $[f(x + \Delta x) - f(x)]/\Delta x$ if (*a*) $f(x)$ is the height of a rocket x seconds after lift-off; (*b*) $f(x)$ is the number of bacteria in a bacterial culture at time x; (*c*) $f(x)$ is the mass of the left-hand x centimeters of a rod; (*d*) $f(x)$ is the position of the image on the linear screen of the point x on the linear slide.

9 A bug is wandering on the x axis. At time t seconds it is at the point $x = t^2 - 2t$. Assume that distance is measured in meters.

(*a*) What is the bug's velocity at time t?

(*b*) What is the bug's velocity at $t = \frac{1}{4}$?

(*c*) What is the bug's speed when $t = \frac{1}{4}$?

(*d*) Is the bug moving to the right or to the left when $t = \frac{1}{4}$?

10 Give an antiderivative of

(*a*) x^5

(*b*) $\sin 2x$

(*c*) $2x \cos x^2$

11 Figure 1 is the graph of a function $f(x)$.

(*a*) Estimate $f(1)$ and $f'(1)$.

(*b*) Let $g(x) = (f(x))^2$. Estimate $g'(3)$.

(*c*) Let $h(x) = 1/f(x)$. Estimate $h'(1)$.

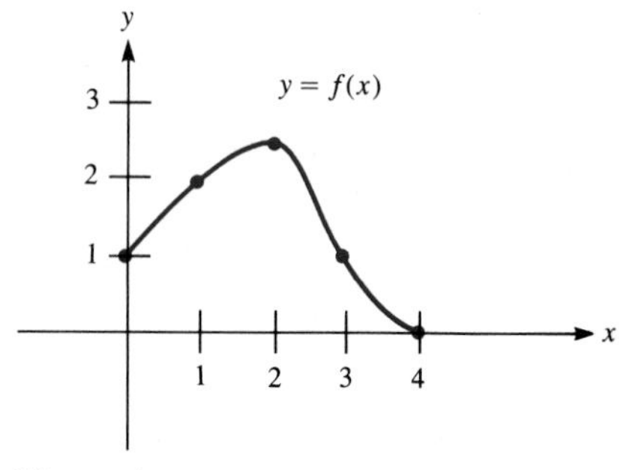

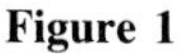

Figure 1

REVIEW EXERCISES FOR CHAP. 3: THE DERIVATIVE

In Exercises 1 to 6 use the definition of the derivative to differentiate the given functions.

1 $5x^3$ **2** $\sqrt{3x}$ **3** $1/(x + 3)$

4 $(2x + 1)^2$ **5** $\cos 3x$ **6** $\sin 5x$

In Exercises 7 to 36 find the derivatives of the given functions.

7 $2x^5 + x^3 - x$ **8** $t^4 - 5t^2 + 2$

9 $\frac{x^2}{4x+1}$ **10** $\frac{(3x+1)^4}{(2x-1)^2}$

11 $\sqrt{3x^2 + 2x + 4}$ **12** $\sqrt{5x^2 - x}$

13 $\sqrt[3]{(2t - 1)^2}$ **14** $(t^2 + 1)^{3/4}$

15 $\sin^2 5x$ **16** $\cos^3 7x$

17 $\frac{(5x+1)^4}{7}$ **18** $\frac{(3x-2)^{-5}}{11}$

19 $x \sin 3x$ **20** $x^2 \cos 4x$

21 $\tan^2 \sqrt[3]{1 + 2x}$ **22** $\left(\frac{\sin 2x}{1 + \tan 3x}\right)^3$

23 $\dfrac{x^3 \cos 2x}{1 + x^2}$

24 $\dfrac{x^2 \sin 5x}{(2x + 1)^3}$

25 $\sqrt[3]{1 + \sqrt{x^2 + 3}}$

26 $\sqrt[4]{x^3 + \sqrt{x + \sin x}}$

27 $\sqrt[3]{(\cot 5x)^7}$

28 $\sqrt[5]{(\csc 3x)^{11}}$

29 $\sqrt{8x + 3}$

30 $\sqrt{5x - 1}$

31 $\dfrac{x^2}{x^3 + 1}$

32 $\dfrac{x^3 + 1}{x^2}$

33 $[(x^2 + 3x)^4 + x]^{-5/7}$

34 $\left(\dfrac{3x + 1}{2x + 1}\right)^4$

35 $x\sqrt{2x + 1} \cos 6x$

36 $\dfrac{\sin 4x}{x\sqrt{2x + 1}}$

In Exercises 37 to 40 the number a is a constant. In each case differentiate the given function and simplify your answer.

37 $\dfrac{1}{a^2} \sin ax - \dfrac{1}{a} x \cos ax$

38 $\dfrac{x^2}{4} - \dfrac{x \sin 2ax}{4a} - \dfrac{\cos 2ax}{8a^2}$

39 $\dfrac{x}{2} - \dfrac{\sin 2ax}{4a}$

40 $-\dfrac{1}{a} \cos ax + \dfrac{1}{3a} \cos^3 ax$

41 The height of a ball thrown straight up is $64t - 16t^2$ feet after t seconds.
(*a*) Show that its velocity after t seconds is $64 - 32t$ feet per second.
(*b*) What is its velocity when $t = 0$? $t = 1$? $t = 2$? $t = 3$?
(*c*) What is its speed when $t = 0$? $t = 1$? $t = 2$? $t = 3$?
(*d*) For what values of t is the ball rising? falling?

42 In the study of the seepage of irrigation water into soil, equations such as $y = \sqrt{t}$ are sometimes used. The equation says that the water penetrates $\sqrt{t}$ feet in t hours.
(*a*) What is the physical significance of $(\sqrt{t})' = 1/(2\sqrt{t})$?
(*b*) What does (*a*) say about the rate at which water penetrates the soil when t is large?

43 Find an equation of the tangent line to the curve $y = x^3 - 2x^2$ at $(1, -1)$.

44 Find an equation of the tangent line to the curve $y = 2x^4 - 6x^2 + 8$ at $(2, 16)$.

45 (*a*) The left-hand x centimeters of a rod have a mass of $3x^4$ grams. What is its density at $x = 1$?
(*b*) Devise a magnification problem mathematically equivalent to (*a*).
(*c*) Devise a velocity problem mathematically equivalent to (*a*).

46 The left-hand x centimeters of a string has a mass of $\sqrt{x}$ grams. What is its density when x is (*a*) $\frac{1}{4}$? (*b*) 1? (*c*) Is its density defined at $x = 0$?

47 A snail crawls $\sqrt{t}$ feet in t seconds. What is its speed when t is (*a*) $\frac{1}{9}$? (*b*) 1? (*c*) 4? (*d*) 9?

48 Sketch a graph of $y = x^3$.
(*a*) Why can the tangent line to this graph at $(0, 0)$ *not* be defined as "the line through $(0, 0)$ that meets the graph just once"?
(*b*) Why can the tangent line to this graph at $(1, 1)$ *not* be defined as "the line through $(1, 1)$ that meets the graph just once"?
(*c*) How is the tangent line at any point on the graph defined?

49 (Contributed by David G. Mead) (*a*) Sketch the curves $y = x^2 + 1$ and $y = -x^2$. (*b*) Find equations of the lines that are tangent to both curves simultaneously.

50 After t hours a certain bacterial population has a mass of $500 + t^3$ grams. Find the rate at which it grows (in grams per hour) when (*a*) $t = 0$, (*b*) $t = 1$, and (*c*) $t = 2$.

51 A lens projects the point x on the x axis onto the point x^3 on a linear screen.
(*a*) How much does it magnify the interval $[2, 2.1]$?
(*b*) How much does it magnify the interval $[1.9, 2]$?
(*c*) What is its magnification at 2?

52 Let f be the function whose value at x is $4x^2$.
(*a*) Compute $[f(2.1) - f(2)]/0.1$.
(*b*) What is the interpretation of the quotient in (*a*) if $f(x)$ denotes the total profit of a firm (in millions of dollars) in its first x years?
(*c*) What is the interpretation of the quotient in (*a*) if $f(x)$ denotes the ordinate in the graph of the parabola $y = 4x^2$?
(*d*) What is the interpretation of the quotient in (*a*) if $f(x)$ is the distance a particle moves in the first x seconds?

53 It costs a certain firm $C(x) = 1000 + 5x + x^2/200$ dollars to produce x calculators, for $x \le 400$.
(*a*) How much does it cost to produce 0 calculators? (This represents start-up costs, which are independent of the number produced.)
(*b*) What is the marginal cost $C'(x)$?
(*c*) What is the marginal cost when $x = 10$?
(*d*) Compute $C(11) - C(10)$, the cost of producing the eleventh calculator.

54 (*a*) If the function f records the trade-in value of a car (dependent on its age), then we may think of the derivative f' as ______.
(*b*) When is the derivative of (*a*) negative? positive? Which is the more usual case?

55 (*Economics*) A certain growing business firm makes a profit of t^2 million dollars in its first t years.
(*a*) How much profit does it make during its third year, that is, from time $t = 2$ to time $t = 3$?
(*b*) How much profit does it make from time $t = 2$ to time $t = 2.5$ (a duration of half a year)?
(*c*) Using (*b*), show that its average rate of profit from time $t = 2$ to time $t = 2.5$ is 4.5 million dollars per year.
(*d*) Find its "rate of profit at time $t = 2$" by considering

short intervals of time from 2 to t, $t > 2$, and letting t approach 2.

56 *(Biology)* A certain increasing bacterial population has a mass of t^2 grams after t hours.

(*a*) By how many grams does the population increase from time $t = 3$ hours to time $t = 4$ hours?

(*b*) By how many grams does it increase from time $t = 3$ hours to $t = 3.01$ hours?

(*c*) By how many grams does the population increase from time $t = 3$ hours to time t hours, where t is larger than 3?

(*d*) Using (*c*), show that the average rate of growth from time 3 to time t, $t > 3$, is $3 + t$ grams per hour. As t approaches 3, the average growth rate approaches 6 grams per hour, which is called "the growth rate at time 3 hours."

57 In each of these functions, y denotes a differentiable function of x. Express the derivative of each with respect to x in terms of y and dy/dx.

(*a*) y^3 (*b*) $\cos y$ (*c*) $1/y$

58 Let $f(t)$ be the height in miles of the cloud top above burst height t minutes after the explosion of a 1-megaton nuclear bomb. Figure 2 is a graph of this function. (Note that the vertical and horizontal scales are different.) Assume that the cloud is not dispersed.

(*a*) What is the physical meaning of $f'(t)$?

(*b*) As t increases, what happens to $f'(t)$?

(*c*) As t increases, what happens to $f(t)$?

(*d*) Estimate how rapidly the cloud is rising at the time of explosion.

(*e*) Estimate how rapidly the cloud is rising 1 minute after the explosion.

In (*d*) and (*e*) make the estimate by drawing a tangent line. The estimate will be in miles per minute.

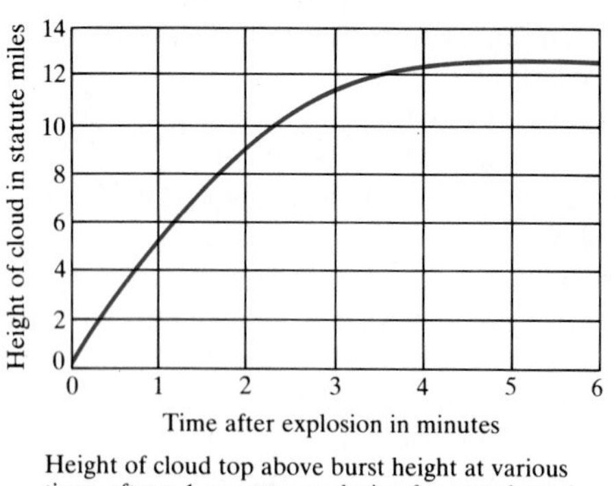

Height of cloud top above burst height at various times after a 1-megaton explosion for a moderately low air burst

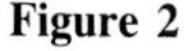

Figure 2

59 Give two antiderivatives for each of these functions:

(*a*) $4x^3$ (*b*) x^3

(*c*) $x^4 + x^3 + \cos x$ (*d*) $x^3 + \sin x$

(*e*) $(x^2 + 1)^2$ *Hint:* For (*e*), first expand $(x^2 + 1)^2$.

60 (*a*) Does $D(x^2 + x^3)$ equal $D(x^2) + D(x^3)$?

(*b*) Does $D(x^2x^3)$ equal $D(x^2)D(x^3)$?

(*c*) Does $D(x^2 - x^3)$ equal $D(x^2) - D(x^3)$?

(*d*) Does $D\left(\frac{x^2}{x^3}\right)$ equal $\frac{D(x^2)}{D(x^3)}$?

61 Figure 3 shows the graph of a function f. (*a*) At which numbers a does $\lim_{x \to a} f(x)$ not exist? (*b*) At which numbers a does $\lim_{x \to a} f(x)$ exist yet f is not continuous at a? (*c*) Where is f continuous but not differentiable?

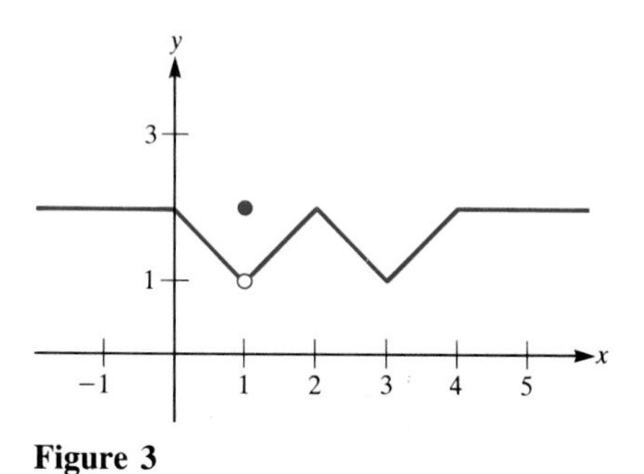

Figure 3

The next exercise briefly introduces a concept (that will later be treated in detail) which might be met early in an elementary physics or chemistry class.

62 In the case of a function of more than one variable we may differentiate with respect to any one of the variables, treating the others as constants. For instance, the derivative of x^3y^2z with respect to x is $3x^2y^2z$; the derivative of x^3y^2z with respect to y is $2x^3yz$; the derivative with respect to z is x^3y^2. Such derivatives are called **partial derivatives** and are denoted with the symbol ∂ as follows:

$$\frac{\partial}{\partial x}(x^3y^2z) = 3x^2y^2z,$$

$$\frac{\partial}{\partial y}(x^3y^2z) = 2x^3yz,$$

$$\frac{\partial}{\partial z}(x^3y^2z) = x^3y^2.$$

Compute the following partial derivatives:

(*a*) $\frac{\partial}{\partial x}\left(\frac{x^2z}{y}\right)$ (*b*) $\frac{\partial}{\partial y}\left(\frac{x^2z}{y}\right)$ (*c*) $\frac{\partial}{\partial z}\left(\frac{x^2z}{y}\right)$

(*d*) $\frac{\partial}{\partial x}(\cos 3x \sin 4y)$ (*e*) $\frac{\partial}{\partial y}(\cos 3x \sin 4y)$

63 Tell what is wrong with this alleged proof that $2 = 1$: Observe that $x^2 = x \cdot x = x + x + \cdots + x$ (x times). Differentiation with respect to x yields the equation $2x = 1 + 1 + \cdots + 1$ (x 1s). Thus $2x = x$. Setting $x = 1$ shows that $2 = 1$.

64 Define $f(x)$ to be $x^2 \sin(1/x)$ if $x \neq 0$, and $f(0)$ to be 0.
(*a*) Show that f has a derivative at 0, namely, 0. *Hint:* Investigate

$$\lim_{\Delta x \to 0} \frac{f(\Delta x) - f(0)}{\Delta x}.$$

(*b*) Show that f has a derivative at $x \neq 0$.
(*c*) Show that the derivative of f is not continuous at $x = 0$.

65 Let $V(r) = 4\pi r^3/3$, the volume of a sphere of radius r. Let $S(r) = 4\pi r^2$, the surface area of a sphere of radius r.
(*a*) Show that $V'(r) = S(r)$.
(*b*) With the aid of a picture showing concentric spheres of radii r and $r + \Delta r$, explain why the result in (*a*) is plausible.

Express the limits in Exercises 66 and 67 as derivatives, and evaluate them.

66 $\displaystyle\lim_{w \to 2} \frac{(1 + w^2)^3 - 125}{w - 2}$

67 $\displaystyle\lim_{\Delta x \to 0} \frac{\sin \sqrt{3 + \Delta x} - \sin \sqrt{3}}{\Delta x}$

68 Find all continuous functions f, whose domains are the x axis, such that $f(0) = 5$ and, for each x, $f(x)$ is an integer.

69 (*a*) Draw a freehand curve indicating a typical function f.
(*b*) Label on it the three points $P_0 = (x, f(x))$, $P_1 = (x + h, f(x + h))$, and $P_2 = (x - h, f(x - h))$.
(*c*) Show that the slope of the line through P_1 and P_2 is

$$\frac{f(x + h) - f(x - h)}{2h}.$$

(*d*) For a differentiable function, what do you think is the value of

$$\lim_{h \to 0} \frac{f(x + h) - f(x - h)}{2h}?$$

(*e*) Compute the limit in (*d*) if $f(x) = x^3$.

70 Let f and g be differentiable functions. Show that

(*a*) $\displaystyle\frac{(fg)'}{fg} = \frac{f'}{f} + \frac{g'}{g}$

(*b*) $\displaystyle\frac{(f/g)'}{f/g} = \frac{f'}{f} - \frac{g'}{g}$

(It is assumed that the denominators are not 0.)
(*c*) Generalize (*a*) to three functions, f, g, and h.

71 The area of a disk of radius r is πr^2 and its circumference is $2\pi r$. Note that $2\pi r$ is the derivative of πr^2. Explain why this is not a coincidence. (*Hint:* Draw two concentric disks of radii r and $r + \Delta r$, and use the definition of the derivative.)

72 * Section 6.1 discusses logarithms. However, even without knowing what they are, you can examine the derivatives of the functions denoted "log" and "ln" on a calculator. This exercise concerns the "ln"-key. Let $f(x) = \ln x$.
(*a*) Estimate $f'(1)$ by computing $\displaystyle\frac{f(1 + \Delta x) - f(1)}{\Delta x}$ for small values of Δx.
(*b*) Estimate $f'(2)$ similarly.
(*c*) Estimate $f'(1/2)$, $f'(3)$, and $f'(4)$ similarly and fill in this table.

x	$\frac{1}{2}$	1	2	3	4
$f'(x)$					

(*d*) Do you see a pattern in the table in (*c*)? Guess the general formula for $f'(x)$.

73 Like Exercise 72, but for $f(x) = \log x$ instead.

74 Pick an "interesting" function on your calculator. Use it to estimate the derivative of that function as various values of x. Make graphs of both f and f'.

75 One interpretation of the derivative is magnification of a projector. Interpret the chain rule in terms of two successive projectors (as in Fig. 3 of Sec. 2.2). Does the interpretation seem reasonable? Explain.

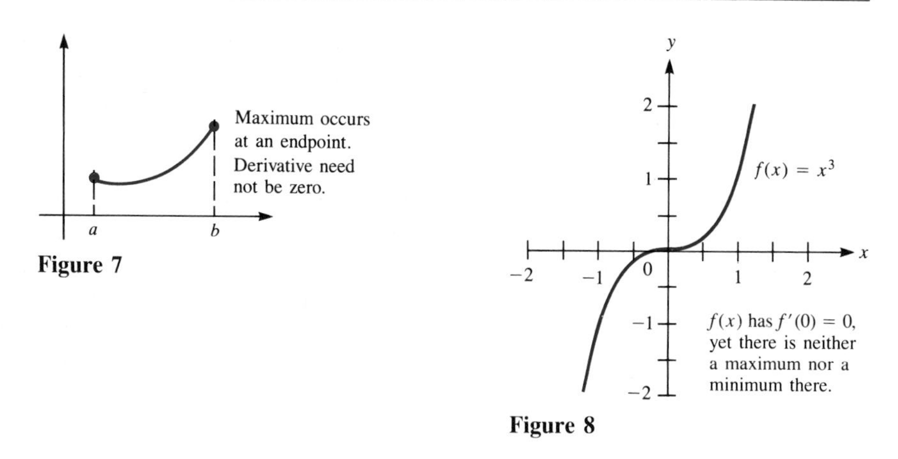

Figure 7

Figure 8

If an extreme value occurs within an open interval and the derivative exists there, the derivative must be 0 there. We will use this theorem beginning in Sec. 4.2 to find maximum and minimum values of a function.

Two warnings must be given right away. First, the theorem isn't necessarily true if the open interval (a, b) is replaced by a closed interval $[a, b]$. A glance at Fig. 7 shows why. If the maximum occurs at an endpoint of the interval, at a or b, as the graph in Fig. 7 illustrates, the derivative at such a point need not be 0. In this graph, the maximum occurs at b, where the derivative is not 0.

Second, the converse of this theorem is not true. Having the derivative equal to 0 at a point does *not* guarantee that there is an extremum at that point. A glance at Fig. 8, which displays the graph of $f(x) = x^3$, shows why. Since $f'(x) = 3x^2, f'(0) = 0$. The tangent is indeed horizontal at $(0, 0)$, but it crosses the curve there. The graph has neither a maximum nor a minimum at the origin.

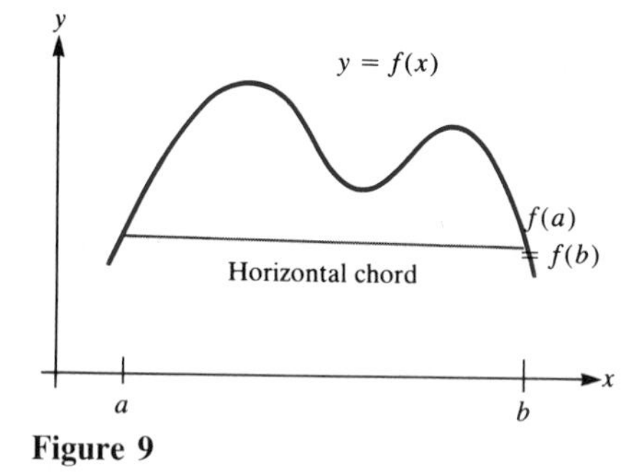

Figure 9

Rolle's Theorem

The next theorem is suggested by the second observation in the special case when the points A and B in Fig. 5 have the same y coordinate. A line segment that joins two points on the graph of a function f is called a *chord* of f. Rolle's theorem relates a horizontal chord and the derivative of f. It says that if the horizontal chord has endpoints $(a, f(a))$ and $(b, f(b))$, then there is at least one number c between a and b such that $f'(c) = 0$; that is, there is a horizontal tangent at $(c, f(c))$. See Figs. 9 and 10.

Specifically, we have

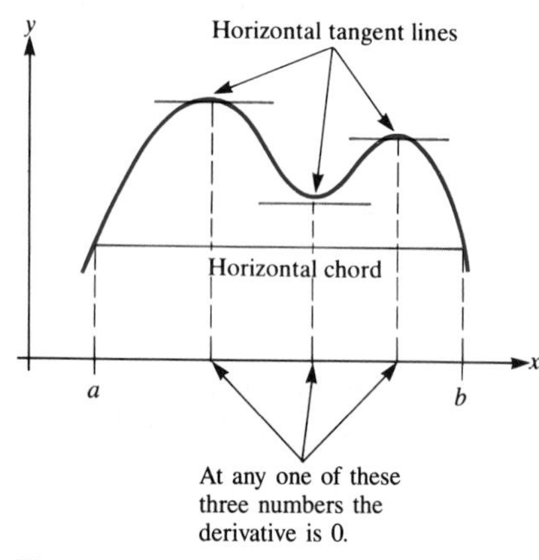

Figure 10

Rolle's Theorem Let f be a continuous function on the closed interval $[a, b]$ and have a derivative at all x in the open interval (a, b). If $f(a) = f(b)$, then there is at least one number c in (a, b) such that $f'(c) = 0$.

EXAMPLE 1 Verify Rolle's theorem for the case $f(x) = \cos x$ and $[a, b] = [\pi, 5\pi]$.

SOLUTION Note that $f(\pi) = -1 = f(5\pi)$. Since $\cos x$ is differentiable for all x, it is continuous on $[\pi, 5\pi]$ and differentiable on $(\pi, 5\pi)$. According to Rolle's theorem, there must be at least one number c in $(\pi, 5\pi)$ for which

$(\cos x)'$ is 0. Now, $(\cos x)' = -\sin x$. Thus there should be at least one solution of the equation

$$-\sin x = 0$$

in the open interval $(\pi, 5\pi)$. As can be checked, the equation has three such solutions, namely, 2π, 3π, and 4π. ■

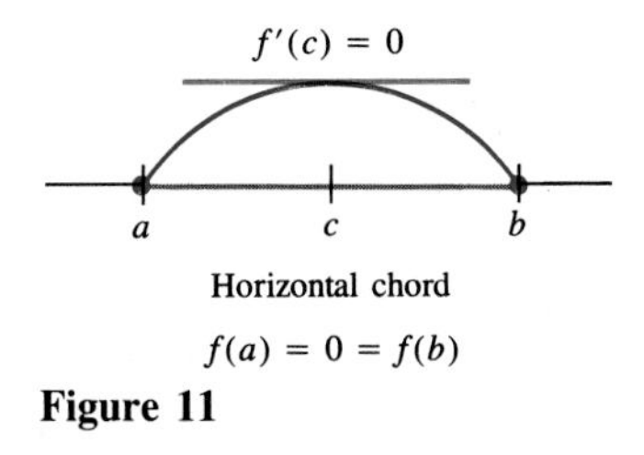

Figure 11

Remark: Assume that $f(x)$ is a differentiable function such that $f'(x)$ is never 0 for x in the interval $[d, e]$. Then the equation $f(x) = 0$ *can have at most one solution in that interval.* If it has two solutions, a and b, then $f(a) = 0$ and $f(b) = 0$. (See Fig. 11.) By Rolle's theorem, there would be a number c between a and b such that $f'(c) = 0$. This violates our assumption that the derivative is never 0 for x in the interval $[d, e]$. In short,

> In an interval in which the derivative $f'(x)$ is never 0, the graph of $y = f(x)$ can have no more than one x intercept.

Example 2 illustrates this observation.

EXAMPLE 2 Use Rolle's theorem to determine how many real roots there are for the equation

$$x^3 - 6x^2 + 15x + 3 = 0. \tag{1}$$

SOLUTION Since $f(x) = x^3 - 6x^2 + 15x + 3$ is a polynomial of odd degree, there is at least one real number r such that $f(r) = 0$. (Recall the argument in Sec. 2.8 based on the intermediate-value theorem.) Could there be another root s? If so, by Rolle's theorem, there would be a number c (between r and s) at which $f'(c) = 0$.

To check, we compute the derivative of $f(x)$ and see if it is ever equal to 0. We have $f'(x) = 3x^2 - 12x + 15$.

To find when $f'(x)$ is 0, we solve the equation $3x^2 - 12x + 15 = 0$ by the quadratic formula, obtaining

$$x = \frac{12 \pm \sqrt{144 - 180}}{6}.$$

Since $144 - 180$ is negative, the equation has no real roots. Since $f'(x)$ is never 0, it follows that the polynomial $x^3 - 6x^2 + 15x + 3$ has only one real root. ■

Example 8 will illustrate a way of estimating the unique solution of Eq. (1).

The Mean-Value Theorem

The third theorem, the "mean-value" theorem, is a generalization of Rolle's theorem, since it concerns any chord of f, not just horizontal ones.

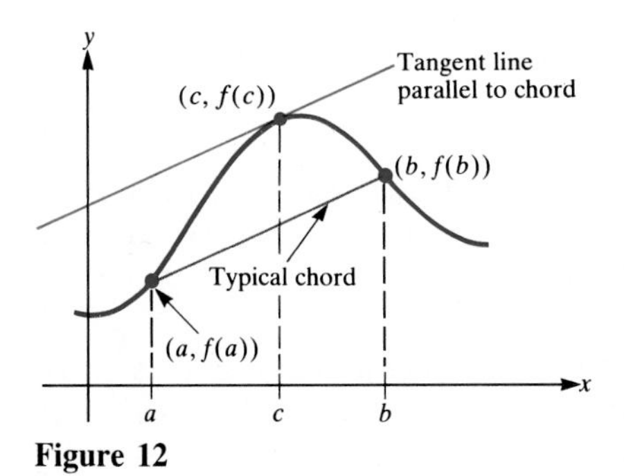

Figure 12

In geometric terms, the theorem asserts that if you draw a chord for the graph of a well-behaved function (as in Fig. 12), then somewhere above or below that chord the graph has at least one tangent line parallel to the chord. (See Fig. 6.)

Let us translate this geometric statement into the language of functions. Call the ends of the chord $(a, f(a))$ and $(b, f(b))$. The slope of the chord is

$$\frac{f(b) - f(a)}{b - a}.$$

Since the tangent line and the chord are parallel, they have the same slopes. If the tangent line is at the point $(c, f(c))$, then

$$f'(c) = \frac{f(b) - f(a)}{b - a}.$$

Specifically, we have

Mean-Value Theorem Let f be a continuous function on the closed interval $[a, b]$ and have a derivative at every x in the open interval (a, b). Then there is at least one number c in the open interval (a, b) such that

$$f'(c) = \frac{f(b) - f(a)}{b - a}.$$

EXAMPLE 3 Verify the mean-value theorem for $f(x) = 2x^3 - 8x + 1$, $a = 1$, and $b = 3$.

SOLUTION

$$f(a) = f(1) = 2 \cdot 1^3 - 8 \cdot 1 + 1 = -5$$

and $$f(b) = f(3) = 2 \cdot 3^3 - 8 \cdot 3 + 1 = 31.$$

According to the mean-value theorem, there is at least one number c between $a = 1$ and $b = 3$ such that

$$f'(c) = \frac{31 - (-5)}{3 - 1} = \frac{36}{2} = 18.$$

Let us find c explicitly. Since $f'(x) = 6x^2 - 8$, we need to solve the equation

$$6x^2 - 8 = 18.$$

that is, $$6x^2 = 26$$

or $$x^2 = \tfrac{26}{6}.$$

The solutions are $\sqrt{\tfrac{13}{3}}$ and $-\sqrt{\tfrac{13}{3}}$. But only $\sqrt{\tfrac{13}{3}}$ is in $(1, 3)$. Hence there is only one number, namely $\sqrt{\tfrac{13}{3}}$, that serves as the c whose existence is guaranteed by the mean-value theorem. ■

The interpretation of the derivative as slope suggested the mean-value theorem. What does the mean-value theorem say when the derivative is interpreted, say, as velocity? This question is considered in Example 4.

EXAMPLE 4 A car moving on the x axis has the x coordinate $f(t)$ at time t. At time a its position is $f(a)$. At some later time b its position is $f(b)$. What does the mean-value theorem assert for this car?

SOLUTION In this case the quotient

$$\frac{f(b) - f(a)}{b - a}$$

equals
$$\frac{\text{Change in position}}{\text{Change in time}};$$

that is,
$$f'(c) = \frac{f(b) - f(a)}{b - a},$$

or "average velocity" for the interval of time $[a, b]$. The mean-value theorem asserts that at some time c during this period the velocity of the car must equal its average velocity. To be specific, if a car travels 210 miles in 3 hours, then at some time its speedometer must read 70 miles per hour. ■

Consequences of the Mean-Value Theorem

There are several ways of writing the mean-value theorem. For example, the equation

$$f'(c) = \frac{f(b) - f(a)}{b - a}$$

is equivalent to
$$f(b) - f(a) = (b - a)f'(c),$$

and hence to
$$f(b) = f(a) + (b - a)f'(c).$$

A different view of the mean-value theorem

In this form, the mean-value theorem asserts that $f(b)$ is equal to $f(a)$ plus a quantity that involves the derivative f' at some number c between a and b. The following important corollaries exploit this alternative view of the mean-value theorem.

Corollary 1 If the derivative of a function is 0 throughout an interval, then the function is constant throughout that interval.

Proof Let a and b be any two numbers in the interval and let the function be denoted by f. To prove the corollary, it suffices to prove that $f(a) = f(b)$, for that forces the function to be constant.

By the mean-value theorem there is a number c between a and b such that

$$f(b) = f(a) + (b - a)f'(c).$$

But $f'(c) = 0$, since $f'(x)$ is 0 for all x in the given interval. Hence

$$f(b) = f(a) + (b - a)(0),$$

which proves that
$$f(b) = f(a). \quad ■$$

When Corollary 1 is interpreted in terms of motion, it is quite plausible. It asserts that if a particle has zero velocity for a period of time, then it does not move during that time.

EXAMPLE 5 Use Corollary 1 to show that $f(x) = \cos^2 3x + \sin^2 3x$ is a constant. Find the constant.

SOLUTION $f'(x) = -6 \cos 3x \sin 3x + 6 \sin 3x \cos 3x = 0$. Corollary 1 says that f is constant. To find the constant, just evaluate f at some specific number of your choice, say at 0. We have $f(0) = \cos^2 (3 \cdot 0) + \sin^2 (3 \cdot 0) = \cos^2 0 + \sin^2 0 = 1^2 + 0^2 = 1$. Thus

$$\cos^2 3x + \sin^2 3x = 1$$

for all x. This should be no surprise since, by the pythagorean theorem, $\cos^2 \theta + \sin^2 \theta = 1$. ■

Corollary 2 If two functions have the same derivatives throughout an interval, then they differ by a constant. That is, if $f'(x) = g'(x)$ for all x in an interval, then there is a constant C such that $f(x) = g(x) + C$.

Proof Define a third function h by the equation

$$h(x) = f(x) - g(x).$$

Then $$h'(x) = f'(x) - g'(x) = 0. \qquad [\text{since } f'(x) = g'(x)]$$

Since the derivative of h is 0, Corollary 1 implies that h is constant, that is, $h(x) = C$ for some fixed number C. Thus

$$f(x) - g(x) = C \qquad \text{or} \qquad f(x) = g(x) + C,$$

and the corollary is proved. ■

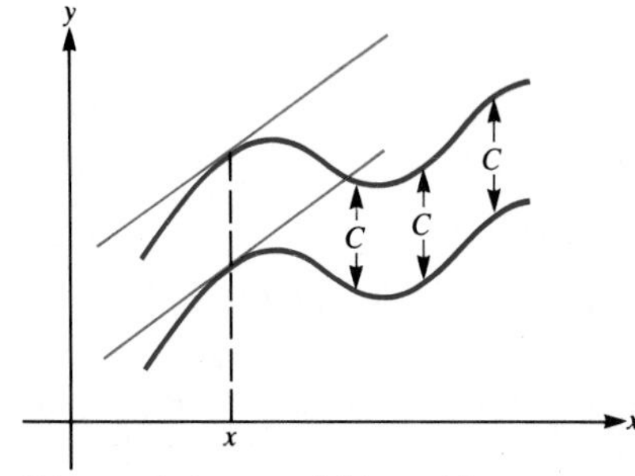

If two graphs have parallel tangent lines at all pairs of points with the same x coordinate, then one graph is obtainable from the other by raising or lowering it.

Figure 13

Is Corollary 2 plausible when the derivative is interpreted as slope? In this case, the corollary asserts that if the graphs of two functions have the property that their tangent lines at points with the same x coordinate are parallel, then one graph can be obtained from the other by raising (or lowering) it by an amount C. If you sketch two such graphs (as in Fig. 13), you will see that the corollary is reasonable.

EXAMPLE 6 What functions have a derivative equal to $2x$ everywhere?

Any antiderivative of $2x$ must be of the form $x^2 + C$.

SOLUTION One such function is x^2; another is $x^2 + 25$. For any constant C, $D(x^2 + C) = 2x$. Are there any other possibilities? Corollary 2 tells us there are not, for if f is a function such that $f'(x) = 2x$, then $f'(x) = (x^2)'$ for all x. Thus the functions f and x^2 differ by a constant, say C, that is,

$$f(x) = x^2 + C.$$

The only functions whose derivatives are $2x$ are of the form $x^2 + C$. ■

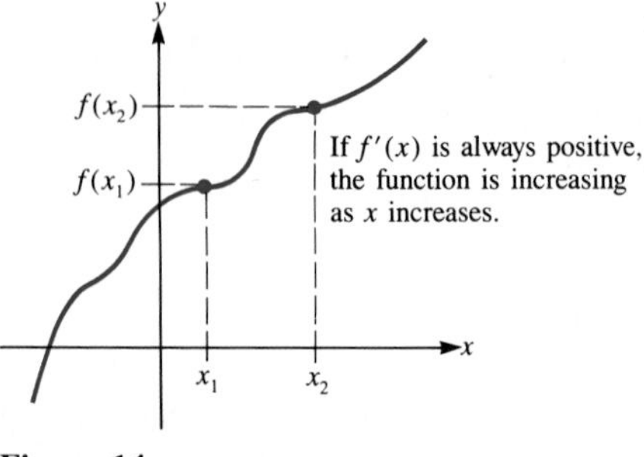

Figure 14

Corollary 1 asserts that if $f'(x) = 0$ for all x, then f is constant. What can be said about f if $f'(x)$ is *positive* for all x? In terms of the graph of f, this assumption implies that all the tangent lines slope upward. It is reasonable to expect that as we move from left to right on the graph in Fig. 14, the y coordinate increases.

A function f is said to be *increasing* if whenever $x_1 < x_2$, one has $f(x_1) < f(x_2)$. A function is *decreasing* if whenever $x_1 < x_2, f(x_1) > f(x_2)$. (See Fig. 14.)

Corollary 3 If f is continuous on $[a, b]$ and has a positive derivative on the open interval (a, b), then f is increasing on the interval $[a, b]$. If f is continuous on $[a, b]$ and has a negative derivative on the open interval (a, b), then f is decreasing on the interval $[a, b]$.

Proof We prove the "increasing" case. Take two numbers x_1 and x_2 such that

$$a \leq x_1 < x_2 \leq b.$$

By the mean-value theorem, there is some number c between x_1 and x_2 such that

$$f(x_2) = f(x_1) + (x_2 - x_1)f'(c).$$

Now, since $x_2 > x_1$, we know $x_2 - x_1$ is positive. Since $f'(c)$ is assumed to be positive, it follows that

$$(x_2 - x_1)f'(c) > 0.$$

(The product of two positive numbers is positive.) Thus $f(x_2) > f(x_1)$, and so $f(x)$ is an increasing function. The corollary is proved. (The "decreasing" case is proved similarly.) ■

EXAMPLE 7 Determine whether $3x + 2 \sin x$ is an increasing function, a decreasing function, or neither.

SOLUTION The function $3x + 2 \sin x$ is made up of $3x$ and $2 \sin x$. The "$3x$" part is an increasing function. The "$2 \sin x$," like $\sin x$, increases as x goes from 0 to $\pi/2$ and decreases as x goes from $\pi/2$ to π. It isn't clear what type of function you will get when you add $3x$ and $2 \sin x$. Let's see what Corollary 3 tells us.

The derivative of $3x + 2 \sin x$ is $3 + 2 \cos x$. Since $\cos x \geq -1$ for all x,

$$3 + 2 \cos x \geq 3 + 2(-1) = 1.$$

Since $(3x + 2 \sin x)'$ is positive, $3x + 2 \sin x$ is an increasing function. It is graphed in Fig. 15, which includes the graphs of $3x$ and $2 \sin x$. ■

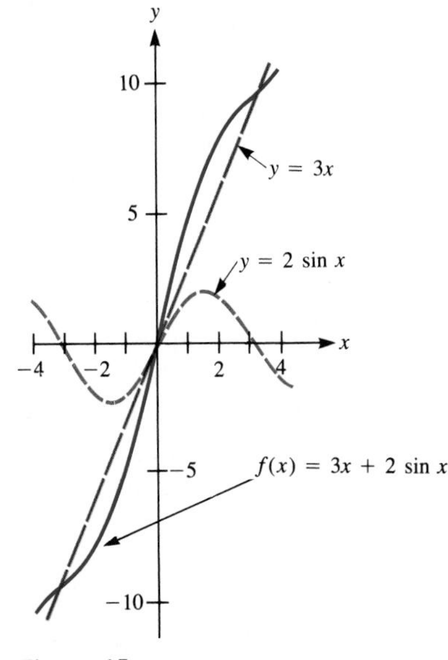

Figure 15

The next example uses Corollary 3 to estimate the root of Eq. (1) in Example 2. The technique is called the "bisection method."

EXAMPLE 8 Estimate the (unique) solution of the equation

$$x^3 - 6x^2 + 15x + 3 = 0.$$

SOLUTION Let $f(x) = x^3 - 6x^2 + 15x + 3$. It was shown in Example 2 that its derivative, $f'(x) = 3x^2 - 12x + 15$, is never 0. Therefore the derivative is always positive or else it is always negative. [If $f'(x)$ were positive somewhere and negative somewhere, then, by the intermediate-value theorem, it would be 0 somewhere in between.] To find whether $f'(x)$ is always positive or is always negative, evaluate $f'(a)$ for any number of your choice, say $a = 2$. Since $f'(2) = 3(2^2) - 12(2) + 15 = 12 - 24 + 15 = 3$, which is positive, $f'(x)$ is *positive* for all x. By Corollary 3, $f(x)$ is an increasing function.

Let r be the (unique) solution of the equation $f(x) = 0$. In order to begin to estimate r, let us find a number a such that $f(a) < 0$ and a number b such that $f(b) > 0$. (Since $f(x)$ is increasing a must be less than b.) Then r must be in the interval $[a, b]$.

To find these numbers a and b, try some number on a calculator, starting with "easy numbers."

The easiest input is 0. We have $f(0) = 0^3 - 6(0^2) + 15(0) + 3 = 3$. Since $f(0)$ is positive and f is increasing, we already know that the solution to $f(x) = 0$ is less than 0, that is, it's negative.

Try the input -1: $f(-1) = (-1)^3 - 6(-1)^2 + 15(-1) + 3 = -1 - 6 - 15 + 3 = -19$. Since $f(-1)$ is negative and $f(0)$ is positive, the root lies somewhere in $[-1, 0]$.

To pin down the root more precisely, compute $f(x)$ *at the midpoint* of the interval $[-1, 0]$, that is, compute $f(-\frac{1}{2})$:

$$\begin{aligned} f(-\tfrac{1}{2}) &= (-\tfrac{1}{2})^3 - 6(-\tfrac{1}{2})^2 + 15(-\tfrac{1}{2}) + 3 \\ &= -\tfrac{1}{8} - \tfrac{6}{4} - \tfrac{15}{2} + 3 \\ &= -\tfrac{49}{8}. \end{aligned}$$

Since $f(-\frac{1}{2})$ is negative and $f(0)$ is positive, the root lies in the interval $[-\frac{1}{2}, 0]$. (Had $f(-\frac{1}{2})$ been positive, the root would lie in $[-1, -\frac{1}{2}]$.)

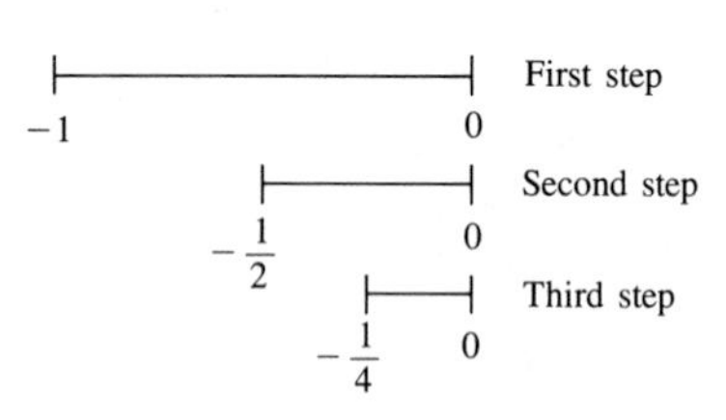

Figure 16

We may continue this bisection process as many times as we please. At the next step we compute $f(-\frac{1}{4})$ since $-\frac{1}{4}$ is the midpoint of the interval $[-\frac{1}{2}, 0]$. You can check that $f(-\frac{1}{4}) = -\frac{73}{64}$ or evaluate it on your calculator. Thus the root lies in the interval $[-\frac{1}{4}, 0]$.

At each step the length of the interval in which we know the root is situated is cut in half. (See Fig. 16.) ■

The method used in Example 8 is called the **bisection technique**.

Proofs of the Three Theorems

Proof of the Interior-Extremum Theorem [$f'(c) = 0$ at maximum or minimum in open interval.] We will prove this for the case of a maximum, the other case being similar. Assume that $f(c)$ is the maximum value of f on $[a, b]$, that c is in (a, b), and that $f'(c)$ exists. It is to be shown that $f'(c) = 0$. We shall prove that $f'(c) \le 0$ and that $f'(c) \ge 0$. From this it will follow that $f'(c)$ must be 0.

Consider the quotient

$$\frac{f(c + \Delta x) - f(c)}{\Delta x}$$

used in defining $f'(c)$. Take Δx so small that $c + \Delta x$ is in the interval $[a, b]$. Since $f(c)$ is the maximum value of $f(x)$ for x in $[a, b]$,

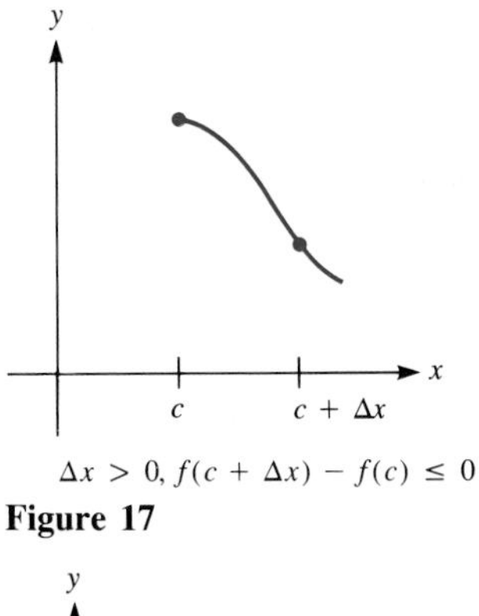

$\Delta x > 0,\ f(c+\Delta x) - f(c) \leq 0$

Figure 17

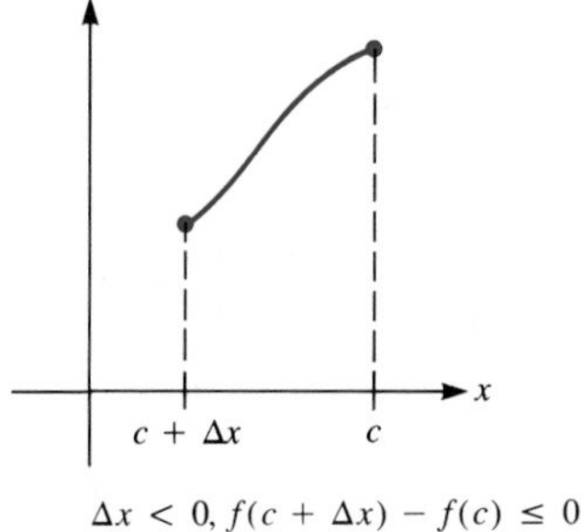

$\Delta x < 0,\ f(c+\Delta x) - f(c) \leq 0$

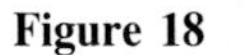

Figure 18

$$f(c + \Delta x) \leq f(c).$$

Hence
$$f(c + \Delta x) - f(c) \leq 0.$$

Therefore, when Δx is positive,

$$\frac{f(c + \Delta x) - f(c)}{\Delta x}$$

is negative or 0. See Fig. 17. Consequently, as $\Delta x \to 0$ through positive values,

$$\frac{f(c + \Delta x) - f(c)}{\Delta x},$$

being negative or 0, cannot approach a positive number. Thus

$$f'(c) = \lim_{\Delta x \to 0} \frac{f(c + \Delta x) - f(c)}{\Delta x} \leq 0.$$

If, on the other hand, Δx is negative, then the denominator of

$$\frac{f(c + \Delta x) - f(c)}{\Delta x}$$

is negative, and the numerator is still less than or equal to 0, since we have $f(c) \geq f(c + \Delta x)$. (See Fig. 18.) Hence, for negative Δx,

$$\frac{f(c + \Delta x) - f(c)}{\Delta x} \geq 0$$

(the quotient of two negative numbers being positive). Thus as $\Delta x \to 0$ through negative values, the quotient approaches a number greater than or equal to 0. Hence $f'(c) \geq 0$.

Since $0 \leq f'(c) \leq 0$, $f'(c)$ must be 0, and the theorem is proved. ■

Proof of Rolle's Theorem (If $f(a) = f(b)$, then $f'(c) = 0$ for at least one number c in $[a, b]$.) Since f is continuous, it has a maximum value M and a minimum value m for x in $[a, b]$. Certainly $m < M$ or perhaps $m = M$.

If $m = M$, f is constant and $f'(x) = 0$ for all x in $[a, b]$. Then any number x in (a, b) will serve as the desired number c.

If $m < M$, then the minimum and maximum cannot both occur at the ends of the interval a and b, since $f(a) = f(b)$. One of them, at least, occurs at a number c, $a < c < b$. And, by the interior-extremum theorem, at that c, $f'(c)$ is 0. This proves Rolle's theorem. ■

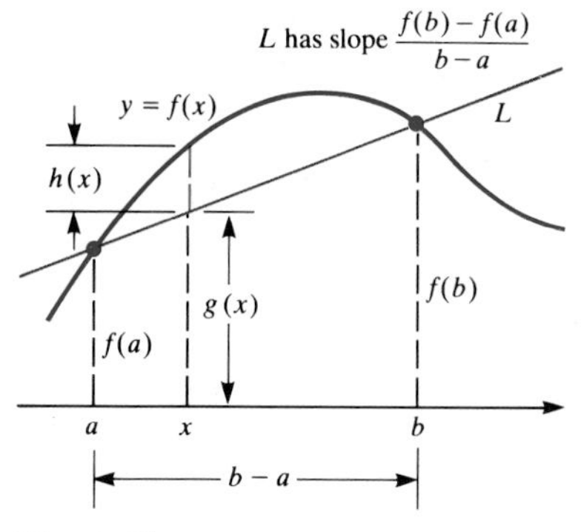

Figure 19

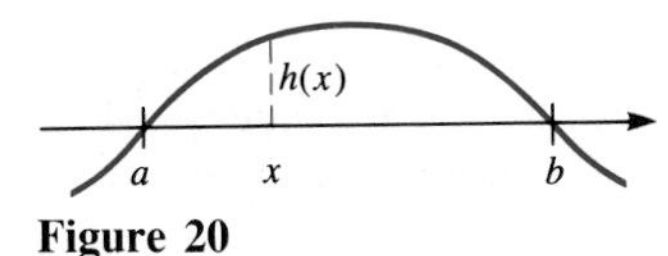

Figure 20

Proof of the Mean-Value Theorem (There is a number c in $[a, b]$ such that $f'(c) = [f(b) - f(a)]/(b - a)$.) We shall prove the theorem by introducing a function to which Rolle's theorem can be applied.

The chord through $(a, f(a))$ and $(b, f(b))$ is part of a line L whose equation is, let us say, $y = g(x)$. Let $h(x) = f(x) - g(x)$, which represents the difference between the graph of f and the line L for a given x. It is clear from Fig. 19 that $h(a) = h(b)$, since both equal 0. (Figure 20 depicts the graph of $h(x)$ corresponding to the curve in Fig. 19.)

By Rolle's theorem, there is at least one number c in the open interval (a, b) such that $h'(c) = 0$. But $h'(c) = f'(c) - g'(c)$. Since $y = g(x)$ is the equation of the line through $(a, f(a))$ and $(b, f(b))$, $g'(x)$ is the slope of the line, that is,

$$g'(x) = \frac{f(b) - f(a)}{b - a}$$

for all x.

Hence $$0 = h'(c) = f'(c) - g'(c)$$

or $$0 = f'(c) - \frac{f(b) - f(a)}{b - a}.$$

In short, $$f'(c) = \frac{f(b) - f(a)}{b - a},$$

and the mean-value theorem is proved. ■

Section Summary

After an informal introduction we stated three theorems, which we restate to serve as a reminder. For brevity, we omit some of the assumptions.

Theorem of the interior extremum. At an extremum in an open interval the derivative is 0.

Rolle's theorem. If $f(a) = f(b)$ and f is differentiable, then there is at least one number c in (a, b) such that $f'(c) = 0$.

Mean-value theorem. If f is differentiable on (a, b), there is at least one number c between a and b such that $f'(c) = (f(b) - f(a))/(b - a)$.

The mean-value theorem implies that

1 If a function has a derivative equal to 0 everywhere, then it is constant.
2 If two functions have the same derivative, they differ by a constant (which might be 0).
3 Where the derivative is positive, the function is increasing; where it is negative, the function is decreasing.

We also described how to use the ideas in this section to show that an equation has only one solution in a given interval and how to estimate this solution by the ''bisection'' technique.

EXERCISES FOR SEC. 4.1: THREE THEOREMS ABOUT THE DERIVATIVE

Exercises 1 and 2 concern the interior extremum.

1 Consider the function $f(x) = x^2$ only for x in $[-1, 2]$.
(*a*) Graph the function $f(x)$ for x in $[-1, 2]$.
(*b*) What is the maximum value of $f(x)$ for x in the interval $[-1, 2]$?
(*c*) Does $f'(x)$ exist at the maximum?
(*d*) Does $f'(x)$ equal 0 at the maximum?
(*e*) Does $f'(x)$ equal 0 at the minimum?

2 Consider the function $f(x) = \sin x$ for x in $[0, \pi]$.
(*a*) Graph the function $f(x)$ for x in $[0, \pi]$.
(*b*) Does $f'(x)$ equal 0 at the maximum value of $f(x)$ for x in $[0, \pi]$?
(*c*) Does $f'(x)$ equal 0 where $f(x)$ has a minimum for x in $[0, \pi]$?

3 (*a*) Graph $y = -x^2 + 3x + 2$ for x in $[0, 2]$.
(*b*) Looking at the graph, estimate the x coordinate where the maximum value of y occurs for x in $[0, 2]$.
(*c*) Find where $dy/dx = 0$.

(*d*) Using (*c*), determine exactly where the maximum occurs.

4 (*a*) Graph $y = 2x^2 - 3x + 1$ for x in $[0, 1]$.
(*b*) Looking at the graph, determine the maximum value of y for x in $[0, 1]$. At which value of x does it occur?
(*c*) Looking at the graph, estimate the x coordinate where the minimum value of y occurs for x in $[0, 1]$.
(*d*) Find where $dy/dx = 0$.
(*e*) Using (*d*), determine exactly where the minimum occurs.

Exercises 5 and 6 concern Rolle's theorem.

5 (*a*) Graph $f(x) = x^{2/3}$ for x in $[-1, 1]$.
(*b*) Show that $f(-1) = f(1)$.
(*c*) Is there a number c in $(-1, 1)$ such that $f'(c) = 0$?
(*d*) Why does this function not contradict Rolle's theorem?

6 (*a*) Graph $f(x) = 1/x^2$ for x in $[-1, 1]$.
(*b*) Show that $f(-1) = f(1)$.
(*c*) Is there a number c in $(-1, 1)$ such that $f'(c) = 0$?
(*d*) Why does this function not contradict Rolle's theorem?

In each of Exercises 7 to 10, verify that the given function satisfies the hypotheses of Rolle's theorem for the given interval. Find all numbers c that satisfy the conclusion of the theorem.

7 $x^2 - 2x - 3$ and $[0, 2]$
8 $x^3 - x$ and $[-1, 1]$
9 $x^4 - 2x^2 + 1$ and $[-2, 2]$
10 $\sin x + \cos x$ and $[0, 4\pi]$

In Exercises 11 to 14, find explicitly all values of c which satisfy the mean-value theorem for the given functions and intervals.

11 $x^2 - 3x$ and $[1, 4]$
12 $2x^2 + x + 1$ and $[-2, 3]$
13 $3x + 5$ and $[1, 3]$
14 $5x - 7$ and $[0, 4]$

15 (*a*) Graph $y = \sin x$ for x in $[\pi/2, 7\pi/2]$.
(*b*) Draw the chord joining $(\pi/2, f(\pi/2))$ and $(7\pi/2, f(7\pi/2))$.
(*c*) Draw all tangents to the graph that are parallel to the chord drawn in (*b*).
(*d*) Using (*c*), determine how many numbers c there are in $[\pi/2, 7\pi/2]$ such that

$$f'(c) = \frac{f(7\pi/2) - f(\pi/2)}{7\pi/2 - \pi/2}.$$

(*e*) Use the graph to estimate the values of the c's.

16 (*a*) Graph $f(x) = \cos x$ for x in $[0, 9\pi/2]$.
(*b*) Draw the chord joining $(0, f(0))$ to $(9\pi/2, f(9\pi/2))$.
(*c*) Draw all tangents to the graph that are parallel to the chord drawn in (*b*).
(*d*) Using (*c*), determine how many numbers c there are in $[0, 9\pi/2]$ such that

$$f'(c) = \frac{f(9\pi/2) - f(0)}{9\pi/2 - 0}.$$

(*e*) Use the graph to estimate the values of the c's.

17 (*a*) Differentiate $\sec^2 x$ and $\tan^2 x$.
(*b*) The derivatives in (*a*) are equal. Corollary 2 then asserts that there exists a constant C such that $\sec^2 x = \tan^2 x + C$. Find that constant.

18 (*a*) Differentiate $\csc^2 x$ and $\cot^2 x$.
(*b*) The derivatives in (*a*) are equal. Find the constant C (promised by Corollary 2) such that $\csc^2 x = \cot^2 x + C$.

In each of Exercises 19 to 24 sketch a graph of a differentiable function that meets the given conditions. (Just draw the graph; there is no need to come up with a formula for the function.)

19 $f'(x) < 0$ for all x.
20 $f'(3) = 0$ but for $x \neq 3$, $f'(x) < 0$.
21 x intercepts 1 and 5; y intercept 2; $f'(x) < 0$ for $x < 4$; $f'(x) > 0$ for $x > 4$.
22 x intercepts 2 and 5; y intercept 3; $f'(x) > 0$ for $x < 1$ and for $x > 3$; $f'(x) < 0$ for x in $(1, 3)$.
23 $f'(x) = 0$ only when $x = 1$ or 4; $f(1) = 3$, $f(4) = 1$; $f'(x) < 0$ for $x < 1$; $f'(x) > 0$ for $x > 4$.
24 $f'(x) = 0$ only when $x = 1$ or 4; $f(1) = 3$, $f(4) = 1$; $f'(x) > 0$ for $x < 1$ and for $x > 4$.

In Exercises 25 to 28 explain why *no* differentiable function satisfies all the conditions.

25 $f(1) = 3$, $f(2) = 4$, $f'(x) < 0$ for all x.
26 $f(2) = 5$, $f(3) = -1$, $f'(x) \geq 0$ for all x.
27 x intercepts only 1 and 2; $f(3) = -1$, $f(4) = 2$.
28 $f(x) = 2$ only when $x = 0$, 1, and 3; $f'(x) = 0$ only when $x = \frac{1}{4}$, $\frac{3}{4}$, and 4.

29 In "*Surely You're Joking, Mr. Feynman!*," Norton, New York, 1985, Nobel laureate Richard P. Feynman writes

> I often liked to play tricks on people when I was at MIT. One time, in mechanical drawing class, some joker picked up a French curve (a piece of plastic for drawing smooth curves—a curly, funny-looking thing) and said, "I wonder if the curves on that thing have some special formula?"
>
> I thought for a moment and said, "Sure they do. The curves are very special curves. Lemme show ya," and I picked up my French curve and began to turn it slowly. "The French curve is made so that at the lowest point on each curve, no matter how you turn it, the tangent is horizontal."
>
> All the guys in the class were holding their French curve up at different angles, holding their pencil up to it at the lowest point and laying it along, and discovering that, sure enough, the tangent is horizontal.

How was Feynman playing a trick on his classmates?

30 (*a*) Show that the equation $5x - 4 \cos x = 0$ has exactly one solution.
(*b*) Find a specific interval which contains the solution.
(*c*) Apply the bisection method twice to this interval.

31 (*a*) Show that the equation $2x^3 - 3x^2 + 12x = 5$ has exactly one solution.
(*b*) Find a specific interval which contains the solution.
(*c*) Apply the bisection method twice to this interval.

32 What can be said about the number of solutions of the equation $f(x) = 3$ for a differentiable function if
(*a*) $f'(x) > 0$ for all x?
(*b*) $f'(x) > 0$ for $x < 7$ and $f'(x) < 0$ for $x > 7$?

In Exercises 33 to 36 the figures depict the graph of $y = f'(x)$ for a differentiable function f. In each case, $f(1) = 2$. Sketch the general shape of the graph of each $y = f(x)$.

33

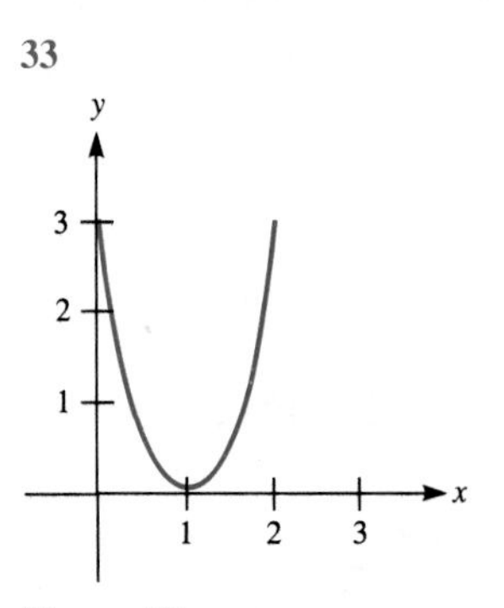

Figure 21

34

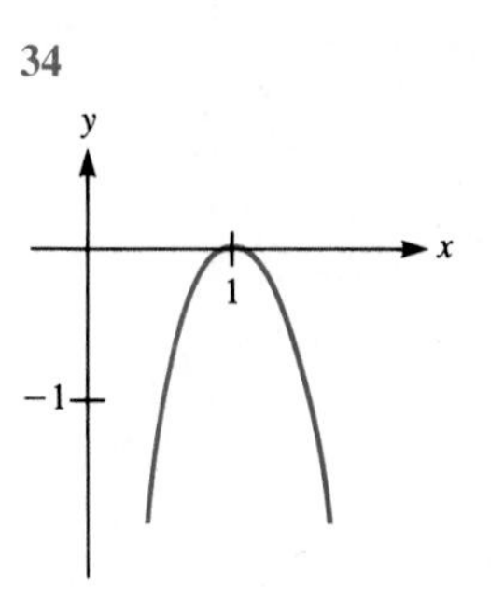

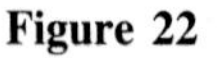
Figure 22

35

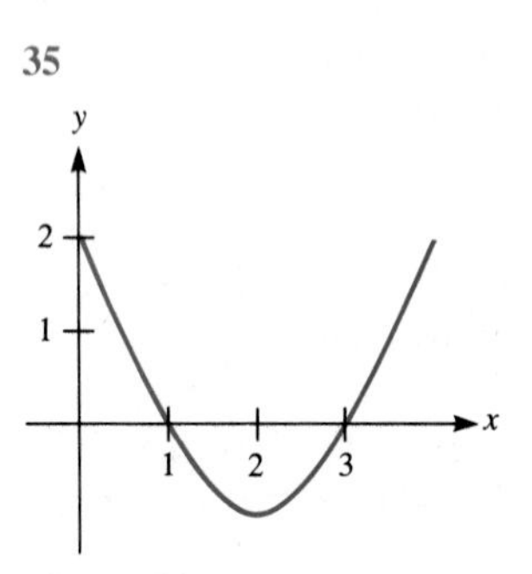

Figure 23

36

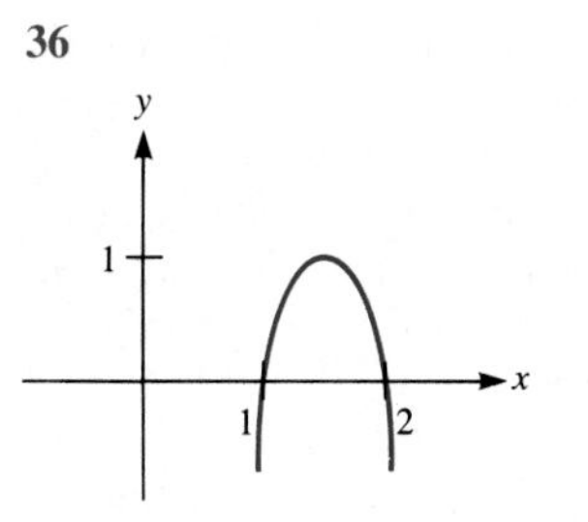

Figure 24

37 For which values of the constant k is the function $7x + k \sin 2x$ always increasing?

38 Consider the function $f(x) = x^3 + ax^2 + c$. Show that if $a < 0$ and $c > 0$, then f has exactly one negative root.

---

39 Let f have a derivative for all x.
(*a*) Is every chord of the graph of f parallel to some tangent to the graph of f?
(*b*) Is every tangent to the graph of f parallel to some chord of the graph of f?

40 (*a*) Show that a polynomial of degree 1, $a_0 + a_1x$, $a_1 \neq 0$, has exactly one root.
(*b*) Using the quadratic formula, show that a polynomial of degree 2, $a_0 + a_1x + a_2x^2$, has at most two distinct roots.
(*c*) Using Rolle's theorem, solve (*b*).
(*d*) Show that a polynomial of degree 3, $a_0 + a_1x + a_2x^2 + a_3x^3$, $a_3 \neq 0$, has at most three distinct roots.
(*e*) Generalize to higher-degree polynomials.
(*f*) Use mathematical induction to prove your generalization in (*e*).

The answers to Exercises 41 and 42 should be phrased in colloquial English. (Section 3.1 introduced the concepts of density and magnification.)

41 State the mean-value theorem in terms of a slide and a screen. [Let x denote the position on the (linear) slide and $f(x)$ denote the position of the image on the screen.] In optical terms, what does the mean-value theorem say?

42 State the mean-value theorem in terms of density and mass. [Let x be the distance from the left end of a string and $f(x)$ the mass of the string from 0 to x.] When stated in these terms, does the mean-value theorem seem reasonable?

43 At time t seconds a thrown ball has the height $f(t) = -16t^2 + 32t + 40$ feet.
(*a*) Show that after 2 seconds it returns to its initial height $f(0)$.
(*b*) What does Rolle's theorem imply about the velocity of the ball?
(*c*) Verify Rolle's theorem in this case by computing the numbers c which it asserts exist.

44 Which of the corollaries to the mean-value theorem implies that (*a*) if two cars on a straight road have the same velocity at every instant, they remain a fixed distance apart? (*b*) If all the tangents to a curve are horizontal, the curve is a horizontal line. Explain in each case.

45 Differentiate for practice:

(*a*) $\sqrt{1-x^2}\sin 3x$ (*b*) $\dfrac{\sqrt[3]{x}}{x^2+1}$ (*c*) $\tan\dfrac{1}{(2x+1)^2}$

46 Is there a differentiable function f whose domain is the x axis such that f is increasing and yet the derivative is *not* positive for all x?

In Exercises 47 and 48 find *all* antiderivatives of the given function. [Recall that $F(x)$ is called an antiderivative of $f(x)$ if $F'(x) = f(x)$.] Check your answer by differentiation.

47 (*a*) $\sqrt{x}$ (*b*) $\sec^2 3x$ (*c*) $\cos 3x$ (*d*) $(2x+1)^{10}$

48 (*a*) $8x^3$ (*b*) $\sin 2x$ (*c*) $1/x^2$ (*d*) $\sqrt[3]{x}$

49 Consider the function f given by the formula $f(x) = x^3 - 3x$.

(*a*) At which numbers x is $f'(x) = 0$?

(*b*) Use the theorem of the interior extremum to show that the maximum value of $x^3 - 3x$ for x in $[1, 5]$ occurs either at 1 or at 5.

(*c*) What is the maximum value of $x^3 - 3x$ for x in $[1, 5]$?

50 (*a*) Recall the definition of $g(x)$ in the proof of the mean-value theorem, and show that

$$g(x) = f(a) + \frac{x-a}{b-a}[f(b) - f(a)].$$

(*b*) Using (*a*), show that

$$g'(x) = \frac{f(b) - f(a)}{b-a}.$$

51 Is this proposed proof of the mean-value theorem correct? *Proof:* Tilt the x and y axes until the x axis is parallel to the given chord. The chord is now "horizontal," and we may apply Rolle's theorem.

4.2 THE FIRST DERIVATIVE AND GRAPHING

In the previous section we saw that where the derivative is positive the function increases and where it is negative the function decreases. In this section we use these facts to help graph functions. In particular we will use them to find high and low points on a graph of a function.

Throughout the section we assume that the functions are differentiable. In the first part the domain of the function is the entire x axis or made up of open intervals. (Thus if a number c is in the domain, an entire open interval around c is also in the domain.) In the second part we consider functions defined on closed intervals $[a, b]$. (In this case the endpoints a and b require special attention.)

A Few Definitions

The graph of a hypothetical function f defined on the entire x axis is shown in Fig. 1. The points P, Q, R, and S are of special interest. S is the highest point on the graph *for all x in the domain*. We call it a *global* (or *absolute*) *maximum*. The point P is higher than all points *near it* on the graph. It is called a *relative maximum* (or local maximum). Similarly, Q is called a *relative minimum* (or local minimum). S, the global maximum, is also a relative maximum. As the theorem of the interior extremum in Sec. 4.1 shows, the tangents at P, Q, and S are horizontal. However, the tangent at R is also horizontal, but R is not a relative extremum (neither a relative maximum nor relative minimum).

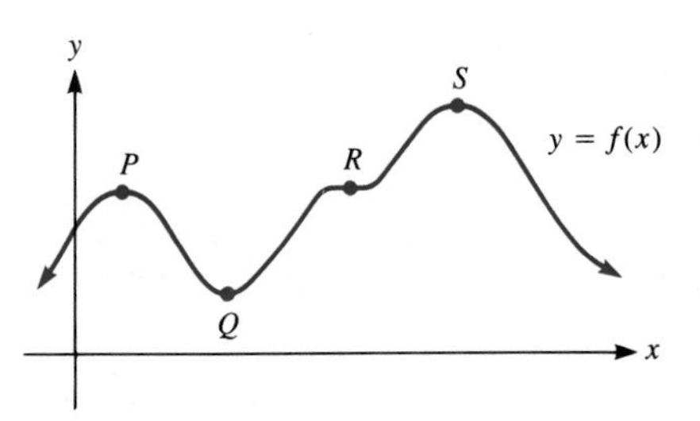

Figure 1

If you were to walk left to right along the graph, you would call P the top of a hill, Q the bottom of a valley, and S the highest point on your walk (it is also the top of a hill). You might not notice R, for you would keep climbing as you pass through it. But near R the walking would be quite easy, almost like walking on a horizontal path.

These important aspects of a function and its graph are made precise in the

following definitions, which are phrased to include functions whose domains may not consist of open intervals.

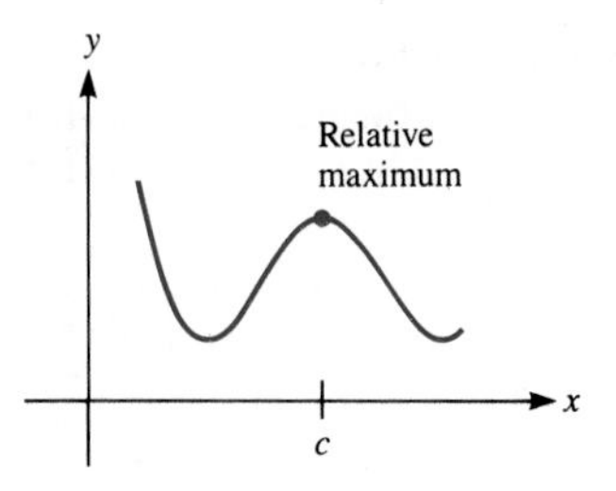

Figure 2

Definition *Relative maximum (local maximum).* The function f has a **relative maximum** (or **local maximum**) at the number c if there is an open interval around c such that $f(c) \geq f(x)$ for all x in that interval that lie in the domain of f. (See Fig. 2.)

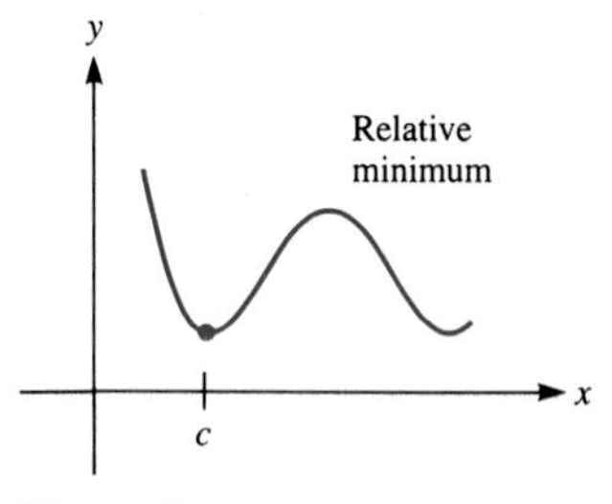

Figure 3

Definition *Relative minimum (local minimum).* The function f has a **relative minimum (local minimum)** at the number c if there is an open interval around c such that $f(c) \leq f(x)$ for all x in that interval that lie in the domain of f. (See Fig. 3.)

Definition *Global maximum.* The function f has a **global maximum** at the number c if $f(c) \geq f(x)$ for all x in the domain of f.

Note that a global maximum is necessarily a local maximum as well. A local maximum is like the summit of a single mountain; a global maximum corresponds to Mount Everest.

Definition *Global minimum.* The function f has a **global minimum** at the number c if $f(c) \leq f(x)$ for all x in the domain of f.

Definition *Critical number and critical point.* A number c at which $f'(c) = 0$ is called a **critical number** for the function f. The corresponding point $(c, f(c))$ on the graph of f is a **critical point** on that graph.

Remark: Some texts define a critical number as a number where the derivative is 0 or else is not defined. Since we emphasize differentiable functions, a critical number is a number where the derivative is 0.

The types of critical points

Figures 4 through 6 show what a graph may look like near a critical point. In each case the tangent is horizontal at the critical point.

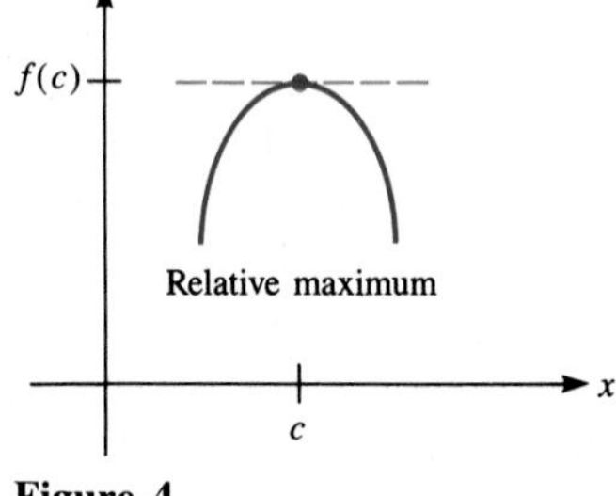

Figure 4

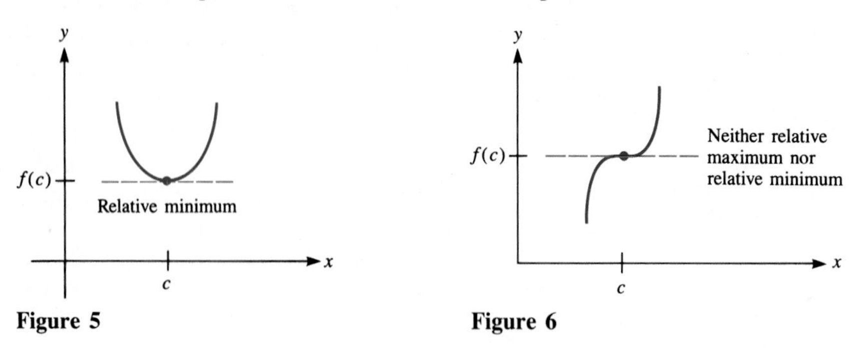

Figure 5

Figure 6

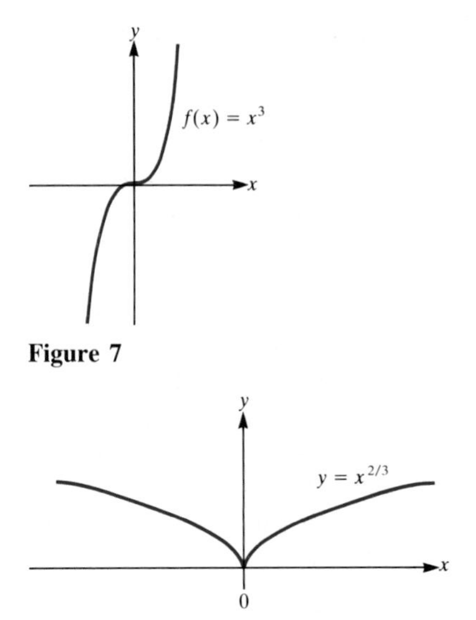

Figure 7

Figure 8

The Connection Between Critical Numbers and Extrema

By the theorem of the interior extremum in Sec. 4.1, there is a close relation between a local maximum (or minimum) and critical points for a differentiable function. If a local maximum occurs at a number c that lies within some open interval within the domain of f, then $f'(c) = 0$. This means that c is a critical number. However, a critical point need not be a local extremum. This is illustrated by the function x^3, whose derivative $3x^2$ is 0 at 0. This $c = 0$ is a critical number of the function x^3. A glance at Fig. 7 shows that x^3 has neither a local maximum nor a local minimum at 0.

Warning: The moral is that to determine whether a function has a local extremum at c, it is not enough to know that $f'(c) = 0$. It is also important to know how the sign of the derivative behaves for inputs near c.

Remark: A function may not be differentiable at a local extremum. For instance, consider $f(x) = x^{2/3}$, which is graphed in Fig. 8. It has a local minimum at $x = 0$. However, $f'(x) = \frac{2}{3}x^{-1/3}$ is not defined at $x = 0$. [This is an unusual situation, but it should be kept in mind as a possibility when dealing with functions that are not differentiable throughout the domain of interest. The curve $y = x^{2/3}$ is said to have a "cusp" at (0, 0).]

How to Test a Critical Point

Say that you have found a number c such that $f'(c) = 0$, and you want to find out whether there may be a local extremum—either a maximum or minimum—at c. The following test describes a way to get the answer. It is an immediate consequence of the fact that when the derivative is positive the function increases and when the derivative is negative the function decreases.

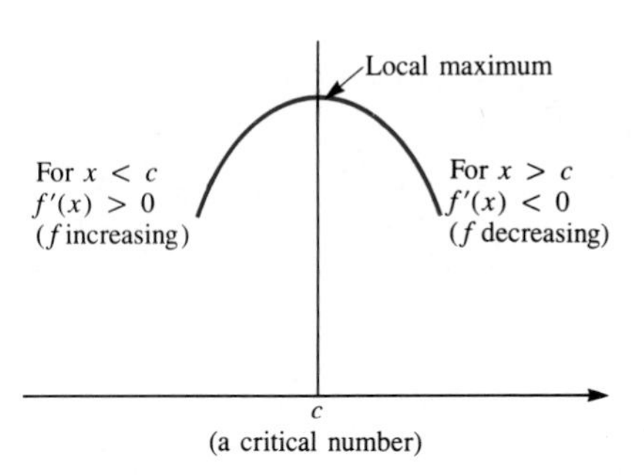

Figure 9

FIRST-DERIVATIVE TEST FOR LOCAL MAXIMUM AT $x = c$

Let f be a function and let c be a number in its domain. Assume that numbers a and b exist such that $a < c < b$, f is continuous on the open interval (a, b), and f is differentiable on the open interval (a, b), except possibly at c.

If $f'(x)$ is positive for all $x < c$ in the interval and is negative for all $x > c$ in the interval, then f has a local maximum at c.

A similar test, with "positive" and "negative" interchanged, holds for a local minimum. (See Figs. 9 and 10.)

Informally, the derivative test says, *"If the derivative changes sign at c, then the function has either a local minimum or a local maximum."* To decide which it is, just make a rough sketch of the graph near $(c, f(c))$ to show on which side of c the function is increasing and on which side it is decreasing.

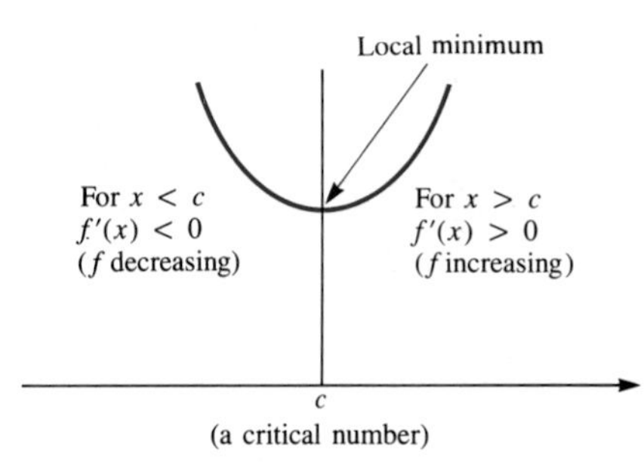

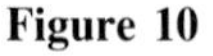
Figure 10

The Derivative and Sketching a Graph

We now illustrate the use of the derivative in graphing.

EXAMPLE 1 Sketch the graph of $f(x) = x^2 - x - 1$.

SOLUTION As in Chap. 2, we first check symmetry, intercepts, and asymptotes.

Symmetry

Since $f(x) = x^2 - x - 1$ and $f(-x) = (-x)^2 - (-x) - 1 = x^2 + x - 1$, f is neither even nor odd. The graph is not symmetric with respect to the y axis or with respect to the origin.

Intercepts

The y intercept is $f(0) = 0^2 - 0 - 1 = -1$. The x intercepts are found by solving the equation

$$x^2 - x - 1 = 0.$$

By the quadratic formula the roots are $(1 \pm \sqrt{5})/2$, which are approximately -0.6 and 1.6. The information gathered so far is displayed in Fig. 11.

Asymptotes

Since $f(x) = x^2 - x - 1$ is a polynomial of degree greater than 1, the graph has no asymptotes. (For large x it resembles $y = x^2$, since x^2 is the dominant term.)

Critical number

Now we make use of the derivative of $x^2 - x - 1$, which is $2x - 1$. To find any critical number, set the derivative equal to 0:

$$2x - 1 = 0$$

or

$$x = \frac{1}{2}.$$

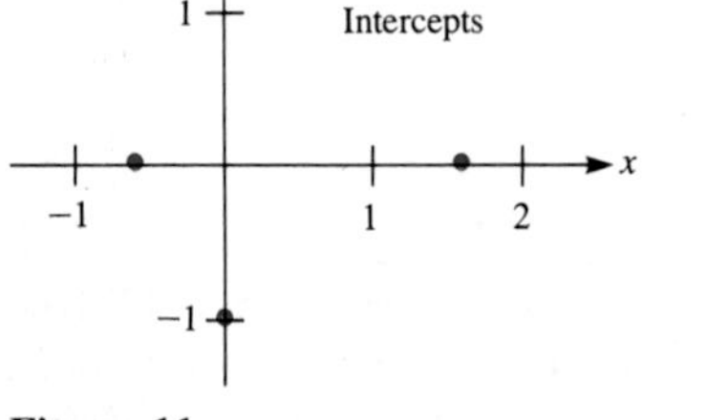

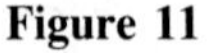
Figure 11

Thus $\frac{1}{2}$ is the only critical number. Since $f(\frac{1}{2}) = (\frac{1}{2})^2 - \frac{1}{2} - 1 = -\frac{5}{4}$, the critical point is $(\frac{1}{2}, -\frac{5}{4})$. See Fig. 12, where the short tangent line at $(\frac{1}{2}, -\frac{5}{4})$ reminds us that $(\frac{1}{2}, -\frac{5}{4})$ is a critical point. (*Stop before reading on.* At this point you could probably sketch the curve, keeping in mind where the derivative must be positive and where it must be negative.)

Testing the critical point

To see whether the critical point is a relative maximum, relative minimum, or neither, check the sign of $f'(x) = 2x - 1$. We know that $2x - 1 = 0$ when $x = \frac{1}{2}$. If $x > \frac{1}{2}$, $2x - 1$ is greater than 0. If $x < \frac{1}{2}$, $2x - 1$ is less than 0. (Test $x = 1$ and $x = 0$, for instance.) This information is recorded in Fig. 13.

Therefore the function is *increasing* for $x > \frac{1}{2}$ and *decreasing* for $x < \frac{1}{2}$. The first-derivative test tells us that the critical point is a relative minimum—in fact, a global minimum. Moreover, it enables us to sketch the rest of the graph—for it is descending to the left of $\frac{1}{2}$ and rising to the right of $\frac{1}{2}$. See Fig. 14. ■

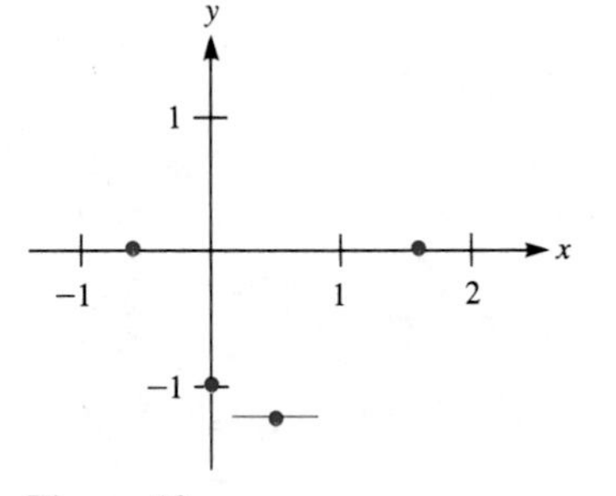

Figure 12

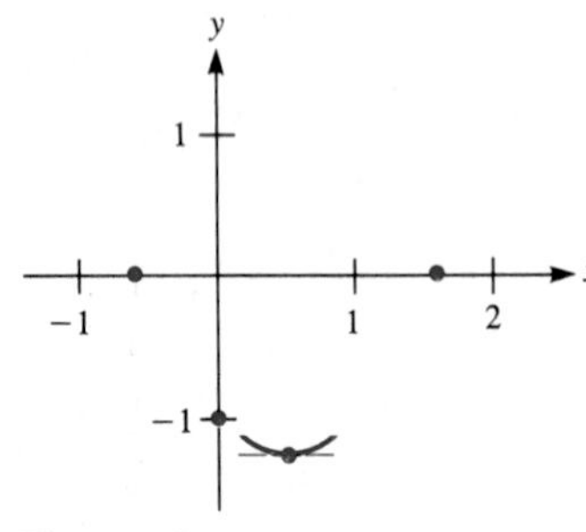

Figure 13

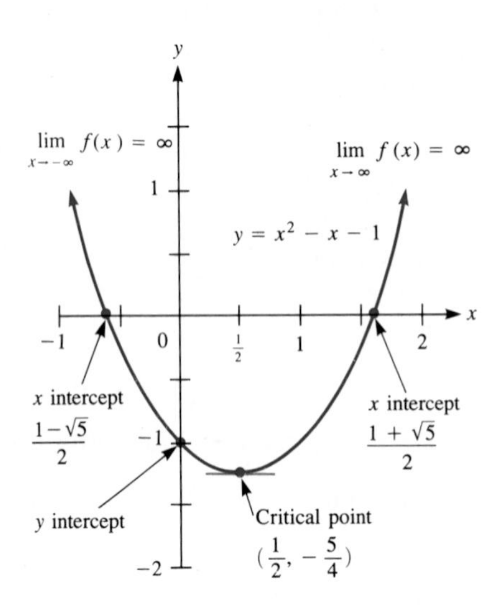

Figure 14

EXAMPLE 2 Sketch the graph of $f(x) = 2x^3 - 3x^2 - 12x$.

Symmetry

SOLUTION Since the polynomial $f(x) = 2x^3 - 3x^2 - 12x$ has terms of even degree and of odd degree, it is neither an even nor odd function. The graph is not symmetric with respect to the y axis nor with respect to the origin.

Intercepts

Since $f(0) = 2 \cdot 0^3 - 3 \cdot 0^2 - 12 \cdot 0 = 0$, the y intercept is 0. To find the x intercepts, it is necessary to solve the equation $f(x) = 0$. In the case of this function, the equation can be solved easily:

$$2x^3 - 3x^2 - 12x = 0$$

so

$$x(2x^2 - 3x - 12) = 0.$$

Either $x = 0$ or $2x^2 - 3x - 12 = 0$. The latter equation can be solved by the quadratic formula:

$$x = \frac{-(-3) \pm \sqrt{(-3)^2 - 4(2)(-12)}}{2 \cdot 2} = \frac{3 \pm \sqrt{9 + 96}}{4} = \frac{3 \pm \sqrt{105}}{4}.$$

These two solutions are approximately -1.8 and 3.3.

The intercepts are recorded in Fig. 15.

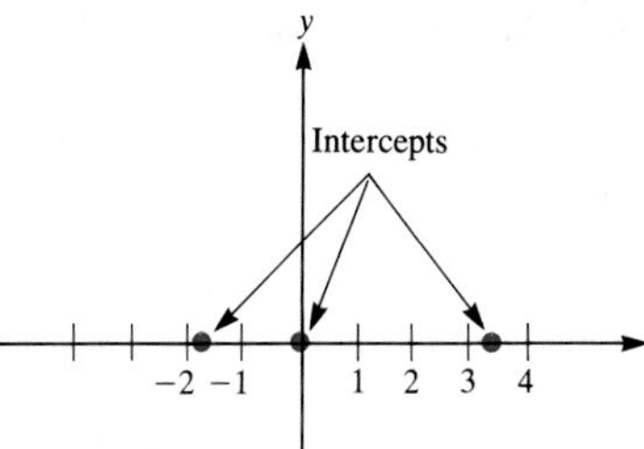
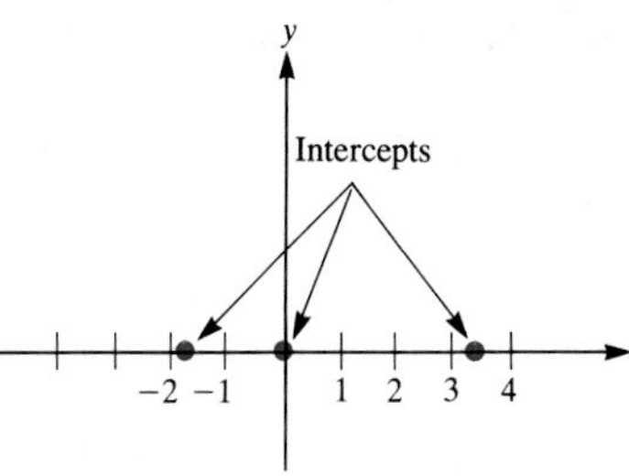

Figure 15

When is $f'(x) = 0$? We have

$$f'(x) = 6x^2 - 6x - 12 = 6(x^2 - x - 2) = 6(x - 2)(x + 1).$$

Thus $f'(x) = 0$ when $6(x - 2)(x + 1) = 0$, that is, when $x = 2$ or $x = -1$. At these critical numbers, the function has the values

$$f(2) = 2 \cdot 2^3 - 3 \cdot 2^2 - 12 \cdot 2 = -20$$

and

$$f(-1) = 2(-1)^3 - 3(-1)^2 - 12(-1) = 7.$$

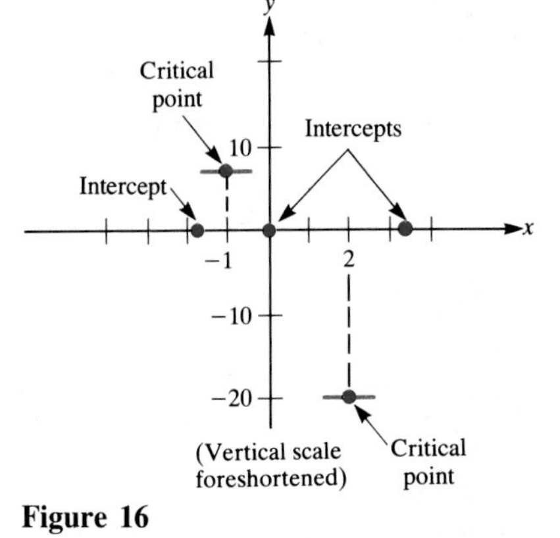

Figure 16

Figure 16 records the data gathered so far. The short segments indicate horizontal tangents.

Next, examine the sign of $f'(x)$ to determine where the function is increasing and where it is decreasing.

In the interval $(-1, 2)$, $f'(x)$ must keep the same sign. Otherwise, it would be positive somewhere and negative somewhere. By the intermediate-value theorem, $f'(x)$ would be 0 somewhere in $(-1, 2)$. However, *there are no critical numbers in that open interval*. Thus $f(x)$ is either increasing throughout $(-1, 2)$ or decreasing throughout $(-1, 2)$. Since $f(2)$ is less than $f(-1)$, the function must *decrease* for x in $(-1, 2)$. Similarly, $f(x)$ increases for $x > 2$ and increases for $x < -1$.

Behavior for x large

To find the shape of the graph for large x, examine $\lim_{x\to\infty} f(x)$ and $\lim_{x\to-\infty} f(x)$. We have

$$\lim_{x\to\infty} (2x^3 - 3x^2 - 12x) = \infty$$

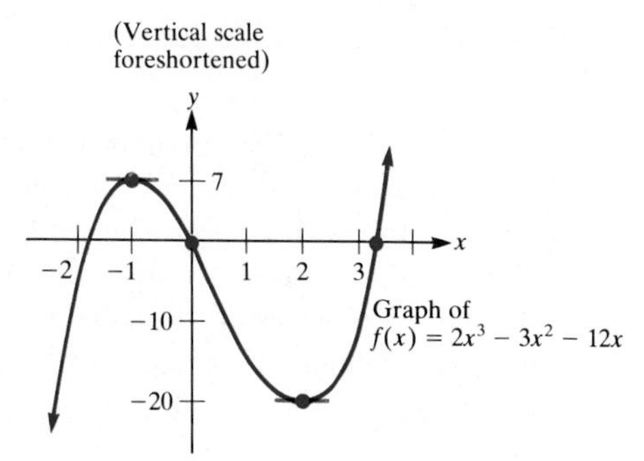

Figure 17 The graph crosses the x axis at $x = 0, \frac{3 - \sqrt{105}}{4} \approx -1.8$, and $\frac{3 + \sqrt{105}}{4} \approx 3.3$.

and $$\lim_{x \to -\infty} (2x^3 - 3x^2 - 12x) = -\infty.$$

There are no asymptotes.

With this last information the curve can be sketched. The graph (with the y axis compressed) appears in Fig. 17.

There is a local maximum at $x = -1$, a local minimum at $x = 2$, but no global maximum or minimum. ■

In the next example a critical point turns out not to be a relative extremum.

EXAMPLE 3 Graph $f(x) = 3x^4 - 4x^3$. Discuss relative maxima and minima.

SOLUTION As in the preceding example the graph is not symmetric with respect to the y axis or the origin. Nor does it have asymptotes. However,

$$\lim_{x \to \infty} f(x) = \infty \quad \text{and} \quad \lim_{x \to -\infty} f(x) = \infty.$$

Intercepts

To find the intercepts, note that $f(0) = 0$ and $3x^4 - 4x^3 = 0$ when we have $x^3(3x - 4) = 0$, that is, when $x = 0$ or $x = \frac{4}{3}$. The derivative is

$$f'(x) = 12x^3 - 12x^2 = 12x^2(x - 1).$$

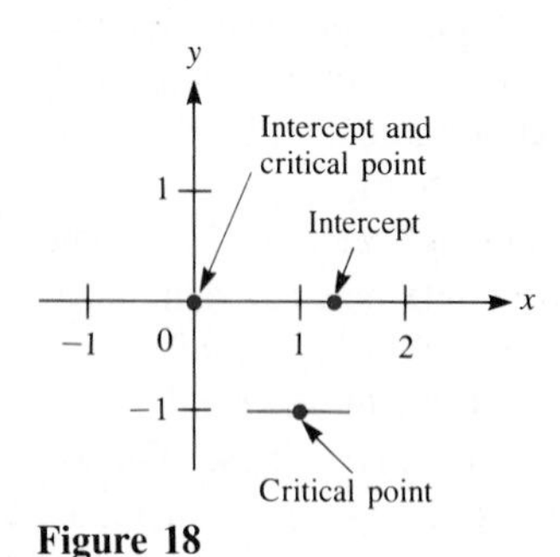

Figure 18

The critical numbers are the solutions of the equation

$$12x^2(x - 1) = 0,$$

namely, 0 and 1. Since $f(0) = 0$, $(0, 0)$ is a critical point. Since $f(1) = 3(1^4) - 4(1^3) = -1$, $(1, -1)$ is a critical point.

The information collected so far is shown in Fig. 18. As in the preceding example, the derivative cannot change sign in an interval that contains no critical numbers. Thus the derivative keeps the same sign in each of these intervals: $(-\infty, 0)$, $(0, 1)$, $(1, \infty)$.

The derivative is negative for x in $(-\infty, 0)$. (Why?) It is negative for x in $(0, 1)$. (Why?) It is positive for x in $(1, \infty)$. (Why?) This information tells us that the function decreases for $x < 1$ and increases for $x > 1$. Since the derivative does not change sign at the critical number 0, 0 is neither a relative maximum nor relative minimum. However, the derivative does change sign at the

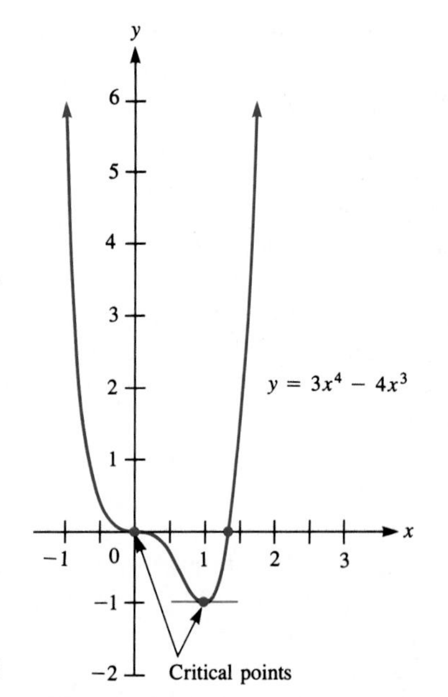

Figure 19

critical number 1, which provides a relative minimum—in fact, a global minimum as well. See Fig. 19. ■

EXAMPLE 4 Graph $f(x) = (2x - 5)/(x - 1)$.

SOLUTION Again the function is neither even nor odd.

Intercepts

Since

$$f(0) = \frac{2 \cdot 0 - 5}{0 - 1} = 5$$

the graph has the y intercept 5. To find x intercepts, solve the equation

$$\frac{2x - 5}{x - 1} = 0.$$

Hence $2x - 5 = 0$ or $x = \frac{5}{2}$. There is one x intercept, $\frac{5}{2}$.

Asymptotes

Since the denominator $x - 1$ is near 0 when x is near 1, there is a vertical asymptote at $x = 1$. To see how the graph looks near the vertical line $x = 1$, consider the behavior of $f(x) = (2x - 5)/(x - 1)$ for x near 1. If x is a little larger than 1, $2x - 5$ is negative and $x - 1$ is a small positive number; their quotient is a large negative number. When x is a little smaller than 1, $2x - 5$ is near -3 and $x - 1$ is a small negative number; their quotient is a large positive number. Furthermore,

$$\lim_{x \to \infty} \frac{2x - 5}{x - 1} = \lim_{x \to \infty} \frac{x\left(2 - \frac{5}{x}\right)}{x\left(1 - \frac{1}{x}\right)} = \frac{2}{1} = 2.$$

Similarly,

$$\lim_{x \to -\infty} \frac{2x - 5}{x - 1} = 2.$$

Therefore $y = 2$ is a horizontal asymptote for x large, positive or negative. The information found up to this point is shown in Fig. 20.

No critical number

Next, *determine the critical numbers of f.* To do this compute $f'(x)$:

$$f'(x) = D\left(\frac{2x-5}{x-1}\right) = \frac{(x-1)\cdot 2 - (2x-5)\cdot 1}{(x-1)^2} = \frac{3}{(x-1)^2}.$$

Since the numerator is never 0, there are no critical numbers.

Always increasing

Where is the function increasing? Decreasing? Since the derivative is $3/(x-1)^2$, it is positive throughout the domain of the function. The function is always increasing on each interval in the domain.

We now complete the graph. For $x > 1$ the graph rises from left to right. It passes through $(\frac{5}{2}, 0)$ and as it rises gets closer and closer to the asymptote $y = 2$. (Since it keeps rising, it must approach this asymptote *from below*.) Similar reasoning obtains the graph for $x < 1$. (See Fig. 21.) ■

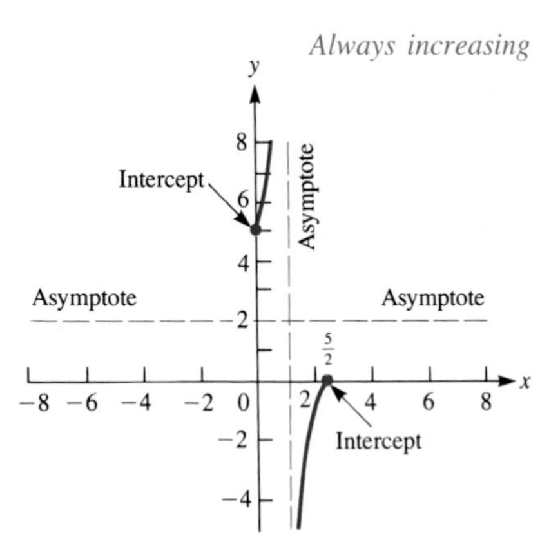

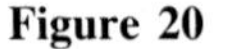
Figure 20

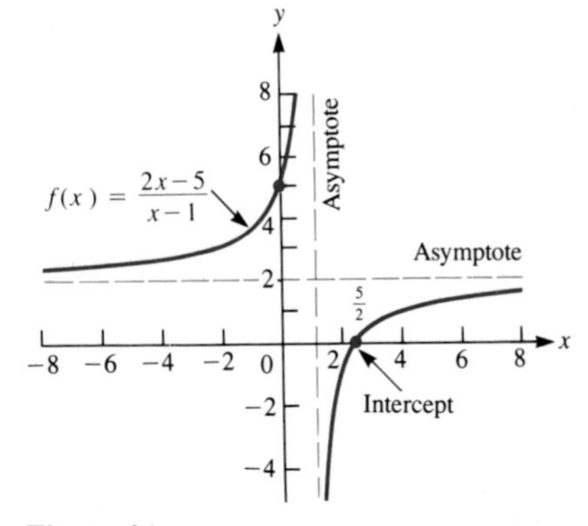

Figure 21

These examples illustrate graphing a function whose domain is the entire x axis or is composed of open intervals. We now turn to graphing a function where the domain is a closed interval.

Graphing a Function on a Closed Interval

In many applied problems we are interested in the behavior of a differentiable function just over some closed interval $[a, b]$. Such a function will have a global maximum for that interval by the maximum-value theorem of Sec. 2.8. That maximum can occur either at an endpoint—a or b—or else at some number c in the open interval (a, b). In the latter case, c must be a critical number, for $f'(c) = 0$ by the interior-extremum theorem of Sec. 4.1.

Figures 22 and 23 show some of the ways in which a relative or global maximum or minimum can occur for a function considered only on a closed interval $[a, b]$. The major point to keep in mind is

> The maximum value of a function f that is differentiable on a closed interval occurs either
>
> **1** at an endpoint of the interval, or
> **2** at a critical number [where $f'(x) = 0$].

In short, *pay special attention to the endpoints of the closed interval.*

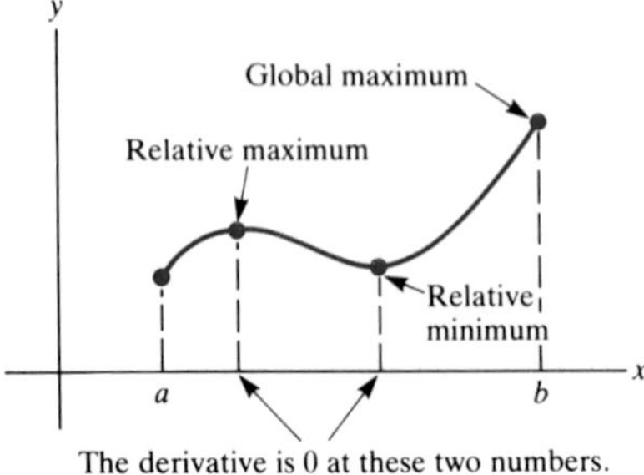

The global maximum occurs at an end. (The derivative need not be 0.)

Figure 22

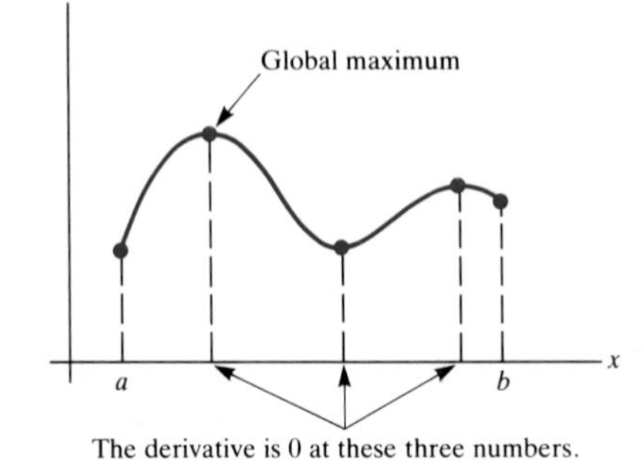

The global maximum occurs at a number other than a or b. (The derivative is 0.)

Figure 23

EXAMPLE 5 Graph $f(x) = x^3 - 3x^2 + 3x$ for x in $[0, 2]$ and find its maximum value.

SOLUTION First compute f at the ends of the interval, 0 and 2:

$$f(0) = 0 \quad \text{and} \quad f(2) = 2.$$

Next, compute $f'(x)$, which is $3x^2 - 6x + 3$. When is $f'(x) = 0$? When

$$3x^2 - 6x + 3 = 0,$$

or

$$3(x^2 - 2x + 1) = 0,$$

or

$$3(x - 1)^2 = 0.$$

Thus 1 is the only critical number, and it lies in the interval $[0, 2]$.

The maximum of f must therefore occur either at an endpoint of the interval (at 0 or 2) or at the only critical number, 1. It is necessary to calculate $f(1)$ to determine where the maximum occurs:

$$f(1) = 1^3 - 3 \cdot 1^2 + 3 \cdot 1 = 1.$$

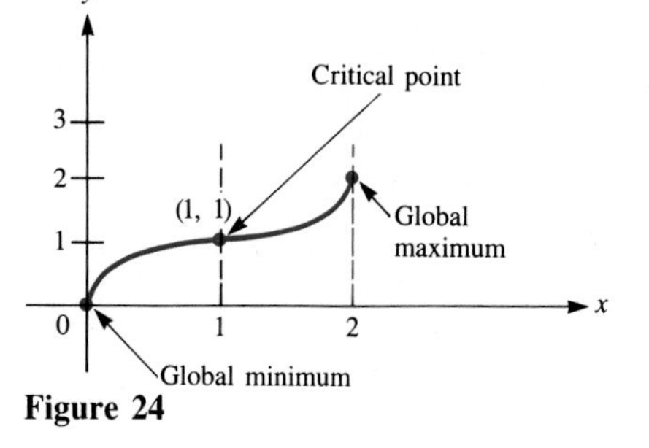

Figure 24

Since $f(0) = 0$, $f(2) = 2$, and $f(1) = 1$, the maximum value is 2, occurring at the endpoint 2.

Now that the problem is solved, it may be instructive to sketch the graph of the function. Since

$$f'(x) = 3(x - 1)^2$$

is positive for all x other than 1, the function is increasing. Figure 24 shows how the graph looks. Observe that the minimum occurs at 0. ■

The following flowchart shows how to maximize a differentiable function on a closed interval. A similar procedure works for finding a minimum for $f(x)$ in $[a, b]$.

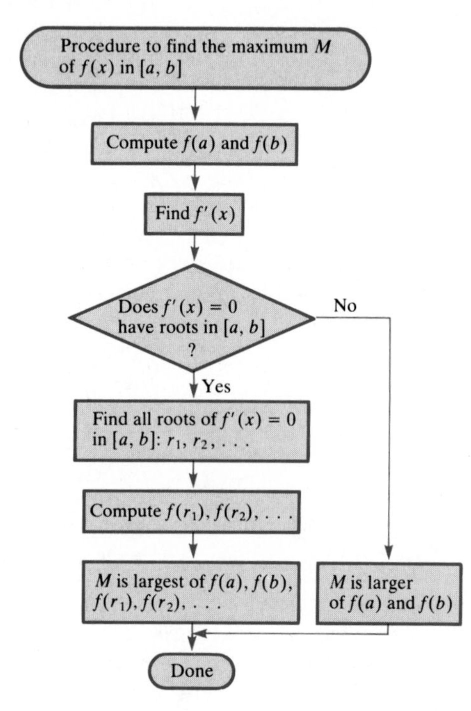

Figure 25

EXERCISES FOR SEC. 4.2: THE FIRST DERIVATIVE AND GRAPHING

In each of Exercises 1 to 8 find all critical numbers of the given function and use the first-derivative test to determine whether a local maximum, a local minimum, or neither, occurs there.

1 x^5 **2** x^6
3 $(x-1)^3$ **4** $(x-1)^4$
5 $3x^4 + x^3$ **6** $2x^3 + 3x^2$
7 $x \sin x + \cos x$ **8** $x \cos x - \sin x$

In Exercises 9 to 16 sketch the general shape of the graph, using the given information. Assume the function and its derivative are defined for all x and are continuous. Explain your reasoning.

9 Critical point $(1, 2)$, $f'(x) < 0$ for $x < 1$ and $f'(x) > 0$ for $x > 1$.

10 Critical point $(1, 2)$, and $f'(x) < 0$ for all x except $x = 1$.

11 x intercept -1; critical points $(1, 3)$ and $(2, 1)$,

$$\lim_{x\to\infty} f(x) = 4, \quad \lim_{x\to-\infty} f(x) = -1.$$

12 y intercept 3; critical point $(1, 2)$,

$$\lim_{x\to\infty} f(x) = \infty, \quad \lim_{x\to-\infty} f(x) = 4.$$

13 x intercept -1; critical points $(1, 5)$ and $(2, 4)$,

$$\lim_{x\to\infty} f(x) = 5, \quad \lim_{x\to-\infty} f(x) = -\infty.$$

14 x intercept 1, y intercept 2, critical points $(1, 0)$ and $(4, 4)$,

$$\lim_{x\to\infty} f(x) = 3, \quad \lim_{x\to-\infty} f(x) = \infty.$$

15 x intercepts 2 and 4, y intercept 2, critical points $(1, 3)$ and $(3, -1)$,

$$\lim_{x\to\infty} f(x) = \infty, \quad \lim_{x\to-\infty} f(x) = 1.$$

16 No x intercepts, y intercept 1, no critical points,

$$\lim_{x\to\infty} f(x) = 2, \quad \lim_{x\to-\infty} f(x) = 0.$$

In Exercises 17 to 32 graph the given functions, showing any intercepts, asymptotes, critical points, or local or global extrema.

17 $x^3 - 3x^2 + 3x$ **18** $x^4 - 4x^3 + 4x^2$
19 $x^4 - 4x + 3$ **20** $x^5 + 5x$
21 $x^2 - 6x + 5$ **22** $2x^2 + 3x + 5$
23 $x^4 + 2x^3 - 3x^2$ **24** $2x^3 + 3x^2 - 6x$
25 $\dfrac{3x+1}{3x-1}$ **26** $\dfrac{x}{x+1}$
27 $\dfrac{x}{x^2+1}$ **28** $\dfrac{x}{x^2-1}$
29 $\dfrac{1}{2x^2 - x}$ **30** $\dfrac{1}{x^2 - 3x + 2}$
31 $\dfrac{x^2+3}{x^2-4}$ **32** $\dfrac{\sqrt{x^2+1}}{x}$

Exercises 33 to 42 concern functions whose domains are restricted to closed intervals. In each find the maximum and the minimum value for the given function over the given interval.

33 $x^2 - x^4$; $[0, 1]$ **34** $4x - x^2$; $[0, 5]$
35 $4x - x^2$; $[0, 1]$ **36** $2x^2 - 5x$; $[-1, 1]$
37 $x^3 - 2x^2 + 5x$; $[-1, 3]$ **38** $x/(x^2+1)$; $[0, 3]$
39 $x^2 + x^4$; $[0, 1]$ **40** $(x+1)/\sqrt{x^2+1}$; $[0, 3]$
41 $\sin x + \cos x$; $[0, \pi]$ **42** $\sin x - \cos x$; $[0, \pi]$

In Exercises 43 to 49 graph the functions.

43 $\dfrac{\sin x}{1 + 2\cos x}$ **44** $\dfrac{\sqrt{x^2-1}}{x}$
45 $\dfrac{1}{(x-1)^2(x-2)}$ **46** $\dfrac{3x^2+5}{x^2-1}$
47 $2x^{1/3} + x^{4/3}$ **48** $\dfrac{3x^2+5}{x^2+1}$

49 $\sqrt{3}\sin x + \cos x$.

50 Let f and g be polynomials without a common root.
(*a*) Show that if the degree of g is odd, the graph of f/g has a vertical asymptote.
(*b*) Show that if f and g have the same degree, the graph of f/g has a horizontal asymptote.
(*c*) Show that if the degree of f is less than the degree of g, the graph of f/g has a horizontal asymptote.

51 (*a*) Graph $y = x$ and $y = \tan x$ relative to the same axes.
(*b*) Use (*a*) to find how many solutions there are to the equation $x = \tan x$.
(*c*) Graph $y = (\sin x)/x$ showing intercepts and asymptotes.
(*d*) Write a short commentary on the critical points of $(\sin x)/x$. [Part (*b*) may come in handy.]

52 A certain differentiable function has $f'(x) < 0$ for $x < 1$ and $f'(x) > 0$ for $x > 1$. Moreover, $f(0) = 3$, $f(1) = 1$, and $f(2) = 2$.
(*a*) What is the minimum value of $f(x)$ for x in $[0, 2]$? Why?
(*b*) What is the maximum value of $f(x)$ for x in $[0, 2]$? Why?

53 In Example 3 of Sec. 2.5 we graphed $y = (x^3 - x)/(x^2 - 4)$. What is the minimum value of y for $x > 2$? (Find the exact corresponding value of x and express y to four decimal places.)

4.3 MOTION AND THE SECOND DERIVATIVE

In Secs. 3.1 and 3.2 we saw that the velocity of an object moving on a line is represented by a derivative. In this section we examine the acceleration mathematically.

Acceleration

Velocity is the rate at which distance changes. The rate at which velocity changes is called **acceleration**. Thus if $y = f(t)$ denotes position on a line at time t, then the derivative dy/dt equals the velocity, and the derivative of the derivative, that is,

$$\frac{d}{dt}\left(\frac{dy}{dt}\right)$$

equals the acceleration.

The second derivative

The derivative of the derivative of a function $y = f(x)$ is called the **second derivative** of the function. It is denoted

$$\frac{d^2y}{dx^2}, \quad D^2y, \quad y'', \quad f'', \quad D^2f, \quad f^{(2)}, \quad \text{or} \quad \frac{d^2f}{dx^2}.$$

If $y = f(t)$, where t denotes time, the second derivative d^2y/dt^2 is also denoted $\ddot{y}$.

For instance, if $y = x^3$,

$$\frac{dy}{dx} = 3x^2 \quad \text{and} \quad \frac{d^2y}{dx^2} = 6x.$$

Other ways of denoting the second derivative of this function are

$$D^2(x^3) = 6x, \qquad \frac{d^2(x^3)}{dx^2} = 6x, \qquad \text{and} \qquad (x^3)'' = 6x.$$

y	$\frac{dy}{dx}$	$\frac{d^2y}{dx^2}$
x^3	$3x^2$	$6x$
$\frac{1}{x}$	$\frac{-1}{x^2}$	$\frac{2}{x^3}$
$\sin 5x$	$5 \cos 5x$	$-25 \sin 5x$

The table in the margin lists dy/dx, the first derivative, and d^2y/dx^2, the second derivative, for a few functions.

Most functions f met in applications of calculus can be differentiated repeatedly in the sense that Df exists, the derivative of Df, namely, D^2f, exists, the derivative of D^2f exists, and so on.

The derivative of the second derivative is called the *third derivative* and is denoted many ways, such as

$$\frac{d^3y}{dx^3}, D^3y, y''', f''', f^{(3)}, \text{or } \frac{d^3f}{dx^3}.$$

The fourth derivative is defined similarly, as the derivative of the third derivative. In the same way we can define the nth derivative for any positive integer n, which is denoted by such symbols as

$$\frac{d^ny}{dx^n}, \qquad D^ny, \qquad f^{(n)}, \qquad \text{or} \qquad \frac{d^nf}{dx^n}.$$

It is read as "the nth derivative with respect to x." For example, if $f(x) = 2x^3 + x^2 - x + 5$, we have

$$f^{(1)}(x) = 6x^2 + 2x - 1$$

$$f^{(2)}(x) = 12x + 2$$

$$f^{(3)}(x) = 12$$

$$f^{(4)}(x) = 0$$

$$f^{(n)}(x) = 0. \qquad \text{for } n \geq 5$$

We will need only the first and second derivatives in this chapter. In Chap. 5 reference will be made to the fourth derivative. All the "higher-order" derivatives will be needed in Chap. 11.

EXAMPLE 1 A falling rock drops $16t^2$ feet in the first t seconds. Find its velocity and acceleration.

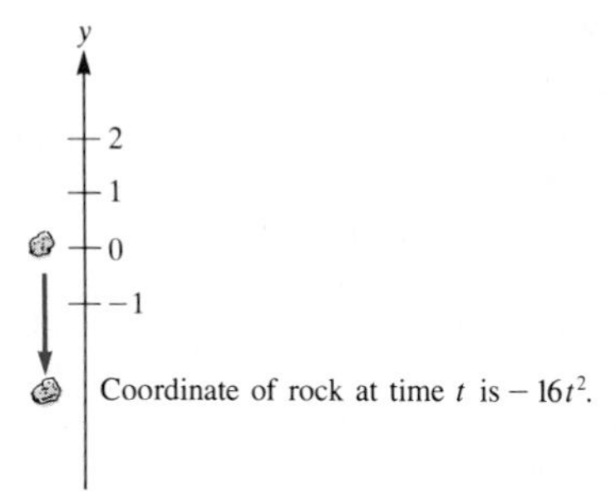

Figure 1

SOLUTION Place the y axis in the usual position, with 0 at the beginning of the fall and the part with positive values above 0, as in Fig. 1. At time t the object has the y coordinate

$$y = -16t^2.$$

The velocity is $(-16t^2)' = -32t$ feet per second, and the acceleration is $(-32t)' = -32$ feet per second per second. The velocity changes at a constant rate. That is, the acceleration is constant. ■

The second derivative represents acceleration in other contexts, as the next example shows.

EXAMPLE 2 Translate this news item into calculus: "The latest unemployment figures can be read as bearing out the forecast that the recession is nearing its peak. Though unemployment continues to increase, it is doing so at a slower rate than before."

SOLUTION Let y be a differentiable function of t that approximates the number of people unemployed at time t. As time changes, so does y: $y = f(t)$. The rate of change in unemployment is the derivative, dy/dt. The news that "unemployment continues to increase" is recorded by the inequality

$$\frac{dy}{dt} > 0.$$

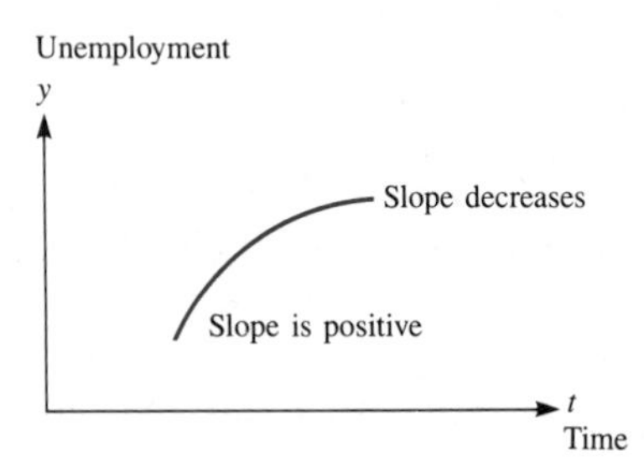

Figure 2

The graph of $y = f(t)$ has a positive slope, indicated in Fig. 2. There is optimism in the article. The rate of increase, dy/dt, is itself declining ("unemployment continues to increase . . . at a *slower rate* than before"). The function dy/dt is decreasing. In terms of the graph of unemployment as a function of time, the slope decreases as you move to the right. That is, the graph becomes less steep as you move to the right. Thus its second derivative

$$\frac{d}{dt}\left(\frac{dy}{dt}\right)$$

is negative: $$\frac{d^2y}{dt^2} < 0.$$

In short, the bad news is that dy/dt is positive. But there is good news: d^2y/dt^2 is negative.

Economics need not be a dismal science. If the first derivative is bleak, maybe the second derivative is promising.

The promise that "the recession is nearing its peak" amounts to the prediction that soon dy/dt will be 0 and then switch sign to become negative. In short, a local maximum in the graph of $y = f(t)$ appears in the economists' crystal ball. ■

Finding Position from Velocity and Acceleration

Knowing the initial position and initial velocity of a moving object and its acceleration at all times, we can calculate its position at all times, as we see in the next two examples, where acceleration is constant. In the first example, the acceleration is 0.

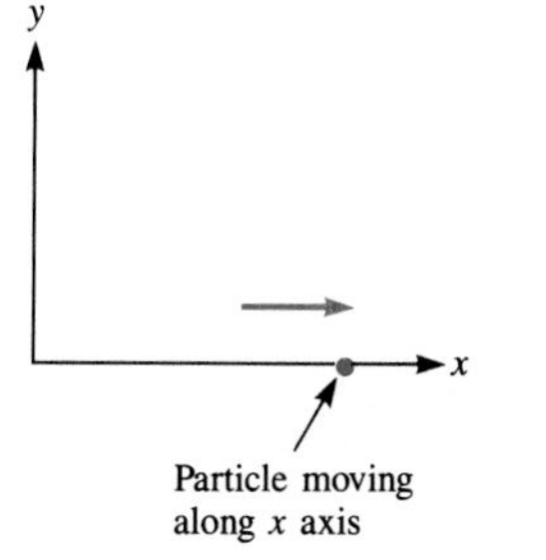

Figure 3

EXAMPLE 3 In the simplest motion, no forces act on a moving particle, hence its acceleration is 0. Assume that a particle is moving on the x axis and no forces act on it. Let its location at time t seconds be $x = f(t)$ feet. See Fig. 3. If at time $t = 0$, $x = 3$ feet and the velocity is 5 feet per second, determine $f(t)$.

SOLUTION The assumption that no force operates on the particle tells us that $d^2x/dt^2 = 0$. Call the velocity v. Then

$$\frac{dv}{dt} = \frac{d}{dt}\left(\frac{dx}{dt}\right) = \frac{d^2x}{dt^2} = 0.$$

Now v is itself a function of time. Since its derivative is 0, v must be constant:

$$v(t) = C$$

for some constant C. Since $v(0) = 5$, the constant C must be 5.

To find the position x as a function of time, note that

$$\frac{dx}{dt} = 5.$$

This equation implies that $x = f(t)$ must be of the form

$$x = 5t + K$$

for some constant K. (Any antiderivative of the constant function 5 is of the form $5t + K$. See Corollary 2 in Sec. 4.1. Since $5t$ is an antiderivative of 5, any other antiderivative must be of the form $5t + K$.) Now, when $t = 0$, $x = 3$. Thus $K = 3$. In short, at any time t seconds, the particle is at $x = 5t + 3$ feet. ■

The next example concerns the case in which the acceleration is constant, but not zero.

EXAMPLE 4 A ball is thrown straight up, with an initial speed of 64 feet per second, from a cliff 96 feet above the ground. Where is the ball t seconds later? When does it reach its maximum height? How high above the ground does the ball rise? When does the ball hit the ground? Assume that there is no air resistance and that the acceleration due to gravity is constant.

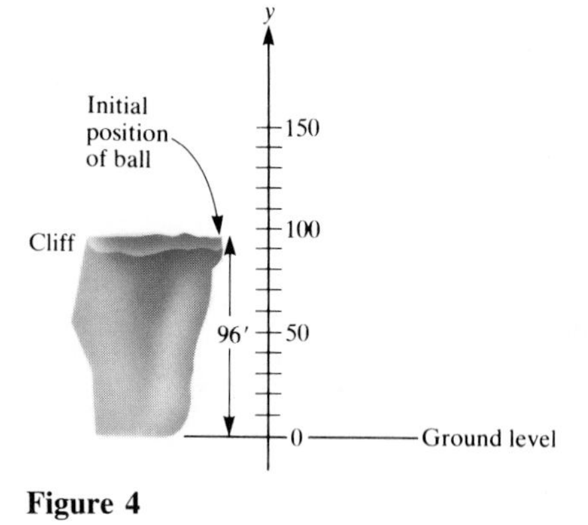

Figure 4

SOLUTION Introduce a vertical coordinate axis to describe the position of the ball. It is more natural to call it the y axis, and so the velocity is dy/dt and acceleration is d^2y/dt^2. Place the origin at ground level and let the positive part of the y axis be above the ground, as in Fig. 4. At time $t = 0$, the velocity $dy/dt = 64$, since the ball is thrown up at a speed of 64 feet per second. (If it had been thrown down, dy/dt would be -64.) As time increases, dy/dt decreases from 64 to 0 (when the ball reaches the top of its path and begins its descent) and continues to decrease through larger and larger negative values as the ball falls down to the ground. Since v is decreasing, the acceleration dv/dt is negative. The (constant) value of dv/dt, gravitational acceleration, is approximately -32 feet per second per second.

From the equation

$$\text{Acceleration} = \frac{dv}{dt} = -32,$$

Velocity is an antiderivative of acceleration.

it follows that

$$v(t) = -32t + C,$$

where C is some constant. [The function $v(t)$ is an antiderivative of the constant function, -32.] To find C, recall that $v = 64$ when $t = 0$. Thus

$$64 = -32 \cdot 0 + C,$$

and $C = 64$. Hence $v = -32t + 64$ for any time t until the ball hits the ground. Now $v(t) = dy/dt$, so

$$\frac{dy}{dt} = v(t) = -32t + 64.$$

Position is an antiderivative of velocity.

Since the position function $y(t)$ is an antiderivative of the velocity, $v(t) = -32t + 64$, we have

$$y(t) = -16t^2 + 64t + K,$$

where K is a constant. To find K, make use of the fact that $y = 96$ when $t = 0$. Thus

$$96 = -16 \cdot 0^2 + 64 \cdot 0 + K,$$

and $K = 96$.

We have obtained a complete description of the position of the ball at any time t while it is in the air:

$$y = -16t^2 + 64t + 96.$$

(As a check, note that when $t = 0$, $y = 96$, the initial height.) This, together with $v = -32t + 64$, provides answers to many questions about the ball's flight.

Maximum height

When does it reach its maximum height? When it is neither rising nor falling. In other words, the velocity is neither positive nor negative, but must be 0. When $v = 0$; that is, when $-32t + 64 = 0$, or when $t = 2$ seconds.

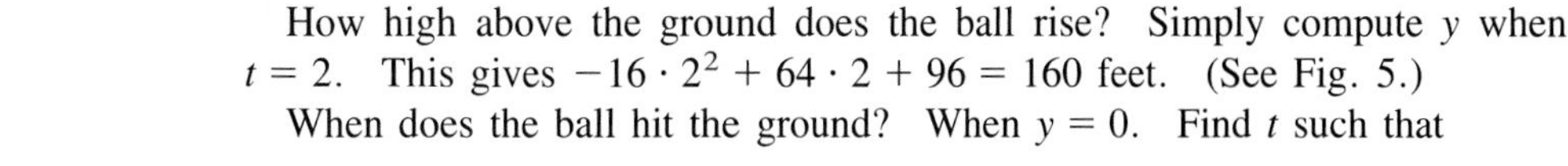

How high above the ground does the ball rise? Simply compute y when $t = 2$. This gives $-16 \cdot 2^2 + 64 \cdot 2 + 96 = 160$ feet. (See Fig. 5.)

When does the ball hit the ground? When $y = 0$. Find t such that

$$y = -16t^2 + 64t + 96 = 0.$$

Division by -16 yields the simpler equation $t^2 - 4t - 6 = 0$, which has the solutions

$$t = \frac{4 \pm \sqrt{16 + 24}}{2} = 2 \pm \sqrt{10}.$$

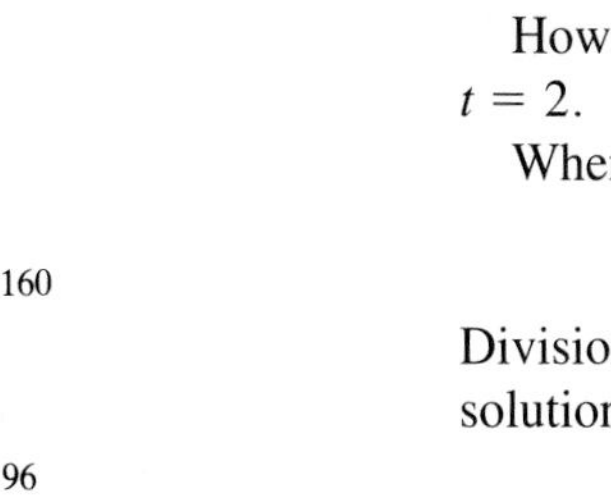

Figure 5

Since $2 - \sqrt{10}$ is negative and the ball cannot hit the ground before it is thrown, the physically meaningful solution is $2 + \sqrt{10}$. The ball lands $2 + \sqrt{10}$ seconds after it is thrown; it is in the air for about 5.2 seconds.

The graphs of position, velocity, and acceleration as functions of time provide another perspective on the motion of the ball, as shown in Figs. 6 to 8.

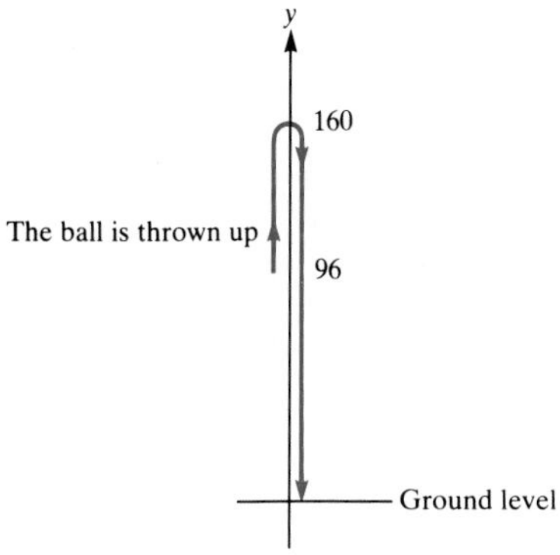

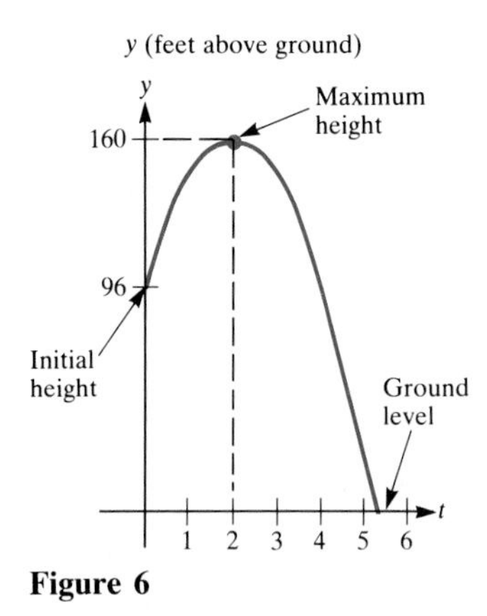

Figure 6

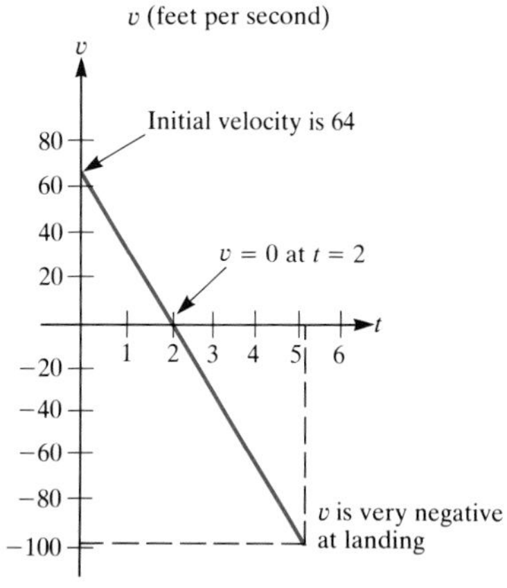

Figure 7

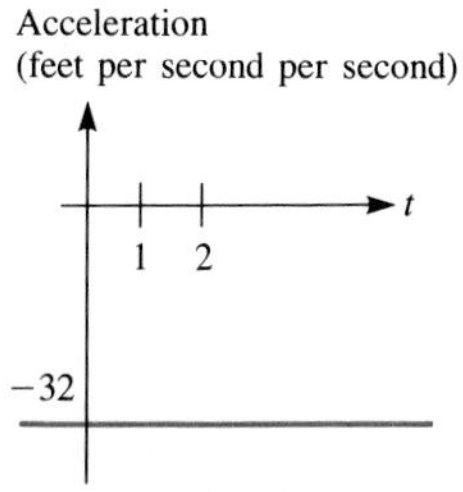

Figure 8

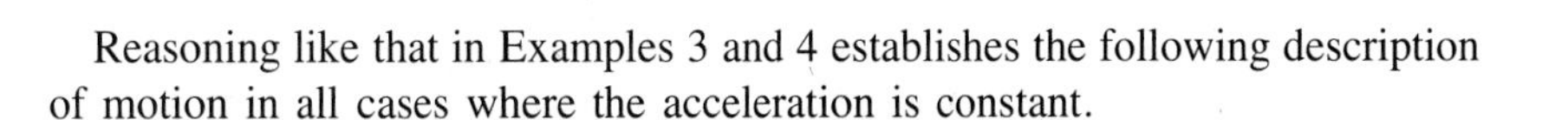

Reasoning like that in Examples 3 and 4 establishes the following description of motion in all cases where the acceleration is constant.

MOTION UNDER CONSTANT ACCELERATION

Assume that a particle moving on the y axis has a constant acceleration a at any time. Assume that at time $t = 0$ it has an initial velocity v_0 and has the initial y coordinate y_0. Then at any time $t \geq 0$ its y coordinate is

$$y = \frac{a}{2}t^2 + v_0 t + y_0.$$

In Example 3, $a = 0$, $v_0 = 5$, and $y_0 = 3$; in Example 4, $a = -32$, $v_0 = 64$, and $y_0 = 96$.

EXERCISES FOR SEC. 4.3: MOTION AND THE SECOND DERIVATIVE

In Exercises 1 to 16 find the first and second derivatives of the given functions.

1 $y = 2x + 3$
2 $y = 5x - 7$
3 $y = x^5$
4 $y = x^6$
5 $y = 2x^3 + x + 2$
6 $y = 4x^3 - x^2 + x$
7 $y = \dfrac{x}{x+1}$
8 $y = \dfrac{x^2}{x-1}$
9 $y = x \cos x$
10 $y = \sec x$
11 $y = \dfrac{\sin x}{x}$
12 $y = \dfrac{x}{\tan x}$
13 $y = (x-2)^4$
14 $y = (x+1)^3$
15 $y = \sin 3x$
16 $y = \tan x^2$

17 Translate into calculus the following news report about the leaning tower of Pisa. "The tower's angle from the vertical was increasing more rapidly." *Suggestion:* Let $\theta = f(t)$ be the angle of deviation from the vertical at time t.

Incidentally, the tower, begun in 1174 and completed in 1350, is 179 feet tall and leans about 14 feet from the vertical. Each day it leans, on the average, another $\frac{1}{5000}$ inch.

18 Translate this news headline into calculus: "Gasoline prices increase more slowly." *Suggestion:* Let $G(t)$ be the price of a gallon of gasoline at time t.

Exercises 19 to 21 concern Example 4.

19 (*a*) How long after the ball in Example 4 is thrown does it pass by the top of the cliff?
(*b*) What are its speed and velocity then?

20 If the ball in Example 4 had simply been dropped from the cliff, what would y be as a function of time? How long would the ball fall?

21 In view of the result of Exercise 20, interpret physically each of the three terms on the right side of the formula $y = -16t^2 + 64t + 96$.

22 Until the Eiffel Tower was completed in 1889, the Washington Monument was the tallest human-created structure.
(*a*) How long would it take a rock to hit the ground if dropped from the top of the Washington Monument, which is 555 feet 5.5 inches high?
(*b*) How long would it take if it were dropped from the top of the Eiffel Tower, which is 300 meters or 984.25 feet tall?

23 A jetliner begins its descent 120 miles from the airport. Its velocity is initially 500 miles per hour and its velocity in landing is 180 miles per hour. Assuming a constant deceleration, how long does the descent take?

24 Let $y = f(t)$ describe the motion on the y axis of an object whose acceleration has the constant value a. Show that

$$y = \frac{a}{2}t^2 + v_0 t + y_0,$$

where v_0 is the velocity when $t = 0$, and y_0 is the position when $t = 0$.

25 At time $t = 0$ a particle is at $y = 3$ feet and has a velocity of -3 feet per second; it has a constant acceleration of 6 feet per second per second. Find its position at any time t.

26 At time $t = 0$ a particle is at $y = 10$ feet and has a velocity of 8 feet per second; it has a constant acceleration of -8 feet per second per second. (*a*) Find its position at any time t. (*b*) What is its maximum y coordinate?

27 At time $t = 0$ a particle is at $y = 0$ and has a velocity of 0 feet per second. Find its position at any time t if its acceleration is always -32 feet per second per second.

28 At time $t = 0$ a particle is at $y = -4$ feet and has a velocity of 6 feet per second; it has a constant acceleration of -32 feet per second per second. (*a*) Find its position at any time t. (*b*) What is its largest y coordinate?

---

29 A car accelerates with constant acceleration from 0 (rest) to 60 miles per hour in 15 seconds. How far does it travel in this period? Be sure to do your computations either all in seconds or all in hours; for instance, 60 miles per hour is 88 feet per second.

30 Show that a ball thrown straight up from the ground takes as long to rise as to fall back to its initial position. How does the velocity with which it strikes the ground compare with its initial velocity? Consider the same question for its speed.

4.4 RELATED RATES

The rate at which a quantity is changing is represented by a derivative. The rate at which that rate is changing is represented by a second derivative. Sometimes the rate at which one quantity is changing is closely related to the rate at which another quantity is changing. This section shows how calculus can be used to determine exactly how one rate of change determines another. Example 1 illustrates the general procedure.

EXAMPLE 1 An angler has a fish at the end of his line, which is reeled in at 2

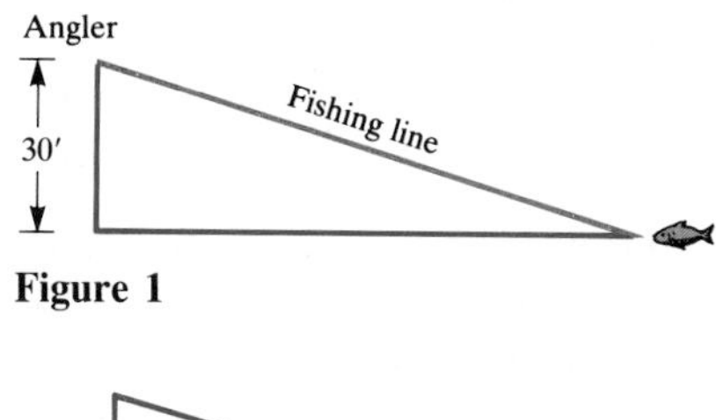

Figure 1

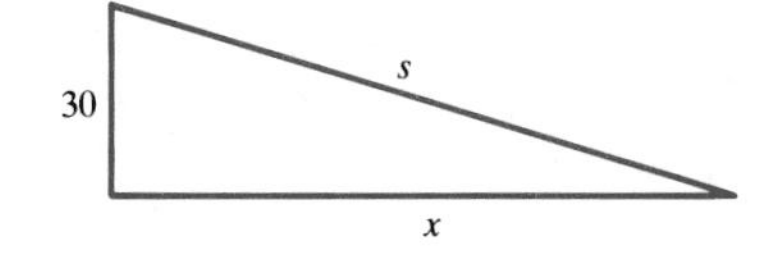

Figure 2

feet per second from a bridge 30 feet above the water. At what speed is the fish moving through the water when the amount of line out is 50 feet? 31 feet? Assume the fish is at the surface of the water. (See Fig. 1.)

SOLUTION Our first impression might be that since the line is reeled in at a constant speed, the fish at the end of the line moves through the water at a constant speed. As we will see, this impression is wrong.

Let s be the length of the line and x the horizontal distance of the fish from the bridge. (See Fig. 2.)

Since the line is reeled in at the rate of 2 feet per second, s is shrinking, and

$$\frac{ds}{dt} = -2.$$

The rate at which the fish moves through the water is given by the derivative, dx/dt. The problem is to find dx/dt when $s = 50$ and also when $s = 31$.

First of all, we need an equation that relates s and x at *any* time, not just when $x = 50$ or $x = 31$. If we consider only $x = 50$ or $x = 31$, there would be no motion, and no chance to use derivatives.

The quantities x and s are related by the pythagorean theorem:

This equation is the heart of the example.

$$x^2 + 30^2 = s^2.$$

Both x and s are functions of time t. Thus both sides of the equation may be differentiated with respect to t, yielding

$$\frac{d(x^2)}{dt} + \frac{d(30^2)}{dt} = \frac{d(s^2)}{dt}$$

or

$$2x\frac{dx}{dt} + 0 = 2s\frac{ds}{dt}.$$

Hence

$$x\frac{dx}{dt} = s\frac{ds}{dt}.$$

This last equation provides the tool for answering the questions.

Since $ds/dt = -2$,

$$x\frac{dx}{dt} = (s)(-2).$$

Hence

$$\frac{dx}{dt} = \frac{-2s}{x}.$$

When $s = 50$,

$$x^2 + 30^2 = 50^2,$$

so $x = 40$. Thus when 50 feet of line is out, the speed is

$$\left|\frac{dx}{dt}\right| = \frac{2s}{x} = \frac{2 \cdot 50}{40} = 2.5 \text{ feet per second.}$$

When $s = 31$,

$$x^2 + 30^2 = 31^2.$$

Hence

$$x = \sqrt{31^2 - 30^2} = \sqrt{961 - 900} = \sqrt{61}.$$

Thus when 31 feet of line is out, the fish is moving at the speed of

$$\frac{2s}{x} = \frac{2 \cdot 31}{\sqrt{61}} = \frac{62}{\sqrt{61}} \approx 7.9 \text{ feet per second.}$$

Let us look at the situation from the fish's point of view. When it is x feet from the point in the water directly below the bridge, its speed is $2s/x$ feet per second. Since s is larger than x, its speed is always greater than 2 feet per second. When x is very large, s/x is near 1, so the fish is moving through the water only a little faster than the line is reeled in. However, when the fish is almost at the point under the bridge, x is very small; then $2s/x$ is huge, and the fish finds itself moving at huge speeds. ■

In Example 1 it would be a tactical mistake to indicate in Fig. 2 that the hypotenuse of the triangle is 50 feet long, for if one leg is 30 feet and the hypotenuse is 50 feet, the triangle is determined; there is nothing left free to vary with time. It is safest to label all the lengths or quantities that can change with letters x, y, s, and so on, even if not all are needed in the solution. Only after you finish differentiating do you determine what the rates are at a specified value of the variable.

The General Procedure

The method used in Example 1 applies to many related rate problems. This is the general procedure, broken into steps:

Warning: Differentiate, then *substitute the specific numbers for the variables. If you reversed the order, you would just be differentiating constants.*

PROCEDURE FOR FINDING A RELATED RATE

1. Find an equation that relates the varying quantities. (If the quantities are geometric, draw a picture and label the varying quantities with letters.)
2. Differentiate both sides of the equation with respect to time, obtaining an equation that relates the various rates of change.
3. Solve the equation obtained in Step 2 for the unknown rate. (Only at this step do you substitute constants for variables.)

EXAMPLE 2 A woman on the ground is watching a jet through a telescope as it approaches at a speed of 10 miles per minute at an altitude of 7 miles. At what rate (in radians per minute) is the angle of the telescope changing when the horizontal distance of the jet from the woman is 24 miles? When the jet is directly above the woman?

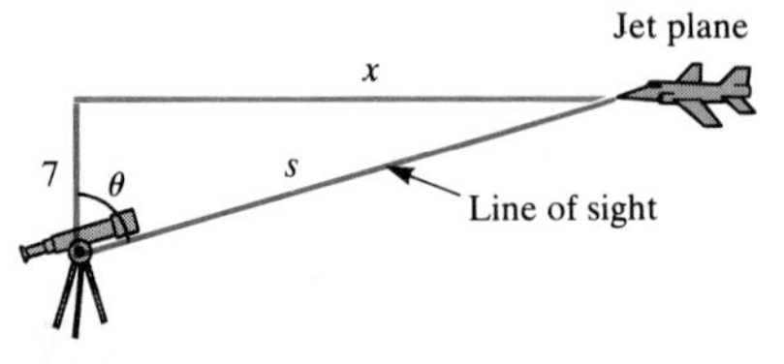

Figure 3

SOLUTION When the plane is far away, the woman sees the angle change very slowly; she moves the telescope slowly. As the plane gets closer, she has to change the direction of the telescope more quickly. Even though the plane keeps a constant speed, the woman must change the direction of the telescope at a varying angular speed. To find that varying angular speed, we follow the procedure just outlined.

Step 1. We make a diagram and label its key parts, as in Fig. 3. We know that $dx/dt = -10$ miles per minute (the minus sign since x is shrinking). We wish to find $d\theta/dt$. The variables θ and x are related by the equation

$$\tan \theta = \frac{x}{7}. \tag{1}$$

We wish to find the rate at which θ changes, $d\theta/dt$.

Step 2. We differentiate both sides of (1) with respect to time.

$$\frac{d}{dt}(\tan \theta) = \frac{d}{dt}\left(\frac{x}{7}\right)$$

or

$$\sec^2 \theta \frac{d\theta}{dt} = \frac{1}{7}\frac{dx}{dt}. \tag{2}$$

(Note the use of the chain rule.)

Step 3. We know that $dx/dt = -10$. Substituting this information into (2) yields

$$\sec^2 \theta \frac{d\theta}{dt} = \frac{-10}{7}. \tag{3}$$

Solving (3) for the unknown rate of change, $d\theta/dt$, yields

$$\frac{d\theta}{dt} = \frac{-10}{7 \sec^2 \theta} \text{ radians per minute.} \tag{4}$$

Equation (4) expresses the rate at which angle θ changes, in terms of the angle θ. However, we want to find $d\theta/dt$ when $x = 24$. So we must rewrite (4) in terms of x. To do this, refer back to Fig. 3. We have

$$\begin{aligned}\sec \theta &= \frac{1}{\cos \theta} \\ &= \frac{1}{7/s} \\ &= \frac{s}{7} \\ &= \frac{\sqrt{x^2 + 7^2}}{7}.\end{aligned}$$

Therefore, Eq. (4) can be written in terms of x as

$$\begin{aligned}\frac{d\theta}{dt} &= \frac{-10}{7\left(\frac{\sqrt{x^2 + 7^2}}{7}\right)^2} \\ &= \frac{-70}{x^2 + 7^2}.\end{aligned}$$

Thus

$$\frac{d\theta}{dt} = \frac{-70}{x^2 + 7^2} \text{ radians per minute.}$$

Only at this point do we substitute a constant for a variable, replacing x by 24, to obtain

$$\frac{d\theta}{dt} = \frac{-70}{24^2 + 7^2} = \frac{-70}{625} \text{ radians per minute.}$$

When the horizontal distance between the plane and the woman is 24 miles, the angle θ of the telescope is changing at the rate of $-70/625$ radians per minute, or about $-6.4°$ per minute. When the jet is directly above the woman $x = 0$, and we obtain $d\theta/dt = -70/49$ radians per minute (which is about $-82°$ per minute). ■

Finding an Acceleration

The method described in Example 1 for determining unknown rates from known ones extends to finding an unknown acceleration. Just differentiate another time. Example 3 illustrates the procedure.

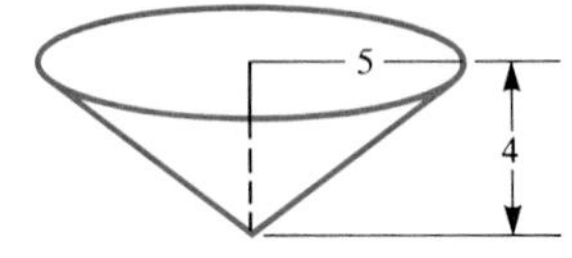

Figure 4

EXAMPLE 3 Water flows into a conical tank at the constant rate of 3 cubic meters per second. The radius of the cone is 5 meters and its height is 4 meters. Let $h(t)$ represent the height of the water above the bottom of the cone at time t. Find dh/dt (the rate at which the water is rising in the tank) and d^2h/dt^2 (the rate at which that rate changes) when the tank is filled to a height of 2 meters. (See Figs. 4 and 5.)

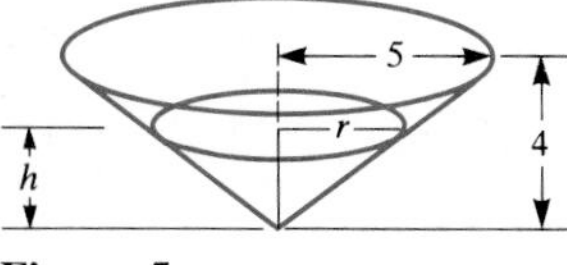

Figure 5

SOLUTION Let $V(t)$ be the volume of water in the tank at time t. The data imply that

$$\frac{dV}{dt} = 3,$$

and hence

$$\frac{d^2V}{dt^2} = 0.$$

To find dh/dt and d^2h/dt^2, first obtain an equation relating V and h.

When the tank is filled to the height h, the water forms a cone of height h and radius r. (See Fig. 5.) By similar triangles,

$$\frac{r}{h} = \frac{5}{4} \quad \text{or} \quad r = \frac{5h}{4}.$$

Thus

$$\begin{aligned} V &= \tfrac{1}{3}\pi r^2 h \\ &= \tfrac{1}{3}\pi(\tfrac{5}{4}h)^2 h \\ &= \tfrac{25}{48}\pi h^3. \end{aligned}$$

The equation relating V and h is

$$V = \frac{25\pi}{48} h^3. \tag{5}$$

From here on, just differentiate as often as needed.

Differentiating both sides of Eq. (5) once (using the chain rule) yields

$$\frac{dV}{dt} = \frac{25\pi}{48} \frac{d(h^3)}{dh} \frac{dh}{dt}$$

or
$$\frac{dV}{dt} = \frac{25\pi}{16} h^2 \frac{dh}{dt}.$$

Since $dV/dt = 3$ all the time,
$$3 = \frac{25\pi h^2}{16} \frac{dh}{dt},$$

from which it follows that
$$\frac{dh}{dt} = \frac{48}{25\pi h^2} \text{ meters per second.} \tag{6}$$

As (6) shows, the larger h is, the slower the water rises. (Why is this to be expected?) Even though water enters the tank at a constant rate, it does not rise at a constant rate.

To find dh/dt when $h = 2$ meters, substitute 2 for h in (6), obtaining
$$\frac{dh}{dt} = \frac{48}{25\pi 2^2} = \frac{12}{25\pi} \text{ meters per second.}$$

Now we turn to the acceleration, d^2h/dt^2. We do not differentiate the equation $dh/dt = 12/(25\pi)$ since this equation holds only when $h = 2$. We must go back to (6), which holds at any time.

Using (6), we have
$$\frac{dh}{dt} = \frac{48}{25\pi h^2}, \tag{7}$$

hence
$$\frac{d^2h}{dt^2} = \frac{48}{25\pi} \frac{d}{dt}\left(\frac{1}{h^2}\right),$$
$$\frac{d^2h}{dt^2} = \frac{48}{25\pi} \frac{-2}{h^3} \frac{dh}{dt},$$

or
$$\frac{d^2h}{dt^2} = \frac{-96}{25\pi h^3} \frac{dh}{dt}. \tag{8}$$

The last equation expresses the acceleration in terms of h and dh/dt. Substituting (7) into (8) gives
$$\frac{d^2h}{dt^2} = \frac{-96}{25\pi h^3} \frac{48}{25\pi h^2},$$

or
$$\frac{d^2h}{dt^2} = \frac{-(96)(48)}{(25\pi)^2 h^5} \text{ meters per second per second.} \tag{9}$$

Equation (9) tells us that, since d^2h/dt^2 is negative, the rate at which the water rises in the tank is decreasing.

The problem also asked for the value of d^2h/dt^2 when $h = 2$. To find that value, replace h by 2 in (9), obtaining
$$\frac{d^2h}{dt^2} = \frac{-(96)(48)}{(25\pi)^2 2^5}$$

or
$$\frac{d^2h}{dt^2} = \frac{-144}{625\pi^2} \text{ meters per second per second.} \quad ■$$

EXERCISES FOR SEC. 4.4: RELATED RATES

Exercises 1 and 2 are related to Example 1.

1 How fast is the fish moving through the water when it is 1 foot horizontally from the bridge?

2 The angler in Example 1 decides to let the line out as the fish swims away. The fish swims away at a constant speed of 5 feet per second relative to the water. How fast is the angler paying out his line when the horizontal distance from the bridge to the fish is (*a*) 1 foot? (*b*) 100 feet?

3 A 10-foot ladder is leaning against a wall. A person pulls the base of the ladder away from the wall at the rate of 1 foot per second.

(*a*) Draw a neat picture of the situation and label the varying lengths by letters and the fixed lengths by numbers.

(*b*) Obtain an equation involving the variables in (*a*).

(*c*) Differentiate it with respect to time.

(*d*) How fast is the top going down the wall when the base of the ladder is 6 feet from the wall? 8 feet from the wall? 9 feet from the wall?

4 A kite is flying at a height of 300 feet in a horizontal wind.

(*a*) Draw a neat picture of the situation and label the varying lengths by letters and the fixed lengths by numbers.

(*b*) When 500 feet of string is out, the kite is pulling the string out at a rate of 20 feet per second. What is the kite's velocity? (Assume the string remains straight.)

5 A beachcomber walks 2 miles per hour along the shore as the beam from a rotating light 3 miles offshore follows him. (See Fig. 6.)

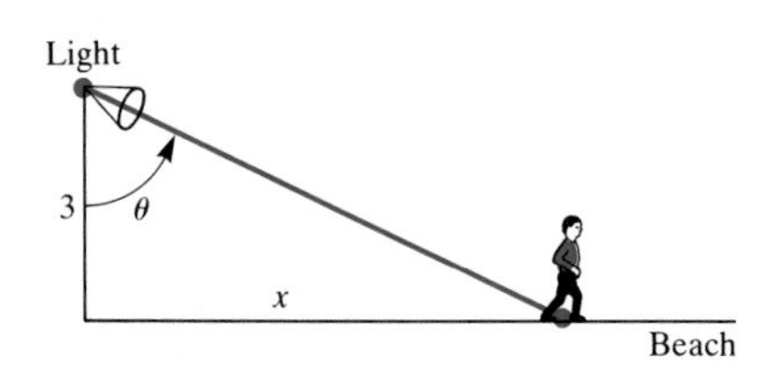

Figure 6

(*a*) Intuitively, what do you think happens to the rate at which the light rotates as the beachcomber walks further and further along the shore away from the lighthouse?

(*b*) Letting x describe the distance of the beachcomber from the point on the shore nearest the light and θ the angle of the light, obtain an equation relating θ and x.

(*c*) With the aid of (*b*), show that $d\theta/dt = 6/(9 + x^2)$ (radians per hour).

(*d*) Does the formula in (*c*) agree with your guess in (*a*)?

6 A man 6 feet tall walks at the rate of 5 feet per second away from a lamp that is 20 feet high. At what rate is his shadow lengthening when he is (*a*) 10 feet from the lamp? (*b*) 100 feet from the lamp?

7 A large spherical balloon is being inflated at the rate of 100 cubic feet per minute. At what rate is the radius increasing when the radius is (*a*) 10 feet? (*b*) 20 feet? (The volume V of a sphere of radius r is $4\pi r^3/3$.)

8 A shrinking spherical balloon loses air at the rate of 1 cubic inch per second. At what rate is its radius changing when the radius is (*a*) 2 inches? (*b*) 1 inch?

9 Bulldozers are moving earth at the rate of 1,000 cubic yards per hour onto a conically shaped hill whose height remains equal to its radius. At what rate is the height of the hill increasing when the hill is (*a*) 20 yards high? (*b*) 100 yards high? (The volume of a cone of radius r and height h is $\pi r^2 h/3$.)

10 The lengths of the two legs of a right triangle depend on time. One leg, whose length is x, increases at the rate of 5 feet per second, while the other, of length y, decreases at the rate of 6 feet per second. At what rate is the hypotenuse changing when $x = 3$ feet and $y = 4$ feet? Is the hypotenuse increasing or decreasing then?

11 Two sides of a triangle and their included angle are changing with respect to time. The angle increases at the rate of 1 radian per second, one side increases at the rate of 3 feet per second, and the other side decreases at the rate of 2 feet per second. Find the rate at which the area is changing when the angle is $\pi/4$, the first side is 4 feet long, and the second side is 5 feet long. Is the area decreasing or increasing then?

12 The length of a rectangle is increasing at the rate of 7 feet per second, and the width is decreasing at the rate of 3 feet per second. When the length is 12 feet and the width is 5 feet, find the rate of change of (*a*) the area, (*b*) the perimeter, (*c*) the length of the diagonal.

Exercises 13 to 16 concern acceleration. The notation $\dot{x}$ for dx/dt, $\dot{\theta}$ for $d\theta/dt$, $\ddot{x}$ for d^2x/dt^2, and $\ddot{\theta}$ for $d^2\theta/dt^2$ is common in physics.

13 What is the acceleration of the fish described in Example 1 when the length of line is (*a*) 300 feet? (*b*) 31 feet?

14 Find $\ddot{\theta}$ in Example 2 when the horizontal distance from the jet is (*a*) 7 miles, (*b*) 1 mile.

15 A particle moves on the parabola $y = x^2$ in such a way that $\dot{x} = 3$ throughout the journey. Find formulas for (*a*) $\dot{y}$ and (*b*) $\ddot{y}$.

16 Call one acute angle of a right triangle θ. The adjacent leg has length x and the opposite leg has length y.

(*a*) Obtain an equation relating x, y, and θ.

(*b*) Obtain an equation involving $\dot{x}$, $\dot{y}$, and $\dot{\theta}$ (and other variables).

(*c*) Obtain an equation involving $\ddot{x}$, $\ddot{y}$, and $\ddot{\theta}$ (and other variables).

---

17 A two-piece extension ladder leaning against a wall is collapsing at the rate of 2 feet per second at the same time as its foot is moving away from the wall at the rate of 3 feet per second. How fast is the top of the ladder moving down the wall when it is 8 feet from the ground and the foot is 6 feet from the wall? (See Fig. 7.)

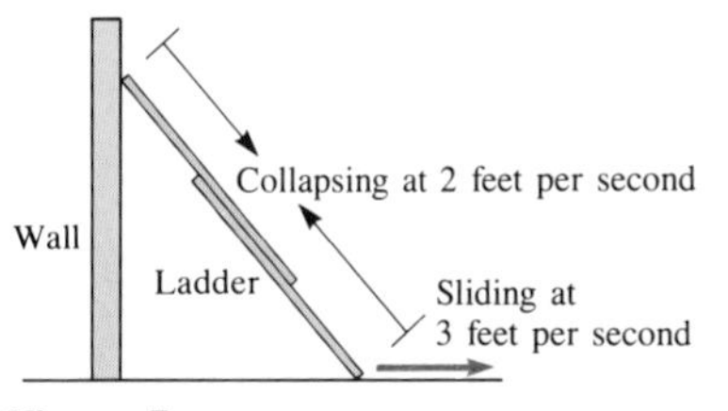

Figure 7

18 At an altitude of x kilometers, the atmospheric pressure decreases at a rate of $128\ (0.88)^x$ millibars per kilometer. A rocket is rising at the rate of 5 kilometers per second vertically. At what rate is the atmospheric pressure changing (in millibars per second) when the altitude of the rocket is (*a*) 1 kilometer? (*b*) 50 kilometers?

19 A woman is walking on a bridge that is 20 feet above a river as a boat passes directly under the center of the bridge (at a right angle to the bridge) at 10 feet per second. At that moment the woman is 50 feet from the center and approaching it at the rate of 5 feet per second. (*a*) At what rate is the distance between the boat and woman changing at that moment? (*b*) Is the rate at which they are approaching or separating increasing or is it decreasing?

20 A spherical raindrop evaporates at a rate proportional to its surface area. Show that the radius shrinks at a constant rate.

21 A couple is on a Ferris wheel when the sun is directly overhead. The diameter of the wheel is 50 feet, and its speed is 0.1 revolution per second. (*a*) What is the speed of their shadows on the ground when they are at a two-o'clock position? (*b*) A one-o'clock position? (*c*) Show that the shadow is moving its fastest when they are at the top or bottom, and its slowest when they are at the three-o'clock position.

22 Water is flowing into a hemispherical bowl of radius 5 feet at the constant rate of 1 cubic foot per minute.

(*a*) At what rate is the top surface of the water rising when its height above the bottom of the bowl is 3 feet? 4 feet? 5 feet?

(*b*) If $h(t)$ is the depth in feet at time t, find $\ddot{h}$ when $h = 3, 4,$ and 5.

23 A man in a hot-air balloon is ascending at the rate of 10 feet per second. How fast is the distance from the balloon to the horizon (that is, the distance the man can see) increasing when the balloon is 1,000 feet high? Assume that the earth is a ball of radius 4,000 miles. (See Fig. 8.)

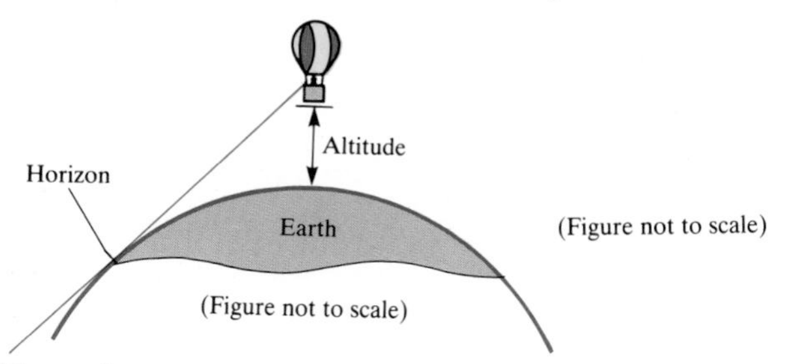

Figure 8

24 (Contributed by Keith Sollers, undergraduate at the University of California at Davis.) We quote from his note to us. "The numbers are ugly, but I think it's a good problem nevertheless. I didn't think it up myself. The Medical Center eye group gave me the problem and asked me to solve it. They were going to put a gas bubble in someone's eye."

The volume of a gas bubble changes from 0.4 cc to 1.6 cc in 74 hours. Assuming that the rate of change of the radius is constant, find

(*a*) The rate at which the radius changes;

(*b*) The rate at which the volume of the bubble is increasing at any volume V;

(*c*) The rate at which the volume is increasing when the volume is 1 cc.

(*Note:* The volume of a ball of radius r is $4\pi r^3/3$. Assume the bubble is spherical.)

4.5 THE SECOND DERIVATIVE AND GRAPHING

The sign of the first derivative tells whether a graph is rising or descending. In this section we examine what the sign of the second derivative tells us about a graph. We will use the information not only to help graph functions but also to provide an additional way to test whether a critical point is a maximum or minimum (or neither).

Concave Up and Concave Down

Assume that $f''(x)$ is positive for all x in the open interval (a, b). Since f'' is the derivative of f', it follows that f' is an increasing function throughout the interval

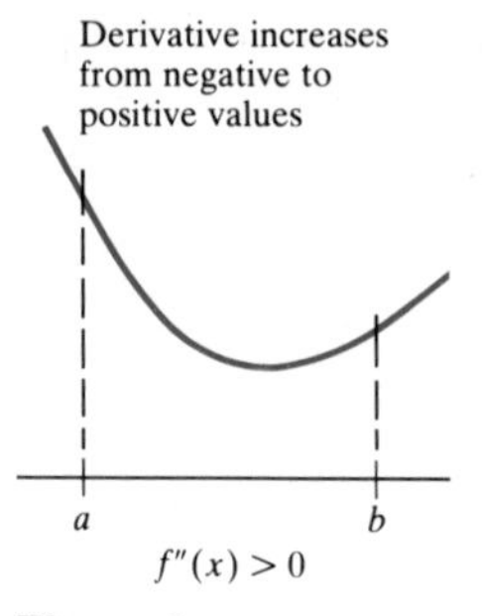

Figure 1

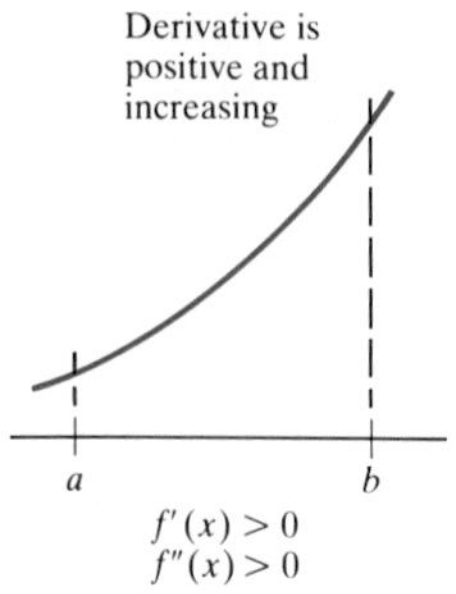

Derivative is
positive and
increasing

a
b

$f'(x) > 0$
$f''(x) > 0$

Figure 2

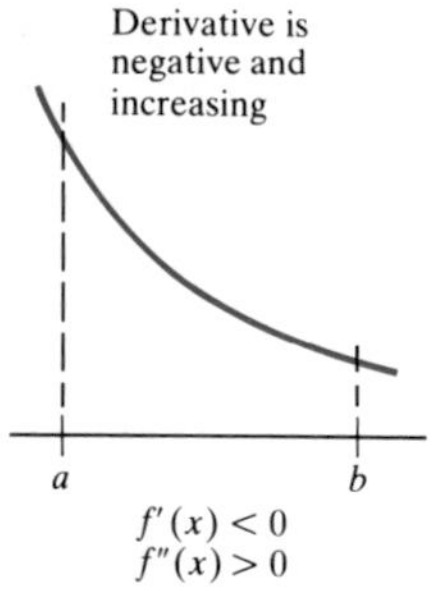

Derivative is
negative and
increasing

a
b

$f'(x) < 0$
$f''(x) > 0$

Figure 3

(a, b). In other words, as x increases, the slope of the graph of $y = f(x)$ increases as we move from left to right on that part of the graph corresponding to the interval (a, b). The slope may increase from negative to positive values, as in Fig. 1. Or the slope may be positive throughout (a, b) and increasing, as in Fig. 2. Or the slope may be negative throughout (a, b) and increasing, as in Fig. 3.

As you drive along such a graph from left to right, your car keeps turning to the left, as in Fig. 4.

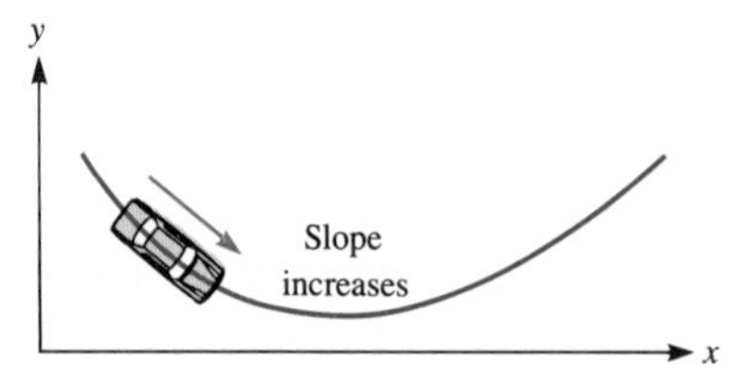

Figure 4

These examples bring us to the definition of a key feature of a function and its graph.

Definition *Concave upward.* A function f whose first derivative is increasing throughout the open interval (a, b) is called **concave upward** in that interval.

Note that when a function is concave upward, it is shaped like part of a *cup* (concave *up*ward).

Where a curve is concave upward, it lies above its tangent lines and below its chords, as shown in Fig. 5.

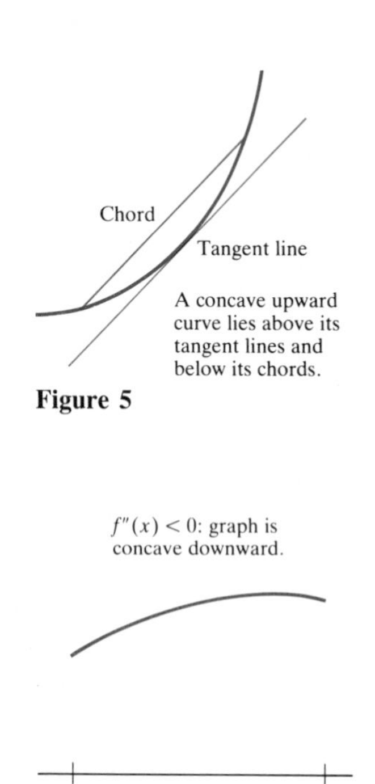

Figure 5

Figure 6

As was observed, in an interval where $f''(x)$ is positive, the function $f'(x)$ is increasing, and so the function f is concave upward. However, if a function is concave upward, $f''(x)$ is not necessarily positive. For instance, $y = x^4$ is concave upward over any interval, since the derivative $4x^3$ is increasing. The second derivative $12x^2$ is not always positive; at $x = 0$ it is 0.

If, on the other hand, $f''(x)$ is negative throughout (a, b), then f' is a decreasing function and the graph of f looks like part of the curve in Fig. 6.

Definition *Concave downward.* A function f whose first derivative is decreasing throughout the open interval (a, b) is called **concave downward** in that interval.

Where a function is concave downward, it lies *below its tangent lines* and

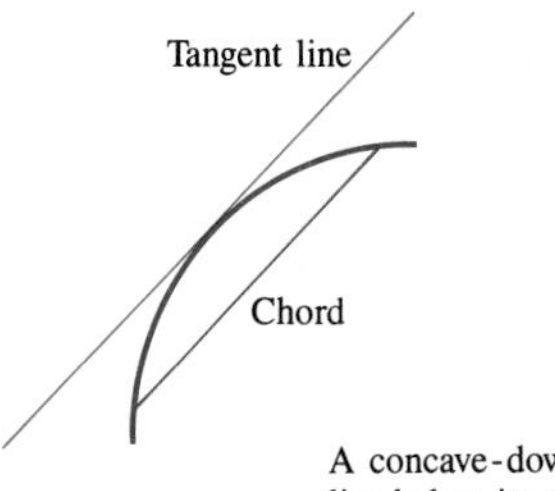

A concave-downward curve lies below its tangent lines and above its chords.

Figure 7

above its chords. (See Fig. 7.) When a car moves from left to right on a concave downward part of a curve, the driver keeps turning it to the right.

EXAMPLE 1 Where is the graph of $f(x) = x^3$ concave upward? Concave downward?

SOLUTION First compute the second derivative. Since $D(x^3) = 3x^2$, $D^2(x^3) = 6x$.

Clearly $6x$ is positive for all positive x and negative for all negative x. The graph, shown in Fig. 8, is concave upward if $x > 0$ and concave downward if $x < 0$. Note that the sense of concavity changes at $x = 0$. When you drive along this curve from left to right, your car turns to the right until you pass through (0, 0). Then it starts turning to the left. ■

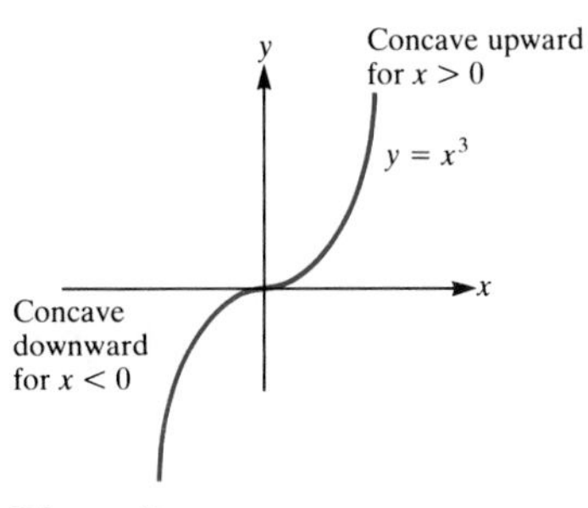

Figure 8

Inflection Points

Of special interest in Example 1 is the point on the graph where the sense of concavity changes. Such a point is called an **inflection point**.

Definition *Inflection point* and *inflection number*. Let f be a function and let a be a number. Assume that there are numbers b and c such that $b < a < c$ and

1. f is continuous on the open interval (b, c).
2. f is concave upward in the interval (b, a) and concave downward in the interval (a, c), or vice versa.

Then the point $(a, f(a))$ is called an **inflection point** or **point of inflection**. The number a is called an **inflection number**.

Observe that if the second derivative changes sign at the number a, then a is an inflection number.

If the second derivative exists at an inflection point, it must be 0. But there can be an inflection point even if f'' is not defined there, as shown by the next example.

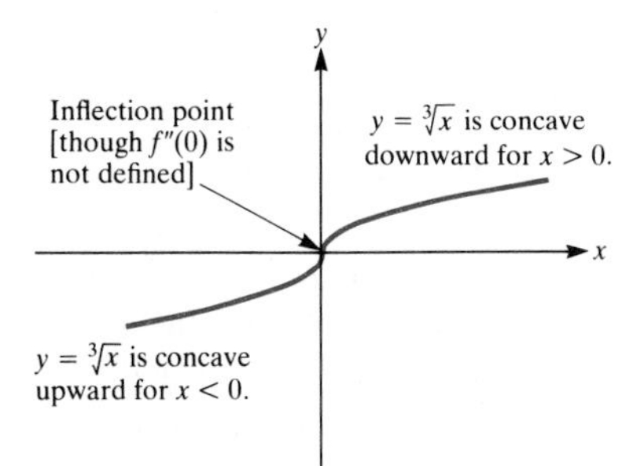

Figure 9

EXAMPLE 2 Examine the concavity of $y = x^{1/3}$.

SOLUTION Here

$$y' = \tfrac{1}{3}x^{-2/3} \qquad \text{and} \qquad y'' = \tfrac{1}{3} \cdot \left(-\tfrac{2}{3}\right) x^{-5/3}.$$

Neither y' nor y'' is defined at 0; however, the sign of y'' changes at 0. When x is negative, y'' is positive; when x is positive, y'' is negative. The concavity switches from upward to downward at $x = 0$. The graph is shown in Fig. 9. There is an inflection point at (0, 0). ■

Finding Inflection Points

The simplest way to look for inflection points is to use the second derivative:

> To find inflection points of $y = f(x)$,
>
> **1** Compute $f''(x)$.
> **2** Look for numbers a such that $f''(a) = 0$ or f'' is not defined at a.
> **3** Check whether $f''(x)$ changes sign at a.

EXAMPLE 3 Find the inflection points of $f(x) = x^4 - 6x^3 + 12x^2$.

SOLUTION We first compute $f''(x)$, as follows:

$$f'(x) = 4x^3 - 18x^2 + 24x$$

$$f''(x) = 12x^2 - 36x + 24$$

$$= 12(x^2 - 3x + 2).$$

Thus $$f''(x) = 12(x - 1)(x - 2).$$

Then we find where this second derivative is 0:

$$0 = 12(x - 1)(x - 2).$$

Hence $x - 1 = 0$ or $x - 2 = 0$, and $x = 1$ or $x = 2$.

The numbers 1 and 2 are candidates for inflection numbers. To see whether they really are inflection numbers, we must check how the sign of $f''(x) = 12(x - 1)(x - 2)$ behaves near each of these two numbers.

If you have a graphing calculator, can you spot the inflection points from the display?

Check 1 first. If x is less than 1, both $x - 1$ and $x - 2$ are negative, hence $f''(x)$ is positive. If x is a little larger than 1, $x - 1$ is positive, but $x - 2$ is negative. (In short, $x - 1$ changes sign at 1, but $x - 2$ does not.) Hence $f''(x) = 12(x - 1)(x - 2)$ changes sign, from positive to negative, at 1. The sense of concavity changes at 1, which tells us that 1 is an inflection number.

Similar reasoning shows that 2 is also an inflection number. The graph has two inflection points, namely $(1, f(1)) = (1, 7)$ and $(2, f(2)) = (2, 16)$. ■

The Second Derivative and Local Extrema

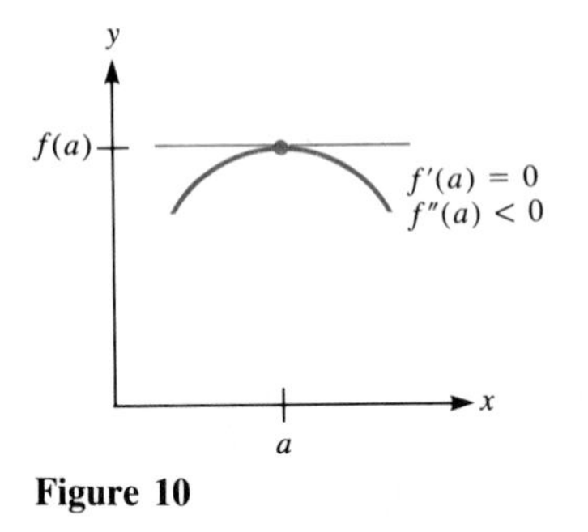

Figure 10

The second derivative is also useful in testing whether at a critical number there is a relative maximum or relative minimum. For instance, let a be a critical number for the function f and assume that $f''(a)$ happens to be negative. If f'' is continuous in some open interval that contains a, then $f''(x)$ remains negative for a suitably small open interval that contains a. This means that the graph of f is concave downward near $(a, f(a))$, hence lies below its tangent lines. In particular, it lies below the horizontal tangent line at the critical point $(a, f(a))$, as illustrated in Fig. 10. Thus the function has a **relative maximum** at the critical number a. However, if $f'(a) = 0$ and $f''(a) > 0$, the critical point $(a, f(a))$ is a relative minimum. This observation suggests the following test for a relative maximum or minimum.

Second-Derivative Test for Relative Maximum or Minimum. Let f be a function such that $f'(x)$ is defined at least on some open interval containing the number a. Assume that $f''(a)$ is defined. If

$$f'(a) = 0 \quad \text{and} \quad f''(a) < 0,$$

then f has a relative maximum at a.

Similarly, if

$$f'(a) = 0 \quad \text{and} \quad f''(a) > 0,$$

then f has a relative minimum at a.

EXAMPLE 4 Find all local extrema of the function $f(x) = x^4 - 2x^3$.

SOLUTION Since f is differentiable throughout its domain, any local extremum can occur only at a critical number. So begin by finding the critical numbers, as follows:

$$f'(x) = (x^4 - 2x^3)' = 4x^3 - 6x^2 = x^2(4x - 6).$$

Setting $f'(x) = 0$ gives $x^2 = 0$ or $4x - 6 = 0$. The critical numbers are therefore

$$x = 0 \quad \text{and} \quad x = \tfrac{3}{2}.$$

Now use the second derivative to determine whether either of these corresponds to a local extremum.

The second derivative is

$$f''(x) = (4x^3 - 6x^2)' = 12x^2 - 12x.$$

At $x = \frac{3}{2}$ we have

$$f''(\tfrac{3}{2}) = 12(\tfrac{3}{2})^2 - 12(\tfrac{3}{2}) = 27 - 18 = 9,$$

which is positive. Since $f'(\frac{3}{2}) = 0$ and $f''(\frac{3}{2}) > 0$, f has a local minimum at $x = \frac{3}{2}$.

How about the other critical number $x = 0$? In this case,

$$f''(0) = 12 \cdot 0^2 - 12 \cdot 0 = 0.$$

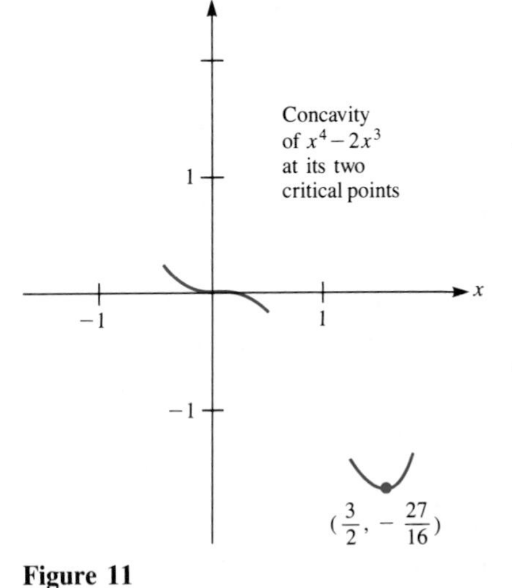

Figure 11

Since $f''(0) = 0$, the second-derivative test tells us nothing about the critical number 0. Instead, we resort to the first-derivative test and examine the sign of $f'(x) = 4x^3 - 6x^2 = x^2(4x - 6)$ for x near 0. For x sufficiently near 0, whether to the right of 0 or to the left, x^2 is positive and $4x - 6$ is negative. Thus $f'(x)$ is negative for x near 0. Since f is a decreasing function near 0, it has neither a local maximum nor a local minimum at 0.

The two critical points are $(\frac{3}{2}, f(\frac{3}{2})) = (\frac{3}{2}, -\frac{27}{16})$ and $(0, f(0)) = (0, 0)$. The first one is a relative minimum, but the second is not a relative extremum. (See Fig. 11.) Example 5 develops the graph of $x^4 - 2x^3$. ■

Using Concavity in Graphing

The final example illustrates how the second derivative, together with the first derivative and other tools of graphing, can help us find the general shape of a graph.

The example continues Example 4.

EXAMPLE 5 Graph $f(x) = x^4 - 2x^3$.

SOLUTION The function $x^4 - 2x^3$ is neither even nor odd; nor does it have asymptotes.

Since $f(0) = 0^4 - 2(0^3)$, its y intercept is 0. To find its x intercepts we solve the equation

$$x^4 - 2x^3 = 0$$
$$x^3(x - 2) = 0.$$

Thus $x = 0$ or $x - 2 = 0$. The x intercepts are 0 and 2.

As we found in Example 4, $f'(x) = 4x^3 - 6x^2 = x^2(4x - 6)$. It is 0 only when $x = 0$ and $x = \frac{3}{2}$, and the critical points are shown in Fig. 11.

Now we examine the concavity of the graph. We have

$$f''(x) = 12x^2 - 12x$$
$$= 12x(x - 1).$$

Thus $f''(x) = 0$ when $x = 0$ and when $x = 1$. We use this information to help determine where the graph is concave up or concave down, and inflection points.

The second derivative $12x(x - 1)$ can change sign only at 0 and 1, which break the x axis into three sections $(-\infty, 0)$, $(0, 1)$, $(1, \infty)$. In each section $f''(x)$ keeps the same sign throughout. To determine that sign choose a convenient sample number in each section. The following table shows how this is done.

Section	$(-\infty, 0)$	$(0, 1)$	$(1, \infty)$
Sample number s	-1	$\frac{1}{2}$	2
$f''(s) = 12s(s - 1)$	24	-3	24
Sign	+	−	+
Concavity	Up	Down	Up

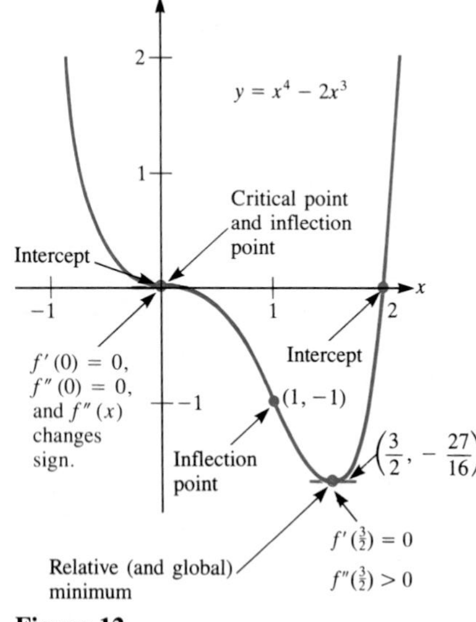

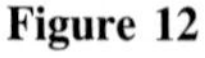
Figure 12

Since the sign of $f''(x)$ changes at $x = 0$ and $x = 1$, 0 and 1 are inflection numbers. The points $(0, f(0)) = (0, 0)$ and $(1, f(1)) = (1, -1)$ are inflection points. The graph of $x^4 - 2x^3$ is shown in Fig. 12. ■

Section Summary

The following chart shows the meaning of the signs of $f(x)$, $f'(x)$, and $f''(x)$ in the graph of $y = f(x)$.

	is positive (>0)	*is negative* (<0)	*changes sign*	*is zero* (=0)
Where the ordinate $f(x)$	the graph is above the x axis.	the graph is below the x axis.	the graph crosses the x axis.	there is an x intercept.
Where the slope $f'(x)$	the graph slopes upward.	the graph slopes downward.	the graph has a horizontal tangent and a relative maximum or minimum.	there is a critical point.
Where $f''(x)$	the graph is concave upward (like a cup).	the graph is concave downward.	the graph has an inflection point.	there may be an inflection point.

Keep in mind that the graph can have an inflection point at a, even though the second derivative is not defined at a (Example 2). Similarly, a graph can have a maximum or minimum at a, even though the first derivative is not defined at a. [Consider $f(x) = |x|$ at $a = 0$.]

EXERCISES FOR SEC. 4.5: THE SECOND DERIVATIVE AND GRAPHING

In Exercises 1 to 16 describe the intervals where the function is concave up or concave down and give any inflection points.

1 $x^3 - 3x^2 + 2$
2 $x^3 - 6x^2 + 1$
3 $x^2 + x + 1$
4 $2x^2 - 5x$
5 x^6
6 x^5
7 $x^4 - 4x^3$
8 $3x^5 - 5x^4$
9 $\dfrac{1}{1 + x^2}$
10 $\dfrac{1}{1 + x^4}$
11 $x^3 - 6x^2 - 15x$
12 $\dfrac{x^2}{2} + \dfrac{1}{x}$
13 $\tan x$
14 $\sin x + \sqrt{3} \cos x$
15 $\cos x$
16 $\cos x + \sin x$

In Exercises 17 to 20 graph the polynomials, showing inflection points, critical points, and intercepts.

17 $x^3 + 3x^2$
18 $2x^3 + 9x^2$
19 $x^4 - 4x^3 + 6x^2$
20 $x^4 + 4x^3 + 6x^2 - 2$

In each of Exercises 21 to 28 sketch the general appearance of the graph of the given function near (1, 1) on the basis of the information given. Assume that f, f', and f'' are continuous.

21 $f(1) = 1, f'(1) = 0, f''(1) = 1$
22 $f(1) = 1, f'(1) = 0, f''(1) = -1$
23 $f(1) = 1, f'(1) = 0, f''(1) = 0$ (Sketch four possibilities.)
24 $f(1) = 1$, $f'(1) = 0$, $f''(1) = 0$, $f''(x) < 0$ for $x < 1$ and $f''(x) > 0$ for $x > 1$
25 $f(1) = 1$, $f'(1) = 0$, $f''(1) = 0$ and $f''(x) < 0$ for x near 1
26 $f(1) = 1, f'(1) = 1, f''(1) = -1$
27 $f(1) = 1, f'(1) = 1, f''(1) = 0$ and $f''(x) < 0$ for $x < 1$ and $f''(x) > 0$ for $x > 1$
28 $f(1) = 1$, $f'(1) = 1$, $f''(1) = 0$ and $f''(x) > 0$ for x near 1
29 Graph $y = 2(x - 1)^{5/3} + 5(x - 1)^{2/3}$, paying particular attention to points where y' does not exist.
30 Graph $y = x + (x + 1)^{1/3}$.

In each of Exercises 31 and 32 sketch a graph of a hypothetical function that meets the given conditions. Assume f' and f'' are continuous. Explain your reasoning.

31 Critical point (2, 4); inflection points (3, 1) and (1, 1);

$$\lim_{x\to\infty} f(x) = 0 \quad \text{and} \quad \lim_{x\to-\infty} f(x) = 0.$$

32 Critical points (−1, 1) and (3, 2); inflection point (4, 1);

$$\lim_{x\to 0^+} f(x) = -\infty, \quad \lim_{x\to 0^-} f(x) = \infty;$$

$$\lim_{x\to\infty} f(x) = 0, \quad \lim_{x\to-\infty} f(x) = \infty.$$

---

33 Figure 13 appeared in "Energy Use in the United States Food System," by John S. Steinhart and Carol E. Steinhart, in *Perspectives on Energy*, edited by Lon C. Ruedisili and Morris W. Firebaugh, Oxford, New York, 1975. The graph shows farm output as a function of energy input.
(*a*) What is the practical significance of the fact that the function has a positive derivative?

(*b*) What is the practical significance of the inflection point?

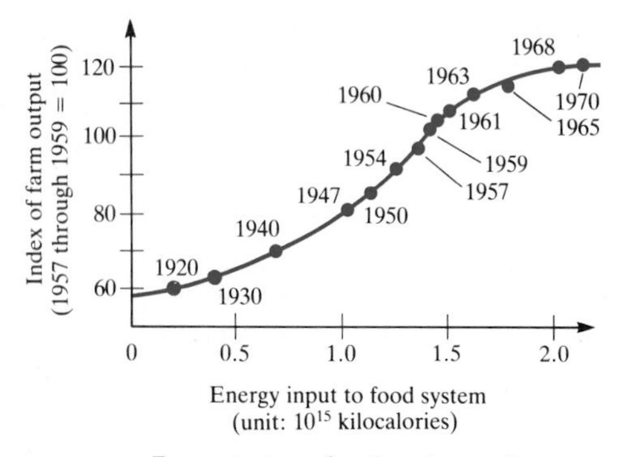

Farm output as a function of energy input to the United States food system, 1920 through 1970.

Figure 13

34 (Contributed by David Hayes.) Let f be a function that is continuous for all x and differentiable for all x other than 0. Figure 14 is the graph of its derivative $f'(x)$ as a function of x.

(*a*) Answer the following questions about f (*not* about f'). Where is f increasing? decreasing? concave upward? concave downward? What are the critical numbers? Where do any relative maxima or relative minima occur? Explain.

(*b*) Assuming that $f(0) = 1$, graph a hypothetical function f that satisfies the conditions given.

(*c*) Graph $f''(x)$.

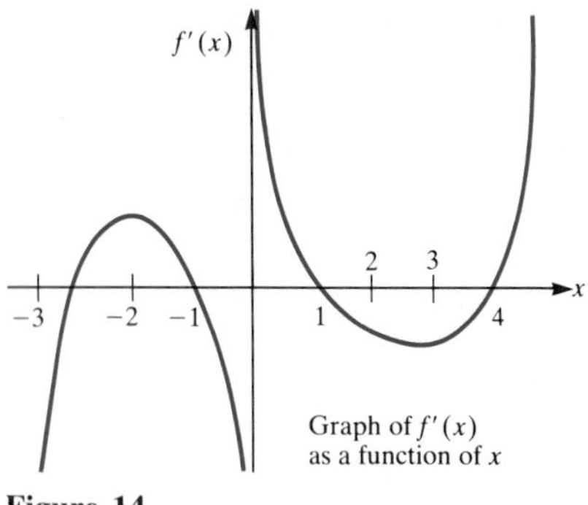

Graph of $f'(x)$ as a function of x

Figure 14

35 Find the critical points and inflection points in $[0, 2\pi]$ of $f(x) = \sin^2 x \cos x$.

36 Can a polynomial of degree 6 have (*a*) no inflection points? (*b*) exactly one inflection point? Explain.

37 Can a polynomial of degree 5 have (*a*) no inflection points? (*b*) exactly one inflection point? Explain.

38 In the theory of **inhibited growth** it is assumed that the growing quantity y approaches some limiting size M. Specifically, one assumes that the rate of growth is proportional both to the amount present and to the amount left to grow:

$$\frac{dy}{dt} = ky(M - y).$$

Prove that the graph of y as a function of time has an inflection point when the amount y is exactly half the amount M.

39 Let f be a function such that $f''(x) = (x - 1)(x - 2)$.

(*a*) For which x is f concave upward?

(*b*) For which x is f concave downward?

(*c*) List its inflection numbers.

(*d*) Find a specific function f whose second derivative is $(x - 1)(x - 2)$.

40 A certain function $y = f(x)$ has the property that

$$y' = \sin y + 2y + x.$$

Show that at a critical number the function has a local minimum.

41 Assume that the domain of $f(x)$ is the entire x axis, and $f'(x)$ and $f''(x)$ are continuous. Assume that (1, 1) is the only critical point and that $\lim_{x\to\infty} f(x) = 0$.

(*a*) Must $f(x)$ be decreasing for $x > 1$?

(*b*) Must $f(x)$ have an inflection point?

42 We stated that if $f(x)$ is defined in an open interval around the critical number a and $f''(a)$ is negative, then $f(x)$ has a relative maximum at a. Explain why this is so, following these steps.

(*a*) Why is $\lim_{\Delta x\to 0} \dfrac{f'(a + \Delta x) - f'(a)}{\Delta x}$ negative?

(*b*) Deduce that if Δx is small and positive, then $f'(a + \Delta x)$ is negative.

(*c*) Show that if Δx is small and negative, then $f'(a + \Delta x)$ is positive.

(*d*) Show that $f'(x)$ changes sign from positive to negative at a. By the first-derivative test for a relative maximum, $f(x)$ has a relative maximum at $x = a$.

4.6 NEWTON'S METHOD FOR SOLVING AN EQUATION

Some equations are very hard to solve exactly; for instance, the equations $x^5 + x - 3 = 0$ or $x + \sin x - 2 = 0$. This section describes an efficient way of estimating the solutions (roots) of such equations, known as *Newton's method*.

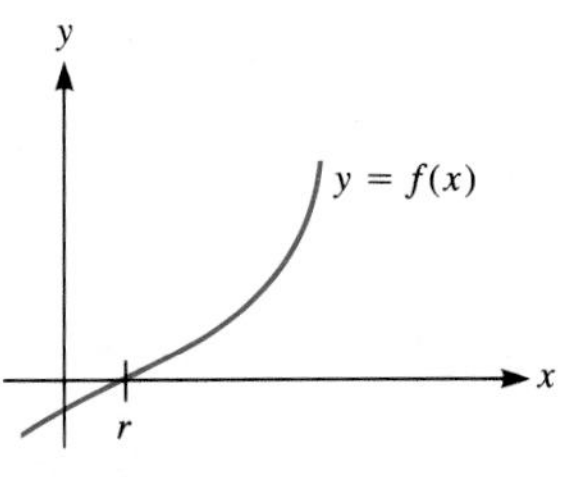

Figure 1

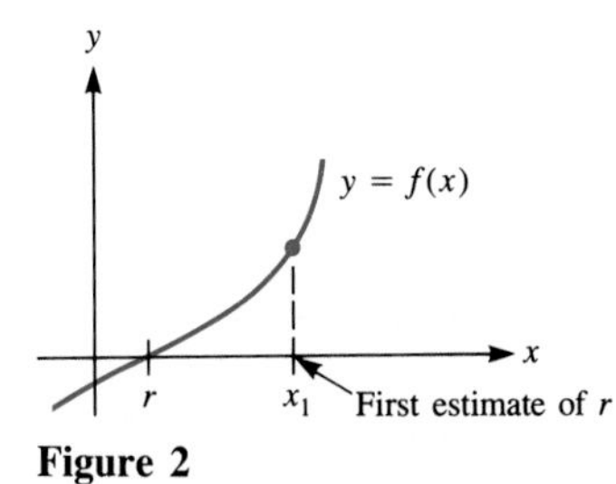

Figure 2

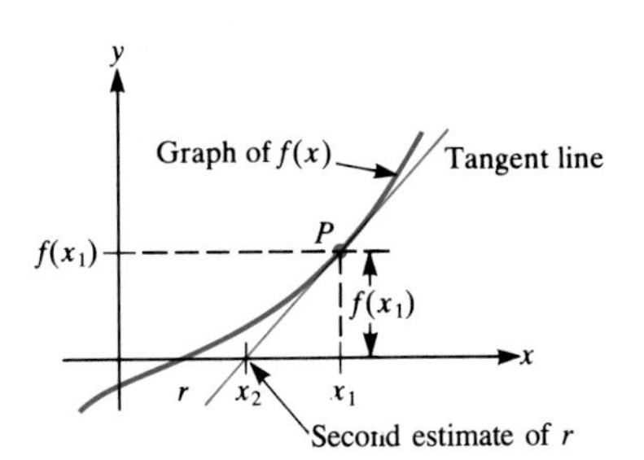

Figure 3

Let $f(x)$ be a function. By a *solution* of the equation $f(x) = 0$ we mean a number r such that $f(r) = 0$. The graph of $y = f(x)$ passes through the point $(r, 0)$, as shown in Fig. 1.

If we don't know r, we make a guess (or estimate) x_1. (How we make this initial guess depends on the particular function f.) Figure 2 shows this guess.

To make a better estimate we use the fact that a short piece of differentiable curve resembles a straight line—namely, its tangent at some point. So draw the tangent at $P = (x_1, f(x_1))$, as in Fig. 3.

If the tangent is not parallel to the x axis, it cuts the x axis at x_2, which is usually closer to r than x is. (Indeed, if the graph of f were a straight line, then x_2 would be *precisely* r.) That, in a nutshell, is Newton's method for getting good estimates of a solution of an equation.

The Formula

To apply Newton's method, we need a formula for x_2 in terms of x_1, the initial guess. To do this, we first find an equation for the tangent line at P. This line passes through the point $(x_1, f(x_1))$ and has slope $f'(x_1)$. Its equation is therefore (by the point-slope formula)

$$y - f(x_1) = f'(x_1)(x - x_1).$$

The x intercept of this line is found by setting $y = 0$:

$$0 - f(x_1) = f'(x_1)(x - x_1). \tag{1}$$

Solving (1), we have, first

$$-\frac{f(x_1)}{f'(x_1)} = x - x_1$$

and finally,

$$x = x_1 - \frac{f(x_1)}{f'(x_1)}. \tag{2}$$

Equation (2) thus gives us

Newton's formula: $x_2 = x_1 - \dfrac{f(x_1)}{f'(x_1)}$

assuming $f'(x_1) \neq 0$.

We can repeat the process using x_2 as our guess. We obtain

$$x_3 = x_2 - \frac{f(x_2)}{f'(x_2)}.$$

Repeating the procedure as often as we wish, we obtain a sequence of estimates $x_1, x_2, x_3, \ldots$ which usually approaches r rapidly.

Advisory: Read x_1 as "x sub 1." *Don't* call it "x to the one." The 1 is a subscript, not an exponent. We mention this because many people incorrectly use exponent terminology to name the estimates.

Examples

Some examples illustrate how you may make the first estimate x_1 efficiently and then how the calculations of $x_2, x_3, \ldots$ are done.

EXAMPLE 1 Use Newton's method to estimate the square root of 3, that is, the positive root of the equation $x^2 - 3 = 0$.

SOLUTION Here $f(x) = x^2 - 3$ and $f'(x) = 2x$. According to Eq. (2), if the first guess is x_1, then the next estimate x_2 should be

$$x_2 = x_1 - \frac{f(x_1)}{f'(x_1)} = x_1 - \frac{x_1^2 - 3}{2x_1} = \frac{x_1 + 3/x_1}{2}.$$

For our initial guess, let us use $x_1 = 2$. Its square is 4, which isn't far from 3. Besides, it's an easy number to work with. Then

$$x_2 = \frac{2 + \frac{3}{2}}{2} = 1.75.$$

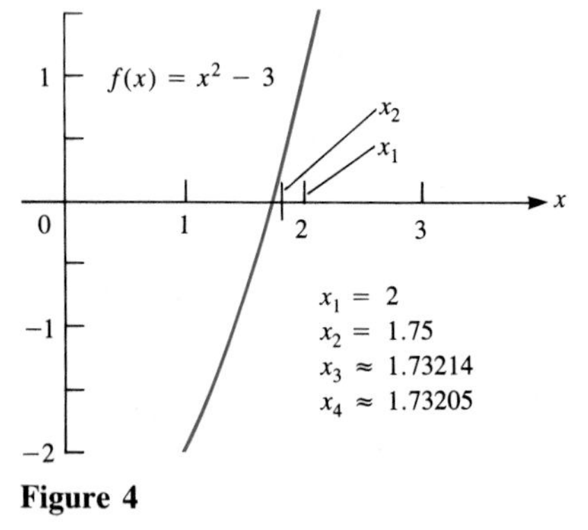

Figure 4

For a better estimate of $\sqrt{3}$, repeat and use 1.75 instead of 2. Thus

$$x_3 = \frac{x_2 + 3/x_2}{2} = \frac{1.75 + 3/1.75}{2} \approx 1.73214$$

is the third estimate, to five decimals. One more repetition of the process yields (to five decimals) $x_4 \approx 1.73205$, which is quite close to $\sqrt{3}$, whose decimal expansion begins 1.7320508. See Fig. 4, which shows x_1, x_2, and the graph of $f(x) = x^2 - 3$. ■

In fact, x_4 agrees with $\sqrt{3}$ to seven decimals.

In Example 1, $x_3 \approx 1.73214$ and $x_4 \approx 1.73205$ agree through the first three decimal places. If we were interested in estimating the root, $\sqrt{3}$, only to three decimal places, we wouldn't bother finding x_4. As a general rule of thumb, when two consecutive estimates agree up to a certain number of decimal places, you may assume that you have obtained an estimate of the root accurate up to that number of places.

In the next example the problem of finding where two curves cross is reduced to solving an equation. We use Newton's method to estimate the solution.

EXAMPLE 2 The line $y = 2x/3$ crosses the curve $y = \sin x$ at a point P, whose x coordinate r is between 0 and π, as shown in Fig. 5. The number r is a solution of the equation $2x/3 = \sin x$, since the graphs have equal y coordinates at $x = r$. Use Newton's method to obtain the first three decimals of r.

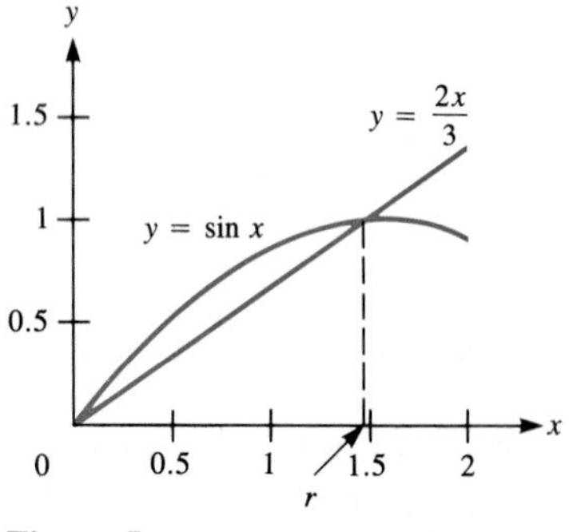

Figure 5

SOLUTION A glance at the graph in Fig. 5 suggests that r is approximately 1.5. The number r satisfies the equation

$$\sin r = \frac{2r}{3}.$$

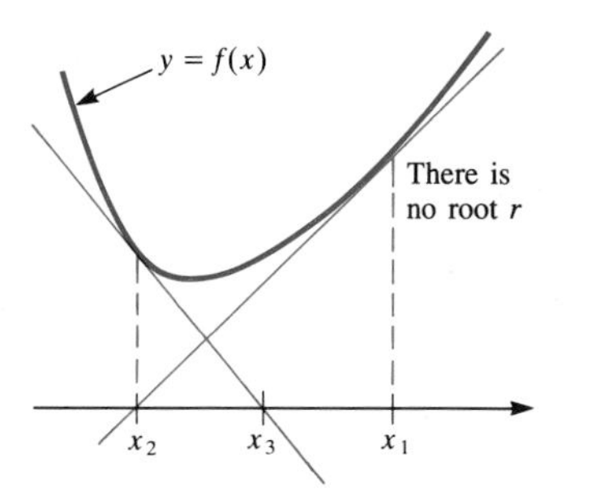

Figure 6

Hence r is a root of the equation

$$\sin x - \frac{2x}{3} = 0.$$

As our first estimate of the solution, we use $x_1 = 1.5$, suggested by Fig. 5. Then we apply Newton's method to the function $f(x) = \sin x - 2x/3$. Since $f'(x) = \cos x - \frac{2}{3}$, we have

$$x_2 = x_1 - \frac{\sin x_1 - \dfrac{2x_1}{3}}{\cos x_1 - \frac{2}{3}}.$$

Hence

$$\begin{aligned} x_2 &= 1.5 - \frac{\sin 1.5 - \dfrac{2(1.5)}{3}}{\cos 1.5 - \frac{2}{3}} \\ &\approx 1.5 - \frac{0.997495 - 1}{0.070737 - 0.666667} \\ &\approx 1.5 - \frac{0.002505}{0.595930} \\ &\approx 1.495796. \end{aligned}$$

Repeating the process, using x_2 to obtain x_3, we have

$$x_3 \approx 1.495782.$$

To three decimal places, $r = 1.496$. ■

Warning

The equation $f(x) = 0$ may not have a solution. In that case the sequence of estimates you produce by Newton's method does not approach a specific number but may "wander all over the place," as in Fig. 6.

Another possibility is that there is a root r, but your first estimate x_1 is so far from it that the sequence of estimates does not approach r. See Fig. 7.

Of course, if you happen to choose x_1 such that $f'(x_1) = 0$, then the Newton recursion, which has $f'(x_1)$ in the denominator, makes no sense. [The tangent at $(x_1, f(x_1))$ does not intersect the x axis.]

However, Newton's method works fine in a situation such as that in Fig. 8. Here the curve is concave up and the successive estimates get closer and closer to the root r, staying between x_1 and r.

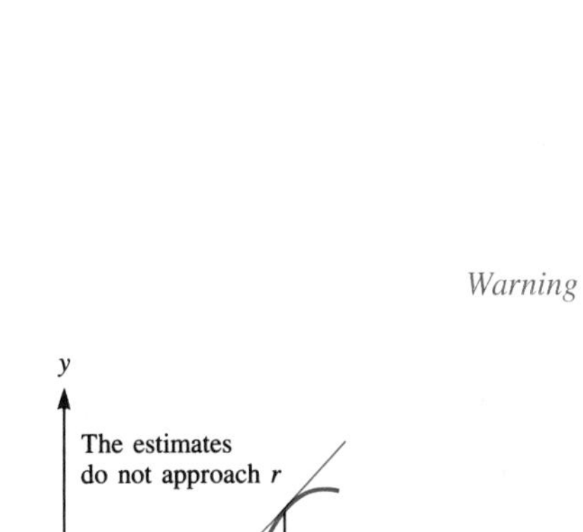

Figure 7

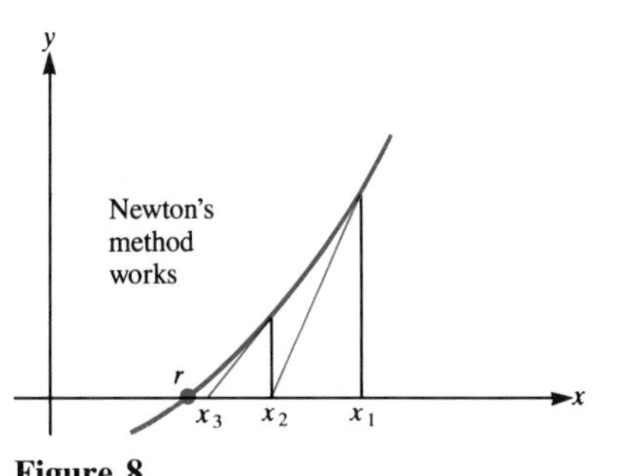

Figure 8

HOW GOOD IS NEWTON'S METHOD?

When you use Newton's method, you produce a sequence of estimates $x_1, x_2, x_3, \ldots$ of a root r. How quickly does the sequence approach r? In other words, how rapidly does the difference between the estimate x_n and the root r, $|x_n - r|$, approach 0?

To get a feel for the rate at which $|x_n - r|$ shrinks as we keep using Newton's method, take the case in Example 1, where we were estimating $\sqrt{3}$ using the recursion

$$x_2 = \frac{x_1 + \dfrac{3}{x_1}}{2}.$$

(continued)

In the following table, we list x_1, x_2, x_3, x_4 to seven decimal places and compare them to $\sqrt{3} \approx 1.7320508$:

Newton's method for solving $x^2 - 3 = 0$.

Estimate	*Value*	*Agreement with $\sqrt{3}$*
x_1	2	
x_2	1.75	First two digits
x_3	1.7321429	First four digits
x_4	1.7320508	First eight digits

At each stage the number of correct digits tends to *double*. (Why this happens is shown in Exercise 17 of Sec. 4.10.)

In the bisection method, described in Sec. 4.1, the error tends to be cut in half at each step. That means that the estimates in the bisection method approach the root much more slowly than do the estimates in Newton's method.

For example, in estimating the positive root of $x^2 - 3 = 0$, we might first say that it's between 1 and 2, since $1^2 - 3 < 0$ and $2^2 - 3 > 0$. Then we would compute $x^2 - 3$ at 1.5, the midpoint of [1, 2], getting $1.5^2 - 3 = -0.75$. Thus the root is between 1.5 and 2. The next table shows a few more steps. (See Fig. 9.)

Bisection method for solving $x^2 - 3 = 0$.

Interval	*Midpoint*
[1, 2]	$m_1 = 1.5$
[1.5, 2]	$m_2 = 1.75$
[1.5, 1.75]	$m_3 = 1.625$
[1.625, 1.75]	$m_4 = 1.6875$
[1.6875, 1.75]	$m_5 = 1.71875$
[1.71875, 1.75]	$m_6 = 1.734375$
[1.71875, 1.734375]	$m_7 = 1.7265625$

Even our seventh midpoint agrees with the actual value of $\sqrt{3}$ in only the first two digits. Newton's method is clearly a much more efficient technique.

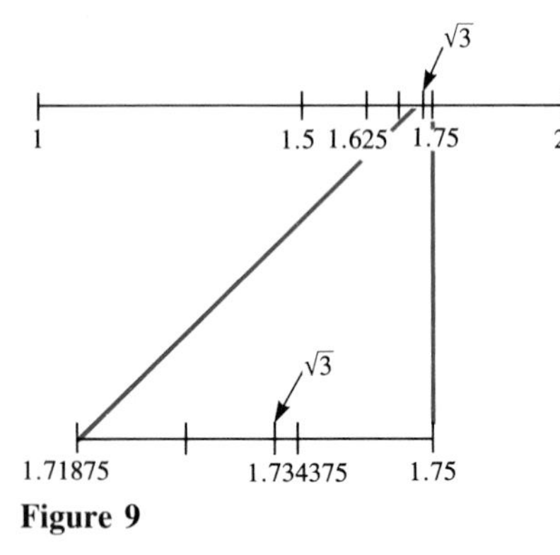

Figure 9

Section Summary

This section developed Newton's recursive formula for estimating a root of an equation, $f(x) = 0$. You start with an estimate x_1 of the root, then compute

$$x_2 = x_1 - \frac{f(x_1)}{f'(x_1)}.$$

Then repeat the process, computing

$$x_3 = x_2 - \frac{f(x_2)}{f'(x_2)},$$

and so on.

The first estimate x_1 should be chosen as well as you can. The number of correct figures in the estimate tends to double at each step.

EXERCISES FOR SEC. 4.6: NEWTON'S METHOD FOR SOLVING AN EQUATION

In Exercises 1 and 2, what does Newton's method give for x_2?

1 $x_1 = 2, f(2) = 0.3, f'(2) = 1.5$

2 $x_1 = 3, f(3) = 0.06, f'(3) = 0.3$

3 Let a be a positive number. Show that the Newton recursion formula for estimating $\sqrt{a}$ is given by

$$x_{i+1} = \frac{x_i + a/x_i}{2}.$$

4 Let a be a number. Show that the Newton recursion formula for estimating $\sqrt[3]{a}$ is given by

$$x_{i+1} = \tfrac{2}{3}x_i + \frac{a}{3x_i^2}.$$

5 Use the formula of Exercise 3 to estimate $\sqrt{15}$. Choose $x_1 = 4$ and compute x_2 and x_3 to three decimals.

6 Use the formula of Exercise 3 to estimate $\sqrt{19}$. Choose $x_1 = 4$ and compute x_2 and x_3 to three decimals.

7 Use the formula of Exercise 4 to estimate $\sqrt[3]{7}$. Choose $x_1 = 2$ and compute x_2 and x_3 to three decimals.

8 Use the formula of Exercise 4 to estimate $\sqrt[3]{25}$. Choose $x_1 = 3$ and compute x_2 and x_3 to three decimals.

9 (*a*) Use the formula in Exercise 3 to estimate $\sqrt{5}$, using $x_1 = 2$ and repeating Newton's method until consecutive estimates agree to four decimal places.
(*b*) Compute $\sqrt{5}$ directly on your calculator. How close is the estimate of (*a*)?

10 (*a*) Use the formula in Exercise 4 to estimate $\sqrt[3]{2}$, using $x_1 = 1$ and repeating Newton's method until consecutive estimates agree to four decimal places.
(*b*) Compute $\sqrt[3]{2}$ directly on your calculator. How close is the estimate of (*a*)?

11 Let $f(x) = x^5 + x - 1$.
(*a*) Show that there is a root of the equation $f(x) = 0$ in the interval $[0, 1]$.
(*b*) Using $x_1 = \frac{1}{2}$ as a first estimate, apply Newton's method to find a second estimate x_2.
(*c*) Why is the root unique?

12 Let $f(x) = x^4 + x - 19$.
(*a*) Show that $f(2) < 0 < f(3)$ and that f must thus have a root r between 2 and 3.
(*b*) Apply Newton's method, starting with $x_1 = 2$. Compute x_2 and x_3.

13 In estimating $\sqrt{3}$ with Newton's method, a student imprudently chooses $x_1 = 10$. What does Newton's method give for x_2, x_3, and x_4?

14 Let $f(x) = 2x^3 - x^2 - 2$.
(*a*) Show that there is exactly one root of the equation $f(x) = 0$ in the interval $[1, 2]$.
(*b*) Using $x_1 = \frac{3}{2}$ as a first estimate, apply Newton's method to find a second estimate x_2.

15 (*a*) Graph $y = x$ and $y = \cos x$ relative to the same axes.
(*b*) Using the graph in (*a*), estimate the positive solution of the equation $x = \cos x$. Is there a negative solution?
(*c*) Using your estimate in (*b*) as x_1, apply Newton's method until consecutive estimates agree to four decimal places.

16 (*a*) Graph $y = \cos x$ and $y = 2 \sin x$ relative to the same axes.
(*b*) Using the graph in (*a*), estimate the solution that lies in $[0, \pi/2]$.
(*c*) Using your estimate in (*b*) as x_1, apply Newton's method until consecutive estimates agree to four decimal places.

Exercises 17 through 19 show that care should be taken in applying Newton's method.

17 Let $f(x) = 2x^3 - 4x + 1$.
(*a*) Show that there must be a root r of $f(x) = 0$ in $[0, 1]$.
(*b*) Take $x_1 = 1$, and apply Newton's method to obtain x_2 and x_3, estimates of r.
(*c*) Graph f, and show what is happening in the sequence of estimates.

18 Let $f(x) = x^2 + 1$.
(*a*) Using Newton's method with $x_1 = 2$, compute x_2, x_3, x_4, and x_5 to two decimal places.
(*b*) Using the graph of f, show geometrically what is happening in (*a*).
(*c*) Using Newton's method with $x_1 = \sqrt{3}/3$, compute x_2 and x_3. What happens to x_n as $n \to \infty$?
(*d*) What happens when you use Newton's method, starting with $x_1 = 1$?

19 Apply Newton's method to the function $x^3 - x$, starting with $x_1 = 1/\sqrt{5}$.
(*a*) Compute x_2 and x_3 exactly (not as decimal approximations).
(*b*) Graph $x^3 - x$ and explain why Newton's method fails in this case.

In Exercise 20 to 22 (Figs. 10 to 12) use Newton's method to estimate θ (to two decimal places). Angles are in radians. Also show that there is only one answer if $0 < \theta < \pi/2$.

20

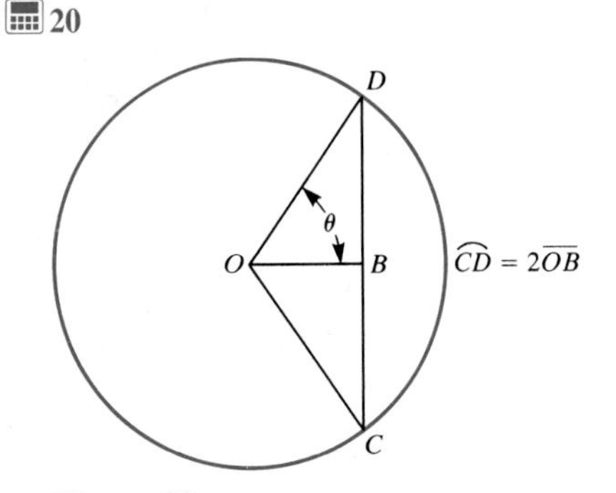

Figure 10

21

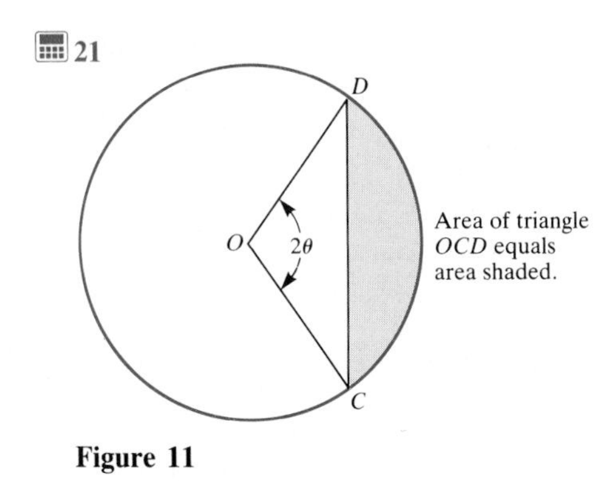

Figure 11

22

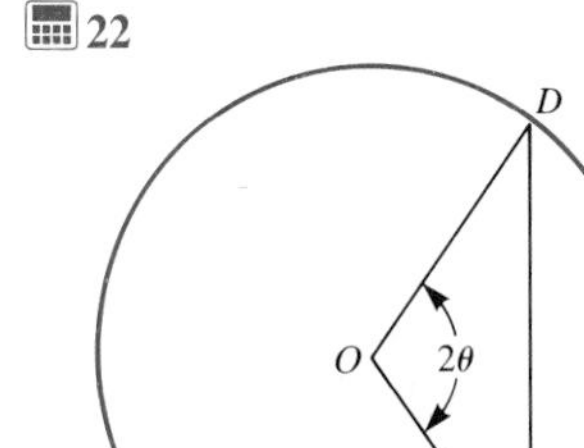

Figure 12

23 The equation $x \tan x = 1$ occurs in the theory of vibrations.
(*a*) How many roots does it have in $[0, \pi/2]$?
(*b*) Find them to two decimal places.

24 (*a*) Show that a critical number of the function $(\sin x)/x$ satisfies the equation $\tan x = x$.
(*b*) Show that $(\sin x)/x$ is an even function. Thus we will consider only positive x.
(*c*) Graph the functions $\tan x$ and x relative to the same axes. How often do they cross for x in $[\pi/2, 3\pi/2]$? for x in $[3\pi/2, 5\pi/2]$? Base your answer on your graphs.
(*d*) Show that $\tan x - x$ is an increasing function for x in $[\pi/2, 3\pi/2]$. What does that tell us about the number of solutions of the equation $\tan x = x$ for x in $[\pi/2, 3\pi/2]$?
(*e*) How many critical numbers does the function $(\sin x)/x$ have?
(*f*) Use Newton's method to estimate the critical number in $[\pi/2, 3\pi/2]$.

25 Examine the solutions of the equation $2x + \sin x = 2$. How many are there? Use Newton's method to evaluate them to two places. Explain the steps in your solution in complete sentences.

26 How many solutions does the equation $\sin x = x$ have? Explain.

27 Use Newton's method to obtain a formula for estimating $\sqrt[5]{a}$.

28 Assume that when you apply Newton's method to estimate a root of $f(x) = 0$ you obtain a sequence of estimates $x_1, x_2, x_3, \ldots, x_n, \ldots$ that approaches a number L. Assume that $f'(L) \neq 0$. Show that L must be a root of $f(x) = 0$.

29 Assume that $f'(x) > 0$ and $f''(x) < 0$ for all x. Assume $f(r) = 0$.
(*a*) Sketch the graph of $y = f(x)$.
(*b*) Describe the behavior of the sequence of Newton's estimates $x_1, x_2, \ldots, x_n, \ldots$ when you choose $x_1 > r$. Include a sketch.
(*c*) Describe the behavior of the sequence if you choose $x_1 < r$. Include a sketch.

4.7 APPLIED MAXIMUM AND MINIMUM PROBLEMS

This section shows how to use the derivative and second derivative in finding extrema in "applied problems." Though the examples are mainly geometric, they illustrate the general procedure. In that large world outside the classroom the problems become messier and more complex. How to set the prices of several products to maximize profit? Too low, and you earn too little; too high, and you lose customers. How to harvest ocean fish to maintain a maximum catch? Too small a harvest, and you starve; too large, and you destroy the fish supply—and you starve. How to set the tax schedules to maximize revenue? Too low, you get little revenue; too high, you destroy the economy.

The Method

In finding the extrema of the functions in Secs. 4.2 and 4.5, you were handed the function $f(x)$. When doing an applied problem, you have to figure out what the function is that you want to maximize or minimize. Once you have the formula for the function, then you just use the techniques discussed earlier in the chapter to find its extrema. The general procedure runs something along these lines.

1 Get a feel for the problem (experiment with particular cases).
2 Devise a formula for the function whose maximum or minimum you want to find.
3 Determine the domain of the function—that is, the inputs *that make sense in the application*.
4 Find the maximum or minimum of the function for inputs that are in the domain discussed in Step 3.

The most important step is finding a formula for the function. To become skillful at doing this, practice.

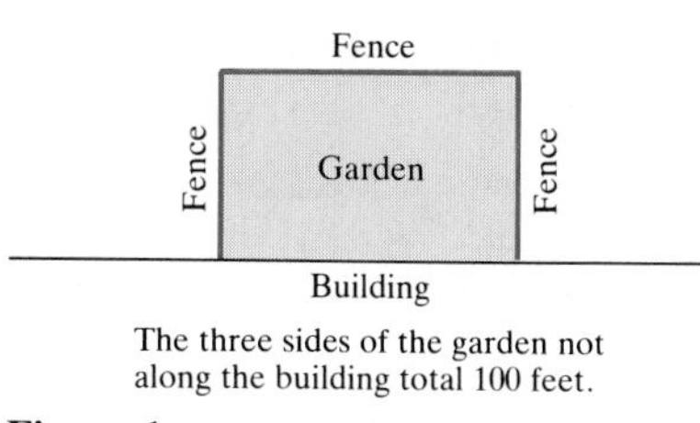

The three sides of the garden not along the building total 100 feet.

Figure 1

A Large Garden

EXAMPLE 1 A couple have enough wire to construct 100 feet of fence. They wish to use it to form three sides of a rectangular garden, one side of which is along a building, as shown in Fig. 1. What shape garden should they choose in order to enclose the largest possible area?

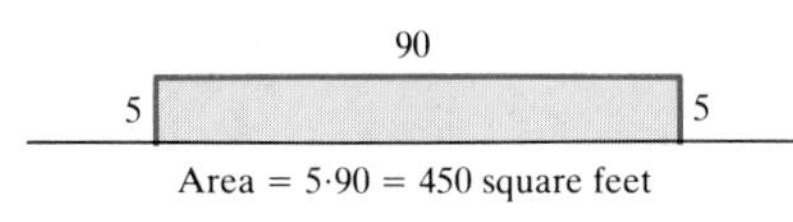

Area = 5·90 = 450 square feet

Figure 2

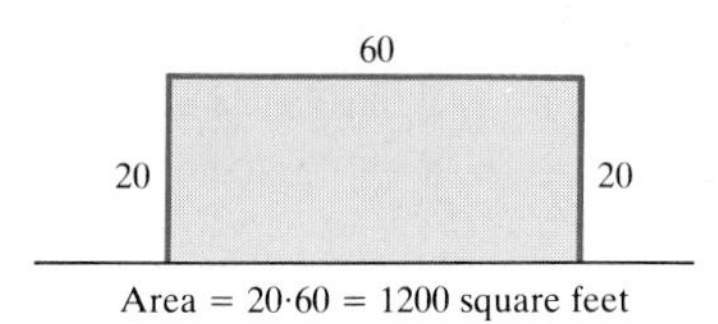

Area = 20·60 = 1200 square feet

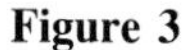

Figure 3

SOLUTION *Step 1.* First we make a few experiments. Figures 2 to 4 show some possible ways of laying out the 100 feet of fence. In the first case the side parallel to the building is very long, in an attempt to make a large area. However, doing this forces the other sides of the garden to be small. The area is $90 \times 5 = 450$ square feet. In the second case, the garden has a larger area, $60 \times 20 = 1200$ square feet. In the third case, the side parallel to the building is only 20 feet long, but the other sides are longer. The area is $20 \times 40 = 800$ square feet.

Clearly, we may think of the area of the garden as a function of the length of the side parallel to the building.

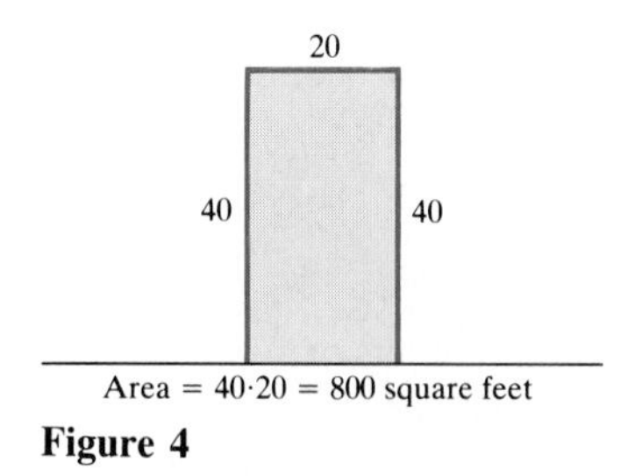

Area = 40·20 = 800 square feet

Figure 4

Step 2. Let $A(x)$ be the area of the garden when the length of the side parallel to the building is x feet, as in Fig. 5. The other sides of the garden have length y. But y is completely determined by x since the fence is 100 feet long:

$$x + 2y = 100.$$

Thus $y = (100 - x)/2$.

Since the area of a rectangle is its length times its width,

$$A(x) = xy$$

$$A(x) = x\left(\frac{100 - x}{2}\right)$$

or

$$A(x) = 50x - \frac{x^2}{2}. \tag{1}$$

(See Fig. 6.)

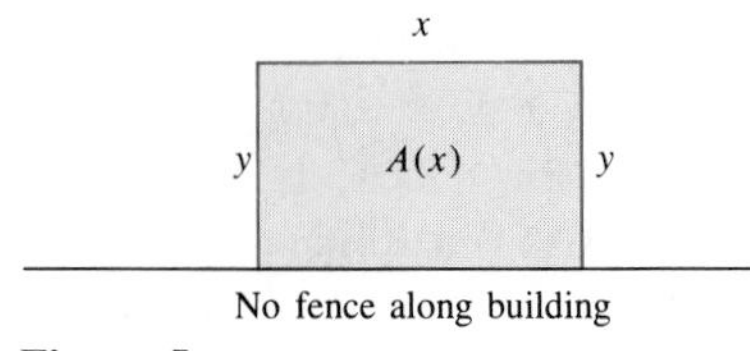

No fence along building

Figure 5

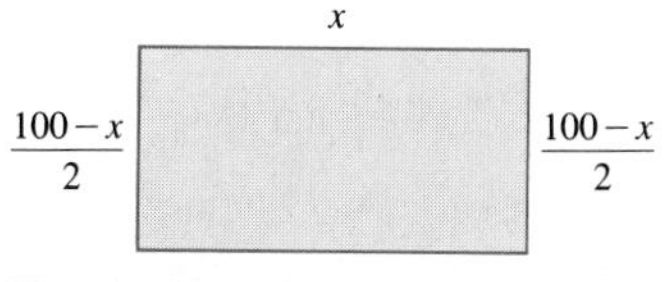

Figure 6

Step 3. Which x in (1) correspond to possible gardens? Since there is only 100 feet of fence, $x \leq 100$. Furthermore, it makes no sense to have a negative

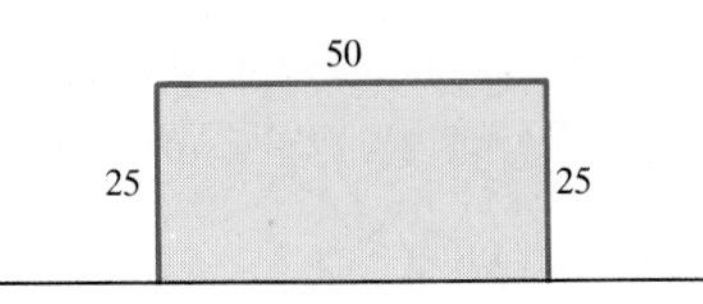

Figure 7

amount of fence; hence $x \geq 0$. Therefore the domain on which we wish to consider the function (1) is the closed interval $[0, 100]$.

Step 4. To maximize $A(x) = 50x - x^2/2$ on $[0, 100]$ we examine $A(0)$, $A(100)$, and the value of $A(x)$ at any critical numbers. To find critical numbers, we differentiate $A(x)$:

$$A(x) = 50x - \frac{x^2}{2}$$

$$A'(x) = 50 - x.$$

Setting $A'(x) = 0$ gives

$$0 = 50 - x$$

or

$$x = 50.$$

There is one critical number, 50.

All that is left is to find the largest of $A(0)$, $A(100)$, and $A(50)$. We have

$$A(0) = 50 \cdot 0 - \frac{0^2}{2} = 0,$$

$$A(100) = 50 \cdot 100 - \frac{100^2}{2} = 0,$$

and

$$A(50) = 50 \cdot 50 - \frac{50^2}{2} = 1250.$$

The maximum possible area is 1250 square feet, and the fence should be laid out as shown in Fig. 7. ■

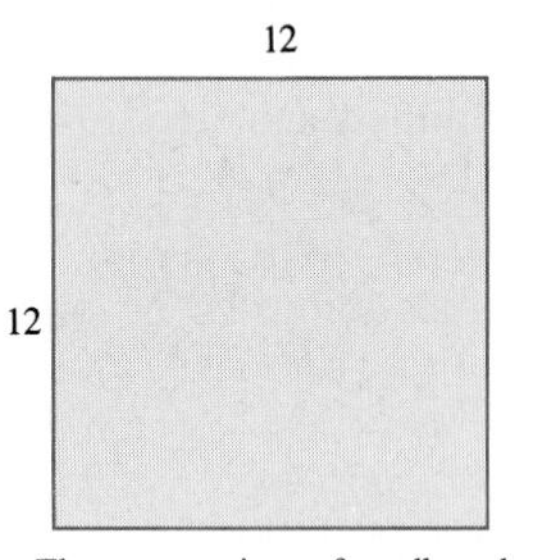

The square piece of cardboard

Figure 8

Four congruent squares removed

Figure 9

A Large Tray

EXAMPLE 2 If we cut four congruent squares out of the corners of a square piece of cardboard 12 inches on each side, we can fold up the four remaining flaps to obtain a tray without a top. What size squares should be cut in order to maximize the volume of the tray? (See Figs. 8 to 10.)

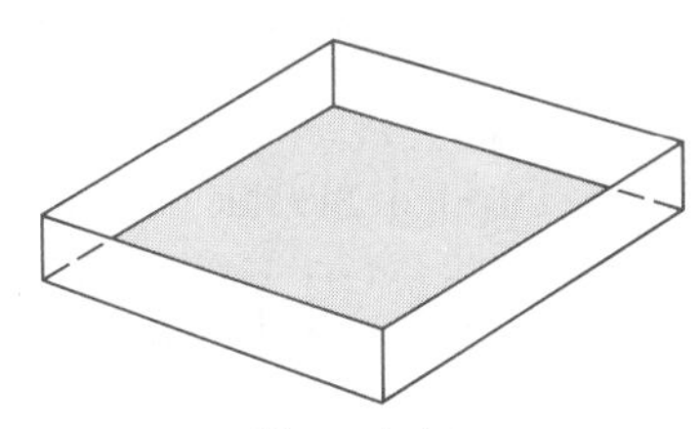
The typical tray

Figure 10

SOLUTION *Step 1.* First we get a feel for the problem. Let us make a couple of experiments.

Say that we remove small squares that are 1 inch by 1 inch, as in Fig. 11. When we fold up the flaps we obtain a tray whose base is a 10-inch by 10-inch square and whose height is 1 inch, as in Fig. 12. The volume of the tray is

$$\text{Area of base} \times \text{height} = \underbrace{10 \times 10}_{\text{base}} \times \underbrace{1}_{\text{height}} = 100 \text{ cubic inches.}$$

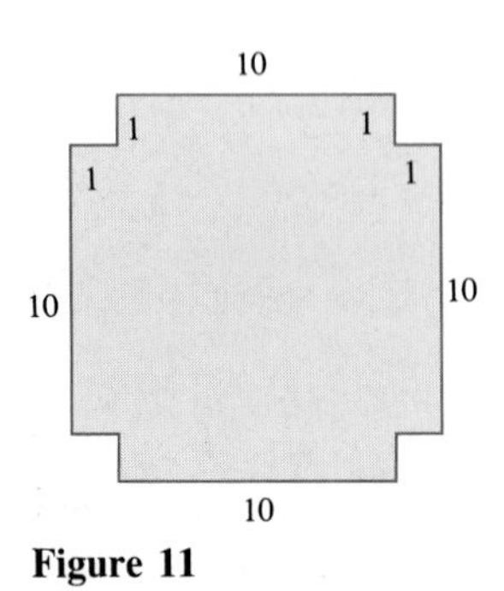

Figure 11

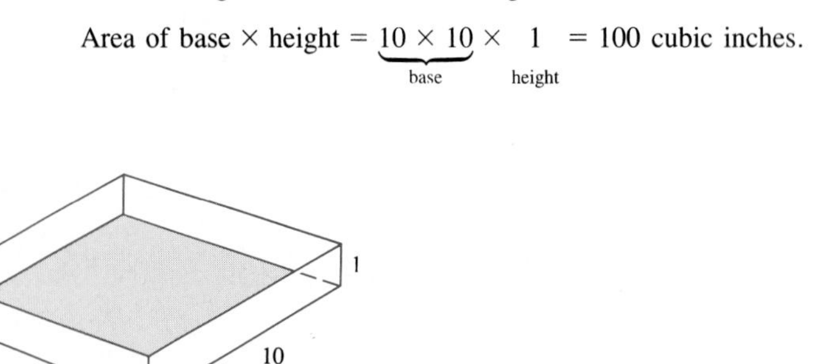

Figure 12

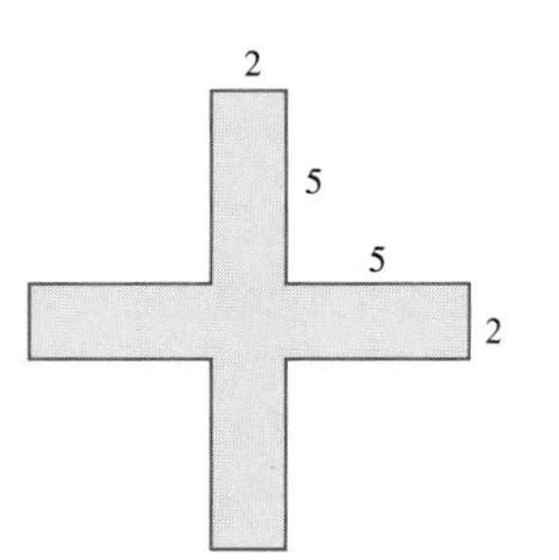

Figure 13

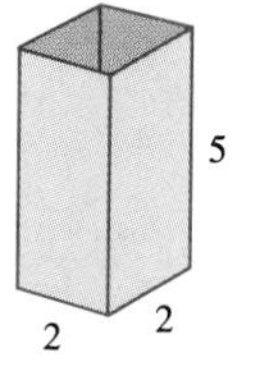

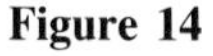
Figure 14

For our second experiment, let's try cutting out a large square, say 5 inches by 5 inches, as in Fig. 13. When we fold up the flaps, we get a very tall tray with a very small base, as in Fig. 14. Its volume is

$$\text{Area of base} \times \text{height} = 2 \times 2 \times 5 = 20 \text{ cubic inches.}$$

Clearly *volume depends on the size of the cut-out squares*. The function we will investigate is of the type $V = f(x)$, where V is the volume of the tray formed by removing four squares whose sides all have length x.

Step 2. To find the formula for $f(x)$ we make a *large*, clear diagram of the typical case, as in Figs. 15 and 16. Now

$$\begin{aligned} \text{Volume of tray} &= \text{length} \cdot \text{width} \cdot \text{height} \\ &= (12 - 2x)(12 - 2x)x \\ &= (12 - 2x)^2 x, \end{aligned}$$

hence

$$V(x) = 4x^3 - 48x^2 + 144x. \tag{2}$$

We have obtained a formula for volume as a function of the length of the sides of the cut-out squares.

Step 3. Next we determine the domain of the function $V(x)$ that is *meaningful* in our problem.

The smallest that x can be is 0. In this case the tray has height 0 and is just a flat piece of cardboard. (The volume is 0.) The size of the cut is not more than 6 inches, since the cardboard has sides of length 12 inches. The cut can be as near 6 inches as we please, and the nearer it is to 6 inches, the smaller is the base of the tray. For convenience of our calculations, let us allow cuts with $x = 6$, when the area of the base is 0 square inches and the height is 6 inches. (The volume is 0 cubic inches.) Therefore the domain of the volume function $V(x)$ is the closed interval $[0, 6]$. *We must find the maximum value of* $4x^3 - 48x^2 + 144x$ *on the closed interval* $[0, 6]$.

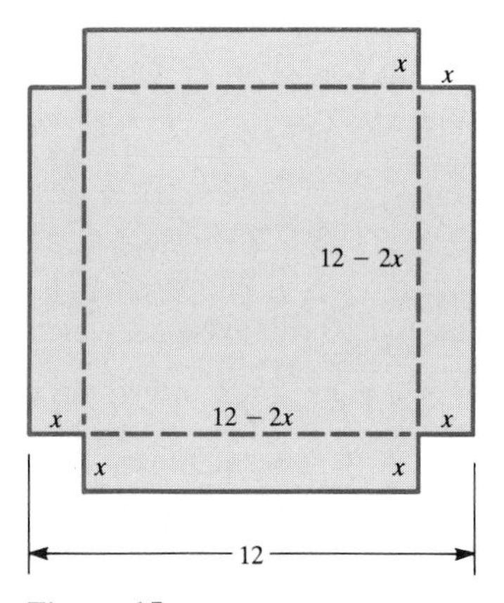

Figure 15

Step 4. To maximize $V(x) = 4x^3 - 48x^2 + 144x$ on $[0, 6]$ we evaluate $V(x)$ at critical numbers in $[0, 6]$ and at the endpoints of $[0, 6]$.

We have

$$\begin{aligned} V'(x) &= 12x^2 - 96x + 144 \\ &= 12(x^2 - 8x + 12) \\ &= 12(x - 2)(x - 6). \end{aligned}$$

A critical number satisfies the equation

$$0 = 12(x - 2)(x - 6).$$

Hence $x - 2 = 0$ or $x - 6 = 0$. The critical numbers are 2 and 6.

The endpoints of the interval $[0, 6]$ are 0 and 6. Therefore the maximum

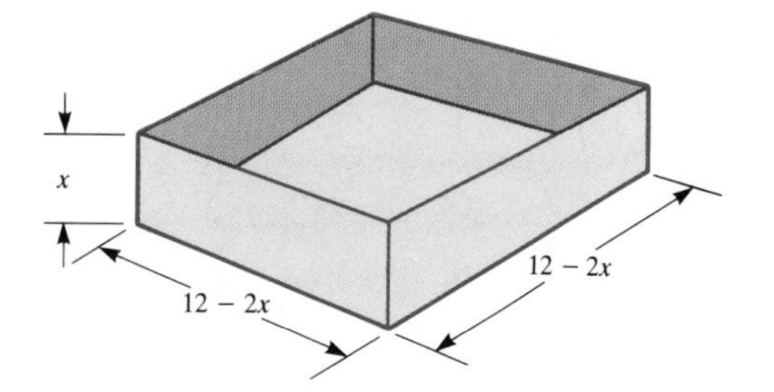

Figure 16

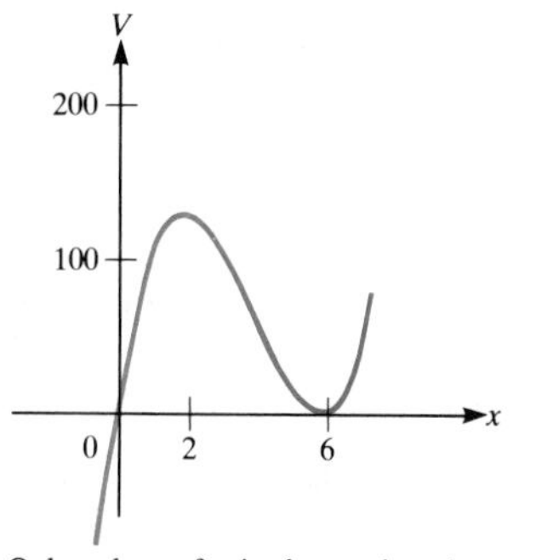

Only values of x in the portion above [0, 6] correspond to physically realizable trays.

Figure 17

value of $V(x)$ for x in $[0, 6]$ is the largest of $V(0)$, $V(2)$, and $V(6)$. Since $V(0) = 0$ and $V(6) = 0$, the largest value is

$$V(2) = 4(2^3) - 48(2^2) + 144 \cdot 2 = 128 \text{ cubic inches.}$$

The cut that produces the tray with the largest volume is $x = 2$ inches. ■

As a matter of interest, let us graph the function V, showing its behavior for all x, not just for values of x significant to the problem. Note in Fig. 17 that at $x = 2$ and $x = 6$ the tangent is horizontal.

Remark: In Example 2 you might say $x = 0$ and $x = 6$ don't really correspond to what you would call a tray. If so, you would restrict the domain of $V(x)$ to the open interval $(0, 6)$. You would then have to examine the behavior of $V(x)$ for x *near* 0 or near 6. By making the domain $[0, 6]$ from the start, you avoid the extra work of examining $V(x)$ for x *near* the ends of the interval.

The key step in each of Examples 1 and 2 and any applied problem is Step 2, finding a mathematical formula for the quantity whose extremum you are seeking. In case the problem is geometrical, the following chart may be of aid.

SETTING UP THE FUNCTION

1 Draw and label the appropriate diagrams. (Make them large enough so there is room for labels.)
2 Label the various quantities by letters, such as x, y, A, V.
3 Identify the quantity to be maximized (or minimized).
4 Express the quantity to be maximized (or minimized) in terms of one or more of the other variables.
5 Finally, express that quantity in terms of only one variable.

An Economical Can

EXAMPLE 3 Of all the tin cans that enclose a volume of 100 cubic inches, which requires the least metal?

SOLUTION The can may be flat or tall. If the can is flat, the side uses little metal, but then the top and bottom bases are large. If the can is shaped like a mailing tube, then the two bases require little metal, but the curved side requires a great deal of metal. (See Fig. 18, where r denotes the radius and h the height

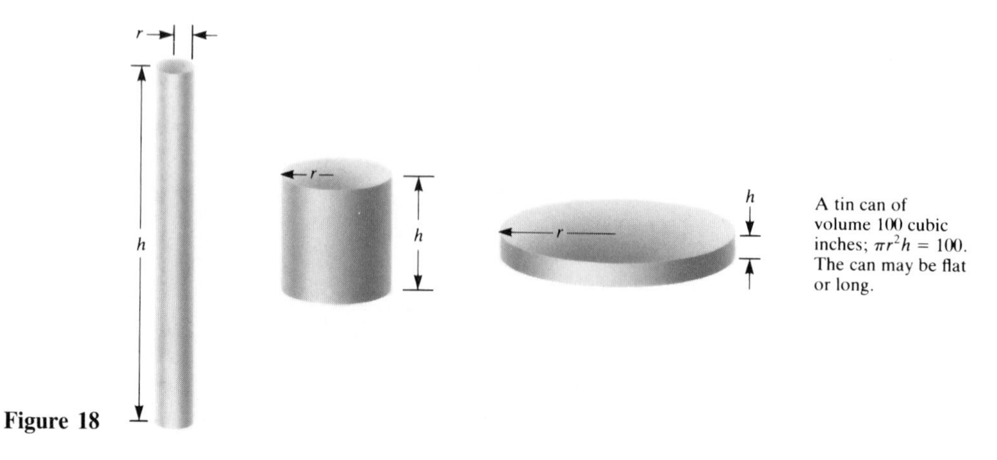

A tin can of volume 100 cubic inches; $\pi r^2 h = 100$. The can may be flat or long.

Figure 18

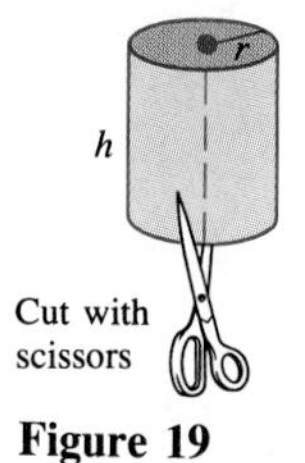

Figure 19

of the can.) What is the ideal compromise between these two extremes?

The surface area S of the can is given by

$$S = 2\pi r^2 + 2\pi rh, \tag{3}$$

which accounts for the two circular bases and the side. Figure 19 shows why the area of the side is $2\pi rh$. Since the amount of metal in the can is proportional to S, it suffices to minimize S.

Equation (3) gives S as a function of two variables, but we can express one of the variables in terms of the other. In the tin cans under consideration, the radius and height are related by the equation

$$\pi r^2 h = 100, \tag{4}$$

since their volume is 100 cubic inches. In order to express S as a function of one variable, use Eq. (4) to eliminate either r or h. Choosing to eliminate h, we solve Eq. (4) for h:

$$h = \frac{100}{\pi r^2}.$$

Substitution into Eq. (3) yields

$$S = 2\pi r^2 + 2\pi r\frac{100}{\pi r^2} \quad \text{or} \quad S = 2\pi r^2 + \frac{200}{r}. \tag{5}$$

Equation (5) expresses S as a function of just one variable, r.

The domain of this function for our purposes is $(0, \infty)$, since the tin can has a positive radius.

Compute dS/dr:

$$\frac{dS}{dr} = 4\pi r - \frac{200}{r^2} = \frac{4\pi r^3 - 200}{r^2}. \tag{6}$$

Setting the derivative equal to 0 to find any critical numbers, we have

$$0 = \frac{4\pi r^3 - 200}{r^2},$$

hence

$$0 = 4\pi r^3 - 200$$

or

$$4\pi r^3 = 200$$

$$r^3 = \frac{200}{4\pi}$$

$$r = \sqrt[3]{\frac{50}{\pi}}.$$

There is only one critical number. Does it provide a minimum? Let's check it two ways, first by the first-derivative test, then by the second-derivative test.

The first derivative is

$$\frac{dS}{dr} = \frac{4\pi r^3 - 200}{r^2}. \tag{7}$$

When $r = \sqrt[3]{50/\pi}$, the numerator in (7) is 0. When $r < \sqrt[3]{50/\pi}$ the numerator is negative and when $r > \sqrt[3]{50/\pi}$ the numerator is positive. (The denominator is always positive.) Since $dS/dr < 0$ for $r < \sqrt[3]{50/\pi}$, and $dS/dr > 0$ for $r > \sqrt[3]{50/\pi}$, the function $S(r)$ decreases for $r < \sqrt[3]{50/\pi}$ and increases for $r >$

Figure 20

Figure 21

Figure 22

$\sqrt[3]{50/\pi}$. That shows that a global minimum occurs at $\sqrt[3]{50/\pi}$. (See Fig. 20.)

Let us instead use the second-derivative test. Differentiation of (6) gives

$$\frac{d^2S}{dr^2} = 4\pi + \frac{400}{r^3}. \tag{8}$$

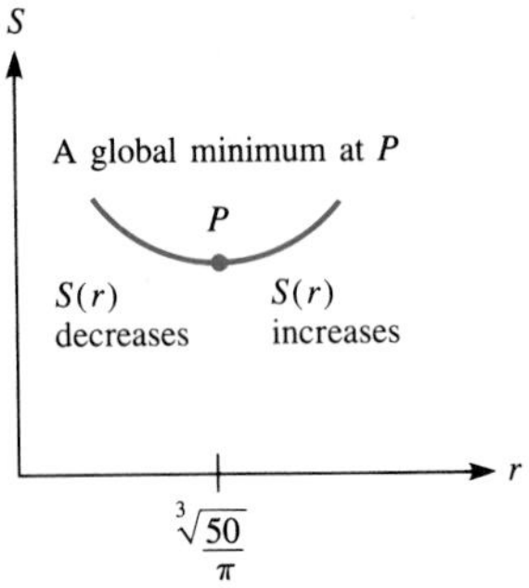

Inspection of (8) shows that for all r in $(0, \infty)$, which is the domain that is meaningful for tin cans, d^2S/dr^2 is positive. (The function is concave upward, as shown in Fig. 21.) Not only is P a relative minimum, it is a global minimum, since the graph lies above its tangents, in particular, the tangent at P.

The minimum of $S(r)$ is shown in Fig. 22.

Incidentally, if you want to know the height of the most economical can, solve (4) for h:

$$h = \frac{100}{\pi r^2} = \frac{100}{\pi(\sqrt[3]{50/\pi})^2}$$

$$= \frac{100}{\pi(\sqrt[3]{50/\pi})^2}\frac{\sqrt[3]{50/\pi}}{\sqrt[3]{50/\pi}} \qquad \text{rationalize the denominator}$$

$$= \frac{100}{\pi(50/\pi)}\sqrt[3]{\frac{50}{\pi}} = 2\sqrt[3]{\frac{50}{\pi}}.$$

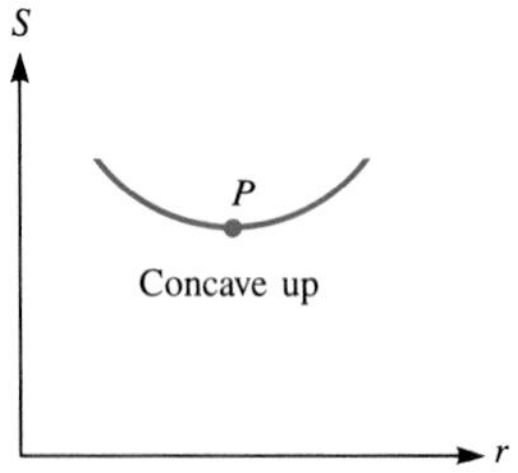

The height of the can is equal to its diameter. ■

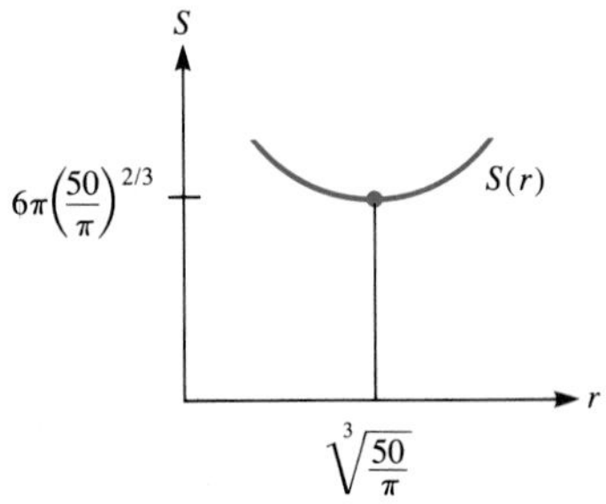

Section Summary

We showed how to use calculus to solve applied problems: experiment, set up a function, find its domain, and critical points. Then test the critical points and endpoints of the domain to determine the extrema.

EXERCISES FOR SEC. 4.7: APPLIED MAXIMUM AND MINIMUM PROBLEMS

(Formulas for various volumes and areas are listed inside the back cover.)

1 Solve Example 1, expressing A in terms of y instead of x.

2 Solve Example 1 if there is enough wire to construct 160 feet of fence.

Exercises 3 to 6 are related to Example 2. In each case find the length of the cut that maximizes the volume of the tray. The dimensions of the cardboard are given.

3 5 inches by 5 inches

4 5 inches by 7 inches

5 4 inches by 8 inches

6 6 inches by 10 inches

7 Solve Example 3, expressing S in terms of h instead of r.

8 Of all cylindrical tin cans *without a top* that contain 100 cubic inches, which requires the least material?

9 Of all enclosed rectangular boxes with square bases that have a volume of 1000 cubic inches, which uses the least material?

10 Of all topless rectangular boxes with square bases that have a volume of 1000 cubic inches, which uses the least material?

11 Find the dimensions of the rectangle of largest area that can be inscribed in a circle of radius a. The typical rectangle is shown in Fig. 23. (*Hint:* Express the area in terms of the angle θ shown.)

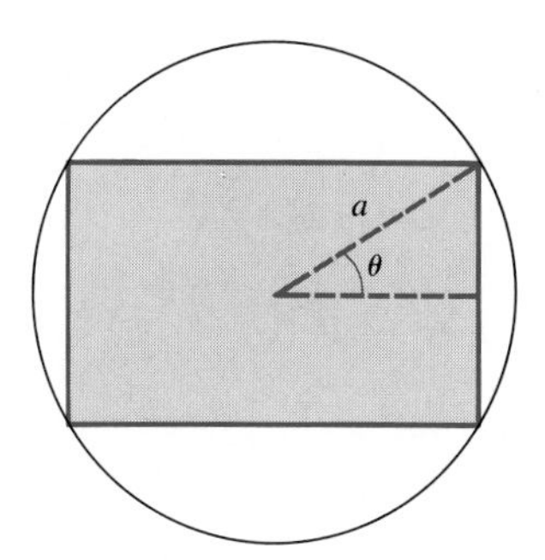

Figure 23

12 Find the dimensions of the rectangle of largest perimeter that can be inscribed in a circle of radius a.

13 Show that of all rectangles of a given perimeter, the square has the largest area. *Suggestion:* Call the fixed perimeter p and keep in mind that it is constant.

14 Show that of all rectangles of a given area, the square has the shortest perimeter. *Suggestion:* Call the fixed area A and keep in mind that it is constant.

15 A rancher wants to construct a rectangular corral. He also wants to divide the corral by a fence parallel to one of the sides. He has 240 feet of fence. What are the dimensions of the corral of largest area he can enclose?

16 A river has a 45° turn, as indicated in Fig. 24. A rancher wants to construct a corral bounded on two sides by the river and on two sides by 1 mile of fence ABC, as shown. Find the dimensions of the corral of largest area.

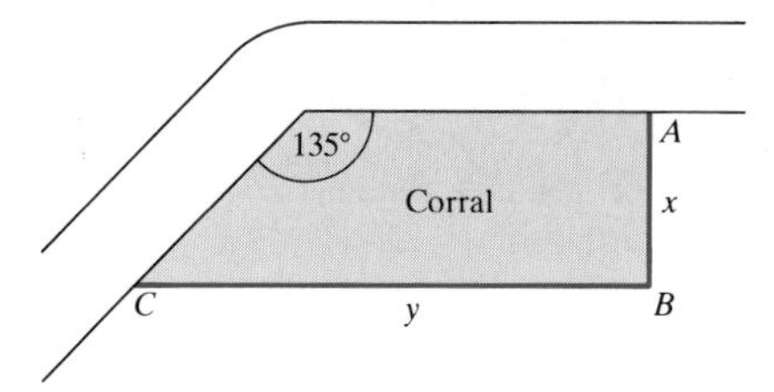

Figure 24

17 (*a*) How should one choose two nonnegative numbers whose sum is 1 in order to maximize the sum of their squares?
(*b*) To minimize the sum of their squares?

18 How should one choose two nonnegative numbers whose sum is 1 in order to maximize the product of the square of one of them and the cube of the other?

19 An irrigation channel made of concrete is to have a cross section in the form of an isosceles trapezoid, three of whose sides are 4 feet long. See Fig. 25. How should the trapezoid be shaped if it is to have the maximum possible area? Consider the area as a function of x and solve.

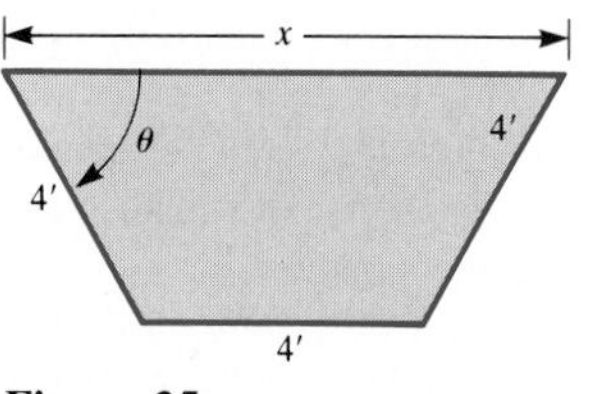

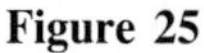
Figure 25

20 (*a*) Solve Exercise 19, expressing the area as a function of θ instead of x.
(*b*) Do the answers in (*a*) and Exercise 19 agree?

In Exercises 21 to 24 use the fact that the combined length and girth (distance around) of a package to be sent through the mail cannot exceed 108 inches.

21 Find the dimensions of the right circular cylinder of largest volume that can be sent through the mail.

22 Find the dimensions of the right circular cylinder of largest surface area that can be sent through the mail.

23 Find the dimensions of the rectangular box with square base of largest volume that can be sent through the mail.

24 Find the dimensions of the rectangular box with square base of largest surface area that can be sent through the mail.

Exercises 25 to 30 concern ''minimal cost'' problems.

25 A cylindrical can is to be made to hold 100 cubic inches. The material for its top and bottom costs twice as much per square inch as the material for its side. Find the radius and height of the most economical can. *Warning:* This is not the same as Example 3.
(*a*) Would you expect the most economical can in this problem to be taller or shorter than the solution to Example 3? (Use common sense, not calculus.)
(*b*) For convenience, call the cost of 1 square inch of the material for the side k cents. Thus the cost of 1 square inch of the material for the top and bottom is $2k$ cents. (The precise value of k will not affect the answer.) Show that a can of radius r and height h costs

$$C = 4k\pi r^2 + 2k\pi rh \qquad \text{cents.}$$

(*c*) Find r that minimizes the function C in (*b*). Keep in mind during any differentiation that k is constant.
(*d*) Find the corresponding h.

26 A rectangular box with a square base is to hold 100 cubic inches. Material for the sides costs twice as much per square inch as the material for the top and bottom.
(*a*) If the base has side x and the height is y, what does the box cost?
(*b*) Find the dimensions of the most economical box.

27 A rectangular box with a square base is to hold 100 cubic inches. Material for the top of the box costs 2 cents per square inch; material for the sides costs 3 cents per square inch; material for the bottom costs 5 cents per square inch. Find the dimensions of the most economical box.

28 The cost of operating a certain truck (for gasoline, oil, and depreciation) is $(20 + s/2)$ cents per mile when it travels at a speed of s miles per hour. A truck driver earns \$18 per hour. What is the most economical speed at which to operate the truck during a 600-mile trip?

(*a*) If you considered only the truck, would you want s to be small or large?

(*b*) If you, the employer, considered only the expense of the driver's wages, would you want s to be small or large?

(*c*) Express cost as a function of s and solve. (Be sure to put the costs all in terms of cents or all in terms of dollars.)

(*d*) Would the answer be different for a 1000-mile trip?

29 A government contractor who is removing earth from a large excavation can route trucks over either of two roads. There are 10,000 cubic yards of earth to move. Each truck holds 10 cubic yards. On one road the cost per truckload is $1 + 2x^2$ cents, when x trucks use that road; the function records the cost of congestion. On the other road the cost is $2 + x^2$ cents per truckload when x trucks use that road. How many trucks should be dispatched to each of the two roads?

30 On one side of a river 1 mile wide is an electric power station; on the other side, s miles upstream, is a factory. (See Fig. 26.) It costs 3 dollars per foot to run cable over land and 5 dollars per foot under water. What is the most economical way to run cable from the station to the factory?

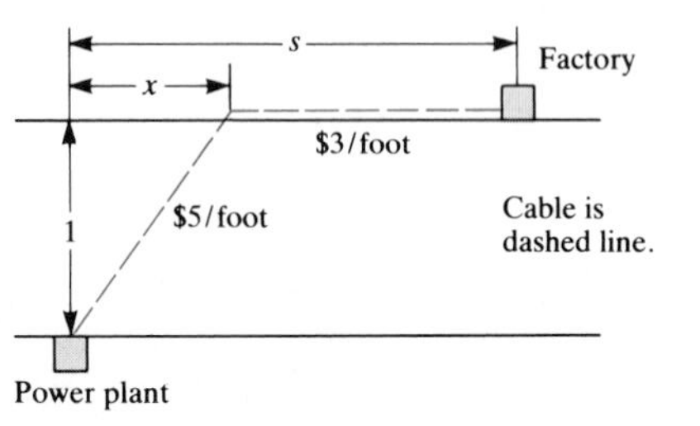

Figure 26

(*a*) Using no calculus, what do you think would be (approximately) the best route if s were very small? if s were very large?

(*b*) Solve with the aid of calculus, and draw the routes for $s = \frac{1}{2}, \frac{3}{4}, 1$, and 2.

(*c*) Solve for arbitrary s.

Warning: Minimizing the length of cable is not the same as minimizing its cost.

31 (From *Dynamics of Airplanes*, by John E. Younger and Baldwin M. Woods, Wiley, New York.) "Recalling that

$$I = A\cos^2\theta + C\sin^2\theta - 2E\cos\theta\sin\theta,$$

we wish to find θ when I is a maximum or a minimum." Show that at an extremum of I,

$$\tan 2\theta = \frac{2E}{C - A}. \qquad \text{(assume that } A \neq C)$$

32 (From *University Physics*, p. 46, by Alvin Hudson and Rex Nelson, Harcourt Brace Jovanovich, New York, 1982.) "By differentiating the equation for the horizontal range,

$$R = \frac{v_0^2 \sin 2\theta}{g},$$

show that the initial elevation angle θ for maximum range is 45°." In the formula for R, v_0 and g are constants. (R is the horizontal distance a baseball covers if you throw it at an angle θ with speed v_0. Air resistance is disregarded.)

(*a*) Using calculus, show that the maximum range occurs when $\theta = 45°$.

(*b*) Solve the same problem without calculus.

33 (*a*) Graph $y = x \sin x$ for x in $[0, \pi]$.

(*b*) Using the first and second derivatives, show that it has a unique relative maximum in the interval $[0, \pi]$.

(*c*) Show that the maximum value of $x \sin x$ occurs when $x \cos x + \sin x = 0$.

(*d*) Use Newton's method, with $x_1 = \pi/2$, to find an estimate x_2 for a root of $x \cos x + \sin x = 0$.

(*e*) Use Newton's method again to find x_3.

34 (*a*) Graph $y = x \cos x$ for x in $[0, \pi/2]$.

(*b*) Using the first and second derivatives, show that it has a unique relative maximum in the interval $[0, \pi/2]$.

(*c*) Show that the maximum value of $x \cos x$ occurs when $\cos x - x \sin x = 0$.

(*d*) Use Newton's method, with $x_1 = \pi/4$, to find an estimate x_2 for a root of $\cos x - x \sin x = 0$.

(*e*) Use Newton's method again to find x_3.

35 Estimate the maximum value of $y = 2 \sin x - x^2$ over the interval $[0, \pi/2]$.

36 Estimate the minimum value of $y = x^3 + \cos x$ over the interval $[0, \pi/2]$.

37 Fencing is to be added to an existing wall of length 20 feet, as shown in Fig. 27. How should the extra fence be added to maximize the area of the enclosed rectangle if the additional fence is

(*a*) 40 feet long? (*b*) 80 feet long? (*c*) 60 feet long?

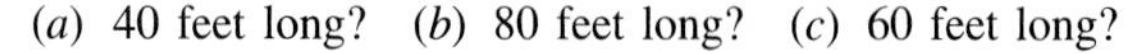

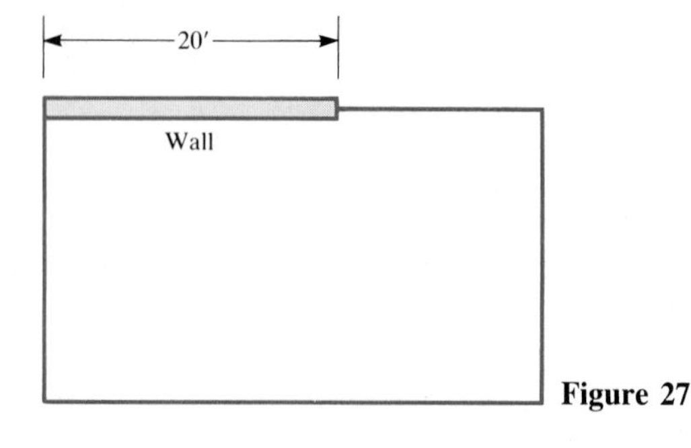

Figure 27

38 Let A and B be constants. Find the maximum and minimum values of $A \cos t + B \sin t$.

39 A spider at corner S of a cube of side 1 inch wishes to capture a fly at the opposite corner F. (See Fig. 28.) The spider, who must walk on the surface of the solid cube, wishes to find the shortest path.

(*a*) Find a shortest path with the aid of calculus.

(*b*) Find a shortest path without calculus.

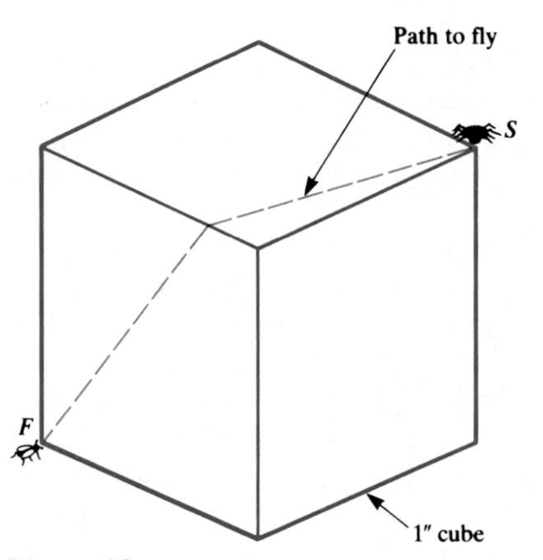

Figure 28

40 A ladder of length b leans against a wall of height a, $a < b$. What is the maximal horizontal distance that the ladder can extend beyond the wall if its base rests on the horizontal ground?

41 A woman can walk 3 miles per hour on grass and 5 miles per hour on sidewalk. She wishes to walk from point A to point B, shown in Fig. 29, in the least time. What route should she follow if s is (*a*) $\frac{1}{2}$? (*b*) $\frac{3}{4}$? (*c*) 1?

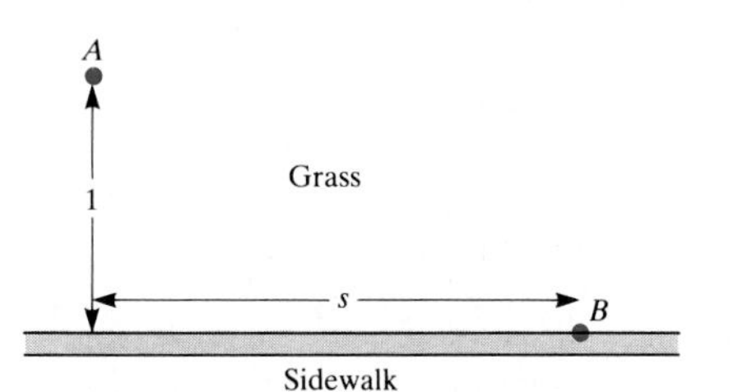

Figure 29

42 The potential energy in a diatomic molecule is given by the formula

$$U(r) = U_0\left[\left(\frac{r_0}{r}\right)^{12} - 2\left(\frac{r_0}{r}\right)^{6}\right],$$

where U_0 and r_0 are constants and r is the distance between the atoms. For which value of r is $U(r)$ a minimum?

43 What are the dimensions of the right circular cylinder of largest volume that can be inscribed in a sphere of radius a?

44 The stiffness of a rectangular beam is proportional to the product of the width and the cube of the height of its cross section. What shape beam should be cut from a log in the form of a right circular cylinder of radius r in order to maximize its stiffness?

45 A rectangular box-shaped house is to have a square floor. Three times as much heat per square foot enters through the roof as through the walls. What shape should the house be if it is to enclose a volume of 12,000 cubic feet and minimize heat entry? (Assume no heat enters through the floor.)

46 (See Fig. 30.) Find the coordinates of the points $P = (x, y)$, with $y \le 1$, on the parabola $y = x^2$, that (*a*) minimize $\overline{PA}^2 + \overline{PB}^2$, (*b*) maximize $\overline{PA}^2 + \overline{PB}^2$.

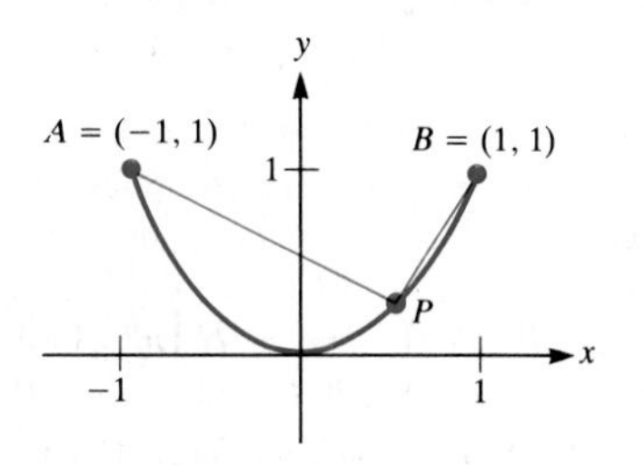

Figure 30

47 The speed of traffic through the Lincoln Tunnel in New York City depends on the amount of traffic. Let S be the speed in miles per hour and let D be the amount of traffic measured in vehicles per mile. The relation between S and D was seen to be approximated closely by the formula

$$S = 42 - \frac{D}{3},$$

for $D \le 100$.

(*a*) Express in terms of S and D the total number of vehicles that enter the tunnel in an hour.

(*b*) What value of D will maximize the flow in (*a*)?

48 When a tract of timber is to be logged, a main logging road is built from which small roads branch off as feeders. The question of how many feeders to build arises in practice. If too many are built, the cost of construction would be prohibitive. If too few are built, the time spent moving the logs to the roads would be prohibitive. The formula for total cost,

$$y = \frac{CS}{4} + \frac{R}{VS},$$

is used in a logger's manual to find how many feeder roads are to be built. R, C, and V are known constants: R is the cost of road at "unit spacing"; C is the cost of moving a log a unit distance; V is the value of timber per acre. S denotes the distance between the regularly spaced feeder roads. (See Fig. 31.) Thus the cost y is a function of S, and the object is to find that value of S that minimizes y. The manual says, "To find the desired S set the two summands equal to each other and solve:

$$\frac{CS}{4} = \frac{R}{VS}.\text{''}$$

Show that the method is valid.

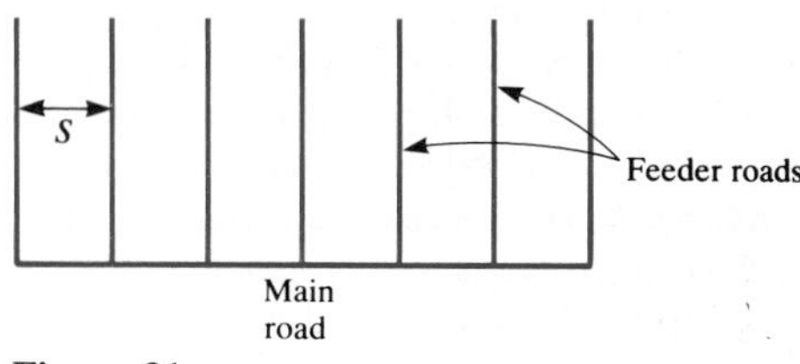

Figure 31

49 A delivery service is deciding how many warehouses to set up in a large city. The warehouses will serve similarly shaped regions of equal area A and, let us assume, an equal number of people.

(*a*) Why would transportation costs per item presumably be proportional to $\sqrt{A}$?

(*b*) Assuming that the warehouse cost per item is inversely proportional to A, show that C, the cost of transportation and storage per item, is of the form $t\sqrt{A} + w/A$, where t and w are appropriate constants.

(*c*) Show that C is a minimum when $A = (2w/t)^{2/3}$.

Exercises 50 and 51 are related.

50 A pipe of length b is carried down a long corridor of width $a < b$ and then around corner C. (See Fig. 32.) During the turn y starts out at 0, reaches a maximum, and then returns to 0. (Try this with a short stick.) Find that maximum in terms of a and b. *Suggestion:* Express y in terms of a, b, and θ; θ is a variable, while a and b are constants.

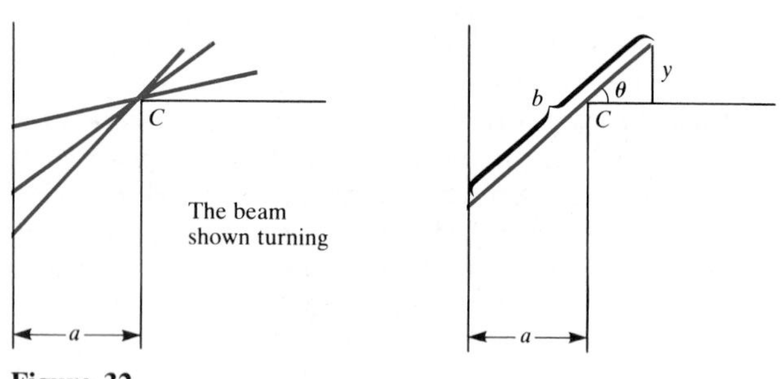

Figure 32

51 Figure 33 shows two corridors meeting at a right angle. One has width 8; the other, width 27. Find the length of the longest pipe that can be carried horizontally from one hall, around the corner and into the other hall. *Suggestion:* Do Exercise 50 first.

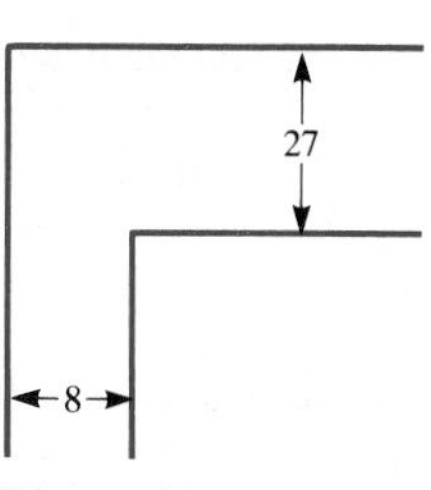

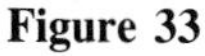
Figure 33

52 Two houses, A and B, are a distance p apart. They are distances q and r, respectively, from a straight road, and on the same side of the road. Find the length of the shortest path that goes from A to the road, and then on to the other house B.

(*a*) Use calculus.

(*b*) Use only elementary geometry. *Hint:* Introduce an imaginary house C such that the midpoint of B and C is on the road and the segment BC is perpendicular to the road; that is, "reflect" B across the road to become C.

53 The base of a painting on a wall is a feet above the eye of an observer, as shown in Fig. 34. The vertical side of the painting is b feet long. How far from the wall should the observer stand to maximize the angle that the painting subtends? *Hint:* It is more convenient to maximize $\tan \theta$ than θ itself. Recall that

$$\tan (A - B) = \frac{\tan A - \tan B}{1 + \tan A \tan B}.$$

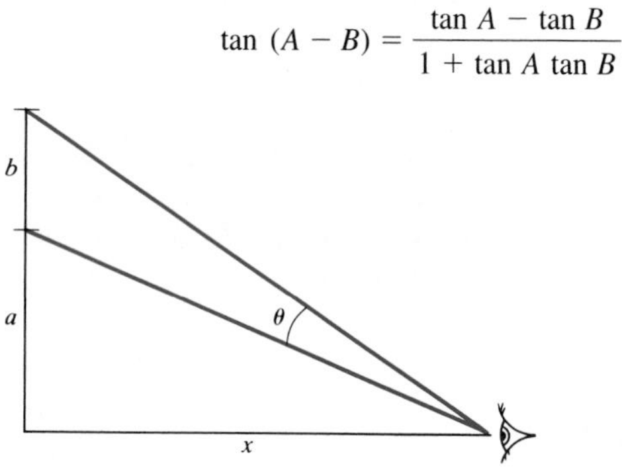

Figure 34

54 Find the point P on the x axis such that the angle APB in Fig. 35 is maximal. *Suggestion:* Note hint in Exercise 53.

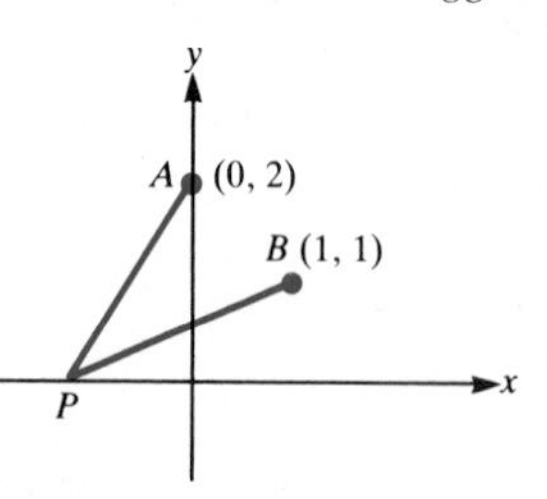

Figure 35

55 *(Economics)* Let p denote the price of some commodity and y the number sold at that price. To be concrete, assume that $y = 250 - p$ for $0 \le p \le 250$. Assume that it costs the producer $100 + 10y$ dollars to manufacture y units. What price p should the producer choose in order to maximize total profit, that is, "revenue minus cost"?

56 *(Leibniz on light)* A ray of light travels from point A to point B in Fig. 36 in minimal time. The point A is in one medium, such as air or a vacuum. The point B is in another medium, such as water or glass. In the first medium light travels at velocity v_1 and in the second at velocity v_2. The media are separated by line L. Show that for the path APB of minimal time,

$$\frac{\sin \alpha}{v_1} = \frac{\sin \beta}{v_2}.$$

Leibniz solved this problem with calculus in a paper published in 1684. (The result is called **Snell's law of refraction.**)

Leibniz then wrote, "other very learned men have sought in many devious ways what someone versed in this calculus can accomplish in these lines as by magic." (See C. H. Edwards Jr., *The Historical Development of the Calculus,* p. 259, Springer-Verlag, New York, 1979.)

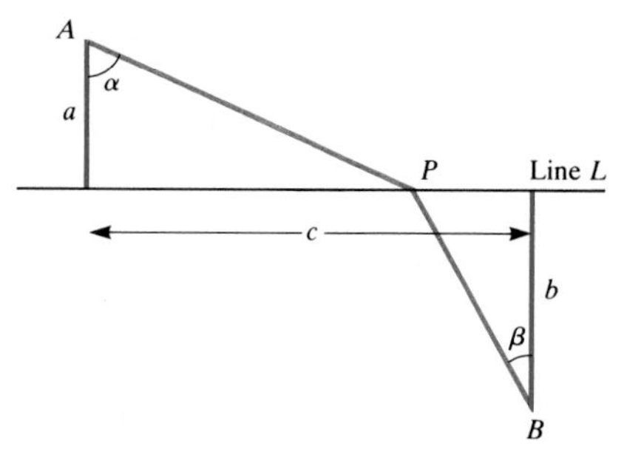

Figure 36

57 The following calculation occurs in an article by Manfred Kochen, "On Determining Optimum Size of New Cities": The net utility to the total client-centered system is

$$U = \frac{RLv}{A} n^{1/2} - nK - \frac{ALc}{v} n^{-1/2}.$$

All symbols except U and n are constant; n is a measure of decentralization. Regarding U as a differentiable function of n, we can determine when $dU/dn = 0$. This occurs when

$$\frac{RLv}{2A} n^{-1/2} - K + \frac{ALc}{2v} n^{-3/2} = 0.$$

This is a cubic equation for $n^{-1/2}$.

(*a*) Check that the differentiation is correct.

(*b*) Of what cubic polynomial is $n^{-1/2}$ a root?

4.8 IMPLICIT DIFFERENTIATION

Sometimes a function $y = f(x)$ is given indirectly by an equation that involves y and x. This section shows how to differentiate y without solving for y explicitly in terms of x.

A Function Given Implicitly

Consider the equation

$$x^2 + y^2 = 25. \tag{1}$$

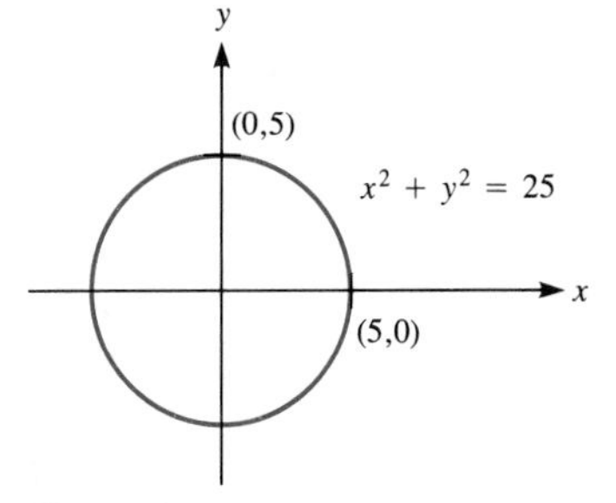

Figure 1

This equation describes a circle of radius 5 and center at the origin, as in Fig. 1. (See Appendix B.) This circle is not the graph of a function, since a vertical line can meet the circle in two points. However, the top half is the graph of a function and so is the bottom half. To find these functions explicitly, solve Eq. (1) for y:

$$y^2 = 25 - x^2$$
$$y = \pm\sqrt{25 - x^2}.$$

So either $y = \sqrt{25 - x^2}$ or $y = -\sqrt{25 - x^2}$. The graph of $y = \sqrt{25 - x^2}$ is

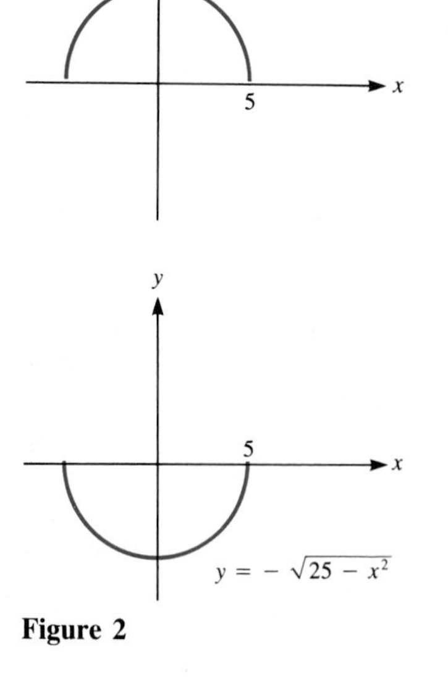

Figure 2

the top semicircle; the graph of $y = -\sqrt{25 - x^2}$ is the bottom semicircle. (See Fig. 2.) There are thus two continuous functions that satisfy Eq. (1).

The equation $x^2 + y^2 = 25$ is said to describe the function $y = f(x)$ **implicitly**. The equations

$$y = \sqrt{25 - x^2} \qquad \text{and} \qquad y = -\sqrt{25 - x^2}$$

describe the function $y = f(x)$ **explicitly**.

Differentiating an Implicit Function

It is possible to differentiate a function given implicitly without having to solve for the function and express it explicitly. An example will illustrate the method, which is simply to differentiate both sides of the equation that defines the function implicitly. This procedure is called **implicit differentiation**.

EXAMPLE 1 Let $y = f(x)$ be the continuous function that satisfies the equation

$$x^2 + y^2 = 25$$

such that $y = 4$ when $x = 3$. Find dy/dx when $x = 3$ and $y = 4$.

SOLUTION (We could, of course, solve for y, $y = \sqrt{25 - x^2}$, and differentiate directly. However, the algebra would be more involved since square roots would appear.) Differentiating both sides of the equation

$$x^2 + y^2 = 25$$

with respect to x yields

$$\frac{d}{dx}(x^2 + y^2) = \frac{d}{dx}(25),$$

or

$$2x + \frac{d(y^2)}{dx} = 0.$$

To differentiate y^2 with respect to x, write $w = y^2$, where y is a function of x. By the chain rule

$$\frac{dw}{dx} = \frac{dw}{dy}\frac{dy}{dx},$$

which gives us

$$\frac{d(y^2)}{dx} = 2y\frac{dy}{dx}.$$

Thus

$$2x + 2y\frac{dy}{dx} = 0,$$

or

$$x + y\frac{dy}{dx} = 0.$$

In particular, when $x = 3$ and $y = 4$,

$$3 + 4\frac{dy}{dx} = 0,$$

Observe that the algebra involves no square roots.

and therefore,

$$\frac{dy}{dx} = -\frac{3}{4}. \quad ■$$

In the next example implicit differentiation is the only way to find the derivative, for in this case there is no formula expressible in terms of trigonometric and algebraic functions giving y explicitly in terms of x.

EXAMPLE 2 Assume that the equation

$$2xy + \pi \sin y = 2\pi$$

defines a function $y = f(x)$. Find dy/dx when $x = 1$ and $y = \pi/2$. (Note that $x = 1$ and $y = \pi/2$ satisfy the equation.)

SOLUTION Implicit differentiation yields

$$\frac{d}{dx}(2xy + \pi \sin y) = \frac{d(2\pi)}{dx},$$

$$\left(2x\frac{dy}{dx} + 2y\frac{dx}{dx}\right) + \pi(\cos y)\frac{dy}{dx} = 0,$$

by the formula for the derivative of a product and the chain rule. Hence

$$2x\frac{dy}{dx} + 2y + \pi(\cos y)\frac{dy}{dx} = 0.$$

Solving for the derivative, dy/dx, we get

$$\frac{dy}{dx} = \frac{-2y}{2x + \pi \cos y}.$$

In particular, when $x = 1$ and $y = \pi/2$,

$$\frac{dy}{dx} = -\frac{2 \cdot \frac{\pi}{2}}{2 \cdot 1 + \pi \cos \frac{\pi}{2}}$$

$$= -\frac{\pi}{2 + \pi \cdot 0} = -\frac{\pi}{2}. \quad ■$$

Implicit Differentiation and Extrema

Example 3 of Sec. 4.7 answered the question, "Of all the tin cans that enclose a volume of 100 cubic inches, which requires the least metal?" The radius of the most economical can is $\sqrt[3]{50/\pi}$. From this and the fact that its volume is 100 cubic inches, its height was found to be $2\sqrt[3]{50/\pi}$, which is exactly twice the radius. In the next example implicit differentiation is used to answer the same question. Not only will the algebra be simpler than before, but the answer will provide the general shape—the proportion between height and radius—quite easily.

EXAMPLE 3 Of all the tin cans that enclose a volume of 100 cubic inches, which requires the least metal?

SOLUTION The height h and radius r of any can of volume 100 cubic inches are related by the equation

$$\pi r^2 h = 100. \tag{2}$$

The surface area S of the can is

$$S = 2\pi r^2 + 2\pi rh. \tag{3}$$

Consider h, and hence S, as functions of r. However, *it is not necessary to find these functions explicitly.*

Differentiation of Eqs. (2) and (3) with respect to r yields

$$\pi r^2 \frac{dh}{dr} + 2\pi rh = \frac{d(100)}{dr} = 0 \tag{4}$$

and

$$\frac{dS}{dr} = 4\pi r + 2\pi r \frac{dh}{dr} + 2\pi h. \tag{5}$$

Since when S is a minimum, $dS/dr = 0$, we have

$$0 = 4\pi r + 2\pi r \frac{dh}{dr} + 2\pi h. \tag{6}$$

Equations (4) and (6) yield, with a little algebra, a relation between h and r, as follows:

Factoring πr out of Eq. (4) and 2π out of Eq. (6) shows that

$$r\frac{dh}{dr} + 2h = 0 \qquad \text{and} \qquad 2r + r\frac{dh}{dr} + h = 0. \tag{7}$$

Elimination of dh/dr from Eqs. (7) yields

$$2r + r\left(\frac{-2h}{r}\right) + h = 0,$$

which simplifies to

$$2r = h. \tag{8}$$

Equation (8) asserts that the height of the most economical can is the same as its diameter. Moreover, this is the ideal shape, no matter what the prescribed volume happens to be. [Equation (4) follows from Eq. (2) merely because 100 is constant.]

The specific dimensions of the most economical can are found by combining the equations

$$2r = h \tag{8}$$

and

$$\pi r^2 h = 100. \tag{2}$$

Elimination of h from these two equations shows that

$$\pi r^2(2r) = 100 \qquad \text{or} \qquad r^3 = \frac{50}{\pi}.$$

Hence

$$r = \sqrt[3]{\frac{50}{\pi}} \qquad \text{and} \qquad h = 2r = 2\sqrt[3]{\frac{50}{\pi}}. \quad \blacksquare$$

As in the case of Example 3, implicit differentiation finds the proportions of a general solution before finding the exact values of the variables. Often it is the proportion, rather than the (perhaps messier) explicit values, that gives more insight into the answer. For instance, Eq. (8) tells that the diameter equals the height for the most economical can.

The procedure illustrated in Example 3 is quite general. It may be of use when maximizing (or minimizing) a quantity that at first is expressed as a function of two variables which are linked by an equation. The equation that links them is called the **constraint**. In Example 3, the constraint is $\pi r^2 h = 100$.

General procedure for using implicit differentiation in an applied maximum problem

USING IMPLICIT DIFFERENTIATION IN AN EXTREMUM PROBLEM

1 Name the various quantities in the problem by letters, such as x, y, A, V.
2 Identify which quantity is to be maximized (or minimized).
3 Express the quantity to be maximized (or minimized) in terms of other quantities, such as x and y.
4 Obtain an equation relating x and y. (This equation is called a constraint.)
5 Differentiate implicitly both the constraint and the expression to be maximized (or minimized), interpreting all the various quantities to be functions of x (or, perhaps, of y).
6 Set the derivative of the expression to be maximized (or minimized) equal to 0 and combine with the derivative of the constraint to obtain an equation relating x and y at a maximum (or minimum).
7 Step 6 gives only a relation or proportion between x and y at an extremum. If the explicit values of x and y are desired, find them by using the fact that x and y also satisfy the constraint.

Warning: Sometimes an extremum occurs where a derivative, such as dy/dx, is not defined. (Exercise 29 illustrates this possibility.)

Section Summary

To differentiate a function given implicitly, differentiate both sides of the defining equation. The chain rule will usually be needed. This technique was then applied to extrema problems in which the quantity to be maximized or minimized is a function of two variables that are related to each other by an equation.

EXERCISES FOR SEC. 4.8: IMPLICIT DIFFERENTIATION

In Exercises 1 to 4 find dy/dx at the indicated values of x and y in two ways: explicitly (solving for y first) and implicitly.

1 $xy = 4$ at $(1, 4)$
2 $x^2 - y^2 = 3$ at $(2, 1)$
3 $x^2y + xy^2 = 12$ at $(3, 1)$
4 $x^2 + y^2 = 100$ at $(6, -8)$

In Exercises 5 to 8 find dy/dx at the given points by implicit differentiation.

5 $\dfrac{2xy}{\pi} + \sin y = 2$ at $(1, \pi/2)$
6 $2y^3 + 4xy + x^2 = 7$ at $(1, 1)$
7 $x^5 + y^3x + yx^2 + y^5 = 4$ at $(1, 1)$
8 $x + \tan xy = 2$ at $(1, \pi/4)$

9 Solve Example 3 by implicit differentiation, but differentiate Eqs. (2) and (3) with respect to h instead of r.
10 What is the shape of the cylindrical can of largest volume that can be constructed with a given surface area? Do not find the radius and height of the largest can; find the ratio between them. *Suggestion:* Call the surface area S and keep in mind that it is constant.

In Exercises 11 to 16, solve by implicit differentiation.

11 Example 1 of Sec. 4.7.
12 Exercise 12 of Sec. 4.7.
13 Exercise 13 of Sec. 4.7.
14 Exercise 14 of Sec. 4.7.

15 Exercise 23 of Sec. 4.7.
16 Exercise 24 of Sec. 4.7.

In Exercises 17 to 20 find dy/dx at a general point (x, y) on the given curve.
17 $xy^3 + \tan(x + y) = 1$
18 $\sec(x + 2y) + \cos(x - 2y) + y = 2$
19 $-7x^2 + 48xy + 7y^2 = 25$
20 $\sin^3(xy) + \cos(x + y) + x = 1$

Exercise 21 shows how to find y'' if y is given implicitly.
21 Assume that $y(x)$ is a differentiable function of x and that $x^3y + y^4 = 2$. Assume that $y(1) = 1$. Find $y''(1)$, following these steps.
(*a*) Show that $x^3y' + 3x^2y + 4y^3y' = 0$.
(*b*) Use (*a*) to find $y'(1)$.
(*c*) Differentiate the equation in (*a*) and thereby show that $x^3y'' + 6x^2y' + 6xy + 4y^3y'' + 12y^2(y')^2 = 0$.
(*d*) Use the equation in (*c*) to find $y''(1)$. [*Hint:* $y(1)$ and $y'(1)$ are known.]
22 Find $y''(1)$ if $y(1) = 2$ and $x^5 + xy + y^5 = 35$.
23 Find $y'(1)$ and $y''(1)$ if $y(1) = 0$ and $\sin y = x - x^3$.
24 Find $y''(2)$ if $y(2) = 1$ and $x^3 + x^2y - xy^3 = 10$.
25 Use implicit differentiation to find the highest and lowest points on the ellipse $x^2 + xy + y^2 = 12$.
26 Does the tangent line to the curve $x^3 + xy^2 + x^3y^5 = 3$ at the point $(1, 1)$ pass through the point $(-2, 3)$?

Exercises 27 and 28 obtain by implicit differentiation the formulas for differentiating $x^{1/n}$ and $x^{m/n}$ with the assumption that they are differentiable functions. Here m and n are integers.
27 Let n be a positive integer. Assume that $y = x^{1/n}$ is a differentiable function of x. From the equation $y^n = x$ deduce by implicit differentiation that $y' = (1/n)x^{1/n-1}$.
28 Let m be a nonzero integer and n a positive integer. Assume that $y = x^{m/n}$ is a differentiable function of x. From the equation $y^n = x^m$ deduce by implicit differentiation that $y' = (m/n)x^{m/n-1}$.
29 (*a*) What difficulty arises when you use implicit differentiation to maximize $x^2 + y^2$ subject to $x^2 + 4y^2 = 16$.
(*b*) Show that a maximum occurs where dy/dx is not defined. What is the maximum of $x^2 + y^2$ subject to $x^2 + 4y^2 = 16$?
(*c*) The problem can be viewed geometrically as "Maximize $x^2 + y^2$ for points on the ellipse $x^2 + 4y^2 = 16$." Sketch the ellipse and interpret (*b*) in terms of it.
30 Differentiate for practice.
(*a*) $\sqrt{4 - 9x^2}$
(*b*) $\sqrt[3]{x}\cos^3 x$
(*c*) $\dfrac{\sec 2x}{1 + \tan 2x}$

4.9 THE DIFFERENTIAL AND LINEARIZATION

A small piece of the graph of a differentiable function looks almost like a straight line, namely a tangent line. This observation suggests a way to estimate the change in the output of a function when the input changes a small amount. The estimate depends on the derivative and introduces the differential.

The Differential

Consider a point $P = (a, f(a))$ on the graph of a differentiable function $f(x)$. The tangent line at P passes through $P = (a, f(a))$ and has slope $f'(a)$. By the point-slope formula (see Appendix B) an equation of the tangent is

$$y - f(a) = f'(a)(x - a),$$

or

$$y = f(a) + f'(a)(x - a).$$

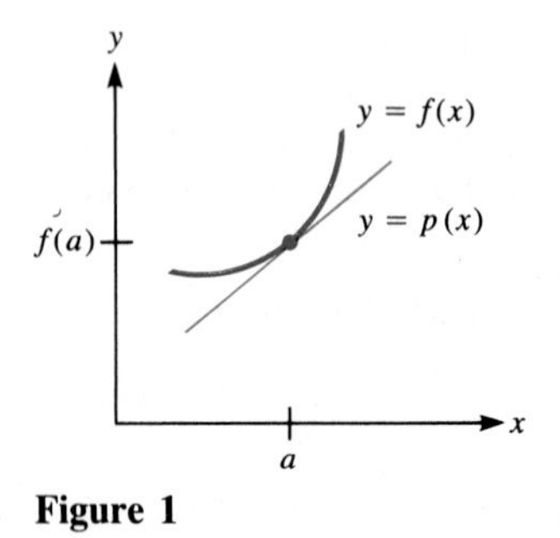

Figure 1

Since $f(a)$ and $f'(a)$ are constants, the function $f(a) + f'(a)(x - a)$ is a polynomial, which we call $p(x)$,

$$p(x) = f(a) + f'(a)(x - a). \tag{1}$$

(The letter "p" reminds us that the function involved is a polynomial.) Figure 1 shows the graph of $y = p(x)$ and $y = f(x)$.

EXAMPLE 1 Find the equation of the tangent line to the graph of $y = x^3$ at the point $(1, 1)$.

SOLUTION In this case, $a = 1$, $f(x) = x^3$, and $f'(x) = 3x^2$. Thus $f(a) = 1^3 = 1$ and $f'(a) = 3 \cdot 1^2 = 3$. By (1), the tangent line at (1, 1) has the equation

$$p(x) = 1 + 3(x - 1).$$

The curves $f(x) = x^3$ and $p(x) = 1 + 3(x - 1)$ are shown in Fig. 2. ■

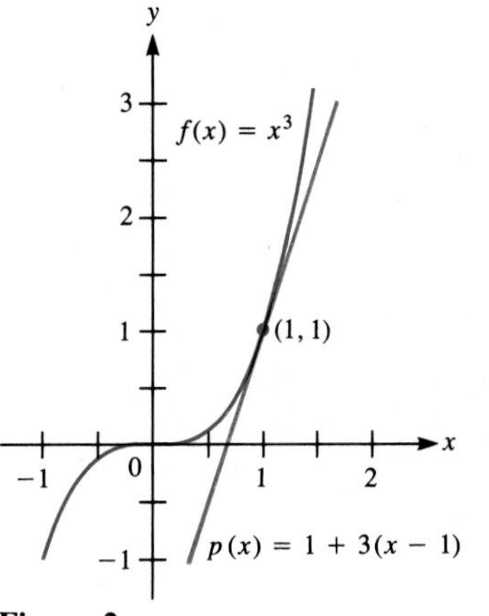

Figure 2

Since the tangent at $P = (a, f(a))$ stays close to the graph of $y = f(x)$ for points on it near P, we may use it as an approximation of that graph.

Now, consider a small number Δx and the input $a + \Delta x$, which is close to a. Then, by (1)

$$p(a + \Delta x) = f(a) + f'(a)[(a + \Delta x) - a]$$

or

$$p(a + \Delta x) = f(a) + f'(a)\,\Delta x. \tag{2}$$

On the other hand, since, by definition, $\Delta f = f(a + \Delta x) - f(a)$, "the change in the function," we have

$$f(a + \Delta x) = f(a) + \Delta f. \tag{3}$$

Figure 3 shows $p(a + \Delta x)$ and $f(a + \Delta x)$.

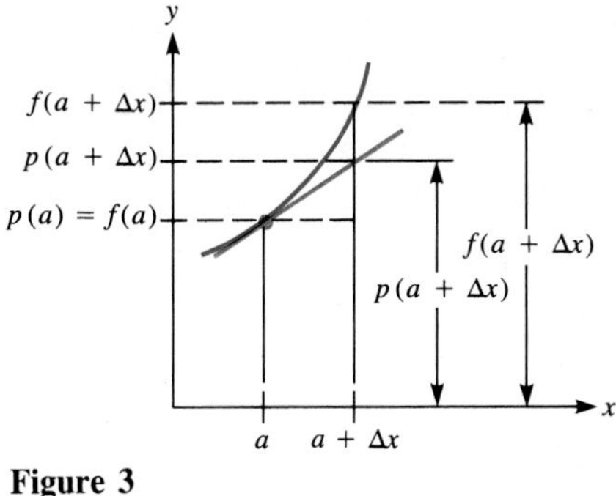

Figure 3

EXAMPLE 2 Compute $f(a + \Delta x)$ and $p(a + \Delta x)$ in case $f(x) = x^3$, $a = 1$, and $\Delta x = 0.1$.

SOLUTION In this case $a + \Delta x = 1 + 0.1 = 1.1$. Thus

$$f(a + \Delta x) = (1.1)^3 = 1.331.$$

Also

$$\begin{aligned} p(a + \Delta x) &= f(a) + f'(a)\Delta x \qquad \text{[by Eq. (2)]} \\ &= 1 + 3(0.1) \\ &= 1.3. \end{aligned}$$

Note that, as expected, $p(1.1)$ provides a good approximation to $f(1.1)$: 1.3 is a good approximation of 1.331. ■

There are two summands in the formula

$$p(a + \Delta x) = f(a) + f'(a)\Delta x.$$

One term is $f(a)$. The other term is $f'(a)\Delta x$, which is small when Δx is small. The small term $f'(a)\Delta x$ is added to $f(a)$ to provide the approximation to $f(a + \Delta x)$, as shown in Fig. 4. Figure 5 contrasts $f'(a)\Delta x$ and Δf.

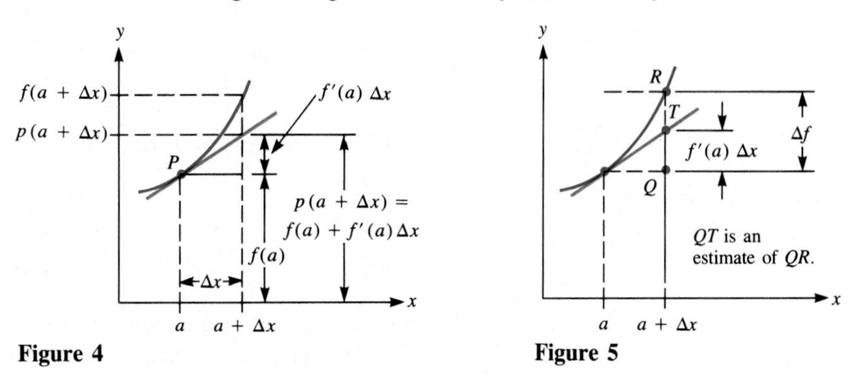

Figure 4 **Figure 5**

The segment QT is a close approximation of QR, which is the actual change in the function $f(x)$, $\Delta f = f(a + \Delta x) - f(a)$. In other words,

$$f'(a)\Delta x \text{ is an estimate of } \Delta f.$$

The expression $f'(a)\Delta x$ is called the "differential"; it is an approximation of the "difference," Δf.

Definition Let $y = f(x)$ be a differentiable function and a be a number in the domain of f. Then $f'(a)\Delta x$ is called the **differential** of f at a and is denoted df or dy.

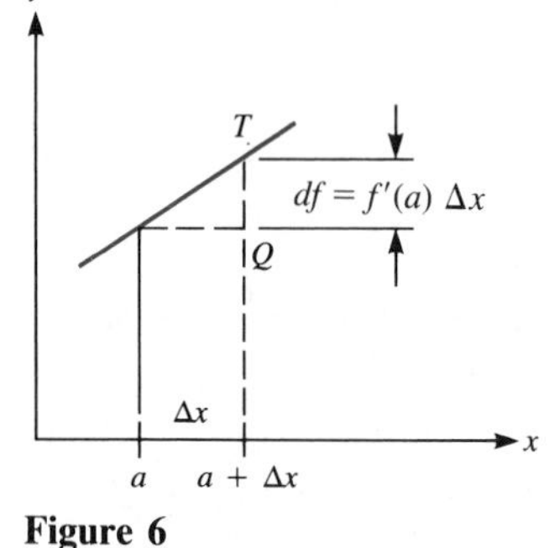

Figure 6

When Δx is small, the differential is small, as shown in Fig. 6. The differential represents *vertical change along the tangent line*. Figure 7 contrasts df and Δf.

EXAMPLE 3 Compute df and Δf for $f(x) = \sqrt{x}$, $a = 9$, and $\Delta x = 0.3$.

SOLUTION Since $f(x) = \sqrt{x}$, $f'(x) = 1/(2\sqrt{x})$. In particular,

$$f'(a) = \frac{1}{2\sqrt{9}} = \frac{1}{6}.$$

Thus

$$df = f'(a)\Delta x$$

$$= \frac{1}{6} \cdot 0.3,$$

so

$$df = 0.05.$$

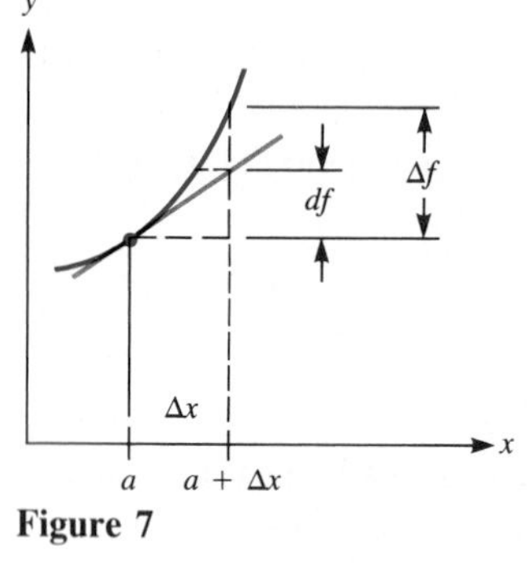

Figure 7

Next, compute

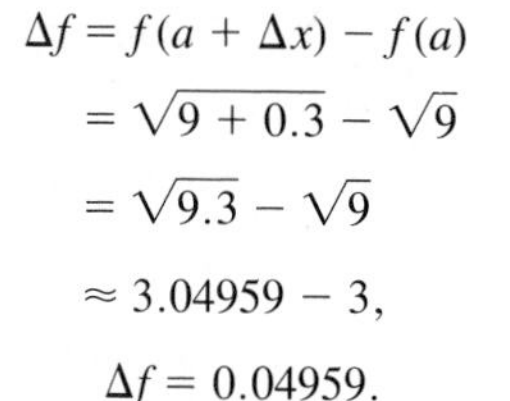

$$\Delta f = f(a + \Delta x) - f(a)$$

$$= \sqrt{9 + 0.3} - \sqrt{9}$$

$$= \sqrt{9.3} - \sqrt{9}$$

$$\approx 3.04959 - 3,$$

so

$$\Delta f = 0.04959.$$

Notice that df is very close to Δf, as was to be expected. ■

Often we will use the symbol x instead of a in the differential. Instead of writing $df = f'(a)\Delta x$ we will write $df = f'(x)\,\Delta x$. For instance, $d(x^3) = 3x^2\,\Delta x$.

If the derivative of a function is known, so is its differential. For example,

$$d(\tan x) = \sec^2 x\,\Delta x,$$

$$d(x^5) = 5x^4\,\Delta x,$$

and

$$d(x) = 1\,\Delta x = \Delta x.$$

$dx = \Delta x$.

Notice that $d(x) = \Delta x$. For this reason it is customary to write Δx also as dx. The differential of f, then, is also written as

$$df = f'(x)dx \qquad \text{or} \qquad dy = f'(x)dx.$$

Thus we can also write that

$$d(\tan x) = \sec^2 x\, dx \qquad \text{and} \qquad d(x^5) = 5x^4\, dx.$$

The origin of the symbol dy/dx

The symbols dy and dx now have meaning individually. It is meaningful to divide both sides of the equation

$$dy = f'(x)\, dx$$

by dx, obtaining

$$dy \div dx = f'(x).$$

This is the origin of the symbol dy/dx for the derivative. It goes back to Leibniz at the end of the seventeenth century when dx denoted a number "vanishingly small."

Using the Differential

As observed earlier, $p(a + \Delta x) = f(a) + f'(a)\, \Delta x$ is a good approximation to $f(a + \Delta x)$ when Δx is small. We have

Approximation formula

$$\boxed{f(a + \Delta x) \approx f(a) + f'(a)\, \Delta x} \tag{4}$$

The output of f at $a + \Delta x$ is approximated by the output of f at a plus the differential $f'(a)\, \Delta x$ (see Fig. 8). Example 4 shows how to use Eq. (4).

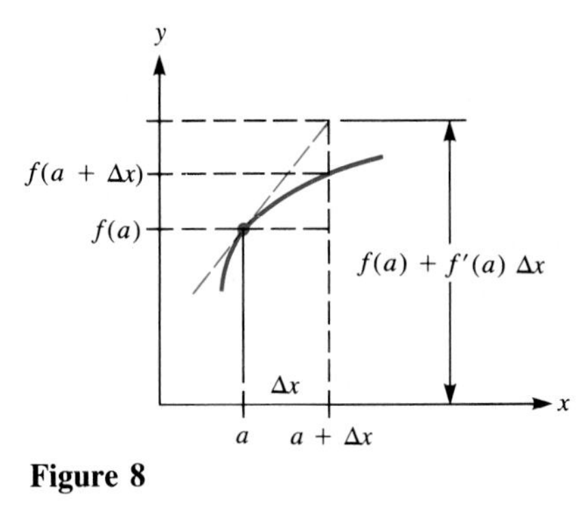

Figure 8

EXAMPLE 4 Use the approximation formula to estimate $\sqrt[3]{29}$.

SOLUTION Because we wish to estimate $\sqrt[3]{29}$, we introduce the cube root function $f(x) = \sqrt[3]{x}$. We know the exact value of $\sqrt[3]{x}$ when $a = 27$, which is near 29, so we use Eq. (4) with $a = 27$ and $\Delta x = 29 - 27 = 2$:

$$\sqrt[3]{27 + 2} \approx \sqrt[3]{27} + f'(27)(29 - 27).$$

Since $f(x) = x^{1/3}$, $f'(x) = \frac{1}{3}x^{-2/3}$. Thus we have

$$\sqrt[3]{29} \approx 3 + \frac{1}{3(27)^{2/3}}(2)$$

or

$$\sqrt[3]{29} \approx 3 + \frac{2}{27} \approx 3 + 0.0741 = 3.0741.$$

As may be checked on a calculator, $\sqrt[3]{29} \approx 3.0723$, rounded to four decimals. ■

The method used in Example 4 amounts to the following general procedure.

How to use a differential to estimate an output of a function

TO ESTIMATE $f(b)$

1 Find a number a near b at which $f(a)$ and $f'(a)$ are easy to calculate.
2 Find $\Delta x = b - a$. (Δx may be positive or negative.)
3 Compute $f(a) + f'(a)\, \Delta x$. This is an estimate of $f(b)$. In short, $f(b) \approx f(a) + (b - a)f'(a)$.

EXAMPLE 5 Use a differential to estimate $\sqrt{61}$.

SOLUTION The object is to estimate the value of the square root function $f(x) = \sqrt{x}$ at the input $x = 61$. (Here $b = 61$.)

In this case $f(64)$ is known. (So use $a = 64$.) We have

$$f(64) = \sqrt{64} = 8 \quad \text{and} \quad f'(64) = \frac{1}{2\sqrt{64}} = \frac{1}{16}.$$

Since $61 = 64 - 3$, Δx is -3. Therefore,

$$\begin{aligned}\sqrt{61} = f(64-3) &\approx f(64) + df \\ &= f(64) + f'(64)(-3) \\ &= 8 + \tfrac{1}{16}(-3) = 7.8125.\end{aligned}$$

Thus $$\sqrt{61} \approx 7.8125.$$

(A calculator shows that, to four decimal places, $\sqrt{61} \approx 7.8102$.) ■

In the next example the differential provides a quick estimate of "relative error." If you make an error E in estimating a quantity Q, the *relative error* is E/Q, which is often expressed as a percent. In this case we are not interested in the error E itself, but only how large it is when compared to the quantity we are measuring.

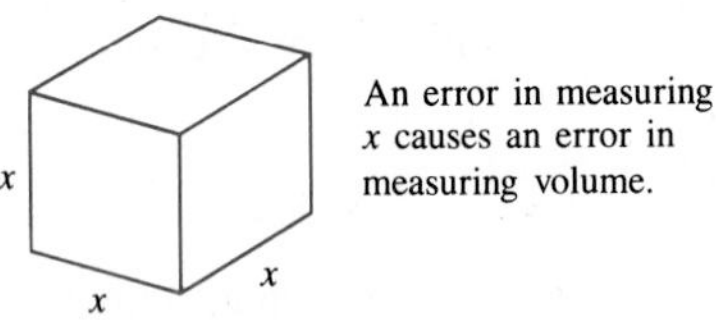

An error in measuring x causes an error in measuring volume.

Figure 9

EXAMPLE 6 The side of a cube is measured with an error of at most 1 percent. What percent error may this cause in calculating the volume of the cube? (See Fig. 9.)

SOLUTION Let x be the length of a side of the cube and V its volume. Let dx denote the possible error in measuring x. The relative error

$$\frac{dx}{x}$$

is at most 0.01 in absolute value. That is, $|dx|/x \le 0.01$.

Estimating relative error

The differential dV is an estimate of the actual error in calculating the volume. Thus

$$\frac{dV}{V}$$

is an estimate of the relative error in the volume.

Since

$$dV = d(x^3) = 3x^2\,dx,$$

it follows that

$$\begin{aligned}\frac{dV}{V} &= \frac{3x^2\,dx}{x^3} \\ &= 3\frac{dx}{x}.\end{aligned}$$

Therefore, the relative error in the volume is about three times the relative error in measuring the side, about 3 percent. ■

The Best Fit

At the beginning of this section we showed that

$$p(x) = f(a) + f'(a)(x - a)$$

is the equation of the tangent line to the graph of $y = f(x)$ at the point $(a, f(a))$. Again, since the tangent line resembles the graph of $y = f(x)$ for x near a, we expect $p(x)$ to be a good approximation of $f(x)$ when x is near a.

In fact at the input a, the functions $p(x)$ and $f(x)$ have the same output.

$$p(a) = f(a) + f'(a)(a - a)$$

$$= f(a) + f'(a) \cdot 0,$$

hence $$p(a) = f(a).$$

Let's also compare the derivatives of $p(x)$ and $f(x)$ at a:

$$p'(x) = [f(a) + f'(a)(x - a)]'$$

$$= 0 + f'(a)(1),$$

hence $$p'(a) = f'(a).$$

Thus $p(x)$ and $f(x)$ have the same first derivative at a.

In short, $p(a) = f(a)$ and $p'(a) = f'(a)$. The two equations give an *algebraic* description of the polynomial $p(x)$. It is the only polynomial of degree at most one that has the same output as f does at a and the same derivative as f does at a. Theorem 1 justifies this assertion.

Theorem 1 Let f be differentiable at a. Let c and k be constants. Assume that the polynomial $q(x) = c + kx$ has the property that $q(a) = f(a)$ and $q'(a) = f'(a)$. Then

$$q(x) = f(a) + f'(a)(x - a).$$

Proof We wish to find c and k such that $q(a) = f(a)$ and $q'(a) = f'(a)$.

First of all, $q(a) = c + ka$ and $q'(a) = k$. Thus

$$c + ka = f(a) \tag{5}$$

and $$k = f'(a). \tag{6}$$

We already have k, $k = f'(a)$.

Substituting (6) into (5), we find that

$$c + f'(a)a = f(a).$$

or $$c = f(a) - f'(a)a. \tag{7}$$

Using Eqs. (6) and (7), we conclude that

$$q(x) = c + kx$$

$$= f(a) - f'(a)a + f'(a)x$$

or $$q(x) = f(a) + f'(a)(x - a).$$

Hence $q(x) = p(x)$. ■

Since a polynomial of degree at most one is called a linear function (its graph is a line), $p(x)$ is called the *linearization* of $f(x)$ at a. On the one hand, it can be viewed geometrically as describing the tangent line to $y = f(x)$ at $(a, f(a))$. On the other hand, it can be viewed algebraically as the unique linear polynomial whose output and first derivative at a coincide with those of $f(x)$ at a. In short, $p(x)$ can be thought of as the linear polynomial whose graph best fits the graph of f near $(a, f(a))$.

EXAMPLE 7 Find the linearization of $f(x) = \tan x$ at $\pi/4$.

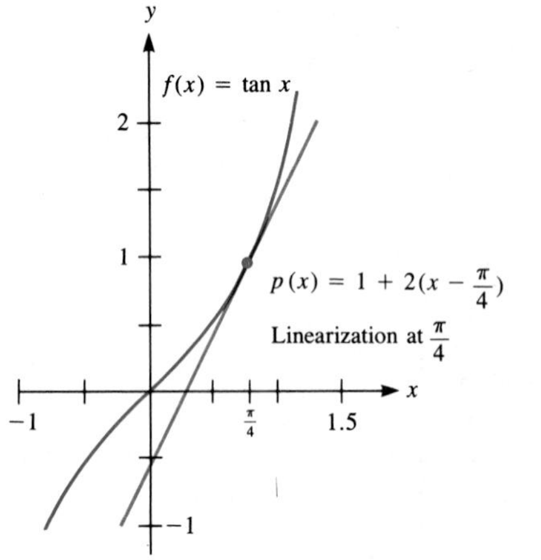

SOLUTION In this case $f'(x) = \sec^2 x$. Since the linearization is at $a = \pi/4$, we compute

$$f\left(\frac{\pi}{4}\right) = \tan\frac{\pi}{4} = 1$$

and

$$f'\left(\frac{\pi}{4}\right) = \sec^2\frac{\pi}{4} = (\sqrt{2})^2 = 2.$$

Thus the linearization of $\tan x$ at $\pi/4$ is

$$p(x) = 1 + 2\left(x - \frac{\pi}{4}\right).$$

It is graphed in Fig. 10. ■

$f(x) = \tan x$; $p(x) = 1 + 2(x - \frac{\pi}{4})$; Linearization at $\frac{\pi}{4}$

Figure 10

Section Summary

This section introduced the differential of a function,

$$df = f'(x)\,dx.$$

Using the fact that the tangent line to the graph of $y = f(x)$ at $(a, f(a))$ resembles the graph itself, at least for points near $(a, f(a))$, we obtained the estimate

$$f(x) \approx f(a) + f'(a)(x - a)$$

for x near a. The polynomial

$$p(x) = f(a) + f'(a)(x - a)$$

is called the linearization of $f(x)$ at a. Its graph is the tangent at $(a, f(a))$. It is the sum of $f(a)$ and $f'(a)(x - a)$, which is small when x is near a.

EXERCISES FOR SEC. 4.9: THE DIFFERENTIAL AND LINEARIZATION

In Exercises 1 to 6 compute df and Δf for the given functions and values of x and dx and represent them on graphs of the functions.

1 x^2 at $x = 1$ and $dx = 0.3$ 2 x^3 at $x = \frac{1}{2}$ and $dx = 0.1$

3 $\sqrt{x}$ at $x = 9$ and $dx = -2$

4 $\sqrt[3]{x}$ at $x = 27$ and $dx = -4$

5 $\tan x$ at $x = \pi/6$ and $dx = \pi/12$

6 $\sin x$ at $x = \pi/3$ and $dx = -\pi/12$

7 Estimate $\sqrt{98}$ using a differential, answering these questions:

(*a*) What function is involved?

(*b*) At what input near 98 is the output of that function (and the output of its derivative) easily computed?

(*c*) What estimate do you get for $\sqrt{98}$?

8 Estimate $\sqrt{26.1}$ using a differential, answering these questions:

(*a*) What function is involved?

(*b*) At what input near 26.1 is the output of that function (and the output of its derivative) easily computed?

(*c*) What estimate do you get for $\sqrt{26.1}$?

In Exercises 9 to 24 use differentials to estimate the given quantities.

9 $\sqrt{119}$ **10** $\sqrt{103}$
11 $\sqrt[3]{25}$ **12** $\sqrt[3]{28}$
13 $\tan\left(\frac{\pi}{4} - 0.01\right)$ **14** $\tan\left(\frac{\pi}{4} + 0.01\right)$
15 $\sin\left(\frac{\pi}{3} - 0.02\right)$ **16** $\sin\left(\frac{\pi}{3} + 0.02\right)$
17 $\sin 0.13$ **18** $\sin 0.3$
19 $1/4.03$ **20** $\sqrt{15.7}$
21 $\sin 32°$ (*Warning:* First translate into radians.)
22 $\sin 2°$ **23** $\cos 28°$ **24** $\tan 3°$

In Exercises 25 to 28 use the approximation formula to establish the given estimates for small values of h.

25 $\frac{1}{1+h} \approx 1 - h$ **26** $(1+h)^{10} \approx 1 + 10h$
27 $\sqrt{1+h} \approx 1 + \frac{h}{2}$ **28** $\sqrt[3]{1+h} \approx 1 + \frac{h}{3}$

In Exercises 29 to 36 calculate the differentials, expressing them in terms of x and dx.

29 $d(1/x^3)$ **30** $d(\sqrt{1+x^2})$
31 $d(\sin 2x)$ **32** $d(\cos 5x)$
33 $d(\csc x)$ **34** $d(\tan x^3)$
35 $d\left(\frac{\cot 5x}{x}\right)$ **36** $d(x^3 \sec^2 5x)$

In Exercises 37 and 38 compute the linearization $p(x)$ for the given functions at the given numbers.

37 $\sqrt{x}$ at (*a*) $a = 1$, (*b*) $a = 2$.
38 $1/x^2$ at (*a*) $a = 2$, (*b*) $a = 3$.
39 Complete the table for $f(x) = \sqrt[3]{x}$ and its linearization at $a = 1$, $p(x)$.

x	$f(x) = \sqrt[3]{x}$	$p(x)$	$f(x) - p(x)$
1.5			
1.1			
1.01			
1.001			
1.0001			

40 Complete the table for $f(x) = 1/x$ and its linearization at $a = 2$, $p(x)$.

x	$f(x) = 1/x$	$p(x)$	$f(x) - p(x)$
1.5			
1.9			
1.95			
1.99			
1.999			

41 The period T of a pendulum is proportional to the square root of its length l. That is, there is a constant k such that $T = k\sqrt{l}$. If the length is measured with an error of at most p percent, use differentials to estimate the possible error in calculating the period.

42 The side of a square is measured with an error of at most 5 percent. Estimate the largest percent error this may induce in the measurement of the area.

---

43 Let $f(x) = x^2$, the area of a square of side x, shown in Fig. 11.
(*a*) Compute df and Δf in terms of x and Δx.
(*b*) In the square in the diagram, shade the part whose area is Δf.
(*c*) Shade the part of the square in (*b*) whose area is df.

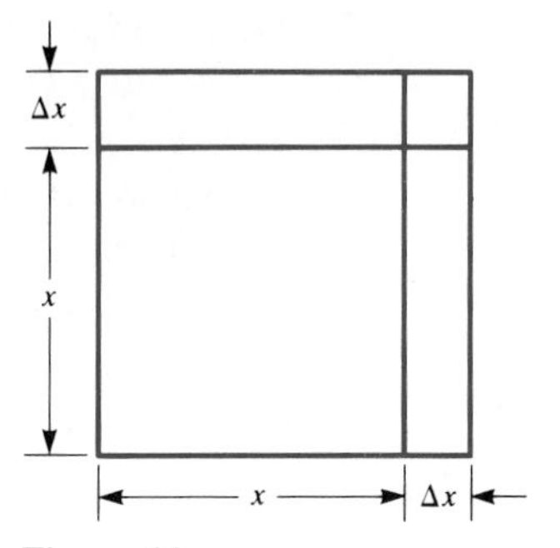

Figure 11

44 (See Exercise 43.) Let $f(x) = x^3$, the volume of a cube of side x.
(*a*) Compute df and Δf in terms of x and Δx.
(*b*) Draw a diagram analogous to Fig. 11 showing cubes of sides x and $x + \Delta x$.
(*c*) Indicate in the diagram in (*b*) the part whose volume is df.

45 Prove that if f and g are two differentiable functions, then
(*a*) $d(f - g) = df - dg$
(*b*) $d(fg) = f\,dg + g\,df$
(*c*) $d\left(\frac{f}{g}\right) = \frac{g\,df - f\,dg}{g^2}$.

46 (*a*) Estimate $\sin\theta$, $\cos\theta$, and $\tan\theta$ for small θ using differentials.
(*b*) Explain why $\cos\theta \approx 1 - (\theta^2/2)$ is a better estimate for $\cos\theta$ than the one in (*a*). [*Hint:* Recall the identity $(1 - \cos\theta)(1 + \cos\theta) = \sin^2\theta$.]

✱ 47 The differential df is an approximation of the actual change Δf. Examine $\lim_{\Delta x \to 0} \Delta f/df$, as follows. Assume that $f'(a) \neq 0$, so that df is not identically 0.
(*a*) Pick a function $f(x)$, a particular number a, and small values of Δx. Using your calculator, compute $\Delta f/df$ for those values.

(b) What do you think happens to $\Delta f/df$ as $\Delta x \to 0$?
(c) Prove your conjecture in (b).

48 The accompanying table lists values of the derivative of a function f at certain inputs.

x	1	1.1	1.2	1.3	1.4	1.5
$f'(x)$	0.7	0.5	0.4	0.2	0.3	0.4

(a) If $f(1) = 3$, estimate $f(1.6)$ as well as you can.
(b) If $f(x)$ is the position of a moving particle at time x, estimate how far the particle moves during the time interval $[1, 1.6]$.

49 Assume that $f(x)$ has continuous first and second derivatives for all x. Let a and x be fixed numbers.
(a) Show that there is a number c_1 between a and x such that $f(x) = f(a) + f'(c_1)\,(x - a)$.
(b) Show that there is a number c_2 between a and c_1 such that

$$f'(c_1) = f'(a) + f''(c_2)(c_1 - a).$$

(c) Show that

$$f(x) = f(a) + f'(a)(x - a) + f''(c_2)(c_1 - a)(x - a).$$

(d) Deduce that $|\Delta f - df| \le |f''(c_2)|(\Delta x)^2$, where $\Delta x = x - a$. This argument suggests that when using df to estimate Δf, the error, $|\Delta f - df|$ approaches 0 as the *square* of Δx.

4.10 THE SECOND DERIVATIVE AND GROWTH OF A FUNCTION

Initially the car is at rest.

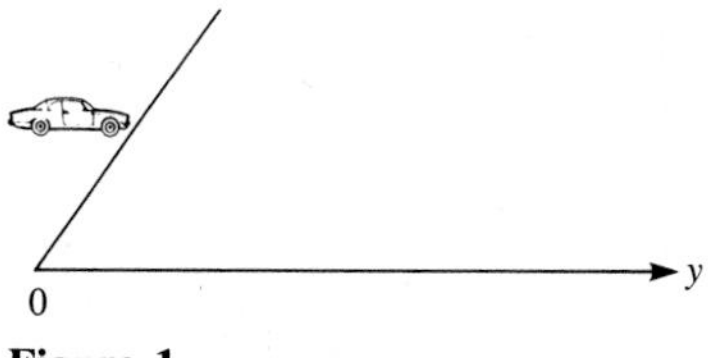

Figure 1

The car has acceleration A.

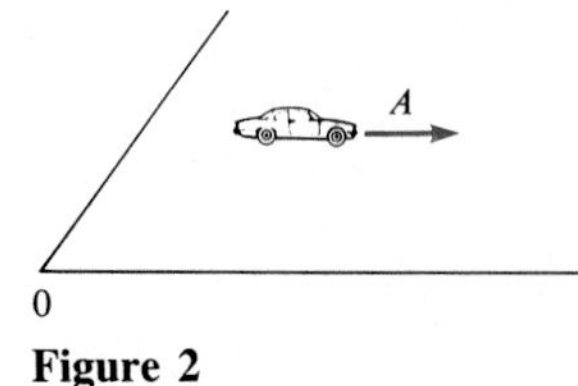

Figure 2

The car after t seconds at constant acceleration A

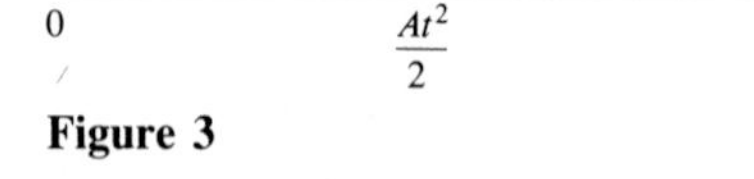

Figure 3

This section shows how the size of the second derivative of a function can tell us something about how rapidly that function can grow. This information will be used to examine the error in using the differential approximation.

The Growth Theorem

The basic idea is this. Say that you know that $f(a) = 0$, $f'(a) = 0$, and $f''(x)$ is small. Since $f''(x)$ tells how quickly $f'(x)$ changes, $f'(x)$ doesn't change very fast. Combining this with the information that $f'(a) = 0$, we see that $f'(x)$ stays small for x near a. But $f'(x)$ is the rate at which $f(x)$ changes, so $f(x)$ changes slowly also. Since $f(a) = 0$, $f(x)$ remains small when x is near a. So, in this way, knowing that $f''(x)$ is small gives us an insight into the function $f(x)$ itself—it can't grow too swiftly. Now we will make this observation precise.

Imagine that you are in the driver's seat of a car. At time $t = 0$ on your stopwatch you are at the origin of the y axis, pointing in the direction of positive y, and at rest. At that moment you start the car and press the accelerator to the floor, giving the car a constant acceleration of A feet per second per second. Where will you be on the y axis t seconds later? (See Figs. 1 and 2, where the y axis is drawn horizontally to save space.)

To answer this, use the general formula from Sec. 4.3,

$$y = \tfrac{1}{2}At^2 + v_0t + y_0.$$

We use A for acceleration because a will be needed elsewhere.

Since your initial velocity is 0, $v_0 = 0$; since you start at the origin, $y_0 = 0$. Thus

$$y = \tfrac{1}{2}At^2.$$

After t seconds you will be at the point on the y axis whose coordinate is $At^2/2$. The distance traveled in the first t seconds equals "one half the acceleration times the square of the time traveled." (See Fig. 3.)

Now, assume that you had neglected to set your stopwatch to 0. Instead, when you started at the origin of the y axis, the watch read t_0 seconds. Then, when it read t seconds, $t > t_0$, you would have traveled

$$y = \tfrac{1}{2}A(t - t_0)^2 \qquad \text{feet}$$

since the time spent traveling is $t - t_0$ seconds rather than t seconds.

Varying acceleration

But what if you don't keep the accelerator fixed throughout the time interval $[t_0, t]$? What if you have a nervous foot?

Imagine that during the time interval your minimum acceleration is m feet per second per second and your maximum acceleration is M feet per second per second. Then it is reasonable to expect that during that time interval you traveled at least

$$\tfrac{1}{2}m(t - t_0)^2 \qquad \text{feet}$$

and at most

$$\tfrac{1}{2}M(t - t_0)^2 \qquad \text{feet.}$$

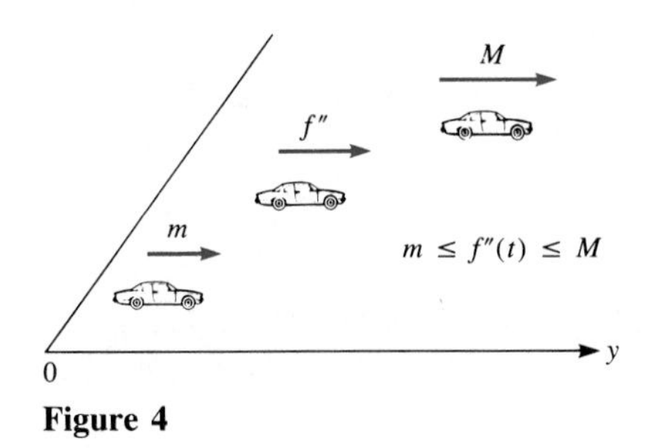

Figure 4

If you were racing against two cars, one with constant acceleration m feet per second per second and another with constant acceleration M feet per second per second, you would come in second. See Fig. 4.

Let's translate these observations into the language of functions.

> **Theorem 1** *The Growth Theorem.* Let $f(t)$ be a function with a continuous second derivative in some closed interval containing the number t_0. Assume that
>
> $$f(t_0) = 0, f'(t_0) = 0,$$
>
> and that there are two constants, m and M, such that
>
> $$m \le f''(t) \le M$$
>
> for all t in the interval. Then for any number t in the interval,
>
> $$\tfrac{1}{2}m(t - t_0)^2 \le f(t) \le \tfrac{1}{2}M(t - t_0)^2.$$

What does this mean about the graph of $y = f(t)$? First, since $f(t_0) = 0$, the graph passes through the point $(t_0, 0)$. Second, since $f'(t_0) = 0$, the tangent is horizontal there. Third, the graph lies between the parabolas,

$$y = \tfrac{1}{2}m(t - t_0)^2 \qquad \text{and} \qquad y = \tfrac{1}{2}M(t - t_0)^2.$$

(See Fig. 5.)

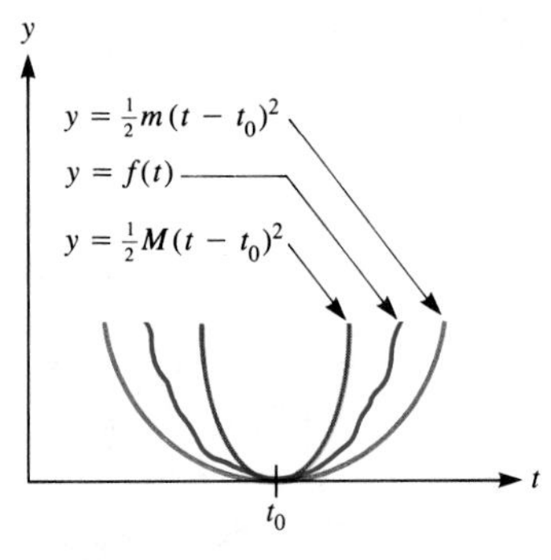

The graph of f lies between two parabolas.
Figure 5

EXAMPLE 1 Let f be a function such that $f(2) = 0, f'(2) = 0$, and $f''(t) \le 3$ for all t in $[2, 5]$. How large can $f(5)$ be?

SOLUTION The formula

$$f(t) \le \tfrac{1}{2}M(t - t_0)^2$$

in this case becomes

$$f(5) \le \tfrac{1}{2} \cdot 3(5 - 2)^2.$$

Thus $f(5) \le 3 \cdot 3^2/2 = \frac{27}{2}$. ■

The important idea in "the second derivative and the growth of a function" is that if $f(a) = 0$ and $f'(a) = 0$ (that is, the function is "flat" near a), then the size of the second derivative can be used to find a bound on the values of the function $f(x)$ itself.

This idea will be generalized in Chap. 11. Its mathematical justification (without reference to automobiles) is found in Exercises 14 and 15.

The Error in Using the Differential

Let $f(x)$ be a function with continuous first and second derivatives. Let a be a number in the domain of $f(x)$. To estimate $f(x)$ when x is near a one uses the linear approximation

$$p(x) = f(a) + f'(a)(x - a).$$

This approximation adds the differential $f'(a)(x - a)$ to $f(a)$. How good is this approximation when x is near a? That is, as x approaches a, *how rapidly does the difference between $f(x)$ and $p(x)$ approach* 0?

To answer this question we examine the *error:*

$$E(x) = f(x) - p(x) = f(x) - [f(a) + f'(a)(x - a)].$$

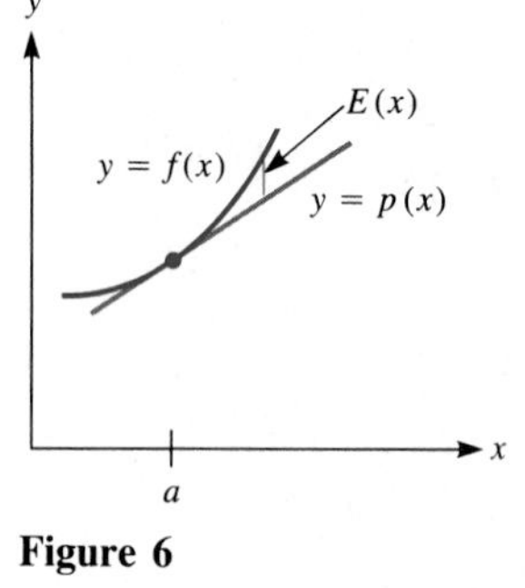

Figure 6

(See Fig. 6.) We want to know how rapidly $E(x)$ approaches 0 as $x \to a$. To put it another way, we want to find out how $E(x)$ grows as $x - a$ increases.

To decide, let us examine the behavior of $E(x)$ and its derivatives at and near a. To begin,

$$E(x) = f(x) - f(a) - f'(a)(x - a),$$

$$E'(x) = f'(x) - f'(a),$$

and

$$E''(x) = f''(x).$$

Thus $\qquad E(a) = 0 \quad \text{and} \quad E'(a) = 0.$

Because $E(a)$ and $E'(a)$ are equal to 0, we can use the growth theorem to say something about the size of $E(x)$ itself, as Theorem 2 will show.

Theorem 2 *Error in using the differential approximation.* Let $f(x)$ be a function with continuous first and second derivatives in some interval I around a. Let the minimum of $f''(x)$ for x in I be m; let the maximum of $f''(x)$ for x in I be M. Then for all x in I,

$$\frac{m(x-a)^2}{2} \le E(x) \le \frac{M(x-a)^2}{2}. \tag{1}$$

Proof We know that $E(a) = 0$ and $E'(a) = 0$. Moreover, $E''(x) = f''(x)$. Thus m and M are also the minimum and maximum of $E''(x)$ for x in I. The growth theorem therefore implies (1). ■

Remark: When x is very near a, $E''(x)$ varies only slightly. So, in a short interval around a, the minimum and maximum of $E''(x)$ are both near $E''(a)$. We would therefore expect the error to be roughly $E''(a)(x - a)^2/2$. This means that the error decreases as the *square* of $x - a$. (If $f''(a) \neq 0$.)

Let's check this assertion for $f(x) = \sqrt{x}$ and $a = 1$. In this case, $f'(x) = 1/(2\sqrt{x})$ and therefore $f'(1) = \frac{1}{2}$. The linear approximation of $f(x)$ near $a = 1$ is

$$p(x) = f(1) + f'(1)(x - 1) = 1 + \tfrac{1}{2}(x - 1)$$

and

$$E(x) = \sqrt{x} - [1 + \tfrac{1}{2}(x - 1)].$$

The following table shows how rapidly $E(x)$ approaches 0 as $x \to 1$ and compares it to $(x - 1)^2$:

x	$E(x)$	$(x-1)^2$	$E(x)/(x-1)^2$
2	$\sqrt{2} - [1 + \frac{1}{2}(2 - 1)] \approx -0.08579$	1	-0.086
1.5	$\sqrt{1.5} - [1 + \frac{1}{2}(1.5 - 1)] \approx -0.0253$	0.25	-0.101
1.1	$\sqrt{1.1} - [1 + \frac{1}{2}(1.1 - 1)] \approx -0.00119$	0.01	-0.119
1.01	$\sqrt{1.01} - [1 + \frac{1}{2}(1.01 - 1)] \approx -0.0000124$	0.0001	-0.124

The final column in the table shows that $E(x)/(x - 1)^2$ is nearly constant. That's what it means when we say "$E(x)$ approaches 0 as the square of $(x - 1)$."

Since $E(x)$ is approximately $f''(1)(x - 1)^2/2$ when x is near 1, $E(x)/(x - 1)^2$ should be near $f''(1)/2$ when x is near 1. Just as a check, let us compute $f''(1)/2$. We have

$$f'(x) = \frac{1}{2\sqrt{x}}$$

and

$$f''(x) = -\tfrac{1}{4}x^{-3/2}.$$

Thus $f''(1)/2 = (-1/4)/2 = -1/8 = -0.125$, which is certainly consistent with the final column of the table.

If you reduce $x - 1$ by a factor of, say, 10, you would expect the error to shrink by a factor of $10^2 = 100$.

Section Summary

Let $f(x)$ have first and second derivatives throughout an open interval that contains a. Assume that $f(a) = 0$, $f'(a) = 0$, and that $m \leq f^{(2)}(x) \leq M$, for two fixed numbers m and M. Then

$$\frac{m(x - a)^2}{2} \leq f(x) \leq \frac{M(x - a)^2}{2}.$$

This "growth theorem" was applied to the error in using the linear approximation, $E(x) = f(x) - p(x)$, to show that this error shrinks as the square of $\Delta x = x - a$.

EXERCISES FOR SEC. 4.10: THE SECOND DERIVATIVE AND GROWTH OF A FUNCTION

1 Initially a car is sitting at rest. The driver starts the car and begins to drive, with an acceleration never permitted to exceed 4 feet per second per second. What can you say about the distance the car travels in the first 30 seconds?

2 A car starts from rest and travels for 100 seconds, its acceleration never exceeding 10 feet per second per second. What can you say about the distance traveled after 100 seconds?

3 A car starts from rest and travels for 4 hours. Its acceleration is always at least 5 miles per hour per hour, but never exceeds 12 miles per hour per hour. What can you say about the distance traveled after 4 hours?

4 A car starts from rest and travels for 6 hours. Its acceleration is always at least 4.1 miles per hour per hour, but never exceeds 15.5 miles per hour per hour. What can you say about the distance traveled after 6 hours?

In Exercises 5 to 8 the functions possess first and second derivatives.

5 What can be said about $f(3)$ if $f(1) = 0$, $f'(1) = 0$, and $f''(x) \le 4$ for all x?

6 What can be said about $f(5)$ if $f(2) = 0$, $f'(2) = 0$, and $f''(x) \le \frac{2}{3}$ for all x?

7 What can be said about $f(2)$ if $f(1) = 0$, $f'(1) = 0$, and $2.5 \le f''(x) \le 2.6$ for all x?

8 What can be said about $f(4)$ if $f(1) = 0$, $f'(1) = 0$, and $2.9 \le f''(x) \le 3.1$ for all x?

9 Let $f(x) = \sqrt{x}$.

(a) What is the linear approximation, $p(x)$, to $\sqrt{x}$ at $x = 4$?

(b) Fill in this table:

x	$E(x) = f(x) - p(x)$	$(x-4)^2$	$E(x)/(x-4)^2$
5			
4.1			
4.01			
3.99			

(c) Compute $f''(4)/2$. (It should be consistent with the fourth column of the table in (b).)

10 Like Exercise 9 for the linear approximation of $\sqrt{x}$ at $x = 3$; use $x = 4, 3.1, 3.01, 2.99$.

In Exercises 11 and 12, assume that f has continuous first and second derivatives.

11 Assume $f(x)$ has the property that $4 \le f^{(2)}(x) \le 5$ for all x.

(a) What can be said about the error in using $p(x) = f(2) + f'(2)(x-2)$ to approximate $f(x)$?

(b) How small should $x - 2$ be to be sure that the absolute value of the error is less than or equal to 0.01?

12 Let $f(x)$ be a certain function. The error in using $p(x) = f(3) + f'(3)(x-3)$ to approximate $f(3.1)$ is 0.02.

(a) What would you expect the error to be when using $p(3.01)$ to approximate $f(3.01)$?

(b) Estimate $f''(3)$.

---

13 Let f have a continuous second derivative for all x. Assume that $f(0) = 0$, $f'(0) = 0$, and $f(3) = 4$.

(a) Show that there is a number c_1 such that $f'(c_1) = \frac{4}{3}$.

(b) Show that there is a number c_2 such that $f''(c_2) = \frac{8}{9}$.

Exercises 14 and 15 outline a mathematical proof of the growth theorem. For the sake of simplicity, we consider only $t_0 = 0$ and $t > 0$. We will show that if $f(0) = 0$, $f'(0) = 0$, and $f''(t) \le M$, then $f(t) \le Mt^2/2$. [A similar argument shows that if $f(0) = 0$, $f'(0) = 0$, and $f''(t) \ge m$, then $f(t) \ge mt^2/2$.]

14 Let $g(t) = f'(t) - Mt$.

(a) Show that $g(0) = 0$ and $g'(t) \le 0$.

(b) Show that for $t \ge 0$, $g(t) \le 0$.

(c) Deduce that for $t \ge 0$, $f'(t) \le Mt$.

15 Let $h(t) = f(t) - Mt^2/2$.

(a) Show that $h(0) = 0$ and $h'(t) \le 0$.

(b) Deduce that for $t \ge 0$, $f(t) \le Mt^2/2$.

16 Let f have a continuous second derivative for all x. Show that if $b > a$ there is a number c in $[a, b]$ such that

$$f(b) = f(a) + f'(a)(b-a) + \frac{f^{(2)}(c)}{2}(b-a)^2.$$

(This equation holds also if $b \le a$, with some c between a and b.)

17 With the aid of the equation in Exercise 16, we can show that the error in Newton's method diminishes rapidly. Let x_1 be an estimate of the root r and let x_2 be the second estimate, obtained by Newton's method. Assume $f'(x_1) \ne 0$.

(a) In the equation for $f(b)$ in Exercise 16, replace b by r and a by x_1 to show that

$$x_2 - r = \frac{f^{(2)}(c)}{2f'(x_1)}(r - x_1)^2,$$

where c is between x_1 and r.

(b) Assume that $x_1 > r$ and that $f'(x)$ and $f''(x)$ are positive for x in $[r, x_1]$. Assume also that $f''(x) \le M_2$ and $f'(x) \ge M_1 > 0$ for all x in $[r, x_1]$. Indicate on a diagram where the numbers $x_2, x_3, \ldots$ are situated. Then use (a) to discuss how the error, $r - x_n$, behaves as n increases.

18 (a) Why is

$$f(a) + f'(a)(x-a) + \frac{f^{(2)}(a)}{2}(x-a)^2$$

probably a better approximation to $f(x)$, when x is near a, than $p(x) = f(a) + f'(a)(x - a)$?

(*b*) Check whether it is, using a function $f(x)$ of your choice and particular numbers a and x, near a.

19 (*a*) Find the polynomial $q(x) = c + dx + ex^2$ that approximates $f(x)$ at $x = a$ in the sense that $q(a) = f(a)$, $q'(a) = f'(a)$, and $q''(a) = f''(a)$.

(*b*) Compare $q(x)$ to the polynomial in Exercise 18(*a*).

4.S SUMMARY

This chapter applied the derivative to graphs, motion, rates of change, extrema, estimates, and growth of a function.

Section 4.1 provided the foundation for the chapter. The theorem of the interior extremum showed that there is a close tie between a maximum or minimum value and a derivative being equal to 0, although the two conditions are not equivalent. The mean-value theorem, which followed from Rolle's theorem, showed that a function whose derivative is positive is increasing and that two functions that have the same derivative differ only by a constant. The diagram on the next page shows how the theorems in Sec. 4.1 fit together.

Section 4.2 described a procedure for graphing functions that involves intercepts, asymptotes, and the first derivative.

The higher derivatives were introduced in Sec. 4.3 and used in the study of motion. The second derivative was shown to represent acceleration, and motion under constant acceleration was described completely. (The position turned out to be a second-degree polynomial in time t.)

Related rates were discussed in Sec. 4.4.

Section 4.5 introduced the concepts of concave upward, concave downward, inflection point, and inflection number.

Newton's method was presented in Sec. 4.6 as a tool for finding approximate solutions of equations. The recursion formula is $x_{i+1} = x_i - f(x_i)/f'(x_i)$. Keep in mind that the process occasionally breaks down.

Section 4.7 applied the techniques for finding extrema to specific problems such as minimizing cost or maximizing volume.

When functions are defined implicitly rather than explicitly, the chain rule can be used to find their derivatives. Section 4.8 explained the technique of implicit differentiation.

The differential was treated in Sec. 4.9. It is related to the linearization of a function at a given point. Its application as a tool for approximation was also discussed.

Section 4.10 explained how the second derivative can be used to place limits on the possible growth of a function. The results were stated in the growth theorem, which helped us analyze the error in using the differential.

Vocabulary and Symbols

theorem of the interior extremum
chord of f
Rolle's theorem
mean-value theorem
increasing (decreasing) function
critical number
critical point
relative (local) maximum or minimum
global maximum or minimum
first-derivative test for local extremum
second derivative d^2y/dx^2, y'', f'', $f^{(2)}$, $D^2(f)$
higher derivatives d^ny/dx^n, $f^{(n)}$, $D^n(f)$
acceleration
related rates
Newton's formula
concave upward, concave downward
inflection number
inflection point
second-derivative test for local extremum
implicit function
implicit differentiation
constraint
differential dy, df
linearization
growth theorem

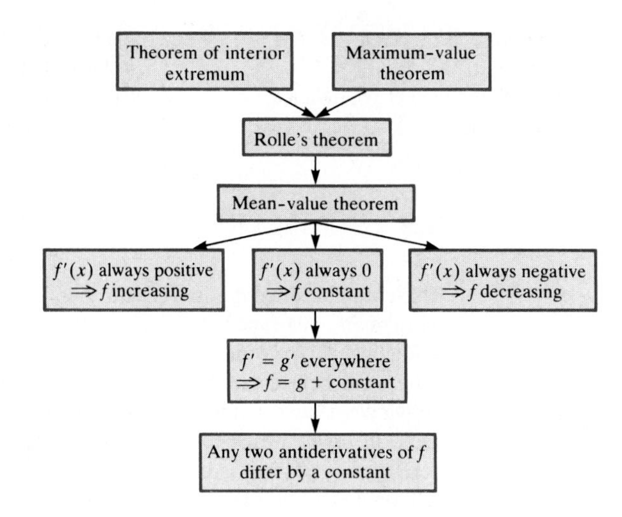

Key Facts

THEOREM OF THE INTERIOR EXTREMUM

("Maximum" case) Let f be defined at least on (a, b). Let c be a number in (a, b) such that $f(c) \geq f(x)$ for all x in (a, b). If f is differentiable at c, then $f'(c) = 0$.

The "minimum" case is similar.

ROLLE'S THEOREM

Let f be continuous on $[a, b]$ and differentiable on (a, b). If $f(a) = f(b)$, then there is at least one number c in (a, b) such that $f'(c) = 0$.

Informally, Rolle's theorem asserts that if a graph of a differentiable function has a horizontal chord, then it has a horizontal tangent line.

MEAN-VALUE THEOREM

Let f be continuous on $[a, b]$ and differentiable on (a, b). Then there is at least one number c in (a, b) such that

$$f'(c) = \frac{f(b) - f(a)}{b - a}.$$

Informally, the mean-value theorem asserts that for any chord on the graph of a differentiable function, there is a tangent line parallel to it.

The conclusion of the mean-value theorem may also be written as

$$f(b) = f(a) + f'(c)(b - a),$$

for some number c in (a, b).

INFORMATION PROVIDED BY f' AND f''

Where f' is positive, f is increasing.
Where f'' is positive, f is concave upward.
Where f' is negative, f is decreasing.
Where f'' is negative, f is concave downward.
Where $f' = 0$, f may have an extremum.
Where $f'' = 0$, f may have an inflection point.
First-derivative test for local maximum at c: $f'(c) = 0$ and f' changes from positive to negative at c.
Second-derivative test for local maximum at c: $f'(c) = 0$ and $f''(c)$ is negative.
First-derivative test for local minimum at c: $f'(c) = 0$ and f' changes from negative to positive at c.
Second-derivative test for local minimum at c: $f'(c) = 0$ and $f''(c)$ is positive.
[If $f'(c)$ or $f''(c)$ does not exist, it is best to study the behavior of $f(x)$ and $f'(x)$ for x near c.]

Two functions, defined over the same interval, with equal derivatives, differ by a constant. [From this it follows, for instance, that if $F'(x) = 2x$, then $F(x)$ must be of the form $x^2 + C$ for some constant C.] This fact will be of use in the next chapter where antiderivatives will be needed.

The second derivative records acceleration, the rate of change of velocity. If the acceleration is constant, the position is given by the formula

$$y = \frac{a}{2}t^2 + v_0t + y_0,$$

where a = acceleration, v_0 = initial velocity, and y_0 = initial position.

The following table summarizes the procedure for graphing f in terms of questions about f, f', and f'':

How to graph using f, f', f''
Is f even or odd?
What are the intercepts?
What are the critical numbers?
Where is the function increasing? decreasing?
Are there any local maxima or minima?
Where is the second derivative positive? negative? zero?
Where is the curve concave upward? concave downward?
Are there any inflection points?
Are there any vertical or horizontal asymptotes?
What happens when $\|x\| \to \infty$?

To find the derivative of a function given implicitly, differentiate the defining equation, remembering that the chain rule may be needed. Differentiation of the resulting equation will then give an equation for the second derivative.

The quantity $df = f'(x)\ \Delta x$ [or $f'(x)\ dx$], the change along a tangent line, is an estimate of Δf, the change along the curve. When Δx is small, Δf and df will be small and their ratio will be near 1.

If you know $f(a)$ and $f'(a)$ and if x is near a, then you can estimate $f(x)$ by

$$f(x) \approx f(a) + f'(a)(x - a).$$

The term $f'(a)(x - a)$ is the differential. The entire expression, $f(a) + f'(a)(x - a)$, is the linearization of the function f at a. It is denoted $p(x)$ and its graph is the tangent line to the graph of f at the point $(a, f(a))$.

This amounts to the same thing as the approximations

$$f(x + \Delta x) \approx f(x) + f'(x)\,\Delta x \qquad \text{or} \qquad f(x + \Delta x) \approx f(x) + df.$$

GUIDE QUIZ ON CHAP. 4: APPLICATIONS OF THE DERIVATIVE

1 (*a*) State all the assumptions in the mean-value theorem.
(*b*) State the conclusion of the mean-value theorem.

2 What does each of the following imply about the graph of a function?
(*a*) As you move from left to right, $f(x)$ changes sign at a from positive to negative?
(*b*) As you move from left to right, $f'(x)$ changes sign at a from positive to negative?
(*c*) As you move from left to right, $f''(x)$ changes sign at a from positive to negative?

3 Find a number c that satisfies the mean-value theorem for the function $f(x) = x^2 - 2x - 8$ and the interval $[0, 6]$.

4 (*a*) Prove that $f(x) = \tan x - x$ is an increasing function of x when $0 \le x < \pi/2$.
(*b*) Deduce that $\tan x > x$ for x in $(0, \pi/2)$.
(*c*) From (*b*) obtain the inequality $x \cos x - \sin x < 0$, if x is in $(0, \pi/2)$.
(*d*) Prove that $(\sin x)/x$ is a decreasing function for x in $(0, \pi/2)$.

5 (*a*) Describe all functions whose derivatives equal $3x^2$.
(*b*) How are you sure that you have found all possibilities?

6 How should one choose two nonnegative numbers whose sum is 1 in order to minimize the sum of the square of one and the cube of the other?

7 A track of length L is to be laid out in the shape of two semicircles at the ends of a rectangle, as shown in Fig. 1. Find the relative proportion of the radius of the circle r and the length of the straight section x if the track is to enclose a maximum area. Discuss also the case of minimum area.

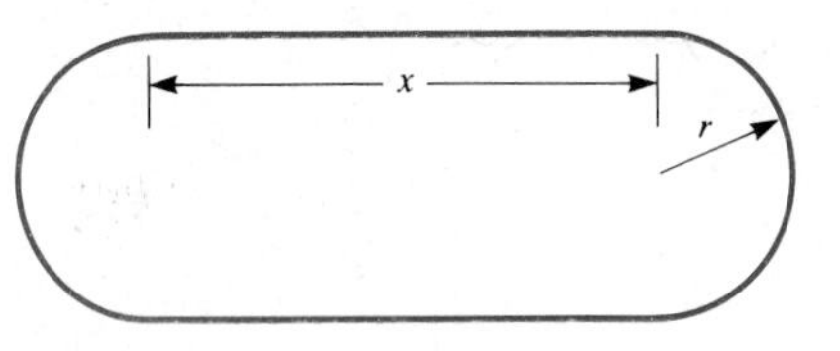

Figure 1

8 The radius of a sphere is measured with an error of at most 2 percent. What percent error may this cause in calculating the volume of the sphere? (The volume of a sphere of radius r is $4\pi r^3/3$.)

9 Graph $y = 1/(x^2 - 3x + 2)$.

10 Graph $f(x) = 3x^4 - 16x^3 + 24x^2$.

11 Graph $y = (x^3 + 1)/(x^2 + 1)$.

12 The volume of a cube of side x is increasing at the rate of 12 cubic meters per minute. What is the rate of change of x when $x = 10$ meters?

13 Find $y'(0)$ and $y''(0)$ if $y(0) = 2$ and $y^2 + x^2y + x^3 = 4$.

14 Find the second derivative of
(*a*) $2x^5 - 1/x$
(*b*) $\cos 2x$
(*c*) $17x^3 - 5x + 2$
(*d*) $\sqrt{x}$
(*e*) $\dfrac{\tan 3x}{1 + 2x}$
(*f*) $\dfrac{\sin 2x}{1 + \sec 2x}$

15 (*a*) Explain how you would find the global maximum of a differentiable function on a closed interval.
(*b*) Describe a test for a relative (local) minimum at a critical point using the first derivative.
(*c*) Describe a test for a relative (local) minimum at a critical point using the second derivative.

16 (*a*) Describe Newton's method of estimating the solution of an equation.
(*b*) Using a diagram, justify the method.

17 A baseball pitcher can throw a ball 90 miles per hour. How high can he throw it? (Disregard air resistance.)

18 (*a*) Estimate $\sqrt{101.5}$ using a differential.
(*b*) Use the second derivative to estimate the error in (*a*).
(*c*) Use a calculator to find the error in (*a*) to five decimal places.

19 Explain why $p(x) = f(a) + f'(a)(x - a)$ should be a good estimate of $f(x)$ for x near a (*a*) geometrically, (*b*) algebraically.

REVIEW EXERCISES FOR CHAP. 4: APPLICATIONS OF THE DERIVATIVE

1 Show that the equation $x^5 + 2x^3 - 2 = 0$ has exactly one solution in the interval $[0, 1]$. (Why does it have at least one? Why is there at most one?)

2 Show that the equation $3 \tan x + x^3 = 2$ has exactly one solution in the interval $[0, \pi/4]$.

3 Let $f(x) = 1/x$.

(*a*) Show that $f'(x)$ is negative for all x in the domain of f.
(*b*) If $x_1 > x_2$, is $f(x_1) < f(x_2)$?

4 A rancher wishes to fence in a rectangular pasture 1 square mile in area, one side of which is along a road. The cost of fencing along the road is higher and equals 5 dollars a foot. The fencing for the other three sides costs 3 dollars a foot. What is the shape of the most economical pasture?

5 Graph $y = \frac{1}{x^2} + \frac{1}{x-1}$.

6 Translate the following excerpt from a news article into the terminology of calculus:

With all the downward pressure on the economy, the first signs of a slowing of inflation seem to be appearing. Some sensitive commodity price indexes are down; the overall wholesale price index is rising at a slightly slower rate.

7 (*a*) Graph $y = \sqrt{x}$ for $0 \le x \le 5$.
(*b*) Compute dy for $x = 4$ and $dx = 1$.
(*c*) Compute Δy for $x = 4$ and $\Delta x = 1$.
(*d*) Using the graph in (*a*), show dy and Δy.

8 Fill in this table:

Interpretation of f(x)	*Interpretation of f'(x)*
The y coordinate in a graph of $y = f(x)$	
Total distance traveled up to time x	
Projection by lens of point x	
Size of population at time x	
Total mass of left x centimeters of string	
Velocity at time x	

9 Explain why each of these two proposed definitions of a tangent line is inadequate:
(*a*) A line L is tangent to a curve at a point P if L meets the curve only at P.
(*b*) A line L is tangent to a curve at a point P if L meets the curve at P and does not cross the curve at P.

10 A window is made of a rectangle and an equilateral triangle, as shown in Fig. 2. What should the dimensions be to maximize the area of the window if its perimeter is prescribed?

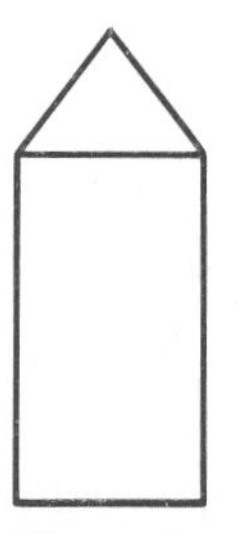

Figure 2

11 A wire of length L is to be cut into two pieces. One piece will be shaped into an equilateral triangle and the other into a square. How should the wire be cut in order to
(*a*) minimize the sum of the areas of the triangle and square?
(*b*) maximize the sum of the areas?

12 A square foot of thin glass is to be melted into two shapes. Some of it will be the thin surface of a cube. The rest will be the thin surface of a sphere. How much of the glass should be used for the cube and how much for the sphere if
(*a*) their total volume is to be a minimum?
(*b*) their total volume is to be a maximum?

13 If f is defined for all x, $f(0) = 0$, and $f'(x) \ge 1$ for all x, what is the most that can be said about $f(3)$? Explain.

14 (*a*) Using differentials, show that $\sqrt[3]{8 + h} \approx 2 + h/12$ when h is small.
(*b*) What is the percent error when $h = 1$? $h = -1$?

15 (*a*) Using a differential, estimate $\sqrt{1 - x^2}$ when x is small.
(*b*) Check how good the estimate is when $x = 0.2$.

16 (*a*) Using a differential, estimate $\frac{1}{\sqrt{1 - x^2}}$ when x is small.
(*b*) Check how good the estimate is when $x = 0.2$.

17 (*a*) Using a differential, estimate $\sqrt{7x + 2}$ when x is small.
(*b*) Check how good the estimate is when $x = 0.1$.

18 (*a*) Using a differential, estimate $\frac{1}{\sqrt{7x + 2}}$ when x is small.
(*b*) Check how good the estimate is when $x = 0.1$.

19 When measuring an angle, a lab assistant might make an error of 1.6°. Estimate the error that might be induced in $\cos\theta$, if θ is measured to be (*a*) 60°, (*b*) 45°, (*c*) 30°.

20 The volume of a ball of radius r is $4\pi r^3/3$.
(*a*) Estimate the change in the volume of a spherical orange as the radius increases from 1 inch to 1.1 inches.
(*b*) Estimate the percentage change in the volume.
(*c*) Estimate the change in volume as the radius increases from 2 to 2.1 inches.
(*d*) Estimate the percentage change in the volume.

21 The surface area of a sphere of radius r is $S = 4\pi r^2$. The volume of a ball of radius r is $V = 4\pi r^3/3$. If you make a 2 percent error in estimating radius, what percent error will this cause in the estimates of S and V?

22 You wish to estimate the volume of a tree trunk that has the form of a right circular cylinder. You want the estimate to be within 3 percent of the correct value. You can measure the height with negligible error. What percentage error is permitted in estimating the circumference, if you will use that estimate in calculating the volume?

23 For what value of the exponent a does the function $y = x^a$ satisfy the equation

$$\frac{dy}{dx} = -y^2?$$

24 Find all functions f such that

$$\frac{d^2f}{dx^2} = x.$$

25 Let $f(x) = \tan x$, $a = \pi/4$. Let $p(x)$ be the linearization of f at a. Let $df = f'(a)\,dx$ and $\Delta f = f(a + \Delta x) - f(a)$. Fill in this table.

Δx	$f(a + \Delta x) - p(a + \Delta x)$	Δf	df	$\Delta f - df$
0.2				
0.1				
0.01				
−0.01				

26 Like Exercise 25 with $f(x) = 1/(1 + x)^2$, $a = 2$.

27 Graph $y = \sqrt{x}/(1 + x)$.

28 Graph $y = x^4 - 12x^3 + 54x^2$.

29 Show that the equation $2x^7 + 3x^5 + 6x + 10 = 0$ has exactly one real solution.

30 For each of the following give an example of a function whose domain is the x axis, such that

(*a*) f has a global minimum at $x = 1$, but 1 is not a critical number.

(*b*) f has an inflection point at (0, 0), but $f''(0)$ is not defined.

(*c*) $f'(2) = 0$, but f does not have a local extremum at $x = 2$.

(*d*) $f''(2) = 0$, but 2 is not an inflection number.

31 Differentiate for practice:

(*a*) $\dfrac{2x^3 - x}{x + 2}$ (*b*) $x^5\sqrt{1 + 3x}$ (*c*) $\dfrac{(2x - 1)^5}{7}$

(*d*) $\sin^4 \sqrt{x}$ (*e*) $\cos(1/x^3)$ (*f*) $\tan \sqrt{1 - x^2}$

32 A rectangular box with a square base is to be constructed. Material for the top and bottom costs a cents per square inch and material for the sides costs b cents per square inch.

(*a*) For a given cost, what shape has the largest volume? (Express your answer in terms of the ratio between height and dimension of base.)

(*b*) For a given volume, what shape is most economical?

33 What is the minimum slope of $y = x^3 - 9x^2 + 15x$?

34 Use a differential to estimate each of the following for small h:

(*a*) $\sec\left(\dfrac{\pi}{3} + h\right)$ (*b*) $\sqrt[3]{1 + h^2}$ (*c*) $\dfrac{1}{(1 - h)^2}$

35 Use a differential to estimate (*a*) $(1.002)^5$ (*b*) $(0.996)^3$.

36 Find $y'(1)$ if $y(1) = 1$ and $\tan\left(\dfrac{\pi}{4}xy\right) + y^3 + x = 3$.

37 Find a tilted asymptote to the graph of $y = \sqrt{x^2 + 2x}$ and graph the function.

38 The derivative of a certain function f is 5 when x is 2.

(*a*) If $f(x)$ is the distance in feet that a rocket travels in x seconds, about how far does it travel from $x = 2$ to $x = 2.1$ seconds?

(*b*) If $f(x)$ is the projection of x on a slide, about how long is the projection of the interval [2, 2.1]?

(*c*) If $f(x)$ is the depth in feet that water penetrates the soil in the first x hours, about how far does the water penetrate in the 6 minutes from 2 to 2.1 hours?

39 A certain function $y = f(x)$ has the property that

$$\frac{dy}{dx} = 3y^2.$$

Show that

$$\frac{d^2y}{dx^2} = 18y^3.$$

40 If dy/dx is proportional to x^2, show that d^2y/dx^2 is proportional to x.

41 If dy/dx is proportional to y^2, show that d^2y/dx^2 is proportional to y^3.

42 Using differentials, show that

$$\sin\left(\frac{\pi}{6} + h\right) \approx \frac{1}{2} + \frac{\sqrt{3}}{2}h$$

for small h.

43 Show that if $dy/dx = 3y^4$, then $d^2y/dx^2 = 36y^7$.

44 Find the maximum value of $\sin^2 \theta \cos \theta$.

---

45 What are the dimensions of the rectangle of largest area that can be inscribed in the ellipse $x^2/a^2 + y^2/b^2 = 1$? (Assume that the sides of the rectangle are parallel to the axes.)

46 Of all squares that can be inscribed in a square of side a what is the side of the one of smallest area?

47 What point on the parabola $y = x^2$ is closest to the point (3, 0)?

48 Let f be a differentiable function and let A be a point not on the graph of f. Show that if B is the point on the graph closest to A, then the segment AB is perpendicular to the tangent line to the curve at B.

Galois, early in the nineteenth century, proved that the root of a fifth-degree polynomial cannot generally be expressed in terms of square roots, cube roots, fourth roots, fifth roots, etc. [His theorem applies to any polynomial with integer coefficients such that (i) its degree is a prime $p \geq 5$, (ii) it is not the product of two polynomials of lower degree with integer coefficients, and (iii) exactly $p - 2$ of its roots are real.] We can, nevertheless, show that roots exist and use Newton's method to calculate them to several decimal places. See Exercises 49 and 50.

49 (*a*) Show that the equation $x^5 - 6x + 3 = 0$ has exactly three real roots.

(*b*) Use Newton's method to find one of the roots to three decimal places.

50 (*a*) Show that the equation $2x^5 - 10x + 5 = 0$ has exactly three real roots.

(*b*) Use Newton's method to find one of the roots to three decimal places.

51 Show that $x^3 - 3x - 3$ has exactly one real root and use Newton's method to estimate it to three decimal places.

52 Show that $x^3 + x - 6$ has exactly one real root and use Newton's method to estimate it to three decimal places.

53 The quantities x and y, which are differentiable functions of t, are related by the equation

$$xy + e^y = e + 1.$$

When $t = 0$, $x = 1$, $y = 1$, $\dot{x} = 2$, and $\ddot{x} = 3$. Find $\dot{y}$ and $\ddot{y}$ when $t = 0$.

54 A load of concrete M hangs from a rope which passes over a pulley B. (See Fig. 3.) A construction worker at C pulls the rope as he walks away at 5 feet per second. The level of the pulley is 10 feet higher than his hand. At what rate is the load rising if the length $\overline{BC}$ is (*a*) 15 feet? (*b*) 100 feet?

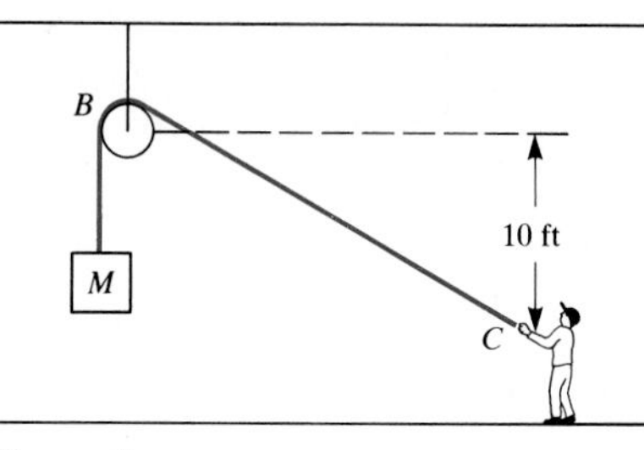

Figure 3

55 Find the volume of the largest right circular cone that can be inscribed in a sphere of radius a.

56 Of all right circular cones with fixed volume V, which shape has the least surface area, including the area of the base? (The area of the curved part of a cone of slant height l and radius r is πrl.)

57 Of all right circular cones with fixed surface area A (including the area of the base), which shape has the largest volume?

58 What point on the line $y = 3x + 7$ is closest to the origin? (Instead of minimizing the distance, it is much more convenient to minimize the square of the distance. Doing so avoids square roots.)

59 (*a*) What is the maximum value of the function $y = 3 \sin t + 4 \cos t$?
(*b*) What is the maximum value of the function $y = A \sin kt + B \cos kt$, where A, B, and k are constants, $k \neq 0$?

60 Let $f(x) = (x - 1)^n(x - 2)$, where n is an integer, $n \geq 2$.
(*a*) Show that $x = 1$ is a critical number.
(*b*) For which values of n will $x = 1$ provide a relative maximum? a relative minimum? neither?

61 Let p and q be constants.
(*a*) Show that, if $p > 0$, the equation $x^3 + px + q = 0$ has exactly one real root.
(*b*) Show that, if $4p^3 + 27q^2 < 0$, the equation $x^3 + px + q = 0$ has three distinct real roots.

62 (*a*) Sketch the graph of $y = 1/x$.
(*b*) Estimate by eye the point (x_0, y_0) on the graph closest to $(3, 1)$.
(*c*) Show that x_0 is a solution of the equation $x^4 - 3x^3 + x - 1 = 0$.
(*d*) Show that the equation in (*c*) has a root between 2.9 and 3.

63 Show that, if $f'(x) \neq 0$, then dy is a good approximation to Δy when dx is small, in the sense that

$$\lim_{dx \to 0} \frac{\Delta y - dy}{dx} = 0.$$

64 A swimmer stands at a point A on the bank of a circular pond of diameter 200 feet. He wishes to reach the diametrically opposite point B by swimming to some point P on the bank and walking the arc PB along the bank. If he swims 100 feet per minute and walks 200 feet per minute, to what point P should he swim in order to reach B in the shortest possible time?

65 Let $f(x) = (x^2 - 4)/(x + \frac{1}{2})$. Observe that $f(2) = f(-2)$.
(*a*) What does Rolle's theorem say about f?
(*b*) For which values of x is $f'(x) = 0$?

66 (*a*) The area of a circle of radius r is πr^2. Use a differential to estimate the change in the area when the radius changes from r to $r + dr$.
(*b*) The circumference of a circle of radius r is $2\pi r$. Explain why $2\pi r$ appears in the answer to (*a*).

67 The graph of a certain function is shown in Fig. 4. List the x coordinates of (*a*) relative maxima, (*b*) relative minima, (*c*) critical points, (*d*) global maximum, (*e*) global minimum.

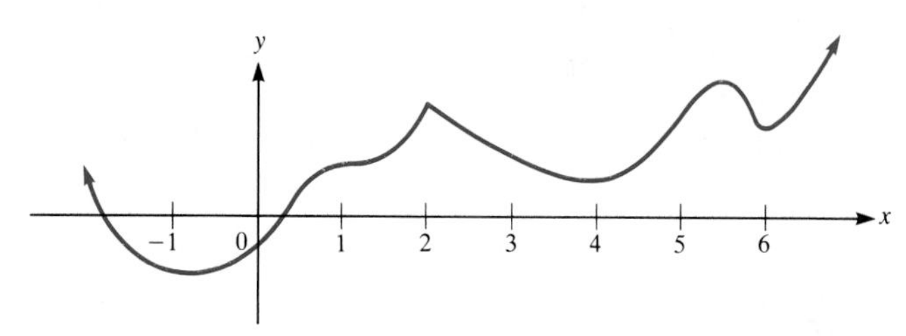

Figure 4

68 How should one choose two positive numbers whose product is 2 in order to minimize the sum of their squares?

69 The left-hand x centimeters of a string 12 centimeters long has a mass of $18x^2 - x^3$ grams.
(*a*) What is its density x centimeters from the left-hand end?
(*b*) Where is its density greatest?

70 Each of Figs. 5 to 8 is the graph of the velocity $v(t)$ of a particle moving on a line. In each case sketch the general shape of (*a*) the graph of acceleration as a function of time, (*b*) the graph of position as a function of time. (Assume that when $t = 0$, the particle's coordinate is 0.)

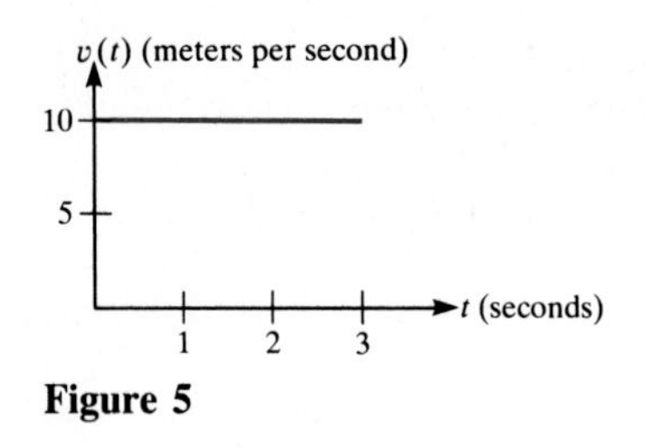

Figure 5

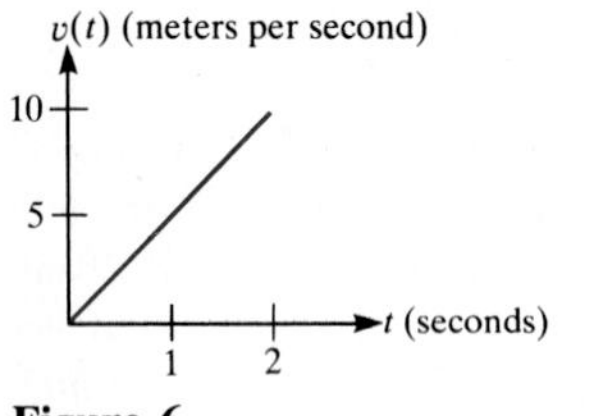

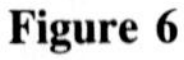

Figure 6

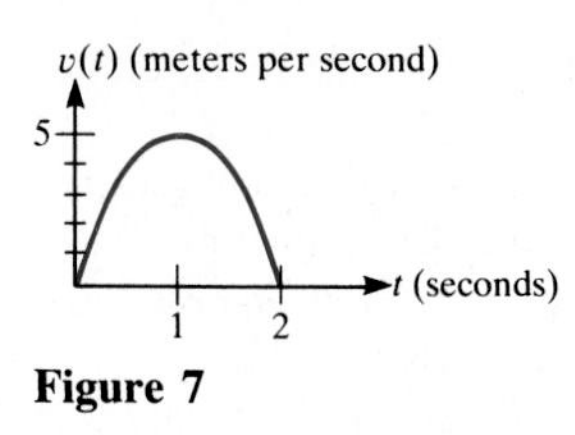

Figure 7

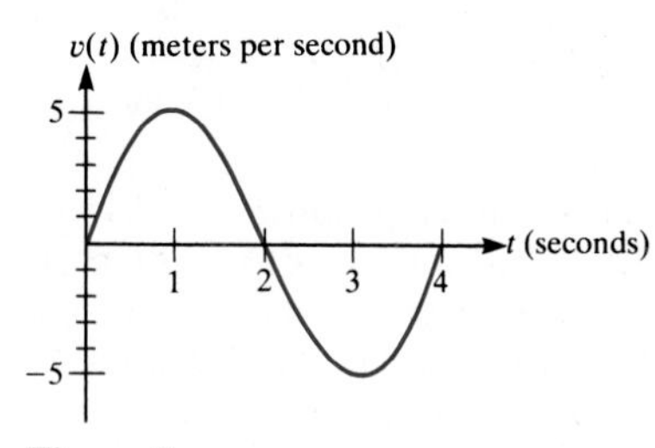

Figure 8

5

THE DEFINITE INTEGRAL

Chapters 3 and 4 were concerned with the derivative, which gives local information, such as the slope at a particular point on a curve or the velocity at a particular time. The present chapter introduces the second major concept of calculus, the definite integral. In contrast to the derivative, the definite integral gives overall global information, such as the area under a curve. The derivative turns out to be one of the tools for evaluating many definite integrals.

Section 5.1 motivates the definite integral through four of its applications.

Section 5.2 introduces a notation for sums and develops formulas for some particular sums. They will be used in Sec. 5.3, which defines the definite integral.

Section 5.4 presents two methods for estimating a definite integral.

Sections 5.5 and 5.6 develop concepts needed in Sec. 5.7, which relates the derivative to the definite integral.

Chapters 2 to 5 constitute the core of calculus. The subsequent chapters are mostly variations or applications of the key ideas in these four chapters.

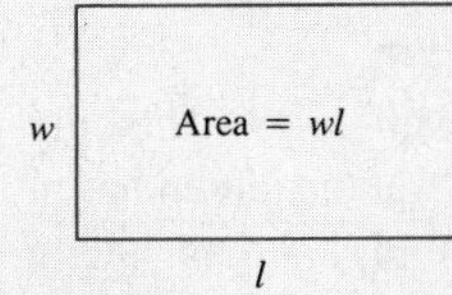

Figure 1

5.1 ESTIMATES IN FOUR PROBLEMS

Just as Chap. 3 introduced the derivative by four problems, this chapter introduces the definite integral by four problems. At first glance these problems may seem unrelated, but by the end of the section it will be clear that they represent one basic problem in various guises. They lead up to the concept of the definite integral, which is defined in Sec. 5.3. The definite integral will help us find the exact values of certain quantities, which we will be able only to estimate in this section.

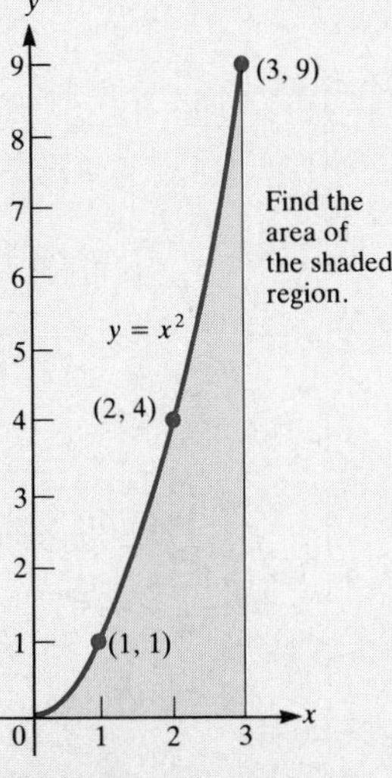

Figure 2

Estimating an Area

It is easy to find the exact area of a rectangle: Just multiply its length by its width. (See Fig. 1.) But how do you find the area of the region in Fig. 2? In this section we will show how to make very accurate *estimates* of the area. The technique we use will lead up to the definition of the definite integral in Sec. 5.3. The definite integral will give the *exact* area.

PROBLEM 1 Estimate the area of the region bounded by the curve $y = x^2$, the x axis, and the vertical line $x = 3$, as shown in Fig. 2.

SOLUTION Since we know how to find the area of a rectangle, we will use rectangles to approximate the region in Fig. 2. Figure 3 shows an approximation by six rectangles whose total area is more than the area under the parabola. Figure 4 shows a similar approximation whose area is less than the area under the parabola.

In each case we break the interval [0, 3] into six shorter intervals, all of width $\frac{1}{2}$. Therefore, the width of each rectangle is $\frac{1}{2}$. In order to find the area of the overestimate and the area of the underestimate, we must find the height of each rectangle. That height is determined by the curve $y = x^2$.

There are six rectangles in the overestimate shown in Fig. 3. The smallest one is shown in Fig. 5. The height of the rectangle in Fig. 5 is equal to the value of x^2 when $x = \frac{1}{2}$. The height is therefore $(\frac{1}{2})^2$. Therefore, its area is $(\frac{1}{2})^2(\frac{1}{2})$, the product of its height and its width. The areas of the other five rectangles can be found similarly, in each case evaluating x^2 at the right end of the rectangle's base in order to find the height. Their total area is

$$(\tfrac{1}{2})^2(\tfrac{1}{2}) + (\tfrac{2}{2})^2(\tfrac{1}{2}) + (\tfrac{3}{2})^2(\tfrac{1}{2}) + (\tfrac{4}{2})^2(\tfrac{1}{2}) + (\tfrac{5}{2})^2(\tfrac{1}{2}) + (\tfrac{6}{2})^2(\tfrac{1}{2}), \tag{1}$$

which is

$$\tfrac{1}{8}(1^2 + 2^2 + 3^2 + 4^2 + 5^2 + 6^2) = \tfrac{91}{8} = 11.375. \tag{2}$$

The area under the parabola is therefore less than 11.375.

The area of the underestimate in Fig. 4 can be computed similarly. The only difference is that the height of each rectangle is the value of x^2 at the left endpoint of the corresponding section. The smallest rectangle in Fig. 4 has height $0^2 = 0$. The largest has height $(\frac{5}{2})^2$. The total area of the six rectangles in Fig. 4 is

$$(0)^2(\tfrac{1}{2}) + (\tfrac{1}{2})^2(\tfrac{1}{2}) + (\tfrac{2}{2})^2(\tfrac{1}{2}) + (\tfrac{3}{2})^2(\tfrac{1}{2}) + (\tfrac{4}{2})^2(\tfrac{1}{2}) + (\tfrac{5}{2})^2(\tfrac{1}{2}), \tag{3}$$

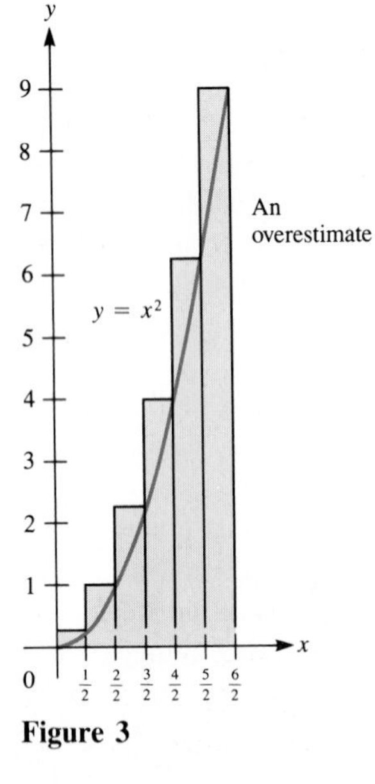

Figure 3

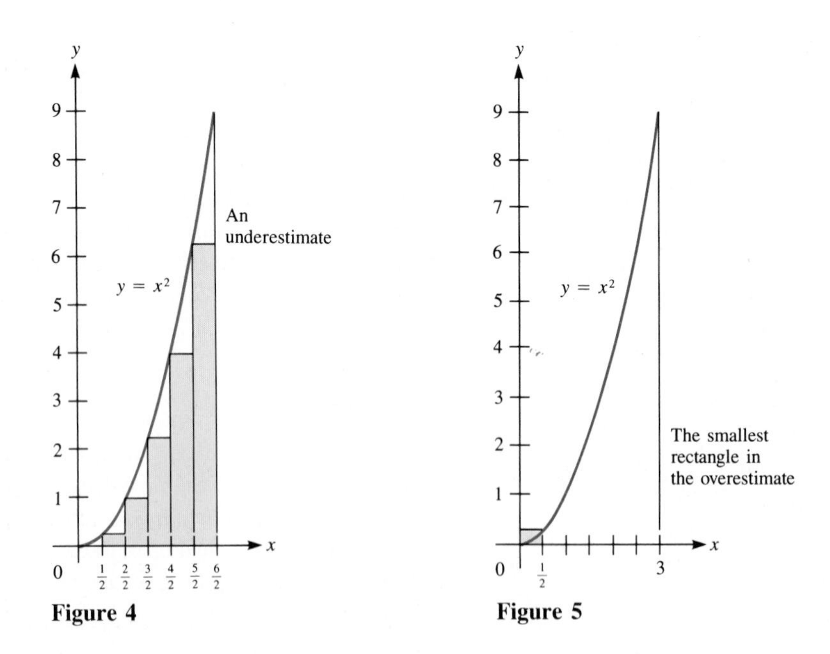

Figure 4

Figure 5

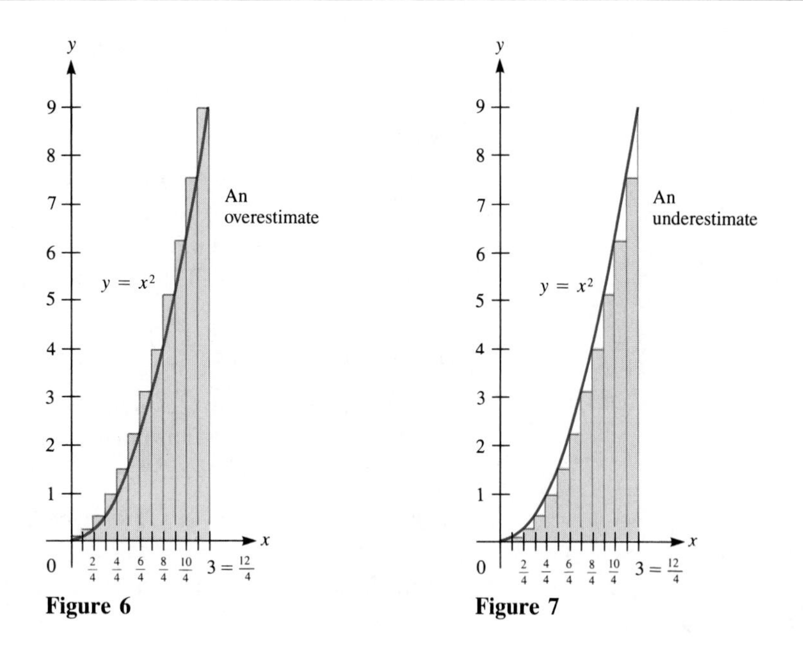

Figure 6 **Figure 7**

which is

$$\tfrac{1}{8}(0^2 + 1^2 + 2^2 + 3^2 + 4^2 + 5^2) = \tfrac{55}{8} = 6.875. \tag{4}$$

We now have trapped the area under the parabola between 6.875 and 11.375.

To get closer estimates we should use more rectangles. Figures 6 and 7 show an overestimate and an underestimate, in each of which there are 12 rectangles. Each has width $\frac{3}{12} = \frac{1}{4}$.

The total area of the overestimate is

$$(\tfrac{1}{4})^2(\tfrac{1}{4}) + (\tfrac{2}{4})^2(\tfrac{1}{4}) + (\tfrac{3}{4})^2(\tfrac{1}{4}) + \cdots + (\tfrac{12}{4})^2(\tfrac{1}{4}), \tag{5}$$

which is

$$\tfrac{1}{64}(1^2 + 2^2 + 3^3 + \cdots + 12^2) = \tfrac{650}{64} = 10.15625. \tag{6}$$

The underestimate provided by the rectangles in Fig. 7 is

$$(\tfrac{0}{4})^2(\tfrac{1}{4}) + (\tfrac{1}{4})^2(\tfrac{1}{4}) + (\tfrac{2}{4})^2(\tfrac{1}{4}) + \cdots + (\tfrac{11}{4})^2(\tfrac{1}{4}) = \tfrac{506}{64} = 7.90625. \tag{7}$$

Now the area of the parabola is trapped between 7.90625 and 10.15625. These numbers are much closer to each other than the numbers 6.875 and 11.375 that we found using only six rectangles.

To get closer estimates we would cut the interval [0, 3] into more sections, maybe 100 or 1,000 or more, and calculate the total area of the corresponding rectangles. (This is an easy computation on a computer.)

In forming our estimates, we chose all the widths of the rectangles to be equal. (That simplified the arithmetic.) Of course, estimates could be made with rectangles of unequal widths.

We determined the heights of the rectangles by evaluating x^2 at either the right endpoint of each section (to get an overestimate) or at the left endpoint of each section (to get an underestimate). We could have determined the height by evaluating x^2 at some other point in each section. For instance, if we divide [0, 3] into six sections of equal widths and determine the height of each rectangle by the value of x^2 at midpoints, we obtain the approximation shown in Fig. 8.

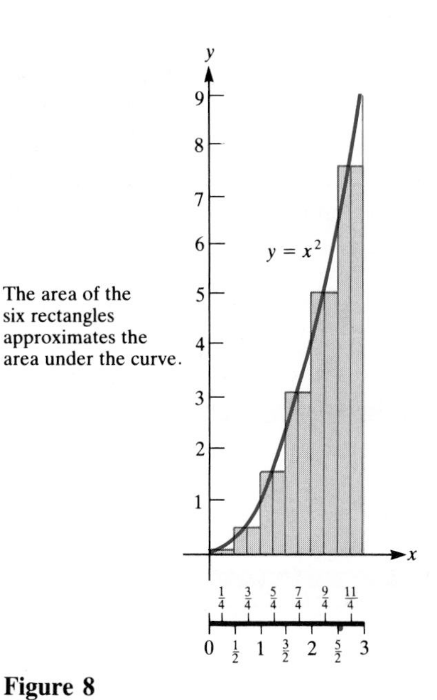

The area of the six rectangles approximates the area under the curve.

Figure 8

The total area of these six rectangles is

$$(\tfrac{1}{4})^2(\tfrac{1}{2}) + (\tfrac{3}{4})^2(\tfrac{1}{2}) + (\tfrac{5}{4})^2(\tfrac{1}{2}) + (\tfrac{7}{4})^2(\tfrac{1}{2}) + (\tfrac{9}{4})^2(\tfrac{1}{2}) + (\tfrac{11}{4})^2(\tfrac{1}{2}), \tag{8}$$

which is $\quad \tfrac{1}{32}(1^2 + 3^2 + 5^2 + 7^2 + 9^2 + 11^2) = \tfrac{286}{32} = 8.9375.$

Though this is probably a better estimate than the earlier ones we made, we do not know whether it is too large or too small.

In any case, we see how to make estimates of the area under the parabola as accurate as we wish—if we had the time. ■

Estimating a Mass

If a string has a density of d grams per centimeter and it is l centimeters long, we can easily compute its total mass. The total mass is just the product of the density and the length:

$$\text{Total mass} = d \cdot l \text{ grams}.$$

See Fig. 9.

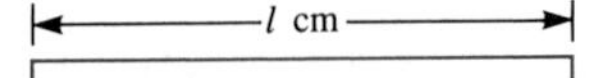

Density d, grams/cm

Figure 9

But how do we find the total mass if the density varies along the string? (Recall the wet string in Sec. 3.1, which is denser toward the end where the water is accumulating.) The next problem shows how to estimate the total mass when the density is not constant.

PROBLEM 2 A thin string 3 centimeters long is very light near one end and gradually gets much heavier toward the other. In fact, at a distance of x centimeters from the light end, its density is x^2 grams per centimeter. Estimate the total mass of the string, shown in Fig. 10.

SOLUTION The maximum density of the string is at its right end, where the density is $3^2 = 9$ grams per centimeter. Since the total length is 3 centimeters, the total mass is at most $9 \cdot 3 = 27$ grams. But we can make better estimates by using the techniques illustrated in Problem 1.

Cut the string into six sections of equal length, as shown in Fig. 11. The density of the string in each of the six pieces varies less than it does over the whole length of the string. Consider the mass in the interval $[\tfrac{3}{2}, \tfrac{4}{2}]$ for instance. (To preserve the pattern, we do not reduce $\tfrac{4}{2}$ to 2.) The density of the string in this interval varies from $(\tfrac{3}{2})^2$ to $(\tfrac{4}{2})^2$ grams per centimeter. The section is $\tfrac{1}{2}$ centimeter long. Its total mass is *at most* $(\tfrac{4}{2})^2(\tfrac{1}{2})$ grams, since that would be its mass if the density were $(\tfrac{4}{2})^2$ grams per centimeter throughout the interval $[\tfrac{3}{2}, \tfrac{4}{2}]$.

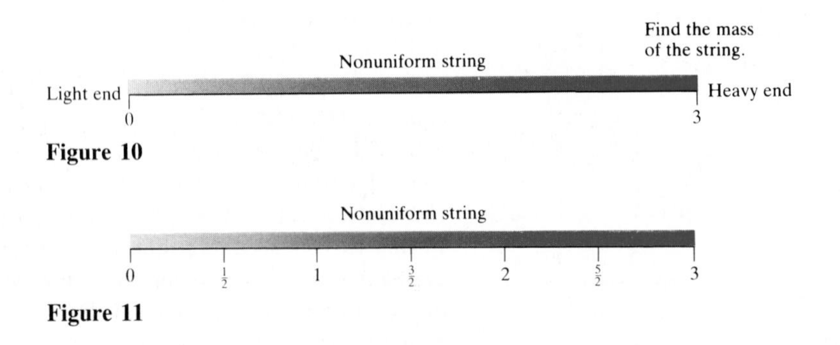

Figure 10

Figure 11

Similarly, the mass in the interval is *at least* $(\frac{3}{2})^2(\frac{1}{2})$ grams. Thus

$$(\tfrac{3}{2})^2(\tfrac{1}{2}) \le \text{Mass in } [\tfrac{3}{2}, \tfrac{4}{2}] \le (\tfrac{4}{2})^2(\tfrac{1}{2}) \text{ grams.} \tag{9}$$

Inequality (9) suggests how to make an overestimate and an underestimate of the total mass of the string.

To find an overestimate, evaluate the density at the right endpoint of each section. Multiply each result by the width of the section and add the products. The overestimate is

$$(\tfrac{1}{2})^2(\tfrac{1}{2}) + (\tfrac{2}{2})^2(\tfrac{1}{2}) + (\tfrac{3}{2})^2(\tfrac{1}{2}) + (\tfrac{4}{2})^2(\tfrac{1}{2}) + (\tfrac{5}{2})^2(\tfrac{1}{2}) + (\tfrac{6}{2})^2(\tfrac{1}{2}). \tag{10}$$

The sum (10) is the very same sum (1) we met in overestimating the area under the parabola. It equals 11.375.

To obtain an underestimate of the total mass of the string, evaluate the density at the left endpoint of each of the six sections. This underestimate is

$$(0)^2(\tfrac{1}{2}) + (\tfrac{1}{2})^2(\tfrac{1}{2}) + (\tfrac{2}{2})^2(\tfrac{1}{2}) + (\tfrac{3}{2})^2(\tfrac{1}{2}) + (\tfrac{4}{2})^2(\tfrac{1}{2}) + (\tfrac{5}{2})^2(\tfrac{1}{2}). \tag{11}$$

This sum also appeared in Problem 1. It equals 6.875. We conclude that

$$6.875 \text{ grams} \le \text{Mass of string} \le 11.375 \text{ grams.}$$

If we wished, we could make better estimates by chopping the string into more sections, not necessarily of equal lengths. Moreover, we could use the density at a midpoint (or any other point) of each section.

We have not found the exact mass of the string, but we do have a method of estimating it as precisely as we wish. ■

Estimating a Distance Traveled

If you drive at a constant speed of v miles per hour for a period of t hours, you would travel vt miles:

$$\begin{aligned} \text{Distance} &= \text{Speed times Time} \\ &= vt \text{ miles.} \end{aligned}$$

But how would you compute the total distance traveled if your speed were not constant? (Imagine that your odometer, which records distance traveled, was broken. However, your speedometer and clock are still working fine, so you know your speed at any instant.) The next problem illustrates how you could make very accurate estimates of the total distance traveled.

PROBLEM 3 An engineer drives a car whose clock and speedometer work, but whose odometer is broken. On a 3-hour trip out of a congested city into the countryside she begins at a snail's pace and, as the traffic thins, gradually speeds up. Indeed, she notices that after traveling t hours her speed is $8t^2$ miles per hour. Thus after the first $\frac{1}{2}$ hour she is crawling along at 2 miles per hour, but after 3 hours she is traveling at 72 miles per hour. Estimate how far the engineer travels in 3 hours.

An estimate for the total distance.

SOLUTION The speed during the 3-hour trip varies from 0 to 72 miles per hour. During shorter time intervals, such a wide fluctuation will not occur.

As in the first two problems, cut the 3 hours of the trip into six equal intervals,

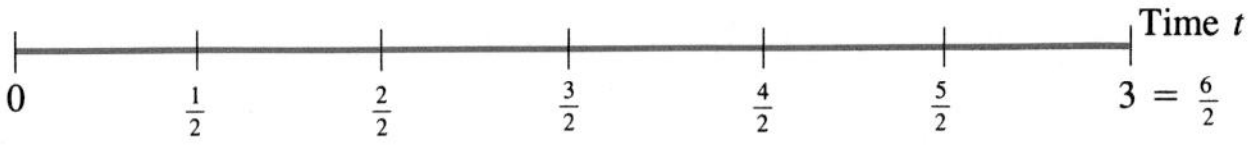

Time t

$0 \quad \frac{1}{2} \quad \frac{2}{2} \quad \frac{3}{2} \quad \frac{4}{2} \quad \frac{5}{2} \quad 3 = \frac{6}{2}$

Figure 12

each $\frac{1}{2}$ hour long, and use them to make an estimate of the total distance covered. Represent time by a line segment cut into six parts of equal length, as in Fig. 12.

Speed increases as t increases.

Consider the distance she travels during one of the six half-hour intervals, say the interval $[\frac{3}{2}, \frac{4}{2}]$. At the beginning of this interval her speed was $8(\frac{3}{2})^2$ miles per hour; at the end she was going $8(\frac{4}{2})^2$ miles per hour. The highest speed during this half hour was $8(\frac{4}{2})^2$ miles per hour. Therefore she traveled at most $8(\frac{4}{2})^2(\frac{1}{2})$ miles during the period. [If her speed were $8(\frac{4}{2})^2$ miles per hour throughout the half hour, she would have traveled exactly $8(\frac{4}{2})^2(\frac{1}{2})$ miles.]

Similarly, since her lowest speed during that half hour was $8(\frac{3}{2})^2$ miles per hour, she traveled at least $8(\frac{3}{2})^2(\frac{1}{2})$ miles. Thus

$$8(\tfrac{3}{2})^2(\tfrac{1}{2}) \leq \text{Distance traveled during time interval } [\tfrac{3}{2}, \tfrac{4}{2}] \leq 8(\tfrac{4}{2})^2(\tfrac{1}{2}) \text{ miles.}$$

Similar reasoning applies to the other five half-hour periods. Adding up the lower and upper estimates for each section of time, we get

$$8(\tfrac{0}{2})^2(\tfrac{1}{2}) + 8(\tfrac{1}{2})^2(\tfrac{1}{2}) + 8(\tfrac{2}{2})^2(\tfrac{1}{2}) + 8(\tfrac{3}{2})^2(\tfrac{1}{2}) + 8(\tfrac{4}{2})^2(\tfrac{1}{2}) + 8(\tfrac{5}{2})^2(\tfrac{1}{2}) \leq \text{Total distance traveled}$$
$$\leq 8(\tfrac{1}{2})^2(\tfrac{1}{2}) + 8(\tfrac{2}{2})^2(\tfrac{1}{2}) + 8(\tfrac{3}{2})^2(\tfrac{1}{2}) + 8(\tfrac{4}{2})^2(\tfrac{1}{2}) + 8(\tfrac{5}{2})^2(\tfrac{1}{2}) + 8(\tfrac{6}{2})^2(\tfrac{1}{2}). \quad (12)$$

The upper and lower estimates in (12) bear a remarkable resemblance to (1) and (3). The only difference is the appearance of the number 8, which is a common factor of all summands in (12). Making the same kind of computations as we did in Problem 1, namely (2) and (4), we conclude that

$$8(6.875) \leq \text{Total distance traveled} \leq 8(11.375) \text{ miles.}$$

Thus the engineer traveled at least 55 miles and at most 91 miles.

To get better estimates, we would cut the 3-hour interval into shorter time periods. ■

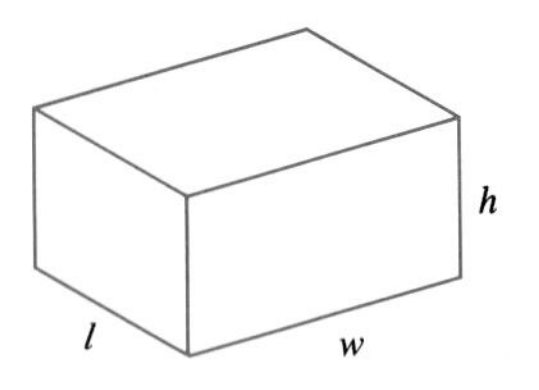

Volume = lwh

Figure 13

Estimating a Volume

The volume of a rectangular box is easy to compute; it is the product of its length, width, and height. See Fig. 13. But finding the volume of a pyramid or ball requires more work. The next example illustrates how we can estimate the volume of a tent, which is a pyramid with a square base. In Sec. 5.3, after we have developed the definite integral, we will obtain its exact volume.

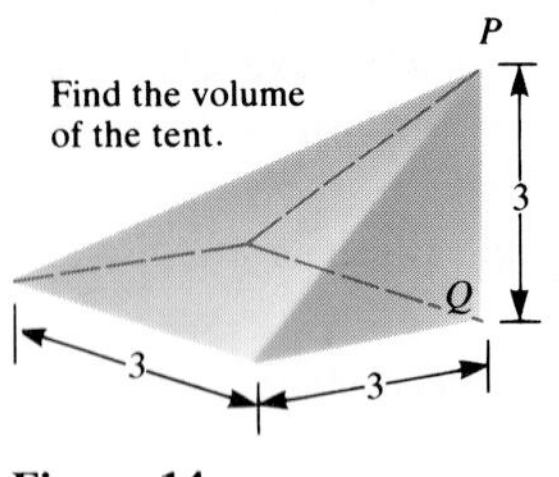

Figure 14

PROBLEM 4 Find the volume inside a tent with a square floor of side 3 feet, whose vertical pole, 3 feet long, rises above a corner of the floor. The tent is shown in Fig. 14.

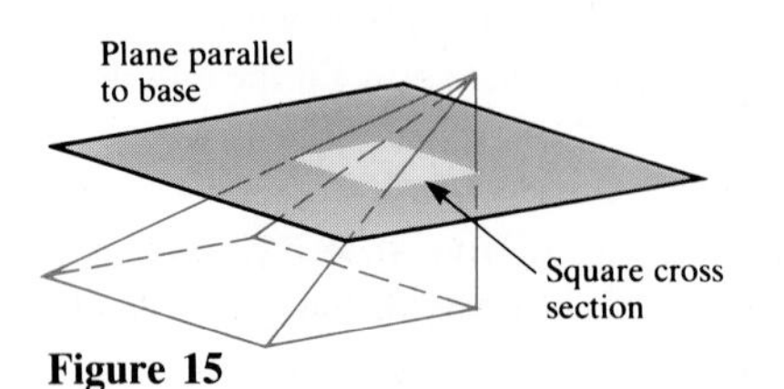

Figure 15

SOLUTION Observe that the cross section of the tent made by any plane parallel to the base is a square, as shown in Fig. 15. Observe that the width of the square equals its distance from the top of the pole. Using this fact we can approximate the tent with boxes.

We begin by cutting a vertical line, representing the pole, into six sections of

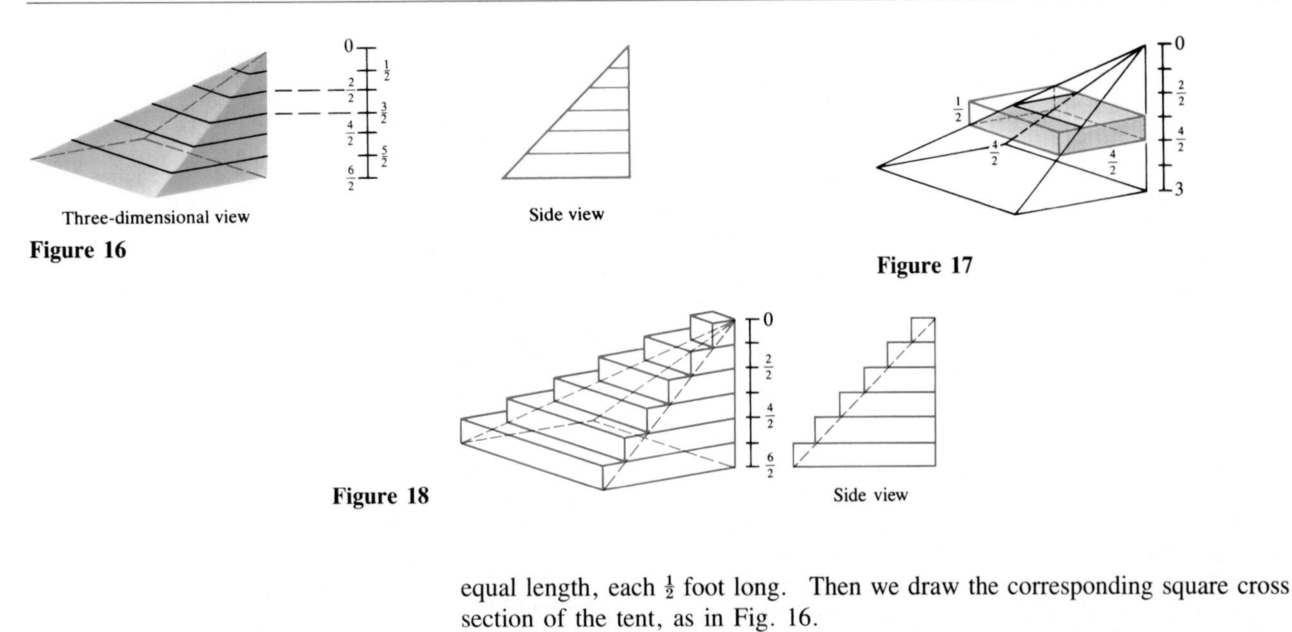

Figure 16

Figure 17

Figure 18

equal length, each $\frac{1}{2}$ foot long. Then we draw the corresponding square cross section of the tent, as in Fig. 16.

We use these square cross sections to form boxes. Consider the part of the tent corresponding to the interval $[\frac{3}{2}, \frac{4}{2}]$ on the pole. The base of this section is a square of side $\frac{4}{2}$ feet. The box with this square as base and height $\frac{1}{2}$ encloses completely the part of the tent corresponding to $[\frac{3}{2}, \frac{4}{2}]$. (See Fig. 17.) The volume of this box is $(\frac{4}{2})^2(\frac{1}{2})$ cubic feet. Figure 18 shows six such boxes, whose total volume is greater than the volume of the tent.

Since the volume of a box is just the area of its base times its height, the total volume of the six boxes is

$$(\tfrac{1}{2})^2(\tfrac{1}{2}) + (\tfrac{2}{2})^2(\tfrac{1}{2}) + (\tfrac{3}{2})^2(\tfrac{1}{2}) + (\tfrac{4}{2})^2(\tfrac{1}{2}) + (\tfrac{5}{2})^2(\tfrac{1}{2}) + (\tfrac{6}{2})^2(\tfrac{1}{2}) \text{ cubic feet.} \tag{13}$$

This sum, which we have met earlier, equals 11.375. It is an *overestimate* of the volume of a tent.

An underestimate can be constructed by considering boxes that are contained in the tent. Figure 19 shows one such box, corresponding to the section $[\frac{3}{2}, \frac{4}{2}]$ on the pole. The total volume of the six boxes formed this way is

$$(\tfrac{0}{2})^2(\tfrac{1}{2}) + (\tfrac{1}{2})^2(\tfrac{1}{2}) + (\tfrac{2}{2})^2(\tfrac{1}{2}) + (\tfrac{3}{2})^2(\tfrac{1}{2}) + (\tfrac{4}{2})^2(\tfrac{1}{2}) + (\tfrac{5}{2})^2(\tfrac{1}{2}) = 6.875 \text{ cubic feet.} \tag{14}$$

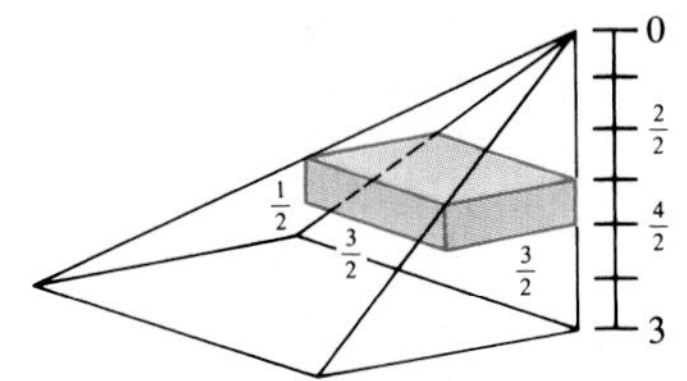

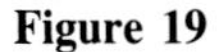

Figure 19

Thus $$6.875 \le \text{Volume of tent} \le 11.375 \text{ cubic feet.}$$

We could get better estimates of the volume by cutting the pole into shorter pieces. ■

Section Summary

We have not found the exact area, mass, distance traveled, or volume in any of our four problems. But we have discovered a way to estimate these quantities as closely as we please. The method would apply to other functions, such as x^3, $\sin x$, 2^x, and so on, and to any interval, not just $[0, 3]$. The arithmetic in each case is the same. The sums formed in three of the problems are identical, while the sums in the distance problem have an additional factor of 8. If we knew what happens to the sums as we cut the interval $[0, 3]$ into smaller sections—if we knew what number they approach—then we could solve all four problems in one blow.

EXERCISES FOR SEC. 5.1: ESTIMATES IN FOUR PROBLEMS

Exercises 1 to 24 concern estimates of areas under curves.

1 In Problem 1 we broke the interval [0, 3] into six sections. Instead break [0, 3] into four sections of equal length and estimate the area under $y = x^2$ and above [0, 3] as follows.
(*a*) Draw the four rectangles whose total area is larger than the area under the curve. The value of x^2 at the right endpoint of each section determines the height of each rectangle.
(*b*) On the diagram in (*a*) show the height and width of each rectangle.
(*c*) Find the total area of the four rectangles.

2 Like Exercise 1, but this time obtain an underestimate of the area by using the value of x^2 at the left endpoint of each section to determine the height of the rectangle.

3 Cutting the interval [0, 3] into five sections of equal length, estimate the area in Problem 1 by finding the sum of the areas of the five rectangles whose heights are determined by (*a*) right endpoints and (*b*) left endpoints. (*c*) Using the information gathered in (*a*) and (*b*), complete this sentence: The area in Problem 1 is certainly less than ________ but larger than ________.

4 Cutting the interval [0, 3] into five sections, estimate the area in Problem 1 by finding the sum of the areas of the five rectangles whose heights are determined by the value of x^2 at the midpoints of the sections.

5 In order to estimate the area in Problem 1 more closely, divide the interval [0, 3] into 15 sections of equal length.
(*a*) What is the width of each section?
(*b*) An overestimate of the area is obtained by using rectangles; the height of each is the value of x^2 at the right endpoint of each section. Draw the rectangles.
(*c*) What is the height of the smallest rectangle? the height of the largest rectangle?
(*d*) Find the total area of the 15 rectangles.

6 Like Exercise 5, but this time form an underestimate using 15 rectangles.

7 Estimate the area under $y = x^2$ and above [1, 2] using the five rectangles in Fig. 20.

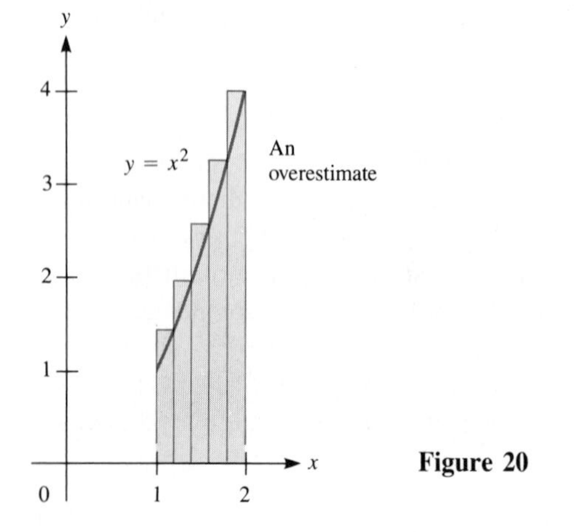

Figure 20

8 Estimate the area under $y = x^2$ and above [1, 2] using the five rectangles in Fig. 21.

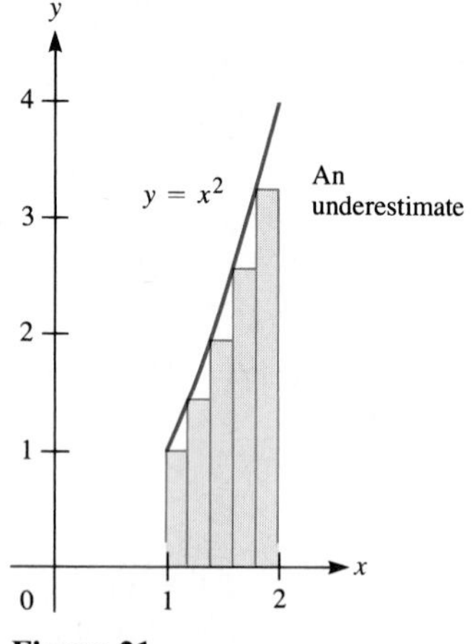

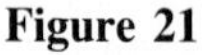

Figure 21

9 Estimate the area under $y = x^2$ and above [1, 2] as follows.
(*a*) Divide [1, 2] into five sections of equal length. (Draw them.)
(*b*) Find the coordinates of the midpoint of each section. (Show the points on your drawing.)
(*c*) On each section construct a rectangle whose height is the value of x^2 at the midpoint. (Draw the five rectangles.)
(*d*) Find the height of each rectangle. (Show the heights on your drawing.)
(*e*) Find the total area of the five rectangles.

10 Like Exercise 9, but with 10 sections of equal length.

11 Estimate the area in Problem 1, using the division of [0, 3] into four sections as shown in Fig. 22. As the points where the height is computed, use $\frac{1}{2}$, $\frac{3}{2}$, 2, and $\frac{14}{5}$ (one of these is in each of the four sections).

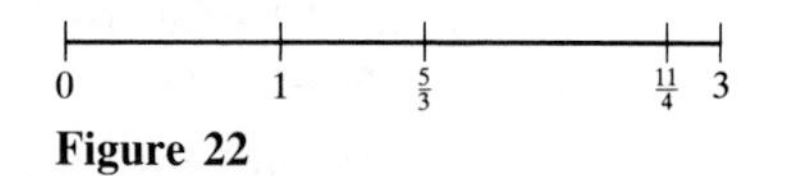

Figure 22

12 Like Exercise 11, but use midpoints to determine the heights of the rectangles.

13 Figure 23 shows the curve $y = 1/x$ for x in [1, 2] and an approximation of the area under the curve by five rectangles of equal width.
(*a*) Make a large copy of Fig. 23.
(*b*) On your diagram show the height and width of each rectangle.
(*c*) Find the total area of the five rectangles.

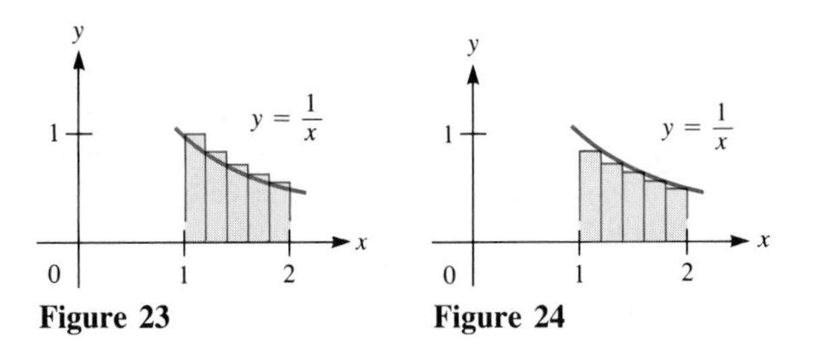

Figure 23 **Figure 24**

14 Like Exercise 13, with the five rectangles in Fig. 24.

In each of Exercises 15 to 20

(*a*) Draw the region.

(*b*) Draw six rectangles of equal width whose total area overestimates the area of the region.

(*c*) On your diagram indicate the height and width of each rectangle.

(*d*) Find the total area of the six rectangles.

15 Under x^2, above [2, 3].

16 Under $1/x$, above [2, 3].

17 Under x^3, above [0, 1].

18 Under $\sqrt{x}$, above [1, 4]. (Give your answer to four decimal places.)

19 Under $\sin x$, above $[0, \pi]$. (Give your answer to four decimal places.)

20 Under $\sqrt{1 + x^3}$, above [0, 1]. (Give your answer to four decimal places.)

21 Estimate the area between the curve $y = x^3$, the x axis, and the vertical line $x = 6$ using a division into (*a*) three sections of equal length with midpoints; (*b*) six sections of equal length with midpoints; (*c*) six sections of equal length with left endpoints; (*d*) six sections of equal length with right endpoints.

22 Estimate the area below the curve $y = 1/x^2$ and above [1, 7] following the directions in Exercise 21.

23 Estimate the area under $y = x^2$ and above [0, 3] by breaking the interval [0, 3] into 50 sections of equal length and (*a*) right endpoints, (*b*) left endpoints.

24 Estimate the area under $y = 1/x$ and above [1, 2] by breaking the interval [1, 2] into 50 sections of equal length and (*a*) right endpoints, (*b*) left endpoints.

25 Estimate the mass of the string in Problem 2 by cutting it into five sections of equal length. For an estimate of the mass of each of these sections use the density at (*a*) the midpoint of each section, (*b*) the right endpoint, and (*c*) the left endpoint. (*d*) On the basis of (*b*) and (*c*), the mass in Problem 2 is less than ________ but larger than ________.

26 Like Exercise 25, with 10 sections of equal length.

27 An electron is being accelerated in such a way that its velocity is t^3 kilometers per second after t seconds, $t \geq 0$. Estimate how far it travels in the first 4 seconds, as follows:

(*a*) Draw the interval [0, 4] as the time axis and cut it into eight sections of equal length.

(*b*) Using the sections in (*a*), make an estimate that is too large.

(*c*) Using the sections in (*a*), make an estimate that is too small.

28 Like Exercise 27, if the velocity is $2^t - 1$ kilometers per second after t seconds, $t \geq 0$.

---

29 A business which now shows no profit is to increase its profit flow gradually in the next 3 years until it reaches a rate of 9 million dollars per year. At the end of the first half year the rate is to be $\frac{1}{4}$ million dollars per year; at the end of 2 years, 4 million dollars per year. In general, at the end of t years, where t is any number between 0 and 3, the rate of profit is to be t^2 million dollars per year. Estimate the total profit during the next 3 years if the plan is successful. Use six intervals of equal length and midpoints.

30 A right circular cone has a height of 3 feet and a radius of 3 feet, as shown in Fig. 25. Estimate its volume by the sum of the volumes of six cylindrical slabs, just as we estimated the volume of the tent with the aid of six rectangular slabs.

(*a*) Make a large and neat diagram that shows the six cylinders used in making an overestimate.

(*b*) Compute the total volume of the six cylinders in (*a*).

(*c*) Make a separate diagram, showing a corresponding underestimate.

(*d*) Compute the total volume of the six cylinders in (*c*). (One has radius 0.)

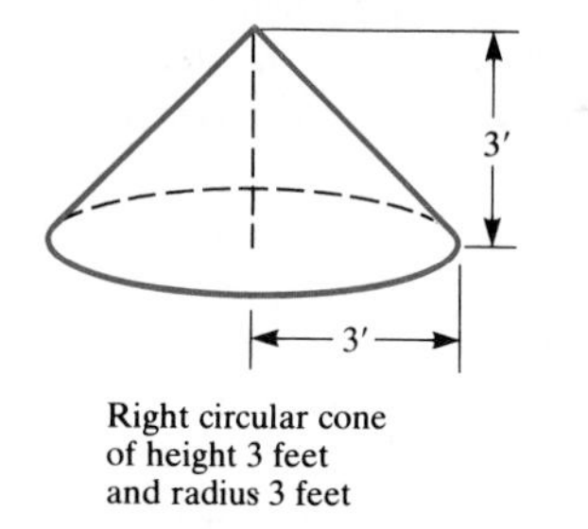

Right circular cone of height 3 feet and radius 3 feet

Figure 25

31 Oil is leaking out of a tank at the rate of 2^{-t} gallons per minute after t minutes. Describe how you would estimate how much oil leaks out during the first 10 minutes. Illustrate your procedure by computing one estimate.

32 Estimate the area of the region under the curve $y = \sin x$ and above the interval $[0, \pi/2]$, cutting the interval as shown in Fig. 26 and using (*a*) left endpoints, (*b*) right endpoints. All but the last section are of the same size.

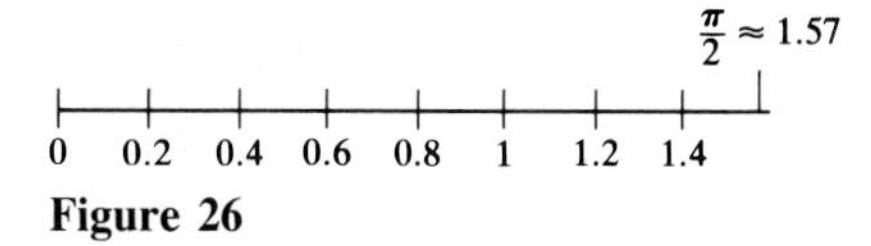

Figure 26

33 Estimate the area under $2^x/x$ and above $[1, 2]$.

34 The kinetic energy of an object, for example, a bullet or car, of mass m and speed v is defined as $mv^2/2$ ergs. (Here mass is measured in grams and speed in centimeters per second.) Now, in a certain machine a uniform rod 3 centimeters long and weighing 32 grams rotates once per second around one of its ends as shown in Fig. 27. Estimate the kinetic energy of this rod by cutting it into six sections, each $\frac{1}{2}$ centimeter long, and taking as the "speed of a section" the speed of its midpoint.

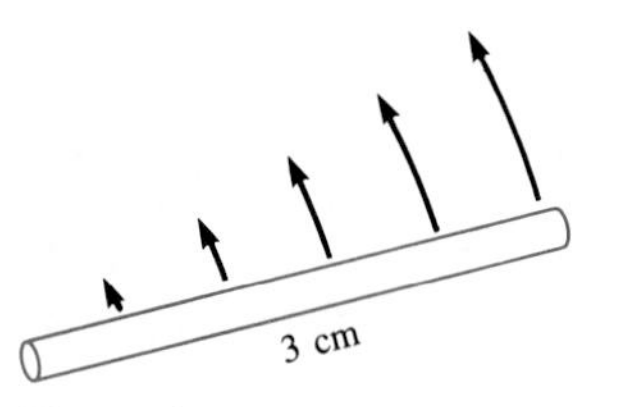

Figure 27

35 (*a*) Draw the region that lies below $y = x$ but above $y = x^2$.
(*b*) Draw six rectangles of equal width whose total area approximates the area of the region in (*a*). (Different students may make different choices.)
(*c*) Indicate on each rectangle its width and height.
(*d*) What is the total area of the six rectangles you chose?

36 Let $y = f(x)$ be a function such that $f(x) \geq 0$, $f'(x) \geq 0$, and $f''(x) > 0$ for all x in $[1, 4]$. An estimate of the area under $y = f(x)$ is made by dividing the interval into sections and forming rectangles. The height of each rectangle is the value of $f(x)$ at the midpoint of the corresponding section. Show that the estimate is less than or equal to the area under the curve. *Hint:* Draw a tangent to the curve at each of the midpoints.

37 Say that to estimate the area of the region below x^2 and above $[0, 3]$ you divide $[0, 3]$ into n sections of equal length, where n is a positive integer. Let $B(n)$ be the value of the overestimate obtained by using right endpoints and $C(n)$ be the value of the underestimate obtained by using left endpoints. Let A be the area we wish to estimate. [We already know that $C(12) = 7.90625 \leq A \leq 10.15625 = B(12)$.]
(*a*) Explain why $B(n) - C(n) = 27/n$.
(*b*) How large should you choose n so that $B(n) - C(n) \leq 0.01$?

38 Differentiate for practice:
(*a*) $(1 + x^2)^{4/3}$
(*b*) $\dfrac{(1 + x^3) \sin 3x}{\sqrt[3]{5x}}$
(*c*) $\dfrac{3x}{8} + \dfrac{3x \sin 4x}{32} + \dfrac{\cos^3 2x \sin 2x}{8}$
(*d*) $\dfrac{3}{8(2x + 3)^2} - \dfrac{1}{4(2x + 3)}$
(*e*) $\dfrac{\cos^3 2x}{6} - \dfrac{\cos 2x}{2}$
(*f*) $x^3\sqrt{x^2 - 1} \tan 5x$

39 Give an example of a function F whose derivative is
(*a*) $(x + 2)^3$ (*b*) $(x^2 + 1)^2$ (*c*) $x \sin x^2$
(*d*) $x^3 + \dfrac{1}{x^3}$ (*e*) $\dfrac{1}{\sqrt{x}}$
(Check by differentiating your answers.)

40 (*a*) Make a large drawing of a cube. Show that it can be divided into three *congruent* pieces, each shaped like the tent in Problem 4.
(*b*) Use (*a*) to find the volume of the tent.
(*c*) What does (*b*) tell us about the area, mass, and distance traveled?

5.2 SUMMATION NOTATION AND APPROXIMATING SUMS

The approximating sums in Sec. 5.1 are cumbersome to write out in full, especially when the number of little sections is large. For this reason we pause to develop a convenient notation that we will use in the next section when we develop the definite integral.

In addition to writing approximating sums in a new form, we will introduce the notion of telescoping sums and use it to develop some formulas needed in the next section.

Sigma Notation

The standard way of describing sums in which all summands have the same general form is called **sigma notation**, named after the Greek letter Σ (sigma), which corresponds to the "s" of "sum." Sigma notation is also called *summation notation,* for obvious reasons.

Definition *Sigma notation.* Let $a_1, a_2, \ldots, a_n$ be n numbers. The sum $a_1 + a_2 + \cdots + a_n$ will be denoted in **sigma notation** by the symbol $\sum_{i=1}^{n} a_i$ or $\Sigma_{i=1}^{n}\, a_i$, which is read as "the sum of a sub i as i goes from 1 to n."

In the sigma notation, the formula for the typical summand is given, as is a description of where the summation starts and ends.

EXAMPLE 1 Write the sum $1^2 + 2^2 + 3^2 + 4^2$ in the sigma notation.

SOLUTION Since the ith summand is the square of i and the summation extends from $i = 1$ to $i = 4$, we have

$$1^2 + 2^2 + 3^2 + 4^2 = \sum_{i=1}^{4} i^2.$$

Simple arithmetic shows that the sum is equal to 30. ■

EXAMPLE 2 Compute $\Sigma_{i=1}^{3}\, 2^i$.

SOLUTION This is short for the sum $2^1 + 2^2 + 2^3$, which is $2 + 4 + 8$, or 14. ■

In the definition of the sigma notation, the letter i (for "index") was used. Any letter, such as j or k, would do just as well. Such an index is sometimes called a summation index or dummy index.

EXAMPLE 3 Compute $\displaystyle\sum_{j=1}^{4} \frac{1}{j}$.

SOLUTION This is short for $\frac{1}{1} + \frac{1}{2} + \frac{1}{3} + \frac{1}{4}$, which is approximately 2.083. ■

Had Example 3 read "Compute $\Sigma_{k=1}^{4} \dfrac{1}{k}$," the result would be the same:

$$\sum_{k=1}^{4} \frac{1}{k} = \frac{1}{1} + \frac{1}{2} + \frac{1}{3} + \frac{1}{4}.$$

The particular letter used to indicate the form of the typical summand is of no special importance.

The summation notation has two properties which will be of use in coming chapters. First of all, if c is a fixed number, then

$$\sum_{i=1}^{n} ca_i = ca_1 + ca_2 + \cdots + ca_n$$

$$= c(a_1 + a_2 + \cdots + a_n) = c \sum_{i=1}^{n} a_i.$$

Thus
$$\sum_{i=1}^{n} ca_i = c \sum_{i=1}^{n} a_i.$$

You can factor a constant factor out of a sum.

This rule is read as "a constant factor can be moved past Σ."

Second,

$$\sum_{i=1}^{n} (a_i + b_i) = (a_1 + b_1) + (a_2 + b_2) + \cdots + (a_n + b_n)$$

$$= (a_1 + a_2 + \cdots + a_n) + (b_1 + b_2 + \cdots + b_n)$$

$$= \sum_{i=1}^{n} a_i + \sum_{i=1}^{n} b_i.$$

Thus
$$\sum_{i=1}^{n} (a_i + b_i) = \sum_{i=1}^{n} a_i + \sum_{i=1}^{n} b_i.$$

This is a direct consequence of the rules of algebra.

EXAMPLE 4 Compute $\sum_{i=1}^{4} \left(i^2 + \frac{1}{i}\right)$.

SOLUTION From the above remark we know that this may be rewritten as

$$\sum_{i=1}^{4} i^2 + \sum_{i=1}^{4} \frac{1}{i}.$$

By Examples 1 and 3, the sum, to three decimals, is $30 + 2.083 = 32.083$. ■

EXAMPLE 5 What is the value of $\Sigma_{i=1}^{5} 3$?

SOLUTION In this case, $a_i = 3$ for each index i. Each summand has the value 3. Thus

$$\sum_{i=1}^{5} 3 = 3 + 3 + 3 + 3 + 3 = 15.$$

$\sum_{i=1}^{n} c = cn.$

More generally, if c is a fixed number not depending on i, then $\Sigma_{i=1}^{n} c = cn$. ■

The next example shows how to interpret the sigma notation when the index does not start at 1.

EXAMPLE 6 Compute $\Sigma_{i=2}^{6} 5i$ (read as "the sum of $5i$ as i goes from 2 to 6").

SOLUTION This is short for $5 \cdot 2 + 5 \cdot 3 + 5 \cdot 4 + 5 \cdot 5 + 5 \cdot 6$, which equals

$$5(2 + 3 + 4 + 5 + 6),$$

or 100. ■

Another useful fact to note is that

Breaking a sum into a "front" end and a "back" end

$$\sum_{i=1}^{m} a_i + \sum_{i=m+1}^{n} a_i = \sum_{i=1}^{n} a_i \qquad 1 \le m < n.$$

For instance,

$$\sum_{i=1}^{4} i^2 + \sum_{i=5}^{10} i^2 = \sum_{i=1}^{10} i^2.$$

Telescoping Sums

Some sums are formed by adding up differences. In certain cases, the differences cancel in a way that makes the sum very easy to evaluate. This notion of a **telescoping sum** is introduced in Example 7.

EXAMPLE 7 Let b_0, b_1, b_2, b_3 be four numbers. Form the three differences

$$a_1 = b_1 - b_0 \qquad a_2 = b_2 - b_1 \qquad a_3 = b_3 - b_2,$$

and compute

$$\sum_{i=1}^{3} a_i.$$

SOLUTION

$$\sum_{i=1}^{3} a_i = a_1 + a_2 + a_3 = (b_1 - b_0) + (b_2 - b_1) + (b_3 - b_2).$$

Cancellations of b_1 and $-b_1$ and of b_2 and $-b_2$ show that

$$a_1 + a_2 + a_3 = b_3 - b_0. \quad ■$$

Example 7 can easily be generalized from four to any finite list of numbers. If $b_0, b_1, \ldots, b_n$ are $n + 1$ numbers and $a_i = b_i - b_{i-1}$, for $i = 1, \ldots, n$, then

$$\sum_{i=1}^{n} a_i = \sum_{i=1}^{n} (b_i - b_{i-1}) = b_n - b_0.$$

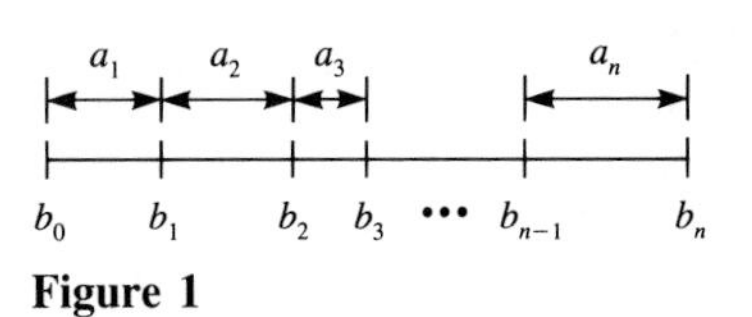

Figure 1

In short, the sum $\Sigma_{i=1}^{n} (b_i - b_{i-1})$ "telescopes" to $b_n - b_0$. (See Fig. 1.)

Two Useful Sums

Using a telescoping sum we will find a short formula for

$$\sum_{i=1}^{n} i = 1 + 2 + 3 + \cdots + n.$$

To begin, consider the telescoping sum $\Sigma_{i=1}^{n} [(i + 1)^2 - i^2]$. [In this case, $b_i = (i + 1)^2$.] Since $b_n = (n + 1)^2$ and $b_0 = (0 + 1)^2 = 1$,

$$\sum_{i=1}^{n} [(i + 1)^2 - i^2] = (n + 1)^2 - 1. \tag{1}$$

On the other hand, we know that

$$(i + 1)^2 - i^2 = i^2 + 2i + 1 - i^2 = 2i + 1.$$

Thus Eq. (1) can be written

$$\sum_{i=1}^{n} (2i + 1) = (n + 1)^2 - 1,$$

which simplifies to

$$2 \sum_{i=1}^{n} i + n = n^2 + 2n.$$

It follows that

Short formula for $\sum_{i=1}^{n} i$

$$\boxed{\sum_{i=1}^{n} i = \frac{n^2}{2} + \frac{n}{2} = \frac{n(n + 1)}{2}.}$$

We will also need a short formula for the sum of the squares of the integers from 1 to n, $\Sigma_{i=1}^{n} i^2 = 1^2 + 2^2 + 3^2 + \cdots + n^2$. This type of sum appeared several times in the preceding section. This time we start with the telescoping sum

$$\sum_{i=1}^{n} [(i + 1)^3 - i^3]. \tag{2}$$

[Here $b_i = (i + 1)^3$.] Hence $b_n = (n + 1)^3$ and $b_0 = (0 + 1)^3 = 1$. Thus, since the sum (2) is telescoping, it "collapses" to $b_n - b_0 = (n + 1)^3 - 1$,

$$\sum_{i=1}^{n} [(i + 1)^3 - i^3] = (n + 1)^3 - 1. \tag{3}$$

By the binomial theorem, $(i + 1)^3 = i^3 + 3i^2 + 3i + 1$. Hence Eq. (3) becomes

$$\sum_{i=1}^{n} (i^3 + 3i^2 + 3i + 1 - i^3) = n^3 + 3n^2 + 3n$$

or

$$3 \sum_{i=1}^{n} i^2 + 3 \sum_{i=1}^{n} i + \sum_{i=1}^{n} 1 = n^3 + 3n^2 + 3n. \tag{4}$$

Since $\Sigma_{i=1}^{n} 1 = n$ and, as was just shown, $\Sigma_{i=1}^{n} i = (n^2/2) + (n/2)$, Eq. (4) provides this equation for $\Sigma_{i=1}^{n} i^2$:

$$3\sum_{i=1}^{n} i^2 + 3\frac{n^2}{2} + 3\frac{n}{2} + n = n^3 + 3n^2 + 3n,$$

from which it follows that

Short formula for $\sum_{i=1}^{n} i^2$

$$\boxed{\sum_{i=1}^{n} i^2 = \frac{n^3}{3} + \frac{n^2}{2} + \frac{n}{6} = \frac{n(n+1)(2n+1)}{6}.} \qquad (5)$$

This formula will be needed in Sec. 5.3. (See Exercise 29 for an application to computer programming.)

Sigma Notation for Approximating Sums

In Sec. 5.1 we used sums of a certain form—a sum of products—to estimate area, mass, distance, and volume. The summation notation will help us to talk about such sums efficiently and in full generality. Consider, for instance, how an approximating sum for the area under x^2 and above the interval $[a, b]$ would be formed.

First, the interval $[a, b]$ is partitioned into smaller sections, perhaps all of equal length, perhaps not. There could be any finite number of sections. Say that there are n sections. These sections are determined by choosing $n - 1$ numbers in (a, b), $x_1, x_2, \ldots, x_{n-1}$,

$$a < x_1 < x_2 < \cdots < x_{n-1} < b.$$

For convenience, introduce

$$x_0 = a \qquad \text{and} \qquad x_n = b.$$

The ith section, $i = 1, 2, \ldots, n$, has the left endpoint x_{i-1} and the right endpoint x_i, as shown in Fig. 2. The typical section is shown in Fig. 3. For instance, the first section is $[x_0, x_1]$, which is $[a, x_1]$, the second section is $[x_1, x_2]$, . . . , and the nth section is $[x_{n-1}, x_n]$, which is $[x_{n-1}, b]$. The length of the ith section is $x_i - x_{i-1}$, which is often denoted Δx_i.

The partition: $x_0 = a$ x_1 x_2 ••• x_{i-1} x_i ••• x_{n-2} x_{n-1} $x_n = b$

Figure 2

After the partition or division into n sections is formed, a number is selected in each section at which to evaluate x^2. The number chosen in $[x_{i-1}, x_i]$ could be its left endpoint x_{i-1}, its right endpoint x_i, its midpoint $(x_{i-1} + x_i)/2$, or any point whatsoever in the section. To allow for the most general possible choice, denote the number chosen in the ith section by c_i. The number c_i is called a **sampling number**. It is shown in Fig. 4. The value of x^2 at c_i, namely c_i^2, is used to determine the height of an approximating rectangle. Figure 5 shows the typical staircase of rectangles formed this way.

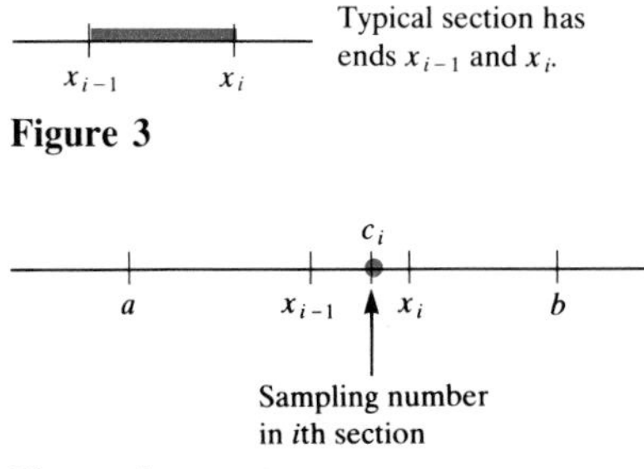

Figure 4

The next step is to evaluate the function x^2 at each c_i and form the sum with n summands:

$$c_1^2(x_1 - x_0) + c_2^2(x_2 - x_1) + \cdots + c_i^2(x_i - x_{i-1}) + \cdots + c_n^2(x_n - x_{n-1}). \qquad (6)$$

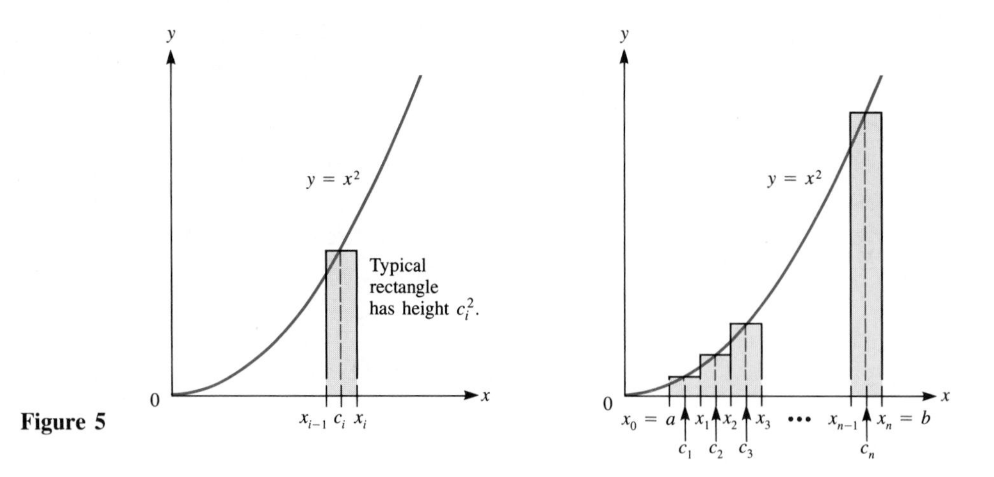

Figure 5

It takes a long time to write down the sum (6) or to read it aloud. The Σ notation compresses it to the simple expression,

$$\sum_{i=1}^{n} c_i^2(x_i - x_{i-1}),$$

which is read, "the sum from 1 to n of $c_i^2(x_i - x_{i-1})$." If the length $x_i - x_{i-1}$ is denoted Δx_i, (6) reduces to

$$\sum_{i=1}^{n} c_i^2\, \Delta x_i. \tag{7}$$

It must be kept in mind that $x_0 = a$ and $x_n = b$. Surely (7) is more convenient than (6).

EXAMPLE 8 The first sum formed in Sec. 5.1 was

$$(\tfrac{1}{2})^2(\tfrac{1}{2}) + (\tfrac{2}{2})^2(\tfrac{1}{2}) + (\tfrac{3}{2})^2(\tfrac{1}{2}) + (\tfrac{4}{2})^2(\tfrac{1}{2}) + (\tfrac{5}{2})^2(\tfrac{1}{2}) + (\tfrac{6}{2})^2(\tfrac{1}{2}). \tag{8}$$

It was formed by cutting [0, 3] into six sections of equal length and evaluating x^2 at the right endpoint of each section. State the values of x_i and c_i and write (8) in sigma notation.

SOLUTION In this case the interval $[a, b]$ is $[0, 3]$. The six sections are formed with the numbers $x_0 = 0$, $x_1 = \frac{1}{2}$, $x_2 = \frac{2}{2}$, $x_3 = \frac{3}{2}$, $x_4 = \frac{4}{2}$, $x_5 = \frac{5}{2}$, and $x_6 = 3$. The sampling numbers were $c_1 = \frac{1}{2}$ (in the first section), $c_2 = \frac{2}{2}$, $c_3 = \frac{3}{2}$, $c_4 = \frac{4}{2}$, $c_5 = \frac{5}{2}$, and $c_6 = \frac{6}{2}$. We have $x_i = i/2$ for $i = 0, 1, 2, \ldots, 6$ and $c_i = i/2$ for $i = 1, 2, \ldots, 6$. Each $x_i - x_{i-1}$ is equal to $\frac{1}{2}$, the width of the intervals. Thus

$$\sum_{i=1}^{n} c_i^2(x_i - x_{i-1}) = \sum_{i=1}^{6} (x_i^2)(\tfrac{1}{2})$$

$$= \sum_{i=1}^{6} \left(\frac{i}{2}\right)^2 (\tfrac{1}{2}).$$

Surely $\Sigma_{i=1}^{6}\, (i/2)^2(\frac{1}{2})$ is easier to write and read than the long sum (8). ■

EXAMPLE 9 Write in sigma notation the typical approximating sum for the area of the region under $f(x) = x^2$ and above $[2, 7]$ using right endpoints as sampling numbers.

SOLUTION In this case, $a = 2$, $b = 7$, and $c_i = x_i$. The typical sum is

$$\sum_{i=1}^{n} x_i^2(x_i - x_{i-1})$$

or

$$\sum_{i=1}^{n} x_i^2 \, \Delta x_i.$$

It is to be understood that

$$2 = x_0 < x_1 < \cdots < x_{n-1} < x_n = 7. \blacksquare$$

But x^2 is just one possible function. Similar approximating sums can be formed for any other function, such as x^3, $1/x$, or $\sin x$. The typical approximating sum for any function $f(x)$ and any interval $[a, b]$ in the domain of the function is formed just like Eq. (6). First, a partition of $[a, b]$ is determined:

$$a = x_0 < x_1 < \cdots < x_{n-1} < x_n = b.$$

Then a sampling number c_i is picked in the ith section, $i = 1, 2, \ldots, n$. The function $f(x)$ is evaluated at each c_i, and finally the approximating sum is formed:

The general approximating sum: long notation

$$f(c_1)(x_1 - x_0) + \cdots + f(c_i)(x_i - x_{i-1}) + \cdots + f(c_n)(x_n - x_{n-1}).$$

In Σ notation this reduces to

$$\sum_{i=1}^{n} f(c_i)(x_i - x_{i-1})$$

The general approximating sum: Σ notation

or

$$\sum_{i=1}^{n} f(c_i) \, \Delta x_i.$$

An approximating sum is also called a Riemann sum.

Such an approximating sum is also called a **Riemann sum** in honor of the nineteenth-century mathematician, Bernhard Riemann, who made many fundamental contributions to various branches of mathematics.

Section Summary

We introduced sigma notation in order to write approximating sums of the type used in Sec. 5.1 conveniently. Using telescoping sums we obtained the formulas

$$\sum_{i=1}^{n} i = \frac{n(n+1)}{2}$$

and

$$\sum_{i=1}^{n} i^2 = \frac{n(n+1)(2n+1)}{6}.$$

EXERCISES FOR SEC. 5.2: SUMMATION NOTATION AND APPROXIMATING SUMS

In Exercises 1 to 4 evaluate the sums.

1 (*a*) $\sum_{i=1}^{3} i$ (*b*) $\sum_{i=1}^{4} 2i$ (*c*) $\sum_{d=1}^{3} d^2$

2 (*a*) $\sum_{i=2}^{4} i^2$ (*b*) $\sum_{j=2}^{4} j^2$ (*c*) $\sum_{i=1}^{3} (i^2 + i)$

3 (*a*) $\sum_{i=1}^{4} 1^i$ (*b*) $\sum_{k=2}^{6} (-1)^k$ (*c*) $\sum_{j=1}^{150} 3$

4 (*a*) $\sum_{i=3}^{5} \frac{1}{i}$ (*b*) $\sum_{i=0}^{4} \cos 2\pi i$ (*c*) $\sum_{i=1}^{3} 2^{-i}$

In Exercises 5 to 8 write in the sigma notation. (Do not evaluate.)

5 (*a*) $1 + 2 + 2^2 + 2^3 + \cdots + 2^{100}$
(*b*) $x^3 + x^4 + x^5 + x^6 + x^7$
(*c*) $\frac{1}{3} + \frac{1}{4} + \cdots + \frac{1}{102}$

6 (*a*) $\frac{1}{2} + \frac{1}{3} + \cdots + \frac{1}{100}$
(*b*) $\frac{1}{3} + \frac{1}{5} + \frac{1}{7} + \frac{1}{9} + \frac{1}{11}$
(*c*) $\frac{1}{1^2} + \frac{1}{3^2} + \frac{1}{5^2} + \cdots + \frac{1}{101^2}$

7 (*a*) ${x_0}^2(x_1 - x_0) + {x_1}^2(x_2 - x_1) + {x_2}^2(x_3 - x_2)$
(*b*) ${x_1}^2(x_1 - x_0) + {x_2}^2(x_2 - x_1) + {x_3}^2(x_3 - x_2)$

8 (*a*) $8{t_0}^2(t_1 - t_0) + 8{t_1}^2(t_2 - t_1) + \cdots + 8{t_{99}}^2(t_{100} - t_{99})$
(*b*) $8{t_1}^2(t_1 - t_0) + 8{t_2}^2(t_2 - t_1) + \cdots + 8{t_n}^2(t_n - t_{n-1})$

In Exercises 9 and 10 evaluate the telescoping sums.

9 (*a*) $\sum_{i=1}^{100} (2^i - 2^{i-1})$ (*b*) $\sum_{i=2}^{100} \left(\frac{1}{i} - \frac{1}{i-1}\right)$

(*c*) $\sum_{i=1}^{50} \left(\frac{1}{2i+1} - \frac{1}{2(i-1)+1}\right)$

10 (*a*) $\sum_{i=1}^{100} \left(\frac{{x_i}^3}{3} - \frac{{x_{i-1}}^3}{3}\right)$ (*b*) $\sum_{i=5}^{70} \left(\frac{1}{x_i} - \frac{1}{x_{i-1}}\right)$

11 Writing out each sum in longhand, show that

(*a*) $\sum_{i=1}^{3} a_i = \sum_{j=1}^{3} a_j = \sum_{k=1}^{3} a_k$

(*b*) $\sum_{i=1}^{3} (a_i + 4) = 12 + \sum_{j=2}^{4} a_{j-1}$

12 Writing out each sum in longhand, show that

(*a*) $\sum_{i=1}^{3} (a_i - b_i) = \sum_{i=1}^{3} a_i - \sum_{i=1}^{3} b_i$

(*b*) $\sum_{i=1}^{2} a_i b_i$ is *not* always equal to $\sum_{i=1}^{2} a_i \cdot \sum_{i=1}^{2} b_i$.

(*c*) $\sum_{i=1}^{3} \left(\sum_{j=1}^{3} b_j\right) a_i = \sum_{j=1}^{3} \left(\sum_{i=1}^{3} a_i\right) b_j$

In Exercises 13 to 18 evaluate the approximating sum $\Sigma_{i=1}^{n} f(c_i)(x_i - x_{i-1})$. In each case the interval is [1, 3] and the partition consists of four sections of equal length. Express answers to two decimal places.

13 $f(x) = 3x$, $c_i = x_i$ **14** $f(x) = 3x$, $c_i = x_{i-1}$
15 $f(x) = 5$, $c_i = x_i$ **16** $f(x) = x^3 + x$, $c_i = x_i$
17 $f(x) = 1/x$, $c_1 = 1.25$, $c_2 = 1.8$, $c_3 = 2.2$, $c_4 = 3$
18 $f(x) = 2^x$, $c_i = x_{i-1}$

19 Evaluate (*a*) $\sum_{i=1}^{100} i$; (*b*) $\sum_{j=1}^{1000} 3j$.

20 Evaluate (*a*) $\sum_{i=1}^{100} i^2$; (*b*) $\sum_{i=1}^{n} (2i^2 + 3i + 6)$.

---

21 Using the method by which Eq. (5) was obtained, show that

$$\sum_{i=1}^{n} i^3 = \frac{n^4}{4} + \frac{n^3}{2} + \frac{n^2}{4}. \tag{9}$$

22 Using the method by which Eq. (5) was obtained, together with the result in Exercise 21, obtain a formula for $\Sigma_{i=1}^{n} i^4$.

23 Let $f(x) = 11$ for all x. Consider the partition $x_0 = a < x_1 < \cdots < x_n = b$ and let c_i be in $[x_{i-1}, x_i]$. Evaluate $\Sigma_{i=1}^{n} f(c_i)(x_i - x_{i-1})$.

24 Write in sigma notation the typical approximating sum for the area of the region under x^3 and above [1, 7] in which the sampling numbers c_i are (*a*) left endpoints, (*b*) right endpoints, (*c*) midpoints.

In Exercises 25 to 27 evaluate $\Sigma_{i=1}^{n} f(c_i)(x_i - x_{i-1})$ for the given data.

25 $f(x) = \sqrt{x}$, $x_0 = 1$, $x_1 = 3$, $x_2 = 5$, $c_1 = 1$, $c_2 = 4$ $(n = 2)$

26 $f(x) = \sqrt[3]{x}$, $x_0 = 0$, $x_1 = 1$, $x_2 = 4$, $x_3 = 10$, $c_1 = 0$, $c_2 = 1$, $c_3 = 8$ $(n = 3)$

27 $f(x) = 1/x$, $x_0 = 1$, $x_1 = 1.25$, $x_2 = 1.5$, $x_3 = 1.75$, $x_4 = 2$, $c_1 = 1$, $c_2 = 1.25$, $c_3 = 1.6$, $c_4 = 2$ $(n = 4)$

28 Write the expression

$$c^{n-1} + c^{n-2}d + c^{n-3}d^2 + \cdots + d^{n-1}$$

in the sigma notation.

29 A computer program designed to evaluate a certain function f takes kn^2 seconds to evaluate $f(n)$, where k is a constant. It takes 3 seconds to evaluate $f(1{,}000)$. About how long will it take to evaluate $f(n)$ for all n from 1 to 1,000,000?

30 The area under $f(x) = \sin x$ and above $[0, \pi]$ is to be estimated by a Riemann sum formed of n sections of equal length and left endpoints of sections as the sampling numbers. Write this sum in sigma notation.

31 Estimate the area under x^2 and above $[0, 3]$ by breaking the interval $[0, 3]$ into 100 sections of equal lengths and taking (*a*) right endpoints, (*b*) left endpoints as sampling points.

32 Estimate the area under 2^{-x} and above $[0, 1]$ by breaking the interval $[0, 1]$ into 100 sections of equal lengths and taking (*a*) right endpoints, (*b*) left endpoints as sampling points.

33 Compute $\Sigma_{i=1}^{n} f(i/n)1/n$ to three decimal places for $n = 5$ and (*a*) $f(x) = 1/x$, (*b*) $f(x) = \sin x$, and (*c*) $f(x) = 2^x$. (*d*) In each case draw the region whose area is being estimated by these sums and the rectangles making up the approximation.

34 Let r be a number other than 1 and $b_i = r^i$, $0 \le i \le n$. Use a telescoping sum to show that

$$1 + r + r^2 + \cdots + r^{n-1} = \frac{1 - r^n}{1 - r}.$$

5.3 THE DEFINITE INTEGRAL

There are two main concepts in calculus: the derivative and the definite integral. This section defines the definite integral and uses it to solve the four problems in Sec. 5.1. In Sec. 5.1 there were only *estimates* of the four quantities; now we will obtain their exact values.

The Definite Integral

In Sec. 5.1 sums of the form

$$\sum_{i=1}^{n} f(c_i)(x_i - x_{i-1})$$

were used to estimate certain quantities such as area, mass, distance, and volume. The larger n is and the shorter the sections $[x_{i-1}, x_i]$ are, the closer we would expect these approximating sums to be to the quantity we are trying to find. We are really interested in *what happens to these approximating sums as all the sections in the partition are chosen smaller and smaller.* This leads to the notion of the definite integral of a function over an interval, which will be defined after we introduce a measure of the "fineness" of a partition.

Definition *Mesh.* The **mesh** of a partition is the length of the longest section (or sections) in the partition.

For instance, the first partition used in Sec. 5.1 has mesh equal to $\frac{1}{2}$. Another term sometimes used for the mesh is *norm.*

If the interval $[a, b]$ is cut into n sections, $[x_0, x_1]$, $[x_1, x_2]$, . . . , $[x_{n-1}, x_n]$, then the mesh is the largest width of any of these sections. Thus the mesh is the maximum $x_i - x_{i-1}$ (or maximum Δx_i). If the mesh is small, all the sections must be short. Hence n must be large. When we say that "the mesh is small," we are simply saying that "all the sections are small."

We are now ready to define the definite integral.

Definition *The definite integral of a function f over an interval* $[a, b]$. If f is a function defined on $[a, b]$ and the sums $\Sigma_{i=1}^{n} f(c_i)(x_i - x_{i-1})$ approach a certain number as the mesh of partitions of $[a, b]$ shrinks toward 0 (no matter how the sampling number c_i is chosen in $[x_{i-1}, x_i]$), that certain number is called the **definite integral of f over $[a, b]$ or the definite integral of f from a to b.** It is denoted

$$\int_a^b f(x)\, dx.$$

The definite integral is a limit of sums.

In short, the definite integral of f over $[a, b]$ is

$$\lim_{\text{mesh}\to 0} \sum_{i=1}^{n} f(c_i)\, \Delta x_i.$$

EXAMPLE 1 In Problem 1 of Sec. 5.1 we formed sums of the form $\Sigma_{i=1}^{n} c_i^2\, \Delta x_i$ to estimate the area of the region under $f(x) = x^2$ and above $[0, 3]$. What is the meaning of $\int_0^3 x^2\, dx$ in that problem?

SOLUTION Any particular sum $\Sigma_{i=1}^{n} c_i^2\, \Delta x_i$ is an *estimate* of that area. As all the Δx_i are chosen smaller, the rectangles become very narrow. Their total area becomes a better approximation of the area under the curve $y = x^2$. The number that these sums approach as $\Delta x_i \to 0$ would be the exact value of the area under the curve. Therefore the definite integral $\int_0^3 x^2\, dx$, the number that the sums $\Sigma_{i=1}^{n} c_i^2\, \Delta x_i$ approach as the mesh approaches 0, equals the area. Thus

$$\text{Area under } x^2 \text{ and above } [0, 3] = \int_0^3 x^2\, dx.$$

If we can find $\int_0^3 x^2\, dx$, we would then know the area under the curve $f(x) = x^2$ above $[0, 3]$. ■

The symbol $\int_a^b x^2\, dx$ is read as "the integral from a to b of x^2." Freeing ourselves from the variable x, we could say, "the integral from a to b of the squaring function." There is nothing special about the symbol x in "x^2." We could just as well have used the letter t—or any other letter. (We would typically pick a letter near the end of the alphabet, since letters near the beginning are customarily used to denote constants.) The notations

$$\int_a^b x^2\, dx, \qquad \int_a^b t^2\, dt, \qquad \int_a^b y^2\, dy, \qquad \int_a^b z^2\, dz$$

all denote the same number, that is, "the definite integral of the squaring function from a to b." For that matter, we might just write

$$\int_a^b (\quad)^2\, d(\quad)$$

to convey the same idea. Usually, however, we find it more convenient to use some letter to name the variable. Since the actual letter chosen to represent the variable has no significance of its own, it is called a *dummy variable*. While x is a typical choice for the dummy variable, we will occasionally find it convenient to use the symbol t. Later in this chapter there will be cases where the interval of integration is $[a, x]$ instead of $[a, b]$. Were we to write $\int_a^x x^2\,dx$, you might think that there is some relation between the x in x^2 and the x in the interval of integration. To avoid possible confusion, we prefer to write $\int_a^x t^2\,dt$ in such cases.

The symbol $\int$ comes from the letter s of sum. The dx traditionally suggests a small section of the x axis and denotes the **variable of integration** (usually x, as in this case). The function $f(x)$ is called the **integrand**, while the numbers a and b are called the **limits of integration**; a is the **lower limit of integration** and b is the **upper limit of integration**. (Limits of integration are not limits in the usual mathematical sense, but the terminology is traditional and difficult to avoid. The left endpoint of the interval of integration is a and the right endpoint is b. Sometimes limits of integration are simply called *ends* or *endpoints of integration,* but these terms are not particularly popular.)

It is important to realize that area, mass, distance traveled, and volume are merely applications of the definite integral. (It is a mistake to link the definite integral too closely with one of its applications, just as it narrows our understanding of the number 2 to link it always with the idea of two fingers.) The definite integral $\int_a^b f(x)\,dx$ is also called the **Riemann integral** in honor of the mathematician who defined it.

Slope, velocity, magnification, and density are particular interpretations or applications of the derivative, which is a purely mathematical concept defined as a limit:

A derivative is a limit.

$$\text{Derivative of } f \text{ at } x = \lim_{\Delta x \to 0} \frac{f(x + \Delta x) - f(x)}{\Delta x}.$$

Similarly, area, total distance, mass, and volume are just particular interpretations of the definite integral, which is also defined as a limit:

A definite integral is also a limit.

$$\text{Definite integral of } f \text{ over } [a, b] = \lim_{\text{mesh} \to 0} \sum_{i=1}^{n} f(c_i)(x_i - x_{i-1}).$$

Remark: If you are visually oriented, you can think of $\Sigma_{i=1}^n f(c_i)\,\Delta x_i$ as the area of a staircase of narrow rectangles and $\int_a^b f(x)\,dx$ as the exact area of the region that the staircase approximates. (See Fig. 1.) But that is only one interpretation.

In advanced calculus it is proved that if f is continuous, then

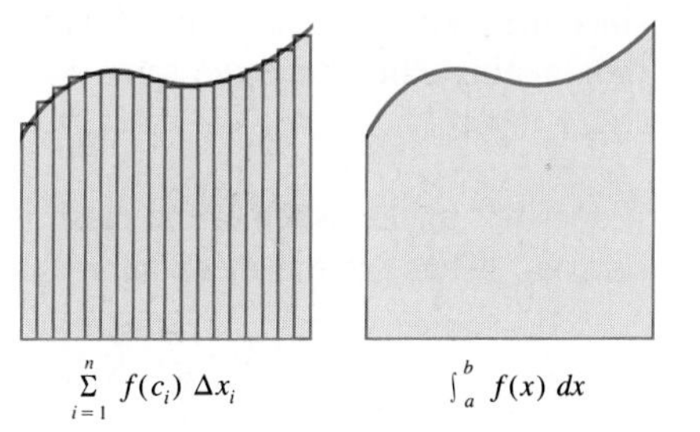

Figure 1

$$\lim_{\text{mesh}\to 0} \sum_{i=1}^{n} f(c_i)(x_i - x_{i-1})$$

exists; that is, a continuous function always has a definite integral. For emphasis, we record this fact, an important result in advanced calculus, as a theorem.

Recall that $x_0 = a$ and $x_n = b$.

Theorem *Existence of the definite integral.* Let f be a continuous function defined on $[a, b]$. Then the approximating sums

$$\sum_{i=1}^{n} f(c_i)(x_i - x_{i-1})$$

approach a single number as the mesh of the partition of $[a, b]$ approaches 0. Hence $\int_a^b f(x)\, dx$ exists.

The Definite Integral of a Constant Function

To bring the definition down to earth, let us use it to evaluate the definite integral of a constant function.

EXAMPLE 2 Let f be the function whose value at any number x is 4; that is, f is the constant function given by the formula $f(x) = 4$. Use only the definition of the definite integral to compute

$$\int_1^3 f(x)\, dx.$$

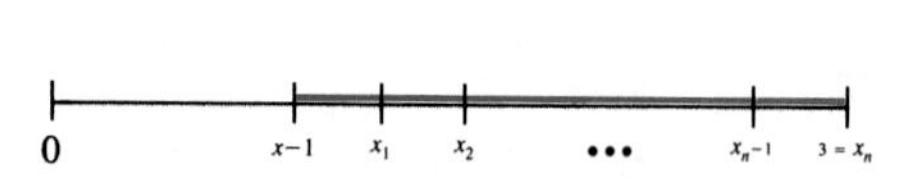

A typical partition of [1, 3]

Figure 2

SOLUTION In this case, a typical partition has $x_0 = 1$ and $x_n = 3$. See Fig. 2. The approximating sum

$$\sum_{i=1}^{n} f(c_i)(x_i - x_{i-1})$$

becomes

$$\sum_{i=1}^{n} 4(x_i - x_{i-1})$$

since, no matter how the sampling number c_i is chosen, $f(c_i) = 4$. Now

$$\sum_{i=1}^{n} 4(x_i - x_{i-1}) = 4 \sum_{i=1}^{n} (x_i - x_{i-1}) = 4(x_n - x_0),$$

since the sum is telescoping. Since $x_n = 3$ and $x_0 = 1$, it follows that all approximating sums have the same value, namely,

$$4(3 - 1) = 8.$$

It does not matter whether the mesh is small or where the c_i are picked in each section. Thus, as the mesh approaches 0,

$$\sum_{i=1}^{n} f(c_i)(x_i - x_{i-1})$$

approaches 8. Indeed, the sums are always 8. Thus

$$\int_1^3 4\,dx = 8. \quad \blacksquare$$

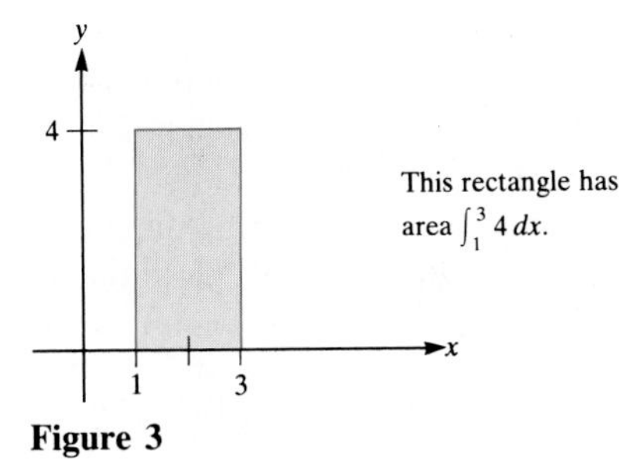

Figure 3

We could have guessed the value of $\int_1^3 4\,dx$ by interpreting the definite integral as an area. To do so, draw a rectangle of height 4 and base coinciding with the interval [1, 3]. (See Fig. 3.) Since the area of a rectangle is its base times its height, it follows again that $\int_1^3 4\,dx = 8$.

Similar reasoning shows that for any constant function that has the fixed value c,

$$\boxed{\int_a^b c\,dx = c(b-a).} \qquad (c \text{ is a constant function.})$$

In Exercise 23 you will be asked to use the area interpretation to show that

$$\int_a^b x\,dx = \frac{b^2}{2} - \frac{a^2}{2}.$$

Although area is the most intuitive of the interpretations, physical scientists, if they want to think of the definite integral concretely, should think of it as giving total mass if we know the density everywhere. This interpretation carries through easily to higher dimensions; the area interpretation does not.

It is the concept of the definite integral that links the four problems of Sec. 5.1, which are summarized below.

PROBLEM 1 The area under the curve $y = x^2$ and above [0, 3] equals the definite integral $\int_0^3 x^2\,dx$.

PROBLEM 2 The mass of the string whose density is x^2 grams per centimeter equals the definite integral $\int_0^3 x^2\,dx$.

PROBLEM 3 The distance traveled by the engineer whose speed is $8t^2$ miles per hour at time t equals the definite integral $\int_0^3 8t^2\,dt$ (the t reminding us of time).

PROBLEM 4 The volume of the tent equals the definite integral $\int_0^3 x^2\,dx$.

It is somewhat satisfying to have reduced all four problems to one problem, that of evaluating the definite integral of x^2 from 0 to 3, $\int_0^3 x^2\,dx$. Once it is found, all four problems are solved.

But it is one thing to define a certain limit; it is quite a different matter to evaluate it. Recall that the derivative was defined in Sec. 3.2, but it took several sections to develop techniques for computing it. Just as we resorted to algebra to differentiate x^2, x^3, and $\sqrt{x}$ in Sec. 3.2, we will use algebra to compute $\int_0^3 x^2\,dx$.

Evaluating $\int_0^b x^2\,dx$

It takes no more work to evaluate $\int_0^b x^2\,dx$ than $\int_0^3 x^2\,dx$, so let us evaluate

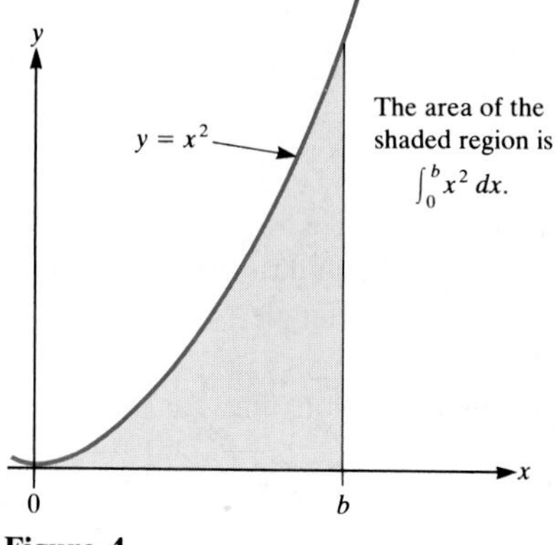

Figure 4

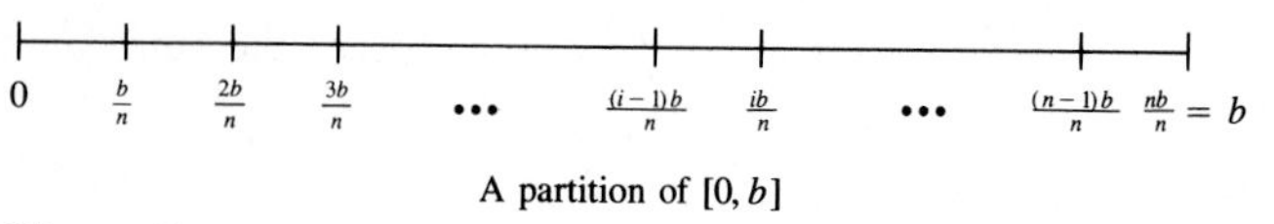

Figure 5

$\int_0^b x^2\,dx$. Although the reasoning will be free of any particular interpretation, the definite integral $\int_0^b x^2\,dx$ can be thought of as the area of the region under the curve $y = x^2$ and above the interval $[0, b]$, as shown in Fig. 4.

We assume that $\int_0^b x^2\,dx$ exists; it remains only to examine certain approximating sums to find that limit. For ease of computation, we use partitions in which all the sections have the same length. As sampling points, we use right-hand endpoints.

So we let n be a positive integer and partition $[0, b]$ by the numbers

$$0, \frac{b}{n}, \frac{2b}{n}, \ldots, \frac{ib}{n}, \ldots, \frac{nb}{n} = b,$$

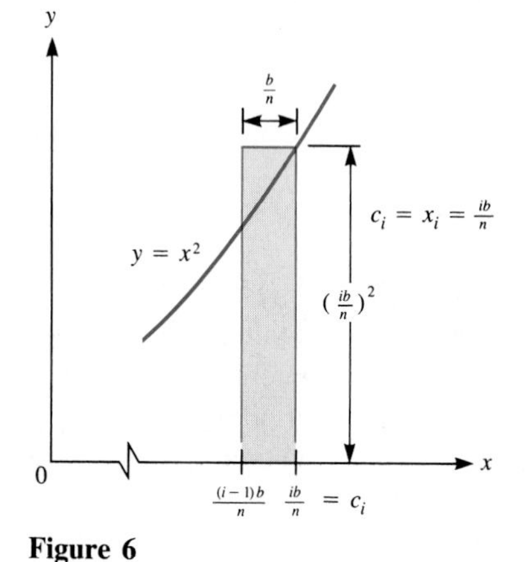

Figure 6

as shown in Fig. 5. There are n sections and each section has length b/n. The ith section is $[(i-1)b/n, ib/n]$; its right endpoint is ib/n. The partition has $x_i = ib/n$, and $x_i - x_{i-1} = b/n$. The sampling point is $c_i = ib/n$. The function is x^2. Since the width of each section is b/n, the mesh is b/n, which approaches 0 as $n \to \infty$.

Figure 6 shows the typical rectangle in the approximation and Fig. 7 depicts the whole collection of rectangles that approximate the area of the region under $f(x) = x^2$ and above $[0, b]$.

If we now add the areas of these rectangles, we obtain a formula for our approximating sum. Thus

$\sum_{i=1}^{n} i^2$ *was found in Sec. 5.2.*

$$\sum_{i=1}^{n} f(c_i)(x_i - x_{i-1}) = \sum_{i=1}^{n} \left(\frac{ib}{n}\right)^2 \frac{b}{n}$$

$$= \frac{b}{n}\frac{b^2}{n^2}\sum_{i=1}^{n} i^2$$

$$= \frac{b^3}{n^3}\left(\frac{n^3}{3} + \frac{n^2}{2} + \frac{n}{6}\right)$$

$$= \frac{b^3}{3} + \frac{b^3}{2n} + \frac{b^3}{6n^2}.$$

As $n \to \infty$, this approaches $b^3/3$. We conclude that

$$\int_0^b x^2\,dx = \frac{b^3}{3}. \tag{1}$$

In particular, when $b = 3$, Eq. (1) tells us that $\int_0^3 x^2\,dx = 3^3/3 = 9$. From this result we immediately conclude that:

Precise answers to the four problems.

The area under $y = x^2$ and above $[0, 3]$ is $\int_0^3 x^2\,dx = 9$ square units.

The mass of the string in Problem 2 of Sec. 5.1 is $\int_0^3 x^2\,dx = 9$ grams.

The engineer in Problem 3 of Sec. 5.1 travels $\int_0^3 8t^2\,dt = 8 \cdot 9 = 72$ miles.

The volume of the tent in Problem 4 of Sec. 5.1 is $\int_0^3 x^2\,dx = 9$ cubic feet.

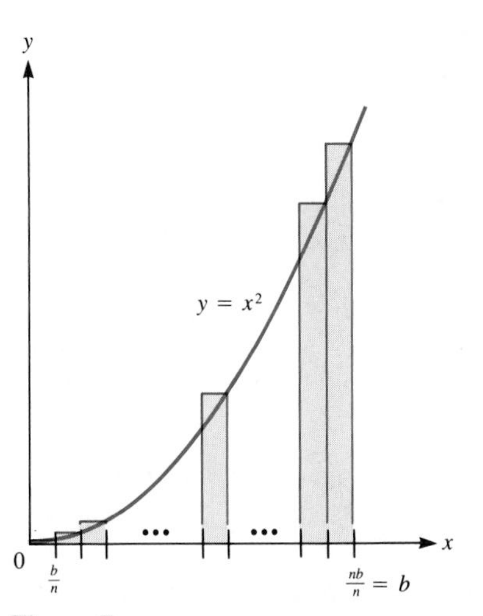

Figure 7

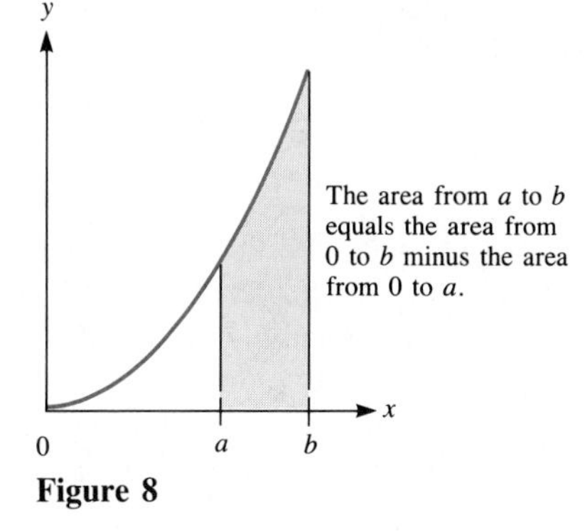

Figure 8

The same technique which showed that $\int_0^b x^2\,dx = b^3/3$ can be used to show that

$$\boxed{\int_a^b x^2\,dx = \frac{b^3}{3} - \frac{a^3}{3},}$$

where $0 \le a \le b$. You might guess this by considering it in terms of area. Figure 8 shows that the area from a to b is equal to the difference between the area from 0 to b and the area from 0 to a. So we expect that

$$\int_a^b x^2\,dx = \int_0^b x^2\,dx - \int_0^a x^2\,dx$$
$$= \frac{b^3}{3} - \frac{a^3}{3}.$$

Exercise 5 presents a formal demonstration that does not depend on areas or sketches.

This is a good time to summarize the four main applications of the definite integral in full generality. Each has already been illustrated by one of the four problems in Sec. 5.1.

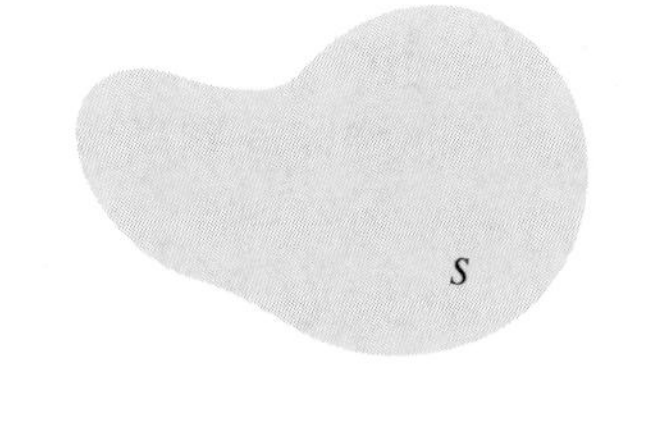

Figure 9

Area of a Plane Region as a Definite Integral Let S be some region in the plane whose area is to be found. Let L be a line in the plane which will be considered to be the x axis. Each line in the plane and perpendicular to L meets S in what shall be called a cross section. (If the line misses S, the cross section is empty.) See Figs. 9 to 11. Let the coordinate on L where the typical line meets

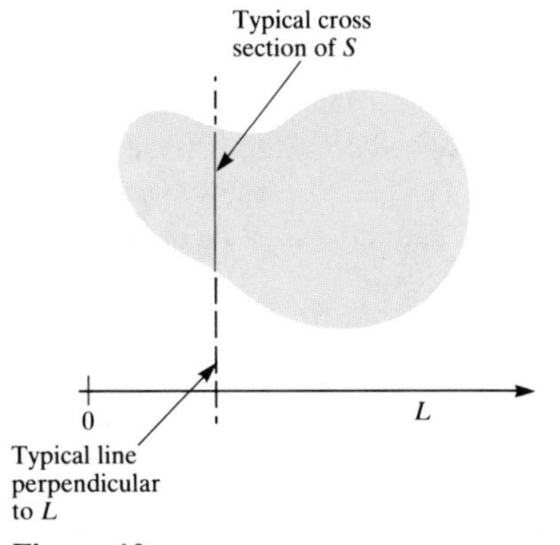

Figure 10

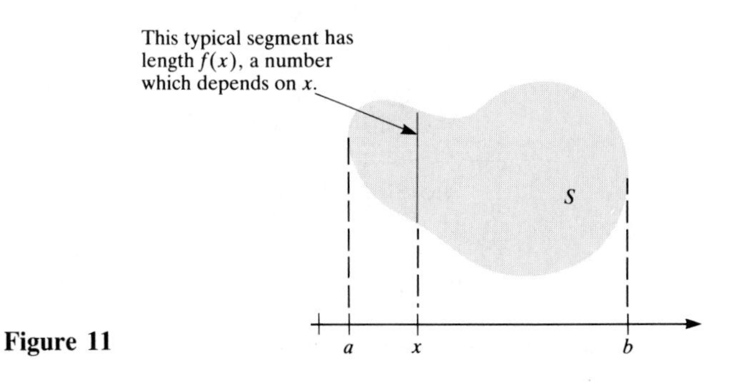

Figure 11

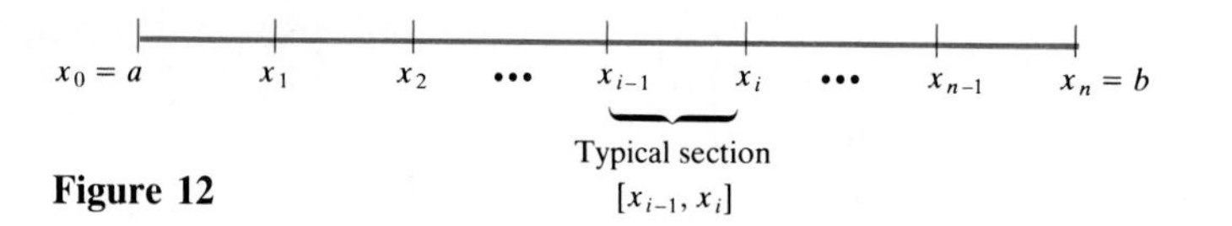

Figure 12

L be x. The length of the typical cross section is denoted by $f(x)$. Assume that the lines which are perpendicular to L and which meet S intersect L in an interval whose ends are a and b. [In Problem 1 of Sec. 5.1, L is the x axis, $f(x) = x^2$, $a = 0$, and $b = 3$.]

To estimate the area of S, proceed just as for the region under $y = x^2$. First cut the interval $[a, b]$ into n sections by means of the numbers $x_0 = a, x_1, x_2, \ldots, x_n = b$, as in Fig. 12.

In each of these sections, select a number c_i at random. In the section $[x_0, x_1]$ select c_1, in $[x_1, x_2]$ select c_2, and so on. Simply stated, in the ith interval, $[x_{i-1}, x_i]$, select c_i.

With this choice of x_i's and c_i's, form a set of rectangles whose typical member is shaded in Fig. 13. Then $\Sigma_{i=1}^n f(c_i)(x_i - x_{i-1})$ is an *estimate* of the area of S. As the lengths $x_i - x_{i-1}$ are chosen to be smaller and smaller, we would expect that these sums tend toward the area of S.

A typical rectangle has base $x_i - x_{i-1}$, height $f(c_i)$, and area $f(c_i)(x_i - x_{i-1})$.

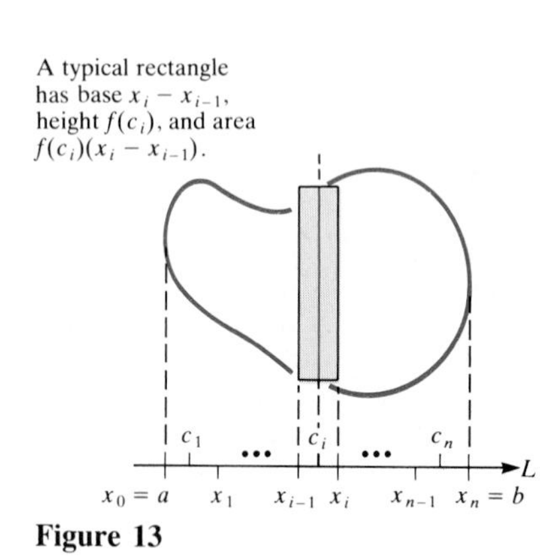

Figure 13

But, by the definition of the definite integral, the sums

$$\sum_{i=1}^{n} f(c_i)(x_i - x_{i-1})$$

approach

$$\int_a^b f(x)\,dx$$

as the mesh, the largest $\Delta x_i = x_i - x_{i-1}$, approaches 0.

Thus

Area of $S = \int_a^b f(x)\,dx$, where $f(x)$ is the length of a cross section of S.

In short, *area is the definite integral of cross-sectional length.* In practical terms, this tells us that if we can compute definite integrals, then we can compute areas.

The Mass of a String as a Definite Integral A string is made of a material whose density may vary from point to point. (Such a string is called nonuniform or **nonhomogeneous.**) How would its total mass be computed if its density at each point is known?

First, place the string somewhere on the x axis and let $f(x)$ be its density at x. The string occupies an interval $[a, b]$ on the axis, as in Fig. 14. Then cut the string into n sections $[x_{i-1}, x_i]$, $i = 1, 2, \ldots, n$, as in Fig. 15. In each section the density is almost constant. (We assume the density is a continuous

Nonuniform string

Figure 14

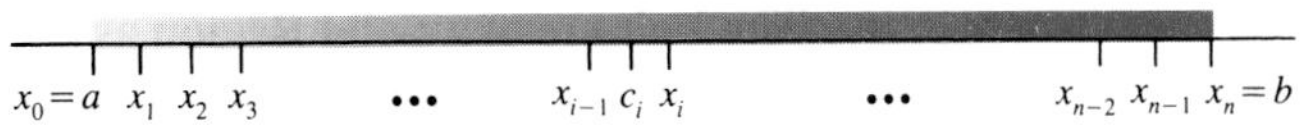

$x_0 = a \quad x_1 \quad x_2 \quad x_3 \quad \cdots \quad x_{i-1} \quad c_i \quad x_i \quad \cdots \quad x_{n-2} \quad x_{n-1} \quad x_n = b$

Figure 15

function. In a short interval it varies very little.) So the mass of the ith section is approximately

$$f(c_i)(x_i - x_{i-1}),$$

where c_i is some point in $[x_{i-1}, x_i]$. (See Fig. 16.) Thus we see that $\sum_{i=1}^{n} f(c_i)(x_i - x_{i-1})$ is an estimate of the total mass. And, what is more important, it seems plausible that as the mesh of the partition approaches 0, the sum $\sum_{i=1}^{n} f(c_i)(x_i - x_{i-1})$ approaches the mass of the string.

The mass in the typical section is approximately $f(c_i)(x_i - x_{i-1})$.

$x_{i-1} \quad c_i \quad x_i$

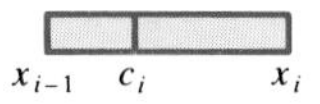

Figure 16

[Problem 2 in Sec. 5.1 is the case where $a = 0$, $b = 3$, and $f(x) = x^2$.] But, by the definition of the definite integral, the sums

$$\sum_{i=1}^{n} f(c_i)(x_i - x_{i-1})$$

approach

$$\int_a^b f(x)\, dx$$

as the mesh approaches 0. Thus

$$\text{Mass} = \int_a^b f(x)\, dx, \text{ where } f(x) \text{ is the density at } x.$$

In short, *mass is the definite integral of density.*

The Distance Traveled as a Definite Integral An engineer takes a trip that begins at time a and ends at time b. At any time t during the trip her velocity is $f(t)$, depending on the time t. How far does she travel? [Problem 3 in Sec. 5.1 is the case in which $a = 0$, $b = 3$, and $f(t) = 8t^2$.]

First, cut the time interval $[a, b]$ into n smaller intervals by a partition and estimate the trip's length by summing the estimates of the distance the engineer travels during each of the time intervals. (See Fig. 17.)

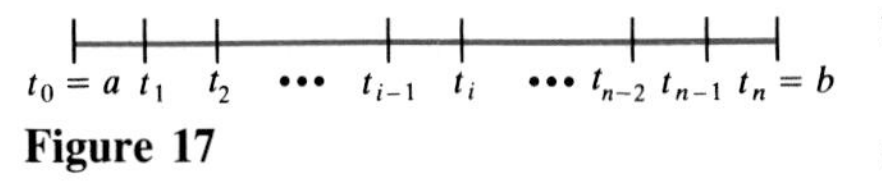

Figure 17

During a small interval of time, the velocity changes little. We thus expect to obtain a reasonable estimate of the distance covered during the ith time interval $[t_{i-1}, t_i]$ by observing the speedometer reading at some instant T_i in that interval, $f(T_i)$, and computing the product $f(T_i)(t_i - t_{i-1})$. Thus $\sum_{i=1}^{n} f(T_i)(t_i - t_{i-1})$ is an estimate of the length of the trip. Moreover, as the mesh of the partition approaches zero, the sum $\sum_{i=1}^{n} f(T_i)(t_i - t_{i-1})$ approaches the length of the trip.

Since these sums also approach the definite integral $\int_a^b f(t)\, dt$, we have

$$\text{Total distance} = \int_a^b f(t)\, dt, \text{ where } f(t) \text{ is the velocity at time } t.$$

In short, *distance is the definite integral of velocity.*

2 Find the mesh of each of these partitions of $[2, 6]$:
(*a*) $x_0 = 2,\ x_1 = 4,\ x_2 = 6$
(*b*) $x_0 = 2,\ x_1 = 3,\ x_2 = 6$
(*c*) $x_0 = 2,\ x_1 = 2.5,\ x_2 = 3,\ x_3 = 3.5,\ x_4 = 4,\ x_5 = 5,\ x_6 = 6$

3 Using the formula for $\int_a^b x^2\,dx$, find the area under the curve $y = x^2$ and above the interval (*a*) $[0, 5]$; (*b*) $[0, 4]$; (*c*) $[4, 5]$.

4 Figure 23 shows the curve $y = x^2$. What is the ratio between the shaded area under the curve and the area of the rectangle *ABCD*?

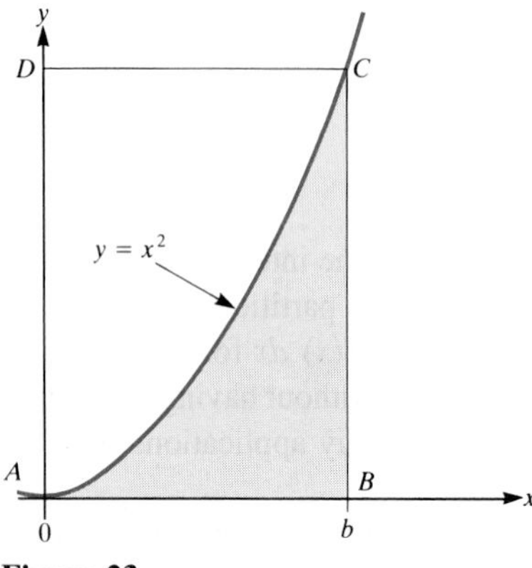

Figure 23

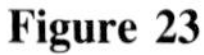

5 (This exercise obtains $\int_a^b x^2\,dx$.) Let n be a positive integer. Partition $[a, b]$ into n sections of equal length using the numbers $x_0, x_1, \ldots, x_n$.
(*a*) What is x_0?
(*b*) What is x_1?
(*c*) What is x_2? x_i?
(*d*) Using the right-hand endpoint of each section as the sampling point, compute $\sum_{i=1}^{n} x_i^2 (x_i - x_{i-1})$.
(*e*) Use (*d*) to find $\int_a^b x^2\,dx$.

6 The same as Exercise 5, except use the left-hand endpoint of each section as the sampling point.

7 A rocket moving with a varying speed travels at $f(t)$ miles per second at time t. Let $t_0, \ldots, t_n$ be a partition of $[a, b]$, and let $T_1, \ldots, T_n$ be sampling numbers. What is the physical interpretation of
(*a*) $t_i - t_{i-1}$? (*b*) $f(T_i)$? (*c*) $f(T_i)(t_i - t_{i-1})$?
(*d*) $\sum_{i=1}^{n} f(T_i)(t_i - t_{i-1})$? (*e*) $\int_a^b f(t)\,dt$?

8 A string occupying the interval $[a, b]$ has the density $f(x)$ grams per centimeter at x. Let $x_0, \ldots, x_n$ be a partition of $[a, b]$, and let $c_1, \ldots, c_n$ be sampling numbers. What is the physical interpretation of
(*a*) $x_i - x_{i-1}$? (*b*) $f(c_i)$?
(*c*) $f(c_i)(x_i - x_{i-1})$? (*d*) $\sum_{i=1}^{n} f(c_i)(x_i - x_{i-1})$?
(*e*) $\int_a^b f(x)\,dx$?

9 What is meant by $\int_1^3 (1/x^2)\,dx$? Describe in detail.

10 What is meant by $\int_1^\pi \sin x\,dx$? Describe in detail.

11 The density of a string is x^2 grams per centimeter at the point x. Find the mass in the interval $[1, 3]$.

12 An object is moving with a velocity of t^2 feet per second at time t seconds. How far does it travel during the time interval $[2, 5]$?

Exercises 13 to 18 concern the definite integral, free of any particular interpretation. Express answers to four decimal places.

13 Estimate $\int_1^3 (1/x)\,dx$, using a partition into four sections of equal length and, as sampling points, (*a*) left endpoints, (*b*) right endpoints.

14 Estimate $\int_0^3 2^x\,dx$, using a partition into three sections of equal length and, as sampling points, (*a*) left endpoints, (*b*) right endpoints.

15 Estimate $\int_0^1 x^3\,dx$, using a partition into five sections of equal length and, as sampling points, (*a*) left endpoints, (*b*) right endpoints.

16 Estimate $\int_0^1 x^3\,dx$, using a partition into 10 sections of equal length and left endpoints as sampling points.

17 Estimate $\int_0^{\pi/2} \sin x\,dx$, using the partition $x_0 = 0$, $x_1 = \pi/6$, $x_2 = \pi/4$, $x_3 = \pi/3$, and $x_4 = \pi/2$ and, as sampling points, (*a*) left endpoints, (*b*) right endpoints.

18 Estimate $\int_0^1 \sqrt{x}\,dx$, using a partition into 10 sections of equal length and, as sampling points, (*a*) left endpoints, (*b*) right endpoints, (*c*) midpoints.

---

19 (Exercise 21 of Sec. 5.2 provides a formula for $\sum_{i=1}^{n} i^3$.)
(*a*) Using the method of this section to find $\int_0^b x^2\,dx$, find $\int_0^b x^3\,dx$.
(*b*) Find the area under the curve $y = x^3$ and above the interval $[1, 2]$.

20 In Exercise 22 of Sec. 5.2 it was shown that

$$\sum_{i=1}^{n} i^4 = \frac{n^5}{5} + \frac{n^4}{2} + \frac{n^3}{3} - \frac{n}{30}.$$

(*a*) Obtain a formula for $\sum_{i=1}^{n} i^5$.
(*b*) Using the method of this section, find $\int_0^b x^5\,dx$.

21 Estimate $\int_1^3 (1/x^2)\,dx$, using a partition into four sections of equal length and left endpoints as sampling points.

22 Estimate $\int_1^4 (2x + 1)\,dx$, using a partition into three sections of equal length and right endpoints as sampling points.

23 Figure 24 shows the graph of $y = x$ and the region below it and above the interval $[a, b]$, $0 \le a < b$. Using the formula for the area of a trapezoid (see inside back cover), show that $\int_a^b x\,dx = b^2/2 - a^2/2$.

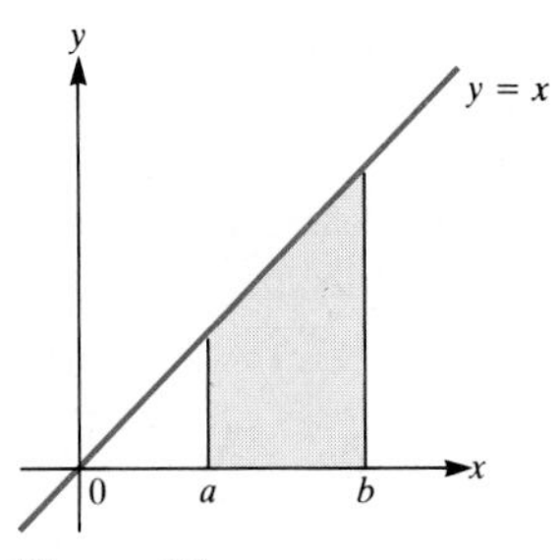

Figure 24

24 (See Exercise 23.)

(*a*) Set up an appropriate definite integral $\int_a^b f(x)\,dx$ which equals the volume of the headlight in Fig. 25, whose cross section by a typical plane perpendicular to the x axis at x is a circle whose radius is $\sqrt{x/\pi}$.

(*b*) Evaluate the definite integral in (*a*) with the aid of Exercise 23.

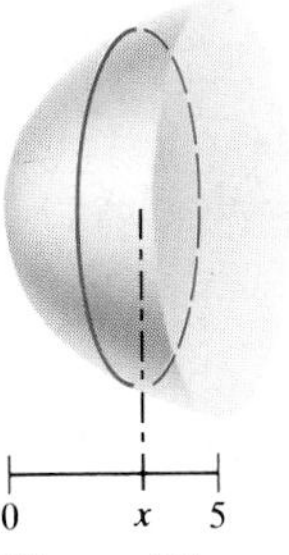

Figure 25

25 By considering Fig. 26, in particular the area of region *ACD*, show that $\int_0^a \sqrt{x}\,dx = \frac{2}{3}a^{3/2}$.

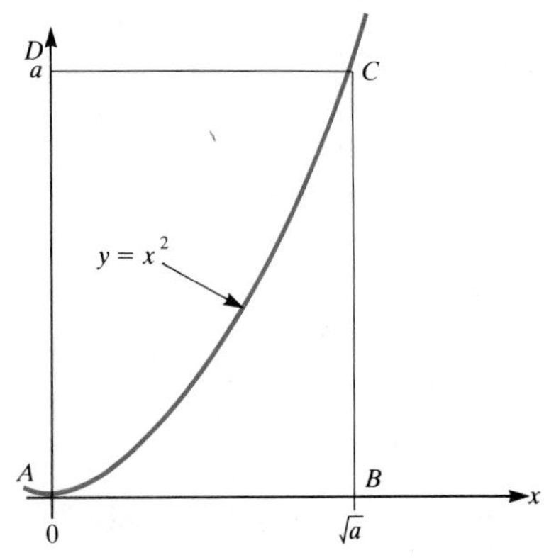

Figure 26

26 Show that the volume of a right circular cone of radius a and height h is $\pi a^2 h/3$. (*Suggestion:* First show that a cross section by a plane perpendicular to the axis of the cone and a distance x from the vertex is a circle of radius ax/h.)

27 (*a*) Sketch a graph of $y = 1/(1 + x^2)$ for $0 \le x \le 1$.

(*b*) Let A be the area under the graph in (*a*) and above $[0, 1]$. Show that $A < 1$.

(*c*) Use elementary geometry to show $\frac{3}{4} < A$.

(*d*) Use a partition of $[0, 1]$ into five sections to obtain lower and upper estimates of A. [It will be shown in Chap. 7 that $\int_0^1 1/(1 + x^2)\,dx = \pi/4$, hence that $A \approx 0.7854$.]

The next two exercises concern a property of the function $1/x$ first noticed in the seventeenth century.

28 (*a*) Write out the typical approximating sum for the area under the curve $y = 1/x$ and above the interval $[1, 2]$, using left endpoints as the c_i and the typical partition $x_0 = 1, x_1, \ldots, x_n = 2$.

(*b*) Show that $3x_0, 3x_1, \ldots, 3x_n$ is a partition of the interval $[3, 6]$. Show that if the left endpoints are used to form the approximating sum for the area under $y = 1/x$ and above $[3, 6]$, then this sum has the same value as the one obtained in (*a*).

(*c*) Show that the area under the curve $y = 1/x$ and above $[1, 2]$ equals the area under the curve $y = 1/x$ and above $[3, 6]$.

29 (See Exercise 28.)

(*a*) Show that if $A(t)$ is the area under the curve $y = 1/x$ and above $[1, t]$, then $A(2) = A(6) - A(3)$.

(*b*) Show that if $x > 1$ and $y > 1$, then $A(x)$ is equal to $A(xy) - A(y)$.

(*c*) By (*b*), $A(xy) = A(x) + A(y)$ for x and y greater than 1. What famous functions f have the property that $f(xy) = f(x) + f(y)$ for all positive x and y?

30 Assume that $f(x) \le -3$ for all x in $[1, 5]$. What can be said about the value of $\int_1^5 f(x)\,dx$? Explain in detail, using the definition of the definite integral.

31 The velocity of an automobile at time t is $v(t)$ feet per second. [Assume $v(t) \ge 0$.] The graph of v for t in $[0, 20]$ is shown in Fig. 27.

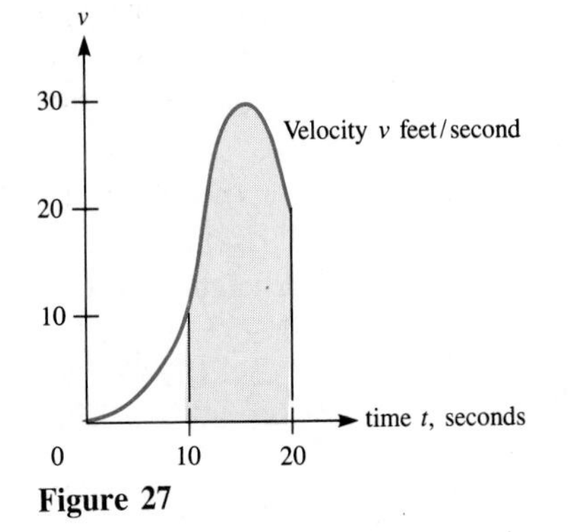

Figure 27

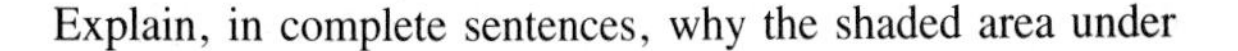
Explain, in complete sentences, why the shaded area under

the curve in Fig. 27 is equal to the distance the automobile goes from time $t = 10$ seconds to $t = 20$ seconds.

32 Let $G(x) = \int_0^x t^2\,dt$. Find dG/dx.

33 A partition of [1, 5] is formed with n sections. The mesh is 0.02. What can be said about n? (Include a diagram.)

34 A partition of [1, 5] is formed with n sections.
(*a*) If the sections all have the same length, what can we say about the mesh of the partition?
(*b*) If the sections are not necessarily all of the same length, what can we say about the mesh of the partition?

5.4 ESTIMATING THE DEFINITE INTEGRAL

A definite integral is a number defined as a limit of sums. So we can estimate it by computing one of those sums. In the previous section we were able to compute $\int_a^b x^2\,dx$ exactly because there is a convenient formula for $\Sigma_{i=1}^n i^2$. But what will we do for such definite integrals as

$$\int_0^1 \sqrt{1 - x^3}\, x^2\,dx \qquad \text{or} \qquad \int_0^1 \sqrt{1 - x^3}\,dx?$$

While Sec. 5.7 describes a way to compute $\int_0^1 \sqrt{1 - x^3}\, x^2\,dx$ exactly, the method won't help us evaluate $\int_0^1 \sqrt{1 - x^3}\,dx$. The present section describes three ways of estimating a definite integral and compares their accuracies.

Approximation by Rectangles

The definite integral $\int_a^b f(x)\,dx$ is, by definition, a limit of sums of the form

$$\sum_{i=1}^{n} f(c_i)(x_i - x_{i-1}). \tag{1}$$

Any such sum consequently provides an estimate of $\int_a^b f(x)\,dx$.

In terms of area, Fig. 1 shows the local approximation by a rectangle of the area under part of the curve. By summing the areas of individual rectangles, we obtain an estimate of the area under the curve. Example 1 illustrates this approach, which is called the **rectangular method**.

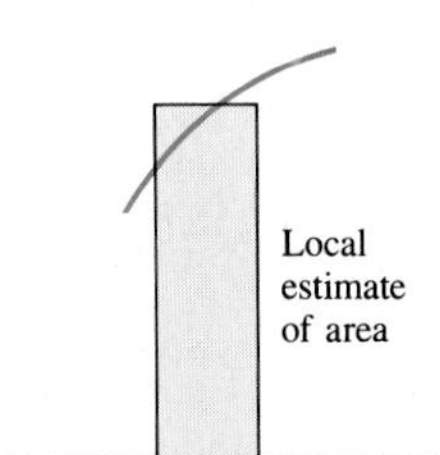

Figure 1

In the rectangular method, divide the interval into sections. (To simplify the arithmetic, choose them to be all the same size.) Then choose a number c_i in the ith section and form a Riemann sum. By the very definition of the definite integral, each Riemann sum approximates it.

EXAMPLE 1 Use four rectangles with equal widths to estimate $\int_0^1 dx/(1 + x^2)$. Use the left endpoint of each section as the sampling number to determine the height of each rectangle.

SOLUTION Since the length of [0, 1] is 1, each of the four sections of equal length has length $\frac{1}{4}$. See Fig. 2. The sum of their areas is

$$\frac{1}{1 + 0^2}\cdot\frac{1}{4} + \frac{1}{1 + (\frac{1}{4})^2}\cdot\frac{1}{4} + \frac{1}{1 + (\frac{2}{4})^2}\cdot\frac{1}{4} + \frac{1}{1 + (\frac{3}{4})^2}\cdot\frac{1}{4},$$

which equals

$$\tfrac{1}{4}(1 + \tfrac{16}{17} + \tfrac{16}{20} + \tfrac{16}{25}).$$

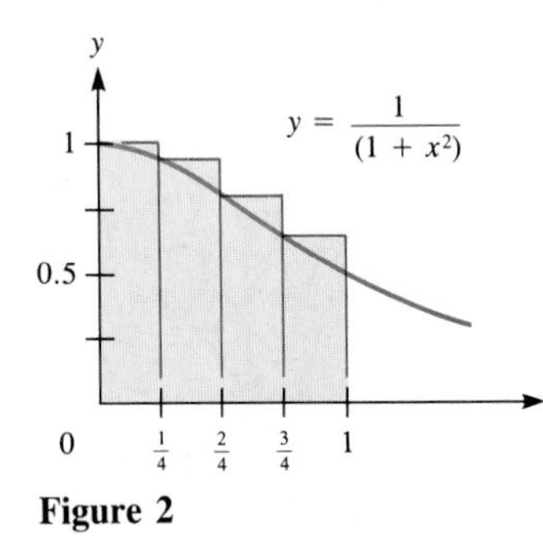

Figure 2

This is approximately

$$\tfrac{1}{4}(1.0000000 + 0.9411765 + 0.8000000 + 0.6400000) = \tfrac{1}{4}(3.3811765) \approx 0.845294.$$

Note that this method was used throughout Sec. 5.1. ■

Approximation by Trapezoids

In the **trapezoidal method**, trapezoids are used instead of rectangles. Recall that the area of a trapezoid of width h and bases b_1 and b_2 is $(b_1 + b_2)h/2$. (See the inside back cover.)

Let n be a positive integer. Divide the interval $[a, b]$ into n sections of equal width $h = (b - a)/n$ with

$$x_0 = a,\ x_1 = a + h,\ x_2 = a + 2h,\ \ldots,\ x_n = b.$$

The sum

$$\frac{f(x_0) + f(x_1)}{2} \cdot h + \frac{f(x_1) + f(x_2)}{2} \cdot h + \cdots + \frac{f(x_{n-1}) + f(x_n)}{2} \cdot h$$

is the **trapezoidal estimate** of $\int_a^b f(x)\,dx$. It is usually written

$$\boxed{\frac{h}{2}[f(x_0) + 2f(x_1) + 2f(x_2) + \cdots + 2f(x_{n-1}) + f(x_n)].} \qquad (2)$$

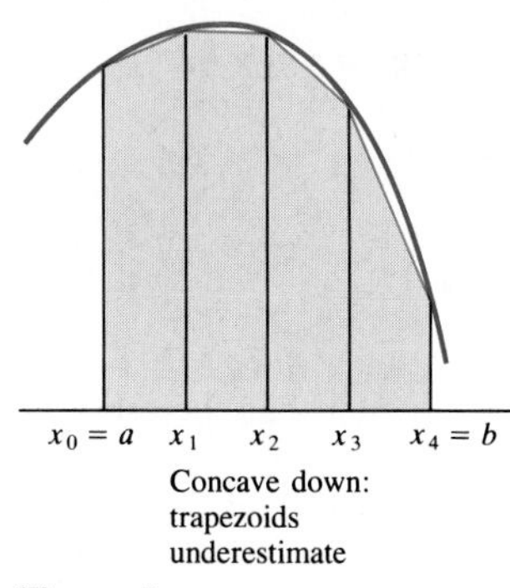

Figure 3

Note that $f(x_0)$ and $f(x_n)$ have coefficient 1, while all other $f(x_i)$'s have coefficient 2. This is due to the double counting of the edges common to two trapezoids.

The diagram in Fig. 3 illustrates the trapezoidal approximation for the case $n = 4$. Note that if f is concave down, the trapezoidal approximation underestimates $\int_a^b f(x)\,dx$. On the other hand, when the curve is concave up, the trapezoids overestimate, as shown in Fig. 4.

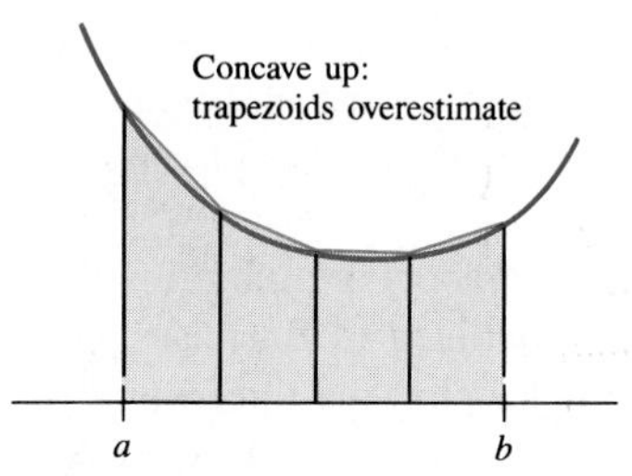

Figure 4

If f is linear—that is, its graph is a line and $f(x)$ is a first-degree polynomial or constant—then the trapezoidal method gives the integral exactly.

Because the trapezoid appears to give a closer approximation than the rectangle generally does, we would expect the trapezoidal method to provide better estimates of a definite integral than we obtain by rectangles.

EXAMPLE 2 Use the trapezoidal method with $n = 4$ to estimate $\int_0^1 dx/(1 + x^2)$.

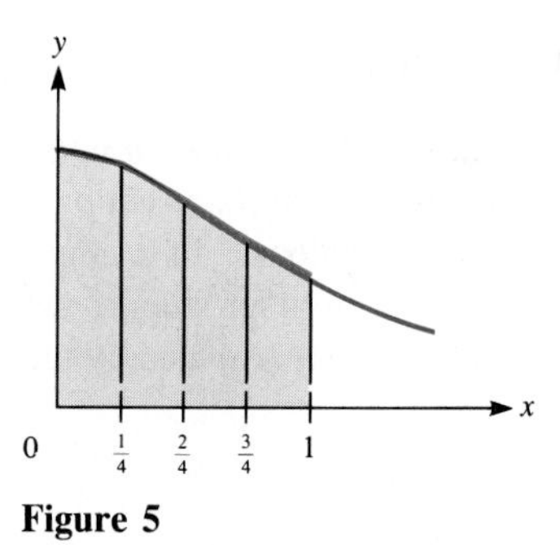

Figure 5

SOLUTION In this case, $a = 0$, $b = 1$, and $n = 4$, so $h = (1 - 0)/4 = \frac{1}{4}$. The four trapezoids are shown in Fig. 5. The trapezoidal estimate is

$$\frac{h}{2}[f(0) + 2f(\tfrac{1}{4}) + 2f(\tfrac{2}{4}) + 2f(\tfrac{3}{4}) + f(1)].$$

Now $h/2 = \frac{1}{4}/2 = 1/8$. To compute the sum in brackets, make a list as shown in Table 1.

Table 1

x_i	$f(x_i)$	*Coefficient*	*Summand*	*Decimal Form*
0	$\frac{1}{1+0^2}$	1	$1 \cdot \frac{1}{1+0}$	1.0000000
$\frac{1}{4}$	$\frac{1}{1+(\frac{1}{4})^2}$	2	$2 \cdot \frac{1}{1+\frac{1}{16}}$	1.8823529
$\frac{2}{4}$	$\frac{1}{1+(\frac{2}{4})^2}$	2	$2 \cdot \frac{1}{1+\frac{1}{4}}$	1.6000000
$\frac{3}{4}$	$\frac{1}{1+(\frac{3}{4})^2}$	2	$2 \cdot \frac{1}{1+\frac{9}{16}}$	1.2800000
$\frac{4}{4}$	$\frac{1}{1+(\frac{4}{4})^2}$	1	$1 \cdot \frac{1}{1+1}$	0.5000000

The trapezoidal sum is therefore approximately

$$\tfrac{1}{8}(1 + 1.8823529 + 1.6 + 1.28 + 0.5) = \tfrac{1}{8}(6.2623529) \approx 0.782794.$$

Thus
$$\int_0^1 \frac{dx}{1+x^2} \approx 0.782794. \quad \blacksquare$$

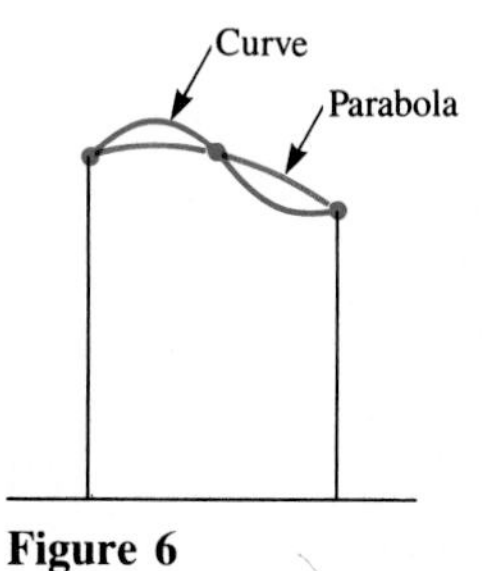

Figure 6

We expect the estimate 0.782794 of Example 2 to be better than the estimate 0.845294 found in Example 1.

Simpson's Method: Approximation by Parabolas

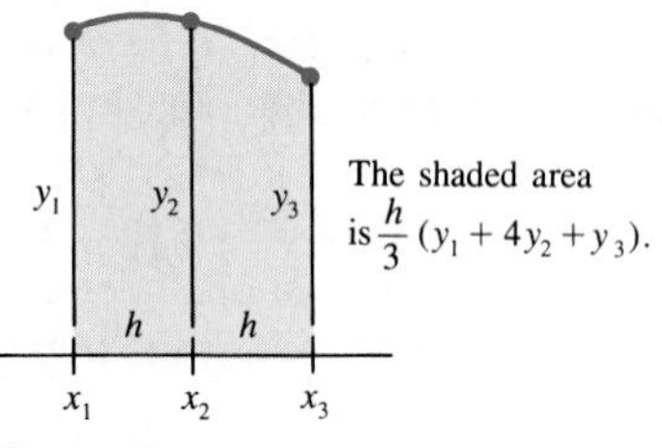

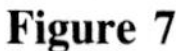

Figure 7

In the trapezoidal method a curve is approximated by chords. In Simpson's method a curve is approximated by parabolas. Given *three* points on a curve, there is a unique parabola (of the form $y = ax^2 + bx + c$) that passes through them, as shown in Fig. 6. The area under the parabola is then used to approximate the area under the curve.

The computations for finding the area under the parabola are more involved than those for the area of a trapezoid. (They are outlined in Exercises 21 to 26.) However, the resulting formula is fairly simple. Let the three points be (x_1, y_1), (x_2, y_2), and (x_3, y_3), with $x_1 < x_2 < x_3$, $x_2 - x_1 = h$, and $x_3 - x_2 = h$, as shown in Fig. 7. The shaded area under the parabola is

$$\frac{h}{3}(y_1 + 4y_2 + y_3). \tag{3}$$

A curve approximated by 3 parabolas

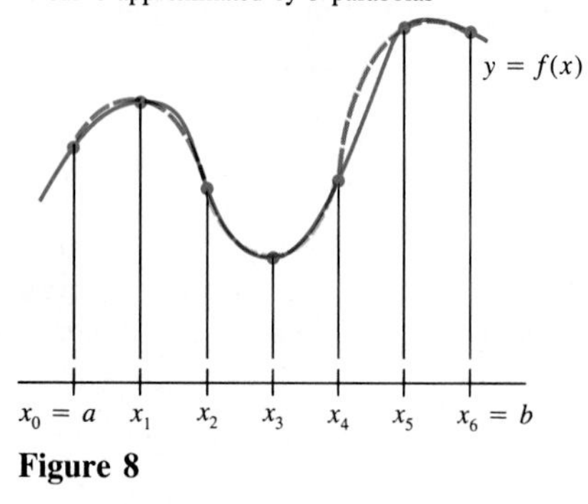

Figure 8

To use (3) to estimate $\int_a^b f(x)\,dx$, break the interval $[a, b]$ into an *even* number n of sections of equal length. Figure 8 illustrates this step for six sections. Formula (3) is applied to each of the $n/2$ pairs of adjacent sections. In the case shown in Fig. 8, $n = 6$ and there are $\frac{6}{2} = 3$ parabolas used to approximate the curve. Then $h = (b - a)/6$ and the area under the curve is approximated by the sum of these three terms:

$$\left\{\frac{(b-a)/6}{3}[f(x_0)+4f(x_1)+f(x_2)]\right\}+\left\{\frac{(b-a)/6}{3}[f(x_2)+4f(x_3)+f(x_4)]\right\}$$
$$+\left\{\frac{(b-a)/6}{3}[f(x_4)+4f(x_5)+f(x_6)]\right\},$$

which a little algebra reduces to

$$\frac{b-a}{18}[f(x_0)+4f(x_1)+2f(x_2)+4f(x_3)+2f(x_4)+4f(x_5)+f(x_6)].$$

This special case ($n = 6$) shows why the coefficients, in general, are 1, 4, 2, 4, . . . , 2, 4, 1, with a 1 at both ends, and 4 and 2 alternating.

In Simpson's method the interval $[a, b]$ is divided into an *even* number of sections of equal length $h = (b - a)/n$ with

$$x_0 = a,\ x_1 = a + h,\ x_2 = a + 2h,\ \dots,\ x_n = b.$$

Then **Simpson's estimate** of $\int_a^b f(x)\,dx$ is

Note that $f(x_0)$ and $f(x_n)$ have coefficient 1, while the coefficients of the other $f(x_i)$'s alternate 4, 2, 4, 2, . . . , 2, 4.

$$\boxed{\frac{h}{3}[f(x_0)+4f(x_1)+2f(x_2)+4f(x_3)+\cdots+2f(x_{n-2})+4f(x_{n-1})+f(x_n)].} \quad (4)$$

EXAMPLE 3 Use Simpson's method with $n = 4$ to estimate $\int_0^1 dx/(1+x^2)$.

SOLUTION Again $h = \frac{1}{4}$. Simpson's formula (4) takes the form

$$\frac{\frac{1}{4}}{3}[f(0)+4f(\tfrac{1}{4})+2f(\tfrac{2}{4})+4f(\tfrac{3}{4})+f(1)].$$

The computations are shown in Table 2. See Fig. 9, where the two parabolas are shown. Note that the parabolas are virtually indistinguishable from the curve itself.

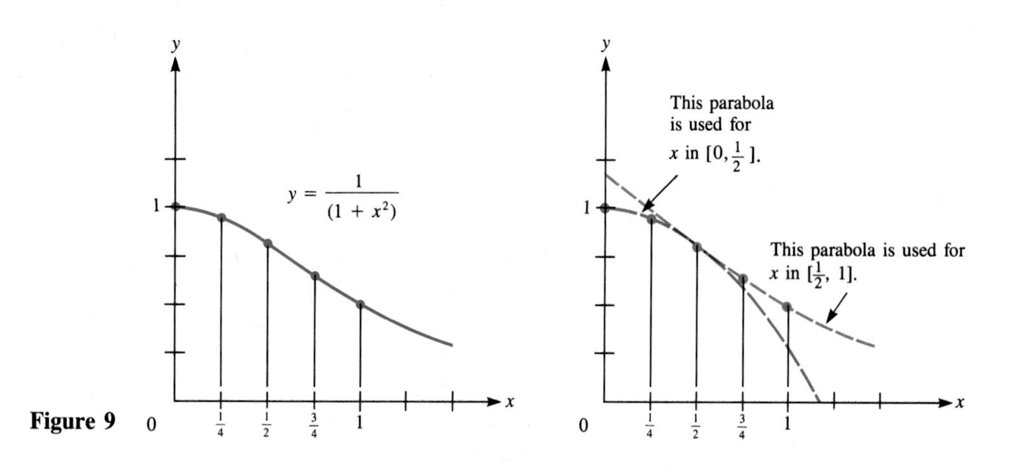

Figure 9

Table 2

x_i	$f(x_i)$	*Coefficient*	*Summand*	*Decimal Form*
0	$\frac{1}{1+0^2}$	1	$1 \cdot \frac{1}{1+0}$	1.0000000
$\frac{1}{4}$	$\frac{1}{1+(\frac{1}{4})^2}$	4	$4 \cdot \frac{1}{1+\frac{1}{16}}$	3.7647059
$\frac{2}{4}$	$\frac{1}{1+(\frac{2}{4})^2}$	2	$2 \cdot \frac{1}{1+\frac{1}{4}}$	1.6000000
$\frac{3}{4}$	$\frac{1}{1+(\frac{3}{4})^2}$	4	$4 \cdot \frac{1}{1+\frac{9}{16}}$	2.5600000
$\frac{4}{4}$	$\frac{1}{1+(\frac{4}{4})^2}$	1	$1 \cdot \frac{1}{1+1}$	0.5000000

The Simpson approximation of $\int_0^1 dx/(1 + x^2)$ is therefore

$$\frac{1}{12}(1 + 3.7647059 + 1.6 + 2.56 + 0.5) = \frac{1}{12}(9.4247059)$$

$$\approx 0.785392.$$

Thus
$$\int_0^1 \frac{dx}{1+x^2} \approx 0.785392. \quad \blacksquare$$

Comparison of the Three Methods

In Chapter 7 it is shown that the value of $\int_0^1 dx/(1 + x^2)$ is $\pi/4$; to six places, this equals 0.785398. Table 3 compares the estimates made in the three examples to this value.

Table 3

Method	*Estimate*	*Error* $= \int_0^1 dx/(1 + x^2) -$ *Estimate*
Rectangles	0.845294	−0.059896
Trapezoids	0.782794	0.002604
Simpson's (parabolas)	0.785392	0.000006

Though each method takes about the same amount of work, the table shows that Simpson's method gives the best estimate. The trapezoidal method is next best. The rectangular method has the largest error. These results should not come as a surprise. Parabolas should fit the curve better than chords do, and chords should fit better than horizontal line segments. Note that the trapezoidal and Simpson's methods in Examples 2 and 3 used the same number of points to evaluate the function; their only difference is in the ''weights'' (coefficients) given the outputs of the functions.

The Error

The size of the error is closely connected to the derivatives of the integrand. For a positive integer k, let M_k be the largest value of $|f^{(k)}(x)|$ for x in $[a, b]$. [Recall that $f^{(k)}(x)$ is the kth derivative of f. For instance, $f^{(2)}(x)$ is the second derivative.] Table 4, developed in upper-division numerical analysis classes, lists upper bounds on the error when $\int_a^b f(x)\,dx$ is estimated by sections of width $h = (b - a)/n$.

Table 4

Method	*Bound on Absolute Value of Error*
Rectangular	$M_1(b-a)h$
Trapezoidal	$\frac{1}{12}M_2(b-a)h^2$
Simpson's	$\frac{1}{180}M_4(b-a)h^4$

The coefficients tell us something. For instance, if $M_4 = 0$, then there is no error in Simpson's method. That is, if $f^{(4)}(x) = 0$ for all x in $[a, b]$, then Simpson's method produces an exact answer. For in this case the error is $M_4(b-a)h^4/180 = 0$. As a consequence, for polynomials of at most degree 3, Simpson's approximation is exact.

But more important is the power of h that appears in the error bound. For instance, if you reduce the width h by a factor of 10 (using 10 times as many points) you expect the error of the rectangular method to shrink by a factor of 10, the error in the trapezoidal method to shrink by a factor of $10^2 = 100$, and the error in Simpson's method by a factor of $10^4 = 10{,}000$. These observations are recorded in Table 5.

Table 5

Method	*Reduction Factor of h*	*Expected Reduction of Error Factor*
Rectangular	10	10
Trapezoidal	10	100
Simpson's	10	10,000

Because the error in the rectangular method approaches 0 so slowly as $h \to 0$, we will not use it for estimating definite integrals. (Rectangular estimates using the midpoint rather than a left or right endpoint are about as accurate as the trapezoidal method. However, this method requires the extra arithmetic of computing midpoints.)

The bounds on the error given in Table 4 are obtained in Exercise 30 and Exercises 9 and 10 of Sec. 11.3.

CALCULATORS AND OTHER TECHNIQUES

Many scientific calculators have a key marked "∫" to compute an estimate of a definite integral defined by the user. Some calculators use Simpson's method to do this, while others have more sophisticated techniques built in. You may even be able to specify the degree of accuracy that you want, although the more decimal places you ask for the longer the calculation will take. If you have such a calculator, you might check to see how rapidly it estimates a definite integral.

The trapezoidal method and Simpson's method are just two examples of what is called **numerical integration**. Such techniques are taught in courses on numerical analysis and are important tools in computing definite integrals that might not otherwise be evaluated. The availability of these techniques means that it is always possible to find out something about the value of a definite integral. Keep this in mind for future reference. We are about to introduce the fundamental theorem of calculus, which is the most popular tool for evaluating definite integrals. There are definite integrals, however, where the fundamental theorem does not help us. In such cases we may fall back on the numerical techniques of this section.

EXERCISES FOR SEC. 5.4: ESTIMATING THE DEFINITE INTEGRAL

In Exercises 1 to 8 estimate the given definite integrals by the trapezoidal method, using the given number of trapezoids.

1 $\int_0^2 \frac{dx}{1+x^2}$, $n = 2$ **2** $\int_0^2 \frac{dx}{1+x^2}$, $n = 4$

3 $\int_0^2 \sin\sqrt{x}\,dx$, $n = 2$ **4** $\int_0^2 \sin\sqrt{x}\,dx$, $n = 3$

5 $\int_1^3 \frac{2^x}{x}\,dx$, $n = 3$ **6** $\int_1^3 \frac{2^x}{x}\,dx$, $n = 6$

7 $\int_1^3 \cos x^2\,dx$, $n = 4$ **8** $\int_1^3 \cos x^2\,dx$, $n = 2$

In Exercises 9 to 12 use Simpson's method to estimate each definite integral with the given n.

9 $\int_0^1 \frac{dx}{x^3+1}$, $n = 2$ **10** $\int_0^1 \frac{dx}{x^3+1}$, $n = 4$

11 $\int_0^1 \frac{dx}{x^4+1}$, $n = 4$ **12** $\int_0^1 \frac{dx}{x^4+1}$, $n = 2$

13 The cross section of a ship's hull is shown in Fig. 10. Estimate its area by (*a*) the trapezoidal method, (*b*) Simpson's method. Dimensions are in feet. (Give your answer to four decimal places.)

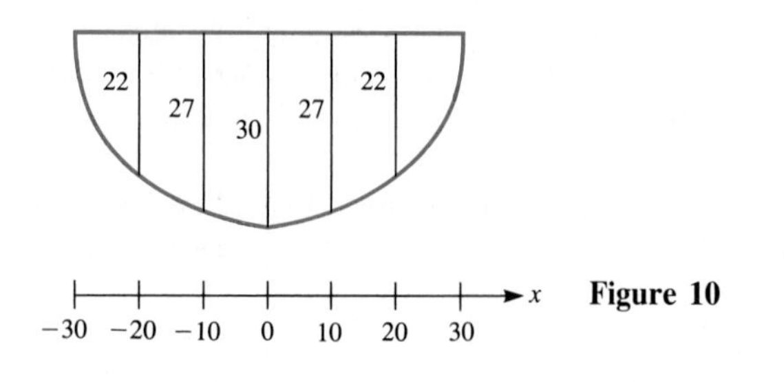

Figure 10

14 A ship is 120 feet long. The area of the cross section of its hull is given at intervals in the table below:

x	0	20	40	60	80	100	120	Feet
Area	0	200	400	450	420	300	150	Square feet

Estimate the volume of the hull in cubic feet by (*a*) the trapezoidal method, (*b*) Simpson's method. Give your answers to four decimal places.

15 A map of Lake Tahoe is shown in Fig. 11. Use Simpson's rule and data from the map to estimate the surface area of the lake. (Each little square is a mile on a side.)

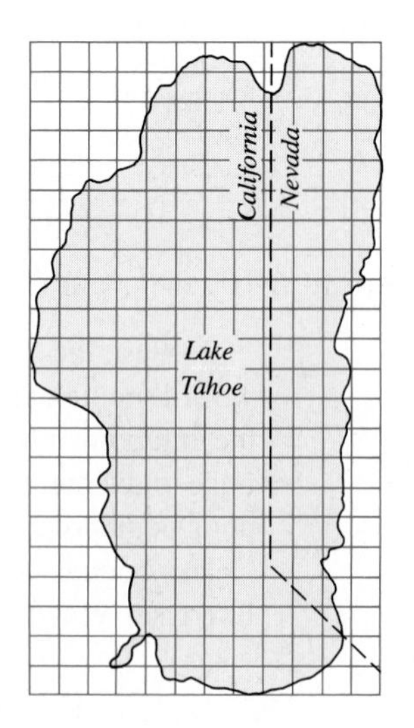

Figure 11

16 The diameter of a tree trunk is given in feet at various heights in the following table:

Height	0	2	4	6	8	10	12	14	16
Diameter	4.0	3.7	3.5	3.1	3.1	2.9	2.7	2.4	2.1

Use Simpson's method to find the volume of the tree trunk between the heights of 0 and 16 feet. Assume that each cross section is circular.

The ''right-point'' estimate of $\int_a^b f(x)\,dx$ is obtained by selecting a positive integer n and dividing $[a, b]$ into n sections of equal width $h = (b - a)/n$. The points of subdivision are $x_0 < x_1 < \cdots < x_n$, with $x_0 = a$, and $x_n = b$. The right-point estimate is the approximating sum

$$h[f(x_1) + f(x_2) + \cdots + f(x_n)].$$

The ''left-point'' estimate is defined similarly; it is given by

$$h[f(x_0) + f(x_1) + \cdots + f(x_{n-1})]$$

and is the method that was used in Example 1.

17 Show that if $f(a) = f(b)$, the left-point, right-point, and trapezoidal estimates for a given value of h are the same.

18 Show that for a given n the average of the left-point estimate and the right-point estimate equals the trapezoidal estimate.

The next two exercises present cases in which the bound of maximum error is actually assumed.

19 Show that if the trapezoidal method with $n = 1$ is used to estimate $\int_0^1 x^2\,dx$, the error equals $(b - a)M_2h^2/12$, where $a = 0$, $b = 1$, $h = 1$, and M_2 is the maximum value of $|d^2(x^2)/dx^2|$ for x in $[0, 1]$.

20 Show that if Simpson's method with $n = 2$ is used to estimate $\int_0^1 x^4\,dx$, the error equals $(b - a)M_4h^4/180$, where $a = 0$, $b = 1$, $h = \frac{1}{2}$, and M_4 is the maximum value of $|d^4(x^4)/dx^4|$ for x in $[0, 1]$.

Exercises 21 to 26 describe the geometric motivation of Simpson's method.

21 Let $f(x) = Ax^2 + Bx + C$. Show that

$$\int_{-h}^{h} f(x)\,dx = \frac{h}{3}[f(-h) + 4f(0) + f(h)].$$

Hint: Just compute both sides.

22 Let f be a function. Show that there is a parabola $y = Ax^2 + Bx + C$ that passes through the three points $(-h, f(-h))$, $(0, f(0))$, and $(h, f(h))$. (See Fig. 12.)

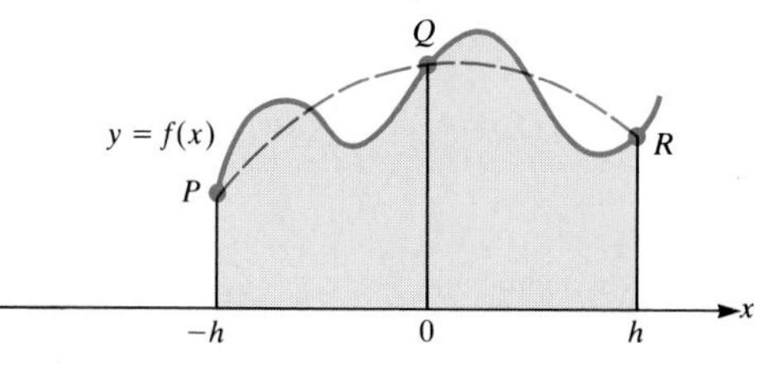

The dashed graph is a parabola, $y = Ax^2 + Bx + C$, through P, Q, and R. The area of the region below the parabola is precisely

$$\frac{h}{3}[f(-h) + 4f(0) + f(h)]$$

and is an approximation of the area of the shaded region.

Figure 12

23 The equation in Exercise 21 (which was known to the Greeks) is called the **prismoidal formula** Use it to compute the volume of
(*a*) A sphere of radius a
(*b*) A right circular cone of radius a and height h

24 Let $f(x) = Ax^2 + Bx + C$. Show that

$$\int_{c-h}^{c+h} f(x)\,dx = \frac{h}{3}[f(c - h) + 4f(c) + f(c + h)].$$

Hint: Use the substitution $x = c + t$ to reduce this to Exercise 21.

25 First, $[a, b]$ is divided into n sections (n even), which are grouped into $n/2$ pairs of adjacent sections. Over each pair the function is approximated by the parabola that passes through the three points of the graph with x coordinates equal to those which determine the two sections of the pair. (See Fig. 8.) The integral of this quadratic function is used as an estimate of the integral of f over each pair of adjacent sections. Show that when these $n/2$ separate estimates are added, Simpson's formula results.

26 Since Simpson's method was designed to be exact when $f(x) = Ax^2 + Bx + C$, one would expect the error associated with it to involve $f^{(3)}(x)$. By a quirk of good fortune, Simpson's method happens to be exact even when $f(x)$ is a *cubic*, $Ax^3 + Bx^2 + Cx + D$. This suggests that the error involves $f^{(4)}(x)$ not $f^{(3)}(x)$.
(*a*) Show that if $f(x) = x^3$,

$$\int_{-h}^{h} f(x)\,dx = \frac{h}{3}[f(-h) + 4f(0) + f(h)].$$

(*b*) Show that Simpson's estimate is exact for cubics.

27 There are many other methods for estimating definite integrals. Some old methods, which had been of only theoretical interest because of their messy arithmetic, have, with the advent of computers, assumed practical importance. This exercise illustrates the simplest of the so-called **Gaussian quadrature** formulas. For simplicity, consider only integrals over $[-1, 1]$.

(*a*) Show that

$$\int_{-1}^{1} f(x)\,dx = f\left(\frac{-1}{\sqrt{3}}\right) + f\left(\frac{1}{\sqrt{3}}\right)$$

for $f(x) = 1$, x, x^2, and x^3.

(*b*) Let a and b be two numbers, $-1 \le a < b \le 1$, such that

$$\int_{-1}^{1} f(x)\,dx = f(a) + f(b)$$

for $f(x) = 1$, x, x^2, and x^3. Show that $a = -1/\sqrt{3}$ and $b = 1/\sqrt{3}$.

(*c*) Show that the Gaussian approximation

$$\int_{-1}^{1} f(x)\,dx \approx f(-1/\sqrt{3}) + f(1/\sqrt{3})$$

has no error when f is a polynomial of degree at most 3.

(*d*) Use the formula in (*a*) to estimate $\int_{-1}^{1} dx/(1+x^2)$.

28 Let f be a function such that $|f^{(2)}(x)| \le 10$ and $|f^{(4)}(x)| \le 50$ for all x in $[1, 5]$. If $\int_1^5 f(x)\,dx$ is to be estimated with an error of at most 0.01, how small must h be in (*a*) the trapezoidal approximation? (*b*) Simpson's approximation?

29 Let T be the trapezoidal estimate of $\int_a^b f(x)\,dx$, using $x_0 = a$, $x_1, \ldots, x_n = b$. Let M be the "midpoint estimate," $\Sigma_{i=1}^{n} f(c_i)(x_i - x_{i-1})$, where $c_i = (x_{i-1} + x_i)/2$. Let S be Simpson's estimate using the $2n+1$ points $x_0, c_1, x_1, c_2, x_2, \ldots, c_n, x_n$. Show that

$$S = \tfrac{2}{3}M + \tfrac{1}{3}T.$$

30 In his *Principia,* published in 1687, Newton examined the error in approximating an area by rectangles. He considered an increasing, differentiable function f defined on the interval $[a, b]$ and drew a figure similar to Fig. 13. All rectangles have the same width h. Let R equal the sum of the areas of the rectangles using right endpoints and let L equal the sum of the areas of the rectangles using left endpoints. Let A be the area under the curve $y = f(x)$ and above $[a, b]$.

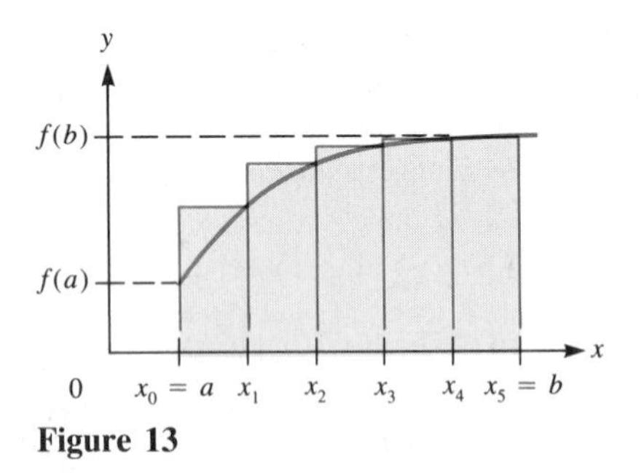

Figure 13

(*a*) Why is $R - L = (f(b) - f(a))h$?

(*b*) Show that any approximating sum for A, formed with rectangles of equal width h and any sampling points, differs from A by at most $(f(b) - f(a))h$.

(*c*) Let M_1 be the maximum value of $|f'(x)|$ for x in $[a, b]$. Show that any approximating sum for A formed with equal widths h differs from A by at most $M_1(b-a)h$. (*Hint:* Use the mean-value theorem.) Since the error involves the first power of h, approximation by rectangles is less efficient than approximation by trapezoids, where the error involves h^2.

(*d*) Newton also considered the case where the rectangles do not necessarily have the same widths. Let h be the largest of their widths. What can be said about the error?

31 Figure 14 shows cross sections of a pond in two directions. Use Simpson's method to estimate the area of the pond using (*a*) the vertical cross sections, (*b*) the horizontal cross sections.

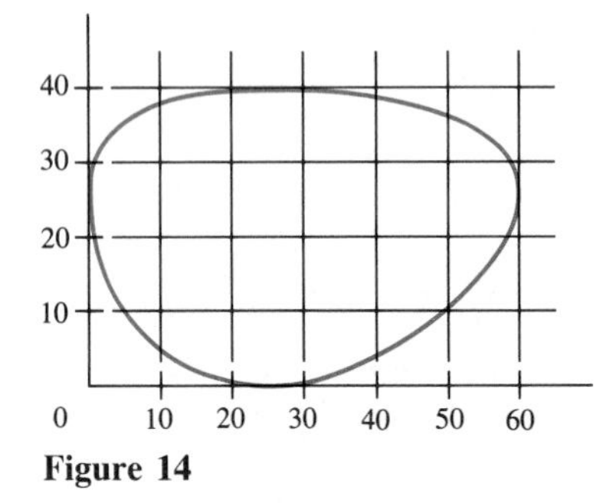

Figure 14

32 The table lists the values of a function f at the given inputs.

x	1	2	3	4	5	6	7
$f(x)$	1	2	1.5	1	1.5	3	3

(*a*) Plot the seven points.

(*b*) Sketch six trapezoids that can be used to estimate $\int_1^7 f(x)\,dx$.

(*c*) Find the trapezoidal estimate of $\int_1^7 f(x)\,dx$.

(*d*) Sketch (by eye) the three parabolas that can be used in Simpson's method to estimate $\int_1^7 f(x)\,dx$.

(*e*) Find the Simpson's estimate of $\int_1^7 f(x)\,dx$.

33 Figure 15 is the graph of a function $f(x)$. Let $g(x) = \int_0^x f(t)\,dt$.

(*a*) Graph $y = g(x)$ as well as you can.

(*b*) Graph $y = f'(x)$ as well as you can.

(*c*) Explain your reasoning.

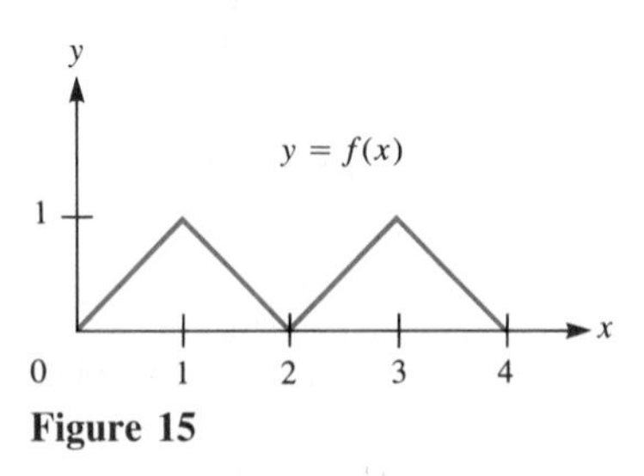

Figure 15

34 Repeat Exercise 33 for the function $f(x)$ in Fig. 16.

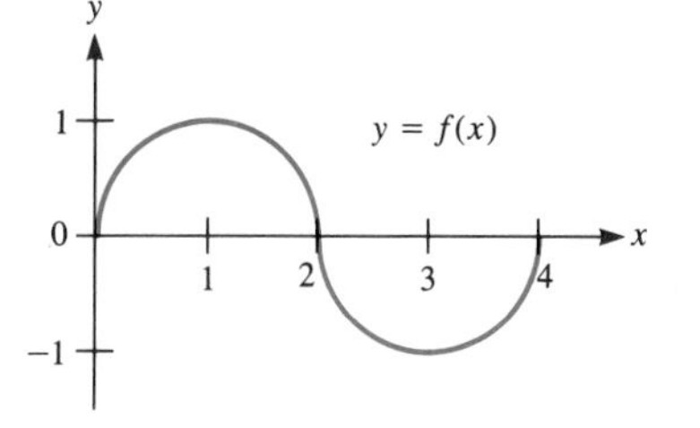

Figure 16

5.5 PROPERTIES OF THE ANTIDERIVATIVE AND THE DEFINITE INTEGRAL

Back in Sec. 3.3 an antiderivative of a function $f(x)$ was defined. It is any function $F(x)$ whose derivative is $f(x)$. For instance x^3 is an antiderivative of $3x^2$. So is $x^3 + 1{,}992$. Keep in mind that *an antiderivative is a function.*

In this section we discuss various properties of antiderivatives and definite integrals. These properties will be needed in Sec. 5.7 where we obtain a relation between antiderivatives and definite integrals. That relation will be a great time-saver in evaluating many (but not all) definite integrals.

Up to this point we have not introduced a symbol for an antiderivative of a function. We will adopt the following standard notation:

Notation Any antiderivative of f is denoted $\int f(x)\,dx$.

For instance, $x^3 = \int 3x^2\,dx$. This equation is read "x^3 is an antiderivative of $3x^2$." That means simply that "the derivative of x^3 is $3x^2$." It is true that $x^3 + 1{,}992 = \int 3x^2\,dx$, since $x^3 + 1{,}992$ is also an antiderivative of $3x^2$. That does *not* mean that the functions x^3 and $x^3 + 1{,}992$ are equal. The symbol $\int 3x^2\,dx$ refers to *any function* whose derivative is $3x^2$.

Warning

If $F'(x) = f(x)$ we write $F(x) = \int f(x)\,dx$. The function $f(x)$ is called the **integrand**. The function $F(x)$ is called an antiderivative of $f(x)$. The symbol for an antiderivative, $\int f(x)\,dx$, is similar to the symbol for a definite integral, $\int_a^b f(x)\,dx$, but they are vastly different concepts. The symbol $\int f(x)\,dx$ denotes a function—any function whose derivative is $f(x)$; the symbol $\int_a^b f(x)\,dx$ denotes a number—one that is defined by a limit of certain sums.

We apologize for the use of such similar notations, $\int f(x)\,dx$ and $\int_a^b f(x)\,dx$, for such distinct concepts. However, it is not for us to undo over three centuries of history. [Some authors have tried, but failed. For instance, one author denoted an antiderivative of $f(x)$ by the symbol $A(f(x))$—thus $A(3x^2) = x^3$. The proposal did not catch on.] Rather, it is up to you to read the symbols $\int f(x)\,dx$ and $\int_a^b f(x)\,dx$ carefully. You read words carefully; for instance, you distinguish between such similar-looking words as "density" and "destiny." Be as careful when reading mathematics.

Properties of Antiderivatives

The tables inside the covers of this book list many antiderivatives. For instance, the third entry is $\int \sin x\,dx = -\cos x$. Of course, $(-\cos x) + 17$ also is an antiderivative of $\sin x$. The following theorem asserts that if you have found an antiderivative $F(x)$ for a function $f(x)$, then any other antiderivative $f(x)$ is of the form $F(x) + C$ for some constant C.

Theorem 1 If F and G are both antiderivatives of f on some interval, then there is a constant C such that

$$F(x) = G(x) + C.$$

Proof The functions F and G have the same derivative f. By Corollary 2 in Sec. 4.1, they must differ by a constant, which we denote C. Thus $F(x) - G(x) = C$ for any x in the interval. Hence $F(x) = G(x) + C$. ■

In tables, C is usually omitted.

When writing down an antiderivative, it is best to add the constant C. (It will be needed in the study of differential equations.) For example,

$$\int 5\,dx = 5x + C,$$

$$\int x^3\,dx = \frac{x^4}{4} + C,$$

and

$$\int \sin 2x\,dx = -\frac{\cos 2x}{2} + C.$$

Observe that

$$\frac{d}{dx}\left(\int x^3\,dx\right) = x^3$$

and

$$\frac{d}{dx}\left(\int \sin 2x\,dx\right) = \sin 2x.$$

Are these two equations profound or trivial? Read them aloud and decide.

The first says, "The derivative of an antiderivative of x^3 is x^3." *It is true simply because that is how we defined an antiderivative.* We know that

$$\frac{d}{dx}\left(\int \frac{x^{17}+1}{\sin^2 x}\,dx\right) = \frac{x^{17}+1}{\sin^2 x}$$

even if we cannot write out a formula for an antiderivative of $(x^{17} + 1)/(\sin^2 x)$. In other words, by the very definition of an antiderivative,

$$\frac{d}{dx}\left(\int f(x)\,dx\right) = f(x).$$

Any property of derivatives implies a corresponding property of antiderivatives. The next theorem records three of the most important properties of antiderivatives.

Properties of antiderivatives

Theorem 2 Assume that f and g are functions with antiderivatives $\int f(x)\,dx$ and $\int g(x)\,dx$. Then the following hold:

(a) $\int cf(x)\,dx = c\int f(x)\,dx$ for any constant c.
(b) $\int (f(x) + g(x))\,dx = \int f(x)\,dx + \int g(x)\,dx$.
(c) $\int (f(x) - g(x))\,dx = \int f(x)\,dx - \int g(x)\,dx$.

Proof (*a*) Befo:e we prove that $\int cf(x)\,dx = c\int f(x)\,dx$, we stop to see what it means. It says that "c times an antiderivative of $f(x)$ is an antiderivative of $cf(x)$." Let $F(x)$ be an antiderivative of $f(x)$. Then the equation says "c times $F(x)$ is an antiderivative of $cf(x)$." To check that this claim is true we must differentiate $cF(x)$ and check whether we do get $cf(x)$. So we compute $(cF(x))'$:

$$\frac{d}{dx}(cF(x)) = c\,\frac{dF}{dx} \qquad (c \text{ is constant})$$

$$= cf(x). \qquad (F \text{ is antiderivative of } f)$$

Thus $cF(x)$ is indeed an antiderivative of $cf(x)$. Therefore, we may write

$$cF(x) = \int cf(x)\,dx.$$

Since $F(x) = \int f(x)\,dx$, we conclude that

$$c\int f(x)\,dx = \int cf(x)\,dx.$$

(*b*) The proof is similar. We show that $\int f(x)\,dx + \int g(x)\,dx$ is an antiderivative of $f(x) + g(x)$. To do this we compute the derivative of the sum $\int f(x)\,dx + \int g(x)\,dx$:

$$\frac{d}{dx}\left(\int f(x)\,dx + \int g(x)\,dx\right) = \frac{d}{dx}\left(\int f(x)\,dx\right) + \frac{d}{dx}\left(\int g(x)\,dx\right) \quad \text{(derivative of a sum)}$$

$$= f(x) + g(x). \qquad \text{(definition of antiderivative)}$$

Thus $\int f(x)\,dx + \int g(x)\,dx = \int (f(x) + g(x))\,dx$.
(*c*) The proof is similar to that of (*b*). ■

EXAMPLE 1 Find (*a*) $\int 6\cos x\,dx$, (*b*) $\int (6\cos x + 3x^2)\,dx$, (*c*) $\int (6\cos x - 3x^2)\,dx$.

SOLUTION (*a*) $\int 6\cos x\,dx = 6\int \cos x\,dx = 6\sin x + C$.
[We used part (*a*) of the theorem to move the "6" past the "$\int$".]
(*b*) $\int (6\cos x + 3x^2)\,dx = \int 6\cos x\,dx + \int 3x^2\,dx$ [part (*b*) of the theorem]
$= 6\sin x + x^3 + C$.
(*c*) $\int (6\cos x - 3x^2)\,dx = \int 6\cos x\,dx - \int 3x^2\,dx$ [part (*c*) of the theorem]
$= 6\sin x - x^3 + C$. ■

The last two parts of Theorem 2 extend to any finite number of functions. For instance,

$$\int (f(x) - g(x) + h(x))\,dx = \int f(x)\,dx - \int g(x)\,dx + \int h(x)\,dx.$$

Theorem 3 Let a be a number other than -1. Then

$$\int x^a\,dx = \frac{x^{a+1}}{a+1} + C.$$

Proof

$$\left(\frac{x^{a+1}}{a+1} + C\right)' = \frac{(a+1)x^{a+1-1}}{a+1} = x^a. \quad ■$$

EXAMPLE 2 Find $\int (2x^5 - 3x^2 + 4)\,dx$.

SOLUTION

$$\int (2x^5 - 3x^2 + 4)\,dx = \int 2x^5\,dx - \int 3x^2\,dx + \int 4\,dx \qquad \text{Theorem 2}$$

$$= 2\int x^5\,dx - 3\int x^2\,dx + \int 4\,dx \qquad \text{Theorem 2}$$

A single constant is enough.

$$= 2\frac{x^6}{6} - 3\frac{x^3}{3} + 4x + C \qquad \text{Theorem 3}$$

$$= \frac{x^6}{3} - x^3 + 4x + C. \quad \blacksquare$$

As Example 2 illustrates, an antiderivative of any polynomial is again a polynomial.

Properties of Definite Integrals

Some of the properties of definite integrals look like similar properties of antiderivatives. However, they are assertions about numbers, not about functions. In the notation for the definite integral $\int_a^b f(x)\,dx$, b is larger than a. It will be useful to be able to speak of "the definite integral from a to b" even if b is less than or equal to a. The following definitions meet this need and will be used in the proofs of the two fundamental theorems of calculus.

Definition *The integral from a to b, where b is less than a.* If b is less than a, then

$$\int_a^b f(x)\,dx = -\int_b^a f(x)\,dx.$$

EXAMPLE 3 Compute $\int_3^0 x^2\,dx$, the integral from 3 to 0 of x^2.

SOLUTION The symbol $\int_3^0 x^2\,dx$ is defined as $-\int_0^3 x^2\,dx$. As was shown in Sec. 5.3, $\int_0^3 x^2\,dx = 9$. Thus

$$\int_3^0 x^2\,dx = -9. \quad \blacksquare$$

Definition *The integral from a to a.* $\int_a^a f(x)\,dx = 0$.

Remark: The definite integral is defined with the aid of partitions. Rather than permit partitions to have sections of length 0, it is simpler just to make the preceding definition.

The point of making these two definitions is that now the symbol $\int_a^b f(x)\,dx$ is defined for any numbers a and b and any continuous function f. It is no longer necessary that a be less than b.

The definite integral has several properties, some of which will be used in this section and some later in the text.

Properties of the Definite Integral Let f and g be continuous functions, and let c be a constant. Then

1 $\int_a^b cf(x)\,dx = c\int_a^b f(x)\,dx$.
2 $\int_a^b (f(x) + g(x))\,dx = \int_a^b f(x)\,dx + \int_a^b g(x)\,dx$.
3 $\int_a^b (f(x) - g(x))\,dx = \int_a^b f(x)\,dx - \int_a^b g(x)\,dx$.
4 If $f(x) \geq 0$ for all x in $[a, b]$, $a < b$, then

$$\int_a^b f(x)\,dx \geq 0.$$

5 If $f(x) \geq g(x)$ for all x in $[a, b]$, $a < b$, then

$$\int_a^b f(x)\,dx \geq \int_a^b g(x)\,dx.$$

6 If a, b, and c are numbers, then

$$\int_a^c f(x)\,dx + \int_c^b f(x)\,dx = \int_a^b f(x)\,dx.$$

7 If m and M are numbers and $m \leq f(x) \leq M$ for all x between a and b, then

$$m(b-a) \leq \int_a^b f(x)\,dx \leq M(b-a) \qquad \text{if } a < b,$$

and

$$m(b-a) \geq \int_a^b f(x)\,dx \geq M(b-a) \qquad \text{if } b < a.$$

Proof of Property 1. Take the case $a < b$. The equation $\int_a^b cf(x)\,dx = c\int_a^b f(x)\,dx$ resembles a theorem about antiderivatives, $\int cf(x)\,dx = c\int f(x)\,dx$. However, its proof is quite different, since $\int_a^b cf(x)\,dx$ is defined as a limit of sums.

We have

$$\int_a^b cf(x)\,dx = \lim_{\Delta x_i \to 0} \sum_{i=1}^n cf(c_i)\,\Delta x_i \qquad \text{definition of definite integral}$$

$$= \lim_{\Delta x_i \to 0} c\sum_{i=1}^n f(c_i)\,\Delta x_i \qquad \text{algebra}$$

$$= c\lim_{\Delta x_i \to 0} \sum_{i=1}^n f(c_i)\,\Delta x_i \qquad \text{properties of limits}$$

$$= c\int_a^b f(x)\,dx. \qquad \text{definition of definite integral} \quad \blacksquare$$

The other properties can be justified by a similar approach. However, we pause only to make them plausible. To do this, interpret each of the properties in terms of areas (assuming the integrands are positive).

Property 5 Property 5 amounts to the assertion that the area of the region under the curve $y = f(x)$ is greater than or equal to the area under the curve $y = g(x)$. (See Fig. 1.)

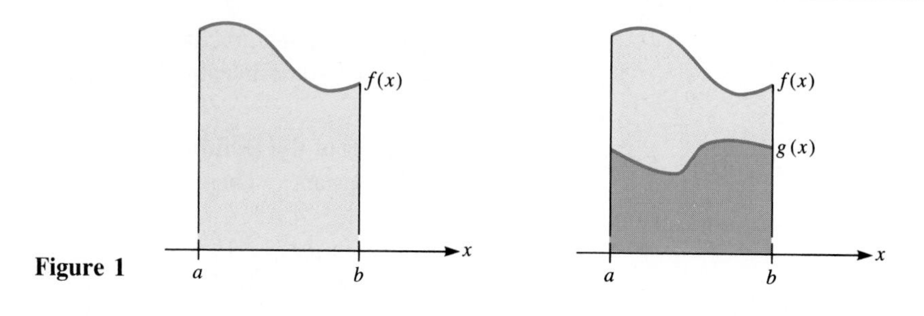

Figure 1

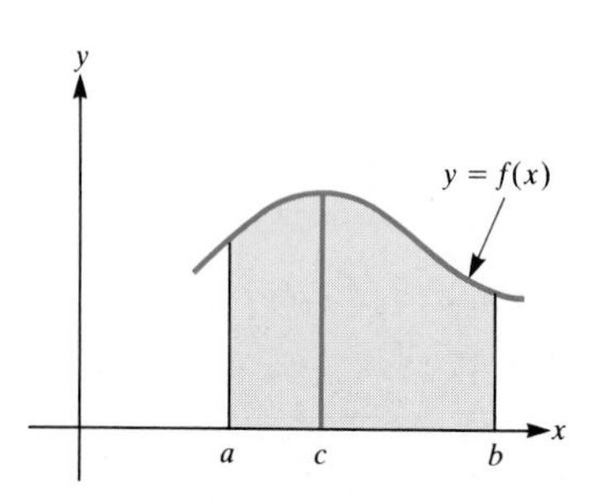

The area from a to c plus the area from c to b is equal to the area from a to b.

Figure 2

Property 6 In the case that $a < c < b$ and $f(x)$ assumes only positive values, property 6 asserts that the area of the region below the graph of f and above the interval $[a, b]$ is the sum of the areas of the regions below the graph and above the smaller intervals $[a, c]$ and $[c, b]$. Figure 2 shows that this is certainly plausible.

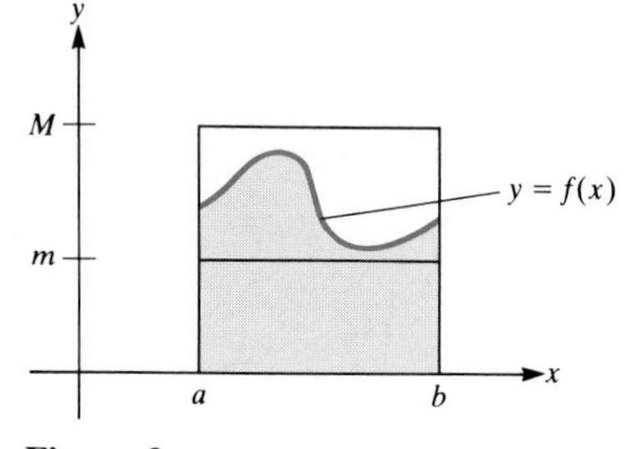

Figure 3

Property 7 The inequalities in property 7 compare the area under $f(x)$ with the areas of two rectangles, one of height M and one of height m. (See Fig. 3.) In the case $a < b$, the area of the larger rectangle is $M(b - a)$ and the area of the smaller rectangle is $m(b - a)$.

The Mean-Value Theorem for Definite Integrals

The mean-value theorem for derivatives says that (under suitable hypotheses) $f(b) - f(a) = f'(c)(b - a)$ for some number c in $[a, b]$. The mean-value theorem for definite integrals has a similar flavor. First we state it geometrically.

If $f(x)$ is positive and $a < b$, then $\int_a^b f(x)\,dx$ can be interpreted as the area of the shaded region in Fig. 4.

Let m be the minimum and M the maximum of $f(x)$ for x in $[a, b]$. The area of the rectangle of height M is larger than the shaded area; the area of the rectangle of height m is smaller than the shaded area. (See Figs. 5 and 6.) Therefore, there is a rectangle whose height h is somewhere between m and M, whose area is the same as the shaded area under the curve $y = f(x)$. (See Fig. 7.) Hence $\int_a^b f(x)\,dx = h(b - a)$.

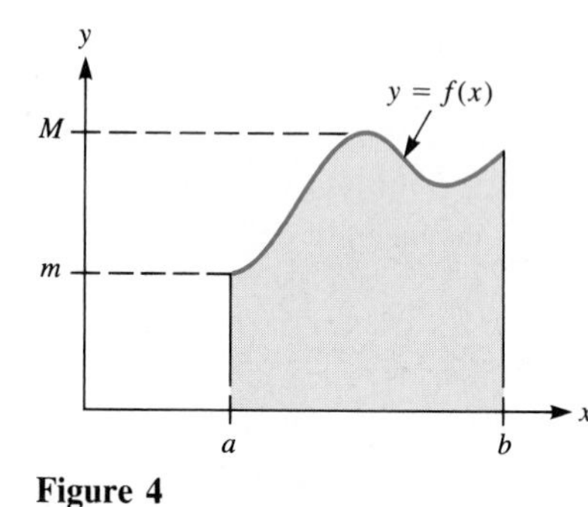

Figure 4

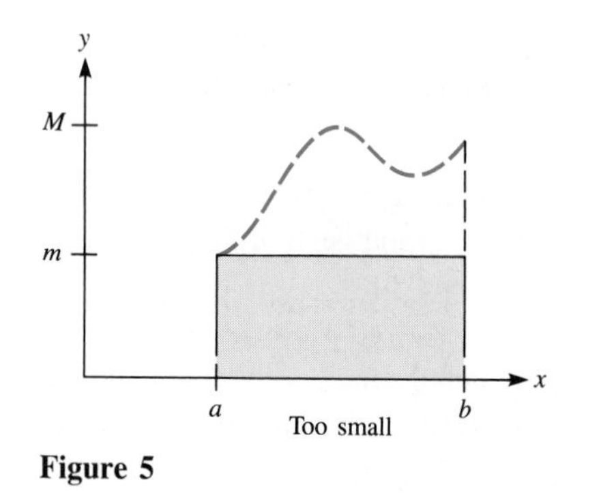

Figure 5

y

M

m

a

b

x

Too large

Figure 6

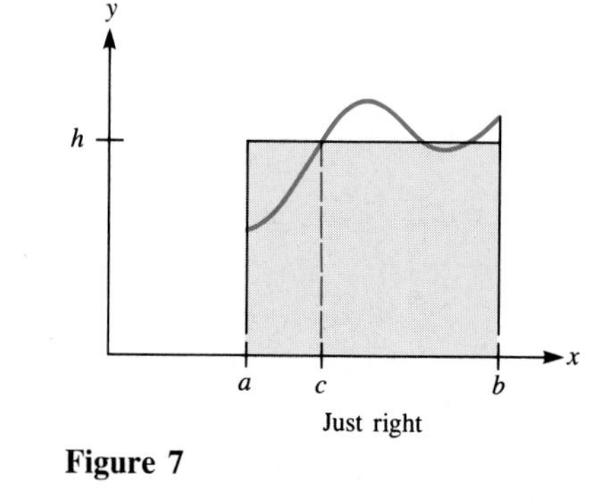

Figure 7

Now h is a number between m and M. By the intermediate-value theorem for continuous functions, there is a number c in $[a, b]$ such that $f(c) = h$. (See Fig. 7.) Hence,

$$\text{Area of shaded region under curve} = f(c)(b - a).$$

Next we state the mean-value theorem for integrals.

Mean-Value Theorem for Definite Integrals Let a and b be numbers, and let f be a continuous function defined for x between a and b. Then there is a number c between a and b such that

$$\int_a^b f(x)\,dx = f(c)(b - a).$$

Proof Consider the case $a < b$. Let M be the maximum and m the minimum of $f(x)$ for x in $[a, b]$. [Recall the maximum- (and minimum-) value theorem of Sec. 2.8.] By property 7,

$$m \le \frac{\int_a^b f(x)\,dx}{b - a} \le M.$$

By the intermediate-value theorem of Sec. 2.8, there is a number c in $[a, b]$ such that

$$f(c) = \frac{\int_a^b f(x)\,dx}{b - a},$$

and the theorem is proved. (The case $b < a$ can be obtained from the case $a < b$.) ■

EXAMPLE 4 Verify the mean-value theorem for $f(x) = x^2$ and $[a, b] = [0, 3]$.

SOLUTION In Sec. 5.3 it was shown that $\int_0^3 x^2\,dx = 9$. Since $f(x) = x^2$, we are looking for c in $[a, b]$ such that

$$\int_0^3 x^2\,dx = 9 = c^2(3 - 0)$$

or

$$9 = 3c^2.$$

Since $c^2 = \frac{9}{3} = 3$, $c = \sqrt{3}$. ($-\sqrt{3}$ is not the interval $[0, 3]$.) See Fig. 8. The rectangle with height $f(\sqrt{3}) = (\sqrt{3})^2 = 3$ and base $[0, 3]$ has the same area as the area of the region under the curve $y = x^2$ and above $[0, 3]$. ■

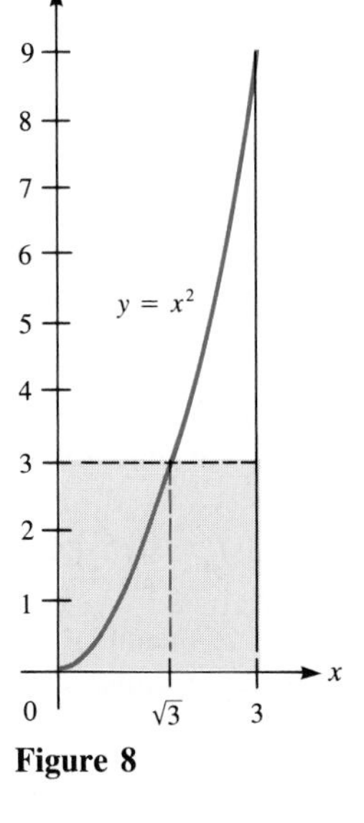

Figure 8

Average Value of a Function

Let $f(x)$ be a continuous function defined on $[a, b]$. What shall we mean by the "average value of $f(x)$ over $[a, b]$"? We cannot add up all the values of $f(x)$ for all x's in $[a, b]$ and divide by the number of x's, since there are an infinite number of such x's. [The average of a finite list of numbers $a_1, a_2, \ldots, a_n$ is their sum divided by n, $(\Sigma_{i=1}^n a_i)/n$. For instance, the average of 1, 2, and 6 is $(1 + 2 + 6)/3 = \frac{9}{3} = 3$.] Instead, we make the following definition:

Definition Let $f(x)$ be defined on the interval $[a, b]$. Assume that $\int_a^b f(x)\,dx$ exists. Then the quotient

$$\frac{\int_a^b f(x)\,dx}{b-a}$$

is called the **average** or **mean value** of the function on this interval.

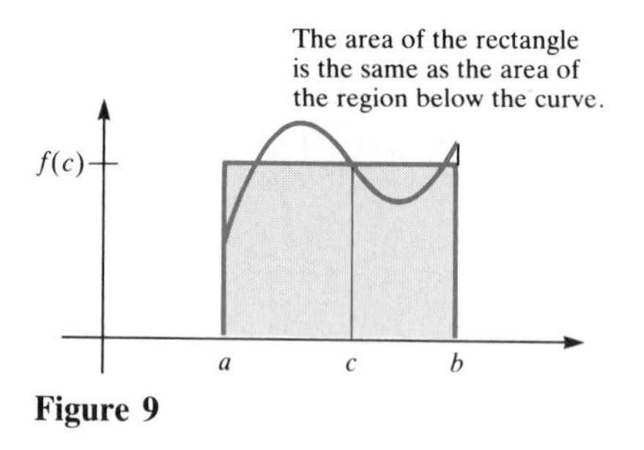

Figure 9

Geometrically speaking [if $f(x)$ is positive], this average value is the height of the rectangle that has the same area as the area of the region under the curve $y = f(x)$, above $[a, b]$. See Fig. 9. Observe that the average value of $f(x)$ over $[a, b]$ is between its maximum and minimum values for x in $[a, b]$. However, it is not necessarily the average of these two numbers.

EXAMPLE 5 Find the average value of x^2 over the interval $[1, 3]$.

SOLUTION The average value of x^2 over $[1, 3]$ by definition equals

$$\frac{\int_1^3 x^2\,dx}{3-1}.$$

In Sec 5.3 it was shown that

$$\int_1^3 x^2\,dx = \frac{3^3}{3} - \frac{1^3}{3}$$

$$= \frac{27}{3} - \frac{1}{3}$$

$$= \frac{26}{3}.$$

Hence $$\text{Average value} = \frac{\left(\frac{26}{3}\right)}{2} = \frac{13}{3} \approx 4.33.$$

[On the other hand, the average of the maximum and minimum values of x^2 on $[1, 3]$ is $(3^2 + 1^2)/2 = 5$.] ■

Section Summary

We introduced the notation $\int f(x)\,dx$ for an antiderivative of $f(x)$. Using this notation we stated various properties of antiderivatives.

We defined the symbol $\int_a^b f(x)\,dx$ in the special case when $b \leq a$, and stated various properties of definite integrals.

The mean-value theorem for definite integrals asserts that $\int_a^b f(x)\,dx$ equals $f(c)$ times $(b-a)$ for at least one value of c in $[a, b]$.

The quantity $(\int_a^b f(x)\,dx)/(b-a)$ is called the average value (or mean value) of $f(x)$ over $[a, b]$. It can be thought of as the height of the rectangle whose area is the same as the area of the region under the curve $y = f(x)$.

EXERCISES FOR SEC. 5.5: PROPERTIES OF THE ANTIDERIVATIVE AND THE DEFINITE INTEGRAL

In Exercises 1 to 10 evaluate the antiderivatives, in each case adding the constant C. Check each answer by differentiating it.

1 $\int 5x^2\,dx$ **2** $\int 7/x^2\,dx$

3 $\int (2x - x^3 + x^5)\,dx$ **4** $\displaystyle\int \left(6x^2 + \frac{1}{\sqrt{x}}\right) dx$

5 (*a*) $\int \cos x\,dx$, (*b*) $\int \cos 2x\,dx$

6 (*a*) $\int \sin x\,dx$, (*b*) $\int \sin 3x\,dx$

7 (*a*) $\int (2 \sin x + 3 \cos x)\,dx$ (*b*) $\int (\sin 2x + \cos 3x)\,dx$

8 $\int \sec x \tan x\,dx$

9 $\int \sec^2 x\,dx$

10 $\int \csc^2 x\,dx$

11 Write a paragraph or two in your own words that tells what the symbols $\int f(x)\,dx$ and $\int_a^b f(x)\,dx$ mean. Include examples.

12 Let $f(x)$ be a differentiable function. Is this equation true or false: "$f(x) = \int (df/dx)\,dx$"?
(*a*) Test some functions of your choice.
(*b*) Answer the question and write a brief justification for your answer.

13 Evaluate: (*a*) $\int_2^5 x^2\,dx$ (*b*) $\int_5^2 x^2\,dx$ (*c*) $\int_5^5 x^2\,dx$

14 Evaluate: (*a*) $\int_1^2 x\,dx$ (*b*) $\int_2^1 x\,dx$ (*c*) $\int_3^3 x\,dx$ ($\int_a^b x\,dx$ was discussed in Sec. 5.3.)

15 Find: (*a*) $\int x\,dx$ (*b*) $\int_3^4 x\,dx$

16 Find: (*a*) $\int 3x^2\,dx$ (*b*) $\int_1^4 3x^2\,dx$

17 If $2 \le f(x) \le 3$, what can be said about $\int_1^6 f(x)\,dx$?

18 If $-1 \le f(x) \le 4$, what can be said about $\int_{-2}^7 f(x)\,dx$?

The mean-value theorem for definite integrals asserts that if $f(x)$ is continuous, then $\int_a^b f(x)\,dx = f(c)(b - a)$ for some number c in $[a, b]$. In each of Exercises 19 to 22 find $f(c)$ and at least one c.

19 $f(x) = 2x$; $[a, b] = [1, 5]$

20 $f(x) = 5x + 2$; $[a, b] = [1, 2]$

21 $f(x) = x^2$; $[a, b] = [0, 4]$

22 $f(x) = x^2 + x$; $[a, b] = [1, 4]$

23 If $\int_1^2 f(x)\,dx = 3$ and $\int_1^5 f(x)\,dx = 7$, find
(*a*) $\int_2^1 f(x)\,dx$ (*b*) $\int_2^5 f(x)\,dx$

24 If $\int_1^3 f(x)\,dx = 4$ and $\int_1^3 g(x)\,dx = 5$, find
(*a*) $\int_1^3 (2f(x) + 6g(x))\,dx$ (*b*) $\int_3^1 [f(x) - g(x)]\,dx$

25 If the maximum value of $f(x)$ on $[a, b]$ is 7 and the minimum value on $[a, b]$ is 3, what can be said about (*a*) $\int_a^b f(x)\,dx$? (*b*) the average value of $f(x)$ in $[a, b]$?

26 Let $f(x) = c$ (constant) for all x in $[a, b]$. Find the average value of $f(x)$ in $[a, b]$.

Exercises 27 to 30 concern the average of a function over an interval. In each case, find the minimum, maximum, and average value of the function over the given interval.

27 $f(x) = x$; $[1, 3]$ **28** $f(x) = x$; $[1, 4]$

29 $f(x) = x^2$; $[2, 3]$ **30** $f(x) = x^2$; $[0, 5]$

In Exercises 31 and 32 use Simpson's method to estimate the average of the function over the given interval. Only the indicated values of the function are known.

31 $[1, 7]$

x	1	2	3	4	5	6	7
$f(x)$	3	1	4	5	2	2	6

32 $[0, 8]$

x	0	1	2	3	4	5	6	7	8
$f(x)$	5	1	2	4	3	3	2	6	1

33 (*a*) Graph $f(x) = x/(x^2 + 1)$ for x in $[0, 2]$.
(*b*) What are the minimum and maximum values of $f(x)$ for x in $[0, 2]$?
(*c*) Using (*b*), what can you say about $\int_0^2 [x/(x^2 + 1)]\,dx$?

34 Is $\int f(x)g(x)\,dx$ equal to $(\int f(x)\,dx)(\int g(x)\,dx)$?

35 (*a*) Show that $(\sin^3 x)/3$ is *not* an antiderivative of $\sin^2 x$.
(*b*) Find an antiderivative of $\sin^2 x$ by using the identity $\sin^2 x = (1 - \cos 2x)/2$.
(*c*) Check your answer in (*b*) by differentiating it.

In Exercises 36 and 37 verify the equations quoted from a table of antiderivatives (integrals). Just differentiate each of the alleged antiderivatives and see whether you obtain the integrand. The number a is constant in each case.

36 $$\int x^2 \sin ax\,dx = \frac{2x}{a^2} \sin ax + \frac{2}{a^3} \cos ax - \frac{x^2}{a} \cos ax + C$$

37 $$\int x \sin^2 ax\,dx = \frac{x^2}{4} - \frac{x \sin 2ax}{4a} - \frac{\cos 2ax}{8a^2} + C$$

38 The average value of a certain function $f(x)$ over $[1, 3]$ is 4. Over $[3, 6]$ it is 5. What is its average value over $[1, 6]$? Explain.

39 Here is another way to define the average of a continuous function $f(x)$ over $[a, b]$. Let n be a positive integer. Pick $n + 1$ equally spaced numbers $x_0 = a, x_2, x_3, \ldots, x_n = b$. Compute the average of $f(x_0), f(x_1), \ldots, f(x_n)$. Then take the limit of this average as $n \to \infty$. Call this limit "the average of $f(x)$ over $[a, b]$." In other words define the average as

$$\lim_{n\to\infty} \frac{\sum_{i=0}^{n} f(x_i)}{n + 1}$$

where the x_i are chosen as described.
(*a*) Use this definition to find the average of x^2 over $[0, 3]$.
(*b*) Does the result in (*a*) agree with the average as defined in the text?
(*c*) Find the relation between the "average" as defined in this exercise and the "average" as defined in the text. (*Hint:* Consider Riemann sums formed with n sections of equal length and left endpoints as sampling points.)

5.6 BACKGROUND FOR THE FUNDAMENTAL THEOREMS OF CALCULUS

This section introduces a function that plays a key role in relating definite integrals to derivatives. This function is the key to the fundamental theorems of calculus, which we will obtain in Sec. 5.7.

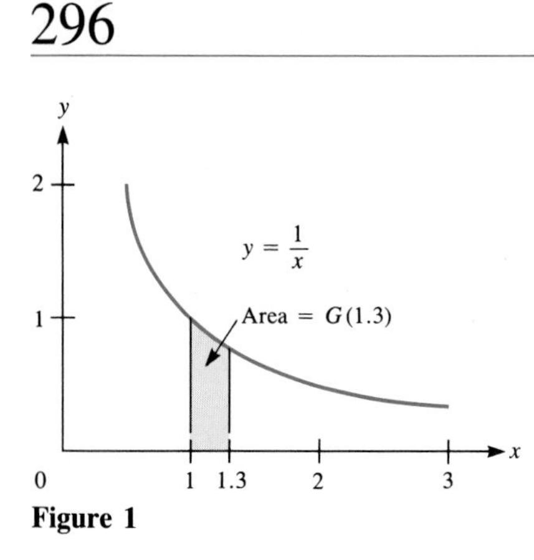

Figure 1

The Function *G*

Since we have efficient ways to make accurate estimates of definite integrals (for instance, Simpson's method), we can examine how $\int_a^b f(x)\,dx$ varies as we change one end of its interval, say b. The answer will turn out to provide another way to evaluate definite integrals, one that permits us to calculate many common definite integrals exactly.

Let us carry out some numerical experiments, using the integrand $f(x) = 1/x$. (Similar experiments could be done with any continuous function of your choice.)

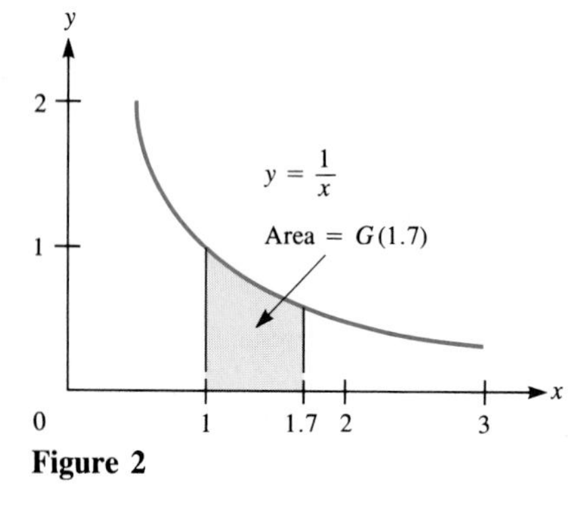

Figure 2

Figures 1 to 5 show areas under the curve $f(x) = 1/x$ and above various intervals. These intervals have the same left endpoint, 1, but different right endpoints: 1.3, 1.7, 2, 2.01, and 2.1. Let the area under the curve and above $[1, x]$ be $G(x)$.

Using integral notation, we might be tempted to write

$$G(x) = \int_1^x \frac{dx}{x}.$$

But x then plays two completely different roles, being used both to describe the interval $[1, x]$ and the integrand $1/x$. Recalling our discussion of dummy variables in Sec. 5.3, we choose t instead of x to describe the integrand. Thus

$$G(x) = \int_1^x \frac{dt}{t}.$$

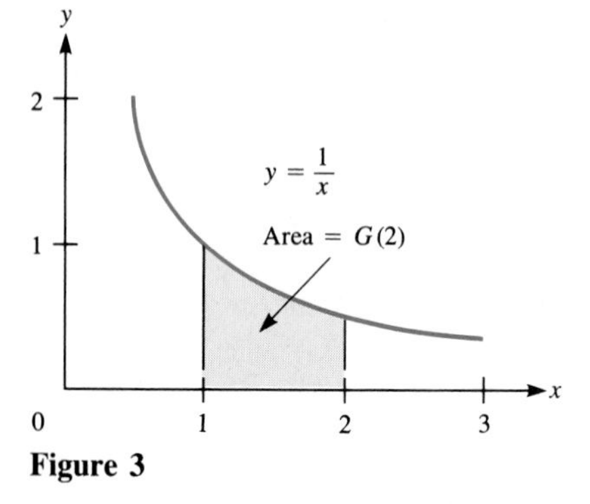

Figure 3

As the right endpoint increases, so does the area. Simpson's method with $n = 8$ gives the estimates of the 5 areas, accurate to 4 places, shown in Table 1. [Using approximating sums to estimate $\int_1^{1.3} (1/t)\,dt$, for instance, would require much more arithmetic.] Figure 6 is a graph of the function G, based on the data in Table 1 and the fact that $G(1) = 0$ (the area from 1 to 1 is 0). You may be able to plot additional points if you have a scientific calculator that can estimate definite integrals.

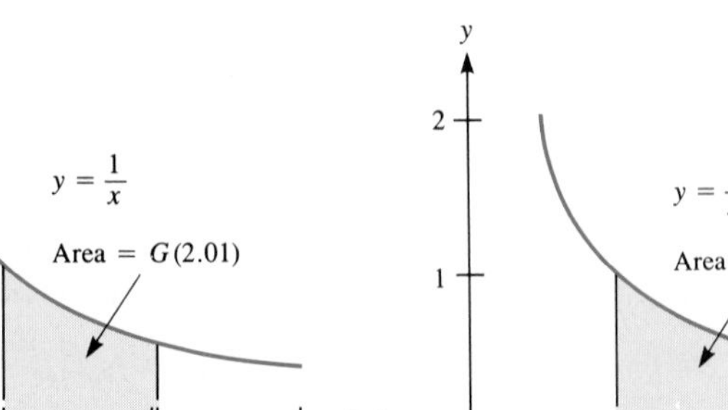

Figure 4

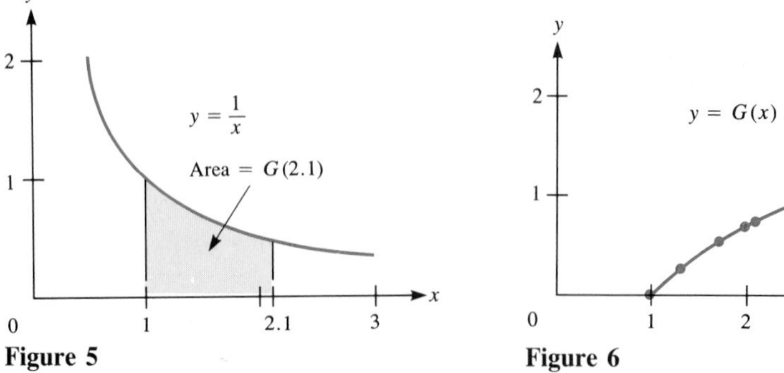

Figure 5

Figure 6

Table 1

x	$G(x)$
1.3	0.2624
1.7	0.5306
2.0	0.6932
2.01	0.6981
2.1	0.7419

The Derivative of G

Watch closely the behavior of $G(x)$ for x *near* 2. It is this behavior that will be important in developing the fundamental theorems of calculus. How rapidly does $G(x)$ change for x near 2? In other words, what is $G'(x)$?

The estimates of $G(2)$ and $G(2.1)$ permit us to estimate $G'(2)$, the rate at which $G(x)$ is changing when x is near 2. By the definition of the derivative,

$$G'(2) = \lim_{\Delta x \to 0} \frac{G(2 + \Delta x) - G(2)}{\Delta x}.$$

Hence,

$$\frac{G(2.1) - G(2)}{0.1}$$

is an approximation of $G'(2)$. We evaluate $[G(2.1) - G(2)]/0.1$:

$$\begin{aligned} \frac{G(2.1) - G(2)}{0.1} &\approx \frac{0.7419 - 0.6932}{0.1} \\ &= \frac{0.0487}{0.1} \qquad (1) \\ &= 0.487. \end{aligned}$$

To get a better estimate of $G'(2)$, we try a similar computation that uses $G(2)$ and $G(2.01)$:

$$\begin{aligned} \frac{G(2.01) - G(2)}{0.01} &\approx \frac{0.6981 - 0.6932}{0.01} \\ &= \frac{0.0049}{0.01} \qquad (2) \\ &= 0.49. \end{aligned}$$

Judging by (1) and (2), we suspect that

$$\lim_{\Delta x \to 0} \frac{G(2 + \Delta x) - G(2)}{\Delta x} = \frac{1}{2}.$$

In other words, we expect that

$$G'(2) = \frac{1}{2},$$

though we cannot be sure, since this is just a guess based on a pair of estimates. *Notice that* $\frac{1}{2}$ *is also the value of the integrand* $1/t$ *when* t *is* 2. In other words, it seems that for the integrand $f(t) = 1/t$,

$$G'(2) = \frac{1}{2} = f(2).$$

Is it a coincidence that our estimate of $G'(2)$ is so close to $f(2)$? Are the two quantities actually equal? In words instead of symbols, it appears that "the derivative of the definite integral with respect to its right-hand endpoint x is equal to the value of the integrand at x." In symbols, our conjecture says that if

$$G(x) = \int_1^x f(t)\,dt$$

then $$G'(x) = f(x).$$

If we can confirm this conjecture, then the definite integral and the derivative are intimately related. To put it in more concrete terms, there would be a relation between the problems of finding areas and finding tangent lines.

Is the conjecture true? That the answer is "yes" is the heart of the fundamental theorems of calculus, which are presented in the next section. The homework exercises give you a chance to gather more numerical data concerning this important question.

Section Summary

Given a continuous function f, we introduced the function $G(x) = \int_a^x f(t)\,dt$, the value of the definite integral of f over the interval $[a, x]$. We estimated $G'(x)$ for $f(t) = 1/t$ and found that it seemed to equal $f(x)$, the value of the integrand at x. Finally, we asked, "If $G(x) = \int_a^x f(t)\,dt$, is $G'(x) = f(x)$?"

EXERCISES FOR SEC. 5.6: BACKGROUND FOR THE FUNDAMENTAL THEOREMS OF CALCULUS

1 Let $f(t) = 3$ for all t. Let $G(x) = \int_1^x f(t)\,dt$.
(*a*) Draw the regions whose areas are $G(2.1)$ and $G(2)$.
(*b*) Compute $[G(2.1) - G(2)]/0.1$.
(*c*) Compute $G(x)$.
(*d*) Is $G'(x)$ equal to $f(x)$?

2 Let $f(t) = t$ for all t. Let $G(x) = \int_2^x f(t)\,dt$.
(*a*) Draw the region whose area is $G(x)$. (It is a trapezoid.)
(*b*) Use (*a*) to find $G(x)$.
(*c*) Is $G'(x) = f(x)$?

In the exercises that require estimates of definite integrals, use either Simpson's method with at least $n = 6$ intervals or a scientific calculator that can evaluate definite integrals; choose at least four decimal place accuracy.

3 Let $f(t) = 1/t$ and let $G(x) = \int_1^x dt/t$.
(*a*) Estimate $G(3)$ and $G(3.1)$.
(*b*) On a graph of $f(t) = 1/t$, shade the regions whose areas are $G(3)$ and $G(3.1)$.
(*c*) Use (*a*) to estimate $G'(3)$.
(*d*) What is $f(3)$? [Compare the result with (*c*).]

4 The same as Exercise 3 with 4 and 4.1 instead of 3 and 3.1.

5 Let $f(t) = 1/(1 + t^3)$ and $G(x) = \int_0^x dt/(1 + t^3)$.
(*a*) Estimate $G(1)$ and $G(1.1)$.
(*b*) Sketch the regions whose areas are $G(1)$ and $G(1.1)$.
(*c*) Use (*a*) to estimate $G'(1)$.
(*d*) What is $f(1)$?

6 The same as Exercise 5 with $G(1.01)$ instead of $G(1.1)$.

7 Let $f(t) = 2^{-t}$ and $G(x) = \int_1^x 2^{-t}\,dt$.
(*a*) Estimate $G(2)$ and $G(1.9)$.
(*b*) Draw the regions whose areas are $G(2)$ and $G(1.9)$.
(*c*) Use (*a*) to estimate $G'(2)$.
(*d*) What is $f(2)$?

8 The same as Exercise 7 with $G(1.99)$ instead of $G(1.9)$.

9 In Sec. 5.3 we learned that $\int_a^b t\,dt = b^2/2 - a^2/2$. Let $G(x) = \int_1^x t\,dt$.
(*a*) Find $G(x)$.
(*b*) Use (*a*) to find $G'(2)$.
(*c*) Is $G'(2)$ equal to the value of the integrand at 2?
(*d*) Is $G'(3)$ equal to the value of the integrand at 3?

10 In Sec. 5.3 it was shown that $\int_a^b t^2\,dt = b^3/3 - a^3/3$. Let $G(x) = \int_1^x t^2\,dt$.
(*a*) Find $G(x)$.
(*b*) Use (*a*) to find $G'(7)$.
(*c*) Is $G'(7)$ equal to the value of the integrand at 7?
(*d*) Is $G'(8)$ equal to the value of the integrand at 8?

It is instructive to see what happens to $\int_a^b f(t)\,dt$ when b is fixed and a changes. Exercises 11 to 13 concern this situation.

11 Let f be a continuous function with positive values. Let $G(x)$ be the area of the region under the graph of f and above the interval $[x, 5]$, $x < 5$.
(*a*) Sketch the region whose area is $G(1)$.
(*b*) Sketch the region whose area is $G(2)$.
(*c*) Is G an increasing function or a decreasing function?
(*d*) Would you expect its derivative to be positive or to be negative?

12 Let $G(x) = \int_x^3 dt/t$, $0 < x < 3$.
(*a*) Draw the regions whose areas are $G(1)$ and $G(1.1)$.
(*b*) Which is larger, $G(1)$ or $G(1.1)$?
(*c*) Is G an increasing or decreasing function?
(*d*) Is G' positive or negative?
(*e*) Estimate $G'(1)$ by estimating $[G(1.01) - G(1)]/0.01$.
(*f*) Does $G'(1)$ seem related to the value of the integrand at 1?

13 Let $G(x) = \int_x^2 dt/(1 + t^3)$, $x < 2$.
(*a*) Draw the regions whose areas are $G(0)$ and $G(0.01)$.
(*b*) Which is larger, $G(0)$ or $G(0.01)$?
(*c*) Is G an increasing or decreasing function?
(*d*) Is G' positive or negative?
(*e*) Estimate $G'(0)$ by estimating $[G(0.01) - G(0)]/0.01$.
(*f*) Does $G'(0)$ seem related to the value of the integrand at 0?

14 Let $G(x) = \int_1^x dt/t$.
(*a*) Draw the little region whose area is $G(2.01) - G(2)$.
(*b*) Using (*a*), discuss why $[G(2.01) - G(2)]/0.01$ is approximately equal to the integrand $1/t$ evaluated at 2.

15 Let $f(t)$ be a continuous function with positive values and let $G(x) = \int_1^x f(t)\, dt$.
(*a*) Sketch the graph of a "typical" such function.
(*b*) On your graph from (*a*), show $G(2)$ and $G(2.01)$.
(*c*) On your graph, shade the region whose area is $G(2.01) - G(2)$.
(*d*) Using (*c*), explain why $[G(2.01) - G(2)]/0.01$ is probably close to $f(2)$.
(*e*) Why would you expect that $G'(2) = f(2)$?

16 Let $f(t)$ be a continuous function with positive values.
(*a*) Let $G(x) = \int_0^x f(t)\, dt$, $x > 0$. Sketch a figure and explain whether $G'(x)$ is positive or negative.
(*b*) Let $G(x) = \int_x^5 f(t)\, dt$, $x < 5$. Sketch a figure and explain whether $G'(x)$ is positive or negative.

5.7 THE FUNDAMENTAL THEOREMS OF CALCULUS

This is the most important section in the whole book, for it develops the two "fundamental theorems of calculus," which provide a connection between the concepts of the derivative and the definite integral. This connection, in turn, gives a way to evaluate some definite integrals exactly, without worrying about summations or estimation errors. This method applies to many, though not all, of the definite integrals that arise in the day-to-day work of engineers, chemists, physicists, economists, biologists, and other scientific professionals.

The first fundamental theorem is of interest to us mainly because it is the basis of the second fundamental theorem. The second fundamental theorem is the one that enables us to calculate many of the definite integrals met daily wherever calculus is used. In particular, it will tell us the value of the definite integral of any polynomial.

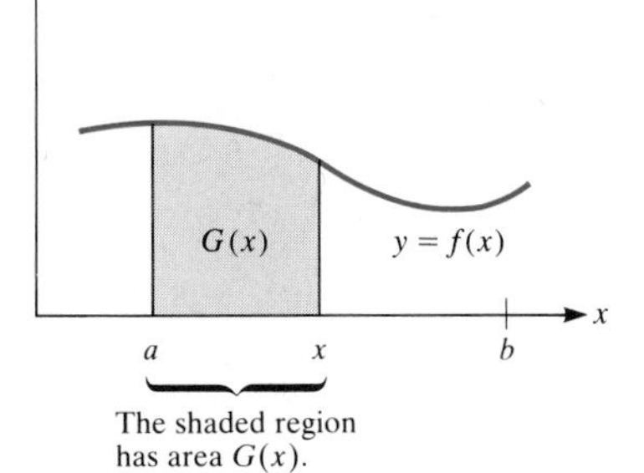

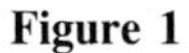
Figure 1

The First Fundamental Theorem of Calculus

The first fundamental theorem of calculus concerns the derivative of the function G introduced in the preceding section.

Let f be a continuous function such that $f(x)$ is positive for x in $[a, b]$. For x in $[a, b]$, let $G(x)$ be the area of the region under the graph of f and above the interval $[a, x]$, as shown in Fig. 1. For instance, $G(a) = 0$.

We will compute the derivative of G, that is,

$$\lim_{\Delta x \to 0} \frac{G(x + \Delta x) - G(x)}{\Delta x}.$$

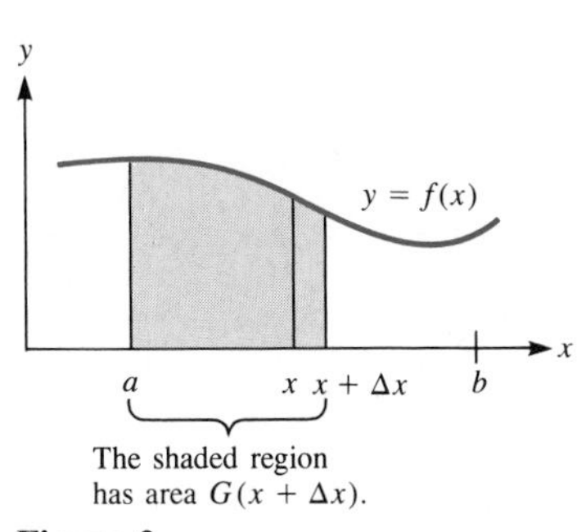

Figure 2

For simplicity, keep Δx positive. Then $G(x + \Delta x)$ is the area under the curve $y = f(x)$ above the interval $[a, x + \Delta x]$. If Δx is small, $G(x + \Delta x)$ is only slightly larger than $G(x)$, as shown in Fig. 2. Then $G(x + \Delta x) - G(x)$ is the area of the thin shaded strip in Fig. 3.

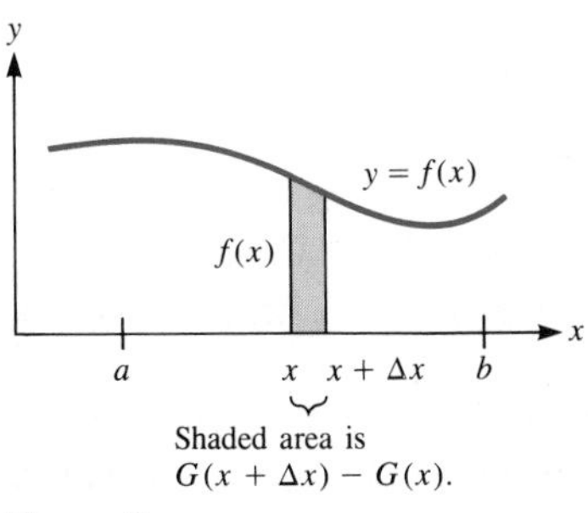

Figure 3

When Δx is small, the narrow shaded strip above $[x, x + \Delta x]$ resembles a rectangle of base Δx and height $f(x)$, with area $f(x)\,\Delta x$. Therefore, it seems reasonable that when Δx is small,

$$\frac{\Delta G}{\Delta x} \approx \frac{f(x)\,\Delta x}{\Delta x} = f(x).$$

In short, it seems plausible that

$$\lim_{\Delta x \to 0} \frac{\Delta G}{\Delta x} = f(x).$$

Briefly,

$$G'(x) = f(x).$$

In words, "the derivative of the area of the region under the graph of f and above $[a, x]$ with respect to x is the value of f at x."

Now we state these observations in terms of definite integrals.

Let f be a continuous function. Let $G(x) = \int_a^x f(t)\,dt$. Then we expect that

$$\frac{d}{dx}\left(\int_a^x f(t)\,dt\right) = f(x).$$

This equation says that "the derivative of the definite integral of f with respect to the right end of the interval is simply f evaluated at that end." This is the substance of the first fundamental theorem of calculus.

FIRST FUNDAMENTAL THEOREM OF CALCULUS

Let f be continuous on an open interval containing the interval $[a, b]$. Let

$$G(x) = \int_a^x f(t)\,dt$$

for $a \le x \le b$. Then G is differentiable on $[a, b]$ and its derivative is f; that is,

$$G'(x) = f(x).$$

As a consequence of this theorem, *every continuous function is the derivative of some function*. There is a similar theorem concerning $H(x) = \int_x^b f(t)\,dt$: $H'(x) = -f(x)$. (See Exercise 50.)

EXAMPLE 1 Give a function whose derivative is $2^x/(1 + x^2)$.

SOLUTION Let $G(x) = \int_0^x 2^t/(1 + t^2)\,dt$. According to the first fundamental theorem

$$G'(x) = \frac{2^x}{1 + x^2}.$$

Thus G is a function whose derivative is $2^x/(1 + x^2)$. ■

You probably would prefer to have a "nice formula" for $G(x)$ in Example 1, the way you have a nice formula for an antiderivative of $3x^2$, namely x^3. Do not try looking for a formula for G in terms of the functions met in algebra, trigonometry, and math analysis—such as polynomials, $\sin x$, $\cos x$, square roots, cube roots, exponentials, and logarithms. Such functions are called *elementary*.

Mathematicians have proved—beyond a shadow of a doubt—that the function G in Example 1 is *not elementary*.

We now turn to the second fundamental theorem of calculus, the one that is such a nice tool for evaluating many common definite integrals.

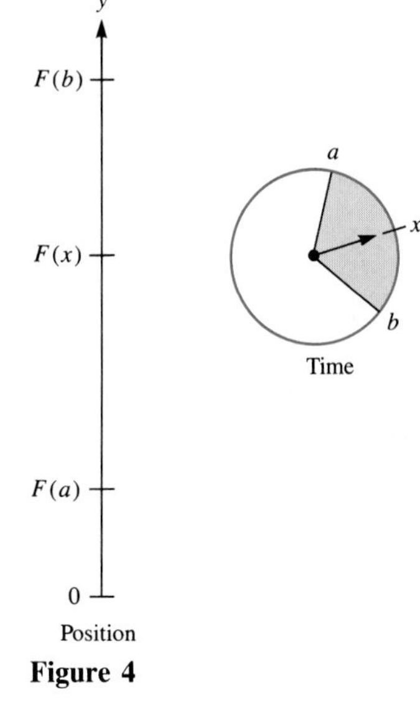

Figure 4

The Second Fundamental Theorem of Calculus

In Chap. 3 it was shown that *velocity is the derivative of the distance*. In Sec. 5.3 it was shown that *the definite integral of velocity is the change in distance*. To make this more precise, consider an object moving upward on the y axis. It starts at time a and stops at time b. At time x it is at $F(x)$ on the y axis. (See Fig 4.) Then the velocity at time x is the derivative $F'(x)$. The change in distance of the moving particle from time a to time b is

$$\text{Final coordinate} - \text{initial coordinate} = F(b) - F(a).$$

This assertion, ''the definite integral of velocity is the change in distance,'' now reads mathematically as

$$\int_a^b F'(x)\,dx = F(b) - F(a). \tag{1}$$

If we give $F'(x)$ the name $f(x)$, then (1) can be stated this way: If $f(x) = F'(x)$, then

$$\int_a^b f(x)\,dx = F(b) - F(a).$$

Our thinking in terms of a moving particle suggests the second fundamental theorem of calculus:

> **SECOND FUNDAMENTAL THEOREM OF CALCULUS**
>
> If f is continuous on $[a, b]$ and if F is an antiderivative of f, then
>
> $$\int_a^b f(x)\,dx = F(b) - F(a).$$

In practical terms this theorem says, ''If you want to evaluate the definite integral of f from a to b, look for an antiderivative of f. Evaluate that antiderivative at b and subtract the value of that antiderivative at a. This difference is the value of the definite integral you are seeking.'' Of course all hinges on finding an antiderivative of the integrand f. For many functions, it is easy to find their antiderivatives. For some it is hard, but they can be found. (Chapter 7 gives some techniques.) For others, the antiderivatives are not elementary: It is impossible to express these antiderivatives in terms of the usual functions met in high school.

To illustrate the power of the second fundamental theorem we will compute $\int_0^3 x^2\,dx$. Recall how hard we worked in Sec. 5.3 to show that it equals 9.

EXAMPLE 2 Use the second fundamental theorem of calculus to evaluate $\int_0^3 x^2\,dx$.

SOLUTION Our search for an antiderivative of x^2 is not hard. $F(x) = x^3/3$ will do. By the second fundamental theorem, with $f(x) = x^2$ and $F(x) = x^3/3$,

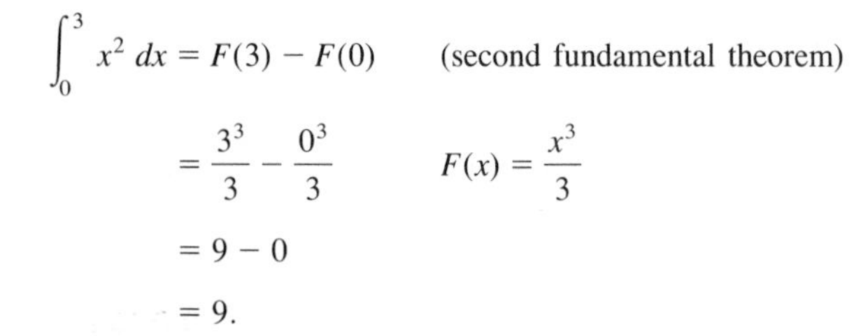

$$\begin{aligned}\int_0^3 x^2\,dx &= F(3) - F(0) && \text{(second fundamental theorem)}\\ &= \frac{3^3}{3} - \frac{0^3}{3} && F(x) = \frac{x^3}{3}\\ &= 9 - 0\\ &= 9.\end{aligned}$$

That is certainly a lot faster than the method in Sec. 5.3. ■

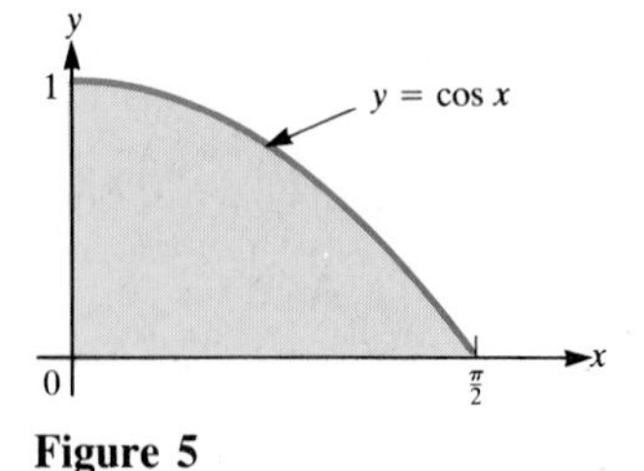

Figure 5

EXAMPLE 3 Find the area of the region under the curve $y = \cos x$, above the x axis, and between $x = 0$ and $x = \pi/2$. (See Fig. 5.)

SOLUTION As was shown in Sec. 5.3, area is the definite integral of the cross-sectional length. In this case,

$$\text{Area} = \int_0^{\pi/2} \cos x\,dx.$$

The fundamental theorem of calculus asserts that if we can find a function F such that

$$F'(x) = \cos x,$$

then the definite integral can be evaluated easily, as $F(\pi/2) - F(0)$. Now, in Chap. 3 it was shown that the derivative of the sine function is the cosine function. So let

$$F(x) = \sin x.$$

Then, by the second fundamental theorem,

$$\begin{aligned}\int_0^{\pi/2} \cos x\,dx &= F\left(\frac{\pi}{2}\right) - F(0)\\ &= \sin\frac{\pi}{2} - \sin 0\\ &= 1 - 0\\ &= 1.\end{aligned}$$

The area is one square unit. ■

Some Terms and Notation

The related processes of computing $\int_a^b f(x)\,dx$ and of finding an antiderivative $\int f(x)\,dx$ are both called **integrating** $f(x)$. Thus integration refers to two separate but related problems: computing a number $\int_a^b f(x)\,dx$ or finding a function $\int f(x)\,dx$. The second fundamental theorem states that the second process may be of use in computing $\int_a^b f(x)\,dx$.

Notation $F(b) - F(a)$ is denoted $F(x)\Big|_a^b$. For instance,

$$\frac{x^3}{3}\Big|_1^3 = \frac{3^3}{3} - \frac{1^3}{3} = \frac{27}{3} - \frac{1}{3} = \frac{26}{3}$$

and
$$\sin x\Big|_0^{\pi/2} = \sin\frac{\pi}{2} - \sin 0 = 1 - 0 = 1.$$

The second fundamental theorem now reads: If $F' = f$ then

$$\int_a^b f(x)\,dx = F(x)\Big|_a^b.$$

EXAMPLE 4 Evaluate $\int_1^2 1/x^2\,dx$ by the second fundamental theorem.

SOLUTION In order to apply the second fundamental theorem, we have to find an antiderivative of $1/x^2$, $\int (1/x^2)\,dx$. In Sec. 3.4 it was observed that

$$\int x^a\,dx = \frac{x^{a+1}}{a+1} + C \qquad a \neq -1.$$

The case $a = -2$ gives us an antiderivative of $1/x^2$:

$$\begin{aligned}\int \frac{1}{x^2}\,dx &= \int x^{-2}\,dx\\ &= \frac{x^{-2+1}}{-2+1} + C\\ &= \frac{x^{-1}}{-1} + C\\ &= -\frac{1}{x} + C.\end{aligned}$$

By the second fundamental theorem

$$\int_1^2 \frac{1}{x^2}\,dx = \left(\frac{-1}{x} + C\right)\Big|_1^2 = \left(\frac{-1}{2} + C\right) - \left(\frac{-1}{1} + C\right) = \frac{-1}{2} + 1 = \frac{1}{2}. \quad ■$$

Note that the C's cancel. We don't need the C when applying the second fundamental theorem.

The second fundamental theorem asserts that

$$\underbrace{\int_1^2 \frac{1}{x^2}\,dx}_{\text{The definite integral: a limit of sums.}} = \underbrace{\int \frac{1}{x^2}\,dx\Big|_1^2}_{\text{The difference between an antiderivative evaluated at 2 and at 1}}$$

The symbols on the right and left of the equal sign are so similar that it is tempting to think that the equation is obvious or says nothing whatever.

Beware: This compact equation is in fact a special instance of the second fundamental theorem.

Remark: Often we write $\int (1/x^2)\,dx$ as $\int (dx/x^2)$, merging the 1 with the dx. More generally, $\int \frac{f(x)}{g(x)}\,dx$ may be written as $\int \frac{f(x)\,dx}{g(x)}$.

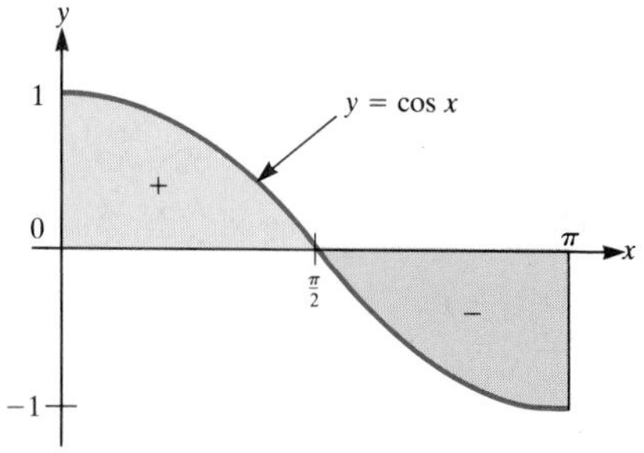

Figure 6

Algebraic Area

When we evaluate $\int_0^\pi \cos x\, dx$, we obtain $\sin \pi - \sin 0 = 0 - 0 = 0$. What does this say about areas? Inspection of Fig. 6 shows what is happening.

For x in $[\pi/2, \pi]$, $\cos x$ is negative and the curve $y = \cos x$ lies *below* the x axis. If we interpret the corresponding area as negative, then we see that it cancels with the area from 0 to $\pi/2$. So let us agree that when we say "$\int_a^b f(x)\, dx$ represents the area under the curve $y = f(x)$," we mean that it represents the area between the curve and the x axis, *with area below the x axis taken as negative*.

A Moral

The next example offers a contrast between two definite integrals, from which we may draw a moral.

EXAMPLE 5 Evaluate (*a*) $\int_0^\pi \sin x\, dx$ and (*b*) $\int_0^\pi \sin x^2\, dx$.

SOLUTION (*a*) It is not hard to find an antiderivative of $\sin x$: $\int \sin x\, dx = -\cos x$. So, by the second fundamental theorem,

$$\int_0^\pi \sin x\, dx = -\cos x \Big|_0^\pi$$
$$= (-\cos \pi) - (-\cos 0)$$
$$= -(-1) - (-1)$$
$$= 1 + 1$$
$$= 2.$$

(*b*) To evaluate $\int_0^\pi \sin x^2\, dx$ we first look for an antiderivative of $\sin x^2$. Let us try $-\cos x^2$. The derivative of $-\cos x^2$ is $(\sin x^2)(2x) = 2x \sin x^2$, by the chain rule. So $-\cos x^2$ is *not* an antiderivative of $\sin x^2$. Too bad. Even if we searched a thousand years for an *elementary* function whose derivative is $\sin x^2$, we would be disappointed. There is no such function. In that case, to evaluate $\int_0^\pi \sin x^2\, dx$, we could use the trapezoidal method or Simpson's method. ■

(The proof that there is no *elementary* function whose derivative is $\sin x^2$, which fills dozens of pages, is reserved for a graduate course.)

The derivative of an elementary function is elementary. This is shown in Chaps. 4 and 6. But an antiderivative of an elementary function need not be elementary. The moral: Evaluating a definite integral cannot always be reduced to calculating $F(b) - F(a)$.

More Examples

EXAMPLE 6 Differentiate the functions

(*a*) $y = \int_0^x \sqrt{1 + t^2}\, dt$,

(*b*) $y = \int_0^{x^3} \sqrt{1 + t^2}\, dt$,

(*c*) $y = \int_x^{10} \sqrt{1 + t^2}\, dt$.

SOLUTION (*a*) By the first fundamental theorem

$$\frac{dy}{dx} = \frac{d}{dx}\left(\int_0^x \sqrt{1+t^2}\, dt\right) = \sqrt{1+x^2}.$$

(*b*) If $y = \int_0^{x^3} \sqrt{1+t^2}\, dt$, the first fundamental theorem does not apply directly since the upper limit of integration is x^3, not x. In this case, let $u = x^3$. Then

$$y = \int_0^u \sqrt{1+t^2}\, dt, \text{ where } u = x^3.$$

By the first fundamental theorem,

$$\frac{dy}{du} = \sqrt{1+u^2}.$$

The chain rule then tells us that

$$\begin{aligned}\frac{dy}{dx} &= \frac{dy}{du}\frac{du}{dx}\\ &= \sqrt{1+u^2}\, 3x^2\\ &= \sqrt{1+(x^3)^2}\, 3x^2\\ &= 3x^2\sqrt{1+x^6}.\end{aligned}$$

(*c*) If $y = \int_x^{10} \sqrt{1+t^2}\, dt$, x is the *lower* limit of integration. As we mentioned right after the first fundamental theorem, the derivative in this case is equal to the *negative* of the integrand at x. That is,

$$\frac{dy}{dx} = -\sqrt{1+x^2}. \quad ■$$

EXAMPLE 7 The density of a string 10 centimeters long is x^3 grams per centimeter at a distance of x centimeters from one end. Find its total mass.

SOLUTION As shown in Sec. 5.3, "mass is the integral of density," that is,

$$\text{Mass} = \int_0^{10} x^3\, dx \text{ grams}.$$

We have reduced the problem to that of evaluating a definite integral. Since $x^4/4$ is an antiderivative of x^3, we can easily apply the second fundamental theorem:

$$\begin{aligned}\text{Mass} = \int_0^{10} x^3\, dx &= \left.\frac{x^4}{4}\right|_0^{10}\\ &= \frac{10^4}{4} - \frac{0^4}{4}\\ &= \frac{10{,}000}{4} = 2{,}500 \text{ grams}. \quad ■\end{aligned}$$

Proof of the First Fundamental Theorem

The first fundamental theorem asserts that the derivative of $G(x) = \int_a^x f(t)\, dt$ is $f(x)$. We gave a convincing argument using areas of regions. However, since definite integrals are defined in terms of approximating sums, not areas, we should include a proof that uses only properties of definite integrals. One advantage of doing this is that the argument also takes care of the case when the integrand is negative.

Proof of the First Fundamental Theorem We wish to show that $G'(x) = f(x)$. To do this, we must make use of the definition of the derivative of a function.

We have

$$G'(x) = \lim_{\Delta x \to 0} \frac{G(x + \Delta x) - G(x)}{\Delta x} \qquad \text{(definition of derivative)}$$

$$= \lim_{\Delta x \to 0} \frac{\int_a^{x+\Delta x} f(t)\, dt - \int_a^x f(t)\, dt}{\Delta x} \qquad \text{(definition of the function } G\text{)}$$

$$= \lim_{\Delta x \to 0} \frac{\int_a^x f(t)\, dt + \int_x^{x+\Delta x} f(t)\, dt - \int_a^x f(t)\, dt}{\Delta x} \qquad \text{(property 6 in Sec. 5.5)}$$

$$= \lim_{\Delta x \to 0} \frac{\int_x^{x+\Delta x} f(t)\, dt}{\Delta x} \qquad \text{(canceling)}$$

$$= \lim_{\Delta x \to 0} \frac{f(c)\, \Delta x}{\Delta x} \qquad \text{(mean-value theorem for integrals; } c \text{ between } x \text{ and } x + \Delta x\text{)}$$

$$= \lim_{\Delta x \to 0} f(c) \qquad \text{(canceling)}$$

$$= f(x). \qquad (f \text{ is continuous})$$

Hence

$$G'(x) = f(x),$$

which was to be proved. ■

Proof of the Second Fundamental Theorem

The second fundamental theorem asserts that if $F' = f$, then $\int_a^b f(x)\, dx = F(b) - F(a)$. We persuaded ourselves that it is true by thinking of f as "velocity" and F as "position." We now prove the theorem, showing that it is an immediate consequence of the first fundamental theorem and the fact that two antiderivatives of the same function differ by a constant.

Proof of the Second Fundamental Theorem We are assuming that $F' = f$ and wish to show that $F(b) - F(a) = \int_a^b f(x)\, dx$. Let $G(x) = \int_a^x f(t)\, dt$. By the first fundamental theorem, G is an antiderivative of f. Since F and G are both antiderivatives of f, they differ by a constant, say C. That is,

$$F(x) = G(x) + C.$$

Thus

$$F(b) - F(a) = (G(b) + C) - (G(a) + C)$$

$$= G(b) - G(a) \qquad C\text{'s cancel}$$

$$= \int_a^b f(x)\,dx - \int_a^a f(x)\,dx \qquad \text{definition of } G$$

$$= \int_a^b f(x)\,dx. \qquad \left(\int_a^a f(x)\,dx = 0\right)$$

Hence
$$F(b) - F(a) = \int_a^b f(x)\,dx. \quad \blacksquare$$

The "Paintbrush" Theorem

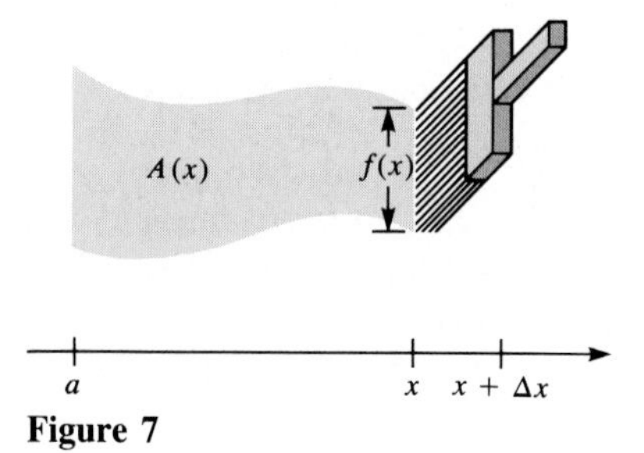

Figure 7

The first fundamental theorem should make sense to anyone who has painted a wall with a paintbrush. The wider the brush, the larger the area you sweep out. Imagine that you hold it as shown in Fig. 7 and move it a distance Δx. The area ΔA swept out is roughly the "width of the brush times Δx." Call the width of the brush $f(x)$. [As you press down, $f(x)$ may increase.] Thus

$$\Delta A \approx f(x)\,\Delta x$$

or

$$\frac{\Delta A}{\Delta x} \approx f(x).$$

Letting Δx approach 0, we get

$$\frac{dA(x)}{dx} = f(x).$$

Since $A(x) = \int_a^x f(t)\,dt$, we have $(d/dx)\int_a^x f(t)\,dt = f(x)$, which is the first fundamental theorem. (Judith Grabiner calls this "the windshield wiper theorem.")

Section Summary

This section concerns the two fundamental theorems of calculus. The first asserts that $(d/dx)(\int_a^x f(t)\,dt) = f(x)$. The second asserts that if $F(x) = \int f(x)\,dx$, an antiderivative of $f(x)$, then $\int_a^b f(x)\,dx = F(b) - F(a)$. Be careful when reading the symbols $\int f(x)\,dx$ and $\int_a^b f(x)\,dx$. They represent quite different concepts, with completely different definitions. Read them as carefully as you read, say, the words "unite" and "untie." They may look similar to the eye, but their meanings are not similar. Antiderivatives are also referred to as *indefinite integrals*.

The second theorem frequently is of use in evaluating a definite integral. For instance, it quickly gives us the value of $\int_0^3 x^2\,dx$. However, it is useless in evaluating $\int_0^1 \sqrt{1+t^3}\,dt$, since there is no elementary function whose derivative is $\sqrt{1+t^3}$.

Because we use the second fundamental theorem much more often than the first, we may refer to it as *the fundamental theorem of calculus,* abbreviated FTC.

We first gave geometric and physical arguments for the two theorems, but concluded with purely mathematical arguments. It turned out that the second theorem followed directly from the first theorem.

EXERCISES FOR SEC. 5.7: THE FUNDAMENTAL THEOREMS OF CALCULUS

In Exercises 1 and 2 evaluate the given expressions.

1 (*a*) $x^3|_1^2$ (*b*) $x^3|_{-1}^2$ (*c*) $\cos x\,|_0^{\pi}$

2 (*a*) $(x + \sec x)|_0^{\pi/4}$ (*b*) $\left.\frac{1}{x}\right|_2^3$ (*c*) $\sqrt{x-1}\,|_5^{10}$

3 State the first fundamental theorem in your own words, using as few mathematical symbols as you can.

4 State the second fundamental theorem in your own words, using as few mathematical symbols as you can.

In Exercises 5 to 16 use the second fundamental theorem to evaluate the given integrals.

5 $\int_1^2 5x^3\,dx$ 6 $\int_{-1}^3 2x^4\,dx$

7 $\int_1^4 (x + 5x^2)\,dx$ 8 $\int_{-1}^2 (6x - 3x^2)\,dx$

9 $\int_{\pi/6}^{\pi/3} 5\cos x\,dx$ 10 $\int_{\pi/4}^{3\pi/4} 3\sin x\,dx$

11 $\int_0^{\pi/2} \sin 2x\,dx$ 12 $\int_0^{\pi/6} \cos 3x\,dx$

13 $\int_4^9 5\sqrt{x}\,dx$ 14 $\int_1^9 \frac{1}{\sqrt{x}}\,dx$

15 $\int_1^8 \sqrt[3]{x^2}\,dx$ 16 $\int_2^4 \frac{4}{x^3}\,dx$

In Exercises 17 to 22 find the average value of the given function over the given interval.

17 x^2; [3, 5]

18 x^4; [1, 2]

19 $\sin x$; $[0, \pi]$

20 $\cos x$; $[0, \pi/2]$

21 $\sec^2 x$; $[\pi/6, \pi/4]$

22 $\sec 2x \tan 2x$; $[\pi/8, \pi/6]$

In Exercises 23 to 32 evaluate the given quantities.

23 The area of the region under the curve $3x^2$ and above [1, 4].

24 The area of the region under the curve $1/x^2$ and above [2, 3].

25 The area of the region under the curve $6x^4$ and above [−1, 1].

26 The area of the region under the curve $\sqrt{x}$ and above [25, 36].

27 The distance an object travels from time $t = 1$ second to time $t = 2$ seconds, if its speed at time t seconds is t^5 feet per second.

28 The distance an object travels from time $t = 1$ second to time $t = 8$ seconds if its speed at time t seconds is $7\sqrt[3]{t}$ feet per second.

29 The total mass of a string in the section [1, 2] if its density at x is $5x^3$ grams per centimeter.

30 The total mass of a string in the section $[\frac{1}{4}, 1]$ if its density at x is $4\sqrt{x}$ grams per centimeter.

31 The volume of a solid located between a plane at $x = 1$ and a plane located at $x = 5$ if the cross-sectional area of the solid by a plane corresponding to x is $6x^3$ square centimeters. (Assume that the planes are all perpendicular to the x axis.)

32 Like Exercise 31, except that the typical cross-sectional area is $1/x^3$ instead of $6x^3$.

33 (*a*) Is $\int x^2\,dx$ a function or is it a number?
(*b*) Is $\int x^2\,dx\,|_1^3$ a function or is it a number?
(*c*) Is $\int_1^3 x^2\,dx$ a function or is it a number?

34 (*a*) Which expression is defined as a limit of sums:

$$\int x^2\,dx\,\Big|^3 \quad \text{or} \quad \int^3 x^2\,dx?$$

(*b*) Why are the two numbers in (*a*) equal?

35 True or false: (*a*) Every elementary function has an elementary derivative. (*b*) Every elementary function has an elementary antiderivative. Explain.

36 True or false: (*a*) $\sin x^2$ has an elementary antiderivative. (*b*) $\sin x^2$ has an antiderivative. Explain.

---

37 Find dy/dx if (*a*) $y = \int \sin x^2\,dx$, (*b*) $y = 3x + \int_{-2}^3 \sin x^2\,dx$, (*c*) $y = \int_{-2}^x \sin t^2\,dt$.

In Exercises 38 to 41 differentiate the given functions.

38 (*a*) $\int_1^x t^4\,dt$ (*b*) $\int_x^1 t^4\,dt$

39 (*a*) $\int_1^x \sqrt[3]{1 + \sin t}\,dt$ (*b*) $\int_1^{x^2} \sqrt[3]{1 + \sin t}\,dt$

40 $\int_{-1}^x 3^{-t}\,dt$

41 $\int_{2x}^{3x} t \tan t\,dt$ *Hint:* First rewrite it as $\int_{2x}^0 t \tan t\,dt + \int_0^{3x} t \tan t\,dt$.

42 Figure 8 shows the graph of a function $f(x)$ for x in [1, 3]. Let $G(x) = \int_1^x f(t)\,dt$. Graph $y = G(x)$ for x in [1, 3] as well as you can. Explain your reasoning.

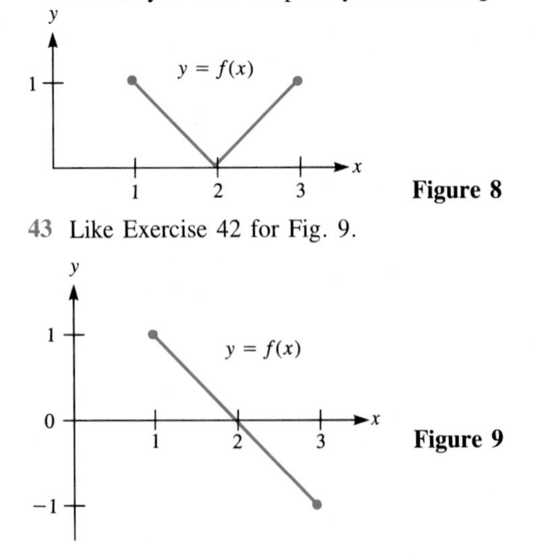

Figure 8

43 Like Exercise 42 for Fig. 9.

Figure 9

44 A plane at a distance x from the center of a sphere of radius r, $0 \le x \le r$, meets the sphere in a circle. (See Fig. 10.)
(*a*) Show that the radius of the circle is $\sqrt{r^2 - x^2}$.
(*b*) Show that the area of the circle is $\pi r^2 - \pi x^2$.
(*c*) Using the fundamental theorem, find the volume of the sphere.

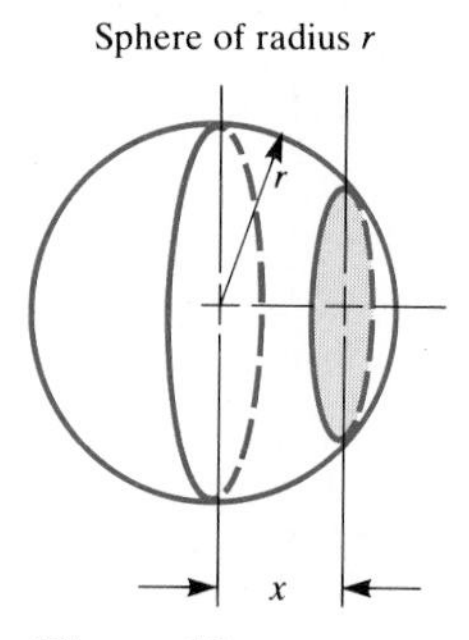

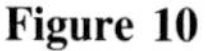

Figure 10

45 Say that you want to find the area of a certain cross-sectional plane of a rock. One way to find it is by sawing the rock in two and measuring the area directly. But suppose you do not want to ruin the rock. However, you do have a very accurate measuring glass, as shown in Fig. 11, which gives you excellent volume measurements. How could you use the glass to get a good estimate of the cross-sectional area?

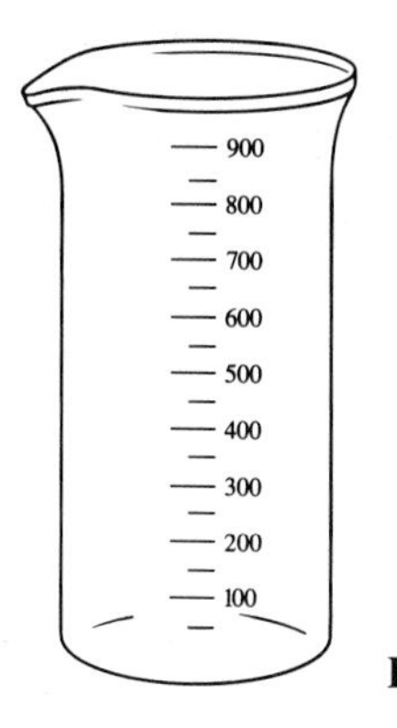

Figure 11

46 Let $v(t)$ be the velocity at time t of an object moving on a straight line. The velocity may be positive or negative.
(*a*) What is the physical meaning of $\int_a^b v(t)\,dt$? Explain.
(*b*) What is the physical meaning of the slope of the graph $y = v(t)$? Explain.

47 Give an example of a function f such that $f(4) = 0$ and $f'(x) = \sqrt[3]{1 + x^2}$. *Hint:* Think of the first fundamental theorem.

48 How often should a machine be overhauled? This depends on the rate $f(t)$ at which it depreciates and the cost A of overhaul. Denote the time interval between overhauls by T.
(*a*) Explain why you would like to minimize $g(T) = [A + \int_0^T f(t)\,dt]/T$.
(*b*) Find dg/dT.
(*c*) Show that when $dg/dT = 0$, $f(T) = g(T)$.
(*d*) Is this reasonable?

49 The integral $\int_0^1 \sqrt{x}\,dx$ gives numerical analysts a pain. The integrand is not differentiable at 0. What is worse, the derivatives (first, second, etc.) of $\sqrt{x}$ all become arbitrarily large for x near 0. It is instructive, therefore, to see how the error in Simpson's method behaves as h is made small.
(*a*) Use the fundamental theorem of calculus to show that $\int_0^1 \sqrt{x}\,dx = \frac{2}{3}$.
(*b*) Fill in this table to at least seven decimal places.

h	*Simpson's Estimate*	*Error*
$\frac{1}{2}$		
$\frac{1}{4}$		
$\frac{1}{8}$		
$\frac{1}{16}$		
$\frac{1}{32}$		
$\frac{1}{64}$		

(*c*) In the typical application of Simpson's method, when you cut h by a factor of 2, you find that the error is cut by a factor of $2^4 = 16$. (That is, the ratio of the two errors would be $\frac{1}{16} = 0.0625$.) Examine the five ratios of consecutive errors in the table in (*b*).
(*d*) Let $E(h)$ be the error in using Simpson's method to estimate $\int_0^1 \sqrt{x}\,dx$ with sections of length h. Assume that $E(h) \approx Ah^k$ for some constants k and A. Estimate k and A.

50 Let $f(x)$ be a continuous function with positive values. Let $H(x) = \int_x^b f(t)\,dt$, $x < b$. Let Δx be positive, $x + \Delta x < b$. Interpreting the definite integral as an area of a region, draw the regions whose areas are
(*a*) $H(x)$;
(*b*) $H(x + \Delta x)$.
(*c*) Is $H(x + \Delta x) - H(x)$ positive or negative?
(*d*) Draw the region whose area is related to $H(x + \Delta x) - H(x)$.
(*e*) When Δx is small, estimate $H(x + \Delta x) - H(x)$ in terms of the integrand f.
(*f*) Use (*e*) to evaluate the derivative $H'(x)$:

$$\frac{dH}{dx} = \lim_{\Delta x \to 0} \frac{H(x + \Delta x) - H(x)}{\Delta x}.$$

51 Let R be a function with continuous second derivative R''. Assume that $R(1) = 2$, $R'(1) = 6$, $R(3) = 5$, and $R'(3) = 8$. Evaluate $\int_1^3 R''(x)\,dx$. (Not all the information given is needed.)

THE LONG ROAD TO CALCULUS

It is often stated that Newton and Leibniz invented calculus in order to solve problems in the physical world. There is no evidence for this claim. Rather, as with their predecessors, Newton and Leibniz were driven by curiosity to solve the "tangent" and "area" problems, that is, to construct a general procedure for finding tangents and areas. Once calculus was available, it was then applied to a variety of fields, notably physics, with spectacular success.

The first five chapters have presented the foundations of calculus in this order: functions, limits and continuity, the derivative, the definite integral, and the fundamental theorem that joins the last two. This bears little relation to the order in which these concepts were actually developed. Nor can we sense in this approach, which follows the standard calculus syllabus, the long struggle that culminated in the creation of calculus.

The origins of calculus go back over 2000 years to the work of the Greeks on areas and tangents. Archimedes (287–212 B.C.) found the area of a section of a parabola, an accomplishment that amounts in our terms to evaluating $\int_0^b x^2\,dx$. He also found the area of an ellipse and both the surface area and the volume of a sphere. Apollonius (around 260–200 B.C.) wrote about tangents to ellipses, parabolas, and hyperbolas, and Archimedes discussed the tangents to a certain spiral-shaped curve. Little did they suspect that the "area" and "tangent" problems were to converge many centuries later.

With the collapse of the Greek world, symbolized by the Emperor Justinian's closing in A.D. 529 of Plato's Academy, which had survived for a thousand years, it was the Arab world that preserved the works of Greek mathematicians. In its liberal atmosphere, Arab, Christian, and Jewish scholars worked together, translating and commenting on the old writings, occasionally adding their own embellishments. For instance, Alhazen (A.D. 965–1039) computed volumes of certain solids, in essence evaluating $\int_0^b x^3\,dx$ and $\int_0^b x^4\,dx$.

It was not until the seventeenth century that several ideas came together to form calculus. In 1637, both Descartes (1596–1650) and Fermat (1601–1665) introduced analytic geometry. Descartes examined a given curve with the aid of algebra, while Fermat took the opposite tack, exploring the geometry hidden in a given equation. For instance, Fermat showed that the graph of $ax^2 + bxy + cy^2 + dx + ey + f = 0$ is always an ellipse, hyperbola, parabola, or one of their degenerate forms.

In this same period, Cavalieri (1598–1647) found the area under the curve $y = x^n$ for $n = 1, 2, 3, \ldots, 9$ by a method the length of whose computations grew rapidly as the exponent increased. Stopping at $n = 9$, he conjectured that the pattern would continue for larger exponents. In the next 20 years, several mathematicians justified his guess. So, even the calculation of the area under $y = x^n$ for a positive integer n, which we take for granted, represented a hard-won triumph.

"What about the other exponents?" we may wonder. Before 1665 there were no other exponents. Nevertheless, it was possible to work with the function which we denote $y = x^{p/q}$ for positive integers p and q by describing it as the function y such that $y^q = x^p$. (For instance, $y = x^{2/3}$ would be the function y that satisfies $y^3 = x^2$.) Wallis (1616–1703) found the area by a method that smacks more of magic than of mathematics. However, Fermat obtained the same result with the aid of an infinite geometric series.

The problem of determining tangents to curves was also in vogue in the first half of the seventeenth century. Descartes showed how to find a line perpendicular to a curve at a point P (by constructing a circle that meets the curve only at P); the tangent was then the line through P perpendicular to that line. Fermat found tangents in a way similar to ours and applied it to maximum-minimum problems.

The stage was set for the union of the "tangent" and "area" techniques. Indeed, Barrow (1630–1677), Newton's teacher at Cambridge, obtained a result equivalent to the fundamental theorem of calculus, but it was not expressed in a useful form.

Newton (1642–1727) arrived in Cambridge in 1661, and during the two years 1665–1666, which he spent at his family's farm to avoid the plague, he developed the essentials of calculus—recognizing that finding tangents and calculating areas are inverse processes. The first integral table ever compiled is to be found in one of his manuscripts of this period. But Newton did not publish his results at that time,

5.S SUMMARY

Four problems led to the concept of the definite integral: area of the region under a curve, total mass of a string of varying density, total distance traveled when the speed is varying, and volume of a certain tent.

All four problems required the same procedure: choosing partitions of some interval, sampling numbers, and then forming approximating sums:

perhaps because of the depression in the book trade after the Great Fire of London in 1665. During those two remarkable years he also introduced negative and fractional exponents, thus demonstrating that such diverse operations as multiplying a number by itself several times, taking its reciprocal, and finding a root of some power of that number are just special cases of a single general exponential function a^x, where x is a positive integer, -1, or a fraction, respectively.

Independently, however, Leibniz (1646–1716) also invented calculus. A lawyer, diplomat, and philosopher, for whom mathematics was a serious avocation, Leibniz established his version in the years 1673–1676, publishing his researches in 1684 and 1686, well before Newton's first publication in 1711. To Leibniz we owe the notations dx and dy, the terms "differential calculus" and "integral calculus," the integral sign, and the word "function." Newton's notation survives only in the symbol $\dot{x}$ for differentiation with respect to time, which is still used in physics.

It was to take two more centuries before calculus reached its present state of precision and rigor. The notion of a function gradually evolved from "curve" to "formula" to any rule that assigns one quantity to another. The great calculus text of Euler, published in 1748, emphasized the function concept by including not even one graph.

In several texts of the 1820s, Cauchy (1789–1857) defined "limit" and "continuous function" much as we do today. He also gave a definition of the definite integral, which with a slight change by Riemann (1826–1866) in 1854 became the definition standard today. So by the mid-nineteenth century the discoveries of Newton and Leibniz were put on a solid foundation.

In 1833, Liouville (1809–1882) demonstrated that the fundamental theorem could not be used to evaluate integrals of all elementary functions. In fact, he showed that the only values of the constant k for which $\int \sqrt{1-x^2}\,\sqrt{1-kx^2}\,dx$ is elementary are 0 and 1.

Still some basic questions remained, such as "What do we mean by area?" (For instance, does the set of points situated within some square and having both coordinates rational have an area? If so, what is this area?) It was as recently as 1887 that Peano (1858–1932) gave a precise definition of area—that quantity which earlier mathematicians had treated as intuitively given.

The history of calculus therefore consists of three periods. First, there was the long stretch when there was no hint that the tangent and area problems were related. Then came the discovery of their intimate connection and the exploitation of this relation from the end of the seventeenth century through the eighteenth century. This was followed by a century in which the loose ends were tied up.

The twentieth century has seen calculus applied in many new areas, for it is the natural language for dealing with continuous processes, such as change with time. In this century mathematicians have also obtained some of the deepest theoretical results about its foundations. Calculus is definitely alive and well and still growing.

References

E. T. Bell, *Men of Mathematics,* Simon and Schuster, New York, 1937. (In particular, chap. 6 on Newton and chap. 7 on Leibniz.)

Carl B. Boyer, *The History of the Calculus and Its Conceptual Development,* Dover, New York, 1959.

Carl B. Boyer, *A History of Mathematics,* Wiley, New York, 1968.

M. J. Crowe, *A History of Vector Analysis,* Dover, New York, 1985. (In particular, chap. 5.)

Philip J. Davis and Reuben Hersh, *The Mathematical Experience,* Birkhäuser Boston, Cambridge, 1981.

C. H. Edwards, Jr., *The Historical Development of the Calculus,* Springer-Verlag, New York, 1979. (In particular, chap. 8, "The Calculus According to Newton," and chap. 9, "The Calculus According to Leibniz.")

Morris Kline, *Mathematical Thought from Ancient* to *Modern Times,* Oxford, New York, 1972. (In particular, chap. 17.)

Pronunciation

Descartes	"Day-CART"
Fermat	"Fair-MA"
Leibniz	"LIBE-nits"
Euler	"OIL-er"
Cauchy	"KOH-shee"
Riemann	"REE-mahn"
Liouville	"LYU-veel"
Peano	"Pay-AHN-oh"

$$\sum_{i=1}^{n} f(c_i)(x_i - x_{i-1}).$$

In Sec. 5.1 the interval was [0, 3], and the function f was the squaring function. The typical sum

$$\sum_{i=1}^{n} f(c_i)(x_i - x_{i-1})$$

is *not* the definite integral, any more than $[f(x + \Delta x) - f(x)]/\Delta x$ is the derivative. A definite integral $\int_a^b f(x)\,dx$ is the limit of the approximating sums as their mesh is chosen smaller and smaller:

$$\int_a^b f(x)\,dx = \lim_{\text{mesh}\to 0} \sum_{i=1}^n f(c_i)(x_i - x_{i-1}).$$

The following table shows at a glance why the definite integral is related to mass, area, volume, and distance. Study this table carefully. It records the core ideas of the chapter.

Function	*Interpretation of Typical Summand*	*Approximating Sum*	*Definite Integral*	*Meaning of Definite Integral*
Density	$f(c_i)(x_i - x_{i-1})$ is estimate of mass in $[x_{i-1}, x_i]$.	$\sum_{i=1}^n f(c_i)(x_i - x_{i-1})$	$\int_a^b f(x)\,dx$	Mass
Length of cross section of a plane region by a line	$f(c_i)(x_i - x_{i-1})$ is area of an approximating rectangle.	$\sum_{i=1}^n f(c_i)(x_i - x_{i-1})$	$\int_a^b f(x)\,dx$	Area
Area of cross section of a solid by a plane	$f(c_i)(x_i - x_{i-1})$ is volume of a thin approximating slab.	$\sum_{i=1}^n f(c_i)(x_i - x_{i-1})$	$\int_a^b f(x)\,dx$	Volume
Speed	$f(T_i)(t_i - t_{i-1})$ is estimate of distance covered from time t_{i-1} to time t_i.	$\sum_{i=1}^n f(T_i)(t_i - t_{i-1})$	$\int_a^b f(t)\,dt$	Distance
Just a function (no application in mind)	$f(c_i)(x_i - x_{i-1})$ is just a product of two numbers.	$\sum_{i=1}^n f(c_i)(x_i - x_{i-1})$	$\int_a^b f(x)\,dx$	Just a number (no application in mind)

Vocabulary and Symbols

sigma notation $\sum_{i=1}^n a_i$
partition
section of a partition
sampling number c_i
mesh
definite integral $\int_a^b f(x)\,dx$
trapezoidal method
Simpson's method
$\int_a^b f(x)\,dx = -\int_b^a f(x)\,dx$
$\int_a^a f(x)\,dx = 0$
first fundamental theorem of calculus
second fundamental theorem of calculus (FTC)

elementary function
integral
antiderivative $\int f(x)\,dx$
average of a function
telescoping sum
$\sum_{i=1}^n (b_i - b_{i-1}) = b_n - b_0$
approximating sum
$\sum_{i=1}^n f(c_i)(x_i - x_{i-1})$
indefinite integral
integrand
$F(x)\Big|_a^b = F(b) - F(a)$

Key Facts

If f is continuous, $\int_a^b f(x)\,dx$ exists.

$$\int x^a\,dx = \frac{x^{a+1}}{a+1} + C, \qquad a \neq -1.$$

PROPERTIES OF ANTIDERIVATIVES

1. $\int cf(x)\,dx = c\int f(x)\,dx$ (c constant).
2. $\int (f(x) + g(x))\,dx = \int f(x)\,dx + \int g(x)\,dx$.
3. $\int (f(x) - g(x))\,dx = \int f(x)\,dx - \int g(x)\,dx$.

PROPERTIES OF DEFINITE INTEGRALS

1. $\int_a^b cf(x)\,dx = c\int_a^b f(x)\,dx$ (c constant).
2. $\int_a^b (f(x) + g(x))\,dx = \int_a^b f(x)\,dx + \int_a^b g(x)\,dx$.
3. $\int_a^b (f(x) - g(x))\,dx = \int_a^b f(x)\,dx - \int_a^b g(x)\,dx$.
4. If $f(x) \geq 0$, then $\int_a^b f(x)\,dx \geq 0$, $(a < b)$.
5. If $f(x) \geq g(x)$, then $\int_a^b f(x)\,dx \geq \int_a^b g(x)\,dx$ $(a < b)$.
6. $\int_a^b f(x)\,dx = \int_a^c f(x)\,dx + \int_c^b f(x)\,dx$.
7. If m and M are constants, $m \leq f(x) \leq M$, then

$$m(b-a) \leq \int_a^b f(x)\,dx \leq M(b-a) \qquad (a < b).$$

The close tie between the derivative and the definite integral is expressed in the two fundamental theorems of calculus:

FIRST FUNDAMENTAL THEOREM

Let f be continuous on an open interval containing the interval $[a, b]$. Then, for $a \leq x \leq b$,

$$\frac{d}{dx}\left(\int_a^x f(t)\,dt\right) = f(x).$$

SECOND FUNDAMENTAL THEOREM

If f is continuous on the interval $[a, b]$ and if F is an antiderivative of f, then

$$\int_a^b f(x)\,dx = F(b) - F(a).$$

The second fundamental theorem is also called the fundamental theorem and is abbreviated by the letters FTC. It provides a tool for computing many definite integrals. If an antiderivative of f is elementary, then FTC is of use. But there are elementary functions, for instance, $\sqrt{1 + x^3}$, which are not derivatives of elementary functions. In these cases, it is necessary to estimate the definite integral, say, by the trapezoidal method or Simpson's method.

TWO WAYS TO ESTIMATE $\int_a^b f(x)\,dx$

Method	*Formula*	*Weights*	*Bound on Error*
Trapezoidal	$\frac{h}{2}[f(x_0) + 2f(x_1) + 2f(x_2) + \cdots + 2f(x_{n-1}) + f(x_n)]$	1, 2, 2, 2, . . . , 2, 1	$\frac{(b-a)M_2h^2}{12}$
Simpson's	$\frac{h}{3}[f(x_0) + 4f(x_1) + 2f(x_2) + \cdots + 4f(x_{n-1}) + f(x_n)]$	1, 4, 2, 4, 2, . . . , 2, 4, 1	$\frac{(b-a)M_4h^4}{180}$

(M_k is maximum of $|f^{(k)}(x)|$ for $a \le x \le b$.) In Simpson's method the number n must be even. In the trapezoidal formula $h/2$ is a factor; in Simpson's method $h/3$ is a factor. In both methods $h = (b - a)/n$. Also in both, the inputs x_i are evenly spaced with successive x_i's a distance h apart: $x_i = a + ih$, $i = 0, 1, 2, \ldots, n$.

Any function whose derivative is the function f is called an **antiderivative** of f and is denoted $\int f(x)\,dx$. Any two antiderivatives of a function over an interval differ by a constant.

The first fundamental theorem implies that every continuous function is the derivative of some function. More specifically, it tells the rate of change of a definite integral as the interval over which the integral is computed changes.

The FTC is *not* a theorem about area or mass. It is a theorem about the limit of the sums $\Sigma_{i=1}^{n} f(c_i)(x_i - x_{i-1})$. In many applications, it is first shown that a certain quantity (area, distance, volume, mass, etc.) is estimated by sums of that type, and then the FTC is called on. However, it may or may not be of use.

MEAN-VALUE THEOREM FOR INTEGRALS

Let a and b be numbers and let f be a continuous function defined for x between a and b. Then there is a number c between a and b such that

$$\int_a^b f(x)\,dx = f(c)(b-a).$$

Although there are formulas for computing some definite integrals, do not forget that a definite integral is a limit of sums. There are two reasons for keeping this fundamental concept clear:

1 In many applications in science the concept of the definite integral is more important than its use as a computational tool.
2 Many definite integrals cannot be evaluated by a formula. Some of the more important of these have been tabulated to several decimal places and published in handbooks of mathematical tables.

To emphasize that the fundamental theorem of calculus does not dispose of all definite integrals, here is a little assortment of elementary functions whose integrals are not elementary:

$$\sqrt{x}\sqrt[n]{1+x}, \qquad \sqrt[n]{1+x^2} \qquad \text{for } n = 3, 4, 5, \ldots$$

$$\sqrt{1+x^n} \qquad \text{for } n = 3, 4, 5, \ldots$$

$$\frac{\sin x}{x} \qquad x \tan x \qquad \frac{2^x}{x} \qquad 2^{x^2}$$

Even when an elementary antiderivative exists, given the easy access to calculators and computers, we may prefer to get a good estimate of the definite integral by the trapezoidal or Simpson's method rather than by struggling to find an antiderivative. (Programmable calculators can provide estimates of definite integrals accurate to several decimal places.)

In the first five chapters we have covered the core of calculus: limits, derivatives, definite integrals, antiderivatives, and the relations between them. The remaining chapters present further applications and generalizations of these ideas.

GUIDE QUIZ ON CHAP. 5: THE DEFINITE INTEGRAL

1 Write out definitions of the symbols $\int_a^b f(x)\,dx$ and $\int f(x)\,dx$ and describe their relation.

2 (*a*) Describe the trapezoidal method of estimating a definite integral.
(*b*) For which class of functions does the trapezoidal method give the exact answer?
(*c*) If you halve h, the length of the small sections used in the approximation, what would you expect to happen to the error in the approximation? [Recall that $h = (b-a)/n$.]
(*d*) If $|f^{(2)}(x)| \le 4$ for x in $[1, 8]$, how large should n (the number of sections) be for you to be sure that the error in the trapezoidal method is less than 0.001?

3 Repeat the previous exercise for Simpson's rule, except that in (*d*) assume instead that $|f^{(4)}(x)| \le 7$.

4 Estimate $\int_0^3 dx/(1+x^3)$ by cutting the interval $[0, 3]$ into six sections of equal length and employing
(*a*) The left endpoint method
(*b*) The trapezoidal method
(*c*) Simpson's method.

5 Let f be a continuous function such that $f(1) = 4$ and $f(3) = 5$. Let $G(x) = \int_1^x f(t)\,dt$.
(*a*) Estimate $G(3.02) - G(3)$ as well as you can.
(*b*) Find $G'(3)$ and $G'(1)$.
(*c*) If f is an increasing function, how large might $G(3)$ be? how small?

6 (*a*) What is the first fundamental theorem of calculus?
(*b*) Why does the first imply the second fundamental theorem of calculus?
(*c*) Explain why the first fundamental theorem is true.

7 (*a*) Differentiate $1/(2x+3)$.
(*b*) Find the area under the curve $y = 1/(2x+3)^2$ and above the interval $[0, 1]$.

8 What is the average of $\sec^2 2x$ for x in $[\pi/12, \pi/8]$?

9 Find $(d^2/dx^2)(\int_{x^2}^{x^3} \cos 3t\,dt)$.

10 Letting $f(x)$ denote the velocity of an object at time x, interpret each step of the proof of the first fundamental theorem in terms of velocity and distance.

11 (*a*) What is meant by an "elementary function"?
(*b*) Give an example of an elementary function whose antiderivative is not elementary.
(*c*) Give an example of a function that is not elementary, yet has an elementary derivative.

REVIEW EXERCISES FOR CHAP. 5: THE DEFINITE INTEGRAL

In each of Exercises 1 to 6 find the area of the region under the given curve and above the given interval.

1 $y = 2x^3$; $[1, 2]$ **2** $y = 6x^2 + 10x^4$; $[-1, 2]$
3 $y = \sin 3x$; $[0, \pi/6]$ **4** $y = 3\cos 2x$; $[\pi/6, \pi/4]$
5 $y = \dfrac{1}{x^3}$; $[2, 3]$ **6** $y = \dfrac{1}{(x+1)^4}$; $[0, 1]$

In each of Exercises 7 to 16 give a formula for the antiderivatives of the given function (including the constant C).

7 $\sec^2 x$ **8** $\sec^2 3x$
9 $\sec x \tan x$ **10** $5 \sec 3x \tan 3x$
11 $4 \csc x \cot x$ **12** $-\csc 5x \cot 5x$
13 $(x^3+1)^2$ (*Suggestion:* First multiply it out.)
14 $\left(x + \dfrac{1}{x}\right)^2$ **15** $100x^{19}$ **16** $\dfrac{24}{x^3}$

17 (*a*) Differentiate $(x^3+1)^6$.
(*b*) Find $\int (x^3+1)^5 x^2\,dx$.

18 (*a*) Differentiate $\sin x^3$.
(*b*) Find $\int x^2 \cos x^3\,dx$.

19 Write these sums without using the sigma notation:
(*a*) $\sum_{j=1}^{3} d^j$ (*b*) $\sum_{k=1}^{4} x^k$
(*c*) $\sum_{i=0}^{3} i2^{-i}$ (*d*) $\sum_{i=2}^{5} \dfrac{i+1}{i}$
(*e*) $\sum_{i=2}^{4} \left(\dfrac{1}{i} - \dfrac{1}{i+1}\right)$ (*f*) $\sum_{i=1}^{4} \sin \dfrac{\pi i}{4}$

20 Write these sums without sigma notation and then evaluate:

(*a*) $\sum_{i=1}^{100} (2^i - 2^{i-1})$ (*b*) $\sum_{i=0}^{100} (2^{i+1} - 2^i)$

(*c*) $\sum_{i=1}^{100} \left(\frac{1}{i} - \frac{1}{i+1}\right)$

21 Evaluate $\int_0^4 4x^2\,dx$ by using its definition as a limit of approximating sums.

22 Evaluate $\int_1^2 (5 + x^2)\,dx$ by using its definition as a limit of approximating sums.

In Exercises 23 to 26 evaluate the given definite integrals by the fundamental theorem of calculus.

23 $\int_1^3 (4x^3 - x^2)\,dx$ 24 $\int_{\pi/4}^{3\pi/4} 4\sin x\,dx$

25 $\int_{1/2}^2 \frac{1}{x^2}\,dx$ 26 $\int_3^{12} \sqrt{3x}\,dx$

In Exercises 27 to 32 differentiate the given functions.

27 $\int_3^x \sqrt{1+t^2}\,dt$ 28 $\int_3^{4x} \sin^3 t\,dt$

29 $\int_{3x}^4 \tan 3t\,dt$ 30 $\int_{3x}^{4x} \frac{1}{1+t^2}\,dt$

31 $\int_4^{\tan x} \cos t\,dt$ 32 $\int_{\sin 3x}^{\cos 3x} 2^t\,dt$

In Exercises 33 to 36 evaluate the expressions.

33 $\int_0^0 2^{x^2}\,dx$ 34 $\int_1^0 x^2\,dx$

35 $\int_2^1 (12x^3 - 2x)\,dx$ 36 $\int_3^2 (x+1)^{-2}\,dx$

37 (*a*) Figure 1 is the graph of a function f. Sketch what the graph of $G(x) = \int_1^x f(t)\,dt$ might look like. Describe your reasoning.

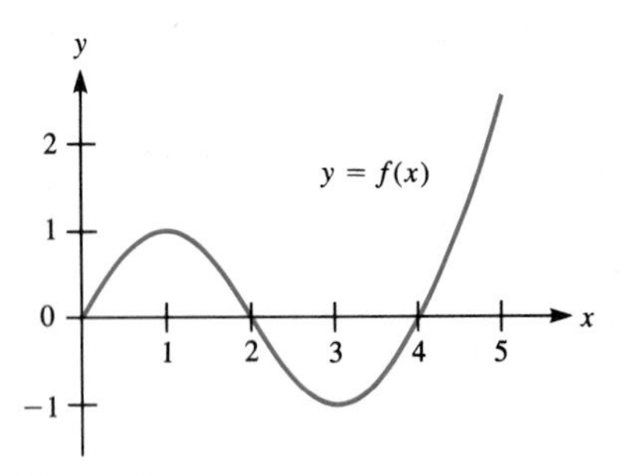

Figure 1

(*b*) Estimate the slope of the curve $y = G(x)$ when $x = 2$ and when $x = 1$.

(*c*) Estimate $G(0)$.

(*d*) Estimate $G(3.01) - G(3)$.

38 If $f(x) = \sin 2x$, how small must h be so that the error in estimating $\int_1^2 f(x)\,dx$ is less than 0.001 (*a*) by the trapezoidal method? (*b*) by Simpson's method?

39 (*a*) Estimate $\int_1^2 x^4\,dx$ by Simpson's method using $h = \frac{1}{2}$ and then $h = \frac{1}{4}$.

(*b*) Find the error in each case.

(*c*) What is the ratio of the error in the second case to the error in the first case?

(*d*) Use a calculator or computer to find the estimate for $h = \frac{1}{20}$ and compare the error in this case to the error in using $h = \frac{1}{2}$.

40 Repeat Exercise 39 using the trapezoidal method instead.

41 Repeat Exercise 39 using the left endpoint method instead.

42 Find

(*a*) $\lim_{\Delta x \to 0} \frac{\int_2^{5+\Delta x} \sin x^2\,dx - \int_2^5 \sin x^2\,dx}{\Delta x}$

(*b*) $\frac{d^2}{dx^2}\left(\int_0^{x^2} \frac{dt}{\sqrt{1-5t^3}}\right)^2$

43 (*a*) Compute $\frac{d}{dx}\left(\int_4^{x^2} \frac{\sqrt{1+u^2}}{2\sqrt{u}}\,du\right)$ for $x > 0$.

(*b*) Compute $\frac{d}{dx}\left(\int_2^x \sqrt{1+t^4}\,dt\right)$ for $x > 0$.

(*c*) In view of (*a*) and (*b*), there is a constant C such that

$$\int_2^x \sqrt{1+t^4}\,dt = \int_4^{x^2} \frac{\sqrt{1+u^2}}{2\sqrt{u}}\,du + C.$$

Find C.

44 An unmanned satellite automatically reports its speed every minute. If a graph is drawn showing speed as a function of time during the flight, what is the physical interpretation of the area under the curve and above the time axis? Explain.

45 We interpreted $f(x)$ as velocity and $F(x)$ as distance in order to motivate the second fundamental theorem of calculus. Instead, let $f(x)$ be density and $F(x)$ be mass along a wire. Specifically, let $F(x)$ be the mass in grams of the left-hand x centimeters of a wire. Then the density of the wire at x is $F'(x)$, which we denote $f(x)$.

(*a*) Express the mass of the section $[a, b]$ in terms of F.

(*b*) Express the same mass in terms of f.

(*c*) Compare (*a*) and (*b*) to obtain another piece of evidence in favor of the second fundamental theorem of calculus.

46 (*a*) Draw the region that is bounded by the curves $y = x^2$ and $y = x^3$.

(*b*) Let $c(x)$ be the length of the cross section of the region made by a line parallel to the y axis and passing through x on the x axis. Find a formula for $c(x)$.

(*c*) Express the area of the region as a definite integral.

(*d*) Find the area of the region.

47 A tiring snail is moving along at the rate of $1/(t+1)^2$ feet per minute t minutes into its journey. How far does it travel in the first 3 minutes?

48 Find the volume of the solid shown in Fig. 2. Cross sections made by planes perpendicular to the x axis are circles.

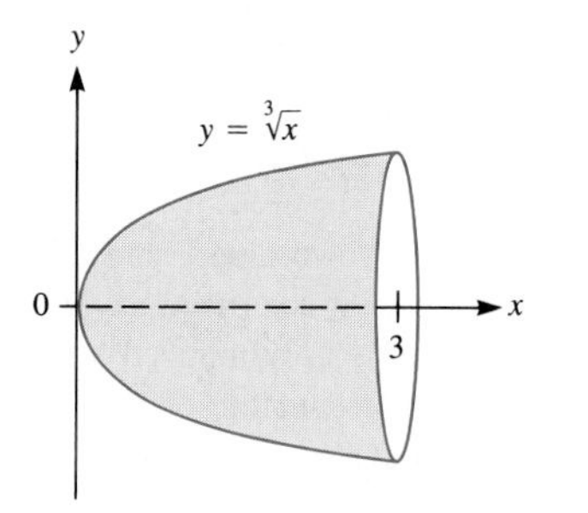

Figure 2

In Exercises 49 to 52 find the areas of the regions shown.

49

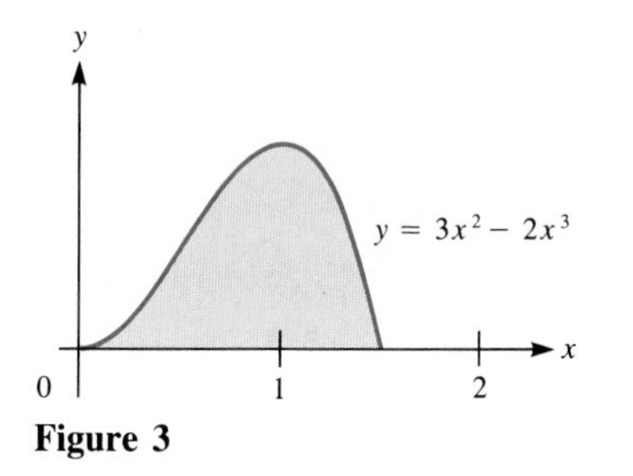

Figure 3

50

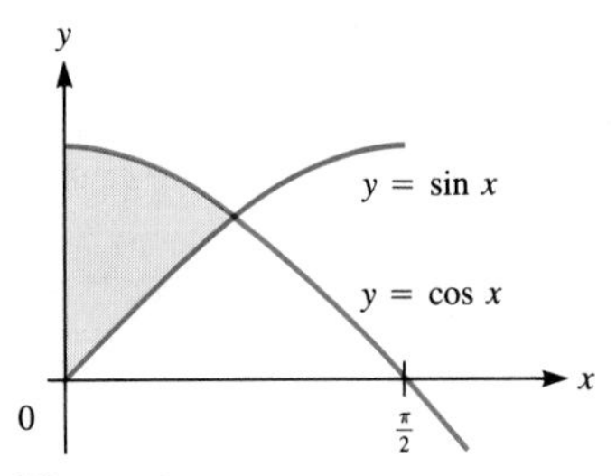

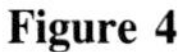
Figure 4

51

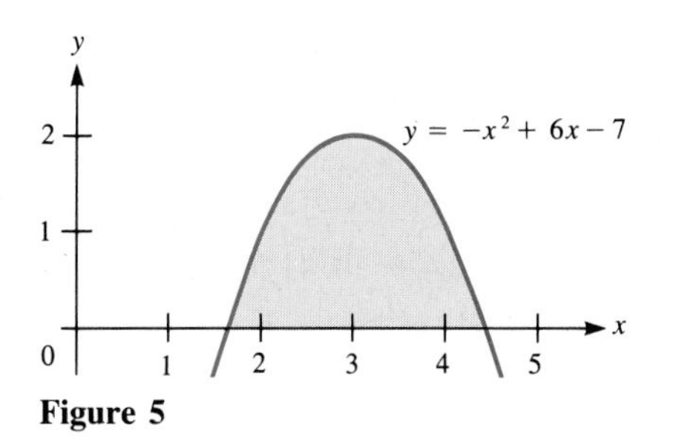

Figure 5

52

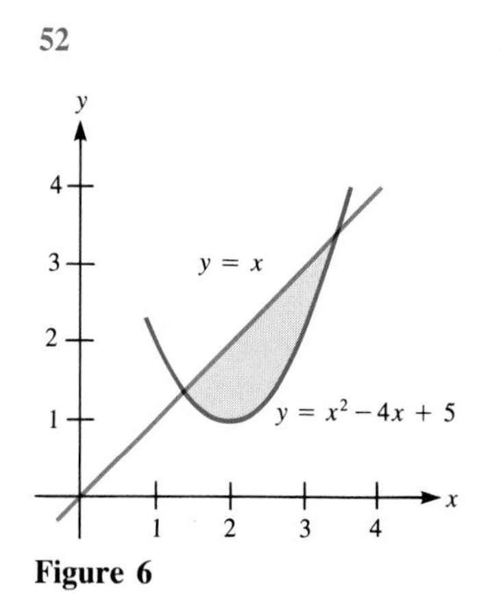

Figure 6

Exercises 53 and 54 each contain three definite integrals; two of them can be evaluated by the fundamental theorem. In each case use the fundamental theorem on the two appropriate integrals and on the third use Simpson's method with $n = 4$.

53 (*a*) $\int_0^1 \sqrt{1 - x}\, dx$ (*b*) $\int_0^1 \sqrt[3]{1 - x^2}\, dx$

(*c*) $\int_0^1 \sqrt[3]{1 + x}\, dx$

54 (*a*) $\int_1^2 \frac{2^x}{x^2}\, dx$ (*b*) $\int_1^2 \frac{1}{x^2}\, dx$ (*c*) $\int_1^2 \frac{x^3 + 1}{x^2}\, dx$

55 Let n be a positive integer and f a function.

(*a*) Show that $\Sigma_{i=1}^{n} f(i/n)(1/n)$ is an approximating sum for the definite integral $\int_0^1 f(x)\, dx$.

(*b*) What is the length of the ith section of the partition in (*a*)?

(*c*) What is the mesh of the partition?

(*d*) Where does the sampling number c_i lie in the ith section?

56 Explain why $\frac{1}{100} \Sigma_{i=1}^{100} f(i/100)$ is an estimate of $\int_0^1 f(x)\, dx$.

57 What definite integrals are estimated by the following sums?

(*a*) $\sum_{i=1}^{200} \left(\frac{i}{100}\right)^3 \frac{1}{100}$ (*b*) $\sum_{i=1}^{100} \left(\frac{i-1}{100}\right)^4 \frac{1}{100}$

(*c*) $\sum_{i=101}^{300} \left(\frac{i}{100}\right)^5 \frac{1}{100}$

58 (See Exercise 57.)

(*a*) Show that $(1/n^3)\, \Sigma_{i=1}^{n} i^2$ is an approximation of $\int_0^1 x^2\, dx$.

(*b*) Compute the sum in (*a*) when $n = 4$.

(*c*) Compute $\int_0^1 x^2\, dx$.

(*d*) Find $\lim_{n\to\infty} (1/n^3)\, \Sigma_{i=1}^{n} i^2$.

59 A man whose jeep has a vertical windshield drives a mile through a vertical rain consisting of drops that are uniformly distributed and falling at a constant rate. (See Fig. 7.) Should he go slow or fast in order to minimize the amount of rain that strikes the windshield?

Figure 7

60 Let f be a function such that $\int_0^x f(t)\,dt = [f(x)]^2$ for $x \geq 0$. Assume that $f(x) > 0$ for $x > 0$.
(*a*) Find $f(0)$. (*b*) Find $f(x)$ for $x > 0$.

61 A particle moves on a line in such a way that its average velocity over any interval of time $[a, b]$ is the same as its velocity at time $(a + b)/2$. Prove that the velocity $v(t)$ must be of the form $ct + d$ for some constants c and d. *Hint:* Differentiate the relation $\int_a^b v(t)\,dt = [v((a + b)/2)](b - a)$ with respect to b and with respect to a.

62 A particle moves on a line in such a way that the average velocity over any interval of time $[a, b]$ is equal to the average of its velocities at the beginning and the end of the interval of time. Prove that the velocity $v(t)$ must be of the form $ct + d$ for some constants c and d.

63 The tank shown in Fig. 8 is being filled with water at the rate of 2 cubic meters per hour. The radius of the tank at a height y above its point is $r = \sqrt[3]{1 + y^2} - 1$. At what rate is the water rising when the depth of the water is $\sqrt{7}$ meters?

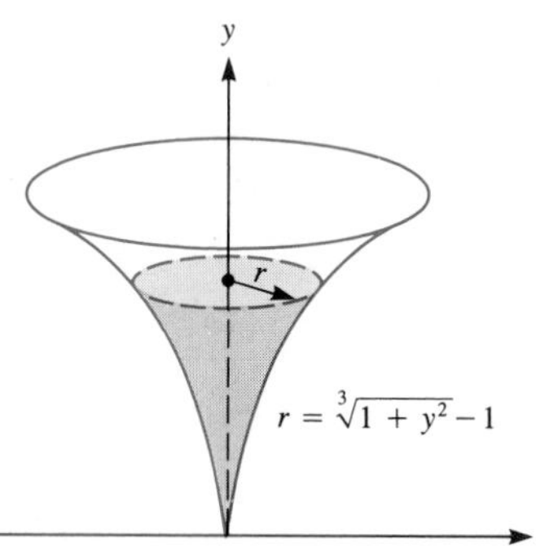

Figure 8

64 (Optimal replacement) The following argument appears in *Optimal Replacement Policy* by D. W. Jorgenson, J. J. McCall, and R. Radner, pp. 92–93, Rand McNally, Chicago, 1967:

The average cost per unit good time, $V(N)$, is

$$V(N) = \frac{N + K}{\int_0^N R(t)\,dt}, \qquad (*)$$

where K is a constant, "the imputed down time," and N is the time of replacement. To determine the optimum preparedness maintenance policy, $V(N)$ is minimized with respect to replacement age N. Differentiation of Eq. (*) with respect to N yields the condition

$$\frac{dV}{dN} = \frac{\int_0^N R(t)\,dt - (N + K)R(N)}{[\int_0^N R(t)\,dt]^2} = 0.$$

Verify that the differentiation is correct.

65 The second fundamental theorem can be proved directly, without referring to the first fundamental theorem. Assume that f is continuous, $f = F'$, and $\int_a^b f(x)\,dx$ exists. The steps are outlined as follows:

(*a*) Given x_{i-1} and x_i, $x_{i-1} < x_i$, in $[a, b]$, show that there is a number c_i in $[x_{i-1}, x_i]$ such that

$$F(x_i) - F(x_{i-1}) = F'(c_i)(x_i - x_{i-1}).$$

(*b*) Given x_{i-1} and x_i in $[a, b]$, show that there is a number c_i in $[x_{i-1}, x_i]$ such that

$$f(c_i)(x_i - x_{i-1}) = F(x_i) - F(x_{i-1}).$$

(*c*) Let $x_0 = a, x_1, \ldots, x_n = b$ determine a partition of $[a, b]$ into n sections. Show that, if the sampling numbers c_i are chosen as in (*b*), then

$$\sum_{i=1}^{n} f(c_i)(x_i - x_{i-1}) = F(b) - F(a).$$

(*d*) Use (*c*) to show that

$$\lim_{\text{mesh}\to 0} \sum_{i=1}^{n} f(c_i)(x_i - x_{i-1}) = F(b) - F(a)$$

[even if the c_i are not chosen as in (*b*)]. This proves the second fundamental theorem directly.

66 A number is dyadic if it can be expressed as the quotient of two integers m/n, where n is a power of 2. (These are the fractions into which an inch is usually divided.) Between any two numbers lies an infinite set of dyadic numbers and also an infinite set of numbers that are not dyadic. With this background, we shall define a function f that does *not* have a definite integral over the interval $[0, 1]$ as follows:

$$\text{Let } f(x) = \begin{cases} 0 & \text{if } x \text{ is dyadic} \\ 3 & \text{if } x \text{ is not dyadic.} \end{cases}$$

(*a*) Show that for any partition of $[0, 1]$ it is possible to choose sampling numbers c_i such that

$$\sum_{i=1}^{n} f(c_i)(x_i - x_{i-1}) = 3.$$

(*b*) Show that for any partition of [0, 1] it is possible to choose sampling numbers c_i such that

$$\sum_{i=1}^{n} f(c_i)(x_i - x_{i-1}) = 0.$$

(*c*) Why does f not have a definite integral over the interval [0, 1]?

67 This exercise concerns the areas of regions under the curve $1/x$.

(*a*) Estimate the area under $y = 1/x$ and above [1, 2], using five sections of equal length and left endpoints.

(*b*) Estimate the area under $y = 1/x$ and above [3, 6], using five sections of equal length and left endpoints.

(*c*) The answers to (*a*) and (*b*) are the same. Would they be the same if you used 100 sections of equal length (instead of 5)? Explain.

(*d*) Write a short paragraph explaining why the area under $y = 1/x$ above [1, 2] equals the area under $y = 1/x$ above [3, 6].

(*e*) Let a and b be numbers greater than 1. Explain why the area under $1/x$ and above $[1, a]$ equals the area under $1/x$ and above $[b, ab]$.

(*f*) For $t > 1$, let $G(t)$ equal the area under the curve $1/x$ and above $[1, t]$. Show that, for a and b greater than 1, $G(ab) = G(a) + G(b)$.

(*g*) What function g studied in precalculus resembles the function G in that $g(ab) = g(a) + g(b)$?

68 As shown in Exercise 67, the function $f(x) = 1/x$ has the remarkable property that for any positive numbers a and b, $\int_1^a f(x)\,dx = \int_b^{ab} f(x)\,dx$. Moreover, $f(1) = 1$. Find all functions $f(x)$ such that $\int_1^a f(x)\,dx = \int_b^{ab} f(x)\,dx$ for all positive numbers a and b and such that $f(1) = 1$. *Hint:* First differentiate with respect to a, holding b fixed. Then think about the equation you get.

69 Assume the function f is defined for all x and has a continuous derivative. Assume $f(0) = 0$ and that $0 < f'(x) \leq 1$.

(*a*) Prove that

$$\left[\int_0^t f(x)\,dx\right]^2 \geq \int_0^t [f(x)]^3\,dx.$$

(*Hint:* Use the first fundamental theorem of calculus.)

(*b*) Give an example of a function $f(x)$ where equality occurs in (*a*).

70 A subway train travels between two stations in 3 minutes. It starts with a constant acceleration until it reaches its maximum speed. It maintains that speed for 2 minutes and then decelerates at the same rate it accelerated. Sketch a possible graph of each of the following and describe your reasoning:

(*a*) Acceleration as a function of time,

(*b*) Velocity as a function of time,

(*c*) Position as a function of time.

71 An accelerometer measures the acceleration of an automobile in miles per hour per second at 1-second intervals for a period of 10 seconds. The automobile starts from rest at time $t = 0$ seconds. The data are recorded in a table:

t	0	1	2	3	4	5	6	7	8	9	10
Acceleration	10	11	12	8	6	4	1	−3	−4	−5	−6

(*a*) Sketch a graph of acceleration as a function of time.

(*b*) Estimate the velocity of the car at the end of each of the first 10 seconds.

(*c*) Estimate how far the automobile travels during the first 5 seconds; during the first 10 seconds.

6

TOPICS IN DIFFERENTIAL CALCULUS

Most of the material in this chapter could be covered before Chap. 5, since it is primarily concerned with derivatives.

Section 6.1 discusses the logarithmic functions. (There is no calculus here.)

Section 6.2 introduces the most important number in calculus,

$$\lim_{x \to 0} (1 + x)^{1/x},$$

which is denoted e and is about 2.718.

Section 6.3 obtains the derivatives of the logarithmic functions.

Section 6.4 describes a type of function—one that assigns distinct outputs to distinct inputs. This concept is needed in Secs. 6.5 and 6.6, where we obtain the derivatives of the exponential functions of the form b^x and the inverse trigonometric functions.

Section 6.7 applies the exponential functions to the study of natural growth and decay. (It depends on Chap. 5.)

Section 6.8 shows how derivatives can be used to evaluate certain limits.

Section 6.9 defines the hyperbolic functions and obtains their properties. These functions are certain simple combinations of exponential functions. (Many calculators have keys for these functions.)

6.1 LOGARITHMS

Exponents are reviewed in App. D.

Consider the question

$$3^? = 9,$$

read as "3 raised to what power equals 9?" The answer, whatever its numerical value might be, is called "the logarithm of 9 to the base 3."

Since

$$3^2 = 9,$$

we say that "the logarithm of 9 to the base 3 is 2" and write "$\log_3 9 = 2$."

Remember: A logarithm is an exponent!

Definition *Logarithm*. If b and c are positive numbers, $b \neq 1$, and $b^x = c$, then the number x is the **logarithm** of c to the base b and we write

$$x = \log_b c.$$

Any exponential equation $b^x = c$ may be translated into a logarithmic equation $x = \log_b c$. Table 1 illustrates some of these translations. Read it over several times, perhaps aloud, until you can, when covering a column, fill in the correct translation of the other column.

Table 1

Exponential Language	*Logarithmic Language*
$3^2 = 9$	$\log_3 9 = 2$
$7^0 = 1$	$\log_7 1 = 0$
$10^3 = 1{,}000$	$\log_{10} 1{,}000 = 3$
$10^{-2} = 0.01$	$\log_{10} 0.01 = -2$
$9^{1/2} = 3$	$\log_9 3 = \frac{1}{2}$
$8^{2/3} = 4$	$\log_8 4 = \frac{2}{3}$
$8^{-1} = \frac{1}{8}$	$\log_8 \frac{1}{8} = -1$
$5^1 = 5$	$\log_5 5 = 1$
$b^1 = b$	$\log_b b = 1$

Since $b^x = c$ is equivalent to $x = \log_b c$, it follows that

$$b^{\log_b c} = c.$$

The equality $b^{\log_b c} = c$ is not deep; it just restates the definition of a logarithm.
The answer to the question "To what power must we raise b to get c?" is "the logarithm of c to the base b."

EXAMPLE 1 Find $\log_5 125$.

SOLUTION Look for an answer to the question

"5 to what power equals 125?"

or, equivalently, for a solution of the equation

$$5^x = 125.$$

Since $5^3 = 125$, the answer is 3; that is,

$$\log_5 125 = 3. \quad ■$$

EXAMPLE 2 Find $\log_{10} \sqrt{10}$.

SOLUTION The question is "What power of 10 equals $\sqrt{10}$?" Now,

$$10^{1/2} = \sqrt{10}.$$

The answer to the question is therefore "$\frac{1}{2}$."

Thus,

$$\log_{10} \sqrt{10} = \tfrac{1}{2}.$$

Since $\sqrt{10} \approx 3.162$, we can conclude that $\log_{10} 3.162 \approx \frac{1}{2}$. ■

In order to get an idea of the logarithm as a function, consider logarithms to the familiar base of 10:

$$y = \log_{10} x.$$

Table 2

x	$\log_{10} x$
100	2
10	1
1	0
0.1	-1
0.01	-2

Begin with a table of values as shown in Table 2. We must restrict ourselves to $x > 0$. A negative number, such as -1, cannot have a logarithm, since there is no power of 10 that equals -1. That is, 10^x is always a positive number, no matter whether x is positive, negative, or zero. Similarly, 0 does not have a logarithm since the equation $10^x = 0$ has no solution. The domain of the function "log to the base 10" consists of the positive real numbers.

With the aid of the data in Table 2, the graph is easy to sketch and is shown in Fig. 1. The graph lies to the right of the y axis. Far to the right it rises slowly; not until x reaches 100 does the y coordinate reach 2. Furthermore, $\log_{10} 1{,}000 = 3$ and $\log_{10} 10^n = n$ for any positive integer n. Although $\log_{10} x$ grows slowly, it does become arbitrarily large as $x \to \infty$: we have $\lim_{x\to\infty} \log_{10} x = \infty$.

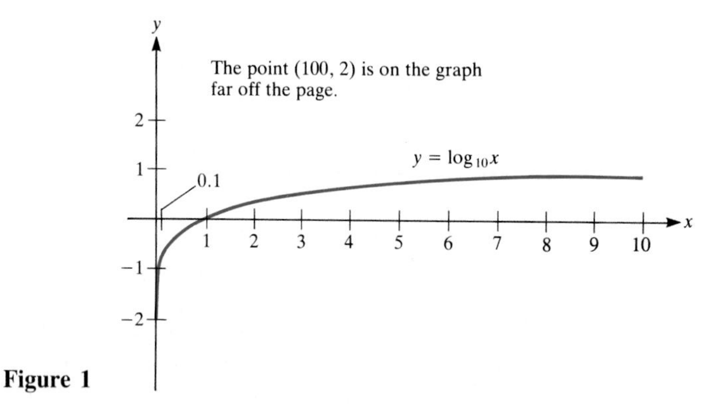

Figure 1

To see how $\log_{10} x$ behaves when x is a small positive number, note that $\log_{10} 0.01 = -2$, $\log_{10} 0.001 = -3$, and $\log_{10} 10^{-n} = -n$ for any positive integer n. We see that $\lim_{x\to 0^+} \log_{10} x = -\infty$. Keeping in mind that $\log_{10} x$ is an increasing function, we conclude that for $x > 0$, $\log_{10} x$ takes on all values, positive and negative. Similarly, for any base $b > 1$, $\lim_{x\to\infty} \log_b x = \infty$ and $\lim_{x\to 0^+} \log_b x = -\infty$.

Logarithms to the base 10 are called **common logarithms**. Many calculators have a $\log_{10}$-key, usually labeled "log." Such a key replaces logarithm tables that filled books, the first of which was published by John Napier in 1614.

Properties of Logarithms

Since each exponential equation $b^x = c$ translates into the corresponding logarithmic equation $x = \log_b c$, every property of exponentials must carry over to some property of logarithms. For instance, the information that

$$b^0 = 1$$

translates, in the language of logarithms, to

$$\log_b 1 = 0.$$

The logarithm of 1 in any base b is 0.

The equation $$b^1 = b$$

amounts to saying that the logarithm of b (the base) in the base b is 1:

$$\log_b b = 1.$$

Thus $\log_{10} 10 = 1$ and $\log_{2.718}(2.718) = 1$.

The following table lists the fundamental properties of exponential functions together with the corresponding properties of logarithms.

Properties of logarithms

Table 3

Exponents	*Logarithms*
$b^0 = 1$	$\log_b 1 = 0$
$b^{1/2} = \sqrt{b}$	$\log_b \sqrt{b} = \frac{1}{2}$
$b^1 = b$	$\log_b b = 1$
$b^{x+y} = b^x b^y$	$\log_b cd = \log_b c + \log_b d$
$b^{-x} = \dfrac{1}{b^x}$	$\log_b \left(\dfrac{1}{c}\right) = -\log_b c$
$b^{x-y} = \dfrac{b^x}{b^y}$	$\log_b \left(\dfrac{c}{d}\right) = \log_b c - \log_b d$
$(b^x)^y = b^{xy}$	$\log_b c^m = m \log_b c$

Of these identities for logarithms, the most fundamental is

$$\log_b cd = \log_b c + \log_b d, \tag{1}$$

which asserts that *the log of the product is the sum of the logs of the factors*. The proof is instructive. (Proofs of the others are left as exercises.)

The log of the product

Theorem 1 For any positive numbers c and d and for any base b,

$$\log_b cd = \log_b c + \log_b d.$$

Proof By the definition of the logarithm as an exponent,

$$c = b^{\log_b c} \quad \text{and} \quad d = b^{\log_b d}.$$

Thus $$cd = b^{\log_b c} b^{\log_b d}. \tag{2}$$

By the basic law of exponents ($b^x b^y = b^{x+y}$),

$$b^{\log_b c} b^{\log_b d} = b^{\log_b c + \log_b d}. \tag{3}$$

Combining Eqs. (2) and (3) shows that

$$cd = b^{\log_b c + \log_b d}.$$

So the exponent to which b must be raised to get cd is

$$\log_b c + \log_b d.$$

In other words, the logarithm of cd to the base b is $\log_b c + \log_b d$. ■

EXAMPLE 3 Use the last line in Table 3 to evaluate $\log_9 (3^7)$ and $\log_5 \sqrt[3]{25^2}$.

SOLUTION

$$\log_9 (3^7) = 7 \log_9 3 = 7(\tfrac{1}{2}) = \tfrac{7}{2}$$

$$\log_5 \sqrt[3]{25^2} = \log_5 (25)^{2/3} = \tfrac{2}{3} \log_5 25 = (\tfrac{2}{3})2 = \tfrac{4}{3}. \quad ■$$

The next example shows how logarithms can be used to solve equations in which the unknown appears in an exponent.

EXAMPLE 4 Find x if $5 \cdot 3^x \cdot 7^{2x} = 2$.

SOLUTION First rewrite the equation as

$$3^x \cdot 7^{2x} = \tfrac{2}{5} = 0.4,$$

and then take logarithms to the base 10 of both sides:

$$\log_{10} (3^x 7^{2x}) = \log_{10} 0.4$$

$$\log_{10} 3^x + \log_{10} 7^{2x} = \log_{10} 0.4 \qquad \text{log of a product}$$

$$x \log_{10} 3 + 2x \log_{10} 7 = \log_{10} 0.4 \qquad \text{log of an exponential}$$

$$x = \frac{\log_{10} 0.4}{\log_{10} 3 + 2 \log_{10} 7} \qquad \text{solving for } x$$

$$x \approx \frac{-0.3979}{0.4771 + 2(0.8451)} \qquad \text{calculator}$$

$$x \approx -0.1836. \quad ■$$

Something your calculator doesn't do.

A calculator or computer program may make $\log_{10} x$ immediately available. In that case, how could you find logarithms with a different base, say, $\log_2 7$? The next example answers this question.

EXAMPLE 5 Express $\log_2 7$ in terms of common logarithms (base 10 logarithms).

SOLUTION Since a logarithm is an exponent, we begin with the equation

$$2^{\log_2 7} = 7.$$

Then we take the logarithm to the base 10 of both sides, obtaining

$$\log_{10}(2^{\log_2 7}) = \log_{10} 7. \tag{4}$$

By the rule for the logarithm of a power $[\log_b (c^m) = m \log_b c]$, transform Eq. (4) into

$$\log_2 7 \log_{10} 2 = \log_{10} 7,$$

and find that

$$\log_2 7 = \frac{\log_{10} 7}{\log_{10} 2}.$$

(A calculator then shows that $\log_{10} 7 \approx 0.8451$, $\log_{10} 2 \approx 0.3010$, and so $\log_2 7 \approx 2.807$. This answer is reasonable, since $\log_2 8 = 3$.) ■

The argument in Example 5 shows that for two different bases, b and c,

$$\log_b x = \log_b c \cdot \log_c x.$$

Since $\log_b c$ is a constant, this equation tells us that $\log_b x$ is proportional to $\log_c x$ and the constant of proportionality is $\log_b c$. [For instance, $\log_3 x = \log_3 9 \log_9 x = 2 \log_9 x$. This means that if (x, y) is on the graph of the $\log_9 x$ function, then $(x, 2y)$ is on the graph of the $\log_3 x$ function.]

So for $b, c > 1$, the graphs of $y = \log_b x$ and $y = \log_c x$ have the same general shape: the first is obtained from the second by an expansion parallel to the y axis by a constant factor, $\log_b c$. (See Fig. 2.)

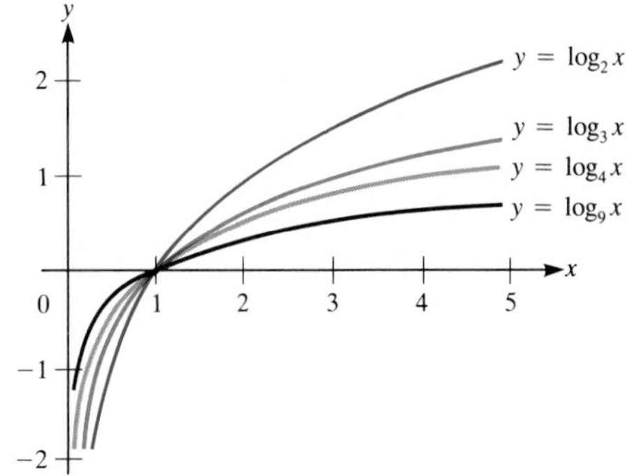

Figure 2

EXAMPLE 6 As every computer buff knows, 2^{10} is approximately 1,000. (It is actually 1,024.) Use this information to estimate $\log_{10} 2$.

SOLUTION

Start with the approximation

$$2^{10} \approx 10^3,$$

and take $\log_{10}$ of both sides, getting

$$\log_{10} (2^{10}) \approx \log_{10} (10^3),$$

hence

$$10 \log_{10} 2 \approx 3.$$

Thus

$$\log_{10} 2 \approx 0.3. \quad \blacksquare$$

What does your calculator give for $\log_{10} 2$?

A Logarithm Grows Slowly

As mentioned earlier, for $b > 1$, $\log_b x \to \infty$ as $x \to \infty$, but very slowly. In fact, it grows much more slowly than any fixed power of x, x^a, $a > 0$, as is illustrated by the following table, which compares $\log_2 x$ with $x^{0.1}$:

Table 4

x	1	2	2^5	2^{10}	2^{100}	2^{200}
$\log_2 x$	0	1	5	10	100	200
$x^{0.1}$	1	1.072	1.414	2	1,024	1,048,576
$\dfrac{\log_2 x}{x^{0.1}}$	0	0.93	3.5	5.0	0.098	0.00019

For large x, $\log_2 x$ is much smaller than $\sqrt[10]{x}$.

Even though at $x = 2^{10}$, $\log_2 x$ is larger than $x^{0.1}$, for large x, the power $x^{0.1}$ takes a commanding lead.

More generally, for any base $b > 1$ and $a > 0$,

$$\lim_{x\to\infty} \frac{\log_b x}{x^a} = 0.$$

Section Summary

We defined the logarithmic functions. The number "$\log_b c$" is the answer to the question, "b to what power equals c?" Therefore $b^{\log_b c} = c$. We also discussed the properties of logarithms, for instance, $\log_b cd = \log_b c + \log_b d$ and $\log_b c^m = m \log_b c$.

EXERCISES FOR SEC. 6.1: LOGARITHMS

1 Translate these equations into the language of logarithms.
(*a*) $2^5 = 32$ (*b*) $3^4 = 81$
(*c*) $10^{-3} = 0.001$ (*d*) $5^0 = 1$
(*e*) $1{,}000^{1/3} = 10$ (*f*) $49^{1/2} = 7$

2 Translate these equations into the language of logarithms.
(*a*) $8^{2/3} = 4$ (*b*) $10^3 = 1{,}000$
(*c*) $10^{-4} = 0.0001$ (*d*) $3^0 = 1$
(*e*) $10^{1/2} = \sqrt{10}$ (*f*) $(\frac{1}{2})^{-2} = 4$

3 (*a*) Fill in this table:

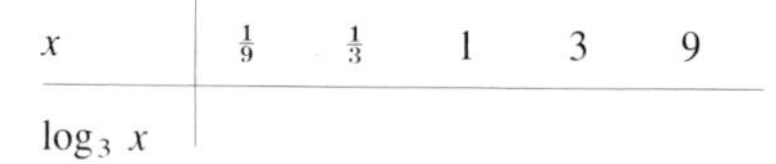

x	$\frac{1}{9}$	$\frac{1}{3}$	1	3	9
$\log_3 x$					

(*b*) Plot the five points in (*a*) and graph $y = \log_3 x$.

4 (*a*) Fill in this table:

x	$\frac{1}{16}$	$\frac{1}{4}$	1	2	4	8	16
$\log_4 x$							

(*b*) Plot the seven points in (*a*) and graph $y = \log_4 x$.

5 Translate these equations into the language of exponents.
(*a*) $\log_2 7 = x$ (*b*) $\log_5 2 = s$
(*c*) $\log_3 \frac{1}{3} = -1$ (*d*) $\log_7 49 = 2$

6 Translate these equations into the language of exponents.
(*a*) $\log_{10} 1{,}000 = 3$ (*b*) $\log_5 \frac{1}{25} = -2$
(*c*) $\log_{1/2} (\frac{1}{4}) = 2$ (*d*) $\log_{64} 128 = \frac{7}{6}$

7 Evaluate
(*a*) $2^{\log_2 16}$ (*b*) $2^{\log_2 (1/2)}$ (*c*) $2^{\log_2 7}$ (*d*) $2^{\log_2 g}$

8 Evaluate
(*a*) $10^{\log_{10} 100}$ (*b*) $10^{\log_{10} 0.01}$
(*c*) $10^{\log_{10} 7}$ (*d*) $10^{\log_{10} p}$

9 Evaluate
(*a*) $\log_3 \sqrt{3}$ (*b*) $\log_3 (3^5)$ (*c*) $\log_3 (\frac{1}{27})$

10 If $\log_4 A = 2.1$, evaluate
(*a*) $\log_4 A^2$ (*b*) $\log_4 (1/A)$ (*c*) $\log_4 16A$

In each of Exercises 11 to 14 solve for x.

11 $2 \cdot 3^x = 7$ **12** $3 \cdot 5^x = 6^x$
13 $3^{5x} = 2^{7x}$ **14** $10^{2x}3^{2x} = 5$

15 (*a*) Evaluate $\log_2 8$, $\log_8 2$, and then their product.
(*b*) Show that, for any two bases a and b,

$$(\log_a b) \times (\log_b a) = 1.$$

Suggestion: Start with $a^{\log_a b} = b$ and take $\log_b$ of both sides.

16 If $\log_3 5 = a$, what is $\log_5 3$?

17 Assume that $\log_{10} 2 \approx 0.30$ and $\log_{10} 3 \approx 0.48$. From this information estimate
(*a*) $\log_{10} 4$ (*b*) $\log_{10} 5$
(*c*) $\log_{10} 6$ (*d*) $\log_{10} 8$

(*e*) $\log_{10} 9$ (*f*) $\log_{10} 1.5$
(*g*) $\log_{10} 1.2$ (*h*) $\log_{10} 1.33$
(*i*) $\log_{10} 20$ (*j*) $\log_{10} 200$
(*k*) $\log_{10} 0.006$

18 Assume that $\log_{10} 2 \approx 0.30$ and $\log_{10} 3 \approx 0.48$. From this information, obtain estimates for
(*a*) $\log_{10} \sqrt{2}$ (*b*) $\log_{10} 0.5$
(*c*) $\log_{10} \frac{2}{3}$ (*d*) $\log_{10} \sqrt[3]{3}$
(*e*) $\log_{10} 18$ (*f*) $\log_{10} 12$
(*g*) $\log_{10} 0.75$ (*h*) $\log_{10} 7.5$
(*i*) $\log_{10} (1/7.5)$ (*j*) $\log_{10} 0.075$
(*k*) $\log_{10} (30\sqrt[3]{2^5})$ (*l*) $\log_{10} \frac{9}{32}$

In Exercises 19 and 20 use logarithms to the base 10 to calculate the given logarithms to two decimal places.

19 (*a*) $\log_3 5$ (*b*) $\log_2 3$

20 (*a*) $\log_{1/2} 3$ (*b*) $\log_7 \frac{1}{2}$

21 (*a*) Express $\log_{1/2} x$ in terms of $\log_2 x$.
(*b*) Sketch the graphs of $y = \log_{1/2} x$ and $y = \log_2 x$.
(*c*) How is one graph in (*b*) obtainable from the other?

22 Let $b > 1$.
(*a*) How are the graphs of $\log_b x$ and $\log_{1/b} x$ related?
(*b*) How are the graphs of $\log_b x$ and $\log_{b^2} x$ related?

23 From the fact that $(b^x)^y = b^{xy}$, deduce that $\log_b c^m = m \log_b c$. (*Hint:* Write c as $b^{\log_b c}$ and consider c^m.)

24 From the fact that $b^{x-y} = b^x/b^y$, deduce that $\log_b (c/d) = \log_b c - \log_b d$. (*Hint:* Write c as $b^{\log_b c}$ and d as $b^{\log_b d}$.)

In Exercises 25 to 28 express $\log_{10} f(x)$ as simply as possible for the given $f(x)$.

25 $f(x) = \dfrac{(\cos^7 x)\sqrt{(x^2+5)^3}}{4 + \tan^2 x}$

26 $f(x) = \sqrt{(1+x^2)^5(3+x)^4\sqrt{1+2x}}$

27 $f(x) = (x\sqrt{2+\cos x})^{x^2}$ **28** $f(x) = \sqrt{\dfrac{x(1+x)}{\sqrt{(1+2x)^3}}}$

29 Find $\log_2 [\log_2 (\log_2 2^{1,024})]$.

30 Is $\log_2 (c + d)$ ever equal to $\log_2 c + \log_2 d$? Explain your answer.

31 Imagine that your calculator fell on the floor and its multiplication and division keys stopped working. However, all the other keys, including the trigonometric, arithmetic, logarithmic, and exponential keys still functioned. Show how you would use your calculator to calculate the product and the quotient of two positive numbers, a and b.

32 (The slide rule) Logarithms are the basis for the design of the slide rule, a device for multiplying or dividing two numbers to three significant figures. Slide rules were common from the early part of the seventeenth century to their recent eclipse by the hand-held calculator. A slide rule consists of two sticks (or circular disks), one of which is fixed and the other is free to slide next to it. A scale is introduced on each stick by placing the number N at a distance $\log_{10} N$ inches from the left end of the stick. (Any base would do as well as 10.) Thus 1 is placed at the left end of each stick and 10 at $\log_{10} 10 = 1$ inch from the left end. The scale of the top stick is at the bottom edge; the scale of the lower stick is at the top edge.
(*a*) Make two such scales on paper or cardboard.
(*b*) Use them to multiply 4 times 25, as follows. Place the 1 of the lower stick at the 4 of the upper stick. Above the 25 of the lower stick appears the product. Check that this works with your two sticks.
(*c*) Explain precisely why the slide rule works.
(*d*) How would you use it to divide two numbers?

33 In 1989, San Francisco and vicinity was struck by an earthquake that measured 7.1 on the Richter scale. The strongest earthquake in recent times had a Richter measure of 8.9 (Columbia-Ecuador in 1906 and Japan in 1933). A "major earthquake" has a measure of at least 7.5.

In his *Introduction to the Theory of Seismology*, Cambridge, 1965, pp. 271–272, K. E. Bullen explains the Richter scale:

"Gutenberg and Richter sought to connect the magnitude M with the energy E of an earthquake by the formula

$$aM = \log_{10}\left(\frac{E}{E_0}\right)$$

and after several revisions arrived in 1956 at the result $a = 1.5$, $E_0 = 2.5 \times 10^{11}$ ergs." (E_0 is the energy of the smallest instrumentally recorded earthquake.)
(*a*) Deduce that $\log_{10} E \approx 11.4 + 1.5M$. (Energy E is measured in ergs. M is the number assigned to the earthquake on the Richter scale.)
(*b*) What is the ratio between the energy of the earthquake that struck Japan in 1933 ($M = 8.9$) and the San Francisco earthquake of 1989 ($M = 7.1$)?
(*c*) What is the ratio between the energy of the San Francisco earthquake of 1906 ($M = 8.3$) and that of the San Francisco earthquake of 1989 ($M = 7.1$)?
(*d*) Find a formula for E in terms of M.
(*e*) If one earthquake has a Richter measure 1 larger than that of another earthquake, what is the ratio of their energies?
(*f*) What is the Richter rating of a 10-megaton H-bomb, that is, of an H-bomb whose energy is equivalent to that in 10 million tons of TNT. (One ton of TNT releases an energy of 4.2×10^6 ergs.)

34 Prove that $\log_3 2$ is irrational. (*Hint:* Assume that it is rational, that is, equal to m/n for some integers m and n, and obtain a contradiction.)

35 If $0 < b < 1$, examine (*a*) $\lim_{x\to\infty} \log_b x$, and (*b*) $\lim_{x\to 0^+} \log_b x$.

36 As of August 1989, the largest known prime was

$$391{,}581 \times 2^{216{,}193} - 1.$$

(*a*) When written in decimal notation, how many digits will it have?

(*b*) How many pages of this book would be needed to print it? (One page can hold about 6,400 digits.)

6.2 THE NUMBER *e*

This section describes a number that is as important in calculus as π is in the study of the circle. We introduce this number, which is always denoted e, by an example involving interest on a bank account.

The Bank

Imagine that you open an account at a bank on January 1 by depositing \$1,000. The bank pays interest monthly at the rate of 5 percent per year. How much will there be in your account at the end of the year? (For simplicity, assume our months have the same length.)

To answer this question, we will find out how much there is in the account at the end of each of the 12 months of the year.

How much is there in the account at the end of January? (At the end of January the account contains the *initial amount*, \$1,000, plus *the interest it earns* during January.) Since there are 12 months, the interest rate for one month is 5%/12. Hence, the interest earned during January is

$$\begin{aligned}(1{,}000)\left(\frac{5\%}{12}\right) &= (1{,}000)\left(\frac{0.05}{12}\right)\\ &= \frac{50}{12}\\ &\approx \$4.17.\end{aligned}$$

In January, the \$1,000 deposit earned \$4.17. So there is \$1,004.17 in the account at the end of January.

How much is there in the account at the end of February?

At the beginning of February, the initial amount is \$1,004.17. It earns interest at the rate of 5%/12. Hence, it earns in the month of February

$$(1{,}004.17)\left(\frac{0.05}{12}\right) \approx \$4.18.$$

Therefore, at the end of February, there is

$$1{,}004.17 + 4.18 = \$1{,}008.35$$

in the account.

In order to compute the amounts at the end of each of the remaining ten months more efficiently on a calculator, consider how the first two calculations were done.

The amount at the end of January is

$$\underbrace{1{,}000}_{\substack{\text{Amount}\\ \text{at beginning}\\ \text{of month}}} + \underbrace{(1{,}000)\left(\frac{0.05}{12}\right)}_{\text{Interest}} = 1{,}000\left(1 + \frac{0.05}{12}\right).$$

In other words, the amount at the end of the month is obtained from the amount at the beginning of the month by multiplying by $[1 + (0.05/12)]$.

The amount at the end of February is

$$\underbrace{\left[1{,}000\left(1+\frac{0.05}{12}\right)\right]}_{\text{Amount at beginning of February}} + \underbrace{\underbrace{\left[1{,}000\left(1+\frac{0.05}{12}\right)\right]}_{\text{Amount at beginning of February}}\left(\frac{0.05}{12}\right)}_{\text{Interest for February}}$$

$$= \left[1{,}000\left(1+\frac{0.05}{12}\right)\right]\left(1+\frac{0.05}{12}\right)$$

$$= 1{,}000\left(1+\frac{0.05}{12}\right)^2.$$

The amount at the end of 2 months is $1{,}000\,[1 + (0.05/12)]^2$.

Similar reasoning shows that at the end of 3 months the account grows to

$$1{,}000\left(1+\frac{0.05}{12}\right)^3 \text{ dollars.} \qquad \text{(Check that this is correct.)}$$

Similarly, at the end of 12 months, the account grows to

$$1{,}000\left(1+\frac{0.05}{12}\right)^{12} \text{ dollars.}$$

In order to compute the amount in the account at the end of each month, store $1 + (0.05/12) \approx 1.0041667$ in the memory of your calculator.

To calculate the amount in the account at the end of, say, 7 months, calculate $(1 + 0.05/12)^7$ using a recall from memory and the y^x key. Then, multiply by 1,000. You should get $1,029.53. Table 1 shows the growth of the account during the year.

At the end of 12 months, the account grows to

$$1{,}000\left(1+\frac{0.05}{12}\right)^{12} \approx \$1{,}051.16.$$

Table 1

Month	*Amount at End of Month*
January	1,004.17
February	1,008.35
March	1,012.55
April	1,016.77
May	1,021.01
June	1,025.26
July	1,029.53
August	1,033.82
September	1,038.13
October	1,042.46
November	1,046.80
December	1,051.16

If, instead, the bank computed interest only once during the year, at the end of December, the interest earned would be simply $(1{,}000)(0.05) = \$50$. The account would grow to only $1,050. When interest is computed monthly, the interest earned each month also earns interest. (During the 12 months shown in the table, the account earned $1{,}051.16 - 1{,}000 = \$51.16$ interest.)

Compound Interest

Simple interest is computed just once a year. Compound interest is computed at regular intervals during the year, monthly, quarterly, or even daily.

Imagine that on January 1 you deposit A dollars in a bank that pays interest at the annual rate r. (In our example, $A = \$1{,}000$ and $r = 0.05$.) The bank "compounds" interest n times a year. (In our example, $n = 12$. If interest is computed daily, $n = 365$.) How much will there be in the account at the end of the year? [In our example, the amount was

$$(1{,}000)[1 + (0.05/12)]^{12}.]$$

At the end of the first interest period, the amount in the account is

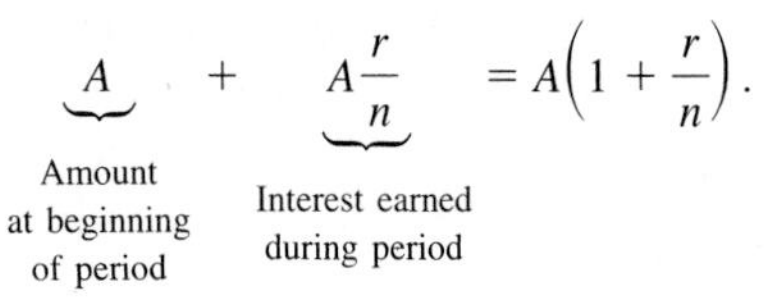

$$\underbrace{A}_{\substack{\text{Amount}\\ \text{at beginning}\\ \text{of period}}} + \underbrace{A\frac{r}{n}}_{\substack{\text{Interest earned}\\ \text{during period}}} = A\left(1 + \frac{r}{n}\right).$$

At the end of the second interest period, the amount in the account is

$$\underbrace{A\left(1 + \frac{r}{n}\right)}_{\substack{\text{Amount}\\ \text{at beginning}\\ \text{of second}\\ \text{interest period}}} + \underbrace{\left[A\left(1 + \frac{r}{n}\right)\right]\frac{r}{n}}_{\substack{\text{Interest earned}\\ \text{during second}\\ \text{period}}} = A\left(1 + \frac{r}{n}\right)\left(1 + \frac{r}{n}\right) = A\left(1 + \frac{r}{n}\right)^2.$$

At the end of the third interest period, the amount is

$$\underbrace{A\left(1 + \frac{r}{n}\right)^2}_{\substack{\text{Amount}\\ \text{at beginning}\\ \text{of third}\\ \text{interest period}}} + \underbrace{\left[A\left(1 + \frac{r}{n}\right)^2\right]\frac{r}{n}}_{\substack{\text{Interest earned}\\ \text{during third}\\ \text{period}}} = A\left(1 + \frac{r}{n}\right)^2\left(1 + \frac{r}{n}\right) = A\left(1 + \frac{r}{n}\right)^3.$$

During each interest period, the amount is "magnified" by the factor $[1 + (r/n)]$.

On December 31, at the end of the n interest periods, the amount in the account grows to

$$\boxed{\text{Final amount} = A\left(1 + \frac{r}{n}\right)^n \text{ dollars.}} \tag{1}$$

The Generous Bank That Pays 100 Percent

Assume that banks pay interest at the rate of 100 percent a year instead of a mere 5 percent a year. That is, assume r = 100 percent = 1. If you open an account on January 1 with \$1 ($A = 1$), how large will it be on December 31?

If your bank pays simple interest (no compounding during the year), your dollar earns \$1 interest and you have \$2 at the end of the year.

A competing bank compounds interest twice a year ($n = 2$). In this case, your dollar would grow to

$$\begin{aligned} A\left(1 + \frac{r}{n}\right)^n &= 1(1 + \tfrac{1}{2})^2 \\ &= (\tfrac{3}{2})^2 \\ &= \$2.25. \end{aligned}$$

Another bank offers to compound quarterly ($n = 4$). One dollar deposited there would grow to

$$\begin{aligned} A\left(1 + \frac{r}{n}\right)^n &= 1(1 + \tfrac{1}{4})^4 \\ &= (\tfrac{5}{4})^4 \\ &\approx \$2.44141. \end{aligned}$$

Another bank promises to compound every day ($n = 365$). In this case, \$1 grows to

$$\left(1 + \frac{1}{365}\right)^{365} \approx \$2.71457.$$

Not to be outdone, another bank says that they will compound every hour. One dollar in this bank would grow to

$$\left(1 + \frac{1}{8{,}760}\right)^{8{,}760} \approx \$2.71813.$$

Table 2 recalls these results:

Table 2

Type	*Simple*	*Semiannual*	*Quarterly*	*Daily*	*Hourly*
n	1	2	4	365	8,760
$\left(1 + \frac{1}{n}\right)^n$	2	2.25	2.44141	2.71457	2.71813

The more often a bank compounds, the larger the value of the account will be at the end of a year. What would happen to an account if a bank compounded every minute? every second?

That question brings us to the number e.

The Number e

No matter how frequently interest is compounded—even if every second—the account does not get arbitrarily large in one year. It turns out that it gets arbitrarily close to a certain amount, which, to five decimal places, is \$2.71828. In other words, we say

$$\lim_{n\to\infty} \left(1 + \frac{1}{n}\right)^n$$

exists and is about 2.71828.

Observe that for large n the expression $(1 + 1/n)^n$ is of the form

$$(1 + \text{small number})^{\text{reciprocal of same small number}}.$$

So we may consider

$$(1 + x)^{1/x}$$

when x is near 0, even if x is not of the form $1/n$, that is, is not the reciprocal of an integer. It can be shown that $\lim_{x\to 0} (1 + x)^{1/x}$ exists and, of course, is about 2.71828.

This brings us to the definition of the number e.

e is not a repeating decimal. The next digit is 4.

Definition *The number e.*

$$e = \lim_{x\to 0} (1 + x)^{1/x} \approx 2.718281828. \qquad (2)$$

Two opposing forces

We assume that the limit in (2) exists, since a proof that it does exist would amount to a big detour. (See Exercise 44 of Sec. 6.3 and Appendix H.) Note that there are two conflicting influences on $(1 + x)^{1/x}$ when x is near 0. For simplicity, consider only $x > 0$. First, the base gets near 1, so there is a chance that $(1 + x)^{1/x}$ gets near 1. But the exponent $1/x$ gets arbitrarily large, so there is also a chance that $(1 + x)^{1/x}$ gets large, since the base $1 + x$ is larger than 1. It turns out that the force pushing $(1 + x)^{1/x}$ toward 1 is not strong enough to do that, but it does manage to keep $(1 + x)^{1/x}$ from exceeding $e \approx 2.71828$.

From the fact that $\lim_{x\to 0} (1 + x)^{1/x} = e$, we can obtain other closely related limits. For instance, $\lim_{h\to 0} (1 + 2h)^{1/(2h)} = e$. (Note that $2h \to 0$ and the exponent is the reciprocal of $2h$.)

EXAMPLE 1 Find $\lim_{h\to 0} (1 + 2h)^{1/h}$.

SOLUTION The expression $(1 + 2h)^{1/h}$ is *not* of the form

$$(1 + \text{small number})^{\text{reciprocal of same small number}}$$

since $1/h$ is not the reciprocal of $2h$. A little algebra gets around this obstacle:

$$\begin{aligned}
\lim_{h\to 0} (1 + 2h)^{1/h} &= \lim_{h\to 0} (1 + 2h)^{2/(2h)} \\
&= \lim_{h\to 0} [(1 + 2h)^{1/(2h)}]^2 \qquad (b^c)^d = b^{cd} \\
&= \left[\lim_{h\to 0} (1 + 2h)^{1/(2h)}\right]^2 \qquad \text{``squaring'' is continuous} \\
&= e^2. \quad \blacksquare
\end{aligned}$$

The next example will be referred to in Sec. 6.3.

EXAMPLE 2 Find $\lim_{\Delta x\to 0} \left(1 + \frac{\Delta x}{x}\right)^{x/\Delta x}$, where x is a fixed number, $x \neq 0$.

SOLUTION Since $\Delta x \to 0$, we may think of $\Delta x/x$ as a small number, s. Thus,

$$\left(1 + \frac{\Delta x}{x}\right)^{x/\Delta x}$$

has the form $(1 + s)^{1/s}$, where s is a small number. Thus,

$$\lim_{\Delta x\to 0} \left(1 + \frac{\Delta x}{x}\right)^{x/\Delta x} = \lim_{s\to 0} (1 + s)^{1/s} = e,$$

by the definition of e. $\blacksquare$

Back to the Bank

The number e was introduced by considering a bank that pays 100 percent interest per year. However, e is useful in analyzing compound interest even when the rate is more modest. Example 3 illustrates this when the rate is 5 percent a year.

EXAMPLE 3 One thousand dollars is deposited in a bank that pays 5 percent interest a year compounded n times a year. If n is very large, how much will there be in the account at the end of the year?

SOLUTION We use the formula $A[1 + (r/n)]^n$ to compute the amount in the account:

$$A\left(1 + \frac{r}{n}\right)^n = 1{,}000\left(1 + \frac{0.05}{n}\right)^n.$$

In order to see how $1{,}000[1 + (0.05/n)]^n$ behaves when n is large, we compute $\lim_{n\to\infty} 1{,}000[1 + (0.05/n)]^n$ as follows:

$$1{,}000\left(1 + \frac{0.05}{n}\right)^n = 1{,}000\left(1 + \frac{0.05}{n}\right)^{\frac{n}{0.05}\cdot n}$$

$$= 1{,}000\left[\left(1 + \frac{0.05}{n}\right)^{\frac{n}{0.05}}\right]^{0.05}.$$

[The expression in brackets is of the form $(1 + x)^{1/x}$, where $x = 0.05/n$, which is small when n is large.]

Thus,

$$\lim_{n\to\infty} 1{,}000\left(1 + \frac{0.05}{n}\right)^n = \lim_{n\to\infty} 1{,}000\left[\left(1 + \frac{0.05}{n}\right)^{\frac{n}{0.05}}\right]^{0.05}$$

$$= 1{,}000e^{0.05}.$$

A calculator gives the value of $e^{0.05}$.

When n is large, the value in the account is approximately $1{,}000e^{0.05} \approx 1{,}000(1.0512711) \approx \$1{,}051.27$. This is not much more than the amount you would get at a bank that compounds monthly. (See Table 1.) ■

Example 3 illustrates what is called "continuously compounded interest." It corresponds to the limiting case of compounding n times a year and letting $n \to \infty$. Under continuous compounding an initial deposit of A dollars, with interest at an annual rate of r, grows in 1 year to

$$\lim_{n\to\infty} A\left(1 + \frac{r}{n}\right)^n \text{ dollars.}$$

The method in Example 3 shows that the limit is Ae^r. In continuous compounding, where the annual interest rate is r, \$1 grows to e^r dollars in 1 year.

The number e appears often in places where you would not expect it. For example, imagine that you write letters to n friends, address the n envelopes, and put the letters randomly in the envelopes, one to an envelope. The probability that all the letters are in wrong envelopes is approximately $1/e$. As $n \to \infty$, the probability approaches $1/e$.

As another example, the nth prime number is approximately equal to $n \log_e n$. If P_n denotes the nth prime number, then $\lim_{n\to\infty} P_n/(n \log_e n) = 1$. The 100th prime number is 541, and $541/(100 \log_e 100) \approx 1.17$. The 664,699th prime number is 10,006,721; the quotient $P_n/(n \log_e n) = 10{,}006{,}721/(664{,}699 \log_e 664{,}699) \approx 1.12$.

Leonhard Euler (pronounced "oiler"), the great Swiss mathematician, introduced the number e (and named it) in his calculus text *Introductio in analisin Infinitorum*, vol. 1, 1748, p. 90, in these words:

$$\text{inventam, } e = 1 + \frac{1}{1} + \frac{1}{1.2} + \frac{1}{1.2.3} + \frac{1}{1.2.3.4} + \&c.,$$

qui termini, ſi in fractiones decimales convertantur atque actu addantur, præbebunt hunc valorem pro e = 2,71828182845904523536028, cujus ultima adhuc nota veritati eſt conſentanea. Quod ſi jam ex hac baſi Logarithmi conſtruantur, ii vocari ſolent Logarithmi *naturales* ſeu *hyperbolici*, quoniam quadratura hyperbolæ per iſtiuſmodi Logarithmos exprimi poteſt. Ponamus autem brevitatis gratia pro numero hoc 2,718281828459 &c. conſtanter litteram e, quæ ergo denotabit baſin Logarithmorum naturalium ſeu hyperbolicorum,

Here it is defined as the limit as $n \to \infty$ of sums of the form

$$1 + \frac{1}{1} + \frac{1}{2!} + \frac{1}{3!} + \frac{1}{4!} + \cdots + \frac{1}{n!}.$$

(See Exercise 42 in Sec. 6.3.) In Chap. 11 it will be shown that Euler's definition of e agrees with our definition.

EXERCISES FOR SEC. 6.2: THE NUMBER e

1 Fill in this table.

x	0.1	0.01	0.001	0.0005
$(1 + x)^{1/x}$				

2 Fill in this table.

x	−0.1	−0.01	−0.001	−0.0005
$(1 + x)^{1/x}$				

In Exercises 3 to 8 evaluate the limits.

3 $\lim_{t \to 0} (1 + t)^{1,000}$

4 $\lim_{x \to \infty} 1.001^x$

5 $\lim_{h \to 0} (1 + 3h)^{1/(4h)}$

6 $\lim_{h \to 0} (1 - h)^{1/h}$

7 $\lim_{\Delta x \to 0} \left(1 + \frac{\Delta x}{2x}\right)^{x/\Delta x}$, $x \neq 0$ is fixed

8 $\lim_{n \to \infty} \left(1 + \frac{3}{n}\right)^{n/2}$

* 9 What do you think happens to $(1 + x)^{1/x}$ as $x \to -1$ from the right? Do some calculations first.

* 10 What do you think happens to $(1 + x)^{1/x}$ as $x \to \infty$? Do some calculations first.

11 A bank pays an annual interest rate of 50 percent. Assume that \$1,000 is deposited at the beginning of the year. How much will there be in the account at the end of the year if interest is (*a*) simple (not compounded), (*b*) compounded every 6 months, (*c*) compounded monthly, (*d*) compounded daily, (*e*) compounded continuously?

12 Like Exercise 11, but with an interest rate of 8 percent per year.

---

13 On the basis of Exercises 9 and 10 and the definition of e, sketch the graph of $y = (1 + x)^{1/x}$ for $x > -1$, $x \neq 0$.

14 Using your calculator, fill in this table and graph $y = e^x$.

x	−3	−2	−1	0	1	2	3
e^x							

15 (*a*) Compute $(2^h - 1)/h$ for $h = 0.1$, 0.001, and -0.001.
(*b*) On the basis of (*a*), estimate the derivative of 2^x at 0.
(*c*) Similarly, estimate the derivative of 3^x at 0.
(*d*) Estimate the derivative of b^x at 0 for your own choices

for b. Experiment with various choices to get the derivative at 0 to be as close to 1 as you can.

16 (*a*) Fill in the table.

n	-2	-10	-100
$\left(1+\frac{1}{n}\right)^n$			

(*b*) On the basis of (*a*), what do you think

$$\lim_{n\to-\infty}\left(1+\frac{1}{n}\right)^n$$

is?

(*c*) Explain, using algebra, why

$$\lim_{n\to-\infty}\left(1+\frac{1}{n}\right)^n = \lim_{n\to\infty}\left(1+\frac{1}{n}\right)^n.$$

[Do *not* assume that $\lim_{x\to0}(1+x)^{1/x}$ exists.]

17 Under continuously compounded interest at the annual rate r, how much will be in an account at the end of t years if the account is opened with A dollars? (The number t need not be an integer.)

6.3 THE DERIVATIVE OF A LOGARITHMIC FUNCTION

In Chaps. 3 and 4, we found the derivatives of a variety of functions, including polynomials and the trigonometric functions. In this section, we obtain the derivative of $\log_b x$ ($b > 0$, $b \neq 1$) and decide which base b to use in calculus. We also obtain an antiderivative for $1/x$.

Most of this section can be covered right after Chap. 4. The small part that depends on Chap. 5 begins with the heading *Logarithms and Integration*. (Exercises that depend on Chap. 5 are so labeled.)

Logarithms and Differentiation

In the proof of Theorem 1 it will be assumed that the function $\log_b x$ is continuous.

Theorem 1 The derivative of the function $\log_b x$ is

$$\frac{\log_b e}{x}$$

for all positive numbers x.

Proof Recall that $f'(x) = \lim_{\Delta x\to0}[f(x+\Delta x) - f(x)]/\Delta x$ by definition of the derivative. Thus, it is necessary to compute

$$\lim_{\Delta x\to0}\frac{\log_b(x+\Delta x) - \log_b(x)}{\Delta x}.$$

There is no simple cancellation that would make this limit "obvious." To find the limit, we first rewrite the quotient using properties of logarithms, as follows:

$$\frac{\log_b (x + \Delta x) - \log_b (x)}{\Delta x} = \frac{\log_b [(x + \Delta x)/x]}{\Delta x} \qquad \log_b c - \log_b d = \log_b \left(\frac{c}{d}\right)$$

$$= \frac{1}{\Delta x} \log_b \left(1 + \frac{\Delta x}{x}\right) \qquad \text{algebra}$$

$$= \log_b \left(1 + \frac{\Delta x}{x}\right)^{1/\Delta x} \qquad \log_b c^m = m \log_b c$$

$$= \log_b \left[\left(1 + \frac{\Delta x}{x}\right)^{x/\Delta x}\right]^{1/x} \qquad \text{power of a power}$$

$$= \frac{1}{x} \log_b \left(1 + \frac{\Delta x}{x}\right)^{x/\Delta x}. \qquad \log_b c^m = m \log_b c$$

After these manipulations, it is easy to take limits:

$$\lim_{\Delta x \to 0} \frac{\log_b (x + \Delta x) - \log_b x}{\Delta x}$$

$$= \lim_{\Delta x \to 0} \frac{1}{x} \log_b \left(1 + \frac{\Delta x}{x}\right)^{x/\Delta x}$$

$$= \frac{1}{x} \lim_{\Delta x \to 0} \log_b \left(1 + \frac{\Delta x}{x}\right)^{x/\Delta x} \qquad 1/x \text{ is fixed}$$

$$= \frac{1}{x} \log_b \left[\lim_{\Delta x \to 0} \left(1 + \frac{\Delta x}{x}\right)^{x/\Delta x}\right] \qquad \log_b \text{ is continuous}$$

$$= \frac{1}{x} \log_b e. \qquad \text{Example 2 of Sec. 6.2}$$

Thus $\log_b x$ has a derivative:

$$(\log_b x)' = \frac{\log_b e}{x}. \qquad \blacksquare$$

In particular,

$$(\log_{10} x)' = \frac{\log_{10} e}{x}.$$

As a table of logarithms or a calculator shows, $\log_{10} e \approx 0.434$. Thus

$$(\log_{10} x)' \approx \frac{0.434}{x}. \tag{1}$$

The Best Base for Logarithms in Calculus

Which is the best of all possible bases b to use? More precisely, for which base b does the formula

$$\frac{\log_b e}{x}$$

take its simplest form? Certainly not $b = 10$, since

$$(\log_{10} x)' = \frac{\log_{10} e}{x} \approx \frac{0.434}{x}.$$

Would $b = 2$ be convenient? By Theorem 1,

$$(\log_2 x)' = \frac{\log_2 e}{x} \approx \frac{1.443}{x}.$$

We would like to avoid such numbers as 0.434 and 1.443. To avoid decimals, we would prefer to use the base b such that

$$\log_b e = 1.$$

Why e is used as a base for logarithms

That is, b^1 must equal e. In this case, b is e. The best of all bases to use for logarithms is e. The derivative of the $\log_e$ function is given by

$$\frac{d}{dx}(\log_e x) = \frac{\log_e e}{x} = \frac{1}{x}.$$

In this case, there is no constant, such as 0.434 or 1.443, to memorize.

The natural logarithm

For this reason, the base e is preferred in calculus. We shall write $\log_e x$ as **ln x**, the **natural logarithm** of x. (Only for purposes of arithmetic, such as multiplying with the aid of logarithms, is base 10 preferable.) Most handbooks of mathematical tables include tables of $\log_{10} x$ (common logarithm) and $\ln x$ (natural logarithm). Scientific calculators usually have an ln-key ($\log_e$) and a log-key ($\log_{10}$). Mathematics texts and some handbooks denote the natural logarithm by "log x," since the common logarithm is seldom used in higher mathematics.

A simple equation summarizes much of this section:

$$\boxed{\frac{d}{dx}(\ln x) = \frac{1}{x}, \qquad x > 0.}$$

The derivative of the natural logarithm function is the function $1/x$.

In earlier examples we have seen that the derivatives of polynomials are polynomials and that the derivatives of trigonometric functions are trigonometric functions. Surprisingly, the derivative of $\ln x$ bears no resemblance to $\ln x$. It is simply the reciprocal of x.

EXAMPLE 1 Find $[\ln (x^2 + 1)]'$.

SOLUTION Let $y = \ln (x^2 + 1)$. Then $y = \ln u$, where $u = x^2 + 1$. By the chain rule,

$$\begin{aligned} \frac{dy}{dx} &= \frac{dy}{du}\frac{du}{dx} \\ &= \frac{d}{du}(\ln u)\frac{d}{dx}(x^2 + 1) \\ &= \frac{1}{u} \cdot 2x = \frac{2x}{x^2 + 1}. \end{aligned}$$

Thus the derivative of $\ln (x^2 + 1)$ is $2x/(x^2 + 1)$. ■

Example 1 illustrates a useful principle:

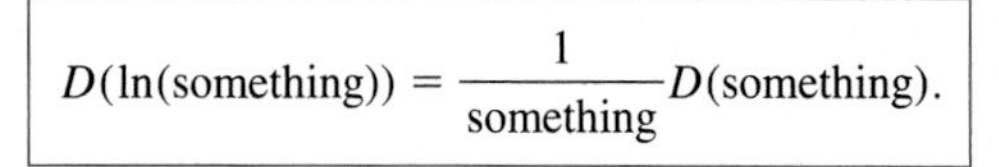

$$D(\ln(\text{something})) = \frac{1}{\text{something}} D(\text{something}).$$

More precisely, if f is a differentiable function, then

$$\frac{d}{dx}[\ln f(x)] = \frac{f'(x)}{f(x)}.$$

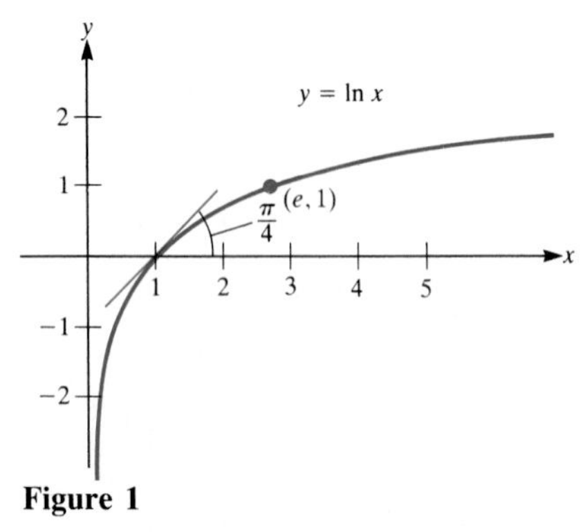

Figure 1

EXAMPLE 2 Graph $y = \ln x$. At what angle does it cross the x axis?

SOLUTION The graph is shown in Fig. 1. Note that at $x = e \approx 2.718$, the y coordinate is 1. The graph crosses the x axis at $(1, 0)$. Since the derivative of $\ln x$ is $1/x$, the slope of the curve $y = \ln x$ at $x = 1$ is $1/1$, which is 1. Since the slope of the tangent line at $(1, 0)$ is 1, the tangent makes an angle of $\pi/4$ radians $(45°)$ with the x axis. ■

EXAMPLE 3 Graph $y = (\ln x)/x$, showing intercepts, extrema, inflection points, and asymptotes.

Domain

SOLUTION Since $\ln x$ is defined only when x is positive, the domain of $\ln x$ is $(0, \infty)$.

y intercept

Since $\ln 0$ is not defined, there is no y intercept.

x intercepts

To find any x intercepts, solve the equation

$$\frac{\ln x}{x} = 0.$$

The numerator must be 0, hence $\ln x = 0$. The only x intercept is $x = 1$.

Critical points

To find the critical points, set the derivative of $(\ln x)/x$ equal to 0:

$$\left(\frac{\ln x}{x}\right)' = 0$$

$$\frac{x\dfrac{1}{x} - (\ln x)1}{x^2} = 0$$

$$\frac{1 - \ln x}{x^2} = 0.$$

Hence

$$1 - \ln x = 0$$

$$\ln x = 1.$$

The only critical number is $x = e$. The critical point is

$$\left(e, \frac{\ln e}{e}\right) = \left(e, \frac{1}{e}\right).$$

Extrema

To determine any extrema, check first whether the first derivative changes sign at the critical number e. We have $y' = (1 - \ln x)/x^2$. If $0 < x < e$, y' is positive. If $x > e$, y' is negative. Since the sign of y' changes from positive to negative, the critical point $(e, 1/e)$ is a relative maximum. Actually, it is even a global maximum since $y = (\ln x)/x$ increases for $0 < x < e$ and decreases for $x > e$.

To find any inflection points, we examine the second derivative:

$$\left(\frac{1 - \ln x}{x^2}\right)' = \frac{x^2\left(-\frac{1}{x}\right) - (1 - \ln x)(2x)}{x^4}$$

$$= \frac{-3 + 2 \ln x}{x^3}.$$

The second derivative is 0 when $-3 + 2 \ln x = 0$, that is, when $2 \ln x = 3$ or $\ln x = \frac{3}{2}$. Hence, at $x = e^{3/2}$, $[(\ln x)/x]'' = 0$. Moreover $\dfrac{-3 + 2 \ln x}{x^3}$ is negative when $0 < x < e^{3/2}$ and positive when $x > e^{3/2}$.

Inflection points

Since y'' changes sign at $x = e^{3/2}$, $e^{3/2}$ is an inflection number and the point $(e^{3/2}, (\ln e^{3/2})/e^{3/2}) = (e^{3/2}, \frac{3}{2}/e^{3/2})$ is an inflection point.

Horizontal asymptotes

To determine if there are any horizontal asymptotes, we consider $\lim_{x\to\infty} (\ln x)/x$. As remarked in Sec. 6.1, this limit is 0. Hence the x axis is an asymptote.

Vertical asymptotes

When x is near 0 (but positive), $\ln x$ is a very large negative number. Thus

$$\frac{\ln x}{x} = (\ln x)\left(\frac{1}{x}\right)$$

is the product of a large negative number and a large positive number. Thus

$$\lim_{x\to 0^+} \frac{\ln x}{x} = -\infty.$$

Therefore, the y axis is a vertical asymptote.

All of this information appears in Fig. 2.

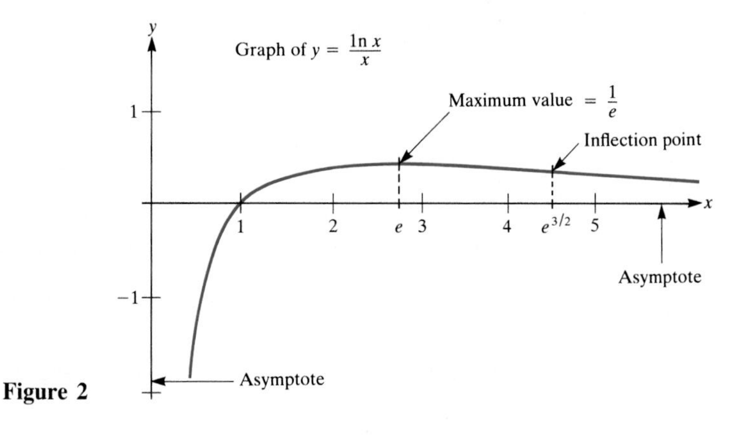

Figure 2

■

Logarithmic Differentiation

The next example presents a special case of implicit differentiation called **logarithmic differentiation.** This is a method for differentiating a function whose logarithm is simpler than the function itself.

EXAMPLE 4 Differentiate $y = [\sqrt[3]{x}\sqrt{(1+x^2)^3}]/x^{4/5}$.

SOLUTION If you were to differentiate directly, you would run into some messy computations. (Do it.) Rather than compute dy/dx directly, take logarithms of both sides of the equation first, obtaining

$$\ln y = \tfrac{1}{3}\ln x + \tfrac{3}{2}\ln(1+x^2) - \tfrac{4}{5}\ln x.$$

Then differentiate this equation implicitly:

$$\frac{1}{y}\frac{dy}{dx} = \frac{1}{3x} + \frac{3}{2}\frac{2x}{1+x^2} - \frac{4}{5}\frac{1}{x}.$$

Solving for dy/dx yields

$$\begin{aligned}\frac{dy}{dx} &= y\left(\frac{1}{3x} + \frac{3x}{1+x^2} - \frac{4}{5x}\right)\\ &= \frac{\sqrt[3]{x}\sqrt{(1+x^2)^3}}{x^{4/5}}\left(\frac{1}{3x} + \frac{3x}{1+x^2} - \frac{4}{5x}\right).\end{aligned}$$

The reader is invited to find dy/dx directly from the explicit formula for y. Doing so will show the advantage of logarithmic differentiation. ■

Logarithms and Integration

If a is not equal to -1, then

$$\int x^a\,dx = \frac{x^{a+1}}{a+1} + C.$$

The formula works, for instance, when $a = -1.01$:

$$\begin{aligned}\int x^{-1.01}\,dx &= \frac{x^{-1.01+1}}{-1.01+1} + C\\ &= \frac{x^{-0.01}}{-0.01} + C\\ &= -100x^{-0.01} + C.\end{aligned}$$

But what if $a = -1$? The formula for $\int x^a\,dx$ then asserts that

$$\int x^{-1}\,dx = \int \frac{x^{-1+1}}{-1+1}\,dx + C,$$

which makes no sense since division by $-1 + 1 = 0$ is totally meaningless.

However, the derivative of the logarithm to the base e is $1/x$, that is,

$$\frac{d}{dx}(\ln x) = \frac{1}{x}.$$

Therefore, when $a = -1$, $\int x^a\,dx$ no longer involves a power of x. Instead, it is a logarithm:

An antiderivative for 1/x.

$$\int \frac{1}{x}\,dx = \ln x + C, \qquad x > 0. \tag{2}$$

One reason that the natural logarithmic function, ln x, is so important is that it serves as an antiderivative of $1/x$.

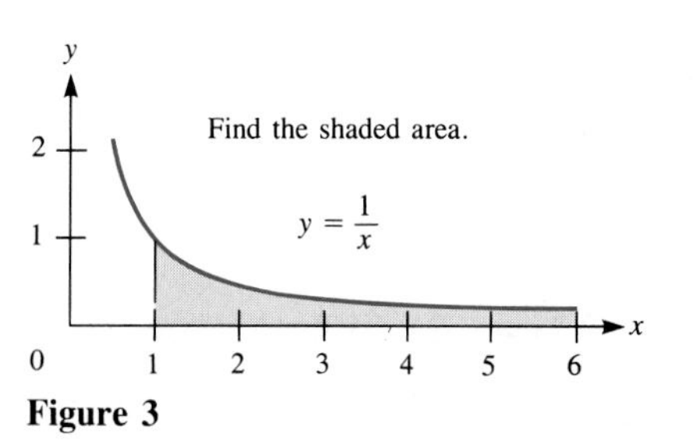

Figure 3

EXAMPLE 5 Find the area under the curve $y = 1/x$ and above the interval [1, 6]. (See Fig. 3.)

SOLUTION The area equals the definite integral

$$\int_1^6 \frac{1}{x}\,dx.$$

By the fundamental theorem of calculus,

$$\int_1^6 \frac{1}{x}\,dx = \ln x \Big|_1^6 = \ln 6 - \ln 1$$

$$= \ln 6 \approx 1.792. \quad \blacksquare$$

Equation (2) makes no sense if x is negative, since only positive numbers have logarithms. However, the next theorem provides an antiderivative for $1/x$ even when x is negative.

Theorem 2

$$\int \frac{1}{x}\,dx = \ln |x| + C \qquad \text{for } x > 0 \text{ or for } x < 0. \tag{3}$$

Proof It is necessary to show that the derivative of $\ln |x|$ is $1/x$. For positive x this has already been done. Now consider x negative.

For $x < 0$, $|x| = -x$. Thus, for negative x,

$$\frac{d}{dx}(\ln |x|) = \frac{d}{dx}(\ln(-x))$$

$$= \frac{1}{(-x)}\frac{d(-x)}{dx} \qquad \text{by chain rule, as in Example 1}$$

$$= \frac{1}{-x}(-1) = \frac{1}{x}.$$

This completes the proof. ■

EXAMPLE 6 Compute $\displaystyle\int_{-3}^{-1} \frac{dx}{x}$.

SOLUTION

$$\int_{-3}^{-1} \frac{dx}{x} \underset{\text{FTC}}{=} \ln |x| \Big|_{-3}^{-1} \qquad \text{by Theorem 2}$$

$$= \ln |-1| - \ln |-3|$$

$$= \ln 1 - \ln 3 = 0 - \ln 3 = -\ln 3. \quad \blacksquare$$

The next theorem shows that the logarithm function enables us to integrate many functions besides $1/x$.

Theorem 3 Let $f(x)$ be a differentiable function. Then, if $f(x) \neq 0$,

$$\int \frac{f'(x)}{f(x)}\, dx = \ln |f(x)| + C.$$

Proof Let $y = \ln |f(x)|$. Then $y = \ln |u|$, where $u = f(x)$. By the chain rule and Theorem 2,

$$\frac{dy}{dx} = \frac{1}{u} f'(x) = \frac{f'(x)}{f(x)}. \quad \blacksquare$$

How to integrate $\dfrac{f'(x)}{f(x)}$

With the aid of the formula in Theorem 3 we can integrate the quotient of two functions if the numerator is exactly the derivative of the denominator, as the next example shows. In fact, if the numerator is a constant times the derivative of the denominator, we can still use the formula, as Example 8 shows.

EXAMPLE 7 Compute $\int 3x^2/(x^3 + 1)\, dx$.

SOLUTION The numerator is exactly the derivative of the denominator. By Theorem 3,

$$\int \frac{3x^2}{x^3 + 1}\, dx = \ln |x^3 + 1| + C. \quad \blacksquare$$

EXAMPLE 8 Compute $\int x/(x^2 + 1)\, dx$.

SOLUTION The numerator is not the derivative of the denominator, which is $2x$. However, $2x$ is a constant times the numerator. A little algebra permits us to use Theorem 3:

$$\int \frac{x}{x^2 + 1}\, dx = \int \frac{\frac{1}{2}(2x)}{x^2 + 1}\, dx$$

$$= \frac{1}{2} \int \frac{2x}{x^2 + 1}\, dx \qquad \textstyle\int cf(x)\, dx = c \int f(x)\, dx,\ c \text{ constant}$$

$$= \frac{1}{2} \ln |x^2 + 1| + C \qquad \text{Theorem 3}$$

$$= \frac{1}{2} \ln (x^2 + 1) + C. \qquad x^2 + 1 \text{ is positive} \quad ■$$

In practical terms, Theorem 3 tells us:

$$\boxed{\int \frac{D(\text{something})}{\text{something}}\, dx = \ln (|\text{something}|) + C.}$$

Section Summary

We found the derivative of $\log_b x$ and introduced the natural logarithm function $\ln x = \log_e x$. ("ln x" stands for *logarithm natural,* a symbol first used in 1893.) The table records derivatives and antiderivatives involving logarithms.

Derivative	*Antiderivative*
$(\ln x)' = \frac{1}{x}$	$\int \frac{dx}{x} = \ln x + C,\ x > 0$
$(\ln \|x\|)' = \frac{1}{x}$	$\int \frac{dx}{x} = \ln \|x\| + C,\ x \neq 0$
$[\ln f(x)]' = \frac{f'(x)}{f(x)}$	$\int \frac{f'(x)}{f(x)}\, dx = \ln f(x) + C,\ f(x) > 0$
$[\ln \|f(x)\|]' = \frac{f'(x)}{f(x)}$	$\int \frac{f'(x)}{f(x)}\, dx = \ln \|f(x)\| + C,\ f(x) \neq 0$
$(\log_b x)' = \frac{\log_b e}{x}$	Not needed as antiderivative

Before differentiating $\ln f(x)$, first simplify by the laws of logarithms. To differentiate a function $y = f(x)$ whose logarithm is much easier to differentiate, first differentiate $\ln y = \ln f(x)$ implicitly and then solve for y'.

EXERCISES FOR SEC. 6.3: THE DERIVATIVE OF A LOGARITHMIC FUNCTION

In Exercises 1 to 10 differentiate the functions.

1 $\ln (1 + x^2)$
2 $\ln (1 + x^3)$
3 $x^2 \ln x$
4 $x \ln x^2$
5 $\frac{\ln x}{x}$
6 $(\ln x)^3$
7 $\ln 5x \sin 2x$
8 $\sec 5x \ln 2x$
9 $\ln (\sin x)$
10 $\cos (\ln x)$

In Exercises 11 to 16 differentiate and simplify your answers.

11 $\ln (2x + 3)$
12 $\frac{x}{3} - \frac{1}{9} \ln (3x + 1)$
13 $\frac{2}{25(5x + 2)} + \frac{1}{25} \ln (5x + 2)$
14 $x + 3 - 6 \ln (x + 3) - \frac{9}{x + 3}$
15 $\ln (x + \sqrt{x^2 - 5})$
16 $\ln (x + \sqrt{x^2 + 1})$

In Exercises 17 to 22 first simplify by using the laws of logarithms; then differentiate.

17 $\frac{1}{5} \ln \frac{x}{3x + 5}$
18 $-\frac{1}{3x} + \frac{5}{9} \ln \left(\frac{3x + 5}{x}\right)$
19 $\frac{1}{10} \ln \left(\frac{5 + x}{5 - x}\right)$
20 $\sqrt{x^2 + 1} \ln \frac{\sqrt{x^2 + 1} - 1}{x}$
21 $\ln [(x^2 + 1)^3(x^5 + 1)^4]$
22 $\ln \frac{\sqrt{2x + 1}\sqrt[3]{3x + 2}}{(x^2 + 1)^5}$
23 Differentiate $\log_{10} \sqrt[3]{x}$.
24 Differentiate $\log_2[(x^2 + 1)^3 \sin 3x]$.

In Exercises 25 and 26 graph the functions as in Example 3.

25 $y = (\ln x)/x^2$
26 $y = (\ln x)/x^3$

In Exercises 27 to 30 differentiate by logarithmic differentiation.

27 $(1 + 3x)^5(\sin 3x)^6$

28 $\sqrt{1 + x^2}\sqrt[3]{(1 + \cos 3x)^5}$

29 $\dfrac{(\sec 4x)^{5/3} \sin^3 2x}{\sqrt{x}}$ 30 $\dfrac{\cot^3 x}{\sqrt[3]{x}\sqrt{(x^3 + 2)^5}}$

Exercises 31 and 32 depend on material from Chap. 5.

31 (*a*) Estimate $\ln 4 = \int_1^4 1/x \, dx$ by Simpson's method with $h = \frac{1}{2}$.
(*b*) How small should h be chosen in order that the error in the estimate would be less than 0.0005?

32 Use Simpson's method to estimate $\ln 2 = \int_1^2 dx/x$ to three decimal places.

33 (*a*) Show that there is a number r in [0, 1] such that $\ln (1 + r) = 1 - r$.
(*b*) Show that there is only one such number r in the interval [0, 1].
(*c*) Use Newton's method with $x_1 = 0.5$ to find x_2, a closer approximation to r.
(*d*) Find r to four decimal places.

34 (*a*) Graph $y = \ln x$ and $y = \sin x$ relative to the same axes.
(*b*) With the aid of the graphs in (*a*), estimate the value of the number x such that $\sin x = \ln x$.
(*c*) Using the estimate in (*b*) as x_1, apply Newton's method to find the value of x to four decimal places.

---

35 Assuming that f and g are differentiable functions with positive values, obtain by logarithmic differentiation the formulas for (*a*) $(fg)'$, (*b*) $(f/g)'$.

36 Use differentials to estimate $\ln (1 + h)$ when h is small.

Exercises 37 to 41 depend on material from Chap. 5. In Exercises 37 and 38 find the indicated antiderivatives.

37 (*a*) $\displaystyle\int \frac{5\,dx}{5x + 1}$ (*b*) $\displaystyle\int \frac{x\,dx}{x^2 + 5}$

(*c*) $\displaystyle\int \frac{\cos x\,dx}{\sin x}$ (*d*) $\displaystyle\int \frac{(1/x)\,dx}{\ln x}$

38 (*a*) $\displaystyle\int \frac{dx}{3x + 2}$ (*b*) $\displaystyle\int \frac{\sin x\,dx}{\cos x}$

(*c*) $\displaystyle\int \frac{(6x + 1)\,dx}{3x^2 + x + 5}$ (*d*) $\displaystyle\int \frac{\ln x\,dx}{x}$

39 Find the area under $y = (\ln x)/x$ and above $[e, e^2]$. [*Hint:* First find the derivative of $(\ln x)^2$.]

40 Find the area under $y = x/(1 + x^2)$ and above [2, 5].

41 Let n be an integer larger than 1.
(*a*) Show that the area of the shaded region in Fig. 4 is equal to

$$\frac{1}{1} + \frac{1}{2} + \frac{1}{3} + \cdots + \frac{1}{n - 1} - \ln n.$$

As $n \to \infty$, the area of the shaded region approaches a number, denoted γ (gamma), called *Euler's constant*. To three decimal places, $\gamma \approx 0.577$. It is not known whether γ is rational. Using geometric intuition (no calculus), show that γ is (*b*) less than 1, (*c*) greater than $\frac{1}{2}$.

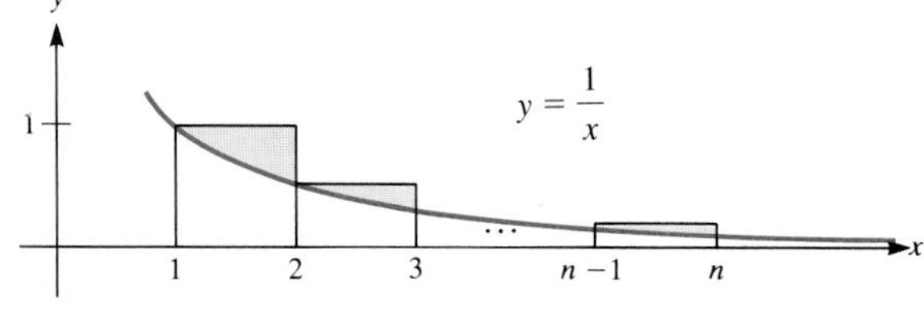

Figure 4

42 (*a*) Show that for all $n > 1$

$$\left(1 + \frac{1}{n}\right)^n < \frac{1}{0!} + \frac{1}{1!} + \frac{1}{2!} + \frac{1}{3!} + \cdots + \frac{1}{n!}.$$

(Recall that $0! = 1$.)
(*b*) Compute $\dfrac{1}{0!} + \dfrac{1}{1!} + \dfrac{1}{2!} + \cdots + \dfrac{1}{6!}$.

Exercises 43 to 45 depend on Chap. 5. Exercises 43 and 44 are a unit.

43 Using approximating sums and the definition of a definite integral, show that

(*a*) $\displaystyle\int_1^2 \frac{dx}{x} < 1$ (*b*) $\displaystyle\int_1^3 \frac{dx}{x} > 1.$

44 This exercise outlines an argument using areas that $\lim_{h \to 0} (1 + h)^{1/h}$ exists. For simplicity, consider only rational $h > 0$.
(*a*) Sketch the curves $y = 1/x$ and $y = 1/x^{1-h}$ relative to the same axes. Note that for $x > 1$ the second curve lies above the first curve.
(*b*) For a given number h (rational) find the number $A(h)$ such that

$$\int_1^{A(h)} \frac{dx}{x^{1-h}} = 1.$$

(*c*) Using Exercise 43, show that there is a number B such that

$$\int_1^B \frac{dx}{x} = 1.$$

Note that $2 < B < 3$.
(*d*) Using (*a*), (*b*), and (*c*), show that, for $h > 0$, $(1 + h)^{1/h} < B$.
(*e*) Why would you expect, on the basis of geometric intuition, that $\lim_{h \to 0} (1 + h)^{1/h} = B$? This number B is, of course, the number e. (A similar argument works for $h < 0$.)

45 (*a*) Find the area of the region under the curve $y = 1/x$ and above $[1, b]$.

(*b*) Using the result in (*a*), show that the area of the region under $y = 1/x$ and above $[1, \infty)$ is infinite.

(*c*) The region below $y = 1/x$ and above $[1, \infty)$ is revolved about the x axis. Find the volume of the resulting solid for $1 \leq x \leq b$.

(*d*) Show that the volume of the unbounded solid in (*c*) is finite.

(*e*) In view of (*b*), it is impossible to paint the region in (*b*) with a finite amount of paint. However, in view of (*d*), we could fill the unbounded solid of revolution with a finite amount of paint, then dip the region in (*b*) into the paint. That would paint the region in (*b*) with a finite amount of paint. What is wrong?

6.4 ONE-TO-ONE FUNCTIONS AND THEIR INVERSE FUNCTIONS

This section develops the notion of an "inverse function," which will be used later in this chapter to obtain the derivatives of b^x and the inverse trigonometric functions.

Most scientific calculators have a $\sqrt{x}$-key and an x^2-key. Each "undoes" the effect of the other. For instance, if you press

1 the 5-key
2 then the $\sqrt{x}$-key
3 then the x^2-key

the calculator will end up displaying 5.

If you reverse the order of the square root and squaring keys, a similar effect results. If you press

1 the 5-key
2 then the x^2-key
3 then the $\sqrt{x}$-key

Why doesn't this work for -5?

the final output, the composition of the two functions, should be 5. (Actually, due to the design of its hardware, the display may be a number very near 5.)

Taking a square root is the "inverse" of squaring a number. Let us examine this idea more generally.

One-to-One Functions

With some functions, "the output determines the input." For instance, the cubing function, $f(x) = x^3$, has this property. If we are told that the output of this function is, say, 64, then we know that the input must have been 4. The cubing function assigns *distinct* outputs to any pair of *distinct* inputs. However, the squaring function, $f(x) = x^2$, does not have this property. If we are told that the output of this function is, say, 25, then we do not know what the input is. It could be 5 or -5, since $5^2 = 25$ and $(-5)^2 = 25$. The squaring function assigns the same output, 25, to the distinct inputs, 5 and -5.

Definition A function f that assigns distinct outputs to distinct inputs is called a **one-to-one function**.

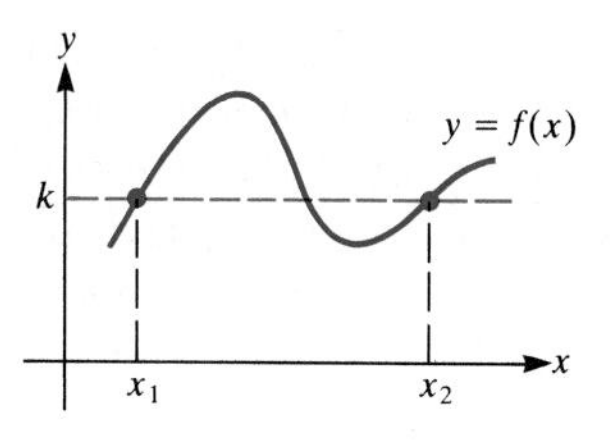

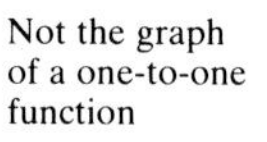
Not the graph of a one-to-one function

Figure 1

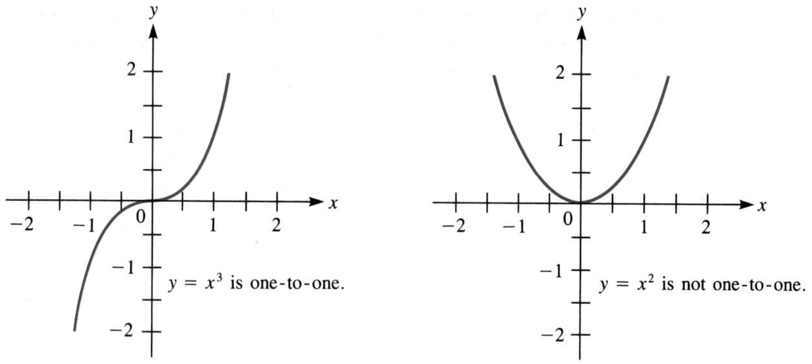

$y = x^3$ is one-to-one.

$y = x^2$ is not one-to-one.

Figure 2

Restricting f to $[a, b]$ produces a one-to-one function.

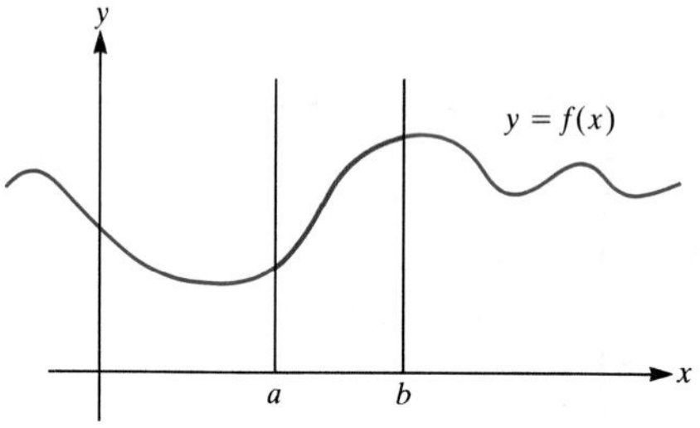

Figure 3

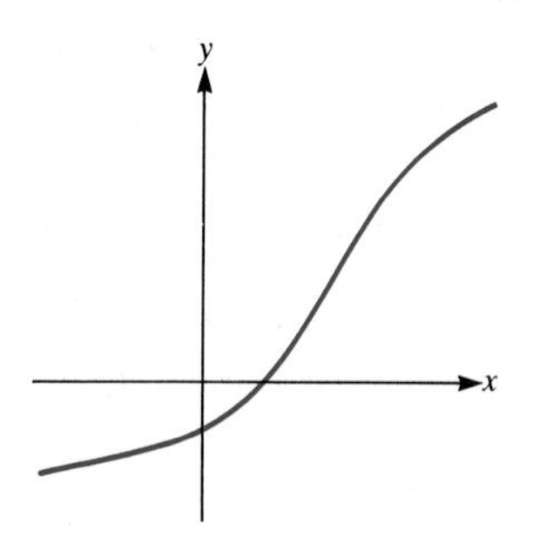

An increasing function (necessarily one-to-one)

Figure 4

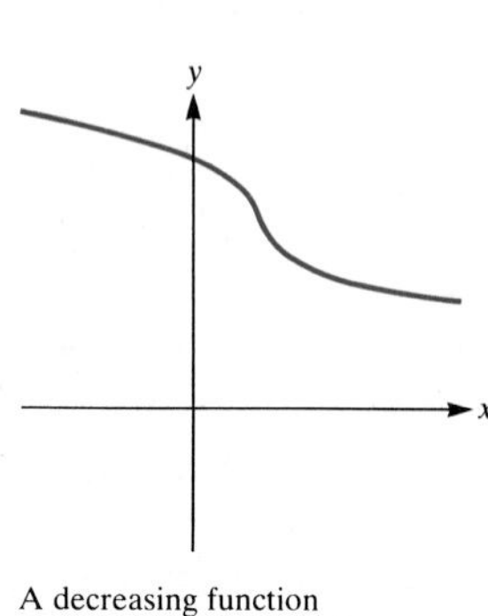

A decreasing function (necessarily one-to-one)

Figure 5

For instance, x^3 is a one-to-one function, but x^2 (with domain taken to be the entire x axis) is *not* one-to-one.

The function that assigns a Social Security number to each U.S. citizen is supposed to be one-to-one. If it were not, different people would have the same Social Security number, a circumstance that could have awkward consequences.

The graph of a one-to-one function has the property that *every horizontal line meets it in at most one point*. To see why, consider the line $y = k$ in Fig. 1. If it meets the graph of a function f in at least two distinct points, say (x_1, k) and (x_2, k), then $f(x_1) = k$ and $f(x_2) = k$. This means that f is not a one-to-one function, since the outputs corresponding to the inputs x_1 and x_2 are equal, namely, k.

On the other hand, if *each horizontal line meets the graph of a function f in at most one point, then f is one-to-one*. (Contrast the graphs of x^2 and x^3 in Fig. 2.)

Though the function x^2 is not one-to-one on the entire x axis, if we restrict its domain to nonnegative numbers, $[0, \infty)$, it is one-to-one. The squares of distinct *positive* numbers are distinct.

It is generally possible to restrict the domain of a function met in calculus to some interval so that the function, considered only on that interval, is one-to-one. (See Fig. 3.)

The simplest way to check whether a continuous function is one-to-one on an interval is to see whether the function is always increasing on that interval (or is always decreasing). (See Figs. 4 and 5.)

The Inverse of a One-to-One Function

Associated with a one-to-one function f is a second function g that records the unique input that yields a given output for the function f.

Definition Let $y = f(x)$ be a one-to-one function. The function g that assigns to each output y of f the corresponding unique input x is called the **inverse** of f. That is, if $y = f(x)$, then $x = g(y)$.

For example, $y = x^3$ is a one-to-one function. Its inverse is found by solving for x in terms of y; that is, $x = \sqrt[3]{y}$.

As the calculator example at the beginning of the section illustrates, x^2 and $\sqrt{x}$ are inverse functions, assuming that the domains are restricted to $x \geq 0$.

Inverse functions come in pairs, each reversing the effect of the other. The table lists some pairs of inversely related functions:

Function f	*Inverse Function g*
Cubing, $y = x^3$	Cube root, $x = \sqrt[3]{y}$
Cube root, $y = \sqrt[3]{x}$	Cubing, $x = y^3$
Squaring, $y = x^2$, $x \geq 0$	Square root, $x = \sqrt{y}$, $y \geq 0$
Square root, $y = \sqrt{x}$, $x \geq 0$	Squaring, $x = y^2$, $y \geq 0$

Inverse Functions on a Calculator

The inverse key might be labeled "2nd F" or have a special color and no label.

To minimize the number of keys, many calculators have a special "inverse" key, sometimes labeled "inv." The inv-key, in combination with some other function key, produces the inverse of that function (if the domain of the function is restricted to make the function one-to-one).

Warning: When you enter a number x in a calculator, then press the squaring key and then the square root key, you may not necessarily get exactly what you started with, x. This discrepancy is due to round-off errors in the calculations performed. Since calculators usually carry out computations to more digits than they display, such a discrepancy is rare.

Some Examples

EXAMPLE 1 Determine the inverse of the "doubling" function f defined by $f(x) = 2x$ and then graph it.

SOLUTION If $y = 2x$, there is only one value of x for each value of y, and it is obtained by solving the equation $y = 2x$ for x: $x = y/2$. Thus f is one-to-one and its inverse function g is the "halving" function: If y is the input in the function g, then the output is $y/2$.

For instance, $f(3) = 6$ and $g(6) = 3$. Thus $(3, 6)$ is on the graph of f, and $(6, 3)$ is on the graph of g. Since it is customary to reserve the x axis for inputs, we should write the formula for g, the "halving" function, as

$$g(x) = \frac{x}{2}.$$

Thus, f has the formula $y = 2x$, doubling, and its inverse, g, has the formula $y = x/2$, halving. The graphs of f and g are lines (see Fig. 6). ■

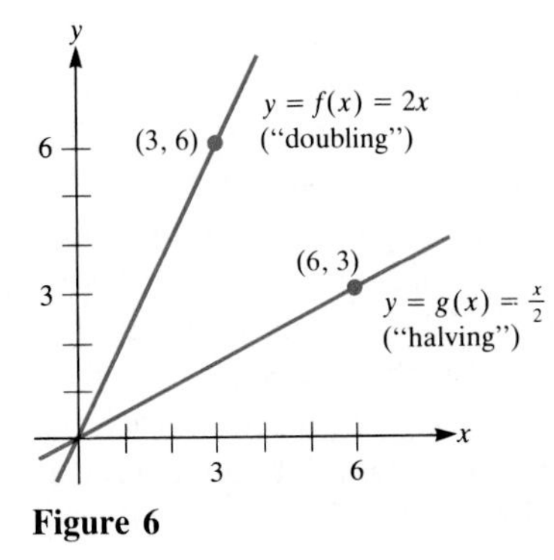

Figure 6

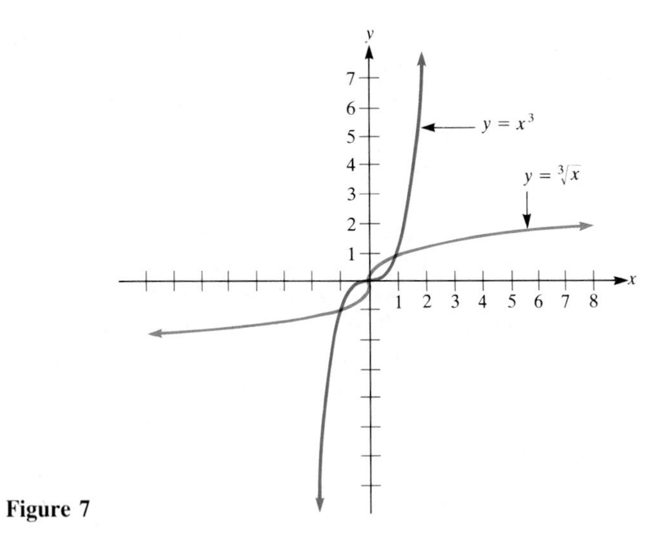

Figure 7

EXAMPLE 2 Graph the cubing function, $y = x^3$, and its inverse (the cube root function, $y = \sqrt[3]{x}$) on the same axes.

SOLUTION We first prepare brief tables:

x	-2	-1	0	1	2
x^3	-8	-1	0	1	8

x	-8	-1	0	1	8
$\sqrt[3]{x}$	-2	-1	0	1	2

These data suggest the general shape of the graphs, as sketched in Fig. 7. ■

The Graph of an Inverse Function

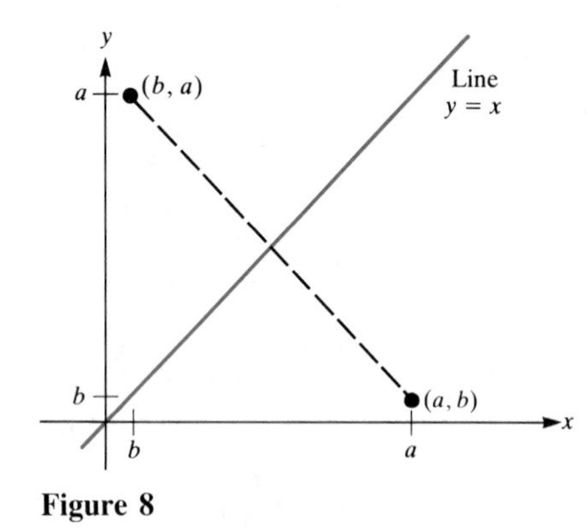

Figure 8

Note the relation between the two tables in Example 2. One is obtained from the other by switching inputs and outputs. This is the case for any one-to-one function and its inverse. If a and b are the respective entries in the input and output columns for one function, then b and a are the respective entries in the input and output columns for the inverse function. Also note the relation between the two graphs in Examples 1 and 2. One graph is obtained from the other by reflecting it across the line $y = x$. This can be done because, if (a, b) is on the graph of one function, then (b, a) is on the graph of the other. If you fold the paper along the line $y = x$, the point (b, a) comes together with the point (a, b), as you will note in Fig. 8. Note that the domain of f is equal to the range of its inverse function, and the range of f is equal to the domain of its inverse. This relation between the graphs holds for any one-to-one function and its inverse.

These examples are typical of the correspondence between a one-to-one function and its inverse. Perhaps the word "reverse" might be more descriptive than "inverse." One final matter of notation: We have used the letter g to denote the inverse of f. It is common to use the symbol f^{-1} (read as "f inverse") to denote the inverse function. We preferred to delay its use because its resemblance to the reciprocal notation might cause confusion. It should be clear from the examples that f^{-1} does *not* mean to divide 1 by f. The symbol inv f

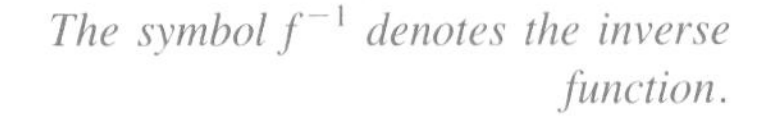
The symbol f^{-1} denotes the inverse function.

would be unambiguous. However, it is longer than the symbol f^{-1} and the weight of tradition is behind f^{-1}.

If f is one-to-one and $y = f(x)$, then we write $x = f^{-1}(y)$.

Section Summary

We defined a one-to-one function. A function that is increasing (or is decreasing) on its domain is one-to-one. A one-to-one function f has an inverse function, f^{-1}. (If f assigns y to x, then f^{-1} assigns x to y.) The graph of f^{-1} is obtained from the graph of f by reflecting it across the line $y = x$.

EXERCISES FOR SEC. 6.4: ONE-TO-ONE FUNCTIONS AND THEIR INVERSES

In Exercises 1 to 8 determine whether the given function is one-to-one on the given domain. If it is, obtain a formula for its inverse.

1 $y = x^4$ (*a*) $[-1, 1]$ (*b*) $[0, 2]$
2 $y = (x - 1)^2$ (*a*) $[0, 2]$ (*b*) $[1, 3]$
3 $y = 1 + x^5$ (*a*) $[0, 1]$ (*b*) $[-100, 100]$
4 $y = (x + 1)^5$ (*a*) $[0, 1]$ (*b*) $[-6, 6]$
5 $y = \sqrt[5]{1 + x^3}$ (*a*) $(-\infty, \infty)$ (*b*) $[0, \infty)$
6 $y = \sqrt[4]{1 + x^2}$ (*a*) $(-\infty\ \infty)$ (*b*) $[0, \infty)$
7 $y = x^{5/3}$ (*a*) $(-\infty, \infty)$ (*b*) $[0, \infty)$
8 $y = x^{2/3}$ (*a*) $(-\infty, \infty)$ (*b*) $[0, \infty)$

In Exercises 9 and 10 decide whether the given graph is the graph of a one-to-one function. If it is, sketch the graph of the inverse function.

9 **10**

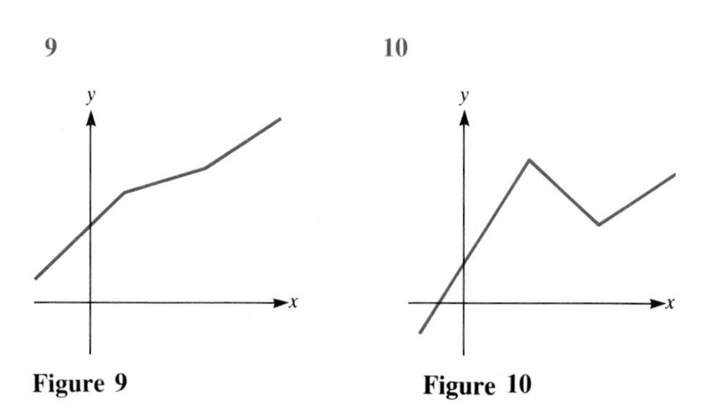

Figure 9 **Figure 10**

11 This table records some of the data about a one-to-one function f:

x	-1	0	1	2
$f(x)$	3	5	6	7

Use the data to plot four points on the graph of (*a*) f and (*b*) g, the inverse of f.

12 This table records some of the data about a one-to-one function f:

x	-5	2	π
$f(x)$	2	1	0

Use the data to plot three points on the graph of (*a*) f and (*b*) g, the inverse of f.

13 What is the inverse of the function $y = 2^x$?

14 What is the inverse of the function $y = \log_2 x$?

---

15 (*a*) Is $y = \sin x$ one-to-one on $[0, \pi/2]$? on $[0, \pi]$?
(*b*) Graph $y = \sin x$ on $[0, \pi/2]$ and its inverse function relative to the same axes.

16 (*a*) Is $y = \cos x$ one-to-one on $[0, \pi/2]$? on $[0, \pi]$?
(*b*) Graph $y = \cos x$ where it is one-to-one and its inverse function.

17 (*a*) Show that $y = \tan x$ is one-to-one on $(-\pi/2, \pi/2)$.
(*b*) Graph $y = \tan x$ for x in $(-\pi/2, \pi/2)$ and its inverse function.

18 (*a*) Show that $y = \ln x$ is one-to-one on $(0, \infty)$.
(*b*) Graph $y = \ln x$ and its inverse function.

19 (*a*) Is the function $y = x^3 + 2x$ one-to-one?
(*b*) Is the function $y = x^4 + 3x$ one-to-one? Explain your answers.

20 (*a*) For which choices of the constants a and b is the function $f(x) = ax + b$ one-to-one?
(*b*) In case a function in (*a*) is one-to-one, give the formula for its inverse.

21 (*a*) For what values of the constant k is the cubic function $x^3 + kx^2 + x$ one-to-one?
(*b*) For what values of k is the function $x^3 + kx^2 - x$ one-to-one?

In Exercises 22 to 24, f and g are functions and the range of g is included in the domain of f. Recalling the definition of a composite function, $(f \circ g)(x) = f(g(x))$ (see Sec. 2.2), what can be said about $f \circ g$ if:

22 f and g are increasing functions? Explain.

23 f and g are decreasing functions? Explain.

24 f is increasing and g is decreasing? Explain.

25 (*a*) For which positive integers n is the polynomial function $y = x^n + 6x^3$ a one-to-one function?

(*b*) For which n that make $x^n + 6x^3$ one-to-one can you solve easily for x as a function of y?

6.5 THE DERIVATIVE OF b^x

In this section we obtain the derivatives of the exponential function b^x, $b > 0$. To obtain these derivatives, we will make use of the fact that the exponential function b^x is the inverse of the logarithmic function $\log_b x$.

b^x Is the Inverse of $\log_b x$

If $y = b^x$, then $x = \log_b y$. This means that the exponential function $y = b^x$ and the logarithm function $x = \log_b y$ for $b \neq 1$ are inverses of each other.

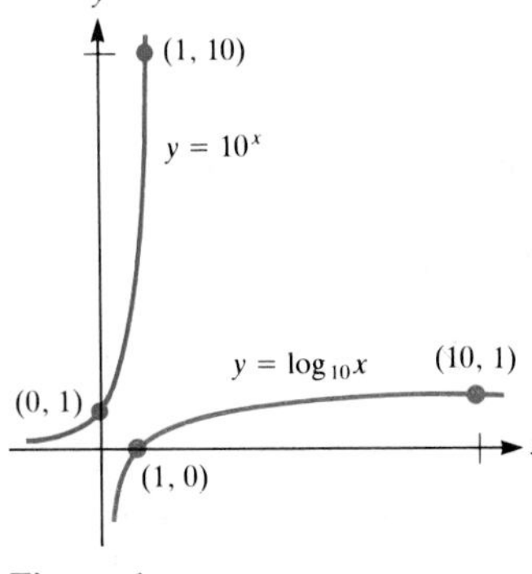

Figure 1

EXAMPLE 1 Graph $y = 10^x$ and its inverse function $y = \log_{10} x$ on the same axes.

SOLUTION In order to graph the logarithm and exponential functions, it is advisable to prepare brief tables first:

x	10	1	0.1
$\log_{10} x$	1	0	−1

x	1	0	−1
10^x	10	1	0.1

The functions are graphed with the aid of the tables, as in Fig. 1. ■

The Derivative of an Inverse Function

We will obtain the derivative of b^x by exploiting the derivative of its inverse function, $\log_b x$. For this reason, let us see how the derivative of a function f is related to the derivative of its inverse function g.

Assume that the function f is differentiable on the interval $[a, b]$ and that its derivative is always positive. Then f is an increasing function, hence one-to-one. Its graph is indicated in Fig. 2.

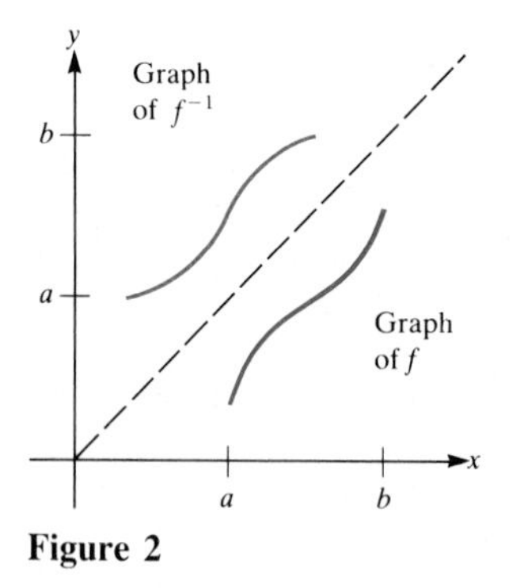

Figure 2

The graph of its inverse function g is the mirror image of that for f in the line $y = x$. Since f is differentiable, its graph locally resembles a line. Since the graph of g is a copy of that of f, we would expect g to be a differentiable function as well. This expectation is formally justified in an advanced calculus course, but we will assume it without proof in the following theorem.

Theorem 1 Let f be a differentiable function with domain $[a, b]$ and range $[c, d]$. Assume that $f'(x)$ is positive for all x in $[a, b]$. Then f is a one-to-one function and its inverse function $g = f^{-1}$ from $[c, d]$ to $[a, b]$ is differentiable. The same conclusion holds if $f'(x)$ is negative for all x in $[a, b]$. ■

We are now in a position to find the derivative of b^x for any fixed base $b > 0$, $b \neq 1$.

The Derivative of b^x

Theorem 2 *The derivative of b^x, $b > 0$, $b \neq 1$.* Let $y = b^x$ be the exponential function with base b. Then

$$\frac{d}{dx}(b^x) = (\ln b)\, b^x.$$

Proof Let $y = b^x$. Since b^x is the inverse of the differentiable function $\log_b x$, we know it has a derivative, dy/dx. To find y', we use logarithmic differentiation:

$$\begin{aligned} y &= b^x && \\ \ln y &= \ln b^x && \text{ln of both sides} \\ \ln y &= x \ln b && \text{ln of a power} \\ \frac{1}{y}\frac{dy}{dx} &= \ln b && \text{derivative with respect to } x \\ \frac{dy}{dx} &= (\ln b)\, y && \text{solving for } \frac{dy}{dx} \\ \frac{dy}{dx} &= (\ln b)\, b^x. && y = b^x \quad \blacksquare \end{aligned}$$

For instance, $D(10^x) = (\ln 10)10^x \approx (2.303)10^x$ and $D(2^x) = (\ln 2)2^x \approx (0.693)2^x$. The derivative of b^x is equal to a constant times b^x, but that constant can be an unpleasant decimal. That coefficient is $\ln b = \log_e b$. If we choose the base b to be e, the coefficient is $\log_e e$, which is 1. This gives us the remarkably simple formula, $D(e^x) = e^x$—the derivative of e^x is e^x itself! A special case then of Theorem 2 is the following result:

Theorem 3 *The derivative of e^x.* Let $y = e^x$ be the exponential function with base e. Then

$$\boxed{\frac{d}{dx}(e^x) = e^x.}$$

Why e is used as a base for exponentials

Because e^x has such a simple derivative, e is the preferred base for the exponential function in calculus. If you ever have to differentiate 10^x or 2^x or some other exponential, just use logarithmic differentiation, as in the proof of Theorem 2.

EXAMPLE 2 Find the derivative of e^{3x}.

SOLUTION Let $y = e^{3x}$. Then $y = e^u$, where $u = 3x$.

Thus

$$\frac{dy}{dx} = \frac{dy}{du}\frac{du}{dx} \quad \text{chain rule}$$

$$= \frac{d(e^u)}{du}\frac{d(3x)}{dx}$$

$$= e^u \cdot 3 \quad \text{Theorem 3}$$

$$= e^{3x} \cdot 3$$

$$= 3e^{3x}. \quad ■$$

EXAMPLE 3 Differentiate $f(x) = \frac{e^{ax}}{a^3}(a^2x^2 - 2ax + 2)$, where a is a nonzero constant.

SOLUTION First bring a^3, which is constant, to the front of the formula by writing the function as

$$f(x) = \frac{1}{a^3}e^{ax}(a^2x^2 - 2ax + 2).$$

Now differentiate:

$$f'(x) = \frac{1}{a^3}[e^{ax}(a^2x^2 - 2ax + 2)]' \quad \text{derivative of constant times a function}$$

$$= \frac{1}{a^3}[e^{ax}(a^2 \cdot 2x - 2a) + (a^2x^2 - 2ax + 2)ae^{ax}]$$

$$= \frac{1}{a^3}[(2a^2x - 2a + a^3x^2 - 2a^2x + 2a)e^{ax}]$$

$$= x^2e^{ax}. \quad \text{after canceling} \quad ■$$

In Sec. 3.2, we stated that the derivative of x^a is ax^{a-1} for any real number a, but proved it only for positive integers, -1, and $\frac{1}{2}$. We now justify the general claim.

Theorem 4 Let a be a fixed real number. Then for $x > 0$,

$$\frac{d(x^a)}{dx} = ax^{a-1}.$$

Proof Let $y = x^a$. Since $x = e^{\ln x}$,

$$y = (e^{\ln x})^a.$$

Hence, by the "power-of-a-power" rule,

$$y = e^{a \ln x}.$$

This can be written as $y = e^u$, where $u = a \ln x$.

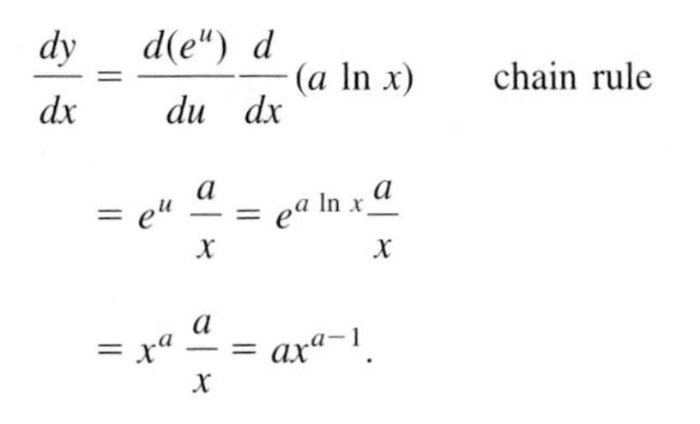

Hence
$$\frac{dy}{dx} = \frac{d(e^u)}{du}\frac{d}{dx}(a \ln x) \qquad \text{chain rule}$$
$$= e^u \frac{a}{x} = e^{a \ln x}\frac{a}{x}$$
$$= x^a \frac{a}{x} = ax^{a-1}.$$

This proves the theorem. ■

For example, $d(x^\pi)/dx = \pi x^{\pi-1}$ and $(x^{\sqrt{2}})' = \sqrt{2}(x^{\sqrt{2}-1})$.

Note that in Theorem 4 the base x is positive. The theorem also holds when $x = 0$ and $a > 1$ and when x is negative and a is a rational number of the form m/n, where m and n are integers and n is odd. (See Exercise 50.) For instance, $D(x^{1/3})$ when $x = -8$ is $(\frac{1}{3})(-8)^{-2/3} = (\frac{1}{3})(-2)^{-2} = (\frac{1}{3})(\frac{1}{4}) = \frac{1}{12}$.

Exponential Functions Grow Rapidly

In Sec. 6.1, it was mentioned that a logarithmic function $\log_b x$, $b > 1$, gets arbitrarily large as $x \to \infty$, but much more slowly than x^a does for any exponent $a > 0$. An exponential function, b^x, $b > 1$, on the other hand, grows much more rapidly than x^a does for any exponent $a > 0$.

The following table compares x^3 and 1.1^x for a few values of x.

Watch 1.1^x catch up.

x	1	10	100	1,000
x^3	1	1,000	1,000,000	10^9
1.1^x	1.1	2.5937	13,780.6	2.4699×10^{41}

The data suggest that $\lim_{x\to\infty} x^3/1.1^x = 0$. In fact, for $a > 0$ and $b > 1$, we have

$$\boxed{\lim_{x\to\infty} \frac{x^a}{b^x} = 0.}$$

(This can be proved with the aid of the binomial theorem. See Exercises 48 and 49.) In particular $\lim_{x\to\infty} x/e^x = 0$.

The next example shows the behavior of x/e^x throughout its domain, which is the entire x axis.

EXAMPLE 4 Graph $f(x) = x/e^x = xe^{-x}$.

SOLUTION

y intercept The y intercept is $f(0) = 0/e^0 = 0$.

x intercepts To find the x intercepts, solve the equation $f(x) = 0$:

$$\frac{x}{e^x} = 0.$$

Thus, $x = 0$ is the only x intercept.

Asymptotes

Since $\lim_{x\to\infty} x/e^x = 0$, the x axis is a horizontal asymptote (for x positive and large). To learn about the behavior of x/e^x for large negative values of x, we examine

$$\lim_{x\to-\infty} \frac{x}{e^x}.$$

When x is a large and negative number, e^x is a small positive number. (Think of $e^{-10} = 1/e^{10}$.) Thus, $\lim_{x\to-\infty} e^x = 0$ and $\lim_{x\to-\infty} x = -\infty$. Consequently,

$$\lim_{x\to-\infty} \frac{x}{e^x} = -\infty.$$

The graph for x negative has no asymptote.

We next examine df/dx:

$$\begin{aligned}\frac{df}{dx} &= \frac{d}{dx}\left(\frac{x}{e^x}\right)\\ &= \frac{e^x - xe^x}{(e^x)^2}\\ &= \frac{1-x}{e^x}.\end{aligned}$$

Critical point

There is a critical point when $x = 1$. When $x < 1$, df/dx is positive, and when $x > 1$, df/dx is negative. Hence a global maximum occurs when $x = 1$.

Next, consider d^2f/dx^2:

$$\begin{aligned}\frac{d^2f}{dx^2} &= \frac{d}{dx}\left(\frac{df}{dx}\right)\\ &= \frac{d}{dx}\left(\frac{1-x}{e^x}\right)\\ &= \frac{e^x(-1) - (1-x)e^x}{(e^x)^2}\\ &= \frac{-e^x - e^x + xe^x}{(e^x)^2}\\ &= \frac{(x-2)e^x}{(e^x)^2}\\ &= \frac{x-2}{e^x}.\end{aligned}$$

The second derivative is 0 at $x = 2$ and changes sign at 2; hence, there is an inflection point there. The graph is concave downward for $x < 2$ and concave upward for $x > 2$.

At the critical number, 1, $f(1) = 1/e^1 = 1/e$. At the inflection number 2, $f(2) = 2/e^2$. The critical point $(1, 1/e)$ and inflection point $(2, 2/e^2)$ are shown in Fig. 3.

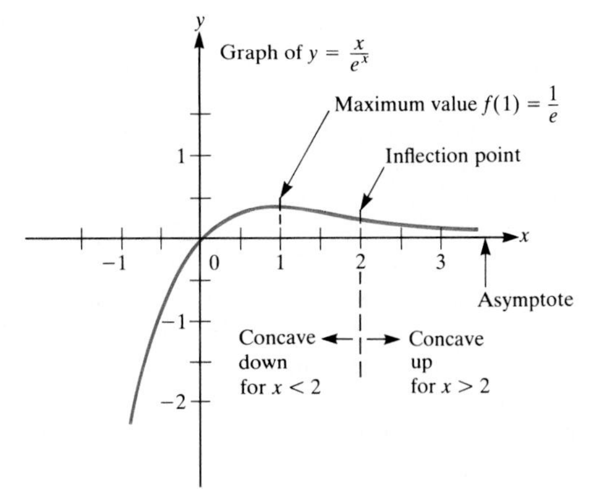

Figure 3

Section Summary

We found that $D(b^x) = (\ln b)b^x$. In particular, $D(e^x) = e^x$. Because of the simplicity of this formula, we will usually use e as the base. By writing x^a as $e^{a \ln x}$, we showed that $D(x^a) = ax^{a-1}$. Finally, we observed that for $b > 1$, b^x grows much faster than any fixed power of x, x^a, $a > 0$. That is, $\lim_{x\to\infty} x^a/b^x = 0$.

EXERCISES FOR SEC. 6.5: THE DERIVATIVE OF b^x

In Exercises 1 to 16 differentiate the given functions. *Hint:* Use logarithmic differentiation in Exercises 7 to 9.

1 e^{x^2}
2 xe^{-4x}
3 x^2e^{2x}
4 $(\sin 2x)(\sin e^{-x})$
5 2^{-x^2}
6 $3^{\sqrt{x}}$
7 $x^{(x^2)}$
8 $(2 + \cos x)^{\sin x}$
9 $x^{\tan 3x}$
10 $(\tan \sqrt{x})e^{-x}$
11 $\dfrac{e^{-4x}}{1 + e^x}$
12 $\dfrac{10^{x^2}}{\ln (1 + x^2)}$
13 $x^{\sqrt{3}} (\sin 3x)e^{x^2}$
14 $\dfrac{x^\pi \tan e^x}{2^x}$
15 $\ln (x + \sqrt{1 + e^{3x}})$
16 $\ln e^{\cos 3\theta}$

In Exercises 17 to 20 differentiate and simplify. (The numbers a, b, and c are positive constants.)

17 $\dfrac{e^{ax}(ax - 1)}{a^2}$

18 $\dfrac{xb^{ax}}{a \ln b} - \dfrac{b^{ax}}{a^2 (\ln b)^2}$

19 $\dfrac{e^{ax}}{a^2 + b^2}(a \sin bx - b \cos bx)$

20 $\dfrac{1}{ac} \ln (b + ce^{ax})$

In Exercises 21 to 28 use a differential to estimate the given quantities. Assume x is small. You may also assume $\ln 10 \approx 2.30$, $e \approx 2.72$, and $\log_{10} e \approx 0.43$.

21 e^x
22 $e^{1.1}$
23 10^x
24 $10^{1.1}$
25 $\ln (1 + x)$
26 $\ln 1.1$
27 $\log_{10} (1 + x)$
28 $\log_{10} 0.98$

In each of Exercises 29 to 34 find (*a*) intercepts, (*b*) critical points, (*c*) local maxima or minima, (*d*) inflection points, and (*e*) asymptotes. (*f*) Graph the function.

29 $f(x) = (1 + x)e^{-x}$
30 $f(x) = x^2e^{-x}$
31 $f(x) = x^3e^{-x}$
32 $f(x) = x\ 2^{-x}$
33 $f(x) = (x - x^2)e^{-x}$
34 $f(x) = xe^x$

In each of Exercises 35 to 38, a region in the plane is described. Find its area. (These exercises depend on material in Chap. 5.)

35 Under e^{3x}, above $[1, 5]$.
36 Under $5e^{-2x}$, above $[0, \ln 2]$.
37 Under 10^x, above $[0, 3]$.
38 Under 2^{-x}, above $[-4, 1]$.
39 Using results of this section and the definition of the derivative, evaluate:

(*a*) $\lim_{h\to 0} \dfrac{e^h - 1}{h}$ (*b*) $\lim_{x\to 1} \dfrac{2^x - 2}{x - 1}$ (*c*) $\lim_{h\to 0} \dfrac{10^h - 1}{h}$

40 Let $f(x) = 5 + (x - x^2)e^x$
(*a*) Show that $f(0) = f(1)$.
(*b*) Find all numbers c in $(0, 1)$ such that $f'(c) = 0$. Rolle's theorem asserts that such a number can be found.

41 (*a*) Graph $y = e^x$ and $y = \tan x$ relative to the same axes.
(*b*) Show that the equation $e^x - \tan x = 0$ has a solution between 0 and $\pi/2$.
(*c*) Choose x_1, an estimate of the solution in (*b*), on the basis of the graph in (*a*). Then determine x_2 by Newton's method.

42 (*a*) Show that the equation $3x + \sin x - e^x = 0$ has a root between 0 and 1.
(*b*) Starting with $x_1 = 0.5$, compute x_2 and x_3, the estimates of the root in (*a*) by Newton's method.

43 (*a*) Graph $y = e^x$ and $y = x + 2$ relative to the same axes.
(*b*) With the aid of (*a*), estimate roots of the equation $e^x - x - 2 = 0$.
(*c*) Use Newton's method and a calculator to estimate the roots to four-decimal accuracy.

44 Let A and k be constants. Show that the derivative of Ae^{kx} is proportional to Ae^{kx}.

45 (*a*) Graph $y = e^x$ and $y = -x$ on the same axes.
(*b*) Using the graphs in (*a*), graph $y = e^x - x$.
(*c*) Find all points on the graph in (*b*) where the tangent line is horizontal.

46 The formula $y = e^{-t} \sin t$ describes a decaying alternating current. Consider $t \geq 0$.
(*a*) For which values of t is $e^{-t} \sin t = 0$? Draw the points where the graph of $y = e^{-t} \sin t$ meets the t axis.
(*b*) Draw the curves $y = e^{-t}$ and $y = -e^{-t}$. Why does the graph of $y = e^{-t} \sin t$ lie between these two graphs?
(*c*) For which t does $e^{-t} \sin t$ equal e^{-t}? What points does this information give on the graph of $y = e^{-t} \sin t$?
(*d*) When does $e^{-t} \sin t$ equal $-e^{-t}$? Use this information to plot more points on the graph of $y = e^{-t} \sin t$.
(*e*) Find all points on the graph of $y = e^{-t} \sin t$ where the tangent line is horizontal and $0 \leq t \leq 4\pi$.

47 (Contributed by Daniel Drucker.) In planning a dam, an engineer estimates that for a dam of height h meters, the cost will be $ae^{h/20}$ dollars, the volume of the lake created bh^3 cubic meters, and the surface area of the lake ch^2 square meters, where a, b, and c are constants.
(*a*) What height maximizes the volume-to-cost ratio?
(*b*) What height maximizes the area-to-cost ratio?

48 This exercise shows that if $b > 1$, then $\lim_{x\to\infty} x/b^x = 0$. (See App. C for the binomial theorem.)
(*a*) Show that for x sufficiently large, x/b^x is a decreasing function.
(*b*) Write $b = 1 + c$, $c > 0$. Using the binomial theorem for $n > 2$, show that

$$b^n > 1 + nc + \frac{n(n-1)}{2}c^2 \qquad \text{if } n > 2.$$

(*c*) From (*b*) deduce that

$$\lim_{n\to\infty} \frac{n}{b^n} = 0$$

(*d*) From (*a*) and (*c*) deduce that $\lim_{x\to\infty} x/b^x = 0$.

49 (This continues Exercise 48.) Show that if $n > 3$ and $b > 1$, then

$$b^n > 1 + nc + \frac{n(n-1)}{2}c^2 + \frac{n(n-1)(n-2)}{3!}c^3.$$

Then, modeling your argument on Exercise 48, show that $\lim_{x\to\infty} x^2/b^x = 0$.

A similar argument shows that for any $b > 1$ and any positive a, $\lim_{x\to\infty} x^a/b^x = 0$. (The case $a \leq 0$ is trivial. Why?)

50 Theorem 4 asserts that $D(x^a) = ax^{a-1}$ for positive x. This exercise treats negative x.
(*a*) To show that $(x^a)' = ax^{a-1}$ when $x < 0$ and $a = m/n$, m an integer, n an odd integer, argue as follows. We take the case when m is odd. Write $x = -t$, so $t > 0$. Then

$$\frac{d}{dx}(x^{m/n}) = \frac{d}{dt}[(-t)^{m/n}]\frac{dt}{dx},$$

by the chain rule. Complete the argument.
(*b*) Treat the case $a = m/n$, where m is even and n is odd.

6.6 THE DERIVATIVES OF THE INVERSE TRIGONOMETRIC FUNCTIONS

We now define the six inverse trigonometric functions and obtain their derivatives. These functions are important in calculus, since they provide antiderivatives of such common functions as $1/(1 + x^2)$ and $1/\sqrt{1 - x^2}$. (This section does not depend on Chap. 5. Only Exercises 58 to 64 refer to Chap. 5 and they are so identified.)

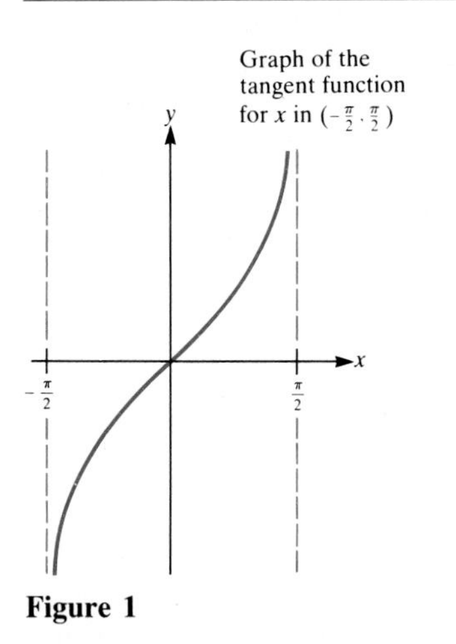

Figure 1

The Inverse Tangent

Consider the function $f(x) = \tan x$ in the open interval $-\pi/2 < x < \pi/2$. (See Fig. 1.)

As x increases in this interval, $\tan x$ increases. Thus the function $\tan x$ is one-to-one if the domain is restricted to be between $-\pi/2$ and $\pi/2$. Note that as $x \to \pi/2$ or $x \to -\pi/2$, $\tan x$ gets very large in absolute value.

The graph of the inverse function f^{-1} is obtained by reflecting the graph of $y = \tan x$ across the line $y = x$. As x gets large, note that $f^{-1}(x) \to \pi/2$, as shown in Fig. 2.

The inverse of the tangent function is called the **arctangent function** and is written $\arctan x$ or $\tan^{-1} x$. [This is not the reciprocal of $\tan x$, which is written $\cot x$, $1/\tan x$, or $(\tan x)^{-1}$ to avoid confusion.] As an example, since the tangent of $\pi/4$ is 1, it follows that $\pi/4$ is the angle whose tangent is 1, that is,

$$\arctan 1 = \frac{\pi}{4} \qquad \text{or} \qquad \tan^{-1} 1 = \frac{\pi}{4}.$$

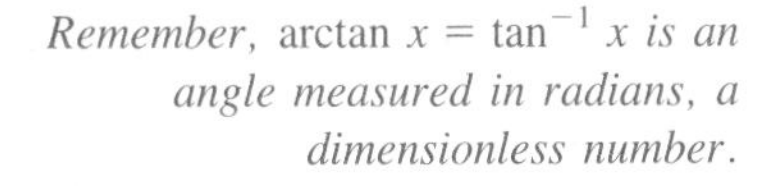

Remember, $\arctan x = \tan^{-1} x$ *is an angle measured in radians, a dimensionless number.*

Observe that the *domain of the arctan function* is the entire x axis and that when $|x|$ is large, $\arctan x$ is near $\pi/2$ or $-\pi/2$. The *range of the arctan function* is the open interval $(-\pi/2, \pi/2)$.

It is frequently useful to picture the tangent and arctangent functions in terms of the unit circle, as shown in Figs. 3 and 4.

The inverse tangent function is built into scientific calculators. Some have a key labeled "$\tan^{-1}$" while others require you to press a key labeled "inv" or "arc" before pressing "tan." To get a feel for this function, calculate $\tan^{-1} 1$, $\tan^{-1} 10$, and $\tan^{-1} 100$, remembering to set the calculator for radians (not degrees). See the graph of $y = \tan^{-1} x$ in Fig. 2.

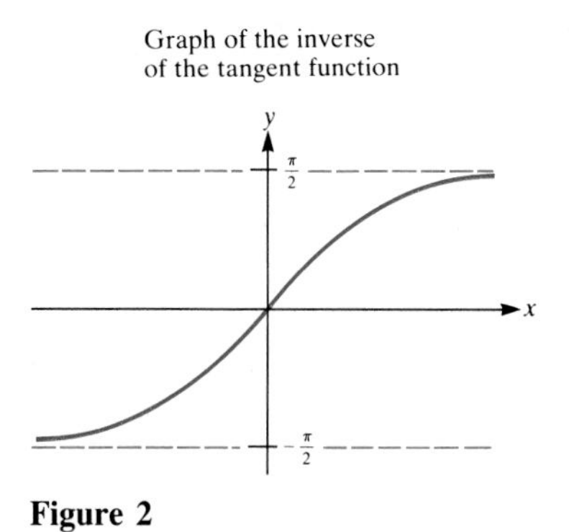

Figure 2

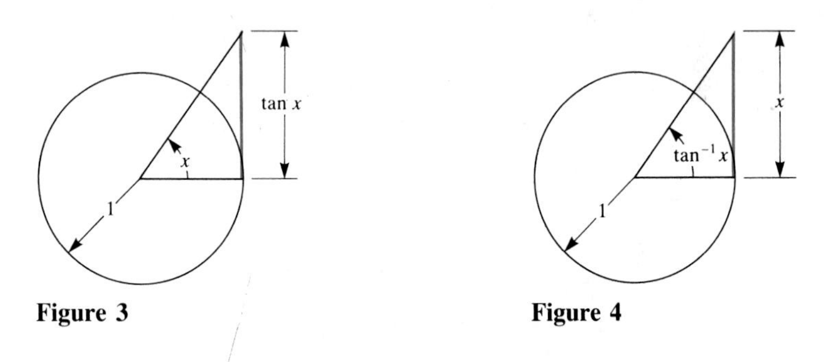

Figure 3 **Figure 4**

The Derivative of the Inverse Tangent

Theorem 1 obtains the derivative of $\tan^{-1} x$. The proof uses the identity $\sec^2 \theta = 1 + \tan^2 \theta$. (To obtain this identity, divide both sides of $\cos^2 \theta + \sin^2 \theta = 1$ by $\cos^2 \theta$.)

Theorem 1 $$\frac{d}{dx}(\tan^{-1} x) = \frac{1}{1 + x^2}$$

Proof Let $y = \tan^{-1} x$. Since $\tan^{-1} x$ is the inverse of a differentiable function, we know that its derivative dy/dx exists. To find dy/dx, we use the fact that $x = \tan y$. Then

$$\frac{d(x)}{dx} = \frac{d}{dx}(\tan y)$$

$$= \frac{d}{dy}(\tan y)\frac{dy}{dx}; \qquad \text{chain rule}$$

hence

$$1 = \sec^2 y \frac{dy}{dx}.$$

Thus we have

$$\frac{dy}{dx} = \frac{1}{\sec^2 y}.$$

To express dy/dx in terms of x, we use the identity $\sec^2 y = 1 + \tan^2 y$, hence $\sec^2 y = 1 + x^2$. Thus

$$\frac{dy}{dx} = \frac{1}{1 + x^2}. \quad ■$$

EXAMPLE 1 Find the derivative of $\tan^{-1} \sqrt{x}$.

SOLUTION Theorem 1 and the chain rule are needed. Let $y = \tan^{-1} \sqrt{x}$. Then $y = \tan^{-1} u$, where $u = \sqrt{x}$. Thus

$$\frac{dy}{dx} = \frac{dy}{du} \cdot \frac{du}{dx} \qquad \text{chain rule}$$

$$= \frac{d}{du}(\tan^{-1} u)\frac{d}{dx}(\sqrt{x})$$

$$= \frac{1}{1 + u^2} \cdot \frac{1}{2\sqrt{x}}$$

$$= \frac{1}{1 + x} \cdot \frac{1}{2\sqrt{x}}$$

$$= \frac{1}{2\sqrt{x}(1 + x)}. \quad ■$$

The Inverse Sine

We turn next to the inverse of the sine function.

The sine function is not one-to-one. (See its graph in Fig. 5. For instance, $\sin 0 = \sin \pi$.) However, if we restrict its domain to $[-\pi/2, \pi/2]$ it is one-to-one. See Fig. 6.

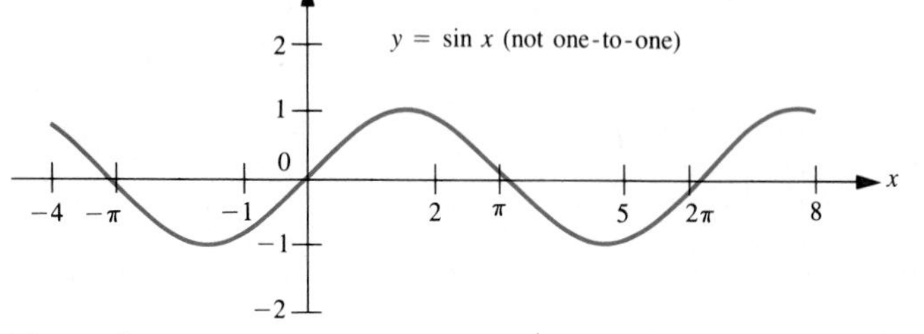

Figure 5

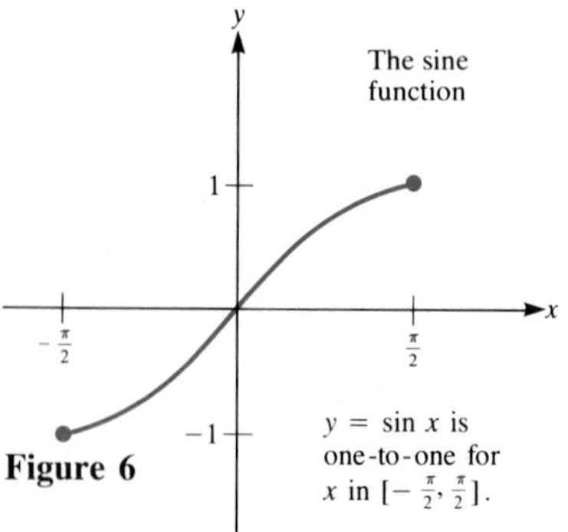

Figure 6

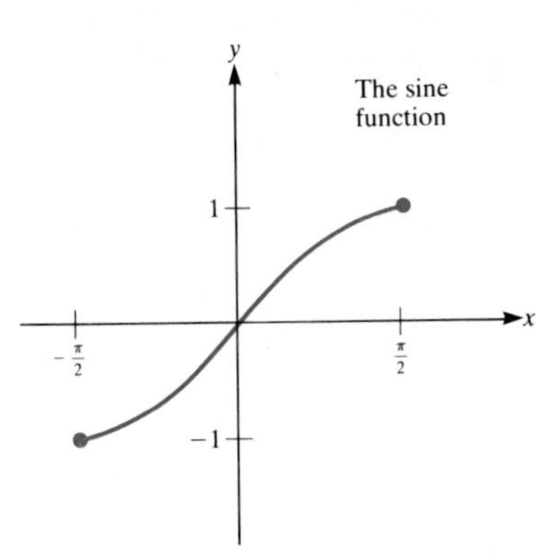

Figure 7

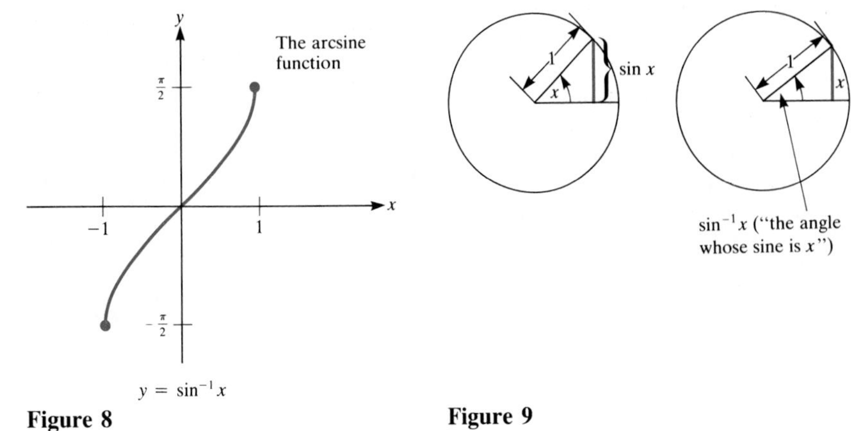

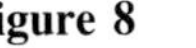

Figure 8

Figure 9

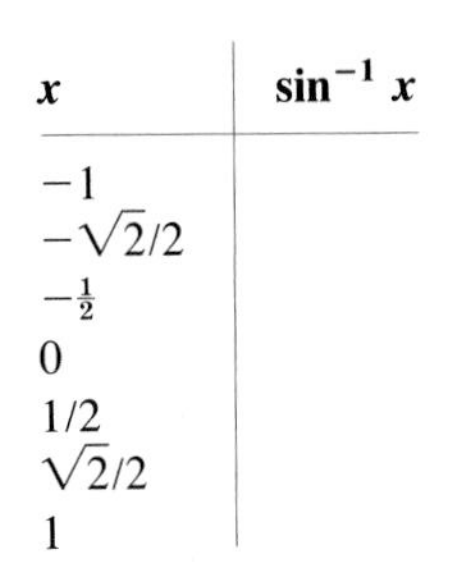

x	$\sin^{-1} x$
-1	
$-\sqrt{2}/2$	
$-\frac{1}{2}$	
0	
$1/2$	
$\sqrt{2}/2$	
1	

As x goes from $-\pi/2$ to $\pi/2$, $\sin x$ increases from -1 to 1. The inverse of the sine function (now restricted to $[-\pi/2, \pi/2]$) is called the *arcsine* function and is written $\arcsin x$ or $\sin^{-1} x$.

For instance, since $\sin(\pi/2) = 1$, it follows that $\arcsin 1 = \pi/2$. Since $\sin(\pi/6) = \frac{1}{2}$, it follows that $\sin^{-1}(\frac{1}{2}) = \pi/6$. Similarly, since $\sin(-\pi/2) = -1$, $\sin^{-1}(-1) = -\pi/2$. (In each case think of "$\sin^{-1} x$" as "the angle whose sine is x.")

The graph of $y = \sin^{-1} x$ can be obtained by flipping the graph in Fig. 7 around the line $y = x$. (See Fig. 8.) You could also get its graph by filling in the table in the margin and plotting the seven points.

It is also useful to visualize these two functions in terms of the unit circle: they are depicted in Fig. 9. Note that

$$-\frac{\pi}{2} \leq \arcsin x \leq \frac{\pi}{2}.$$

Note also that the domain of the arcsine function is $[-1, 1]$ and its range is $[-\pi/2, \pi/2]$.

The Derivative of the Inverse Sine

The proof of the next theorem is similar to that of Theorem 1.

Theorem 2 $$\frac{d}{dx}(\sin^{-1} x) = \frac{1}{\sqrt{1 - x^2}}$$

Proof Let $y = \sin^{-1} x$; hence $x = \sin y$. Thus

$$\frac{dx}{dx} = \frac{d}{dx}(\sin y) \qquad \text{implicit differentiation}$$

$$= \frac{d}{dy}(\sin y)\frac{dy}{dx} \qquad \text{chain rule}$$

or

$$1 = \cos y \frac{dy}{dx}.$$

Hence

$$\frac{dy}{dx} = \frac{1}{\cos y}.$$

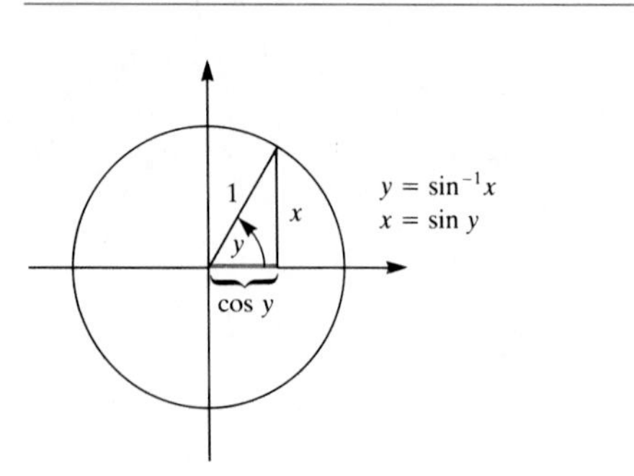

Figure 10

To express $\cos y$ in terms of x, draw the unit circle and indicate on it that $x = \sin y$. Inspection of Fig. 10 shows that $\cos y$ is *positive* (since $-\pi/2 \le y \le \pi/2$) and by the pythagorean theorem,

$$\cos^2 y + x^2 = 1$$

Thus $\cos y = \sqrt{1 - x^2}$, the *positive* square root, and dy/dx takes the form

$$\frac{dy}{dx} = \frac{1}{\sqrt{1 - x^2}}.$$

This proves that $$\frac{d}{dx}(\sin^{-1} x) = \frac{1}{\sqrt{1 - x^2}}. \quad ■$$

EXAMPLE 2 Find the derivative of $\sin^{-1}(3x/4)$.

SOLUTION Let $y = \sin^{-1}(3x/4)$. This is a composite function, with

$$y = \sin^{-1} u \qquad \text{where } u = \frac{3x}{4}.$$

By the chain rule, $$\frac{dy}{dx} = \frac{dy}{du} \cdot \frac{du}{dx}.$$

Now, $$\frac{dy}{du} = \frac{1}{\sqrt{1 - u^2}} \qquad \text{and} \qquad \frac{du}{dx} = \frac{3}{4}.$$

Thus
$$\begin{aligned} \frac{dy}{dx} &= \frac{1}{\sqrt{1 - u^2}} \cdot \frac{3}{4} \\ &= \frac{1}{\sqrt{1 - (3x/4)^2}} \cdot \frac{3}{4} \\ &= \frac{1}{\sqrt{1 - 9x^2/16}} \cdot \frac{3}{4} \\ &= \frac{3}{\sqrt{16 - 9x^2}}. \quad ■ \end{aligned}$$

EXAMPLE 3 Differentiate $x\sqrt{1 - x^2} + \sin^{-1} x$.

SOLUTION

$$\begin{aligned} \frac{d}{dx}(x\sqrt{1 - x^2} + \sin^{-1} x) &= \frac{d}{dx}(x\sqrt{1 - x^2}) + \frac{d}{dx}(\sin^{-1} x) \\ &= \left[x\frac{d}{dx}(\sqrt{1 - x^2}) + \sqrt{1 - x^2}\frac{dx}{dx}\right] + \frac{1}{\sqrt{1 - x^2}} \\ &= x \cdot \frac{1}{2} \cdot \frac{-2x}{\sqrt{1 - x^2}} + \sqrt{1 - x^2} + \frac{1}{\sqrt{1 - x^2}} \\ &= \frac{-x^2}{\sqrt{1 - x^2}} + \sqrt{1 - x^2} + \frac{1}{\sqrt{1 - x^2}} \\ &= \frac{1 - x^2}{\sqrt{1 - x^2}} + \sqrt{1 - x^2} \\ &= \sqrt{1 - x^2} + \sqrt{1 - x^2} \\ &= 2\sqrt{1 - x^2}. \quad ■ \end{aligned}$$

So what is an antiderivative of $\sqrt{1 - x^2}$*?*

The Inverse Cosine

The cosine function is not one-to-one. For instance, $\cos 0 = 1 = \cos 2\pi$. (See Fig. 11.) However, if we restrict its domain to $[0, \pi]$, then it is one-to-one. As x goes from 0 to π, $\cos x$ decreases from $\cos 0 = 1$ to $\cos \pi = -1$. (See Fig. 12.) Its domain is $[0, \pi]$ and its range is $[-1, 1]$. Therefore it has an inverse function, called *arccosine* and denoted $\cos^{-1} x$. In short, $\cos^{-1} x$ denotes the angle between 0 and π whose cosine is x. Figure 13 shows $\cos^{-1} x$ for positive x; Fig. 14 shows it for negative x, when $\cos^{-1} x$ is a second-quadrant angle. For instance, since $\cos(\pi/3) = \frac{1}{2}$, it follows that $\cos^{-1}(\frac{1}{2}) = \pi/3$. Both equations record the fact that $\cos(\pi/3) = \frac{1}{2}$. The graph of $\cos^{-1} x$ is shown in Fig. 15.

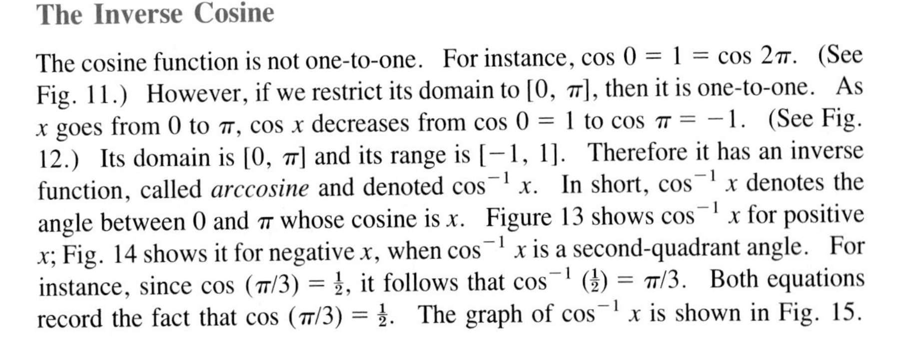

Figure 11

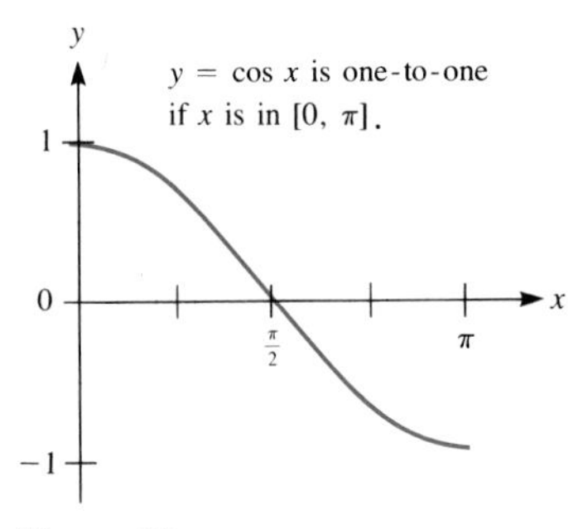

Figure 12

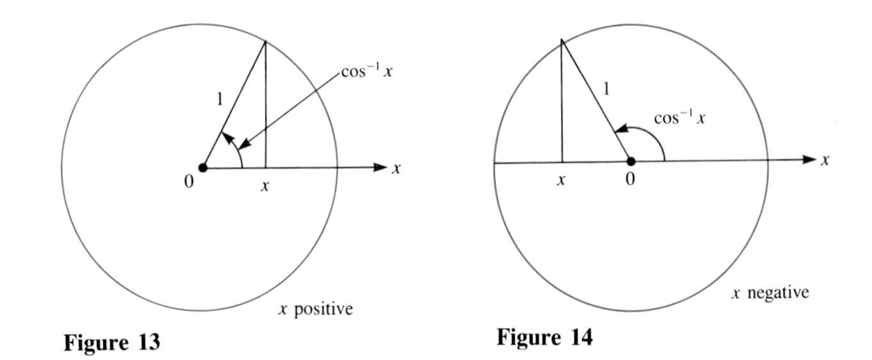

Figure 13

Figure 14

The Derivative of the Inverse Cosine

Reasoning like that in the proof of Theorem 2 shows that

$$(\cos^{-1} x)' = \frac{-1}{\sqrt{1 - x^2}}.$$

(The negative sign reflects the fact that $\cos^{-1} x$ is a decreasing function.)

Theorem 3 $\quad \dfrac{d}{dx}(\cos^{-1} x) = \dfrac{-1}{\sqrt{1 - x^2}}$

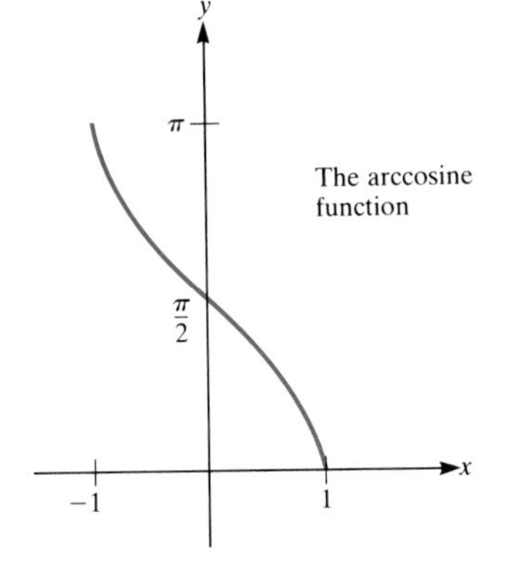

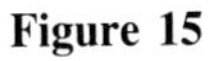
Figure 15

Since the derivative of $\cos^{-1} x$ differs from the derivative of $\sin^{-1} x$ only by the constant factor -1, it is not needed for finding antiderivatives. In most integral tables, $\sin^{-1} x$ is used in preference to $\cos^{-1} x$.

The Inverse Secant

After $\tan^{-1} x$ and $\sin^{-1} x$, the next most important function for the computation of antiderivatives is the inverse of the secant function. Recall that $\sec x = 1/\cos x$; it is graphed in Sec. 2.6. Since $|\sec x| \geq 1$, the inverse of the secant function is defined only for inputs of absolute value ≥ 1. For $|x| \geq 1$, define

Section Summary

The key points in this section are the definitions of $\tan^{-1} x$, $\sin^{-1} x$, $\cos^{-1} x$, and $\sec^{-1} x$ and the calculations of their derivatives.

Function	*Derivative*	*Domain*	*Range*
$\tan^{-1} x$	$\dfrac{1}{1+x^2}$	All real numbers	$(-\pi/2, \pi/2)$
$\sin^{-1} x$	$\dfrac{1}{\sqrt{1-x^2}}$	$[-1, 1]$	$[-\pi/2, \pi/2]$
$\cos^{-1} x$	$\dfrac{-1}{\sqrt{1-x^2}}$	$[-1, 1]$	$[0, \pi]$
$\sec^{-1} x$	$\dfrac{1}{\lvert x\rvert\sqrt{x^2-1}}$	$\lvert x\rvert \geq 1$	$[0, \pi/2)$ and $(\pi/2, \pi]$

With each derivative comes a corresponding antiderivative. In the notation of Chap. 5:

$$\int \frac{dx}{1+x^2} = \tan^{-1} x + C$$

$$\int \frac{dx}{\sqrt{1-x^2}} = \sin^{-1} x + C \qquad |x| < 1$$

$$\int \frac{dx}{|x|\sqrt{x^2-1}} = \sec^{-1} x + C \qquad |x| > 1.$$

I didn't like the symbols for sine, cosine, tangent, and so on. To me, "sin *f*" looked like *s* times *i* times *n* times *f*! So I invented another symbol, like a square root sign, that was a sigma with a long arm sticking out of it, and I put the *f* underneath. . . .

Now the inverse sine was the same sigma, but left-to-right reflected so that it started with the horizontal line with the value underneath, and then the sigma. *That* was the inverse sine, NOT $\sin^{-1} f$—that was crazy! They had that in books! To me, $\sin^{-1}$ meant 1/sine, the reciprocal. So my symbols were better.

I didn't like $f(x)$—that looked to me like *f* times *x*. I also didn't like dy/dx—you have a tendency to cancel the *d*'s—so I made a different sign, something like an & sign. For logarithms it was a big L extended to the right, with the thing you take the log of inside, and so on.

I thought my symbols were just as good, if not better, than the regular symbols—it doesn't make any difference *what* symbols you use—but I discovered later that it *does* make a difference. Once when I was explaining something to another kid in high school, without thinking I started to make these symbols, and he said, "What the hell are those?" I realized then that if I'm going to talk to anybody else, I'll have to use the standard symbols, so I eventually gave up my own symbols.

From Richard P. Feynman, *"Surely You're Joking, Mr. Feynman,"* Bantam, New York, 1986.

Feynman won the Nobel prize in physics in 1965.

EXERCISES FOR SEC. 6.6: THE DERIVATIVES OF THE INVERSE TRIGONOMETRIC FUNCTIONS

(Exercises 58 to 64 rely on Chap. 5.)

1 Draw the unit circle used in defining the trigonometric functions and the tangent line at (1, 0). On your diagram draw the following angles:
(*a*) $\tan^{-1} 1$ (*b*) $\tan^{-1} 2$ (*c*) $\tan^{-1}(-1)$

2 Like Exercise 1 for (*a*) $\tan^{-1} 0.5$, (*b*) $\tan^{-1}(-3)$, (*c*) $\tan^{-1} 3$.

3 Using your calculator, fill in this table and use the data to graph $y = \tan^{-1} x$.

x	−4	−3	−2	−1	0	1	2	3	4
$\tan^{-1} x$									

4 Examine (*a*) $\lim_{x\to\infty} \tan^{-1} x$ and (*b*) $\lim_{x\to-\infty} \tan^{-1} x$.

5 Draw the unit circle used in defining the trigonometric functions. On your diagram draw the following angles:
(*a*) $\sin^{-1}(\frac{1}{2})$ (*b*) $\sin^{-1} 1$ (*c*) $\sin^{-1}(-1)$.

6 Like Exercise 5 for (*a*) $\sin^{-1}(\frac{1}{3})$, (*b*) $\sin^{-1}(0)$, (*c*) $\sin^{-1}(-\frac{1}{2})$.

7 Using your calculator, fill in this table and use the data to graph $y = \sin^{-1} x$.

x	−1	−0.8	−0.6	−0.4	−0.2	0	0.2	0.4	0.6	0.8	1
$\sin^{-1} x$											

8 Like Exercise 7 for $y = \cos^{-1} x$.

9 Using your calculator fill in this table and use the data to graph $y = \sec^{-1} x$.

x	1	2	3	4	−1	−2	−3	−4
$\sec^{-1} x$								

10 Like Exercise 9 for $\csc^{-1} x$.

11 Evaluate without a calculator
(*a*) $\sin^{-1}(\frac{1}{2})$ (*b*) $\tan^{-1}(1/\sqrt{3})$
(*c*) $\sin^{-1}(-\sqrt{3}/2)$ (*d*) $\tan^{-1}(-\sqrt{3})$
(*e*) $\sec^{-1}\sqrt{2}$

12 Which of these are *meaningless?* (Try each on your calculator.)
(*a*) $\cos^{-1} 1.5$ (*b*) $\sec^{-1} 1.5$ (*c*) $\tan^{-1} 1.5$
(*d*) $\sec^{-1} 0.3$ (*e*) $\sin^{-1} 2.4$

In Exercises 13 to 20 evaluate the expressions without recourse to a calculator. A sketch of the unit circle or an appropriate right triangle may help.

13 $\sin(\tan^{-1} 1)$ **14** $\tan(\sec^{-1} 2)$
15 $\tan[\sin^{-1}(-\sqrt{2}/2)]$ **16** $\tan[\sin^{-1}(\sqrt{3}/2)]$
17 $\sin(\sin^{-1} 0.3)$ **18** $\sin(\tan^{-1} 0)$
19 $\sin^{-1}(\sin \pi)$ **20** $\cos^{-1}[\cos(-\pi/4)]$

In Exercises 21 to 50 differentiate the given functions.

21 $\sin^{-1} 5x$ **22** $\tan^{-1} 3x$
23 $\sec^{-1} 3x$ **24** $\sin^{-1} e^{-x}$
25 $\tan^{-1}\sqrt[3]{x}$ **26** $-\dfrac{1}{3}\sin^{-1}\dfrac{3}{x}$
27 $x^2 \sec^{-1}\sqrt{x}$ **28** $\dfrac{1}{\sin^{-1} 2x}$
29 $\sin 3x \sin^{-1} 3x$ **30** $x^3 \tan^{-1} 2x$
31 $\dfrac{x \sec^{-1} 3x}{e^{2x}}$ **32** $\arcsin(2x - 3)$
33 $\arctan\sqrt{x}$ **34** $\operatorname{arcsec}\sqrt{x}$
35 $\ln \sec^{-1}\sqrt{x}$ **36** $\ln[(\sin^{-1} 5x)^2]$
37 $\dfrac{x}{\tan^{-1} 10^x}$ **38** $10^{\sec^{-1} 2x}$
39 $\sin^{-1} x - \sqrt{1 - x^2}$ **40** $2^x \cdot \log_3 x \cdot \sec 3x$
41 $(\tan^{-1} 2x)^3$ **42** $(\sin^{-1}\sqrt{x} - 1)^4$
43 $\dfrac{x}{2}\sqrt{2 - x^2} + \sin^{-1}\dfrac{x}{\sqrt{2}}$
44 $\sqrt{3x^2 - 1} - \tan^{-1}\sqrt{3x^2 - 1}$
45 $\frac{2}{3}\sec^{-1}\sqrt{3x^5}$

In Exercises 46 to 50 differentiate and simplify your answer.

46 $\dfrac{1}{2}\left[(x - 3)\sqrt{6x - x^2} + 9\sin^{-1}\dfrac{x - 3}{3}\right]$

47 $\sqrt{1 + x}\sqrt{2 - x} - 3\sin^{-1}\sqrt{\dfrac{2 - x}{3}}$

48 $x \sin^{-1} 3x + \frac{1}{3}\sqrt{1 - 9x^2}$

49 $x(\sin^{-1} 2x)^2 - 2x + \sqrt{1 - 4x^2}\sin^{-1} 2x$

50 $x \tan^{-1} 5x - \frac{1}{10}\ln(1 + 25x^2)$

In Exercises 51 and 52 differentiate the given functions. Note that quite different functions may have very similar derivatives.

51 (*a*) $\ln(x + \sqrt{x^2 - 9})$ (*b*) $\sin^{-1}\dfrac{x}{3}$

52 (*a*) $-\dfrac{1}{5}\ln\dfrac{5 + \sqrt{25 - x^2}}{x}$ (*b*) $-\dfrac{1}{5}\sin^{-1}\dfrac{5}{x}$

53 Assume that your scientific calculator does not have a special key for π. Describe what keys you could press in order to have the calculator display π to as many decimal places as the display accommodates. (There may be more than one answer.)

Exercises 54 to 56 are related. Use $\pi \approx 3.14$.

54 (*a*) Compute the differential $d(\tan^{-1} x)$.
(*b*) Use (*a*) to estimate $\tan^{-1} 1.1$. (*Suggestion:* $\tan^{-1} 1$ is known.)

55 (*a*) Compute the differential $d(\sin^{-1} x)$.
(*b* Use (*a*) to estimate $\sin^{-1} 0.47$.

56 (*a*) Compute the differential $d(\sec^{-1} x)$.
(*b*) Use (*a*) to estimate $\sec^{-1} 2.08$.

57 Show that $\tan^{-1} \frac{1}{2} + \tan^{-1} \frac{1}{3} = \pi/4$. *Hint:* Use the identity for $\tan(A + B)$.

Exercises 58 to 64 rely on material from Chap. 5.

58 (*a*) Show that

$$\int \frac{dx}{\sqrt{1 - a^2x^2}} = \frac{1}{a} \sin^{-1} ax + C \qquad (a > 0).$$

Use (*a*) to find

(*b*) $\int \frac{dx}{\sqrt{1 - 25x^2}}$ (*c*) $\int \frac{dx}{\sqrt{1 - 3x^2}}$

59 (*a*) Show that

$$\int \frac{dx}{\sqrt{a^2 - x^2}} = \sin^{-1} \frac{x}{a} + C \qquad (a > 0).$$

Use (*a*) to find

(*b*) $\int \frac{dx}{\sqrt{25 - x^2}}$ (*c*) $\int \frac{dx}{\sqrt{5 - x^2}}$

60 Find the area below $y = 1/(1 + x^2)$ and above the interval (*a*) $[-1, 1]$, (*b*) $(0, \sqrt{3})$, (*c*) $[0, 10]$.

61 (*a*) Why is $\pi/4 = \int_0^1 \frac{dx}{1 + x^2}$?
(*b*) Use Simpson's rule with $n = 6$ to obtain an estimate of $\pi/4$ and, consequently, of π.

62 (*a*) Why is $\pi/4 = \int_0^1 \sqrt{1 - x^2}\, dx$? (*Hint:* Why is the graph part of the unit circle?)
(*b*) Use Simpson's rule with $n = 6$ to obtain an estimate of $\pi/4$ and, consequently, of π.

63 Find the area below $y = 1/(x\sqrt{x^2 - 1})$ and above $[\sqrt{2}, 2]$.

64 (*a*) Differentiate $\tan^{-1} ax$, where a is a constant.
(*b*) Find the area under $y = 1/(1 + 3x^2)$ and above $[0, 1]$.

65 For which numbers x is (*a*) $\sin(\arcsin x) = x$? (*b*) $\arcsin(\sin x) = x$?

6.7 THE DIFFERENTIAL EQUATION OF NATURAL GROWTH AND DECAY

This section depends on Chap. 5. It introduces the concept of a differential equation and applies it to the study of growth and decay.

Differential Equations

The abbreviation for "differential equation" is D.E.

An equation that involves one or more of the derivatives of a function is called a **differential equation**. For example, Sec. 4.3 examined the differential equation for motion under constant acceleration,

$$\frac{d^2y}{dt^2} = a. \qquad a \text{ constant} \tag{1}$$

Finding an antiderivative $F(x)$ for a function $f(x)$ amounts to solving the differential equation

$$\frac{dF}{dx} = f(x) \tag{2}$$

for the unknown function $F(x)$.

Solution of a D.E.

A **solution** of a differential equation is any function that satisfies the equation. To solve a differential equation means to find all its solutions. In Sec. 4.3 it was shown that the most general solution of Eq. (1) is

$$y = \frac{at^2}{2} + v_0t + y_0,$$

A solution of a D.E. is a function, not a number.

where v_0 and y_0 are constants. The most general solution of

$$\frac{dF}{dx} = x^2 \tag{3}$$

is

$$F(x) = \frac{x^3}{3} + C,$$

where C represents an arbitrary constant.

Order of a D.E.

The order of a differential equation is the highest order of the derivatives that appear in it. Thus Eq. (1) is of order 2, and Eq. (2) is of order 1.

This section examines a special and important type of first-order differential equation, called **separable**. After showing how to solve it, we will apply it to the study of natural growth and decay.

Separable D.E.

A **separable differential equation** is one that can be written in the form

$$\frac{dy}{dx} = \frac{f(x)}{g(y)} \quad \text{or} \quad \frac{dy}{dx} = \frac{g(y)}{f(x)} \tag{4}$$

where $f(x)$ and $g(y)$ are differentiable functions. Such an equation can be solved by *separating the variables,* that is, bringing all the x's to one side and all the y's to the other side.

For instance, the first equation in (4) can be changed to the following equation in differentials:

$$g(y)\, dy = f(x)\, dx.$$

This is solved by integrating both sides:

$$\int g(y)\, dy = \int f(x)\, dx + C.$$

The second equation in (4) can be rewritten as $dy/g(y) = dx/f(x)$, and is solved by integrating:

$$\int \frac{dy}{g(y)} = \int \frac{dx}{f(x)} + C.$$

The solution of a separable differential equation (in fact, any first-order differential equation) will generally involve one arbitrary constant. Each choice of that constant determines a specific function that satisfies the differential equation.

Example 1 illustrates the method by solving the differential equation that turns out to be the key to the study of natural growth and decay. The solution fits any situation where dP/dt, the rate of change of a quantity, is proportional to the quantity P itself.

EXAMPLE 1 Solve the differential equation

$$\frac{dP}{dt} = kP \tag{5}$$

where k is a constant and P has only positive values.

SOLUTION Bring all expressions involving P to one side and all those involving t to the other, obtaining

$$\frac{dP}{P} = k\, dt. \tag{6}$$

Then

$$\int \frac{dP}{P} = \int k\, dt$$

Since $P > 0$, $\ln |P| = \ln P$.

or

$$\ln P = kt + C, \tag{7}$$

where C is some constant. Thus

$$P(t) = e^{kt+C}$$
$$= e^{kt}\, e^{C}.$$

Denote the arbitrary positive constant e^C by the letter A. Then

$$P(t) = Ae^{kt},$$

where A is any positive constant, is the most general solution of $dP/dt = kP$. ■

Natural Growth

The change in the size of the world's population is determined by the relation between the birth and death rates. If they are equal, the size remains constant. If they are unequal, the population either grows or shrinks. Improvements in nutrition and preventive medicine, among other scientific advances, have lowered the death rate.

Consequently, for the past 2 centuries the world population has increased dramatically, as shown in Fig. 1. The size of the world population in 1650 is estimated to have been about 0.5 billion; in 1750 about 0.7 billion; in 1850 about 1.1 billion; in 1980 about 4.5 billion; in 1990 about 5.3 billion. The population has been growing at an accelerating pace. If back in 1850 you had moved every human being to China, then that total number would just match the present population of China.

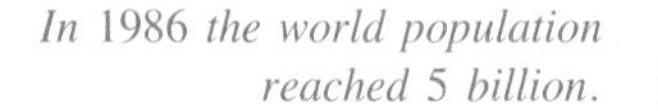

It took a million years for the world population to reach 1 billion. It reached 5 billion a mere 140 years later, in 1986.

What will the population be in the year 2000 if the present rate of growth continues? To answer this question we must describe the size of the population mathematically.

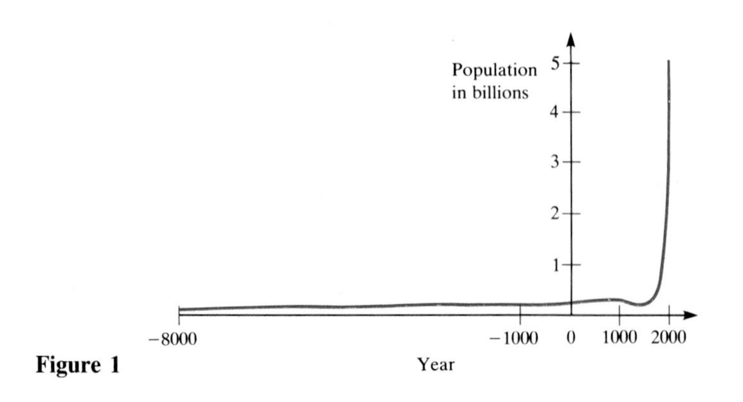

Figure 1

The Mathematics of Natural Growth

Let $P(t)$ denote the size of the population at time t. Actually, $P(t)$ is an integer, and the graph of P has "jumps" whenever someone is born or dies. However, assume that P is a "smooth" (differentiable) function that approximates the size of the population.

The derivative P' then records the rate of change of the population. It is reasonable to expect the rate of growth $P'(t)$ to be proportional to the size of the population $P(t)$: A large population will produce more babies in a year than a small population.

More precisely, we assume that there is a fixed positive number k independent of time, such that

The D.E. of natural growth

$$P'(t) = kP(t) \qquad k > 0. \tag{8}$$

This is the differential equation of **natural growth**. The constant k is called the **growth constant**.

To forecast the population, it is necessary to solve Eq. (8). By Example 1, the general solution is

$$P(t) = Ae^{kt}, \tag{9}$$

where A is any positive constant.

What is the meaning of A? To find out, set $t = 0$ in Eq. (9), obtaining

$$P(0) = Ae^{k \cdot 0}$$

or

$$P(0) = A.$$

Thus A is the size of the *initial population,* that is, the population at time $t = 0$. Hence

$$P(t) = P(0)e^{kt}. \tag{10}$$

Equation (10) describes the growth of any quantity that changes at a rate proportional to the quantity present. Because natural growth involves the exponential function e^{kt}, it is also called **exponential growth**.

EXAMPLE 2 The size of the world population at the beginning of 1988 was approximately 5.14 billion. At the beginning of 1989 it was 5.23 billion. Assume that the growth rate remains constant.

(a) What is the growth constant k?
(b) What will the population be in 2000?
(c) When will the population double?

SOLUTION Let $P(t)$ be the population in billions at time t. For convenience, measure time starting in the year 1988; that is, $t = 0$ corresponds to 1988 and $t = 1$ to 1989. Thus $P(0) = 5.14$ and $P(1) = 5.23$. The equation describing the population in billions at time t is

$$P(t) = 5.14e^{kt}. \tag{11}$$

(*a*) To find k, we note that

$$P(1) = 5.14e^{k \cdot 1},$$

so

$$\begin{aligned} 5.14e^k &= 5.23 \\ e^k &= \frac{5.23}{5.14} \\ k &= \ln \frac{5.23}{5.14} \\ &\approx 0.0174. \end{aligned}$$

Hence Eq. (11) takes the form

$$P(t) = 5.14e^{0.0174t}. \tag{12}$$

This equation is all that we need to answer the remaining questions.

(*b*) The year 2000 corresponds to $t = 12$, so in the year 2000 the population in billions will be

$$\begin{aligned} P(12) &= 5.14e^{0.0174 \cdot 12} \\ &\approx 5.14e^{0.21} \\ &\approx 5.14(1.23) \\ &\approx 6.33. \end{aligned}$$

The population will be approximately 6.33 billion in the year 2000 (unless there are some major changes in the birth or death rates).

(*c*) The population will double when it reaches $2(5.14) = 10.28$ billion. We need to solve for t in the equation $P(t) = 10.28$. We have

$$\begin{aligned} 5.14e^{kt} &= 10.28 \\ e^{kt} &= 2 \\ kt &= \ln 2 \\ t &= \frac{\ln 2}{k} \\ &\approx \frac{0.6931}{0.0174} \\ &\approx 39.83. \end{aligned}$$

The population will double in approximately 40 years, which corresponds to the year 2028. ■

The time it takes for a population to double in size is called the *doubling time* and is denoted t_2. Benjamin Franklin estimated a doubling time of 20 years for the population of the United States, starting in 1751 when the colonies' population was about 1.3 million. Exponential growth is often described by its doubling time t_2 rather than by its growth constant k. However, if you know either t_2 or k you can figure out the other, as we now show.

By the definition of the doubling time, t_2, $P(t_2) = 2\,P(0)$. As shown in Example 2,

$$t_2 = \frac{\ln 2}{k}. \tag{13}$$

Since $\ln 2 \approx 0.693$, we often write

Doubling time from growth constant

$$t_2 \approx \frac{0.693}{k}.$$

Knowing the doubling time, we can get the growth constant k by rewriting Eq. (13) as

Growth constant from doubling time

$$k = \frac{\ln 2}{t_2}.$$

Exponential growth may also be described in terms of an annual percentage increase, such as "The population is growing 6 percent per year." That is, each year the population is multiplied by a factor of 1.06: $P(t+1) = P(t)(1.06)$. On the other hand, from the exponential growth function, we see that

$$\begin{aligned} P(t+1) &= P(0)e^{k(t+1)} \\ &= P(0)e^{kt}e^{k} \\ &= P(t)e^{k}. \end{aligned}$$

That is, during each unit of time the population increases by a factor of e^k. Now, when k is small, $e^k \approx 1 + k$. (See Exercise 36.) Consequently we can approximate 6 percent annual growth by letting $k = 0.06$. This approximation is valid whenever the growth rate is only a few percent. Since population figures are themselves only an approximation, setting the growth constant k equal to the annual percentage growth rate does not significantly affect the accuracy of our computations.

EXAMPLE 3 Find the doubling time if the growth rate is 2 percent per year.

SOLUTION The growth rate is 2 percent, so we set $k = 0.02$. Then

$$\begin{aligned} t_2 &\approx \frac{0.693}{0.02} \\ &= 34.65 \text{ years.} \end{aligned}$$ ■

The Mathematics of Natural Decay

As Glen Seaborg observes in the conversation given on p. 374, some radioactive elements decay at a rate proportional to the amount present. The time it takes for half the initial amount to decay is denoted $t_{1/2}$ and is called the element's **half-life**

Similarly, in medicine one speaks of the half-life of a drug administered to a patient: the time required for half the drug to be removed from the body. This half-life depends both on the drug and the patient, and can vary from 20 minutes for penicillin to 2 weeks for quinacrine, an antimalarial drug. This half-life is critical in determining how frequently a drug can be administered. Several elderly patients died from overdoses before it was realized that the half-life is longer in the elderly than in the young.

Letting $P(t)$ again represent the amount present at time t, we have

Now k is negative.

$$P'(t) = kP(t) \qquad k < 0 \tag{14}$$

where k is the **decay constant**. This is the same equation as (8), so

$$P(t) = P(0)e^{kt},$$

as before, except that now k is a *negative number*. Since k is negative, the factor e^{kt} is a decreasing function of t. (See Fig. 2.)

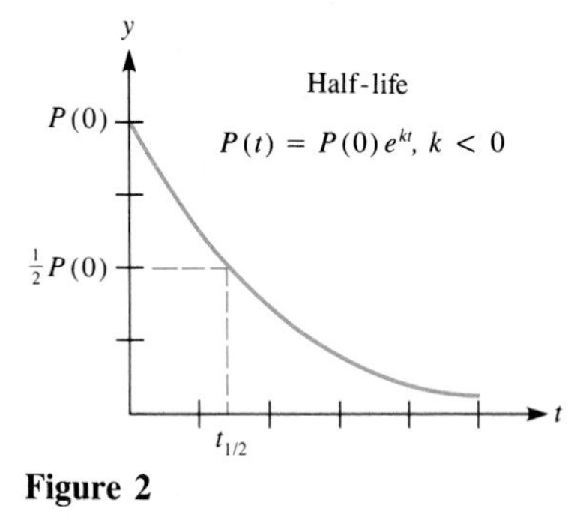

Figure 2

To relate $t_{1/2}$ and k, begin with the information

$$P(t_{1/2}) = \frac{1}{2}P(0).$$

Thus

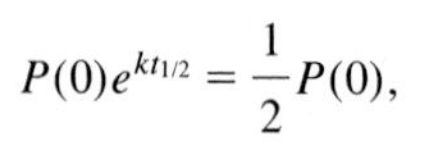

$$P(0)e^{kt_{1/2}} = \frac{1}{2}P(0),$$

so

$$e^{kt_{1/2}} = \frac{1}{2}.$$

Consequently,

$$kt_{1/2} = \ln\frac{1}{2}.$$

Recalling that $\ln\left(\frac{1}{2}\right) = -\ln 2$, we have

$$kt_{1/2} = -\ln 2,$$

and finally

Half-life from the decay constant

$$\boxed{t_{1/2} = \frac{-\ln 2}{k}.} \tag{15}$$

Since both the numerator and denominator in (15) are negative, $t_{1/2}$ is positive, as we would expect. The half-life also determines k:

The decay constant from half-life

$$\boxed{k = \frac{-\ln 2}{t_{1/2}}.} \tag{16}$$

EXAMPLE 4 The Chernobyl nuclear reactor accident, in April 1986, released radioactive cesium 137 into the air. The half-life of ^{137}Cs is 27.9 years.

(*a*) Find the decay constant k of ^{137}Cs.

(*b*) When will only one-fourth of an initial amount remain?

(*c*) When will only 20 percent of an initial amount remain?

SOLUTION

(*a*) By Eq. (16),

$$k \approx \frac{-0.693}{27.9} \approx -0.0248.$$

(*b*) This can be done without the aid of any formulas. Since $\frac{1}{4} = \frac{1}{2} \cdot \frac{1}{2}$, in two half-lives only one-quarter of an initial amount remains. The answer is $2(27.9) = 55.8$ years.

(*c*) We want to find t such that only 20 percent remains. While we know the answer is greater than 55.8 years (since 20 percent is even less than 25 percent), finding the exact time requires a calculation.

We want

$$P(t) = 0.20P(0).$$

That is, we want to solve

$$P(0)e^{kt} = 0.20P(0).$$

Then

$$e^{kt} = 0.20$$

$$kt = \ln 0.20$$

$$t = \frac{\ln 0.20}{k}.$$

Since $k \approx -0.0248$, this gives

$$t \approx \frac{-1.609}{-0.0248} \approx 64.9 \text{ years}.$$

After 64.9 years only 20 percent of the original amount remains. ■

Section Summary

A separable equation is one that can be written in the form

$$\frac{dy}{dx} = \frac{f(x)}{g(y)} \quad \text{or} \quad \frac{dy}{dx} = \frac{g(y)}{f(x)}.$$

It is solved by bringing the x's to one side and the y's to the other and then integrating.

The differential equation of natural growth or decay,

$$\frac{dP}{dt} = kP,$$

can be solved as a separable differential equation. Its solutions are of the form

$$P(t) = P(0)e^{kt},$$

where $P(0)$ is the amount at time $t = 0$.

Natural Growth	*Natural Decay*
$P(t) = P(0)e^{kt}$	$P(t) = P(0)e^{kt}$
$k > 0$: growth constant	$k < 0$: decay constant
t_2: doubling time	$t_{1/2}$: half-life
$t_2 = \frac{\ln 2}{k} \approx \frac{0.69}{k}$	$t_{1/2} = \frac{-\ln 2}{k} \approx \frac{-0.69}{k}$
$k = \frac{\ln 2}{t_2}$	$k = \frac{-\ln 2}{t_{1/2}}$

THE SCIENTIST, THE SENATOR, AND HALF-LIFE

During the hearings in 1963 before the Senate Foreign Relations Committee on the nuclear test ban treaty, this exchange took place between Glen Seaborg, winner of the Nobel prize for chemistry in 1951, and Senator James W. Fulbright.

Seaborg: Tritium is used in a weapon, and it decays with a half-life of about 12 years. But the plutonium and uranium have such long half-lives that there is no detectable change in a human lifetime.
Fulbright: I am sure this seems to be a very naive question, but why do you refer to half-life rather than whole life? Why do you measure by half-lives?
Seaborg: Here is something that I could go into a very long discussion on.
Fulbright: I probably wouldn't benefit adequately from a long discussion. It seems rather odd that you should call it a half-life rather than its whole life.
Seaborg: Well, I will try. If we have, let us say, one million atoms of a material like tritium, in 12 years half of those will be transformed into a decay product and you will have 500,000 atoms.

Then, in another 12 years, half of what remains transforms, so you have 250,000 atoms left. And so forth.

On that basis it never all decays, because half is always left, but of course you finally get down to where your last atom is gone.

AIDS AND THE EXPONENTIAL CURVE

In 1984, John Platt, a biophysicist at the University of Chicago, wrote a short piece for private circulation. At a time when most of us ignored AIDS, Platt recognized that the limited data on its spread in America suggested a frightening prospect: we are all susceptible to AIDS and the disease has been spreading in a simple exponential manner.

> Exponential growth is a geometric increase. Remember the old kiddy problem: if you place a penny on square one of a checkerboard and double the number of coins on each subsequent square—2, 4, 8, 16, 32 . . . —how big is the stack by the 64th square? The answer: about as high as the universe is wide. Nothing in the external environment inhibits this increase, thus giving to exponential processes their relentless character. In the real, noninfinite world, of course, some limit will eventually arise, and the process slows down, reaches a steady state, or destroys the entire system: the stack of pennies falls over, the bacterial cells exhaust their supply of nutrients.
>
> Platt noticed that data for the initial spread of AIDS fell right on an exponential curve. He then followed the simplest possible procedure of extrapolating the curve unabated into the 1990's. Most of us were incredulous, accusing Platt of the mathematical gamesmanship that scientists call "curve fitting." After all aren't exponential models unrealistic?
>
> Well, hello 1987—world-wide data still match Platt's extrapolated curve. This will not, of course, go on forever.

Stephen Jay Gould, The exponential spread of AIDS,
New York Times Magazine, April 19, 1987.

EXERCISES FOR SEC. 6.7: THE DIFFERENTIAL EQUATION OF NATURAL GROWTH AND DECAY

In Exercises 1 to 8, solve the differential equations by separating the variables.

1 $\dfrac{dy}{dx} = \dfrac{x^3}{y^2}$ **2** $\dfrac{dy}{dx} = \dfrac{x^2}{y+3}$

3 $\dfrac{dy}{dx} = \dfrac{y+1}{x+2}$ **4** $\dfrac{dy}{dx} = \dfrac{y^2+1}{2x}$

5 $\dfrac{dy}{dx} = \dfrac{\sin 3x}{\cos 2y}$ **6** $\dfrac{dy}{dx} = \dfrac{\sec^2 x}{\sin y}$

7 $\dfrac{dy}{dx} = \dfrac{e^y}{1+x^2}$ **8** $\dfrac{dy}{dx} = \dfrac{5+y}{e^x}$

9 The amount of a certain growing substance increases at the rate of 10 percent per hour. Find (*a*) k, (*b*) t_2.

10 A quantity is increasing according to the law of natural growth. The amount present at time $t = 0$ is A. It will double when $t = 10$.
(*a*) Express the amount in the form Ae^{kt} for suitable k.
(*b*) Express the amount in the form Ab^t for suitable b.

11 The mass of a certain bacterial culture after t hours is $10 \cdot 3^t$ grams.
(*a*) What is the initial amount?
(*b*) What is the growth constant k?
(*c*) What is the percent increase in any period of 1 hour?

12 Let $f(t) = 3 \cdot 2^t$.
(*a*) Solve the equation $f(t) = 12$.
(*b*) Solve the equation $f(t) = 5$.
(*c*) Find k such that $f(t) = 3e^{kt}$.

13 In 1988 the world population was about 5.1 billion and was increasing at the rate of 1.7 percent per year. If it continues to grow at that rate, when will it (*a*) double? (*b*) quadruple? (*c*) reach 100 billion?

14 The population of Latin America has a doubling time of 27 years. Estimate the percent it grows per year.

15 At 1 P.M. a bacterial culture weighed 100 grams. At 4:30 P.M. it weighed 250 grams. Assuming that it grows at a rate proportional to the amount present, find (*a*) at what time it will grow to 400 grams, (*b*) its growth constant.

16 A bacterial culture grows from 100 to 400 grams in 10 hours according to the law of natural growth.
(*a*) How much was present after 3 hours?
(*b*) How long will it take the mass to double? quadruple? triple?

17 A radioactive substance disintegrates at the rate of 0.05 gram per day when its mass is 10 grams.
(*a*) How much of the substance will remain after t days if the initial amount is A?
(*b*) What is its half-life?

18 A disintegrating radioactive substance decreases from 12 to 11 grams in 1 day. Find its half-life.

19 In 1990 the population of Mexico was 89 million and of the United States 250 million. If the population of Mexico increases at 1.8% per year and the population of the United States at 0.7% per year, when would the two nations have the same size population?

20 The size of the population in India was 689 million in 1980 and 832 million in 1990. What is its doubling time t_2?

21 Carbon 14 (chemical symbol ^{14}C), an isotope of carbon, is radioactive and has a half-life of approximately 5,730 years. If the ^{14}C concentration in a piece of wood of unknown age is half of the concentration in a present-day live specimen, then it is about 5,730 years old. (This assumes that ^{14}C concentrations in living objects remain about the same.) This gives a way of estimating the age of an undated specimen. Show that if A_c is the concentration of ^{14}C in a live (contemporary) specimen and A_u is the concentration of ^{14}C in a specimen of unknown age, then the age of the undated material is about $8{,}300 \ln (A_c/A_u)$ years. (This method, called *radiocarbon dating*, is reliable up to around 70,000 years.) See Radiocarbon Dating in the *Encyclopedia of Science and Technology*, 6th ed., McGraw-Hill, New York, 1987.

22 This newspaper article illustrates the rapidity of exponential growth:

> **U.S. HIT FOR 200-YR. DEBT**
>
> LAS VEGAS—An autograph dealer is demanding the U.S. government pay off a 200-year-old note. At seven percent interest, the debt amounts to $14 billion.
>
> It was issued to Haym Salomon on March 27, 1782, by finance chief Robert Morse in return for a $30,000 loan.
>
> The note is payable to the bearer—but the statute of limitations ran out on it some 150 years ago.

(*a*) Is the figure of $14 billion correct? Assume that the interest is compounded annually.
(*b*) What interest rate would be required to produce an account of $14 billion if interest were compounded once a year?
(*c*) Answer (*b*) for "continuous compounding," which is another term for natural growth (a bank account increases at a rate proportional to the amount in the account at any instant).

23 From a newspaper article:

'Rule of 72'

I've been hearing bankers and investment advisers talk about something called the "rule of 72." Could you explain what it means? — B. H.

How quickly would you like to double your money? That's what the "rule of 72" will tell you. To find out how fast your money will double at any given interest rate or yield, simply divide that yield into 72. This will tell you how many years doubling will take.

Let's say you have a long-term certificate of deposit paying 12 percent. At that rate your money would double in six years. A money-market fund paying 10 percent would take 7.2 years to double your investment.

(*a*) Explain the rule of 72 and what number should be used instead of 72.

(*b*) Why do you think 72 is used?

24 As mentioned in the text, Benjamin Franklin conjectured that the population of the United States would double every 20 years, beginning in 1751, when the population was 1.3 million.

(*a*) If Franklin's conjecture were right, what would the population of the United States be in 1990?

(*b*) In 1990 the population was 250 million. Assuming natural growth, what would the doubling time be?

---

25 The differential equation

$$L \frac{di}{dt} + Ri = E$$

occurs in the study of electric circuits. L, R, and E are constants that describe the inductance, resistance, and voltage, i is the current, and t is time. L and R are positive. Assume that di/dt is positive. Then $E - Ri$ is positive as well.

(*a*) Solve for di/dt in terms of i.

(*b*) The equation in (*a*) is separable. Solve it and express the answer in terms of R, L, E, and i_0, the initial current.

26 Find all functions $y = f(t)$ such that

$$\frac{dy}{dt} = k(y - A),$$

where k and A are constants. For negative k, this is Newton's law of cooling; y is the temperature of some heated object at time t. The room temperature is A. The differential equation $dy/dt = k(y - A)$ says, "The object cools at a rate proportional to the difference between its temperature and the room temperature."

(*a*) Solve by separating the variables.

(*b*) Solve by first rewriting it as $(d/dt)\,(y - A) = k(y - A)$. (Why is this step legal?)

27 A company is founded with a capital investment A. The plan is to have its rate of investment proportional to its total investment at any time. Let $f(t)$ denote the rate of investment at time t.

(*a*) Show that there is a constant k such that $f(t) = k[A + \int_0^t f(x)\,dx]$ for any $t \geq 0$.

(*b*) Find a formula for f.

28 (Doomsday equation) A differential equation of the form $dP/dt = kP^{1.01}$ is called a **doomsday equation**. The rate of growth is just slightly higher than that for natural growth. Solve the differential equation to find $P(t)$. How does $P(t)$ behave as t increases? Does $P(t)$ increase forever?

29 If the population of the western hemisphere is growing exponentially and the population of the eastern hemisphere is growing exponentially, does it follow that the population of the world is growing exponentially?

30 The following situations are all mathematically the same:

(1) A drug is administered in a dose of A grams to a patient and gradually leaves the system through excretion.

(2) Initially there is an amount A of smoke in a room. The air conditioner is turned on and gradually the smoke is removed. (Assume that the smoke is always thoroughly mixed.)

(3) Initially there is an amount A of some pollutant in a lake, when further dumping of toxic materials is prohibited. The rate at which water enters the lake equals the rate at which it leaves. (Assume the pollutant is thoroughly mixed.)

In each case, let $P(t)$ be the initial amount present (whether drug, smoke, or pollutant).

(*a*) Why is it reasonable to assume that there is a constant k such that for small intervals of time, Δt, $\Delta P \approx kP(t)\Delta t$?

(*b*) From (*a*) deduce that $P(t) = Ae^{kt}$.

(*c*) Is k positive or negative?

31 A certain fish increases in number at a rate proportional to the size of the population. In addition, it is being harvested at a constant rate. Let $P(t)$ be the size of the fish population at time t.

(*a*) Show that there are positive constants h and k such that for small Δt, $\Delta P \approx kP\,\Delta t - h\,\Delta t$.

(*b*) Find a formula for $P(t)$ in terms of $P(0)$, h, and k.

(*c*) Describe the behavior of $P(t)$ in the three cases $h = kP(0)$, $h > kP(0)$, and $h < kP(0)$.

32 If each of two functions grows exponentially, does their product? their quotient? their sum? their difference?

33 Find all functions $f(x)$ such that $f(x) = 3 \int_0^x f(t)\,dt$.

34 Let $I(x)$ be the intensity of sunlight at a depth of x meters in the ocean. As x increases, $I(x)$ decreases.

(*a*) Why is it reasonable to assume that there is a constant k (negative) such that $\Delta I \approx kI(x)\,\Delta x$ for small Δx?

(*b*) Deduce that $I(x) = I(0)e^{kx}$, where $I(0)$ is the intensity of sunlight at the surface. Incidentally, sunlight at a depth of 1 meter is only one-fourth as intense as at the surface.

35 A salesman, trying to persuade a tycoon to invest in Standard Coagulated Mutual Fund, shows him the accompanying graph which records the value of a similar investment made in the fund in 1965. ''Look! In the first 5 years the investment increased \$1,000,'' the salesman observed, ''but in the past 5 years it increased by \$2,000. It's really improving. Look at the slope of the graph from 1985 to 1990, which you can see clearly in Fig. 3.''

The tycoon replied, ''Hogwash. Though your graph is steeper from 1985 to 1990, in fact, the rate of return is less than from 1965 to 1970. Indeed, that was your best period.''

(*a*) If the percentage return on the accumulated investment remains the same over each 5-year period as the first 5-year period, sketch the graph.

(*b*) Explain the tycoon's reasoning.

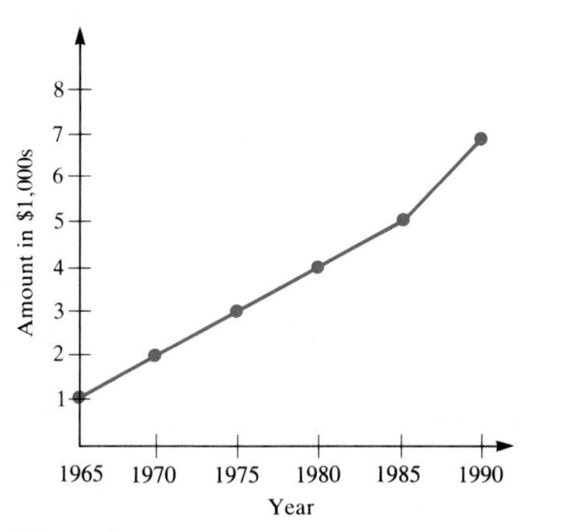

Figure 3

36 In the text we used the approximation $e^k \approx 1 + k$ to permit us to set k equal to percentage growth in unit time. Here we justify that approximation. Let $P(t) = Ae^{kt}$.

(*a*) In words, what does the quantity $[P(t+1) - P(t)]/P(t)$ represent?

(*b*) Show that $[P(t+1) - P(t)]/P(t) = e^k - 1$.

(*c*) Using a differential, show that $e^k - 1 \approx k$, when k is small.

(So if there is a 2 percent growth in 1 year, it is reasonable to set k equal to 0.02.)

37 If the doubling time is t_2, express $P(t) = P(0)e^{kt}$ in terms of $P(0)$ and an exponential involving t and t_2. (The base of the exponential should be chosen to be convenient; it will *not* be e.)

38 A particle moving through a liquid meets a ''drag'' proportional to the velocity; that is, its acceleration is proportional to its velocity. Let x denote its position and v its velocity at time t. Assume $v > 0$.

(*a*) Show that there is a positive constant k such that

$$\frac{dv}{dt} = -kv.$$

(*b*) Show that there is a constant A such that

$$v = Ae^{-kt}.$$

(*c*) Show that there is a constant B such that

$$x = -\frac{1}{k}Ae^{-kt} + B.$$

(*d*) How far does the particle travel as t goes from 0 to ∞?

39 In many cases of growth there is obviously a finite upper bound M which the population size must approach. Why is it reasonable to assume (or to take as a model) that

$$\frac{dP}{dt} = kP(t)[M - P(t)] \qquad 0 < P(t) < M \qquad (17)$$

for some constant k? This type of growth is called **inhibited** or **logistic**.

40 (*a*) Solve the differential equation in Exercise 39. *Hints:* You will need the identities

$$\frac{1}{P(M-P)} = \frac{1}{M}\left(\frac{1}{P} + \frac{1}{M-P}\right)$$

and $$\ln A - \ln B = \ln \frac{A}{B}.$$

After simplification, your answer should have the form

$$P(t) = \frac{M}{1 + ae^{-Mkt}}$$

for a suitable constant a.

(*b*) Find $\lim_{t\to\infty} P(t)$. Is this reasonable?

(*c*) Express a in terms of $P(0)$, M, and k.

41 By considering Eq. (17) in Exercise 39 directly (not the explicit formula in Exercise 40), show that

(*a*) P is an increasing function.

(*b*) The maximum rate of change of P occurs when $P(t) = M/2$.

(*c*) The graph of $P(t)$ has an inflection point.

42 Using AIDS data from 1982 to 1988, epidemiologists have noticed that the number of cases appears to be a cubic polynomial, not an exponential. James M. Hyman observed, ''I don't know of any other epidemic that's growing polynomially. Usually it's a very quick exponential, saturates out, then dies back down.'' It should be kept in mind that data on AIDS has a wide margin of error. See *Modeling the AIDS Epidemic,* Allyn Jackson, Notices of the American Mathematical Society, October 1989, vol. 36, no. 8, pp. 981–983. The accompanying table shows the number of deaths from AIDS in the United States in the given years. (*Source:* U.S. Centers for Disease Control).

Year	1982	1983	1984	1985	1986	1987
Deaths	434	1416	3196	6242	10,620	13,933

(*a*) What exponential curve $y = Ae^{kt}$ matches the data for the years 1982 and 1985? (Set $t = 0$ for the year 1982 and find A and k.)

(*b*) Graph the data in the table and the curve $y = Ae^{kt}$ that you obtained in (*a*). Use the same axes.

(*c*) If the number of deaths continued to grow exponentially, how many deaths would there have been in the year 1988, when $t = 6$? (The actual number of deaths in 1988 was 15,463, showing that the exponential phase of the epidemic had passed. However, if a cubic polynomial now provides the best model for the number of deaths, the growth rate is still very large.)

(For a discussion of more sophisticated models of the AIDS epidemic, see James R. Thompson, *Empirical Model Building,* Wiley, New York, 1989; in particular, pp. 79–81.)

6.8 L'HÔPITAL'S RULE

Back in Sec. 2.4 we distinguished between determinate and indeterminate limits. In this section we describe a technique for evaluating many indeterminate limits, known as l'Hôpital's rule.

Zero-over-Zero Limits

If f and g are functions and a is a number such that

$$\lim_{x\to a} f(x) = 2 \qquad \text{and} \qquad \lim_{x\to a} g(x) = 3,$$

then

$$\lim_{x\to a} \frac{f(x)}{g(x)} = \frac{2}{3}.$$

This problem presents no difficulty; no more information is needed about the functions f and g. But if

$$\lim_{x\to a} f(x) = 0 \qquad \text{and} \qquad \lim_{x\to a} g(x) = 0,$$

then finding

$$\lim_{x\to a} \frac{f(x)}{g(x)}$$

may present a serious problem. If you know *only* that

$$\lim_{x\to a} f(x) = 0 \qquad \text{and} \qquad \lim_{x\to a} g(x) = 0,$$

then you do not have enough information to determine

$$\lim_{x\to a} \frac{f(x)}{g(x)}.$$

In such cases, simply substituting the limits of the numerator and denominator produces 0/0, a meaningless expression traditionally called an **indeterminate form**

Theorem 1 describes a general technique for dealing with the troublesome quotient

$$\frac{f(x)}{g(x)}$$

when $f(x) \to 0$ and $g(x) \to 0$. It is known as the **zero-over-zero case** of l'Hôpital's rule.

Theorem 1 *L'Hôpital's rule (zero-over-zero case).* Let a be a number and let f and g be differentiable over some open interval that contains a. Assume also that $g'(x)$ is not 0 for any x in that interval except perhaps at a. If

$$\lim_{x\to a} f(x) = 0, \qquad \lim_{x\to a} g(x) = 0, \qquad \text{and} \qquad \lim_{x\to a} \frac{f'(x)}{g'(x)} = L,$$

then
$$\lim_{x\to a} \frac{f(x)}{g(x)} = L. \quad ■$$

Before worrying about *why* this theorem is true, we illustrate its use by an example.

EXAMPLE 1 Find $\lim_{x\to 1} (x^5 - 1)/(x^3 - 1)$.

SOLUTION In this case,

$$a = 1,\ f(x) = x^5 - 1, \text{ and } g(x) = x^3 - 1.$$

First checking that the assumptions of l'Hôpital's rule hold

All the assumptions of l'Hôpital's rule are satisfied. In particular,

$$\lim_{x\to 1} (x^5 - 1) = 0 \qquad \text{and} \qquad \lim_{x\to 1} (x^3 - 1) = 0.$$

According to l'Hôpital's rule,

$$\lim_{x\to 1} \frac{x^5 - 1}{x^3 - 1} = \lim_{x\to 1} \frac{(x^5 - 1)'}{(x^3 - 1)'}$$

if the latter limit exists. Now,

$$\lim_{x\to 1} \frac{(x^5 - 1)'}{(x^3 - 1)'} = \lim_{x\to 1} \frac{5x^4}{3x^2} \qquad \text{differentiation of numerator and denominator}$$

$$= \lim_{x\to 1} \tfrac{5}{3}x^2 \qquad \text{algebra}$$

$$= \tfrac{5}{3}.$$

Thus
$$\lim_{x\to 1} \frac{x^5 - 1}{x^3 - 1} = \frac{5}{3}. \quad ■$$

Argument for a special case of Theorem 1

A complete proof of Theorem 1 may be found in Exercises 158 and 159 of Sec. 6.S. Let us pause long enough here to make the theorem plausible. To do so, consider the *special case* where f, f', g, and g' are all continuous throughout an open interval containing a. Assume that $g'(x) \neq 0$ throughout the interval. Since we have $\lim_{x\to a} f(x) = 0$ and $\lim_{x\to a} g(x) = 0$, it follows by continuity that $f(a) = 0$ and $g(a) = 0$.

Assume that $\lim_{x\to a} f'(x)/g'(x) = L$. Then

$$\lim_{x\to a} \frac{f(x)}{g(x)} = \lim_{x\to a} \frac{f(x) - f(a)}{g(x) - g(a)} \qquad \text{since } f(a) = 0 \text{ and } g(a) = 0$$

$$= \lim_{x\to a} \frac{\dfrac{f(x) - f(a)}{x - a}}{\dfrac{g(x) - g(a)}{x - a}} \qquad \text{algebra}$$

$$= \frac{\lim_{x \to a} \frac{f(x) - f(a)}{x - a}}{\lim_{x \to a} \frac{g(x) - g(a)}{x - a}} \qquad \text{limit of quotients}$$

$$= \frac{f'(a)}{g'(a)} \qquad \text{by definition of } f'(a) \text{ and } g'(a)$$

$$= \frac{\lim_{x \to a} f'(x)}{\lim_{x \to a} g'(x)} \qquad f' \text{ and } g' \text{ are continuous}$$

$$= \lim_{x \to a} \frac{f'(x)}{g'(x)} \qquad \text{``limit of quotient'' property}$$

$$= L. \qquad \text{by assumption}$$

Consequently, $$\lim_{x \to a} \frac{f(x)}{g(x)} = L.$$

Sometimes it may be necessary to apply l'Hôpital's rule more than once, as in the next example.

EXAMPLE 2 Find $\lim_{x \to 0} (\sin x - x)/x^3$.

SOLUTION As $x \to 0$, both numerator and denominator approach 0. By l'Hôpital's rule,

$$\lim_{x \to 0} \frac{\sin x - x}{x^3} = \lim_{x \to 0} \frac{(\sin x - x)'}{(x^3)'},$$

$$= \lim_{x \to 0} \frac{\cos x - 1}{3x^2}.$$

Repeated application of l'Hôpital's rule

But as $x \to 0$, both $\cos x - 1 \to 0$ and $3x^2 \to 0$. So use l'Hôpital's rule again:

$$\lim_{x \to 0} \frac{\cos x - 1}{3x^2} = \lim_{x \to 0} \frac{(\cos x - 1)'}{(3x^2)'}$$

$$= \lim_{x \to 0} \frac{-\sin x}{6x}.$$

Or recall from Sec. 2.7 that $\lim_{x \to 0} \frac{\sin x}{x} = 1.$

Both $\sin x$ and $6x$ approach 0 as $x \to 0$. Use l'Hôpital's rule yet another time:

$$\lim_{x \to 0} \frac{-\sin x}{6x} = \lim_{x \to 0} \frac{(-\sin x)'}{(6x)'}$$

$$= \lim_{x \to 0} \frac{-\cos x}{6}$$

$$= \frac{-1}{6}.$$

So after three applications of l'Hôpital's rule we find that

$$\lim_{x \to 0} \frac{\sin x - x}{x^3} = -\frac{1}{6}. \quad ■$$

Sometimes a limit may be simplified before l'Hôpital's rule is applied. For instance, consider

$$\lim_{x\to 0} \frac{(\sin x - x)\cos^5 x}{x^3}.$$

Since $\lim_{x\to 0} \cos^5 x = 1$, we have

$$\lim_{x\to 0} \frac{(\sin x - x)\cos^5 x}{x^3} = \left(\lim_{x\to 0} \frac{\sin x - x}{x^3}\right) \cdot 1,$$

which, by Example 2, is $-\frac{1}{6}$. This shortcut saves a lot of work, as may be checked by finding the limit using l'Hôpital's rule without separating $\lim_{x\to 0} \cos^5 x$.

Replacing a by ∞ in Theorem 1, etc.

Theorem 1 concerns limits as $x \to a$. L'Hôpital's rule also applies if $x \to \infty$, $x \to -\infty$, $x \to a^+$, or $x \to a^-$. In the first case, we would assume that $f(x)$ and $g(x)$ are differentiable in some interval (c, ∞) and $g'(x)$ is not zero there. In the case of $x \to a^+$, assume that $f(x)$ and $g(x)$ are differentiable in some open interval (a, b) and $g'(x)$ is not 0 there.

Infinity-over-Infinity Limits

Theorem 1 concerns the problem of finding the limit of $f(x)/g(x)$ when both $f(x)$ and $g(x)$ approach 0. But a similar problem arises when both $f(x)$ and $g(x)$ get arbitrarily large as $x \to a$ or as $x \to \infty$. The behavior of the quotient $f(x)/g(x)$ will be influenced by how rapidly $f(x)$ and $g(x)$ become large.

In short, if $\lim_{x\to a} f(x) = \infty$ and $\lim_{x\to a} g(x) = \infty$, then $\lim_{x\to\infty} [f(x)/g(x)]$ is an indeterminate limit. "Infinity-over-infinity" is indeterminate. (See Sec. 2.4.)

Another indeterminate limit

The next theorem presents a form of l'Hôpital's rule that covers the case in which $f(x) \to \infty$ and $g(x) \to \infty$.

L'Hôpital's rule for the infinity-over-infinity case

Theorem 2 *L'Hôpital's rule (infinity-over-infinity case).* Let f and g be defined and differentiable for all x larger than some fixed number. Then, if

$$\lim_{x\to\infty} f(x) = \infty, \qquad \lim_{x\to\infty} g(x) = \infty, \qquad \text{and} \qquad \lim_{x\to\infty} \frac{f'(x)}{g'(x)} = L,$$

it follows that $$\lim_{x\to\infty} \frac{f(x)}{g(x)} = L.$$

A similar result holds for $x \to a$, $x \to a^-$, $x \to a^+$, or $x \to -\infty$. Moreover, $\lim_{x\to\infty} f(x)$ and $\lim_{x\to\infty} g(x)$ could both be $-\infty$, or one could be ∞ and the other $-\infty$.

EXAMPLE 3 Find $\lim_{x\to\infty} \frac{\ln x}{x^2}$.

SOLUTION First note that $\ln x \to \infty$ and $x^2 \to \infty$ as $x \to \infty$. We may therefore use l'Hôpital's rule in the "infinity-over-infinity" form.

We have
$$\lim_{x\to\infty} \frac{\ln x}{x^2} = \lim_{x\to\infty} \frac{(\ln x)'}{(x^2)'}$$
$$= \lim_{x\to\infty} \frac{1/x}{2x}$$
$$= \lim_{x\to\infty} \frac{1}{2x^2}$$
$$= 0.$$

Hence $\lim_{x\to\infty} [(\ln x)/x^2] = 0$. This says that $\ln x$ grows much more slowly than x^2 does as x gets large. ■

The next example conveys a warning.

EXAMPLE 4 Find $\lim_{x\to\infty} \dfrac{x - \cos x}{x}$. (1)

SOLUTION Both numerator and denominator approach ∞ as $x \to \infty$. Trying l'Hôpital's rule, we obtain

$$\lim_{x\to\infty} \frac{x - \cos x}{x} = \lim_{x\to\infty} \frac{(x - \cos x)'}{x'}$$
$$= \lim_{x\to\infty} \frac{1 + \sin x}{1}.$$

But $\lim_{x\to\infty} (1 + \sin x)$ does not exist, since $\sin x$ oscillates back and forth from -1 to 1 as $x \to \infty$.

L'Hôpital's rule may fail to provide an answer.

What can we conclude about (1)? Nothing at all. L'Hôpital's rule says that *if* $\lim_{x\to\infty} f'/g'$ exists, then $\lim_{x\to\infty} f/g$ exists. It says nothing about the case when $\lim_{x\to\infty} f'/g'$ does not exist.

It is not difficult to evaluate (1) directly, as follows:

$$\lim_{x\to\infty} \frac{x - \cos x}{x} = \lim_{x\to\infty} \left(1 - \frac{\cos x}{x}\right) \qquad \text{algebra}$$
$$= 1 - 0 \qquad \text{since } |\cos x| \le 1$$
$$= 1. \quad ■$$

Moral: Look carefully at a limit before you decide to use l'Hôpital's rule.

The proof of Theorem 2 is left to an advanced calculus course. However, it is easy to see why it is plausible. Imagine that $f(t)$ and $g(t)$ describe the locations on the x axis of two cars at time t. Call the cars the f-car and the g-car. See Fig. 1. Their velocities are therefore $f'(t)$ and $g'(t)$. These two cars are on endless journeys. But let us assume that as time $t \to \infty$ the f-car tends to travel at a speed closer and closer to L times the speed of the g-car. That is, assume that

$$\lim_{t\to\infty} \frac{f'(t)}{g'(t)} = L.$$

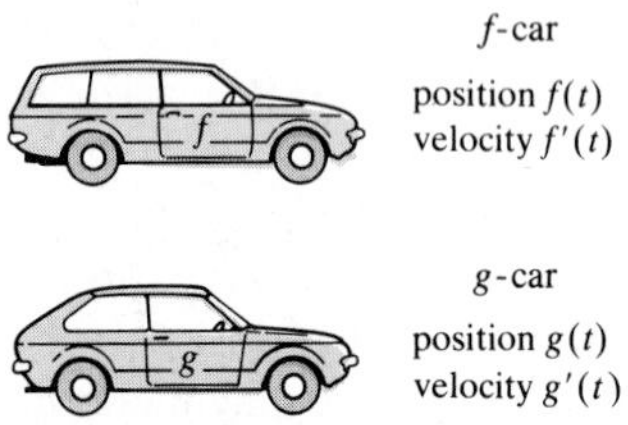

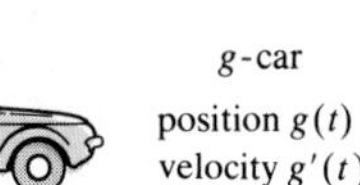

Figure 1

No matter how the two cars move in the short run, it seems reasonable that in the long run the f-car will tend to travel about L times as far as the g-car; that is,

$$\lim_{t\to\infty} \frac{f(t)}{g(t)} = L.$$

Transforming Limits to Make Them Fit the Rule

Many problems can be transformed to limits to which l'Hôpital's rule applies. For instance, the problem of finding

The zero-times-infinity case

$$\lim_{x\to 0^+} x \ln x$$

does not fit into l'Hôpital's rule, since it does not involve the quotient of two functions. As $x \to 0^+$, one factor, x, approaches 0 and the other factor, $\ln x$, approaches $-\infty$. So this is another type of indeterminate limit, a small number times a large number (''zero times infinity''). It is not obvious how this product, $x \ln x$, behaves as $x \to 0^+$. (In fact, such a limit can turn out to be ''small, medium, or large.'') A little algebraic manipulation, however, transforms the zero-times-infinity case into a problem to which l'Hôpital's rule applies, as we demonstrate in the next example.

EXAMPLE 5 Find $\lim_{x\to 0^+} x \ln x$.

SOLUTION Rewrite $x \ln x$ as a quotient, $\ln x/(1/x)$. Note that

$$\lim_{x\to 0^+} \ln x = -\infty \qquad \text{and} \qquad \lim_{x\to 0^+} \frac{1}{x} = \infty.$$

By l'Hôpital's rule,

$$\begin{aligned} \lim_{x\to 0^+} \frac{\ln x}{1/x} &= \lim_{x\to 0^+} \frac{1/x}{-1/x^2} \\ &= \lim_{x\to 0^+} (-x) \\ &= 0. \end{aligned}$$

Thus

$$\lim_{x\to 0^+} \frac{\ln x}{1/x} = 0,$$

from which it follows that $\lim_{x\to 0^+} x \ln x = 0$. (The factor x, which approaches 0, dominates the factor $\ln x$.) ■

The final example illustrates another type of limit that can be found by first relating it to problems to which l'Hôpital's rule applies.

Try this on your calculator first.

EXAMPLE 6 $\lim_{x\to 0^+} x^x$.

SOLUTION Since this limit involves an exponential, not a quotient, it does not fit directly into l'Hôpital's rule. But a little algebraic manipulation will change the problem to one covered by l'Hôpital's rule.

Let $$y = x^x.$$

Then $$\ln y = \ln x^x = x \ln x.$$

By Example 5, $$\lim_{x\to 0^+} x \ln x = 0.$$

Therefore

$$\lim_{x\to 0^+} \ln y = 0,$$

and

$$\begin{aligned} \lim_{x\to 0^+} y &= \lim_{x\to 0} e^{\ln y} \\ &= e^{\lim_{x\to 0} \ln y} \qquad e^x \text{ is continuous} \\ &= e^0 \\ &= 1. \end{aligned}$$

Hence $x^x \to 1$ as $x \to 0^+$. ■

Section Summary

We described l'Hôpital's rule, which is a technique for dealing with limits of the form "zero-over-zero" and "infinity-over-infinity." In both of these cases it asserts that

$$\lim_{x\to a} \frac{f(x)}{g(x)} = \lim_{x\to a} \frac{f'(x)}{g'(x)},$$

if the latter limit exists. Note that it concerns the quotient of two derivatives, *not* the derivative of the quotient.

The following table shows how some limits not of the zero-over-zero or infinity-over-infinity forms can be brought into those forms:

Form	*Name*	*Method*
$f(x)g(x)$; $f(x) \to 0$, $g(x) \to \infty$	Zero-times-infinity	Rewrite as $\dfrac{g(x)}{1/f(x)}$.
$f(x)^{g(x)}$; $f(x) \to 1$, $g(x) \to \infty$ $f(x)^{g(x)}$; $f(x) \to 0$, $g(x) \to 0$	One-to-the-infinity Zero-to-the-zero	Write $y = f(x)^{g(x)}$; take $\ln y$, find limit of $\ln y$ and then the limit of $y = e^{\ln y}$.

EXERCISES FOR SEC. 6.8: L'HÔPITAL'S RULE

In Exercises 1 to 12 check that l'Hôpital's rule applies and use it to find the limits.

1 $\lim_{x\to 2} \dfrac{x^3 - 8}{x^2 - 4}$

2 $\lim_{x\to 1} \dfrac{x^7 - 1}{x^3 - 1}$

3 $\lim_{x\to 0} \dfrac{\sin 3x}{\sin 2x}$

4 $\lim_{x\to 0} \dfrac{\sin x^2}{(\sin x)^2}$

5 $\lim_{x\to\infty} \dfrac{x^3}{e^x}$

6 $\lim_{x\to\infty} \dfrac{x^5}{3^x}$

7 $\lim_{x\to 0} \dfrac{1 - \cos x}{x^2}$

8 $\lim_{x\to 0} \dfrac{\sin x - x}{(\sin x)^3}$

9 $\lim_{x\to 0} \dfrac{\tan 3x}{\ln (1 + x)}$

10 $\lim_{x\to 1} \dfrac{\cos (\pi x/2)}{\ln x}$

11 $\lim_{x\to\infty} \dfrac{(\ln x)^2}{x}$ **12** $\lim_{x\to 0} \dfrac{\sin^{-1} x}{e^{2x} - 1}$

In each of Exercises 13 to 18 transform the problem into one to which l'Hôpital's rule applies; then find the limit.

13 $\lim_{x\to 0} (1 - 2x)^{1/x}$ **14** $\lim_{x\to 0} (1 + \sin 2x)^{\csc x}$

15 $\lim_{x\to 0^+} (\sin x)^{(e^x - 1)}$ **16** $\lim_{x\to 0^+} x^2 \ln x$

17 $\lim_{x\to 0^+} (\tan x)^{\tan 2x}$ **18** $\lim_{x\to 0^+} (e^x - 1) \ln x$

In Exercises 19 to 46 find the limits. Use l'Hôpital's rule if it applies. *Warning:* l'Hôpital's rule, carelessly applied, may give a wrong answer or no answer.

19 $\lim_{x\to\infty} \dfrac{2^x}{3^x}$ **20** $\lim_{x\to\infty} \dfrac{2^x + x}{3^x}$

21 $\lim_{x\to\infty} \dfrac{\log_2 x}{\log_3 x}$ **22** $\lim_{x\to 1} \dfrac{\log_2 x}{\log_3 x}$

23 $\lim_{x\to\infty} \left(\dfrac{1}{x} - \dfrac{1}{\sin x}\right)$

24 $\lim_{x\to\infty} (\sqrt{x^2 + 3} - \sqrt{x^2 + 4x})$

25 $\lim_{x\to\infty} \dfrac{x^2 + 3 \cos 5x}{x^2 - 2 \sin 4x}$ **26** $\lim_{x\to\infty} \dfrac{e^x - 1/x}{e^x + 1/x}$

27 $\lim_{x\to 0} \dfrac{3x^3 + x^2 - x}{5x^3 + x^2 + x}$ **28** $\lim_{x\to\infty} \dfrac{3x^3 + x^2 - x}{5x^3 + x^2 + x}$

29 $\lim_{x\to\infty} \dfrac{\sin x}{4 + \sin x}$ **30** $\lim_{x\to\infty} x \sin 3x$

31 $\lim_{x\to 1^+} (x - 1) \ln (x - 1)$ **32** $\lim_{x\to \pi/2} \dfrac{\tan x}{x - (\pi/2)}$

33 $\lim_{x\to 0} (\cos x)^{1/x}$ **34** $\lim_{x\to 0^+} x^{1/x}$

35 $\lim_{x\to\infty} \dfrac{\sin 2x}{\sin 3x}$ **36** $\lim_{x\to 1} \dfrac{x^2 - 1}{x^3 - 1}$

37 $\lim_{x\to 0} \dfrac{xe^x(1 + x)^3}{e^x - 1}$ **38** $\lim_{x\to 0} \dfrac{xe^x \cos^2 6x}{e^{2x} - 1}$

39 $\lim_{x\to 0} (\csc x - \cot x)$ **40** $\lim_{x\to 0} \dfrac{\csc x - \cot x}{\sin x}$

41 $\lim_{x\to 0} \dfrac{5^x - 3^x}{\sin x}$ **42** $\lim_{x\to 0} \dfrac{\tan^5 x - \tan^3 x}{1 - \cos x}$

43 $\lim_{x\to 2} \dfrac{x^3 + 8}{x^2 + 5}$ **44** $\lim_{x\to \pi/4} \dfrac{\sin 5x}{\sin 3x}$

45 $\lim_{x\to 0} \left(\dfrac{1}{1 - \cos x} - \dfrac{2}{x^2}\right)$ **46** $\lim_{x\to 0} \dfrac{\sin^{-1} x}{\tan^{-1} 2x}$

---

47 In Fig. 2 the unit circle is centered at O, BQ is a vertical tangent line, and the length of BP is the same as the length of BQ. What happens to the point E as $Q \to B$?

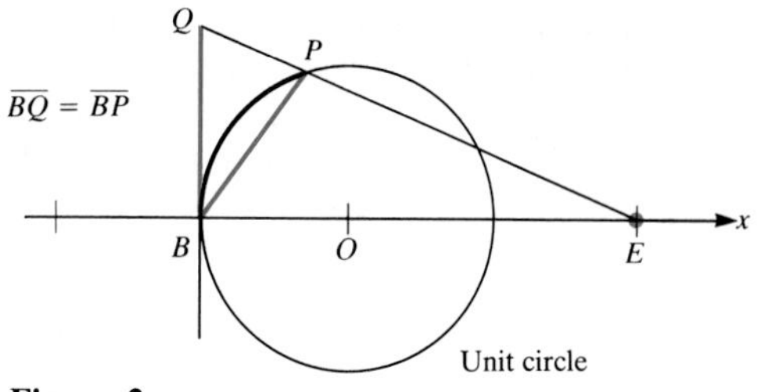

Figure 2

48 In Fig. 3 the unit circle is centered at the origin, BQ is a vertical tangent line, and the length of BQ is the same as the arc length $\widehat{BP}$. Prove that the x coordinate of R approaches -2 as $P \to B$.

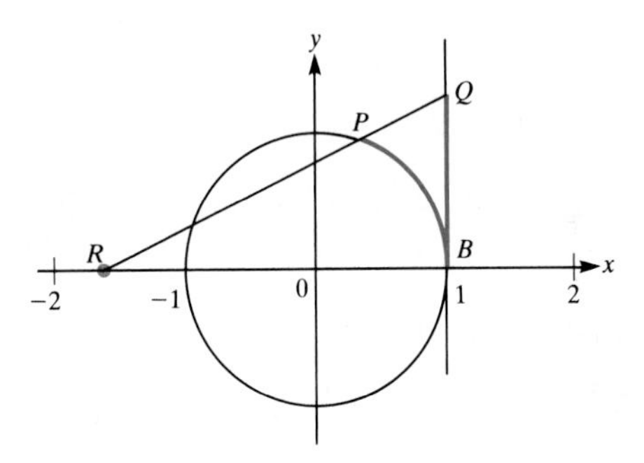

Figure 3

49 *Warning:* As Albert Einstein observed, "Common sense is the deposit of prejudice laid down in the mind before the age of 18." Exercise 20 of Sec. 2.7 asked the reader to guess a certain limit. Now that limit will be computed. In Fig. 4, which shows a circle, let $f(\theta)$ = area of triangle ABC and let $g(\theta)$ = area of the shaded region formed by deleting triangle OAC from the sector OBC. Clearly, $0 < f(\theta) < g(\theta)$.
(*a*) What would you guess is the value of $\lim_{\theta\to 0} f(\theta)/g(\theta)$?
(*b*) Find $\lim_{\theta\to 0} f(\theta)/g(\theta)$.

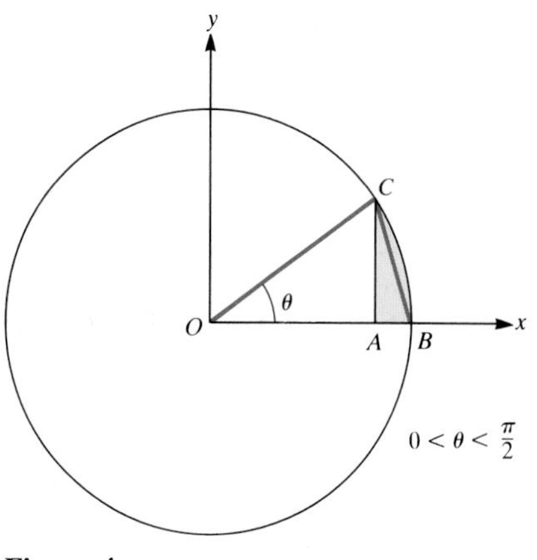

Figure 4

50 Figure 5 shows a triangle ABC and a shaded region cut from the parabola $y = x^2$ by a horizontal line. Find the limit, as $x \to 0$, of the ratio between the area of the triangle and the area of the shaded region. (This depends on Chap. 5.)

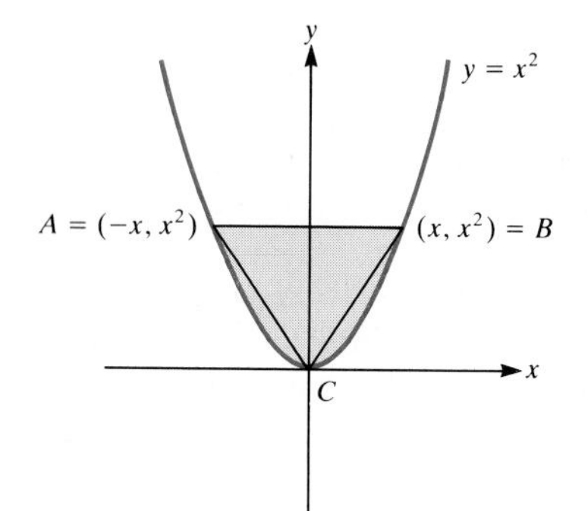

Figure 5

51 *(Economics)* In Eugene Silberberg, *The Structure of Economics,* McGraw-Hill, New York, 1978, this argument appears:

Consider the production function

$$y = k[\alpha x_1^{-\rho} + (1 - \alpha)x_2^{-\rho}]^{-1/\rho},$$

where k, α, x_1, x_2 are positive constants and $\alpha < 1$. Taking limits as $\rho \to 0^+$, we find that

$$\lim_{\rho \to 0^+} y = k\, x_1^{\alpha}\, x_2^{1-\alpha},$$

which is the Cobb-Douglas function, as expected.

Fill in the details.

52 Linus proposes this proof for Theorem 1: ''Since

$$\lim_{x \to a^+} f(x) = 0 \qquad \text{and} \qquad \lim_{x \to a^+} g(x) = 0,$$

I will define $f(a) = 0$ and $g(a) = 0$. Next I consider $x > a$ but near a. I now have continuous functions f and g defined on the closed interval $[a, x]$ and differentiable on the open interval (a, x). So, using the mean-value theorem, I conclude that there is a number c, $a < c < x$, such that

$$\frac{f(x) - f(a)}{x - a} = f'(c) \qquad \text{and} \qquad \frac{g(x) - g(a)}{x - a} = g'(c).$$

Since $f(a) = 0$ and $g(a) = 0$, these equations tell me that

$$f(x) = (x - a)f'(c) \qquad \text{and} \qquad g(x) = (x - a)g'(c).$$

Thus
$$\frac{f(x)}{g(x)} = \frac{f'(c)}{g'(c)}.$$

Hence
$$\lim_{x \to a^+} \frac{f(x)}{g(x)} = \lim_{x \to a^+} \frac{f'(c)}{g'(c)}$$
$$= L.$$

Alas, Linus made one error. What is it?

53 Find $\displaystyle\lim_{x \to 0} \left(\frac{1 + 2^x}{2}\right)^{1/x}$.

54 In R. P. Feynman, *Lectures on Physics,* Addison-Wesley, Reading, Mass., 1963, this remark appears: ''Here is the quantitative answer of what is right instead of kT. This expression

$$\frac{\hbar\omega}{e^{\hbar\omega/kT} - 1}$$

should, of course, approach kT as $\omega \to 0$. . . . See if you can prove that it does—learn how to do the mathematics.'' Do the mathematics.

55 Graph $y = x^x$ for $0 < x \le 1$, showing its minimum point.

56 Graph $y = (1 + x)^{1/x}$ for $x > -1$, $x \ne 0$, showing (*a*) where y is decreasing, (*b*) asymptotes, and (*c*) behavior of y for x near 0.

57 Graph $y = x \ln x$.

58 Graph $y = x^2 \ln x$.

6.9 THE HYPERBOLIC FUNCTIONS AND THEIR INVERSES

Certain combinations of the exponential functions e^x and e^{-x} occur often enough in differential equations and engineering—for instance, in the study of electric transmission and suspension cables—to be given names. This section defines these so-called **hyperbolic functions** and obtains their basic properties. Since the letter x will be needed later for another purpose, we will use the letter t when writing the two preceding exponentials, namely, e^t and e^{-t}.

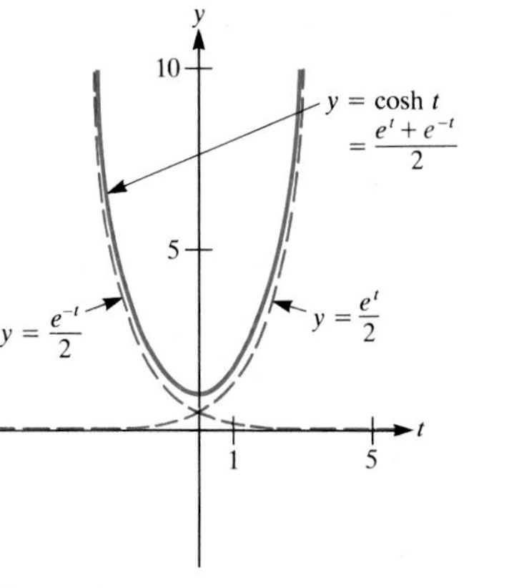

Figure 1

The Hyperbolic Functions

Definition *The hyperbolic cosine.* Let t be a real number. The **hyperbolic cosine** of t, denoted $\cosh t$, is given by the formula

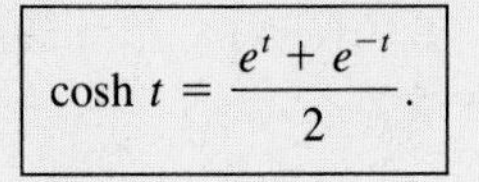

$$\cosh t = \frac{e^t + e^{-t}}{2}.$$

To graph $\cosh t$, note first that

$$\cosh(-t) = \frac{e^{-t} + e^{-(-t)}}{2} = \frac{e^{-t} + e^{t}}{2} = \cosh t.$$

Pronounced as written, "cosh," rhyming with "gosh"

Since $\cosh(-t) = \cosh t$, the cosh function is even, and so its graph is symmetric with respect to the vertical axis. Furthermore, $\cosh t$ is the sum of two terms:

$$\cosh t = \frac{e^t}{2} + \frac{e^{-t}}{2}.$$

As $t \to \infty$, the second term, $e^{-t}/2$, approaches 0. Thus for $t > 0$ and large, the graph of $\cosh t$ is just a little above that of $e^t/2$. This information, together with the fact that $\cosh 0 = (e^0 + e^{-0})/2 = 1$, is the basis for Fig. 1.

For $|t| \to \infty$, the graph of $y = \cosh t$ is asymptotic to the graph of $y = e^t/2$ or $y = e^{-t}/2$.

The curve $y = \cosh t$ in Fig. 1 is called a **catenary** (from the Latin *catena* meaning "chain"). It describes the shape of a freely hanging chain.

Definition *The hyperbolic sine.* Let t be a real number. The **hyperbolic sine** of t, denoted $\sinh t$, is given by the formula

$$\sinh t = \frac{e^t - e^{-t}}{2}.$$

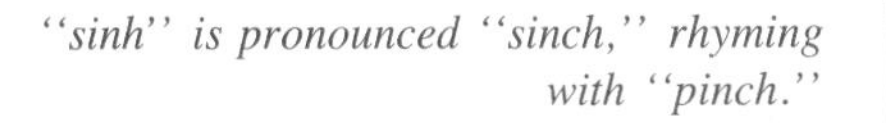

"sinh" is pronounced "sinch," rhyming with "pinch."

It is a simple matter to check that $\sinh 0 = 0$ and $\sinh(-t) = -\sinh t$, so that the graph of $\sinh t$ is symmetric with respect to the origin. Moreover, it lies below the graph of $e^t/2$. However, as $t \to \infty$, the two graphs approach each other since $e^{-t}/2 \to 0$ as $t \to \infty$. Figure 2 shows the graph of $\sinh t$.

Note the contrast between $\sinh t$ and $\sin t$. As t becomes large, the hyperbolic sine becomes large, $\lim_{t\to\infty} \sinh t = \infty$ and $\lim_{t\to-\infty} \sinh t = -\infty$. There is a similar contrast between $\cosh t$ and $\cos t$. While the trigonometric functions are periodic, the hyperbolic functions are not.

Example 1 shows why the functions $(e^t + e^{-t})/2$ and $(e^t - e^{-t})/2$ are called **hyperbolic**

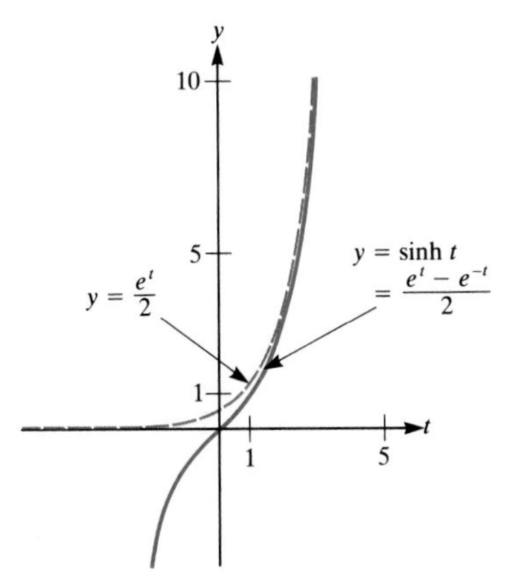

Figure 2

Why they are called hyperbolic functions

EXAMPLE 1 Show that for any real number t the point with coordinates

$$x = \cosh t, \qquad y = \sinh t$$

lies on the hyperbola $x^2 - y^2 = 1.$

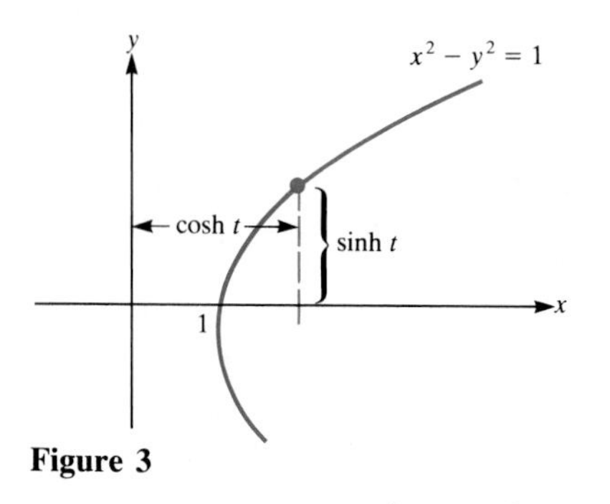

Figure 3

SOLUTION Compute $\cosh^2 t - \sinh^2 t$ and see whether it equals 1. We have

$$\cosh^2 t - \sinh^2 t = \left(\frac{e^t + e^{-t}}{2}\right)^2 - \left(\frac{e^t - e^{-t}}{2}\right)^2$$

$$= \frac{e^{2t} + 2e^te^{-t} + e^{-2t}}{4} - \frac{e^{2t} - 2e^te^{-t} + e^{-2t}}{4}$$

$$= \frac{2+2}{4} \qquad \text{cancellation}$$

$$= 1.$$

$\cosh^2 t - \sinh^2 t = 1$

Observe that since $\cosh t \geq 1$, the point $(\cosh t, \sinh t)$ is on the right half of the hyperbola $x^2 - y^2 = 1$, as shown in Fig. 3. ■

[Since $(\cos \theta, \sin \theta)$ lies on the circle $x^2 + y^2 = 1$, the trigonometric functions are called circular functions.]

The four other hyperbolic functions, namely, the hyperbolic tangent, the hyperbolic secant, the hyperbolic cotangent, and the hyperbolic cosecant, are defined as follows:

$$\tanh t = \frac{\sinh t}{\cosh t} \qquad \operatorname{sech} t = \frac{1}{\cosh t} \qquad \coth t = \frac{\cosh t}{\sinh t} \qquad \operatorname{csch} t = \frac{1}{\sinh t}.$$

Each can be expressed in terms of exponentials. For instance,

$$\tanh t = \frac{(e^t - e^{-t})/2}{(e^t + e^{-t})/2} = \frac{e^t - e^{-t}}{e^t + e^{-t}}.$$

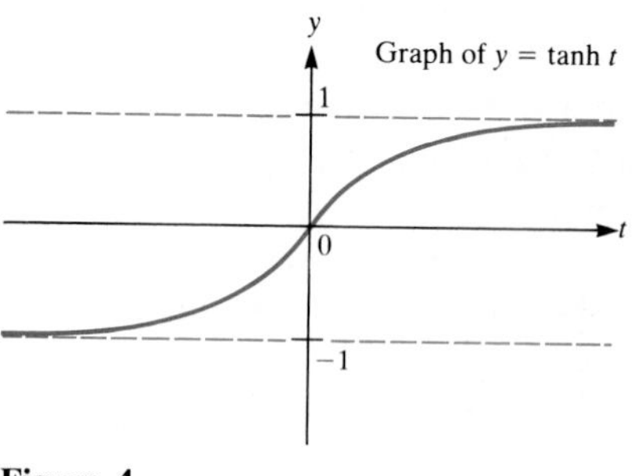

Figure 4

$-1 < \tanh t < 1$, so tanh is bounded. and cosh and sinh are unbounded (the exact opposite of their trig analogs).

As $t \to \infty$, $e^t \to \infty$ and $e^{-t} \to 0$. Thus $\lim_{t\to\infty} \tanh t = 1$. Similarly, $\lim_{t\to-\infty} \tanh t = -1$. Figure 4 is a graph of $y = \tanh t$.

The Derivatives of the Hyperbolic Functions

The derivatives of the six hyperbolic functions can be computed directly. For instance,

$$(\cosh t)' = \frac{(e^t + e^{-t})'}{2} = \frac{e^t - e^{-t}}{2} = \sinh t.$$

Table 1 lists the six derivatives. Notice that the formulas, except for the minus signs, are like those for the derivatives of the trigonometric functions.

Table 1

Function	*Derivative*
$\cosh t$	$\sinh t$
$\sinh t$	$\cosh t$
$\tanh t$	$\operatorname{sech}^2 t$
$\coth t$	$-\operatorname{csch}^2 t$
$\operatorname{sech} t$	$-\operatorname{sech} t \tanh t$
$\operatorname{csch} t$	$-\operatorname{csch} t \coth t$

The Inverses of the Hyperbolic Functions

Inverse hyperbolic functions appear on some calculators and in integral tables. Just as the hyperbolic functions are expressed in terms of the exponential function, each inverse hyperbolic function can be expressed in terms of a logarithm. They provide useful antiderivatives as well as solutions to some differential equations.

Consider the inverse of sinh t first. Since sinh t is increasing, it is one-to-one; there is no need to restrict its domain. To find its inverse, it is necessary to solve the equation

$$x = \sinh t$$

Finding the inverse of the hyperbolic sine

for t as a function of x. The steps are straightforward:

$$x = \frac{e^t - e^{-t}}{2},$$

$$2x = e^t - \frac{1}{e^t},$$

$$2xe^t = (e^t)^2 - 1,$$

or

$$(e^t)^2 - 2xe^t - 1 = 0. \tag{1}$$

Equation (1) is quadratic in the unknown e^t. By the quadratic formula,

$$e^t = \frac{2x \pm \sqrt{(2x)^2 + 4}}{2}$$

$$= x \pm \sqrt{x^2 + 1}.$$

Since $e^t > 0$ and $\sqrt{x^2 + 1} > x$, the plus sign is kept and the minus sign is rejected. Thus

$$e^t = x + \sqrt{x^2 + 1} \qquad \text{and} \qquad t = \ln (x + \sqrt{x^2 + 1}).$$

Consequently, the inverse of the function sinh t is given by the formula

Formula for $\sinh^{-1} x$

$$\boxed{\sinh^{-1} x = \ln (x + \sqrt{x^2 + 1}).}$$

Computation of $\tanh^{-1} x$ is a little different. Since the derivative of tanh t is $\text{sech}^2\, t$, the function tanh t is increasing and has an inverse. However, $|\tanh t| < 1$, and so the inverse function will be defined only for $|x| < 1$.

We find the inverse of $x = \tanh t$ as follows:

$$x = \tanh t = \frac{e^t - e^{-t}}{e^t + e^{-t}} = \frac{e^t - (1/e^t)}{e^t + (1/e^t)} = \frac{e^{2t} - 1}{e^{2t} + 1}.$$

Thus

$$xe^{2t} + x = e^{2t} - 1$$

$$1 + x = e^{2t}(1 - x)$$

$$e^{2t} = \frac{1 + x}{1 - x}$$

$$2t = \ln \left(\frac{1 + x}{1 - x}\right)$$

$$t = \frac{1}{2} \ln \left(\frac{1 + x}{1 - x}\right),$$

Formula for $\tanh^{-1} x$

so

$$\boxed{\tanh^{-1} x = \frac{1}{2} \ln \left(\frac{1 + x}{1 - x}\right)} \qquad |x| < 1.$$

Inverses of the other four hyperbolic functions are computed similarly. The functions $\cosh^{-1} x$ and $\text{sech}^{-1} x$ are chosen to be positive. Their formulas are included in Table 2.

The derivatives are found by differentiating the formulas in the second column.

$\cosh x$ and $\text{sech}\, x$ are one-to-one for $x \geq 0$.

Table 2

Function	*Formula*	*Derivative*	*Domain*
$\cosh^{-1} x$	$\ln (x + \sqrt{x^2 - 1})$	$\dfrac{1}{\sqrt{x^2 - 1}}$	$x \geq 1$
$\sinh^{-1} x$	$\ln (x + \sqrt{x^2 + 1})$	$\dfrac{1}{\sqrt{x^2 + 1}}$	x axis
$\tanh^{-1} x$	$\dfrac{1}{2} \ln \left(\dfrac{1 + x}{1 - x}\right)$	$\dfrac{1}{1 - x^2}$	$\|x\| < 1$
$\coth^{-1} x$	$\dfrac{1}{2} \ln \left(\dfrac{x + 1}{x - 1}\right)$	$\dfrac{1}{1 - x^2}$	$\|x\| > 1$
$\text{sech}^{-1} x$	$\ln \left(\dfrac{1 + \sqrt{1 - x^2}}{x}\right)$	$\dfrac{-1}{x\sqrt{1 - x^2}}$	$0 < x \leq 1$
$\text{csch}^{-1} x$	$\ln \left(\dfrac{1}{x} + \sqrt{1 + \dfrac{1}{x^2}}\right)$	$\dfrac{-1}{\|x\|\sqrt{1 + x^2}}$	$x \neq 0$

This example depends on Chap. 5.

EXAMPLE 2 Verify the formula

$$\int \frac{dx}{a^2 - x^2} = \frac{1}{a} \tanh^{-1} \frac{x}{a} + C \qquad a > 0,\ |x| < a,$$

included in many integral tables.

SOLUTION Let

$$\begin{aligned} y &= \frac{1}{a} \tanh^{-1} \frac{x}{a} + C \\ &= \frac{1}{a} \frac{1}{2} \ln \left(\frac{1 + x/a}{1 - x/a}\right) + C && \text{from preceding table} \\ &= \frac{1}{2a} \ln \left(\frac{a + x}{a - x}\right) + C && \text{algebra} \\ &= \frac{1}{2a} [\ln (a + x) - \ln (a - x)] + C. && \ln (A/B) = \ln A - \ln B \end{aligned}$$

Differentiation then yields

$$y' = \frac{1}{2a}\left(\frac{1}{a + x} + \frac{1}{a - x}\right) = \frac{1}{2a} \cdot \frac{2a}{a^2 - x^2} = \frac{1}{a^2 - x^2}. \quad \blacksquare$$

EXERCISES FOR SEC. 6.9: THE HYPERBOLIC FUNCTIONS AND THEIR INVERSES

Only Exercises 19 to 22, and 34 depend on Chap. 5. Exercises 1 to 14 concern the hyperbolic functions.

1 Show that $\cosh t + \sinh t = e^t$.

2 Show that $\cosh t - \sinh t = e^{-t}$.

In Exercises 3 to 6 differentiate the given functions and express the derivatives in terms of hyperbolic functions.

3 $\sinh t$ **4** $\tanh t$

5 $\operatorname{sech} t$ **6** $\coth t$

In Exercises 7 to 14 differentiate the functions. Be sure to use the chain rule.

7 $\cosh 3x$ **8** $\sinh 5x$

9 $\tanh \sqrt{x}$ **10** $\operatorname{sech} (\ln x)$

11 $e^{3x} \sinh x$ **12** $\dfrac{(\tanh 2x)(\operatorname{sech} 3x)}{\sqrt{1 + 2x}}$

13 $(\cosh 4x)(\coth 5x)(\operatorname{csch} x^2)$

14 $\dfrac{(\coth 5x)^{5/2}(\tanh 3x)^{1/3}}{3x + 4 \cos x}$

Exercises 15 to 22 concern the inverse hyperbolic functions. In Exercises 15 to 18 obtain the given formulas using the technique illustrated in the text.

15 $\cosh^{-1} x = \ln (x + \sqrt{x^2 - 1}) \qquad x \geq 1$

16 $\operatorname{sech}^{-1} x = \ln \left(\dfrac{1 + \sqrt{1 - x^2}}{x}\right) \qquad 0 < x \leq 1$

17 $\coth^{-1} x = \dfrac{1}{2} \ln \left(\dfrac{x + 1}{x - 1}\right) \qquad |x| > 1$

18 $\operatorname{csch}^{-1} x = \ln \left(\dfrac{1}{x} + \sqrt{1 + \dfrac{1}{x^2}}\right) \qquad x \neq 0$

In Exercises 19 to 22 verify the integration formulas by differentiation (as in Example 2).

19 $\displaystyle\int \frac{dx}{\sqrt{x^2 - 1}} = \cosh^{-1} x + C \qquad x > 1$

20 $\displaystyle\int \frac{dx}{\sqrt{x^2 + 1}} = \sinh^{-1} x + C$

21 $\displaystyle\int \frac{dx}{x\sqrt{1 - x^2}} = -\operatorname{sech}^{-1} x + C \qquad 0 < x < 1$

22 $\displaystyle\int \frac{dx}{x\sqrt{1 + x^2}} = -\operatorname{csch}^{-1} x + C \qquad x > 0$

In Exercises 23 to 28 use the definitions of the hyperbolic functions to verify the given identities.

23 (*a*) $\cosh (x + y) = \cosh x \cosh y + \sinh x \sinh y$
(*b*) $\sinh (x + y) = \sinh x \cosh y + \cosh x \sinh y$

24 $\tanh (x + y) = \dfrac{\tanh x + \tanh y}{1 + \tanh x \tanh y}$

25 (*a*) $\cosh (x - y) = \cosh x \cosh y - \sinh x \sinh y$
(*b*) $\sinh (x - y) = \sinh x \cosh y - \cosh x \sinh y$

26 (*a*) $\cosh 2x = \cosh^2 x + \sinh^2 x$
(*b*) $\sinh 2x = 2 \sinh x \cosh x$

27 (*a*) $2 \sinh^2 (x/2) = \cosh x - 1$
(*b*) $2 \cosh^2 (x/2) = \cosh x + 1$

28 $\operatorname{sech}^2 x + \tanh^2 x = 1$

29 At what angle does the graph of $y = \tanh x$ cross the x axis?

30 At what angle does the graph of $y = \sinh x$ cross the x axis?

---

31 (*a*) Compute $\cosh t$ and $\sinh t$ for $t = -3, -2, -1, 0, 1, 2,$ and 3.
(*b*) Plot the seven points $(\cosh t, \sinh t)$ given by (*a*). They should lie on the hyperbola $x^2 - y^2 = 1$.

32 (*a*) Compute $\cosh t$ and $e^t/2$ for $t = 0, 1, 2, 3,$ and 4.
(*b*) Using the data in (*a*), graph $\cosh t$ and $e^t/2$ relative to the same axes.

33 (*a*) Compute $\tanh x$ for $x = 0, 1, 2,$ and 3.
(*b*) Using the data in (*a*) and the fact that $\tanh (-x) = -\tanh x$, graph $y = \tanh x$.

34 Some integral tables contain the formulas

$$\int \frac{dx}{\sqrt{ax + b}\sqrt{cx + d}} = \begin{cases} \dfrac{2}{\sqrt{-ac}} \tan^{-1} \sqrt{\dfrac{-c(ax + b)}{a(cx + d)}} & a > 0,\ c < 0 \\[2ex] \dfrac{2}{\sqrt{ac}} \tanh^{-1} \sqrt{\dfrac{c(ax + b)}{a(cx + d)}} & a,\ c > 0 \end{cases}$$

The first formula is used if a and c have opposite signs, the second if they have the same signs. Check each of the formulas by differentiation.

35 Find the inflection points on the curve $y = \tanh x$.

36 Graph $y = \sinh x$ and $y = \sinh^{-1} x$ relative to the same axes and show any inflection points.

37 One of the applications of hyperbolic functions is to the study of motion in which the resistance of the medium is proportional to the square of the velocity. Suppose that a body starts from rest and falls x meters in t seconds. Let g (a constant) be the acceleration due to gravity. It can be shown that there is a constant $V > 0$ such that

$$x = \frac{V^2}{g} \ln \cosh \frac{gt}{V}.$$

(*a*) Find the velocity $v(t) = dx/dt$ as a function of t.
(*b*) Show that $\lim_{t \to \infty} v(t) = V$.
(*c*) Compute the acceleration dv/dt as a function of t.
(*d*) Show that the acceleration equals $g - g(v/V)^2$.
(*e*) What is the limit of the acceleration as $t \to \infty$?

38 Two electrons *repel* each other with a force proportional to the distance x between them. There is thus a positive constant k such that

$$\frac{d^2x}{dt^2} = kx.$$

(a) Show that, for any constants A and B, the function $x = A \cosh \sqrt{k}t + B \sinh \sqrt{k}t$ satisfies the given differential equation.

(b) If at time $t = 0$ the electrons are a distance a apart and motionless, show that $A = a$ and $B = 0$. (Thus $x = a \cosh \sqrt{k}\, t$.)

6.S SUMMARY

This chapter completed the work of finding the derivatives of all the elementary functions. It began in Sec. 6.1 with a discussion of logarithms ("a logarithm is an exponent"), defined e in Sec. 6.2, and obtained the derivatives of the logarithmic functions in Sec. 6.3.

Section 6.5 used the notion of an inverse function (defined in Sec. 6.4) to find the derivative of b^x and Sec. 6.6 did the same for the inverse trigonometric functions.

Separable differential equations were solved in Sec. 6.7 and applied to natural growth and decay.

Section 6.8 presented l'Hôpital's rule, which can deal with many limits that are of the form $\lim_{x \to a} f(x)/g(x)$, where $f(x)$ and $g(x)$ both approach 0 or both approach ∞. In such instances $f(x)/g(x)$ is called an indeterminate form of the zero-over-zero case or infinity-over-infinity case. Certain other indeterminate forms can be transformed to the type covered by l'Hôpital's rule.

Section 6.9 discussed the hyperbolic functions and their inverses, which are a kind of shorthand for functions that can otherwise be described in terms of exponential and logarithmic functions. Exercises in this review that require the hyperbolic functions are so labeled and should be omitted if Sec. 6.9 was not covered.

Vocabulary and Symbols

$e = \lim_{x \to 0} (1 + x)^{1/x} \approx 2.718$
$e = \lim_{n \to \infty} (1 + 1/n)^n$
$\ln x$ $(= \log_e x)$
one-to-one
inverse function
inverse trigonometric functions: $\arcsin x$, $\sin^{-1} x$, etc.
elementary function
algebraic function
transcendental function
differential equation (D.E.)
solution of a differential equation
order of a differential equation
separable differential equation
natural growth or decay $dP/dt = kP$
growth constant k (k positive for growth, negative for decay)
doubling-time t_2
half-life $t_{1/2}$
l'Hôpital's rule
indeterminate form
zero-over-zero case
infinity-over-infinity case
hyperbolic functions
inverse hyperbolic functions

Key Facts

The following table lists the derivatives found in Chap. 3 and in this chapter:

Formulas

Function	*Derivative*	*Function*	*Derivative*	*Function*	*Derivative*
x^a	ax^{a-1}	$\sin x$	$\cos x$	$\sin^{-1} x$	$\dfrac{1}{\sqrt{1-x^2}}$
$\sqrt{x}$	$\dfrac{1}{2\sqrt{x}}$	$\cos x$	$-\sin x$		
e^x	e^x	$\tan x$	$\sec^2 x$	$\tan^{-1} x$	$\dfrac{1}{1+x^2}$
b^x	$(\ln b)\, b^x$	$\cot x$	$-\csc^2 x$		
$\ln x$	$\dfrac{1}{x}, x > 0$	$\sec x$	$\sec x \tan x$	$\sec^{-1} x$	$\dfrac{1}{\lvert x\rvert\sqrt{x^2-1}}$
$\ln\lvert x\rvert$	$\dfrac{1}{x}, x \neq 0$	$\csc x$	$-\csc x \cot x$		

Remember where the minus signs go.

It is not necessary to memorize the formulas for the derivatives of $\cos^{-1}$, $\cot^{-1}$, and $\csc^{-1}$. (They are obtained by putting minus signs in front of the last three formulas in the list.)

A separable D.E. has the form

$$\frac{dy}{dx} = \frac{g(x)}{h(y)}.$$

[It may also have the form $dy/dx = g(x)h(y)$ or $dy/dx = h(y)/g(x)$.] The equation is solved by bringing all x's to one side and all y's to the other (separating the variables), then integrating:

$$\int h(y)\, dy = \int g(x)\, dx + C.$$

L'Hôpital's rule concerns the behavior of $f(x)/g(x)$ in case numerator and denominator both approach 0 or both approach ∞. If

$$\lim_{x \to a} \frac{f'(x)}{g'(x)} \text{ exists,}$$

then

$$\lim_{x \to a} \frac{f(x)}{g(x)} = \lim_{x \to a} \frac{f'(x)}{g'(x)}.$$

The forms "zero times infinity," "one to the infinity," and "zero to the zero" can be reduced to l'Hôpital's form. In the last two cases, take logs, find the limits of the logs, and then be sure to complete the problem by finding the original limits.

GUIDE QUIZ ON CHAP. 6: TOPICS IN DIFFERENTIAL CALCULUS

1 Differentiate:
(*a*) $\sin^{-1}(x^3)$ (*b*) $e^{\sin^{-1} 3x}$
(*c*) $\sin[(\tan^{-1} x)^2]$ (*d*) $(x^2 + 5)x^{\sqrt{x}}$
(*e*) $\ln(\sin x)$ (*f*) 7^{4x-1}
(*g*) $(\tan x)/(\sec^{-1} 2x)$ (*h*) $\ln(x^2 - 5x + 4)$
(*i*) $\{x^{-5}[\sqrt[3]{\cos 4x}]^8\}^{1/4}$

2 Differentiate and simplify your answer.
(*a*) $\dfrac{e^{ax}}{a^2 + b^2}(a \cos bx + b \sin bx)$
(*b*) $x \sin^{-1} ax + \dfrac{1}{a}\sqrt{1 - a^2x^2}$
(*c*) $x \tan^{-1} ax - \dfrac{1}{2a} \ln(1 + a^2x^2)$
(*d*) $x \sec^{-1} ax - \dfrac{1}{a} \ln(ax + \sqrt{a^2x^2 - 1})$, $ax > 0$

3 Solve:
(*a*) $\dfrac{dy}{dx} = e^{-2y}x^3$ (*b*) $\dfrac{dy}{dx} = \dfrac{4y^2 + 1}{y}$

4 Radon has a half-life of 3.825 days. How long does it take for radon to diminish to only 10 percent of its original amount? 50 percent of its original amount?

5 Examine these limits:
(*a*) $\lim_{x \to 0} \dfrac{\sin 2x}{\tan^{-1} 3x}$ (*b*) $\lim_{x \to \infty} \dfrac{\sin 2x}{\tan^{-1} 3x}$
(*c*) $\lim_{x \to (\pi/2)^-} (\sec x - \tan x)$ (*d*) $\lim_{x \to 0} (1 - \cos 2x)^x$
(*e*) $\lim_{x \to (\pi/2)^-} \dfrac{\int_0^x \tan^2 \theta\, d\theta}{\tan x}$ (uses Chap. 5)

6 Arrange these functions in order of increasing size for large x: x^3, $\ln x$, 1.001^x, 2^x, $\log_{10} x$.

7 Let $f(x) = (\ln x)/x$.
(*a*) What is the domain of f?
(*b*) Find any x intercepts of f.
(*c*) Find any critical numbers of f.
(*d*) Use the second-derivative test to find whether there is a local maximum or minimum at any critical number.
(*e*) For which x is f increasing? decreasing?
(*f*) Does f have a global minimum? a global maximum?
(*g*) Find $\lim_{x \to 0^+} f(x)$.
(*h*) Find $\lim_{x \to \infty} f(x)$.
(*i*) Graph f.

8 (*a*) What is meant by e?
(*b*) What is e to three decimal places?

9 Assume that a calculator has a log-key ($\log_{10}$). How could you use the log-key to find $\log_2 3$?

10 Determine whether $f(x) = x^2 - 2x$ is one-to-one on the following domains. If it is, find its inverse function.
(*a*) $[0, 2]$ (*b*) $[1, 2]$ (*c*) $[-2, 0]$

Exercise 11 uses the hyperbolic functions from Sec. 6.9.

11 Differentiate the functions.
(*a*) $\cosh 2x$ (*b*) $\sinh \sqrt{x}$
(*c*) $\tanh 4x^3$ (*d*) $\sinh^{-1} e^x$

REVIEW EXERCISES FOR CHAP. 6: TOPICS IN DIFFERENTIAL CALCULUS

The next chapter, which treats the problem of finding an antiderivative, assumes a mastery of the formal computation of derivatives. Do as many of the first 78 exercises as you can find time for.

In Exercises 1 to 78 differentiate and simplify. (Assume that the input for any logarithm is positive.)

1 $\sqrt{1+x^3}$ 2 5^{x^2}

3 $\sqrt{x}$ 4 $\dfrac{1}{\sqrt{x}}$

5 $\cos^2 3x$ 6 $\sin^{-1} 3x$

7 $\sqrt{x^3}$ 8 $\dfrac{1}{3}\tan^{-1}\dfrac{x}{3}$

9 $\sqrt{\sin x}$ 10 $\dfrac{\cos 5x}{x^2}$

11 $\cot x^2$ 12 $x^2 \sin 3x$

13 $x^{5/6}\sin^{-1} x$ 14 $e^{\sqrt{x}}$

15 $x^2 e^{3x}$ 16 $\sin^2 2x$

17 $\ln(\sec 3x + \tan 3x)$ 18 $\dfrac{1}{5}\tan^{-1}\dfrac{x}{5}$

19 $\cos\sqrt{x}$ 20 $e^{-x}\tan x^2$

21 $\ln(\sec x + \tan x)$ 22 $3\cot 5x + 5\csc 3x$

23 $\dfrac{1}{\sqrt{6+3x^2}}$ 24 $\ln(\sin 2x)$

25 $\sqrt{\frac{2}{15}(5x+7)^3}$ 26 $x\cos 5x + \sin 5x$

27 $\dfrac{x}{3} - \dfrac{4}{9}\ln(3x+4)$ 28 $\dfrac{\sqrt{4-9x^2}}{x}$

29 $(1+x^2)^5 \sin 3x$ 30 $\cos[\log_{10}(3x+1)]$

31 $(1+2x)^5\cos 3x$ 32 $\dfrac{\sin 3x}{(x^2+1)^5}$

33 $\cos 3x \sin 4x$ 34 $\tan\sqrt{x}$

35 $\csc 3x^2$ 36 $\dfrac{1}{\sqrt{1-x^2}}$

37 $x\left(\dfrac{x^2}{1+x}\right)^3$ 38 $\left(\dfrac{1+2x}{1+3x}\right)^4$

39 $x\sec 3x$ 40 $[\ln(x^2+1)]^3$

41 $\ln(x+\sqrt{x^2+1})$ 42 $\dfrac{\sin^3 2x}{x^2+x}$

43 $e^{-x}\tan^{-1} x^2$ 44 $\log_{10}(x^2+1)$

45 $2e^{\sqrt{x}}(\sqrt{x}-1)$ 46 $\sin^{-1}\dfrac{x}{5}$

47 $\dfrac{\ln x}{x^2}$ 48 $(x^2+1)\sin 2x$

49 $\dfrac{\sin^2 x}{\cos x}$ 50 $x^5 - 2x + \ln(2x+3)$

51 $(\sec^{-1} 3x)x^2\ln(1+x^2)$ 52 $\sin^3(1+x^2)$

53 $x - 2\ln(x-1) + \dfrac{1}{x+1}$

54 $\ln\sqrt{\dfrac{1+x^2}{1+x^3}}$ 55 $\ln\left(\dfrac{1}{6x^2+3x+1}\right)$

56 $\ln(\sqrt{4+x}\,\sqrt[3]{x^2+1})$

57 $\ln\left[\dfrac{(5x+1)^3(6x+1)^2}{(2x+1)^4}\right]$

58 $\dfrac{1}{8}\left[\ln(2x+1) + \dfrac{2}{2x+1} - \dfrac{1}{2(2x+1)^2}\right]$

59 $\dfrac{1}{2}\left(x\sqrt{9-x^2} + 9\sin^{-1}\dfrac{x}{3}\right)$

60 $\dfrac{1}{8}\left[2x+3 - 6\ln(2x+3) - \dfrac{9}{2x+3}\right]$

61 $\tan 3x - 3x$ 62 $\dfrac{1}{\sqrt{6}}\tan^{-1}\left(x\sqrt{\dfrac{2}{3}}\right)$

63 $\ln(x+\sqrt{x^2+25})$ 64 $\dfrac{e^x(\sin 2x - 2\cos 2x)}{5}$

65 $x\sin^{-1} x + \sqrt{1-x^2}$

66 $x\tan^{-1} x - \frac{1}{2}\ln(1+x^2)$

67 $\frac{1}{3}\ln(\tan 3x + \sec 3x)$

68 $\frac{1}{6}\tan^2 3x + \frac{1}{3}\ln\cos 3x$

69 $-\frac{1}{3}\cos 3x + \frac{1}{9}\cos^3 3x$ 70 $\frac{1}{8}e^{-2x}(4x^2+4x+2)$

71 $\dfrac{x}{2}\sqrt{4x^2+3} + \dfrac{3}{4}\ln\left(2x + \sqrt{4x^2+3}\right)$

72 $\sqrt{1+\sqrt[3]{x}}$

73 $\dfrac{x^3(x^4 - x + 3)}{(x+1)^2}$ **74** $\dfrac{1}{1 + \csc 5x}$

75 $(1 + 3x)^{x^2}$ **76** $2^x 5^{x^2} 7^{x^3}$

77 $\dfrac{x \ln x}{(1+x^2)^5}$ **78** $x^3 \cot^3 \sqrt{x^3}$

79 (*a*) Fill in this table for the function $\log_2 x$:

x	$\frac{1}{8}$	$\frac{1}{4}$	$\frac{1}{2}$	1	2	4	8
$\log_2 x$							

(*b*) Plot the seven points in (*a*) and graph $\log_2 x$.
(*c*) What is $\lim_{x\to\infty} \log_2 x$?
(*d*) What is $\lim_{x\to 0^+} \log_2 x$?

Exercises 80 to 92 depend on material from Chap. 5. In Exercises 80 to 90 check the equations by differentiation. The letters a and b denote constants.

80 $\displaystyle\int \ln ax\,dx = x \ln ax - x + C$

81 $\displaystyle\int x \ln ax\,dx = \frac{x^2}{2}\ln ax - \frac{x^2}{4} + C$

82 $\displaystyle\int x^2 \ln ax\,dx = \frac{x^3}{3}\ln ax - \frac{x^3}{9} + C$

83 $\displaystyle\int \frac{dx}{x \ln ax} = \ln(\ln ax) + C$

84 $\displaystyle\int \frac{dx}{\sin x} = \ln|\csc x - \cot x| + C$

85 $\displaystyle\int \tan ax\,dx = -\frac{1}{a}\ln|\cos ax| + C$

86 $\displaystyle\int \tan^3 ax\,dx = \frac{1}{2a}\tan^2 ax + \frac{1}{a}\ln|\cos ax| + C$

87 $\displaystyle\int \frac{dx}{\sin ax \cos ax} = \frac{1}{a}\ln|\tan ax| + C$

88 $\displaystyle\int \frac{dx}{\sqrt{ax^2+b}} = \frac{1}{\sqrt{a}}\ln\left|x\sqrt{a} + \sqrt{ax^2+b}\right| + C \quad (a > 0)$

89 $\displaystyle\int \frac{x\,dx}{(ax+b)^2} = \frac{b}{a^2(ax+b)} + \frac{1}{a^2}\ln|ax+b| + C$

90 $\displaystyle\int (\ln ax)^2\,dx = x(\ln ax)^2 - 2x\ln ax + 2x + C$

91 (*a*) Differentiate $\ln\left|\dfrac{1+x}{1-x}\right|$.

(*b*) Find the area of the region under the curve $y = 1/(1-x^2)$ and above $[0, \frac{1}{2}]$.

92 Find the area of the region under the curve $y = x^3/(x^4+5)$ and above $[1, 2]$.

93 Differentiate and then simplify your answer:

$$\ln\left|\frac{\sqrt{ax+b} - b}{\sqrt{ax+b} + b}\right|.$$

94 Show that, for b and $c > 1$, $\log_b x/\log_c x$ is constant by differentiating it.

Exercises 95 and 96 depend on Chap. 5.

95 Find these integrals:

(*a*) $\displaystyle\int \frac{3x^2+1}{x^3+x-6}\,dx$ (*b*) $\displaystyle\int \frac{\cos 2x}{\sin 2x}\,dx$

(*c*) $\displaystyle\int \frac{dx}{5x+3}$ (*d*) $\displaystyle\int \frac{dx}{(5x+3)^2}$

96 Compute:

(*a*) $\displaystyle\int \frac{dx}{(3x+2)^2}$ (*b*) $\displaystyle\int \frac{dx}{\sqrt{3x+2}}$ (*c*) $\displaystyle\int \frac{dx}{3x+2}$

97 Differentiate:

(*a*) $\ln\left[\dfrac{\sqrt[3]{\tan 4x}\,(1-2x)^5}{\sqrt[3]{(1-3x)^2}}\right]$

(*b*) $\dfrac{(1+x^2)^3\sqrt{1+x}}{\sin 3x}$

98 The graph in Fig. 1 shows the tangent line to the curve $y = \ln x$ at a point (x_0, y_0). Prove that AB has length 1, independent of the choice of (x_0, y_0).

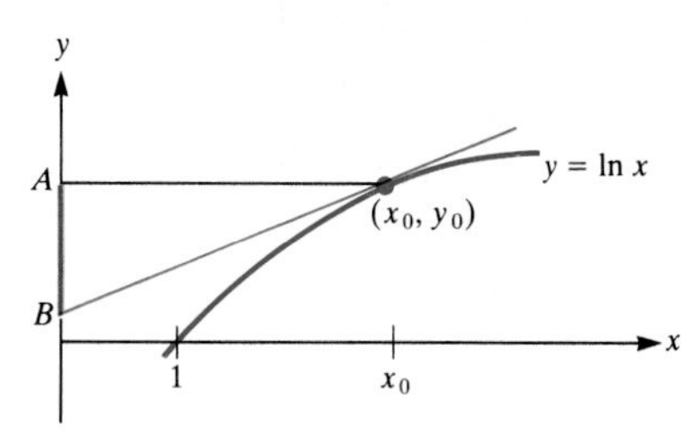

Figure 1

99 Examine (*a*) $\lim_{x\to 1} \log_x 2$, (*b*) $\log_{x\to\infty} \log_x 2$.

100 The **information content** or **entropy** of a binary source (such as a telegraph that transmits dots and dashes) whose two values occur with probabilities p and $1-p$ is defined as $H(p) = -p\ln p - (1-p)\ln(1-p)$, where $0 < p < 1$. Show that H has a maximum at $p = \frac{1}{2}$. The practical significance of this result is that for maximum flow of information per unit time, the two values should, in the long run, appear in equal proportions.

101 (See Exercise 100.) Let p be fixed so that $0 < p < 1$. Define $M(q) = -p \ln q - (1 - p) \ln (1 - q)$. Show that $H(p) \leq M(q)$ for $0 < q < 1$ and that equality holds if and only if $p = q$.

In each of Exercises 102 to 104, evaluate the limit by first showing that it is the derivative of a certain function at a specific number.

102 $\lim_{x \to 3} \frac{\ln (1 + 2x) - \ln 7}{x - 3}$

103 $\lim_{x \to 0} \frac{\ln (2 + x) - \ln 2}{x}$

104 $\lim_{x \to 0} \frac{\ln (1 + x)}{x}$

105 (*a*) Evaluate $(1 - 2h)^{1/h}$ for $h = 0.01$.
(*b*) Find $\lim_{h \to 0} (1 - 2h)^{1/h}$.

106 Find the coordinates of the point P on the curve $y = \ln x$ such that the line through (0, 0) and P is tangent to the curve.

In Exercises 107 to 137 examine the limits.

107 $\lim_{x \to \infty} x^{1/\log_2 x}$

108 $\lim_{x \to \pi/4} \frac{\sin x - \sqrt{2}/2}{x - \pi/4}$

109 $\lim_{h \to 0} \frac{e^{3+h} - e^3}{h}$

110 $\lim_{x \to 1} \frac{\sin \pi x}{x - 1}$

111 $\lim_{x \to 0} \frac{\cos \sqrt{x} - 1}{\tan x}$

112 $\lim_{x \to \infty} 2^x e^{-x}$

113 $\lim_{x \to 0} (1 + 2x^2)^{1/x^2}$

114 $\lim_{x \to 0} (1 + 3x)^{1/x}$

115 $\lim_{x \to -\infty} \frac{e^x - e^{-x}}{e^x + e^{-x}}$

116 $\lim_{x \to \infty} \frac{x^2 + 5}{2x^2 + 6x}$

117 $\lim_{x \to 0} \frac{1 - \cos x}{x + \tan x}$

118 $\lim_{x \to 0^+} (\sin x)^{\sin x}$

119 $\lim_{x \to 0} \frac{\sin 2x}{e^{3x} - 1}$

120 $\lim_{x \to \infty} \frac{(x^2 + 1)^5}{e^x}$

121 $\lim_{x \to \infty} \frac{3x - \sin x}{x + \sqrt{x}}$

122 $\lim_{x \to 0} \frac{xe^x}{e^x - 1}$

123 $\lim_{x \to 2} \frac{x^3 - 8}{x^2 - 4}$

124 $\lim_{x \to 0} \frac{\sin x^2}{x \sin x}$

125 $\lim_{x \to \pi/2} \frac{\sin x}{1 + \cos x}$

126 $\lim_{x \to 1} \frac{e^x + 1}{e^x - 1}$

127 $\lim_{x \to 3^+} \frac{\ln (x - 3)}{x - 3}$

128 $\lim_{x \to 0^+} \frac{1 - \cos x}{x - \tan x}$

129 $\lim_{x \to \infty} \frac{\cos x}{x}$

130 $\lim_{x \to 2} \frac{5^x + 3^x}{x}$

131 $\lim_{x \to 0} \frac{5^x - 3^x}{x}$

132 $\lim_{x \to \infty} \frac{e^{-x}}{x^2}$

133 $\lim_{x \to \infty} \frac{\ln (x^2 + 1)}{\ln (x^2 + 8)}$

134 $\lim_{x \to \infty} e^{-x} \ln x$

135 $\lim_{x \to \infty} (2^x - x^{10})^{1/x}$

136 $\lim_{x \to 0^+} \left(\frac{\sin x}{x}\right)^{1/x}$

137 $\lim_{h \to 0} \frac{e^{3+h} - e^3}{1 - h}$

138 Graph $y = xe^{-x}/(x + 1)$, showing intercepts, critical points, relative maxima and minima, and asymptotes.

139 (*a*) Why does calculus use radian measure?
(*b*) Why does calculus use the base e for logarithms?
(*c*) Why does calculus use the base e for exponentials?

140 Using the e^x-key, the ln-key, and the × (multiplication) but not the y^x-key, how would you calculate 3^{80} on a calculator?

141 In which cases below is it possible to determine $\lim_{x \to a} f(x)^{g(x)}$ without further information about the functions?
(*a*) $\lim_{x \to a} f(x) = 0$; $\lim_{x \to a} g(x) = 7$
(*b*) $\lim_{x \to a} f(x) = 2$; $\lim_{x \to a} g(x) = 0$
(*c*) $\lim_{x \to a} f(x) = 0$; $\lim_{x \to a} g(x) = 0$
(*d*) $\lim_{x \to a} f(x) = 0$; $\lim_{x \to a} g(x) = \infty$
(*e*) $\lim_{x \to a} f(x) = \infty$; $\lim_{x \to a} g(x) = 0$
(*f*) $\lim_{x \to a} f(x) = \infty$; $\lim_{x \to a} g(x) = -\infty$

142 In which cases below is it possible to determine $\lim_{x \to a} f(x)/g(x)$ without further information about the functions?
(*a*) $\lim_{x \to a} f(x) = 0$; $\lim_{x \to a} g(x) = \infty$
(*b*) $\lim_{x \to a} f(x) = 0$; $\lim_{x \to a} g(x) = 1$
(*c*) $\lim_{x \to a} f(x) = 0$; $\lim_{x \to a} g(x) = 0$
(*d*) $\lim_{x \to a} f(x) = \infty$; $\lim_{x \to a} g(x) = -\infty$

143 (*a*) State the assumptions in the zero-over-zero case of l'Hôpital's rule.
(*b*) State the conclusion.

144 Prove that

(*a*) $(\sin^{-1} x)' = \frac{1}{\sqrt{1 - x^2}}$ (*b*) $(e^x)' = e^x$

(*c*) $(\tan^{-1} x)' = \frac{1}{1 + x^2}$

145 What is the inverse of each of these functions?
(*a*) $\ln x$ (*b*) e^x (*c*) x^3
(*d*) $3x$ (*e*) $\sqrt[3]{x}$ (*f*) $\sin^{-1} x$

146 (*a*) Let $f(x) = 5^{7x}6^{8x+3}$. Show that $f'(x)$ is proportional to $f(x)$.

(*b*) Does (*a*) contradict the theorem that asserts that the only functions whose derivatives are proportional to the functions are of the form Ae^{kx}?

147 (*a*) Use a differential to show that $\log_{10}(1 + x) \approx 0.434x$ for small values of x.

(*b*) Use (*a*) to estimate $\log_{10} 1.05$.

148 Use logarithmic differentiation to find the derivative of

(*a*) $\dfrac{x^{3/5}(1 + 2x)^4 \sin^3 2x}{\tan^2 5x}$ (*b*) $\dfrac{x^3}{\sqrt[3]{x^3 + x^2} \cos 4x}$

149 Find $y'(0)$ if $y = f(x)$ is a function satisfying the equation

$$\ln(1 + y) + xy = \ln 2.$$

150 Does the graph of $y = x^4 - 4 \ln x$ have (*a*) any local maxima? (*b*) local minima? (*c*) inflection points?

151 The graph in Fig. 2 shows the tangent line to the curve $y = e^x$ at a point (x_0, y_0). Find the length of the segment AB.

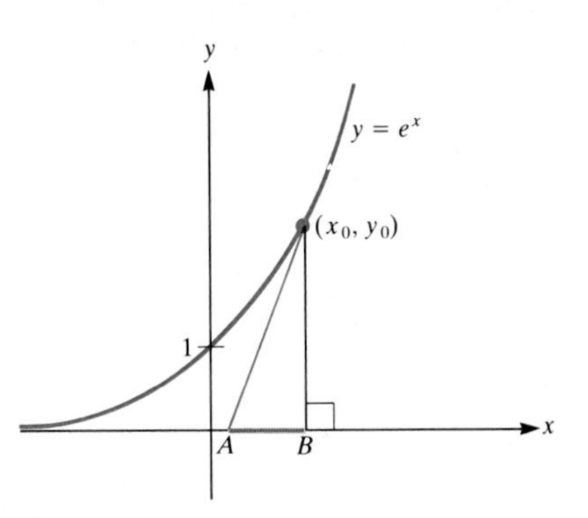

Figure 2

152 Find all positive integer solutions of the equation $x^y = y^x$ where $x \neq y$. [*Hint:* First take logarithms of both sides of the equation, then study the graph of $(\ln x)/x$.]

153 Let $f(x) = (1 + x)^{1/x}$ for $x > -1$, $x \neq 0$. Let $f(0) = e$.

(*a*) Show that f is continuous at $x = 0$.

(*b*) Show that f is differentiable at $x = 0$. What is $f'(0)$?

154 Newton computed ln 0.8, ln 0.9, ln 1.1, and ln 1.2. Then, using these values and laws of logarithms, he found ln 2, ln 3, and ln 5. How could that be done?

155 It was once conjectured that the speed of a ball falling from rest is proportional to the distance s that it drops.

(*a*) Show that, if this conjecture were correct, s would grow exponentially as a function of time.

(*b*) With the aid of (*a*) show that the speed would also grow exponentially.

(*c*) Recalling that the initial speed is 0, show that (*b*) leads to an absurd conclusion.

In fact, ds/dt is proportional to time t rather than to distance s.

156 According to a Chinese riddle, the lilies in a pond doubled in number each day. At the end of a month of 30 days the lilies finally covered the entire pond. On what day was exactly half the pond covered?

157 Find the limit of $(1^x + 2^x + 3^x)^{1/x}$ as (*a*) $x \to 0$, (*b*) $x \to \infty$, and (*c*) $x \to -\infty$.

158 The proof of Theorem 1 in Sec. 6.8, to be outlined in Exercise 159, depends on the following generalized mean-value theorem.

Generalized mean-value theorem. Let f and g be two functions that are continuous on $[a, b]$ and differentiable on (a, b). Furthermore, assume that $g'(x)$ is never 0 for x in (a, b). Then there is a number c in (a, b) such that

$$\frac{f(b) - f(a)}{g(b) - g(a)} = \frac{f'(c)}{g'(c)}.$$

(*a*) During a given time interval one car travels twice as far as another car. Use the generalized mean-value theorem to show that there is at least one instant when the first car is traveling exactly twice as fast as the second car.

(*b*) To prove the generalized mean-value theorem, introduce a function h:

$$h(x) = f(x) - f(a) - \frac{f(b) - f(a)}{g(b) - g(a)}[g(x) - g(a)].$$

Show that $h(b) = 0$ and $h(a) = 0$. Then apply Rolle's theorem to h.

Remarks: The function h is geometrically quite similar to the function h used in the proof of the mean-value theorem in Sec. 4.1. It is easy to check that $h(x)$ is the vertical distance between the point $(g(x), f(x))$ and the line through $(g(a), f(a))$ and $(g(b), f(b))$.

159 This exercise proves Theorem 1 of Sec. 6.8, l'Hôpital's rule in the zero-over-zero case. Assume the hypotheses of that theorem. Define $f(a) = 0$ and $g(a) = 0$, so that f and g are continuous at a. Note that

$$\frac{f(x)}{g(x)} = \frac{f(x) - f(a)}{g(x) - g(a)},$$

and apply the generalized mean-value theorem from Exercise 158.

160 If $$\lim_{t\to\infty} f(t) = \infty = \lim_{t\to\infty} g(t)$$

and $$\lim_{t\to\infty} \frac{\ln f(t)}{\ln g(t)} = 1,$$

must $$\lim_{t\to\infty} \frac{f(t)}{g(t)} = 1?$$

161 If

$$\lim_{t\to\infty} f(t) = \infty = \lim_{t\to\infty} g(t)$$

and

$$\lim_{t\to\infty} \frac{f(t)}{g(t)} = 3,$$

what can be said about

$$\lim_{t\to\infty} \frac{\ln f(t)}{\ln g(t)}?$$

(Do not assume that f and g are differentiable.)

162 Give an example of a pair of functions f and g such that we have $\lim_{x\to 0} f(x) = 1$, $\lim_{x\to 0} g(x) = \infty$, and $\lim_{x\to 0} f(x)^{g(x)} = 2$.

163 Figure 3 shows the graph of the velocity $v(t)$ of a particle moving along the x axis.

(*a*) Approximately how far does the particle move during the time interval [0, 1]? During the time interval [1, 2]? During the time interval [2, 4]?

(*b*) Let $x(t)$ be its coordinate at time t. Assume that $x(0) = 0$. Graph $x(t)$ as a function of t.

(*c*) Graph the acceleration as a function of t.

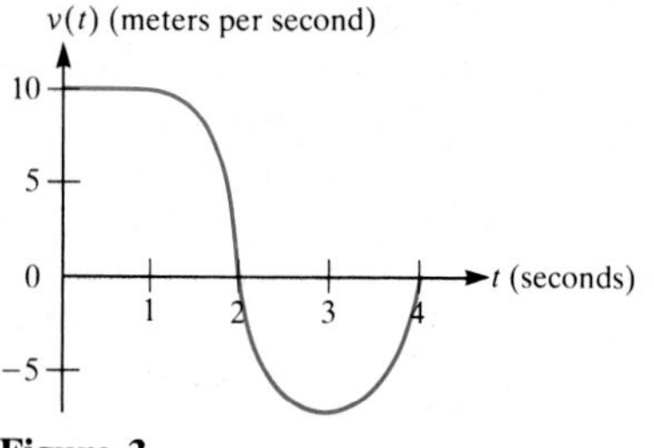

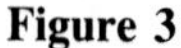

Figure 3

164 This table records the velocity $v(t)$ in meters per second of an object moving on the x axis during its first second of motion:

t	0	0.1	0.2	0.3	0.4	0.5	0.6	0.7	0.8	0.9	1.0
$v(t)$	0	2.0	2.5	2.0	3.0	3.1	3.2	3.5	3.2	3.0	3.1

(*a*) Sketch a rough graph of velocity as a function of time.

(*b*) Sketch a rough graph of acceleration as a function of time.

(*c*) Let $x(t)$ be the position of the object at time t. Assuming that $x(0) = 0$, sketch a rough graph of $x(t)$ as a function of t.

165 We said in Sec. 6.2 that the more often a bank compounds in a year, the more there will be in the account at the end of the year. This exercise justifies that statement.

(*a*) Show that the claim is true if, when r is a fixed positive number, the function $(1 + rx)^{1/x}$ is a decreasing function of x for $x > 0$.

(*b*) Show that $f(x) = (1 + rx)^{1/x}$ decreases for $x > 0$. *Hint:* Show that $\ln f(x)$ is a decreasing function.

7

COMPUTING ANTIDERIVATIVES

In Chap. 5 we found that such varied concepts as area, mass, distance traveled, and volume can be represented by definite integrals, $\int_a^b f(x)\, dx$. For this reason we may want to evaluate a definite integral.

One approach is to use some approximation technique, such as Simpson's method, discussed in Sec. 5.4.

Another approach is to use a calculator that has a built-in program for estimating a definite integral to any desired degree of accuracy. (See the discussion on page 437.)

A third approach is to use the fundamental theorem of calculus. If we can find an antiderivative $F(x)$ of the integrand $f(x)$, then $\int_a^b f(x)\, dx$ is simply $F(b) - F(a)$.

The problem of finding an antiderivative differs from that of finding a derivative in two important ways. First, the antiderivatives of some elementary functions, such as e^{x^2}, are not elementary. On the other hand, as we saw in Chaps. 3, 4, and 6, the derivatives of all elementary functions are elementary. Second, a slight change in the form of a function can cause a great change in the form of its antiderivative. For instance,

$$\int \frac{dx}{x^2+1} = \tan^{-1} x + C \qquad \text{while} \qquad \int \frac{x\, dx}{x^2+1} = \frac{1}{2} \ln\,(x^2+1) + C,$$

as you may check by differentiating $\tan^{-1} x$ and $\frac{1}{2} \ln\,(x^2+1)$.

On the other hand, a slight change in the form of an elementary function produces only a slight change in the form of its derivative.

There are three ways to find an antiderivative:

1 By hand, using techniques described in this chapter
2 By integral tables
3 By computer

Section 7.1 gives a few shortcuts, describes how to use integral tables, and discusses the strengths and weaknesses of computers.

Section 7.2 presents the most important technique for finding an antiderivative, "substitution."

Section 7.3 describes "integration by parts," a technique that has many uses (such as in differential equations) besides finding antiderivatives.

Sections 7.4 and 7.5 form a unit, which shows how to integrate any rational function.

Section 7.6 describes some substitutions to use on integrands of special types.

Section 7.7 offers an opportunity to practice the techniques when there is no clue as to which technique is the best to use.

7.1 SHORTCUTS, INTEGRAL TABLES, AND MACHINES

In this section we list antiderivatives of some common functions and some general shortcuts, then describe integral tables and the computation of antiderivatives by computers.

Some Common Integrands

Every formula for a derivative provides a corresponding formula for an antiderivative or integral. For instance, since $(x^3/3)' = x^2$, it follows that

$$\int x^2\,dx = \frac{x^3}{3} + C.$$

The following miniature integral table lists a few formulas that should be memorized. Each can be checked by differentiating the right-hand side of the equation.

$$\int x^a\,dx = \frac{x^{a+1}}{a+1} + C \qquad \text{for } a \neq -1$$

$$\int \frac{1}{x}\,dx = \ln|x| + C \qquad \left(\text{This is } \int x^a\,dx \text{ for } a = -1\right)$$

$$\int \frac{f'(x)}{f(x)}\,dx = \ln|f(x)| + C$$

$$\int u^n\,du = \frac{u^{n+1}}{n+1} + C \qquad n \neq -1,\ u = f(x)$$

$$\int e^x\,dx = e^x + C$$

$$\int \sin x\,dx = -\cos x + C \qquad \text{(Remember the } -\text{)}$$

$$\int \cos x\,dx = \sin x + C$$

$$\int \frac{1}{\sqrt{1-x^2}}\,dx = \sin^{-1} x + C$$

$$\int \frac{1}{1+x^2}\,dx = \tan^{-1} x + C$$

$$\int \frac{1}{|x|\sqrt{x^2-1}}\,dx = \sec^{-1} x + C$$

Antiderivative of a polynomial

EXAMPLE 1 Find $\int (2x^4 - 3x + 2)\,dx$.

SOLUTION
$$\int (2x^4 - 3x + 2)\,dx = \int 2x^4\,dx - \int 3x\,dx + \int 2\,dx$$
$$= 2\int x^4\,dx - 3\int x\,dx + 2\int 1\,dx$$
$$= 2\frac{x^5}{5} - 3\frac{x^2}{2} + 2x + C. \blacksquare$$

One constant of integration is enough.

EXAMPLE 2 Find $\int \frac{4x^3}{x^4 + 1}\,dx$.

SOLUTION The numerator is precisely the derivative of the denominator. Hence

Antiderivative of f'/f

$$\int \frac{4x^3}{x^4+1}\,dx = \ln|x^4 + 1| + C.$$

Since $x^4 + 1$ is always positive, the absolute-value sign is not needed, and
$$\int \frac{4x^3}{x^4+1}\,dx = \ln(x^4 + 1) + C. \blacksquare$$

EXAMPLE 3 Find $\int \sqrt{x}\,dx$.

Antiderivative of x^a

SOLUTION
$$\int \sqrt{x}\,dx = \int x^{1/2}\,dx$$
$$= \frac{x^{1/2+1}}{\frac{1}{2}+1} + C$$
$$= \tfrac{2}{3}x^{3/2} + C$$
$$= \tfrac{2}{3}(\sqrt{x})^3 + C. \blacksquare$$

EXAMPLE 4 Find $\int \frac{1}{x^3}\,dx$.

SOLUTION
$$\int \frac{1}{x^3}\,dx = \int x^{-3}\,dx$$
$$= \frac{x^{-3+1}}{-3+1} + C$$
$$= -\tfrac{1}{2}x^{-2} + C$$
$$= -\frac{1}{2x^2} + C. \blacksquare$$

EXAMPLE 5 Find $\int \left(3\cos x - 4\sin x + \frac{1}{x^2}\right) dx$.

SOLUTION
$$\int\left(3\cos x - 4\sin x + \frac{1}{x^2}\right) dx = 3\int \cos x\,dx - 4\int \sin x\,dx + \int \frac{1}{x^2}\,dx$$
$$= 3\sin x + 4\cos x - \frac{1}{x} + C. \blacksquare$$

EXAMPLE 6 Find $\int \frac{x}{1+x^2}\,dx$.

SOLUTION If the numerator were exactly $2x$, then the numerator would be the derivative of the denominator and we would have the case $\int [f'(x)/f(x)]\,dx$. In that case, the antiderivative would be $\ln(1+x^2)$. But the numerator can be multiplied by 2 if we simultaneously divide by 2:

Multiplying the integrand by a constant

$$\int \frac{x}{1+x^2}\,dx = \frac{1}{2}\int \frac{2x}{1+x^2}\,dx.$$

This step depends on the fact that a constant can be moved past the integral sign:

$$\frac{1}{2}\int \frac{2x}{1+x^2}\,dx = \frac{1}{2}\cdot 2\int \frac{x}{1+x^2}\,dx = \int \frac{x}{1+x^2}\,dx.$$

Thus
$$\int \frac{x}{1+x^2}\,dx = \frac{1}{2}\int \frac{2x}{1+x^2}\,dx$$
$$= \tfrac{1}{2}\ln(1+x^2) + C. \quad ■$$

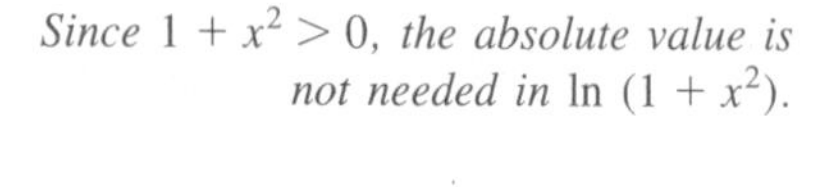

Since $1 + x^2 > 0$, the absolute value is not needed in $\ln(1+x^2)$.

We present three shortcuts for evaluating some special but fairly common definite integrals.

SHORTCUT 1 If f is an odd function, then

$$\int_{-a}^{a} f(x)\,dx = 0. \tag{1}$$

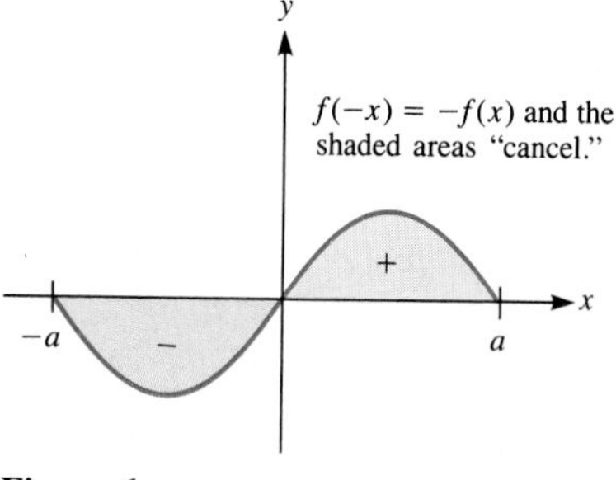

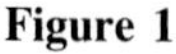

Figure 1

EXPLANATION Recall that for an odd function $f(-x) = -f(x)$. Figure 1 suggests why (1) holds. The shaded area left of the y axis equals the shaded area to the right. As integrals, however, these two areas represent quantities of opposite sign:

$$\int_{-a}^{0} f(x)\,dx = -\int_{0}^{a} f(x)\,dx.$$

Therefore, the definite integral over the entire interval is 0. ■

EXAMPLE 7 Find $\int_{-2}^{2} x^3\sqrt{4-x^2}\,dx$.

SOLUTION The function $f(x) = x^3\sqrt{4-x^2}$ is odd. (Check it.) By Shortcut 1,

$$\int_{-2}^{2} x^3\sqrt{4-x^2} = 0. \quad ■$$

SHORTCUT 2 $\int_{0}^{a} \sqrt{a^2-x^2}\,dx = \frac{1}{4}\pi a^2$.

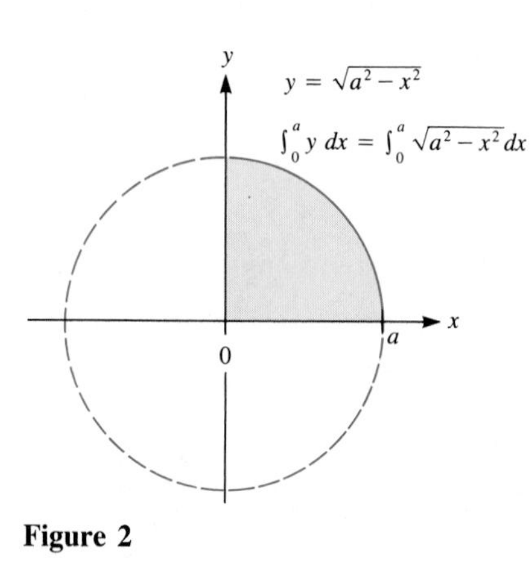

Figure 2

EXPLANATION The graph of $y = \sqrt{a^2-x^2}$ is part of a circle of radius a. The definite integral $\int_0^a \sqrt{a^2-x^2}\,dx$ is a quarter of the area of that circle. (See Fig. 2.) ■

EXAMPLE 8 Find $\int_0^1 \sqrt{1-x^2}\,dx$.

SOLUTION Use Shortcut 2, with $a = 1$, to get

$$\int_0^1 \sqrt{1-x^2}\,dx = \frac{\pi}{4}. \quad \blacksquare$$

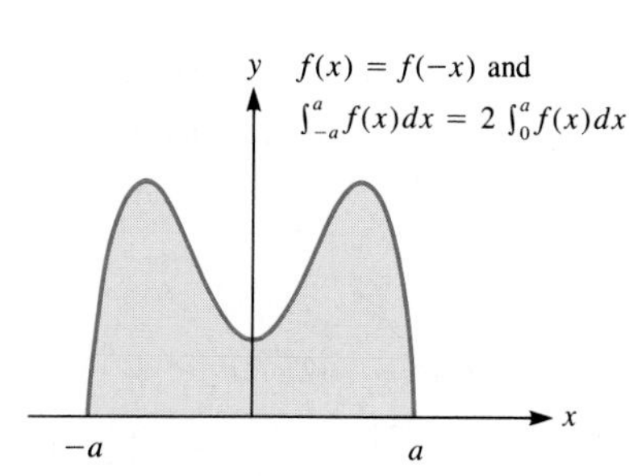

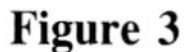

Figure 3

SHORTCUT 3 If f is an even function,

$$\int_{-a}^{a} f(x)\,dx = 2\int_0^a f(x)\,dx.$$

EXPLANATION A glance at Fig. 3 suggests why this shortcut is valid. $\blacksquare$

EXAMPLE 9 Find $\int_{-1}^{1} \sqrt{1-x^2}\,dx$.

SOLUTION Since $\sqrt{1-x^2}$ is an even function, by Shortcut 3,

$$\int_{-1}^{1} \sqrt{1-x^2}\,dx = 2\int_0^1 \sqrt{1-x^2}\,dx.$$

So, by Example 8,

$$\int_{-1}^{1} \sqrt{1-x^2}\,dx = 2 \cdot \frac{\pi}{4} = \frac{\pi}{2}. \quad \blacksquare$$

Using an Integral Table

Inside the covers of this book you will find a list of antiderivatives. This list constitutes a short integral table. *Burington's Handbook of Mathematical Tables and Formulas,* 5th edition, McGraw-Hill, 1973, lists over 300 integrals in 33 pages. (Usually integral tables use "log" to denote "ln"; it is understood that e is the base of the logarithms.)

The best way to use an integral table is to browse through one (buy one or check one out from the library). Notice how the formulas are grouped. First might come the forms that are so common that everyone will use them often. Then may come "forms containing $ax + b$," then "forms containing $a^2 \pm x^2$," then "forms containing $ax^2 + bx + c$," and so on, running through many different algebraic forms. Trigonometric forms are often next, followed by sections on logarithmic and exponential functions. Our integral table from the inside front cover is similarly grouped. We use Formulas 21 and 22 from the section "Expressions containing $ax + b$" in the next two examples.

EXAMPLE 10 Use the integral table to integrate

$$\int \frac{dx}{x\sqrt{3x+2}}$$

SOLUTION Search until you find Formula 21,

$$\int \frac{dx}{x\sqrt{ax+b}} = \frac{1}{\sqrt{b}} \ln \left| \frac{\sqrt{ax+b} - \sqrt{b}}{\sqrt{ax+b} + \sqrt{b}} \right| \qquad b > 0,$$

and replace $ax + b$ by $3x + 2$ and b by 2. Thus

$$\int \frac{dx}{x\sqrt{3x+2}} = \frac{1}{\sqrt{2}} \ln \left| \frac{\sqrt{3x+2} - \sqrt{2}}{\sqrt{3x+2} + \sqrt{2}} \right| + C. \quad ■$$

Contrast this example with the next one. The integrands look much alike, but their integrals do not.

EXAMPLE 11 Use the integral table to integrate $\int \frac{dx}{x\sqrt{3x-2}}$.

SOLUTION This time we need Formula 22:

$$\int \frac{dx}{x\sqrt{ax+b}} = \frac{2}{\sqrt{-b}} \tan^{-1} \sqrt{\frac{ax+b}{-b}} \qquad b < 0.$$

Thus
$$\int \frac{dx}{x\sqrt{3x-2}} = \frac{2}{\sqrt{2}} \tan^{-1} \sqrt{\frac{3x-2}{2}} + C. \quad ■$$

When using a table of integrals be very cautious and keep a cool head. There is no need to make a big fuss about integral tables. Just match the patterns carefully, including any conditions on the variables and their coefficients. Note that some formulas are expressed in terms of an integral of a different integrand. In these cases you will have to search through the table more than once. (Exercises 34 and 35 illustrate this situation.)

COMPUTERS

Using an integral table is an exercise in "pattern matching," where you hunt for the formula that fits the particular integral you have. Computers are good at pattern matching, so it is not surprising to learn that for many years computers have been used to find antiderivatives. MACSYMA is probably the best known of the computer-based mathematics programs that can do calculus. In addition to matching problems with formulas from its large internal table of integrals, MACSYMA also performs various substitutions and computations to turn integrals into forms it can solve. Originally confined to large computer systems, MACSYMA now operates on personal computers. Wolfram Research's Mathematica, Prescience's Theorist, and the Soft Warehouse's Derive also find antiderivatives for personal computer users. With this kind of computer power increasingly available, calculus users do not need to rely as much on formal integration techniques or tables of integrals.

EXERCISES FOR SEC. 7.1: SHORTCUTS, INTEGRAL TABLES, AND MACHINES

In Exercises 1 to 14 find the integrals. Use the short list at the beginning of the section.

1 $\int 5x^3\, dx$

2 $\int (8 + 11x)\, dx$

3 $\int x^{1/3}\, dx$

4 $\int \sqrt[3]{x^2}\, dx$

5 $\int \frac{6\, dx}{x^2}$

6 $\int \frac{dx}{x^3}$

7 $\int 5e^{-2x}\, dx$

8 $\int \frac{5\, dx}{1 + x^2}$

9 $\int \frac{6\,dx}{|x|\sqrt{x^2-1}}$ **10** $\int \frac{5\,dx}{\sqrt{1-x^2}}$

11 $\int \frac{4x^3\,dx}{1+x^4}$ **12** $\int \frac{e^x\,dx}{1+e^x}$

13 $\int \frac{\sin x\,dx}{1+\cos x}$ **14** $\int \frac{dx}{1+3x}$

In Exercises 15 to 20, change the integrand into an easier one by algebra and find the antiderivatives.

15 $\int \frac{1+2x}{x^2}\,dx$ **16** $\int \frac{1+2x}{1+x^2}\,dx$ $\left(\textit{Hint: } \frac{a+b}{c} = \frac{a}{c} + \frac{b}{c}\right)$

17 $\int (x^2+3)^2\,dx$ **18** $\int (1+e^x)^2\,dx$ (*Hint:* First multiply out the integrand.)

19 $\int (1+3x)x^2\,dx$ **20** $\int \frac{1+\sqrt{x}}{x}\,dx$

In Exercises 21 to 26 use a shortcut to evaluate the integral.

21 $\int_{-1}^{1} x^5\sqrt{1+x^2}\,dx$

22 $\int_{-\pi/2}^{\pi/2} \sin 3x \cos 5x\,dx$

23 $\int_{-1}^{1} x^5\sqrt[4]{1-x^2}\,dx$

24 $\int_{-\pi}^{\pi} \sin^3 x\,dx$

25 $\int_{-3}^{3} \sqrt{9-x^2}\,dx$

26 $\int_{-3}^{3} (x^3\sqrt{9-x^2} + 10\sqrt{9-x^2})\,dx$

In Exercises 27 to 32 find the antiderivative with the aid of a table of integrals. For example, the one inside the front cover will do.

27 (*a*) $\int \frac{dx}{(3x+2)^2}$ (*b*) $\int \frac{dx}{x(3x+2)}$

28 (*a*) $\int \frac{dx}{x\sqrt{3x+4}}$ (*b*) $\int \frac{dx}{x^2\sqrt{3x+4}}$

29 (*a*) $\int \frac{dx}{x\sqrt{3x-4}}$ (*b*) $\int \frac{dx}{x^2\sqrt{3x-4}}$

30 (*a*) $\int \frac{dx}{4x^2+9}$ (*b*) $\int \frac{dx}{4x^2-9}$

31 (*a*) $\int \frac{dx}{x^2+3x+5}$ (*b*) $\int \frac{dx}{x^2+2x+5}$

32 (*a*) $\int \frac{dx}{\sqrt{11-x^2}}$ (*b*) $\int \frac{dx}{\sqrt{11+x^2}}$

33 (A shortcut for $\int_0^{\pi/2} \sin^2\theta\,d\theta$)
(*a*) Why would you expect $\int_0^{\pi/2} \cos^2\theta\,d\theta$ to equal $\int_0^{\pi/2} \sin^2\theta\,d\theta$?
(*b*) Why is $\int_0^{\pi/2} \sin^2\theta\,d\theta + \int_0^{\pi/2} \cos^2\theta\,d\theta = \pi/2$?
(*c*) Conclude that $\int_0^{\pi/2} \sin^2\theta\,d\theta = \pi/4$.

34 Using the integral table on the inside front cover of the book, find $\int x\,dx/\sqrt{2x^2+x+5}$. [Use Formula 42 first, followed by Formula 40.]

35 Using the integral table in the front of the book, find

(*a*) $\int \frac{dx}{\sqrt{3x^2+x+2}}$ (*b*) $\int \frac{dx}{\sqrt{-3x^2+x+2}}$.

7.2 THE SUBSTITUTION METHOD

This section describes the **substitution method** that changes the form of an integrand, preferably to one that we can integrate more easily. Sometimes we can use a substitution to transform an integral not listed in an integral table to one that is listed. Several examples will illustrate the technique, which is really the chain rule in disguise. After the examples, Theorem 1 provides the basis of the substitution method. The technique uses differentials, defined in Sec. 4.9. Recall that if $y = f(x)$, then $dy = f'(x)\,dx$.

Examples of the Substitution Method

EXAMPLE 1 Find $\int (\sin x^2)\,2x\,dx$.

SOLUTION Note that $2x$ is the derivative of x^2. Make the substitution $u = x^2$. Then

$$du = 2x\,dx \qquad \text{and} \qquad \int (\sin x^2)\, 2x\, dx = \int \sin u\, du.$$

Now it is easy to find $\int \sin u\, du$:

$$\int \sin u\, du = -\cos u + C.$$

Replacing u by x^2 in $-\cos u$ yields $-\cos x^2$. Thus

$$\int (\sin x^2)\, 2x\, dx = -\cos x^2 + C.$$

This answer can be checked by differentiation (using the chain rule):

$$\begin{aligned}\frac{d}{dx}(-\cos x^2 + C) &= \sin x^2 \frac{d}{dx}(x^2) + \frac{d}{dx}(C)\\ &= (\sin x^2)\, 2x. \quad \blacksquare\end{aligned}$$

EXAMPLE 2 Find $\int 5e^{x^5} x^4\, dx$.

SOLUTION Introduce $u = x^5$. Then $du = 5x^4\, dx$ and

$$\begin{aligned}\int 5e^{x^5}x^4\, dx &= \int e^u\, du\\ &= e^u + C\\ &= e^{x^5} + C.\end{aligned}$$

Check:

$$\begin{aligned}\frac{d}{dx}(e^{x^5} + C) &= e^{x^5}\frac{d}{dx}(x^5) + \frac{d}{dx}(C) \qquad \text{chain rule}\\ &= e^{x^5}(5x^4)\\ &= 5e^{x^5}x^4. \quad \blacksquare\end{aligned}$$

EXAMPLE 3 Find $\int \sin^2 \theta \cos \theta\, d\theta$.

SOLUTION Note that $\cos \theta$ is the derivative of $\sin \theta$, and introduce $u = \sin \theta$. Hence

$$du = \cos \theta\, d\theta.$$

Then

$$\begin{aligned}\int \underbrace{\sin^2 \theta}_{u^2}\, \underbrace{\cos \theta\, d\theta}_{du} &= \int u^2\, du\\ &= \frac{u^3}{3} + C\\ &= \frac{\sin^3 \theta}{3} + C.\end{aligned}$$

Check:

$$\begin{aligned}\frac{d}{d\theta}\left(\frac{\sin^3 \theta}{3} + C\right) &= \frac{d}{d\theta}\left(\frac{\sin^3 \theta}{3}\right) + \frac{dC}{d\theta}\\ &= \frac{3 \sin^2 \theta}{3}\frac{d}{d\theta}(\sin \theta) + 0 \qquad \text{chain rule}\\ &= \sin^2 \theta \cos \theta. \quad \blacksquare\end{aligned}$$

EXAMPLE 4 Find $\int (1 + x^3)^5 x^2\,dx$.

SOLUTION The derivative of $1 + x^3$ is $3x^2$, which differs from the x^2 in the integrand only by the constant factor 3. So let $u = 1 + x^3$. Hence

$$du = 3x^2\,dx \quad \text{and} \quad \frac{du}{3} = x^2\,dx.$$

Then

$$\int (1 + x^3)^5 x^2\,dx = \int u^5 \frac{du}{3}$$

$$= \frac{1}{3}\int u^5\,du = \frac{1}{3}\frac{u^6}{6} + C$$

$$= \frac{(1 + x^3)^6}{18} + C.$$

(You could check this answer by differentiating it.) ■

In Example 4 note that if the x^2 were not present in the integrand, the substitution method would not work. To find $\int (1 + x^3)^5\,dx$, it would be necessary to multiply out $(1 + x^3)^5$ first, a most unpleasant chore.

Description of the Substitution Method

In each of Examples 1 to 3 the integrand $f(x)$ could be written in the form

$$(\text{function of } h(x)) \times (\text{derivative of } h(x)),$$

for some function $h(x)$. To put it another way, $f(x)\,dx$ could be written as

$$(\text{function of } h(x)) \times (\text{differential of } h(x)).$$

Whenever this is the case, the substitution of u for $h(x)$ and du for $h'(x)\,dx$ transforms $\int f(x)\,dx$ to another integral, one involving u instead of x, $\int g(u)\,du$.

If you can find an antiderivative $G(u)$ of $g(u)$, replace u by $h(x)$. The resulting function, $G(h(x))$, is an antiderivative of $f(x)$. (This claim will be justified at the end of the section.)

More Examples

EXAMPLE 5 Compare the problems of finding these antiderivatives:

$$\int \frac{dx}{\sqrt{1 + x^3}} \quad \text{and} \quad \int \frac{x^2\,dx}{\sqrt{1 + x^3}}.$$

SOLUTION It turns out that the first antiderivative is *not* an elementary function. That means you could compute until the next ice age and never find a function expressible in terms of polynomials, square roots, trigonometric functions, logarithms, exponentials, etc., whose derivative is $1/\sqrt{1 + x^3}$. However the second integral is elementary and not hard to find.

Since x^2 differs from the derivative of $1 + x^3$ only by a constant factor 3, use the substitution $u = 1 + x^3$. Hence

$$du = 3x^2\,dx \quad \text{and} \quad \frac{du}{3} = x^2\,dx.$$

SOLUTION Let $u = 1 + x^3$. Then $du = 3x^2\,dx$. Furthermore, as x goes from 1 to 2, $u = 1 + x^3$ goes from $1 + 1^3 = 2$ to $1 + 2^3 = 9$. Thus

This is the last you see of x.

$$\int_1^2 3(1 + x^3)^5\,x^2\,dx = \int_2^9 u^5\,du = \left.\frac{u^6}{6}\right|_2^9 = \frac{9^6 - 2^6}{6}.$$

Once you make the substitution, you work only with expressions involving u. There is no need to bring back x again. ■

You might go back and apply the technique illustrated in Example 8 to Example 6.

A Radical Substitution

We have so far limited our substitutions to functions that are already part of the integrand. Sometimes it helps to be more creative. The next example takes advantage of the identity $1 - \sin^2\theta = \cos^2\theta$ to eliminate a radical in the integrand.

EXAMPLE 9 Find $\displaystyle\int \frac{\sqrt{1 - x^2}}{x^2}\,dx$.

SOLUTION We let $x = \sin\theta$. Then $dx = \cos\theta\,d\theta$ and

Trouble arises when $\cos\theta < 0$. *See Exercise 48.*

$$\sqrt{1 - x^2} = \sqrt{1 - \sin^2\theta} = \sqrt{\cos^2\theta} = \cos\theta.$$

Therefore,

$$\int \frac{\sqrt{1 - x^2}}{x^2}\,dx = \int \frac{\cos\theta\,(\cos\theta\,d\theta)}{\sin^2\theta} = \int \frac{\cos^2\theta}{\sin^2\theta}\,d\theta = \int \cot^2\theta\,d\theta = \int (\csc^2\theta - 1)\,d\theta = -\cot\theta - \theta + C.$$

Since $x = \sin\theta$, $\theta = \sin^{-1} x$. We also have $\cos\theta = \sqrt{1 - \sin^2\theta} = \sqrt{1 - x^2}$. Therefore,

$$\cot\theta = \frac{\cos\theta}{\sin\theta} = \frac{\sqrt{1 - x^2}}{x}.$$

Our antiderivative is thus

$$-\cot\theta - \theta + C = -\frac{\sqrt{1-x^2}}{x} - \sin^{-1} x + C. \quad ■$$

Example 9 illustrates a technique known as *trigonometric substitution*. We will return to this topic in Sec. 7.6.

Why Substitution Works

Theorem 1 Assume that f and g are continuous functions and $u = h(x)$ is differentiable. Suppose that $f(x)$ can be written as $g(u)\,(du/dx)$ and that G is an antiderivative of g. Then $G(u(x))$ is an antiderivative of $f(x)$.

Proof We differentiate $G(u(x))$ and check that the result is $f(x)$, as follows:

$$\frac{d}{dx}G(u(x)) = \frac{dG}{du}\frac{du}{dx} \qquad \text{(chain rule)}$$

$$= g(u)\frac{du}{dx} \qquad \text{(by definition of } G\text{)}$$

$$= f(x). \qquad \text{(by assumption)} \quad ■$$

Theorem 2 *(Substitution in a definite integral)* Under the same assumptions as in Theorem 1,

$$\int_a^b f(x)\,dx = \int_{u(a)}^{u(b)} g(u)\,du.$$

Proof Let $F(x) = G(u(x))$, where G is defined in the previous proof.

$$\int_a^b f(x)\,dx = F(b) - F(a) \qquad \text{(fundamental theorem of calculus)}$$

$$= G(u(b)) - G(u(a)) \qquad \text{(definition of } F\text{)}$$

$$= \int_{u(a)}^{u(b)} g(u)\,du. \qquad \text{(fundamental theorem, again)} \quad ■$$

Section Summary

This section introduced the most commonly used integration technique, "substitution," which replaces $\int f(x)\,dx$ by $\int g(u)\,du$ and $\int_a^b f(x)\,dx$ by $\int_{u(a)}^{u(b)} g(u)\,du$.

It is to be hoped that the problem of finding $\int g(u)\,du$ is easier than that of finding $\int f(x)\,dx$. If it is not, try another substitution or one of the methods presented in the rest of the chapter. It is important to keep in mind that there is no simple routine method for antidifferentiation of elementary functions. Practice in integration pays off in the quick recognition of which technique is most promising.

EXERCISES FOR SEC. 7.2: THE SUBSTITUTION METHOD

In Exercises 1 to 14 use the given substitutions to find the antiderivatives or definite integrals.

1 $\int (1 + 3x)^5\, 3\, dx;\quad u = 1 + 3x$

2 $\int e^{\sin\theta} \cos\theta\, d\theta;\quad u = \sin\theta$

3 $\int_0^1 \frac{x}{\sqrt{1 + x^2}}\, dx;\quad u = 1 + x^2$

4 $\int_{\sqrt{8}}^{\sqrt{15}} \sqrt{1 + x^2}\, x\, dx;\quad u = 1 + x^2$

5 $\int \sin 2x\, dx;\quad u = 2x$

6 $\int \frac{e^{2x}}{(1 + e^{2x})^2}\, dx;\quad u = 1 + e^{2x}$

7 $\int_{-1}^{2} e^{3x}\, dx;\quad u = 3x$

8 $\int_2^3 \frac{e^{1/x}}{x^2}\, dx;\quad u = \frac{1}{x}$

9 $\int \frac{1}{\sqrt{1 - 9x^2}}\, dx;\quad u = 3x$

10 $\int \frac{t\, dt}{\sqrt{2 - 5t^2}};\quad u = 2 - 5t^2$

11 $\int_{\pi/6}^{\pi/4} \tan\theta \sec^2\theta\, d\theta;\quad u = \tan\theta$

12 $\int_{\pi^2/16}^{\pi^2/4} \frac{\sin\sqrt{x}}{\sqrt{x}}\, dx;\quad u = \sqrt{x}$

13 $\int \frac{(\ln x)^4}{x}\, dx;\quad u = \ln x$

14 $\int \frac{\sin(\ln x)}{x}\, dx;\quad u = \ln x$

In Exercises 15 to 36 use appropriate substitutions to find the antiderivatives.

15 $\int (1 - x^2)^5\, x\, dx$

16 $\int \frac{x\, dx}{(x^2 + 1)^3}$

17 $\int \sqrt[3]{1 + x^2}\, x\, dx$

18 $\int \frac{\sin\theta}{\cos^2\theta}\, d\theta$

19 $\int \frac{e^{\sqrt{t}}}{\sqrt{t}}\, dt$

20 $\int e^x \sin e^x\, dx$

21 $\int \sin 3\theta\, d\theta$

22 $\int \frac{dx}{\sqrt{2x + 5}}$

23 $\int (x - 3)^{5/2}\, dx$

24 $\int \frac{dx}{(4x + 3)^3}$

25 $\int \frac{2x + 3}{x^2 + 3x + 2}\, dx$

26 $\int \frac{2x + 3}{(x^2 + 3x + 5)^4}\, dx$

27 $\int e^{2x}\, dx$

28 $\int \frac{dx}{\sqrt{x}(1 + \sqrt{x})^3}$

29 $\int x^4 \sin x^5\, dx$

30 $\int \frac{\cos(\ln x)\, dx}{x}$

31 $\int \frac{x}{1 + x^4}\, dx$

32 $\int \frac{x^3}{1 + x^4}\, dx$

33 $\int \frac{x\, dx}{(1 + x)^3}$

34 $\int \frac{x^2\, dx}{(1 + x)^3}$

35 $\int \frac{\ln 3x\, dx}{x}$

36 $\int \frac{\ln x^2\, dx}{x}$

In Exercises 37 to 42 find the area of the region under the graph of the given function and above the given interval.

37 $e^{x^3} x^2$; $[1, 2]$

38 $\sin^3\theta \cos\theta$; $[0, \pi/2]$

39 $\frac{x^2 + 3}{(x + 1)^4}$; $[0, 1]$

40 $\frac{x^2 - x}{(3x + 1)^2}$; $[1, 2]$

41 $\frac{(\ln x)^3}{x}$; $[1, e]$

42 $\tan^5\theta \sec^2\theta$; $\left[0, \frac{\pi}{3}\right]$

In Exercises 43 to 46 use substitution to evaluate the integral.

43 $\int \frac{x^2\, dx}{ax + b};\quad a \neq 0$

44 $\int \frac{x\, dx}{(ax + b)^2};\quad a \neq 0$

45 $\int \frac{x^2\, dx}{(ax + b)^2};\quad a \neq 0$

46 $\int x(ax+b)^n\,dx$; for (a) $n=-1$, (b) $n=-2$, (c) $n \neq -1, -2$. Assume $a \neq 0$.

---

47 Jack (using the substitution $u = \cos\theta$) claims that $\int 2\cos\theta\sin\theta\,d\theta = -\cos^2\theta$, while Jill (using the substitution $u = \sin\theta$) claims that the answer is $\sin^2\theta$. Who is right? Explain.

48 Jill says, "$\int_0^\pi \cos^2\theta\,d\theta$ is obviously positive." Jack claims, "No, it's zero. Just make the substitution $u = \sin\theta$; hence $du = \cos\theta\,d\theta$. Then I get

$$\int_0^\pi \cos^2\theta\,d\theta = \int_0^\pi \cos\theta\cos\theta\,d\theta$$
$$= \int_0^0 \sqrt{1-u^2}\,du = 0.$$

Simple."

(a) Who is right? What is the mistake?

(b) Use the identity $\cos^2\theta = (1+\cos 2\theta)/2$ to evaluate the integral without substitution.

49 Jill asserts that $\int_{-2}^1 2x^2\,dx$ is obviously positive. "After all, the integrand is never negative and $-2 < 1$. It equals the area under $y = 2x^2$ and above $[-2, 1]$." "You're wrong again," Jack replies, "It's negative. Here are my computations. Let $u = x^2$; hence $du = 2x\,dx$. Then

$$\int_{-2}^1 2x^2\,dx = \int_{-2}^1 x\cdot 2x\,dx$$
$$= \int_4^1 \sqrt{u}\,du = -\int_1^4 \sqrt{u}\,du,$$

which is obviously negative." Who is right? Explain.

50 Show that if f is an odd function then $\int_{-a}^a f(x)\,dx = 0$. Suggestion: First show that $\int_{-a}^0 f(x)\,dx = -\int_0^a f(x)\,dx$ by using the substitution $u = -x$. (Do not refer to "areas.")

51 Show that if f is an even function, then $\int_{-a}^a f(x)\,dx = 2\int_0^a f(x)\,dx$. Suggestion: First show that $\int_{-a}^0 f(x)\,dx = \int_0^a f(x)\,dx$ by using the substitution $u = -x$. (Do not refer to "areas.")

52 (a) Graph $y = (\ln x)/x$.
(b) Find the area under the curve in (a) and above the interval $[e, e^2]$.

53 (a) Graph $y = xe^{-x^2}$.
(b) Find the area of the region bounded by $y = xe^{-x^2}$, the line $x + y = 1$, and the x axis. (You will need Newton's method of estimating a solution of an equation.)

54 The velocity of a particle at time t seconds is $e^{-t}\sin\pi t$ meters per second. Find how far it travels in the first second, from time $t = 0$ to $t = 1$, (a) using the integral table in the front of the book, (b) using Simpson's method with $n = 4$, expressing your answer to four decimal places.

7.3 INTEGRATION BY PARTS

Recall that u' is the derivative of u.

This section develops a way of expressing $\int uv'\,dx$ in terms of $\int vu'\,dx$ for any two differentiable functions $u(x)$ and $v(x)$. We hope that finding $\int vu'\,dx$ is easier than finding $\int uv'\,dx$. The method is called **integration by parts**. It works especially well when the integrand has the form x^ne^x, $x^n\ln x$, $x^n\sin x$, or is an inverse trigonometric function.

Integration by Parts

Integration by parts is based on the equation

$$\boxed{\int u\,dv = uv - \int v\,du,} \tag{1}$$

where u and v are both differentiable functions of x.

We will first illustrate the method by an example. Right after the example we will explain why (1) is valid.

EXAMPLE 1 Find $\int xe^x\,dx$.

SOLUTION To use the formula $\int u\,dv = uv - \int v\,du$ we must write $xe^x\,dx$ as $u\,dv$. One way to do this is

$$u = x \qquad dv = e^x\,dx.$$

Of course we must find v itself. Since $dv = e^x\,dx$,

$$\frac{dv}{dx} = e^x,$$

that is, v is an antiderivative of e^x, $v = \int e^x\,dx$. But $\int e^x\,dx = e^x$. Hence we may choose $v = e^x$. (We could choose $v = e^x + C$, for any constant C, but we chose the simplest v whose derivative is e^x.) We therefore have

$$u = x \qquad dv = e^x\,dx$$

$$du = dx \qquad v = e^x.$$

By integration by parts, Eq. (1),

$$\int \underbrace{x}_{u}\,\underbrace{e^x\,dx}_{dv} = \underbrace{xe^x}_{uv} - \int \underbrace{e^x}_{v}\,\underbrace{dx.}_{du}$$

Hence

$$\int xe^x\,dx = xe^x - e^x + C.$$

(You may check that the derivative of $xe^x - e^x + C$ is xe^x.) ■

The Basis of Integration by Parts

Just as the chain rule is the basis for integration by substitution, the formula for the derivative of a product is the basis for integration by parts.

Let $u(x)$ and $v(x)$ have continuous derivatives. Then

$$\frac{d}{dx}(u(x)v(x)) = u(x)v'(x) + v(x)u'(x).$$

This tells us that $u(x)v(x)$ is an antiderivative of $u(x)v'(x) + v(x)u'(x)$. So we may write

$$u(x)v(x) = \int (u(x)v'(x) + v(x)u'(x))\,dx. \tag{2}$$

Rewriting (2) tells us that

$$\int u(x)v'(x)\,dx = u(x)v(x) - \int v(x)u'(x)\,dx. \tag{3}$$

Rewriting (3) in terms of the differentials of u and v, we have

$$\int u\,dv = uv - \int v\,du,$$

which is (1).

Examples of Integration by Parts

EXAMPLE 2 Find $\int x \ln x\,dx$.

SOLUTION Setting $dv = \ln x\, dx$ is not a wise move, since $v = \int \ln x\, dx$ is not immediately apparent. But setting $u = \ln x$ is promising because $du = d(\ln x) = dx/x$ is much easier than $\ln x$. This second approach goes through smoothly:

$$u = \ln x \qquad du = \frac{dx}{x}$$

$$dv = x\, dx \qquad v = \frac{x^2}{2}.$$

(Note that we needed to find $v = \int x\, dx$.) Thus

$$\int x \ln x\, dx = \int \underbrace{\ln x}_{u}\, \underbrace{x\, dx}_{dv} = \underbrace{\ln x}_{u}\, \underbrace{\frac{x^2}{2}}_{v} - \int \underbrace{\frac{x^2}{2}}_{v}\, \underbrace{\frac{dx}{x}}_{du}$$

$$= \frac{x^2 \ln x}{2} - \int \frac{x\, dx}{2}$$

$$= \frac{x^2 \ln x}{2} - \frac{x^2}{4} + C.$$

You may check the result by differentiation. ■

The key to applying integration by parts is the labeling of u and dv. Usually three conditions should be met:

1. v can be found by integrating and should not be too messy.
2. du should not be messier than u.
3. $\int v\, du$ should be easier than the original $\int u\, dv$.

The next example shows how to integrate any inverse trigonometric function.

EXAMPLE 3 Find $\int \tan^{-1} x\, dx$.

SOLUTION Recall that the derivative of $\tan^{-1} x$ is $1/(1 + x^2)$, a much simpler function than $\tan^{-1} x$. This suggests the following approach:

$$u = \tan^{-1} x \qquad du = \frac{dx}{1 + x^2}$$

$$dv = dx \qquad v = x$$

Integrating an inverse trig function by parts

$$\int \underbrace{\tan^{-1} x}_{u}\, \underbrace{dx}_{dv} = \underbrace{(\tan^{-1} x)}_{u}\, \underbrace{x}_{v} - \int \underbrace{x}_{v}\, \underbrace{\frac{dx}{1 + x^2}}_{du}$$

$$= x \tan^{-1} x - \int \frac{x}{1 + x^2}\, dx.$$

It is easy to compute $\int x\, dx/(1 + x^2)$, since the numerator is a constant times the derivative of the denominator:

$$\int \frac{x\, dx}{1 + x^2} = \frac{1}{2} \int \frac{2x}{1 + x^2}\, dx = \tfrac{1}{2} \ln (1 + x^2).$$

Hence

Check by differentiation

$$\int \tan^{-1} x\,dx = x \tan^{-1} x - \tfrac{1}{2} \ln (1 + x^2) + C. \quad ■$$

To check that you understand the idea in Example 3, find $\int \sin^{-1} x\,dx$ by exactly the same method. (Compare your answer with the integral table in the front of the book.)

EXAMPLE 4 Find $\int x \sin x\,dx$.

SOLUTION There are two approaches. We could choose $u = \sin x$ and $dv = x\,dx$:

$$\int x \sin x\,dx = \int \underbrace{\sin x}_{u} \underbrace{(x\,dx)}_{dv}.$$

Then $du = \cos x\,dx$, which is not any worse than $u = \sin x$. However, since $dv = x\,dx$, $v = x^2/2$. Thus,

$$\int \underbrace{\sin x}_{u} \underbrace{(x\,dx)}_{dv} = \underbrace{\sin x}_{u} \underbrace{\frac{x^2}{2}}_{v} - \int \underbrace{\frac{x^2}{2}}_{v} \underbrace{\cos x\,dx}_{du}.$$

We have replaced the problem of finding $\int x \sin x\,dx$ with the harder problem of finding $\frac{1}{2} \int x^2 \cos x\,dx$. That is *not* progress: we have *raised* the exponent of x in the integrand from 1 to 2.

Let us take a second approach to $\int x \sin x\,dx$. This time let $u = x$ and $dv = \sin x\,dx$. Hence,

$$u = x \qquad du = dx$$
$$dv = \sin x\,dx \qquad v = -\cos x.$$

What happens? Integration by parts goes through smoothly!

$$\int \underbrace{x}_{u} \underbrace{\sin x\,dx}_{dv} = \underbrace{x}_{u} \underbrace{(-\cos x)}_{v} - \int \underbrace{(-\cos x)}_{v} \underbrace{dx}_{du}$$
$$= -x \cos x + \int \cos x\,dx$$
$$= -x \cos x + \sin x + C. \quad ■$$

EXAMPLE 5 Find $\int x^2 e^x\,dx$.

SOLUTION If we let $u = x^2$, then $du = 2x\,dx$. This is good, for it *lowers* the exponent of x. Hence, try $u = x^2$ and therefore $dv = e^x\,dx$:

$$u = x^2 \qquad du = 2x\,dx$$
$$dv = e^x\,dx \qquad v = e^x.$$

Thus

$$\int \underbrace{x^2}_{u} \underbrace{e^x\,dx}_{dv} = \underbrace{x^2}_{u} \underbrace{e^x}_{v} - \int \underbrace{e^x}_{v} \underbrace{2x\,dx}_{du}$$
$$= x^2 e^x - 2 \int e^x x\,dx$$
$$= x^2 e^x - 2[xe^x - e^x + C] \qquad \text{Example 1}$$
$$= x^2 e^x - 2xe^x + 2e^x - 2C.$$

We may rename $-2C$, the arbitrary constant, as K, obtaining

$$\int x^2 e^x \, dx = x^2 e^x - 2xe^x + 2e^x + K. \quad \blacksquare$$

Example 5 generalizes.

The idea behind Example 5 applies to integrals of the form $\int P(x)g(x)\,dx$, where $P(x)$ is a polynomial and $g(x)$ is a function—such as $\sin x$, $\cos x$, or e^x—that can be repeatedly integrated. Let $u = P(x)$ and $dv = g(x)\,dx$.

Definite Integrals and Integration by Parts

Evaluation by integration by parts of a definite integral $\int_a^b f(x)\,dx$, where $f(x) = u(x)v'(x)$, takes the form

$$\int_a^b f(x)\,dx = \int_a^b u\,dv = uv\Big|_a^b - \int_a^b v\,du$$

$$= u(b)v(b) - u(a)v(a) - \int_a^b v(x)u'(x)\,dx.$$

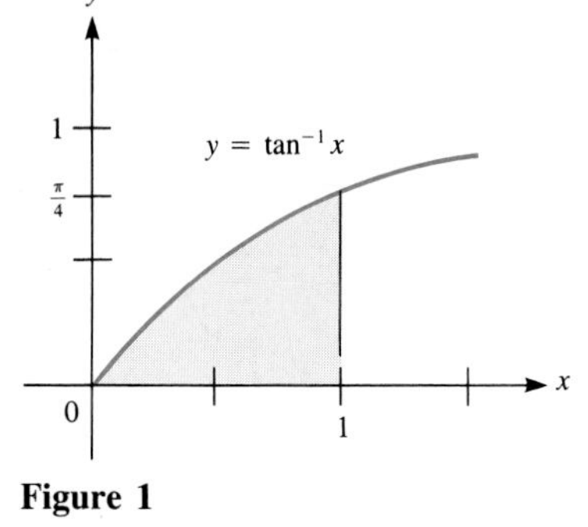

Figure 1

EXAMPLE 6 Find the area under the curve $y = \tan^{-1} x$, above $[0, 1]$. (See Fig. 1.)

SOLUTION By Sec. 5.3, the area is $\int_0^1 \tan^{-1} x\,dx$. Now, by Example 3,

$$\int \tan^{-1} x\,dx = x \tan^{-1} x - \tfrac{1}{2} \ln (1 + x^2) + C.$$

Since only one antiderivative is needed in order to apply the fundamental theorem of calculus, we may choose $C = 0$. Then

$$\int_0^1 \tan^{-1} x\,dx = x \tan^{-1} x \Big|_0^1 - \frac{1}{2} \ln (1 + x^2) \Big|_0^1$$

$$= 1 \tan^{-1} 1 - 0 \tan^{-1} 0 - \tfrac{1}{2} \ln (1 + 1^2) + \tfrac{1}{2} \ln (1 + 0^2)$$

$$= \frac{\pi}{4} - \frac{1}{2} \ln 2. \quad \blacksquare$$

Reduction Formulas

Many formulas in a table of integrals express the integral of a function that involves the nth power of some expression in terms of the integral of a function that involves the $(n - 1)$st or lower power of the same expression. These are **reduction formulas** or **recursion formulas**. Usually they are obtained by an integration by parts.

An example of a reduction formula is

$$\int \sin^n x\,dx = -\frac{\sin^{n-1} x \cos x}{n} + \frac{n-1}{n} \int \sin^{n-2} x\,dx. \qquad \text{integer } n \geq 2 \tag{4}$$

We illustrate how to use it by finding $\int \sin^5 x\,dx$. In this case $n = 5$. By (4),

$$\int \sin^5 x\,dx = -\frac{\sin^4 x \cos x}{5} + \frac{4}{5} \int \sin^3 x\,dx. \tag{5}$$

Use (4) again to dispose of $\int \sin^3 x\,dx$. In this case $n = 3$:

$$\begin{aligned}\int \sin^3 x\,dx &= -\frac{\sin^2 x \cos x}{3} + \frac{2}{3}\int \sin x\,dx \\ &= -\frac{\sin^2 x \cos x}{3} - \frac{2}{3}\cos x\end{aligned} \tag{6}$$

(since $\int \sin x\,dx = -\cos x$). Combining (5) and (6) gives

$$\int \sin^5 x\,dx = -\frac{\sin^4 x \cos x}{5} + \frac{4}{5}\left(\frac{-\sin^2 x \cos x}{3} - \frac{2}{3}\cos x\right) + C.$$

Every time you use (4) you lower the exponent of $\sin x$ by 2. If you keep applying (4), you eventually run into the exponent 1 (as we did) or, if n is even, into the exponent 0.

The next example shows how (4) can be obtained by integration by parts.

EXAMPLE 7 Obtain the reduction formula (4).

SOLUTION First write $\int \sin^n x\,dx$ as $\int \sin^{n-1} x \sin x\,dx$. Then let $u = \sin^{n-1} x$ and $dv = \sin x\,dx$. Thus

$$u = \sin^{n-1} x \qquad du = (n-1)\sin^{n-2} x \cos x\,dx$$

$$dv = \sin x\,dx \qquad v = -\cos x.$$

Integration by parts yields

$$\int \underbrace{\sin^{n-1} x}_{u}\,\underbrace{\sin x\,dx}_{dv} = \underbrace{(\sin^{n-1} x)}_{u}\underbrace{(-\cos x)}_{v} - \int \underbrace{(-\cos x)}_{v}\underbrace{(n-1)\sin^{n-2} x \cos x\,dx}_{du}.$$

The integral on the right of the preceding equation is equal to

$$\begin{aligned}-\int (n-1)\cos^2 x \sin^{n-2} x\,dx &= -(n-1)\int (1 - \sin^2 x)\sin^{n-2} x\,dx \\ &= -(n-1)\int \sin^{n-2} x\,dx + (n-1)\int \sin^n x\,dx.\end{aligned}$$

Thus

$$\int \sin^n x\,dx = -\sin^{n-1} x \cos x - \left[-(n-1)\int \sin^{n-2} x\,dx + (n-1)\int \sin^n x\,dx\right]$$

or

$$\int \sin^n x\,dx = -\sin^{n-1} x \cos x + (n-1)\int \sin^{n-2} x\,dx - (n-1)\int \sin^n x\,dx.$$

Rather than being dismayed by the reappearance of $\int \sin^n x\,dx$, collect like terms:

$$n\int \sin^n x\,dx = -\sin^{n-1} x \cos x + (n-1)\int \sin^{n-2} x\,dx,$$

from which (4) follows. ■

The reduction formula for $\int \cos^n x\,dx$ is obtained similarly. (It is Formula 50 in the table on the front cover with $a = 1$.)

An Unusual Example

In the next example one integration by parts appears at first to be useless, but two in succession find the integral.

EXAMPLE 8 Find $\int e^x \cos x\, dx$.

SOLUTION Proceed as follows:

$$\int \underbrace{e^x}_{u} \underbrace{\cos x\, dx}_{dv} = \underbrace{e^x}_{u} \underbrace{\sin x}_{v} - \int \underbrace{\sin x}_{v} \underbrace{e^x\, dx}_{du} \tag{7}$$

$$du = e^x\, dx \qquad v = \sin x.$$

It may seem that nothing useful has been accomplished; $\cos x$ is replaced by $\sin x$. But watch closely as the new integral is treated by an integration by parts. Capital letters U and V, instead of u and v, are used to distinguish this computation from the preceding one.

Repeated integration by parts

$$\int \underbrace{e^x}_{U} \underbrace{\sin x\, dx}_{dV} = \underbrace{e^x}_{U} \underbrace{(-\cos x)}_{V} - \int \underbrace{(-\cos x)}_{V} \underbrace{e^x\, dx}_{dU} \qquad (dU = e^x\, dx,\ V = -\cos x)$$

$$= -e^x \cos x + \int e^x \cos x\, dx. \tag{8}$$

Combining (7) and (8) yields

$$\int e^x \cos x\, dx = e^x \sin x - \left(-e^x \cos x + \int e^x \cos x\, dx\right)$$

$$= e^x (\sin x + \cos x) - \int e^x \cos x\, dx.$$

Bringing $-\int e^x \cos x\, dx$ to the left side of the equation gives

$$2 \int e^x \cos x\, dx = e^x (\sin x + \cos x),$$

and we conclude that

$$\int e^x \cos x\, dx = \tfrac{1}{2} e^x (\sin x + \cos x).$$

See Exercise 47.

The most general antiderivative is $\frac{1}{2}e^x (\sin x + \cos x) + C$. ■

Section Summary

Integration by parts is described by the formula

$$\int u\, dv = uv - \int v\, du.$$

When you break up the original integral into the parts u and dv, try to make your choices so that

1 You can find v and it is not too messy.
2 The derivative of u is nicer than u.
3 You can integrate $\int v\, du$.

Sometimes you have to apply integration by parts more than once, for instance, in finding $\int e^x \cos x\, dx$. Integration by parts is also the way to develop recursion formulas, such as the one for $\int \sin^n x\, dx$.

EXERCISES FOR SEC. 7.3: INTEGRATION BY PARTS

In Exercises 1 to 20 evaluate the integrals by integration by parts.

1 $\int xe^{2x}\, dx$ **2** $\int (x+3)e^{-x}\, dx$ **3** $\int x \sin 2x\, dx$ **4** $\int (x+3) \cos 2x\, dx$

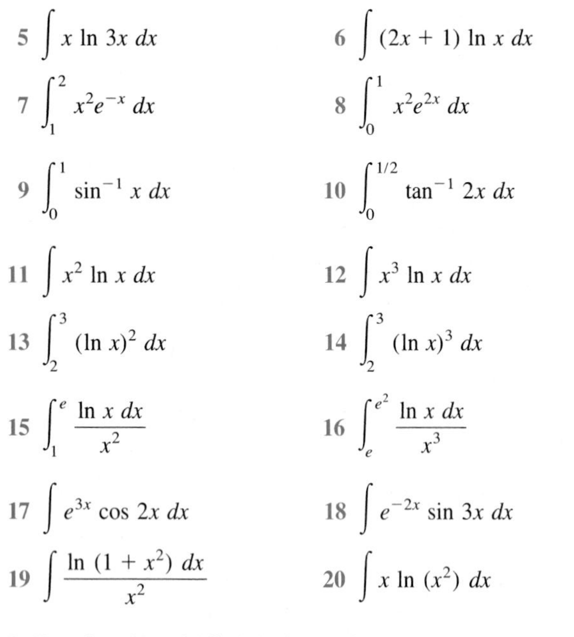

5 $\int x \ln 3x \, dx$

6 $\int (2x + 1) \ln x \, dx$

7 $\int_1^2 x^2 e^{-x} \, dx$

8 $\int_0^1 x^2 e^{2x} \, dx$

9 $\int_0^1 \sin^{-1} x \, dx$

10 $\int_0^{1/2} \tan^{-1} 2x \, dx$

11 $\int x^2 \ln x \, dx$

12 $\int x^3 \ln x \, dx$

13 $\int_2^3 (\ln x)^2 \, dx$

14 $\int_2^3 (\ln x)^3 \, dx$

15 $\int_1^e \frac{\ln x \, dx}{x^2}$

16 $\int_e^{e^2} \frac{\ln x \, dx}{x^3}$

17 $\int e^{3x} \cos 2x \, dx$

18 $\int e^{-2x} \sin 3x \, dx$

19 $\int \frac{\ln (1 + x^2) \, dx}{x^2}$

20 $\int x \ln (x^2) \, dx$

In Exercises 21 to 24 find the integrals two ways: (*a*) by substitution, (*b*) by integration by parts.

21 $\int x\sqrt{3x + 7} \, dx$

22 $\int \frac{x \, dx}{\sqrt{2x + 7}}$

23 $\int x(ax + b)^3 \, dx$

24 $\int \frac{x \, dx}{\sqrt[3]{ax + b}}, \qquad a \neq 0$

25 Use the recursion in Example 7 to find

(*a*) $\int \sin^2 x \, dx$ (*b*) $\int \sin^4 x \, dx$ (*c*) $\int \sin^6 x \, dx$.

26 Use the recursion in Example 7 to find

(*a*) $\int \sin^3 x \, dx$ (*b*) $\int \sin^5 x \, dx$.

27 Explain how you would go about finding

$$\int x^{10} (\ln x)^{18} \, dx.$$

(Do not say, ''I'd use integral tables or a computer.'') Explain why your approach would work, but include only enough calculation to convince the reader that it would succeed.

28 Let $P(x)$ be a polynomial.

(*a*) Check by differentiation that $(P(x) - P'(x) + P''(x) - \cdots)e^x$ is an antiderivative of $P(x)e^x$. (Note that the signs alternate and that the derivatives are taken to successively higher orders until they are 0.)

(*b*) Use (*a*) to find $\int (3x^3 - 2x - 2)e^x \, dx$.

(*c*) Apply integration by parts to $\int P(x)e^x \, dx$ to show how the formula in (*a*) could be obtained.

29 (*a*) Graph $y = e^x \sin x$ for x in $[0, \pi]$, showing extrema and inflection points.

(*b*) Find the area of the region below the graph and above the interval $[0, \pi]$.

30 (*a*) Graph $y = e^{-x} \sin x$ for x in $[0, \pi]$, showing extrema and inflection points.

(*b*) Find the area of the region under the graph and above the interval $[0, \pi]$.

31 Figure 2 shows a shaded region whose cross sections by planes perpendicular to the x axis are squares. Find its volume.

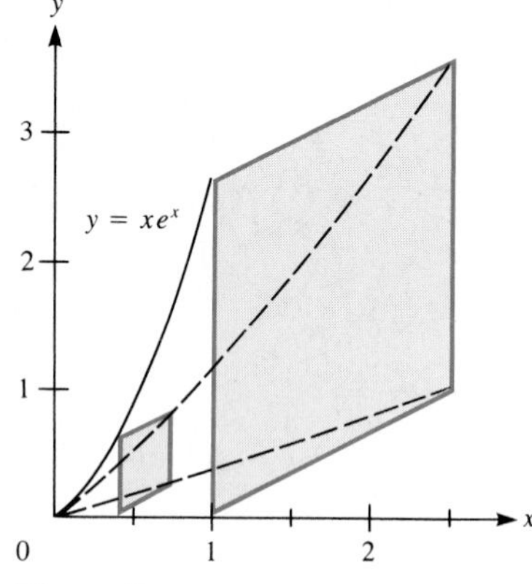

Figure 2

32 Figure 3 shows a solid whose cross sections by planes perpendicular to the x axis are circles. The solid meets the x axis in the interval $[1, e]$. Find its volume.

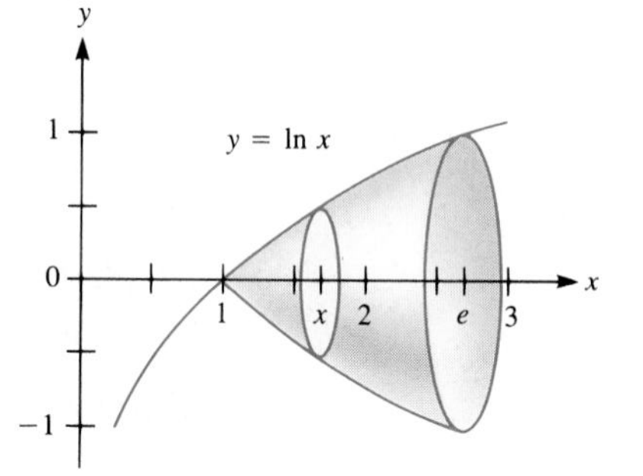

Figure 3

In Exercises 33 to 36 find the integrals. In each case a substitution is required before integration by parts can be used. In Exercises 35 and 36 the notation exp (u) is used for e^u. This notation is often used for clarity.

33 $\int \sin \sqrt{x} \, dx$

34 $\int \sin \sqrt[3]{x} \, dx$

35 $\int \exp (\sqrt{x}) \, dx$

36 $\int \exp (\sqrt[3]{x}) \, dx$

37 Given that $\int [(\sin x)/x] \, dx$ is not elementary, deduce that $\int \cos x \ln x \, dx$ is not elementary.

38 Given that $\int x \tan x \, dx$ is not elementary, deduce that $\int (x/\cos x)^2 \, dx$ is not elementary.

39 Let I_n denote $\int_0^{\pi/2} \sin^n \theta \, d\theta$, where n is a nonnegative integer.

(*a*) Evaluate I_0 and I_1.
(*b*) Using the recursion in Example 7, show that

$$I_n = \frac{n-1}{n} I_{n-2}, \qquad \text{for } n \geq 2.$$

(*c*) Use (*b*) to evaluate I_2 and I_3.
(*d*) Use (*b*) to evaluate I_4 and I_5.

(*e*) Explain why $I_n = \dfrac{2 \cdot 4 \cdot 6 \cdot \cdots \cdot (n-1)}{3 \cdot 5 \cdot 7 \cdot \cdots \cdot n}$ when n is odd.

(*f*) Explain why $I_n = \dfrac{1 \cdot 3 \cdot 5 \cdot \cdots \cdot (n-1)}{2 \cdot 4 \cdot 6 \cdot \cdots \cdot n} \cdot \dfrac{\pi}{2}$ when n is even.

(*g*) Explain why $\int_0^{\pi/2} \sin^n \theta \, d\theta = \int_0^{\pi/2} \cos^n \theta \, d\theta$. *Suggestion:* Use the substitution $u = \pi/2 - \theta$.

40 Find $\int \ln(x+1)\, dx$ using
(*a*) $u = \ln(x+1)$, $dv = dx$, $v = x$
(*b*) $u = \ln(x+1)$, $dv = dx$, $v = x + 1$
(*c*) Which is easier?

In Exercises 41 to 44 obtain recursion formulas for the integrals.

41 $\int x^n e^{ax}\, dx$, n an integer > 0, a a nonzero constant
42 $\int (\ln x)^n\, dx$, n an integer > 0
43 $\int x^n \sin x\, dx$, n an integer > 0
44 $\int \cos^n ax\, dx$, integer $n > 0$
45 Find $\int_{-1}^{1} x^3 \sqrt{1 + x^{20}}\, dx$.
46 Find $\int_{-\pi/4}^{\pi/4} \tan x\, (1 + \cos x)^{3/2}\, dx$.
47 According to the reasoning in Example 8, it appears that $\int e^x \cos x\, dx$ must equal $\frac{1}{2} e^x (\sin x + \cos x)$. This would contradict the fact that for any constant C, $\frac{1}{2} e^x (\sin x + \cos x) + C$ is also an antiderivative of $e^x \cos x$. Resolve the paradox.
48 (*a*) What does the graph of $y = \cos ax$ look like when $a = 1$? $a = 2$? $a = 3$? a is very large? Here a is a constant. Include graphs and a written description in your answer.
(*b*) Let $f(x)$ be a function with a continuous derivative. Assume that $f(x)$ is positive. What does the graph of $y = f(x) \cos ax$ look like when a is large? Express your response in terms of the graph of $y = f(x)$. Include a sketch of $y = f(x) \cos ax$ to give an idea of its shape.
(*c*) On the basis of (*b*), what do you think happens to

$$\int_0^1 f(x) \cos ax\, dx$$

as $a \to \infty$? Give an intuitive explanation.
(*d*) Use integration by parts to justify your answer in (*c*).

7.4 HOW TO INTEGRATE CERTAIN RATIONAL FUNCTIONS

This section, together with Sec. 7.5, shows how to integrate any rational function, $P(x)/Q(x)$, when $P(x)$ and $Q(x)$ are polynomials. It turns out that the antiderivative is a sum of a rational function, logarithms, and arctangents.

The Basic Idea

Which of the following two expressions would you rather integrate?

$$\int \frac{2x^3 + 5x^2 + 3x + 6}{(x^2+2)(x+1)^2}\, dx \tag{1}$$

or

$$\int \left(\frac{1}{x+1} + \frac{2}{(x+1)^2} + \frac{x}{x^2+2} \right) dx. \tag{2}$$

The first integral looks hopeless, while the second can be done by the methods of Secs. 7.1 and 7.2:

$$\int \left(\frac{1}{x+1} + \frac{2}{(x+1)^2} + \frac{x}{x^2+2} \right) dx = \int \frac{dx}{x+1} + \int \frac{2\,dx}{(x+1)^2} + \int \frac{x\,dx}{x^2+2}$$

$$= \ln|x+1| - \frac{2}{x+1} + \frac{1}{2} \ln(x^2+2) + C.$$

If you put the integrand in (2) over a common denominator, you will find that

$$\frac{1}{x+1} + \frac{2}{(x+1)^2} + \frac{x}{x^2+2} = \frac{2x^3+5x^2+3x+6}{(x^2+2)(x+1)^2}.$$

By evaluating (2), we had also evaluated (1), with its forbidding integrand.

In Sec. 7.5 we will see how to rewrite an integral such as (1) in a much simpler form [such as (2)]. That section is completely algebraic. It turns out that any rational function can be expressed as the sum of a polynomial and constant multiples of these three types of rational functions:

$$\frac{1}{(ax+b)^n} \qquad \frac{1}{(ax^2+bx+c)^n} \qquad \text{and} \qquad \frac{x}{(ax^2+bx+c)^n}. \tag{3}$$

In this section we focus on integrating the three types of functions in (3). Once we know how to integrate them, we can integrate any rational function.

Computing $\int \frac{dx}{(ax+b)^n}$

The first type, $\int dx/[(ax+b)^n]$, was found in Example 7 in Sec. 7.2, provided that $n \neq 1$. The result was

$$\int \frac{dx}{(ax+b)^n} = \frac{1}{a(1-n)(ax+b)^{n-1}} + C.$$

If, instead, we have $n = 1$, the antiderivative is

$$\int \frac{dx}{ax+b} = \frac{1}{a} \ln |ax+b| + C.$$

These two integrations are accomplished by the substitution $u = ax + b$, as the following example illustrates.

EXAMPLE 1 Find $\int \frac{dx}{(3x+2)^5}$ and $\int \frac{dx}{3x+2}$.

SOLUTION Use the substitution $u = 3x + 2$. Then $du = 3\,dx$ and $dx = \frac{1}{3}\,du$. We now have

$$\begin{aligned}\int \frac{dx}{(3x+2)^5} &= \int u^{-5} \frac{1}{3}\,du = \frac{1}{3} \cdot \frac{u^{-4}}{-4} + C \\ &= \frac{-1}{12}(3x+2)^{-4} + C \\ &= \frac{-1}{12(3x+2)^4} + C.\end{aligned}$$

We have

$$\begin{aligned}\int \frac{dx}{3x+2} &= \int \frac{1}{u}\frac{1}{3}\,du \\ &= \frac{1}{3} \ln |u| + C \\ &= \frac{1}{3} \ln |3x+2| + C.\end{aligned}$$

The integral $\int dx/(3x+2)$ could also be evaluated by noticing that the numerator is almost the derivative of the denominator. ■

More generally, any integral of the form

$$\int \frac{P(x)\,dx}{(ax + b)^n},$$

where $P(x)$ is a polynomial, can be computed by making the substitution $u = ax + b$.

Let us turn our attention to the integrands in (3) that involve some power of a quadratic factor in the denominator.

Computing $\int \frac{dx}{ax^2 + bx + c}$

It will be assumed that the polynomial $ax^2 + bx + c$ is irreducible; that is, it cannot be factored into two first-degree polynomials. This is the case when the discriminant $b^2 - 4ac$ is negative. (See Exercises 39 to 41.) We illustrate how to find

$$\int \frac{dx}{ax^2 + bx + c}$$

with some examples.

EXAMPLE 2 Find $\int \frac{dx}{4x^2 + 1}$.

SOLUTION This resembles

$$\int \frac{dx}{x^2 + 1} = \tan^{-1} x + C.$$

$u = -2x$ would work too.

For this reason, make a substitution so that $u^2 = 4x^2$. To do this, let $u = 2x$; hence

$$du = 2\,dx \qquad \text{and} \qquad \frac{du}{2} = dx.$$

Then

$$\int \frac{dx}{4x^2 + 1} = \int \frac{1}{u^2 + 1}\frac{du}{2} = \frac{1}{2}\int \frac{du}{u^2 + 1} = \frac{1}{2}\tan^{-1} u + C$$
$$= \tfrac{1}{2}\tan^{-1} 2x + C. \quad ■$$

EXAMPLE 3 Find $\int \frac{dx}{4x^2 + 9}$.

SOLUTION Again the motivation is provided by the fact that

$$\int \frac{dx}{x^2 + 1} = \tan^{-1} x + C.$$

This time choose u such that $9u^2 = 4x^2$. This substitution is suggested by the equation

$$\frac{1}{4x^2 + 9} = \frac{1}{9u^2 + 9} = \frac{1}{9}\frac{1}{u^2 + 1}.$$

So choose u such that $3u = 2x$; hence

$$3\,du = 2\,dx \quad \text{and} \quad \tfrac{3}{2}\,du = dx.$$

Thus
$$\int \frac{dx}{4x^2+9} = \int \frac{1}{9u^2+9}\,\frac{3}{2}\,du$$
$$= \frac{3}{18}\int \frac{du}{u^2+1}$$
$$= \frac{1}{6}\tan^{-1} u + C$$
$$= \frac{1}{6}\tan^{-1}\frac{2x}{3} + C.$$

This can be checked by differentiation or integral tables.

(Note that only at the end is it necessary to solve for u; $u = 2x/3$.) ■

Remark: As Examples 2 and 3 illustrate, to find $\int dx/(ax^2+c)$, where $a > 0$ and $c > 0$, use the substitution u such that $ax^2 = cu^2$.

The next example uses the algebraic technique "completing the square." It is based on the identity

$$x^2 + bx + c = \left(x + \frac{b}{2}\right)^2 + c - \frac{b^2}{4}.$$

(See Appendix C for further discussion.)

EXAMPLE 4 Find $\displaystyle\int \frac{dx}{x^2+4x+13}$.

SOLUTION First, as mentioned before Example 2, $ax^2 + bx + c$ is irreducible, if $b^2 - 4ac < 0$. In $x^2 + 4x + 13$, we have $a = 1$, $b = 4$, and $c = 13$; hence $b^2 - 4ac = 4^2 - 4 \cdot 1 \cdot 13 = 16 - 52 = -36 < 0$. Therefore $x^2 + 4x + 13$ is irreducible.

Completing the square to "get rid of 4x."

Begin by completing the square in the denominator:

$$x^2 + 4x + 13 = x^2 + 4x + 2^2 + 13 - 2^2$$
$$= (x+2)^2 + 9.$$

Thus
$$\int \frac{dx}{x^2+4x+13} = \int \frac{dx}{(x+2)^2+9},$$

an integral reminiscent of those in Examples 2 and 3.

To complete the integration, introduce a function u such that

$$9u^2 = (x+2)^2.$$

To do this, let $3u = x + 2$; that is, $u = (x+2)/3$. It follows that $3\,du = dx$. Thus

$$\int \frac{dx}{(x+2)^2+9} = \int \frac{3\,du}{9u^2+9} = \frac{3}{9}\int \frac{du}{u^2+1} = \frac{1}{3}\tan^{-1} u + C.$$

Consequently,

You can check this by differentiation.

$$\int \frac{dx}{x^2+4x+13} = \frac{1}{3}\tan^{-1}\frac{x+2}{3} + C. \quad ■$$

In the next example the coefficient of x^2 is not 1.

EXAMPLE 5 Find $\int \frac{dx}{4x^2 + 8x + 13}$.

Check that $4x^2 + 8x + 13$ is irreducible.

SOLUTION First make the coefficient of x^2 equal to 1 by factoring out the 4:

$$\int \frac{dx}{4x^2 + 8x + 13} = \frac{1}{4} \int \frac{dx}{x^2 + 2x + \frac{13}{4}}.$$

Then complete the square:

$$\begin{aligned} x^2 + 2x + \frac{13}{4} &= x^2 + 2x \qquad + \frac{13}{4} \\ &= x^2 + 2x + 1 + \frac{13}{4} - 1 \\ &= (x + 1)^2 + \frac{9}{4}. \end{aligned}$$

So

$$\int \frac{dx}{4x^2 + 8x + 13} = \frac{1}{4} \int \frac{dx}{(x + 1)^2 + \frac{9}{4}}. \tag{4}$$

Choose a substitution such that

$$(x + 1)^2 = \tfrac{9}{4}u^2.$$

To do this, let

$$x + 1 = \tfrac{3}{2}u.$$

Then

$$dx = \tfrac{3}{2}\,du$$

and

$$u = \tfrac{2}{3}(x + 1).$$

This substitution turns the right side of (4) into

$$\begin{aligned} \frac{1}{4} \int \frac{\frac{3}{2}\,du}{\frac{9}{4}u^2 + \frac{9}{4}} &= \frac{1}{6} \int \frac{du}{u^2 + 1} \\ &= \frac{1}{6} \tan^{-1} u + C \\ &= \frac{1}{6} \tan^{-1} \frac{2}{3}(x + 1) + C. \end{aligned}$$

If you are suspicious of the answer, just differentiate it. ■

As these examples show, to compute

$$\int \frac{dx}{ax^2 + bx + c} \qquad (b^2 - 4ac < 0),$$

complete the square and then make a substitution. The integral will involve an arctangent, so we see that the inverse tangent function is important in integrating rational functions.

Integrating $\int dx/[(ax^2 + bx + c)^n]$ when $n > 1$ involves a recursion formula that is usually listed in integral tables. It is seldom needed. (See Exercise 46 for a derivation of the recursion formula.)

Computing $\int \frac{x\,dx}{ax^2 + bx + c}$

The computation of $\int (x\,dx)/(ax^2 + bx + c)$ can be reduced to the integration of $\int dx/(ax^2 + bx + c)$, as shown in Example 6.

EXAMPLE 6 Find $\int \frac{x\,dx}{4x^2 + 8x + 13}$.

SOLUTION If the numerator were $8x + 8$, it would be the derivative of the denominator. The problem would then be covered by the formula

$$\int \frac{f'(x)}{f(x)}\,dx = \ln|f(x)| + C. \tag{5}$$

For this reason we write $x = \frac{1}{8}(8x) = \frac{1}{8}[(8x + 8) - 8]$. We then have

$$\begin{aligned}
\int \frac{x\,dx}{4x^2 + 8x + 13} &= \frac{1}{8}\int \frac{(8x + 8) - 8}{4x^2 + 8x + 13}\,dx \\
&= \frac{1}{8}\left(\int \frac{8x + 8}{4x^2 + 8x + 13}\,dx - \int \frac{8}{4x^2 + 8x + 13}\,dx\right) \\
&= \frac{1}{8}\left[\ln(4x^2 + 8x + 13) - \frac{4}{3}\tan^{-1}\frac{2(x + 1)}{3}\right] + C,
\end{aligned}$$

by Example 5 and Eq. (5). ■

Section Summary

Integrand	*Method of Integration*
$\frac{1}{(ax + b)^n}$	Substitute $u = ax + b$.
$\frac{1}{ax^2 + c}$, $a, c > 0$	Substitute so $cu^2 = ax^2$: $u = \sqrt{\frac{a}{c}}\,x$.
$\frac{1}{ax^2 + bx + c}$, $b^2 - 4ac < 0$	Factor out a, complete the square, then substitute.
$\frac{x}{ax^2 + bx + c}$, $b^2 - 4ac < 0$	First write x in numerator as $\frac{1}{2a}(2ax + b) - \frac{b}{2a}$, then break into two parts.

Recursive formulas for dealing with the integrands

$$\frac{1}{(ax^2 + bx + c)^n} \quad \text{and} \quad \frac{x}{(ax^2 + bx + c)^n} \qquad b^2 - 4ac < 0,\ n \geq 2$$

are included in most integral tables.

EXERCISES FOR SEC. 7.4: HOW TO INTEGRATE CERTAIN RATIONAL FUNCTIONS

Compute the integrals in Exercises 1 to 36.

1 $\int \frac{dx}{3x - 4}$

2 $\int \frac{2\,dx}{3x + 6}$

3 $\int \frac{5\,dx}{(2x + 7)^2}$

4 $\int \frac{dx}{(4x + 1)^3}$

5 $\int \frac{dx}{x^2 + 9}$

6 $\int \frac{dx}{9x^2 + 1}$

7 $\int_0^3 \frac{x\,dx}{x^2 + 9}$

8 $\int_0^1 \frac{x\,dx}{x^2 + 2}$

9 $\int \frac{2x + 3}{x^2 + 9}\,dx$

10 $\int \frac{3x - 5}{x^2 + 9}\,dx$

11 $\int \frac{dx}{16x^2 + 25}$

12 $\int \frac{dx}{9x^2 + 4}$

13 $\int \frac{x\,dx}{16x^2 + 25}$

14 $\int \frac{x\,dx}{9x^2 + 4}$

15 $\int \frac{x + 2}{9x^2 + 4}\,dx$

16 $\int \frac{2x - 1}{9x^2 + 4}\,dx$

17 $\int \frac{dx}{2x^2 + 3}$

18 $\int \frac{x\,dx}{2x^2 + 3}$

19 $\int \frac{dx}{x^2 + 2x + 3}$

20 $\int \frac{dx}{x^2 + 2x + 5}$

21 $\int \frac{dx}{x^2 - 2x + 3}$

22 $\int \frac{x\,dx}{x^2 - 2x + 3}$

23 $\int \frac{dx}{2x^2 + x + 3}$

24 $\int \frac{dx}{3x^2 - 12x + 13}$

25 $\int \frac{dx}{x^2 + 4x + 7}$

26 $\int \frac{dx}{x^2 + 4x + 9}$

27 $\int \frac{dx}{2x^2 + 4x + 7}$

28 $\int \frac{dx}{2x^2 + 6x + 5}$

29 $\int \frac{2x\,dx}{x^2 + 2x + 3}$

30 $\int \frac{2x\,dx}{x^2 + 2x + 5}$

31 $\int \frac{3x\,dx}{5x^2 + 3x + 2}$

32 $\int \frac{x\,dx}{5x^2 - 3x + 2}$

33 $\int \frac{x + 1}{x^2 + x + 1}\,dx$

34 $\int \frac{x + 3}{x^2 + x + 1}\,dx$

35 $\int \frac{3x + 5}{3x^2 + 2x + 1}\,dx$

36 $\int \frac{x + 5}{2x^2 + 3x + 5}\,dx$

---

37 Find the area of the region under $y = (x + 1)/(x^2 + x + 1)$ and above $[0, 1]$.

38 Find the area of the region under $y = 1/(2x^2 + x + 1)$ and above $[1, 2]$.

39 Show that the equation $ax^2 + bx + c = 0$, $a \neq 0$, has
(a) no real roots if $b^2 - 4ac < 0$. (Use the quadratic formula.)
(b) two distinct roots if $b^2 - 4ac > 0$.
(c) one root if $b^2 - 4ac = 0$.

40 (a) Show that if $b^2 - 4ac > 0$, then $ax^2 + bx + c = a(x - r_1)(x - r_2)$, where r_1 and r_2 are the distinct roots of $ax^2 + bx + c$.
(b) Show that if $b^2 - 4ac = 0$, then $ax^2 + bx + c = a(x - r)(x - r)$, with r the only root of $ax^2 + bx + c = 0$.
Taken together, (a) and (b) show that if $b^2 - 4ac \geq 0$, then $ax^2 + bx + c$ is reducible.

41 (a) Show that if $ax^2 + bx + c$ is reducible, then it can be written in the form $a(x - s_1)(x - s_2)$, for some real numbers s_1 and s_2.
(b) Deduce that s_1 and s_2 are roots of $ax^2 + bx + c = 0$.
(c) Deduce that $b^2 - 4ac \geq 0$.
From (a), (b), and (c), it follows that if $ax^2 + bx + c$ is reducible, then $b^2 - 4ac \geq 0$. (Compare with the conclusions in the preceding exercise.)

In Exercises 42 and 43 determine which polynomials are irreducible and which are reducible. Write reducible polynomials as the product of first-degree factors.

42 (a) $x^2 - 9$ (b) $x^2 - 5$
(c) $x^2 + 9$ (d) $x^2 + 3x + 2$
(e) $x^2 + 6x + 9$ (f) $2x^2 + 3x + 2$
(g) $x^2 + 5x + 2$

43 (a) $x^2 - 4$ (b) $x^2 - 3$
(c) $x^2 + 3$ (d) $2x^2 + 3x + 1$
(e) $2x^2 + 3x + 7$ (f) $2x^2 + 3x - 7$
(g) $49x^2 + 25$

44 Find $\int \frac{dx}{(4x^2 + 1)^2}$ by using the integral table in the front of the book.

45 Find $\int \frac{x\,dx}{(4x^2 + 8x + 13)^2}$ by using the integral table in the front of the book.

46 This exercise shows how Formula 38 in the integral table can be obtained. For simplicity we assume that completing the square has been done already. Instead of $ax^2 + bx + c$, we consider therefore only $x^2 + p$, where p is a positive constant.

(*a*) Use an integration by parts to show that

$$\int \frac{dx}{(x^2+p)^n} = \frac{x}{(x^2+p)^n} + 2n \int \frac{x^2}{(x^2+p)^{n+1}}\,dx.$$

(*b*) Write the numerator of the integrand on the right side of the equation in (*a*) as $(x^2 + p) - p$ and continue.

7.5 INTEGRATION OF RATIONAL FUNCTIONS BY PARTIAL FRACTIONS

In this section we show how to represent any rational function, $A(x)/B(x)$ [where $A(x)$ and $B(x)$ are polynomials] as the sum of three types of functions:

1 Polynomials (perhaps 0).

2 $\dfrac{k}{(px+q)^n}$,

where k, p, and q are constants and n is a positive integer.

3 $\dfrac{rx+s}{(ax^2+bx+c)^m}$,

where r, s, a, b, and c are constants and m is a positive integer.

We know how to integrate each of these three types. To integrate type 2 use the substitution $u = px + q$. To integrate type 3 write it as

$$\frac{rx}{(ax^2+bx+c)^m} + \frac{s}{(ax^2+bx+c)^m}$$

and use the methods of Sec. 7.4 or an integral table on each of the two summands. Therefore we can integrate any rational function.

For instance, we will see in Example 7 that

$$\frac{3x^3+2x^2+x-3}{x^2-1}$$

can be expressed as the sum

$$3x + 2 + \frac{5/2}{x+1} + \frac{3/2}{x-1}.$$

(To check that they are equal, put the second expression above a common denominator.) Therefore,

$$\begin{aligned}
\int \frac{3x^3+2x^2+x-3}{x^2-1}\,dx &= \int \left(3x + 2 + \frac{\frac{5}{2}}{x+1} + \frac{\frac{3}{2}}{x-1}\right) dx \\
&= \int (3x+2)\,dx + \frac{5}{2}\int \frac{dx}{x+1} + \frac{3}{2}\int \frac{dx}{x-1} \\
&= \tfrac{3}{2}x^2 + 2x + \tfrac{5}{2}\ln|x+1| + \tfrac{3}{2}\ln|x-1| + C.
\end{aligned}$$

The representation of $A(x)/B(x)$ as a sum of a polynomial and functions of the type $k/(px+q)^n$ and $(rx+s)/(ax^2+bx+c)^m$ is called the **partial-fraction**

representation of $A(x)/B(x)$. It is shown in advanced algebra that such a representation is always possible and that it is unique. If you take a course in differential equations, you may see it used for purposes other than integration. (See, for instance, W. E. Boyce and R. C. DiPrima, *Elementary Differential Equations,* Wiley, New York, 1986, in particular, pp. 288–293.)

Algebraic Preliminaries

We state some facts about polynomials with real coefficients that we will need in finding partial-fraction representations.

1 The *degree* of a polynomial $a_nx^n + a_{n-1}x^{n-1} + \cdots + a_1x + a_0$, where $a_n \neq 0$, is n.

For example, the degree of $3x + 5$ is 1 and the degree of $-x^2 + x + 1$ is 2.

2 A nonconstant polynomial is called *reducible* if it can be written as the product of two polynomials of smaller degree. Otherwise it is called *irreducible*.

For example, $x^2 - 5$ is reducible since $x^2 - 5 = (x - \sqrt{5})(x + \sqrt{5})$. On the other hand, $x^2 + 5$ is irreducible. Any polynomial of degree 1, such as $2x + 7$ is irreducible. A polynomial of degree 2, $ax^2 + bx + c$ is irreducible *only* when $b^2 - 4ac < 0$. (See Exercises 39 to 41 of Sec. 7.4.)

In upper-division algebra it is shown that any polynomial of degree at least 3 is reducible. (See Exercise 39 in Sec. 2.8.) This is far from obvious.

3 Let $A(x)$ be a polynomial and r a number such that $A(r) = 0$. Then $x - r$ is a factor of $A(x)$. For example, let $A(x) = x^3 - 8$ and $r = 2$. Then $2^3 - 8 = 0$ and $x^3 - 8 = (x - 2)(x^2 + 2x + 4)$. (See Exercise 56.)

4 If the degree of $A(x)$ is at least as large as the degree of $B(x)$, then $A(x)/B(x)$ can be written in the form

$$Q(x) + \frac{R(x)}{B(x)}$$

where $Q(x)$ is a polynomial and the degree of $R(x)$ is *less* than the degree of $B(x)$. To find $Q(x)$ and $R(x)$ use long division. For instance, consider

$$\frac{A(x)}{B(x)} = \frac{2x^3 - x^2 + 3x}{x^2 + 1}.$$

Since $3 \geq 2$, the degree of the numerator is at least as large as the degree of the denominator. Long division is possible:

$$\begin{array}{rrl}
 & 2x - 1 & \leftarrow \text{quotient } Q(x) \\
B(x) \rightarrow x^2 + 0x + 1 \,\big) & 2x^3 - x^2 + 3x + 0 & \leftarrow A(x) \\
 & 2x^3 + 0x^2 + 2x & \\
 & -x^2 + x + 0 & \\
 & -x^2 - 0x - 1 & \\
\hline
 & x + 1 & \leftarrow \text{remainder } R(x)
\end{array}$$

Thus

$$\frac{2x^3 - x^2 + 3x}{x^2 + 1} = 2x - 1 + \frac{x + 1}{x^2 + 1}. \tag{1}$$

You may check by putting the right-hand side of (1) all over the denominator $x^2 + 1$.

A rational function $A(x)/B(x)$ in which the degree of $A(x)$ is *less than* the degree of $B(x)$ is called *proper*. Otherwise it is *improper*. As mentioned, an improper rational function can be expressed as the sum of a polynomial and a proper rational function.

We will concentrate on the representation of a proper rational function and treat an improper rational function afterward.

Representing a Proper Rational Function $A(x)/B(x)$

The general procedure for representing a proper rational function consists of four steps:

1 Write $B(x)$ as a product of first-degree polynomials and irreducible second-degree polynomials.

2 If $px + q$ appears exactly n times in the factorization of $B(x)$, form

List summands of the form $\dfrac{k_i}{(px+q)^i}$.

$$\frac{k_1}{px+q} + \frac{k_2}{(px+q)^2} + \cdots + \frac{k_n}{(px+q)^n},$$

where the constants $k_1, k_2, \ldots, k_n$ are to be determined later.

3 If $ax^2 + bx + c$ appears exactly m times in the factorization of $B(x)$, then form the sum

List summands of the form $\dfrac{r_jx + s_j}{(ax^2+bx+c)^j}$.

$$\frac{r_1x+s_1}{ax^2+bx+c} + \frac{r_2x+s_2}{(ax^2+bx+c)^2} + \cdots + \frac{r_mx+s_m}{(ax^2+bx+c)^m},$$

where the constants $r_1, r_2, \ldots, r_m$ and $s_1, s_2, \ldots, s_m$ are to be determined later.

4 Find all the constants (k_i's, r_j's, and s_j's) mentioned in steps 2 and 3 so that the sum of all the expressions formed in steps 2 and 3 equals $A(x)/B(x)$.

In our examples we will assume that the factorization of $B(x)$ is known. Exercises 53 to 56 discuss the factorization of polynomials of degrees 2 and 3.

Remarks: Steps 2 and 3 above refer to the number of times a factor occurs in the denominator. If you factor $2x^2 + 4x + 2$, you may obtain $(x+1)(2x+2)$. Note that $2x + 2$ is a constant times $x + 1$. The factorization may be written as $2(x+1)^2$, where $x + 1$ is a repeated factor. We say that "$x + 1$ appears exactly two times in the factorization." Always collect factors that are constants times each other.

Step 4 requires the most work: both algebra and arithmetic.

EXAMPLE 1 Carry out steps 2 and 3 for

$$\frac{A(x)}{B(x)} = \frac{x^3 - 2x + 1}{(x+2)^3(x+1)(x^2+x+1)^2}.$$

SOLUTION The degree of the numerator is 3; the degree of the denominator (if multiplied out) is 8. Thus, $A(x)/B(x)$ is proper. The polynomial $x^2 + x + 1$ is irreducible, by the "$b^2 - 4ac < 0$" criterion. Thus $B(x)$ is already in factored form.

In steps 2 and 3 there are eight constants we must find for the partial-fraction representation of $A(x)/B(x)$. For convenience we will label these constants c_1 through c_8 (rather than using k_i's, r_j's, and s_j's). Therefore, we have

$$\frac{x^3 - 2x + 1}{(x+2)^3(x+1)(x^2+x+1)^2} = \underbrace{\frac{c_1}{x+2} + \frac{c_2}{(x+2)^2} + \frac{c_3}{(x+2)^3}}_{\text{since } (x+2)^3 \text{ is a factor of } B(x)} + \underbrace{\frac{c_4}{x+1}}_{\substack{\text{since } (x+1)^1 \text{ is} \\ \text{a factor of } B(x)}} + \underbrace{\frac{c_5x + c_6}{x^2+x+1} + \frac{c_7x + c_8}{(x^2+x+1)^2}}_{\substack{\text{since } (x^2+x+1)^2 \\ \text{is a factor of } B(x)}}.$$

■

A simple check

Remark: In Example 1 the number of unknown constants—8—equals the degree of $B(x)$. This will always be the case and serves as a check on your algebra.

In the examples we will use fairly simple denominators $B(x)$ for three reasons. First, to keep the algebra manageable. Second, because many applications involve only simple $B(x)$. Third, there are computer programs for dealing with more complicated $B(x)$. You can learn the technique of partial fractions well enough from our basic examples without grinding through the tedious problems that are best left to machine calculation.

First consider the case where $B(x)$ is the product of first-degree factors (no second-degree factor appears).

$B(x)$ Involves Only Linear Factors

EXAMPLE 2 Express $\dfrac{4x - 1}{x^2 + x - 2}$ in partial fractions.

SOLUTION The denominator $x^2 + x - 2$ is reducible since $b^2 - 4ac = 1^2 - 4 \cdot 1 \cdot (-2) = 9$ is greater than 0. Its factorization is $x^2 + x - 2 = (x+2)(x-1)$. Thus

$$\frac{4x-1}{x^2+x-2} = \frac{4x-1}{(x+2)(x-1)} = \frac{c_1}{x+2} + \frac{c_2}{x-1}$$

for suitable constants c_1 and c_2.

To find c_1 and c_2, clear the denominators by multiplying by $(x+2)(x-1)$, obtaining

$$4x - 1 = c_1(x-1) + c_2(x+2). \qquad (2)$$

Equation (2) holds for *all* values of x. In particular it holds for $x = 1$ and $x = -2$, the roots of the polynomials $x - 1$ and $x + 2$. Therefore

Substituting

$$4(1) - 1 = c_1(1-1) + c_2(1+2) \qquad \text{substitute 1 for } x \text{ in Eq. (2)}$$
$$4(-2) - 1 = c_1(-2-1) + c_2(-2+2). \qquad \text{substitute } -2 \text{ for } x$$

These equations reduce to

Solving

$$\begin{cases} 3 = 0c_1 + 3c_2 \\ -9 = -3c_1 + 0c_2. \end{cases}$$

(The 0's come from our particular choices of x.) In short

$$\begin{cases} 3 = 3c_2 \\ -9 = -3c_1, \end{cases}$$

from which we deduce that

$$c_1 = 3 \quad \text{and} \quad c_2 = 1.$$

Then

$$\frac{4x - 1}{x^2 + x - 2} = \frac{3}{x + 2} + \frac{1}{x - 1}. \tag{3}$$

You may check that (3) is correct by putting the right-hand side over a common denominator. ■

Example 2 illustrated substitution for x. If there are, say, n unknown constants $c_1, c_2, \ldots, c_n$, then substitute n different choices of x. You will then have n equations in n unknowns to solve. In particular, if there is a linear factor $px + q$, substitute its root $-q/p$, for that will produce a 0 (and cause terms to "disappear").

Example 2 can be solved by a completely different method, called **comparison of coefficients** (or **equating coefficients**). It depends on the fact that if two polynomials are equal, then their corresponding coefficients are equal.

To illustrate this approach, rewrite (2), collecting terms of like degree:

$$4x - 1 = \underbrace{(c_1 + c_2)x}_{\text{first-degree}} + \underbrace{2c_2 - c_1}_{\text{constant term}}. \tag{4}$$

Comparing coefficients on both sides of (4) gives two equations:

$$\begin{cases} 4 = c_1 + c_2 \\ -1 = 2c_2 - c_1. \end{cases} \tag{5}$$

The equations in (5) can be solved several ways. For instance, if we add them, the c_1's cancel; then

$$3 = 3c_2$$

or $c_2 = 1$. Since $4 = c_1 + c_2$, it follows that $c_1 = 3$.

The next example illustrates a *shortcut* for finding the partial fractions when the factorization of the denominator $B(x)$ involves *only linear factors* and *none of them is repeated*.

EXAMPLE 3 Find the partial-fraction representation of

$$\frac{6x^2 - 7x - 1}{(x - 1)(x + 1)(x - 2)}.$$

SOLUTION There are constants c_1, c_2, c_3 such that

$$\frac{6x^2 - 7x - 1}{(x - 1)(x + 1)(x - 2)} = \frac{c_1}{x - 1} + \frac{c_2}{x + 1} + \frac{c_3}{x - 2}. \tag{6}$$

To find c_1 multiply both sides of (6) by $x - 1$, obtaining

Multiply (6) by $x - 1$.

$$\frac{6x^2 - 7x - 1}{(x+1)(x-2)} = c_1 + (x-1)\left(\frac{c_2}{x+1} + \frac{c_3}{x-2}\right). \tag{7}$$

Now replace x in (7) by the solution of $x - 1 = 0$, namely 1, obtaining

Substitute 1 for x.

$$\frac{6(1)^2 - 7(1) - 1}{(1+1)(1-2)} = c_1 + 0.$$

Hence $$c_1 = \frac{6 - 7 - 1}{(2)(-1)} = 1.$$

Note that we did not clear the whole denominator in (7).

To obtain c_2 multiply (6) by $x + 1$, obtaining

Multiply (6) by $x + 1$.

$$\frac{6x^2 - 7x - 1}{(x-1)(x-2)} = c_2 + (x+1)\left(\frac{c_1}{x-1} + \frac{c_3}{x-2}\right). \tag{8}$$

Replacing x in (8) by -1, the solution of $x + 1 = 0$, gives

Substitute -1 for x.

$$\frac{6(-1)^2 - 7(-1) - 1}{(-1-1)(-1-2)} = c_2 + 0.$$

Hence $$c_2 = 2.$$

Multiply (6) by $x - 2$ and substitute 2 for x.

To obtain c_3 multiply (6) by $x - 2$ and replace x by 2. As you may check, this gives

$$c_3 = 3. \quad ■$$

If a linear factor is repeated you may use either substitution or comparison of coefficients. The next example illustrates both methods.

EXAMPLE 4 Find the partial-fraction representation of $\dfrac{-x^2 + 2x + 4}{(x+1)^2(2x+3)}$.

SOLUTION *(First solution: Substitution)*

$$\frac{-x^2 + 2x + 4}{(x+1)^2(2x+3)} = \frac{c_1}{x+1} + \frac{c_2}{(x+1)^2} + \frac{c_3}{2x+3}.$$

Clearing the denominator gives

$$-x^2 + 2x + 4 = c_1(x+1)(2x+3) + c_2(2x+3) + c_3(x+1)^2. \tag{9}$$

Since there are three unknowns, we make three different substitutions.

First we use -1, the root of the linear factor $x + 1$. Then we use $-\frac{3}{2}$, the root of $2x + 3$. For the third substitution, let us use 0, which will not lead to complex arithmetic. (The substitution of 1992 for x, say, would invite error.) Making these three substitutions in (9) gives

$$-(-1)^2 + 2(-1) + 4 = 0 + c_2(2(-1) + 3) + 0 \qquad (x = -1)$$

$$-(-\tfrac{3}{2})^2 + 2(-\tfrac{3}{2}) + 4 = 0 + 0 + c_3(-\tfrac{3}{2} + 1)^2 \qquad (x = -\tfrac{3}{2})$$

$$-(0^2) + 2(0) + 4 = c_1(0+1)(2\cdot 0 + 3) + c_2(2\cdot 0 + 3) + c_3(0+1)^2. \qquad (x = 0)$$

These three equations reduce to

$$1 = c_2$$
$$-\tfrac{5}{4} = \tfrac{1}{4}c_3 \qquad (10)$$
$$4 = 3c_1 + 3c_2 + c_3.$$

Hence $c_2 = 1$, $c_3 = -5$, and therefore

$$4 = 3c_1 + 3(1) + (-5),$$

from which it follows that $c_1 = 2$.

Hence

$$\frac{-x^2 + 2x + 4}{(x+1)^2(2x+3)} = \frac{2}{x+1} + \frac{1}{(x+1)^2} - \frac{5}{2x+3}.$$

(Second solution: Comparison of coefficients)

The solution begins like the first solution, reaching (9). But this time we multiply the products and collect terms of like degree. Equation (9) becomes

$$-x^2 + 2x + 4 = (2c_1 + c_3)x^2 + (5c_1 + 2c_2 + 2c_3)x + 3c_1 + 3c_2 + c_3.$$

Comparing coefficients on both sides gives three equations for the unknowns:

$$-1 = 2c_1 + c_3 \qquad x^2 \text{ coefficient}$$
$$2 = 5c_1 + 2c_2 + 2c_3 \qquad x \text{ coefficient} \qquad (11)$$
$$4 = 3c_1 + 3c_2 + c_3. \qquad \text{constant term}$$

Solving these simultaneous equations gives $c_1 = 2$, $c_2 = 1$, and $c_3 = -5$. (See Exercise 44.) ■

Either the substitution method or the comparison of coefficients method can be used even when second-degree factors are present in $B(x)$. In the next example, which concerns a rational function that appears in the differential equations text mentioned earlier, we use a mix of the two methods.

EXAMPLE 5 Obtain the partial-fraction representation of $\dfrac{x^2}{x^4 - 1}$.

SOLUTION First factor the denominator: $x^4 - 1 = (x^2 + 1)(x^2 - 1) = (x^2 + 1)(x + 1)(x - 1)$. Then

$$\frac{x^2}{x^4 - 1} = \frac{c_1}{x+1} + \frac{c_2}{x-1} + \frac{c_3x + c_4}{x^2 + 1}.$$

Clear the denominator:

$$x^2 = c_1(x-1)(x^2+1) + c_2(x+1)(x^2+1) + (c_3x + c_4)(x-1)(x+1). \qquad (12)$$

Substitute 1 and -1, the roots of the linear factors:

$$1 = 0 + 4c_2 + 0 \qquad \text{substituting } 1$$
$$1 = -4c_1 + 0 + 0. \qquad \text{substituting } -1$$

Already we see that $c_1 = -\tfrac{1}{4}$ and $c_2 = \tfrac{1}{4}$.

Next substitute 0, obtaining

$$0 = -c_1 + c_2 - c_4. \qquad \text{substituting } 0$$

Hence $c_4 = \tfrac{1}{2}$.

We still have to find c_3. We could substitute another number, say 2, or compare coefficients in (12). Let us compare coefficients of just the highest degree, x^3. Without going to the bother of multiplying (12) out in full, we can read off the coefficient of x^3 on both sides by sight, getting

$$0 = c_1 + c_2 + c_3.$$

Since $c_1 = -\frac{1}{4}$ and $c_2 = \frac{1}{4}$, it follows that $c_3 = 0$. Hence

$$\frac{x^2}{x^4 - 1} = \frac{-\frac{1}{4}}{x + 1} + \frac{\frac{1}{4}}{x - 1} + \frac{\frac{1}{2}}{x^2 + 1}. \quad \blacksquare$$

Representing an Improper Rational Function

If $A(x)/B(x)$ is improper, divide $B(x)$ into $A(x)$ first so that you obtain

$$\frac{A(x)}{B(x)} = Q(x) + \frac{R(x)}{B(x)},$$

where $Q(x)$ is a polynomial and $R(x)/B(x)$ is proper (or may even be 0).

EXAMPLE 6 Find the partial-fraction representation of

$$\frac{A(x)}{B(x)} = \frac{2x^5 + 11x^4 + 28x^3 + 39x^2 + 32x + 13}{(x + 1)^2(2x + 3)}. \tag{13}$$

SOLUTION The degree of the numerator in (13) is 5, while the degree of the denominator is 3. Since the degree of the numerator is greater than or equal to the degree of the denominator, (13) is improper. To divide, we first multiply out the denominator to obtain $B(x) = 2x^3 + 7x^2 + 8x + 3$. Then divide:

$$\begin{array}{r}
x^2 + 2x + 3 \leftarrow Q(x) \\
B(x) \rightarrow 2x^3 + 7x^2 + 8x + 3 \,\overline{)\, 2x^5 + 11x^4 + 28x^3 + 39x^2 + 32x + 13} \leftarrow A(x) \\
\underline{2x^5 + 7x^4 + 8x^3 + 3x^2} \\
4x^4 + 20x^3 + 36x^2 + 32x \\
\underline{4x^4 + 14x^3 + 16x^2 + 6x} \\
6x^3 + 20x^2 + 26x + 13 \\
\underline{6x^3 + 21x^2 + 24x + 9} \\
-x^2 + 2x + 4 \leftarrow R(x)
\end{array}$$

Thus
$$\frac{A(x)}{B(x)} = x^2 + 2x + 3 + \frac{-x^2 + 2x + 4}{(x + 1)^2(2x + 3)}.$$

We then find the partial-fraction representation of

$$\frac{-x^2 + 2x + 4}{(x + 1)^2(2x + 3)}.$$

Luckily that has been done in Example 4. Hence

$$\frac{2x^5 + 11x^4 + 28x^3 + 39x^2 + 32x + 13}{(x + 1)^2(2x + 3)} = x^2 + 2x + 3 + \frac{2}{x + 1} + \frac{1}{(x + 1)^2} - \frac{5}{2x + 3}.$$

$\blacksquare$

EXAMPLE 7 Find the partial-fraction representation of

$$\frac{3x^3 + 2x^2 + x - 3}{x^2 - 1}. \tag{14}$$

SOLUTION Since (14) is improper, we first carry out a long division:

$$\begin{array}{r} 3x + 2 \\ x^2 + 0x - 1 \overline{) 3x^3 + 2x^2 + x - 3} \\ \underline{3x^3 + 0x^2 - 3x } \\ 2x^2 + 4x - 3 \\ \underline{2x^2 + 0x - 2} \\ 4x - 1. \end{array}$$

Thus
$$\frac{3x^3 + 2x^2 + x - 3}{x^2 - 1} = 3x + 2 + \frac{4x - 1}{x^2 - 1}.$$

Next we work on $(4x - 1)/(x^2 - 1)$, first factoring the denominator $x^2 - 1 = (x + 1)(x - 1)$. There are constants c_1 and c_2 such that

$$\frac{4x - 1}{(x + 1)(x - 1)} = \frac{c_1}{x + 1} + \frac{c_2}{x - 1}. \tag{15}$$

Since there are only linear factors and they are not repeated, we use the technique illustrated in Example 3.

To find c_1 multiply (15) by $x + 1$, obtaining

$$\frac{4x - 1}{x - 1} = c_1 + (x + 1)\frac{c_2}{x - 1}.$$

Replace x by -1, the root of $x + 1$. This gives

$$\frac{4(-1) - 1}{(-1) - 1} = c_1.$$

Hence $c_1 = 5/2$.

To find c_2 multiply (15) by $x - 1$ and replace x by 1, the root of $x - 1$. You may check that $c_2 = 3/2$.

All told,

$$\frac{3x^3 + 2x^2 + x - 3}{x^2 - 1} = 3x + 2 + \frac{\frac{5}{2}}{x + 1} + \frac{\frac{3}{2}}{x - 1}. \quad ■$$

Section Summary

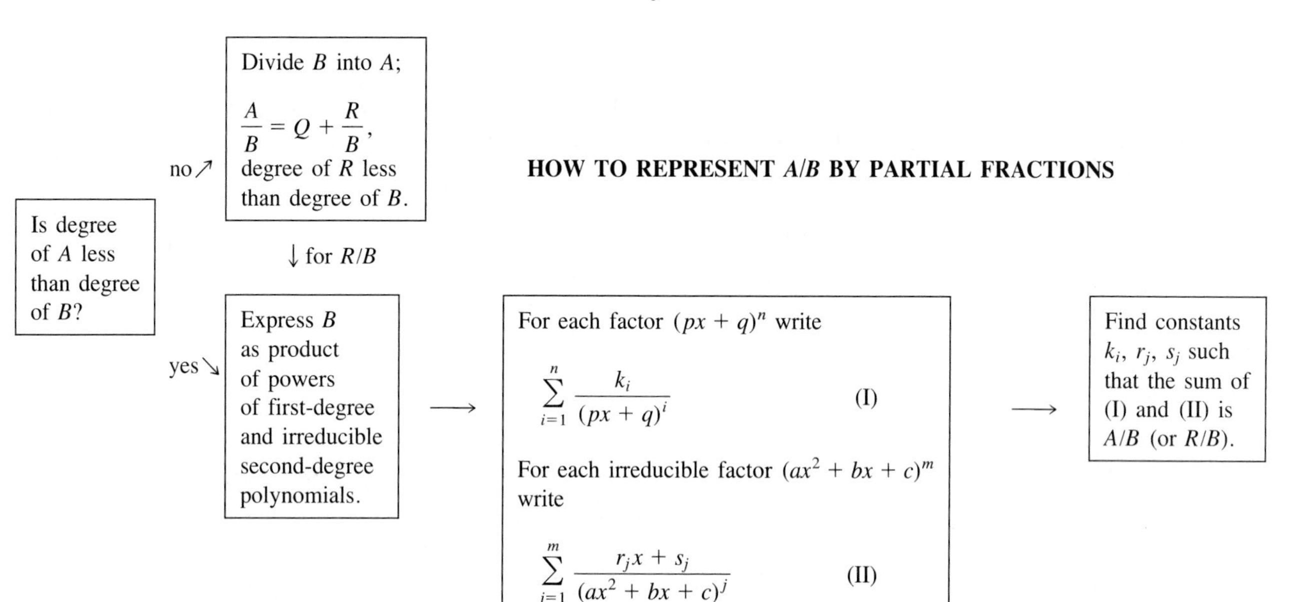

Useful Facts The number of unknown constants equals the degree of B. (Check that you have the right number.)

If $b^2 - 4ac < 0$, $ax^2 + bx + c$ is irreducible. If $b^2 - 4ac \geq 0$, $ax^2 + bx + c$ is reducible.

The constants can be found in two ways: comparing coefficients or substituting numbers for x. In case $B(x)$ has no second-degree irreducible factors and its linear factors are not repeated, there is a special technique. (Multiply by a factor $px + q$ and then replace x by the root of the factor.)

THE REAL WORLD

Say that you wanted to compute the definite integral

$$\int_1^2 \frac{x + 3}{x^3 + x^2 + 2x + 1}\, dx.$$

One way is by partial fractions, but this can be tedious. You would probably prefer to estimate the definite integral by one of the approximation techniques in Sec. 5.4. Alternatively, computers and many scientific calculators can be programmed to estimate a definite integral. With the Hewlett-Packard 28S, for example, you would enter the integrand, the variable of integration, the limits of integration, and how small you want the error to be. It takes the HP-28S about 10 seconds to find that the answer is 0.49353 with an error less than 0.00001.

As noted in Chap. 5, in some cases computers and calculators can even give the exact value of a definite integral by first producing an antiderivative. In practical applications, however, formal antidifferentiation is not that important. The present example could theoretically be computed exactly by partial fractions, but modern computational tools can evaluate it accurately to as many decimal places as may be required.

EXERCISES FOR SEC. 7.5: INTEGRATION OF RATIONAL FUNCTIONS BY PARTIAL FRACTIONS

In Exercises 1 to 10 indicate the form of the partial-fraction representation of the proper rational function listed, but do *not* find the unknown constants. Remember that the number of constants equals the degree of the denominator.

1 $\dfrac{x+3}{(x+1)(x+2)}$ 2 $\dfrac{5}{(x-1)(x+3)}$

3 $\dfrac{1}{(x-1)^2(x+2)}$ 4 $\dfrac{3}{(x+1)(x+5)(x+4)}$

5 $\dfrac{6x^2-2}{(x-1)(x-2)(2x-3)}$ 6 $\dfrac{x^3}{(x+1)^2(x+2)^2}$

7 $\dfrac{x}{(x+1)(x^2+x+1)^2}$ 8 $\dfrac{x^3+2}{(x-1)^2(x^2+2x+3)^2}$

9 $\dfrac{x+3}{(x^2-1)(3x+5)^3}$ 10 $\dfrac{x^2+2}{(x+1)^2(2x+2)(3x+3)}$

Exercises 11 to 14 concern improper rational functions. In each case express the given function as the sum of a polynomial and a proper rational function.

11 $\dfrac{x^2}{x^2+x+1}$ 12 $\dfrac{x^3}{(x+1)(x+2)}$

13 $\dfrac{x^5-2x+1}{(x+1)(x^2+1)}$ 14 $\dfrac{x^5+x}{(x+1)^2(x-2)}$

In Exercises 15 to 32 express the rational function in terms of partial fractions. (For practice in solving simultaneous equations, see Exercises 37 to 44 and the discussion preceding them.)

15 $\dfrac{4x-3}{x(x-1)}$ 16 $\dfrac{x+2}{x(x+1)}$

17 $\dfrac{5x^2-x-1}{x^2(x-1)}$ 18 $\dfrac{2x^2+4x+3}{x(x+1)^2}$

19 $\dfrac{x}{(x+1)(x+2)}$ 20 $\dfrac{x+4}{(x-1)(x+3)}$

21 $\dfrac{2x}{x^2-1}$ 22 $\dfrac{8}{x^2-4}$

23 $\dfrac{2x^2+3}{x(x+1)(x+2)}$ 24 $\dfrac{5x^2-2x-2}{x(x^2-1)}$

25 $\dfrac{6x^2-7x-1}{(x+1)(x+1)(x+2)}$ 26 $\dfrac{2x^2-10x+14}{(x-2)(x-3)(x-4)}$

27 $\dfrac{5x^2+9x+6}{(x+1)(x^2+2x+2)}$ 28 $\dfrac{5x^2+2x+3}{x(x^2+x+1)}$

29 $\dfrac{x^3+5x+1}{x(x+1)}$ 30 $\dfrac{x^3-3x^2+3x-3}{x^2-3x+2}$

31 $\dfrac{3x^3+2x^2+3x+1}{x(x^2+1)}$ 32 $\dfrac{x^5+2x^4+4x^3+2x^2+x-2}{x^4-1}$

Exercises 33 to 36 are suggested by examples or exercises in the differential equations text mentioned earlier. In each case find the partial-fraction representation.

33 $\dfrac{x-1}{x^2-x-2}$ 34 $\dfrac{x^2+6}{(x^2+1)(x^2+4)}$

35 $\dfrac{2}{x^2+3x-4}$ 36 $\dfrac{2x-3}{x^2+7x+10}$

Exercises 37 to 44 concern solving simultaneous equations. By way of illustration we solve the equations

$$2c_1 - 3c_2 = 5$$
$$3c_1 + 4c_2 = 6$$

in two different ways. In one approach we solve for one of the unknowns in terms of the other unknown (using one equation). Then we substitute the result in the other equation. Thus $c_1 = (5+3c_2)/2$, using the first equation. Substitution in the second equation gives $3(5+3c_2)/2 + 4c_2 = 6$, an equation in only one unknown. Solve it for c_2, then get c_1.

In another approach we multiply each equation by a constant so that the coefficients of, say, c_1 become equal. Then subtract one equation from another. Thus

$$3(2c_1 - 3c_2) = 3 \cdot 5$$
$$2(3c_1 + 4c_2) = 2 \cdot 6$$

or

$$6c_1 - 9c_2 = 15$$
$$6c_1 + 8c_2 = 12.$$

Subtracting gives $-17c_2 = 3$, hence $c_2 = -\frac{3}{17}$. Then obtain c_1 using any of the equations. Both approaches apply to three equations in three unknowns.

In Exercises 37 to 44 solve the simultaneous equations and check that your answers satisfy the equations.

37 $\begin{aligned} 3c_1 - 2c_2 &= 3 \\ c_1 + c_2 &= 4 \end{aligned}$ 38 $\begin{aligned} 2c_1 + 5c_2 &= -3 \\ 3c_1 - 4c_2 &= 2 \end{aligned}$

39 $\begin{aligned} c_1 + 5c_2 &= 6 \\ 2c_1 - 3c_2 &= -2 \end{aligned}$ 40 $\begin{aligned} 5c_1 + 2c_2 &= 2 \\ -3c_1 + 4c_2 &= 1 \end{aligned}$

41 $\begin{aligned} c_1 + 2c_2 + c_3 &= 9 \\ c_1 - c_2 &= -1 \\ c_1 + c_3 &= 3 \end{aligned}$ 42 $\begin{aligned} 2c_1 - c_2 + c_3 &= -7 \\ 3c_1 - c_2 - 2c_3 &= 5 \\ c_1 + c_2 + c_3 &= -2 \end{aligned}$

43 $\begin{aligned} c_1 + c_3 &= 4 \\ c_2 - c_3 &= -6 \\ c_1 + c_2 + c_3 &= -1 \end{aligned}$

44 Solve the simultaneous equations (11) in Example 4.

In Exercises 45 and 46 find the areas of the given regions.

45 Under $y = 1/(x^3 + x)$ and above $[1, 2]$.

46 Under $y = (x^3 + 2x + 2)/(x^2 + 1)$ and above $[0, 1]$.

47 The region under $y = (x + 2)/(x^2 + x)$ and above $[1, 2]$ is revolved about the x axis to produce a solid. Find its volume.

48 The region under

$$y = \frac{1}{\sqrt{(x^2 - 4)(x - 1)}}$$

and above $[3, 5]$ is revolved about the x axis to produce a solid. Find its volume.

In Exercises 49 to 52 evaluate the integrals.

49 $\displaystyle\int_0^1 \frac{dx}{x^2 + 3x + 2}$ **50** $\displaystyle\int_0^1 \frac{dx}{x^2 + 2x + 2}$

51 $\displaystyle\int_1^2 \frac{x^3\,dx}{x^3 + 1}$ **52** $\displaystyle\int_0^1 \frac{x^3 + x}{x^2 + 2}\,dx$

We did not discuss the problem of factoring a polynomial $B(x)$ into linear and irreducible quadratic polynomials. Exercises 53 to 59 concern this problem when the degree of $B(x)$ is 2, 3, or 4.

53 In Exercise 40 of Sec. 7.4 it was shown that if $b^2 - 4ac \geq 0$, the polynomial $ax^2 + bx + c$ is reducible. Indeed, it equals $a(x - r_1)(x - r_2)$ where r_1 and r_2 are its roots, which can be found by the quadratic formula. Factor each of these polynomials: (*a*) $x^2 + 6x + 5$, (*b*) $x^2 - 5$, (*c*) $2x^2 + 6x + 3$.

54 (*a*) Show that $x^2 + 3x - 5$ is reducible.
(*b*) Using (*a*), find $\int dx/(x^2 + 3x - 5)$ by partial fractions.
(*c*) Find $\int dx/(x^2 + 3x - 5)$ by using an integral table.

55 Let $P(x)$ be a polynomial with integer coefficients. If $P(r) = 0$, then $x - r$ is a factor of $P(x)$. You may search for a root r by Newton's method. There is an algebraic technique for determining any *rational* roots of $P(x) = 0$. Let $r = p/q$, where p and q are integers with no common divisor larger than 1. We may assume that q is positive. The rational-root test asserts that if p/q is a root of $a_nx^n + a_{n-1}x^{n-1} + \cdots + a_0$, then q must divide a_n and p must divide a_0.

For instance, consider $P(x) = 3x^3 + x^2 + x - 2$. If $P(p/q) = 0$, then p divides -2 and q divides 3. Then p must be $1, 2, -1$, or -2 and q must be 1 or 3. There are 8 combinations of p and q to check. For example, consider $p = 1, q = 1$, that is, $p/q = 1$. Note that $P(1) = 3$, so $1/1$ is *not* a root. It turns out that the choice $p = 2, q = 3$ produces a root $\frac{2}{3}$. [Check that $P(\frac{2}{3}) = 0$.] Of course, a polynomial of degree greater than 1 need not have a rational root. Determine all rational roots of the following polynomials:
(*a*) $x^2 + x - 12$
(*b*) $2x^3 - 11x^2 + 17x - 6$
(*c*) $x^4 + x^3 + x^2 + x + 1$
(*d*) $3x^3 - 2x^2 - 4x - 1$

56 To factor a cubic $P(x) = ax^3 + bx^2 + cx + d$ first find or estimate a root r. Then divide $x - r$ into $P(x)$, obtaining a quotient $Q(x)$, that is, a quadratic polynomial such that $P(x) = Q(x)(x - r)$. If $Q(x)$ is irreducible, we are done. If $Q(x)$ is reducible, factoring it completes the factorization of $P(x)$. Illustrate this procedure for
(*a*) $4x^3 + 4x^2 - 13x - 3$
(*b*) $2x^3 - x^2 - x - 3$
(*c*) $x^3 + x + 1$
(*d*) $x^3 - 8$

57 Since the only polynomials that are irreducible over the real numbers are the first-degree polynomials and the irreducible second-degree polynomials, the polynomial $x^4 + 1$ must be reducible. Factor it. [*Suggestion:* Find constants a and b such that $x^4 + 1 = (x^2 + ax + 1)(x^2 + bx + 1)$.]

58 (See Exercise 57.) Compute as easily as possible.

(*a*) $\displaystyle\int \frac{x^3\,dx}{x^4 + 1}$ (*b*) $\displaystyle\int \frac{x\,dx}{x^4 + 1}$ (*c*) $\displaystyle\int \frac{dx}{x^4 + 1}$

59 (*a*) Write $x^4 + x^2 + 1$ as the product of irreducible polynomials of second degree. (See the suggestion in Exercise 57.)

(*b*) Compute $\displaystyle\int \frac{dx}{x^4 + x^2 + 1}$.

7.6 SPECIAL TECHNIQUES

So far in this chapter you have met three techniques for computing integrals. The first, substitution, and the second, integration by parts, are used most often. Partial fractions applies only to a special class of integrands, the rational functions. In this section we compute some integrals such as $\int \sin^2 \theta\, d\theta$, $\int \sin mx \cos nx\, dx$, and $\int \sec \theta\, d\theta$, which you may meet in applications outside of a calculus course. Then we describe some substitutions that deal with special classes of integrands, including the case of trigonometric substitution. You may or may not have occasion to use them. At least you will be aware that there are such techniques.

Computing $\int \sin mx \sin nx\, dx$

m and n are constants.

The integrals $\int \sin mx \sin nx\, dx$, $\int \cos mx \sin nx\, dx$, and $\int \cos mx \cos nx\, dx$ are needed in the study of Fourier series, an important tool in such varied areas as heat, sound, and signal processing. These three integrals can be computed with the aid of these identities:

$$\sin A \sin B = \tfrac{1}{2}\cos(A-B) - \tfrac{1}{2}\cos(A+B);$$

$$\sin A \cos B = \tfrac{1}{2}\sin(A+B) + \tfrac{1}{2}\sin(A-B);$$

$$\cos A \cos B = \tfrac{1}{2}\cos(A-B) + \tfrac{1}{2}\cos(A+B).$$

These identities can be checked using the well-known identities for $\sin(A \pm B)$ and $\cos(A \pm B)$.

EXAMPLE 1 Find $\int_0^{\pi/4} \sin 3x \sin 2x\, dx$.

SOLUTION

$$\begin{aligned}\int_0^{\pi/4} \sin 3x \sin 2x\, dx &= \int_0^{\pi/4} \left(\tfrac{1}{2}\cos x - \tfrac{1}{2}\cos 5x\right) dx \\ &= \left(\tfrac{1}{2}\sin x - \tfrac{1}{10}\sin 5x\right)\Big|_0^{\pi/4} \\ &= \left(\frac{\sqrt{2}}{4} + \frac{\sqrt{2}}{20}\right) - \left(\frac{0}{2} - \frac{0}{10}\right) \\ &= \frac{3\sqrt{2}}{10}. \quad \blacksquare\end{aligned}$$

Computing $\int \sin^2 x\, dx$ and $\int \cos^2 x\, dx$

These integrals can be computed with the aid of the identities

$$\sin^2 x = \frac{1 - \cos 2x}{2} \quad \text{and} \quad \cos^2 x = \frac{1 + \cos 2x}{2}.$$

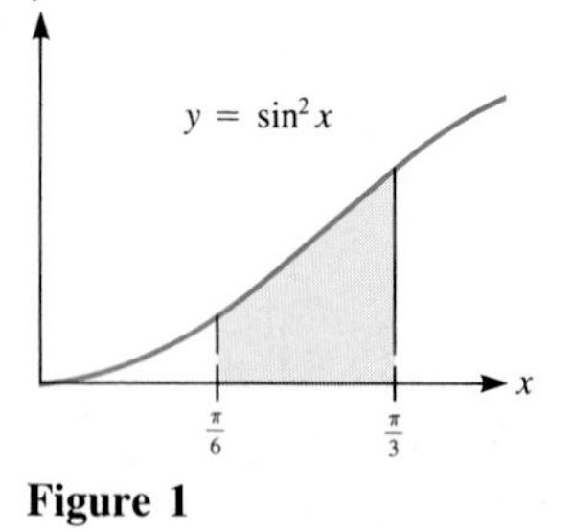

Figure 1

EXAMPLE 2 Find the area of the region under the curve $y = \sin^2 x$ and above the interval $[\pi/6, \pi/3]$ shown in Fig. 1.

SOLUTION We first find an antiderivative of $\sin^2 x$:

$$\begin{aligned}\int \sin^2 x\, dx &= \int \frac{1 - \cos 2x}{2}\, dx \\ &= \int \frac{dx}{2} - \int \frac{\cos 2x}{2}\, dx \\ &= \frac{x}{2} - \frac{\sin 2x}{4} + C.\end{aligned}$$

Thus, by the fundamental theorem of calculus, the area $\int_{\pi/6}^{\pi/3} \sin^2 x\, dx$ equals

$$\left(\frac{x}{2}-\frac{\sin 2x}{4}\right)\Bigg|_{\pi/6}^{\pi/3}=\left[\frac{\pi/3}{2}-\frac{\sin 2(\pi/3)}{4}\right]-\left[\frac{\pi/6}{2}-\frac{\sin 2(\pi/6)}{4}\right]$$
$$=\frac{\pi}{6}-\frac{\sin 2\pi/3}{4}-\frac{\pi}{12}+\frac{\sin \pi/3}{4}$$
$$=\frac{\pi}{12}-\frac{\sqrt{3}/2}{4}+\frac{\sqrt{3}/2}{4}=\frac{\pi}{12}.$$ ■

Computing $\int \sec\theta\, d\theta$

A century before the invention of calculus, the cartographer Gerhardus Mercator had to estimate $\int_0^\alpha \sec\theta\, d\theta$ in order to determine where to place the lines of latitude on his maps. On a Mercator map, a straight line corresponds to a voyage with a constant compass heading, a property of great use to navigators. (Figure 2 is part of a Mercator map.) In 1645, Henry Bond conjectured that, on the basis of numerical evidence, $\int_0^\alpha \sec\theta\, d\theta = \ln\tan(\alpha/2+\pi/4)$ but offered no proof. In 1666, Nicolaus Mercator (no relation to Gerhardus) offered the royalties on one of his inventions to the mathematician who could prove Bond's conjecture was right. Within two years James Gregory provided the missing proof, well before the tools of calculus were available.

Today we have several ways to find $\int_0^\alpha \sec\theta\, d\theta$, all of which involve a trick.

EXAMPLE 3 Find $\displaystyle\int \sec\theta\, d\theta$.

SOLUTION

$$\int \sec\theta\, d\theta = \int \frac{1}{\cos\theta}\, d\theta$$
$$= \int \frac{\cos\theta}{\cos^2\theta}\, d\theta$$
$$= \int \frac{\cos\theta}{1-\sin^2\theta}\, d\theta.$$

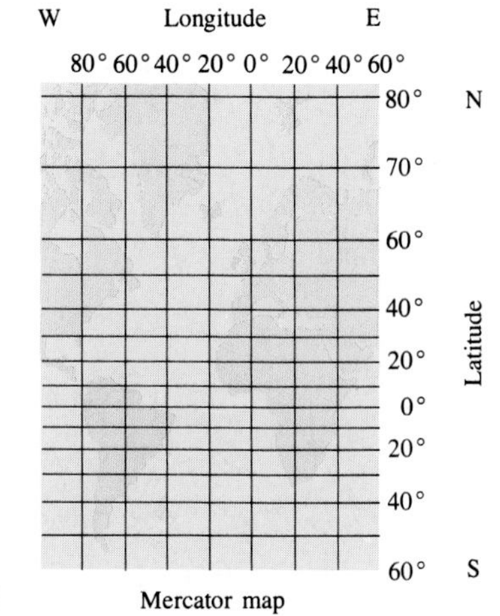

Figure 2

The substitution $u = \sin \theta$ and $du = \cos \theta \, d\theta$ transforms this last integral into the integral of a rational function:

$$\int \frac{du}{1 - u^2} = \frac{1}{2} \int \left(\frac{1}{1 + u} + \frac{1}{1 - u} \right) du$$

$$= \tfrac{1}{2}[\ln (1 + u) - \ln (1 - u)] + C$$

$$= \frac{1}{2} \ln \frac{1 + u}{1 - u} + C.$$

Since $u = \sin \theta$,
$$\frac{1}{2} \ln \frac{1 + u}{1 - u} = \frac{1}{2} \ln \frac{1 + \sin \theta}{1 - \sin \theta}.$$

Thus

$$\int \sec \theta \, d\theta = \frac{1}{2} \ln \left(\frac{1 + \sin \theta}{1 - \sin \theta} \right) + C. \quad ■ \tag{1}$$

Most integral tables have the formula

Another formula for $\int \sec \theta \, d\theta$

$$\int \sec \theta \, d\theta = \ln |\sec \theta + \tan \theta| + C.$$

Exercise 36 shows that this formula agrees with (1).

Remark: Using (1) we find that

$$\int_0^\alpha \sec \theta \, d\theta = \frac{1}{2} \ln \left(\frac{1 + \sin \theta}{1 - \sin \theta} \right) \Bigg|_0^\alpha$$

$$= \frac{1}{2} \ln \left(\frac{1 + \sin \alpha}{1 - \sin \alpha} \right) - \frac{1}{2} \ln \left(\frac{1 + \sin 0}{1 - \sin 0} \right)$$

$$= \frac{1}{2} \ln \left(\frac{1 + \sin \alpha}{1 - \sin \alpha} \right).$$

This result does not look like Bond's formula, $\ln \tan(\alpha/2 + \pi/4)$. Exercise 36 invites you to show that the two formulas both represent the same function.

In contrast to Example 3, $\int \sec^2 \theta \, d\theta$ is easy, since it is simply $\tan \theta + C$.

Computing $\int \tan \theta \, d\theta$

The integration of $\tan \theta$ is much more direct than the integration of $\sec \theta$.

EXAMPLE 4 Find $\int \tan \theta \, d\theta$.

SOLUTION

$$\int \tan \theta \, d\theta = \int \frac{\sin \theta}{\cos \theta} \, d\theta$$

$$= \int \frac{-du}{u} \qquad \text{(substitute } u = \cos \theta\text{)}$$

$$= -\ln u + C$$

$$= -\ln |\cos \theta| + C. \quad ■$$

Finding $\int \tan^2 \theta \, d\theta$ is comparatively easy. Using the trigonometric identity $\tan^2 \theta = \sec^2 \theta - 1$, we obtain

$$\begin{aligned}\int \tan^2 \theta \, d\theta &= \int (\sec^2 \theta - 1) \, d\theta \\ &= \tan \theta - \theta + C.\end{aligned}$$

The Substitution $u = \sqrt[n]{ax + b}$

The next example illustrates the use of the substitution $u = \sqrt[n]{ax + b}$. After the example we describe the integrands for which the substitution is appropriate.

EXAMPLE 5 Find $\int_4^7 x^2\sqrt{3x + 4} \, dx$.

SOLUTION Let $u = \sqrt{3x + 4}$, hence $u^2 = 3x + 4$. Thus $x = (u^2 - 4)/3$ and $dx = (2u \, du)/3$. Moreover, as x goes from 4 to 7, u goes from $\sqrt{16} = 4$ to $\sqrt{25} = 5$. Thus

$$\begin{aligned}\int_4^7 x^2\sqrt{3x + 4} \, dx &= \int_4^5 \left(\frac{u^2 - 4}{3}\right)^2 u \, \frac{2u \, du}{3} \\ &= \frac{2}{27} \int_4^5 (u^2 - 4)^2 \, u^2 \, du \\ &= \frac{2}{27} \int_4^5 (u^6 - 8u^4 + 16u^2) \, du \\ &= \frac{2}{27} \left(\frac{u^7}{7} - \frac{8u^5}{5} + \frac{16u^3}{3}\right)\Bigg|_4^5 \\ &= \frac{2}{27}\left[\left(\frac{5^7}{7} - \frac{8 \cdot 5^5}{5} + \frac{16 \cdot 5^3}{3}\right) - \left(\frac{4^7}{7} - \frac{8 \cdot 4^5}{5} + \frac{16 \cdot 4^3}{3}\right)\right]. \quad \blacksquare\end{aligned}$$

The substitution in Example 5 illustrates the substitution $u = \sqrt[n]{ax + b}$, where the integer n is greater than or equal to 2. It may be used to integrate any "rational functions of x and $\sqrt[n]{ax + b}$," which we define as follows.

A **polynomial in two variables x and y** is a sum of terms of the form

$$a_{mn}x^m y^n,$$

where m and n are nonnegative integers and a_{mn} is a real number. (The symbol a_{mn} is short for $a_{m,n}$. The comma is usually omitted.) For instance, the expression $2x^3 - \sqrt{2}xy^7 + xy$ is a polynomial in x and y. The quotient of two such polynomials is called a **rational function of x and y**

Let $R(x, y)$ be a rational function of x and y. Let n be an integer greater than or equal to 2. Replacing y by $\sqrt[n]{ax + b}$ creates what is called a "rational function of x and $\sqrt[n]{ax + b}$," denoted $R(x, \sqrt[n]{ax + b})$. For instance, if

$$R(x, y) = \frac{x + y^2}{2x - y},$$

then replacing y by $\sqrt[3]{4x + 5}$ yields

$$R(x, \sqrt[3]{4x + 5}) = \frac{x + (\sqrt[3]{4x + 5})^2}{2x - \sqrt[3]{4x + 5}},$$

a rational function of x and $\sqrt[3]{4x + 5}$.

To integrate $R(x, \sqrt[n]{ax+b})$

To integrate $R(x, \sqrt[n]{ax+b})$, let $u = \sqrt[n]{ax+b}$. Then $u^n = ax + b$, $x = (u^n - b)/a$ and $dx = nu^{n-1}\,du/a$. The integrand is now a *rational function of u* and can be integrated by partial fractions.

There are many other substitutions, and we describe four of them. Since you may see them used occasionally, it is good to know what they are.

Three Trigonometric Substitutions

Any rational function of x and $\sqrt{a^2 - x^2}$, where a is a constant, is transformed into a rational function of $\cos\theta$ and $\sin\theta$ by the substitution $x = a\sin\theta$. (Recall Example 9 in Sec. 7.2.) Similar substitutions are possible in situations involving $\sqrt{a^2 + x^2}$ or $\sqrt{x^2 - a^2}$. In each case, one of the trigonometric identities $1 - \sin^2\theta = \cos^2\theta$, $\tan^2\theta + 1 = \sec^2\theta$, or $\sec^2\theta - 1 = \tan^2\theta$ is used to convert a sum or difference of squares into a perfect square.

If the integrand is a rational function of x and:

How to integrate $R(x, \sqrt{a^2 - x^2})$ — *Case 1* $\sqrt{a^2 - x^2}$; let $x = a\sin\theta$ $\quad \left(a > 0,\ -\dfrac{\pi}{2} \le \theta \le \dfrac{\pi}{2}\right)$.

$R(x, \sqrt{a^2 + x^2})$ — *Case 2* $\sqrt{a^2 + x^2}$; let $x = a\tan\theta$ $\quad \left(a > 0,\ -\dfrac{\pi}{2} < \theta < \dfrac{\pi}{2}\right)$.

$R(x, \sqrt{x^2 - a^2})$ — *Case 3* $\sqrt{x^2 - a^2}$; let $x = a\sec\theta$ $\quad \left(a > 0,\ 0 \le \theta \le \pi,\ \theta \ne \dfrac{\pi}{2}\right)$.

The motivation behind this general procedure is quite simple. Consider case 1, for instance. If you replace x in $\sqrt{a^2 - x^2}$ by $a\sin\theta$, you obtain

How to make the square root sign in $\sqrt{a^2 - x^2}$ *disappear*

$$\begin{aligned}\sqrt{a^2 - x^2} = \sqrt{a^2 - (a\sin\theta)^2} &= \sqrt{a^2(1 - \sin^2\theta)} \\ &= \sqrt{a^2\cos^2\theta} \\ &= a\cos\theta.\end{aligned}$$

(Keep in mind that a and $\cos\theta$ are positive.) The important thing is that *the square root sign disappears*.

Case 3 raises a fine point. We have $a > 0$. However, whenever x is negative, θ is a second-quadrant angle, so $\tan\theta$ is *negative*. In that case,

$$\begin{aligned}\sqrt{x^2 - a^2} &= \sqrt{(a\sec\theta)^2 - a^2} \\ &= a\sqrt{\sec^2\theta - 1} \\ &= a\sqrt{\tan^2\theta} \\ &= a(-\tan\theta). \qquad \text{since } -\tan\theta \text{ is positive}\end{aligned}$$

If $c < 0$, $\sqrt{c^2} = -c$.

In the examples and exercises involving case 3 it will be assumed that x varies through nonnegative values, so that θ remains in the first quadrant and $\sqrt{\sec^2\theta - 1} = \tan\theta$.

Note that for $\sqrt{a^2 - x^2}$ to be meaningful, $|x|$ must be no larger than a. On the other hand, for $\sqrt{x^2 - a^2}$ to be meaningful, $|x|$ must be at least as large as a. The quantity $\sqrt{a^2 + x^2}$ is meaningful for all values of x.

EXAMPLE 6 Compute $\displaystyle\int \sqrt{1 + x^2}\, dx$.

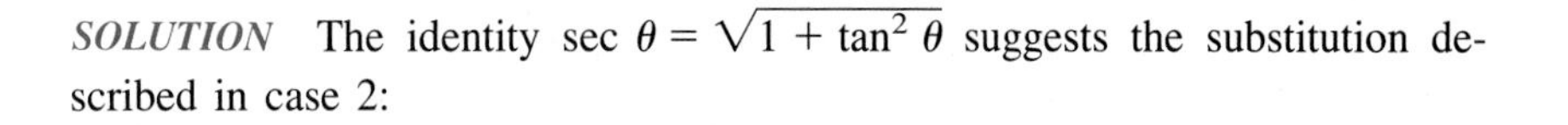

SOLUTION The identity $\sec\theta = \sqrt{1+\tan^2\theta}$ suggests the substitution described in case 2:

$$x = \tan\theta.$$

Hence $$dx = \sec^2\theta\, d\theta.$$

(See Fig. 3 for the geometry of this substitution.)

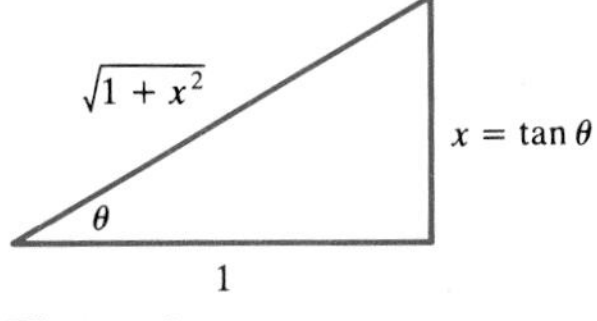

Figure 3

Thus $$\int \sqrt{1+x^2}\, dx = \int \sec\theta \sec^2\theta\, d\theta = \int \sec^3\theta\, d\theta.$$

By Formula 55 from the integral table,

$$\int \sec^3\theta\, d\theta = \frac{\sec\theta\tan\theta}{2} + \frac{1}{2}\ln|\sec\theta + \tan\theta| + C.$$

To express the antiderivative just obtained in terms of x rather than θ, it is necessary to express $\tan\theta$ and $\sec\theta$ in terms of x. Starting with the definition $x = \tan\theta$, find $\sec\theta$ by means of the relation $\sec\theta = \sqrt{1+\tan^2\theta} = \sqrt{1+x^2}$, as in Fig. 3. Thus

$$\int \sqrt{1+x^2}\, dx = \frac{x\sqrt{1+x^2}}{2} + \frac{1}{2}\ln(\sqrt{1+x^2} + x) + C. \quad ■$$

EXAMPLE 7 Compute $\displaystyle\int_4^5 \frac{dx}{\sqrt{x^2-9}}$.

SOLUTION Let $x = 3\sec\theta$; hence $dx = 3\sec\theta\tan\theta\, d\theta$. (See Fig. 4.) Thus, letting $\alpha = \sec^{-1}(\frac{4}{3})$ and $\beta = \sec^{-1}(\frac{5}{3})$, we obtain

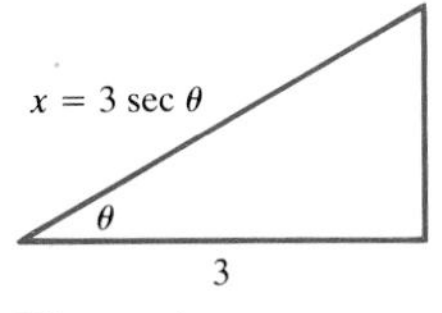

Figure 4

$$\begin{aligned}\int_4^5 \frac{dx}{\sqrt{x^2-9}} &= \int_\alpha^\beta \frac{3\sec\theta\tan\theta\, d\theta}{\sqrt{9\sec^2\theta - 9}} \\ &= \int_\alpha^\beta \frac{\sec\theta\tan\theta\, d\theta}{\tan\theta} \\ &= \int_\alpha^\beta \sec\theta\, d\theta \\ &= \ln|\sec\theta + \tan\theta| \Big|_\alpha^\beta \\ &= \ln\left(\frac{5}{3} + \frac{4}{3}\right) - \ln\left(\frac{4}{3} + \frac{\sqrt{7}}{3}\right) \qquad \text{using Fig. 5 to find } \tan\theta \text{ at the limits of integration} \\ &= \ln 3 - \ln\left(\frac{4+\sqrt{7}}{3}\right) \\ &= \ln\left(\frac{9}{4+\sqrt{7}}\right). \quad ■\end{aligned}$$

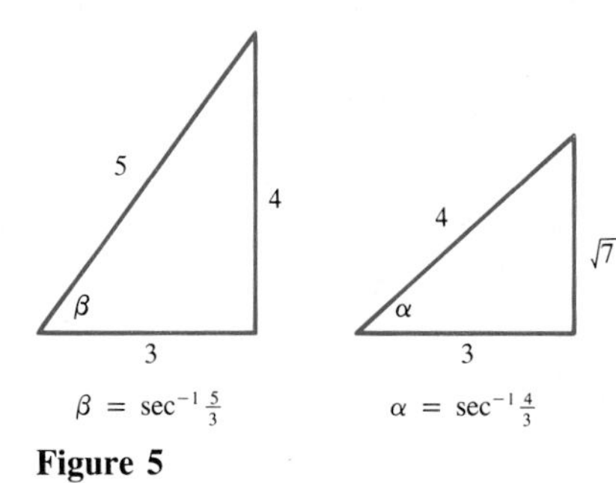

Figure 5

A Half-Angle Substitution for $R(\cos\theta, \sin\theta)$

Any rational function of $\cos\theta$ and $\sin\theta$ is transformed into a rational function of u by the substitution $u = \tan(\theta/2)$. See Exercises 53 and 54. This is sometimes useful after one of the three basic trigonometric substitutions has been used, leaving the integrand in terms of $\cos\theta$ and $\sin\theta$. The $u = \tan(\theta/2)$ substitution then yields an integral that can be done by partial fractions.

Section Summary

We discussed some special integrals and integration techniques. First we saw how to compute

$$\int \sin mx \sin nx\, dx, \int \sin mx \cos nx\, dx, \int \cos mx \cos nx\, dx,$$

$$\int \sin^2 x\, dx, \int \cos^2 x\, dx,$$

$$\int \sec \theta\, d\theta, \int \sec^2 \theta\, d\theta, \int \tan \theta\, d\theta, \text{ and } \int \tan^2 \theta\, d\theta.$$

The integration of higher powers of the trigonometric functions is discussed in the exercises.

We also pointed out that the substitution $u = \sqrt[n]{ax+b}$ transforms $R(x, \sqrt[n]{ax+b})$ into a rational function of u, which can be treated by partial fractions.

$R(x, \sqrt{a^2 - x^2})$, $R(x, \sqrt{x^2 - a^2})$, and $R(x, \sqrt{a^2 + x^2})$ can be transformed into rational functions of $\cos \theta$ and $\sin \theta$ by trigonometric substitutions. Any $R(\cos \theta, \sin \theta)$ can be transformed into a rational function of u by the substitution $u = \tan(\theta/2)$, which can then be evaluated by partial fractions.

EXERCISES FOR SEC. 7.6: SPECIAL TECHNIQUES

Exercises 1 to 18 are related to Examples 1 to 4. In Exercises 1 to 14 find the integrals.

1 $\int \sin 5x \sin 3x\, dx$

2 $\int \sin 5x \cos 2x\, dx$

3 $\int \cos 3x \sin 2x\, dx$

4 $\int \cos 2\pi x \sin 5\pi x\, dx$

5 $\int \sin^2 3x\, dx$

6 $\int \cos^2 5x\, dx$

7 $\int (3 \sin 2x + 4 \sin^2 5x)\, dx$

8 $\int (5 \cos 2x + \cos^2 7x)\, dx$

9 $\int (3 \sin^2 \pi x + 4 \cos^2 \pi x)\, dx$

10 $\int \sec 3\theta\, d\theta$

11 $\int \tan 2\theta\, d\theta$

12 $\int \sec^2 4x\, dx$

13 $\int \tan^2 5x\, dx$

14 $\int \frac{dx}{\cos^2 3x}$

15 Show that $\sin A \sin B = \frac{1}{2} \cos (A - B) - \frac{1}{2} \cos (A + B)$.

16 Show that $\sin A \cos B = \frac{1}{2} \sin (A + B) + \frac{1}{2} \sin (A - B)$.

In a Mercator map, as shown below, the distance between the line representing the equator and the parallel line representing the latitude α is proportional to $\int_0^\alpha \sec \theta\, d\theta$. Exercises 17 and 18 concern $F(\alpha) = \int_0^\alpha \sec \theta\, d\theta$.

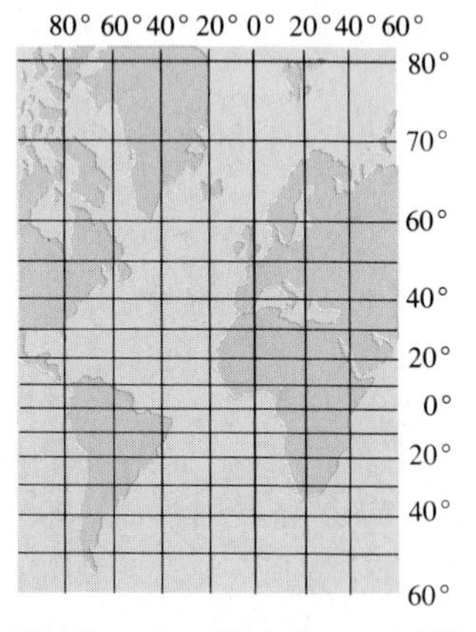

17 Compute $F(\alpha)$ for $\alpha = 20°, 40°, 60°, 80°$, first expressing these angles in radians.

18 How large would a Mercator map have to be to depict all points on the earth except the north and south poles?

Exercises 19 to 28 concern the substitution $u = \sqrt[n]{ax+b}$. In each case evaluate the integral.

19 $\int x^2\sqrt{2x+1}\, dx$

20 $\int \frac{x^2\, dx}{\sqrt[3]{x+1}}$

21 $\int \frac{dx}{\sqrt{x}+3}$

22 $\int \frac{\sqrt{2x+1}}{x}\,dx$

23 $\int x\sqrt[3]{3x+2}\,dx$

24 $\int \frac{\sqrt{x}+3}{\sqrt{x}-2}\,dx$

25 $\int \frac{x\,dx}{\sqrt{x}+3}$

26 $\int x\,(3x+2)^{5/3}\,dx$

27 $\int \frac{dx}{\sqrt[3]{x}+\sqrt{x}}$ *Hint:* Let $u = \sqrt[6]{x}$.

28 $\int (x+2)\sqrt[5]{x-3}\,dx$

Exercises 29 and 30 concern recursion formulas for $\tan^n \theta$ and $\sec^n \theta$.

29 In the text we found $\int \tan\theta\,d\theta$ and $\int \tan^2\theta\,d\theta$.

(*a*) Obtain the recursion

$$\int \tan^n \theta\,d\theta = \frac{\tan^{n-1}\theta}{n-1} - \int \tan^{n-2}\theta\,d\theta.$$

Begin by writing

$$\tan^n\theta = \tan^{n-2}\theta\,\tan^2\theta$$
$$= \tan^{n-2}\theta\,(\sec^2\theta - 1).$$

(*b*) Use the recursion to find $\int \tan^3\theta\,d\theta$.

(*c*) Find $\int \tan^4\theta\,d\theta$.

30 In the text we found $\int \sec\theta\,d\theta$ and $\int \sec^2\theta\,d\theta$.

(*a*) Obtain the recursion

$$\int \sec^n\theta\,d\theta = \frac{\sec^{n-2}\theta\,\tan\theta}{n-1} + \frac{n-2}{n-1}\int \sec^{n-2}\theta\,d\theta.$$

Begin by writing $\sec^n\theta = \sec^{n-2}\theta\,\sec^2\theta$, and integrating by parts. After the integration, $\tan^2\theta$ will appear in an integrand. Write it as $\sec^2\theta - 1$.

(*b*) Evaluate $\int \sec^3\theta\,d\theta$.

(*c*) Evaluate $\int \frac{d\theta}{\cos^4\theta}$.

(*d*) Evaluate $\int \sec^2 2x\,dx$.

31 Find (*a*) $\int \csc\theta\,d\theta$, (*b*) $\int \csc^2\theta\,d\theta$.

32 Find (*a*) $\int \cot\theta\,d\theta$, (*b*) $\int \cot^2\theta\,d\theta$.

33 Consider $\int \sin^n\theta\cos^m\theta\,d\theta$, where m and n are nonnegative integers, m odd. To evaluate $\int \sin^n\theta\cos^m\theta\,d\theta$, write it as $\int \sin^n\theta\cos^{m-1}\theta\cos\theta\,d\theta$. Then rewrite $\cos^{m-1}\theta$ as $(1-\sin^2\theta)^{(m-1)/2}$ and use the substitution $u = \sin\theta$. Using this technique, find

(*a*) $\int \sin^3\theta\cos^3\theta\,d\theta$

(*b*) $\int \sin^4\theta\cos\theta\,d\theta$

(*c*) $\int_0^{\pi/2} \sin^4\theta\cos^3\theta\,d\theta$

(*d*) $\int \cos^5\theta\,d\theta$

34 (See Exercise 33.) How would you integrate $\int \sin^n\theta\cos^m\theta\,d\theta$, where m and n are nonnegative integers, n odd? Illustrate your technique by three examples.

35 (See Exercises 33 and 34.) The techniques in Exercises 33 and 34 apply to $\int \sin^n\theta\cos^m\theta\,d\theta$ only when at least one of m and n is odd. If both are even, first use the identities

$$\sin^2\theta = \frac{1-\cos 2\theta}{2} \quad\text{and}\quad \cos^2\theta = \frac{1+\cos 2\theta}{2}.$$

You will get a polynomial in $\cos 2\theta$. If $\cos 2\theta$ appears only to odd powers, the technique of Exercise 33 suffices. To treat an even power of $\cos 2\theta$, use the identity $\cos^2 2\theta = (1+\cos 4\theta)/2$ and continue. Using this method find

(*a*) $\int \cos^2\theta\sin^4\theta\,d\theta$ (*b*) $\int_0^{\pi/4}\cos^2\theta\sin^2\theta\,d\theta$

36 (*a*) Show that

$$\frac{1}{2}\ln\left(\frac{1+\sin\alpha}{1-\sin\alpha}\right) = \ln\tan\left(\frac{\alpha}{2}+\frac{\pi}{4}\right).$$

(*b*) Show that

$$\frac{1}{2}\ln\left(\frac{1+\sin\alpha}{1-\sin\alpha}\right) = \ln|\sec\alpha+\tan\alpha|.$$

These identities show that Bond's conjecture for $\int \sec\theta\,d\theta$ agrees with the result in Example 3.

37 The region R under $y = \sin x$ and above $[0, \pi]$ is revolved about the x axis to produce a solid S.

(*a*) Draw R.

(*b*) Draw S.

(*c*) Set up a definite integral for the area of R.

(*d*) Set up a definite integral for the volume of S.

(*e*) Evaluate the integrals in (*c*) and (*d*).

38 An arbitrary periodic sound wave can be approximated by a sum of simpler sound waves that correspond to pure pitches. This suggests representing a function $y = f(x)$ as the sum of cosine and sine functions:

$$f(x) = \frac{a_0}{2} + \sum_{k=1}^{n} a_k\cos kx + \sum_{k=1}^{n} b_k\sin kx,$$

for an integer n and some constants $a_0, a_1, \ldots, a_n, b_1, \ldots, b_n$. Show that if the representation for $f(x)$ is valid, then (*a*) $a_m = (1/\pi)\int_0^{2\pi} f(x)\cos mx\,dx$, and (*b*) $b_m = (1/\pi)\int_0^{2\pi} f(x)\sin mx\,dx$. [*Hint:* In the case of (*a*), evaluate $\int_0^{2\pi} f(x)\cos mx\,dx$.] (In the theory of Fourier series such representations, with n replaced by ∞, are studied.)

39 When a voltage V is introduced across a resistance R, the power dissipated is V^2/R. If the voltage varies with time but R is constant, $V = V(t)$, then the average power during the time interval $[a, b]$ is defined as

$$\frac{1}{b-a}\,\frac{1}{R}\int_a^b [V(t)]^2\,dt.$$

In case of sinusoidal voltage, $V(t) = M\sin t$, where M is the maximum voltage.

(*a*) Show that the average power over the time interval $[0, \pi/2]$ is $\frac{1}{2}M^2/R$.

(*b*) Show that the constant voltage $M/\sqrt{2}$ would produce the same power.

When we say we have 110 volts in our homes, we are referring to $M/\sqrt{2}$. The maximum voltage is actually $\sqrt{2}(110) \approx 156$ volts.

The **root mean square** (RMS) of a function $f(x)$ is defined as the square root of $\int_a^b [f(x)]^2\,dx/(b-a)$. A stereo system is often described as "producing 50 watts RMS," even though this is a simple average, with no squares and no square roots.

In Exercises 40 to 50 find the integrals using trigonometric substitution. (a is a positive constant.)

40 $\int \sqrt{4-x^2}\,dx$ **41** $\int \dfrac{dx}{\sqrt{9+x^2}}$

42 $\int \dfrac{x^2\,dx}{\sqrt{x^2-9}}$ **43** $\int x^3\sqrt{1-x^2}\,dx$

44 $\int \dfrac{\sqrt{4+x^2}}{x}\,dx$ **45** $\int \sqrt{a^2-x^2}\,dx$

46 $\int \dfrac{dx}{\sqrt{a^2-x^2}}$ **47** $\int \sqrt{a^2+x^2}\,dx$

48 $\int \sqrt{x^2-a^2}\,dx$ **49** $\int \dfrac{dx}{\sqrt{25x^2-16}}$

50 $\int_{\sqrt{2}}^{2} \sqrt{x^2-1}\,dx$

51 Transform the following integrals into integrals of rational functions of $\cos\theta$ and $\sin\theta$. Do *not* try to evaluate the integrals.

(*a*) $\int \dfrac{x+\sqrt{9-x^2}}{x^3}\,dx$ (*b*) $\int \dfrac{x^3\sqrt{5-x^2}}{1+\sqrt{5-x^2}}\,dx$

52 Transform the following integrals into integrals of rational functions of $\cos\theta$ and $\sin\theta$. Do *not* try to evaluate the integrals.

(*a*) $\int \dfrac{x^2+\sqrt{x^2-9}}{x}\,dx$ (*b*) $\int \dfrac{x^3\sqrt{5+x^2}}{x+2}\,dx$

Exercises 53 to 55 concern $\int R(\cos\theta, \sin\theta)\,d\theta$.

53 Let $-\pi < \theta < \pi$ and $u = \tan(\theta/2)$. (See Fig. 6 for the case $u \ge 0$.) The following steps show that this substitution transforms $\int R(\sin\theta, \cos\theta)\,d\theta$ into the integral of a rational function of u (which can be integrated by partial fractions).

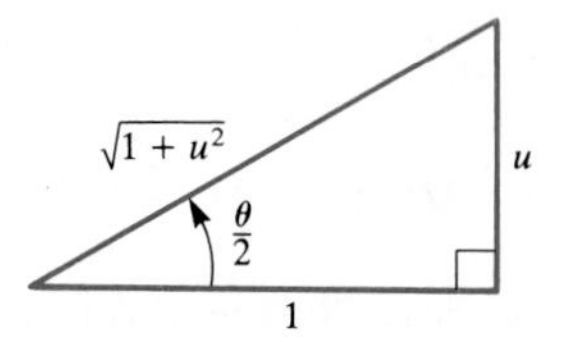

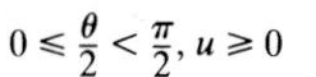

Figure 6

(*a*) Show that $\cos\dfrac{\theta}{2} = \dfrac{1}{\sqrt{1+u^2}}$ and $\sin\dfrac{\theta}{2} = \dfrac{u}{\sqrt{1+u^2}}$.

(*b*) Show that $\cos\theta = \dfrac{1-u^2}{1+u^2}$.

(*c*) Show that $\sin\theta = \dfrac{2u}{1+u^2}$.

(*d*) Show that $d\theta = \dfrac{2\,du}{1+u^2}$. (*Hint:* Note that $\theta = 2\tan^{-1} u$.)

Combining (*b*), (*c*), and (*d*) shows that the substitution $u = \tan(\theta/2)$ transforms $\int R(\sin\theta, \cos\theta)\,d\theta$ into the integral of a rational function of u.

54 Using the substitution $u = \tan(\theta/2)$, transform the following integrals into integrals of rational functions. (Do not evaluate them.)

(*a*) $\int \dfrac{1+\sin\theta}{1+\cos^2\theta}\,d\theta$

(*b*) $\int \dfrac{5+\cos\theta}{(\sin\theta)^2+\cos\theta}\,d\theta$

(*c*) $\int_0^{\pi/2} \dfrac{5\,d\theta}{2\cos\theta+3\sin\theta}$ (Be sure to transform the limits of integration also.)

55 Compute $\int_0^{\pi/2} \dfrac{d\theta}{4\sin\theta+3\cos\theta}$.

56 Explain why any rational function of $\tan\theta$ and $\sec\theta$ has an elementary antiderivative.

57 Let $0 \le \theta < \pi/2$.

(*a*) Show that $\int \sec\theta\,d\theta = \ln|\sec\theta+\tan\theta| + C$, by differentiating $\ln|\sec\theta+\tan\theta|$.

(*b*) Does (*a*) contradict the formula given in Example 3?

58 Show that any rational function of x, $\sqrt{x+a}$, $\sqrt{x+b}$ has an elementary antiderivative. *Hint:* Use the substitution $u = \sqrt{x+a}$. However, it is not the case that every rational function of $\sqrt{x+a}$, $\sqrt{x+b}$, and $\sqrt{x+c}$ has an elementary antiderivative. For instance,

$$\int \frac{dx}{\sqrt{x}\sqrt{x+1}\sqrt{x-1}} = \int \frac{dx}{\sqrt{x^3-x}}$$

is not an elementary function.

59 Every rational function of x and $\sqrt[n]{(ax+b)/(cx+d)}$ has an elementary antiderivative. Explain why.

60 In a Mercator map the meridians (vertical lines of longitude) are spaced at equal distances, but the lines of latitude (horizontal lines) are not. Instead they are placed so that "locally the map shrinks horizontal and vertical distances by the same factor." Figure 7 shows the spherical earth, with radius 1 for convenience and a corresponding section of the map. (*a*) Show that if $\overline{A'B'} = \widehat{AB}$, then $\overline{C'D'} = \sec\theta\,\widehat{CD}$, where

$$\int \underbrace{\ln x}_{u} \underbrace{\frac{dx}{x}}_{dv} = \underbrace{(\ln x)}_{u} \underbrace{(\ln x)}_{v} - \int \underbrace{\ln x}_{v} \underbrace{\frac{dx}{x}}_{du}.$$

Bringing $\int \ln x \, dx/x$ all to one side produces the equation

$$2 \int \ln x \frac{dx}{x} = (\ln x)^2,$$

from which it follows that

$$\int \ln x \frac{dx}{x} = \frac{(\ln x)^2}{2} + C.$$

The method worked, but it is not the easiest one to use. Since $1/x$ is the derivative of $\ln x$, we could have used the substitution $u = \ln x$, $du = dx/x$; thus

$$\int \frac{\ln x \, dx}{x} = \int u \, du = \frac{u^2}{2} + C = \frac{(\ln x)^2}{2} + C. \quad \blacksquare$$

EXAMPLE 9 $\int_0^{3/5} \sqrt{9 - 25x^2} \, dx.$

DISCUSSION This integral reminds us of $\int_0^a \sqrt{a^2 - x^2} \, dx = \pi a^2/4$, the area of a quadrant of a circle of radius a. This resemblance suggests a substitution u such that $9u^2 = 25x^2$ or $u = \frac{5}{3} x$, hence $dx = \frac{3}{5} du$. Then substitution gives

$$\begin{aligned} \int_0^{3/5} \sqrt{9 - 25x^2} \, dx &= \int_0^1 \sqrt{9 - 9u^2} \, \tfrac{3}{5} \, du \\ &= \frac{9}{5} \int_0^1 \sqrt{1 - u^2} \, du \\ &= \frac{9}{5} \frac{\pi}{4} = \frac{9\pi}{20}. \quad \blacksquare \end{aligned}$$

EXAMPLE 10 $\int \sin^5 2x \cos 2x \, dx.$

DISCUSSION We could try integration by parts with $u = \sin^5 2x$ and $dv = \cos 2x \, dx$. (Check that it works.)

However $\cos 2x$ is almost the derivative of $\sin 2x$. For this reason make the substitution

$$u = \sin 2x \qquad du = 2 \cos 2x \, dx;$$

hence

$$\cos 2x \, dx = \frac{du}{2}.$$

Then

$$\begin{aligned} \int \sin^5 2x \cos 2x \, dx &= \int u^5 \frac{du}{2} \\ &= \frac{1}{2} \frac{u^6}{6} + C \\ &= \frac{\sin^6 2x}{12} + C. \quad \blacksquare \end{aligned}$$

EXAMPLE 11 $\int_{-3}^{3} x^3 \cos x \, dx.$

DISCUSSION Since the integrand is of the form $P(x) \cos x$, where P is a polynomial, repeated integration by parts would work. On the other hand, x^3 is an odd function and $\cos x$ is an even function. The integrand is therefore an odd function and the integral over $[-3, 3]$ is 0. ■

EXAMPLE 12 $\int \sin^2 3x \, dx.$

Or use $\int \sin mx \sin nx \, dx$

DISCUSSION You could rewrite this integral as $\int \sin 3x \sin 3x \, dx$ and use integration by parts. However, it is easier to use the trigonometric identity $\sin^2 \theta = (1 - \cos 2\theta)/2$:

$$\int \sin^2 3x \, dx = \int \frac{1 - \cos 6x}{2} \, dx$$

$$= \int \frac{dx}{2} - \int \frac{\cos 6x}{2} \, dx$$

$$= \frac{x}{2} - \frac{\sin 6x}{12} + C. \quad ■$$

EXAMPLE 13 $\int_1^2 \frac{x^3 - 1}{(x + 2)^2} \, dx.$

DISCUSSION Partial fractions would certainly work. (The first step would be division of $x^3 - 1$ by $x^2 + 4x + 4$.) However, the substitution $u = x + 2$ is easier. (It makes the denominator simply u^2.) We have

$$u = x + 2 \qquad du = dx \qquad \text{and} \qquad x = u - 2.$$

Note the new limits for u.

Thus

$$\int_1^2 \frac{x^3 - 1}{(x + 2)^2} \, dx = \int_3^4 \frac{(u - 2)^3 - 1}{u^2} \, du$$

$$= \int_3^4 \frac{u^3 - 6u^2 + 12u - 8 - 1}{u^2} \, du$$

$$= \int_3^4 \left(u - 6 + \frac{12}{u} - \frac{9}{u^2} \right) du$$

$$= \left(\frac{u^2}{2} - 6u + 12 \ln |u| + \frac{9}{u} \right) \Big|_3^4$$

$$= -\frac{13}{4} + 12 \ln 4 - 12 \ln 3. \quad ■$$

EXERCISES FOR SEC. 7.7: WHAT TO DO IN THE FACE OF AN INTEGRAL

All the integrals in Exercises 1 to 59 are elementary. In each case, list the technique or techniques that would be of use. If there is a preferred technique, state what it is. Do *not* evaluate the integrals. (For practice in evaluating integrals, see Sec. 7.S, Exercises 43 to 162.)

1 $\int \frac{1 + x}{x^2} \, dx$

2 $\int \frac{x^2}{1 + x} \, dx$

3 $\int \frac{dx}{x^2 + x^3}$

4 $\int \frac{x + 1}{x^2 + x^3} \, dx$

5 $\int \tan^{-1} 2x \, dx$

6 $\int \sin^{-1} 2x \, dx$

7 $\int x^{10} e^x \, dx$

8 $\int \frac{\ln x}{x^2} \, dx$

9 $\int \frac{\sec^2 \theta \, d\theta}{\tan \theta}$

10 $\int \frac{\tan \theta \, d\theta}{\sin^2 \theta}$

11 $\int \frac{x^3}{\sqrt[3]{x+2}} \, dx$

12 $\int \frac{x^2}{\sqrt[3]{x^3+2}} \, dx$

13 $\int \frac{2x+1}{(x^2+x+1)^5} \, dx$

14 $\int \sqrt{\cos \theta} \sin \theta \, d\theta$

15 $\int \tan^2 \theta \, d\theta$

16 $\int \frac{d\theta}{\sec^2 \theta}$

17 $\int e^{\sqrt{x}} \, dx$

18 $\int \sin \sqrt{x} \, dx$

19 $\int \frac{dx}{(x^2-4x+3)^2}$

20 $\int \frac{x+1}{x^5} \, dx$

21 $\int \frac{x^5}{x+1} \, dx$

22 $\int \frac{\ln x}{x(1+\ln x)} \, dx$

23 $\int \frac{e^{3x} \, dx}{1+e^x+e^{2x}}$

24 $\int \frac{\cos x \, dx}{(3+\sin x)^2}$

25 $\int \ln (e^x) \, dx$

26 $\int \ln (\sqrt[3]{x}) \, dx$

27 $\int \frac{x^4-1}{x+2} \, dx$

28 $\int \frac{x+2}{x^4-1} \, dx$

29 $\int \frac{dx}{\sqrt{x}(3+\sqrt{x})^2}$

30 $\int \frac{dx}{(3+\sqrt{x})^3}$

31 $\int (1+\tan \theta)^3 \sec^2 \theta \, d\theta$

32 $\int \frac{e^{2x}+1}{e^x-e^{-x}} \, dx$

33 $\int \frac{e^x+e^{-x}}{e^x-e^{-x}} \, dx$

34 $\int \frac{(x+3)(\sqrt{x+2}+1)}{\sqrt{x+2}-1} \, dx$

35 $\int \frac{(\sqrt[3]{x+2}-1) \, dx}{\sqrt{x+2}+1}$

36 $\int \frac{dx}{x^2-9}$

37 $\int \frac{x+7}{(3x+2)^{10}} \, dx$

38 $\int \frac{x^3 \, dx}{(3x+2)^7}$

39 $\int \frac{2^x+3^x}{4^x} \, dx$

40 $\int \frac{2^x}{1+2^x} \, dx$

41 $\int \frac{(x+\sin^{-1} x) \, dx}{\sqrt{1-x^2}}$

42 $\int \frac{x+\tan^{-1} x}{1+x^2} \, dx$

43 $\int x^3 \sqrt{1+x^2} \, dx$

44 $\int x(1+x^2)^{3/2} \, dx$

45 $\int \frac{x \, dx}{\sqrt{x^2-1}}$

46 $\int \frac{x^3}{\sqrt{x^2-1}} \, dx$

47 $\int \frac{x \, dx}{(x^2-9)^{3/2}}$

48 $\int \frac{\tan^{-1} x}{1+x^2} \, dx$

49 $\int \frac{\tan^{-1} x}{x^2} \, dx$

50 $\int \frac{\sin (\ln x)}{x} \, dx$

51 $\int \cos x \ln (\sin x) \, dx$

52 $\int \frac{x \, dx}{\sqrt{x^2+4}}$

53 $\int \frac{dx}{x^2+x+5}$

54 $\int \frac{x \, dx}{x^2+x+5}$

55 $\int \frac{x+3}{(x+1)^5} \, dx$

56 $\int \frac{x^5+x+\sqrt{x}}{x^3} \, dx$

57 $\int (x^2+9)^{10} x \, dx$

58 $\int (x^2+9)^{10} x^3 \, dx$

59 $\int \frac{x^4 \, dx}{(x+1)^2(x-2)^3}$

In Exercises 60 to 62, (*a*) determine which positive integers n yield integrals you can evaluate and (*b*) evaluate them.

60 $\int \sqrt{1+x^n} \, dx$

61 $\int (1+x^2)^{1/n} \, dx$

62 $\int (1+x)^{1/n} \sqrt{1-x} \, dx$

63 Find $\int \frac{dx}{\sqrt{x+2}-\sqrt{x-2}}$.

64 Find $\int \sqrt{1-\cos x} \, dx$.

In Exercises 65 to 71, evaluate the integrals. Trigonometric substitution may be helpful.

65 $\int \frac{x \, dx}{(\sqrt{9-x^2})^5}$

66 $\int \frac{dx}{\sqrt{9-x^2}}$

67 $\int \frac{dx}{(\sqrt{9-x^2})^5}$

68 $\int \frac{dx}{x\sqrt{x^2+9}}$

69 $\int \frac{x \, dx}{\sqrt{x^2+9}}$

70 $\int \frac{dx}{x+\sqrt{x^2+25}}$

71 $\int (x^3+x^2) \sqrt{x^2-5} \, dx$

7.S SUMMARY

Method	*Description*
Substitution (Sec. 7.2)	Introduce $u = h(x)$. If $f(x)\,dx = g(u)\,du$, then $\int f(x)\,dx = \int g(u)\,du$.
Substitution in the definite integral (Sec. 7.2)	If, in the above substitution, $u = A$ when $x = a$ and $u = B$ when $x = b$, then $\int_a^b f(x)\,dx = \int_A^B g(u)\,du$.
Table of integrals (Sec. 7.1)	Obtain and become familiar with a table of integrals. Substitution, together with integral tables, will usually be adequate.
Integration by parts (Sec. 7.3)	$\int u\,dv = uv - \int v\,du$. Choose u and v so $u\,dv = f(x)\,dx$ and $\int v\,du$ is easier than $\int u\,dv$.
Partial fractions (applies to any rational function of x) (Secs. 7.4 and 7.5)	This is an algebraic method. Write the integrand as a sum of a polynomial (if the degree of the numerator is greater than or equal to the degree of the denominator) plus terms of the type $\frac{k_i}{(ax+b)^i}$ and $\frac{r_j x + s_j}{(ax^2+bx+c)^j}$. The number of unknown constants is the same as the degree of the denominator. A table of integrals treats the integrals of these two types. (For the first type, use the substitution $u = ax + b$; for the second, complete the square.)
Certain trigonometric functions (Sec. 7.6)	$\int \sin mx \cos nx\,dx$, etc. $\int \sin^2\theta\,d\theta$, $\int \cos^2\theta\,d\theta$ $\int \tan\theta\,d\theta$, $\int \tan^2\theta\,d\theta$ $\int \sec\theta\,d\theta$, $\int \sec^2\theta\,d\theta$
To integrate any rational function of $\cos\theta$ and $\sin\theta$ (Sec. 7.6, Exercises 53 to 55)	Let $u = \tan(\theta/2)$. Then $\cos\theta = \frac{1-u^2}{1+u^2}$ $\sin\theta = \frac{2u}{1+u^2}$ $d\theta = \frac{2\,du}{1+u^2}$, and the new integrand is a rational function of u.
To integrate rational functions of x and one of $\sqrt{a^2-x^2}$, $\sqrt{a^2+x^2}$, $\sqrt{x^2-a^2}$ (Sec. 7.6)	For $\sqrt{a^2-x^2}$ or a^2-x^2, let $x = a\sin\theta$. For $\sqrt{a^2+x^2}$ or a^2+x^2, let $x = a\tan\theta$. For $\sqrt{x^2-a^2}$, let $x = a\sec\theta$.
To integrate rational functions of x and $\sqrt[n]{ax+b}$ (Sec. 7.6)	Let $u = \sqrt[n]{ax+b}$, hence $u^n = ax + b$, $nu^{n-1}\,du = a\,dx$, and $x = (u^n - b)/a$. The new integrand is a rational function of u.

The fundamental theorem of calculus, proved in Chap. 5, raised the problem of finding antiderivatives. Now, some very simple and important functions do not have elementary antiderivatives; for instance,

$$\int \frac{\sin x\,dx}{x} \qquad \int e^{x^2}\,dx \qquad \int \frac{dx}{\ln x} \qquad \int x\tan x\,dx$$

$$\int \frac{\ln x}{x+1}\,dx \qquad \int \sqrt{1 - \frac{\sin^2 x}{4}}\,dx \qquad \text{and} \qquad \int \sqrt[3]{x - x^2}\,dx$$

are not elementary. If the definite integral $\int_0^1 e^{x^2}\,dx$ is needed, an estimate must be made, for example, by an approximating sum (after all, a definite integral is defined as a limit of such sums), the trapezoidal method, or Simpson's method. The "elliptic integral" $\int_0^{\pi/2} \sqrt{1 - k^2 \sin^2 x}\,dx$, $0 \le k \le 1$, frequently used in engineering, is tabulated for various values of k to four decimal places in most handbooks.

Some definite integrals over intervals of the form $[-a, a]$ can be simplified before evaluation. If $f(x)$ is an even function, then $\int_{-a}^a f(x)\,dx = 2\int_0^a f(x)\,dx$; if it is an odd function, then $\int_{-a}^a f(x)\,dx = 0$. (For instance, $\int_{-1}^1 xe^{x^4}\,dx = 0$.)

GUIDE QUIZ ON CHAP. 7: COMPUTING ANTIDERIVATIVES

In Exercises 1 to 25 evaluate the integrals.

1 $\int \frac{x^3\,dx}{1+x^4}$ 2 $\int \sqrt{4x+3}\,dx$

3 $\int \sec^5 2x \tan 2x\,dx$ 4 $\int \tan^3 x \sec^2 x\,dx$

5 $\int \frac{dx}{x^4-1}$ 6 $\int \frac{x^4\,dx}{x^4-1}$

7 $\int \frac{\tan(\ln x)}{x}\,dx$ 8 $\int x^2 \cos^2 x^3\,dx$

9 $\int x^2 \sec x^3\,dx$ 10 $\int \frac{x^4-\sqrt{x}}{x^3}\,dx$

11 $\int \frac{dx}{(x+3)\sqrt{x+3}}$ 12 $\int_0^1 \frac{x^3}{x^3+1}\,dx$

13 $\int x \cos 3x\,dx$ 14 $\int \tan^{-1} 5x\,dx$

15 $\int_{\sqrt{3}}^{2} \frac{x}{7-x^2}\,dx$ 16 $\int_0^{\sqrt{7}/4} \frac{dx}{\sqrt{7-x^2}}$

17 $\int \frac{x^3\,dx}{1+x^8}$ 18 $\int x^3 e^{-2x}\,dx$

19 $\int \frac{x^2+2}{(x-3)(x+4)(x-1)}\,dx$ 20 $\int \frac{x^3}{x^3-1}\,dx$

The following exercises may require techniques from Sec. 7.6.

21 $\int \sqrt{4-9x^2}\,dx$ 22 $\int \frac{dx}{\sqrt{9x^2+16}}$

23 $\int \frac{dx}{\sin^5 3x}$ 24 $\int \frac{x^2\,dx}{\sqrt{x^2-9}}$

25 $\int \frac{dx}{3+\cos x}$

REVIEW EXERCISES FOR CHAP. 7: COMPUTING ANTIDERIVATIVES

1 (*a*) By an appropriate substitution, transform this definite integral into a simpler definite integral:

$$\int_0^{\pi/2} \sqrt{(1+\cos\theta)^3}\,\sin\theta\,d\theta.$$

(*b*) Evaluate the new definite integral in (*a*).

2 Two of these antiderivatives are elementary functions; evaluate them.

(*a*) $\int \ln x\,dx$ (*b*) $\int \frac{\ln x\,dx}{x}$ (*c*) $\int \frac{dx}{\ln x}$

3 Evaluate

(*a*) $\int_1^2 (1+x^3)^2\,dx$ (*b*) $\int_1^2 (1+x^3)^2 x^2\,dx$

4 Compute with the aid of a table of integrals:

(*a*) $\int \frac{e^x\,dx}{5e^{2x}-3}$ (*b*) $\int \frac{dx}{\sqrt{x^2-3}}$

5 Compute

(*a*) $\int \frac{dx}{x^3}$ (*b*) $\int \frac{dx}{\sqrt{x+1}}$ (*c*) $\int \frac{e^x\,dx}{1+5e^x}$

6 Compute

$$\int \frac{5x^4-5x^3+10x^2-8x+4}{(x^2+1)(x-1)}\,dx$$

7 Transform the definite integral

$$\int_0^3 \frac{x^3}{\sqrt{x+1}}\,dx$$

to another definite integral in two different ways (and evaluate):

(*a*) By the substitution $u = x + 1$.
(*b*) By the substitution $u = \sqrt{x+1}$.

8 (*a*) Transform the definite integral

$$\int_{-1}^{4} \frac{x+2}{\sqrt{x+3}}\,dx$$

into an easier definite integral by a substitution.
(*b*) Evaluate the integral obtained in (*a*).

9 Compute $\int x^2 \ln(1+x)\,dx$ (*a*) without an integral table, (*b*) with an integral table.

10 Verify that the following factorizations into irreducible polynomials are correct.

(*a*) $x^3 - 1 = (x-1)(x^2+x+1)$
(*b*) $x^4 - 1 = (x-1)(x+1)(x^2+1)$
(*c*) $x^3 + 1 = (x+1)(x^2-x+1)$

Express Exercises 11 to 17 as a sum of partial fractions. (Do not integrate.) Exercise 10 may be helpful.

11 $\dfrac{2x^2 + 3x + 1}{x^3 - 1}$ **12** $\dfrac{x^4 + 2x^2 - 2x + 2}{x^3 - 1}$

13 $\dfrac{2x - 1}{x^3 + 1}$ **14** $\dfrac{x^4 + 3x^3 - 2x^2 + 3x - 1}{x^4 - 1}$

15 $\dfrac{2x + 5}{x^2 + 3x + 2}$ **16** $\dfrac{5x^3 + 11x^2 + 6x + 1}{x^2 + x}$

17 $\dfrac{5x^3 + 6x^2 + 8x + 5}{(x^2 + 1)(x + 1)}$

18 The fundamental theorem can be used to evaluate one of these definite integrals, but not the other. Evaluate one of them.

(*a*) $\int_0^1 \sqrt[3]{x}\,\sqrt{x}\,dx$ (*b*) $\int_0^1 \sqrt[3]{1 - x}\,\sqrt{x}\,dx$

19 Compute $\int \dfrac{x^3\,dx}{(x - 1)^2}$

(*a*) Using partial fractions.
(*b*) Using the substitution $u = x - 1$.
(*c*) Which method is easier?

20 (*a*) Compute $\int \dfrac{x^{2/3}\,dx}{x + 1}$.

(*b*) What does a table of integrals say about $\int \dfrac{x^{2/3}\,dx}{x + 1}$?

21 Compute $\int x\,\sqrt[3]{x + 1}\,dx$ using
(*a*) The substitution $u = \sqrt[3]{x + 1}$.
(*b*) The substitution $u = x + 1$.

In Exercises 22 to 25 evaluate the integrals.

22 $\int_0^1 (e^x + 1)^3 e^x\,dx$ **23** $\int_0^1 (x^4 + 1)^5\,x^3\,dx$

24 $\int_1^e \dfrac{\sqrt{\ln x}}{x}\,dx$ **25** $\int_0^{\pi/2} \dfrac{\cos\theta\,d\theta}{\sqrt{1 + \sin\theta}}$

26 (*a*) Without an integral table, evaluate

$$\int \sin^5\theta\,d\theta \quad \text{and} \quad \int \tan^6\theta\,d\theta.$$

(*b*) Evaluate them with an integral table.

27 Two of these three antiderivatives are elementary. Find them.

(*a*) $\int \sqrt{1 - 4\sin^2\theta}\,d\theta$ (*b*) $\int \sqrt{4 - 4\sin^2\theta}\,d\theta$

(*c*) $\int \sqrt{1 + \cos\theta}\,d\theta$

28 Find $\int \cot 3\theta\,d\theta$.

29 Find $\int \csc 5\theta\,d\theta$.

30 Compute

(*a*) $\int \sec^5 x \tan x\,dx$ (*b*) $\int \dfrac{\sin x}{(\cos x)^3}\,dx$

31 Compute

$$\int \frac{x^3\,dx}{(1 + x^2)^4}$$

in two different ways:
(*a*) By the substitution $u = 1 + x^2$.
(*b*) By the substitution $x = \tan\theta$.

32 Find $\int \dfrac{x\,dx}{\sqrt{9x^4 + 16}}$ (*a*) without an integral table, (*b*) with an integral table.

33 Transform $\int (x^2/\sqrt{1 + x})\,dx$ by each of the substitutions (*a*) $u = \sqrt{1 + x}$, (*b*) $u = 1 + x$, (*c*) $x = \tan^2\theta$. (*d*) Solve the easiest of the resulting problems.

34 Compute $\int x\sqrt{1 + x}\,dx$ in three ways:
(*a*) Let $u = \sqrt{1 + x}$.
(*b*) Let $x = \tan^2\theta$.
(*c*) By parts, with $u = x$, $dv = \sqrt{1 + x}\,dx$.

35 Find $\int x\sqrt{(1 - x^2)^5}\,dx$ using the substitutions (*a*) $u = x^2$, (*b*) $u = 1 - x^2$, (*c*) $x = \sin\theta$.

In each of Exercises 36 to 39 use an appropriate trigonometric substitution to find the antiderivative.

36 $\int \dfrac{dx}{x^2\sqrt{x^2 + 25}}$ **37** $\int (9 - x^2)^{3/2}\,dx$

38 $\int \dfrac{dx}{x\sqrt{4 + x^2}}$ **39** $\int \sqrt{4x^2 - 9}\,dx$

40 Transform the problem of finding $\int (x^3/\sqrt{1 + x^2})\,dx$ to a different problem, using (*a*) integration by parts with $dv = (x\,dx)/\sqrt{1 + x^2}$, (*b*) the substitution $x = \tan\theta$, (*c*) the substitution $u = \sqrt{1 + x^2}$.

41 Two of these three integrals are elementary. Evaluate them.

(*a*) $\int \sin^2 x\,dx$ (*b*) $\int \sin\sqrt{x}\,dx$ (*c*) $\int \sin x^2\,dx$

42 (*a*) Show that

$$\int \frac{dx}{\sqrt{1 - a^2x^2}} = \frac{1}{a}\sin^{-1} ax + C. \qquad a > 0$$

Use (*a*) to find

(*b*) $\int \dfrac{dx}{\sqrt{1 - 25x^2}}$ (*c*) $\int \dfrac{dx}{\sqrt{1 - 3x^2}}$

In Exercises 43 to 162 find the integrals. Exercises 43 to 70 rely only on the techniques of Secs. 7.1 to 7.5. Some of the exercises from 71 to 162 require the techniques in Sec. 7.6; you may also use a table of integrals in such cases (possibly with a suitable substitution).

43 $\int \frac{x^3}{(x^4+1)^3}\,dx$

44 $\int \frac{x^2\,dx}{x^4-1}$

45 $\int \frac{x^4+x^2+1}{x^3}\,dx$

46 $\int x^{1/4}(1+x^{1/5})\,dx$

47 $\int 10^x\,dx$

48 $\int \frac{dx}{x^3+4x}$

49 $\int \frac{\sin x\,dx}{3+\cos x}$

50 $\int \tan x\,dx$

51 $\int x^2\sqrt{x^3-1}\,dx$

52 $\int \frac{3\,dx}{x^2+4x-5}$

53 $\int \frac{\tan^{-1} x}{x^2}\,dx$

54 $\int \ln \sqrt{2x-1}\,dx$

55 $\int e^x \sin 3x\,dx$

56 $\int x^3 \tan^{-1} x\,dx$

57 $\int x\sqrt{x^2+4}\,dx$

58 $\int \frac{x+2}{x^2+1}\,dx$

59 $\int \frac{x^2\,dx}{1+x^6}$

60 $\int \sqrt[3]{4x+7}\,dx$

61 $\int x^2 \sin x^3\,dx$

62 $\int \frac{\ln x^4}{x}\,dx$

63 $\int x^4 \ln x\,dx$

64 $\int \frac{\tan^{-1} 3x}{1+9x^2}\,dx$

65 $\int \frac{e^{\sqrt{x}}}{\sqrt{x}}\,dx$

66 $\int \sin(\ln x)\,dx$

67 $\int \frac{x\,dx}{\sqrt{(x^2+1)^3}}$

68 $\int \frac{2+\sqrt[3]{x}}{x}\,dx$

69 $\int \frac{dx}{\sqrt{(x+1)^3}}$

70 $\int \frac{2x+3}{x^2+3x+5}\,dx$

71 $\int \frac{x\,dx}{x^4-2x^2-3}$

72 $\int \frac{3\,dx}{\sqrt{1-5x^2}}$

73 $\int \frac{x^2\,dx}{\sqrt[3]{x-1}}$

74 $\int \ln(4+x^2)\,dx$

75 $\int \frac{\sqrt{x^2+4}}{x}\,dx$

76 $\int \sqrt{\tan\theta}\sec^2\theta\,d\theta$

77 $\int \sec^5\theta \tan\theta\,d\theta$

78 $\int \tan^6\theta\,d\theta$

79 $\int \frac{dx}{x\sqrt{x^2+9}}$

80 $\int (e^x+1)^2\,dx$

81 $\int \frac{(1-x)^2}{\sqrt[3]{x}}\,dx$

82 $\int (1+\sqrt{x})x\,dx$

83 $\int \frac{\sin x\,dx}{1+3\cos^2 x}$

84 $\int (e^{2x})^3 e^x\,dx$

85 $\int \left(e^x - \frac{1}{e^x}\right)^2 dx$

86 $\int \frac{dx}{(\sqrt{x}+1)(\sqrt{x})}$

87 $\int x \sin^{-1} x^2\,dx$

88 $\int x \sin^{-1} x\,dx$

89 $\int \frac{dx}{e^{2x}+5e^x}$

90 $\int \frac{x^7\,dx}{\sqrt{x^2+1}}$

91 $\int (2x+1)\sqrt{3x+2}\,dx$

92 $\int x\sqrt{x^4-1}\,dx$

93 $\int \frac{x^2\,dx}{(x-1)^3}$

94 $\int \frac{dx}{\sqrt{9+x^2}}$

95 $\int \frac{e^x+1}{e^x-1}\,dx$

96 $\int \frac{dx}{4x^2+1}$

97 $\int (1+3x^2)^2\,dx$

98 $\int \frac{x\,dx}{x^3+1}$

99 $\int \frac{x^3\,dx}{x^3+1}$

100 $\int \frac{x^2\,dx}{\sqrt{2x+1}}$

101 $\int \frac{dx}{\sqrt{2x+1}}$

102 $\int (x+\sin x)^2\,dx$

103 $\int \sin^2 3x \cos^2 3x\,dx$

104 $\int \sin^3 3x \cos^2 3x\,dx$

105 $\int \tan^4 3\theta\,d\theta$

106 $\int \cos^3 x \sin^2 x\,dx$

107 $\int \cos^2 x\,dx$

108 $\int \frac{dx}{(4+x^2)^2}$

109 $\int \frac{dx}{(4-x^2)^{3/2}}$

110 $\int \frac{x^3\,dx}{x^4-1}$

111 $\int \frac{e^x\,dx}{1+e^{2x}}$

112 $\int \frac{dx}{x^2+5x-6}$

113 $\int \frac{dx}{x^2+5x+6}$

114 $\int \frac{\sqrt{x^2+1}}{x^4}\,dx$

115 $\int \frac{4x+10}{x^2+5x+6}\,dx$

116 $\int \sqrt{4x^2+1}\,dx$

117 $\int \frac{dx}{2x^2+5x+6}$

118 $\int \sqrt{-4x^2+1}\,dx$

119 $\int \frac{dx}{2x^2+5x-6}$

120 $\int \frac{dx}{2+3\sin x}$

121 $\int \frac{dx}{\sin^2 x}$

122 $\int \frac{dx}{3+2\sin x}$

123 $\int \ln(x^2+5)\,dx$

124 $\int x^3 e^{-5x}\,dx$

125 $\int \sqrt{(1+2x)(1-2x)}\,dx$

126 $\int x \sin 3x\,dx$

127 $\int \frac{2x\,dx}{\sqrt{x^2+1}}$

128 $\int \frac{2\,dx}{\sqrt{x^2+1}}$

129 $\int \frac{x^4 + 4x^3 + 6x^2 + 4x - 3}{x^4 - 1}\,dx$

130 $\int \frac{x^2 - 3x}{(x+1)(x-1)^2\,dx}$ **131** $\int \frac{12x^2 + 2x + 3}{4x^3 + x}\,dx$

132 $\int \frac{6x^3 + 2x + \sqrt{3}}{1 + 3x^2}\,dx$ **133** $\int \sqrt{\frac{1}{x^2} + \frac{1}{x^4}}\,dx$

134 $\int \frac{x\,dx}{\sqrt{1 - 9x^2}}$ **135** $\int \frac{dx}{\sqrt{1 - 9x^2}}$

136 $\int \frac{dx}{x\sqrt{3x^2 - 5}}$ **137** $\int \frac{dx}{(3x^2 + 2)^{3/2}}$

138 $\int \frac{dx}{\sin 5x}$ **139** $\int \frac{dx}{\cos 4x}$

140 $\int \frac{x^2\,dx}{1 + 3x^3 + 2x^6}$ **141** $\int e^x \sin^2 x\,dx$

142 $\int x^3\sqrt{1 - 3x^2}\,dx$ **143** $\int \sqrt{1 + \sqrt{1 + \sqrt{x}}}\,dx$

144 $\int \frac{x^3\,dx}{1 - 4x^2}$ **145** $\int \sec^4 x\,dx$

146 $\int \frac{dx}{\cos^3 x}$ **147** $\int \cos^3 x\,dx$

148 $\int x^2 \ln (x^3 + 1)\,dx$ **149** $\int \frac{\ln x + \sqrt{x}}{x}\,dx$

150 $\int \frac{(3 + x^2)^2\,dx}{x}$ **151** $\int \frac{dx}{e^x}$

152 $\int \frac{(1 + 3\cos x)^2\,dx}{\sin x}$ **153** $\int (e^{2x} + 1)e^{-x}\,dx$

154 $\int \sqrt{9x^2 - 4}\,dx$ **155** $\int \frac{2x^2 + 4x + 3}{x^3 + 2x^2 + 3x}\,dx$

156 $\int \frac{dx}{x^4 + 3x^2 + 1}$ **157** $\int \frac{\sec^2 \theta\,d\theta}{\sqrt{\sec^2 \theta - 1}}$

158 $\int \ln (2x + x^2)\,dx$ **159** $\int x \sin^2 x\,dx$

160 $\int \frac{dx}{1 + 2e^{3x}}$ **161** $\int x \tan^2 x\,dx$

162 $\int \sqrt{1 + \cos 3\theta}\,d\theta$

163 Compute

(*a*) $\int \frac{dx}{x^2 + 4x + 3}$ (*b*) $\int \frac{dx}{x^2 + 4x + 4}$

(*c*) $\int \frac{dx}{x^2 + 4x + 5}$ (*d*) $\int \frac{dx}{x^2 + 4x - 2}$

Exercise 164 is the basis of Exercises 165 to 174.

164 Let p and q be rational numbers. Prove $\int x^p(1 - x)^q\,dx$ is an elementary function (*a*) if p is an integer (*Hint:* If $q = s/t$, let $1 - x = v^t$), (*b*) if q is an integer, (*c*) if $p + q$ is an integer.

Chebyshev proved that these are the only cases for which the antiderivative in question is elementary. In particular,

$$\int \sqrt{x}\sqrt[3]{1 - x}\,dx \quad \text{and} \quad \int \sqrt[3]{x - x^2}\,dx$$

are not elementary. Chebyshev's theorem also holds for $\int x^p(1 + x)^q\,dx$.

165 Deduce from Exercise 164 that $\int \sqrt{1 - x^3}\,dx$ is not elementary.

166 Deduce from Exercise 164 that $\int (1 - x^n)^{1/m}\,dx$, where m and n are positive integers, is elementary if and only if $m = 1$, $n = 1$, or $m = 2 = n$.

167 Deduce from Exercise 164 that $\int \sqrt{\sin x}\,dx$ is not elementary. *Hint:* Let $u = \sin^2 x$.

168 Deduce from Exercise 164 that $\int \sin^a x\,dx$, where a is rational, is elementary if and only if a is an integer.

169 Deduce from Exercise 164 that $\int \sin^p x \cos^q x\,dx$, where p and q are rational, is elementary if and only if p or q is an odd integer or $p + q$ is an even integer.

170 Deduce from Exercise 169 that $\int \sec^p x \tan^q x\,dx$, where p and q are rational, is elementary only if $p + q$ or q is odd, or if p is even.

171 (*a*) Deduce from Exercise 164 that $\int (x/\sqrt{1 + x^n})\,dx$, where n is a positive integer, is elementary only when $n = 1$, 2, or 4.
(*b*) Evaluate the integral for $n = 1$, 2, and 4.

172 (*a*) Deduce from Exercise 164 that $\int (x^2/\sqrt{1 + x^n})\,dx$, where n is a positive integer, is elementary only when $n = 1$, 2, 3, or 6.
(*b*) Evaluate the integral for $n = 1$, 2, 3, and 6.

173 (*a*) Using Exercise 164, determine for which positive integers n the integral $\int (x^n/\sqrt{1 + x^4})\,dx$ is elementary.
(*b*) Evaluate the integral for $n = 3$ and $n = 5$.

174 The following is an excerpt from an engineering text:
The last equation may be written

$$\theta = \int \frac{c\,dr}{r\sqrt{r^6 - c^2}}$$

where c is a constant. The integral is easily evaluated by the substitution. . . .
(*a*) What substitution did the text recommend?
(*b*) Using Exercise 164, determine for which positive integers, n, $\int (c/(r^n \sqrt{r^6 - c^2}))\,dr$ is elementary.

175 Consider the problem of finding the area under $y = e^{x^2}$ from $x = 0$ to $x = 1$.
(*a*) Why is the FTC useless in determining this area?
(*b*) Estimate $\int_0^1 e^{x^2}\,dx$ by using Simpson's method with $n = 6$.

176 Assuming that $\int (e^x/x)\,dx$ is not elementary (a theorem of Liouville), prove that $\int 1/(\ln x)\,dx$ is not elementary.

177 Evaluate $\int_{-1}^{1} xe^{x^4}\,dx$.

178 The factor theorem asserts that if r is a root of polynomial $P(x)$, then $x - r$ is a divisor of $P(x)$. Use this theorem to factor each of the following polynomials into irreducible factors:
(*a*) $x^3 - 1$ (*b*) $x^3 + 1$ (*c*) $x^3 - 5$
(*d*) $x^3 + 8$ (*e*) $x^4 - 1$ (*f*) $x^3 - 3x + 2$

179 (See Exercise 178.) Represent in partial fractions:

(*a*) $\dfrac{x}{x^3 - 1}$ (*b*) $\dfrac{1}{x^3 - 3x + 2}$

(*c*) $\dfrac{x^5}{x^3 + 1}$ (*d*) $\dfrac{1}{x^3 + 8}$.

180 Show these integrals are elementary:
(*a*) $\int x^{1/3}(1 + x)^{5/3}\,dx$, (*b*) $\int \sqrt[4]{x(1 + x)^3}\,dx$. (*Hint:* Let $x = 1/t$.)

181 One integral table lists the antiderivative $\int (\sqrt{x^2 + a^2}/x)\,dx$ as

$$\sqrt{x^2 + a^2} - a \ln\left(\frac{a + \sqrt{x^2 + a^2}}{x}\right),$$

while another lists it as

$$\sqrt{x^2 + a^2} + a \ln\left(\frac{\sqrt{x^2 + a^2} - a}{x}\right).$$

Is there an error in one?

182 One of these integrals is elementary; the other is not. Evaluate the one that is elementary.
(*a*) $\int \ln(\cos x)\,dx$ (*b*) $\int \cos(\ln x)\,dx$

183 From the fact that $\int x \tan x\,dx$ is not elementary, deduce that the following are not elementary:

(*a*) $\displaystyle\int x^2 \sec^2 x\,dx$ (*b*) $\displaystyle\int x^2 \tan^2 x\,dx$

(*c*) $\displaystyle\int \frac{x^2\,dx}{1 + \cos x}$

184 Three of these six antiderivatives are elementary. Compute them.

(*a*) $\displaystyle\int x \cos x\,dx$ (*b*) $\displaystyle\int \frac{\cos x}{x}\,dx$ (*c*) $\displaystyle\int \frac{x\,dx}{\ln x}$

(*d*) $\displaystyle\int \frac{\ln x^2}{x}\,dx$ (*e*) $\displaystyle\int \sqrt{x - 1}\sqrt{x}\sqrt{x + 1}\,dx$

(*f*) $\displaystyle\int \sqrt{x - 1}\sqrt{x + 1}\,x\,dx$

185 From the fact that $\int (\sin x)/x\,dx$ is not elementary, deduce that the following are not elementary:

(*a*) $\displaystyle\int (\cos^2 x)/x^2\,dx$ (*b*) $\displaystyle\int (\sin^2 x)/x^2\,dx$

(*c*) $\displaystyle\int \sin e^x\,dx$ (*d*) $\displaystyle\int \cos x \ln x\,dx$

186 There are two values of a for which $\int \sqrt{1 + a \sin^2\theta}\,d\theta$ is elementary. What are they?

187 From Exercise 186 deduce that there are two values of a for which

$$\int \frac{\sqrt{1 + ax^2}}{\sqrt{1 - x^2}}\,dx$$

is elementary. What are they?

188 There are three values of b for which $\int \sqrt{1 + b\cos\theta}\,d\theta$ is elementary. What are they?

189 From Exercise 188 deduce that there are three values of b for which

$$\int \frac{\sqrt{1 + bx}}{\sqrt{1 - x^2}}\,dx$$

is elementary.

190 Using the fact that $\int e^{x^2}\,dx$ is not elementary, deduce that the following are not elementary:

(*a*) $\displaystyle\int \frac{e^x}{\sqrt{x}}\,dx$ (*b*) $\displaystyle\int \sqrt{x}\,e^x\,dx$ (*c*) $\displaystyle\int x^2 e^{x^2}\,dx$

8

APPLICATIONS OF THE DEFINITE INTEGRAL

This chapter presents various geometric and physical applications of the definite integral. It is based on Chap. 5 and can be covered before Chap. 7. None of the examples depend on the material in Chap. 7; the few exercises that do are set apart.

In Chap. 5 we showed that area and volume can both be represented as definite integrals: ''Area is the integral of the length of cross sections made by lines'' and ''Volume is the integral of the areas of cross sections made by planes.'' Section 8.1 gives further practice in computing areas and Sec. 8.4 in computing volumes.

Because clear diagrams of objects in space are essential in computing volumes and in the work of several later chapters, we devote Sec. 8.2 to some advice on how to draw.

Section 8.3 describes a shorthand method for setting up a definite integral. It saves us the trouble of forming partitions, choosing sampling numbers, and writing limits of approximating sums. The shorthand method is applied in Sec. 8.4 to finding volumes of solids.

Section 8.5 introduces the shell technique, a method for finding the volume of certain solids.

Section 8.6 uses the definite integral to define and compute centroids of plane regions, while Sec. 8.7 uses it to compute the work done by a varying force that moves an object along a straight line. (In Chap. 16 we study the work done in moving an object along a curve.)

Section 8.8 extends the notion of a definite integral to infinite intervals and to unbounded functions—cases that do not fit into the definition of a definite integral, but do arise in applications in areas such as statistics, economics, and physics.

8.1 COMPUTING AREA BY PARALLEL CROSS SECTIONS

''c'' is short for ''cross section.''

In Sec. 5.3 it was shown that the area of a plane region is equal to the integral of its cross-sectional length. Let $c(x)$ denote the length of the intersection with the given region of the vertical line through $(x, 0)$. (See Fig. 1.) Then the area of

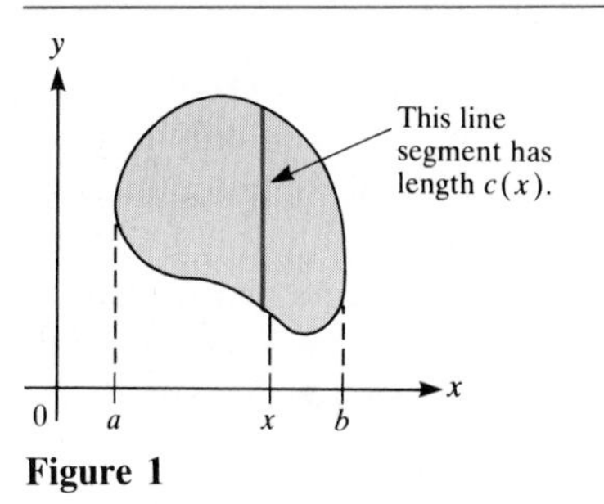

Figure 1

the region is equal to $\int_a^b c(x)\,dx$, where a and b are shown in Fig. 1. Note that x need not refer to the x axis of the xy plane; it may refer to any conveniently chosen line in the plane. It may even refer to the y axis; in this case, the cross-sectional length would be denoted $c(y)$.

To compute an area:

1 Find a, b and the cross-sectional length $c(x)$.
2 Evaluate $\int_a^b c(x)\,dx$ by the fundamental theorem of calculus if $\int c(x)\,dx$ is elementary.

Chapter 7 showed how to accomplish step 2. The present section is concerned primarily with step 1, how to find the cross-sectional length $c(x)$.

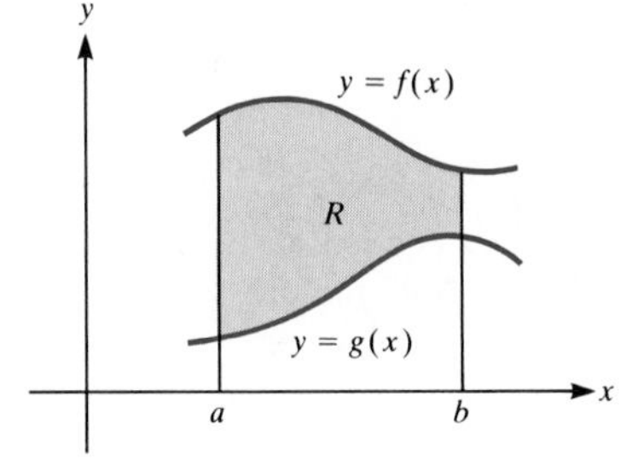

Figure 2

The Area of the Region between Two Curves

Let f and g be two continuous functions such that $f(x) \geq g(x)$ for all x in the interval $[a, b]$. Let R be the region between the curve $y = f(x)$ and the curve $y = g(x)$ for x in $[a, b]$, as shown in Fig. 2. In these circumstances, the length of the cross section of R made by a line perpendicular to the x axis can be computed in terms of f and g. Inspection of Fig. 3 shows that

$$c(x) = f(x) - g(x).$$

In short, to find $c(x)$, subtract the smaller value, $g(x)$, from the larger value, $f(x)$. It follows that

$$\text{Area of } R = \int_a^b [f(x) - g(x)]\,dx.$$

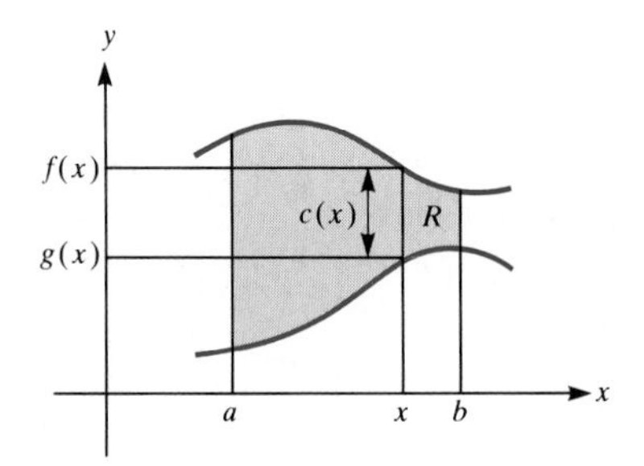

Figure 3

EXAMPLE 1 Find the area of the region shown in Fig. 4. The region is bounded by the curve $y = x^2$, the line $y = -3x/2$, and the line $x = 2$. (Since we are interested in the actual geometric area, the area of the region below the x axis is positive.)

SOLUTION In this case, $f(x) = x^2$ and $g(x) = -\frac{3}{2}x$. For x in $[0, 2]$, the cross-sectional length is $c(x) = x^2 - (-\frac{3}{2}x)$. (See Fig. 5.) Thus the area of the region is

$$\int_0^2 \left[x^2 - \left(-\frac{3x}{2}\right)\right] dx = \int_0^2 \left(x^2 + \frac{3x}{2}\right) dx.$$

This definite integral can be evaluated by the fundamental theorem of calculus:

$$\int_0^2 \left(x^2 + \frac{3x}{2}\right) dx = \left(\frac{x^3}{3} + \frac{3x^2}{4}\right)\Bigg|_0^2$$

$$= \left(\frac{2^3}{3} + \frac{3 \cdot 2^2}{4}\right) - \left(\frac{0^3}{3} + \frac{3 \cdot 0^2}{4}\right) = \frac{17}{3}. \quad \blacksquare$$

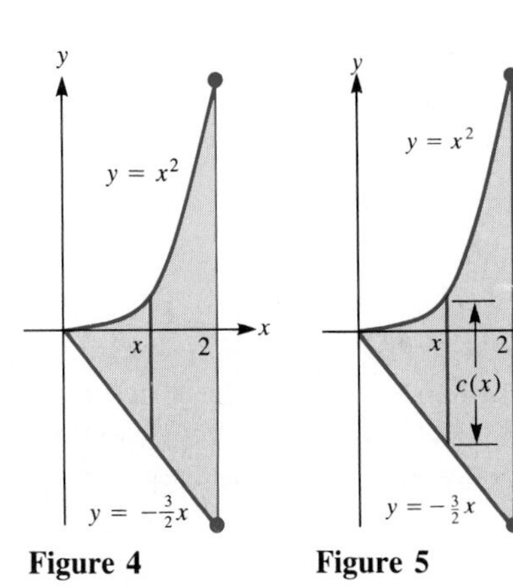

Figure 4

Figure 5

EXAMPLE 2 Find the area of the region in Fig. 4, but this time use cross sections parallel to the x axis.

SOLUTION Since the cross-sectional length is to be expressed in terms of y, first express the equations of the curves bounding the region in terms of y. The curve $y = x^2$ may be written as $x = \sqrt{y}$, since we are interested only in positive x. The curve $y = -\frac{3}{2}x$ can be expressed as $x = -\frac{2}{3}y$ by solving for x in terms of y. The line $x = 2$ also bounds the region. (See Fig. 6.)

For each number y in a certain interval $[a, b]$ that will be determined, the line with y coordinate equal to y meets the region in Fig. 6 in a line segment of length $c(y)$. It is necessary to determine the numbers a and b as well as the formula for $c(y)$.

The point P in Fig. 6 lies on the parabola $y = x^2$ (or $x = \sqrt{y}$) and has the x coordinate 2. Thus $P = (2, 2^2) = (2, 4)$. The point Q lies on the line $y = -\frac{3}{2}x$ and has x coordinate 2. Thus $Q = (2, -\frac{3}{2} \cdot 2) = (2, -3)$. Consequently, a cross section of the region is determined for each number y in the interval $[-3, 4]$. The area of the region is therefore

y varies from the y coordinate of Q to the y coordinate of P.

$$\int_{-3}^{4} c(y)\, dy. \qquad (1)$$

Next, we must find a formula for $c(y)$. For $0 \leq y \leq 4$, the cross section is determined by the line $x = 2$ and the parabola $x = \sqrt{y}$. Thus for $0 \leq y \leq 4$,

$$c(y) = 2 - \sqrt{y},$$

the larger minus the smaller. (See Fig. 6.) For $-3 \leq y \leq 0$, the cross section is determined by the line $x = 2$ and the line $x = -\frac{2}{3}y$. Thus for $-3 \leq y \leq 0$,

$$c(y) = 2 - \left(-\frac{2}{3}y\right) = 2 + \frac{2y}{3}.$$

The integral (1) breaks into two separate integrals, each of which is easily evaluated by the fundamental theorem of calculus:

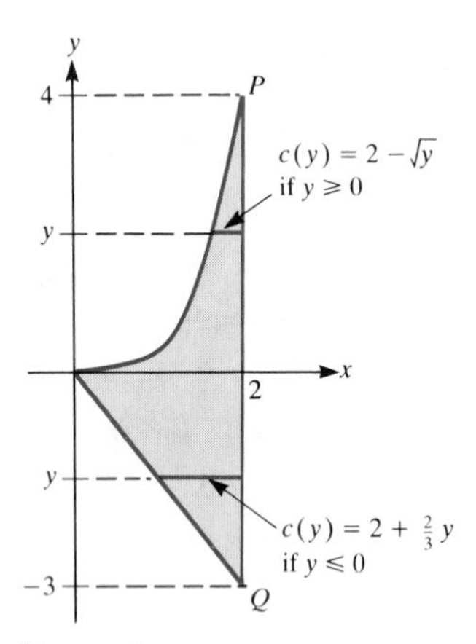

Figure 6

$$\int_{-3}^{0} c(y)\, dy + \int_{0}^{4} c(y)\, dy = \int_{-3}^{0} \left(2 + \frac{2y}{3}\right) dy + \int_{0}^{4} (2-\sqrt{y})\, dy$$
$$= \left(2y + \frac{y^2}{3}\right)\Bigg|_{-3}^{0} + \left(2y - \frac{2}{3}y^{3/2}\right)\Bigg|_{0}^{4}$$
$$= \left(2 \cdot 0 + \frac{0^2}{3}\right) - \left(2(-3) + \frac{(-3)^2}{3}\right)$$
$$+ \left(2 \cdot 4 - \frac{2}{3}4^{3/2}\right) - \left(2 \cdot 0 - \frac{2}{3}0^{3/2}\right)$$
$$= 0 - (-3) + \frac{8}{3} - 0 = \frac{17}{3}. \quad \blacksquare$$

Example 1 needed only one integral, but Example 2 needed two. Moreover, in Example 2 the formula for the cross-sectional length when $0 \leq y \leq 4$ involved $\sqrt{y}$, which is a little harder to handle than x^2, which appeared in the corresponding formula in Example 1. Although both approaches to finding the area of the region in Fig. 4 are valid, the one with cross sections parallel to the y axis is more convenient.

Choose the direction for cross sections to make life easy.

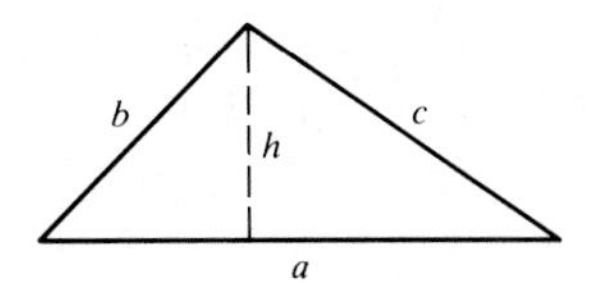

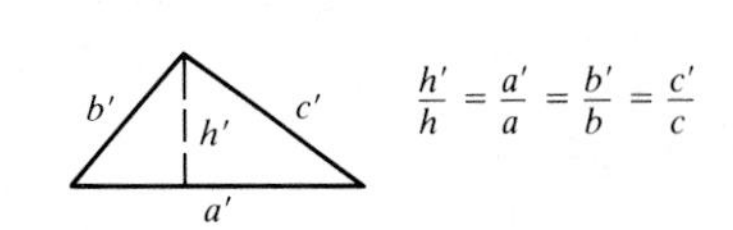

Figure 7

Computing Cross Sections

Perhaps a region is described not by formulas but geometrically, maybe as a circle, triangle, trapezoid, or some other polygon. Two geometric facts are often of use in finding the cross-sectional length in these circumstances:

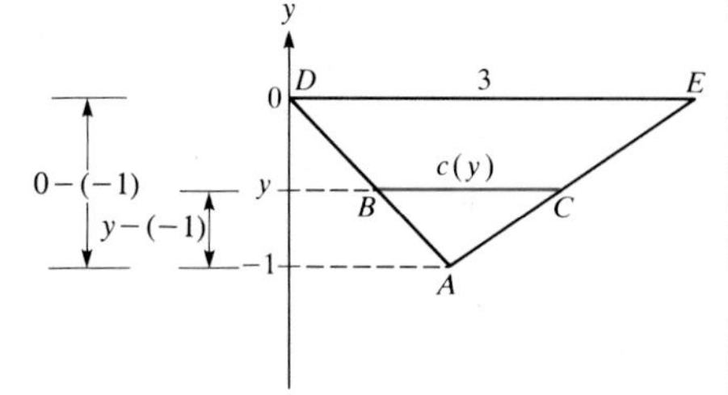

Figure 8

> **The Pythagorean Theorem** In a right triangle whose legs have lengths a and b and whose hypotenuse has length c, $c^2 = a^2 + b^2$.

> **Corresponding Parts of Similar Triangles are Proportional** If a, b, and c are the lengths of the sides of one triangle and a', b', and c' are the lengths of the corresponding sides of a similar triangle, then $a'/a = b'/b = c'/c$.

Figure 9

In addition, corresponding altitudes of the two triangles are also in the same proportion. (In Fig. 7 they have lengths h and h'.)

EXAMPLE 3 Find the formula for $c(y)$ as shown in the triangles in Figs. 8 and 9.

SOLUTION In the first figure, ΔABC is similar to ΔADE. Since corresponding parts of similar triangles are proportional, the sides BC and DE are in the same ratio as the corresponding altitudes perpendicular to them; that is,

$$\frac{\overline{BC}}{\overline{DE}} = \frac{y-(-1)}{0-(-1)}.$$

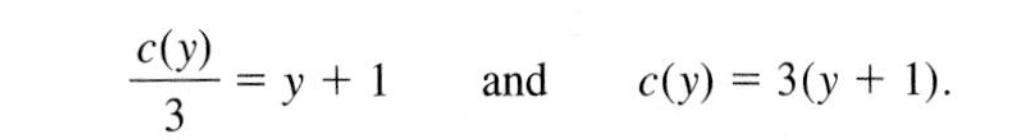

Hence $$\frac{c(y)}{3} = y + 1 \quad \text{and} \quad c(y) = 3(y+1).$$

In the case of Fig. 8, we have, again by similar triangles,

$$\frac{c(y)}{3} = \frac{y}{1} \quad \text{and} \quad c(y) = 3y.$$

Placing the y coordinate system as in Fig. 9 provides the simpler formula for cross-sectional length. ■

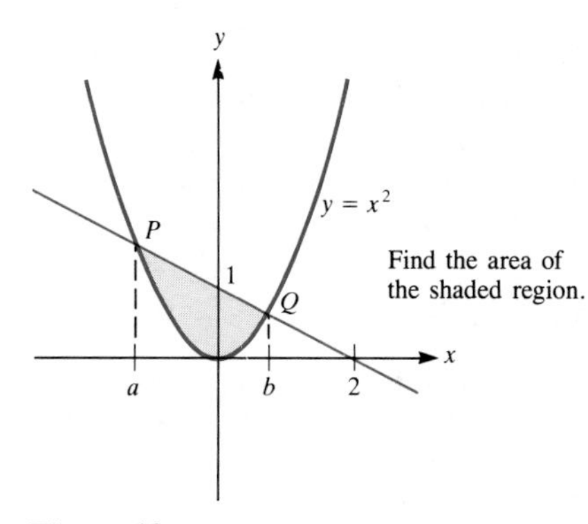

Figure 10

EXAMPLE 4 Set up a definite integral for the area of the region above the parabola $y = x^2$ and below the line through (2, 0) and (0, 1) shown in Fig. 10.

SOLUTION Since the x intercept of the line is 2 and the y intercept is 1, an equation for the line is

Appendix B describes the intercept equation of a line.

$$\frac{x}{2} + \frac{y}{1} = 1.$$

Hence $y = 1 - x/2$. The length $c(x)$ of a cross section of the region taken parallel to the y axis is therefore

$$c(x) = \left(1 - \frac{x}{2}\right) - x^2 = 1 - \frac{x}{2} - x^2.$$

To find the interval $[a, b]$ of integration, we must find the x coordinates of the points P and Q in Fig. 10. For these values of x,

$$x^2 = 1 - \frac{x}{2},$$

so

$$2x^2 + x - 2 = 0. \qquad (2)$$

The solutions of Eq. (2) are

$$x = \frac{-1 \pm \sqrt{17}}{4}.$$

Hence

$$\text{Area} = \int_{(-1-\sqrt{17})/4}^{(-1+\sqrt{17})/4} \left(1 - \frac{x}{2} - x^2\right) dx. \quad ■$$

Section Summary

The big idea in this section was already developed in Chap. 5: The area of a region is the definite integral of cross-sectional length. To compute this area, find the cross-sectional length $c(x)$ or $c(y)$ and set up a definite integral.

EXERCISES FOR SEC. 8.1: COMPUTING AREA BY PARALLEL CROSS SECTIONS

In each of Exercises 1 to 6 (*a*) draw the region, (*b*) compute the lengths of vertical cross sections ($c(x)$), (*c*) compute the lengths of horizontal cross sections ($c(y)$).

1 The finite region bounded by $y = \sqrt{x}$ and $y = x^2$
2 The finite region bounded by $y = x^2$ and $y = x^3$
3 The triangle bounded by $y = 2x$, $y = 3x$, and $x = 1$
4 The region bounded by $y = x^2$, $y = 2x$, and $x = 1$
5 The triangle with vertices (0, 0), (3, 0), (0, 4)
6 The triangle with vertices (1, 0), (3, 0), (2, 1)

In Exercises 7 to 12 sketch the finite regions bounded by the given curves. Then find their areas by (*a*) vertical cross sections, (*b*) horizontal cross sections.

7 $y = x^2$ and $y = 3x - 2$
8 $y = 2x^2$ and $y = x + 1$
9 $y = 4x$ and $y = 2x^2$
10 $y = x^2$ and $y = 4$
11 $y = 1/x^2$, $y = 0$, $x = 1$, $x = 3$
12 $x = y^2$ and $x = 3y - 2$

In each of Exercises 13 to 18 find the area of the region between the two curves and above (or below) the given interval.

13 $y = x^2$ and $y = x^3$; [0, 1]
14 $y = x^2$ and $y = x^3$; [1, 2]
15 $y = x^2$ and $y = \sqrt{x}$; [0, 1]
16 $y = x^3$ and $y = -x$; [1, 2]
17 $y = \sin x$ and $y = \cos x$; $[0, \pi/4]$
18 $y = \sin x$ and $y = \cos x$; $[\pi/2, \pi]$

Exercises 19 to 28 use techniques in Chap. 7.
In Exercises 19 and 20 find the area of the region between the two curves and above the given interval.

19 $y = x^3$ and $y = \sqrt[3]{2x - 1}$; [1, 2]
20 $y = 1 + x$ and $y = \ln x$; $[1, e]$
21 Find the area under the curve $y = \tan^{-1} 2x$ and above the interval $[1/2, 1/\sqrt{3}]$.
22 (*a*) Find the area $A(b)$ under the curve $y = e^{-x} \cos^2 x$ and above the interval $[0, b]$.
(*b*) Find $\lim_{b\to\infty} A(b)$.
23 Find the area of the region in the first quadrant below $y = -7x + 29$ and above the portion of $y = 8/(x^2 - 8)$ that lies in the first quadrant.
24 Find the area of the region below $y = 10^x$ and above $y = \log_{10} x$ for x in [1, 10].
25 Find the area under the curve $y = x/(x^2 + 5x + 6)$ and above the interval [1, 2].
26 Find the area of the region below $y = (2x + 1)/(x^2 + x)$ and above the interval [2, 3].
27 Find the area of the region bounded by $y = \tan x$, $y = 0$,

and $x = \pi/4$ (consider only $x \geq 0$), (*a*) by vertical cross sections (*b*) by horizontal cross sections.

28 Find the area of the region bounded by $y = \sin x$, $y = 0$, $x = 0$, and $x = \pi/2$ by (*a*) vertical cross sections (*b*) horizontal cross sections.

---

29 Let $A(t)$ be the area of the region in the first quadrant between $y = x^2$ and $y = 2x^2$ and inside the rectangle bounded by $x = t$, $y = t^2$, and the coordinate axes. (See the shaded region in Fig. 11.) If $R(t)$ is the area of the rectangle, find (*a*) $\lim_{t\to 0} A(t)/R(t)$ and (*b*) $\lim_{t\to\infty} A(t)/R(t)$.

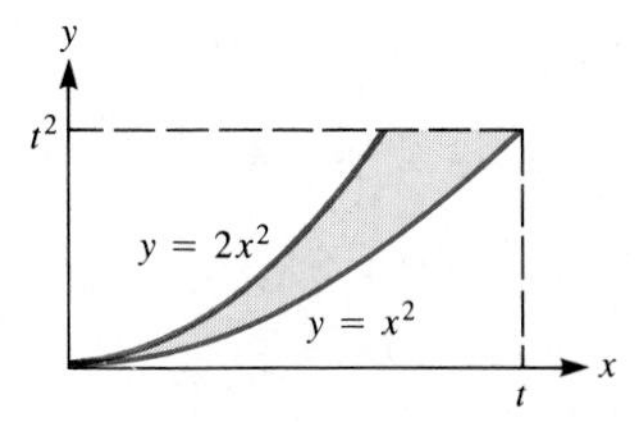

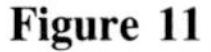

Figure 11

30 (*a*) Draw the curve $y = e^x/x$ for $x > 0$, showing any asymptotes or critical points.
(*b*) Find the number t such that the area below $y = e^x/x$ and above the interval $[t, t + 1]$ is a minimum.

31 Estimate the area of the region bounded by $y = x^3$, $y = x + 2$, and the y axis. Remark: You may need to estimate the solution of an equation. What method do we have?

32 Estimate the area of the region between $y = 3$ and $y = e^x/x$. *Remark*: You will meet an equation which you cannot solve exactly. What method do we have for estimating a solution? Also, $\int e^x/x \, dx$ is not elementary. What methods do we have for estimating a definite integral?

33 What fraction of the rectangle whose vertices are $(0, 0)$, $(a, 0)$, (a, a^4), and $(0, a^4)$, with a positive, is occupied by the region under the curve $y = x^4$ and above $[0, a]$?

34 Figure 12 shows a right triangle ABC.
(*a*) Find equations for the lines parallel to each edge, AC, BC, and AB, that cut the triangle into two pieces of equal area.
(*b*) Are the three lines in (*a*) concurrent; that is, do they meet at a single point?

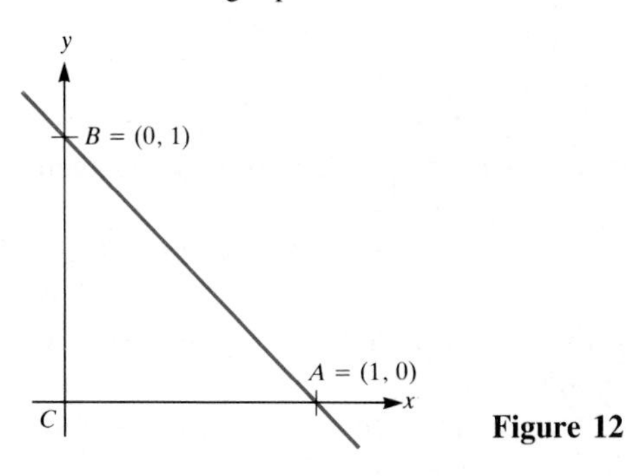

Figure 12

35 Lef f be an increasing function with $f(0) = 0$, and assume that it has an elementary antiderivative. Then f^{-1} is an increasing function, and $f^{-1}(0) = 0$. Prove that if f^{-1} is elementary, then it also has an elementary antiderivative. *Hint:* See Fig. 13.

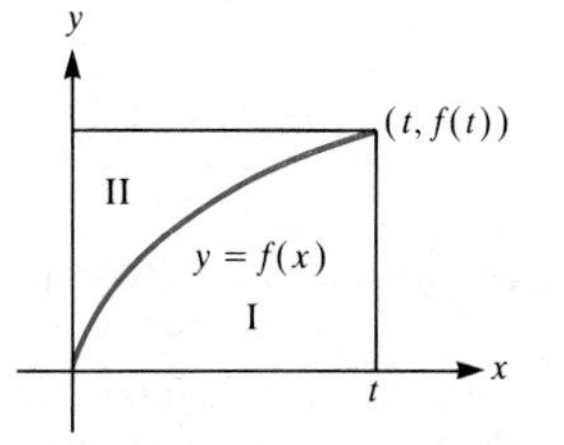

Figure 13

36 Show that the shaded area in Fig. 14 is two-thirds the area of the parallelogram $ABCD$. This is an illustration of a theorem of Archimedes concerning sectors of parabolas.

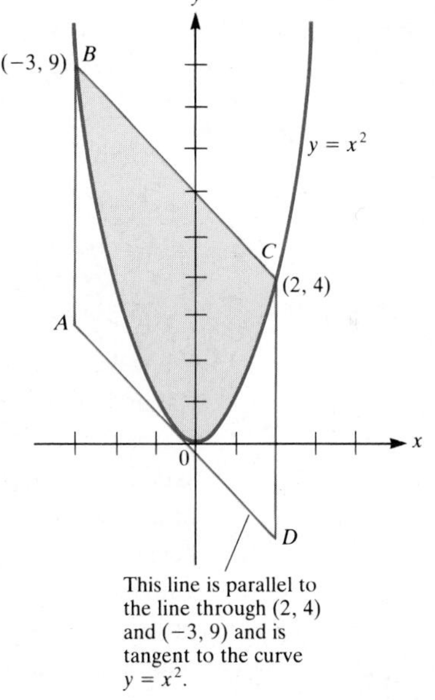

Figure 14

37 Figure 15 shows the graph of an increasing function $y = f(x)$ such that $f(0) = 0$. Assume that $f'(x)$ is continuous and $f'(0) > 0$. Do not assume $f''(x)$ exists. Investigate

$$\lim_{t\to 0^+} \frac{\text{Shaded area under curve}}{\text{Area of triangle } ABC}. \tag{3}$$

(*a*) Experiment with various functions, including some trigonometric functions and polynomials. [Make sure that $f'(0) \neq 0$.]
(*b*) Make a conjecture about (3) and explain why it is true.

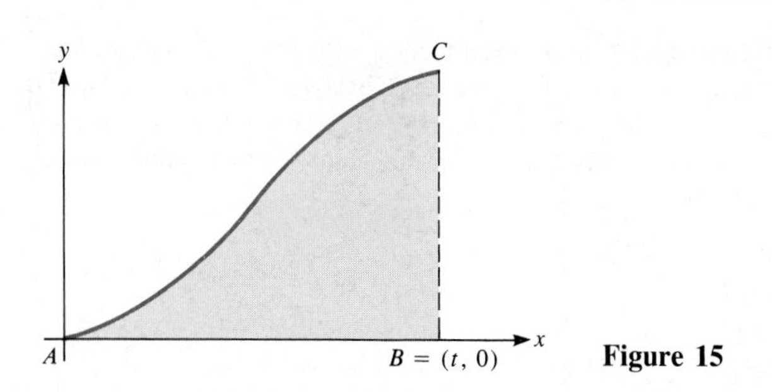

Figure 15

*38 Repeat the previous exercise, but now assume that $f'(0) = 0$, f'' is continuous, and $f''(0) \neq 0$.

39 A region R inside a square of side a consists of all points that are closer to the center of the square than to its border. (See Fig. 16.) Find the area of R.

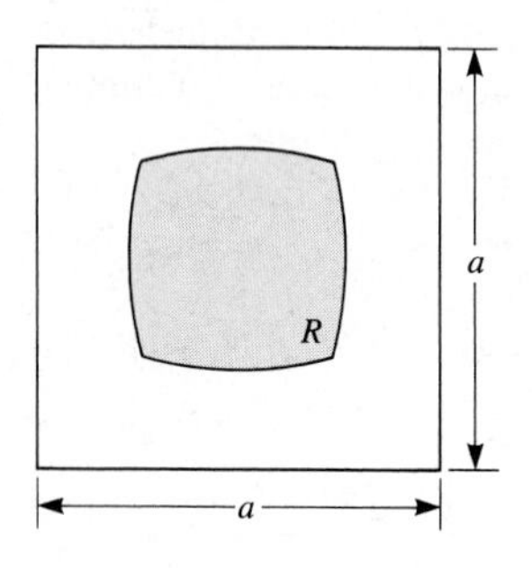

Figure 16

Exercises 40 and 41 use techniques from Chap. 7.

40 Find the area of the region above the interval [1, 2] and below the curve $y = x^2/(x^3 + x^2 + x + 1)$.

41 (*a*) Draw the region inside the ellipse

$$\frac{x^2}{a^2} + \frac{y^2}{b^2} = 1.$$

(*b*) Find $c(x)$, the vertical cross-sectional length.

(*c*) Find $c(y)$, the horizontal cross-sectional length.

(*d*) Find the area of R. (If you do not know how to compute the integral, use an integral table.)

8.2 SOME POINTERS ON DRAWING

We weren't born knowing how to draw solids. As we grew up, we lived in flatland: the surface of the earth. Few high school math courses cover solid geometry, so calculus is often the first place where students have to think and sketch in terms of three dimensions. That is why we pause now for a few words of advice on how to draw. Too often the student cannot work a problem simply because his or her diagrams are so primitive that they confuse even the person who drew them. The following guidelines are not based on any profound artistic principles. Instead, they derive from years of experience in attempting to create diagrams that do more good than harm.

A Few Words of Advice

1 *Draw large.* Most students tend to draw diagrams that are so small that there is no room to place labels or sketch cross sections.

2 *Draw neatly.* Use a straightedge to make straight lines that are actually straight. Use a compass to make circles that look like circles. (A jar lid or the base of a soda can will do just fine.) Draw each line or curve slowly. You may want to add a second color.

3 *Avoid clutter.* If you end up with too many labels or the cross section

doesn't show up well, add separate diagrams for important parts of the figure.

4 *Practice.*

EXAMPLE 1 Draw a diagram of a ball of radius a that shows the circular cross section made by a plane at a distance x from the center of the ball. Use the diagram to help find the radius of the cross section as a function of x.

Figure 1 is a terrible solution. Is it a potato or a ball? What segment has length r? What's x? What does the cross section look like?

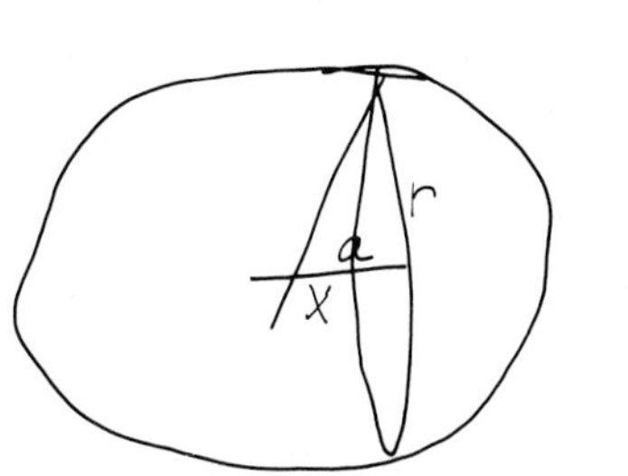

Terrible drawing

Figure 1

REASONABLE SOLUTION First, draw the ball carefully, as in Fig. 2. The equator is drawn to give it perspective. Add a little shading.

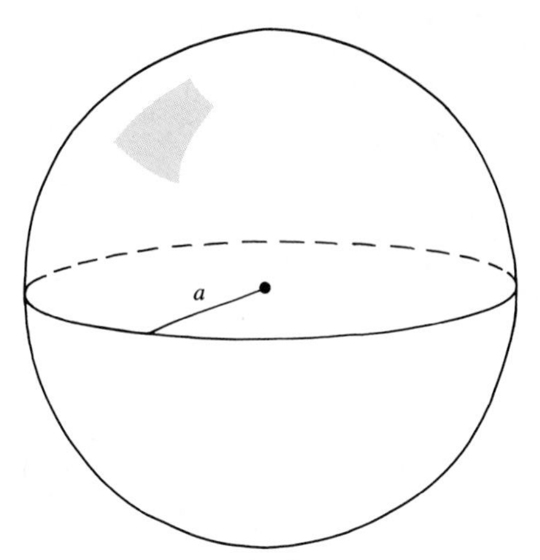

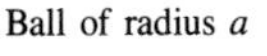
Ball of radius a

Figure 2

Next show a typical cross section at a distance x from the center, as in Fig. 3.

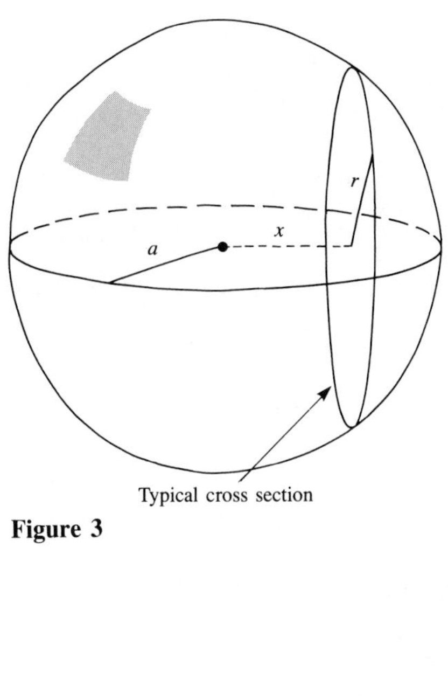

Figure 3

Shading the cross section helps.

To find r, the radius of the cross section, in terms of x, we sketch a companion diagram. The radius we want is part of a right triangle. In order to avoid clutter, we draw only the part of interest in a convenient side view, as in Fig. 4. (Try putting this information into Fig. 1. It will probably become cluttered and unreadable.)

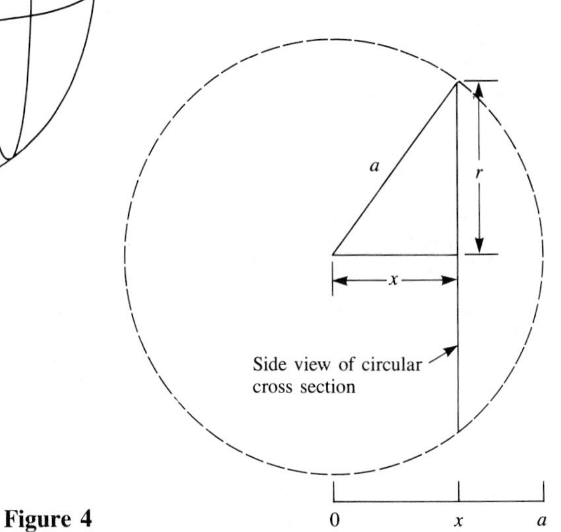

Figure 4

Inspection of the right triangle in Fig. 4 shows that

$$r^2 + x^2 = a^2,$$

hence that

$$r = \sqrt{a^2 - x^2}. \quad \blacksquare$$

EXAMPLE 2 A pyramid has a square base with a side of length a. The top of the pyramid is above the center of the base at a height h. Draw the pyramid and its cross sections by planes parallel to the base. Then find the area of the cross section in terms of its distance x from the top.

Figure 5 shows a terrible solution. The diagram is too small; there's no room

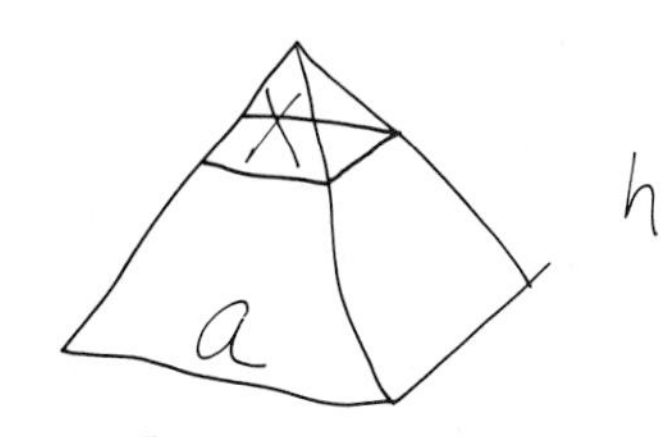

Figure 5

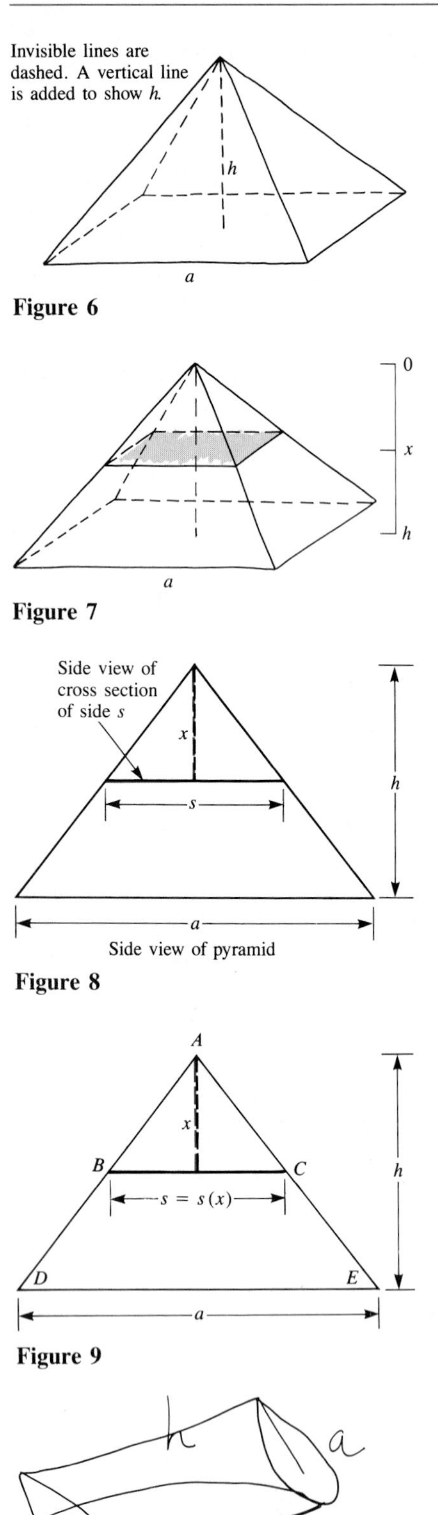

Figure 6

Figure 7

Figure 8

Figure 9

Figure 10

for the symbols and it's unclear to what h refers. Also, one of the edges of the pyramid is hidden behind another.

REASONABLE SOLUTION First draw a large pyramid, as in Fig. 6. Note that opposite edges of the base are drawn as parallel lines. While artists draw parallel lines as meeting in a point to enhance the sense of perspective, for our purposes it is more useful to use parallel lines to depict lines that are parallel. Then show a typical cross section in perspective and side views, as in Figs. 7 and 8. Note the x axis, which is drawn separate from the pyramid.

As x increases, so does s, the width of the square cross section. Thus s is a function of x, which we could call $s(x)$ [or $f(x)$, if you prefer]. A glance at Fig. 8 shows that $s(0) = 0$ and $s(h) = a$. To find $s(x)$ for all x in $[0, h]$, use the similar triangles ABC and ADE, shown in Fig. 9. These triangles show that

$$\frac{x}{s} = \frac{h}{a};$$

hence

$$s = \frac{ax}{h}. \tag{1}$$

As a check on (1), replace x by 0 and by h; we get 0 and a for the respective values of s, as expected. Finally, the area A of the cross sections is given by

$$A = s^2 = \left(\frac{ax}{h}\right)^2. \quad \blacksquare$$

EXAMPLE 3 A cylindrical drinking glass of height h and radius a is full of water. It is tilted until the remaining water covers exactly half the base.

(*a*) Draw a diagram of the glass and water.

(*b*) Show a cross section of the water that is a triangle.

(*c*) Find the area of the triangle in terms of the distance x of the cross section from the axis of the glass.

Figure 10 shows a terrible solution. The diagram is too small. It's unclear what has length a. The cross section is unclear. What does x refer to? Of course the student gives up in despair.

REASONABLE SOLUTION Draw a *neat, large* diagram first, as in Fig. 11. Don't put in too much data at first. When showing the cross section, we draw only the water. Figures 12 to 16 show various views. Let u and v be the lengths of the two legs of the cross section, as shown in Fig. 16.

Comparing Figs. 14 and 16, we have, by similar triangles, the relation

$$\frac{u}{a} = \frac{v}{h};$$

hence

$$v = \frac{h}{a}u.$$

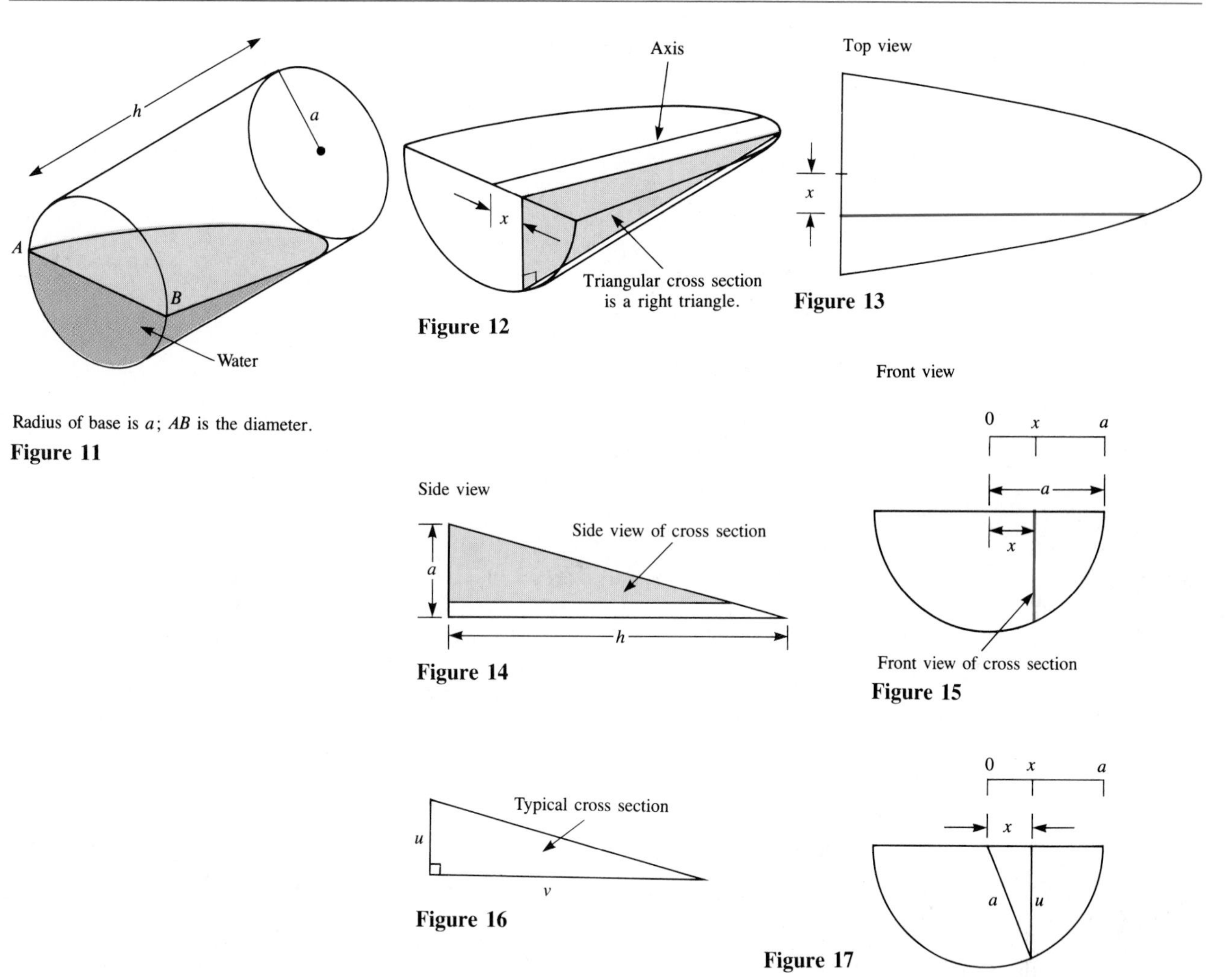

Radius of base is a; AB is the diameter.

Figure 11

Figure 12

Figure 13

Figure 14

Figure 15

Figure 16

Figure 17

Let $A(x)$ be the area of the cross section at a distance x from the center of the base. If we can find u and v as functions of x, we will be able to write down a formula for $A(x) = \frac{1}{2}uv$ in terms of x.

Figure 15 suggests how to find u. Copy it over and draw in the necessary radius, as in Fig. 17. By the pythagorean theorem,

$$u = \sqrt{a^2 - x^2}.$$

All told,

$$\begin{aligned} A(x) &= \frac{1}{2}uv \\ &= \frac{1}{2}u\left(\frac{h}{a}u\right) \\ &= \frac{h}{2a}u^2 \\ &= \frac{h}{2a}(a^2 - x^2). \end{aligned}$$

As a check, note that

$$A(a) = \frac{h}{2a}(a^2 - a^2) = 0,$$

which makes sense. Also the formula gives

$$A(0) = \frac{h}{2a}a^2$$
$$= \frac{1}{2}ah,$$

again agreeing with the geometry of, say, Fig. 12. ■

When you look back at these three examples, you will see that most of the work is spent on making clear diagrams. If you can't draw a straight line, use a straightedge. If you can't draw a circle, use a compass.

EXERCISES FOR SEC. 8.2: SOME POINTERS ON DRAWING

1 (See Example 2.) Cross sections of the pyramid in Example 2 are made by using planes perpendicular to the base and parallel to an edge of the base. What is the area of the cross section made by a plane that is a distance x from the top of the pyramid?
(*a*) Draw a large perspective view of the pyramid.
(*b*) Copy the diagram in (*a*) and show the typical cross section shaded.
(*c*) Draw a side view that clearly shows the shape of the cross section.
(*d*) Draw a different side view.
(*e*) Put necessary labels, such as x, a, and h, on the diagrams, where appropriate. You will need to introduce more labels.
(*f*) Find the area of the cross section as a function of x, $A(x)$.

2 (See Example 3.) Cross sections of the water in Example 3 are made by using planes parallel to the plane that passes through the horizontal diameter of the base and the axis of the glass. What is the area of the cross section made by a plane that is at a distance x from the center of the base?
(*a*) Draw a large perspective view of the water and glass.
(*b*) Copy the diagram in (*a*) and show the typical cross section shaded.
(*c*) Draw a side view that clearly shows the shape of the cross section.
(*d*) Draw a different side view.
(*e*) Put necessary labels, such as x, a, and h, on the diagrams, where appropriate. You will need to introduce more labels.
(*f*) Find the area of the cross section as a function of x, $A(x)$.

3 (See Example 3.) Cross sections of the water in Example 3 are made by using planes perpendicular to the axis of the glass. Make clear diagrams, including perspective and side views, that show the typical cross section. Do not find its area.

4 A cylindrical glass is full of water. The glass is tilted until the remaining water just covers the base of the glass. (Try it.) The radius of the glass is a and its height is h. Consider parallel planes such that cross sections of the water are rectangles.
(*a*) Make clear diagrams that show the situation. (You may want to include a top view to show the cross sections.)
(*b*) Obtain a formula for the area of the cross section. *Advice:* The two planes at a distance x from the axis of the glass cut out cross sections of different areas. So introduce an x axis with 0 at the center of the base and extending from $-a$ to a in a convenient direction.

5 The same as Exercise 4, but this time the cross sections are trapezoids.

6 A right circular cone has a radius a and height h, as shown in Fig. 18. Consider cross sections made by planes parallel to the base of the cone. (*a*) Draw perspective and side views of the situation. (*b*) Drawing as many diagrams as necessary, find the area of the cross section made by the plane at a distance x from the vertex of the cone.

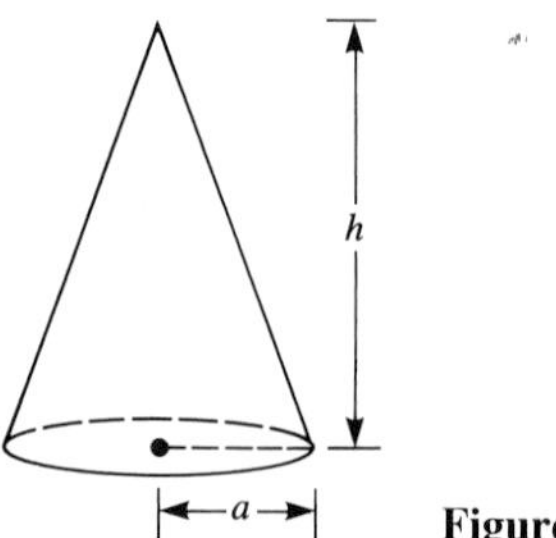

Figure 18

7 See Exercise 6. Draw the typical cross section made by a plane parallel to the axis of the cone. Draw perspective and side views of the situation, but do not find a formula for the area of the cross section.

8 Draw a cross section of a right circular cylinder that is (*a*) a circle, (*b*) an ellipse that is not a circle, (*c*) a rectangle.

9 Figure 19 indicates an unbounded, solid right circular cone. Draw a cross section that is bounded by (*a*) a circle, (*b*) an ellipse (but not a circle), (*c*) a parabola, (*d*) a hyperbola.

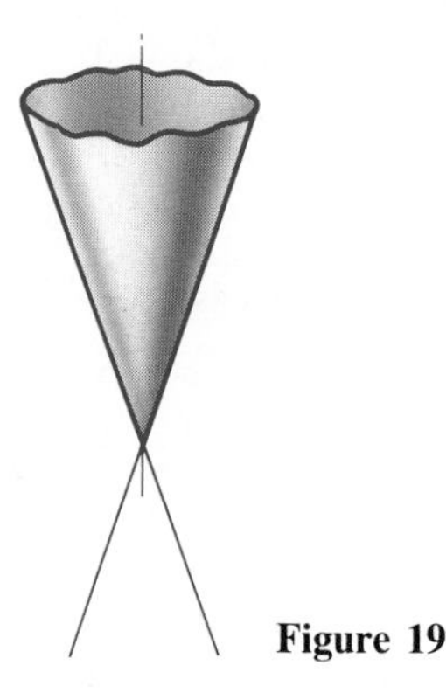

Figure 19

10 A lumberjack saws a wedge out of a cylindrical tree of radius a. His first cut is parallel to the ground and stops at the axis of the tree. His second cut makes an angle θ with the first cut and meets it along a diameter.
(*a*) Draw a typical cross section that is a triangle.
(*b*) Find the area of the triangle as a function of x, the distance of the plane from the axis of the tree.
(*c*) Draw a typical cross section that is a rectangle.
(*d*) Find the area of the rectangle as a function of x, the distance of the plane from the axis of the tree.

11 The plane region between the curves $y = x$ and $y = x^2$ is spun around the x axis to produce a solid resembling the bell of a trumpet.
(*a*) Draw the plane region.
(*b*) Draw the solid region produced by spinning this region around the x axis.
(*c*) Draw the typical cross section made by the plane perpendicular to the x axis. Show this in both perspective and side views.
(*d*) Find the area of the cross section in terms of the distance x of the plane from the origin of the x axis.

12 Draw a cross section of a solid cube by a plane that is (*a*) a square, (*b*) an equilateral triangle, (*c*) a five-sided polygon, (*d*) a regular hexagon. [*Hint* for (*d*): The vertices of the hexagon are midpoints of edges of the cube.]

13 Obtain a circular stick such as a broom handle or a dowel. Saw off a piece, making one cut perpendicular to the axis and the second cut at an angle to the axis. Mark on the piece you cut out the borders of cross sections that are (*a*) rectangles, (*b*) trapezoids.

8.3 SETTING UP A DEFINITE INTEGRAL

This section presents an informal shortcut for setting up a definite integral to evaluate some quantity. First, the formal and informal approaches are contrasted in the case of setting up the definite integral for area. Then the informal approach will be illustrated as commonly applied in a variety of fields.

The Complete Approach

Recall how the formula $A = \int_a^b f(x)\,dx$ was obtained in Sec. 5.3. The interval $[a, b]$ was partitioned by the numbers $x_0 < x_1 < x_2 < \cdots < x_n$ with $x_0 = a$ and $x_n = b$. A sampling number was chosen in each section $[x_{i-1}, x_i]$. For convenience, let us use x_{i-1} as that sampling number. The sum

$$\sum_{i=1}^{n} f(x_{i-1})(x_i - x_{i-1})$$

is then formed. It equals the total area of the rectangular approximation in Fig. 1.

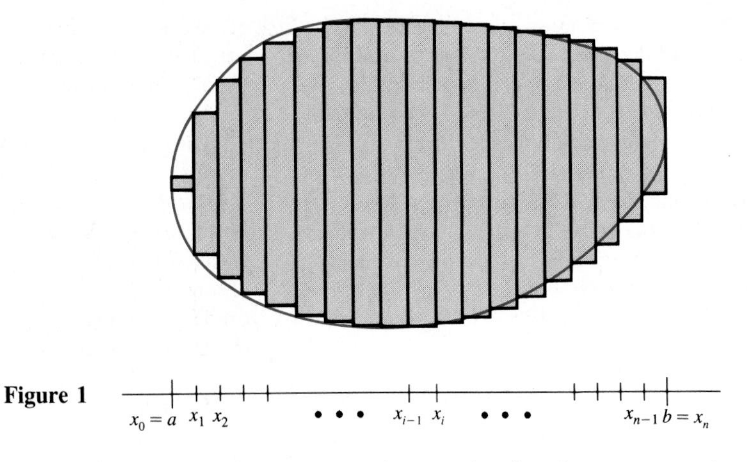

Figure 1

For brevity, we sometimes write Δx_i for $x_i - x_{i-1}$, in which case we can express the approximating sum as

$$\sum_{i=1}^{n} f(x_{i-1})\ \Delta x_i. \tag{1}$$

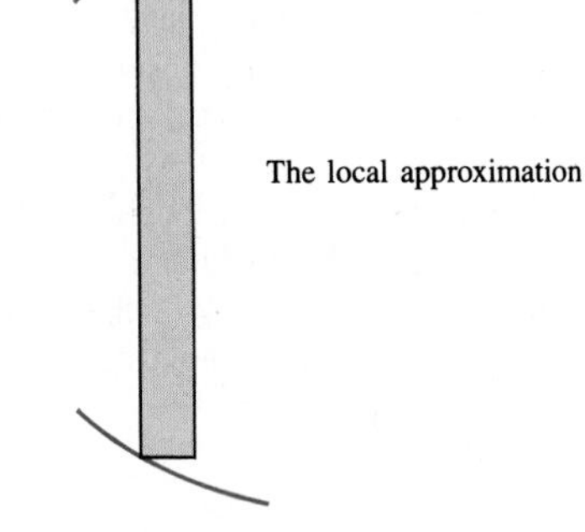

As all of the Δx_i approach 0, the sum (1) approaches the area of the region under consideration. But, by the definition of the definite integral, the sum (1) approaches

$$\int_a^b f(x)\ dx$$

as the mesh of the partition approaches 0. Thus

$$\text{Area} = \int_a^b f(x)\ dx. \tag{2}$$

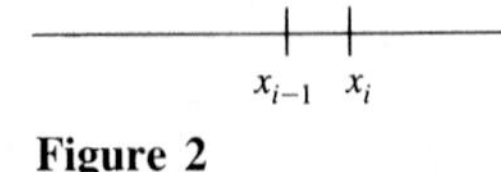

Figure 2

That is the complete or "formal" approach to obtain the formula (2). Now consider the "informal" approach, which is just a shorthand for the complete approach.

The Shorthand Approach

The heart of the complete approach is the *local estimate* $f(x_{i-1})(x_i - x_{i-1})$, the area of a rectangle of height $f(x_{i-1})$ and width $x_i - x_{i-1}$, which is shown in Fig. 2.

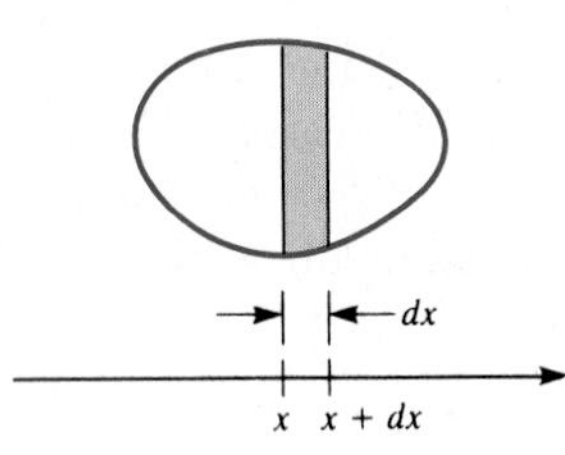

Figure 3

In the shorthand approach to setting up a definite integral attention is focused on the *local approximation*. No mention is made of the partition or the sampling numbers or the mesh approaching 0. We illustrate this shorthand approach by obtaining formula (2) informally. This is *not* a new method of integration, but just a way to save time when setting up an integral—finding out the integrand and the interval of integration.

For example, consider a small positive number dx. What would be a good estimate of the area of the region corresponding to the short interval $[x, x + dx]$ of width dx shown in Fig. 3? The area of the rectangle of width dx and height $f(x)$ shown in Fig. 4 would seem to be a plausible estimate. The area of this thin rectangle is

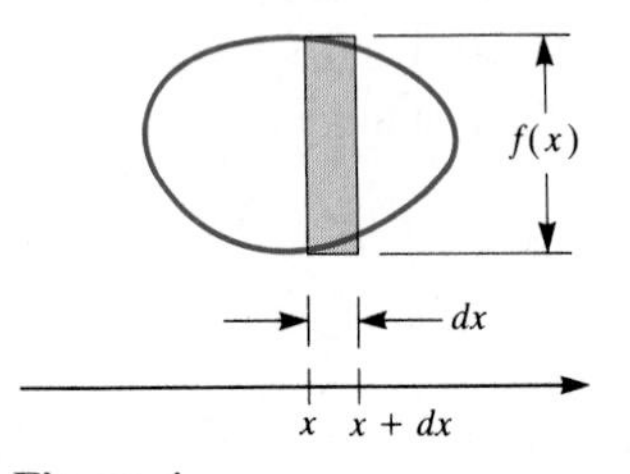

Figure 4

$$f(x)\ dx. \tag{3}$$

Without further ado, we then write

$$\text{Area} = \int_a^b f(x)\ dx, \tag{4}$$

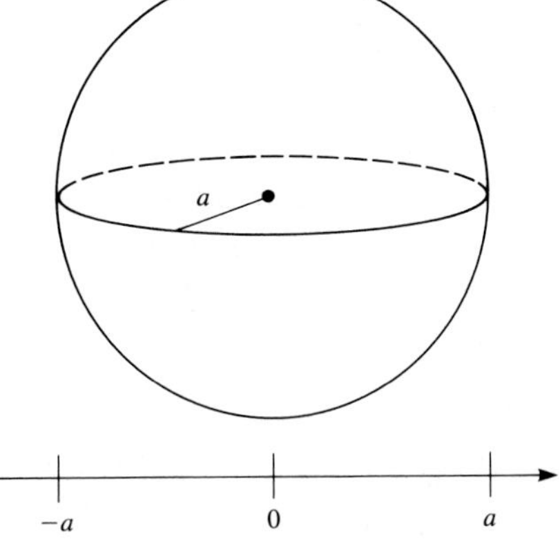

Figure 5

which is formula (2). The leap from the local approximation (3) to the definite integral (4) omits many steps of the complete approach. This informal approach is the shorthand commonly used in applications of calculus. It's the way engineers, physicists, and mathematicians set up integrals.

It should be emphasized that it is only an abbreviation of the formal approach, which deals with approximating sums.

The Volume of a Ball

EXAMPLE 1 Find the volume of a ball of radius a. First use the complete approach. Then use the shorthand approach.

SOLUTION Both approaches require good diagrams. In the complete approach we show an x axis, a partition, sampling numbers, and the approximating disks. See Figs. 5 and 6. The thickness of the ith disk is $x_i - x_{i-1}$, as shown in the side view of Fig. 7, while its radius is labeled r_i, as shown in the end view of Fig. 8. The volume of this typical disk is

$$\pi r_i^2(x_i - x_{i-1}). \tag{5}$$

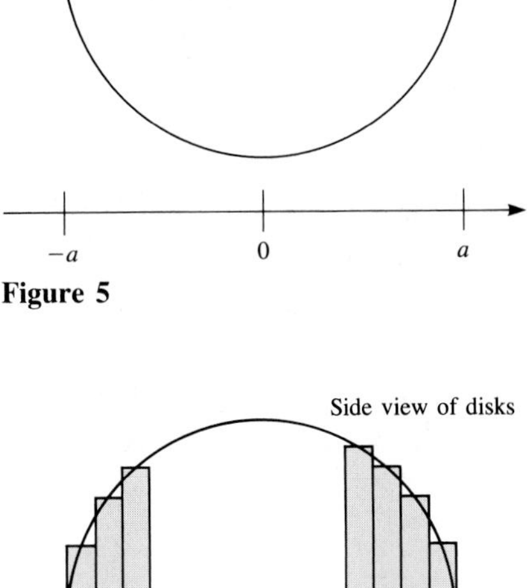

Figure 6

All that remains is to determine r_i. Figure 9 helps us do that. By the pythagorean theorem,

$$r_i^2 = a^2 - c_i^2. \tag{6}$$

Combining (1), (5), and (6) gives the typical estimate of the volume of a sphere of radius a:

$$\sum_{i=1}^{n} \pi(a^2 - c_i^2)(x_i - x_{i-1}). \tag{7}$$

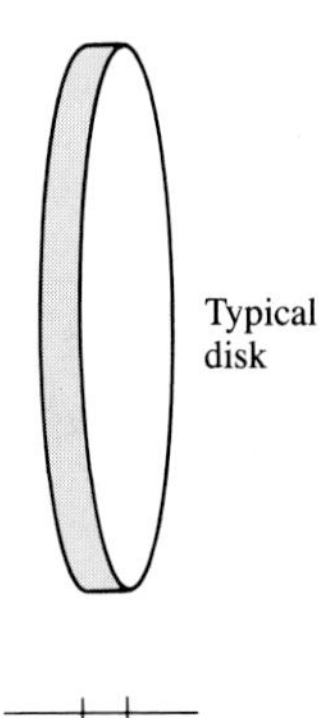

Figure 7

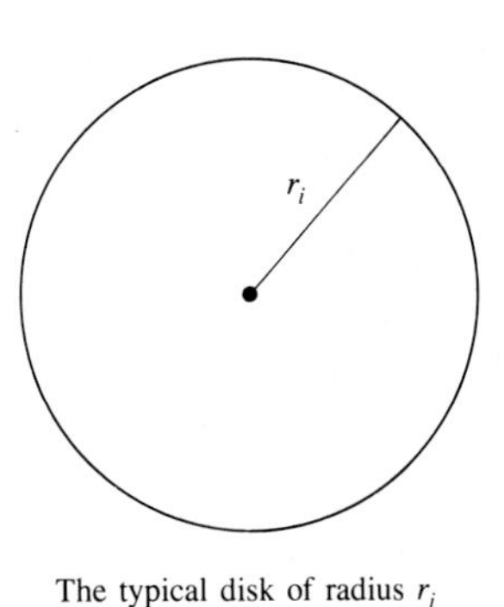

The typical disk of radius r_i (side view).

Figure 8

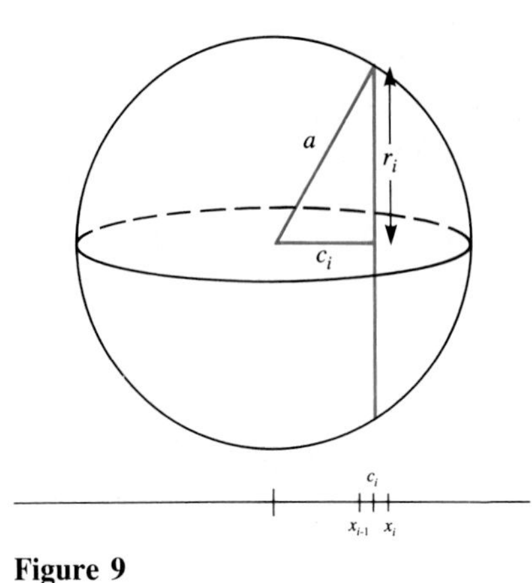

Figure 9

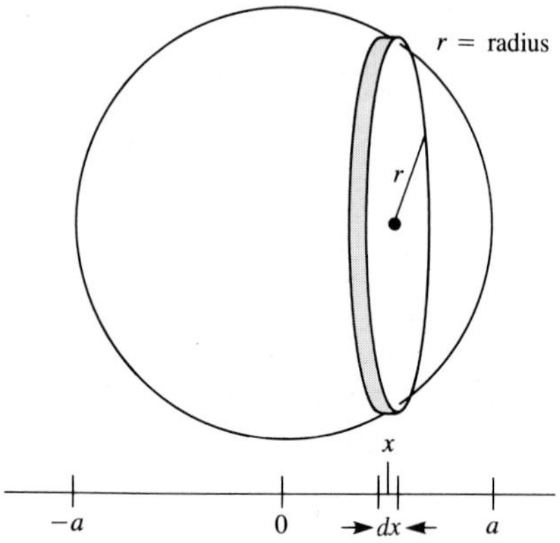

Figure 10

By the definition of the definite integral,

$$\lim_{\text{mesh}\to 0} \sum_{i=1}^{n} \pi(a^2 - c_i^2)(x_i - x_{i-1}) = \int_{-a}^{a} \pi(a^2 - x^2)\, dx.$$

Hence

$$\text{Volume of ball of radius } a = \int_{-a}^{a} \pi(a^2 - x^2)\, dx.$$

(By the fundamental theorem of calculus, the integral equals $4\pi a^3/3$.)

Now for the shorthand approach. We draw only a short section of an x axis and label its length dx. Then we draw an approximating disk, whose radius we label r, as in Fig. 10. Since the disk has a base of area πr^2 and thickness dx, its volume is $\pi r^2\, dx$. Moreover, as Fig. 11 shows, $r^2 = a^2 - x^2$. Hence the local approximation is

$$\pi(a^2 - x^2)\, dx. \tag{8}$$

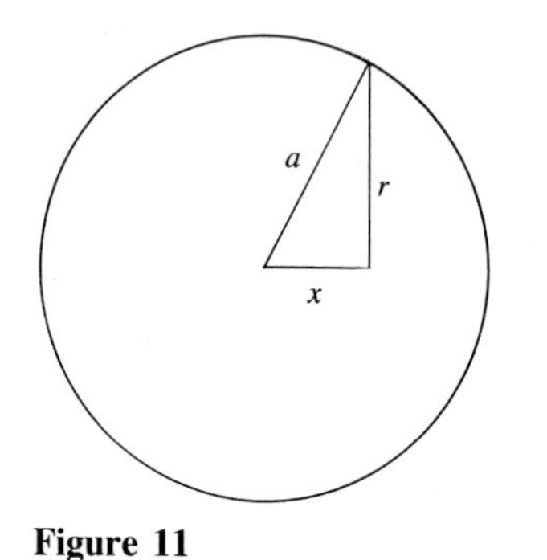

Figure 11

Then, without further ado, without writing any c_i, $x_i - x_{i-1}$, or approximating sum, we have

$$\text{Volume of ball of radius } a = \int_{-a}^{a} \pi(a^2 - x^2)\, dx.$$

The key to this bookkeeping is the local approximation (8) in differential form, which gives the necessary integrand. The limits of integration are determined separately. ■

Volcanic Ash

EXAMPLE 2 After the explosion of a volcano, ash gradually settles from the atmosphere and falls on the ground. The depth diminishes with distance from the volcano. Assume that the depth of the ash at a distance x feet from the volcano is Ae^{-kx} feet, where A and k are positive constants. Set up a definite integral for the total volume of ash that falls within a distance b of the volcano.

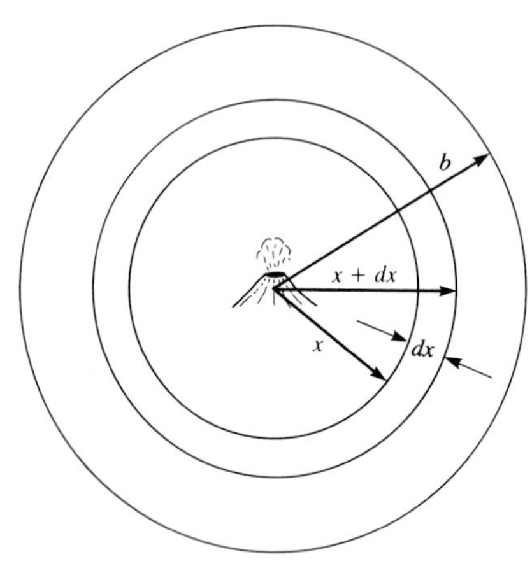

Figure 12

SOLUTION First estimate the volume of ash that falls on a very narrow ring of width dx and inner radius x centered at the volcano. (See Fig. 12.) This estimate can be made since the depth of the ash depends only on the distance from the volcano. On this ring the depth is almost constant.

The area of this ring is approximately that of a rectangle of length $2\pi x$ and width dx. (See Fig. 13.) So the area of the ring is approximately

$$2\pi x\, dx.$$

[Exercise 5 shows that its area is $2\pi x\, dx + \pi(dx)^2$.]

Although the depth of the ash on this narrow ring is not constant, it does not vary much. A good estimate of the depth throughout the ring is Ae^{-kx}. Thus the volume of ash that falls on the typical ring of inner radius x and outer radius $x + dx$ is approximately

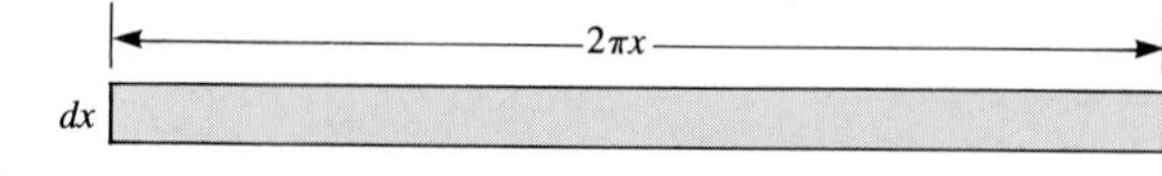

Figure 13 An approximation of the shaded band in Fig. 12.

The local approximation

$$Ae^{-kx}\, 2\pi x\, dx \text{ cubic feet.} \tag{9}$$

Once we have the key local estimate (9), we immediately write down the definite integral for the total volume of ash that falls within a distance b of the volcano:

$$\text{Total volume} = \int_0^b Ae^{-kx}\, 2\pi x\, dx.$$

(The limits of integration must be determined just as in the formal approach.) This completes the shorthand setting up of the definite integral. (It could be evaluated by a technique in Chap. 7 or by Formula 62 inside the front cover of this book.) ■

Kinetic Energy

The next example of the informal approach to setting up definite integrals concerns kinetic energy. The kinetic energy associated with an object of mass m kilograms and velocity v meters per second is defined as

$$\text{Kinetic energy} = \frac{mv^2}{2} \text{ joules.}$$

If the various parts of the objects are not all moving at the same speed, an integral is needed to express the total kinetic energy.

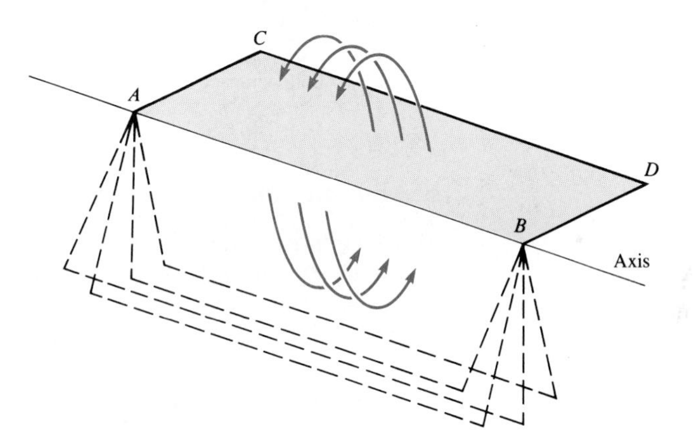

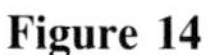

Figure 14

EXAMPLE 3 A thin rectangular piece of sheet metal is spinning around one of its longer edges 3 times per second, as shown in Fig. 14. The length of its shorter edge is 6 meters and the length of its longer edge is 10 meters. The density of the sheet metal is 4 kilograms per square meter. Find the kinetic energy of the spinning rectangle.

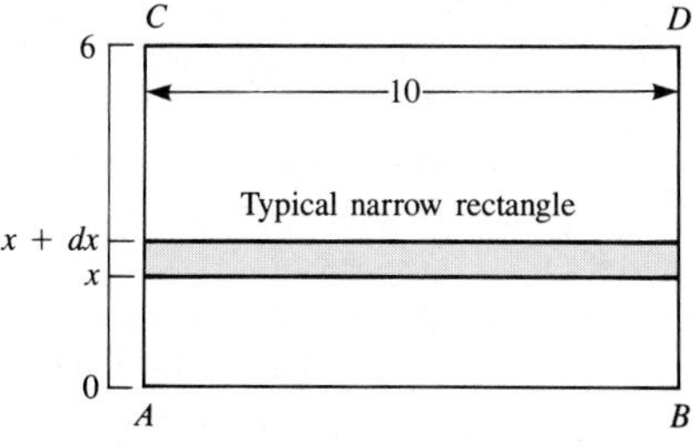

Figure 15

SOLUTION The farther a mass is from the axis, the faster it moves, and therefore the larger its kinetic energy. To find the total kinetic energy of the rotating piece of sheet metal, imagine it divided into narrow rectangles of length 10 meters and width dx meters parallel to the edge AB; a typical one is shown in Fig. 15. (Introduce an x axis parallel to edge AC with the origin corresponding to A.) Since all points of this typical narrow rectangle move at roughly the same speed, we will be able to estimate its kinetic energy. That estimate will provide the key local approximation in the informal approach to setting up a definite integral.

First of all, the mass of the typical rectangle is

$$4 \cdot 10\ dx \text{ kilograms,}$$

since its area is 10 dx square meters and the density is 4 kilograms per square meter.

Second, we must estimate its velocity. The narrow rectangle is spun 3 times per second around a circle of radius x. In 1 second each point in it covers a distance of about

$$3 \cdot 2\pi x = 6\pi x \text{ meters.}$$

Consequently, the velocity of the typical rectangle is

$$6\pi x \text{ meters per second.}$$

The local estimate of the kinetic energy associated with the typical rectangle is therefore

$$\frac{1}{2} \underbrace{40\ dx}_{\text{mass}} \underbrace{(6\pi x)^2}_{\text{velocity squared}} \text{ joules}$$

The local approximation

or simply

$$720\pi^2 x^2\ dx \text{ joules.} \tag{10}$$

Having obtained the local estimate (10), we jump directly to the definite integral and conclude that

$$\text{Total energy of spinning rectangle} = \int_0^6 720\pi^2 x^2\ dx \text{ joules.} \quad ■$$

Section Summary

This section presented a shorthand approach to setting up a definite integral for a quantity Q. In this method we estimate how much of the quantity Q corresponds to a very short section $[x, x + dx]$ of the x axis, say $f(x)\ dx$. Then $Q = \int_a^b f(x)\ dx$, where a and b are determined by the particular situation.

EXERCISES FOR SEC. 8.3: SETTING UP A DEFINITE INTEGRAL

1 In Sec. 5.3 we showed that "total mass is the definite integral of density." That is, if the density of a wire b centimeters long is $f(x)$ grams per centimeter at a distance of x centimeters from one end then the mass of the wire is $\int_0^b f(x)\ dx$ grams. Develop this fact in the informal style of this section.

2 In Sec. 5.3 we showed that if $f(t)$ is the velocity at time t of an object moving along the x axis, then $\int_a^b f(t)\ dt$ is the change in position during the time interval $[a, b]$. Develop this fact in the informal style of this section. Keep in mind that $f(t)$ may be positive or negative.

3 The depth of rain at a distance r feet from the center of a storm is $g(r)$ feet.

(*a*) Estimate the total volume of rain that falls between a distance r feet and a distance $r + dr$ feet from the center of the storm. (Assume that dr is a small positive number.)

(*b*) Using (*a*), set up a definite integral for the total volume of rain that falls between 1,000 and 2,000 feet from the center of the storm.

4 The following analysis of primitive agriculture is taken from *Is There an Optimum Level of Population?*, edited by S. Fred Singer, McGraw-Hill, New York, 1971:

Consider a circular range of radius a with the home base of production at the center. Let $G(r)$ denote the density of foodstuffs (in calories per square meter) at radius r meters from the home base. Then the total number of calories produced in the range is given by the definite integral ________.

Using the informal approach, set up the definite integral that appeared in the blank.

5 In Example 2 the area of the ring with inner radius x and outer radius $x + dx$ was informally estimated to be approximately $2\pi x\, dx$.

(*a*) Using the formula for the area of a circle, show that the area of the ring is $2\pi x\, dx + \pi(dx)^2$.

(*b*) Show that the ring has the same area as a trapezoid of height dx and bases of lengths $2\pi x$ and $2\pi(x + dx)$.

6 Think of a circular disk of radius a as being composed of concentric circular rings, as in Fig. 16.

A circular disk composed of rings

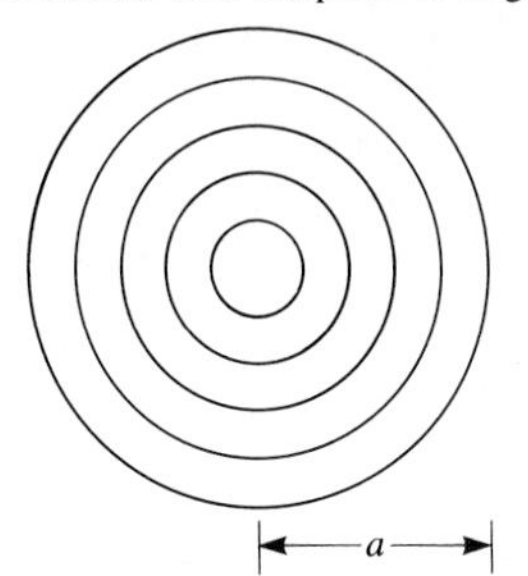

Figure 16

(*a*) Using the shorthand approach, set up a definite integral for the area of the disk. (Draw a good picture of the local approximation.)

(*b*) Evaluate the integral in (*a*).

Exercises 7 to 9 concern the volumes of the given solids. In each case (*a*) draw a good picture of the local approximation of width dx, (*b*) set up the appropriate definite integral, and (*c*) evaluate the integral.

7 A right circular cone of radius a and height h.

8 A pyramid with a square base of side a and of height h. Its top vertex is above one corner of the base. (Use square cross sections.)

9 A pyramid with a triangular base of area A and of height h. (The triangle need not be equilateral. See Fig. 17.)

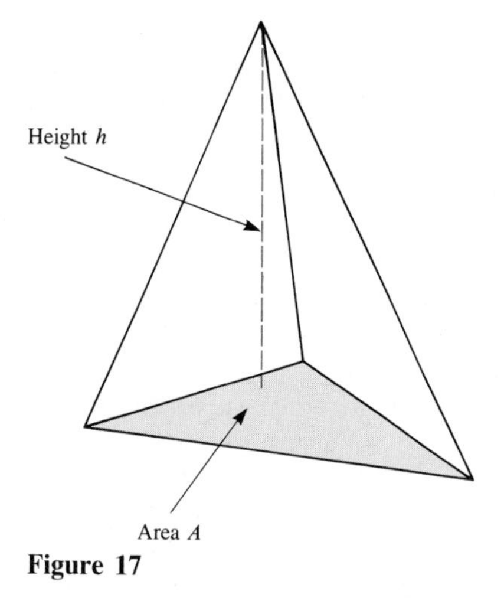

Figure 17

10 (*Surface area of a sphere*) Find the surface area of a sphere of radius a. *Suggestion*: Begin by estimating the area of the narrow band shown in Fig. 18.

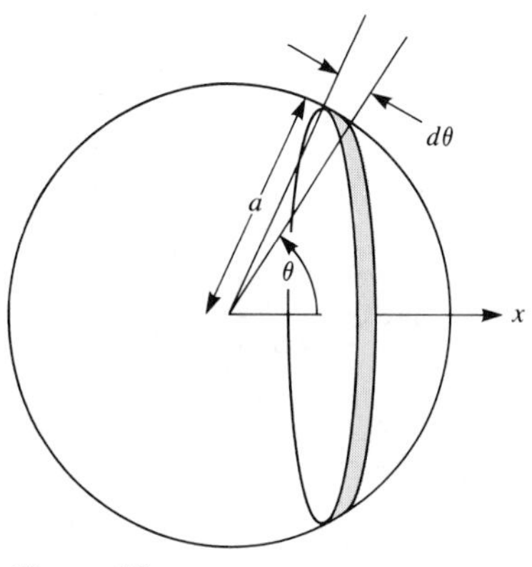

Figure 18

11 (*Actuarial tables*) Let $F(t)$ be the fraction of people born in 1900 who are alive t years later. [The number t need not be an integer. For instance, $F(14.5)$ is the fraction who reach the age of 14 years 6 months.]

(*a*) What is $F(150)$, probably?

(*b*) What is $F(0)$?

(*c*) Sketch the general shape of the graph of $y = F(t)$.

(*d*) Let $f(t) = F'(t)$. (Assume F is differentiable.) Is $f(t)$ positive or negative?

(*e*) What fraction of the people born in 1900 die during the time interval $[t, t + dt]$? (Express your answer in terms of F.)

(*f*) Answer (*e*), but express your answer in terms of f.

(*g*) Evaluate $\int_0^{150} f(t)\, dt$.

(*h*) What integral would you propose to call "the average life span of the people born in 1900"? Why?

12 Let $F(t)$ be the fraction of ball bearings that wear out during the first t hours of use. Thus $F(0) = 0$ and $F(t) \leq 1$.
(*a*) As t increases, what would you think happens to $F(t)$?
(*b*) Show that during the short interval of time $[t, t + dt]$, the fraction of ball bearings that wear out is approximately $F'(t)\,dt$. (Assume F is differentiable.)
(*c*) Assume all wear out in at most 1,000 hours. What is $F(1{,}000)$?
(*d*) Using the assumptions in (*b*) and (*c*), devise a definite integral for the average life of the ball bearings.

13 At time t hours, $0 \leq t \leq 24$, a firm uses electricity at the rate of $e(t)$ joules per hour. The rate schedule indicates that the cost per joule at time t is $c(t)$ dollars. Assume that both e and c are continuous functions.
(*a*) Estimate the cost of electricity consumed between times t and $t + dt$, where dt is a small positive number.
(*b*) Using (*a*), set up a definite integral for the total cost of electricity for the 24-hour period.

14 (*Present value*) The **present value** of a promise to pay one dollar t years from now is $g(t)$ dollars.
(*a*) What is $g(0)$?
(*b*) Why is it reasonable to assume that $g(t) \leq 1$ and that g is a decreasing function of t?
(*c*) What is the present value of a promise to pay q dollars t years from now?
(*d*) Assume that an investment made now will result in an income flow at the rate of $f(t)$ dollars per year t years from now. (Assume that f is a continuous function.) Estimate informally the present value of the income to be earned between time t and time $t + dt$, where dt is a small positive number.
(*e*) On the basis of the local estimate made in (*d*), set up a definite integral for the present value of all the income to be earned from now to time b years in the future.

15 (*Population*) Let the number of females in a certain population in the age range from x years to $x + dx$ years, where dx is a small positive number, be approximately $f(x)\,dx$. Assume that, on average, women of age x produce $m(x)$ offspring during the year before they reach age $x + 1$. Assume that both f and m are continuous functions.
(*a*) What definite integral represents the number of women between ages a and b years?
(*b*) What definite integral represents the total number of offspring during the calendar year produced by women whose ages at the beginning of the calendar year were between a and b years?

Exercises 16 to 21 concern kinetic energy. They are all based on the concept that a particle of mass M moving with velocity V has the kinetic energy $MV^2/2$. (See Example 3.) An object whose density is the same at all its points is called **homogeneous**. If the object is planar, such as a square or disk, and has mass M kilograms and area A square meters, its density is M/A kilograms per square meter.

16 The piece of sheet metal in Example 3 is rotated around the line midway between the edges AB and CD at the rate of 5 revolutions per second.
(*a*) Using the informal approach, obtain a local approximation for the kinetic energy of a narrow strip of the metal.
(*b*) Using (*a*), set up a definite integral for the kinetic energy of the piece of sheet metal.
(*c*) Evaluate the integral in (*b*).

17 A circular piece of metal of radius 7 meters has a density of 3 kilograms per square meter. It rotates 5 times per second around an axis perpendicular to the circle and passing through the center of the circle.
(*a*) Devise a local approximation for the kinetic energy of a narrow ring in the circle.
(*b*) With the aid of (*a*), set up a definite integral for the kinetic energy of the rotating metal.
(*c*) Evaluate the integral in (*b*).

18 The density of a rod x centimeters from its left end is $g(x)$ grams per centimeter. The rod has a length of b centimeters. The rod is spun around its left end 7 times per second.
(*a*) Estimate the mass of the rod in the section that is between x and $x + dx$ centimeters from the left end. (Assume that dx is small.)
(*b*) Estimate the kinetic energy of the mass in (*a*).
(*c*) Set up a definite integral for the kinetic energy of the rotating rod.

19 A homogeneous square of mass M kilograms and side a meters rotates around an edge 5 times per second.
(*a*) Obtain a "local estimate" of the kinetic energy. What part of the square would you use? Why? Draw it.
(*b*) What is the local estimate?
(*c*) What definite integral represents the total kinetic energy of the square?
(*d*) Evaluate it.

20 Like Exercise 19, but this time the square is spun around a line through its center and parallel to an edge.

21 Like Exercise 19 for a disk of radius a and mass M spinning around a line through its center and perpendicular to it. It is spinning at the rate of ω radians per second. (See Fig. 19.)

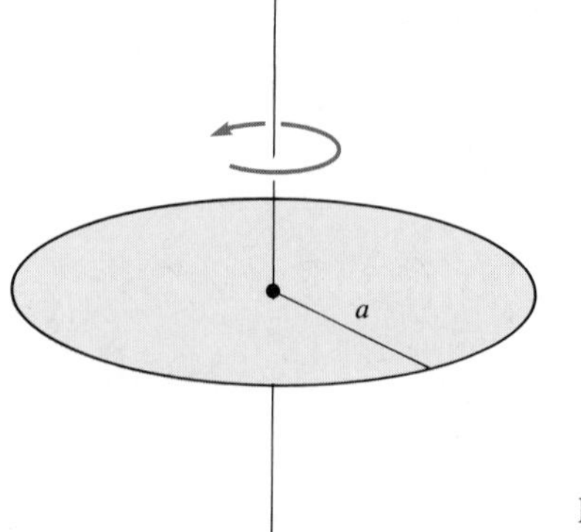

Figure 19

In Exercises 22 and 23 you will meet definite integrals that cannot be evaluated by the fundamental theorem of calculus (since the desired antiderivative is not elementary). Use (*a*) the trapezoidal and (*b*) Simpson's method with six sections to estimate the definite integrals.

22 A homogeneous object of mass M occupies the region under $y = e^{x^2}$ and above $[0, 1]$. It is spun at the rate of ω radians per second around the y axis. Estimate its kinetic energy.

23 A homogeneous object of mass M occupies the region under $y = (\sin x)/x$ and above $[\pi/2, \pi]$. It is spun around the line $x = 1$ at the rate of ω radians per second. Estimate its kinetic energy.

The solutions of Exercises 24 to 27 require techniques of Chap. 7 or else the use of integral tables. In each case find the kinetic energy of a planar homogeneous object that occupies the given region, has mass M, and is spun around the y axis ω radians per second.

24 The region under $y = e^x$ and above the interval $[1, 2]$.

25 The region under $y = \tan^{-1} x$ and above the interval $[0, 1]$.

26 The region under $y = 1/(1 + x)$ and above $[2, 4]$.

27 The region under $y = \sqrt{1 + x^2}$ and above $[0, 2]$.

28 A solid homogeneous right circular cylinder of radius a, height h, and mass M is spun at the rate of ω radians per second around its axis. Find its kinetic energy. (Include a good picture on which your local approximation is based.)

29 A solid homogeneous ball of radius a and mass M is spun at the rate of ω radians per second around a diameter. Find its kinetic energy. (Include a good picture on which your local approximation is based.)

30 (*Beware*) Consider the following argument: "Approximate the surface area of the sphere of radius a shown in Fig. 20 as follows. To approximate the surface area between x and $x + dx$, let us try using the area of the narrow curved part of the cylinder used to approximate the volume between x and $x + dx$. (This part is shaded in Fig. 20.) This local approximation can be pictured (when unrolled and laid flat) as a rectangle of width dx and length $2\pi r$. The surface area of a sphere is $\int_{-a}^{a} 2\pi r\, dx = 4\pi \int_0^a \sqrt{a^2 - x^2}\, dx$. But $\int_0^a \sqrt{a^2 - x^2}\, dx = \pi a^2/4$, since it equals the area of a quadrant of a disk. Hence the area of the sphere is $\pi^2 a^2$." This does not agree with the correct value, $4\pi a^2$, which was discovered by Archimedes in the third century B.C. What is wrong with the argument?

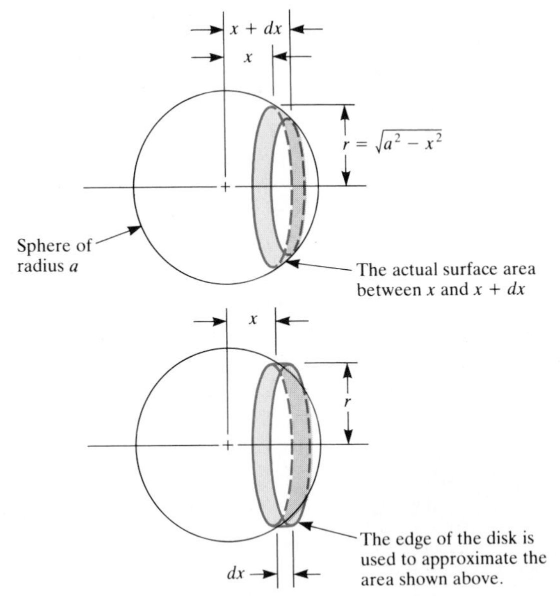

Figure 20

31 (*Poiseuille's law of blood flow*) A fluid flowing through a pipe does not all move at the same velocity. The velocity of any part of the fluid depends on its distance from the center of the pipe. The fluid at the center of the pipe moves fastest, whereas the fluid near the wall of the pipe moves slowest. Assume that the velocity of the fluid at a distance x centimeters from the axis of the pipe is $g(x)$ centimeters per second.
(*a*) Estimate the flow of fluid (in cubic centimeters per second) through a thin ring of inner radius r and outer radius $r + dr$ centimeters centered at the axis of the pipe and perpendicular to the axis.
(*b*) Using (*a*), set up a definite integral for the flow (in cubic centimeters per second) of fluid through the pipe. (Let the radius of the pipe be b centimeters.)
(*c*) Poiseuille (1797–1869), studying the flow of blood through arteries, used the function $g(r) = k(b^2 - r^2)$, where k is a constant. Show that in this case the flow of blood through an artery is proportional to the fourth power of the radius of the artery.

32 The density of the earth at a distance of r miles from its center is $g(r)$ pounds per cubic mile. Set up a definite integral for the total mass of the earth. (Take the radius of the earth to be 4,000 miles.)

8.4 COMPUTING VOLUMES

In Chap. 5 we found that "volume is the integral of cross-sectional area," $V = \int_a^b A(x)\, dx$. Some volumes were computed in Chaps. 5, 7, and earlier in this chapter. This section offers further practice in setting up a definite integral for the volume of a solid.

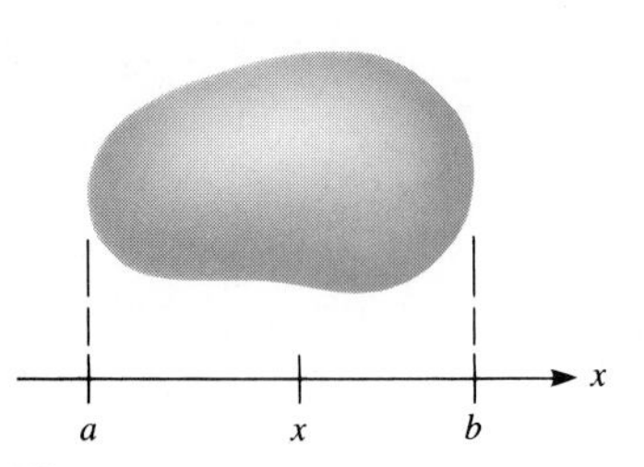

Figure 1

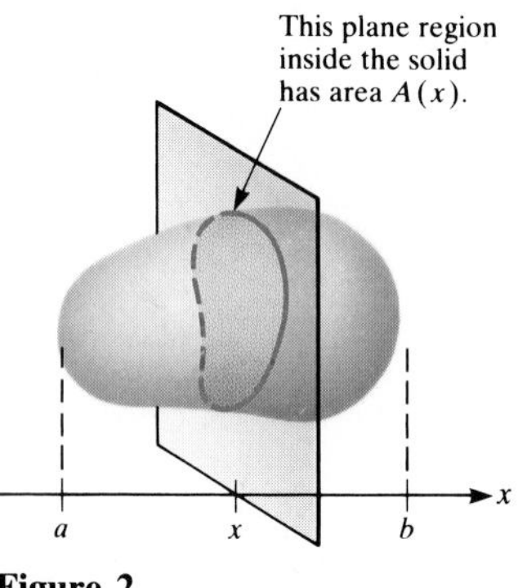

Figure 2

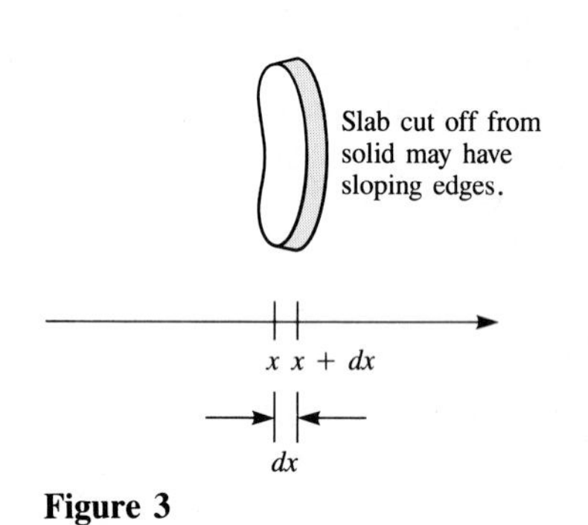

Figure 3

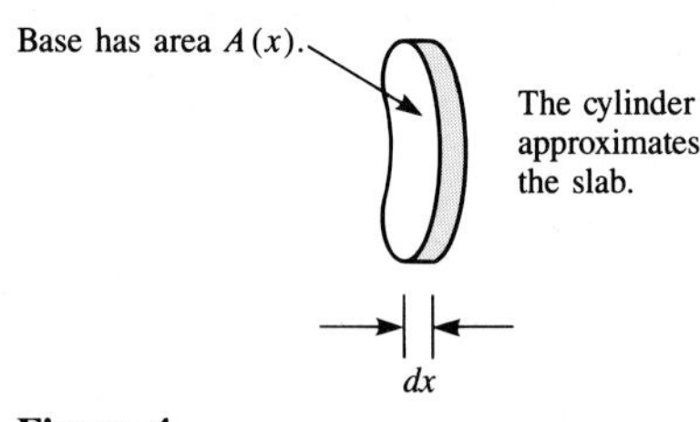

Figure 4

Figure 1 shows a typical solid whose volume we might wish to express as a definite integral. Say that we could find a formula for $A(x)$, the area of the general cross section shown in Fig. 2.

We could then make a local approximation of the volume of the solid slab contained between parallel planes corresponding to x and $x + dx$ on the axis shown in Fig. 3. The slab is not a cylinder, since its shaded rim is not perpendicular to the cross section of area $A(x)$, but it resembles the cylinder of base $A(x)$ and thickness dx shown in Fig. 4.

The volume of the cylinder is precisely the area of its base times its thickness, $A(x)\,dx$. Taking this as the local approximation, we can then get the total volume of the solid by integration:

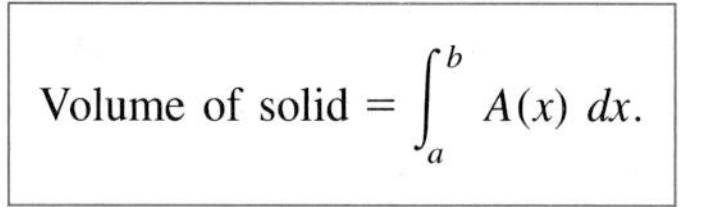

$$\text{Volume of solid} = \int_a^b A(x)\,dx.$$

Given a particular solid, one just has to find a, b, and the cross-sectional area $A(x)$ in order to construct a definite integral for the volume of the solid. These are the steps for finding the volume:

1 Choose a line to serve as an x axis.
2 For each plane perpendicular to that axis, find the area of the cross section of the solid made by the plane. Call this area $A(x)$.
3 Determine the limits of integration, a and b, for the region.
4 Evaluate the definite integral $\int_a^b A(x)\,dx$.

Most of the effort is usually spent in finding the integrand $A(x)$.

In addition to the pythagorean theorem and the properties of similar triangles, formulas for the areas of familiar plane figures may be needed. (See inside back cover.) Also keep in mind that if corresponding dimensions of similar figures have the ratio k, then their areas have the ratio k^2; that is, the area is proportional to the square of the ratios of corresponding line segments.

EXAMPLE 1 Find the volume of a solid triangular pyramid whose base is a right triangle of sides 3, 4, and 5. The altitude of the pyramid is above the vertex of the right angle and has length 2. (See Fig. 5.)

SOLUTION There are three convenient directions in which to define cross sections by planes, namely, parallel to each of the three right-triangular faces. Choose, say, planes parallel to the base, as shown in Fig. 6.

Introduce an x axis perpendicular to the base triangle and with origin in the plane of that triangle. The typical cross section is a triangle T. Let b be the

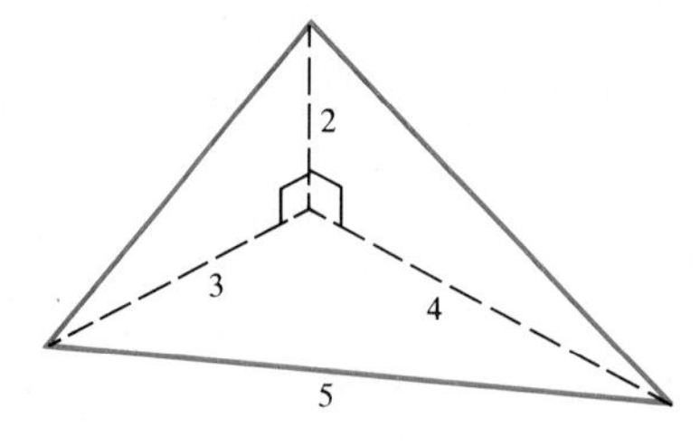

Figure 5

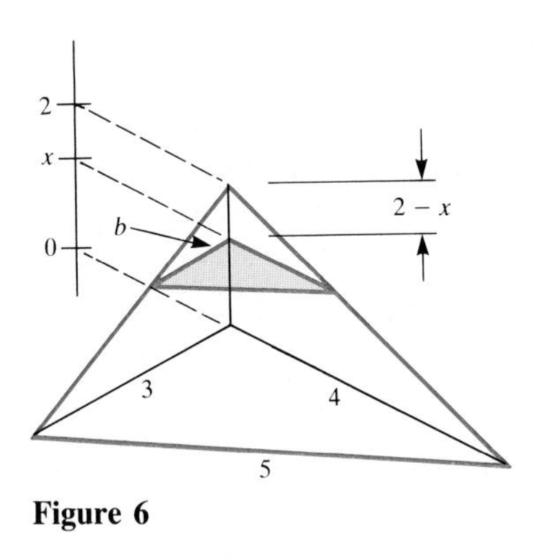

Figure 6

length of the side of T parallel to the side of length 3 in the base. We find the area of T by noting that T is similar to the base triangle, which has area $\frac{1}{2}\cdot 3\cdot 4 = 6$. Recall that the areas of similar figures have the ratio k^2, where k is the ratio of their corresponding sides. By similar triangles, the corresponding sides of lengths b and 3 have the ratio

$$\frac{b}{3} = \frac{2-x}{2},$$

so

$$\frac{\text{Area of } T}{\text{Area of base}} = \left(\frac{2-x}{2}\right)^2$$

or

$$\frac{A(x)}{6} = \left(\frac{2-x}{2}\right)^2.$$

Solving this last equation for $A(x)$ gives

$$A(x) = \tfrac{3}{2}(2-x)^2.$$

Thus the volume of the pyramid is

$$\int_0^2 \tfrac{3}{2}(2-x)^2\,dx = -\frac{3}{2}\frac{(2-x)^3}{3}\Big|_0^2$$

$$= \left[-\frac{3}{2}\frac{(2-2)^3}{3}\right] - \left[-\frac{3}{2}\frac{(2-0)^3}{3}\right]$$

$$= 0 + 4 = 4.$$

The volume of the pyramid is 4 cubic units. ■

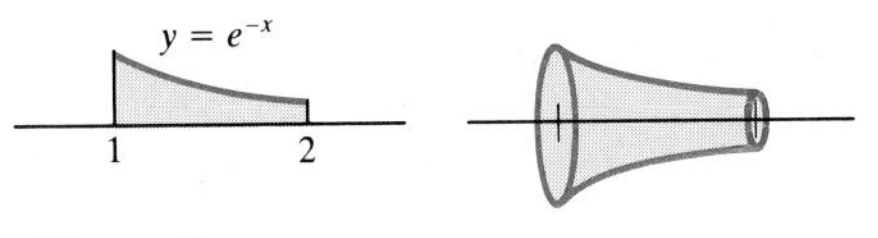

The solid formed by revolving a region in a plane about a line in that plane is called a **solid of revolution**. In the next example we use the slab technique to find the volume of such a solid.

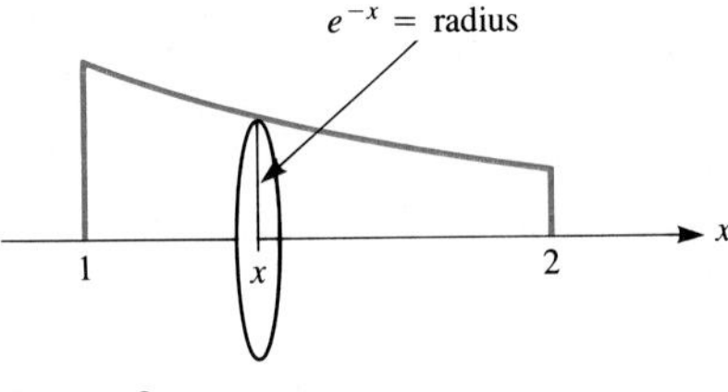

EXAMPLE 2 The region under $y = e^{-x}$ and above $[1, 2]$ is revolved about the x axis. Find the volume of the resulting solid of revolution. (See Fig. 7.)

SOLUTION The typical cross section by a plane perpendicular to the x axis is a disk of radius e^{-x}, as shown in Fig. 8. The local approximation of the volume is shown in Fig. 9. This approximation equals

$$\pi(e^{-x})^2\,dx$$

or

$$\pi e^{-2x}\,dx.$$

The volume of the solid is therefore

$$\int_1^2 \pi e^{-2x}\,dx.$$

Recall that $\int e^{ax}\,dx = (1/a)e^{ax}$. Hence

$$\int_1^2 \pi e^{-2x}\,dx = \frac{\pi e^{-2x}}{-2}\Big|_1^2$$

$$= -\frac{\pi}{2}(e^{-4} - e^{-2})$$

$$= \frac{\pi}{2}(e^{-2} - e^{-4}). \quad ■$$

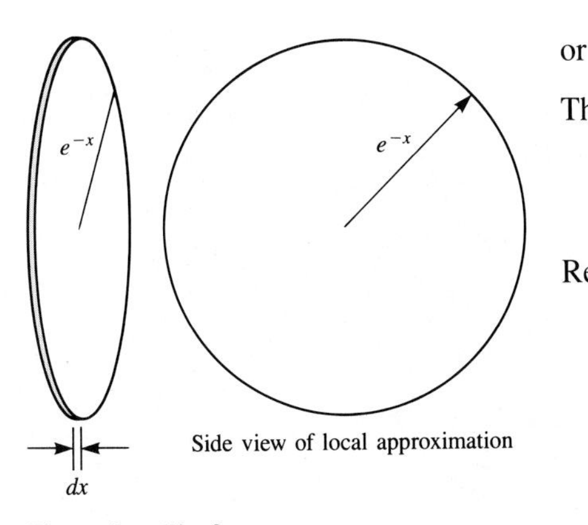

Blowup from Fig. 8.

Figure 9

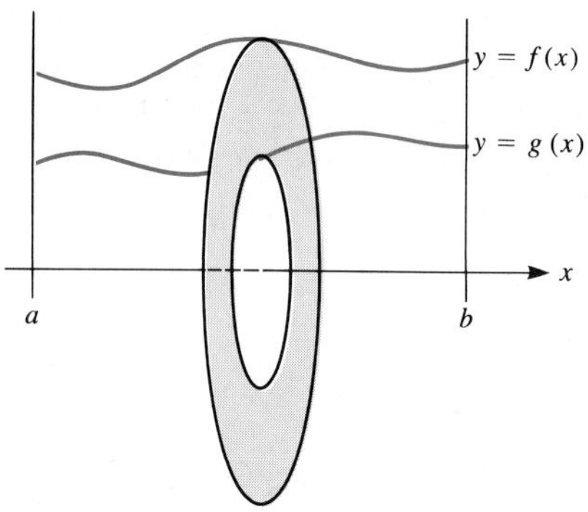

Rings as Cross Sections: The Washer Technique

If the region being revolved around the x axis is bounded by two curves $y = f(x)$ and $y = g(x)$, $f(x) \geq g(x) \geq 0$, then we have to take into account the hole in the resulting solid of revolution. The cross sections perpendicular to the x axis are no longer disks but "washers." Figure 10 shows that the cross section has area

$$\pi(f(x))^2 - \pi(g(x))^2,$$

the difference of the areas of the disks. Hence the volume is

$$\int_a^b \pi[(f(x))^2 - (g(x))^2]\, dx.$$

Figure 10

It is not necessary to memorize this formula. It's just a special case of parallel cross sections.

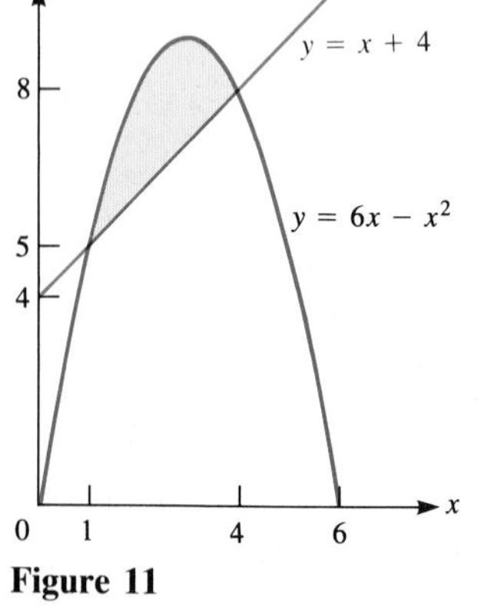

Figure 11

EXAMPLE 3 The region shown in Fig. 11 is revolved about the x axis to form a solid of revolution. Express the volume of the solid as a definite integral.

SOLUTION We first draw a local approximation to the volume. (See Fig. 12.) The area of the typical cross section is

$$\pi(6x - x^2)^2 - \pi(x + 4)^2.$$

Hence the local approximation of volume is

$$[\pi(6x - x^2)^2 - \pi(x + 4)^2]\, dx.$$

We have the integrand. Next we find the interval of integration $[a, b]$. The ends of the interval are determined by where the curves cross, that is, when

$$x + 4 = 6x - x^2$$

or

$$x^2 - 5x + 4 = 0.$$

Hence

$$(x - 1)(x - 4) = 0,$$

$$x = 1 \qquad \text{or} \qquad x = 4,$$

so the volume of the solid is

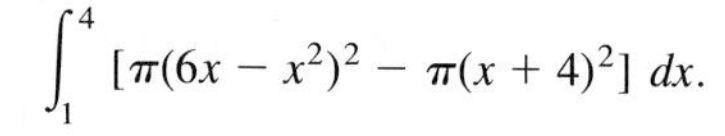

$$\int_1^4 [\pi(6x - x^2)^2 - \pi(x + 4)^2]\, dx.$$

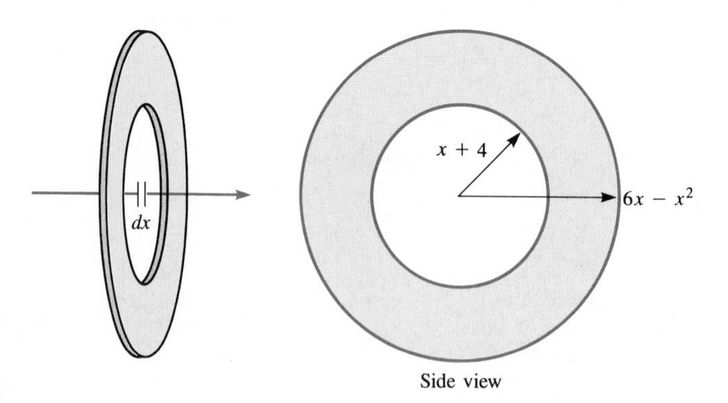

Figure 12 ■

Section Summary

This section reviewed the calculation of volumes by using parallel cross sections. When calculating a volume, first draw a clear picture to help find the local approximation.

EXERCISES FOR SEC. 8.4: COMPUTING VOLUMES

In each of Exercises 1 to 7, (*a*) draw the solid, (*b*) draw the typical cross section, (*c*) draw the local approximation, (*d*) find the volume of the local approximation, (*e*) set up the definite integral, and (*f*) evaluate the integral.

1 Find the volume of a cone of radius a and height h.

2 The base of a solid is the region bounded by $y = x^2$, the line $x = 1$ and the x and y axes. Each cross section perpendicular to the x axis is a square. (See Fig. 13.) Find the volume of the solid.

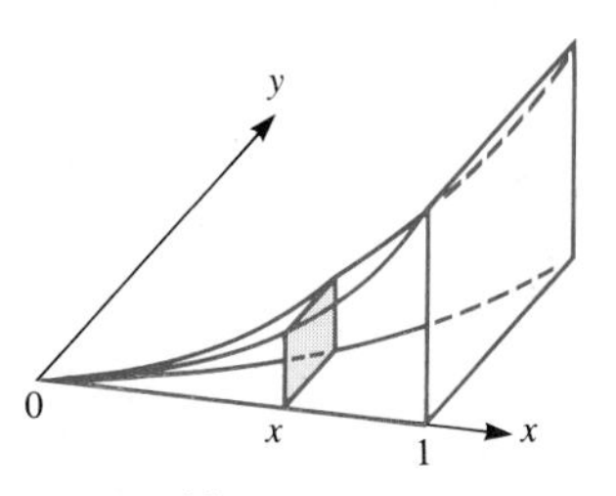

Figure 13

3 The base of a solid is a disk of radius 3. Each plane perpendicular to a given diameter meets the solid in a square, one side of which is in the base of the solid. (See Fig. 14.) Find its volume.

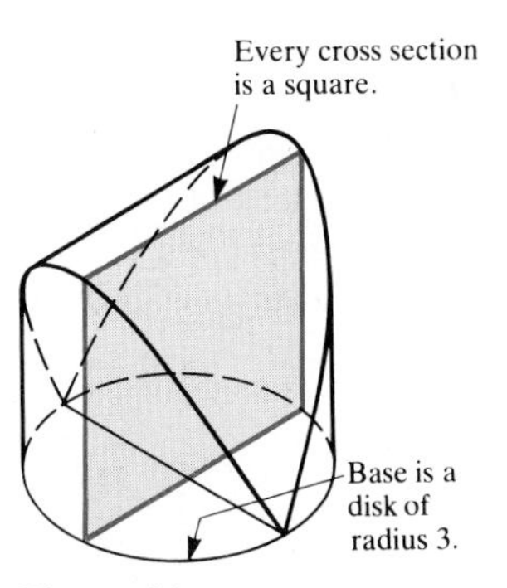

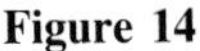

Figure 14

4 Repeat Exercise 2 except that the cross sections are equilateral triangles.

5 Find the volume of a pyramid with a square base of side a and height h, using square cross sections. The top of the pyramid is above the center of the base.

6 Repeat Exercise 5, but using trapezoidal cross sections.

7 Find the volume of the solid whose base is the disk of radius 5 and whose cross sections perpendicular to the x axis are equilateral triangles. (See Fig. 15.)

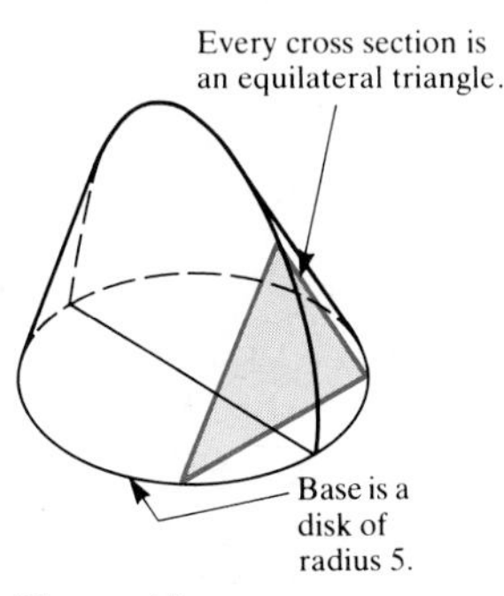

Figure 15

8 (*a*) Find the volume of the pyramid shown in Fig. 16 by using cross sections perpendicular to the edge of length c.

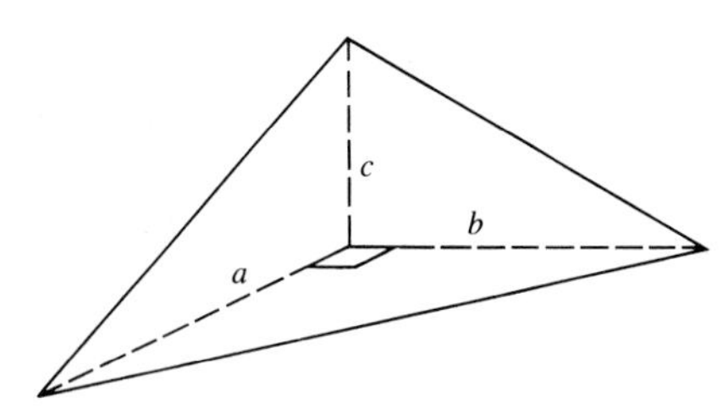

Figure 16

(*b*) Find the volume of the pyramid shown in Fig. 16 by using cross sections perpendicular to the edge of length b.

In Exercises 9 to 14 find the volume of the solid formed by revolving the given region R about the given axis.

9 R is bounded by $y = \sqrt{x}$, $x = 1$, $x = 2$, and the x axis, about the x axis.

10 R is bounded by $y = 1/\sqrt{1 + x^2}$, $x = 0$, $x = 1$, and the x axis, about the x axis.

11 R is bounded by $y = 1/\sqrt{x}$, $y = 1/x$, $x = 1$, and $x = 2$, about the x axis.

12 R is bounded by $y = x^2$ and $y = x^3$, about the y axis.

In Exercises 13 and 14 use an integral table or techniques from Chap. 7.

13 R is bounded by $y = \tan x$, $y = \sin x$, $x = 0$, and $x = \pi/4$, about the x axis.

14 R is bounded by $y = \sec x$, $y = \cos x$, $x = \pi/6$, and $x = \pi/3$, about the x axis.

15 A cylindrical drinking glass of height h and radius a, full of water, is tilted until the water just covers the base. How much water is left? Use rectangular cross sections. Refer to Fig. 17 and follow the directions preceding Exercise 1.

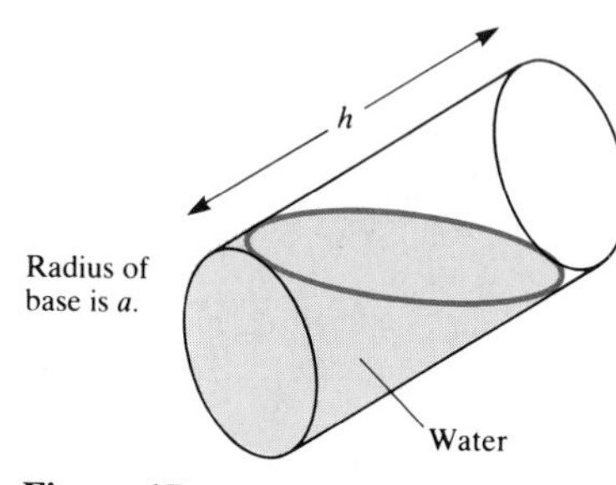

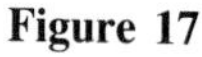
Figure 17

16 Repeat Exercise 15, but use trapezoidal cross sections.

17 Solve Exercise 15 by common sense. Don't use any calculus at all.

18 A cylindrical drinking glass of height h and radius a, full of water, is tilted until the water remaining just covers half the base. Find the volume of water, using triangular cross sections.

19 Repeat Exercise 18, using rectangular cross sections.

20 A solid is formed in the following manner. A plane region R and a point P not in that plane are given. The solid consists of all line segments joining P to points in R. If R has area A and P is a distance h from the plane of R, show that the volume of the solid is $Ah/3$. (See Fig. 18.)

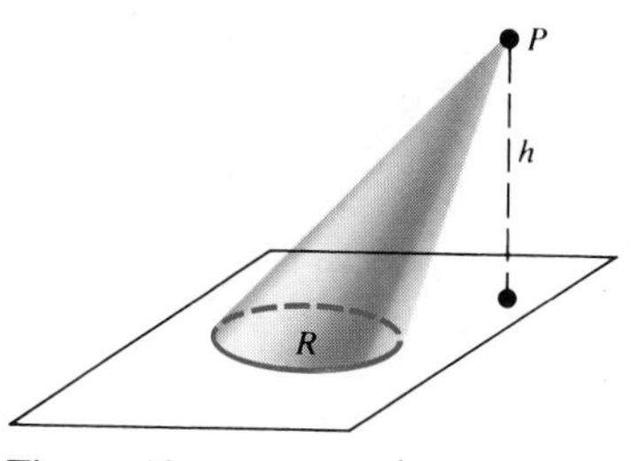

Figure 18

21 A drill of radius 3 inches bores a hole through a sphere of radius 5 inches, passing symmetrically through the center of the sphere.

(*a*) Draw the part of the sphere removed by the drill.

(*b*) Find $A(x)$, the area of a cross section of the region in (*a*) made by a plane perpendicular to the axis of the drill and at a distance x from the center of the sphere.

(*c*) Find the volume removed.

22 What fraction of the volume of a sphere is contained between parallel planes that trisect the diameter to which they are perpendicular?

23 The disk bounded by the circle $(x - b)^2 + y^2 = a^2$, where $0 < a < b$, is revolved around the y axis. Find the volume of the doughnut (torus) produced.

24 (Contributed by Archimedes.) Find the volume of one octant of the region common to two right circular cylinders of radius 1 whose axes intersect at right angles, as shown in Fig. 19.

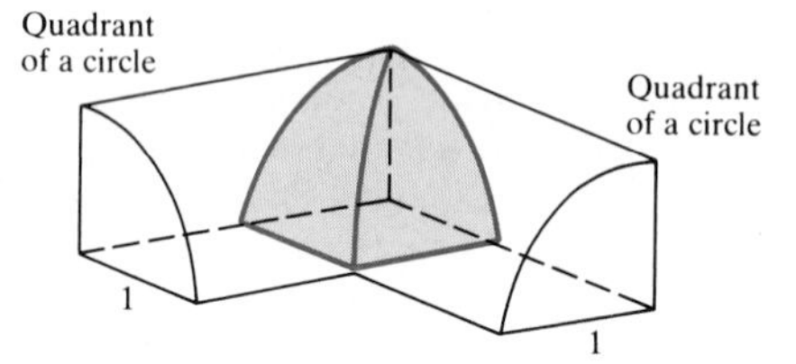

Figure 19

25 When a region R of area A situated to the right of the y axis is revolved around the y axis, the resulting solid of revolution has volume V. When R is revolved around the line $x = -k$, the volume of the resulting solid is V^*. Express V^* in terms of k, A, and V.

8.5 THE SHELL TECHNIQUE

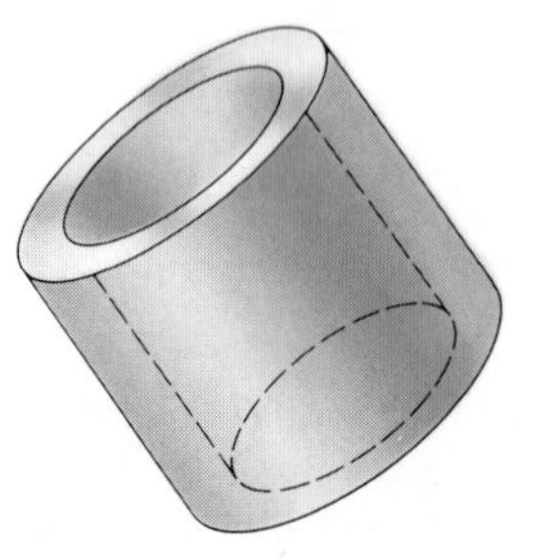

Figure 1

As we have seen, the volume of any solid can be computed by approximating by slabs cut off by parallel planes. However, for a solid of revolution, there is also a second way of viewing it—as built up of concentric pipes or tubes of the type shown in Fig. 1. This approach leads to integrals which in some cases are much simpler than those you get in the "slab" approach.

The Shell Technique

Let R be a region in the plane and L a line in the plane that does not meet R or meets R only on its border, as in Fig. 2.

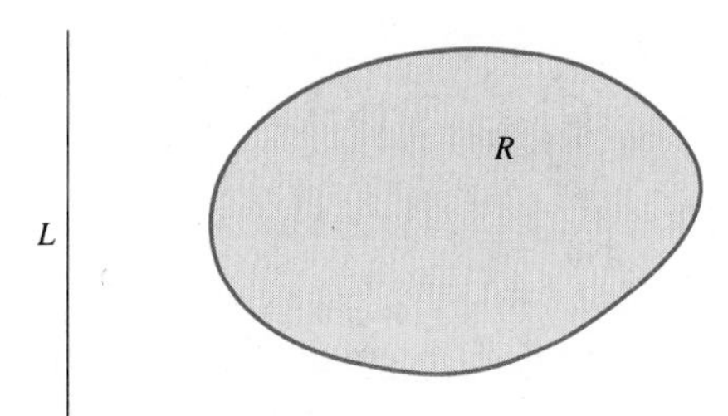

Figure 2

Cut R into narrow pieces by lines parallel to L, as in Fig. 3, and approximate each piece by a rectangle. Then the solid of revolution S formed by revolving R about L is approximated by the collection of solids of revolution formed by revolving the rectangles about L. (See Fig. 4.) Each rectangle sweeps out a cylindrical shell or tube. In Fig. 5 these shells are pulled out so that each one is visible. (The sections of some hand-held collapsible telescopes are made of tubes like these shown in Fig. 5.)

To set up a definite integral for the volume of S we make a local approximation by estimating the volume of a typical tube with a thin wall. (Think of a tin can without its top and bottom.)

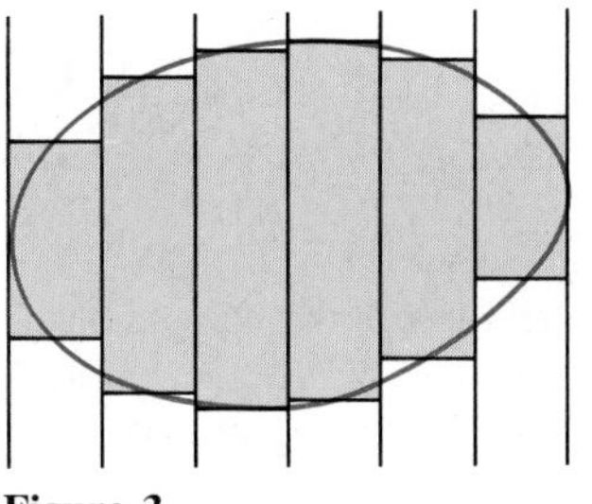

Figure 3

The Local Approximation

Introduce an x axis in the plane of R and perpendicular to L. Assume that L lies to the left of R and cuts the x axis at $x = k$ and that R lies above the interval $[a, b]$ as in Fig. 6.

We estimate the volume of the solid of revolution formed by revolving about L the part of R between those lines parallel to L which meet the x axis at x and $x + dx$. (Assume dx is a small positive number. See Fig. 7.)

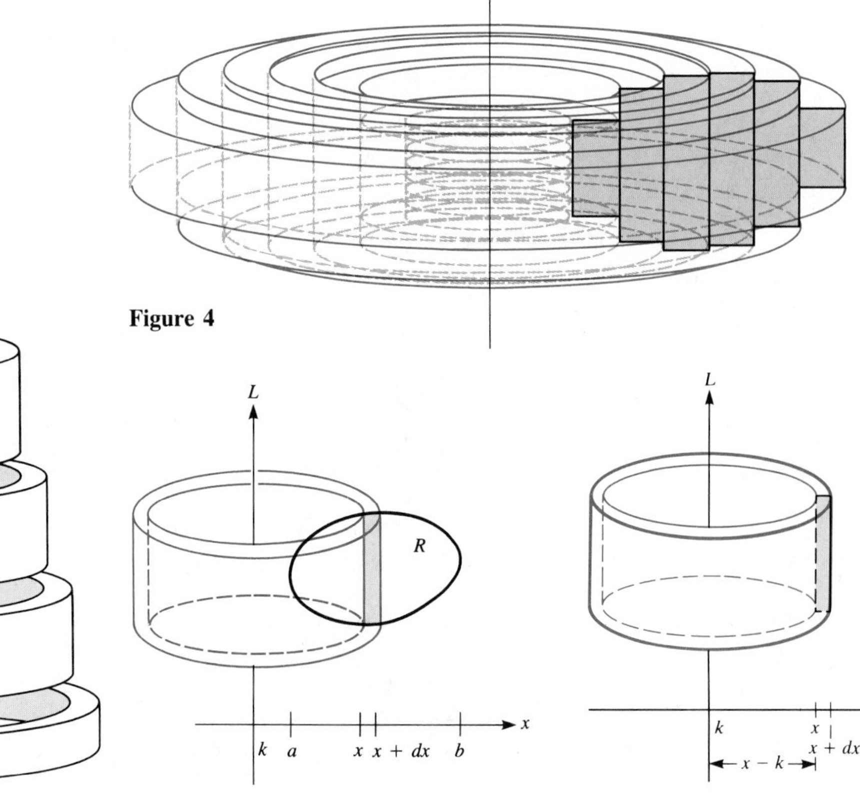

Figure 4

Figure 6

Figure 7

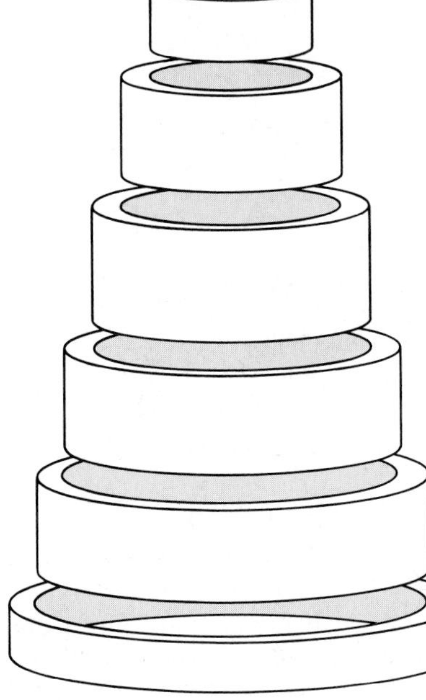

Figure 5

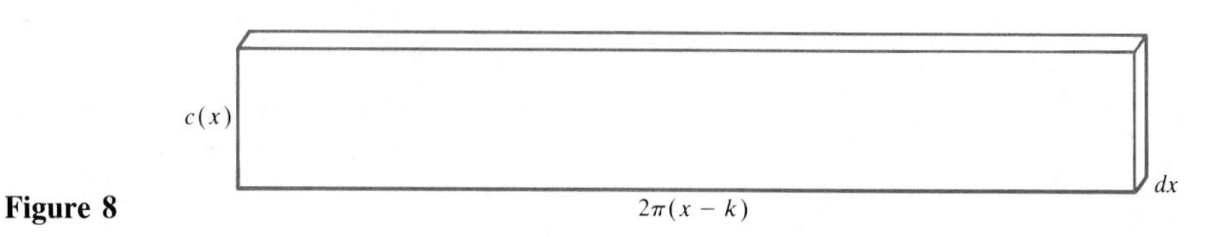

Figure 8

To estimate the volume of the shell or tube in Fig. 7, begin by letting $c(x)$ be the length of the cross section of R made by a line parallel to L and meeting the coordinate axis at x. The radius of the shell is $x - k$. Imagine cutting the shell along a direction parallel to L and then laying it flat as though it were a carpet. When laid flat, the shell resembles a thin slab of thickness dx, width $c(x)$, and length $2\pi(x - k)$, as shown in Fig. 8. The volume of the shell, therefore, is presumably about

$$2\pi(x - k)c(x)\,dx. \tag{1}$$

With the aid of this local estimate (1), we then conclude that the volume of the solid of revolution is

$$\text{Volume} = \int_a^b 2\pi(x - k)c(x)\,dx. \tag{2}$$

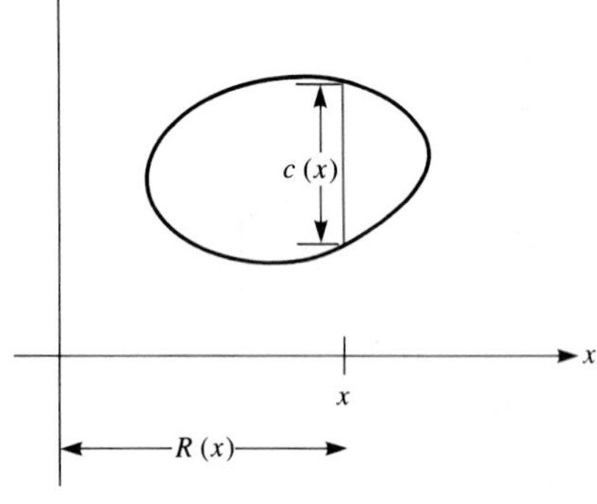

Figure 9

This is the formula for computing volumes by the shell technique. If $x - k$ is denoted $R(x)$, the "radius of the shell" as in Fig. 9, then

$$\text{Volume} = \int_a^b 2\pi R(x)c(x)\,dx.$$

Rather than memorizing these formulas, draw a typical thin shell and picture it rolled out like a carpet.

Examples

The first example, which would be very difficult by the "parallel slab" approach, is quite easy by the "concentric shell" approach.

EXAMPLE 1 The region under $y = 1 + x + x^5$ and above the interval $[0, 1]$ is revolved about the y axis. Find the volume of the resulting solid.

SOLUTION The region and solid are shown in Fig. 10. The typical local approximation is shown in Fig. 11, where the typical shell is presented as if it were rolled flat.

By inspection we see that x is the radius of the shell and $1 + x + x^5$ is the height of the shell. Hence the local approximation to the volume is

$$\underbrace{2\pi x}_{\text{width}}\ \underbrace{(1 + x + x^5)}_{\text{height}}\ \underbrace{dx.}_{\text{thickness}}$$

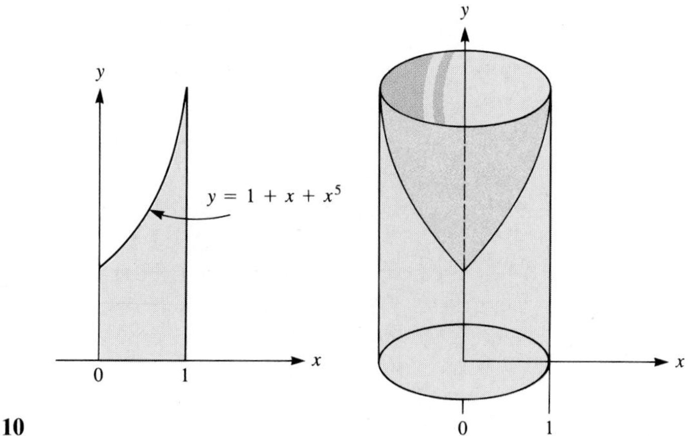

Figure 10

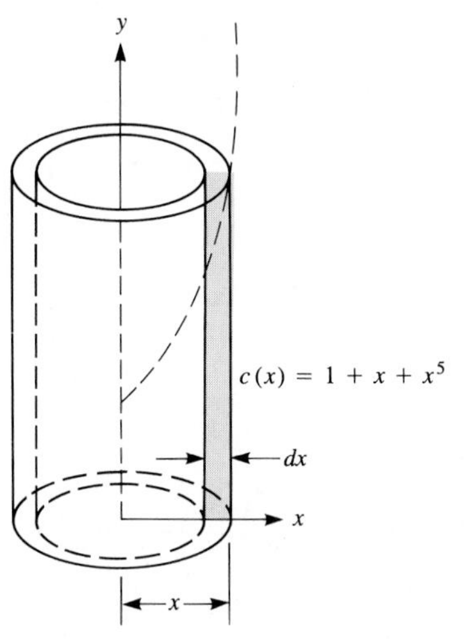

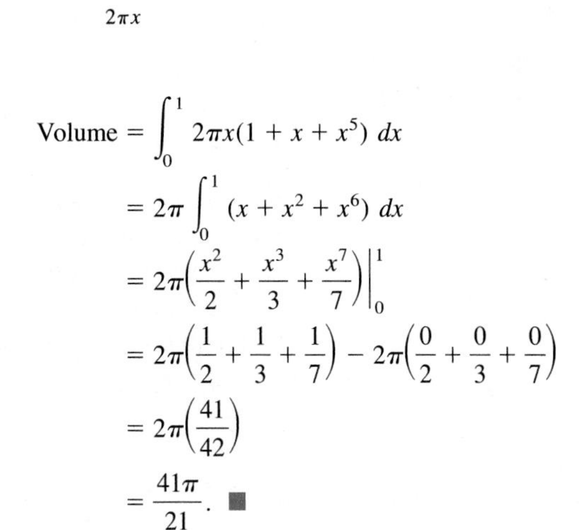

Figure 11

Hence

$$\begin{aligned}
\text{Volume} &= \int_0^1 2\pi x(1 + x + x^5)\, dx \\
&= 2\pi \int_0^1 (x + x^2 + x^6)\, dx \\
&= 2\pi\left(\frac{x^2}{2} + \frac{x^3}{3} + \frac{x^7}{7}\right)\Bigg|_0^1 \\
&= 2\pi\left(\frac{1}{2} + \frac{1}{3} + \frac{1}{7}\right) - 2\pi\left(\frac{0}{2} + \frac{0}{3} + \frac{0}{7}\right) \\
&= 2\pi\left(\frac{41}{42}\right) \\
&= \frac{41\pi}{21}. \quad \blacksquare
\end{aligned}$$

Remark: To solve Example 1 by the slab method you would take slabs perpendicular to the y axis. Each slab would be a disk. To find the radius of the disk you would want to solve for x as a function of y. Try it. (See Exercise 22.)

EXAMPLE 2 Set up an integral for the volume of a ball of radius a by the shell technique.

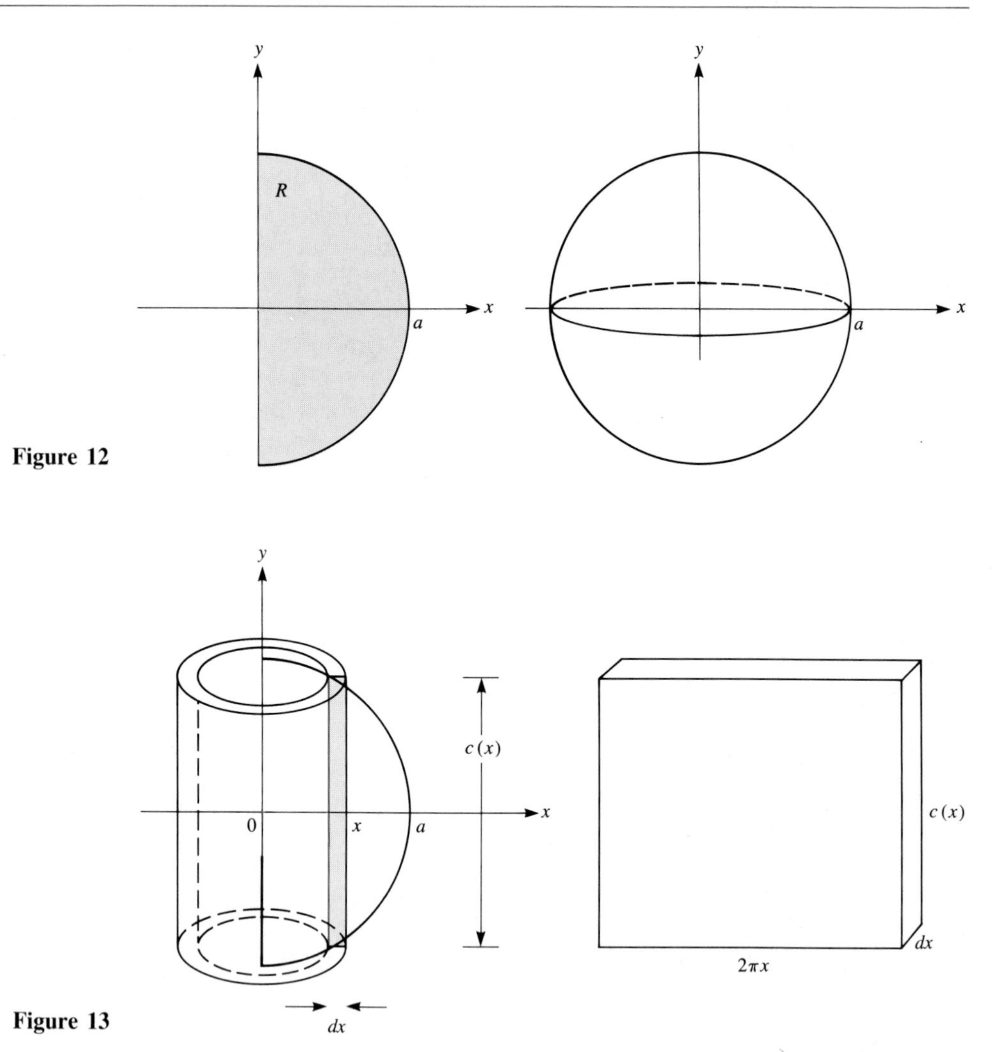

Figure 12

Figure 13

SOLUTION A ball of radius a can be thought of as the solid obtained by revolving a semicircle R of radius a about its diameter. (See Fig. 12.) The typical shell is shown in Fig. 13.

As the radius of the shell we may use x. The height of the shell can be found by the pythagorean theorem and Fig. 14. We have $c(x) = 2\sqrt{a^2 - x^2}$. Thus

$$\text{Volume} = \int_0^a 2\pi x(2\sqrt{a^2 - x^2})\, dx$$

$$= 4\pi \int_0^a x\sqrt{a^2 - x^2}\, dx.$$

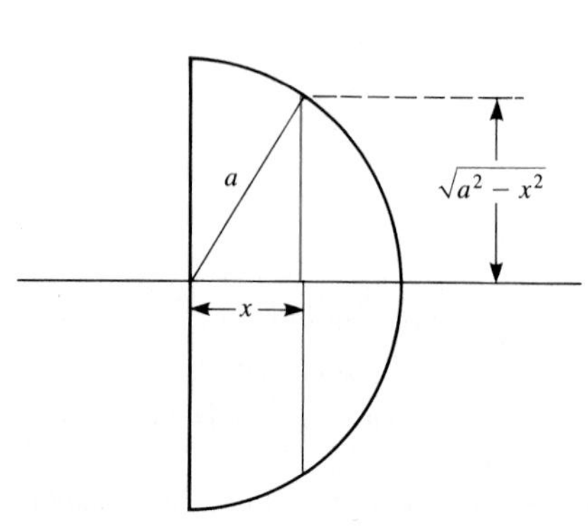

Figure 14

This integral can be evaluated by an integral table or by using the substitution $u = a^2 - x^2$ described in Sec. 7.2. (The volume is $4\pi a^3/3$.) ■

Section Summary

The volume of a solid of revolution may be found by approximating the solid by concentric thin shells. The volume of such a shell is approximately

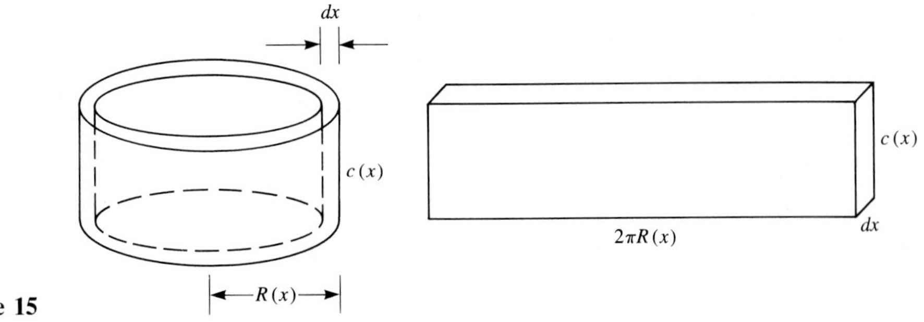

Figure 15

$2\pi R(x)\, c(x)\, dx$. (See Fig. 15.) The shell technique is often useful when "slabs" or "washers" are difficult to set up.

EXERCISES FOR SEC. 8.5: THE SHELL TECHNIQUE

In Exercises 1 to 4 draw the typical approximating cylindrical shell for the solid described, and determine the volume of this shell.

1 The trapezoid bounded by $y = x$, $x = 1$, $x = 2$, and the x axis is revolved about the y axis.

2 The trapezoid in Exercise 1 is revolved about the line $x = -3$.

3 The triangle with vertices (0, 0), (1, 0), (0, 2) is revolved about the y axis.

4 The triangle in Exercise 3 is revolved about the x axis.

5 Find the volume of the solid produced by revolving about the y axis the finite region bounded by $y = x^2$ and $y = x^3$.

6 Repeat Exercise 5, except the region is revolved about the line $x = -2$.

7 Find the volume of the solid produced by revolving about the x axis the finite region bounded by $y = \sqrt{x}$ and $y = \sqrt[3]{x}$.

8 Repeat Exercise 7, except the region is revolved about the y axis.

9 Find the volume of a right circular cone of radius a and height h by the shell method.

10 (Integral tables or Chap. 7 needed.) Find the volume of the doughnut (ring, torus) produced by revolving the disk of radius a about a line L at a distance $b > a$ from its center. (See Fig. 16.)

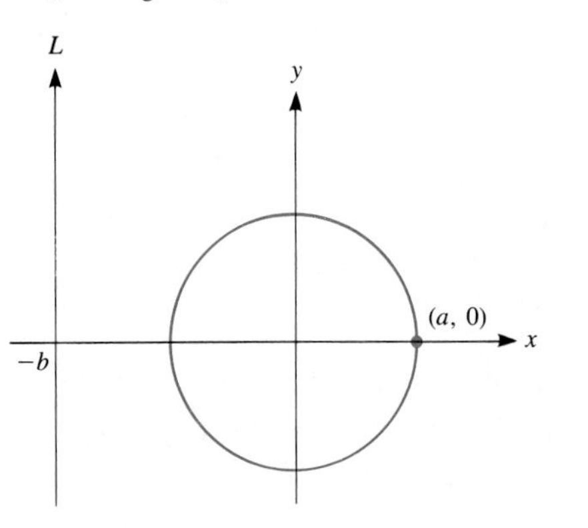

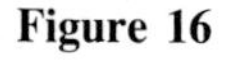
Figure 16

11 Let R be the region bounded by $y = x + x^3$, $x = 1$, $x = 2$, and the x axis. Find the volume of the solid produced by revolving R about (*a*) the y axis, (*b*) the line $x = 3$.

12 Find the volume of the solid produced by revolving the region R in Exercise 11 about (*a*) the x axis, (*b*) the line $y = -2$.

Exercises 13 to 20 may require integral tables or techniques in Chap. 7.

13 Find the volume of the solid of revolution formed by revolving the region bounded by $y = 2 + \cos x$, $x = \pi$, $x = 10\pi$, and the x axis around (*a*) the y axis, (*b*) the x axis.

14 The region below $y = \cos x$, above the x axis, and between $x = 0$ and $x = \pi/2$ is revolved around the x axis. Find the volume of the resulting solid of revolution by (*a*) parallel cross sections, (*b*) concentric shells.

15 Let R be the region below $y = 1/(1 + x^2)^2$ and above [0, 1]. Find the volume of the solid produced by revolving R about the y axis.

16 The region between $y = e^{x^2}$, the x axis, $x = 0$, and $x = 1$ is revolved about the y axis. (It is interesting to note that the fundamental theorem of calculus is of no use in evaluating the area of this region.) Find the volume of the solid produced.

17 The region R below $y = e^x(1 + \sin x)/x$ and above $[\pi, 10\pi]$ is revolved about the y axis to produce a solid of revolution. (*a*) Try to find the volume of this solid by parallel cross sections and by the shell technique. (*b*) Which is easier? (*c*) Why?

18 Let R be the region below $y = \ln x$ and above [1, *e*]. Find the volume of the solid produced by revolving R about (*a*) the x axis, (*b*) the y axis.

19 Let R be the region below $y = 1/(x^2 + 4x + 1)$ and above [0, 1]. Find the volume of the solid produced by revolving R about the line $x = -2$.

20 Let R be the region below $y = 1/\sqrt{2 + x^2}$ and above

$[\sqrt{3}, \sqrt{8}]$. Find the volume of the solid produced by revolving R about (*a*) the x axis, (*b*) the y axis.

---

21 When a region R in the first quadrant is revolved about the y axis, a solid of volume 24 is produced. When R is revolved about the line $x = -3$, a solid of volume 82 is produced. What is the area of R?

In Exercises 22 to 25 use an integral table or techniques from Chap. 7.

22 Solve Example 1 by the slab approach. View the solid as a cylinder from which a small part has been deleted. First find the volume of the deleted part. *Hint:* Don't try to solve for x as a function of y. Instead change the variable of integration from y to x.

23 Let R be a region in the first quadrant. When it is revolved around the x axis, a solid of revolution is produced. When it is revolved around the y axis, another solid of revolution is produced. Give an example of such a region R with the property that the volume of the first solid *cannot* be evaluated by the fundamental theorem of calculus, but the volume of the second solid can be.

24 Let a and b be positive numbers and $y = f(x)$ be a decreasing differentiable function of x such that $f(0) = b$ and $f(a) = 0$. Prove that $\int_0^a 2xy\,dx = \int_0^b x^2\,dy$, (*a*) by considering the volume of a certain solid, (*b*) by integration by parts.

25 Let f in Exercise 24 be elementary. (*a*) Show that, if x^2f' has an elementary integral, so does xf, and conversely. (*b*) Consider the solid obtained by rotating the region bounded by the curve $y = f(x)$, the x axis, and the y axis around the y axis. Show that its volume expressed by the shell technique involves an elementary integral only when its volume by the slab technique involves an elementary integral.

8.6 THE CENTROID OF A PLANE REGION

This section introduces the centroid of a plane region and shows various ways of computing it.

The Center of Mass of n Point Masses

We disregard the weight of the seesaw.

A small boy on one side of a seesaw (which we regard as weightless) can balance a bigger boy on the other side. For example, the two boys in Fig. 1 balance. (According to physical laws, each boy exerts a force on the seesaw, due to gravitational attraction, proportional to his mass.)

The small mass with the long lever arm balances the large mass with the small lever arm. Each boy contributes the same tendency to turn—but in opposite directions.

This tendency is called the **moment**:

$$\text{Moment} = (\text{Mass}) \cdot (\text{Lever arm}),$$

where the lever arm can be positive or negative. To be more precise, introduce on the seesaw an x axis with its origin 0 at the fulcrum, the point on which the seesaw rests. Define the moment about 0 of a mass m located at the point x on the x axis to be the product mx. Then the bigger boy has a moment (90)(4), while the smaller boy has a moment (40)(−9). The total moment of the lever-mass system is 0, and the masses balance. (See Fig. 2.)

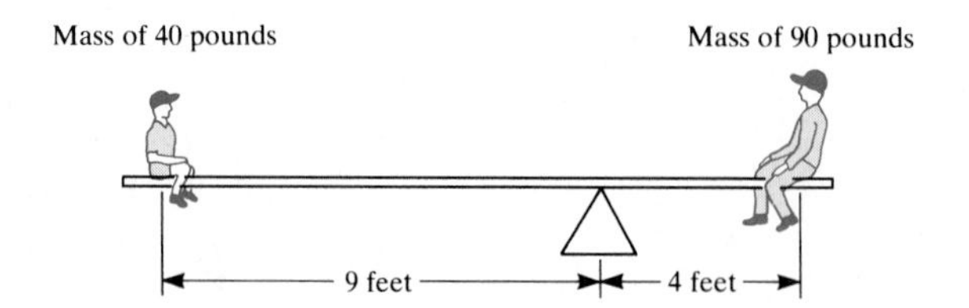

Figure 1

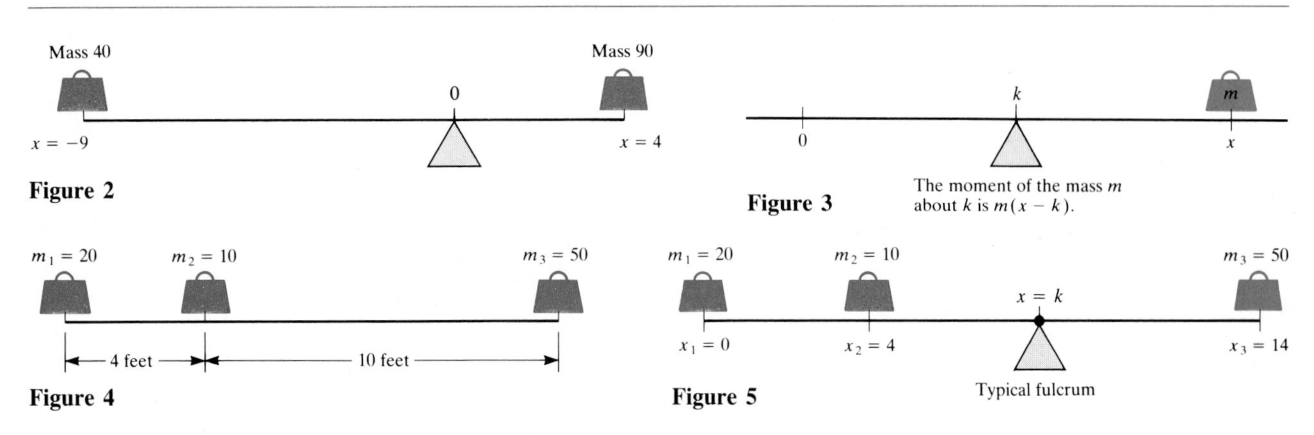

Figure 2

Figure 3

Figure 4

Figure 5

If a mass m is located on a line at coordinate x, define its moment about the point having coordinate k as the product $m(x - k)$. (See Fig. 3.)

Now consider several point masses $m_1, m_2, \ldots, m_n$. If mass m_i is located at x_i, with $i = 1, 2, \ldots, n$, then $\Sigma_{i=1}^{n} m_i (x_i - k)$ is the total moment of all the masses about the point k. If a fulcrum is placed at k, then the seesaw rotates clockwise if the total moment is greater than 0, rotates counterclockwise if it is less than 0, and is in equilibrium if the total moment is 0.

EXAMPLE 1 Where should the fulcrum be placed so that the three masses in Fig. 4 will be in equilibrium?

SOLUTION Introduce an x axis with origin at mass m_1 and compute the moments about a typical fulcrum having coordinate k; then select k to make the total moment 0. (See Fig. 5.)

Let the total moment about k be M. Then

$$M = 20(0 - k) + 10(4 - k) + 50(14 - k).$$

We seek k such that M is equal to 0, or equivalently,

$$0 - 20k + 40 - 10k + 700 - 50k = 0$$

or

$$80k = 740.$$

Hence

$$k = \frac{740}{80} = 9.25.$$

The fulcrum is to the right of the midpoint, which was to be expected. ■

As illustrated by the example, the balancing point of masses $m_1, m_2, \ldots, m_n$, located respectively at $x_1, x_2, \ldots, x_n$ on an x axis, is found by solving the equation

$$\sum_{i=1}^{n} m_i(x_i - k) = 0 \tag{1}$$

for the number k. Expanding Eq. (1), we obtain

$$\sum_{i=1}^{n} m_i x_i - \sum_{i=1}^{n} m_i k = 0$$

or
$$\sum_{i=1}^{n} m_i x_i = \sum_{i=1}^{n} m_i k.$$

Thus
$$k \sum_{i=1}^{n} m_i = \sum_{i=1}^{n} m_i x_i,$$

and finally,
$$\boxed{k = \frac{\sum_{i=1}^{n} m_i x_i}{\sum_{i=1}^{n} m_i}.} \tag{2}$$

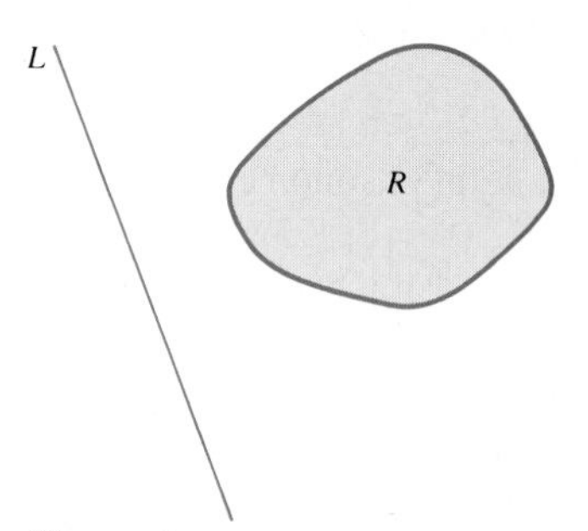

Figure 6

The number k given by Eq. (2) is called the **center of mass** of the system of masses. It is the point about which all the masses balance. *The center of mass is found by dividing the total moment about* 0 *by the total mass.* It is usually denoted $\bar{x}$.

Finding the center of mass of a finite number of "point masses" involves only arithmetic, no calculus. Now let us turn our attention to finding the center of mass of a continuous distribution of matter in the plane. For this purpose, definite integrals will be needed.

The Moment and Centroid of a Plane Region

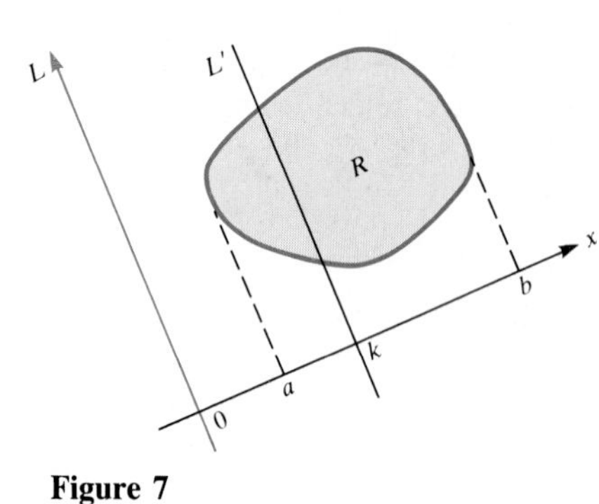

Figure 7

Figure 8

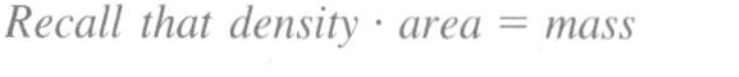
Recall that density · area = mass

Let R be a region in the plane. Imagine that R is occupied by a thin piece of metal that has a density of σ (sigma) grams per square centimeter. Throughout this section σ will be assumed to be constant (that is, the metal is "homogeneous"). Let L be a line in the plane. (See Fig. 6.) Is there a line parallel to L on which R balances?

Let L' be *any* line parallel to L. We will compute the moment about L' and then see how to choose L' to make that moment equal to 0. To compute the moment of R about L', introduce an x axis perpendicular to L with its origin at its intersection with L. Assume that L' passes through the x axis at the point $x = k$, as in Fig. 7. In addition, assume that each line parallel to L meets R either in a line segment or at a point on the boundary of R. The lever arm of the mass distributed throughout R varies from point to point. However, the length of the lever arm is almost constant for the mass located between two lines parallel to L and close to each other.

Consider the moment about L' of the mass in R located between the lines parallel to L and passing through the points on the x axis with coordinates x and $x + dx$, where dx is a small positive number. (See Fig. 8.)

Let $c(x)$ be the length of the cross section of R at x. Then the area of that portion of R between the lines passing through x and $x + dx$ is approximately $c(x)\,dx$. The mass of that portion is consequently about $\sigma c(x)\,dx$. The lever arm around L' of this mass is about $x - k$, which may be positive or negative. The local approximation of the moment of the mass in R about L' is therefore

$$\underbrace{(x-k)}_{\text{lever arm}}\ \underbrace{\sigma\ \underbrace{c(x)\,dx}_{\text{area}}}_{\text{mass}}.$$

Following the informal approach, we conclude that

The moment of a plane homogeneous mass about a line

$$\text{Moment of the mass in } R \text{ about } L' = \int_a^b (x-k)\sigma c(x)\,dx. \tag{3}$$

We will use Eq. (3) as the formal definition of the moment of a plane distribution of matter about a line.

The moment around L' may or may not be 0. Let us determine k, hence the line L', such that the moment given by formula (3) is 0. To do this, solve the equation

$$0 = \int_a^b (x-k)\sigma c(x)\,dx$$

for k.

We then have

$$0 = \int_a^b x\sigma c(x)\,dx - \int_a^b k\sigma c(x)\,dx$$

k and σ are constants.

$$0 = \sigma\int_a^b xc(x)\,dx - k\sigma\int_a^b c(x)\,dx$$

$$k\sigma\int_a^b c(x)\,dx = \sigma\int_a^b xc(x)\,dx$$

σ cancels.

$$k\int_a^b c(x)\,dx = \int_a^b xc(x)\,dx$$

Formula for balancing line $x = k$ (σ constant)

$$k = \frac{\int_a^b xc(x)\,dx}{\int_a^b c(x)\,dx}. \tag{4}$$

The numerator is called **the moment of R about L.** The density σ does not appear in this moment, since it is assumed to be constant. The denominator $\int_a^b c(x)\,dx$ is the area of R. Formula (4) shows that there is a unique balancing line $x = k$ parallel to L. Its coordinate is given by

$$k = \frac{\text{Moment of } R \text{ about } L}{\text{Area of } R}.$$

Assume now that the plane is furnished with an xy coordinate system. There is a unique balancing line parallel to the y axis. Its x coordinate equals

$$\bar{x} = \frac{\int_a^b xc(x)\,dx}{\text{Area of } R}. \tag{5}$$

Similarly, there is a unique balancing line parallel to the x axis. Its y coordinate is given by

$$\bar{y} = \frac{\int_c^d yc(y)\,dy}{\text{Area of } R}. \tag{6}$$

where $c(y)$ is the cross-section function of R for lines parallel to the x axis and $[c, d]$ is the interval of integration. (See Fig. 9.)

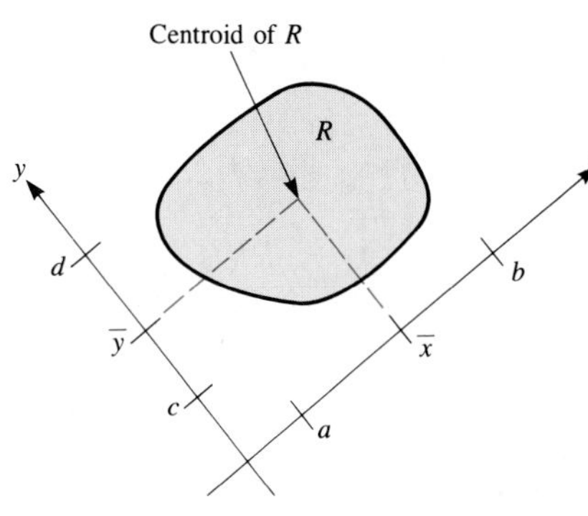

Figure 9

Definition *Centroid of R.* The **centroid of** R is defined as the point $(\bar{x}, \bar{y})$, where

$$\bar{x} = \frac{\text{Moment of } R \text{ about } y \text{ axis}}{\text{Area of } R} \quad \text{and} \quad \bar{y} = \frac{\text{Moment of } R \text{ about } x \text{ axis}}{\text{Area of } R},$$

Formula for the centroid

or

$$\bar{x} = \frac{\int_a^b xc(x)\,dx}{\text{Area of } R} \quad \text{and} \quad \bar{y} = \frac{\int_c^d yc(y)\,dy}{\text{Area of } R}. \tag{7}$$

It can be shown that the region R *balances on any line through its centroid.* Moreover, if R is suspended motionless from a string attached at its centroid, it remains in equilibrium.

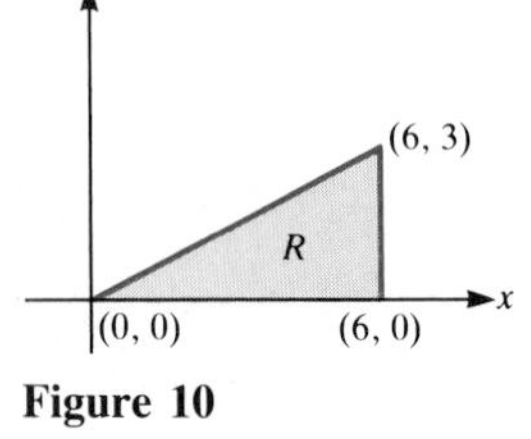

Figure 10

EXAMPLE 2 Let R be the triangle in the xy plane with vertices at (0, 0), (6, 0), and (6, 3), as shown in Fig. 10. Find (*a*) $\bar{x}$, and (*b*) $\bar{y}$.

SOLUTION

(*a*) To find $\bar{x}$, we first compute the moment of R around the y axis,

$$\int_0^6 xc(x)\,dx,$$

where $c(x)$ is the cross-section function for R. To find $c(x)$, use the equation of the line through (0, 0) and (6, 3), namely, $y = x/2$. Thus $c(x) = x/2$. Consequently, the moment of R around the y axis is

$$\int_0^6 x \cdot \frac{x}{2}\,dx = \frac{1}{2}\int_0^6 x^2\,dx = \frac{1}{2}\frac{x^3}{3}\bigg|_0^6 = \frac{1}{2}\frac{216}{3} = 36.$$

To use formula (7), we divide 36 by the area of the triangle, which is $\frac{1}{2} \cdot 6 \cdot 3 = 9$. Thus

$$\bar{x} = \tfrac{36}{9} = 4.$$

The balancing line parallel to the y axis has the equation $x = 4$.

(*b*) To find $\bar{y}$, first compute the moment $\int_0^3 yc(y)\,dy$. Since the line through (0, 0) and (6, 3) has the equation $y = x/2$, on that line $x = 2y$. Inspection of Fig. 10 shows that $c(y) = 6 - 2y$. Thus

$$\int_0^3 yc(y)\,dy = \int_0^3 y(6-2y)\,dy = \int_0^3 (6y - 2y^2)\,dy$$

$$= \left(3y^2 - \frac{2y^3}{3}\right)\bigg|_0^3 = (27 - 18) - 0 = 9.$$

Hence

$$\bar{y} = \frac{9}{\text{Area of } R} = \frac{9}{9} = 1.$$

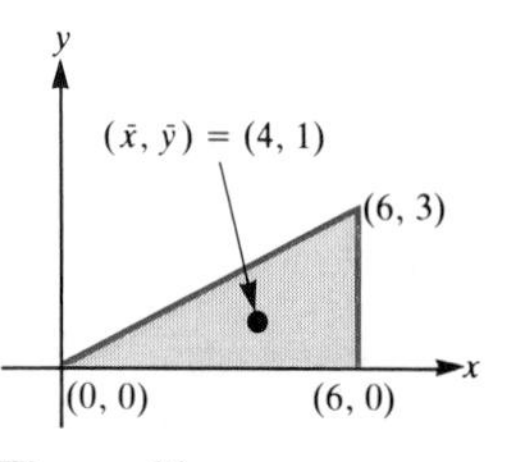

Figure 11

Parts (*a*) and (*b*) tell us that the centroid is at (4, 1),

$$(\bar{x}, \bar{y}) = (4, 1),$$

as shown in Fig. 11. ■

A Shortcut for $\bar{y}$

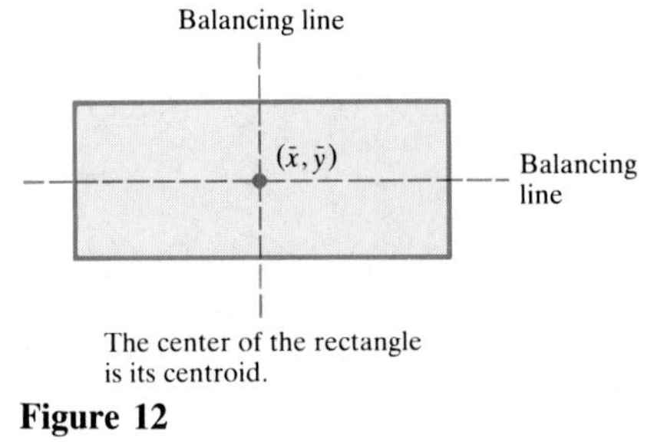

The center of the rectangle is its centroid.

Figure 12

Using formula (7), we can show that the centroid of a rectangle is its center. This fact is plausible, since, by symmetry, a rectangle would balance on a line through its center and parallel to an edge, as shown in Fig. 12.

Consider a rectangle resting on the x axis, as in Fig. 13. Let $\bar{y}$ be the y coordinate of its center of mass. By the definition of $\bar{y}$,

$$\bar{y} = \frac{\text{Moment of rectangle about } x \text{ axis}}{\text{Area of rectangle}}$$

or

$$\begin{pmatrix}\text{Moment of rectangle}\\ \text{about } x \text{ axis}\end{pmatrix} = \begin{pmatrix} y \text{ coordinate of}\\ \text{centroid of rectangle}\end{pmatrix} \cdot \begin{pmatrix}\text{Area of}\\ \text{rectangle}\end{pmatrix}. \tag{8}$$

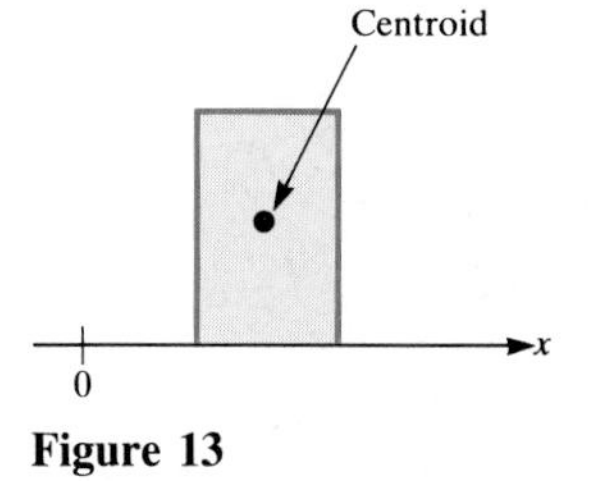

Figure 13

Formula (8) is the basis of a shortcut for computing $\bar{y}$ of a region under a curve and above the x axis. We now develop this shortcut.

Let $y = f(x)$ be a continuous function such that $f(x) \geq 0$ for x in $[a, b]$. Let R be the region below the curve $y = f(x)$ and above $[a, b]$. The moment around the x axis of the portion of R between two lines parallel to the y axis, one with x coordinate x and one with x coordinate $x + dx$, where dx is a small positive number, can be estimated easily. (See Fig. 14, where this narrow band is shaded.) The narrow band has area approximately $f(x)\ dx$, and the y coordinate of its centroid is approximately $f(x)/2$. By formula (8), its moment around the x axis is approximately $\frac{1}{2}f(x) \cdot f(x)\ dx$. Hence

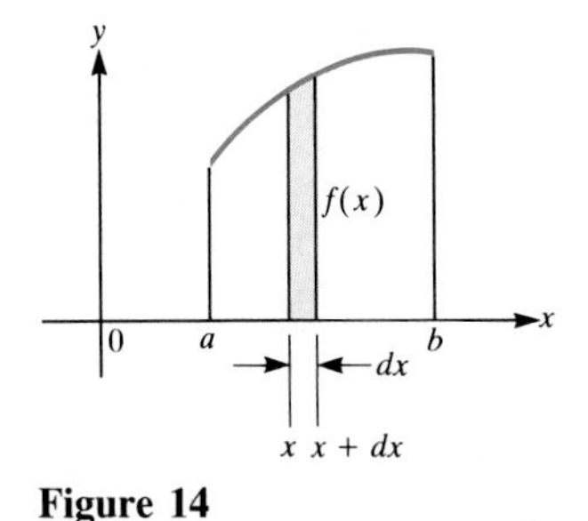

Figure 14

$$\text{Moment of } R \text{ about } x \text{ axis} = \int_a^b \tfrac{1}{2}[f(x)]^2\ dx. \tag{9}$$

Consequently, $\bar{y}$, the y coordinate of the centroid of R, equals

$$\boxed{\bar{y} = \frac{\displaystyle\int_a^b \frac{[f(x)]^2}{2}\ dx}{\text{Area of } R}.} \tag{10}$$

This formula is preferable to formula (7) if R is given as a region under a curve, $y = f(x)$.

EXAMPLE 3 Find the centroid of the semicircular region of radius a shown in Fig. 15.

SOLUTION By symmetry, $\bar{x} = 0$.

To find $\bar{y}$, use (10). The function f in this case is given by the formula $f(x) = \sqrt{a^2 - x^2}$, an even function. Thus the moment of R about the x axis is

$$\int_{-a}^{a} \frac{(\sqrt{a^2 - x^2})^2}{2}\ dx = \int_{-a}^{a} \frac{a^2 - x^2}{2}\ dx = 2\int_0^a \frac{a^2 - x^2}{2}\ dx$$

$$= \int_0^a (a^2 - x^2)\ dx = \left(a^2 x - \frac{x^3}{3}\right)\Bigg|_0^a$$

$$= \left(a^3 - \frac{a^3}{3}\right) - 0 = \tfrac{2}{3}a^3.$$

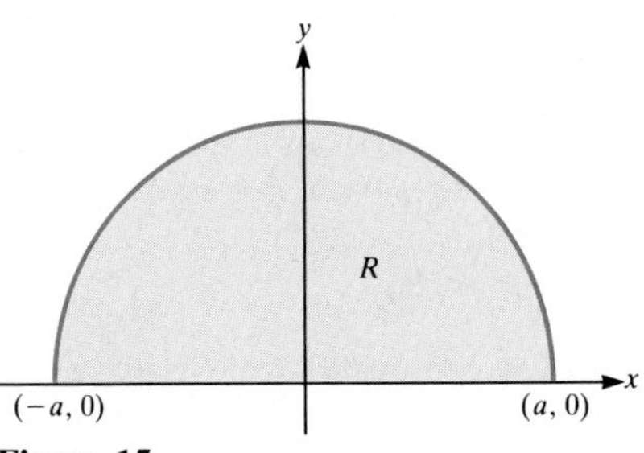

Figure 15

Thus
$$\bar{y} = \frac{\frac{2}{3}a^3}{\text{Area of } R} = \frac{\frac{2}{3}a^3}{\frac{1}{2}\pi a^2} = \frac{4a}{3\pi}.$$

[Since $4/(3\pi) \approx 0.42$, the center of gravity of R is at a height of about $0.42a$.] ■

Section Summary

We began with the notion of the moment of a point mass, "mass × lever," or $m(x - k)$. This definition suggested how to define the moment of a region R with density σ, namely $M = \int_a^b (x - k)\sigma c(x)\,dx$. The value of k that makes this integral equal to 0 is denoted $\bar{x}$. ($\bar{y}$ is defined similarly.) The point $(\bar{x}, \bar{y})$ is called the centroid of R.

Don't memorize these formulas; be able to derive them.

In particular, if R is the region below the curve $y = f(x)$ and above $[a, b]$, then

$$\bar{x} = \frac{\int_a^b xf(x)\,dx}{\text{Area of } R} \quad \text{and} \quad \bar{y} = \frac{\int_a^b \frac{1}{2}(f(x))^2\,dx}{\text{Area of } R}.$$

EXERCISES FOR SEC. 8.6: THE CENTROID OF A PLANE REGION

In Exercises 1 to 8 find the centroid of the given region R, with constant density σ. (Exercises 5 to 8 require integral tables or techniques of Chap. 7.)

1 R is bounded by $y = x^2$ and $y = 4$.

2 R is bounded by $y = x^4$ and $y = 1$.

3 R is bounded by $y = 4x - x^2$ and the x axis.

4 R is bounded by $y = x$, $x + y = 1$, and the x axis.

5 The region bounded by $y = e^x$ and the x axis, between the lines $x = 1$ and $x = 2$.

6 The region bounded by $y = \sin 2x$ and the x axis, between the lines $x = 0$ and $x = \pi/2$.

7 The region bounded by $y = \sqrt{1 + x}$ and the x axis, between the lines $x = 0$ and $x = 3$.

8 The region bounded by $y = \ln x$ and the x axis, between the lines $x = 1$ and $x = e$.

Exercises 9 to 11 concern Pappus's theorem, which relates the volume of a solid of revolution to the centroid of the planar region R that is revolved.

9 (*a*) Prove Pappus's theorem: Let R be a region in the plane and L a line in the plane that does not cross R (though it can touch R at its border). Then the volume of the solid formed by revolving R about L is equal to the product:

(Distance centroid of R is revolved) · (Area of R).

(*b*) Use Pappus's theorem to find the volume of the torus or "doughnut" formed by revolving a circle of radius 3 inches about a line 5 inches from its center.

10 Use Pappus's theorem to find the centroid of the half disk R of radius a.

11 Use Pappus's theorem to find the centroid of the right triangle in Fig. 16.

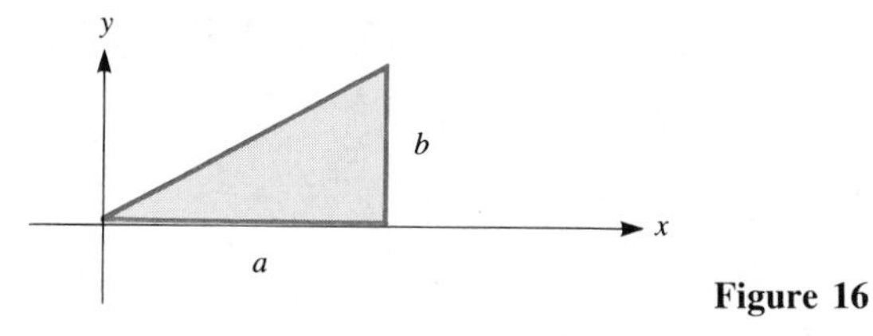

Figure 16

Exercise 12 is used in Exercises 13 to 16.

12 Let f and g be continuous functions such that $f(x) \geq g(x) \geq 0$ for x in $[a, b]$. Let R be the region above $[a, b]$ which is bounded by the curves $y = f(x)$ and $y = g(x)$.
(*a*) Set up a definite integral (in terms of f and g) for the moment of R about the y axis.
(*b*) Set up a definite integral with respect to x (in terms of f and g) for the moment of R about the x axis.

In Exercises 13 to 16 find (*a*) the moment of the given region R about the y axis, (*b*) the moment of R about the x axis, (*c*) the area of R, (*d*) $\bar{x}$, (*e*) $\bar{y}$. Assume the density is 1. (See Exercise 12.)

13 R is bounded by the curves $y = x^2$, $y = x^3$.

14 R is bounded by $y = x$, $y = 2x$, $x = 1$, and $x = 2$.

15 R is bounded by the curves $y = 3^x$ and $y = 2^x$ between $x = 1$ and $x = e$.

16 (Use a table of integrals or techniques from Chap. 7.) R is bounded by the curves $y = x - 1$ and $y = \ln x$, between $x = 1$ and $x = e$.

17 In a letter of 1680 Leibniz wrote:

> Huygens, as soon as he had published his book on the pendulum, gave me a copy of it; and at that time I was

quite ignorant of Cartesian algebra and also of the method of indivisibles, indeed I did not know the correct definition of the center of gravity. For, when by chance I spoke of it to Huygens, I let him know that I thought that a straight line drawn through the center of gravity always cut a figure into two equal parts; since that clearly happened in the case of a square, or a circle, an ellipse, and other figures that have a center of magnitude, I imagined that it was the same for all other figures. Huygens laughed when he heard this, and told me that nothing was further from the truth. (Quoted in C. H. Edwards, *The Historical Development of the Calculus,* p. 239, Springer-Verlag, New York, 1979.)

Give an example showing that "nothing is further from the truth."

18 Let a be a constant ≥ 1. Let R be the region below $y = x^a$, above the x axis, between the lines $x = 0$ and $x = 1$.
(*a*) Sketch R for large a.
(*b*) Compute the centroid $(\bar{x}, \bar{y})$ of R.
(*c*) Find $\lim_{a\to\infty} \bar{x}$ and $\lim_{a\to\infty} \bar{y}$.
(*d*) For large a, does the centroid of R lie in R?

19 (Contributed by Jeff Lichtman.) Let f and g be two continuous functions such that $f(x) \geq g(x) \geq 0$ for x in $[0, 1]$. Let R be the region under $y = f(x)$ and above $[0, 1]$; let R^* be the region under $y = g(x)$ and above $[0, 1]$.
(*a*) Do you think the center of mass of R is at least as high as the center of mass of R^*? (An opinion only.)
(*b*) Let $g(x) = x$. Define $f(x)$ to be $\frac{1}{3}$ for $0 \leq x \leq \frac{1}{3}$ and $f(x)$ to be x if $\frac{1}{3} \leq x \leq 1$. (Note that f is continuous.) Find $\bar{y}$ for R and also for R^*. (Which is larger?)
(*c*) Let a be a constant, $0 \leq a \leq 1$. Let $f(x) = a$ for $0 \leq x \leq a$, and let $f(x) = x$ for $a \leq x \leq 1$. Find $\bar{y}$ for R.
(*d*) Show that the number a for which $\bar{y}$ defined in part (*c*) is a minimum is a root of the equation $x^3 + 3x - 1 = 0$.
(*e*) Show that the equation in (*d*) has only one real root q.
(*f*) Find q to four decimal places.

20 This exercise shows that the three medians of a triangle meet at the centroid of the triangle. (A **median** of a triangle is a line that passes through a vertex and the midpoint of the opposite edge.)

Let R be a triangle with vertices A, B, and C. It suffices to show that the centroid of R lies on the median through C and the midpoint M of the edge AB. Introduce an xy coordinate system such that the origin is at A, and B lies on the x axis, as in Fig. 17.
(*a*) Compute $(\bar{x}, \bar{y})$.
(*b*) Find the equation of the median through C and M.
(*c*) Verify that the centroid lies on the median computed in (*b*).
(*d*) Why would you expect the centroid to lie on each median? (Just use physical intuition.)

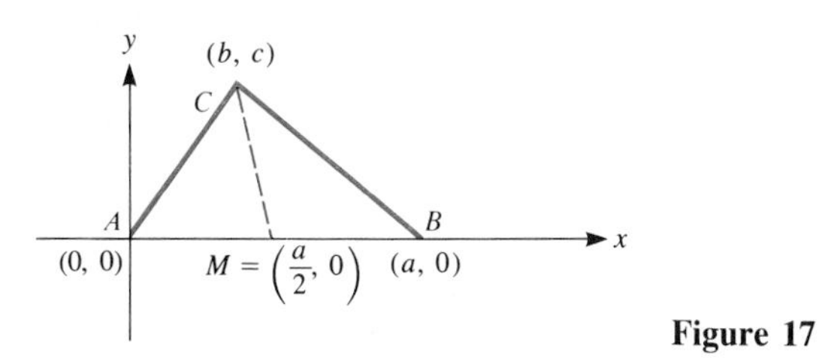

Figure 17

21 Cut an irregular shape out of cardboard and find three balancing lines for it experimentally. Are they concurrent; that is, do they pass through a common point? (It can be shown that each line through the centroid is a balancing line.)

8.7 WORK

In this section we treat the work accomplished by a force operating along a line, for example the work done when you stretch a spring. If the force has the *constant* value F and it operates over a distance s in the *direction* of the force, then the work W accomplished is simply

$$\text{Work} = \text{Force} \cdot \text{Distance}$$

or

$$W = F \cdot s.$$

If force is measured in newtons and distance in meters, work is measured in newton-meters or joules. For example, the force needed to lift a mass of m kilograms at the surface of the earth is about $9.8m$ newtons.)

A weightlifter who raises 100 kilograms a distance of 0.5 meter accomplishes $9.8(100)(0.5) = 490$ joules of work. On the other hand, the weightlifter who just carries the barbell from one place to another in the weightlifting room, without raising or lowering it, accomplishes no work because the barbell was moved a distance zero in the direction of the force.

The Stretched Spring

Hooke's law says a spring's force is proportional to the distance it is stretched.

As you stretch a spring (or rubber band) from its rest position, the further you stretch it the harder you have to pull. According to Hooke's law, the force you must exert is proportional to the distance that the spring is stretched, as shown in Fig. 1.

Because the force is *not* constant, we cannot compute the work accomplished just by multiplying force times distance. Instead, an integration is required, as the next example illustrates.

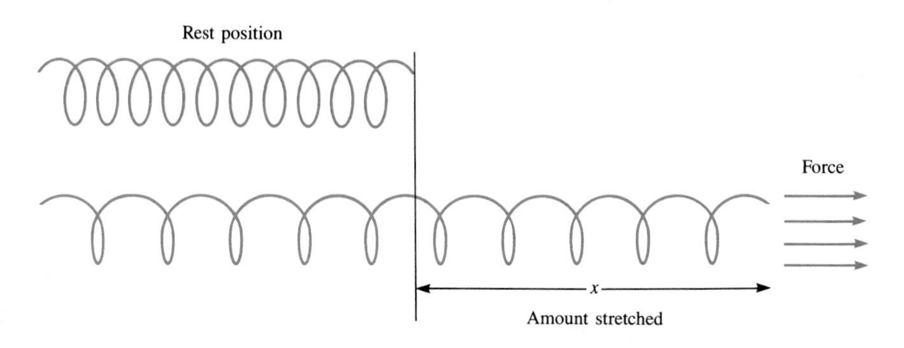

Figure 1

EXAMPLE 1 A spring is stretched 0.5 meter longer than its rest length. The force required to keep it at that length is 3 newtons. Find the total work accomplished in stretching the spring 0.5 meter from its rest position.

SOLUTION Let us estimate the work involved in stretching the spring from x to $x + dx$. (See Fig. 2.)

The distance dx is small. As the end of the spring is stretched from x to $x + dx$, the force is almost constant. Since the force is proportional to x, it is of the form kx for some constant k. We know that $F = 3$ when $x = 0.5$, so

$$F = kx$$

$$3 = k(0.5)$$

$$6 = k.$$

The work accomplished in stretching the spring from x to $x + dx$ is then approximately

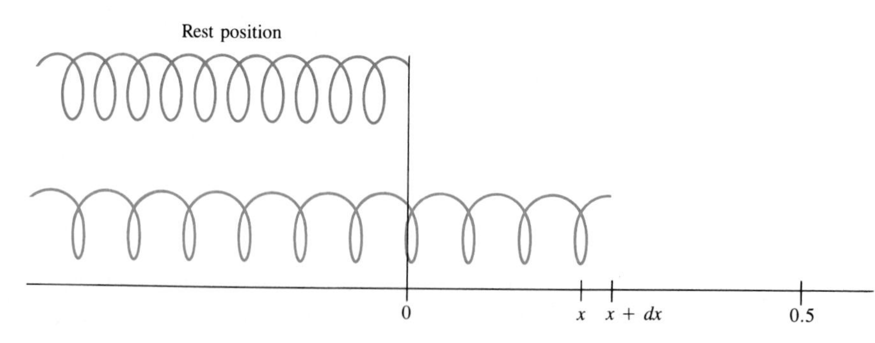

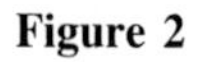

Figure 2

$$\underbrace{kx}_{\text{force}} \cdot \underbrace{dx}_{\text{distance}} \quad \text{joules.}$$

Hence the total work is

$$\int_a^b kx\,dx = \int_0^{0.5} 6x\,dx$$

$$= 3x^2 \Big|_0^{0.5}$$

$$= 0.75 \text{ joule.} \quad \blacksquare$$

Work in Launching a Rocket

The force of gravity that the earth exerts on an object diminishes as the object gets further and further away from the earth. The work required to lift an object 1 foot at sea level is greater than the work required to lift the same object the same distance at the top of Mt. Everest. However, the difference in altitudes is so small in comparison to the radius of the earth that the difference in work is negligible. On the other hand, when an object is rocketed into space, the fact that the force of gravity diminishes with distance from the center of the earth is critical.

According to Newton, the force of gravity on a given mass is proportional to the reciprocal of the square of the distance of that mass from the center of the earth. That is, there is a constant k such that the gravitational force at distance r from the center of the earth, $F(r)$, is given by

$$F(r) = \frac{k}{r^2}.$$

(See Fig. 3.)

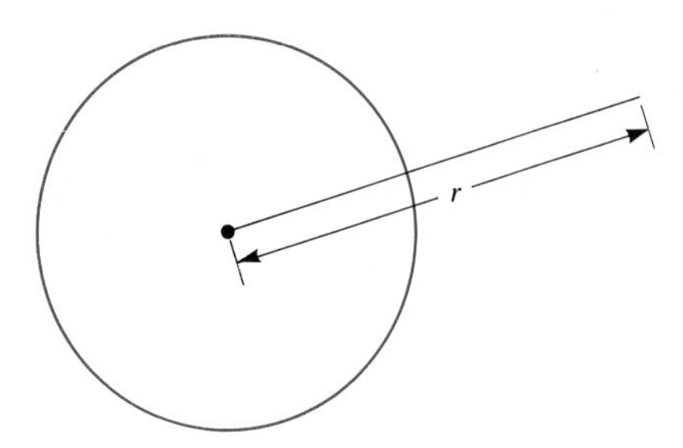

Figure 3

EXAMPLE 2 How much work is required to lift a 1-pound payload from the surface of the earth to the moon, which is about 240,000 miles away? (The earth's surface is at a distance of 4,000 miles from its center.)

SOLUTION The work W necessary to lift an object a distance x against a constant vertical force F is the product of force times distance:

$$W = F \cdot x.$$

Since the gravitational pull of the earth on the payload *changes* with distance from the earth, an integral will be needed to express the total work required to lift the load.

The payload weighs 1 pound at the surface of the earth. The farther it is from the center of the earth, the less it weighs, for the force of the earth on the mass is inversely proportional to the square of the distance of the mass from the center of the earth. Thus the force on the payload is given by k/r^2 pounds, where k is a constant, which will be determined in a moment, and r is the distance in miles from the payload to the center of the earth. When $r = 4{,}000$ (miles), the force is 1 pound; thus

$$1 = \frac{k}{4{,}000^2}.$$

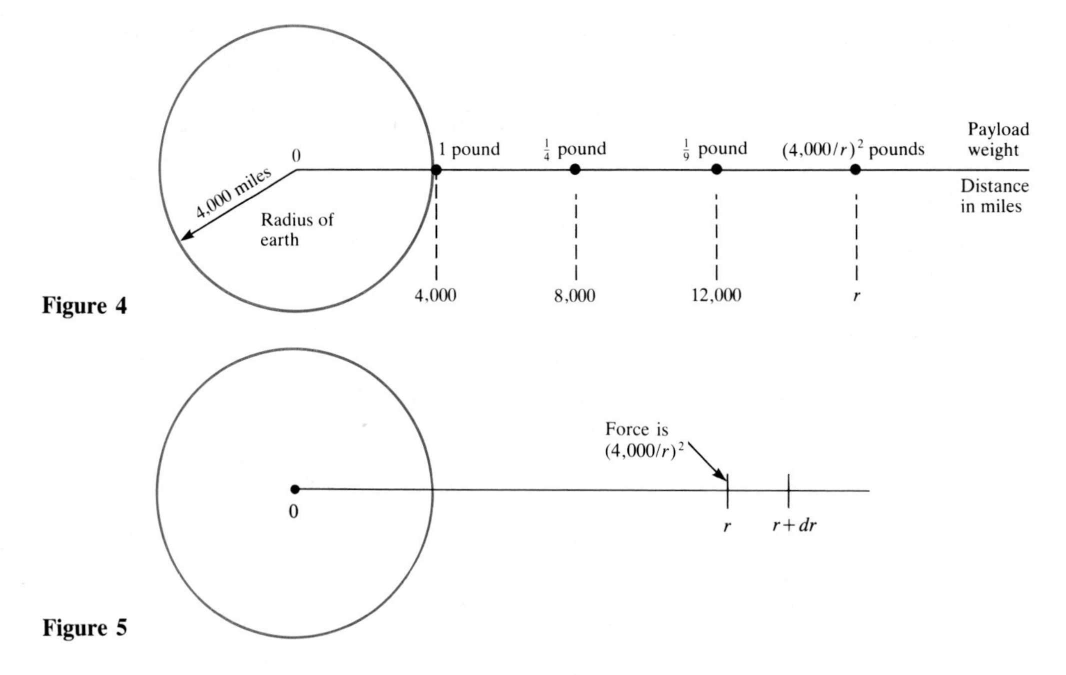

Figure 4

Figure 5

From this it follows that $k = 4{,}000^2$, and therefore the gravitational force on a 1-pound mass is, in general, $(4{,}000/r)^2$ pounds. As the payload recedes from the earth, it loses weight (but not mass), as recorded in Fig. 4. The work done in lifting the payload from point r to point $r + dr$ is approximately

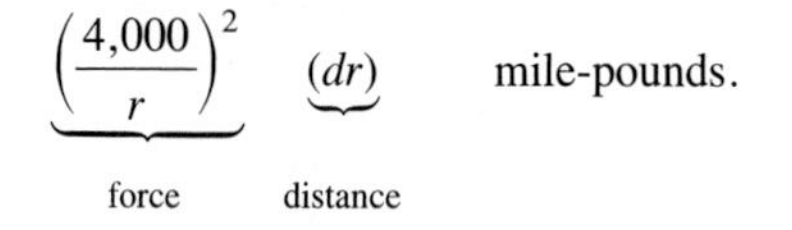

$$\underbrace{\left(\frac{4{,}000}{r}\right)^2}_{\text{force}} \underbrace{(dr)}_{\text{distance}} \qquad \text{mile-pounds.}$$

(See Fig. 5.)

Hence the work required to move the 1-pound mass from the surface of the earth to the moon is given by the integral

$$\int_{4{,}000}^{240{,}000} \left(\frac{4{,}000}{r}\right)^2 dr = -\frac{4{,}000^2}{r}\Bigg|_{4{,}000}^{240{,}000}$$

$$= -4{,}000^2\left(\frac{1}{240{,}000} - \frac{1}{4{,}000}\right)$$

$$= -\frac{4{,}000}{60} + 4{,}000$$

$$\approx 3{,}933 \text{ mile-pounds.}$$

The work is just a little less than if the payload were lifted 4,000 miles against a constant gravitational force equal to that at the surface of the earth. ■

EXERCISES FOR SEC. 8.7: WORK

The springs in Exercises 1 and 2 obey Hooke's law, as in Example 1.

1 A spring is stretched 0.20 meter from its rest length. The force required to keep it at that length is 5 newtons. Assuming that the force of the spring is proportional to the distance it is stretched,
(*a*) find the work accomplished in stretching the spring 0.20 meter from its rest length;
(*b*) find the work accomplished in stretching the spring 0.30 meter from its rest length.

2 A spring is stretched 3 meters from its rest length. The force required to keep it at that length is 24 newtons. Assuming that the force of the spring is proportional to the distance it is stretched,
(*a*) find the work accomplished in stretching the spring 3 meters from its rest length;
(*b*) find the work accomplished in stretching the spring 4 meters from its rest length.

3 Suppose a spring does not obey Hooke's law. Instead, the force it exerts when stretched x meters from its rest length is $F(x) = 3x^2$ newtons. Find the work done in stretching the spring 0.80 meter from its rest length.

4 Suppose a spring does not obey Hooke's law. Instead, the force it exerts when stretched x meters from its rest length is $F(x) = 2\sqrt{x}$ newtons. Find the work done in stretching the spring 0.50 meter from its rest length.

5 (See Example 2.) How much work is done in lifting the 1-pound payload the first 4,000 miles of its journey to the moon?

6 If a mass which weighs 1 pound at the surface of the earth were launched from a position 20,000 miles from the center of the earth, how much work would be required to send it to the moon (240,000 miles from the center of the earth)?

7 Assume that the force of gravity obeys an inverse cube law, so that the force on a 1-pound payload a distance r miles from the center of the earth ($r \geq 4{,}000$) is $(4{,}000/r)^3$ pounds. How much work would be required to lift a 1-pound payload from the surface of the earth to the moon?

8 Geologists, when considering the origin of mountain ranges, estimate the energy required to lift a mountain up from sea level. Assume that two mountains are composed of the same type of matter, which weighs k pounds per cubic foot. Both are right circular cones in which the height is equal to the radius. One mountain is twice as high as the other. The base of each is at sea level. If the work required to lift the matter in the smaller mountain above sea level is W, what is the corresponding work for the larger mountain?

9 (See Exercise 8.) Assume that Mt. Everest has the shape of a right circular cone of height 30,000 feet and radius 150,000 feet, with uniform density of 200 pounds per cubic foot.
(*a*) How much work was required to lift the material in Mt. Everest if it was initially all at sea level?
(*b*) How does this work compare with the energy of a 1-megaton H bomb? (One megaton is the energy in a million tons of TNT: about 3×10^{14} foot-pounds.)

10 A town in a flat valley made a conical hill out of its rubbish, as shown in Fig. 6. The work required to lift all the rubbish was W. Happy with the result, the town decided to make another hill with twice the volume, but of the same shape. How much work will be required to build this hill? Explain.

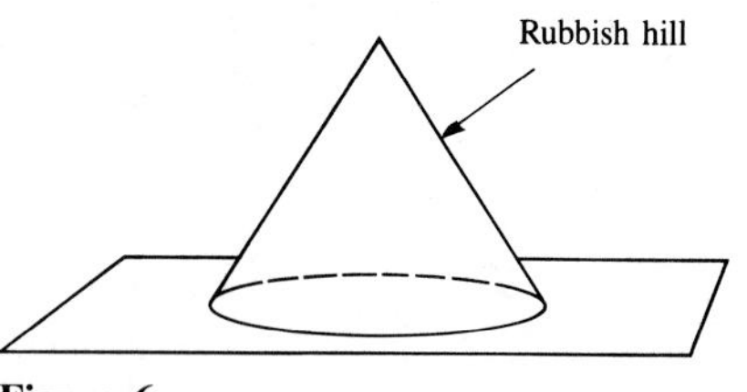

Figure 6

---

11 A container is full of water which weighs 62.4 pounds per cubic foot. All the water is pumped out of an opening at the top of the container. Develop a definite integral for the work accomplished. [The integral involves only a, b, and $A(x)$, the cross-sectional area shown in Fig. 7.]

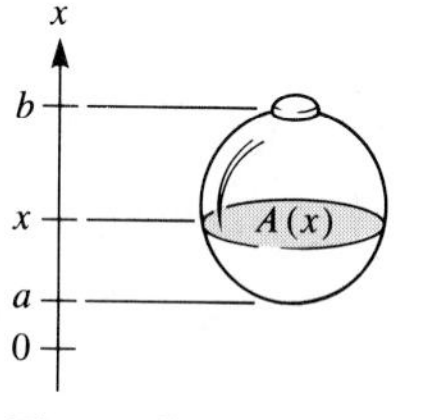

Figure 7

12 A horizontal tank in the form of a cylinder with base R is full of water. The cylinder has height h feet. (See Fig. 8.) Develop a definite integral for the total work accomplished when all the water is pumped out an opening at the top.

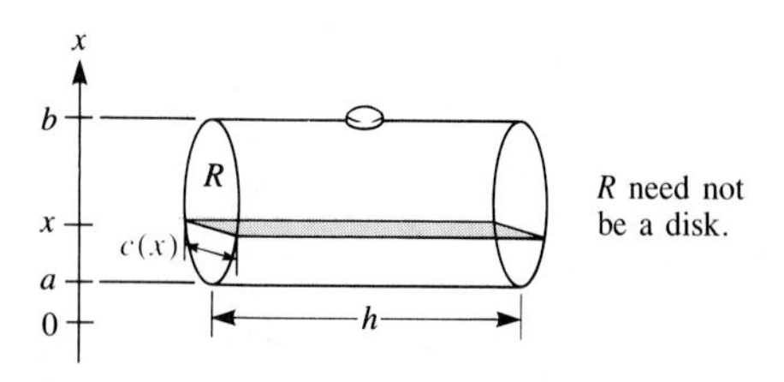

Figure 8

[Express the integral in terms of a, b, $c(x)$, and h.]

13 Show that the work in Exercise 12 is the same as it would be if all the water were located at the centroid of the base R.

8.8 IMPROPER INTEGRALS

This section develops the analog of a definite integral when the interval of integration is infinite or the integrand becomes arbitrarily large in the interval of integration. The definition of a definite integral does not cover these cases.

Improper Integrals: Interval Unbounded

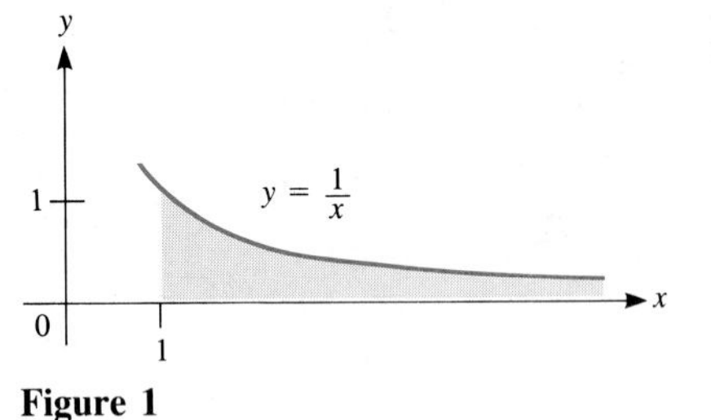

Figure 1

A question about areas will introduce the notion of an "improper integral." Figure 1 shows the region under $y = 1/x$ and above the interval $[1, \infty)$. Figure 2 shows the region under $y = 1/x^2$ and above the same interval.

Let us compute the areas of the two regions. We might be tempted to say that the area in Fig. 1 is $\int_1^\infty (1/x)\,dx$. Unfortunately, the symbol $\int_a^\infty f(x)\,dx$ has not been given any meaning so far in this book. The definition of the definite integral $\int_a^b f(x)\,dx$ involves a limit of sums of the form

$$\sum_{i=1}^{n} f(c_i)(x_i - x_{x-1}),$$

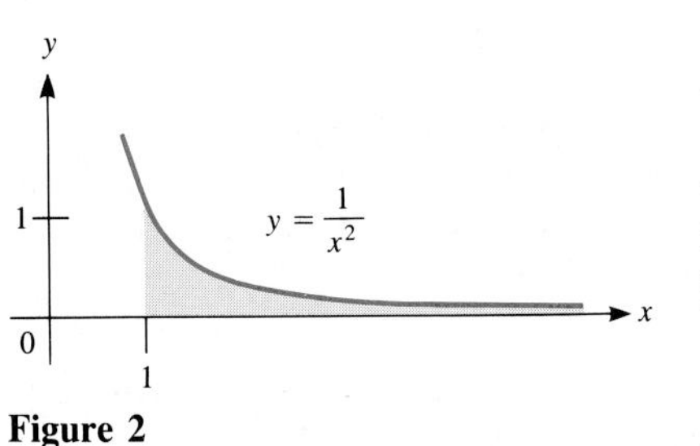

Figure 2

where each $x_i - x_{i-1}$ is the length of an interval $[x_{i-1}, x_i]$. If you cut the interval $[1, \infty]$ into a finite number of intervals, then at least one section has infinite length, and such a sum is meaningless.

It does make sense, however, to find the area of that part of the region in Fig. 1 from $x = 1$ to $x = b$, where $b > 1$, and determine what happens to that number as $b \to \infty$. To do this, first find $\int_1^b (1/x)\,dx$, as follows:

$$\int_1^b \frac{dx}{x} = \ln x \Big|_1^b = \ln b - \ln 1 = \ln b.$$

Then,

$$\lim_{b\to\infty} \int_1^b \frac{dx}{x} = \lim_{b\to\infty} \ln b = \infty.$$

We conclude that the area of the region in Fig. 1 is infinite.

Next, examine the area of the region in Fig. 2. We first find

$$\int_1^b \frac{dx}{x^2} = -\frac{1}{x}\Big|_1^b = -\frac{1}{b} - \left(-\frac{1}{1}\right) = 1 - \frac{1}{b}.$$

Thus,

$$\lim_{b\to\infty} \int_1^b \frac{dx}{x^2} = \lim_{b\to\infty} \left(1 - \frac{1}{b}\right) = 1.$$

In this case the area is finite. Though the regions in Figs. 1 and 2 look a lot alike, one has an infinite area, and the other, a finite area. This contrast suggests the following definitions.

Definition *Convergent improper integral* $\int_a^\infty f(x)\,dx$. Let f be continuous for $x \geq a$. If $\lim_{b\to\infty} \int_a^b f(x)\,dx$ exists, the function f is said to have a **convergent improper integral** from a to ∞. The value of the limit is denoted by $\int_a^\infty f(x)\,dx$:

$$\int_a^\infty f(x)\,dx = \lim_{b\to\infty} \int_a^b f(x)\,dx.$$

$\int_1^\infty dx/x^2$ is convergent.

We saw that $\int_1^\infty dx/x^2$ is a convergent improper integral with value 1.

Definition *Divergent improper integral* $\int_a^\infty f(x)\,dx$. Let f be a continuous function. If $\lim_{b\to\infty} \int_a^b f(x)\,dx$ does not exist, the function f is said to have a **divergent improper integral** from a to ∞.

$\int_1^\infty dx/x$ is divergent.

As we saw, $\int_1^\infty dx/x$ is a divergent improper integral.

The improper integral $\int_1^\infty dx/x$ is divergent because $\int_1^b dx/x \to \infty$ as $b \to \infty$. But an improper integral $\int_a^\infty f(x)\,dx$ can be divergent without being infinite. Consider, for instance, $\int_0^\infty \cos x\,dx$. We have

$$\int_0^b \cos x\,dx = \sin x \Big|_0^b = \sin b.$$

Divergence due to integral oscillating

$\int_0^\infty \cos x\,dx$ is divergent.

As $b \to \infty$, $\sin b$ does not approach a limit, nor does it become arbitrarily large. As $b \to \infty$, $\sin b$ just keeps going up and down in the range -1 to 1 infinitely often. Thus $\int_0^\infty \cos x\,dx$ is divergent.

The improper integral $\int_{-\infty}^b f(x)\,dx$

The improper integral $\int_{-\infty}^b f(x)\,dx$ is defined similarly:

$$\int_{-\infty}^b f(x)\,dx = \lim_{a\to-\infty} \int_a^b f(x)\,dx.$$

If the limit exists, $\int_{-\infty}^b f(x)\,dx$ is a *convergent* improper integral. If the limit does not exist, it is a *divergent* improper integral.

The improper integral $\int_{-\infty}^\infty f(x)\,dx$

To deal with improper integrals over the entire x axis, define

$$\int_{-\infty}^\infty f(x)\,dx$$

to be the sum

$$\int_{-\infty}^0 f(x)\,dx + \int_0^\infty f(x)\,dx,$$

which will be called **convergent** if both

$$\int_{-\infty}^0 f(x)\,dx \qquad \text{and} \qquad \int_0^\infty f(x)\,dx$$

are convergent. [If at least one of the two is divergent, $\int_{-\infty}^\infty f(x)\,dx$ will be called **divergent**.]

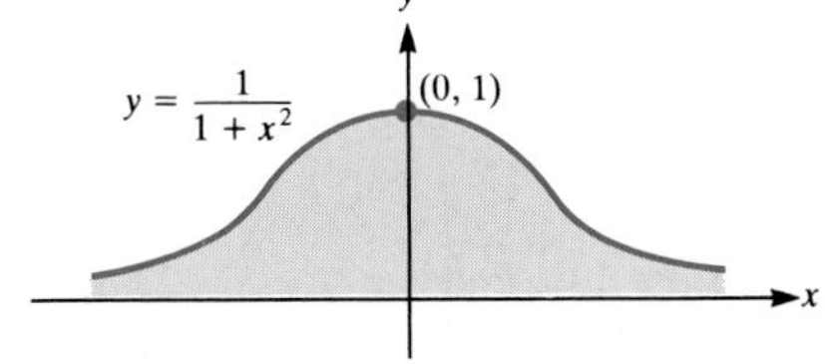

Figure 3

EXAMPLE 1 Determine the area of the region bounded by the curve $y = 1/(1 + x^2)$ and the x axis, as indicated in Fig. 3.

SOLUTION The area in question equals $\int_{-\infty}^{\infty} dx/(1+x^2)$.

Now,
$$\int_0^{\infty} \frac{dx}{1+x^2} = \lim_{b\to\infty} \int_0^b \frac{dx}{1+x^2}$$
$$= \lim_{b\to\infty} (\tan^{-1} b - \tan^{-1} 0) = \frac{\pi}{2}.$$

By symmetry,
$$\int_{-\infty}^{0} \frac{dx}{1+x^2} = \frac{\pi}{2}.$$

Hence
$$\int_{-\infty}^{\infty} \frac{dx}{1+x^2} = \frac{\pi}{2} + \frac{\pi}{2},$$

and the area in question is π. ■

Comparison Test for $\int_a^{\infty} f(x)\,dx$, $f(x) \geq 0$

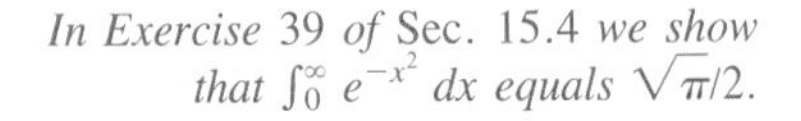
In Exercise 39 of Sec. 15.4 we show that $\int_0^{\infty} e^{-x^2}\,dx$ equals $\sqrt{\pi}/2$.

Is $\int_0^{\infty} e^{-x^2}\,dx$ convergent or divergent? (This integral is important in statistical theory.) We cannot evaluate $\int_0^b e^{-x^2}\,dx$ by the fundamental theorem since e^{-x^2} does not have an elementary antiderivative. Even so, there is a way of showing that $\int_0^{\infty} e^{-x^2}\,dx$ is in fact convergent without determining its exact value. The method is described in Theorem 1, and depends on the following principle:

Let $h(x)$ be defined for all $x \geq a$ and have the property that $x_1 > x_2$ implies $h(x_1) \geq h(x_2)$. Assume that there is a number B such that $h(x) \leq B$ for all $x \geq a$. Then as $x \to \infty$, $h(x)$ approaches a limit L, and $L \leq B$. [We will assume this result, which is a fundamental property of the real number system. Figure 4 suggests that it is plausible. The graph of the function $y = h(x)$ stays below the line $y = B$ and must have a horizontal asymptote.]

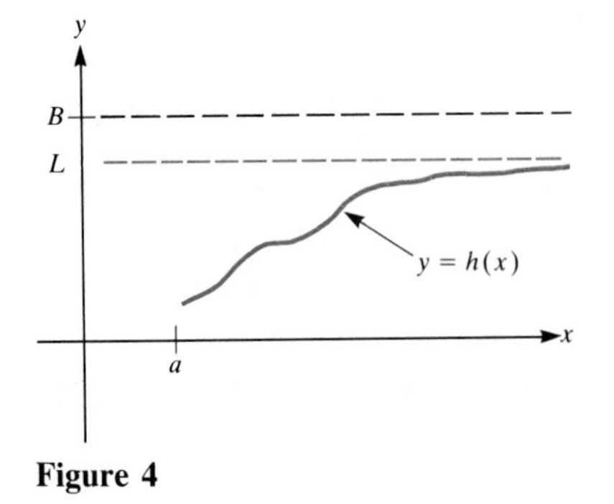

Figure 4

> **Theorem 1** *Comparison test for convergence of improper integrals.* Let $f(x)$ and $g(x)$ be continuous functions for $x \geq a$. Assume that $0 \leq f(x) \leq g(x)$ and that $\int_a^{\infty} g(x)\,dx$ is convergent. Then $\int_a^{\infty} f(x)\,dx$ is convergent and
> $$\int_a^{\infty} f(x)\,dx \leq \int_a^{\infty} g(x)\,dx.$$

Proof Let $h(b) = \int_a^b f(x)\,dx$ for $b \geq a$. Since $f(x) \geq 0$, it follows that $x_1 > x_2$ implies $h(x_1) \geq h(x_2)$. Moreover,
$$h(b) = \int_a^b f(x)\,dx \leq \int_a^b g(x)\,dx \leq \int_a^{\infty} g(x)\,dx.$$

Since $h(b)$ never exceeds $B = \int_a^{\infty} g(x)\,dx$, it follows that $\lim_{b\to\infty} h(b)$ exists and is not larger than $\int_a^{\infty} g(x)\,dx$. Thus $\int_a^{\infty} f(x)\,dx$ exists, and
$$\int_a^{\infty} f(x)\,dx \leq \int_a^{\infty} g(x)\,dx. \quad ■$$

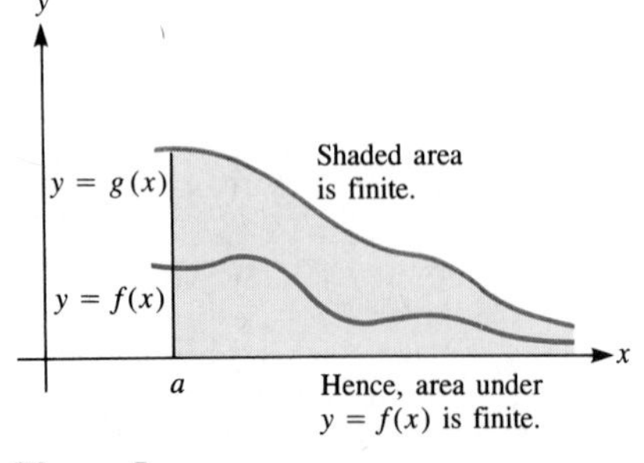

Figure 5

In geometric terms, Theorem 1 asserts that if the area under $y = g(x)$ is finite, so is the area under $y = f(x)$. (See Fig. 5.)

A similar convergence test holds for $g(x) \leq f(x) \leq 0$. If $\int_a^{\infty} g(x)\,dx$ converges, so does $\int_a^{\infty} f(x)\,dx$.

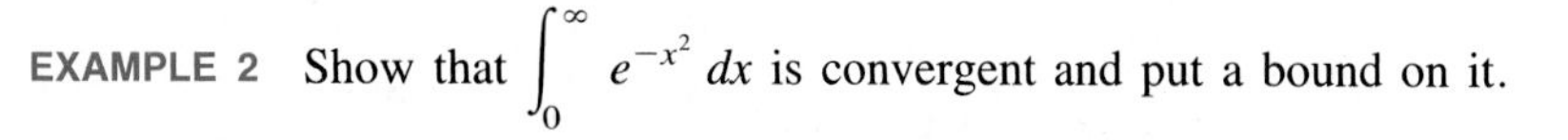

EXAMPLE 2 Show that $\int_0^\infty e^{-x^2}\, dx$ is convergent and put a bound on it.

SOLUTION Since e^{-x^2} does not have an elementary antiderivative, we cannot evaluate $\int_0^b e^{-x^2}\, dx$ and use the result to determine the behavior of $\int_0^b e^{-x^2}\, dx$ as $b \to \infty$.

However, we can compare $\int_0^\infty e^{-x^2}\, dx$ to an improper integral that we know converges.

For $x \geq 1$, $x^2 \geq x$; hence $e^{-x^2} \leq e^{-x}$. (See Fig. 6.) Now,

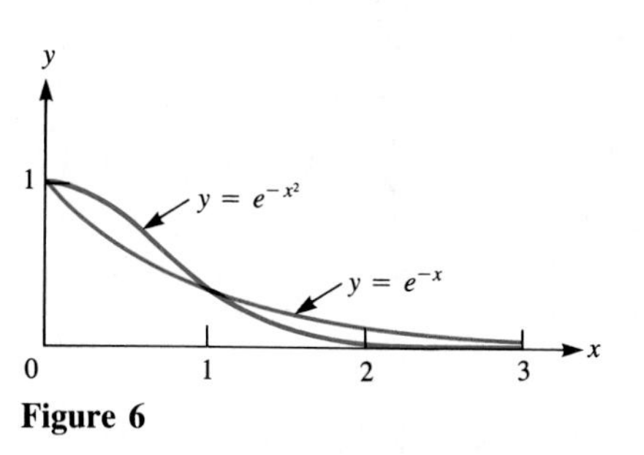

Figure 6

$$\int_1^b e^{-x}\, dx = -e^{-x}\Big|_1^b = e^{-1} - e^{-b}.$$

Thus $$\lim_{b\to\infty} \int_1^b e^{-x}\, dx = \frac{1}{e}$$

and $\int_1^\infty e^{-x}\, dx$ is convergent.

Since $0 < e^{-x^2} \leq e^{-x}$ for $x \geq 1$, the comparison test tells us that $\int_1^\infty e^{-x^2}\, dx$ is convergent. Furthermore,

$$\int_1^\infty e^{-x^2}\, dx \leq \int_1^\infty e^{-x}\, dx = \frac{1}{e}.$$

Thus $$\int_0^\infty e^{-x^2}\, dx = \int_0^1 e^{-x^2}\, dx + \int_1^\infty e^{-x^2}\, dx \leq \int_0^1 e^{-x^2}\, dx + \frac{1}{e}.$$

Since $e^{-x^2} \leq 1$ for $0 < x \leq 1$, we conclude that

$$\int_0^\infty e^{-x^2}\, dx \leq 1 + \frac{1}{e}. \quad ■$$

There is a similar comparison test for divergence:

Theorem 2 *Comparison test for divergence of improper integrals.* Let $f(x)$ and $g(x)$ be continuous functions for $x \geq a$. Assume that $0 \leq g(x) \leq f(x)$ and that $\int_a^\infty g(x)\, dx$ is divergent. Then $\int_a^\infty f(x)\, dx$ is also divergent.

Proof $$\int_a^\infty f(x)\, dx = \lim_{b\to\infty} \int_a^b f(x)\, dx \geq \lim_{b\to\infty} \int_a^b g(x)\, dx = \infty. \quad ■$$

EXAMPLE 3 Show that $\int_1^\infty (x^2+1)/x^3\, dx$ is divergent.

SOLUTION For $x > 0$,

$$\frac{x^2+1}{x^3} > \frac{x^2}{x^3} = \frac{1}{x}.$$

Since $\int_1^\infty dx/x = \infty$, it follows that $\int_1^\infty (x^2+1)/x^3\, dx = \infty$. ■

Convergence of $\int_a^\infty f(x)\, dx$ if $\int_a^\infty |f(x)|\, dx$ Converges

The next theorem provides a way to establish the convergence of $\int_a^\infty f(x)\, dx$ when $f(x)$ is a function that takes on both positive and negative values. It says that if $\int_a^\infty |f(x)|\,dx$ converges, so does $\int_a^\infty f(x)\,dx$. The argument for this depends on showing that the "negative and positive parts of the function" both lead to convergent integrals.

Theorem 3 *The absolute-convergence test.* If $f(x)$ is continuous and $\int_a^\infty |f(x)|\,dx$ converges to the number L, then $\int_a^\infty f(x)\,dx$ is convergent and converges to a number between L and $-L$.

Proof We will break the function $f(x)$ into two functions that do not change sign. That will enable us to use Theorem 2.

Figure 7 shows the graphs of $y = f(x)$ and four functions closely related to $f(x)$.

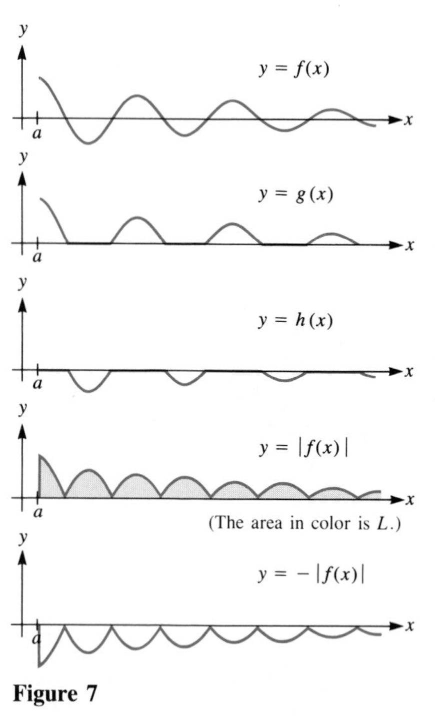

Figure 7

$g(x)$ equals $f(x)$ if $f(x)$ is positive; otherwise $g(x) = 0$.

$h(x)$ equals $f(x)$ if $f(x)$ is negative; otherwise $h(x) = 0$.

Note that $f(x) = g(x) + h(x)$. We will show that $\int_a^\infty g(x)\,dx$ and $\int_a^\infty h(x)\,dx$ both converge.

First, since $\int_a^\infty |f(x)|\,dx$ converges and $0 \le g(x) \le |f(x)|$, $\int_a^\infty g(x)\,dx$ converges to a number A, where

$$0 \le A \le \int_a^\infty |f(x)|\,dx = L.$$

Second, since $\int_a^\infty (-|f(x)|)\,dx$ converges and $0 \ge h(x) \ge -|f(x)|$, $\int_a^\infty h(x)\,dx$ converges to a number B, where

$$0 \ge B \ge -\int_a^\infty |f(x)|\,dx = -L.$$

Thus $\int_a^\infty f(x)\,dx = \int_a^\infty [g(x) + h(x)]\,dx$ converges to $A + B$, which is a number in the interval $[-L, L]$. Hence $\int_a^\infty f(x)\,dx$ is convergent. ■

EXAMPLE 4 Show that $\int_0^\infty e^{-x^2} \sin x\,dx$ is convergent.

SOLUTION Since $|\sin x| \le 1$, $|e^{-x^2} \sin x| \le e^{-x^2}$. Since $\int_0^\infty e^{-x^2}\,dx$ converges (by Example 2), the comparison test tells us that $\int_0^\infty |e^{-x^2} \sin x|\,dx$ converges. Theorem 3 then implies that $\int_0^\infty e^{-x^2} \sin x\,dx$ converges. ■

Improper Integrals: Integrand Unbounded

There is a second type of improper integral, in which the function is unbounded in the interval $[a, b]$. If $f(x)$ becomes arbitrarily large in the interval $[a, b]$, then it is possible to have arbitrarily large approximating sums $\sum_{i=1}^n f(c_i)(x_i - x_{i-1})$ no matter how fine the partition may be by choosing a c_i that makes $f(c_i)$ large. The next example shows how to get around this difficulty.

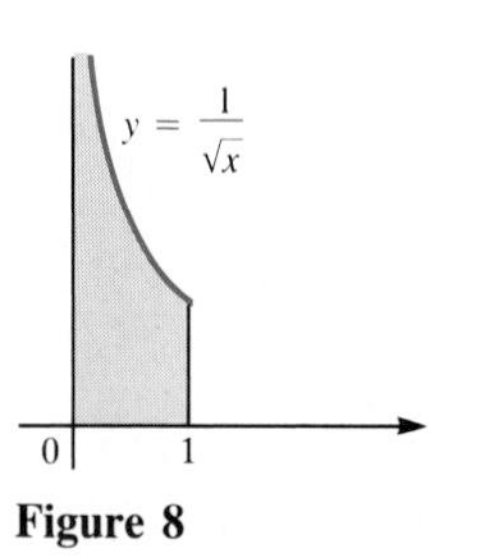

Figure 8

EXAMPLE 5 Determine the area of the region bounded by $y = 1/\sqrt{x}$, $x = 1$, and the coordinate axes shown in Fig. 8.

SOLUTION Resist for the moment the temptation to write "Area =

$\int_0^1 1/\sqrt{x}\,dx$,'' for $\int_0^1 1/\sqrt{x}\,dx$ does not exist according to the definition of the definite integral given in Chap. 5, since its integrand is unbounded in [0, 1]. (Note also that the integrand is not defined at 0.) Instead, consider the behavior of $\int_t^1 1/\sqrt{x}\,dx$ as t approaches 0 from the right. Since

$$\int_t^1 \frac{1}{\sqrt{x}}\,dx = 2\sqrt{x}\,\Big|_t^1 = 2\sqrt{1} - 2\sqrt{t} = 2(1 - \sqrt{t}),$$

it follows that

$$\lim_{t\to 0^+} \int_t^1 \frac{dx}{\sqrt{x}} = 2.$$

The area in question is 2.

Check to see that this is the same value for the area that can be obtained by taking horizontal cross sections and evaluating an improper integral from 0 to ∞. ■

The reasoning in Example 5 motivates the definition of the second type of improper integral, in which the function rather than the interval is unbounded.

Definition *Convergent and divergent improper integrals* $\int_a^b f(x)\,dx$. Let f be continuous at every number in $[a, b]$ except a. If $\lim_{t\to a^+} \int_t^b f(x)\,dx$ exists, the function f is said to have a **convergent improper integral** from a to b. The value of the limit is denoted $\int_a^b f(x)\,dx$. If $\lim_{t\to a^+} \int_t^b f(x)\,dx$ does not exist, the function f is said to have a **divergent improper integral** from a to b; in brief, $\int_a^b f(x)\,dx$ does not exist.

In a similar manner, if f is not defined at b, define $\int_a^b f(x)\,dx$ as $\lim_{t\to b^-} \int_a^t f(x)\,dx$, if this limit exists.

As Example 5 showed, the improper integral $\int_0^1 1/\sqrt{x}\,dx$ is convergent and has the value 2.

More generally, if a function $f(x)$ is not defined at certain isolated numbers, break the domain of $f(x)$ into intervals $[a, b]$ for which $\int_a^b f(x)\,dx$ is either improper or ''proper''—that is, an ordinary definite integral.

A ''proper'' integral is a definite integral.

For instance, the improper integral $\int_{-\infty}^{\infty} 1/x^2\,dx$ is troublesome for four reasons: $\lim_{x\to 0^-} 1/x^2 = \infty$, $\lim_{x\to 0^+} 1/x^2 = \infty$, and the range extends infinitely to the left and also to the right. (See Fig. 9.) To treat the integral, write it as the sum of four improper integrals of the two basic types:

$$\int_{-\infty}^{\infty} \frac{1}{x^2}\,dx = \int_{-\infty}^{-1} \frac{1}{x^2}\,dx + \int_{-1}^{0} \frac{1}{x^2}\,dx + \int_{0}^{1} \frac{1}{x^2}\,dx + \int_{1}^{\infty} \frac{1}{x^2}\,dx.$$

All four of the integrals on the right have to be convergent for $\int_{-\infty}^{\infty} 1/x^2\,dx$ to be convergent. As a matter of fact, only the first and last are, so $\int_{-\infty}^{\infty} 1/x^2\,dx$ is divergent.

We conclude with two applications of improper integrals.

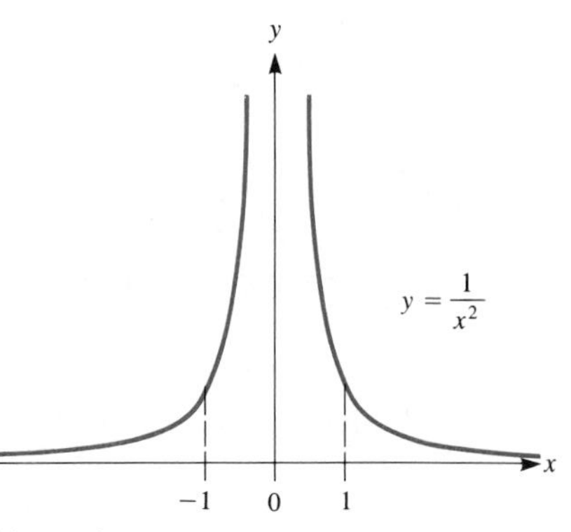

Figure 9

An Improper Integral in Economics

EXAMPLE 6 (*Present value of future income.*) Both business and government frequently face the question: ''What is 1 dollar t years in the future worth today?'' Implicit in this question are such considerations as the present value of a business being dependent on its future profit and the cost of a dam being weighed against its future revenue. Determine the present value of a business whose rate of profit t years in the future is $f(t)$ dollars per year.

SOLUTION To begin the analysis, assume that the annual interest rate r remains constant and that 1 dollar deposited today is worth e^{rt} dollars t years from now. This assumption corresponds to continuously compounded interest or to natural growth. Thus A dollars today will be worth Ae^{rt} dollars t years from now. (See Sec. 6.2.) What is the present value of the promise of 1 dollar t years from now? In other words, what amount A invested today will be worth 1 dollar t years from now? To find out, solve the equation $Ae^{rt} = 1$ for A. The solution is

The present value of \$1 t years from now is \$$e^{-rt}$.

$$A = e^{-rt}. \tag{1}$$

Now consider the present value of the future profit of a business (or future revenue of a dam). Assume that the profit flow t years from now is $f(t)$ dollars per year. This rate may vary within the year; consider f to be a continuous function of time. The profit in the small interval of time dt, from time t to time $t + dt$, would be approximately $f(t)\,dt$. The total future profit, $F(T)$, from now, when $t = 0$, to some time T in the future is therefore

$$F(T) = \int_0^T f(t)\,dt. \tag{2}$$

But the **present value** of the future profit is *not* given by Eq. (2). It is necessary to consider the present value of the profit earned in a typical short interval of time from t to $t + dt$. According to Eq. (1), its present value is approximately

$$e^{-rt} f(t)\,dt.$$

Hence the present value of future profit from $t = 0$ to $t = T$ is given by

$$\int_0^T e^{-rt} f(t)\,dt. \tag{3}$$

The present value of all future profit is therefore the improper integral $\int_0^\infty e^{-rt} f(t)\,dt$. ■

To see what influence the interest rate r has, denote by $P(r)$ the present value of all future revenue when the interest rate is r; that is,

$$P(r) = \int_0^\infty e^{-rt} f(t)\,dt. \tag{4}$$

If the interest rate r is raised, then according to Eq. (4) the present value of a business declines. An investor choosing between investing in a business or placing her money in a bank account finds the bank account more attractive when r is raised.

Laplace transform

Equation (4) assigns to a profit function f (which is a function of time t) a present-value function P, which is a function of r, the interest rate. In the theory of differential equations, P is called the **Laplace transform of** f. (See Exercises 36 to 42.)

An Improper Integral and Escape Velocity

In Sec. 8.7 we saw that the total work required to lift a 1-pound payload from the surface of the earth to the moon is 3,933 mile-pounds. The work required to launch the payload on an endless journey is given by the improper integral

$$\int_{4,000}^{\infty} \left(\frac{4,000}{r}\right)^2 dr = 4,000 \text{ mile-pounds.}$$

Since the integral is convergent, only a finite amount of energy is needed to send a payload on an endless journey—as the Voyager spacecraft has demonstrated. It takes only a little more energy than is required to lift the payload to the moon.

That the work required for the endless journey is finite raises the question "With what initial velocity must we launch the payload so that it never falls back, but continues to rise forever away from the earth?" If the initial velocity is too small, the payload will rise for a while, then fall back, as anyone who has thrown a ball straight up knows quite well.

The energy we supply the payload is kinetic energy. The force of gravity slows the payload and reduces its kinetic energy. We do not want the kinetic energy to shrink to zero. If it were ever zero, then the velocity of the payload would be zero. At that point the payload would start to fall back to earth.

As we will show, the kinetic energy of the payload is reduced by *exactly* the amount of work done on the payload by gravity. If v_{esc} is the minimal velocity needed for the payload to "escape" and not fall back, then

$$\tfrac{1}{2}m(v_{\text{esc}})^2 = 4,000 \text{ mile-pounds,} \tag{5}$$

where m is the mass of the payload. Equation (5) can be solved for v_{esc}, the **escape velocity**.

In order to solve (5) for v_{esc}, we must calculate the mass of a payload that weighs 1 pound at the surface of the earth. The weight of 1 pound is the gravitational force of the earth pulling on the payload. Newton's equation

$$\text{Force} = \text{Mass} \times \text{Acceleration}, \tag{6}$$

known as his "second law of motion," provides the relationship we need among force, mass, and the acceleration of that mass.

The acceleration of an object at the surface of the earth is 32 feet per second per second, or 0.0061 mile per second per second. Then (6), for the 1-pound payload, becomes

$$1 = m(0.0061) \tag{7}$$

Combining (5) and (7) gives

$$\frac{1}{2}\frac{1}{0.0061}(v_{\text{esc}})^2 = 4,000,$$

or

$$(v_{\text{esc}})^2 = (8,000)(0.0061) = 48.8$$

Hence $v_{\text{esc}} \approx 7$ miles per second, which is about 25,000 miles per hour, a speed first attained by human beings when Apollo 8 traveled to the moon in December 1968. All that remains is to justify the claim that the change in kinetic energy equals the work done by the force.

Let $v(r)$ be the velocity of the payload when it is r miles from the center of the earth. Let $F(r)$ be the force on the payload when it is r miles from the center of the earth. Since the force is in the opposite direction from the motion, we will define $F(r)$ to be negative.

Let a and b be numbers, $4{,}000 \le a < b$. (See Fig. 10.) We wish to show that

$$\underbrace{\frac{1}{2}m(v(b))^2 - \frac{1}{2}m(v(a))^2}_{\text{Change in kinetic energy}} = \underbrace{\int_a^b F(r)\,dr}_{\text{Work done by gravity}}. \tag{8}$$

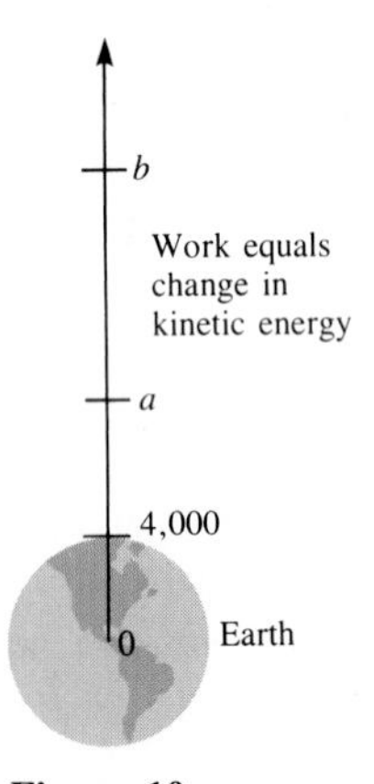

Figure 10

In this equation m is payload mass. Note that both sides of (8) are negative.

Equation (8) resembles the fundamental theorem of calculus. If we could show that $m(v(r))^2/2$ is an antiderivative of $F(r)$, then (8) would follow immediately. Let us find the derivative of $m(v(r))^2/2$ and show that it equals $F(r)$:

$$\begin{aligned}\frac{d}{dr}\left[\frac{1}{2}m(v(r))^2\right] &= mv(r)\frac{dv}{dr} = mv(r)\frac{dv/dt}{dr/dt} \qquad (t \text{ is time})\\ &= mv(r)\frac{d^2r/dt^2}{v(r)} = m\frac{d^2r}{dt^2}\\ &= \text{mass} \times \text{acceleration}\\ &= F(r). \qquad \text{Newton's law}\end{aligned}$$

Hence (8) is valid and we have justified our calculation of escape velocity.

Section Summary

We introduced two types of integrals that are not definite integrals, but are defined as limits of definite integrals. The "improper integral" $\int_a^\infty f(x)\,dx$ is defined as $\lim_{b\to\infty} \int_a^b f(x)\,dx$. If $f(x)$ is continuous in $[a, b]$ except at a, then $\int_a^b f(x)\,dx$ is defined as $\lim_{t\to a^+} \int_t^b f(x)\,dx$. A similar definition holds if $f(x)$ is not defined at b. We also developed two comparison tests for convergence or divergence of $\int_a^\infty f(x)\,dx$, where the integrand keeps a constant sign. In the case where the integrand $f(x)$ may have both positive and negative values, we showed that if $\int_a^\infty |f(x)|\,dx$ converges, so does $\int_a^\infty f(x)\,dx$.

EXERCISES FOR SEC. 8.8: IMPROPER INTEGRALS

In Exercises 1 to 22 determine whether the improper integral is convergent or divergent. Evaluate the convergent ones if possible.

1 $\displaystyle\int_1^\infty \frac{dx}{x^3}$ **2** $\displaystyle\int_1^\infty \frac{dx}{\sqrt[3]{x}}$

3 $\displaystyle\int_0^\infty e^{-x}\,dx$ **4** $\displaystyle\int_0^\infty \frac{dx}{x+100}$

5 $\displaystyle\int_0^\infty \frac{x^3\,dx}{x^4+1}$ **6** $\displaystyle\int_1^\infty x^{-1.01}\,dx$

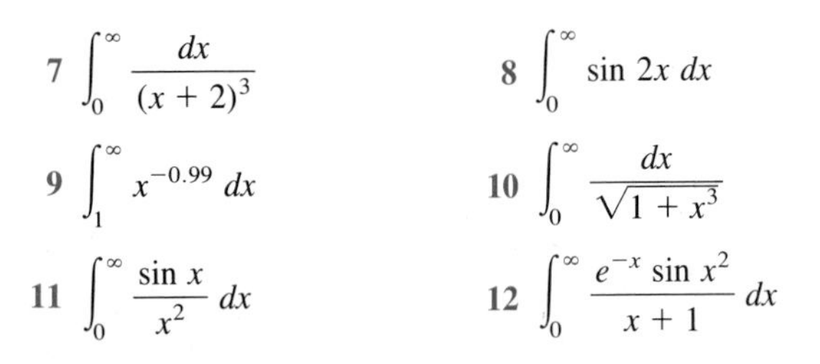

7 $\displaystyle\int_0^\infty \frac{dx}{(x+2)^3}$ **8** $\displaystyle\int_0^\infty \sin 2x\,dx$

9 $\displaystyle\int_1^\infty x^{-0.99}\,dx$ **10** $\displaystyle\int_0^\infty \frac{dx}{\sqrt{1+x^3}}$

11 $\displaystyle\int_0^\infty \frac{\sin x}{x^2}\,dx$ **12** $\displaystyle\int_0^\infty \frac{e^{-x}\sin x^2}{x+1}\,dx$

Exercises 13 to 18 require an integral table or techniques from Chap. 7.

13 $\displaystyle\int_1^\infty \frac{\ln x\,dx}{x}$

14 $\displaystyle\int_0^\infty \frac{dx}{x^2+4}$

15 $\displaystyle\int_0^\infty \frac{x\,dx}{x^4+1}$

16 $\displaystyle\int_0^\infty e^{-2x}\sin 3x\,dx$

17 $\displaystyle\int_0^1 \frac{dx}{\sqrt{x}\sqrt{1-x}}$ (Observe that the integrand is undefined at both 0 and 1.)

18 $\displaystyle\int_0^\infty \frac{dx}{(x+1)(x+2)(x+3)}$

19 $\displaystyle\int_0^1 \frac{dx}{\sqrt[3]{x}}$

20 $\displaystyle\int_0^\infty \frac{x\,dx}{\sqrt{1+x^4}}$

21 $\displaystyle\int_0^\infty \frac{e^{-x}}{\sqrt{x}}\,dx$

22 $\displaystyle\int_0^1 \frac{dx}{(x-1)^2}$

23 Let R be the region between the curves $y = 1/x$ and $y = 1/(x+1)$ to the right of the line $x = 1$. Is the area of R finite or infinite? If it is finite, evaluate it.

24 Let R be the region between the curves $y = 1/x$ and $y = 1/x^2$ to the right of $x = 1$. Is the area of R finite or infinite? If it is finite, evaluate it.

25 Describe how you would go about estimating $\int_0^\infty e^{-x^2}\,dx$ with an error less than 0.01. (Do not do the arithmetic.)

26 Describe how you would go about estimating $\displaystyle\int_0^\infty \frac{dx}{\sqrt{1+x^4}}$ with an error less than 0.01. (Do not do the arithmetic.)

---

27 The function $f(x) = (\sin x)/x$ for $x \neq 0$ and $f(0) = 1$ occurs in communication theory. Show that the energy E of the signal represented by f is finite, where

$$E = \int_{-\infty}^{\infty} [f(x)]^2\,dx.$$

28 Plankton are small football-shaped organisms. The resistance they meet when falling through water is proportional to the integral

$$\int_0^\infty \frac{dx}{\sqrt{(a^2+x)(b^2+x)(c^2+x)}},$$

where a, b, and c describe the dimensions of the plankton. Is this improper integral convergent or divergent?

29 In R. P. Feynman, *Lectures on Physics,* Addison-Wesley, Reading, Mass., 1963, appears this remark: " . . . the expression becomes

$$\frac{U}{V} = \frac{(kT)^4}{\hbar^3\pi^2c^3}\int_0^\infty \frac{x^3\,dx}{e^x-1}.$$

This integral is just some number that we can get, approximately, by drawing a curve and taking the area by counting squares. It is roughly 6.5. The mathematicians among us can show that the integral is exactly $\pi^4/15$."

Show at least that the integral is convergent.

30 For which positive constants p is $\int_0^1 dx/x^p$ convergent? divergent?

31 For which positive constants p is $\int_1^\infty dx/x^p$ convergent? divergent?

32 Let $f(x)$ be a positive function and let R be the region under $y = f(x)$ and above $[1, \infty]$. Assume that the area of R is infinite. Does it follow that the volume of the solid of revolution formed by revolving R about the x axis is infinite?

The remaining exercises require techniques from Chap. 7 or an integral table.

33 If the profit flow in Example 6 remains constant, say, $f(t) = k > 0$, the total future profit is obviously infinite.
(*a*) Show that the present value is k/r, which is finite.
(*b*) Why is the result in (*a*) reasonable? (Explain without calculus.)

34 (*a*) Show that $\int_1^\infty (\cos x)/x^2\,dx$ is convergent.
(*b*) Show that $\int_1^\infty (\sin x)/x\,dx$ is convergent. (*Hint:* Start with integration by parts.)
(*c*) Show that $\int_0^\infty (\sin x)/x\,dx$ is convergent.
(*d*) Show that $\int_0^\infty \sin e^x\,dx$ is convergent.

35 Find the error in the following computations: The substitution $x = y^2$, $dx = 2y\,dy$, yields

$$\int_0^1 \frac{1}{x}\,dx = \int_0^1 \frac{2y}{y^2}\,dy = \int_0^1 \frac{2}{y}\,dy$$

$$= 2\int_0^1 \frac{1}{y}\,dy = 2\int_0^1 \frac{1}{x}\,dx.$$

Hence $$\int_0^1 \frac{1}{x}\,dx = 2\int_0^1 \frac{1}{x}\,dx,$$

from which it follows that $\int_0^1 dx/x = 0$.

Let $f(t)$ be a continuous function defined for $t \geq 0$. Assume that, for certain fixed positive numbers r, $\int_0^\infty e^{-rt}f(t)\,dt$ converges and that $e^{-rt}f(t) \to 0$ as $t \to \infty$. Define $P(r)$ to be $\int_0^\infty e^{-rt}f(t)\,dt$. The function P is called the **Laplace transform** of the function f. It is an important tool for solving differential equations. In Exercises 36 to 40 find the Laplace transforms of the given functions.

36 $f(t) = t$

37 $f(t) = t^2$

38 $f(t) = e^t$ (assume $r > 1$)

39 $f(t) = \sin t$

40 $f(t) = \cos t$

41 Let f and its derivative f' both have Laplace transforms. Let P be the Laplace transform of f, and let Q be the Laplace transform of f'. Show that

$$Q(r) = -f(0) + rP(r).$$

42 Let P be the Laplace transform of f. Let a be a positive constant, and let $g(t) = f(at)$. Let P be the Laplace transform of f, and let Q be the Laplace transform of g. Show that $Q(r) = (1/a)P(r/a)$.

8.S SUMMARY

The following table summarizes most of the applications of the definite integral treated in this chapter:

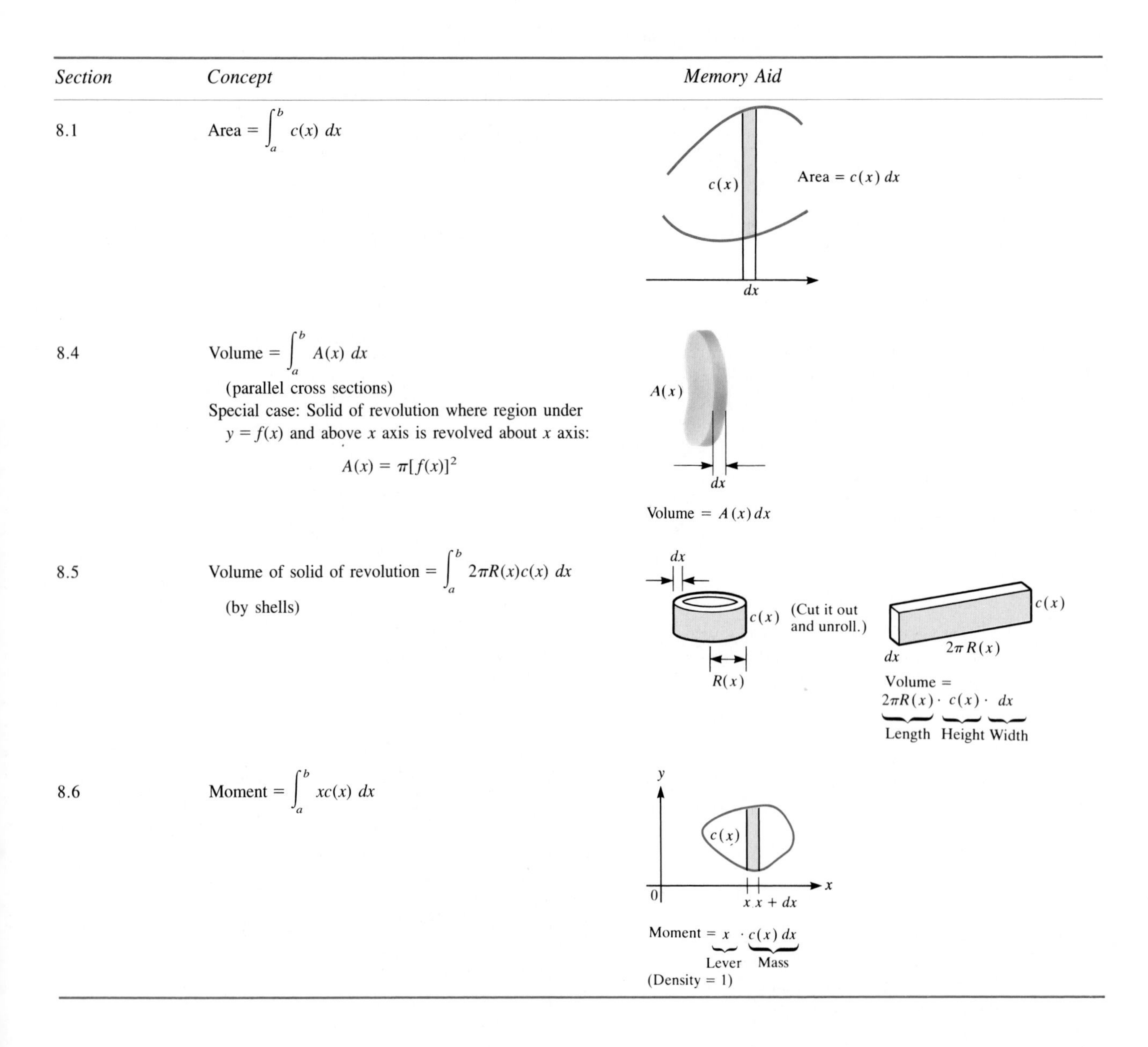

Section	*Concept*	*Memory Aid*
8.1	Area $= \int_a^b c(x)\, dx$	$c(x)$; Area $= c(x)\, dx$; dx
8.4	Volume $= \int_a^b A(x)\, dx$ (parallel cross sections) Special case: Solid of revolution where region under $y = f(x)$ and above x axis is revolved about x axis: $A(x) = \pi[f(x)]^2$	$A(x)$; dx; Volume $= A(x)\, dx$
8.5	Volume of solid of revolution $= \int_a^b 2\pi R(x)c(x)\, dx$ (by shells)	dx; $c(x)$; $R(x)$; (Cut it out and unroll.) $c(x)$; dx; $2\pi R(x)$; Volume $= \underbrace{2\pi R(x)}_{\text{Length}} \cdot \underbrace{c(x)}_{\text{Height}} \cdot \underbrace{dx}_{\text{Width}}$
8.6	Moment $= \int_a^b xc(x)\, dx$	y; $c(x)$; 0; x; x $x + dx$; Moment $= \underbrace{x}_{\text{Lever}} \cdot \underbrace{c(x)\, dx}_{\text{Mass}}$ (Density = 1)

Vocabulary and Symbols

cross-sectional length $c(x)$, $c(y)$
cross-sectional area $A(x)$
solid of revolution
shell technique
lever arm
moment about a line
centroid, center of mass, $(\bar{x}, \bar{y})$
work
improper integral (convergent and divergent)

Key Facts

The centroid $(\bar{x}, \bar{y})$ of a plane region R is given by

$$\bar{x} = \frac{\int_a^b xc(x)\,dx}{\text{Area of } R} \quad \text{and} \quad \bar{y} = \frac{\int_c^d yc(y)\,dy}{\text{Area of } R}.$$

If R is the region below $y = f(x)$ and above $[a, b]$, then

$$\bar{x} = \frac{\int_a^b xf(x)\,dx}{\text{Area of } R} \quad \text{and} \quad \bar{y} = \frac{\frac{1}{2}\int_a^b [f(x)]^2\,dx}{\text{Area of } R}.$$

If $0 \le f(x) \le g(x)$ and $\int_a^\infty g(x)\,dx$ is convergent, so is $\int_a^\infty f(x)\,dx$ and $\int_a^\infty f(x)\,dx \le \int_a^\infty g(x)\,dx$.

If $0 \le g(x) \le f(x)$ and $\int_a^\infty g(x)\,dx$ is divergent, so is $\int_a^\infty f(x)\,dx$.

If $\int_a^\infty |f(x)|\,dx$ is convergent, so is $\int_a^\infty f(x)\,dx$.

GUIDE QUIZ ON CHAP. 8: APPLICATIONS OF THE DEFINITE INTEGRAL

1 (*a*) Draw the region R bounded by $y = (2 - x)/2$ and $y = x^2$.
(*b*) Set up a definite integral for the area of R using vertical cross sections.
(*c*) Like (*b*), with horizontal cross sections.
(*d*) Find the area of R.

2 (*a*) Draw a solid right circular cone.
(*b*) Draw the cross section of the cone made by a plane that includes the axis of the cone.
(*c*) Draw the cross section of the cone made by a plane that does not meet the axis of the cone.
(*d*) Draw the typical cross section made by a plane that meets the axis of the cone in just one point.

3 Let $f(x)$ and $g(x)$ be functions such that $0 \le g(x) \le f(x)$. The region between the curves $y = g(x)$ and $y = f(x)$ and above the interval $[a, b]$ is revolved about the line $y = -3$ to obtain a solid of revolution S. Drawing a local approximation, obtain a definite integral for the volume of S.

4 Let R be the region shown in Fig. 1.

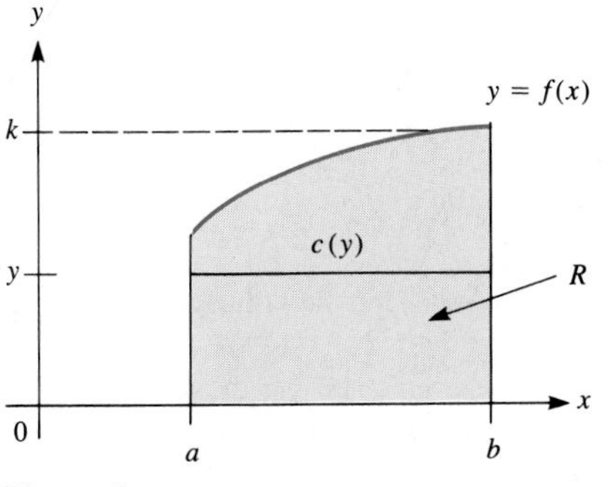

Figure 1

(*a*) Why is it reasonable to use $\int_0^k yc(y)\,dy$ to measure the moment of R about the x axis? (Sketch the local estimate.)
(*b*) Why is the moment in (*a*) equal to $\int_a^b \frac{1}{2}(f(x))^2\,dx$?

5 Find the centroid of the trapezoid shown in Fig. 2.

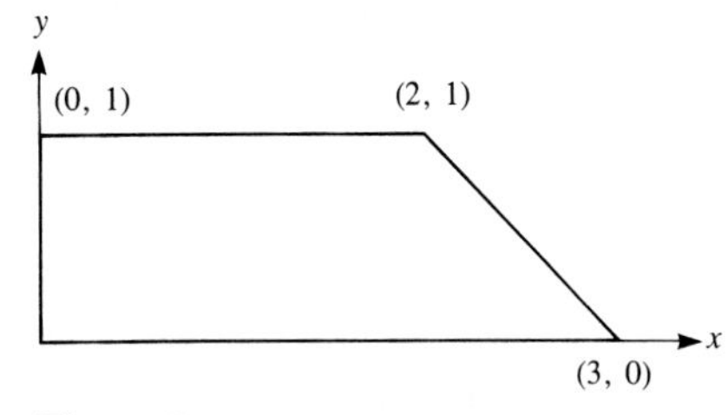

Figure 2

6 If the work required to stretch a spring 6 centimeters from its rest position is W, how much work is required to stretch it 3 centimeters from its rest position?

7 Let R be the region beneath $y = 1/x^2$ and above $[1, \infty)$.
(*a*) Obtain an (improper) integral for the area of R using horizontal cross sections.
(*b*) Evaluate the integral in (*a*)
(*c*) Obtain an (improper) integral for the area of R using vertical cross sections.
(*d*) Evaluate the integral in (*c*).
(*e*) Let S be the solid of revolution obtained by revolving R about the x axis. Set up an (improper) integral for the volume of S by the slab method.
(*f*) Evaluate the integral in (*e*).
(*g*) Set up an (improper) integral for the volume of S using the shell technique.
(*h*) Evaluate the integral in (*g*).

8 (*a*) With the aid of diagrams, explain why $\int_a^\infty f(x)\,dx$ is convergent if $\int_a^\infty |f(x)|\,dx$ is convergent.
(*b*) If $\int_a^\infty |f(x)|\,dx = 15$, what can be said about $\int_a^\infty f(x)\,dx$?

REVIEW EXERCISES FOR CHAP. 8: APPLICATIONS OF THE DEFINITE INTEGRAL

In Exercises 1 to 4 set up integrals for the given quantities; do not evaluate them.

1 The area of the region above the parabola $y = x^2$ and below the line $y = 2x$, using (*a*) vertical cross sections, (*b*) horizontal cross sections.

2 The volume of the wedge cut from a right circular cylinder of height 5 inches and radius 3 inches by a plane that bisects one base and touches the other base at one point.

3 The volume of the solid obtained by revolving the triangle whose vertices are (2, 0), (2, 1), and (3, 2) about the x axis. (Use the shell technique.)

4 The moment of the region in the first quadrant bounded by $y = x^2$ and $y = x^3$, about the line $y = -2$.

In Exercises 5 to 9

(*a*) Find the area of R.

(*b*) Find the volume of the solid of revolution formed by revolving R about the x axis.

(*c*) The same as (*b*), but around the y axis.

(*d*) The same as (*b*), but around the line $y = -1$.

5 R is the region below the curve $y = x/(1 + x)$ and above $[1, 2]$.

6 R is the region below $y = 1/(1 + x)^2$ and above $[0, 1]$.

7 R is the region below $y = \sin 2x$ and above $[0, \pi/2]$.

8 R is the region below $y = \sqrt{x^2 - 9}$ and above $[3, 4]$.

9 R is the region below $y = 1/(2x + 1)$ and above $[0, 1]$.

In Exercises 10 to 13 find the moments of the given regions R about the given lines L. In each case the density is 1.

10 R: below $y = \sec x$, above $[\pi/6, \pi/4]$; L: the x axis.

11 R: below $y = (\sin x)/x$, above $[\pi/2, \pi]$; L: the y axis.

12 R: below $y = 1/\sqrt{x^2 + 1}$, above $[0, 1]$; L: the x axis.

13 R: below $y = 1/\sqrt{x^2 + 1}$, above $[0, 1]$; L: the y axis.

In Exercises 14 to 18 find the area of the given regions R.

14 R is the region below $y = 1/(x^2 + 3x + 2)$ and above $y = 1/(x^2 + 3x + 4)$, between $x = 0$ and $x = 1$.

15 R is below $y = x$ and above $y = \tan^{-1} x$, between $x = 0$ and $x = 1$.

16 R is below $y = x\sqrt{2x + 1}$ and above $[0, 4]$.

17 R is below $y = \cos^3 x$ and above $y = \sin^3 x$, between $x = 0$ and $x = \pi/4$.

18 R is below $y = 1/\sqrt{4 - x^2}$ and above $y = x/\sqrt{4 - x^2}$, between $x = 0$ and $x = 1$.

19 Find the area of the region between the curves $y = 1/(x^2 - x)$ and $y = 1/(x^3 - x)$ (*a*) between $x = 2$ and $x = 3$, (*b*) to the right of $x = 3$.

20 Let R be the region below $y = \tan x$ and above $[0, \pi/4]$.

(*a*) Find the area of R.

(*b*) Find the moment of R about the x axis.

(*c*) Find $\bar{y}$.

(*d*) Set up an integral for the moment of R about the y axis. (Don't try to evaluate it; the fundamental theorem of calculus is useless here.)

(*e*) Find the volume of the solid of revolution formed by revolving R around the line $y = -1$.

21 Let R be the region below $y = \sin^2 x$ and above $y = \sin^3 x$, between $x = 0$ and $x = \pi/2$.

(*a*) Find the area of R.

(*b*) Find the moment of R about the x axis. (See Formula 73 in the integral table for $\int_0^{\pi/2} \sin^n x\, dx$.)

(*c*) Find $\bar{y}$.

22 Is $\int_0^\infty dx/(\sqrt{x}\sqrt{x+1}\sqrt{x+2})$ convergent or divergent?

23 A drill of radius a inches bores a hole through the center of a sphere of radius b inches, leaving a ring whose height is 2 inches. Find the volume of the ring.

24 A barrel is made by rotating an ellipse around one of its axes and then cutting off equal caps, top and bottom. It is 3 feet high and 3 feet wide at its midsection. Its top and bottom have a diameter of 2 feet. What is its volume?

25 Let R be the region to the right of the y axis, below $y = e^{-x}$ and above the x axis.

(*a*) Find the area of R.

(*b*) Find the volume of the region obtained by revolving R about the x axis.

(*c*) Find the volume of the region obtained by revolving R about the y axis.

26 Let l be a line which intersects the triangle ABC and is parallel to BC. Suppose that l is twice as far from the point A as from the line BC. (See Fig. 3.) Show that the centroid of ABC is on l.

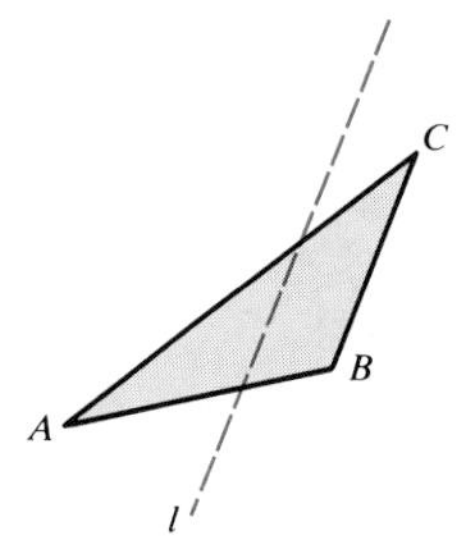

Figure 3

27 (*a*) Develop the shell formula for the volume of a solid of revolution.

(*b*) What is the device for remembering the formula?

28 Show that $\int_0^\infty e^{-rx} \sin ax\, dx$ equals $a/(a^2 + r^2)$, where $r > 0$ and a are constants.

29 Show that $\int_0^\infty e^{-rx} \cos ax\, dx$ equals $r/(a^2 + r^2)$, where $r > 0$ and a are constants.

30 Let f be a continuous function such that $f(0) = 2$ and $f(x) \to 3$ as $x \to \infty$. Find the limit of $(1/b) \int_0^b f(x)\, dx$ as (*a*) $b \to 0$, (*b*) $b \to \infty$.

31 By interpreting these improper integrals as expressions for the area of a certain region, show that

$$\int_0^\infty \frac{dx}{1+x^2} = \int_0^1 \sqrt{\frac{1-y}{y}}\, dy.$$

32 Define $G(a) = \int_0^\infty a/(1+a^2x^2)\, dx$.
(*a*) Compute $G(0)$.
(*b*) Compute $G(a)$ if a is negative.
(*c*) Compute $G(a)$ if a is positive.
(*d*) Graph G.

33 Prove that $\int_0^\infty (\sin x^2)/x\, dx = \frac{1}{2}\int_0^\infty (\sin x)/x\, dx$.

34 Is $\int_0^\infty dx/(x-1)^2$ convergent or divergent?

35 Evaluate $\int_0^\infty e^{-x} \sin(2x+3)\, dx$.

36 (*a*) Sketch $y = e^{-x}(1+\sin x)$ for $x \ge 0$.
(*b*) The region beneath the curve in (*a*) and above the positive x axis is revolved around the y axis. Find the volume of the resulting solid.

37 From the fact that $\int (e^x/x)\, dx$ is not elementary, deduce that $\int e^x \ln x\, dx$ is not elementary.

38 Is this computation correct?

$$\int_{-2}^{1} \frac{dx}{2x+1} = \frac{1}{2}\ln|2x+1| \Big|_{-2}^{1} = \frac{1}{2}\ln 3 - \frac{1}{2}\ln 3 = 0.$$

39 Find the error in the following computations:

$$\int_{-1}^{1} \frac{1}{x^2}\, dx = \frac{-1}{x}\Big|_{-1}^{1} = \frac{-1}{1} - \frac{-1}{-1} = -2.$$

(The integrand is positive, yet the integral is negative.)

40 It can be proved that $\int_0^\infty x^{n-1}/(1+x)\, dx = \pi \csc n\pi$ for $0 < n < 1$. Verify that this equation is correct for $n = \frac{1}{2}$.

41 Compute $\int_0^1 x^4 \ln x\, dx$.

42 Show that $\int_0^\infty dx/(1+x^4) = \int_0^\infty x^2\, dx/(1+x^4)$. *Hint:* Let $x = 1/y$.

43 Show that $\int_0^1 (-\ln x)^3\, dx = \int_0^\infty x^3 e^{-x}\, dx$.

44 A solid is formed by revolving the region below $y = e^{2x} \sin 3x$ and above $[0, \pi/3]$ around the x axis. Find its volume.

45 Find the centroid of the region bounded by the parabola $y = x^2$ and the line $y = 3x - 2$.

46 Find the centroid of the finite region bounded by $y = 2^x$ and $y = x^2$, to the right of the y axis.

47 Is $\displaystyle\int_0^\infty \frac{x^2 - 5x^3}{x^6+1} \sin 3x\, dx$ convergent or divergent?

48 Is $\displaystyle\int_0^1 \frac{\ln x}{1-x^2}\, dx$ convergent or divergent?

49 (*The gamma function*: A generalization of $n!$ from integers to any positive n.) For a real number $n > 0$, define $\Gamma(n)$ to be $\int_0^\infty e^{-x}x^{n-1}\, dx$.
(*a*) Evaluate $\Gamma(1)$.
(*b*) Show that $\Gamma(n+1) = n\Gamma(n)$.
(*c*) Using (*a*) and (*b*), evaluate $\Gamma(2)$, $\Gamma(3)$, $\Gamma(4)$, and $\Gamma(5)$.
(*d*) What is the relationship between $n!$ and $\Gamma(n)$?

The **gamma function** generalizes the factorial, which is defined only at nonnegative integers, to all positive real numbers. Handbooks of mathematical tables usually include values of the gamma function.

50 At time t, $0 \le t \le \pi/2$, a particle is at the point $x = A \sin t$ on the x axis.
(*a*) Find the average of the square of its speed with respect to distance.
(*b*) Find the average of the square of its speed with respect to time.

51 Water flows out of a hole in the bottom of a cylindrical tank of radius r and height h at the rate of $\sqrt{y}$ cubic feet per second when the depth of the water is y feet. Initially the tank is full. (See Fig. 4.)
(*a*) How long will it take to become half full?
(*b*) How long will it take to empty?

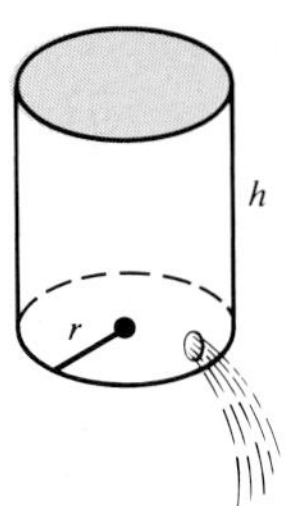

Figure 4

52 Is the area under the curve $y = (\ln x)/x^2$, above the x axis and to the right of the line $x = 1$, finite or infinite?

53 (*a*) Let $G(a) = \int_0^\infty 1/[(1+x^a)(1+x^2)]\, dx$. Evaluate $G(0)$, $G(1)$, $G(2)$.
(*b*) Show, using the substitution $x = 1/y$, that

$$G(a) = \int_0^\infty \frac{x^a\, dx}{(1-x^a)(1+x^2)}.$$

(*c*) From (*b*), show that $G(a) = \pi/4$, independent of a.

54 (*a*) Show that the region R under $y = 1/x$ and above $[1, \infty)$ has infinite area. When you revolve R about the x axis you obtain a solid S. Show that S has finite volume.
(*b*) If you filled S with paint, you would be covering R with a finite amount of paint, even though it has infinite area. How would you resolve this apparent paradox?

55 Sketch the region common to two circles of radius 1 whose centers are a distance 1 apart. Find the area of this region, using (*a*) vertical cross sections, (*b*) horizontal cross sections, (*c*) elementary geometry, but no calculus.

56 When computing the internal energy of a crystal, Claude Garrod, in *Twentieth Century Physics*, Faculty Publishing, Davis, Calif., 1984, p. 326, states that the integral

$$\int_0^{\pi/2} \frac{\sin x}{e^{0.26 \sin x} - 1}\, dx$$

"cannot be evaluated analytically. However, it can easily be computed numerically using Simpson's rule. The result is 5.56."

(*a*) Is the integral proper or improper?
(*b*) How does the integrand behave when x is near 0?
(*c*) What does "cannot be evaluated analytically" mean?
(*d*) Use Simpson's rule with $n = 6$ to estimate the integral.

57 Find the error in the following computations: Using the substitution $u = 1/x$, $du = -1/x^2\, dx$, we have

$$\int_{-1}^{1} \frac{1}{1+x^2}\, dx = \int_{-1}^{1} \frac{1}{1+1/u^2}\left(-\frac{1}{u^2}\, du\right)$$
$$= -\int_{-1}^{1} \frac{1}{1+u^2}\, du.$$

58 Let f be a function such that $f(x) > 0$. Assume that f has derivatives of all orders and that $\ln f(x) = f(x) \int_0^x f(t)\, dt$. Find (*a*) $f(0)$, (*b*) $f^{(1)}(0)$, (*c*) $f^{(2)}(0)$.

59 Find the number a, $0 \le a \le 2\pi$ that maximizes the function

$$f(a) = \int_0^{2\pi} \sin x \sin (x + a)\, dx.$$

A nonnegative function $f(x)$, such that $\int_{-\infty}^{\infty} f(x)\, dx = 1$, is called a **probability distribution**. [The probability that a certain variable observed in an experiment is between x and $x + \Delta x$ is approximately $f(x)\,\Delta x$.] The improper integral $\int_{-\infty}^{\infty} xf(x)\, dx$, if it exists, is called the **mean** of the distribution and is denoted μ. The improper integral $\int_{-\infty}^{\infty} (x-\mu)^2 f(x)\, dx$, if it exists, is called the **variance** of the distribution and is denoted μ_2. The square root of μ_2 is called the **standard deviation** of the distribution and is denoted σ.

60 Let k be a positive constant. Define $f(x)$ to be ke^{-kx} if $x > 0$ and 0 if $x \le 0$.
(*a*) Show that $\int_{-\infty}^{\infty} f(x)\, dx = 1$.
(*b*) Find μ. (*c*) Find μ_2. (*d*) Find σ.

61 Let $f(x) = e^{-x^2/2k^2}/(\sqrt{2\pi}\, k)$, where k is a positive constant. This is a **normal distribution**.
(*a*) Assume that $\int_0^{\infty} e^{-x^2}\, dx = \sqrt{\pi}/2$. Show that $\int_{-\infty}^{\infty} f(x)\, dx = 1$.
(*b*) Find μ. (*c*) Find μ_2. (*d*) Find σ.

62 Estimate $\int_0^{\infty} e^{-x^2}\, dx$ to four decimal places.

63 The following quote is taken from an article on the energy stored in solar ponds:

> The effect of the free surface of the pond on the temperature at the bottom of the pond is given by an expression involving the integral
>
> $$\int_a^{\infty} \left(1 - \frac{k}{v^2}\right) e^{-v^2}\, dv.$$

Show that the integral is convergent. (The constants a and k are positive.)

64 If the force of gravity varied as $1/r$ instead of $1/r^2$, how high would a 1-pound payload rise if you launched it at a speed of 7 miles per second. (See discussion of escape velocity in Sec. 8.8.)

65 It follows from Exercise 49 that $\int_0^{\infty} x^n e^{-x}\, dx = n!$. Use this to find $\int_0^{\infty} x^n e^{-ax}\, dx$, where a is a positive constant.

66 A particle is drawn toward a fixed particle by the force of gravitational attraction. Show that if the force is proportional to r^{-2}, where r is the distance between the particles, then the total work accomplished when the particles meet is infinite. (This result, which violates common sense, suggests that at atomic distances the force of gravity must vary according to some other formula. This paradox is resolved in quantum mechanics.)

9

PLANE CURVES AND POLAR COORDINATES

This chapter presents further applications of the derivative and integral. Section 9.1 describes polar coordinates. (There is no calculus here.) Section 9.2 shows how to compute the area of a flat region that has a convenient description in polar coordinates. Section 9.3 introduces a method of describing a curve that is especially useful in the study of motion. (Instead of y being a function of x, x and y are both functions of a third variable, which is often time t or angle θ.)

Section 9.4 shows how to compute the speed of an object moving along a curved path. It also shows how to express the length of a curve as a definite integral. Section 9.5 shows how to express the area of a "surface of revolution" as a definite integral. The sphere is an instance of such a surface.

Section 9.6 shows how the derivative and second derivative provide tools for measuring how curvy a curve is at each of its points. This measure, called "curvature," will be needed in Chap. 13 in the study of motion along a curve.

Section 9.7 develops the reflecting properties of the parabola, ellipse, and hyperbola, illustrating their many applications, from microwave reflectors to ovens.

9.1 POLAR COORDINATES

Rectangular coordinates are only one of the ways to describe points in the plane by pairs of numbers. In this section another system called polar coordinates is described.

Polar Coordinates

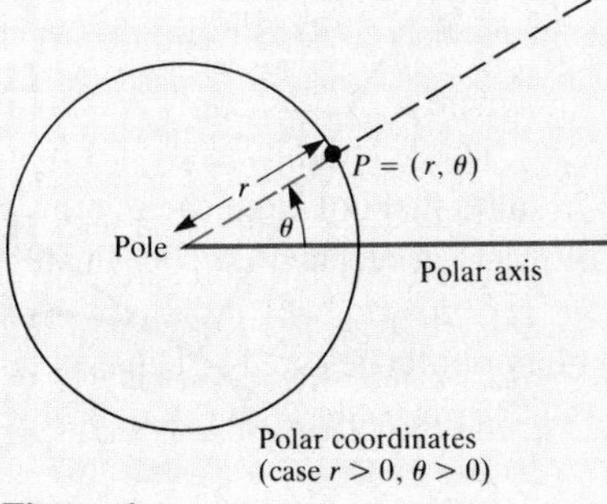

Figure 1

The rectangular coordinates x and y describe a point P in the plane as the intersection of a vertical line and a horizontal line. Polar coordinates describe a point P as the intersection of a circle and a ray from the center of that circle. They are defined as follows.

Select a point in the plane and a ray emanating from this point. The point is called the **pole**, and the ray the **polar axis**. (See Fig. 1.) Measure positive angles θ counterclockwise from the polar axis and negative angles clockwise.

Now let r be a number. To plot the point P that corresponds to the pair of numbers r and θ, proceed as follows:

If r is positive, P is the intersection of the circle of radius r whose center is at the pole and the ray of angle θ emanating from the pole.

If r is 0, P is the pole, no matter what θ is.

If r is negative, P is at a distance $|r|$ from the pole on the ray directly opposite the ray of angle θ, that is, on the ray of angle $\theta + \pi$.

In each case P is denoted (r, θ), and the pair r and θ are called **polar coordinates** of P. Note that the point (r, θ) is on the circle of radius $|r|$ whose center is the pole. Also observe that the pole is the midpoint of the points (r, θ) and $(-r, \theta)$. Notice that the point $(-r, \theta + \pi)$ is the same as the point (r, θ). Moreover, changing the angle by 2π does not change the point; that is, $(r, \theta) = (r, \theta + 2\pi) = (r, \theta + 4\pi) = \cdots = (r, \theta + 2k\pi)$ for any integer k.

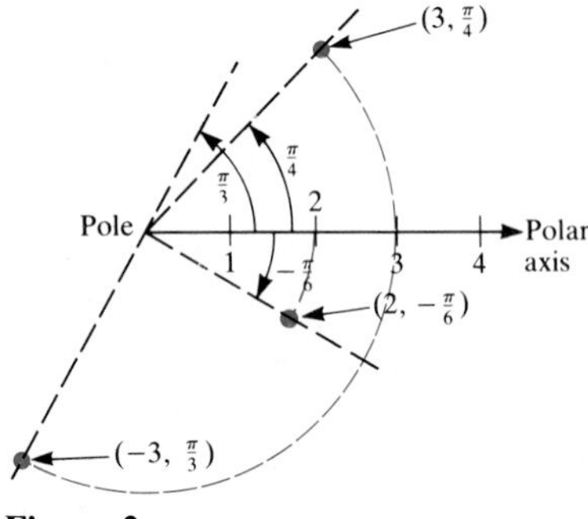

Figure 2

EXAMPLE 1 Plot the points $(3, \pi/4)$, $(2, -\pi/6)$, $(-3, \pi/3)$ in polar coordinates.

SOLUTION To plot $(3, \pi/4)$, go out a distance 3 on the ray of angle $\pi/4$ (shown in Fig. 2). To plot $(2, -\pi/6)$, go out a distance 2 on the ray of angle $-\pi/6$. To plot $(-3, \pi/3)$, draw the ray of angle $\pi/3$, and then go a distance 3 in the *opposite* direction from the pole. (See Fig. 2.) ■

The relation between polar and rectangular coordinates

It is customary to have the polar axis coincide with the positive x axis as in Fig. 3. In that case, inspection of the diagram shows the following relation between the rectangular coordinates (x, y) and the polar coordinates of the point P:

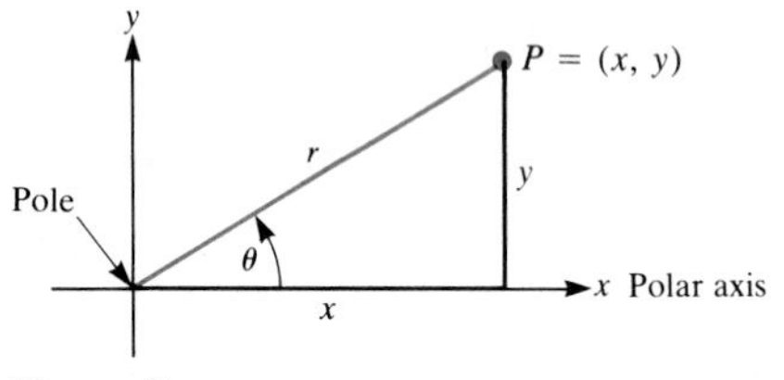

Figure 3

$$x = r\cos\theta \qquad y = r\sin\theta,$$
$$r^2 = x^2 + y^2 \qquad \tan\theta = \frac{y}{x}.$$

These equations hold even if r is negative. If r is positive, then $r = \sqrt{x^2 + y^2}$. Furthermore, if $-\pi/2 < \theta < \pi/2$, then $\theta = \tan^{-1}(y/x)$.

Graphing $r = f(\theta)$

Just as we may graph the set of points (x, y), where x and y satisfy a certain equation, so may we graph the set of points (r, θ), where r and θ satisfy a certain equation. It is important, however, to keep in mind that although each point in the plane is specified by a unique ordered pair (x, y) in rectangular coordinates, there are *many ordered pairs (r, θ) in polar coordinates which specify each point*. For instance, the point whose rectangular coordinates are $(1, 1)$ has polar coordinates $(\sqrt{2}, \pi/4)$ or $(\sqrt{2}, \pi/4 + 2\pi)$ or $(\sqrt{2}, \pi/4 + 4\pi)$ or $(-\sqrt{2}, \pi/4 + \pi)$ and so on.

The simplest equation in polar coordinates has the form $r = k$, where k is a positive constant. Its graph is the circle of radius k, centered at the pole. (See

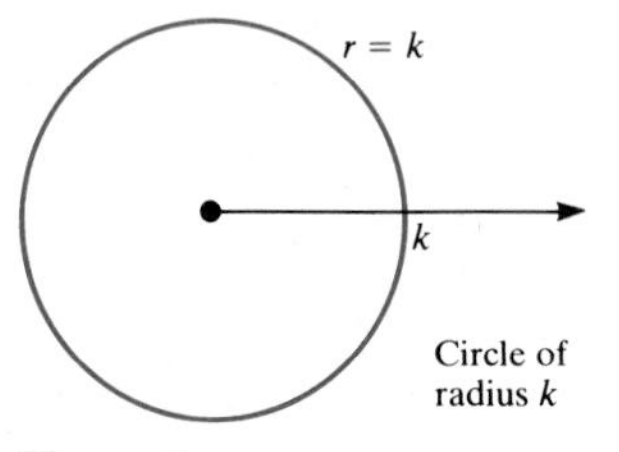

Figure 4

SOLUTION Since $y = r \sin \theta$, $r \sin \theta = 2$,

or
$$r = \frac{2}{\sin \theta} = 2 \csc \theta.$$

This is more complicated than the original equation, but it is still sometimes useful. ■

EXAMPLE 7 Transform the equation $r = 2 \cos \theta$ into rectangular coordinates and graph it.

SOLUTION Since $r^2 = x^2 + y^2$ and $r \cos \theta = x$, first multiply the equation $r = 2 \cos \theta$ by r, obtaining

$$r^2 = 2r \cos \theta.$$

Hence
$$x^2 + y^2 = 2x.$$

To graph this curve, rewrite the equation as

$$x^2 - 2x + y^2 = 0$$

and complete the square, obtaining

$$(x - 1)^2 + y^2 = 1.$$

The graph is a circle of radius 1 and center at (1, 0) in rectangular coordinates. It is graphed in Fig. 13. ■

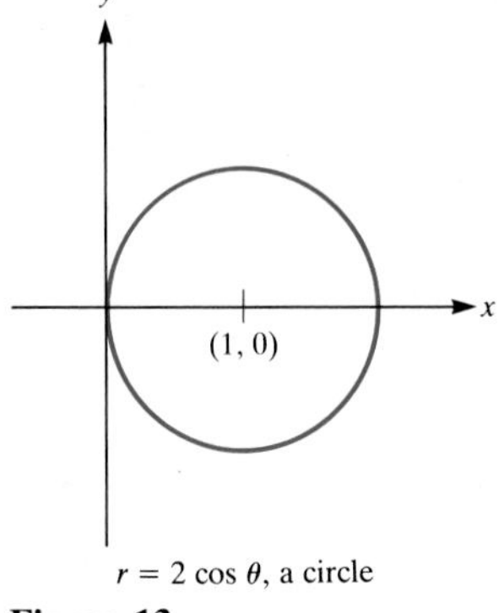

$r = 2 \cos \theta$, a circle

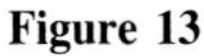
Figure 13

Remark: The step in Example 7 where we multiply by r deserves some attention. If $r = 2 \cos \theta$, then certainly $r^2 = 2r \cos \theta$. However, if $r^2 = 2r \cos \theta$, it does not follow that $r = 2 \cos \theta$. We can "cancel the r" only when r is not 0. If $r = 0$, it is true that $r^2 = 2r \cos \theta$, but it is not necessarily true that $r = 2 \cos \theta$. Since $r = 0$ satisfies the equation $r^2 = 2r \cos \theta$, the pole is on the curve $r^2 = 2r \cos \theta$. Luckily, it is also on the original curve $r = 2 \cos \theta$, since $\theta = \pi/2$ makes $r = 0$. Hence the graphs of $r^2 = 2r \cos \theta$ and $r = 2 \cos \theta$ are the same. [However, as you may check, the graphs of $r = 2 + \cos \theta$ and $r^2 = r(2 + \cos \theta)$ are *not* the same. The origin lies on the second curve, but not on the first.]

The Intersection of Two Curves

Finding the intersection of two curves in polar coordinates is complicated by the fact that a given point has many descriptions in polar coordinates. Example 8 illustrates how to find the intersection.

EXAMPLE 8 Find the intersection of the curve $r = 1 - \cos \theta$ and the circle $r = \cos \theta$.

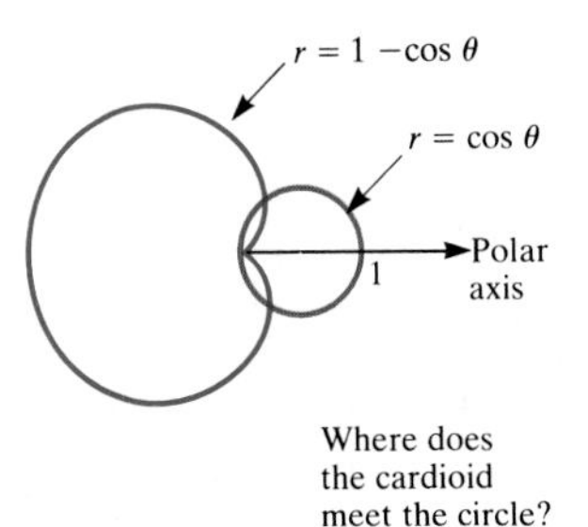

Where does the cardioid meet the circle?

Figure 14

SOLUTION First graph the curves. The curve $r = \cos \theta$ is a circle half the size of the one in Example 7. The curve $r = 1 - \cos \theta$ is shown in Fig. 14. (It is a cardioid, being congruent to $r = 1 + \cos \theta$.) It appears that there are three points of intersection.

If a point of intersection is produced because the same value of θ yields the same value of r in both equations, we would have

$$1 - \cos\theta = \cos\theta.$$

Hence $\cos\theta = \frac{1}{2}$. Thus $\theta = \pi/3$ or $\theta = -\pi/3$ (or any angle differing from these by $2n\pi$, n an integer). This gives two of the three points, but it fails to give the origin. Why?

How does the origin get to be on the circle $r = \cos\theta$? Because, when $\theta = \pi/2$, $r = 0$. How does it get to be on the cardioid $r = 1 - \cos\theta$? Because, when $\theta = 0$, $r = 0$. The origin lies on both curves, but we would not learn this by simply equating $1 - \cos\theta$ and $\cos\theta$. ■

When checking for the intersection of two curves, $r = f(\theta)$ and $r = g(\theta)$ in polar coordinates, examine the origin separately. The curves may also intersect at other points not obtainable by setting $f(\theta) = g(\theta)$. This possibility is due to the fact that a given point P has an infinite number of descriptions in polar coordinates; that is, (r, θ) is the same as the points $(r, \theta + 2n\pi)$ and $(-r, \theta + (2n+1)\pi)$ for any integer n. The safest procedure is to graph the two curves first and then see why they intersect at the points suggested by the graphs.

Section Summary

We introduced polar coordinates and showed how to graph curves given in the form $r = f(\theta)$. Some of the more common polar curves are listed below.

Equation	*Curve*
$r = a,\ a > 0$	circle of radius a, center at pole
$r = 1 + \cos\theta$	cardioid
$r = a\theta,\ a > 0$	archimedean spiral
$r = \sin 3\theta$	3-leafed rose (one loop symmetric about $\theta = \pi/6$)
$r = \sin n\theta$, n odd	n-leafed rose
$r = \sin n\theta$, n even	$2n$-leafed rose
$r = \cos n\theta$, n odd	n-leafed rose (one loop symmetric about $\theta = 0$)
$r = \cos n\theta$, n even	$2n$-leafed rose
$r = a\csc\theta$	the line $y = a$
$r = a\sec\theta$	the line $x = a$
$r = a\cos\theta,\ a > 0$	circle of radius $a/2$ through pole and $(a, 0)$
$r = a\sin\theta,\ a > 0$	circle of radius $a/2$ through pole and $(a, \pi/2)$

We also mentioned that to find the intersection of two curves in polar coordinates, first graph them.

EXERCISES FOR SEC. 9.1: POLAR COORDINATES

1 Plot the points whose polar coordinates are
(*a*) $(1, \pi/6)$ (*b*) $(2, \pi/3)$
(*c*) $(2, -\pi/3)$ (*d*) $(-2, \pi/3)$
(*e*) $(2, 7\pi/3)$ (*f*) $(0, \pi/4)$

2 Find the rectangular coordinates of the points in Exercise 1.

3 Give at least three pairs of polar coordinates (r, θ) for the point $(3, \pi/4)$, (*a*) with $r > 0$, (*b*) with $r < 0$.

4 Find the polar coordinates (r, θ) with $0 \le \theta < 2\pi$ and r positive, for the point whose rectangular coordinates are
(*a*) $(\sqrt{2}, \sqrt{2})$ (*b*) $(-1, \sqrt{3})$

(c) $(-5, 0)$ (d) $(-\sqrt{2}, -\sqrt{2})$
(e) $(0, -3)$ (f) $(1, 1)$

In Exercises 5 to 8 transform the equation into one in rectangular coordinates.

5 $r = \sin\theta$
6 $r = \csc\theta$
7 $r = 3/(4\cos\theta + 5\sin\theta)$
8 $r = 4\cos\theta + 5\sin\theta$

In Exercises 9 to 12 transform the equation into one in polar coordinates.

9 $x = -2$
10 $y = x^2$
11 $xy = 1$
12 $x^2 + y^2 = 4x$

In Exercises 13 to 22 graph the given equations.

13 $r = 1 + \sin\theta$
14 $r = 3 + 2\cos\theta$
15 $r = 2^{-\theta/\pi}$
16 $r = 4^{\theta/\pi}$, $\theta > 0$
17 $r = \cos 3\theta$
18 $r = \sin 2\theta$
19 $r = 2$
20 $r = 3$
21 $r = 3\sin\theta$
22 $r = -2\cos\theta$

23 (a) Graph $r = 1/\theta$, $\theta > 0$.
(b) What happens to the y coordinate of (r, θ) as $\theta \to \infty$?

24 (a) Graph $r = 1/\sqrt{\theta}$, $\theta > 0$.
(b) What happens to the y coordinate of (r, θ) as $\theta \to \infty$?

In Exercises 25 to 30, find the intersections of the curves.

25 $r = 1 + \cos\theta$ and $r = \cos\theta - 1$
26 $r = \sin 2\theta$ and $r = 1$
27 $r = \sin 3\theta$ and $r = \cos 3\theta$
28 $r = 2\sin 2\theta$ and $r = 1$
29 $r = \sin\theta$ and $r = \cos 2\theta$
30 $r = \cos\theta$ and $r = \cos 2\theta$

The curve $r = 1 + a\cos\theta$ (or $r = 1 + a\sin\theta$) is called a *limaçon* (pronounced lee′ · ma · son). Its shape depends on the choice of the constant a. For $a = 1$ we have the cardioid of Example 2. Exercises 31 to 33 concern other choices of a.

31 Graph $r = 1 + 2\cos\theta$. (If $|a| > 1$, then the graph of $r = 1 + a\cos\theta$ crosses itself and forms a loop.)

32 Graph $r = 1 + \frac{1}{2}\cos\theta$.

33 Consider the curve $r = 1 + a\cos\theta$, where $0 \le a \le 1$.
(a) Relative to the same polar axis, graph the curves corresponding to $a = 0, \frac{1}{4}, \frac{1}{2}, \frac{3}{4}, 1$.
(b) For $a = \frac{1}{4}$ the graph in (a) is convex, but not for $a = 1$. Show that for $\frac{1}{2} < a \le 1$ the curve is not convex. ("Convex" is defined on p. 85.) *Hint:* Find the points on the curve farthest to the left and compare them to the point on the curve corresponding to $\theta = \pi$.

34 (a) Graph $r = 3 + \cos\theta$.
(b) Find the point on the graph in (a) that has the maximum y coordinate.

35 Find the y coordinate of the highest point on the right-hand leaf of the four-leaved rose $r = \cos 2\theta$.

36 Graph $r^2 = \cos 2\theta$. Note that, if $\cos 2\theta$ is negative, r is not defined and that, if $\cos 2\theta$ is positive, there are two values of r, $\sqrt{\cos 2\theta}$ and $-\sqrt{\cos 2\theta}$. This curve is called a *lemniscate*.

37 Where do the spirals $r = \theta$ and $r = 2\theta$, for $\theta \ge 0$, intersect?

In Appendix G it is shown that the graph of $r = 1/(1 + e\cos\theta)$ is a parabola if $e = 1$, an ellipse if $0 \le e < 1$, and a hyperbola if $e > 1$. (e here is not related to $e \approx 2.718$.) Exercises 38 and 39 concern such graphs.

38 (a) Graph $r = \dfrac{1}{1 + \cos\theta}$.
(b) Find an equation in rectangular coordinates for the curve in (a).

39 (a) Graph $r = \dfrac{1}{1 - \frac{1}{2}\cos\theta}$.
(b) Find an equation in rectangular coordinates for the curve in (a).

40 The spiral $r = \theta$ meets the circle $r = 2\sin\theta$ at a point other than the origin. Use Newton's method to estimate the coordinates of that point. (Give both the polar and rectangular coordinates.)

9.2 AREA IN POLAR COORDINATES

Section 5.3 showed how to compute the area of a region if the lengths of parallel cross sections are known. Sums based on estimating rectangles led to the formula

$$\text{Area} = \int_a^b c(x)\,dx,$$

where $c(x)$ denotes the cross-sectional length. Now we consider quite a different situation, in which sectors of a circle, not rectangles, provide an estimate of the area.

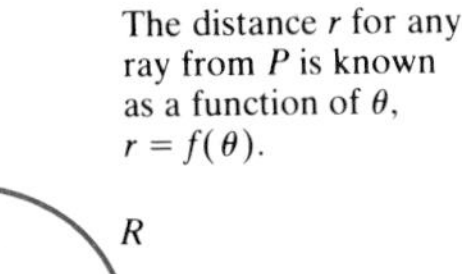

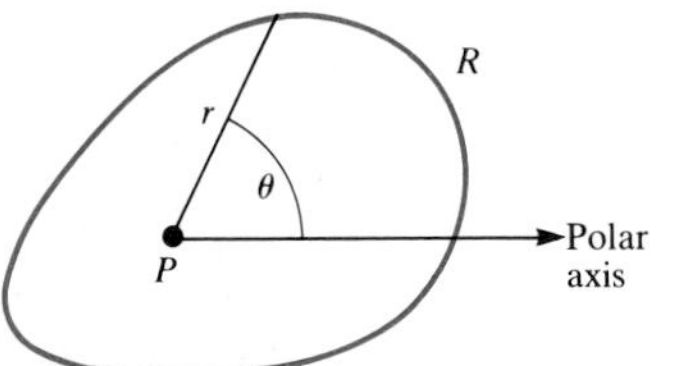

Figure 1

Let R be a region in the plane and P a point inside it. Assume that the distance r from P to any point on the boundary of R is known as a function $r = f(\theta)$. Assume that any ray from P meets the boundary of R just once, as in Fig. 1.

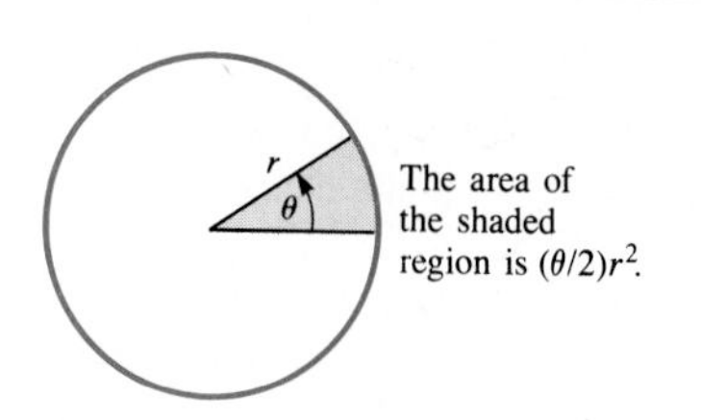

Figure 2

The cross sections made by the rays from P are *not* parallel. Instead, like spokes in a wheel, they all meet at the point P. It would be unnatural to use rectangles to estimate the area, but it is reasonable to use sectors of circles that have P as a common vertex.

Begin by recalling that in a circle of radius r a sector of central angle θ has area $(\theta/2)r^2$. (See Fig. 2.) This formula plays the same role now as the formula for the area of a rectangle did in Sec. 5.3.

Area in Polar Coordinates (informal approach)

Assume $f(\theta) \geq 0$.

Let R be the region bounded by the rays $\theta = \alpha$ and $\theta = \beta$ and by the curve $r = f(\theta)$, as shown in Fig. 3. To obtain a **local estimate** for the area of R, consider the portion of R between the rays corresponding to the angles θ and $\theta + d\theta$, where $d\theta$ is a small positive number. (See Fig. 4.) The area of the narrow wedge which is shaded in Fig. 4 is approximately that of a sector of a circle of radius $r = f(\theta)$ and angle $d\theta$, shown in Fig. 5. The area of the sector in Fig. 5 is

$$\frac{[f(\theta)]^2\, d\theta}{2}. \tag{1}$$

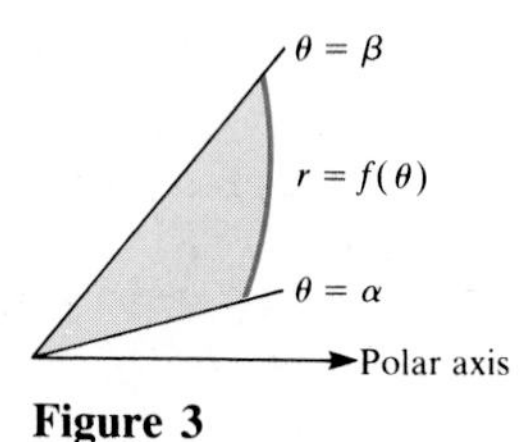

Figure 3

Having found the local estimate of area (1), we conclude that the area of R is

$$\int_\alpha^\beta \frac{[f(\theta)]^2\, d\theta}{2}.$$

How to find area in polar coordinates.

The area of the region bounded by the rays $\theta = \alpha$ and $\theta = \beta$ and by the curve $r = f(\theta)$ is

Formula (2) is applied in Sec. 13.5 to the motion of satellites and planets.

$$\boxed{\int_\alpha^\beta \frac{[f(\theta)]^2}{2}\, d\theta \quad \text{or} \quad \int_\alpha^\beta \frac{r^2\, d\theta}{2}.} \tag{2}$$

It is assumed that no ray from the origin between α and β crosses the curve twice.

It may seem surprising to find $[f(\theta)]^2$, not just $f(\theta)$, in the integrand. But remember that area has the dimension "length times length." Since θ, given in radians, is dimensionless, being defined as "length of circular arc divided by length of radius," $d\theta$ is also dimensionless. Hence $f(\theta)\, d\theta$, having the dimension of length, not of area, could *not* be correct. But $\frac{1}{2}[f(\theta)]^2\, d\theta$, having the dimension of area (length times length), is plausible. For rectangular coordinates, in the expression $f(x)\, dx$, both $f(x)$ and dx have the dimension of length, one along the y axis, the other along the x axis; thus $f(x)\, dx$ has the dimension of area.

Memory device

As an aid in remembering the area of the narrow sector in Fig. 5, note that it

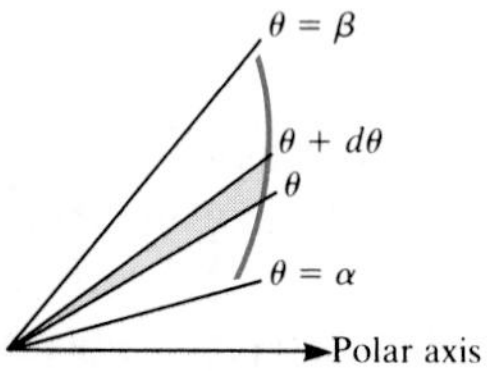

θ = β
θ + dθ
θ
θ = α
Polar axis

Figure 4

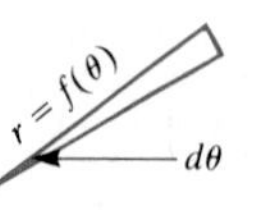

This narrow sector of a circle approximates the narrow wedge in Fig. 4.

Figure 5

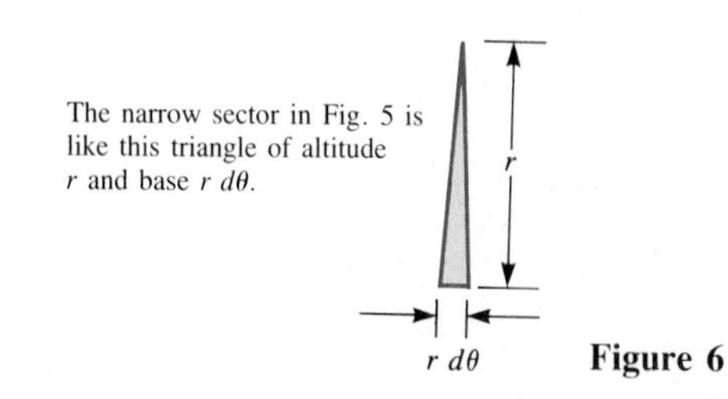

Figure 6

resembles a triangle of height r and base $r\,d\theta$, as shown in Fig. 6. Its area is

$$\frac{1}{2}\cdot r \cdot r\,d\theta = \frac{r^2\,d\theta}{2}.$$

EXAMPLE 1 Find the area of the region bounded by the curve $r = 3 + 2\cos\theta$, shown in Fig. 7.

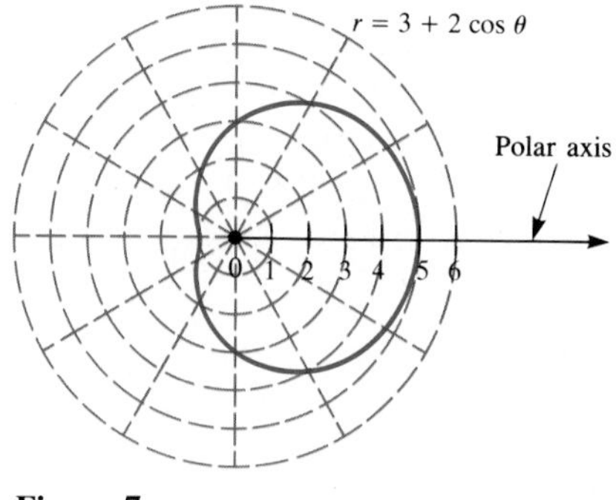

Figure 7

SOLUTION By the formula just obtained, this area is

$$\int_0^{2\pi} \tfrac{1}{2}(3 + 2\cos\theta)^2\,d\theta = \frac{1}{2}\int_0^{2\pi}(9 + 12\cos\theta + 4\cos^2\theta)d\theta$$

$$= \tfrac{1}{2}(9\theta + 12\sin\theta + 2\theta + \sin 2\theta)\Big|_0^{2\pi} = 11\pi. \quad ■$$

Observe that any line through the origin intersects the region of Example 1 in a segment of length 6, since $(3 + 2\cos\theta) + [3 + 2\cos(\theta + \pi)] = 6$ for any θ. Also, any line through the center of a circle of radius 3 intersects the circle in a segment of length 6. Thus two sets in the plane can have equal corresponding cross-sectional lengths through a fixed point and yet have different areas: the set in Example 1 has area 11π, while the circle of radius 3 has area 9π. *Knowing the lengths of all the cross sections of a region through a given point is not enough to determine the area of the region!*

EXAMPLE 2 Find the area of the region inside one of the eight loops of the eight-leaved rose $r = \cos 4\theta$.

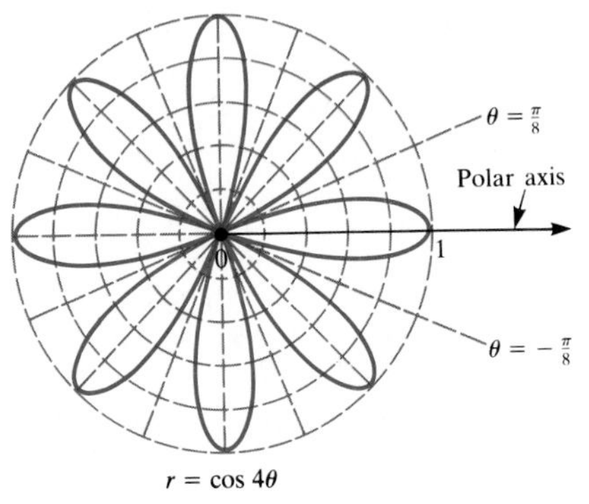

Figure 8

SOLUTION To graph one of the loops, start with $\theta = 0$. For that angle, $r = \cos(4\cdot 0) = \cos 0 = 1$. The point $(r, \theta) = (1, 0)$ is the outer tip of a loop. As θ increases from 0 to $\pi/8$, $\cos 4\theta$ decreases from $\cos 0 = 1$ to $\cos(\pi/2) = 0$. One of the eight loops is therefore bounded by the rays $\theta = \pi/8$ and $\theta = -\pi/8$. It is shown in Fig. 8.

The area of the loop which is bisected by the polar axis is

$$\int_{-\pi/8}^{\pi/8}\frac{r^2}{2}\,d\theta = \int_{-\pi/8}^{\pi/8}\frac{\cos^2 4\theta}{2}\,d\theta$$

$$= \int_{-\pi/8}^{\pi/8}\frac{1 + \cos 8\theta}{4}\,d\theta$$

$$= \left(\frac{\theta}{4} + \frac{\sin 8\theta}{32}\right)\Bigg|_{-\pi/8}^{\pi/8}$$

$$= \left(\frac{\pi}{32} + \frac{\sin\pi}{32}\right) - \left[\frac{-\pi}{32} + \frac{\sin(-\pi)}{32}\right]$$

$$= \frac{\pi}{16}. \quad ■$$

The Area between Two Curves

Assume that $r = f(\theta)$ and that $r = g(\theta)$ describe two curves in polar coordinates and that $f(\theta) \geq g(\theta) \geq 0$ for θ in $[\alpha, \beta]$. Let R be the region between these two curves and the rays $\theta = \alpha$ and $\theta = \beta$, as shown in Fig. 9.

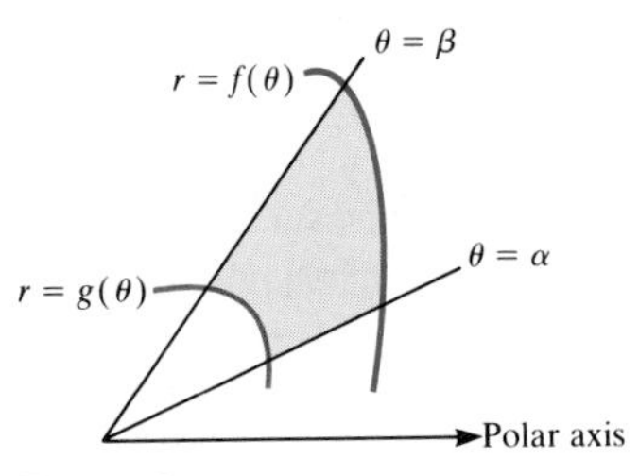

Figure 9

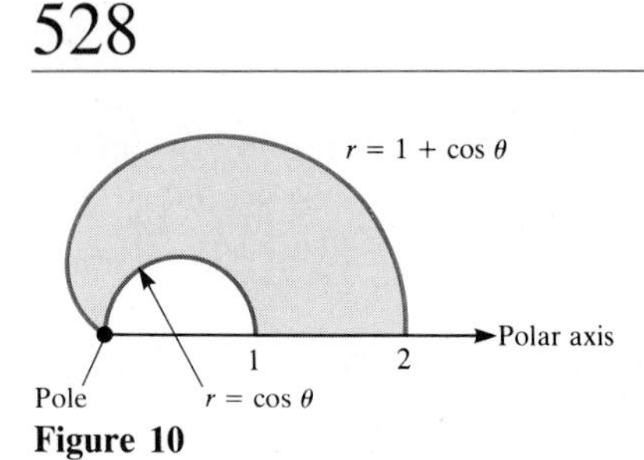

Figure 10

The area of R is obtained by subtracting the area within the inner curve, $r = g(\theta)$, from the area within the outer curve, $r = f(\theta)$.

EXAMPLE 3 Find the area of the top half of the region inside the cardioid $r = 1 + \cos\theta$ and outside the circle $r = \cos\theta$.

We must integrate over two different intervals to find the two areas.

SOLUTION The region is shown in Fig. 10. The top half of the circle $r = \cos\theta$ is swept out as θ goes from 0 to $\pi/2$. The top half of the cardioid is swept out as θ goes from 0 to π.

The area of the top half of the cardioid is

$$\begin{aligned}\frac{1}{2}\int_0^{\pi}(1+\cos\theta)^2\,d\theta &= \frac{1}{2}\int_0^{\pi}(1+2\cos\theta+\cos^2\theta)\,d\theta\\ &= \frac{1}{2}\int_0^{\pi}\left(1+2\cos\theta+\frac{1+\cos 2\theta}{2}\right)d\theta\\ &= \frac{1}{2}\int_0^{\pi}\left(\frac{3}{2}+2\cos\theta+\frac{\cos 2\theta}{2}\right)d\theta\\ &= \frac{1}{2}\left[\frac{3\theta}{2}+2\sin\theta+\frac{\sin 2\theta}{4}\right]_0^{\pi}\\ &= \frac{3\pi}{4}.\end{aligned}$$

The area of the top half of the circle $r = \cos\theta$ is

Or note that it's half the area of a circle of radius $\frac{1}{2}$.

$$\frac{1}{2}\int_0^{\pi/2}\cos^2\theta\,d\theta = \frac{\pi}{8}.$$

Thus the area in question is

$$\frac{3\pi}{4} - \frac{\pi}{8} = \frac{5\pi}{8}. \quad \blacksquare$$

Area in Polar Coordinates (formal approach)

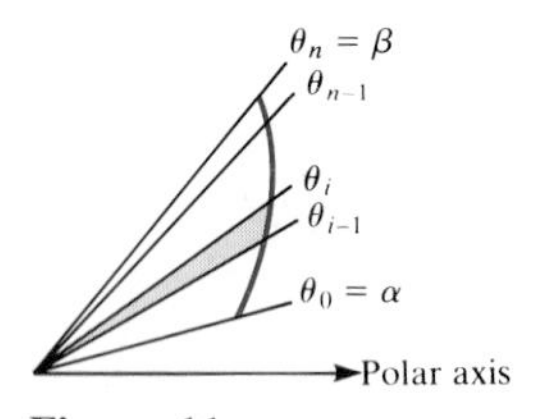

Figure 11

We wish to estimate the area of the shaded region in Fig. 11. To do so, introduce a partition of $[\alpha, \beta]$:

$$\alpha = \theta_0 < \theta_1 < \cdots < \theta_n = \beta.$$

Now estimate the area of R bounded between the rays whose angles are

$$\theta_{i-1} \qquad \text{and} \qquad \theta_i,$$

as shown in Fig. 12.

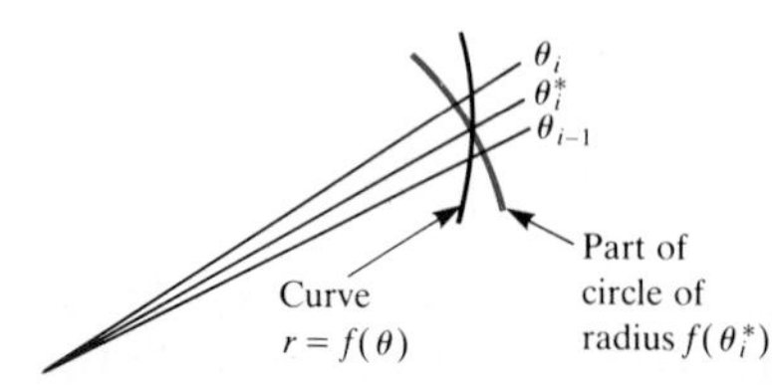

Figure 12

Pick any angle θ_i^* in the interval $[\theta_{i-1}, \theta_i]$ and use a sector of a circle whose radius is $f(\theta_i^*)$ to approximate the area of the shaded region. The approximating sector has radius $f(\theta_i^*)$ and angle $\theta_i - \theta_{i-1}$; hence its area is

$$\frac{[f(\theta_i^*)]^2}{2}(\theta_i - \theta_{i-1}).$$

The sum

$$\sum_{i=1}^{n} \frac{[f(\theta_i^*)]^2}{2}(\theta_i - \theta_{i-1})$$

is an estimate of the area of R.

As the mesh of the partition of $[\alpha, \beta]$ approaches 0, these sums become better and better approximations to the area of R and also approach the definite integral

$$\int_\alpha^\beta \frac{[f(\theta)]^2}{2}\,d\theta,$$

which, therefore, is the area of R. This establishes formula (2).

Section Summary

In this section we determined how to find the area within a curve $r = f(\theta)$ and the rays $\theta = \alpha$ and $\theta = \beta$. The heart of the method is the local approximation by a narrow sector of radius r and angle $d\theta$, which has area $r^2\,d\theta/2$. (It resembles a triangle of height r and base $r\,d\theta$.) This approximation leads to the formula,

$$\text{Area} = \int_\alpha^\beta \frac{r^2}{2}\,d\theta.$$

EXERCISES FOR SEC. 9.2: AREA IN POLAR COORDINATES

In each of Exercises 1 to 6, draw the bounded region enclosed by the indicated curve and rays and then find its area.

1 $r = 2\theta,\ \alpha = 0,\ \beta = \dfrac{\pi}{2}$

2 $r = \sqrt{\theta},\ \alpha = 0,\ \beta = \pi$

3 $r = \dfrac{1}{1+\theta},\ \alpha = \dfrac{\pi}{4},\ \beta = \dfrac{\pi}{2}$

4 $r = \sqrt{\sin\theta},\ \alpha = 0,\ \beta = \dfrac{\pi}{2}$

5 $r = \tan\theta,\ \alpha = 0,\ \beta = \dfrac{\pi}{4}$

6 $r = \sec\theta,\ \alpha = \dfrac{\pi}{6},\ \beta = \dfrac{\pi}{4}$

In each of Exercises 7 to 16 draw the region bounded by the indicated curve and then find its area.

7 $r = 2\cos\theta$

8 $r = e^\theta,\ 0 \le \theta \le 2\pi$

9 Inside the cardioid $r = 3 + 3\sin\theta$ and outside the circle $r = 3$

10 $r = \sqrt{\cos 2\theta}$

11 One loop of $r = \sin 3\theta$

12 One loop of $r = \cos 2\theta$

13 Inside one loop of $r = 2\cos 2\theta$ and outside $r = 1$

14 Inside $r = 1 + \cos\theta$ and outside $r = \sin\theta$

15 Inside $r = \sin\theta$ and outside $r = \cos\theta$

16 Inside $r = 4 + \sin\theta$ and outside $r = 3 + \sin\theta$

---

17 Sketch the graph of $r = 4 + \cos\theta$. Is it a circle?

18 (*a*) Show that the area of the triangle in Fig. 13 is $\int_0^\beta \frac{1}{2}\sec^2\theta\,d\theta$.

(*b*) From (*a*) and the fact that the area of a triangle is $\frac{1}{2}$(base)(height), show that $\tan\beta = \int_0^\beta \sec^2\theta\,d\theta$.

(*c*) With the aid of this equation, obtain another proof that $(\tan x)' = \sec^2 x$.

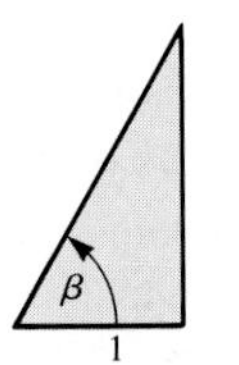

Figure 13

19 Show that the area of the shaded crescent between the two circular arcs is equal to the area of square $ABCD$. (See Fig. 14.) This type of result encouraged mathematicians from the time of the Greeks to try to find a method using only straightedge and compass for constructing a square whose area equals that of a given circle. This was proved impossible at the end of the nineteenth century.

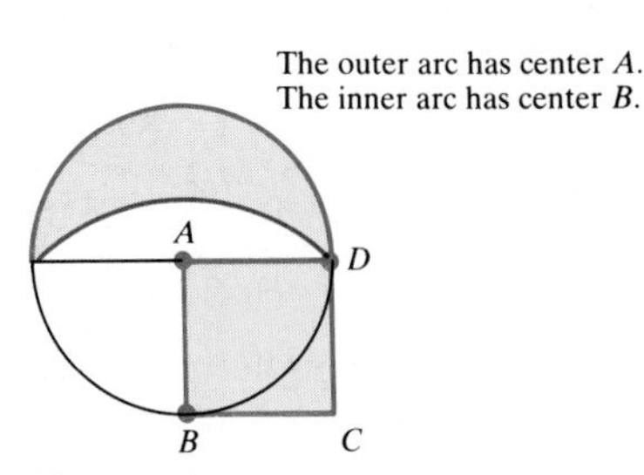

Figure 14

20 (*a*) Graph $r = 1/\theta$ for $0 < \theta \leq \pi/2$.
(*b*) Is the area of the region bounded by the curve drawn in (*a*) and the rays $\theta = 0$ and $\theta = \pi/2$ finite or infinite?

21 (*a*) Sketch the curve $r = 1/(1 + \cos\theta)$.
(*b*) What is the equation of the curve in (*a*) in rectangular coordinates?
(*c*) Find the area of the region bounded by the curve in (*a*) and the rays $\theta = 0$ and $\theta = 3\pi/4$, using polar coordinates.
(*d*) Solve (*c*) using rectangular coordinates and the equation in (*b*).

22 Find the area of the region bounded by $r = e^{\theta}$, $r = 2\cos\theta$ and $\theta = 0$. (You may need Newton's method to estimate a limit of integration.)

23 Estimate the area of the bounded region between $r = \sqrt[3]{1 + \theta^2}$, $\theta = 0$, and $\theta = \pi/2$, using Simpson's method.

24 Figure 15 shows a point P inside a convex region R.

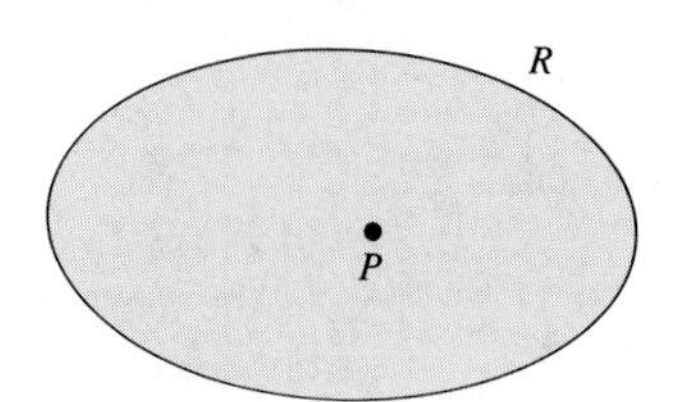

Figure 15

(*a*) Assume that P cuts each chord through P into two intervals of equal length. Must each chord through P cut R into two regions of equal areas?
(*b*) Assume that each chord through P cuts R into two regions of equal areas. Must P cut each chord through P into two intervals of equal lengths?

* 25 Let R be a region in the plane bounded by a loop whose equation is $r = f(\theta)$ in polar coordinates. Assume that every chord of R that passes through the pole has length at least 1.
(*a*) Draw several examples of such an R.
(*b*) Make a general conjecture about the area of R.
(*c*) Prove it.

* 26 Like Exercise 25, except that each chord through the pole has length at most 1.

9.3 PARAMETRIC EQUATIONS

Up to this point we have considered curves described in three forms: "y is a function of x," "x and y are related implicitly," and "r is a function of θ." But a curve is often described by giving both x and y as functions of a third variable. We introduce this situation as it arises naturally in the study of motion.

Two Examples

If a ball is thrown horizontally with a speed of 32 feet per second, it falls in a curved path. Air resistance disregarded, its position t seconds later is given by $x = 32t$, $y = -16t^2$ relative to the coordinate system in Fig. 1. Here the curve is completely described, not by expressing y as a function of x, but by expressing both x and y as functions of a third variable t. The third variable is called a **parameter** (*para* meaning "together," *meter* meaning "measure"). The equations $x = 32t$, $y = -16t^2$ are called **parametric equations** for the curve.

In this example it is easy to eliminate t and so find a direct relation between x and y:

$$t = \frac{x}{32}.$$

Hence

$$y = -16\left(\frac{x}{32}\right)^2 = -\frac{16}{(32)^2}x^2 = -\frac{1}{64}x^2.$$

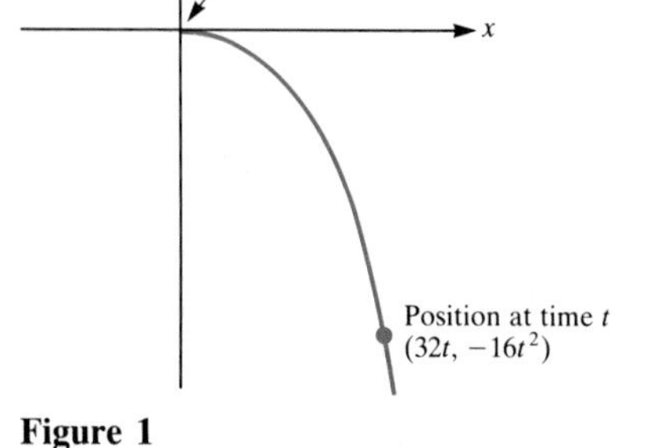

Figure 1

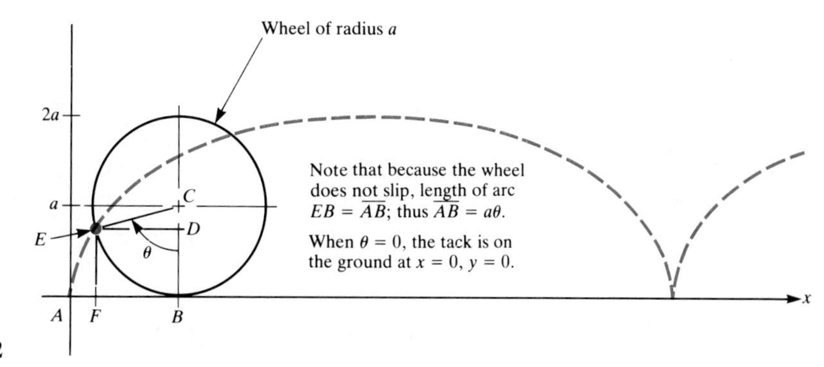

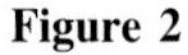
Figure 2

The path of the falling ball is part of the parabola $y = -\frac{1}{64}x^2$.

In Example 1 elimination of the parameter would lead to a complicated equation involving x and y. One advantage of parametric equations is that they can provide a simple description of a curve, although it may be impossible to find an equation in x and y which describes the curve.

EXAMPLE 1 As a bicycle wheel of radius a rolls along, a tack stuck in its circumference traces out a curve called a **cycloid**, which consists of a sequence of arches, one arch for each revolution of the wheel. (See Fig. 2.)

Find the position of the tack as a function of the angle θ through which the wheel turns.

SOLUTION The x coordinate of the tack, corresponding to θ, is

$$\overline{AF} = \overline{AB} - \overline{ED} = a\theta - a\sin\theta,$$

and the y coordinate is

$$\overline{EF} = \overline{BC} - \overline{CD} = a - a\cos\theta.$$

Then the position of the tack, as a function of the parameter θ, is

$$\begin{cases} x = a\theta - a\sin\theta \\ y = a - a\cos\theta. \end{cases}$$

In this case, eliminating θ would lead to a complicated relation between x and y. ■

Any curve $y = f(x)$ can be given parametrically: $x = t$, $y = f(t)$.

Any curve can be described parametrically. For instance, consider the curve $y = e^x + x$. It is perfectly legal to introduce a parameter t equal to x and write

$$\begin{cases} x = t \\ y = e^t + t. \end{cases}$$

This device may seem a bit artificial, but it will be useful in the next section in order to apply results for curves expressed by means of parametric equations to curves given in the form $y = f(x)$.

How to Find dy/dx and d^2y/dx^2

How can we find the slope of a curve which is described parametrically by the equations

$$x = g(t) \qquad y = h(t)?$$

An often difficult, perhaps impossible, approach is to solve the equation $x = g(t)$ for t as a function of x and substitute the result into the equation $y = h(t)$, thus expressing y explicitly in terms of x; then differentiate the result to find dy/dx. Fortunately, there is a very easy way, which we will now describe.

Assume that y is a differentiable function of x. Then, by the chain rule,

$$\frac{dy}{dt} = \frac{dy}{dx}\frac{dx}{dt},$$

from which it follows that

Formula for the slope of a parameterized curve

$$\boxed{\frac{dy}{dx} = \frac{\dfrac{dy}{dt}}{\dfrac{dx}{dt}}.} \tag{1}$$

It is assumed that in formula (1) dx/dt is not 0. To obtain d^2y/dx^2 just replace y in (1) by dy/dx, obtaining

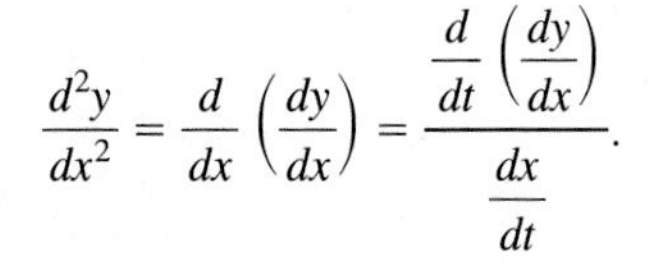

$$\frac{d^2y}{dx^2} = \frac{d}{dx}\left(\frac{dy}{dx}\right) = \frac{\dfrac{d}{dt}\left(\dfrac{dy}{dx}\right)}{\dfrac{dx}{dt}}.$$

EXAMPLE 2 At what angle does the arch of the cycloid shown in Example 1 meet the x axis at the origin?

SOLUTION The parametric equations of the cycloid are

$$x = a\theta - a \sin \theta \qquad \text{and} \qquad y = a - a \cos \theta.$$

Here θ is the parameter. Then

$$\frac{dx}{d\theta} = a - a \cos \theta \qquad \text{and} \qquad \frac{dy}{d\theta} = a \sin \theta.$$

Consequently,

$$\frac{dy}{dx} = \frac{dy/d\theta}{dx/d\theta} = \frac{a \sin \theta}{a - a \cos \theta} = \frac{\sin \theta}{1 - \cos \theta}.$$

When θ is near 0, (x, y) is near the origin. How does the slope, which is $\sin \theta/(1 - \cos \theta)$, behave as $\theta \to 0^+$? L'Hôpital's rule applies, and we have

$$\lim_{\theta \to 0^+} \frac{\sin \theta}{1 - \cos \theta} = \lim_{\theta \to 0^+} \frac{\cos \theta}{\sin \theta} = \infty.$$

Thus the cycloid comes in vertically at the origin. ■

EXAMPLE 3 Find d^2y/dx^2 for the cycloid of Example 1.

SOLUTION From Example 2 we know that

$$\frac{dy}{dx} = \frac{\sin\theta}{1-\cos\theta}.$$

Hence

$$\frac{d^2y}{dx^2} = \frac{\dfrac{d}{d\theta}\left(\dfrac{dy}{dx}\right)}{\dfrac{dx}{d\theta}} = \frac{\dfrac{d}{d\theta}\left(\dfrac{\sin\theta}{1-\cos\theta}\right)}{\dfrac{dx}{d\theta}}.$$

As shown in Example 2, $dx/d\theta = a - a\cos\theta$. Thus

$$\frac{d^2y}{dx^2} = \frac{\left[\dfrac{(1-\cos\theta)(\cos\theta)-(\sin\theta)(\sin\theta)}{(1-\cos\theta)^2}\right]}{a-a\cos\theta} \qquad \text{derivative of quotient}$$

$$= \frac{\left[\dfrac{\cos\theta-\cos^2\theta-\sin^2\theta}{(1-\cos\theta)^2}\right]}{a-a\cos\theta}$$

$$= \frac{\cos\theta-1}{a(1-\cos\theta)^3} \qquad (\sin^2\theta+\cos^2\theta=1)$$

$$= \frac{-1}{a(1-\cos\theta)^2}.$$

Since the denominator is positive (or 0), the quotient, when defined, is negative. This agrees with Fig. 2, which shows a concave-down arch. ■

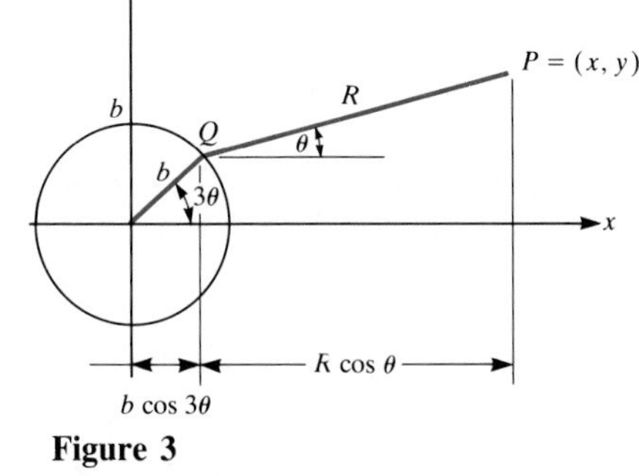

Figure 3

The Rotary Engine

The next two examples use parametric equations to describe the geometric principles of the rotary engine recognized by Felix Wankel in 1954. He found that it is possible for an equilateral triangle to revolve in a certain curve in such a way that its corners maintain contact with the curve and its centroid sweeps out a circle.

EXAMPLE 4 Let b and R be fixed positive numbers and consider the curve given parametrically by

$$x = b\cos 3\theta + R\cos\theta \qquad \text{and} \qquad y = b\sin 3\theta + R\sin\theta.$$

Show that an equilateral triangle can revolve in this curve while its centroid describes a circle of radius b.

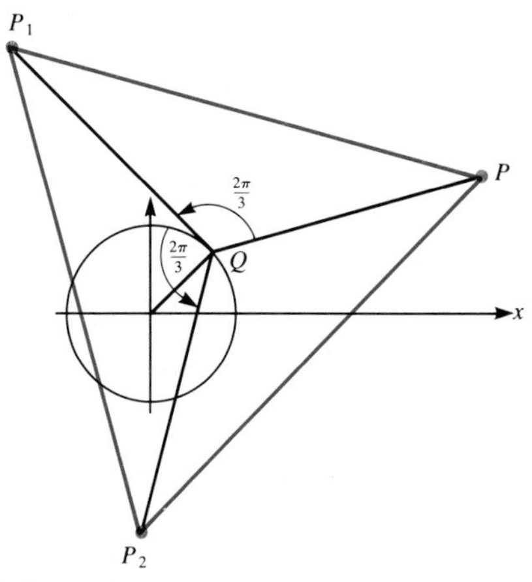

Figure 4

SOLUTION Figure 3 shows the typical point $P = (x, y)$ that corresponds to the parameter value θ. As θ increases by $2\pi/3$ from any given angle, the point Q goes once around the circle of radius b and returns to its initial position. During this revolution of Q the point P moves to a point P_1 whose angle, instead of being θ, is $\theta + 2\pi/3$. Thus, if P is on the curve, so are the points P_1 and P_2 shown in Fig. 4; these form an equilateral triangle.

Consequently, each vertex of the equilateral triangle sweeps out the curve once, while the centroid Q goes three times around the circle of radius b. ■

What does the curve described in Example 4 look like? Wankel graphed it without knowing that mathematicians had met it long before in a different setting, described in Example 5, which provides a way of graphing the curve.

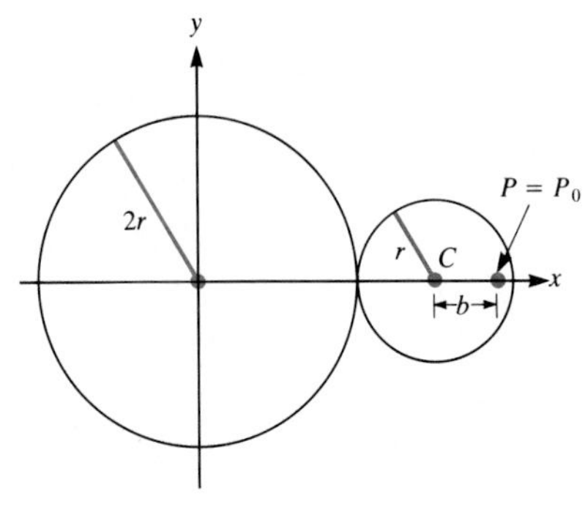

Figure 5

EXAMPLE 5 A circle of radius r rolls without slipping around a fixed circle of radius $2r$. Describe the path swept out by a point P located at a distance b from the center of the moving circle, $0 \leq b \leq r$.

SOLUTION Place the rolling circle as shown in Fig. 5. Note that the center C of the rolling circle traces out a circle of radius $3r$. Let $R = 3r$.

As the little circle rolls counterclockwise around the fixed circle without slipping, the point P traces out a path whose initial point P_0 is shown in Fig. 5. The typical point P on the path as the circle rolls around the larger circle is shown in Fig. 6. Since the radius of the rolling circle is half that of the fixed circle (and there is no slipping), angle ACB is 2θ. Thus the angle that CP makes with the x axis is the sum of θ and 2θ, which is 3θ. Consequently, $P = (x, y)$ has coordinates given parametrically as

Why is angle ACB equal to 2θ?

$$x = b \cos 3\theta + R \cos \theta \qquad \text{and} \qquad y = b \sin 3\theta + R \sin \theta.$$

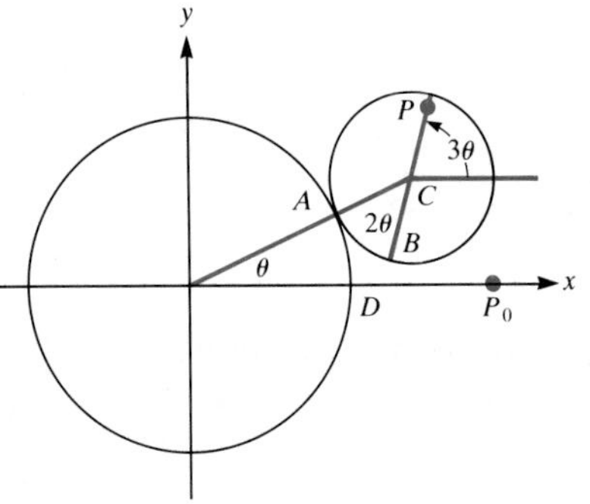

Figure 6

Thus the curve swept out by P is precisely the curve Wankel studied.

Long known to mathematicians as an **epitrochoid**, it is shown in Fig. 7. ■

In order that the moving rotor in the rotary engine can turn the drive shaft, teeth are placed in it along a circle of radius $2b$ which engage teeth in the drive shaft, which has radius b. (See Fig. 8.) For each complete rotation of the rotor, the drive shaft completes three rotations.

It was a Stuttgart professor, Othmar Baier, who showed that Wankel's curve was an epitrochoid. This insight was of aid in simplifying the machining of the working surface of the motor.

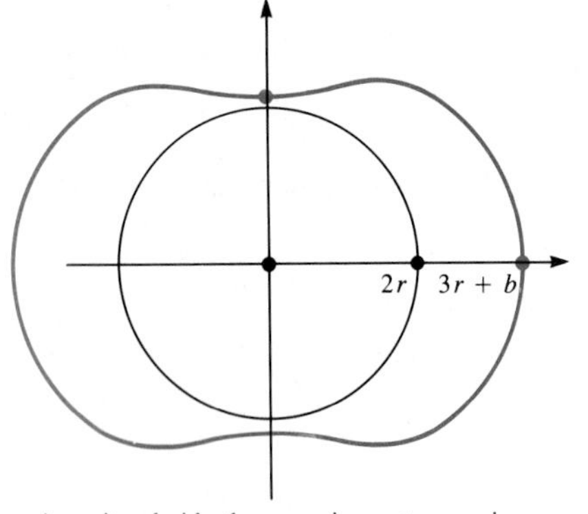

An epitrochoid: the curve in a rotary engine
Figure 7

Section Summary

This section described parametric equations, where x and y are given as functions of a third variable, often time (t) or angle (θ). We also showed how to compute dy/dx and d^2y/dx^2:

$$\frac{dy}{dx} = \frac{dy/dt}{dx/dt}$$

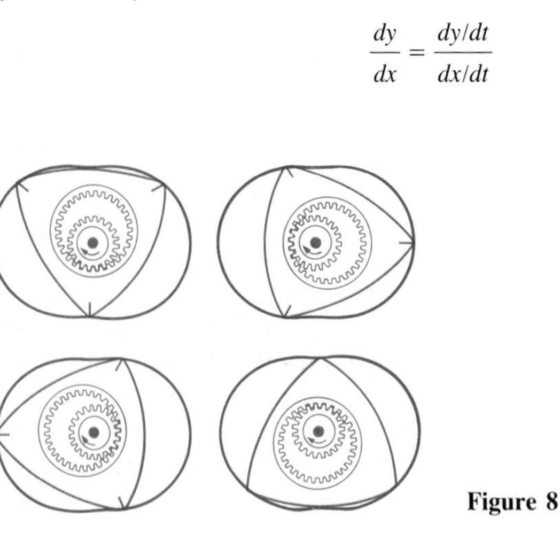
Figure 8

and, replacing y by dy/dx,

$$\frac{d^2y}{dx^2} = \frac{\dfrac{d}{dt}\left(\dfrac{dy}{dx}\right)}{\dfrac{dx}{dt}}.$$

EXERCISES FOR SEC. 9.3: PARAMETRIC EQUATIONS

1 Consider the parametric equations $x = 2t + 1$, $y = t - 1$.
(*a*) Fill in this table:

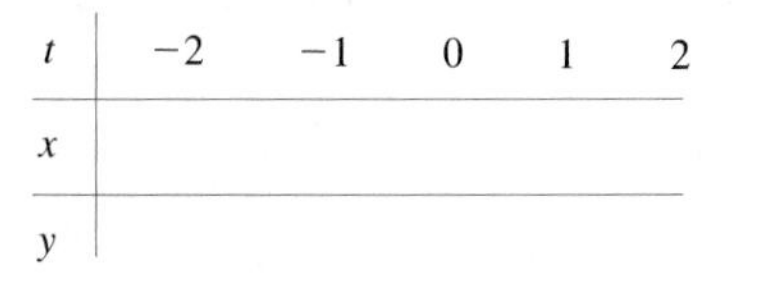

t	-2	-1	0	1	2
x					
y					

(*b*) Plot the five points (x, y) obtained in (*a*).
(*c*) Graph the curve given by the parametric equations $x = 2t + 1$, $y = t - 1$.
(*d*) Eliminate t to find an equation for the graph involving only x and y.

2 Consider the parametric equations $x = t + 1$, $y = t^2$.
(*a*) Fill in this table:

t	-2	-1	0	1	2
x					
y					

(*b*) Plot the five points (x, y) obtained in (*a*).
(*c*) Graph the curve.
(*d*) Find an equation in x and y that describes the curve.

3 Consider the parametric equations $x = t^2$, $y = t^2 + t$.
(*a*) Fill in this table:

t	-3	-2	-1	0	1	2	3
x							
y							

(*b*) Plot the seven points (x, y) obtained in (*a*).
(*c*) Graph the curve given by $x = t^2$, $y = t^2 + t$.
(*d*) Eliminate t and find an equation for the graph in terms of x and y.

4 Consider the parametric equations $x = 2 \cos t$, $y = 3 \sin t$.
(*a*) Fill in this table, expressing the entries decimally.

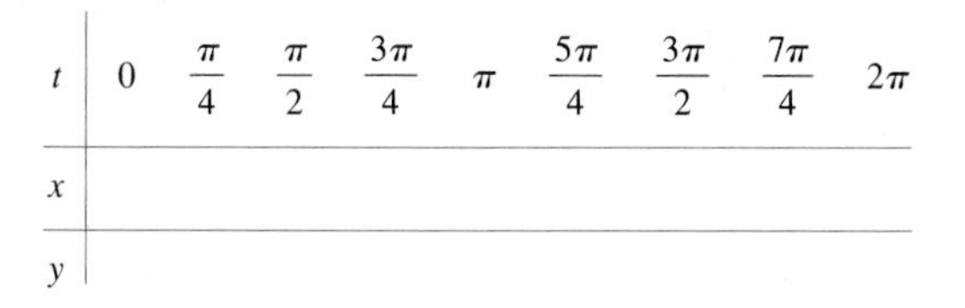

t	0	$\frac{\pi}{4}$	$\frac{\pi}{2}$	$\frac{3\pi}{4}$	π	$\frac{5\pi}{4}$	$\frac{3\pi}{2}$	$\frac{7\pi}{4}$	2π
x									
y									

(*b*) Plot the eight distinct points in (*a*).
(*c*) Graph the curve given by $x = 2 \cos t$, $y = 3 \sin t$.
(*d*) Using the identity $\cos^2 t + \sin^2 t = 1$, eliminate t.

In Exercises 5 to 8 express the curves parametrically with parameter t.

5 $y = \sqrt{1 + x^3}$ **6** $y = \tan^{-1} 3x$
7 $r = \cos 2\theta$ **8** $r = 3 + \cos \theta$

In Exercises 9 to 14 find dy/dx and d^2y/dx^2 for the given curves.

9 $x = t^3 + t$, $y = t^7 + t + 1$
10 $x = \sin 3t$, $y = \cos 4t$
11 $x = 1 + \ln t$, $y = t \ln t$
12 $x = e^{t^2}$, $y = \tan t$
13 $r = \cos 3\theta$
14 $r = 2 + 3 \sin \theta$

In Exercises 15 to 16 find the equation of the tangent line to the given curve at the given point.

15 $x = t^3 + t^2$; $y = t^5 + t$; $(2, 2)$

16 $x = \dfrac{t^2 + 1}{t^3 + t^2 + 1}$, $y = \sec 3t$; $(1, 1)$

In Exercises 17 and 18 find d^2y/dx^2.

17 $x = t^3 + t + 1$, $y = t^2 + t + 2$
18 $x = e^{3t} + \sin 2t$, $y = e^{3t} + \cos t^2$
19 For which values of t is the curve in Exercise 17 concave up? concave down?
20 Let $x = t^3 + 1$ and $y = t^2 + t + 1$. For which values of t is the curve concave up? concave down?
21 Find the slope of the three-leaved rose, $r = \sin 3\theta$, at the point $(r, \theta) = (\sqrt{2}/2, \pi/12)$.
22 (*a*) Find the slope of the cardioid $r = 1 + \cos \theta$ at the point (r, θ).
(*b*) What happens to the slope in (*a*) as $\theta \to \pi^-$?
(*c*) What does (*b*) tell us about the graph of the cardioid? (Show it on the graph.)

---

23 Obtain parametric equations for the circle of radius a and center (h, k), using as parameter the angle θ shown in Fig. 9.

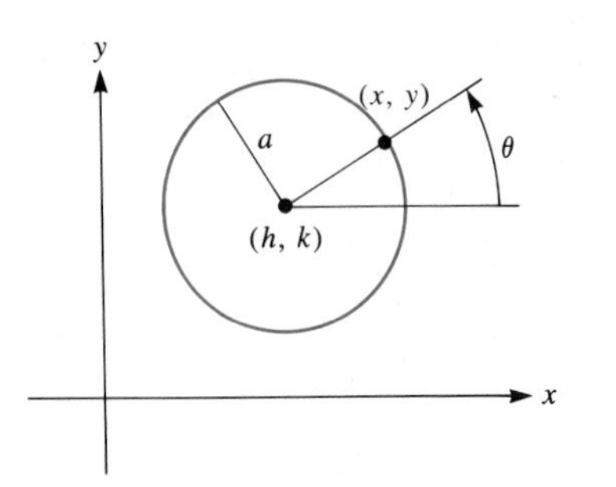

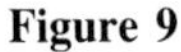
Figure 9

24 At time $t \geq 0$ a ball is at the point $(24t, -16t^2 + 5t + 3)$.
(*a*) Where is it at time $t = 0$?
(*b*) What is its horizontal speed at that time?
(*c*) What is its vertical speed at that time?

25 A ball is thrown at an angle α and initial velocity v_0, as sketched in Fig. 10. It can be shown that if time is in seconds and distance in feet, then t seconds later the ball is at the point

$$\begin{cases} x = (v_0 \cos \alpha)t \\ y = (v_0 \sin \alpha)t - 16t^2. \end{cases}$$

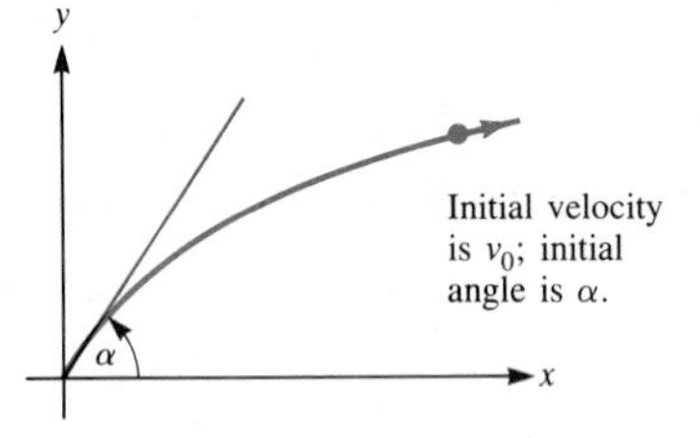

Figure 10

(*a*) Eliminate t.
(*b*) In view of (*a*), what type of curve does the ball follow?
(*c*) Find the coordinates of its highest point.

26 (*a*) The spiral $r = e^{2\theta}$ meets the ray $\theta = \alpha$ at an infinite number of points. Show that at all of these points the curve has the same slope.
(*b*) Show that the analog of (*a*) is not true for the spiral $r = \theta$.

27 The spiral $r = \theta$, $\theta > 0$ meets the ray $\theta = \alpha$ at an infinite number of points (α, α), $(\alpha + 2\pi, \alpha)$, $(\alpha + 4\pi, \alpha)$, ... What happens to the angle between the spiral and the ray at the point $(\alpha + 2\pi n, \alpha)$ as $n \to \infty$?

28 Let a and b be positive numbers. Consider the curve given parametrically by the equations

$$x = a \cos t \qquad y = b \sin t.$$

(*a*) Show that the curve is the ellipse

$$\frac{x^2}{a^2} + \frac{y^2}{b^2} = 1.$$

(*b*) Find the area of the region bounded by the ellipse in (*a*) by making a substitution that expresses $4 \int_0^a y\,dx$ in terms of an integral in which the variable is t and the range of integration is $[0, \pi/2]$.

29 Consider the curve given parametrically by

$$x = t^2 + e^t \qquad y = t + e^t$$

for t in $[0, 1]$.
(*a*) Plot the points corresponding to $t = 0$, $1/2$, and 1.
(*b*) Find the slope of the curve at the point $(1, 1)$.
(*c*) Find the area of the region under the curve and above the interval $[1, e + 1]$. [See Exercise 28(*b*).]

30 What is the slope of the cycloid in Fig. 2 at the first point on it to the right of the y axis at the height a?

31 The region under the arch of the cycloid

$$\begin{cases} x = a\theta - a \sin \theta \\ y = a - a \cos \theta \end{cases} \qquad (0 \leq \theta \leq 2\pi)$$

and above the x axis is revolved around the x axis. Find the volume of the solid of revolution produced.

32 The same as the preceding exercise, except the region is revolved around the y axis instead of the x axis.

33 L'Hôpital's rule in Sec. 6.8 asserts that if $\lim_{t \to 0} f(t) = 0$, $\lim_{t \to 0} g(t) = 0$, and $\lim_{t \to 0} [f'(t)/g'(t)]$ exists, then $\lim_{t \to 0} [f(t)/g(t)] = \lim_{t \to 0} [f'(t)/g'(t)]$. Interpret that rule in terms of the parameterized curve $x = g(t)$, $y = f(t)$. [Make a sketch of the curve near $(0, 0)$ and show on it the geometric meaning of the quotients $f(t)/g(t)$ and $f'(t)/g'(t)$.]

34 Let a be a positive constant. Consider the curve given parametrically by the equations $x = a \cos^3 t$, $y = a \sin^3 t$.
(*a*) Sketch the curve.
(*b*) Find the slope of the curve at the point corresponding to the parameter value t.

35 Consider a tangent line to the curve in Exercise 34 at a point P in the first quadrant. Show that the length of the segment of that line intercepted by the coordinate axes is a.

36 A circle of radius a is situated above the xy plane in a plane that is inclined at an angle to the xy plane. The shadow of the circle cast on the xy plane by light perpendicular to the xy plane is an oval curve. Find parametric equations for this curve and use them to show that the shadow is an ellipse.

9.4 ARC LENGTH AND SPEED ON A CURVE

In Sec. 4.3 we studied the motion of an object moving on a line. If at time t its position is $x(t)$, then its velocity is dx/dt and its speed is $|dx/dt|$. Now we will examine the speed of an object moving along a curved path.

Arc Length

The path of some particle is given parametrically:

$$\begin{cases} x = g(t) \\ y = h(t). \end{cases}$$

(Think of t as time.) A physicist might ask the following questions: How far does the particle travel from time $t = a$ to time $t = b$? What is the speed of the particle at time t?

Consider the "distance traveled" question first. The second will then be easy to answer, making use of the derivative. (In our reasoning we shall call the parameter t and think of it as time, but the results apply to any parameter.) We give an informal argument. At the end of this section there is a more detailed argument, using approximating sums.

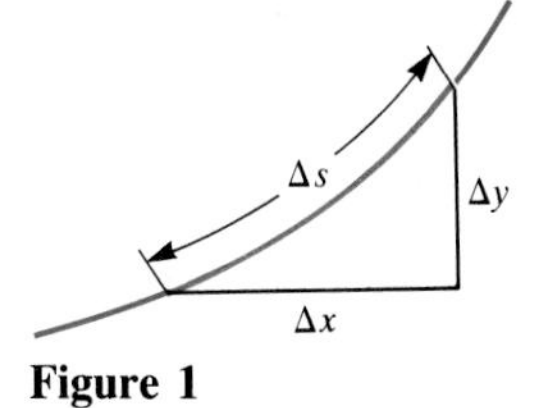

Figure 1

Assume that $x = g(t)$ and $y = h(t)$ have continuous derivatives. Let us make a local estimate of the arc length swept out on the path during the short interval of time from t to $t + dt$.

s denotes arc length.

Let s denote arc length along the path. During the time dt the change in the x coordinate, Δx, is approximately $dx = g'(t)\,dt$ and the change in the y coordinate, Δy, is approximately $dy = h'(t)\,dt$. The corresponding change in arc length, Δs, is approximately ds. Now, a very small piece of the curve resembles a straight line. A relation between dx, dy, and ds is therefore suggested by Fig. 1, which is a "right triangle" whose longest side is almost straight.

In view of the pythagorean theorem, it is reasonable to suspect that

This formula is the key device for dealing with arc length.

$$(ds)^2 = (dx)^2 + (dy)^2$$

or

$$ds = \sqrt{(dx)^2 + (dy)^2}. \tag{1}$$

Rewriting Eq. (1) in the form

$$ds = \sqrt{\left(\frac{dx}{dt}\right)^2 + \left(\frac{dy}{dt}\right)^2}\,dt \tag{2}$$

gives us the local estimate of arc length. From this we conclude that the arc length of the curve corresponding to t in $[a, b]$ is given by the integral

Formula for arc length of a parameterized curve

$$\boxed{\text{Arc length} = \int_a^b \sqrt{\left(\frac{dx}{dt}\right)^2 + \left(\frac{dy}{dt}\right)^2}\,dt.} \tag{3}$$

This formula holds for a curve given parametrically and if the derivatives dx/dt and dy/dt are continuous. If the curve is given in the form $y = f(x)$, it may be put in parametric form:

$$y = f(t) \qquad x = t.$$

Since $dx/dt = 1$ and $x = t$, formula (3) for the arc length of the curve $y = f(x)$ for x in $[a, b]$ takes the following form:

Formula for arc length of curve $y = f(x)$

$$\text{Arc length} = \int_a^b \sqrt{1 + \left(\frac{dy}{dx}\right)^2}\, dx. \tag{4}$$

(It is assumed that the derivative dy/dx is continuous.)

Three examples will show how these formulas are applied. The first goes back to the year 1657, when the 20-year-old Englishman, William Neil, found the length of an arc on the graph of $y = x^{3/2}$. His method was much more complicated. Earlier in that century, Thomas Harnot had found the length of an arc of the spiral $r = e^{\theta}$, but his work was not well publicized.

EXAMPLE 1 Find the arc length of the curve $y = x^{3/2}$ for x in $[0, 1]$. (See Fig. 2.)

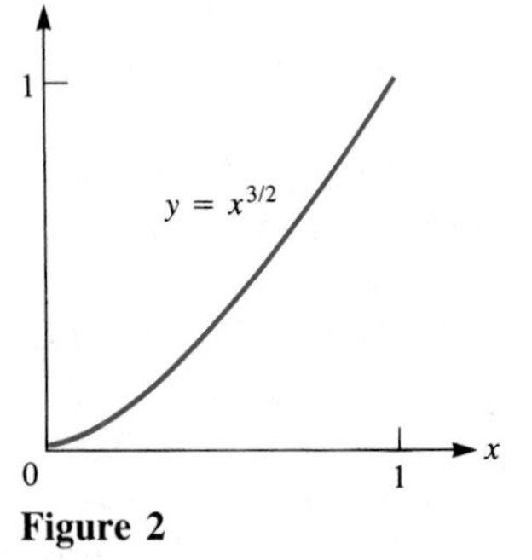

Figure 2

SOLUTION By formula (4),

$$\text{Arc length} = \int_0^1 \sqrt{1 + \left(\frac{dy}{dx}\right)^2}\, dx.$$

Since $y = x^{3/2}$, $dy/dx = \frac{3}{2}x^{1/2}$. Thus

$$\begin{aligned}
\text{Arc length} &= \int_0^1 \sqrt{1 + \left(\frac{3}{2}x^{1/2}\right)^2}\, dx \\
&= \int_0^1 \sqrt{1 + \frac{9x}{4}}\, dx \\
&= \int_1^{13/4} \sqrt{u} \cdot \frac{4}{9}\, du \qquad \text{where } u = 1 + (9x/4),\ du = \frac{9}{4}dx \\
&= \frac{4}{9} \cdot \frac{2}{3} u^{3/2} \Big|_1^{13/4} = \frac{8}{27}\left[\left(\frac{13}{4}\right)^{3/2} - 1^{3/2}\right] \\
&= \frac{8}{27}\left(\frac{13^{3/2}}{8} - 1\right) = \frac{13^{3/2} - 8}{27}. \qquad \blacksquare
\end{aligned}$$

Incidentally, the length of the curve $y = x^a$, where a is a rational number, usually *cannot* be computed with the aid of the fundamental theorem. The only cases in which it can be computed by the fundamental theorem are $a = 1$ (the graph is the line $y = x$) and $a = 1 + 1/n$, where n is an integer. Exercise 29 treats this question.

EXAMPLE 2 Find the distance s which the ball described at the beginning of Sec. 9.3 travels during the first b seconds.

SOLUTION Here

$$x = 32t \qquad \text{and} \qquad y = -16t^2.$$

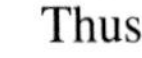

Thus

$$\frac{dx}{dt} = 32 \qquad \text{and} \qquad \frac{dy}{dt} = -32t.$$

By formula (3),

$$s = \int_0^b \sqrt{(32)^2 + (-32t)^2}\, dt = 32 \int_0^b \sqrt{1 + t^2}\, dt,$$

See Formula 31 *in the integral table.*

a definite integral that can be evaluated with the aid of a table or the substitution $t = \tan \theta$; its value is

$$16b\sqrt{1 + b^2} + 16 \ln (b + \sqrt{1 + b^2}). \quad \blacksquare$$

EXAMPLE 3 Find the length of one arch of the cycloid in Example 1 of Sec. 9.3.

SOLUTION Here the parameter is θ, and we compute $dx/d\theta$ and $dy/d\theta$:

$$\frac{dx}{d\theta} = \frac{d}{d\theta}(a\theta - a \sin \theta) = a - a \cos \theta,$$

and

$$\frac{dy}{d\theta} = \frac{d}{d\theta}(a - a \cos \theta) = a \sin \theta.$$

To complete one arch, θ varies from 0 to 2π. By formula (3), the length of one arch is $\int_0^{2\pi} \sqrt{(a - a \cos \theta)^2 + (a \sin \theta)^2}\, d\theta$. Thus

$$\begin{aligned}
\text{Length of arch} &= a \int_0^{2\pi} \sqrt{(1 - \cos \theta)^2 + (\sin \theta)^2}\, d\theta \\
&= a \int_0^{2\pi} \sqrt{1 - 2 \cos \theta + (\cos^2 \theta + \sin^2 \theta)}\, d\theta \\
&= a \int_0^{2\pi} \sqrt{2 - 2 \cos \theta}\, d\theta \\
&= a\sqrt{2} \int_0^{2\pi} \sqrt{1 - \cos \theta}\, d\theta \\
&= a\sqrt{2} \int_0^{2\pi} \sqrt{2} \sin \frac{\theta}{2}\, d\theta \qquad \text{trigonometry} \\
&= 2a \int_0^{2\pi} \sin \frac{\theta}{2}\, d\theta \\
&= 2a \left(-2 \cos \frac{\theta}{2}\right)\Bigg|_0^{2\pi} \\
&= 2a[-2(-1) - (-2)(1)] = 8a.
\end{aligned}$$

This step works since $\sin (\theta/2) \geq 0$ *for* $0 \leq \theta \leq 2\pi$.

This means that while θ varies from 0 to 2π, a bicycle travels a distance $2\pi a \approx 6.28a$, and the tack in the tire travels a distance $8a$. $\blacksquare$

Speed of a Particle Moving on a Curve

In practice we are not interested as much in the length of the path as in the speed of the particle as it moves along the path. The work done so far in this section helps us find this speed easily.

Consider a particle which at time t is at the point $(x, y) = (g(t), h(t))$. Choose a point B on the curve from which to measure distance along the curve, as shown

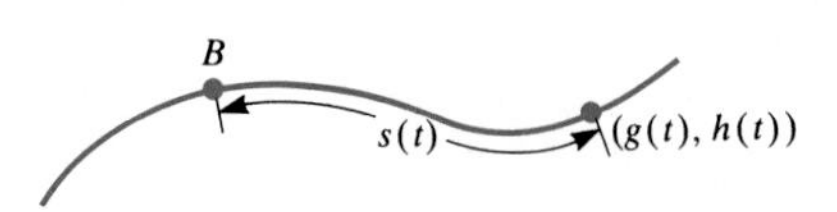

Figure 3

in Fig. 3. Let $s(t)$ denote the distance from B to $(g(t), h(t))$. We shall always assume that B has been chosen in such a way that $s(t)$ is an *increasing* function of t, hence $ds/dt \geq 0$.

Definition *Speed on a curved path.* If ds/dt exists, it is called the **speed** of the particle.

Since $s(t)$ is assumed to be an increasing function, speed is not negative. As early as Sec. 3.1 we were able to treat the speed of a particle moving in a straight path. Now it is possible to compute the speed of a particle moving on a curved path.

Speed of a particle on a curved path

If a particle at time t is at the point $(x, y) = (g(t), h(t))$, where g and h are functions having continuous derivatives, then its speed at time t is equal to

$$\sqrt{[g'(t)]^2 + [h'(t)]^2}.$$

The argument is short. Let $s(t)$ denote the arc length along the curve from some base point $B = (g(a), h(a))$ to the particle at time t.

Now,
$$s(t) = \int_a^t \sqrt{[g'(T)]^2 + [h'(T)]^2}\, dT.$$

(The letter T is introduced, since t is already used to describe the interval of integration.) Differentiation of this relation with respect to t (using the first fundamental theorem of calculus) yields

$$\frac{ds}{dt} = \sqrt{[g'(t)]^2 + [h'(t)]^2}.$$

EXAMPLE 4 Find the speed at time t of the ball described at the beginning of Sec. 9.3.

SOLUTION At time t the ball is at the point

$$(x, y) = (32t, -16t^2).$$

Newton's dot notation: $\dot{x} = \frac{dx}{dy}$, $\dot{y} = \frac{dy}{dt}$

Thus dx/dt, usually written $\dot{x}$, is 32, and $\dot{y}$ is $-32t$. The speed of the ball is

$$\frac{ds}{dt} = \sqrt{\dot{x}^2 + \dot{y}^2} = \sqrt{32^2 + (-32t)^2}$$

$$= 32\sqrt{1 + t^2} \quad \text{feet per second.} \quad \blacksquare$$

Arc Length and Speed in Polar Coordinates

So far in this section curves have been described in rectangular coordinates. Next consider a curve given in polar coordinates by the equation $r = f(\theta)$.

HOW TO FIND THE ARC LENGTH OF $r = f(\theta)$

The length of the curve $r = f(\theta)$ for θ in $[\alpha, \beta]$ is equal to

$$\int_\alpha^\beta \sqrt{[f(\theta)]^2 + [f'(\theta)]^2}\, d\theta$$

or

$$\int_\alpha^\beta \sqrt{r^2 + (r')^2}\, d\theta.$$

(Assume that f has a continuous derivative.)

This formula can be derived from that for the arc length of a parameterized curve in rectangular coordinates, as follows. Find the rectangular coordinates of the point whose polar coordinates are

$$(r, \theta) = (f(\theta), \theta).$$

They are

$$\begin{cases} x = f(\theta) \cos \theta \\ y = f(\theta) \sin \theta. \end{cases}$$

The curve is now given in rectangular form with parameter θ. Thus its length is

$$\int_\alpha^\beta \sqrt{\left(\frac{dx}{d\theta}\right)^2 + \left(\frac{dy}{d\theta}\right)^2}\, d\theta.$$

Now,

$$\frac{dx}{d\theta} = f(\theta)(-\sin \theta) + f'(\theta) \cos \theta,$$

and

$$\frac{dy}{d\theta} = f(\theta) \cos \theta + f'(\theta) \sin \theta.$$

Hence

$$\left(\frac{dx}{d\theta}\right)^2 + \left(\frac{dy}{d\theta}\right)^2 = [f(\theta)]^2 \sin^2 \theta - 2f(\theta)f'(\theta) \sin \theta \cos \theta + [f'(\theta)]^2 \cos^2 \theta$$

$$+ [f(\theta)]^2 \cos^2 \theta + 2f(\theta)f'(\theta) \sin \theta \cos \theta + [f'(\theta)]^2 \sin^2 \theta,$$

which, by the identity $\sin^2 \theta + \cos^2 \theta = 1$, simplifies to $[f(\theta)]^2 + [f'(\theta)]^2$. This justifies the formula.

EXAMPLE 5 Find the length of the spiral $r = e^{-3\theta}$ for θ in $[0, 2\pi]$.

SOLUTION First compute

$$r' = \frac{dr}{d\theta} = -3e^{-3\theta},$$

and then use the formula

$$\text{Arc length} = \int_\alpha^\beta \sqrt{r^2 + (r')^2}\, d\theta = \int_0^{2\pi} \sqrt{(e^{-3\theta})^2 + (-3e^{-3\theta})^2}\, d\theta$$

$$= \int_0^{2\pi} \sqrt{e^{-6\theta} + 9e^{-6\theta}}\, d\theta = \sqrt{10} \int_0^{2\pi} \sqrt{e^{-6\theta}}\, d\theta$$

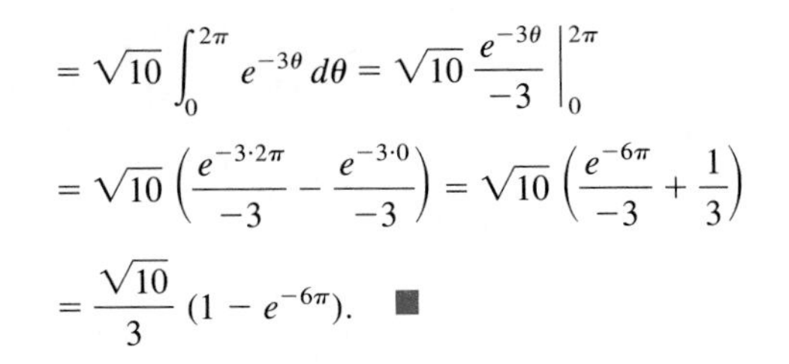

$$= \sqrt{10}\int_0^{2\pi} e^{-3\theta}\,d\theta = \sqrt{10}\,\frac{e^{-3\theta}}{-3}\Bigg|_0^{2\pi}$$

$$= \sqrt{10}\left(\frac{e^{-3\cdot 2\pi}}{-3} - \frac{e^{-3\cdot 0}}{-3}\right) = \sqrt{10}\left(\frac{e^{-6\pi}}{-3} + \frac{1}{3}\right)$$

$$= \frac{\sqrt{10}}{3}(1 - e^{-6\pi}). \quad \blacksquare$$

Arc Length (a more formal approach) Partition the time interval $[a, b]$ and use this partition to inscribe a polygon in the curve of the moving particle, as shown in Fig. 4.

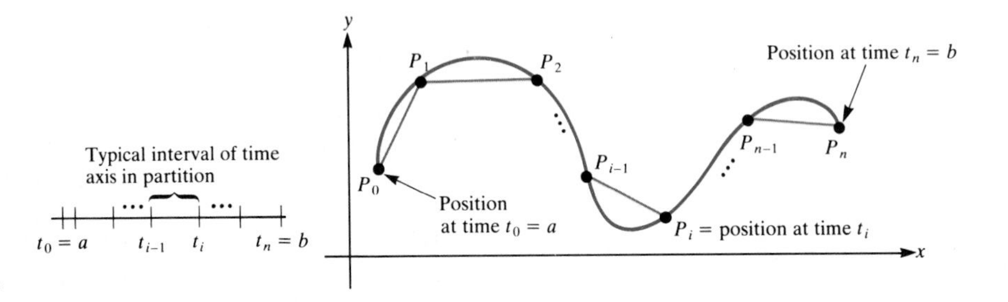

Figure 4

We are assuming $x = g(t)$ and $y = h(t)$ have continuous derivatives.

The length of such a polygon should approach the arc length as the mesh of the partition of $[a, b]$ shrinks toward 0 since the points P_i along the curve will get closer and closer together. The length of the typical straight-line segment $P_{i-1}P_i$, where $P_{i-1} = (g(t_{i-1}), h(t_{i-1}))$ and $P_i = (g(t_i), h(t_i))$, is (by the distance formula)

$$\sqrt{[g(t_i) - g(t_{i-1})]^2 + [h(t_i) - h(t_{i-1})]^2},$$

and so the length of the polygon is the sum

$$\sum_{i=1}^{n} \sqrt{[g(t_i) - g(t_{i-1})]^2 + [h(t_i) - h(t_{i-1})]^2}. \tag{5}$$

We shall relate this sum to sums of the type appearing in the definition of a definite integral over $[a, b]$.

By the mean-value theorem there exist numbers T_i^* and T_i^{**}, both in the interval $[t_{i-1}, t_i]$, such that we have $g(t_i) - g(t_{i-1}) = g'(T_i^*)(t_i - t_{i-1})$ and $h(t_i) - h(t_{i-1}) = h'(T_i^{**})(t_i - t_{i-1})$. Thus the sum (5) can be rewritten

$$\sum_{i=1}^{n} \sqrt{[g'(T_i^*)]^2 + [h'(T_i^{**})]^2}\,(t_i - t_{i-1}). \tag{6}$$

If T_i^{**} were equal to T_i^*, then this sum (6) would be an approximating sum used in defining

$$\int_a^b \sqrt{[g'(t)]^2 + [h'(t)]^2}\,dt.$$

To get around this difficulty, notice that since h' is continuous, $h'(T_i^*)$ is near $h'(T_i^{**})$ when the mesh of the partition of $[a, b]$ is small. If the sum (6) is a good approximation to the arc length, then presumably so is the sum

$$\sum_{i=1}^{n} \sqrt{[g'(T_i^*)]^2 + [h'(T_i^*)]^2}\,(t_i - t_{i-1}).$$

This step is justified in advanced calculus.

In other words, it is reasonable to expect that

$$\text{Arc length} = \lim_{\text{mesh}\to 0} \sum_{i=1}^{n} \sqrt{[g'(T_i^*)]^2 + [h'(T_i^*)]^2}\,(t_i - t_{i-1}).$$

But that limit is precisely the definition of the definite integral

$$\int_a^b \sqrt{[g'(t)]^2 + [h'(t)]^2}\,dt.$$

This shows why this definite integral should yield the arc length.

Section Summary

We first obtained a formula for the arc length of a curve given parametrically, $s = \int_a^b \sqrt{(dx/dt)^2 + (dy/dt)^2}\,dt$. Using the first fundamental theorem of calculus, we deduced that the speed, ds/dt, equals $\sqrt{(dx/dt)^2 + (dy/dt)^2}$. Both the informal and formal approaches are based on a small right triangle whose legs are parallel to the x and y axes.

If the curve is given in polar coordinates, then $s = \int_\alpha^\beta \sqrt{r^2 + (dr/d\theta)^2}\,d\theta$.

EXERCISES FOR SEC. 9.4: ARC LENGTH AND SPEED ON A CURVE

In Exercises 1 to 8 find the arc lengths of the given curves over the given intervals.

1 $y = x^{3/2}$, x in $[1, 2]$
2 $y = x^{2/3}$, x in $[0, 1]$
3 $y = (e^x + e^{-x})/2$, x in $[0, b]$
4 $y = x^2/2 - (\ln x)/4$, x in $[2, 3]$
5 $x = \cos^3 t$, $y = \sin^3 t$, t in $[0, \pi/2]$
6 $r = e^\theta$, θ in $[0, 2\pi]$
7 $r = 1 + \cos\theta$, θ in $[0, \pi]$
8 $r = \cos^2(\theta/2)$, θ in $[0, \pi]$

In each of Exercises 9 to 12 find the speed of the particle at time t, given the parametric description of its path.

9 $x = 50t$, $y = -16t^2$
10 $x = \sec 3t$, $y = \sin^{-1} 4t$
11 $x = t + \cos t$, $y = 2t - \sin t$
12 $x = \csc 3t$, $y = \tan^{-1}\sqrt{t}$

13 (*a*) Graph $x = t^2$, $y = t$ for $0 \le t \le 3$.
(*b*) Estimate its arc length from (0, 0) to (9, 3) by an inscribed polygon whose vertices have x coordinates 0, 1, 4, and 9.
(*c*) Set up a definite integral for the arc length in question.
(*d*) Estimate the definite integral in (*c*) by using a partition of [0, 3] into three sections, each of length 1, and the trapezoidal method.
(*e*) Estimate the definite integral in (*c*) by Simpson's method with six sections.
(*f*) If your calculator has a program to evaluate definite integrals, use it to evaluate the definite integral in (*c*) to four decimal places.

14 (*a*) Graph $y = 1/x^2$ for x in $[1, 2]$
(*b*) Estimate the length of the arc in (*a*) by using an inscribed polygon with vertices at (1, 1), $(\frac{5}{4}, (\frac{4}{5})^2)$, $(\frac{3}{2}, (\frac{2}{3})^2)$, and $(2, \frac{1}{4})$.
(*c*) Set up a definite integral for the length in question.
(*d*) Estimate the definite integral in (*c*) by the trapezoidal method, using four sections.
(*e*) Estimate the definite integral in (*c*) by Simpson's method, using four sections.
(*f*) If your calculator has a program to evaluate definite integrals, use it to estimate the definite integral in (*c*) to four decimal places.

15 How long is the spiral $r = e^{-3\theta}$, $\theta \ge 0$?

16 How long is the spiral $r = 1/\theta$, $\theta \ge 2\pi$?

17 Assume that a curve is described in rectangular coordinates in the form $x = f(y)$. Show that

$$\text{Arc length} = \int_c^d \sqrt{1 + \left(\frac{dx}{dy}\right)^2}\,dy$$

where y ranges in the interval $[c, d]$.

18 (See Exercise 17.) Consider the arc length of the curve $y = x^{2/3}$ for x in the interval $[1, 8]$.

(*a*) Set up a definite integral for this arc length using x as the parameter.
(*b*) Set up a definite integral for this arc length using y as the parameter.
(*c*) Evaluate the easier of the integrals in (*a*) and (*b*).

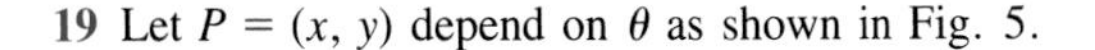

19 Let $P = (x, y)$ depend on θ as shown in Fig. 5.

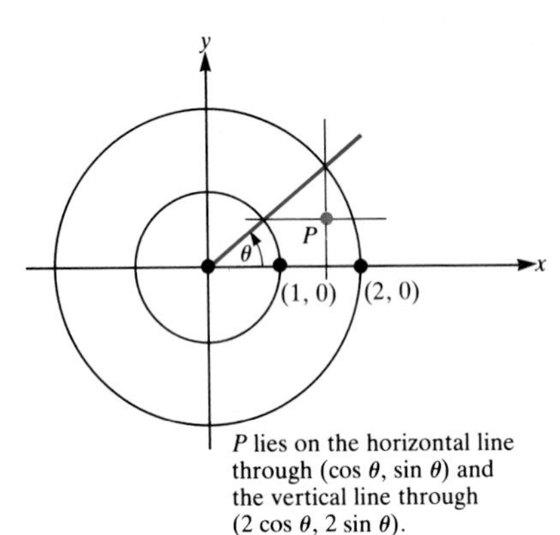

P lies on the horizontal line through $(\cos \theta, \sin \theta)$ and the vertical line through $(2 \cos \theta, 2 \sin \theta)$.

Figure 5

(*a*) Sketch the curve that P sweeps out.
(*b*) Show that $P = (2 \cos \theta, \sin \theta)$.
(*c*) Set up a definite integral for the length of the curve described in P. (Do not evaluate it.)
(*d*) Eliminate θ and show that P is on the ellipse

$$\frac{x^2}{4} + \frac{y^2}{1} = 1.$$

20 (*a*) At time t a particle has polar coordinates $r = g(t)$, $\theta = h(t)$. How fast is it moving?
(*b*) Use the formula in (*a*) to find the speed of a particle which at time t is at the point $(r, \theta) = (e^t, 5t)$.

21 (*a*) How far does a bug travel from time $t = 1$ to time $t = 2$ if at time t it is at the point $(\cos \pi t, \sin \pi t)$?
(*b*) How fast is it moving at time t?
(*c*) Graph its path relative to an xy coordinate system. Where is it at time $t = 1$? At $t = 2$?
(*d*) Eliminate t to find a relation between x and y.

22 Find the arc length of the archimedean spiral $r = a\theta$ for θ in $[0, 2\pi]$.

23 Consider the cardioid $r = 1 + \cos \theta$ for θ in $[0, \pi]$. We may consider r as a function of θ or as a function of s, arc length along the curve, measured, say, from $(2, 0)$.
(*a*) Find the average of r with respect to θ.
(*b*) Find the average of r with respect to s. *Hint:* Express all quantities appearing in this average in terms of θ.

24 We obtained the formula for arc length in polar coordinates from the one in terms of rectangular coordinates. After a somewhat complicated computation, we arrived at a fairly simple integrand. Obtain that formula, $\int_\alpha^\beta \sqrt{r^2 + (r')^2}\, d\theta$, by making a local estimate of the arc length corresponding to a small change in θ, $d\theta$.

25 Let $r = f(\theta)$ describe a curve in polar coordinates. Assume that $df/d\theta$ is continuous. Let θ be a function of time t. Let $s(t)$ be the length of the curve corresponding to the time interval $[a, t]$.
(*a*) What definite integral is equal to $s(t)$?
(*b*) What is the speed ds/dt?

26 The function $r = f(\theta)$ describes, for θ in $[0, 2\pi]$, a curve in polar coordinates. Assume r' is continuous and $f(\theta) > 0$. Prove that the average of r as a function of arc length is at least as large as the quotient $2A/s$, where A is the area swept out by the radius and s is the arc length of the curve. When is the average equal to $2A/s$?

27 The equations $x = \cos t$, $y = 2 \sin t$, t in $[0, \pi/2]$ describe a quarter of an ellipse. Draw this arc and examine its length. That is, describe various ways of estimating the length and compare their efficiencies.

28 Let $y = f(x)$ for x in $[0, 1]$ describe a curve that starts at $(0, 0)$, ends at $(1, 1)$, and lies in the square with vertices $(0, 0)$, $(1, 0)$, $(1, 1)$, and $(0, 1)$. Assume f has a continuous derivative.
(*a*) What can be said about the arc length of the curve? How small and how large can it be?
(*b*) Answer (*a*) if it is assumed also that $f'(x) \geq 0$ for x in $[0, 1]$.

29 Consider the length of the curve $y = x^m$, where m is a rational number. Show that the fundamental theorem of calculus is of aid in computing this length only if $m = 1$ or if m is of the form $1 + 1/n$ for some integer n. *Hint:* Chebyshev proved that $\int x^p(1 + x)^q\, dx$ is elementary for rational numbers p and q only when at least one of p, q, and $p + q$ is an integer.

30 If one convex polygon P_1 lies inside another convex polygon P_2, is the perimeter of P_1 necessarily less than the perimeter of P_2?

9.5 THE AREA OF A SURFACE OF REVOLUTION

In this section we develop a formula for expressing the surface area of a solid of revolution as a definite integral. In particular, we will show that the surface area of a sphere is four times the area of a cross section through its center. (See Fig. 1.) This was one of the great discoveries of Archimedes in the third century B.C.

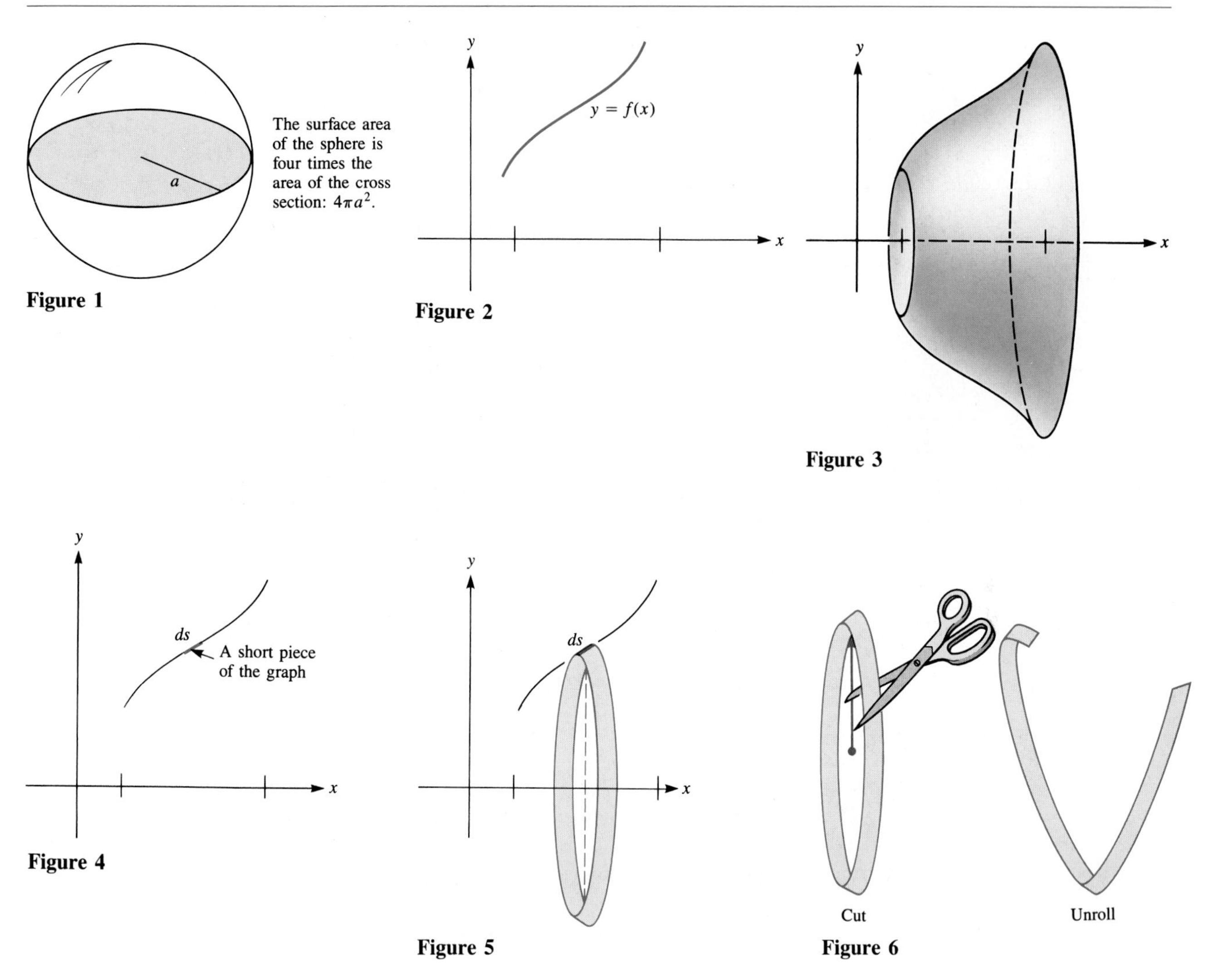

Figure 1

Figure 2

Figure 3

Figure 4

Figure 5

Figure 6

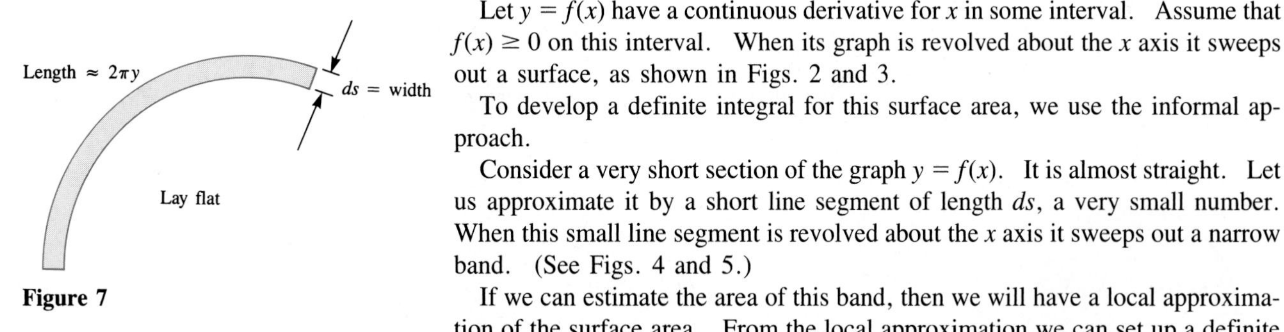

Figure 7

$2\pi y$

ds

Rectangle of area $2\pi y\ ds$

Figure 8

Let $y = f(x)$ have a continuous derivative for x in some interval. Assume that $f(x) \geq 0$ on this interval. When its graph is revolved about the x axis it sweeps out a surface, as shown in Figs. 2 and 3.

To develop a definite integral for this surface area, we use the informal approach.

Consider a very short section of the graph $y = f(x)$. It is almost straight. Let us approximate it by a short line segment of length ds, a very small number. When this small line segment is revolved about the x axis it sweeps out a narrow band. (See Figs. 4 and 5.)

If we can estimate the area of this band, then we will have a local approximation of the surface area. From the local approximation we can set up a definite integral for the entire surface area.

Imagine cutting the band with scissors and laying it flat, as in Figs. 6 and 7. It seems reasonable that the area of the flat band in Fig. 7 is close to the area of a flat rectangle of length $2\pi y$ and width ds, as in Fig. 8. (See Exercises 29 and 30.)

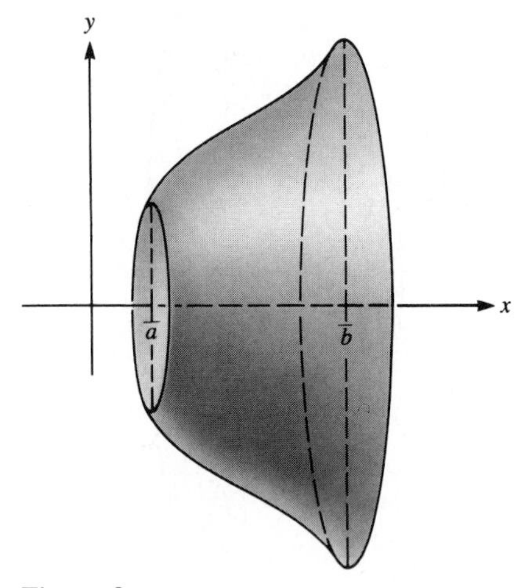

Figure 9

The local approximation of the surface area is therefore given by the formula

$$\text{Local approximation} = 2\pi y\, ds.$$

It leads to the formula

$$\boxed{\text{Surface area} = \int_{s_0}^{s_1} 2\pi y\, ds,} \tag{1}$$

where $[s_0, s_1]$ describes the appropriate interval on the "s axis." Since s is a clumsy parameter, for computations we will use substitutions to change (1) into more convenient integrals.

Say that the section of the graph that was revolved corresponds to the interval $[a, b]$ on the x axis, as in Fig. 9. Then the integral $\int_{s_0}^{s_1} 2\pi y\, ds$ becomes

$$\int_a^b 2\pi y \frac{ds}{dx}\, dx.$$

Since

$$\frac{ds}{dx} = \sqrt{1 + \left(\frac{dy}{dx}\right)^2},$$

this gives us the formula

$$\boxed{\text{Surface area} = \int_a^b 2\pi y \sqrt{1 + \left(\frac{dy}{dx}\right)^2}\, dx.} \tag{2}$$

(It is assumed that $y \geq 0$ and that dy/dx is continuous.)

EXAMPLE 1 Find the surface area of a sphere of radius a.

SOLUTION The circle of radius a has the equation $x^2 + y^2 = a^2$. The top half has the equation $y = \sqrt{a^2 - x^2}$. The sphere of radius a is formed by revolving the top half about the x axis. (See Fig. 10.) We have

$$\text{Surface area of sphere} = \int_{-a}^{a} 2\pi y \frac{ds}{dx}\, dx.$$

Now $ds/dx = \sqrt{1 + (dy/dx)^2}$, where we have $dy/dx = -x/\sqrt{a^2 - x^2}$. Thus

$$\begin{aligned}
\text{Surface area of sphere} &= \int_{-a}^{a} 2\pi y \sqrt{1 + \left(\frac{-x}{\sqrt{a^2 - x^2}}\right)^2}\, dx \\
&= \int_{-a}^{a} 2\pi\sqrt{a^2 - x^2} \sqrt{1 + \frac{x^2}{a^2 - x^2}}\, dx \\
&= \int_{-a}^{a} 2\pi\sqrt{a^2 - x^2} \sqrt{\frac{a^2}{a^2 - x^2}}\, dx \\
&= \int_{-a}^{a} 2\pi a\, dx = 2\pi a x \Big|_{-a}^{a} \\
&= 4\pi a^2.
\end{aligned}$$

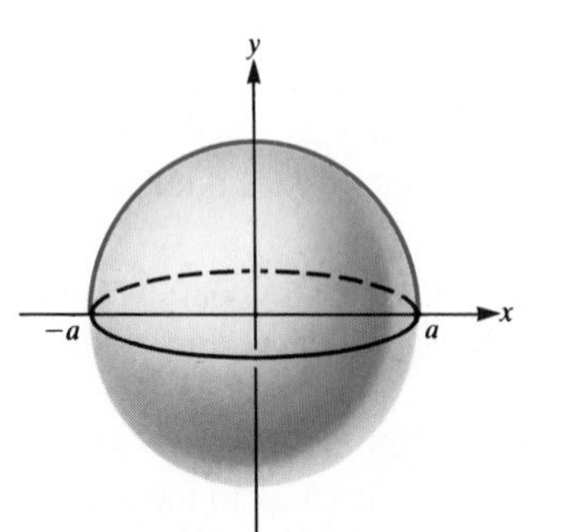

Figure 10

The surface area of a sphere is 4 times the area of its equatorial cross section. ■

If the graph is given parametrically, $x = g(t)$, $y = h(t)$, where g and h have continuous derivatives and $h(t) \geq 0$, then it is natural to express the integral $\int_{s_0}^{s_1} 2\pi y \, ds$ as an integral over an interval on the t axis. If t varies in the interval $[a, b]$, then

$$\int_{s_0}^{s_1} 2\pi y \, ds = \int_a^b 2\pi y \frac{ds}{dt} \, dt = \int_a^b 2\pi y \sqrt{\left(\frac{dx}{dt}\right)^2 + \left(\frac{dy}{dt}\right)^2} \, dt.$$

So we have

> Surface area for a curve given parametrically
>
> $$= \int_a^b 2\pi y \sqrt{\left(\frac{dx}{dt}\right)^2 + \left(\frac{dy}{dt}\right)^2} \, dt. \qquad (3)$$

Formula (2) is just the special case of the formula when the parameter is chosen to be x.

As the formulas are stated, they seem to refer only to surfaces obtained by revolving a curve about the x axis. In fact, they refer to revolution about any line. The factor y in the integrand,

$$2\pi y \sqrt{\left(\frac{dx}{dt}\right)^2 + \left(\frac{dy}{dt}\right)^2},$$

is the distance from the typical point on the curve to the axis of revolution. Replace y by R (for *radius*) to free ourselves from coordinate systems. (Use capital R to avoid confusion with polar coordinates.) The expression

$$\sqrt{\left(\frac{dx}{dt}\right)^2 + \left(\frac{dy}{dt}\right)^2} \, dt$$

is simply ds, since

$$\frac{ds}{dt} = \sqrt{\left(\frac{dx}{dt}\right)^2 + \left(\frac{dy}{dt}\right)^2}.$$

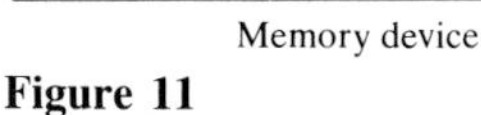

Figure 11

The simplest way to write the formula for surface area of revolution is then

> $$\text{Surface area} = \int_c^d 2\pi R \, ds,$$

where the interval $[c, d]$ refers to the parameter s. However, in practice s is seldom used as the parameter. Instead, x, y, t, or θ is used and the interval of integration describes the interval through which the parameter varies.

To remember this formula, think of a narrow circular band of width ds and radius R as analogous to the rectangle shown in Fig. 11.

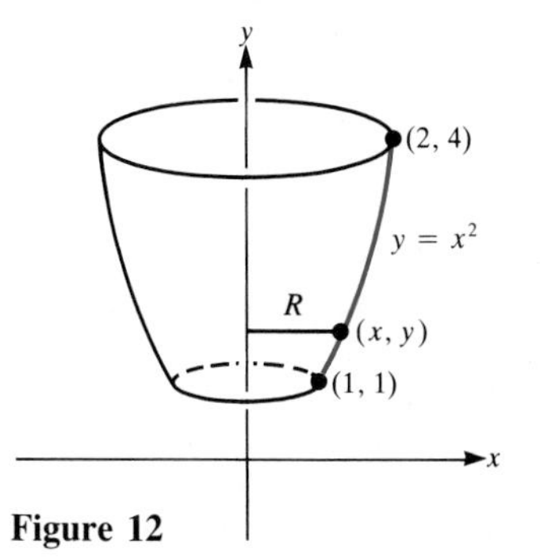

Figure 12

EXAMPLE 2 Find the area of the surface obtained by revolving around the y axis the part of the parabola $y = x^2$ that lies between $x = 1$ and $x = 2$. (See Fig. 12.)

R is found by inspection of a diagram.

SOLUTION The surface area is $\int_a^b 2\pi R\, ds$. Since the curve is described as a function of x, choose x as the parameter. By inspection of Fig. 12, $R = x$. Next, note that

$$ds = \frac{ds}{dx}\, dx = \sqrt{1 + \left(\frac{dy}{dx}\right)^2}\, dx.$$

The surface area is therefore

$$\int_1^2 2\pi x \sqrt{1 + 4x^2}\, dx.$$

To evaluate the integral, use the substitution

$$u = 1 + 4x^2 \qquad du = 8x\, dx.$$

Hence $x\, dx = du/8$. The new limits of integration are $u = 5$ and $u = 17$. Thus

$$\text{Surface area} = \int_5^{17} 2\pi\sqrt{u}\, \frac{du}{8} = \frac{\pi}{4}\int_5^{17} \sqrt{u}\, du$$

$$= \frac{\pi}{4} \cdot \frac{2}{3}\, u^{3/2} \Big|_5^{17} = \frac{\pi}{6}(17^{3/2} - 5^{3/2}). \quad ■$$

Section Summary

This section developed a definite integral for the area of a surface of revolution. It rests on the local estimate of the area swept out by a short segment of length ds revolved around a line L at a distance R from the segment: $2\pi R\, ds$. We gave an informal argument for this estimate; Exercises 29 and 30 develop it formally.

EXERCISES FOR SEC. 9.5: THE AREA OF A SURFACE OF REVOLUTION

In each of Exercises 1 to 4 set up a definite integral for the area of the indicated surface using the suggested parameter. Show the radius R on a diagram; do *not* evaluate the definite integrals.

1 The curve $y = x^3$; x in $[1, 2]$; revolved about the x axis; parameter x.

2 The curve $y = x^3$; x in $[1, 2]$; revolved about the line $y = -1$; parameter x.

3 The curve $y = x^3$; x in $[1, 2]$; revolved about the y axis; parameter y.

4 The curve $y = x^3$; x in $[1, 2]$; revolved about the y axis; parameter x.

5 Find the area of the surface obtained by rotating about the x axis that part of the curve $y = e^x$ that lies above $[0, 1]$.

6 Find the area of the surface formed by rotating one arch of the curve $y = \sin x$ about the x axis.

7 One arch of the cycloid given parametrically by the formula $x = \theta - \sin\theta$, $y = 1 - \cos\theta$ is revolved around the x axis. Find the area of the surface produced.

8 The curve given parametrically by $x = e^t \cos t$, $y = e^t \sin t$, t in $[0, \pi/2]$, is revolved around the x axis. Find the area of the surface produced.

In each of Exercises 9 to 16 find the area of the surface formed by revolving the indicated curve about the indicated axis. Leave the answer as a definite integral, but indicate how it could be evaluated by the fundamental theorem of calculus.

9 $y = 2x^3$ for x in $[0, 1]$; about the x axis.

10 $y = 1/x$ for x in $[1, 2]$; about the x axis.

11 $y = x^2$ for x in $[1, 2]$; about the x axis.

12 $y = x^{4/3}$ for x in $[1, 8]$; about the y axis.

13 $y = x^{2/3}$ for x in $[1, 8]$; about the line $y = 1$.

14 $y = x^3/6 + 1/(2x)$ for x in $[1, 3]$; about the y axis.

15 $y = x^3/3 + 1/(4x)$ for x in $[1, 2]$; about the line $y = -1$.

16 $y = \sqrt{1 - x^2}$ for x in $[-1, 1]$; about the line $y = -1$.

17 Consider the smallest tin can that contains a given sphere. (The height and diameter of the tin can equal the diameter of the sphere.)

(*a*) Compare the volume of the sphere with the volume of the tin can. Archimedes, who obtained the solution about 2200 years ago, considered it his greatest accomplishment. Cicero wrote, about two centuries after Archimedes' death:

I shall call up from the dust [the ancient equivalent of a blackboard] and his measuring-rod an obscure, insignificant person belonging to the same city [Syracuse], who lived many years after, Archimedes. When I was quaestor I tracked out his grave, which was unknown to the Syracusans (as they totally denied its existence), and found it enclosed all round and covered with brambles and thickets; for I remembered certain doggerel lines inscribed, as I had heard, upon his tomb, which stated that a sphere along with a cylinder had been set up on the top of his grave. Accordingly, after taking a good look all around (for there are a great quantity of graves at the Agrigentine Gate), I noticed a small column rising a little above the bushes, on which there was the figure of a sphere and a cylinder. And so I at once said to the Syracusans (I had their leading men with me) that I believed it was the very thing of which I was in search. Slaves were sent in with sickles who cleared the ground of obstacles, and when a passage to the place was opened we approached the pedestal fronting us; the epigram was traceable with about half the lines legible, as the latter portion was worn away.

(Cicero, *Tusculan Disputations,* vol. 23, translated by J. E. King, Loeb Classical Library, Harvard University, Cambridge, 1950.) Archimedes was killed by a Roman soldier in 212 B.C. Cicero was quaestor in 75 B.C.

(*b*) Compare the surface area of the sphere with the area of the curved side of the can.

18 (*a*) Compute the area of the portion of a sphere of radius a that lies between two parallel planes at distances c and $c + h$ from the center of the sphere ($0 \le c \le c + h \le a$).

(*b*) The result in (*a*) depends only on h, not on c. What does this mean geometrically? (See Fig. 13.)

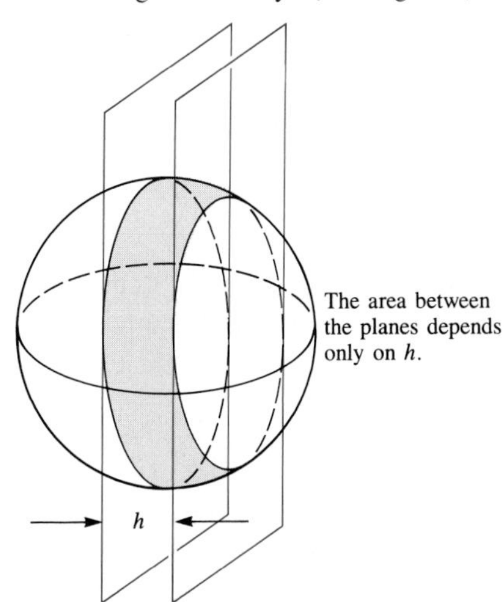

Figure 13

In Exercises 19 and 20 estimate the surface area obtained by revolving the given arc about the given line. Use either Simpson's method with six sections or a programmable calculator or computer.

19 $y = x^{1/4}$, x in [1, 3], about the x axis.

20 $y = x^{1/5}$, x in [1, 3], about the line $y = -1$.

Exercises 21 to 25 are concerned with the area of a surface obtained by revolving a curve given in polar coordinates.

21 Show that the area of the surface obtained by revolving the curve $r = f(\theta)$, $\alpha \le \theta \le \beta$, around the polar axis is

$$\int_\alpha^\beta 2\pi r \sin\theta \sqrt{r^2 + (r')^2}\, d\theta.$$

[Use formula (3).]

22 Solve Exercise 21 by making an informal local estimate first.

23 Use the formula in Exercise 21 to find the surface area of a sphere of radius a.

24 Find the area of the surface formed by revolving the portion of the curve $r = 1 + \cos\theta$ in the first quadrant about (*a*) the x axis, (*b*) the y axis. [The identity $1 + \cos\theta = 2\cos^2(\theta/2)$ may help in (*b*).]

25 The curve $r = \sin 2\theta$, θ in $[0, \pi/2]$, is revolved around the polar axis. Set up an integral for the surface area.

26 The portion of the curve $x^{2/3} + y^{2/3} = 1$ situated in the first quadrant is revolved around the x axis. Find the area of the surface produced.

27 Although the fundamental theorem of calculus is of no use in computing the perimeter of the ellipse $x^2/a^2 + y^2/b^2 = 1$, it is useful in computing the surface area of the "football" formed when the ellipse is rotated about one of its axes. Assuming that $a > b$ and that the ellipse is revolved around the x axis, find that area. Does your answer give the correct formula for the surface area of a sphere of radius a, $4\pi a^2$? (Let $b \to a^-$.)

28 The region bounded by $y = 1/x$ and the x axis and situated to the right of $x = 1$ is revolved around the x axis.

(*a*) Show that its volume is finite but its surface area is infinite.

(*b*) Does this mean that an infinite area can be painted by pouring a finite amount of paint into this solid?

Exercises 29 and 30 develop an exact formula for the area of the surface obtained by revolving a line segment about a line that does not meet it.

29 A right circular cone has slant height L and radius r, as shown in Fig. 14. If this cone is cut along a line through its vertex and laid flat, it becomes a sector of a circle of radius L, as shown in Fig. 15. By comparing Fig. 15 to a complete disk of radius L, find the area of the sector and thus the area of the cone in Fig. 14.

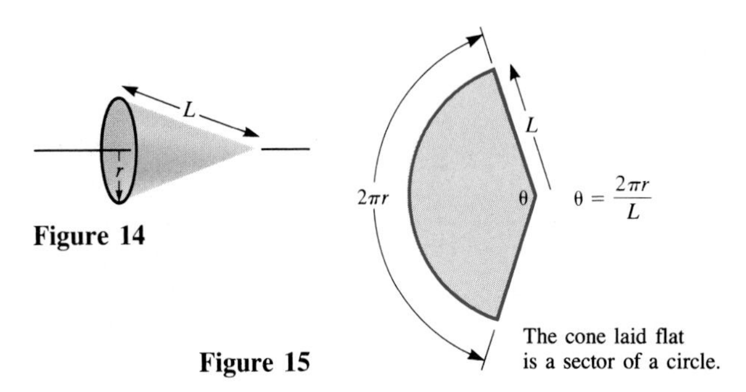

Figure 14

Figure 15

30 Consider a line segment of length L in the plane which does not meet a certain line in the plane, called the axis. (See Fig. 16.) When the line segment is revolved around the axis, it sweeps out a curved surface. Show that the area of this surface equals

$$2\pi rL,$$

where r is the distance from the midpoint of the line segment to the axis. The surface in Fig. 16 is called a **frustum of a cone**. Follow these steps:

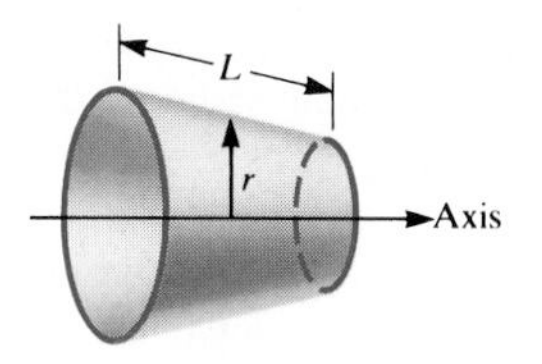

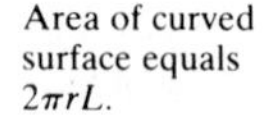
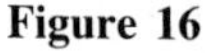
Area of curved surface equals $2\pi rL$.

Figure 16

(*a*) Complete the cone by extending the frustum as shown in Fig. 17. Label the radii and lengths as in that figure.

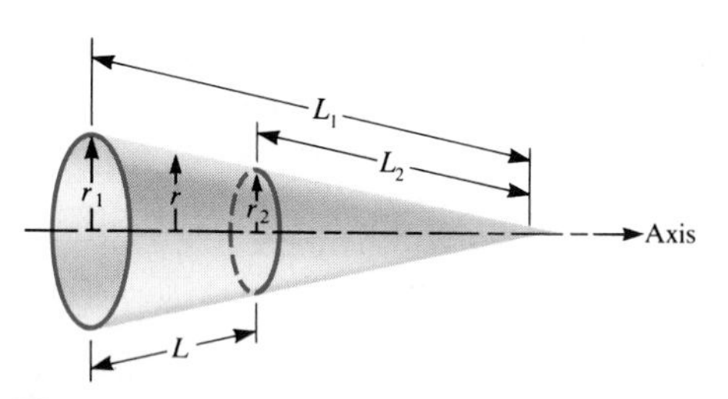

Figure 17

Show that $\dfrac{r_1}{r_2} = \dfrac{L_1}{L_2}$, hence $r_1L_2 = r_2L_1$.

(*b*) Show that the surface area of the frustum is $\pi r_1L_1 - \pi r_2L_2$.

(*c*) Express L_1 as $L_2 + L$ and, using the result of (*a*), show that

$$\begin{aligned}\pi r_1L_1 - \pi r_2L_2 &= \pi r_2(L_1 - L_2) + \pi r_1L \\ &= \pi r_2L + \pi r_1L.\end{aligned}$$

(*d*) Show that the surface area of the frustum is $2\pi rL$, where $r = (r_1 + r_2)/2$.

31 The derivative of the volume of a sphere, $4\pi r^3/3$, is $4\pi r^2$, its surface area. Is this simply a coincidence?

32 Define the moment of a curve around the x axis to be $\int_a^b y\,ds$, where a and b refer to the range of the arc length s. The moment of the curve around the y axis is defined as $\int_a^b x\,ds$. The centroid of the curve, $(\bar{x}, \bar{y})$, is defined by setting

$$\bar{x} = \frac{\displaystyle\int_a^b x\,ds}{\text{Length of curve}} \qquad \bar{y} = \frac{\displaystyle\int_a^b y\,ds}{\text{Length of curve}}.$$

Find the centroid of the top half of the circle $x^2 + y^2 = a^2$.

33 (See Exercise 32.) Show that the area of the surface obtained by revolving about the x axis a curve that lies above it is equal to the length of the curve times the distance that the centroid of the curve moves.

34 Use Exercise 33 to find the surface area of the doughnut formed by revolving a circle of radius a around a line a distance b from its center, $b \geq a$.

35 Use Exercise 33 to find the area of the curved part of a cone of radius a and height h.

9.6 CURVATURE

In this section we use calculus to obtain a measure of the ''curviness'' or ''curvature'' of a graph at points on its curve. This concept is used in Sec. 13.4 in the study of motion along a curved path.

Curvature

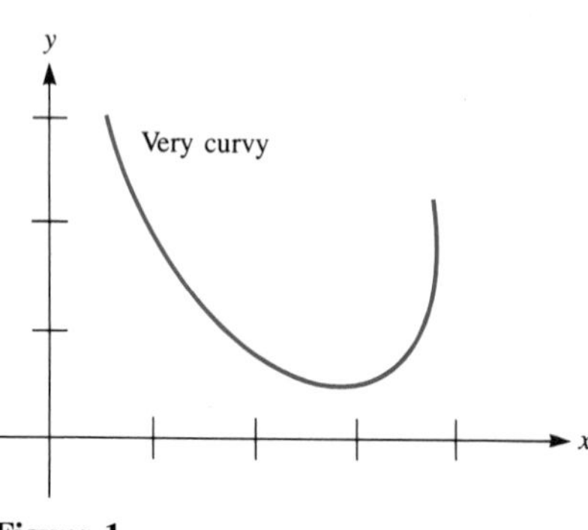

Figure 1

The curve in Fig. 1 is certainly ''curvier'' than the curve in Fig. 2. If you were to drive a car at a fixed speed along the curve in Fig. 1, the direction of the car changes rapidly. But as you travel along the curve in Fig. 2, the direction changes more slowly. (See Figs. 3 and 4.)

This contrast suggests that ''curvature'' can be measured by the direction's rate of change. We have a way of assigning a numerical value to direction, namely, the angle of the tangent line. The rate of change of this angle with respect to arc length could then be our measure of curvature.

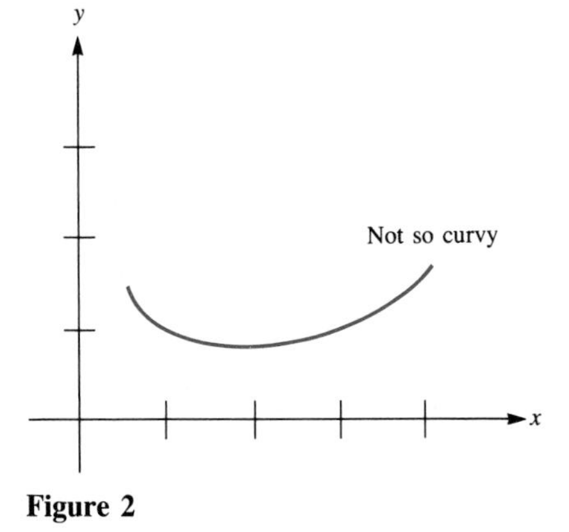

Figure 2

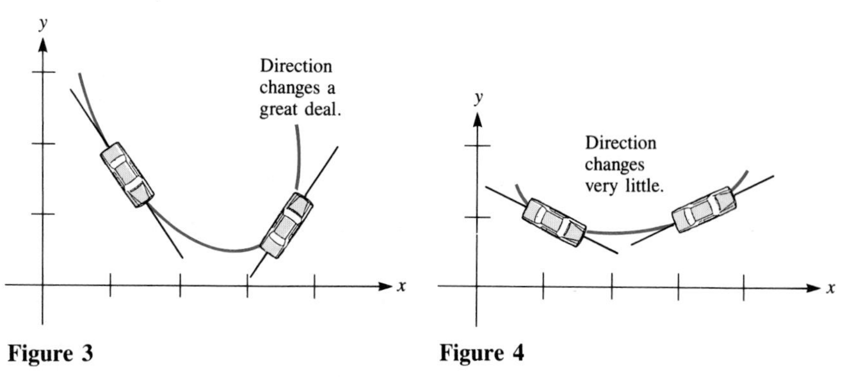

Figure 3

Figure 4

κ is the Greek letter "kappa."

Definition *Curvature*. Assume that a curve is given parametrically, with the parameter of the typical point P being s, the distance along the curve from a fixed point P_0 to P. Let ϕ be the angle between the tangent line at P and the positive part of the x axis. The **curvature** κ at P is the absolute value of the derivative, $d\phi/ds$:

$$\boxed{\kappa = \left|\frac{d\phi}{ds}\right|}$$

(if the derivative exists). (See Fig. 5.)

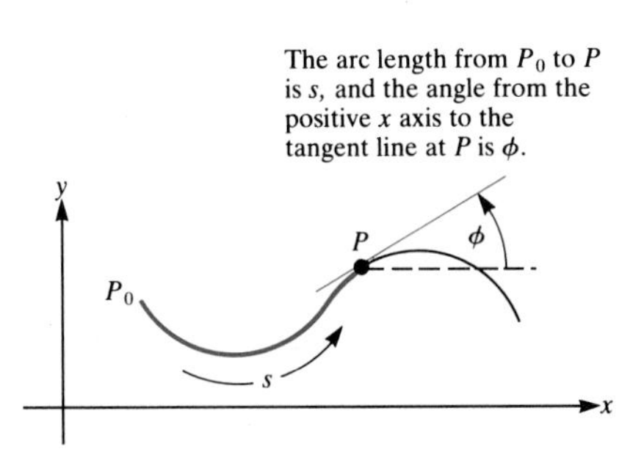

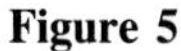
Figure 5

Observe that a straight line has zero curvature everywhere, since ϕ is constant.

What is the curvature of a circle? A small circle should have large curvature; a large circle should have small curvature. The next theorem shows that curvature, as given in the definition, meets that demand. The curvature κ of a circle of radius a turns out to be $1/a$, the reciprocal of the radius.

Theorem 1 For a circle of radius a, the curvature $|d\phi/ds|$ is constant and equals $1/a$, the reciprocal of the radius.

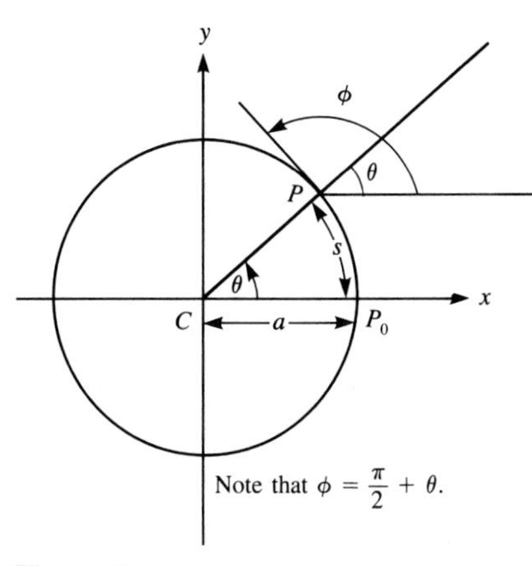

Figure 6

Proof It is necessary to express ϕ as a function of arc length s on a circle of radius a. Measure s counterclockwise from the point P_0, as shown in Fig. 6. (When $s = 0$, choose $\phi = \pi/2$.) Then $\phi = \pi/2 + \theta$, since an exterior angle of triangle PCP_0 is the sum of the two opposite angles of the triangle. By definition of radian measure,

$$\theta = \frac{s}{a}.$$

Hence

$$\phi = \frac{s}{a} + \frac{\pi}{2}.$$

Thus
$$\frac{d\phi}{ds} = \frac{1}{a} + 0$$
$$= \frac{1}{a},$$

as claimed. ■

Computing Curvature

When a curve is given in the form $y = f(x)$, the curvature can be expressed in terms of the first and second derivatives dy/dx and d^2y/dx^2.

The curvature of $y = f(x)$

Theorem 2 Let arc length s be measured along the curve $y = f(x)$ from a fixed point P_0. Assume that x increases as s increases. Assume that y' and y'' are continuous. Then

$$\kappa = \text{curvature} = \frac{|d^2y/dx^2|}{[1 + (dy/dx)^2]^{3/2}}.$$

Proof By the chain rule,

Same as $\frac{d\phi}{dx} = \frac{d\phi}{ds}\frac{ds}{dx}$

$$\frac{d\phi}{ds} = \frac{d\phi/dx}{ds/dx}.$$

As was shown in Sec. 9.4,

$$\frac{ds}{dx} = \left[1 + \left(\frac{dy}{dx}\right)^2\right]^{1/2}.$$

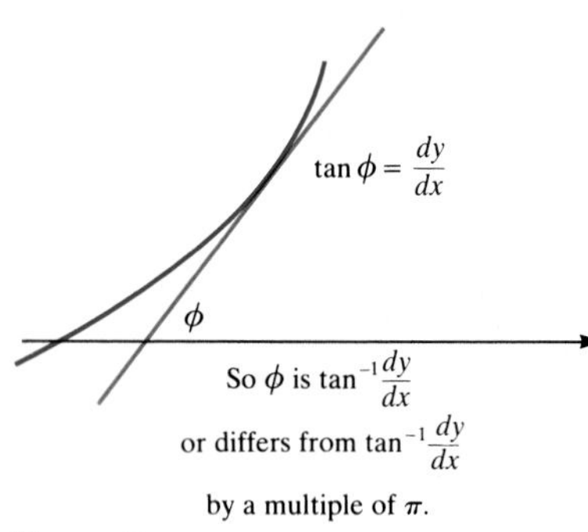

Figure 7

All that remains is to express $d\phi/dx$ in terms of dy/dx and d^2y/dx^2. Note that, in Fig. 7,

$$\frac{dy}{dx} = \tan\phi \qquad \text{(slope of tangent line).} \tag{1}$$

We find $d\phi/dx$ by differentiating both sides of (1) with respect to x.

Thus $$\frac{d^2y}{dx^2} = \sec^2\phi\,\frac{d\phi}{dx} = (1 + \tan^2\phi)\frac{d\phi}{dx} = \left[1 + \left(\frac{dy}{dx}\right)^2\right]\frac{d\phi}{dx}.$$

and we have

$$\frac{d\phi}{dx} = \frac{d^2y/dx^2}{1 + (dy/dx)^2}.$$

Consequently, $$\frac{d\phi}{ds} = \frac{d\phi/dx}{ds/dx} = \frac{d^2y/dx^2}{[1 + (dy/dx)^2]\sqrt{1 + (dy/dx)^2}},$$

and the theorem is proved. ■

EXAMPLE 1 Find the curvature at a typical point (x, y) on the curve $y = x^2$.

SOLUTION In this case, $dy/dx = 2x$ and $d^2y/dx^2 = 2$. Thus the curvature at (x, y) is

$$\kappa = \frac{|d^2y/dx^2|}{[1 + (dy/dx)^2]^{3/2}} = \frac{2}{[1 + (2x)^2]^{3/2}}.$$

The maximum curvature occurs when $x = 0$. As $|x|$ increases, the curvature approaches 0, and the curve gets straighter. ■

Theorem 2 holds for curves given parametrically.

Theorem 2 tells how to find the curvature if y is given as a function of x. But it holds as well when the curve is described parametrically, where x and y are functions of some parameter such as t or θ. Just use the fact that

$$\frac{dy}{dx} = \frac{\dfrac{dy}{dt}}{\dfrac{dx}{dt}}$$

and

$$\frac{d^2y}{dx^2} = \frac{\dfrac{d}{dt}\left(\dfrac{dy}{dx}\right)}{\dfrac{dx}{dt}}.$$

Example 2 illustrates the procedure.

EXAMPLE 2 The cycloid determined by a wheel of radius 1 has the parametric equations

$$x = \theta - \sin\theta \qquad \text{and} \qquad y = 1 - \cos\theta,$$

as shown in Sec. 9.3. (See Fig. 8.) Find the curvature at a typical point on this curve.

SOLUTION First find dy/dx in terms of θ. We have

$$\frac{dx}{d\theta} = 1 - \cos\theta \qquad \text{and} \qquad \frac{dy}{d\theta} = \sin\theta.$$

Thus

$$\frac{dy}{dx} = \frac{\sin\theta}{1 - \cos\theta}.$$

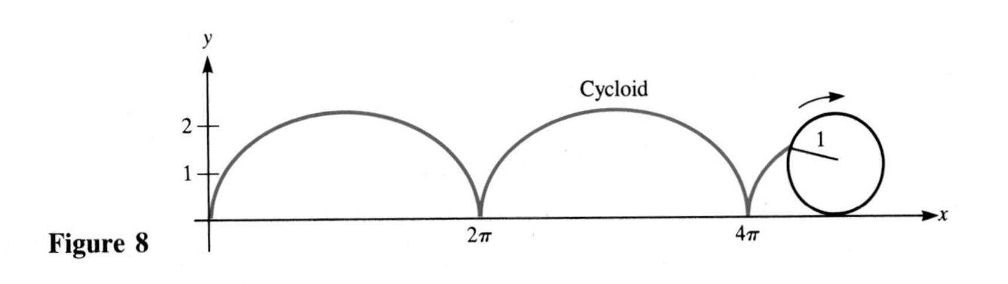

Figure 8

Next find d^2y/dx^2. We have

$$\frac{d^2y}{dx^2} = \frac{d}{dx}\left(\frac{dy}{dx}\right)$$

$$= \frac{\dfrac{d}{d\theta}\left(\dfrac{dy}{dx}\right)}{\dfrac{dx}{d\theta}}$$

$$= \frac{\dfrac{d}{d\theta}\left(\dfrac{\sin\theta}{1-\cos\theta}\right)}{1-\cos\theta}$$

$$= \frac{\dfrac{(1-\cos\theta)\cos\theta - \sin\theta\sin\theta}{(1-\cos\theta)^2}}{1-\cos\theta}$$

$$= \frac{\cos\theta - \cos^2\theta - \sin^2\theta}{(1-\cos\theta)^3}$$

$$= \frac{\cos\theta - 1}{(1-\cos\theta)^3}$$

$$= \frac{-1}{(1-\cos\theta)^2}.$$

Thus the curvature is

$$\kappa = \frac{\left|\dfrac{-1}{(1-\cos\theta)^2}\right|}{\left[1+\left(\dfrac{\sin\theta}{1-\cos\theta}\right)^2\right]^{3/2}}$$

$$= \frac{\dfrac{1}{(1-\cos\theta)^2}}{\left[\dfrac{(1-\cos\theta)^2+\sin^2\theta}{(1-\cos\theta)^2}\right]^{3/2}}$$

$$= \frac{1-\cos\theta}{[1-2\cos\theta+\cos^2\theta+\sin^2\theta]^{3/2}}$$

$$= \frac{1-\cos\theta}{[2-2\cos\theta]^{3/2}}$$

$$= \frac{1}{2^{3/2}\sqrt{1-\cos\theta}}.$$

Since $y = 1 - \cos\theta$ and $2^{3/2} = \sqrt{8}$, the curvature equals $1/\sqrt{8y}$. ■

As Example 2 shows, the computations in computing curvature can be tedious. When a curve is given in parametric form, it is often simpler to use an alternative formula for curvature, which is described in the next theorem.

Theorem 3 Assume that g and h are functions with continuous second derivatives. If, as we move along the parameterized curve $x = g(t)$, $y = h(t)$, both x and the arc length s from a point P_0 increase as t increases, then

$$\kappa = \frac{|\dot{x}\ddot{y} - \dot{y}\ddot{x}|}{[\dot{x}^2 + \dot{y}^2]^{3/2}}.$$

(The dot notation for derivatives shortens the formula: $\dot{x} = dx/dt$, $\ddot{x} = d^2x/dt^2$, $\dot{y} = dy/dt$, and $\ddot{y} = d^2y/dt^2$.)

Proof We already know that

$$\frac{dy}{dx} = \frac{dy/dt}{dx/dt} = \frac{\dot{y}}{\dot{x}}.$$

Furthermore,

$$\frac{d^2y}{dx^2} = \frac{d}{dx}\left(\frac{\dot{y}}{\dot{x}}\right) = \frac{\frac{d}{dt}(\dot{y}/\dot{x})}{\frac{dx}{dt}} = \frac{1}{\dot{x}}\left(\frac{\dot{x}\ddot{y} - \dot{y}\ddot{x}}{\dot{x}^2}\right) = \frac{\dot{x}\ddot{y} - \dot{y}\ddot{x}}{\dot{x}^3}.$$

By Theorem 2,

$$\kappa = \frac{|d^2y/dx^2|}{[1 + (dy/dx)^2]^{3/2}} = \frac{|(\dot{x}\ddot{y} - \dot{y}\ddot{x})/\dot{x}^3|}{[1 + (\dot{y}/\dot{x})^2]^{3/2}} = \frac{|\dot{x}\ddot{y} - \dot{y}\ddot{x}|}{[\dot{x}^2 + \dot{y}^2]^{3/2}}. \quad \blacksquare$$

EXAMPLE 3 Find the curvature for the cycloid of Example 2 using the parametric formula for κ.

SOLUTION We have $x = \theta - \sin\theta$ and $y = 1 - \cos\theta$, so

$$\frac{dx}{d\theta} = 1 - \cos\theta,$$

$$\frac{d^2x}{d\theta^2} = \sin\theta,$$

$$\frac{dy}{d\theta} = \sin\theta,$$

and $$\frac{d^2y}{d\theta^2} = \cos\theta.$$

Hence
$$\kappa = \frac{|(1-\cos\theta)\cos\theta - \sin\theta\sin\theta|}{[(1-\cos\theta)^2 + \sin^2\theta]^{3/2}}$$
$$= \frac{|\cos\theta - \cos^2\theta - \sin^2\theta|}{[1 - 2\cos\theta + \cos^2\theta + \sin^2\theta]^{3/2}}$$
$$= \frac{1-\cos\theta}{[2 - 2\cos\theta]^{3/2}}$$
$$= \frac{1}{2^{3/2}\sqrt{1-\cos\theta}}.$$

The result agrees with that in Example 2. ■

Radius of Curvature

As Theorem 1 shows, a circle with curvature κ has radius $1/\kappa$. This suggests the following definition.

A large radius of curvature implies a small curvature.

Definition *Radius of curvature.* The **radius of curvature** of a curve at a point is the reciprocal of the curvature:

$$\text{Radius of curvature} = \frac{1}{\text{Curvature}} = \frac{1}{\kappa}.$$

As can easily be checked, the radius of curvature of a circle of radius a is, fortunately, a.

The cycloid in Example 2 has radius of curvature at the point (x, y) equal to

$$\frac{1}{1/\sqrt{8y}} = \sqrt{8y}.$$

In particular, at the top of an arch, the radius of curvature is $\sqrt{8 \cdot 2} = 4$.

The Osculating Circle

At a given point P on a curve, the osculating circle at P is defined to be that circle which (a) passes through P, (b) has the same slope at P as the curve does, and (c) has the same second derivative there. By Theorem 2, the osculating circle and the curve have the same curvature at P. Hence they have the same radius of curvature.

For instance, consider the parabola $y = x^2$ of Example 1. When $x = 1$, the curvature is $2/5^{3/2}$ and the radius of curvature is $5^{3/2}/2 \approx 5.6$. The osculating circle at (1, 1) is shown in Fig. 9.

Observe that the osculating circle in Fig. 9 *crosses the parabola* as it passes through the point (1, 1). Although this may be surprising, a little reflection will show why it is to be expected.

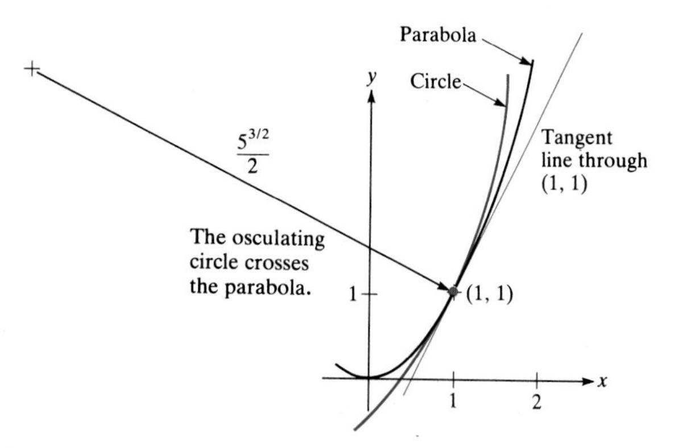

Figure 9

Think of driving along the parabola $y = x^2$. If you start at $(1, 1)$ and drive up along the parabola, the curvature diminishes. It is smaller than that of the circle of curvature at $(1, 1)$. Hence you would be turning your steering wheel to the left and would be traveling *outside* the circle of curvature at $(1, 1)$. On the other hand, if you start at $(1, 1)$ and move to the left on the parabola, the curvature increases and is greater than that of the osculating circle at $(1, 1)$, so you would be driving *inside* the osculating circle at $(1, 1)$. This informal argument shows why the osculating circle crosses the curve in general. At a point where the curvature is neither a local maximum nor a local minimum, the osculating circle crosses the curve. In the case of the curve $y = x^2$, the only osculating circle that does *not* cross the curve at its point of tangency is the one that is tangent at $(0, 0)$, where the curvature is a maximum.

Section Summary

We defined the curvature of a curve as the absolute value of the rate at which the angle between the tangent and the x axis changes as a function of arc length. If the curve is given parametrically then its curvature is

$$\frac{|\dot{x}\ddot{y} - \dot{y}\ddot{x}|}{[\dot{x}^2 + \dot{y}^2]^{3/2}}.$$

If the parameter is x then $\dot{x} = 1$, $\ddot{x} = 0$ and this formula becomes

$$\frac{\left|\dfrac{d^2y}{dx^2}\right|}{\left[1 + \left(\dfrac{dy}{dx}\right)^2\right]^{3/2}}.$$

Radius of curvature is defined as the reciprocal of the curvature.

EXERCISES FOR SEC. 9.6: CURVATURE

In each of Exercises 1 to 6 find the curvature and radius of curvature of the given curve at the given point.

1 $y = x^2$ at $(1, 1)$ **2** $y = \cos x$ at $(0, 1)$
3 $y = e^{-x}$ at $(1, 1/e)$ **4** $y = \ln x$ at $(e, 1)$
5 $y = \tan x$ at $(\pi/4, 1)$ **6** $y = \sec 2x$ at $(\pi/6, 2)$

In Exercises 7 to 10 find the curvatures of the given curves for the given values of the parameter.

7 $\begin{cases} x = 2\cos 3t \\ y = 2\sin 3t \end{cases}$ at $t = 0$

8 $\begin{cases} x = 1 + t^2 \\ y = t^3 + t^4 \end{cases}$ at $t = 2$

9 $\begin{cases} x = e^{-t} \cos t \\ y = e^{-t} \sin t \end{cases}$ at $t = \pi/6$

10 $\begin{cases} x = \cos^3 \theta \\ y = \sin^3 \theta \end{cases}$ at $\theta = \pi/3$

11 (*a*) Compute the curvature and radius of curvature for the curve $y = (e^x + e^{-x})/2$.
(*b*) Show that the radius of curvature at (x, y) is y^2.

12 Find the radius of curvature along the curve $y = \sqrt{a^2 - x^2}$, where a is a constant. (Since the curve is part of a circle of radius a, the answer should be a.)

13 For what value of x is the radius of curvature of $y = e^x$ smallest? *Hint:* How does one find the minimum of a function?

14 For what value of x is the radius of curvature of $y = x^2$ smallest?

---

15 (*a*) Show that where a curve has its tangent parallel to the x axis its curvature is simply the absolute value of the second derivative d^2y/dx^2.
(*b*) Show that the curvature is never larger than the absolute value of d^2y/dx^2.

16 An engineer lays out a railroad track as indicated in Fig. 10. BC is part of a circle. AB and CD are straight and tangent to the circle. After the first train runs over this track, the engineer is fired because the curvature is not a continuous function. Why should it be?

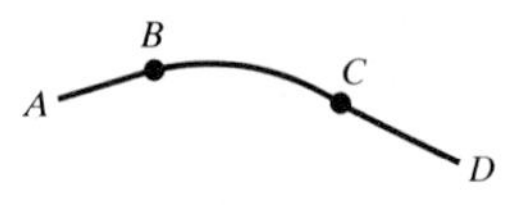

Figure 10

17 Railroad curves are banked to reduce wear on the rails and flanges. The greater the radius of curvature, the less the curve must be banked. The best bank angle A satisfies the equation $\tan A = v^2/(32R)$, where v is speed in feet per second and R is radius of curvature in feet. A train travels in the elliptical track

$$\frac{x^2}{1{,}000^2} + \frac{y^2}{500^2} = 1$$

(where x and y are measured in feet) at 60 miles per hour (equals 88 feet per second). Find the best angle A at the points (1,000, 0) and (0, 500).

18 The flexure formula in the theory of beams asserts that the bending moment M required to bend a beam is proportional to the desired curvature, $M = c/R$, where c is a constant depending on the beam and R is the radius of curvature. A beam is bent to form the parabola $y = x^2$. What is the ratio between the moments required at (0, 0) and at (2, 4)? (See Fig. 11.)

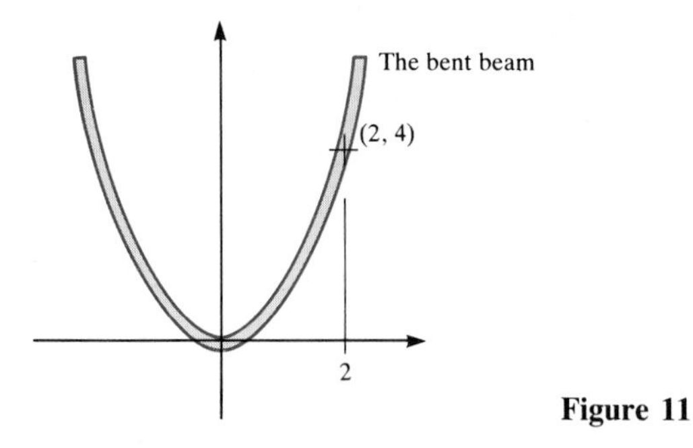

Figure 11

Exercises 19 to 21 are related.

19 Find the radius of curvature at a typical point on the ellipse

$$\begin{cases} x = a \cos \theta \\ y = b \sin \theta. \end{cases}$$

20 (*a*) Show, by eliminating θ, that the curve in Exercise 19 is the ellipse

$$\frac{x^2}{a^2} + \frac{y^2}{b^2} = 1.$$

(*b*) What is the radius of curvature of this ellipse at $(a, 0)$? at $(0, b)$?

21 An ellipse has a major diameter of length 6 and a minor diameter of length 4. Draw the circles that most closely approximate this ellipse at the four points which lie at the extremities of its diameters. (See Exercises 19 and 20.)

In each of Exercises 22 to 24 a curve is given in polar coordinates. To find its curvature write it in rectangular coordinates with parameter θ, using the equations $x = r \cos \theta$ and $y = r \sin \theta$.

22 Find the curvature of $r = a \cos \theta$.

23 Show that at the point (r, θ) the cardioid $r = 1 + \cos \theta$ has curvature $3\sqrt{2}/(4\sqrt{r})$.

24 Find the curvature of $r = \cos 2\theta$.

25 If, on a curve, $dy/dx = y^3$, express the curvature in terms of y.

26 As is shown in physics, the larger the radius of curvature of a turn, the faster a given car can travel around that turn. The radius of curvature required is proportional to the square of the maximum speed. Or conversely, the maximum speed around a turn is proportional to the square root of the radius of curvature. If a car moving on the path $y = x^3$ (x and y measured in miles) can go 30 miles per hour at (1, 1) without sliding off, how fast can it go at (2, 8)?

27 At the top of the cycloid in Example 2 the radius of curvature is twice the diameter of the rolling circle. What would you have guessed the radius of curvature to be at this point? Why is it not simply the diameter of the wheel, since the wheel at each moment is rotating about its point of contact with the ground?

9.7 THE REFLECTION PROPERTIES OF THE CONIC SECTIONS

This section obtains the reflection properties of the parabola, ellipse, and hyperbola and describes their many applications. The main tool is a formula that determines the angle between two lines in terms of their slopes.

The Angle between Two Lines

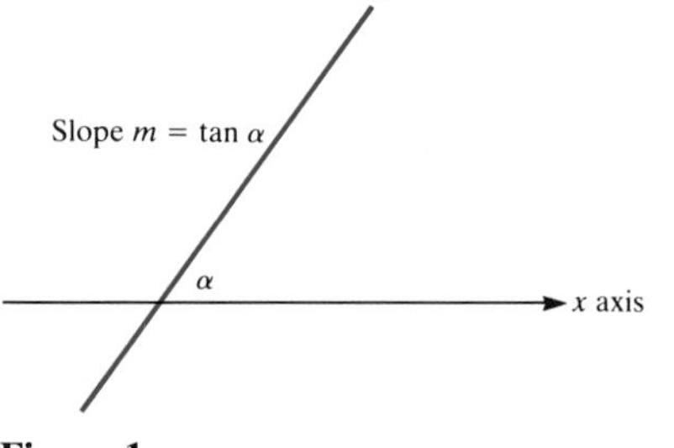

Figure 1

Consider a line L in the xy plane. It forms an **angle of inclination** α, $0 \le \alpha < \pi$, with the positive x axis. The slope m of L is $\tan\alpha$. (If $\alpha = \pi/2$, the slope is not defined.) See Fig. 1.

Consider two lines L and L' with angles of inclination α and α' and slopes m and m', respectively, as in Fig. 2. There are two (supplementary) angles between the two lines. The following definition serves to distinguish one of these two angles as *the* angle between L and L'.

Definition *Angle between two lines.* Let L and L' be two lines in the xy plane so named that L has the *larger* angle of inclination, $\alpha > \alpha'$. The angle θ between L and L' is defined to be

$$\theta = \alpha - \alpha'.$$

If L and L' are parallel, define θ to be 0.

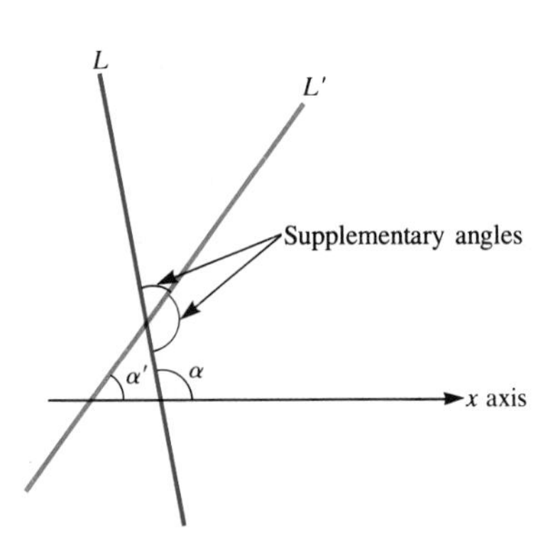

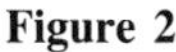

Figure 2

Figure 3 shows θ for some typical L and L'. In each case, θ is the counterclockwise angle from L' to L. Note that $0 \le \theta < \pi$.

The tangent of θ is easily expressed in terms of the slopes m and m'. We have

$$\tan\theta = \tan(\alpha - \alpha')$$

$$= \frac{\tan\alpha - \tan\alpha'}{1 + \tan\alpha\tan\alpha'} \qquad \text{by the identity for } \tan(A - B)$$

$$= \frac{m - m'}{1 + mm'}.$$

Formula for tangent of angle between two lines in terms of their slopes

Thus

$$\boxed{\tan\theta = \frac{m - m'}{1 + mm'}}, \qquad (1)$$

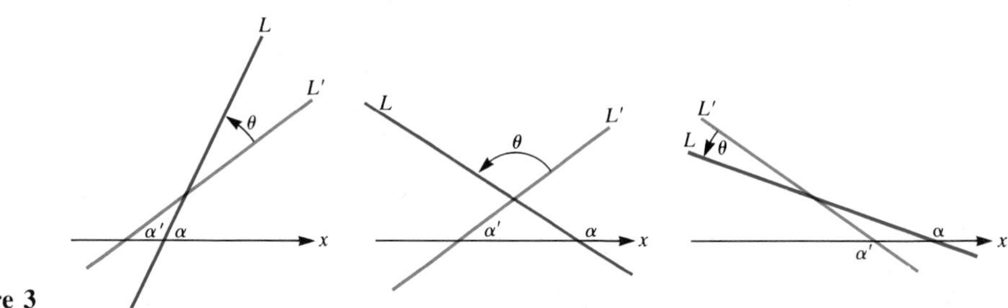

Figure 3

where m is the slope of the line with larger angle of inclination. If $mm' = -1$, then $\theta = \pi/2$; this corresponds to the fact that as mm' approaches -1, $|\tan\theta| \to \infty$.

The Reflecting Property of the Parabola

A satellite dish receiver and a microwave reflector are parabolic in shape. The reason is that all radio waves parallel to the axis of the parabola, after bouncing off the parabola, pass through a common point. This point is called the focus of the parabola. (See Fig. 4.) Similarly, the reflector behind a flashlight bulb is parabolic.

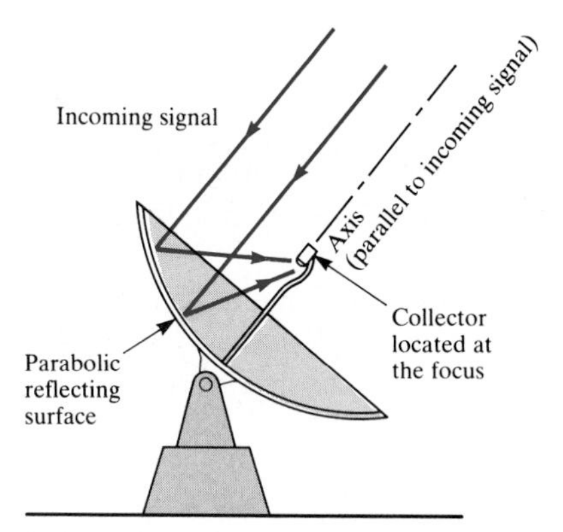

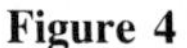
Figure 4

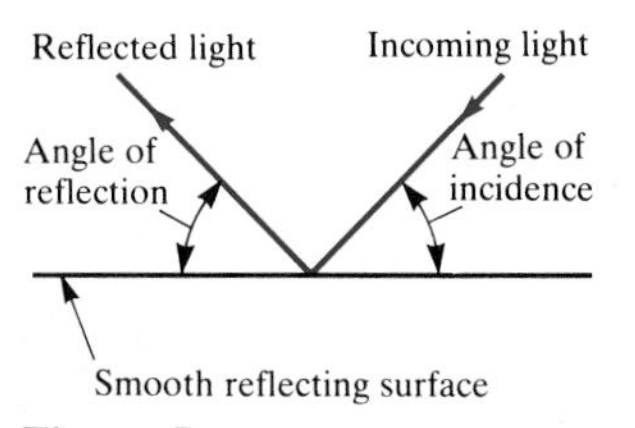

Figure 5

Now we will show why parabolas have this property. Our argument uses the fact that the angle of reflection of a light ray equals its angle of incidence, as shown in Fig. 5.

Consider a parabola with focus $F = (c, 0)$, $c > 0$, and directrix $x = -c$. Its equation is $y^2 = 4cx$. The top half has the equation $y = 2\sqrt{cx}$. (See Fig. 6.)

Let P be an arbitrary point on the curve $y = 2\sqrt{cx}$. We must show that the angle α between the line FP and the tangent line to the parabola at P equals the angle β between the tangent line and the line through P parallel to the x axis. To do this, we will show that $\tan\alpha = \tan\beta$.

The slope of the tangent line at a point on the parabola $y = 2\sqrt{cx}$ is $\sqrt{c}/\sqrt{x}$. The slope of the line FP is, by the two-point formula,

$$\frac{y-0}{x-c} = \frac{y}{x-c}.$$

By formula (1),

$$\tan\alpha = \frac{\dfrac{y}{x-c} - \dfrac{\sqrt{c}}{\sqrt{x}}}{1 + \left(\dfrac{y}{x-c}\right)\left(\dfrac{\sqrt{c}}{\sqrt{x}}\right)}.$$

Also,

$$\tan\beta = \frac{\sqrt{c}}{\sqrt{x}}, \tag{2}$$

since it is the slope of the tangent line at P.

Some algebra, together with the equation $y = 2\sqrt{cx}$, simplifies the formula for $\tan\alpha$, as follows:

$$\tan\alpha = \frac{y\sqrt{x} - \sqrt{c}(x-c)}{(x-c)\sqrt{x} + y\sqrt{c}} = \frac{2\sqrt{cx}\sqrt{x} - \sqrt{c}x + \sqrt{c}c}{x\sqrt{x} - c\sqrt{x} + 2\sqrt{cx}\sqrt{c}}.$$

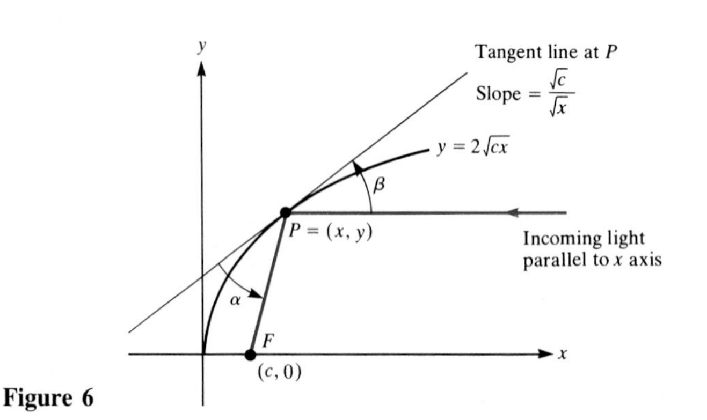

Figure 6

Some further algebra simplifies this expression and shows that

$$\tan \alpha = \frac{\sqrt{c}}{\sqrt{x}}. \tag{3}$$

By Eqs. (2) and (3), $\tan \alpha = \tan \beta$, and the reflection property of the parabola is established. (See Exercise 11.)

Diocles, in his book *On Burning Mirrors,* written around the year 190 B.C., studied both spherical and parabolic reflectors, both of which had been considered by earlier scientists. Some had thought that a spherical reflector focuses incoming light at a single point. This is false, and Diocles showed that a spherical reflector subtending an angle of 60° reflects light that is parallel to its axis of symmetry to points on this axis that occupy about one-thirteenth of the radius. He proposed an experiment, "Perhaps you would like to make two examples of a burning-mirror, one spherical, one parabolic, so that you can measure the burning power of each." *On Burning Mirrors* contains the first known proof that a parabola has the reflecting property described in this section. (See Diocles, *On Burning Mirrors,* edited by G. J. Toomer, Springer, New York, 1976.) [It is a simple exercise in trigonometry to show that a spherical mirror of radius r and subtending an angle of 60° causes light parallel to its axis of symmetry to reflect and meet the axis in an interval of length $r\,(1/\sqrt{3} - \frac{1}{2}) \approx r/12.9$. So even a spherical reflector makes an effective solar oven. After all, a potato or hamburger is not a point.]

The reflecting property of the parabola was known to Greek mathematicians before Diocles. Legend has it that Archimedes (287–212 B.C.) arranged mirrors in a parabolic arc to burn the ships of an invader.

In the science museum in Paris two parabolic reflectors about 2 meters in diameter are set up some 20 meters apart. A person whispering at the focus of one reflector can be heard clearly by a person listening at the other focus—but by nobody else.

The Reflecting Property of the Ellipse

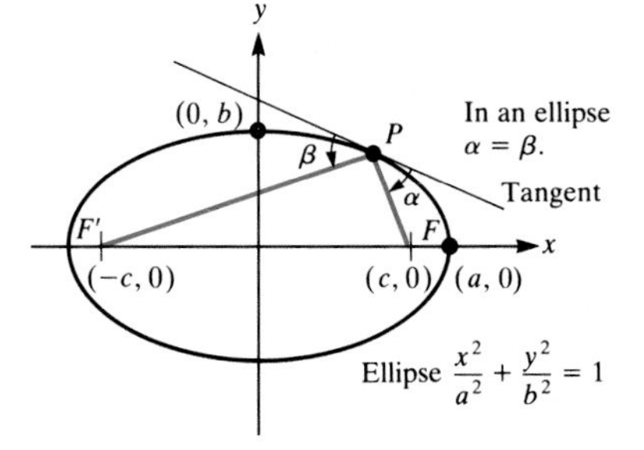

Figure 7

An ellipse with foci F and F' has the property that sound (or light) starting at F passes through F' after bouncing off the ellipse. This follows from the following geometric property. For any point P on the ellipse, the lines PF and PF' make equal angles α and β with the tangent line to the ellipse at P. (See Fig. 7.)

The argument follows the same general lines. First, write the equation of the ellipse in the form

$$\frac{x^2}{a^2} + \frac{y^2}{b^2} = 1.$$

The foci are at $(c, 0)$ and $(-c, 0)$, where $c^2 = a^2 - b^2$. Then compute the slopes of PF and PF' by the two-point formula and the slope of the tangent line at P by differentiation (either explicit or implicit). Finally, use formula (1) and check that $\tan \alpha = \tan \beta$. The details are left as Exercise 12.

For a geometric (noncalculus) proof of the reflection property of the ellipse, see the high school text *Geometry: A Guided Inquiry,* by G. D. Chakerian, C. Crabill, and S. Stein, Sunburst, Pleasantville, N.Y., 1986.

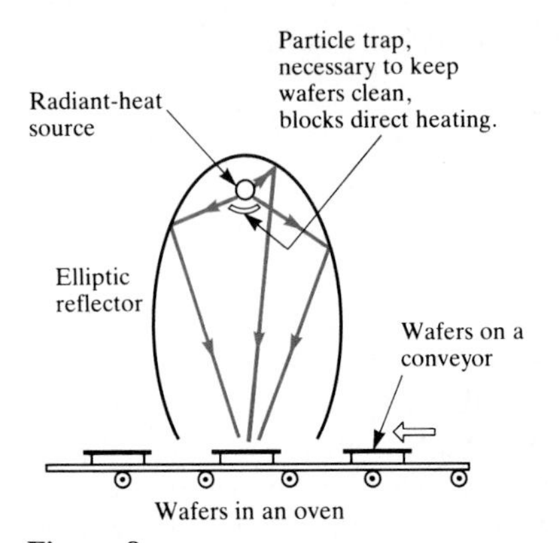

Figure 8

The reflecting property of the ellipse has diverse applications. For instance, in the construction of chips for computers it is necessary to bake a photomask onto the surface of a silicon wafer, a process that requires focusing heat at the mask. This is accomplished by placing a heat source at one focus of an ellipse and the wafer at the other focus, as in Fig. 8.

The reflection property is used in wind tunnel tests of aircraft noise. The test is run in an elliptical chamber, with the aircraft model at one focus and a microphone at the other.

Whispering rooms, such as the rotunda in the Capitol in Washington, D.C., are based on the same principle. A person talking quietly at one focus can be heard easily at the other focus. (Fortunately, since all the paths of the sound from F to F' have the same length, a whisper at F arrives at F' at one time.)

An ellipsoidal reflector cup is used for crushing kidney stones. (An ellipsoid is formed by rotating an ellipse about the line through its foci.) An electrode is placed at one focus and an ellipsoid positioned so that the stone is at the other focus. Shock waves generated at the electrode bounce off the ellipsoid, concentrate at the other focus, and pulverize the stones without damaging other parts of the body. The patient recovers in a few days instead of the few months required after surgery. (See COMAP newsletter #20, November 1986.)

Apollonius, a contemporary of Diocles, proved that the lines from a point P on an ellipse to the foci make equal angles with the tangent at P. However, there is no evidence that this property was put to any use in ancient times.

This is typical of scientific research. The discoveries motivated simply by a desire to know what is not known often become the basis of surprising applications, sooner or later. That's why it is difficult and dangerous to distinguish "applied" from "pure" research. What is "applied" today may be useless tomorrow. What is "pure" today may turn out to be quite useful tomorrow.

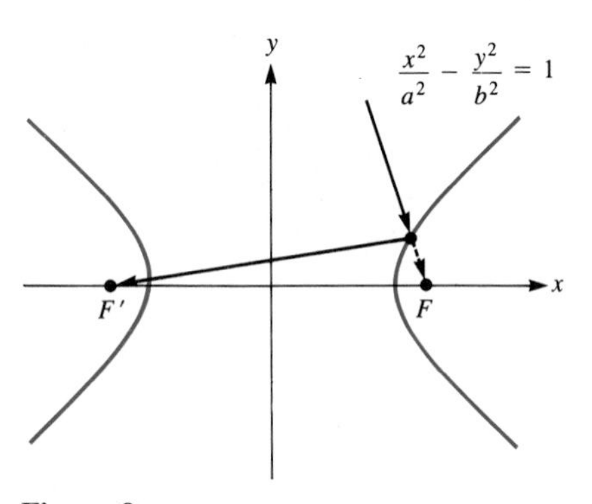

Figure 9

The Reflecting Property of the Hyperbola

The hyperbola with foci at F and F' shown in Fig. 9 also has a reflection property. Light arriving on a line through one focus reflects off the hyperbola and passes through the other focus. (See Exercise 13.)

One design for a reflecting telescope contains parabolic and hyperbolic mirrors as shown in Fig. 10. F is the focus of both mirrors; F' is the second focus of the hyperbola. A ray of light parallel to FF' bounces off the parabola and goes toward F. Before reaching F it bounces off the hyperbolic mirror and goes toward F', where the eyepiece is placed.

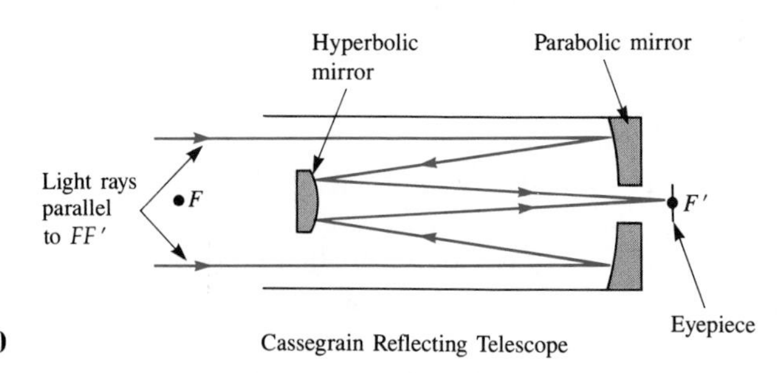

Figure 10 Cassegrain Reflecting Telescope

EXERCISES FOR SEC. 9.7: THE REFLECTION PROPERTIES OF THE CONICS

In Exercises 1 and 2 find the angle between lines with the given inclinations.

1 $\pi/4$ and $3\pi/4$ **2** $5\pi/6$ and $\pi/6$

In Exercises 3 to 6 find the tangent of the angle between lines with the given slopes.

3 2 and -3 **4** 2 and $-1/2$

5 -2 and -3 **6** 1 and $-\sqrt{3}$

In each of Exercises 7 to 10 find the tangent of the angle between the two curves at the indicated point of intersection.

7 $y = \sin x$ and $y = \cos x$ at $(\pi/4, \sqrt{2}/2)$

8 $y = x^2$ and $y = x^3$ at $(1, 1)$

9 $y = e^x$ and $y = e^{-x}$ at $(0, 1)$

10 $y = \sec x$ and $y = \sqrt{2} \tan x$ at $(\pi/4, \sqrt{2})$

---

11 Complete the algebra needed to obtain (3).

12 Establish the reflection property of the ellipse. (See Appendix G for the coordinates of the foci of the ellipse $x^2/a^2 + y^2/b^2 = 1$.)

13 Establish the reflection property of the hyperbola. (See Appendix G for the coordinates of the foci of the hyperbola $x^2/a^2 - y^2/b^2 = 1$.)

The remaining exercises concern the angle γ shown in Fig. 11. The curve is the graph of $r = f(\theta)$ in polar coordinates. P is a point on the curve. The angle between the ray from the pole O through P and the tangent line at P is called γ.

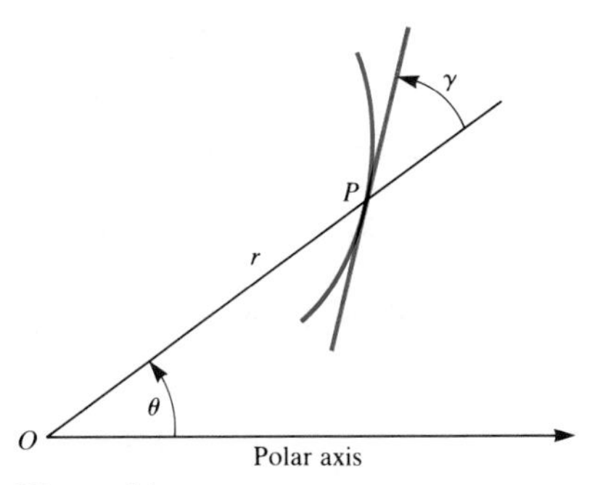

Figure 11

14 Find $\tan \gamma$ as follows:

(*a*) Express the slope of the ray OP in terms of θ.

(*b*) Obtain the slope of the tangent line by expressing the curve parametrically as

$$\begin{cases} x = f(\theta) \cos \theta \\ y = f(\theta) \sin \theta. \end{cases}$$

[This slope will be expressed in terms of $f(\theta)$, $f'(\theta)$, $\cos \theta$, and $\sin \theta$.]

(*c*) Using (*a*), (*b*), and formula (1), show that

$$\tan \gamma = \frac{f(\theta)}{f'(\theta)},$$

if $f'(\theta) \neq 0$.

(*d*) The answer in (*c*) is quite simple. Draw a picture showing why it is to be expected.

15 Find γ for the spiral $r = e^\theta$.

16 Show that for the cardioid $r = 1 - \cos \theta$, $\gamma = \theta/2$.

17 If for the curve $r = f(\theta)$, γ always equals θ, what are all the possibilities for f?

18 (*a*) For the cardioid $r = 1 + \cos \theta$ find $\lim_{\theta \to \pi^-} \gamma$.

(*b*) Sketch $r = 1 + \cos \theta$ using the information obtained in (*a*).

19 If for the curve $r = f(\theta)$, γ is independent of θ, what are the possibilities for f?

20 Four dogs are chasing each other counterclockwise at the same speed. Initially they are at the four vertices of a square of side a. As they chase each other, each running directly toward the dog in front, they approach the center of the square in spiral paths. How far does each dog travel? (See Fig. 12.)

(*a*) Use calculus.

(*b*) Answer the question without calculus.

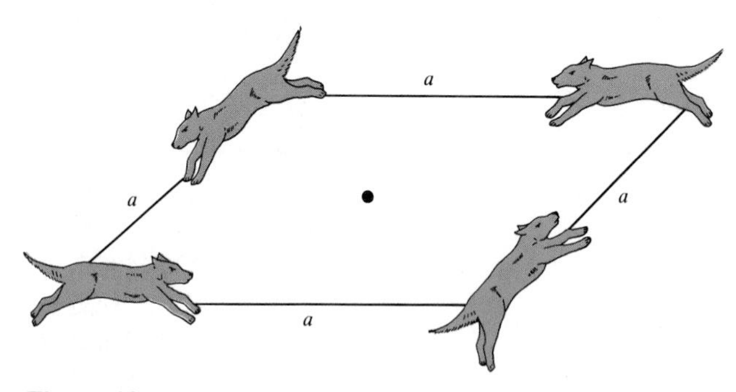

Figure 12

9.S SUMMARY

The following table provides memory aids for most of the formulas in the chapter:

Section	Concept	Key	Memory Aid
9.2	Area $= \int_{\alpha}^{\beta} \frac{r^2}{2}\, d\theta$	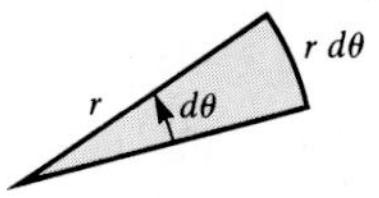 r, $d\theta$, $r\, d\theta$	Area $= \frac{1}{2}r \cdot r\, d\theta$
9.4	Arc length $= \int_a^b \sqrt{\left(\frac{dx}{dt}\right)^2 + \left(\frac{dy}{dt}\right)^2}\, dt$	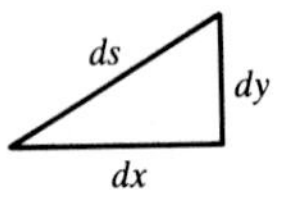 ds, dy, dx	$(ds)^2 = (dx)^2 + (dy)^2$
	Arc length $= \int_a^b \sqrt{1 + \left(\frac{dy}{dx}\right)^2}\, dx$		
	Arc length $= \int_{\alpha}^{\beta} \sqrt{r^2 + (r')^2}\, d\theta$	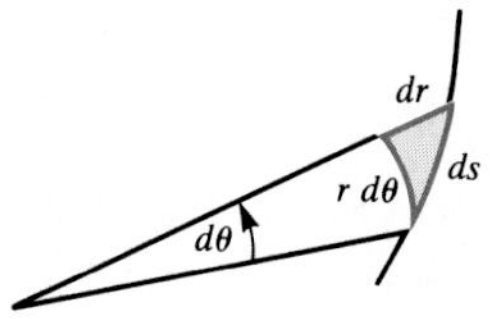 dr, ds, $r\, d\theta$, $d\theta$	
	Speed $= \sqrt{\left(\frac{dx}{dt}\right)^2 + \left(\frac{dy}{dt}\right)^2}$		$(ds)^2 = (r\, d\theta)^2 + (dr)^2$
9.5	Area of surface of revolution $= \int_c^d 2\pi R\, ds$	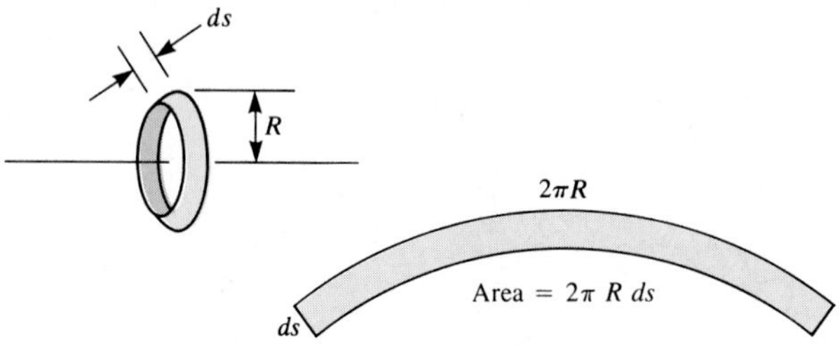 ds, R, $2\pi R$, Area $= 2\pi\, R\, ds$, ds	

Vocabulary and Symbols

polar coordinates (r, θ)
pole, polar axis
parametric representation
arc length s
surface of revolution
curvature κ
radius of curvature
angle between two lines

Key Facts

Polar and rectangular coordinates are related through the equations $x = r \cos \theta$, $y = r \sin \theta$, $r^2 = x^2 + y^2$, $\tan \theta = y/x$.

The slope of a curve given parametrically as $x = g(t)$, $y = h(t)$ is

$$\frac{dy}{dx} = \frac{dy/dt}{dx/dt} = \frac{\dot{y}}{\dot{x}}.$$

One can find d^2y/dx^2 by noting that

$$\frac{d}{dx}\left(\frac{dy}{dx}\right) = \frac{\frac{d}{dt}\left(\frac{dy}{dx}\right)}{\frac{dx}{dt}}.$$

The area $\int_a^b y\, dx$ under a curve given parametrically is obtained by a substitution in which y, dx, and the limits of integration are all expressed in terms of the parameter t. [See Exercise 28(*b*) of Sec. 9.3.] The parameter may be denoted by a different letter, such as θ, for instance.

The angle between two lines in the plane equals the larger inclination minus the smaller inclination. The tangent of this angle is

$$\frac{m - m'}{1 + mm'},$$

where m is the slope of the line of larger inclination and m' is the slope of the other line.

The curvature of a curve is the absolute value of the rate at which the angle of inclination of the tangent line changes with respect to arc length: $|d\phi/ds|$. The curvature of $y = f(x)$ is

$$\frac{|d^2y/dx^2|}{[1 + (dy/dx)^2]^{3/2}}.$$

Curvature can also be computed for curves given parametrically, and thus in polar coordinates, since $r = f(\theta)$ implies that $x = f(\theta) \cos \theta$ and $y = f(\theta) \sin \theta$.

The radius of curvature is the reciprocal of the curvature.

GUIDE QUIZ ON CONCEPTS: CHAPS. 3 TO 9

1 (*a*) What is meant by "the derivative of a function $f(x)$"?
(*b*) Use the definition to find the derivative of x^3.

2 Describe at least three applications or interpretations of the derivative.

3 Explain why $(fg)' = fg' + gf'$, where f and g are differentiable functions.

4 Describe at least two ways of determining whether a critical point is a relative maximum.

5 Of what use is the second derivative in (*a*) graphing, (*b*) the study of motion, (*c*) estimating error in the linear approximation of a function, (*d*) finding curvature.

6 (*a*) What is meant by $\int_a^b f(x)\, dx$, the definite integral of $f(x)$ from a to b?
(*b*) Use the definition in (*a*) to find $\int_0^1 3\, dx$.

7 Describe at least three applications or interpretations of the definite integral.

8 (*a*) What is meant by an antiderivative?
(*b*) How is it used to evaluate a definite integral?

9 (*a*) State the first fundamental theorem of calculus (the one concerning the derivative of a definite integral with respect to an end of the interval).
(*b*) Give a persuasive argument for it.
(*c*) Explain why it implies the second fundamental theorem, the one used to evaluate some definite integrals.

10 (*a*) What is meant by an elementary function?
(*b*) Give an example of a function that is not elementary.
(*c*) Is the derivative of an elementary function necessarily elementary? Explain.
(*d*) Is the antiderivative of an elementary function necessarily elementary? Explain.

11 If using the fundamental theorem of calculus is useless or cumbersome to evaluate $\int_a^b f(x)\, dx$, how might you estimate $\int_a^b f(x)\, dx$? Discuss the accuracy of your method.

12 What is meant by "$\int_a^\infty f(x)\,dx$"?
13 Explain why the convergence of $\int_a^\infty |f(x)|\,dx$ implies the convergence of $\int_a^\infty f(x)\,dx$.
14 Develop informally the formula for
(*a*) area in polar coordinates.
(*b*) arc length of $y = f(x)$.
(*c*) area of a surface of revolution.
(*d*) volume by concentric tubes.
(*e*) the centroid of a plane region.
15 (*a*) State the method of integration by parts.
(*b*) Illustrate it by an example.
(*c*) For what types of integrands is it especially appropriate?
(*d*) Explain why the formula in (*a*) is correct.
16 Describe the washer and shell techniques for finding the volume of a solid of revolution and discuss the advantages of each.
17 Explain why an improper integral cannot meet the definition of a definite integral. How does one get around this problem?
18 What is the distinction between centroid and center of mass?
19 Describe polar coordinates (r, θ) and discuss the complications of converting between polar coordinates and rectangular coordinates.
20 Develop the expressions for ds, the differential of arc length, in terms of rectangular, polar, and parametric equations.

GUIDE QUIZ ON CHAP. 9: PLANE CURVES AND POLAR COORDINATES

In Exercises 1 to 7 set up definite integrals for the given quantities, but do not evaluate them.

1 The area of the region within one loop of the curve $r = \cos 5\theta$.
2 The length of the curve $y = x^4$ from $x = 1$ to $x = 2$.
3 The area of the surface obtained by revolving the curve in Exercise 2 around the line $y = 20$.
4 The area of the surface obtained by revolving the curve in Exercise 2 around the line $x = 1$.
5 The length of the curve $r = 2 + \sin\theta$, $0 \le \theta \le 2\pi$.
6 The area of the surface obtained by revolving about the polar axis the part of the curve in Exercise 5 for $0 \le \theta \le \pi$.
7 The length of the curve $x = 5t^2$, $y = \sqrt{t}$, $0 \le t \le 1$.
8 At time t a moving particle is at the point $x = 12t$, $y = -16t^2 + 5$. What is its speed when $t = 1$?
9 A curve is given parametrically as $x = \tan t$, $y = \sec t$. Eliminate t to find an equation in x and y for the curve.
10 (*a*) Find the inclination of the tangent line to the curve $r = \cos 2\theta$ at the point for which $\theta = \pi/8$.
(*b*) Graph the curve and check that the answer in (*a*) is reasonable.
11 What is the radius of the circle that best approximates the curve $y = 1/x$ at $(1, 1)$ in the sense that it has the same radius of curvature as the curve does at $(1, 1)$?
12 State the reflection properties of the parabola, ellipse, and hyperbola.

REVIEW EXERCISES FOR CHAP. 9: PLANE CURVES AND POLAR COORDINATES

1 (*a*) Develop the formula for area in polar coordinates.
(*b*) What is the device for remembering the formula?
2 (*a*) Develop the formula for arc length with parameter t.
(*b*) What is the device for remembering the formula?
3 (*a*) Develop the formula for the area of a surface of revolution with parameter x.
(*b*) What is the device for remembering the formula?
4 See Exercise 14 in Sec. 9.7.
(*a*) Develop the formula for $\tan\gamma$, where γ is the angle between the ray from the pole to the point P on the curve $r = f(\theta)$ and the tangent line to the curve at P.
(*b*) Draw the memory device for the formula in (*a*).
5 (*a*) Define "curvature."
(*b*) From (*a*), obtain the formula for the curvature of the curve $y = f(x)$.
6 Consider the curve $y = e^x$ for x in $[0, 1]$.
(*a*) Set up integrals for its arc length and for the areas of the surfaces obtained by rotating the curve around the x axis and also about the y axis.
(*b*) Two of the three integrals are elementary. Evaluate them.
7 Consider the curve $y = \sin x$ for x in $[0, \pi]$. Proceed as in Exercise 6. This time, however, only one of the three integrals is elementary. Evaluate it.
8 Graph $r = 3/(\cos\theta + 2\sin\theta)$ after first finding the rectangular form of the equation.
9 Find the maximum y coordinate of the curve $r = 1 + \cos\theta$.
10 Find the minimum x coordinate of the curve $r = 1 + \cos\theta$.
11 Is the total length of the curve $r = 1/(1 + \theta)$, $\theta \ge 0$, finite or infinite?
12 Is the total length of the curve $r = 1/(1 + \theta^2)$, $\theta \ge 0$, finite or infinite?
13 (*a*) Graph $r = \sin 2\theta$ for $0 \le \theta \le \pi/2$.
Find the point on the graph with the largest (*b*) x coordinate, (*c*) y coordinate, (*d*) r coordinate.
14 Assume that x and y are functions of t. Obtain the formula for d^2y/dx^2 in terms of the derivatives $\dot{x}$, $\ddot{x}$, $\dot{y}$, $\ddot{y}$.

15 If

$$\begin{cases} x = \cos 2t \\ y = \sin 3t, \end{cases}$$

express dy/dx and d^2y/dx^2 in terms of t.

16 Find the radius of curvature of the curve $y = \ln x$ at $(e, 1)$.

17 Let $r = e^{\theta}$, $0 \le \theta \le \pi/4$, describe a curve in polar coordinates. In parts (*b*) to (*f*) set up definite integrals for the quantities, and show that they could be evaluated by the fundamental theorem of calculus, but do not evaluate them.

(*a*) Sketch the curve.

(*b*) The area of the region R below the curve and above the interval $[1, e^{\pi/4}\sqrt{2}/2]$ on the x axis.

(*c*) The volume of the solid obtained by revolving R, defined in (*b*), about the x axis.

(*d*) The volume of the solid obtained by revolving R about the y axis.

(*e*) The area of the surface obtained by revolving the curve in (*a*) around the x axis.

(*f*) The area of the surface obtained by revolving the curve in (*a*) around the y axis.

Rare is the curve for which the corresponding five integrals are all elementary.

18 Find the length of the curve $y = \ln x$ from $x = 1$ to $x = \sqrt{3}$. An integral table will save a lot of work.

---

19 Find an equation of the tangent line to the spiral $r = \theta$ at the point for which $\theta = \pi/3$.

20 A set R in the plane bounded by a curve is **convex** if, whenever P and Q are points in R, the line segment PQ also lies in R. A curve is convex if it is the boundary of a convex set. Consider a convex curve C with no line segments on it.

(*a*) Show why the average radius of curvature with respect to angle ϕ as you traverse C equals (Length of C)$/2\pi$. (See p. 551.)

(*b*) Deduce from (*a*) that a convex curve of length L has a radius of curvature equal to $L/2\pi$ somewhere on the curve.

21 Prove that the average value of the curvature as a function of arc length s as you sweep out a convex curve C in the counterclockwise direction is 2π/Length of C. (See Exercise 20 for the definition of convex curve.)

22 A girl is standing at the base of a hill that makes the angle β with the horizontal. At what angle θ should she throw a baseball to have it land as high up the hill as possible? (See Fig. 1.) *Suggestion:* Rather than maximizing GB, maximize GA. Also, the parametric equations of a ball thrown at an angle θ and velocity v are

$$\begin{cases} x = v(\cos \theta)t \\ y = -16t^2 + v(\sin \theta)t, \end{cases}$$

where distance is measured in feet and time in seconds. Simplify your answer by trigonometry. (It will simplify to $\pi/4 + \beta/2$.)

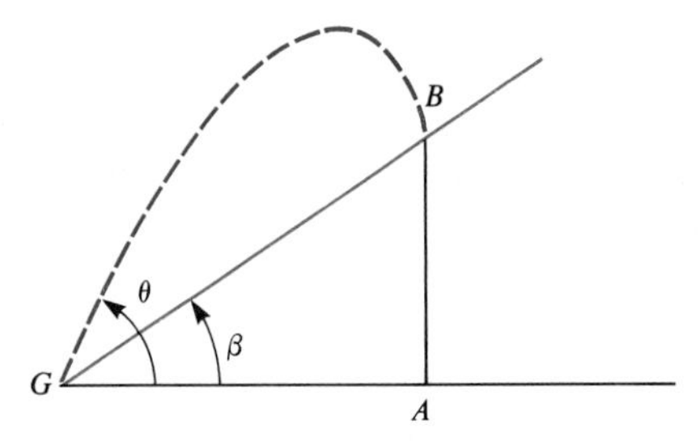

Figure 1

23 Find the centroid of the circular arc AB in Fig. 2. (*Hint:* How should the centroid of a curve be defined?)

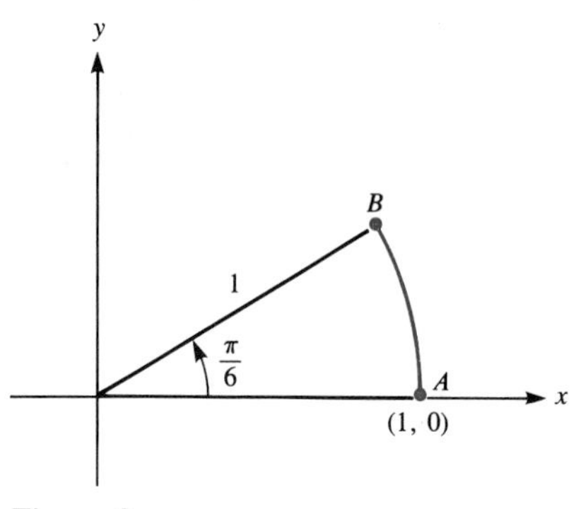

Figure 2

24 Let a be a number, and let $x = g(t)$ and $y = h(t)$, $t \ge a$, describe a curve. Let $P = (g(a), h(a))$, and let $Q(t) = (g(t), h(t))$, $t > a$.

(*a*) Sketch the chord $PQ(t)$ and the tangent line $T(t)$ at $Q(t)$.

(*b*) What is the slope of $PQ(t)$? What is the slope of $T(t)$?

(*c*) Under what assumptions will the two slopes in (*b*) have the same limit as $t \to a^+$?

25 Read the generalized mean-value theorem stated in Exercise 158 of Sec. 6.S.

(*a*) What does it say about the curve given parametrically by the equations $x = g(t)$, $y = f(t)$? (Here t plays the role of the x in Exercise 158.)

(*b*) Let $h(t)$, for t in $[a, b]$, be defined as the vertical distance from $(g(t), f(t))$ to the line that passes through $(g(a), f(a))$ and $(g(b), h(b))$, as in the proof of the mean-value theorem in Sec. 4.1. Use the function h to prove the generalized mean-value theorem.

26 (*a*) How far is the point (a, b) from the line $x + y = 1$?

(*b*) The portion of the curve $y = 5 - (x - 2)^2$ that lies above the line $x + y = 1$ is revolved around that line to produce a surface of revolution. Set up a definite integral for the area of this surface.

27 Let a be a rational number. Consider the curve $y = x^a$ for x in the interval $[1, 2]$. Show that the area of the surface obtained by rotating this curve around the x axis can be evaluated by the fundamental theorem of calculus in the cases $a = 1$ and $1 + 2/n$, where n is any nonzero integer. (These are the only rational a for which the pertinent integral is elementary.) Assume Chebyshev's theorem, which asserts that if p and q are rational numbers then $\int x^p(1 + x)^q\, dx$ is elementary only when p is an integer, q is an integer, or $p + q$ is an integer.

28 A particle is moving along the cardioid $r = 1 + \cos\theta$. At the point for which $\theta = \pi/2$ its "horizontal velocity" dx/dt is 5.
(*a*) Find its "vertical velocity" dy/dt at that point.
(*b*) Find the rate at which its distance from the origin is changing.

29 A circle of radius b rolls without slipping on the inside of a circle of radius a, $a > b$. A point P on the circumference of the moving circle traces out a curve called a **hypocycloid**. Assume that initially the point P is at the point of contact of the two circles and that the coordinates of the center of the moving circle are initially $(a - b, 0)$. Using Fig. 3, which shows the angles θ and ϕ, obtain parametric equations for the hypocycloid, as follows:
(*a*) Show that $b\phi = a\theta$.
(*b*) Show that $x = (a - b)\cos\theta + b\cos(\phi - \theta)$ and $y = (a - b)\sin\theta - b\sin(\phi - \theta)$.

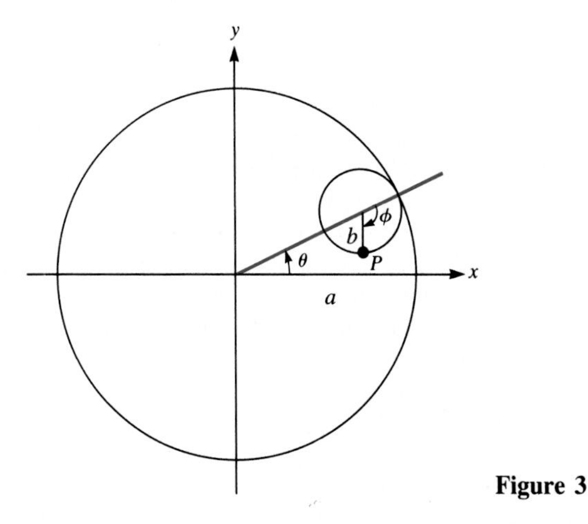

Figure 3

30 (See Exercise 29.) When $b = a/4$, the equations for the hypocycloid take a very simple form, obtained as follows:
(*a*) Show that $\sin 3\theta = 3\sin\theta - 4\sin^3\theta$ and $\cos 3\theta = 4\cos^3\theta - 3\cos\theta$.
(*b*) Using part (*a*) and also Exercise 29, show that if $f(x,y)$ is on the hypocycloid then $x = a\cos^3\theta$ and $y = a\sin^3\theta$.
(*c*) From (*b*) deduce that the hypocycloid in the special case $b = a/4$ has the equation $x^{2/3} + y^{2/3} = a^{2/3}$.
(*d*) Graph the equation $x^{2/3} + y^{2/3} = 1$.

10

SERIES

Trying to find $\sin\theta = b/c$

How is $\sin\theta$ computed? One approach might be to draw a right triangle with one angle θ, as in Fig. 1. Then measure the length of the opposite side b, the length of the hypotenuse c, and calculate b/c (''opposite'' over ''hypotenuse''). (Try it.) You would be lucky to get even two decimal places correct. Clearly this method cannot give the many decimal places a calculator displays for $\sin\theta$.

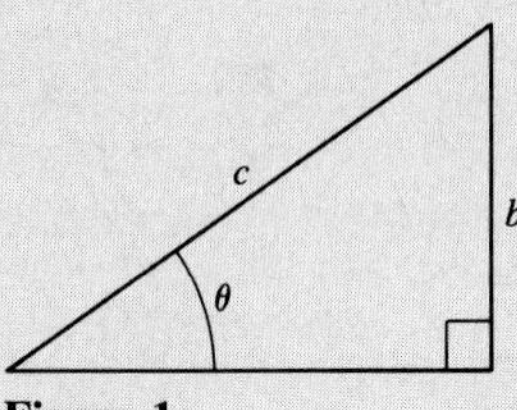

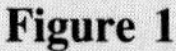
Figure 1

Chapter 11 describes one way of obtaining such accuracy. More generally, it will show how to use polynomials to evaluate such important functions as $\sin x$, $\cos x$, $\tan^{-1} x$, e^x, and $\ln x$ to as many decimal places as we please. For instance, when $|x| \leq 1$, the polynomial

$$x - \frac{x^3}{6} + \frac{x^5}{120}$$

approximates $\sin x$ with an error less than 0.0002 (provided angle x is given in radians).

Such an estimate has other uses than simply evaluating the function. For instance, it also provides a way of estimating integrals, such as

$$\int_0^1 \frac{\sin x}{x}\,dx.$$

[The fundamental theorem of calculus is useless here since $(\sin x)/x$ does not have an elementary antiderivative.] Just evaluate

$$\int_0^1 \frac{x - \frac{x^3}{6} + \frac{x^5}{120}}{x}\,dx = \int_0^1 \left(1 - \frac{x^2}{6} + \frac{x^4}{120}\right) dx$$

instead. It will provide a good estimate of $\int_0^1 (\sin x)/x\,dx$. Since the integrand is now a polynomial, the fundamental theorem of calculus can be used to obtain the estimate

$$x - \frac{x^3}{18} + \frac{x^5}{600}\Bigg|_0^1 = 1 - \frac{1}{18} + \frac{1}{600} \approx 0.946,$$

which gives $\int_0^1 (\sin x)/x\,dx$ to three decimal places.

10.1 AN INFORMAL INTRODUCTION TO SERIES

The main goal of this chapter and the next is to show how polynomials can be used to approximate functions that are not polynomials. The accompanying table anticipates some of the formulas we will obtain. In the table $(-1)^{n-1}$ appears. It is a bookkeeping device to give $+1$ when n is odd and -1 when n is even. Sometimes we use $(-1)^n$ instead, whenever we need $+1$ when n is even and -1 when n is odd. (In this table n is a positive integer and $n!$ denotes the product of all the integers from n down to 1. For instance, $4! = 4 \cdot 3 \cdot 2 \cdot 1 = 24$ and $5! = 5 \cdot 4 \cdot 3 \cdot 2 \cdot 1 = 120$.)

Approximating Some Functions with Polynomials

Function	*Approximating Polynomial*
e^x	$1 + x + \frac{x^2}{2!} + \frac{x^3}{3!} + \cdots + \frac{x^n}{n!}$
$\ln(1 + x)$	$x - \frac{x^2}{2} + \frac{x^3}{3} - \cdots + (-1)^{n-1}\frac{x^n}{n}, \quad \lvert x\rvert \le 1$
$\sin x$	$x - \frac{x^3}{3!} + \frac{x^5}{5!} - \cdots + (-1)^n \frac{x^{2n+1}}{(2n+1)!}$

The larger n is, the better the approximation, as long as we keep x in some restricted interval.

Example 1 illustrates the use of such polynomials.

EXAMPLE 1 Use the approximations in the table to estimate $\sqrt{e} = e^{1/2}$.

SOLUTION By the first row of the table, for each positive integer n,

$$1 + \frac{1}{2} + \frac{(\frac{1}{2})^2}{2!} + \frac{(\frac{1}{2})^3}{3!} + \cdots + \frac{(\frac{1}{2})^n}{n!}$$

is an estimate of $e^{1/2}$. Let us compute some of these estimates, keeping in mind that as n increases the estimates improve.

n	$1 + \frac{1}{2} + \frac{(\frac{1}{2})^2}{2!} + \frac{(\frac{1}{2})^3}{3!} + \cdots + \frac{(\frac{1}{2})^n}{n!}$	*Decimal Form*	*Sum*
1	$1 + \frac{1}{2}$	$1 + 0.5$	1.5
2	$1 + \frac{1}{2} + \frac{(\frac{1}{2})^2}{2!}$	$1 + 0.5 + 0.125$	1.625
3	$1 + \frac{1}{2} + \frac{(\frac{1}{2})^2}{2!} + \frac{(\frac{1}{2})^3}{3!}$	$1 + 0.5 + 0.125 + 0.02083\ldots$	1.64583 . . .
4	$1 + \frac{1}{2} + \frac{(\frac{1}{2})^2}{2!} + \frac{(\frac{1}{2})^3}{3!} + \frac{(\frac{1}{2})^4}{4!}$	$1 + 0.5 + 0.125 + 0.02083 + 0.00260\ldots$	1.6484375

As a calculator shows, $e^{1/2} \approx 1.64872$ to five decimal places. So the estimate with $n = 4$ is quite close. ■

Example 1 raises two closely related questions:

1 How do we choose n to achieve a prescribed accuracy, say, to 10 decimal places?

2 How can we estimate the "error"—the difference between an estimate and the number we are estimating? After all, there's little point in making an estimate if we have no idea about the size of its error.

Example 1 also depicts a battle between two forces. On the one hand, the individual summands are getting very small—shrinking toward 0; so their sums may not get very large. On the other hand, there are more and more of them; so their sums might become arbitrarily large.

In Example 1, the first force was stronger, and the sums—no matter how many summands we take—stay less than $\sqrt{e} \approx 1.64872$. But in Example 2 the sums behave quite differently.

EXAMPLE 2 What happens to sums of the form

$$\frac{1}{\sqrt{1}} + \frac{1}{\sqrt{2}} + \cdots + \frac{1}{\sqrt{n}} \tag{1}$$

as n gets larger and larger? Will they stay less than some fixed number or will they get arbitrarily large, eventually passing 100, then 1,000, and so on?

SOLUTION The table lists values of (1) for n up through 5.

n	$\frac{1}{\sqrt{1}} + \frac{1}{\sqrt{2}} + \cdots + \frac{1}{\sqrt{n}}$	*Decimal Form (7 places)*
1	$\frac{1}{\sqrt{1}}$	1.0000000
2	$\frac{1}{\sqrt{1}} + \frac{1}{\sqrt{2}}$	1.7071068
3	$\frac{1}{\sqrt{1}} + \frac{1}{\sqrt{2}} + \frac{1}{\sqrt{3}}$	2.2844571
4	$\frac{1}{\sqrt{1}} + \frac{1}{\sqrt{2}} + \frac{1}{\sqrt{3}} + \frac{1}{\sqrt{4}}$	2.7844571
5	$\frac{1}{\sqrt{1}} + \frac{1}{\sqrt{2}} + \frac{1}{\sqrt{3}} + \frac{1}{\sqrt{4}} + \frac{1}{\sqrt{5}}$	3.2316706

These computations don't answer the question: What will happen to the sums when n becomes arbitrarily large? In fact, even if we calculated the values of $1/\sqrt{1} + 1/\sqrt{2} + \cdots + 1/\sqrt{n}$ all the way up to $n = 1{,}000{,}000$, we still wouldn't know the answer. Why? Because we can't be sure what happens to the sums when n is a billion or a trillion or even larger. Do the sums get arbitrarily large or do they stay below some fixed number? No computer, even a supercomputer operating at billions of computations per second, can answer that question.

However, an algebraic insight can help us answer the question. Observe that

$$\frac{1}{\sqrt{1}} + \frac{1}{\sqrt{2}} + \cdots + \frac{1}{\sqrt{n}}$$

has n summands and that the smallest of them is $1/\sqrt{n}$. Therefore (1) is at least as large as

$$\underbrace{\frac{1}{\sqrt{1}} + \frac{1}{\sqrt{2}} + \cdots + \frac{1}{\sqrt{n}}}_{n \text{ summands}} \geq \underbrace{\frac{1}{\sqrt{n}} + \frac{1}{\sqrt{n}} + \cdots + \frac{1}{\sqrt{n}}}_{n \text{ summands}} = n\left(\frac{1}{\sqrt{n}}\right) = \sqrt{n}.$$

Thus $1/\sqrt{1} + 1/\sqrt{2} + \cdots + 1/\sqrt{n}$ is at least as large as $\sqrt{n}$. (In fact, when $n \geq 2$, the sum is larger than $\sqrt{n}$.)

Now, as n gets larger and larger, $\sqrt{n}$ grows arbitrarily large. For $n =$ 1,000,000, for instance, we have

$$\frac{1}{\sqrt{1}} + \frac{1}{\sqrt{2}} + \cdots + \frac{1}{\sqrt{1{,}000{,}000}} \geq \sqrt{1{,}000{,}000} = 1{,}000.$$

So the sums of form $1/\sqrt{1} + 1/\sqrt{2} + \cdots + 1/\sqrt{n}$ also become arbitrarily large. They do *not* stay less than some fixed number. ■

Traveler's Advisory In both Examples 1 and 2, the individual summands approach 0 as n becomes large. Yet in the first case, the sums stay less than $\sqrt{e}$, while in the second they grow arbitrarily large. This contrast shows that we must be careful when dealing with such sums, especially since they may play a role in approximating important functions.

THINGS TO COME

In most of this chapter the summands are constant; in Chap. 11 they involve a variable.

Sec. 10.2	introduces the notion of a "sequence" of numbers (such as $1/\sqrt{1}, 1/\sqrt{2}, 1/\sqrt{3}, \ldots$).
Sec. 10.3	introduces the notion of a "series," which involves adding up more and more terms from a sequence of numbers. (Example: $1/\sqrt{1}, 1/\sqrt{1} + 1/\sqrt{2}, 1/\sqrt{1} + 1/\sqrt{2} + 1/\sqrt{3}, \ldots$.)
Secs. 10.4–7	develop methods for determining when these sums approach a number and, if they do, how big the error is when you use a particular sum to estimate that number.
Sec. 11.1	answers the question: Given a function, how do we find polynomials whose graphs resemble the graph of the function, at least over some interval? These polynomials give us a way to approximate e^x and $\sin x$ to as many decimal places as we please. We will see the importance of series involving powers of x or $x - a$.
Sec. 11.2	gives a formula for estimating the error in these approximations.
Sec. 11.3	explains where the formula came from.
Secs. 11.4–5	explore series further, showing how they can be used to evaluate the indeterminate limit "zero-over-zero."
Sec. 11.6	is not about calculus at all. It develops complex numbers, which are needed in Sec. 11.7 (and in many sophomore science courses).
Sec. 11.7	uses series and complex numbers to show that the functions $\sin x$ and $\cos x$ are intimately related to the exponential function e^x. This relation is used in many undergraduate classes in physics, engineering, and mathematics.

As you go through Chaps. 10 and 11, check back to this outline from time to time. It will help you keep track of what you're doing and why.

EXERCISES FOR SEC. 10.1: AN INFORMAL INTRODUCTION TO SERIES

1 Estimate $\sqrt[3]{e} = e^{1/3}$ by using the following approximations with $x = \frac{1}{3}$.

(*a*) $1 + x + \dfrac{x^2}{2!} + \dfrac{x^3}{3!}$

(*b*) $1 + x + \dfrac{x^2}{2!} + \dfrac{x^3}{3!} + \dfrac{x^4}{4!}$

2 Estimate $1/e = e^{-1}$ using the following approximations with $x = -1$.

(*a*) $1 + x + \dfrac{x^2}{2!} + \dfrac{x^3}{3!} + \dfrac{x^4}{4!}$

(*b*) $1 + x + \dfrac{x^2}{2!} + \dfrac{x^3}{3!} + \dfrac{x^4}{4!} + \dfrac{x^5}{5!}$

3 It will be shown in Sec. 11.2 that the polynomial $x - x^3/6$ is an excellent approximation to $\sin x$ (angle in radians) for $|x| \le \frac{1}{2}$. Using a calculator, fill in this table to seven decimal places.

x	$x - \dfrac{x^3}{6}$	$\sin x$	$\sin x - \left(x - \dfrac{x^3}{6}\right)$
0.1			
0.2			
0.3			
0.4			
0.5			

This illustrates that the estimate is accurate to at least three decimal places.

4 The polynomial $x - x^3/3! + x^5/5!$ is an excellent approximation of $\sin x$ (angle in radians) for $|x| \le 1$. Using a calculator, compute to at least seven decimal places:

(*a*) $\sin 1$,

(*b*) $x - \dfrac{x^3}{3!} + \dfrac{x^5}{5!}$ for $x = 1$.

5 Estimate $\int_{1/2}^{1} (e^x - 1)/x \, dx$ by approximating e^x by the polynomial

(*a*) $1 + x + \dfrac{x^2}{2!}$,

(*b*) $1 + x + \dfrac{x^2}{2!} + \dfrac{x^3}{3!}$.

6 Estimate $\int_{1/4}^{1/2} (\sin x)/x \, dx$ by approximating $\sin x$ by the polynomial

(*a*) x,

(*b*) $x - \dfrac{x^3}{3!}$,

(*c*) $x - \dfrac{x^3}{3!} + \dfrac{x^5}{5!}$.

7 (*a*) The polynomial $x - x^2/2 + x^3/3 - \cdots \pm x^n/n$, $|x| \le 1$, is a good estimate of $\ln(1 + x)$ when n is large. So, to estimate $\ln(1.5)$, which is $\ln(1 + 0.5)$, we use the polynomial with x replaced by $\frac{1}{2}$. Fill in the table with the use of a calculator:

n	$\dfrac{1}{2} - \dfrac{(\frac{1}{2})^2}{2} + \dfrac{(\frac{1}{2})^3}{3} - \cdots + (-1)^{n-1}\dfrac{(\frac{1}{2})^n}{n}$	*Decimal Form*
1		
2		
3		
4		
5		

(*b*) Look up $\ln(1.5)$ on your calculator. What is the error when the result in the fifth row of the table is used as an approximation?

8 (*a*) (See Exercise 7.) To estimate $\ln(0.5)$ write it as $\ln(1 + (-0.5))$. Fill in the table.

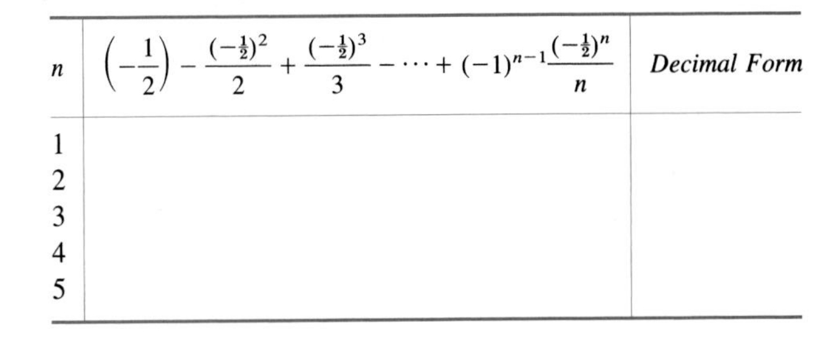

n	$\left(-\dfrac{1}{2}\right) - \dfrac{(-\frac{1}{2})^2}{2} + \dfrac{(-\frac{1}{2})^3}{3} - \cdots + (-1)^{n-1}\dfrac{(-\frac{1}{2})^n}{n}$	*Decimal Form*
1		
2		
3		
4		
5		

(*b*) By how much does the estimate in the fifth row differ from $\ln(0.5)$?

9 One way to estimate ln 2 is to write it as ln (1 + 1) and use a polynomial in Exercise 7 that approximates ln (1 + x) with $x = 1$. Another way is to note that $\ln 2 = -\ln (0.5)$ and use the approach in Exercise 8.

(*a*) Using the polynomial approximation of degree 5 ($n = 5$) in both cases, which gives the better estimate?

(*b*) Write a paragraph that gives your opinion on which approach is better and why you think so.

∗10 What happens to sums of the form

$$\frac{1}{\sqrt[3]{1}} + \frac{1}{\sqrt[3]{2}} + \frac{1}{\sqrt[3]{3}} + \cdots + \frac{1}{\sqrt[3]{n}}$$

as n gets larger? Explore and explain.

11 In the text it is asserted that $1/\sqrt{1} + 1/\sqrt{2} + \cdots + 1/\sqrt{n}$ is larger than $\sqrt{n}$ for $n \geq 2$. Check that this is true for (*a*) $n = 2$, (*b*) $n = 4$.

12 (*a*) Check these factorings by multiplying out the right-hand side.

$$1 - x^2 = (1 - x)(1 + x)$$

$$1 - x^3 = (1 - x)(1 + x + x^2)$$

$$1 - x^4 = (1 - x)(1 + x + x^2 + x^3)$$

(*b*) For every positive integer n

$$1 - x^n = (1 - x)(1 + x + \cdots + x^{n-1}).$$

Use this to show that

$$1 + x + \cdots + x^{n-1} = \frac{1}{1 - x} - \frac{x^n}{1 - x} \qquad (x \neq 1).$$

Now assume that $|x| < 1$. Then x^n approaches 0 as n increases (as will be shown in the next section). Thus, for $|x| < 1$ and large n, $1 + x + x^2 + \cdots + x^{n-1}$ is a polynomial approximation of the function $1/(1 - x)$.

(*c*) Compute $1 + x + x^2 + \cdots + x^{n-1}$ for $n = 6$ and $x = 0.3$. How much does it differ from $1/(1 - x)$ for $x = 0.3$?

(*d*) The same as (*c*), with $x = -0.9$.

13 This exercise obtains the polynomial approximation of ln (1 + x) given in the table on p. 570. It begins with the algebraic identity

$$\frac{1}{1 - t} = 1 + t + t^2 + t^3 + \cdots + t^{n-1} + \frac{t^n}{1 - t} \qquad t \neq 1.$$

Replace t by $-t$, getting

$$\frac{1}{1 + t} = 1 - t + t^2 - t^3 + \cdots + (-1)^{n-1}t^{n-1} + \frac{(-1)^n t^n}{1 + t} \qquad t \neq -1. \qquad (2)$$

(*a*) Integrate both sides of (2) over the interval from 0 to x, $x > 0$, to show that

$$\ln (1 + x) = x - \frac{x^2}{2} + \frac{x^3}{3} - \frac{x^4}{4} + \cdots + \frac{(-1)^{n-1}x^n}{n} + (-1)^n \int_0^x \frac{t^n}{1 + t}\, dt.$$

(*b*) Show that for $0 \leq x \leq 1$, $\int_0^x [t^n/(1 + t)]\, dt$ approaches 0 as n increases. Here x is fixed. [*Hint:* $1/(1 + t) \leq 1$ for $t \geq 0$.]

14 This exercise obtains polynomials that approximate $\tan^{-1} x$ for $|x| \leq 1$ and shows one way of computing π. In (2) in the preceding exercise, replace t by t^2 to obtain

$$\frac{1}{1 + t^2} = 1 - t^2 + t^4 - t^6 + \cdots + (-1)^{n-1}t^{2n-2} + (-1)^n \frac{t^{2n}}{1 + t^2}. \qquad (3)$$

(*a*) Consider only $0 \leq x \leq 1$. Integrate both sides of (3) over $[0, x]$ to show that

$$\tan^{-1} x = x - \frac{x^3}{3} + \frac{x^5}{5} - \frac{x^7}{7} + \cdots + (-1)^{n-1}\frac{x^{2n-1}}{2n - 1} + (-1)^n \int_0^x \frac{t^{2n}}{1 + t^2}\, dt. \qquad (4)$$

(*b*) Show that for fixed x, $0 < x \leq 1$, the integral in (4) approaches 0 as n increases.

(*c*) Use the polynomial in (*a*), with $n = 5$ (so its degree is 9) to estimate $\tan^{-1} 1$.

(*d*) Use the result in (*c*) to estimate π. [Recall that $\tan^{-1} 1 = \pi/4$.]

15 ∗ What happens to sums of the form

$$\frac{1}{1 \cdot 2} + \frac{1}{2 \cdot 3} + \frac{1}{3 \cdot 4} + \cdots + \frac{1}{n(n + 1)}$$

as n increases? Do they get arbitrarily large or do they approach some number? The table will help you begin your exploration.

n	$\frac{1}{1 \cdot 2} + \frac{1}{2 \cdot 3} + \frac{1}{3 \cdot 4} + \cdots + \frac{1}{n(n+1)}$	*Sum as Decimal*	*Sum as Fraction*
1			
2			
3			
4			
5			

(*a*) Fill in at least the five rows of the table, and more if you wish.

(*b*) On the basis of your computations, what do you think

happens to the sums as n increases? (If you don't see a pattern, go up to $n = 10$.)
(*c*) Justify your opinion in (*b*).

16 (*a*) Use the polynomial in (4), with $n = 5$, to estimate $\tan^{-1}(\frac{1}{2})$.
(*b*) Use the result in (*a*) to estimate $\tan^{-1} 2$. [*Hint:* For positive x, what is the relation between $\tan^{-1}(1/x)$ and $\tan^{-1} x$?]
(*c*) Draw a right triangle with one leg 20 cm long and the other 10 cm; use a protractor to estimate $\tan^{-1} 2$.
(*d*) What does your calculator give as an estimate of $\tan^{-1} 2$?

10.2 SEQUENCES

This section introduces the notion of a sequence of numbers and the limit of such a sequence. The rest of Chaps. 10 and 11 deals with a special type of sequence, formed by adding terms from a given sequence.

Sequences

A **sequence** of real numbers,

$$a_1, a_2, a_3, \ldots, a_n, \ldots,$$

is a function that assigns to each positive integer n a number a_n. The number a_n is called the ***n*th term** of the sequence. For example, the sequence

$$\left(1+\frac{1}{1}\right)^1, \left(1+\frac{1}{2}\right)^2, \left(1+\frac{1}{3}\right)^3, \ldots, \left(1+\frac{1}{n}\right)^n, \ldots$$

was considered in Sec. 6.2 in the study of the number e. In this case,

$$a_n = \left(1+\frac{1}{n}\right)^n.$$

As another example, Newton's method of estimating a root of $f(x) = 0$ involves the construction of a sequence $x_1, x_2, \ldots,$ where

$$x_{n+1} = x_n - \frac{f(x_n)}{f'(x_n)}.$$

Sometimes the notation $\{a_n\}$ is used as an abbreviation of the sequence $a_1, a_2, \ldots, a_n, \ldots$. For instance, e involves the sequence $\{(1 + 1/n)^n\}$.

The limit of a sequence

If, as n gets larger, a_n approaches a number L, then L is called the **limit** of the sequence. If the sequence $a_1, a_2, \ldots$ has a limit L, we write

$$\lim_{n\to\infty} a_n = L.$$

For instance, we write

$$\lim_{n\to\infty}\left(1+\frac{1}{n}\right)^n = e.$$

If a_n becomes and remains arbitrarily large and positive as n gets larger, we write $\lim_{n\to\infty} a_n = \infty$. (The limit does not exist in this case.) For instance, $\lim_{n\to\infty} 2^n = \infty$. Similarly, we write $\lim_{n\to\infty}(-2^n) = -\infty$.

To show that $\lim_{n\to\infty} a_n = 0$, show that $\lim_{n\to\infty} |a_n| = 0$.

The assertion that "a_n approaches the number L" is equivalent to the assertion that $|a_n - L| \to 0$ as $n \to \infty$. In particular, if $L = 0$, the assertion that $a_n \to 0$ as

tan(1) = 6 √2

$n \to \infty$ is equivalent to the assertion that $|a_n| \to 0$ as $n \to \infty$. This means that if the absolute value of a_n approaches 0 as $n \to \infty$, then a_n approaches 0 as $n \to \infty$.

A sequence need not begin with the term a_1. Later, sequences of the form a_0, a_1, a_2, . . . will be considered. In such a case, a_0 is called the **zeroth term**. Or we may consider a ''tail end'' of a sequence, a sequence that begins with the term a_k: a_k, a_{k+1}, a_{k+2}, ...

The Sequence $\{r^n\}$

The next example introduces a simple but important sequence.

EXAMPLE 1 A certain device depreciates in value over the years. In fact, at the end of any year it has only 80 percent of the value it had at the beginning of the year. What happens to its value in the long run if its value when new is \$1?

SOLUTION Let a_n be the value of the device at the end of the nth year. Thus $a_1 = 0.8$ and $a_2 = (0.8)(0.8) = 0.8^2 = 0.64$. Similarly, $a_3 = 0.8^3$. The question concerns the sequence $\{0.8^n\}$.

This table lists a few values of 0.8^n, rounded off to four decimal places:

n	1	2	3	4	5	10	20
0.8^n	0.8	0.64	0.512	0.4096	0.3277	0.1074	0.0115

The entries in the table suggest that

$$\lim_{n \to \infty} 0.8^n = 0.$$

In the long run the device will be worth less than a nickel, then less than a penny, etc. ■

Even if the device in Example 1 lost only 1 percent of this value each year, in the long run it would still be worth less than a nickel, then less than a penny, etc. The accompanying table lists some values of 0.99^n to illustrate this fact.

n	1	2	3	4	5	10	20	40	80	100	200	400
0.99^n	0.99	0.9801	0.9703	0.9606	0.9510	0.9044	0.8179	0.6690	0.4475	0.3660	0.1340	0.01795

As the table indicates, 0.99^n approaches 0 as $n \to \infty$, but much more slowly than 0.8^n does. Our experience with 0.8^n and 0.99^n suggests the following theorem.

Theorem 1 If r is a number in the open interval $(-1, 1)$, then

$$\lim_{n\to\infty} r^n = 0.$$

(See Example 4 and Exercises 26 and 27.)

An important fact about sequences to keep in mind is that the terms of the sequence $\{a_n\}$ usually never equal their limit L but merely approach it arbitrarily closely.

Furthermore, not every sequence has a limit, as we saw in the case $\{2^n\}$. However, a sequence may fail to have a limit without approaching infinity, as the next example illustrates.

EXAMPLE 2 Let $a_n = (-1)^n$ for $n = 1, 2, 3, \ldots$. What happens to a_n when n is large? Does the sequence have a limit?

SOLUTION The first four terms of the sequence are

$$a_1 = (-1)^1 = -1,$$

$$a_2 = (-1)^2 = 1,$$

$$a_3 = (-1)^3 = -1,$$

$$a_4 = (-1)^4 = 1.$$

The numbers of this sequence continue to alternate $-1, 1, -1, 1, \ldots$. This sequence does not approach a single number. Therefore, it does not have a limit. ■

Definition *Convergent and divergent sequences.* A sequence that has a limit is said to **converge** or to be **convergent**. A sequence that does not have a limit is said to **diverge** or to be **divergent**.

The sequences $\{0.8^n\}$ and $\{0.99^n\}$ converge to 0. The sequence $\{(-1)^n\}$ of Example 2 is divergent.

Calculate 1.0006^n for some large n.

For $r > 1$ and $r \le -1$, the sequence $\{r^n\}$ is divergent. For instance,

$$\lim_{n\to\infty} 1.0006^n = \infty.$$

The Sequence $\{k^n/n!\}$

Example 3 introduces a type of sequence that occurs in the study of $\sin x$, $\cos x$, and e^x.

$n!$ is defined in Appendix C.

EXAMPLE 3 Does the sequence defined by $a_n = 3^n/n!$ converge or diverge?

SOLUTION The first terms of this sequence are computed (to two decimal places) with the aid of the table below.

a_n eventually decreases.

n	1	2	3	4	5	6	7	8
3^n	3	9	27	81	243	729	2,187	6,561
$n!$	1	2	6	24	120	720	5,040	40,320
$a_n = \frac{3^n}{n!}$	3.00	4.50	4.50	3.38	2.03	1.01	0.43	0.16

Although a_2 is larger than a_1 and a_3 is equal to a_2, from a_4 through a_8, as the table shows, the terms decrease.

The numerator 3^n becomes large as $n \to \infty$, influencing a_n to grow large. But the denominator $n!$ also becomes large as $n \to \infty$, influencing the quotient a_n to shrink toward 0. For $n = 1$ and 2 the first influence dominates, but then, as the table shows, the denominator $n!$ seems to grow faster than the numerator 3^n, forcing a_n toward 0.

To see why
$$\frac{3^n}{n!} \to 0$$
as $n \to \infty$, consider, for instance, a_{10}. Express a_{10} as the product of 10 fractions:
$$a_{10} = \tfrac{3}{1}\tfrac{3}{2}\tfrac{3}{3}\tfrac{3}{4}\tfrac{3}{5}\tfrac{3}{6}\tfrac{3}{7}\tfrac{3}{8}\tfrac{3}{9}\tfrac{3}{10}.$$
The first three fractions are greater than or equal to 1, but all the seven remaining fractions are less than or equal to $\frac{3}{4}$. Thus
$$a_{10} < \tfrac{3}{1}\tfrac{3}{2}\tfrac{3}{3}(\tfrac{3}{4})^7.$$
Similarly,
$$a_{100} < \tfrac{3}{1}\tfrac{3}{2}\tfrac{3}{3}(\tfrac{3}{4})^{97}.$$
More generally, for $n > 4$,
$$a_n < \tfrac{3}{1}\tfrac{3}{2}\tfrac{3}{3}(\tfrac{3}{4})^{n-3}.$$
By Theorem 1,
$$\lim_{n\to\infty} (\tfrac{3}{4})^n = 0.$$
Thus
$$\lim_{n\to\infty} a_n = 0. \blacksquare$$

Reasoning like that in Example 3 shows that for any fixed number k,

This limit will be used often.

$$\boxed{\lim_{n\to\infty} \frac{k^n}{n!} = 0.}$$

This means that the factorial function grows faster than any exponential k^n.

The following theorem will be used several times to show that a sequence converges. It says that if a sequence is nondecreasing ($a_1 \le a_2 \le a_3 \le \cdots$) but does not get arbitrarily large, then it must be convergent.

Theorem 2 Let $\{a_n\}$ be a nondecreasing sequence with the property that there is a number B such that $a_n \le B$ for all n. That is, $a_1 \le a_2 \le a_3 \le a_4 \le \cdots \le a_n \le \cdots$ and $a_n \le B$ for all n. Then the sequence $\{a_n\}$ is convergent and a_n approaches a number L less than or equal to B.

Similarly, if $\{a_n\}$ is a nonincreasing sequence and there is a number B such that $a_n \ge B$ for all n, then the sequence $\{a_n\}$ is convergent and its limit is greater than or equal to B.

The proof, which is given in advanced calculus, is omitted. Figure 1 shows that Theorem 2 is at least plausible. (The a_n's increase but, being less than B, approach some number L, and that number is not larger than B.) The next example applies Theorem 2 to our opening example.

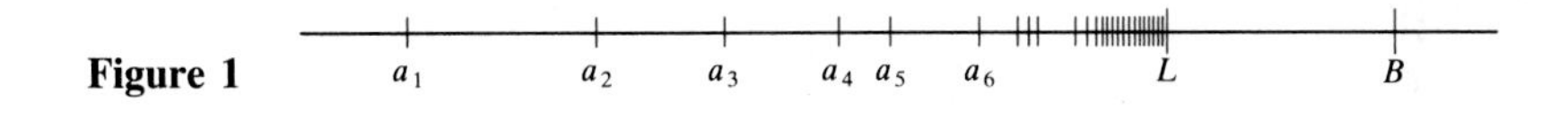

Figure 1

EXAMPLE 4 Use the second half of Theorem 2 to show that $0.8^n \to 0$ as $n \to \infty$.

SOLUTION The sequence $0.8, 0.8^2, 0.8^3, \ldots, 0.8^n, \ldots$ is decreasing. Furthermore, $0.8^n \ge 0$ for all n. (See Fig. 2.) Thus $\lim_{n\to\infty} 0.8^n$ exists and is greater than or equal to 0. Call this limit L. We wish to show that $L = 0$.

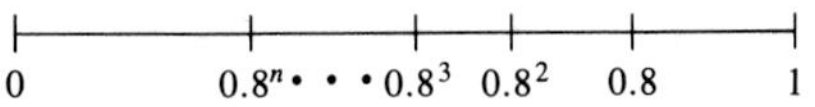

0 $0.8^n \cdot \cdot \cdot 0.8^3$ 0.8^2 0.8 1

Figure 2

Note that the sequence $0.8^2, 0.8^3, \ldots, 0.8^{n+1}$ also has the limit L. The nth term of this sequence is 0.8^{n+1}. Thus

$$\begin{aligned} L &= \lim_{n\to\infty} (0.8)^{n+1} \\ &= \lim_{n\to\infty} (0.8)(0.8)^n \\ &= 0.8 \lim_{n\to\infty} (0.8)^n \\ &= 0.8L. \end{aligned}$$

Since $L = 0.8L$, it follows that $L = 0$. ■

An argument similar to that in Example 4 proves Theorem 1. (See Exercises 26 and 27.)

Properties of Limits of Sequences

The limits of sequences $\{a_n\}$ behave like the limits of functions $f(x)$, as discussed in Sec. 2.4. The following theorem will be given without proof.

Remember that A and B are real numbers (not "infinity").

Theorem 3 If $\lim_{n\to\infty} a_n = A$ and $\lim_{n\to\infty} b_n = B$, then

(*a*) $\lim_{n\to\infty} (a_n + b_n) = A + B.$

(*b*) $\lim_{n\to\infty} (a_n - b_n) = A - B.$

(*c*) $\lim_{n\to\infty} a_n b_n = AB.$

(*d*) $\lim_{n\to\infty} \frac{a_n}{b_n} = \frac{A}{B} \quad (B \neq 0).$

(*e*) If k is a constant, $\lim_{n\to\infty} ka_n = kA.$ ■

For instance,

$$\lim_{n\to\infty}\left[\frac{3}{n} + \left(\frac{1}{2}\right)^n\right] = 3 \lim_{n\to\infty} \frac{1}{n} + \lim_{n\to\infty}\left(\frac{1}{2}\right)^n$$
$$= 3 \cdot 0 + 0 = 0.$$

Note that (*e*) was used in Example 4, where it was asserted that $\lim_{n\to\infty} (0.8)(0.8)^n = 0.8 \lim_{n\to\infty} (0.8)^n$.

Techniques for dealing with $\lim_{x\to\infty} f(x)$ can often be applied to determining $\lim_{n\to\infty} a_n$. The essential point is

$$\text{if } \lim_{x\to\infty} f(x) = L, \qquad \text{then} \qquad \lim_{n\to\infty} f(n) = L.$$

Example 5 illustrates this observation.

EXAMPLE 5 Find $\lim_{n\to\infty} (n/2^n)$.

SOLUTION Consider the function $f(x) = x/2^x$. By l'Hôpital's rule (infinity-over-infinity case),

$$\lim_{x\to\infty} \frac{x}{2^x} = \lim_{x\to\infty} \frac{1}{2^x \ln 2} = 0.$$

Thus

$$\lim_{n\to\infty} \frac{n}{2^n} = 0. \quad ■$$

The converse of the statement "if $\lim_{x\to\infty} f(x) = L$, then $\lim_{n\to\infty} f(n) = L$" is not true. For example, take $f(x) = \sin \pi x$. Then $\lim_{n\to\infty} f(n) = 0$, but $\lim_{x\to\infty} f(x)$ does not exist.

The Precise Definition of $\lim_{n\to\infty} a_n = L$

In Secs. 2.9 and 2.10 various limit concepts were given precise (as opposed to informal) definitions. The following definition is in the same spirit.

Precise definition of limit of a sequence

Definition *Limit of a sequence.* The number L is the **limit of the sequence** $\{a_n\}$ if for each $\epsilon > 0$ there is an integer N such that

$$|a_n - L| < \epsilon$$

for all integers $n > N$.

EXAMPLE 6 Use the precise definition to show that $\lim_{n\to\infty} 1/n = 0$.

SOLUTION Given $\epsilon > 0$ we want to show that there is an integer N such that

$$\left|\frac{1}{n} - 0\right| < \epsilon$$

for all integers $n > N$.

For instance, if $\epsilon = 0.01$, we want

$$\left|\frac{1}{n} - 0\right| < 0.01$$

or simply

$$\frac{1}{n} < 0.01.$$

Multiplying by n gives

$$1 < 0.01\, n$$

and then

$$n > \frac{1}{0.01} = 100.$$

Hence $N = 100$ suffices.

The general case is similar. We wish

$$\left|\frac{1}{n} - 0\right| < \epsilon,$$

or

$$\frac{1}{n} < \epsilon.$$

Hence

$$1 < n\epsilon$$

and finally

$$n > \frac{1}{\epsilon}.$$

Any integer $N > 1/\epsilon$ will suffice. ■

k^n AND ENERGY FROM THE ATOM

In a nuclear chain reaction, when a neutron strikes the nucleus of an atom of uranium or plutonium, on the average a certain number of neutrons split off. Call this number k. These k neutrons then strike further atoms. Since each splits off k neutrons, in this second generation there are k^2 neutrons. In the third generation there are k^3 neutrons, and so on. Each generation is born in a fraction of a second and produces energy.

If k is less than 1, then the chain reaction dies out, since $k^n \to 0$ as $n \to \infty$. A successful chain reaction—whether in a nuclear reactor or an atomic bomb—requires that k be greater than 1, since then $k^n \to \infty$ as $n \to \infty$.

In September 1941, Enrico Fermi and Leo Szilard achieved $k = 0.87$ with a uranium pile at Columbia University. In April 1942, they obtained an encouraging $k = 0.918$. In the meantime, Samuel Allison at the University of Chicago measured $k = 0.94$ in his own atomic pile. Finally, at 3:53 P.M. on December 2, 1942, in an experiment set up in a squash court at the University of Chicago, Fermi and Szilard attained $k = 1.0006$. With this k the neutron intensity doubled every 2 minutes. They had achieved the first controlled, sustained, chain reaction, producing energy from the atom. Fermi let the pile run for 4.5 minutes. Had he let it go on much longer, the atomic pile, the squash court, the university, and part of Chicago might have disappeared.

Eugene Wigner, one of the scientists present, wrote, "We felt as, I presume, everyone feels who has done something that he knows will have very far-reaching consequences which he cannot foresee."

Szilard had a different reaction: "There was a crowd there and then Fermi and I stayed there alone. I shook hands with Fermi and I said I thought this day would go down as a black day in the history of mankind."

However it may be regarded, December 2, 1942, is a historic date. Before that date k was less than 1, and $\lim_{n\to\infty} k^n = 0$. After that date k was larger than 1 and $\lim_{n\to\infty} k^n = \infty$.

Based on Richard Rhodes,
The Making of the Atomic Bomb,
Simon and Schuster, New York, 1986

Section Summary

We defined sequences and their limits if they converge. The sequence r^n ($|r| < 1$) and $k^n/n!$ will be used often in Chaps. 10 and 11. We have

$$\lim_{n\to\infty} r^n = 0 \qquad (|r| < 1) \qquad \text{and} \qquad \lim_{n\to\infty} \frac{k^n}{n!} = 0. \qquad (k \text{ any constant})$$

We pointed out that if $\{a_n\}$ is increasing, but remains below a number B, then $\{a_n\}$ is convergent and its limit is not larger than B. Similarly, if $\{a_n\}$ is decreasing, but remains above a number B, then $\{a_n\}$ is convergent and its limit is not less than B.

EXERCISES FOR SEC. 10.2: SEQUENCES

In Exercises 1 to 12 write out the first three terms of the given sequences and state whether the sequence converges or diverges. If it converges, give its limit.

1 $\{0.999^n\}$

2 $\{1.01^n\}$

3 $\{1^n\}$

4 $\{(-0.8)^n\}$

5 $\{n!\}$

6 $\left\{\frac{10^n}{n!}\right\}$

7 $\left\{\frac{3n+5}{5n-3}\right\}$

8 $\left\{\frac{(-1)^n}{n}\right\}$

9 $\left\{\frac{\cos n}{n}\right\}$

10 $\left\{\frac{n}{2^n}+\frac{3n+1}{4n+2}\right\}$

11 $\left\{\left(1+\frac{2}{n}\right)^n\right\}$ (*Hint:* Review Sec. 6.2.)

12 $\left\{\left(\frac{n-1}{n}\right)^n\right\}$

13 (*a*) Fill in this table:

n	1	2	3	4	5	6	7	8
$6^n/n!$								

(*b*) Plot the points obtained in (*a*), with n along the horizontal axis.
(*c*) What is the largest value of $6^n/n!$?
(*d*) What is $\lim_{n\to\infty} 6^n/n!$?

14 What is the largest value of $(11.8)^n/n!$? Explain.

15 Find a value of n such that 0.999^n is less than 0.0001
(*a*) by experimenting with the aid of your calculator;
(*b*) by solving the equation $0.999^x = 0.0001$.

16 Find a value of n such that 1.0006^n is larger than 2
(*a*) by experimenting with the aid of your calculator;
(*b*) by solving the equation $1.0006^x = 2$.

---

17 Assume that each year inflation eats away 5 percent of the value of a dollar. Let a_n be the value of a dollar after n years of such inflation.
(*a*) Find a_4. (*b*) Find $\lim_{n\to\infty} a_n$.

18 The binomial theorem asserts that if n is a positive integer then $(1+x)^n$ is equal to $1+nx$ plus other terms that are positive if $x>0$. Use this to show that if $r>1$, then $\lim_{n\to\infty} r^n = \infty$.

19 The Fibonacci sequence was first considered in the *Liber Abaci*, by Fibonacci, around the year 1200. (This was the book that introduced decimal notation to Europe.) The nth term of this sequence is usually denoted F_n, with $F_1 = 1$, $F_2 = 1$, and $F_{n+2} = F_{n+1} + F_n$, for $n \geq 1$. For instance, $F_3 = 1 + 1 = 2$, $F_4 = F_3 + F_2 = 2 + 1 = 3$, $F_5 = 5$, and $F_6 = 8$. (Each term from the third on is the sum of the two preceding terms.)
(*a*) Compute F_7, F_8, F_9, and F_{10}.
(*b*) Compute F_{n+1}/F_n, for $n = 1, 2, \ldots, 10$.
(*c*) What do you think happens to the sequence $a_n = F_{n+1}/F_n$ as $n \to \infty$?

20 Define a sequence $\{a_n\}$ as follows: $a_1 = 1$, $a_2 = 3$, and $a_{n+2} = (1 + a_{n+1})/a_n$, for $n \geq 1$. Examine the behavior of a_n as $n\to\infty$. Try other choices of a_1 and a_2.

21 For each integer $n \geq 1$, let

$$a_n = \frac{1}{n} + \frac{1}{n+1} + \cdots + \frac{1}{2n} = \sum_{i=n}^{2n} \frac{1}{i}.$$

For instance, $a_3 = \frac{1}{3} + \frac{1}{4} + \frac{1}{5} + \frac{1}{6} = 0.95$. Examine the sequence $\{a_n\}$ by performing some numerical experiments. Write up your data, conjectures, and arguments.

22 Let $f(x)$ be a function with a continuous derivative. Let $x_1, x_2, \ldots, x_n, \ldots$ be a sequence of numbers obtained by Newton's method,

$$x_{n+1} = x_n - \frac{f(x_n)}{f'(x_n)}.$$

Show that if $\{x_n\}$ converges to L and $f'(L) \neq 0$, then L is a solution of the equation $f(x) = 0$.

In Exercises 23 and 24 determine the given limits by first showing that each limit is a definite integral $\int_a^b f(x)\,dx$ for a suitable interval $[a, b]$ and function f. (Review Sec. 5.3.)

23 $\lim_{n\to\infty} \sum_{i=1}^{n} \left(\frac{i}{n}\right)^2 \frac{1}{n}$

24 $\lim_{n\to\infty} \sum_{i=1}^{n} \frac{n}{n^2+i^2}$

25 This exercise contrasts two sequences that approach the solution r of the equation $x = \cos x$.
(*a*) Draw $y = x$ and $y = \cos x$ and estimate the x coordinate of the point where they cross.
(*b*) Use Newton's method, and record the sequence of estimates. [Let x_1 be your estimate in (*a*).]
(*c*) Let x_1 be your estimate in (*a*) and obtain a sequence $x_2 = \cos x_1$, $x_3 = \cos x_2$, $\ldots$, $x_n = \cos x_{n-1}$, $\ldots$. (Just keep pressing the cosine key on your calculator.) Keep a record of the results.
(*d*) Contrast the sequence in (*b*) and (*c*). Which requires less work? Which approaches the solution more quickly? Discuss the speed with which the error goes toward 0.

26 Prove Theorem 1 in the style of Example 4.

27 (*a*) Use the precise definition of $\lim_{n\to\infty} a_n = L$ to prove Theorem 1 for $0 < r < 1$.
(*b*) Use (*a*) to prove Theorem 1 for $-1 < r < 0$.

28 Use the precise definition of $\lim_{n\to\infty} a_n = L$ to prove that $\lim_{n\to\infty} (\sin n)/n = 0$.

29 Use the precise definition of $\lim_{n\to\infty} a_n = L$ to prove that $\lim_{n\to\infty} 3/n^2 = 0$.

30 Use the precise definition of $\lim_{n\to\infty} a_n = L$ to prove that the statement "$\lim_{n\to\infty} (-1)^n = 0$" is false.

10.3 SERIES

This section examines the convergence of the sequence formed by adding up more and more terms of a given sequence.

Series

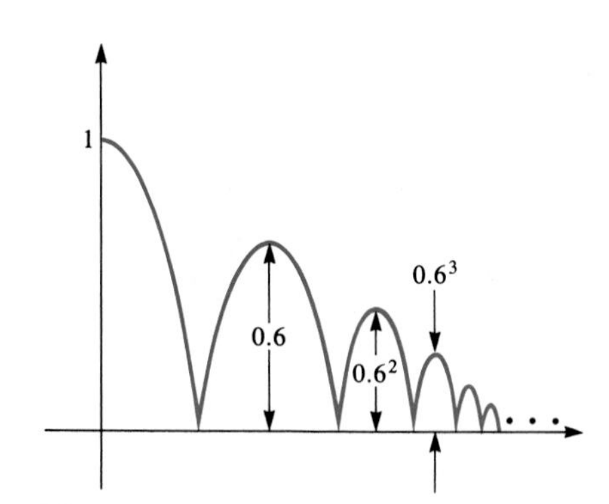

Figure 1

Consider a tennis ball that is dropped from a height of 1 meter. It rebounds 0.6 meter. It continues to bounce, and each fall is 60 percent as high as the previous fall. (See Fig. 1.) What is the total distance the ball falls?

The third fall is $(0.6)^2$ meter, the next $(0.6)^3$ meter, and so on. We are asking, "What happens to the sum

$$1 + 0.6 + 0.6^2 + \cdots + (0.6)^n$$

as $n \to \infty$?" Example 1 explores this question. (Similar sums arise in many areas. See Exercise 28 for an example in medicine and Exercises 29 and 30 for examples in economics.)

EXAMPLE 1 Given the sequence $1, 0.6, 0.6^2, 0.6^3, \ldots$, form a new sequence $\{S_n\}$ as follows:

S is short for "sum."

$$S_1 = 1,$$

$$S_2 = 1 + 0.6,$$

$$S_3 = 1 + 0.6 + 0.6^2,$$

and, in general, $\quad S_n = 1 + 0.6 + 0.6^2 + \cdots + (0.6)^{n-1}.$

Each S_n is the sum of the first n terms of the sequence, $1, 0.6, (0.6)^2, (0.6)^3, \ldots$. Examine the behavior of S_n as $n \to \infty$.

SOLUTION

$$S_1 = 1$$

$$S_2 = 1 + 0.6 = 1.6$$

$$S_3 = 1 + 0.6 + 0.6^2 = 1 + 0.6 + 0.36 = 1.96$$

$$S_4 = 1 + 0.6 + 0.6^2 + 0.6^3 = 1 + 0.6 + 0.36 + 0.216 = 2.176$$

$$S_5 = 1 + 0.6 + 0.6^2 + 0.6^3 + 0.6^4 = 1 + 0.6 + 0.36 + 0.216 + 0.1296 = 2.3056$$

Continuing in this way, we fill in the following table to four decimal place accuracy:

n	1	2	3	4	5	6	7	8	9	10	11	12	13
S_n	1	1.6	1.96	2.176	2.3056	2.3834	2.4300	2.4580	2.4748	2.4849	2.4909	2.4946	2.4967

Figure 2 displays these data.

Two influences on S_n

Two influences affect the growth of S_n as n increases. On the one hand, the number of summands increases, causing S_n to get larger. On the other hand, the summands approach 0, so that S_n grows more and more slowly as n increases.

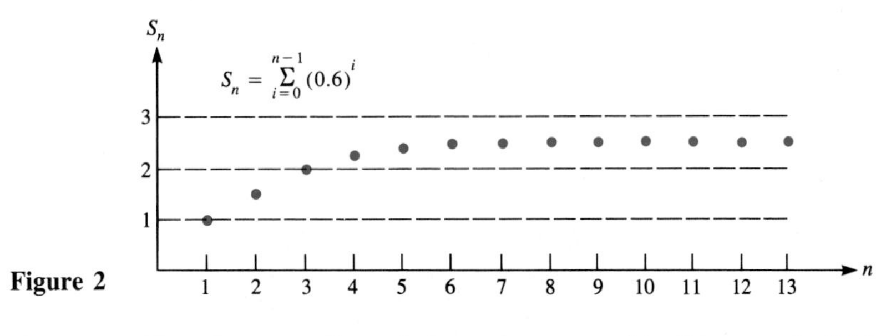

Figure 2

How these two forces balance is answered by Theorem 1 of this section, which shows that the sequence $\{S_n\}$ converges and that its limit is 2.5:

$$\lim_{n\to\infty} S_n = 2.5. \quad \blacksquare$$

The rest of this section extends the ideas introduced in Example 1.

Let $a_1, a_2, a_3, \ldots, a_n, \ldots$ be a sequence. From this sequence a new sequence $S_1, S_2, S_3, \ldots, S_n, \ldots$ can be formed:

$$S_1 = a_1,$$

$$S_2 = a_1 + a_2,$$

$$S_3 = a_1 + a_2 + a_3,$$

$$\cdots\cdots\cdots\cdots\cdots\cdots$$

Summation notation was introduced in Sec. 5.2.

$$S_n = a_1 + a_2 + a_3 + \cdots + a_n = \sum_{i=1}^{n} a_i.$$

The sequence of sums, $S_1, S_2, S_3, \ldots, S_n$, is called the **series** obtained from the sequence $a_1, a_2, a_3, \ldots, a_n, \ldots$.

Traditionally, it is referred to as "the series whose nth term is a_n." Common notations for the sequence $\{S_n\}$ are $\Sigma_{n=1}^{\infty} a_n$ and $a_1 + a_2 + a_3 + \cdots + a_n + \cdots$. The sum

$$S_n = a_1 + a_2 + \cdots + a_n = \sum_{i=1}^{n} a_i$$

is called a **partial sum** or the ***n*th partial sum**. If the sequence of partial sums of a series converges to L, then L is called the **sum** of the series and the series is said to be **convergent**. We write

$$\lim_{n\to\infty} S_n = L.$$

Only finitely many summands are ever added up.

Frequently one writes $L = a_1 + a_2 + \cdots + a_n + \cdots$. Remember, however, that we do not add an infinite number of terms; we take the limit of finite sums. A series that is not convergent is called **divergent**.

A Note on Notation Starting with the sequence $a_1, a_2, \ldots, a_n, \ldots$, we form a new sequence, $S_1, S_2, \ldots, S_n, \ldots$, whose terms are the partial sums $S_1 = a_1$, $S_2 = a_1 + a_2$, $\ldots$, $S_n = a_1 + a_2 + \cdots + a_n$. The symbol

$$\sum_{n=1}^{\infty} a_n$$

is shorthand for this sequence $S_1, S_2, \ldots, S_n, \ldots$. If the sequence of these sums, $S_1, S_2, \ldots, S_n, \ldots$, converges to a number L, we also write

$$\sum_{n=1}^{\infty} a_n = L.$$

The symbol $\sum_{n=1}^{\infty} a_n$ has two meanings.

So the symbol $\Sigma_{n=1}^{\infty} a_n$ stands for two different concepts: a sequence of partial sums and also, if that sequence converges, for its limit. This limit is called the "sum" of the series.

So, in Example 1 we investigated the *series*

$$\sum_{n=1}^{\infty} 0.6^{n-1},$$

namely, the sequence of partial sums 1, $1 + 0.6$, $1 + 0.6 + 0.6^2, \ldots, 1 + 0.6 + 0.6^2 + \cdots + (0.6)^{n-1}$. This sequence converges to 2.5. That permits us to write

$$\sum_{n=1}^{\infty} (0.6)^{n-1} = 2.5,$$

which says, "The series $\Sigma_{n=1}^{\infty} (0.6)^{n-1}$ converges to the number 2.5." We also say, for the sake of brevity, "Its sum is 2.5."

Geometric Series

Example 1 concerns the series whose nth term is $(0.6)^{n-1}$:

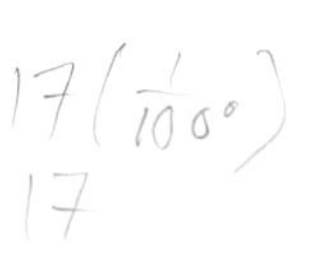

$$S_n = 1 + 0.6 + 0.6^2 + \cdots + 0.6^{n-1}.$$

It is a special case of a geometric series, which will now be defined.

Appendix C treats geometric series with a finite number of terms.

Definition *Geometric series.* Let a and r be real numbers. The series

$$a + ar + ar^2 + \cdots + ar^{n-1} + \cdots$$

is called the **geometric series with initial term a and ratio r**.

The series in Example 1 is a geometric series with initial term 1 and ratio 0.6.

Theorem 1 If $-1 < r < 1$, the geometric series

$$a + ar + \cdots + ar^{n-1} + \cdots$$

converges to $a/(1 - r)$.

Proof Let S_n be the sum of the first n terms:

$$S_n = a + ar + \cdots + ar^{n-1}.$$

By the formula in Appendix C for the sum of a finite geometric series,

$$S_n = \frac{a(1 - r^n)}{1 - r}.$$

See Exercise 26.

By Theorem 1 in Sec. 10.2, $\lim_{n \to \infty} r^n = 0$.

Thus
$$\lim_{n \to \infty} S_n = \frac{a}{1 - r},$$

proving the theorem. ■

In particular, if $a = 1$ and $r = 0.6$, as in Example 1, the geometric series has the sum

$$\frac{1}{1 - 0.6} = \frac{1}{0.4} = 2.5.$$

The *n*th-Term Test

Theorem 1 says nothing about geometric series in which $r \geq 1$ or $r \leq -1$. The next theorem, which concerns series in general, not just geometric series, will be useful in settling this case.

Theorem 2 *The nth-term test for divergence.* If $\lim_{n \to \infty} a_n \neq 0$, then the series $a_1 + a_2 + \cdots + a_n + \cdots$ diverges. (The same conclusion holds if $\{a_n\}$ has no limit.)

We take an indirect approach.

Proof Assume that the series $a_1 + a_2 + \cdots$ converges. Since S_n is the sum $a_1 + a_2 + \cdots + a_n$, while S_{n-1} is the sum of the first $n - 1$ terms, it follows that $S_n = S_{n-1} + a_n$, or

$$a_n = S_n - S_{n-1}.$$

Let
$$S = \lim_{n \to \infty} S_n.$$

Then we also have
$$S = \lim_{n \to \infty} S_{n-1},$$

since S_{n-1} runs through the same numbers as S_n. Thus

$$\begin{aligned}\lim_{n \to \infty} a_n &= \lim_{n \to \infty} (S_n - S_{n-1}) \\ &= \lim_{n \to \infty} S_n - \lim_{n \to \infty} S_{n-1} \\ &= S - S \\ &= 0.\end{aligned}$$

So if a series converges, its nth term must approach 0.

This proves the theorem. ■

Theorem 2 implies that if $a \neq 0$ and $r \geq 1$, the geometric series

$$a + ar + \cdots + ar^{n-1} + \cdots$$

diverges. For instance, if $r = 1$,

$$\lim_{n\to\infty} ar^n = \lim_{n\to\infty} a1^n = a,$$

which is not 0. If $r > 1$, then r^n gets arbitrarily large as n increases; hence $\lim_{n\to\infty} ar^n$ does not exist. Similarly, if $r \le -1$, $\lim_{n\to\infty} ar^n$ does not exist. The above results and Theorem 1 can be summarized by this statement: The geometric series

$$\sum_{n=1}^{\infty} ar^{n-1} = a + ar + ar^2 + \cdots + ar^{n-1} + \cdots,$$

for $a \neq 0$, converges if and only if $|r| < 1$.

Warning: Even if the nth term approaches 0, the series can diverge!

This is a good time to read Appendix F.

Theorem 2 tells us that *if* the series $a_1 + a_2 + a_3 + \cdots$ converges, *then* a_n approaches 0 as $n \to \infty$. The converse of this statement is *not true*. If a_n approaches 0 as $n \to \infty$, it does *not* follow that the series $a_1 + a_2 + a_3 + \cdots$ converges. Be careful to make this distinction.

Recall the series

$$\frac{1}{\sqrt{1}} + \frac{1}{\sqrt{2}} + \cdots + \frac{1}{\sqrt{n}} + \cdots$$

discussed in Sec. 10.1. Even though its nth term approaches 0 as $n \to \infty$, the sums get arbitrarily large. The nth term approaches 0 so ''slowly'' that the sums S_n get arbitrarily large.

The harmonic series was so named by the Greeks because of the role of 1/n in musical harmony.

In the next example, the nth term approaches 0 much faster than $1/\sqrt{n}$ does. Still the series diverges. The series in this example is called the **harmonic series**. The argument that it diverges is due to the French mathematician Nicolas of Oresme, who presented it about the year 1360.

EXAMPLE 2 Show that the harmonic series $1/1 + 1/2 + \cdots + 1/n + \cdots$ diverges.

SOLUTION Collect the summands in longer and longer groups in the manner indicated below (each group from the third on has twice the number of summands as the previous group):

$$\underbrace{\tfrac{1}{1}} + \underbrace{\tfrac{1}{2}} + \underbrace{\tfrac{1}{3} + \tfrac{1}{4}} + \underbrace{\tfrac{1}{5} + \tfrac{1}{6} + \tfrac{1}{7} + \tfrac{1}{8}} + \underbrace{\tfrac{1}{9} + \tfrac{1}{10} + \cdots + \tfrac{1}{16}} + \underbrace{\tfrac{1}{17} + \cdots}.$$

The sum of the terms in each group is at least $\frac{1}{2}$. For instance,

$$\tfrac{1}{5} + \tfrac{1}{6} + \tfrac{1}{7} + \tfrac{1}{8} > \tfrac{1}{8} + \tfrac{1}{8} + \tfrac{1}{8} + \tfrac{1}{8} = \tfrac{4}{8} = \tfrac{1}{2},$$

and

$$\tfrac{1}{9} + \tfrac{1}{10} + \cdots + \tfrac{1}{16} > \tfrac{1}{16} + \tfrac{1}{16} + \cdots + \tfrac{1}{16} = \tfrac{8}{16} = \tfrac{1}{2}.$$

Since the repeated addition of $\frac{1}{2}$'s produces sums as large as we please, the series diverges. ■

An important moral: The nth-term test is only a test for divergence.

If the series $a_1 + a_2 + \cdots + a_n + \cdots$ converges, it *follows* that $a_n \to 0$. However, if $a_n \to 0$, it *does not necessarily follow* that $a_1 + a_2 + \cdots + a_n + \cdots$ converges. Indeed, there is no general, practical rule for determining whether a series converges or diverges. Fortunately, a few rules suffice to decide on the convergence or divergence of the most common series; they will be presented in this chapter.

Two basic properties of series, similar to those of definite integrals, are recorded in the next theorem. Exercise 35 asks for the proof.

Theorem 3
(*a*) If $\Sigma_{n=1}^{\infty} a_n$ is a convergent series with sum L, and if c is a number, then $\Sigma_{n=1}^{\infty} ca_n$ is convergent and has the sum cL.
(*b*) If, furthermore, $\Sigma_{n=1}^{\infty} b_n$ is a convergent series with sum M, then $\Sigma_{n=1}^{\infty} (a_n + b_n)$ is a convergent series with sum $L + M$.

Front ends don't affect convergence.

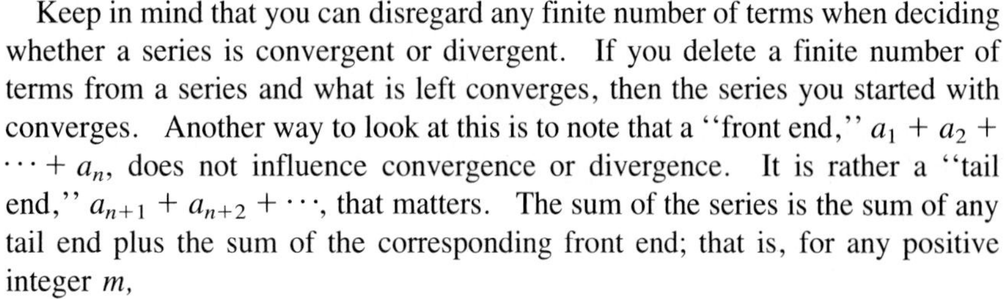

Keep in mind that you can disregard any finite number of terms when deciding whether a series is convergent or divergent. If you delete a finite number of terms from a series and what is left converges, then the series you started with converges. Another way to look at this is to note that a "front end," $a_1 + a_2 + \cdots + a_n$, does not influence convergence or divergence. It is rather a "tail end," $a_{n+1} + a_{n+2} + \cdots$, that matters. The sum of the series is the sum of any tail end plus the sum of the corresponding front end; that is, for any positive integer m,

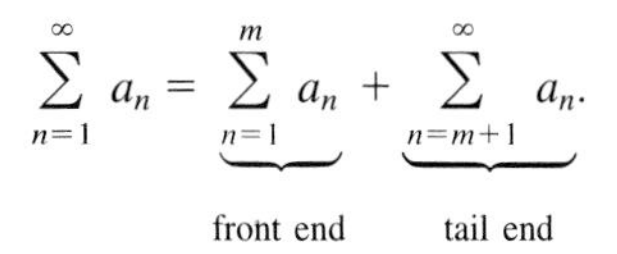

$$\sum_{n=1}^{\infty} a_n = \underbrace{\sum_{n=1}^{m} a_n}_{\text{front end}} + \underbrace{\sum_{n=m+1}^{\infty} a_n}_{\text{tail end}}.$$

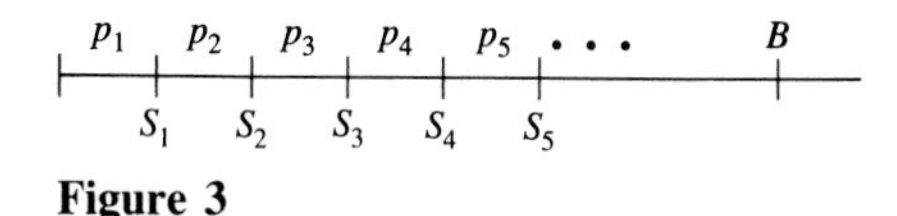

Figure 3

Suppose that $\Sigma_{n=1}^{\infty} p_n$ is a series with nonnegative terms and you can show that there is a number B such that every partial sum $S_1 = p_1$, $S_2 = p_1 + p_2$, . . . , $S_n = p_1 + p_2 + \cdots + p_n$, is less than or equal to B. By Theorem 2 of Sec. 10.2, they have a limit L, which is less than or equal to B. (See Fig. 3.) This means that $\Sigma_{n=1}^{\infty} p_n$ is convergent (and its sum is less than or equal to B). This observation will be useful in establishing the convergence of a series of nonnegative terms, even though it does not tell us what the sum of the series is.

A similar statement holds for a series $\Sigma_{n=1}^{\infty} a_n$ in which $a_n \le 0$ for all n. If there is a number A such that each partial sum is greater than or equal to A, then the series converges and its sum is greater than or equal to A.

Section Summary

Starting with any sequence $\{a_n\}$ we can form a new sequence $\{S_n\}$, where S_n is the sum of the first n terms of $\{a_n\}$, $S_n = a_1 + a_2 + \cdots + a_n$. The new sequence is called the "series" derived from the original sequence $\{a_n\}$. If the series converges, then a_n must approach 0 as $n \to \infty$. (The converse is not true.) It follows that if a_n does not approach 0 as $n \to \infty$, then the series $a_1 + a_2 + \cdots + a_n + \cdots$ diverges.

If $a_n = ar^{n-1}$, where $|r| < 1$, we obtain the geometric series a, $a + ar$, $a + ar + ar^2$, . . . , denoted simply as $a + ar + ar^2 + \cdots + ar^{n-1} + \cdots$. It converges to $a/(1 - r)$.

If a_n is nonnegative and $a_1 + a_2 + \cdots + a_n \le B$ for some fixed number B for all n, then $\Sigma_{n=1}^{\infty} a_n$ is convergent and approaches a number no larger than B. This principle will be used in the next section.

EXERCISES FOR SEC. 10.3: SERIES

Exercises 1 to 3 are based on suggestions of James T. Vance, Jr. Each one concerns the series $\Sigma_{n=1}^{\infty} a_n$ and the sequence of its partial sums $\{S_n\}$.

1 Suppose you know that $a_n \to 0$ as $n \to \infty$. Of the following four statements, only one is true. Which is it?
(*a*) The series definitely converges, but we need more information to determine its sum.
(*b*) The series definitely converges, and $\Sigma_{n=1}^{\infty} a_n = 0$.
(*c*) The series definitely diverges.
(*d*) There's not enough information to decide whether the series converges or diverges.

2 Suppose you know that $a_n \to 6$ as $n \to \infty$. Identify the one true statement:
(*a*) The series definitely converges, but we need more information to determine its sum.
(*b*) The series definitely converges, and its sum is 6.
(*c*) The series definitely diverges.
(*d*) There is not enough information to decide whether the series converges or diverges.

3 Suppose you know that $S_n \to 3$ as $n \to \infty$. Which of the following statements are true? (More than one may be true.)
(*a*) The series definitely diverges.
(*b*) $a_n \to 3$ as $n \to \infty$.
(*c*) $a_n \to 0$ as $n \to \infty$.
(*d*) The series definitely converges.
(*e*) The sum of the series is 3.
(*f*) We need more information to decide what happens to a_n as $n \to \infty$.

4 This exercise concerns the series $\Sigma_{n=1}^{\infty} 5(-1/2)^n$.
(*a*) Express the fourth term of this series as a decimal.
(*b*) Express the fourth partial sum of this series as a decimal.
(*c*) Find the limit as $n \to \infty$ of the nth term of the series.
(*d*) Find the limit as $n \to \infty$ of the nth partial sum of this series.
(*e*) Does the series converge? If so, what is its sum?

In each of Exercises 5 to 12 determine whether the given geometric series converges. If it does, find its sum.

5 $1 + \frac{1}{2} + \frac{1}{4} + \frac{1}{8} + \cdots + (\frac{1}{2})^{n-1} + \cdots$

6 $1 - \frac{1}{3} + \frac{1}{9} - \frac{1}{27} + \cdots + (-\frac{1}{3})^{n-1} + \cdots$

7 $\sum_{n=1}^{\infty} 10^{-n}$

8 $\sum_{n=1}^{\infty} 10^{n}$

9 $\sum_{n=1}^{\infty} 5(0.99)^{n}$

10 $\sum_{n=1}^{\infty} 7(-1.01)^{n}$

11 $\sum_{n=1}^{\infty} 4\left(\frac{2}{3}\right)^{n}$

12 $-\frac{3}{2} + \frac{3}{4} - \frac{3}{8} + \cdots + \frac{3}{(-2)^n} + \cdots$

In Exercises 13 to 20 determine whether the given series converge or diverge. Find the sums of the convergent series.

13 $-5 + 5 - 5 + 5 - \cdots + (-1)^n\, 5 + \cdots$

14 $\sum_{n=1}^{\infty} \frac{1}{[1 + (1/n)]^n}$

15 $\sum_{n=1}^{\infty} \frac{2}{n}$

16 $\sum_{n=1}^{\infty} \frac{n}{2n+1}$

17 $\sum_{n=1}^{\infty} 6\left(\frac{4}{5}\right)^{n}$

18 $\sum_{n=1}^{\infty} 100\left(\frac{-8}{9}\right)^{n}$

19 $\sum_{n=1}^{\infty} (2^{-n} + 3^{-n})$

20 $\sum_{n=1}^{\infty} (4^{-n} + n^{-1})$

21 What is the total distance traveled—both up and down—by the ball described in the opening paragraph of this section?

22 A rubber ball, when dropped on concrete, rebounds 90 percent of the distance it falls. If it is dropped from a height of 6 feet, how far does it travel—both up and down—before coming to rest?

---

23 The repeating decimal

$$3.171717\ldots,$$

where the 17's continue forever, can be viewed as 3 plus a geometric series:

$$3 + \frac{17}{100} + \frac{17}{100^2} + \frac{17}{100^3} + \cdots.$$

Using the formula for the sum of a geometric series, evaluate the decimal.

24 (See Exercise 23.) Evaluate the repeating decimal 0.3333 . . . .

25 (See Exercise 23.) Evaluate the repeating decimal 4.1256256256 . . . .

26 Show that if $|r| < 1$, the sum of the geometric series $a + ar + ar^2 + \cdots$ differs from S_n by $ar^n/(1-r)$.

27 This is a quote from an economics text: "The present value of the land, if a new crop is planted at time t, $2t$, $3t$, etc., is

$$P = g(t)e^{-rt} + g(t)e^{-2rt} + g(t)e^{-3rt} + \cdots.$$

Note that each term is the previous term multiplied by e^{-rt}. By the formula for the sum of a geometric series,

$$P = \frac{g(t)e^{-rt}}{1 - e^{-rt}}."$$

Check that the missing step, which simplified the formula for P, was correct.

28 A patient takes A grams of a certain medicine every 6 hours. The amount of each dose active in the body t hours later is Ae^{-kt} grams, where k is a positive constant and time is measured in hours.

(a) Show that immediately after taking the medicine for the nth time, the amount active in the body is

$$S_n = A + Ae^{-6k} + Ae^{-12k} + \cdots + Ae^{-6(n-1)k}.$$

(b) If, as $n \to \infty$, $S_n \to \infty$, the patient would be in danger. Does $S_n \to \infty$? If not, what is $\lim_{n\to\infty} S_n$?

29 (*How banks, with the assistance of the public, create money.*) If a deposit of A dollars is made at a bank, the bank can lend out most of this amount. However, it cannot lend out all the amount, for it must keep a reserve to meet the demands of depositors who may withdraw money from their accounts. The government stipulates what this reserve is, usually between 10 and 20 percent of the amount deposited. Assume that a bank is allowed to lend 80 percent of the amount deposited. If a person deposits \$1,000, then the bank can lend another person \$800. Assume that this borrower deposits all the amount; then the bank can lend a third person \$640 of that deposit. This process can go on indefinitely, through a fourth person, a fifth, and so on. If this process continues indefinitely, how large will all the deposits total in the long run? Note that this total is much larger than the initial deposit. The banks have created money. *Note:* By manipulating the reserve requirement, the government can heat up or cool off the economy. In 1990 the Federal Reserve Board required a 12 percent reserve on checking-account deposits. At the end of the year, fearing a recession, it removed any reserve requirements on certain other accounts, in order to encourage banks to make loans.

30 Deficit spending by the federal government inflates the nation's money supply. However, much of the money paid out by the government is spent in turn by those who receive it, thereby producing additional spending. This produces a chain reaction, called by economists the **multiplier effect**. It results in much greater total spending than the government's original expenditure. To be specific, suppose the government spends 1 billion dollars and that the recipients of that expenditure in turn spend 80 percent while retaining 20 percent. Let S_n be the *total* spending generated after n transactions in the chain, 80 percent of receipts being expended at each step.

(a) Show that $S_n = 1 + 0.8 + 0.8^2 + \cdots + 0.8^{n-1}$ billion dollars.

(b) Show that as n increases, the total spending approaches 5 billion dollars. (The number 5 is called the **multiplier**.)

(c) What would be the total spending if 90 percent of receipts is spent at each step instead of 80 percent?

31 How long does the ball in Exercise 22 bounce?

32 A gambler tosses a penny until a head appears. On the average, how many times does she toss a penny to get a head? Parts (a) and (b) concern this question.

(a) Experiment with a penny on 10 runs. Each run consists of tossing a penny until a head appears. Average the lengths of the 10 runs.

(b) The probability of a run of length one is $\frac{1}{2}$, since a head must appear on the first toss. The probability of a run of length two is $(\frac{1}{2})^2$. The probability of having a head appear for the first time on the nth toss is $(\frac{1}{2})^n$. It is shown in probability theory that the average number of tosses to get a head is $\sum_{n=1}^{\infty} n/2^n$. (This is a theoretical average approached as the experiment is repeated many times.) Compute $\sum_{n=1}^{8} n/2^n$. (The next exercise sums the infinite series.)

33 Oresme, around the year 1360, summed the series $\sum_{n=1}^{\infty} n/2^n$ by drawing the endless staircase shown in Fig. 4, in which each stair has width 1 and is twice as high as the stair immediately to its right.

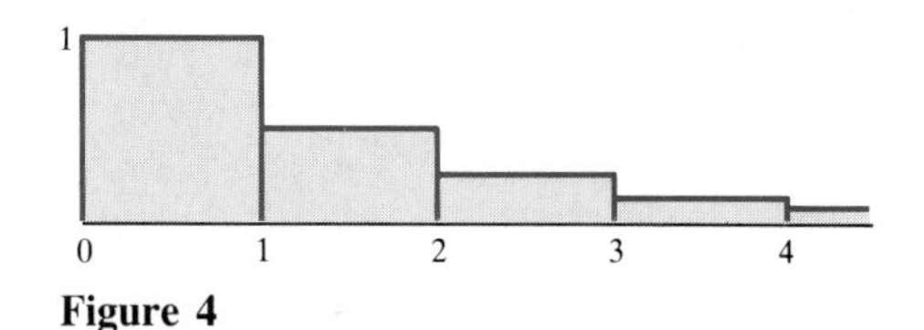

Figure 4

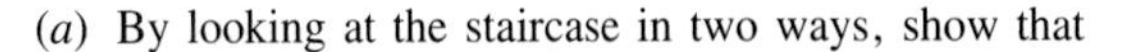

(a) By looking at the staircase in two ways, show that

$$1 + \tfrac{1}{2} + \tfrac{1}{4} + \tfrac{1}{8} + \cdots = \tfrac{1}{2} + \tfrac{2}{4} + \tfrac{3}{8} + \cdots.$$

(b) Use (a) to sum $\sum_{n=1}^{\infty} n/2^n$.

(c) Use the same idea to find $\sum_{n=1}^{\infty} np^n$, when $0 < p < 1$.

34 (a) Using your calculator compute enough partial sums of the series $\sum_{n=1}^{\infty} n\, 3^{-n}$ to offer an opinion as to whether it converges or diverges.

(b) If it converges, what do you think its sum is?

(c) Show that it converges. *Hint:* The coefficient n is less than 2^n.

35 (a) If c is a number and $\sum_{n=1}^{\infty} a_n$ is a convergent series with sum L, show that $\sum_{n=1}^{\infty} ca_n$ is a convergent series with sum cL. (Use the precise definition of limit in Sec. 10.2.)

(b) If $\sum_{n=1}^{\infty} a_n$ and $\sum_{n=1}^{\infty} b_n$ are convergent series with sums L and M, respectively, show that $\sum_{n=1}^{\infty} (a_n + b_n)$ is a convergent series with sum $L + M$.

10.4 THE INTEGRAL TEST

In this section we use improper integrals of the form $\int_a^\infty f(x)\,dx$ to establish convergence or divergence of series whose terms are positive and decreasing. Furthermore, we obtain a way of estimating the error when we use a partial sum to estimate the sum of the series.

The Integral Test

Let $f(x)$ be a decreasing function such that $f(x) > 0$. We can obtain a sequence from $f(x)$ by defining a_n to be $f(n)$. For instance, the sequence 1/1, 1/2, 1/3, . . . , $1/n$, . . . can be obtained from the function $f(x) = 1/x$ that way. It turns out that the convergence (or divergence) of the series $\Sigma_{n=1}^\infty a_n$ is closely connected with the convergence (or divergence) of the improper integral $\int_1^\infty f(x)\,dx$. This connection is expressed in the following theorem.

> **Theorem 1** *Integral test.* Let $f(x)$ be a continuous decreasing function such that $f(x) > 0$ for $x \geq 1$. Let $a_n = f(n)$. Then
> (*a*) If $\int_1^\infty f(x)\,dx$ is convergent, so is the series $\Sigma_{n=1}^\infty a_n$.
> (*b*) If $\int_1^\infty f(x)\,dx$ is divergent, so is the series $\Sigma_{n=1}^\infty a_n$.

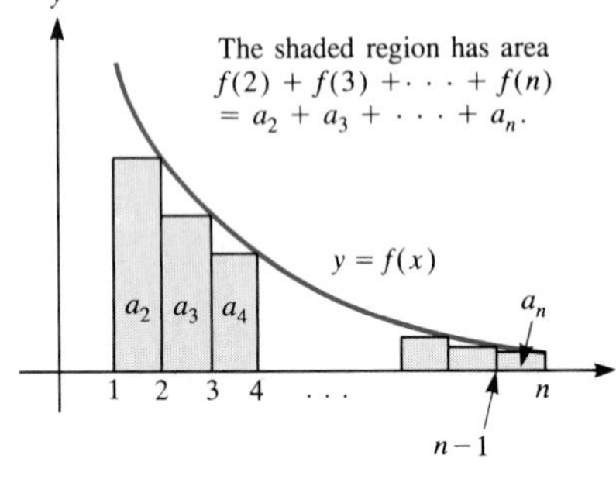

Figure 1

Proof Figures 1 and 2 are the key to the proof. Note how the rectangles are constructed in each case.

In Fig. 1 the rectangles lie below the curve $y = f(x)$. Each rectangle has width 1. Comparing the "staircase" area with the area under the curve gives the inequality

$$a_2 + a_3 + \cdots + a_n < \int_1^n f(x)\,dx,$$

and therefore

$$a_1 + a_2 + a_3 + \cdots + a_n < a_1 + \int_1^n f(x)\,dx. \tag{1}$$

If $\int_1^\infty f(x)\,dx$ is convergent, with value I, then

$$a_1 + a_2 + \cdots + a_n < a_1 + I.$$

Since the partial sums of the series $\Sigma_{n=1}^\infty a_n$ are all bounded by the number $a_1 + I$, the series $\Sigma_{n=1}^\infty a_n$ converges and its sum is less than or equal to $a_1 + I$.

Now, Fig. 2 shows that

$$a_1 + a_2 + \cdots + a_n > \int_1^{n+1} f(x)\,dx. \tag{2}$$

It follows that if $\int_1^\infty f(x)\,dx$ diverges, so must the series $\Sigma_{n=1}^\infty a_n$. ■

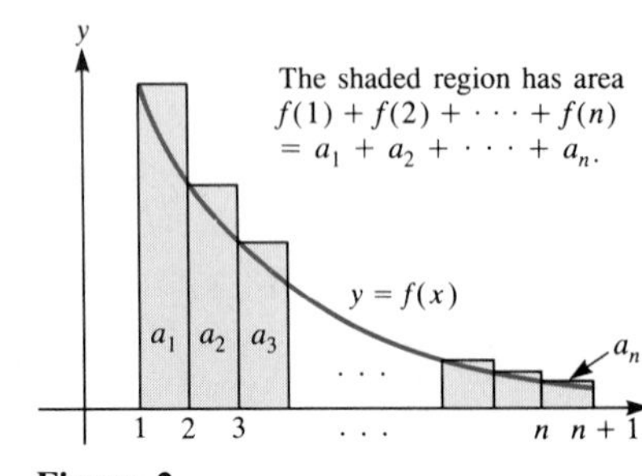

Figure 2

EXAMPLE 1 Use the integral test to determine the convergence or divergence of

(*a*) $\dfrac{1}{1} + \dfrac{1}{2} + \cdots + \dfrac{1}{n} + \cdots = \displaystyle\sum_{n=1}^{\infty} \frac{1}{n}$

(b) $\dfrac{1}{1^{1.01}} + \dfrac{1}{2^{1.01}} + \cdots + \dfrac{1}{n^{1.01}} + \cdots = \sum_{n=1}^{\infty} \dfrac{1}{n^{1.01}}.$

SOLUTION (a) Let $f(x) = 1/x$, which is a decreasing positive function for $x > 0$. Then $a_n = f(n) = 1/n$. We have

$$\int_1^{\infty} \frac{dx}{x} = \lim_{b\to\infty} \int_1^b \frac{dx}{x} = \lim_{b\to\infty} [\ln b - \ln 1] = \infty.$$

Since $\int_1^{\infty} dx/x$ is divergent, so is the series $\sum_{n=1}^{\infty} 1/n$.

(b) Let $f(x) = 1/x^{1.01}$, which is a decreasing positive function. Then $a_n = f(n) = 1/n^{1.01}$. We have

$$\int_1^{\infty} \frac{dx}{x^{1.01}} = \lim_{b\to\infty} \int_1^b \frac{dx}{x^{1.01}} = \lim_{b\to\infty} \left. \frac{x^{-1.01+1}}{-1.01+1} \right|_1^b = \lim_{b\to\infty} \left. \frac{x^{-0.01}}{-0.01} \right|_1^b = \lim_{b\to\infty} \left[\frac{b^{-0.01}}{-0.01} - \left(\frac{1^{-0.01}}{-0.01} \right) \right] = 100.$$

Since $\int_1^{\infty} dx/x^{1.01}$ is convergent, so is $\sum_{n=1}^{\infty} 1/n^{1.01}$. By (1), its sum is less than $a_1 + 100 = 101$. ■

The argument in Example 1 extends to a family of series known as p-series.

Definition For a positive number p, the series

$$\sum_{n=1}^{\infty} \frac{1}{n^p}$$

is called a **p-series.**

For example, when $p = 1$ we obtain the harmonic series $\sum_{n=1}^{\infty} 1/n$ and for $p = 1.01$, the series $\sum_{n=1}^{\infty} 1/n^{1.01}$.

An argument similar to those in Example 1 establishes the following theorem.

Theorem 2 If $p \leq 1$, the p-series $\sum_{n=1}^{\infty} 1/n^p$ diverges. If $p > 1$, the p-series $\sum_{n=1}^{\infty} 1/n^p$ converges.

Note that there is a p-series for each positive number p. A negative exponent

p would not give a series of interest. For instance, when $p = -1$, we obtain $\Sigma_{n=1}^{\infty} 1/n^{-1} = \Sigma_{n=1}^{\infty} n$, which is clearly divergent since its nth term does not approach 0 as $n \to \infty$. (For any negative p, $\lim_{n\to\infty} 1/n^p = \infty$.)

Controlling the Error

When we use a front end of a series (a partial sum) to estimate the sum of the whole series, there will be an error, namely, the sum of the corresponding tail end. For the sum of a front end to be a good estimate of the sum of the whole series, we must be sure that the sum of the corresponding tail end is small. Otherwise, we would be like the carpenter who measures a board as "5 feet long with an error of perhaps as much as 5 feet." That is why we wish to be sure that the sum of the tail end is small.

Let S_n be the sum of the first n terms of a convergent series $\Sigma_{n=1}^{\infty} a_n$ whose sum is S. The difference

$$R_n = S - S_n = a_{n+1} + a_{n+2} + a_{n+3} + \cdots$$

is called the **remainder** or **error** in using the sum of the first n terms to approximate the sum of the series. That is,

$$\underbrace{a_1 + a_2 + \cdots + a_n}_{\text{partial sum } S_n} + \underbrace{a_{n+1} + a_{n+2} + \cdots}_{\text{tail end } R_n} = \underbrace{a_1 + a_2 + \cdots + a_n + a_{n+1} + \cdots}_{\text{sum of series } S}$$

so

$$S_n + R_n = S.$$

For the series of the special type considered in this section, it is possible to use an improper integral to estimate the error. The reasoning depends once again on comparing a staircase of rectangles with the area under a curve.

Recall that $f(x)$ is a continuous decreasing positive function. The error in using $S_n = f(1) + f(2) + \cdots + f(n) = \Sigma_{i=1}^{n} f(i)$ to approximate $\Sigma_{i=1}^{\infty} f(i)$ is the sum $\Sigma_{i=n+1}^{\infty} f(i)$. This sum is the area of the endless staircase of rectangles shown in Fig. 3. Comparing the rectangles with the region under the curve $y = f(x)$, we conclude that

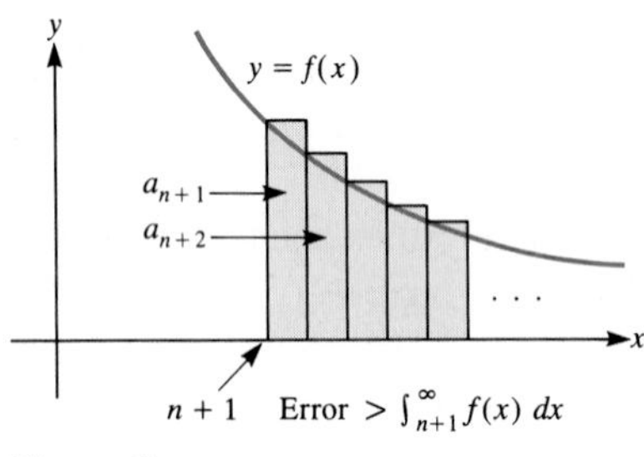

Figure 3

$$R_n = a_{n+1} + a_{n+2} + \cdots = f(n+1) + f(n+2) + \cdots > \int_{n+1}^{\infty} f(x)\,dx. \tag{3}$$

Inequality (3) gives a *lower* estimate of the error.

The staircase in Fig. 4, which lies below the curve, gives an *upper* estimate of the error. Inspection of Fig. 4 shows that

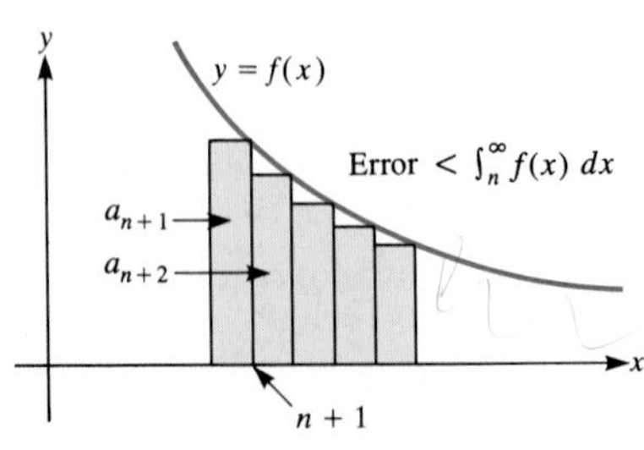

Figure 4

$$R_n = a_{n+1} + a_{n+2} + \cdots = f(n+1) + f(n+2) + \cdots < \int_{n}^{\infty} f(x)\,dx.$$

Putting these observations together yields the following estimate of the error.

Estimating the error

Theorem 3 *A bound on the error.* Let $f(x)$ be a continuous decreasing positive function such that $\int_1^{\infty} f(x)\,dx$ is convergent. Then the error R_n in using $f(1) + f(2) + \cdots + f(n)$ to estimate $\Sigma_{i=1}^{\infty} f(i)$ satisfies the inequality

$$\int_{n+1}^{\infty} f(x)\,dx < R_n < \int_{n}^{\infty} f(x)\,dx. \tag{4}$$

EXAMPLE 2 The first five terms of the series $1/1^2 + 1/2^2 + \cdots + 1/n^2 + \cdots$ are used to estimate the sum of the series.
(*a*) Estimate the error in using just the first five terms.
(*b*) Estimate $\Sigma_{n=1}^{\infty} 1/n^2$.

SOLUTION
(*a*) By inequality (4), the error R_5 satisfies the inequality

$$\int_6^\infty \frac{dx}{x^2} < R_5 < \int_5^\infty \frac{dx}{x^2}.$$

Now, $$\int_5^\infty \frac{dx}{x^2} = -\frac{1}{x}\bigg|_5^\infty = 0 - \left(-\frac{1}{5}\right) = \frac{1}{5}.$$

Similarly, $$\int_6^\infty \frac{dx}{x^2} = \frac{1}{6}.$$

Thus $$\tfrac{1}{6} < R_5 < \tfrac{1}{5}.$$

(*b*) The sum of the first five terms of the series is

$$\frac{1}{1^2} + \frac{1}{2^2} + \frac{1}{3^2} + \frac{1}{4^2} + \frac{1}{5^2} \approx 1.4636.$$

"Round down" lower bounds and "round up" upper bounds.

Since the sum of the remaining terms (the "tail end") is between $\frac{1}{6}$ and $\frac{1}{5}$, the sum of the series is between $1.463 + \frac{1}{6}$ and $1.464 + \frac{1}{5}$, hence between 1.629 and 1.664. (In the 18th century Euler proved that the sum is $\pi^2/6$.) ■

Estimating a Partial Sum S_n

We still restrict our attention to series that satisfy the hypotheses of the integral test in Theorem 1. That is, there is a positive, decreasing function $f(x)$ such that $f(n) = a_n$.

Just as we can use an (improper) integral to estimate the sum of a tail end of such a series, we can also use a (definite) integral to estimate a partial sum $S_n = a_1 + a_2 + \cdots + a_n$.

In the course of proving Theorem 1, we obtained equations (1) and (2). Taken together, they give us the inequality

$$\boxed{\int_1^{n+1} f(x)\,dx < a_1 + a_2 + \cdots + a_n < a_1 + \int_1^n f(x)\,dx.} \tag{5}$$

Since we may be able to evaluate $\int_1^{n+1} f(x)\,dx$ and $\int_1^n f(x)\,dx$ by the fundamental theorem, we may be able to use (5) to put an upper and lower bound on $S_n = \Sigma_{k=1}^n a_k$.

EXAMPLE 3 Use (5) to estimate the sum of the first million terms of the harmonic series.

SOLUTION By (5)

$$\int_1^{1,000,001} \frac{dx}{x} < \sum_{n=1}^{1,000,000} \frac{1}{n} < 1 + \int_1^{1,000,000} \frac{dx}{x},$$

hence

$$\ln(1{,}000{,}001) < \sum_{n=1}^{1{,}000{,}000} \frac{1}{n} < 1 + \ln(1{,}000{,}000).$$

Evaluating the logarithm with a calculator, we conclude that

$$13.81 < \sum_{n=1}^{1{,}000{,}000} \frac{1}{n} < 14.82. \quad ■$$

Section Summary

We developed a test for convergence or divergence for series whose terms a_n are of the form $f(n)$ for a continuous, positive, decreasing function $f(x)$. The series converges if $\int_1^\infty f(x)\,dx$ converges, and diverges if $\int_1^\infty f(x)\,dx$ diverges.

We also used integrals to estimate the error in using a partial sum S_n of such a series as an estimate of the series' sum. Furthermore, we saw how to use integrals to estimate a partial sum. (Rather than try to memorize the formulas, just draw the appropriate staircase diagrams.)

We assumed that $f(x)$ is decreasing for $x \geq 1$. Actually, Theorem 1 holds if we assume that $f(x)$ is decreasing from some point on, that is, there is some number a such that $f(x)$ is decreasing for $x \geq a$. (The argument for this involves similar "staircase" diagrams.)

EXERCISES FOR SEC. 10.4: THE INTEGRAL TEST

In Exercises 1 to 8 use the integral test to determine whether the series diverge or converge.

1 $\sum_{n=1}^{\infty} \frac{1}{n^{1.1}}$

2 $\sum_{n=1}^{\infty} \frac{1}{n^{0.9}}$

3 $\sum_{n=1}^{\infty} \frac{n}{n^2 + 1}$

4 $\sum_{n=1}^{\infty} \frac{1}{n^2 + 1}$

5 $\sum_{n=2}^{\infty} \frac{1}{n \ln n}$

6 $\sum_{n=1}^{\infty} \frac{1}{n + 1{,}000}$

7 $\sum_{n=1}^{\infty} \frac{\ln n}{n}$

8 $\sum_{n=1}^{\infty} \frac{n^3}{e^n}$

In Exercises 9 to 12 use Theorem 2 to determine whether the series diverge or converge.

9 $\sum_{n=1}^{\infty} \frac{1}{\sqrt[3]{n}}$

10 $\sum_{n=1}^{\infty} \frac{1}{n^3}$

11 $\sum_{n=1}^{\infty} \frac{1}{\sqrt{n}}$

12 $\sum_{n=1}^{\infty} \frac{1}{n^{0.999}}$

13 (*a*) Prove that if $p > 1$, the p-series converges.
(*b*) Give two numbers between which its sum lies.

14 Prove that if $p \leq 1$, the p-series diverges.

In each of Exercises 15 to 18 (*a*) compute the sum of the first four terms of the series to four decimal places, (*b*) give upper and lower bounds on the error R_4, (*c*) combine (*a*) and (*b*) to estimate the sum of the series.

15 $\sum_{n=1}^{\infty} \frac{1}{n^3}$

16 $\sum_{n=1}^{\infty} \frac{1}{n^4}$

17 $\sum_{n=1}^{\infty} \frac{1}{n^2 + 1}$

18 $\sum_{n=1}^{\infty} \frac{1}{n^2 + n}$

19 (*a*) If you used S_{100} to estimate $\sum_{n=1}^{\infty} 1/n^2$, what could you say about the error R_{100}?
(*b*) How large should you choose n to be sure that the error R_n is less than 0.0001?

20 (*a*) If you use S_{1000} to estimate $\sum_{n=1}^{\infty} 1/n^3$, what could you say about the error R_{1000}?
(*b*) How large should you choose n to be sure that the error R_n is less than 0.0001?

21 (*a*) How many terms of the series $\sum_{n=1}^{\infty} 1/n^4$ should you use to be sure that the remainder is less than 0.0001?
(*b*) Estimate $\sum_{n=1}^{\infty} 1/n^4$ to three decimal places.

22 Like Exercise 21 for the series $\sum_{n=1}^{\infty} 1/n^5$.

23 What does the integral test say about the geometric series $\Sigma_{n=1}^{\infty} k^n$, when $0 < k < 1$?

24 Let $f(x)$ be a positive continuous function that is decreasing for $x \geq a$. Let $a_n = f(n)$. Show in detail (with appropriate diagrams and exposition) why $\int_a^{\infty} f(x)\,dx$ and $\Sigma_{n=1}^{\infty} a_n$ both converge or both diverge. Use your own words. Don't just mimic the book's treatment of the case $a = 1$.

25 (See Exercise 24.) Show that $\Sigma_{n=1}^{\infty} n^3 e^{-n}$ converges.

26 Show that for $n \geq 2$,

$$2\sqrt{n+1} - 2 < \sum_{i=1}^{n} \frac{1}{\sqrt{i}} < 2\sqrt{n} - 1.$$

27 (*a*) By comparing the sum with integrals, show that

$$\ln \tfrac{201}{100} < \tfrac{1}{100} + \tfrac{1}{101} + \tfrac{1}{102} + \cdots + \tfrac{1}{200} < \ln \tfrac{200}{99}.$$

(*b*) Show that $\lim_{n\to\infty} \Sigma_{i=n}^{2n} 1/i = \ln 2$.

28 (*a*) Let $f(x)$ be a decreasing continuous positive function for $x \geq 1$ such that $\int_1^{\infty} f(x)\,dx$ is convergent. Show that

$$\int_1^{\infty} f(x)\,dx < \sum_{n=1}^{\infty} f(n) < f(1) + \int_1^{\infty} f(x)\,dx.$$

(*b*) Use (*a*) to estimate $\Sigma_{n=1}^{\infty} 1/n^2$.

29 In Example 1 we showed that the p-series for $p = 1$ diverges but the p-series for $p = 1.01$ converges. This contrast occurs even though the corresponding terms of the two series seem to resemble each other so closely. (For instance, $1/7^{1.01} \approx 0.140$ and $1/7^1 \approx 0.143$.) What happens to the ratio $(1/n^{1.01})/(1/n)$ as $n \to \infty$?

Exercises 30 and 31 concern products, rather than sums, of numbers.

30 Let $\{a_n\}$ be a sequence of positive numbers. Denote the product $(1 + a_1)(1 + a_2) \cdots (1 + a_n)$ by $\Pi_{i=1}^{n} (1 + a_i)$.

(*a*) Show that $\Sigma_{i=1}^{n} a_i \leq \Pi_{i=1}^{n} (1 + a_i)$.

(*b*) Show that if $\lim_{n\to\infty} \Pi_{i=1}^{n} (1 + a_i)$ exists, then $\Sigma_{n=1}^{\infty} a_n$ is convergent.

31 (See Exercise 30.)

(*a*) Show that $1 + a_i \leq e^{a_i}$. *Hint:* Show that $1 + x \leq e^x$ for $x \geq 0$.

(*b*) Show that if the series $\Sigma_{n=1}^{\infty} a_n$ is convergent, then $\lim_{n\to\infty} \Pi_{i=1}^{n} (1 + a_i)$ exists.

32 Here is an argument that there is an infinite number of primes. Assume that there is only a finite number of primes, $p_1, p_2, \ldots, p_m$.

(*a*) Show that

$$\frac{1}{1 - (1/p_i)} = 1 + \frac{1}{p_i} + \frac{1}{p_i^2} + \frac{1}{p_i^3} + \cdots.$$

(*b*) Show then that

$$\frac{1}{1 - 1/p_1}\,\frac{1}{1 - 1/p_2} \cdots \frac{1}{1 - 1/p_m} = \sum_{n=1}^{\infty} \frac{1}{n}.$$

(Assume the series can be multiplied term by term.)

(*c*) From (*b*) obtain a contradiction.

10.5 COMPARISON TESTS

So far in this chapter three tests for convergence (or divergence) of a series have been presented. The first concerned a special type of series, a geometric series. The second, the nth-term test for divergence, asserts that if the nth term of a series does *not* approach 0, the series diverges. The third, the integral test, applies to certain series of positive terms. In this section two further tests are developed: the comparison and limit-comparison tests. This section concerns only series with all terms positive.

Comparison Tests

The first test is similar to the comparison test for improper integrals in Sec. 8.8.

Theorem 1 *The comparison tests for convergence and divergence.*
(*a*) If $0 \leq p_n \leq c_n$ for each n and $\Sigma_{n=1}^{\infty} c_n$ converges, so does $\Sigma_{n=1}^{\infty} p_n$.
(*b*) If $0 \leq d_n \leq p_n$ for each n and $\Sigma_{n=1}^{\infty} d_n$ diverges, so does $\Sigma_{n=1}^{\infty} p_n$.

Proof

(*a*) Let the sum of the series $c_1 + c_2 + \cdots$ be C. Let S_n denote the partial sum $p_1 + p_2 + \cdots + p_n$. Then, for each n,

$$S_n = p_1 + p_2 + \cdots + p_n \leq c_1 + c_2 + \cdots + c_n \leq C.$$

Since the p_n's are nonnegative,

$$S_1 \leq S_2 \leq \cdots \leq S_n \leq \cdots.$$

$S_1 \leq S_2 \leq \cdots < C.$

Since each S_n is less than or equal to C, Theorem 2 of Sec. 10.2 assures us that the sequence

$$S_1, S_2, \ldots, S_n, \ldots$$

converges to a number L (less than or equal to C). In other words, the series $p_1 + p_2 + \cdots$ converges (and its sum is less than or equal to the sum $c_1 + c_2 + \cdots$).

(*b*) The divergence test follows immediately from the convergence test. If the series $p_1 + p_2 + \cdots$ converged, so would the series $d_1 + d_2 + \cdots$, which is assumed to diverge. ■

If the unshaded staircase has finite area, so does the shaded lower staircase.
If the shaded staircase has infinite area, so does the unshaded staircase.

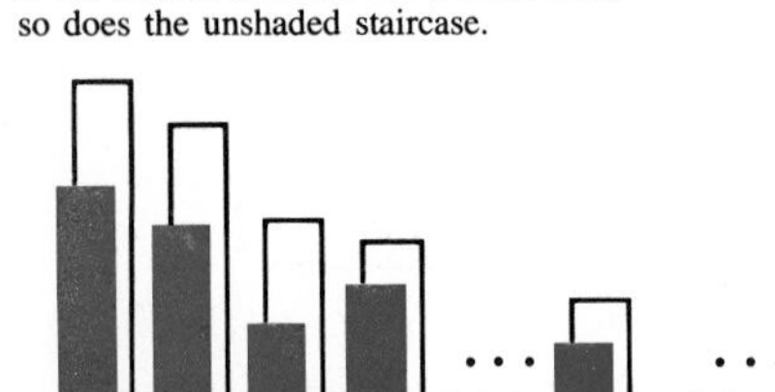

Figure 1

Figure 1 presents the two comparison tests in Theorem 1 in terms of endless staircases.

In order to apply the comparison test to a series of positive terms you have to compare it to a series whose convergence or divergence is already known. What series can we use for comparison? We know the p-series converges for $p > 1$ and diverges for $p \leq 1$. Also a geometric series $\Sigma_{n=1}^{\infty} r^n$ with positive terms converges for $r < 1$ but diverges for $r \geq 1$. Moreover, when we multiply one of these series by a nonzero constant, we don't affect its convergence or divergence.

EXAMPLE 1 Does the series

$$\sum_{n=1}^{\infty} \frac{n+1}{n+2}\frac{1}{n^2} = \frac{2}{3}\cdot\frac{1}{1^2} + \frac{3}{4}\cdot\frac{1}{2^2} + \frac{4}{5}\cdot\frac{1}{3^2} + \cdots$$

converge or diverge?

SOLUTION The coefficients $\frac{2}{3}, \frac{3}{4}, \frac{4}{5}, \ldots$ approach 1 as $n \to \infty$, so they are a minor influence. The series resembles the series

$$\frac{1}{1^2} + \frac{1}{2^2} + \cdots + \frac{1}{n^2} + \cdots,$$

which was shown by the integral test to be convergent. Since the fraction $(n+1)/(n+2)$ is less than 1,

$$\frac{n+1}{n+2}\frac{1}{n^2} < \frac{1}{n^2}.$$

Thus, by the comparison test for convergence, the series

$$\frac{2}{3}\frac{1}{1^2} + \frac{3}{4}\frac{1}{2^2} + \cdots + \frac{n+1}{n+2}\frac{1}{n^2} + \cdots$$

also converges. ■

But the comparison test does not tell what the series converges to.

EXAMPLE 2 Does the series

$$\frac{2}{3} \cdot \frac{1}{1} + \frac{3}{4} \cdot \frac{1}{2} + \cdots + \frac{n+1}{n+2} \cdot \frac{1}{n} + \cdots$$

converge or diverge?

SOLUTION Again the coefficient $(n + 1)/(n + 2)$ is a minor influence. We suspect that $1/n$ is the main influence and that the series diverges.

Unfortunately, the terms in this series are *less* than the terms of the harmonic series, $\Sigma_{n=1}^{\infty} 1/n$. So the divergence test does not directly apply. But the series

$$\frac{1}{2} \cdot \frac{1}{1} + \frac{1}{2} \cdot \frac{1}{2} + \frac{1}{2} \cdot \frac{1}{3} + \cdots + \frac{1}{2} \cdot \frac{1}{n} + \cdots$$

is also divergent, since it's just a multiple of a divergent series. Now the divergence test can be directly applied: the series

$$\sum_{n=1}^{\infty} \frac{n+1}{n+2} \cdot \frac{1}{n}$$

is, term by term, larger than the terms of the divergent series

$$\sum_{n=1}^{\infty} \frac{1}{2} \cdot \frac{1}{n}.$$

Hence it's divergent. ■

Limit-Comparison Tests

There is a variation of the comparison test that produces a much quicker solution of Example 2. It is the **limit-comparison test**.

Theorem 2 *The limit-comparison test.* Let $\Sigma_{n=1}^{\infty} p_n$ be a series of positive terms to be tested for convergence or divergence.

(*a*) Let $\Sigma_{n=1}^{\infty} c_n$ be a convergent series of positive terms. If $\lim_{n\to\infty} p_n/c_n$ exists, then $\Sigma_{n=1}^{\infty} p_n$ also converges.

(*b*) Let $\Sigma_{n=1}^{\infty} d_n$ be a divergent series of positive terms. If $\lim_{n\to\infty} p_n/d_n$ exists and is not 0 or if it is infinite, then $\Sigma_{n=1}^{\infty} p_n$ also diverges.

Proof We shall prove part (*a*). Let $\lim_{n\to\infty} p_n/c_n = a$. Since as $n \to \infty$, $p_n/c_n \to a$, there must be an integer N such that, for all $n \geq N$, p_n/c_n remains less than, say, $a + 1$. Thus

$$p_n < (a+1)c_n \qquad n \geq N.$$

Now the series

$$(a+1)c_N + (a+1)c_{N+1} + \cdots + (a+1)c_n + \cdots,$$

being $a + 1$ times the tail end of a convergent series, is itself convergent. By the comparison test,

$$p_N + p_{N+1} + \cdots + p_n + \cdots$$

is convergent. Hence $p_1 + p_2 + \cdots + p_n + \cdots$ is convergent.
Part (*b*) can be proved similarly. ■

Note that in Theorem 2(*b*) nothing is said about the case $\lim_{n\to\infty} p_n/d_n = 0$. In this circumstance the series $\sum_{n=1}^{\infty} p_n$ can either converge or diverge. For instance, take $\sum_{n=1}^{\infty} d_n$ to be the divergent series $\sum_{n=1}^{\infty} 1/\sqrt{n}$. The series $\sum_{n=1}^{\infty} 1/n^2$ is convergent and $\lim_{n\to\infty} (1/n^2)/(1/\sqrt{n}) = 0$. Contrarily, the harmonic series $\sum_{n=1}^{\infty} 1/n$ is divergent and again $\lim_{n\to\infty} (1/n)/(1/\sqrt{n}) = 0$.
The next example shows how convenient the limit-comparison test is. Contrast the solution there with that in Example 2.

EXAMPLE 3 Does the series

$$\sum_{n=1}^{\infty} \frac{n+1}{n+2} \cdot \frac{1}{n} = \frac{2}{3} \cdot \frac{1}{1} + \frac{3}{4} \cdot \frac{1}{2} + \frac{4}{5} \cdot \frac{1}{3} + \cdots$$

converge or diverge?

SOLUTION As with Example 2, we expect this series to behave like the harmonic series. For this reason we examine the ratio between corresponding terms:

$$\lim_{n\to\infty} \frac{\dfrac{n+1}{n+2} \cdot \dfrac{1}{n}}{\dfrac{1}{n}} = \lim_{n\to\infty} \frac{n+1}{n+2} = 1.$$

Since the limit is not 0, and the harmonic series diverges, the limit-comparison test tells us that $\sum_{n=1}^{\infty} \frac{n+1}{n+2} \cdot \frac{1}{n}$ diverges. ■

EXAMPLE 4 Does

$$\sum_{n=1}^{\infty} \frac{(1 + 1/n)^n[1 + (-1/2)^n]}{2^n}$$

converge or diverge?

SOLUTION Note that as $n \to \infty$, $(1 + 1/n)^n \to e$ and $1 + (-\frac{1}{2})^n \to 1$. The major influence is the 2^n in the denominator. So use the limit-comparison test, with the convergent geometric series $1 + \frac{1}{2} + \frac{1}{4} + \cdots + 1/2^n + \cdots$, which the given series resembles. Then

$$\lim_{n\to\infty} \frac{\dfrac{\left(1 + \dfrac{1}{n}\right)^n \left[1 + \left(-\dfrac{1}{2}\right)^n\right]}{2^n}}{\dfrac{1}{2^n}} = \lim_{n\to\infty} \left(1 + \frac{1}{n}\right)^n \left[1 + \left(-\frac{1}{2}\right)^n\right] = e \cdot 1 = e.$$

Since $\sum_{n=1}^{\infty} 2^{-n}$ is convergent, so is the given series. ■

EXAMPLE 5 Does $\Sigma_{n=1}^{\infty} n^3/2^n$ converge or diverge?

SOLUTION The typical term $n^3/2^n$ reminds us of $1/2^n$, so we suspect that the series $\Sigma_{n=1}^{\infty} n^3/2^n$ might converge. We try the limit-comparison test, obtaining

$$\lim_{n\to\infty} \frac{\frac{n^3}{2^n}}{\frac{1}{2^n}} = \lim_{n\to\infty} n^3 = \infty.$$

Since the limit is not finite, the test gives no information. So we start all over and look at $n^3/2^n$ again.

The numerator n^3 approaches ∞ much more slowly than 2^n, so we still suspect that $\Sigma_{n=1}^{\infty} n^3/2^n$ converges. Now, n^3 approaches ∞ *more slowly than any exponential* b^n, $b > 1$. For example, for large n, n^3 is less than $(1.5)^n$. This means that for large n

$$\frac{n^3}{2^n} < \frac{(1.5)^n}{2^n} = (0.75)^n.$$

The geometric series $\Sigma_{n=1}^{\infty} (0.75)^n$ converges. Since $n^3/2^n < 0.75^n$ for all but a finite number of values of n, the comparison test tells us that $\Sigma_{n=1}^{\infty} n^3/2^n$ converges. ■

Section Summary

We developed two tests for convergence or divergence of a series with positive terms, $\Sigma_{n=1}^{\infty} p_n$. If, for each n, p_n is less than the corresponding term of a convergent series, then $\Sigma_{n=1}^{\infty} p_n$ converges. If p_n is larger than the corresponding term of a divergent series of positive terms, then $\Sigma_{n=1}^{\infty} p_n$ diverges. This comparison test is the basis of the limit-comparison test, which is often much easier to apply. This test depends only on the limit of the ratio of p_n to the corresponding term of a series of positive terms known to converge or diverge.

EXERCISES FOR SEC. 10.5: COMPARISON TESTS

In Exercises 1 to 4 use the comparison test to determine whether the series converge or diverge.

1 $\displaystyle\sum_{n=1}^{\infty} \frac{1}{n^2+3}$

2 $\displaystyle\sum_{n=1}^{\infty} \frac{n+2}{(n+1)\sqrt{n}}$

3 $\displaystyle\sum_{n=1}^{\infty} \frac{\sin^2 n}{n^2}$

4 $\displaystyle\sum_{n=1}^{\infty} \frac{1}{n2^n}$

In Exercises 5 to 8 use the limit-comparison test to determine whether the series converge or diverge.

5 $\displaystyle\sum_{n=1}^{\infty} \frac{5n+1}{(n+2)n^2}$

6 $\displaystyle\sum_{n=1}^{\infty} \frac{2^n+n}{3^n}$

7 $\displaystyle\sum_{n=1}^{\infty} \frac{n+1}{(5n+2)\sqrt{n}}$

8 $\displaystyle\sum_{n=1}^{\infty} \frac{(1+1/n)^n}{n^2}$

In Exercises 9 to 28 use any test discussed so far in this chapter to determine whether the series converges or diverges.

9 $\displaystyle\sum_{n=1}^{\infty} \frac{n^2}{3^n}$

10 $\displaystyle\sum_{n=1}^{\infty} \frac{2^n}{n^2}$

11 $\displaystyle\sum_{n=1}^{\infty} \frac{1}{n^n}$

12 $\displaystyle\sum_{n=1}^{\infty} \frac{1}{n!}$

13 $\displaystyle\sum_{n=1}^{\infty} \frac{4n+1}{(2n+3)n^2}$

14 $\displaystyle\sum_{n=1}^{\infty} \frac{n^2(2^n+1)}{3^n+1}$

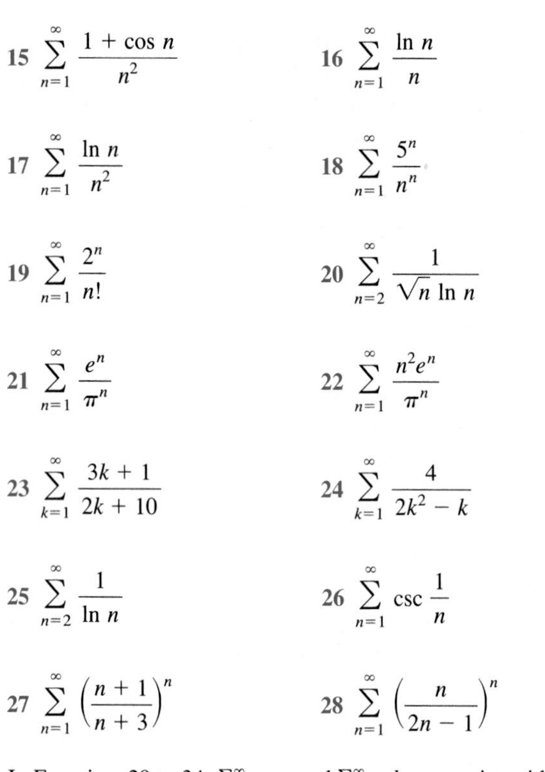

15 $\sum_{n=1}^{\infty} \frac{1 + \cos n}{n^2}$

16 $\sum_{n=1}^{\infty} \frac{\ln n}{n}$

17 $\sum_{n=1}^{\infty} \frac{\ln n}{n^2}$

18 $\sum_{n=1}^{\infty} \frac{5^n}{n^n}$

19 $\sum_{n=1}^{\infty} \frac{2^n}{n!}$

20 $\sum_{n=2}^{\infty} \frac{1}{\sqrt{n} \ln n}$

21 $\sum_{n=1}^{\infty} \frac{e^n}{\pi^n}$

22 $\sum_{n=1}^{\infty} \frac{n^2 e^n}{\pi^n}$

23 $\sum_{k=1}^{\infty} \frac{3k + 1}{2k + 10}$

24 $\sum_{k=1}^{\infty} \frac{4}{2k^2 - k}$

25 $\sum_{n=2}^{\infty} \frac{1}{\ln n}$

26 $\sum_{n=1}^{\infty} \csc \frac{1}{n}$

27 $\sum_{n=1}^{\infty} \left(\frac{n + 1}{n + 3}\right)^n$

28 $\sum_{n=1}^{\infty} \left(\frac{n}{2n - 1}\right)^n$

In Exercises 29 to 34, $\Sigma_{n=1}^{\infty} a_n$ and $\Sigma_{n=1}^{\infty} b_n$ are series with positive terms. What, if anything, can we conclude about the convergence or divergence of $\Sigma_{n=1}^{\infty} a_n$ if:

29 $\Sigma_{n=1}^{\infty} b_n$ is divergent and $\lim_{n\to\infty} a_n/b_n = 0$?

30 $\Sigma_{n=1}^{\infty} b_n$ is convergent and $\lim_{n\to\infty} a_n/b_n = \infty$?

31 $\Sigma_{n=1}^{\infty} b_n$ is convergent and $3b_n \le a_n \le 5b_n$?

32 $\Sigma_{n=1}^{\infty} b_n$ is divergent and $3b_n \le a_n \le 5b_n$?

33 $\Sigma_{n=1}^{\infty} b_n$ is convergent and $a_n < b_n^2$?

34 $\Sigma_{n=1}^{\infty} b_n$ is divergent, $b_n \to 0$ as $n \to \infty$, and $a_n < b_n^2$?

---

35 For which values of the positive number x does the series $\Sigma_{n=1}^{\infty} x^n/(n2^n)$ converge? diverge?

36 For which values of the positive exponent k does the series $\Sigma_{n=2}^{\infty} 1/(n^k \ln n)$ converge? diverge?

37 Prove part (*b*) of Theorem 2.

38 For which constants p does $\Sigma_{n=1}^{\infty} n^p e^{-n}$ converge?

39 (*a*) Show that $\Sigma_{n=1}^{\infty} 1/(1 + 2^n)$ converges.
(*b*) Show that the sum of the series in (*a*) is between 0.64 and 0.77. (Use the first three terms and control the sum of the rest of the series by comparing it to the sum of a geometric series.)

40 (*a*) Show that $\Sigma_{k=n+1}^{\infty} 1/k!$ is less than the sum of the geometric series whose first term is $1/(n + 1)!$ and whose ratio is $1/(n + 2)$.
(*b*) Use (*a*) with $n = 4$ to show that

$$1 + 1 + \frac{1}{2!} + \frac{1}{3!} + \frac{1}{4!} < \sum_{k=0}^{\infty} \frac{1}{k!} < 1 + 1 + \frac{1}{2!} + \frac{1}{3!} + \frac{1}{4!} + \frac{1}{5!} \cdot \frac{1}{1 - \frac{1}{6}}.$$

(*c*) From (*b*) deduce that

$$2.71 < \sum_{k=0}^{\infty} \frac{1}{k!} < 2.72.$$

(*d*) Find a value of n such that $\Sigma_{k=n+1}^{\infty} 1/k! < 0.0005$.
(*e*) Use (*d*) to estimate $\Sigma_{k=0}^{\infty} 1/k!$ to three decimal places.

41 Prove the following result, which is used in the statistical theory of stochastic processes: Let $\{a_n\}$ and $\{c_n\}$ be two sequences of nonnegative numbers such that $\Sigma_{n=1}^{\infty} a_n c_n$ converges and $\lim_{n\to\infty} c_n = 0$. Then $\Sigma_{n=1}^{\infty} a_n c_n^2$ converges.

42 Find a specific number B, expressed as a decimal, such that

$$\sum_{n=1}^{\infty} \frac{\ln n}{n^2} < B.$$

43 Find a specific number B, expressed as a decimal, such that

$$\sum_{n=1}^{\infty} \frac{n + 2}{n + 1} \cdot \frac{1}{n^3} < B.$$

44 Estimate $\Sigma_{n=1}^{\infty} 1/(n2^n)$ to three decimal places.

10.6 RATIO TESTS

The next test is suggested by the test for the convergence of a geometric series. In a geometric series the ratio between consecutive terms is constant. The "ratio test" concerns series where this ratio is "almost constant."

The Ratio Test

Theorem 1 *The ratio test.* Let $p_1 + p_2 + \cdots + p_n + \cdots$ be a series of positive terms.
(*a*) If $\lim_{n\to\infty} p_{n+1}/p_n$ exists and is less than 1, the series converges.
(*b*) If $\lim_{n\to\infty} p_{n+1}/p_n$ exists and is greater than 1 or is infinite, the series diverges.

Proof
(*a*) Let

$$\lim_{n\to\infty} \frac{p_{n+1}}{p_n} = s < 1.$$

Select a number r such that $s < r < 1$. Then there is an integer N such that for all $n \geq N$,

$$\frac{p_{n+1}}{p_n} < r,$$

and, therefore,

$$p_{n+1} < rp_n.$$

Thus

$$p_{N+1} < rp_N$$

$$p_{N+2} < rp_{N+1} < r(rp_N) = r^2 p_N$$

$$p_{N+3} < rp_{N+2} < r(r^2 p_N) = r^3 p_N,$$

and so on.

Thus the terms of the series

$$p_N + p_{N+1} + p_{N+2} + \cdots$$

are less than the corresponding terms of the geometric series

$$p_N + rp_N + r^2 p_N + \cdots$$

(except for the first term p_N, which equals the first term of the geometric series). Since $r < 1$, the latter series converges. By the comparison test, $p_N + p_{N+1} + p_{N+2} + \cdots$ converges. Adding in the front end,

$$p_1 + p_2 + \cdots + p_{N-1},$$

still results in a convergent series.

(*b*) If $\lim_{n\to\infty} p_{n+1}/p_n$ is greater than 1 or is infinite, then for all n from some point on, p_{n+1} is larger than p_n. Thus the nth term of the series $p_1 + p_2 + \cdots$ cannot approach 0. By the nth-term test for divergence the series diverges. ■

No information if ratio approaches 1

No mention is made in Theorem 1 of the case $\lim_{n\to\infty} p_{n+1}/p_n = 1$. The reason for this omission is that anything can happen; the series may diverge or it may converge. (Exercise 28 illustrates these possibilities.) Also, $\lim_{n\to\infty} p_{n+1}/p_n$ may not exist. In these cases, one must look to other tests to determine whether the series diverges or converges.

The ratio test is a natural one to try if the nth term of a series involves powers of a fixed number, as the next example shows.

EXAMPLE 1 Show that the series $p + 2p^2 + 3p^3 + \cdots + np^n + \cdots$ converges for any fixed number p for which $0 < p < 1$.

SOLUTION Let a_n denote the nth term of the series. Then

$$a_n = np^n \qquad \text{and} \qquad a_{n+1} = (n+1)p^{n+1}.$$

The ratio between consecutive terms is

$$\frac{a_{n+1}}{a_n} = \frac{(n+1)p^{n+1}}{np^n} = \frac{n+1}{n}p.$$

Thus

$$\lim_{n\to\infty} \frac{a_{n+1}}{a_n} = p < 1,$$

and the series converges. ■

To see what its sum is, look at Exercise 33 in Sec. 10.3.

EXAMPLE 2 Find for which positive values of x the series

$$\frac{1}{0!} + \frac{x}{1!} + \frac{x^2}{2!} + \frac{x^3}{3!} + \cdots + \frac{x^n}{n!} + \cdots$$

converges and for which it diverges. (Each choice of x determines a specific series with constant terms.)

SOLUTION If we start the series at $n = 0$, then the nth term, a_n, is $x^n/n!$ Thus

$$a_{n+1} = \frac{x^{n+1}}{(n+1)!},$$

and therefore

$$\frac{a_{n+1}}{a_n} = \frac{\dfrac{x^{n+1}}{(n+1)!}}{\dfrac{x^n}{n!}} = x\frac{n!}{(n+1)!} = \frac{x}{n+1}.$$

Since x is fixed,

$$\lim_{n\to\infty} \frac{x}{n+1} = 0.$$

By the ratio test, the series converges for all positive x. ■

In the next section, it will be shown to converge for negative x too.

The next example uses the ratio test to establish divergence.

EXAMPLE 3 Show that the series $2/1 + 2^2/2 + \cdots + 2^n/n + \cdots$ diverges.

SOLUTION In this case, $a_n = 2^n/n$ and

$$\frac{a_{n+1}}{a_n} = \frac{\dfrac{2^{n+1}}{n+1}}{\dfrac{2^n}{n}}$$

$$= \frac{2^{n+1}}{n+1}\frac{n}{2^n} = 2\frac{n}{n+1}.$$

Thus

$$\lim_{n\to\infty} \frac{a_{n+1}}{a_n} = 2,$$

which is larger than 1. By the ratio test, the series diverges. ■

So the series is like a geometric series with ratio 2.

It is not really necessary to call on the powerful ratio test to establish the divergence of the series in Example 3. Since $\lim_{n\to\infty} 2^n/n = \infty$, its nth term gets arbitrarily large; by the nth-term test, the series diverges. (Comparison with the harmonic series also demonstrates divergence.)

The Root Test

The next test, closely related to the ratio test, is of use when the nth term contains only nth powers, such as n^n or 3^n. It is not useful if factorials such as $n!$ are present.

Theorem 2 *The root test.* Let $\Sigma_{n=1}^{\infty} p_n$ be a series of positive terms. Then
(*a*) if $\lim_{n\to\infty} \sqrt[n]{p_n}$ exists and is less than 1, $\Sigma_{n=1}^{\infty} p_n$ converges;
(*b*) if $\lim_{n\to\infty} \sqrt[n]{p_n}$ exists and is greater than 1 or is infinite, $\Sigma_{n=1}^{\infty} p_n$ diverges.
(*c*) if $\lim_{n\to\infty} \sqrt[n]{p_n} = 1$, no conclusion can be drawn ($\Sigma_{n=1}^{\infty} p_n$ may converge or may diverge).

The proof of the root test is outlined in Exercises 26 and 29.

EXAMPLE 4 Use the root test to determine whether $\Sigma_{n=1}^{\infty} 3^n/n^n$ converges or diverges.

SOLUTION We have

$$\lim_{n\to\infty} \sqrt[n]{\frac{3^n}{n^n}} = \lim_{n\to\infty} \frac{3}{n} = 0.$$

By the root test, the series converges. ■

Section Summary

We developed two tests for convergence or divergence of a series $\Sigma_{n=1}^{\infty} p_n$ with positive terms, both motivated by geometric series. In the ratio test, we examine $\lim_{n\to\infty} p_{n+1}/p_n$, and in the root test, $\lim_{n\to\infty} \sqrt[n]{p_n}$. The ratio test is convenient to use when the terms involve powers and factorials. The root test is convenient when only powers appear.

EXERCISES FOR SEC. 10.6: RATIO TESTS

In Exercises 1 to 6 try the ratio test to decide whether the series converges or diverges. If the ratio test gives no information, use another test to decide.

1 $\sum_{n=1}^{\infty} \frac{n^2}{3^n}$

2 $\sum_{n=1}^{\infty} \frac{(n+1)^2}{n2^n}$

3 $\sum_{n=1}^{\infty} \frac{n \ln n}{3^n}$

4 $\sum_{n=1}^{\infty} \frac{n!}{3^n}$

5 $\sum_{n=1}^{\infty} \frac{(2n+1)(2^n+1)}{3^n+1}$

6 $\sum_{n=1}^{\infty} \frac{n!}{n^n}$

7 Determine whether the series $\Sigma_{n=1}^{\infty} 2^n/(n+1)^n$ converges (*a*) by the ratio test, (*b*) by the root test, (*c*) by the comparison test.

8 Determine whether the series $\Sigma_{n=1}^{\infty} 2^n/n^5$ converges (*a*) by the ratio test, (*b*) by the *n*th-term test.

9 For which positive numbers x does the series $\Sigma_{n=1}^{\infty} nx^n$ (*a*) converge? (*b*) diverge?

10 For which positive numbers x does the series $\Sigma_{n=1}^{\infty} x^n/n$ (*a*) converge? (*b*) diverge?

11 For which positive numbers x does the series $\Sigma_{n=1}^{\infty} x^n/2^n$ (*a*) converge? (*b*) diverge?

12 For which positive numbers x does the series $\Sigma_{n=1}^{\infty} 2^n x^n/n!$ (*a*) converge? (*b*) diverge?

In Exercises 13 and 14 use the root test to determine whether the series converge or diverge.

13 $\displaystyle\sum_{n=1}^{\infty} \frac{n^n}{3^{n^2}}$

14 $\displaystyle\sum_{n=1}^{\infty} \frac{(1+1/n)^n(2n+1)^n}{(3n+1)^n}$

In Exercises 15 to 20 find by any legal means a number B in decimal form that is larger than the sum of the series, each of which is convergent.

15 $\displaystyle\sum_{n=1}^{\infty} \frac{n^2}{2^n}$

16 $\displaystyle\sum_{n=1}^{\infty} \frac{\ln n}{n^2}$

17 $\displaystyle\sum_{n=1}^{\infty} \frac{n^3}{n!}$

18 $\displaystyle\sum_{n=1}^{\infty} \frac{n+2}{2^n}$

19 $\displaystyle\sum_{n=2}^{\infty} \frac{1}{n^2-1}$

20 $\displaystyle\sum_{n=1}^{\infty} \frac{n}{n^3+1}$

In Exercises 21 to 24 find by any legal means a number m such that the mth partial sum of the series exceeds 1,000. Each series is divergent.

21 $\displaystyle\sum_{n=1}^{\infty} \frac{\ln n}{n}$

22 $\displaystyle\sum_{n=1}^{\infty} \frac{n}{n^2+1}$

23 $\displaystyle\sum_{n=1}^{\infty} (1.01)^n$

24 $\displaystyle\sum_{n=1}^{\infty} \frac{(n+2)^2}{n+1} \cdot \frac{1}{\sqrt{n}}$

25 Use the result of Example 2 to show that, for $x > 0$, $\lim_{n\to\infty} x^n/n! = 0$. (This fact was established directly in Sec. 10.2.)

26 This exercise shows that the root test gives no information if $\lim_{n\to\infty} \sqrt[n]{p_n} = 1$.
(*a*) Show that for $p_n = 1/n$, $\Sigma_{n=1}^{\infty} p_n$ diverges and $\lim_{n\to\infty} \sqrt[n]{p_n} = 1$.
(*b*) Show that for $p_n = 1/n^2$, $\Sigma_{n=1}^{\infty} p_n$ converges and $\lim_{n\to\infty} \sqrt[n]{p_n} = 1$.

27 Solve Example 3 using the root test.

28 This exercise shows that the ratio test is useless when $\lim_{n\to\infty} p_{n+1}/p_n = 1$.
(*a*) Show that if $p_n = 1/n$, then $\Sigma_{n=1}^{\infty} p_n$ diverges and $\lim_{n\to\infty} p_{n+1}/p_n = 1$.
(*b*) Show that if $p_n = 1/n^2$, then $\Sigma_{n=1}^{\infty} p_n$ converges and $\lim_{n\to\infty} p_{n+1}/p_n = 1$.

29 (Proof of the root test, Theorem 2)
(*a*) Assume that $\lim_{n\to\infty} \sqrt[n]{p_n} = L < 1$. Pick any r, $L < r < 1$, and then N such that $\sqrt[n]{p_n} < r$ for $n > N$. Show that $p_n < r^n$ for $n > N$ and compare a tail end of $\Sigma_{n=1}^{\infty} p_n$ to a geometric series.
(*b*) Assume that $\displaystyle\lim_{n\to\infty} \sqrt[n]{p_n} = L > 1$. Pick any number r, $1 < r < L$, and then N such that $\sqrt[n]{p_n} > r$ for $n > N$. Show that $p_n > r^n$ for $n > N$. From this conclude that $\Sigma_{n=1}^{\infty} p_n$ diverges.

10.7 TESTS FOR SERIES WITH BOTH POSITIVE AND NEGATIVE TERMS

The tests for convergence or divergence in Secs. 10.4 to 10.6 concern series whose terms are positive. This section examines series which may have both positive and negative terms. Two tests for the convergence of such a series are presented. The alternating-series test applies to series whose terms alternate in sign, +, −, +, −, . . . and decrease in absolute value. In the absolute-convergence test the signs may vary in any way.

Alternating Series

Definition *Alternating series.* If $p_1, p_2, \ldots, p_n, \ldots$ is a sequence of positive numbers, then the series

$$\sum_{n=1}^{\infty} (-1)^{n+1} p_n = p_1 - p_2 + p_3 - p_4 + \cdots + (-1)^{n+1} p_n + \cdots$$

and the series

$$\sum_{n=1}^{\infty} (-1)^n p_n = -p_1 + p_2 - p_3 + p_4 - \cdots + (-1)^n p_n + \cdots$$

are called **alternating series.**

For instance,

$$1 - \frac{1}{3} + \frac{1}{5} - \frac{1}{7} + \cdots + (-1)^{n+1} \frac{1}{2n-1} + \cdots$$

and

$$-1 + 1 - 1 + 1 - \cdots + (-1)^n + \cdots$$

are alternating series.

By the nth-term test, the second series diverges. The following theorem implies that the first series converges.

Theorem 1 *The alternating-series test.* If $p_1, p_2, \ldots, p_n, \ldots$ is a decreasing sequence of positive numbers such that $\lim_{n \to \infty} p_n = 0$, then the series whose nth term is $(-1)^{n+1} p_n$,

$$\sum_{n=1}^{\infty} (-1)^{n+1} p_n = p_1 - p_2 + p_3 - \cdots + (-1)^{n+1} p_n + \cdots,$$

converges.

Proof The idea of the proof is easily conveyed by a specific case. For the sake of concreteness and simplicity, consider the series in which $p_n = 1/n$, that is, the "alternating harmonic series"

$$1 - \frac{1}{2} + \frac{1}{3} - \frac{1}{4} + \cdots + (-1)^{n+1} \frac{1}{n} + \cdots.$$

Consider first the partial sums of an *even* number of terms, $S_2, S_4, S_6, \ldots$. For clarity, group the summands in pairs:

$$S_2 = (1 - \tfrac{1}{2})$$

$$S_4 = (1 - \tfrac{1}{2}) + (\tfrac{1}{3} - \tfrac{1}{4}) = S_2 + (\tfrac{1}{3} - \tfrac{1}{4})$$

$$S_6 = (1 - \tfrac{1}{2}) + (\tfrac{1}{3} - \tfrac{1}{4}) + (\tfrac{1}{5} - \tfrac{1}{6}) = S_4 + (\tfrac{1}{5} - \tfrac{1}{6})$$

. . . .

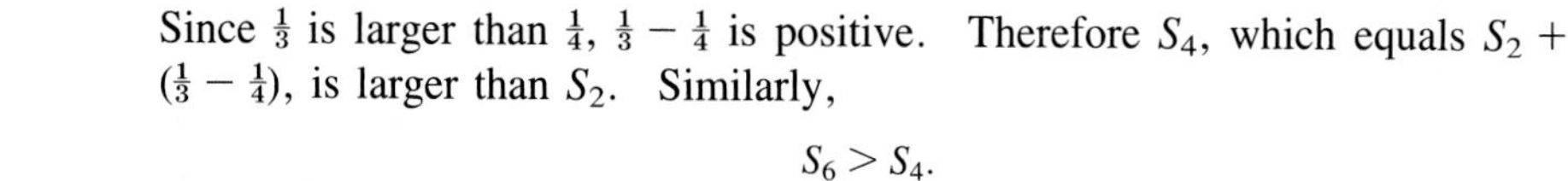

Since $\frac{1}{3}$ is larger than $\frac{1}{4}$, $\frac{1}{3} - \frac{1}{4}$ is positive. Therefore S_4, which equals $S_2 + (\frac{1}{3} - \frac{1}{4})$, is larger than S_2. Similarly,

$$S_6 > S_4.$$

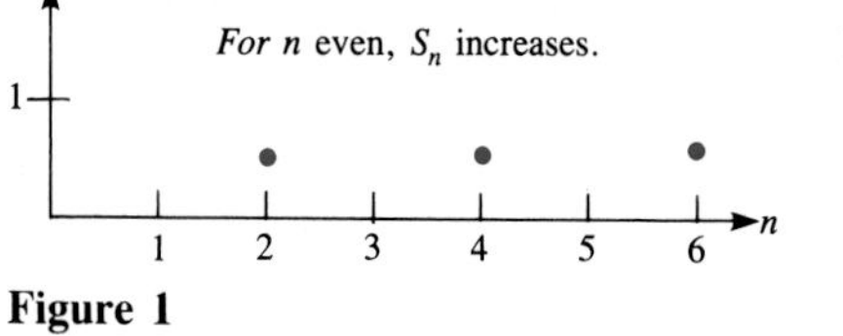

Figure 1

More generally, then, $S_2 < S_4 < S_6 < S_8 < \cdots$.

The sequence $\{S_{2n}\}$ is increasing. (See Fig. 1.)

Next, it will be shown that S_{2n} is less than 1, the first term of the given sequence. First of all,

$$S_2 = 1 - \tfrac{1}{2} < 1.$$

Next, consider S_4:

$$\begin{aligned} S_4 &= 1 - \tfrac{1}{2} + \tfrac{1}{3} - \tfrac{1}{4} \\ &= 1 - (\tfrac{1}{2} - \tfrac{1}{3}) - \tfrac{1}{4} \\ &< 1 - (\tfrac{1}{2} - \tfrac{1}{3}). \end{aligned}$$

Since $\frac{1}{2} - \frac{1}{3}$ is positive, this shows that

$$S_4 < 1.$$

Similarly,

$$\begin{aligned} S_6 &= 1 - (\tfrac{1}{2} - \tfrac{1}{3}) - (\tfrac{1}{4} - \tfrac{1}{5}) - \tfrac{1}{6} \\ &< 1 - (\tfrac{1}{2} - \tfrac{1}{3}) - (\tfrac{1}{4} - \tfrac{1}{5}) \\ &< 1. \end{aligned}$$

In general then,

$$S_{2n} < 1$$

for all n.

The sequence

$$S_2, S_4, S_6, \ldots$$

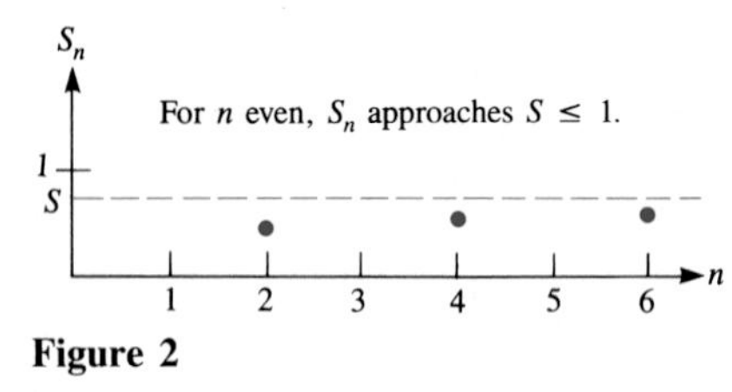

Figure 2

is therefore increasing and yet bounded by the number 1, as indicated in Fig. 2. By Theorem 2 of Sec. 10.2, $\lim_{n\to\infty} S_{2n}$ exists. Call this limit S, which is less than or equal to 1. (See Fig. 2.)

All that remains to be shown is that the numbers

$$S_1, S_3, S_5, \ldots$$

also converge to S.

Note that

$$S_3 = 1 - \tfrac{1}{2} + \tfrac{1}{3} = S_2 + \tfrac{1}{3}$$

$$S_5 = 1 - \tfrac{1}{2} + \tfrac{1}{3} - \tfrac{1}{4} + \tfrac{1}{5} = S_4 + \tfrac{1}{5}.$$

The term $1/(2n+1)$ will be p_{2n+1} in the general case.

In general,

$$S_{2n+1} = S_{2n} + \frac{1}{2n+1}.$$

Thus

$$\begin{aligned} \lim_{n\to\infty} S_{2n+1} &= \lim_{n\to\infty} \left(S_{2n} + \frac{1}{2n+1}\right) \\ &= \lim_{n\to\infty} S_{2n} + \lim_{n\to\infty} \frac{1}{2n+1} \\ &= S + 0 \\ &= S. \end{aligned}$$

Since the partial sums

$$S_2, S_4, S_6, \ldots$$

and the partial sums

$$S_1, S_3, S_5, \ldots$$

both have the same limit S, it follows that

$$\lim_{n\to\infty} S_n = S.$$

Thus the series

$$1 - \tfrac{1}{2} + \tfrac{1}{3} - \tfrac{1}{4} + \tfrac{1}{5} - \cdots$$

converges.

A similar argument applies to any alternating series whose nth term approaches 0 and whose terms decrease in absolute value. ■

Decreasing alternating series

An alternating series whose terms decrease in absolute value as n increases will be called a **decreasing alternating series**. Theorem 1 shows that a decreasing alternating series whose nth term approaches 0 as $n \to \infty$ converges.

EXAMPLE 1 Estimate the sum S of the series $1 - \tfrac{1}{2} + \tfrac{1}{3} - \tfrac{1}{4} + \cdots$.

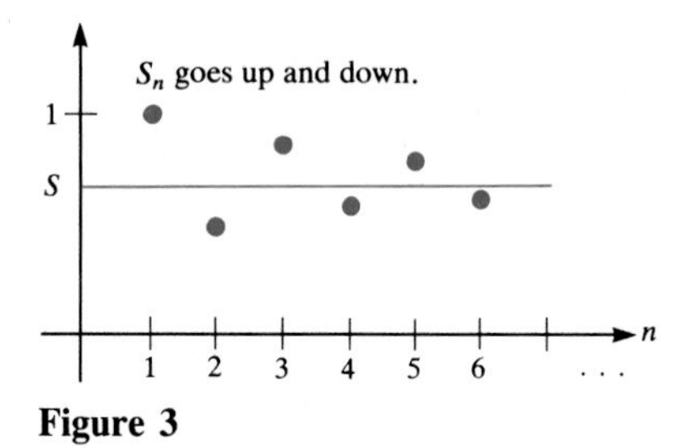

Figure 3

SOLUTION These are the first five partial sums:

$$S_1 = 1 = 1.00$$

$$S_2 = 1 - \tfrac{1}{2} = 0.500$$

$$S_3 = 1 - \tfrac{1}{2} + \tfrac{1}{3} \approx 0.500 + 0.3333 = 0.8333$$

$$S_4 = S_3 - \tfrac{1}{4} \approx 0.8333 - 0.250 = 0.5833$$

$$S_5 = S_4 + \tfrac{1}{5} \approx 0.5833 + 0.200 = 0.7833.$$

Figure 3 is a graph of S_n as a function of n. The sums $S_1, S_3, \ldots$ approach S from above. The sums $S_2, S_4, \ldots$ approach S from below. For instance,

$$S_4 < S < S_5$$

gives the information that $0.583 < S < 0.784$. (See Fig. 4.) ■

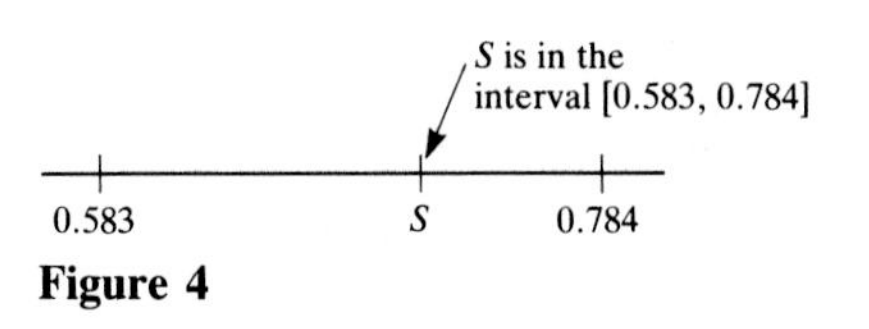

Figure 4

As Fig. 3 suggests, any partial sum of a series satisfying the hypothesis of the alternating-series test differs from the sum of the series by less than the absolute value of the first omitted term. That is, if S_n is the sum of the first n terms of the series and S is the sum of the series, then the error

$$R_n = S - S_n$$

The error in estimating the sum of a decreasing alternating series

has absolute value at most p_{n+1}, which is the absolute value of the first omitted term. Moreover, S is between S_n and S_{n+1} for every n.

EXAMPLE 2 Does the series

$$\frac{3}{1!} - \frac{3^2}{2!} + \frac{3^3}{3!} - \frac{3^4}{4!} + \frac{3^5}{5!} - \cdots + (-1)^{n+1}\frac{3^n}{n!} + \cdots$$

converge or diverge?

SOLUTION This is an alternating series. By Example 3 of Sec. 10.2, its nth term approaches 0. Let us see whether the absolute values of the terms decrease in size, term by term. The first few absolute values are

At first the terms increase.

$$\frac{3}{1!} = 3$$

$$\frac{3^2}{2!} = \frac{9}{2} = 4.5$$

$$\frac{3^3}{3!} = \frac{27}{6} = 4.5$$

$$\frac{3^4}{4!} = \frac{81}{24} = 3.375.$$

At first they increase. However, the fourth term is less than the third. Let us show that the rest of the terms decrease in size. For instance,

But then they decrease.

$$\frac{3^5}{5!} = \frac{3}{5}\frac{3^4}{4!} < \frac{3^4}{4!},$$

and, for $n \geq 3$,
$$\frac{3^{n+1}}{(n+1)!} = \frac{3}{n+1}\frac{3^n}{n!} < \frac{3^n}{n!}.$$

By the alternating-series test, the series that begins

$$\frac{3^3}{3!} - \frac{3^4}{4!} + \frac{3^5}{5!} - \frac{3^6}{6!} + \cdots$$

converges. Call its sum S. If the two terms

$$\frac{3}{1!} - \frac{3^2}{2!}$$

are added on, we obtain the original series, which therefore converges and has the sum

$$\frac{3}{1!} - \frac{3^2}{2!} + S. \quad ■$$

In the alternating-series test the absolute values of the terms must eventually be decreasing.

As Example 2 illustrates, the alternating-series test works as long as the nth term approaches 0 and the terms decrease in size from some point on in the series.

It may seem that any alternating series whose nth term approaches 0 converges. *This is not the case,* as is shown by this series:

$$\frac{2}{1} - \frac{1}{1} + \frac{2}{2} - \frac{1}{2} + \frac{2}{3} - \frac{1}{3} + \frac{2}{4} - \frac{1}{4} + \cdots, \tag{1}$$

whose terms alternate $2/n$ and $-1/n$.

Let S_n be the sum of the first n terms of (1). Then

$$S_2 = \frac{2}{1} - \frac{1}{1} = \frac{1}{1},$$

$$S_4 = \left(\frac{2}{1} - \frac{1}{1}\right) + \left(\frac{2}{2} - \frac{1}{2}\right) = \frac{1}{1} + \frac{1}{2},$$

$$S_6 = \left(\frac{2}{1} - \frac{1}{1}\right) + \left(\frac{2}{2} - \frac{1}{2}\right) + \left(\frac{2}{3} - \frac{1}{3}\right) = \frac{1}{1} + \frac{1}{2} + \frac{1}{3},$$

and more generally,

$$S_{2n} = \frac{1}{1} + \frac{1}{2} + \frac{1}{3} + \cdots + \frac{1}{n}.$$

Since S_{2n} gets arbitrarily large as $n \to \infty$ (the harmonic series diverges), the series (1) diverges.

Also, an alternating series whose terms decrease in size from some point on need not converge. Consider, for instance, the series

$$\frac{2}{1} - \frac{3}{2} + \frac{4}{3} - \frac{5}{4} + \cdots + (-1)^{n+1}\left(\frac{n+1}{n}\right) + \cdots.$$

Since the absolute value of the nth term approaches 1, the nth term does not approach 0. By the nth-term test for divergence, the series diverges.

Absolute Convergence

Consider a series $\qquad a_1 + a_2 + \cdots + a_n + \cdots,$

whose terms may be positive, negative, or zero. It is reasonable to expect it to behave at least as "nicely" as the series

$$|a_1| + |a_2| + \cdots + |a_n| + \cdots,$$

since by making all the terms positive we give the series more chance to diverge. This is similar to the case with improper integrals in Sec. 8.8, where it was shown that if $\int_a^\infty |f(x)|\,dx$ converges, then so does $\int_a^\infty f(x)\,dx$. The next theorem (and its proof) is similar to Theorem 3 in Sec. 8.8. (Reread it.)

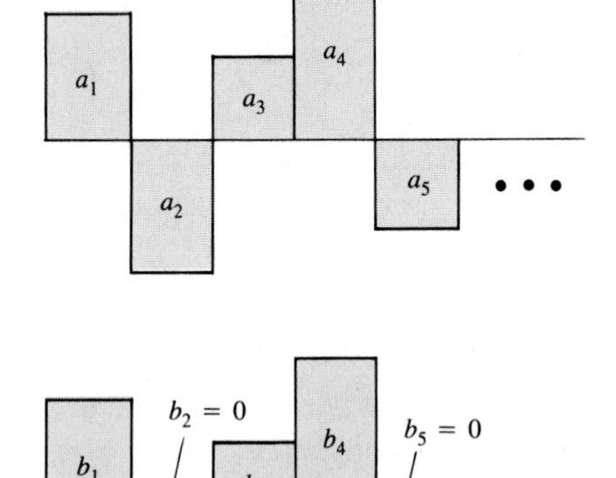

> **Theorem 2** *Absolute-convergence test.* If the series $\Sigma_{n=1}^\infty |a_n|$ converges, then so does the series $\Sigma_{n=1}^\infty a_n$. Furthermore, if $\Sigma_{n=1}^\infty |a_n| = S$ then $\Sigma_{n=1}^\infty a_n$ is between S and $-S$.

Proof We introduce two series in order to record the behavior of the positive and negative terms in $\Sigma_{n=1}^\infty a_n$ separately.

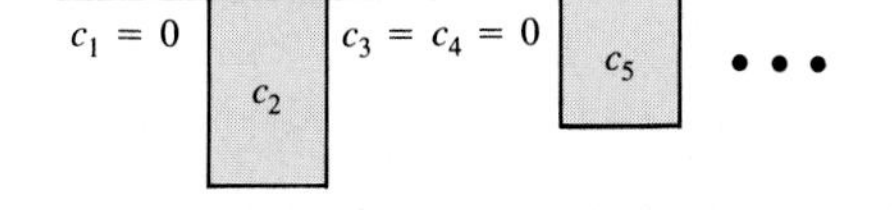

Let $\qquad b_n = a_n$ if $a_n > 0$, and let $b_n = 0$ otherwise.

Let $\qquad c_n = a_n$ if $a_n < 0$, and let $c_n = 0$ otherwise.

Figure 5 indicates these series as well as the series $\Sigma_{n=1}^\infty a_n$ and $\Sigma_{n=1}^\infty |a_n|$.

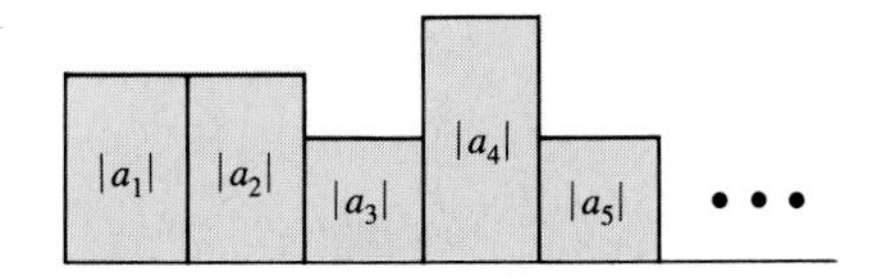

Note that $a_n = b_n + c_n$. To establish the convergence of $\Sigma_{n=1}^\infty a_n$ we show that both $\Sigma_{n=1}^\infty b_n$ and $\Sigma_{n=1}^\infty c_n$ converge. First of all, since b_n is nonnegative and $b_n \le |a_n|$, $\Sigma_{n=1}^\infty b_n$ converges by the comparison test. In fact, it converges to a number $A \le S$.

Since c_n is nonpositive, and $c_n \ge -|a_n|$, $\Sigma_{n=1}^\infty c_n$ converges to a number $B \ge -S$. Thus $\Sigma_{n=1}^\infty a_n = \Sigma_{n=1}^\infty (b_n + c_n)$ converges to $A - B$, which is between S and $-S$. ■

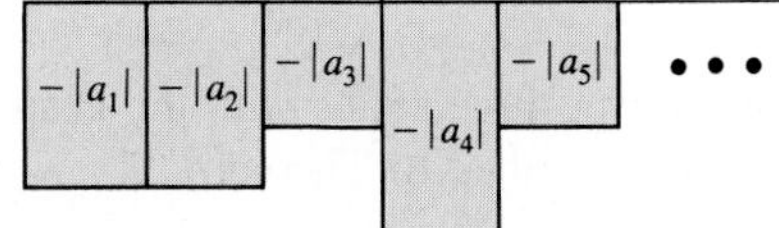

Figure 5

EXAMPLE 3 Examine the series

$$\frac{\cos x}{1^2} + \frac{\cos 2x}{2^2} + \frac{\cos 3x}{3^2} + \cdots + \frac{\cos nx}{n^2} + \cdots \tag{2}$$

for convergence or divergence.

SOLUTION The number x is fixed. The numbers $\cos nx$ may be positive, negative, or zero, in an irregular manner. However, for all n, $|\cos nx| \le 1$.
Recall that the series

$$\frac{1}{1^2} + \frac{1}{2^2} + \frac{1}{3^2} + \cdots + \frac{1}{n^2} + \cdots$$

converges, as shown in Sec. 10.4. Since $|\cos nx|/n^2 \le 1/n^2$, the series

$$\frac{|\cos x|}{1^2} + \frac{|\cos 2x|}{2^2} + \frac{|\cos 3x|}{3^2} + \cdots + \frac{|\cos nx|}{n^2} + \cdots \tag{3}$$

Advanced calculus shows that for $0 \le x \le 2\pi$, series (2) sums to $\dfrac{3x^2 - 6\pi x + 2\pi^2}{12}$.

converges by the comparison test. Since (3) converges, Theorem 2 tells us that (2) converges. ■

The **alternating harmonic series**

$$1 - \tfrac{1}{2} + \tfrac{1}{3} - \tfrac{1}{4} + \cdots$$

converges, as shown by Theorem 1. However, when all the terms are replaced by their absolute values, the resulting series, the harmonic series, does not converge; that is,

$$1 + \tfrac{1}{2} + \tfrac{1}{3} + \tfrac{1}{4} + \cdots$$

diverges. Thus the converse of Theorem 2 is false.
The following definitions are frequently used in describing these various cases of convergence or divergence.

Definition *Absolute convergence.* A series $a_1 + a_2 + \cdots$ is said to **converge absolutely** if the series $|a_1| + |a_2| + \cdots$ converges.

Theorem 2 can be stated simply: "If a series converges absolutely, then it converges."

Definition *Conditional convergence.* A series $a_1 + a_2 + \cdots$ is said to **converge conditionally** if it converges but does *not* converge absolutely.

$1 - \frac{1}{2} + \frac{1}{3} - \frac{1}{4} + \cdots$ converges conditionally.

For instance, the alternating harmonic series $1 - \frac{1}{2} + \frac{1}{3} - \frac{1}{4} + \cdots$ is conditionally convergent.

The Absolute-Limit-Comparison Test

When you combine the limit-comparison test for positive series with the absolute-convergence test, you obtain a single test, described in Theorem 3.

Theorem 3 *The absolute-limit-comparison test.* Let $\Sigma_{n=1}^{\infty} a_n$ be a series whose terms may be negative or positive. Let $\Sigma_{n=1}^{\infty} c_n$ be a convergent series of positive terms. If

$$\lim_{n\to\infty} \left| \frac{a_n}{c_n} \right|$$

exists, then $\Sigma_{n=1}^{\infty} a_n$ is absolutely convergent, hence convergent.

Proof Note that $|a_n/c_n| = |a_n|/c_n$, since c_n is positive. The limit-comparison test tells us that $\Sigma_{n=1}^{\infty} |a_n|$ converges. Then the absolute-convergence test assures us that $\Sigma_{n=1}^{\infty} a_n$ converges. ■

One advantage of the absolute-limit-comparison test over the limit-comparison test is that we don't have to follow it by the absolute-convergence test. Another is that we don't have to worry about the arithmetic of negative numbers.

EXAMPLE 4 Show that

$$\frac{3}{1}\left(\frac{1}{2}\right) - \frac{5}{2}\left(\frac{1}{2}\right)^2 + \frac{7}{3}\left(\frac{1}{2}\right)^3 - \cdots + (-1)^{n+1}\frac{2n+1}{n}\left(\frac{1}{2}\right)^n + \cdots \tag{4}$$

converges.

SOLUTION Consider the series of positive terms

$$\frac{3}{1}\left(\frac{1}{2}\right) + \frac{5}{2}\left(\frac{1}{2}\right)^2 + \frac{7}{3}\left(\frac{1}{2}\right)^3 - \cdots + \frac{2n+1}{n}\left(\frac{1}{2}\right)^n + \cdots.$$

The fact that $(2n + 1)/n \to 2$ as $n \to \infty$ suggests use of the limit-comparison test, comparing the series to the convergent geometric series $\Sigma_{n=1}^{\infty} (\frac{1}{2})^n$. We have

$$\lim_{n\to\infty} \frac{\dfrac{2n+1}{n}\left(\dfrac{1}{2}\right)^n}{(\frac{1}{2})^n} = 2.$$

Thus $\Sigma_{n=1}^{\infty} [(2n + 1)/n](\frac{1}{2})^n$ converges. Consequently, the given series (4), with positive and negative terms, converges absolutely. Thus it converges. ■

The Absolute-Ratio Test

The ratio test of Sec. 10.6 also has an analog when absolute values are introduced.

Theorem 4 *The absolute-ratio test.* Let $\Sigma_{n=1}^{\infty} a_n$ be a series such that

$$\lim_{n\to\infty} \left| \frac{a_{n+1}}{a_n} \right| = L < 1.$$

Then $\Sigma_{n=1}^{\infty} a_n$ converges. If $L > 1$ or if $\lim_{n\to\infty} |a_{n+1}/a_n| = \infty$, then $\Sigma_{n=1}^{\infty} a_n$ diverges.

Proof Take the case $L < 1$. By the ratio test, $\sum_{n=1}^{\infty} |a_n|$ converges. Since $\sum_{n=1}^{\infty} |a_n|$ converges, it follows that $\sum_{n=1}^{\infty} a_n$ converges also. The case $L > 1$ is treated in Exercise 34. The case $L = \infty$ can be treated as follows. If $\lim_{n \to \infty} |a_{n+1}/a_n| = \infty$, the ratio $|a_{n+1}|/|a_n|$ gets arbitrarily large as $n \to \infty$. So from some point on the numbers $|a_n|$ increase. By the nth-term test for divergence, $\sum_{n=1}^{\infty} a_n$ is divergent. ■

The absolute-ratio test simplifies work with minus signs.

Theorem 4 establishes the convergence of the series in Example 4 as follows. Let $a_n = (-1)^{n+1}[(2n+1)/n](\frac{1}{2})^n$. Then

$$\left|\frac{a_{n+1}}{a_n}\right| = \left|\frac{(-1)^{n+2}\dfrac{2n+3}{n+1}\left(\dfrac{1}{2}\right)^{n+1}}{(-1)^{n+1}\dfrac{2n+1}{n}\left(\dfrac{1}{2}\right)^{n}}\right| = \frac{2n+3}{2n+1}\cdot\frac{n}{n+1}\cdot\frac{1}{2},$$

which approaches $\frac{1}{2}$ as $n \to \infty$. Thus $\sum_{n=1}^{\infty} a_n$ converges (in fact, absolutely).

EXAMPLE 5 For which values of x does $\sum_{n=1}^{\infty} x^n/n$ converge? converge absolutely? diverge?

SOLUTION Inspired by the absolute-ratio test, we examine

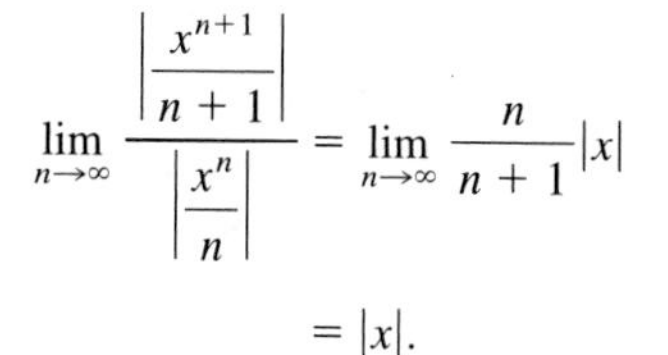

$$\lim_{n \to \infty} \frac{\left|\dfrac{x^{n+1}}{n+1}\right|}{\left|\dfrac{x^n}{n}\right|} = \lim_{n \to \infty} \frac{n}{n+1}|x|$$

$$= |x|.$$

If $|x| > 1$, the series diverges. If $|x| < 1$, the series converges absolutely, hence converges. If $|x| = 1$, the absolute-ratio test gives no information. We must look at this case—when x is 1 or -1—separately.

When $x = 1$, the series $\sum_{n=1}^{\infty} x^n/n$ becomes $\sum_{n=1}^{\infty} 1/n$, the harmonic series, which is divergent. When $x = -1$, the series becomes the alternating harmonic series, which is convergent, but only conditionally. (See Fig. 6.) ■

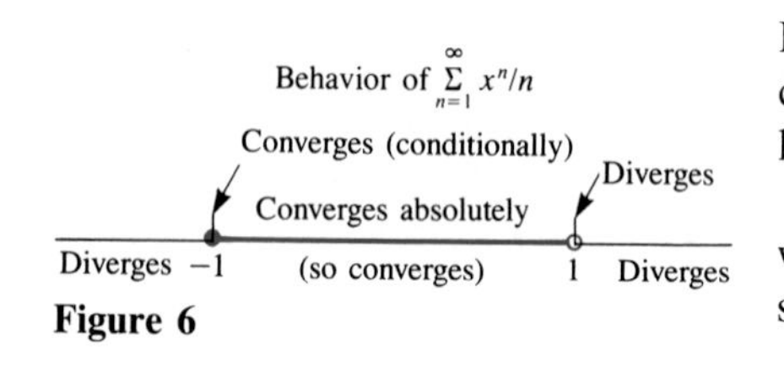

Figure 6

A series that converges absolutely has the property that no matter how the terms are rearranged, the new series converges and has the same sum as the original series. It might be expected that all convergent series have this property, but this is not the case. For instance, the alternating harmonic series

$$\frac{1}{1} - \frac{1}{2} + \frac{1}{3} - \frac{1}{4} + \frac{1}{5} - \cdots \tag{5}$$

does not. To show this, rearrange the summands so that two positive summands alternate with one negative summand, as follows:

$$\frac{1}{1} + \frac{1}{3} - \frac{1}{2} + \frac{1}{5} + \frac{1}{7} - \frac{1}{4} + \cdots. \tag{6}$$

Rearranging a conditionally convergent series is dangerous.

The positive summands in (6) have much more influence than the negative summands. In the battle between the positives and the negatives, the positives will win by a bigger margin in (6) than in (5). In fact, it can be shown that the sum of (6) is $\frac{3}{2}\ln 2$. [But the sum of (5) is just ln 2, as shown in Exercise 24 of Sec. 11.5.]

In advanced calculus it is demonstrated that a conditionally convergent series can be rearranged to converge to any preassigned sum or even to diverge to ∞ or $-\infty$.

Section Summary

Earlier in this chapter we described ways to test for the convergence or divergence of series whose terms are all positive. This section describes several tests for series that may be a mix of positive and negative terms.

If the signs alternate and the absolute value of the terms decreases and approaches 0, the series converges.

If the series converges when "all its terms are made positive," then it converges.

This absolute-convergence test in combination with the limit-comparison test gives us a single test, called the absolute-limit-comparison test. The absolute-convergence test in combination with the ratio test gives us the absolute-ratio test. (This will be the most important test in Chap. 11.)

EXERCISES FOR SEC. 10.7: TESTS FOR SERIES WITH BOTH POSITIVE AND NEGATIVE TERMS

In Exercises 1 to 8, which concern alternating series, determine which series converge and which diverge. Explain your answers.

1 $\frac{1}{2} - \frac{2}{3} + \frac{3}{4} - \frac{4}{5} + \cdots + (-1)^{n+1}\frac{n}{n+1} + \cdots$

2 $-\frac{1}{1+\frac{1}{2}} + \frac{1}{1+\frac{1}{4}} - \frac{1}{1+\frac{1}{8}} + \cdots + (-1)^n\frac{1}{1+2^{-n}} + \cdots$

3 $\frac{1}{\sqrt{1}} - \frac{1}{\sqrt{2}} + \frac{1}{\sqrt{3}} - \frac{1}{\sqrt{4}} + \cdots + (-1)^{n+1}\frac{1}{\sqrt{n}} + \cdots$

4 $\frac{5}{1!} - \frac{5^2}{2!} + \frac{5^3}{3!} - \frac{5^4}{4!} + \cdots + (-1)^{n+1}\frac{5^n}{n!} + \cdots$

5 $\frac{3}{\sqrt{1}} - \frac{2}{\sqrt{1}} + \frac{3}{\sqrt{2}} - \frac{2}{\sqrt{2}} + \frac{3}{\sqrt{3}} - \frac{2}{\sqrt{3}} + \cdots$

6 $\sqrt{1} - \sqrt{2} + \sqrt{3} - \sqrt{4} + \cdots + (-1)^{n+1}\sqrt{n} + \cdots$

7 $\frac{1}{3} - \frac{2}{5} + \frac{3}{7} - \frac{4}{9} + \frac{5}{11} - \cdots + (-1)^{n+1}\frac{n}{2n+1} + \cdots$

8 $\frac{1}{1^2} - \frac{1}{2^2} + \frac{1}{3^2} - \frac{1}{4^2} + \cdots + (-1)^{n+1}\frac{1}{n^2} + \cdots$

9 Consider the alternating harmonic series

$$\sum_{n=1}^{\infty} \frac{(-1)^{n+1}}{n}.$$

(*a*) Compute S_5 and S_6 to five decimal places.
(*b*) Is the estimate S_5 smaller or larger than the sum of the series?
(*c*) Use (*a*) and (*b*) to find two numbers between which the sum of the series must lie.

10 Consider the series $\sum_{n=1}^{\infty} (-1)^{n+1}2^{-n}/n$.
(*a*) Estimate the sum of the series using S_6.
(*b*) Estimate the error R_6.

In Exercises 11 to 26 determine which series diverge, converge absolutely, or converge conditionally. Explain your answers.

11 $\sum_{n=1}^{\infty} \frac{(-1)^n}{\sqrt[3]{n^2}}$

12 $\sum_{n=1}^{\infty} (-1)^n \ln\frac{1}{n}$

13 $\sum_{n=2}^{\infty} \frac{(-1)^n}{n \ln n}$

14 $\sum_{n=1}^{\infty} \frac{\sin n}{n^{1.01}}$

15 $\sum_{n=1}^{\infty} \left(1 - \cos\frac{\pi}{n}\right)$

16 $\sum_{n=1}^{\infty} (-1)^n \cos\left(\frac{\pi}{n^2}\right)$

17 $\sum_{n=1}^{\infty} \sin\left(\frac{\pi}{n^2}\right)$

18 $\sum_{n=1}^{\infty} \frac{(-2)^n}{n!}$

19 $\frac{1}{1^2} + \frac{1}{2^2} - \frac{1}{3^2} - \frac{1}{4^2} + \frac{1}{5^2} + \frac{1}{6^2} - \cdots$

(Two +'s alternating with two −'s.)

20 $\sum_{n=1}^{\infty} \frac{(-3)^n(1+n^2)}{n!}$

21 $\sum_{n=1}^{\infty} \frac{\cos n\pi}{2n+1}$

22 $\sum_{n=1}^{\infty} \frac{(-1)^n(n+5)}{n^2}$

23 $\sum_{n=1}^{\infty} \frac{(-9)^n}{10^n + n}$

24 $\sum_{n=1}^{\infty} \frac{(-1)^n}{\sqrt[3]{n}}$

25 $\sum_{n=1}^{\infty} \frac{(-1.01)^n}{n!}$

26 $\sum_{n=1}^{\infty} \frac{(-\pi)^{2n+1}}{(2n+1)!}$

---

27 For which values of x does $\sum_{n=1}^{\infty} x^n/n!$ converge?

28 The series $\sum_{n=1}^{\infty} (-1)^{n+1}2^{-n}$ is both a geometric series and a decreasing alternating series whose nth term approaches 0.

(a) Compute S_6 to three decimal places.
(b) Using the fact that the series is a decreasing alternating series, put a bound on R_6.
(c) Using the formula for the sum of a geometric series, compute R_6 exactly.

29 (a) How many terms of the series $\Sigma_{n=1}^{\infty} (\sin n)/n^2$ must you take to be sure the error is less than 0.005? Explain.
(b) Estimate $\Sigma_{n=1}^{\infty} (\sin n)/n^2$ to two decimal places.

30 Estimate $\Sigma_{n=0}^{\infty} (-1)^n/n! = 1 - 1 + 1/2! - 1/3! + \cdots$ to two decimal places. Explain your reasoning.

31 (a) Show $\Sigma_{n=1}^{\infty} 2^n/n!$ converges.
(b) Estimate the sum of the series in (a) to two decimal places.

32 Let $P(x)$ and $Q(x)$ be two polynomials of degree at least one. Assume that for $n \geq 1$, $Q(n) \neq 0$. What relation must there be between the degrees of $P(x)$ and $Q(x)$ if
(a) $P(n)/Q(n) \to 0$ as $n \to \infty$?
(b) $\Sigma_{n=1}^{\infty} P(n)/Q(n)$ converges absolutely?
(c) $\Sigma_{n=1}^{\infty} (-1)^n P(n)/Q(n)$ converges conditionally?

33 Prove Theorem 1.

34 This exercise treats the second half of the absolute-ratio test.
(a) Show that if

$$\lim_{n\to\infty} \left|\frac{a_{n+1}}{a_n}\right| = L > 1,$$

then $|a_n| \to \infty$ as $n \to \infty$. *Suggestion:* First show that there is a number r, $r > 1$, such that for some integer N, $|a_{n+1}| > r|a_n|$ for all $n \geq N$.
(b) From (a) deduce that a_n does not approach 0 as $n \to \infty$.

35 Consider the series $\Sigma_{n=1}^{\infty} nx^n/(2n+1)$. For which values of x does it converge? converge absolutely? diverge? Record your conclusions in a diagram on the x axis.

36 Repeat Exercise 35 for the series (a) $\Sigma_{n=1}^{\infty} x^n/n!$, (b) $\Sigma_{n=1}^{\infty} x^n/n^2$.

37 Is this argument okay? Add the alternating harmonic series to half of itself:

$$\begin{array}{l} 1 - \frac{1}{2} + \frac{1}{3} - \frac{1}{4} + \frac{1}{5} - \frac{1}{6} + \frac{1}{7} - \frac{1}{8} + \frac{1}{9} - \frac{1}{10} + \frac{1}{11} - \frac{1}{12} + \cdots = S \\ \quad \frac{1}{2} \quad - \frac{1}{4} \quad + \frac{1}{6} \quad - \frac{1}{8} \quad + \frac{1}{10} \quad - \frac{1}{12} + \cdots = \frac{1}{2}S \\ \hline 1 \quad + \frac{1}{3} - \frac{1}{2} + \frac{1}{5} \quad + \frac{1}{7} - \frac{1}{4} + \frac{1}{9} \quad + \frac{1}{11} - \frac{1}{6} + \cdots = \frac{3}{2}S \end{array}$$

Rearranging the last line produces the alternating harmonic series, whose sum is S. Thus $S = \frac{3}{2}S$, from which it follows that $S = 0$.

38 Is this alleged proof of Theorem 2 correct? Consider the sequence whose nth term is $a_n + |a_n|$. Since its terms are nonnegative and

$$\sum_{n=1}^{\infty} (a_n + |a_n|) \leq 2 \sum_{n=1}^{\infty} |a_n|,$$

$\Sigma_{n=1}^{\infty} (a_n + |a_n|)$ is convergent. Thus $\Sigma_{n=1}^{\infty} [(a_n + |a_n|) - |a_n|]$ is convergent. This shows $\Sigma_{n=1}^{\infty} a_n$ is convergent.

39 If $\Sigma_{n=1}^{\infty} a_n$ converges, $a_n > 0$, what, if anything, can we say about the convergence or divergence of

(a) $\displaystyle\sum_{n=1}^{\infty} \sin a_n$? (b) $\displaystyle\sum_{n=1}^{\infty} \cos a_n$?

10.S SUMMARY

This chapter opened with sequences and then turned to the study of series formed by adding the terms of a given sequence. That is, if $a_1, a_2, a_3, \ldots$ is a sequence, we consider the partial sums $S_1 = a_1$, $S_2 = a_1 + a_2$, $S_3 = a_1 + a_2 + a_3, \ldots$, which form a sequence $S_1, S_2, S_3, \ldots$ called a series.

Of particular importance is the geometric series

$$a + ar + ar^2 + \cdots + ar^{n-1} + \cdots,$$

which converges to

$$\frac{a}{1-r}$$

when $|r| < 1$. Also of importance is the p-series,

$$\frac{1}{1^p} + \frac{1}{2^p} + \cdots + \frac{1}{n^p} + \cdots,$$

which converges when $p > 1$ and diverges when $p \leq 1$.

The map shows the relation of the tests for convergence and divergence. The absolute-ratio test is the one used most often in the next chapter. The following chart surveys the various tests.

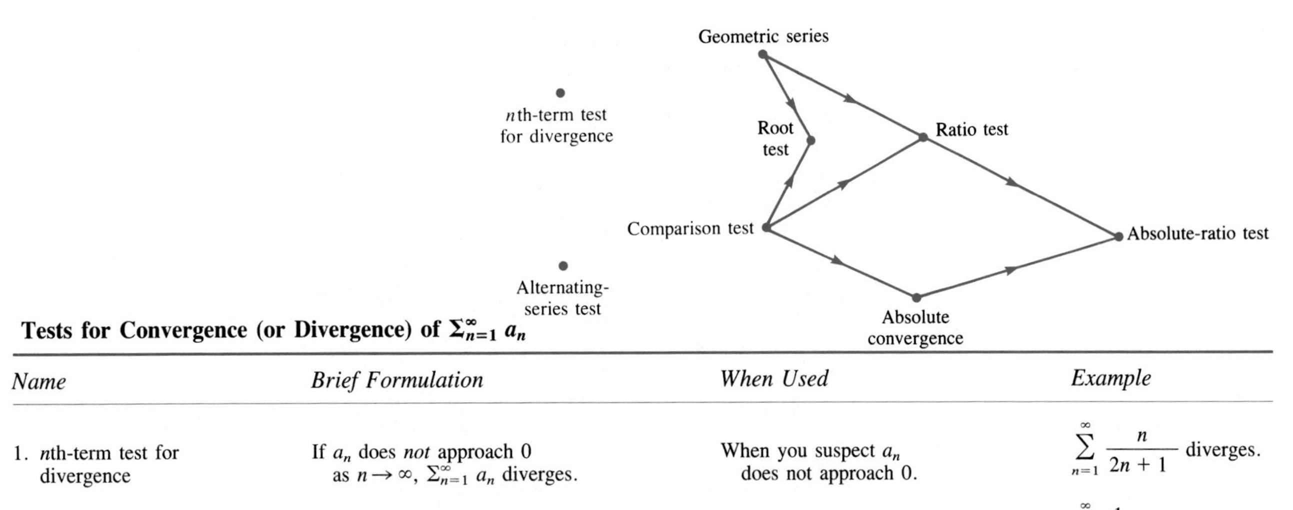

Tests for Convergence (or Divergence) of $\Sigma_{n=1}^{\infty} a_n$

Name	*Brief Formulation*	*When Used*	*Example*
1. nth-term test for divergence	If a_n does *not* approach 0 as $n \to \infty$, $\Sigma_{n=1}^{\infty} a_n$ diverges.	When you suspect a_n does not approach 0.	$\displaystyle\sum_{n=1}^{\infty} \frac{n}{2n+1}$ diverges.
2. Integral test	If $f(x) > 0$ decreases and $\int_1^{\infty} f(x)\, dx$ converges, $\Sigma_{n=1}^{\infty} a_n$ converges. If $\int_1^{\infty} f(x)\, dx$ diverges, $\Sigma_{n=1}^{\infty} a_n$ diverges.	When you have a *positive decreasing* series $a_n = f(n)$ and $\int f(x)\, dx$ is easy to calculate.	$\displaystyle\sum_{n=1}^{\infty} \frac{1}{n^2}$ converges, since $\int_1^{\infty} x^{-2}\, dx$ is convergent.
3. Comparison test	If $0 \le a_n \le c_n$ and $\Sigma_{n=1}^{\infty} c_n$ converges, so must $\Sigma_{n=1}^{\infty} a_n$. If $a_n \ge d_n \ge 0$ and $\Sigma_{n=1}^{\infty} d_n$ diverges, so must $\Sigma_{n=1}^{\infty} a_n$.	When you have a *positive* series that can be compared to a series known to converge or diverge.	$\displaystyle\sum_{n=1}^{\infty} \frac{n+1}{n+2} \cdot \frac{1}{2^n}$ converges, since $\Sigma_{n=1}^{\infty} 1/2^n$ does.
4. Limit-comparison test	If $a_n/c_n \to$ nonzero limit, then $\Sigma_{n=1}^{\infty} a_n$ converges if $\Sigma_{n=1}^{\infty} c_n$ does and diverges if $\Sigma_{n=1}^{\infty} c_n$ does. (For a more general statement, see Theorem 2, Sec. 10.5.)	When you have a positive series very much like a series known to converge or diverge.	$\displaystyle\sum_{n=1}^{\infty} \frac{1 + (-\frac{1}{2})^n}{2^n}$ converges, since $\Sigma_{n=1}^{\infty} 1/2^n$ converges.
5. Ratio test		See ''absolute-ratio'' test, which is more useful.	
6. Decreasing-alternating-series test	A decreasing alternating series whose nth term $\to 0$ converges.	When you have an *alternating series* whose terms diminish in absolute value from some point on and approach 0.	$\displaystyle\sum_{n=1}^{\infty} \frac{(-1)^n}{n}$ converges.
7. Absolute-convergence test	If $\Sigma_{n=1}^{\infty} \lvert a_n \rvert$ converges, so does $\Sigma_{n=1}^{\infty} a_n$.	When you feel that the series would converge even if its terms were all made positive.	$\displaystyle\sum_{n=1}^{\infty} \frac{\cos n}{n^2}$ converges.
8. Absolute-ratio test	If $\lvert a_{n+1}/a_n \rvert \to L < 1$, $\Sigma_{n=1}^{\infty} a_n$ converges (absolutely). If $\lvert a_{n+1}/a_n \rvert \to L > 1$, $\Sigma_{n=1}^{\infty} a_n$ diverges. If $L = 1$, no information.	Especially suitable for series involving factorials or powers, such as x^n.	$\displaystyle\sum_{n=0}^{\infty} \frac{x^n}{n!}$ converges absolutely for all x.
9. Root test	If $\sqrt[n]{\lvert a_n \rvert} \to L < 1$, $\Sigma_{n=1}^{\infty} a_n$ converges.	If something like n^n appears, try root test.	$\displaystyle\sum_{n=1}^{\infty} \frac{3^n}{n^n}$ converges.

Vocabulary and Symbols

sequence a_n or $\{a_n\}$
convergent, divergent sequence
limit of a sequence
series
partial sum S_n
convergent, divergent series
nth term of series
geometric series
nth-term test for divergence
integral test
p-series
harmonic series
comparison test
limit-comparison test
ratio test
root test
alternating series
decreasing alternating series
alternating harmonic series
alternating-series test
absolute convergence
conditional convergence
absolute-convergence test
absolute-ratio test
root test

Key Facts

If $|r| < 1$, $r^n \to 0$ as $n \to \infty$.
$x^n/n! \to 0$ as $n \to \infty$.
A geometric series $\Sigma_{n=0}^{\infty} ax^n$ converges to $a/(1-x)$ if $|x| < 1$.
A p-series $\Sigma_{n=1}^{\infty} 1/n^p$ converges if $p > 1$ and diverges if $p \le 1$.

Estimating the Error

Assume that $\Sigma_{n=1}^{\infty} a_n = S$. Let $\Sigma_{i=1}^{n} a_i = S_n$. The error, or remainder, R_n is defined to be the difference $S - S_n$. $(S_n + R_n = S)$

If $\Sigma_{n=1}^{\infty} a_n$ is a decreasing alternating series whose nth term approaches 0, then $|R_n| < |a_{n+1}|$.

If $\Sigma_{n=1}^{\infty} a_n$ is a positive series to which the integral test applies, then $\int_{n+1}^{\infty} f(x)\,dx < R_n < \int_n^{\infty} f(x)\,dx$.

If $\Sigma_{n=1}^{\infty} a_n$ is a convergent geometric series, $a_n = ar^{n-1}$, then $R_n = ar^n/(1-r)$.

An error in other series may sometimes be estimated if the terms from some point on are less than those of a geometric series or one to which the integral test applies.

GUIDE QUIZ ON CHAP. 10: SERIES

1 Explain in your own words
  (a) Why the comparison test for convergence works.
  (b) Why the ratio test for convergence works.
  (c) Why the alternating-series test works.
  (d) Why the absolute-convergence test works.

2 If $\Sigma_{n=1}^{\infty} a_n$ is convergent, must $\Sigma_{n=1}^{\infty} a_n^2$ be convergent?

3 How many terms of the series $\Sigma_{n=1}^{\infty} (-1)^{n+1}(1/n^2)$ should be used to estimate its sum to three-decimal-place accuracy?
  (a) First, solve by noticing that it is an alternating, decreasing series.
  (b) Then solve by noticing that $|R_k| \le \Sigma_{n=k+1}^{\infty} 1/n^2$, which can be estimated by an integral.

4 For which type of series does each of these tests imply convergence:
  (a) Alternating-series test;
  (b) Integral test;
  (c) Comparison test;
  (d) Ratio test;
  (e) Absolute-convergence test;
  (f) Absolute-ratio test?

5 Estimate $\Sigma_{n=1}^{\infty} n/e^n$ to two-decimal-place accuracy in two ways:
  (a) By controlling the error by comparing it to an integral.
  (b) By controlling the error by comparing it to a geometric series.

6 Test for convergence or divergence.

  (a) $\displaystyle\sum_{n=1}^{\infty} \frac{n!}{n^n}$

  (b) $\displaystyle\sum_{n=1}^{\infty} \tan\frac{1}{n}$

  (c) $\displaystyle\sum_{n=1}^{\infty} \frac{n^2+1}{n}\cdot\frac{x^n}{n+1}$ (Answer may depend on x.)

(d) $\sum_{n=1}^{\infty} \frac{2^n}{\left(1+\frac{1}{n}\right)^{n^2}}$

7 Given that $a_n \le b_n$ for $n \ge 1$ and that $\Sigma_{n=1}^{\infty} b_n = 9$, which of the following assertions is correct?
(a) $\Sigma_{n=1}^{\infty} a_n$ definitely converges and its sum is at most 9.
(b) $\Sigma_{n=1}^{\infty} a_n$ definitely converges, but we can say nothing about its sum.
(c) $\Sigma_{n=1}^{\infty} a_n$ definitely diverges.
(d) $\Sigma_{n=1}^{\infty} a_n$ may converge or diverge.

8 Assume that $|a_n| \le 1/2^n$ for $n \ge 1$.
(a) Must $\Sigma_{n=1}^{\infty} |a_n|$ converge? If so, what can we say about its sum?
(b) Must $\Sigma_{n=1}^{\infty} a_n$ converge? If so, what can we say about its sum?

9 (a) If $\Sigma_{n=1}^{9} 1/n^2$ is used to approximate $\Sigma_{n=1}^{\infty} 1/n^2$, is it correct to say the error is not more than $1/10^2$, the first omitted term? If not, what can be said about the error?
(b) If $\Sigma_{n=1}^{9} (-1)^n/n^2$ is used to approximate $\Sigma_{n=1}^{\infty} (-1)^n/n^2$, is it correct to say the absolute value of the error is not more than $1/10^2$, the first omitted term? If not, what can be said about the error?

REVIEW EXERCISES FOR CHAP. 10: SERIES

Sometimes convergence or divergence of a series can be established by more than one of the tests developed in this chapter. In Exercises 1 to 32 determine the convergence or divergence of the given series by as many tests as can be applied in each case.

1 $\sum_{n=1}^{\infty} \frac{(-1)^n}{n^2}$

2 $\sum_{n=1}^{\infty} \frac{(-1)^n}{3^n}$

3 $\sum_{k=1}^{\infty} \frac{\sqrt{k}}{k^2+1}$

4 $\sum_{k=1}^{\infty} \frac{\sqrt{k}}{k^2-2}$

5 $\sum_{n=1}^{\infty} \left[\frac{3+1/n}{2+1/n}\right]^n$

6 $\sum_{n=1}^{\infty} \left[\frac{2}{3+(1/n)}\right]^n$

7 $\sum_{n=1}^{\infty} \frac{1}{2^n-3}$

8 $\sum_{n=1}^{\infty} \frac{10^n}{n!}$

9 $\sum_{k=1}^{\infty} \frac{\cos^2 k}{2^k}$

10 $\sum_{k=1}^{\infty} \frac{\sin^2 k}{k^2}$

11 $\sum_{n=1}^{\infty} \frac{n^n}{n!}$

12 $\sum_{n=1}^{\infty} (-1)^n \frac{(n+1)^2}{n!}$

13 $\sum_{n=1}^{\infty} \frac{1}{n\sqrt{n}}$

14 $\sum_{n=0}^{\infty} (-1)^n \frac{\pi^{2n+1}}{2^{2n+1}(2n+1)!}$

15 $\sum_{n=1}^{\infty} (-1)^n \ln\left(\frac{n+1}{n}\right)$

16 $\sum_{n=1}^{\infty} \cos \frac{1}{n}$

17 $\sum_{n=1}^{\infty} n \sin \frac{1}{n}$

18 $\sum_{n=1}^{\infty} \frac{(-2)^n}{n}$

19 $\sum_{n=0}^{\infty} \frac{5n^3+6n+1}{n^5+n^3+2}$

20 $\sum_{n=0}^{\infty} \frac{5n^2-3n+1}{2n^3+n^2-1}$

21 $\sum_{n=1}^{\infty} \frac{2^{-n}}{n}$

22 $\sum_{n=1}^{\infty} \ln\left(\frac{n+1}{n}\right)$

23 $\sum_{n=0}^{\infty} \frac{(-1)^n(\frac{1}{2})^n}{n!}$

24 $\sum_{n=0}^{\infty} (-1)^n \frac{\pi^{2n}}{(2n)!}$

25 $\sum_{n=0}^{\infty} \frac{n+2}{n+1}\left(\frac{2}{3}\right)^n$

26 $\sum_{n=1}^{\infty} \frac{\ln n}{n}$

27 $\sum_{n=1}^{\infty} \frac{n \cos n}{1+n^4}$

28 $\sum_{n=1}^{\infty} \frac{\sqrt{n+1}-\sqrt{n}}{n}$

29 $\sum_{n=1}^{\infty} \frac{n-3}{n\sqrt{n}}$

30 $\sum_{n=0}^{\infty} \sin \frac{1}{n}$

31 $\sum_{n=0}^{\infty} \frac{10^n}{n!}$

32 $\sum_{n=1}^{\infty} \frac{(-1)^n n^2}{(2n)!}$

In Exercises 33 to 36 approximate the sum to two-decimal-place accuracy.

33 $\sum_{n=1}^{\infty} \frac{1}{3^n}$

34 $\sum_{n=1}^{\infty} \frac{n}{3^n}$

35 $\sum_{n=1}^{\infty} (-1)^n \frac{n}{5^n}$

36 $\sum_{n=1}^{\infty} \frac{(\frac{1}{2})^n}{n!}$

37 Let $\{a_n\}$ and $\{b_n\}$ be sequences of positive terms. Assume that for all n

$$\frac{a_{n+1}}{a_n} \le \frac{b_{n+1}}{b_n}.$$

(a) Prove that if $\Sigma_{n=1}^{\infty} b_n$ converges, so does $\Sigma_{n=1}^{\infty} a_n$. (*Hint:* Rewrite the inequality as $a_{n+1}/b_{n+1} \le a_n/b_n$.)
(b) Use the result in (a) to prove that if $\lim_{n\to\infty} a_{n+1}/a_n = r < 1$, then $\Sigma_{n=1}^{\infty} a_n$ converges.

38 (*a*) Show that

$$\sum_{n=0}^{\infty} \frac{\cos\,[(2n+1)t]}{(2n+1)^2}$$

converges for all t.

(*b*) In the theory of Fourier series it is shown that for $0 \le t < \pi$, the sum of the series in (*a*) is $(\pi^2 - 2\pi t)/8$. Deduce that

$$\frac{\pi^2}{8} = \frac{1}{1^2} + \frac{1}{3^2} + \frac{1}{5^2} + \cdots.$$

39 (*a*) Show that if $\{a_n\}$ is a sequence of positive terms and $\Sigma_{n=1}^{\infty} a_n$ converges, so does $\Sigma_{n=1}^{\infty} a_n^2$.

(*b*) Give an example of a sequence $\{a_n\}$ such that $\Sigma_{n=1}^{\infty} a_n$ converges but $\Sigma_{n=1}^{\infty} a_n^2$ does not.

40 The zeta function $\zeta(p)$ is defined for $p > 1$ as $\zeta(p) = \Sigma_{n=1}^{\infty} n^{-p}$.

(*a*) Examine $\lim_{p \to 1^+} \zeta(p)$.

(*b*) Show that $(p-1)^{-1} < \zeta(p) < p(p-1)^{-1}$.

(*c*) Show that $\lim_{p \to 1^+} \zeta(p)(p-1) = 1$.

41 (*a*) Show that if $\Sigma_{n=1}^{\infty} a_n^2$ and $\Sigma_{n=1}^{\infty} b_n^2$ converge, so does $\Sigma_{n=1}^{\infty} a_n b_n$.

(*b*) Show that if $\Sigma_{n=1}^{\infty} a_n^2$ converges, so does $\Sigma_{n=1}^{\infty} a_n/n$.

In Exercises 42 to 47 a short formula for estimating $n!$ is obtained.

42 Let f have the properties that for $x \ge 1$, $f(x) \ge 0$, $f'(x) > 0$, and $f''(x) < 0$. Let a_n be the area of the region below the graph of $y = f(x)$ and above the line segment that joins $(n, f(n))$ with $(n+1, f(n+1))$.

(*a*) Draw a large-scale version of Fig. 1. The individual regions of areas a_1, a_2, a_3, and a_4 should be clear and not too narrow.

(*b*) Using geometry, show that the series $a_1 + a_2 + a_3 + \cdots$ converges and has a sum no larger than the area of the triangle with vertices $(1, f(1))$, $(2, f(2))$, $(1, f(2))$.

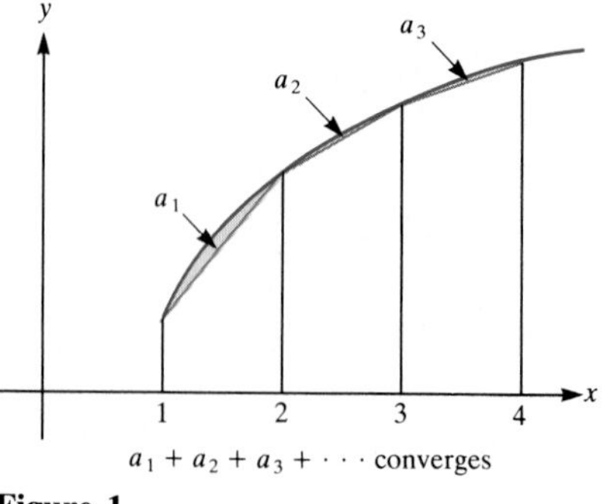

Figure 1

43 Let $y = \ln x$.

(*a*) Using Exercise 42, show that as $n \to \infty$,

$$\int_1^n \ln x\,dx - \left[\frac{\ln 1 + \ln 2}{2} + \frac{\ln 2 + \ln 3}{2} + \cdots + \frac{\ln(n-1) + \ln n}{2}\right]$$

has a limit; denote this limit by C.

(*b*) Show that (*a*) is equivalent to the assertion

$$\lim_{n \to \infty} (n \ln n - n + 1 - \ln n! + \ln \sqrt{n}) = C.$$

44 From Exercise 43(*b*), deduce that there is a constant k such that

$$\lim_{n \to \infty} \frac{n!}{k(n/e)^n \sqrt{n}} = 1.$$

Exercises 45 and 46 are related. Review Exercise 39 of Sec. 7.3 first.

45 Let $I_n = \int_0^{\pi/2} \sin^n x\,dx$.

(*a*) Evaluate I_0 and I_1.

(*b*) Show that

$$I_{2n} = \frac{2n-1}{2n}\,\frac{2n-3}{2n-2} \cdots \frac{3}{4}\,\frac{1}{2}\,\frac{\pi}{2}$$

and

$$I_{2n+1} = \frac{2n}{2n+1}\,\frac{2n-2}{2n-1} \cdots \frac{4}{5}\,\frac{2}{3}.$$

(*c*) Show that

$$\frac{I_7}{I_6} = \frac{6}{7}\,\frac{6}{5}\,\frac{4}{5}\,\frac{4}{3}\,\frac{2}{3}\,\frac{2}{1}\,\frac{2}{\pi}.$$

(*d*) Show that

$$\frac{I_{2n+1}}{I_{2n}} = \frac{2n}{2n+1}\,\frac{2n}{2n-1}\,\frac{2n-2}{2n-1} \cdots \frac{2}{3}\,\frac{2}{1}\,\frac{2}{\pi}.$$

(*e*) Show that

$$\frac{2n}{2n+1} I_{2n} < \frac{2n}{2n+1} I_{2n-1} = I_{2n+1} < I_{2n},$$

and thus

$$\lim_{n \to \infty} \frac{I_{2n+1}}{I_{2n}} = 1.$$

(*f*) From (*d*) and (*e*), deduce that

$$\lim_{n \to \infty} \frac{2 \cdot 2}{1 \cdot 3}\,\frac{4 \cdot 4}{3 \cdot 5}\,\frac{6 \cdot 6}{5 \cdot 7} \cdots \frac{(2n)(2n)}{(2n-1)(2n+1)} = \frac{\pi}{2}.$$

This is **Wallis's formula**, usually written in shorthand as

$$\frac{2 \cdot 2}{1 \cdot 3}\,\frac{4 \cdot 4}{3 \cdot 5}\,\frac{6 \cdot 6}{5 \cdot 7} \cdots = \frac{\pi}{2}.$$

46 (*a*) Show that $2 \cdot 4 \cdot 6 \cdot 8 \cdots 2n = 2^n n!$.

(*b*) Show that $1 \cdot 3 \cdot 5 \cdots (2n-1) = (2n)!/(2^n n!)$.

(*c*) From Exercise 45 deduce that

$$\lim_{n\to\infty} \frac{(n!)^2 4^n}{(2n)!\sqrt{2n+1}} = \sqrt{\frac{\pi}{2}}.$$

47 (*a*) Using Exercise 46(*c*), show that k in Exercise 44 equals $\sqrt{2\pi}$. Thus a good estimate of $n!$ is provided by the formula

$$n! \approx \sqrt{2\pi n}\left(\frac{n}{e}\right)^n.$$

This is known as **Stirling's formula**.

(*b*) Using the factorial key on a calculator, compute 20!. Then compute the ratio $\sqrt{2\pi n}(n/e)^n/n!$ for $n = 20$.

* **48** A sequence $a_1, a_2, a_3, \ldots, a_n, \ldots$ is defined as follows. First one picks a_1 and a_2. Then

$$a_3 = \frac{2 + a_2}{a_1},$$

and, more generally, for $n \geq 3$,

$$a_n = \frac{2 + a_{n-1}}{a_{n-2}}.$$

(*a*) Choosing various values for a_1 and a_2, examine the sequences you obtain.

(*b*) Write up your observations and conjectures.

(*c*) If $\{a_n\}$ converges, what can be said about the limit?

49 If $\lim_{n\to\infty} x^{2n}/(1 + x^{2n})$ exists, call it $f(x)$. (x is fixed; n varies.)

(*a*) Compute $f(\frac{1}{2}), f(2), f(1)$.

(*b*) For which x is $f(x)$ defined? Graph $y = f(x)$.

(*c*) Where is f continuous?

50 Use elementary geometry to show that the area of that part of the endless staircase shown in Fig. 2 above the curve $y = 1/x$ is between $\frac{1}{2}$ and 1.

The area of the shaded region above the curve is known as **Euler's constant** γ, whose decimal representation begins 0.577. Thus, if a_n is defined by $a_n = \sum_{i=1}^{n} 1/i - \ln n$, then $a_n \to \gamma$ as $n \to \infty$. It is not known whether γ is rational or irrational.

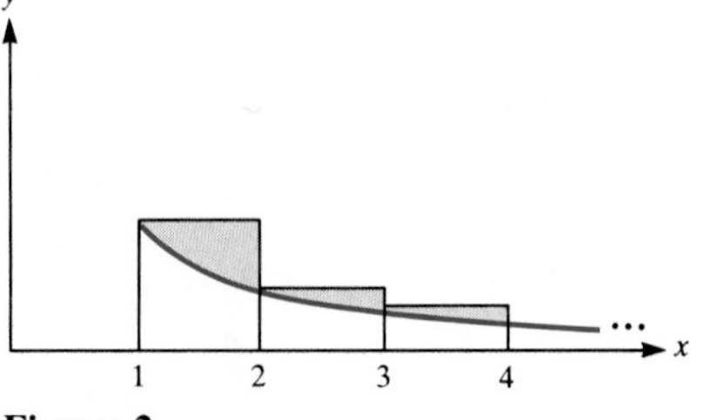

Figure 2

51 For which positive numbers p and q

(*a*) does $\sum_{n=1}^{\infty} (\ln n)^p/n^q$ converge?

(*b*) does $\sum_{n=1}^{\infty} n^q/(\ln n)^p$ converge?

52 Does $\sum_{n=1}^{\infty} (e^{1/n} - 1)$ converge or diverge?

53 Does $\sum_{n=1}^{\infty} (e^{1/n} - 1)/n$ converge or diverge?

11

POWER SERIES AND COMPLEX NUMBERS

The preceding chapter developed the notion of an infinite series and tests for their convergence and divergence. This chapter applies infinite series to find ways of computing the values of functions such as e^x and $\sin x$, evaluating integrals, calculating limits of indeterminate "zero-over-zero" form, and obtaining a close link between the trigonometric and exponential functions.

11.1 TAYLOR SERIES

In this section we develop a method of constructing polynomial approximations to functions that are not polynomials. The higher derivatives of all orders will play a pivotal role.

Taylor Polynomials

We spend years learning how to add, subtract, multiply, and divide. These same operations are built into any calculator or computer. Both we and machines can evaluate a polynomial, such as

$$a_0 + a_1x + a_2x^2 + \cdots + a_nx^n,$$

when x and the coefficients $a_0, a_1, \ldots, a_n$ are given numbers, since only multiplication and addition are required. But how do we evaluate e^x or $\sin x$? We resort to our calculator or look in a table that lists values of e^x. If e^x were a polynomial in disguise, then it would be easy to evaluate e^x by finding the polynomial and evaluating it instead. But e^x cannot be a polynomial, since it grows much more rapidly than any polynomial as $x \to \infty$. This follows from the fact that

$$\lim_{x\to\infty} \frac{x^n}{e^x} = 0,$$

for any positive integer n.

Might e^x nevertheless coincide with some polynomial—at least over a very short interval? No, since e^x equals its own derivative and no polynomial equals its own derivative (other than the polynomial that has constant value 0). After all, when you differentiate a nonconstant polynomial, you get a polynomial of lower degree. These observations are recorded in Fig. 1.

Since we cannot write e^x as a polynomial, we settle for the next best thing. Let's look for a polynomial that closely *approximates* e^x. However, no polynomial can be a good approximation of e^x for *all* x, since e^x grows too fast as $x \to \infty$. Let us search, instead, for a polynomial that is close to e^x for x in some short interval.

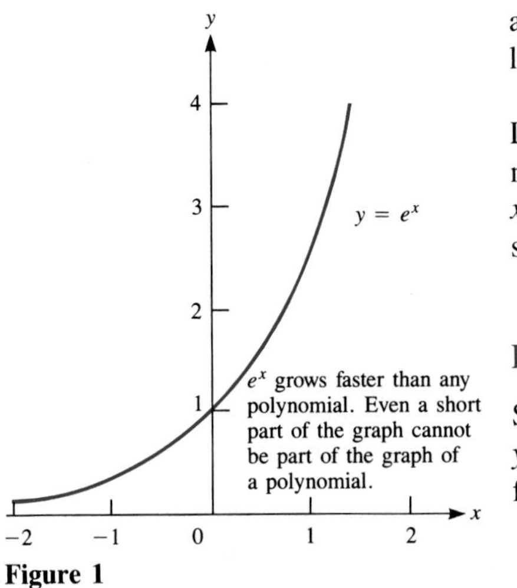

Figure 1

Fitting a Polynomial Locally

Suppose that we want to find a polynomial that closely approximates a function $y = f(x)$ for x near the number 0. For instance, what polynomial $p(x)$ of the form $a_0 + a_1x + a_2x^2 + a_3x^3$ would be the best fit?

First, we insist that

$$p(0) = f(0),$$

so the approximation is exact when $x = 0$.

Second, we would like the slope of the graph of $p(x)$ to be the same as that of $f(x)$ when x is 0. Therefore we require

$$p'(0) = f'(0).$$

These are only two equations so far, while we have four numbers to determine: a_0, a_1, a_2, and a_3. To have four equations in the four unknowns, we will also insist that

$$p^{(2)}(0) = f^{(2)}(0)$$

and

$$p^{(3)}(0) = f^{(3)}(0).$$

We would expect the graphs of $f(x)$ and such a polynomial $p(x)$ to resemble each other for x close to 0, as in Fig. 2.

To find the unknowns a_0, a_1, a_2, and a_3 we first compute $p(x)$, $p'(x)$, $p^{(2)}(x)$, and $p^{(3)}(x)$ at 0. Table 1 displays the computations that express the a_k's in terms of $f(x)$ and its derivatives. For example, note how we compute $p^{(2)}(x) = 2a_2 + 3 \cdot 2a_3x$ and evaluate it at 0 to obtain $p^{(2)}(0) = 2a_2 + 3 \cdot 2a_3 \cdot 0 = 2a_2$. Then we obtain an equation for a_2 by equating $p^{(2)}(0)$ and $f^{(2)}(0)$; that is, $2a_2 = f^{(2)}(0)$, so $a_2 = f^{(2)}(0)/2$.

Near the y axis the two graphs are close.

y = f(x), some function

y = p(x), a polynomial

Figure 2

Table 1

p(x) and Its Derivatives	*Their Values at 0*	*Equation for a_k*	*Formula for a_k*
$p(x) = a_0 + a_1x + a_2x^2 + a_3x^3$	$p(0) = a_0$	$a_0 = f(0)$	$a_0 = f(0)$
$p^{(1)}(x) = a_1 + 2a_2x + 3a_3x^2$	$p^{(1)}(0) = a_1$	$a_1 = f^{(1)}(0)$	$a_1 = f^{(1)}(0)$
$p^{(2)}(x) = 2a_2 + 3 \cdot 2a_3x$	$p^{(2)}(0) = 2a_2$	$2a_2 = f^{(2)}(0)$	$a_2 = \dfrac{f^{(2)}(0)}{2}$
$p^{(3)}(x) = 3 \cdot 2a_3$	$p^{(3)}(0) = 3 \cdot 2a_3$	$3 \cdot 2a_3 = f^{(3)}(0)$	$a_3 = \dfrac{f^{(3)}(0)}{3 \cdot 2}$

Factorials appear in the denominator.

We can write a general formula for a_k if we let $f^{(0)}(x)$ denote $f(x)$ and recall that $0! = 1$ (by definition), $1! = 1$, $2! = 2 \cdot 1 = 2$, and $3! = 3 \cdot 2$. According to Table 1,

$$a_k = \frac{f^{(k)}(0)}{k!}, \qquad k = 0, 1, 2, 3.$$

Therefore

$$p(x) = f(0) + f'(0)x + \frac{f^{(2)}(0)}{2!}x^2 + \frac{f^{(3)}(0)}{3!}x^3.$$

The coefficient of x^k is *completely determined* by the kth derivative of f evaluated at 0. It equals the kth derivative of f at 0 divided by $k!$. In Example 1 we compute $p(x)$ for $f(x) = e^x$.

EXAMPLE 1 Find the polynomial of degree 3 that best approximates e^x near $x = 0$.

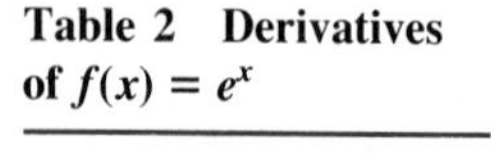

Table 2 Derivatives of $f(x) = e^x$

At x	*At 0*
$f^{(0)}(x) = e^x$	$f^{(0)}(0) = 1$
$f^{(1)}(x) = e^x$	$f^{(1)}(0) = 1$
$f^{(2)}(x) = e^x$	$f^{(2)}(0) = 1$
$f^{(3)}(x) = e^x$	$f^{(3)}(0) = 1$

SOLUTION All we need to do is compute e^x and its first three derivatives. Dividing them by a suitable factorial will give us the coefficients of the polynomial. Table 2 records the computations, which are especially simple because the derivative of e^x is e^x itself.

So the third-degree approximating polynomial is

$$p(x) = 1 + x + \frac{x^2}{2!} + \frac{x^3}{3!}.$$

Figure 3 contrasts e^x and $p(x)$. ■

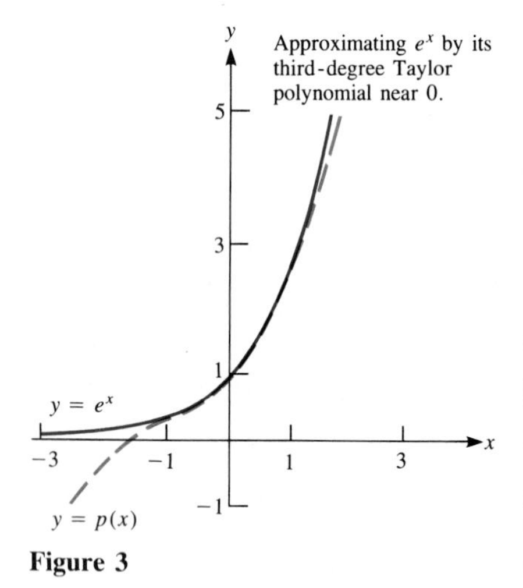

Figure 3

The Taylor Polynomials at 0

Example 1 illustrates the general procedure for finding polynomials that behave much like a given function near 0. These approximating polynomials are given a name in the following definition.

> **Definition** *Taylor polynomial.* Let f be a function with derivatives of all orders at 0 and let n be a nonnegative integer. Then the polynomial
>
> $$P_n(x; 0) = f(0) + f^{(1)}(0)x + \frac{f^{(2)}(0)}{2!}x^2 + \cdots + \frac{f^{(n)}(0)}{n!}x^n \tag{1}$$
>
> is called the ***n*th-order Taylor polynomial** associated with f at 0.

The degree of $P_n(x; 0)$ is at most n. Whether $P_n(x; 0)$ actually approximates $f(x)$ for x near 0 is not evident. Later in this chapter we will show that the Taylor polynomials for e^x and $\sin x$ do provide good approximations of the functions for x near 0. [We already have a clue in Fig. 3 that $P_3(x; 0)$ does a fairly good job near 0 for e^x.] The bigger n is, the better the approximation and the longer the interval where the approximation is good. Notice that $P_1(x; 0)$, which can be written as $f(0) + f'(0)(x - 0)$, is our old friend from Sec. 4.9, the linearization of f near 0. *The Taylor polynomials generalize the linear approximation to polynomials of higher degree.*

Table 3 Derivatives of $f(x) = \sin x$

At x	*At 0*
$f^{(0)}(x) = \sin x$	$f^{(0)}(0) = \sin 0 = 0$
$f^{(1)}(x) = \cos x$	$f^{(1)}(0) = \cos 0 = 1$
$f^{(2)}(x) = -\sin x$	$f^{(2)}(0) = -\sin 0 = 0$
$f^{(3)}(x) = -\cos x$	$f^{(3)}(0) = -\cos 0 = -1$
$f^{(4)}(x) = \sin x$	$f^{(4)}(0) = \sin 0 = 0$
$f^{(5)}(x) = \cos x$	$f^{(5)}(0) = \cos 0 = 1$

EXAMPLE 2 Find the Taylor polynomial $P_5(x; 0)$ associated with the function $f(x) = \sin x$.

SOLUTION Again we make a table for computing the coefficients of the polynomial. (See Table 3.)

Thus

$$P_5(x; 0) = f(0) + f^{(1)}(0)x + \frac{f^{(2)}(0)}{2!}x^2 + \frac{f^{(3)}(0)}{3!}x^3 + \frac{f^{(4)}(0)}{4!}x^4 + \frac{f^{(5)}(0)}{5!}x^5$$

$$= 0 + x + \frac{0}{2!}x^2 - \frac{1}{3!}x^3 + \frac{0}{4!}x^4 + \frac{1}{5!}x^5$$

$$= x - \frac{x^3}{3!} + \frac{x^5}{5!}$$

$$= x - \frac{x^3}{6} + \frac{x^5}{120}.$$

Figure 4 contrasts the graphs of $x - x^3/3! + x^5/5!$ and $\sin x$ near 0. ■

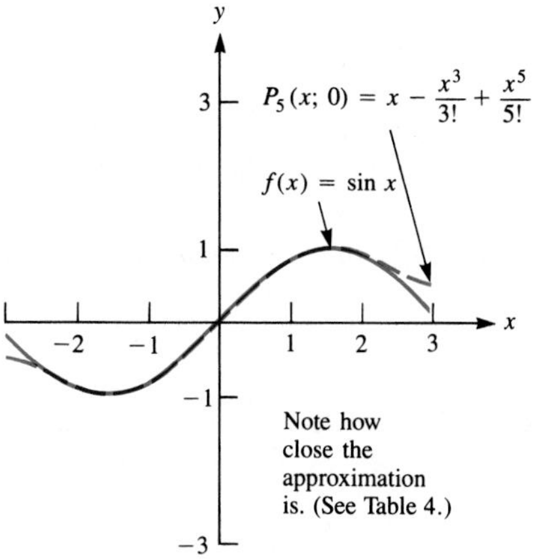

Figure 4

Having found the fifth-order Taylor polynomial for $\sin x$, let us see how good an approximation it is of $\sin x$. Table 4 compares their values to six decimal place accuracy for inputs both near 0 and far from 0. As we see, the closer we are to 0, the better the Taylor approximation is. When x is large, $P_5(x; 0)$ gets very large, but $\sin x$ stays no larger than 1.

Table 4

x	$\sin x$	$P_5(x; 0)$
0	0	0
0.1	0.099833	0.099833
0.5	0.479426	0.479427
1	0.841471	0.841667
2	0.909297	0.933333
π	0	0.524044
2π	0	46.546732

Taylor Polynomials in General

Up to this point we have constructed polynomials to approximate a given function near 0. But we can do the same thing near any number a. Instead of expressing the approximating polynomial in terms of powers of x, it will be convenient to write it in terms of powers of $x - a$. To find the formula for these polynomials, we will imitate for a what we did at 0.

For example, what polynomial of the form

$$p(x) = a_0 + a_1(x - a) + a_2(x - a)^2 + a_3(x - a)^3$$

shall we use to approximate a function near a? This time we demand that

$$p(a) = f(a)$$

$$p^{(1)}(a) = f^{(1)}(a)$$

$$p^{(2)}(a) = f^{(2)}(a)$$

and $$p^{(3)}(a) = f^{(3)}(a).$$

So we compute $p(a)$, $p^{(1)}(a)$, $p^{(2)}(a)$, and $p^{(3)}(a)$ with the aid of Table 5.

Table 5

$p(x)$ and Its Derivatives	*Their Values at a*	*Equation for a_k*	*Formula for a_k*
$p(x) = a_0 + a_1(x-a) + a_2(x-a)^2 + a_3(x-a)^3$	$p(a) = a_0$	$a_0 = f(a)$	$a_0 = f(a)$
$p^{(1)}(x) = a_1 + 2a_2(x-a) + 3a_3(x-a)^2$	$p^{(1)}(a) = a_1$	$a_1 = f^{(1)}(a)$	$a_1 = f^{(1)}(a)$
$p^{(2)}(x) = 2a_2 + 3 \cdot 2a_3(x-a)$	$p^{(2)}(a) = 2a_2$	$2a_2 = f^{(2)}(a)$	$a_2 = \dfrac{f^{(2)}(a)}{2}$
$p^{(3)}(x) = 3 \cdot 2a_3$	$p^{(3)}(a) = 3 \cdot 2a_3$	$3 \cdot 2a_3 = f^{(3)}(a)$	$a_3 = \dfrac{f^{(3)}(a)}{3 \cdot 2}$

Thus

$$p(x) = f(a) + f^{(1)}(a)(x-a) + \frac{f^{(2)}(a)}{2!}(x-a)^2 + \frac{f^{(3)}(a)}{3!}(x-a)^3. \qquad (2)$$

If you compare (2) with (1), you will see that in the general case, at any number a, the derivatives of f are evaluated at a (instead of at 0) and there are powers of $x - a$ (instead of powers of x).

This leads us to the definition of the nth-order Taylor polynomial associated with a function $f(x)$ at the number a.

The nth-order Taylor polynomial has degree at most n.

Definition *Taylor polynomial of degree n, $P_n(x; a)$.* If the function f has derivatives through order n at a, then the **nth-order Taylor polynomial** of f at a is defined as

$$P_n(x; a) = f(a) + f^{(1)}(a)(x-a) + \frac{f^{(2)}(a)}{2!}(x-a)^2 + \cdots + \frac{f^{(n)}(a)}{n!}(x-a)^n.$$

The polynomial is denoted $P_n(x; a)$.

Table 6 Derivatives of $f(x) = 1/x$

At x	*At 1*
$f^{(0)}(x) = 1/x$	$f^{(0)}(1) = 1$
$f^{(1)}(x) = -1/x^2$	$f^{(1)}(1) = -1$
$f^{(2)}(x) = 2/x^3$	$f^{(2)}(1) = 2$
$f^{(3)}(x) = -6/x^4$	$f^{(3)}(1) = -6$
$f^{(4)}(x) = 24/x^5$	$f^{(4)}(1) = 24$

EXAMPLE 3 Compute the fourth-order Taylor polynomial at $a = 1$ for the function $f(x) = 1/x$.

SOLUTION Table 6 shows the values of the derivatives of $1/x$ at 1. Thus

$$\begin{aligned} P_4(x; 1) &= f(1) + f^{(1)}(1)(x-1) + \frac{f^{(2)}(1)}{2!}(x-1)^2 + \frac{f^{(3)}(1)}{3!}(x-1)^3 + \frac{f^{(4)}(1)}{4!}(x-1)^4 \\ &= 1 - (x-1) + \frac{2}{2}(x-1)^2 - \frac{6}{3!}(x-2)^3 + \frac{24}{4!}(x-1)^4 \\ &= 1 - (x-1) + (x-1)^2 - (x-1)^3 + (x-1)^4. \quad \blacksquare \end{aligned}$$

Table 7

x	$1/x$	$P_4(x; 1)$
1	1	1
1.1	0.909091	0.909100
1.5	0.666667	0.687500
2	0.500000	1.000000
0.5	2.000000	1.937500

Table 7 compares the polynomial $P_4(x; 1)$ found in Example 3 with the function $f(x) = 1/x$ for a few inputs near 1.

The graphs of the two functions for x near 1 are shown in Fig. 5.

The Taylor Series

A Taylor polynomial for f at a is a partial sum of the series

$$\sum_{n=0}^{\infty} \frac{f^{(n)}(a)(x-a)^n}{n!} = f(a) + f^{(1)}(a)(x-a) + \frac{f^{(2)}(a)}{2!}(x-a)^2 + \cdots + \frac{f^{(n)}(a)}{n!}(x-a)^n + \cdots.$$

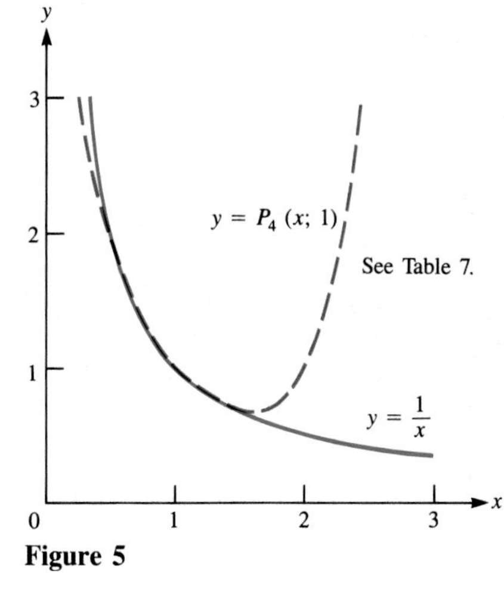

Figure 5

This series is called the **Taylor series at $x = a$** associated with the function $f(x)$. When $a = 0$, the series is also called the **Maclaurin series** associated with $f(x)$.

Computations like those in Example 1 show that the Maclaurin series associated with e^x is

$$\sum_{n=0}^{\infty} \frac{x^n}{n!} = 1 + x + \frac{x^2}{2!} + \frac{x^3}{3!} + \cdots + \frac{x^n}{n!} + \cdots.$$

Furthermore, computations similar to those in Example 2 show that the Maclaurin series associated with $\sin x$ is

$$\sum_{n=0}^{\infty} (-1)^n \frac{x^{2n+1}}{(2n+1)!} = x - \frac{x^3}{3!} + \frac{x^5}{5!} - \cdots + (-1)^n \frac{x^{2n+1}}{(2n+1)!} + \cdots$$

and with $\ln(1+x)$ is

$$x - \frac{x^2}{2} + \frac{x^3}{3} - \frac{x^4}{4} + \cdots + \frac{(-1)^{n+1}x^n}{n} + \cdots.$$

There are important questions to ask about such series:

For what values of x does it converge?

If it converges, does it approach $f(x)$?

If it does approach $f(x)$, how quickly?

To answer these questions, we will examine how the "error" $f(x) - P_n(x;\ a)$ behaves as n increases. In practical terms, we may want to know how large to choose n to estimate $f(x)$ within some prescribed error, such as 0.0005. We show how to do this in the next section.

Section Summary

We introduced the nth-order Taylor polynomial at a:

$$P_n(x;\ a) = \sum_{k=0}^{n} \frac{f^{(k)}(a)(x-a)^k}{k!}$$

It is the unique polynomial of degree at most n such that it and its first n derivatives at a coincide with $f(a)$ and the first n derivatives of $f(x)$ at a. The Taylor polynomials are the partial sums of the Taylor series at a:

$$\sum_{k=0}^{\infty} \frac{f^{(k)}(a)(x-a)^k}{k!}.$$

When $a = 0$, the Taylor series reduces to

$$\sum_{k=0}^{\infty} \frac{f^{(k)}(0)x^k}{k!},$$

which is the Maclaurin series associated with $f(x)$.

EXERCISES FOR SEC. 11.1: TAYLOR SERIES

In Exercises 1 to 12 compute the given Taylor polynomials $P_n(x; a)$ and graph them and the given function relative to the same axes for x near a. (A scientific calculator that computes Taylor series and plots graphs would be helpful.)

1 $P_1(x; 0)$, $P_2(x; 0)$ for $1/(1 + x)$

2 $P_1(x; 1)$, $P_2(x; 1)$ for $1/(1 + x)$

3 $P_1(x; 0)$, $P_2(x; 0)$, $P_3(x; 0)$ for $\ln (1 + x)$

4 $P_1(x; 1)$, $P_2(x; 1)$, $P_3(x; 1)$ for $\ln (1 + x)$

5 $P_1(x; 0)$, $P_2(x; 0)$, $P_3(x; 0)$, $P_4(x; 0)$ for e^x

6 $P_1(x; 2)$, $P_2(x; 2)$, $P_3(x; 2)$, $P_4(x; 2)$ for e^x

7 $P_1(x; 0)$, $P_2(x; 0)$, $P_3(x; 0)$ for $\tan^{-1} x$

8 $P_1(x; -1)$, $P_2(x; -1)$, $P_3(x; -1)$ for $\tan^{-1} x$

9 $P_2(x; 0)$, $P_4(x; 0)$ for $\cos x$

10 $P_7(x; 0)$ for $\sin x$

11 $P_4(x; \pi/4)$ for $\cos x$

12 $P_3(x; \pi/4)$ for $\sin x$

In Exercises 13 to 18 compute the Maclaurin series associated with the given function.

13 $\ln (1 + x)$

14 $\ln (1 - x)$

15 $\cos x$

16 $\sin x$

17 e^{-x}

18 $1/(1 - x)$

Some calculators are programmed to compute Taylor polynomials. Do Exercises 19 and 20 with such a calculator (or with a computer).

19 (*a*) Graph $P_{10}(x; 0)$ for e^x.
(*b*) Compute $P_{10}(x; 0)$ and e^x at $x = 1$, $x = 2$, and $x = 4$.

20 (*a*) Graph $P_{10}(x; 1)$ for $\ln x$.
(*b*) Compute $P_{10}(x; 1)$ and $\ln x$ at $x = 1/2$, $x = 2$, $x = 4$.

21 Let $f(x) = 2 + 3x - 4x^2$.
(*a*) Find $P_1(x; 0)$, $P_2(x; 0)$, and $P_3(x; 0)$.
(*b*) What is the Maclaurin series associated with $f(x)$?

22 What can be said about the degree of the polynomial $P_n(x; 0)$?

23 Let $f(x) = (1 + x)^3$. (*a*) Find $P_3(x; 0)$. (*b*) Check that your answer to (*a*) is correct by expanding $(1 + x)^3$.

24 Let $f(x) = (1 + x)^4$. (*a*) Find $P_4(x; 0)$. (*b*) Check that your answer to (*a*) is correct by expanding $(1 + x)^4$.

25 Find $P_6(x; \pi/6)$ for $\sin x$.

26 Find $P_6(x; \pi/4)$ for $\cos x$.

27 (*a*) Find $P_4(x; 0)$ for $\sqrt{1 + x}$.
(*b*) Graph $P_4(x; 0)$ and $\sqrt{1 + x}$ for x in $[1, -1]$.
(*c*) Fill in this table (four decimal places).

x	-1	-0.5	-0.1	0	0.1	0.5	1
$\sqrt{1+x}$							
$P_4(x; 0)$							

28 (*a*) Find $P_4(x; 1)$ for $1/x$.
(*b*) Graph $P_4(x; 1)$ and $1/x$ for x in $(0, 3]$.
(*c*) Fill in this table (four decimal places).

x	0.1	0.5	0.9	1	1.1	2	3
$1/x$							
$P_4(x; 1)$							

---

29 Can there be a polynomial $p(x)$ such that $\sin x = p(x)$ for all x in the interval $[1, 1.0001]$? Explain.

30 Can there be a polynomial $p(x)$ such that $\ln x = p(x)$ for all x in the interval $[1, 1.0001]$? Explain.

31 (*a*) Which polynomials are even functions?
(*b*) If f is an even function, is $P_n(x; 0)$ necessarily an even function? Explain.

32 (*a*) Which polynomials are odd functions?
(*b*) If f is an odd function, is $P_n(x; 0)$ necessarily an odd function? Explain.

33 (*a*) Is $\tan^{-1} x$ an odd function? an even function?
(*b*) What powers of x will have a coefficient of 0 in the Maclaurin series associated with $\tan^{-1} x$?
(*c*) Find $P_3(x; 0)$ for $\tan^{-1} x$.
(*d*) Using a calculator, fill in the table below for $P_3(x; 0)$ and $\tan^{-1} x$ for x in [0, 2]. (Write your entries to four decimal places.)

x	0	0.1	0.2	0.3	0.5	1	2
$P_3(x; 0)$							
$\tan^{-1} x$							

34 Find the Maclaurin series associated with $(1 + x)^n$, where n is a positive integer.

11.2 THE ERROR IN TAYLOR SERIES

If a function $f(x)$ has derivatives of all orders, we can grind them out at some number a and compute the associated Taylor series:

$$\sum_{n=0}^{\infty} \frac{f^{(n)}(a)}{n!}(x - a)^n.$$

But does that series converge to $f(x)$? Does it really "represent" $f(x)$? In other words, does the nth partial sum—the nth Taylor polynomial at a—approach $f(x)$ as $n \to \infty$?

To answer the question, we need to know what happens to the difference between $f(x)$ and $P_n(x; a)$ as $n \to \infty$. We hope that for "nice" functions,

$$f(x) - P_n(x; a) \to 0$$

as $n \to \infty$. This measure of the discrepancy between $f(x)$ and $P_n(x; a)$ is important enough to have a name.

The Remainder $R_n(x; a)$

Definition *The remainder (error)* $R_n(x; a)$. Let f be a function and let $P_n(x; a)$ be the associated nth-order Taylor polynomial at a. The number $R_n(x; a)$ defined by the equation

$$f(x) = P_n(x; a) + R_n(x; a)$$

is called the **remainder** or **error** in using the Taylor polynomial $P_n(x; a)$ to approximate $f(x)$.

If the remainder $R_n(x; a)$ does approach 0 as $n \to \infty$, then the function $f(x)$ is represented by its Taylor series. The following theorem expresses $R_n(x; a)$ in terms of a derivative (the "derivative" or "Lagrange" form of the remainder). This formula enables us to show that for $f(x) = e^x$, $\sin x$, or $\cos x$, $R_n(x; a)$ indeed approaches 0 as $n \to \infty$. (In Sec. 11.3 we will derive the formula.)

Theorem 1 *Lagrange's form of the remainder.* Assume that a function $f(x)$ has continuous derivatives of orders through $n + 1$ in an interval that includes the numbers a and x. Let $P_n(x; a)$ be the nth Taylor polynomial associated with $f(x)$ in powers of $x - a$. Then there is a number c_n between a and x such that

$$R_n(x; a) = \frac{f^{(n+1)}(c_n)}{(n+1)!}(x-a)^{n+1}.$$

To relate this theorem to our previous experience, take the case $n = 1$. The Taylor polynomial $P_1(x; a)$ is $f(a) + f'(a)(x - a)$, the linear approximation discussed in Secs. 4.9 and 4.10. Theorem 1 asserts that the error in using the linear approximation is equal to

$$\frac{f^{(2)}(c_2)(x-a)^2}{2!}$$

for some number c_2 between a and x. That is what we showed in Sec. 4.10, reasoning in terms of automobiles.

Theorem 1 does not give a formula for c_n. All that it says is that there is a number c_n between a and x such that $R_n(x; a)$ is equal to

$$\frac{f^{(n+1)}(c_n)}{(n+1)!}(x-a)^{n+1},$$

where c_n depends on a, x, and n.

In the case of some famous functions, the Lagrange form of the remainder is enough to let us show that $R_n(x; a) \to 0$ as $n \to \infty$.

EXAMPLE 1 Show that the Maclaurin series associated with $f(x) = e^x$ represents $f(x)$ for all x.

SOLUTION Take the case $x > 0$. We must show that $R_n(x; 0) \to 0$ as $n \to \infty$. By Theorem 1, there is for each positive integer n a number c_n between 0 and x such that

$$R_n(x; 0) = \frac{f^{(n+1)}(c_n)}{(n+1)!}x^{n+1}.$$

All the derivatives of the function e^x are simply e^x itself. Hence $f^{(n+1)}(c_n) = e^{c_n}$. Since $0 \le c_n \le x$ and $f(x)$ is increasing, we have

$$1 = e^0 \le e^{c_n} \le e^x.$$

Thus

Once we know that $R_n(x; 0) \to 0$ as $n \to \infty$, we know $f(x) = \sum_{n=0}^{\infty} f^{(n)}(0)x^n/n!$.

$$\frac{1}{(n+1)!}x^{n+1} \le R_n(x; 0) \le \frac{e^x}{(n+1)!}x^{n+1}.$$

Since x is fixed and $n \to \infty$, $R_n(x; 0) \to 0$ as $n \to \infty$.

For $x < 0$, the reasoning is similar to the case $x > 0$, the main difference being that we now have $x \le c_n \le 0$. The reader may carry out the details.

Since the series for e^x was already found in Sec. 11.1, we can therefore write

$$e^x = 1 + x + \frac{x^2}{2!} + \frac{x^3}{3!} + \cdots + \frac{x^n}{n!} + \cdots = \sum_{n=0}^{\infty} \frac{x^n}{n!}.$$

This equation clearly holds when $x = 0$, since it then reduces to the equation $e^0 = 1$. ■

EXAMPLE 2 Show that the Maclaurin series associated with $f(x) = \sin x$ represents $f(x)$ for all x.

SOLUTION All that is needed is to show that $R_n(x; 0) \to 0$ as $n \to \infty$. Now, by the Lagrange form of the remainder,

$$R_n(x; 0) = \frac{f^{(n+1)}(c_n)x^{n+1}}{(n+1)!},$$

where c_n is between 0 and x.

If $f(x) = \sin x$, then $f^{(1)}(x) = \cos x$, $f^{(2)}(x) = -\sin x$, $f^{(3)}(x) = -\cos x$, $f^{(4)}(x) = \sin x$, and so on. The higher derivatives are either $\pm\sin x$ or $\pm\cos x$. Thus, for any nonnegative integer n and real number c,

$$|f^{(n+1)}(c)| \le 1.$$

Consequently, $$|R_n(x; 0)| = \frac{|f^{(n+1)}(c_n)x^{n+1}|}{(n+1)!} \le \frac{|x|^{n+1}}{(n+1)!},$$

which approaches 0 as $n \to \infty$. (We showed in Sec. 10.2 that $\lim_{n\to\infty} k^n/n! = 0$ for any number k.)

Hence the Maclaurin series associated with $\sin x$ represents $\sin x$ for all x. Since that series is $\Sigma_{n=0}^{\infty} (-1)^n x^{2n+1}/(2n+1)!$, we have

$$\sin x = x - \frac{x^3}{3!} + \frac{x^5}{5!} - \cdots + (-1)^n \frac{x^{2n+1}}{(2n+1)!} + \cdots.$$

(Terms in the Maclaurin series with value 0 are not shown.) ■

Using a front end of a power series to estimate $\sin x$

Example 2 provides a remarkably efficient way to estimate $\sin x$ for x in the range, say, 0° to 45°: Use the polynomial $x - x^3/3! + x^5/5!$ or $x - x^3/3!$ (x is the radian measure of the angle). The error in using the first estimate is less than $(\pi/4)^7/7! < 0.00004$. (Why?) The error in using the second estimate, $x - x^3/6$, is less than $(\pi/4)^5/5! < 0.003$.

The front end of the power-series representation of a function $f(x)$ is often taken as an approximation of the function. The next example illustrates this use of power series.

EXAMPLE 3 Estimate $\sqrt{e} = e^{1/2}$ using the first four terms of the Maclaurin series for e^x. Discuss the error.

SOLUTION We have $f(x) = e^x = 1 + x + \dfrac{x^2}{2!} + \dfrac{x^3}{3!} + \cdots$.

Thus $e^{1/2}$ is approximated by

$$1+\left(\frac{1}{2}\right)+\frac{(\frac{1}{2})^2}{2!}+\frac{(\frac{1}{2})^3}{3!}=1+\frac{1}{2}+\frac{1}{8}+\frac{1}{48}\approx 1.64583.$$

By Lagrange's form of the remainder, the error can be written as

$$\frac{f^{(4)}(c)(\frac{1}{2})^4}{4!}$$

for some number c in $[0, \frac{1}{2}]$. Since $f(x)=e^x, f^{(4)}(x)=e^x$. Thus the error is of the form

$$\frac{e^c(\frac{1}{16})}{24}=\frac{e^c}{384} \qquad \text{for } c \text{ in } [0, \tfrac{1}{2}].$$

Since $e^c \le e^{1/2} < 2$ (because $e < 3$), we see that the error is positive but less than $\frac{2}{384}=\frac{1}{192}\approx 0.0052$. Thus, to two decimal places, $e^{1/2}\approx 1.65$. ■

In statistics, the integral $\int_{-\infty}^{b}(1/\sqrt{2\pi})\,e^{-x^2/2}\,dx$ is of major importance. Since $e^{-x^2/2}$ does not have an elementary antiderivative, the integral must be estimated by other means and its values tabulated. For instance, Burington's *Handbook of Mathematical Tables and Formulas,* 5th ed. (McGraw-Hill, 1972), lists it for b in the range $[0, 4]$ at intervals of 0.01.

Using a front end of a power series to estimate $\int_a^b f(x)\,dx$

The next example shows how to estimate $\int_a^b f(x)\,dx$ when $f(x)$ is representable by a power series.

EXAMPLE 4 Use the Maclaurin series for e^x to estimate $\displaystyle\int_0^1 e^{-x^2}\,dx$.

SOLUTION

$$e^x = 1 + x + \frac{x^2}{2!}+\frac{x^3}{3!}+\cdots.$$

Replacing x by $-x^2$ yields

$$e^{-x^2}=1-x^2+\frac{x^4}{2!}-\frac{x^6}{3!}+\cdots. \qquad (1)$$

For $0<|x|\le 1$, series (1) is a decreasing alternating series. Thus

$$1-x^2+\frac{x^4}{2!}-\frac{x^6}{3!}<e^{-x^2}<1-x^2+\frac{x^4}{2!}-\frac{x^6}{3!}+\frac{x^8}{4!}.$$

Hence

$$\int_0^1\left(1-x^2+\frac{x^4}{2!}-\frac{x^6}{3!}\right)dx<\int_0^1 e^{-x^2}\,dx<\int_0^1\left(1-x^2+\frac{x^4}{2!}-\frac{x^6}{3!}+\frac{x^8}{4!}\right)dx,$$

or

$$1-\frac{1}{3}+\frac{1}{5\cdot 2!}-\frac{1}{7\cdot 3!}<\int_0^1 e^{-x^2}\,dx<1-\frac{1}{3}+\frac{1}{5\cdot 2!}-\frac{1}{7\cdot 3!}+\frac{1}{9\cdot 4!}.$$

From this it follows that

Supply the omitted arithmetic.

$$0.742<\int_0^1 e^{-x^2}\,dx<0.748. \quad ■$$

The next example shows how to use a Maclaurin series to estimate a number to a prescribed number of decimal places.

EXAMPLE 5 Use the Maclaurin series for e^x to estimate e to four places.

SOLUTION To get four-place accuracy the estimate must be within 0.00005 of e. The error in using $P_n(1; 0)$ to approximate $e = e^1$ has the form

$$\frac{f^{(n+1)}(c_n)}{(n+1)!}(1-0)^{n+1},$$

where $f(x) = e^x$ and $0 \le c_n \le 1$. Since all higher derivatives of e^x are e^x itself, the error has the form

$$\frac{e^{c_n}}{(n+1)!}.$$

We wish to find n such that

$$\frac{e^{c_n}}{(n+1)!} \le 0.00005.$$

Even though we do not know the exact value of c_n, we can say that $c_n \le 1$, so $e^{c_n} \le e$. Furthermore, though we do not know the value of e (that is what we are trying to find!), we do know that it is less than 3. Hence $e^{c_n} \le 3$. Therefore, let us find n such that

$$\frac{3}{(n+1)!} \le 0.00005. \tag{2}$$

A calculator shows that

$$\frac{3}{8!} \approx 0.000074$$

and

$$\frac{3}{9!} \approx 0.000008.$$

The smallest n that satisfies (2) is therefore 8. We can use $P_8(1; 0)$ to estimate e. Using $P_8(1; 0)$, we get the estimate

$$1 + 1 + \frac{1}{2!} + \frac{1}{3!} + \frac{1}{4!} + \frac{1}{5!} + \frac{1}{6!} + \frac{1}{7!} + \frac{1}{8!},$$

which a calculator shows is approximately 2.718278770. Thus $e \approx 2.7183$ to four decimal places. ■

Beyond Lagrange

The Maclaurin series for $1/(1 + x)$ is

$$1 - x + x^2 - x^3 + \cdots + (-1)^n x^n + \cdots.$$

This is a geometric series with first term 1 and ratio $-x$. Since it converges to $1/(1 + x)$ when $|x| < 1$, we know that $R_n(x; 0) \to 0$ as $n \to \infty$. (See Sec. 10.3.)

Let us see what happens when we use Lagrange's formula in the case $x = -\frac{1}{2}$. It says that there is a number c_n in $[-\frac{1}{2}, 0]$ such that

$$R_n(-\tfrac{1}{2}; 0) = \frac{f^{(n+1)}(c_n)}{(n+1)!}\left(-\frac{1}{2}\right)^{n+1}$$

Carry out the computation. A straightforward computation shows that

$$f^{(n+1)}(c_n) = (-1)^{n+1}(n+1)!(1+c_n)^{-n-2}.$$

Hence
$$\left|R_n\left(-\frac{1}{2}; 0\right)\right| = \frac{1}{(1+c_n)^{n+2}}\left(\frac{1}{2}\right)^{n+1} = \frac{1}{(2+2c_n)^{n+1}}\,\frac{1}{1+c_n}. \tag{3}$$

All that we know about c_n is that it is in $[-\frac{1}{2}, 0]$. If it is close to $-\frac{1}{2}$, then (3) could be very near 2. So the Lagrange formula for the remainder fails to show that $R_n(-\frac{1}{2}; 0)$ approaches 0 as $n \to \infty$.

The formula fails to give any information when $-1 < x \leq -\frac{1}{2}$, but it does show that $R_n(x; 0) \to 0$ for $-\frac{1}{2} < x < 1$, as may be checked.

There is another form for the remainder, which expresses it as an integral rather than as a derivative. That form, which does show that $R_n(x; 0) \to 0$ as $n \to \infty$ for all x such that $|x| < 1$, is included in any advanced calculus text.

A SURPRISING POLYNOMIAL

The basic idea of Taylor's series is to approximate a function by a polynomial whose derivatives up through order n at some point coincide with those of the function.

We might expect that the following procedure would also produce a good polynomial approximation to a given function $f(x)$ throughout an interval $[a, b]$. Divide the interval into n sections of equal length by $n + 1$ points, of which the leftmost is a and the rightmost is b. There is a unique polynomial $P(x)$ of degree at most n that coincides with the given function at these $n + 1$ values. We would expect that when n is large, $|P(x) - f(x)|$ would be small for all x in $[a, b]$.

This is not the case, even for such a pleasant function as $f(x) = 1/(1 + x^2)$ and the interval $[-5, 5]$. In numerical analysis it is proved that for large n the polynomial you get this way does not stay near $1/(1 + x^2)$. Some of its values become arbitrarily large as n increases. In fact, if $n = 5m + 1$ and m is odd the polynomial $P(x)$ differs from $1/(1 + x^2)$ at $x = 4.875$ by more than $1.8^m/451$, a quantity that grows exponentially as m increases. It is surprising phenomena like this that show why intuition is no substitute for a rigorous, careful proof.

Section Summary

We gave a formula for the error in using a Taylor polynomial, namely

$$R_n(x; a) = \frac{f^{(n+1)}(c_n)}{(n+1)!}(x-a)^{n+1}.$$

With the aid of this formula we can show that the Maclaurin series for e^x, $\sin x$, and $\cos x$ represent the functions. In addition, the formula can be used to control the error in estimating an integral when Taylor polynomials are used as approximations of integrands.

EXERCISES FOR SEC. 11.2: THE ERROR IN TAYLOR SERIES

1 Show that the Maclaurin series associated with $\cos x$ represents $\cos x$ for all x.

2 Show that the Maclaurin series associated with e^{-x} represents e^{-x} for all x.

3 Show that the Maclaurin series associated with a polynomial $f(x)$ represents the polynomial for all x. [Examine $R_n(x; 0)$.]

4 Show that the Maclaurin series associated with $1/(1 + x)$ represents $1/(1 + x)$ for all $-\frac{1}{2} < x < 1$. [Examine $R_n(x; 0)$. Actually the representation holds for $-1 < x < 1$.]

5 Show that the Taylor series in powers of $x - a$ for e^x represents e^x for all x.

6 Show that the Taylor series in powers of $x - a$ for $\cos x$ represents $\cos x$ for all x.

7 Estimate e^{-1} to three places, using the Maclaurin series for e^x.

8 Estimate e^2 to three places, using the Maclaurin series for e^x.

9 Estimate $\cos 20°$ to three decimal places. (First replace $20°$ by $20\pi/180 = \pi/9$.) For π use 3.14159 so that round-off errors are kept small.

10 Estimate $\sin 40°$ with the aid of the Taylor series in powers of $x - \pi/4$. Use 3.14159 for π and obtain an estimate accurate to three decimal places.

11 Let $f(x) = \sqrt{x}$.
(*a*) Use a differential to estimate $\sqrt{23}$.
(*b*) Find $P_1(x; 25)$ and use it to estimate $\sqrt{23}$.
(*c*) Calculate the error in the approximation in (*b*).
(*d*) Use $P_2(x; 25)$ to estimate $\sqrt{23}$.
(*e*) Calculate the error in (*d*).
(*f*) Compare the error calculated in (*e*) with a bound given by Theorem 1.

12 Let $f(x) = \sin x$.
(*a*) Use a differential to estimate $\sin 47°$. (Work in radians.)
(*b*) Find $P_1(x; \pi/4)$ and use it to estimate $\sin 47°$.
(*c*) Calculate the error in (*b*).
(*d*) Use $P_2(x; \pi/4)$ to estimate $\sin 47°$.
(*e*) Calculate the error in (*d*).
(*f*) Compare the error calculated in (*e*) with a bound given by Theorem 1.

13 (*a*) Show that for x in $[0, 2]$

$$x + \frac{x^2}{2!} + \frac{x^3}{3!} + \cdots + \frac{x^n}{n!} \le e^x - 1$$

$$\le x + \frac{x^2}{2!} + \frac{x^3}{3!} + \cdots + \frac{x^n}{n!} + \frac{e^2 x^{n+1}}{(n+1)!}.$$

(*b*) Use (*a*) to find $\int_0^2 \frac{e^x - 1}{x}\,dx$ to three decimal places.

14 Find $\int_0^1 \frac{1 - \cos x}{x}\,dx$ to three decimal places, using an approach like that in Exercise 13.

15 Estimate $\int_0^\infty e^{-5x^2}\,dx$ following these steps:
(*a*) Find a number b such that

$$\int_b^\infty e^{-5x^2}\,dx < 0.0005.$$

(Use the fact that $e^{-5x^2} < e^{-5x}$ for $x > 1$.)
(*b*) Let b be the number you found in (*a*). Estimate $\int_0^b e^{-5x^2}\,dx$ with an error of less than 0.0005. (Use the Maclaurin series for e^{-5x^2}.)
(*c*) Continue (*a*) and (*b*) to get a two-place estimate of $\int_0^\infty e^{-5x^2}\,dx$.

16 Estimate $\int_0^\infty \frac{\sin(x^6/100)}{x^6}\,dx$, following these steps:
(*a*) Find a number b such that

$$\left|\int_b^\infty \frac{\sin(x^6/100)}{x^6}\,dx\right| < 0.001.$$

(Use the fact that $|\sin x| \le 1$.)
(*b*) Let b be the number you found in (*a*). Estimate

$$\int_0^b \frac{\sin(x^6/100)}{x^6}\,dx,$$

with an error less than 0.001. (Use the Maclaurin series for $\sin x$.)
(*c*) Combine (*a*) and (*b*) to get a two-place estimate of $\int_0^\infty \frac{\sin(x^6/100)}{x^6}\,dx$.

---

17 Let f be a function that has derivatives of all orders for all x. Assume that $|f^{(n)}(x)| \le n$ for all n. Show why $f(x)$ is represented by its Maclaurin series.

18 Let f be a function that has derivatives of all orders for all x. Assume that $|f^{(n)}(x)| \le 2^n$ for all n. Does $R_n(x; 0)$ necessarily approach 0 as $n \to \infty$?

19 Evaluate (or estimate) $\int_0^1 dx/(1 + x^2)$ by
(*a*) the fundamental theorem of calculus,
(*b*) Simpson's method (six sections),
(*c*) the trapezoidal method (six sections),
(*d*) using a Taylor series (six nonzero terms).

20 Evaluate (or estimate) $\int_0^1 dx/(1 + x^3)$ as in Exercise 19.

Exercises 21 to 25 develop a function that is *not* represented by its Maclaurin series except at $x = 0$.

21 Graph $y = e^{-1/x^2}$ for x in $[-1, 1]$.

22 Show that $\lim_{x\to 0} (e^{-1/x^2}/x^n) = 0$, where n is any integer. (*Hint:* Let $t = 1/x^2$ and express the limit in terms of a limit as $t \to \infty$.)

23 Let $P(x)$ and $Q(x)$ be polynomials. Show that

$$\lim_{x\to 0} \frac{P(x)e^{-1/x^2}}{Q(x)} = 0.$$

24 Let $f(x) = e^{-1/x^2}$ if $x \neq 0$ and $f(0) = 0$.
(*a*) Find $f'(0)$. (*Hint:* Recall the definition of a derivative.)
(*b*) Find $f^{(2)}(0)$.
(*c*) Show why $f^{(n)}(0) = 0$ for all n.

25 Let $g(x) = e^{-1/x^2} + e^x$ if $x \neq 0$ and $g(0) = 1$.
(*a*) What is the Maclaurin series for $g(x)$?
(*b*) For which x does the Maclaurin series for $g(x)$ converge to $g(x)$? (This is a classic example of a function whose Maclaurin series converges for all x, yet represents the function at only one number.)

26 Assume that $f(x)$ has a continuous fourth derivative. Let M_4 be the maximum of $|f^{(4)}(x)|$ for x in $[-1, 1]$. Show that

$$\left|\int_{-1}^{1} f(x)\,dx - f\left(\frac{1}{\sqrt{3}}\right) - f\left(-\frac{1}{\sqrt{3}}\right)\right| \le \frac{7M_4}{270}.$$

Suggestion: Use the representation $f(x) = f(0) + f^{(1)}(0)x + f^{(2)}(0)x^2/2 + f^{(3)}(0)x^3/6 + f^{(4)}(c)x^4/24$, where c depends on x.

27 Assume that the Maclaurin series associated with $f(x)$ converges to $f(x)$ and that it is an alternating, decreasing series.
(*a*) Using the fact that it is an alternating, decreasing series, what can you say about the error in using $P_n(x; 0)$ to approximate $f(x)$?
(*b*) Using Lagrange's form of the remainder, what can you say about the error in using $P_n(x; 0)$?

11.3 WHY THE ERROR IN TAYLOR SERIES IS CONTROLLED BY A DERIVATIVE

This section obtains Lagrange's formula for the error in using a Taylor polynomial to approximate a function. At the same time it shows how information about the higher derivatives of a function can tell us a good deal about the function itself. Before reading further, review Sec. 4.10, which discusses the growth of a function and the error in the linear approximation, $p(x) = P_1(x; a)$.

The Derivatives of the Error Function

The Taylor polynomial $p(x) = P_n(x; a)$ is constructed in such a way that it and its first n derivatives at a coincide with $f(x)$ and its first n derivatives at a:

$$p(a) = f(a),\ p'(a) = f'(a),\ \ldots,\ p^{(n)}(a) = f^{(n)}(a).$$

Let $R(x) = R_n(x; a) = f(x) - p(x)$, the error in using $p(x)$ to approximate $f(x)$. Then $R(a) = f(a) - p(a) = 0$, $R'(a) = f'(a) - p'(a) = 0$, ..., $R^{(n)}(a) = f^{(n)}(a) - p^{(n)}(a) = 0$. Consequently

$$\boxed{R(a) = 0,\ R^{(1)}(a) = 0,\ \ldots,\ R^{(n)}(a) = 0.} \tag{1}$$

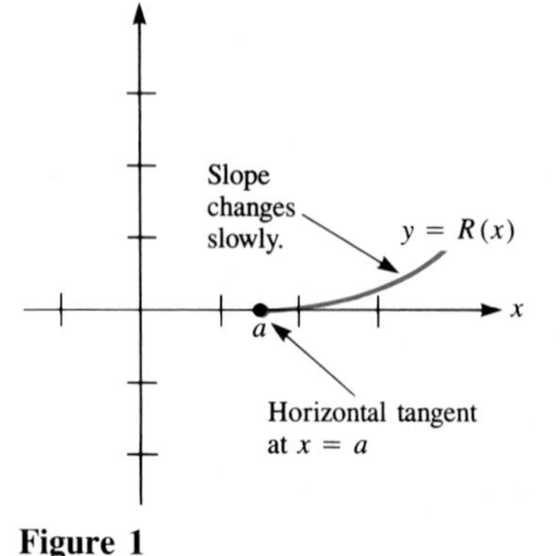

Figure 1

Consider the graph of the remainder function $R(x)$ near $x = a$. Because $R(a) = 0$, the graph passes through the point $(a, 0)$. Because $R^{(1)}(a) = 0$, the tangent at $(a, 0)$ is horizontal. The fact that $R^{(2)}(a) = 0$ suggests that $R^{(1)}(x)$ changes slowly for x near a. We expect the graph of $R(x)$ to be fairly flat near $x = a$, as suggested in Fig. 1.

We can also say something about $R^{(n+1)}(x)$. Recall that the $(n + 1)$st derivative of a polynomial of degree n is 0 for all x. Thus

$$R^{(n+1)}(x) = f^{(n+1)}(x) - p^{(n+1)}(x)$$
$$= f^{(n+1)}(x) - 0;$$

hence

$$\boxed{R^{(n+1)}(x) = f^{(n+1)}(x).} \tag{2}$$

The key facts about $R(x)$

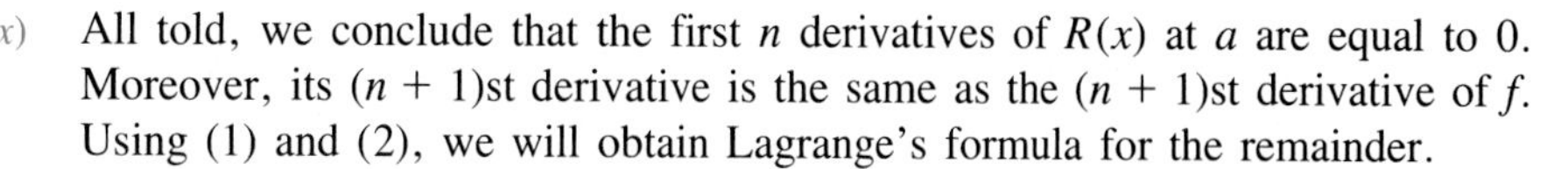

All told, we conclude that the first n derivatives of $R(x)$ at a are equal to 0. Moreover, its $(n + 1)$st derivative is the same as the $(n + 1)$st derivative of f. Using (1) and (2), we will obtain Lagrange's formula for the remainder.

The Growth of $R(x) = R_n(x; a)$

Let b be a number. We wish to show that

$$R(b) = \frac{f^{(n+1)}(c_n)}{(n+1)!}(b-a)^{n+1} \tag{3}$$

for some number c_n between a and b. The key is Theorem 1. The argument for Theorem 1 depends on two properties of definite integrals:

1 If $f(x) \le g(x)$, then $\int_a^b f(x)\,dx \le \int_a^b g(x)\,dx$ for $a < b$. (Sec. 5.5)
2 $\int_a^b h'(x)\,dx = h(b) - h(a)$. (fundamental theorem of calculus)

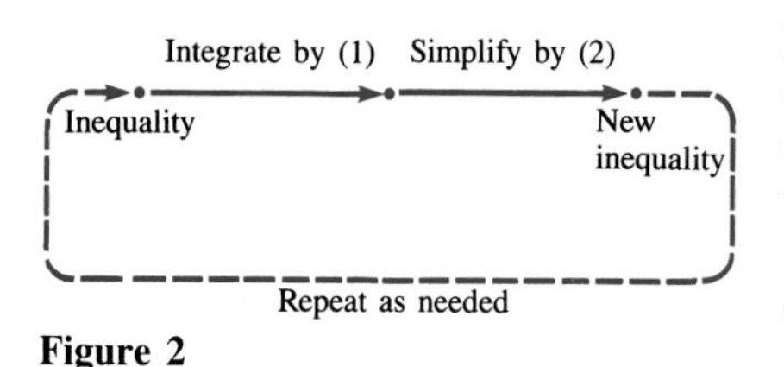

Figure 2

We sketch the idea of the proof: We start with an inequality and integrate it by property 1 to get a new inequality. Then we use property 2 to simplify one side of the inequality. Then we repeat the process as often as needed: integrate, simplify, integrate, simplify, ... Figure 2 is a flowchart of the reasoning, which is simpler than it may seem when you read the proof the first time.

We repeat the process n times to establish the following theorem, from which (3) will quickly follow.

A similar theorem holds for $b < a$.

Theorem 1 Let $R(x)$ have derivatives of all orders through the $(n + 1)$st for x in $[a, b]$. Assume that

$$R(a) = 0,\ R^{(1)}(a) = 0,\ \ldots,\ R^{(n)}(a) = 0$$

and that there are constants m and M such that

$$m \le R^{(n+1)}(x) \le M$$

for all x between a and b. Then

$$\frac{m(b-a)^{n+1}}{(n+1)!} \le R(b) \le \frac{M(b-a)^{n+1}}{(n+1)!}.$$

Proof We take the case $n = 2$. (The argument when n is larger is similar.) That is, we assume that

$$R(a) = 0, \qquad R'(a) = 0, \qquad R''(a) = 0$$

and that

$$m \le R'''(x) \le M,$$

where m and M are constants. We wish to show that

$$\frac{m(b-a)^3}{3!} \le R(b) \le \frac{M(b-a)^3}{3!}.$$

We shall prove that

$$R(b) \le \frac{M(b-a)^3}{3!}.$$

The proof that

$$\frac{m(b-a)^3}{3!} \le R(b)$$

is similar.

We start with

$$R'''(x) \le M$$

and obtain, by property 1,

Integrate an inequality.

$$\int_a^t R'''(x)\,dx \le \int_a^t M\,dx. \tag{4}$$

By property 2,

Fundamental theorem of calculus

$$\int_a^t R'''(x)\,dx = R''(t) - R''(a)$$

$$= R''(t) - 0, \text{ since } R''(a) = 0.$$

Thus $$\int_a^t R'''(x)\,dx = R''(t).$$

Now simplify the right side of (4):

$$\int_a^t M\,dx = Mx \Big|_a^t = M(t-a).$$

So (4) becomes

$$R''(t) \le M(t-a).$$

Using x again as our usual name for a variable, we have

$$R''(x) \le M(x-a). \tag{5}$$

We now have an inequality on $R''(x)$.

So information about the size of R''' has told us something about the size of R''.

Just as we integrated the inequality $R'''(x) \le M$, we integrate (5) by property 1 to obtain

Integrate again

$$\int_a^t R''(x)\,dx \le \int_a^t M(x-a)\,dx. \tag{6}$$

Fundamental theorem of calculus

Now $$\int_a^t R''(x)\,dx = R'(t) - R'(a) = R'(t) - 0 = R'(t),$$

and $$\int_a^t M(x-a)\,dx = \frac{M(x-a)^2}{2}\Big|_a^t = \frac{M(t-a)^2}{2} - 0.$$

Hence (6) becomes

$$R'(t) \le \frac{M(t-a)^2}{2}.$$

Returning to our customary variable x, we have

We now have an inequality on $R'(x)$.

$$R'(x) \le \frac{M(x-a)^2}{2}. \tag{7}$$

At this point we have a statement about the size of the first derivative R'. We repeat the process, integrating (7) to get information about the size of the function R itself.

Integration of (7) gives

Integrate again

$$\int_a^t R'(x)\,dx \le \int_a^t \frac{M(x-a)^2}{2}\,dx.$$

By calculations similar to those already done,

Fundamental theorem of calculus

$$\int_a^t R'(x)\,dx = R(t)$$

and

$$\int_a^t \frac{M(x-a)^2}{2}\,dx = \frac{M(t-a)^3}{3!}.$$

(At this stage we see why a factorial appears in the denominator.)

Replacing t by x, we have

$$R(x) \le \frac{M(x-a)^3}{3!}.$$

In particular, when x is b,

We finally have an inequality on $R(x)$ itself.

$$R(b) \le \frac{M(b-a)^3}{3!}.$$

A similar argument works for any value of n. It involves n integrations and simplifications in a row. ■

The Lagrange Formula

Theorem 1 provides a short proof of Theorem 2, ''Lagrange's form of the remainder.''

Theorem 2 Assume that a function f has continuous derivatives of orders through $n+1$ in an interval that includes a and b. Let $P_n(x; a)$ be the nth Taylor polynomial associated with f in powers of $x-a$. Then there is a number c_n between a and b such that

$$R_n(b; a) = \frac{f^{(n+1)}(c_n)(b-a)^{n+1}}{(n+1)!}.$$

Proof Let M be the maximum value and m the minimum value of $f^{(n+1)}(x)$ for x in the closed interval with endpoints a and b. Let

$$R(x) = f(x) - P_n(x; a).$$

Then

$$R(a) = R^{(1)}(a) = \cdots = R^{(n)}(a) = 0.$$

Moreover, since $P_n(x; a)$ is a polynomial of degree at most n, its $(n + 1)$st derivative is 0 for all x. Thus

$$R^{(n+1)}(x) = f^{(n+1)}(x).$$

So $R(x)$ satisfies the assumptions of Theorem 1. Consequently,

$$\frac{m(x - a)^{n+1}}{(n + 1)!} \leq R(x) \leq \frac{M(x - a)^{n+1}}{(n + 1)!}$$

for x in the interval with endpoints a and b. Thus

$$R(x) = \frac{q(x - a)^{n+1}}{(n + 1)!},$$

for some number q between m and M. Since $f^{(n+1)}$ is continuous and q is between the maximum and minimum values M and m of $f^{(n+1)}$, there is a number c_n between a and x such that

$$f^{(n+1)}(c_n) = q.$$

Thus

$$R(x) = \frac{f^{(n+1)}(c_n)(x - a)^{n+1}}{(n + 1)!}.$$

In particular, when x is b,

$$R(b) = \frac{f^{(n+1)}(c_n)(b - a)^{n+1}}{(n + 1)!}$$

for some number c_n between a and b. ■

EXERCISES FOR SEC. 11.3: WHY THE ERROR IN TAYLOR SERIES IS CONTROLLED BY A DERIVATIVE

1 In the case we examined in Theorem 1 ($n = 2$) we proved that $R(b) \leq M(b - a)^3/3!$ Establish the other inequality, $m(b - a)^3/3! \leq R(b)$.

2 Prove Theorem 1 for $n = 1$.

3 Prove Theorem 1 for $n = 3$.

4 Prove Theorem 1 for $n = 4$.

5 Let f be a function with continuous derivatives f', f'', and f'''. Assume that $f^{(3)}(x) \leq M$ and that $f(a) = 2$, $f'(a) = 1$, and $f''(a) = 3$. Integrating the inequality on $f^{(3)}(x)$ three times, obtain an inequality on $f(x)$.

6 Let f be a function with continuous derivatives through order $n + 1$. Assume that $m \leq f^{(n+1)}(x) \leq M$ for two constants m and M. Integrating this inequality n times over the interval $[a, x]$, what inequalities do you obtain about $f(x)$? Investigate for (*a*) $n = 1$, (*b*) $n = 2$, (*c*) $n = 3$. (*d*) What do you think the general result will be?

The following exercises use the method of this section to show how the higher derivatives control the size of the error in the various methods for estimating a definite integral: the left-point method, the trapezoidal method, the midpoint method, and Simpson's method. In each case $h = (b - a)/n$, where the interval $[a, b]$ is divided into n sections of equal lengths. $E(h)$ denotes the error for an individual section $[c, c + h]$ ($[c - h/2, c + h/2]$ in the midpoint method, $[c - h, c + h]$ in Simpson's method). (See Fig. 3.) The arguments are outlined in full. In each case a function $E(t)$ is introduced, its derivatives analyzed, and then a higher derivative is used to estimate the size of $E(h)$. Since there are n sections, the total error $E = \int_a^b f(x)\,dx -$ Estimate is at most $n\,E(h)$.

Local error in left endpoint method

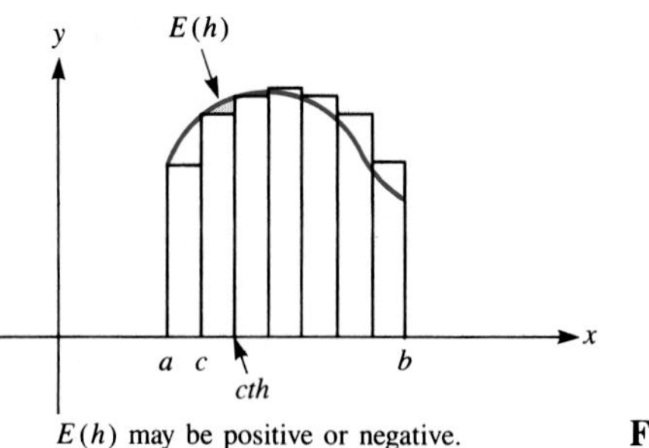

$E(h)$ may be positive or negative.

Figure 3

7 In the left-point method, the interval $[a, b]$ is divided into n sections of equal length and an approximating sum is formed with the left endpoints as sampling numbers.
(*a*) Show that one wishes to estimate

$$E(h) = \int_c^{c+h} f(x)\,dx - f(c)h.$$

(*b*) Introduce the function

$$E(t) = \int_c^{c+t} f(x)\,dx - f(c)t.$$

Show that $E(0) = 0$, $E^{(1)}(0) = 0$, and $E^{(2)}(t) = f^{(1)}(c+t)$.

(*c*) Let the maximum of $f^{(1)}(x)$ be M_1 and the minimum be m_1 for x in $[a, b]$. Integrating twice, deduce that

$$\frac{1}{2}m_1h^2 \le E(h) \le \frac{1}{2}M_1h^2.$$

(*d*) Show that the total error E in using the method satisfies the inequalities

$$\frac{m_1(b-a)h}{2} \le E \le \frac{M_1(b-a)h}{2}.$$

(Recall that there are n sections.)

8 In the midpoint method, the interval $[a, b]$ is divided into n sections of equal length and an approximating sum is formed with the midpoints as sampling numbers.
(*a*) Show that one wishes to estimate

$$E(h) = \int_{c-h/2}^{c+h/2} f(x)\,dx - f(c)h.$$

(*b*) Introduce the function

$$E(t) = \int_{c-t/2}^{c+t/2} f(x)\,dx - f(c)t.$$

Show that $E(0) = 0$, $E^{(1)}(0) = 0$, and that

$$E^{(2)}(t) = \frac{1}{4}[f^{(1)}(c+t/2) - f^{(1)}(c-t/2)].$$

(*c*) Use the mean-value theorem to show that

$$E^{(2)}(t) = \frac{t}{4}f^{(2)}(T)$$

for some number T between $c - t/2$ and $c + t/2$.

(*d*) Let the maximum value of $f^{(2)}(x)$ be M_2, and the minimum value be m_2 for x in $[a, b]$. From (*b*) and (*c*) deduce that

$$\frac{m_2h^3}{24} \le E(h) \le \frac{M_2h^3}{24}.$$

(*e*) Show that the total error E satisfies the inequalities

$$\frac{m_2(b-a)h^2}{24} \le E \le \frac{M_2(b-a)h^2}{24}.$$

9 Review the trapezoidal method of Sec. 5.4.
(*a*) Show that one wishes to estimate

$$E(h) = \int_c^{c+h} f(x)\,dx - \frac{h}{2}[f(c) + f(c+h)].$$

(*b*) Introduce the function

$$E(t) = \int_c^{c+t} f(x)\,dx - \frac{t}{2}[f(c) + f(c+t)].$$

Show that $E(0) = 0$, $E'(0) = 0$, and that

$$E''(t) = -\frac{t}{2}f''(c+t).$$

(*c*) Let the maximum value of $f''(x)$ be M_2 and the minimum value be m_2 for x in $[a, b]$. Deduce that

$$-\frac{m_2h^3}{12} \ge E(h) \ge -\frac{M_2h^3}{12}.$$

(*d*) Deduce that E, the total error in the trapezoidal method, satisfies the inequalities

$$\frac{-m_2(b-a)\,h^2}{12} \ge E \ge \frac{-M_2(b-a)\,h^2}{12}.$$

10 Review Simpson's method in Sec. 5.4, in particular Exercise 20 of that section. In this method the interval $[a, b]$ is divided into n sections of length $h = (b-a)/n$, and n is even. The approximation is based on $n/2$ sections of length $2h$ (a parabola being drawn through three points).
(*a*) Show that one wishes to estimate

$$E(h) = \int_{c-h}^{c+h} f(x)\,dx - \frac{h}{3}[f(c-h) + 4f(c) + f(c+h)].$$

(*b*) Introduce the function

$$E(t) = \int_{c-t}^{c+t} f(x)\,dx - \frac{t}{3}[f(c-t) + 4f(c) + f(c+t)]$$

and show that

$$E^{(1)}(t) = \frac{2}{3}[f(c+t) + f(c-t)] - \frac{4}{3}f(c) - \frac{t}{3}[f^{(1)}(c+t) - f^{(1)}(c-t)].$$

(*c*) Show that

$$E^{(2)}(t) = \frac{1}{3}[f^{(1)}(c+t) - f^{(1)}(c-t)] - \frac{t}{3}[f^{(2)}(c+t) + f^{(2)}(c-t)].$$

(*d*) Show that

$$E^{(3)}(t) = \frac{-t}{3}[f^{(3)}(c+t) - f^{(3)}(c-t)].$$

(*e*) Show that

$$E^{(3)}(t) = \frac{-2t^2}{3} f^{(4)}(T)$$

for some T in $(c-t, c+t)$.

(*f*) Show that

$$E(0) = 0, \qquad E^{(1)}(0) = 0, \qquad \text{and} \qquad E^{(2)}(0) = 0.$$

(*g*) Let the maximum of $f^{(4)}(x)$ be M_4 and the minimum be m_4 for x in $[a, b]$. Deduce that

$$-\frac{m_4 h^5}{90} \geq E(h) \geq -\frac{M_4 h^5}{90}.$$

(*h*) Deduce that the total error in Simpson's method E satisfies the inequalities

$$-\frac{m_4(b-a)h^4}{180} \geq E \geq -\frac{M_4(b-a)h^4}{180}.$$

11.4 POWER SERIES AND RADIUS OF CONVERGENCE

Our use of Taylor polynomials to approximate a function led us to consider series of the form

$$\sum_{n=0}^{\infty} a_n(x-a)^n = a_0 + a_1(x-a) + a_2(x-a)^2 + \cdots + a_n(x-a)^n + \cdots.$$

Such a series is called a **power series** in $x - a$. If $a = 0$, we obtain a series in powers of x:

$$\sum_{n=0}^{\infty} a_n x^n = a_0 + a_1 x + a_2 x^2 + \cdots a_n x^n + \cdots.$$

We will now look at some properties of power series and will see that they behave very much like polynomials.

The Convergence of a Power Series

For each fixed choice of x, a power series becomes a series with constant terms.

The power series $a_0 + a_1 x + a_2 x^2 + \cdots$ certainly converges when $x = 0$. It may or may not converge for other choices of x. However, as Theorem 1 will show, if the series converges at a certain value c, it converges at any number x whose absolute value is less than $|c|$. Since the proof of Theorem 1 uses the comparison test and the absolute-convergence test, it offers a nice review of important concepts from Chap. 10.

Theorem 1 Let c be a nonzero number. Assume that $\Sigma_{n=0}^{\infty} a_n c^n$ converges. Then, if $|x| < |c|$, $\Sigma_{n=0}^{\infty} a_n x^n$ converges. In fact, it converges absolutely.

Proof Since $\Sigma_{n=0}^{\infty} a_n c^n$ converges, the nth term $a_n c^n$ approaches 0 as $n \to \infty$. Thus there is an integer N such that for $n \geq N$, $|a_n c^n| \leq 1$. From now on in the proof, consider only $n \geq N$. Now,

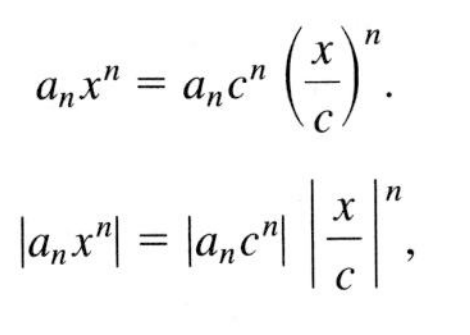

$$a_n x^n = a_n c^n \left(\frac{x}{c}\right)^n.$$

Since $$|a_n x^n| = |a_n c^n| \left|\frac{x}{c}\right|^n,$$

it follows that for $n \geq N$,

$$|a_n x^n| \leq \left|\frac{x}{c}\right|^n. \qquad (\text{since } |a_n c^n| \leq 1 \text{ for } n \geq N)$$

The series $$\sum_{n=N}^{\infty} \left|\frac{x}{c}\right|^n$$

is a geometric series with the ratio $|x/c| < 1$. Hence it converges.

Since $$|a_n x^n| \leq \left|\frac{x}{c}\right|^n$$

for $n \geq N$, the series $$\sum_{n=N}^{\infty} |a_n x^n|$$

converges (by the comparison test). Thus $\sum_{n=N}^{\infty} a_n x^n$ converges (in fact, absolutely). Putting in the front end $\sum_{n=0}^{N-1} a_n x^n$, we conclude that the series $\sum_{n=0}^{\infty} a_n x^n$ converges absolutely if $|x| < |c|$. ■

The Radius of Convergence of $\sum_{n=0}^{\infty} a_n x^n$

The x's for which the series converges form an unbroken set.

By Theorem 1, the set of numbers x such that $\sum_{n=0}^{\infty} a_n x^n$ converges has no holes. In other words, the set of such x consists of one unbroken piece, which includes the number 0. Moreover if c is in that set, so is the entire open interval $(-|c|, |c|)$.

There are two possibilities. In the first case, there are arbitrarily large c's such that the series converges for x in $(-c, c)$. This means that the series converges for all x. In the second case, there is a bound on the numbers c such that the series converges for x in $(-c, c)$. It is shown in advanced calculus that there is then a smallest bound on the c's; call it R. Consequently,

1 Either $a_0 + a_1 x + a_2 x^2 + \cdots$ converges for all x,
2 Or there is a number R such that $a_0 + a_1 x + a_2 x^2 + \cdots$ converges for all x such that $|x| < R$ but diverges when $|x| > R$.

See Fig. 1.

In the second case, R is called the **radius of convergence** of the series. In case 1, the radius of convergence is said to be infinite, $R = \infty$. For the geometric series $1 + x + x^2 + \cdots + x^n + \cdots$, $R = 1$, since the series converges when $|x| < 1$ and diverges when $|x| > 1$. (It also diverges when $x = 1$ and $x = -1$.) A power series with radius of convergence R may or may not converge at $x = R$ and at $x = -R$. For convenient reference, these observations are stated as Theorem 2.

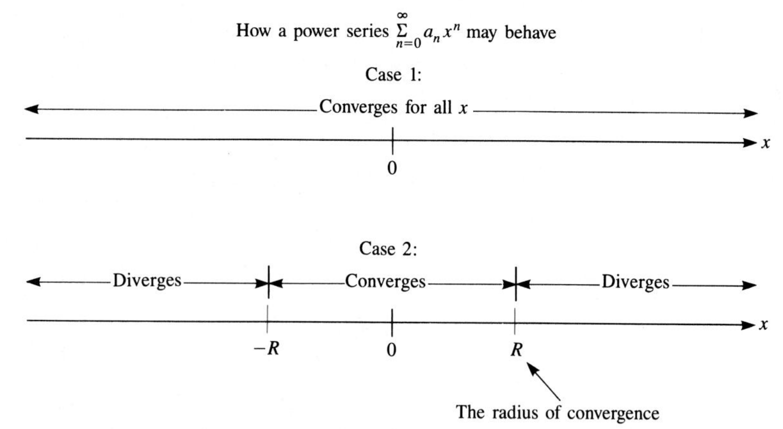

Figure 1

> **Theorem 2** Associated with the power series $\Sigma_{n=0}^{\infty} a_n x^n$ is a radius of convergence R. If $R = 0$, the series converges only for $x = 0$. If R is a positive real number, the series converges for $|x| < R$ and diverges for $|x| > R$. If R is ∞, the series converges for all x.

EXAMPLE 1 Find all values of x for which $x - \frac{x^2}{2} + \frac{x^3}{3} - \frac{x^4}{4} + \cdots + \frac{(-1)^{n+1} x^n}{n} + \cdots$ converges.

SOLUTION Because of the presence of x^n, use the absolute-ratio test. The absolute value of the ratio of successive terms is

$$\left| \frac{\dfrac{(-1)^{n+2}x^{n+1}}{n+1}}{\dfrac{(-1)^{n+1}x^n}{n}} \right| = |x| \frac{n}{n+1}.$$

As $n \to \infty$, $n/(n+1) \to 1$. Thus,

if $|x| < 1$, $$\lim_{n\to\infty} |x| \frac{n}{n+1} = |x| < 1;$$

if $|x| > 1$, $$\lim_{n\to\infty} |x| \frac{n}{n+1} = |x| > 1.$$

The absolute-ratio test takes care of $|x| < 1$ and $|x| > 1$.

By the absolute-ratio test, the series converges for $|x| < 1$ and diverges for $|x| > 1$. It remains to see what happens when $x = 1$ or $x = -1$.

Checking the behavior at $x = 1$

For $x = 1$, we obtain the alternating harmonic series:

$$1 - \tfrac{1}{2} + \tfrac{1}{3} - \tfrac{1}{4} + \cdots.$$

This series converges, by the alternating-series test. Thus $x - x^2/2 + x^3/3 - x^4/4 + \cdots$ converges when $x = 1$. Hence the series $x - x^2/2 + x^3/3 - x^4/4 + \cdots$ converges for $-1 < x \leq 1$.

Checking the behavior at $x = -1$

What about $x = -1$? The series then becomes

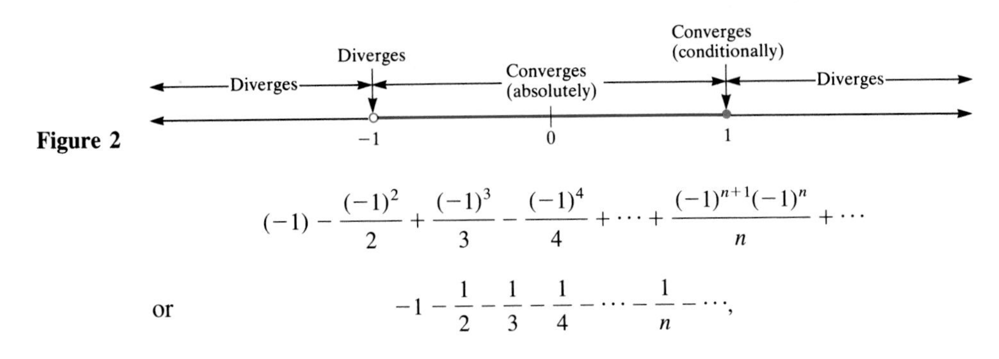

Figure 2

$$(-1) - \frac{(-1)^2}{2} + \frac{(-1)^3}{3} - \frac{(-1)^4}{4} + \cdots + \frac{(-1)^{n+1}(-1)^n}{n} + \cdots$$

or

$$-1 - \frac{1}{2} - \frac{1}{3} - \frac{1}{4} - \cdots - \frac{1}{n} - \cdots,$$

which, being the negative of the harmonic series, diverges.

The radius of convergence is $R = 1$. Figure 2 records the information obtained about the series. ■

EXAMPLE 2 Find the radius of convergence of

$$\sum_{n=0}^{\infty} \frac{x^n}{n!} = 1 + x + \frac{x^2}{2!} + \frac{x^3}{3!} + \cdots + \frac{x^n}{n!} + \cdots,$$

the Maclaurin series for e^x.

SOLUTION Because of the presence of the power x^n and the factorial $n!$ and the fact that x may be negative, the absolute-ratio test is the logical test to use. The absolute value of the ratio between successive terms is

$$\left| \frac{\dfrac{x^{n+1}}{(n+1)!}}{\dfrac{x^n}{n!}} \right| = |x| \frac{n!}{(n+1)!} = \frac{|x|}{n+1}.$$

Since

$$\lim_{n \to \infty} \frac{|x|}{n+1} = 0,$$

the limit of the ratio between successive terms is 0, which is less than 1. Consequently, the series converges for all x. That is, the radius of convergence R is infinite. ■

A case where $R = \infty$

The next example represents the opposite extreme, $R = 0$.

EXAMPLE 3 Find the radius of convergence of the series

$$\sum_{n=1}^{\infty} n^n x^n = 1x + 2^2x^2 + 3^3x^3 + \cdots + n^n x^n + \cdots.$$

SOLUTION The series converges for $x = 0$.

If $x \neq 0$, consider the nth term $n^n x^n$, which can be written as $(nx)^n$. As $n \to \infty$, $|nx| \to \infty$. Thus the nth term does not approach 0 as $n \to \infty$. By the nth-term test, the series diverges. In short, the series converges only when $x = 0$. The radius of convergence in this case is $R = 0$. ■

A case where $R = 0$

Radius of Convergence of $\sum_{n=0}^{\infty} a_n(x-a)^n$

Just as a power series in x has an associated radius of convergence, so does a power series in $x - a$. To see this, consider any such power series,

$$\sum_{n=0}^{\infty} a_n(x-a)^n = a_0 + a_1(x-a) + a_2(x-a)^2 + \cdots. \tag{1}$$

Let $u = x - a$. Then series (1) becomes

$$\sum_{n=0}^{\infty} a_n u^n = a_0 + a_1 u + a_2 u^2 + \cdots. \tag{2}$$

Series (2) has a certain radius of convergence R. That is, (2) converges for $|u| < R$ and diverges for $|u| > R$. Consequently (1) converges for $|x - a| < R$ and diverges for $|x - a| > R$. The number R is called the radius of convergence of the series (1). (R may be infinite.) As Fig. 3 suggests, the series $\Sigma_{n=0}^{\infty} a_n(x-a)^n$ converges in an interval whose midpoint is a. The question marks in Fig. 3 indicate that the series may converge or may diverge at the numbers $a - R$ and $a + R$. These cases must be looked at separately.

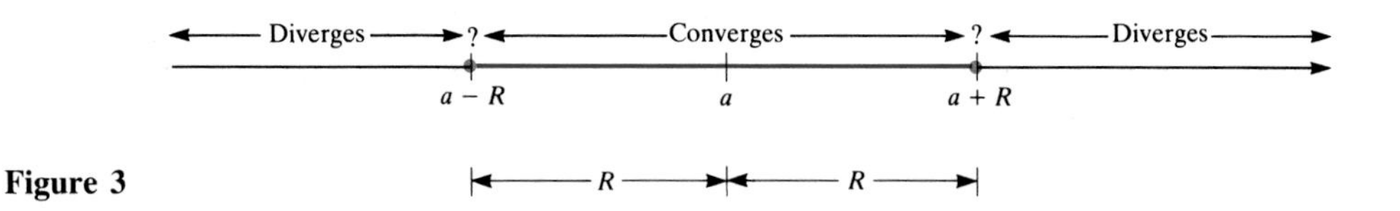

Figure 3

These observations are summarized in the following theorem.

> **Theorem 3** Associated with the power series $\Sigma_{n=0}^{\infty} a_n(x-a)^n$ is a radius of convergence R. If $R = 0$, the series converges only for $x = a$. If R is a positive real number, the series converges for $|x - a| < R$ and diverges for $|x - a| > R$. If $R = \infty$, the series converges for all x.

EXAMPLE 4 Find all values of x for which

The series $\sum_{n=1}^{\infty} (-1)^{n+1} \frac{(x-1)^n}{n}$

$$(x-1) - \frac{(x-1)^2}{2} + \frac{(x-1)^3}{3} - \frac{(x-1)^4}{4} + \cdots \tag{3}$$

converges.

SOLUTION Note that this is Example 1 with x replaced by $x - 1$. Thus $x - 1$ plays the role that x played in Example 1. Consequently, series (3) converges for $-1 < x - 1 \le 1$, that is, for $0 < x \le 2$, and diverges for all other values of x. Its radius of convergence is $R = 1$. The set of values where the series converges is an interval whose midpoint is $x = 1$. The convergence of (3) is recorded in Fig. 4.

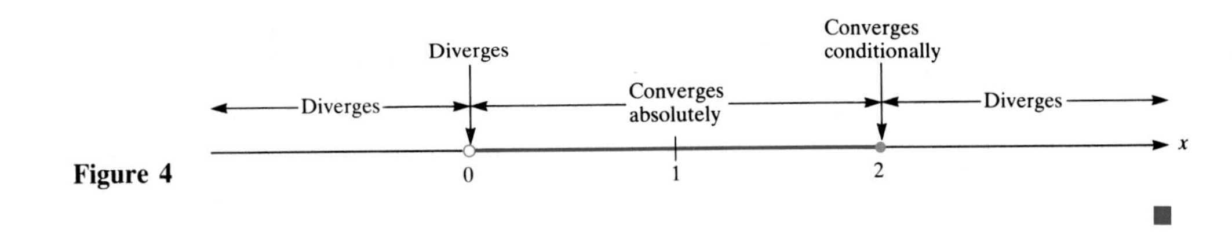

Figure 4

■

The General Binomial Theorem

Appendix C reviews the binomial theorem.

If r is 0 or a positive integer, $(1 + x)^r$ is a polynomial of degree r. Its Maclaurin series has only a finite number of nonzero terms, the one of highest degree being x^r. The formula

$$(1 + x)^r = \sum_{k=0}^{r} \frac{r!}{k!(r-k)!} x^k$$

is known as the binomial theorem. It can also be written

$$(1 + x)^r = \sum_{k=0}^{r} \frac{r(r-1)\cdots(r-k+1)}{1 \cdot 2 \cdots k} x^k.$$

Newton generalized the binomial theorem to all exponents, as illustrated in Example 5.

EXAMPLE 5 Find the Maclaurin series associated with $f(x) = (1 + x)^r$, where r is not 0 or a positive integer and determine its radius of convergence.

SOLUTION The following table will help in computing $f^{(n)}(0)$:

n	$f^{(n)}(x)$	$f^{(n)}(0)$
0	$(1 + x)^r$	1
1	$r(1 + x)^{r-1}$	r
2	$r(r-1)(1 + x)^{r-2}$	$r(r-1)$
3	$r(r-1)(r-2)(1 + x)^{r-3}$	$r(r-1)(r-2)$
...	...	...
n	$r(r-1)\cdots(r-n+1)(1 + x)^{r-n}$	$r(r-1)(r-2)\cdots(r-n+1)$

Consequently, the Maclaurin series associated with $(1 + x)^r$ is

$$1 + rx + \frac{r(r-1)}{1 \cdot 2} x^2 + \frac{r(r-1)(r-2)}{1 \cdot 2 \cdot 3} x^3 + \cdots. \tag{4}$$

Note that the series does not stop, for r is not a positive integer or 0.

For $x = 0$, the series clearly converges. So consider $x \neq 0$. Let b_n be the term containing the power x^n. Then

$$b_n = \frac{r(r-1)(r-2)\cdots(r-n+1)}{1 \cdot 2 \cdot 3 \cdots n} x^n,$$

and

$$b_{n+1} = \frac{r(r-1)(r-2)\cdots(r-n)}{1 \cdot 2 \cdot 3 \cdots (n+1)} x^{n+1}.$$

Thus
$$\left|\frac{b_{n+1}}{b_n}\right| = \left|\frac{\dfrac{r(r-1)(r-2)\cdots(r-n)}{1\cdot 2\cdot 3\cdots(n+1)}x^{n+1}}{\dfrac{r(r-1)(r-2)\cdots(r-n+1)}{1\cdot 2\cdot 3\cdots n}x^n}\right|$$
$$= \left|\frac{r-n}{n+1}x\right|.$$

Since r is fixed,
$$\lim_{n\to\infty}\left|\frac{b_{n+1}}{b_n}\right| = |x|.$$

By the absolute-ratio test, series (4) converges when $|x| < 1$ and diverges when $|x| > 1$. ■

In Example 5 it was shown that for $|x| < 1$ the Maclaurin series associated with $(1 + x)^r$ converges to something, but does it actually converge to $(1 + x)^r$?

Let us check the case $r = -1$. When $r = -1$, series (4) becomes
$$1 + (-1)x + \frac{(-1)(-2)}{1\cdot 2}x^2 + \frac{(-1)(-2)(-3)}{1\cdot 2\cdot 3}x^3 + \cdots,$$
or
$$1 - x + x^2 - x^3 + \cdots.$$
This series converges for $|x| < 1$. Moreover, it does represent the function $(1 + x)^r = (1 + x)^{-1}$, for it is a geometric series with first term 1 and ratio $-x$.

But, as was pointed out at the end of Sec. 11.2, Lagrange's formula for the remainder $R_n(x; 0)$ does not show that $R_n(x; 0) \to 0$ as $n \to \infty$. It is true that for $|x| < 1$ series (4) does converge to $(1 + x)^r$, but we leave the proof to an advanced calculus course. The fact that $(1 + x)^r$ is equal to the series (4) is known as the **general binomial theorem** or, simply, the **binomial theorem**. Series (4) is called the **binomial expansion** of $(1 + x)^r$.

Section Summary

Motivated by Taylor series, we investigated series of the form $\Sigma_{n=0}^{\infty} a_n x^n$ and, more generally, $\Sigma_{n=0}^{\infty} a_n(x - a)^n$. Associated with each such series is a radius of convergence R. (If the series converges for all x, we take R to be infinite.) If $\Sigma_{n=0}^{\infty} a_n x^n$ has radius of convergence R, then it converges for all x in $(-R, R)$, but diverges for all x such that $|x| > R$. Similarly, if $\Sigma_{n=0}^{\infty} a_n(x - a)^n$ has radius of convergence R, it converges for all x such that x is in $(a - R, a + R)$ but diverges if $|x - a| > R$. Convergence or divergence at the two endpoints of the interval of convergence must be checked separately.

EXERCISES FOR SEC. 11.4: POWER SERIES AND RADIUS OF CONVERGENCE

In Exercises 1 to 12 draw the appropriate diagrams (like Fig. 4) showing where the series converge or diverge. Explain your work.

1 $\displaystyle\sum_{n=1}^{\infty} \frac{x^n}{n^2}$

2 $\displaystyle\sum_{n=1}^{\infty} \frac{x^n}{\sqrt{n}}$

3 $\displaystyle\sum_{n=0}^{\infty} \frac{x^n}{3^n}$

4 $\displaystyle\sum_{n=1}^{\infty} n^2 e^{-n} x^n$

5 $\displaystyle\sum_{n=0}^{\infty} \frac{2n^2 + 1}{n^2 - 5}x^n$

6 $\displaystyle\sum_{n=1}^{\infty} \frac{x^n}{n}$

7 $\sum_{n=0}^{\infty} \frac{x^n}{(2n)!}$

8 $\sum_{n=0}^{\infty} \frac{2^n x^n}{n!}$

9 $\sum_{n=0}^{\infty} \frac{x^n}{(2n+1)!}$

10 $\sum_{n=0}^{\infty} n! x^n$

11 $\sum_{n=1}^{\infty} \frac{(-1)^{n+1} x^n}{n}$

12 $\sum_{n=1}^{\infty} \frac{2^n x^n}{n}$

13 Assume that $\Sigma_{n=0}^{\infty} a_n x^n$ converges when $x = 9$ and diverges when $x = -12$. What, if anything, can be said about
(*a*) convergence when $x = 7$?
(*b*) absolute convergence when $x = -7$?
(*c*) absolute convergence when $x = 9$?
(*d*) convergence when $x = -9$?
(*e*) divergence when $x = 10$?
(*f*) divergence when $x = -15$?
(*g*) divergence when $x = 15$?

14 Assume that $\Sigma_{n=0}^{\infty} a_n x^n$ converges when $x = -5$ and diverges when $x = 8$. What, if anything, can be said about
(*a*) convergence at $x = 4$?
(*b*) absolute convergence at $x = 4$?
(*c*) convergence at $x = 7$?
(*d*) absolute convergence at $x = -5$?
(*e*) convergence at $x = -9$?
(*f*) convergence at $x = -8$?

15 If the series $\Sigma_{n=0}^{\infty} a_n x^n$ converges whenever x is positive, must it converge whenever x is negative?

16 If $\Sigma_{n=0}^{\infty} a_n 6^n$ converges, what can be said about the convergence of

(*a*) $\Sigma_{n=0}^{\infty} a_n(-6)^n$? (*b*) $\Sigma_{n=0}^{\infty} a_n 5^n$? (*c*) $\Sigma_{n=0}^{\infty} a_n(-5)^n$?

In Exercises 17 to 28 draw the appropriate diagrams showing where the series converge and diverge.

17 $\sum_{n=0}^{\infty} \frac{(x-2)^n}{n!}$

18 $\sum_{n=1}^{\infty} \frac{(x-1)^n}{n3^n}$

19 $\sum_{n=0}^{\infty} \frac{(x-1)^n}{n+3}$

20 $\sum_{n=0}^{\infty} \frac{(x-4)^n}{2n+1}$.

21 $\sum_{n=1}^{\infty} \frac{n(x-2)^n}{2n+3}$

22 $\sum_{n=2}^{\infty} \frac{(x-5)^n}{n \ln n}$

23 $\sum_{n=0}^{\infty} \frac{(x+3)^n}{5^n}$

24 $\sum_{n=1}^{\infty} n(x+1)^n$

25 $\sum_{n=1}^{\infty} \frac{(x-5)^n}{n^2}$

26 $\sum_{n=0}^{\infty} (-1)^n \frac{(x+4)^n}{n+2}$

27 $\sum_{n=0}^{\infty} n!(x-1)^n$

28 $\sum_{n=0}^{\infty} \frac{(n^2+1)}{n^3+1}(x+2)^n$

In Exercises 29 to 32 write out the first five terms of the binomial expansions of the given functions.

29 $(1+x)^{1/2}$

30 $(1+x)^{1/3}$

31 $(1+x)^{-3}$

32 $(1+x)^{-4}$

33 (*a*) If the power series $\Sigma_{n=0}^{\infty} a_n x^n$ diverges when $x = 3$, at which x must it diverge?
(*b*) If the power series $\Sigma_{n=0}^{\infty} a_n(x+5)^n$ diverges when $x = -2$, at which x must it diverge?

34 If $\Sigma_{n=0}^{\infty} a_n(x-3)^n$ converges for $x = 7$, at what other values of x must the series necessarily converge?

35 Find the radius of convergence of $\Sigma_{n=0}^{\infty} x^{2n+1}/(2n+1)!$.

36 If $\Sigma_{n=0}^{\infty} a_n x^n$ has a radius of convergence 3 and $\Sigma_{n=0}^{\infty} b_n x^n$ has a radius of convergence 5, what can be said about the radius of convergence of $\Sigma_{n=0}^{\infty} (a_n + b_n)x^n$?

37 (*a*) Using the first four nonzero terms of the Maclaurin series for $\sqrt{1+x^3}$, estimate $\int_0^1 \sqrt{1+x^3}\, dx$, an integral that cannot be evaluated by the fundamental theorem of calculus.
(*b*) Evaluate the integral in (*a*) to three decimal places by Simpson's method.

38 (*a*) Write out the first four terms of the Maclaurin series associated with $f(x) = (1+x)^{-2}$.
(*b*) Find a formula for the typical term.
(*c*) Replace x by $-x$ to obtain the Maclaurin series for $(1-x)^{-2}$. (Give the first four terms.)

39 What is the radius of convergence for the Maclaurin series for (*a*) e^x, (*b*) $\sin x$, (*c*) $\cos x$, (*d*) $\ln(1+x)$, (*e*) $\tan^{-1} x$?

---

40 In R. P. Feynman, *Lectures on Physics*, Addison-Wesley, Reading, Mass., 1963, this statement appears in sec. 15.8 of vol. 1:

> An approximate formula to express the increase of mass, for the case when the velocity is small, can be found by expanding $m_0/\sqrt{1 - v^2/c^2} = m_0(1 - v^2/c^2)^{-1/2}$ in a power series, using the binomial theorem. We get
>
> $$m_0\left(1 - \frac{v^2}{c^2}\right)^{-1/2} = m_0\left(1 + \frac{1}{2}\frac{v^2}{c^2} + \frac{3}{8}\frac{v^4}{c^4} + \cdots\right).$$
>
> We see clearly from the formula that the series converges rapidly when v is small and the terms after the first two or three are negligible.

Check the expansion and justify the equation.

41 In *Introduction to Fluid Mechanics*, by Stephen Whitaker, Krieger, New York, 1981, the following argument appears in the discussion of flow through a nozzle:

> The pressure p equals
>
> $$\left(1 + \frac{\gamma - 1}{2}M^2\right)^{\gamma/(1-\gamma)}$$

By the binomial theorem and the fact that $v^2 = M^2\gamma RT$,

$$p = 1 - \frac{1}{2}\frac{v^2}{RT} + \frac{\gamma(2\gamma - 1)}{8}M^4 + \cdots.$$

Fill in the steps. (γ is specific heat, which is about 1.4, and M is a Mach number, which is in the range 1 to 2.)

42 (*a*) The ellipse $x^2/a^2 + y^2/b^2 = 1$ for $a \le b$ has the parameterization

$$x = a \cos t, \qquad y = b \sin t.$$

Show that the arc length of one quadrant of an ellipse is

$$\int_0^{\pi/2} b\sqrt{1 - \left[1 - \left(\frac{a}{b}\right)^2\right]\sin^2 t}\, dt.$$

The integrand does not have an elementary antiderivative.

(*b*) Assume that in (*a*) $a < b$. Then the arc length integral has the form $\int_0^{\pi/2} b\sqrt{1 - k^2 \sin^2 t}\, dt$, where $0 < k < 1$. The "elliptic integral"

$$E = \int_0^{\pi/2} \sqrt{1 - k^2 \sin^2 \theta}\, d\theta$$

is tabulated in mathematical handbooks for many values of k in $[0, 1]$. Using the binomial theorem and the formula for $\int_0^{\pi/2} \sin^n \theta\, d\theta$ (Formula 73 in the table of integrals), obtain the first six terms for E as a series in powers of k^2.

11.5 MANIPULATING POWER SERIES

Where they converge, power series behave like polynomials. You can differentiate or integrate them term by term. You can add, subtract, multiply, and divide them. We will state these properties precisely and apply them. Proofs are to be found in any advanced calculus text.

Differentiating a Power Series

In Sec. 3.4 we showed that you can differentiate the sum of a finite number of functions by adding their derivatives. Theorem 1 generalizes this to power series in x. [A similar theorem holds for power series in $(x - a)$.]

Theorem 1 *Differentiating a power series.* Assume that $R > 0$ and that $\sum_{n=0}^{\infty} a_n x^n$ converges to $f(x)$ for $|x| < R$. Then for $|x| < R$, f is differentiable, $\sum_{n=1}^{\infty} n a_n x^{n-1}$ converges, and

$$f'(x) = a_1 + 2a_2 x + 3a_3 x^2 + \cdots.$$

This theorem is *not* covered by the fact that the derivative of the sum of a *finite* number of functions is the sum of their derivatives.

EXAMPLE 1 Obtain a power series for the function $1/(1 - x)^2$ from that for $1/(1 - x)$.

SOLUTION From the formula for the sum of a geometric series, we know that

$$\frac{1}{1 - x} = 1 + x + x^2 + x^3 + \cdots \qquad \text{for } |x| < 1.$$

According to Theorem 1, if we differentiate both sides of this equation, we obtain a true equation, namely,

$$\frac{1}{(1 - x)^2} = 0 + 1 + 2x + 3x^2 + \cdots \qquad \text{for } |x| < 1.$$

Note that the series can also be written as $\Sigma_{n=0}^{\infty}(n+1)x^n$.

Thus $$\frac{1}{(1-x)^2} = 1 + 2x + 3x^2 + \cdots = \sum_{n=1}^{\infty} nx^{n-1} \qquad \text{for } |x| < 1. \blacksquare$$

Suppose that $f(x)$ has a power-series representation $a_0 + a_1x + a_2x^2 + \cdots$; Theorem 1 enables us to find what the coefficients $a_0, a_1, a_2, \ldots$ must be. The formula for a_n appears in Theorem 2, whose proof is similar to the way we found the coefficients of the Taylor polynomials in Sec. 11.1.

Theorem 2 *Formula for a_n.* Let R be a positive number and suppose that $f(x)$ is represented by the power series $\Sigma_{n=0}^{\infty} a_nx^n$ for $|x| < R$; that is,

$$f(x) = a_0 + a_1x + \cdots + a_nx^n + \cdots \qquad \text{for } |x| < R.$$

Then $$a_n = \frac{f^{(n)}(0)}{n!}. \tag{1}$$

Proof When $x = 0$ we obtain $f(0) = a_0 + a_10 + a_20^2 + \cdots$. Hence

$$f(0) = a_0,$$

Getting a_0

which is (1) for $n = 0$. To obtain a_1, differentiate $f(x)$ and get

$$f^{(1)}(x) = a_1 + 2a_2x + 3a_3x^2 + \cdots + na_nx^{n-1} + \cdots. \tag{2}$$

Getting a_1

Set $x = 0$ in (2) and obtain

$$f^{(1)}(0) = a_1.$$

This establishes (1) for $n = 1$.

Getting a_2

To obtain a_2, differentiate (2) and get

$$f^{(2)}(x) = 2a_2 + 3 \cdot 2a_3x + \cdots + n(n-1)a_nx^{n-2} + \cdots. \tag{3}$$

Letting $x = 0$ gives

$$f^{(2)}(0) = 2a_2.$$

Hence $$a_2 = \frac{f^{(2)}(0)}{2},$$

and (1) is established for $n = 2$.

Getting a_3

To obtain a_3, differentiate (3) as follows:

$$f^{(3)}(x) = 3 \cdot 2a_3 + 4 \cdot 3 \cdot 2a_4x + \cdots + n(n-1)(n-2)a_nx^{n-3} + \cdots. \tag{4}$$

Set $x = 0$, obtaining

$$f^{(3)}(0) = 3 \cdot 2a_3,$$

or $$a_3 = \frac{f^{(3)}(0)}{3!}.$$

This establishes (1) for $n = 3$ and also shows why the factorial appears in the denominator of (1). You may differentiate (4) and verify (1) for $n = 4$. The argument applies for all n and can be completed by induction. ■

Just keep on differentiating.

Theorem 2 also tells us that there can be only one series of the form $\Sigma_{n=0}^{\infty} a_nx^n$ that represents $f(x)$, for the coefficients a_n are completely determined by the function $f(x)$ and its derivatives. That series must be the Maclaurin series we obtained in Sec. 11.1.

Integrating a Power Series

Just as we may differentiate a power series term by term, we can integrate it term by term.

Theorem 3 *Integrating a power series.* Assume that $R > 0$ and

$$f(x) = a_0 + a_1x + a_2x^2 + \cdots + a_nx^n + \cdots \qquad \text{for } |x| < R.$$

Then
$$a_0x + \frac{a_1x^2}{2} + \frac{a_2x^3}{3} + \cdots + \frac{a_nx^{n+1}}{n+1} + \cdots$$

converges for $|x| < R$, and

$$\int_0^x f(t)\,dt = a_0x + \frac{a_1x^2}{2} + \frac{a_2x^3}{3} + \cdots.$$

Note that the t is used to avoid writing $\int_0^x f(x)\,dx$, an expression in which x describes both the interval of integration $[0, x]$ and the independent variable of the function. The next example shows the power of Theorem 3.

EXAMPLE 2 Integrate the power series for $1/(1 + x)$ to obtain a power series for $\ln(1 + x)$.

SOLUTION Start with the series $1/(1 - x) = 1 + x + x^2 + \cdots$ for $|x| < 1$. Replace x by $-x$ and obtain

$$\frac{1}{1+x} = 1 - x + x^2 - x^3 + x^4 - \cdots \qquad \text{for } |x| < 1.$$

By Theorem 3,
$$\int_0^x \frac{dt}{1+t} = x - \frac{x^2}{2} + \frac{x^3}{3} - \frac{x^4}{4} + \cdots \qquad \text{for } |x| < 1.$$

Now,
$$\int_0^x \frac{dt}{1+t} = \ln(1+t)\Big|_0^x$$
$$= \ln(1+x) - \ln(1+0)$$
$$= \ln(1+x).$$

The power series for $\ln(1 + x)$

Thus
$$\ln(1+x) = x - \frac{x^2}{2} + \frac{x^3}{3} - \frac{x^4}{4} + \cdots \qquad \text{for } |x| < 1. \quad ■$$

The Algebra of Power Series

In addition to differentiating and integrating power series, we may also add, subtract, multiply, and divide them just like polynomials. Theorem 4 states the rules for these operations. Two illustrative examples then follow.

Theorem 4 *The algebra of power series.* Assume that

$$f(x) = a_0 + a_1x + a_2x^2 + \cdots \quad \text{and} \quad g(x) = b_0 + b_1x + b_2x^2 + \cdots$$

for $|x| < R$. Then for $|x| < R$,

(*a*) $f(x) + g(x) = \Sigma_{n=0}^{\infty} (a_n + b_n)x^n$

(*b*) $f(x) - g(x) = \Sigma_{n=0}^{\infty} (a_n - b_n)x^n$

(*c*) $f(x)g(x) = a_0b_0 + (a_0b_1 + a_1b_0)x + (a_0b_2 + a_1b_1 + a_2b_0)x^2 + \cdots$
(This says "multiply two power series the way you multiply polynomials, term by term: start with constant terms and work up.")

(*d*) $f(x)/g(x)$ is obtainable by long division, if $g(x) \neq 0$ for $|x| < R$.

EXAMPLE 3 Find the first four terms of the Maclaurin series for $e^x/(1 - x)$.

SOLUTION

$$\begin{aligned} e^x\frac{1}{1-x} &= \left(1 + x + \frac{x^2}{2!} + \frac{x^3}{3!} + \cdots\right)(1 + x + x^2 + x^3 + \cdots) \\ &= 1 \cdot 1 + (1 \cdot 1 + 1 \cdot 1)x + \left(1 \cdot 1 + 1 \cdot 1 + \frac{1}{2!} \cdot 1\right)x^2 \\ &\quad + \left(1 \cdot 1 + 1 \cdot 1 + \frac{1}{2!} \cdot 1 + \frac{1}{3!} \cdot 1\right)x^3 + \cdots \\ &= 1 + 2x + \frac{5}{2}x^2 + \frac{8}{3}x^3 + \cdots, \qquad |x| < 1. \end{aligned}$$

This way is a lot easier than using the formula for the series in terms of the higher derivatives of $e^x/(1 - x)$ evaluated at 0. ■

EXAMPLE 4 Find the first four terms of the power series in x for $e^x/\cos x$.

SOLUTION Write down the power series in x for e^x and $\cos x$ up through the terms of degree 3 and arrange the long division as follows:

$$\begin{array}{r|l} & 1 + x + x^2 + \dfrac{2x^3}{3} + \cdots \\ \hline 1 + 0x - \dfrac{x^2}{2} + 0x^3 + \cdots & 1 + x + \dfrac{x^2}{2} + \dfrac{x^3}{6} + \cdots \\ & 1 + 0x - \dfrac{x^2}{2} + 0x^3 + \cdots \\ \cline{2-2} & \quad x + x^2 + \dfrac{x^3}{6} + \cdots \\ & \quad x + 0x^2 - \dfrac{x^3}{2} + \cdots \\ \cline{2-2} & \qquad x^2 + \dfrac{2x^3}{3} + \cdots \\ & \qquad x^2 + 0x^3 - \cdots \\ \cline{2-2} & \qquad\quad \dfrac{2x^3}{3} + \cdots \\ & \qquad\quad \dfrac{2x^3}{3} + \cdots \\ \cline{2-2} & \qquad\qquad \cdots \end{array}$$

Thus the power series in x for $e^x/\cos x$ begins

$$1 + x + x^2 + \frac{2x^3}{3}. \blacksquare$$

Limits by Power Series

Section 6.8 presented l'Hôpital's rule as a way to deal with $\lim_{x\to a} f(x)/g(x)$ when both numerator and denominator approach 0 as $x \to a$. The next example shows how to use power series to calculate such limits. In fact, many people prefer this approach since most common functions are represented by power series.

EXAMPLE 5 Find $\lim_{x\to 0} \dfrac{1 - \cos x}{1 + x - e^x}$.

SOLUTION Using the Maclaurin series for $\cos x$ and e^x, we obtain

$$\frac{1-\cos x}{1+x-e^x} = \frac{1 - \left(1 - \dfrac{x^2}{2} + \dfrac{x^4}{24} - \cdots\right)}{1 + x - \left(1 + x + \dfrac{x^2}{2} + \dfrac{x^3}{6} + \cdots\right)}$$

$$= \frac{\dfrac{x^2}{2} - \dfrac{x^4}{24} + \cdots}{\dfrac{-x^2}{2} - \dfrac{x^3}{6} - \cdots}$$

$$= \frac{\dfrac{1}{2} - \dfrac{x^2}{24} + \cdots}{-\dfrac{1}{2} - \dfrac{x}{6} - \cdots}. \quad \text{algebra}$$

Hence
$$\lim_{x\to 0}\frac{1-\cos x}{1+x-e^x} = \lim_{x\to 0}\frac{\dfrac{1}{2} - \dfrac{x^2}{24} + \cdots}{-\dfrac{1}{2} - \dfrac{x}{6} - \cdots} = \frac{\dfrac{1}{2}}{-\dfrac{1}{2}} = -1. \blacksquare$$

Power Series Around a

Power series in $x - a$

The various theorems and methods of this section were stated for power series in x. But analogous theorems hold for power series in $x - a$. Such series may be differentiated and integrated inside the interval in which they converge. For instance, Theorem 2 generalizes to the following assertion.

Theorem 5 *Formula for a_n.* Let R be a positive number and suppose that $f(x)$ is represented by $\Sigma_{n=0}^{\infty} a_n(x-a)^n$ for $|x - a| < R$. Then

$$a_n = \frac{f^{(n)}(a)}{n!}.$$

The proof is similar to that of Theorem 2: Differentiate n times and replace x by a.

Section Summary

We showed how to operate with power series to obtain new power series—by differentiation, by integration, or by an algebraic operation, such as multiplying or dividing two series. For instance, from the geometric series for $1/(1 + x)$, you can obtain the series for $\ln(1 + x)$ by integration, or the series for $-1/(1 + x)^2$ by differentiating.

HOW SOME CALCULATORS FIND e^x

The power series in x for e^x is

$$1 + x + \frac{x^2}{2!} + \frac{x^3}{3!} + \cdots + \frac{x^n}{n!} + \cdots. \tag{5}$$

For $x = 10$, this would give

$$e^{10} = 1 + 10 + \frac{10^2}{2!} + \frac{10^3}{3!} + \frac{10^4}{4!} + \cdots + \frac{10^n}{n!} + \cdots.$$

Although the terms eventually become very small, the first few terms are quite large. (For instance, the fifth term, $10^4/4!$, is about 417.) So when x is large, series (5) provides a time-consuming procedure for calculating e^x.

Some calculators use the following method instead.

The values of e^x at certain inputs are built into the memory:

$$\begin{aligned} e^1 &\approx 2.718281828459, \\ e^{10} &\approx 22{,}026.46579, \\ e^{100} &\approx 2.6881171 \times 10^{43}, \\ e^{0.1} &\approx 1.1051709181, \\ e^{0.01} &\approx 1.0100501671, \\ e^{0.001} &\approx 1.0010005002. \end{aligned}$$

[This may be found with the aid of series (5).] Then, to find $e^{315.425}$, say, the calculator computes

$$(e^{100})^3(e^{10})^1(e^1)^5(e^{0.1})^4(e^{0.01})^2(e^{0.001})^5.$$

Here it makes use of the identities $e^{x+y} = e^x e^y$ and $(e^x)^y = e^{xy}$.

EXERCISES FOR SEC. 11.5: MANIPULATING POWER SERIES

1 Differentiate the Maclaurin series for $\sin x$ to obtain the Maclaurin series for $\cos x$.

2 Differentiate the Maclaurin series for e^x to show that $D(e^x) = e^x$.

3 (*a*) Show that for $|t| < 1$, $1/(1 + t^2) = 1 - t^2 + t^4 - t^6 + \cdots$.

(*b*) Use Theorem 3 to show that for $|x| < 1$,

$$\tan^{-1} x = x - \frac{x^3}{3} + \frac{x^5}{5} - \frac{x^7}{7} + \cdots.$$

(*c*) Give the formula for the nth term of the series in (*b*).

(*d*) Use the formula in (*b*) to estimate $\tan^{-1} \frac{1}{2}$ to three decimal places.

4 (*a*) Using Theorem 3, show that for $|x| < 1$, we have $\int_0^x dt/(1 + t^3) = x - x^4/4 + x^7/7 - x^{10}/10 + \cdots$.

(*b*) Use (*a*) to express $\int_0^{0.7} dt/(1 + t^3)$ as a series of numbers.

(*c*) Use (*b*) to evaluate $\int_0^{0.7} dt/(1 + t^3)$ to three decimal places.

(*d*) Describe how you would evaluate $\int_0^{0.7} dt/(1+t^3)$ using the fundamental theorem of calculus. (Don't carry out the details.)

*(*e*) Use a symbolic math program to compute the exact value of $\int_0^{0.7} dt/(1+t^3)$.

5 (*a*) Find the first three nonzero terms of the Maclaurin series for $\tan x$ by dividing the series for $\sin x$ by the series for $\cos x$.

(*b*) Find the first two nonzero terms by using the formula for the nth term, $f^{(n)}(0)x^n/n!$.

6 Find the first four terms of the Maclaurin series for $(1-\cos x)/(1-x^2)$ by division of series.

In Exercises 7 and 8 obtain the first three nonzero terms in the power series in x for the indicated functions by algebraic operations with known series.

7 $e^x \sin x$ 8 $\dfrac{x}{\cos x}$

In Exercises 9 to 14 use power series to determine the limits.

9 $\lim_{x\to 0} \dfrac{1-\cos x}{x^2}$ 10 $\lim_{x\to 0} \dfrac{\sin 3x}{\sin 2x}$

11 $\lim_{x\to 0} \dfrac{\sin^2 x^3}{(1-\cos x^2)^3}$ 12 $\lim_{x\to 0} \left[\dfrac{1}{\sin x} - \dfrac{1}{\ln(1+x)}\right]$

13 $\lim_{x\to 0} \dfrac{(e^x-1)^2}{\sin x^2}$ 14 $\lim_{x\to 0} \dfrac{\sin x\,(1-\cos x)}{e^{x^3}-1}$

15 (*a*) Give a numerical series equal to $\int_0^1 \sqrt{x}\sin x\,dx$.

(*b*) Use it to evaluate the integral to four decimal places.

16 Estimate $\int_0^{1/2} \sqrt{x}\,e^{-x}\,dx$ to four decimal places.

17 Let $f(x) = \sum_{n=0}^{\infty} n^2x^n$.

(*a*) What is the domain of $f(x)$?

(*b*) Find $f^{(100)}(0)$.

18 Let $f(x) = \tan^{-1} x$. Making use of the Maclaurin series for $\tan^{-1} x$, find (*a*) $f^{(100)}(0)$, (*b*) $f^{(101)}(0)$.

---

19 (*a*) From the Maclaurin series for $\cos x$ in powers of x, obtain the Maclaurin series for $\cos 2x$.

(*b*) Exploiting the identity $\sin^2 x = (1-\cos 2x)/2$, obtain the Maclaurin series for $(\sin^2 x)/x^2$.

(*c*) Estimate $\int_0^1 [(\sin x)/x]^2\,dx$ using the first three nonzero terms of the series in (*b*).

(*d*) Find a bound on the error in the estimate in (*c*).

20 What theorems in the text justify the assertion that

$$\lim_{x\to 0}(a_0 + a_1x + a_2x^2 + a_3x^3 + \cdots) = a_0?$$

Assume the series has a nonzero radius of convergence.

21 Let $\sum_{n=0}^{\infty} a_nx^n$ and $\sum_{n=0}^{\infty} b_nx^n$ converge for $|x|<1$. If they converge to the same limit for each x in $(-1, 1)$, must $a_n = b_n$ for all n?

22 This exercise outlines a way to compute logarithms of numbers larger than 1. It depends on the fact that a number $y > 1$ can be written in the form $(1+x)/(1-x)$ for a number x in $(0, 1)$. For instance, when $x = \frac{1}{2}$, $y = 3$. (Check this.) Then $\ln y = \ln(1+x) - \ln(1-x)$. In what follows, we assume $|x| < 1$.

We have shown that

$$\ln(1+x) = x - \frac{x^2}{2} + \frac{x^3}{3} - \cdots + (-1)^{n-1}\frac{x^n}{n} + \cdots.$$

Replacing x by $-x$ gives

$$\ln(1-x) = -x - \frac{x^2}{2} - \frac{x^3}{3} - \cdots - \frac{x^n}{n} - \cdots.$$

Subtracting gives

$$\ln y = \ln\left(\frac{1+x}{1-x}\right) = 2\left(x + \frac{x^3}{3} + \cdots + \frac{x^{2n+1}}{2n+1} + \cdots\right).$$

(*a*) Estimate ln 3 to two decimal places. (*Hint:* Control the error by comparing the tail end of the series to a geometric series.)

(*b*) Is the error in (*a*) less than the first omitted term?

23 Use the method of Exercise 22 to estimate ln 5 to two decimal places. Include a description of your procedures.

24 Here are five ways to compute ln 2. Which seems to be the most efficient? least efficient? Explain.

(*a*) The series for $\ln(1+x)$ when $x = 1$.

(*b*) The series for $\ln(1+x)$ when $x = -\frac{1}{2}$. (This gives $\ln\frac{1}{2} = -\ln 2$.)

(*c*) The series for $\ln[(1+x)/(1-x)]$ when $x = \frac{1}{3}$.

(*d*) Simpson's method applied to the integral $\int_1^2 dx/x$.

(*e*) The root of $e^x = 2$. (Use Newton's method.)

25 (*a*) For which x does $\sum_{n=0}^{\infty} n^2x^n$ converge?

(*b*) Starting with the Maclaurin series for $x^2/(1-x)$, sum the series in (*a*).

(*c*) Does your formula seem to give the correct answer when $x = \frac{1}{3}$?

26 This exercise uses power series to give a new perspective on l'Hôpital's rule. Assume that f and g can be represented by power series in some open interval containing 0:

$$f(x) = \sum_{n=0}^{\infty} a_nx^n \quad\text{and}\quad g(x) = \sum_{n=0}^{\infty} b_nx^n.$$

Assume that $f(0) = 0$, $g(0) = 0$, and $g'(0) \neq 0$. Under these assumptions explain why

$$\lim_{x\to 0}\frac{f(x)}{g(x)} = \lim_{x\to 0}\frac{f'(x)}{g'(x)}.$$

27 In R. P. Feynman, *Lectures on Physics,* Addison-Wesley, Reading, Mass., 1963, appears this remark:

Thus the average energy is

$$\langle E\rangle = \frac{\hbar\omega(0 + x + 2x^2 + 3x^3 + \cdots)}{1 + x + x^2 + \cdots}.$$

Now the two sums which appear here we shall leave for the reader to play with and have some fun with. When we are all finished summing and substituting for x in the sum, we should get—if we make no mistakes in the sum—

$$\langle E\rangle = \frac{\hbar\omega}{e^{\hbar\omega/kT} - 1}.$$

This, then, was the first quantum-mechanical formula ever known, or ever discussed, and it was the beautiful culmination of decades of puzzlement.

Have the aforementioned fun, given that $x = e^{-\hbar\omega/kT}$.

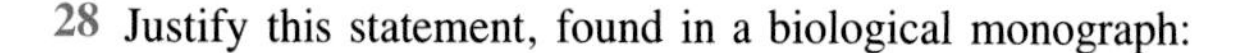

28 Justify this statement, found in a biological monograph:

Expanding the equation

$$a \cdot \ln (x + p) + b \cdot \ln (y + q) = M,$$

we obtain

$$a\left(\ln p + \frac{x}{p} - \frac{x^2}{2p^2} + \frac{x^3}{3p^3} - \cdots\right) + b\left(\ln q + \frac{y}{q} - \frac{y^2}{2q^2} + \frac{y^3}{3q^3} - \cdots\right) = M.$$

11.6 COMPLEX NUMBERS

Let us think of the number line of real numbers as coinciding with the x axis of an xy coordinate system. This number line, with its addition, subtraction, multiplication, and division, is but a small part of a number system that occupies the plane and which obeys the usual rules of arithmetic. This section describes that system, known as the **complex numbers**. One of the important properties of the complex numbers is that any nonconstant polynomial has a root; in particular, the equation $x^2 = -1$ has two solutions.

The Complex Numbers

By a complex number z we shall mean an expression of the form $x + iy$ or $x + yi$, where x and y are real numbers and i is a symbol such that $i^2 = -1$. This expression will be identified with the point (x, y) in the xy plane, as in Fig. 1. Every point in the plane may therefore be thought of as a complex number.

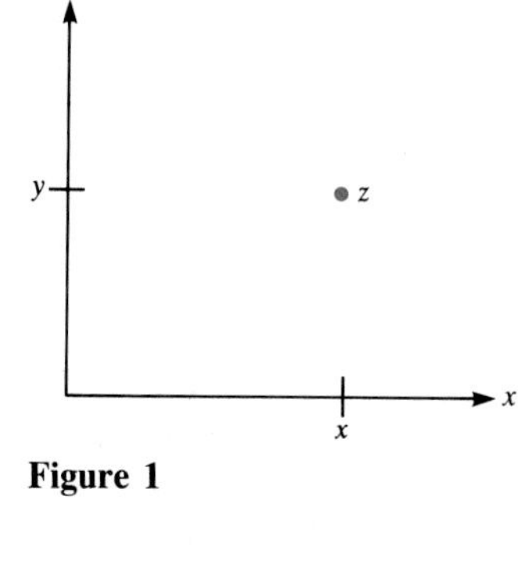

Figure 1

To add or multiply two complex numbers, follow the usual rules of arithmetic of real numbers, with one new proviso:

Whenever you see i^2, replace it by -1.

For instance, to add the complex numbers $3 + 2i$ and $-4 + 5i$, just collect like terms:

$$(3 + 2i) + (-4 + 5i) = [3 + (-4)] + (2i + 5i) = -1 + 7i.$$

Addition does not make use of the fact that $i^2 = -1$. However, multiplication does, as Example 1 shows.

EXAMPLE 1 Compute the product $(2 + i)(3 + i)$.

SOLUTION We can multiply the complex numbers just as we would multiply binomials. (Recall the mnemonic FOIL for "first, outer, inner, last.") We have

$$(2 + i)(3 + i) = 2 \cdot 3 + 2i + 3i + i^2 = 6 + 5i - 1 = 5 + 5i.$$

Figure 2 shows the complex numbers $2 + i$, $3 + i$, and their product, $5 + 5i$. ■

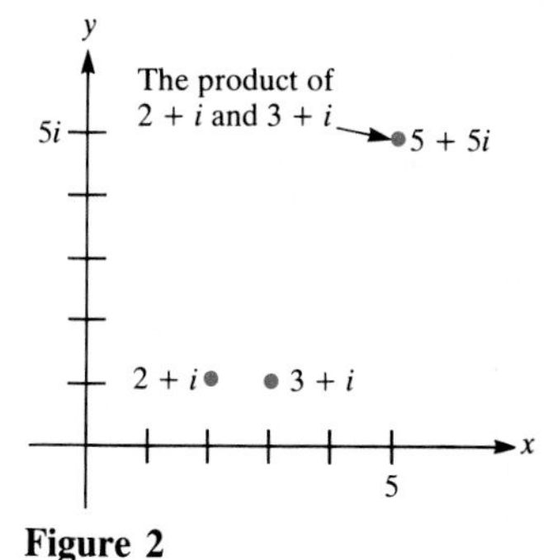

Figure 2

-1 has two square roots.

Note that $(-i)(-i) = i^2 = -1$. Both i and $-i$ are square roots of -1. The symbol $\sqrt{-1}$ traditionally denotes i rather than $-i$.

Real numbers are on the x axis, imaginary on the y axis.

A complex number that lies on the y axis is called **imaginary**. Every complex number z is the sum of a real number and an imaginary number, $z = x + iy$. The number x is called the **real part of** z, and y is called the **imaginary part**. One writes "Re $z = x$" and "Im $z = y$."

We have seen how to add and multiply complex numbers. Subtraction is straightforward. For instance,

$$(3 + 2i) - (4 - i) = (3 - 4) + [2i - (-i)]$$
$$= -1 + 3i.$$

Division of complex numbers requires rationalizing the denominator, as Example 2 illustrates.

EXAMPLE 2 Compute $\dfrac{1 + 5i}{3 + 2i}$.

SOLUTION To divide, "rationalize the denominator":

$$\frac{1 + 5i}{3 + 2i} \cdot \frac{3 - 2i}{3 - 2i} = \frac{3 - 2i + 15i + 10}{9 - 4i^2} = \frac{13 + 13i}{13} = 1 + i. \quad ■$$

Conjugate of z

The solution of Example 2 used the **conjugate** of a complex number. The conjugate of the complex number $z = x + yi$ is the complex number $x - yi$, which is denoted $\bar{z}$. Note that

$$z\bar{z} = (x + yi)(x - yi) = x^2 + y^2$$

and

$$z + \bar{z} = (x + yi) + (x - yi) = 2x.$$

Thus both $z\bar{z}$ and $z + \bar{z}$ are real.

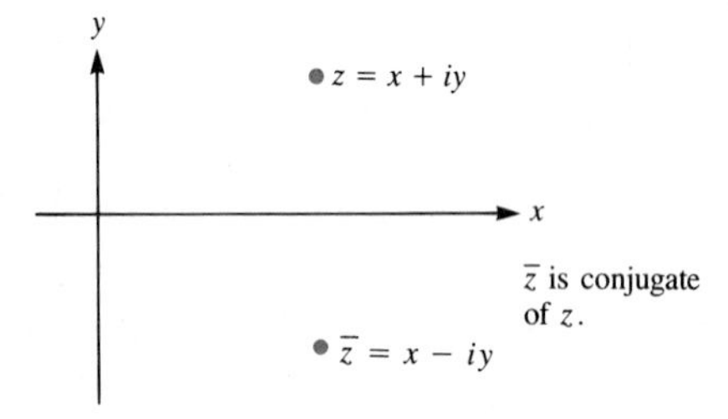

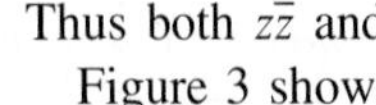
$\bar{z}$ is conjugate of z.

Figure 3

Figure 3 shows the relation between z and $\bar{z}$, which is the mirror image of z in the x axis.

Now All Polynomials Have Roots

Every polynomial has a root in the complex numbers.

The complex numbers provide the equation $x^2 + 1 = 0$ with two solutions, i and $-i$. This illustrates an important property of the complex numbers: If $f(x) = a_nx^n + a_{n-1}x^{n-1} + \cdots + a_0$ is any polynomial of degree $n \geq 1$, with real or complex coefficients, then there is a complex number z such that $f(z) = 0$. This fact is known as the **fundamental theorem of algebra**. Its proof requires advanced mathematics. Example 3 illustrates this theorem.

EXAMPLE 3 Solve the quadratic equation $z^2 - 4z + 5 = 0$.

SOLUTION By the quadratic formula, the solutions are

$$z = \frac{-(-4) \pm \sqrt{(-4)^2 - 4 \cdot 1 \cdot 5}}{2}$$

$$= \frac{4 \pm \sqrt{-4}}{2} = \frac{4 \pm 2i}{2} = 2 \pm i.$$

The solutions are $2 + i$ and $2 - i$.

These solutions can be checked by substitution in the original equation. For instance,

$$(2 + i)^2 - 4(2 + i) + 5 = (4 + 4i + i^2) - 8 - 4i + 5$$
$$= 4 + 4i - 1 - 8 - 4i + 5 = 0 + 0i = 0.$$

Yes, it checks. The solution $2 - i$ can be checked similarly. ■

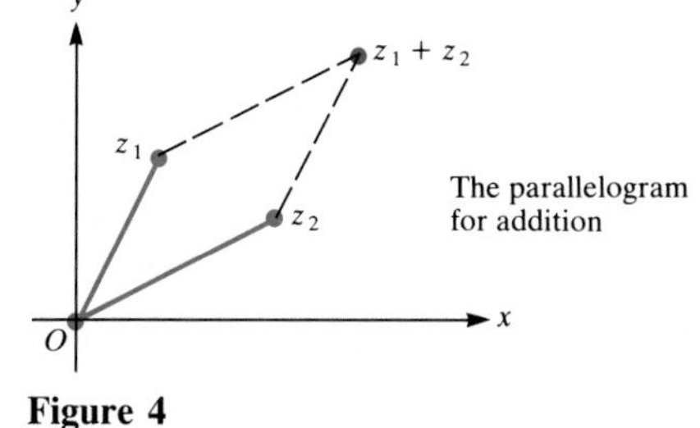

Figure 4

The sum of the complex numbers z_1 and z_2 is the fourth vertex (opposite O) in a parallelogram determined by the origin O and the points z_1 and z_2, as shown in Fig. 4. The geometry of the product of z_1 and z_2 is more involved.

The Geometry of the Product

The geometric relation between z_1, z_2, and their product z_1z_2 is easily described in terms of the magnitude and argument of a complex number. Each complex number z other than the origin is at some distance r from the origin and has a polar angle θ relative to the positive x axis. The distance r is called the **magnitude of z**, and θ is called an **argument of z**. A complex number has an infinity of arguments differing from each other by an integer multiple of 2π. The complex number 0, which lies at the origin, has magnitude 0 and any angle as argument. In short, we may think of magnitude and argument as polar coordinates r and θ of z, with the restriction that r is nonnegative. The magnitude of z is denoted $|z|$. The symbol arg z denotes any of the arguments of z, it being understood that if θ is an argument of z, then so is $\theta + 2n\pi$ for any integer n.

The symbols $|z|$ and arg z

EXAMPLE 4

(*a*) Draw all complex numbers of magnitude 3.

(*b*) Draw the complex number z of magnitude 3 and argument $\pi/6$.

SOLUTION

(*a*) The complex numbers of magnitude 3 form a circle of radius 3 with center at O. (See Fig. 5.)

(*b*) The complex number of magnitude 3 and argument $\pi/6$ is shown in Fig. 5. ■

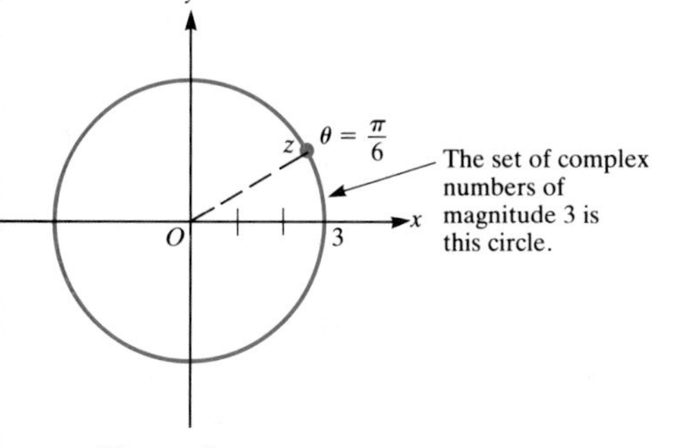

Figure 5

Note that $|x + iy| = \sqrt{x^2 + y^2}$, by the pythagorean theorem. Each complex number $z = x + iy$ other than 0 can be written as the product of a positive real number and a complex number of magnitude 1. To show this, let $z = x + iy$ have magnitude r and argument θ. Recalling the relation between polar and rectangular coordinates, we conclude that

$$z = r \cos \theta + i\, r \sin \theta$$
$$= r (\cos \theta + i \sin \theta).$$

The number r is a positive real number; the number $\cos \theta + i \sin \theta$ has magnitude 1 since $\sqrt{\cos^2 \theta + \sin^2 \theta} = 1$. Figure 6 shows the number r and $\cos \theta + i \sin \theta$, whose product is z. (The quantity $\cos \theta + i \sin \theta$ occurs so frequently in working with complex numbers that the shorthand notation cis $\theta = \cos \theta + i \sin \theta$ is sometimes used. This is convenient, provided you don't confuse "cis" with "cos.")

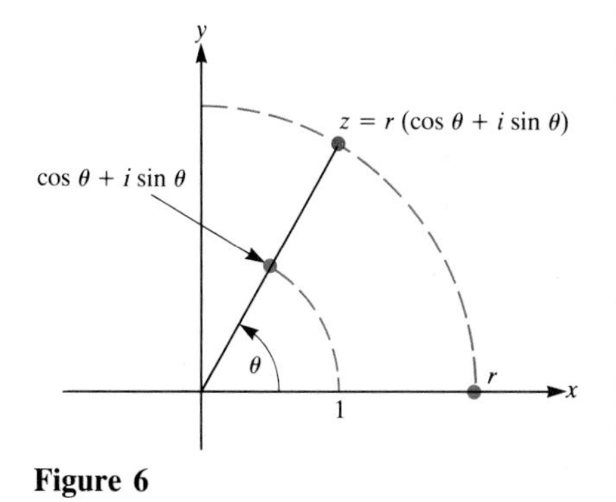

Figure 6

The following theorem describes *how to multiply two complex numbers if they are given in polar form,* that is, in terms of their magnitudes and arguments.

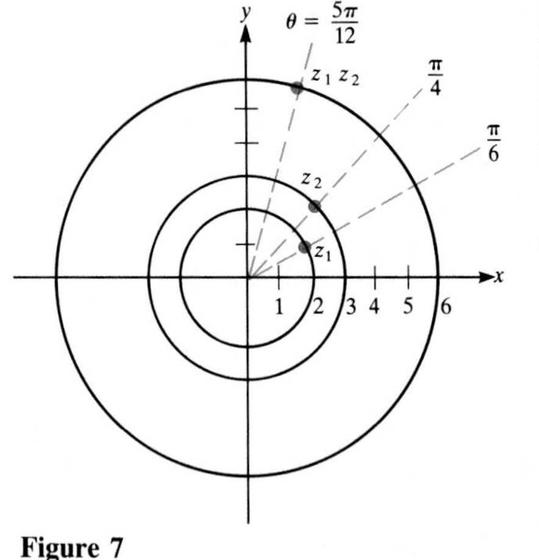

Figure 7

Theorem Assume that z_1 has magnitude r_1 and argument θ_1 and that z_2 has magnitude r_2 and argument θ_2. Then the product z_1z_2 has magnitude r_1r_2 and argument $\theta_1 + \theta_2$.

Proof

$$\begin{aligned} z_1z_2 &= r_1(\cos\,\theta_1 + i\sin\,\theta_1)r_2(\cos\,\theta_2 + i\sin\,\theta_2) \\ &= r_1r_2(\cos\,\theta_1 + i\sin\,\theta_1)(\cos\,\theta_2 + i\sin\,\theta_2) \\ &= r_1r_2[\cos\,\theta_1\cos\,\theta_2 - \sin\,\theta_1\sin\,\theta_2 + i(\sin\,\theta_1\cos\,\theta_2 + \cos\,\theta_1\sin\,\theta_2)] \\ &= r_1r_2[\cos\,(\theta_1 + \theta_2) + i\sin\,(\theta_1 + \theta_2)]. \qquad \text{(by trigonometric identities)} \end{aligned}$$

Thus the argument of z_1z_2 is $\theta_1 + \theta_2$ and the magnitude of z_1z_2 is r_1r_2. This proves the theorem. ■

In practical terms, the theorem says, *"to multiply two complex numbers just add their arguments and multiply their magnitudes."*

EXAMPLE 5 Find z_1z_2 for z_1 and z_2 in Fig. 7.

SOLUTION z_1 has magnitude 2 and argument $\pi/6$; z_2 has magnitude 3 and argument $\pi/4$. Thus z_1z_2 has magnitude $2 \times 3 = 6$ and argument $\pi/6 + \pi/4 = 5\pi/12$. It is shown in Fig. 7. ■

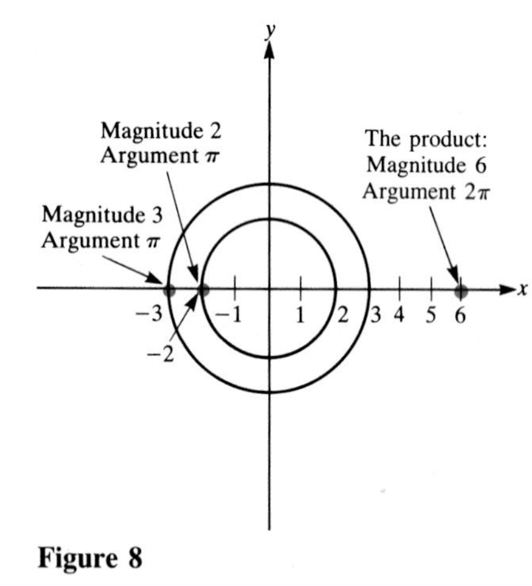

Figure 8

EXAMPLE 6 Using the geometric description of multiplication, find the product of the real numbers -2 and -3.

SOLUTION The number -2 has magnitude 2 and argument π. The number -3 has magnitude 3 and argument π. Therefore $(-2) \times (-3)$ has magnitude $2 \times 3 = 6$ and argument $\pi + \pi = 2\pi$. The complex number with magnitude 6 and argument 2π is just our old friend, the real number 6. Thus $(-2) \times (-3) = 6$ in agreement with the customary product of two negative real numbers. (See Fig. 8.) ■

Division of Complex Numbers

See Exercise 17.

Division of complex numbers given in polar form is similar, except that the magnitudes are divided and the arguments subtracted:

$$\frac{r_1(\cos\,\theta_1 + i\sin\,\theta_1)}{r_2(\cos\,\theta_2 + i\sin\,\theta_2)} = \frac{r_1}{r_2}[\cos\,(\theta_1 - \theta_2) + i\sin\,(\theta_1 - \theta_2)].$$

EXAMPLE 7 Let $z_1 = 6(\cos\,\pi/2 + i\sin\,\pi/2)$ and $z_2 = 3(\cos\,\pi/6 + i\sin\,\pi/6)$. Find (*a*) z_1z_2 and (*b*) z_1/z_2.

SOLUTION

(*a*)

$$z_1z_2 = 6 \cdot 3\left[\cos\left(\frac{\pi}{2}+\frac{\pi}{6}\right) + i\sin\left(\frac{\pi}{2}+\frac{\pi}{6}\right)\right]$$

$$= 18\left(\cos\frac{2\pi}{3} + i\sin\frac{2\pi}{3}\right)$$

$$= 18\left(-\frac{1}{2} + i\frac{\sqrt{3}}{2}\right) = -9 + 9\sqrt{3}i.$$

(*b*)

$$\frac{z_1}{z_2} = \frac{6\left(\cos\frac{\pi}{2} + i\sin\frac{\pi}{2}\right)}{3\left(\cos\frac{\pi}{6} + i\sin\frac{\pi}{6}\right)}$$

$$= 2\left(\cos\frac{\pi}{3} + i\sin\frac{\pi}{3}\right)$$

$$= 2\left(\frac{1}{2} + \frac{\sqrt{3}}{2}i\right) = 1 + \sqrt{3}i. \quad ■$$

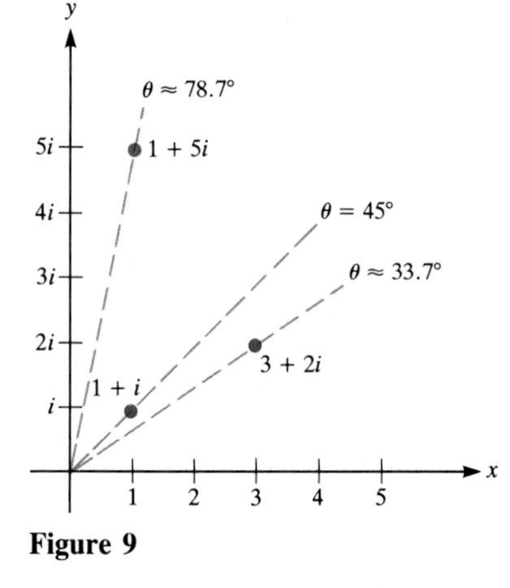

Figure 9

EXAMPLE 8 Compute $(1 + i)(3 + 2i)$ and check the answer in terms of magnitudes and arguments.

SOLUTION

$$(1 + i)(3 + 2i) = 3 + 2i + 3i + 2i^2$$

$$= 3 + 2i + 3i - 2$$

$$= 1 + 5i.$$

As a check, let us see if $|1 + 5i| = |1 + i|\,|3 + 2i|$. We have

$$|1 + 5i| = \sqrt{1^2 + 5^2} = \sqrt{26},$$

$$|1 + i| = \sqrt{1^2 + 1^2} = \sqrt{2},$$

$$|3 + 2i| = \sqrt{3^2 + 2^2} = \sqrt{13}.$$

Since $\sqrt{26} = \sqrt{2}\sqrt{13}$, the magnitude of $1 + 5i$ is the product of the magnitudes of $1 + i$ and $3 + 2i$.

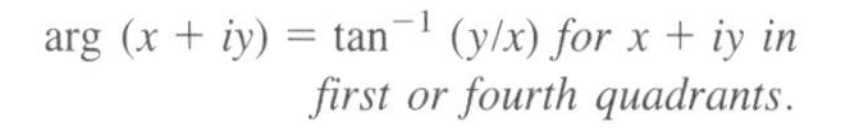

Next consider arguments. First of all, $\arg(1 + 5i) = \tan^{-1} 5 \approx 1.3734$. Similarly, $\arg(1 + i) = \tan^{-1}1 \approx 0.7854$ and $\arg(3 + 2i) = \tan^{-1}\frac{2}{3} \approx 0.5880$. Note that $0.7854 + 0.5880 = 1.3734$. (See also Fig. 9.) ■

Powers of z

When the polar coordinates of z are known, it is easy to compute the powers z^2, z^3, z^4, ... Let z have magnitude r and argument θ. Then $z^2 = z \cdot z$ has magnitude $r \cdot r = r^2$ and argument $\theta + \theta = 2\theta$. So, to square a complex number, just square its magnitude and double its angle.

More generally, to compute z^n for any positive integer n, find $|z|^n$ and multiply the argument of z by n. In short, we have **DeMoivre's law**:

$$[r(\cos\theta + i\sin\theta)]^n = r^n(\cos n\theta + i\sin n\theta).$$

Example 9 illustrates this geometric view of computing powers.

EXAMPLE 9 Let z have magnitude 1 and argument $2\pi/5$. Compute and sketch z, z^2, z^3, z^4, z^5, and z^6.

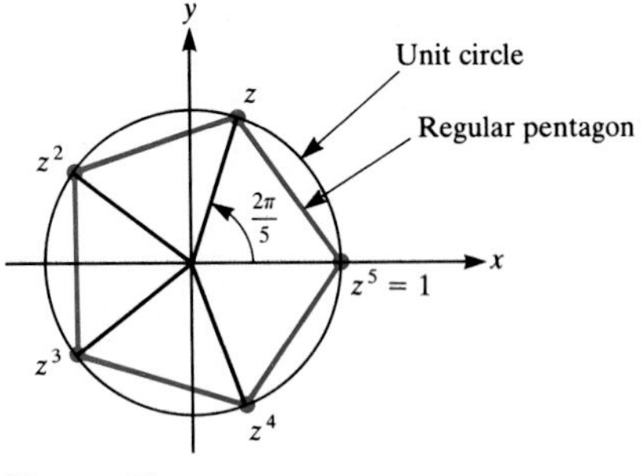

Figure 10

SOLUTION Since $|z| = 1$, it follows that $|z^2| = |z|^2 = 1^2 = 1$. Similarly, for all positive integers n, $|z^n| = 1$; that is, z^n is a point on the unit circle with center 0. All that remains is to examine the arguments of z^2, z^3, etc.

The argument of z^2 is $2(2\pi/5) = 4\pi/5$. Similarly, $\arg z^3 = 6\pi/5$, $\arg z^4 = 8\pi/5$, $\arg z^5 = 10\pi/5 = 2\pi$, and $\arg z^6 = 12\pi/5$. Observe that $z^5 = 1$, since it has magnitude 1 and argument 2π. Similarly, $z^6 = z$, since both z and z^6 have magnitude 1 and arguments that differ by an integer multiple of 2π. (Or, algebraically, $z^6 = z^5 \cdot z = 1z = z$.) The powers of z form the vertices of a regular pentagon, as shown in Fig. 10. ■

The equation $x^5 = 1$ has only one real root, namely, 1. However, it has five complex roots. For instance, the number z shown in Fig. 10 is a solution of $x^5 = 1$, since $z^5 = 1$. Another root is z^2, since $(z^2)^5 = z^{10} = (z^5)^2 = 1^2 = 1$. Similarly, z^3 and z^4 are roots of $x^5 = 1$. The roots are 1, z, z^2, z^3, and z^4.

The powers of i

The powers of i will be needed in the next section. They are $i^2 = -1$, $i^3 = i^2 \cdot i = (-1)i = -i$, $i^4 = i^3 \cdot i = (-i)i = -i^2 = 1$, $i^5 = i^4 \cdot i = i$, and so on. They repeat in blocks of four: for any integer n, $i^{n+4} = i^n$.

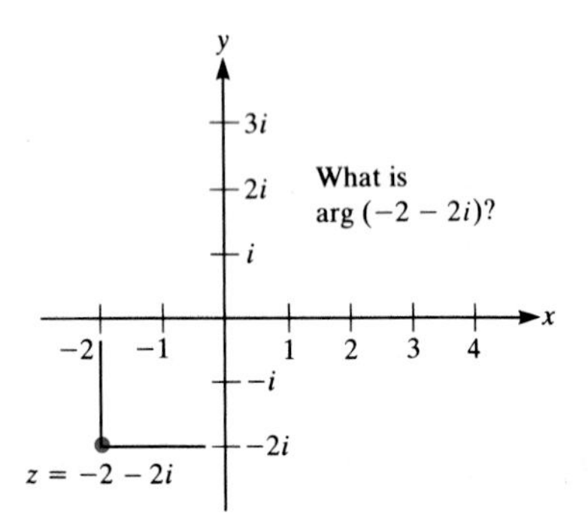

Figure 11

It is often useful to express a complex number $z = x + iy$ in polar form. Recall that $|z| = \sqrt{x^2 + y^2}$. To find θ, it is best to sketch z in order to see in which quadrant it lies. Although $\tan\theta = y/x$, we cannot say that $\theta = \tan^{-1}(y/x)$, since $\tan^{-1} u$ lies between $\pi/2$ and $-\pi/2$ for any real number u. However, θ may be a second- or third-quadrant angle. For instance, to put $z = -2 - 2i$ in polar form, first sketch z, as in Fig. 11. We have $|z| = \sqrt{(-2)^2 + (-2)^2} = \sqrt{8}$ and $\arg z = 5\pi/4$. Thus

$$z = \sqrt{8}\left(\cos\frac{5\pi}{4} + i\sin\frac{5\pi}{4}\right).$$

Note that $\tan^{-1}[-2/(-2)]$, which is $\pi/4$, is *not* an argument of z.

Roots of z

Each complex number z, other than 0, has exactly n nth roots for each positive integer n. These roots can be found by expressing z in polar coordinates. If $z = r(\cos\theta + i\sin\theta)$, that is, has magnitude r and argument θ, then one nth root is

$$r^{1/n}\left(\cos\frac{\theta}{n} + i\sin\frac{\theta}{n}\right).$$

To check that this is an nth root of z, just raise it to the nth power.

To find the other nth roots of z, change the argument of z from θ to $\theta + 2k\pi$,

ALTERNATING CURRENT AND COMPLEX NUMBERS

As early as their sophomore year, many students will take a course in electric circuits, where they will find the complex numbers used in the analysis of alternating currents. They will use a different notation, which is described in this table:

Standard	*Electrical Engineering*
i	j
$\bar{z}$	z^*
$r(\cos\theta + i\sin\theta)$	$r \angle \theta$

The symbol j is used in the following discussion of alternating current. (i is used for current.)

The complex numbers, introduced by mathematicians in the course of their pure research, were accepted by them as a legitimate structure early in the nineteenth century. At that time, electricity was only an object of laboratory interest; it had little practical significance. Yet before the century was over, the discoveries concerning electricity and magnetism were to transform our world, and complex numbers were to serve as a tool in that transformation by simplifying the algebra of alternating currents. The details are worth sketching, not only to show the importance of complex numbers, but also to demonstrate how the "pure" knowledge of one generation can become the "practical" technique of a later generation.

In the case of a *direct current,* such as that provided by a battery, the voltage E is constant. This constant voltage, working against a resistance R, produces a constant current I. The three real numbers E, I, and R are related by the equation

$$E = IR,$$

which says, "current is proportional to voltage."

For the long-distance transmission of electric power, *alternating currents* are far more efficient than direct currents. An alternating current is produced by rotating a coiled wire in a fixed magnetic field. Charles Proteus Steinmetz, an engineer at General Electric when the United States was starting to bring electricity to the cities, found the algebra of alternating currents unwieldy. As he wrote in 1893,

> The current rises from zero to a maximum; then decreases again to nothing, reverses and rises to a maximum in the opposite direction; decreases to zero, again reverses and rises to a maximum in the first direction—and so on.
>
> Thus in all calculations with alternating current, instead of a simple mechanical value of direct current theory, the investigator had to use a complicated function of time to represent the alternating current. The theory of alternating current apparatus thereby became so complicated that the investigator never got very far. . . .
>
> The idea suggested itself at length of representing the alternating current by a single complex number. . . . This proved the solution of the alternating current calculation.
>
> It gave to the alternating current a single numerical value, just as to the direct current, instead of the complicated function of time of the previous theory; and thereby it made alternating current calculations as simple as direct current calculations.
>
> The introduction of the complex number has eliminated the function of time from the alternating current theory, and has made the alternating current theory the simple algebra of the complex number, just as the direct current theory leads to the simple algebra of the real number.

Steinmetz, it has been said, "generated electricity out of the square root of -1." He used three complex numbers: E to describe the electromotive force (voltage), I to describe the current, and R to describe the impedance in the circuit (resistance). They are linked by the equation $E = IR$.

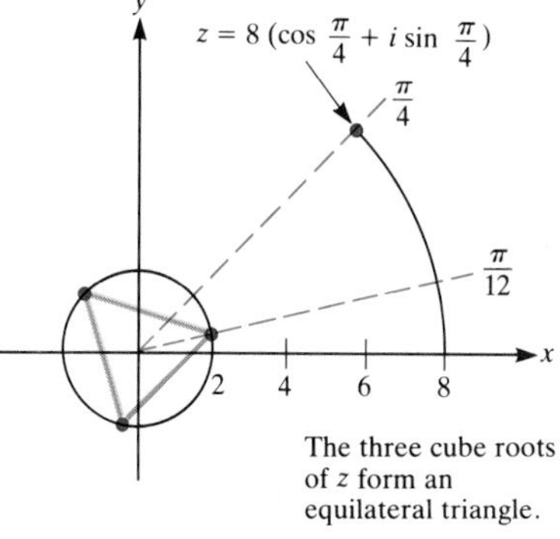

The three cube roots of z form an equilateral triangle.

Figure 12

where $k = 1, 2, \ldots, n-1$. Then

$$r^{1/n}\left(\cos\frac{\theta + 2k\pi}{n} + i\sin\frac{\theta + 2k\pi}{n}\right)$$

is also an nth root of z. (Why?)

For instance, let $z = 8[\cos(\pi/4) + i\sin(\pi/4)]$. Then the three cube roots of z all have magnitude $8^{1/3} = 2$. Their arguments are

$$\frac{\pi/4}{3} = \frac{\pi}{12}, \qquad \frac{\pi/4 + 2\pi}{3} = \frac{\pi}{12} + \frac{2\pi}{3}, \qquad \frac{\pi/4 + 4\pi}{3} = \frac{\pi}{12} + \frac{4\pi}{3}.$$

These three roots are shown in Fig. 12, along with z.

Section Summary

The real numbers, with which we all grew up, are just a small part of the complex numbers, which fill up the xy plane. We add them by a "parallelogram rule." To multiply them "we multiply their magnitudes and add their angles." Using the complex numbers we can see that "negative real times negative real is positive," since $180° + 180° = 360°$, which describes the positive x axis. We also saw how to raise a complex number to a power and how to take its roots.

EXERCISES FOR SEC. 11.6: COMPLEX NUMBERS

In Exercises 1 and 2 compute the given quantities:

1 (a) $(2 + 3i) + (5 - 2i)$ (b) $(2 + 3i)(2 - 3i)$

(c) $\dfrac{1}{2 - i}$ (d) $\dfrac{3 + 2i}{4 - i}$

2 (a) $(2 + 3i)^2$ (b) $\dfrac{4}{3 - i}$

(c) $(1 + i)(3 - i)$ (d) $\dfrac{1 + 5i}{2 - 3i}$

3 Let z_1 have magnitude 2 and argument $\pi/6$, and let z_2 have magnitude 3 and argument $\pi/3$.
(a) Plot z_1 and z_2.
(b) Find z_1z_2 using the polar form.
(c) Write z_1 and z_2 in the form $x + iy$.
(d) With the aid of (c) compute z_1z_2.

4 Let z_1 have magnitude 2 and argument $\pi/4$, and let z_2 have magnitude 1 and argument $3\pi/4$.
(a) Plot z_1 and z_2.
(b) Find their product using the polar form.
(c) Write z_1 and z_2 in the form $x + iy$.
(d) With the aid of (c), compute z_1z_2.

5 The complex number z has argument $\pi/3$ and magnitude 1. Find and plot (a) z^2, (b) z^3, (c) z^4.

6 Find (a) i^3, (b) i^4, (c) i^5, (d) i^{73}.

7 If z has magnitude 2 and argument $\pi/6$, what are the magnitude and argument of (a) z^2? (b) z^3? (c) z^4? (d) z^n? (e) Sketch z, z^2, z^3, z^4.

8 Let z have magnitude 0.9 and argument $\pi/4$. (a) Find and plot z^2, z^3, z^4, z^5, z^6. (b) What happens to z^n as $n \to \infty$?

9 Find and plot all solutions of the equation $z^5 = 32[\cos(\pi/4) + i\sin(\pi/4)]$.

10 Find and plot all solutions of $z^4 = 8 + 8\sqrt{3}\,i$. (*Hint:* First draw $8 + 8\sqrt{3}i$.)

11 Let z have magnitude r and argument θ. Let w have magnitude $1/r$ and argument $-\theta$. Show that $zw = 1$. (w is called the **reciprocal of** z, denoted z^{-1} or $1/z$.)

12 Find z^{-1} if $z = 4 + 4i$. (Use the definition in Exercise 11.)

13 (a) By substitution, verify that $2 + 3i$ is a solution of the equation $x^2 - 4x + 13 = 0$.
(b) Use the quadratic formula to find all solutions of the equation $x^2 - 4x + 13 = 0$.

14 (a) Use the quadratic formula to find the solutions of the equation $x^2 + x + 1 = 0$.
(b) Plot the solutions in (a).

15 Write in polar form: (a) $5 + 5i$, (b) $-\dfrac{1}{2} - \dfrac{\sqrt{3}}{2}i$, (c) $-\dfrac{\sqrt{2}}{2} + \dfrac{\sqrt{2}}{2}i$, (d) $3 + 4i$.

16 Write in rectangular form as simply as possible:

(a) $3\left(\cos\dfrac{3\pi}{4} + i\sin\dfrac{3\pi}{4}\right)$,

(b) $2\left(\cos\dfrac{\pi}{6} + i\sin\dfrac{\pi}{6}\right)$,

(c) $10(\cos\pi + i\sin\pi)$,

(d) $\frac{1}{5}(\cos 22° + i\sin 22°)$. (Express the answer to three decimal places.)

17 Let z_1 have magnitude r_1 and argument θ_1, and let z_2 have magnitude r_2 and argument θ_2.
(a) Explain why the magnitude of z_1/z_2 is r_1/r_2.
(b) Explain why the argument of z_1/z_2 is $\theta_1 - \theta_2$.

18 Compute

$$\frac{\cos\dfrac{5\pi}{4} + i\sin\dfrac{5\pi}{4}}{\cos\dfrac{3\pi}{4} + i\sin\dfrac{3\pi}{4}}$$

two ways: (a) by the result in Exercise 17, (b) by conjugating the denominator.

19 Compute

(a) $(2 + 3i)(1 + i)$ (b) $\dfrac{2 + 3i}{1 + i}$

(c) $(7 - 3i)\overline{(7 - 3i)}$

(d) $3(\cos 42° + i\sin 42°) \cdot 5(\cos 168° + i\sin 168°)$

(e) $\dfrac{\sqrt{8}\,(\cos 147^\circ + i \sin 147^\circ)}{\sqrt{2}\,(\cos 57^\circ + i \sin 57^\circ)}$

(f) $1/(3 - i)$

(g) $[3(\cos 52^\circ + i \sin 52^\circ)]^{-1}$

(h) $\left(\cos \dfrac{\pi}{6} + i \sin \dfrac{\pi}{6}\right)^{12}$

20 Compute

(a) $(3 + 4i)(3 - 4i)$ (b) $\dfrac{3 + 5i}{-2 + i}$

(c) $\dfrac{1}{2 + i}$ (d) $\left(\cos \dfrac{\pi}{12} + i \sin \dfrac{\pi}{12}\right)^{20}$

(e) $[r(\cos \theta + i \sin \theta)]^{-1}$

(f) Re $((r\,(\cos \theta + i \sin \theta))^{10})$

(g) $\dfrac{3\left(\cos \dfrac{\pi}{6} + i \sin \dfrac{\pi}{6}\right)}{5 - 12i}$

21 Find and plot all solutions of $z^3 = i$.

22 Sketch all complex numbers z such that (a) $z^6 = 1$, (b) $z^6 = 64$.

23 Using the fact that

$$(\cos \theta + i \sin \theta)^n = \cos n\theta + i \sin n\theta$$

find formulas for $\cos 3\theta$ and $\sin 3\theta$ in terms of $\cos \theta$ and $\sin \theta$.

24 (a) If $|z_1| = 1$ and $|z_2| = 1$, how large can $|z_1 + z_2|$ be? (*Hint:* Draw some pictures.)
(b) If $|z_1| = 1$ and $|z_2| = 1$, what can be said about $|z_1 z_2|$?

25 Show that (a) $\overline{z_1 z_2} = \bar{z}_1 \bar{z}_2$, (b) $\overline{z_1 + z_2} = \bar{z}_1 + \bar{z}_2$.

26 If arg z is θ, what is the argument of (a) $\bar{z}$, (b) $1/z$?

27 For which complex numbers z is $\bar{z} = 1/z$?

28 Let $z = \dfrac{1}{\sqrt{2}} + \dfrac{i}{\sqrt{2}}$.

(a) Compute z^2 algebraically.
(b) Compute z^2 by putting z into polar form.

29 Let $z = \dfrac{1}{2} + \dfrac{i}{2}$.

(a) Sketch the numbers z^n for $n = 1, 2, 3, 4$, and 5.
(b) What happens to z^n as $n \to \infty$?

30 Let $z = 1 + i$.
(a) Sketch the numbers $z^n/n!$ for $n = 1, 2, 3, 4$, and 5.
(b) What happens to $z^n/n!$ as $n \to \infty$?

31 As the complex number $z = t + it$ moves along the line $y = x$, z^2 sweeps out a curve.
(a) Plot five points on that curve.
(b) Give parametric equations for that curve.
(c) What type of curve is it?

32 Like Exercise 31 for $z = 1 + it$ on the line $x = 1$.

33 Like Exercise 31 for $z = t + i$ on the line $y = 1$.

34 (a) Graph $r = \cos \theta$ in polar coordinates.
(b) Pick five points on the curve in (a). Viewing each as a complex number z, plot z^2.
(c) As z runs through the curve in (a), what curve does z^2 sweep out? (Give its polar equation.)

35 Let a, b, and c be complex numbers such that $a \neq 0$ and $b^2 - 4ac \neq 0$. Show that $ax^2 + bx + c = 0$ has two roots.

36 Find and plot the roots of $x^2 + ix + 3 - i = 0$.

37 Compute the roots of the following equations and plot them relative to the same axes:
(a) $x^2 - 3x + 2 = 0$
(b) $x^2 - 3x + 2.25 = 0$
(c) $x^2 - 3x + 2.5 = 0$
(d) $x^2 - 3x + 1.5 = 0$

38 The partial-fraction representation of a rational function is much simpler when we have complex numbers. No second-degree polynomial $ax^2 + bx + c$ is needed. This exercise indicates why this is the case.

Let z_1 and z_2 be the roots of $ax^2 + bx + c = 0$, $a \neq 0$.
(a) Using the quadratic formula (or by other means), show that $z_1 + z_2 = -b/a$ and $z_1 z_2 = c/a$.
(b) From (a) deduce that

$$ax^2 + bx + c = a(x - z_1)(x - z_2).$$

(c) With the aid of (b) show that

$$\frac{1}{ax^2 + bx + c} = \frac{1}{a(z_1 - z_2)}\left(\frac{1}{x - z_1} - \frac{1}{x - z_2}\right).$$

Part (c) shows that the theory of partial fractions, described in Sec. 7.5, becomes much simpler when complex numbers are allowed as the coefficients of the polynomials. Only partial fractions of the type $k/(ax + b)^n$ are needed.

39 Let $f(x) = a_0 + a_1 x + a_2 x^2 + a_3 x^3 + a_4 x^4$, where each coefficient a_i is real.
(a) Show that if c is a root of $f(x) = 0$, then so is $\bar{c}$.
(b) Show that if c is a root of $f(x) = 0$ and is not real, then $(x - c)(x - \bar{c})$ divides $f(x)$.
(c) Using the fundamental theorem of algebra, show that any fourth-degree polynomial with real coefficients can be expressed as the product of polynomials of degree at most 2 with real coefficients.

11.7 THE RELATION BETWEEN THE EXPONENTIAL AND THE TRIGONOMETRIC FUNCTIONS

With the aid of complex numbers, Euler in 1743 discovered that the trigonometric functions can be expressed in terms of the exponential function e^z, where z is complex. This section will retrace his discovery. In particular, it will show that

Surprise: Pulling trigonometry out of e^x

$$e^{i\theta} = \cos\theta + i\sin\theta, \qquad \cos\theta = \frac{e^{i\theta} + e^{-i\theta}}{2}, \qquad \text{and} \qquad \sin\theta = \frac{e^{i\theta} - e^{-i\theta}}{2i}.$$

Complex Series

In order to relate the exponential function to the trigonometric functions, we will use infinite series such as $\Sigma_{n=0}^{\infty} z_n$, where z_n's are complex numbers. Such a series is said to converge to S if its nth partial sum S_n approaches S in the sense that $|S - S_n| \to 0$ as $n \to \infty$. It is shown in advanced calculus that if $\Sigma_{n=0}^{\infty} |z_n|$ converges, so does $\Sigma_{n=0}^{\infty} z_n$, and the series $\Sigma_{n=0}^{\infty} z_n$ is said to converge absolutely. If a series converges absolutely, we may rearrange its terms in any order without changing the sum.

Let $z_n = x_n + iy_n$, x_n and y_n are real. If $\Sigma_{n=0}^{\infty} z_n$ converges, so do $\Sigma_{n=0}^{\infty} x_n$ and $\Sigma_{n=0}^{\infty} y_n$. If $\Sigma_{i=0}^{n} z_n = S = a + bi$, then $\Sigma_{n=0}^{\infty} x_n = a$ and $\Sigma_{n=0}^{\infty} y_n = b$. $\Sigma_{n=0}^{\infty} x_n$ is called the real part of $\Sigma_{n=0}^{\infty} z_n$, denoted $\mathrm{Re}(\Sigma_{n=0}^{\infty} z_n)$ and $\Sigma_{n=0}^{\infty} y_n$ is the imaginary part, denoted $\mathrm{Im}(\Sigma_{n=0}^{\infty} z_n)$.

EXAMPLE 1 Determine for which complex numbers z, $\Sigma_{n=0}^{\infty} z^n/n!$ converges.

SOLUTION We will examine absolute convergence, that is, the convergence of $\Sigma_{n=0}^{\infty} |z|^n/n!$. This series has real terms. In fact, it is the Maclaurin series for $e^{|z|}$, which converges for all real numbers (and $|z|$ is real). Since $\Sigma_{n=0}^{\infty} z^n/n!$ converges absolutely for all z, it converges for all z. ■

Defining e^z

The Maclaurin series for e^x when x is real suggests the following definition:

> **Definition** *e^z for complex z.* Let z be a complex number. Define e^z to be the sum of the convergent series $\Sigma_{n=0}^{\infty} z^n/n!$.

It can be shown by multiplying the series for e^{z_1} and e^{z_2} that $e^{z_1+z_2} = e^{z_1}e^{z_2}$ in accordance with the basic law of exponents. When the expression for z is complicated, we sometimes write e^z as $\exp z$. For example, in exp notation the law of exponents becomes $\exp(z_1 + z_2) = (\exp z_1)(\exp z_2)$.

The Link between $e^{i\theta}$, $\cos\theta$, and $\sin\theta$

The following theorem, from Euler, provides the key link between the exponential function e^z and the trigonometric functions $\cos\theta$ and $\sin\theta$.

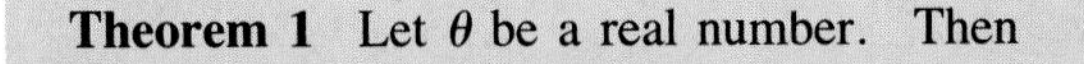

Theorem 1 Let θ be a real number. Then

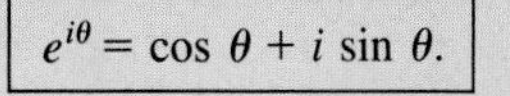

$$\boxed{e^{i\theta} = \cos\theta + i\sin\theta.}$$

Proof By definition of e^z for any complex number,

$$e^{i\theta} = 1 + i\theta + \frac{(i\theta)^2}{2!} + \frac{(i\theta)^3}{3!} + \frac{(i\theta)^4}{4!} + \cdots.$$

Thus

$$e^{i\theta} = 1 + i\theta + \frac{i^2\theta^2}{2!} + \frac{i^3\theta^3}{3!} + \frac{i^4\theta^4}{4!} + \cdots$$

$$= 1 + i\theta - \frac{\theta^2}{2!} - \frac{i\theta^3}{3!} + \frac{\theta^4}{4!} + \cdots$$

$$= \left(1 - \frac{\theta^2}{2!} + \frac{\theta^4}{4!} - \cdots\right) + i\left(\theta - \frac{\theta^3}{3!} + \cdots\right)$$

$$= \cos\theta + i\sin\theta.$$

Recall that $i^2 = -1$, $i^3 = -i$, $i^4 = 1$, $i^5 = i$, ...

Figure 1 shows $e^{i\theta}$, which lies on the standard unit circle. ■

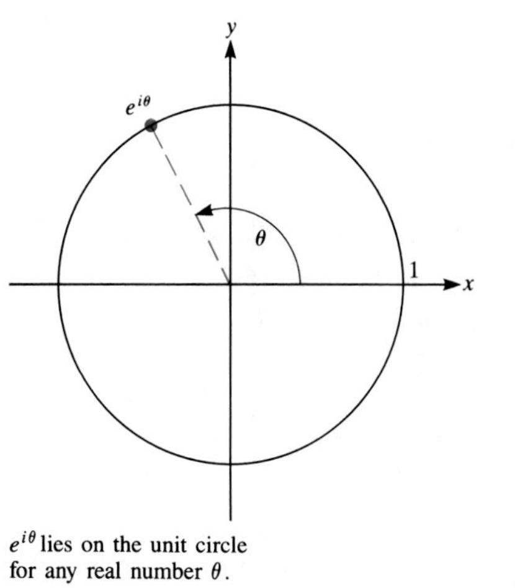

$e^{i\theta}$ lies on the unit circle for any real number θ.

Figure 1

Theorem 1 asserts, for instance, that

$$e^{\pi i} = \cos\pi + i\sin\pi = -1 + i\cdot 0 = -1,$$

or

$$e^{\pi i} = -1,$$

an equation that links e (the fundamental number in calculus), π (the fundamental number in trigonometry), i (the fundamental complex number), and the negative number -1. The history of that short equation would recall the struggles of hundreds of mathematicians to create the number system that we now take for granted.

With the aid of Theorem 1, both $\cos\theta$ and $\sin\theta$ may be expressed in terms of the exponential function.

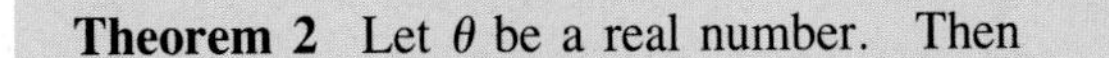

Theorem 2 Let θ be a real number. Then

$$\boxed{\cos\theta = \frac{e^{i\theta} + e^{-i\theta}}{2} \quad \text{and} \quad \sin\theta = \frac{e^{i\theta} - e^{-i\theta}}{2i}.}$$

Proof By Theorem 1,

$$e^{i\theta} = \cos\theta + i\sin\theta. \tag{1}$$

Replacing θ by $-\theta$ in Eq. (1), we obtain

$$e^{-i\theta} = \cos\theta - i\sin\theta. \tag{2}$$

Addition of Eqs. (1) and (2) yields

$$e^{i\theta} + e^{-i\theta} = 2 \cos \theta.$$

Hence $$\cos \theta = \frac{e^{i\theta} + e^{-i\theta}}{2}.$$

Subtraction of Eq. (2) from Eq. (1) yields

$$e^{i\theta} - e^{-i\theta} = 2i \sin \theta.$$

Hence $$\sin \theta = \frac{e^{i\theta} - e^{-i\theta}}{2i}.$$

This establishes the theorem. ■

Recall Sec. 6.9, where cosh x *and* sinh x *were defined.*

The hyperbolic functions cosh x and sinh x were defined in terms of the exponential function by

$$\cosh x = \frac{e^x + e^{-x}}{2} \qquad \text{and} \qquad \sinh x = \frac{e^x - e^{-x}}{2}.$$

Old saying: "God created the complex numbers; anything less is the work of man."

Theorem 2 shows that the trigonometric functions could be similarly defined in terms of the exponential function—if complex numbers were available.

Indeed, from the complex numbers and e^z we could even obtain the derivative formulas for $\sin \theta$ and $\cos \theta$. For instance,

$$(\sin \theta)' = \left(\frac{e^{i\theta} - e^{-i\theta}}{2i}\right)' = \frac{ie^{i\theta} + ie^{-i\theta}}{2i} = \cos \theta.$$

(That the familiar rules for differentiation extend to complex functions is justified in a course in complex variables.)

Sketching e^z

The magnitude and argument of e^{x+iy}

If $z = x + iy$, the evaluation of e^z can be carried out as follows:

$$e^z = e^{x+iy} = e^x e^{iy} = e^x(\cos y + i \sin y).$$

Thus the magnitude of e^{x+iy} is e^x and the argument of e^{x+iy} is y.

EXAMPLE 2 Compute and sketch (*a*) $e^{2+(\pi/6)i}$, (*b*) $e^{2+\pi i}$, and (*c*) $e^{2+3\pi i}$.

SOLUTION (*a*) $e^{2+(\pi/6)i}$ has magnitude e^2 and argument $\pi/6$. (*b*) $e^{2+\pi i}$ has

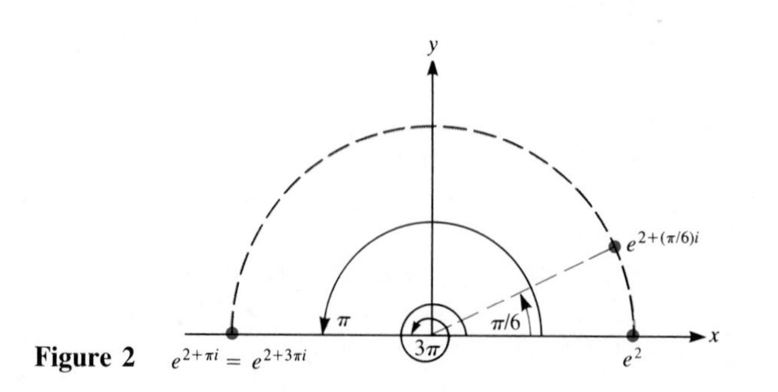

Figure 2

magnitude e^2 and argument π; it equals $-e^2$. (c) $e^{2+3\pi i}$ has magnitude e^2 and argument 3π, so is the same number as the number in (b). The results are sketched in Fig. 2. ■

The next example illustrates a typical computation in alternating currents.

EXAMPLE 3 Find the real part of $100\, e^{j(\pi/6)}e^{j\omega t}$. Here t refers to time, ω is a real constant related to frequency, and j is the electrical engineers' symbol for $\sqrt{-1}$.

SOLUTION

$$\begin{aligned} 100\, e^{j(\pi/6)}e^{j\omega t} &= 100\, e^{j(\pi/6)+j\omega t} \\ &= 100\, e^{j[(\pi/6)+\omega t]} \\ &= 100\left[\cos\left(\frac{\pi}{6}+\omega t\right) + j\sin\left(\frac{\pi}{6}+\omega t\right)\right]. \end{aligned}$$

Thus

$$\text{Re}(100\, e^{j(\pi/6)}e^{j\omega t}) = 100\cos\left(\frac{\pi}{6}+\omega t\right). \quad ■$$

It is often convenient to think of $\cos\theta$ as $\text{Re}(e^{i\theta})$. The next example exploits this point of view.

EXAMPLE 4 Evaluate $\displaystyle\sum_{n=0}^{\infty} \frac{\cos n\theta}{n!}$.

SOLUTION Recall that $e^{in\theta} = \cos n\theta + i \sin n\theta$. Hence $\cos n\theta = \text{Re}(e^{in\theta})$, and we have

Recall the definition of e^z.

$$\begin{aligned} \sum_{n=0}^{\infty} \frac{\cos n\theta}{n!} &= \text{Re} \sum_{n=0}^{\infty} \frac{e^{in\theta}}{n!} = \text{Re} \sum_{n=0}^{\infty} \frac{(e^{i\theta})^n}{n!} \\ &= \text{Re} \sum_{n=0}^{\infty} \frac{(\exp i\theta)^n}{n!} \\ &= \text{Re}\,(e^{\exp i\theta}) \\ &= \text{Re}\,(e^{\cos\theta + i\sin\theta}) \\ &= \text{Re}\,(e^{\cos\theta} e^{i\sin\theta}) \\ &= e^{\cos\theta}\,\text{Re}\,(e^{i\sin\theta}) \\ &= e^{\cos\theta}\,\text{Re}\,[\cos(\sin\theta) + i\sin(\sin\theta)] \\ &= e^{\cos\theta}\cos(\sin\theta). \end{aligned}$$

Hence

$$\sum_{n=0}^{\infty} \frac{\cos n\theta}{n!} = e^{\cos\theta}\cos(\sin\theta). \quad ■$$

Section Summary

Using power series, we obtained the fundamental relation $e^{i\theta} = \cos\theta + i\sin\theta$ and showed that $\cos\theta$ and $\sin\theta$ can be expressed in terms of the exponential function. Since $\ln x$ is the inverse of e^x, it too is obtained from the exponential function. We may define even x^n, $x > 0$, in terms of the exponential function as $e^{n \ln x}$. Similarly, a^x, $a > 0$, can be defined as $e^{x \ln a}$. These observations suggest that the most fundamental function in calculus is e^x.

EXERCISES FOR SEC. 11.7: THE RELATION BETWEEN THE EXPONENTIAL AND THE TRIGONOMETRIC FUNCTIONS

In Exercises 1 to 6 sketch the numbers given and state their real and imaginary parts.

1 $e^{(5\pi i/4)}$
2 $5e^{(\pi i/4)}$
3 $2e^{(\pi i/4)} + 3e^{(\pi i/6)}$
4 e^{2+3i}
5 $e^{(\pi i/6)}e^{(3\pi i/4)}$
6 $2e^{\pi i} \cdot 3e^{-(\pi i/3)}$

In Exercises 7 to 10 express the given numbers in the form $re^{i\theta}$ for a positive real number r and argument θ, $-\pi < \theta \leq \pi$.

7 $\dfrac{e^2}{\sqrt{2}} - \dfrac{e^2}{\sqrt{2}}i$
8 $3\left(\cos \dfrac{\pi}{4} + i \sin \dfrac{\pi}{4}\right)$

9 $5\left(\cos \dfrac{\pi}{6} + i \sin \dfrac{\pi}{6}\right) \cdot 3\left(\cos \dfrac{\pi}{2} + i \sin \dfrac{\pi}{2}\right)$

10 $7\left(\cos \dfrac{7\pi}{3} + i \sin \dfrac{7\pi}{3}\right)$

In Exercises 11 to 14 plot the given numbers.

11 $\exp(\pi i/4 + 3\pi i)$
12 $\exp(1 + 9\pi i/4)$
13 $\exp(2 - \pi i/3)$
14 $\exp(-1 + 17\pi i/6)$

15 Let $z = e^{a+bi}$. Find (*a*) $|z|$, (*b*) $\bar{z}$, (*c*) z^{-1}, (*d*) Re z, (*e*) Im z, (*f*) arg z. [In (*f*) assume a and b are positive.]

16 For which values of a and b is $\lim_{n\to\infty} (e^{a+ib})^n = 0$?

17 Find all z such that $e^z = 1$.

18 Find all z such that $e^z = -1$.

19 (*a*) Find $|e^{3+4i}|$.
(*b*) Plot the complex number e^{3+4i}.

20 (*a*) Plot all complex numbers of the form e^{x+4i}, x real.
(*b*) Plot all complex numbers of the form e^{3+iy}, y real.

21 In Claude Garrod's *Twentieth Century Physics*, Faculty Publishing, Davis, Calif., p. 107, there is the remark: "Using the fact that

$$(e^{-i\omega_0 t})^*(e^{-i\omega_0 t}) = 1,$$

we can easily evaluate the probability density for these standing waves." In this text, z^* denotes the conjugate of z and ω_0 is real. Justify the equation.

22 Use the fact that $1 + \cos\theta + \cos 2\theta + \cdots + \cos (n-1)\theta$ is the real part of $1 + e^{\theta i} + e^{2\theta i} + \cdots + e^{(n-1)\theta i}$ to find a short formula for that trigonometric sum.

23 Find all z such that $e^z = 3 + 4i$.

24 Assuming that $e^{z_1+z_2} = e^{z_1}e^{z_2}$ for complex numbers z_1 and z_2, obtain the trigonometric identities for $\cos(A + B)$ and $\sin(A + B)$.

25 Evaluate
$$\sum_{n=0}^{\infty} \frac{\cos n\theta}{2^n}.$$
(First show that it converges.)

26 Evaluate
$$\sum_{n=0}^{\infty} \frac{\sin n\theta}{n!}.$$
(First show that it converges.)

Exercises 27 and 28 treat the complex logarithms of a complex number. They show that $z = \ln w$ is not single-valued.

27 Let w be a nonzero complex number. Show that there are an infinite number of z such that $e^z = w$.

28 (See Exercise 27.) If $e^z = w$, write $z = \ln w$, although $\ln w$ is not a uniquely defined number. If b is a nonzero complex number and q is a complex number, define b^q to be $e^{q \ln b}$. Since $\ln b$ is not unique, b^q is usually not unique. List all possible values of (*a*) $(-1)^i$, (*b*) $10^{1/2}$, (*c*) 10^3.

29 Show that $\sin z = -i \sinh (iz)$.

30 When z is real, $|\sin z| \leq 1$ and $|\cos z| \leq 1$. Do these inequalities hold for all complex z?

31 Let
$$z = \frac{1+i}{\sqrt{2}}.$$
(*a*) Plot z, $z^2/2!$, $z^3/3!$, $z^4/4!$.
(*b*) Plot $1 + z + z^2/2! + z^3/3! + z^4/4!$, which is an estimate of $\exp[(1 + i)/\sqrt{2}]$.
(*c*) Plot $\exp[(1 + i)/\sqrt{2}]$ relative to the same axes.

32 An integral table lists $\int xe^{ax}\,dx = e^{ax}(ax - 1)/a^2$. At first glance, finding the integral of $xe^{ax} \cos bx$ may appear to be a much harder problem. However, by noticing that $\cos bx =$ Re (e^{ibx}), we can reduce it to a simpler problem. Following this approach, find $\int xe^{ax} \cos bx\,dx$. *Suggestion*: The formula for $\int xe^{ax}\,dx$ holds when a is complex.

33 In Sec. 6.9 we define $\cosh x = (e^x + e^{-x})/2$ and $\sinh x = (e^x - e^{-x})/2$. We can use the same definitions when x is complex. In view of Theorem 2, let us define sine and cosine for complex z by $\sin z = (e^{iz} - e^{-iz})/(2i)$ and $\cos z = (e^{iz} + e^{-iz})/2$. Establish the following links between the hyperbolic and trigonometric functions:
(*a*) $\cosh z = \cos (iz)$
(*b*) $\sinh z = -i \sin(iz)$.

34 Let z be a complex number and θ a real number. What is the geometric relation between z and $e^{i\theta} z$? Experiment, conjecture, and explain.

11.S SUMMARY

The motivation for this chapter was our search for a way to approximate a function $f(x)$ by a polynomial for x near a number a. We chose the unique polynomial of degree at most n whose first n derivatives at a are the same as those of the function. This polynomial turned out to be

$$\sum_{k=0}^{n} \frac{f^{(k)}(a)(x-a)^k}{k!},$$

which we called the nth Taylor polynomial for $f(x)$ at a, denoted $P_n(x; a)$. When $a = 0$, the polynomial takes the simpler form

$$P_n(x; 0) = \sum_{k=0}^{n} \frac{f^{(k)}(0)x^k}{k!}.$$

Keep in mind that $0! = 1$, $1! = 1$, and $f^{(0)}(x) = f(x)$.

These polynomials are the "front ends" or partial sums of the power series

$$\sum_{k=0}^{\infty} \frac{f^{(k)}(a)(x-a)^k}{k!},$$

which is the Taylor series associated with $f(x)$ at a. If $a = 0$, it is also called the Maclaurin series associated with $f(x)$.

The question, "Does $P_n(x; a)$ become arbitrarily close to $f(x)$ as $n \to \infty$?" can be rephrased "Does the Taylor series for $f(x)$ at a converge to $f(x)$?" Sometimes it does not, even though it may converge. But for such important functions as e^x, $\sin x$, and $\cos x$, it does.

The key to showing that the Taylor series does converge to $f(x)$ for certain functions is Lagrange's theorem, which expresses

$$R_n(x; a) = f(x) - P_n(x; a)$$

conveniently in terms of the function $f(x)$:

$$R_n(x; a) = \frac{f^{(n+1)}(c_n)(x-a)^{n+1}}{(n+1)!},$$

where c_n is some number between a and x, and depends on a, x, and n.

With the aid of Lagrange's theorem we can estimate the error in using Taylor polynomials to approximate $f(x)$. Such approximations permit us to evaluate e^x, $\sin x$, $\cos x$, and $\ln x$ to as many decimal places as we have time for. Such approximations also help in approximating definite integrals.

The proof of Lagrange's formula in Sec. 11.3 illustrates the significance of higher order derivatives in estimating the growth of a function. Exercises in that section use the same method to obtain the formula for the errors in the trapezoidal and Simpson's methods.

Taylor series led us to consider power series in general. Associated with each such series is a radius of convergence R, which can be finite or infinite. Where they converge, power series can be differentiated and integrated just like polynomials. We can also use them to evaluate indeterminate limits of the zero-over-zero form.

With the aid of the power series for e^x we defined e^z for complex numbers. Once we enter the realm of complex numbers, we discover that the trigonometric functions are simple combinations of exponentials:

$$\sin \theta = \frac{e^{i\theta} - e^{-i\theta}}{2i} \quad \text{and} \quad \cos \theta = \frac{e^{i\theta} + e^{-i\theta}}{2},$$

and that

$$e^{i\theta} = \cos \theta + i \sin \theta.$$

The number e^{x+iy}, where x and y are real, is at a distance e^x from the origin and on the ray from the origin of angle y.

Vocabulary and Symbols

power series
Maclaurin series
radius of convergence
Taylor series $\Sigma_{n=0}^{\infty} f^{(n)}(a)(x-a)^n/n!$
Taylor polynomial $P_n(x; a)$
remainder (error) in Taylor series $R_n(x; a) = f(x) - P_n(x; a)$
Lagrange's theorem
complex number
magnitude and argument of a complex number

Function	*Maclaurin Series*	*Page*
e^x	$1 + x + \frac{x^2}{2!} + \frac{x^3}{3!} + \cdots = \sum_{n=0}^{\infty} \frac{x^n}{n!}$	627
$\sin x$	$x - \frac{x^3}{3!} + \frac{x^5}{5!} - \frac{x^7}{7!} + \cdots = \sum_{n=0}^{\infty} (-1)^n \frac{x^{2n+1}}{(2n+1)!}$	627
$\cos x$	$1 - \frac{x^2}{2!} + \frac{x^4}{4!} - \frac{x^6}{6!} + \cdots = \sum_{n=0}^{\infty} (-1)^n \frac{x^{2n}}{(2n)!}$	635
$\frac{1}{1-x}$	$1 + x + x^2 + \cdots = \sum_{n=0}^{\infty} x^n, \quad \lvert x\rvert < 1$	574
$\ln(1+x)$	$x - \frac{x^2}{2} + \frac{x^3}{3} - \cdots = \sum_{n=1}^{\infty} (-1)^{n+1} \frac{x^n}{n}, \quad -1 < x \le 1$	627
$\tan^{-1} x$	$x - \frac{x^3}{3} + \frac{x^5}{5} - \cdots = \sum_{n=0}^{\infty} (-1)^n \frac{x^{2n+1}}{2n+1}, \quad \lvert x\rvert \le 1$	574
$\sin^{-1} x$	$x + \frac{1}{2}\frac{x^3}{3} + \frac{1 \cdot 3}{2 \cdot 4}\frac{x^5}{5} + \frac{1 \cdot 3 \cdot 5}{2 \cdot 4 \cdot 6}\frac{x^7}{7} + \cdots, \quad \lvert x\rvert \le 1$	675
$(1+x)^r$	$1 + rx + \frac{r(r-1)}{2!}x^2 + \frac{r(r-1)(r-2)}{3!}x^3 + \cdots, \quad \lvert x\rvert < 1$	647

GUIDE QUIZ ON CHAP. 11: POWER SERIES AND COMPLEX NUMBERS

1 A Maclaurin series converges for $x = 2$ and diverges for $x > 2$. For which other values of x must it converge? diverge?

2 The series $\Sigma_{n=0}^{\infty} a_n(x-2)^n$ diverges for $x = 5$. For which other values of x must it diverge?

3 Give an example of a Maclaurin series that converges for $|x| < 1$ and no other values of x.

4 Estimate $\int_0^1 \sin x^2 \, dx$ to two decimal places.

5 Obtain Lagrange's formula for $R_4(x; a)$ by repeatedly "integrating and simplifying."

6 Give the Maclaurin series and their radii of convergence for the following functions:

(*a*) $\frac{1}{1+x^2}$ (*b*) e^{-x} (*c*) $\cos x$ (*d*) $\ln(1-x)$

(*e*) $\frac{1}{1-2x}$ (*f*) $1 + 3x + 5x^2$ (*g*) $\tan^{-1} 2x$

7 Assume that $f(x)$ is represented by a Maclaurin series for some open interval $(-R, R)$, $f(x) = \Sigma_{n=0}^{\infty} a_n x^n$. Explain why $a_n = f^{(n)}(0)/n!$.

8 Determine for which x the series converge or diverge:

(*a*) $\sum_{n=1}^{\infty} \frac{n}{1+n^2}(x-2)^n$ (*b*) $\sum_{n=0}^{\infty} \frac{2^n}{n!}(x+3)^n$

9 (*a*) Which terms $a_n x^n$ in the Maclaurin series for sec x will have $a_n = 0$?
(*b*) Obtain the first three nonzero terms for the Maclaurin series for sec x by dividing the series for cos x into 1.

10 (*a*) Explain why $e^{i\theta} = \cos\theta + i\sin\theta$.
(*b*) Plot all solutions of the equation $e^z = 2$.
(*c*) Plot all solutions of the equation $z^6 = i$.
(*d*) Plot all solutions of the equation $z^2 = 3 + 4i$.

REVIEW EXERCISES FOR CHAP. 11: POWER SERIES AND COMPLEX NUMBERS

1 Suppose $\Sigma_{k=0}^{\infty} a_k$ is a convergent series with sum S. Let $R_n = S - \Sigma_{k=0}^{n} a_k$. What can be said about R_n if

(*a*) $\Sigma_{k=0}^{\infty} a_k$ is alternating and a_k is decreasing in absolute value?
(*b*) $\Sigma_{k=0}^{\infty} a_k$ is geometric and $a_k = ar^k$?
(*c*) $a_k = f(k)$, where $f(x)$ is a decreasing continuous function?
(*d*) $\Sigma_{k=n+1}^{\infty} |a_k| = 27$?
(*e*) $a_k = f^{(k)}(0)/k!$ for some function $f(x)$.

2 Let $f(x) = \int_0^x (\sin t)/t\, dt$.
(*a*) Write out the first five terms (including terms equal to 0) for the Maclaurin series for $f(x)$.
(*b*) Give a formula for the nth nonzero term.
(*c*) Give $\int_0^1 (\sin t)/t\, dt$ to three decimal places.

3 Give the formula for the nth nonzero term of the Maclaurin series for
(*a*) $\tan^{-1} 3x$ (*b*) $\ln(1 + x^2)$ (*c*) e^{-x}
(*d*) $\sin x^2$ (*e*) $\cos x^2$ (*f*) $\cos^2 x$
In some cases it may be simpler to start at $n = 0$.

4 A text on the economics of equipment maintenance contains this argument:

Adding the discounted net benefits over all the cycles yields

$$B(N) = \sum_{n=1}^{\infty} e^{-(n-1)(N+K)\alpha}\left[\int_0^N e^{-\alpha t}b(t)\, dt - e^{-\alpha N}c(N)\right].$$

Summing the geometric series gives

$$B(N) = \frac{\int_0^N e^{-\alpha t}b(t)\, dt - e^{-\alpha N}c(N)}{1 - e^{-\alpha(N+K)}}.$$

Sum the geometric series to check that the missing algebra is correct.

5 Obtain the first three nonzero terms of the Maclaurin series for $\sin 2x$:
(*a*) by replacing x by $2x$ in the Maclaurin series for $\sin x$,
(*b*) by using the formula $a_n = f^{(n)}(0)/n!$,
(*c*) by using the identity $\sin 2x = 2 \sin x \cos x$ and the Maclaurin series for $\sin x$ and $\cos x$.

6 Obtain the first three nonzero terms of the Maclaurin series for $\sin^2 x$:
(*a*) by using the formula $a_n = f^{(n)}(0)/n!$,
(*b*) by using the identity $\sin^2 x = (1 - \cos 2x)/2$ and the series for $\cos 2x$.

7 An engineer wishes to use a partial sum of the Maclaurin series for e^x to approximate e^x. How many terms of the series should be used to be sure that the error is less than
(*a*) 0.01 for $|x| \le 1$? (*b*) 0.001 for $|x| \le 1$?
(*c*) 0.01 for $|x| \le 2$? (*d*) 0.001 for $|x| \le 2$?

In Exercises 8 to 11 graph the given functions and the Taylor polynomials $P_0(x; 0)$, $P_1(x; 0)$, $P_2(x; 0)$, and $P_3(x; 0)$. In each exercise graph the functions relative to the same axes.

8 $1/(1 + x)$ **9** $\ln(1 + x)$
10 $\sin x$ **11** $\cos x$

In Exercises 12 to 17 show for the given functions that $R_n(x; 0) \to 0$ as $n \to \infty$, using the derivative form for $R_n(x; 0)$.

12 e^x **13** e^{-x}
14 $\sin 2x$ **15** $\cos x$

16 $\dfrac{1}{1 + x}$, $-\frac{1}{2} < x < 1$

17 $\ln(1 + x)$, $-\frac{1}{2} < x < 1$

In each of Exercises 18 to 21 use the first three nonzero terms of a Maclaurin series to estimate the quantity. Also put an upper bound on the error.

18 $\sqrt[10]{e}$ **19** $\cos \frac{1}{3}$
20 $\sin 28°$ **21** $1/\sqrt{e}$

In each of Exercises 22 to 29 determine for which x the series diverges, converges absolutely, and converges conditionally. Give the radius of convergence in each case and the sum of the series if it is easily determined.

22 $\displaystyle\sum_{n=1}^{\infty} \frac{2^n x^n}{n}$ **23** $\displaystyle\sum_{n=1}^{\infty} nx^{n-1}$

24 $\displaystyle\sum_{n=0}^{\infty} \frac{(x-3)^n}{n!}$ **25** $\displaystyle\sum_{n=0}^{\infty} \frac{x^{2n}}{n!}$

26 $\displaystyle\sum_{n=1}^{\infty} (-n)^n x^n$ **27** $\displaystyle\sum_{n=1}^{\infty} \frac{x^n}{n}$

28 $\displaystyle\sum_{n=0}^{\infty} \frac{3^n(x - \frac{2}{3})^n}{4^n}$ **29** $\displaystyle\sum_{n=0}^{\infty} \frac{n^5 + 2}{n^3 + 1}(x + 1)^n$

30 Estimate or compute exactly

$$\sum_{n=1}^{\infty} (-2)^{-n} - \sum_{n=1}^{10} (-2)^{-n}$$

(*a*) by noticing that the series is alternating decreasing, (*b*) by noticing that the absolute value of the error is less than $\Sigma_{n=11}^{\infty} 2^{-n}$, (*c*) by noticing that $\Sigma_{n=1}^{\infty} (-2)^{-n}$ is a geometric series, (*d*) by considering $f(x) = (1+x)^{-1}$ for $x = \frac{1}{2}$ and using Lagrange's form of the remainder.

31 Estimate the positive root of the equation $e^x = 2x + 1$
(*a*) using a Taylor polynomial of degree 3 to approximate e^x,
(*b*) using Newton's method for the function $e^x - 2x - 1$.

32 Express $x^2 + x + 2$ as a polynomial in powers of $x - 5$.

33 In each of the following integrals,
(1) Write the given integral as an infinite series of numbers.
(2) Estimate the error if you use the first three nonzero terms from (1) to approximate the integral.

(*a*) $\int_0^{1/2} \cos x^3\,dx$ (*b*) $\int_0^1 \sin x^2\,dx$

(*c*) $\int_0^{1/2} \sqrt[3]{1+x^3}\,dx$ (*d*) $\int_1^2 e^{-x^3}\,dx$

34 Write a few paragraphs on the subject, "What is a Taylor series and what good is it?"

In each of Exercises 35 to 44 determine the radius of convergence and for which x the series converges or diverges.

35 $\sum_{n=1}^{\infty} \frac{(n!)^2 x^n}{n^{2n}}$ **36** $\sum_{n=1}^{\infty} \frac{2^n (x-1)^n}{[1+(1/n)]^n}$

37 $\sum_{n=1}^{\infty} (n^2 x)^n$ **38** $\sum_{n=0}^{\infty} \frac{(2n+1)x^n}{n!}$

39 $\sum_{n=1}^{\infty} \frac{(-1)^n n^2 (x-3)^n}{n^3+1}$ **40** $\sum_{n=2}^{\infty} \frac{(x-1)^n}{\ln n}$

41 $\sum_{n=2}^{\infty} (\ln n)(x+2)^n$ **42** $\sum_{n=1}^{\infty} \frac{(x+3)^n}{\sqrt[3]{n}}$

43 $\sum_{n=1}^{\infty} \frac{(-1)^n x^{2n}}{\sqrt{n}}$ **44** $\sum_{n=1}^{\infty} \left(\frac{n+1}{n+3}\right)^{n^2} x^n$

45 Prove in your own words that if $\Sigma_{n=0}^{\infty} a_n x^n$ converges at the number d, then it also converges at any number x in $(-|d|, |d|)$.

46 Suppose $\Sigma_{n=0}^{\infty} a_n (x-2)^n$ converges for $x = 7$ and diverges for $x = -3$. What can be said about its radius of convergence?

In Exercises 47 to 49 determine the limits using power series. It might be instructive to solve the problems by l'Hôpital's rule also.

47 $\lim_{x\to 0} \frac{\ln(1+x^2) - \sin^2 x}{\tan x^2}$

48 $\lim_{x\to 0} \frac{(e^{x^2}-1)^2}{1 - x^2/2 - \cos x}$ **49** $\lim_{x\to 0} \frac{(1-\cos x^2)^5}{(x - \sin x)^{20}}$

50 For $|x| \le 1$, $\tan^{-1} x = \Sigma_{n=0}^{\infty} (-1)^n x^{2n+1}/(2n+1)$. Thus $\pi/4 = 1 - \frac{1}{3} + \frac{1}{5} - \frac{1}{7} + \cdots$, which converges slowly. The identity

$$\frac{\pi}{4} = 4\tan^{-1}\frac{1}{5} - \tan^{-1}\frac{1}{70} + \tan^{-1}\frac{1}{99} \tag{1}$$

provides a faster way to estimate π.
(*a*) Using the first three nonzero terms of the Maclaurin series for $\tan^{-1} x$ and identity (1), estimate $\pi/4$ and hence π.
(*b*) Discuss the size of the error in (*a*) in the estimate of $\pi/4$.
(*c*) How would you show that identity (1) is true? (Describe a method, but do not carry out the computations.)

51 Let $f(x) = \Sigma_{n=0}^{\infty} 2^n x^n$. Find $f^{(33)}(0)$.

52 Give an example of a Maclaurin series whose radius of convergence is 1 and which
(*a*) converges at 1 and -1,
(*b*) diverges at 1 and -1,
(*c*) converges at 1 and diverges at -1.

53 Show that if $\Sigma_{n=0}^{\infty} a_n 3^n$ converges, then so does $\Sigma_{n=0}^{\infty} n a_n 2^n$.

54 Find $f^{(99)}(0)$, $f^{(100)}(0)$, and $f^{(101)}(0)$ if $f(x)$ is
(*a*) $\tan^{-1} x$, (*b*) e^{x^2}.

55 Find the first four nonzero terms of the Maclaurin series for $e^{-x} \sin x$ by
(*a*) multiplying the series for e^{-x} by the one for $\sin x$,
(*b*) dividing the series for $\sin x$ by the one for e^x.

56 The integral $\int_0^1 e^{-x^2}\,dx$ cannot be evaluated by the fundamental theorem of calculus.
(*a*) Replacing x in the power series $e^x = \Sigma_{n=0}^{\infty} x^n/n!$ by $-x^2$, obtain the power series for e^{-x^2}.
(*b*) Show that

$$\int_0^1 e^{-x^2}\,dx = 1 - \frac{1}{3} + \frac{1}{5\cdot 2!} - \frac{1}{7\cdot 3!} + \cdots.$$

(*c*) Use (*b*) to estimate $\int_0^1 e^{-x^2}\,dx$ to three-decimal-place accuracy.
(*d*) Use Simpson's method to estimate $\int_0^1 e^{-x^2}\,dx$ to three decimal places.

57 Estimate $\int_0^{1/2} x \tan x\,dx$ using the first three nonzero terms of the Maclaurin series for $\tan x$.

58 Assume that $f^{(2)}$ is continuous and $f^{(2)}(a) \ne 0$. By the mean-value theorem, $f(a+h) = f(a) + hf'(a+\theta h)$ for some θ in $[0, 1]$.
(*a*) When h is small, why is θ unique?
(*b*) Prove that $\theta \to \frac{1}{2}$ as $h \to 0$.

59 Although $f'(a)$ is the limit of $[f(a+\Delta x) - f(a)]/\Delta x$, there is a better way to estimate $f'(a)$ than by that quotient. Assume that $f^{(3)}$ is continuous. Show that

(*a*) $$\frac{f(a+\Delta x) - f(a)}{\Delta x} = f'(a) + \frac{f^{(2)}(c_1)}{2}\Delta x$$

for some c_1 between a and $a + \Delta x$, and

(b) $$\frac{f(a + \Delta x) - f(a - \Delta x)}{2\Delta x} = f'(a) + \frac{f^{(3)}(c_2)}{6}(\Delta x)^2,$$

where c_2 is in $[a - \Delta x, a + \Delta x]$. [Since the error in using the quotient in (b) involves $(\Delta x)^2$, while the error in using the standard quotient involves Δx, the quotient in (b) is more accurate when Δx is small.] Test this observation on the function $y = x^3$ at $a = 2$.

60 Consider $\int_0^b xe^{-x}\,dx$, when b is a small positive number. Since e^{-x} is then close to $1 - x$, the definite integral behaves like $\int_0^b (x - x^2)\,dx = b^2/2 - b^3/3$, hence approximately like $b^2/2$. On the other hand, $\int_0^b xe^{-x}\,dx = 1 - e^{-b}(1 + b)$, and, since $e^{-b} \approx 1 - b$, we have $1 - e^{-b}(1 + b)$ approximately equal to $1 - (1 - b)(1 + b) = b^2$. Hence $\int_0^b xe^{-x}\,dx$ behaves like b^2. Which is correct, $b^2/2$ or b^2? Find the error.

61 In advanced mathematics a certain function $E(x)$ is *defined* as the sum $\sum_{n=0}^{\infty} x^n/n!$. Pretending that you have never heard of e or e^x, solve the following problems:
(a) Show that $E(0) = 1$.
(b) Show that $E'(x) = E(x)$.
(c) Show that $E(x)E(-x) = 1$. *Hint:* Differentiate $E(x)E(-x)$ and use (a) and (b).
(d) Deduce that $E(x + y)/E(x)$ is independent of x.
(e) Deduce that $E(x + y) = E(x)E(y)$.

62 A certain function f has $f(0) = 3$, $f^{(1)}(0) = 2$, $f^{(2)}(0) = 5$, $f^{(3)}(0) = \frac{1}{2}$, and $f^{(j)}(x) = 0$ if $j > 3$. Give an explicit formula for $f(x)$.

Exercises 63 and 64 illustrate the use of power series in solving differential equations.

63 Say that all you knew about a function $f(x)$ is that $f'(x) = f(x)$ for all x. [In Sec. 6.7 it was shown that $f(x) = Ae^x$ for some constant A. However, do not make use of this fact.]
(a) Show that $R_n(x; 0) \to 0$ as $n \to \infty$.
(b) If $f(0) = A$, find $f'(0)$, $f''(0)$, and $f'''(0)$.
(c) Find the Maclaurin series for $f(x)$.
(d) Use (c) to show $f(x) = Ae^x$.

64 Say that all you know about a function $f(x)$ is that $f^{(2)}(x) = -f(x)$, $f(0) = A$, and $f'(0) = B$.
(a) Find $f^{(2)}(0)$.
(b) Find $f^{(3)}(0)$.
(c) Find $f^{(n)}(0)$ for any positive integer n.
(d) Compute the Maclaurin series for $f(x)$.
(e) Show that it represents $f(x)$ for all x. *Hint:* Examine $R^{(n)}(x; 0)$.
(f) Show that $f(x) = A\cos x + B\sin x$.
(g) Check that $A\cos x + B\sin x$ satisfies the differential equation $f^{(2)}(x) = -f(x)$.

Exercises 65 to 71 outline an argument due to Euler that

$$\frac{1}{1^2} + \frac{1}{2^2} + \frac{1}{3^2} + \cdots = \frac{\pi^2}{6}.$$

65 Show that if

$$\frac{1}{1^2} + \frac{1}{3^2} + \frac{1}{5^2} + \frac{1}{7^2} + \cdots + \frac{1}{(2n-1)^2} + \cdots = \frac{\pi^2}{8},$$

then
$$\frac{1}{1^2} + \frac{1}{2^2} + \frac{1}{3^2} + \cdots + \frac{1}{n^2} + \cdots = \frac{\pi^2}{6}.$$

Hint: Break the second series into its odd and even terms.

66 Show that

$$\int_0^1 \frac{\sin^{-1} x}{\sqrt{1 - x^2}}\,dx = \frac{\pi^2}{8}.$$

67 Use the binomial theorem to show that if $|t| < 1$, then

$$\frac{1}{\sqrt{1 - t^2}} = 1 + \frac{1}{2}t^2 + \frac{1 \cdot 3}{2 \cdot 4}t^4 + \frac{1 \cdot 3 \cdot 5}{2 \cdot 4 \cdot 6}t^6 + \cdots.$$

68 (See Exercise 67.) Show that

$$\sin^{-1} x = x + \frac{1}{2}\frac{x^3}{3} + \frac{1 \cdot 3}{2 \cdot 4}\frac{x^5}{5} + \frac{1 \cdot 3 \cdot 5}{2 \cdot 4 \cdot 6}\frac{x^7}{7} + \cdots$$

for $|x| < 1$. This equation is also valid when $x = 1$ or -1.

69 Use the substitution $x = \sin\theta$ to show that

$$\int_0^1 \frac{x^{2n+1}}{\sqrt{1 - x^2}}\,dx = \int_0^{\pi/2} \sin^{2n+1}\theta\,d\theta.$$

70 Assuming that it is safe to integrate term by term, even in the case of an improper integral, show that

$$\int_0^1 \frac{\sin^{-1} x}{\sqrt{1 - x^2}}\,dx = \frac{1}{1^2} + \frac{1}{3^2} + \cdots + \frac{1}{(2n-1)^2} + \cdots.$$

71 Deduce that

$$\sum_{n=1}^{\infty} n^{-2} = \frac{\pi^2}{6}.$$

No one knows whether $\sum_{n=1}^{\infty} n^{-3}$ is related to π^3. All that is known is that the sum is not a rational number. This was proved by Roger Apéry in 1979. Euler found $\sum_{n=1}^{\infty} n^{-k}$ for each positive even number k; the sum is a rational multiple of π^k.

12

ALGEBRAIC OPERATIONS ON VECTORS

In this chapter we discuss the algebra of vectors in the plane and in space. This algebra was developed primarily in response to James Clerk Maxwell's *Treatise on Electricity and Magnetism,* published in 1873. Josiah Gibbs, who in 1863 earned the first doctorate in engineering awarded in the United States and became a mathematical physicist, put vector analysis in its present form. His *Elements of Vector Analysis,* published in 1881, introduced the notation used in this chapter.

Section 12.1 introduces vectors (which are represented by arrows) and their arithmetic.

Section 12.2 examines the notion of a projection of a line segment, vector, or planar region on a line or plane. (A projection is related to the shadow cast by parallel rays of light.)

Section 12.3 defines the dot product of two vectors, which is a number. The dot product is used in Sec. 12.4 to obtain equations of lines and planes. Section 12.5 summarizes the properties of determinants used in Sec. 12.6 to construct a vector perpendicular to two given vectors (the cross product). Section 12.7 then applies the cross product.

12.1 THE ALGEBRA OF VECTORS

This section introduces the notion of vectors in the plane or in space and their basic algebra. It also describes rectangular coordinates in space.

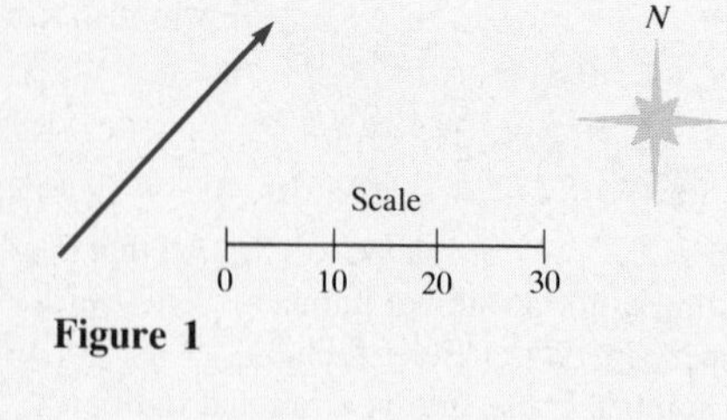

Figure 1

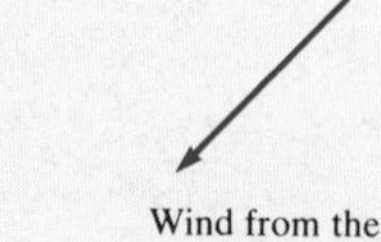

Wind from the south at 10 miles per hour

Wind from the northeast at 20 miles per hour

Figure 2

Common Examples of Vectors

An adequate description of the wind indicates both its speed and its direction. One way to describe a wind of 30 miles per hour from the southwest is to draw an arrow aimed in the direction in which the window blows, scaled so that its length represents a magnitude of 30, as in Fig. 1.

Relative to this same scale, some more wind arrows are shown in Fig. 2. Similarly, the flow of water on the surface of a stream is best indicated by a few sample arrows, as in Fig. 3. Of course, associated with *each* point on the surface is an arrow representing the velocity of the water at that point.

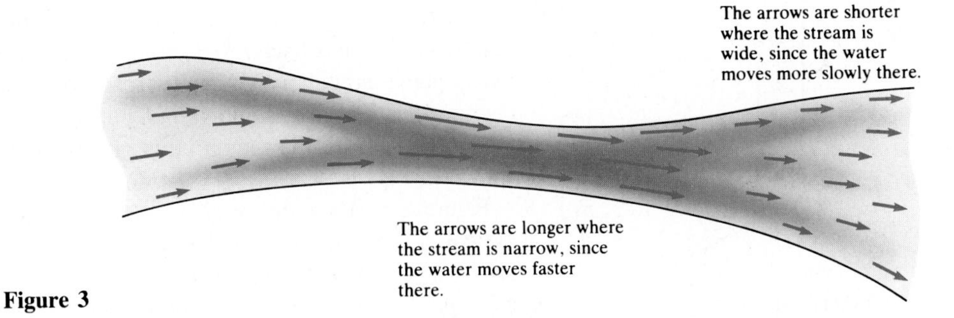

Figure 3

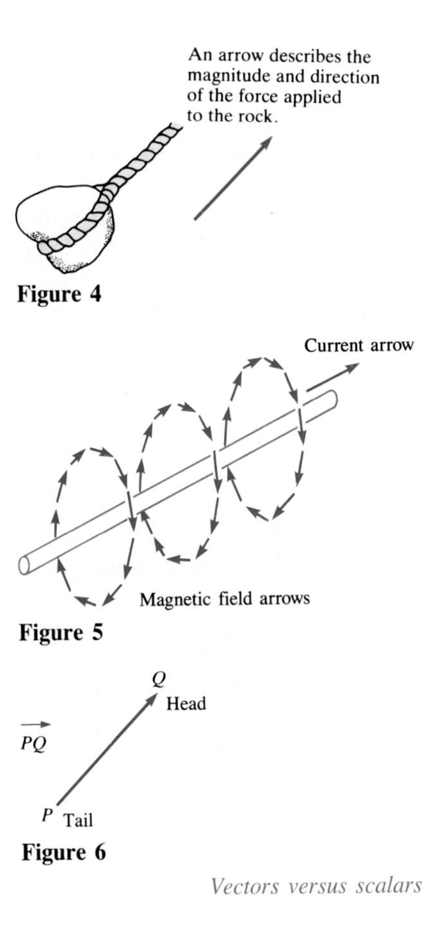

Figure 4

Figure 5

Figure 6

When we pull a heavy rock by a rope, the force we exert has both a magnitude and a direction, as shown in Fig. 4. The magnitude describes how hard we pull; the direction of our pull is along the rope. An electric current flowing in a wire may be represented by an arrow; the magnetic field it creates around the wire is also represented by arrows, as in Fig. 5.

The Notion of a Vector

In each of these examples a concept is represented by a magnitude and a direction. Vectors were introduced to describe such concepts. A vector may be represented by an arrow that records both a magnitude and a direction.

To be more precise, any two distinct points P and Q in the plane determine a **directed line segment** $\overrightarrow{PQ}$ from P to Q, as shown in Fig. 6. If $\overrightarrow{P_1Q_1}$ and $\overrightarrow{P_2Q_2}$ are two directed line segments that have both the same length and the same direction, then we say that they represent the same vector. This means that the two line segments are parallel, have the same length, and point in the same direction. (See Fig. 7.) A directed line segment has a particular location. A vector does not.

The arrows in Fig. 7 all represent the same vector. Our choice of where to place a vector is influenced by the context. Sometimes we may put the base at the origin; sometimes, as in Fig. 3, we put the base elsewhere. In a weather map, where a vector indicates wind, the tail is put at the place where the wind was observed.

In print, boldface letters such as $\mathbf{A}$, $\mathbf{B}$, $\mathbf{F}$, $\mathbf{r}$, and $\mathbf{v}$ are used to denote vectors. In handwriting, the symbols $\mathbb{A}$ and $\vec{A}$ are used. The length of $\mathbf{A}$ is denoted by $\|\mathbf{A}\|$. The length of $\mathbf{A}$ is also called the **norm** of $\mathbf{A}$ or **magnitude** of $\mathbf{A}$.

Vectors versus scalars

Lowercase letters, such as a, b, x, y, and z, will be used to name numbers. Numbers will also be called **scalars** to distinguish them from vectors. Thus $\mathbf{A}$ is a vector, but $\|\mathbf{A}\|$ is a scalar.

Coordinates in Space

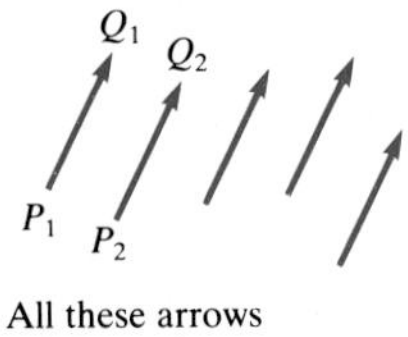

In order to describe and to draw vectors that may not lie in the xy plane, we introduce a coordinate system in space.

First, pick a pair of perpendicular, intersecting lines to serve as the x and y axes. The positive parts of these axes are indicated by arrows. These two lines determine the **xy plane**. The line perpendicular to the xy plane and meeting the x and y axes will be called the z axis. The point where the three axes meet is called the **origin**. The 0 of the z axis will be put at the origin. But which half of

Q_1 Q_2

P_1 P_2

All these arrows represent the same vector.

Figure 7

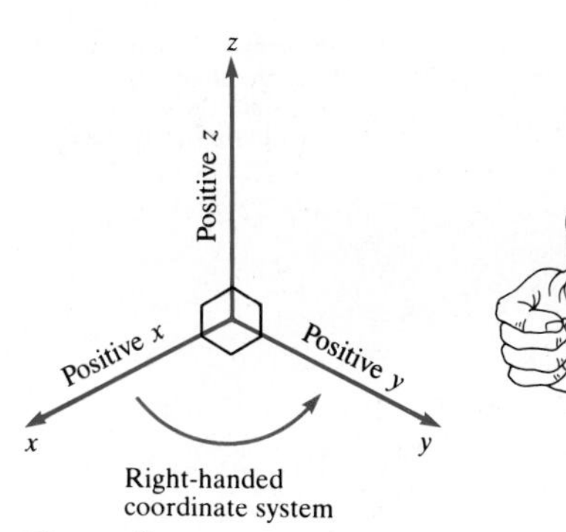

Figure 8

the z axis will have positive numbers and which half will have the negative numbers? It is customary to determine this by the **right-hand rule**. Moving in the xy plane through a right angle from the positive x axis to the positive y axis determines a sense of rotation around the z axis. If the fingers of the right hand curl in that sense, the thumb points in the direction of the *positive* z axis, as shown in Fig. 8.

Figure 8 portrays one way of drawing the three axes, called the **isometric** system. The three right angles appear on the page as 120° angles. To emphasize that they depict right angles, you may wish to add right-angle symbols, as shown in Fig. 8. Another way of drawing the axes is shown in Fig. 9, where the y and z axes make a right angle.

Any point Q in space is now described by three numbers: First, two numbers specify the x and y coordinates of the point P in the xy plane directly below (or above) Q; then the height of Q above (or below) the xy plane is recorded by the z coordinate of the point R where the plane through Q and parallel to the xy plane meets the z axis. The point Q is then denoted (x, y, z). See Fig. 10.

The points (x, y, z) for which $z = 0$ lie in the xy plane. There are an infinite number of these points. The points (x, y, z) for which $x = 0$ lie entirely in the plane determined by the y and z axes, which is called the **yz plane**. Similarly, the equation $y = 0$ describes the **xz plane**. The xy, xz, and yz planes are called the **coordinate planes**.

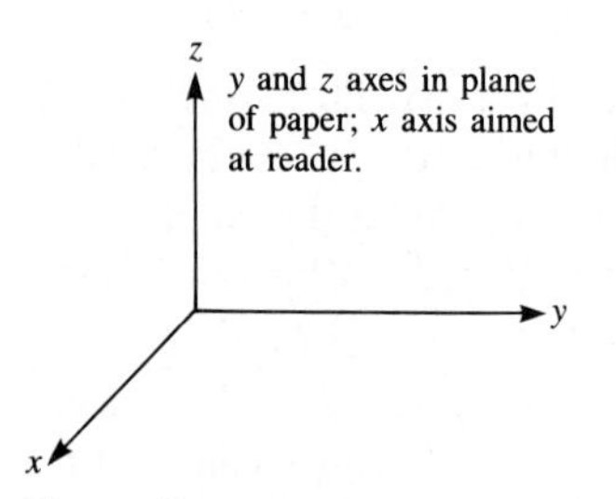

Figure 9

EXAMPLE 1 Plot the point (1, 2, 3).

SOLUTION One way is to first plot the point (1, 2) in the xy plane. Then, on a line perpendicular to the xy plane at that point, show the point (1, 2, 3) as done in Fig. 11.

Another way is to draw a box whose edges are parallel to the axes and which has the origin (0, 0, 0) and (1, 2, 3) as opposite corners, as shown in Fig. 12. (This time, the y and z axes make a right angle.) ■

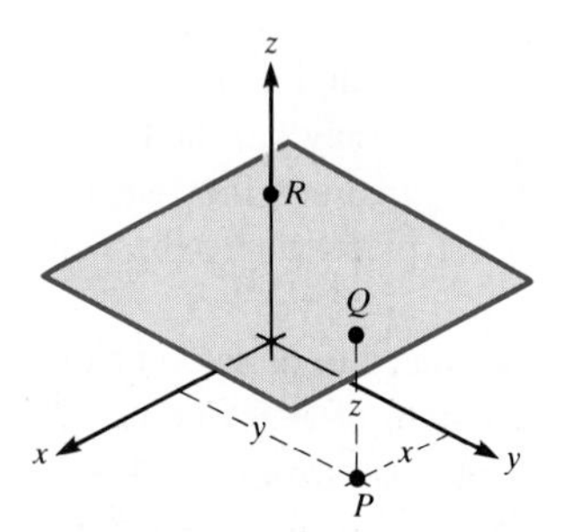

Figure 10

You were not born plotting points, but a few minutes' practice should make up for this omission. Use a ruler to draw the lines; doing this will inspire you to be careful and clear. (Then your diagrams will do good, not harm.)

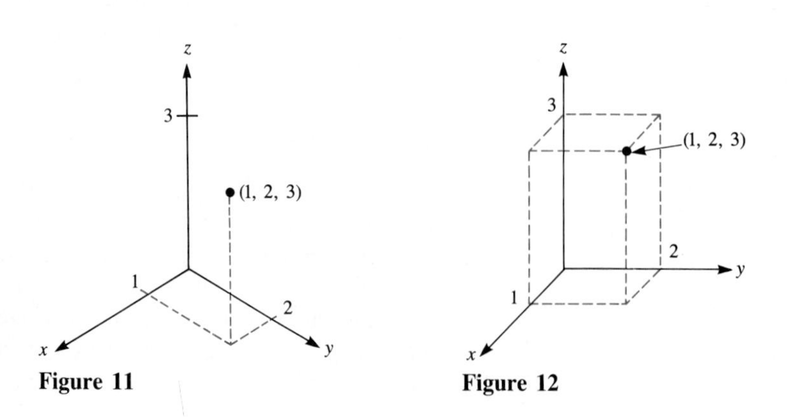

Figure 11 **Figure 12**

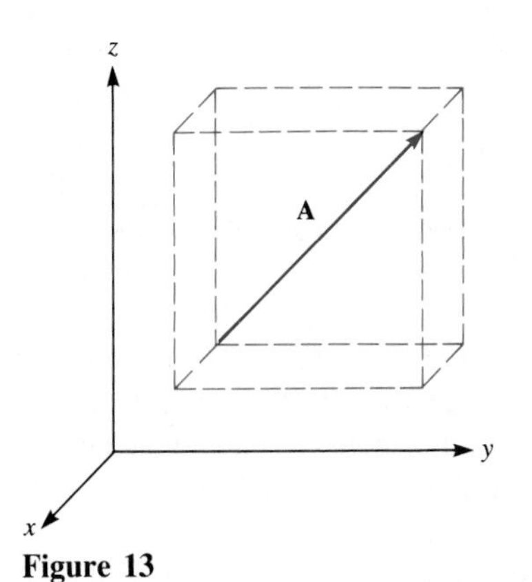

Figure 13

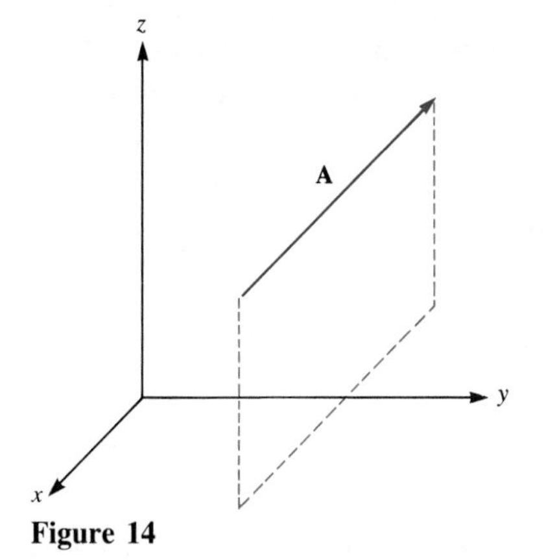

Figure 14

Figure 15

Drawing Vectors

If we are interested only in vectors that lie in a given plane, then it is easy to draw them. A page in this book or a piece of paper can be used to serve as the plane. We then simply display the vector as an arrow on the paper.

But how do we indicate vectors that do not lie in the plane of this page? Figures 13 to 15 show some ways of adding extra lines to show how a vector **A** is situated in space.

Note that parallel lines are represented by parallel lines. (An artist would represent parallel lines by lines that meet at a point to show perspective.)

Adding and Subtracting Vectors

The **sum of two vectors A** and **B** is defined as follows. Place **B** in such a way that its tail is at the head of **A**. Then the vector **A** + **B** goes from the tail of **A** to the head of **B**. Observe that **B** + **A** = **A** + **B**, since both sums lie on the diagonal of a parallelogram, as shown in Fig. 16. (For simplicity the diagrams show vectors in the plane of the page; the definitions hold for all vectors.)

For example, if **W** is a wind vector (describing the motion of the air relative to the earth) and **A** is a vector describing the motion of an airplane relative to the air, then **W** + **A** is the vector describing the motion of the airplane relative to the earth. (See Fig. 17.)

The concept of the sum of two vectors is also important in the study of force. If $\mathbf{F}_1$ and $\mathbf{F}_2$ describe the forces in two ropes lifting a heavy rock, as shown in Fig. 18, then a single rope with the force $\mathbf{F}_1 + \mathbf{F}_2$ pulling at the same point has the same effect on the rock.

The **difference** of **A** and **B**, denoted **A** − **B**, is the vector that you add to **B** to

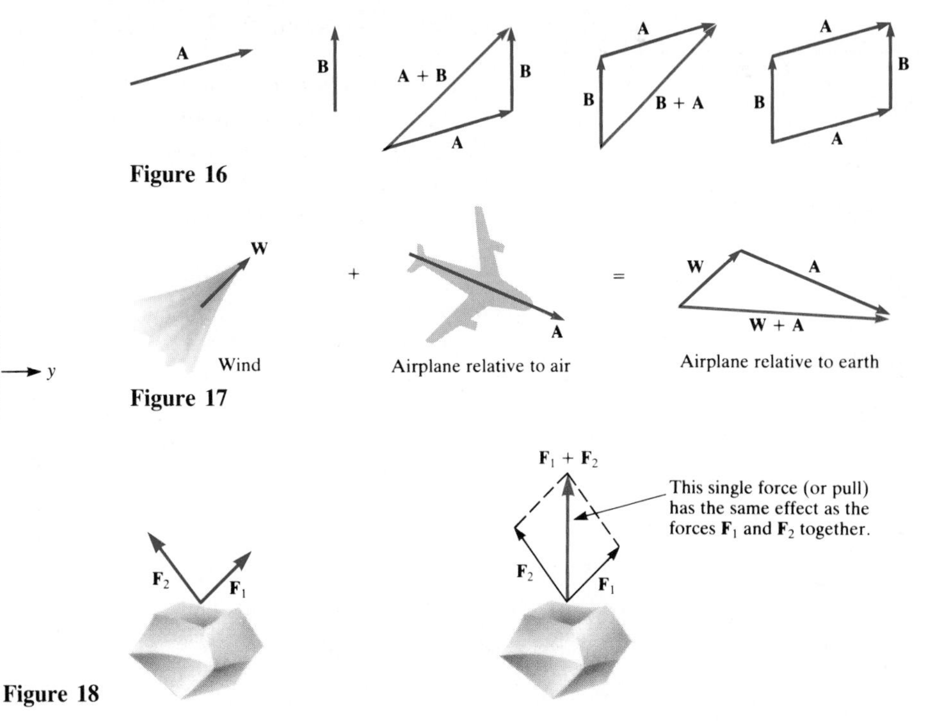

Figure 16

Figure 17

Figure 18

get **A**. Thus

$$\mathbf{B} + (\mathbf{A} - \mathbf{B}) = \mathbf{A},$$

as shown in Fig. 19.

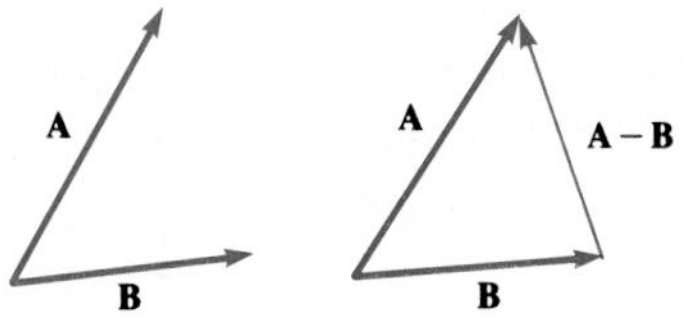

What must you add to **B** to get **A**?

Figure 19

The zero vector

It will be useful to define the vector $\overrightarrow{PP}$, whose head and tail are the same point. It has norm 0, but is not assigned a direction. This vector is called the **zero vector** and denoted **0**. Note that $\|\mathbf{0}\| = 0$.

In the arithmetic of numbers $(a + b) + c = a + (b + c)$. This means that whether we interpret $a + b + c$ as $(a + b) + c$ or as $a + (b + c)$, we get the same answer. [We may "associate" the b with a or with c; hence the equation $(a + b) + c = a + (b + c)$ is called the **associative law**.] The same law holds for vectors, as indicated by Fig. 20.

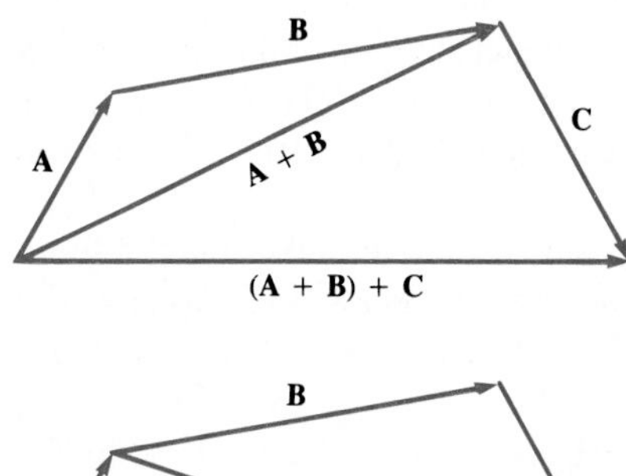

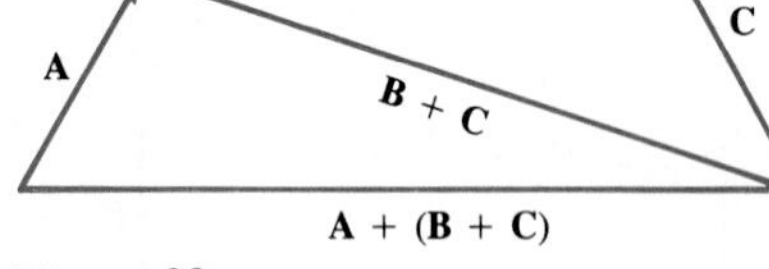

Figure 20

Thus the symbol $\mathbf{A} + \mathbf{B} + \mathbf{C}$ denotes a well-defined vector. In particular, if **A**, **B**, and **C** form the three sides of a triangle, as in Fig. 21, $\mathbf{A} + \mathbf{B} + \mathbf{C} = \mathbf{0}$.

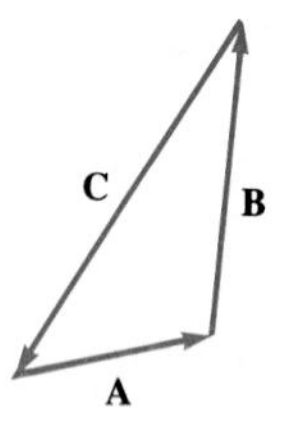

Figure 21

The negative of a vector **A**, other than **0**, is defined as the vector with the same length as **A** but with the opposite direction. It is denoted $-\mathbf{A}$. See Fig. 22. If $\mathbf{A} = \overrightarrow{PQ}$, then $-\mathbf{A} = \overrightarrow{QP}$. A quick sketch shows that $\mathbf{A} + (-\mathbf{A}) = \mathbf{0}$ and that $\mathbf{A} - \mathbf{B} = \mathbf{A} + (-\mathbf{B})$. The negative of **0** is defined to be **0**.

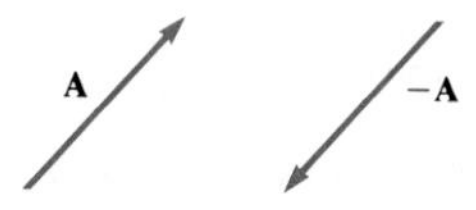

Figure 22

The Product of a Scalar and a Vector

So far the algebra of vectors has just duplicated the algebra of ordinary addition and subtraction of scalars. Now we introduce an operation in which a scalar and a vector combine to form a vector. The scalar "operates" on the vector to produce a new vector, according to the following definition:

> **Definition** *The product of a scalar and a vector.* Let c be a scalar and **A** a vector. The **product** $c\mathbf{A}$ is defined as follows:
>
> **1** If $c = 0$ or $\mathbf{A} = \mathbf{0}$, then $c\mathbf{A} = \mathbf{0}$.
> **2** If c is positive and **A** is not **0**, then $c\mathbf{A}$ is the vector c times as long as **A** and pointing in the same direction as **A**.
> **3** If c is negative and **A** is not **0**, then $c\mathbf{A}$ is the vector $|c|$ times as long as **A** and pointing in the direction opposite the direction of **A**.

For instance, $7\mathbf{0} = \mathbf{0}$ and $(-1)\mathbf{A} = -\mathbf{A}$. Figure 23 displays $(-1)\mathbf{A}$, $\frac{1}{2}\mathbf{A}$, and $2\mathbf{A}$. Note that for any vector **A**,

$$\boxed{\|c\mathbf{A}\| = |c|\,\|\mathbf{A}\|.}$$

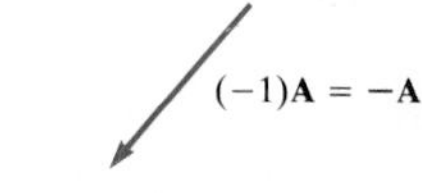

A ½A (−2)A

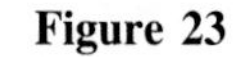

Figure 23

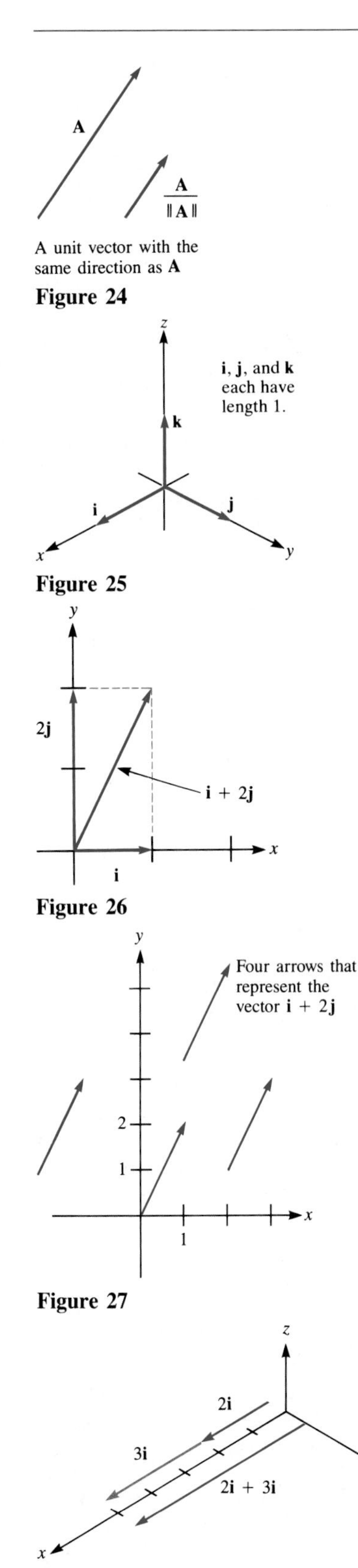

Figure 24

Figure 25

Figure 26

Figure 27

Figure 28

Remark *Test for parallel vectors.* Observe that two nonzero vectors **A** and **B** are parallel if and only if there is a scalar c such that $\mathbf{B} = c\mathbf{A}$.

Division by a Scalar

Division of a vector **A** by a nonzero scalar c, denoted $\mathbf{A}/c$, is defined as the product $(1/c)\mathbf{A}$. For instance, $\mathbf{A}/2 = \frac{1}{2}\mathbf{A}$ and $\mathbf{A}/0.1 = 10\mathbf{A}$.

EXAMPLE 2 Let **A** be a nonzero vector. Find the length and direction of the vector

$$\frac{\mathbf{A}}{\|\mathbf{A}\|}.$$

SOLUTION By definition,

$$\frac{\mathbf{A}}{\|\mathbf{A}\|} = \frac{1}{\|\mathbf{A}\|}\mathbf{A},$$

a vector $1/\|\mathbf{A}\|$ times as long as **A**. Since the length of **A** is $\|\mathbf{A}\|$, the length of $(1/\|\mathbf{A}\|)\mathbf{A}$ is

$$\frac{1}{\|\mathbf{A}\|}\|\mathbf{A}\| = 1.$$

Also, since $1/\|\mathbf{A}\|$ is positive, $\mathbf{A}/\|\mathbf{A}\|$ has the same direction as **A**. ■

Example 2 tells us that if we divide a vector **A** by its norm we obtain a vector of length 1 that points in the same direction as **A** does. This fact will be used often. Any vector of length 1 is called a **unit vector**. (See Fig. 24.) Unit vectors are used to record directions. There is exactly one unit vector for each direction in space. Usually, a unit vector is denoted by a lowercase bold letter, such as **u**. Note that **u** and $-\mathbf{u}$ point in opposite directions.

The three most important unit vectors are those that indicate the directions of the positive x, y, and z axes. They will be denoted **i**, **j**, and **k**, respectively, as in Fig. 25. (Engineers and physicists usually denote them $\hat{x}$, $\hat{y}$, and $\hat{z}$. The notations $\mathbf{e}_1$, $\mathbf{e}_2$, and $\mathbf{e}_3$ are also used.)

EXAMPLE 3 Draw the vector $\mathbf{i} + 2\mathbf{j}$.

SOLUTION First draw **i** and $2\mathbf{j}$, then their sum, as in Fig. 26.

Of course we do not have to put the tail of $\mathbf{i} + 2\mathbf{j}$ at the origin. The three other arrows in Fig. 27 also show the same vector $\mathbf{i} + 2\mathbf{j}$. ■

Figure 28 shows $2\mathbf{i}$, $3\mathbf{i}$, and their sum. As Fig. 28 suggests, $2\mathbf{i} + 3\mathbf{i} = 5\mathbf{i}$. This illustrates a useful fact: For any scalars c and d and any vector **A**,

$$(c + d)\mathbf{A} = c\mathbf{A} + d\mathbf{A}.$$

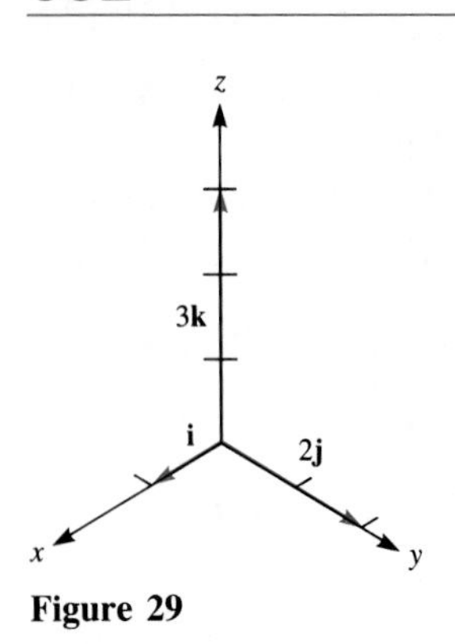

Figure 29

Another useful "distributive rule" is

$$c\mathbf{A} + c\mathbf{B} = c(\mathbf{A} + \mathbf{B})$$

for any scalar c and vectors **A** and **B**. (Its proof is outlined in Exercise 27.)

EXAMPLE 4 Draw the vector $\mathbf{i} + 2\mathbf{j} + 3\mathbf{k}$.

SOLUTION First draw **i**, $2\mathbf{j}$, and $3\mathbf{k}$, as in Fig. 29. The sum is formed by joining "the tail to the head," as in Fig. 30.

The box determined by $\mathbf{i} + 2\mathbf{j} + 3\mathbf{k}$, shown in Fig. 31, helps show its position. ■

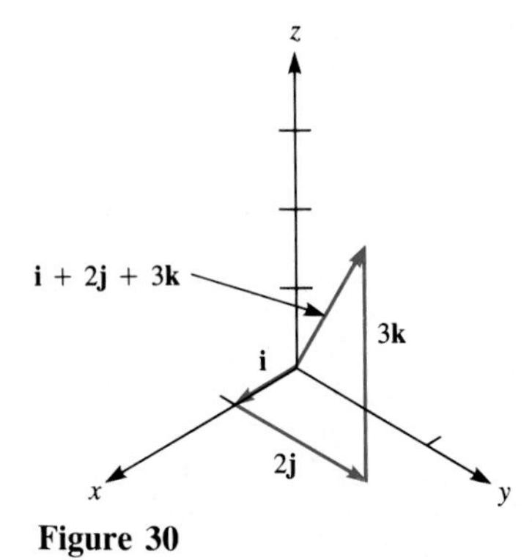

Figure 30

Components of a Vector

The vector **A** whose tail is at the origin and whose head is at the point (x, y, z) is equal to

$$x\mathbf{i} + y\mathbf{j} + z\mathbf{k}.$$

The numbers x, y, and z are called the **scalar components** of **A** along the x, y, and z axes, respectively. These three scalars completely describe **A**, and we may denote **A** as

$$\overrightarrow{(x, y, z)} \quad \text{or} \quad \langle x, y, z \rangle.$$

When we are interested only in vectors in the xy plane, we write

$$\mathbf{A} = x\mathbf{i} + y\mathbf{j}, \qquad \mathbf{A} = \overrightarrow{(x, y)}, \qquad \text{or} \qquad \mathbf{A} = \langle x, y \rangle.$$

EXAMPLE 5 Find the scalar components of $\mathbf{A} + \mathbf{B}$ if $\mathbf{A} = 2\mathbf{j} + 3\mathbf{k}$ and $\mathbf{B} = 2\mathbf{j} + \mathbf{k}$.

SOLUTION We draw **A** and **B**, as in Fig. 32.

Inspection of Fig. 32 shows that $\mathbf{A} + \mathbf{B} = \langle 0, 2 + 2, 3 + 1 \rangle = 4\mathbf{j} + 4\mathbf{k}$. So the scalar components of $\mathbf{A} + \mathbf{B}$ are 0, 4, and 4, respectively. ■

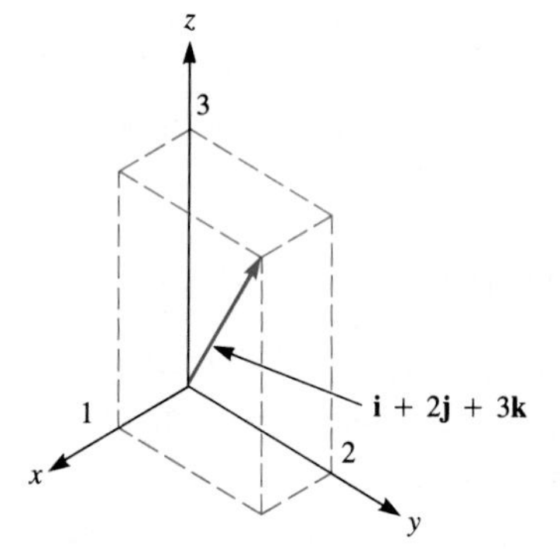

Figure 31

Generally the notation $x\mathbf{i} + y\mathbf{j} + z\mathbf{k}$ is preferable to $\langle x, y, z \rangle$. First of all, it is more geometric, leading us to think in terms of arrows rather than in terms of numbers. Second, when x, y, and z are messy expressions, the notation $x\mathbf{i} + y\mathbf{j} + z\mathbf{k}$ is easier to read.

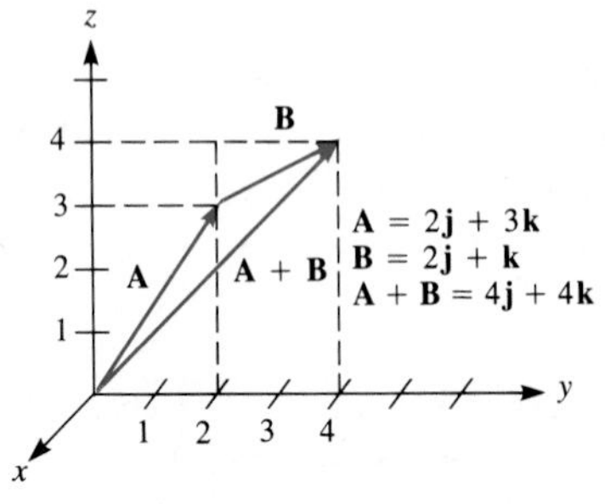

Figure 32

Computing the Length of a Vector

How long is the vector $\mathbf{A} = x\mathbf{i} + y\mathbf{j} + z\mathbf{k}$? To find out, draw the box with edges parallel to the axes that **A** determines, as in Fig. 33.

The three sides of the box have lengths $|x|$, $|y|$, and $|z|$ and the length of the longest diagonal is $\|\mathbf{A}\|$. Let d be the length of the diagonal of the rectangular base of the box, as shown in Fig. 33. By two applications of the pythagorean theorem, we have

$$|x|^2 + |y|^2 = d^2 \qquad \text{and} \qquad d^2 + |z|^2 = \|\mathbf{A}\|^2.$$

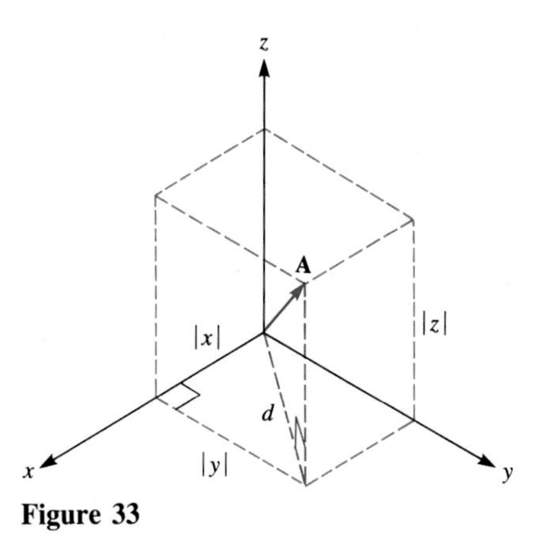

Figure 33

These two equations imply that $\|\mathbf{A}\|^2 = |x|^2 + |y|^2 + |z|^2$. Now, $|x|^2 = x^2$, whether x is positive, negative, or zero. Similarly, $|y|^2 = y^2$ and $|z|^2 = z^2$. Hence

$$\|\mathbf{A}\|^2 = x^2 + y^2 + z^2 \qquad \text{or} \qquad \|\mathbf{A}\| = \sqrt{x^2 + y^2 + z^2}.$$

EXAMPLE 6 Find $\|2\mathbf{i} + 3\mathbf{j} - 4\mathbf{k}\|$.

SOLUTION

$$\begin{aligned}\|2\mathbf{i} + 3\mathbf{j} - 4\mathbf{k}\| &= \sqrt{2^2 + 3^2 + (-4)^2}\\ &= \sqrt{4 + 9 + 16}\\ &= \sqrt{29} \quad \blacksquare\end{aligned}$$

Computing A + B, A − B, −A, and cA

The definitions of $\mathbf{A} + \mathbf{B}$, $\mathbf{A} - \mathbf{B}$, $-\mathbf{A}$, and $c\mathbf{A}$ are all geometric. The following theorem tells how to compute them in terms of components.

Theorem 1 Let $\mathbf{A} = x_1\mathbf{i} + y_1\mathbf{j} + z_1\mathbf{k}$ and $\mathbf{B} = x_2\mathbf{i} + y_2\mathbf{j} + z_2\mathbf{k}$. Then

$$\mathbf{A} + \mathbf{B} = (x_1 + x_2)\mathbf{i} + (y_1 + y_2)\mathbf{j} + (z_1 + z_2)\mathbf{k} \qquad \text{(add components)}$$

$$-\mathbf{A} = -x_1\mathbf{i} - y_1\mathbf{j} - z_1\mathbf{k} \qquad \text{(change signs)}$$

$$\mathbf{A} - \mathbf{B} = (x_1 - x_2)\mathbf{i} + (y_1 - y_2)\mathbf{j} + (z_1 - z_2)\mathbf{k} \qquad \text{(subtract components)}$$

$$c\mathbf{A} = cx_1\mathbf{i} + cy_1\mathbf{j} + cz_1\mathbf{k}. \qquad \text{(multiply components by } c\text{)}$$

Proof The sum of **A** and **B** is

$$\mathbf{A} + \mathbf{B} = (x_1\mathbf{i} + y_1\mathbf{j} + z_1\mathbf{k}) + (x_2\mathbf{i} + y_2\mathbf{j} + z_2\mathbf{k}).$$

As with ordinary arithmetic, since vector addition is commutative and associative, we may rearrange the summands:

$$\begin{aligned}\mathbf{A} + \mathbf{B} &= (x_1\mathbf{i} + x_2\mathbf{i}) + (y_1\mathbf{j} + y_2\mathbf{j}) + (z_1\mathbf{k} + z_2\mathbf{k})\\ &= (x_1 + x_2)\mathbf{i} + (y_1 + y_2)\mathbf{j} + (z_1 + z_2)\mathbf{k}.\end{aligned}$$

The formula for $-\mathbf{A}$ can be checked by verifying that

$$\mathbf{A} + (-x_1\mathbf{i} - y_1\mathbf{j} - z_1\mathbf{k}) = \mathbf{0}.$$

The calculation is straightforward:

$$\begin{aligned}(x_1\mathbf{i} + y_1\mathbf{j} + z_1\mathbf{k}) + (-x_1\mathbf{i} - y_1\mathbf{j} - z_1\mathbf{k}) &= (x_1 + (-x_1))\mathbf{i} + (y_1 + (-y_1))\mathbf{j} + (z_1 + (-z_1))\mathbf{k}\\ &= 0\mathbf{i} + 0\mathbf{j} + 0\mathbf{k}\\ &= \mathbf{0}.\end{aligned}$$

The formula for $\mathbf{A} - \mathbf{B}$ follows immediately, since $\mathbf{A} - \mathbf{B} = \mathbf{A} + (-\mathbf{B})$.

The formula for $c\mathbf{A}$ can be obtained as follows. For simplicity assume that c and the components of **A** are all positive. The box determined by **A** has sides x_1,

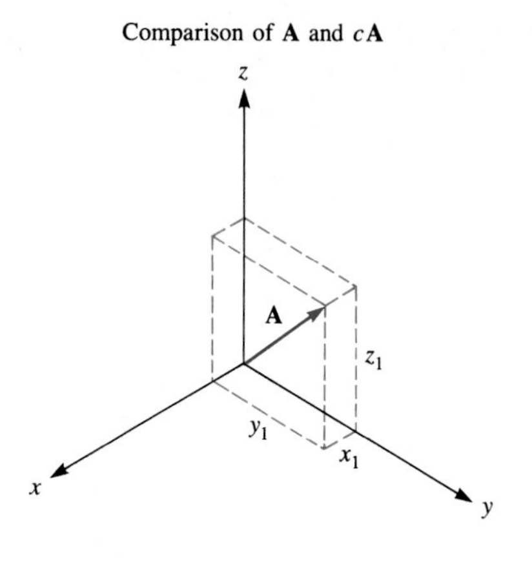

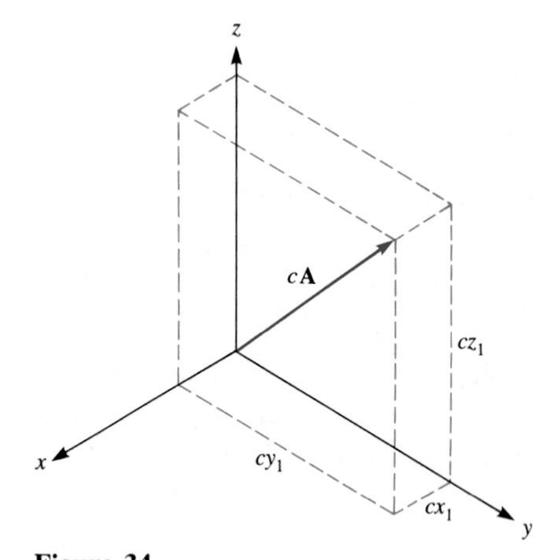

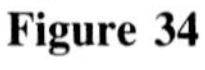
Figure 34

y_1, and z_1. Magnify this box by the factor c. The new box determines a vector which is $c\mathbf{A}$. (See Fig. 34.)

Thus

$$c\mathbf{A} = cx_1\mathbf{i} + cy_1\mathbf{j} + cz_1\mathbf{k}. \quad \blacksquare$$

The following calculations illustrate the theorem.

$$(2\mathbf{i} + 3\mathbf{j} + 4\mathbf{k}) + (\mathbf{i} - 7\mathbf{j} + 8\mathbf{k}) = 3\mathbf{i} - 4\mathbf{j} + 12\mathbf{k}$$

$$-(2\mathbf{i} + 3\mathbf{j} + 4\mathbf{k}) = -2\mathbf{i} - 3\mathbf{j} - 4\mathbf{k}$$

$$(2\mathbf{i} + 3\mathbf{j} + 4\mathbf{k}) - (\mathbf{i} - 7\mathbf{j} + 8\mathbf{k}) = \mathbf{i} + 10\mathbf{j} - 4\mathbf{k}$$

$$2.3(2\mathbf{i} + 3\mathbf{j} + 4\mathbf{k}) = 4.6\mathbf{i} + 6.9\mathbf{j} + 9.2\mathbf{k}$$

In short, there are no surprises.

EXAMPLE 7 If $\mathbf{A} = 2\mathbf{i} - 3\mathbf{j} + 4\mathbf{k}$, compute the length of $\mathbf{A}/\|\mathbf{A}\|$, using the components of $\mathbf{A}$. (The result should be 1.)

SOLUTION First of all,

$$\|\mathbf{A}\| = \sqrt{2^2 + (-3)^2 + 4^2} = \sqrt{29}.$$

Then

$$\frac{\mathbf{A}}{\sqrt{29}} = \frac{1}{\sqrt{29}}(2\mathbf{i} - 3\mathbf{j} + 4\mathbf{k})$$

$$= \frac{2}{\sqrt{29}}\mathbf{i} - \frac{3}{\sqrt{29}}\mathbf{j} + \frac{4}{\sqrt{29}}\mathbf{k},$$

whose length is

$$\sqrt{\left(\frac{2}{\sqrt{29}}\right)^2 + \left(\frac{-3}{\sqrt{29}}\right)^2 + \left(\frac{4}{\sqrt{29}}\right)^2} = \sqrt{\frac{4}{29} + \frac{9}{29} + \frac{16}{29}}$$

$$= \sqrt{\frac{29}{29}}$$

$$= 1. \quad \blacksquare$$

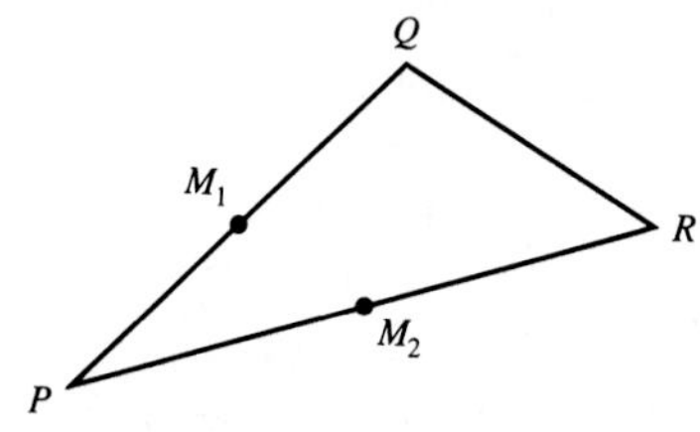

Figure 35

Example 8 shows how vectors can be used to establish geometric properties.

EXAMPLE 8 Prove that the line which joins the midpoints of two sides of a triangle is parallel to the third side and half as long.

SOLUTION Let the triangle have vertices P, Q, and R. Let the midpoint of side PQ be M_1, and the midpoint of side PR be M_2, as in Fig. 35.

Introduce an xy coordinate system in the plane of the triangle. Though its origin could be anywhere in the plane, we should put it at P in order to simplify the calculations. (See Fig. 36.)

We wish to show that the vector $\overrightarrow{M_1M_2}$ is $\frac{1}{2}\overrightarrow{QR}$. To do so, we compute $\overrightarrow{M_1M_2}$ and $\overrightarrow{QR}$ in terms of vectors involving P, Q, and R.

First of all, $\overrightarrow{PM_1} = \frac{1}{2}\overrightarrow{PQ}$ and $\overrightarrow{PM_2} = \frac{1}{2}\overrightarrow{PR}$.

Thus

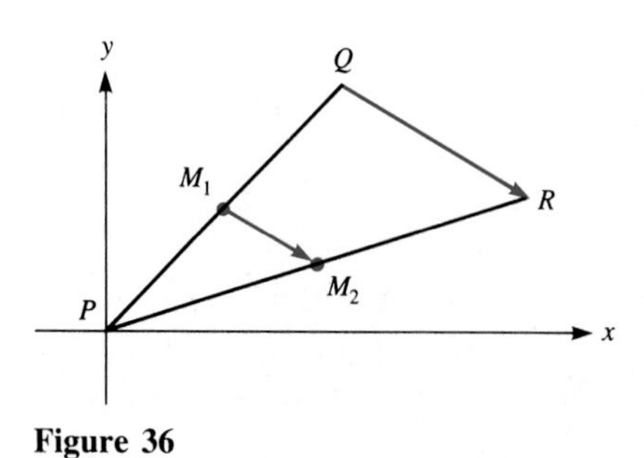

Figure 36

$$\overrightarrow{M_1M_2} = \tfrac{1}{2}\overrightarrow{PR} - \tfrac{1}{2}\overrightarrow{PQ}$$
$$= \tfrac{1}{2}(\overrightarrow{PR} - \overrightarrow{PQ})$$
$$= \tfrac{1}{2}(\overrightarrow{QR}). \quad ■$$

Remark: Engineers and physicists commonly denote components of a vector by subscripts. They would write

$$\mathbf{A} = A_x\mathbf{i} + A_y\mathbf{j} + A_z\mathbf{k} \qquad \text{or} \qquad \mathbf{A} = A_x\hat{x} + A_y\hat{y} + A_z\hat{z}.$$

Section Summary

We introduced several basic terms: vector, scalar, length (norm), components along axes, and unit vector. The symbols and operations we used include: $\mathbf{A}$, $\|\mathbf{A}\|$, $\mathbf{0}$, $\mathbf{A} + \mathbf{B}$, $\mathbf{A} - \mathbf{B}$, $-\mathbf{A}$, $c\mathbf{A}$, $\mathbf{A}/c$, $\mathbf{i}$, $\mathbf{j}$, $\mathbf{k}$, $\langle x, y, z\rangle$.

The various concepts were defined geometrically. Then we saw how to represent them or compute them in terms of components. Note in particular that two nonzero vectors are parallel if (and only if) one is a scalar times the other.

EXERCISES FOR SEC. 12.1: THE ALGEBRA OF VECTORS

In Exercises 1 and 2 use the plane of your paper as the xy plane.

1 Draw the vector $2\mathbf{i} + 3\mathbf{j}$, placing its tail at (*a*) (0, 0), (*b*) (−1, 2), (*c*) (1, 1).

2 Draw the vector $-\mathbf{i} + 2\mathbf{j}$, placing its tail at (*a*) (0, 0), (*b*) (3, 0), (*c*) (−2, 2).

In Exercises 3 to 6 draw the vector $\mathbf{A}$ and enough extra lines to show how it is situated in space.

3 $\mathbf{A} = 2\mathbf{i} + \mathbf{j} + 3\mathbf{k}$, (*a*) tail at (0, 0, 0), (*b*) tail at (1, 1, 1).

4 $\mathbf{A} = \mathbf{i} + \mathbf{j} + \mathbf{k}$, (*a*) tail at (0, 0, 0), (*b*) tail at (2, 3, 4).

5 $\mathbf{A} = -\mathbf{i} - 2\mathbf{j} + 2\mathbf{k}$, (*a*) tail at (0, 0, 0), (*b*) tail at (1, 1, −1).

6 $\mathbf{A} = \mathbf{j} + \mathbf{k}$, (*a*) tail at (0, 0, 0), (*b*) tail at (−1, −1, −1).

In Exercises 7 to 10 plot the points P and Q, draw the vector $\overrightarrow{PQ}$, express it in the form $x\mathbf{i} + y\mathbf{j} + z\mathbf{k}$, and find its length.

7 $P = (0, 0, 0)$, $Q = (1, 3, 4)$

8 $P = (1, 2, 3)$, $Q = (2, 5, 4)$

9 $P = (2, 5, 4)$, $Q = (1, 2, 2)$

10 $P = (1, 1, 1)$, $Q = (-1, 3, -2)$

In Exercises 11 and 12 express the vector $\mathbf{A}$ in the form $x\mathbf{i} + y\mathbf{j}$. North is along the positive y axis and east is along the positive x axis.

11 (*a*) $\|\mathbf{A}\| = 10$ and $\mathbf{A}$ points northwest;
(*b*) $\|\mathbf{A}\| = 6$ and $\mathbf{A}$ points south;
(*c*) $\|\mathbf{A}\| = 9$ and $\mathbf{A}$ points southeast;
(*d*) $\|\mathbf{A}\| = 5$ and $\mathbf{A}$ points east.

12 (*a*) $\|\mathbf{A}\| = 1$ and $\mathbf{A}$ points southwest;
(*b*) $\|\mathbf{A}\| = 2$ and $\mathbf{A}$ points west;
(*c*) $\|\mathbf{A}\| = \sqrt{8}$ and $\mathbf{A}$ points northeast;
(*d*) $\|\mathbf{A}\| = \tfrac{1}{2}$ and $\mathbf{A}$ points south.

13 The wind is 30 miles per hour to the northeast. An airplane is traveling 100 miles per hour relative to the air, and the vector from the tail of the plane to its front tip points to the southeast. (See Fig. 37.)
(*a*) What is the speed of the plane relative to the ground?
(*b*) What is the direction of the flight relative to the ground?

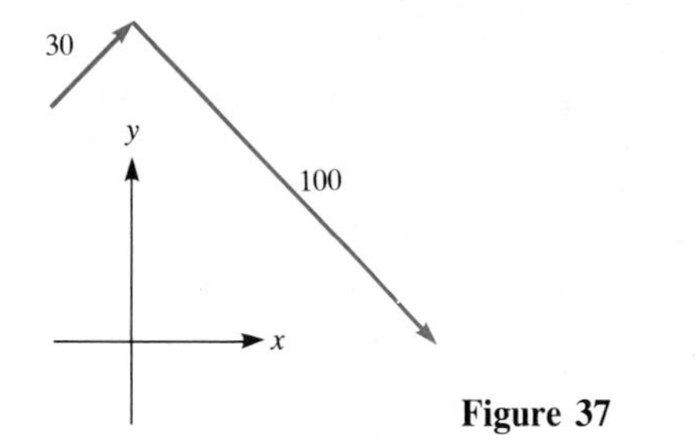

Figure 37

14 (See Exercise 13.) The jet stream is moving 200 miles per hour to the southeast. A plane with a speed of 550 miles per hour relative to the air is aimed to the northwest.
(*a*) Draw the vectors representing the wind and the plane relative to the air. (Choose a scale and make an accurate drawing.)
(*b*) Using your drawing, estimate the speed of the plane relative to the ground.
(*c*) Compute the speed in (*b*) exactly.

15 Compute $\mathbf{A} + \mathbf{B}$ and $\mathbf{A} - \mathbf{B}$ if
(*a*) $\mathbf{A} = \langle -1, 2, 3\rangle$ and $\mathbf{B} = \langle 7, 0, 2\rangle$.
(*b*) $\mathbf{A} = 3\mathbf{j} + 4\mathbf{k}$ and $\mathbf{B} = 6\mathbf{i} + 7\mathbf{j}$.

16 Compute $\mathbf{A} + \mathbf{B}$ and $\mathbf{A} - \mathbf{B}$ if
(*a*) $\mathbf{A} = \langle \tfrac{1}{2}, \tfrac{1}{3}, \tfrac{1}{6}\rangle$ and $\mathbf{B} = \langle 2, 3, -\tfrac{1}{3}\rangle$.
(*b*) $\mathbf{A} = 2\mathbf{i} + 3\mathbf{j} + 4\mathbf{k}$ and $\mathbf{B} = -\mathbf{i} + 5\mathbf{j} + 6\mathbf{k}$.

17 Compute and sketch $c\mathbf{A}$ if $\mathbf{A} = 2\mathbf{i} + 3\mathbf{j} + \mathbf{k}$ and c is
(*a*) 2, (*b*) -2, (*c*) $\frac{1}{2}$, (*d*) $-\frac{1}{2}$.

18 Express each of the following vectors in the form $c(2\mathbf{i} + 3\mathbf{j} + 4\mathbf{k})$ for suitable c:
(*a*) $\overrightarrow{(4, 6, 8)}$
(*b*) $-2\mathbf{i} - 3\mathbf{j} - 4\mathbf{k}$
(*c*) $\mathbf{0}$
(*d*) $\frac{2}{11}\mathbf{i} + \frac{3}{11}\mathbf{j} + \frac{4}{11}\mathbf{k}$

19 If $\|\mathbf{A}\| = 6$, what is the length of (*a*) $-2\mathbf{A}$, (*b*) $\mathbf{A}/3$, (*c*) $\mathbf{A}/\|\mathbf{A}\|$, (*d*) $-\mathbf{A}$, (*e*) $\mathbf{A} + 2\mathbf{A}$?

20 If $\|\mathbf{A}\| = 3$, what is the length of (*a*) $-4\mathbf{A}$, (*b*) $13\mathbf{A} - 7\mathbf{A}$, (*c*) $\mathbf{A}/\|\mathbf{A}\|$, (*d*) $\mathbf{A}/0.05$, (*e*) $\mathbf{A} - \mathbf{A}$?

21 (*a*) Find a unit vector $\mathbf{u}$ that has the same direction as $\mathbf{A} = \mathbf{i} + 2\mathbf{j} + 3\mathbf{k}$.
(*b*) Draw $\mathbf{A}$ and $\mathbf{u}$, with their tails at the origin.

22 (*a*) Find a unit vector $\mathbf{u}$ that has the same direction as $\mathbf{A} = 2\mathbf{i} - 2\mathbf{j} + \mathbf{k}$.
(*b*) Draw $\mathbf{A}$ and $\mathbf{u}$, with their tails at the origin.

23 (*Midpoint formula*) Let A and B be two points in space. Let M be their midpoint. Let $\mathbf{A} = \overrightarrow{OA}$, $\mathbf{B} = \overrightarrow{OB}$, and $\mathbf{M} = \overrightarrow{OM}$.
(*a*) Show that $\mathbf{M} = \mathbf{A} + \frac{1}{2}(\mathbf{B} - \mathbf{A})$.
(*b*) Deduce that $\mathbf{M} = (\mathbf{A} + \mathbf{B})/2$. *Hint:* Draw a picture.

24 Let A and B be two distinct points in space. Let C be the point on the line segment AB that is twice as far from A as it is from B. Let $\mathbf{A} = \overrightarrow{OA}$, $\mathbf{B} = \overrightarrow{OB}$, and $\mathbf{C} = \overrightarrow{OC}$. Show that $\mathbf{C} = \frac{1}{3}\mathbf{A} + \frac{2}{3}\mathbf{B}$. *Hint:* Draw a picture.

25 Show that $2\mathbf{i} + 3\mathbf{j} + 4\mathbf{k}$ and $6\mathbf{i} + 9\mathbf{j} + 12\mathbf{k}$ are parallel.

26 Show that $\mathbf{i} - 3\mathbf{j} + 6\mathbf{k}$ and $-2\mathbf{i} + 6\mathbf{j} - 12\mathbf{k}$ are parallel.

27 This exercise outlines two different proofs of the distributive rule, $c(\mathbf{A} + \mathbf{B}) = c\mathbf{A} + c\mathbf{B}$.
(*a*) (*Algebraic proof*) Write $\mathbf{A}$ and $\mathbf{B}$ in components, and obtain the rule by expressing both $c(\mathbf{A} + \mathbf{B})$ and $c\mathbf{A} + c\mathbf{B}$ in components.
(*b*) (*Geometric proof*) Draw the triangle whose sides are $\mathbf{A}$, $\mathbf{B}$, and $\mathbf{A} + \mathbf{B}$. Magnify this triangle by c and interpret the vectors that form the sides of the new triangle. (Take c positive first, then negative.)

28 If $\|\mathbf{A}\| = 3$ and $\|\mathbf{B}\| = 5$, (*a*) how large can $\|\mathbf{A} + \mathbf{B}\|$ be? (*b*) how small?

---

29 (*a*) Show that the vectors $\mathbf{u}_1 = \frac{1}{2}\mathbf{i} + (\sqrt{3}/2)\,\mathbf{j}$ and $\mathbf{u}_2 = (\sqrt{3}/2)\,\mathbf{i} - \frac{1}{2}\mathbf{j}$ are perpendicular unit vectors. *Hint:* What angles do they make with the x axis?
(*b*) Find scalars x and y such that $\mathbf{i} = x\mathbf{u}_1 + y\mathbf{u}_2$.

30 (*a*) Show that the vectors $\mathbf{u}_1 = (\sqrt{2}/2)\,\mathbf{i} + (\sqrt{2}/2)\,\mathbf{j}$ and $\mathbf{u}_2 = (-\sqrt{2}/2)\,\mathbf{i} + (\sqrt{2}/2)\,\mathbf{j}$ are perpendicular unit vectors. *Hint:* Draw them.
(*b*) Express $\mathbf{i}$ in the form $x\mathbf{u}_1 + y\mathbf{u}_2$. *Hint:* Draw $\mathbf{i}$, $\mathbf{u}_1$, and $\mathbf{u}_2$.
(*c*) Express $\mathbf{j}$ in the form $x\mathbf{u}_1 + y\mathbf{u}_2$.
(*d*) Express $-2\mathbf{i} + 3\mathbf{j}$ in the form $x\mathbf{u}_1 + y\mathbf{u}_2$.

31 (*a*) Draw a unit vector $\mathbf{u}$ tangent to the curve $y = \sin x$ at $(0, 0)$.
(*b*) Express $\mathbf{u}$ in the form $x\mathbf{i} + y\mathbf{j}$.

32 (*a*) Draw a unit vector $\mathbf{u}$ tangent to the curve $y = x^3$ at $(1, 1)$.
(*b*) Express $\mathbf{u}$ in the form $x\mathbf{i} + y\mathbf{j}$.

33 (*a*) Draw the vectors $\mathbf{A} = 2\mathbf{i} + \mathbf{j}$, $\mathbf{B} = 4\mathbf{i} - \mathbf{j}$, and $\mathbf{C} = 5\mathbf{i} + 2\mathbf{j}$.
(*b*) With the aid of the drawing show that there are scalars x and y such that $\mathbf{C} = x\mathbf{A} + y\mathbf{B}$.
(*c*) Using the drawing in (*a*), estimate x and y.
(*d*) Find x and y exactly.

34 (See Exercise 33.) Let $\mathbf{A}$ and $\mathbf{B}$ be two nonzero and nonparallel vectors in the xy plane. Let $\mathbf{C}$ be any vector in the xy plane. Show with the aid of a sketch that there are scalars x and y such that $\mathbf{C} = x\mathbf{A} + y\mathbf{B}$.

35 Let $\mathbf{A}$, $\mathbf{B}$, and $\mathbf{C}$ be three vectors that do not all lie in one plane. Let $\mathbf{D}$ be any vector in space. Show with the aid of a sketch that there are scalars x, y, and z such that $\mathbf{D} = x\mathbf{A} + y\mathbf{B} + z\mathbf{C}$.

36 Let A, B, and C be the vertices of a triangle. Let $\mathbf{A} = \overrightarrow{OA}$, $\mathbf{B} = \overrightarrow{OB}$, and $\mathbf{C} = \overrightarrow{OC}$.
(*a*) Let P be the point that is on the line segment joining A to the midpoint of the edge BC and twice as far from A as from the midpoint. Show that $\overrightarrow{OP} = (\mathbf{A} + \mathbf{B} + \mathbf{C})/3$.
(*b*) Use (*a*) to show that the three medians of a triangle are concurrent.

37 (*a*) What is the sum of the five vectors shown in Fig. 38?
(*b*) Sketch the figure corresponding to the sum $\mathbf{A} + \mathbf{C} + \mathbf{D} + \mathbf{E} + \mathbf{B}$.

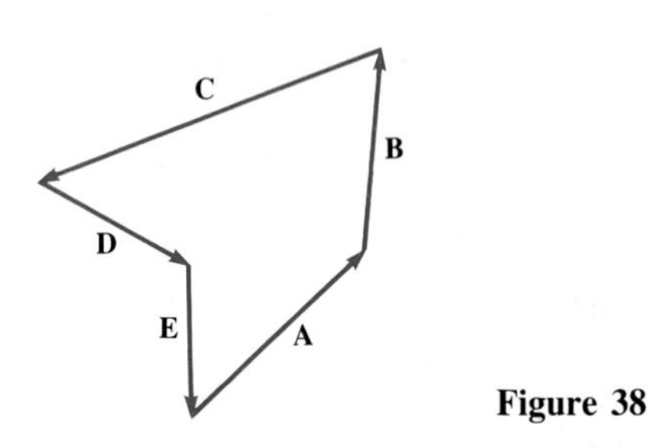

Figure 38

38 The midpoints of a quadrilateral in space are joined to form another quadrilateral. Prove that this second quadrilateral is a parallelogram.

39 (*a*) Using an appropriate diagram, explain why $\|\mathbf{A} + \mathbf{B}\| \le \|\mathbf{A}\| + \|\mathbf{B}\|$. (This is called the **triangle inequality**.)
(*b*) For which pairs of vectors $\mathbf{A}$ and $\mathbf{B}$ is $\|\mathbf{A} + \mathbf{B}\| = \|\mathbf{A}\| + \|\mathbf{B}\|$?

40 From Exercise 39 deduce that for any four real numbers x_1, y_1, x_2, and y_2,

$$x_1x_2 + y_1y_2 \le \sqrt{x_1^2 + y_1^2}\sqrt{x_2^2 + y_2^2}.$$

When does equality hold?

12.2 PROJECTIONS

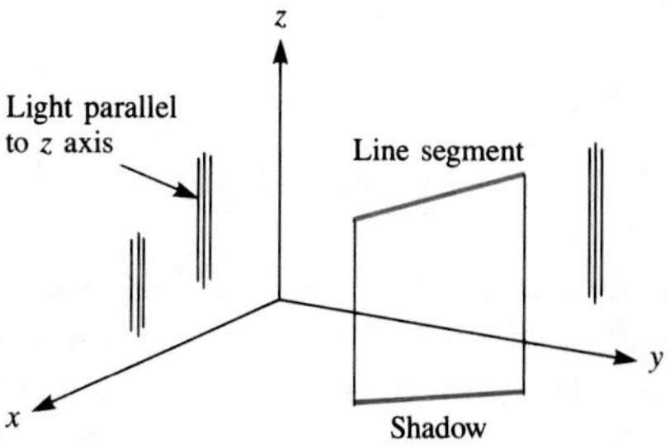

This section develops the notion of a projection of a vector and other geometric objects. This concept is illustrated by the shadow cast on a plane by parallel rays of light that are perpendicular to that plane.

Figure 1 shows the shadow of a line segment cast on the xy plane by light parallel to the z axis. If the line segment is parallel to the z axis, its shadow is just a point. If it is parallel to the xy plane, its shadow is as long as itself. The length of the shadow clearly depends on the direction in which the line segment is placed.

The line segment's shadow is only one of several projections that will be examined in this and later chapters.

Light parallel to z axis; Line segment; Shadow

Figure 1

Projection of a Line Segment on a Line

Let AB be a line segment and L a line through A. Drop a perpendicular from B to L, meeting L in the point D. The segment AD, shown in Fig. 2, is called the **projection of AB on L**. (We may think of it as the shadow of AB cast on line L by light parallel to BD.)

$AD \approx$ projection of AB on line L

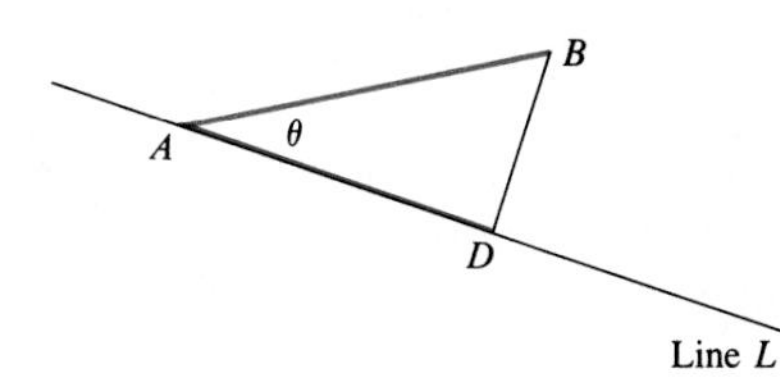

Figure 2

The length of the projection depends on the angle θ in the right triangle ADB shown in Fig. 2. Since

$$\cos\theta = \frac{\text{Adjacent}}{\text{Hypotenuse}} = \frac{\overline{AD}}{\overline{AB}},$$

we have

$$\boxed{\overline{AD} = \overline{AB}\cos\theta.}$$

The cosine function will appear several times in connection with projections.

EXAMPLE 1 A line segment of length 5 makes an angle of $\pi/3$ with a certain line. What is the length of the projection of the segment on the line? (See Fig. 3.)

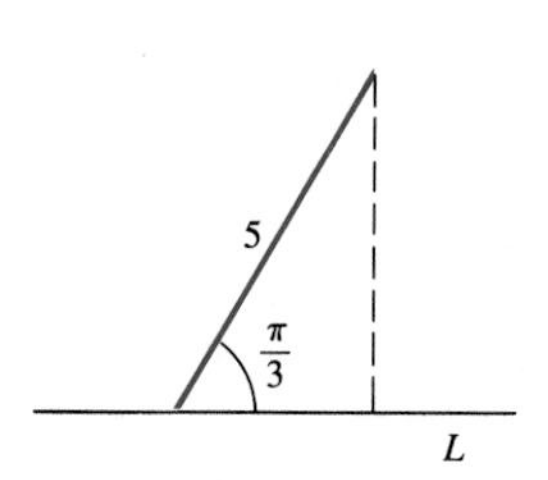

Figure 3

SOLUTION The length of the projection is $5\cos(\pi/3) = \frac{5}{2}$. ■

In defining the projection of a segment on a line, we assumed that the line passes through an end of the segment. If line L does not pass through either end of AB, the projection of AB on L is defined as follows.

Draw the planes perpendicular to L that pass through A and B. The planes meet L at the points C and D, respectively. The segment CD is called the **projection of AB on line L**. (See Fig. 4.)

The projection of AB on L is CD.

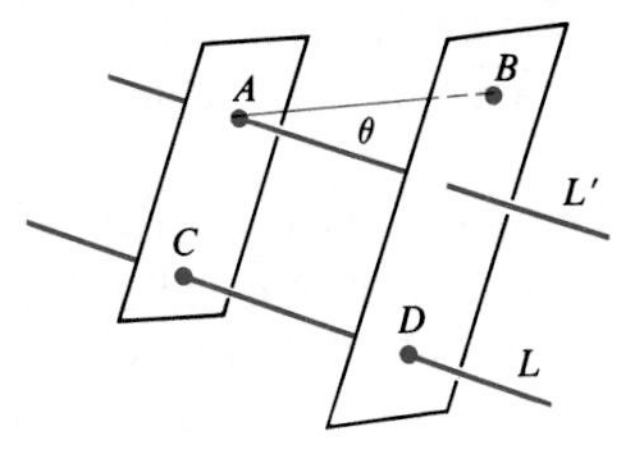

Figure 4

To find the length of CD, draw a line L' through A and parallel to L. Let the angle between AB and L' be θ, $0 \le \theta \le \pi/2$, as in Fig. 4. Then

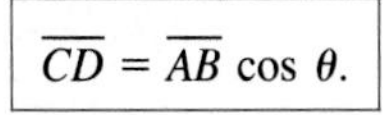

$$\boxed{\overline{CD} = \overline{AB}\cos\theta.}$$

As before, the length of the projection turns out to be the length of the projected line segment multiplied by the cosine of the appropriate angle.

Projection of a Line Segment on a Plane

AC and BD are perpendicular to $\mathcal{P}$.

Figure 5

Let AB be a line segment and $\mathcal{P}$ be a plane. Drop perpendiculars from A and B to $\mathcal{P}$, as shown in Fig. 5.

The two lines meet in $\mathcal{P}$ in points C and D, respectively. The segment CD is called the **projection of AB on the plane $\mathcal{P}$**. To find the length of CD, draw a line L parallel to CD and passing through A. Let θ be the angle between AB and L, $0 \leq \theta \leq \pi/2$. (See Fig. 6.) This angle θ is called the **angle between AB and the plane $\mathcal{P}$**.

Once again we have

θ = angle between AB and L
= angle between AB and $\mathcal{P}$.

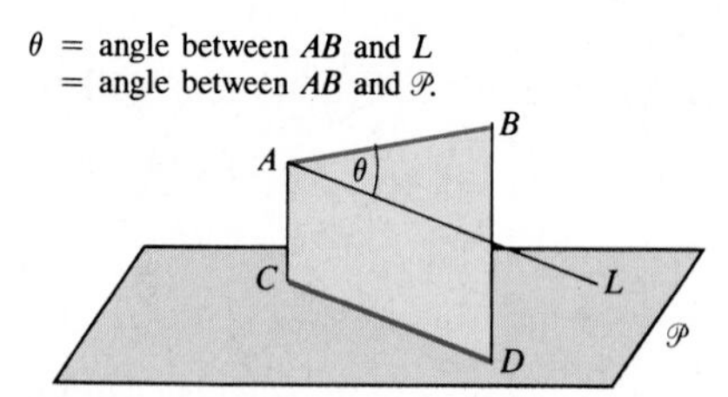

Figure 6

$$\boxed{\overline{CD} = \overline{AB} \cos \theta.}$$

Projection of a Flat Region on a Plane

Any line perpendicular to a given plane is called a **normal** to the plane. The angle between two planes is defined as the angle θ between their normals, where θ is chosen to lie in $[0, \pi/2]$. See Fig. 7.

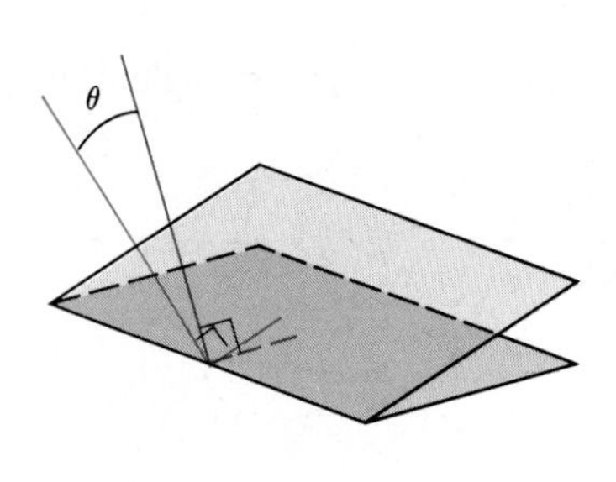

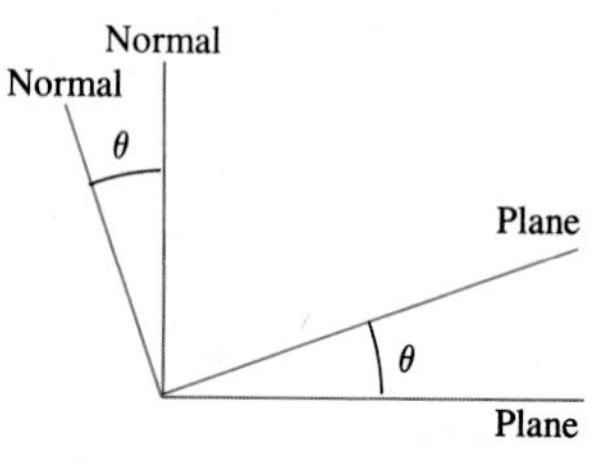

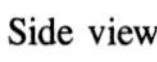

Figure 7

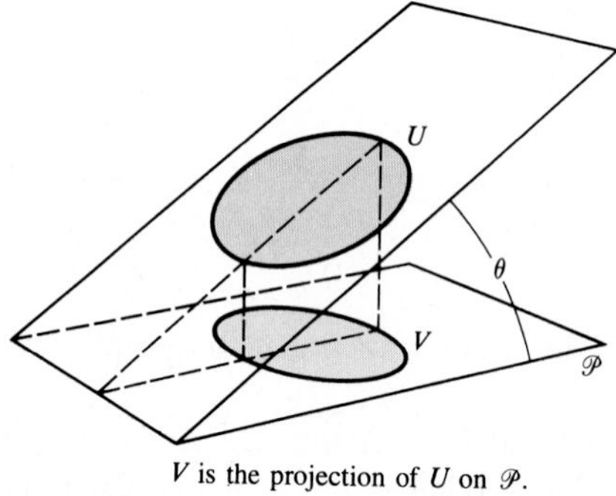

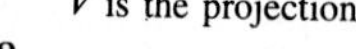

V is the projection of U on $\mathcal{P}$.

Figure 8

The angle between two planes is called a **dihedral angle** [from the Greek, *di* (two) and *hedra* (seat)].

Now consider a region U in a plane tilted at some angle θ to a plane $\mathcal{P}$. The projection of U on $\mathcal{P}$ consists of all the points where lines through U and perpendicular to $\mathcal{P}$ meet $\mathcal{P}$. (See Fig. 8, where the projection is the region labeled V.) The area of V depends on the area of U and the angle between the plane of U and the plane $\mathcal{P}$. To find the area of V we will use the fact that "area is the integral of cross-sectional length," as shown in Sec. 5.3.

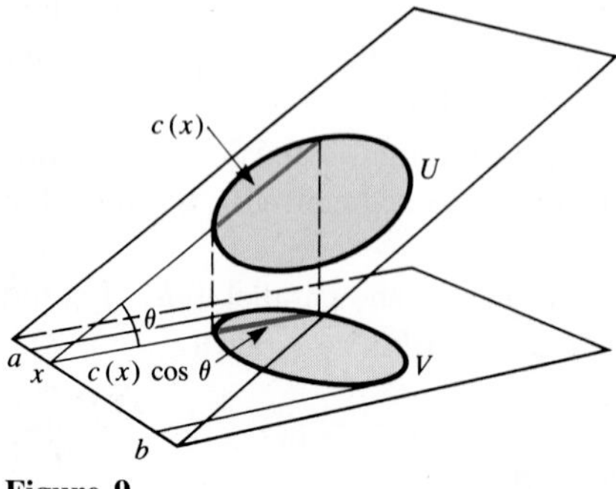

Figure 9

In Fig. 9 the line of intersection of the two planes is used as an axis. Let the cross section of U corresponding to a number x on this axis have length $c(x)$. Then the corresponding cross section of V has length $c(x) \cos \theta$, from the formula for the projection of a line segment on a plane.

Thus

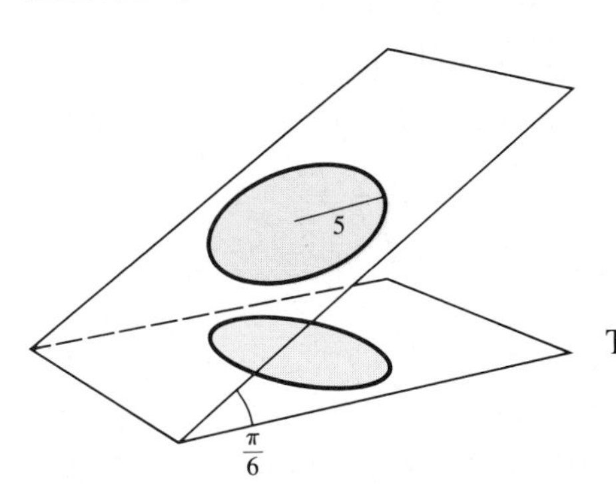

Figure 10

$$\text{Area of } V = \int_a^b c(x) \cos\theta \, dx$$

$$= \cos\theta \int_a^b c(x)\, dx$$

$$= \cos\theta \text{ (Area of } U).$$

This time, $\cos\theta$ tells us how area behaves under projection:

$$\boxed{\text{Area of projection } V = \cos\theta \text{ (Area of } U).} \qquad (1)$$

As a check, take $\theta = 0$ and $\theta = \pi/2$. Does formula (1) give the correct areas?

EXAMPLE 2 A disk of radius 5 lies in a plane tilted at an angle $\pi/6$ to the xy plane. Find the area of its projection on the xy plane. (See Fig. 10.)

SOLUTION Since the area of the disk is $\pi 5^2 = 25\pi$, the area of the projection is $25\pi \cos(\pi/6) = 25\pi\sqrt{3}/2$. ■

Projection of a Vector on a Line

Up to this moment we were concerned only with lengths and areas. Now we consider the projections of vectors, which record not only lengths but directions as well.

Let **A** be a vector and let L be a line. Represent **A** by an arrow whose tail is on L, as in Fig. 11. Let C be the tail of **A** and D be the head of **A**, that is, $\mathbf{A} = \overrightarrow{CD}$. The line through D that meets L and is perpendicular to L meets L in a point E. The vector $\overrightarrow{CE}$ is called the projection of **A** on L and is denoted

$$\mathbf{proj}_L\, \mathbf{A} \qquad \text{or} \qquad \mathbf{proj}(\mathbf{A} \text{ on } L).$$

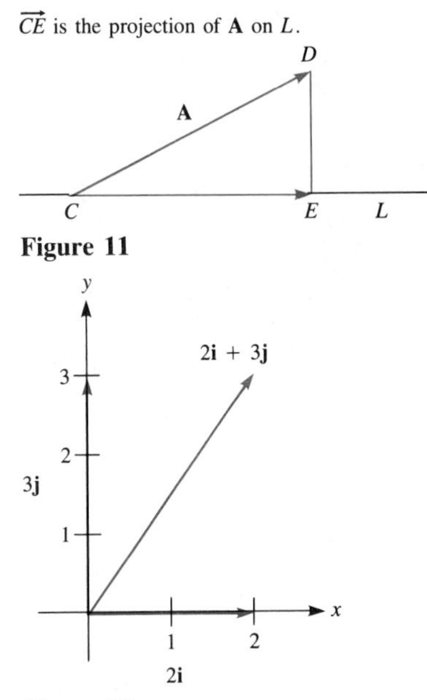

$\overrightarrow{CE}$ is the projection of **A** on L.

Figure 11

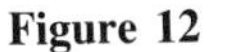

Figure 12

Note that $\mathbf{proj}_L\, \mathbf{A}$ is a vector. Observe that $\mathbf{A} = \overrightarrow{CE} + \overrightarrow{ED}$, and Fig. 11 shows that **A** is the sum of a vector parallel to L and a vector perpendicular to L.

EXAMPLE 3 Find $\mathbf{proj}_L(2\mathbf{i} + 3\mathbf{j})$, where L is (a) the x axis, (b) the y axis.

SOLUTION Figure 12 is the necessary diagram.

(a) Inspection of Fig. 12 shows that the projection of $2\mathbf{i} + 3\mathbf{j}$ on the x axis is $2\mathbf{i}$.

(b) Inspection of Fig. 12 shows that the projection of $2\mathbf{i} + 3\mathbf{j}$ on the y axis is $3\mathbf{j}$. ■

In the next section we will see how to compute the projection of a vector in any direction, not just along the axes.

EXAMPLE 4 Find the projection of $2\mathbf{i} + 3\mathbf{j} + 4\mathbf{k}$ on the y axis.

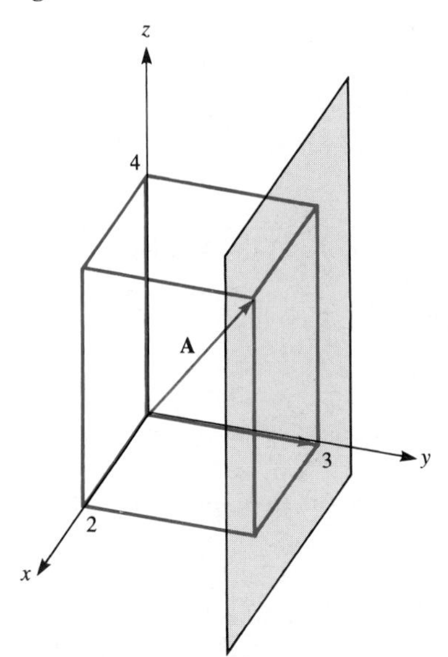

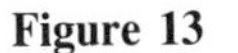

Figure 13

SOLUTION To project $2\mathbf{i} + 3\mathbf{j} + 4\mathbf{k}$ on the y axis, place its tail at the origin and draw a plane perpendicular to the y axis through its head, as in Fig. 13. The plane cuts off the vector $3\mathbf{j}$ on the y axis. Hence

$$\mathbf{proj}_L(2\mathbf{i} + 3\mathbf{j} + 4\mathbf{k}) = 3\mathbf{j}.$$

From Fig. 13 we can also see that the projection of $2\mathbf{i} + 3\mathbf{j} + 4\mathbf{k}$ on the x axis is $2\mathbf{i}$ and on the z axis is $4\mathbf{k}$. ■

Projection of a Vector on a Vector

We now define the projection of a vector on a vector.

> **Definition** *Projection of* **A** *on* **B**. Let **A** be a vector and **B** be a nonzero vector. The **projection of A on B** is defined to be the projection of **A** on any line parallel to **B**. It is denoted
>
> $$\mathbf{proj_B\,A} \qquad \text{or} \qquad \mathbf{proj}(\mathbf{A} \text{ on } \mathbf{B}).$$

Note that $\mathbf{proj_B\,A}$ does *not* depend on the magnitude of **B**. Example 5 reinforces this observation.

EXAMPLE 5 Find the projection of $2\mathbf{i} + 3\mathbf{j}$ on (*a*) **i**, (*b*) $-\mathbf{i}$, (*c*) $3\mathbf{i}$.

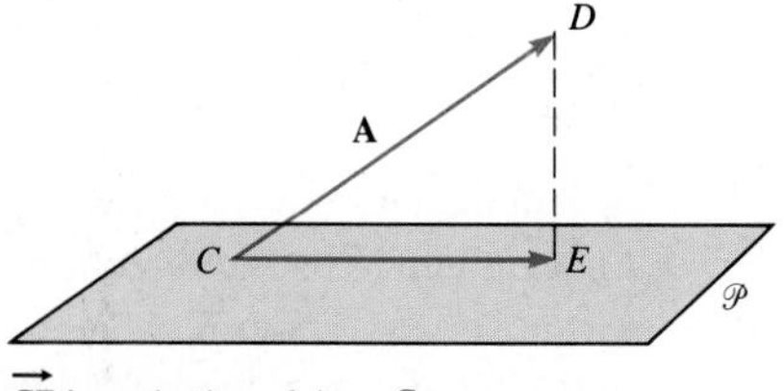

$\overrightarrow{CE}$ is projection of **A** on $\mathscr{P}$.
Figure 14

SOLUTION In each case the x axis can be used as the line parallel to the vector on which $2\mathbf{i} + 3\mathbf{j}$ is projected. Thus
(*a*) $\mathbf{proj_i}(2\mathbf{i} + 3\mathbf{j}) = 2\mathbf{i}$
(*b*) $\mathbf{proj_{-i}}(2\mathbf{i} + 3\mathbf{j}) = 2\mathbf{i}$
(*c*) $\mathbf{proj_{3i}}(2\mathbf{i} + 3\mathbf{j}) = 2\mathbf{i}$. ■

As Example 5 shows, if **B** and **C** are parallel, then $\mathbf{proj_B\,A}$ is equal to $\mathbf{proj_C\,A}$.

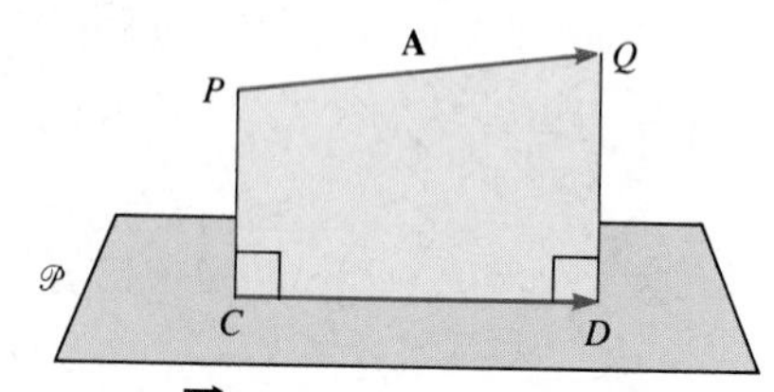

$\overrightarrow{CD}$ is the projection of **A** on $\mathscr{P}$
Figure 15

Projection of a Vector on a Plane

Next we define the projection of a vector **A** on a plane $\mathscr{P}$. Let $\mathbf{A} = \overrightarrow{CD}$. For convenience, we place **A** so that its tail C is in $\mathscr{P}$, as in Fig. 14. The line through D that is perpendicular to the plane $\mathscr{P}$ meets $\mathscr{P}$ in a point E. Then the **projection of *A* on $\mathscr{P}$** is $\overrightarrow{CE}$ and is denoted $\mathbf{proj}_{\mathscr{P}}\,\mathbf{A}$.

If we choose to represent a vector **A** by an arrow where the tail is not in the plane $\mathscr{P}$, $\mathbf{proj}_{\mathscr{P}}\,\mathbf{A}$ would appear as in Fig. 15.

EXAMPLE 6 Find the projection of $\mathbf{A} = \mathbf{i} + 3\mathbf{j} + 2\mathbf{k}$ on the xy plane.

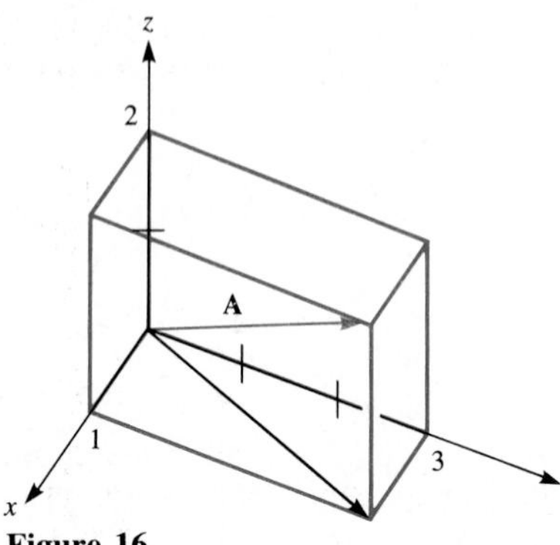

Figure 16

SOLUTION Figure 16 shows that the projection of **A** on the xy plane is $\mathbf{i} + 3\mathbf{j}$. ■

Note that the projection of a vector on a plane is again a vector.

EXAMPLE 7 What relation is there between the vectors $\mathbf{proj}_{\mathscr{P}}\,\mathbf{A}$, $\mathbf{proj}_{\mathscr{P}}\,\mathbf{B}$, and $\mathbf{proj}_{\mathscr{P}}(\mathbf{A} + \mathbf{B})$?

SOLUTION Figure 17 shows $\mathscr{P}$, **A**, **B**, $\mathbf{A} + \mathbf{B}$, and the projection of these three vectors on $\mathscr{P}$. Inspection of Fig. 17 shows that $\mathbf{proj}_{\mathscr{P}}\,\mathbf{A} + \mathbf{proj}_{\mathscr{P}}\,\mathbf{B} =$

$\mathbf{proj}_{\mathcal{P}}(\mathbf{A} + \mathbf{B})$. ■

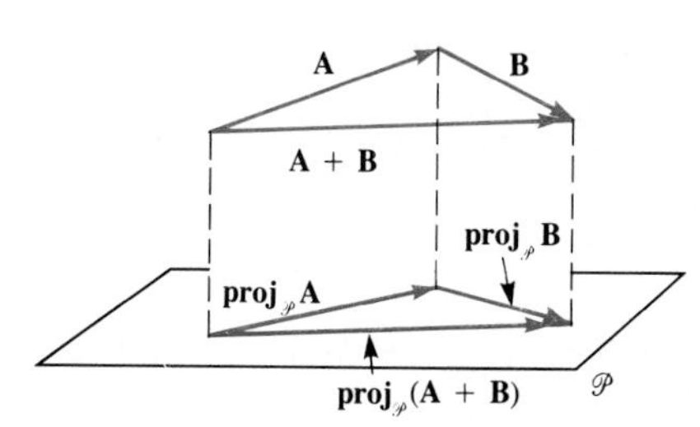

Figure 17

One final remark about the projection of vectors. Let **A** be a vector and **B** be a nonzero vector. Let $\mathcal{P}$ be a plane perpendicular to **B**, as in Fig. 18. Then $\mathbf{A} = \overrightarrow{CE} + \overrightarrow{ED} = \mathbf{proj_B\ A} + \mathbf{proj}_{\mathcal{P}}\ \mathbf{A}$. This says that **A** is the sum of a vector parallel to **B** and a vector perpendicular to **B**. Both vectors are of use, and we will denote the vector perpendicular to **B** by the symbol $\mathbf{orth_B\ A}$. "Orth" is short for "orthogonal" (in Greek, *ortho* is "right" and *gonal* is "angle"). We may now write

$$\mathbf{A} = \mathbf{proj_B\ A} + \mathbf{orth_B\ A}.$$

(We use "orth" rather than "perp" because "perp" looks too much like "proj.")

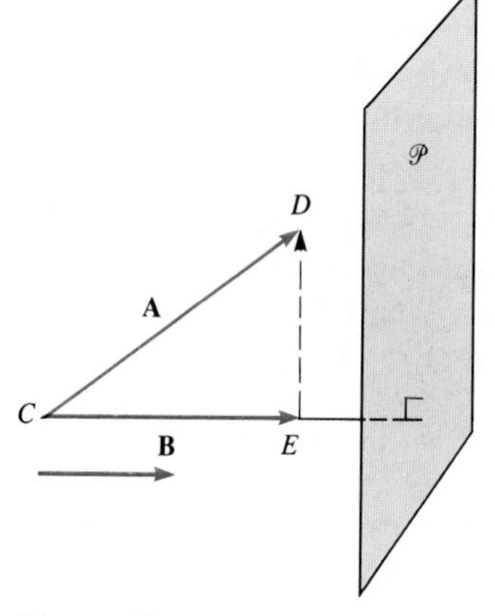

Figure 18

EXAMPLE 8 Let $\mathbf{A} = \mathbf{i} + 2\mathbf{j} + 3\mathbf{k}$ and $\mathbf{B} = -\mathbf{k}$. Find $\mathbf{proj_B\ A}$ and $\mathbf{orth_B\ A}$.

SOLUTION Figure 19 shows **A**, **B**, $\mathbf{proj_B\ A}$, and $\mathbf{orth_B\ A}$. Since **B** is parallel to the z axis, we may use the xy plane as the plane perpendicular to **B**. From Fig. 19 we see that $\mathbf{proj_B\ A} = 3\mathbf{k}$ and $\mathbf{orth_B\ A} = \mathbf{i} + 2\mathbf{j}$. ■

Section Summary

We defined several projections:

1 A line segment on a line or on a plane;
2 A flat region on a plane;
3 A vector on a line, or on a vector, or on a plane.

We observed the key role of "$\cos\theta$" in determining lengths or areas of projections of line segments or planar regions. Also we showed that, given a nonzero vector **B**, any vector is the sum of a vector parallel to **B** and a vector perpendicular (orthogonal) to **B**:

$$\mathbf{A} = \mathbf{proj_B\ A} + \mathbf{orth_B\ A}.$$

The vector $\mathbf{orth_B\ A}$ is the projection of **A** on a plane $\mathcal{P}$ perpendicular to **B**: $\mathbf{orth_B\ A} = \mathbf{proj}_{\mathcal{P}}\ \mathbf{A}$.

At this point we were able to calculate only the simplest projections, the ones that could be done by eye. The next section develops a tool for computing $\mathbf{proj_B\ A}$ for general **A** and **B**.

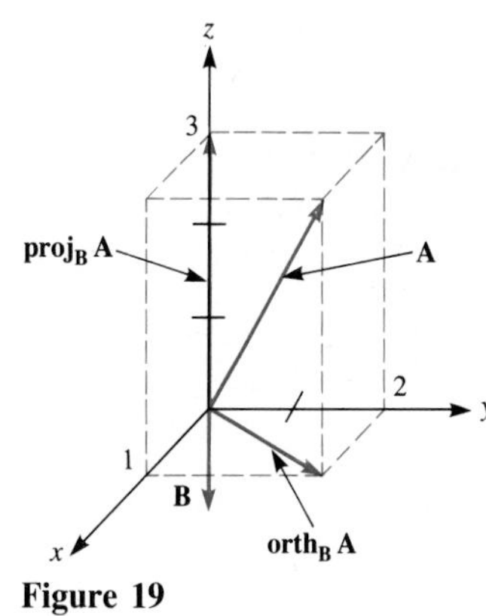

Figure 19

EXERCISES FOR SEC. 12.2: PROJECTIONS

1 The line segment AB joins the points $A = (0, 0, 0)$ and $B = (2, 1, 3)$.
(*a*) Draw AB and its projections on the three axes.
(*b*) Find the lengths of the projections in (*a*).
(*c*) Find the length of AB.

2 The line segment AB joins the points $A = (0, 1, 2)$ and

$B = (2, 1, 3)$.
(*a*) Draw AB and its projections on the three axes.
(*b*) Find the lengths of the projections in (*a*).
(*c*) Find the length of AB.

3 The line segment AB joins the points $A = (0, 0, 0)$ and $B = (2, 1, 3)$. Find the lengths of its projections on the xy, yz, and xz planes. Include a clear diagram.

4 The line segment AB joins the points $A = (1, 1, 1)$ and $B = (2, 0, 3)$. Find the lengths of its projections on the xy, yz, and xz planes. Include a clear diagram.

5 A line segment AB has a projection of length 2 on the y axis and a projection of length 3 on the xz plane.
(*a*) Draw a clear diagram that shows this information.
(*b*) Find the length of AB.

6 A line segment AB has a projection of length 1 on the x axis and a projection of length 2 on the yz plane.
(*a*) Draw a clear diagram that shows this information.
(*b*) Find the length of AB.

7 In our argument for the formula "Area of projection $V = \cos\theta$ (Area of U)," the diagram shows the line of intersection of the two planes as coming out of the page.
(*a*) Draw a diagram for the proof in which the line of intersection of the two planes is parallel to the y axis of an xyz coordinate system.
(*b*) Write up the proof, using the diagram in (*a*).

8 A flat region U lies in a plane whose angle with the xz plane is $\pi/3$. The projection of U on the xz plane has area 5.
(*a*) Draw a clear diagram that shows this information.
(*b*) Find the area of U.

9 (*a*) Make a clear diagram showing the vector $\mathbf{i} + \mathbf{j} + 2\mathbf{k}$ and its projections on the x axis, on the y axis, and on the xy plane.
(*b*) Find the three projections in (*a*).

10 (*a*) Make a clear diagram of the vector $\mathbf{i} + 2\mathbf{j} + 3\mathbf{k}$ and its projections on the x axis, on the y axis, and on the xy plane.
(*b*) Find the three projections in (*a*).

11 (*a*) Make a clear diagram of the vector $2\mathbf{i} + 3\mathbf{j}$ and the line $y = x$ in the xy plane.
(*b*) Draw the projection of $2\mathbf{i} + 3\mathbf{j}$ on the line $y = x$.
(*c*) Write the projection in (*b*) in the form $x\mathbf{i} + y\mathbf{j}$, estimating x and y from your diagram.

12 (*a*) Make a clear diagram of the vectors $2\mathbf{i} + 3\mathbf{j}$ and $\mathbf{i} - 2\mathbf{j}$ in the xy plane.
(*b*) Draw $\mathbf{proj}_{\mathbf{i}-2\mathbf{j}}(2\mathbf{i} + 3\mathbf{j})$.
(*c*) Write the projection in (*b*) in the form $x\mathbf{i} + y\mathbf{j}$, estimating x and y from your diagram.

13 What is the projection of $\mathbf{i} - 2\mathbf{j} + 4\mathbf{k}$ on (*a*) $\mathbf{i}$, (*b*) $-\mathbf{i}$, (*c*) the z axis, (*d*) $\mathbf{j}$, (*e*) the yz plane?

14 What is the projection of $2\mathbf{i} - \mathbf{j} - \mathbf{k}$ on (*a*) $\mathbf{j}$, (*b*) $-\mathbf{k}$, (*c*) $3\mathbf{k}$, (*d*) $\mathbf{j} + \mathbf{k}$, (*e*) the xz plane?

15 Let $\mathbf{A}$ be a nonzero vector. For which vectors $\mathbf{B}$ is $\mathbf{proj_B\,A}$ equal to $\mathbf{A}$? *Hint:* Draw some pictures.

16 Let $\mathbf{A}$ be a nonzero vector. For which vectors $\mathbf{B}$ is $\mathbf{proj_B\,A}$ equal to $\mathbf{0}$?

17 (*a*) What relation is there between $\mathbf{proj_B\,A}$ and $\mathbf{proj_B}(-\mathbf{A})$?
(*b*) What relation is there between $\mathbf{proj_B\,A}$ and $\mathbf{proj_{-B}\,A}$?

18 Let $\mathbf{B}$ be a nonzero vector. For which vectors $\mathbf{A}$ is $\|\mathbf{proj_B\,A}\| = \|\mathbf{A}\|$? Explain, with the aid of a diagram.

19 Find $\mathbf{proj_k}(2\mathbf{i} - 3\mathbf{j} + 2\mathbf{k})$ and $\mathbf{orth_k}(2\mathbf{i} - 3\mathbf{j} + 2\mathbf{k})$.

20 Find $\mathbf{proj_{-i}}(\mathbf{i} + 3\mathbf{j} + \mathbf{k})$ and $\mathbf{orth_{-i}}(\mathbf{i} + 3\mathbf{j} + \mathbf{k})$.

---

21 Let L_1 and L_2 be perpendicular lines in the xy plane. Let $\mathbf{A}$ be a vector in the xy plane. What relation is there between $\mathbf{A}$, $\mathbf{proj}_{L_1}\,\mathbf{A}$, and $\mathbf{proj}_{L_2}\,\mathbf{A}$? Explain, with the aid of a diagram.

22 A vector $\mathbf{D} = x\mathbf{i} + y\mathbf{j} + z\mathbf{k}$ has the projection $\mathbf{A}$ on the xy plane, $\mathbf{B}$ on the yz plane, and $\mathbf{C}$ on the xz plane.
(*a*) Draw a clear diagram that shows $\mathbf{A}$, $\mathbf{B}$, $\mathbf{C}$, and $\mathbf{D}$.
(*b*) Express $\mathbf{A}$, $\mathbf{B}$, and $\mathbf{C}$ in terms of $\mathbf{i}$, $\mathbf{j}$, and $\mathbf{k}$.
(*c*) Express $\mathbf{D}$ in terms of $\mathbf{A}$, $\mathbf{B}$, and $\mathbf{C}$.

23 A line segment has projections of lengths a, b, and c on the coordinate axes. What, if anything, can be said about its length?

24 A line segment has projections of lengths d, e, and f on the coordinate planes. What, if anything, can be said about its length?

25 Find a relation between $\mathbf{proj}_L\,\mathbf{A}$, $\mathbf{proj}_L\,\mathbf{B}$, and $\mathbf{proj}_L(\mathbf{A} + \mathbf{B})$. Explain. *Hint:* Draw some pictures.

26 Explain why the projection of a circle is an ellipse. *Hint:* Set up coordinate systems in the plane of the circle and in the plane of its shadow (which might as well be taken to be the xy plane). Choose the axes for these coordinate systems to be as convenient as possible. Then express the equation of the shadow in terms of x and y by utilizing the equation of the circle.

12.3 THE DOT PRODUCT OF TWO VECTORS

This section introduces the "dot product" or "scalar product," a number that is defined for every pair of vectors.

Consider a rock being pulled along level ground by a rope inclined at a fixed angle to the ground. Let the force applied to the rock be represented by the vector $\mathbf{F}$. The force $\mathbf{F}$ can be expressed as the sum of a vertical force $\mathbf{F}_2$ and a horizontal force $\mathbf{F}_1$, as shown in Fig. 1.

We may replace $\mathbf{F}$ with $\mathbf{F}_1$ and $\mathbf{F}_2$, which together have the same effect on the rock as $\mathbf{F}$.

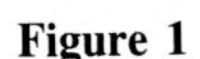

Figure 1

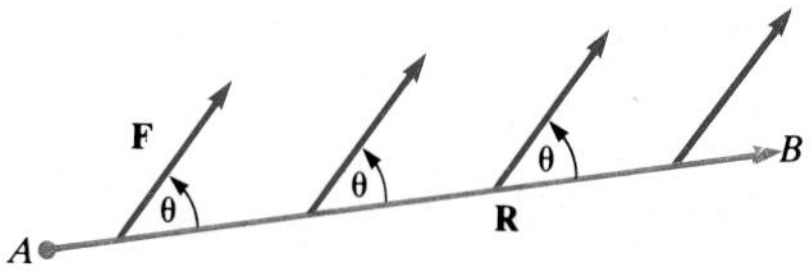

A constant force $\mathbf{F}$ moves an object from A to B, perhaps against gravity, air resistance, friction, etc.

Figure 2

How much work is done by the force $\mathbf{F}$ in moving the rock along the ground? The physicist defines the work accomplished by a constant force $\mathbf{F}$ (whatever direction it may have) as the product of the component of $\mathbf{F}$ in the direction of motion and the distance traveled. Say that the force $\mathbf{F}$, as shown in Fig. 2, moves an object along a straight line from the tail to the head of $\mathbf{R}$.

By definition

$$\text{Work} = \underbrace{\|\mathbf{F}\| \cos \theta}_{\substack{\text{Force in}\\ \text{direction}\\ \text{of } \mathbf{R}}} \cdot \underbrace{\|\mathbf{R}\|}_{\substack{\text{Distance}\\ \text{traveled}}}$$

where θ is the angle between $\mathbf{R}$ and $\mathbf{F}$.

The force $\mathbf{F}_2$ in Fig. 1 accomplishes no work. The work accomplished by $\mathbf{F}$ in pulling the rock is the same as that accomplished by $\mathbf{F}_1$.

The Dot Product

This important physical concept illustrates the dot product of two vectors, which will be introduced after the following definition.

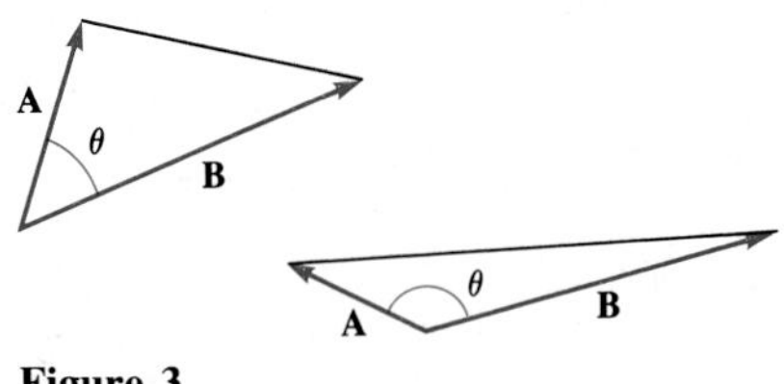

Figure 3

Definition *Angle between two nonzero vectors.* Let $\mathbf{A}$ and $\mathbf{B}$ be two nonparallel and nonzero vectors. They determine a triangle and an angle θ, shown in Fig. 3. The **angle between A and B** is θ. Note that

$$0 < \theta < \pi.$$

If $\mathbf{A}$ and $\mathbf{B}$ are parallel, the angle between them is 0 (if they have the same direction) or π (if they have opposite directions). The angle between $\mathbf{0}$ and any other vector is not defined.

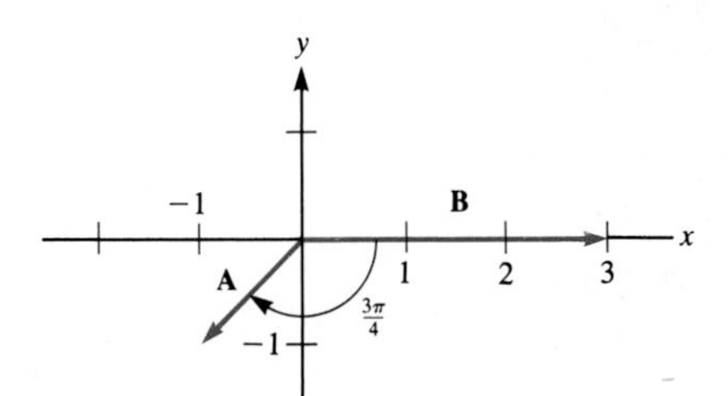

Figure 4

The angle between $\mathbf{i}$ and $\mathbf{j}$ is $\pi/2$. The angle between $\mathbf{A} = -\mathbf{i} - \mathbf{j}$ and $\mathbf{B} = 3\mathbf{i}$ is $3\pi/4$, as Fig. 4 shows. The angle between $\mathbf{k}$ and $-\mathbf{k}$ is π; the angle between $2\mathbf{i}$ and $5\mathbf{i}$ is 0.

The dot product can now be defined.

Definition *Dot product.* Let $\mathbf{A}$ and $\mathbf{B}$ be two nonzero vectors. Their **dot product** is the number

$$\|\mathbf{A}\| \, \|\mathbf{B}\| \cos \theta,$$

where θ is the angle between $\mathbf{A}$ and $\mathbf{B}$. If $\mathbf{A}$ or $\mathbf{B}$ is $\mathbf{0}$, their dot product is 0. The dot product is denoted $\mathbf{A} \cdot \mathbf{B}$. It is a scalar and is also called the **scalar product** of $\mathbf{A}$ and $\mathbf{B}$.

The dot product satisfies several useful identities, which follow from the definition:

$$\mathbf{A} \cdot \mathbf{B} = \mathbf{B} \cdot \mathbf{A} \qquad \text{(the dot product is commutative)}$$

$$\mathbf{A} \cdot \mathbf{A} = \|\mathbf{A}\|^2$$

$$(c\mathbf{A}) \cdot \mathbf{B} = c(\mathbf{A} \cdot \mathbf{B}) = \mathbf{A} \cdot (c\mathbf{B}) \qquad (c \text{ is a scalar})$$

and

$$\mathbf{0} \cdot \mathbf{A} = 0.$$

For instance, to establish that $\mathbf{A} \cdot \mathbf{A} = \|\mathbf{A}\|^2$, we calculate $\mathbf{A} \cdot \mathbf{A}$:

$$\begin{aligned} \mathbf{A} \cdot \mathbf{A} &= \|\mathbf{A}\| \, \|\mathbf{A}\| \cos \theta \\ &= \|\mathbf{A}\|^2, \end{aligned}$$

since the angle θ between $\mathbf{A}$ and $\mathbf{A}$ is 0, and $\cos 0 = 1$.

EXAMPLE 1 Find the dot product $\mathbf{A} \cdot \mathbf{B}$ if $\mathbf{A} = 3\mathbf{i} + 3\mathbf{j}$ and $\mathbf{B} = -5\mathbf{i}$.

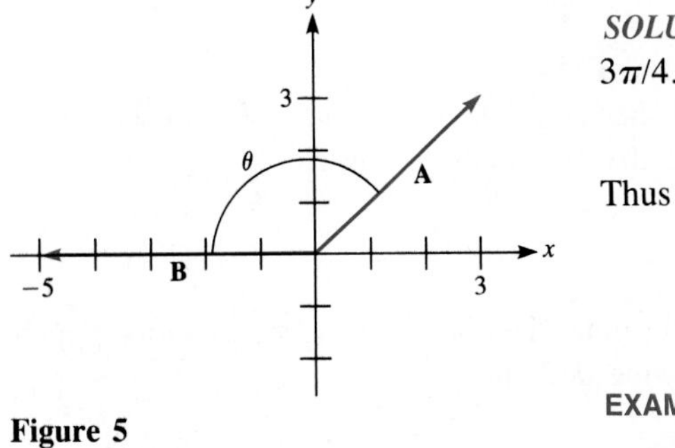

Figure 5

SOLUTION Inspection of Fig. 5 shows that θ, the angle between $\mathbf{A}$ and $\mathbf{B}$, is $3\pi/4$. Also,

$$\|\mathbf{A}\| = \sqrt{3^2 + 3^2} = \sqrt{18} \qquad \text{and} \qquad \|\mathbf{B}\| = \sqrt{5^2 + 0^2} = 5.$$

Thus

$$\begin{aligned} \mathbf{A} \cdot \mathbf{B} &= \|\mathbf{A}\| \, \|\mathbf{B}\| \cos \theta \\ &= \sqrt{18} \cdot 5\left(\frac{-\sqrt{2}}{2}\right) = -15. \quad \blacksquare \end{aligned}$$

EXAMPLE 2 Find (*a*) $\mathbf{i} \cdot \mathbf{j}$, (*b*) $\mathbf{i} \cdot \mathbf{i}$, (*c*) $2\mathbf{k} \cdot (-3\mathbf{k})$.

SOLUTION (*a*) The angle between $\mathbf{i}$ and $\mathbf{j}$ is $\pi/2$. Thus

$$\begin{aligned} \mathbf{i} \cdot \mathbf{j} &= \|\mathbf{i}\| \, \|\mathbf{j}\| \cos \frac{\pi}{2} \\ &= 1 \cdot 1 \cdot 0 \\ &= 0. \end{aligned}$$

(*b*) The angle between $\mathbf{i}$ and $\mathbf{i}$ is 0. Thus

$$\begin{aligned} \mathbf{i} \cdot \mathbf{i} &= \|\mathbf{i}\| \, \|\mathbf{i}\| \cos 0 \\ &= 1 \cdot 1 \cdot 1 \\ &= 1. \end{aligned}$$

(This is a special case of the fact that $\mathbf{A} \cdot \mathbf{A} = \|\mathbf{A}\|^2$.)

(*c*) The angle between $2\mathbf{k}$ and $-3\mathbf{k}$ is π. Thus

$$\begin{aligned} 2\mathbf{k} \cdot (-3\mathbf{k}) &= \|2\mathbf{k}\| \, \|-3\mathbf{k}\| \cos \pi \\ &= 2 \cdot 3 \cdot (-1) \\ &= -6. \quad \blacksquare \end{aligned}$$

Computations like those in Example 2 show that

$$a\mathbf{i} \cdot b\mathbf{i} = ab, \qquad a\mathbf{j} \cdot b\mathbf{j} = ab, \qquad \text{and} \qquad a\mathbf{k} \cdot b\mathbf{k} = ab,$$

while $a\mathbf{i} \cdot b\mathbf{j} = 0, \quad a\mathbf{i} \cdot b\mathbf{k} = 0, \quad \text{and} \quad a\mathbf{j} \cdot b\mathbf{k} = 0.$

In particular, $\mathbf{i} \cdot \mathbf{i} = \mathbf{j} \cdot \mathbf{j} = \mathbf{k} \cdot \mathbf{k} = 1$, while $\mathbf{i} \cdot \mathbf{j} = \mathbf{i} \cdot \mathbf{k} = \mathbf{j} \cdot \mathbf{k} = 0$.

The Geometry of the Dot Product

Let **A** and **B** be nonzero vectors and θ the angle between them. Their dot product is

$$\mathbf{A} \cdot \mathbf{B} = \|\mathbf{A}\| \, \|\mathbf{B}\| \cos \theta.$$

The quantities $\|\mathbf{A}\|$ and $\|\mathbf{B}\|$, being the lengths of vectors, are positive. However, $\cos \theta$ can be positive, zero, or negative. Note that $\cos \theta = 0$ only when $\theta = \pi/2$, that is, when **A** and **B** are perpendicular. So the dot product provides a way of telling whether **A** and **B** are perpendicular:

A test for perpendicularity

> Let **A** and **B** be nonzero vectors. If $\mathbf{A} \cdot \mathbf{B} = 0$, then **A** and **B** are perpendicular. Conversely, if **A** and **B** are perpendicular, then $\mathbf{A} \cdot \mathbf{B} = 0$.

As we noted in the discussion following Example 2, **i**, **j**, and **k** have dot products of 0 with each other. Since they point in the directions of the coordinate axes, they are perpendicular to each other.

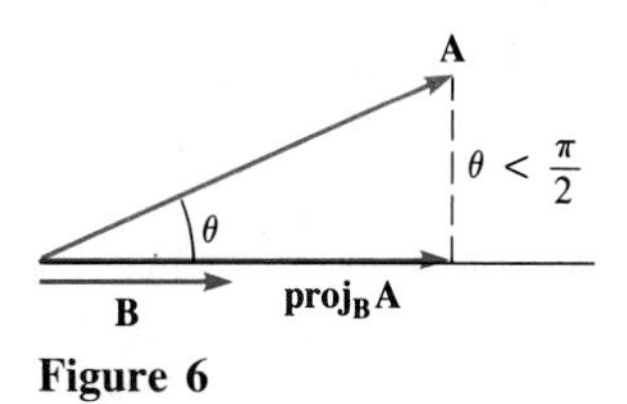

Figure 6

Now let us examine what it means for $\mathbf{A} \cdot \mathbf{B}$ to be positive or negative. If $\mathbf{A} \cdot \mathbf{B}$ is positive, then $\cos \theta$ is positive. This tells us that the angle between **A** and **B** is acute (less than $\pi/2$). Figure 6 shows this situation. As Fig. 6 illustrates, in this case the projection of **A** on **B** points in the *same* direction as **B**.

We summarize this for emphasis:

> Let **A** and **B** be vectors. If $\mathbf{A} \cdot \mathbf{B}$ is positive, then the angle between the vectors is less than $\pi/2$. In this case $\mathbf{proj_B}\ \mathbf{A}$ points in the same direction as **B**.

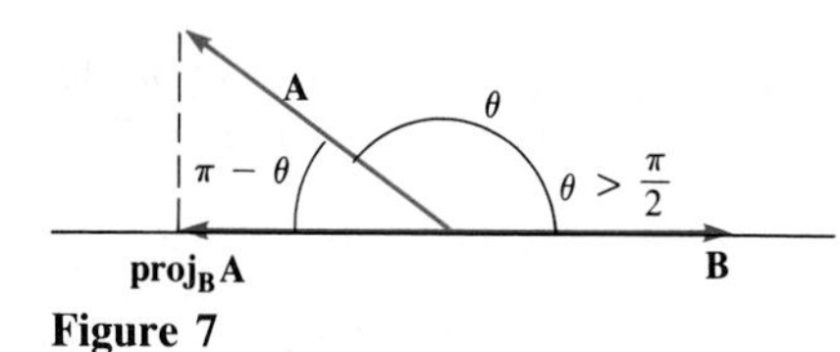

Figure 7

If $\mathbf{A} \cdot \mathbf{B}$ is negative, then the angle between **A** and **B** is obtuse (greater than $\pi/2$). Figure 7 shows this situation. As Fig. 7 illustrates, $\mathbf{proj_B}\ \mathbf{A}$ points in the direction opposite that of **B**.

> Let **A** and **B** be vectors. If $\mathbf{A} \cdot \mathbf{B}$ is negative, then the angle between the vectors is greater than $\pi/2$. In this case $\mathbf{proj_B}\ \mathbf{A}$ points in the direction opposite to that of **B**.

The Component of A on B

The quantity $\|\mathbf{A}\| \cos \theta$ that occurs in the definition of the dot product $\mathbf{A} \cdot \mathbf{B}$ is a scalar, which can be 0, positive, or negative. It is called the **component of A on B**, denoted $\text{comp}_{\mathbf{B}}\ \mathbf{A}$. If $\theta \le \pi/2$, $\text{comp}_{\mathbf{B}}\ \mathbf{A}$ is the length of the projection of **A** on **B**. But if $\theta > \pi/2$, $\text{comp}_{\mathbf{B}}\ \mathbf{A}$ is the negative of the length of $\mathbf{proj_B}\ \mathbf{A}$. In short:

$$\text{comp}_{\mathbf{B}}\,\mathbf{A} = \|\mathbf{A}\|\cos\theta = \pm\|\mathbf{proj}_{\mathbf{B}}\,\mathbf{A}\|$$

(+ if the angle θ between **A** and **B** is acute, − otherwise)

Note that $\mathbf{A}\cdot\mathbf{B} = (\text{comp}_{\mathbf{B}}\,\mathbf{A})\|\mathbf{B}\| = (\text{comp}_{\mathbf{A}}\,\mathbf{B})\|\mathbf{A}\|$.

Sometimes $\mathbf{proj}_{\mathbf{B}}\,\mathbf{A}$, which is a vector, is called the **vector component of A on B** to contrast it with $\text{comp}_{\mathbf{B}}\,\mathbf{A}$, which is just a number. (In such cases $\text{comp}_{\mathbf{B}}\,\mathbf{A}$ is usually called the **scalar component of A on B**, to emphasize the distinction.) Keep in mind that $\text{comp}_{\mathbf{B}}\,\mathbf{A}$ is a *number*, which is plus or minus the length of $\mathbf{proj}_{\mathbf{B}}\,\mathbf{A}$.

The components of $\mathbf{A} = x\mathbf{i} + y\mathbf{j} + z\mathbf{k}$ on **i**, **j**, and **k** are x, y, and z, respectively. (Why?) The projections of **A** on **i**, **j**, and **k** are $x\mathbf{i}$, $y\mathbf{j}$, and $z\mathbf{k}$, respectively.

EXAMPLE 3 Find the projection and component of $3\mathbf{i} + 2\mathbf{j}$ on **i** and on $-2\mathbf{i}$.

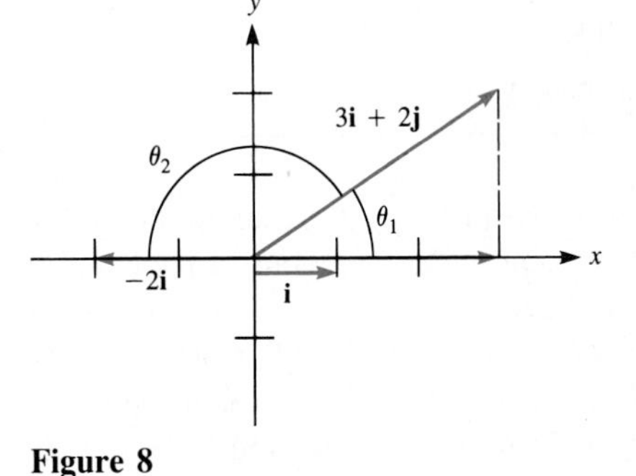

Figure 8

SOLUTION Figure 8 shows $3\mathbf{i} + 2\mathbf{j}$ and its projection on the x axis. Inspection of Fig. 8 shows that $\mathbf{proj}_{\mathbf{i}}(3\mathbf{i} + 2\mathbf{j}) = 3\mathbf{i}$ and that $\mathbf{proj}_{-2\mathbf{i}}(3\mathbf{i} + 2\mathbf{j}) = 3\mathbf{i}$. (The projection of **A** on **B** does not change when the direction of **B** is switched.)

Since θ_1, the angle between $3\mathbf{i} + 2\mathbf{j}$ and **i**, is less than $\pi/2$,

$$\begin{aligned}\text{comp}_{\mathbf{i}}(3\mathbf{i} + 2\mathbf{j}) &= +\|\mathbf{proj}_{\mathbf{i}}(3\mathbf{i} + 2\mathbf{j})\| \\ &= \|3\mathbf{i}\| \\ &= 3.\end{aligned}$$

However, the component of $3\mathbf{i} + 2\mathbf{j}$ on $-2\mathbf{i}$ involves the angle θ_2 in Fig. 8, which is larger than $\pi/2$. In this case

$$\begin{aligned}\text{comp}_{-2\mathbf{i}}(3\mathbf{i} + 2\mathbf{j}) &= -\|\mathbf{proj}_{-2\mathbf{i}}(3\mathbf{i} + 2\mathbf{j})\| \\ &= -\|3\mathbf{i}\| \\ &= -3.\end{aligned}$$

That -3 is negative tells us that $\mathbf{proj}_{-2\mathbf{i}}(3\mathbf{i} + 2\mathbf{j})$ and $-2\mathbf{i}$ point in opposite directions. (The component of **A** on **B** changes sign when the direction of **B** is switched.) ■

Components have a property that we will need in a moment. We present it in Theorem 1.

Theorem 1 Let **A**, **B**, and **C** be vectors, where **C** is not the zero vector. Then

$$\text{comp}_{\mathbf{C}}(\mathbf{A} + \mathbf{B}) = \text{comp}_{\mathbf{C}}\,\mathbf{A} + \text{comp}_{\mathbf{C}}\,\mathbf{B}. \tag{1}$$

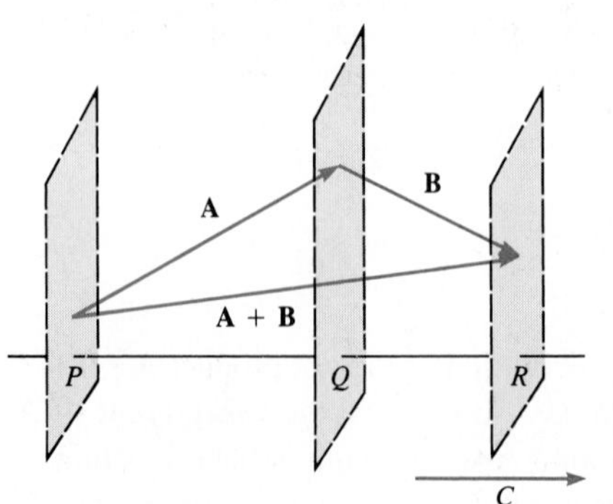

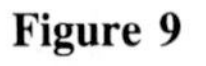
Figure 9

Proof Figure 9 illustrates the case when the three components in (1) are positive. In Fig. 9 the planes are perpendicular to **C** and pass through the tails and heads of **A** and **B**. P, Q, and R are the points where these planes meet a line parallel to **C**. Inspection of Fig. 9 shows that

$$\text{comp}_{\mathbf{C}}\,\mathbf{A} = \|\mathbf{proj}_{\mathbf{C}}\,\mathbf{A}\| = \overline{PQ}$$

$$\text{comp}_{\mathbf{C}}\,\mathbf{B} = \|\mathbf{proj}_{\mathbf{C}}\,\mathbf{B}\| = \overline{QR}$$

$$\text{comp}_{\mathbf{C}}(\mathbf{A}+\mathbf{B}) = \|\mathbf{proj}_{\mathbf{C}}(\mathbf{A}+\mathbf{B})\| = \overline{PR}.$$

Since $\overline{PQ} + \overline{QR} = \overline{PR}$, the theorem follows. ■

Theorem 1 implies the next theorem, which asserts that the dot product is distributive.

Theorem 2

$$(\mathbf{A}+\mathbf{B})\cdot\mathbf{C} = \mathbf{A}\cdot\mathbf{C} + \mathbf{B}\cdot\mathbf{C}. \tag{2}$$

$$\mathbf{C}\cdot(\mathbf{A}+\mathbf{B}) = \mathbf{C}\cdot\mathbf{A} + \mathbf{C}\cdot\mathbf{B}. \tag{3}$$

Proof We prove (2) by expressing the dot product in terms of components.

$$\begin{aligned}(\mathbf{A}+\mathbf{B})\cdot\mathbf{C} &= \text{comp}_{\mathbf{C}}(\mathbf{A}+\mathbf{B})\|\mathbf{C}\| \\ &= (\text{comp}_{\mathbf{C}}\,\mathbf{A} + \text{comp}_{\mathbf{C}}\,\mathbf{B})\|\mathbf{C}\| \qquad \text{(Theorem 1)} \\ &= (\text{comp}_{\mathbf{C}}\,\mathbf{A})\|\mathbf{C}\| + (\text{comp}_{\mathbf{C}}\,\mathbf{B})\|\mathbf{C}\| \\ &= \mathbf{A}\cdot\mathbf{C} + \mathbf{B}\cdot\mathbf{C}.\end{aligned}$$

The other distributive rule follows by the commutativity of the dot product:

$$\begin{aligned}\mathbf{C}\cdot(\mathbf{A}+\mathbf{B}) &= (\mathbf{A}+\mathbf{B})\cdot\mathbf{C} = \mathbf{A}\cdot\mathbf{C} + \mathbf{B}\cdot\mathbf{C} \\ &= \mathbf{C}\cdot\mathbf{A} + \mathbf{C}\cdot\mathbf{B}. \quad ■\end{aligned}$$

Computing A · B in Terms of i, j, k Components

Up to this point we have evaluated **A · B** only for fairly simple vectors. We will now obtain a formula for **A · B** in terms of the vectors' components. We state the formula, illustrate it by an example, then prove it.

Theorem 3 *Formula for dot product in terms of components.*

$$(x_1\mathbf{i} + y_1\mathbf{j} + z_1\mathbf{k})\cdot(x_2\mathbf{i} + y_2\mathbf{j} + z_2\mathbf{k}) = x_1x_2 + y_1y_2 + z_1z_2.$$

For plane vectors this reduces to

$$(x_1\mathbf{i} + y_1\mathbf{j})\cdot(x_2\mathbf{i} + y_2\mathbf{j}) = x_1x_2 + y_1y_2.$$

EXAMPLE 4 Find $(2\mathbf{i} + 3\mathbf{j} - 4\mathbf{k})\cdot(\mathbf{i} + 2\mathbf{j} + 3\mathbf{k})$.

SOLUTION By Theorem 3, the dot product is

$$\begin{aligned}2\cdot 1 + 3\cdot 2 + (-4)\cdot 3 &= 2 + 6 - 12 \\ &= -4. \quad ■\end{aligned}$$

Proof of Theorem 3 We prove the theorem in the case of plane vectors. The argument for vectors in space is similar, but longer. Using the distributive rules, we have

$$\begin{aligned}(x_1\mathbf{i} + y_1\mathbf{j}) \cdot (x_2\mathbf{i} + y_2\mathbf{j}) &= (x_1\mathbf{i} + y_1\mathbf{j}) \cdot x_2\mathbf{i} + (x_1\mathbf{i} + y_1\mathbf{j}) \cdot y_2\mathbf{j} \\ &= x_1\mathbf{i} \cdot x_2\mathbf{i} + y_1\mathbf{j} \cdot x_2\mathbf{i} + x_1\mathbf{i} \cdot y_2\mathbf{j} + y_1\mathbf{j} \cdot y_2\mathbf{j} \\ &= x_1x_2 + 0 + 0 + y_1y_2 \\ &= x_1x_2 + y_1y_2. \quad \blacksquare\end{aligned}$$

Finding the Angle between Two Vectors

Now that we can compute dot products of vectors in terms of their **i**, **j**, and **k** components, we can easily test whether two vectors are perpendicular. All we do is check whether their dot product is 0.

EXAMPLE 5 Show that the vectors $\mathbf{A} = 2\mathbf{i} + \mathbf{j} - 3\mathbf{k}$ and $\mathbf{B} = 4\mathbf{i} - 5\mathbf{j} + \mathbf{k}$ are perpendicular.

SOLUTION

$$\begin{aligned}\mathbf{A} \cdot \mathbf{B} &= 2 \cdot 4 + (1)(-5) + (-3)(1) \\ &= 8 - 5 - 3 = 0.\end{aligned}$$

Therefore **A** and **B** are perpendicular. ■

EXAMPLE 6 Are the vectors $3\mathbf{i} + 7\mathbf{j}$ and $9\mathbf{i} - 4\mathbf{j}$ perpendicular? (Sketch them before reading the solution; do they seem perpendicular?)

SOLUTION Two nonzero vectors are perpendicular if and only if their dot product is 0. The product in this case is

$$\begin{aligned}(3\mathbf{i} + 7\mathbf{j}) \cdot (9\mathbf{i} - 4\mathbf{j}) &= 3 \cdot 9 + 7(-4) \\ &= 27 - 28 = -1.\end{aligned}$$

Since the dot product is *not* 0, the vectors are not perpendicular. ■

Not only can the dot product tell us when the angle between two vectors is $\pi/2$, but it can also help us find the angle between two vectors in general.

To see why, consider the definition of the dot product:

$$\mathbf{A} \cdot \mathbf{B} = \|\mathbf{A}\| \, \|\mathbf{B}\| \cos \theta.$$

(**A** and **B** are nonzero vectors.) Solving for $\cos \theta$, we obtain

$$\boxed{\cos \theta = \frac{\mathbf{A} \cdot \mathbf{B}}{\|\mathbf{A}\| \, \|\mathbf{B}\|}.} \tag{4}$$

Equation (4) tells us how to find the cosine of the angle between two vectors. With the aid of a calculator, we then can find the angle itself. Note that if $\cos \theta > 0$, then $0 < \theta < \pi/2$, and when $\cos \theta < 0$, then $\pi/2 < \theta \leq \pi$.

EXAMPLE 7 Find the angle θ between the two vectors $\mathbf{A} = 2\mathbf{i} - \mathbf{j} + 3\mathbf{k}$ and $\mathbf{B} = \mathbf{i} + \mathbf{j} + 2\mathbf{k}$.

SOLUTION

$$\cos\theta = \frac{(2\mathbf{i} - \mathbf{j} + 3\mathbf{k}) \cdot (\mathbf{i} + \mathbf{j} + 2\mathbf{k})}{\|2\mathbf{i} - \mathbf{j} + 3\mathbf{k}\| \, \|\mathbf{i} + \mathbf{j} + 2\mathbf{k}\|}$$

$$= \frac{2 \cdot 1 + (-1) \cdot 1 + 3 \cdot 2}{\sqrt{2^2 + (-1)^2 + 3^2}\,\sqrt{1^2 + 1^2 + 2^2}}$$

$$= \frac{2 - 1 + 6}{\sqrt{14}\,\sqrt{6}}$$

$$= \frac{7}{\sqrt{84}} \approx 0.764.$$

A calculator or trigonometric table shows that θ is about 40.2° or 0.702 radian. ■

Computing $\mathbf{proj_B\,A}$ and $\mathbf{comp_B\,A}$

In Sec. 12.2 we were able to find $\mathbf{proj_B\,A}$ only for the simplest cases. Now, with the help of the dot product, we will be able to compute $\mathbf{proj_B\,A}$ in general, without having to draw the vectors.

First of all, recall that $\mathbf{proj_B\,A}$ does not depend on the length of $\mathbf{B}$. It will be convenient to assume that $\mathbf{B}$ is a unit vector, which we denote $\mathbf{u}$. (There are two unit vectors parallel to $\mathbf{B}$, namely $\mathbf{B}/\|\mathbf{B}\|$ and $-\mathbf{B}/\|\mathbf{B}\|$.)

Let us find the projection of a vector $\mathbf{A}$ on a unit vector $\mathbf{u}$. Figure 10 shows the situation when the angle between $\mathbf{A}$ and $\mathbf{u}$ is less than $\pi/2$.

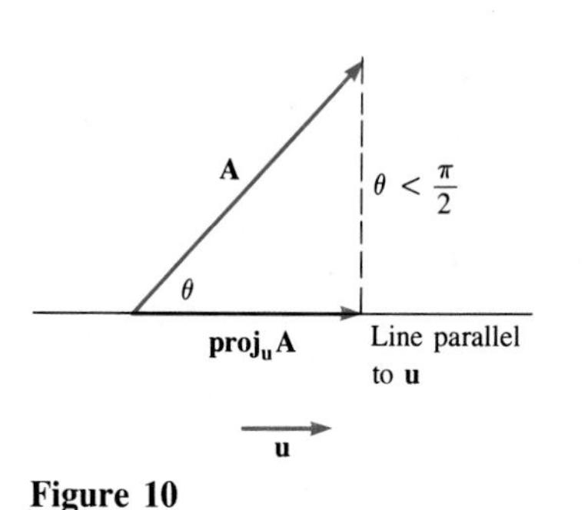

Figure 10

The length of $\mathbf{proj_u\,A}$ is equal to $\|\mathbf{A}\| \cos\theta$, since $\theta < \pi/2$. Moreover, $\mathbf{proj_u\,A}$ points in the same direction as $\mathbf{u}$. Thus

$$\mathbf{proj_u\,A} = \|\mathbf{A}\| \, (\cos\theta)\mathbf{u}. \tag{5}$$

On the other hand,

$$\mathbf{A} \cdot \mathbf{u} = \|\mathbf{A}\| \, \|\mathbf{u}\| \cos\theta,$$

hence

$$\mathbf{A} \cdot \mathbf{u} = \|\mathbf{A}\| \cos\theta, \tag{6}$$

since $\mathbf{u}$ is a unit vector. Combining (5) and (6) tells us that

(This formula also holds when the angle between **A** *and* **u** *is greater than* $\pi/2$.*)*

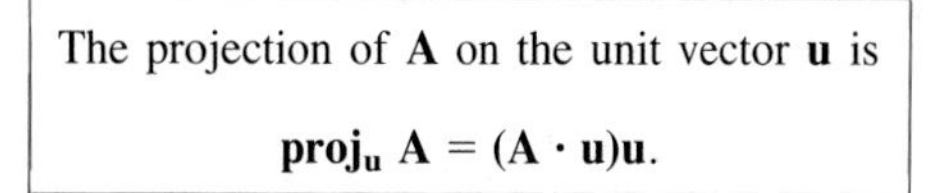
The projection of **A** on the unit vector **u** is

$$\mathbf{proj_u\,A} = (\mathbf{A} \cdot \mathbf{u})\mathbf{u}.$$

The projection of **A** on any nonzero vector **B** is the same as the projection of **A** on the unit vector $\mathbf{B}/\|\mathbf{B}\|$. Thus

$$\mathbf{proj_B\,A} = \left(\mathbf{A} \cdot \frac{\mathbf{B}}{\|\mathbf{B}\|}\right)\frac{\mathbf{B}}{\|\mathbf{B}\|}.$$

Note also that

> The component of **A** on the unit vector **u** is
>
> $$\text{comp}_{\mathbf{u}}\, \mathbf{A} = \mathbf{A} \cdot \mathbf{u}.$$

The component of **A** on any nonzero vector **B** is therefore

$$\text{comp}_{\mathbf{B}}\, \mathbf{A} = \mathbf{A} \cdot \frac{\mathbf{B}}{\|\mathbf{B}\|}.$$

EXAMPLE 8 Find the component and projection of $2\mathbf{i} + 3\mathbf{j}$ on $\mathbf{i} - 2\mathbf{j}$.

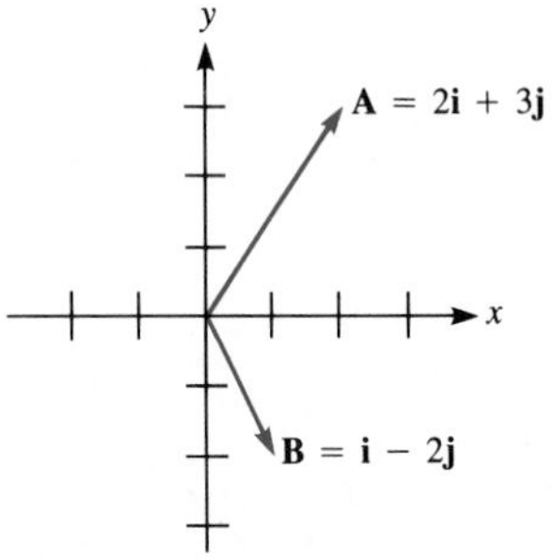

SOLUTION In this case $\mathbf{A} = 2\mathbf{i} + 3\mathbf{j}$ and $\mathbf{B} = \mathbf{i} - 2\mathbf{j}$. (See Fig. 11.)
The unit vector in the direction of **B** is

$$\mathbf{u} = \frac{\mathbf{B}}{\|\mathbf{B}\|} = \frac{\mathbf{i} - 2\mathbf{j}}{\sqrt{1^2 + (-2)^2}} = \frac{\mathbf{i} - 2\mathbf{j}}{\sqrt{5}}.$$

y
A = 2i + 3j
x
B = i − 2j

Figure 11

Therefore the component of **A** on **B** is

$$\begin{aligned}
\text{comp}_{\mathbf{B}}\, \mathbf{A} &= \text{comp}_{\mathbf{u}}\, \mathbf{A} \\
&= \text{comp}\left(2\mathbf{i} + 3\mathbf{j} \text{ on } \frac{\mathbf{i} - 2\mathbf{j}}{\sqrt{5}}\right) \\
&= (2\mathbf{i} + 3\mathbf{j}) \cdot \left(\frac{\mathbf{i}}{\sqrt{5}} - \frac{2\mathbf{j}}{\sqrt{5}}\right) \\
&= \frac{2}{\sqrt{5}} - \frac{6}{\sqrt{5}} \\
&= -\frac{4}{\sqrt{5}}.
\end{aligned}$$

(That the component is negative shows that the angle between $2\mathbf{i} + 3\mathbf{j}$ and $\mathbf{i} - 2\mathbf{j}$ is obtuse, which agrees with Fig. 11.)
Finally,

$$\begin{aligned}
\mathbf{proj}_{\mathbf{B}}\, \mathbf{A} &= \mathbf{proj}_{\mathbf{u}}\, \mathbf{A} \\
&= (\text{comp}_{\mathbf{u}}\, \mathbf{A})\mathbf{u} \\
&= -\frac{4}{\sqrt{5}}\left(\frac{\mathbf{i}}{\sqrt{5}} - \frac{2\mathbf{j}}{\sqrt{5}}\right) \\
&= -\frac{4\mathbf{i}}{5} + \frac{8\mathbf{j}}{5},
\end{aligned}$$

which is shown in Fig. 12. ■

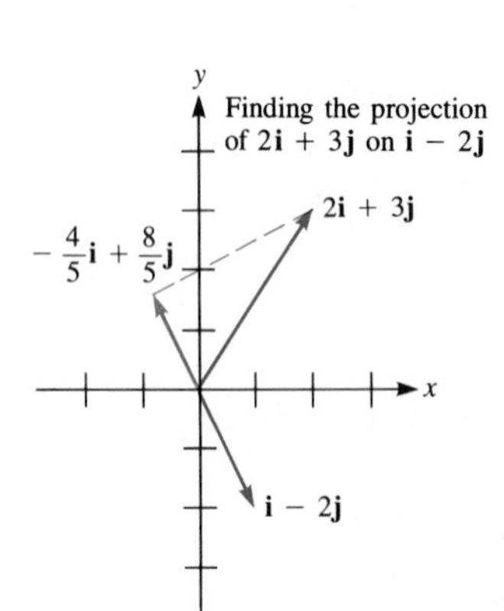

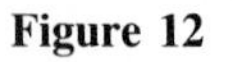
Figure 12

EXAMPLE 9 A line segment joins the points $P = (1, 2, 3)$ and $Q = (3, 1, 4)$. What is the length of its projection on the line L through $R = (5, 2, 7)$ and $S = (-1, 2, 3)$?

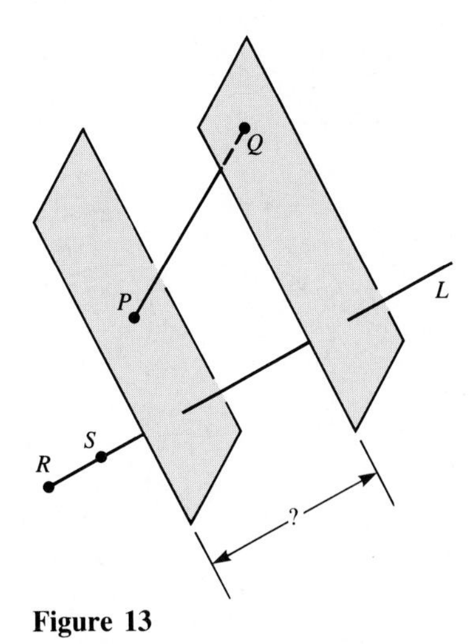

Figure 13

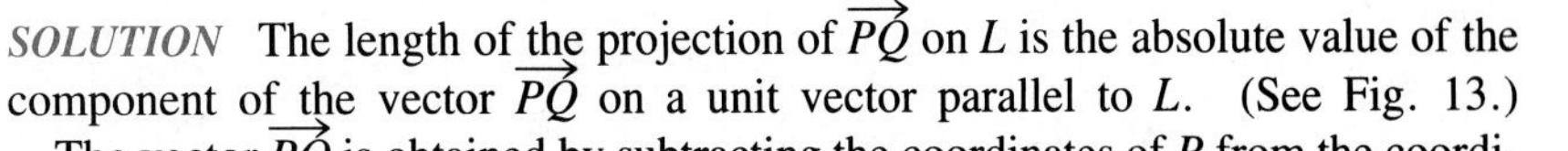

SOLUTION The length of the projection of $\overrightarrow{PQ}$ on L is the absolute value of the component of the vector $\overrightarrow{PQ}$ on a unit vector parallel to L. (See Fig. 13.)

The vector $\overrightarrow{PQ}$ is obtained by subtracting the coordinates of P from the coordinates of Q:

$$\begin{aligned}\overrightarrow{PQ} &= \langle 3-1, 1-2, 4-3\rangle \\ &= \langle 2, -1, 1\rangle \\ &= 2\mathbf{i} - \mathbf{j} + \mathbf{k}.\end{aligned}$$

To obtain a unit vector parallel to L, we first obtain a vector parallel to L:

$$\begin{aligned}\overrightarrow{RS} &= \langle -1-5, 2-2, 3-7\rangle \\ &= \langle -6, 0, -4\rangle \\ &= -6\mathbf{i} + 0\mathbf{j} - 4\mathbf{k}.\end{aligned}$$

A unit vector parallel to L is

$$\begin{aligned}\mathbf{u} = \frac{\overrightarrow{RS}}{\|\overrightarrow{RS}\|} &= \frac{-6\mathbf{i} + 0\mathbf{j} - 4\mathbf{k}}{\sqrt{(-6)^2 + 0^2 + (-4)^2}} \\ &= \frac{-6\mathbf{i} + 0\mathbf{j} - 4\mathbf{k}}{\sqrt{52}}.\end{aligned}$$

The length of $\mathbf{proj}_L\, \overrightarrow{PQ}$ is therefore

$$\begin{aligned}|\text{comp}_{\mathbf{u}}\, \overrightarrow{PQ}| &= |\overrightarrow{PQ} \cdot \mathbf{u}| \\ &= \left|(2\mathbf{i} - \mathbf{j} + \mathbf{k}) \cdot \left(\frac{-6\mathbf{i} + 0\mathbf{j} - 4\mathbf{k}}{\sqrt{52}}\right)\right| \\ &= \left|\frac{-12 + 0 - 4}{\sqrt{52}}\right| \\ &= \frac{16}{\sqrt{52}}. \quad \blacksquare\end{aligned}$$

In Sec. 12.2 it was shown that if $\mathbf{A}$ is a vector and $\mathbf{B}$ is a nonzero vector, then $\mathbf{A}$ is the sum of a vector parallel to $\mathbf{B}$ and a vector perpendicular to $\mathbf{B}$:

$$\mathbf{A} = \underbrace{\mathbf{proj_B\, A}}_{\text{parallel to } \mathbf{B}} + \underbrace{\mathbf{orth_B\, A}}_{\text{perpendicular to } \mathbf{B}}.$$

Computing $\mathbf{orth_B}$ A

Since we now can compute $\mathbf{proj_B\, A}$, we can also compute $\mathbf{orth_B\, A}$ for it equals $\mathbf{A} - \mathbf{proj_B\, A}$. The next example illustrates this observation.

EXAMPLE 10 Let $\mathbf{A} = 2\mathbf{i} + 3\mathbf{j} - \mathbf{k}$ and $\mathbf{B} = 8\mathbf{i} - 4\mathbf{j} + \mathbf{k}$. Find $\mathbf{proj_B\, A}$ and $\mathbf{orth_B\, A}$.

SOLUTION For convenience replace $\mathbf{B}$ by the unit vector $\mathbf{u} = \mathbf{B}/\|\mathbf{B}\|$:

$$\mathbf{u} = \frac{8\mathbf{i} - 4\mathbf{j} + \mathbf{k}}{\sqrt{8^2 + (-4)^2 + 1^2}} = \frac{8\mathbf{i} - 4\mathbf{j} + \mathbf{k}}{\sqrt{81}} = \frac{8\mathbf{i} - 4\mathbf{j} + \mathbf{k}}{9}.$$

Hence

$$\mathbf{proj_B\, A} = \mathbf{proj_u\, A} = (\mathbf{A} \cdot \mathbf{u})\mathbf{u}$$

$$\begin{aligned}\mathbf{proj_B A} &= (2\mathbf{i} + 3\mathbf{j} - \mathbf{k}) \cdot \frac{(8\mathbf{i} - 4\mathbf{j} + \mathbf{k})}{9}\mathbf{u} \\ &= \frac{16 - 12 - 1}{9}\mathbf{u} \\ &= \frac{3}{9}\,\frac{8\mathbf{i} - 4\mathbf{j} + \mathbf{k}}{9} \\ &= \frac{8\mathbf{i} - 4\mathbf{j} + \mathbf{k}}{27}.\end{aligned}$$

Thus
$$\begin{aligned}\mathbf{orth_B\ A} = \mathbf{orth_u\ A} &= \mathbf{A} - \mathbf{proj_u\ A} \\ &= (2\mathbf{i} + 3\mathbf{j} - \mathbf{k}) - \frac{8\mathbf{i} - 4\mathbf{j} + \mathbf{k}}{27} \\ &= \frac{46}{27}\mathbf{i} + \frac{85}{27}\mathbf{j} - \frac{28}{27}\mathbf{k}.\end{aligned}$$

(Check: See whether the dot product of $\mathbf{proj_B\ A}$ and $\mathbf{orth_B\ A}$ is 0.) ■

THE DOT PRODUCT IN BUSINESS AND STATISTICS

Imagine that a fast food restaurant sells 30 hamburgers, 20 salads, 15 soft drinks, and 13 orders of french fries. This is recorded by the four-dimensional "vector" $\langle 30, 20, 15, 13\rangle$. A hamburger sells for \$1.99, a salad for \$1.50, a soft drink for \$1.00, and an order of french fries for \$1.10. The "price vector" is $\langle 1.99, 1.50, 1.00, 1.10\rangle$. The dot product of these two vectors, $30(1.99) + 20(1.50) + 15(1.00) + 13(1.10)$, would be the total amount paid for all items. Descriptions of the economy use "production vectors," "cost vectors," "price vectors," and "profit vectors" with many more than the four components of our restaurant example.

In statistics the coefficient of correlation is defined in terms of a dot product. For instance, you may determine the height and weight of n persons. Let the height of the ith person be h_i and the weight be w_i. Let h be the average of the n heights and w be the average of the n weights. Let $\mathbf{H} = \langle h_1 - h, h_2 - h, \ldots, h_n - h\rangle$ and $\mathbf{W} = \langle w_1 - w, w_2 - w, \ldots, w_n - w\rangle$. Then the coefficient of correlation between the heights and weights is defined to be

$$\frac{\mathbf{H} \cdot \mathbf{W}}{\|\mathbf{H}\|\,\|\mathbf{W}\|}.$$

In analogy with vectors in the plane or space,

$$\mathbf{H} \cdot \mathbf{W} = \sum_{i=1}^{n} (h_i - h)(w_i - w), \qquad \|\mathbf{H}\| = \sqrt{\sum_{i=1}^{n} (h_i - h)^2},$$
$$\|\mathbf{W}\| = \sqrt{\sum_{i=1}^{n} (w_i - w)^2}.$$

It turns out that the coefficient of correlation is simply the cosine of the angle between the points $\langle h_1 - h, h_2 - h, \ldots, h_n - h\rangle$ and $\langle w_1 - w, w_2 - w, \ldots, w_n - w\rangle$ in n-dimensional space.

EXAMPLE 11 Let $PQRS$ be a parallelogram. Show that if the two diagonals of the parallelogram are perpendicular, then the four sides have the same length (forming a rhombus).

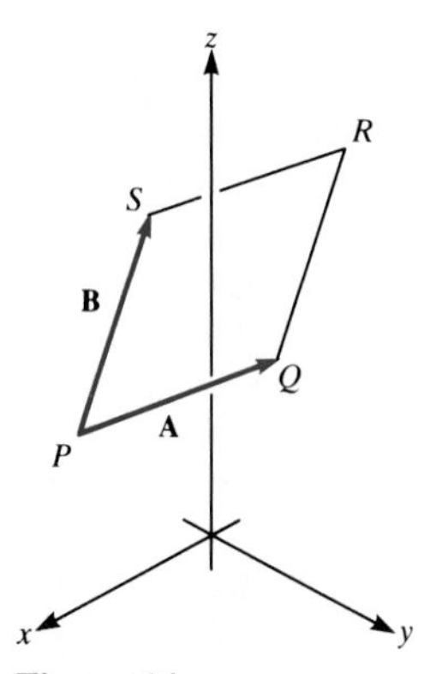

Figure 14

SOLUTION Let $\mathbf{A} = \overrightarrow{PQ}$ and $\mathbf{B} = \overrightarrow{PS}$, as shown in Fig. 14. One diagonal is $\mathbf{A} + \mathbf{B}$ and the other is $\mathbf{A} - \mathbf{B}$. (Draw them.) If these diagonals are perpendicular, then

$$(\mathbf{A} + \mathbf{B}) \cdot (\mathbf{A} - \mathbf{B}) = 0$$

or

$$\mathbf{A} \cdot \mathbf{A} + \mathbf{B} \cdot \mathbf{A} - \mathbf{A} \cdot \mathbf{B} - \mathbf{B} \cdot \mathbf{B} = 0 \qquad \text{(expanding)}$$

$$\mathbf{A} \cdot \mathbf{A} - \mathbf{B} \cdot \mathbf{B} = 0 \qquad (\mathbf{A} \cdot \mathbf{B} = \mathbf{B} \cdot \mathbf{A})$$

$$\mathbf{A} \cdot \mathbf{A} = \mathbf{B} \cdot \mathbf{B}$$

$$\|\mathbf{A}\|^2 = \|\mathbf{B}\|^2$$

$$\|\mathbf{A}\| = \|\mathbf{B}\|.$$

This shows that the sides PQ and PS are of equal length. Hence all the sides are of the same length. ■

Section Summary

We have defined the dot product and developed several of its properties, which we list for convenient reference:

$$\mathbf{A} \cdot \mathbf{B} = \|\mathbf{A}\|\,\|\mathbf{B}\| \cos\theta \qquad \text{(definition)}$$

$$= (\text{comp}_{\mathbf{B}}\, \mathbf{A})\|\mathbf{B}\|$$

$$= (\text{comp}_{\mathbf{A}}\, \mathbf{B})\|\mathbf{A}\|$$

$\mathbf{A} \cdot \mathbf{B} = 0$ is a test for perpendicularity.

$$\mathbf{A} \cdot \mathbf{A} = \|\mathbf{A}\|^2$$

$$\mathbf{A} \cdot \mathbf{B} = \mathbf{B} \cdot \mathbf{A} \qquad \text{(commutativity)}$$

$$\mathbf{A} \cdot (\mathbf{B} + \mathbf{C}) = \mathbf{A} \cdot \mathbf{B} + \mathbf{A} \cdot \mathbf{C}. \qquad \text{(distributivity)}$$

$$(x_1\mathbf{i} + y_1\mathbf{j} + z_1\mathbf{k}) \cdot (x_2\mathbf{i} + y_2\mathbf{j} + z_2\mathbf{k}) = x_1x_2 + y_1y_2 + z_1z_2.$$

For a unit vector $\mathbf{u}$

$$\text{comp}_{\mathbf{u}}\, \mathbf{A} = \mathbf{A} \cdot \mathbf{u} \qquad \text{and} \qquad \textbf{proj}_{\mathbf{u}}\, \mathbf{A} = (\mathbf{A} \cdot \mathbf{u})\mathbf{u}.$$

To find $\text{comp}_{\mathbf{B}}\, \mathbf{A}$ and $\textbf{proj}_{\mathbf{B}}\, \mathbf{A}$, replace $\mathbf{B}$ by the unit vector $\mathbf{B}/\|\mathbf{B}\|$. While $\textbf{proj}_{\mathbf{B}}\, \mathbf{A}$ is a *vector*, $\text{comp}_{\mathbf{B}}\, \mathbf{A}$ is a *number*. They are related, since $\text{comp}_{\mathbf{B}}\, \mathbf{A} = \pm\|\textbf{proj}_{\mathbf{B}}\, \mathbf{A}\|$ (positive if $\textbf{proj}_{\mathbf{B}}\, \mathbf{A}$ and $\mathbf{B}$ point in the same direction, negative if they point in opposite directions).

Three main uses of the dot product:

1 To test whether two vectors are perpendicular ($\mathbf{A} \cdot \mathbf{B} = 0$)

2 To find the angle between vectors:

$$\cos\theta = \frac{\mathbf{A} \cdot \mathbf{B}}{\|\mathbf{A}\|\,\|\mathbf{B}\|}$$

3 To compute projections and components

EXERCISES FOR SEC. 12.3: THE DOT PRODUCT OF TWO VECTORS

In Exercises 1 to 4 compute $\mathbf{A} \cdot \mathbf{B}$.

1 $\mathbf{A}$ has length 3, $\mathbf{B}$ has length 4, and the angle between $\mathbf{A}$ and $\mathbf{B}$ is $\pi/4$.

2 $\mathbf{A}$ has length 2, $\mathbf{B}$ has length 3, and the angle between $\mathbf{A}$ and $\mathbf{B}$ is $3\pi/4$.

3 $\mathbf{A}$ has length 5, $\mathbf{B}$ has length $\frac{1}{2}$, and the angle between $\mathbf{A}$ and $\mathbf{B}$ is $\pi/2$.

4 $\mathbf{A}$ is the zero vector $\mathbf{0}$, and $\mathbf{B}$ has length 5.

In Exercises 5 to 8 compute $\mathbf{A} \cdot \mathbf{B}$ using the formula in terms of components.

5 $\mathbf{A} = -2\mathbf{i} + 3\mathbf{j}$, $\mathbf{B} = 3\mathbf{i} + 4\mathbf{j}$

6 $\mathbf{A} = 0.3\mathbf{i} + 0.5\mathbf{j}$, $\mathbf{B} = 2\mathbf{i} - 1.5\mathbf{j}$

7 $\mathbf{A} = 2\mathbf{i} - 3\mathbf{j} - \mathbf{k}$, $\mathbf{B} = 3\mathbf{i} + 4\mathbf{j} - \mathbf{k}$

8 $\mathbf{A} = \mathbf{i} + \mathbf{j} + \mathbf{k}$, $\mathbf{B} = 2\mathbf{i} + 3\mathbf{j} - 5\mathbf{k}$

9 (*a*) Draw the vectors $7\mathbf{i} + 12\mathbf{j}$ and $9\mathbf{i} - 5\mathbf{j}$.
(*b*) Do they seem to be perpendicular?
(*c*) Determine whether they are perpendicular by examining their dot product.

10 (*a*) Draw the vectors $\mathbf{i} + 2\mathbf{j} + 3\mathbf{k}$ and $\mathbf{i} + \mathbf{j} - \mathbf{k}$.
(*b*) Do they seem to be perpendicular?
(*c*) Determine whether they are perpendicular by examining their dot product.

11 (*a*) Estimate the angle between $\mathbf{A} = 3\mathbf{i} + 4\mathbf{j}$ and $\mathbf{B} = 5\mathbf{i} + 12\mathbf{j}$ by drawing them.
(*b*) Find the angle between $\mathbf{A}$ and $\mathbf{B}$.

12 Let $P = (6, 1)$, $Q = (3, 2)$, $R = (1, 3)$, and $S = (4, 5)$.
(*a*) Draw the vectors $\overrightarrow{PQ}$ and $\overrightarrow{RS}$.
(*b*) Using the diagram in (*a*) estimate the angle between $\overrightarrow{PQ}$ and $\overrightarrow{RS}$.
(*c*) Using the dot product, find the cosine of the angle between $\overrightarrow{PQ}$ and $\overrightarrow{RS}$.
(*d*) Using (*c*) and a calculator, find the angle in (*b*).

13 Find the angle between $2\mathbf{i} - 4\mathbf{j} + 6\mathbf{k}$ and $\mathbf{i} + 2\mathbf{j} + 3\mathbf{k}$.

14 Find the angle between $\mathbf{i} + \mathbf{j} + 3\mathbf{k}$ and $3\mathbf{i} + 6\mathbf{j} - 3\mathbf{k}$.

15 Find the angle between $\overrightarrow{AB}$ and $\overrightarrow{CD}$ if $A = (1, 3)$, $B = (7, 4)$, $C = (2, 8)$, and $D = (1, -5)$.

16 Find the angle between $\overrightarrow{AB}$ and $\overrightarrow{CD}$ if $A = (1, 2, -5)$, $B = (1, 0, 1)$, $C = (0, -1, 3)$, and $D = (2, 1, 4)$.

17 Find the component and projection of $3\mathbf{i} - 4\mathbf{j}$ on $\mathbf{i} + 2\mathbf{j}$.

18 Find the component and projection of $2\mathbf{i} - 4\mathbf{j} + 5\mathbf{k}$ on $3\mathbf{i} - \mathbf{j} - \sqrt{2}\mathbf{k}$.

19 Find the vectors $\mathbf{C}$ and $\mathbf{D}$ in Fig. 15.

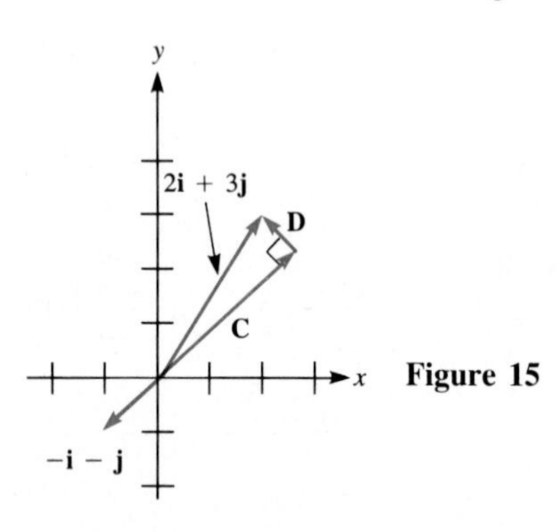

Figure 15

20 Find the length of the projection of $-4\mathbf{i} + 5\mathbf{j}$ on the line through $(2, -1)$ and $(6, 1)$
(*a*) By making a drawing and estimating the length by eye.
(*b*) By using the dot product.

21 (*a*) Find a vector $\mathbf{C}$ parallel to $\mathbf{i} + 2\mathbf{j}$ and a vector $\mathbf{D}$ perpendicular to $\mathbf{i} + 2\mathbf{j}$ such that $-3\mathbf{i} + 4\mathbf{j} = \mathbf{C} + \mathbf{D}$.
(*b*) Draw the vectors in (*a*) to check that your answer is reasonable.

22 (*a*) Find a vector $\mathbf{C}$ parallel to $2\mathbf{i} - \mathbf{j}$ and a vector $\mathbf{D}$ perpendicular to $2\mathbf{i} - \mathbf{j}$ such that $3\mathbf{i} + 4\mathbf{j} = \mathbf{C} + \mathbf{D}$.
(*b*) Draw the vectors in (*a*) to check that your answer is reasonable.

23 If $\text{comp}_{\mathbf{B}}\, \mathbf{A}$ is 15, what is the component of $\mathbf{A}$ on (*a*) $2\mathbf{B}$, (*b*) $-\mathbf{B}$?

24 The projection of the vector $\mathbf{A}$ on the vector $\mathbf{B}$ is the vector $\mathbf{i} - \mathbf{j} + 4\mathbf{k}$. What is the projection of the vector $\mathbf{A}$ on (*a*) $2\mathbf{B}$, (*b*) $-\mathbf{B}$?

25 For which nonzero vectors is $\mathbf{proj}_{\mathbf{B}}\, \mathbf{A}$ equal to $\mathbf{proj}_{\mathbf{A}}\, \mathbf{B}$?

26 For which nonzero vectors is $\text{comp}_{\mathbf{B}}\, \mathbf{A}$ equal to $\text{comp}_{\mathbf{A}}\, \mathbf{B}$?

27 Find the component of $3\mathbf{i} - \mathbf{j} + 2\mathbf{k}$ on (*a*) $3\mathbf{j} + 2\mathbf{k}$, (*b*) $-3\mathbf{j} - 2\mathbf{k}$.

28 Find the component of $3\mathbf{i} + 4\mathbf{j} + 2\mathbf{k}$ on (*a*) $\mathbf{i} + \mathbf{j} + 3\mathbf{k}$, (*b*) $-\mathbf{i} - \mathbf{j} - 3\mathbf{k}$, (*c*) $2\mathbf{i} + 2\mathbf{j} + 6\mathbf{k}$.

In Exercises 29 and 30 find and draw $\mathbf{proj}_{\mathbf{B}}\, \mathbf{A}$ and $\mathbf{orth}_{\mathbf{B}}\, \mathbf{A}$.

29 $\mathbf{A} = 3\mathbf{i} + 2\mathbf{j}$, $\mathbf{B} = 4\mathbf{i} - 3\mathbf{j}$.

30 $\mathbf{A} = \mathbf{i} + 2\mathbf{j} + 3\mathbf{k}$, $\mathbf{B} = \mathbf{i} + \mathbf{j}$.

31 How far is the point $(1, 5)$ from the line through $(4, 2)$ and $(3, 7)$? *Hint:* Draw a picture and think in terms of vectors.

32 How far is the point $(1, 2, -3)$ from the line through $(2, 1, 4)$ and $(1, 5, -2)$?

33 Give an example of a vector in the xy plane that is perpendicular to $3\mathbf{i} - 2\mathbf{j}$.

34 Give an example of a vector that is perpendicular to $5\mathbf{i} - 3\mathbf{j} + 4\mathbf{k}$.

Exercises 35 to 39 refer to the cube in Fig. 16.

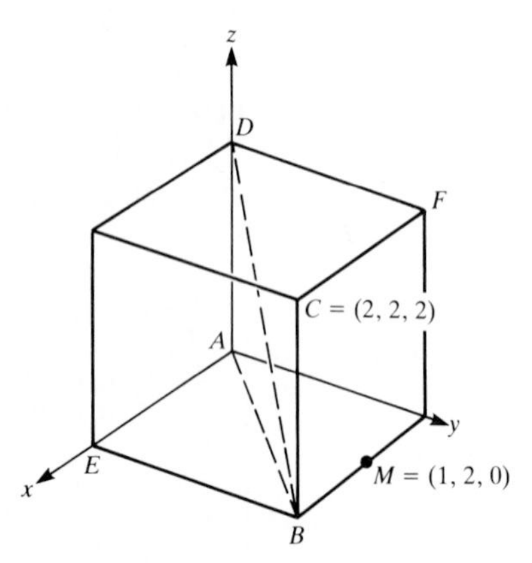

Figure 16

35 Find the cosine of the angle between $\overrightarrow{AC}$ and $\overrightarrow{BD}$.
36 Find the cosine of the angle between $\overrightarrow{AF}$ and $\overrightarrow{BD}$.
37 Find the cosine of the angle between $\overrightarrow{AC}$ and $\overrightarrow{AM}$.
38 Find the cosine of the angle between $\overrightarrow{MD}$ and $\overrightarrow{MF}$.
39 Find the cosine of the angle between $\overrightarrow{EF}$ and $\overrightarrow{BD}$.
40 How far is the point (1, 2, 3) from the line through the points (1, 4, 2) and (2, 1, −4)?

---

41 If $\mathbf{A} \cdot \mathbf{B} = \mathbf{A} \cdot \mathbf{C}$ and $\mathbf{A}$ is not $\mathbf{0}$, must $\mathbf{B} = \mathbf{C}$?
42 By considering the dot product of the two unit vectors $\mathbf{u}_1 = \cos\theta_1\,\mathbf{i} + \sin\theta_1\,\mathbf{j}$ and $\mathbf{u}_2 = \cos\theta_2\,\mathbf{i} + \sin\theta_2\,\mathbf{j}$, prove that

$$\cos(\theta_1 - \theta_2) = \cos\theta_1\cos\theta_2 + \sin\theta_1\sin\theta_2.$$

43 Prove Theorem 3 for the two vectors in space, $\mathbf{A} = x_1\mathbf{i} + y_1\mathbf{j} + z_1\mathbf{k}$ and $\mathbf{B} = x_2\mathbf{i} + y_2\mathbf{j} + z_2\mathbf{k}$.
44 Consider a tetrahedron (not necessarily regular). It has six edges. Show that the line segment joining the midpoints of two opposite edges is perpendicular to the line segment joining another pair of opposite edges if and only if the remaining two edges are of the same length.
45 Let $\mathbf{u}_1$, $\mathbf{u}_2$, and $\mathbf{u}_3$ be unit vectors such that each two are perpendicular. Let $\mathbf{A}$ be a vector.
(*a*) Draw a picture that shows that there are scalars x, y, and z such that $\mathbf{A} = x\mathbf{u}_1 + y\mathbf{u}_2 + z\mathbf{u}_3$.
(*b*) Express x as a dot product.
46 (*a*) Let $\mathbf{A}$ be a vector in the xy plane and $\mathbf{u}_1$ and $\mathbf{u}_2$ perpendicular unit vectors in that plane. If $\mathbf{A} \cdot \mathbf{u}_1 = 0$ and $\mathbf{A} \cdot \mathbf{u}_2 = 0$, must $\mathbf{A} = \mathbf{0}$?
(*b*) Let $\mathbf{v}_1$ and $\mathbf{v}_2$ be nonparallel unit vectors in the xy plane. If $\mathbf{A} \cdot \mathbf{v}_1 = 0$ and $\mathbf{A} \cdot \mathbf{v}_2 = 0$, must $\mathbf{A} = \mathbf{0}$?
47 A firm sells x chairs at C dollars per chair and y desks at D dollars per desk. It costs the firm c dollars to make a chair and d dollars to make a desk. What is the economic interpretation of
(*a*) Cx?
(*b*) $(x\mathbf{i} + y\mathbf{j}) \cdot (C\mathbf{i} + D\mathbf{j})$?
(*c*) $(x\mathbf{i} + y\mathbf{j}) \cdot (c\mathbf{i} + d\mathbf{j})$?
(*d*) $(x\mathbf{i} + y\mathbf{j}) \cdot (C\mathbf{i} + D\mathbf{j}) > (x\mathbf{i} + y\mathbf{j}) \cdot (c\mathbf{i} + d\mathbf{j})$?
48 The output of a firm that manufactures x_1 washing machines, x_2 refrigerators, x_3 dishwashers, x_4 stoves, and x_5 clothes dryers is recorded by the five-dimensional production vector $\mathbf{P} = \langle x_1, x_2, x_3, x_4, x_5\rangle$. Similarly, the cost vector $\mathbf{C} = \langle y_1, y_2, y_3, y_4, y_5\rangle$ records the cost of producing each item; for instance, each refrigerator costs the firm y_2 dollars.
(*a*) What is the economic significance of $\mathbf{P} \cdot \mathbf{C} = \langle 20, 0, 7, 9, 15\rangle \cdot \langle 50, 70, 30, 20, 10\rangle$?
(*b*) If the firm doubles its production of all items in (*a*), what is its new production vector?
49 Let P_1 be the profit from selling a washing machine and P_2, P_3, P_4, and P_5 be defined analogously for the firm of Exercise 48. (Some of the P's may be negative.) What does it mean to the firm to have $\langle P_1, P_2, P_3, P_4, P_5\rangle$ "perpendicular" to $\langle x_1, x_2, x_3, x_4, x_5\rangle$?
50 Use Theorem 1 to prove that $x\mathbf{A} \cdot y\mathbf{B} = xy\,\mathbf{A} \cdot \mathbf{B}$.
51 Theorem 3 can be proved without using the distributive rule. The following proof is based on the law of cosines (see Sec. 2.6). Let $\mathbf{A} = x_1\mathbf{i} + y_1\mathbf{j}$ and $\mathbf{B} = x_2\mathbf{i} + y_2\mathbf{j}$. Let the angle between $\mathbf{A}$ and $\mathbf{B}$ be θ. Consider the triangle whose sides are $\mathbf{A}$, $\mathbf{B}$, and $\mathbf{A} - \mathbf{B}$.
(*a*) Why is $\|\mathbf{A} - \mathbf{B}\|^2 = \|\mathbf{A}\|^2 + \|\mathbf{B}\|^2 - 2\mathbf{A} \cdot \mathbf{B}$?
(*b*) Compute $\|\mathbf{A} - \mathbf{B}\|$, $\|\mathbf{A}\|$, and $\|\mathbf{B}\|$ and deduce Theorem 3.

---

52 A force $\mathbf{F}$ of 10 newtons has the direction of the vector $2\mathbf{i} + 3\mathbf{j} + \mathbf{k}$. This force pushes an object on a ramp in a straight line from the point (3, 1, 5) to the point (4, 3, 7), where coordinates are measured in meters. How much work does the force accomplish?
53 Establish Theorem 1 when (*a*) both $\text{comp}_\mathbf{C}\,\mathbf{A}$ and $\text{comp}_\mathbf{C}\,\mathbf{B}$ are negative, (*b*) $\text{comp}_\mathbf{C}\,\mathbf{A}$ is positive and $\text{comp}_\mathbf{C}\,\mathbf{B}$ is negative.
54 Write a program to compute
(*a*) The dot product of $\langle x_1, y_1, z_1\rangle$ and $\langle x_2, y_2, z_2\rangle$.
(*b*) The angle between the vectors in (*a*).
55 Write a program to compute
(*a*) The component of $\langle x_1, y_1, z_1\rangle$ on $\langle x_2, y_2, z_2\rangle$.
(*b*) The projection of $\langle x_1, y_1, z_1\rangle$ on $\langle x_2, y_2, z_2\rangle$.

12.4 LINES AND PLANES

This section develops the basic geometry of lines and planes and, in particular, formulas for determining the distance from a point to a line or plane.

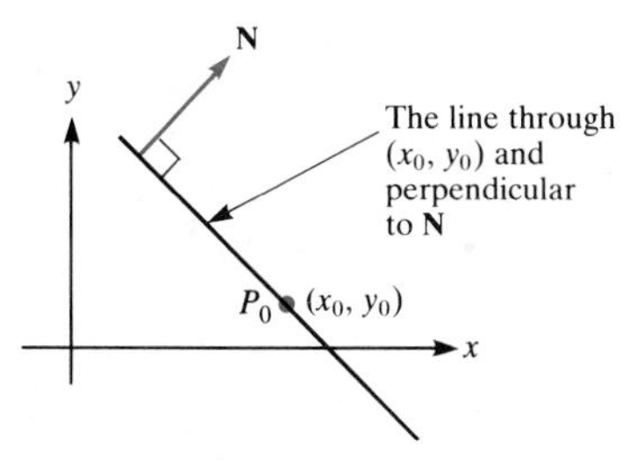

Figure 1

Lines in the Plane

Let $\mathbf{N} = A\mathbf{i} + B\mathbf{j}$ be a nonzero vector and $P_0 = (x_0, y_0)$ be a point in the xy plane. There is a unique line through P_0 that is perpendicular to $\mathbf{N}$, as shown in Fig. 1. $\mathbf{N}$ is called a *normal* to the line. The next theorem provides an algebraic criterion for determining whether the point (x, y) lies on this line.

Theorem 1 An equation of the line (in the xy plane) passing through $P_0 = (x_0, y_0)$ and perpendicular to the nonzero vector $\mathbf{N} = A\mathbf{i} + B\mathbf{j}$ is given by

$$\boxed{A(x - x_0) + B(y - y_0) = 0.}$$

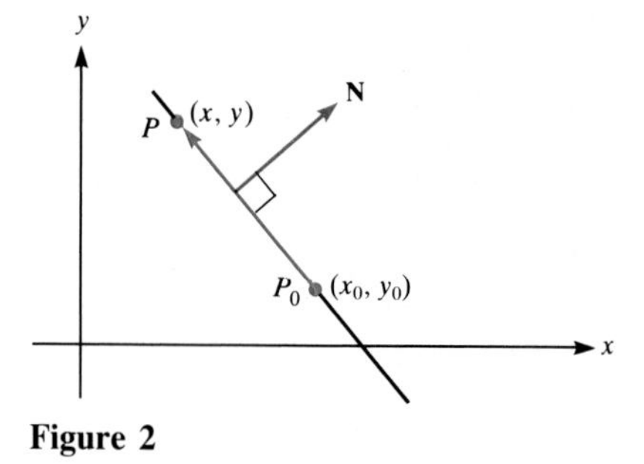

Figure 2

Proof Let $P = (x, y)$ be a point on the line perpendicular to **N**. (See Fig. 2.) Then the vector $\overrightarrow{P_0P} = (x - x_0)\mathbf{i} + (y - y_0)\mathbf{j}$ is perpendicular to **N**. Hence

$$\begin{aligned} 0 &= [(x - x_0)\mathbf{i} + (y - y_0)\mathbf{j}] \cdot \mathbf{N} \\ &= A(x - x_0) + B(y - y_0). \end{aligned}$$

Conversely, it must be shown that if $A(x - x_0) + B(y - y_0) = 0$, then (x, y) is on the line through P_0 perpendicular to **N**. But $A(x - x_0) + B(y - y_0)$ is the scalar product of the vectors **N** and $(x - x_0)\mathbf{i} + (y - y_0)\mathbf{j}$. If this scalar product is zero, then the two vectors **N** and $(x - x_0)\mathbf{i} + (y - y_0)\mathbf{j}$ are perpendicular. Thus (x, y) lies on the line through (x_0, y_0) perpendicular to **N**. ■

EXAMPLE 1 Find an equation of the line through $(2, -7)$ and perpendicular to the vector $4\mathbf{i} + 3\mathbf{j}$.

SOLUTION By Theorem 1, an equation is

$$4(x - 2) + 3(y + 7) = 0,$$

which, when expanded, is

$$4x + 3y + 13 = 0. \quad ■$$

As Theorem 1 and Example 1 show,

> To find a vector perpendicular to a given line $Ax + By + C = 0$, read off the coefficients of x and y in order, A and B, and form the vector $A\mathbf{i} + B\mathbf{j}$.

$A\mathbf{i} + B\mathbf{j}$ is perpendicular to the line $Ax + By + C = 0$. The constant term C plays no role in determining the direction of the line or of a vector perpendicular to it.

Theorem 2 The distance from the point $P_1 = (x_1, y_1)$ to the line L whose equation is $Ax + By + C = 0$ is

$$\boxed{\frac{|Ax_1 + By_1 + C|}{\sqrt{A^2 + B^2}}.}$$

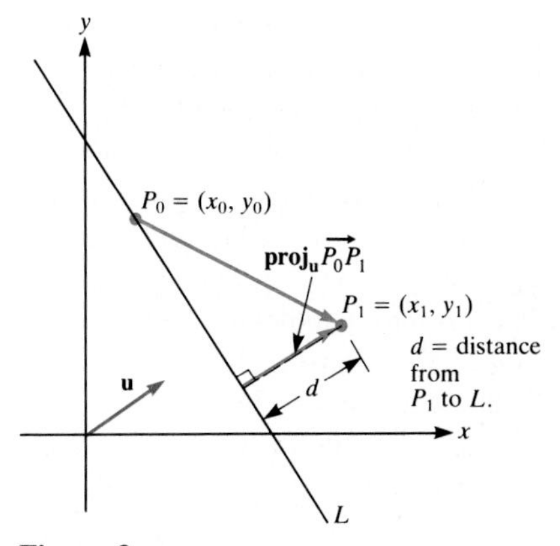

Figure 3

Proof Select a point $P_0 = (x_0, y_0)$ on the line L. Let $\mathbf{u}$ be a unit vector perpendicular to L. Inspection of the right triangle in Fig. 3 shows that the distance from P_1 to the line L is the length of $\mathbf{proj_u}\ \overrightarrow{P_0P_1}$. This length is the absolute value of $\text{comp}_\mathbf{u}\ \overrightarrow{P_0P_1}$. Hence

$$\text{Distance from } P_1 \text{ to } L = |\text{comp}_\mathbf{u}\ \overrightarrow{P_0P_1}| = |\mathbf{u} \cdot \overrightarrow{P_0P_1}|.$$

Since $A\mathbf{i} + B\mathbf{j}$ is perpendicular to L,

$$\mathbf{u} = \frac{A\mathbf{i} + B\mathbf{j}}{\sqrt{A^2 + B^2}}$$

is a unit normal to L. Thus

$$\begin{aligned}\text{Distance from } P_1 \text{ to } L &= \left| \frac{A\mathbf{i} + B\mathbf{j}}{\sqrt{A^2 + B^2}} \cdot \overrightarrow{P_0P_1} \right| \\ &= \left| \frac{A\mathbf{i} + B\mathbf{j}}{\sqrt{A^2 + B^2}} \cdot [(x_1 - x_0)\mathbf{i} + (y_1 - y_0)\mathbf{j}] \right| \\ &= \frac{|A(x_1 - x_0) + B(y_1 - y_0)|}{\sqrt{A^2 + B^2}} \\ &= \frac{|Ax_1 + By_1 - (Ax_0 + By_0)|}{\sqrt{A^2 + B^2}}.\end{aligned}$$

Since $P_0 = (x_0, y_0)$ is on L, $Ax_0 + By_0 + C = 0$; that tells us that $Ax_0 + By_0 = -C$. Thus

$$\text{Distance from } P_1 \text{ to } L = \frac{|Ax_1 + By_1 - (-C)|}{\sqrt{A^2 + B^2}} = \frac{|Ax_1 + By_1 + C|}{\sqrt{A^2 + B^2}}$$

and the theorem is established. ■

EXAMPLE 2 Find the distance from the line $4x + y = 1$ to the origin.

SOLUTION The line has the equation

$$4x + y - 1 = 0.$$

In this case, $A = 4$, $B = 1$, $C = -1$, $x_1 = 0$, and $y_1 = 0$. By Theorem 2, the distance from the line to the origin is

$$\frac{|4 \cdot 0 + 1 \cdot 0 - 1|}{\sqrt{4^2 + 1^2}} = \frac{1}{\sqrt{17}}. \quad ■$$

EXAMPLE 3 Find the distance from point (3, 7) to the line $2x - 4y + 5 = 0$.

SOLUTION By Theorem 2, the distance is

$$\frac{|2 \cdot 3 - 4 \cdot 7 + 5|}{\sqrt{2^2 + 4^2}} = \frac{|6 - 28 + 5|}{\sqrt{20}} = \frac{|-17|}{\sqrt{20}} = \frac{17}{\sqrt{20}}. \quad ■$$

Planes

With the aid of the dot product we can deal with planes in space as easily as we dealt with lines in a plane.

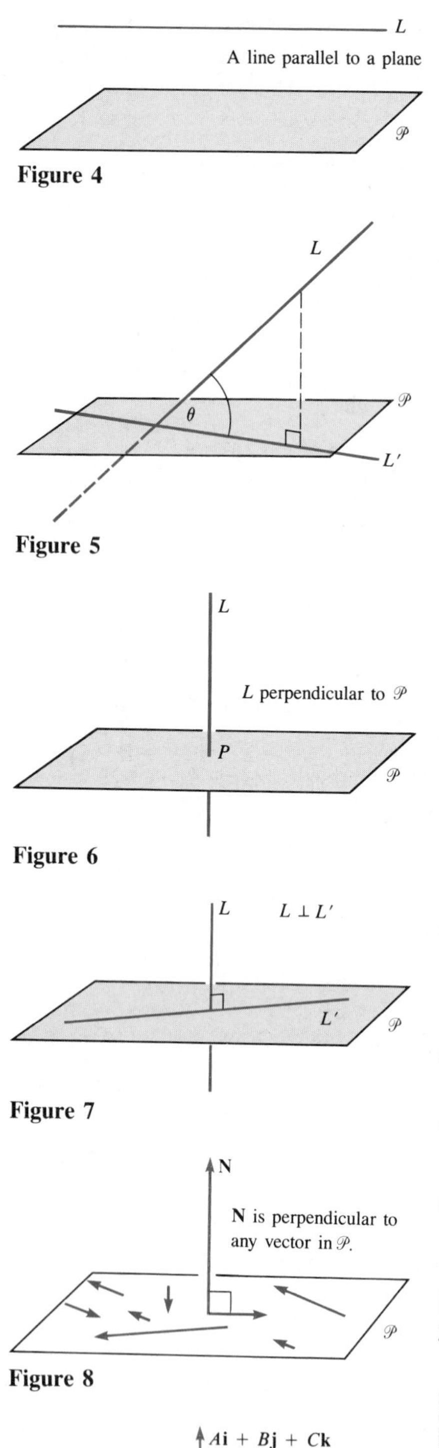

Figure 4

Figure 5

Figure 6

Figure 7

Figure 8

First, we need the notion of the angle between a line L and a plane $\mathcal{P}$. If L does not meet $\mathcal{P}$, the angle between L and $\mathcal{P}$ is defined to be 0. (See Fig. 4.) L is said to be **parallel to** $\mathcal{P}$. If L meets $\mathcal{P}$, let L' be the projection of L on $\mathcal{P}$, as in Fig. 5. In general, L' is a line. We define the angle between L and $\mathcal{P}$ to be the angle θ between L and L', $0 \le \theta < \pi/2$. If the projection of L onto $\mathcal{P}$ is just a point P, we say that L is perpendicular to $\mathcal{P}$ and that the angle between L and $\mathcal{P}$ is $\pi/2$. (See Fig. 6.) In this case L is perpendicular to every line L' in $\mathcal{P}$ that passes through P, as shown in Fig. 7. If $\mathbf{N}$ is a vector parallel to L, we say that $\mathbf{N}$ is perpendicular to $\mathcal{P}$ and call it a **normal to the plane** $\mathcal{P}$. If $\mathbf{A}$ is a vector parallel to a line L' in $\mathcal{P}$, then $\mathbf{N}$ is perpendicular to $\mathbf{A}$. (See Fig. 8.) Therefore $\mathbf{N}$ is perpendicular to any vector that can be drawn in the plane $\mathcal{P}$.

With this geometry in mind, we are ready to treat the equation of a plane in space.

Theorem 3 An equation of the plane passing through (x_0, y_0, z_0) and perpendicular to the nonzero vector $A\mathbf{i} + B\mathbf{j} + C\mathbf{k}$ is given by

$$\boxed{A(x - x_0) + B(y - y_0) + C(z - z_0) = 0.}$$

Proof Let $P = (x, y, z)$ be a point on the plane and let $P_0 = (x_0, y_0, z_0)$. Then the vectors $A\mathbf{i} + B\mathbf{j} + C\mathbf{k}$ and $\overrightarrow{P_0P}$ are perpendicular. (See Fig. 9.)

Thus
$$\begin{aligned} 0 &= (A\mathbf{i} + B\mathbf{j} + C\mathbf{k}) \cdot \overrightarrow{P_0P} \\ &= (A\mathbf{i} + B\mathbf{j} + C\mathbf{k}) \cdot ((x - x_0)\mathbf{i} + (y - y_0)\mathbf{j} + (z - z_0)\mathbf{k}) \\ &= A(x - x_0) + B(y - y_0) + C(z - z_0). \end{aligned}$$

As in the proof of Theorem 1, the steps are reversible. ■

EXAMPLE 4 Find an equation of the plane through $(1, -2, 4)$ that is perpendicular to the vector $5\mathbf{i} + 3\mathbf{j} + 6\mathbf{k}$.

SOLUTION By Theorem 3, an equation of the plane is

$$5(x - 1) + 3[y - (-2)] + 6(z - 4) = 0,$$

which simplifies to
$$5x + 3y + 6z = 23. \quad ■$$

Example 4 shows that the equation $5x + 3y + 6z = 23$ describes a plane. [The plane passes through $(1, -2, 4)$ and is perpendicular to the vector $5\mathbf{i} + 3\mathbf{j} + 6\mathbf{k}$.] The next theorem generalizes this result.

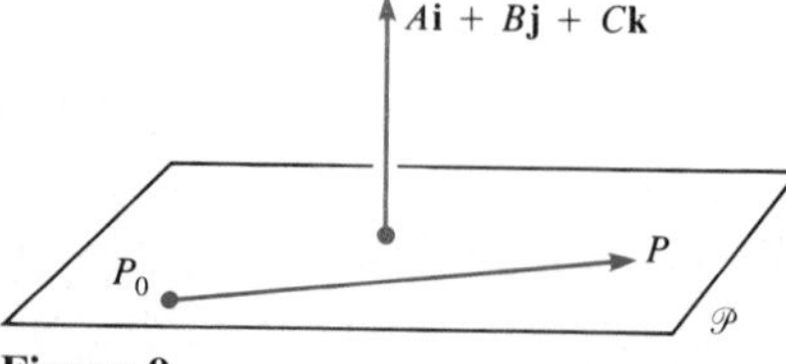

Figure 9

Theorem 4 Let A, B, C, and D be constants such that not all A, B, and C are 0. Then the equation $Ax + By + Cz + D = 0$ describes a plane. Moreover, the vector $A\mathbf{i} + B\mathbf{j} + C\mathbf{k}$ is perpendicular to this plane.

Proof Choose a point $P_0 = (x_0, y_0, z_0)$ that satisfies the equation $Ax + By + Cz + D = 0$. Let $P = (x, y, z)$ be any other point that satisfies the equation.

We will show that $\overrightarrow{P_0P}$ is perpendicular to the vector $A\mathbf{i} + B\mathbf{j} + C\mathbf{k}$. First,

$$\overrightarrow{P_0P} = (x - x_0)\mathbf{i} + (y - y_0)\mathbf{j} + (z - z_0)\mathbf{k}.$$

Recall that $Ax + By + Cz = -D$ and $Ax_0 + By_0 + Cz_0 = -D$.

Then
$$\begin{aligned}(A\mathbf{i} + B\mathbf{j} + C\mathbf{k}) \cdot \overrightarrow{P_0P} &= A(x - x_0) + B(y - y_0) + C(z - z_0)\\ &= (Ax + By + Cz) - (Ax_0 + By_0 + Cz_0)\\ &= -D - (-D)\\ &= 0.\end{aligned}$$

The steps are reversible. Thus the set of points that satisfy the equation $Ax + By + Cz + D = 0$ is the plane perpendicular to $A\mathbf{i} + B\mathbf{j} + C\mathbf{k}$ and passing through the point (x_0, y_0, z_0). ■

EXAMPLE 5 Find a normal to the plane $2x - y + 3z + 4 = 0$. Also, find a point on the plane.

SOLUTION By Theorem 4, $\mathbf{N} = 2\mathbf{i} - \mathbf{j} + 3\mathbf{k}$ is a normal to the plane. To find a point on the plane, pick x and y as you please, say $x = 3$, $y = 2$, and solve for z:

$$2(3) - (2) + 3z + 4 = 0.$$

Hence $8 + 3z = 0$, and $z = -\frac{8}{3}$. The point $(3, 2, -\frac{8}{3})$ lies on the plane. ■

The next theorem is a consequence of Theorem 4. Its proof, which is practically the same as that of Theorem 2, is omitted.

Theorem 5 The distance from the point (x_1, y_1, z_1) to the plane

$$Ax + By + Cz + D = 0$$

is

$$\frac{|Ax_1 + By_1 + Cz_1 + D|}{\sqrt{A^2 + B^2 + C^2}}.$$

EXAMPLE 6 What is the distance from the point $(2, 1, 5)$ to the plane that has the equation $x - 3y + 4z + 8 = 0$?

SOLUTION By Theorem 5, the desired distance is

$$\frac{|1 \cdot 2 - 3 \cdot 1 + 4 \cdot 5 + 8|}{\sqrt{1^2 + (-3)^2 + 4^2}} = \frac{|2 - 3 + 20 + 8|}{\sqrt{26}} = \frac{27}{\sqrt{26}}.$$ ■

Lines in Space

So far in this section we have been concerned with lines in the xy plane and planes in space. Vectors also provide a neat way to treat the geometry of lines in space, as will now be shown.

Consider the line L through the point $P_0 = (x_0, y_0, z_0)$ and parallel to the vector $\mathbf{A} = a_1\mathbf{i} + a_2\mathbf{j} + a_3\mathbf{k}$ shown in Fig. 10. A point $P = (x, y, z)$ is on this line if and only if the vector $\overrightarrow{P_0P}$ is parallel to $\mathbf{A}$. One way to express that $\overrightarrow{P_0P}$ is parallel to $\mathbf{A}$ is to assert that there is a scalar t such that

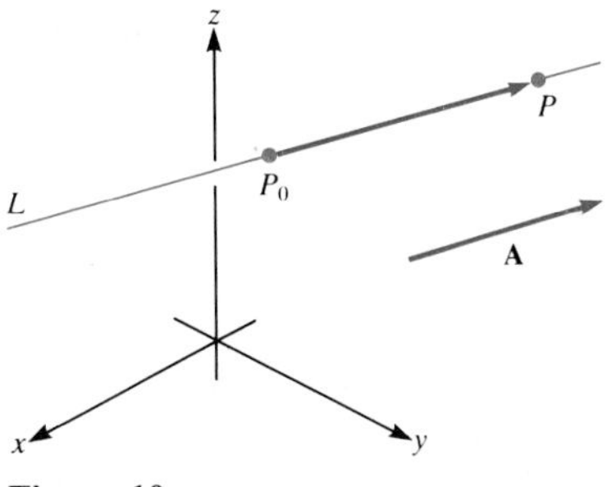

Figure 10

$$\boxed{\overrightarrow{P_0P} = t\mathbf{A};} \tag{1}$$

that is, $(x - x_0)\mathbf{i} + (y - y_0)\mathbf{j} + (z - z_0)\mathbf{k} = ta_1\mathbf{i} + ta_2\mathbf{j} + ta_3\mathbf{k}.$

In short, $x - x_0 = ta_1, \quad y - y_0 = ta_2, \quad z - z_0 = ta_3.$

Consequently, we have these **parametric equations** for the line through (x_0, y_0, z_0) parallel to $\mathbf{A} = a_1\mathbf{i} + a_2\mathbf{j} + a_3\mathbf{k}$:

Parametric equations

$$\boxed{\begin{cases} x = x_0 + a_1 t \\ y = y_0 + a_2 t \\ z = z_0 + a_3 t. \end{cases}}$$

These equations can also be expressed vectorially. Let $\mathbf{P}$ be the vector $\overrightarrow{OP}$ and $\mathbf{P}_0$ be the vector $\overrightarrow{OP_0}$. The three parametric equations reduce to the single vector equation

Vector equation

$$\boxed{\mathbf{P} = \mathbf{P}_0 + t\mathbf{A},}$$

called a **vector equation** of the line.

An equation of the line through (x_0, y_0, z_0) and parallel to $\mathbf{A} = a_1\mathbf{i} + a_2\mathbf{j} + a_3\mathbf{k}$ can also be given in the form

Symmetric equations

$$\boxed{\frac{x - x_0}{a_1} = \frac{y - y_0}{a_2} = \frac{z - z_0}{a_3},}$$

if none of a_1, a_2, a_3 is 0. These are called **symmetric equations** of the line. (This says that the components of $\overrightarrow{P_0P}$ are proportional to the corresponding components of $\mathbf{A}$.) These symmetric equations describe the line as the intersection of two planes, namely,

$$\frac{x - x_0}{a_1} = \frac{y - y_0}{a_2} \quad \text{and} \quad \frac{y - y_0}{a_2} = \frac{z - z_0}{a_3}.$$

The parametric, vector, and symmetric equations of a line are just different versions of Eq. (1).

EXAMPLE 7 Write parametric equations for the line through (1, 2, 2) and parallel to the vector $3\mathbf{i} - \mathbf{j} + 5\mathbf{k}$. Does the point (10, −1, 16) lie on the line?

SOLUTION Parametric equations of the line are given by

$$\begin{cases} x = 1 + 3t \\ y = 2 - t \\ z = 2 + 5t. \end{cases}$$

Does (10, −1, 16) lie on this line? To find out, determine if there is a number t such that these three equations are *simultaneously* satisfied:

$$10 = 1 + 3t$$
$$-1 = 2 - t$$
$$16 = 2 + 5t.$$

The first equation, $10 = 1 + 3t$, has the solution $t = 3$. This value, 3, does satisfy the second equation, since $-1 = 2 - 3$. But it does *not* satisfy the third equation, since $16 \neq 2 + 5 \cdot 3$. Hence $(10, -1, 16)$ is *not* on the line. ■

EXAMPLE 8 Find where the line L through $P_0 = (2, 1, 3)$ and $P_1 = (4, -2, 5)$ meets the plane $\mathcal{P}$ whose equation is $2x + y - 4z + 5 = 0$.

SOLUTION The line L has the vector equation $\mathbf{P} = \mathbf{P}_0 + t(\overrightarrow{P_0P_1})$, which reads

$$\langle x, y, z\rangle = \langle 2, 1, 3\rangle + t\langle 4 - 2, -2 - 1, 5 - 3\rangle$$

or

$$\langle x, y, z\rangle = \langle 2, 1, 3\rangle + t\langle 2, -3, 2\rangle. \tag{2}$$

Equation (2) is short for the three parametric equations

$$\begin{cases} x = 2 + 2t \\ y = 1 - 3t \\ z = 3 + 2t. \end{cases} \tag{3}$$

Equation (3) describes an arbitrary point on the line L. Now we will find the value of t such that the point lies on the plane $\mathcal{P}$. Once we have t, we then calculate the point that lies on both L and $\mathcal{P}$.

Substituting (3) into the equation of $\mathcal{P}$, we have

$$2(2 + 2t) + (1 - 3t) - 4(3 + 2t) + 5 = 0,$$

which reduces to $-2 - 7t = 0$. Hence $t = -\frac{2}{7}$. Substituting $t = -\frac{2}{7}$ into Eq. (3) gives us the point where L meets $\mathcal{P}$:

$$(x, y, z) = \left(2 + 2\left(-\frac{2}{7}\right), 1 - 3\left(-\frac{2}{7}\right), 3 + 2\left(-\frac{2}{7}\right)\right) = \left(\frac{10}{7}, \frac{13}{7}, \frac{17}{7}\right). \quad ■$$

Describing the Direction of Vectors and Lines

The direction of a vector in the plane is described by a single angle, the angle it makes with the positive x axis. The direction of a vector in space involves three angles, two of which almost determine the third.

Definition *Direction angles of a vector.* Let $\mathbf{A}$ be a nonzero vector in space. The angle between

$\mathbf{A}$ and $\mathbf{i}$ is denoted α,

$\mathbf{A}$ and $\mathbf{j}$ is denoted β,

$\mathbf{A}$ and $\mathbf{k}$ is denoted γ.

The angles α, β, and γ are called the **direction angles of A**. (See Fig. 11.)

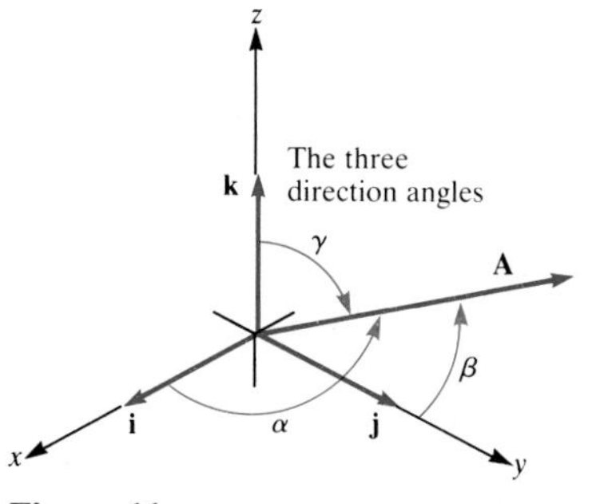

Figure 11

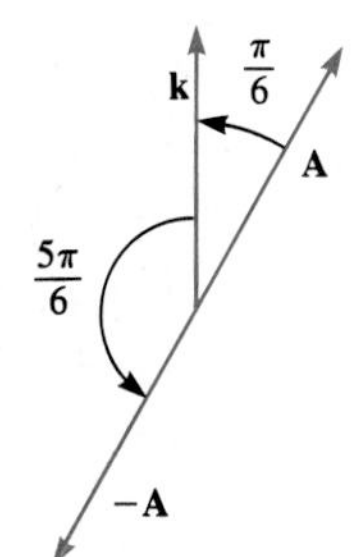

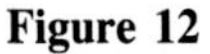
Figure 12

> **Definition** *Direction cosines of a vector.* The **direction cosines** of a vector are the cosines of its direction angles, $\cos \alpha$, $\cos \beta$, and $\cos \gamma$.

EXAMPLE 9 The angle between a vector **A** and **k** is $\pi/6$. Find γ and $\cos \gamma$ for (*a*) **A**, (*b*) $-$**A**.

SOLUTION (*a*) By definition, the direction angle γ for **A** is $\pi/6$. It follows that $\cos \gamma = \cos (\pi/6) = \sqrt{3}/2$. (*b*) To find γ and $\cos \gamma$ for $-$**A**, we draw Fig. 12. For $-$**A**, $\gamma = 5\pi/6$ and $\cos \gamma = \cos (5\pi/6) = -\sqrt{3}/2$. ■

As Example 9 illustrates, if the direction angles of **A** are α, β, and γ, then the direction angles of $-$**A** are $\pi - \alpha$, $\pi - \beta$, and $\pi - \gamma$. The direction cosines of $-$**A** are the negatives of the direction cosines of **A**.

The three direction angles are not independent of each other, as is shown by the next theorem. Two of them determine the third up to sign.

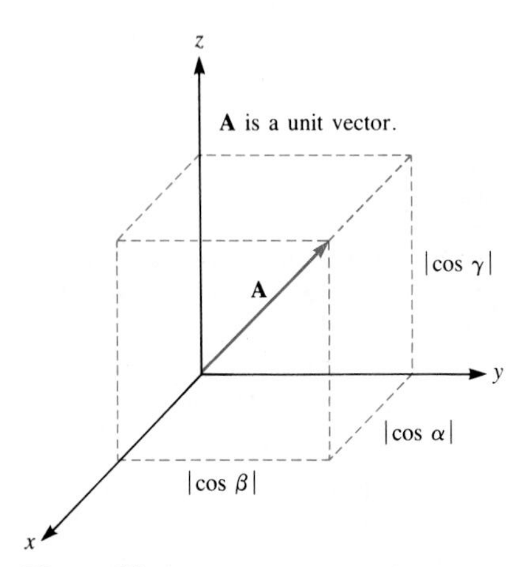

Figure 13

> **Theorem 6** If α, β, and γ are the direction angles of the vector **A**, then
> $$\cos^2 \alpha + \cos^2 \beta + \cos^2 \gamma = 1.$$

Proof It is no loss of generality to assume that **A** is a unit vector. Its component on the x axis, $\text{comp}_{\mathbf{i}}\, \mathbf{A}$, is

$$\begin{aligned} \mathbf{A} \cdot \mathbf{i} &= \|\mathbf{A}\|\, \|\mathbf{i}\| \cos \alpha \\ &= 1 \cdot 1 \cdot \cos \alpha \\ &= \cos \alpha. \end{aligned}$$

Similarly, $\text{comp}_{\mathbf{j}}\, \mathbf{A} = \cos \beta$ and $\text{comp}_{\mathbf{k}}\, \mathbf{A} = \cos \gamma$. Thus the three sides of the box determined by **A** have lengths $|\cos \alpha|$, $|\cos \beta|$, and $|\cos \gamma|$. (See Fig. 13.) Since **A** is a unit vector, we have

$$1 = \|\mathbf{A}\|^2 = \cos^2 \alpha + \cos^2 \beta + \cos^2 \gamma. \quad ■$$

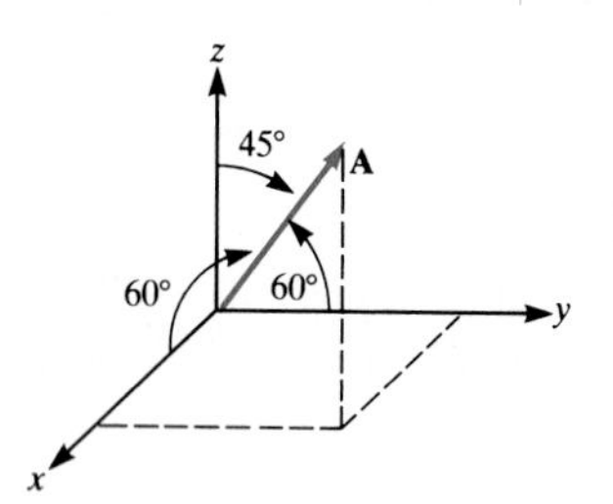

Figure 14

EXAMPLE 10 The vector **A** makes an angle of 60° with the x and y axes. What angle does it make with the z axis?

SOLUTION Here $\alpha = 60°$ and $\beta = 60°$; hence

$$\cos \alpha = \tfrac{1}{2} \quad \text{and} \quad \cos \beta = \tfrac{1}{2}.$$

Since
$$\cos^2 \alpha + \cos^2 \beta + \cos^2 \gamma = 1,$$
it follows that
$$(\tfrac{1}{2})^2 + (\tfrac{1}{2})^2 + \cos^2 \gamma = 1,$$
$$\cos^2 \gamma = \tfrac{1}{2}.$$
Thus
$$\cos \gamma = \frac{\sqrt{2}}{2} \quad \text{or} \quad \cos \gamma = -\frac{\sqrt{2}}{2}.$$
Hence
$$\gamma = 45° \quad \text{or} \quad 135°.$$

Figures 14 and 15 show the two possibilities for **A**. ■

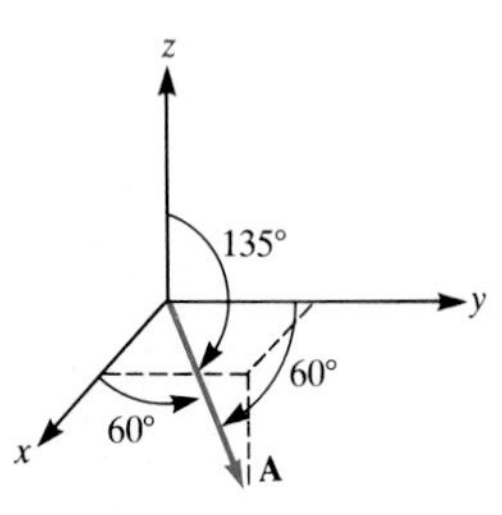

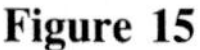
Figure 15

Definition *Direction numbers of a line.* If the vector $a_1\mathbf{i} + a_2\mathbf{j} + a_3\mathbf{k}$ is parallel to the line L, then the numbers a_1, a_2, and a_3 are called **direction numbers** of L. (Note that direction numbers are not unique.)

If a vector $\mathbf{A}$ is parallel to L, so is $-\mathbf{A}$. This means that if a_1, a_2, a_3 are direction numbers for L, so are $-a_1$, $-a_2$, $-a_3$.

EXAMPLE 11 Let $P_0 = (3, 1, 4)$ and $P_1 = (2, 5, 1)$. Find (*a*) the direction cosines of $\overrightarrow{P_0P_1}$, (*b*) the direction angles of $\overrightarrow{P_0P_1}$, and (*c*) the direction numbers for the line through P_0 and P_1.

SOLUTION (*a*) $\overrightarrow{P_0P_1} = \langle 2-3, 5-1, 1-4\rangle = \langle -1, 4, -3\rangle$. Hence

$$\frac{\overrightarrow{P_0P_1}}{|\overrightarrow{P_0P_1}|} = \frac{-\mathbf{i} + 4\mathbf{j} - 3\mathbf{k}}{\sqrt{(-1)^2 + 4^2 + (-3)^2}} = \frac{-\mathbf{i} + 4\mathbf{j} - 3\mathbf{k}}{\sqrt{26}}.$$

The direction cosines of $\overrightarrow{P_0P_1}$ are therefore

$$\cos\alpha = -\frac{1}{\sqrt{26}}, \qquad \cos\beta = \frac{4}{\sqrt{26}}, \qquad \cos\gamma = -\frac{3}{\sqrt{26}}.$$

(*b*) Using a calculator we find that (in radians)

$$\alpha \approx 1.77, \qquad \beta \approx 0.67, \qquad \text{and} \qquad \gamma \approx 2.20.$$

(*c*) Direction numbers come directly from $\overrightarrow{P_0P_1}$. Hence -1, 4, -3 (in that order) are direction numbers for the line through P_0 and P_1. ■

Section Summary

We used the dot product to obtain the equations of lines and planes. A line in the *xy* plane has an equation $Ax + By + C = 0$. The vector $A\mathbf{i} + B\mathbf{j}$ is a normal to that line. Similarly, a plane in space has an equation $Ax + By + Cz + D = 0$, and $A\mathbf{i} + B\mathbf{j} + C\mathbf{k}$ is a normal to that plane. We also saw how to find the distance from a point in the *xy* plane to a line in the *xy* plane, as well as the distance from a point in space to any plane.

A line in space is described by the condition $\overrightarrow{P_0P} = t\mathbf{A}$. This equation can be expressed in the form $\mathbf{P} = \mathbf{P}_0 + t\mathbf{A}$, and rewritten as parametric equations. A line can also be given by symmetric equations, which assert that the vectors $\overrightarrow{P_0P}$ and $\mathbf{A}$ are parallel.

We concluded by defining direction angles and direction numbers.

EXERCISES FOR SEC. 12.4: LINES AND PLANES

In each of Exercises 1 to 4 find an equation of the line through the given point and perpendicular to the given vector.

1 $(2, 3)$, $4\mathbf{i} + 5\mathbf{j}$
2 $(1, 0)$, $2\mathbf{i} - \mathbf{j}$
3 $(4, 5)$, $2\mathbf{i} + 3\mathbf{j}$
4 $(2, -1)$, $\mathbf{i} + 3\mathbf{j}$

In each of Exercises 5 to 8 find a vector in the *xy* plane that is perpendicular to the given line.

5 $2x - 3y + 8 = 0$
6 $\pi x - \sqrt{2}y = 7$
7 $y = 3x + 7$
8 $2(x - 1) + 5(y + 2) = 0$

In each of Exercises 9 and 10 find the distance from the given point to the given line.

9 The point (0, 0) to $3x + 4y - 10 = 0$
10 The point $(\frac{3}{2}, \frac{2}{3})$ to $2x - y + 5 = 0$

In Exercises 11 and 12 find a normal and a unit normal to the given planes.
11 $2x - 3y + 4z + 11 = 0$
12 $z = 2x - 3y + 4$

In each of Exercises 13 to 16 find the distance from the given point to the given plane.
13 The point (0, 0, 0) to the plane $2x - 4y + 3z + 2 = 0$
14 The point (1, 2, 3) to the plane $x + 2y - 3z + 5 = 0$
15 The point (2, 2, −1) to the plane that passes through (1, 4, 3) and has a normal $2\mathbf{i} - 7\mathbf{j} + 2\mathbf{k}$
16 The point (0, 0, 0) to the plane that passes through (4, 1, 0) and is perpendicular to the vector $\mathbf{i} + \mathbf{j} + \mathbf{k}$
17 Find the direction cosines of the vector $2\mathbf{i} + 3\mathbf{j} + 4\mathbf{k}$.
18 Find the direction cosines of the vector from (1, 3, 2) to (4, −1, 5).
19 (*a*) Are the direction cosines of a vector unique?
(*b*) Are the direction numbers of a line unique?
20 Let $P_0 = (2, 1, 5)$ and $P_1 = (3, 0, 4)$. Find the direction cosines and direction angles of (*a*) $\overrightarrow{P_0P_1}$ and (*b*) $\overrightarrow{P_1P_0}$.
21 Give parametric equations for the line through $(\frac{1}{2}, \frac{1}{3}, \frac{1}{2})$ and with direction numbers 2, −5, and 8 in (*a*) scalar form, (*b*) vector form.
22 Give parametric equations for the line through (1, 2, 3) and (4, 5, 7) in (*a*) scalar form, (*b*) vector form.
23 Give symmetric equations for the line through the points (1, 0, 3) and (2, 1, −1).
24 Give symmetric equations for the line through the points (7, −1, 5) and (4, 3, 2).
25 A vector **A** has direction angles $\alpha = 70°$ and $\beta = 80°$. Find the third direction angle γ and show the possibilities on a diagram.
26 Suppose that the three direction angles of a vector are equal. What can they be? Draw the cases.
27 Find the angle between the line through (3, 2, 2) and (4, 3, 1) and the line through (3, 2, 2) and (5, 2, 7).
28 Find the angle between the planes $2x + 3y + 4z = 11$ and $3x - y + 2z = 13$.
29 Find where the line through (1, 2) and (3, 5) meets the line through (1, −1) and (2, 3).
30 Find where the line through (1, 2, 1) and (2, 1, 3) meets the plane that is perpendicular to the vector $2\mathbf{i} + 5\mathbf{j} + 7\mathbf{k}$ and passes through the point (1, −2, −3).
31 Are the three points (1, 2, −3), (1, 6, 2), and (7, 14, 11) on a single line?
32 Where does the line through (1, 2, 4) and (2, 1, −1) meet the plane $x + 2y + 5z = 0$?
33 Give parametric equations for the line through (1, 3, −5) that is perpendicular to the plane $2x - 3y + 4z = 11$.
34 Give parametric equations for the line through (1, 3, 4) that is parallel to the line through (2, 4, 6) and (5, 3, −2).

---

35 A square of side a lies in the plane $2x + 3y + 2z = 8$. What is the area of its projection (*a*) on the xy plane? (*b*) on the yz plane? (*c*) on the xz plane?
36 Prove Theorem 5.
37 If α, β, and γ are direction angles of a vector, what is $\sin^2 \alpha + \sin^2 \beta + \sin^2 \gamma$?
38 Find the angle between the line through (1, 3, 2) and (4, 1, 5) and the plane $x - y - 2z + 15 = 0$.
39 A disk of radius a is situated in the plane $x + 3y + 4z = 5$. What is the area of its projection in the plane $2x + y - z = 6$?
40 What point on the line through (1, 2, 5) and (3, 1, 1) is closest to the point (2, −1, 5)?
41 Does the line through (5, 7, 10) and (3, 4, 5) meet the line through (1, 4, 0) and (3, 6, 4)? If so, where?
Warning: Use parametric equations but give the parameters of the lines different names, such as t and s.
42 Develop a general formula for determining the distance from the point $P_1 = (x_1, y_1, z_1)$ to the line through the point $P_0 = (x_0, y_0, z_0)$ and parallel to the vector $\mathbf{A} = a_1\mathbf{i} + a_2\mathbf{j} + a_3\mathbf{k}$. The formula should be expressed in terms of the vectors $\overrightarrow{P_0P_1}$ and **A**.
43 How far is the point (1, 2, −1) from the line through (1, 3, 5) and (2, 1, −3)?
(*a*) Solve by calculus, minimizing a certain function.
(*b*) Solve by vectors.
44 Find the direction cosines of the vector **A** shown in Fig. 16.
Hint: First draw a large diagram.

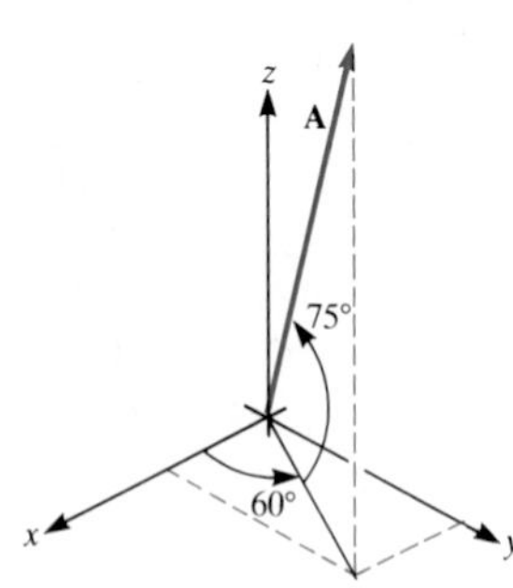

Figure 16

45 How small can the largest of three directions angles ever be?

12.5 DETERMINANTS

This section develops a small part of the extensive field of linear algebra. Keep in mind that the few definitions and theorems we present have broad generalizations that only a full linear algebra course could adequately explore. We introduce here only ideas needed in the next section.

Determinants

Definition *Matrix.* An array of four numbers a_1, a_2, b_1, b_2 forming a square is called a **2 by 2 matrix** and is denoted

$$\begin{pmatrix} a_1 & a_2 \\ b_1 & b_2 \end{pmatrix}.$$

Similarly, an array of nine numbers arranged in a square is called a **3 by 3 matrix**:

$$\begin{pmatrix} a_1 & a_2 & a_3 \\ b_1 & b_2 & b_3 \\ c_1 & c_2 & c_3 \end{pmatrix}.$$

For instance,

$$\begin{pmatrix} 2 & 3 \\ 4 & -1 \end{pmatrix} \quad \text{and} \quad \begin{pmatrix} 1 & 5 & 0 \\ 0 & 2 & 3 \\ 8 & 4 & -1 \end{pmatrix}$$

are 2 by 2 and 3 by 3 matrices, respectively. The set of entries of a matrix in a horizontal line is called a **row** of the matrix. The set of entries in a vertical line is called a **column**. A 3 by 3 matrix thus consists of three rows; it also has three columns. For convenience, number the rows from top to bottom and the columns from left to right:

$$\begin{pmatrix} \text{row 1} \\ \text{row 2} \\ \text{row 3} \end{pmatrix} \quad \text{or} \quad \begin{pmatrix} \text{c} & \text{c} & \text{c} \\ \text{o} & \text{o} & \text{o} \\ \text{l} & \text{l} & \text{l} \\ \text{u} & \text{u} & \text{u} \\ \text{m} & \text{m} & \text{m} \\ \text{n} & \text{n} & \text{n} \\ 1 & 2 & 3 \end{pmatrix}.$$

In the matrix

$$\begin{pmatrix} 2 & 3 \\ 4 & -1 \end{pmatrix},$$

the first row is (2, 3), and the second column is $\begin{pmatrix} 3 \\ -1 \end{pmatrix}$.

Associated with each matrix is an important number called its **determinant**

Definition *Determinant of a 2 by 2 matrix.* The **determinant** of the matrix

$$\begin{pmatrix} a_1 & a_2 \\ b_1 & b_2 \end{pmatrix}$$

is the number $a_1b_2 - a_2b_1$.

It is denoted $\begin{vmatrix} a_1 & a_2 \\ b_1 & b_2 \end{vmatrix}$.

EXAMPLE 1 Compute the determinants:

(a) $\begin{vmatrix} 2 & 3 \\ 1 & 4 \end{vmatrix}$, (b) $\begin{vmatrix} 0 & -5 \\ 2 & 8 \end{vmatrix}$.

SOLUTION

(a) $\begin{vmatrix} 2 & 3 \\ 1 & 4 \end{vmatrix} = 2 \cdot 4 - 3 \cdot 1 = 8 - 3 = 5.$

(b) $\begin{vmatrix} 0 & -5 \\ 2 & 8 \end{vmatrix} = 0 \cdot 8 - (-5) \cdot 2 = 0 + 10 = 10.$ ■

The determinant of a 3 by 3 matrix may be defined with the aid of 2 by 2 matrices as follows.

Definition *Determinant of a 3 by 3 matrix.* The **determinant** of the matrix

$$\begin{pmatrix} a_1 & a_2 & a_3 \\ b_1 & b_2 & b_3 \\ c_1 & c_2 & c_3 \end{pmatrix}$$

Note the minus sign.

is the number $a_1 \begin{vmatrix} b_2 & b_3 \\ c_2 & c_3 \end{vmatrix} - a_2 \begin{vmatrix} b_1 & b_3 \\ c_1 & c_3 \end{vmatrix} + a_3 \begin{vmatrix} b_1 & b_2 \\ c_1 & c_2 \end{vmatrix}$.

It is denoted $\begin{vmatrix} a_1 & a_2 & a_3 \\ b_1 & b_2 & b_3 \\ c_1 & c_2 & c_3 \end{vmatrix}$.

EXAMPLE 2 Compute $\begin{vmatrix} 3 & 2 & -4 \\ 1 & 5 & 0 \\ 2 & -7 & 3 \end{vmatrix}$.

SOLUTION By definition, this determinant equals

$$3 \begin{vmatrix} 5 & 0 \\ -7 & 3 \end{vmatrix} - 2 \begin{vmatrix} 1 & 0 \\ 2 & 3 \end{vmatrix} + (-4) \begin{vmatrix} 1 & 5 \\ 2 & -7 \end{vmatrix}.$$

Now $\begin{vmatrix} 5 & 0 \\ -7 & 3 \end{vmatrix} = 15,$ $\begin{vmatrix} 1 & 0 \\ 2 & 3 \end{vmatrix} = 3,$ and $\begin{vmatrix} 1 & 5 \\ 2 & -7 \end{vmatrix} = -17.$

Hence the 3 by 3 determinant equals

$$3 \cdot 15 - 2 \cdot 3 + (-4)(-17) = 45 - 6 + 68 = 107. \quad ■$$

$$\begin{pmatrix} \not{a}_1 & \not{a}_2 & \not{a}_3 \\ \not{b}_1 & b_2 & b_3 \\ \not{c}_1 & c_2 & c_3 \end{pmatrix}$$

Note that in the definition of a 3 by 3 determinant a_2 has a negative sign. Also observe that the 2 by 2 matrix associated with a_1 is obtained by blotting out of the 3 by 3 matrix the row and column on which a_1 lies, as shown in the margin. A similar procedure works for a_2 and a_3 (but remember the minus sign that goes with a_2).

Properties of Determinants

We now obtain some useful theorems about determinants that are true of 2 by 2 and 3 by 3 matrices. However, the proofs, which are straightforward computations, will be given in only one of the two cases.

Theorem 1 If two rows (or two columns) of a matrix are identical, then the determinant of the matrix is 0.

Proof Let us show that when two rows are identical, the matrix has determinant 0; for instance,

$$\begin{vmatrix} a_1 & a_2 & a_3 \\ a_1 & a_2 & a_3 \\ c_1 & c_2 & c_3 \end{vmatrix} = 0.$$

This determinant equals

$$\begin{aligned} a_1 \begin{vmatrix} a_2 & a_3 \\ c_2 & c_3 \end{vmatrix} - a_2 \begin{vmatrix} a_1 & a_3 \\ c_1 & c_3 \end{vmatrix} + a_3 \begin{vmatrix} a_1 & a_2 \\ c_1 & c_2 \end{vmatrix} \\ &= a_1(a_2c_3 - a_3c_2) - a_2(a_1c_3 - a_3c_1) + a_3(a_1c_2 - a_2c_1) \\ &= a_1a_2c_3 - a_1a_3c_2 - a_2a_1c_3 + a_2a_3c_1 + a_3a_1c_2 - a_3a_2c_1 \\ &= 0. \qquad \text{(the terms cancel in pairs)} \end{aligned}$$

The other cases may be proved similarly. ■

Theorem 2 If two rows (or two columns) of a matrix are switched with each other, the determinant of the resulting matrix is the determinant of the original matrix with its sign changed.

Proof Take the case, for instance, in which the second and third columns of a 3 by 3 matrix are switched. We will prove that

$$\begin{vmatrix} a_1 & a_3 & a_2 \\ b_1 & b_3 & b_2 \\ c_1 & c_3 & c_2 \end{vmatrix} = - \begin{vmatrix} a_1 & a_2 & a_3 \\ b_1 & b_2 & b_3 \\ c_1 & c_2 & c_3 \end{vmatrix}. \tag{1}$$

Don't read ahead; do the computations yourself.

Simply calculate both determinants and compare the results:

$$\begin{vmatrix} a_1 & a_3 & a_2 \\ b_1 & b_3 & b_2 \\ c_1 & c_3 & c_2 \end{vmatrix} = a_1 \begin{vmatrix} b_3 & b_2 \\ c_3 & c_2 \end{vmatrix} - a_3 \begin{vmatrix} b_1 & b_2 \\ c_1 & c_2 \end{vmatrix} + a_2 \begin{vmatrix} b_1 & b_3 \\ c_1 & c_3 \end{vmatrix}, \tag{2}$$

while

$$\begin{vmatrix} a_1 & a_2 & a_3 \\ b_1 & b_2 & b_3 \\ c_1 & c_2 & c_3 \end{vmatrix} = a_1 \begin{vmatrix} b_2 & b_3 \\ c_2 & c_3 \end{vmatrix} - a_2 \begin{vmatrix} b_1 & b_3 \\ c_1 & c_3 \end{vmatrix} + a_3 \begin{vmatrix} b_1 & b_2 \\ c_1 & c_2 \end{vmatrix}. \tag{3}$$

Since, for example,

$$\begin{vmatrix} b_3 & b_2 \\ c_3 & c_2 \end{vmatrix} = b_3c_2 - b_2c_3 \quad \text{and} \quad \begin{vmatrix} b_2 & b_3 \\ c_2 & c_3 \end{vmatrix} = b_2c_3 - b_3c_2,$$

direct comparison of the three summands in (2) and in (3) establishes (1). The other cases are proved similarly. ■

The Geometry of a 2 by 2 Determinant

Let $\mathbf{A} = a_1\mathbf{i} + a_2\mathbf{j}$ and $\mathbf{B} = b_1\mathbf{i} + b_2\mathbf{j}$. These two vectors determine a parallelogram, as shown in Fig. 1. The vectors **A** and **B** are said to **span** the parallelogram. (If **A** and **B** are parallel, the parallelogram collapses to a line segment.) We will show that the area of this parallelogram is the absolute value of the determinant whose rows are formed from the components of **A** and **B**,

abs *is short for "absolute value of."*

$$\text{Area of parallelogram} = \text{abs} \begin{vmatrix} a_1 & a_2 \\ b_1 & b_2 \end{vmatrix}.$$

That means that we can think of a 2 by 2 determinant as plus or minus the area of a certain parallelogram. In the next section we will obtain a geometric interpretation of a 3 by 3 determinant as the volume of a solid.

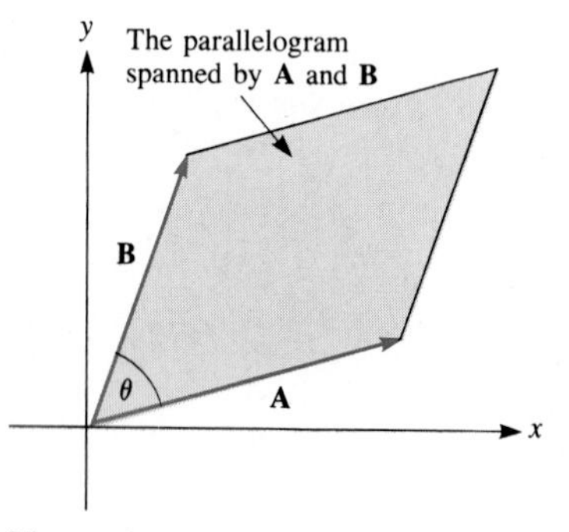

Figure 1

> **Theorem 3** The area of the parallelogram spanned by $\mathbf{A} = a_1\mathbf{i} + a_2\mathbf{j}$ and $\mathbf{B} = b_1\mathbf{i} + b_2\mathbf{j}$ is
>
> $$\pm \begin{vmatrix} a_1 & a_2 \\ b_1 & b_2 \end{vmatrix}.$$

Proof The area of the parallelogram in Fig. 1 is the length of its base **A** times its height h as shown in Fig. 2. Thus,

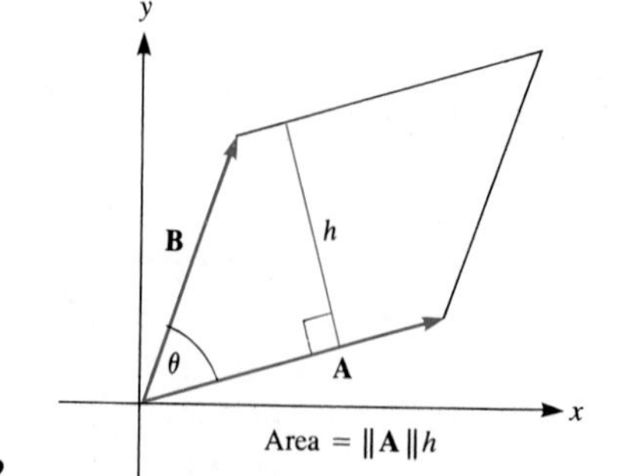

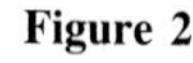

Figure 2

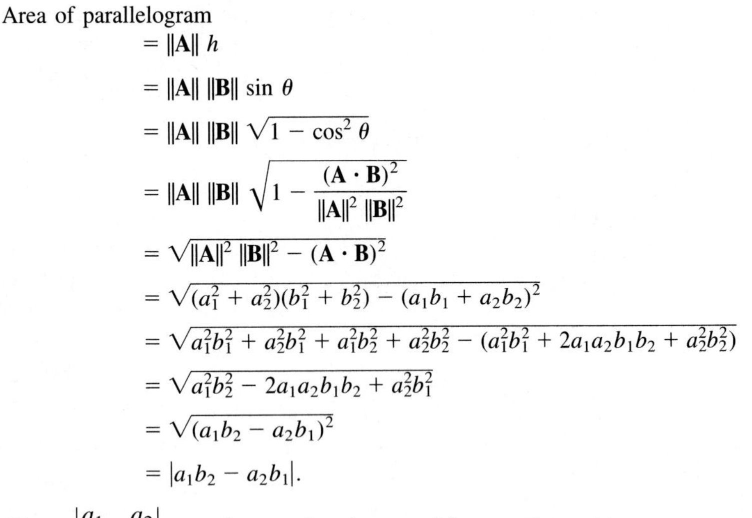

Area of parallelogram

$$
\begin{aligned}
&= \|\mathbf{A}\|\, h \\
&= \|\mathbf{A}\|\, \|\mathbf{B}\| \sin\theta \\
&= \|\mathbf{A}\|\, \|\mathbf{B}\| \sqrt{1-\cos^2\theta} \\
&= \|\mathbf{A}\|\, \|\mathbf{B}\| \sqrt{1-\frac{(\mathbf{A}\cdot\mathbf{B})^2}{\|\mathbf{A}\|^2\,\|\mathbf{B}\|^2}} \\
&= \sqrt{\|\mathbf{A}\|^2\,\|\mathbf{B}\|^2-(\mathbf{A}\cdot\mathbf{B})^2} \\
&= \sqrt{(a_1^2+a_2^2)(b_1^2+b_2^2)-(a_1b_1+a_2b_2)^2} \\
&= \sqrt{a_1^2b_1^2+a_2^2b_1^2+a_1^2b_2^2+a_2^2b_2^2-(a_1^2b_1^2+2a_1a_2b_1b_2+a_2^2b_2^2)} \\
&= \sqrt{a_1^2b_2^2-2a_1a_2b_1b_2+a_2^2b_1^2} \\
&= \sqrt{(a_1b_2-a_2b_1)^2} \\
&= |a_1b_2-a_2b_1|.
\end{aligned}
$$

Since $\begin{vmatrix} a_1 & a_2 \\ b_1 & b_2 \end{vmatrix} = a_1b_2 - a_2b_1$, the proof is complete. ■

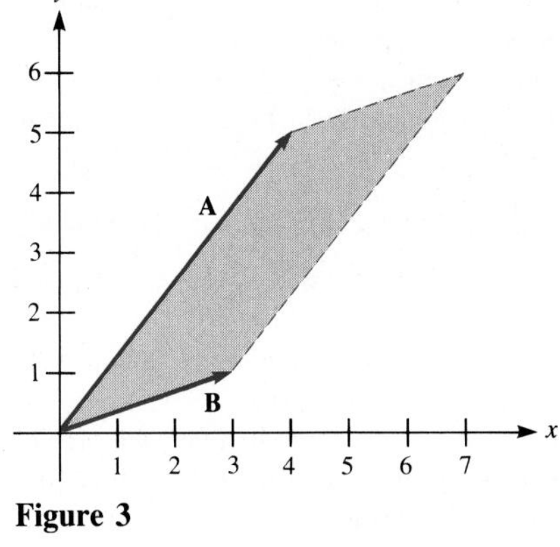

Figure 3

EXAMPLE 3 Find the area of the parallelogram spanned by

$$\mathbf{A} = \langle 4, 5\rangle \text{ and } \mathbf{B} = \langle 3, 1\rangle,$$

shown in Fig. 3.

SOLUTION The area is the absolute value of

$$\begin{vmatrix} 4 & 5 \\ 3 & 1 \end{vmatrix} = 4 - 15 = -11.$$

Hence the area is $|-11| = 11$. ■

Section Summary

In this section we defined the determinants of 2 by 2 and 3 by 3 matrices. We showed that if two rows (or two columns) of a matrix are the same, then the determinant is 0.

We also showed that a 2 by 2 determinant can be interpreted geometrically: The absolute value of the determinant $\begin{vmatrix} a_1 & a_2 \\ b_1 & b_2 \end{vmatrix}$ is the area of the parallelogram spanned by the vectors $\mathbf{A} = \langle a_1, a_2\rangle$ and $\mathbf{B} = \langle b_1, b_2\rangle$.

EXERCISES FOR SEC. 12.5: DETERMINANTS

In Exercises 1 to 10 evaluate the determinants. Use a shortcut if there is one.

1 $\begin{vmatrix} 3 & 4 \\ 7 & 2 \end{vmatrix}$

2 $\begin{vmatrix} 3 & 3 \\ 7 & 7 \end{vmatrix}$

3 $\begin{vmatrix} 4 & 2 & 0 \\ 5 & 6 & -1 \\ 1 & -1 & 2 \end{vmatrix}$

4 $\begin{vmatrix} 1 & 3 & 1 \\ 2 & 1 & 2 \\ 4 & 5 & 4 \end{vmatrix}$

5 $\begin{vmatrix} 2 & 1 & 4 \\ 2 & 1 & 4 \\ 3 & 5 & 7 \end{vmatrix}$

6 $\begin{vmatrix} 0 & 0 & 0 \\ 1 & 5 & 9 \\ 3 & -1 & 2 \end{vmatrix}$

7 $\begin{vmatrix} 1 & 3 & 4 \\ 2 & 5 & 7 \\ 1 & 3 & 4 \end{vmatrix}$

8 $\begin{vmatrix} 1 & 0 & 2 \\ 2 & 3 & -5 \\ 2 & 3 & -5 \end{vmatrix}$

9 $\begin{vmatrix} 1 & 1 & 1 \\ 3 & 4 & 3 \\ 5 & 2 & -1 \end{vmatrix}$ **10** $\begin{vmatrix} 1 & 2 & 3 \\ -1 & 4 & -2 \\ 5 & 1 & 0 \end{vmatrix}$

In Exercises 11 to 14 find the area of the parallelogram spanned by the given vectors.

11 $\mathbf{A} = 3\mathbf{i} + 4\mathbf{j}$, $\mathbf{B} = 2\mathbf{i} - \mathbf{j}$
12 $\mathbf{A} = 2\mathbf{i} + 3\mathbf{j}$, $\mathbf{B} = 4\mathbf{i} + 6\mathbf{j}$
13 $\mathbf{A} = 3\mathbf{i} + 5\mathbf{j}$, $\mathbf{B} = 5\mathbf{i} - 3\mathbf{j}$
14 $\mathbf{A} = -\mathbf{i} - \mathbf{j}$, $\mathbf{B} = 2\mathbf{i} + 3\mathbf{j}$

---

15 (*a*) Let (x_1, y_1) and (x_2, y_2) be two distinct points in the xy plane. Show that the equation

$$\begin{vmatrix} x & y & 1 \\ x_1 & y_1 & 1 \\ x_2 & y_2 & 1 \end{vmatrix} = 0$$

is an equation for the line determined by the two given points.

(*b*) Use the formula in (*a*) to find an equation of the line through the points (2, 3) and (1, −4).

16 Prove that if the first and third columns of a 3 by 3 matrix are identical, then the determinant of the matrix is 0.

17 Show that Theorem 2 implies Theorem 1.

18 Prove that if the first and third rows of a 3 by 3 matrix are switched, the determinant of the resulting matrix is the determinant of the original matrix with its sign changed.

19 Prove that

$$\begin{vmatrix} ka_1 & ka_2 \\ b_1 & b_2 \end{vmatrix} = k \begin{vmatrix} a_1 & a_2 \\ b_1 & b_2 \end{vmatrix}.$$

This illustrates the general theorem: If the entries in a single row (or single column) are all multiplied by the number k, the determinant of the resulting matrix is k times the determinant of the original matrix.

20 Prove that

$$\begin{vmatrix} a_1 & a_2 & a_3 \\ b_1 & b_2 & b_3 \\ c_1 & c_2 & c_3 \end{vmatrix} = \begin{vmatrix} a_1 & b_1 & c_1 \\ a_2 & b_2 & c_2 \\ a_3 & b_3 & c_3 \end{vmatrix}.$$

This says that if you spin a matrix around the diagonal that stretches from the top left corner to the bottom right, the determinant of the resulting matrix is the same as that of the original matrix. The second matrix is called the **transpose** of the first.

21 Prove that

$$\begin{vmatrix} a_1 + kb_1 & a_2 + kb_2 \\ b_1 & b_2 \end{vmatrix} = \begin{vmatrix} a_1 & a_2 \\ b_1 & b_2 \end{vmatrix}.$$

This illustrates the general theorem: If you multiply a row by a scalar and add the result to a different row, the resulting matrix has the same determinant as the original one. A similar theorem holds for columns.

22 Using the geometric interpretation of a determinant, explain why the equation in Exercise 21 is to be expected.

23 Is the determinant $\begin{vmatrix} a_1 & a_2 \\ b_1 & b_2 \end{vmatrix}$ related to the area of the parallelogram spanned by the "column vectors," $\mathbf{C} = \langle a_1, b_1 \rangle$ and $\mathbf{D} = \langle a_2, b_2 \rangle$?

24 Develop a formula for the area of the parallelogram spanned by $\mathbf{A} = \langle a_1, a_2, a_3 \rangle$ and $\mathbf{B} = \langle b_1, b_2, b_3 \rangle$.

25 (See Exercise 24.) Write a program to compute the area of the parallelogram spanned by $\mathbf{A} = \langle a_1, a_2, a_3 \rangle$ and $\mathbf{B} = \langle b_1, b_2, b_3 \rangle$.

26 Write a program for evaluating an arbitrary 3 by 3 determinant.

12.6 THE CROSS PRODUCT OF TWO VECTORS

It is frequently necessary in applications in space to construct a nonzero vector perpendicular to two given vectors **A** and **B**. This section provides a formula for finding such a vector.

The Cross Product of Two Vectors

Given two vectors, $\mathbf{A} = a_1\mathbf{i} + a_2\mathbf{j} + a_3\mathbf{k}$ and $\mathbf{B} = b_1\mathbf{i} + b_2\mathbf{j} + b_3\mathbf{k}$, we will give an algebraic definition of a certain vector **C**, which we will show is perpendicular to both **A** and **B**. Then we will obtain a geometric interpretation of **C**. (See Fig. 1.)

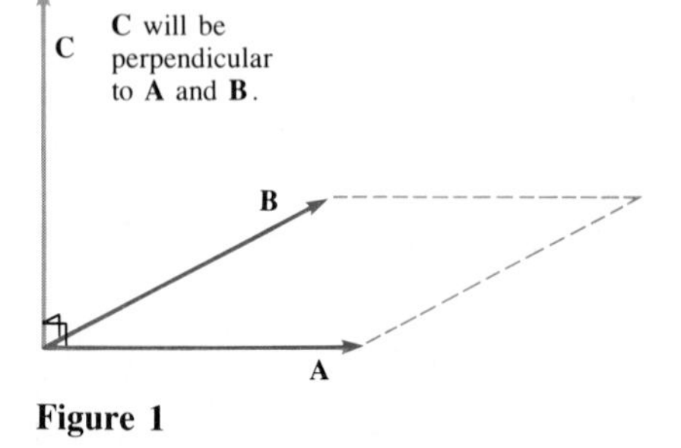

Figure 1

In the following definition, vectors appear as entries in a determinant. Although such a determinant does not fit into the general definition in the preceding section (where all entries were numbers), the notation is so convenient that we shall not hesitate to employ it.

Algebraic Definition of Cross Product

Definition *Cross product (vector product).* Let

$$\mathbf{A} = a_1\mathbf{i} + a_2\mathbf{j} + a_3\mathbf{k} \quad \text{and} \quad \mathbf{B} = b_1\mathbf{i} + b_2\mathbf{j} + b_3\mathbf{k}.$$

The vector

$$\begin{vmatrix} \mathbf{i} & \mathbf{j} & \mathbf{k} \\ a_1 & a_2 & a_3 \\ b_1 & b_2 & b_3 \end{vmatrix} = \mathbf{i}\begin{vmatrix} a_2 & a_3 \\ b_2 & b_3 \end{vmatrix} - \mathbf{j}\begin{vmatrix} a_1 & a_3 \\ b_1 & b_3 \end{vmatrix} + \mathbf{k}\begin{vmatrix} a_1 & a_2 \\ b_1 & b_2 \end{vmatrix}$$

$$= (a_2b_3 - a_3b_2)\mathbf{i} - (a_1b_3 - a_3b_1)\mathbf{j} + (a_1b_2 - a_2b_1)\mathbf{k}$$

is called the **cross product** (or **vector product**) of **A** and **B**. It is denoted

$$\mathbf{A} \times \mathbf{B}.$$

Theorem 1 shows that $\mathbf{A} \times \mathbf{B}$ *is perpendicular to* **A** *and to* **B**.

The determinant for $\mathbf{A} \times \mathbf{B}$ is expanded along its first row:

$$\begin{pmatrix} \mathbf{i} & \mathbf{j} & \mathbf{k} \\ a_1 & a_2 & a_3 \\ b_1 & b_2 & b_3 \end{pmatrix} \qquad \begin{pmatrix} \mathbf{i} & \mathbf{j} & \mathbf{k} \\ a_1 & a_2 & a_3 \\ b_1 & b_2 & b_3 \end{pmatrix} \qquad \begin{pmatrix} \mathbf{i} & \mathbf{j} & \mathbf{k} \\ a_1 & a_2 & a_3 \\ b_1 & b_2 & b_3 \end{pmatrix}$$

Delete the two lines through **i**. The determinant of the remaining square is the coefficient of **i** in $\mathbf{A} \times \mathbf{B}$.

Delete the two lines through **j**. The negative of the determinant of the remaining square is the coefficient of **j** in $\mathbf{A} \times \mathbf{B}$.

Delete the two lines through **k**. The determinant of the remaining square is the coefficient of **k** in $\mathbf{A} \times \mathbf{B}$.

EXAMPLE 1 Compute $\mathbf{A} \times \mathbf{B}$ if $\mathbf{A} = 2\mathbf{i} - \mathbf{j} + 3\mathbf{k}$ and $\mathbf{B} = 3\mathbf{i} + 4\mathbf{j} + \mathbf{k}$.

SOLUTION By definition,

$$\mathbf{A} \times \mathbf{B} = \begin{vmatrix} \mathbf{i} & \mathbf{j} & \mathbf{k} \\ 2 & -1 & 3 \\ 3 & 4 & 1 \end{vmatrix} = \mathbf{i}\begin{vmatrix} -1 & 3 \\ 4 & 1 \end{vmatrix} - \mathbf{j}\begin{vmatrix} 2 & 3 \\ 3 & 1 \end{vmatrix} + \mathbf{k}\begin{vmatrix} 2 & -1 \\ 3 & 4 \end{vmatrix}$$

$$= -13\mathbf{i} + 7\mathbf{j} + 11\mathbf{k}. \quad \blacksquare$$

Note that $\mathbf{A} \times \mathbf{B}$ is a vector, while $\mathbf{A} \cdot \mathbf{B}$ is a scalar. The most important property of the cross product is expressed in the following theorem.

Theorem 1 $\mathbf{A} \times \mathbf{B}$ is a vector perpendicular to both **A** and **B**.

Proof Let us show, for instance, that $\mathbf{A} \cdot (\mathbf{A} \times \mathbf{B})$ is 0. Now, $\mathbf{A} = a_1\mathbf{i} + a_2\mathbf{j} + a_3\mathbf{k}$, $\mathbf{B} = b_1\mathbf{i} + b_2\mathbf{j} + b_3\mathbf{k}$, and

$$\mathbf{A} \times \mathbf{B} = \mathbf{i}\begin{vmatrix} a_2 & a_3 \\ b_2 & b_3 \end{vmatrix} - \mathbf{j}\begin{vmatrix} a_1 & a_3 \\ b_1 & b_3 \end{vmatrix} + \mathbf{k}\begin{vmatrix} a_1 & a_2 \\ b_1 & b_2 \end{vmatrix}.$$

Thus

$$\mathbf{A} \cdot (\mathbf{A} \times \mathbf{B}) = a_1\begin{vmatrix} a_2 & a_3 \\ b_2 & b_3 \end{vmatrix} - a_2\begin{vmatrix} a_1 & a_3 \\ b_1 & b_3 \end{vmatrix} + a_3\begin{vmatrix} a_1 & a_2 \\ b_1 & b_2 \end{vmatrix},$$

which is precisely the definition of the determinant

$$\begin{vmatrix} a_1 & a_2 & a_3 \\ a_1 & a_2 & a_3 \\ b_1 & b_2 & b_3 \end{vmatrix}.$$

Since two rows are identical, the determinant equals 0. Thus $\mathbf{A} \cdot (\mathbf{A} \times \mathbf{B}) = 0$.

A similar argument shows that $\mathbf{B} \cdot (\mathbf{A} \times \mathbf{B}) = 0$. This completes the proof. ■

EXAMPLE 2 Verify Theorem 1 in the case of the vectors **A**, **B**, and $\mathbf{A} \times \mathbf{B}$ of Example 1.

SOLUTION In the case of Example 1,

$$\begin{aligned} \mathbf{A} \cdot (\mathbf{A} \times \mathbf{B}) &= (2\mathbf{i} - \mathbf{j} + 3\mathbf{k}) \cdot (-13\mathbf{i} + 7\mathbf{j} + 11\mathbf{k}) \\ &= -26 - 7 + 33 \\ &= 0, \end{aligned}$$

and

$$\begin{aligned} \mathbf{B} \cdot (\mathbf{A} \times \mathbf{B}) &= (3\mathbf{i} + 4\mathbf{j} + \mathbf{k}) \cdot (-13\mathbf{i} + 7\mathbf{j} + 11\mathbf{k}) \\ &= -39 + 28 + 11 \\ &= 0. \end{aligned}$$

These results verify the theorem. ■

EXAMPLE 3 Compute $\mathbf{i} \times \mathbf{j}$ and $\mathbf{j} \times \mathbf{i}$.

SOLUTION

$$\mathbf{i} \times \mathbf{j} = \begin{vmatrix} \mathbf{i} & \mathbf{j} & \mathbf{k} \\ 1 & 0 & 0 \\ 0 & 1 & 0 \end{vmatrix} = 0\mathbf{i} - 0\mathbf{j} + \mathbf{k} = \mathbf{k},$$

and

$$\mathbf{j} \times \mathbf{i} = \begin{vmatrix} \mathbf{i} & \mathbf{j} & \mathbf{k} \\ 0 & 1 & 0 \\ 1 & 0 & 0 \end{vmatrix} = 0\mathbf{i} - 0\mathbf{j} - \mathbf{k} = -\mathbf{k}.$$

Thus $\mathbf{i} \times \mathbf{j}$ is the negative of $\mathbf{j} \times \mathbf{i}$. ■

Computations like those in Example 3 show that the cross products of the basic unit vectors are:

$$\begin{array}{lll} \mathbf{i} \times \mathbf{i} = \mathbf{0} & \mathbf{j} \times \mathbf{j} = \mathbf{0} & \mathbf{k} \times \mathbf{k} = \mathbf{0} \\ \mathbf{i} \times \mathbf{j} = \mathbf{k} & \mathbf{j} \times \mathbf{k} = \mathbf{i} & \mathbf{k} \times \mathbf{i} = \mathbf{j} \\ \mathbf{j} \times \mathbf{i} = -\mathbf{k} & \mathbf{k} \times \mathbf{j} = -\mathbf{i} & \mathbf{i} \times \mathbf{k} = -\mathbf{j}. \end{array}$$

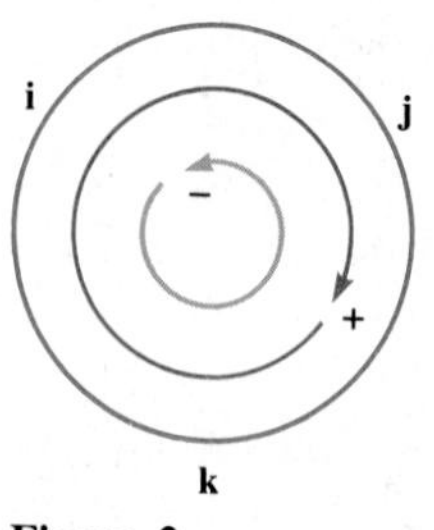

Figure 2

Figure 2 is a way to remember the last six formulas. The cross product of any two of **i**, **j**, **k** is ± the third, + if the direction from the first to the second is clockwise, − if counterclockwise.

The order of the factors in the vector product is critical.

Example 3 shows that $\mathbf{A} \times \mathbf{B}$ may be different from $\mathbf{B} \times \mathbf{A}$. Indeed, it is easy to show that for all vectors **A** and **B**,

$$\mathbf{B} \times \mathbf{A} = -(\mathbf{A} \times \mathbf{B}).$$

This property corresponds to the fact that when two rows of a matrix are interchanged, its determinant changes sign.

Another surprising property of the operation $\times$ is that for all vectors $\mathbf{A}$,

$$\mathbf{A} \times \mathbf{A} = \mathbf{0}.$$

This corresponds to the fact that if two rows of a matrix are identical, then its determinant is 0. More generally, if $\mathbf{A}$ and $\mathbf{B}$ are parallel,

$$\mathbf{A} \times \mathbf{B} = \mathbf{0}.$$

After these shocks, it may be comforting to know that

$$\mathbf{A} \times (\mathbf{B} + \mathbf{C}) = \mathbf{A} \times \mathbf{B} + \mathbf{A} \times \mathbf{C},$$

which is reminiscent of the arithmetic of numbers. This distributive law can be established by a straightforward computation. (See Exercise 40 for an alternate geometric approach.)

On the other hand, the associative law does not hold. Usually $\mathbf{A} \times (\mathbf{B} \times \mathbf{C})$ does *not* equal $(\mathbf{A} \times \mathbf{B}) \times \mathbf{C}$. For instance,

$$\mathbf{i} \times (\mathbf{i} \times \mathbf{j}) = \mathbf{i} \times \mathbf{k} = -\mathbf{j}$$

but

$$(\mathbf{i} \times \mathbf{i}) \times \mathbf{j} = \mathbf{0} \times \mathbf{j} = \mathbf{0}.$$

A scalar can be factored out of a cross product:

$$(c\mathbf{A}) \times \mathbf{B} = c(\mathbf{A} \times \mathbf{B})$$

$$\mathbf{A} \times (c\mathbf{B}) = c(\mathbf{A} \times \mathbf{B}).$$

We can summarize these facts in Theorem 2, which can be proved using the definition of the cross product.

Theorem 2 For any vectors $\mathbf{A}$, $\mathbf{B}$, and $\mathbf{C}$,

(*a*) $\mathbf{B} \times \mathbf{A} = -(\mathbf{A} \times \mathbf{B})$
(*b*) $\mathbf{A} \times \mathbf{A} = \mathbf{0}$
(*c*) If $\mathbf{A}$ is parallel to $\mathbf{B}$, then $\mathbf{A} \times \mathbf{B} = \mathbf{0}$.
(*d*) $\mathbf{A} \times (\mathbf{B} + \mathbf{C}) = \mathbf{A} \times \mathbf{B} + \mathbf{A} \times \mathbf{C}$
$(\mathbf{B} + \mathbf{C}) \times \mathbf{A} = \mathbf{B} \times \mathbf{A} + \mathbf{C} \times \mathbf{A}$
(*e*) $(c\mathbf{A}) \times \mathbf{B} = \mathbf{A} \times (c\mathbf{B}) = c(\mathbf{A} \times \mathbf{B})$.

Geometric Description of the Cross Product

The definition of $\mathbf{A} \times \mathbf{B}$ is algebraic. For many applications it is important to have a geometric description of $\mathbf{A} \times \mathbf{B}$, one that expresses the direction and magnitude of $\mathbf{A} \times \mathbf{B}$ in terms of those of $\mathbf{A}$ and $\mathbf{B}$.

Let $\mathbf{A}$ and $\mathbf{B}$ be nonzero vectors that are not parallel. We already know that $\mathbf{A} \times \mathbf{B}$ is perpendicular to both $\mathbf{A}$ and $\mathbf{B}$. However, there are two directions in which $\mathbf{A} \times \mathbf{B}$ may point. Which is the correct one? The clue is given by the case

$$\mathbf{i} \times \mathbf{j} = \mathbf{k},$$

shown in Fig. 3.

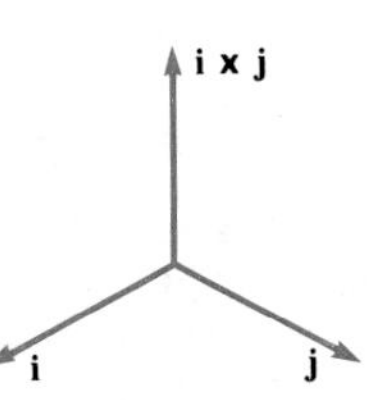
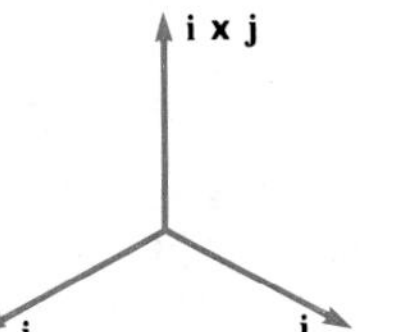

Figure 3

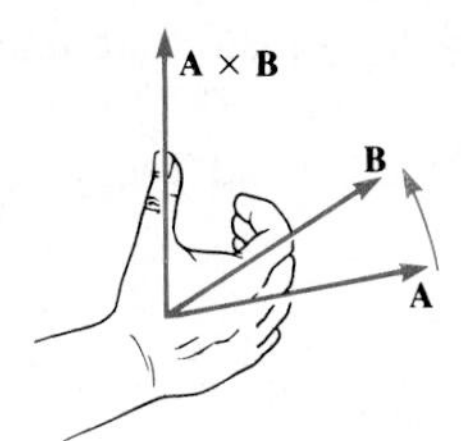

Since the positive x, y, and z axes were chosen to match the right-hand rule, this rule describes in general the direction of $\mathbf{A} \times \mathbf{B}$. If the fingers of the right hand curl from $\mathbf{A}$ to $\mathbf{B}$ through an angle less than 180°, the thumb points in the direction of $\mathbf{A} \times \mathbf{B}$. (See Fig. 4.)

The direction of $\mathbf{A} \times \mathbf{B}$ is completely determined. But what is the length of $\mathbf{A} \times \mathbf{B}$? It is given by the following theorem.

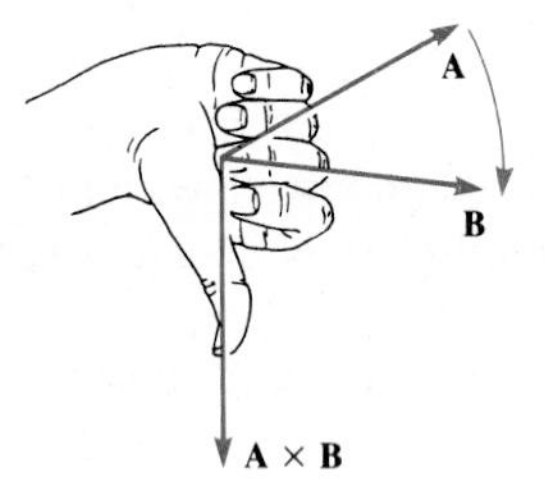

Figure 4

Theorem 3 The magnitude of $\mathbf{A} \times \mathbf{B}$ is equal to the area of the parallelogram spanned by $\mathbf{A}$ and $\mathbf{B}$.

Proof Let θ be the angle between $\mathbf{A}$ and $\mathbf{B}$. Then the area of the parallelogram spanned by $\mathbf{A}$ and $\mathbf{B}$ (shown in Fig. 5) is

$$\|\mathbf{A}\|\,\|\mathbf{B}\| \sin\theta.$$

To avoid square roots, consider the square of the area of the parallelogram. We wish to show that

$$\|\mathbf{A}\|^2\|\mathbf{B}\|^2 \sin^2\theta = \|\mathbf{A} \times \mathbf{B}\|^2;$$

that is,

$$\|\mathbf{A}\|^2\|\mathbf{B}\|^2 \sin^2\theta = \begin{vmatrix} a_2 & a_3 \\ b_2 & b_3 \end{vmatrix}^2 + \begin{vmatrix} a_1 & a_3 \\ b_1 & b_3 \end{vmatrix}^2 + \begin{vmatrix} a_1 & a_2 \\ b_1 & b_2 \end{vmatrix}^2. \tag{1}$$

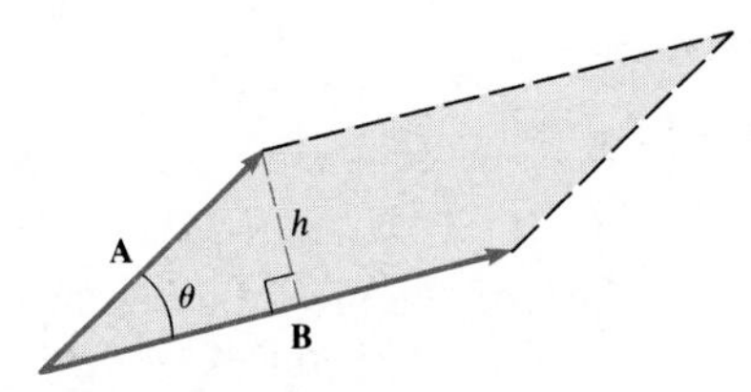

The parallelogram has area $\|\mathbf{B}\|h = \|\mathbf{B}\|\|\mathbf{A}\|\sin\theta = \|\mathbf{A}\|\|\mathbf{B}\|\sin\theta$.

Figure 5

Now,

$$\begin{aligned} \sin^2\theta &= 1 - \cos^2\theta \\ &= 1 - \frac{(\mathbf{A}\cdot\mathbf{B})^2}{\|\mathbf{A}\|^2\|\mathbf{B}\|^2} \\ &= \frac{\|\mathbf{A}\|^2\|\mathbf{B}\|^2 - (\mathbf{A}\cdot\mathbf{B})^2}{\|\mathbf{A}\|^2\|\mathbf{B}\|^2}. \end{aligned}$$

Thus,

$$\sin^2\theta = \frac{(a_1^2 + a_2^2 + a_3^2)(b_1^2 + b_2^2 + b_3^2) - (a_1b_1 + a_2b_2 + a_3b_3)^2}{\|\mathbf{A}\|^2\|\mathbf{B}\|^2}. \tag{2}$$

Comparison of Eqs. (1) and (2) with the aid of elementary algebra completes the proof. ■

$\mathbf{A} \times \mathbf{B}$ *described geometrically*

In short, $\mathbf{A} \times \mathbf{B}$ *is that vector perpendicular to both* $\mathbf{A}$ *and* $\mathbf{B}$, *whose direction is obtained by the right-hand rule and whose length is the area of the parallelogram spanned by* $\mathbf{A}$ *and* $\mathbf{B}$. If $\mathbf{A}$ and $\mathbf{B}$ are not $\mathbf{0}$, and θ is the angle between $\mathbf{A}$ and $\mathbf{B}$, this area is $\|\mathbf{A}\|\,\|\mathbf{B}\|\sin\theta$. If $\mathbf{A}$ or $\mathbf{B}$ is $\mathbf{0}$, or if $\mathbf{A}$ is parallel to $\mathbf{B}$, then $\mathbf{A} \times \mathbf{B}$ is the vector $\mathbf{0}$.

If $\mathbf{n}$ is a unit vector perpendicular to both $\mathbf{A}$ and $\mathbf{B}$ and determined by the right-hand rule, then

$$\mathbf{A} \times \mathbf{B} = \|\mathbf{A}\|\,\|\mathbf{B}\|(\sin\theta)\mathbf{n},$$

which is "the area of the parallelogram spanned by $\mathbf{A}$ and $\mathbf{B}$ times the unit vector $\mathbf{n}$."

Some Applications of the Cross Product

In the next example the cross product is used to obtain the geometric interpretation of a 2 by 2 determinant discussed in Sec. 12.5.

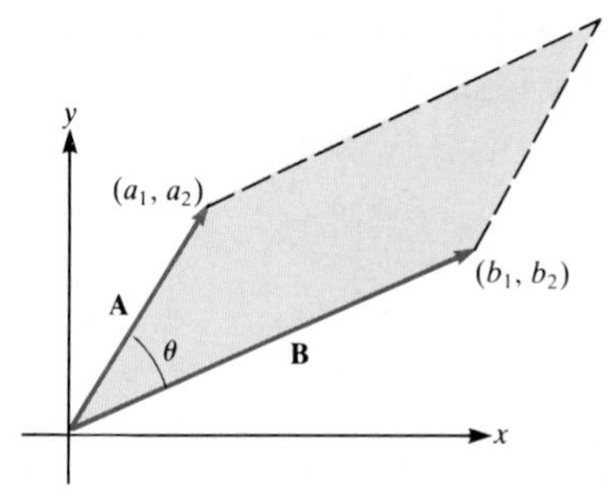

Figure 6

EXAMPLE 4 A parallelogram in the plane has the vertices $(0, 0)$, (a_1, a_2), (b_1, b_2), and $(a_1 + b_1, a_2 + b_2)$. Find its area. (See Fig. 6.)

SOLUTION The parallelogram is spanned by the vectors

$$\mathbf{A} = a_1\mathbf{i} + a_2\mathbf{j} + 0\mathbf{k} \qquad \text{and} \qquad \mathbf{B} = b_1\mathbf{i} + b_2\mathbf{j} + 0\mathbf{k}.$$

Consequently, its area is the magnitude of the vector

$$\begin{vmatrix} \mathbf{i} & \mathbf{j} & \mathbf{k} \\ a_1 & a_2 & 0 \\ b_1 & b_2 & 0 \end{vmatrix} = 0\mathbf{i} - 0\mathbf{j} + \mathbf{k}\begin{vmatrix} a_1 & a_2 \\ b_1 & b_2 \end{vmatrix}.$$

Thus

$$\text{Area of parallelogram spanned by } \mathbf{A} \text{ and } \mathbf{B} = \pm\begin{vmatrix} a_1 & a_2 \\ b_1 & b_2 \end{vmatrix}. \quad ■$$

Remark: It can be shown that

$$\begin{vmatrix} a_1 & a_2 \\ b_1 & b_2 \end{vmatrix}$$

is positive if **A** is the initial side of angle θ and **B** is the terminal side, where θ is traversed in a counterclockwise direction. (We can also describe this by saying that the orientation "from **A** to **B**" is counterclockwise. Check this for $\mathbf{A} = \mathbf{i}$ and $\mathbf{B} = \mathbf{j}$.) The determinant is negative when the orientation from **A** to **B** is clockwise. (Figure 6 illustrates this case.)

The next example is typical of the geometric applications of the cross product.

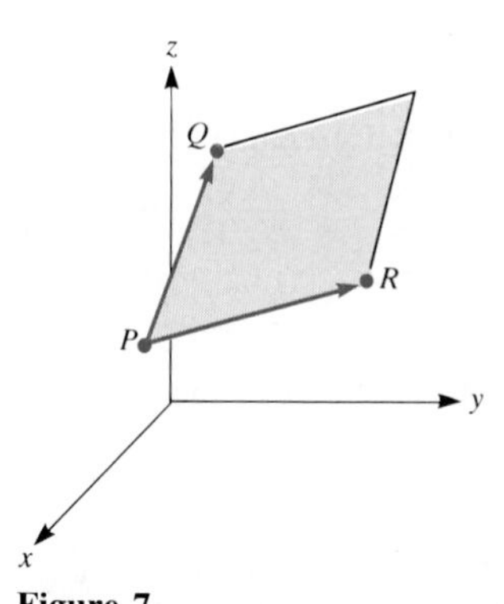

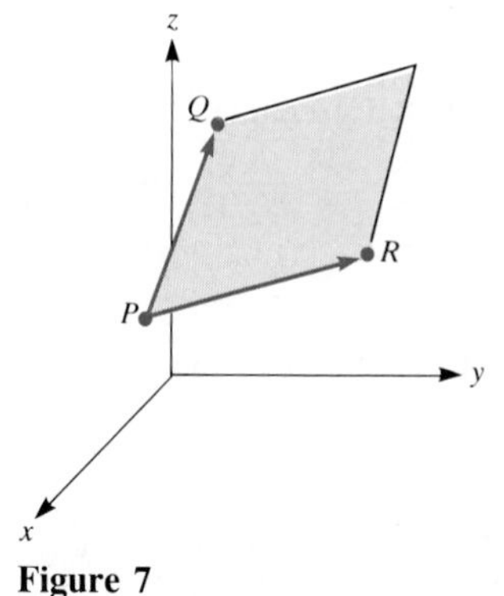
Figure 7

EXAMPLE 5 Find a vector perpendicular to the plane determined by the three points

$$P = (1, 3, 2), \qquad Q = (4, -1, 1), \qquad \text{and} \qquad R = (3, 0, 2).$$

SOLUTION The vectors $\overrightarrow{PQ}$ and $\overrightarrow{PR}$ lie in a plane (see Fig. 7). The vector $\mathbf{N} = \overrightarrow{PQ} \times \overrightarrow{PR}$, being perpendicular to both $\overrightarrow{PQ}$ and $\overrightarrow{PR}$, is perpendicular to the plane. Now,

$$\overrightarrow{PQ} = 3\mathbf{i} - 4\mathbf{j} - \mathbf{k} \qquad \text{and} \qquad \overrightarrow{PR} = 2\mathbf{i} - 3\mathbf{j} + 0\mathbf{k}.$$

Thus

$$\mathbf{N} = \begin{vmatrix} \mathbf{i} & \mathbf{j} & \mathbf{k} \\ 3 & -4 & -1 \\ 2 & -3 & 0 \end{vmatrix} = -3\mathbf{i} - 2\mathbf{j} - \mathbf{k}. \quad ■$$

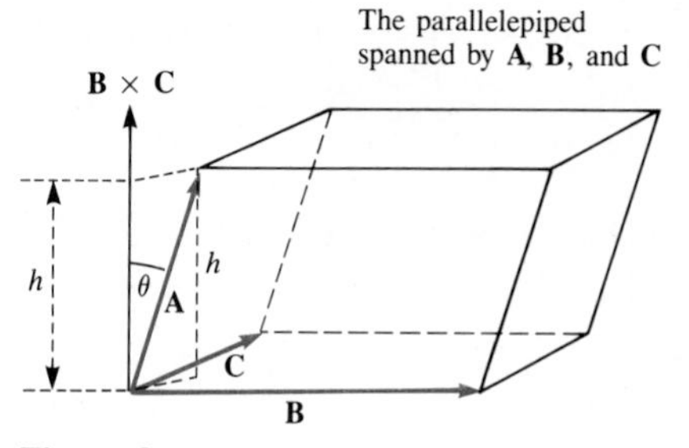

Figure 8

The Scalar Triple Product

Let **A**, **B**, and **C** be vectors placed so that their tails coincide. They span a parallelepiped, shown in Fig. 8.

The volume of the parallelepiped is the product of the area of the base and the height (h in the diagram). The area of the base is $\|\mathbf{B} \times \mathbf{C}\|$. If θ is the angle between $\mathbf{B} \times \mathbf{C}$ and $\mathbf{A}$, then

$$h = \|\mathbf{A}\| \cos \theta.$$

Thus the volume of the parallelepiped is

$$\underbrace{\|\mathbf{B} \times \mathbf{C}\|}_{\text{Area of base}} \; \underbrace{\|\mathbf{A}\| \cos \theta}_{\text{Height}}.$$

Hence

$$\boxed{\text{Volume of parallelepiped} = |\mathbf{A} \cdot (\mathbf{B} \times \mathbf{C})|.}$$

Similar arguments show that the volume is also equal to the absolute value of $\mathbf{B} \cdot (\mathbf{A} \times \mathbf{C})$ or $\mathbf{C} \cdot (\mathbf{A} \times \mathbf{B})$. In each case the volume is the absolute value of the dot product of one vector with the cross product of the other two vectors. Any of these expressions, such as $\mathbf{A} \cdot (\mathbf{B} \times \mathbf{C})$, is called a **scalar triple product**. Using properties of determinants, we can show that

$$\boxed{\mathbf{A} \cdot (\mathbf{B} \times \mathbf{C}) = (\mathbf{A} \times \mathbf{B}) \cdot \mathbf{C}.}$$

Moreover, interchanging any two vectors in $\mathbf{A} \cdot (\mathbf{B} \times \mathbf{C})$ changes the sign. For instance $\mathbf{C} \cdot (\mathbf{B} \times \mathbf{A}) = -\mathbf{A} \cdot (\mathbf{B} \times \mathbf{C})$.

The Geometry of a 3 by 3 Determinant

The parallelepiped spanned by $\mathbf{A}$, $\mathbf{B}$, and $\mathbf{C}$ has volume equal to the absolute value of $\mathbf{A} \cdot (\mathbf{B} \times \mathbf{C})$. This fact is the basis of the following theorem.

Theorem 4 The absolute value of the determinant

$$\begin{vmatrix} a_1 & a_2 & a_3 \\ b_1 & b_2 & b_3 \\ c_1 & c_2 & c_3 \end{vmatrix}$$

is equal to the volume of the parallelepiped spanned by the three "row vectors," $\mathbf{A} = \langle a_1, a_2, a_3 \rangle$, $\mathbf{B} = \langle b_1, b_2, b_3 \rangle$, and $\mathbf{C} = \langle c_1, c_2, c_3 \rangle$.

Proof We know that the volume of the parallelepiped is the absolute value of

$$\mathbf{A} \cdot (\mathbf{B} \times \mathbf{C}).$$

That is,

$$\begin{aligned} \text{Volume} &= \pm(a_1\mathbf{i} + a_2\mathbf{j} + a_3\mathbf{k}) \cdot ((b_1\mathbf{i} + b_2\mathbf{j} + b_3\mathbf{k}) \times (c_1\mathbf{i} + c_2\mathbf{j} + c_3\mathbf{k})) \\ &= \pm(a_1\mathbf{i} + a_2\mathbf{j} + a_3\mathbf{k}) \cdot \left(\begin{vmatrix} b_2 & b_3 \\ c_2 & c_3 \end{vmatrix} \mathbf{i} - \begin{vmatrix} b_1 & b_3 \\ c_1 & c_3 \end{vmatrix} \mathbf{j} + \begin{vmatrix} b_1 & b_2 \\ c_1 & c_2 \end{vmatrix} \mathbf{k} \right) \end{aligned}$$

$$= \pm\left(a_1 \begin{vmatrix} b_2 & b_3 \\ c_2 & c_3 \end{vmatrix} - a_2 \begin{vmatrix} b_1 & b_3 \\ c_1 & c_3 \end{vmatrix} + a_3 \begin{vmatrix} b_1 & b_2 \\ c_1 & c_2 \end{vmatrix}\right)$$

$$= \pm \begin{vmatrix} a_1 & a_2 & a_3 \\ b_1 & b_2 & b_3 \\ c_1 & c_2 & c_3 \end{vmatrix}. \quad \text{(by definition of a 3 by 3 determinant)}$$ ■

EXAMPLE 6 Find the volume of the parallelepiped spanned by $\mathbf{i} + 2\mathbf{j} - 3\mathbf{k}$, $2\mathbf{i} - 2\mathbf{j} - 3\mathbf{k}$, and $\mathbf{i} + 3\mathbf{j} + 2\mathbf{k}$.

SOLUTION The volume is the absolute value of

$$\begin{vmatrix} 1 & 2 & -3 \\ 2 & -2 & -3 \\ 1 & 3 & 2 \end{vmatrix} = 1 \begin{vmatrix} -2 & -3 \\ 3 & 2 \end{vmatrix} - 2 \begin{vmatrix} 2 & -3 \\ 1 & 2 \end{vmatrix} - 3 \begin{vmatrix} 2 & -2 \\ 1 & 3 \end{vmatrix}$$

$$= (1)(5) - 2(7) - (3)(8)$$

$$= -33.$$

The volume is $|-33| = 33$. ■

The Vector Triple Product

The expression $\mathbf{A} \times (\mathbf{B} \times \mathbf{C})$ is called the **vector triple product**. If $\mathbf{B}$ and $\mathbf{C}$ are nonzero and nonparallel vectors, $\mathbf{B} \times \mathbf{C}$ is not $\mathbf{0}$. Then $\mathbf{A} \times (\mathbf{B} \times \mathbf{C})$ is perpendicular to $\mathbf{B} \times \mathbf{C}$. (Why?) Therefore it can be written in the form $x\mathbf{B} + y\mathbf{C}$ for suitable scalars x and y. (Why?) That is,

$$\mathbf{A} \times (\mathbf{B} \times \mathbf{C}) = x\mathbf{B} + y\mathbf{C}.$$

In Exercise 38 it is shown that $x = \mathbf{A} \cdot \mathbf{C}$ and $y = -\mathbf{A} \cdot \mathbf{B}$. Thus:

Formula for the vector triple product
$\mathbf{A} \times (\mathbf{B} \times \mathbf{C}) = (\mathbf{A} \cdot \mathbf{C})\mathbf{B} - (\mathbf{A} \cdot \mathbf{B})\mathbf{C}$.

Some physicists call this the most important vector identity, write it as

$$\mathbf{A} \times (\mathbf{B} \times \mathbf{C}) = \mathbf{B}(\mathbf{A} \cdot \mathbf{C}) - \mathbf{C}(\mathbf{A} \cdot \mathbf{B}),$$

and call it the ''BAC-CAB rule.''

TORQUE AND THE CROSS PRODUCT

Let L be a line in space, P a point not on L, and $\mathbf{F}$ a force applied at P. This force produces a turning tendency about L. How do we compute this turning tendency? (See Fig. 9.)

Consider a special case. Let O be the point such that $\mathbf{r} = \overrightarrow{OP}$ is perpendicular to L and $\mathbf{F}$ is perpendicular to L and $\mathbf{r}$, as in Fig. 10. In this simple case we define the turning tendency to be the product of the lever arm $\|\mathbf{r}\|$ and the magnitude of the force $\|\mathbf{F}\|$. Thus the turning tendency is $\|\mathbf{r} \times \mathbf{F}\|$.

Consider the general case, shown in Fig. 11. Let O be any point on L and let $\mathbf{r} = \overrightarrow{OP}$, as in Fig. 11. Let $\mathbf{n}$ be a unit vector parallel to L. Define $\mathbf{r}_1 = \mathbf{orth_n}\, \mathbf{r}$ and $\mathbf{r}_2 = \mathbf{proj_n}\, \mathbf{r}$.

Then $\mathbf{F} = \mathbf{F}_1 + \mathbf{F}_2 + \mathbf{F}_3$, where $\mathbf{F}_1$ is parallel to $\mathbf{r}_1$, $\mathbf{F}_2$ is parallel to $\mathbf{n}$, and $\mathbf{F}_3$ is

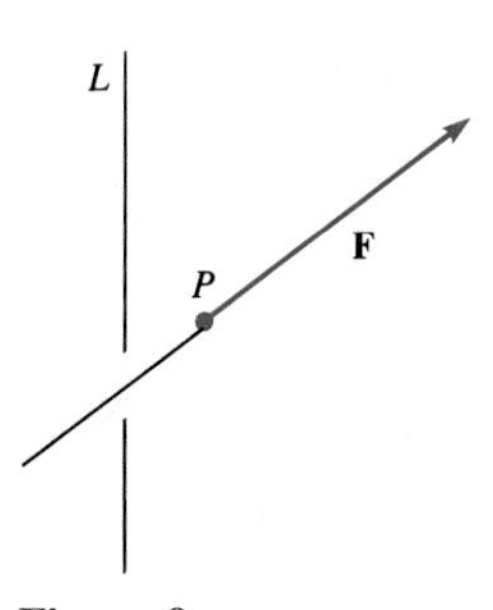

Figure 9

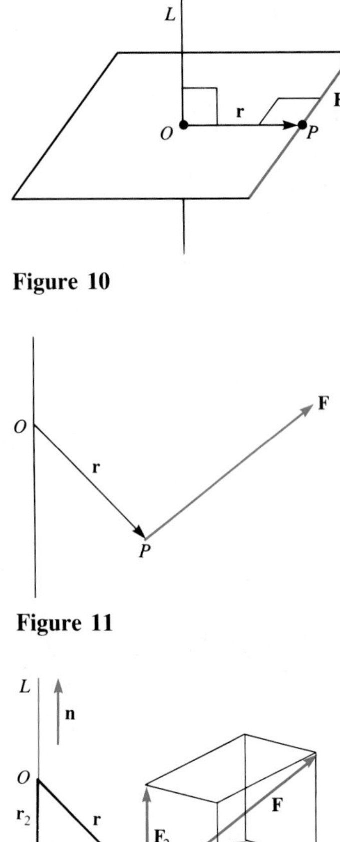

Figure 10

Figure 11

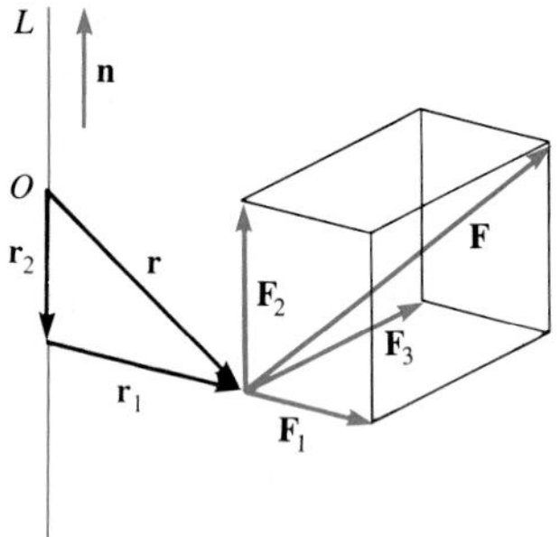

Figure 12

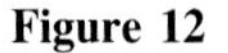

perpendicular to $\mathbf{r}_1$ and $\mathbf{n}$, as shown in Fig. 12. Since $\mathbf{F}_1$ and $\mathbf{F}_2$ act in the plane of L and $\mathbf{r}$, they contribute no turning tendency about L. Therefore the turning tendency of $\mathbf{F}$ is that of $\mathbf{F}_3$. But, as we saw, this turning tendency is $\|\mathbf{r}_1 \times \mathbf{F}_3\|$. Note that $\|\mathbf{r}_1 \times \mathbf{F}_3\|$ is the same as the absolute value of $\mathbf{n} \cdot (\mathbf{r}_1 \times \mathbf{F}_3)$. We now show that $\mathbf{n} \cdot (\mathbf{r}_1 \times \mathbf{F}_3) = \mathbf{n} \cdot (\mathbf{r} \times \mathbf{F})$, and hence we can express the turning tendency directly in terms of the original vectors, $\mathbf{n}$, $\mathbf{r}$, and $\mathbf{F}$.

We have

$$\begin{aligned} \mathbf{n} \cdot (\mathbf{r} \times \mathbf{F}) &= \mathbf{n} \cdot [(\mathbf{r}_1 + \mathbf{r}_2) \times (\mathbf{F}_1 + \mathbf{F}_2 + \mathbf{F}_3)] \\ &= \mathbf{n} \cdot (\mathbf{r}_1 \times \mathbf{F}_1) + \mathbf{n} \cdot (\mathbf{r}_1 \times \mathbf{F}_2) + \mathbf{n} \cdot (\mathbf{r}_1 \times \mathbf{F}_3) \\ &\quad + \mathbf{n} \cdot (\mathbf{r}_2 \times \mathbf{F}_1) + \mathbf{n} \cdot (\mathbf{r}_2 \times \mathbf{F}_2) + \mathbf{n} \cdot (\mathbf{r}_2 \times \mathbf{F}_3) \\ &= \mathbf{n} \cdot (\mathbf{r}_1 \times \mathbf{F}_3), \end{aligned}$$

since the other summands are 0. (Why?)

Therefore the turning tendency of $\mathbf{F}$, applied at P, around L, is the absolute value of $\mathbf{n} \cdot (\mathbf{r} \times \mathbf{F})$. The vector $\mathbf{r} \times \mathbf{F}$ is called the **torque** of $\mathbf{F}$ about O. To find the turning tendency of $\mathbf{F}$ applied at P, about any line L through O, compute the dot product of $\mathbf{r} \times \mathbf{F}$ with a unit vector that is parallel to L, getting $\mathbf{n} \cdot (\mathbf{r} \times \mathbf{F})$. Then $|\mathbf{n} \cdot (\mathbf{r} \times \mathbf{F})|$ is the magnitude of that turning tendency.

Thus, while $\mathbf{r} \times \mathbf{F}$ is not meaningful in itself, $\mathbf{n} \cdot (\mathbf{r} \times \mathbf{F})$ is meaningful for each unit vector $\mathbf{n}$.

Section Summary

We defined the ''vector product'' or ''cross product,'' $\mathbf{A} \times \mathbf{B}$, algebraically:

$$\mathbf{A} \times \mathbf{B} = \begin{vmatrix} \mathbf{i} & \mathbf{j} & \mathbf{k} \\ a_1 & a_2 & a_3 \\ b_1 & b_2 & b_3 \end{vmatrix}.$$

We then showed that $\mathbf{A} \times \mathbf{B}$ is perpendicular to both $\mathbf{A}$ and $\mathbf{B}$ and that $\|\mathbf{A} \times \mathbf{B}\|$ is (plus or minus) the area of the parallelogram spanned by $\mathbf{A}$ and $\mathbf{B}$. The direction of $\mathbf{A} \times \mathbf{B}$ is given by the right-hand rule. (If $\mathbf{A}$ and $\mathbf{B}$ are parallel, $\mathbf{A} \times \mathbf{B} = \mathbf{0}$.)

For any vectors $\mathbf{A}$, $\mathbf{B}$, and $\mathbf{C}$:

$$\begin{aligned} \mathbf{A} \times \mathbf{A} &= \mathbf{0}, \\ \mathbf{B} \times \mathbf{A} &= -(\mathbf{A} \times \mathbf{B}), \\ \mathbf{A} \times (\mathbf{B} + \mathbf{C}) &= \mathbf{A} \times \mathbf{B} + \mathbf{A} \times \mathbf{C}, \\ (c\mathbf{A}) \times \mathbf{B} &= c(\mathbf{A} \times \mathbf{B}) = \mathbf{A} \times c\mathbf{B}. \end{aligned}$$

Moreover the scalar triple product $\mathbf{A} \cdot (\mathbf{B} \times \mathbf{C})$ equals the determinant whose three rows from top to bottom are the components of $\mathbf{A}$, $\mathbf{B}$, and $\mathbf{C}$, respectively.

The vector triple product $\mathbf{A} \times (\mathbf{B} \times \mathbf{C})$ equals $(\mathbf{A} \cdot \mathbf{C})\mathbf{B} - (\mathbf{A} \cdot \mathbf{B})\mathbf{C}$.

EXERCISES FOR SEC. 12.6: THE CROSS PRODUCT OF TWO VECTORS

In Exercises 1 to 4 compute and sketch $\mathbf{A} \times \mathbf{B}$.

1 $\mathbf{A} = \mathbf{k}$, $\mathbf{B} = \mathbf{j}$

2 $\mathbf{A} = \mathbf{i} + \mathbf{j}$, $\mathbf{B} = \mathbf{i} - \mathbf{j}$

3 $\mathbf{A} = \mathbf{i} + \mathbf{j} + \mathbf{k}$, $\mathbf{B} = \mathbf{i} + \mathbf{j}$

4 $\mathbf{A} = \mathbf{k}$, $\mathbf{B} = \mathbf{i} + \mathbf{j}$

In Exercises 5 and 6 compute $\mathbf{A} \times \mathbf{B}$ and check that it is perpendicular to both $\mathbf{A}$ and to $\mathbf{B}$.

5 $\mathbf{A} = 2\mathbf{i} - 3\mathbf{j} + \mathbf{k}$, $\mathbf{B} = \mathbf{i} + \mathbf{j} + 2\mathbf{k}$

6 $\mathbf{A} = \mathbf{i} - \mathbf{j}$, $\mathbf{B} = \mathbf{j} + 4\mathbf{k}$

In Exercises 7 to 10 use the cross product.

7 Find the area of a parallelogram three of whose vertices are (0, 0, 0), (1, 5, 4), and (2, −1, 3).

8 Find the area of a parallelogram three of whose vertices are (1, 2, −1), (2, 1, 4), and (3, 5, 2).

9 Find the area of the triangle two of whose sides are $\mathbf{i} + \mathbf{j}$ and $3\mathbf{i} - \mathbf{j}$.

10 Find the area of the triangle two of whose sides are $\mathbf{i} + 2\mathbf{j} + 3\mathbf{k}$ and $2\mathbf{i} - \mathbf{j} + 2\mathbf{k}$.

In Exercises 11 to 14 find the volumes of the parallelepipeds spanned by the given vectors.

11 $\langle 2, 1, 3\rangle$, $\langle 3, -1, 2\rangle$, $\langle 4, 0, 3\rangle$.

12 $\underline{3\mathbf{i}} + \underline{4\mathbf{j}} + \underline{2\mathbf{k}}$, $2\mathbf{i} + 3\mathbf{j} + 4\mathbf{k}$, $\mathbf{i} - \mathbf{j} - \mathbf{k}$.

13 $\overrightarrow{PQ}$, $\overrightarrow{PR}$, $\overrightarrow{PS}$, where $P = (1, 1, 1)$, $Q = (2, 1, -2)$, $R = (3, 5, 2)$, and $S = (1, -1, 2)$.

14 $\overrightarrow{PQ}$, $\overrightarrow{PR}$, $\overrightarrow{PS}$, where $P = (0, 0, 0)$, $Q = (3, 3, 2)$, $R = (1, 4, -1)$, $S = (1, 2, 3)$.

15 Find a vector perpendicular to the plane determined by the points (1, 2, 1), (2, 1, −3), and (0, 1, 5).

16 Find a vector perpendicular to the plane determined by the points (1, 3, −1), (2, 1, 1), and (1, 3, 4).

17 Find a vector that is perpendicular to the line through the points (3, 6, 1) and (2, 7, 2) and also to the line through the points (2, 1, 4) and (1, −2, 3).

18 Find a vector perpendicular to the line through (1, 2, 1) and (4, 1, 0) and also to the line through (3, 5, 2) and (2, 6, −3).

19 Evaluate $\mathbf{A} \cdot (\mathbf{A} \times \mathbf{B})$.

20 If $\mathbf{A} \times \mathbf{B} \neq \mathbf{0}$, find (*a*) $\mathbf{proj}_{\mathbf{A}\times\mathbf{B}}\, \mathbf{A}$, (*b*) $\mathbf{orth}_{\mathbf{A}\times\mathbf{B}}\, \mathbf{A}$.

21 Prove that $\mathbf{B} \times \mathbf{A} = -(\mathbf{A} \times \mathbf{B})$ in two ways:
(*a*) using the algebraic definition of the cross product;
(*b*) using the geometric description of the cross product.

22 Show that if $\mathbf{B} = c\mathbf{A}$, then $\mathbf{A} \times \mathbf{B} = \mathbf{0}$:
(*a*) using the algebraic definition of the cross product;
(*b*) using the geometric description of the cross product.

23 In the proof of Theorem 1 it was shown that $\mathbf{A}$ is perpendicular to $\mathbf{A} \times \mathbf{B}$. Show that $\mathbf{B}$ is perpendicular to $\mathbf{A} \times \mathbf{B}$.

24 Complete the algebra in the proof of Theorem 3 that shows that Eq. (1) holds.

25 Evaluate (*a*) $\mathbf{i} \times 2\mathbf{i}$ (*b*) $\mathbf{i} \times \mathbf{k}$ (*c*) $\mathbf{j} \times \mathbf{i}$.

26 Show that $\mathbf{A} \cdot (\mathbf{B} \times \mathbf{C}) = (\mathbf{A} \times \mathbf{B}) \cdot \mathbf{C} = -\mathbf{A} \cdot (\mathbf{C} \times \mathbf{B})$.

✎ **27** Let $\mathbf{A}$ and $\mathbf{B}$ be nonzero vectors. As we know, if $\mathbf{A} \cdot \mathbf{B} = 0$, then $\mathbf{A}$ and $\mathbf{B}$ are perpendicular. What, if anything, can we conclude if, instead, $\mathbf{A} \times \mathbf{B} = \mathbf{0}$?

✎ **28** Show that the points (0, 0, 0), (x_1, y_1, z_1), (x_2, y_2, z_2), and (x_3, y_3, z_3) lie on a plane if and only if

$$\begin{vmatrix} x_1 & y_1 & z_1 \\ x_2 & y_2 & z_2 \\ x_3 & y_3 & z_3 \end{vmatrix} = 0.$$

✎ **29** (*a*) If $\mathbf{B}$ is parallel to $\mathbf{C}$, is $\mathbf{A} \times \mathbf{B}$ parallel to $\mathbf{A} \times \mathbf{C}$?
(*b*) If $\mathbf{B}$ is perpendicular to $\mathbf{C}$, is $\mathbf{A} \times \mathbf{B}$ perpendicular to $\mathbf{A} \times \mathbf{C}$?

✎ **30** Let $\mathbf{A}$ be a nonzero vector. If $\mathbf{A} \times \mathbf{B} = \mathbf{0}$ and $\mathbf{A} \cdot \mathbf{B} = 0$, must $\mathbf{B} = \mathbf{0}$?

31 Show that $\mathbf{A} \times (\mathbf{A} \times \mathbf{B}) = (\mathbf{A} \cdot \mathbf{B})\mathbf{A} - (\mathbf{A} \cdot \mathbf{A})\mathbf{B}$.

32 Show that $(\mathbf{A} \times \mathbf{B}) \times (\mathbf{C} \times \mathbf{D}) = [(\mathbf{A} \times \mathbf{B}) \cdot \mathbf{D}]\mathbf{C} - [(\mathbf{A} \times \mathbf{B}) \cdot \mathbf{C}]\mathbf{D}$. *Hint:* Think of $\mathbf{A} \times \mathbf{B}$ as a single vector, $\mathbf{E}$.

33 Using the vector triple product, show that

$$\|\mathbf{A} \times \mathbf{B}\|^2 = \|\mathbf{A}\|^2\, \|\mathbf{B}\|^2 - (\mathbf{A} \cdot \mathbf{B})^2.$$

34 (*a*) Give an example of a vector perpendicular to the vector $3\mathbf{i} - \mathbf{j} + \mathbf{k}$.
(*b*) Give an example of a unit vector perpendicular to the vector $3\mathbf{i} - \mathbf{j} + \mathbf{k}$.

35 Let $\mathbf{u}$ be a unit vector and $\mathbf{B}$ be a vector. What happens as you keep "crossing by $\mathbf{u}$," that is, as you form the sequence $\mathbf{B}$, $\mathbf{u} \times \mathbf{B}$, $\mathbf{u} \times (\mathbf{u} \times \mathbf{B})$, and so on?

36 (*Crystallography*) A crystal is described by three vectors $\mathbf{v}_1$, $\mathbf{v}_2$, and $\mathbf{v}_3$. They span a "fundamental" parallelepiped, whose copies fill out the crystal lattice. (See Fig. 13.) The atoms are at the corners. In order to study the diffraction of x-rays and light through a crystal, crystallographers work with the "reciprocal lattice," as follows. Its fundamental parallelepiped is spanned by three vectors, $\mathbf{k}_1$, $\mathbf{k}_2$, and $\mathbf{k}_3$. The vector $\mathbf{k}_1$ is perpendicular to the parallelogram spanned by $\mathbf{v}_2$ and $\mathbf{v}_3$ and has a length equal to the reciprocal of the distance between that parallelogram and the opposite parallelogram of the fundamental parallelepiped. The vectors $\mathbf{k}_2$ and $\mathbf{k}_3$ are defined similarly in terms of the other four faces of the fundamental parallelepiped.
(*a*) Show that $\mathbf{k}_1$, $\mathbf{k}_2$, $\mathbf{k}_3$ may be chosen to be

$$\mathbf{k}_1 = \frac{\mathbf{v}_2 \times \mathbf{v}_3}{\mathbf{v}_1 \cdot (\mathbf{v}_2 \times \mathbf{v}_3)}, \qquad \mathbf{k}_2 = \frac{\mathbf{v}_3 \times \mathbf{v}_1}{\mathbf{v}_1 \cdot (\mathbf{v}_2 \times \mathbf{v}_3)},$$

$$\mathbf{k}_3 = \frac{\mathbf{v}_1 \times \mathbf{v}_2}{\mathbf{v}_1 \cdot (\mathbf{v}_2 \times \mathbf{v}_3)}.$$

(*b*) Show that the volume of the fundamental parallelepiped determined by $\mathbf{k}_1$, $\mathbf{k}_2$, and $\mathbf{k}_3$ is the reciprocal of the volume of the one determined by $\mathbf{v}_1$, $\mathbf{v}_2$, and $\mathbf{v}_3$.

(*c*) Is the reciprocal of the reciprocal lattice the original lattice? For instance, is

$$\mathbf{v}_1 = \frac{\mathbf{k}_2 \times \mathbf{k}_3}{\mathbf{k}_1 \cdot (\mathbf{k}_2 \times \mathbf{k}_3)}?$$

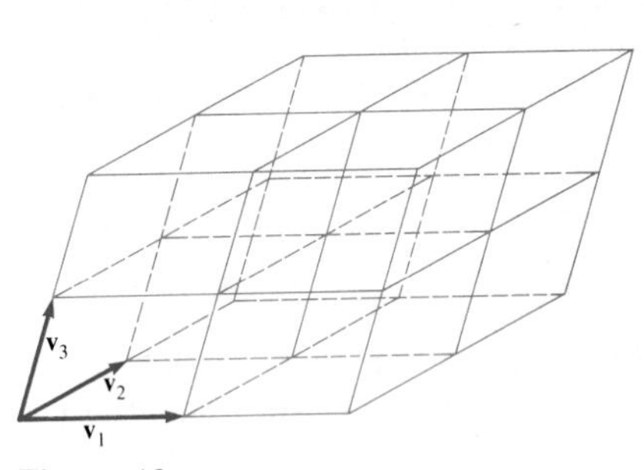

Figure 13

37 Is this argument correct? We *start* with the geometric definition of $\mathbf{A} \times \mathbf{B}$ in terms of the area of a parallelogram and the right-hand rule. We then obtain the algebraic formula for $\mathbf{A} \times \mathbf{B}$ in terms of the components of $\mathbf{A}$ and $\mathbf{B}$ as follows: First, we compute the cross products of the basic vectors ($\mathbf{i} \times \mathbf{i} = \mathbf{0}$, $\mathbf{i} \times \mathbf{j} = \mathbf{k}$, etc.). Then we compute the cross product of $\mathbf{A} = x_1\mathbf{i} + y_1\mathbf{j} + z_1\mathbf{k}$ and $\mathbf{B} = x_2\mathbf{i} + y_2\mathbf{j} + z_2\mathbf{k}$ by expanding $\mathbf{A} \times \mathbf{B} = (x_1\mathbf{i} + y_1\mathbf{j} + z_1\mathbf{k}) \times (x_2\mathbf{i} + y_2\mathbf{j} + z_2\mathbf{k})$.

(*a*) Multiply out, obtaining nine products, and collect.

(*b*) Does the result agree with the definition of cross product given in this section?

(*c*) Is the argument correct?

38 Let $\mathbf{B}$ and $\mathbf{C}$ be nonzero, nonparallel vectors and $\mathbf{A}$ a vector that is perpendicular neither to $\mathbf{B}$ nor $\mathbf{C}$.

(*a*) Why are there scalars x and y such that

$$\mathbf{A} \times (\mathbf{B} \times \mathbf{C}) = x\mathbf{B} + y\mathbf{C}?$$

(*b*) Why is $0 = x(\mathbf{A} \cdot \mathbf{B}) + y(\mathbf{A} \cdot \mathbf{C})$?

(*c*) Using (*b*), show that there is a scalar z such that

$$\mathbf{A} \times (\mathbf{B} \times \mathbf{C}) = z[(\mathbf{A} \cdot \mathbf{C})\mathbf{B} - (\mathbf{A} \cdot \mathbf{B})\mathbf{C}].$$

(*d*) It would be nice if there were a simple geometric way to show that z is a constant and equals 1. Of course we could show that $z = 1$ by writing $\mathbf{A}$, $\mathbf{B}$, and $\mathbf{C}$ in components and grinding out a tedious calculation. But that would hardly be instructive. Can you figure out why $z = 1$ in a simpler way?

In this section $\mathbf{A} \times \mathbf{B}$ was defined in terms of components, and then its geometric description was obtained. This is the opposite of the way we dealt with the dot product. Exercises 39 to 41 outline a different approach to the cross product. We define $\mathbf{A} \times \mathbf{B}$ as follows. If $\mathbf{A}$ or $\mathbf{B}$ is $\mathbf{0}$ or if $\mathbf{A}$ is parallel to $\mathbf{B}$, we define $\mathbf{A} \times \mathbf{B}$ to be $\mathbf{0}$. Otherwise, $\mathbf{A} \times \mathbf{B}$ is the vector whose length is the area of the parallelogram that they span and whose direction is given by the right-hand rule.

39 Let $\mathbf{A}$ be a nonzero vector and $\mathbf{B}$ be a vector. Let $\mathbf{B}_1$ be the projection of $\mathbf{B}$ on a plane perpendicular to $\mathbf{A}$. Let $\mathbf{B}_2$ be obtained by rotating $\mathbf{B}_1$ 90° in the direction given by the right-hand rule with thumb pointing in the same direction as $\mathbf{A}$.

(*a*) Show that $\mathbf{A} \times \mathbf{B} = \mathbf{A} \times \mathbf{B}_1$. (Draw a clear diagram.)

(*b*) Show that $\mathbf{A} \times \mathbf{B} = \|\mathbf{A}\|\mathbf{B}_2$.

40 Using Exercise 39(*b*), show that for $\mathbf{A}$ not $\mathbf{0}$, $\mathbf{A} \times (\mathbf{B} + \mathbf{C}) = \mathbf{A} \times \mathbf{B} + \mathbf{A} \times \mathbf{C}$. (*Hint:* Draw a large, clear picture.)

41 (*a*) From the distributive law $\mathbf{A} \times (\mathbf{B} + \mathbf{C}) = \mathbf{A} \times \mathbf{B} + \mathbf{A} \times \mathbf{C}$, and the fact that $\mathbf{D} \times \mathbf{E} = -\mathbf{E} \times \mathbf{D}$, deduce the distributive law $(\mathbf{B} + \mathbf{C}) \times \mathbf{A} = \mathbf{B} \times \mathbf{A} + \mathbf{C} \times \mathbf{A}$.

(*b*) From the distributive law $\mathbf{A} \times (\mathbf{B} + \mathbf{C}) = \mathbf{A} \times \mathbf{B} + \mathbf{A} \times \mathbf{C}$, deduce that $\mathbf{A} \times (\mathbf{B} + \mathbf{C} + \mathbf{D}) = \mathbf{A} \times \mathbf{B} + \mathbf{A} \times \mathbf{C} + \mathbf{A} \times \mathbf{D}$. (*Hint:* Think of $\mathbf{B} + \mathbf{C}$ as a single vector $\mathbf{E}$.)

(*c*) Repair the proof in Exercise 37.

42 (*a*) Devise a procedure for determining whether the point $P = (x, y)$ is inside the triangle whose three vertices are $P_1 = (x_1, y_1)$, $P_2 = (x_2, y_2)$, and $P_3 = (x_3, y_3)$.

(*b*) Write a program that implements (*a*).

43 (*a*) Devise a procedure for determining whether the point $P = (x, y, z)$ is inside the tetrahedron whose four vertices are $P_1 = (x_1, y_1, z_1)$, $P_2 = (x_2, y_2, z_2)$, $P_3 = (x_3, y_3, z_3)$, and $P_4 = (x_4, y_4, z_4)$.

(*b*) Write a program that implements (*a*).

This problem arises in creating hidden-line algorithms for computer graphics software.

12.7 MORE ON LINES AND PLANES

With the aid of projections, the dot product, and the cross product, we can answer just about anything anyone might want to know about vectors, lines, and planes. As a sample, here are seven questions that our seven examples will answer:

1. How do we find a normal vector **N** to the plane $\mathcal{P}$ through the given points P, Q, and R?
2. How far is the given point P from the plane $\mathcal{P}$ through the given points Q, R, and S?
3. What is the projection of a given vector **A** on a given plane $\mathcal{P}$?
4. How far is a given point P from the line L through points Q and R?
5. How do we find parametric equations of the line L that passes through a given point P and is perpendicular to a given plane $\mathcal{P}$?
6. How far apart are two skew lines L_1 and L_2? (Two lines are **skew** if they do not intersect and they are not parallel.)
7. How do we find the point in line L_1 that is nearest line L_2?

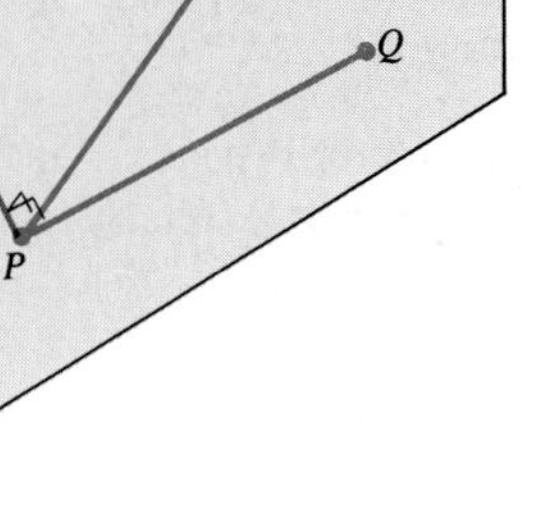

Figure 1

EXAMPLE 1 Find a vector **N** normal to the plane $\mathcal{P}$ through $P = (1, 1, 1)$, $Q = (3, 1, -1)$, and $R = (5, 6, 4)$. (See Fig. 1.)

SOLUTION The cross product $\overrightarrow{PQ} \times \overrightarrow{PR}$ provides a normal to $\mathcal{P}$. Now

$$\overrightarrow{PQ} = \langle 3-1, 1-1, -1-1 \rangle = \langle 2, 0, -2 \rangle$$

and

$$\overrightarrow{PR} = \langle 5-1, 6-1, 4-1 \rangle = \langle 4, 5, 3 \rangle.$$

Thus

$$\overrightarrow{PQ} \times \overrightarrow{PR} = \begin{vmatrix} \mathbf{i} & \mathbf{j} & \mathbf{k} \\ 2 & 0 & -2 \\ 4 & 5 & 3 \end{vmatrix} = 10\mathbf{i} - 14\mathbf{j} + 10\mathbf{k}.$$

Hence $\mathbf{N} = 10\mathbf{i} - 14\mathbf{j} + 10\mathbf{k}$ is normal to $\mathcal{P}$. ■

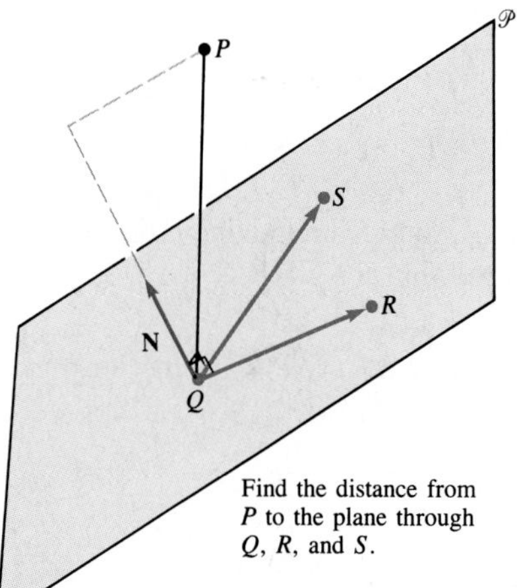

Find the distance from P to the plane through Q, R, and S.

Figure 2

EXAMPLE 2 Find the distance from the point $P = (2, 1, 3)$ to the plane $\mathcal{P}$ through the points $Q = (1, 1, 2)$, $R = (2, 2, 3)$, and $S = (3, 4, -1)$. (See Fig. 2.)

SOLUTION The distance from P to the plane $\mathcal{P}$ is the absolute value of $\text{comp}_{\mathbf{N}} \overrightarrow{QP}$, where **N** is a normal to $\mathcal{P}$.

To produce such a normal **N**, we use the cross product $\overrightarrow{QR} \times \overrightarrow{QS}$:

$$\begin{aligned} \mathbf{N} = \overrightarrow{QR} \times \overrightarrow{QS} &= \langle 2-1, 2-1, 3-2 \rangle \times \langle 3-1, 4-1, -1-2 \rangle \\ &= \langle 1, 1, 1 \rangle \times \langle 2, 3, -3 \rangle \\ &= \langle -6, 5, 1 \rangle. \end{aligned}$$

Also,

$$\begin{aligned} \overrightarrow{QP} &= \langle 2-1, 1-1, 3-2 \rangle \\ &= \langle 1, 0, 1 \rangle. \end{aligned}$$

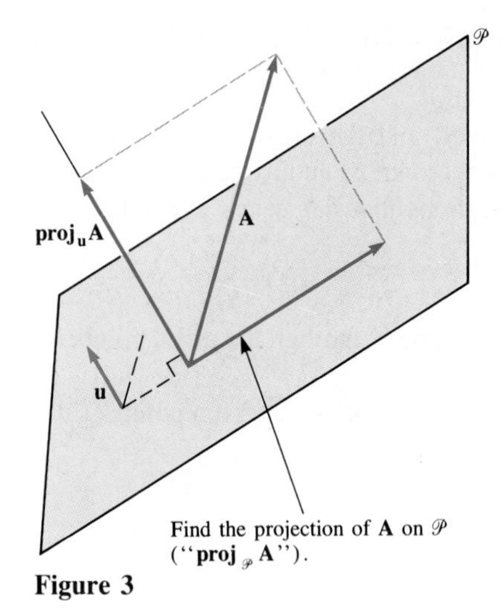

Figure 3

Thus

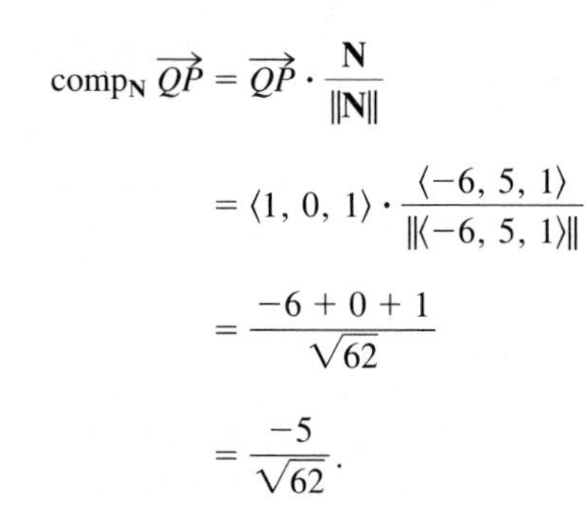

$$\text{comp}_{\mathbf{N}}\,\overrightarrow{QP} = \overrightarrow{QP}\cdot\frac{\mathbf{N}}{\|\mathbf{N}\|}$$

$$= \langle 1, 0, 1\rangle\cdot\frac{\langle -6, 5, 1\rangle}{\|\langle -6, 5, 1\rangle\|}$$

$$= \frac{-6+0+1}{\sqrt{62}}$$

$$= \frac{-5}{\sqrt{62}}.$$

The distance from P to $\mathscr{P}$ is $5/\sqrt{62}$. ■

EXAMPLE 3 Find the projection of the vector $\mathbf{A} = \mathbf{i} + 2\mathbf{j} + 3\mathbf{k}$ in the plane $\mathscr{P}$ whose equation is $2x - 5y + 6z + 2 = 0$. (See Fig. 3.)

SOLUTION We will first find the projection of $\mathbf{A}$ on a unit vector $\mathbf{u}$ perpendicular to $\mathscr{P}$. Then we will find

$$\mathbf{proj}_{\mathscr{P}}\,\mathbf{A} = \mathbf{A} - \mathbf{proj_u}\,\mathbf{A}.$$

The coefficients of $2x - 5y + 6z + 2$ give us a vector $\mathbf{B} = \langle 2, -5, 6\rangle$ normal to $\mathscr{P}$. Let $\mathbf{u}$ be the unit vector $\mathbf{B}/\|\mathbf{B}\| = (2\mathbf{i} - 5\mathbf{j} + 6\mathbf{k})/\sqrt{65}$. Then

$$\mathbf{proj_u}\,\mathbf{A} = (\mathbf{A}\cdot\mathbf{u})\mathbf{u}$$

$$= \left((\mathbf{i} + 2\mathbf{j} + 3\mathbf{k})\cdot\frac{2\mathbf{i} - 5\mathbf{j} + 6\mathbf{k}}{\sqrt{65}}\right)\frac{2\mathbf{i} - 5\mathbf{j} + 6\mathbf{k}}{\sqrt{65}}$$

$$= \frac{2 - 10 + 18}{\sqrt{65}}\,\frac{2\mathbf{i} - 5\mathbf{j} + 6\mathbf{k}}{\sqrt{65}}$$

$$= \tfrac{10}{65}(2\mathbf{i} - 5\mathbf{j} + 6\mathbf{k})$$

$$= \tfrac{4}{13}\mathbf{i} - \tfrac{10}{13}\mathbf{j} + \tfrac{12}{13}\mathbf{k}.$$

Hence

$$\mathbf{proj}_{\mathscr{P}}\,\mathbf{A} = \mathbf{A} - \mathbf{proj_u}\,\mathbf{A}$$

$$= (\mathbf{i} + 2\mathbf{j} + 3\mathbf{k}) - (\tfrac{4}{13}\mathbf{i} - \tfrac{10}{13}\mathbf{j} + \tfrac{12}{13}\mathbf{k})$$

$$= \tfrac{9}{13}\mathbf{i} + \tfrac{36}{13}\mathbf{j} + \tfrac{27}{13}\mathbf{k}. \quad ■$$

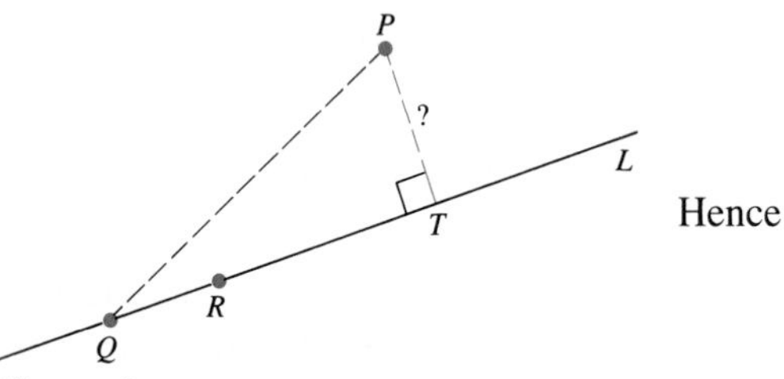

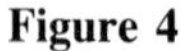
Figure 4

EXAMPLE 4 Find how far the point $P = (2, 1, 3)$ is from the line L through the points $Q = (1, 5, 2)$ and $R = (2, 3, 4)$. (See Fig. 4.)

SOLUTION Inspection of Fig. 4 shows what to do. We want the length of TP. Introducing vectors, we have Fig. 5.

Now

$$\overrightarrow{QT} = \mathbf{proj}_{\overrightarrow{QR}}\,\overrightarrow{QP}$$

and

$$\overrightarrow{TP} = \mathbf{orth}_{\overrightarrow{QR}}\,\overrightarrow{QP} = \overrightarrow{QP} - \mathbf{proj}_{\overrightarrow{QR}}\,\overrightarrow{QP}. \tag{1}$$

Using (1) we will find $\overrightarrow{TP}$ and, finally, $\|\overrightarrow{TP}\|$.

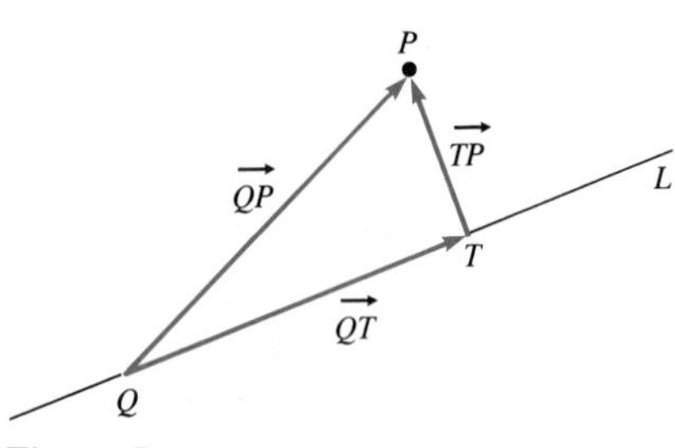

Figure 5

To begin,

$$\overrightarrow{QP} = \langle 2-1, 1-5, 3-2\rangle = \langle 1, -4, 1\rangle$$

and

$$\overrightarrow{QR} = \langle 2-1, 3-5, 4-2\rangle = \langle 1, -2, 2\rangle.$$

Let $\mathbf{u} = \overrightarrow{QR}/\|\overrightarrow{QR}\| = \langle 1, -2, 2\rangle/\sqrt{9} = \langle 1, -2, 2\rangle/3$.

Then

$$\begin{aligned}\mathbf{proj}_{\overrightarrow{QR}}\,\overrightarrow{QP} &= \mathbf{proj}_{\mathbf{u}}\,\overrightarrow{QP}\\ &= (\overrightarrow{QP}\cdot\mathbf{u})\mathbf{u}\\ &= \left(\langle 1, -4, 1\rangle\cdot\frac{\langle 1, -2, 2\rangle}{3}\right)\frac{\langle 1, -2, 2\rangle}{3}\\ &= \frac{11}{9}\langle 1, -2, 2\rangle\\ &= \left\langle\frac{11}{9}, -\frac{22}{9}, \frac{22}{9}\right\rangle.\end{aligned}$$

Thus

$$\begin{aligned}\overrightarrow{TP} &= \overrightarrow{QP} - \mathbf{proj}_{\overrightarrow{QR}}\,\overrightarrow{QP}\\ &= \langle 1, -4, 1\rangle - \left\langle\frac{11}{9}, -\frac{22}{9}, \frac{22}{9}\right\rangle\\ &= \left\langle -\frac{2}{9}, -\frac{14}{9}, -\frac{13}{9}\right\rangle.\end{aligned}$$

Hence

$$\|\overrightarrow{TP}\| = \left\|\left\langle -\frac{2}{9}, -\frac{14}{9}, -\frac{13}{9}\right\rangle\right\| = \frac{\sqrt{369}}{9} = \frac{\sqrt{41}}{3}. \blacksquare$$

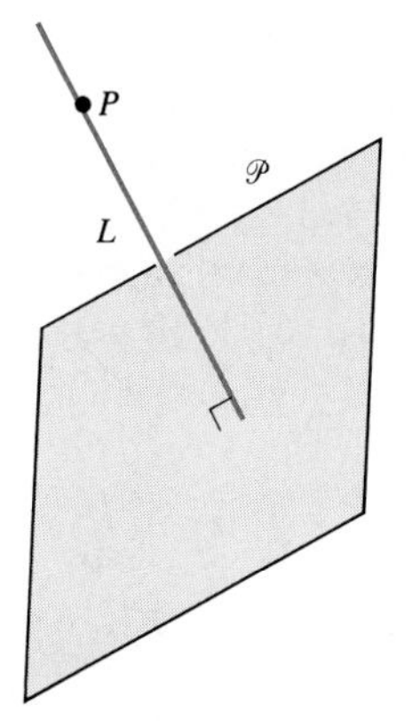

Figure 6

EXAMPLE 5 Find parametric equations of the line through $P = (2, 1, 5)$ and perpendicular to the plane $2x - 4y + 3z = 8$. (See Fig. 6.)

SOLUTION The vector $\mathbf{N} = \langle 2, -4, 3\rangle$ is normal to $\mathcal{P}$, hence provides direction numbers for L. A vector equation for L is determined by the point $P = (2, 1, 5)$ and the vector $\mathbf{N} = \langle 2, -4, 3\rangle$. It is

$$P = 2\mathbf{i} + \mathbf{j} + 5\mathbf{k} + t(2\mathbf{i} - 4\mathbf{j} + 3\mathbf{k}).$$

This equation breaks into three scalar equations,

$$\begin{cases} x = 2 + 2t\\ y = 1 - 4t\\ z = 5 + 3t.\end{cases}$$

These form the parametric equations of the line L. ■

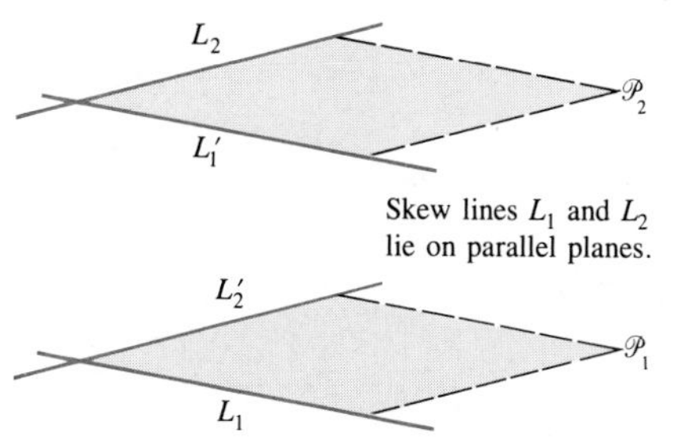

Skew lines L_1 and L_2 lie on parallel planes.

Figure 7

The next example concerns skew lines L_1 and L_2, as in Fig. 7. To draw such lines first draw line L_1' parallel to L_1 and intersecting L_2. Then draw L_2' parallel to L_2 and intersecting L_1. (See Fig. 7.) L_1 and L_2' determine a plane $\mathcal{P}_1$; L_2 and L_1' determine a plane $\mathcal{P}_2$. Note that $\mathcal{P}_1$ is parallel to $\mathcal{P}_2$. By "the distance between the lines L_1 and L_2" we mean the perpendicular distance between the planes $\mathcal{P}_1$ and $\mathcal{P}_2$.

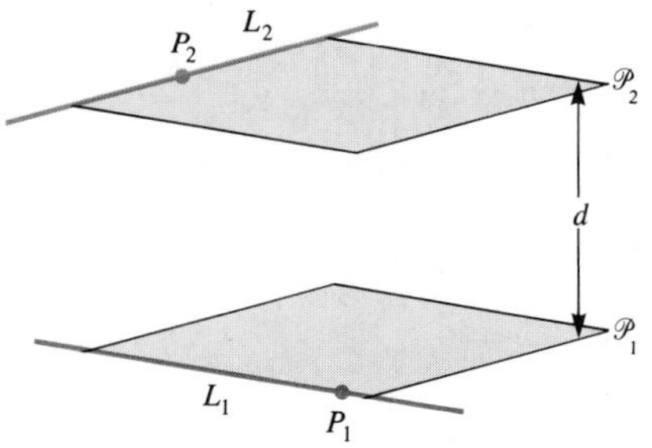

Figure 8

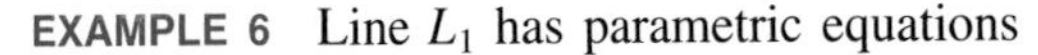

EXAMPLE 6 Line L_1 has parametric equations

$$\begin{cases} x = 2 + 3t \\ y = 1 - 5t \\ z = -3 + 2t. \end{cases} \tag{2}$$

Line L_2 has parametric equations

$$\begin{cases} x = 3 + 2s \\ y = 1 + 6s \\ z = 5 + 7s. \end{cases} \tag{3}$$

How far apart are they?

SOLUTION The distance between L_1 and L_2 is the distance d between the parallel planes $\mathcal{P}_1$ and $\mathcal{P}_2$ in which they lie, as in Fig. 8. Choose any point P_1 on L_1 and any point P_2 on L_2. Let $\mathbf{N}$ be a vector perpendicular to the plane $\mathcal{P}_1$ (and, of course, to $\mathcal{P}_2$ as well). Then

$$d = \|\mathbf{proj_N}\, \overrightarrow{P_1P_2}\|.$$

(See Fig. 9.)

We first choose specific points P_1 on L_1 and P_2 on L_2. Equation (2) for any t gives a suitable P_1. The easiest occurs when $t = 0$. Doing this gives us $P_1 = (2, 1, -3)$. Similarly, letting $s = 0$ in (3), we choose $P_2 = (3, 1, 5)$.

Next, we construct a vector $\mathbf{N}$ perpendicular to $\mathcal{P}_1$. Such a vector is perpendicular to both L_1 and L_2'. Equation (2) gives us the vector $3\mathbf{i} - 5\mathbf{j} + 2\mathbf{k}$ parallel to L_1. Equation (3) gives us $2\mathbf{i} + 6\mathbf{j} + 7\mathbf{k}$ parallel to L_2.

Hence

$$\begin{vmatrix} \mathbf{i} & \mathbf{j} & \mathbf{k} \\ 3 & -5 & 2 \\ 2 & 6 & 7 \end{vmatrix} = -47\mathbf{i} - 17\mathbf{j} + 28\mathbf{k}$$

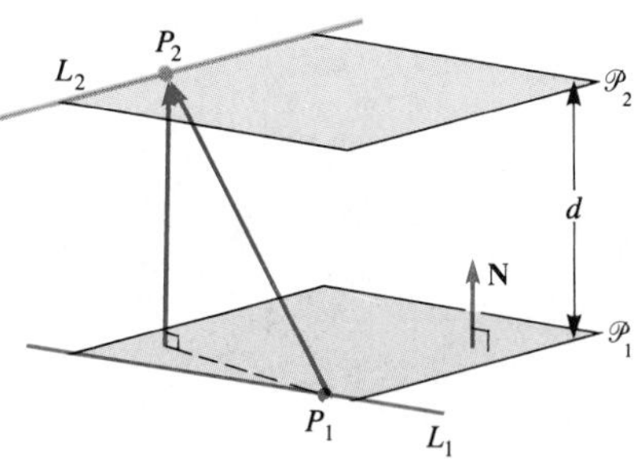

Figure 9

serves as $\mathbf{N}$.

Next, we compute

$$\begin{aligned}
\mathbf{proj_N}\, \overrightarrow{P_1P_2} &= \overrightarrow{P_1P_2} \cdot \frac{\mathbf{N}}{\|\mathbf{N}\|}\, \frac{\mathbf{N}}{\|\mathbf{N}\|} \\
&= \langle 3-2,\ 1-1,\ 5-(-3)\rangle \cdot \frac{\langle -47, -17, 28\rangle}{\|\langle -47, -17, 28\rangle\|}\, \frac{\langle -47, -17, 28\rangle}{\|\langle -47, -17, 28\rangle\|} \\
&= \frac{-47 + 0 + 224}{(-47)^2 + (-17)^2 + 28^2} \langle -47, -17, 28\rangle \\
&= \frac{177}{3282} \langle -47, -17, 28\rangle.
\end{aligned}$$

Finally,

$$\begin{aligned}
\|\mathbf{proj_N}\, \overrightarrow{P_1P_2}\| &= \frac{177}{3282} \|\langle -47, -17, 28\rangle\| \\
&= \frac{177}{3282}\sqrt{3282} \\
&= \frac{177}{\sqrt{3282}}.
\end{aligned}$$

The distance between L_1 and L_2 is $177/\sqrt{3282}$. ■

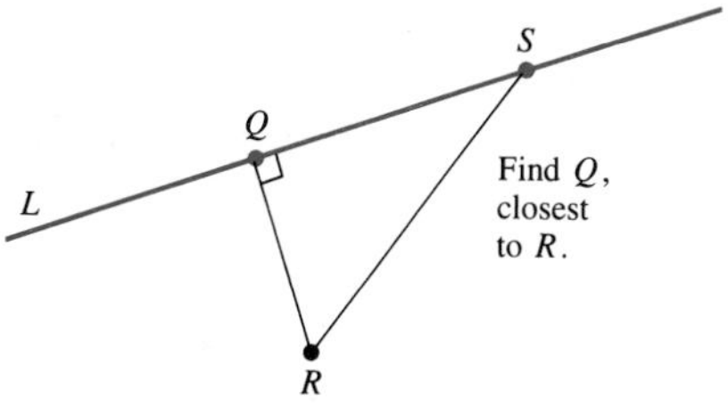

Find Q, closest to R.

Figure 10

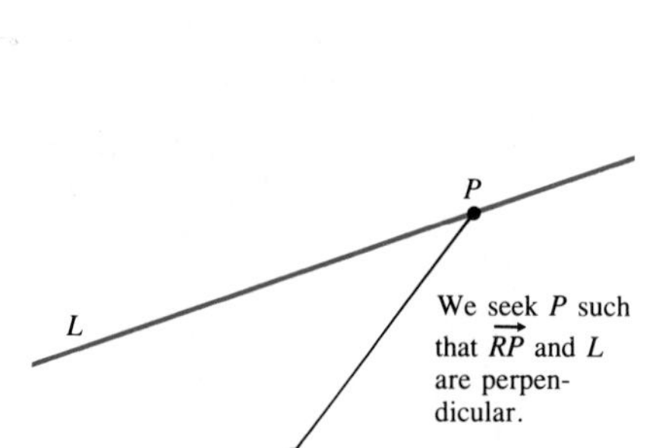

Figure 11

EXAMPLE 7 The line L has parametric equations

$$\begin{cases} x = 3 + 2t \\ y = 4 - t \\ z = 5 + 6t. \end{cases} \tag{4}$$

Find the point Q on L that is closest to the point $R = (4, 1, 7)$.

SOLUTION Figure 10 shows the geometry. If Q is the point on L closest to R, then QR is perpendicular to L. To see this, let S be any other point on L. The triangle RQS is a right triangle, and its hypotenuse RS is longer than its leg QR. Conversely, if QR is perpendicular to L, then Q is the point on L that is closest to R.

Therefore we seek the point Q on L such that QR is perpendicular to L. From (4) we obtain the vector $\langle 2, -1, 6\rangle$ parallel to L. Equation (4) describes an arbitrary P on L. (See Fig. 11.)

Since $R = (4, 1, 7)$

$$\begin{aligned} \overrightarrow{RP} &= \langle (3 + 2t) - 4, (4 - t) - 1, (5 + 6t) - 7\rangle \\ &= \langle 2t - 1, 3 - t, 6t - 2\rangle. \end{aligned}$$

We first find t such that $\overrightarrow{RP} \cdot \langle 2, -1, 6\rangle = 0$.

We have

$$\begin{aligned} 0 = \overrightarrow{RP} \cdot \langle 2, -1, 6\rangle &= (2t - 1)(2) + (3 - t)(-1) + (6t - 2)(6) \\ &= 41t - 17, \end{aligned}$$

hence

$$41t - 17 = 0$$

or

$$t = \frac{17}{41}.$$

Substituting 17/41 for t in (4) gives us the point on L that is closest to R, namely

$$\begin{aligned} (x, y, z) &= \left[3 + 2\left(\frac{17}{41}\right), 4 - \left(\frac{17}{41}\right), 5 + 6\left(\frac{17}{41}\right)\right] \\ &= \left(\frac{157}{41}, \frac{147}{41}, \frac{307}{41}\right). \quad \blacksquare \end{aligned}$$

Section Summary

We used seven examples to show how to deal with the geometry of lines and planes. It is often easier to solve the general problem, using letters, and then substitute the numbers at the end. Rather than memorize the formulas, keep in mind the basic ideas: projections, normals to planes, parametric equations of lines, dot products, and cross products. The exercises include applications to the design of a solar collector, computer graphics, and the construction of hoppers.

EXERCISES FOR SEC. 12.7: MORE ON LINES AND PLANES

1 Find a vector perpendicular to the plane through (2, 1, 3), (4, 5, 1), and (−2, 2, 3).

2 How far is the point (1, 2, 2) from the plane through (0, 0, 0), (3, 5, −2), and (2, −1, 3)?

3 Find the projection of $3\mathbf{i} + 4\mathbf{j} + 5\mathbf{k}$ on the plane $x + 2y + 3z + 5 = 0$.

4 How far is the point (1, 2, 3) from the line through (−2, −1, 3) and (4, 1, 2)?

5 Find parametric equations of the line through (1, 1, 2) and perpendicular to the plane $3x - y + z = 6$.

6 How far apart are the lines whose vector equations are $2\mathbf{i} + 4\mathbf{j} + \mathbf{k} + t(\mathbf{i} + \mathbf{j} + \mathbf{k})$ and $\mathbf{i} + 3\mathbf{j} + 2\mathbf{k} + s(2\mathbf{i} - \mathbf{j} - \mathbf{k})$?

7 Find the point on the line through (1, 2, 1) and (2, −1, 3) that is closest to the line through (3, 0, 3) and parallel to the vector $\mathbf{i} + 2\mathbf{j} + 5\mathbf{k}$.

8 (*a*) How would you find an equation for the plane through points $P_1 = (x_1, y_1, z_1)$, $P_2 = (x_2, y_2, z_2)$, and $P_3 = (x_3, y_3, z_3)$?
(*b*) Find an equation for the plane through (2, 2, 1), (0, 1, 5), and (2, −1, 0).

9 (*a*) How would you decide whether the line through $P_1 = (x_1, y_1, z_1)$ and $P_2 = (x_2, y_2, z_2)$ is parallel to the line through $P_3 = (x_3, y_3, z_3)$ and $P_4 = (x_4, y_4, z_4)$?
(*b*) Is the line through (1, 2, −3) and (5, 9, 4) parallel to the line through (−1, −1, 2) and (1, 3, 5)?

10 (*a*) How would you decide whether the line through $P_1 = (x_1, y_1, z_1)$ and $P_2 = (x_2, y_2, z_2)$ is parallel to the plane $Ax + By + Cz + D = 0$?
(*b*) Is the line through (1, −2, 3) and (5, 3, 0) parallel to the plane $2x - y + z + 3 = 0$?

11 (*a*) How would you decide whether the line through P_1 and P_2 is parallel to the plane through Q_1, Q_2, and Q_3?
(*b*) Is the line through (0, 0, 0) and (1, 1, −1) parallel to the plane through (1, 0, 1), (2, 1, 0), and (1, 3, 4)?

12 (*a*) How would you decide whether the plane through P_1, P_2, and P_3 is parallel to the plane through Q_1, Q_2, and Q_3?
(*b*) Is the plane through (1, 2, 3), (4, 1, −1), and (2, 0, 1) parallel to the plane through (2, 3, 4), (5, 2, 0), and (3, 1, 2)?

13 (*a*) How would you find the angle between the planes $A_1x + B_1y + C_1z + D_1 = 0$ and $A_2x + B_2y + C_2z + D_2 = 0$?
(*b*) Find the angle between $x - y - z - 1 = 0$ and $x + y + z + 2 = 0$.

14 Assume that the planes $A_1x + B_1y + C_1z + D_1 = 0$ and $A_2x + B_2y + C_2z + D_2 = 0$ meet in a line L.
(*a*) How would you find direction numbers for L?
(*b*) How would you find a point on L?
(*c*) Find parametric equations for the line that is the intersection of the planes $2x - y + 3z + 4 = 0$ and $3x + 2y + 5z + 2 = 0$.

15 (*a*) How would you decide whether the four points $P_1 = (x_1, y_1, z_1)$, $P_2 = (x_2, y_2, z_2)$, $P_3 = (x_3, y_3, z_3)$, and $P_4 = (x_4, y_4, z_4)$ lie in a plane?
(*b*) Do the points (1, 2, 3), (4, 1, −5), (2, 1, 6), and (3, 5, 3) lie in a plane?

16 What is the angle between the line through (1, 2, 1) and (−1, 3, 0) and the plane $x + y - 2z = 0$?

17 (*a*) If you know the coordinates of point P and parametric equations of line L, how would you find an equation of the plane that contains P and L? (Assume P is not on L.)
(*b*) Find an equation for the plane through (1, 1, 1) that contains the line

$$\begin{cases} x = 2 + t \\ y = 3 - t \\ z = 4 + 2t. \end{cases}$$

18 (*a*) How many unit vectors are perpendicular to the plane $Ax + By + Cz + D = 0$?
(*b*) How would you find one of them?
(*c*) Find a unit vector perpendicular to the plane $3x - 2y + 4z + 6 = 0$.

19 (*a*) How many unit vectors are perpendicular to the line

$$\begin{cases} x = x_0 + a_1t \\ y = y_0 + a_2t \\ z = z_0 + a_3t? \end{cases}$$

(*b*) How would you find one of them?
(*c*) Find a unit vector perpendicular to the line through the points (1, 0, −3) and (2, 1, −1) and parallel to the plane $4x + 5y + 6z = 0$.

20 (*a*) How would you go about producing a specific point on the plane $Ax + By + Cz + D = 0$?
(*b*) Give the coordinates of a specific point that lies on the plane $3x - y + z + 10 = 0$.

21 (*a*) How would you go about producing a specific point that lies on both planes $A_1x + B_1y + C_1z + D_1 = 0$ and $A_2x + B_2y + C_2z + D_2 = 0$?
(*b*) Find a point that lies on both planes $3x + z + 2 = 0$ and $x - y - z + 5 = 0$.

22 The planes $A_1x + B_1y + C_1z + D_1 = 0$ and $A_2x + B_2y + C_2z + D_2 = 0$ intersect in a line L. Find direction cosines of a vector parallel to L.

---

23 Find the area of the projection of the parallelogram spanned by $2\mathbf{i} + 3\mathbf{j} + 4\mathbf{k}$ and $\mathbf{i} - \mathbf{j} + \mathbf{k}$ on the plane $x + y + z = 62$.

24 (*a*) Let $\mathbf{A}$ and $\mathbf{B}$ be vectors in space. How would you find the area of the parallelogram they span?
(*b*) Find the area of the parallelogram spanned by $\langle 2, 3, 1\rangle$ and $\langle 4, -1, 5\rangle$.

25 A disk of radius a lies in the plane $2x + 4y + 3z = 5$. Find the area of its projection in the plane $x - y + 2z = 6$.

26 A ray of light parallel to the vector **A** is reflected off a mirror. Let **n** be a unit normal to the plane of the mirror. Show that the vector $\mathbf{B} = \mathbf{A} - 2(\mathbf{A} \cdot \mathbf{n})\mathbf{n}$ is in the direction of the reflected ray, as shown in Fig. 12. (Assume that the angle of incidence equals the angle of reflection, that is, $\alpha = \beta$ in Fig. 12. **A** and **B** lie in a plane perpendicular to the mirror.)

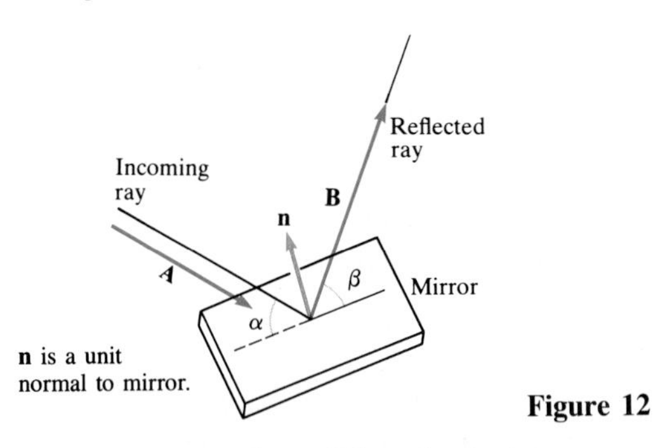

Figure 12

27 *(Computer graphics)* When developing programs that display the brightness of a surface, computer scientists face problems of which the following is a simple example. A light source at $P = (-1, -2, 2)$ illuminates a sphere of radius 1 and center at $O = (0, 0, 0)$. The eyes of the observer are at $Q = (2, 2, 3)$. (See Fig. 13.) When light from P strikes the sphere at point $R = (a, b, c)$, it is reflected in such a way that the incoming and reflected rays make equal angles with the radius vector $\overrightarrow{OP}$. Find the point R such that the reflected ray passes through Q. (If the sphere were perfectly smooth, the light source were just at a point, and our eyes just a point, we would see only that point R.) *Hint:* See Exercise 26.

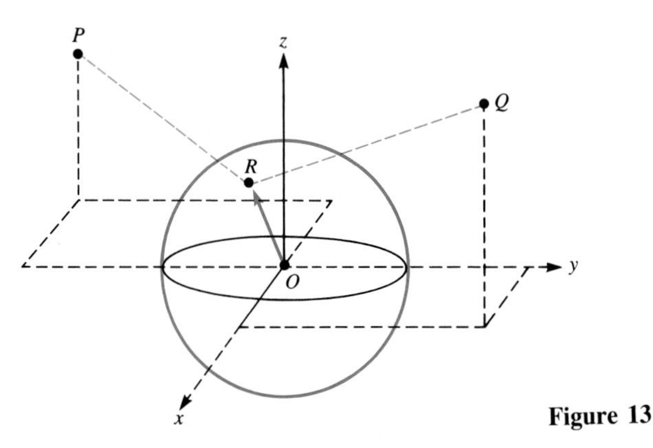

Figure 13

28 (Contributed by Mark O'Donnell.) The following problem arose in the design of a solar collector. The front of a house is aligned 60° from south, as shown in Fig. 14. The roof is inclined at 22° from the horizontal. At noon the sun has an elevation 75° above the horizon. (See Fig. 14.)

(*a*) Find the cosine of the angle between the vector $\overrightarrow{OS}$, which points to the sun, and the vector **N**, a normal to the roof.

(*b*) Find the angle between **N** and $\overrightarrow{OS}$.

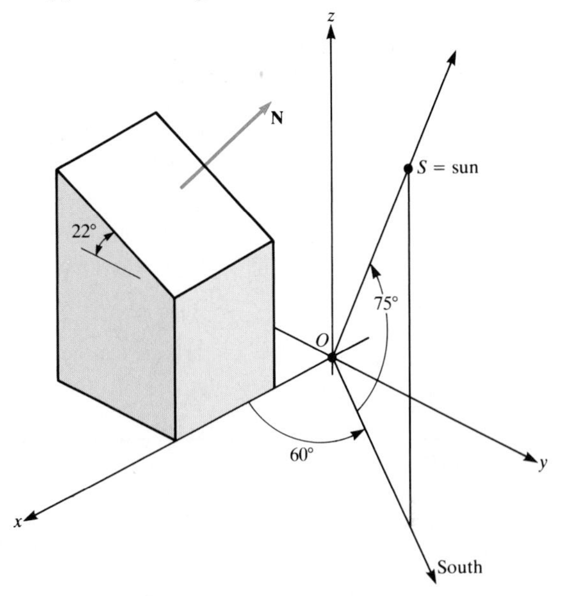

Figure 14

29 (Contributed by Melvyn Kopald Stein.) An industrial hopper is shaped as shown in Fig. 15. Its top and bottom are squares of different sizes. The angle between the plane ABD and plane BDC is 70°. The angle between the plane ABD and plane ABC is 80°. What is the angle between plane ABC and plane BCD? (This angle is needed during fabrication of the hopper, since the planes ABC and BCD are made from a single piece of heavy-gauge sheet metal bent along the edge BC.)

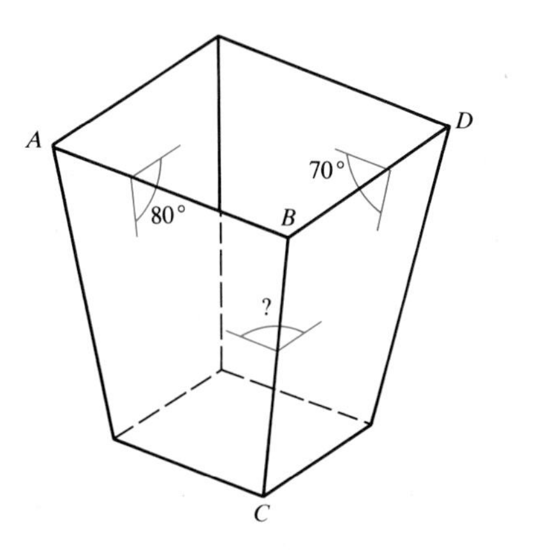

Figure 15

30 (*a*) Let L_1 be the line through P_1 and Q_1 and let L_2 be the line through P_2 and Q_2. Assume that L_1 and L_2 are skew lines. How would you find the point R_1 on L_1 and point

R_2 on L_2 such that $\overrightarrow{R_1R_2}$ is perpendicular to both L_1 and L_2?

(b) Find R_1 and R_2 when $P_1 = (3, 2, 1)$, $Q_1 = (1, 1, 1)$, $P_2 = (0, 2, 0)$, $R_2 = (2, 1, -1)$.

31 Here is another way to solve Example 4. It depends on computing the area of the parallelogram spanned by $\overrightarrow{QR}$ and $\overrightarrow{QP}$ in two ways.

(a) Show that

$$\|\overrightarrow{QR} \times \overrightarrow{QP}\| = \|\overrightarrow{QR}\|\,\|\overrightarrow{PT}\|.$$

(b) Use (a) to find the distance from P to L.

12.S SUMMARY

This chapter concerns mainly the algebra of vectors, which will be applied in the following chapters.

A vector may be pictured as an arrow. Two arrows that point in the same direction and have the same length represent the same vector. The following table summarizes the basic concepts of vectors in space. (For plane vectors disregard the third component.)

Symbol	*Name*	*Geometric Description*	*Algebraic Formula if* $\mathbf{A} = a_1\mathbf{i} + a_2\mathbf{j} + a_3\mathbf{k}$ *and* $\mathbf{B} = b_1\mathbf{i} + b_2\mathbf{j} + b_3\mathbf{k}$
A	Vector	Direction and magnitude (A)	$a_1\mathbf{i} + a_2\mathbf{j} + a_3\mathbf{k}$
$\|\mathbf{A}\|$	Length (norm, magnitude)	Length of **A**	$\sqrt{a_1^2 + a_2^2 + a_3^2}$
$-\mathbf{A}$	Negative or opposite of **A**	(A, −A)	$-a_1\mathbf{i} - a_2\mathbf{j} - a_3\mathbf{k}$
$\mathbf{A} + \mathbf{B}$	Sum of **A** and **B**	(A + B, B, A)	$(a_1 + b_1)\mathbf{i} + (a_2 + b_2)\mathbf{j} + (a_3 + b_3)\mathbf{k}$
$\mathbf{A} - \mathbf{B}$	Difference of **A** and **B**	(A − B, B, A)	$(a_1 - b_1)\mathbf{i} + (a_2 - b_2)\mathbf{j} + (a_3 - b_3)\mathbf{k}$
$c\mathbf{A}$	Scalar multiple of **A**	(A, cA)	$ca_1\mathbf{i} + ca_2\mathbf{j} + ca_3\mathbf{k}$
$\mathbf{A} \cdot \mathbf{B}$	Dot or scalar product	$\|\mathbf{A}\|\,\|\mathbf{B}\| \cos\theta$	$a_1b_1 + a_2b_2 + a_3b_3$
$\mathbf{A} \times \mathbf{B}$	Cross or vector product	Magnitude: area of parallelogram spanned by **A** and **B**, $\|\mathbf{A}\|\,\|\mathbf{B}\| \sin\theta$ Direction: perpendicular to **A** and **B**, direction by right-hand rule	$\begin{vmatrix} \mathbf{i} & \mathbf{j} & \mathbf{k} \\ a_1 & a_2 & a_3 \\ b_1 & b_2 & b_3 \end{vmatrix}$
$\mathrm{proj}_{\mathbf{B}}\,\mathbf{A}$	(Vector) projection of **A** on **B**	(A, B, $\mathrm{proj}_{\mathbf{B}}\mathbf{A}$)	$(\mathbf{A} \cdot \mathbf{u})\mathbf{u}$, where $\mathbf{u} = \mathbf{B}/\|\mathbf{B}\|$

Symbol	Name	Geometric Description	Algebraic Formula if $\mathbf{A} = a_1\mathbf{i} + a_2\mathbf{j} + a_3\mathbf{k}$, $\mathbf{B} = b_1\mathbf{i} + b_2\mathbf{j} + b_3\mathbf{k}$, and $\mathbf{C} = c_1\mathbf{i} + c_2\mathbf{j} + c_3\mathbf{k}$
$\text{comp}_{\mathbf{B}}\,\mathbf{A}$	(Scalar) projection of **A** on **B**; Component of **A** on **B**	$\lVert\mathbf{proj_B\,A}\rVert$ if angle between **A** and **B** is less than or equal to $\pi/2$; $-\lVert\mathbf{proj_B\,A}\rVert$ if angle between **A** and **B** is greater than $\pi/2$	$\mathbf{A}\cdot\mathbf{u}$, where $\mathbf{u} = \mathbf{B}/\lVert\mathbf{B}\rVert$
$\mathbf{orth_B\,A}$	(Vector) projection of **A** on plane perpendicular to **B**	orth$_{\mathbf{B}}$A; A; B	$\mathbf{A} - (\mathbf{A}\cdot\mathbf{u})\mathbf{u}$, where $\mathbf{u} = \mathbf{B}/\lVert\mathbf{B}\rVert$
$\mathbf{A}\cdot(\mathbf{B}\times\mathbf{C})$	Scalar triple product	$\pm$ volume of parallelepiped spanned by **A**, **B**, and **C**	$\begin{vmatrix} a_1 & a_2 & a_3 \\ b_1 & b_2 & b_3 \\ c_1 & c_2 & c_3 \end{vmatrix}$
$\mathbf{A}\times(\mathbf{B}\times\mathbf{C})$	Vector triple product		$(\mathbf{A}\cdot\mathbf{C})\mathbf{B} - (\mathbf{A}\cdot\mathbf{B})\mathbf{C}$

Vocabulary and Symbols

vector **A**, **B**, $\langle x, y\rangle$, $\overrightarrow{(x, y)}$
$\langle x, y, z\rangle$, $\overrightarrow{(x, y, z)}$
length, norm, magnitude, $\lVert\mathbf{A}\rVert$
sum and difference of vectors, $\mathbf{A} + \mathbf{B}$, $\mathbf{A} - \mathbf{B}$
scalar
product of scalar and vector, $c\mathbf{A}$
unit vector, **u**
basic unit vectors, **i**, **j**, **k**
dot product (scalar product), $\mathbf{A}\cdot\mathbf{B}$
projection of a vector
component of a vector
matrix
determinant
cross product (vector product), $\mathbf{A}\times\mathbf{B}$
normal to line or plane
parametric equations of line
symmetric equations of line
direction angles, α, β, γ
direction cosines $\cos\alpha$, $\cos\beta$, $\cos\gamma$
direction numbers

Key Facts

If **A** is not **0**, then $\mathbf{A}/\lVert\mathbf{A}\rVert$ is a unit vector.
If $\mathbf{A}\cdot\mathbf{B} = 0$, then $\mathbf{A} = \mathbf{0}$, $\mathbf{B} = \mathbf{0}$, or **A** is perpendicular to **B**.
The cosine of the angle θ between **A** and **B** is given by

$$\cos\theta = \frac{\mathbf{A}\cdot\mathbf{B}}{\lVert\mathbf{A}\rVert\,\lVert\mathbf{B}\rVert}.$$

$$\mathbf{A}\cdot\mathbf{A} = \lVert\mathbf{A}\rVert^2 \qquad \mathbf{A}\times\mathbf{A} = \mathbf{0}$$

$$\mathbf{A}\cdot\mathbf{B} = \mathbf{B}\cdot\mathbf{A} \qquad \mathbf{A}\times\mathbf{B} = -\mathbf{B}\times\mathbf{A}$$

$$\mathbf{A}\cdot(\mathbf{B}+\mathbf{C}) = \mathbf{A}\cdot\mathbf{B} + \mathbf{A}\cdot\mathbf{C} \qquad \mathbf{A}\times(\mathbf{B}+\mathbf{C}) = \mathbf{A}\times\mathbf{B} + \mathbf{A}\times\mathbf{C}$$

The projection of a flat region U onto a plane at an angle θ to the plane of U has area equal to the area of U times $\cos\theta$.

An equation of the line through the point (x_0, y_0) and perpendicular to the vector $A\mathbf{i} + B\mathbf{j}$ is

$$A(x - x_0) + B(y - y_0) = 0.$$

The vector $A\mathbf{i} + B\mathbf{j}$ is perpendicular to the line

$$Ax + By + C = 0.$$

The distance from the point (x_1, y_1) to the line $Ax + By + C = 0$ is

$$\frac{|Ax_1 + By_1 + C|}{\sqrt{A^2 + B^2}}.$$

An equation of the plane through the point (x_0, y_0, z_0) and perpendicular to the vector $A\mathbf{i} + B\mathbf{j} + C\mathbf{k}$ is

$$A(x - x_0) + B(y - y_0) + C(z - z_0) = 0.$$

The vector $A\mathbf{i} + B\mathbf{j} + C\mathbf{k}$ is perpendicular to the plane

$$Ax + By + Cz + D = 0.$$

The distance from the point (x_1, y_1, z_1) to the plane $Ax + By + Cz + D = 0$ is

$$\frac{|Ax_1 + By_1 + Cz_1 + D|}{\sqrt{A^2 + B^2 + C^2}}.$$

The direction angles α, β, and γ of a vector are the angles it makes with $\mathbf{i}$, $\mathbf{j}$, and $\mathbf{k}$; $\cos\alpha$, $\cos\beta$, and $\cos\gamma$ are the direction cosines of the vector (or of a line parallel to the vector). They are related by the equation

$$\cos^2\alpha + \cos^2\beta + \cos^2\gamma = 1.$$

The line through $P_0 = (x_0, y_0, z_0)$ parallel to $\mathbf{A} = a_1\mathbf{i} + a_2\mathbf{j} + a_3\mathbf{k}$ is given parametrically as

$$\begin{cases} x = x_0 + a_1 t \\ y = y_0 + a_2 t \\ z = z_0 + a_3 t, \end{cases}$$

or vectorially as

$$\overrightarrow{OP} = \overrightarrow{OP_0} + t\mathbf{A}.$$

Also, the line has the description in the symmetric form (if a_1, a_2, a_3 are not 0)

$$\frac{x - x_0}{a_1} = \frac{y - y_0}{a_2} = \frac{z - z_0}{a_3}.$$

An arrangement

$$\begin{pmatrix} a_1 & a_2 \\ b_1 & b_2 \end{pmatrix}$$

is a 2 by 2 matrix. A 2 by 2 determinant is defined by

$$\begin{vmatrix} a_1 & a_2 \\ b_1 & b_2 \end{vmatrix} = a_1b_2 - a_2b_1.$$

The absolute value of the determinant is the area of the parallelogram spanned by $\mathbf{A} = a_1\mathbf{i} + a_2\mathbf{j}$ and $\mathbf{B} = b_1\mathbf{i} + b_2\mathbf{j}$. A 3 by 3 determinant is defined as

$$\begin{vmatrix} a_1 & a_2 & a_3 \\ b_1 & b_2 & b_3 \\ c_1 & c_2 & c_3 \end{vmatrix} = a_1\begin{vmatrix} b_2 & b_3 \\ c_2 & c_3 \end{vmatrix} - a_2\begin{vmatrix} b_1 & b_3 \\ c_1 & c_3 \end{vmatrix} + a_3\begin{vmatrix} b_1 & b_2 \\ c_1 & c_2 \end{vmatrix}.$$

Its absolute value is the volume of the parallelepiped spanned by $\langle a_1, a_2, a_3\rangle$, $\langle b_1, b_2, b_3\rangle$, and $\langle c_1, c_2, c_3\rangle$.

GUIDE QUIZ ON CHAP. 12: ALGEBRAIC OPERATIONS ON VECTORS

1 Given $\mathbf{A} = \mathbf{i} + 2\mathbf{j} - \mathbf{k}$ and $\mathbf{B} = 2\mathbf{i} - \mathbf{j} + 3\mathbf{k}$, find
(*a*) $\mathbf{A} \cdot \mathbf{B}$,
(*b*) $\|\mathbf{A}\|$,
(*c*) a unit vector in the direction of **A**,
(*d*) the component of **B** on **A**,
(*e*) the projection of **B** on **A**,
(*f*) the component of **A** on **B**,
(*g*) the projection of **B** perpendicular to **A**,
(*h*) the cosine of the angle between **A** and **B**,
(*i*) the angle between **A** and **B**,
(*j*) $\mathbf{A} \times \mathbf{B}$,
(*k*) $\mathbf{B} \times \mathbf{A}$,
(*l*) a unit vector perpendicular to both **A** and **B**,
(*m*) the area of the parallelogram spanned by **A** and **B**.

2 Draw the necessary diagrams and explain why the distance from the point (x_1, y_1) to the line $Ax + By + C = 0$ is

$$\frac{|Ax_1 + By_1 + C|}{\sqrt{A^2 + B^2}}.$$

3 Find the direction cosines of a vector normal to the plane $x - 2y + 2z = 16$.

4 Prove that

$$\begin{vmatrix} a_1 & a_2 & a_3 \\ b_1 & b_2 & b_3 \\ c_1 + b_1 & c_2 + b_2 & c_3 + b_3 \end{vmatrix} = \begin{vmatrix} a_1 & a_2 & a_3 \\ b_1 & b_2 & b_3 \\ c_1 & c_2 & c_3 \end{vmatrix}.$$

5 (*a*) Find the volume of the parallelepiped shown in Fig. 1.
(*b*) Justify the general formula you used in (*a*).

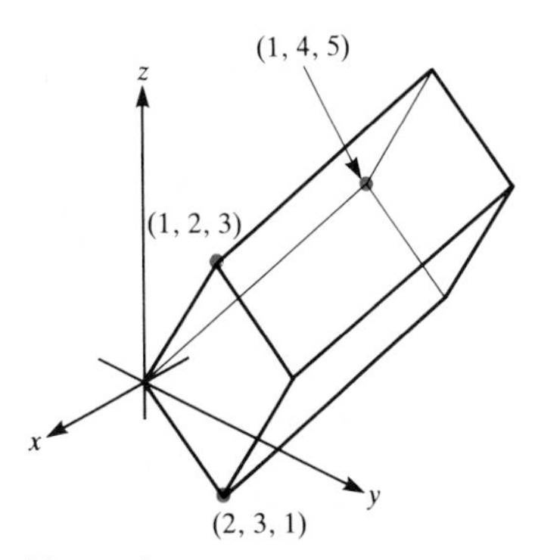

Figure 1

6 Let **A** be a vector in space and P, Q, and R three points that determine a plane.
(*a*) Using a picture, show that there are vectors $\mathbf{A}_1$ and $\mathbf{A}_2$ such that $\mathbf{A} = \mathbf{A}_1 + \mathbf{A}_2$ and $\mathbf{A}_1$ is parallel to the plane and $\mathbf{A}_2$ is perpendicular to the plane.
(*b*) Find formulas for $\mathbf{A}_1$ and $\mathbf{A}_2$.

7 Where does the line through (1, 2, 1) and (3, 1, 1) meet the plane determined by the three points (2, −1, 1), (5, 2, 3), and (4, 1, 3)?

8 (*a*) Show that $\text{comp}_\mathbf{C}\, \mathbf{A} + \text{comp}_\mathbf{C}\, \mathbf{B} = \text{comp}_\mathbf{C}(\mathbf{A} + \mathbf{B})$. (Draw a picture.)
(*b*) Define the dot product of vectors **D** and **E**.
(*c*) Using (*a*) show that $(\mathbf{A} + \mathbf{B}) \cdot \mathbf{C} = \mathbf{A} \cdot \mathbf{C} + \mathbf{B} \cdot \mathbf{C}$.
(*d*) Obtain the formula for dot product in terms of components.

9 In Fig. 2, **n** is a unit normal to the plane $\mathcal{P}$. Develop formulas for $\mathbf{proj_n}\, \mathbf{A}$ and $\mathbf{proj}_\mathcal{P}\, \mathbf{A}$ in terms of **A** and **n**.

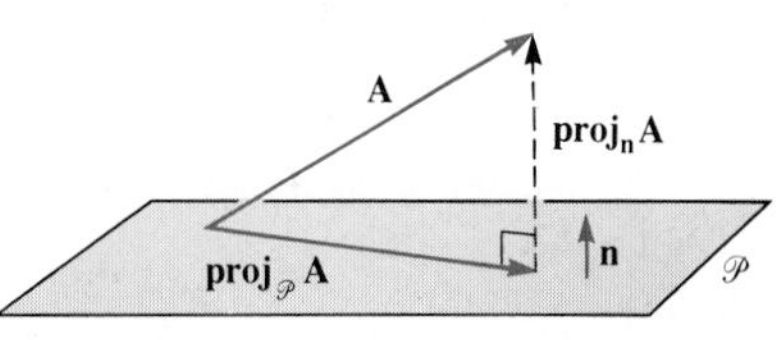

Figure 2

10 Explain in detail why the projection of a flat region of area A on a plane inclined at an angle θ to its plane has area $A \cos \theta$.

11 Where does the line of intersection of the planes $x + 2y + z = 4$ and $2x - y + z = 1$ meet the plane $3x + 2y + z = 6$?

12 Let L be the line in which the planes $x + y + 3z = 5$ and $2x - y + z = 2$ intersect.
(*a*) Find a vector parallel to L.
(*b*) Find a point on L.
(*c*) Find parametric equations for L.

13 The planes $x + 2y + 3z = 6$ and $2x - 3y + 4z = 8$ intersect in a line L.
(*a*) Find a vector parallel to L.
(*b*) Find a point on L.
(*c*) Find parametric equations for L.
(*d*) Find symmetric equations for L.

14 A plane contains the points (1, 1, 1), (2, 3, 2), and (1, 4, 5).
(*a*) Find two vectors that lie in the plane.
(*b*) Construct a normal to the plane.
(*c*) Find an equation of the plane.

15 A plane contains the line through (1, 1, 2) with direction numbers 2, 3, 4. It also contains the line through (1, 1, 2) with direction numbers 3, 1, 5.
(*a*) Find a normal to the plane.
(*b*) What angle does the normal in (*a*) make with the y axis?
(*c*) Find an equation of the plane.

16 (*a*) Explain why the area of the parallelogram spanned by $\langle a_1, a_2\rangle$ and $\langle b_1, b_2\rangle$ is the absolute value of the determinant

$$\begin{vmatrix} a_1 & a_2 \\ b_1 & b_2 \end{vmatrix}.$$

(*b*) Explain why the volume of the parallelepiped spanned by $\langle a_1, a_2, a_3\rangle$, $\langle b_1, b_2, b_3\rangle$, and $\langle c_1, c_2, c_3\rangle$ is the absolute value of the determinant

$$\begin{vmatrix} a_1 & a_2 & a_3 \\ b_1 & b_2 & b_3 \\ c_1 & c_2 & c_3 \end{vmatrix}.$$

17 Explain why the equation $\mathbf{A} \times (\mathbf{B} \times \mathbf{C}) = (\mathbf{A} \cdot \mathbf{C})\mathbf{B} - (\mathbf{A} \cdot \mathbf{B})\mathbf{C}$ is "to be expected."

18 A square whose sides have length 4 lies in the plane $z = x$.
(*a*) Draw a clear diagram that shows the plane and the square in perspective.
(*b*) Draw the projection of the square on the xy plane.
(*c*) Must the projection in (*b*) be a square? a rectangle? a parallelogram? Explain.
(*d*) What is the area of the projection in (*b*)?
(*e*) What is the area of the square's projection on the plane $2x - y + 3z = 8$?

19 The vector **A** in Fig. 3 has magnitude 4. Find
(*a*) The projection of **A** on the xy plane.
(*b*) The projection of **A** on each axis.
(*c*) The direction cosines of **A**.
(*d*) The direction angles of **A**.
(*e*) The projection of **A** on the yz plane.

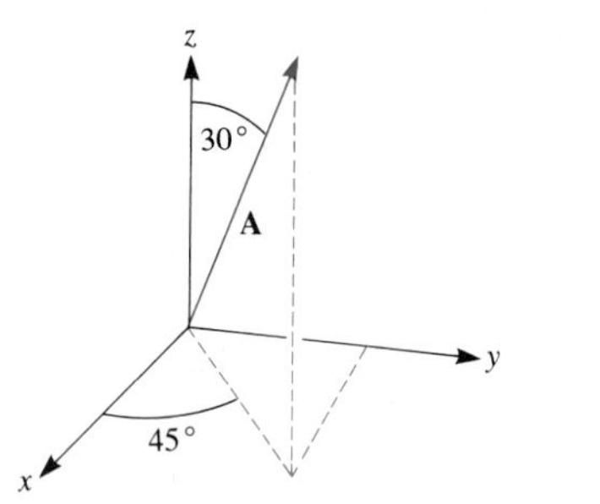

Figure 3

REVIEW EXERCISES FOR CHAP. 12: ALGEBRAIC OPERATIONS ON VECTORS

1 (*a*) Define $\mathbf{A} \cdot \mathbf{B}$, where **A** and **B** are vectors in space.
(*b*) What is the formula for $\mathbf{A} \cdot \mathbf{B}$ in terms of the components of **A** and **B**?
(*c*) Prove that if $\mathbf{A} \cdot \mathbf{B} = 0$ and neither **A** nor **B** is the zero vector, then **A** is perpendicular to **B**.

2 What is the component of $\mathbf{A} = -2\mathbf{i} + 3\mathbf{j}$ on the vector
(*a*) **i**? (*b*) **j**?
(*c*) $0.6\mathbf{i} + 0.8\mathbf{j}$? (*d*) $4\mathbf{i} - 5\mathbf{j}$?

3 What is the projection of $\mathbf{A} = -2\mathbf{i} + 3\mathbf{j}$ on the vector
(*a*) **i**? (*b*) **j**?
(*c*) $0.6\mathbf{i} + 0.8\mathbf{j}$? (*d*) $4\mathbf{i} - 5\mathbf{j}$?

4 (*a*) Give an application of the dot product in physics and one in economics.
(*b*) Give an application of the cross product in physics.

5 Prove that $(a_1\mathbf{i} + a_2\mathbf{j} + a_3\mathbf{k}) \cdot (b_1\mathbf{i} + b_2\mathbf{j} + b_3\mathbf{k}) = a_1b_1 + a_2b_2 + a_3b_3$.

6 Let **A** be a nonzero vector. (*a*) How would you produce a vector **B** perpendicular to **A**? (*b*) How would you produce a vector perpendicular to both **A** and **B**?

7 (*a*) Define $\mathbf{A} \times \mathbf{B}$.
(*b*) Describe $\mathbf{A} \times \mathbf{B}$ geometrically.

8 Find the point on the plane $2x - y + 3z + 12 = 0$ that is nearest the origin. Use vectors, not calculus.

9 Let **A** and **B** be two nonparallel nonzero vectors in space.
(*a*) Explain why there are scalars x and y such that

$$\mathbf{A} \times (\mathbf{A} \times \mathbf{B}) = x\mathbf{A} + y\mathbf{B}.$$

(*b*) Show that $\mathbf{A} \times (\mathbf{A} \times \mathbf{B})$ is not $\mathbf{0}$.
(*c*) Show that $(\mathbf{A} \times \mathbf{A}) \times \mathbf{B} = \mathbf{0}$.

10 Find scalars x and y such that

$$x(2\mathbf{i} + \mathbf{j}) + y(\mathbf{i} + 3\mathbf{j}) = 4\mathbf{i} - 2\mathbf{j}.$$

11 Find scalars x, y, and z such that $x(3\mathbf{i} + \mathbf{j} + 2\mathbf{k}) + y(\mathbf{i} - \mathbf{j} + 2\mathbf{k}) + z(2\mathbf{i} + 2\mathbf{j} + \mathbf{k})$ is equal to $3\mathbf{i} + 3\mathbf{j} + 3\mathbf{k}$.

12 A determinant in which some column has only 0's is 0. Prove this for the determinant

$$\begin{vmatrix} a_1 & 0 & a_3 \\ b_1 & 0 & b_3 \\ c_1 & 0 & c_3 \end{vmatrix}.$$

13 Show that

$$\begin{vmatrix} 0 & a & b \\ -a & 0 & c \\ -b & -c & 0 \end{vmatrix} = 0$$

for all numbers a, b, c.

14 Suppose that a vector from (1, 3, 3) to the plane $x - 4y + 5z + 4 = 0$ makes an angle of 45° with that plane. Find the length of the vector.

15 Figure 4 shows a pyramid with a square base. Find the cosine of the angle between
(*a*) $\overrightarrow{CA}$ and $\overrightarrow{CB}$, (*b*) $\overrightarrow{EA}$ and $\overrightarrow{EB}$, (*c*) $\overrightarrow{AD}$ and $\overrightarrow{AC}$.

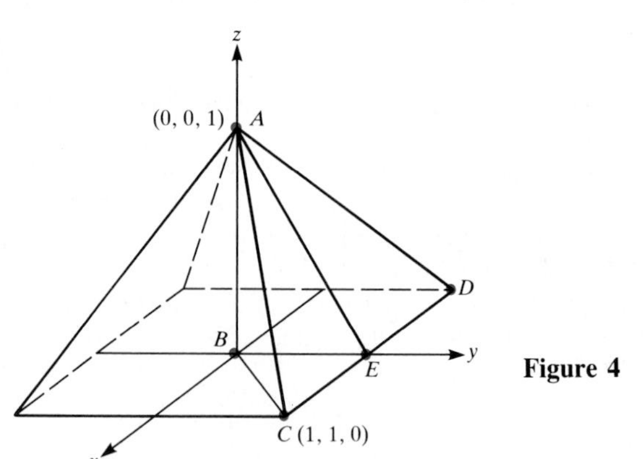

Figure 4

16 The planes $2x + 5y + z = 10$ and $3x - y + 4z = 11$ meet in a line. For this line find
(*a*) direction numbers,
(*b*) direction cosines,
(*c*) direction angles,
(*d*) a point on the line.

17 How far apart are the planes parallel to the plane $Ax + By + Cz + D = 0$ that pass through the points (x_1, y_1, z_1) and (x_2, y_2, z_2)?

18 Find the angle between the planes $x - 3y + 4z = 10$ and $2x + y + z = 11$.

19 Find parametric equations of the line through $(1, 1, 2)$ that is parallel to the planes $x + 2y + 3z = 0$ and $2x - y + 3z + 4 = 0$.

20 Does the plane through $(1, 1, -1)$, perpendicular to $2\mathbf{i} + 4\mathbf{j} + 5\mathbf{k}$, pass through the point $(4, 5, -7)$?

21 Is the line through $(1, 4, 7)$ and $(5, 10, 15)$ perpendicular to the plane $2x + 3y + 4z = 17$?

22 Let $\mathbf{A} = 2\mathbf{i} + 3\mathbf{j} + 4\mathbf{k}$. Find
(*a*) $\text{comp}_{\mathbf{i}}\, \mathbf{A}$
(*b*) $\text{comp}_{-\mathbf{i}}\, \mathbf{A}$
(*c*) $\mathbf{proj}_{\mathbf{i}}\, \mathbf{A}$
(*d*) $\mathbf{proj}_{-\mathbf{i}}\, \mathbf{A}$
(*e*) $\mathbf{proj}_{\mathbf{i}+\mathbf{j}}\, \mathbf{A}$.

23 Find the point on the plane

$$\frac{x}{2} + \frac{y}{3} + \frac{z}{4} = 1$$

(*a*) nearest the origin,
(*b*) nearest the point $(1, 2, 3)$.

24 Let Q, R, and S be distinct points such that S is not on the line L through Q and R.
(*a*) Make a sketch that indicates there are two points P on L such that the angle between PS and L is $\pi/3$.
(*b*) How would you find the points P?
(*c*) Find the points P when $Q = (1, 0, 1)$, $R = (2, -1, 2)$, and $S = (3, 2, -4)$.

25 How would you determine whether the points (x_1, y_1) and (x_2, y_2) are on the same side or on opposite sides of the line $Ax + By + C = 0$? (Don't say "I'd draw them." Develop an algebraic procedure that could be programmed.)

26 How would you determine whether the points (x_1, y_1, z_1) and (x_2, y_2, z_2) are on the same side or on opposite sides of the plane $Ax + By + Cz + D = 0$?

27 Express in terms of cross products and dot products the statement that the line through P_1 and P_2 is parallel to the plane through P_3, P_4, and P_5.

28 A parallelepiped is spanned by the three vectors $\mathbf{A} = \mathbf{i} + 2\mathbf{j} + 3\mathbf{k}$, $\mathbf{B} = 2\mathbf{i} + \mathbf{j} + \mathbf{k}$, $\mathbf{C} = 3\mathbf{i} + 3\mathbf{j} + \mathbf{k}$.
(*a*) Find the volume of the parallelepiped.
(*b*) Find the area of the face spanned by $\mathbf{A}$ and $\mathbf{C}$.
(*c*) Find the angle between $\mathbf{A}$ and the face spanned by $\mathbf{B}$ and $\mathbf{C}$.

29 Find the angle between the line through $(0, 0, 0)$ and $(1, 1, 1)$ and the plane through $(1, 2, 3)$, $(4, 1, 5)$, and $(2, 0, 6)$.

30 Let $\mathbf{u}_1$ and $\mathbf{u}_2$ be perpendicular unit vectors in space. Find $(\mathbf{u}_1 \times \mathbf{u}_2) \times \mathbf{u}_1$ and $\mathbf{u}_1 \times (\mathbf{u}_1 \times \mathbf{u}_2)$.

31 Let L_1 and L_2 be two lines in the xy plane through $(0, 0)$, one of slope m_1 and one of slope m_2.
(*a*) Show that the point $(1, m_1)$ is on L_1 and that $\mathbf{i} + m_1\mathbf{j}$ is a vector parallel to L_1.
(*b*) Show that the point $(1, m_2)$ is on L_2 and that $\mathbf{i} + m_2\mathbf{j}$ is a vector parallel to L_2.
(*c*) Using (*a*) and (*b*), prove that L_1 is perpendicular to L_2 if and only if $m_1m_2 = -1$.

32 Express in terms of cross products and dot products the assertion that the points P_4 and P_5 are situated on the same side of the plane through P_1, P_2, and P_3.

33 Show that

$$\mathbf{A} \times (\mathbf{B} \times \mathbf{C}) + \mathbf{B} \times (\mathbf{C} \times \mathbf{A}) + \mathbf{C} \times (\mathbf{A} \times \mathbf{B}) = \mathbf{0}.$$

34 Let $\mathbf{B}$ be a nonzero vector. Assume that $\mathbf{A}_1$ and $\mathbf{A}_1'$ are parallel to $\mathbf{B}$, $\mathbf{A}_2$ and $\mathbf{A}_2'$ are perpendicular to $\mathbf{B}$, and $\mathbf{A}_1 + \mathbf{A}_2 = \mathbf{A}_1' + \mathbf{A}_2'$. Must $\mathbf{A}_1 = \mathbf{A}_1'$ and $\mathbf{A}_2 = \mathbf{A}_2'$? Explain both with and without pictures.

35 Express in terms of dot products and cross products the assertion that the plane through the points P, Q, and R is perpendicular to the plane through the points S, T, and U.

36 Express in terms of dot products and cross products the assertion that the intersection of the planes $A_1x + B_1y + C_1z + D_1 = 0$ and $A_2x + B_2y + C_2z + D_2 = 0$ is perpendicular to the intersection of $A_3x + B_3y + C_3z + D_3 = 0$ and $A_4x + B_4y + C_4z + D_4 = 0$.

37 Write a program that finds the distance between the point $P_1 = (x_1, y_1, z_1)$ and the line through $P_2 = (x_2, y_2, z_2)$ and $P_3 = (x_3, y_3, z_3)$.

38 Write a program to determine whether the line through $P_1 = (x_1, y_1, z_1)$ and $P_2 = (x_2, y_2, z_2)$ intersects the line through $P_3 = (x_3, y_3, z_3)$ and $P_4 = (x_4, y_4, z_4)$.

39 (*Gram-Schmidt procedure*) Let $\mathbf{A}$ and $\mathbf{B}$ be perpendicular nonzero vectors and $\mathbf{C}$ an arbitrary vector.
(*a*) Draw a picture which shows that there are vectors $\mathbf{C}_1$ and $\mathbf{C}_2$ such that $\mathbf{C}_1$ is of the form $x\mathbf{A} + y\mathbf{B}$ and $\mathbf{C}_2$ is perpendicular to the plane of $\mathbf{A}$ and $\mathbf{B}$ such that $\mathbf{C} = \mathbf{C}_1 + \mathbf{C}_2$.
(*b*) Find a formula for $\mathbf{C}_2$ in terms of dot and cross products of $\mathbf{A}$, $\mathbf{B}$, and $\mathbf{C}$.

40 Find the point P on the line L through $P_0 = (2, 1, 1)$ and $P_1 = (1, 3, 2)$ such that the line through it and the point $P_2 = (3, 1, 5)$ is perpendicular to L.

41 Find a point P on the line L in Exercise 40 such that the angle between L and the line through P and $(4, 1, 1)$ is $\pi/3$.

42 A ray of light parallel to the vector $2\mathbf{i} + 3\mathbf{j} + 4\mathbf{k}$ is reflected off a flat mirror. The equation for the plane of the mirror is $x - y + 2z = 0$. Find a vector parallel to the reflected ray.

43 The bulb of a flashlight is located at the point $(1, 2, 3)$. The other end of the flashlight is at $(3, 4, 6)$. The light shines on a flat mirror lying in the plane whose equation is $x + 3y +$

$3z = 0$. (See Fig. 5.)

(*a*) At what point does the light strike the plane?

(*b*) At what point does the reflected light strike the plane $y + z = 4$?

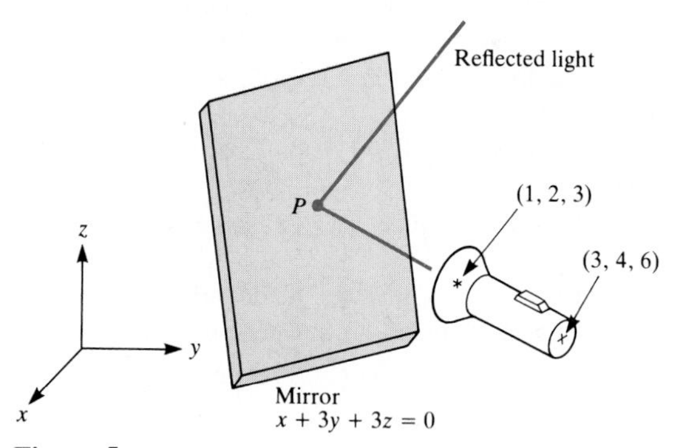

Figure 5

44 *(Contributed by Melvyn Kopald Stein.)* A symmetrically designed chute joins a 5-foot-square intake to a 3-foot-square opening, 10 feet lower, as shown in Fig. 6. Find the angle between two adjacent sides of the chute.

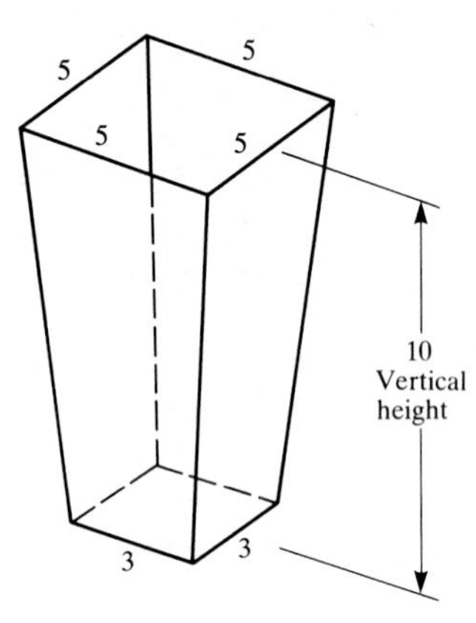

Figure 6

45 Let **A**, **B**, and **C** be noncoplanar vectors and $\mathbf{D} = x\mathbf{A} + y\mathbf{B} + z\mathbf{C}$. Show that

$$z = \frac{\mathbf{D} \cdot (\mathbf{B} \times \mathbf{C})}{\mathbf{A} \cdot (\mathbf{B} \times \mathbf{C})}.$$

Hint: Draw a picture.

46 Let **A**, **B**, and **C** be noncoplanar vectors. Let **N** be a vector perpendicular to the plane determined by **A** and **B**.

(*a*) Find $\mathbf{proj}_{\mathbf{N}}\ \mathbf{C}$.

(*b*) Find the projection of **C** on the plane determined by **A** and **B**.

47 Figure 7 shows a triangle *PQR* and three vectors, each going from one of the vertices to the midpoint of the opposite side.

(*a*) Prove that the three vectors can be placed to form the sides of a triangle.

(*b*) What relation is there between the area of the triangle formed in (*a*) and the area of triangle *PQR*?

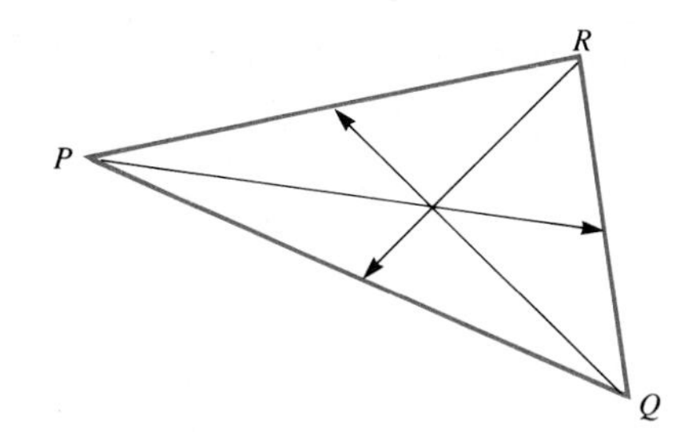

Figure 7

48 New York has latitude 41° and longitude 74°. San Francisco has latitude 38° and longitude 122°. Assuming that the radius of the Earth is 4,000 miles, find

(*a*) the great circle distance between the two cities;

(*b*) the straight line distance between them.

49 A polygonal region of area A lies in some plane in space (that is, it is a flat region, not a twisted one). The projections of the region on the coordinate planes have areas A_1, A_2, and A_3. If you are given values for A_1, A_2, and A_3, what can you say about A? (See Fig. 8.)

(*a*) If you know A_1, A_2, and A_3, can you find A?

(*b*) Show that $A \leq A_1 + A_2 + A_3$.

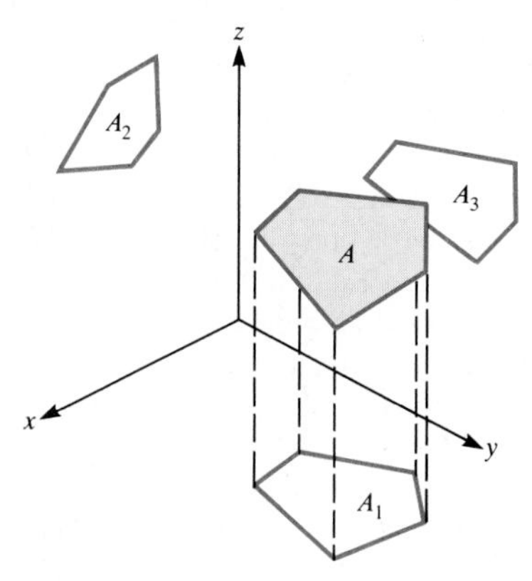

Figure 8

13

THE DERIVATIVE OF A VECTOR FUNCTION

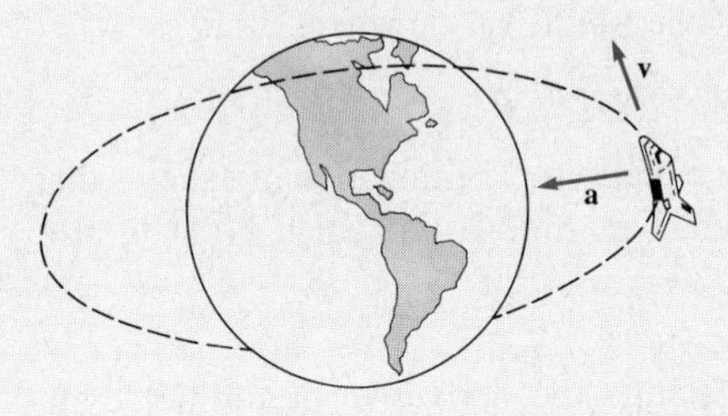

The velocity and acceleration of an object traveling on a curve are most easily described by vectors. At any moment in the flight of the space shuttle, a vector **v** records its speed and direction and a vector **a** the magnitude and direction of the acceleration, whether due to the pull of the earth or the firing of rockets, as shown in the figure. Both **v** and **a** are functions of time.

Sections 13.1 and 13.2, motivated by the study of motion, develop the notion of a derivative of a vector function. This derivative is similar to the derivative defined in Sec. 3.2. Then Secs. 13.3 and 13.4 examine the acceleration of an object that is moving along a curve. Section 13.5 shows that Newton's law of gravitation implies Kepler's three laws of planetary motion.

Only Secs. 13.1 and 13.2, and the vector **T** in Sec. 13.4, are needed in the remaining chapters.

13.1 THE DERIVATIVE OF A VECTOR FUNCTION

Consider an object moving in a plane. It might be a mass at the end of a rope, a ball, a satellite, a comet, a raindrop, or an astronaut's spacecraft. Call this object a "particle" and assume that all its mass is located at a single point. Denote the position of the particle at time t relative to an xy-coordinate system by (x, y). We shall describe its position with the position vector $\mathbf{r}$, whose tail is at $(0, 0)$ and whose head is at (x, y), as in Fig. 1. Thus $\mathbf{r} = x\mathbf{i} + y\mathbf{j}$, where x and y depend on time t. Therefore, $\mathbf{r}$ depends on t and may be written as $\mathbf{r} = \mathbf{G}(t)$ or as $\mathbf{r}(t)$.

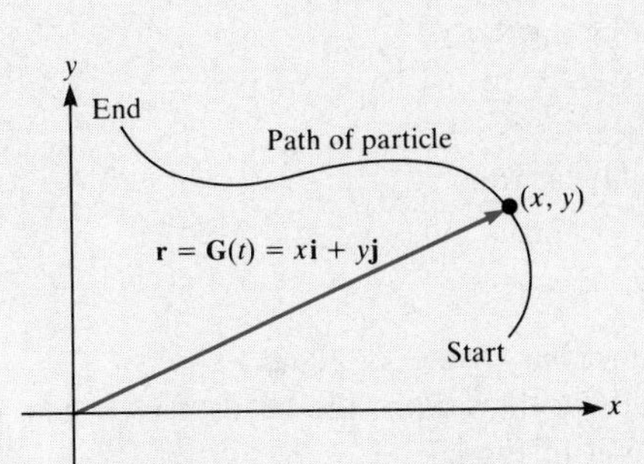

Figure 1

If the particle is moving on a curve in space, its position at time t is described by a position vector $\mathbf{r} = \mathbf{G}(t) = x\mathbf{i} + y\mathbf{j} + z\mathbf{k}$, where x, y, and z are functions of time. Note that $x = x(t)$, $y = y(t)$, and $z = z(t)$ are each scalar functions of t. Thus a single vector function is a bookkeeping device that records three scalar functions.

Vector Functions

This brings us to an important definition, which calls attention to this new type of function.

Definition *Vector function.* A function whose inputs are scalars and whose outputs are vectors is called a **vector function** or **vector-valued function**. It is usually denoted by a boldface letter, such as **F** or **G**.

Frequently **r** *is a function of time t or arc length s.*

Usually the scalar input may be thought of as time, and $\mathbf{G}(t)$ as the position vector at time t. The scalar input might sometimes be arc length along the curve, and $\mathbf{G}(s)$ would be the position vector when the particle has swept out a distance s along the curve.

EXAMPLE 1 Sketch the path of a particle that has the position vector $\mathbf{G}(t) = 2t\mathbf{i} + 4t^2\mathbf{j}$ at time t.

SOLUTION At time t the particle is at the point (x, y), where

$$x = 2t \qquad \text{and} \qquad y = 4t^2.$$

The path is given parametrically by these equations. Elimination of t shows that $y = x^2$. The path is a parabola, sketched in Fig. 2, in which $\mathbf{G}(1)$ is also depicted. ■

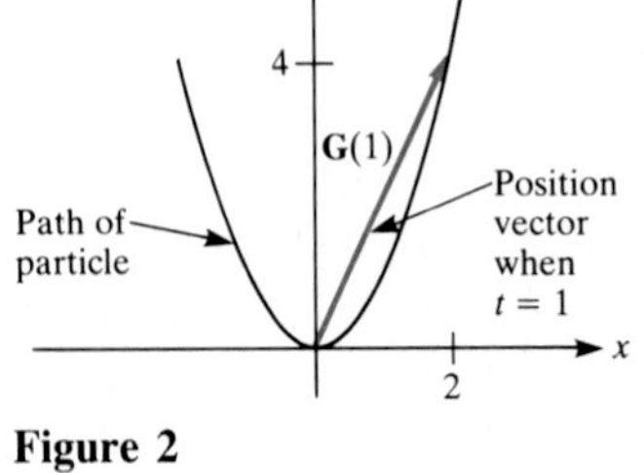
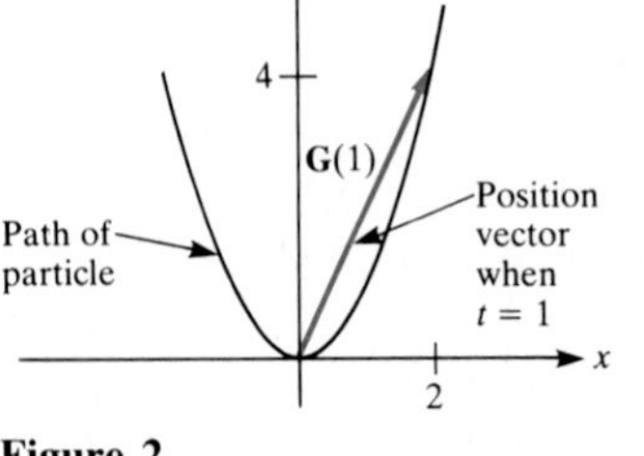

Figure 2

In the case of motion on a horizontal line the derivative of position with respect to time is sufficient to describe the motion of the particle. If the derivative is positive, the particle is moving to the right. If the derivative is negative, the particle is moving to the left. The speed is simply the absolute value of the derivative. But the study of motion in the plane or in space depends on the concept of the derivative of a vector function.

In order to define the derivative of a vector function we need the concept of a limit of a vector function.

Limit of a Vector Function

Let **G** be a vector function, $\mathbf{G}(t) = x(t)\mathbf{i} + y(t)\mathbf{j} + z(t)\mathbf{k}$. We define the limit of $\mathbf{G}(t)$ as $t \to a$ in terms of the limits of its components.

Definition *Limit of a vector function.* Let $\mathbf{G}(t) = x(t)\mathbf{i} + y(t)\mathbf{j} + z(t)\mathbf{k}$ be a vector function. Let a be a number and $\mathbf{A} = a_1\mathbf{i} + a_2\mathbf{j} + a_3\mathbf{k}$ a vector. We say that

$$\lim_{t \to a} \mathbf{G}(t) = \mathbf{A}$$

if

$$\lim_{t \to a} x(t) = a_1, \quad \lim_{t \to a} y(t) = a_2, \quad \text{and} \quad \lim_{t \to a} z(t) = a_3.$$

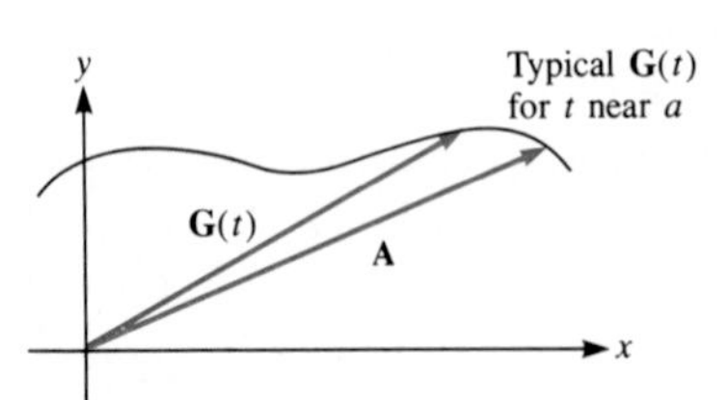

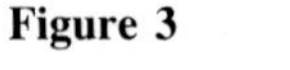
Figure 3

(A similar definition holds for vector functions in the xy plane.)

If $\mathbf{G}(t)$ and **A** are drawn with their tails at the origin, the head of $\mathbf{G}(t)$ gets close to the head of **A** as $t \to a$, as shown in Fig. 3.

EXAMPLE 2 Find $\lim_{t\to 1} \mathbf{G}(t)$, where $\mathbf{G}(t) = t^2\mathbf{i} + 2t\mathbf{j}$.

SOLUTION By the definition of the limit of a vector function,

$$\lim_{t\to 1} \mathbf{G}(t) = (\lim_{t\to 1} t^2)\mathbf{i} + (\lim_{t\to 1} 2t)\mathbf{j} = \mathbf{i} + 2\mathbf{j}. \quad ■$$

The limits of vector functions behave just like limits of scalar functions. For instance,

$$\lim_{t\to a} (\mathbf{G}(t) + \mathbf{H}(t)) = \lim_{t\to a} \mathbf{G}(t) + \lim_{t\to a} \mathbf{H}(t),$$

if the limits on the right-hand side of the equation exist. Moreover, if $f(t)$ is a scalar function then $f(t)\mathbf{G}(t)$ is a vector function and

$$\lim_{t\to a} f(t)\,\mathbf{G}(t) = \lim_{t\to a} f(t) \lim_{t\to a} \mathbf{G}(t),$$

if the limits on the right-hand side exist.

Continuity of a vector function is defined the same way as continuity of a scalar function.

Definition *Continuity of a vector function.* The vector function $\mathbf{G}(t)$ is **continuous** at a if

1 $\mathbf{G}(a)$ is defined,
2 $\lim_{t\to a} \mathbf{G}(t)$ exists, and
3 $\lim_{t\to a} \mathbf{G}(t) = \mathbf{G}(a)$.

Continuity of a vector function $\mathbf{G}(t)$ is equivalent to the continuity of its component scalar functions $x(t)$, $y(t)$, and $z(t)$.

Derivative of a Vector Function

Now it is possible to define the derivative of a vector function. The definition will be modeled after that of the derivative of a scalar function,

$$f'(x) = \lim_{\Delta x\to 0} \frac{f(x + \Delta x) - f(x)}{\Delta x}.$$

After the definition is given, its geometric meaning will be examined.

Definition *Derivative of a vector function.* Let $\mathbf{G}$ be a vector function. The limit

$$\lim_{\Delta t\to 0} \frac{\mathbf{G}(t + \Delta t) - \mathbf{G}(t)}{\Delta t} \tag{1}$$

(if it exists) is called the **derivative of G** at t. It is denoted $\mathbf{G}'(t)$.

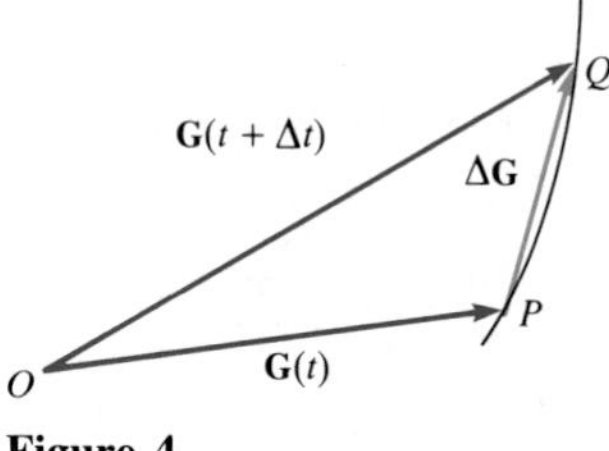

Figure 4

What does the quotient in (1) mean geometrically? The numerator, which may be called $\Delta\mathbf{G}$, is a vector. The denominator Δt is a scalar. The quotient $\Delta\mathbf{G}/\Delta t$ is a vector.

If the derivative exists, then when Δt is small, $\Delta\mathbf{G}$ is a short vector. (See Fig. 4.) It is the vector from the head of $\mathbf{G}(t)$ to the head of $\mathbf{G}(t + \Delta t)$. That is, if $\mathbf{G}(t) = \overrightarrow{OP}$ and $\mathbf{G}(t + \Delta t) = \overrightarrow{OQ}$, then $\Delta\mathbf{G} = \overrightarrow{PQ}$. The vector $\Delta\mathbf{G}$ lies on a chord of the path; when Δt is small, the direction of $\Delta\mathbf{G}$ is close to that of the tangent line to the curve at P. Since Δt is a scalar (for simplicity, consider it to be positive), the vector $\Delta\mathbf{G}/\Delta t$ is parallel to the vector $\Delta\mathbf{G}$ and points in the same direction as $\Delta\mathbf{G}$. Therefore, the vector $\Delta\mathbf{G}/\Delta t$, when Δt is small and positive, would presumably point almost along a tangent line at P.

Thus $\mathbf{G}'(t)$ is presumably parallel to the tangent line at P. The direction of $\mathbf{G}'(t)$ is the direction in which the particle is moving as it passes through P. These observations suggest the following definition.

Definition *Tangent vector to a curve.* Let $\mathbf{r} = \mathbf{G}(t)$ describe a curve in the plane or space. Assume that $\mathbf{G}'(t)$ exists and is not $\mathbf{0}$. Then the vector $\mathbf{G}'(t)$ is called a **tangent vector** to the curve.

To compute $\mathbf{G}'(t)$, just differentiate the scalar components of $\mathbf{G}(t)$. For if $\mathbf{G}(t) = x(t)\mathbf{i} + y(t)\mathbf{j} + z(t)\mathbf{k}$, then

$$\begin{aligned}
\lim_{\Delta t\to 0} \frac{\Delta\mathbf{G}}{\Delta t} &= \lim_{\Delta t\to 0} \frac{\Delta x\mathbf{i} + \Delta y\mathbf{j} + \Delta z\mathbf{k}}{\Delta t} \\
&= \lim_{\Delta t\to 0} \left(\frac{\Delta x}{\Delta t}\mathbf{i} + \frac{\Delta y}{\Delta t}\mathbf{j} + \frac{\Delta z}{\Delta t}\mathbf{k}\right) \\
&= \lim_{\Delta t\to 0} \frac{\Delta x}{\Delta t}\mathbf{i} + \lim_{\Delta t\to 0} \frac{\Delta y}{\Delta t}\mathbf{j} + \lim_{\Delta t\to 0} \frac{\Delta z}{\Delta t}\mathbf{k} \\
&= \frac{dx}{dt}\mathbf{i} + \frac{dy}{dt}\mathbf{j} + \frac{dz}{dt}\mathbf{k}.
\end{aligned}$$

(Physicists often choose to write this in Newton's dot notation as $\dot{x}\mathbf{i} + \dot{y}\mathbf{j} + \dot{z}\mathbf{k}$.)

For the record, we state this useful fact as a theorem.

Theorem 1 Let $\mathbf{G}(t) = x(t)\mathbf{i} + y(t)\mathbf{j} + z(t)\mathbf{k}$. If the derivatives x', y', and z' exist at $t = a$, then $\mathbf{G}(t)$ has a derivative at $t = a$ and

$$\mathbf{G}'(a) = x'(a)\mathbf{i} + y'(a)\mathbf{j} + z'(a)\mathbf{k}.$$

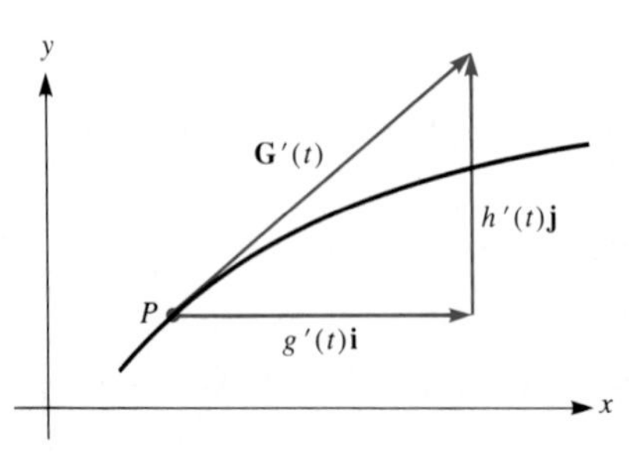

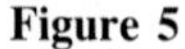
Figure 5

Back in Sec. 9.3 we considered plane curves given parametrically, $(x, y) = (g(t), h(t))$. Such a curve is described by the single vector function $\mathbf{G}(t) = g(t)\mathbf{i} + h(t)\mathbf{j}$. The tangent vector defined in this section is simply $\mathbf{G}'(t) = g'(t)\mathbf{i} + h'(t)\mathbf{j}$, shown in Fig. 5.

According to Fig. 5, the slope of the tangent line to the curve at the point P is $h'(t)/g'(t)$. Fortunately, this formula agrees with the formula in Sec. 9.3 for finding the slope of a curve given parametrically:

$$\frac{dy}{dx} = \frac{dy/dt}{dx/dt} = \frac{h'(t)}{g'(t)}.$$

The vector approach to the definition of a tangent line is more general, for it also applies to curves in space. The next example illustrates this for a particular curve.

EXAMPLE 3 At time t a particle has the position vector $\mathbf{r} = \mathbf{G}(t) = 3 \cos 2\pi t\, \mathbf{i} + 3 \sin 2\pi t\, \mathbf{j} + 5t\mathbf{k}$. Describe its path and find the tangent vector $\mathbf{G}'(t)$.

SOLUTION At time t the particle is at the point

$$\begin{cases} x = 3 \cos 2\pi t \\ y = 3 \sin 2\pi t \\ z = 5t. \end{cases}$$

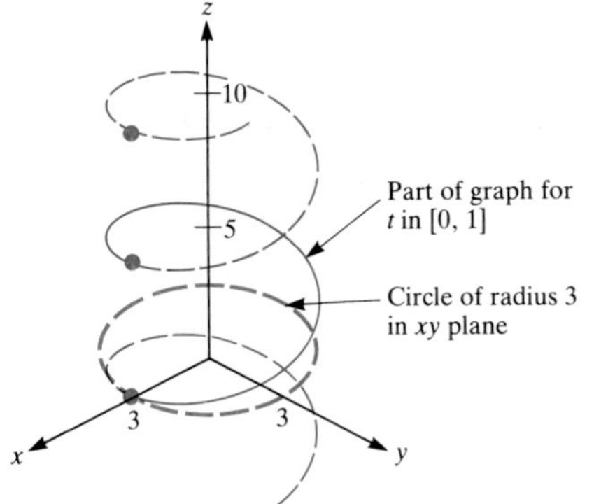

Figure 6

Notice that $x^2 + y^2 = (3 \cos 2\pi t)^2 + (3 \sin 2\pi t)^2 = 9$. Thus the point is always above or below the circle

$$x^2 + y^2 = 9.$$

Moreover, as t increases, $z = 5t$ increases.

The path is thus the spiral spring sketched in Fig. 6. When t increases by 1, the angle $2\pi t$ increases by 2π, and the particle goes once around the spiral. This type of corkscrew path is called a **helix**. At time t,

$$\mathbf{G}'(t) = -6\pi \sin 2\pi t\, \mathbf{i} + 6\pi \cos 2\pi t\, \mathbf{j} + 5\mathbf{k}. \quad \blacksquare$$

$\mathbf{G}'(t)$ and Motion along a Curve

We next interpret $\mathbf{G}'(t)$ in terms of motion along a curve $\mathbf{r} = \mathbf{G}(t) = g(t)\mathbf{i} + h(t)\mathbf{j}$ in the xy plane. The length of the vector $\mathbf{G}'(t) = g'(t)\mathbf{i} + h'(t)\mathbf{j}$ is $\sqrt{[g'(t)]^2 + [h'(t)]^2}$ or, equivalently,

$$\sqrt{\left(\frac{dx}{dt}\right)^2 + \left(\frac{dy}{dt}\right)^2}.$$

The magnitude of $\mathbf{G}'(t)$ *is the speed.*

This is the formula obtained in Sec. 9.4 for the speed of a particle moving on a plane curve.

Similar reasoning shows that the speed of a particle moving on a curve in space is

$$\sqrt{\left(\frac{dx}{dt}\right)^2 + \left(\frac{dy}{dt}\right)^2 + \left(\frac{dz}{dt}\right)^2}.$$

If
$$\mathbf{r} = \mathbf{G}(t) = x(t)\mathbf{i} + y(t)\mathbf{j} + z(t)\mathbf{k},$$

then
$$\|\mathbf{G}'(t)\| = \sqrt{\left(\frac{dx}{dt}\right)^2 + \left(\frac{dy}{dt}\right)^2 + \left(\frac{dz}{dt}\right)^2} = \text{Speed}.$$

Thus the length of the vector $\mathbf{G}'(t)$ may be interpreted as the speed of the particle:

$$\|\mathbf{r}'\| = \|\mathbf{G}'(t)\| = \text{Speed of the particle at time } t.$$

$\mathbf{G}'(t)$ *is also called the* **velocity vector**.

For this reason, $\mathbf{r}'$ or $\mathbf{G}'(t)$ is called the **velocity vector**. It points in the direction the particle is moving at a given instant, and its length is the speed of the particle. Frequently the velocity vector will be denoted $\mathbf{v}$.

EXAMPLE 4 Find the speed at time t of the particle described in Example 3.

SOLUTION

$$\begin{aligned}\text{Speed} = \|\mathbf{G}'(t)\| &= \sqrt{(-6\pi \sin 2\pi t)^2 + (6\pi \cos 2\pi t)^2 + 5^2} \\ &= \sqrt{36\pi^2(\sin^2 2\pi t + \cos^2 2\pi t) + 25} \\ &= \sqrt{36\pi^2 + 25}.\end{aligned}$$

The particle travels at a constant speed along its helical path. In t units of time it travels the distance $\sqrt{36\pi^2 + 25}\ t$.

Note that the velocity vector *is not constant;* its *direction always changes.* However, its length in this example remains constant, for the speed is constant. ■

EXAMPLE 5 Sketch the path of a particle whose position vector at time $t \geq 0$ is $\mathbf{G}(t) = \cos t^2\ \mathbf{i} + \sin t^2\ \mathbf{j}$. Find its speed at time t.

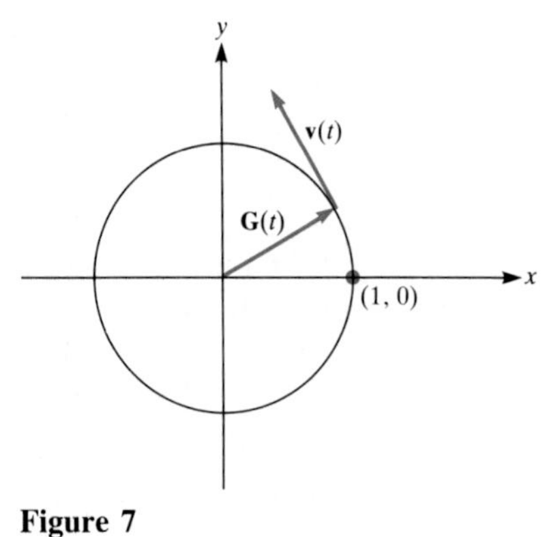

Figure 7

SOLUTION Note that

$$\|\mathbf{G}(t)\| = \sqrt{\cos^2 t^2 + \sin^2 t^2} = 1.$$

So the path of the particle is on the circle of radius 1 and center (0, 0). (See Fig. 7.) The speed of the particle is

$$\begin{aligned}\|\mathbf{v}(t)\| = \|\mathbf{G}'(t)\| &= \|-2t \sin t^2\ \mathbf{i} + 2t \cos t^2\ \mathbf{j}\| \\ &= \sqrt{(-2t \sin t^2)^2 + (2t \cos t^2)^2} \\ &= 2t\sqrt{\sin^2 t^2 + \cos^2 t^2} = 2t.\end{aligned}$$

The particle travels faster and faster around a circle of radius 1. ■

Section Summary

We introduced the notion of a vector function, $\mathbf{G}(t)$. Such a function assigns to a scalar input t a vector output $\mathbf{G}(t)$. We introduced limits, continuity, and derivatives of vector functions. If we have $\mathbf{G}(t)$ in the form $x(t)\mathbf{i} + y(t)\mathbf{j} + z(t)\mathbf{k}$, the derivative of $\mathbf{G}$ is simply $x'(t)\mathbf{i} + y'(t)\mathbf{j} + z'(t)\mathbf{k}$.

When $\mathbf{r} = \mathbf{G}(t)$ describes the position of a particle moving along a curve, $\mathbf{G}'(t)$ is called the velocity vector. It points in the direction of motion and its magnitude is the speed.

EXERCISES FOR SEC. 13.1: THE DERIVATIVE OF A VECTOR FUNCTION

1 At time t a particle has the position vector $\mathbf{r} = \mathbf{G}(t) = t\mathbf{i} + t^2\mathbf{j}$.
(*a*) Compute and draw $\mathbf{G}(1)$, $\mathbf{G}(2)$, and $\mathbf{G}(3)$.
(*b*) Show that the path is a parabola.

2 At time t a particle has the position vector $\mathbf{r} = \mathbf{G}(t) = (2t + 1)\mathbf{i} + 4t\mathbf{j}$.
(*a*) Compute and draw $\mathbf{G}(0)$, $\mathbf{G}(1)$, and $\mathbf{G}(2)$.
(*b*) Show that the path is a straight line.

3 Let $\mathbf{G}(t) = 2t\mathbf{i} + t^2\mathbf{j}$.
(*a*) Compute and draw $\mathbf{G}(1.1)$, $\mathbf{G}(1)$, and their difference $\Delta\mathbf{G} = \mathbf{G}(1.1) - \mathbf{G}(1)$.
(*b*) Compute and draw $\Delta\mathbf{G}/0.1$, where $\Delta\mathbf{G}$ is defined in part (*a*).
(*c*) Compute $\mathbf{G}'(1)$.

4 Let $\mathbf{G}(t) = 3t\mathbf{i} + 2t^2\mathbf{j}$.
(*a*) Compute and draw $\Delta\mathbf{G} = \mathbf{G}(2.01) - \mathbf{G}(2)$.
(*b*) Compute and draw $\Delta\mathbf{G}/0.01$.
(*c*) Compute $\mathbf{G}'(2)$.

5 At time t the position vector of a thrown ball is $\mathbf{r}(t) = 32t\mathbf{i} - 16t^2\mathbf{j}$.
(a) Draw $\mathbf{r}(1)$ and $\mathbf{r}(2)$.
(b) Sketch the path.
(c) Compute and draw $\mathbf{v}(0)$, $\mathbf{v}(1)$, and $\mathbf{v}(2)$. In each case place the tail of the vector at the head of the corresponding position vector.

6 At time $t \geq 0$ a particle is at the point $x = 2t$, $y = 4t^2$.
(a) What is the position vector $\mathbf{r}(t)$ at time t?
(b) Sketch the path.
(c) How fast is the particle moving when $t = 1$?
(d) Draw $\mathbf{v}(1)$ with its tail at the head of $\mathbf{r}(1)$.

7 Let $\mathbf{G}(t)$ describe the path of a particle moving in the xy plane. If $\mathbf{G}(1) = 2.3\mathbf{i} + 4.1\mathbf{j}$ and $\mathbf{G}(1.2) = 2.31\mathbf{i} + 4.05\mathbf{j}$, estimate
(a) how far the particle moves during the time interval $[1, 1.2]$,
(b) the slope of the tangent vector to the path at $\mathbf{G}(1)$,
(c) the velocity vector $\mathbf{G}'(1)$,
(d) the speed of the particle at time $t = 1$.

8 Let $\mathbf{G}(t)$ describe the path of a particle moving in space. If $\mathbf{G}(2) = 1.7\mathbf{i} + 3.6\mathbf{j} + 8\mathbf{k}$ and $\mathbf{G}(2.01) = 1.73\mathbf{i} + 3.59\mathbf{j} + 8.02\mathbf{k}$, estimate
(a) how far the particle moves during the time interval $[2, 2.01]$,
(b) the velocity vector $\mathbf{G}'(2)$,
(c) the speed of the particle at time $t = 2$.

In Exercises 9 to 12 compute the velocity vectors and speeds for the given paths.

9 $\mathbf{r}(t) = \cos 3t\ \mathbf{i} + \sin 3t\ \mathbf{j} + 6t\ \mathbf{k}$

10 $\mathbf{r}(t) = 3\cos 5t\ \mathbf{i} + 2\sin 5t\ \mathbf{j} + t^2\ \mathbf{k}$

11 $\mathbf{r}(t) = \ln(1 + t^2)\mathbf{i} + e^{3t}\mathbf{j} + \dfrac{\tan t}{1 + 2t}\mathbf{k}$

12 $\mathbf{r}(t) = \sec^2 3t\ \mathbf{i} + \sqrt{1 + t^2}\mathbf{j}$

13 Let $\mathbf{G}(t) = t^2\mathbf{i} + t^3\mathbf{j}$.
(a) Sketch the vector $\Delta\mathbf{G} = \mathbf{G}(1.1) - \mathbf{G}(1)$.
(b) Sketch the vector $\Delta\mathbf{G}/\Delta t$, where $\Delta\mathbf{G}$ is given in (a) and $\Delta t = 0.1$.
(c) Sketch $\mathbf{G}'(1)$.
(d) Find $\|\Delta\mathbf{G}/\Delta t - \mathbf{G}'(1)\|$, where $\Delta\mathbf{G}$ is given in (a) and $\Delta t = 0.1$.

14 At time t the position vector of a particle is

$$\mathbf{G}(t) = 2\cos 4\pi t\ \mathbf{i} + 2\sin 4\pi t\ \mathbf{j} + t\mathbf{k}.$$

(a) Sketch its path.
(b) Find its speed.
(c) Find a unit tangent vector to the path at time t.

15 At time t the position vector of a particle is

$$\mathbf{G}(t) = t\cos 2\pi t\ \mathbf{i} + t\sin 2\pi t\ \mathbf{j} + t\mathbf{k}.$$

Sketch the path of the particle.

16 At time t a particle is at $(4t, 16t^2)$.
(a) Show that the particle moves on the curve $y = x^2$.
(b) Draw $\mathbf{r}(t)$ and $\mathbf{v}(t)$ for $t = 0, \frac{1}{4}, \frac{1}{2}$.
(c) What happens to $\|\mathbf{v}(t)\|$ and the direction of $\mathbf{v}(t)$ for large t?

17 At time $t \geq 1$ a particle is at the point (t, t^{-1}).
(a) Draw the path of the particle.
(b) Draw $\mathbf{r}(1)$, $\mathbf{r}(2)$, and $\mathbf{r}(3)$.
(c) Draw $\mathbf{v}(1)$, $\mathbf{v}(2)$, and $\mathbf{v}(3)$.
(d) As times goes on, what happens to dx/dt, dy/dt, $\|\mathbf{v}\|$, and $\mathbf{v}$?

18 At time t a particle is at $(2\cos t^2, \sin t^2)$.
(a) Show that it moves on an ellipse.
(b) Compute $\mathbf{v}(t)$.
(c) How does $\|\mathbf{v}(t)\|$ behave for large t? What does this say about the particle?

---

19 An electron travels at constant speed clockwise in a circle of radius 100 feet 200 times a second. At time $t = 0$ it is at $(100, 0)$.
(a) Compute $\mathbf{r}(t)$ and $\mathbf{v}(t)$.
(b) Draw $\mathbf{r}(0)$, $\mathbf{r}(\frac{1}{800})$, $\mathbf{v}(0)$, $\mathbf{v}(\frac{1}{800})$.
(c) How do $\|\mathbf{r}(t)\|$ and $\|\mathbf{v}(t)\|$ behave as time goes on?

20 A ball is thrown up at an initial speed of 200 feet per second and at an angle of 60° from the horizontal. If we disregard air resistance, then at time t it is at $(100t, 100\sqrt{3}t - 16t^2)$, as long as it is in flight. Compute and draw $\mathbf{r}(t)$ and $\mathbf{v}(t)$ (a) when $t = 0$, (b) when the ball reaches its maximum height, (c) when the ball strikes the ground.

21 Instead of time t, use arc length s along the path as a parameter, $\mathbf{r} = \mathbf{G}(s)$.
(a) Show that $d\mathbf{r}/ds$ is a unit vector.
(b) Sketch $\Delta\mathbf{r}$ and the arc of length Δs. Why is it reasonable that $\|\Delta\mathbf{r}/\Delta s\|$ is near 1 when Δs is small?

22 A particle at time $t = 0$ is at the point (x_0, y_0, z_0). It moves on a line through that point in the direction of the unit vector $\mathbf{u} = \cos\alpha\ \mathbf{i} + \cos\beta\ \mathbf{j} + \cos\gamma\ \mathbf{k}$. It travels at the constant speed of 3 feet per second.
(a) Give a formula for its position vector $\mathbf{r} = \mathbf{G}(t)$.
(b) Find its velocity vector $\mathbf{v} = \mathbf{G}'(t)$.

23 A particle moves in a circular orbit of radius a. At time t its position vector is

$$\mathbf{r}(t) = a\cos 2\pi t\ \mathbf{i} + a\sin 2\pi t\ \mathbf{j}.$$

(a) Draw its position vector when $t = 0$ and when $t = \frac{1}{4}$.
(b) Draw its velocity vector when $t = 0$ and when $t = \frac{1}{4}$.
(c) Show that its velocity vector is always perpendicular to its position vector.

24 (a) Show that the paths $\mathbf{G}(t) = t\mathbf{i} + t^2\mathbf{j} + t^3\mathbf{k}$ and $\mathbf{H}(t) = t^2\mathbf{i} + t^3\mathbf{j} + t^4\mathbf{k}$ intersect when $t = 1$.
(b) At what angle do they intersect?

25 A rock is thrown up at an angle θ from the horizontal and at a speed v_0.
(a) Show that

$$\mathbf{r}(t) = (v_0\cos\theta)t\mathbf{i} + [(v_0\sin\theta)t - 16t^2]\mathbf{j}.$$

[At time $t = 0$, the rock is at $(0, 0)$; the x axis is horizontal. Time is in seconds and distance is in feet.]
(*b*) Show that the horizontal distance that the rock travels by the time it reaches its initial height is the same whether the angle is θ or its complement $(\pi/2) - \theta$.
(*c*) What value of θ maximizes the horizontal distance traveled?

26 A spaceship outside any gravitational field is on the path $\mathbf{G}(t) = t^2\mathbf{i} + 3t\mathbf{j} + 4t^3\mathbf{k}$. At time $t = 1$ it shuts off its rockets and coasts along the tangent line to the curve at that point.
(*a*) Where is it at time $t > 1$?
(*b*) Does it pass through the point (9, 15, 50)?
(*c*) If not, how close does it get to that point?

27 A particle traveling on the curve $\mathbf{r}(t) = \ln t\mathbf{i} + \cos 3t\mathbf{j}$, $t \geq 1$, leaves the curve when $t = 2$ and travels along the tangent to the curve at $\mathbf{r}(2)$. Where is it when $t = 3$?

28 Drawing a picture of $\mathbf{G}(t)$, $\mathbf{G}(t + \Delta t)$, and $\mathbf{G}(t + \Delta t) - \mathbf{G}(t)$, explain why

$$\left\| \frac{\Delta \mathbf{G}}{\Delta t} \right\|$$

is an estimate of the speed of a particle moving on the curve $\mathbf{r} = \mathbf{G}(t)$.

29 The moment a ball is dropped straight down from a tall tree, you shoot an arrow directly at it. Assume that there is no air resistance. Show that the arrow will hit the ball. (Assume that the ball does not hit the ground first.)
(*a*) Solve with the aid of the formulas in Exercise 25.
(*b*) Solve with a maximum of intuition and a minimum of computation.

In Exercises 30 to 36 $\mathbf{v}(t)$ is the velocity vector at time t for a moving particle and $\mathbf{r}(0)$ is the particle's position at time $t = 0$. Find $\mathbf{r}(t)$, the position vector of the particle at time t. (This reviews the integration techniques of Chap. 7.)

30 $\mathbf{v}(t) = \sin^2 3t\ \mathbf{i} + \dfrac{t}{3t^2 + 1}\mathbf{j}$; $\mathbf{r}(0) = \mathbf{j}$

31 $\mathbf{v}(t) = \dfrac{t}{t^2 + t + 1}\mathbf{i} + \tan^{-1} 3t\ \mathbf{j}$; $\mathbf{r}(0) = \mathbf{i} + \mathbf{j}$

32 $\mathbf{v}(t) = \dfrac{t^3}{t^4 + 1}\mathbf{i} + \ln (t + 1)\ \mathbf{j}$; $\mathbf{r}(0) = \mathbf{0}$

33 $\mathbf{v}(t) = e^{2t} \sin 3t\ \mathbf{i} + \dfrac{t^3}{3t + 2}\mathbf{j}$; $\mathbf{r}(0) = \mathbf{i} + 3\mathbf{j}$

34 $\mathbf{v}(t) = \dfrac{t}{(t + 1)(t + 2)(t + 3)}\mathbf{i} + \dfrac{t^2}{(t + 2)^3}\mathbf{j}$; $\mathbf{r}(0) = \mathbf{i} - \mathbf{j}$

35 $\mathbf{v}(t) = \dfrac{[\ln (t + 1)]^3}{t + 1}\mathbf{i} + \dfrac{1}{\sqrt{1 - 4t^2}}\mathbf{j} + \sec^2 3t\ \mathbf{k}$; $\mathbf{r}(0) = \mathbf{i} + \mathbf{j} + \mathbf{k}$

36 $\mathbf{v}(t) = t^3 e^{-t}\ \mathbf{i} + (1 + t)(2 + t)\mathbf{j}$; $\mathbf{r}(0) = 2\mathbf{i} - \mathbf{j}$

37 Use computer software to graph $\mathbf{r} = \mathbf{G}(t) = (3 \cos t + \cos 3t)\mathbf{i} + (3 \sin t + \sin 3t)\mathbf{j}$, $0 \leq \theta \leq 2\pi$.

13.2 PROPERTIES OF THE DERIVATIVE OF A VECTOR FUNCTION

Back in Chap. 3 we obtained various properties of the derivatives of functions $y = f(x)$, for instance $(f + g)' = f' + g'$ and the chain rule. In this section we obtain properties of the derivatives of vector functions. Throughout this section and the rest of the text, we assume that the various functions have derivatives.

The product of a scalar function and a vector function

A vector function $\mathbf{G}$ assigns to a number t a vector $\mathbf{G}(t)$. A numerical function f assigns to a number t a number $f(t)$; for emphasis we call such a function a **scalar function**. The function that assigns to the number t the vector $f(t)\mathbf{G}(t)$ is a *vector function*. Thus the product of a scalar function and a vector function is a vector function. It is denoted $f\mathbf{G}$. The notation $\mathbf{G}f$, where the scalar appears second, is seldom used. The first theorem concerns the derivative of $f\mathbf{G}$. Note its similarity to the theorem about the derivative of the product of two scalar functions in Sec. 3.4. Its proof resembles the proof of that theorem.

Theorem 1 If f is a scalar function and $\mathbf{G}$ is a vector function, the derivative of the vector function $f\mathbf{G}$ is

$$f\mathbf{G}' + f'\mathbf{G}.$$

Proof First work with $\Delta(f\mathbf{G})$, as follows:

$$\begin{aligned}\Delta(f\mathbf{G}) &= f(t+\Delta t)\mathbf{G}(t+\Delta t) - f(t)\mathbf{G}(t)\\ &= [f(t)+\Delta f][\mathbf{G}(t)+\Delta\mathbf{G}] - f(t)\mathbf{G}(t)\\ &= f(t)\,\mathbf{G}(t) + f(t)\,\Delta\mathbf{G} + \Delta f\,\mathbf{G}(t) + \Delta f\,\Delta\mathbf{G} - f(t)\mathbf{G}(t)\\ &= f(t)\Delta\mathbf{G} + \Delta f\,\mathbf{G}(t) + \Delta f\,\Delta\mathbf{G}.\end{aligned}$$

Thus

$$\begin{aligned}\lim_{\Delta t\to 0}\frac{\Delta(f\mathbf{G})}{\Delta t} &= \lim_{\Delta t\to 0}\left[f(t)\frac{\Delta\mathbf{G}}{\Delta t} + \frac{\Delta f}{\Delta t}\mathbf{G}(t) + \Delta f\frac{\Delta\mathbf{G}}{\Delta t}\right]\\ &= f(t)\mathbf{G}'(t) + f'(t)\mathbf{G}(t) + 0\mathbf{G}'(t).\end{aligned}$$

This proves the theorem. ■

EXAMPLE 1 Find the derivative of the vector function $t^2(\cos t\,\mathbf{i} + \sin t\,\mathbf{j})$.

SOLUTION By Theorem 1, the derivative is

$$\begin{aligned}t^2(\cos t\,\mathbf{i} + \sin t\,\mathbf{j})' + (t^2)'(\cos t\,\mathbf{i} + \sin t\,\mathbf{j})\\ &= t^2(-\sin t\,\mathbf{i} + \cos t\,\mathbf{j}) + 2t(\cos t\,\mathbf{i} + \sin t\,\mathbf{j})\\ &= (-t^2\sin t + 2t\cos t)\mathbf{i} + (t^2\cos t + 2t\sin t)\mathbf{j}.\end{aligned}$$

Actually, Theorem 1 is not needed for this particular example. We could have written the function as

$$t^2\cos t\,\mathbf{i} + t^2\sin t\,\mathbf{j}$$

and differentiated each scalar component as in the preceding section. ■

The dot product of two vector functions

From two vector functions $\mathbf{G}$ and $\mathbf{H}$ we can obtain a scalar function $\mathbf{G}\cdot\mathbf{H}$ by defining the value of $\mathbf{G}\cdot\mathbf{H}$ at t to be the dot product

$$\mathbf{G}(t)\cdot\mathbf{H}(t).$$

Theorem 2 If $\mathbf{G}$ and $\mathbf{H}$ are vector functions, then

$$(\mathbf{G}\cdot\mathbf{H})' = \mathbf{G}\cdot\mathbf{H}' + \mathbf{H}\cdot\mathbf{G}'.$$

The proof of Theorem 2, which is similar to the proof of Theorem 1, is outlined in Exercise 8.

EXAMPLE 2 Use Theorem 2 to show that if the length of $\mathbf{G}(t)$ is constant, then $\mathbf{G}'(t)$ is perpendicular to $\mathbf{G}(t)$.

SOLUTION Recall that for any vector $\mathbf{A}$, $\mathbf{A}\cdot\mathbf{A} = \|\mathbf{A}\|^2$. Thus $\mathbf{G}(t)\cdot\mathbf{G}(t) = \|\mathbf{G}(t)\|^2$. The assumption that $\mathbf{G}(t)$ has constant length is equivalent to the statement that $\mathbf{G}(t)\cdot\mathbf{G}(t)$ is constant. Thus

$$\begin{aligned}0 &= (\mathbf{G}\cdot\mathbf{G})' && \text{(Derivative of a constant is 0)}\\ &= \mathbf{G}\cdot\mathbf{G}' + \mathbf{G}\cdot\mathbf{G}' && \text{(Theorem 2 with } \mathbf{G}=\mathbf{H})\\ &= 2(\mathbf{G}\cdot\mathbf{G}').\end{aligned}$$

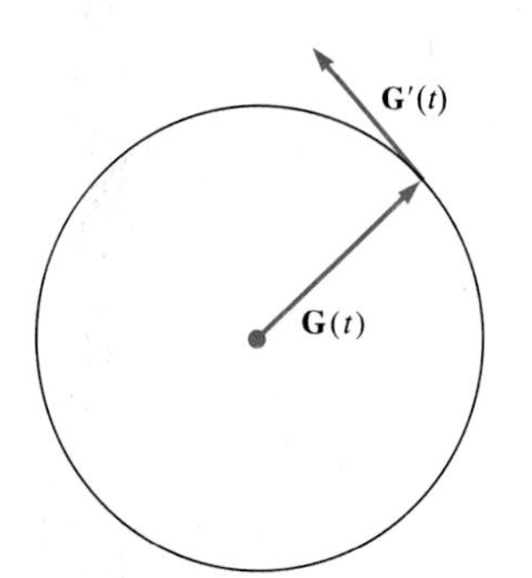

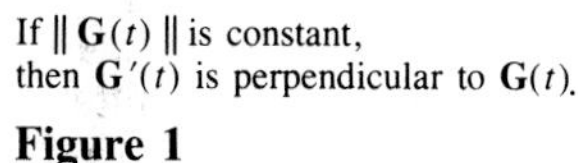
If $\|\mathbf{G}(t)\|$ is constant, then $\mathbf{G}'(t)$ is perpendicular to $\mathbf{G}(t)$.

Figure 1

Thus $\mathbf{G}(t) \cdot \mathbf{G}'(t) = 0$ for each t. That implies that $\mathbf{G}(t)$ and $\mathbf{G}'(t)$ are perpendicular (or else one of them is $\mathbf{0}$). ■

The result obtained in Example 2 also follows from high school geometry. If $\|\mathbf{G}(t)\|$ is constant, the particle moves on a sphere (or circle). $\mathbf{G}'(t)$ is tangent to the sphere and therefore perpendicular to $\mathbf{G}(t)$, since any ray drawn from the center of a sphere is perpendicular to the tangent at the point where the ray meets the sphere. (See Fig. 1.)

For emphasis we honor the substance of Example 2 with a box.

> If $\mathbf{G}(t)$ has constant length, then the vectors $\mathbf{G}(t)$ and $\mathbf{G}'(t)$ are perpendicular for each input t.

Theorem 3 If $\mathbf{G}$ and $\mathbf{H}$ are vector functions then

$$(\mathbf{G} \times \mathbf{H})' = \mathbf{G} \times \mathbf{H}' + \mathbf{G}' \times \mathbf{H}.$$

The proof is requested in Exercise 9. Note the resemblance of Theorem 3 to the formula for the derivative a product of scalar functions, $(gh)' = gh' + hg'$. However, there is an important difference. In the vector case, each term has $\mathbf{G}$ to the left of $\mathbf{H}$. The order is important here, whereas before it was not.

The Chain Rule

Let $\mathbf{G}$ be a vector function of the scalar s, $\mathbf{r} = \mathbf{G}(s)$, and let s in turn be a function of the scalar t, $s = f(t)$. Then we may consider $\mathbf{r}$ to be a function of t:

$$\mathbf{r} = \mathbf{G}(f(t)).$$

(Think of s as arc length along a curve and t as time.) Denote this (composite) function $\mathbf{G} \circ f$.

Theorem 4 *Chain rule*. If $\mathbf{G}$ is a vector function and f is a scalar function, then $\mathbf{G} \circ f$ is a vector function, and its derivative at t is

$$\mathbf{G}'(f(t))f'(t).$$

In other words, if $\mathbf{r} = \mathbf{G}(s)$ and $s = f(t)$, then

$$\frac{d\mathbf{r}}{dt} = \frac{d\mathbf{r}}{ds}\frac{ds}{dt}.$$

The proof of Theorem 4 is outlined in Exercise 11.

EXAMPLE 3 Let $\mathbf{r} = s^2\mathbf{i} + s^3\mathbf{j}$ where $s = e^{2t}$. Compute $d\mathbf{r}/dt$.

SOLUTION

$$\frac{d\mathbf{r}}{ds} = 2s\mathbf{i} + 3s^2\mathbf{j} \quad \text{and} \quad \frac{ds}{dt} = 2e^{2t}.$$

Thus

$$\frac{d\mathbf{r}}{dt} = (2s\mathbf{i} + 3s^2\mathbf{j})2e^{2t},$$

You could also compute $d\mathbf{r}/dt$ by first expressing $\mathbf{r}$ in terms of t.

which is usually written with the scalar coefficient in front:

$$\frac{d\mathbf{r}}{dt} = 2e^{2t}(2s\mathbf{i} + 3s^2\mathbf{j}) = 2e^{2t}(2e^{2t}\mathbf{i} + 3e^{4t}\mathbf{j})$$
$$= 4e^{4t}\mathbf{i} + 6e^{6t}\mathbf{j}. \quad \blacksquare$$

Section Summary

Derivatives of vector functions behave much like derivatives of scalar functions. If f is a scalar function and $\mathbf{G}$ and $\mathbf{H}$ vector functions, then:

$$(f\mathbf{G})' = f'\mathbf{G} + f\mathbf{G}'$$

$$(\mathbf{G} \cdot \mathbf{H})' = \mathbf{G} \cdot \mathbf{H}' + \mathbf{H} \cdot \mathbf{G}'$$

$$(\mathbf{G} \times \mathbf{H})' = \mathbf{G} \times \mathbf{H}' + \mathbf{G}' \times \mathbf{H} \qquad \text{(note the order of } \mathbf{G} \text{ and } \mathbf{H}\text{)}$$

$$(\mathbf{G}(f(t)))' = \mathbf{G}'(f(t))f'(t). \qquad \text{(chain rule)}$$

We also showed that if $\mathbf{G}(t)$ has constant length, then $\mathbf{G}'(t)$ is perpendicular to $\mathbf{G}(t)$.

EXERCISES FOR SEC. 13.2: PROPERTIES OF THE DERIVATIVE OF A VECTOR FUNCTION

In Exercises 1 to 4 compute $(\mathbf{G} \cdot \mathbf{H})'$ in two ways: First with the aid of the identity $(\mathbf{G} \cdot \mathbf{H})' = \mathbf{G} \cdot \mathbf{H}' + \mathbf{H} \cdot \mathbf{G}'$, then by computing $\mathbf{G} \cdot \mathbf{H}$ and differentiating the result.

1 $\mathbf{G}(t) = t^2\mathbf{i} + t^3\mathbf{j}$, $\mathbf{H}(t) = 3t\mathbf{i} + 4t^2\mathbf{j}$

2 $\mathbf{G}(t) = 3t\mathbf{i} + \cos 3t\ \mathbf{j}$, $\mathbf{H}(t) = t^2\mathbf{i} + \sin 3t\ \mathbf{j}$

3 $\mathbf{G}(t) = e^{2t}\mathbf{i} + t^2\mathbf{j}$, $\mathbf{H}(t) = t\mathbf{i} + \sin 3t\mathbf{j}$

4 $\mathbf{G}(t) = \tan^{-1} 3t\ \mathbf{i} + (1/t)\mathbf{j}$, $\mathbf{H}(t) = \tan 2t\ \mathbf{i} + \ln (1 + t)\ \mathbf{j}$

5 Suppose $\mathbf{G}(1) = \mathbf{i} + \mathbf{j}$ and $\mathbf{G}'(1) = 2\mathbf{i} - 3\mathbf{j}$. Find $[t^2\mathbf{G}(t)]'$ at $t = 1$.

6 Suppose $\mathbf{G}(0) = 3\mathbf{i}$ and $\mathbf{G}'(0) = 2\mathbf{i} - 3\mathbf{j} + 4\mathbf{k}$. Find $[t^3\mathbf{G}(t)]'$ at $t = 0$.

7 Theorem 1 was proved without referring to scalar components. Prove Theorem 1 with the aid of scalar components as follows: First write $\mathbf{G}(t) = g(t)\mathbf{i} + h(t)\mathbf{j}$ and then differentiate $f(t)\mathbf{G}(t) = f(t)g(t)\mathbf{i} + f(t)h(t)\mathbf{j}$. (A similar argument would work for vectors in space.)

8 This outlines a proof of Theorem 2.

(*a*) Express $\Delta(\mathbf{G} \cdot \mathbf{H})$ as simply as possible with the aid of the two equations $\mathbf{G}(t + \Delta t) = \mathbf{G}(t) + \Delta\mathbf{G}$ and

$$\mathbf{H}(t + \Delta t) = \mathbf{H}(t) + \Delta\mathbf{H}.$$

(*b*) Using (*a*), find

$$\lim_{\Delta t \to 0} \frac{\Delta(\mathbf{G} \cdot \mathbf{H})}{\Delta t}.$$

9 (*a*) Prove that

$$(\mathbf{G} \times \mathbf{H})' = \mathbf{G} \times \mathbf{H}' + \mathbf{G}' \times \mathbf{H}.$$

Suggestion: Proceed as in the proof of Theorem 1.

(*b*) Would it be correct to assert, as in the case of a derivative of the product of scalar functions, that $(\mathbf{G} \times \mathbf{H})'$ is equal to $\mathbf{G} \times \mathbf{H}' + \mathbf{H} \times \mathbf{G}'$?

10 Solve Exercise 9(*a*) by writing $\mathbf{G}(t) = g_1\mathbf{i} + g_2\mathbf{j} + g_3\mathbf{k}$ and $\mathbf{H}(t) = h_1\mathbf{i} + h_2\mathbf{j} + h_3\mathbf{k}$, expressing $\mathbf{G}(t) \times \mathbf{H}(t)$ in components, and then differentiating. Compare the result with $\mathbf{G}(t) \times \mathbf{H}'(t) + \mathbf{G}'(t) \times \mathbf{H}(t)$, also expressed in scalar components.

11 To prove Theorem 4, write $\mathbf{G}(s) = g(s)\mathbf{i} + h(s)\mathbf{j}$ and work with components. Carry out the details.

12 Let $\mathbf{G}$ and $\mathbf{H}$ be two vector functions. Define $\mathbf{G} + \mathbf{H}$ by defining $(\mathbf{G} + \mathbf{H})(t)$ to be $\mathbf{G}(t) + \mathbf{H}(t)$. Show that $(\mathbf{G} + \mathbf{H})' = \mathbf{G}' + \mathbf{H}'$.

13 Let $\mathbf{G}(t) = \cos 2t\ \mathbf{i} + \sin 2t\ \mathbf{j}$.
(*a*) Show that $\mathbf{G}(t)$ is a unit vector.
(*b*) Is $\mathbf{G}'(t)$ a unit vector?

14 A particle at time $t \geq 0$ is at the point (t, t^2, t).
(*a*) Find its speed at time t.
(*b*) Show that its path lies above a parabola in the xy plane.
(*c*) Show that its path lies in a plane.
(*d*) Plot its path.

---

15 A particle moves in a path such that $\mathbf{r}(t) = e^t \cos t\ \mathbf{i} + e^t \sin t\ \mathbf{j}$. Show that the angle between its position vector and velocity vector has the constant value $\pi/4$.

16 Assume that $\mathbf{G}'(t)$ is always perpendicular to $\mathbf{G}(t)$. Show that the path $\mathbf{r} = \mathbf{G}(t)$ lies on the surface of a ball.

17 Let $\mathbf{G}$ be a vector function such that $\mathbf{G}(t)$ is never $\mathbf{0}$. Show that $\mathbf{G}$ can be written in the form $f\mathbf{H}$, where $f(t) > 0$ and $\|\mathbf{H}(t)\| = 1$ for all t.

18 If the velocity vector $\mathbf{v}(t)$ is constant, show that the path lies on a straight line.

19 A particle moves on the path $\mathbf{r} = \mathbf{G}(t)$, which does not pass through the origin. Prove that if P is a point on the path closest to the origin O, then the position vector $\overrightarrow{OP}$ is perpendicular to the velocity vector at P.

20 (Practice in differentiation) Find $\mathbf{G}'(t)$ if $\mathbf{G}(t)$ is
(*a*) $\sqrt{1 + \sqrt{3t}}\ \mathbf{i} + \dfrac{\sin^2 4t}{1 + t^3}\ \mathbf{j}$
(*b*) $3^{t^2} \ln (1 + \sqrt{t})\ \mathbf{i} + \sin^{-1} 3t\ \mathbf{j}$
(*c*) $\sec^{-1} 5t\ \mathbf{i} + \sec^3 2t\ \mathbf{j}$

13.3 THE ACCELERATION VECTOR

If $\mathbf{r} = \mathbf{G}(t)$ is the position vector at time t, then $\mathbf{v} = \mathbf{G}'(t)$ is the velocity vector at time t. The definition of the acceleration vector is motivated by the definition of acceleration in the case of a particle moving on a line.

Definition *Acceleration vector.* The derivative of the velocity vector is called the **acceleration vector** and is denoted $\mathbf{a}$:

$$\mathbf{a} = \frac{d\mathbf{v}}{dt}.$$

EXAMPLE 1 Let $\mathbf{G}(t) = 32t\mathbf{i} - 16t^2\mathbf{j}$ be the position of a thrown ball at time t. Compute $\mathbf{v}(t)$ and $\mathbf{a}(t)$.

SOLUTION

$$\mathbf{v}(t) = 32\mathbf{i} - 32t\mathbf{j} \qquad \text{and} \qquad \mathbf{a}(t) = -32\mathbf{j}.$$

In this case, the acceleration vector is constant in direction and length. It points directly downward, as does the vector that represents the force of gravity. Figure 1 displays $\mathbf{v}$ and $\mathbf{a}$ at two points on the path. ■

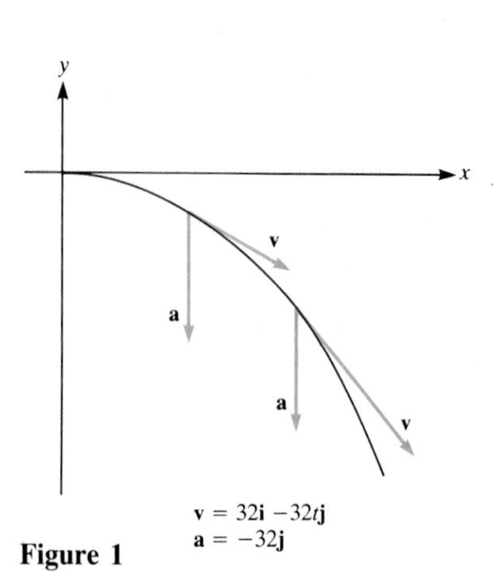

Figure 1

As this example shows, it is a simple matter to compute the velocity vector and the acceleration vector $\mathbf{a}$ when the components of the position vector $\mathbf{r}(t)$ are given. If

$$\mathbf{r}(t) = x(t)\mathbf{i} + y(t)\mathbf{j} + z(t)\mathbf{k},$$

then

$$\mathbf{v}(t) = \frac{dx}{dt}\mathbf{i} + \frac{dy}{dt}\mathbf{j} + \frac{dz}{dt}\mathbf{k}$$

and

$$\mathbf{a}(t) = \frac{d^2x}{dt^2}\mathbf{i} + \frac{d^2y}{dt^2}\mathbf{j} + \frac{d^2z}{dt^2}\mathbf{k}.$$

If no forces act on a moving particle, $\mathbf{v}$ is constant in direction and length; hence $\mathbf{a} = \mathbf{0}$. That is, if the vector $\mathbf{F}$, representing the forces, is $\mathbf{0}$, then $\mathbf{a} = \mathbf{0}$. Newton's second law asserts universally that $\mathbf{F}$, $\mathbf{a}$, and the mass m of the particle are related by the vector equation

The relation between force and acceleration

$$\mathbf{F} = m\mathbf{a}.$$

This little equation says several things: (1) The direction of the acceleration vector $\mathbf{a}$ is the same as the direction of $\mathbf{F}$. (2) A force $\mathbf{F}$ applied to a heavy mass produces a shorter acceleration vector $\mathbf{a}$ than the same force applied to a light mass. (3) For a given mass, the magnitude of $\mathbf{a}$ is proportional to the magnitude of $\mathbf{F}$.

We may always think of the acceleration vector as representing the effect of a force on the particle, since the acceleration vector $\mathbf{a}$ and the force vector $\mathbf{F}$ point in the same direction. If the mass of the particle is 1, then $\mathbf{F} = \mathbf{a}$.

Uniform Circular Motion

Consider now a particle moving in a circular orbit at constant speed v. It may be, perhaps, a heavy mass at the end of a rope or a satellite in a circular orbit around the earth. The following theorem describes the acceleration vector associated with this motion.

> **Theorem** If a particle moves in a circular path of radius r at a constant speed v, its acceleration vector is directed toward the center of the circle and has magnitude v^2/r.

$r = \|\mathbf{r}\|$

Proof Introduce an xy-coordinate system such that (0, 0) is at the center of the orbit and the particle is at $(r, 0)$ at time 0. Let $\mathbf{r} = \mathbf{r}(t)$ be the position vector of the particle at time $t \geq 0$. Assume that the particle travels counterclockwise; then θ is positive, as shown in Fig. 2.

Since the particle moves at the constant speed v, it sweeps out an arc length vt up to time t. By definition of radian measure,

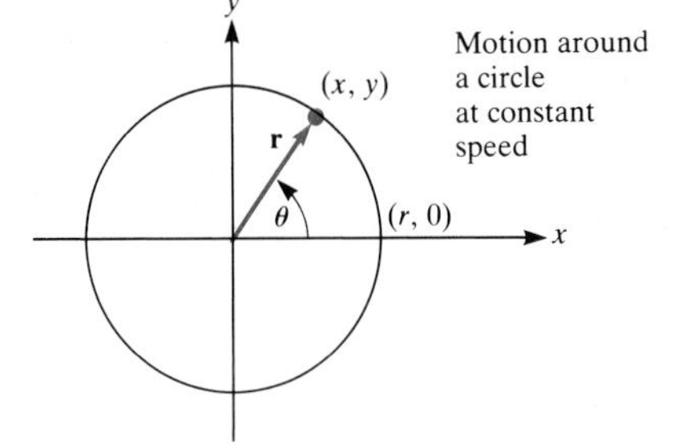

Figure 2

$$\theta = \frac{vt}{r},$$

where θ is the angle of $\mathbf{r}$ at time t. Thus

$$\mathbf{r} = x\mathbf{i} + y\mathbf{j} = r\cos\theta\,\mathbf{i} + r\sin\theta\,\mathbf{j} = r\cos\frac{vt}{r}\,\mathbf{i} + r\sin\frac{vt}{r}\,\mathbf{j}.$$

Now $\mathbf{v}$ and $\mathbf{a}$ can be computed explicitly. Remembering that r is constant, we have then

$v = \|\mathbf{v}\|$

$$\begin{aligned}\mathbf{v} &= \frac{d\mathbf{r}}{dt}\\ &= \frac{-rv}{r}\sin\frac{vt}{r}\,\mathbf{i} + \frac{rv}{r}\cos\frac{vt}{r}\,\mathbf{j}\\ &= -v\sin\frac{vt}{r}\,\mathbf{i} + v\cos\frac{vt}{r}\,\mathbf{j}.\end{aligned}$$

Hence
$$\mathbf{a} = \frac{d\mathbf{v}}{dt}$$
$$= \frac{-v^2}{r}\cos\frac{vt}{r}\,\mathbf{i} - \frac{v^2}{r}\sin\frac{vt}{r}\,\mathbf{j}$$
$$= \frac{v^2}{r}(-\cos\theta\,\mathbf{i} - \sin\theta\,\mathbf{j}).$$

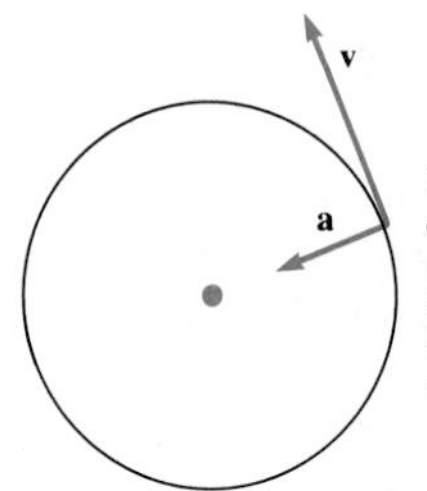

In circular motion at constant speed the acceleration vector is directed toward the center of the circle.

Figure 3

From this last equation we can read off the direction and length of **a**. First of all,
$$-\cos\theta\,\mathbf{i} - \sin\theta\,\mathbf{j}$$
is a unit vector pointing in the direction opposite that of
$$\mathbf{r} = r\cos\theta\,\mathbf{i} + r\sin\theta\,\mathbf{j}.$$
Hence **a** points toward the center of the circle. (See Fig. 3.) Since **a** is v^2/r times a unit vector, its magnitude is v^2/r. ■

This theorem was discovered in 1657 by Huygens while developing a theory of clock mechanisms. Anyone who has spun a pail of water at the end of a rope should find the theorem plausible. First of all, to hold the pail in its orbit one must pull on the rope. Thus the force of the rope on the pail is directed toward the center of its circular orbit. Second, the faster one spins the pail (keeping the rope at fixed length), the harder one must pull on the rope. Hence the appearance of v^2 in the numerator is reasonable. Third, if the same speed is maintained but the radius of the circle is decreased, more force is required.

The Speed of a Satellite

Determining the orbit speed of a satellite.

With the aid of the theorem we can determine the speed that a satellite requires to achieve a circular orbit around the earth. Imagine the satellite being swung around the earth like a pail at the end of a rope 4,000 miles long, the radius of the earth, as in Fig. 4. Instead of the tension on a rope, the force of gravity pulls the satellite in toward the earth from a linear path. Now, if a particle moves in a circle of radius 4,000 miles with speed v miles per second, it has an acceleration toward the center of the earth of $v^2/4{,}000$ miles per second per second. This acceleration must coincide with the acceleration of 32 feet per second per second, approximately 0.0061 mile per second per second, which gravity imparts to any object at the surface of the earth. Thus $0.0061 \approx v^2/4{,}000$, and

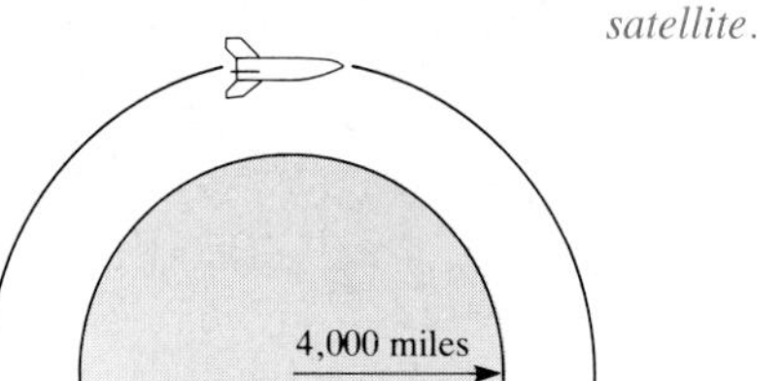

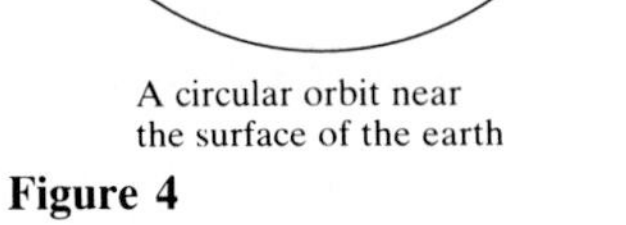

A circular orbit near the surface of the earth

Figure 4

$$\text{Orbital velocity} = v \approx \sqrt{(4{,}000)(0.0061)}$$
$$= \sqrt{24.4}$$
$$\approx 5 \text{ miles per second.}$$

How many miles per hour is 5 miles per second?

The velocity necessary to maintain an object in orbit at the surface of the earth is 5 miles per second, which is less than the escape velocity, 7 miles per second, that was computed in Sec. 8.8.

The Ultracentrifuge

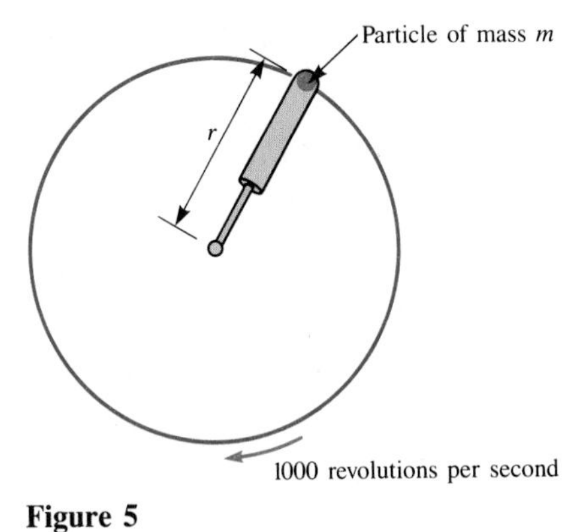

Figure 5

An ultracentrifuge can spin a test tube at the rate of 1,000 revolutions per second. Such a centrifuge is used to separate particles, such as macromolecules, from a liquid. (See Fig. 5.)

Assume that the particle is r meters from the center of the centrifuge. In one second it travels $(1{,}000)2\pi r$ meters. Hence its speed is $2{,}000\pi r$ meters per second. The magnitude of its acceleration vector is $v^2/r = (2{,}000\pi r)^2/r = 4{,}000{,}000\pi^2 r$ meters per second per second. If its mass is m, then a force of $4{,}000{,}000\pi^2 rm$ newtons directed toward the center of the centrifuge is needed to keep it in its circular orbit. (Even for $r = 0.5$ meter, this force is much greater than the force of gravity at sea level on the molecule, namely, $9.8m$ newtons.) If the liquid cannot provide such a force, the macromolecule will move toward the bottom of the test tube. In this way a centrifuge ''sediments'' materials that do not separate just under the much weaker force of gravity. Of course, the base of the test tube must be quite strong, since it must provide sufficient force to keep its contents in a circular orbit.

EXERCISES FOR SEC. 13.3: THE ACCELERATION VECTOR

1 Let $\mathbf{r}(t) = t^2\mathbf{i} + t^3\mathbf{j}$. Compute and sketch $\mathbf{r}(1)$, $\mathbf{v}(1)$, and $\mathbf{a}(1)$.

2 Let

$$\mathbf{r}(t) = \cos\frac{\pi e^t}{2}\,\mathbf{i} + \sin\frac{\pi e^t}{2}\,\mathbf{j}.$$

Compute and sketch $\mathbf{r}(0)$, $\mathbf{v}(0)$, and $\mathbf{a}(0)$.

3 Let $\mathbf{r}(t) = 4t\mathbf{i} - 16t^2\mathbf{j}$ describe a falling ball. Compute and draw $\mathbf{r}$, $\mathbf{v}$, and $\mathbf{a}$ for $t = 0$, $t = 1$, and $t = 2$.

4 Let $\mathbf{r}(t) = 10\cos 2\pi t\,\mathbf{i} + 10\sin 2\pi t\,\mathbf{j}$ denote the position vector of a particle at time t.
(*a*) What is the shape of the path this particle follows?
(*b*) Compute $\mathbf{v}$ and $\mathbf{a}$.
(*c*) Draw $\mathbf{r}$, $\mathbf{v}$, and $\mathbf{a}$ for $t = \frac{1}{4}$.
(*d*) Show that $\mathbf{a}$ is always in the direction opposite that of $\mathbf{r}$.

5 Let $\mathbf{r}(t) = t\mathbf{i} + (1/t)\mathbf{j}$.
(*a*) Sketch the path of the particle for $t > 0$.
(*b*) Compute and sketch $\mathbf{r}(1)$, $\mathbf{v}(1)$, and $\mathbf{a}(1)$.
(*c*) What happens to $\mathbf{v}(t)$ and $\mathbf{a}(t)$ when t is large?
(*d*) What happens to the speed when t is large?

6 Suppose at time t a particle has the position vector $\mathbf{r}(t) = (t + \cos t)\mathbf{i} + (t - \sin t)\mathbf{j}$.
(*a*) Show that $\mathbf{a}$ has constant magnitude.
(*b*) Sketch the path corresponding to t in $[0, 4\pi]$.
(*c*) Sketch $\mathbf{r}$, $\mathbf{v}$, and $\mathbf{a}$ for $t = 0$, $\pi/2$, π, $3\pi/2$, and 2π.

7 Let $\mathbf{r}(t) = 5\cos 3t\,\mathbf{i} + 5\sin 3t\,\mathbf{j} + 6t\mathbf{k}$ describe the position at time t of a particle moving on a helical path.
(*a*) Compute $\mathbf{v}$ and $\mathbf{a}$ at time t.
(*b*) How would you interpret in terms of the motion the fact that $\mathbf{a}$ is parallel to the xy plane?

8 Let $\mathbf{r}(t) = 2t\mathbf{i} + 3t\mathbf{j} + 13t^2\mathbf{k}$.
(*a*) Compute $\mathbf{v}$ at time t.
(*b*) Compute $\mathbf{a}$ at time t.

9 (*a*) If $\mathbf{v}$ has constant direction, must $\mathbf{a}$ have constant direction? Explain.
(*b*) If $\mathbf{v}$ has constant length, must $\mathbf{a}$ have constant length? Explain.

10 The momentum of a particle of mass m and velocity vector $\mathbf{v}$ is the vector $m\mathbf{v}$. Newton stated his second law in the form $\mathbf{F} = (m\mathbf{v})'$.
(*a*) Prove that $\mathbf{F} = m\mathbf{a} + m'\mathbf{v}$.
(*b*) Deduce that, if m is constant, then $\mathbf{F} = m\mathbf{a}$. (In relativity theory m is not necessarily constant.)

11 Prove that if $\mathbf{a}$ is always perpendicular to $\mathbf{v}$ on a certain path, the speed of the particle is constant.

12 (*a*) Prove that if the speed of a particle is constant, $\mathbf{a}$ is perpendicular to $\mathbf{v}$.
(*b*) Does this make sense physically?

13 (*a*) Prove that if a particle moves in a circular orbit [centered at (0, 0)], then $\mathbf{r}\cdot\mathbf{a} + \mathbf{v}\cdot\mathbf{v} = 0$.
(*b*) From (*a*) deduce that $\mathbf{r}\cdot\mathbf{a} \le 0$.
(*c*) What does (*b*) say about the direction of the force vector $\mathbf{F}$?

14 A particle moves in the circular orbit given by $\mathbf{r} = \cos t^2\,\mathbf{i} + \sin t^2\,\mathbf{j}$.
(*a*) Compute $\mathbf{v}$ and $\mathbf{a}$.
(*b*) Verify that $\mathbf{r}\cdot\mathbf{a} \le 0$.

15 Two cars make the same circular turn around a corner. The

first is traveling at 15 miles per hour; the second at 30 miles per hour. How many times as large must the force be to keep the second car from skidding as compared to the force required to keep the first car from skidding? (Assume the cars have equal mass.)

16 A girl is spinning a pail of water at the end of a rope.
(*a*) If she doubles the speed of the pail, how many times as hard must she pull on the rope?
(*b*) If instead she doubles the length of the rope, but keeps the speed of the pail the same, will she have to pull more or less? How much?

Exercises 17 to 21 concern satellites.

17 The acceleration that gravity imparts to an object decreases with the square of the distance of the object from the center of the earth.
(*a*) Show that if an object is r miles from the center of the earth, it has an acceleration of $(0.0061)(4{,}000/r)^2$ miles per second per second.
(*b*) With the aid of (*a*), find the velocity of a satellite in orbit at an altitude of 1,000 miles.

18 How long would it take a satellite to orbit the earth just above the earth's surface? One hundred miles above the earth's surface? One thousand miles above the surface?

19 What altitude must an orbiting satellite have in order to stay directly above a fixed spot on the equator? This type of orbit is called "geosynchronous," from the Greek, *geo* (earth), *syn* (same), *chrono* (time). It is also called a **stationary orbit**.

20 A certain satellite in circular orbit goes around the earth once every 92 minutes. How high is it above the earth?

21 Show that the orbital velocity at the surface of the earth is $1/\sqrt{2}$ times the escape velocity.

22 (*a*) If a curve in the plane is given parametrically, x and y being functions of t, how are $\dot{x}$, $\dot{y}$, and dy/dx related? Express d^2y/dx^2 in terms of $\dot{x}$, $\dot{y}$, $\ddot{x}$, and $\ddot{y}$.
(*b*) If at a certain instant $\mathbf{v} = 2\mathbf{i} + 3\mathbf{j}$ and $\mathbf{a} = \mathbf{i} + 4\mathbf{j}$, find dy/dx, d^2y/dx^2, and the radius of curvature at that instant.

23 A particle in space moves under the influence of a force that is always directed toward the origin. (For instance, the particle may be moving under the influence of the gravitational field of the sun.) Prove that its path lies in a plane:
(*a*) Let $\mathbf{r}$, $\mathbf{v}$, and $\mathbf{v}' = \mathbf{a}$ be the position, velocity, and acceleration vectors. Show that $\mathbf{a}(t) = f(t)\mathbf{r}(t)$, where f is a scalar function.
(*b*) Show that $(\mathbf{r} \times \mathbf{v})' = \mathbf{0}$.
(*c*) From (*b*) it follows that $\mathbf{r} \times \mathbf{v}$ is a constant vector $\mathbf{C}$. Show that if $\mathbf{C}$ is not $\mathbf{0}$, then the particle travels in a plane perpendicular to $\mathbf{C}$.
(*d*) If $\mathbf{C}$ in part (*c*) is $\mathbf{0}$, what is the path of the particle?

In Exercises 24 to 26 the acceleration vector $\mathbf{a}(t)$ at time t is given, as are the initial position $\mathbf{r}(0)$ and initial velocity $\mathbf{v}(0)$. Determine $\mathbf{r}(t)$ by two integrations.

24 $\mathbf{a}(t) = \dfrac{t}{(1-t^2)^{3/2}}\,\mathbf{i} - \dfrac{4}{(1+2t)^2}\,\mathbf{j}$, $0 \le t < 1$; $\mathbf{r}(0) = 2\mathbf{i}$, $\mathbf{v}(0) = 4\mathbf{i} + 2\mathbf{j}$

25 $\mathbf{a}(t) = 2\sec^2 t \tan t\ \mathbf{i} + \sec 2t \tan 2t\ \mathbf{j}$, $0 \le t < \dfrac{\pi}{4}$; $\mathbf{r}(0) = 5\mathbf{j}$, $\mathbf{v}(0) = 3\mathbf{i}$

26 $\mathbf{a}(t) = 2^t(\ln 2)^2\,\mathbf{i} - \dfrac{16t}{(1+4t^2)^2}\,\mathbf{j}$, $0 \le t$; $\mathbf{r}(0) = 2\mathbf{i}$, $\mathbf{v}(0) = (3 + \ln 2)\mathbf{i} + 2\mathbf{j}$

27 This table records the position vector of a moving object at these instants.

t	2	2.01	2.02
$\mathbf{G}(t)$	$\langle 3, 4, 5\rangle$	$\langle 3.02, 3.99, 5.02\rangle$	$\langle 3.0403, 3.9698, 5.0404\rangle$

(*a*) Estimate $\mathbf{v}(2)$ and $\mathbf{v}(2.01)$.
(*b*) Estimate $\mathbf{a}(2)$.

28 We showed using vectors that the acceleration vector of a particle moving in a circle of radius r at constant speed v has magnitude v^2/r. Figure 6 suggests an intuitive basis for this result.

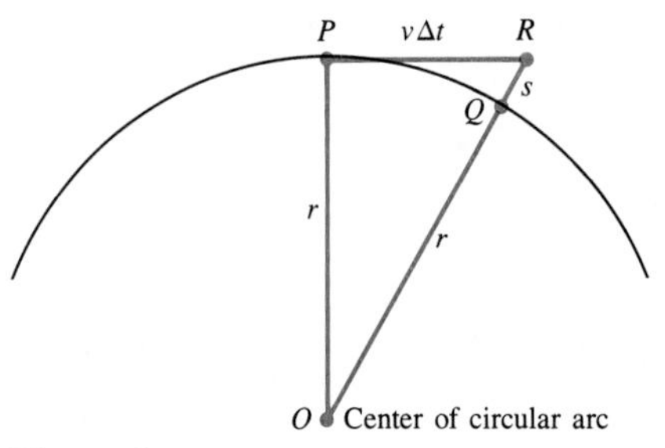

Figure 6

During a very short time interval Δt, the particle moved from P to Q. During this time, the particle "fell toward O" a distance s. (Had it simply followed a straight path along the tangent at P, it would have reached the point R, a distance $r + s$ from O.)
(*a*) Show that s is approximately $(v\,\Delta t)^2/(2r)$.
(*b*) Why does (*a*) suggest that the magnitude of the acceleration is v^2/r?

29 The following table lists the acceleration vector of a moving particle at various times.

t	1	1.05	1.10	1.20
$\mathbf{a}(t)$	$3\mathbf{i} + 4\mathbf{j}$	$4\mathbf{i} + 5\mathbf{j}$	$5\mathbf{i} + 7\mathbf{j}$	$5\mathbf{i} + 8\mathbf{j}$

Assume that $\mathbf{r}(1) = \mathbf{i} + \mathbf{j}$ and $\mathbf{v}(1) = 3\mathbf{j}$.
(*a*) Estimate $\mathbf{v}(1.05)$, $\mathbf{v}(1.10)$, and $\mathbf{v}(1.20)$.
(*b*) Estimate $\mathbf{r}(1.05)$, $\mathbf{r}(1.10)$, and $\mathbf{r}(1.20)$.

30 If you have mass m kilograms and are sitting down, the seat

of the chair exerts a force of $9.8m$ newtons on you (at sea level). This is called a force of "one g." To experience a larger force you may go to a carnival where you can be spun in a cylinder (a "rotor-ride") or take a ride on a roller coaster. Or you could make a sharp turn in a jet plane. Say that you want to experience the force of $7g$, that is, have the seat of the plane push against you with a force of $7(9.8m)$ newtons. If you are traveling at the speed of 1,000 kilometers per hour what radius turn should you make? First express your speed in meters per second.

13.4 THE COMPONENTS OF ACCELERATION

In this section we determine the projections of the acceleration vector along a tangent and along a line perpendicular to the tangent. It turns out that the component perpendicular to the tangent involves the radius of curvature. It is for this reason that the radius of curvature plays a key role in the design of turns in highways and railroads.

The Unit Vectors T and N

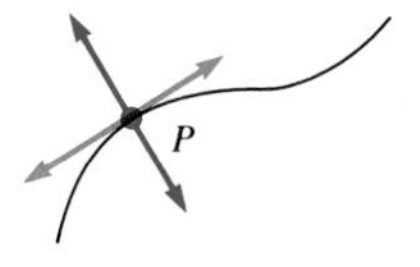

There are two unit vectors tangent to the curve at P and two unit vectors perpendicular to the curve at P.

Figure 1

Consider a particle moving along a curve in the plane. At each point P on the curve there are two unit tangent vectors and two unit vectors perpendicular to the curve, as shown in Fig. 1.

We choose one of the two tangent vectors and one of the two normal vectors, as follows.

Let the position vector of the particle be $\mathbf{r}(s)$, where the parameter is arc length along the curve. The vector

$$\frac{d\mathbf{r}}{ds}$$

is tangent to the curve. Its magnitude is

$$\sqrt{\left(\frac{dx}{ds}\right)^2 + \left(\frac{dy}{ds}\right)^2}.$$

But for any parameterization, we have

$$\frac{ds}{dt} = \sqrt{\left(\frac{dx}{dt}\right)^2 + \left(\frac{dy}{dt}\right)^2}.$$

In particular, since $ds/ds = 1$, we have

$$1 = \sqrt{\left(\frac{dx}{ds}\right)^2 + \left(\frac{dy}{ds}\right)^2}.$$

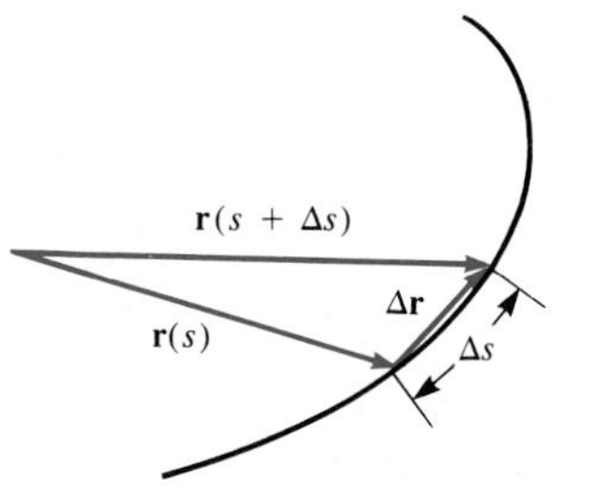

Figure 2

Therefore $d\mathbf{r}/ds$ is a unit vector.

(Figure 2 shows geometrically why $\|d\mathbf{r}/ds\| = 1$. For small Δs the ratio between the chord length $\|\Delta\mathbf{r}\|$ and arc length Δs is near 1, that is, $\|\Delta\mathbf{r}/\Delta s\| \approx 1$.)

Definition *Unit tangent* **T**. Let $\mathbf{r} = \mathbf{G}(s)$ describe a curve, where s is arc length. Then $d\mathbf{r}/ds$ is a **unit tangent vector** which will be denoted $\mathbf{T}$.

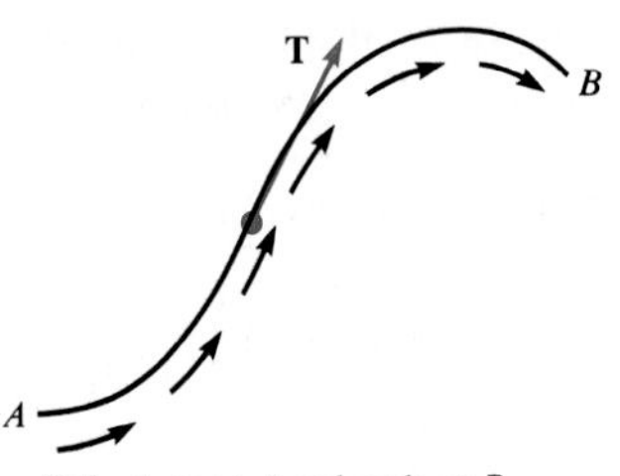

Path starts at A and ends at B.

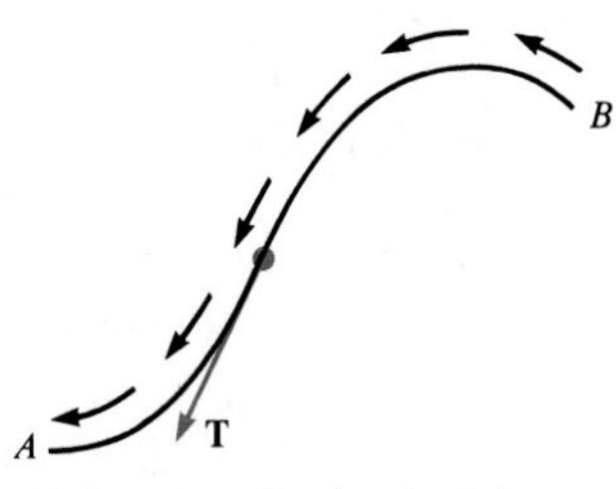

Path starts at B and ends at A.

Figure 3

Note that **T** depends on the direction in which the curve is swept out. It points in the direction of the velocity vector, which depends on the direction in which the curve is traveled. (See Fig. 3.)

If the curve is described using a parameter other than arc length s, **T** can be computed easily. It is equal to

$$\frac{d\mathbf{r}/dt}{\|d\mathbf{r}/dt\|} = \frac{\mathbf{v}}{\|\mathbf{v}\|},$$

where **v** is the velocity vector. First, $\mathbf{v}/\|\mathbf{v}\|$ points in the same direction as **v**, hence in the same direction as **T**. Second, it is a unit vector.

EXAMPLE 1 A particle moves along a curve given by

$$\mathbf{r}(t) = t\mathbf{i} + t^2\mathbf{j}.$$

Find $\mathbf{T}(0)$ and $\mathbf{T}(1)$ and sketch them on the curve.

SOLUTION

$$\mathbf{v}(t) = \mathbf{r}'(t) = \mathbf{i} + 2t\mathbf{j}.$$

Thus,

$$\mathbf{T} = \frac{\mathbf{v}(t)}{\|\mathbf{v}(t)\|}$$

$$= \frac{\mathbf{i} + 2t\mathbf{j}}{\|\mathbf{i} + 2t\mathbf{j}\|}$$

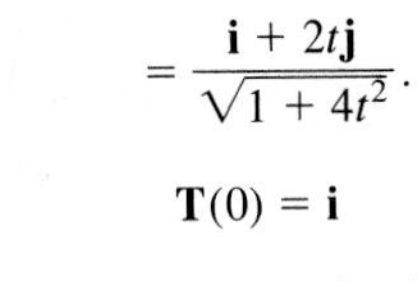

$$= \frac{\mathbf{i} + 2t\mathbf{j}}{\sqrt{1 + 4t^2}}.$$

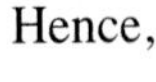

Hence,

$$\mathbf{T}(0) = \mathbf{i}$$

and

$$\mathbf{T}(1) = \frac{\mathbf{i} + 2\mathbf{j}}{\sqrt{5}}.$$

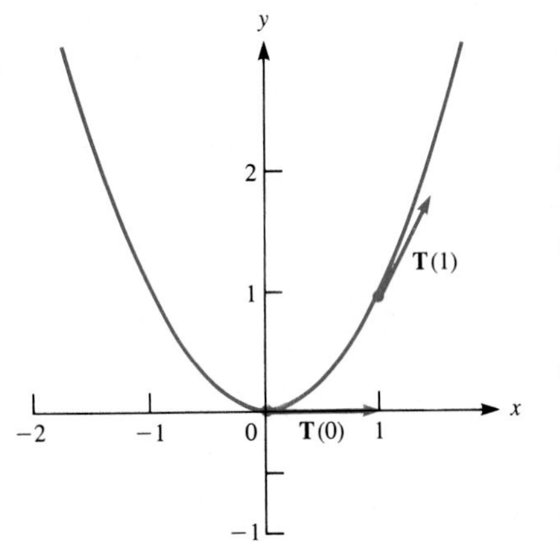

Figure 4

To describe the curve, recall that the formula $\mathbf{r}(t) = t\mathbf{i} + t^2\mathbf{j}$ records the parametric equations:

$$\begin{cases} x = t \\ y = t^2. \end{cases}$$

Hence $y = x^2$. The curve is a parabola. Figure 4 shows the curve, $\mathbf{T}(0)$, and $\mathbf{T}(1)$. ■

We have thus picked out one of the two tangent vectors, and the choice records the direction of motion. As the particle traverses the path, **T** remains of unit length, but may change in direction, as shown in Fig. 4.

A unit vector perpendicular to **T** will also be needed. There are two choices. The one we shall take indicates the direction in which the particle is veering—to the right or to the left. The sense of concavity determines the choice of this normal vector as indicated in Fig. 5. It tells us on which side of the tangent line the nearby section of the curve lies.

This description of the choice of vector perpendicular to **T** is intuitive and physical. For computations later in this section we need a more formal definition. We now give this definition and then show that it agrees with the informal choice.

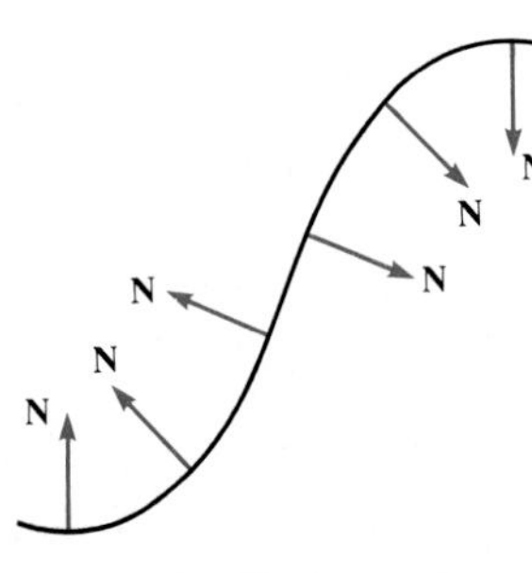

The choice of unit normal vector

Figure 5

Definition *Principal unit normal vector* **N**. Let s denote arc length on a curve. If $\mathbf{T}'(s) = d\mathbf{T}/ds$ is not $\mathbf{0}$, the vector

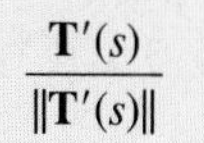

$$\frac{\mathbf{T}'(s)}{\|\mathbf{T}'(s)\|}$$

is called the **principal unit normal vector** and is denoted $\mathbf{N}(s)$.

We must show three things:

(*i*) **N** is a unit vector.
(*ii*) **N** is perpendicular to **T**.
(*iii*) **N** points in the direction toward which $\mathbf{T}(s)$ is turning.

The argument is brief:

(*i*) Since any vector divided by its length is a unit vector, **N** is a unit vector.
(*ii*) Since $\mathbf{T}(s)$ has constant length, its derivative $\mathbf{T}'(s)$ is perpendicular to it. (See Sec. 13.2.) Thus, $\mathbf{N}(s) = \mathbf{T}'(s)/\|\mathbf{T}'(s)\|$ is perpendicular to $\mathbf{T}(s)$.
(*iii*) To see that **N** points in the direction **T** is turning requires looking at Fig. 6. It compares $\mathbf{T}(s + \Delta s)$ and $\mathbf{T}(s)$ for small Δs and shows that $\Delta\mathbf{T}$ indicates the direction of turning (to the right in this case). Since

$$\mathbf{T}'(s) = \lim_{\Delta s \to 0} \frac{\Delta \mathbf{T}}{\Delta s},$$

$\mathbf{T}'(s)$ points in the direction of turning. Therefore $\mathbf{N}(s) = \mathbf{T}'(s)/\|\mathbf{T}'(s)\|$ does also.

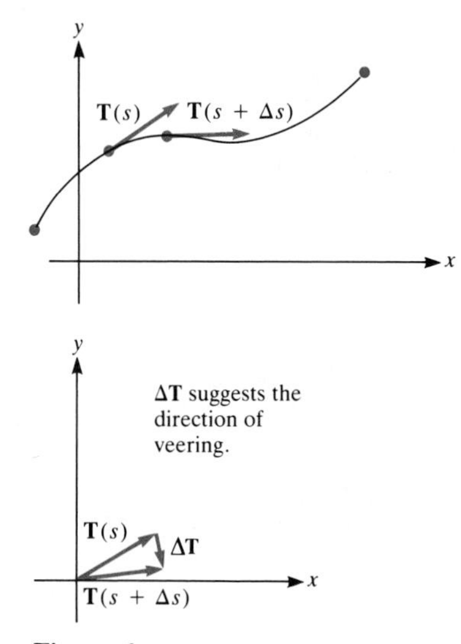

Figure 6

Although arc length s appears in the definition of the principal unit normal, it plays no particular role in our argument that **N** points in the direction we want. So, for any parameterization in which $\mathbf{T}'(t)$ is not $\mathbf{0}$, the vector

$$\frac{\mathbf{T}'(t)}{\|\mathbf{T}'(t)\|}$$

also is the principal unit normal.

EXAMPLE 2 (*a*) Find $\mathbf{T}(t)$ and $\mathbf{N}(t)$ for the curve

$$\mathbf{r}(t) = 3 \cos 2t\, \mathbf{i} + 3 \sin 2t\, \mathbf{j}.$$

(*b*) Draw the curve and show $\mathbf{T}(\pi/6)$ and $\mathbf{N}(\pi/6)$.

SOLUTION (*a*) First, differentiation gives

$$\mathbf{v}(t) = -6 \sin 2t\, \mathbf{i} + 6 \cos 2t\, \mathbf{j}$$

and

$$\|\mathbf{v}(t)\| = \sqrt{(-6 \sin 2t)^2 + (6 \cos 2t)^2}$$
$$= \sqrt{36(\sin^2 2t + \cos^2 2t)}$$
$$= 6.$$

Thus

$$\mathbf{T}(t) = \frac{\mathbf{v}(t)}{\|\mathbf{v}(t)\|}$$

$$= \frac{-6 \sin 2t\ \mathbf{i} + 6 \cos 2t\ \mathbf{j}}{6};$$

hence
$$\mathbf{T}(t) = -\sin 2t\ \mathbf{i} + \cos 2t\ \mathbf{j}.$$

To find $\mathbf{N}(t) = \mathbf{T}'(t)/\|\mathbf{T}'(t)\|$, we differentiate $\mathbf{T}(t)$:

$$\mathbf{T}'(t) = -2 \cos 2t\ \mathbf{i} - 2 \sin 2t\ \mathbf{j}.$$

Then we compute

$$\|\mathbf{T}'(t)\| = \sqrt{(-2 \cos 2t)^2 + (-2 \sin 2t)^2}$$
$$= 2.$$

Thus
$$\mathbf{N}(t) = \frac{\mathbf{T}'(t)}{\|\mathbf{T}'(t)\|}$$
$$= \frac{-2 \cos 2t\ \mathbf{i} - 2 \sin 2t\ \mathbf{j}}{2}$$
$$= -\cos 2t\ \mathbf{i} - \sin 2t\ \mathbf{j}.$$

(*b*) Since

$$\|\mathbf{r}(t)\| = \sqrt{(3 \cos 2t)^2 + (3 \sin 2t)^2}$$
$$= 3,$$

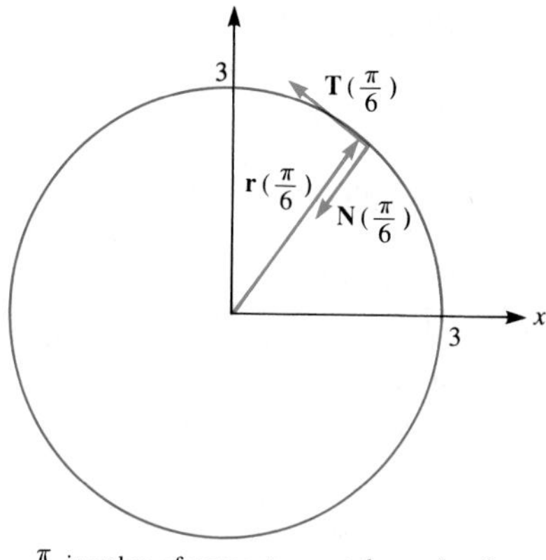

$\frac{\pi}{6}$ is value of parameter, not the angle of $\mathbf{r}$.

Figure 7

the curve is a circle. Figure 7 shows the curve, $\mathbf{r}(\pi/6)$, $\mathbf{T}(\pi/6)$, and $\mathbf{N}(\pi/6)$, which we compute as follows:

$$\mathbf{r}\left(\frac{\pi}{6}\right) = 3 \cos 2\left(\frac{\pi}{6}\right) \mathbf{i} + 3 \sin 2\left(\frac{\pi}{6}\right) \mathbf{j}$$
$$= 3 \cos \frac{\pi}{3}\ \mathbf{i} + 3 \sin \frac{\pi}{3}\ \mathbf{j} = \left(\frac{3}{2}\right) \mathbf{i} + \left(\frac{3}{2}\right)\sqrt{3}\ \mathbf{j}$$
$$\mathbf{T}\left(\frac{\pi}{6}\right) = -\sin 2\left(\frac{\pi}{6}\right) \mathbf{i} + \cos 2\left(\frac{\pi}{6}\right) \mathbf{j} = -\left(\frac{\sqrt{3}}{2}\right) \mathbf{i} + \left(\frac{1}{2}\right) \mathbf{j}$$
$$\mathbf{N}\left(\frac{\pi}{6}\right) = -\cos 2\left(\frac{\pi}{6}\right) \mathbf{i} - \sin 2\left(\frac{\pi}{6}\right) \mathbf{j} = -\left(\frac{1}{2}\right) \mathbf{i} - \left(\frac{\sqrt{3}}{2}\right) \mathbf{j}.$$

Note that $\mathbf{N}(\pi/6) = (-1/3)\mathbf{r}(\pi/6)$. ■

There is another natural geometric parameter associated with a plane curve, namely, the angle that $\mathbf{T}$ makes with the vector $\mathbf{i}$. This is the angle ϕ in Fig. 8. (This angle ϕ is used in Sec. 9.6 for the study of curvature.) The next theorem is concerned with $d\mathbf{T}/d\phi$. It is assumed that the curve is not a straight line. (Otherwise ϕ is constant.)

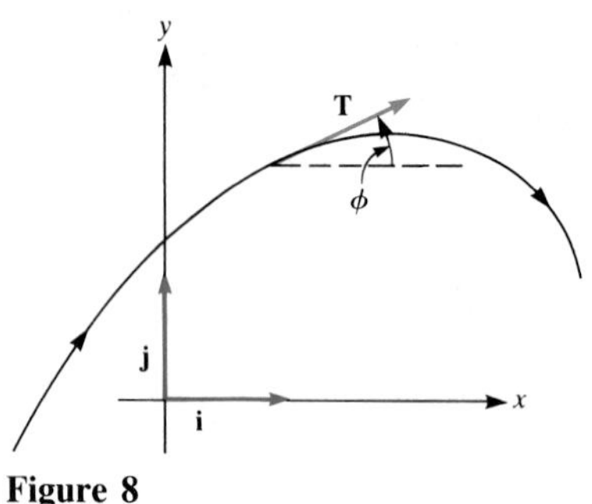

Figure 8

Theorem 1 Let $\mathbf{r}(t)$ be the position vector on a curve at time t. Let ϕ denote the angle that

$$\mathbf{T}(t) = \frac{\mathbf{v}(t)}{\|\mathbf{v}(t)\|}$$

makes with the vector $\mathbf{i}$. Then $\mathbf{T} = \cos \phi\ \mathbf{i} + \sin \phi\ \mathbf{j}$ and

$$\left\|\frac{d\mathbf{T}}{d\phi}\right\| = 1.$$

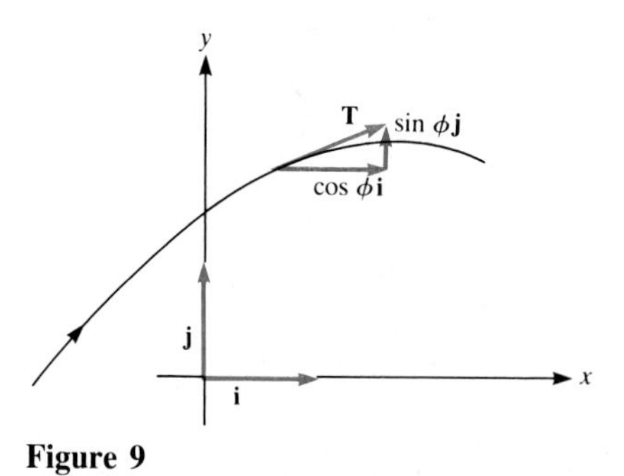

Figure 9

Proof Since **T** is a unit vector and makes an angle ϕ with **i**, the vector components of **T** along **i** and **j** are cos ϕ **i** and sin ϕ **j**. (See Fig. 9.) Thus

$$\mathbf{T} = \cos\phi\ \mathbf{i} + \sin\phi\ \mathbf{j}.$$

Hence

$$\frac{d\mathbf{T}}{d\phi} = -\sin\phi\ \mathbf{i} + \cos\phi\ \mathbf{j}$$

and

$$\left\|\frac{d\mathbf{T}}{d\phi}\right\| = \sqrt{(-\sin\phi)^2 + (\cos\phi)^2} = 1. \quad \blacksquare$$

Writing a in Terms of T and N

Let $\mathbf{r}(t)$ be the position vector of a particle at time t. The velocity vector is easily expressed in terms of the speed and **T**, $\mathbf{v} = v\mathbf{T}$. In the study of motion, it is useful to express **a** in terms of **T** and **N**, in the form

$$\mathbf{a}(t) = a_T\mathbf{T} + a_N\mathbf{N};$$

a_T is called the **tangential component** and a_N the **normal component** of the acceleration vector. For instance, in the case of motion at constant speed v around a circle of radius r

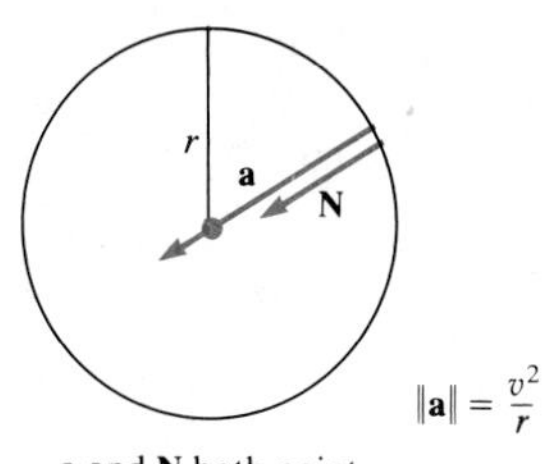

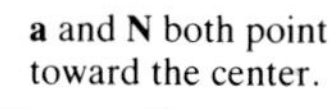
a and **N** both point toward the center.

Figure 10

$$\mathbf{a}(t) = 0\mathbf{T} + \frac{v^2}{r}\mathbf{N},$$

so $a_T = 0$ and $a_N = v^2/r$. (See Fig. 10.)

Our goal is to obtain a formula for a_T and a_N for motion on any curve and with *speed not necessarily constant*. The key is Theorem 2. Before reading it, review the definition of radius of curvature from Sec. 9.6.

Since $\mathbf{v} = v\mathbf{T}$,

$$\mathbf{a} = \frac{d\mathbf{v}}{dt} = v\frac{d\mathbf{T}}{dt} + \frac{dv}{dt}\mathbf{T} = v\frac{d\mathbf{T}}{dt} + \frac{d^2s}{dt^2}\mathbf{T}.$$

All that remains is to find $d\mathbf{T}/dt$.

Theorem 2 Let s represent arc length along the path of a moving particle. Let $v = ds/dt$ be the speed (which is always assumed to be positive). Then

$$\frac{d\mathbf{T}}{dt} = \frac{v}{r}\mathbf{N},$$

where r is the radius of curvature, which is assumed not to be 0.

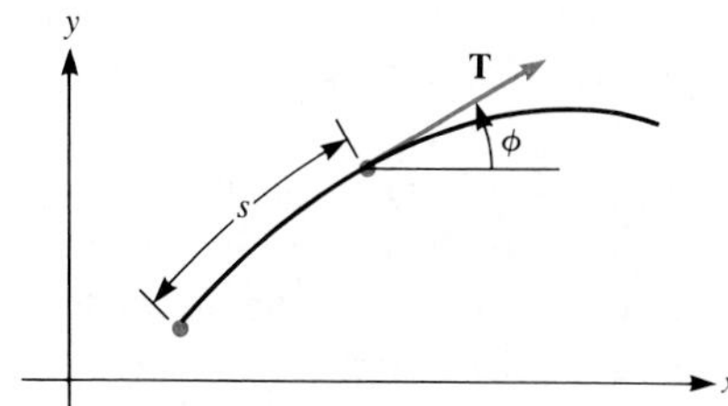

Figure 11

Proof We will use the angle ϕ shown in Fig. 11. Recall that curvature κ is defined as $|d\phi/ds|$.

We have

$$\frac{d\mathbf{T}}{dt} = \left\|\frac{d\mathbf{T}}{dt}\right\|\mathbf{N} \qquad \text{(definition of N)}$$

$$= \left\|\frac{d\mathbf{T}}{d\phi}\right\| \left|\frac{d\phi}{dt}\right|\mathbf{N} \qquad \text{(chain rule)}$$

$$= \left|\frac{d\phi}{dt}\right| \mathbf{N} \qquad \text{(Theorem 1)}$$

$$= \left|\frac{d\phi}{ds}\frac{ds}{dt}\right| \mathbf{N} \qquad \text{(chain rule)}$$

$$= \kappa v \, \mathbf{N}$$

$$= \frac{v}{r}\mathbf{N}. \qquad \text{(definition of radius of curvature)} \blacksquare$$

We are now ready to prove the following theorem, which provides formulas for a_T and a_N for a parameterized plane curve. For simplicity, the theorem is stated in terms of a moving particle.

Theorem 3 If a particle moves in a plane curve and arc length s is measured in such a way that $v = ds/dt$ is positive, and the radius of curvature r is not 0, then

$$\mathbf{a} = \frac{d^2s}{dt^2}\mathbf{T} + \frac{v^2}{r}\mathbf{N}.$$

In other words,

$$a_T = \frac{d^2s}{dt^2} \quad \text{and} \quad a_N = \frac{v^2}{r}.$$

Proof

$$\mathbf{a} = \frac{d\mathbf{v}}{dt}$$

$$= \frac{d}{dt}(v\mathbf{T})$$

$$= v\frac{d\mathbf{T}}{dt} + \frac{dv}{dt}\mathbf{T} \qquad \text{(derivative of scalar times vector function)}$$

$$= v\frac{v}{r}\mathbf{N} + \frac{dv}{dt}\mathbf{T} \qquad \text{(Theorem 2)}$$

$$= \frac{v^2}{r}\mathbf{N} + \frac{d^2s}{dt^2}\mathbf{T}. \qquad \left(v = \frac{ds}{dt}\right) \blacksquare$$

a_T is determined by change of speed, a_N by speed and radius of curvature.

Observe that a_T is determined by the "scalar" acceleration, the rate at which the speed is changing. On the other hand, a_N depends only on the speed and the radius of curvature; it is the normal component of the acceleration of a particle moving in a circle of radius r with constant speed v. Note that the scalar component of $\mathbf{a}$ along $\mathbf{N}$, being v^2/r, is positive.

a_N is positive; a_T need not be.

The fact that a_N is positive is reasonable, since $\mathbf{a}$ points in the same direction as the external force $\mathbf{F}$; $\mathbf{F}$ causes the particle to veer, and $\mathbf{N}$ records the direction in which the particle veers. a_T can be positive or negative depending on whether the particle is speeding up or slowing down, as shown in Fig. 12.

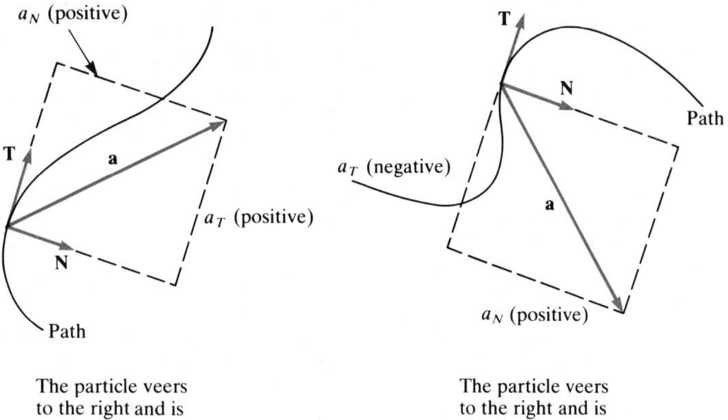

Figure 12

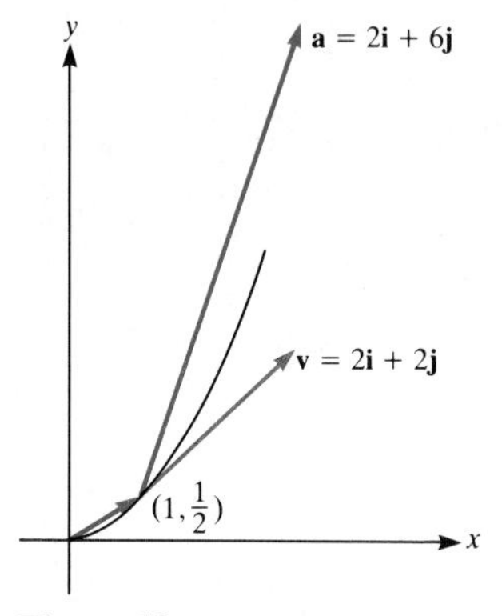

Figure 13

EXAMPLE 3 At time t a particle is at $(t^2, t^4/2)$ on the curve $2y = x^2$. Determine $\mathbf{r}$, $\mathbf{v}$, $\mathbf{a}$, a_N, a_T, and curvature κ when $t = 1$.

SOLUTION First, since $\mathbf{r} = t^2\,\mathbf{i} + \frac{1}{2}t^4\,\mathbf{j}$, we have $\mathbf{v} = 2t\mathbf{i} + 2t^3\mathbf{j}$. Thus

$$\mathbf{a} = 2\mathbf{i} + 6t^2\mathbf{j}.$$

From this it follows that $\|\mathbf{a}\| = \sqrt{4 + 36t^4}$.

Let us sketch $\mathbf{a}$, $\mathbf{v}$, and $\mathbf{r}$ at $t = 1$. At that time,

$$\mathbf{a} = 2\mathbf{i} + 6\mathbf{j},\ \mathbf{v} = 2\mathbf{i} + 2\mathbf{j},\ \text{and } \mathbf{r} = \mathbf{i} + \tfrac{1}{2}\mathbf{j}.$$

They are shown in Fig. 13. Since $a_T = \mathbf{a} \cdot \mathbf{T} = \mathbf{a} \cdot (\mathbf{v}/\|\mathbf{v}\|)$, we have

$$a_T = (2\mathbf{i} + 6\mathbf{j}) \cdot \frac{2\mathbf{i} + 2\mathbf{j}}{\|2\mathbf{i} + 2\mathbf{j}\|} = \frac{4 + 12}{\sqrt{8}} = 4\sqrt{2}.$$

Finding a_N without using the formula v^2/r

Rather than compute a_N by the formula $a_N = v^2/r$, which involves computation of the radius of curvature, use the pythagorean relation

$$a_N^2 + a_T^2 = \|\mathbf{a}\|^2.$$

At $t = 1$ this becomes

$$a_N^2 + (4\sqrt{2})^2 = (\sqrt{4 + 36})^2 = 40.$$

Hence $a_N^2 = 40 - 32 = 8$, and $a_N = 2\sqrt{2}$. Figure 14 shows $\mathbf{a}$, $a_T\mathbf{T}$, and $a_N\mathbf{N}$ at $t = 1$.

Finally, we find the curvature κ by using the equation

$$a_N = \frac{v^2}{r} = \kappa v^2.$$

Hence
$$\kappa = \frac{a_N}{v^2} = \frac{2\sqrt{2}}{\|\mathbf{v}\|^2} = \frac{2\sqrt{2}}{\|2\mathbf{i} + 2\mathbf{j}\|^2} = \frac{\sqrt{2}}{4}. \quad \blacksquare$$

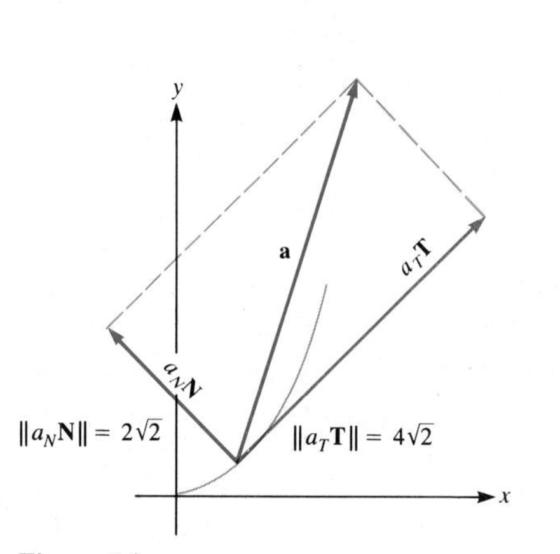

Figure 14

Curvature in Terms of v and a

In Example 3 the curvature κ was determined using only the velocity $\mathbf{v}$ and acceleration $\mathbf{a}$. There is a simple formula for κ in terms of $\mathbf{v}$ and $\mathbf{a}$, which we obtain as follows:

$$\begin{aligned} \|\mathbf{v} \times \mathbf{a}\| &= \left\| \mathbf{v} \times \left(\frac{d^2s}{dt^2}\mathbf{T} + \frac{v^2}{r}\mathbf{N} \right) \right\| \\ &= \left\| \frac{v^2}{r}\mathbf{v} \times \mathbf{N} \right\| && (\mathbf{v} \text{ and } \mathbf{T} \text{ are parallel}) \\ &= \frac{v^2}{r} v && (\mathbf{v} \text{ and } \mathbf{N} \text{ are perpendicular, } \|\mathbf{v}\| = v,\ \|\mathbf{N}\| = 1) \\ &= v^3\kappa. \end{aligned}$$

Therefore

$$\boxed{\kappa = \frac{\|\mathbf{v} \times \mathbf{a}\|}{\|\mathbf{v}\|^3}.} \tag{1}$$

EXAMPLE 4 Find the curvature at a point on a curve where $\mathbf{v} = 2\mathbf{i} + 2\mathbf{j}$ and $\mathbf{a} = 2\mathbf{i} + 6\mathbf{j}$. (These are the same vectors that appear in Example 3.)

SOLUTION First, $\|\mathbf{v}\| = \|2\mathbf{i} + 2\mathbf{j}\| = \sqrt{8}$. Second,

$$\mathbf{v} \times \mathbf{a} = \begin{vmatrix} \mathbf{i} & \mathbf{j} & \mathbf{k} \\ 2 & 2 & 0 \\ 2 & 6 & 0 \end{vmatrix} = 8\mathbf{k}.$$

Thus $\|\mathbf{v} \times \mathbf{a}\| = 8$ and we have

$$\kappa = \frac{\|\mathbf{v} \times \mathbf{a}\|}{\|\mathbf{v}\|^3} = \frac{8}{(\sqrt{8})^3} = \frac{1}{\sqrt{8}}.$$

(You may check that this answer agrees with the one in Example 3, $\sqrt{2}/4$.) ■

Section Summary

We defined two unit vectors $\mathbf{T}$ and $\mathbf{N}$ for a plane curve, $\mathbf{T} = d\mathbf{r}/ds$, and $\mathbf{N} = (d\mathbf{T}/ds)/\|d\mathbf{T}/ds\|$. $\mathbf{T}$ points in the direction of the velocity vector. If the parameter is t rather than arc length s, we have $\mathbf{T} = (d\mathbf{r}/dt)/\|d\mathbf{r}/dt\| = \mathbf{v}/\|\mathbf{v}\|$. Similarly, $\mathbf{N} = (d\mathbf{T}/dt)/\|d\mathbf{T}/dt\|$. We showed that $d\mathbf{T}/dt = (v/r)\mathbf{N}$, where r is the radius of curvature. This fact is the basis of the equation

$$\mathbf{a} = \frac{d^2s}{dt^2}\mathbf{T} + \frac{v^2}{r}\mathbf{N}.$$

We also showed how to find the normal and tangential components of $\mathbf{a}$ if we have all the components of $\mathbf{a}$ and $\mathbf{v}$ along $\mathbf{i}$ and $\mathbf{j}$. Finally, we showed that the curvature κ is given by

$$\kappa = \frac{\|\mathbf{v} \times \mathbf{a}\|}{\|\mathbf{v}\|^3}.$$

Exercises 37 to 42 extend this discussion to curves in space.

EXERCISES FOR SEC. 13.4: THE COMPONENTS OF ACCELERATION

In each of Exercises 1 to 6 compute and draw $\mathbf{T}(1)$ and $\mathbf{N}(1)$ for the given path at the point $\mathbf{r}(1)$. Also sketch the path for $t > 0$.

1 $\mathbf{r}(t) = t\mathbf{i} - t^2\mathbf{j}$
2 $\mathbf{r}(t) = 2t\mathbf{i} - 16t^2\mathbf{j}$
3 $\mathbf{r}(t) = \cos t\ \mathbf{i} + \sin t\ \mathbf{j}$
4 $\mathbf{r}(t) = t^2\mathbf{i} + t^3\mathbf{j}$
5 $\mathbf{r}(t) = e^t\mathbf{i} + e^{2t}\mathbf{j}$
6 $\mathbf{r}(t) = (t - \sin t)\ \mathbf{i} + (1 - \cos t)\ \mathbf{j}$ (a cycloid)
7 True or false: At a given point P on a curve, $\mathbf{T}$ depends on the direction in which the curve is being swept out.
8 True or false: At a given point P on a curve, $\mathbf{N}$ depends on the direction in which the curve is being swept out.
9 Let $\mathbf{r}(t) = t\mathbf{i} + t^2\mathbf{j}$ describe the path of a particle.
(*a*) Graph the path.
(*b*) Compute $\mathbf{r}(1)$, $\mathbf{v}(1)$, $\mathbf{T}(1)$, and $\mathbf{N}(1)$.
(*c*) Draw the vectors in (*b*), placing the tails of the last three at the head of $\mathbf{r}(1)$.
10 Let $\mathbf{r}(t) = \cos 2\pi t\ \mathbf{i} - \sin 2\pi t\ \mathbf{j}$.
(*a*) Sketch the path corresponding to $0 \le t \le 1$.
(*b*) Compute $\mathbf{r}(0)$, $\mathbf{v}(0)$, $\mathbf{T}(0)$, and $\mathbf{N}(0)$.
(*c*) Draw the four vectors in (*b*).
(*d*) Check that $\mathbf{T}(0)$ points in the direction of motion when the particle is at $\mathbf{r}(0)$.
11 A particle has the position vector $\mathbf{r}(t)$ at time t. Its speed when $t = 1$ is 3 meters per second. Evaluate $\mathbf{v}(1) \cdot \mathbf{T}(1)$.
12 If $\mathbf{v}$ is the velocity vector and $\mathbf{N}$ is the principal normal vector, evaluate $\mathbf{v} \cdot \mathbf{N}$ and $\mathbf{v} \cdot \mathbf{T}$.

In Exercises 13 to 16 find a_N and a_T, given that at a certain value of t:

13 $\dfrac{dv}{dt} = 3,\ v = 4,\ r = 5$ (r = radius of curvature)

14 $\dfrac{d^2s}{dt^2} = -2,\ v = 3,\ \kappa = \dfrac{1}{2}$

15 $\dfrac{d^2s}{dt^2} = 0,\ v = 1,\ r = 3$

16 $\dfrac{d^2s}{dt^2} = 3,\ v = 2,\ \|\mathbf{a}\| = 5$

In Exercises 17 and 18 note that the path is a circle. Find a_T, a_N, and κ in each case when $t = 1$, if

17 $\mathbf{r}(t) = 5 \cos 3\pi t\ \mathbf{i} + 5 \sin 3\pi t\ \mathbf{j}$
18 $\mathbf{r}(t) = 5 \cos \pi t^2\ \mathbf{i} + 5 \sin \pi t^2\ \mathbf{j}$
19 Show that (*a*) $\mathbf{a} \cdot \mathbf{T} = d^2s/dt^2$, (*b*) $\mathbf{a} \cdot \mathbf{N} = v^2/r$.
20 What can be said about $\mathbf{a} \cdot \mathbf{v}$ if the particle is (*a*) slowing down? (*b*) speeding up?
21 A moving particle has the position vector $\mathbf{r}(t) = t^2\mathbf{i} + t^3\mathbf{j}$ at time t. Find a_T, a_N, and κ.
22 A moving particle has the position vector $\mathbf{r}(t) = 3 \cos t\ \mathbf{i} + 4 \sin t\ \mathbf{j}$ at time t. Find a_T, a_N, and the radius of curvature.
23 At a certain instant, a particle headed northeast is speeding up and veering to the left. Taking north as the positive y axis and east as the positive x axis, draw (*a*) $\mathbf{T}$, (*b*) $\mathbf{N}$, (*c*) enough typical $\mathbf{a}$'s to indicate the directions possible for $\mathbf{a}$.
24 Repeat Exercise 23 for a particle headed north, slowing down, and veering to the left.
25 At a certain instant $\mathbf{r} = \mathbf{i} + \mathbf{j}$, $\mathbf{v} = 3\mathbf{i} + 4\mathbf{j}$, and $\mathbf{a} = 3\mathbf{i} - 3\mathbf{j}$.
(*a*) Draw $\mathbf{r}$, $\mathbf{v}$, and $\mathbf{a}$.
(*b*) Is the particle speeding up or slowing down? Explain.
(*c*) Estimate a_T and a_N graphically.
(*d*) Compute a_T, a_N, and κ.
26 At a certain instant $\mathbf{v} = 2\mathbf{i} + 3\mathbf{j}$ and $\mathbf{a} = 3\mathbf{i} - 4\mathbf{j}$.
(*a*) Draw $\mathbf{v}$, $\mathbf{a}$, $\mathbf{T}$, and $\mathbf{N}$.
(*b*) Find v and d^2s/dt^2.
(*c*) Find the radius of curvature.
27 Let $\mathbf{r}$ describe the journey presented in Example 3.
(*a*) Compute and draw $\mathbf{a}$ for $t = 1/\sqrt{6}$.
(*b*) From your drawing estimate a_T and a_N at $t = 1/\sqrt{6}$.
(*c*) Compute a_T and a_N at $t = 1/\sqrt{6}$.
(*d*) Compute (*b*) and (*c*) in decimal form.
28 At time $t \ge 0$ a particle is at the point $(\cos t + t \sin t, \sin t - t \cos t)$.
(*a*) Show that $\|\mathbf{v}\| = t$ and $\|\mathbf{a}\| = \sqrt{1 + t^2}$.
(*b*) Show that $a_T = 1$ and $a_N = t$.
(*c*) Show that the radius of curvature equals t.
29 Let $\mathbf{r}(t) = e^t\mathbf{i} + e^{2t}\mathbf{j}$.
(*a*) Compute $\mathbf{v}(t)$ and $\mathbf{a}(t)$ and express them in terms of $\mathbf{i}$ and $\mathbf{j}$.
(*b*) Compute $v(t)$ and dv/dt.
(*c*) Find a_T, a_N, and κ.
30 At a certain instant $\mathbf{v} = 5\mathbf{i} + 12\mathbf{j}$ and $\mathbf{a} = \mathbf{i} + 2\mathbf{j}$.
(*a*) Draw $\mathbf{v}$ and $\mathbf{a}$.
(*b*) On the basis of the sketch for (*a*) estimate a_T and a_N.
(*c*) Compute a_T and a_N.
31 If a particle travels at a constant speed what can be said about $\mathbf{a} \cdot \mathbf{v}$?
32 Let $\mathbf{r}(t) = t\mathbf{i} + t^3\mathbf{j}$.
(*a*) Compute $\mathbf{T}(t)$.
(*b*) Show that $\mathbf{N}$ is not defined when $t = 0$.
(*c*) Sketch the path. What property of the path causes $\mathbf{N}$ not to be defined when $t = 0$?

---

33 Show that for motion on a curve in the xy plane there is a scalar c, not necessarily constant, such that $\mathbf{N}' = c\mathbf{T}$.
34 Theorem 1 asserts that $d\mathbf{T}/d\phi$ has length 1. Explain why this is just a special case of the fact that when arc length s is used as a parameter for a curve, $\mathbf{r} = \mathbf{G}(s)$, then $d\mathbf{r}/ds$ has length 1.

35 Deduce from Theorem 3 this formula for obtaining r, the radius of curvature:

$$\frac{v^4}{r^2} = \left(\frac{d^2x}{dt^2}\right)^2 + \left(\frac{d^2y}{dt^2}\right)^2 - \left(\frac{d^2s}{dt^2}\right)^2.$$

36 According to Sec. 9.6, the radius of curvature of the curve $y = f(x)$ is given by the formula

$$r = \frac{[1 + (dy/dx)^2]^{3/2}}{|d^2y/dx^2|}.$$

(*a*) Obtain this formula from the result in Exercise 35.
(*b*) Obtain this formula from Eq. (1).

In the case of a curve in space $\mathbf{r} = \mathbf{r}(s)$, where s is arc length, three mutually perpendicular unit vectors are needed, not just **T** and **N**. Exercises 37 to 42 develop these vectors and extend the notion of curvature to curves that need not lie in a plane.

37 Show that $\mathbf{T} = d\mathbf{r}/ds$ and $\mathbf{N} = (d\mathbf{T}/ds)/\|d\mathbf{T}/ds\|$ are unit vectors. (Near any point P on it, the curve is nearly planar, almost lying in the plane through P parallel to **N** and **T**.)

38 Why is it reasonable to *define* curvature κ for a curve in space as $\|d\mathbf{T}/ds\|$? (*Hint:* Review Theorem 2.) Radius of curvature is then defined as $1/\kappa$. With these definitions, the proof of Theorem 3 goes through without change.

39 Define **B** to be $\mathbf{T} \times \mathbf{N}$. (See Fig. 15.)

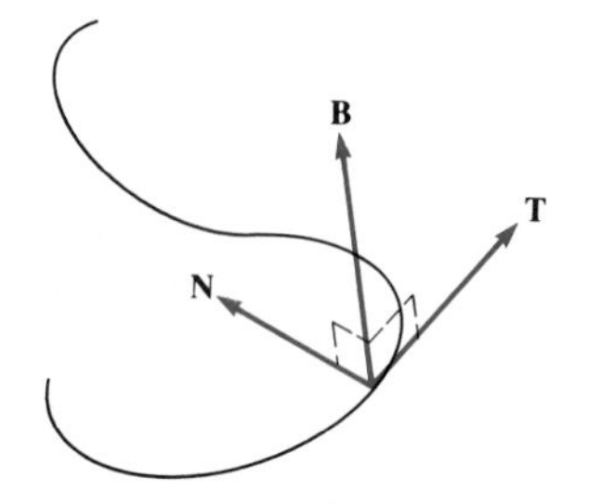

Figure 15

(*a*) Compute **T**, **N**, and **B** for the helix $\mathbf{r}(t) = 3\cos 2t\,\mathbf{i} + 3\sin 2t\,\mathbf{j} + 5t\mathbf{k}$. (*Note:* t is not arc length.)
(*b*) Draw the helix and the vectors in (*a*) at $\mathbf{r}(\pi/4)$.

40 (*a*) Show that $d\mathbf{B}/dt$ is perpendicular to **B**.
(*b*) Why are there scalars x and y such that $d\mathbf{B}/dt = x\mathbf{T} + y\mathbf{N}$?
(*c*) By differentiating $\mathbf{B} \cdot \mathbf{T} = 0$, show that $d\mathbf{B}/dt$ is perpendicular to **T**, hence $d\mathbf{B}/dt = y\mathbf{N}$.

41 Assume $\mathbf{B}(t) = \mathbf{c}$, a constant.
(*a*) Deduce that $\mathbf{c} \cdot \mathbf{v} = 0$, where **v** is the velocity vector.
(*b*) Deduce that the path lies in a plane.

42 By Exercise 40 there is a scalar τ (tau) such that $d\mathbf{B}/dt = -\tau\mathbf{N}$. By Exercise 41 if τ is always 0, the curve lies in a plane. Hence τ is a measure of the tendency of the plane determined by **T** and **N** to change direction. Thus τ serves to measure the tendency of the curve to twist out of the plane determined by **T** and **N**. For this reason τ is called the **torsion** of the curve.
(*a*) Show that $d\mathbf{N}/ds = p\mathbf{T} + q\mathbf{B}$ for some scalars p and q.
(*b*) By differentiating the equation $\mathbf{B} \cdot \mathbf{N} = 0$, show that $q = \tau$.
(*c*) By differentiating the equation $\mathbf{N} \cdot \mathbf{T} = 0$, show that $p = -\kappa$.

Thus $d\mathbf{N}/ds = -\kappa\mathbf{T} + \tau\mathbf{B}$. The formulas for $d\mathbf{T}/ds$, $d\mathbf{B}/ds$, and $d\mathbf{N}/ds$ are known as the **Frenet formulas**: $d\mathbf{T}/ds = \kappa\mathbf{N}$, $d\mathbf{N}/ds = -\kappa\mathbf{T} + \tau\mathbf{B}$, $d\mathbf{B}/ds = \tau\mathbf{N}$.

43 (*a*) Sketch the curve in space given by

$$\mathbf{r} = 3\cos 2\pi t\,\mathbf{i} + 3\sin 2\pi t\,\mathbf{j} + 3t\mathbf{k}.$$

(*b*) Find **v** and **a**.
(*c*) Find the curvature κ. (See Exercise 38.)
(*d*) Compute a_T and a_N.

13.5 NEWTON'S LAW IMPLIES KEPLER'S THREE LAWS

After hundreds of pages of computation based on observations by the astronomer Tycho Brahe in the last 30 years of the sixteenth century, plus lengthy detours and lucky guesses, Kepler arrived at these three laws of planetary motion:

KEPLER'S THREE LAWS

1. Every planet travels around the sun in an elliptical orbit such that the sun is situated at one focus (discovered 1605, published 1609).
2. The velocity of a planet varies in such a way that the line joining the planet to the sun sweeps out equal areas in equal times (discovered 1602, published 1609).
3. The square of the time required by a planet for one revolution around the sun is proportional to the cube of its mean distance from the sun (discovered 1618, published 1619).

The work of Kepler shattered the crystal spheres which for 2,000 years had carried the planets. Before him astronomers admitted only circular motion and motion compounded of circular motions. Copernicus, for instance, used five circles to describe the motion of Mars.

The ellipse got a cold reception.

The ellipse was not welcomed. In 1605 Kepler complained to a skeptical astronomer:

> You have disparaged my oval orbit. . . . If you are enraged because I cannot take away oval flight how much more you should be enraged by the motions assigned by the ancients, which I did take away. . . . You disdain my oval, a single cart of dung, while you endure a whole stable. (If indeed my oval is a cart of dung.)

But the astronomical tables that Kepler based on his theories, and published in 1627, proved to be more accurate than any other, and the ellipse gradually gained acceptance.

The three laws stood as mysteries alongside a related question: If there are no crystal spheres, what propels the planets? Bullialdus, a French mathematician, suggested in 1645:

The inverse square law was conjectured.

> That force with which the sun seizes or pulls the planets, a physical force which serves as hands for it, is sent out in straight lines into all the world's space . . . ; since it is physical it is decreased in greater space; . . . the ratio of this decrease is the same as that for light, namely as the reciprocal of the square of the distance.

In 1666, Hooke, more of an experimental scientist than a mathematician, wondered

> why the planets should move about the sun . . . being not included in any solid orbs . . . nor tied to it . . . by any visible strings. . . . I cannot imagine any other likely cause besides these two: The first may be from an unequal density of the medium . . . ; if we suppose that part of the medium, which is farthest from the centre, or sun, to be more dense outward, than that which is more near, it will follow, that the direct motion will be always deflected inwards, by the easier yielding of the inwards. . . .
>
> But the second cause of inflecting a direct motion into a curve may be from an attractive property of the body placed in the centre; whereby it continually endeavours to attract or draw it to itself. For if such a principle be supposed all the phenomena of the planets seem possible to be explained by the common principle of mechanic motions. . . . By this hypothesis, the phenomena of the comets as well as of the planets may be solved.

In 1674, Hooke, in an announcement to the Royal Society, went further:

> All celestial bodies have an attraction towards their own centres, whereby they attract not only their own parts but also other celestial bodies that are within the sphere of their activity. . . . All bodies that are put into direct simple motion will so continue to move forward in a straight line till they are, by some other effectual powers, deflected and bent into a motion describing a circle, ellipse, or some other more compound curve. . . . These attractive powers are much more powerful in operating by how much the nearer the body wrought upon is to their own centres. . . . It is a notion which if fully prosecuted as it ought to be, will mightily assist the astronomer to reduce all the celestial motions to a certain rule. . . .

Hooke pressed Newton to work on the problem.

Trying to interest Newton in the question, Hooke wrote on November 24, 1679: "I shall take it as a great favor if . . . you will let me know your thoughts of that of compounding the celestial motion of planets of a direct motion by the tangent and an attractive motion toward the central body." But four days later, Newton replied:

> My affection to philosophy [science] being worn out, so that I am almost as little concerned about it as one tradesman uses to be about another man's trade or a countryman about learning, I must acknowledge myself averse from spending that time in writing about it which I think I can spend otherwise more to my own content and the good of others. . . .

In a letter to Newton on January 17, 1680, Hooke returned to the problem of planetary motion:

> It now remains to know the properties of a curved line (not circular . . .) made by a central attractive power which makes the velocities of descent from the tangent line or equal straight motion at all distances in a duplicate proportion to the distances reciprocally taken. I doubt not that by your excellent method you will easily find out what that curve must be, and its properties, and suggest a physical reason of this proportion.

Hooke succeeded in drawing Newton back to science, as Newton himself admitted in his *Principia,* published in 1687: "I am beholden to him only for the diversion he gave me from my other studies to think on these things and for his dogmaticalness in writing as if he had found the motion in the ellipse, which inclined me to try it."

It seems that Newton then obtained a proof—perhaps containing a mistake (the history is not clear)—that if the motion is elliptical, the force varies as the inverse square. In 1684, at the request of the astronomer Halley, Newton provided a correct proof. With Halley's encouragement, Newton spent the next year and a half writing the *Principia*.

Halley, of Halley's comet, paid for publication of the Principia.

In the *Principia,* which develops the science of mechanics and applies it to celestial motions, Newton begins with two laws:

1 Every body continues in its state of rest, or of uniform motion in a straight line, unless it is compelled to change this state by forces impressed upon it.
2 The change of momentum is proportional to the motive force impressed, and is made in the direction of the straight line in which that force is impressed.

To state these in the language of vectors, let $\mathbf{v}$ be the velocity of the body, $\mathbf{F}$ the impressed force, and m the mass of the body. The first law asserts that $\mathbf{v}$ is constant if $\mathbf{F}$ is $\mathbf{0}$. **Momentum** is defined as $m\mathbf{v}$; the second law asserts that

$$\mathbf{F} = \frac{d}{dt}(m\mathbf{v}).$$

If m is constant, this reduces to

$$\mathbf{F} = m\mathbf{a},$$

where $\mathbf{a}$ is the acceleration vector.

Newton assumed a universal **law of gravitation**: Any particle P exerts an attractive force on any other particle Q, and the direction of the force is from Q

toward P. Then *assuming* that the orbit of a planet moving about the sun (both treated as points) is an ellipse, he *deduced* that this force is inversely proportional to the square of the distance between the particles P and Q.

Nowhere in the *Principia* does he deduce from the inverse-square law of gravity that the planets' orbits are ellipses. (However, there are general theorems in the *Principia* on the basis of which this deduction could have been made.) In the *Principia* he showed that Kepler's second law (concerning areas) was equivalent to the assumption that the force acting on a planet is directed toward the sun. Finally, he deduced Kepler's third law.

Newton's universal law of gravitation asserts that any particle, of mass M, exerts a force on any other particle, of mass m, and that the magnitude of this force is proportional to the product of the two masses, mM, inversely proportional to the square of the distance between them, and is directed toward the particle of mass M.

We will assume that the sun is fixed at O.

Assume that the sun has mass M and is located at point O and that the planet has mass m and is located at point P. (See Fig. 1.) Let $\mathbf{r} = \overrightarrow{OP}$ and $r = \|\mathbf{r}\|$. Then the sun exerts a force $\mathbf{F}$ on the planet given by the formula

$$\mathbf{F} = -\frac{GMm}{r^3}\mathbf{r}, \tag{1}$$

where G is a universal constant. It is convenient to introduce the unit vector $\mathbf{u} = \mathbf{r}/|\mathbf{r}|$, which points in the direction of $\mathbf{r}$. Then Eq. (1) reads

$$\mathbf{F} = -\frac{GMm}{r^2}\mathbf{u}.$$

Now, $\mathbf{F} = m\mathbf{a}$, where $\mathbf{a}$ is the acceleration vector of the planet. Thus

$$m\mathbf{a} = -\frac{GMm}{r^2}\mathbf{u},$$

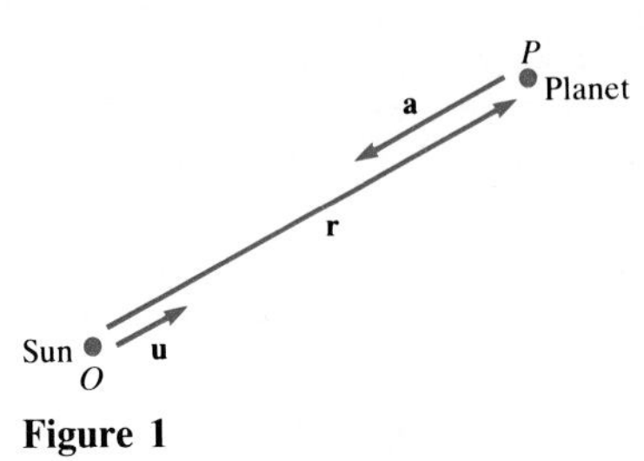

Figure 1

from which it follows that

$$\mathbf{a} = -\frac{q\mathbf{u}}{r^2}, \tag{2}$$

where $q = GM$ is independent of the planet.

The vectors $\mathbf{u}$, $\mathbf{r}$, and $\mathbf{a}$ are indicated in Fig. 1.

The following exercises show how to obtain Kepler's three laws from the single law of Newton, $\mathbf{a} = -q\mathbf{u}/r^2$. Complete solutions are in the *Student's Solutions Manual*.

EXERCISES FOR SEC. 13.5: NEWTON'S LAW IMPLIES KEPLER'S THREE LAWS

Exercises 1–3 obtain Kepler's "area" law.

1 Let $\mathbf{r}(t)$ be the position vector of a given planet at time t. Let $\Delta\mathbf{r} = \mathbf{r}(t + \Delta t) - \mathbf{r}(t)$. Show that for small Δt,

$$\frac{1}{2}\|\mathbf{r} \times \Delta\mathbf{r}\|$$

approximates the area swept out by the position vector during the small interval of time Δt. *Hint:* Draw a picture.

2 From Exercise 1 deduce that $\frac{1}{2}\|\mathbf{r} \times d\mathbf{r}/dt\|$ is the rate at which the position vector $\mathbf{r}$ sweeps out area. (See Fig. 2.)

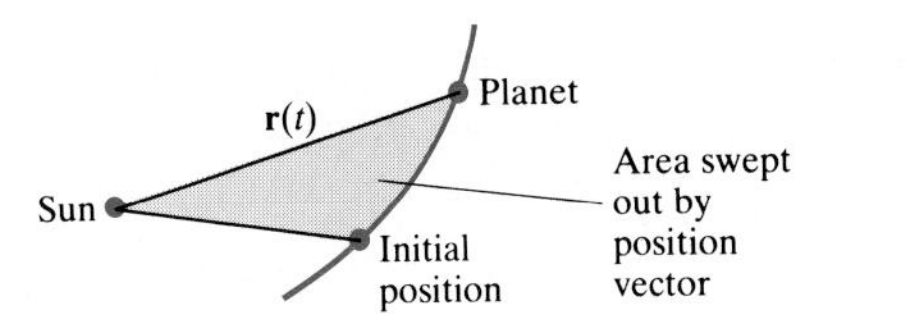

Figure 2

Let $\mathbf{v} = d\mathbf{r}/dt$. The vector $\mathbf{r} \times \mathbf{v}$, introduced in Exercise 2, will play a central role in the argument.

3 With the aid of Eq. (2), show that the vector $\mathbf{r} \times \mathbf{v}$ is constant, independent of time.

Since $\mathbf{r} \times \mathbf{v}$ is constant, $\frac{1}{2}\|\mathbf{r} \times \mathbf{v}\|$ is constant. In view of Exercise 2, it follows that the radius vector of a given planet sweeps out area at a constant rate. To put it another way, the radius vector sweeps out equal areas in equal times. This is Kepler's second law.

Introduce an xyz-coordinate system such that the unit vector $\mathbf{k}$, which points in the direction of the positive z axis, has the same direction as the constant vector $\mathbf{r} \times \mathbf{v}$. Thus there is a positive constant h such that

$$\mathbf{r} \times \mathbf{v} = h\mathbf{k}. \tag{3}$$

Exercises 4–13 obtain Kepler's "ellipse" law.

4 Show that h in Eq. (3) is twice the rate at which the position vector of the planet sweeps out area.

5 Show that the planet remains in the plane perpendicular to $\mathbf{k}$ that passes through the sun.

By Exercise 5, the orbit of the planet is planar. We may assume that the orbit lies in the xy plane; for convenience, locate the origin of the xy coordinates at the sun. Also introduce polar coordinates in this plane, with the pole at the sun and the polar axis along the positive x axis, as in Fig. 3.

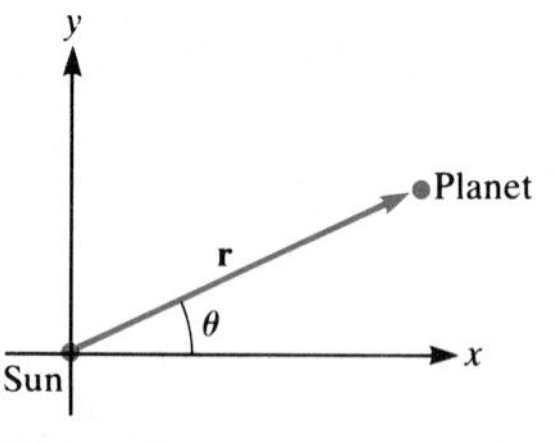

Figure 3

6 (a) Show that during the time interval $[t_0, t]$ the position vector of the planet sweeps out the area

$$\frac{1}{2}\int_{t_0}^{t} r^2 \frac{d\theta}{dt}\, dt.$$

(b) From (a) deduce that the radius vector sweeps out area at the rate $\frac{1}{2}r^2\, d\theta/dt$.

Henceforth use the dot notation for differentiation with respect to time. Thus $\dot{\mathbf{r}} = \mathbf{v}$, $\dot{\mathbf{v}} = \mathbf{a}$, and $\dot{\theta} = d\theta/dt$.

7 Show that $\mathbf{r} \times \mathbf{v} = r^2\dot{\theta}\,\mathbf{k}$.

8 Show that $\dot{\mathbf{u}} = (d\mathbf{u}/d\theta)\dot{\theta}$ and is perpendicular to $\mathbf{u}$. Recall that $\mathbf{u}$ is defined as $\mathbf{r}/\|\mathbf{r}\|$.

9 Recalling that $\mathbf{r} = r\mathbf{u}$, show that $h\mathbf{k} = r^2(\mathbf{u} \times \dot{\mathbf{u}})$.

10 Using Eq. (2) and Exercise 9, show that $\mathbf{a} \times h\mathbf{k} = q\dot{\mathbf{u}}$. [*Hint:* What is the vector identity for $\mathbf{A} \times (\mathbf{B} \times \mathbf{C})$?]

11 Deduce from Exercise 10 that $\mathbf{v} \times h\mathbf{k}$ and $q\mathbf{u}$ differ by a constant vector.

By Exercise 11, there is a constant vector $\mathbf{C}$ such that

$$\mathbf{v} \times h\mathbf{k} = q\mathbf{u} + \mathbf{C}. \tag{4}$$

Rotate the coordinate system shown in Fig. 3 in such a way that the polar axis points in the direction of $\mathbf{C}$. Then the angle between $\mathbf{r}$ and $\mathbf{C}$ is the angle θ of polar coordinates.

The next exercise requires the vector identity $(\mathbf{A} \times \mathbf{B}) \cdot \mathbf{C} = \mathbf{A} \cdot (\mathbf{B} \times \mathbf{C})$, which is valid for any three vectors $\mathbf{A}$, $\mathbf{B}$, and $\mathbf{C}$.

12 (a) Show that $(\mathbf{r} \times \mathbf{v}) \cdot h\mathbf{k} = h^2$.
(b) Show that $\mathbf{r} \cdot (\mathbf{v} \times h\mathbf{k}) = rq + \mathbf{r} \cdot \mathbf{C}$.
(c) Combining (a) and (b), deduce that $h^2 = rq + rc \cos \theta$, where $c = \|\mathbf{C}\|$.

It follows from Exercise 12 that the polar equation for the orbit of the planet is given by

$$r(\theta) = \frac{h^2}{q + c \cos \theta}. \tag{5}$$

13 By expressing Eq. (5) in rectangular coordinates, show that it describes a conic section.

Since the orbit of a planet is bounded and is also a conic section, it must be an ellipse. This establishes Kepler's first law.

Kepler's third law asserts that the square of the time required for a planet to complete one orbit is proportional to the cube of its mean distance from the sun.

First the term "mean distance" must be defined. For Kepler this meant the average of the shortest distance and the longest distance from the planet to the sun in its orbit. Let us compute this average for the ellipse of semimajor axis a and semiminor axis b, shown in Fig. 4. The sun is at the focus F, which is also the pole of the polar coordinate system we are using. The line through the two foci contains the polar axis.

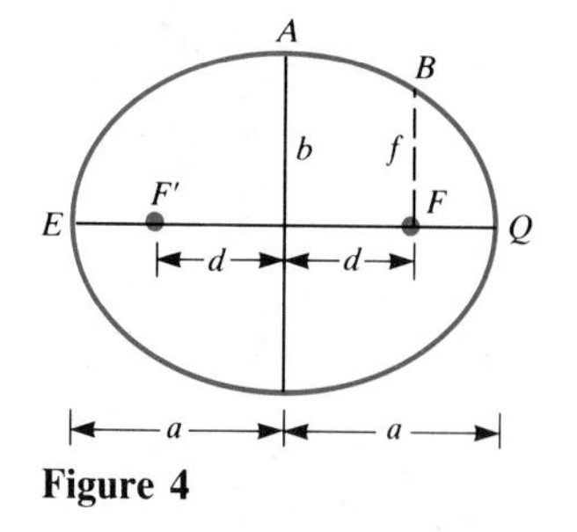

Figure 4

Recall that an ellipse is the set of points P such that the sum of the distances from P to the two foci F and F' is constant, $2a$. The shortest distance from the planet to the sun is $\overline{FQ} = a - d$ and the longest distance is $\overline{EF} = a + d$. Thus Kepler's mean distance is

$$\frac{(a-d)+(a+d)}{2} = a.$$

Now let T be the time required by the given planet to complete one orbit. Kepler's third law asserts that T^2 is proportional to a^3. Exercises 14 to 18 establish this law by showing that T^2/a^3 is the same for all planets.

14 Using the fact that the area of the ellipse in Fig. 4 is πab, show that $Th/2 = \pi ab$, hence that

$$T = \frac{2\pi ab}{h}. \tag{6}$$

The rest of the argument depends only on Eqs. (5) and (6) and the "fixed sum of two distances" property of an ellipse.

15 Using Eq. (5), show that f in Fig. 4 equals h^2/q.

16 Show that $b^2 = af$, as follows:
  (a) From the fact that $\overline{F'A} + \overline{FA} = 2a$, deduce that $a^2 = b^2 + d^2$.
  (b) From the fact that $\overline{F'B} + \overline{FB} = 2a$, deduce that $d^2 = a^2 - af$.
  (c) From (a) and (b), deduce that $b^2 = af$.

17 From Exercises 15 and 16, deduce that $b^2 = ah^2/q$.

18 Combining Eq. (6) and Exercise 17, show that

$$\frac{T^2}{a^3} = \frac{4\pi^2}{q}.$$

Since $4\pi^2/q$ is a constant, the same for all planets, Kepler's third law is established.

FOR FURTHER READING

Isaac Newton, *Principia,* University of California, Berkeley, 1947.

V. Frederick Reckey, "Isaac Newton: Man, Myth, and Mathematician," *College Mathematics Journal,* Vol. 18, 362–388, 1987.

J. L. Russell, "Kepler's Laws of Planetary Motion 1609–1666," *British Journal for the History of Science,* Vol. 2, 1–24, 1964.

D. T. Whiteside, "Newton's Early Thoughts on Planetary Motion: A Fresh Look," *British Journal for the History of Science,* Vol. 2, 117–137, 1964.

A. Koestler, *The Watershed,* Doubleday, New York, 1960.

M. Caspar, *Kepler,* Hellman, New York, 1959.

J. Kepler, "Copernican astronomy," *Great Books of the Western World,* Encyclopaedia Britannica, Chicago, 1952.

13.S SUMMARY

In this chapter we defined and applied the derivative of a vector function **G,** which assigns to a scalar t a vector $\mathbf{G}(t)$. $\mathbf{G}(t)$ may be thought of as the position vector **r** in the plane or in space of a moving particle at time t.

The derivative of **G** is defined as

$$\lim_{\Delta t \to 0} \frac{\Delta \mathbf{G}}{\Delta t}.$$

The mathematician denotes it $\mathbf{G}'$, but a physicist denotes it **v** and calls it the velocity vector, for its length is the speed of the moving particle and its direction is that in which the particle is moving at a given instant.

The derivative of a vector function has many useful properties that resemble those of the ordinary derivative. For instance

$$(\mathbf{G} \cdot \mathbf{H})' = \mathbf{G} \cdot \mathbf{H}' + \mathbf{H} \cdot \mathbf{G}'.$$

In particular, if $\mathbf{G} = \mathbf{H}$, then

$$(\mathbf{G} \cdot \mathbf{G})' = 2\mathbf{G} \cdot \mathbf{G}'.$$

From this it follows that if $\|\mathbf{G}\|$ is constant, then $\mathbf{G}'(t)$ is perpendicular to $\mathbf{G}(t)$ for all t.

If arc length s is used as the parameter then $\mathbf{G}'(s)$ has length 1 ($\|\mathbf{v}\| = 1$).

The acceleration vector is the second derivative of $\mathbf{G}$, $\mathbf{a} = (\mathbf{G}')' = \mathbf{v}'$. If $\mathbf{G}$ is given in components relative to the basic unit vectors $\mathbf{i}$ and $\mathbf{j}$,

$$\mathbf{G}(t) = x(t)\mathbf{i} + y(t)\mathbf{j},$$

then

$$\mathbf{G}'(t) = \frac{dx}{dt}\mathbf{i} + \frac{dy}{dt}\mathbf{j}$$

and

$$\mathbf{G}''(t) = \frac{d^2x}{dt^2}\mathbf{i} + \frac{d^2y}{dt^2}\mathbf{j}.$$

In the study of a particle moving in the xy plane, two other perpendicular unit vectors are of interest: $\mathbf{T}$, pointing straight ahead in the direction of motion, and $\mathbf{N}$, pointing in the direction in which the particle is turning. $\mathbf{T}$ and $\mathbf{N}$ depend on time. It turns out that

$$\mathbf{a} = \frac{d^2s}{dt^2}\mathbf{T} + \frac{v^2}{r}\mathbf{N},$$

where s is arc length and r is radius of curvature.

From this equation we obtain an expression for curvature in terms of $\mathbf{a}$ and $\mathbf{v}$:

$$\kappa = \frac{\|\mathbf{v} \times \mathbf{a}\|}{\|\mathbf{v}\|^3}.$$

The optional section on Newton and Kepler is not reviewed in this summary.

Vectors are used for two purposes in this chapter. First, they are a bookkeeping device. Instead of writing the three parametric equations $x = f(t)$, $y = g(t)$, $z = h(t)$, we write one vector equation $\mathbf{r} = \mathbf{G}(t)$.

Second, vectors provide a language in which the symbols closely correspond to intuitive concepts. For instance, instead of saying "The x coordinate changes at the rate of 3 feet per second and the y coordinate changes at the rate of 4 feet per second," we say simply "The velocity vector $\mathbf{v}$ is $3\mathbf{i} + 4\mathbf{j}$."

This table records some of the concepts in both mathematical and physical terms:

Mathematical Formulation	*Physical Interpretation*
Vector function $\mathbf{G}$	Parameterized path of moving particle
Derivative of $\mathbf{G}$	Velocity vector $\mathbf{v}$
Second derivative of $\mathbf{G}$	Acceleration vector $\mathbf{a}$
$\mathbf{T} = \dfrac{\mathbf{v}}{\|\mathbf{v}\|}$	Unit vector pointing straight ahead
$\mathbf{N} = \dfrac{d\mathbf{T}/dt}{\|d\mathbf{T}/dt\|}$	Unit vector perpendicular to path and pointing in the direction the path is veering
$\mathbf{a} = \dfrac{d^2s}{dt^2}\mathbf{T} + \dfrac{v^2}{r}\mathbf{N}$	Acceleration vector, sum of two vectors: one, the acceleration if the particle were moving in a straight line; the other, the acceleration if the particle were moving at a constant speed on a circle whose radius is the radius of curvature

Vocabulary and Symbols

vector function $\mathbf{G}$, $\mathbf{r}$
limit of a vector function
continuity of a vector function
position vector $\mathbf{r}$
derivative of $\mathbf{G}$, $\mathbf{G}'$
velocity vector $\mathbf{G}'$, $\mathbf{v}$
acceleration vector $\mathbf{G}''$, $\mathbf{a}$
unit tangent vector $\mathbf{T}$
principal unit normal vector $\mathbf{N}$
scalar components of $\mathbf{a}$, a_T, and a_N, along $\mathbf{T}$ and $\mathbf{N}$

GUIDE QUIZ ON CHAP. 13: THE DERIVATIVE OF A VECTOR FUNCTION

1 Let $\mathbf{G}(t)$ be a vector function.
(a) Define $\mathbf{G}'(t)$.
(b) Assuming that $\mathbf{G}(1) = 3\mathbf{i} + 4\mathbf{j}$ and $\mathbf{G}(1.01) = 3.02\mathbf{i} + 4.03\mathbf{j}$, estimate $\mathbf{G}'(1)$.

2 Let $\mathbf{r} = \mathbf{G}(t)$ describe the position of a particle at time t. By drawing $\Delta\mathbf{G}$ for small positive Δt, explain why $\|\mathbf{G}'(t)\|$ is the speed of the particle.

3 Let $\mathbf{G}(s)$ be a vector function, where s is arc length along the curve. Explain why $d\mathbf{G}/ds$ is a unit vector
(a) using formulas in Chap. 9;
(b) using pictures, showing $\Delta\mathbf{G}$ and Δs for small positive Δs.

4 Define the vectors $\mathbf{T}$ and $\mathbf{N}$ and explain why $d\mathbf{T}/dt = (v/r)\mathbf{N}$, where r is the radius of curvature and v is the speed.

5 Explain why $\mathbf{a} = (d^2s/dt^2)\mathbf{T} + (v^2/r)\mathbf{N}$.

6 A particle moving along a curve has $\mathbf{v} = 3\mathbf{i} + 4\mathbf{j}$ and $\mathbf{a} = 2\mathbf{i} + \mathbf{j}$ at a certain instant. Find (a) the curvature κ and (b) the radius of curvature at that instant. (c) Find the tangential component of acceleration at that instant. (d) Is the particle speeding up or slowing down? (e) Find the normal component of acceleration.

7 Let $\mathbf{r} = t^2\mathbf{i} + t\,\mathbf{j}$. When $t = 1$, find
(a) $\mathbf{T}$
(b) $\mathbf{N}$
(c) $\mathbf{v}$
(d) d^2s/dt^2
(e) κ
(f) the radius of curvature.

8 Explain why $\kappa = \dfrac{\|\mathbf{v} \times \mathbf{a}\|}{\|\mathbf{v}\|^3}$.

REVIEW EXERCISES FOR CHAP. 13: THE DERIVATIVE OF A VECTOR FUNCTION

You may also use the exercises in Sec. 13.5 for review.

1 Let

$$\mathbf{G}(t) = \frac{e^t + e^{-t}}{2}\mathbf{i} + \frac{e^t - e^{-t}}{2}\mathbf{j}.$$

(a) Show that the particle moves on the curve $x^2 - y^2 = 1$.
(b) Find a_T and a_N when $t = 1$.

2 An astronaut traveling on the path $\mathbf{G}(t) = t\mathbf{i} + t^2\mathbf{j} + t^3\mathbf{k}$ shuts off her engine when at the point (1, 1, 1).
(a) Show that she passes through the point (3, 5, 7).
(b) How near does she get to the point (5, 8, 9)?

3 At time t the position vector of a particle is

$$\mathbf{G}(t) = t\mathbf{i} + 3t\mathbf{j} + 4t\mathbf{k}.$$

(a) Show that any point in the path lies in the plane $3x = y$ and also in the plane $4y = 3z$.
(b) Sketch the planes in (a) and indicate their intersection, which is the path of the particle.

4 At a certain moment a particle moving on a curve has $\mathbf{v} = 2\mathbf{i} + 3\mathbf{j}$ and $\mathbf{a} = -\mathbf{i} + 2\mathbf{j}$.
(a) What is the speed of the particle?
(b) Find $\mathbf{T}$ and $\mathbf{N}$, expressing each in the form $x\mathbf{i} + y\mathbf{j}$, for suitable scalars x and y.
(c) Draw $\mathbf{v}$, $\mathbf{a}$, $\mathbf{T}$, and $\mathbf{N}$.
(d) Compute $\mathbf{a} \cdot \mathbf{N}$.
(e) From (a) and (d) obtain the radius of curvature.
(f) Compute $\mathbf{a} \cdot \mathbf{T}$.
(g) Is the particle speeding up or slowing down? Compute d^2s/dt^2.
(h) Find dy/dt.
(i) Find dy/dx.
(j) Find d^2y/dt^2.
(k) Obtain the curvature directly from $\mathbf{v}$ and $\mathbf{a}$.

5 At time $t = 0$ the velocity vector of a certain particle is $0\mathbf{i} + 1\mathbf{j}$. If the acceleration vector $\mathbf{a}(t)$ is $6t\mathbf{i} + e^t\mathbf{j}$, find (a) $\mathbf{v}$ and $\mathbf{a}$ when $t = 1$; (b) v, a_T, a_N, and r (the radius of curvature) when $t = 1$.

6 (a) Show that, when $t = 1$, the particles on the paths $\mathbf{G}(t) = t^2\mathbf{i} + 6t\mathbf{j} + \mathbf{k}$ and $\mathbf{H}(t) = t^3\mathbf{i} + (t + 5)\mathbf{j} + \mathbf{k}$ collide.
(b) At what angle do they collide?

7 Let $\mathbf{G}(t) = e^t \cos t\ \mathbf{i} + e^t \sin t\ \mathbf{j} + e^t\ \mathbf{k}$.
(a) Find the speed at time t.
(b) Find the distance traveled during the time interval [0, 1].

8 A particle moves on a straight line at a constant speed of 4 feet per second in the direction of the unit vector

$$\frac{\mathbf{i}}{3} + \frac{2\mathbf{j}}{3} + \frac{2\mathbf{k}}{3}.$$

At time $t = 0$ the particle is at the point (2, 5, 7). Find its position vector at time t.

9 A particle moves along a straight line at a constant speed. At time $t = 0$ it is at the point (1, 2, 1), and at time $t = 2$ it is at (4, 1, 5).
(a) Find its speed.
(b) Find the direction cosines of the line.
(c) Find the position vector of the particle at time t.

10 An engineer wishes to add a curved track to join the tracks shown in Fig. 1.
(a) Show that she could add an arc of a circle in such a way that it has the same slope at (1, 0) and at (−1, 0) as the tracks there.
(b) If a train traveled from left to right on these tracks, what would a passenger experience passing through (−1, 0)?

Take two cases: a passenger standing up and a passenger leaning against the left side of the railway car.

(*c*) To avoid the consequences in (*b*) the engineer joins the two straight sections by a curved track given by the equation $y = A + Bx^2 + Cx^4$. How should she choose A, B, and C so that the passengers will have a smooth ride?

(Tables have been published that describe how to join straight sections of track.)

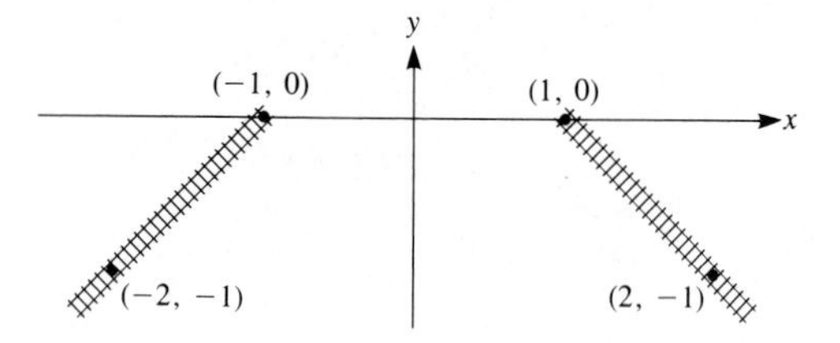

Figure 1

11 At time t an astronaut has the position vector $\mathbf{r}(t) = \sin 3t\ \mathbf{i} + \tan t\ \mathbf{j} + e^{2t}\ \mathbf{k}$. When $t = \pi/4$ he shuts off his rockets and coasts off along the tangent line. Where will he meet the plane $x + 2y + 3z = 100$?

12 One astronaut is traveling along the curve given by $\mathbf{G}(t) = (t^2 - 1)\mathbf{i} + t\mathbf{j} + t\mathbf{k}$ and another along $\mathbf{F}(t) = \cos \pi t\ \mathbf{i} + \sin \pi t\ \mathbf{j} + t\mathbf{k}$. When $t = 1$ the first astronaut shuts off her rockets and coasts along the tangent line. When $t = 2$, the second astronaut shuts off his rockets and coasts along the tangent line. What is the closest that the astronauts are to each other when they are coasting?

13 Let $\mathbf{G}(t)$ be a vector function. Assume that $\mathbf{G}'(t) = 0$, for all t. Show that $\mathbf{G}(t)$ is constant.

14 A particle is moving along the curve $\mathbf{r} = \mathbf{G}(t)$. Assume that the acceleration vector is always pointed at the origin $(0, 0, 0)$.

(*a*) Show that $\mathbf{r} \times \mathbf{v}$ is constant.

(*b*) Show that the path of the particle lies in a plane.

15 Express a_N in terms of $\mathbf{a} \times \mathbf{v}$ and $\mathbf{v}$.

16 (Contributed by Michael Stora.) Express $\mathbf{N}$ in terms of $\mathbf{a}$ and $\mathbf{v}$, using dot products but not cross products. *Hint:* Draw a picture.

14

PARTIAL DERIVATIVES

Throughout the first 13 chapters we were concerned with functions of one variable, their derivatives, and their integrals. But many quantities are functions of more than one variable. For instance, the volume of a cylindrical can of radius r and height h is $\pi r^2 h$, a function of two variables r and h. The volume of a rectangular box of sides l, w, and h is the product lwh, a function of three variables. In the present chapter, we discuss functions of several variables. Throughout we will assume that the functions are "well behaved," that is, they have partial derivatives of all orders, a concept defined in Sec. 14.5.

This chapter generalizes ideas we met in Chaps. 2, 3, and 4 concerning functions of one variable to functions of more than one variable.

Sections 14.1, 14.2, and 14.3 concern the graphs of equations and functions involving several variables. There is no calculus here, just practice in drawing surfaces in space.

Section 14.4 defines limits and continuity for functions of more than one variable.

Section 14.5 defines the partial derivatives of such functions. It is the analog of Sec. 3.2.

Section 14.6 develops the chain rule; it is the analog of Sec. 3.6.

Section 14.8 defines the tangent plane to a surface and the differential dz. This is the analog of Sec. 4.9, which relates a tangent line to a curve and the differential dy.

Section 14.7 generalizes partial derivatives, while Secs. 14.9 and 14.10 are concerned with finding extrema. These latter two sections generalize the tests for extrema discussed in Chap. 4.

Section 14.11 treats a special, but important, case of the chain rule.

14.1 GRAPHS

As early as Chap. 1 we graphed equations in two variables. For instance, the graph of $y = x^2$ is a parabola, the graph of $x^2 + y^2 = 1$ is a circle, and, as we saw in Chap. 12, the graph of $Ax + By + C = 0$ (not both A and B zero) is a line. In Chap. 12 we also graphed an equation in three variables, namely $Ax + By + Cz + D = 0$ (not all A, B, and C zero); the graph is a plane. In this section we graph some other equations, partly to become familiar with certain surfaces, partly to develop our ability to draw geometric objects in space.

Planes

The set of points (x, y, z) that satisfy some given equation in x, y, and z is called the **graph** of that equation. For instance, the graph of $Ax + By + Cz + D = 0$, where A, B, and C are not all 0, is a plane. A few examples show how to draw planes.

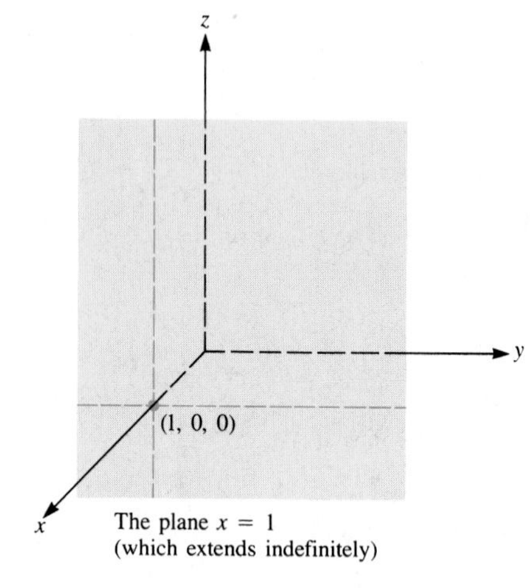

Figure 1

EXAMPLE 1 Sketch the plane $x = 1$.

SOLUTION A point (x, y, z) satisfies the equation $x = 1$ if its x coordinate is 1. Such a point is one unit away from the yz plane and in front of it, as we view it. It is sketched in Fig. 1. Note carefully the way hidden parts of the axes are dashed. Also note the inclusion of the point $(1, 0, 0)$ and the extra lines through it to show where the plane is situated. ■

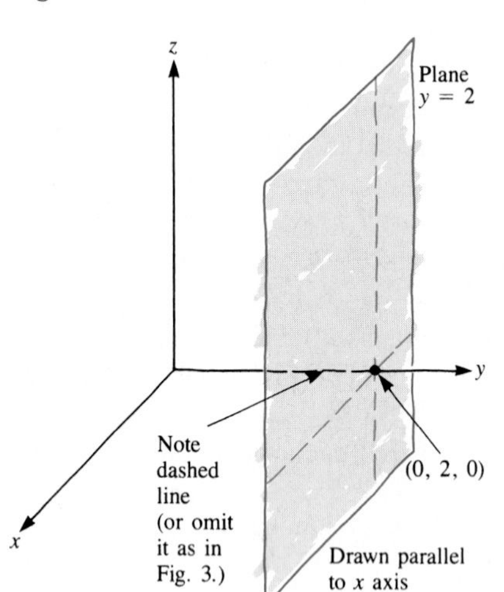

Figure 2

EXAMPLE 2 Sketch the plane $y = 2$.

SOLUTION This plane is parallel to the xz plane and two units to the right of it. It is shown in Figs. 2 and 3. ■

EXAMPLE 3 Sketch the plane $x = y$.

SOLUTION Note that if $(x, y, 0)$ is on the graph, so is (x, y, z) for any z. For instance, $(2, 2, 0)$ is on the graph; so are $(2, 2, 1)$ and $(2, 2, -1)$. In fact, $(2, 2, z)$ is on the graph for any z. (See Fig. 4.)

The equation $x = y$ puts no demands on z, only on x and y. This suggests that we first look at those points (x, y, z) on the graph of $x = y$ for which $z = 0$. This is the line $x = y$ in the xy plane. Then the graph consists of all points in vertical lines through this line, as shown in Fig. 5. Figure 5 depicts only the part of the plane $y = x$ that lies in the "first octant," where x, y, and z are positive. But that small part is enough to indicate the location of the plane. ■

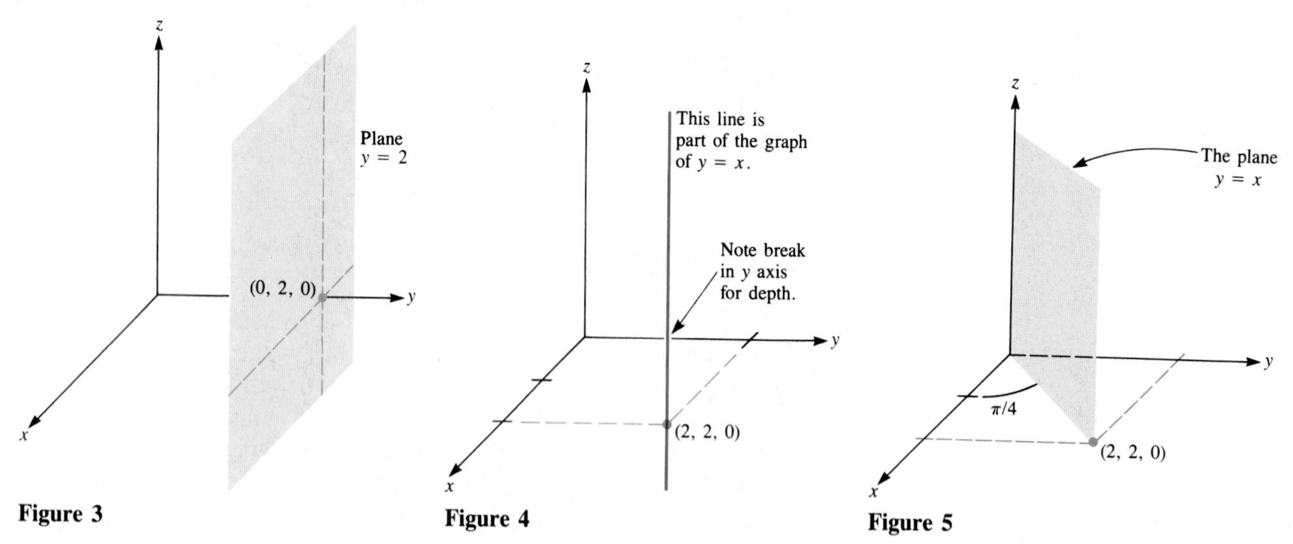

Figure 3 **Figure 4** **Figure 5**

EXAMPLE 4 Graph the plane $x/1 + y/2 + z/3 = 1$.

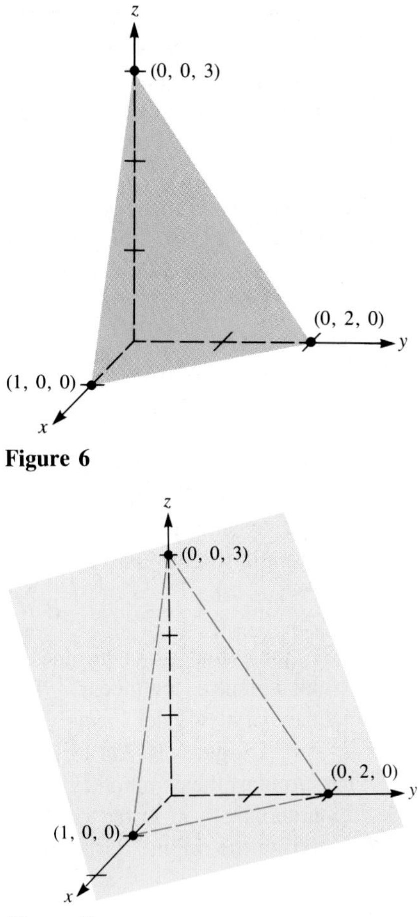

Figure 6

Figure 7

SOLUTION This plane turns out not to be parallel to any of the axes. Since one tilted plane looks much like any other, special care is needed when drawing it. One way is to show where it cuts the three axes. These intercepts are found by setting two of the variables x, y, and z in the equation $x/1 + y/2 + z/3 = 1$ equal to zero, and solving for the third.

To find the x intercept of the above equation, set y and z equal to zero, getting

$$\frac{x}{1} + \frac{0}{2} + \frac{0}{3} = 1$$

or $x = 1$. Therefore the point $(1, 0, 0)$ lies in the tilted plane. Similarly, so do the points $(0, 2, 0)$ and $(0, 0, 3)$. The sketch of the plane in Fig. 6 emphasizes these three points. Figure 7 shows another way to draw the plane. Notice in both figures broken lines or dashes to distinguish the visible from the hidden lines. ■

The Equation of a Sphere

The set of all points that are a fixed distance r from a given point (x_0, y_0, z_0) is a **sphere** of radius r and center (x_0, y_0, z_0). To sketch this sphere, show the horizontal equator as is done in Fig. 8. Draw it large enough so that there is room to label its center and radius. Add a little shading if you like.

A point (x, y, z) is on this sphere when the distance between it and (x_0, y_0, z_0) is r, that is, when

$$\sqrt{(x - x_0)^2 + (y - y_0)^2 + (z - z_0)^2} = r.$$

The distance from $P_0 = (x_0, y_0, z_0)$ to $P = (x, y, z)$ is r, so

Equation of sphere of radius r and center (x_0, y_0, z_0)

$$(x - x_0)^2 + (y - y_0)^2 + (z - z_0)^2 = r^2.$$

Often the origin $(0, 0, 0)$ of the xyz coordinate system is placed at the center of the sphere. The equation of a sphere of radius r and center $(0, 0, 0)$ is simply

$$x^2 + y^2 + z^2 = r^2.$$

For instance, the equation

$$x^2 + y^2 + z^2 = 25$$

describes a sphere of radius 5 and center $(0, 0, 0)$. As another example, the equation

$$x^2 + y^2 + z^2 = 3$$

describes a sphere of radius $\sqrt{3}$ and center $(0, 0, 0)$.

Note that the sphere is a *surface*. The solid that it bounds is called a **ball**.

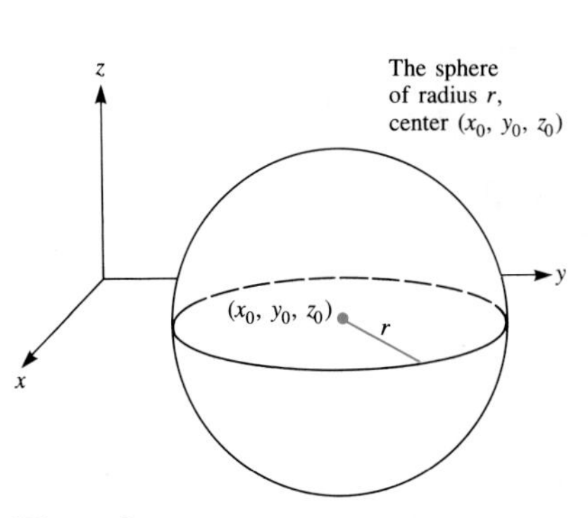

Figure 8

Cylinders

The graph of an equation that involves only one or two of the variables x, y, z is fairly easy to draw. For instance, say the equation involves only x and y. (There is no restriction on z.) This was the situation with the equation in Example 3. The point (x, y, z) is on the graph if and only if the point $(x, y, 0)$ is on the graph. This suggests the following procedure:

1 Graph the equation in x and y in the xy plane. (The graph is usually a curve.)
2 Graph the surface formed by all the lines parallel to the z axis that meet the graph drawn in step 1. (This is how we drew the graph of $x = y$.)

Figure 9 shows the typical case.
Graphs of the type formed in steps 1 and 2 are called right cylinders.

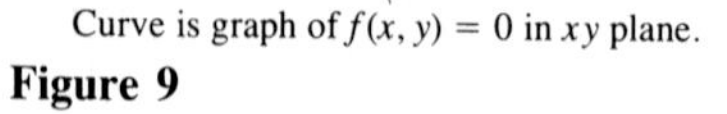

Figure 9

> **Definition** *Right cylinder.* Let R be a set in a plane. The set formed by all lines that are perpendicular to the given plane and that meet R is called the **right cylinder** determined by R. (Usually we say simply "cylinder," it being understood that the lines are at a right angle to the plane.)

For instance, if R is the line $x = y$ in the xy plane, the cylinder it determines is the plane $x = y$ in space. It may seem strange to call a plane a "cylinder." The **right circular cylinder** of daily life is the special case of a cylinder, namely the graph in space of an equation of the form $x^2 + y^2 = r^2$. Figures 10 and 11 show the graphs of two cylinders. The one in Fig. 11 is circular, the common cylinder of daily life. To draw Fig. 10, first sketch the parabola $z = x^2$ in the xz plane (lightly, in pencil). It should be tangent to the x axis at the origin. Use a ruler to draw the lines.

There is computer software for drawing surfaces. Figure 12 was obtained with the aid of Mathematica (Wolfram Research, Inc.).

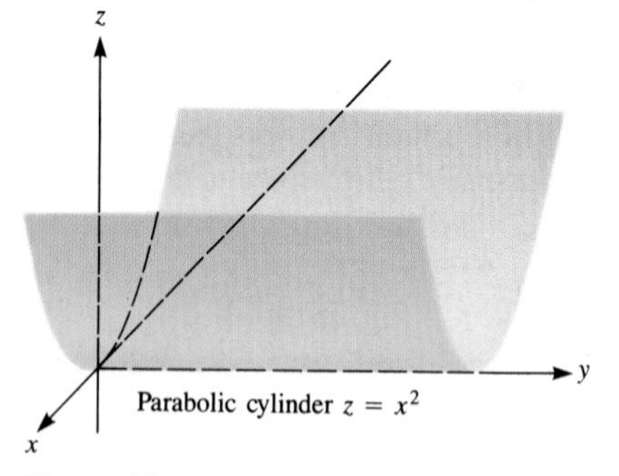

Figure 10

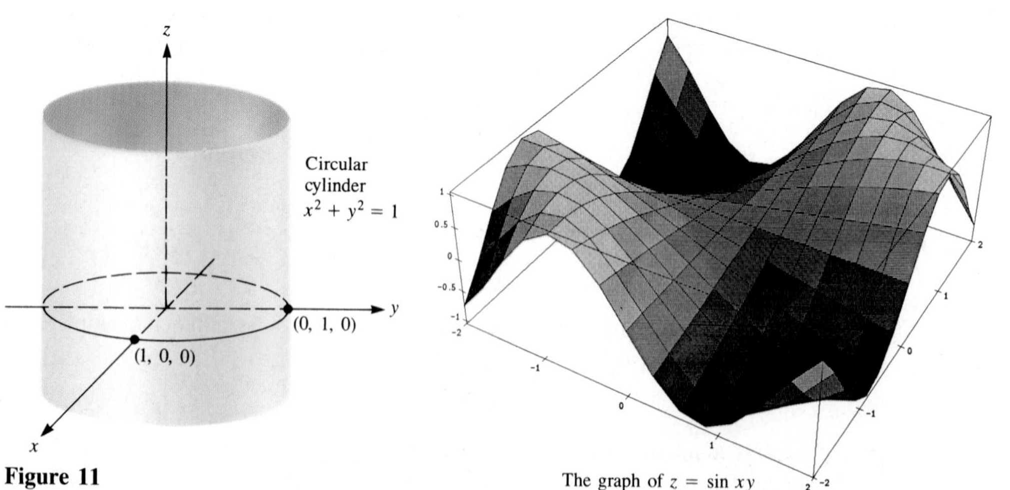

Figure 11

Figure 12

Surfaces of Revolution

Some surfaces are formed by revolving a planar curve around a line in the plane of the curve. For instance, a spherical surface can be obtained by revolving a semicircle about its diameter. Such a surface is called a **surface of revolution**. In Sec. 9.5 we found how to compute the area of such a surface. Now we will see how to find its equation.

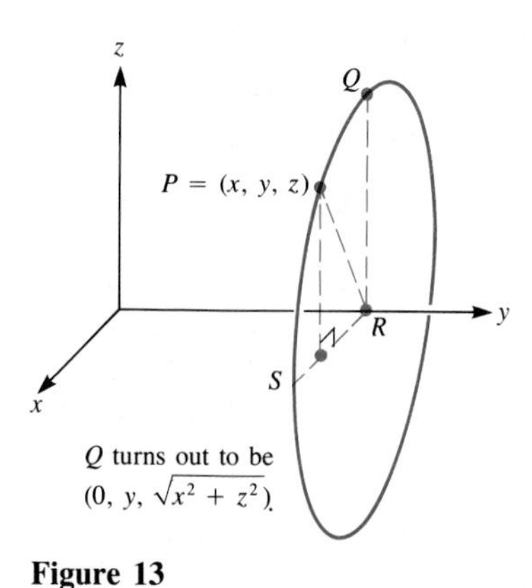

Figure 13

EXAMPLE 5 Consider the part of the line $z = y/2$ that lies in the first quadrant of the yz plane. Find the equation of the cone obtained when this line is revolved about the y axis.

SOLUTION Call the line that is revolved L. We wish to find an equation that tells us when the point $P = (x, y, z)$ lies on the cone. The point P on the cone comes from revolving a point Q on the line L. We will find the coordinates of Q. By using the fact that those coordinates satisfy a known equation ("the z coordinate is one-half the y coordinate") we then obtain an equation in x, y, and z. That equation tells when P lies on the cone.

Figure 13 shows $P = (x, y, z)$ and the point Q whose coordinates we wish to determine.

The y coordinate of Q is the same as that of P. The x coordinate of Q is 0. To find the z coordinate of Q, we use the fact that $\overline{QR} = \overline{PR}$. By right triangle PSR,

$$\overline{PR} = \sqrt{x^2 + z^2}.$$

Hence the z coordinate of Q is $\sqrt{x^2 + z^2}$, and

$$Q = (0, y, \sqrt{x^2 + z^2}).$$

Since the point Q lies on the line L,

$$\sqrt{x^2 + z^2} = \tfrac{1}{2}y. \tag{1}$$

Equation (1) is the equation of the cone, shown in Fig. 14. ■

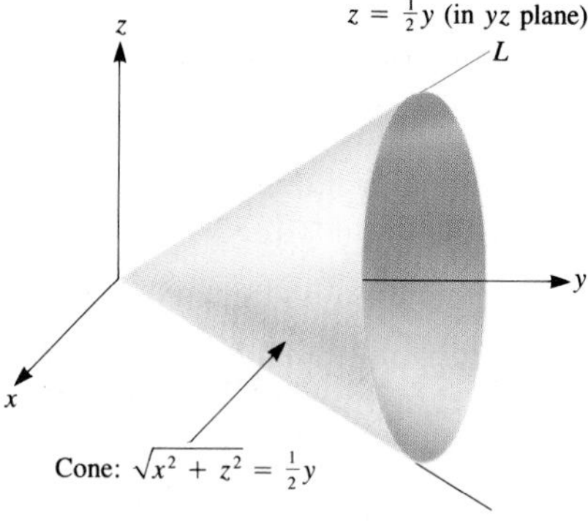

Figure 14

The method used in Example 5 applies to a surface of revolution about any of the axes:

> Draw a typical point $P = (x, y, z)$ on the surface. Then draw the point Q on the curve that is revolved to produce P. Find the coordinates of Q in terms of the coordinates of P. Substituting these coordinates into the equation that Q satisfies produces an equation that P satisfies.

Section Summary

We discussed the equations of spheres, generalized cylinders, and surfaces of revolution. We also suggested ways to graph these surfaces:

1. Use a ruler.
2. Make the graphs large enough.
3. Indicate hidden parts by dashed, broken, or omitted lines.

It is a good idea to make your sketch first lightly in pencil. Then complete it in ink, colored pencil, etc.

EXERCISES FOR SEC. 14.1: GRAPHS

In Exercises 1 to 12 sketch the given planes.

1 $x = 2$
2 $x = -1$
3 $y = 1$
4 $y = -1$
5 $z = -2$
6 $z = 2$
7 $y = 2x$
8 $y = -x$
9 $z = x$
10 $-x + 2y - 3z = 6$
11 $x + y - z = 1$
12 $3x - 2y + z = 6$

In Exercises 13 to 16 write an equation of the given sphere and sketch the sphere.

13 Center (0, 0, 0), radius 3
14 Center (1, 1, 0), radius $\sqrt{2}$
15 Center (1, 2, 3), radius 1
16 Center (1, 2, −1), radius 3

In Exercises 17 to 26 graph the given cylinders.

17 $x^2 + y^2 = 5$
18 $(x - 1)^2 + (y - 2)^2 = 9$
19 $x + y = 1$
20 $z = -x^2$
21 $z = y^2$
22 $z + y = 1$
23 $y = x^2$
24 $x = y^2$
25 $y = 2^{-x}$
26 $z = y^3$

In Exercises 27 to 30 graph the given curve, the surface obtained by revolving the curve about the given axis, and find an equation of the surface. Show the circle and right triangle used in obtaining that equation.

27 The part of the line $z = 2y$ in the first quadrant of the yz plane, about the y axis.
28 The part of the line $z = y + 2$ in the first quadrant of the yz plane, about the z axis.
29 The part of the circle $x^2 + y^2 = 1$ in the xy plane for which $y \geq 0$, about the x axis.
30 The part of the parabola $z = y^2$ in the first quadrant of the yz plane, (*a*) about the z axis, (*b*) about the y axis.

31 Let S be the set of all points $P = (x, y, z)$ such that the distance from P to (1, 2, 3) equals the distance from P to (4, −1, 2).
(*a*) Obtain an equation involving x, y, and z that tells when the point $P = (x, y, z)$ is in S.
(*b*) Using (*a*), show that S is a plane.
(*c*) Draw it.

32 Draw the part of the plane $x = y - 2$ that lies inside the cylinder $(y - 1)^2 + (z - 2)^2 = 4$.

33 S is the solid region that lies within both the cylinder $y^2 + z^2 = 1$ and the cylinder $x^2 + z^2 = 1$. Draw the part of S that lies in the first octant (where all coordinates are positive).

34 The line $z = 2y + 1$ in the yz plane is revolved around the line $y + z = 1$ in the yz plane.
(*a*) Sketch the double cone produced.
(*b*) Find an equation of the double cone.

35 The curve $z = e^{-y^2}$ in the yz plane is revolved about the z axis to produce a surface.
(*a*) Find an equation of this surface and graph it.
(*b*) Estimate the volume of the region below the surface, above the xy plane, and between the cylinders $x^2 + y^2 = r^2$ and $x^2 + y^2 = (r + dr)^2$, shown in Fig. 15. Assume that dr is small.
(*c*) Express the total volume under the surface and above the xy plane as an integral of the form $\int_0^\infty g(r)\,dr$ for a suitable function.
(*d*) Evaluate the integral in (*c*).

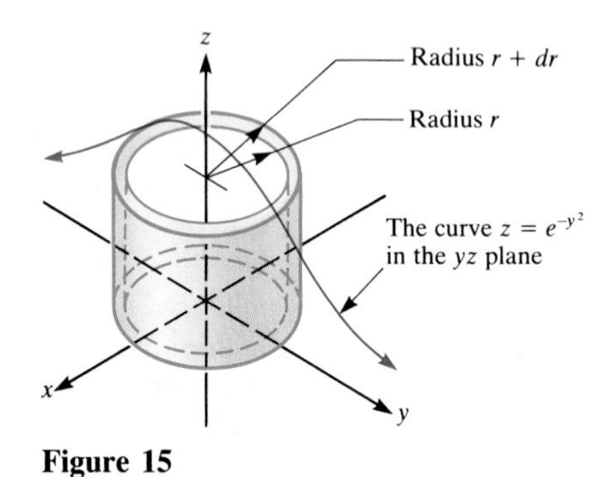

Figure 15

14.2 QUADRIC SURFACES

The graph of $Ax^2 + Bxy + Cy^2 + Dx + Ey + F = 0$ in the xy plane is a conic section. (See Appendix G for details.) By moving the axes, we can transform the equation to one of the forms shown in Fig. 1, depending on whether the graph is an ellipse, hyperbola, or parabola. (We disregard such degenerate cases as $x^2 + y^2 + 4 = 0$, whose graph is empty.)

In this section we examine the graph of a second-degree polynomial in *three* variables:

$$Ax^2 + By^2 + Cz^2 + Dxy + Eyz + Fxz + Gx + Hy + Iz + J = 0. \tag{1}$$

The graph is called a **quadric surface** (also **quadratic surface**).

We assume that at least one of the coefficients A through F is not 0. Otherwise we would be examining the equation $Gx + Hy + Iz + J = 0$, whose graph would be a plane. Graphs of (1) can be of several types, such as a sphere or cone, and even consist of two pieces (like a hyperbola). See Fig. 2, where the various possibilities are shown.

By moving the xyz axes, we can transform (1) to a simpler form (where most of the coefficients are 0). We will discuss the graphs of these simpler forms, namely,

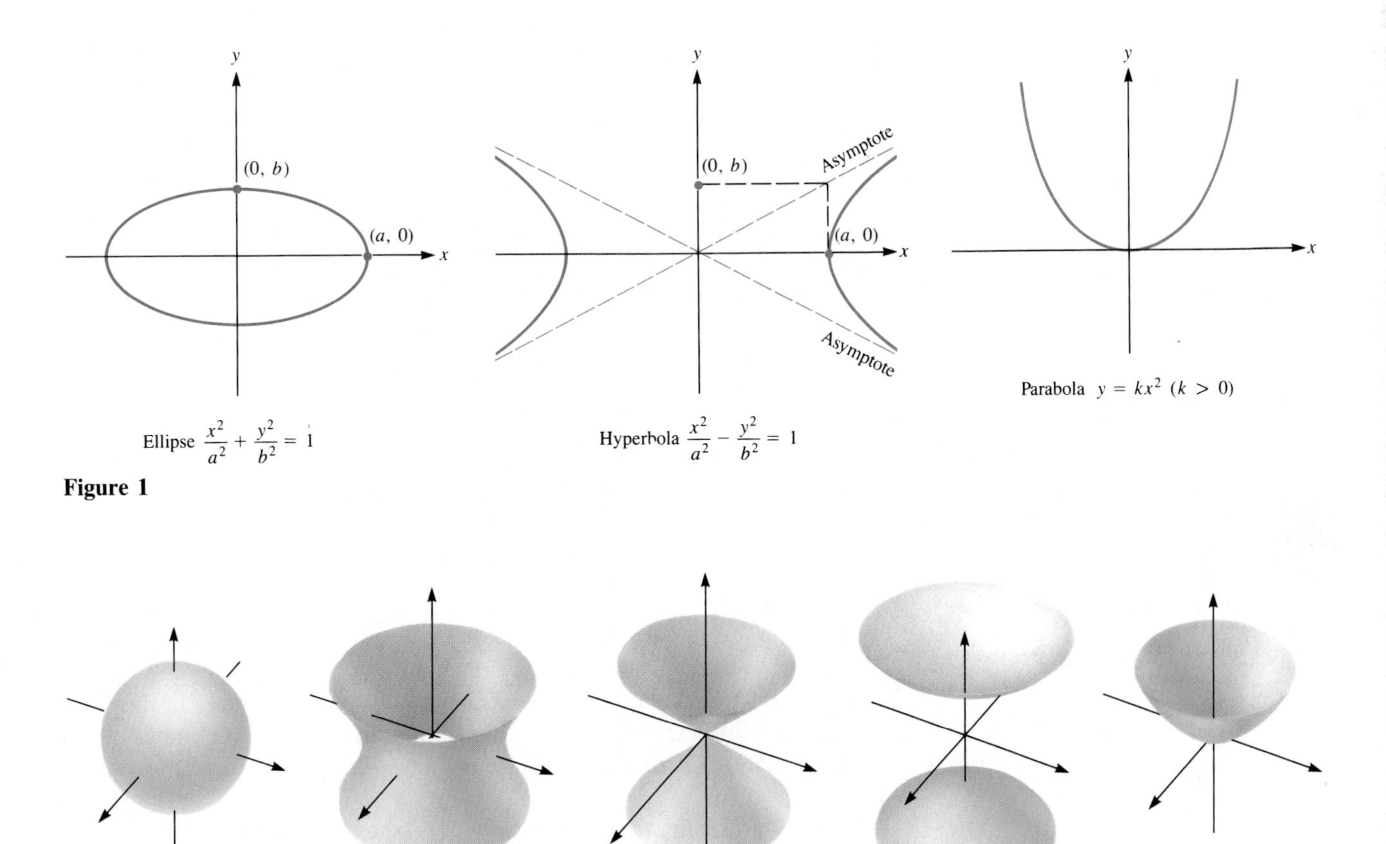

Figure 1

Figure 2

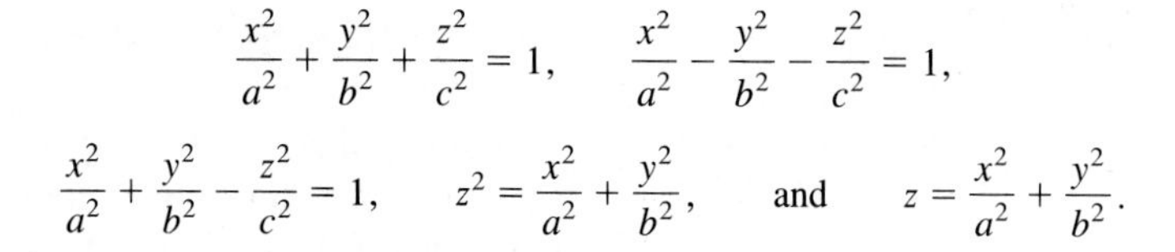

$$\frac{x^2}{a^2} + \frac{y^2}{b^2} + \frac{z^2}{c^2} = 1, \qquad \frac{x^2}{a^2} - \frac{y^2}{b^2} - \frac{z^2}{c^2} = 1,$$

$$\frac{x^2}{a^2} + \frac{y^2}{b^2} - \frac{z^2}{c^2} = 1, \qquad z^2 = \frac{x^2}{a^2} + \frac{y^2}{b^2}, \qquad \text{and} \qquad z = \frac{x^2}{a^2} + \frac{y^2}{b^2}.$$

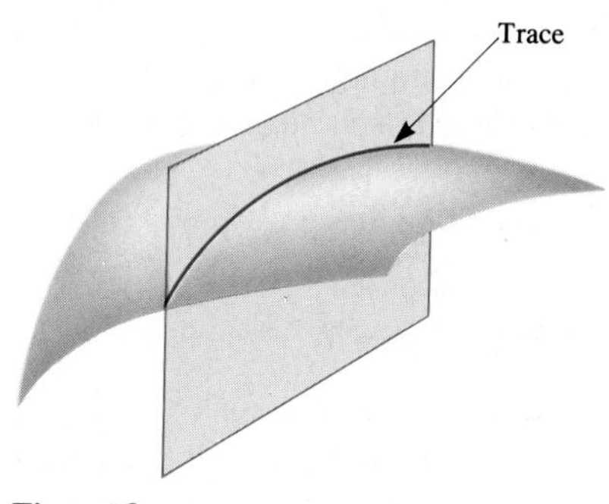

Figure 3

Throughout we assume a, b, and c are positive.

To sketch the surfaces we will examine their intersections with various planes. The intersection of a plane with a surface is called a **trace** of the surface. (See Fig. 3.)

The Ellipsoid

The graph of $x^2 + y^2 + z^2 = a^2$, $a > 0$, is a sphere of radius a and center $(0, 0, 0)$. Hence the graph of

$$\frac{x^2}{a^2} + \frac{y^2}{a^2} + \frac{z^2}{a^2} = 1$$

is a sphere. Now, let a, b, and c be positive numbers. The graph of

$$\frac{x^2}{a^2} + \frac{y^2}{b^2} + \frac{z^2}{c^2} = 1 \tag{2}$$

is called an **ellipsoid**. (When $a = b = c$, the ellipsoid is a sphere.) To graph an ellipsoid, first find its intercepts with the axes. For instance, the x intercepts are found by setting y and z equal to zero:

$$\frac{x^2}{a^2} + \frac{0^2}{b^2} + \frac{0^2}{c^2} = 1,$$

hence $x = a$ and $-a$. Similarly, the y intercepts are b and $-b$, and the z intercepts are c and $-c$.

What is the trace of the ellipsoid in the xy plane? To find out, set $z = 0$ in (2), obtaining

$$\frac{x^2}{a^2} + \frac{y^2}{b^2} = 1.$$

Therefore the trace is an ellipse. The trace in the plane $z = k$, where k is any number in $(-c, c)$ is found by replacing z in (2) by k:

$$\frac{x^2}{a^2} + \frac{y^2}{b^2} + \frac{k^2}{c^2} = 1$$

or

$$\frac{x^2}{a^2} + \frac{y^2}{b^2} = 1 - \frac{k^2}{c^2}. \tag{3}$$

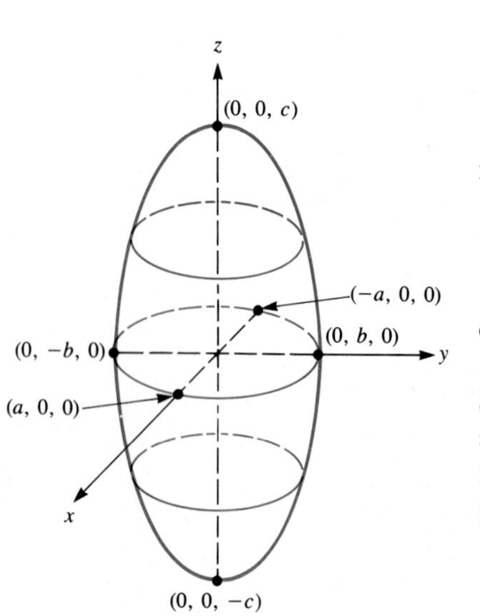

Figure 4

Since $1 - k^2/c^2$ is a constant, (3) describes an ellipse. Each trace of the ellipsoid (2) in a plane parallel to the xy plane is an ellipse. The graph of (2) resembles an egg or football. Figure 4 shows some traces parallel to the xy plane.

If $a = b$ in (2), the trace in the plane $z = k$ has the equation

$$\frac{x^2}{a^2} + \frac{y^2}{a^2} + \frac{k^2}{c^2} = 1$$

or

$$x^2 + y^2 = a^2\left(1 - \frac{k^2}{c^2}\right). \tag{4}$$

Since $a^2(1 - k^2/c^2)$ is simply a constant, (4) describes a circle of radius $a\sqrt{1 - k^2/c^2}$. In this case the ellipsoid is a surface of revolution. However, if a, b, and c are all different, then the ellipsoid (2) is not a surface of revolution.

Hyperboloids of Two Sheets

When the hyperbola $y^2 - z^2 = 1$ in the yz plane is revolved around the y axis, we obtain the surface shown in Fig. 5. This surface, which consists of two separate pieces, is an example of a hyperboloid of two sheets. A point $P = (x, y, z)$ lies on the surface of revolution if $Q = (0, y, \sqrt{x^2 + z^2})$ lies on the given hyperbola, whose equation is

$$(y \text{ coordinate})^2 - (z \text{ coordinate})^2 = 1.$$

(See Fig. 6.)

Hence the equation of the surface is

$$y^2 - (\sqrt{x^2 + z^2})^2 = 1,$$

or

$$y^2 - x^2 - z^2 = 1. \tag{5}$$

Therefore the surface in Fig. 5 is a quadric surface.

More generally, a **hyperboloid of two sheets** is defined as the graph of one of these equations:

$$\frac{x^2}{a^2} - \frac{y^2}{b^2} - \frac{z^2}{c^2} = 1,$$
$$-\frac{x^2}{a^2} + \frac{y^2}{b^2} - \frac{z^2}{c^2} = 1, \tag{6}$$
$$-\frac{x^2}{a^2} - \frac{y^2}{b^2} + \frac{z^2}{c^2} = 1.$$

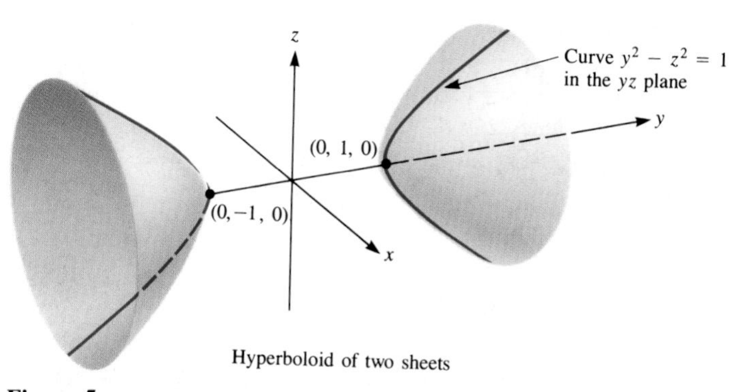

Figure 5

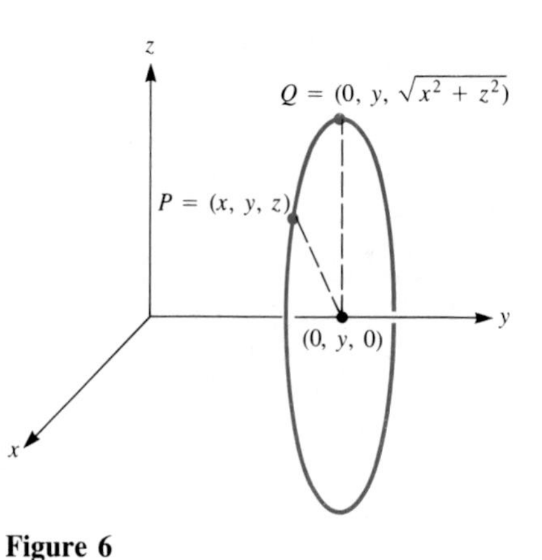

Figure 6

(There are *two* minuses and one plus, and the surface comes in *two* pieces, called **sheets** or **branches**.)

If no two of a^2, b^2, and c^2 are equal, then the surface is not a surface of revolution. Equation (5) is a special case of the second equation in (6) when $a^2 = b^2 = c^2 = 1$.

In Example 1 we examine one hyperboloid of two sheets in detail.

EXAMPLE 1 Graph $z^2 = 1 + x^2/4 + y^2/9$.

SOLUTION Rewriting the equation as

$$-\frac{x^2}{4} - \frac{y^2}{9} + \frac{z^2}{1} = 1, \tag{7}$$

we see that its graph is a hyperboloid of two sheets.

Consider traces in planes of the form $z = k$. If $k = 0$, (7) gives

$$-\frac{x^2}{4} - \frac{y^2}{9} + \frac{0^2}{1} = 1. \tag{8}$$

The left side of (8) is never positive. Thus the trace is empty. That means that the surface does not meet the xy plane. Similarly, if $-1 < k < 1$, the trace in the plane $z = k$ is empty.

Consider the plane $z = 1$. Substituting $z = 1$ into (7) gives

$$-\frac{x^2}{4} - \frac{y^2}{9} + \frac{1^2}{1} = 1$$

or

$$0 = \frac{x^2}{4} + \frac{y^2}{9}. \tag{9}$$

Equation (9) has only one solution: (0, 0). So the plane $z = 1$ meets the surface (7) in only one point, (0, 0, 1).

If $k > 1$ (or $k < -1$), the trace in the plane $z = k$ is found by examining the equation

$$-\frac{x^2}{4} - \frac{y^2}{9} + \frac{k^2}{1} = 1,$$

which can be rewritten

$$\frac{x^2}{4} + \frac{y^2}{9} = k^2 - 1. \tag{10}$$

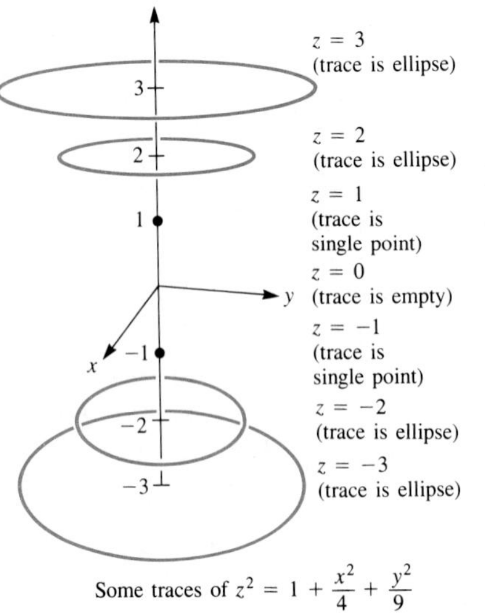

Some traces of $z^2 = 1 + \frac{x^2}{4} + \frac{y^2}{9}$

Figure 7

Equation (10) describes an ellipse. The larger k is, the larger the ellipse is. Figure 7 shows traces in the six planes $z = 1$, $z = -1$, $z = 2$, $z = -2$, $z = 3$, and $z = -3$.

The surface (7) is made of ellipses. How do they fit together? To find out, examine the trace in the yz plane, $x = 0$.

Substituting 0 for x in (7) gives

$$-\frac{0^2}{4} - \frac{y^2}{9} + \frac{z^2}{1} = 1$$

or

$$z^2 - \frac{y^2}{9} = 1. \tag{11}$$

The graph of (11) is a hyperbola in the yz plane. It is shown in Fig. 8, where we use it to complete our graph of the surface (7).

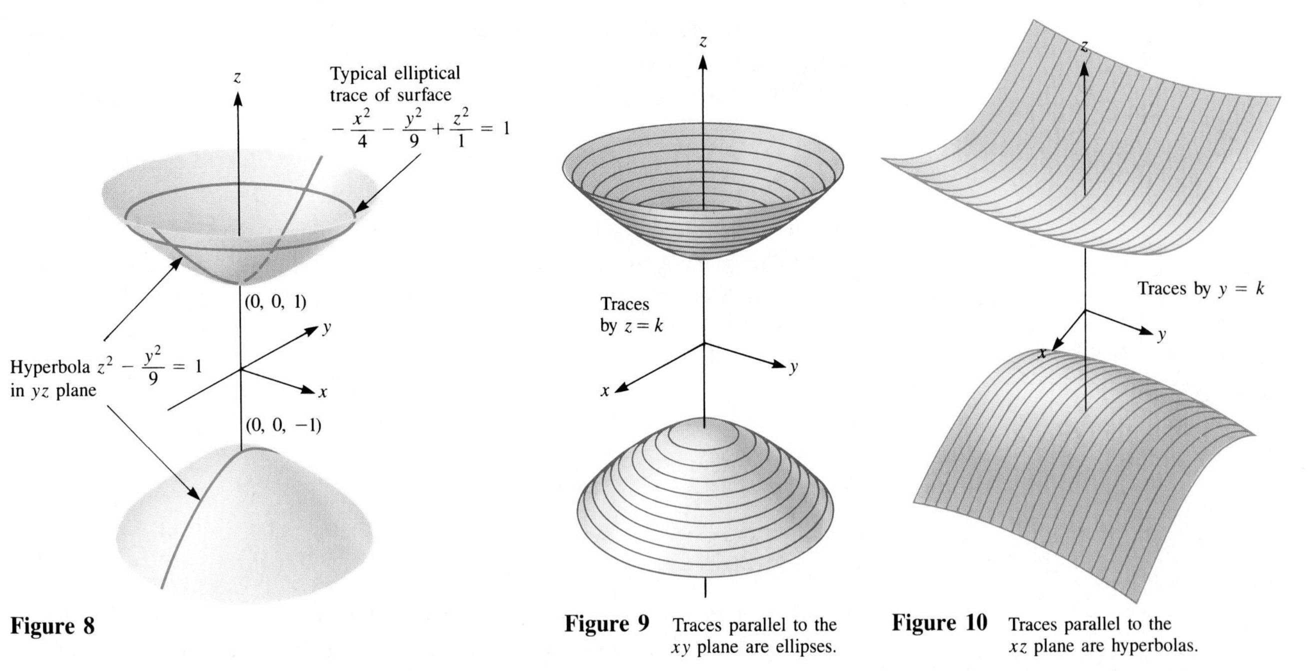

Figure 8

Figure 9 Traces parallel to the xy plane are ellipses.

Figure 10 Traces parallel to the xz plane are hyperbolas.

Drawing a couple of elliptical traces helps add perspective. (This hyperboloid is *not* a surface of revolution.) Computer graphics software usually shows many more traces, such as in Figs. 9 to 11, where a variety of intersections are displayed. ■

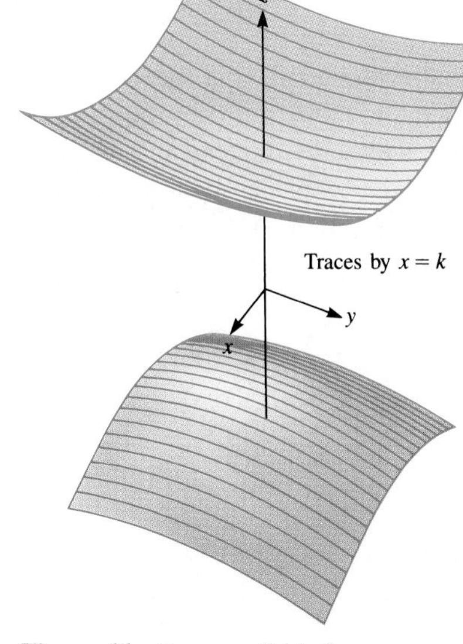

Figure 11 Traces parallel to the yz plane are hyperbolas.

Hyperboloids of One Sheet

When you revolve the hyperbola $y^2 - z^2 = 1$ in the yz plane about the z axis instead of the y axis, you obtain a surface that consists of *one* piece. This surface resembles the cooling tower at a power plant. (See Fig. 12.) It is an example of a hyperboloid of one sheet.

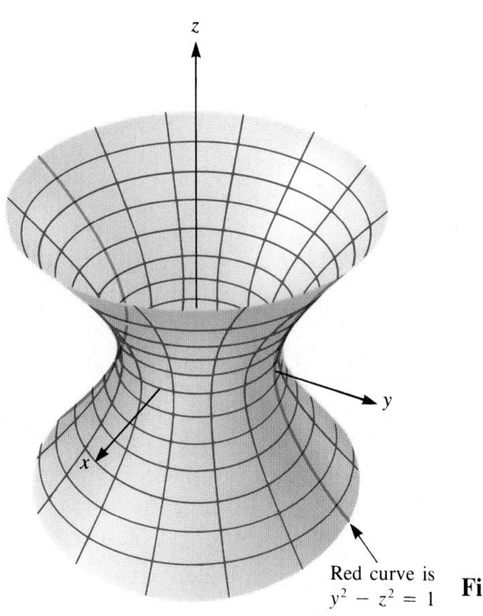

Figure 12

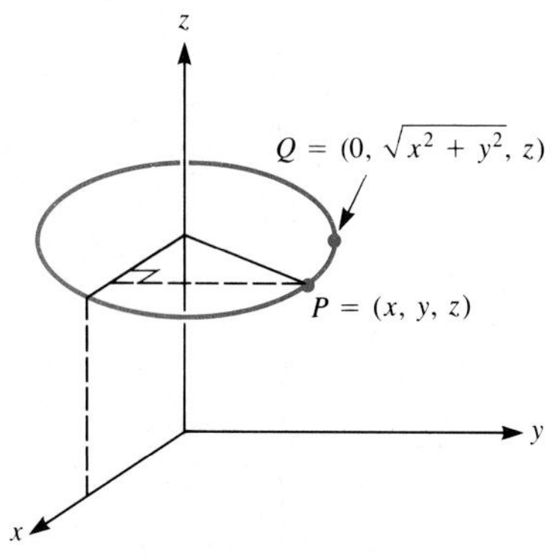

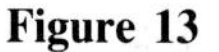
Figure 13

To obtain the equation of this surface, let $P = (x, y, z)$ be a typical point on the surface. Then $Q = (0, \sqrt{x^2 + y^2}, z)$ is on the hyperbola $y^2 - z^2 = 1$ in the yz plane. See Fig. 13. Since Q is on the hyperbola $y^2 - z^2 = 1$,

$$(\sqrt{x^2 + y^2})^2 - z^2 = 1$$

or

$$x^2 + y^2 - z^2 = 1. \tag{12}$$

The main difference between (12) and the previous example $y^2 - x^2 - z^2 = 1$ is that now there is only one minus, whereas before there were two minuses. This leads to the definition of a **hyperboloid of one sheet**, namely the graph of any one of the equations

$$-\frac{x^2}{a^2} + \frac{y^2}{b^2} + \frac{z^2}{c^2} = 1,$$

$$\frac{x^2}{a^2} - \frac{y^2}{b^2} + \frac{z^2}{c^2} = 1,$$

or

$$\frac{x^2}{a^2} + \frac{y^2}{b^2} - \frac{z^2}{c^2} = 1.$$

As we noted, there is *one* minus sign and only *one* sheet. The equation of a hyperboloid of *two* sheets has *two* minuses. Traces of these surfaces in planes parallel to the coordinate planes are either ellipses or hyperbolas.

The Cone

The cone is another example of a surface that is the graph of a second-degree polynomial.

Figure 14 shows a double right circular cone with a half-vertex angle α. It is obtained by revolving the line $(\tan \alpha)y = z$ in the yz plane about the y axis. The point (x, y, z) is on the double cone if the point $(0, y, \sqrt{x^2 + z^2})$ is on the line $(\tan \alpha)y = z$ in the yz plane; that is

$$y \tan \alpha = \sqrt{x^2 + z^2}. \tag{13}$$

Squaring both sides of (13) gives

$$(\tan^2 \alpha)y^2 = x^2 + z^2.$$

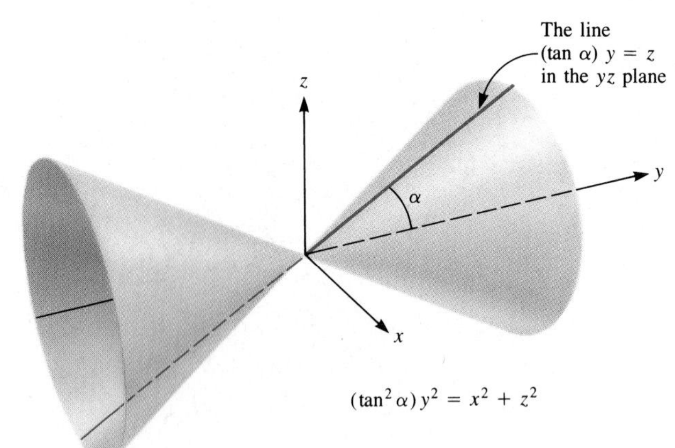

Figure 14

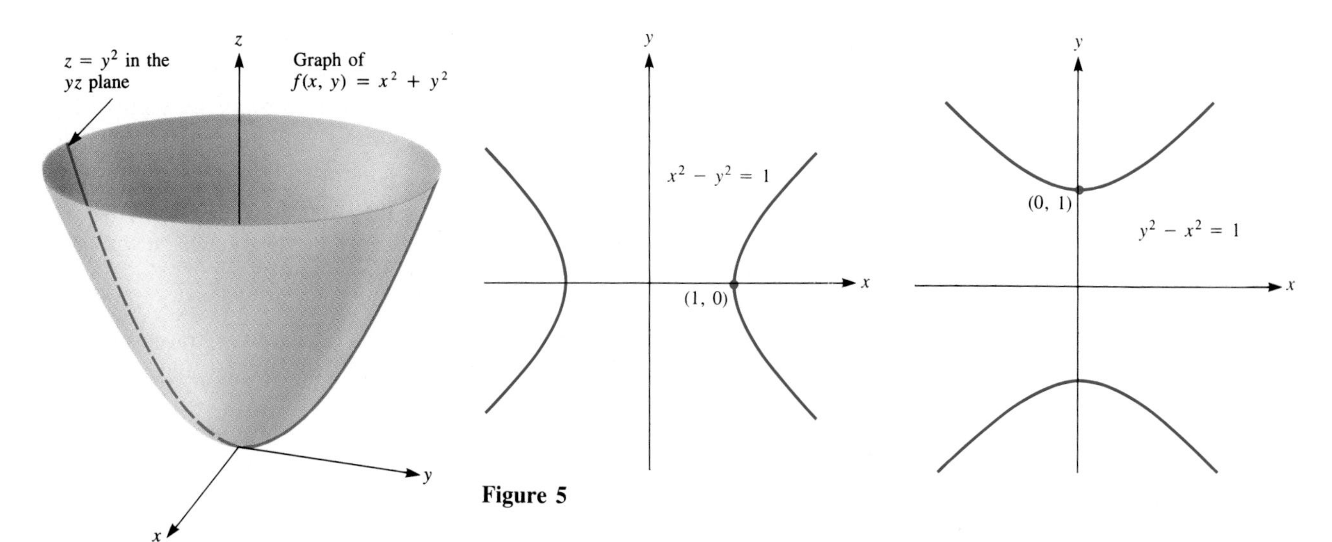

Figure 5

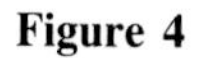

Figure 4

Next we examine traces in plane $z = k$, for k positive. Such a trace is described by the equation

$$k = y^2 - x^2.$$

When $k = 0$, we obtain the equation $0 = y^2 - x^2$, or $0 = (y + x)(y - x)$. In this case the trace consists of the two lines $y = -x$ and $y = x$. When $k = 1$, we have the hyperbola $1 = y^2 - x^2$. The trace in $z = 1$ is distinguished in color. As k increases, the traces become larger hyperbolas. Figure 7 shows the part of the graph *above* the xy plane.

Next consider the part of the graph below the xy plane. For negative k the trace in the plane $z = k$ is described by the equation

$$k = y^2 - x^2. \tag{1}$$

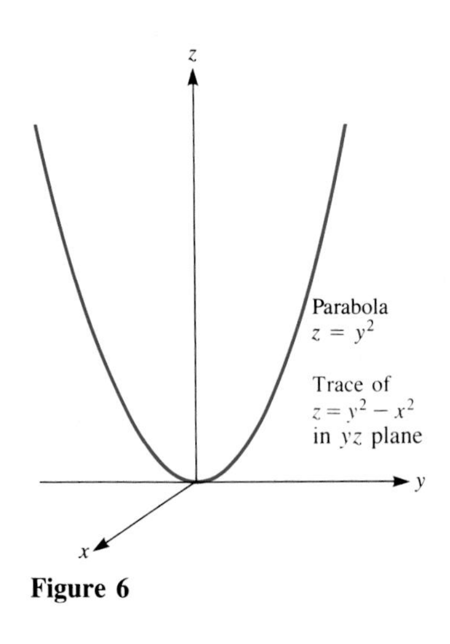

Figure 6

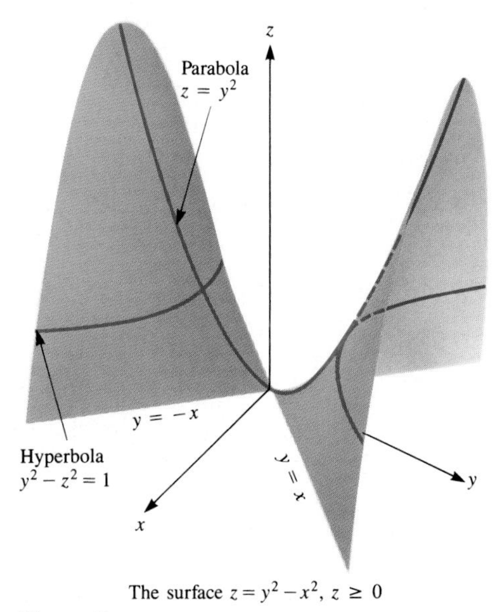

The surface $z = y^2 - x^2$, $z \geq 0$

Figure 7

For instance, when $k = -1$, we have

$$-1 = y^2 - x^2$$

or

$$x^2 - y^2 = 1.$$

Equation (1) describes a hyperbola. To see how these hyperbolas fit together, we find the trace of the surface $z = y^2 - x^2$ in the xz plane by setting $y = 0$. The trace is $z = -x^2$, a parabola. With all this information, we sketch the part of the surface that lies below the xy plane, in Fig. 8.

Piecing the two figures together gives us Fig. 9.

Figures 10 and 11 were drawn by computer. They show traces in planes parallel to the xy plane and traces in planes parallel to the xz plane. ■

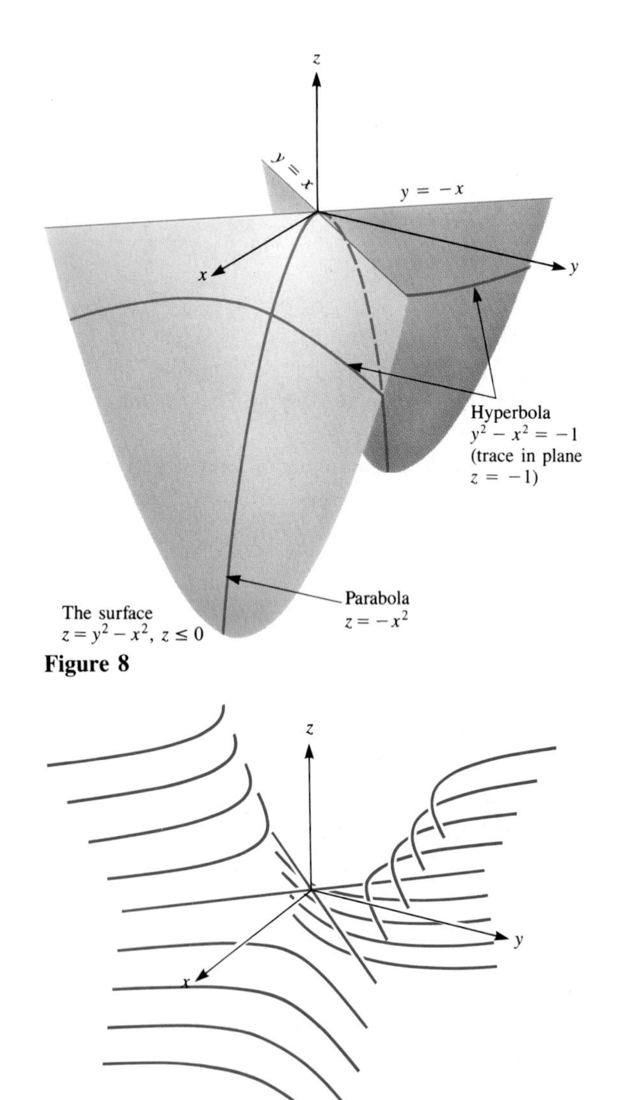

The surface $z = y^2 - x^2$, $z \leq 0$

Figure 8

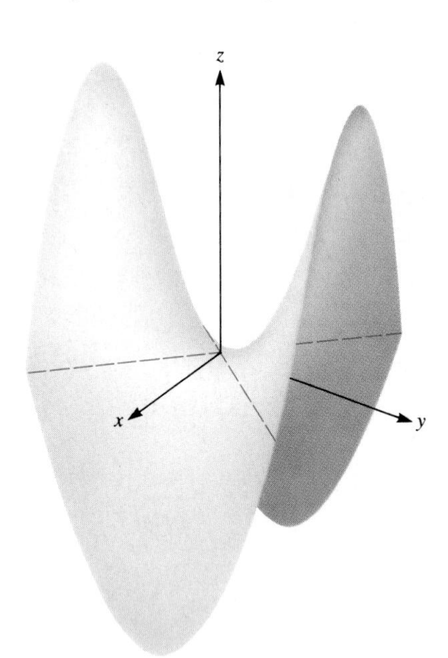

Figure 9 Saddle $z = y^2 - x^2$

Traces of $z = y^2 - x^2$ parallel to the xy plane are hyperbolas.

Figure 10

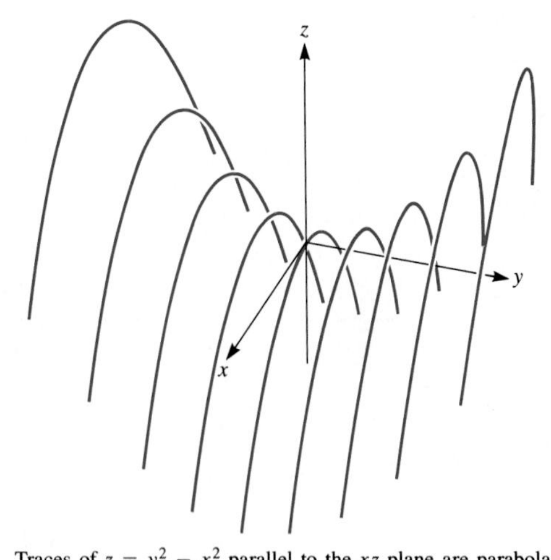

Traces of $z = y^2 - x^2$ parallel to the xz plane are parabolas.

Figure 11

In Example 4 we graphed the surface $z = y^2 - x^2$. It is an example of an equation of the form

$$z = \frac{y^2}{b^2} - \frac{x^2}{a^2}$$

or $z = x^2/a^2 - y^2/b^2$. The graphs of these equations are called **hyperbolic paraboloids**. They complete the roster of quadric surfaces.

The surface in Example 4 resembles a saddle or the pass between two hills. It is usually called a **saddle**.

Level Curves of $z = f(x, y)$

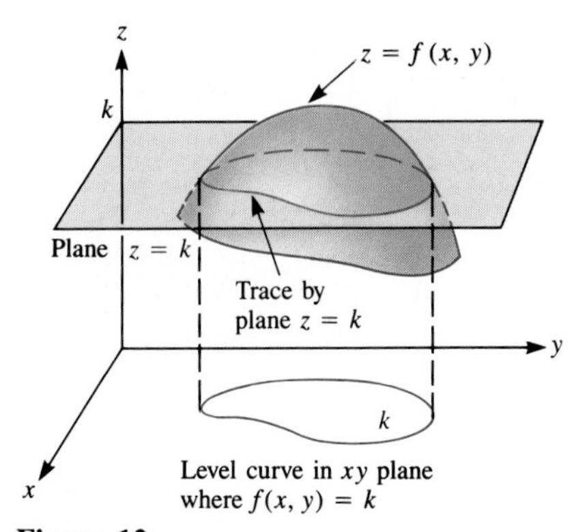

Figure 12

Even such a "simple" function as $z = y^2 - x^2$ has a fairly complicated graph. There is another way to display the behavior of a function $z = f(x, y)$—geometrically. Instead of drawing the actual trace in a plane $z = k$, we draw the projection of that trace on the xy plane. The projection is called a **level curve** of the function $f(x, y)$. At all points (x, y) on that level curve, the function has the value k. We put a k next to it to indicate the value of f on the curve. (See Fig. 12.)

When you walk on the surface $z = f(x, y)$, staying above the level curve $f(x, y) = k$, your path would be level. Your altitude above the xy plane would stay constant, namely k.

The next example illustrates the use of level curves. It presents a different way of depicting the function $f(x, y) = y^2 - x^2$ of Example 4.

EXAMPLE 5 Describe the behavior of $f(x, y) = y^2 - x^2$ by sketching some of its level curves.

SOLUTION The level curve when $f(x, y)$ has the fixed value 0 has the equation

$$y^2 - x^2 = 0.$$

It consists of the lines $y = x$ and $y = -x$.

The level curve where $f(x, y) = 1$ has the equation

$$y^2 - x^2 = 1.$$

It is a hyperbola. Figure 13 shows these two level curves as well as those corresponding to $k = 2, -1$, and -2. In Fig. 13 the level curve for $k = 0$ (two lines) separates the plane into four pieces. In two opposite sections, the function is negative; in the other two, the function is positive. It also indicates that as you go out on the x axis, the function becomes more negative. Also, as you go out along the y axis, the function becomes more positive. At the origin the function has the value 0. But near the origin there are points where the function is positive and points where it is negative. ■

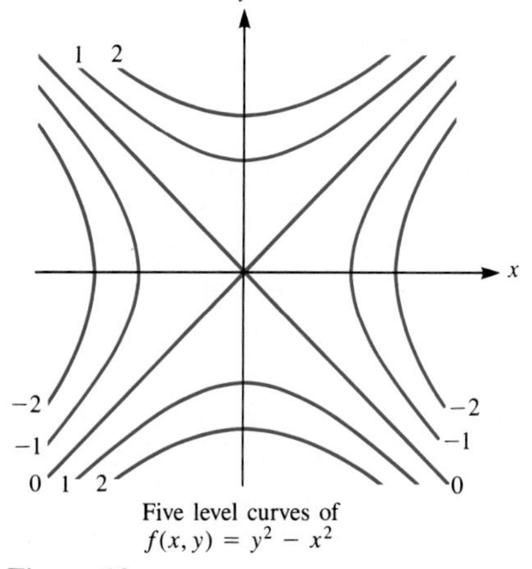

Figure 13

Level Curves in Daily Life

Level curves go by many names, as the following table shows:

Function f	*Name of Level Curve*
Altitude of land	Contour line
Air pressure	Isobar
Temperature	Isotherm
Utility (in economic theory)	Indifference curve
Gravitational potential	Equipotential curve

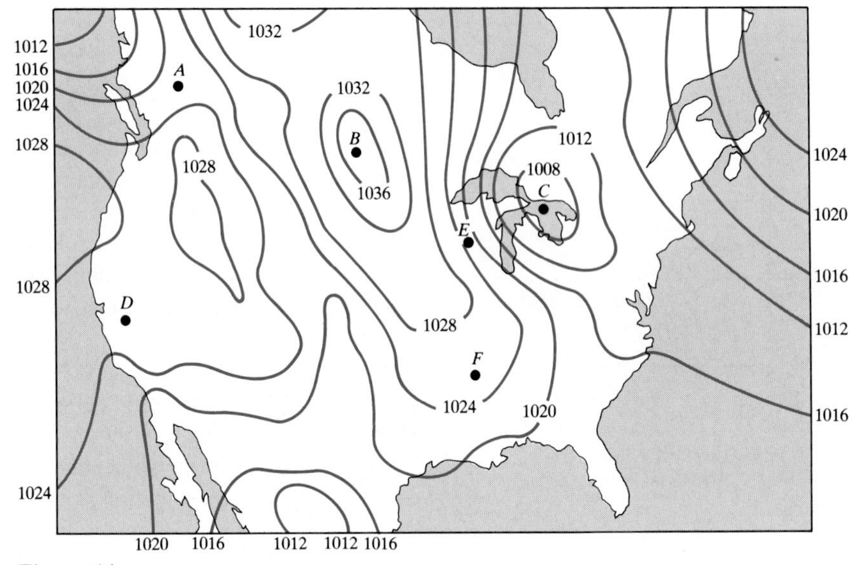

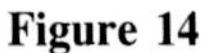
Figure 14

The daily weather map in newspapers often indicates the barometric pressure function by a few well-chosen level curves (called *isobars*), as in Fig. 14. In this case, the function is "pressure at (x, y)" in some barometric unit (such as *torr*). By studying the map, you can see that the highest pressure occurs near point B (more than 1036, but less than 1040 since the isobars are made at intervals of 4). The minimum pressure occurs near C (less than 1008, but more than 1004). Given some clay and enough time, we could even construct the surface that is the graph of the pressure function. A perspective drawing of the surface would look something like Fig. 15, with B as the highest point and C as the lowest point.

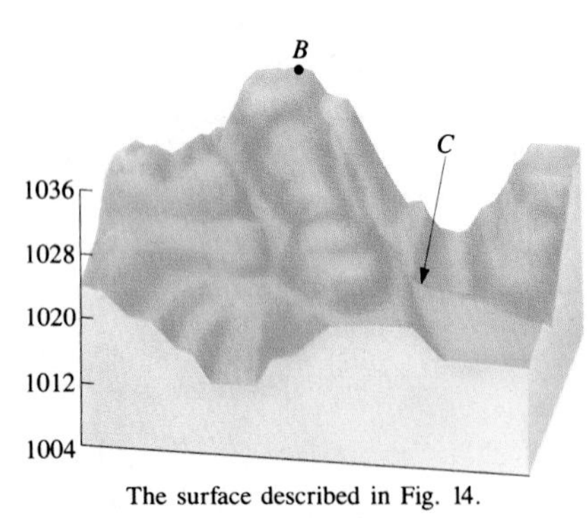

The surface described in Fig. 14.

Figure 15

The distribution of radioactive fallout from the explosion of a 15-megaton H-bomb in 1954 can best be described in terms of level curves. In this case, $f(P)$ is the total radiation dose at point P accumulated within 4 days after the explosion. (This dose is measured in rads; a dose of 700 rads in 4 days is lethal.) Figure 16 shows some of the level curves of this function. Note that they are not all at the same intervals.

A U.S. Geological Survey map indicates height by contour lines. For instance, one curve may show all the points at the 300-foot altitude. This is the 300-foot contour. The map in Fig. 17 depicts Mount Shasta by using contour lines at intervals of 400 feet. A hiker walking along that part of the mountain which corresponds to a contour line neither rises nor descends.

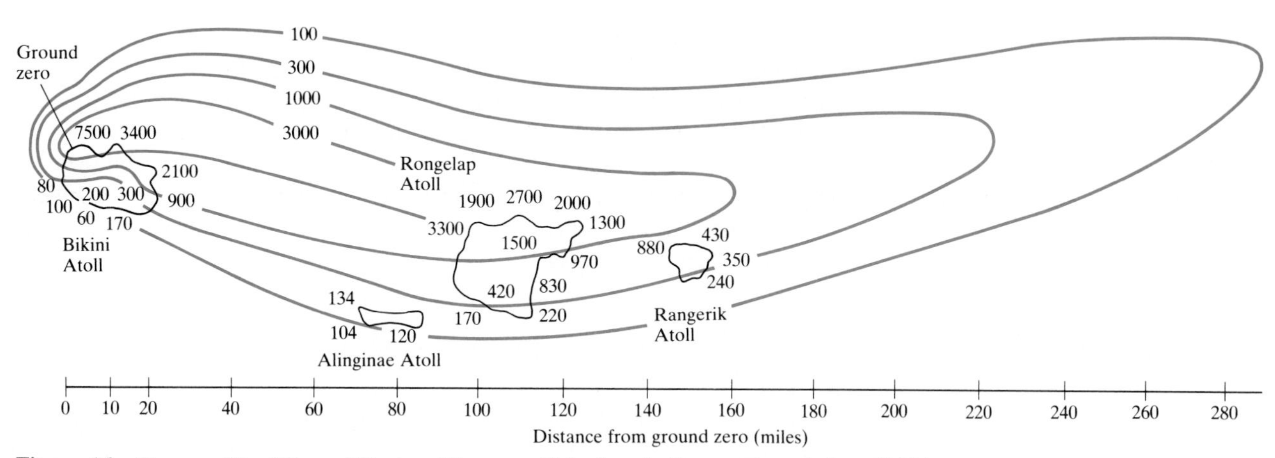

Figure 16 (*Source: The Effect of Nuclear Weapons, U.S. Atomic Energy Commission, 1962.*)

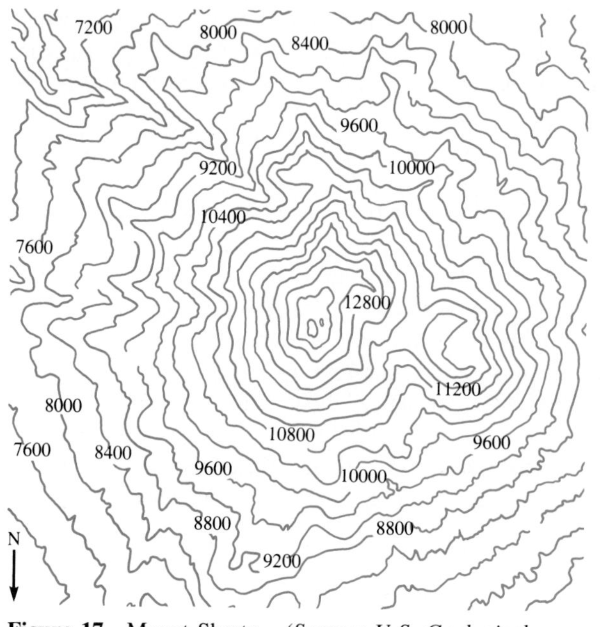

Figure 17 Mount Shasta. (*Source: U.S. Geological Survey.*)

Figure 18 North face of Mount Shasta. (*Photograph by Tim Feldman.*)

Mount Shasta is shown in Fig. 18. Note that where the contour lines are close together the slope is steeper.

Let $f(x, y)$ be the altitude of the mountain corresponding to the point (x, y) on the map. Then the 9600-foot contour consists of those points where $f(x, y) = 9600$. On the 9600-foot contour the altitude function f is constant: Its value is 9600.

Level Surfaces of $u = f(x, y, z)$

The graph of $y = f(x)$ consists of certain points in the xy plane. The graph of $z = f(x, y)$ consists of certain points in the xyz space. But what if we have a function of three variables, $u = f(x, y, z)$? (The volume V of a box of sides x,

y, z is given by the equation $V = xyz$; this is an example of a function of three variables.) We cannot graph the set of points (x, y, z, u) where $u = f(x, y, z)$ since we live in space of only three dimensions. What we could do is pick a constant k and draw the "level surfaces," the set of points where $f(x, y, z) = k$. Varying k may give an idea of this function's behavior, just as varying the k of $f(x, y) = k$ yields information about the behavior of a function of two variables.

For example, let $T = f(x, y, z)$ be the temperature (Fahrenheit) at the point (x, y, z). Then the level surface

$$68 = f(x, y, z)$$

consists of all points where the temperature is 68°.

EXAMPLE 6 Describe the level surfaces of the function $u = x^2 + y^2 + z^2$.

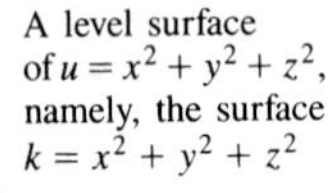
A level surface of $u = x^2 + y^2 + z^2$, namely, the surface $k = x^2 + y^2 + z^2$

Figure 19

SOLUTION For each k we examine the equation $k = x^2 + y^2 + z^2$. If k is negative, there are no points in the "level surface." If $k = 0$, there is only one point, the origin $(0, 0, 0)$. If $k = 1$, the equation is $1 = x^2 + y^2 + z^2$, which describes a sphere of radius 1, center $(0, 0, 0)$. If k is positive, the level surface $f(x, y, z) = k$ is a sphere of radius $\sqrt{k}$, center $(0, 0, 0)$. See Fig. 19. ■

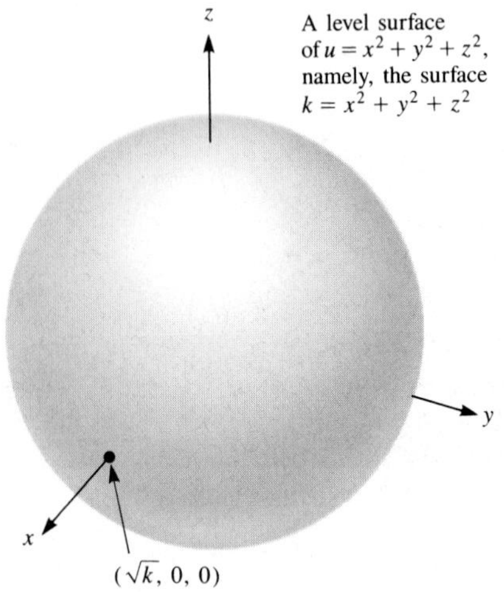

Section Summary

We introduced the idea of a function of two variables $z = f(P)$, where P is in some region in the xy plane. The graph of $z = f(P)$ is usually a surface. But it is often more useful to sketch a few of its level curves than to sketch that surface. Each level curve is the projection of a trace of the surface in a plane of the form $z = k$. Note that at all points (x, y) on a level curve the function has the same value. In other words, the function f is constant on a level curve.

In particular, we used level curves to analyze the function $z = y^2 - x^2$, whose graph is a saddle.

For functions of three variables $u = f(x, y, z)$, we defined level surfaces. When considered on a level surface, $k = f(x, y, z)$, such a function is constant, with value k.

EXERCISES FOR SEC. 14.3: FUNCTIONS AND THEIR LEVEL CURVES

In Exercises 1 to 10, graph the given functions.

1 $f(x, y) = y$
2 $f(x, y) = x + 1$
3 $f(x, y) = 3$
4 $f(x, y) = -2$
5 $f(x, y) = x^2$
6 $f(x, y) = y^2$
7 $f(x, y) = x + y + 1$
8 $f(x, y) = 2x - y + 1$
9 $f(x, y) = x^2 + 2y^2$
10 $f(x, y) = \sqrt{x^2 + y^2}$

In Exercises 11 to 14 draw for the given functions the level curves corresponding to the values -1, 0, 1, and 2 (if they are not empty).

11 $f(x, y) = x + y$
12 $f(x, y) = x + 2y$
13 $f(x, y) = x^2 + 2y^2$
14 $f(x, y) = x^2 - 2y^2$

In Exercises 15 to 18 draw the level curves for the given functions that pass through the given points.

15 $f(x, y) = x^2 + y^2$ through $(1, 1)$. *Hint:* First compute $f(1, 1)$.
16 $f(x, y) = x^2 + 3y^2$ through $(1, 2)$.
17 $f(x, y) = x^2 - y^2$ through $(3, 2)$.
18 $f(x, y) = x^2 - y^2$ through $(2, 3)$.

---

19 (*a*) Draw the level curves for the function $f(x, y) = x^2 + y^2$ corresponding to the values $k = 0, 1, \ldots, 9$.
(*b*) By inspection of the curves in (*a*), decide where the function is changing most rapidly. Explain why you think so.

20 Let $f(P)$ be the average daily solar radiation at the point P (measured in langleys). The level curves corresponding to 350, 400, 450, and 500 langleys are shown in Fig. 20.

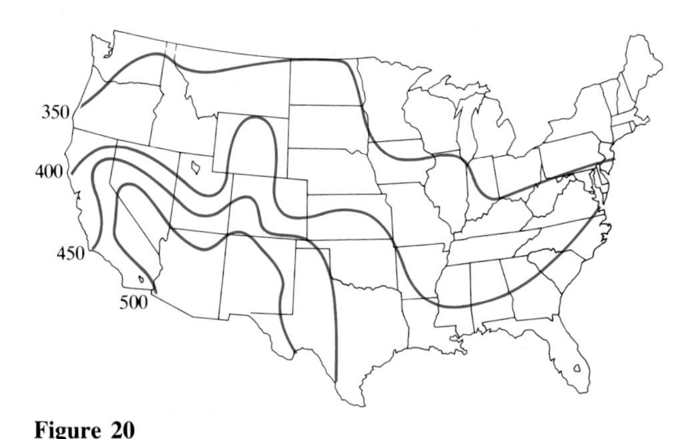

Figure 20

(*a*) What can be said about the ratio between the maximum and minimum solar radiation at points in the United States?

(*b*) Why are there rather sharp bends in the level curves in two areas?

21 At which point in Fig. 14 would you expect the wind to be the strongest?

22 Let $u = g(x, y, z)$ be a function of three variables. Describe the level surface $g(x, y, z) = 1$ if $g(x, y, z)$ is

(*a*) $x + y + z$
(*b*) $x^2 + y^2 + z^2$
(*c*) $x^2 + y^2 - z^2$
(*d*) $x^2 - y^2 - z^2$

14.4 LIMITS AND CONTINUITY

In the next section we will be differentiating functions of two or more variables. Before that we should discuss limits and continuity of such functions. In particular, we describe the domains of the functions commonly met in applications.

Domains

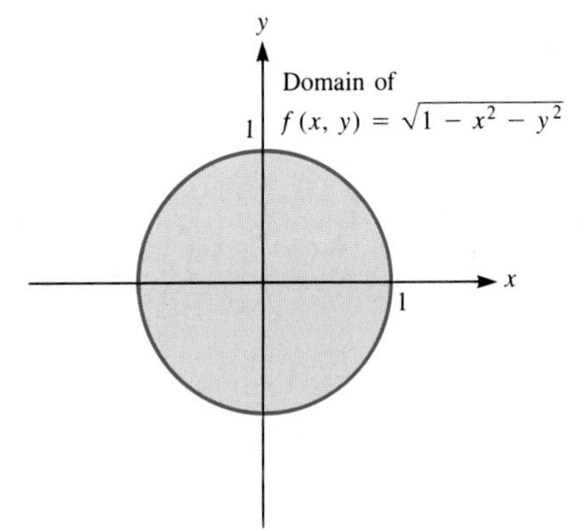

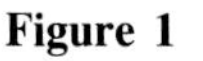

Figure 1

The **domain** of a function $f(x, y)$ is the set of points where it is defined. The domain of $f(x, y) = x + y$ is the entire xy plane. However, the domain of $f(x, y) = \sqrt{1 - x^2 - y^2}$ is much smaller. In order for the square root of $1 - x^2 - y^2$ to be defined, $1 - x^2 - y^2$ must not be negative. In other words, we must have $x^2 + y^2 \leq 1$. The domain is the disk bounded by the circle $x^2 + y^2 = 1$, shown in Fig. 1. The domain includes the boundary $x^2 + y^2 = 1$.

The domain of $f(x, y) = 1/\sqrt{1 - x^2 - y^2}$ is even smaller. Now we must not let $1 - x^2 - y^2$ be 0 or negative. The domain of $1/\sqrt{1 - x^2 - y^2}$ consists of the points (x, y) such that $x^2 + y^2 < 1$. It is the disk in Fig. 1 *without* its boundary.

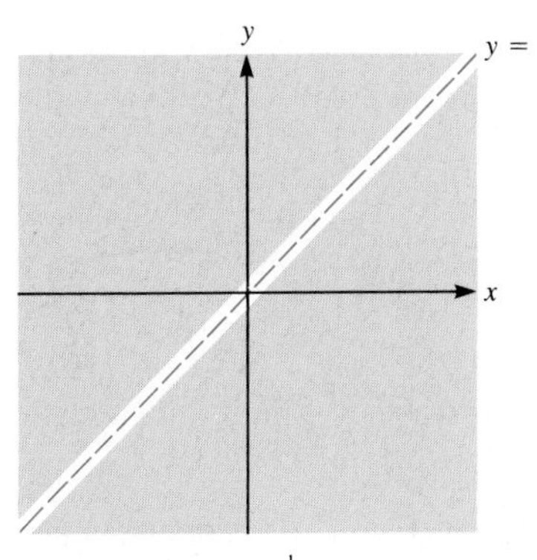

Domain of $f(x, y) = \frac{1}{y - x}$ consists of all points not on the line $y = x$.

Figure 2

The function $f(x, y) = 1/(y - x)$ is defined everywhere except on the line $y - x = 0$. Its domain is the xy plane from which the line $y = x$ is removed. (See Fig. 2.)

The domain of a function of interest to us will either be the entire xy plane or some region bordered by curves or lines, or perhaps such a region with a few points omitted.

We say a point P_0 is on the *boundary* of the region R if every disk with center P_0, no matter how small, contains points in R and points not in R. By a disk we mean all points within a circle of positive radius. (See Fig. 3.) The collection of all the boundary points of R is called the **boundary** of R. If R is bordered by some curves or lines, those curves form the boundary of R. If R consists of all points in the xy plane except (0, 0), the point (0, 0) is the only point on its boundary.

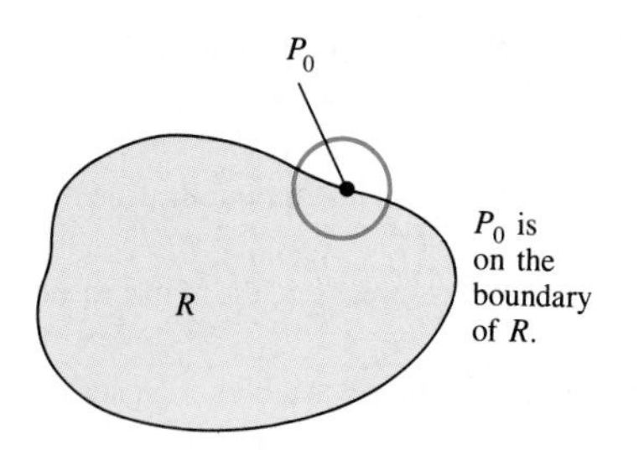

Figure 3

Let P_0 be a point in the domain of a function f. If there is a disk with center P_0 that lies within the domain of f, we call P_0 an **interior** point of the domain. (See Fig. 4.) When P_0 is an interior point of the domain of f, we know that $f(P)$ is defined for all points P sufficiently near P_0. Every point P_0 in the domain not on its boundary is an interior point. A set R is called **open** if each point P of R is an interior point of R. The entire xy plane is open. So is any disk without its circumference. More generally, the set of points inside some closed curve but not on it forms an open set.

P_0 is an interior point.

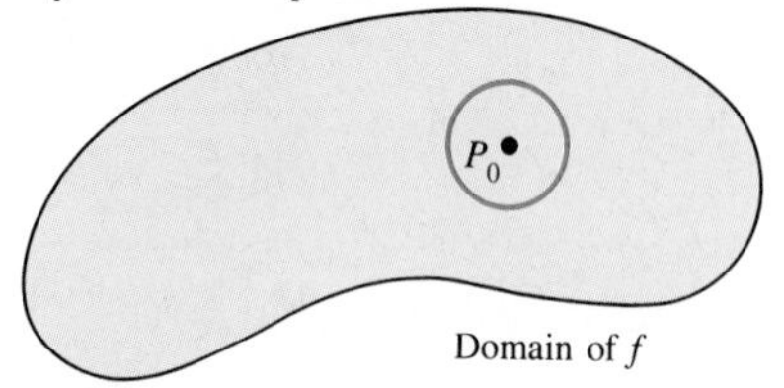

Figure 4

Limits

In Sec. 2.3 we defined $\lim_{x \to a} f(x) = L$. We assumed that $f(x)$ is defined at least on open intervals (c, a) and (a, b). It did not need to be defined at a. If $f(x)$ approaches L as $x \to a$ both from the right and the left, then we say that "the limit of $f(x)$ as x approaches a is L." We now define the limit of a function whose inputs are points P in the xy plane.

Definition *Limit of $f(x, y)$ at $P_0 = (a, b)$.* Let f be a function defined at least at every point in some disk with center P_0, except perhaps at P_0. If there is a number L such that $f(P)$ approaches L whenever P approaches P_0, we call L the **limit of $f(P)$ as P approaches P_0**. We write

$$\lim_{P \to P_0} f(P) = L$$

or

$$f(P) \to L \quad \text{as} \quad P \to P_0.$$

We also write

$$\lim_{(x,y) \to (a,b)} f(x, y) = L.$$

EXAMPLE 1 Let $f(x, y) = \dfrac{x^2 - y^2}{x^2 + y^2}$. Determine whether $\lim_{P \to (1,1)} f(P)$ exists.

SOLUTION If $P = (x, y)$ is near $(1, 1)$, $x^2 - y^2$ is near $1^2 - 1^2 = 0$ and $x^2 + y^2$ is near $1^2 + 1^2 = 2$. Thus $f(x, y)$ is near $0/2 = 0$. We may write

$$\lim_{(x,y) \to (1,1)} \frac{x^2 - y^2}{x^2 + y^2} = 0.$$

(The limit exists and is 0.) ■

EXAMPLE 2 Let $f(x, y) = \dfrac{x^2 - y^2}{x^2 + y^2}$. Determine whether $\lim_{P \to (0,0)} f(P)$ exists.

SOLUTION When (x, y) is near $(0, 0)$, both the numerator and denominator of $(x^2 - y^2)/(x^2 + y^2)$ are small numbers. There are, as in Chap. 2, two influences. The numerator is pushing the quotient toward 0 while the denominator is influencing the quotient to be large. We must be careful.

We try a few inputs near (0, 0). For instance, (0.01, 0) is near (0, 0) and

$$f(0.01, 0) = \frac{(0.01)^2 - 0^2}{(0.01)^2 + 0^2} = 1.$$

Also, (0, 0.01) is near (0, 0) and

$$f(0, 0.01) = \frac{0^2 - (0.01)^2}{0^2 + (0.01)^2} = -1.$$

More generally for $x \neq 0$,

$$f(x, 0) = 1;$$

for $y \neq 0$, $$f(0, y) = -1.$$

Since x can be as near 0 as we please and y can be as near 0 as we please, it is *not* the case that $\lim_{P \to (0,0)} f(P)$ exists. ■

If P_0 is not an interior point of the domain of f, we modify the definition of limit slightly. Let P_0 be a point on the boundary of the domain of f. If $f(P) \to L$ as P approaches P_0 through points in the domain of f, we say that "L is the limit of $f(P)$ as $P \to P_0$."

The Precise Definition of $\lim_{P \to P_0} f(P) = L$

In Sec. 2.10 we gave a precise definition of $\lim_{x \to a} f(x) = L$. The following definition is made in the same spirit.

Definition *Precise definition of limit of $f(x, y)$ at $P_0 = (a, b)$.* Let f be a function defined at least at every point in some disk with center P_0 except perhaps at P_0. Let L be a number. Assume that for each positive number ϵ there is a positive number δ such that

$$|f(P) - L| < \epsilon$$

whenever P is in the disk of radius δ and center P_0 but is not P_0. In that case we call L **the limit of $f(P)$ as P approaches P_0.**

Figure 5 shows the "challenge" ϵ and the "response" disk of radius δ.

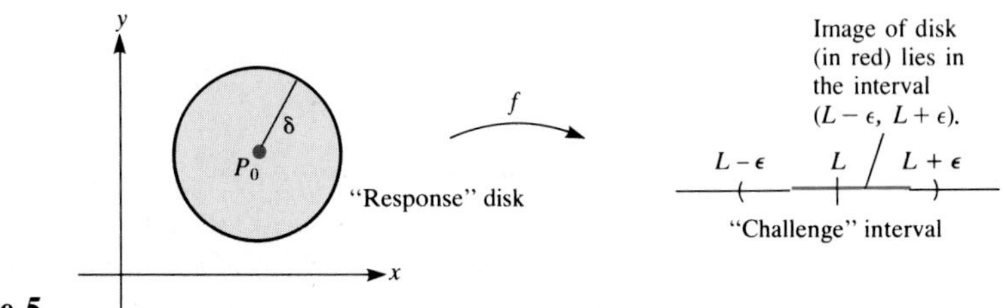

Figure 5

For every point P in the shaded disk (except perhaps P_0) $f(P)$ lies in the open interval $(L - \epsilon, L + \epsilon)$.

Continuity

Once we have the notion of a limit of a function of two variables, we are in a position to define a continuous function $f(x, y)$.

Recall how continuity at a number a of $y = f(x)$ was defined in Sec. 2.8. There were three conditions:

1 $f(a)$ is defined.
2 $\lim_{x \to a} f(x)$ exists.
3 $\lim_{x \to a} f(x) = f(a)$.

We define the continuity of $f(x, y)$ similarly.

Definition *Continuity of $f(x, y)$ at $P_0 = (a, b)$.* Assume that $f(P)$ is defined throughout some disk with center P_0. Then f is **continuous** at P_0 if $\lim_{P \to P_0} = f(P_0)$.

This means

1 $f(P_0)$ is defined (that is, P_0 is in the domain of f).
2 $\lim_{P \to P_0} f(P)$ exists.
3 $\lim_{P \to P_0} f(P) = f(P_0)$.

Continuity at a point on the boundary of the domain can be defined similarly. A function $f(P)$ is **continuous** if it is continuous at every point in its domain.

EXAMPLE 3 Determine whether $f(x, y) = \dfrac{x^2 - y^2}{x^2 + y^2}$ is continuous at $(1, 1)$.

SOLUTION This is the same function explored in Examples 1 and 2. First, $f(1, 1)$ is defined. (It equals 0.) Second, $\lim_{(x,y) \to (1,1)} \dfrac{x^2 - y^2}{x^2 + y^2}$ exists. (It is 0.) Third, $\lim_{(x,y) \to (1,1)} f(x, y) = f(1, 1)$.

Hence $f(x, y)$ is continuous at $(1, 1)$. ■

In fact, the function of Example 3 is continuous at every point (x, y) in its domain. We do not need to worry about the behavior of $f(x, y)$ when (x, y) is near $(0, 0)$ because $(0, 0)$ is *not* in the domain. Since $f(x, y)$ is continuous at every point *in its domain,* it is a continuous function.

Remark. Note the similarity of the behavior of the function in Example 3 to the behavior of the function $f(x) = 1/x$. Though $1/x$ "explodes" near $x = 0$, it, too, is a continuous function.

EXERCISES FOR SEC. 14.4: LIMITS AND CONTINUITY

In Exercises 1 to 8 evaluate the limits, if they exist.

1 $\lim_{(x,y)\to(2,3)} \dfrac{x+y}{x^2+y^2}$

2 $\lim_{(x,y)\to(1,1)} \dfrac{x^2}{x^2+y^2}$

3 $\lim_{(x,y)\to(0,0)} \dfrac{x^2}{x^2+y^2}$

4 $\lim_{(x,y)\to(0,0)} \dfrac{xy}{x^2+y^2}$

5 $\lim_{(x,y)\to(2,3)} x^y$

6 $\lim_{(x,y)\to(0,0)} (x^2)^y$

7 $\lim_{(x,y)\to(0,0)} (1+xy)^{1/(xy)}$

8 $\lim_{(x,y)\to(0,0)} (1+x)^{1/y}$

In Exercises 9 to 14, (*a*) describe the domain of the given functions and (*b*) state whether the functions are continuous.

9 $f(x, y) = 1/(x + y)$

10 $f(x, y) = 1/(x^2 + 2y^2)$

11 $f(x, y) = 1/(9 - x^2 - y^2)$

12 $f(x, y) = \sqrt{x^2 + y^2 - 25}$

13 $f(x, y) = \sqrt{16 - x^2 - y^2}$

14 $f(x, y) = 1/\sqrt{49 - x^2 - y^2}$

In Exercises 15 to 20 find the boundary of the given region R.

15 R consists of all points (x, y) such that $x^2 + y^2 \le 1$.

16 R consists of all points (x, y) such that $x^2 + y^2 < 1$.

17 R consists of all points (x, y) such that $1/(x^2 + y^2)$ is defined.

18 R consists of all points (x, y) such that $1/(x + y)$ is defined.

19 R consists of all points (x, y) such that $y < x^2$.

20 R consists of all points (x, y) such that $y \le x$.

Exercises 21 to 24 concern the precise definition of $\lim_{(x,y)\to P_0} f(x, y)$.

21 Let $f(x, y) = x + y$.

(*a*) Show that if $P = (x, y)$ lies within a distance 0.01 of (1, 2), then $|x - 1| < 0.01$ and $|y - 2| < 0.01$. (See Fig. 6.)

(*b*) Show that if $|x - 1| < 0.01$ and $|y - 2| < 0.01$, then $|f(x, y) - 3| < 0.02$.

(*c*) Find a number $\delta > 0$ such that if $P = (x, y)$ is in the disk of center (1, 2) and radius δ, then $|f(x, y) - 3| < 0.001$.

(*d*) Show that for any positive number ϵ, no matter how small, there is a positive number δ such that when $P = (x, y)$ is in the disk of radius δ and center (1, 2), then $|f(x, y) - 3| < \epsilon$. (Give δ as a function of ϵ.)

(*e*) What may we conclude on the basis of (*d*)?

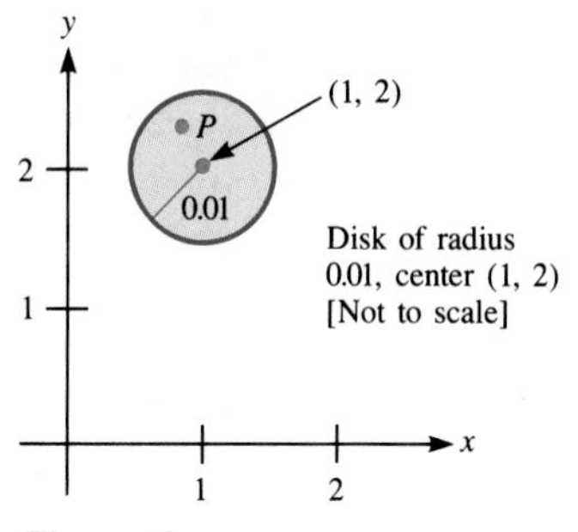

Figure 6

22 Let $f(x, y) = 2x + 3y$.

(*a*) Find a disk with center (1, 1) such that whenever P is in that disk, $|f(P) - 5| < 0.01$.

(*b*) Let ϵ be any positive number. Show that there is a disk with center (1, 1) such that whenever P is in that disk, $|f(P) - 5| < \epsilon$. (Give δ as a function of ϵ.)

(*c*) What may we conclude on the basis of (*b*)?

23 Let $f(x, y) = x^2y/(x^4 + 2y^2)$.

(*a*) What is the domain of f?

(*b*) Fill in this table:

(x, y)	(0.01, 0.01)	(0.01, 0.02)	(0.001, 0.003)
$f(x, y)$			

(*c*) On the basis of (*b*), do you think $\lim_{P\to(0,0)} f(P)$ exists? If so, what is its value?

(*d*) Fill in this table:

(x, y)	(0.5, 0.25)	(0.1, 0.01)	(0.001, 0.000001)
$f(x, y)$			

(*e*) On the basis of (*d*), do you think $\lim_{P\to(0,0)} f(P)$ exists? If so, what is its value?

(*f*) Does $\lim_{P\to(0,0)} f(P)$ exist? If so, what is it? Explain.

24 Let $f(x, y) = 5x^2y/(2x^4 + 3y^2)$.
(*a*) What is the domain of f?
(*b*) As P approaches $(0, 0)$ on the line $y = 2x$, what happens to $f(P)$?
(*c*) As P approaches $(0, 0)$ on the line $y = 3x$, what happens to $f(P)$?
(*d*) As P approaches $(0, 0)$ on the parabola $y = x^2$, what happens to $f(P)$?
(*e*) Does $\lim_{P \to (0,0)} f(P)$ exist? If so, what is it? Explain.

14.5 PARTIAL DERIVATIVES

This section generalizes the notion of a derivative to functions of two or more variables.

Recall the definition of the derivative from Chap. 3. The derivative of $f(x)$ at the number a is defined as

$$\lim_{\Delta x \to 0} \frac{f(a + \Delta x) - f(a)}{\Delta x}.$$

We may interpret this derivative as the slope of the tangent to the curve $y = f(x)$ at the point $(a, f(a))$. We may also think of it as telling how rapidly the function $f(x)$ changes for x near a. Both of these interpretations will generalize to functions of several variables.

Partial Derivatives

Consider the function $f(x, y) = x^2y^3$. If we hold y constant and differentiate with respect to x, we obtain $d(x^2y^3)/dx = 2xy^3$. This derivative is called the "partial derivative" of x^2y^3 with respect to x. We could hold x fixed instead and find the derivative of x^2y^3 with respect to y, that is, $d(x^2y^3)/dy = 3x^2y^2$. This derivative is called the "partial derivative" of x^2y^3 with respect to y. This example introduces the general idea of a partial derivative. First we define them. Then we will see what they mean in terms of slope and rate of change.

Definition *Partial derivatives*. Assume that the domain of $f(x, y)$ includes the region within some disk with center (a, b). If

$$\lim_{\Delta x \to 0} \frac{f(a + \Delta x, b) - f(a, b)}{\Delta x}$$

exists, this limit is called the **partial derivative of f with respect to x** at (a, b). Similarly, if

$$\lim_{\Delta y \to 0} \frac{f(a, b + \Delta y) - f(a, b)}{\Delta y}$$

exists, it is called the **partial derivative of f with respect to y** at (a, b).

The following notations are used for the partial derivative of $z = f(x, y)$ with respect to x:

Notations for partial derivatives

$$\frac{\partial z}{\partial x}, \quad \frac{\partial f}{\partial x}, \quad f_x, \quad f_1, \quad \text{or} \quad z_x.$$

And the following are used for the partial derivative of $z = f(x, y)$ with respect to y:

$$\frac{\partial z}{\partial y}, \quad \frac{\partial f}{\partial y}, \quad f_y, \quad f_2, \quad \text{or} \quad z_y.$$

Since physicists and engineers use the subscript notation in the study of vectors, they prefer to use

$$\frac{\partial f}{\partial x} \quad \text{and} \quad \frac{\partial f}{\partial y}$$

to denote the two partial derivatives. (As previously noted, they write $\mathbf{A} = a\mathbf{i} + b\mathbf{j} + c\mathbf{k}$ as $\mathbf{A} = A_x\hat{x} + A_y\hat{y} + A_z\hat{z}$, where the subscripts do *not* represent partial derivatives.) We will tend to use the notation of the physicists and engineers for partial derivatives (although not for vectors). The symbol $\partial f/\partial x$ may be viewed as "the rate at which the function $f(x, y)$ changes when x varies and y is kept fixed." The symbol $\partial f/\partial y$ records "the rate at which the function $f(x, y)$ changes when y varies and x is kept fixed."

The value of $\partial f/\partial x$ at (a, b) is denoted

$$\frac{\partial f}{\partial x}(a, b) \quad \text{or} \quad \left.\frac{\partial f}{\partial x}\right|_{(a,b)}.$$

(We prefer the first form.) In the middle of a sentence, we will write it as $\partial f/\partial x(a, b)$.

EXAMPLE 1 If $f(x, y) = \sin(x^2y)$, find (*a*) $\partial f/\partial x$, (*b*) $\partial f/\partial y$, and (*c*) $\partial f/\partial y$ at $(1, \pi/4)$.

SOLUTION

(*a*) To find $(\partial/\partial x)(\sin x^2y)$, differentiate with respect to x, keeping y constant:

$$\begin{aligned}\frac{\partial}{\partial x}(\sin x^2y) &= \cos(x^2y)\frac{\partial}{\partial x}(x^2y) && \textit{chain rule}\\ &= \cos(x^2y)(2xy) && \textit{y is constant}\\ &= 2xy\cos(x^2y).\end{aligned}$$

(*b*) To find $(\partial/\partial y)(\sin x^2y)$, differentiate with respect to y, keeping x constant:

$$\begin{aligned}\frac{\partial}{\partial y}(\sin x^2y) &= \cos(x^2y)\frac{\partial}{\partial y}(x^2y) && \textit{chain rule}\\ &= \cos(x^2y)(x^2) && x^2 \textit{ is constant}\\ &= x^2\cos(x^2y).\end{aligned}$$

(*c*) By (*b*)

$$\begin{aligned}\frac{\partial f}{\partial y}(1, \pi/4) &= x^2\cos(x^2y)\big|_{(1,\pi/4)}\\ &= 1^2\cos\left(1^2\frac{\pi}{4}\right)\\ &= \frac{\sqrt{2}}{2}. \quad ■\end{aligned}$$

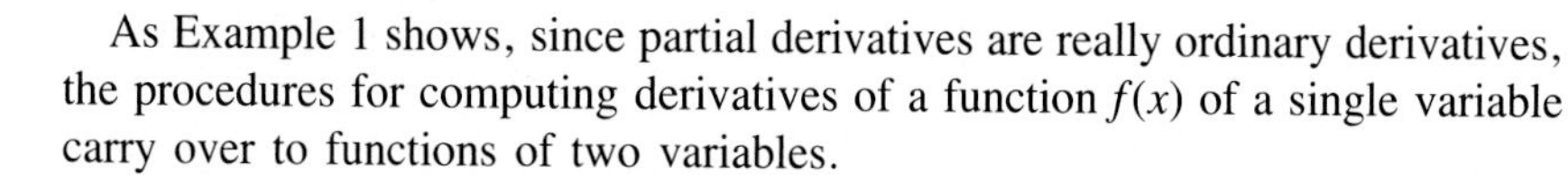
As Example 1 shows, since partial derivatives are really ordinary derivatives, the procedures for computing derivatives of a function $f(x)$ of a single variable carry over to functions of two variables.

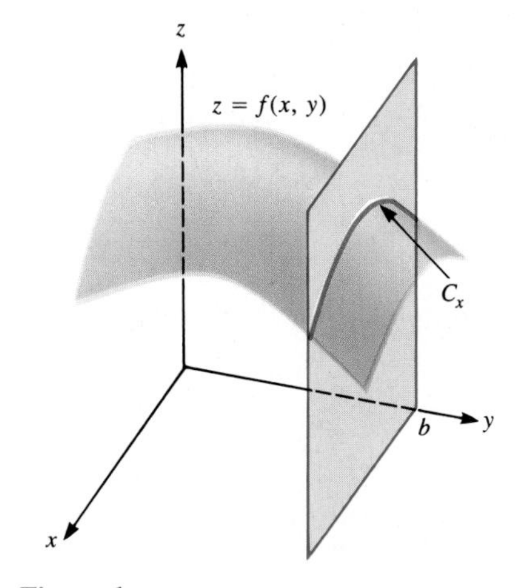

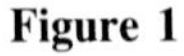
Figure 1

Partial Derivatives and Slope

Let $f(x, y)$ be defined at a point (a, b) and at least throughout some disk with center (a, b). Figure 1 shows the trace of the surface $z = f(x, y)$ in the plane $y = b$. Call this curve C_x, since x is free to vary on this curve. Figure 2 shows the trace of the surface in the plane $x = a$. Call this curve C_y, since y is free to vary on this curve. Let $P = (a, b, f(a, b))$. The curve C_x is the graph of $z = f(x, y)$ where y has the fixed value b. Its slope at P is the derivative of f with respect to x, where y is fixed at the value b. In other words

$$\frac{\partial f}{\partial x}(a, b)$$

is the slope of the curve C_x at the point P. Similarly

$$\frac{\partial f}{\partial y}(a, b)$$

is the slope of the curve C_y at the point P.

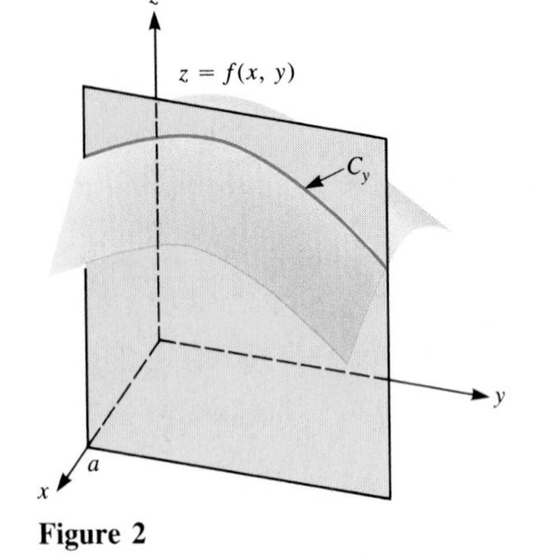

Figure 2

EXAMPLE 2 Let $f(x, y) = xy^2/6$. (*a*) Compute $(\partial f/\partial y)(1, 1)$. (*b*) Use (*a*) to draw the tangent to the trace C_y of the surface $z = xy^2/6$ in the plane $x = 1$.

SOLUTION

(*a*)

$$\frac{\partial}{\partial y}\left(\frac{xy^2}{6}\right) = \frac{2xy}{6} = \frac{xy}{3}.$$

Thus
$$\left.\frac{\partial}{\partial y}\left(\frac{xy^2}{6}\right)\right|_{(1,1)} = \frac{1}{3}.$$

(*b*) By (*a*) the slope of the tangent to the curve C_y in Fig. 3 is $\frac{1}{3}$. This slope represents a rise of $\frac{1}{3}$ in a horizontal run of 1.

Note that on the curve C_y, $z = 1 \cdot y^2/6 = y^2/6$. Thus the trace of the surface in the plane $x = 1$ is a parabola. ■

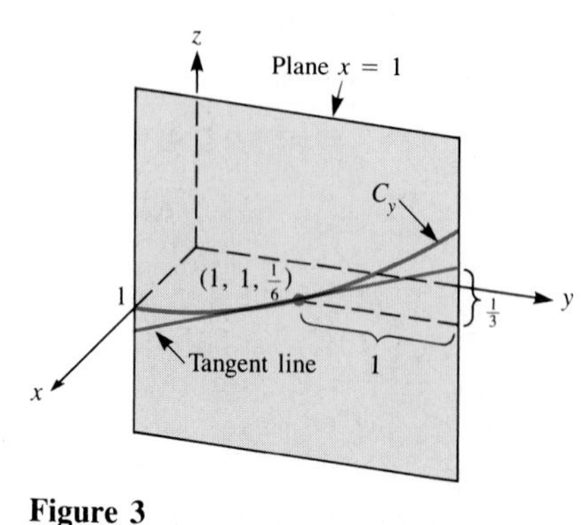

Figure 3

Higher-Order Partial Derivatives

Just as there are derivatives of derivatives, so are there partial derivatives of partial derivatives. For instance, if

$$z = 2x + 5x^4y^7,$$

then
$$\frac{\partial z}{\partial x} = 2 + 20x^3y^7 \quad \text{and} \quad \frac{\partial z}{\partial y} = 35x^4y^6.$$

We may go on and compute partial derivatives of $\partial z/\partial x$ and $\partial z/\partial y$:

$$\frac{\partial}{\partial x}\left(\frac{\partial z}{\partial x}\right) = 60x^2y^7 \qquad \frac{\partial}{\partial y}\left(\frac{\partial z}{\partial x}\right) = 140x^3y^6$$

$$\frac{\partial}{\partial x}\left(\frac{\partial z}{\partial y}\right) = 140x^3y^6 \qquad \frac{\partial}{\partial y}\left(\frac{\partial z}{\partial y}\right) = 210x^4y^5.$$

There are four partial derivatives of the second order:

$$\frac{\partial}{\partial x}\left(\frac{\partial z}{\partial x}\right), \quad \frac{\partial}{\partial y}\left(\frac{\partial z}{\partial x}\right), \quad \frac{\partial}{\partial y}\left(\frac{\partial z}{\partial y}\right), \quad \frac{\partial}{\partial x}\left(\frac{\partial z}{\partial y}\right).$$

These are usually denoted, in order,

$$\frac{\partial^2 z}{\partial x^2}, \quad \frac{\partial^2 z}{\partial y\,\partial x}, \quad \frac{\partial^2 z}{\partial y^2}, \quad \frac{\partial^2 z}{\partial x\,\partial y}.$$

To compute $\partial^2 z/\partial x\,\partial y$, you first differentiate with respect to y, then with respect to x. To compute $\partial^2 z/\partial y\,\partial x$, you first differentiate with respect to x, then with respect to y. In both cases "differentiate from right to left in the order that the variables occur."

Equality of the mixed partials

In the computations just done, the two mixed partials $\partial^2 z/\partial x\,\partial y$ and $\partial^2 z/\partial y\,\partial x$ are equal. For the functions commonly encountered, the two mixed partials are equal. (For a proof, see Appendix K.)

EXAMPLE 3 Compute $\dfrac{\partial^2 z}{\partial x^2}$, $\dfrac{\partial^2 z}{\partial y\,\partial x}$, and $\dfrac{\partial^2 z}{\partial x\,\partial y}$ for $z = y \cos xy$.

SOLUTION First compute

$$\frac{\partial^2 z}{\partial x^2} = \frac{\partial}{\partial x}\left(\frac{\partial z}{\partial x}\right) = \frac{\partial}{\partial x}(-y^2 \sin xy) = -y^3 \cos xy.$$

Then

$$\frac{\partial^2 z}{\partial y\,\partial x} = \frac{\partial}{\partial y}\left(\frac{\partial z}{\partial x}\right) = \frac{\partial}{\partial y}(-y^2 \sin xy) = -2y \sin xy - xy^2 \cos xy.$$

Finally,

$$\begin{aligned}\frac{\partial^2 z}{\partial x\,\partial y} &= \frac{\partial}{\partial x}\left(\frac{\partial z}{\partial y}\right)\\ &= \frac{\partial}{\partial x}(-yx \sin xy + \cos xy)\\ &= -y\frac{\partial}{\partial x}(x \sin xy) + \frac{\partial}{\partial x}(\cos xy)\\ &= -y(xy \cos xy + \sin xy) - y \sin xy\end{aligned}$$

$$= -xy^2 \cos xy - y \sin xy - y \sin xy$$
$$= -2y \sin xy - xy^2 \cos xy.$$

As expected, the two mixed partials are equal. ■

Functions of More Than Two Variables

A quantity may depend on more than two variables. For instance, the volume of a box depends on three variables: the length l, width w, and height h, $V = lwh$. The "chill factor" depends on the temperature, humidity, and wind velocity. The temperature T at any point in the atmosphere is a function of the three space coordinates, x, y, and z: $T = f(x, y, z)$.

To differentiate, hold all variables constant except one.

The notions and notations of partial derivatives carry over to functions of more than two variables. If $u = f(x, y, z, t)$, there are four first-order partial derivatives. For instance, the partial derivative of u with respect to x, holding y, z, and t fixed, is denoted

$$\frac{\partial u}{\partial x}, \qquad \frac{\partial f}{\partial x}, \qquad u_x, \qquad \text{etc.}$$

Also, we may list the variables held constant. For instance,

$$\left(\frac{\partial u}{\partial x}\right)_{y,z,t}$$

denotes the partial derivative of u with respect to x when y, z, and t are held constant. Higher-order partial derivatives are defined and denoted similarly.

Many basic problems in chemistry and physics are examined in terms of equations involving partial derivatives (known as PDEs). The next example illustrates this.

EXAMPLE 4 Let $T(x, y, z)$ be the temperature at the point (x, y, z) in a metal object. If the temperature does not vary with time, then it is known that

$$\frac{\partial^2 T}{\partial x^2} + \frac{\partial^2 T}{\partial y^2} + \frac{\partial^2 T}{\partial z^2} = 0. \tag{1}$$

(This will be shown in Chap. 17.) Show that the function $T(x, y, z) = 1/\sqrt{x^2 + y^2 + z^2}$ satisfies Eq. (1).

SOLUTION We compute $\partial^2 T/\partial x^2$, $\partial^2 T/\partial y^2$, $\partial^2 T/\partial z^2$, substitute them into (1), and see if we get a true equation. To begin,

$$\frac{\partial T}{\partial x} = \frac{\partial}{\partial x}(x^2 + y^2 + z^2)^{-1/2}$$

$$= -\frac{1}{2}(2x)(x^2 + y^2 + z^2)^{-3/2}$$

$$= \frac{-x}{(x^2 + y^2 + z^2)^{3/2}}.$$

Thus $$\frac{\partial^2 T}{\partial x^2} = \frac{\partial}{\partial x}\left(\frac{-x}{(x^2 + y^2 + z^2)^{3/2}}\right)$$

$$= \frac{(x^2 + y^2 + z^2)^{3/2}(-1) - (-x)(3/2)(x^2 + y^2 + z^2)^{1/2}\,2x}{(x^2 + y^2 + z^2)^3}$$

$$= \frac{-(x^2 + y^2 + z^2)^{3/2} + 3x^2(x^2 + y^2 + z^2)^{1/2}}{(x^2 + y^2 + z^2)^3}$$

$$= \frac{-(x^2 + y^2 + z^2) + 3x^2}{(x^2 + y^2 + z^2)^{5/2}}.$$

Similar computations (or symmetry in the roles of x, y, and z) show that

$$\frac{\partial^2 T}{\partial y^2} = \frac{-(x^2 + y^2 + z^2) + 3y^2}{(x^2 + y^2 + z^2)^{5/2}}$$

and $$\frac{\partial^2 T}{\partial z^2} = \frac{-(x^2 + y^2 + z^2) + 3z^2}{(x^2 - y^2 + z^2)^{5/2}}.$$

Thus $$\frac{\partial^2 T}{\partial x^2} + \frac{\partial^2 T}{\partial y^2} + \frac{\partial^2 T}{\partial z^2} = \frac{-3(x^2 + y^2 + z^2) + 3x^2 + 3y^2 + 3z^2}{(x^2 + y^2 + z^2)^{5/2}}$$

$$= 0.$$

The function $T(x, y, z) = 1/\sqrt{x^2 + y^2 + z^2}$ satisfies (1). ■

It is quite rare for a function $u = f(x, y, z)$ to satisfy (1). For instance, $u = x^2yz$ does *not* satisfy (1), as the following calculations show:

$$\frac{\partial u}{\partial x} = 2xyz, \qquad \frac{\partial u}{\partial y} = x^2 z, \qquad \frac{\partial u}{\partial z} = x^2 y;$$

hence $$\frac{\partial^2 u}{\partial x^2} = 2yz, \qquad \frac{\partial^2 u}{\partial y^2} = 0, \qquad \frac{\partial^2 u}{\partial z^2} = 0.$$

It follows that for all (x, y, z),

$$\frac{\partial^2 u}{\partial x^2} + \frac{\partial^2 u}{\partial y^2} + \frac{\partial^2 u}{\partial z^2} = 2yz.$$

Therefore $u = x^2yz$ does *not* satisfy (1). That tells us that x^2yz cannot describe a temperature distribution in space that does not vary in time.

Partial Derivatives in Economics

Functions of several variables are common in the study of economics. For instance, the total cost of production C may depend on the number of workers w, the hourly wage h, the amount of raw materials used r, and the cost per pound of the raw materials p. We would have $C = wh + rp$, a function of four variables.

EXAMPLE 5 Assume that $C = 21h + 32p$, where h is the hourly wage and p is the cost per pound of raw materials. Compute $\partial C/\partial h$ and $\partial C/\partial p$ and interpret them.

SOLUTION

$$\frac{\partial C}{\partial h} = 21;$$

the rate of change of total cost is \$21 for each \$1 rise in the hourly wage, assuming that the price of raw materials is held constant. Also

$$\frac{\partial C}{\partial p} = 32;$$

the rate of change of total cost is \$32 for each \$1 rise in the price per pound of raw materials, assuming that the hourly wage is held fixed. ■

Partial Derivatives and a Vibrating String

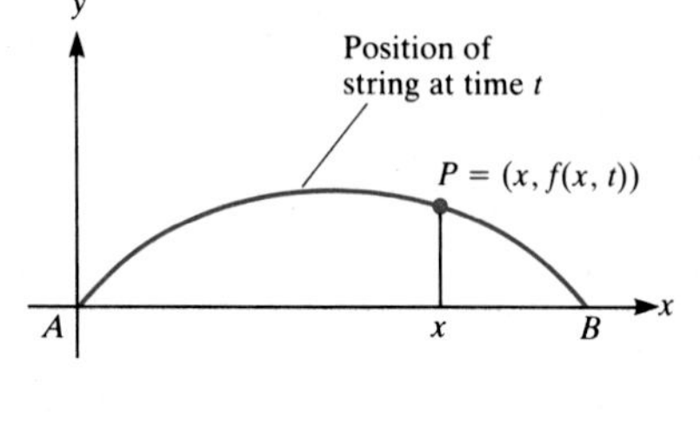

Figure 4

Figure 4 shows the position of a vibrating string at time t. The string is fixed at points A and B and moves parallel to the y axis. Let $y = f(x, t)$ be the height of the string at the point with abscissa x at time t, as shown in the figure. In this case, the partial derivatives are denoted $\partial y/\partial x$ and $\partial y/\partial t$. The first one, $\partial y/\partial x$, is the slope of the curve in Fig. 4 at the point P. The second, $\partial y/\partial t$, is the velocity of the string along the vertical line with abscissa x. In other words, $\partial y/\partial t$ tells how fast the point P is moving.

Section Summary

We introduced the partial derivatives of functions of several variables. For instance, if $u = f(x, y, z, t)$, then the partial derivative $\partial u/\partial x$ is the derivative of u with respect to x, keeping the remaining variables y, z, and t constant. Higher-order partial derivatives are defined much like higher-order derivatives. Two corresponding "mixed partials," such as

$$\frac{\partial}{\partial y}\left(\frac{\partial u}{\partial x}\right) \quad \text{and} \quad \frac{\partial}{\partial x}\left(\frac{\partial u}{\partial y}\right),$$

are equal for the functions usually met in applications.

EXERCISES FOR SEC. 14.5: PARTIAL DERIVATIVES

In Exercises 1 to 14 compute $\partial f/\partial x$ and $\partial f/\partial y$ for the given functions.

1 $3x + 2y$
2 $x^2y + 4$
3 x^3y^4
4 $6x - 7y$
5 $x \cos xy$
6 $\ln(x + 2y)$
7 $\tan^{-1} xy$
8 $\tan^{-1} y/x$
9 $x \sin^2(x + y)$
10 $(x^2 + \cos 3y)/(1 + x)$
11 $\sqrt{x} \sec x^2y$
12 $\sin^{-1} (x + 3y)$
13 e^x/y^3
14 $1/\sqrt{x^2 + y^2}$

In Exercises 15 to 20 compute the four partial derivatives of order two:

$$\frac{\partial^2 f}{\partial x^2}, \quad \frac{\partial^2 f}{\partial y^2}, \quad \frac{\partial^2 f}{\partial x\, \partial y}, \quad \text{and} \quad \frac{\partial^2 f}{\partial y\, \partial x}$$

and check that the last two are equal.

15 $f(x, y) = 5x^2 - 3xy + 6y^2$
16 $f(x, y) = x^4y^7$
17 $f(x, y) = \dfrac{1}{\sqrt{x^2 + y^2}}$
18 $f(x, y) = \sin (x^2y)$
19 $f(x, y) = \tan (x + 3y)$
20 $f(x, y) = \dfrac{x}{y}$

In each of Exercises 21 to 24 find the slope of the trace of the given surface in a plane perpendicular to the y axis and passing through the given point.

21 $z = xy^2$ at $(1, 2)$

22 $z = \cos(x + 2y)$ at $(\pi/4, \pi/2)$
23 $z = x/y$ at $(1, 1)$
24 $z = x^2 e^{xy}$ at $(1, 0)$

In each of Exercises 25 and 26 find the slope of the trace of the given surface in a plane perpendicular to the x axis and passing through the given point.
25 $z = ye^{xy}$ at $(1, 1)$
26 $z = e^{x/y}$ at $(0, 1)$
27 (*a*) Draw the trace of the surface $z = x^2 + y^2$ in the plane $x = 2$.
(*b*) Find the slope of the curve in (*a*) at the point $(2, 1, 5)$.
(*c*) Draw the tangent line to the curve in (*a*) at the point $(2, 1, 5)$. [Make use of the slope found in (*b*)].
28 (*a*) Draw the trace of the surface $z = x^2 y$ in the plane $y = \frac{1}{2}$.
(*b*) Find the slope of the curve in (*a*) at the point $(1, \frac{1}{2}, \frac{1}{2})$.
(*c*) Draw the tangent line to the curve in (*a*) at the point $(1, \frac{1}{2}, \frac{1}{2})$. [Make use of the slope found in (*b*).]
29 Assume $f(1, 1) = 3$, $f(1.02, 1) = 3.05$, and $f(1, 0.97) = 2.4$. Use this information to estimate $\partial f/\partial x$ and $\partial f/\partial y$ at $(1, 1)$.
30 Assume $f(2, 3) = 1$, $f(1.98, 3) = 1.03$, and $f(2, 3.04) = 0.98$. Use this information to estimate $\partial f/\partial x$ and $\partial f/\partial y$ at $(2, 3)$.

---

31 The temperature $T(x, y)$ in a flat piece of metal in the xy plane is recorded at the points listed in this table:

Point	(1, 2)	(1.01, 2)	(1, 2.02)	(1.01, 2.02)
Temp.	5	5.025	5.06	5.09

On the basis of this information, estimate $\partial T/\partial x$ and $\partial T/\partial y$ at $(1, 2)$.
32 Find $(\partial/\partial x)(x^2 y)$ at (a, b) using the definition of the partial derivative $\partial f/\partial x$ as a certain limit.
33 The function f defined for all points (x, y) has the property that $\partial f/\partial x = 0$ and $\partial f/\partial y = 0$ for all (x, y). Moreover, $f(1, 1) = 3$.
(*a*) Give an example of such a function.
(*b*) Find all such functions, explaining your reasoning.
34 Does $T(x, y) = 1/\sqrt{x^2 + y^2}$ satisfy the differential equation

$$\frac{\partial^2 T}{\partial x^2} + \frac{\partial^2 T}{\partial y^2} = 0?$$

(Note the similarity to Example 4.)

35 Is there a function f such that

$$\frac{\partial f}{\partial x} = e^x \cos y \quad \text{and} \quad \frac{\partial f}{\partial y} = e^x \sin y?$$

36 Let $f(x, y)$ be a continuous function. For each fixed value of y, define $g(y) = \int_a^b f(x, y)\, dx$.
(*a*) Using the definition of $g'(y)$, explain why the equation

$$\frac{dg}{dy} = \int_a^b \frac{\partial f}{\partial y}\, dx$$

is plausible.
(*b*) Verify that the equation in (*a*) holds when $f(x, y) = x^3 y^2$.
Note: The equation in (*a*) is justified in Appendix K.
37 Let $f(x, y) = \int_0^x \sqrt{y + t}\, dt$. Find (*a*) $\partial f/\partial x$, (*b*) $\partial f/\partial y$.
38 (This exercise suggests what the Taylor polynomials for functions of two variables look like.) Let $f(x, y) = a + bx + cy + dx^2 + exy + ky^2$, where a, b, c, d, e, and k are constants. Show that
(*a*) $a = f(0, 0)$
(*b*) $b = \dfrac{\partial f}{\partial x}(0, 0)$
(*c*) $c = \dfrac{\partial f}{\partial y}(0, 0)$
(*d*) $d = \dfrac{1}{2}\dfrac{\partial^2 f}{\partial x^2}(0, 0)$
(*e*) $e = \dfrac{\partial^2 f}{\partial x\, \partial y}(0, 0)$
(*f*) $k = \dfrac{1}{2}\dfrac{\partial^2 f}{\partial y^2}(0, 0)$
39 A function $z = f(x, t)$, where z denotes temperature, x position, and t time, is said to justify the heat equation if

$$a^2 \frac{\partial^2 f}{\partial x^2} = \frac{\partial f}{\partial t},$$

where a is a constant. Show that the function $f(x, t) = e^{-\pi^2 a^2 t} \sin \pi x$ satisfies the heat equation.
40 Let $u = f(x, y)$. Assume that $u(1, 2)$ is 3, $\partial u/\partial x$ at $(1, 2)$ is 2, and $\partial u/\partial y$ at $(1, 2)$ is 1.2.
(*a*) Estimate $u(1, 2.01)$.
(*b*) Estimate $u(0.98, 2)$.
(*c*) Estimate $u(1.02, 2.03)$.
In each part describe your reasoning.

14.6 THE CHAIN RULE

The chain rule for functions of one variable asserts that if $y = f(u)$ and $u = g(x)$, then

$$\frac{dy}{dx} = \frac{dy}{du}\frac{du}{dx}. \tag{1}$$

(See Sec. 3.6.) In this section we develop the chain rule for functions of several variables. The formula involves a *sum* of terms similar to the right-hand side of (1).

The Change Δf

Let $z = f(x, y)$ be a function of two variables with continuous partial derivatives at least throughout a disk centered at the point (x, y). We will express $\Delta f = f(x + \Delta x, y + \Delta y) - f(x, y)$ in terms of $\partial f/\partial x$ and $\partial f/\partial y$. Then we will be able to obtain the chain rule.

As Fig. 1 shows, the change in the value of the function f as the argument changes from (x, y) to $(x + \Delta x, y + \Delta y)$, labeled c, can be expressed as the sum of two changes. First, we keep y fixed, but change x to $x + \Delta x$ to get the change labeled a. Then we keep the x coordinate fixed at $x + \Delta x$ and change y to $y + \Delta y$ to get the change labeled b. As the figure shows, $c = a + b$. That is,

$$\begin{aligned}\Delta f &= f(x + \Delta x, y + \Delta y) - f(x, y)\\ &= a + b.\end{aligned}$$

Hence

$$\Delta f = [f(x + \Delta x, y + \Delta y) - f(x + \Delta x, y)] + [f(x + \Delta x, y) - f(x, y)]. \tag{2}$$

[Equation (2) is simply a special case of the algebraic identity $r - t = (r - s) + (s - t)$.] All that remains is to approximate each of the bracketed quantities in Eq. (2) in terms of partial derivatives of f.

Take the quantity $f(x + \Delta x, y) - f(x, y)$ first. This is the change in the func-

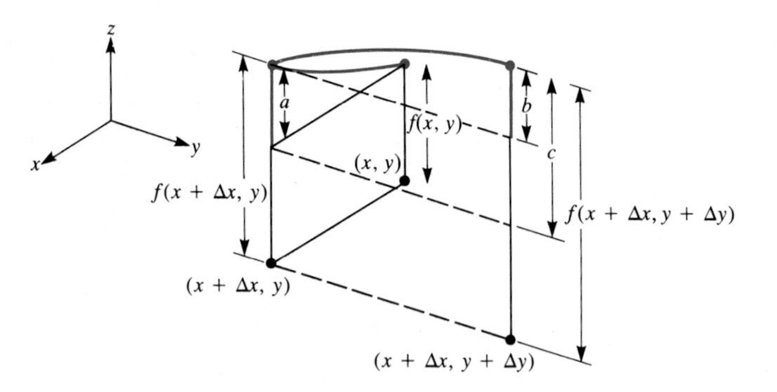

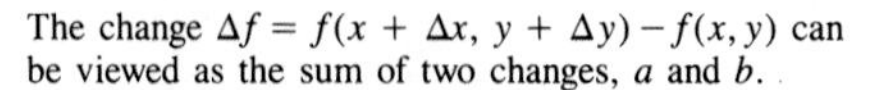

Figure 1 The change $\Delta f = f(x + \Delta x, y + \Delta y) - f(x, y)$ can be viewed as the sum of two changes, a and b.

tion as we move from (x, y) to $(x + \Delta x, y)$ in Fig. 2. On this path y is constant. By the mean-value theorem, there is a number c_1 between x and $x + \Delta x$ such that

$$f(x + \Delta x, y) - f(x, y) = \frac{\partial f}{\partial x}(c_1, y)\,\Delta x. \tag{3}$$

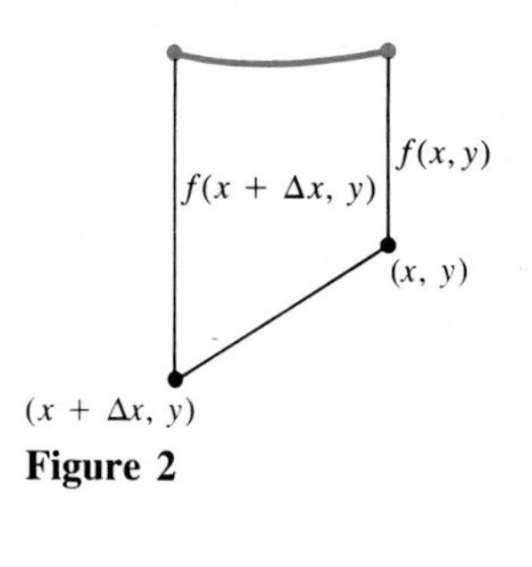

Figure 2

Next treat the difference $f(x + \Delta x, y + \Delta y) - f(x + \Delta x, y)$, the other bracketed quantity in Eq. (2). This quantity is the change in the function as we go from $(x + \Delta x, y)$ to $(x + \Delta x, y + \Delta y)$, a path on which the x coordinate is constant, with value $x + \Delta x$. By the mean-value theorem, there is a number c_2 between y and $y + \Delta y$ such that

$$f(x + \Delta x, y + \Delta y) - f(x + \Delta x, y) = \frac{\partial f}{\partial y}(x + \Delta x, c_2)\,\Delta y. \tag{4}$$

(See Fig. 3.)

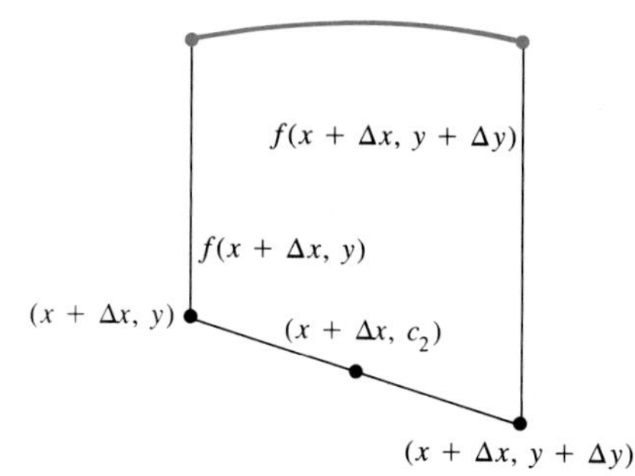

Figure 3

Combining Eqs. (2), (3), and (4) we obtain

$$\boxed{\Delta f = \frac{\partial f}{\partial x}(c_1, y)\,\Delta x + \frac{\partial f}{\partial y}(x + \Delta x, c_2)\,\Delta y.} \tag{5}$$

When both Δx and Δy are small, the points (c_1, y) and $(x + \Delta x, c_2)$ are near the point (x, y). If we assume that the partial derivatives $\partial f/\partial x$ and $\partial f/\partial y$ are continuous at $P = (x, y)$, then we may conclude that

$$\frac{\partial f}{\partial x}(c_1, y) = \frac{\partial f}{\partial x}(x, y) + \epsilon_1 \quad \text{and} \quad \frac{\partial f}{\partial y}(x + \Delta x, c_2) = \frac{\partial f}{\partial y}(x, y) + \epsilon_2, \tag{6}$$

where both ϵ_1 and ϵ_2 approach 0 as Δx and Δy both approach 0.

Combining Eqs. (5) and (6) gives the key to the chain rule. For emphasis, we state it as a theorem.

Theorem 1 Let f have continuous partial derivatives $\partial f/\partial x$ and $\partial f/\partial y$ for all points within some disk with center at the point (x, y). Then Δf, which is the change $f(x + \Delta x, y + \Delta y) - f(x, y)$, can be written

This equation is the core of this section.

$$\boxed{\Delta f = \frac{\partial f}{\partial x}(x, y)\,\Delta x + \frac{\partial f}{\partial y}(x, y)\,\Delta y + \epsilon_1\,\Delta x + \epsilon_2\,\Delta y,} \tag{7}$$

where ϵ_1 and ϵ_2 approach 0 as Δx and Δy approach 0. (Both ϵ_1 and ϵ_2 are functions of the four variables x, y, Δx, and Δy.)

The Chain Rule

Theorems 2 and 3 are two cases of the chain rule for functions of more than one variable. Afterward we will state the chain rule for functions of any number of variables.

Theorem 2 *Chain rule.* Let $z = f(x, y)$ have continuous partial derivatives $\partial f/\partial x$ and $\partial f/\partial y$, and let $x = x(t)$ and $y = y(t)$ be differentiable functions of t. Then z is a differentiable function of t and

$$\frac{dz}{dt} = \frac{\partial z}{\partial x}\frac{dx}{dt} + \frac{\partial z}{\partial y}\frac{dy}{dt}. \tag{8}$$

Proof By definition,

$$\frac{dz}{dt} = \lim_{\Delta t \to 0} \frac{\Delta z}{\Delta t}.$$

Now, Δt induces changes Δx and Δy in x and y, respectively. According to Theorem 1,

$$\Delta z = \frac{\partial f}{\partial x}(x, y)\,\Delta x + \frac{\partial f}{\partial y}(x, y)\,\Delta y + \epsilon_1\,\Delta x + \epsilon_2\,\Delta y,$$

where $\epsilon_1 \to 0$ and $\epsilon_2 \to 0$ as Δx and Δy approach 0. (Keep in mind that x and y are fixed.) Thus

$$\frac{\Delta z}{\Delta t} = \frac{\partial f}{\partial x}(x, y)\,\frac{\Delta x}{\Delta t} + \frac{\partial f}{\partial y}(x, y)\,\frac{\Delta y}{\Delta t} + \epsilon_1\,\frac{\Delta x}{\Delta t} + \epsilon_2\,\frac{\Delta y}{\Delta t}.$$

and

$$\frac{dz}{dt} = \lim_{\Delta t \to 0}\frac{\Delta z}{\Delta t} = \frac{\partial f}{\partial x}(x, y)\,\frac{dx}{dt} + \frac{\partial f}{\partial y}(x, y)\frac{dy}{dt} + 0\frac{dx}{dt} + 0\frac{dy}{dt}.$$

This proves the theorem. ■

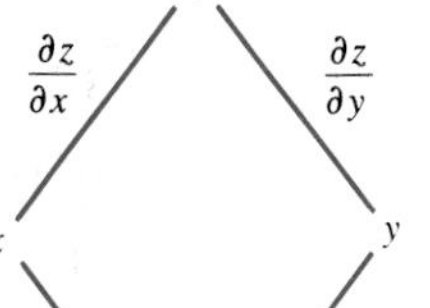
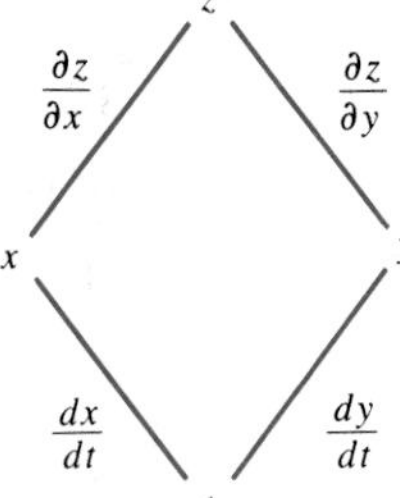

Figure 4

The two summands on the right-hand side of (8) remind us of the chain rule for functions of one variable. Why is there a "+" in (8)? The "+" first appears in Eq. (5) and you can trace it back to Figs. 2 and 3.

The diagram in Fig. 4 helps in using the chain rule of Theorem 2.

There are two paths from the top variable z down to the bottom variable t. Label each edge with the appropriate partial derivative (or derivative). For each path there is a summand in the chain rule. The left-hand path gives us the summand

$$\frac{\partial z}{\partial x}\frac{dx}{dt}.$$

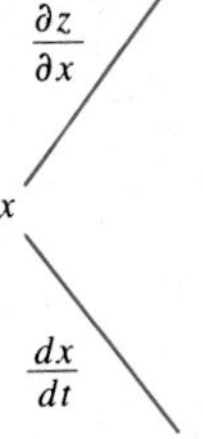

Figure 5

(See Fig. 5.) The right-hand path gives us the summand

$$\frac{\partial z}{\partial y}\frac{dy}{dt}.$$

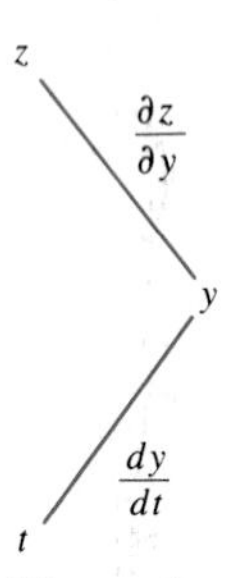

Figure 6

(See Fig. 6.) Then dz/dt is the sum of those two summands.

EXAMPLE 1 Let $z = x^2y^3$, $x = 3t^2$, and $y = t/3$. Find dz/dt when $t = 1$.

SOLUTION In order to apply Theorem 2, compute $\partial z/\partial x$, $\partial z/\partial y$, dx/dt, and dy/dt:

$$\frac{\partial z}{\partial x} = 2xy^3, \qquad \frac{\partial z}{\partial y} = 3x^2y^2$$

$$\frac{dx}{dt} = 6t, \qquad \frac{dy}{dt} = \frac{1}{3}.$$

By Theorem 2,

$$\frac{dz}{dt} = 2xy^3 \cdot 6t + 3x^2y^2 \cdot \frac{1}{3}.$$

In particular, when $t = 1$, x is 3 and y is $\frac{1}{3}$. Therefore, when $t = 1$,

$$\frac{dz}{dt} = 2 \cdot 3\left(\frac{1}{3}\right)^3 6 \cdot 1 + 3 \cdot 3^2\left(\frac{1}{3}\right)^2 \frac{1}{3}$$

$$= \frac{36}{27} + \frac{27}{27} = \frac{7}{3}. \quad ■$$

In Example 1, the derivative dz/dt can be found without using Theorem 2. To do this, express z explicitly in terms of t:

$$z = x^2y^3 = (3t^2)^2\left(\frac{t}{3}\right)^3 = \frac{t^7}{3}.$$

Then
$$\frac{dz}{dt} = \frac{7t^6}{3}.$$

When $t = 1$, this gives
$$\frac{dz}{dt} = \frac{7}{3},$$

in agreement with the first computation.

The proof of the next chain rule is almost identical to the proof of Theorem 2. (See Exercise 24.)

Theorem 3 *Chain rule.* Let $z = f(x, y)$ have continuous partial derivatives, $\partial f/\partial x$ and $\partial f/\partial y$. Let $x = x(t, u)$ and $y = y(t, u)$ have continuous partial derivatives

$$\frac{\partial x}{\partial t}, \quad \frac{\partial x}{\partial u}, \quad \frac{\partial y}{\partial t}, \quad \frac{\partial y}{\partial u}.$$

Then
$$\frac{\partial z}{\partial t} = \frac{\partial z}{\partial x}\frac{\partial x}{\partial t} + \frac{\partial z}{\partial y}\frac{\partial y}{\partial t}$$

and
$$\frac{\partial z}{\partial u} = \frac{\partial z}{\partial x}\frac{\partial x}{\partial u} + \frac{\partial z}{\partial y}\frac{\partial y}{\partial u}.$$

Figure 7 lists the variables:

	z		Top variable
x		y	Middle variables
t		u	Bottom variables

Figure 7

To find $\partial z/\partial t$, draw all the paths from z down to t. Label the edges by the appropriate partial derivative, as in Fig. 8.

Each path from the top variable down to the bottom variable contributes a summand in the chain rule. The only difference between Figs. 4 and 8 is that ordinary derivatives dx/dt and dy/dt appear in Fig. 4, while partial derivatives $\partial x/\partial t$ and $\partial y/\partial t$ appear in Fig. 8.

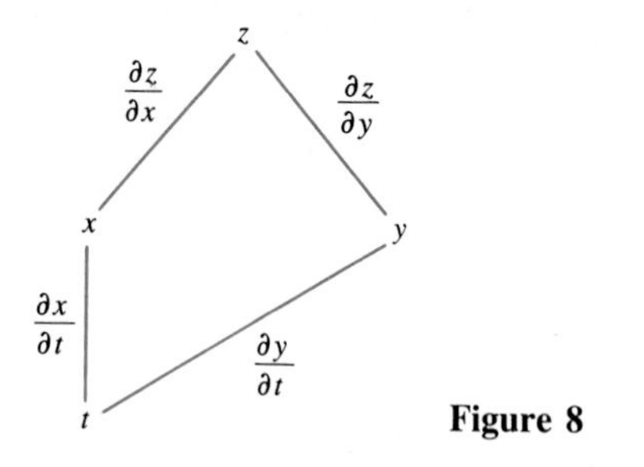

Figure 8

In the first chain rule (Theorem 2) there are two middle variables and one bottom variable. In the second chain rule (Theorem 3) there are two middle variables and two bottom variables. The chain rule holds for any number of middle variables and any number of bottom variables. For instance, there may be three middle variables and, say, four bottom variables. Call the middle variables x_1, x_2, x_3. (We use the subscript notation, for if we use letters, such as u, v, w, we will soon run out of the alphabet.) Call the bottom variables t_1, t_2, t_3, t_4 as in Fig. 9.

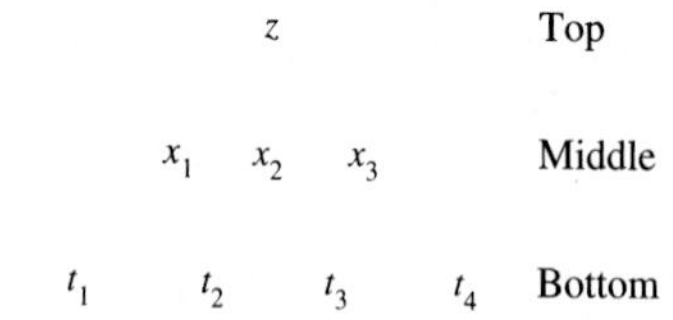

z Top

x_1 x_2 x_3 Middle

t_1 t_2 t_3 t_4 Bottom

Figure 9

In this situation z is a function of x_1, x_2, and x_3. Furthermore, each x_i is a function of t_1, t_2, t_3, and t_4. Hence z is a (composite) function of t_1, t_2, t_3, t_4. There are four partial derivatives of interest,

$$\frac{\partial z}{\partial t_1}, \frac{\partial z}{\partial t_2}, \frac{\partial z}{\partial t_3}, \frac{\partial z}{\partial t_4}.$$

(For instance, $\partial z/\partial t_1$ is the partial derivative of z with respect to t_1, holding t_2, t_3, and t_4 constant.) To find $\partial z/\partial t_1$, use Fig. 10.

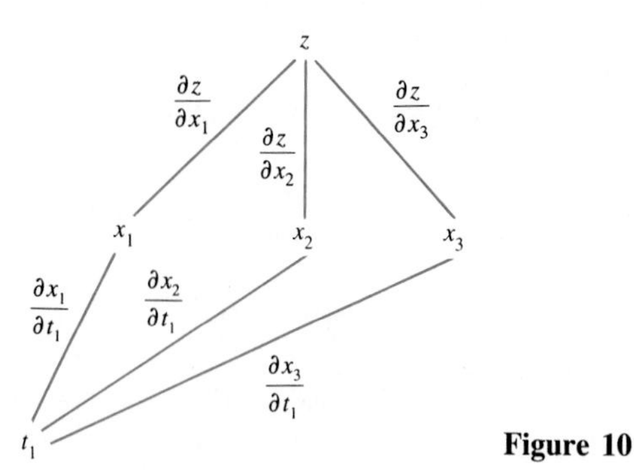

Figure 10

Then $\partial z/\partial t_1$ is the sum of three products—one product for each middle variable. Each such product is the product of the two partial derivatives on a path from the top variable z to the bottom variable t. In short,

$$\boxed{\frac{\partial z}{\partial t_1} = \frac{\partial z}{\partial x_1}\frac{\partial x_1}{\partial t_1} + \frac{\partial z}{\partial x_2}\frac{\partial x_2}{\partial t_1} + \frac{\partial z}{\partial x_3}\frac{\partial x_3}{\partial t_1}.} \tag{9}$$

For each bottom variable, the chain rule gives a formula for the partial derivative of the "top" variable z with respect to that variable.

In the general chain rule the top variable is a function of some "middle" variables, $x_1, x_2, \ldots, x_m$. Each middle variable x_i is a function of some "bottom" variables, $t_1, t_2, \ldots, t_n$. Then the partial derivative of the top variable with respect to a bottom variable t_j is a sum:

Chain rule in general form

$$\boxed{\frac{\partial z}{\partial t_j} = \sum_{i=1}^{m} \frac{\partial z}{\partial x_i}\frac{\partial x_i}{\partial t_j}.}$$

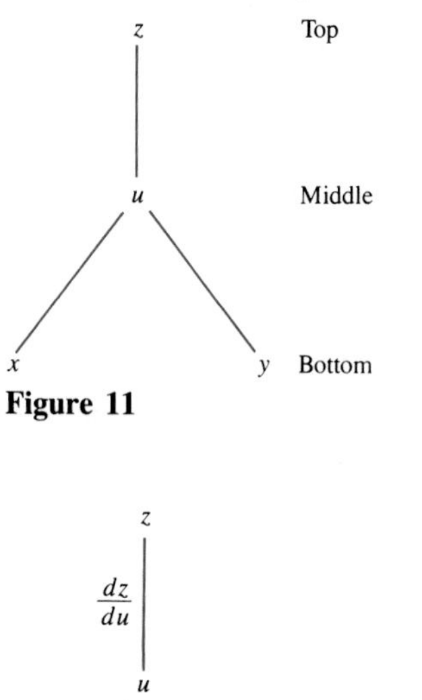

Figure 11

In the general chain rule there is a summand for each middle variable.

In the next example there is only one middle variable and two bottom variables.

EXAMPLE 2 Let $z = f(u)$ be a function of a single variable. Let $u = 2x + 3y$. Then z is a composite function of x and y. Show that

$$2\frac{\partial z}{\partial y} = 3\frac{\partial z}{\partial x}. \tag{10}$$

SOLUTION The situation is shown in Fig. 11. We will evaluate both $\partial z/\partial x$ and $\partial z/\partial y$ by the chain rule and then check whether (10) is true.

To find $\partial z/\partial x$ we consider all paths from z down to x in Fig. 11. (Since there is only one middle variable, there is only one path; it is shown in Fig. 12.) Since $u = 2x + 3y$, $\partial u/\partial x = 2$. Thus

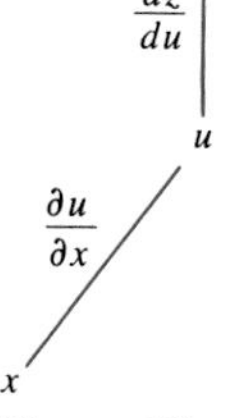

Figure 12

$$\frac{\partial z}{\partial x} = \frac{dz}{du}\frac{\partial u}{\partial x} = \frac{dz}{du}\cdot 2 = 2\frac{dz}{du}. \tag{11}$$

(Note that one derivative is ordinary, while the other is a partial derivative.)

To find $\partial z/\partial y$, consider all paths from z down to y in Fig. 11. (Again, there is only one; it is shown in Fig. 13.) Since $u = 2x + 3y$, $\partial u/\partial y = 3$. Thus

$$\frac{\partial z}{\partial y} = \frac{dz}{du}\frac{\partial u}{\partial y} = \frac{dz}{du}\cdot 3 = 3\frac{dz}{du}. \tag{12}$$

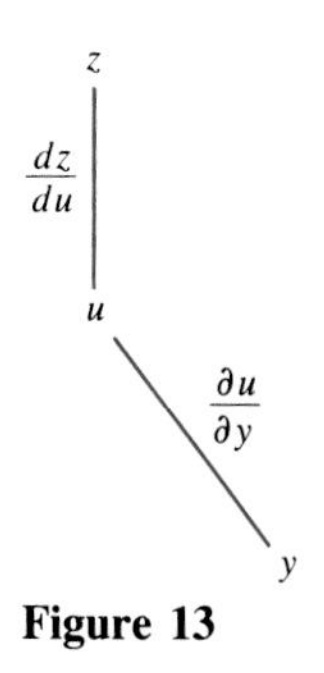

Figure 13

Figures 12 and 13 tell us that $\partial z/\partial x = 2\ dz/du$ and $\partial z/\partial y = 3\ dz/du$. Substitute these into the equation

$$2\frac{\partial z}{\partial y} = 3\frac{\partial z}{\partial x}$$

to see whether we obtain a true equation:

$$2\left(3\frac{dz}{du}\right) \stackrel{?}{=} 3\left(2\frac{dz}{du}\right). \tag{13}$$

Since (13) is true, we have verified (10). ■

The Main Use of the Chain Rule

There is a fundamental difference between Examples 1 and 2. In the first example, we were dealing with explicitly given functions. We did not really need to use the chain rule to find the derivative, dz/dt. As remarked after the example, we could have shown that $z = t^7/3$ and easily found that $dz/dt = (7t^6)/3$. But in Example 2, we were dealing with a general type of function formed in a certain way: We showed that (10) holds for *every* differentiable function $f(u)$. For instance, if $f(u) = u^2$, then $z = u^2$ where $u = 2x + 3y$. That means that $z = (2x + 3y)^2$. When $f(u) = \sin u$, then $z = \sin(2x + 3y)$. No matter what $f(u)$ we choose, we know that $2\ \partial z/\partial y = 3\ \partial z/\partial x$.

Example 2 shows why the chain rule is important. It enables us to make *general statements* about the partial derivatives of an infinite number of functions, all of which are formed the same way. The next example illustrates this use again.

Equation (10) is an example of a partial differential equation, an equation that involves partial derivatives. The first partial differential equation ever studied was

$$\frac{\partial^2 y}{\partial t^2} = \frac{\partial y}{\partial x} + x\frac{\partial^2 y}{\partial x^2}.$$

D'Alembert introduced it in 1743 when studying the vibrations in a hanging chain. He was not able to solve it, that is, to find y as a function of x and t. However, in 1746 he obtained the partial differential equation for a vibrating string:

$$\frac{\partial^2 y}{\partial t^2} = k^2\frac{\partial^2 y}{\partial x^2}.$$

(See Fig. 4 in Sec. 14.5.) This "wave equation" created a great deal of excitement, especially since d'Alembert showed that any function of the form

$$g(x + kt) + h(x - kt)$$

is a solution.

EXAMPLE 3 If $u = f(x - y, y - z, z - x)$, show that

$$\frac{\partial u}{\partial x} + \frac{\partial u}{\partial y} + \frac{\partial u}{\partial z} = 0. \tag{14}$$

SOLUTION First of all, f is a function of three variables. These variables are being replaced by $x - y$, $y - z$, and $z - x$. Introducing three letters not used already, we have $u = f(r, s, t)$, where

$$r = x - y, \qquad s = y - z, \qquad \text{and} \qquad t = z - x. \tag{15}$$

Therefore, u is a composite function of x, y, and z. (See Fig. 14.)

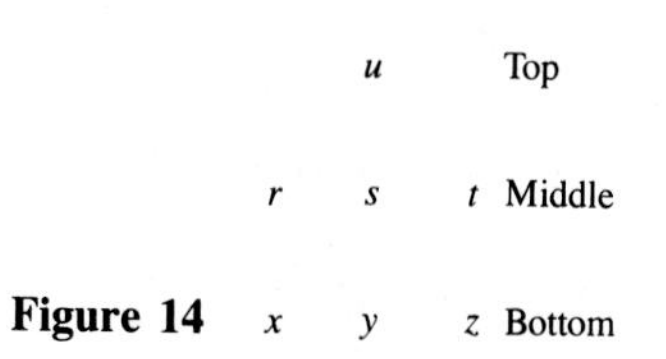

Figure 14

We will compute $\partial u/\partial x$, $\partial u/\partial y$, and $\partial u/\partial z$ separately, add them, and check whether their sum is 0.

To compute $\partial u/\partial x$ we use Fig. 15.

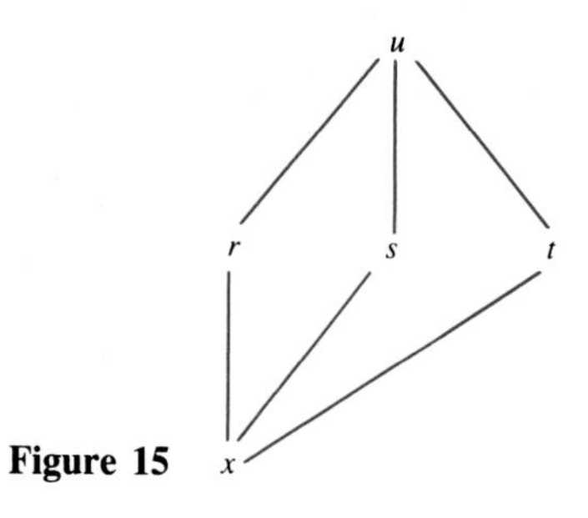

Figure 15

We see by Fig. 15 that the chain rule for $\partial u/\partial x$ involves three summands:

$$\frac{\partial u}{\partial x} = \frac{\partial u}{\partial r}\frac{\partial r}{\partial x} + \frac{\partial u}{\partial s}\frac{\partial s}{\partial x} + \frac{\partial u}{\partial t}\frac{\partial t}{\partial x}. \tag{16}$$

We cannot simplify $\partial u/\partial r$, $\partial u/\partial s$, $\partial u/\partial t$. However,

$$\frac{\partial r}{\partial x} = \frac{\partial}{\partial x}(x - y) = 1, \qquad \frac{\partial s}{\partial x} = \frac{\partial}{\partial x}(y - z) = 0, \qquad \frac{\partial t}{\partial x} = \frac{\partial}{\partial x}(z - x) = -1. \tag{17}$$

Combining (16) and (17) gives

$$\frac{\partial u}{\partial x} = \frac{\partial u}{\partial r}(1) + \frac{\partial u}{\partial s}(0) + \frac{\partial u}{\partial t}(-1) = \frac{\partial u}{\partial r} - \frac{\partial u}{\partial t}.$$

Similar computations show that

$$\frac{\partial u}{\partial y} = \frac{-\partial u}{\partial r} + \frac{\partial u}{\partial s} \tag{18}$$

and

$$\frac{\partial u}{\partial z} = \frac{-\partial u}{\partial s} + \frac{\partial u}{\partial t}. \tag{19}$$

Then $$\frac{\partial u}{\partial x} + \frac{\partial u}{\partial y} + \frac{\partial u}{\partial z} = \left(\frac{\partial u}{\partial r} - \frac{\partial u}{\partial t}\right) + \left(\frac{-\partial u}{\partial r} + \frac{\partial u}{\partial s}\right) + \left(\frac{-\partial u}{\partial s} + \frac{\partial u}{\partial t}\right) = 0,$$

which we were to show. ■

Some advice

You do not have to be a great mathematician to apply the chain rule. However, you must do careful bookkeeping. First, display the top, middle, and bottom variables. Second, keep in mind that the number of middle variables determines the number of summands.

EXAMPLE 4 In the study of light or sound, engineers and physicists meet the **wave equation**

$$\frac{\partial^2 z}{\partial t^2} = k^2 \frac{\partial^2 z}{\partial x^2}, \tag{20}$$

where k is a constant. Here z is a function of x and t. Show that any function of the form $z = g(x + kt)$ satisfies the wave equation.

SOLUTION In order to find the partial derivatives $\partial^2 z/\partial x^2$ and $\partial^2 z/\partial t^2$ we express $z = g(x + kt)$ as a composition of functions:

$$z = g(u) \qquad \text{where } u = x + kt.$$

Note that g is a function of just one variable. Figure 16 lists the variables.

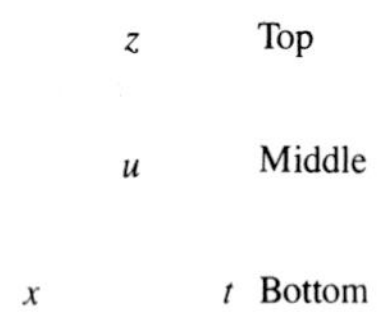

z Top

u Middle

x t Bottom

Figure 16

We will compute $\partial^2 z/\partial x^2$ and $\partial^2 z/\partial t^2$ in terms of derivatives of g and then check whether (20) holds. We first compute $\partial^2 z/\partial x^2$. First of all,

Recall that $u = x + kt$.

$$\frac{\partial z}{\partial x} = \frac{dz}{du}\frac{\partial u}{\partial x} = \frac{dz}{du} \cdot 1 = \frac{dz}{du}. \tag{21}$$

(There is only one path from z down to x. See Fig. 16.) In Eq. (21) dz/du is viewed as a function of x and t; that is, u is replaced by $x + kt$. Next,

$$\frac{\partial^2 z}{\partial x^2} = \frac{\partial}{\partial x}\left(\frac{\partial z}{\partial x}\right) = \frac{\partial}{\partial x}\left(\frac{dz}{du}\right).$$

Now, dz/du, viewed as a function of x and t, may be expressed as a composite function. Letting $w = dz/du$, we have

$$w = f(u), \qquad \text{where } u = x + kt.$$

(See Fig. 17.)

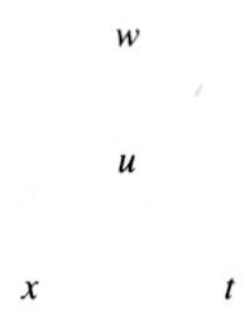

w

u

x t

Figure 17

Therefore

$$\frac{\partial^2 z}{\partial x^2} = \frac{\partial}{\partial x}\left(\frac{\partial z}{\partial x}\right) = \frac{\partial w}{\partial x}$$

$$= \frac{dw}{du} \cdot \frac{\partial u}{\partial x} \qquad \text{(only one path down to } x\text{)}$$

$$= \frac{d}{du}\left(\frac{dz}{du}\right)\frac{\partial u}{\partial x}$$

$$= \frac{d^2 z}{du^2} \cdot 1;$$

hence

$$\frac{\partial^2 z}{\partial x^2} = \frac{d^2 z}{du^2}. \tag{22}$$

Then we also express $\partial^2 z/\partial t^2$ in terms of d^2z/du^2, as follows. First of all,

$$\frac{\partial z}{\partial t} = \frac{dz}{du}\frac{\partial u}{\partial t} = \frac{dz}{du} \cdot k = k\frac{dz}{du}.$$

(See Fig. 18.)

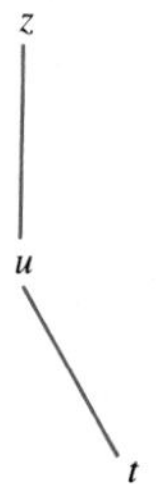

Figure 18

Then

$$\frac{\partial^2 z}{\partial t^2} = \frac{\partial}{\partial t}\left(\frac{\partial z}{\partial t}\right) = \frac{\partial}{\partial t}\left(k\frac{dz}{du}\right)$$

$$= k\frac{d}{du}\left(\frac{dz}{du}\right) \cdot \frac{\partial u}{\partial t} \qquad \text{(only one path down to } t\text{)}$$

$$= k\frac{d^2 z}{du^2} \cdot k;$$

hence

$$\frac{\partial^2 z}{\partial t^2} = k^2\frac{d^2 z}{du^2}. \tag{23}$$

Comparing (22) and (23) shows that

$$\frac{\partial^2 z}{\partial t^2} = k^2\,\frac{\partial^2 z}{\partial x^2}. \quad \blacksquare$$

The Chain Rule Rescues the Economists

Eugene Silberberg writes in *The Structure of Economics,*

> The development of marginal productivity theory . . . led to the conclusion that the factors of production would be paid the value of their marginal product. Roughly speaking, factors would be hired until their contributions to the output of the firm just equaled the cost of acquiring additional units of the factor . . . [Would] the firm be capable of making these payments? . . . Would enough output be produced (or perhaps would too much be produced) . . . ?
>
> A theorem developed by the great Swiss mathematician Euler . . . came to the rescue of this analysis.

We will show how Euler rescued the economists. Let $f(x, y)$ be the production (measured in dollars) of x units of labor and y units of capital. If the amounts of labor and capital are, say, doubled, we could expect that the output might double, that is, $f(2x, 2y) = 2f(x, y)$. More generally, we would expect that for any positive number k,

$$f(kx, ky) = kf(x, y). \tag{24}$$

A function satisfying Eq. (24) is called **homogeneous**.

The **marginal product of labor** is defined as $\partial f/\partial x$, and the **marginal product of capital** as $\partial f/\partial y$.

The economists wanted "the total production $f(x, y)$ to equal cost of labor plus cost of capital if each is paid at the rate of its marginal product." That is, they expected the following equation to be true:

$$f(x, y) = x\frac{\partial f}{\partial x} + y\frac{\partial f}{\partial y}. \tag{25}$$

Euler derived Eq. (25) from Eq. (24), as follows. Start with Eq. (24):

$$f(kx, ky) = kf(x, y). \tag{26}$$

Both sides of (26) are functions of k, x, and y. Differentiate both sides of (26) with respect to k (holding x and y fixed).

Differentiating the right side of Eq. (26) first gives

$$\left(\frac{\partial}{\partial k}(kf(x, y))\right)_{x,\,y} = f(x, y). \tag{27}$$

Differentiating the left side of Eq. (26) requires the chain rule. Let $z = f(u, v)$ where $u = kx$ and $v = ky$. Then z is a composite function of k, x, and y, namely $z = f(kx, ky)$. (See Fig. 19.)

We compute

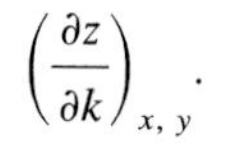

$$\left(\frac{\partial z}{\partial k}\right)_{x,\,y}.$$

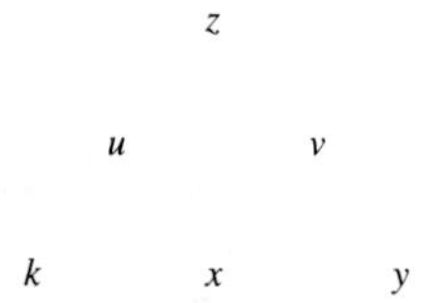

z

u v

k x y

Figure 19

By the chain rule:

We use the notations f_1 and f_2 for the partial derivatives with respect to the first and second variables.

$$\left(\frac{\partial z}{\partial k}\right)_{x,\,y} = f_1(kx,\, ky)\,\frac{\partial(kx)}{\partial k} + f_2(kx,\, ky)\frac{\partial(ky)}{\partial k}.$$

But
$$\frac{\partial(kx)}{\partial k} = x \qquad \text{and} \qquad \frac{\partial(ky)}{\partial k} = y.$$

So
$$\left(\frac{\partial z}{\partial k}\right)_{x,\,y} = xf_1(kx,\, ky) + yf_2(kx,\, ky). \tag{28}$$

Since (27) and (28) represent the partial derivatives of the two sides of (24) with respect to k, we have

$$f(x,\, y) = xf_1(kx,\, ky) + yf_2(kx,\, ky). \tag{29}$$

Since Eq. (29) holds for all $k > 0$, it holds for $k = 1$. Substituting $k = 1$ into Eq. (29) gives

$$f(x,\, y) = xf_1(x,\, y) + yf_2(x,\, y);$$

that is,
$$f(x,\, y) = x\,\frac{\partial f}{\partial x} + y\,\frac{\partial f}{\partial y},$$

which is Euler's theorem on homogeneous functions, and the economists are saved.

Section Summary

If z is a function of $x_1, x_2, \ldots, x_m$ and each x_i is a function of $t_1, t_2, \ldots, t_n$, then there are n partial derivatives $\partial z/\partial t_j$. Each is a sum of m products of the form $(\partial z/\partial x_i)(\partial x_i/\partial t_j)$. To do the bookkeeping, first make a roster as shown in Fig. 20.

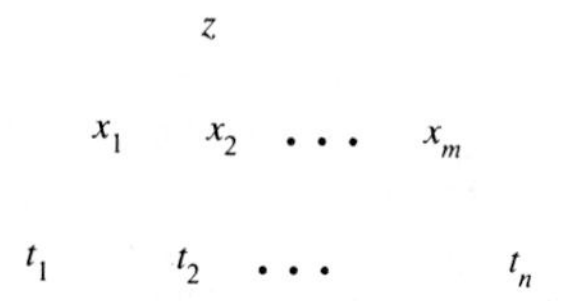

z

$x_1 \quad x_2 \quad \cdots \quad x_m$

$t_1 \quad t_2 \quad \cdots \quad t_n$

Figure 20

To compute $\partial z/\partial t_j$ list all paths from z down to t_j, as shown in Fig. 21. Each path "contributes" a product.

The chain rule for functions of two variables is based on the fact that $f(x + \Delta x,\, y + \Delta y) - f(x,\, y)$ is approximated very well by

$$\frac{\partial f}{\partial x}\,\Delta x + \frac{\partial f}{\partial y}\,\Delta y.$$

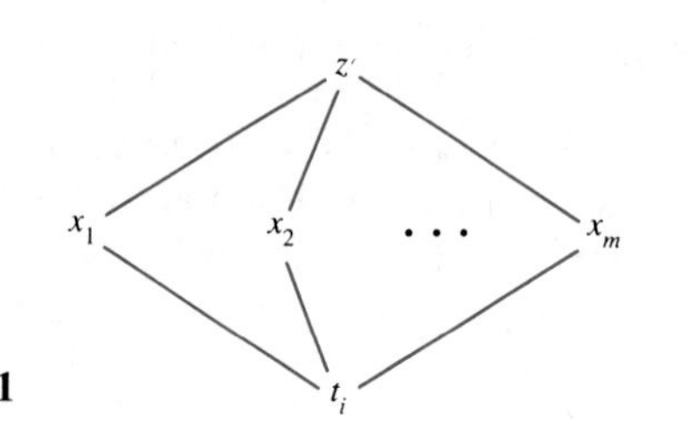

Figure 21

EXERCISES FOR SEC. 14.6: THE CHAIN RULE

In Exercises 1 to 4 verify the chain rule in Theorem 2 by computing dz/dt two ways: (*a*) with the chain rule, (*b*) without the chain rule (by writing z as a function of t).

1 $z = x^2y^3$, $x = t^2$, $y = t^3$

2 $z = xe^y$, $x = t$, $y = 1 + 3t$

3 $z = \cos(xy^2)$, $x = e^{2t}$, $y = \sec 3t$

4 $z = \ln(x + 3y)$, $x = t^2$, $y = \tan 3t$.

In Exercises 5 and 6 verify the chain rule in Theorem 3 by computing $\partial z/\partial t$ two ways: (*a*) with the chain rule, (*b*) without the chain rule (by writing z as a function of t and u).

5 $z = x^2y$, $x = 3t + 4u$, $y = 5t - u$

6 $z = \sin(x + 3y)$, $x = \sqrt{t/u}$, $y = \sqrt{t} + \sqrt{u}$

7 Assume that $z = f(x_1, x_2, x_3, x_4, x_5)$ and that each x_i is a function of t_1, t_2, t_3.

(*a*) List all the variables, showing top, middle, and bottom variables.

(*b*) Draw the paths involved in expressing $\partial z/\partial t_3$ in terms of the chain rule.

(*c*) Express $\partial z/\partial t_3$ in terms of a sum of products of partial derivatives.

(*d*) When computing $\partial z/\partial x_2$, which variables are constant?

(*e*) When computing $\partial z/\partial t_3$, which variables are constant?

8 If $z = f(g(t_1, t_2, t_3), h(t_1, t_2, t_3))$

(*a*) How many middle variables are there?

(*b*) How many bottom variables?

(*c*) What does the chain rule say about $\partial z/\partial t_3$? (Include a diagram showing the paths.)

9 Find dz/dt if $\partial z/\partial x = 4$, $\partial z/\partial y = 3$, $dx/dt = 4$, and $dy/dt = 1$.

10 Find dz/dt if $\partial z/\partial x = 3$, $\partial z/\partial y = 2$, $dx/dt = 4$, and $dy/dt = -3$.

11 Let $z = f(x, y)$, $x = u + v$, and $y = u - v$.

(*a*) Show that $(\partial z/\partial x)^2 - (\partial z/\partial y)^2 = (\partial z/\partial u)(\partial z/\partial v)$. (Include diagrams.)

(*b*) Verify (*a*) when $f(x, y) = x^2 + 2y^3$.

12 Let $z = f(x, y)$, $x = u^2 - v^2$, and $y = v^2 - u^2$.

(*a*) Show that

$$u\frac{\partial z}{\partial v} + v\frac{\partial z}{\partial u} = 0.$$

(Include diagrams.)

(*b*) Verify (*a*) when $f(x, y) = \sin(x + 2y)$.

13 Let $z = f(t - u, -t + u)$.

(*a*) Show that $\dfrac{\partial z}{\partial t} + \dfrac{\partial z}{\partial u} = 0$. (Include diagrams.)

(*b*) Verify (*a*) when $f(x, y) = x^2y$.

14 Let $w = f(x - y, y - z, z - x)$.

(*a*) Show that $\dfrac{\partial w}{\partial x} + \dfrac{\partial w}{\partial y} + \dfrac{\partial w}{\partial z} = 0$. (Include diagrams.)

(*b*) Verify (*a*) in the case $f(s, t, u) = s^2 + t^2 - u$.

15 Let $z = f(u, v)$ where $u = ax + by$, $v = cx + dy$, and a, b, c, d are constants. Show that

(*a*) $$\frac{\partial^2 z}{\partial x^2} = a^2\frac{\partial^2 f}{\partial u^2} + 2ac\frac{\partial^2 f}{\partial u\,\partial v} + c^2\frac{\partial^2 f}{\partial v^2}$$

(*b*) $$\frac{\partial^2 z}{\partial y^2} = b^2\frac{\partial^2 f}{\partial u^2} + 2bd\frac{\partial^2 f}{\partial u\,\partial v} + d^2\frac{\partial^2 f}{\partial v^2}$$

(*c*) $$\frac{\partial^2 z}{\partial x\,\partial y} = ab\frac{\partial^2 f}{\partial u^2} + (ad + bc)\frac{\partial^2 f}{\partial u\,\partial v} + cd\frac{\partial^2 f}{\partial v^2}.$$

16 Let a, b, and c be given constants and consider the partial differential equation

$$a\frac{\partial^2 z}{\partial x^2} + b\frac{\partial^2 z}{\partial x\,\partial y} + c\frac{\partial^2 z}{\partial y^2} = 0.$$

Assume a solution of the form $z = f(y + mx)$, where m is a constant. Show that for this function to be a solution, $am^2 + bm + c$ must be 0.

17 (*a*) Show that any function of the form $z = f(x + y)$ is a solution of the partial differential equation

$$\frac{\partial^2 z}{\partial x^2} - 2\frac{\partial^2 z}{\partial x\,\partial y} + \frac{\partial^2 z}{\partial y^2} = 0.$$

(*b*) Verify (*a*) for $z = (x + y)^3$.

18 Let $u(x, t)$ be the temperature at point x along a rod at time t. The function u satisfies the one-dimensional heat equation for a constant k:

$$\frac{\partial u}{\partial t} = k\frac{\partial^2 u}{\partial x^2}.$$

(*a*) Show that $u(x, t) = e^{kt}g(x)$ satisfies the heat equation if $g(x)$ is any function such that $g''(x) = g(x)$.

(*b*) Show that if $g(x) = 3e^{-x} + 4e^x$, then $g''(x) = g(x)$.

19 (*a*) Show that any function of the form $z = f(x + y) + e^y g(x - y)$ is a solution of the partial differential equation

$$\frac{\partial^2 z}{\partial x^2} - \frac{\partial^2 z}{\partial y^2} - \frac{\partial z}{\partial x} + \frac{\partial z}{\partial y} = 0.$$

(*b*) Check (*a*) for $z = (x + y)^2 + e^y \sin(x - y)$.

20 Let $z = f(x, y)$ denote the temperature at the point (x, y) in the first quadrant. If polar coordinates are used, then we would write $z = g(r, \theta)$.

(*a*) Express $\partial z/\partial r$ in terms of $\partial z/\partial x$ and $\partial z/\partial y$. [*Hint:* What is the relation between rectangular coordinates (x, y) and polar coordinates (r, θ)?]

(*b*) Express $\partial z/\partial \theta$ in terms of $\partial z/\partial x$ and $\partial z/\partial y$.

(*c*) Show that

$$\left(\frac{\partial z}{\partial x}\right)^2 + \left(\frac{\partial z}{\partial y}\right)^2 = \left(\frac{\partial z}{\partial r}\right)^2 + \frac{1}{r^2}\left(\frac{\partial z}{\partial \theta}\right)^2.$$

21 Let $u = f(r)$ and $r = (x^2 + y^2 + z^2)^{1/2}$. Show that

$$\frac{\partial^2 u}{\partial x^2} + \frac{\partial^2 u}{\partial y^2} + \frac{\partial^2 u}{\partial z^2} = \frac{d^2 u}{dr^2} + \frac{2}{r}\frac{du}{dr}.$$

---

22 At what rate is the volume of a rectangular box changing when its width is 3 feet and increasing at the rate of 2 feet per second, its length is 8 feet and decreasing at the rate of 5 feet per second, and its height is 4 feet and increasing at the rate of 2 feet per second?

23 The temperature T at (x, y, z) in space is $f(x, y, z)$. An astronaut is traveling in such a way that his x and y coordinates increase at the rate of 4 miles per second and his z coordinate decreases at the rate of 3 miles per second. Compute the rate dT/dt at which the temperature changes at a point where

$$\frac{\partial T}{\partial x} = 4, \qquad \frac{\partial T}{\partial y} = 7, \qquad \text{and} \qquad \frac{\partial T}{\partial z} = 9.$$

24 We proved the chain rule when there are two middle variables and one bottom variable. Prove Theorem 3, where there are two middle variables and two bottom variables.

25 To prove the general chain rule when there are three middle variables, we need an analog of Theorem 1 concerning Δf when f is a function of three variables.

(*a*) Let $u = f(x, y, z)$ be a function of three variables. Show that

$$\begin{aligned}\Delta f &= f(x + \Delta x, y + \Delta y, z + \Delta z) - f(x, y, z)\\ &= [f(x + \Delta x, y, z) - f(x, y, z)]\\ &\quad + [f(x + \Delta x, y + \Delta y, z) - f(x + \Delta x, y, z)]\\ &\quad + [f(x + \Delta x, y + \Delta y, z + \Delta z) - f(x + \Delta x, y + \Delta y, z)].\end{aligned}$$

(*b*) Using (*a*) show that

$$\begin{aligned}\Delta f = \frac{\partial f}{\partial x}(x, y, z)\,\Delta x + \frac{\partial f}{\partial y}(x, y, z)\,\Delta y\\ + \frac{\partial f}{\partial z}(x, y, z)\,\Delta z + \epsilon_1\,\Delta x + \epsilon_2\,\Delta y + \epsilon_3\,\Delta z,\end{aligned}$$

where $\epsilon_1, \epsilon_2, \epsilon_3 \to 0$ as $\Delta x, \Delta y, \Delta z \to 0$.

(*c*) Obtain the general chain rule in the case of three middle variables and any number of bottom variables.

26 Let (r, θ) be polar coordinates for the point (x, y) given in rectangular coordinates.

(*a*) From the relation $r = \sqrt{x^2 + y^2}$, show that $\partial r/\partial x = \cos\theta$.

(*b*) From the relation $r = x/\cos\theta$, show that $\partial r/\partial x = 1/(\cos\theta)$.

(*c*) Explain why (*a*) and (*b*) are not contradictory.

27 Let $z = f(x, y)$, where $x = r\cos\theta$ and $y = r\sin\theta$. Show that

$$\frac{\partial^2 z}{\partial r^2} = \cos^2\theta\,\frac{\partial^2 f}{\partial x^2} + 2\cos\theta\sin\theta\,\frac{\partial^2 f}{\partial x\,\partial y} + \sin^2\theta\,\frac{\partial^2 f}{\partial y^2}.$$

28 Let $u = f(x, y)$, where $x = r\cos\theta$ and $y = r\sin\theta$. Verify the following equation, which appears in electromagnetic theory,

$$\frac{1}{r}\frac{\partial}{\partial r}\left(r\frac{\partial u}{\partial r}\right) + \frac{1}{r^2}\frac{\partial^2 u}{\partial \theta^2} = \frac{\partial^2 u}{\partial x^2} + \frac{\partial^2 u}{\partial y^2}.$$

29 Let u be a function of x and y, where x and y are both functions of s and t. Show that

$$\begin{aligned}\frac{\partial^2 u}{\partial s^2} = \frac{\partial^2 u}{\partial x^2}\left(\frac{\partial x}{\partial s}\right)^2 + 2\frac{\partial^2 u}{\partial x\,\partial y}\frac{\partial x}{\partial s}\frac{\partial y}{\partial s} + \frac{\partial^2 u}{\partial y^2}\left(\frac{\partial y}{\partial s}\right)^2\\ + \frac{\partial u}{\partial x}\frac{\partial^2 x}{\partial s^2} + \frac{\partial u}{\partial y}\frac{\partial^2 y}{\partial s^2}.\end{aligned}$$

30 In developing Eq. (7), we used the path that started at (x, y), went to $(x + \Delta x, y)$, and ended at $(x + \Delta x, y + \Delta y)$. Could we have used the path from (x, y), through $(x, y + \Delta y)$, to $(x + \Delta x, y + \Delta y)$ instead? If "no," explain why. If "yes," write out the argument, using this path.

Exercises 31 to 35 concern homogeneous functions, defined in the economics discussion.

31 Verify that each of the following functions is homogeneous and also verify that each satisfies the conclusion of Euler's theorem:

(*a*) $3x + 4y$

(*b*) x^3y^{-2}

(*c*) $xe^{x/y}$

32 A function $f(x, y)$ is homogeneous of degree r if $f(kx, ky) = k^r f(x, y)$ for all $k > 0$. Show that each of the following functions is homogeneous of some degree r:

(*a*) $f(x, y) = x^2(\ln x - \ln y)$

(*b*) $f(x, y) = 1/\sqrt{x^2 + y^2}$

(*c*) $f(x, y) = \sin\left(\dfrac{y}{x}\right)$

33 (See Exercise 32.) Show that if f is homogeneous of degree r, then $x\,\partial f/\partial x + y\,\partial f/\partial y = rf$. This is the general form of Euler's theorem.

34 (See Exercise 33.) Verify Euler's theorem for the functions in Exercise 32.

35 (See Exercise 32.) Show that if f is homogeneous of degree r, then $\partial f/\partial x$ is homogeneous of degree $r - 1$.

14.7 DIRECTIONAL DERIVATIVES AND THE GRADIENT

In this section we generalize the notion of a partial derivative to that of a directional derivative. Then we introduce a vector, called "the gradient," to provide a short formula for the directional derivative. The gradient will have other uses later in this chapter and in Chaps. 16 and 17.

Directional Derivatives

If $z = f(x, y)$, the partial derivative $\partial f/\partial x$ tells us how rapidly z changes as we move the input point (x, y) in a direction parallel to the x axis. Similarly, $\partial f/\partial y$ tells how fast z changes as we move parallel to the y axis. But we can ask, "How rapidly does z change when we move the input point (x, y) in any fixed direction in the xy plane?" The answer is given by the directional derivative.

Consider a function $z = f(x, y)$, let's say the temperature at (x, y). Let (a, b) be a point and let $\mathbf{u}$ be a unit vector in the xy plane. Draw a line through (a, b) and parallel to $\mathbf{u}$. Call it the t axis and let its positive part point in the direction of $\mathbf{u}$. Place the 0 of the t axis at (a, b). (See Fig. 1.) Each value of t determines a point (x, y) on the t axis and thus a value of z. Along the t axis, z can therefore be viewed as a function of t, $z = g(t)$. The derivative dg/dt, evaluated at $t = 0$, is called the **directional derivative** of $z = f(x, y)$ at (a, b) in the direction $\mathbf{u}$. It is denoted $D_{\mathbf{u}}f$. The directional derivative is the slope of the tangent line to the curve $z = g(t)$ at $t = 0$. (See Fig. 2.)

When $\mathbf{u} = \mathbf{i}$, we obtain the directional derivative $D_{\mathbf{i}}f$, which is simply $\partial f/\partial x$. When $\mathbf{u} = \mathbf{j}$, we obtain $D_{\mathbf{j}}f$, which is $\partial f/\partial y$.

The directional derivative generalizes the two partial derivatives $\partial f/\partial x$ and $\partial f/\partial y$. After all, we can ask for the rate of change of $z = f(x, y)$ in any direction in the xy plane, not just the directions indicated by the vectors $\mathbf{i}$ and $\mathbf{j}$.

The following theorem shows how to compute a directional derivative.

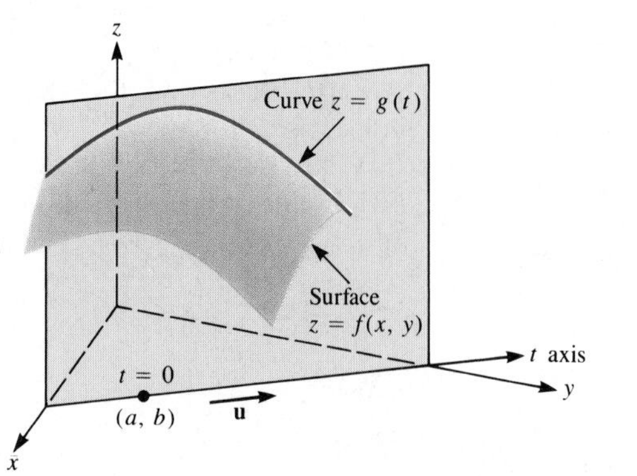

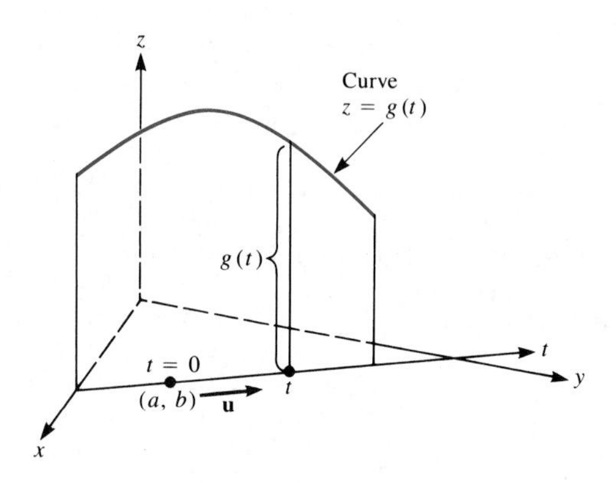

Figure 1

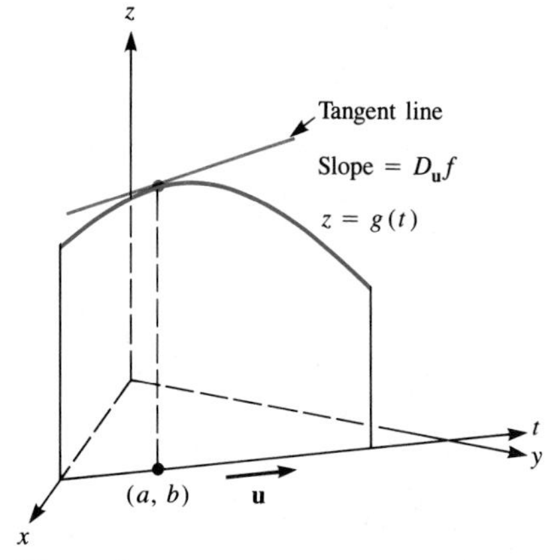

Figure 2

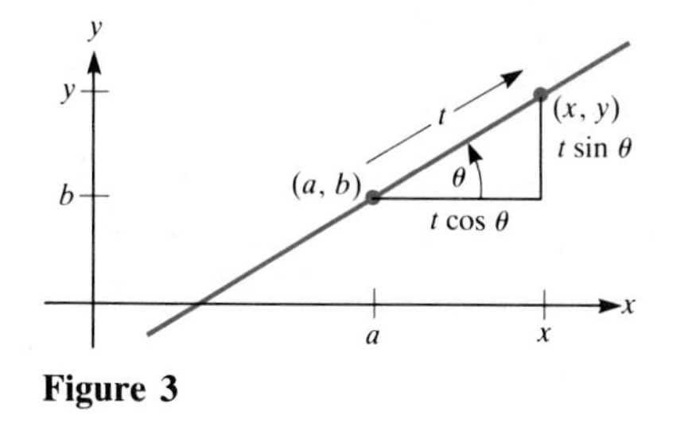

Figure 3

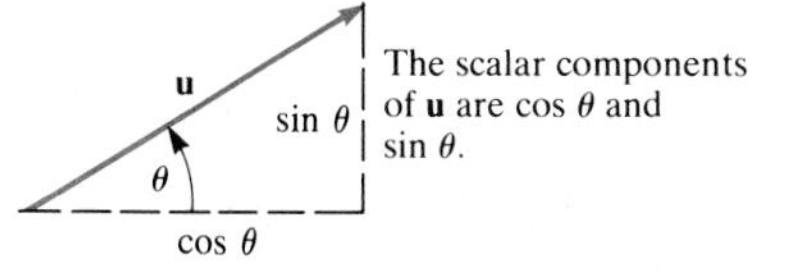

Figure 4

> **Theorem 1** If $f(x, y)$ has continuous partial derivatives $\partial f/\partial x$ and $\partial f/\partial y$, then the directional derivative of f at (a, b) in the direction $\mathbf{u} = \cos\theta\,\mathbf{i} + \sin\theta\,\mathbf{j}$ is
>
> $$\frac{\partial f}{\partial x}(a, b)\cos\theta + \frac{\partial f}{\partial y}(a, b)\sin\theta. \tag{1}$$

Proof The directional derivative of f at (a, b) in the direction θ is the derivative of the function

$$g(t) = f(a + t\cos\theta,\ b + t\sin\theta)$$

when $t = 0$. (See Figs. 3 and 4.)

Now g is a composite function

$$g(t) = f(x, y) \qquad \text{where } \begin{cases} x = a + t\cos\theta \\ y = b + t\sin\theta\,. \end{cases}$$

The chain rule tells us that

$$g'(t) = \frac{\partial f}{\partial x}\frac{dx}{dt} + \frac{\partial f}{\partial y}\frac{dy}{dt}.$$

Moreover, $$\frac{dx}{dt} = \cos\theta \qquad \text{and} \qquad \frac{dy}{dt} = \sin\theta.$$

Thus $$g'(0) = \frac{\partial f}{\partial x}(a, b)\cos\theta + \frac{\partial f}{\partial y}(a, b)\sin\theta,$$

and the theorem is proved. ■

Checking the formula when $\theta = 0$

When $\theta = 0$, that is, $\mathbf{u} = \mathbf{i}$, the formula given by Theorem 1 becomes

$$\frac{\partial f}{\partial x}(a, b)\cos 0 + \frac{\partial f}{\partial y}(a, b)\sin 0 = \frac{\partial f}{\partial x}(a, b)(1) + \frac{\partial f}{\partial y}(a, b)(0)$$
$$= \frac{\partial f}{\partial x}(a, b).$$

Checking the formula when $\theta = \pi$

When $\theta = \pi$, that is, $\mathbf{u} = -\mathbf{i}$, the formula given by Theorem 1 becomes

$$\frac{\partial f}{\partial x}(a, b)\cos\pi + \frac{\partial f}{\partial y}(a, b)\sin\pi = \frac{\partial f}{\partial x}(a, b)(-1) + \frac{\partial f}{\partial y}(a, b)(0)$$
$$= -\frac{\partial f}{\partial x}(a, b).$$

(This makes sense: If the temperature increases as you walk east, then it decreases as you walk west.)

Checking when $\theta = \pi/2$

When $\theta = \pi/2$, that is, $\mathbf{u} = \mathbf{j}$, Theorem 1 asserts that the directional derivative is

$$\frac{\partial f}{\partial x}(a, b)\cos\frac{\pi}{2} + \frac{\partial f}{\partial y}(a, b)\sin\frac{\pi}{2} = \frac{\partial f}{\partial x}(a, b)(0) + \frac{\partial f}{\partial y}(a, b)(1) = \frac{\partial f}{\partial y}(a, b),$$

which also is expected.

EXAMPLE 1 Compute the derivative of $f(x, y) = x^2y^3$ at $(1, 2)$ in the direction given by the angle $\pi/3$. [$\mathbf{u} = \cos(\pi/3)\,\mathbf{i} + \sin(\pi/3)\,\mathbf{j}$.] Interpret the results if f describes a temperature distribution.

SOLUTION First of all,

$$\frac{\partial f}{\partial x} = 2xy^3 \quad \text{and} \quad \frac{\partial f}{\partial y} = 3x^2y^2.$$

Hence
$$\frac{\partial f}{\partial x}(1, 2) = 16 \quad \text{and} \quad \frac{\partial f}{\partial y}(1, 2) = 12.$$

Second
$$\cos\frac{\pi}{3} = \frac{1}{2} \quad \text{and} \quad \sin\frac{\pi}{3} = \frac{\sqrt{3}}{2}.$$

Thus the derivative of f in the direction given by $\theta = \pi/3$ is

$$16\left(\frac{1}{2}\right) + 12\left(\frac{\sqrt{3}}{2}\right) = 8 + 6\sqrt{3}.$$

If x^2y^3 is the temperature in degrees at the point (x, y), where x and y are measured in centimeters, then the rate at which the temperature changes at $(1, 2)$, in the direction given by $\theta = \pi/3$, is $(8 + 6\sqrt{3})$ degrees per centimeter. ■

The Gradient

Formula (1) resembles the formula for the dot product. To exploit this similarity, it is useful to introduce the vector whose scalar components are $\partial f/\partial x(a, b)$ and $\partial f/\partial y(a, b)$.

Definition *The gradient of f(x, y).* The vector

$$\frac{\partial f}{\partial x}(a, b)\mathbf{i} + \frac{\partial f}{\partial y}(a, b)\mathbf{j}$$

is the **gradient** of f at (a, b) and is denoted ∇f. (It is also called "del f," because of the upside-down delta ∇.)

The del symbol is in boldface to emphasize that the gradient of f is a vector.

For instance, let $f(x, y) = x^2 + y^2$. We compute and draw ∇f at a few points, listed in the following table:

(x, y)	$\frac{\partial f}{\partial x} = 2x$	$\frac{\partial f}{\partial y} = 2y$	∇f
$(1, 2)$	2	4	$2\mathbf{i} + 4\mathbf{j}$
$(3, 0)$	6	0	$6\mathbf{i}$
$(2, -1)$	4	-2	$4\mathbf{i} - 2\mathbf{j}$

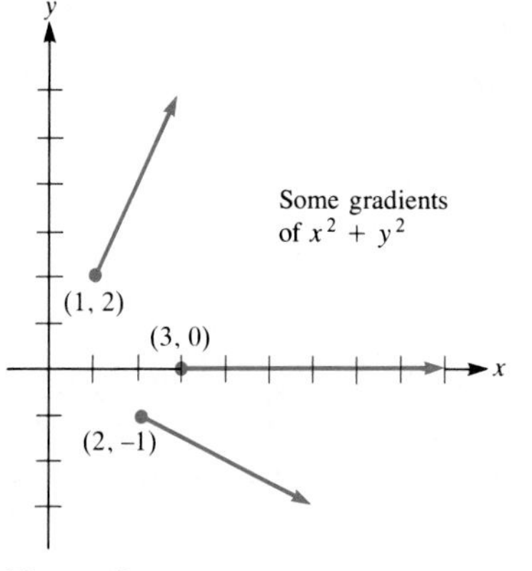

Figure 5

Figure 5 shows ∇f in each case, with the tail of ∇f placed at the point where ∇f is computed.

In vector notation, Theorem 1 reads as follows:

Theorem 1 *(Rephrased)* If $z = f(x, y)$ has continuous partial derivatives $\partial f/\partial x$ and $\partial f/\partial y$, then

$$D_{\mathbf{u}}f = \nabla f \cdot \mathbf{u}.$$

The gradient is introduced not merely to simplify the statement of Theorem 1. Its importance is made clear in the next theorem.

A Different View of the Gradient

The meaning of $\|\nabla f\|$ and the direction of ∇f

Theorem 2 *Significance of* ∇f. Let $z = f(x, y)$ have continuous partial derivatives $\partial f/\partial x$ and $\partial f/\partial y$. Let (a, b) be a point in the plane where ∇f is not $\mathbf{0}$. Then the length of ∇f at (a, b) is the largest directional derivative of f at (a, b); the direction of ∇f is the direction in which the directional derivative at (a, b) has its largest value.

Proof By Theorem 1 (rephrased), if $\mathbf{u}$ is a unit vector, then, at (a, b),

$$D_{\mathbf{u}}f = \nabla f \cdot \mathbf{u}.$$

By the definition of the dot product

$$\nabla f \cdot \mathbf{u} = \|\nabla f\| \, \|\mathbf{u}\| \cos \alpha,$$

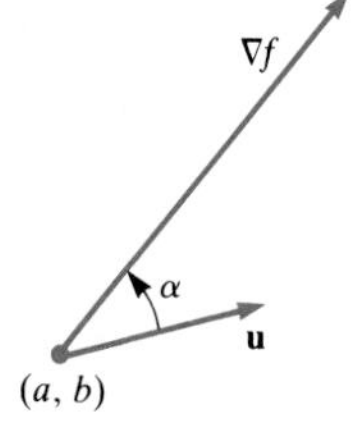

Figure 6

where α is the angle between ∇f and $\mathbf{u}$, as shown in Fig. 6. Since $\|\mathbf{u}\| = 1$,

$$D_{\mathbf{u}}f = \|\nabla f\| \cos \alpha. \tag{2}$$

The largest value of $\cos \alpha$, for $0 \le \alpha \le \pi$, occurs when $\alpha = 0$; that is, when $\cos \alpha = 1$. Thus, by Eq. (2), the largest directional derivative of $f(x, y)$ at (a, b) occurs when the direction is that of ∇f at (a, b). For that choice of $\mathbf{u}$, $D_{\mathbf{u}}f = \|\nabla f\|$. This proves the theorem. ■

What does Theorem 2 tell a bug wandering about a flat piece of metal? If it is at the point (a, b) and wishes to get warmer as quickly as possible, it should compute the gradient of the temperature function and then go in the direction indicated by the gradient. If, instead, it wishes to cool off as quickly as possible, it should go in the direction opposite the gradient.

EXAMPLE 2 What is the largest directional derivative of $f(x, y) = x^2y^3$ at $(2, 3)$? In what direction does this maximum directional derivative occur?

SOLUTION At the point (x, y),

$$\nabla f = 2xy^3\mathbf{i} + 3x^2y^2\mathbf{j}.$$

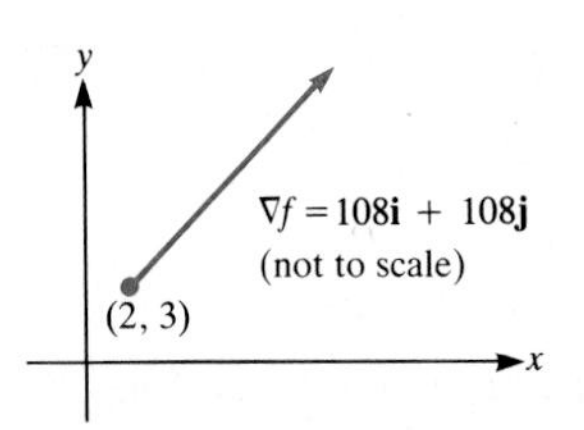

Figure 7

Thus, at (2, 3), $\qquad \nabla f = 108\mathbf{i} + 108\mathbf{j},$

which is sketched in Fig. 7 (not to scale). Note that its angle θ is $\pi/4$. The maximal directional derivative of x^2y^3 at (2, 3) is $\|\nabla f\| = 108\sqrt{2}$. This is achieved at the angle $\theta = \pi/4$, relative to the x axis, that is, for

$$\mathbf{u} = \cos\frac{\pi}{4}\mathbf{i} + \sin\frac{\pi}{4}\mathbf{j} = \frac{\sqrt{2}}{2}\mathbf{i} + \frac{\sqrt{2}}{2}\mathbf{j}. \quad \blacksquare$$

Incidentally, if $f(x, y)$ denotes the temperature at (x, y), the gradient ∇f helps to indicate the direction in which heat flows. It tends to flow "toward the coldest," which boils down to the mathematical assertion, "Heat tends to flow in the direction of $-\nabla f$."

The gradient and directional derivative have been interpreted in terms of a temperature distribution in the plane and a wandering bug. It is also instructive to interpret these concepts in terms of a hiker on the surface of a mountain.

Consider a mountain above the xy plane. The altitude of the point on the surface above the point (x, y) will be denoted by $f(x, y)$. The directional derivative

$$D_{\mathbf{u}}f$$

indicates the rate at which altitude changes per unit change in *horizontal* distance. The gradient ∇f at (a, b) points in the compass direction the hiker should choose to climb in the direction of steepest ascent. The length of ∇f tells the hiker the steepest slope available. (See Fig. 8.)

Generalization to $f(x, y, z)$

The notions of directional derivative and gradient can be generalized with little effort to functions of three (or more) variables. It is easiest to interpret the

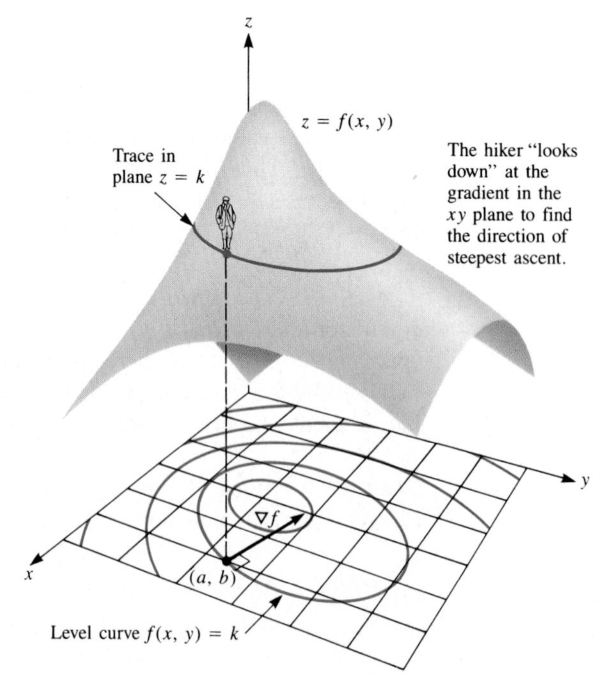

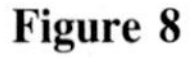
Figure 8

directional derivative of $f(x, y, z)$ in a particular direction in space as indicating the rate of change of the function in that direction in space. A useful interpretation is how fast the temperature changes in a given direction.

Let $\mathbf{u}$ be a unit vector in space, with direction angles α, β, and γ. Then $\mathbf{u} = \cos\alpha\,\mathbf{i} + \cos\beta\,\mathbf{j} + \cos\gamma\,\mathbf{k}$. We now define the derivative of $f(x, y, z)$ in the direction $\mathbf{u}$.

Definition *Directional derivative of $f(x, y, z)$.* The **directional derivative** of f at (a, b, c) in the direction of the unit vector $\mathbf{u} = \cos\alpha\,\mathbf{i} + \cos\beta\,\mathbf{j} + \cos\gamma\,\mathbf{k}$ is $g'(0)$, where g is defined by

$$g(t) = f(a + t\cos\alpha,\ b + t\cos\beta,\ c + t\cos\gamma).$$

It is denoted $D_{\mathbf{u}}f$.

Note that t is the measure of length along the line through (a, b, c) with direction angles α, β, and γ. Therefore $D_{\mathbf{u}}f$ is just a derivative along the t axis.

The proof of the following theorem is like that of Theorem 1.

Theorem 3 If $f(x, y, z)$ has continuous partial derivatives $\partial f/\partial x$, $\partial f/\partial y$, and $\partial f/\partial z$, then the directional derivative of f at (a, b, c) in the direction of the unit vector $\mathbf{u} = \cos\alpha\,\mathbf{i} + \cos\beta\,\mathbf{j} + \cos\gamma\,\mathbf{k}$ is

$$\frac{\partial f}{\partial x}(a, b, c)\cos\alpha + \frac{\partial f}{\partial y}(a, b, c)\cos\beta + \frac{\partial f}{\partial z}(a, b, c)\cos\gamma.$$

Definition *The gradient of $f(x, y, z)$.* The vector

$$\frac{\partial f}{\partial x}(a, b, c)\,\mathbf{i} + \frac{\partial f}{\partial y}(a, b, c)\,\mathbf{j} + \frac{\partial f}{\partial z}(a, b, c)\,\mathbf{k}$$

is the **gradient** of f at (a, b, c) and is denoted ∇f.

Theorem 3 thus asserts that the derivative of $f(x, y, z)$ in the direction of the unit vector $\mathbf{u}$ equals the dot product of $\mathbf{u}$ and the gradient of f:

$$\boxed{D_{\mathbf{u}}f = \nabla f \cdot \mathbf{u}.}$$

Just as in the case of a function of two variables, ∇f, evaluated at (a, b, c), points in the direction $\mathbf{u}$ that produces the largest directional derivative at (a, b, c). Moreover, $\|\nabla f\|$ is that largest directional derivative. The proof is practically identical with that of Theorem 2.

EXAMPLE 3 The temperature at the point (x, y, z) in a solid piece of metal is given by the formula $f(x, y, z) = e^{2x+y+3z}$ degrees. In what direction at the point $(0, 0, 0)$ does the temperature increase most rapidly?

SOLUTION First compute

$$\frac{\partial f}{\partial x} = 2e^{2x+y+3z}, \qquad \frac{\partial f}{\partial y} = e^{2x+y+3z}, \qquad \frac{\partial f}{\partial z} = 3e^{2x+y+3z}.$$

Then form the gradient vector:

$$\nabla f = 2e^{2x+y+3z}\mathbf{i} + e^{2x+y+3z}\mathbf{j} + 3e^{2x+y+3z}\mathbf{k}.$$

At (0, 0, 0), $$\nabla f = 2\mathbf{i} + \mathbf{j} + 3\mathbf{k}.$$

Consequently, the direction of most rapid increase in temperature is that given by the vector $2\mathbf{i} + \mathbf{j} + 3\mathbf{k}$. The rate of increase is then

$$\|2\mathbf{i} + \mathbf{j} + 3\mathbf{k}\| = \sqrt{14} \text{ degrees per unit length.}$$

If the line through (0, 0, 0) parallel to $2\mathbf{i} + \mathbf{j} + 3\mathbf{k}$ is given a coordinate system so that it becomes the t axis, with $t = 0$ at the origin and the positive part in the direction of $2\mathbf{i} + \mathbf{j} + 3\mathbf{k}$, then $df/dt = \sqrt{14}$ at 0. ■

> The gradient was denoted $\boldsymbol{\Delta}$ by Hamilton in 1846. By 1870 it was denoted ∇, an upside-down delta, and therefore called "atled." In 1871 Maxwell wrote, "The quantity ∇P is a vector. I venture, with much diffidence, to call it the *slope* of P." The name "slope" is no longer used, having been replaced by "gradient." "Gradient" goes back to the word "grade," the slope of a road or surface. The name "del" first appeared in print in 1901, in *Vector Analysis, A text-book for the use of students of mathematics and physics founded upon the lectures of J. Willard Gibbs,* by E. B. Wilson.

Section Summary

We defined the derivative of $f(x, y)$ at (a, b) in the direction of the unit vector $\mathbf{u}$ in the xy plane and the derivative of $f(x, y, z)$ at (a, b, c) in the direction of the unit vector $\mathbf{u}$ in space. Then we introduced the gradient vector ∇f in terms of its components and obtained the formula

$$D_{\mathbf{u}} f = \nabla f \cdot \mathbf{u}.$$

By examining this formula we saw that the length and direction of ∇f at a given point are significant:

∇f points in the direction $\mathbf{u}$ that maximizes $D_{\mathbf{u}} f$ at the given point.
$\|\nabla f\|$ is the maximum directional derivative of f at the given point.

EXERCISES FOR SEC. 14.7: DIRECTIONAL DERIVATIVES AND THE GRADIENT

As usual, we assume that all functions mentioned have continuous partial derivatives. In Exercises 1 and 2 compute the directional derivatives of x^4y^5 at (1, 1) in the indicated directions.

1 (*a*) $\mathbf{i}$, (*b*) $-\mathbf{i}$, (*c*) $\cos(\pi/4)\,\mathbf{i} + \sin(\pi/4)\,\mathbf{j}$

2 (*a*) $\mathbf{j}$, (*b*) $-\mathbf{j}$, (*c*) $\cos(\pi/3)\,\mathbf{i} + \sin(\pi/3)\,\mathbf{j}$

In Exercises 3 and 4 compute the directional derivatives of x^2yz^3 in the directions of the given vectors.

3 (*a*) $\mathbf{j}$, (*b*) $\mathbf{k}$, (*c*) $-\mathbf{i}$

4 (*a*) $\mathbf{i} + \mathbf{j} + \mathbf{k}$, (*b*) $2\mathbf{i} - \mathbf{j} + 2\mathbf{k}$, (*c*) $\mathbf{i} + \mathbf{k}$ (Note that these are not unit vectors. First construct a unit vector with the same direction.)

5 Assume that at the point (2, 3) $\partial f/\partial x = 4$ and $\partial f/\partial y = 5$.
(*a*) Draw ∇f at (2, 3).
(*b*) What is the maximal directional derivative of f at (2, 3)?
(*c*) For which $\mathbf{u}$ is $D_{\mathbf{u}}f$ at (2, 3) maximal? (Write $\mathbf{u}$ in the form $x\mathbf{i} + y\mathbf{j}$.)

6 Assume that at the point (1, 1) $\partial f/\partial x = 3$ and $\partial f/\partial y = -3$.
(*a*) Draw ∇f at (1, 1).
(*b*) What is the maximal directional derivative of f at (1, 1)?
(*c*) For which $\mathbf{u}$ is $D_{\mathbf{u}}f$ at (1, 1) maximal? (Write $\mathbf{u}$ in the form $x\mathbf{i} + y\mathbf{j}$.)

In Exercises 7 and 8 compute and draw ∇f at the indicated points for the given functions.

7 $f(x, y) = x^2y$ at (*a*) (2, 5), (*b*) (3, 1).

8 $f(x, y) = 1/\sqrt{x^2 + y^2}$ at (*a*) (1, 2), (*b*) (3, 0).

9 If the maximal directional derivative of f at (a, b) is 5, what is the minimal directional derivative there? Explain.

10 For a given function $f(x, y)$ at a given point (a, b) is there always a direction in which the directional derivative is 0? Explain.

11 If $(\partial f/\partial x)(a, b) = 2$ and $(\partial f/\partial y)(a, b) = 3$, in what direction should a directional derivative at (a, b) be computed in order that it be (*a*) 0? (*b*) as large as possible? (*c*) as small as possible?

12 If at the point (a, b, c) $\partial f/\partial x = 2$, $\partial f/\partial y = 3$, and $\partial f/\partial z = 4$, what is the largest directional derivative of f at (a, b, c)?

13 Assume that $f(1, 2) = 2$ and $f(0.99, 2.01) = 1.98$.
(*a*) Which directional derivatives $D_{\mathbf{u}}f$ at (1, 2) can be estimated with this information? (Give $\mathbf{u}$.)
(*b*) Estimate the directional derivatives in (*a*).

14 Assume that $f(1, 1, 1) = 3$ and $f(1.1, 1.2, 1.1) = 3.1$.
(*a*) Which directional derivatives $D_{\mathbf{u}}f$ at (1, 1, 1) can be estimated with this information? (Give $\mathbf{u}$.)
(*b*) Estimate the directional derivatives in (*a*).

15 When a bug crawls east, it discovers that the temperature increases at the rate of 0.02° per centimeter. When it crawls north, the temperature decreases at the rate of −0.03° per centimeter.
(*a*) If the bug crawls south, at what rate does the temperature change?
(*b*) If the bug crawls 30° north of east, at what rate does the temperature change?
(*c*) If the bug is happy with its temperature, in what direction should it crawl to try to keep the temperature the same?

16 A bird is very sensitive to the temperature. It notices that when it flies in the direction $\mathbf{i}$, the temperature increases at the rate of 0.03° per centimeter. When it flies in the direction $\mathbf{j}$, the temperature decreases at the rate of 0.02° per centimeter. When it flies in the direction $\mathbf{k}$, the temperature increases at the rate of 0.05° per centimeter. It decides to fly off in the direction of the vector $\langle 2, 5, 1\rangle$. Will it be getting warmer or colder?

17 Assume that $f(1, 2) = 3$ and that the directional derivative of f at (1, 2) in the direction of the (nonunit) vector $\mathbf{i} + \mathbf{j}$ is 0.7. Use this information to estimate $f(1.1, 2.1)$.

18 Assume that $f(1, 1, 2) = 4$ and that the directional derivative of f at (1, 1, 2) in the direction of the vector from (1, 1, 2) to (1.01, 1.02, 1.99) is 3. Use this information to estimate $f(0.99, 0.98, 2.01)$.

In Exercises 19 to 24 find the directional derivative of the function in the given direction and the maximum directional derivative.

19 xyz^2 at (1, 0, 1); $\mathbf{i} + \mathbf{j} + \mathbf{k}$

20 x^3yz at (2, 1, −1); $2\mathbf{i} - \mathbf{k}$

21 $e^{xy \sin z}$ at (1, 1, $\pi/4$); $\mathbf{i} + \mathbf{j} + 3\mathbf{k}$

22 $\tan^{-1} \sqrt{x^2 + y + z}$ at (1, 1, 1); $-\mathbf{i}$

23 $\ln(1 + xyz)$ at (2, 3, 1); $-\mathbf{i} + \mathbf{j}$

24 $x^xye^{z^2}$ at (1, 1, 0); $\mathbf{i} - \mathbf{j} + \mathbf{k}$

25 Let $f(x, y, z) = 2x + 3y + z$.
(*a*) Compute ∇f at (0, 0, 0) and at (1, 1, 1).
(*b*) Draw ∇f for the two points in (*a*), in each case putting its tail at the point.

26 Let $f(x, y, z) = x^2 + y^2 + z^2$.
(*a*) Compute ∇f at (2, 0, 0), (0, 2, 0), and (0, 0, 2).
(*b*) Draw ∇f for the three points in (*a*), in each case putting its tail at the point.

27 Assume that ∇f at (a, b) is not $\mathbf{0}$. Show that there are two unit vectors $\mathbf{u}_1$ and $\mathbf{u}_2$, such that the directional derivatives of f at (a, b) in the direction of $\mathbf{u}_1$ and $\mathbf{u}_2$ are 0.

28 Assume that ∇f at (a, b, c) is not $\mathbf{0}$. How many unit vectors $\mathbf{u}$ are there such that $D_{\mathbf{u}}f = 0$? Explain.

29 Let $T(x, y, z)$ be the temperature at the point (x, y, z). Assume that ∇T at (1, 1, 1) is $2\mathbf{i} + 3\mathbf{j} + 4\mathbf{k}$.
(*a*) Find $D_{\mathbf{u}}T$ at (1, 1, 1) if $\mathbf{u}$ is in the direction of the vector $\mathbf{i} - \mathbf{j} + 2\mathbf{k}$.
(*b*) Estimate the change in temperature as you move from the point (1, 1, 1) a distance 0.2 in the direction of the vector $\mathbf{i} - \mathbf{j} + 2\mathbf{k}$.
(*c*) Find three unit vectors $\mathbf{u}$ such that $D_{\mathbf{u}}T = 0$ at (1, 1, 1).

30 A bug at the point (1, 2) is very sensitive to temperature and observes that if it moves in the direction $\mathbf{i}$ the temperature increases at the rate of 2° per centimeter. If it moves in the direction $\mathbf{j}$, the temperature decreases at the rate of 2° per centimeter. In what direction should it move if it wants
(*a*) to warm up most rapidly?
(*b*) to cool off most rapidly?
(*c*) to change the temperature as little as possible?

---

31 Let $f(x, y) = 1/\sqrt{x^2 + y^2}$; the function f is defined everywhere except at (0, 0). Let $\mathbf{r} = \langle x, y\rangle$.
(*a*) Show that $\nabla f = -\mathbf{r}/\|\mathbf{r}\|^3$.
(*b*) Show that $\|\nabla f\| = 1/\|\mathbf{r}\|^2$.

32 Let $f(x, y, z) = 1/\sqrt{x^2 + y^2 + z^2}$, which is defined everywhere except at (0, 0, 0). (This function is related to the potential in a gravitational field due to a point-mass.) Let $\mathbf{r} = x\mathbf{i} + y\mathbf{j} + z\mathbf{k}$. Express ∇f in terms of $\mathbf{r}$.

33 Let $f(x, y) = x^2 + y^2$. Prove that if (a, b) is an arbitrary point on the curve $x^2 + y^2 = 9$, then ∇f computed at (a, b) is perpendicular to the tangent line to that curve at (a, b).

34 Let $f(x, y, z)$ equal temperature at (x, y, z). Let $P = (a, b, c)$ and Q be a point very near (a, b, c). Show that $\nabla f \cdot \overrightarrow{PQ}$ is a good estimate of the change in temperature from point P to point Q.

35 (*a*) If $(\partial f/\partial x)(a, b, c) = 2$, $(\partial f/\partial y)(a, b, c) = 3$, and $(\partial f/\partial z)(a, b, c) = 1$, find three different unit vectors $\mathbf{u}$ such that $D_{\mathbf{u}}f$ at (a, b, c) is 0.
(*b*) How many unit vectors $\mathbf{u}$ are there such that $D_{\mathbf{u}}f$ at (a, b, c) is 0?

36 Let $f(x, y) = xy$.
(*a*) Draw the level curve $xy = 4$ carefully.
(*b*) Compute ∇f at three convenient points on that level curve and draw it with its tail at the point where it is evaluated.
(*c*) What angle does ∇f seem to make with the curve at the point where it is evaluated?
(*d*) Prove that the angle is what you think it is.

37 Let $f(x, y)$ be the temperature at (x, y). Assume that ∇f at (1, 1) is $2\mathbf{i} + 3\mathbf{j}$. A bug is crawling northwest at the rate of 3 centimeters per second. Let $g(t)$ be the temperature at the point where the bug is at time t seconds. Then dg/dt is the rate at which temperature changes on the bug's journey (degrees per second). Find dg/dt when the bug is at (1, 1).

38 If $f(P)$ is the electric potential at the point P, then the electric field $\mathbf{E}$ at P is given by $-^2\nabla f$. Calculate $\mathbf{E}$ if $f(x, y) = \sin \alpha x \cos \beta y$, where α and β are constants.

39 The equality $\partial^2 f/\partial x\, \partial y = \partial^2 f/\partial y\, \partial x$ can be written as $D_{\mathbf{i}}(D_{\mathbf{j}}f) = D_{\mathbf{j}}(D_{\mathbf{i}}f)$. Show for any two unit vectors $\mathbf{u}_1$ and $\mathbf{u}_2$ that $D_{\mathbf{u}_2}(D_{\mathbf{u}_1}f) = D_{\mathbf{u}_1}(D_{\mathbf{u}_2}f)$. (Assume that all partial derivatives of f of all orders are continuous.)

40 Prove the first part of Theorem 2 without the aid of vectors. That is, prove that the maximum value of

$$g(\theta) = \partial f/\partial x(a, b) \cos \theta + \partial f/\partial y(a, b) \sin \theta$$

is $\sqrt{[\partial f/\partial x(a, b)]^2 + [\partial f/\partial y(a, b)]^2}$.

41 Figure 9 shows two level curves of a function $f(x, y)$ near the point (1, 2), namely $f(x, y) = 3$ and $f(x, y) = 3.02$. Use the diagram to estimate
(*a*) $D_{\mathbf{i}}f$ at (1, 2),
(*b*) $D_{\mathbf{j}}f$ at (1, 2).
(*c*) Draw ∇f at (1, 2).

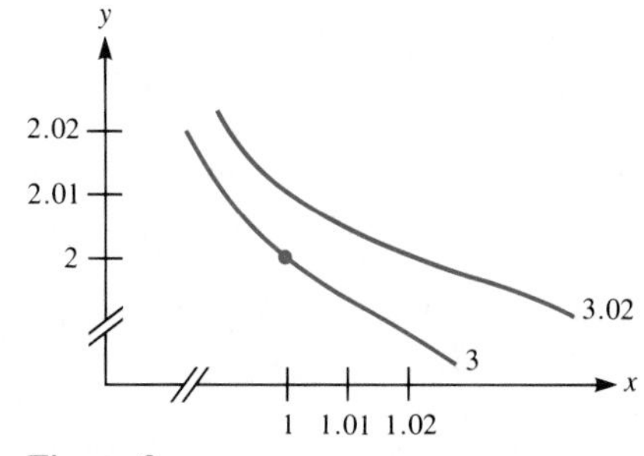

Figure 9

14.8 NORMALS AND THE TANGENT PLANE

In this section we first find how to obtain a normal vector to a curve given implicitly, as a level curve $f(x, y) = k$. Then we find how to obtain a normal to a surface given implicitly, as a level surface $f(x, y, z) = k$. With the aid of this vector we define the tangent plane to a surface at a given point on the surface.

Normals to a Curve in the *xy* Plane

We saw in Sec. 13.4 how to find a normal vector to a curve when the curve is given parametrically, $\mathbf{r} = \mathbf{G}(t)$. Now we will see how to find a normal when the curve is given implicitly, as a level curve $f(x, y) = k$.

Theorem 1 The gradient ∇f at (a, b) is a normal to the level curve of f passing through (a, b).

Proof Let $\mathbf{G}(t) = x(t)\mathbf{i} + y(t)\mathbf{j}$ be a parameterization of the level curve of f that passes through the point (a, b). On this curve, $f(x, y)$ is constant and has the value $f(a, b)$. Let $\mathbf{G}'(t_0)$ be the tangent vector to the curve at (a, b) and let the gradient of f at (a, b) be $\nabla f = (\partial f/\partial x)(a, b)\mathbf{i} + (\partial f/\partial y)(a, b)\mathbf{j}$. We wish to show that

$$\nabla f \cdot \mathbf{G}'(t_0) = 0;$$

that is,

$$\frac{\partial f}{\partial x}(a, b)\frac{dx}{dt}(t_0) + \frac{\partial f}{\partial y}(a, b)\frac{dy}{dt}(t_0) = 0. \tag{1}$$

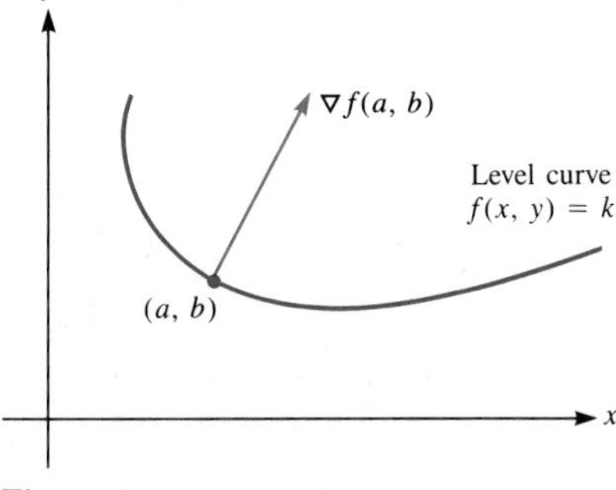

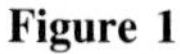
Figure 1

The left side of Eq. (1) has the form of a chain rule. To make use of this fact, introduce the function $u(t)$ defined as

$$u(t) = f(x(t), y(t)).$$

Note that $u(t)$ is the value of f at a point on the level curve that passes through (a, b). Hence $u(t) = f(a, b)$. What is more important is that $u(t)$ is a constant function. Therefore, $du/dt = 0$.

Now, $u = f(x, y)$, where x and y are functions of t. The chain rule asserts that

$$\frac{du}{dt} = \frac{\partial f}{\partial x}\frac{dx}{dt} + \frac{\partial f}{\partial y}\frac{dy}{dt}.$$

Since $du/dt = 0$, Eq. (1) follows. Hence ∇f, evaluated at (a, b), is a normal to the level curve of f that passes through (a, b). ■

Figure 1 shows a typical level curve and gradient. The gradient is perpendicular to the level curve. Moreover, as we saw in Sec. 14.7, the gradient points in the direction in which the function increases most rapidly.

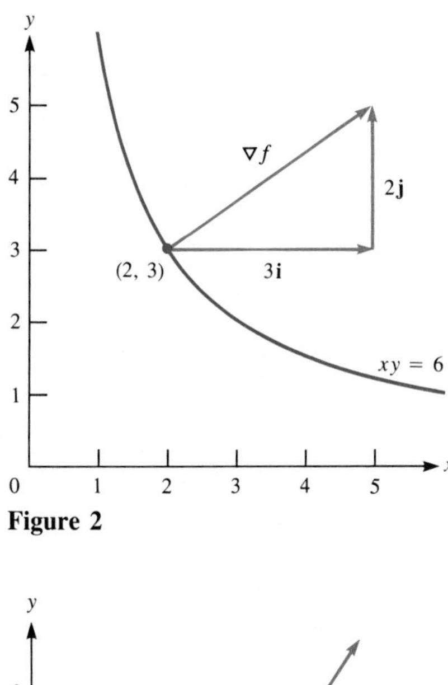

Figure 2

EXAMPLE 1 Find and draw a normal vector to the hyperbola $xy = 6$ at the point $(2, 3)$.

SOLUTION Let $f(x, y) = xy$. Then $\partial f/\partial x = y$ and $\partial f/\partial y = x$. Hence

$$\nabla f = y\mathbf{i} + x\mathbf{j}.$$

In particular

$$\nabla f(2, 3) = 3\mathbf{i} + 2\mathbf{j}.$$

This gradient and the level curve $xy = 6$ are shown in Fig. 2. ■

EXAMPLE 2 Find an equation of the tangent line to the ellipse $x^2 + 3y^2 = 7$ at the point $(2, 1)$.

SOLUTION As we saw in Sec. 12.4, we may write the equation of a line in the plane if we know a point on the line and a vector normal to the line. We know that $(2, 1)$ lies on the line. We use a gradient to produce a normal.

The ellipse $x^2 + 3y^2 = 7$ is a level curve of the function $f(x, y) = x^2 + 3y^2$. Since $\partial f/\partial x = 2x$ and $\partial f/\partial y = 6y$, $\nabla f = 2x\mathbf{i} + 6y\mathbf{j}$. In particular

$$\nabla f(2, 1) = 4\mathbf{i} + 6\mathbf{j}.$$

Hence the tangent line at $(2, 1)$ has an equation

$$4(x - 2) + 6(y - 1) = 0,$$

or

$$4x + 6y = 14.$$

It is shown in Fig. 3. ■

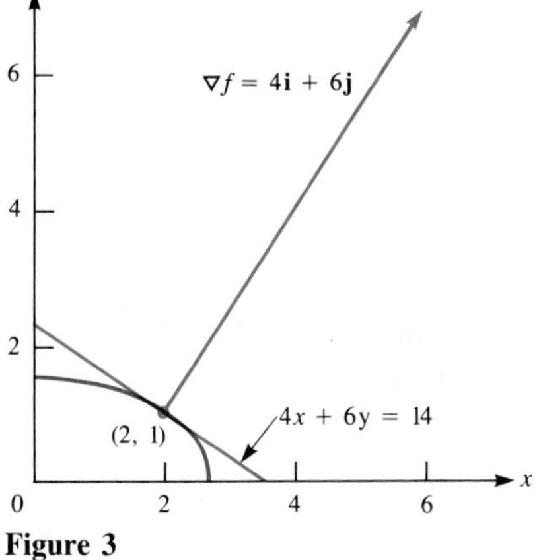

Figure 3

Normals to a Surface

We can construct a vector perpendicular to a surface $f(x, y, z) = k$ at a given point $P = (a, b, c)$ as easily as we constructed a vector perpendicular to a planar curve. It turns out that the gradient vector ∇f, evaluated at (a, b, c), is perpendicular to the surface $f(x, y, z) = k$. The proof is similar to the proof of Theorem 1.

Before proceeding, it is necessary to state what is meant by a vector being perpendicular to a surface.

Definition *Normal vector to a surface.* A vector is perpendicular to a surface at the point (a, b, c) on this surface if the vector is perpendicular to each curve on the surface through the point (a, b, c). [A vector is perpendicular to a curve at a point (a, b, c) on the curve if the vector is perpendicular to a tangent vector to the curve at (a, b, c).] Such a vector is called a **normal vector**.

Theorem 2 The gradient ∇f at (a, b, c) is a normal to the level surface of f passing through (a, b, c).

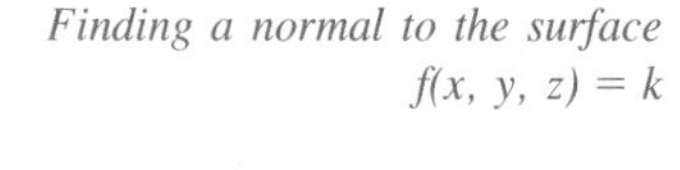

Proof Let $\mathbf{G}(t) = x(t)\mathbf{i} + y(t)\mathbf{j} + z(t)\mathbf{k}$ be the parameterization of a curve in the level surface of f that passes through the point (a, b, c). Assume $\mathbf{G}(t_0) = \langle a, b, c\rangle$. Then $\mathbf{G}'(t_0)$ is the tangent vector to the curve at (a, b, c) and the gradient at (a, b, c) is

$$\nabla f = \frac{\partial f}{\partial x}(a, b, c)\mathbf{i} + \frac{\partial f}{\partial y}(a, b, c)\mathbf{j} + \frac{\partial f}{\partial z}(a, b, c)\mathbf{k}.$$

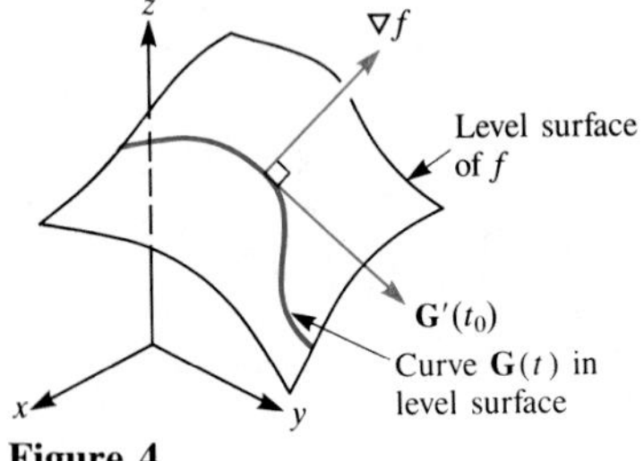

Figure 4

We wish to show that

$$\nabla f \cdot \mathbf{G}'(t_0) = 0;$$

that is,

$$\frac{\partial f}{\partial x}(a, b, c)x'(t_0) + \frac{\partial f}{\partial y}(a, b, c)y'(t_0) + \frac{\partial f}{\partial z}(a, b, c)z'(t_0) = 0. \tag{2}$$

(See Fig. 4.) Introduce a function $u(t)$ defined by

$$u(t) = f(x(t), y(t), z(t)).$$

Once again the chain rule comes to our aid.

By the chain rule,

$$\left.\frac{du}{dt}\right|_{t_0} = \frac{\partial f}{\partial x}(a, b, c)x'(t_0) + \frac{\partial f}{\partial y}(a, b, c)y'(t_0) + \frac{\partial f}{\partial z}(a, b, c)z'(t_0). \tag{3}$$

However, *since the curve* $\mathbf{G}(t)$ *lies on a level surface of* f, $u(t)$ *is constant.* [In fact, $u(t) = f(a, b, c)$.] Thus $du/dt = 0$, and the right side of Eq. (3) is 0, as required. ■

We can check Theorem 2 by seeing whether it is correct when the level surfaces are just planes. In this case we consider $f(x, y, z) = Ax + By + Cz + D$. The plane $Ax + By + Cz + D = 0$ is the level surface $f(x, y, z) = 0$. According to Theorem 2, ∇f is perpendicular to this surface. Now, $\partial f/\partial x = A$, $\partial f/\partial y = B$, and $\partial f/\partial z = C$. Hence

$$\begin{aligned}\nabla f &= \frac{\partial f}{\partial x}\mathbf{i} + \frac{\partial f}{\partial y}\mathbf{j} + \frac{\partial f}{\partial z}\mathbf{k} \\ &= A\mathbf{i} + B\mathbf{j} + C\mathbf{k}.\end{aligned}$$

This agrees with the fact that $A\mathbf{i} + B\mathbf{j} + C\mathbf{k}$ is a normal to the plane $Ax + By + Cz + D = 0$, as we saw in Sec. 12.4.

EXAMPLE 3 Find a normal vector to the ellipsoid $x^2 + y^2/4 + z^2/9 = 3$ at the point $(1, 2, 3)$.

SOLUTION The ellipsoid is a level surface of the function

$$f(x, y, z) = x^2 + \frac{y^2}{4} + \frac{z^2}{9}.$$

The gradient of f is

$$\nabla f = 2x\mathbf{i} + \frac{y}{2}\mathbf{j} + \frac{2z}{9}\mathbf{k}.$$

At $(1, 2, 3)$, $$\nabla f = 2\mathbf{i} + \mathbf{j} + \tfrac{2}{3}\mathbf{k}.$$

This vector is normal to the ellipsoid at $(1, 2, 3)$. ■

Tangent Planes to a Surface

We now can define a tangent plane to a surface at a point on the surface.

Definition *Tangent plane to a surface*. Consider a surface that is a level surface of a function $u = f(x, y, z)$. Let (a, b, c) be a point on this surface where ∇f is not 0. The tangent plane to the surface at the point (a, b, c) is that plane through (a, b, c) that is perpendicular to the vector ∇f evaluated at (a, b, c).

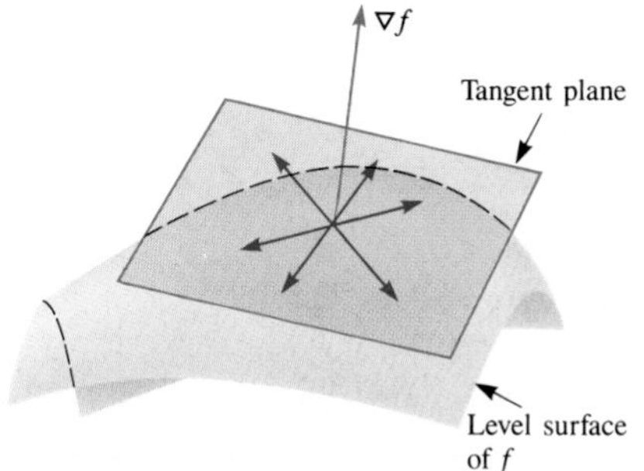

Figure 5

The tangent plane at (a, b, c) is the plane that best approximates the surface near (a, b, c). It consists of all the tangent lines at (a, b, c) to curves in the surface that pass through the point (a, b, c). See Fig. 5.

Note that an equation of the tangent plane to the surface $f(x, y, z) = k$ at (a, b, c) is

$$\frac{\partial f}{\partial x}(x - a) + \frac{\partial f}{\partial y}(y - b) + \frac{\partial f}{\partial z}(z - c) = 0,$$

where the partial derivatives are evaluated at (a, b, c).

EXAMPLE 4 Find an equation of the tangent plane to the ellipsoid $x^2 + y^2/4 + z^2/9 = 3$ at the point $(1, 2, 3)$.

SOLUTION By Example 3, the vector $2\mathbf{i} + \mathbf{j} + \frac{2}{3}\mathbf{k}$ is normal to the surface at the point (1, 2, 3). The tangent plane consequently has an equation

$$2(x - 1) + 1(y - 2) + \tfrac{2}{3}(z - 3) = 0. \quad \blacksquare$$

Normals and Tangent Planes to $z = f(x, y)$

A surface may be described explicitly in the form $z = f(x, y)$ rather than implicitly in the form $f(x, y, z) = k$. The techniques already developed enable us to find the normal and tangent plane in the case $z = f(x, y)$ as well.

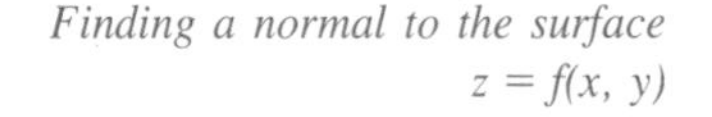

Finding a normal to the surface $z = f(x, y)$

We need only rewrite the equation $z = f(x, y)$ in the form $z - f(x, y) = 0$ [or, if you prefer, $f(x, y) - z = 0$]. Then define $g(x, y, z)$ to be $z - f(x, y)$. The surface $z = f(x, y)$ is simply the particular level surface of g given by $g(x, y, z) = 0$. There is no need to memorize an extra formula for a vector normal to the surface $z = f(x, y)$. The next example illustrates this advice.

EXAMPLE 5 Find a vector perpendicular to the saddle $z = y^2 - x^2$ at the point (1, 2, 3).

SOLUTION In this case, rewrite $z = y^2 - x^2$ as $z + x^2 - y^2 = 0$. The surface in question is a level surface of $g(x, y, z) = z + x^2 - y^2$. Hence $\nabla g = 2x\mathbf{i} - 2y\mathbf{j} + \mathbf{k}$ is perpendicular to the surface at (x, y, z). In particular, the vector $2\mathbf{i} - 4\mathbf{j} + \mathbf{k}$ is perpendicular to the surface at the point (1, 2, 3).

This surface, which looks like a saddle near the origin, was graphed in Sec. 14.2. The surface and the normal vector $2\mathbf{i} - 4\mathbf{j} + \mathbf{k}$ are shown in Fig. 6. ■

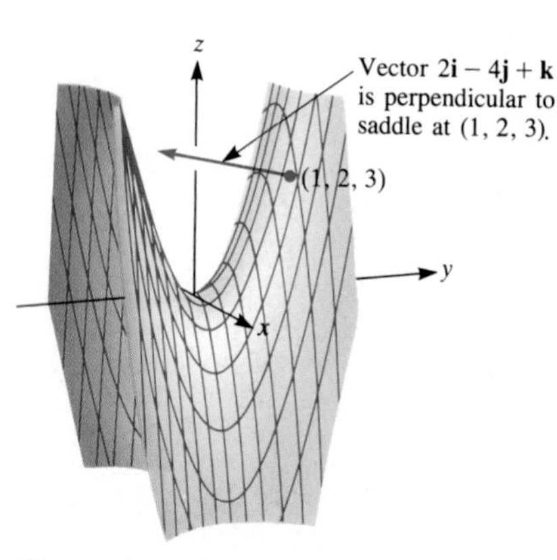

Figure 6

The Differential of $z = f(x, y)$

In the case of a function of one variable, $y = f(x)$, the tangent line at $(a, f(a))$ closely approximates the graph of $y = f(x)$. The equation of the tangent line $y = f(a) + f'(a)(x - a)$ gives us a linear approximation $p(x)$ of $f(x)$. (See Sec. 4.9.)

We can use the tangent plane to the surface $z = f(x, y)$ similarly. To find the equation of the plane tangent at $(a, b, f(a, b))$, we first rewrite the equation of the surface as

$$g(x, y, z) = f(x, y) - z = 0.$$

Then ∇g is a normal to the surface at $(a, b, f(a, b))$. Now,

$$\nabla g = \frac{\partial f}{\partial x}\mathbf{i} + \frac{\partial f}{\partial y}\mathbf{j} - \mathbf{k},$$

where the partial derivatives are evaluated at (a, b).

The equation of the tangent plane at $(a, b, f(a, b))$ is therefore

$$\frac{\partial f}{\partial x}(a, b)(x - a) + \frac{\partial f}{\partial y}(a, b)(y - b) - (z - f(a, b)) = 0.$$

We can rewrite this equation as

$$z = f(a, b) + \frac{\partial f}{\partial x}(a, b)(x - a) + \frac{\partial f}{\partial y}(a, b)(y - b). \tag{4}$$

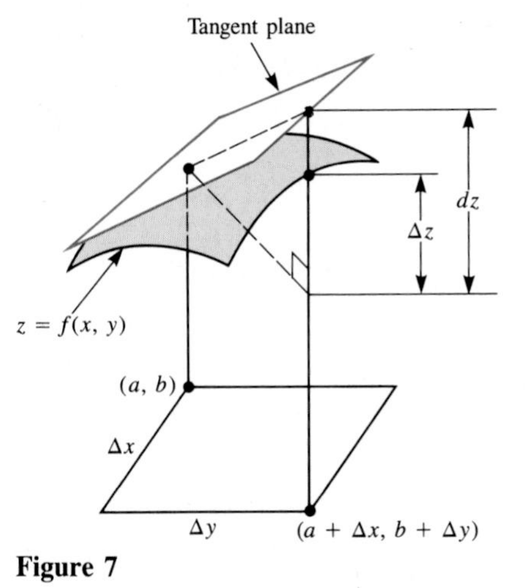

Figure 7

Equation (4) gives the z coordinate on the tangent plane as a function of x and y. It is an estimate of the z coordinate on the surface $z = f(x, y)$. [It is exact when $(x, y) = (a, b)$.] We define the **differential** of z, dz, as

$$dz = \frac{\partial f}{\partial x}(a, b)(x - a) + \frac{\partial f}{\partial y}(a, b)(y - b).$$

[This is the analog of the differential of $y = f(x)$, namely, $f'(a)\,(x - a)$.]

Thus

$$f(a, b) + dz$$

is an estimate of $f(x, y)$. Setting $\Delta x = x - a$ and $\Delta y = y - b$ gives

$$dz = \frac{\partial f}{\partial x}(a, b)\,\Delta x + \frac{\partial f}{\partial y}(a, b)\,\Delta y. \tag{5}$$

(See Fig. 7.) On the other hand

$$\Delta z = f(a + \Delta x, b + \Delta y) - f(a, b).$$

While dz is "change in z along the tangent plane," Δz is the "change in z on the surface $z = (x, y)$."

Recall the expression we developed in Sec. 14.6 for $\Delta f(=\Delta z)$. It is

$$\Delta z = \frac{\partial f}{\partial x}(a, b)\,\Delta x + \frac{\partial f}{\partial y}(a, b)\,\Delta y + \epsilon_1\,\Delta x + \epsilon_2\,\Delta y, \tag{6}$$

where ϵ_1 and $\epsilon_2 \to 0$ as Δx and $\Delta y \to 0$. Comparison of (5) and (6) shows why dz is a good approximation of Δz when Δx and Δy are small.

EXAMPLE 6 Let $z = f(x, y) = x^2y$. Let $\Delta z = f(1.01, 2.02) - f(1, 2)$ and let

$$dz = \left.\frac{\partial f}{\partial x}\right|_{(1,\,2)}(0.01) + \left.\frac{\partial f}{\partial y}\right|_{(1,\,2)}(0.02).$$

Compute Δz and dz.

SOLUTION

$$\Delta z = (1.01)^2(2.02) - 1^2 2 = 2.060602 - 2 = 0.060602$$

Since $\partial f/\partial x = 2xy$ and $\partial f/\partial y = x^2$, we have $\partial f/\partial x = 4$ and $\partial f/\partial y = 1$ at $(1, 2)$. Hence

$$dz = (4)(0.01) + (1)(0.02) = 0.06.$$

Note that dz is a good approximation of Δz. ■

EXAMPLE 7 In an experiment to find an average speed, an engineer uses the formula $v = s/t$, where s is the distance traveled, t is the elapsed time, and v is the average speed. If there may be a 1 percent error in measuring s, and a 2 percent error in measuring t, how large a percent error may there be in estimating v?

SOLUTION Let the error in measuring s be Δs and the error in measuring t be Δt. Then

$$\left|\frac{\Delta s}{s}\right| \le 0.01 \qquad \text{and} \qquad \left|\frac{\Delta t}{t}\right| \le 0.02.$$

We wish to find out how large $|\Delta v/v|$ might be. Since dv is a good approximation of Δv when Δs and Δt are small, and dv is easier to compute than is Δv, we estimate $|dv/v|$.

Now,
$$dv = \frac{\partial v}{\partial s}\,\Delta s + \frac{\partial v}{\partial t}\,\Delta t$$
$$= \frac{1}{t}\,\Delta s - \frac{s}{t^2}\,\Delta t.$$

Hence
$$\frac{dv}{v} = \frac{dv}{s/t} = \frac{t}{s}\left(\frac{1}{t}\,\Delta s - \frac{s}{t^2}\,\Delta t\right)$$
$$= \frac{\Delta s}{s} - \frac{\Delta t}{t}.$$

Thus
$$\left|\frac{dv}{v}\right| = \left|\frac{\Delta s}{s} - \frac{\Delta t}{t}\right| \le \left|\frac{\Delta s}{s}\right| + \left|\frac{\Delta t}{t}\right| = 0.01 + 0.02 = 0.03.$$

We may expect an error as large as 3 percent in estimating v. ■

Section Summary

We found that when f is a function of two variables, ∇f is normal to a level curve of f. If f is a function of three variables, then ∇f is a normal to a level surface of f. The gradient then helped us define the tangent plane to a surface at a point (a, b, c): it is the plane through (a, b, c) perpendicular to the gradient at (a, b, c). The tangent plane to a surface $z = f(x, y)$ can be found by viewing the surface as a level surface of the function $z - f(x, y)$, which is a function of three variables.

We then related the tangent plane to the differential $\partial f/\partial x\ \Delta x + \partial f/\partial y\ \Delta y$, which is an approximation of Δf.

EXERCISES FOR SEC. 14.8: NORMALS AND THE TANGENT PLANE

In Exercises 1 to 4 find a vector perpendicular to the curve at the given point and draw the curve and the vector.

1 $xy = 8$ at $(2, 4)$

2 $x^2 + y^2 = 25$ at $(-3, 4)$

3 $\dfrac{x^2}{4} + \dfrac{y^2}{9} = 2$ at $(2, 3)$

4 $x^2 - y^2 = 1$ at $(3, -\sqrt{8})$

In Exercises 5 to 8, find a normal vector to the given surface at the given point.

5 $x^2 + 2y^2 + 3z^2 = 6$ at $(1, 1, 1)$

6 $xy + xz + yz = 11$ at $(1, 2, 3)$

7 $z = x^2 + y^2$ at $(1, 1, 2)$

8 $z = x^2$ at $(1, 3, 1)$

In Exercises 9 to 12 draw the given surface, a normal vector, and the tangent plane at the given point.

9 $x^2 + y^2 + z^2 = 3$ at $(1, 1, 1)$

10 $z = y^2$ at $(1, 3, 1)$

11 $z = x^2 + y^2$ at $(1, 2, 5)$

12 $x^2 + y^2 - z^2 = 7$ at $(2, 2, 1)$

In Exercises 13 to 16 find an equation of the tangent plane to the given surface at the given point.

13 $xyz = 6$ at $(1, 2, 3)$

14 $x^2 + 3y^2 - z^2 = 8$ at $(3, 1, 2)$

15 $\dfrac{e^{xy}}{1 + z^2} = \dfrac{1}{2}$ at $(0, 2, 1)$

16 $\sin^{-1}\left(\dfrac{x}{\sqrt{yz}}\right) = \dfrac{\pi}{6}$ at $(1, 2, 2)$

In Exercises 17 and 18 compute Δz and the corresponding dz.

17 $z = \cos(x + y^2)$, $\Delta z = \cos(1.01 + (2.03)^2) - \cos(1 + 2^2)$

18 $z = \sqrt{x^2 + y^2}$, $\Delta z = \sqrt{(3.02)^2 + (3.99)^2} - \sqrt{3^2 + 4^2}$.

19 In estimating the volume of a right circular cylindrical tree trunk, a lumberjack may make a 5 percent error in estimating the diameter and a 3 percent error in measuring the height. How large an error may he make in estimating the volume?

20 Let T denote the time it takes for a pendulum to complete a back-and-forth swing. If the length of the pendulum is L and g the acceleration due to gravity, then

$$T = 2\pi\sqrt{\frac{L}{g}}.$$

A 3 percent error may be made in measuring L and a 2 percent error in measuring g. How large an error may we make in estimating T?

---

21 Let $A(x, y) = xy$ be the area of a rectangle of sides x and y. Compute ΔA and dA and show them in Fig. 8.

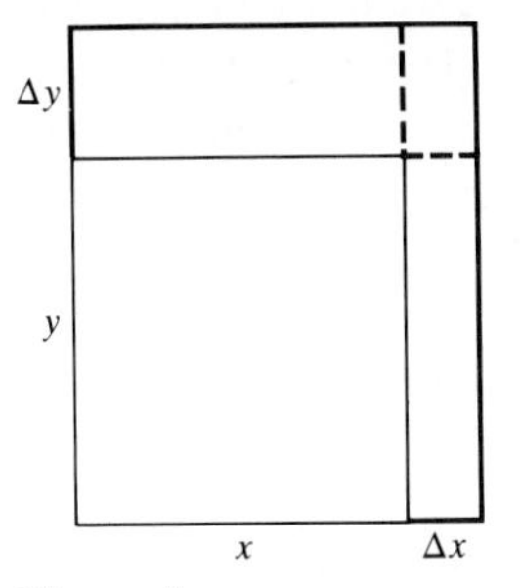

Figure 8

The differential of a function $u = f(x, y, z)$ is defined to be $\partial f/\partial x\ \Delta x + \partial f/\partial y\ \Delta y + \partial f/\partial z\ \Delta z$, in analogy with the differential of a function of two variables.

22 Let $V(x, y, z) = xyz$ be the volume of a box of sides x, y, and z. Compute ΔV and dV and show them in Fig. 9.

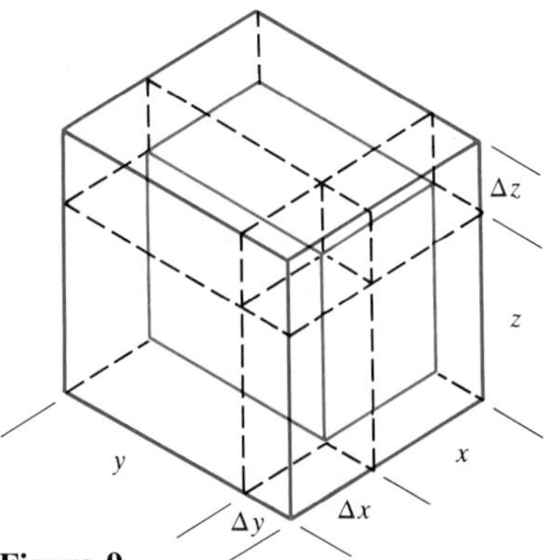

Figure 9

23 Let $u = f(x, y, z)$ and $\mathbf{r} = \mathbf{G}(t)$. Then u is a composite function of t. Show that

$$\frac{du}{dt} = \nabla f \cdot \mathbf{G}'(t),$$

where ∇f is evaluated at $\mathbf{G}(t)$. For instance, let $u = f(x, y, z)$ be the temperature at the point (x, y, z) and let $\mathbf{G}$ describe the journey of a bug. Then the rate of change in the temperature as observed by the bug is the dot product of the temperature gradient ∇f and the velocity vector $\mathbf{v} = \mathbf{G}'$.

24 We have found a way to find a normal and a tangent plane to a surface. How would you find a *tangent line* to a surface? Illustrate your method by finding a line that is tangent to the surface $z = xy$ at $(2, 3, 6)$.

25 Suppose you are at the point (a, b, c) on the level surface $f(x, y, z) = k$. At that point $\nabla f = 2\mathbf{i} + 3\mathbf{j} - 4\mathbf{k}$.
(*a*) If $\mathbf{u}$ is tangent to the surface at (a, b, c), what would $D_{\mathbf{u}}f$ equal?
(*b*) If $\mathbf{u}$ is normal to the level surface at (a, b, c), what would $D_{\mathbf{u}}f$ equal? (There are two such normals.)

26 (*a*) Draw three level curves of the function f defined by $f(x, y) = xy$. Include the curve through $(1, 1)$ as one of them.
(*b*) Draw three level curves of the function g defined by $g(x, y) = x^2 - y^2$. Include the curve through $(1, 1)$ as one of them.
(*c*) Prove that each level curve of f intersects each level curve of g at a right angle.
(*d*) If we think of f as air pressure, how may we interpret the level curves of g?

27 (*a*) Draw a level curve for the function $2x^2 + y^2$.
(*b*) Draw a level curve for the function y^2/x.
(*c*) Prove that any level curve of $2x^2 + y^2$ crosses any level curve of y^2/x at a right angle.

28 The surfaces $x^2yz = 1$ and $xy + yz + zx = 3$ both pass through the point $(1, 1, 1)$. The tangent planes to these surfaces meet in a line. Find parametric equations for this line.

29 Let $T(x, y, z)$ be the temperature at the point (x, y, z), where ∇T is not $\mathbf{0}$. A level surface $T(x, y, z) = k$ is called an *isotherm*. Show that if you are at the point (a, b, c) and wish to move in the direction in which the temperature changes most rapidly, you would move in a direction perpendicular to the isotherm that passes through (a, b, c).

30 Two surfaces $f(x, y, z) = 0$ and $g(x, y, z) = 0$ both pass through the point (a, b, c). Their intersection is a curve. How would you find a tangent vector to that curve at (a, b, c)?

31 How far is it from point $(2, 1, 3)$ to the tangent plane to $z = xy$ at $(3, 4, 12)$?

32 Write a short essay on the wonders of the chain rule. Include a description of how it was used to show that $D_{\mathbf{u}}f = \nabla f \cdot \mathbf{u}$ and in showing that ∇f is a normal to the level surface of f at the point where it is evaluated.

The angle between two surfaces that pass through (a, b, c) is defined as the angle between the two lines through (a, b, c) that are perpendicular to the two surfaces at the point (a, b, c). This angle is taken to be acute. Use this definition in Exercises 33 to 35.

33 (*a*) Show that the point (1, 1, 2) lies on the surfaces $xyz = 2$ and $x^3yz^2 = 4$.
(*b*) Find the angle between the surfaces in (*a*) at the point (1, 1, 2).

34 (*a*) Show that the point (1, 2, 3) lies on the plane

$$2x + 3y - z = 5$$

and the sphere

$$x^2 + y^2 + z^2 = 14.$$

(*b*) Find the angle between them at the point (1, 2, 3).

35 (*a*) Show that the surfaces $z = x^2y^3$ and $z = 2xy$ pass through the point (2, 1, 4).
(*b*) At what angle do they cross at that point?

36 Let $z = f(x, y)$ describe a surface. Assume that at (3, 5), $z = 7$, $\partial z/\partial x = 2$, and $\partial z/\partial y = 3$.
(*a*) Find two vectors that are tangent to the surface at (3, 5, 7).
(*b*) Find a normal to the surface at (3, 5, 7).
(*c*) Estimate $f(3.02, 4.99)$.

14.9 CRITICAL POINTS AND EXTREMA

In the case of a function of one variable, $y = f(x)$, the first and second derivatives were of use in searching for relative extrema. First, we looked for critical numbers, that is, solutions of the equation $f'(x) = 0$. Then we checked the value of $f''(x)$ at each such point. If $f''(x)$ were positive, the critical number gave a relative minimum. If $f''(x)$ were negative, the critical number gave a relative maximum. If $f''(x)$ were 0, then anything might happen: a relative minimum or maximum or neither. (For instance, at 0 the functions x^4, $-x^4$, and x^3 have both first and second derivatives equal to 0, but the first function has a relative minimum there, the second has a relative maximum, and the third has neither.) In such a case, we would resort to other tests.

This section develops the idea of a critical point for functions $f(x, y)$ of two variables and shows how to use the second-order partial derivatives $\partial^2 f/\partial x^2$, $\partial^2 f/\partial y^2$, and $\partial^2 f/\partial x\, \partial y$ to see whether the critical point provides a relative maximum, relative minimum, or neither.

Extrema of $f(x, y)$

The number M is called the **maximum** (or **global maximum**) of f over a set R in the plane if it is the largest value of $f(x, y)$ for (x, y) in R. A **relative maximum** (or **local maximum**) of f occurs at a point (a, b) in R if there is a disk around (a, b) such that $f(a, b) \geq f(x, y)$ for all points (x, y) in the disk. **Minimum** and **relative** (or **local**) **minimum** are defined similarly.

Let us look closely at the surface above a point (a, b) where a relative maximum of f occurs. Assume that f is defined for all points within some circle around (a, b) and possesses partial derivatives at (a, b). Let L_1 be the line $y = b$ in the xy plane; let L_2 be the line $x = a$ in the xy plane. (See Fig. 1. Assume, for convenience, that the values of f are positive.)

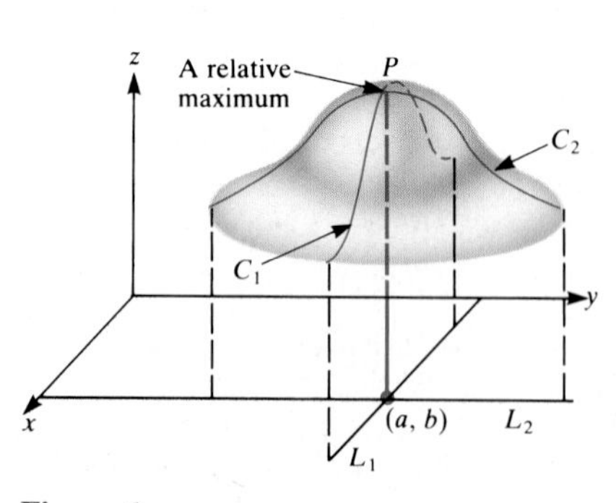

Figure 1

Let C_1 be the curve in the surface directly above the line L_1. Let C_2 be the curve in the surface directly above the line L_2. Let P be the point on the surface directly above (a, b).

Since f has a relative maximum at (a, b), no point on the surface near P is higher than P. Thus P is a highest point on the curve C_1 and on the curve C_2 (for points near P). The study of functions of one variable showed that both these curves have horizontal tangents at P. In other words, at (a, b) both partial derivatives of f must be 0:

$$\frac{\partial f}{\partial x}(a, b) = 0 \quad \text{and} \quad \frac{\partial f}{\partial y}(a, b) = 0.$$

This conclusion is summarized in the following theorem.

Theorem 1 Let f be defined on a domain that includes the point (a, b) and all points within some circle whose center is (a, b). If f has a relative maximum (or relative minimum) at (a, b) and $\partial f/\partial x$ and $\partial f/\partial y$ exist at (a, b), then both these partial derivatives are 0 at (a, b); that is,

$$\frac{\partial f}{\partial x}(a, b) = 0 = \frac{\partial f}{\partial y}(a, b).$$

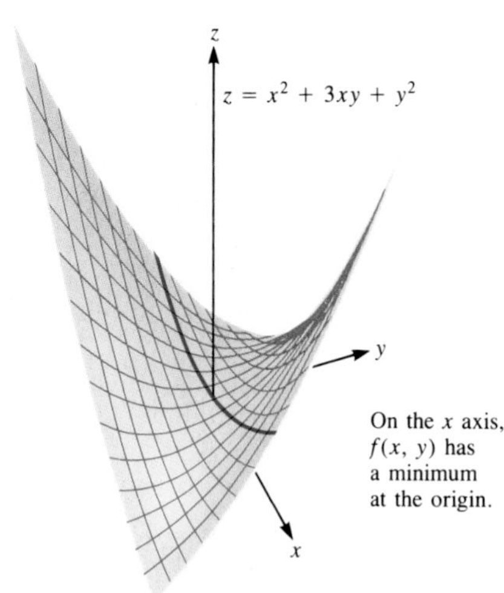

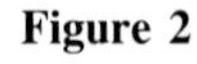
Figure 2

A point (a, b) where both partial derivatives $\partial f/\partial x$ and $\partial f/\partial y$ are 0 is clearly of importance. The following definition is analogous to that of a critical point of a function of one variable.

Definition *Critical point.* If $(\partial f/\partial x)(a, b) = 0$ and $(\partial f/\partial y)(a, b) = 0$, the point (a, b) is a **critical point** of the function $f(x, y)$.

You might expect that if (a, b) is a critical point of f and the two second partial derivatives $\partial^2 f/\partial x^2$ and $\partial^2 f/\partial y^2$ are both positive at (a, b), then f necessarily has a relative minimum at (a, b). The next example shows that *the situation is not that simple*.

EXAMPLE 1 Find the critical points of $f(x, y) = x^2 + 3xy + y^2$ and determine whether there is an extremum there.

SOLUTION First, find any critical points by setting $\partial f/\partial x$ and $\partial f/\partial y$ both equal to 0. This gives the simultaneous equations

$$\begin{cases} 2x + 3y = 0 \\ 3x + 2y = 0. \end{cases}$$

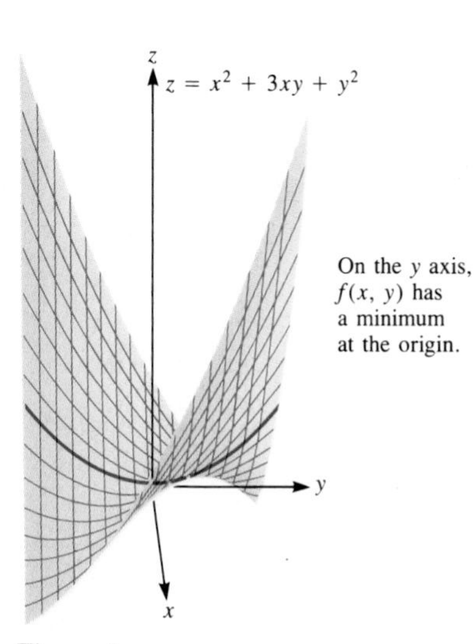

Figure 3

Since the only solution of these equations is $(x, y) = (0, 0)$, the function has one critical point, namely $(0, 0)$.

Now look at the graph of f for (x, y) near $(0, 0)$.

First, consider how f behaves for points on the x axis. We have $f(x, 0) = x^2 + 3 \cdot x \cdot 0 + 0^2 = x^2$. Therefore, considered *only as a function of* x, the function has a minimum at the origin. (See Fig. 2.)

On the y axis, the function reduces to $f(0, y) = y^2$, whose graph is another parabola with a minimum at the origin. (See Fig. 3.) Note also that $\partial^2 f/\partial x^2 = 2$ and $\partial^2 f/\partial y^2 = 2$, so both are positive at $(0, 0)$.

So far, the evidence suggests that f has a relative minimum at $(0, 0)$. However, consider its behavior on the line $y = -x$. For points (x, y) on this line

$$f(x, y) = f(x, -x) = x^2 + 3x(-x) + (-x)^2 = -x^2.$$

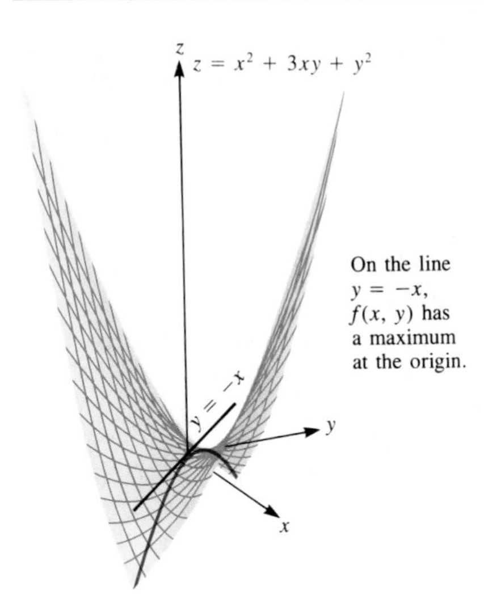

Figure 4

On this line the function assumes negative values, and its graph is a parabola opening downward, as shown in Fig. 4.

Thus $f(x, y)$ has neither a relative maximum nor minimum at the origin. Its graph resembles a saddle. ■

Example 1 shows that to determine whether a critical point of $f(x, y)$ provides an extremum, it is not enough to look at $\partial^2 f/\partial x^2$ and $\partial^2 f/\partial y^2$ [for they describe the behavior of $f(x, y)$ only on lines parallel to the x axis and y axis, respectively]. The criteria are more complicated and involve the mixed partial derivative $\partial^2 f/\partial x\, \partial y$ as well. Appendix K proves the following theorem. At the end of this section a proof is presented in the special case when $f(x, y)$ is a polynomial of the form $Ax^2 + 2Bxy + Cy^2$, where A, B, and C are constants.

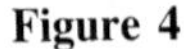

Theorem 2 *The second-partial-derivative test for f(x, y).* Let (a, b) be a critical point of the function $f(x, y)$. Assume that the partial derivatives $\partial f/\partial x$, $\partial f/\partial y$, $\partial^2 f/\partial x^2$, $\partial^2 f/\partial x\, \partial y$, and $\partial^2 f/\partial y^2$ are continuous at and near (a, b). Let

In subscript notation, $D = f_{xx}f_{yy} - (f_{xy})^2$.

$$D = \frac{\partial^2 f}{\partial x^2}(a, b)\frac{\partial^2 f}{\partial y^2}(a, b) - \left(\frac{\partial^2 f}{\partial x\, \partial y}(a, b)\right)^2.$$

1 If $D > 0$ and $(\partial^2 f/\partial x^2)(a, b) > 0$, then f has a relative minimum at (a, b).

2 If $D > 0$ and $(\partial^2 f/\partial x^2)(a, b) < 0$, then f has a relative maximum at (a, b).

3 If $D < 0$, then f has neither a relative minimum nor a relative maximum at (a, b). [There is a saddle point at (a, b).]

If $D = 0$, then anything can happen; there may be a relative minimum, a relative maximum, or a saddle. These possibilities are illustrated in Exercise 42.

The quantity D in Theorem 2 is called the **discriminant of f**. Note that it also equals the determinant

$$\begin{vmatrix} \dfrac{\partial^2 f}{\partial x^2} & \dfrac{\partial^2 f}{\partial x\, \partial y} \\ \dfrac{\partial^2 f}{\partial y\, \partial x} & \dfrac{\partial^2 f}{\partial y^2} \end{vmatrix},$$

since $\partial^2 f/(\partial x\, \partial y) = \partial^2 f/(\partial y\, \partial x)$. This determinant is called the **Hessian of f**. To see what the theorem says, consider case 1, the test for a relative minimum. It says that $(\partial^2 f/\partial x^2)(a, b) > 0$ (which is to be expected) and that

$$\frac{\partial^2 f}{\partial x^2}(a, b)\frac{\partial^2 f}{\partial y^2}(a, b) - \left[\frac{\partial^2 f}{\partial x\, \partial y}(a, b)\right]^2 > 0,$$

Or equivalently,
$$\left[\frac{\partial^2 f}{\partial x\, \partial y}(a, b)\right]^2 < \frac{\partial^2 f}{\partial x^2}(a, b)\frac{\partial^2 f}{\partial y^2}(a, b). \tag{1}$$

Memory aid regarding size of $\partial^2 f/\partial x\, \partial y$

Since the square of a real number is never negative, and $(\partial^2 f/\partial x^2)(a, b)$ is positive, it follows that $(\partial^2 f/\partial y^2)(a, b) > 0$, which was to be expected. But inequality (1) says more. It says that the *mixed partial* $(\partial^2 f/\partial x\, \partial y)(a, b)$ *must not be too large*. For a relative maximum or minimum, inequality (1) must hold. This may be easier to remember than "$D > 0$."

EXAMPLE 2 Examine each of these functions for relative extrema:
(*a*) $f(x, y) = x^2 + 3xy + y^2$,
(*b*) $g(x, y) = x^2 + 2xy + y^2$,
(*c*) $h(x, y) = x^2 + xy + y^2$.

SOLUTION

(*a*) The case $x^2 + 3xy + y^2$ is Example 1. The origin is the only critical point, and it provides neither a relative maximum nor a relative minimum. We can check this by the use of the discriminant. We have

$$\frac{\partial^2 f}{\partial x^2}(0, 0) = 2, \qquad \frac{\partial^2 f}{\partial x\, \partial y}(0, 0) = 3, \qquad \text{and} \qquad \frac{\partial^2 f}{\partial y^2}(0, 0) = 2.$$

Hence $D = 2 \cdot 2 - 3^2 = -5$ is negative. By the second-partial-derivative test, there is neither a relative maximum nor a relative minimum at the origin. Instead, there is a saddle there.

(*b*) It is a straightforward matter to show that all the points on the line $x + y = 0$ are critical points of $g(x, y) = x^2 + 2xy + y^2$. Moreover,

$$\frac{\partial^2 g}{\partial x^2}(x, y) = 2, \qquad \frac{\partial^2 g}{\partial x\, \partial y}(x, y) = 2, \qquad \text{and} \qquad \frac{\partial^2 g}{\partial y^2}(x, y) = 2.$$

Thus the discriminant $D = 2 \cdot 2 - 2^2 = 0$. Since $D = 0$, the discriminant gives no information.

Note, however, that $x^2 + 2xy + y^2 = (x + y)^2$ and so, being the square of a real number, is always greater than or equal to 0. Hence the origin provides a relative minimum of $x^2 + 2xy + y^2$. [In fact, any point on the line $x + y = 0$ provides a relative minimum. Since $g(x, y) = (x + y)^2$, the function is constant on each line $x + y = c$, for any choice of the constant c. See Fig. 5.]

(*c*) For $h(x, y) = x^2 + xy + y^2$, again the origin is the only critical point and we have

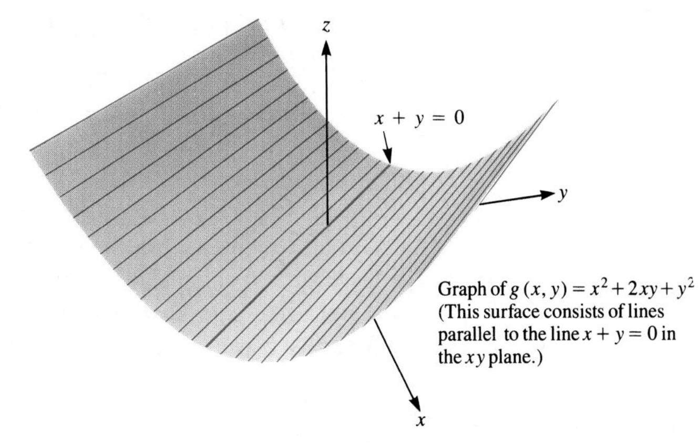

Figure 5

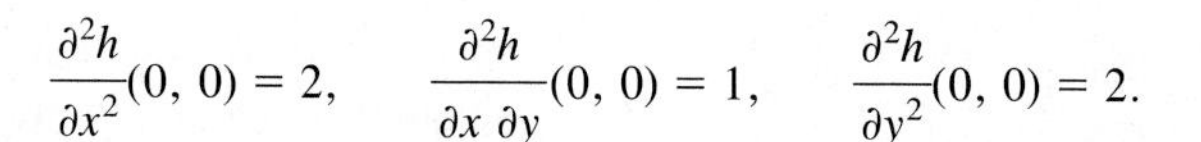

$$\frac{\partial^2 h}{\partial x^2}(0, 0) = 2, \qquad \frac{\partial^2 h}{\partial x\,\partial y}(0, 0) = 1, \qquad \frac{\partial^2 h}{\partial y^2}(0, 0) = 2.$$

In this case, $D = 2 \cdot 2 - 1^2 = 3$ is positive and $(\partial^2 h/\partial x^2)(0, 0) > 0$. Hence $x^2 + xy + y^2$ has a relative minimum at the origin.

The graph of h is shown in Fig. 6. ■

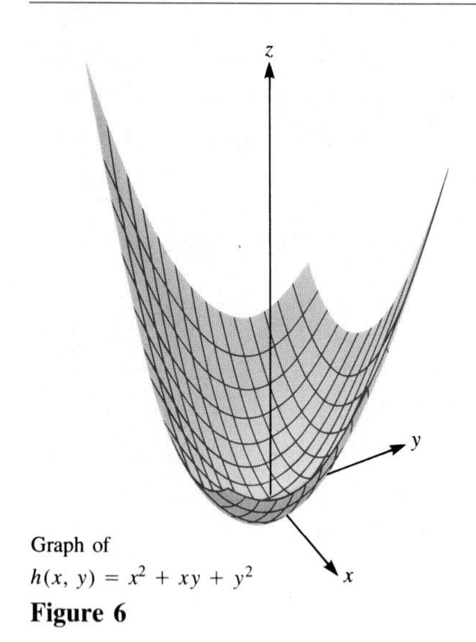

Graph of
$h(x, y) = x^2 + xy + y^2$
Figure 6

EXAMPLE 3 Examine $f(x, y) = x + y + 1/(xy)$ for global and relative extrema.

SOLUTION When x and y are both large positive numbers or small positive numbers, then $f(x, y)$ may be arbitrarily large. There is therefore no global maximum. By allowing x and y to be negative numbers of large absolute values, we see that there is no global minimum.

Let us look for critical points. We have

$$\frac{\partial f}{\partial x} = 1 - \frac{1}{x^2 y} \quad \text{and} \quad \frac{\partial f}{\partial y} = 1 - \frac{1}{xy^2}.$$

Setting these derivatives equal to 0 gives

$$\frac{1}{x^2 y} = 1 \quad \text{and} \quad \frac{1}{xy^2} = 1. \tag{2}$$

Hence $x^2 y = xy^2$. Since the function f is not defined when x or y is 0, we may assume $xy \neq 0$. Dividing both sides of $x^2 y = xy^2$ by xy gives $x = y$. By (2), $1/x^3 = 1$; hence $x = 1$. Thus there is only one critical point, namely, (1, 1).

To find whether it is a relative extremum, use Theorem 2. We have

$$\frac{\partial^2 f}{\partial x^2} = \frac{2}{x^3 y}, \qquad \frac{\partial^2 f}{\partial x\,\partial y} = \frac{1}{x^2 y^2}, \qquad \frac{\partial^2 f}{\partial y^2} = \frac{2}{xy^3}.$$

Thus at (1, 1),

$$\frac{\partial^2 f}{\partial x^2} = 2, \qquad \frac{\partial^2 f}{\partial x\,\partial y} = 1, \qquad \text{and} \qquad \frac{\partial^2 f}{\partial y^2} = 2.$$

Therefore,

$$D = 2 \cdot 2 - 1^2 = 3 > 0.$$

Since $D > 0$ and $(\partial^2 f/\partial x^2)(1, 1) > 0$, the point (1, 1) provides a relative minimum. ■

Extrema on a Bounded Region

In Sec. 4.7, we saw how to find a maximum of a differentiable function, $y = f(x)$, on an interval $[a, b]$. The procedure is as follows:

1. First find any numbers x in $[a, b]$ (other than a or b) where $f'(x) = 0$. Such a number is called a critical number. If there are no critical numbers, the maximum occurs at a or b.
2. If there are critical numbers, evaluate f at them. Also find the maximum of $f(a)$ and $f(b)$. The maximum of f in $[a, b]$ is the largest of the numbers: $f(a)$, $f(b)$, and the values of f at critical numbers.

We can similarly find the maximum of $f(x, y)$ in a region R in the plane

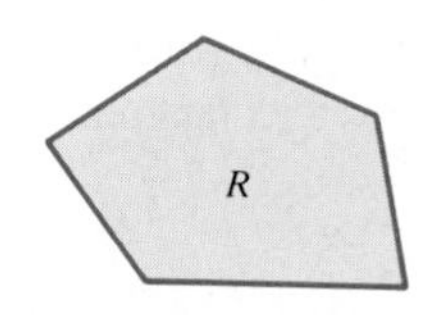

A continuous function on R (which includes the border) has a maximum value at some point in R.

Figure 7

bounded by some polygon or curve. (See Fig. 7.) It is assumed that R includes its border and is a finite region in the sense that it lies within some disk. (In advanced calculus, it is proved that a continuous function defined on such a domain has a maximum—and a minimum—value.) If f has continuous partial derivatives, the procedure for finding a maximum is similar to that for maximizing a function on a closed interval.

1. First find any points that are in R but not on the boundary of R where both $\partial f/\partial x$ and $\partial f/\partial y$ are 0. These are called **critical points**. (If there are no critical points, the maximum occurs on the boundary.)
2. If there are critical points, evaluate f at them. Also find the maximum of f on the boundary. The maximum of f on R is the largest value of f on the boundary and at critical points.

A similar procedure finds the minimum value on a bounded region.

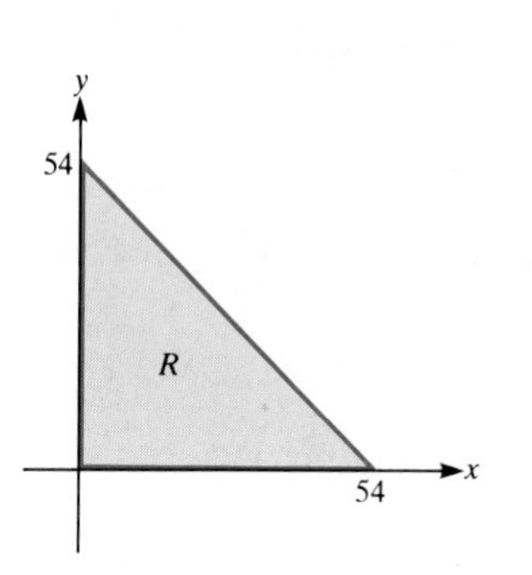

Figure 8

EXAMPLE 4 Maximize the function $f(x, y) = xy(108 - 2x - 2y) = 108xy - 2x^2y - 2xy^2$ on the triangle R bounded by the x axis, the y axis, and the line $x + y = 54$. (See Fig. 8.)

SOLUTION First find any critical points. We have

$$\frac{\partial f}{\partial x}(x, y) = 108y - 4xy - 2y^2 = 0$$

$$\frac{\partial f}{\partial y}(x, y) = 108x - 2x^2 - 4xy = 0$$

which give the simultaneous equations

$$2y(54 - 2x - y) = 0,$$
$$2x(54 - x - 2y) = 0.$$

Since a critical point in R but not on its boundary has $x \neq 0$ and $y \neq 0$, we solve the simultaneous equations

$$54 - 2x - y = 0,$$
$$54 - x - 2y = 0.$$

By the first equation, $y = 54 - 2x$. Substitution of this into the second equation gives: $54 - x - 2(54 - 2x) = 0$, or $-54 + 3x = 0$. Hence $x = 18$ and therefore $y = 54 - 2 \cdot 18 = 18$.

Evaluate f at critical points.

The point (18, 18) lies in the interior of R, since it lies above the x axis, to the right of the y axis, and below the line $x + y = 54$. Furthermore, $f(18, 18) = 18 \cdot 18\,(108 - 2 \cdot 18 - 2 \cdot 18) = 11{,}664$.

Evaluate f on boundary.

Next we examine the function $f(x, y) = xy(108 - 2x - 2y)$ on the boundary of the triangle R. On the base of R, $y = 0$, so $f(x, y) = 0$. On the left edge of R, $x = 0$, so again $f(x, y) = 0$. On the slanted edge, which lies on the line $x + y = 54$, we have $108 - 2x - 2y = 0$, so $f(x, y) = 0$ on this edge also. Thus $f(x, y) = 0$ on the entire boundary.

Therefore, the maximum occurs at the critical point (18, 18) and has the value 11,664. ■

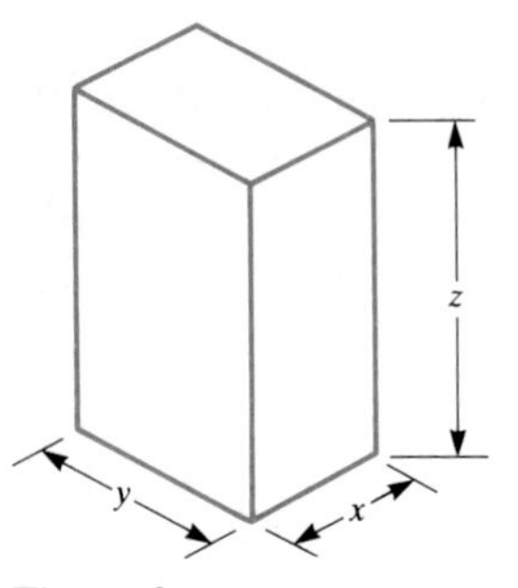

Figure 9

EXAMPLE 5 The combined length and girth (distance around) of a package sent through the mail cannot exceed 108 inches. If the package is a rectangular box, how large can its volume be?

SOLUTION Introduce letters to name the quantities of interest. We label its length (a longest side) z and the other sides x and y, as in Fig. 9. The volume $V = xyz$ is to be maximized, subject to girth plus length at most 108, that is,

$$2x + 2y + z \leq 108.$$

Since we want the largest box, we might as well restrict our attention to boxes for which

$$2x + 2y + z = 108. \tag{3}$$

By Eq. (3), $z = 108 - 2x - 2y$. Thus $V = xyz$ can be expressed as a function of two variables:

$$V = f(x, y) = xy(108 - 2x - 2y).$$

This function is to be maximized on the triangle described by $x \geq 0$, $y \geq 0$, $2x + 2y \leq 108$, that is, $x + y \leq 54$.

Why is $2x + 2y \leq 108$?

These are the same function and region as in the previous example. Hence, the largest box has $x = y = 18$ and $z = 108 - 2x - 2y = 108 - 2 \cdot 18 - 2 \cdot 18 = 36$; its dimensions are 18 inches by 18 inches by 36 inches and its volume is 11,664 cubic inches. ■

Remark: In Example 5 we let z be the length of a longest side, an assumption that was never used. So if the Postal Service regulations read ''The length of one edge plus the girth around the other edges shall not exceed 108 inches,'' the effect would be the same. You would not be able to send a larger box by, say, measuring the girth around the base formed by its largest edges.

EXAMPLE 6 Find the maximum and minimum values of $f(x, y) = x^2 + y^2 - 2x - 4y$ on the disk R of radius 3 and center $(0, 0)$.

SOLUTION First, find any critical points. We have

$$\frac{\partial f}{\partial x} = 2x - 2 \quad \text{and} \quad \frac{\partial f}{\partial y} = 2y - 4.$$

The equations

$$2x - 2 = 0$$
$$2y - 4 = 0$$

have the solutions $x = 1$ and $y = 2$. This point lies in R (since its distance from the origin is $\sqrt{1^2 + 2^2} = \sqrt{5}$, which is less than 3). At the critical point $(1, 2)$, the value of the function is $1^2 + 2^2 - 2(1) - 4(2) = 5 - 2 - 8 = -5$.

Second, find the behavior of f on the boundary, which is a circle of radius 3. We parameterize this circle:

$$x = 3 \cos \theta$$
$$y = 3 \sin \theta.$$

On this circle,

$$\begin{aligned} f(x, y) &= x^2 + y^2 - 2x - 4y \\ &= (3 \cos \theta)^2 + (3 \sin \theta)^2 - 2(3 \cos \theta) - 4(3 \sin \theta) \\ &= 9 \cos^2 \theta + 9 \sin^2 \theta - 6 \cos \theta - 12 \sin \theta \\ &= 9 - 6 \cos \theta - 12 \sin \theta. \end{aligned}$$

We now must find the maximum and minimum of the single-variable function $g(\theta) = 9 - 6 \cos \theta - 12 \sin \theta$ for θ in $[0, 2\pi]$.

To do this, find $g'(\theta)$:

$$g'(\theta) = 6 \sin \theta - 12 \cos \theta.$$

Setting $g'(\theta) = 0$ gives

$$0 = 6 \sin \theta - 12 \cos \theta$$

or

$$\sin \theta = 2 \cos \theta. \tag{4}$$

To solve (4), divide by $\cos \theta$ (which will not be 0), getting

$$\frac{\sin \theta}{\cos \theta} = 2$$

or

$$\tan \theta = 2.$$

There are two angles θ in $[0, 2\pi]$ such that $\tan \theta = 2$. One is in the first quadrant, $\theta = \tan^{-1} 2$, and the other is in the third quadrant, $\pi + \tan^{-1} 2$. To evaluate $g(\theta) = 9 - 6 \cos \theta - 12 \sin \theta$ at these angles, we must compute their cosine and sine. The right triangle in Fig. 10 helps us do this.

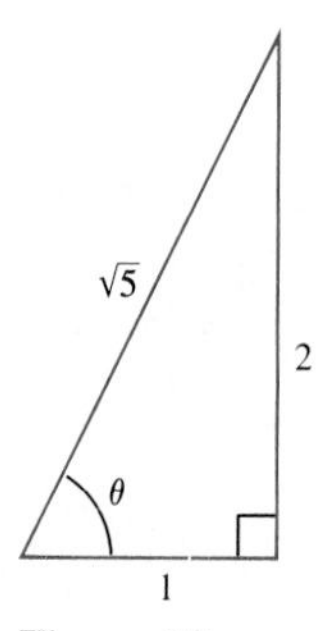

Figure 10

Inspection of Fig. 10 shows that for $\theta = \tan^{-1} 2$,

$$\cos \theta = \frac{1}{\sqrt{5}} \quad \text{and} \quad \sin \theta = \frac{2}{\sqrt{5}}.$$

For this angle

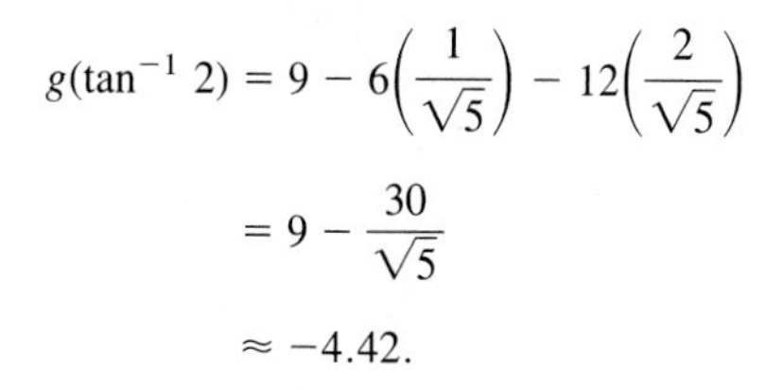

$$\begin{aligned} g(\tan^{-1} 2) &= 9 - 6\left(\frac{1}{\sqrt{5}}\right) - 12\left(\frac{2}{\sqrt{5}}\right) \\ &= 9 - \frac{30}{\sqrt{5}} \\ &\approx -4.42. \end{aligned}$$

When $\theta = \pi + \tan^{-1} 2$,

$$\cos \theta = \frac{-1}{\sqrt{5}} \quad \text{and} \quad \sin \theta = \frac{-2}{\sqrt{5}}.$$

So

$$\begin{aligned} g(\pi + \tan^{-1} 2) &= 9 - 6\left(\frac{-1}{\sqrt{5}}\right) - 12\left(\frac{-2}{\sqrt{5}}\right) \\ &= 9 + \frac{30}{\sqrt{5}} \\ &\approx 22.42. \end{aligned}$$

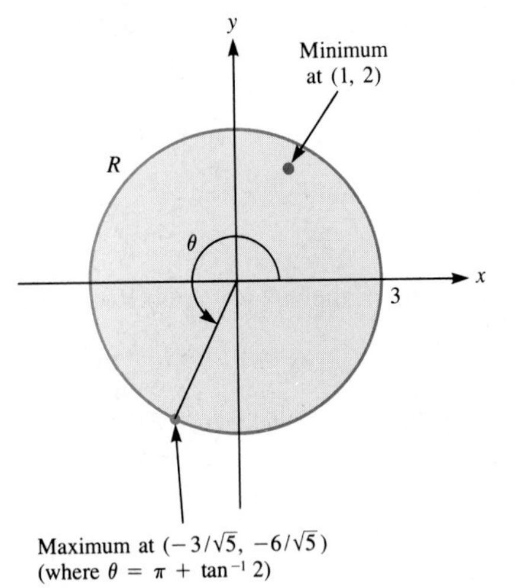

Figure 11

Since $g(2\pi) = g(0) = 9 - 6(1) - 12(0) = 3$, the maximum of f on the border of R is about 22.42 and the minimum is about -4.42. (Recall that at the critical point the value of f is -5.)

We conclude that the maximum value of f on R is about 22.42 and the minimum value is -5 [and it occurs at the point (1, 2), which is not on the boundary]. See Fig. 11. ■

Proof of Theorem 2 in a Special Case

We will prove Theorem 2 in case $f(x, y)$ is a second-degree polynomial of the form

$$f(x, y) = Ax^2 + 2Bxy + Cy^2.$$

(We call the coefficient of xy "$2B$" rather than B, in order to simplify the proof.)

Theorem 3 Let $f(x, y) = Ax^2 + 2Bxy + Cy^2$, where A, B, and C are constants. Then (0, 0) is a critical point. Let

$$D = \frac{\partial^2 f}{\partial x^2}(0, 0)\frac{\partial^2 f}{\partial y^2}(0, 0) - \left(\frac{\partial^2 f}{\partial x\,\partial y}(0, 0)\right)^2.$$

1 If $D > 0$ and $(\partial^2 f/\partial x^2)(0, 0) > 0$, then f has a relative minimum at (0, 0).
2 If $D > 0$ and $(\partial^2 f/\partial x^2)(0, 0) < 0$, then f has a relative maximum at (0, 0).
3 If $D < 0$, then f has neither a relative minimum nor a relative maximum at (0, 0).

Proof We prove case 1, leaving cases 2 and 3 as exercises.

First, compute the partial derivatives of f:

$$\frac{\partial f}{\partial x} = 2Ax + 2By, \qquad \frac{\partial f}{\partial y} = 2Bx + 2Cy,$$

$$\frac{\partial^2 f}{\partial x^2} = 2A, \qquad \frac{\partial^2 f}{\partial x\,\partial y} = 2B, \qquad \frac{\partial^2 f}{\partial y^2} = 2C.$$

Note that both $\partial f/\partial x$ and $\partial f/\partial y$ are 0 at (0, 0). Hence (0, 0) is a critical point and $f(0, 0) = 0$. We must show that $f(x, y) \geq 0$ for (x, y) near (0, 0). [In fact, we will show that $f(x, y) \geq 0$ for all (x, y).]

Expressing case 1 in terms of A, B, and C, we have

$$D = (2A)(2C) - (2B)^2 = 4(AC - B^2) > 0$$

and $(\partial^2 f/\partial x^2)(0, 0) = 2A > 0$. In short, we are assuming that $AC - B^2 > 0$ and $A > 0$ and want to deduce that $f(x, y) = Ax^2 + 2Bxy + Cy^2 \geq 0$. Since A is positive, this amounts to showing that

$$A(Ax^2 + 2Bxy + Cy^2) \geq 0. \tag{5}$$

To establish (5), we complete the square:

$$A(Ax^2 + 2Bxy + Cy^2) = A^2x^2 + 2ABxy + ACy^2$$
$$= A^2x^2 + 2ABxy + B^2y^2 - B^2y^2 + ACy^2$$
$$= (Ax + By)^2 + (AC - B^2)y^2.$$

Now, $(Ax + By)^2 \geq 0$ and $y^2 \geq 0$ since they are squares of real numbers. Moreover, $AC - B^2$ is positive, by assumption. Thus (5) holds and the theorem is proved. ■

Section Summary

We defined a critical point of $f(x, y)$ as a point where both partial derivatives $\partial f/\partial x$ and $\partial f/\partial y$ are 0. Even if $\partial^2 f/\partial x^2$ and $\partial^2 f/\partial y^2$ are negative there, such a point need not provide a relative maximum. We must also know that $\partial^2 f/\partial x\,\partial y$ is not too large in absolute value. If $\partial^2 f/\partial x^2 < 0$ and

$$\left(\frac{\partial^2 f}{\partial x\,\partial y}\right)^2 < \frac{\partial^2 f}{\partial x^2}\frac{\partial^2 f}{\partial y^2},$$

then there is indeed a relative maximum at the critical point. (Note that the two inequalities imply $\partial^2 f/\partial y^2 < 0$.) Similar criteria hold for a relative maximum. When

$$\left(\frac{\partial^2 f}{\partial x\,\partial y}\right)^2 = \frac{\partial^2 f}{\partial x^2}\frac{\partial^2 f}{\partial y^2},$$

the critical point may be a relative maximum, relative minimum, or neither.
We also saw how to find extrema of a function defined on a bounded region.

EXERCISES FOR SEC. 14.9: CRITICAL POINTS AND EXTREMA

Use Theorems 1 and 2 to determine any relative maxima or minima of the functions in Exercises 1 to 10.

1 $x^2 + 3xy + y^2$ **2** $x^2 - y^2$
3 $x^2 - 2xy + 2y^2 + 4x$ **4** $x^4 + 8x^2 + y^2 - 4y$
5 $x^2 - xy + y^2$ **6** $x^2 + 2xy + 2y^2 + 4x$
7 $2x^2 + 2xy + 5y^2 + 4x$ **8** $-4x^2 - xy - 3y^2$
9 $4/x + 2/y + xy$ **10** $x^3 - y^3 + 3xy$

Let f be a function of x and y such that at (a, b) both $\partial f/\partial x$ and $\partial f/\partial y$ equal 0. In each of Exercises 11 to 16 values are specified for $\partial^2 f/\partial x^2$, $\partial^2 f/\partial x\,\partial y$, and $\partial^2 f/\partial y^2$ at (a, b). Assume that all these partial derivatives are continuous. On the basis of the given information decide whether (*a*) f has a relative maximum at (a, b), (*b*) f has a relative minimum at (a, b), (*c*) neither (*a*) nor (*b*) occurs, (*d*) there is inadequate information.

11 $\partial^2 f/\partial x\,\partial y = 4$, $\partial^2 f/\partial x^2 = 2$, $\partial^2 f/\partial y^2 = 8$
12 $\partial^2 f/\partial x\,\partial y = -3$, $\partial^2 f/\partial x^2 = 2$, $\partial^2 f/\partial y^2 = 4$
13 $\partial^2 f/\partial x\,\partial y = 3$, $\partial^2 f/\partial x^2 = 2$, $\partial^2 f/\partial y^2 = 4$
14 $\partial^2 f/\partial x\,\partial y = 2$, $\partial^2 f/\partial x^2 = 3$, $\partial^2 f/\partial y^2 = 4$
15 $\partial^2 f/\partial x\,\partial y = -2$, $\partial^2 f/\partial x^2 = -3$, $\partial^2 f/\partial y^2 = -4$
16 $\partial^2 f/\partial x\,\partial y = -2$, $\partial^2 f/\partial x^2 = 3$, $\partial^2 f/\partial y^2 = -4$

In Exercises 17 to 24 find the critical points and the relative extrema of the given functions.

17 $x + y - \dfrac{1}{xy}$ **18** $3xy - x^3 - y^3$
19 $12xy - x^3 - y^3$ **20** $6xy - x^2y - xy^2$
21 $\exp(x^3 + y^3)$ **22** 2^{xy}
23 $3x + xy + x^2y - 2y$ **24** $x + y + \dfrac{8}{xy}$

25 Find the dimensions of the open rectangular box of volume 1 of smallest surface area. Use Theorem 2 as a check that the critical point provides a minimum.

26 The material for the top and bottom of a rectangular box costs 3 cents per square foot, and that for the sides 2 cents per square foot. What is the least expensive box that has a volume of 1 cubic foot? Use Theorem 2 as a check that the critical point provides a minimum.

27 Let $P_1 = (x_1, y_1)$, $P_2 = (x_2, y_2)$, $P_3 = (x_3, y_3)$, and $P_4 = (x_4, y_4)$. Find the coordinates of the point P that minimizes the sum of the squares of the distances from P to the four points.

28 Find the dimensions of the rectangular box of largest volume if its total surface area is to be 12 square meters.

29 Three nonnegative numbers x, y, and z have the sum 1.
(*a*) How small can $x^2 + y^2 + z^2$ be?
(*b*) How large can it be?

30 Each year a firm can produce r radios and t television sets at a cost of $2r^2 + rt + 2t^2$ dollars. It sells a radio for \$600 and a television set for \$900.
(*a*) What is the profit from the sale of r radios and t television sets, that is, the revenue less the cost?
(*b*) Find the combination of r and t that maximizes profit. Use the discriminant as a check.

31 Find the dimensions of the rectangular box of largest volume that can be inscribed in a sphere of radius 1.

32 For which values of the constant k does $x^2 + kxy + 3y^2$ have a relative minimum at $(0, 0)$?

33 For which values of the constant k does the function $kx^2 + 5xy + 4y^2$ have a relative minimum at $(0, 0)$?

34 Let $f(x, y) = (2x^2 + y^2)\, e^{-x^2-y^2}$.
(*a*) Find all critical points of f.
(*b*) Examine the behavior of f when $x^2 + y^2$ is large.
(*c*) What is the minimum value of f?
(*d*) What is the maximum value of f?

35 Find the maximum and minimum values of the function in Exercise 34 on the circle (*a*) $x^2 + y^2 = 1$, (*b*) $x^2 + y^2 = 4$. *Hint:* Express the function in terms of θ.

36 Find the maximum value of $f(x, y) = 3x^2 - 4y^2 + 2xy$ for points (x, y) in the square region whose vertices are $(0, 0)$, $(0, 1)$, $(1, 0)$, and $(1, 1)$.

37 Find the maximum value of $f(x, y) = xy$ for points in the triangular region whose vertices are $(0, 0)$, $(1, 0)$, and $(0, 1)$.

38 Maximize the function $-x + 3y + 6$ on the quadrilateral whose vertices are $(1, 1)$, $(4, 2)$, $(0, 3)$, and $(5, 6)$.

---

39 (*a*) Show that $z = x^2 - y^2 + 2xy$ has no maximum and no minimum.
(*b*) Find the minimum and maximum of z if we consider only (x, y) on the circle of radius 1 and center $(0, 0)$, that is, all (x, y) such that $x^2 + y^2 = 1$.
(*c*) Find the minimum and maximum of z if we consider all (x, y) in the disk of radius 1 and center $(0, 0)$, that is, all (x, y) such that $x^2 + y^2 \le 1$.

40 If, at the point (x_0, y_0), $\partial z/\partial x = 0 = \partial z/\partial y$, $\partial^2 z/\partial x^2 = 3$, and $\partial^2 z/\partial y^2 = 12$, for what values of $\partial^2 z/\partial x\, \partial y$ is it certain that z has a relative minimum at (x_0, y_0)?

41 Let $U(x, y, z) = x^{1/2}y^{1/3}z^{1/6}$ be the "utility" or "desirability" to a given consumer of the amounts x, y, and z of three different commodities. Their prices are, respectively, 2 dollars, 1 dollar, and 5 dollars, and the consumer has 60 dollars to spend. How much of each product should he buy to maximize the utility?

42 This exercise shows that if the discriminant D is 0, then any of the three outcomes mentioned in Theorem 2 are possible.
(*a*) Let $f(x, y) = x^2 + 2xy + y^2$. Show that at $(0, 0)$ both $\partial f/\partial x$ and $\partial f/\partial y$ are 0, $\partial^2 f/\partial x^2$ and $\partial^2 f/\partial y^2$ are positive, $D = 0$, and f has a relative minimum.
(*b*) Let $f(x, y) = x^2 + 2xy + y^2 - x^4$. Show that at $(0, 0)$ both $\partial f/\partial x$ and $\partial f/\partial y$ are 0, $\partial^2 f/\partial x^2$ and $\partial^2 f/\partial y^2$ are positive, $D = 0$, and f has neither a relative maximum nor a relative minimum at $(0, 0)$.
(*c*) Give an example of a function $f(x, y)$ for which $(0, 0)$ is a critical point and $D = 0$ there, but f has a relative maximum at $(0, 0)$.

43 Let $(x_1, y_1), (x_2, y_2), \ldots, (x_n, y_n)$ be n points in the plane. Statisticians define the **line of regression** as the line that minimizes the sum of the squares of the differences between y_i and the ordinates of the line at x_i. (See Fig. 12.) Let the typical line in the plane have the equation $y = mx + b$.
(*a*) Show that the line of regression minimizes the sum $\Sigma_{i=1}^n [y_i - (mx_i + b)]^2$ considered as a function of m and b.
(*b*) Let $f(m, b) = \Sigma_{i=1}^n [y_i - (mx_i + b)]^2$. Compute $\partial f/\partial m$ and $\partial f/\partial b$.
(*c*) Show that when $\partial f/\partial m = 0 = \partial f/\partial b$, we have

$$m \sum_{i=1}^{n} x_i^2 + b \sum_{i=1}^{n} x_i = \sum_{i=1}^{n} x_i y_i$$

and

$$m \sum_{i=1}^{n} x_i + nb = \sum_{i=1}^{n} y_i.$$

(*d*) When do the simultaneous equations in (*c*) have a unique solution for m and b?
(*e*) Find the regression line for the points $(1, 1)$, $(2, 3)$, and $(3, 5)$.

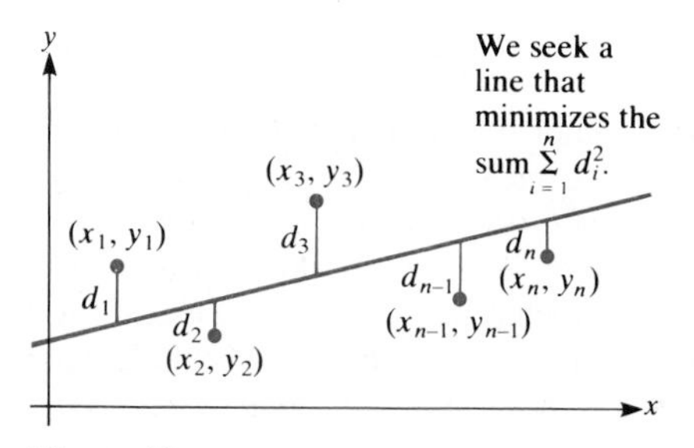

Figure 12

44 If your calculator is programmed to compute lines of regression, find and draw the line of regression for the points $(1, 1)$, $(2, 1.5)$, $(3, 3)$, $(4, 2)$, and $(5, 3.5)$.

45 Let $f(x, y) = (y - x^2)(y - 2x^2)$.

(*a*) Show that f has neither a local minimum nor a local maximum at (0, 0).
(*b*) Show that f has a local minimum at (0, 0) when considered only on any fixed line through (0, 0).
Suggestion for (b): Graph $y = x^2$ and $y = 2x^2$ and show where $f(x, y)$ is positive and where it is negative.

46 Two nonoverlapping rectangles are placed in the triangle whose vertices are (0, 0), (1, 0), and (0, 1). If their sides are parallel to the coordinate axes, how large can their total area be?

47 Let $f(x, y) = ax + by + c$, for constants a, b, and c. Let R be a polygon in the xy plane. Show that the maximum and minimum values of $f(x, y)$ on R are assumed at vertices of the polygon.

48 Find (*a*) the minimum value of xyz, and (*b*) the maximum value of xyz, for all triplets of nonnegative real numbers x, y, z such that $x + y + z = 1$.

49 (*a*) Deduce from Exercise 48 that for any three nonnegative numbers a, b, and c, $\sqrt[3]{abc} \le (a + b + c)/3$. This asserts that the "geometric mean" of three numbers is not larger than the "arithmetic mean." (*b*) Obtain a corresponding result for four numbers.

50 Prove case 2 of Theorem 3.

51 Prove case 3 of Theorem 3.

14.10 LAGRANGE MULTIPLIERS

In Sec. 4.7 we discussed the problem of finding the minimal surface area of a right circular can of volume 100. The problem was to minimize

$$2\pi r^2 + 2\pi rh, \tag{1}$$

subject to the constraint that

$$\pi r^2 h - 100 = 0. \tag{2}$$

The solution began by using Eq. (2) to eliminate h. Then the expression $2\pi r^2 + 2\pi rh$ was written as a function of r alone.

In Sec. 4.8 the same problem was solved with the aid of implicit differentiation. Although Eq. (2) was not solved to give h explicitly as a function of r, it was clear that h could be considered a function of r. Thus the derivative dh/dr made sense.

In both solutions the variables r and h assumed quite different roles. We singled out one of them, r, to be the independent variable and the other, h, to be the dependent variable. In the method of Lagrange multipliers, *all the variables are treated the same*. None is distinguished from the others. Variables that play similar roles in the assumptions will play similar roles in the details of the solution. Furthermore, the method of Lagrange multipliers generalizes easily to several variables and several constraints.

An Example

First let us illustrate the method of Lagrange multipliers by using it to solve the can problem just cited. The explanation of why it works will follow.

The first step is to form a certain function L of the variables r, h, and λ:

λ ("lambda") is the Greek letter corresponding to our letter l.

$$L(r, h, \lambda) = 2\pi r^2 + 2\pi rh - \lambda(\pi r^2 h - 100);$$

that is, $L(r, h, \lambda) = (\text{function to be minimized}) - \lambda(\text{constraint})$.

(λ is called a **Lagrange multiplier**.) Then compute the partial derivatives

$$\frac{\partial L}{\partial r}, \quad \frac{\partial L}{\partial h}, \quad \text{and} \quad \frac{\partial L}{\partial \lambda}$$

and find where they are all simultaneously 0:

$$0 = \frac{\partial L}{\partial r} = 4\pi r + 2\pi h - 2\pi\lambda rh,$$

$$0 = \frac{\partial L}{\partial h} = 2\pi r - \lambda\pi r^2,$$

$$0 = \frac{\partial L}{\partial \lambda} = -(\pi r^2 h - 100).$$

This gives three equations in three unknowns, r, h, and λ:

$$0 = 4\pi r + 2\pi h - 2\pi\lambda rh, \tag{3}$$

$$0 = 2\pi r - \lambda\pi r^2, \tag{4}$$

$$0 = \pi r^2 h - 100. \tag{5}$$

Note that Eq. (5) is just the given constraint (2).

Since $r = 0$ does not satisfy Eq. (5), we may divide Eq. (4) by πr, obtaining

$$0 = 2 - \lambda r.$$

Hence
$$\lambda r = 2. \tag{6}$$

Combining Eq. (6) with Eq. (3) yields

$$0 = 4\pi r + 2\pi h - (2\pi)(2)h$$

or
$$0 = 4\pi r - 2\pi h.$$

Hence
$$2r = h. \tag{7}$$

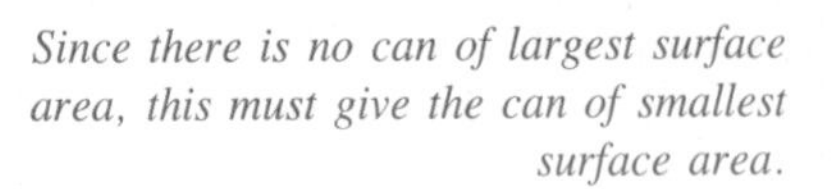

Since there is no can of largest surface area, this must give the can of smallest surface area.

Equation (7) already shows that the can of smallest surface area has its diameter equal to its height.

To find r and h explicitly, combine Eqs. (5) and (7), obtaining

$$0 = \pi r^2(2r) - 100.$$

Hence
$$2\pi r^3 = 100,$$

so
$$r = \left(\frac{50}{\pi}\right)^{1/3}$$

It is then possible to solve for h. This illustrates the method of Lagrange multipliers.

Why the Method Works

Let us see why Lagrange's method works. Consider the following problem:

Maximize or minimize $u = f(x, y)$, given the constraint $g(x, y) = 0$.

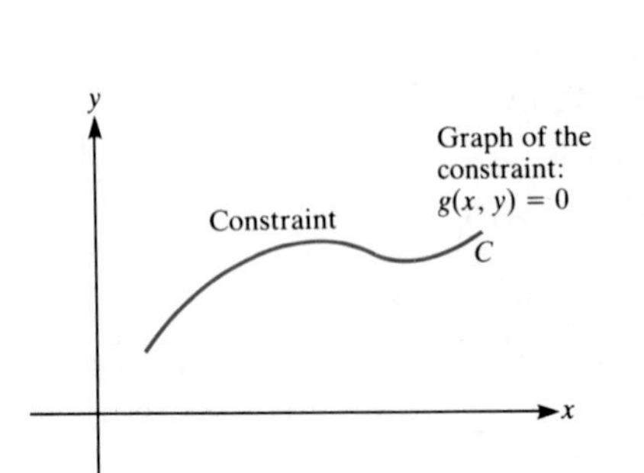

Figure 1

The graph of $g(x, y) = 0$ is in general a curve C, as shown in Fig. 1. Assume that f, *considered only on points of* C, takes a maximum (or minimum) value at the point P_0. Let C be parameterized by the vector function $\mathbf{G}(t) = x(t)\mathbf{i} + y(t)\mathbf{j}$. Let $\mathbf{G}(t_0) = \overrightarrow{OP_0}$. Then u is a function of t:

$$u = f(x(t), y(t)),$$

and, as shown in the proof of Theorem 1 of Sec. 14.8,

$$\frac{du}{dt} = \nabla f \cdot \mathbf{G}'(t_0). \tag{8}$$

Since f, considered only on C, has a maximum at $\mathbf{G}(t_0)$,

$$\frac{du}{dt} = 0$$

at t_0. Thus, by Eq. (8),

$$\nabla f \cdot \mathbf{G}'(t_0) = 0.$$

This means that ∇f is perpendicular to $\mathbf{G}'(t_0)$ at P_0. But ∇g, evaluated at P_0, is also perpendicular to $\mathbf{G}'(t_0)$, since the gradient ∇g is perpendicular to the level curve $g(x, y) = 0$. (We assume that ∇g is not $\mathbf{0}$.) (See Fig. 2.) Thus

$$\nabla f \text{ is parallel to } \nabla g.$$

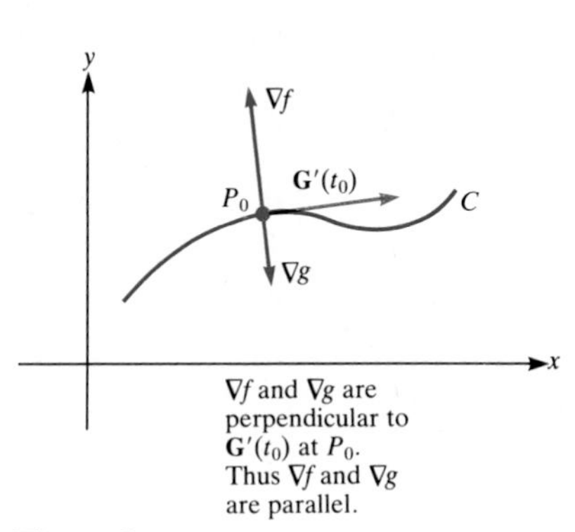

∇f and ∇g are perpendicular to $\mathbf{G}'(t_0)$ at P_0. Thus ∇f and ∇g are parallel.

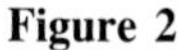
Figure 2

In other words, there is a scalar λ such that $\nabla f = \lambda \nabla g$.

Thus at a maximum or minimum of f, subject to the constraint $g(x, y) = 0$, there is a scalar λ such that

$$\frac{\partial f}{\partial x}\mathbf{i} + \frac{\partial f}{\partial y}\mathbf{j} = \lambda\left(\frac{\partial g}{\partial x}\mathbf{i} + \frac{\partial g}{\partial y}\mathbf{j}\right)$$

or, equivalently,

$$\frac{\partial f}{\partial x} = \lambda\frac{\partial g}{\partial x}$$

and

$$\frac{\partial f}{\partial y} = \lambda\frac{\partial g}{\partial y}.$$

Consequently, at such an extremum, occurring at $P_0 = (x_0, y_0)$, there is a scalar λ such that these three conditions hold:

$$\begin{cases} \dfrac{\partial f}{\partial x}(x_0, y_0) = \lambda\dfrac{\partial g}{\partial x}(x_0, y_0), \\[2ex] \dfrac{\partial f}{\partial y}(x_0, y_0) = \lambda\dfrac{\partial g}{\partial y}(x_0, y_0), \end{cases} \tag{9}$$

and

$$g(x_0, y_0) = 0. \tag{10}$$

Conditions (9) and (10), which provide three equations for three unknowns x_0, y_0, and λ, have a simple description in terms of the function L, defined as follows:

$$L(x, y, \lambda) = f(x, y) - \lambda g(x, y).$$

Conditions (9) are equivalent to

$$\frac{\partial L}{\partial x} = 0 \quad \text{and} \quad \frac{\partial L}{\partial y} = 0.$$

Condition (10) is equivalent to

$$\frac{\partial L}{\partial \lambda} = 0,$$

since this partial derivative is $-g(x, y)$.

EXAMPLE 1 Maximize the function x^2y for points (x, y) on the circle $x^2 + y^2 = 1$.

First put the constraint in the form "something = 0."

SOLUTION First, put the constraint in the form

$$x^2 + y^2 - 1 = 0.$$

Then the function L is easy to write down. In this case, L is given by

$$L(x, y, \lambda) = x^2y - \lambda(x^2 + y^2 - 1).$$

The three partial derivatives, which are set equal to 0, are

$$\frac{\partial L}{\partial x} = 2xy - 2\lambda x = 0,$$

$$\frac{\partial L}{\partial y} = x^2 - 2\lambda y = 0,$$

$$\frac{\partial L}{\partial \lambda} = -(x^2 + y^2 - 1) = 0.$$

Since at a maximum of x^2y the number x is not 0, $2x$ can be canceled from the first of the three equations. These equations then simplify to

$$y - \lambda = 0, \tag{11}$$

$$x^2 - 2\lambda y = 0, \tag{12}$$

$$x^2 + y^2 = 1. \tag{13}$$

By Eqs. (11) and (12), $$x^2 = 2y^2. \tag{14}$$

By Eqs. (13) and (14), $$2y^2 + y^2 = 1.$$

Hence $$y^2 = \tfrac{1}{3}.$$

Thus $$y = \frac{\sqrt{3}}{3} \quad \text{or} \quad y = -\frac{\sqrt{3}}{3}.$$

By Eq. (14), $$x = \sqrt{2}\, y \quad \text{or} \quad x = -\sqrt{2}\, y.$$

There are only four points to be considered on the circle:

$$\left(\frac{\sqrt{6}}{3}, \frac{\sqrt{3}}{3}\right), \quad \left(\frac{-\sqrt{6}}{3}, \frac{\sqrt{3}}{3}\right), \quad \left(\frac{-\sqrt{6}}{3}, \frac{-\sqrt{3}}{3}\right), \quad \left(\frac{\sqrt{6}}{3}, \frac{-\sqrt{3}}{3}\right).$$

At the first and second points x^2y is positive, while at the third and fourth x^2y is negative. The first two points provide the maximum value of x^2y on the circle $x^2 + y^2 = 1$, namely,

$$\left(\frac{\sqrt{6}}{3}\right)^2 \frac{\sqrt{3}}{3} = \frac{2\sqrt{3}}{9}.$$

The third and fourth points provide the minimum value of x^2y, namely,

$$\frac{-2\sqrt{3}}{9}. \quad \blacksquare$$

More Variables

The same method applies to finding a maximum or minimum of $f(x, y, z)$ subject to the constraint $g(x, y, z) = 0$. In this case, form the function $L(x, y, z) = f(x, y, z) - \lambda g(x, y, z)$ and set the four partial derivatives $\partial L/\partial x$, $\partial L/\partial y$, $\partial L/\partial z$, and $\partial L/\partial \lambda$ equal to 0.

More Constraints

Lagrange multipliers can also be used to maximize $f(x, y, z)$ subject to more than one constraint; for instance, the constraints may be

$$g(x, y, z) = 0 \quad \text{and} \quad h(x, y, z) = 0. \tag{15}$$

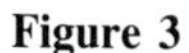

Figure 3

The two surfaces (15) in general meet in a curve C, as shown in Fig. 3. Assume that C is parameterized by the function $\mathbf{G}$. Then at a maximum (or minimum) of f at a point $P_0 = (x_0, y_0, z_0)$ on C,

$$\nabla f \cdot \mathbf{G}'(t_0) = 0.$$

Thus ∇f, evaluated at P_0, is perpendicular to $\mathbf{G}'(t_0)$. But ∇g and ∇h, being normal vectors at P_0 to the level surfaces $g(x, y, z) = 0$ and $h(x, y, z) = 0$, respectively, are both perpendicular to $\mathbf{G}'(t_0)$. Thus

$$\nabla f, \qquad \nabla g, \qquad \text{and} \qquad \nabla h$$

are all perpendicular to $\mathbf{G}'(t_0)$ at (x_0, y_0, z_0). (See Fig. 4.) Consequently, ∇f lies in the plane determined by the vectors ∇g and ∇h (which we assume are not parallel). Hence there are scalars λ and μ such that

$$\nabla f = \lambda \nabla g + \mu \nabla h.$$

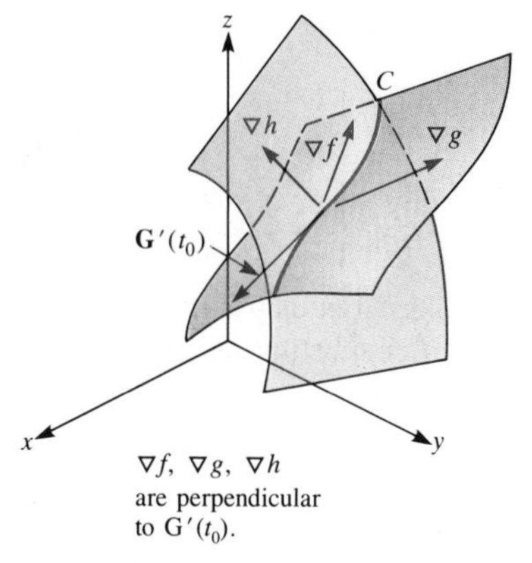

Figure 4

This equation asserts that at P_0 there are scalars λ and μ such that

$$\begin{cases} \dfrac{\partial f}{\partial x} = \lambda \dfrac{\partial g}{\partial x} + \mu \dfrac{\partial h}{\partial x}, \\[2mm] \dfrac{\partial f}{\partial y} = \lambda \dfrac{\partial g}{\partial y} + \mu \dfrac{\partial h}{\partial y}, \\[2mm] \dfrac{\partial f}{\partial z} = \lambda \dfrac{\partial g}{\partial z} + \mu \dfrac{\partial h}{\partial z}. \end{cases} \tag{16}$$

Equations (15) and (16) together are equivalent to the brief assertion that the five partial derivatives of the function L defined by

$$L(x, y, z, \lambda, \mu) = f(x, y, z) - \lambda g(x, y, z) - \mu h(x, y, z) \tag{17}$$

are 0. Hence to maximize (or minimize) f subject to the two constraints $g(x, y, z) = 0$ and $h(x, y, z) = 0$, form the function L in Eq. (17) and proceed as in Example 1.

A rigorous development of the material in this section belongs in an advanced calculus course. If a maximum occurs at an endpoint of the curves in question or if the two surfaces do not meet in a curve, this method does not apply. We will content ourselves by illustrating the method with an example in which there are two constraints.

EXAMPLE 2 Minimize the quantity $x^2 + y^2 + z^2$ subject to the constraints $x + 2y + 3z = 6$ and $x + 3y + 9z = 9$.

SOLUTION There are three variables and two constraints. Each of the two constraints mentioned describes a plane. Thus the two constraints together describe a *line*. The function $x^2 + y^2 + z^2$ is the square of the distance from (x, y, z) to the origin. So the problem can be rephrased as ''How far is the origin from a certain line?'' (It could be solved by vector algebra.) When viewed in this perspective, the problem certainly has a solution; that is, there is clearly a minimum.

Before forming the Lagrange function L, write the constraints with 0 on one side of the equation:

Rewrite the constraints.

$$x + 2y + 3z - 6 = 0 \quad \text{and} \quad x + 3y + 9z - 9 = 0.$$

The Lagrange function is

Form the Lagrange function.

$$L(x, y, z, \lambda, \mu) = x^2 + y^2 + z^2 - \lambda(x + 2y + 3z - 6) - \mu(x + 3y + 9z - 9).$$

Set the partial derivatives equal to 0.

Setting the five partial derivatives of L equal to 0 gives

$$\left.\begin{aligned} \frac{\partial L}{\partial x} = 0&: \quad 2x - \lambda - \mu = 0 \\ \frac{\partial L}{\partial y} = 0&: \quad 2y - 2\lambda - 3\mu = 0 \\ \frac{\partial L}{\partial z} = 0&: \quad 2z - 3\lambda - 9\mu = 0 \end{aligned}\right\} \tag{18}$$

$$\left.\begin{aligned} \frac{\partial L}{\partial \lambda} = 0&: \quad x + 2y + 3z - 6 = 0 \\ \frac{\partial L}{\partial \mu} = 0&: \quad x + 3y + 9z - 9 = 0 \end{aligned}\right\} \tag{19}$$

There are many ways to solve these five equations. Let us use the first three to express x, y, and z in terms of λ and μ. Then, after substituting the results in the last two equations, we will find λ and μ.

By (18),

$$x = \frac{\lambda + \mu}{2}, \qquad y = \frac{2\lambda + 3\mu}{2}, \qquad z = \frac{3\lambda + 9\mu}{2}.$$

Equations (19) then become

$$\frac{\lambda + \mu}{2} + \frac{2(2\lambda + 3\mu)}{2} + \frac{3(3\lambda + 9\mu)}{2} - 6 = 0$$

and

$$\frac{\lambda + \mu}{2} + \frac{3(2\lambda + 3\mu)}{2} + \frac{9(3\lambda + 9\mu)}{2} - 9 = 0,$$

which yield

$$\begin{aligned} 14\lambda + 34\mu &= 12 \\ 34\lambda + 91\mu &= 18. \end{aligned} \tag{20}$$

Solving (20) gives $\lambda = \dfrac{240}{59}$ and $\mu = -\dfrac{78}{59}$.

Thus

$$x = \frac{\lambda + \mu}{2} = \frac{81}{59},$$

$$y = \frac{2\lambda + 3\mu}{2} = \frac{123}{59},$$

$$z = \frac{3\lambda + 9\mu}{2} = \frac{9}{59}.$$

Since there is no maximum, this must be a minimum.

The minimum of $x^2 + y^2 + z^2$ is thus

$$\left(\frac{81}{59}\right)^2 + \left(\frac{123}{59}\right)^2 + \left(\frac{9}{59}\right)^2 = \frac{21{,}771}{3{,}481} = \frac{369}{59}. \quad \blacksquare$$

In Example 2 there were three variables, x, y, and z, and two constraints. There may, in some cases, be many variables, $x_1, x_2, \ldots, x_n$, and many constraints. Write each constraint as "something = 0" and introduce Lagrange multipliers $\lambda_1, \lambda_2, \ldots, \lambda_m$, one for each constraint.

THE INCREDIBLE SAILBOAT

A sailboat can move "against the wind." Even if the wind is blowing from the north, it is still possible to make some northward progress, though the boat cannot go directly north. We will show that the desire to maximize northward progress leads to an extremum problem. The reader may solve it by Lagrange multipliers, in Exercise 19.

Figure 5 is the top view of the sailboat. The wind vector **W** is pointed down along the y axis. The angle between the y axis and the sail is t. The angle between the sail and the line of symmetry of the boat is u. The angle between the boat and the x axis is v.

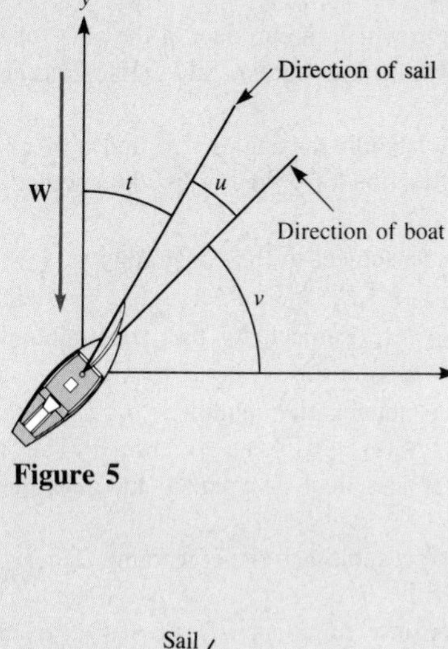

Figure 5

The vector **W** is the sum of a vector $\mathbf{W}_1$ perpendicular to the sail and a vector $\mathbf{W}_2$ parallel to the sail. (See Fig. 6.) The force against the sail is proportional to $\|\mathbf{W}_1\|$, which is $\|\mathbf{W}\| \sin t$.

The vector $\mathbf{W}_1$ is the sum of a vector $\mathbf{W}_3$ in the direction of the boat and a vector $\mathbf{W}_4$ perpendicular to that direction. (See Fig. 7.)

Note that the magnitude of the force in the direction of the boat is proportional to $\|\mathbf{W}_3\| = \|\mathbf{W}_1\| \sin u = \|\mathbf{W}\| \sin t \sin u$.

The northward component of this force is proportional to $\|\mathbf{W}_3\| \sin v$. All told, the northward component of the force is proportional to

$$\|\mathbf{W}\| \sin t \sin u \sin v. \tag{21}$$

The sailor wishes to maximize the product $\sin t \sin u \sin v$ subject to the constraint $t + u + v = \pi/2$. In Exercise 19 you are asked to find the angles t, u, and v that maximize the quantity (21) and therefore the northerly motion of the boat.

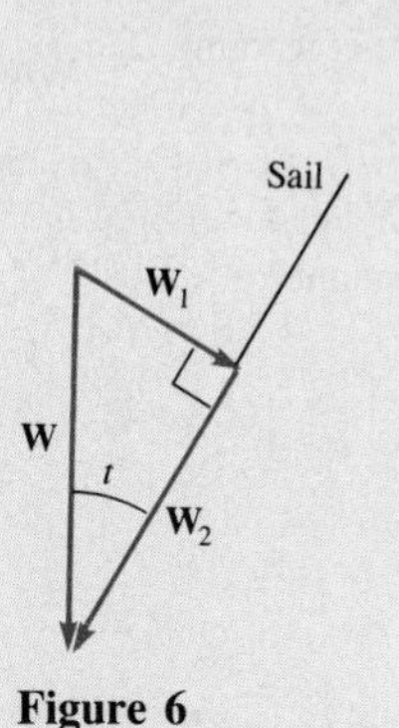

Figure 6

Figure 7

Section Summary

The method of Lagrange multipliers can be used to maximize (or minimize) a function when the variables are subject to one or more constraints. For instance, if you wish to maximize $f(x, y, z)$ subject to the constraints $g(x, y, z) = 0$ and $h(x, y, z) = 0$, form

$$L(x, y, z, \lambda, \mu) = f(x, y, z) - \lambda g(x, y, z) - \mu h(x, y, z)$$

and set the partial derivatives of L with respect to each of the five variables equal to 0.

For a rigorous treatment of the method of Lagrange multipliers and the conditions under which it is valid, see, for instance, M. Protter and C. B. Morrey, *Intermediate Calculus*, 2d ed., pp. 266–271, Springer, New York, 1985.

EXERCISES FOR SEC. 14.10: LAGRANGE MULTIPLIERS

In the exercises use Lagrange multipliers unless otherwise suggested.

1 Maximize xy for points on the circle $x^2 + y^2 = 4$.

2 Minimize $x^2 + y^2$ for points on the line

$$2x + 3y - 6 = 0.$$

3 Minimize $2x + 3y$ on the portion of the hyperbola $xy = 1$ in the first quadrant.

4 Maximize $x + 2y$ on the ellipse $x^2 + 2y^2 = 8$.

5 Find the largest area of all rectangles whose perimeters are 12 centimeters.

6 A rectangular box is to have a volume of 1 cubic meter. Find its dimensions if its surface area is minimal.

7 Find the point on the plane $x + 2y + 3z = 6$ that is closest to the origin. *Suggestion:* Minimize the square of the distance in order to avoid square roots.

8 Maximize $x + y + 2z$ on the sphere $x^2 + y^2 + z^2 = 9$.

9 Minimize the distance from (x, y, z) to $(1, 3, 2)$ for points on the plane $2x + y + z = 5$.

10 Find the dimensions of the box of largest volume whose surface area is to be 6 square inches.

11 Maximize $x^2y^2z^2$ subject to the constraint

$$x^2 + y^2 + z^2 = 1.$$

12 Find the points on the surface $xyz = 1$ closest to the origin.

13 Minimize $x^2 + y^2 + z^2$ on the line common to the two planes $x + 2y + 3z = 0$ and $2x + 3y + z = 4$.

14 The plane $2y + 4z - 5 = 0$ meets the cone $z^2 = 4(x^2 + y^2)$ in a curve. Find the point on this curve nearest the origin.

In Exercises 15 to 18 solve the given exercise in Sec. 14.9 by Lagrange multipliers.

15 Exercise 25

16 Exercise 26

17 Exercise 28

18 Exercise 29

---

19 (*a*) To advise the sailor how to set the sail and direct the boat, maximize $\sin t \sin u \sin v$ subject to $t + u + v = \pi/2$, where t, u, and v are all greater than or equal to 0. [See Eq. (21).]
(*b*) Sketch a diagram of the boat and sail when the boat is proceeding north as fast as possible.
(*c*) When the boat is moving north most rapidly, what fraction of the force of the wind pushes the boat north?

20 Solve Example 2 by vector algebra.

21 Solve Exercise 13 by vector algebra.

22 (*a*) Sketch the elliptical paraboloid $z = x^2 + 2y^2$.
(*b*) Sketch the plane $x + y + z = 1$.
(*c*) Sketch the intersection of the surfaces in (*a*) and (*b*).
(*d*) Find the highest point on the intersection in (*c*).

23 (*a*) Sketch the ellipsoid $x^2 + y^2/4 + z^2/9 = 1$ and the point $P(2, 1, 3)$.
(*b*) Find the point Q on the ellipsoid that is nearest P.
(*c*) What is the angle between PQ and the tangent plane at Q?

24 (*a*) Sketch the hyperboloid $x^2 - y^2/4 - z^2/9 = 1$. (How many sheets does it have?)
(*b*) Sketch the point $(1, 1, 1)$. (Is it "inside" or "outside" the hyperboloid?)
(*c*) Find the point on the hyperboloid nearest P.

25 Maximize $x^3 + y^3 + 2z^3$ on the intersection of the surfaces $x^2 + y^2 + z^2 = 4$ and $(x - 3)^2 + y^2 + z^2 = 4$.

26 Show that a triangle in which the product of the sines of the three angles is maximized is equilateral. (Use Lagrange multipliers.)

27 Solve Exercise 26 by labeling the angles x, y, and $\pi - x - y$ and minimizing a function of x and y by the method of Sec. 14.9.

28 Maximize $x + 2y + 3z$ subject to the constraints $x^2 + y^2 + z^2 = 1$ and $x + y + z = 0$.

29 (*a*) Maximize $x_1x_2 \cdots x_n$ subject to the constraint that $\Sigma_{i=1}^n x_i = 1$ and all $x_i \geq 0$.
(*b*) Deduce that for nonnegative numbers $a_1, a_2, \ldots, a_n$, $\sqrt[n]{a_1a_2 \cdots a_n} \leq (a_1 + a_2 + \cdots + a_n)/n$. (The **geometric mean** is less than or equal to the **arithmetic mean**.)

30 (*a*) Maximize $\Sigma_{i=1}^n x_iy_i$ subject to the constraints $\Sigma_{i=1}^n x_i^2 = 1$ and $\Sigma_{i=1}^n y_i^2 = 1$.
(*b*) Deduce that for any numbers $a_1, a_2, \ldots, a_n$ and $b_1, b_2, \ldots, b_n$, $\Sigma_{i=1}^n a_ib_i \leq (\Sigma_{i=1}^n a_i^2)^{1/2}(\Sigma_{i=1}^n b_i^2)^{1/2}$, which is called the Schwarz inequality.

Hint: Let $$x_i = \frac{a_i}{(\Sigma_{i=1}^n a_i^2)^{1/2}}$$

and $$y_i = \frac{b_i}{(\Sigma_{i=1}^n b_i^2)^{1/2}}.$$

(*c*) How would you justify the inequality in (*b*), for $n = 3$, by vectors?

31 Let $a_1, a_2, \ldots, a_n$ be fixed nonzero numbers. Maximize $\Sigma_{i=1}^n a_ix_i$ subject to $\Sigma_{i=1}^n x_i^2 = 1$.

32 Let p and q be positive numbers that satisfy the equation $1/p + 1/q = 1$. Obtain Hölder's inequality for nonnegative numbers a_i and b_i,

$$\sum_{i=1}^n a_ib_i \leq \left(\sum_{i=1}^n a_i^p\right)^{1/p}\left(\sum_{i=1}^n b_i^q\right)^{1/q},$$

as follows.
(*a*) Maximize $\Sigma_{i=1}^n x_iy_i$ subject to $\Sigma_{i=1}^n x_i^p = 1$ and $\Sigma_{i=1}^n y_i^q = 1$.

(*b*) By letting $x_i = \dfrac{a_i}{(\sum_{i=1}^n a_i^p)^{1/p}}$

and $y_i = \dfrac{b_i}{(\sum_{i=1}^n b_i^q)^{1/q}}$,

obtain Hölder's inequality.

Note that Hölder's inequality, with $p = 2$ and $q = 2$, reduces to the Schwarz inequality in Exercise 30.

33 A consumer has a budget of B dollars and may purchase n different items. The price of the ith item is p_i dollars. When the consumer buys x_i units of the ith item, the total cost is $\sum_{i=1}^n p_i x_i$. Assume that $\sum_{i=1}^n p_i x_i = B$ and that the consumer wishes to maximize her utility $u(x_1, x_2, \ldots, x_n)$.

(*a*) Show that when $x_1, \ldots, x_n$ are chosen to maximize utility, then

$$\frac{\partial u/\partial x_i}{p_i} = \frac{\partial u/\partial x_j}{p_j}.$$

(*b*) Explain the result in (*a*) using just economic intuition. *Hint:* Consider a slight change in x_i and x_j, with the other x_k's held fixed.

34 The following is quoted from Colin W. Clark in *Mathematical Bioeconomics,* Wiley, New York, 1976:

[S]uppose there are N fishing grounds. Let $H^i = H^i(R^i, E^i)$ denote the production function for the total harvest H^i on the ith ground as a function of the recruited stock level R^i and effort E^i on the ith ground. The problem is to determine the least total cost $\sum_{i=1}^N c_i E^i$ at which a given total harvest $H = \sum_{i=1}^N H^i$ can be achieved. This problem can be easily solved by Lagrange multipliers. The result is simply

$$\frac{1}{c_i}\frac{\partial H^i}{\partial E^i} = \text{constant}$$

[independent of i].

Verify his assertion. The c_i's are constants. The superscripts name the functions; they are not exponents.

35 (*Computer science*) This exercise is based on J. D. Ullman, *Principles of Database Systems,* pp. 82–83, Computer Science Press, Potomac, Md., 1980. It arises in the design of efficient "bucket" sorts. (A *bucket sort* is a particular way of rearranging information in a database.) Let $p_1, p_2, \ldots, p_k$ and B be positive constants. Let $b_1, b_2, \ldots, b_k$ be k nonnegative variables satisfying $\sum_{j=1}^k b_j = B$. The quantity $\sum_{j=1}^k p_j \cdot 2^{B-b_j}$ represents the expected search time. What values of $b_1, b_2, \ldots, b_k$ does the method of Lagrange multipliers suggest provide the minimum expected search time?

14.11 THE CHAIN RULE REVISITED

In the chain rule of Sec. 14.6, z is a function of $x_1, x_2, \ldots, x_m$ and each x_i is a function of $t_1, t_2, \ldots, t_n$. In some applications, some of the t's may be the same as some of the x's. We illustrate this situation, which is met in elementary thermodynamics, and describe how to deal with it.

Overlapping Variables

Let $z = f(x, y)$ and consider a level curve $f(x, y) = k$. Assume that this level curve is also described explicitly as $y = g(x)$. Therefore, z is a function of x and y, but y is a function of x. Therefore we may view z as a function of x. The following diagram records this situation:

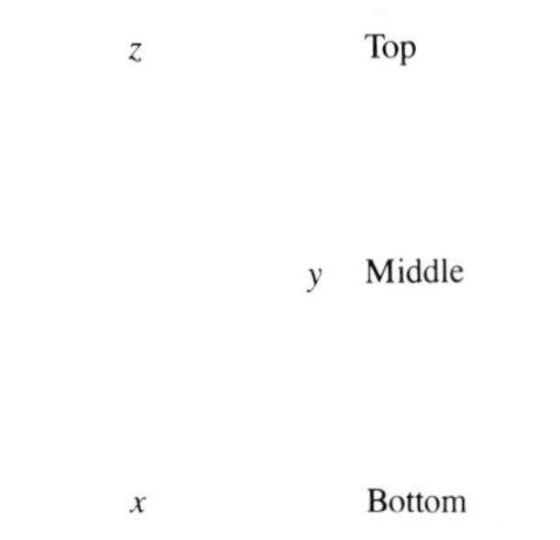

Note that x appears both as a middle variable and a bottom variable.

SOLUTION Equation (14) looks a little bit like a chain rule, but in order to decide which particular form it is, we must determine the top variable, middle variables, and bottom variables.

On the left side of (14), E is viewed as a function of T and V. E is therefore the top variable and T and V the bottom variables.

Since both

$$\left(\frac{\partial E}{\partial T}\right)_P \quad \text{and} \quad \left(\frac{\partial E}{\partial P}\right)_T$$

occur on the right side of (14), we expect P and T to be the middle variables. The diagram is

	E		Top
P		T	Middle
T		V	Bottom

Since the left side of (14) is $(\partial E/\partial T)_V$, we draw the paths from E down to the bottom variable T:

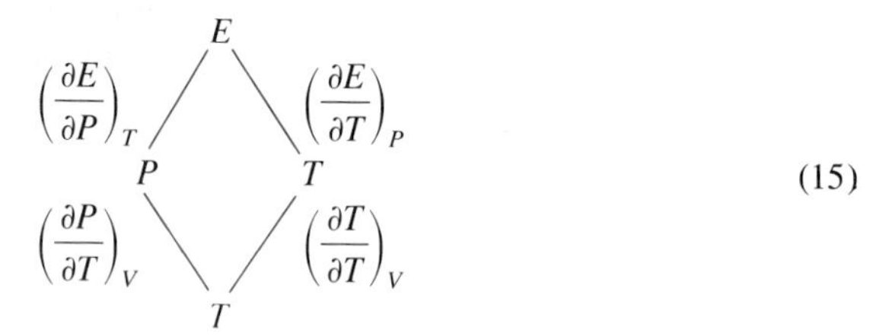

Each of the two paths in (15) provides a summand in the chain rule. Therefore,

$$\left(\frac{\partial E}{\partial T}\right)_V = \left(\frac{\partial E}{\partial P}\right)_T\left(\frac{\partial P}{\partial T}\right)_V + \left(\frac{\partial E}{\partial T}\right)_P\left(\frac{\partial T}{\partial T}\right)_V. \tag{16}$$

Since $(\partial T/\partial T)_V = 1$, (16) reduces to

$$\left(\frac{\partial E}{\partial T}\right)_V = \left(\frac{\partial E}{\partial P}\right)_T\left(\frac{\partial P}{\partial T}\right)_V + \left(\frac{\partial E}{\partial T}\right)_P. \tag{17}$$

Comparison of (17) and (14) completes the argument. ■

Note that if subscripts were not used, then (14) would read

$$\frac{\partial E}{\partial T} = \frac{\partial E}{\partial T} + \frac{\partial E}{\partial P}\frac{\partial P}{\partial T},$$

which is nonsense.

Section Summary

This section illustrated how to use the chain rule when a variable is both a middle and bottom variable. In short, use subscripts on the partial derivatives in order to avoid confusion in the bookkeeping.

EXERCISES FOR SEC. 14.11: THE CHAIN RULE REVISITED

1 Let $u = f(x, y, z)$. Assume that $x = h(y, z)$ satisfies the equation $f(h(y, z), y, z) = k$. Using the chain rule, obtain the formula for $(\partial x/\partial z)_y$ analogous to Theorem 1.

2 (*This continues Exercise 1.*) Obtain the formula for $(\partial x/\partial y)_z$ analogous to Theorem 1.

3 Verify Theorem 1 for $f(x, y, z) = x^2 + 2y^2 + z^2$, $k = 4$, and $z = \sqrt{4 - x^2 - 2y^2}$.

4 Verify the analog of Theorem 1 for $(\partial x/\partial z)_y$ when $f(x, y, z) = 5x + 2y + 3z$ and $k = 6$.

Exercises 5 to 8 continue the thermodynamics example.

5 Show that $(\partial P/\partial T)_V = \dfrac{-(\partial V/\partial T)_P}{(\partial V/\partial P)_T}$.

6 Show that

(*a*) $\left(\dfrac{\partial E}{\partial V}\right)_P = \left(\dfrac{\partial E}{\partial T}\right)_P \left(\dfrac{\partial T}{\partial V}\right)_P$

(*b*) $\left(\dfrac{\partial E}{\partial P}\right)_V = \left(\dfrac{\partial E}{\partial T}\right)_P \left(\dfrac{\partial T}{\partial P}\right)_V + \left(\dfrac{\partial E}{\partial P}\right)_T$

7 Show that $(\partial P/\partial T)_V(\partial T/\partial P)_V = 1$.

Hint: Express each of the partial derivatives as a quotient of partial derivatives, as in Exercise 5.

8 Show that $(\partial P/\partial T)_V(\partial T/\partial V)_P(\partial V/\partial P)_T = -1$.

9 Let $z = r^2 + s^2 + t^2$ and let $t = rsu$.

(*a*) The symbol $\partial z/\partial r$ has two interpretations. What are they?

(*b*) Evaluate $\partial z/\partial r$ in both cases in (*a*).

10 Let $u = F(x, y, z)$ and $z = f(x, y)$. Thus u is a (composite) function of x and y: $u = G(x, y) = F(x, y, f(x, y))$. Assume that $G(x, y) = x^2y$. Obtain a formula for $\partial f/\partial x$ in terms of $\partial F/\partial x$, $\partial F/\partial y$, and $\partial F/\partial z$. (All three need not appear in your answer.)

11 Two functions u and v of the variables x and y are defined implicitly by the two simultaneous equations

$$F(u, v, x, y) = 0$$

and

$$G(u, v, x, y) = 0.$$

Assuming all necessary differentiability, find a formula for $\partial u/\partial x$ in terms of partial derivatives of F and G.

14.S SUMMARY

This chapter developed properties of functions of several variables analogous to properties of functions of one variable. The basis of the chapter is the equation

$$\Delta f = \frac{\partial f}{\partial x}\Delta x + \frac{\partial f}{\partial y}\Delta y + \epsilon_1 \Delta x + \epsilon_2 \Delta y.$$

Here $\Delta f = f(x + \Delta x, y + \Delta y) - f(x, y)$, the partial derivatives are evaluated at (x,y), and ϵ_1 and ϵ_2 approach 0 as both Δx and Δy approach 0. (Similar equations hold for functions of more variables.)

The following table summarizes most of the chapter.

The Plane	*Space*
Distance formula: $\sqrt{(x_2 - x_1)^2 + (y_2 - y_1)^2}$	Distance formula: $\sqrt{(x_2 - x_1)^2 + (y_2 - y_1)^2 + (z_2 - z_1)^2}$
Circle of radius r and center (x_0, y_0): $(x - x_0)^2 + (y - y_0)^2 = r^2$	Sphere of radius r and center (x_0, y_0, z_0): $(x - x_0)^2 + (y - y_0)^2 + (z - z_0)^2 = r^2$
Ellipse: $\dfrac{x^2}{a^2} + \dfrac{y^2}{b^2} = 1$	Ellipsoid: $\dfrac{x^2}{a^2} + \dfrac{y^2}{b^2} + \dfrac{z^2}{c^2} = 1$
Hyperbola: $\dfrac{x^2}{a^2} - \dfrac{y^2}{b^2} = 1$	Hyperboloid of one sheet: $\pm\dfrac{x^2}{a^2} \pm \dfrac{y^2}{b^2} \pm \dfrac{z^2}{c^2} = 1$ (one −, two +'s)
	Hyperboloid of two sheets: $\pm\dfrac{x^2}{a^2} \pm \dfrac{y^2}{b^2} \pm \dfrac{z^2}{c^2} = 1$ (two −'s, one +)
Parabola: $y = kx^2$	Elliptical paraboloid: $z = \dfrac{x^2}{a^2} + \dfrac{y^2}{b^2}$

Derivative f measures rate of change of $f(x)$	Partial derivatives $\partial f/\partial x$ and $\partial f/\partial y$ measure rate of change of $f(x, y)$ in the x and y directions. Directional derivative $D_{\mathbf{u}}f$ measures rate of change of f in direction $\mathbf{u}$.
Chain rule: $\dfrac{dy}{dx} = \dfrac{dy}{du}\dfrac{du}{dx}$	Chain rule: If y is a function of $u_1, u_2, \ldots, u_n$, and each u_i is a function of $x_1, x_2, \ldots, x_m$, then $\dfrac{\partial y}{\partial x_i} = \sum_{j=1}^{n} \dfrac{\partial y}{\partial u_j}\dfrac{\partial u_j}{\partial x_i}$
Higher derivatives: $\dfrac{d^2y}{dx^2}, \ldots$	Higher partial derivatives: $\dfrac{\partial^2 f}{\partial x^2}, \dfrac{\partial^2 f}{\partial y^2}, \dfrac{\partial^2 f}{\partial x\,\partial y}, \ldots$
$\Delta f = f'(x)\,\Delta x + \epsilon \Delta x$ ($\epsilon \to 0$ as $\Delta x \to 0$)	$\Delta z = \Delta f = \dfrac{\partial f}{\partial x}\Delta x + \dfrac{\partial f}{\partial y}\Delta y + \epsilon_1 \Delta x + \epsilon_2 \Delta y$ ($\epsilon_1, \epsilon_2 \to 0$ as $\Delta x, \Delta y \to 0$)
dy defined as $f'(x)\,\Delta x$	dz defined as $\dfrac{\partial f}{\partial x}\Delta x + \dfrac{\partial f}{\partial y}\Delta y$
Tangent line	Tangent plane
dy is change along tangent line	dz is change along tangent plane
Critical number a: $f'(a) = 0$	Critical point (a, b): $\dfrac{\partial f}{\partial x}(a, b) = 0 = \dfrac{\partial f}{\partial y}(a, b)$
Second derivative test for local extremum: $f'(a) = 0$ $f''(a) > 0$: local minimum $f'(a) = 0$ $f''(a) < 0$: local maximum	Second-partial-derivative test for local extremum: $\dfrac{\partial f}{\partial x}(a, b) = 0 = \dfrac{\partial f}{\partial y}(a, b)$ $\left(\dfrac{\partial^2 f}{\partial x\,\partial y}\right)^2 < \dfrac{\partial^2 f}{\partial x^2}\dfrac{\partial^2 f}{\partial y^2}$, and $\dfrac{\partial^2 f}{\partial x^2} > 0$: local minimum $\dfrac{\partial^2 f}{\partial x^2} < 0$: local maximum

The gradient was introduced to give a short expression for a directional derivative. The gradient of $f(x, y)$ is $\nabla f = \partial f/\partial x\mathbf{i} + \partial f/\partial y\mathbf{j}$. The gradient of $f(x, y, z)$ is $\nabla f = \partial f/\partial x\mathbf{i} + \partial f/\partial y\mathbf{j} + \partial f/\partial z\mathbf{k}$.

The gradient has many uses:

1 Describing a directional derivative: $D_{\mathbf{u}} f = \nabla f \cdot \mathbf{u}$.
2 Providing a normal to a curve in the plane or to a surface. [∇f is normal to the level curve $f(x, y) = k$ or to the level surface $f(x, y, z) = k$.]
3 Rephrasing a chain rule; for instance:

$$\frac{du}{dt} = \frac{\partial u}{\partial x}\frac{dx}{dt} + \frac{\partial u}{\partial y}\frac{dy}{dt} + \frac{\partial u}{\partial z}\frac{dz}{dt}$$

can be written

$$\frac{du}{dt} = \nabla u \cdot \mathbf{G}'(t),$$

which shows that ∇u is perpendicular to any curve in a surface where u is constant.

Lagrange multipliers is a method of finding an extremum of a function of several variables subject to one or more constraints.

Vocabulary

graph	chain rule
trace	critical point
quadric surface	directional derivative
level curve	gradient
level surface	differential
limit, continuity	discriminant
partial derivative	tangent plane

GUIDE QUIZ ON CHAP. 14: PARTIAL DERIVATIVES

1 Give an example of an equation of the following and sketch it:

(*a*) An ellipsoid

(*b*) A hyperboloid of one sheet

(*c*) A hyperboloid of two sheets

(*d*) An elliptical paraboloid

2 (*a*) Explain in detail why

$$\Delta f = f(x + \Delta x, y + \Delta y) - f(x, y) = \frac{\partial f}{\partial x}\Delta x + \frac{\partial f}{\partial y}\Delta y + \epsilon_1 \Delta x + \epsilon_2 \Delta y,$$

where ϵ_1 and $\epsilon_2 \to 0$ as Δx and $\Delta y \to 0$. (Include a diagram of the curves involved.) What assumptions about f are used?

(*b*) Use (*a*) to explain why

$$\frac{\partial u}{\partial t} = \frac{\partial u}{\partial x}\frac{\partial x}{\partial t} + \frac{\partial u}{\partial y}\frac{\partial y}{\partial t}.$$

3 Sketch the surface whose level curves are shown in Fig. 1. Show the corresponding traces.

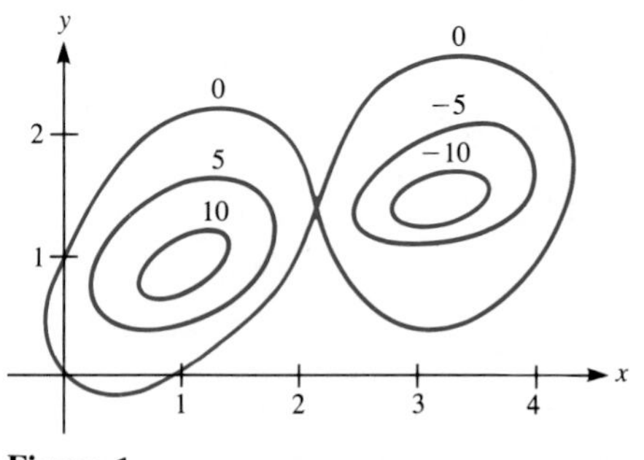

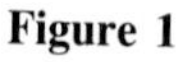
Figure 1

4 Figure 2 shows four level curves of $f(x, y)$ near $(0, 0)$.

(*a*) Estimate $\partial f/\partial x$ at $(0, 0)$.

(*b*) Estimate $\partial f/\partial y$ at $(0, 0)$.

(*c*) Draw ∇f at $(0, 0)$.

(*d*) What angle does ∇f make with the level curve that passes through $(0, 0)$?

(*e*) Estimate $D_{\mathbf{u}} f$ at $(0, 0)$, where $\mathbf{u} = \cos 37° \, \mathbf{i} + \sin 37° \, \mathbf{j}$.

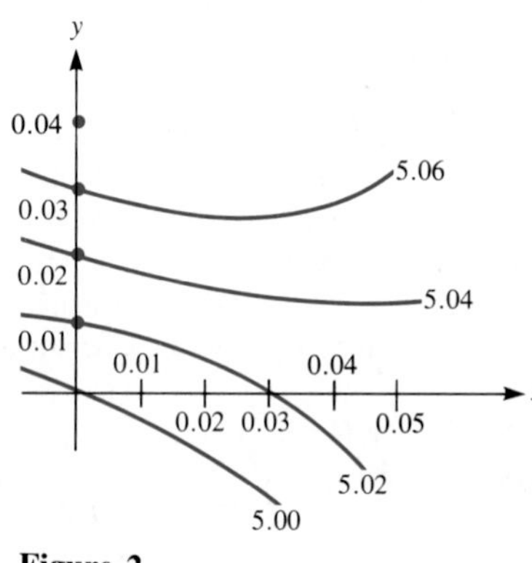

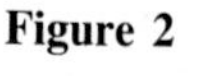
Figure 2

5 A function f defined for all points (x, y) has the property that $\partial f/\partial x = 2$ and $\partial f/\partial y = 3$ for all (x, y). Moreover, $f(0, 0) = 0$.

(*a*) Give an example of such a function.

(*b*) Find all such functions, explaining your reasoning.

6 Evaluate the following partial derivatives:

(*a*) $\dfrac{\partial}{\partial y}[\sec^3(x + 2y) \ln (1 + 2xy)]$

(*b*) $\dfrac{\partial}{\partial x}(x \tan^{-1} 3xy)$

(*c*) $\dfrac{\partial}{\partial x}(ye^{x^3 y})$

(*d*) $\dfrac{\partial^2}{\partial x \, \partial y}[\cos (2x + 3y)]$

7 Suppose $w = f(u, v)$ has continuous partial derivatives with respect to u and v, and $u = x + y$ and $v = x - y$. Show that

(a) $\dfrac{\partial w}{\partial x} \cdot \dfrac{\partial w}{\partial y} = \left(\dfrac{\partial f}{\partial u}\right)^2 - \left(\dfrac{\partial f}{\partial v}\right)^2$

(b) $\dfrac{\partial^2 w}{\partial x\, \partial y} = \dfrac{\partial^2 f}{\partial u^2} - \dfrac{\partial^2 f}{\partial v^2}$

8 Graph

(a) $x^2 + y^2 + z^2 = 9$

(b) $x^2 + 4y^2 + 9z^2 = 36$

(c) $-x^2 - \dfrac{y^2}{4} + z^2 = 1$

(d) $x^2 - 2y^2 + z^2 = 8$

(e) $z = y^2 - 2x^2$

(f) $z^2 = 6(x^2 + y^2)$

9 (a) Let $y(x, t) = y_0 \sin(kx - kvt)$, where y_0, k, and $v \neq 0$ are constants. Show that $y(x, t)$ satisfies the equation

$$\frac{\partial^2 y}{\partial x^2} = \frac{1}{v^2} \frac{\partial^2 y}{\partial t^2}.$$

(b) Let $f(x)$ have first and second derivatives. Show that $f(x - vt)$ and $f(x + vt)$ both satisfy the equation in (a), if v is a constant.

10 Give an example of a function f such that it has a critical point at (0, 0), $\partial^2 f/\partial x^2(0, 0) > 0$, $\partial^2 f/\partial y^2(0, 0) > 0$, and
(a) (0, 0) is a local minimum.
(b) (0, 0) is not a local minimum.

11 A house in the form of a box is to hold 10,000 cubic feet. The glass walls admit heat at the rate of 5 units per minute per square foot, the roof admits heat at the rate of 3 units per minute per square foot, and the floor admits heat at a rate of 1 unit per minute per square foot. What should the shape of the house be in order to minimize the rate at which heat enters?

12 (a) Using a calculator, evaluate $(1.1)^2 \ln(1.2)$.
(b) Using the differential of the function $f(x, y) = x^2 \ln y$ and making use of $f(1, 1)$, estimate $(1.1)^2 \ln(1.2)$.

13 The kinetic energy of a particle of mass m and velocity v is given by $K = \frac{1}{2}mv^2$. If the maximum error in measuring m is 1 percent and in measuring v is 3 percent, estimate the maximum error in measuring K.

14 The temperature $T(x, y)$ in a flat plate is known at three points: $T(1, 2) = 3$, $T(1.01, 2) = 3.03$, and $T(1, 2.01) = 2.98$. A bug at time t is at the point $(x, y) = (t, 2t^2)$. As it passes through the point (1, 2), at what rate is the temperature changing as a function of time?

15 (a) Sketch the ellipsoid $x^2 + y^2/4 + z^2/9 = 1$ and its trace in the plane $x + y + z = 1$.
(b) Using Lagrange multipliers, find the points on that trace that are closest to the origin.

16 The surface $z = f(x, y)$ passes through the point (1, 2, 3), where $\partial f/\partial x = 1$ and $\partial f/\partial y = -1$.
(a) Draw two lines that lie in the tangent plane to the surface at (1, 2, 3).
(b) Find an equation of that tangent plane.
(c) Where does the line through the origin and (4, 1, 2) meet that plane?

17 You arrive early in the chemistry auditorium and see this left on the blackboard from the previous lecture:

$$\left(\frac{\partial H}{\partial T}\right)_P \left(\frac{\partial T}{\partial P}\right)_H + \left(\frac{\partial H}{\partial P}\right)_T = 0.$$

You have no idea what the letters refer to. Nevertheless, show that the formula is correct.

18 Figure 3 shows the level curve $f(x, y) = 1$ and several gradients ∇f at points on this curve. Draw as well as you can the level curves (a) $f(x, y) = 2$, (b) $f(x, y) = 0$. (c) Explain your reasoning.

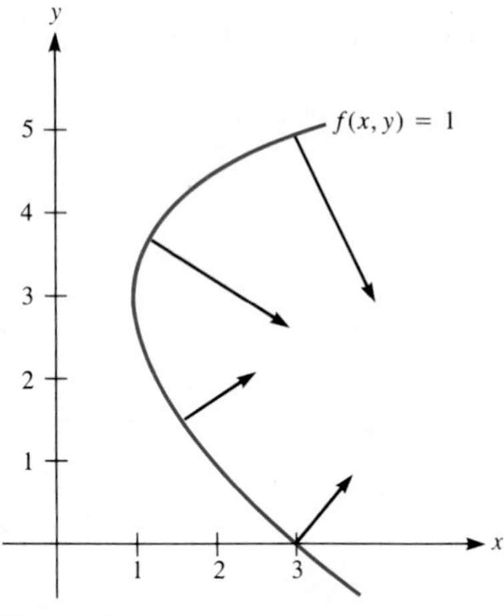

Figure 3

REVIEW EXERCISES FOR CHAP. 14: PARTIAL DERIVATIVES

1 Consider the part of the hyperbola $zy = 1$ in the yz plane for which y is positive. (a) Find the equation of the surface obtained by revolving this curve around the y axis. (b) Is the surface a quadric?

2 (a) Draw the hyperbolas $y^2 - (z^2/4) = 1$ and $y^2 - (z^2/4) = -1$ and their asymptotes.
(b) Draw the surfaces obtained by revolving the curves and lines in (a) around the z axis. (Draw them relative to the same axes, in order to show how they fit together.)
(c) Give equations for the surfaces in (b).

3 (a) Draw $z = x^2 + 2y^2$ and a vector normal to this surface at (1, 1, 3).

(*b*) How far is the point (1, 1, 1) from the tangent plane at (1, 1, 3)?

4 (*a*) Graph the surface $z = 2x^2y$ for $x, y \geq 0$.
(*b*) Find and draw a vector tangent to the surface at (1, 1, 2) and parallel to the xz plane.
(*c*) Find and draw a vector tangent to the surface at (1, 1, 2) and parallel to the yz plane.
(*d*) Find and draw a vector perpendicular to the surface at (1, 1, 2).
(*e*) Determine the derivative of z at (1, 1, 2) in the direction of the vector in (*c*).
(*f*) Determine the derivative of z at (1, 1, 2) in the direction of the vector in (*d*).
(*g*) Determine the derivative of z at (1, 1, 2) in the direction of the vector $\langle 2, 3, 4\rangle$.

5 Sketch the intersection of the cylinder $x^2 + y^2 = 4$ with (*a*) the plane $z = y + 1$, (*b*) the plane $z = x$.

6 Sketch the intersection of the cylinders $x^2 + y^2 = 1$ and $x^2 + z^2 = 1$.

7 Assume that the function $z = f(x, y)$ satisfies the equation $xyz + x + y + z^5 + 3 = 0$.
(*a*) Find $(\partial z/\partial x)_y$, by differentiating both sides of the equation.
(*b*) Similarly, find $(\partial x/\partial y)_z$.
(*c*) Similarly, find $(\partial y/\partial z)_x$.
(*d*) Show that $(\partial z/\partial x)_y\,(\partial x/\partial y)_z\,(\partial y/\partial z)_x = -1$.

8 Let $z = x^2y$, where $y = e^{3x}u$. Thus z may be considered a function of x and y or of x and u.
(*a*) Compute $\partial z/\partial x$, considering z to be a function of x and y.
(*b*) Compute $\partial z/\partial x$, considering z to be a function of x and u.
(*c*) Use subscript notation to distinguish the partial derivatives in (*a*) and (*b*).

9 Let θ be fixed and $f(x, y)$ be a function, and define

$$g(u, v) = f(u \cos\theta - v \sin\theta,\ u \sin\theta + v \cos\theta).$$

Show that

$$\left(\frac{\partial f}{\partial x}\right)^2 + \left(\frac{\partial f}{\partial y}\right)^2 = \left(\frac{\partial g}{\partial u}\right)^2 + \left(\frac{\partial g}{\partial v}\right)^2.$$

10 Let $u = f(x, y)$ and $v = g(x, y)$. Assume that u and v determine x and y, $x = h(u, v)$ and $y = j(u, v)$. Show that

$$\frac{\partial u}{\partial x}\frac{\partial x}{\partial u} + \frac{\partial u}{\partial y}\frac{\partial y}{\partial u} = 1$$

and

$$\frac{\partial u}{\partial x}\frac{\partial x}{\partial v} + \frac{\partial u}{\partial y}\frac{\partial y}{\partial v} = 0.$$

11 A fence perpendicular to the xy plane has as its base the line segment whose ends are (1, 0, 0) and (0, 1, 0). It is bounded on the top by the parabolic cylinder $z = y^2$.
(*a*) Draw the fence.
(*b*) Find its area.

12 (*a*) Sketch the portion of the cylinder $x^2 + y^2 = 1$ that lies below the paraboloid $z = x^2 + 2y^2$ and above the xy plane.
(*b*) Find the area of the surface sketched in (*a*).

13 (*a*) Sketch the surface $z = x^2/4 + y^2/9$ and the point $P(2, 3, 2)$ that lies on it.
(*b*) Sketch the curve in the surface whose slope at P is equal to $\partial z/\partial x$ evaluated at (2, 3).
(*c*) Sketch the curve in the surface whose slope at P is equal to $\partial z/\partial y$ evaluated at (2, 3).
(*d*) Sketch the curve in the surface whose slope at P is $D_{\mathbf{u}}(z)$ evaluated at (2, 3), when $\mathbf{u} = \dfrac{1}{2}\mathbf{i} + \dfrac{\sqrt{3}}{2}\mathbf{j}$.
(*e*) Find and draw a vector normal to the surface at P.
(*f*) Sketch the trace in the plane $z = -2$.
(*g*) Sketch the level curve that passes through the point (2, 3).
(*h*) Find and draw a vector normal to the curve in (*g*) at the point (2, 3).

14 (*a*) Sketch the surfaces $z = 2x^2 + y^2$ and $x^2 + y^2 + 2z^2 = 20$ and their intersection.
(*b*) Find the point P in the first octant where the plane $y = x$ meets the intersection in (*a*).
(*c*) Find and sketch a vector perpendicular to the paraboloid in (*a*) at P.
(*d*) Find and sketch a vector perpendicular to the ellipsoid in (*a*) at P.
(*e*) Find and draw a vector tangent to the intersection in (*a*) at P.

15 Explain why the gradient of $F(x, y, z)$ is perpendicular to the level surface $F(x, y, z) = k$.

16 Find a unit vector perpendicular to (*a*) the surface $z = x^2 + y^2$ at (1, 1, 2), (*b*) the surface $x + y^2 + z^3 = 3$ at (1, 1, 1), (*c*) the sphere $x^2 + y^2 + z^2 = 9$ at (x, y, z).

17 The temperature T at the point (x, y, z) is $x^2 + y^2 + z^2$. As a bird flies on the path $t\mathbf{i} + t^2\mathbf{j} + t^3\mathbf{k}$, it notices that the air is getting hotter. Find the rate of change in the temperature along the bird's flight at the point (1, 1, 1) (*a*) with respect to time, (*b*) with respect to distance.

18 (*a*) If a plane curve is given in the form $\mathbf{r} = \mathbf{G}(t)$, how would you find a vector parallel to the curve? Perpendicular to the curve?
(*b*) If a plane curve is given in the form $f(x, y) = 0$, how would you find a vector perpendicular to the curve?
(*c*) If a curve is given as the intersection of two surfaces, how could you find a vector perpendicular to the curve? parallel to the curve?
(*d*) If a space curve is given in the form $\mathbf{r} = \mathbf{G}(t)$, how would you find a vector parallel to the curve? perpendicular to the curve?

19 (*a*) Using a calculator, draw the curve $x^2 + xy + y^2 = 3$. (Appendix G shows that this curve is an ellipse.)
(*b*) Using Lagrange multipliers find the points on the ellipse nearest the origin and farthest from the origin.

20 (*a*) Show that $\partial^2(xy)/\partial x\,\partial y = 1$.

(*b*) If $\partial^2 f/\partial x \partial y = 1$ for all (x, y), must $f(x, y)$ be of the form $xy + C$, where C is a constant?

21 Let $z = f(x, y)$, $x = g(u, v)$, and $y = h(u, v)$. Thus $z = p(u, v)$. Assume that $\partial g/\partial u = \partial h/\partial v$ and $\partial g/\partial v = -\partial h/\partial u$. Show that

$$\frac{\partial^2 p}{\partial u^2} + \frac{\partial^2 p}{\partial v^2} = \left(\frac{\partial^2 f}{\partial x^2} + \frac{\partial^2 f}{\partial y^2}\right)\left[\left(\frac{\partial g}{\partial u}\right)^2 + \left(\frac{\partial g}{\partial v}\right)^2\right].$$

22 (See Exercises 31 to 35 of Sec. 14.6.) Prove that if $f(x, y)$ is homogeneous of degree n and has continuous second-order partial derivatives, then

$$x^2\frac{\partial^2 f}{\partial x^2} + 2xy\frac{\partial^2 f}{\partial x\, \partial y} + y^2\frac{\partial^2 f}{\partial y^2} = n(n-1)f.$$

23 The partial differential equation

$$\frac{\partial^2 y}{\partial t^2} = a^2\frac{\partial^2 y}{\partial x^2} \qquad (a \text{ constant})$$

describes a vibrating string. Assume that $y(x, t)$ has the form $f(x)g(t)$, that is, of a product of a function of x and a function of t.

(*a*) Show that $fg'' = a^2 f'' g$.

(*b*) From (*a*) it follows that

$$\frac{f''}{f} = \frac{g''}{a^2 g}.$$

The left side of this equation is a function of x and the right side is a function of t. Deduce that there is a constant k such that $f''/f = k$ and $g''/(a^2 g) = k$.

(*c*) Assume that k in (*b*) is positive. Show that any function of the form $f(x) = c_1 e^{\sqrt{k}x} + c_2 e^{-\sqrt{k}x}$ satisfies the equation $f''/f = k$.

(*d*) Show that any function $g(t)$ of the form $c_3 e^{a\sqrt{k}t} + c_4 e^{-a\sqrt{k}t}$ satisfies the equation $g''/(a^2 g) = k$.

Thus any function of the form

$$(c_1 e^{\sqrt{k}x} + c_2 e^{-\sqrt{k}x})(c_3 e^{a\sqrt{k}t} + c_4 e^{-a\sqrt{k}t})$$

satisfies the given partial differential equation.

24 The heat-flow equation for $u(x, t)$ is

$$\frac{\partial u}{\partial t} = a^2\frac{\partial^2 u}{\partial x^2}.$$

Following the approach in the preceding exercise, we try $u(x, t) = f(x)g(t)$.

(*a*) Show that there is a constant k such that $f''/f = k$ and $g'/(a^2 g) = k$.

(*b*) (Assume that k is negative.) What form must g have?

(*c*) Show that $c_1 \sin(\sqrt{-k}x) + c_2 \cos(\sqrt{-k}x)$ satisfies the equation $f''/f = k$.

Thus $[c_1 \sin(\sqrt{-k}x) + c_2 \cos(\sqrt{-k}x)]e^{a^2 kt}$ is a solution of the heat equation.

25 Let $z = f(u, v)$, where u and v are functions of x and y. Then, indirectly, $z = g(x, y)$. Show that if $du = \partial u/\partial x\, dx + \partial u/\partial y\, dy$ and $dv = \partial v/\partial x\, dx + \partial v/\partial y\, dy$, then the two expressions for dz,

$$dz = \frac{\partial z}{\partial u}du + \frac{\partial z}{\partial v}dv \qquad \text{and} \qquad dz = \frac{\partial z}{\partial x}dx + \frac{\partial z}{\partial y}dy,$$

are equal.

26 Tell what is wrong with this "proof" that if $z = f(x, y)$ and x and y are functions of t, then dz/dt is just $(\partial z/\partial x)(dx/dt)$: Since dz/dt is $\lim_{\Delta t \to 0} \Delta z/\Delta t$, it follows that $dz/dt = \lim_{\Delta t \to 0} (\Delta z/\Delta x)(\Delta x/\Delta t)$, where Δx is the change in x induced by the change in t. Thus $dz/dt = (\partial z/\partial x)(dx/dt)$.

27 Let (x, y) be rectangular coordinates in the plane, and (X, Y, Z) in space. Assume that F is a one-to-one correspondence between the plane and space such that x and y depend continuously on X, Y, and Z and have continuous partial derivatives with respect to them. Similarly, assume that through the inverse function F^{-1}, X, Y, and Z are continuous functions of x and y and have continuous partial derivatives with respect to them. From this deduce that $2 = 3$. *Hint:* $2 = dx/dx + dy/dy$ and $3 = dX/dX + dY/dY + dZ/dZ$. Use the chain rule. Incidentally, there is a one-to-one correspondence between the plane and space, but it does not have the specified properties of continuity and differentiability.

28 When studying the motion of the membrane of a circular drum, Euler in 1759 met the equation

$$\frac{1}{c^2}\frac{\partial^2 z}{\partial t^2} = \frac{\partial^2 z}{\partial x^2} + \frac{\partial^2 z}{\partial y^2}. \tag{1}$$

[z is the vertical displacement of the point (x, y) on the membrane at time t.] Because the drum is circular, he decided to use polar coordinates (r, θ) instead of rectangular coordinates (x, y). (*a*) Show that Eq. (1) then becomes

$$\frac{1}{c^2}\frac{\partial^2 z}{\partial t^2} = \frac{1}{r}\frac{\partial z}{\partial r} + \frac{\partial^2 z}{\partial r^2} + \frac{1}{r^2}\frac{\partial^2 z}{\partial \theta^2}. \tag{2}$$

(*b*) He guessed that the solution to (2) must have the form

$$z(t, r, \theta) = u(r)\sin(\alpha t + a)\sin(\beta\theta + b).$$

Show that u must satisfy the equation

$$u'' + \frac{1}{r}u' + \left(\frac{\alpha^2}{c^2} - \frac{\beta^2}{r^2}\right)u = 0,$$

later to be called Bessel's equation.

29 In the ideal gas $P = nRT/V$, where P is pressure, V is volume, T is temperature, R is a constant, and n is the number of moles, assumed constant.

(*a*) Compute $(\partial P/\partial V)_T$ and $(\partial P/\partial T)_V$.

(*b*) From the fact that there is a function $f(x, y, z)$ such that $f(P, V, T) = 0$, deduce that $(\partial T/\partial V)_P = -(\partial P/\partial V)_T/(\partial P/\partial T)_V$.

(*c*) Solve the equation $P = nRT/V$ for T and compute

$(\partial T/\partial V)_P$ directly. The result should agree with that found in (*b*).

30 The surfaces $2x^2 + 3y^2 + z^2 = 6$ and $x^3 + y^3 + z^3 = 3$ pass through the point (1, 1, 1). At what angle do they cross there?

In Exercises 31 to 37 compute Δf and df to three decimal places for the indicated f, x, y, Δx, and Δy.

31 $f(x, y) = \ln \dfrac{x^2 + y^2}{xy}$, $x = 1$, $y = 2$, $\Delta x = 0.01$, $\Delta y = 0.02$

32 $f(x, y) = e^{xy} \cos^{-1} x$, $x = 0$, $y = 2$, $\Delta x = 0.01$, $\Delta y = -0.01$

33 $f(x, y) = x \tan^{-1} xy$, $x = 1$, $y = 1$, $\Delta x = 0.3$, $\Delta y = 0.1$

34 $f(x, y) = x^3y^4$, $x = 1$, $y = 2$, $\Delta x = 0.1$, $\Delta y = -0.1$

35 $f(x, y) = \sqrt{x^2 + y^2}$, $x = 3$, $y = 4$, $\Delta x = 0.02$, $\Delta y = 0.03$

36 $f(x, y) = \ln(x + 2y)$, $x = 2$, $y = 3$, $\Delta x = -0.1$, $\Delta y = 0.2$

37 $f(x, y) = 3x - 4y + 2$, $x = 7$, $y = 9$, $\Delta x = 0.2$, $\Delta y = -0.3$

In Exercises 38 to 41 x is measured with a possible error of 3 percent and y with a possible error of 4 percent. Estimate the maximum possible percentage error in measuring the indicated quantity.

38 x^3y^2

39 x^3/y^2

40 x^5y

41 x^my^n (where m and n are constants)

42 Let $P_0 = (a, b, c)$ be a point not on the smooth curve given as the intersection of the surfaces $F(x, y, z) = 0$ and $G(x, y, z) = 0$. Let P_1 be a nearest point to P_0 on the curve. Prove that $\overrightarrow{P_0P_1}$ is perpendicular to the curve at P_1. (Use Lagrange multipliers.)

43 (*a*) Use a differential to estimate the volume of a rectangular box whose sides have lengths 1.03, 2.02, and 3.99 meters.
(*b*) Use a calculator to compute the volume.

44 (*a*) Use a differential to estimate the surface area of a rectangular box whose sides have lengths 4.01, 1.97, and 2.98 meters.
(*b*) Use a calculator to compute the area.

45 Two sides of a triangle are measured to be 8 centimeters and 11 centimeters and the included angle to be 45°. An error of 0.2 centimeter may be made in measuring length and an error of 3° in measuring angle.
(*a*) Use a differential to estimate the maximum possible error in the computed area of the triangle.
(*b*) Use a calculator to evaluate the maximum possible error.

46 Consider a curve given as the intersection of the surfaces $g(x, y, z) = 0$ and $h(x, y, z) = 0$. The function $u = f(x, y, z)$, considered only on this curve, has a maximum at (a, b, c). Why might we expect there to be, in general, scalars λ and μ such that

$$\nabla f = \lambda \nabla g + \mu \nabla h,$$

where the gradients are evaluated at (a, b, c)? (Include a sketch of the curve and the gradients.)

47 Find the point on the surface $z = x^2 + y^2$ closest to the plane $x + y - z = 1$, as follows:
(*a*) Sketch the surface and the plane.
(*b*) Prove that they do not meet.
(*c*) Find the point on the surface $z = x^2 + y^2$ where the normal is also perpendicular to the plane. Explain.

48 (*a*) Sketch the elliptic paraboloid $z = 2x^2 + 3y^2$.
(*b*) Sketch the line through (1, 0, 0) and (0, 1, 0).
(*c*) Sketch the planes through the points in (*b*) that are tangent to the paraboloid. (Estimate by eye.)
(*d*) Find where the planes in (*c*) touch the paraboloid.

49 Show that if $f(x, y)$ satisfies the equation

$$\frac{\partial^2 u}{\partial x^2} + \frac{\partial^2 u}{\partial y^2} = 0$$

so does the function

$$g(x, y) = f\left(\frac{x}{x^2 + y^2}, \frac{y}{x^2 + y^2}\right).$$

List the top, middle, and bottom variables you use in your calculations.

50 Figure 4 shows a smooth surface $\mathcal{S}$ and a point P outside $\mathcal{S}$. Show that if Q is the point on $\mathcal{S}$ closest to P, then $\overrightarrow{PQ}$ is perpendicular to $\mathcal{S}$ at the point Q. [Assume that $\mathcal{S}$ is the graph of the function $f(x, y, z) = 0$.]

Figure 4

51 Refer to Fig. 5.
(*a*) Estimate the gradient at P and explain your reasoning.
(*b*) Where is the gradient longer, at P or at Q?
(*c*) Estimate $(\partial f/\partial x)(0.02, 0.05)$.
(*d*) Estimate $D_{\mathbf{u}}f(0.02, 0.05)$ for $\mathbf{u} = \cos 20°\,\mathbf{i} + \sin 20°\,\mathbf{j}$.

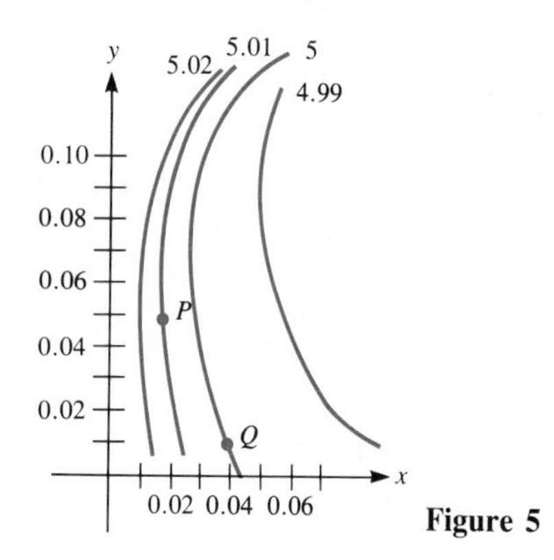

Figure 5

52 Where the level curves $f(x, y) = 1$ and $f(x, y) = 2$ are close to each other, would you expect the gradient ∇f to be short or long?

53 A cylindrical tank with hemispherical ends is to hold c cubic feet. (See Fig. 6.) What should the ratio between h and r be in order to minimize the surface area of the tank?
(*a*) Solve by Lagrange multipliers.
(*b*) Solve by implicit differentiation.
(*c*) Solve by the method of Sec. 4.7.

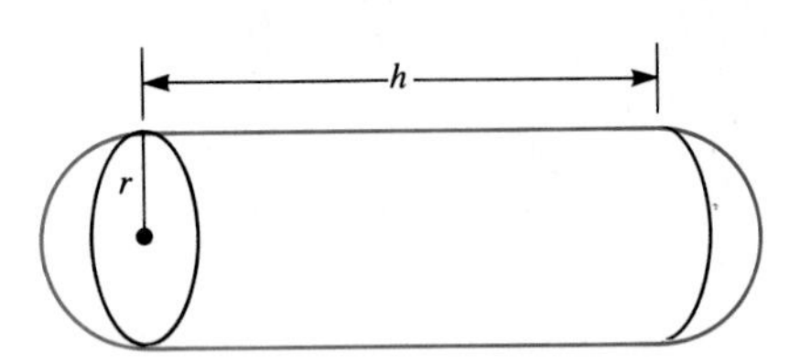

Figure 6

54 Find the maximum and minimum values of $x + 2y + 3z$ on the curve where the plane $x + y + 2z = 2$ meets the cylinder $x^2 + y^2 = 9$.

55 Figure 7 shows level curves of $f(x, y)$ and a curve C.
(*a*) Find the maximum and minimum values of f on C.
(*b*) At which points do they occur?

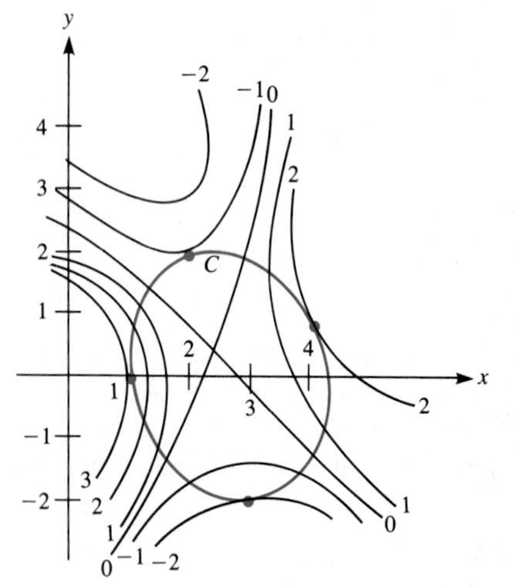

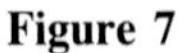
Figure 7

56 Let $f(x, y, z) = \int_x^y g(z, t)\, dt$.
(*a*) What is $\partial f/\partial y$?
(*b*) What is $\partial f/\partial x$?
(*c*) What is $\partial f/\partial z$? (See Appendix K.)

57 Let $f(x, y) = \int_{x+3}^{x+4} g(y, t)\, dt$. Find $\partial f/\partial x$.

58 Let z be a function of x and y and let $x = e^u$ and $y = e^v$. Show that

$$\frac{\partial^2 z}{\partial u^2} + \frac{\partial^2 z}{\partial v^2} = x^2\frac{\partial^2 z}{\partial x^2} + y^2\frac{\partial^2 z}{\partial y^2} + x\frac{\partial z}{\partial x} + y\frac{\partial z}{\partial y}.$$

59 Let $f(x, y)$ be defined for all (x, y). Assume that $\partial f/\partial x = 0$ for all points (x, y) and $\partial f/\partial y = 1$ for all points $(0, y)$, and $f(0, 0) = 0$. Find $f(x, y)$ and explain your reasoning.

60 Let r and θ be polar coordinates of the point (x, y) in the first quadrant. Then $x = r\cos\theta$, $y = r\sin\theta$, $\theta = \tan^{-1}(y/x)$, and $r = \sqrt{x^2 + y^2}$.
(*a*) Show that $\partial x/\partial\theta = -r\sin\theta$.
(*b*) Show that $\partial\theta/\partial x = -\dfrac{\sin\theta}{r}$.
(*c*) If $y = f(x)$ and $x = f^{-1}(y)$, functions of one variable, dy/dx is the reciprocal of dx/dy. Do (*a*) and (*b*) contradict this fact?

61 The wave equations for string waves, sound waves, and electromagnetic waves are all of the form

$$\frac{\partial^2}{\partial x^2}\psi(x, t) - \alpha\frac{\partial^2}{\partial t^2}\psi(x, t) = 0,$$

where α is a constant. Show that if $\alpha\omega^2 = k^2$, then the function $\psi(x, t) = A\cos(kx - \omega t) + B\sin(kx - \omega t)$ satisfies the equation. (k and ω are constants.)

62 Show that $z = \ln(1 + x^r + y^r)$, where r is constant, satisfies the partial differential equation

$$\frac{\partial^2 f}{\partial x\,\partial y} + \frac{\partial f}{\partial x}\frac{\partial f}{\partial y} = 0.$$

63 (*a*) Explain why ∇F is perpendicular to the level surface $F(x, y, z) = 0$.
(*b*) Explain why ∇f is perpendicular to the level curve $f(x, y) = 0$.

64 Find a tangent vector to the curve of intersection of the surfaces $x^2 + 2y^2 + 3z^2 = 36$ and $2x^2 - y^2 + z^2 = 7$ at the point $(1, 2, 3)$.

65 (*a*) Sketch the level curve of the function $5x^2 + 3y^2$ that passes through $(1, 1)$.
(*b*) Do the same for the function y^5/x^3.
(*c*) Show that the two curves cross at a right angle at the point $(1, 1)$.
(*d*) Prove that each level curve of $5x^2 + 3y^2$ crosses each level curve of y^5/x^3 at a right angle.

66 Let $\mathbf{G}(s)$, where s denotes arc length, parameterize a curve in space. Let $u = F(x, y, z)$ be a function of x, y, and z. Then u may be considered a composite function of s. Show that (*a*) $du/ds = \nabla F \cdot \mathbf{G}'(s)$, where ∇F is evaluated at $\mathbf{G}(s)$. (*b*) $du/ds = D_{\mathbf{T}}(u)$, where $\mathbf{T}$ is a unit tangent vector to the curve at $\mathbf{G}(s)$. [Recall that $D_{\mathbf{T}}(u)$ is the directional derivative along $\mathbf{T}$.]

67 Sketch the surfaces:
(*a*) $\dfrac{x^2}{4} + \dfrac{y^2}{4} - z^2 = 1$ (*b*) $x^2 - 4y^2 - 9z^2 = 36$
(*c*) $z^2 = x^2 + 2y^2$

68 The pressure P, volume V, and temperature T of a gas are related by the equation $(P + a/V^2)(V - b) = cT$, where a, b, and c are constants. Thus any two of P, V, and T determine the third.
(*a*) Compute $\partial V/\partial T$, $\partial T/\partial P$, and $\partial P/\partial V$.
(*b*) Show that the product of the three partial derivatives in (*a*) is -1.

69 Consider Euler's partial differential equation

$$a\frac{\partial^2 z}{\partial x^2} + 2b\frac{\partial^2 z}{\partial x \partial y} + c\frac{\partial^2 z}{\partial y^2} = 0,$$

where a, b, and c are constants and $b^2 \neq ac$. Show that

$$z = f(x + r_1 y) + g(x + r_2 y),$$

where r_1 and r_2 are the roots of $a + 2bx + cx^2 = 0$, is a solution of the differential equation. The functions f and g are differentiable.

70 Determine the minimum value of the function $f(x, y) = x^4 - x^2y^2 + y^4$.

71 If $u = x^4 f(y/x, z/x)$ show that $x(\partial u/\partial x) + y(\partial u/\partial y) + z(\partial u/\partial z) = 4u$.

72 A wire of length 1 is to be cut into three pieces which will be bent into a square, a circle, and an equilateral triangle. How should this be done to (*a*) minimize their total area? (*b*) maximize their total area?

73 Find the maximum and the minimum of $f(x, y) = 4x^2 - 3y^2 + 2xy$ on the square $0 \le x \le 1$, $0 \le y \le 1$.

74 (*a*) Show that if you rotate the xy plane clockwise by the angle θ, the point (x, y) moves to the point $(x \cos\theta - y \sin\theta, x \sin\theta + y\cos\theta)$. [Either use Appendix G or apply complex numbers, multiplying out $(\cos\theta + i\sin\theta)(x + iy)$.]
(*b*) The line through $(\frac{1}{2}, 0, 0)$ and $(\frac{1}{2}, 1, 2)$ is revolved about the z axis. Find the equation of the surface produced.
(*c*) What kind of surface is it?

75 Though defined at first algebraically, the gradient turned out to be an intrinsic geometric property of the function, for its direction and magnitude are expressible in terms of the function, without reference to any particular coordinate system. Consider, then, a function $f(P)$, where P runs over a portion of the plane. If f is given in rectangular coordinates, then

$$\nabla f = \frac{\partial f}{\partial x}\mathbf{i} + \frac{\partial f}{\partial y}\mathbf{j}.$$

What if f is described in polar coordinates, $f(r, \theta)$? How would ∇f be expressed? The natural unit vectors to use, instead of $\mathbf{i}$ and $\mathbf{j}$, are $\mathbf{u}_r$ and $\mathbf{u}_\theta$ shown in Fig. 8. (In physics $\mathbf{u}_\theta$ is denoted $\hat{\boldsymbol{\theta}}$ and $\mathbf{u}_r$, $\hat{\mathbf{r}}$.)
(*a*) If $\nabla f = A\mathbf{u}_r + B\mathbf{u}_\theta$ for scalars A and B, show that $A = D_{\mathbf{u}_r}(f)$ and $B = D_{\mathbf{u}_\theta}(f)$.
(*b*) Show that $D_{\mathbf{u}_r}(f) = \partial f/\partial r$.
(*c*) Why would $D_{\mathbf{u}_\theta}(f) = \dfrac{1}{r}\dfrac{\partial f}{\partial \theta}$? $\left(\text{Not } \dfrac{\partial f}{\partial \theta}!\right)$
Give a persuasive argument, not necessarily a rigorous proof.
(*d*) From (*a*), (*b*), and (*c*) deduce that

$$\nabla f = \frac{\partial f}{\partial r}\mathbf{u}_r + \frac{1}{r}\frac{\partial f}{\partial \theta}\mathbf{u}_\theta.$$

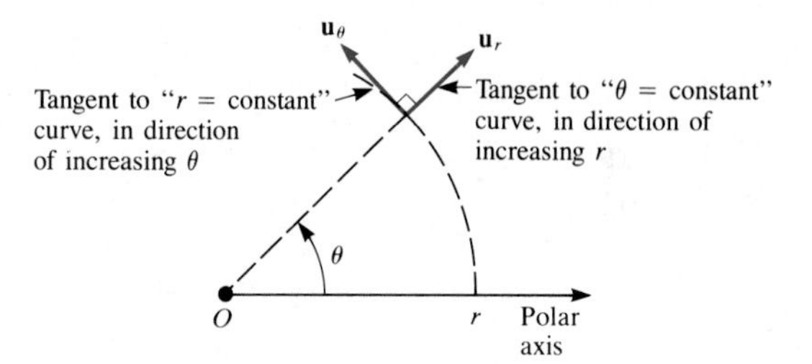

Figure 8

76 (See Exercise 75.) Let $f(P)$ be the reciprocal of the distance from P in the plane to the origin.
(*a*) In rectangular coordinates f has the formula $f(x, y) = 1/\sqrt{x^2 + y^2}$. Calculate ∇f using rectangular coordinates.
(*b*) In polar coordinates f has the formula $f(r, \theta) = 1/r$. Calculate ∇f using polar coordinates.
(*c*) Sketch ∇f as calculated in (*a*) and in (*b*). Show that the results are the same.

77 Solve Exercise 75 by expressing $\partial f/\partial x\mathbf{i} + \partial f/\partial y\mathbf{j}$ in terms of $\partial f/\partial\theta$, $\partial f/\partial r$, $\mathbf{u}_r$, and $\mathbf{u}_\theta$. That is, express $\mathbf{i}$ and $\mathbf{j}$ in terms of $\mathbf{u}_r$ and $\mathbf{u}_\theta$ and express $\partial f/\partial x$ and $\partial f/\partial y$ in terms of $\partial f/\partial r$ and $\partial f/\partial\theta$. For convenience, assume that (x, y) is in the first quadrant.

15

DEFINITE INTEGRALS OVER PLANE AND SOLID REGIONS

The two basic concepts in calculus are the derivative and the definite integral. Chapter 14 generalized the derivative to functions of more than one variable. The present chapter generalizes the definite integral.

In Chap. 5 the definite integral of $f(x)$ over an interval $[a, b]$, $\int_a^b f(x)\,dx$, was defined as follows. First partition the interval $[a, b]$ into n shorter intervals, select a sampling point c_i in each of the intervals, and form the sum $\Sigma_{i=1}^n f(c_i)L_i$ where L_i is the length of the ith interval. (See the figure below.)

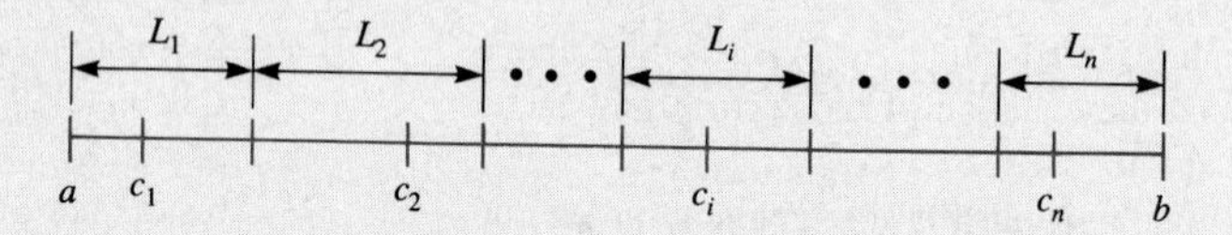

(We use L_i instead of $x_i - x_{i-1}$, used in Chap. 5, to emphasize the similarity of this definition with the definitions to be made in this chapter.) The limit of such sums as the length of all the intervals approaches 0 is called "the definite integral of f over $[a, b]$":

$$\int_a^b f(x)\,dx = \lim_{\text{mesh}\to 0} \sum_{i=1}^{n} f(c_i)L_i.$$

In this chapter we will first work with functions defined on a plane region, such as a disk, rectangle, or triangle. Instead of chopping up an interval we will be chopping up such a region and forming sums similar to those used to define $\int_a^b f(x)\,dx$. This will be done in Secs. 15.1 to 15.4.

Sections 15.5 to 15.7 develop and apply the notion of a definite integral over a spatial region in a similar manner.

It turns out, as we will see, that these so-called "multiple integrals" can often be evaluated by integrating two or three times over intervals.

15.1 THE DEFINITE INTEGRAL OF A FUNCTION OVER A REGION IN THE PLANE

Two problems will introduce the definite integral of a function over a region in the plane.

A region in the plane shall be a set of points in the plane enclosed by curves or polygons. When a region is partitioned into subsets, the smaller regions shall also be bounded by curves or polygons.

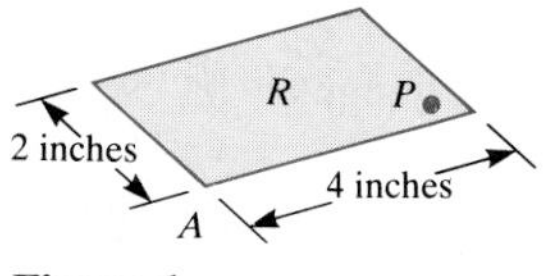

Figure 1

A Volume Problem

PROBLEM 1 Estimate the volume of the solid S which we now describe. Above each point P in a 4-inch by 2-inch rectangle R erect a line segment whose length, in inches, is the square of the distance from P to the corner A. (R is shown in Fig. 1.) These segments form a solid S which is shown in Fig. 2. Note that the highest point of S is above the corner of R opposite A; there its height, by the pythagorean theorem, is $4^2 + 2^2 = 20$ inches.

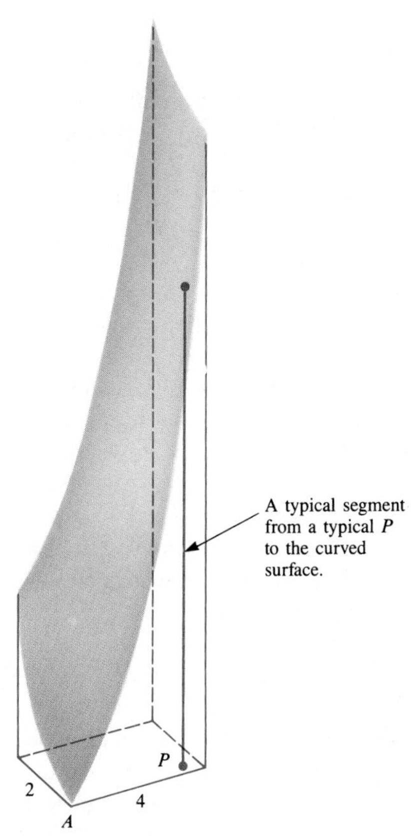

Figure 2

APPROACH Observe that the volume of S is certainly less than $4 \cdot 2 \cdot 20 = 160$ cubic inches, since S can be put into a box whose base has area $4 \cdot 2$ square inches and whose height is 20 inches.

In order to make more accurate estimates, cut the rectangular base into smaller pieces. For convenience, cut it into four congruent rectangles R_1, R_2, R_3, and R_4, as in Fig. 3. To estimate the volume of S, estimate the volume of that portion of S above each of the rectangles R_1, R_2, R_3, and R_4, and add these estimates. To do this, select a point in each of the four rectangles, say, the center of each, and above each rectangle form a box whose height is the height of S above the center of the corresponding rectangle. (See Figs. 4 and 5.)

For instance, the height of the box above R_1 is $1^2 + (\frac{3}{2})^2 = \frac{13}{4}$ inches, and the height of the box above R_2 is $1^2 + (\frac{1}{2})^2 = \frac{5}{4}$ inches. Adding the volumes of the four boxes, we obtain the following estimate of the volume of S:

$$\left(\tfrac{13}{4}\right)2 + \left(\tfrac{5}{4}\right)2 + \left(\tfrac{45}{4}\right)2 + \left(\tfrac{37}{4}\right)2 = 50 \text{ cubic inches.}$$

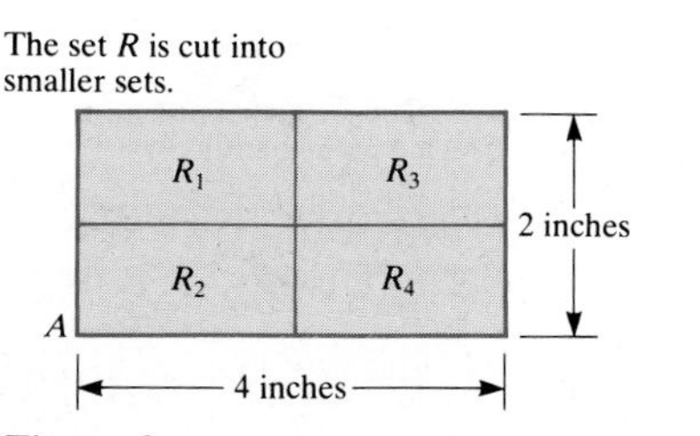

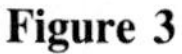
Figure 3

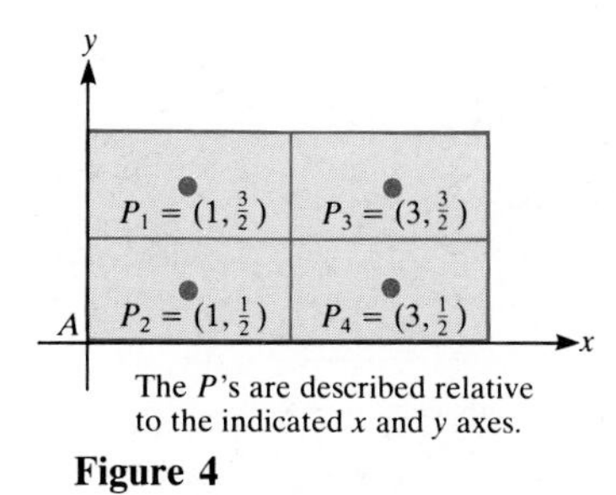

Figure 4

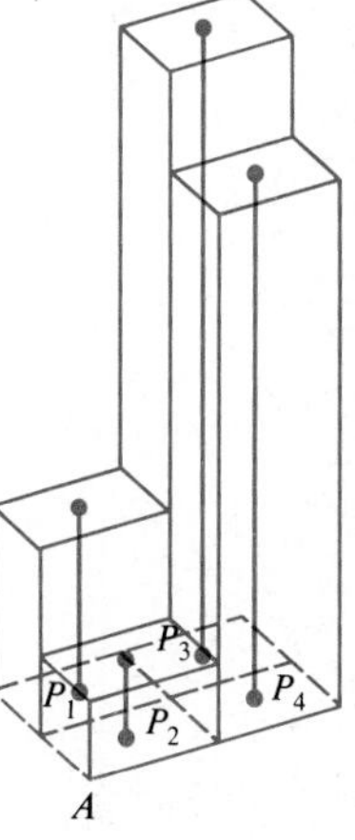

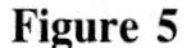
Figure 5

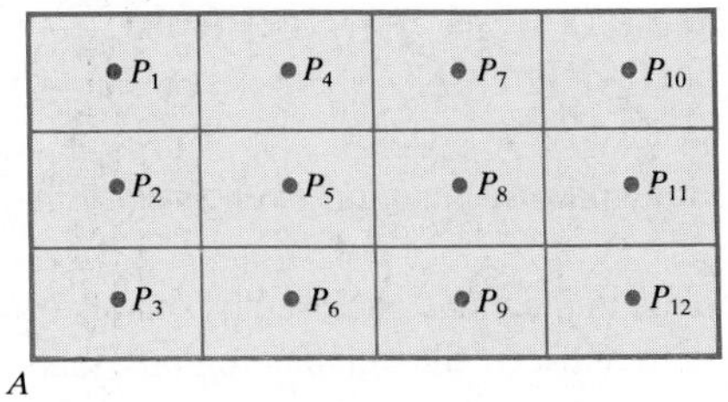

Figure 6

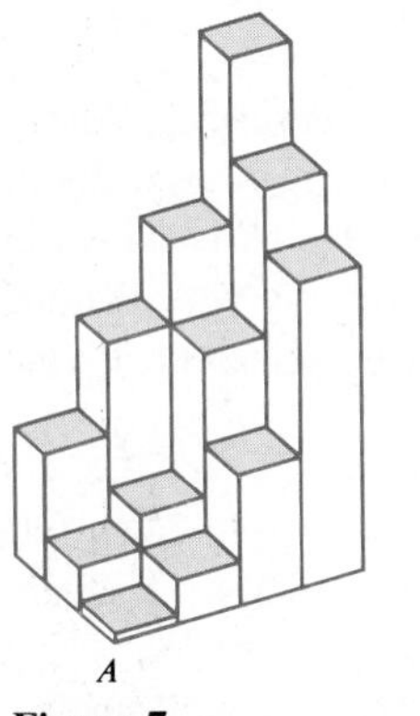

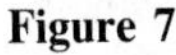

Figure 7

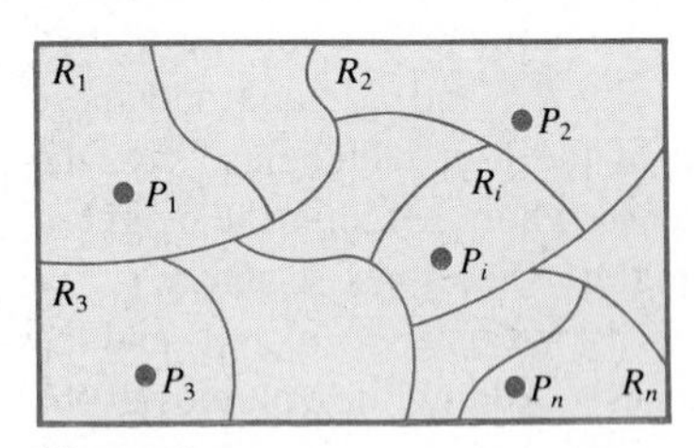

Figure 8

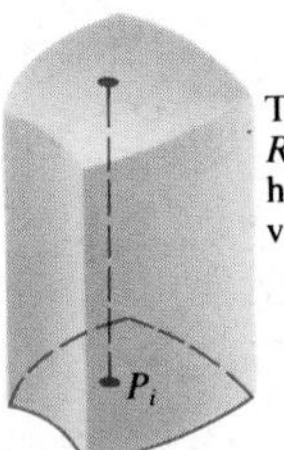

The cylinder has base R_i with area A_i, height $c(P_i)$, and volume $c(P_i)A_i$.

Figure 9

This is only an estimate. With the same partition of R we could make other estimates by choosing other P's to determine the heights of approximating boxes.

Cutting the rectangle R into 12 rectangles as shown in Fig. 6 and again using the centers as the "sampling points," we obtain a presumably more accurate estimate. The result this time is $\frac{1414}{27} \approx 52.4$.

The approximating 12 boxes, solid cylinders with rectangular bases, are shown in Fig. 7 (at a reduced scale).

In general, to estimate the volume of S, begin by partitioning R into smaller subsets $R_1, R_2, \ldots, R_n$ and selecting points P_1 in R_1, P_2 in $R_2, \ldots, P_n$ in R_n. (See Fig. 8.)

Denote the height of S above a typical point P_i in R_i by $c(P_i)$, and the area of R_i by A_i. Then

$$c(P_1)A_1 + c(P_2)A_2 + \cdots + c(P_n)A_n$$

is an estimate of the volume of S by a sum of the volumes of n solid cylinders, one cylinder above each of the small regions $R_1, R_2, \ldots, R_n$. A typical one may look like the one in Fig. 9. ■

A Mass Problem

PROBLEM 2 Estimate the mass of the rectangular sheet R described as follows. Its dimensions are 4 centimeters by 2 centimeters. The material is sparse near the corner A and dense far from A. Indeed, assume that the density in the vicinity of any point P is numerically equal to the square of the distance from P to A (in grams per square centimeter). See Fig. 10. Note that it is densest at the corner opposite A, where its density is $4^2 + 2^2 = 20$ grams per square centimeter.

APPROACH To begin, observe that the total mass is certainly less than $4 \cdot 2 \cdot 20 = 160$ grams, since the area of R is $4 \cdot 2$ square centimeters and the maximum density is 20 grams per square centimeter.

In order to make more accurate estimates, cut the rectangle into smaller pieces, to be specific, into the four congruent rectangles shown in Fig. 11. Estimate the total mass by estimating the mass in each of the rectangles R_1, R_2, R_3, and R_4. To do this, select a point, say, the center, in each of the four rectangles and compute the density at each of these four points. (See Fig. 12.) The density at P_1 is $1^2 + (\frac{3}{2})^2 = \frac{13}{4}$ grams per square centimeter. An estimate of the mass in R_1 is $\frac{13}{4} \cdot 2$ grams, since the area of R_1 is 2 square centimeters. The sum of the estimates for each of the four rectangles,

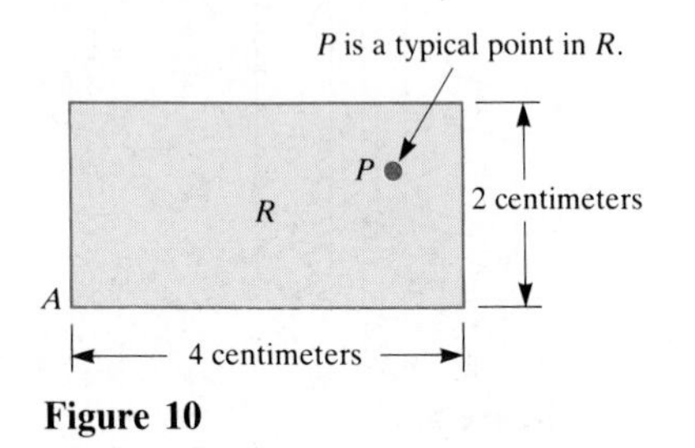

Figure 10

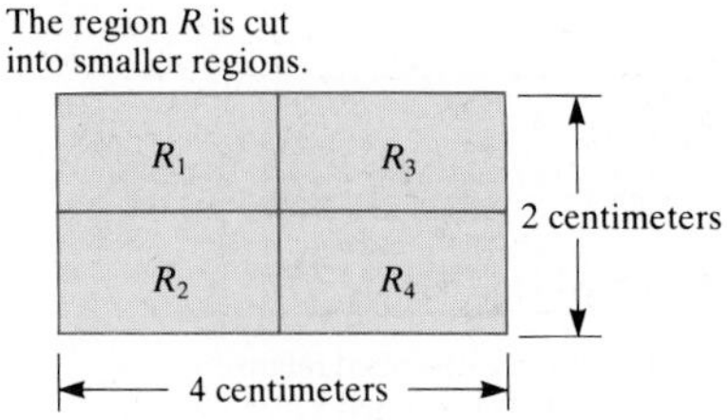

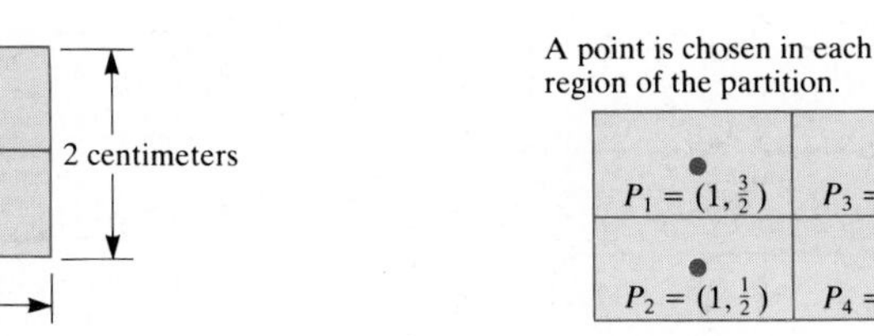

Figure 11

Figure 12

$$(\tfrac{13}{4})2 + (\tfrac{5}{4})2 + (\tfrac{45}{4})2 + (\tfrac{37}{4})2 = 50 \text{ grams},$$

is an estimate of the total mass in R.

This is only an estimate. Just as for the volume in Problem 1 other estimates can be made in the same way. To do so, partition R into small subsets R_1, R_2, . . . , R_n and select points P_1 in R_1, P_2 in R_2, . . . , P_n in R_n. Denote the density at P_i by $f(P_i)$ and the area of R_i by A_i. Then

$$f(P_1)A_1 + f(P_2)A_2 + \cdots + f(P_n)A_n$$

is an estimate of the total mass. ■

Although only estimates have been made, they show the similarity of the two problems.

Even though we have found neither the volume in Problem 1 nor the mass in Problem 2, it is clear that if we know the answer to one, we have the answer to the other: The arithmetic for calculating any estimate for the volume is the same as that for an estimate of the mass.

The similarity of the sums formed for both problems to the sums met in Chap. 5 suggests that the idea of the definite integral can be generalized from intervals $[a, b]$ to regions in the plane. First, in order to speak of fine partitions on the plane, we need two definitions.

Definition *Diameter of a region.* Let S be a region in the plane bounded by a curve or polygon. The **diameter** of S is the largest distance between two points of S.

Note that the diameter of a square of side s is $s\sqrt{2}$, and the diameter of a circle whose radius is r is $2r$, its usual diameter.

Definition *Mesh of a partition in the plane.* Let R_1, R_2, . . . , R_n be a partition of a region R in the plane into smaller regions. The **mesh** of this partition is the largest of the diameters of the regions R_1, R_2, . . . , R_n.

For example, the mesh of the partition used in both Problems 1 and 2 is $\sqrt{5}$. Now the definite integral over a plane region can be defined.

Definition *Definite integral of a function f over a region R in the plane.* Let f be a function that assigns to each point P in a region R in the plane a number $f(P)$. Consider the typical sum

$$f(P_1)A_1 + f(P_2)A_2 + \cdots + f(P_n)A_n$$

formed from a partition of R, where A_i is the area of R_i, and P_i is in R_i. If these sums approach a certain number as the mesh of the partitions shrinks toward 0 (no matter how P_i is chosen in R_i), that number is called the **definite integral** of f over the set R and is written

$$\int_R f(P)\, dA.$$

That is, in shorthand

$$\lim_{\text{mesh}\to 0} \sum_{i=1}^{n} f(P_i)A_i = \int_R f(P)\, dA.$$

It is illuminating to compare this definition with the one for the definite integral over an interval. Both are numbers that are approached by certain sums of products. The sums are formed in a similar manner, as the following table shows:

This table contrasts integration over an interval with integration over a planar region.

Given	*For each subset in a partition compute*	*Select in each of the subsets*	*Take the limit of sums of the form*
An interval and a function defined there	Its length L_i	A point (described by its coordinate c_i)	$\sum_{i=1}^{n} f(c_i)L_i$
A set in the plane and a function defined there	Its area A_i	A point P_i	$\sum_{i=1}^{n} f(P_i)A_i$

The definite integral is not defined as a sum formed in this table, but rather as the number approached by these sums when the mesh approaches 0. The definite integral of f over R is sometimes called the "integral of f over R" or the "integral of $f(P)$ over R."

It is proved in advanced calculus that if $f(P)$ is continuous and the region R is bounded by curves and lines, and lies within some disk, then $\int_R f(P)\, dA$ exists.

For example, the volume in Problem 1 and the mass in Problem 2 are both given by the definite integral

$$\int_R (x^2 + y^2)\, dA,$$

where R is the rectangle that has vertices (0, 0), (4, 0), (0, 2), and (4, 2).

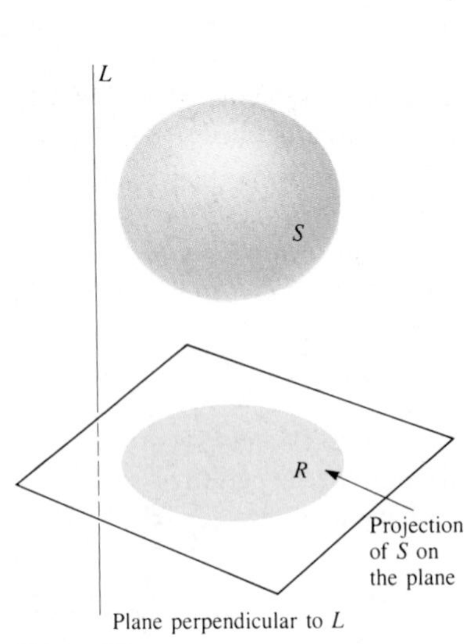

Figure 13

Applications of $\int_R f(P)\, dA$

The two illustrations given of the definite integral over a plane region are quite important, and we emphasize them by stating them in full generality.

Volume of a Solid Expressed as an Integral over a Plane Region Consider a solid set S and pick a line L in space. Assume that all lines parallel to L that meet S intersect S in a line segment or a point. Pick a plane perpendicular to L. Let R be the projection of S on that plane, that is, the set of all points where lines parallel to L that meet S intersect the plane. For each point P in R let $c(P)$ be the length of the intersection with S of the line through P parallel to L. (See Figs. 13 and 14.) Partition R into smaller regions $R_1, R_2, \ldots, R_n$ and pick a sampling

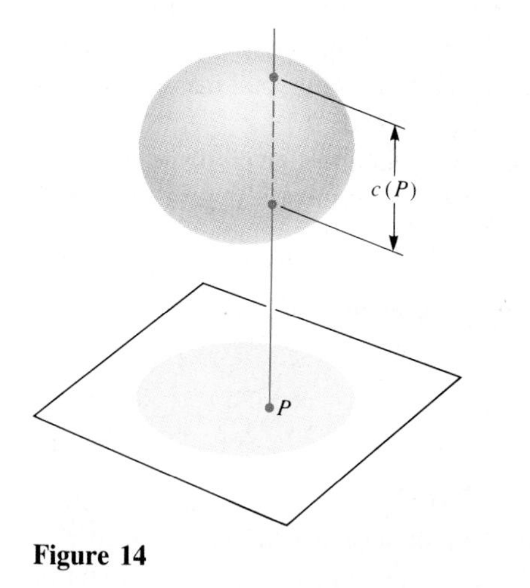

Figure 14

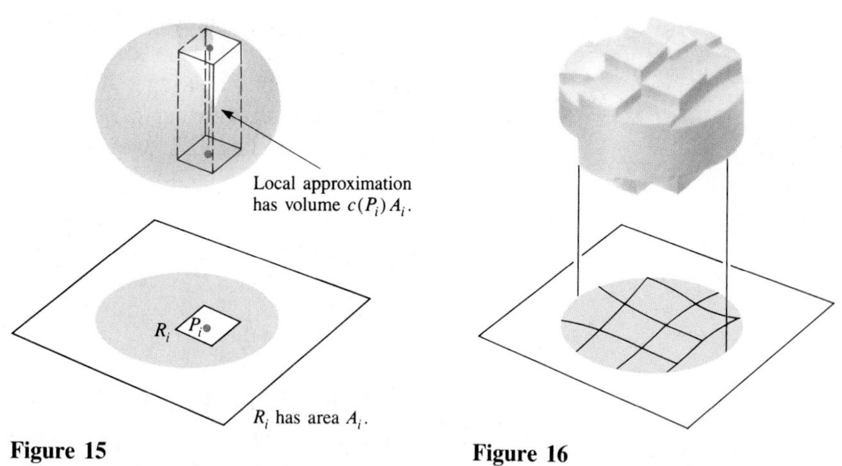

Figure 15

Figure 16

point P_i in R_i for each $i = 1, 2, \ldots, n$. Let the area of R_i be A_i. Then approximate the volume of S above R_i by a cylinder of height $c(P_i)$ and base congruent to R_i, as shown in Fig. 15. The volume of the ith cylinder is $c(P_i)A_i$. Hence

$$\sum_{i=1}^{n} c(P_i)A_i$$

is an estimate of the volume of S. (See Fig. 16.) Thus

$$\text{Volume of } S = \int_R c(P)\, dA.$$

Mass of a Flat Region (Lamina) Expressed as an Integral over a Plane Region *R* Consider a plane distribution of mass through a region R, as shown in Fig. 17. The density may vary throughout the region. Denote the density at P by $\sigma(P)$ (in grams per square centimeter). To estimate the total mass in the region R, partition R into small regions $R_1, R_2, \ldots, R_n$ and pick a sampling point P_i in R_i for each $i = 1, 2, \ldots, n$. Then the mass in R_i is approximately $\sigma(P_i)A_i$, since density times area gives mass if the density is constant. Thus

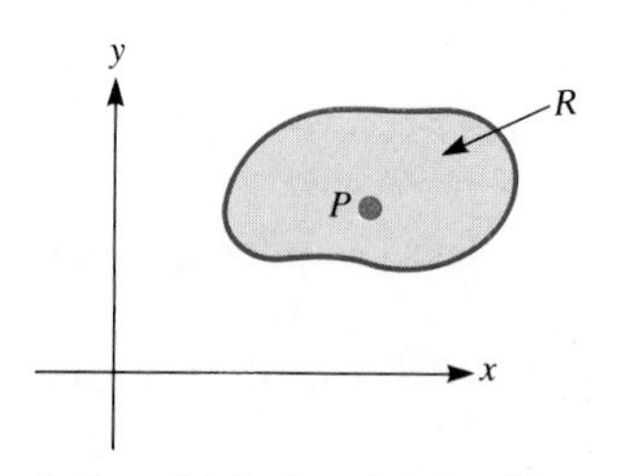

A plane distribution of matter. The density at P is $\sigma(P)$ (grams per square centimeter).

Figure 17

$$\sum_{i=1}^{n} \sigma(P_i)A_i$$

is an estimate of the mass in R. Consequently,

σ is a standard symbol for density in the plane.

$$\text{Mass in } R = \int_R \sigma(P)\, dA.$$

For engineers and physicists this is an important interpretation of the two-dimensional integral. The definite integral of density equals mass.

The definite integral of density over an interval gives the total mass for matter distributed along a string. (This was shown in Chap. 5.)

To emphasize further the similarity between integrals over plane sets and integrals over intervals, we define the average of a function over a plane set.

Definition *Average value.* The **average value** of f over the region R is

$$\frac{\int_R f(P)\, dA}{\text{Area of } R}.$$

If $f(P)$ is positive for all P in R, there is a simple geometric interpretation of the average of f over R. Let S be the solid situated below the graph of f (a surface) and above the region R. The average value of f over R is the height of the cylinder whose base is R and whose volume is the same as the volume of S. (See Fig. 18.)

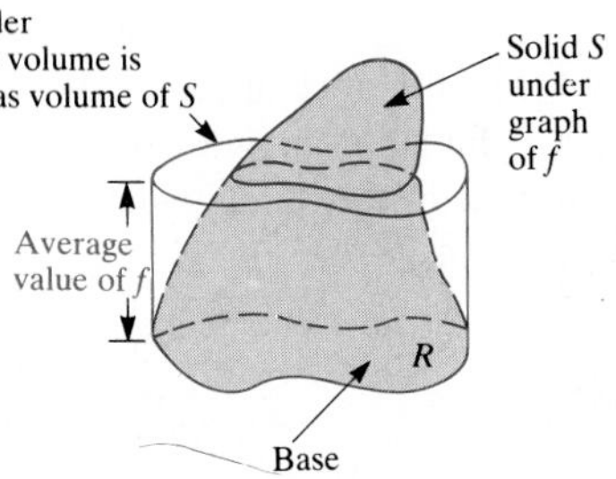

Figure 18

The integral $\int_R f(P)\, dA$ is called "an integral over a plane region" to distinguish it from $\int_a^b f(x)\, dx$, which, for contrast, is called "an integral over an interval."

Often $\int_R f(P)\, dA$ is denoted $\iint_R f(P)\, dA$, with the two integral signs emphasizing that the integral is over a plane set. However, the symbol dA, which calls to mind areas, is an adequate reminder.

The integral of the function $f(P) = 1$ over a region is of special interest. The typical approximating sum $\sum_{i=1}^n f(P_i)A_i$ then equals $\sum_{i=1}^n 1 \cdot A_i = A_1 + A_2 + \cdots + A_n$, which is the area of the region R that is being partitioned. Since *every* approximating sum has this same value, it follows that

$$\lim_{\text{mesh}\to 0} \sum_{i=1}^{n} f(P_i)A_i = \text{Area of } R.$$

Consequently

The integral of the constant function, 1, gives area.

$$\boxed{\int_R 1\, dA = \text{Area of } R.}$$

This formula will come in handy on several occasions. The 1 is often omitted, in which case we write $\int_R dA = \text{Area of } R$.

This table summarizes some of the main applications of the integral $\int_R f(P)\, dA$:

Integral	*Interpretation*
$\int_R 1\, dA$	Area of R
$\int_R \sigma(P)\, dA,\quad \sigma(P) = \text{density}$	Mass of R
$\int_R c(P)\, dA,\quad c(P) = \text{length of cross section of solid}$	Volume of the solid

Properties of Integrals Integrals over plane regions have properties similar to those of integrals over intervals:

1 $\int_R cf(P)\,dA = c\int_R f(P)\,dA$ for any constant c.
2 $\int_R [f(P) + g(P)]\,dA = \int_R f(P)\,dA + \int_R g(P)\,dA$.
3 If $f(P) \le g(P)$ for all points P in R, then $\int_R f(P)\,dA \le \int_R g(P)\,dA$.
4 If R is broken into two regions, R_1 and R_2, overlapping at most on their boundaries, then

$$\int_R f(P)\,dA = \int_{R_1} f(P)\,dA + \int_{R_2} f(P)\,dA.$$

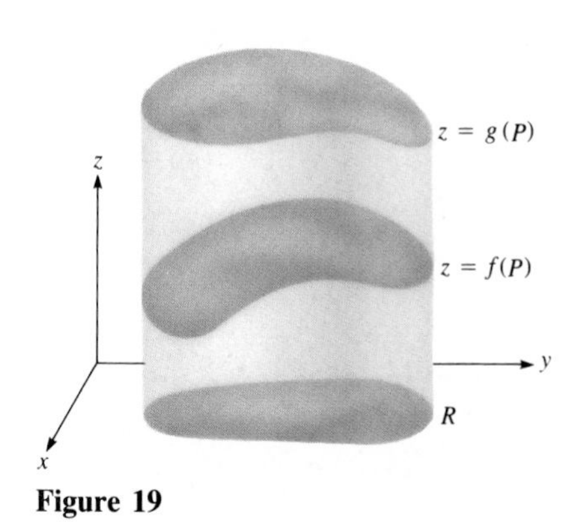

Figure 19

For instance, consider (3) when $f(P)$ and $g(P)$ are positive. Then $\int_R f(P)\,dA$ is the volume under the surface $z = f(P)$ and above R in the xy plane. Similarly $\int_R g(P)\,dA$ is the volume under $z = g(P)$ and above R. Then (3) asserts that the volume of a solid is not larger than the volume of a solid that contains it. (See Fig. 19.)

EXERCISES FOR SEC. 15.1: THE DEFINITE INTEGRAL OF A FUNCTION OVER A REGION IN THE PLANE

1 In the estimates for the volume in Problem 1, the centers of the subrectangles were used as the P_i's. Make an estimate for the volume in Problem 1 by using the same partition but taking as P_i (*a*) the lower left corner of each R_i, (*b*) the upper right corner of each R_i. (*c*) What do (*a*) and (*b*) tell about the volume of the solid?

2 Estimate the mass in Problem 2 using a partition of R into eight congruent squares and taking as the P_i's (*a*) centers, (*b*) upper right corners, (*c*) lower left corners.

3 Let R be a set in the plane whose area is A. Let f be the function such that $f(P) = 5$ for every point P in R.
(*a*) What can be said about any approximating sum $\Sigma_{i=1}^{n} f(P_i)A_i$ formed for this R and this f?
(*b*) What is the value of $\int_R f(P)\,dA$?

4 Let R be the square with vertices (1, 1), (5, 1), (5, 5), and (1, 5). Let $f(P)$ be the distance from P to the y axis.
(*a*) Estimate $\int_R f(P)\,dA$ by partitioning R into four squares and using midpoints as sampling points.
(*b*) Show that $16 \le \int_R f(P)\,dA \le 80$.

5 (*a*) Let f and R be as in Problem 2. Use the estimate of $\int_R f(P)\,dA$ obtained in the text to estimate the average of f over R.
(*b*) Using the information from Exercise 2, show that the average is between 4 and 10.

6 Assume that for all P in R, $m \le f(P) \le M$, where m and M are constants. Let A be the area of R. By examining approximating sums, show that

$$mA \le \int_R f(P)\,dA \le MA.$$

7 (*a*) Let R be the rectangle with vertices (0, 0), (2, 0), (2, 3), and (0, 3). Let $f(x, y) = \sqrt{x + y}$. Estimate $\int_R \sqrt{x + y}\,dA$ by partitioning R into six squares and choosing the sampling points to be their centers.
(*b*) Use (*a*) to estimate the average value of f over R.

8 (*a*) Let R be the square with vertices (0, 0), (0.8, 0), (0.8, 0.8), and (0, 0.8). Let $f(P) = f(x, y) = e^{xy}$. Estimate $\int_R e^{xy}\,dA$ by partitioning R into 16 squares and choosing the sampling points to be their centers.
(*b*) Use (*a*) to estimate the average value of $f(P)$ over R.
(*c*) Show that $0.64 \le \int_R f(P)\,dA \le 0.64e^{0.64}$.

9 (*a*) Let R be the triangle with vertices (0, 0), (4, 0), and (0, 4) shown in Fig. 20. Let $f(x, y) = x^2y$. Use the partition into four triangles and sampling points shown in the diagram to estimate $\int_R f(P)\,dA$.
(*b*) What is the maximum value of $f(x, y)$ in R?
(*c*) From (*b*) obtain an upper bound on $\int_R f(P)\,dA$.

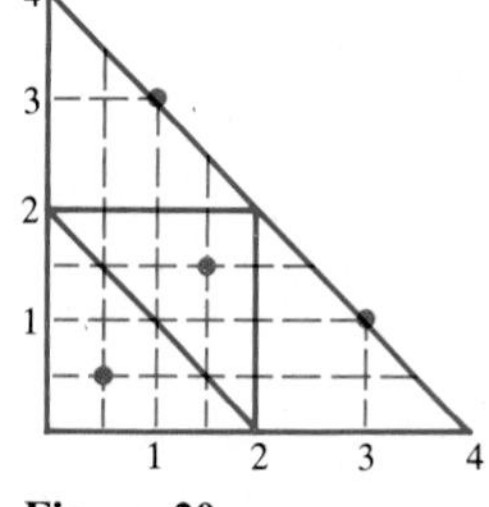

Figure 20

10 (*a*) Sketch the surface $z = \sqrt{x^2 + y^2}$.
(*b*) Let $\mathcal{V}$ be the region in space below the surface in (*a*) and above the square R with vertices (0, 0), (1, 0), (1, 1), and (0, 1). Let V be the volume of $\mathcal{V}$. Show that $V \le \sqrt{2}$.

(*c*) Using a partition of R with 16 squares, find an estimate for V that is too large.

(*d*) Using the partition in (*c*), find an estimate for V that is too small.

11 The amount of rain that falls at point P during 1 year is $f(P)$ inches. Let R be some geographic region, and assume areas are measured in square inches.

(*a*) What is the meaning of $\int_R f(P)\,dA$?

(*b*) What is the meaning of

$$\frac{\int_R f(P)\,dA}{\text{Area of } R}?$$

12 A region R in the plane is divided into two regions R_1 and R_2. The function $f(P)$ is defined throughout R. Assume that you know the areas of R_1 and R_2 (they are A_1 and A_2) and the average of f over R_1 and the average of f over R_2 (they are f_1 and f_2). Find the average of f over R. (See Fig. 21.)

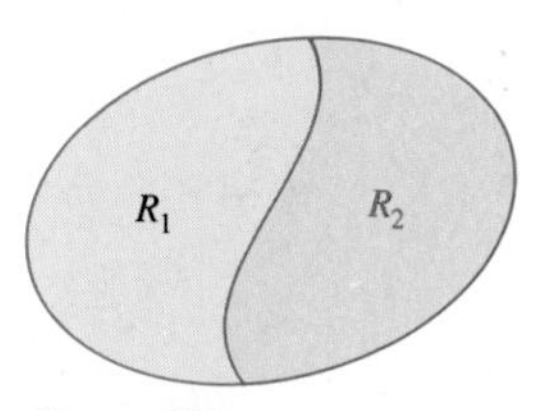

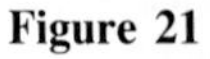
Figure 21

13 (*a*) Let R be a disk of radius 1. Let $f(P)$, for P in R, be the distance from P to the center of the disk. By cutting R into narrow circular rings with center at the center of the disk, evaluate $\int_R f(P)\,dA$.

(*b*) Find the average of $f(P)$ over R.

14 A point Q on the xy plane is at a distance b from the center of a disk R of radius a ($a < b$) in the xy plane. For P in R let $f(P) = 1/\overline{PQ}$. Find positive numbers c and d such that:

$$c < \int_R f(P)\,dA < d.$$

(The numbers c and d depend on a and b.) See Fig. 22.

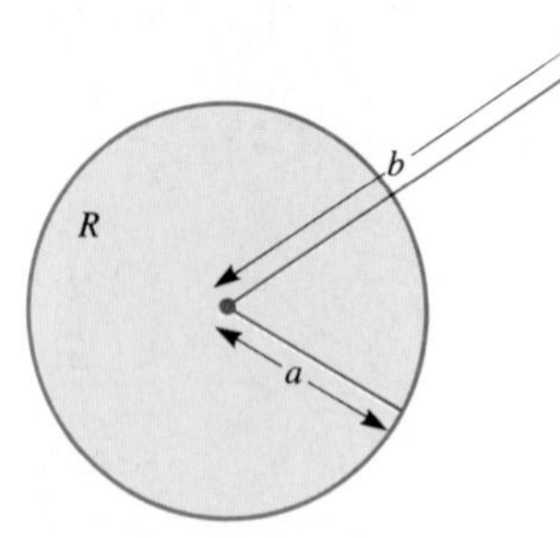

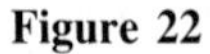
Figure 22

15 Figure 23 shows the parts of some level curves of a function $z = f(x, y)$ and a square R. Estimate $\int_R f(P)\,dA$, and describe your reasoning.

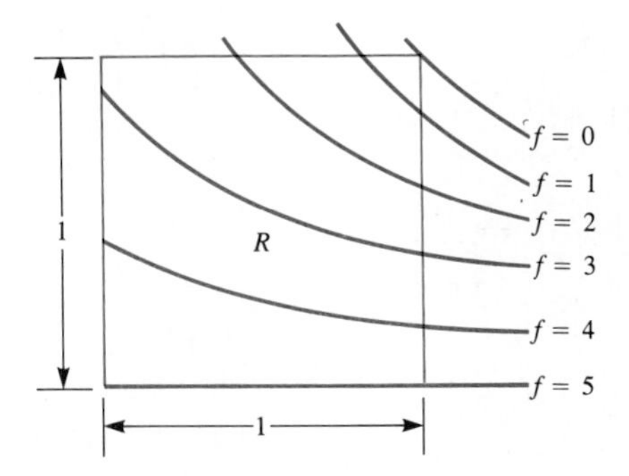

Figure 23

16 This exercise involves estimating an integral by choosing points randomly. A computing machine can be used to generate random numbers and thus random points in the plane which can be used to estimate definite integrals, as we now show. Say that a complicated region R lies in the square whose vertices are (0, 0), (2, 0), (2, 2), and (0, 2), and a complicated function f is defined in R. The machine generates 100 random points (x, y) in the square. Of these, 73 lie in R. The average value of f for these 73 points is 2.31.

(*a*) What is a reasonable estimate of the area of R?

(*b*) What is a reasonable estimate of $\int_R f(P)\,dA$?

Techniques such as this one which utilize randomness are called **Monte Carlo methods**. These methods are not very efficient, since the error decreases on the order of $1/\sqrt{n}$, where n is the number of random points. That is a slow rate.

17 (This exercise illustrates the Monte Carlo method described in Exercise 16.) Let R be the disk bounded by the unit circle $x^2 + y^2 = 1$ in the xy plane. Let $f(x, y) = e^{x^2y}$ be the temperature at (x, y). (*a*) Estimate the average value of f over R by evaluating $f(x, y)$ at twenty random points in R. (Adjust your program to select each of x and y randomly in the interval $[-1, 1]$. In this way you construct a random point (x, y) in the square whose vertices are $(1, 1)$, $(-1, 1)$, $(-1, -1)$, $(1, -1)$. Consider only those points that lie in R.)

(*b*) Use (*a*) to estimate $\int_R f(P)\,dA$.

(*c*) Show why $\pi/e \le \int_R f(P)\,dA \le \pi e$.

***18** If a region has diameter d, how large do you think its area can be?

(*a*) Experiment with various shapes.

(*b*) Make a conjecture.

(*c*) Can a region of diameter d always be placed inside some circle of diameter d?

(*d*) If a region has diameter d, how small can its area be?

19 If a square of side 1 is partitioned into n regions, what is the largest that the mesh can be? (It is not known in general how small the mesh can be. For $n = 9$, it can be as small as $\frac{5}{11}$.)

15.2 COMPUTING $\int_R f(P)\,dA$ USING RECTANGULAR COORDINATES

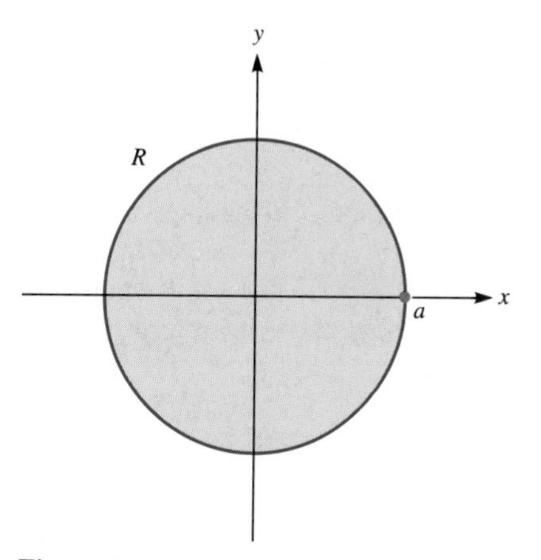

Figure 1

In this section, we will show how to use rectangular coordinates to evaluate $\int_R f(P)\,dA$. This method requires that both R and f be described in rectangular coordinates. In particular, it requires a precise description of the cross sections of R. We first show how to describe plane regions R in rectangular coordinates.

Describing *R* in Rectangular Coordinates

Some examples illustrate how to describe planar regions by their cross sections in terms of rectangular coordinates.

EXAMPLE 1 Describe a disk R of radius a in rectangular coordinates.

SOLUTION Introduce an xy coordinate system with its origin at the center of the disk, as in Fig. 1. A glance at the figure shows that x ranges from $-a$ to a. All that remains is to tell how y varies for each x in $[-a, a]$.

Figure 2 shows a typical x in $[-a, a]$ and corresponding cross section. The circle has the equation $x^2 + y^2 = a^2$. The top half has the description $y = \sqrt{a^2 - x^2}$ and the bottom half, $y = -\sqrt{a^2 - x^2}$. So, for each x in $[-a, a]$, y varies from $-\sqrt{a^2 - x^2}$ to $\sqrt{a^2 - x^2}$. (As a check, test $x = 0$. Does y vary from $-\sqrt{a^2 - 0^2} = -a$ to $\sqrt{a^2 - 0^2} = a$? It does, as an inspection of Fig. 2 shows.)

All told, this is the description of R by vertical cross sections:

$$-a \leq x \leq a, \qquad -\sqrt{a^2 - x^2} \leq y \leq \sqrt{a^2 - x^2}. \quad \blacksquare$$

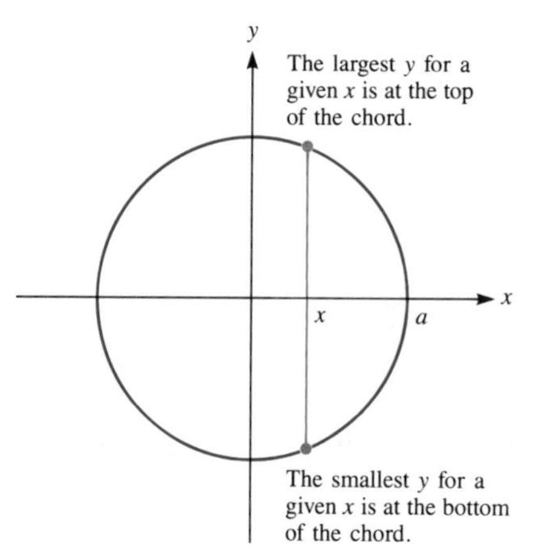

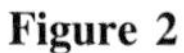
Figure 2

EXAMPLE 2 Let R be the region bounded by $y = x^2$, the x axis, and the line $x = 2$. Describe R in terms of cross sections parallel to the y axis.

SOLUTION A glance at R in Fig. 3 shows that for points (x, y) in R, x ranges from 0 to 2. To describe R completely, we shall describe the behavior of y for any x in the interval $[0, 2]$.

Hold x fixed and consider only the cross section above the point $(x, 0)$. It extends from the x axis to the curve $y = x^2$; for any x, the y coordinate varies from 0 to x^2. This is a complete description of R by vertical cross sections, written in compact notation:

$$0 \leq x \leq 2, \qquad 0 \leq y \leq x^2. \quad \blacksquare$$

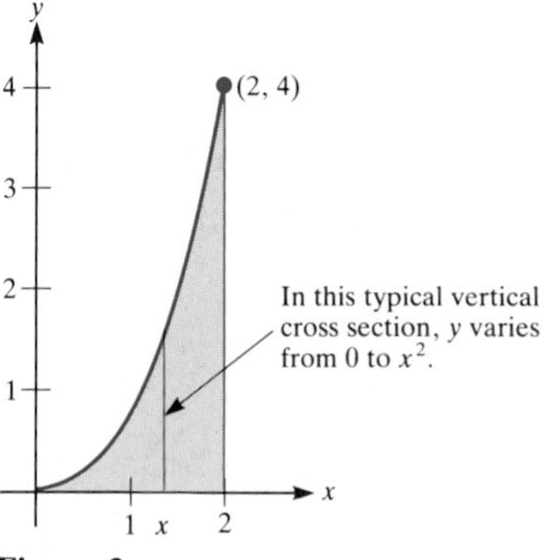

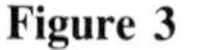
Figure 3

EXAMPLE 3 Describe the region R of Example 2 by cross sections parallel to the x axis, that is, by horizontal cross sections.

SOLUTION A glance at R in Fig. 4 shows that y varies from 0 to 4. For any y in the interval $[0, 4]$, x varies from a smallest value $x_1(y)$ to a largest value $x_2(y)$. Note that $x_2(y) = 2$ for each value of y in $[0, 4]$. To find $x_1(y)$, utilize the fact that the point $(x_1(y), y)$ is on the curve $y = x^2$, that is,

$$x_1(y) = \sqrt{y}.$$

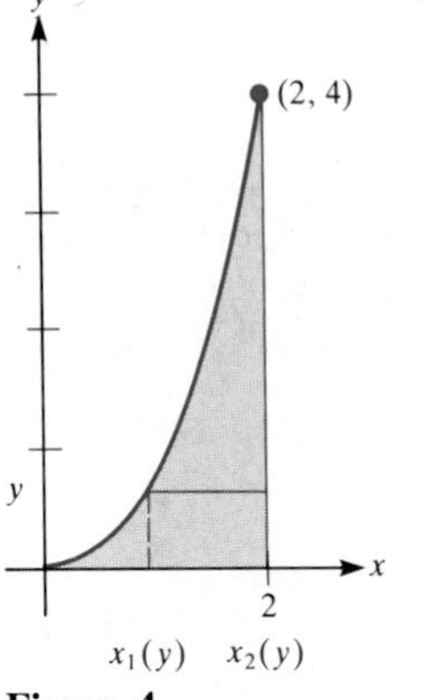

Figure 4

The description of R in terms of horizontal cross sections is

$$0 \leq y \leq 4, \qquad \sqrt{y} \leq x \leq 2. \quad \blacksquare$$

EXAMPLE 4 Describe the region R whose vertices are (0, 0), (6, 0), (4, 2), and (0, 2) by vertical cross sections and then by horizontal cross sections. (See Fig. 5.)

SOLUTION Clearly, x varies between 0 and 6. For any x in the interval [0, 4], y ranges from 0 to 2 (independently of x). For x in [4, 6], y ranges from 0 to the value of y on the line through (4, 2), and (6, 0). This line has the equation $y = 6 - x$. The description of R by vertical cross sections therefore requires two separate statements:

$$0 \leq x \leq 4, \qquad 0 \leq y \leq 2$$

and

$$4 \leq x \leq 6, \qquad 0 \leq y \leq 6 - x.$$

Use of horizontal cross sections provides a simpler description. First, y goes from 0 to 2. For each y in [0, 2], x goes from 0 to the value of x on the line $y = 6 - x$. Solving this equation for x yields $x = 6 - y$.

The description is much shorter:

$$0 \leq y \leq 2, \qquad 0 \leq x \leq 6 - y. \quad \blacksquare$$

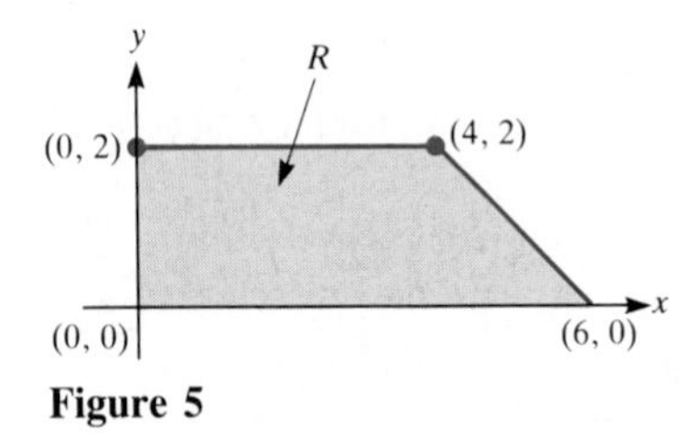

Figure 5

These examples are typical. First, determine the range of one coordinate, and then see how the other coordinate varies for any fixed value of the first coordinate.

Evaluating $\int_R f(P)\, dA$ by Repeated Integrals

We will offer an intuitive development of a formula for computing definite integrals over plane regions.

We first develop a way for computing a definite integral over a rectangle. After applying this formula in Example 5, we make the slight modification needed to evaluate integrals over more general regions.

Consider a rectangular region R whose description by cross sections is

$$a \leq x \leq b, \qquad c \leq y \leq d,$$

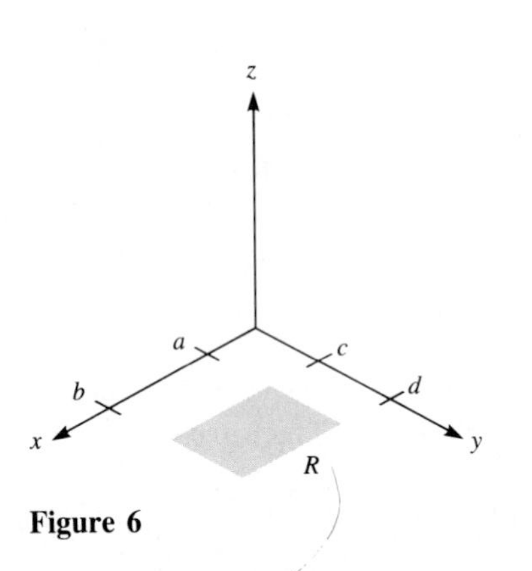

Figure 6

as shown in Fig. 6. If $f(P) \geq 0$ for all P in R, then $\int_R f(P)\, dA$ is the volume V of the solid whose base is R and which has, above P, height $f(P)$. (See Fig. 7.) Let $A(x)$ be the area of the cross section made by a plane perpendicular to the x axis and having abscissa x, as in Fig. 8. As was shown in Sec. 5.3,

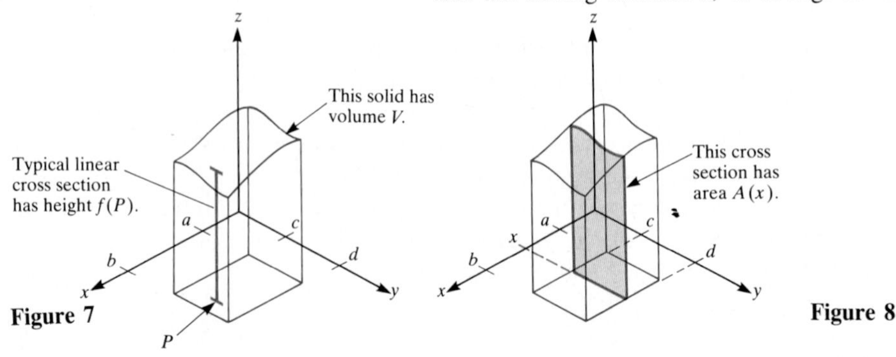

Figure 7

Figure 8

$$V = \int_a^b A(x)\,dx.$$

But the area $A(x)$ is itself expressible as a definite integral:

$$A(x) = \int_c^d f(x, y)\,dy.$$

Note that x is held fixed throughout the integration. This reasoning provides a repeated integral whose value is $V = \int_R f(P)\,dA$, namely,

$$\begin{aligned}\int_R f(P)\,dA &= V \\ &= \int_a^b A(x)\,dx \\ &= \int_a^b \left[\int_c^d f(x, y)\,dy\right] dx.\end{aligned}$$

In short

An integral over a rectangle expressed as a repeated integral

$$\boxed{\int_R f(P)\,dA = \int_a^b \left[\int_c^d f(x, y)\,dy\right] dx.}$$

Of course, cross sections by planes perpendicular to the y axis could be used. Then similar reasoning shows that

$$\int_R f(P)\,dA = \int_c^d \left[\int_a^b f(x, y)\,dx\right] dy.$$

The quantities $\int_a^b [\int_c^d f(x, y)\,dy]\,dx$ and $\int_c^d [\int_a^b f(x, y)\,dx]\,dy$ are called **double integrals, repeated integrals**, or **iterated integrals**. Usually the brackets are omitted and the integrals are written $\int_a^b \int_c^d f(x, y)\,dy\,dx$ and $\int_c^d \int_a^b f(x, y)\,dx\,dy$.

The order of dx and dy matters; the differential that is on the left tells which integration is performed first.

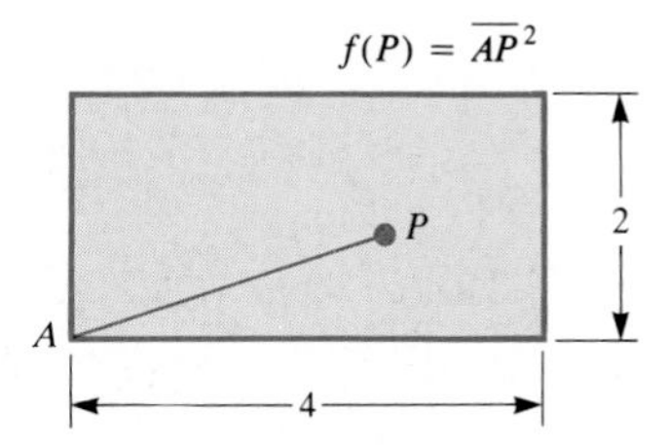

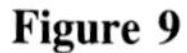
Figure 9

EXAMPLE 5 Compute the definite integral $\int_R f(P)\,dA$, where R is the rectangle shown in Fig. 9 and the function f is defined by $f(P) = \overline{AP}^2$. (This integral was estimated in Sec. 15.1.)

SOLUTION Introduce xy coordinates in the convenient manner depicted in Fig. 10. Then f has this description in rectangular coordinates:

$$f(x, y) = \overline{AP}^2 = x^2 + y^2.$$

To describe R, observe that x takes all values from 0 to 4 and that for each x the number y takes all values between 0 and 2. Thus

$$\int_R f(P)\,dA = \int_0^4 \left[\int_0^2 (x^2 + y^2)\,dy\right] dx.$$

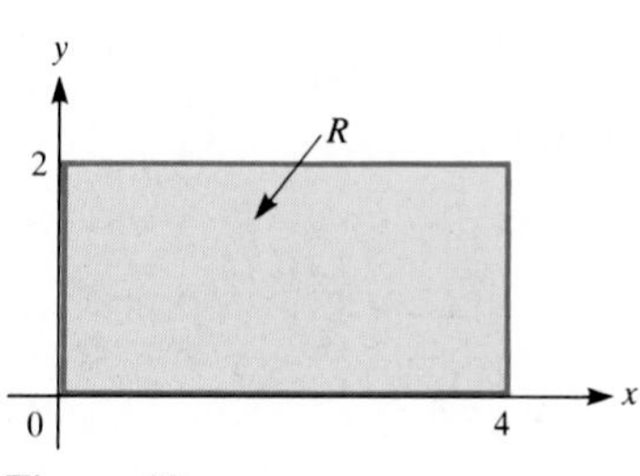

Figure 10

We must first compute $\displaystyle\int_0^2 (x^2 + y^2)\,dy,$

where x is fixed in [0, 4]. [This is the cross-sectional area $A(x)$.] To apply the fundamental theorem of calculus, first find a function $F(x, y)$ such that

$$\frac{\partial F}{\partial y} = x^2 + y^2.$$

Keep in mind that x is constant during this first integration.

$$F(x, y) = x^2 y + \frac{y^3}{3}$$

is such a function. The appearance of x in its formula should not disturb us, since x is fixed for the time being. By the fundamental theorem of calculus,

The notation $\big|_{y=0}^{y=2}$ tells which of the two variables is replaced by 0 and 2.

$$\int_0^2 (x^2 + y^2)\,dy = \left(x^2 y + \frac{y^3}{3}\right)\Bigg|_{y=0}^{y=2}$$

$$= \left(x^2 \cdot 2 + \frac{2^3}{3}\right) - \left(x^2 \cdot 0 + \frac{0^3}{3}\right)$$

$$= 2x^2 + \tfrac{8}{3}.$$

[The integral $2x^2 + \frac{8}{3}$ is the area $A(x)$ discussed earlier in this section.]

Now compute
$$\int_0^4 (2x^2 + \tfrac{8}{3})\,dx.$$

By the fundamental theorem of calculus,

$$\int_0^4 \left(2x^2 + \frac{8}{3}\right) dx = \left(\frac{2x^3}{3} + \frac{8x}{3}\right)\Bigg|_0^4 = \frac{160}{3}.$$

How close were the estimates in Sec. 15.1?

Hence the two-dimensional definite integral has the value $\frac{160}{3}$. The volume of the region in Problem 1 of Sec. 15.1 is $\frac{160}{3}$ cubic inches. The mass in Problem 2 is $\frac{160}{3}$ grams. ■

If R is not a rectangle, the repeated integral that equals $\int_R f(P)\,dA$ differs from that for the case where R is a rectangle only in the intervals of integration. If R has the description

$$a \le x \le b, \qquad y_1(x) \le y \le y_2(x),$$

by cross sections parallel to the y axis, then

$$\boxed{\int_R f(P)\,dA = \int_a^b \left[\int_{y_1(x)}^{y_2(x)} f(x, y)\,dy\right] dx.}$$

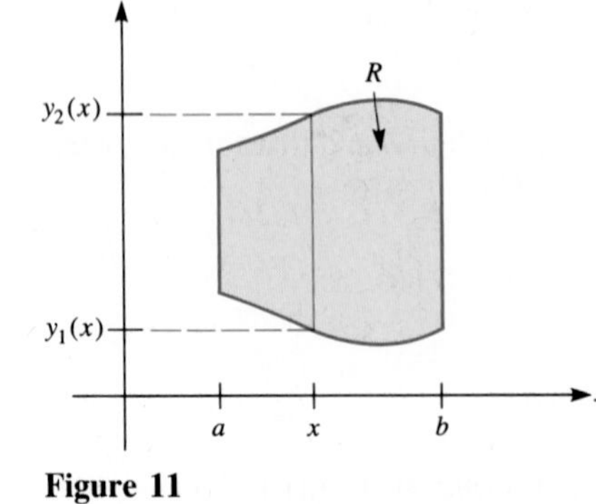

Figure 11

See Fig. 11.

Similarly, if R has the description

$$c \le y \le d, \qquad x_1(y) \le x \le x_2(y),$$

by cross sections parallel to the x axis, then

$$\boxed{\int_R f(P)\,dA = \int_c^d \left[\int_{x_1(y)}^{x_2(y)} f(x, y)\,dx\right] dy.}$$

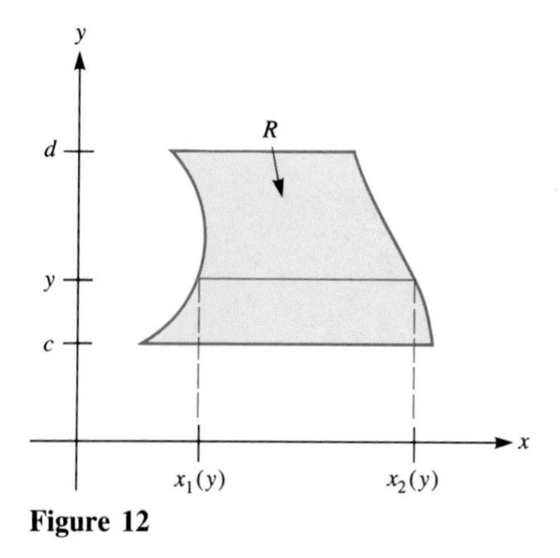

Figure 12

The intervals of integration are determined by R; the function f influences only the integrand. (See Fig. 12.)

In the next example R is the region bounded by $y = x^2$, $x = 2$, and $y = 0$; the function is $f(x, y) = 3xy$. The integral $\int_R 3xy\,dA$ has at least three interpretations:

1. If at each point $P = (x, y)$ in R we erect a line segment above P of length $3xy$, then the integral is the volume of the resulting solid. (See Fig. 13.)
2. If the density of matter at (x, y) in R is $3xy$, then $\int_R 3xy\,dA$ is the total mass in R.
3. If the temperature at (x, y) in R is $3xy$ then $\int_R 3xy\,dA/\text{Area of } R$ is the average temperature in R.

EXAMPLE 6 Evaluate $\int_R 3xy\,dA$ over the region R shown in Fig. 14.

R is discussed in Examples 2 and 3.

SOLUTION If cross sections parallel to the y axis are used, then R is described by

$$0 \le x \le 2, \qquad 0 \le y \le x^2.$$

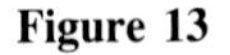
Figure 13

Thus

$$\int_R 3xy\,dA = \int_0^2 \left(\int_0^{x^2} 3xy\,dy \right) dx,$$

which is easy to compute. First, with x fixed,

$$\int_0^{x^2} 3xy\,dy = \left(3x\frac{y^2}{2}\right)\Bigg|_{y=0}^{y=x^2}$$

$$= 3x\frac{(x^2)^2}{2} - 3x\frac{0^2}{2}$$

$$= \frac{3x^5}{2}.$$

Then,

$$\int_0^2 \frac{3x^5}{2}\,dx = \frac{3x^6}{12}\Bigg|_0^2 = 16.$$

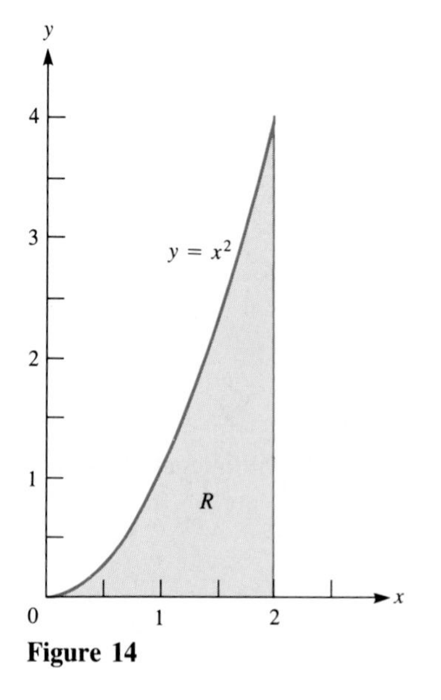

Figure 14

Figure 15 shows which integration is performed first.

R can also be described in terms of cross sections parallel to the x axis:

$$0 \le y \le 4, \qquad \sqrt{y} \le x \le 2.$$

Then,

$$\int_R 3xy\,dA = \int_0^4 \left(\int_{\sqrt{y}}^2 3xy\,dx \right) dy,$$

which, as the reader may verify, equals 16. See Fig. 16. ■

In Example 6 we could evaluate $\int_R f(P)\,dA$ by cross sections in either direction. In the next example we don't have that choice.

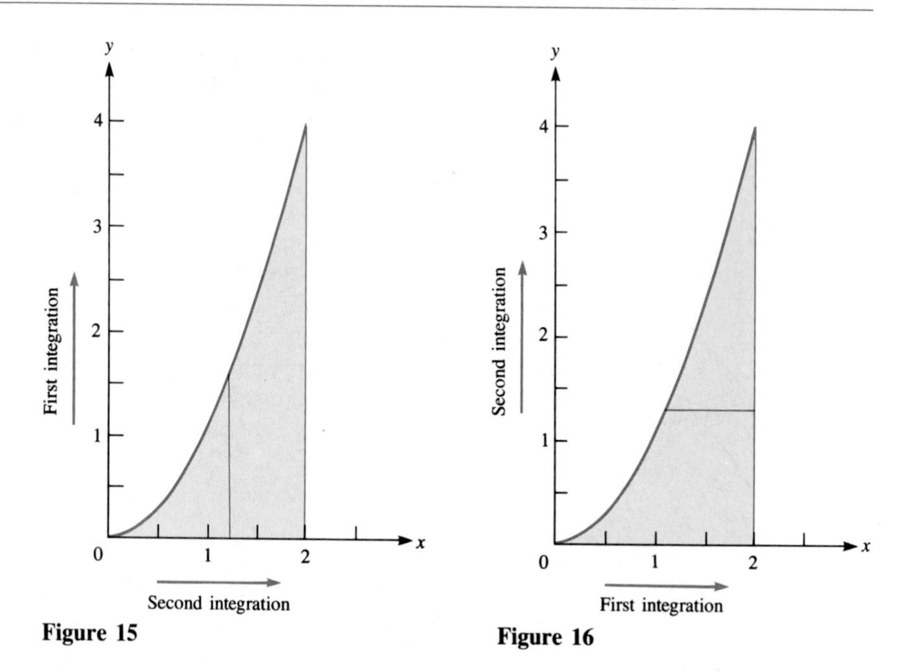

Figure 15

Figure 16

Figure 17

EXAMPLE 7 A triangular lamina is located as in Fig. 17. Its density at (x, y) is e^{y^2}. Find its mass, that is,

$$\int_R f(P)\, dA,$$

where $f(x, y) = e^{y^2}$.

SOLUTION The description of R by vertical cross sections is

$$0 \le x \le 2, \qquad \frac{x}{2} \le y \le 1.$$

Hence
$$\int_R f(P)\, dA = \int_0^2 \left(\int_{x/2}^1 e^{y^2}\, dy \right) dx.$$

Unfortunately, the fundamental theorem of calculus is useless in computing

$$\int_{x/2}^1 e^{y^2}\, dy,$$

so we try horizontal cross sections instead.

The description of R is now

$$0 \le y \le 1, \qquad 0 \le x \le 2y.$$

Thus
$$\int_R f(P)\, dA = \int_0^1 \left(\int_0^{2y} e^{y^2}\, dx \right) dy.$$

The first integration, $\int_0^{2y} e^{y^2}\, dx$, is easy, since y is fixed; the integrand is constant. Thus

$$\begin{aligned} \int_0^{2y} e^{y^2}\, dx &= e^{y^2} \int_0^{2y} 1\, dx \\ &= e^{y^2} x \Big|_{x=0}^{x=2y} \\ &= e^{y^2} 2y. \end{aligned}$$

The second definite integral in the repeated integral is thus $\int_0^1 e^{y^2} 2y\, dy$, which can be evaluated by the fundamental theorem of calculus, since $d(e^{y^2})/dy = e^{y^2} 2y$:

$$\int_0^1 e^{y^2} 2y\, dy = e^{y^2}\Big|_0^1 = e^{1^2} - e^{0^2} = e - 1.$$

The total mass is $e - 1$. ■

Notice that computing a definite integral over R involves, first, a wise choice of an xy-coordinate system; second, a description of R and f relative to this coordinate system; and finally, the computation of two successive definite integrals over intervals. The order of these integrations should be considered carefully since computation may be much simpler in one than in the other. This order is determined by the description of R by cross sections.

Section Summary

We showed that the integral of $f(P)$ over a plane region R can be evaluated by a repeated integral, where the limits of integration are determined by R (not by f). If each line parallel to the y axis meets R in at most two points then

$$\int_R f(P)\, dA = \int_a^b \left[\int_{y_1(x)}^{y_2(x)} f(x, y)\, dy\right] dx,$$

where R has the description

$$a \le x \le b, \qquad y_1(x) \le y \le y_2(x).$$

If each line parallel to the x axis meets R in at most two points, then, similarly,

$$\int_R f(P)\, dA = \int_c^d \left[\int_{x_1(y)}^{x_2(y)} f(x, y)\, dx\right] dy.$$

EXERCISES FOR SEC. 15.2: COMPUTING $\int_R f(P)\, dA$ USING RECTANGULAR COORDINATES

Exercises 1 to 12 give practice in describing plane regions by cross sections in rectangular coordinates. In Exercises 1 to 12 describe the regions by (*a*) vertical cross sections, (*b*) horizontal cross sections.

1 The triangle whose vertices are (0, 0), (2, 1), (0, 1).
2 The triangle whose vertices are (0, 0), (2, 0), (1, 1).
3 The parallelogram with vertices (0, 0), (1, 0), (2, 1), (1, 1).
4 The parallelogram with vertices (2, 1), (5, 1), (3, 2), (6, 2).
5 The disk of radius 5 and center (0, 0).
6 The trapezoid with vertices (1, 0), (3, 2), (3, 3), (1, 6).
7 The triangle bounded by the lines $y = x$, $x + y = 2$, and $x + 3y = 8$.
8 The region bounded by the ellipse $4x^2 + y^2 = 4$.
9 The triangle bounded by the lines $x = 0$, $y = 0$, and $2x + 3y = 6$.
10 The region bounded by the curves $y = e^x$, $y = 1 - x$, and $x = 1$.
11 The quadrilateral bounded by the lines $y = 1$, $y = 2$, $y = x$, $y = x/3$.
12 The quadrilateral bounded by the lines $x = 1$, $x = 2$, $y = x$, $y = 5 - x$.

In Exercises 13 to 16 draw the regions and describe them by horizontal cross sections.

13 $0 \le x \le 2$, $2x \le y \le 3x$
14 $1 \le x \le 2$, $x^3 \le y \le 2x^2$
15 $0 \le x \le \pi/4$, $0 \le y \le \sin x$ and $\pi/4 \le x \le \pi/2$, $0 \le y \le \cos x$
16 $1 \le x \le e$, $(x - 1)/(e - 1) \le y \le \ln x$

In Exercises 17 to 22 evaluate the repeated integrals.

17 $\int_0^1 \left(\int_0^x (x + 2y)\, dy\right) dx$

18 $\int_1^2 \left(\int_x^{2x} dy\right) dx$

19 $\int_0^2 \left(\int_0^{x^2} xy^2\, dy\right) dx$

20 $\int_1^2 \left(\int_0^y e^{x+y}\, dx\right) dy$

21 $\int_1^2 \left(\int_0^{\sqrt{y}} yx^2\, dx \right) dy$ 22 $\int_0^1 \left(\int_0^x y \sin \pi x\, dy \right) dx$

23 (*a*) Sketch the solid region S below the plane $z = 1 + x + y$ and above the triangle R in the xy plane with vertices $(0, 0)$, $(1, 0)$, $(0, 2)$.
(*b*) Describe R in terms of coordinates.
(*c*) Set up an iterated integral for the volume of S.
(*d*) Evaluate the expression in (*c*), and show in the manner of Fig. 15 or 16 which integration you performed first.
(*e*) Carry out (*c*) and (*d*) in the other order of integration.

24 Let S be the solid region below the paraboloid $z = x^2 + 2y^2$ and above the rectangle in the xy plane with vertices $(0, 0)$, $(1, 0)$, $(1, 2)$, $(0, 2)$. Carry out the steps of Exercise 23 in this case.

25 Let S be the solid region below the saddle $z = xy$ and above the triangle in the xy plane with vertices $(1, 1)$, $(3, 1)$, and $(1, 4)$. Carry out the steps of Exercise 23 in this case.

26 Let S be the solid region below the saddle $z = xy$ and above the region in the first quadrant of the xy plane bounded by the parabolas $y = x^2$ and $y = 2x^2$ and the line $y = 2$. Carry out the steps of Exercise 23 in this case.

27 Find the mass of a thin lamina occupying the finite region bounded by $y = 2x^2$ and $y = 5x - 3$ and whose density at (x, y) is xy.

28 Find the mass of a thin lamina occupying the triangle whose vertices are $(0, 0)$, $(1, 0)$, $(1, 1)$, and whose density at (x, y) is $1/(1 + x^2)$.

29 The temperature at (x, y) is $T(x, y) = \cos(x + 2y)$. Find the average temperature in the triangle with vertices $(0, 0)$, $(1, 0)$, $(0, 2)$.

30 The temperature at (x, y) is $T(x, y) = e^{x-y}$. Find the average temperature in the region in the first quadrant bounded by the triangle with vertices $(0, 0)$, $(1, 1)$, and $(3, 1)$.

In each of Exercises 31 to 34 replace the given iterated integral by an equivalent one with the order of integration reversed. First sketch the region R of integration.

31 $\int_0^2 \left(\int_0^{x^2} x^3 y\, dy \right) dx$ 32 $\int_0^{\pi/2} \left(\int_0^{\cos x} x^2\, dy \right) dx$

33 $\int_0^1 \left(\int_{x/2}^x xy\, dy \right) dx + \int_1^2 \left(\int_{x/2}^1 xy\, dy \right) dx$

34 $\int_{-1/\sqrt{2}}^0 \left(\int_{-x}^{\sqrt{1-x^2}} x^3 y\, dy \right) dx + \int_0^1 \left(\int_0^{\sqrt{1-x^2}} x^3 y\, dy \right) dx$

In Exercises 35 to 38 evaluate the iterated integrals. First sketch the region of integration.

35 $\int_0^1 \left(\int_x^1 \sin y^2\, dy \right) dx$ 36 $\int_0^1 \left(\int_{\sqrt{x}}^1 \frac{dy}{\sqrt{1 + y^3}} \right) dx$

37 $\int_0^1 \left(\int_{\sqrt[3]{y}}^1 \sqrt{1 + x^4}\, dx \right) dy$

38 $\int_1^2 \left(\int_1^y \frac{\ln x}{x}\, dx \right) dy + \int_2^4 \left(\int_{y/2}^2 \frac{\ln x}{x}\, dx \right) dy$

39 Let $f(x, y) = y^2 e^{y^2}$ and let R be the triangle bounded by $y = a$, $y = x/2$, and $y = x$. Assume that a is positive.
(*a*) Set up two repeated integrals for $\int_R f(P)\, dA$.
(*b*) Evaluate the easier one.

40 Let R be the finite region bounded by the curve $y = \sqrt{x}$ and the line $y = x$. Let $f(x, y) = (\sin y)/y$ if $y \neq 0$ and $f(x, 0) = 1$. Compute $\int_R f(P)\, dA$.

41 In Sec. 8.6 it was shown that if the density is 1, then the moment about the x axis of the region under $y = f(x)$ and above $[a, b]$ is given by the integral $\int_a^b \{[f(x)]^2/2\}\, dx$. Obtain that result by evaluating an appropriate plane integral by an iterated integral.

42 The temperature at the point (x, y) at time t is $T(x, y, t) = e^{-tx} \sin(x + 3y)$. Let $f(t)$ be the average temperature in the rectangle $0 \le x \le \pi$, $0 \le y \le \pi/2$ at time t. Find df/dt.

43 Let f be a function such that $f(-x, y) = -f(x, y)$.
(*a*) Give some examples of such functions.
(*b*) For what type regions R in the xy plane is $\int_R f(x, y)\, dA$ certainly equal to 0?

44 Find $\int_R (2x^3y^2 + 7)\, dA$ where R is the square with vertices $(1, 1)$, $(-1, 1)$, $(-1, -1)$, and $(1, -1)$. Do this with as little work as possible.

45 Let $f(x, y)$ be a continuous function. Define $g(x)$ to be $\int_R f(P)\, dA$, where R is the rectangle with vertices $(3, 0)$, $(3, 5)$, $(x, 0)$, and $(x, 5)$, $x > 3$. Express dg/dx as a suitable integral.

15.3 MOMENTS AND CENTERS OF MASS

Review the definitions in Sec. 8.6.

In Sec. 8.6 moments, balancing lines, and the center of mass for a plane region furnished with a *constant density* (a homogeneous distribution of mass) were studied. Now with the aid of integrals over plane sets, these concepts can be extended to cases where the density is not constant.

Let L be a line in the plane of the region R, which is furnished with a distribution of matter that has density $\sigma(P)$ at the point P. (Density may be measured in grams per square centimeter.) Introduce an x axis perpendicular to L with its

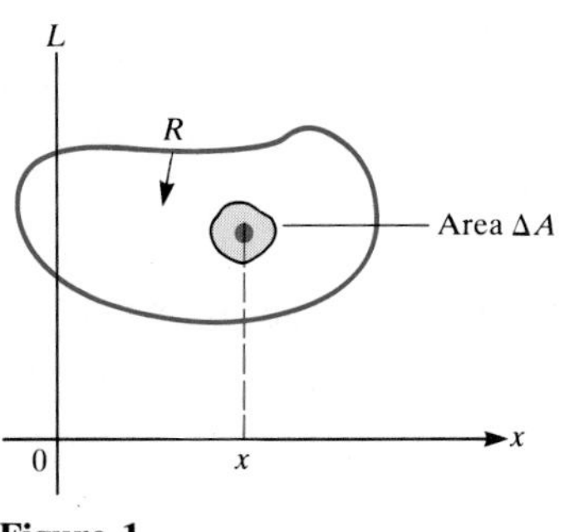

Figure 1

origin on L, as in Fig. 1. Consider the moment of the mass in a small region of area ΔA. Let P be a point in the small region and let its x coordinate be x.

The mass in the small region is approximately

$$\underbrace{\sigma(P)}_{\text{Density}}\ \underbrace{\Delta A}_{\text{Area}}.$$

The lever arm of this mass around L is approximately x. So the moment about L of this small bit of mass is approximately

$$\boxed{\underbrace{x}_{\text{Lever}}\ \underbrace{\sigma(P)\,\Delta A}_{\text{Mass}}.} \tag{1}$$

The local estimate (1) suggests the following definition.

Definition *First moment.* The **first moment** of a plane distribution of matter around a line L in the plane is the integral

$$\int_R x\,\sigma(P)\,dA.$$

Here $\sigma(P)$ is the density of the matter at P, L is the y axis, and x is the coordinate of P measured on an axis perpendicular to L.

Various (first) moments

Frequently the term "first moment" is shortened to "moment." The first moment is usually computed around either the x axis or the y axis. The moment around the x axis is denoted M_x:

$$M_x = \int_R y\,\sigma(P)\,dA.$$

M_x has a y in the integrand; M_y has an x.

(In this case, L is the x axis and the lever arms are measured along the y axis.) The moment around the y axis is denoted M_y:

$$M_y = \int_R x\,\sigma(P)\,dA.$$

The center of mass for any continuous density function $\sigma(P)$ is defined as in the case of a constant density, discussed in Sec. 8.6.

Definition *Center of mass.* Let R be a region in the xy plane furnished with a density function $\sigma(P)$. The **center of mass** $(\bar{x}, \bar{y})$ is defined by the formulas

$$\bar{x} = \frac{\int_R x\,\sigma(P)\,dA}{\int_R \sigma(P)\,dA} = \frac{\text{Moment around } y \text{ axis}}{\text{Total mass}}$$

and

$$\bar{y} = \frac{\int_R y\,\sigma(P)\,dA}{\int_R \sigma(P)\,dA} = \frac{\text{Moment around } x \text{ axis}}{\text{Total mass}}.$$

If the density is constant, the center of mass is often called the **centroid** of R. (In this case, σ cancels in the formulas for $\bar{x}$ and $\bar{y}$, since a constant can be pulled outside the integrals in both numerator and denominator.) The center of mass is also called "the center of gravity."

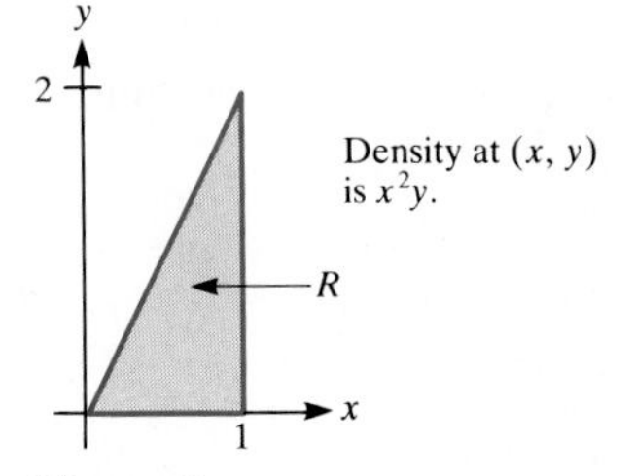

Figure 2

EXAMPLE 1 A lamina occupies the triangle whose vertices are (0, 0), (1, 0), and (1, 2) and has density $\sigma = x^2y$ at the point (x, y). (See Fig. 2.) Find its center of mass.

SOLUTION In this case it will be necessary to compute three definite integrals:

$$\text{Mass} = \int_R x^2y\, dA,$$

and the first moments,

$$M_y = \int_R x\, \sigma\, dA = \int_R x \cdot x^2y\, dA = \int_R x^3y\, dA,$$

and

$$M_x = \int_R y\, \sigma\, dA = \int_R y \cdot x^2y\, dA = \int_R x^2y^2\, dA.$$

To evaluate these integrals, describe R by, say, vertical cross sections:

$$0 \le x \le 1, \qquad 0 \le y \le 2x.$$

Then

$$\text{Mass} = \int_R x^2y\, dA = \int_0^1 \left(\int_0^{2x} x^2y\, dy \right) dx.$$

The first integration goes as follows:

$$\int_0^{2x} x^2y\, dy = \left. \frac{x^2y^2}{2} \right|_{y=0}^{y=2x} = \frac{x^2(2x)^2}{2} - \frac{x^2 0^2}{2} = 2x^4.$$

The second integration gives

$$\int_0^1 2x^4\, dx = \left. \frac{2x^5}{5} \right|_0^1 = \frac{2}{5}.$$

Thus

$$\text{Mass} = \int_R x^2y\, dA = \frac{2}{5}.$$

Next compute, $M_y = \int_R x^3y\, dA = \int_0^1 (\int_0^{2x} x^3y\, dy)\, dx$. The inner integral (the one in parentheses) is

$$\int_0^{2x} x^3y\, dy = \left. \frac{x^3y^2}{2} \right|_{y=0}^{y=2x} = \frac{x^3(2x)^2}{2} - \frac{x^3 0^2}{2} = 2x^5.$$

The outer integral is then $\displaystyle\int_0^1 2x^5\, dx = \left. \frac{2x^6}{6} \right|_0^1 = \frac{1}{3}$.

Thus $$M_y = \int_R x^3 y \, dA = \frac{1}{3}.$$

Consequently, $$\bar{x} = \frac{\frac{1}{3}}{\frac{2}{5}} = \frac{1}{3}\frac{5}{2} = \frac{5}{6}.$$

Next, compute $M_x = \int_R x^2y^2 \, dA$. This equals $\int_0^1 (\int_0^{2x} x^2y^2 \, dy) \, dx$. The inner integral is

$$\int_0^{2x} x^2y^2 \, dy = \left. \frac{x^2y^3}{3} \right|_{y=0}^{y=2x} = \frac{8x^5}{3}.$$

The outer integral is $$M_x = \int_0^1 \frac{8x^5}{3} \, dx = \left. \frac{8x^6}{18} \right|_0^1 = \frac{4}{9}.$$

Hence $$\bar{y} = \frac{\frac{4}{9}}{\frac{2}{5}} = \frac{4}{9}\frac{5}{2} = \frac{10}{9}.$$

The center of mass is therefore

$$(\bar{x}, \bar{y}) = (\tfrac{5}{6}, \tfrac{10}{9}). \quad \blacksquare$$

Moment of Inertia

The kinetic energy of a particle of mass M kilograms moving along a line at the speed of v meters per second is defined to be $Mv^2/2$ joules. In the next example we compute the kinetic energy of a flat object spinning around a line.

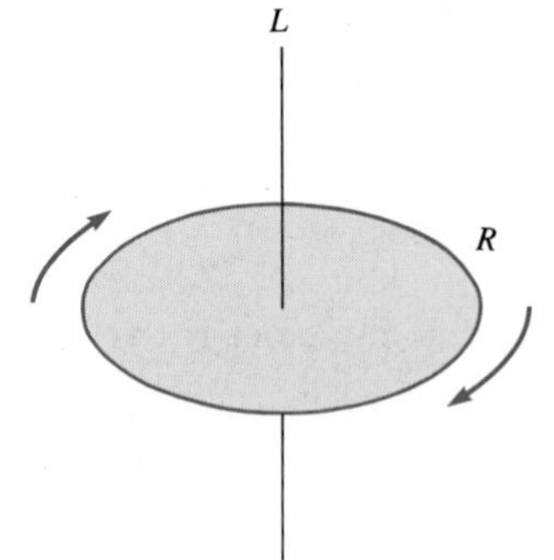

Figure 3

EXAMPLE 2 A thin piece of metal spins around the line L perpendicular to it at the rate of ω radians per second. The density of the metal at point P is $\sigma(P)$ kilograms per square meter. Find its kinetic energy.

SOLUTION Figure 3 shows the metal, which occupies the region R.

To estimate the total kinetic energy, we make a local estimate. Partition R into n small patches $R_1, R_2, \ldots, R_n$. Consider a typical small patch R_i of the metal with area A_i. Choose a point P_i in R_i. Let the distance from P_i to L be r_i, as shown in Fig. 4. The mass in this small patch is approximately $\sigma(P_i) \, A_i$. Furthermore, since the patch is small, all points in it are moving at about the same speed. To find that speed, recall that if s is the length of arc on a circle of radius r_i, subtended by an angle θ, then $s = r_i\theta$. Thus

$$\frac{ds}{dt} = r_i \frac{d\theta}{dt}.$$

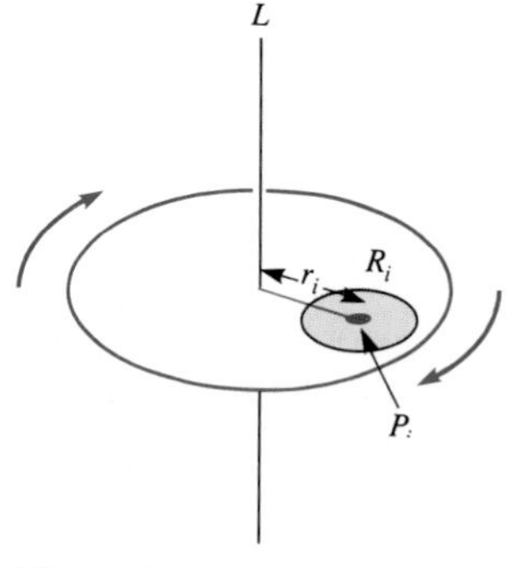

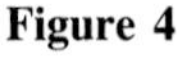
Figure 4

Since $d\theta/dt$ is given as ω radians per second, we see that the speed of the particle is $r_i\omega$.

Particles in R_i are all moving at approximately the speed $r_i\omega$ meters per second. The kinetic energy of this small patch is therefore approximately

$$\tfrac{1}{2} \underbrace{\sigma(P_i) \, A_i}_{\text{Mass}} \underbrace{(r_i \, \omega)^2}_{\text{Speed}^2}.$$

Thus $$\sum_{i=1}^{n} \tfrac{1}{2} r_i^2 \, \sigma(P_i) \, \omega^2 \, A_i \tag{2}$$

is an estimate of the total kinetic energy. Taking the limit of (2) as the mesh approaches 0 gives

$$\text{Total kinetic energy} = \int_R \tfrac{1}{2}r^2\,\sigma(P)\,\omega^2\,dA \quad \text{joules.} \quad \blacksquare$$

The kinetic energy in Example 2 can be rewritten as

$$\text{Kinetic energy} = \tfrac{1}{2}\left(\int_R r^2\,\sigma(P)\,dA\right)\omega^2 \quad \text{joules.}$$

The quantity $\int_R r^2\,\sigma(P)\,dA$ is determined by the position of the axis of rotation and the piece of metal. It is called the **moment of inertia** of the object about L and denoted I. We have then

$$\text{Kinetic energy of circular motion} = \tfrac{1}{2}I\omega^2. \tag{3}$$

Compare (3) with

$$\text{Kinetic energy of linear motion} = \tfrac{1}{2}Mv^2, \tag{4}$$

where M is the mass and v is the speed.

Comparison of (3) with (4) shows that the moment of inertia I plays the same role in circular motion around an axis that mass M plays in motion along a line. By moment of inertia about a point in a plane, we mean the moment of inertia about the line through the point and perpendicular to the plane.

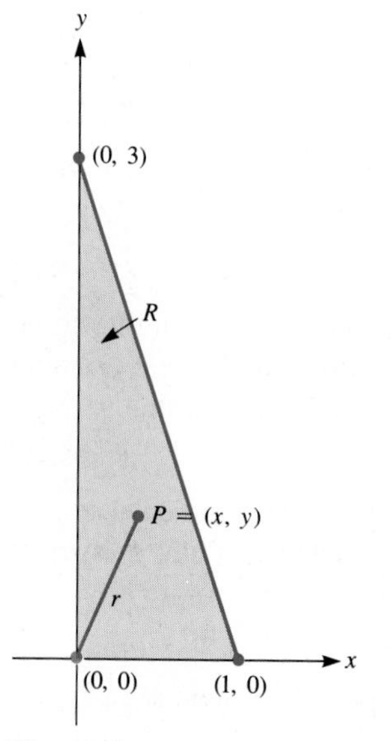

Figure 5

EXAMPLE 3 A homogeneous triangular piece of metal of mass M has vertices (0, 0), (1, 0), and (0, 3). Find its moment of inertia about (0, 0).

SOLUTION The metal is shown in Fig. 5. Its density at any point is $\sigma(P) = M/\text{Area} = M/(\frac{3}{2}) = 2M/3$.

(*a*) The moment of inertia about (0, 0) is

$$\int_R r^2\,\sigma(P)\,dA = \int_R (\sqrt{x^2+y^2})^2\,\frac{2M}{3}\,dA$$

$$= \frac{2}{3}M\int_R (x^2+y^2)\,dA. \tag{5}$$

To evaluate the integral in (5) we first describe R in rectangular coordinates. The line joining (0, 3) and (1, 0) has the equation $y = 3 - 3x$. Hence R has the description

$$0 \le x \le 1, \qquad 0 \le y \le 3 - 3x.$$

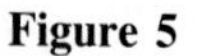

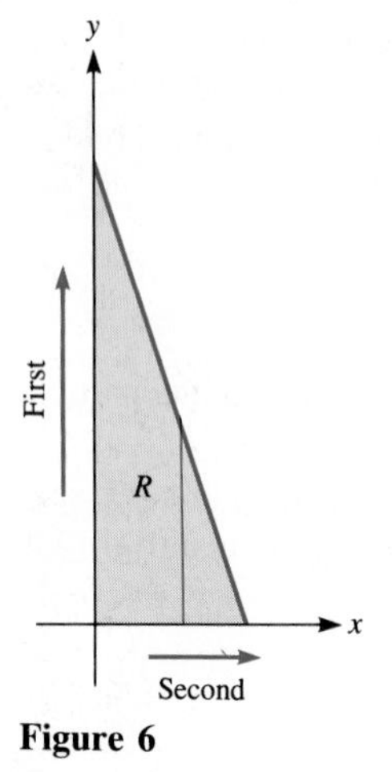

Figure 6

Therefore, $\int_R (x^2+y^2)\,dA = \int_0^1 \int_0^{3-3x} (x^2+y^2)\,dy\,dx$. (We are computing an iterated integral in the order "first y, second x," as shown in Fig. 6.) The first integration gives

$$\int_0^{3-3x} (x^2+y^2)\,dy = \left(x^2y + \frac{y^3}{3}\right)\Bigg|_0^{3-3x}$$

$$= \left[x^2(3-3x) + \frac{(3-3x)^3}{3}\right] - (0)$$

$$= -12x^3 + 30x^2 - 27x + 9.$$

The second integration gives

$$\int_0^1 (-12x^3 + 30x^2 - 27x + 9)\,dx = \left(\frac{-12x^4}{4} + \frac{30x^3}{3} - \frac{27x^2}{2} + 9x\right)\Bigg|_0^1$$

$$= -3 + 10 - \tfrac{27}{2} + 9$$

$$= \tfrac{5}{2}.$$

Thus the moment of inertia about (0, 0) is $(2M/3)(5/2) = 5M/3$. ■

Section Summary

We defined the first moment (or simply ''moment'') of a plane distribution of matter around a line in the plane: $\int_R x\,\sigma(P)\,dA$. [x is the distance—which may be positive or negative—to the line; $\sigma(P)$ is density.] Using it, we defined the centroid $(\bar{x}, \bar{y})$.

Then we defined a ''second moment,'' the moment of inertia around a line perpendicular to the plane: $\int_R r^2\,\sigma(P)\,dA$ [r is the distance to the line; $\sigma(P)$ is density].

EXERCISES FOR SEC. 15.3: MOMENTS AND CENTERS OF MASS

In each of Exercises 1 to 8 find the center of mass of the lamina occupying the given region and having the given density.

1 The triangle with vertices (0, 0), (1, 0), (0, 1); density at (x, y) is $x + y$.

2 The triangle with vertices (0, 0), (2, 0), (1, 1); density at (x, y) is y.

3 The square with vertices (0, 0), (1, 0), (1, 1), (0, 1); density at (x, y) equal to $y \tan^{-1} x$.

4 The finite region bounded by $y = 1 + x$ and $y = 2^x$; density at (x, y) is $x + y$.

5 The triangle with vertices (0, 0), (1, 2), (1, 3); density at (x, y) is xy.

6 The finite region bounded by $y = x^2$, the x axis, and $x = 2$; density at (x, y) is e^x.

7 The finite region bounded by $y = x^2$ and $y = x + 6$, situated to the right of the y axis; density at (x, y) is $2x$.

8 The trapezoid with vertices (0, 0), (3, 0), (2, 1), (0, 1); density at (x, y) is $\sin x$.

In each of Exercises 9 to 14 find the moment of inertia of the lamina occupying the given region. The moment of inertia is to be calculated around the line perpendicular to the lamina that passes through the given point.

9 A homogeneous equilateral triangle of side a and mass M about a line through a vertex.

10 A homogeneous equilateral triangle of side a and mass M about a line through its centroid.

11 A homogeneous square of side a and mass M about a line through a vertex.

12 A homogeneous square of side a and mass M about a line through its centroid.

13 The triangle with vertices (0, 0), (2, 0), (1, 1); density at (x, y) is xy; about the origin.

14 The region bounded by $y = x^2$ and $y = x^3$ in the first quadrant; density at (x, y) is ye^{-x}; about the origin.

---

15 We defined the moment of inertia around a line L perpendicular to the plane of the lamina.

(a) How would you define the moment of inertia of a lamina around a line L in the plane of the lamina? Give the definition in terms of an integral.

(b) Let I_x be the moment of inertia of a lamina in the xy plane around the x axis, I_y the moment of inertia around the y axis, and I the moment of inertia about the z axis. Express I in terms of I_x and I_y.

16 Let L_1 be the line perpendicular to the lamina and passing through its center of mass. Let L_2 be parallel to L_1, and a distance a from L_1. Let I_1 be the moment of inertia of the lamina about L_1 and I_2 be the moment of inertia about L_2. Show that $I_2 = I_1 + a^2M$, where M is the mass of the lamina. (That equation tells us that it takes less energy to spin the lamina around a line through the center of mass than around any other line.)

17 A plane distribution of matter occupies the region R. It is cut into two pieces, occupying regions R_1 and R_2, as in Fig. 7. The part in R_1 has mass M_1 and centroid $(\bar{x}_1, \bar{y}_1)$. The part in R_2 has mass M_2 and centroid $(\overline{x_2}, \overline{y_2})$. Find the centroid $(\bar{x}, \bar{y})$ of the entire mass, which occupies R. [Express $(\bar{x}, \bar{y})$ in terms of M_1, M_2, $\overline{x_1}$, $\overline{x_2}$, $\overline{y_1}$, and $\overline{y_2}$.]

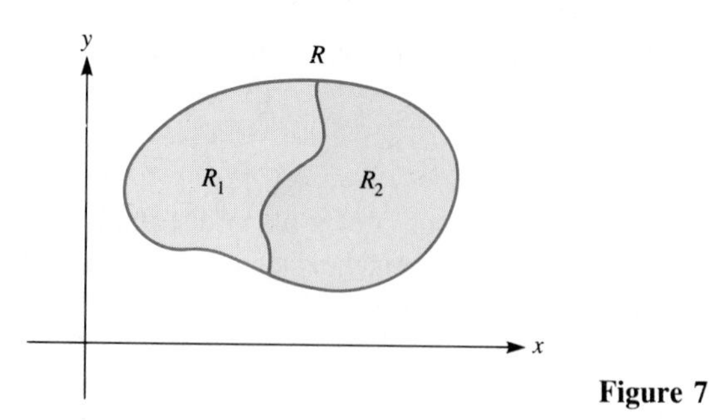

Figure 7

18 Use the formula in Exercise 17 to find the center of mass of the homogeneous lamina shown in Fig. 8.

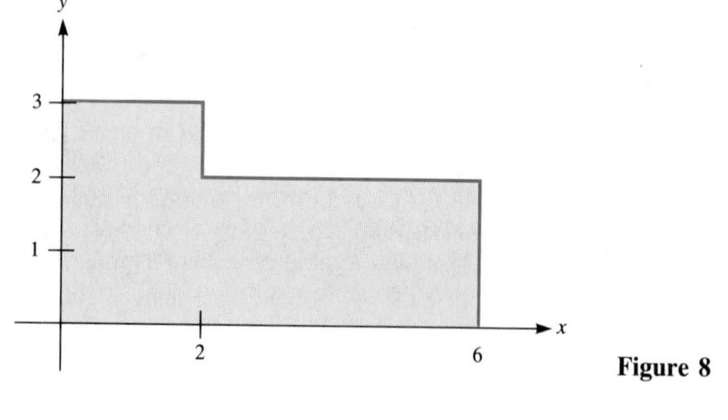

Figure 8

19 Consider a distribution of mass in a plane region R with density $\sigma(P)$ at P. Let $(\bar{x}, \bar{y})$ be its center of mass relative to an xy-coordinate system. Let L be the line $y = \bar{y}$.
(*a*) What is the moment of R about L?
(*b*) Explain.

20 Consider a distribution of mass in a plane region R with density $\sigma(P)$ at P. Show that any line in the plane that passes through the center of the mass is a balancing line, following these steps.
(*a*) For convenience, place the origin of the xy-coordinate system at the center of mass. Show that $\int_R x\,\sigma(P)\,dA = 0$ and $\int_R y\,\sigma(P)\,dA = 0$.
(*b*) Let L be any line $ax + by = 0$ through the origin. Show that the moment of the mass about L is

$$\int_R \frac{ax + by}{\sqrt{a^2 + b^2}}\,\sigma(P)\,dA.$$

(*c*) From (*a*) and (*b*) deduce that the moment of the mass about L is 0. Thus all balancing lines for the mass pass through a single point. Any two of them therefore determine that point, which is called the center of mass. It is customary to use the two lines parallel to the x and y axes to determine that point.

15.4 COMPUTING $\int_R f(P)\,dA$ USING POLAR COORDINATES

This section shows how to evaluate $\int_R f(P)\,dA$ by using polar coordinates. This method is especially appropriate when the region R has a simple description in polar coordinates, for instance, if it is a disk or cardioid. First, we examine how to describe cross sections in polar coordinates. Then we describe the iterated integral in polar coordinates that equals $\int_R f(P)\,dA$.

Describing R in Polar Coordinates

In describing a region R in polar coordinates, we first determine the range of θ and then see how r varies for any fixed value of θ. (The reverse order is seldom useful.) Some examples show how to find how r varies for each θ.

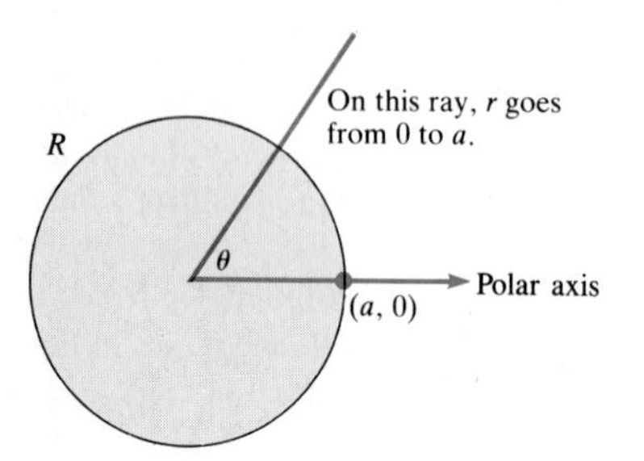

Figure 1

EXAMPLE 1 Let R be the disk of radius a and center at the pole of a polar coordinate system. (See Fig. 1.) Describe R in terms of cross sections by rays emanating from the pole.

SOLUTION To sweep out R, θ goes from 0 to 2π. Hold θ fixed and consider the behavior of r on the ray of angle θ. Clearly, r goes from 0 to a, independently of θ. (See Fig. 1.) The complete description is

$$0 \le \theta \le 2\pi, \qquad 0 \le r \le a. \quad ■$$

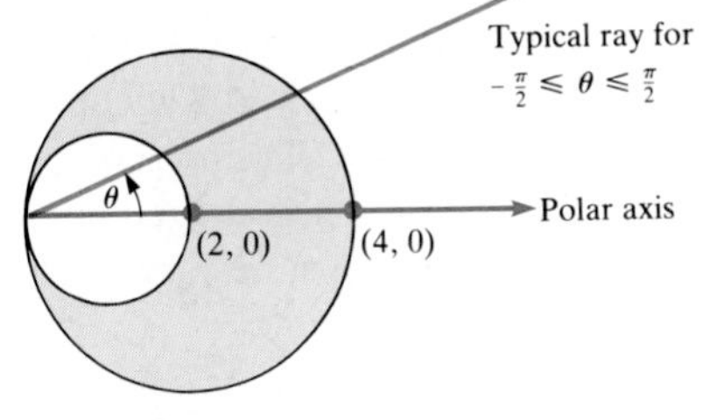

Figure 2

EXAMPLE 2 Let R be the region between the circles $r = 2 \cos \theta$ and $r = 4 \cos \theta$. Describe R in terms of cross sections by rays from the pole. (See Fig. 2.)

SOLUTION To sweep out this region, use the rays from $\theta = -\pi/2$ to $\theta = \pi/2$. For each such θ, r varies from $2 \cos \theta$ to $4 \cos \theta$. The complete description is

$$-\frac{\pi}{2} \le \theta \le \frac{\pi}{2}, \qquad 2 \cos \theta \le r \le 4 \cos \theta. \quad ■$$

As Examples 1 and 2 suggest, polar coordinates provide simple descriptions for regions bounded by circles. The next example shows that polar coordinates may also provide simple descriptions of regions bounded by straight lines, especially if some of the lines pass through the origin.

EXAMPLE 3 Let R be the triangular region whose vertices, in rectangular coordinates, are (0, 0), (1, 1), and (0, 1). Describe R in polar coordinates.

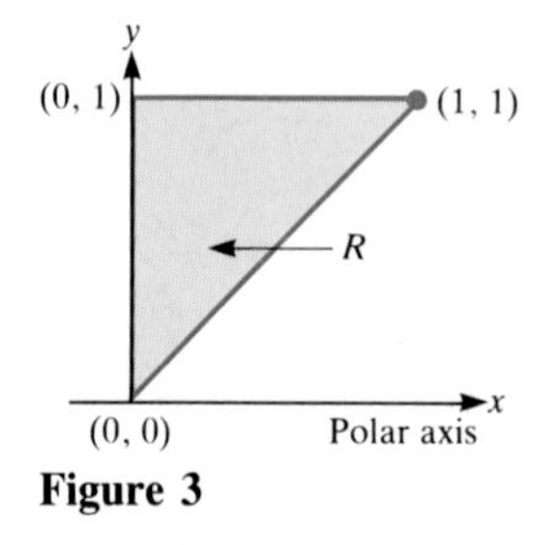

Figure 3

SOLUTION Inspection of R in Fig. 3 shows that θ varies from $\pi/4$ to $\pi/2$. For each θ, r goes from 0 until the point (r, θ) is on the line $y = 1$, that is, on the line $r \sin \theta = 1$. Thus the upper limit of r for each θ is $1/(\sin \theta)$. The description of R is

$$\frac{\pi}{4} \le \theta \le \frac{\pi}{2}, \qquad 0 \le r \le \frac{1}{\sin \theta}. \quad ■$$

In general, cross sections by rays lead to descriptions of plane regions of the form:

$$\alpha \le \theta \le \beta, \qquad r_1(\theta) \le r \le r_2(\theta).$$

A Basic Difference Between Rectangular and Polar Coordinates

Before we can set up an iterated integral in polar coordinates for $\int_R f(P)\, dA$ we must contrast certain properties of rectangular and polar coordinates.

Consider all points (x, y) in the plane that satisfy the inequalities

$$x_0 \le x \le x_0 + \Delta x \qquad \text{and} \qquad y_0 \le y \le y_0 + \Delta y,$$

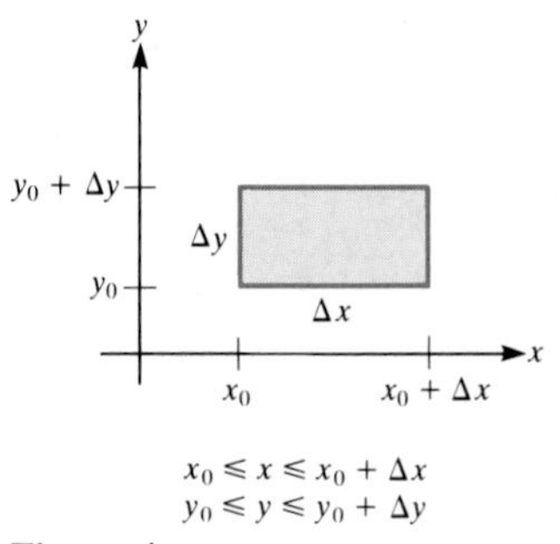

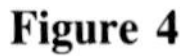

Figure 4

where x_0, Δx, y_0, and Δy are fixed numbers with Δx and Δy positive. The set is a rectangle of sides Δx and Δy shown in Fig. 4. The area of this rectangle is simply the product of Δx and Δy; that is,

$$\text{Area} = \Delta x\, \Delta y. \tag{1}$$

This will be contrasted with the case of polar coordinates.

Consider the set in the plane consisting of the points (r, θ) such that

$$r_0 \le r \le r_0 + \Delta r \qquad \text{and} \qquad \theta_0 \le \theta \le \theta_0 + \Delta\theta,$$

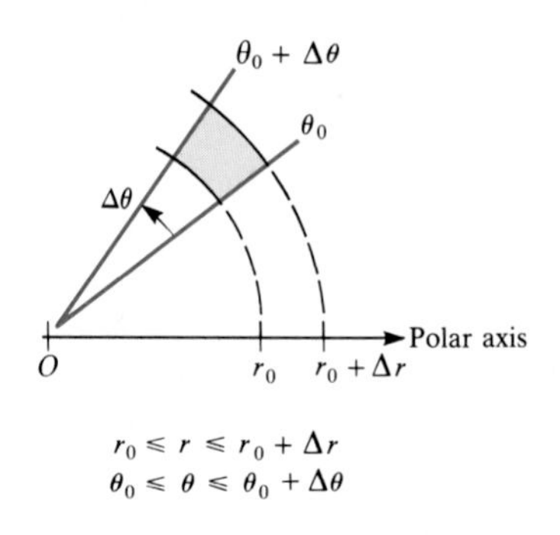

Figure 5

where r_0, Δr, θ_0, and $\Delta\theta$ are fixed numbers, with r_0, Δr, and $\Delta\theta$ all positive, as shown in Fig. 5.

When Δr and $\Delta\theta$ are small, the set is approximately a rectangle, one side of which has length Δr and the other, $r_0\,\Delta\theta$. So its area is approximately $r_0\,\Delta r\,\Delta\theta$. (For its exact value, see Exercise 37.) In this case,

$$\text{Area} \approx r_0\,\Delta r\,\Delta\theta. \tag{2}$$

The area is *not* the product of Δr and $\Delta\theta$. (It couldn't be since $\Delta\theta$ is in radians, a dimensionless quantity—''arc length subtended on a circle divided by length of radius''—so $\Delta r\,\Delta\theta$ has the dimension of length, not of area.) The presence of this extra factor r_0 will be reflected in the integrand we use when integrating in polar coordinates. It is necessary to replace dA by $r\,dr\,d\theta$, not simply by $dr\,d\theta$.

How to Evaluate $\int_R f(P)\,dA$ with Polar Coordinates

The method for computing $\int_R f(P)\,dA$ with polar coordinates involves an iterated integral where the dA is replaced by $r\,dr\,d\theta$. (A more detailed explanation of why the r must be added is outlined in Exercise 50. See also App. L.)

EVALUATING $\int_R f(P)\,dA$ IN POLAR COORDINATES

1 Express $f(P)$ in terms of r and θ: $f(r, \theta)$.

2 Describe the region R in polar coordinates:

$$\alpha \le \theta \le \beta, \qquad r_1(\theta) \le r \le r_2(\theta).$$

3 Evaluate the repeated integral:

$$\int_\alpha^\beta \int_{r_1(\theta)}^{r_2(\theta)} f(r, \theta)\,r\,dr\,d\theta.$$

Notice the factor r in the integrand.

EXAMPLE 4 Let R be the semicircle of radius a shown in Fig. 6. Let $f(P)$ be the distance from a point P to the x axis. Evaluate $\int_R f(P)\,dA$ by a repeated integral in polar coordinates.

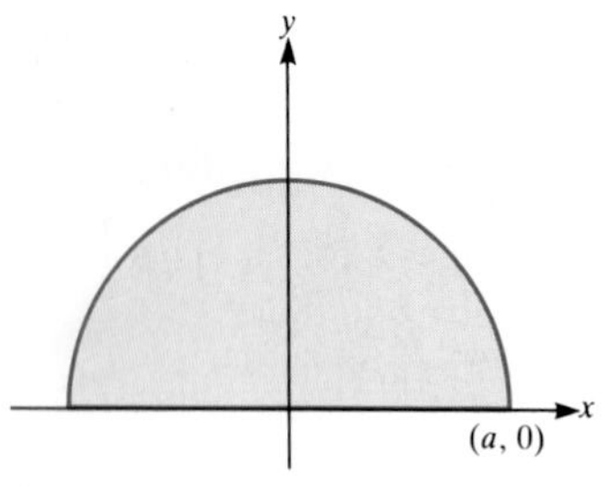

Figure 6

SOLUTION In polar coordinates, R has the description

$$0 \le \theta \le \pi, \qquad 0 \le r \le a.$$

The distance from P to the x axis is, in rectangular coordinates, y. Since $y = r\sin\theta$, $f(P) = r\sin\theta$. Thus,

Notice this r, needed when using polar coordinates.

$$\int_R f(P)\,dA = \int_0^\pi \left[\int_0^a (r\sin\theta)\,r\,dr\right] d\theta.$$

From here on the calculations are like those in the preceding section.

The calculation of the repeated integral is like that for a repeated integral in rectangular coordinates. First, evaluate the inside integral:

$$\int_0^a r^2\sin\theta\,dr = \sin\theta\int_0^a r^2\,dr = \sin\theta\left(\frac{r^3}{3}\right)\bigg|_0^a = \frac{a^3\sin\theta}{3}.$$

The outer integral is therefore

$$\int_0^{\pi} \frac{a^3 \sin\theta}{3}\,d\theta = \frac{a^3}{3}\int_0^{\pi} \sin\theta\,d\theta = \frac{a^3}{3}(-\cos\theta)\Big|_0^{\pi}$$

$$= \frac{a^3}{3}[(-\cos\pi) - (-\cos 0)] = \frac{a^3}{3}(1+1) = \frac{2a^3}{3}.$$

Thus
$$\int_R y\,dA = \frac{2a^3}{3}. \quad ■$$

Example 5 refers to a ball of radius a. Generally, we will distinguish between a **ball**, which is a solid region, and a **sphere**, which is only the surface of a ball.

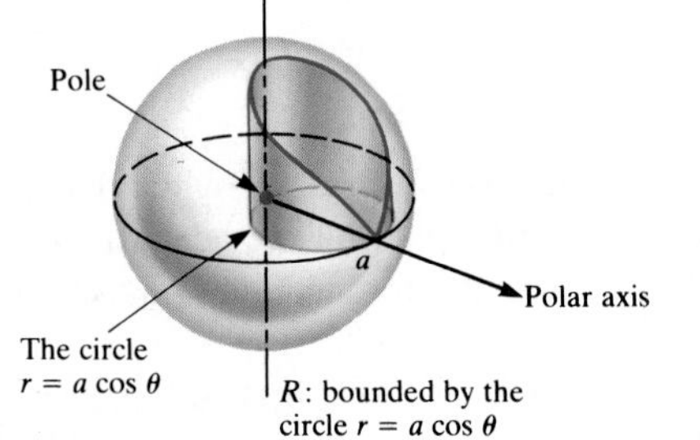

Figure 7

EXAMPLE 5 A ball of radius a has its center at the pole of a polar coordinate system. Find the volume of the part of the ball that lies above the plane region R bounded by the curve $r = a\cos\theta$. (See Fig. 7.)

SOLUTION It is necessary to describe R and f in polar coordinates, where $f(P)$ is the length of a cross section of the solid made by a vertical line through P. R is described as follows: r goes from 0 to $a\cos\theta$ for each θ in $[-\pi/2, \pi/2]$, that is,

$$-\frac{\pi}{2} \le \theta \le \frac{\pi}{2}, \qquad 0 \le r \le a\cos\theta.$$

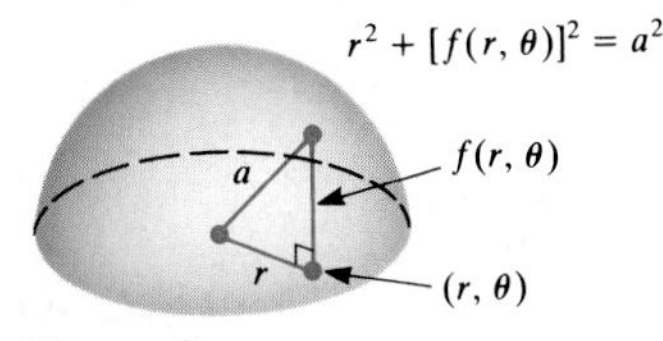

Figure 8

To express $f(P)$ in polar coordinates, consider Fig. 8, which shows the top half of a ball of radius a. By the pythagorean theorem,

$$r^2 + [f(r,\theta)]^2 = a^2.$$

Thus
$$f(r,\theta) = \sqrt{a^2 - r^2}.$$

Consequently,

$$\text{Volume} = \int_R f(P)\,dA = \int_{-\pi/2}^{\pi/2}\left(\int_0^{a\cos\theta} \sqrt{a^2 - r^2}\,r\,dr\right)d\theta.$$

Remember to double.

Exploiting symmetry, compute half the volume, keeping θ in $[0, \pi/2]$, and then double the result:

$$\int_0^{a\cos\theta} \sqrt{a^2 - r^2}\,r\,dr = \frac{-(a^2 - r^2)^{3/2}}{3}\Big|_0^{a\cos\theta}$$

$$= -\left[\frac{(a^2 - a^2\cos^2\theta)^{3/2}}{3} - \frac{(a^2)^{3/2}}{3}\right]$$

$$= \frac{a^3}{3} - \frac{(a^2 - a^2\cos^2\theta)^{3/2}}{3}$$

$$= \frac{a^3}{3} - \frac{a^3(1 - \cos^2\theta)^{3/2}}{3}$$

$$= \frac{a^3}{3}(1 - \sin^3\theta).$$

(The trigonometric formula used above, $\sin\theta = \sqrt{1 - \cos^2\theta}$, is true when $0 \le \theta \le \pi/2$ but not when $-\pi/2 \le \theta < 0$.)

The second integration is then carried out:

$$\int_0^{\pi/2} \frac{a^3}{3}(1 - \sin^3 \theta)\, d\theta = \frac{a^3}{3} \int_0^{\pi/2} [1 - (1 - \cos^2 \theta) \sin \theta]\, d\theta$$
$$= \frac{a^3}{3}\left(\theta + \cos \theta - \frac{\cos^3 \theta}{3}\right)\Bigg|_0^{\pi/2}$$
$$= \frac{a^3}{3}\left[\frac{\pi}{2} - \left(1 - \frac{1}{3}\right)\right]$$
$$= a^3\left(\frac{3\pi - 4}{18}\right).$$

We remembered. The total volume is twice as large:

$$a^3\left(\frac{3\pi - 4}{9}\right). \quad ■$$

EXAMPLE 6 A circular disk of radius a is formed of a material which has a density at each point equal to the distance from that point to the center.
(*a*) Set up a repeated integral in rectangular coordinates for the total mass of the disk.
(*b*) Set up a repeated integral in polar coordinates for the total mass of the disk.
(*c*) Compute the easier one.

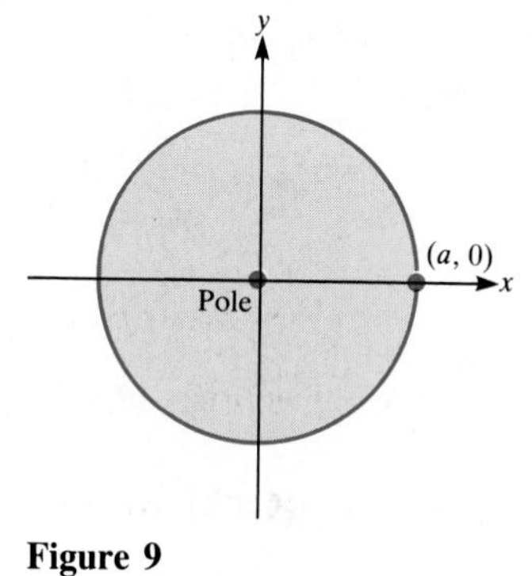

Figure 9

SOLUTION The disk is shown in Fig. 9.
(*a*) (Rectangular coordinates) The density $\sigma(P)$ at the point $P = (x, y)$ is $\sqrt{x^2 + y^2}$. The disk has the description

$$-a \leq x \leq a, \qquad -\sqrt{a^2 - x^2} \leq y \leq \sqrt{a^2 - x^2}.$$

Thus $$\text{Mass} = \int_R \sigma(P)\, dA = \int_{-a}^{a} \left(\int_{-\sqrt{a^2-x^2}}^{\sqrt{a^2-x^2}} \sqrt{x^2 + y^2}\, dy\right)dx.$$

(*b*) (Polar coordinates) The density $\sigma(P)$ at $P = (r, \theta)$ is r. The disk has the description

$$0 \leq \theta \leq 2\pi, \qquad 0 \leq r \leq a.$$

Thus $$\text{Mass} = \int_R \sigma(P)\, dA = \int_0^{2\pi} \left(\int_0^{a} r \cdot r\, dr\right) d\theta$$
$$= \int_0^{2\pi} \left(\int_0^{a} r^2\, dr\right) d\theta.$$

(*c*) Even the first integration in the repeated integral in (*a*) would be tedious. However, the repeated integral in (*b*) is a delight: The first integration gives

$$\int_0^a r^2\, dr = \frac{r^3}{3}\Bigg|_0^a = \frac{a^3}{3}.$$

The second integration gives

$$\int_0^{2\pi} \frac{a^3}{3}\, d\theta = \frac{a^3\theta}{3}\Bigg|_0^{2\pi} = \frac{2\pi a^3}{3}.$$

The total mass is $2\pi a^3/3$. ■

Section Summary

We saw how to calculate an integral $\int_R f(P)\,dA$ by introducing polar coordinates. In this case

$$\int_R f(P)\,dA = \int_\alpha^\beta \int_{r_1(\theta)}^{r_2(\theta)} f(r,\theta)\,r\,dr\,d\theta.$$

The extra r in the integrand is due to the fact that a small region corresponding to changes dr and $d\theta$ has area approximately $r\,dr\,d\theta$ (not $dr\,d\theta$). Polar coordinates are convenient when either the function f or the region R has a simple description in terms of r and θ.

EXERCISES FOR SEC. 15.4: COMPUTING $\int_R f(P)\,dA$ USING POLAR COORDINATES

In Exercises 1 to 6 draw and describe the given regions in the form $\alpha \le \theta \le \beta$, $r_1(\theta) \le r \le r_2(\theta)$.

1 The region inside the curve $r = 3 + \cos\theta$.

2 The region between the curve $r = 3 + \cos\theta$ and the curve $r = 1 + \sin\theta$.

3 The triangle whose vertices have the rectangular coordinates $(0, 0)$, $(1, 1)$, and $(1, \sqrt{3})$.

4 The circle bounded by the curve $r = 3\sin\theta$.

5 The region shown in Fig. 10.

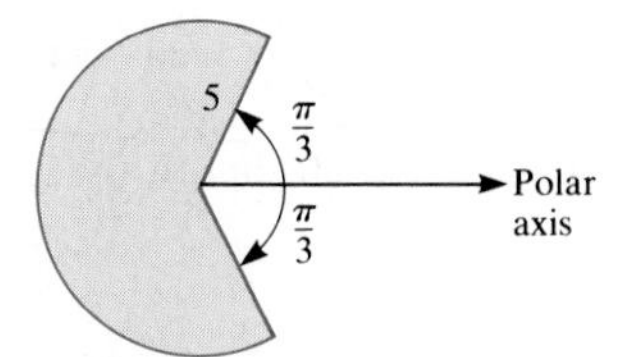

Figure 10

6 The region in the loop of the three-leaved rose, $r = \sin 3\theta$, that lies in the first quadrant.

7 (*a*) Draw the region R bounded by the lines $y = 1$, $y = 2$, $y = x$, $y = x/\sqrt{3}$. Describe R in terms of (*b*) horizontal cross sections, (*c*) vertical cross sections, (*d*) cross sections by polar rays.

8 (*a*) Draw the region R whose description is given by

$$-2 \le y \le 2, \qquad -\sqrt{4 - y^2} \le x \le \sqrt{4 - y^2}.$$

(*b*) Describe R by vertical cross sections.
(*c*) Describe R by cross sections by polar rays.

9 Describe in polar coordinates the square whose vertices have rectangular coordinates $(0, 0)$, $(1, 0)$, $(1, 1)$, $(0, 1)$.

10 Describe the trapezoid whose vertices have rectangular coordinates $(0, 1)$, $(1, 1)$, $(2, 2)$, $(0, 2)$
(*a*) in polar coordinates,
(*b*) by horizontal cross sections,
(*c*) by vertical cross sections.

In Exercises 11 to 14 draw the regions and evaluate $\int_R r^2\,dA$ for the given regions R.

11 $-\pi/2 \le \theta \le \pi/2$, $0 \le r \le \cos\theta$

12 $0 \le \theta \le \pi/2$, $0 \le r \le \sin^2\theta$

13 $0 \le \theta \le 2\pi$, $0 \le r \le 1 + \cos\theta$

14 $0 \le \theta \le 0.3$, $0 \le r \le \sin 2\theta$

In Exercises 15 to 18 draw R and evaluate $\int_R y^2\,dA$ for the given regions R.

15 The circle of radius a, center at the pole.

16 The circle of radius a, center at $(a, 0)$ in polar coordinates.

17 The region within the cardioid $r = 1 + \sin\theta$.

18 The region within one leaf of the four-leaved rose $r = \sin 2\theta$.

In Exercises 19 and 20 use iterated integrals in polar coordinates to find the given point.

19 The center of mass of the region within the cardioid $r = 1 + \cos\theta$.

20 The center of mass of the region within the leaf of $r = \cos 3\theta$ that lies along the polar axis.

The average of a function $f(P)$ over a region R in the plane is defined as $\int_R f(P)\,dA/(\text{Area of } R)$. In each of Exercises 21 to 24 find the average of the given function over the given region.

21 $f(P)$ is the distance from P to the pole; R is one leaf of the three-leaved rose, $r = \sin 3\theta$.

22 $f(P)$ is the distance from P to the x axis; R is the region between the rays $\theta = \pi/6$, $\theta = \pi/4$ and the circles $r = 2$, $r = 3$.

23 $f(P)$ is the distance from P to a fixed point on the border of a disk R of radius a. (*Hint:* Choose the pole wisely.)

24 $f(P)$ is the distance from P to the x axis; R is the region within the cardioid $r = 1 + \cos\theta$.

In Exercises 25 to 28 evaluate the given iterated integrals using polar coordinates. Note that in Exercises 25 and 28 polar coordinates are convenient because of the form of the integrand.

25 $\int_0^1 \left(\int_0^x \sqrt{x^2 + y^2}\, dy \right) dx$

26 $\int_0^1 \left(\int_0^{\sqrt{1-x^2}} x^3\, dy \right) dx$

27 $\int_0^1 \left(\int_x^{\sqrt{1-x^2}} xy\, dy \right) dx$

28 $\int_1^2 \left[\int_{x/\sqrt{3}}^{\sqrt{3}x} (x^2 + y^2)^{3/2}\, dy \right] dx$

29 Evaluate the integrals over the given regions.
(*a*) $\int_R \cos(x^2 + y^2)\, dA$; R is the portion in the first quadrant of the disk of radius a centered at the origin.
(*b*) $\int_R \sqrt{x^2 + y^2}\, dA$; R is the triangle bounded by the line $y = x$, the line $x = 2$, and the x axis.

30 Find the volume of the region above the paraboloid $z = x^2 + y^2$ and below the plane $z = x + y$.

31 Let R be a plane lamina in the shape of the region bounded by the graph of the equation $r = 2a \sin \theta$ $(a > 0)$. If the variable density of the lamina is given by $\sigma(r, \theta) = \sin \theta$, find the center of mass R.

32 The area of a region R is equal to $\int_R 1\, dA$. Use this to find the area of a disk of radius a. (Use an iterated integral in polar coördinates.)

In Exercises 33 to 36 find the moment of inertia of a homogeneous lamina of mass M of the given shape, around the given line.

33 A disk of radius a, about the line perpendicular to it through its center.

34 A disk of radius a, about a line perpendicular to it through a point on the circumference.

35 A disk of radius a, about a diameter.

36 A disk of radius a, about a tangent.

---

37 Find the area of the shaded region in Fig. 5 as follows:
(*a*) Find the area of the ring between two circles, one of radius r_0, the other of radius $r_0 + \Delta r$.
(*b*) What fraction of the area in (*a*) is included between two rays whose angles differ by $\Delta\theta$?
(*c*) Show that the area of the shaded region in Fig. 5 is precisely

$$\left(r_0 + \frac{\Delta r}{2} \right) \Delta r\, \Delta\theta.$$

38 In Example 5 we computed half the volume and doubled the result. Evaluate the repeated integral

$$\int_{-\pi/2}^{\pi/2} \left(\int_0^{a\cos\theta} \sqrt{a^2 - r^2}\, r\, dr \right) d\theta$$

directly. The result should still be $a^3(3\pi - 4)/9$. *Caution:* Use trigonometric formulas with care.

S. P. Thompson, in *Life of Lord Kelvin* (Macmillan, London, 1910), wrote,

> Once when lecturing to a class he [the physicist Lord Kelvin] used the word "mathematician" and then interrupting himself asked his class: "Do you know what a mathematician is?" Stepping to his blackboard he wrote upon it: $\int_{-\infty}^{\infty} e^{-x^2}\, dx = \sqrt{\pi}$. Then putting his finger on what he had written, he turned to his class and said, "A mathematician is one to whom that is as obvious as that twice two makes four is to you."

On the other hand, the mathematician Littlewood wrote,

> Many things are not accessible to intuition at all, the value of $\int_0^\infty e^{-x^2}\, dx$ for instance.

From J. E. Littlewood, "Newton and the Attraction of the Sphere," *Mathematical Gazette,* vol. 63, 1948. Now consider Exercise 39.

39 This exercise shows that $\int_0^\infty e^{-x^2}\, dx = \sqrt{\pi}/2$. Let R_1, R_2, R_3 be the three regions indicated in Fig. 11, and $f(P) = e^{-r^2}$, where r is the distance from P to the origin. Hence $f(r, \theta) = e^{-r^2}$ in polar coordinates and in rectangular coordinates $f(x, y) = e^{-x^2-y^2}$. (Observe that R_1 is inside R_2 and R_2 is inside R_3.)
(*a*) Show that $\int_{R_1} f(P)\, dA = (\pi/4)(1 - e^{-a^2})$ and that $\int_{R_3} f(P)\, dA = (\pi/4)(1 - e^{-2a^2})$.
(*b*) By considering $\int_{R_2} f(P)\, dA$ and the results in (*a*), show that

$$\frac{\pi}{4}(1 - e^{-a^2}) < \left(\int_0^a e^{-x^2}\, dx \right)^2 < \frac{\pi}{4}(1 - e^{-2a^2}).$$

(*c*) Show that $\int_0^\infty e^{-x^2}\, dx = \sqrt{\pi}/2$.

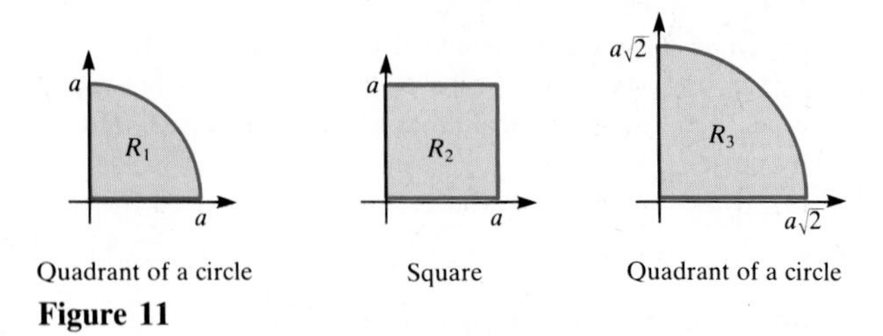

Figure 11

40 Figure 12 shows the "bell curve" or "normal curve" often used to assign grades in large classes. Using the fact established in Exercise 39 that $\int_0^\infty e^{-x^2}\, dx = \sqrt{\pi}/2$, show that the area under the curve in Fig. 12 is 1.

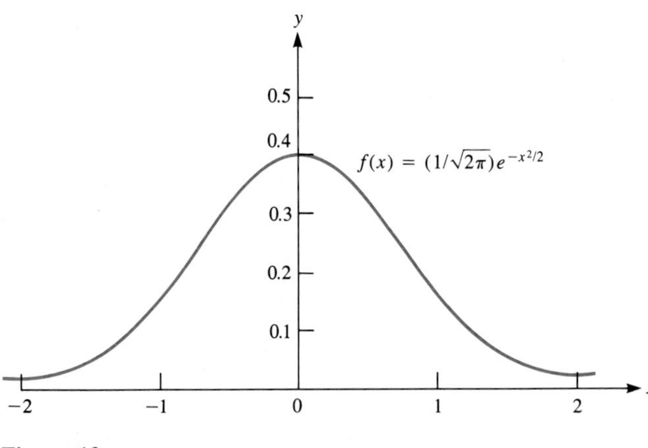

Figure 12

41 Find the center of mass of a homogeneous semicircular region of radius a.

42 What is wrong with this reasoning?

We shall obtain a repeated integral in polar coordinates equal to $\int_R f(P)\,dA$, where R has the description $\alpha \le \theta \le \beta$, $r_1(\theta) \le r \le r_2(\theta)$. Assuming that $f(P) \ge 0$, construct the solid whose base is R and whose height at P is $f(P)$. The cross section of this solid by a plane perpendicular to R and passing through the ray of angle θ is a plane section of area $A(\theta) = \int_{r_1(\theta)}^{r_2(\theta)} f(r, \theta)\,dr$. Since $V = \int_\alpha^\beta A(\theta)\,d\theta$,

$$\int_R f(P)\,dA = \int_\alpha^\beta \left[\int_{r_1(\theta)}^{r_2(\theta)} f(r, \theta)\,dr\right] d\theta.$$

This is a *wrong formula*, since r is not present in the integrand. Find the error.

43 (The spread of epidemics.) In the theory of a spreading epidemic it is assumed that the probability that a contagious individual infects an individual D miles away depends only on D. Consider a population that is uniformly distributed in a circular city whose radius is 1 mile. Assume that the probability we mentioned is proportional to $2 - D$. For a fixed point Q let $f(P) = 2 - \overline{PQ}$. Let R be the region occupied by the city.

(*a*) Why is the exposure of a person residing at Q proportional to $\int_R f(P)\,dA$, assuming that contagious people are uniformly distributed throughout the city?

(*b*) Compute this definite integral when Q is the center of town and when Q is on the edge of town.

(*c*) In view of (*b*), which is the safer place?

Transportation problems lead to integrals over plane sets, as Exercises 44 to 49 illustrate.

44 Show that the average travel distance from the center of a disk of area A to points in the disk is precisely $2\sqrt{A}/(3\sqrt{\pi}) \approx 0.376\sqrt{A}$.

45 Show that the average travel distance from the center of a regular hexagon of area A to points in the hexagon is

$$\frac{\sqrt{2A}}{3^{3/4}}\left(\frac{1}{3} + \frac{\ln 3}{4}\right) \approx 0.377\sqrt{A}.$$

46 Show that the average travel distance from the center of a square of area A to points in the square is $(\sqrt{2} + \ln \tan 3\pi/8)\sqrt{A}/6 \approx 0.383\sqrt{A}$.

47 Show that the average travel distance from the centroid of an equilateral triangle of area A to points in the triangle is

$$\frac{\sqrt{A}}{3^{9/4}}\left(2\sqrt{3} + \ln \tan \frac{5\pi}{12}\right) \approx 0.404\sqrt{A}.$$

In Exercises 44 to 47 the distance is the ordinary straight-line distance. In cities the usual street pattern suggests that the "metropolitan" distance between the points (x_1, y_1) and (x_2, y_2) should be measured by $|x_1 - x_2| + |y_1 - y_2|$.

48 Show that if in Exercise 44 metropolitan distance is used, then the average is $8\sqrt{A}/(3\pi^{3/2}) \approx 0.479\sqrt{A}$.

49 Show that if in Exercise 46 metropolitan distance is used, then the average is $\sqrt{A}/2$. In most cities the metropolitan average tends to be about 25 percent larger than the direct-distance average.

50 We asserted that the extra factor r must be inserted into the integrand because the area of a small patch is approximately $r\,\Delta r\,\Delta\theta$, not $\Delta r\,\Delta\theta$. This exercise justifies this claim. Let R be the region bounded by $\theta = \alpha$, $\theta = \beta$, $r = a$, and $r = b$. Divide it into n^2 small pieces by rays $\theta_0 = \alpha, \theta_1, \ldots, \theta_n = \beta$ and by parts of circles of radii $r_0 = a, r_1, r_2, \ldots, r_n = b$. (See Fig. 13.) The patch determined by θ_{i-1}, θ_i, r_{j-1}, and r_j has area precisely

$$\frac{r_{j-1} + r_j}{2}(r_j - r_{j-1})(\theta_i - \theta_{i-1})$$

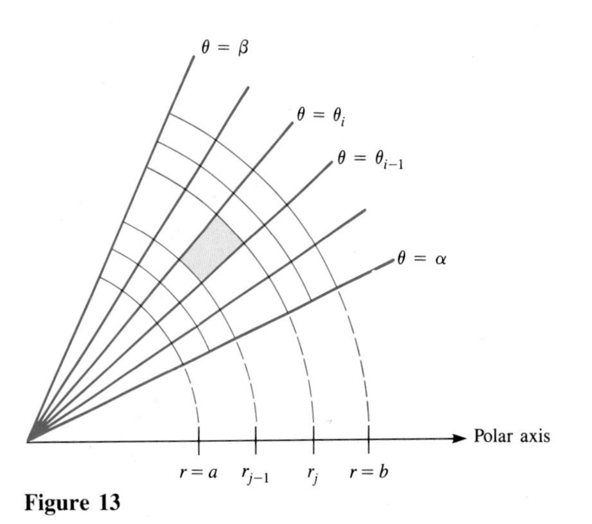

Figure 13

as shown in Exercise 37. As a sampling point use

$$P_{ij} = \left(\frac{r_{j-1} + r_j}{2}, \theta_i\right).$$

Then

$$\sum_{i=1}^{n}\left(\sum_{j=1}^{n} f(P_{ij})\left(\frac{r_{j-1} + r_j}{2}\right)(r_j - r_{j-1})(\theta_i - \theta_{i-1})\right) \qquad (3)$$

is an approximation of $\int_R f(P)\, dA$.

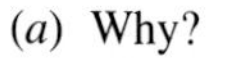

(*a*) Why?

(*b*) Why is $\sum_{j=1}^{n} f(P_{ij})\left(\frac{r_{j-1} + r_j}{2}\right)(r_j - r_{j-1})$ an approximation of $\int_a^b f(r, \theta_i)\, r\, dr$?

(*c*) Let $g(\theta) = \int_a^b f(r, \theta)\, r\, dr$. Show that (3) is approximately equal to $\Sigma_{i=1}^{n}\, g(\theta_i)(\theta_i - \theta_{i-1})$.

(*d*) Explain why (3) is an approximation of the repeated integral $\int_\alpha^\beta (\int_a^b f(r, \theta)\, r\, dr)\, d\theta$.

(*e*) Explain why $\int_R f(P)\, dA = \int_\alpha^\beta (\int_a^b f(r, \theta)\, r\, dr)\, d\theta$.

15.5 THE DEFINITE INTEGRAL OF A FUNCTION OVER A REGION IN SPACE

The notion of a definite integral over an interval in the line or over a plane region generalizes to integrals over solids located in space. (These solids will be assumed to be bounded by smooth surfaces or planes.) Rather than plunge directly into the definition, let us first illustrate the idea with a problem.

A Mass Problem

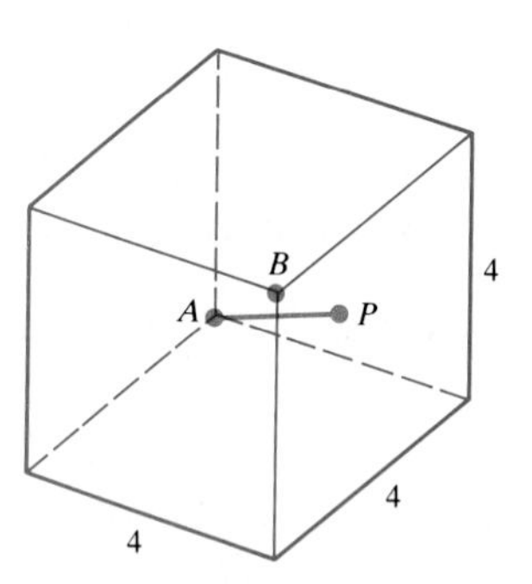

The density at P is the square of the distance $\overline{AP}$. P is a typical point in the cube.

Figure 1

PROBLEM A cube of side 4 centimeters is made of a material of varying density. Near one corner A it is very light; at the opposite corner it is very dense. In fact, the density $f(P)$ (in grams per cubic centimeter) at any point P in the cube is the square of the distance from A to P (in centimeters). How do we estimate the mass of the cube? (See Fig. 1.)

APPROACH Proceed exactly as in the case of the string of Sec. 5.1 and the rectangular plate of Sec. 15.1. First, partition the cube into regions $R_1, R_2, \ldots, R_n$; then compute the density at a selected point P_i in each R_i and form the sum

$$f(P_1)V_1 + f(P_2)V_2 + \cdots + f(P_n)V_n,$$

where V_i is the volume of R_i. As the R_i's become smaller, we obtain more reliable estimates of the total mass of the cube.

Observe first of all that the maximum density is the square of the length of the longest diagonal. This density is $\overline{AB}^2 = 4^2 + 4^2 + 4^2 = 48$ grams per cubic centimeter. Since the total volume is $4 \cdot 4 \cdot 4 = 64$ cubic centimeters, the total mass is less than $48 \cdot 64 = 3{,}072$ grams.

The arithmetic in evaluating even the simplest approximating sum is tedious. It may be of value, though, to go through the drudgery of computing one such sum. The following is a sample.

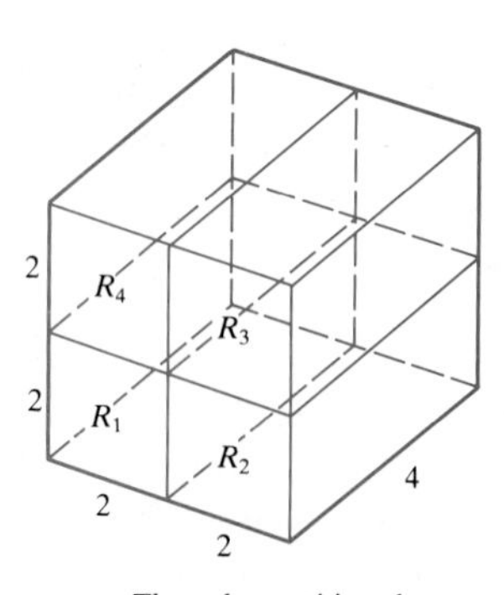

The cube partitioned into four boxes, R_1, R_2, R_3, R_4.

Figure 2

Partition the cube into four 2- by 2- by 4-centimeter boxes, as shown and labeled in Fig. 2. This table displays the computation:

Region	*Volume*	*Minimum Density*	*Maximum Density*	*Mass Is Between*
R_1	16	0	$2^2 + 2^2 + 4^2 = 24$	0 and 384 grams
R_2	16	$2^2 = 4$	$4^2 + 2^2 + 4^2 = 36$	64 and 576 grams
R_3	16	$2^2 + 2^2 = 8$	$4^2 + 4^2 + 4^2 = 48$	128 and 768 grams
R_4	16	$2^2 = 4$	$2^2 + 4^2 + 4^2 = 36$	64 and 576 grams

Thus the mass of the cube is between

$$0 + 64 + 128 + 64 = 256 \text{ grams}$$

and

$$384 + 576 + 768 + 576 = 2{,}304 \text{ grams}.$$

This is more precise information than the fact that the mass is less than 3,072 grams. ■

If the cube is cut into smaller regions, perhaps sixty-four 1- by 1- by 1-centimeter cubes, more accurate estimates can be made. The important idea is that the procedure for making an approximation is practically the same as the one that led to the sums

$$\sum_{i=1}^{n} f(c_i)(x_i - x_{i-1})$$

of Chap. 5 and to the sums

$$\sum_{i=1}^{n} f(P_i)A_i$$

of Sec. 15.1.

Definite Integrals over Spatial Regions

Two definitions are needed before defining the definite integral of a function over a region R in space.

Definition *Diameter of a region in space.* Let S be a set of points in space bounded by some surface. The **diameter** of S is the largest distance between two points of S.

For instance, the diameter of a cube of side s is $s\sqrt{3}$, the length of its longest diagonal. The diameter of a ball is its customary diameter.

Definition *Mesh of a partition in space.* Let $R_1, R_2, \ldots, R_n$ be a partition of a region R in space. The **mesh** of this partition is the largest of the diameters of the regions $R_1, R_2, \ldots, R_n$.

The typical function of interest will have some region R in space as its domain. A function f will assign to each point P in R a number, denoted $f(P)$. For the sake of concreteness, think of $f(P)$ as the density at P, or the temperature at P.

Definition *The definite integral of a function f over a set R in space.* Let f be a function that assigns to each point P of a region R in space a number

$f(P)$. Consider the typical sum

$$f(P_1)V_1 + f(P_2)V_2 + \cdots + f(P_n)V_n,$$

formed from a partition $R_1, R_2, \ldots, R_n$ of R, where V_i is the volume of R_i, and P_i is in R_i. If these sums approach a certain number as the mesh of the partition shrinks toward 0 (no matter how P_i is chosen in R_i), we call that certain number the **definite integral** of f over the region R. The definite integral of f over R is denoted

$$\int_R f(P)\, dV.$$

If $f(P)$ is thought of as the density at P of some solid matter, the definite integral can be interpreted as the total mass of the solid.

The notation $\iiint_R f(P)dV$ is also used, to emphasize the three dimensions.

EXAMPLE 1 If $f(P) = 1$ for each point P in a solid region R, compute $\int_R f(P)\, dV$.

SOLUTION Each approximating sum $\Sigma_{i=1}^n f(P_i)V_i$ has the value

$$\sum_{i=1}^n 1 \cdot V_i = V_1 + V_2 + \cdots + V_n = \text{Volume of } R.$$

Hence
$$\int_R f(P)\, dV = \text{Volume of } R,$$

a fact that will be useful for computing volumes. ■

The average value of a function f defined on a region R in space is defined as

Average of a function

$$\frac{\int_R f(P)\, dV}{\text{Volume of } R}.$$

This is the analog of the definition of the average of a function over an interval (Sec. 5.5) or the average of a function over a plane region (Sec. 15.1). If f describes the density of matter in R, then the average value of f is the density of a *homogeneous* solid occupying R and having the same total mass as the given solid. [For if the number

$$\frac{\int_R f(P)\, dV}{\text{Volume of } R}$$

is multiplied by the volume of R, the result is

$$\int_R f(P)\, dV,$$

which is the total mass.]

δ usually denotes density in a solid.

Moments, centers of mass, and moments of inertia are defined for spatial regions very much as they are for planar regions. If $\delta(P)$ is the density of the material at P, then we have:

Name	Formula
M = Mass	$\int_R \delta(P)\,dV$
M_{xy} = Moment with respect to the xy plane	$\int_R z\,\delta(P)\,dV$
M_{xz} = Moment with respect to the xz plane	$\int_R y\,\delta(P)\,dV$
M_{yz} = Moment with respect to the yz plane	$\int_R x\,\delta(P)\,dV$
$(\bar{x}, \bar{y}, \bar{z})$ = Center of mass (also called **center of gravity**)	$\bar{x} = \dfrac{\int_R x\,\delta(P)\,dV}{\text{Mass in } R}$, $\bar{y} = \dfrac{\int_R y\,\delta(P)\,dV}{\text{Mass in } R}$, $\bar{z} = \dfrac{\int_R z\,\delta(P)\,dV}{\text{Mass in } R}$
$(\bar{x}, \bar{y}, \bar{z})$ = Centroid	$\bar{x} = \dfrac{\int_R x\,dV}{\text{Volume of } R}$, $\bar{y} = \dfrac{\int_R y\,dV}{\text{Volume of } R}$, $\bar{z} = \dfrac{\int_R z\,dV}{\text{Volume of } R}$
I_x = Moment of inertia with respect to x axis	$\int (y^2 + z^2)\,\delta(P)\,dV$
I_y = Moment of inertia with respect to y axis	$\int_R (x^2 + z^2)\,\delta(P)\,dV$
I_z = Moment of inertia with respect to z axis	$\int_R (x^2 + y^2)\,\delta(P)\,dV$
Moment of inertia with respect to any line L	$\int_R f(P)\,\delta(P)\,dV$, where $f(P)$ = square of distance from P to L

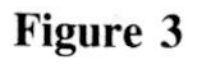

Figure 3

Describing a Solid Region

In order to evaluate definite integrals over spatial regions, it is necessary to describe these regions in terms of a coordinate system.

A description of a solid region in rectangular coordinates will have the form

$$a \le x \le b, \qquad y_1(x) \le y \le y_2(x), \qquad z_1(x, y) \le z \le z_2(x, y).$$

(This is the order x, y, then z. There are six possible orders, as you may check.) The inequalities on x and y describe the ''shadow'' or projection of the region on the xy plane. The inequalities for z then tell how z varies on a line parallel to the z axis and passing through the point (x, y) in the projection. (See Fig. 3.)

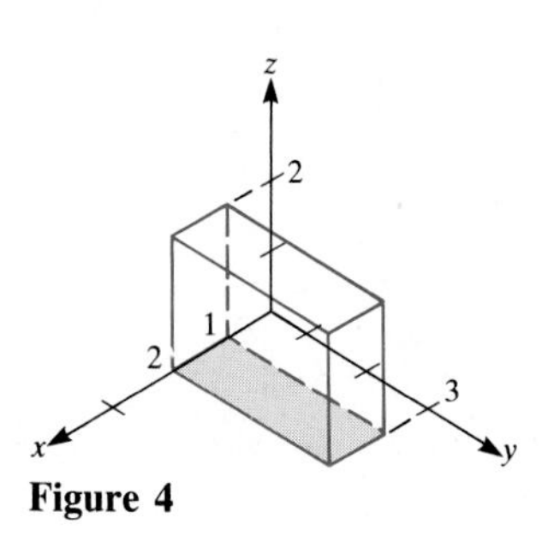

Figure 4

EXAMPLE 2 Describe in terms of x, y, and z the rectangular box shown in Fig. 4.

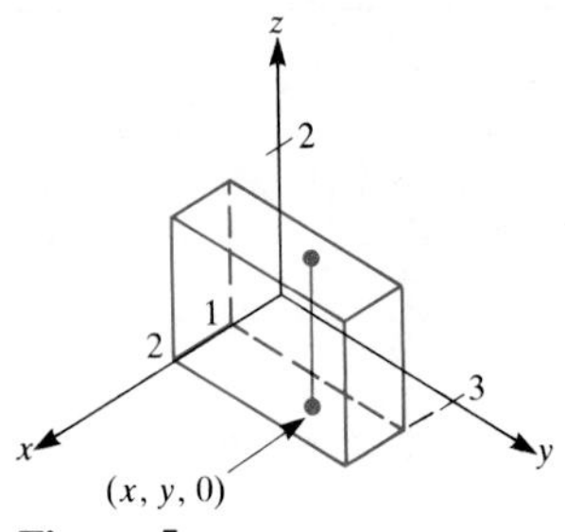

Figure 5

SOLUTION The shadow of the box on the xy plane has a description $1 \le x \le 2$, $0 \le y \le 3$. For each point in this shadow, z varies from 0 to 2, as shown in Fig. 5. So the description of the box is

$$1 \le x \le 2, \qquad 0 \le y \le 3, \qquad 0 \le z \le 2,$$

which is read from left to right as "x goes from 1 to 2; for each such x, the variable y goes from 0 to 3; for each such x and y, the variable z goes from 0 to 2."

Of course, we could have changed the order of x and y in the description of the shadow or projected the box on one of the other two coordinate planes. (All told, there are six possible descriptions.) ■

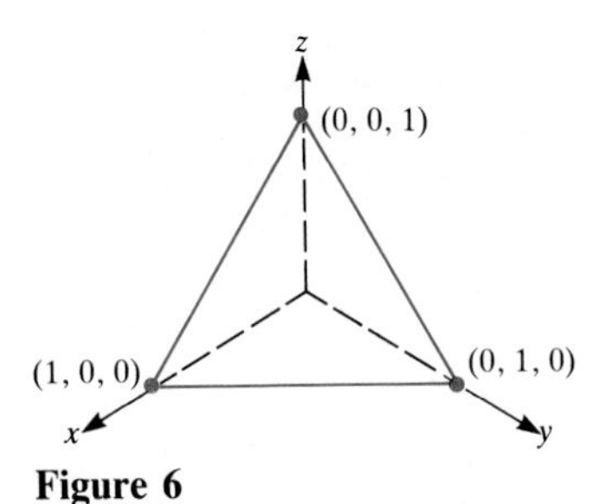

Figure 6

EXAMPLE 3 Describe by cross sections the tetrahedron bounded by the planes $x = 0$, $y = 0$, $z = 0$, and $x + y + z = 1$, as shown in Fig. 6.

SOLUTION For the sake of variety, project the tetrahedron onto the xz plane. The shadow is shown in Fig. 7. A description of the shadow is

$$0 \le x \le 1, \qquad 0 \le z \le 1 - x,$$

since the slanted edge has the equation $x + z = 1$. For each point (x, z) in this shadow, y ranges from 0 up to the value of y that satisfies the equation $x + y + z = 1$, that is, up to $y = 1 - x - z$. (See Fig. 8.) A description of the tetrahedron is

$$0 \le x \le 1, \qquad 0 \le z \le 1 - x, \qquad 0 \le y \le 1 - x - z.$$

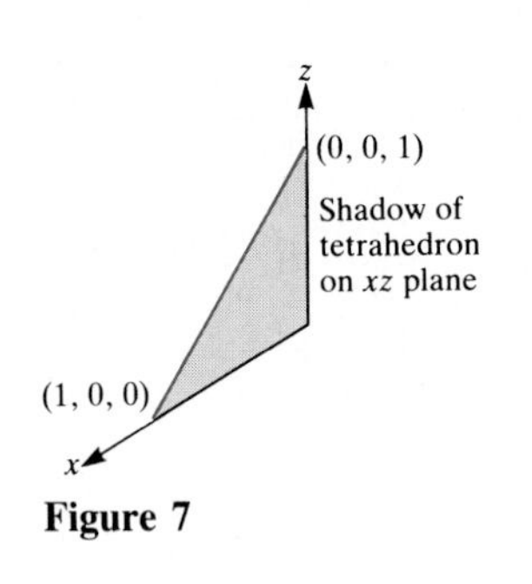

Figure 7

That is, x goes from 0 to 1; for each x, z goes from 0 to $1 - x$; for each x and z, y goes from 0 to $1 - x - z$. ■

EXAMPLE 4 Describe in rectangular coordinates the ball of radius 4 whose center is at the origin.

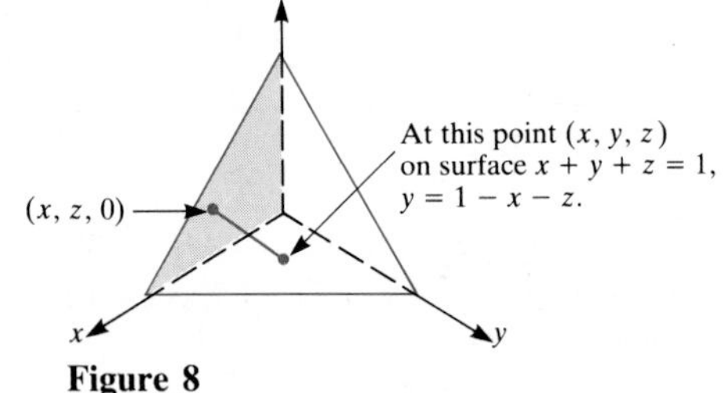

Figure 8

SOLUTION The shadow of the ball on the xy plane is the disk of radius 4 and center (0, 0). Its description is

$$-4 \le x \le 4, \qquad -\sqrt{16 - x^2} \le y \le \sqrt{16 - x^2}.$$

Hold (x, y) fixed in the xy plane and consider the way z varies on the line parallel to the z axis that passes through the point $(x, y, 0)$. Since the sphere that bounds the ball has the equation

$$x^2 + y^2 + z^2 = 16,$$

for each appropriate (x, y), z varies from

$$-\sqrt{16 - x^2 - y^2} \qquad \text{to} \qquad \sqrt{16 - x^2 - y^2}.$$

This describes the line segment shown in Fig. 9.

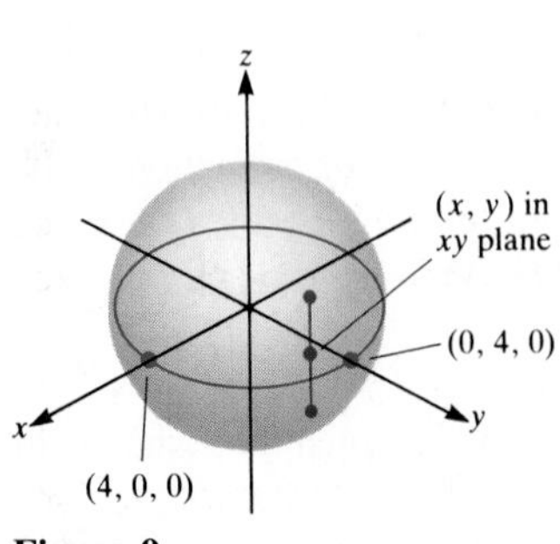

Figure 9

The ball, therefore, has a description

$$-4 \le x \le 4,$$
$$-\sqrt{16 - x^2} \le y \le \sqrt{16 - x^2},$$
$$-\sqrt{16 - x^2 - y^2} \le z \le \sqrt{16 - x^2 - y^2}. \quad ■$$

Repeated Integrals for $\int_R f(P)\, dV$

The repeated integral in rectangular coordinates for $\int_R f(P)\, dV$ is similar to that for evaluating integrals over plane sets. It involves three integrations instead of two. The limits of integration are determined by the description of R in rectangular coordinates. If R has the description

$$a \le x \le b, \qquad y_1(x) \le y \le y_2(x), \qquad z_1(x, y) \le z \le z_2(x, y),$$

then
$$\int_R f(P)\, dV = \int_a^b \left\{ \int_{y_1(x)}^{y_2(x)} \left[\int_{z_1(x, y)}^{z_2(x, y)} f(x, y, z)\, dz \right] dy \right\} dx.$$

An example illustrates how this formula is applied. In Exercise 40 an argument for its plausibility is presented.

EXAMPLE 5 Compute $\int_R z\, dV$, where R is the tetrahedron in Example 3.

SOLUTION A description of the tetrahedron is

$$0 \le y \le 1, \qquad 0 \le x \le 1 - y, \qquad 0 \le z \le 1 - x - y.$$

Hence
$$\int_R z\, dV = \int_0^1 \left[\int_0^{1-y} \left(\int_0^{1-x-y} z\, dz \right) dx \right] dy.$$

Compute the inner integral first, treating x and y as constants. By the fundamental theorem,

$$\int_0^{1-x-y} z\, dz = \left. \frac{z^2}{2} \right|_{z=0}^{z=1-x-y} = \frac{(1 - x - y)^2}{2}.$$

The next integration, where y is fixed, is

$$\int_0^{1-y} \frac{(1 - x - y)^2}{2}\, dx = \left. -\frac{(1 - x - y)^3}{6} \right|_{x=0}^{x=1-y}$$
$$= -\frac{0^3}{6} + \frac{(1 - y)^3}{6} = \frac{(1 - y)^3}{6}.$$

The third integration is

$$\int_0^1 \frac{(1 - y)^3}{6}\, dy = \left. -\frac{(1 - y)^4}{24} \right|_0^1 = -\frac{0^4}{24} + \frac{1^4}{24} = \frac{1}{24}. \quad \blacksquare$$

Finding $\bar{z}$ for a tetrahedron

Having computed $\int_R z\, dV$, we can find the z coordinate of the centroid of the tetrahedron in Example 3 with just a little more effort. By definition,

$$\bar{z} = \frac{\int_R z\, dV}{\text{Volume of } R}$$

The volume of the tetrahedron is $\frac{1}{3} \cdot \text{Height} \cdot \text{Area of base} = \frac{1}{3} \cdot 1 \cdot \frac{1}{2} = \frac{1}{6}$.

Thus
$$\bar{z} = \frac{\frac{1}{24}}{\frac{1}{6}} = \frac{1}{4},$$

which is one-fourth the distance from the base to the opposite vertex.

Section Summary

We defined $\int_R f(P)\, dV$, where R is a region in space. For instance, if $f(P)$ is the density of matter near P, then $\int_R f(P)\, dV$ is the total mass. We also showed how to evaluate these integrals by introducing rectangular coordinates. First, describe R, for instance, as

$$a \le x \le b, \qquad y_1(x) \le y \le y_2(x), \qquad z_1(x, y) \le z \le z_2(x, y).$$

(There are six possible orders.) Then

$$\int_R f(P)\, dV = \int_a^b \left\{ \int_{y_1(x)}^{y_2(x)} \left[\int_{z_1(x, y)}^{z_2(x, y)} f(x, y, z)\, dz \right] dy \right\} dx.$$

EXERCISES FOR SEC. 15.5: THE DEFINITE INTEGRAL OF A FUNCTION OVER A REGION IN SPACE

Exercises 1 to 8 concern the definition of $\int_R f(P)\, dV$.

1 Find upper and lower estimates for the mass of the cube in the opening problem of this section by partitioning it into eight cubes. (See Fig. 1.)

2 Using the same partition as in the text, estimate the mass of the cube, but select as the P_i's the centers of the four rectangular boxes.

3 If R is a ball of radius r and $f(P) = 5$ for each point in R, compute $\int_R f(P)\, dV$ by examining approximating sums. Recall that the ball has volume $\frac{4}{3}\pi r^3$.

4 How would you define the average distance from points of a certain set in space to a fixed point P_0?

5 Estimate the mass of the cube described in the opening problem by cutting it into eight congruent cubes and using their centers as the P_i's.

6 If R is a three-dimensional set and $f(P)$ is never more than 8 for all P in R,
(*a*) what can we say about the maximum possible value of $\int_R f(P)\, dV$?
(*b*) what can we say about the average of f over R?

7 What is the mesh of the partition of the cube used in the text?

8 What is the mesh of the partition used in Exercise 1?

In Exercises 9 to 14 draw the solids described.

9 $1 \le x \le 3,\ 0 \le y \le 2,\ 0 \le z \le x$

10 $0 \le x \le 1,\ 0 \le y \le 1,\ 1 \le z \le 1 + x + y$

11 $0 \le y \le 1,\ 0 \le x \le y^2,\ y \le z \le 2y$

12 $0 \le y \le 1,\ y^2 \le x \le y,\ 0 \le z \le x + y$

13 $-1 \le z \le 1,\ -\sqrt{1 - z^2} \le x \le \sqrt{1 - z^2},$
$-\frac{1}{2} \le y \le \sqrt{1 - x^2 - z^2}$

14 $0 \le z \le 3,\ 0 \le y \le \sqrt{9 - z^2},\ 0 \le x \le \sqrt{9 - y^2 - z^2}$

In Exercises 15 to 18 evaluate the repeated integrals.

15 $\displaystyle\int_0^1 \left[\int_0^2 \left(\int_0^x z\, dz \right) dy \right] dx.$

16 $\displaystyle\int_0^1 \left[\int_{x^3}^{x^2} \left(\int_0^{x+y} z\, dz \right) dy \right] dx.$

17 $\displaystyle\int_2^3 \left\{ \int_x^{2x} \left[\int_0^1 (x + z)\, dz \right] dy \right\} dx.$

18 $\displaystyle\int_0^1 \left\{ \int_0^x \left[\int_0^3 (x^2 + y^2)\, dz \right] dy \right\} dx.$

19 Describe the solid cylinder of radius a and height h shown in Fig. 10 in rectangular coordinates
(*a*) in the order first x, then y, then z,
(*b*) in the order first x, then z, then y.

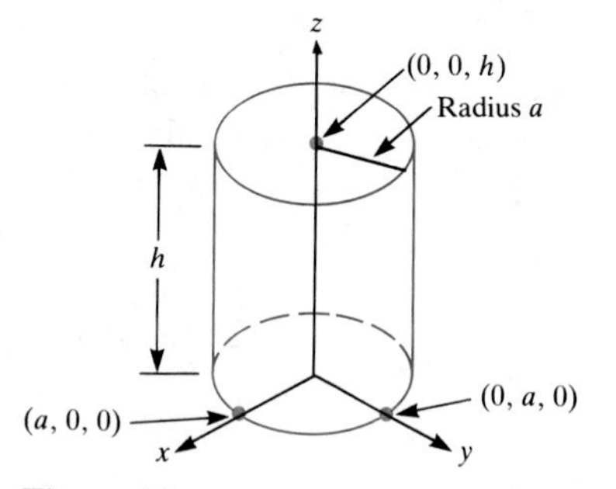

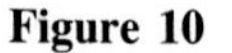
Figure 10

20 Describe the prism shown in Fig. 11 in rectangular coordinates, in two ways:
(*a*) First project it onto the xy plane.
(*b*) First project it onto the xz plane.

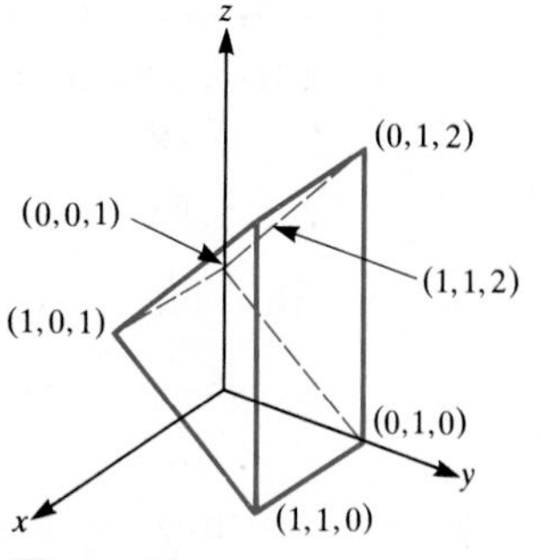

Figure 11

21 Describe the tetrahedron shown in Fig. 12 in rectangular coordinates in two ways:

(*a*) First project it onto the xy plane.
(*b*) First project it onto the xz plane.

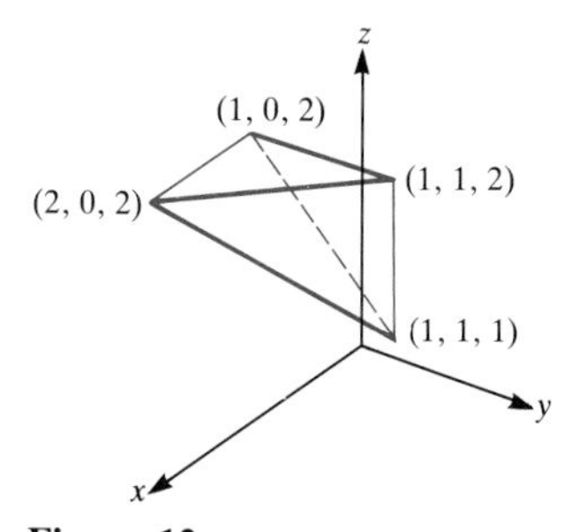

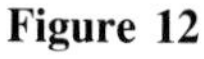
Figure 12

22 Describe the tetrahedron whose vertices are given in Fig. 13 in rectangular coordinates as follows:
(*a*) Draw its shadow on the xy plane.
(*b*) Obtain equations of its top and bottom planes.
(*c*) Describe the region.

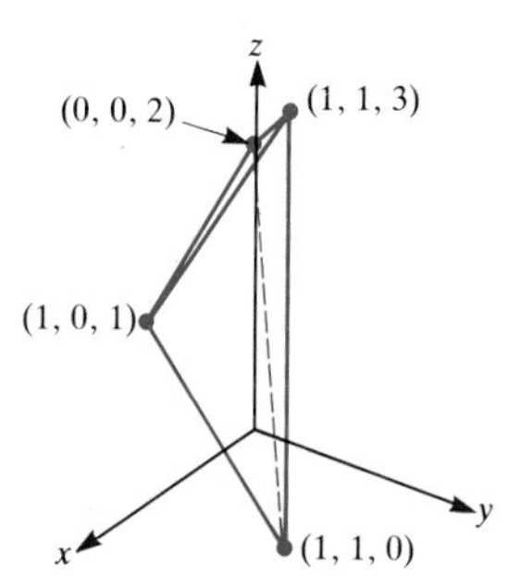

Figure 13

23 Let R be the tetrahedron whose vertices are $(0, 0, 0)$, $(a, 0, 0)$, $(0, b, 0)$, and $(0, 0, c)$, where a, b, and c are positive.
(*a*) Sketch the tetrahedron.
(*b*) Find the equation of its top surface.
(*c*) Compute $\int_R z\, dV$.
(*d*) Find $\bar{z}$, the z coordinate of the centroid of R.

24 Compute $\int_R z\, dV$, where R is the region above the rectangle whose vertices are $(0, 0, 0)$, $(2, 0, 0)$, $(2, 3, 0)$, and $(0, 3, 0)$ and below the plane $z = x + 2y$.

25 Find the mass of the cube in the opening problem. (See Fig. 1.)

26 Find the average value of the square of the distance from a corner of a cube of side a to points in the cube.

27 Find the average of the square of the distance from a point P in a cube of side a to the center of the cube.

28 A solid consists of all points below the surface $z = xy$ that are above the triangle whose vertices are $(0, 0, 0)$, $(1, 0, 0)$, and $(0, 2, 0)$. If the density at (x, y, z) is $x + y$, find the total mass.

29 Compute $\int_R xy\, dV$ for the tetrahedron of Example 3.

30 (*a*) Describe in rectangular coordinates the right circular cone of radius r and height h if its axis is on the positive z axis and its vertex is at the origin. Draw the cross sections for fixed x and fixed x and y.
(*b*) Find the z coordinate of its centroid.

31 The temperature at the point (x, y, z) is e^{-x-y-z}. Find the average temperature in the tetrahedron whose vertices are $(0, 0, 0)$, $(1, 1, 0)$, $(0, 0, 2)$, and $(1, 0, 0)$.

32 The temperature at the point (x, y, z), $y > 0$, is $e^{-x}/\sqrt{y}$. Find the average temperature in the region bounded by the cylinder $y = x^2$, the plane $y = 1$, and the plane $z = 2y$.

Exercises 33 to 36 concern the moment of inertia. Note that if the object is homogeneous, has mass M and volume V, its density $\delta(P)$ is M/V.

33 A homogeneous rectangular solid box has mass M and sides of lengths a, b, and c. Find its moment of inertia about an edge of length a.

34 A rectangular homogeneous box of mass M has dimensions a, b, and c. Show that the moment of inertia of the box about a line through its center and parallel to the side of length a is $M(b^2 + c^2)/12$.

35 A right solid circular cone has altitude h, radius a, constant density, and mass M.
(*a*) Why is its moment of inertia about its axis less than Ma^2?
(*b*) Show that its moment of inertia about its axis is $3Ma^2/10$.

36 Let P_0 be a fixed point in a solid of mass M. Show that for all choices of three mutually perpendicular lines that meet at P_0 the sum of the moments of inertia of the solid about the lines is the same.

---

37 (*Parallel axis theorem*) The center of mass of a solid of mass M is located at $(0, 0, 0)$. Let its moment of inertia about the x axis be I.
(*a*) Find the moment of inertia of the solid about a line parallel to the x axis and a distance k from it.
(*b*) About which line parallel to the x axis is the moment of inertia of the solid least?

38 Without using a repeated integral, evaluate $\int_R x\, dV$, where R is a spherical ball whose center is $(0, 0, 0)$ and whose radius is a.

39 The work done in lifting a weight of w pounds a vertical distance of x feet is wx foot-pounds. Imagine that through geological activity a mountain is formed consisting of material originally at sea level. Let the density of the material near point P in the mountain be $g(P)$ pounds per cubic foot and the height of P be $h(P)$ feet. What definite integral represents the total work expended in forming the mountain? This type of problem is important in the geological theory of mountain formation.

40 In Sec. 15.2 an intuitive argument was presented for the equality

$$\int_R f(P)\, dA = \int_a^b \left[\int_{y_1(x)}^{y_2(x)} f(x, y)\, dy\right] dx.$$

Here is an intuitive argument for the equality

$$\int_R f(P)\, dV = \int_{x_1}^{x_2} \left\{\int_{y_1(x)}^{y_2(x)} \left[\int_{z_1(x, y)}^{z_2(x, y)} f(x, y, z)\, dz\right] dy\right\} dx.$$

To start, interpret $f(P)$ as "density."

(*a*) Let $R(x)$ be the plane cross section consisting of all points in R with abscissa x. Show that the average density in $R(x)$ is

$$\frac{\displaystyle\int_{y_1(x)}^{y_2(x)} \left[\int_{z_1(x, y)}^{z_2(x, y)} f(x, y, z)\, dz\right] dy}{\text{Area of } R(x)}$$

(*b*) Show that the mass of R between the plane sections $R(x)$ and $R(x + \Delta x)$ is approximately

$$\int_{y_1(x)}^{y_2(x)} \left[\int_{z_1(x, y)}^{z_2(x, y)} f(x, y, z)\, dz\right] dy\, \Delta x.$$

(*c*) From (*b*) obtain a repeated integral in rectangular coordinates for $\int_R f(P)\, dV$.

15.6 COMPUTING $\int_R f(P)\, dV$ USING CYLINDRICAL COORDINATES

So far in this chapter we have worked with rectangular coordinates. In this section we introduce and apply cylindrical coordinates, which combine polar coordinates in the xy plane with the z coordinate of an xyz system.

Cylindrical Coordinates

Cylindrical coordinates combine polar coordinates in the plane with the z of rectangular coordinates in space. Each point P in space receives the name (r, θ, z), as in Fig. 1. We are free to choose the direction of the polar axis; usually it will coincide with the x axis of an (x, y, z) system. Note that (r, θ, z) is directly above (or below) $P^* = (r, \theta)$ in the $r\theta$ plane. Since the set of all points $P = (r, \theta, z)$ for which r is some constant, is a circular cylinder, this coordinate system is especially convenient for describing such cylinders. Just as with polar coordinates, cylindrical coordinates of a point are not unique.

Figure 2 shows the surfaces $\theta = k$, $r = k$, and $z = k$, where k is a positive number.

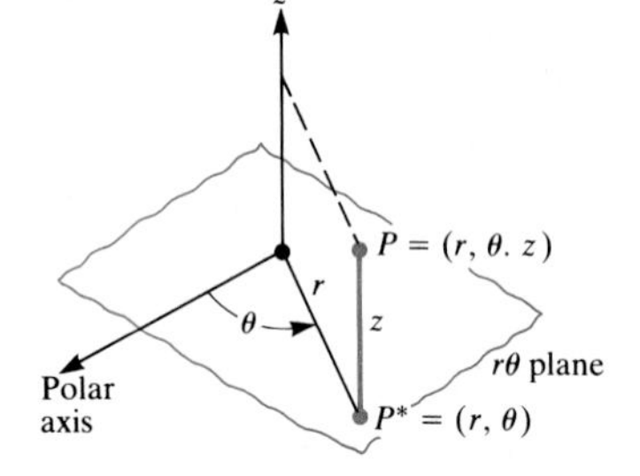

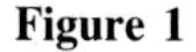
Figure 1

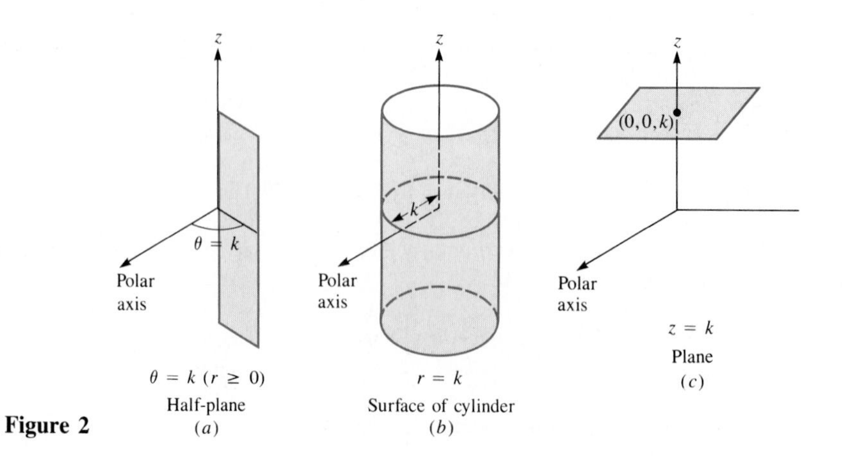

Figure 2

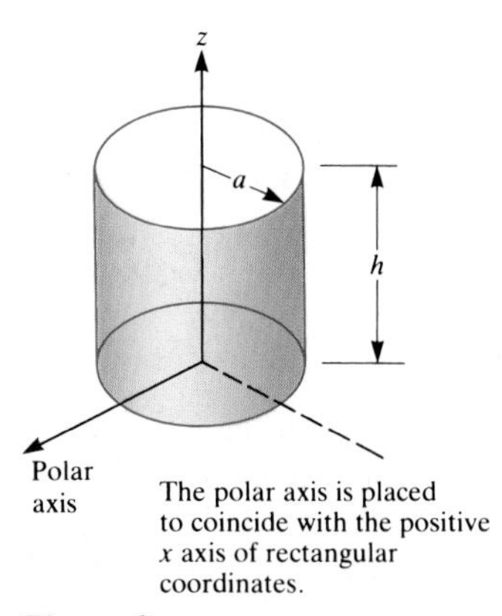

Figure 3

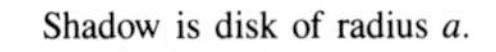

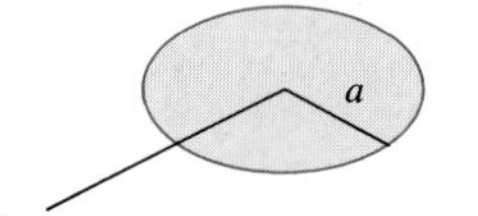

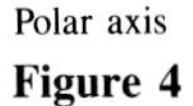

Figure 4

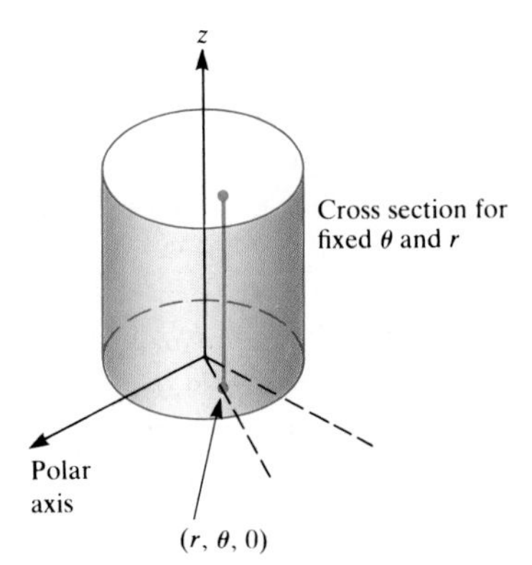

Figure 5

EXAMPLE 1 Describe a solid cylinder of radius a and height h in cylindrical coordinates. Assume that the axis of the cylinder is on the positive z axis and the lower base has its center at the pole, as in Fig. 3.

SOLUTION The shadow of the cylinder on the $r\theta$ plane is the disk of radius a with center at the pole shown in Fig. 4. Its description is

$$0 \le \theta \le 2\pi, \qquad 0 \le r \le a.$$

For each point (r, θ) in the shadow the line through the point parallel to the z axis intersects the cylinder in a line segment. On this segment z varies from 0 to h for every (r, θ). (See Fig. 5.) Thus a description of the cylinder is

$$0 \le \theta \le 2\pi, \qquad 0 \le r \le a, \qquad 0 \le z \le h. \quad ■$$

EXAMPLE 2 Describe in cylindrical coordinates the region in space formed by the intersection of a solid cylinder of radius 3 with a ball of radius 5 whose center is on the axis of the cylinder. Locate the cylindrical coordinate system as shown in Fig. 6.

SOLUTION Note that the point $P = (r, \theta, z)$ is a distance $\sqrt{r^2 + z^2}$ from the origin O, for, by the pythagorean theorem, $r^2 + z^2 = \overline{OP}^2$. (See Fig. 7.) We will use this fact in a moment.

Now consider the description of the solid. First of all, θ varies from 0 to 2π and r from 0 to 3, bounds determined by the cylinder. For fixed θ and r, the cross section of the solid is a line segment determined by the sphere that bounds the ball, as shown in Fig. 8. Now, since the sphere has radius 5, for any point (r, θ, z) on it,

$$r^2 + z^2 = 25 \qquad \text{or} \qquad z = \pm\sqrt{25 - r^2}.$$

Thus, on the line segment determined by fixed r and θ, z varies from $-\sqrt{25 - r^2}$ to $\sqrt{25 - r^2}$.

The solid has this description:

$$0 \le \theta \le 2\pi, \qquad 0 \le r \le 3, \qquad -\sqrt{25 - r^2} \le z \le \sqrt{25 - r^2}. \quad ■$$

Integrating in Cylindrical Coordinates

The set of all points (r, θ, z) whose r coordinates are between r and $r + \Delta r$, whose θ coordinates are between θ and $\theta + \Delta\theta$, and whose z coordinates are

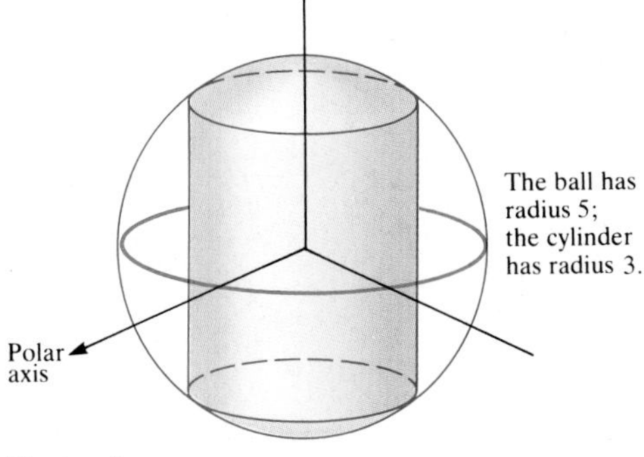

Figure 6

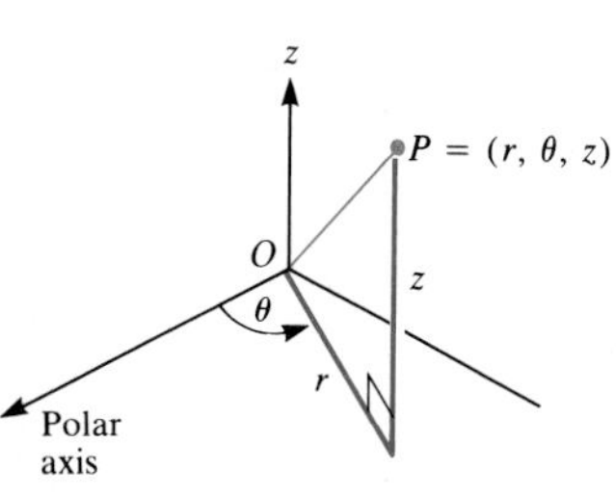

Figure 7

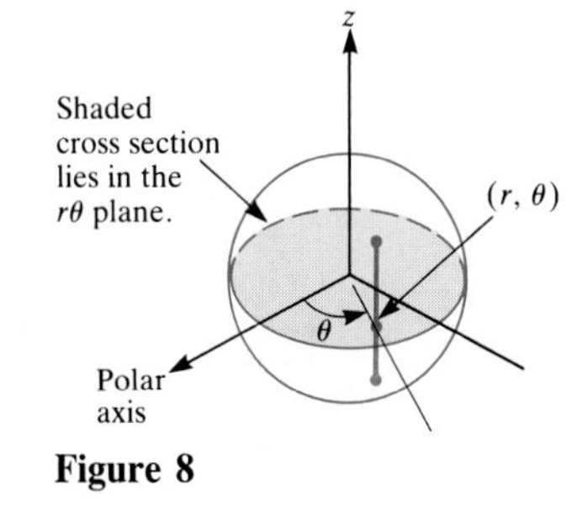

Figure 8

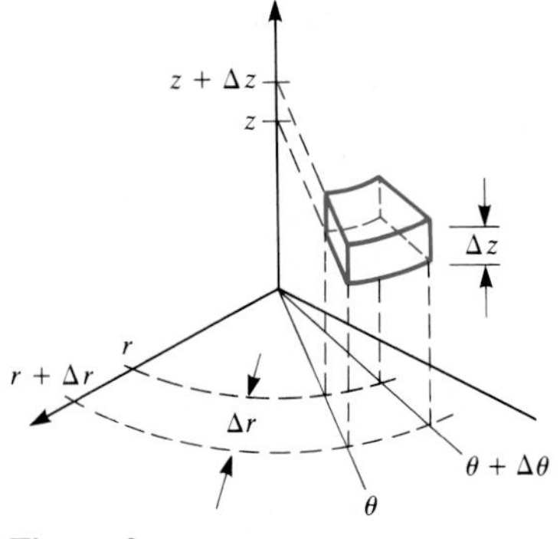

Figure 9

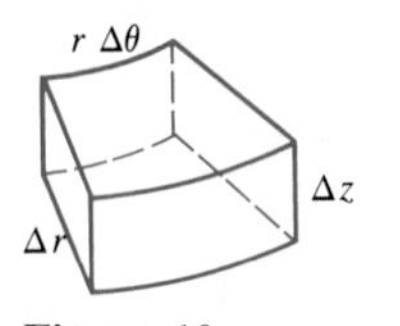

Figure 10

between z and $z + \Delta z$ is shown in Figs. 9 and 10. It is a solid with four flat surfaces and two curved surfaces.

When Δr is small, the area of the base of the solid is approximately $r\ \Delta r\ \Delta\theta$, as shown in Sec. 15.4. Thus, when Δr, $\Delta\theta$, and Δz are small, the volume ΔV of the solid in Fig. 10 is

$$\Delta V = (\text{Area of base})(\text{height}) \approx r\ \Delta r\ \Delta\theta\ \Delta z.$$

That is,

$$\Delta V \approx r\ \Delta r\ \Delta\theta\ \Delta z.$$

To evaluate $\int_R f(P)\ dV$ in cylindrical coordinates we express the integrand in cylindrical coordinates and describe the region R in cylindrical coordinates. It must be kept in mind that dV is replaced by $r\ dz\ dr\ d\theta$. There are six possible orders of integration, but the most common one is: z varies first, then r, finally θ:

$$\boxed{\int_R f(P)\ dV = \int_\alpha^\beta \left\{ \int_{r_1(\theta)}^{r_2(\theta)} \left[\int_{z_1(r,\ \theta)}^{z_2(r,\ \theta)} f(r,\ \theta,\ z)\ r\ dz \right] dr \right\} d\theta.}$$

EXAMPLE 3 Find the volume of a ball R of radius a using cylindrical coordinates.

SOLUTION Place the origin of a cylindrical coordinate system at the center of the ball, as in Fig. 11.

The volume of the ball is $\int_R 1\ dV$. The description of R in cylindrical coordinates is

$$0 \le \theta \le 2\pi, \qquad 0 \le r \le a, \qquad -\sqrt{a^2 - r^2} \le z \le \sqrt{a^2 - r^2}.$$

The repeated integral for the volume is thus

Note, as with polar coordinates, the extra factor r.

$$\int_R 1\ dV = \int_0^{2\pi} \left[\int_0^a \left(\int_{-\sqrt{a^2-r^2}}^{\sqrt{a^2-r^2}} 1 \cdot r\ dz \right) dr \right] d\theta.$$

(Note the r in the integrand.)

Evaluation of the first integral, where r and θ are fixed, yields

Note that the order of integration is determined by the order in describing R.

$$\int_{-\sqrt{a^2-r^2}}^{\sqrt{a^2-r^2}} r\ dz = rz \Big|_{z=-\sqrt{a^2-r^2}}^{z=\sqrt{a^2-r^2}} = 2r\sqrt{a^2 - r^2}.$$

Evaluation of the second integral, where θ is fixed, yields

$$\int_0^a 2r\sqrt{a^2 - r^2}\ dr = \frac{-2(a^2 - r^2)^{3/2}}{3}\Big|_{r=0}^{r=a} = \frac{2a^3}{3}.$$

Finally, evaluation of the third integral gives

$$\int_0^{2\pi} \frac{2a^3}{3}\ d\theta = \frac{2a^3}{3}\int_0^{2\pi} d\theta = \frac{2a^3}{3} \cdot 2\pi = \frac{4}{3}\pi a^3. \quad \blacksquare$$

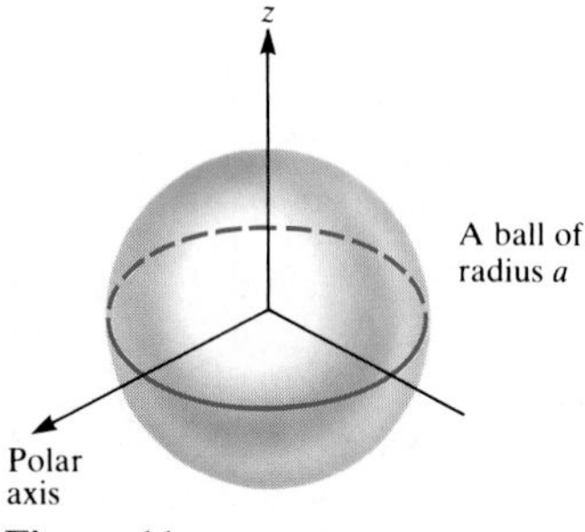

Figure 11

Section Summary

We introduced cylindrical coordinates (r, θ, z). They can be used to evaluate an integral $\int_R f(P)\,dV$ by an iterated integral, such as

$$\int_\alpha^\beta \left\{ \int_{r_1(\theta)}^{r_2(\theta)} \left[\int_{z_1(r,\theta)}^{z_2(r,\theta)} f(r, \theta, z)\, r\, dz \right] dr \right\} d\theta.$$

EXERCISES FOR SEC. 15.6: COMPUTING $\int_R f(P)\,dV$ USING CYLINDRICAL COORDINATES

Exercises 1 to 24 concern describing regions in cylindrical coordinates.

In Exercises 1 to 6 sketch graphs of all points (r, θ, z) such that

1 $r = 1$

2 $\theta = \pi/4$ (consider only nonnegative r)

3 $z = 1$

4 $r = z$

5 $r^2 + z^2 = 9$

6 $r = 2\cos\theta$

In Exercises 7 and 8 draw the cross sections corresponding to fixed θ of the regions described. (Restrict r to nonnegative values.)

7 The region R consists of all points within a distance a of the origin of the $r\theta z$-coordinate system.

8 The region R consists of all points of the solid of Exercise 7 that are within the cylinder whose equation is $r = a\cos\theta$.

9 Describe the solid in Exercise 7 in cylindrical coordinates.

10 Describe the solid in Exercise 8 in cylindrical coordinates.

11 (*a*) What are the cylindrical coordinates of the point $P = (x, y, z)$? Assume (x, y) is in the first quadrant.
(*b*) What are the rectangular coordinates of the point $P = (r, \theta, z)$?

12 Describe in cylindrical coordinates the solid cone shown in Fig. 12.

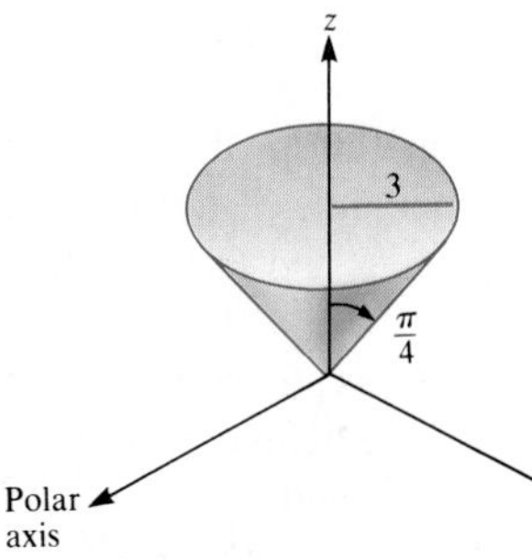

Figure 12

13 Give the equation in cylindrical coordinates of (*a*) the xy plane, (*b*) the part of the plane $x = y$ for which $x \geq 0$, (*c*) the z axis.

14 Describe in cylindrical coordinates the solid that lies above the first-quadrant leaf of $r = \sin 3\theta$ and inside the hyperboloid $x^2 - y^2 + z^2 = 1$.

In Exercises 15 to 20 draw the solids described.

15 $0 \leq \theta \leq 2\pi,\ 0 \leq r \leq 1,\ r \leq z \leq 1$

16 $0 \leq \theta \leq \pi/2,\ 1 \leq z \leq 2,\ 1 \leq r \leq z$

17 $0 \leq \theta \leq \pi/2,\ 0 \leq r \leq 1,\ 0 \leq z \leq r\cos\theta$

18 $0 \leq \theta \leq \pi/2,\ 0 \leq r \leq \cos\theta,\ 1 \leq z \leq 2$

19 $0 \leq \theta \leq \pi/2,\ 1 \leq r \leq 1 + \sin\theta,\ r \leq z \leq 2$

20 $0 \leq \theta \leq \pi/2,\ 0 \leq r \leq \cos\theta,\ r \leq z \leq \sqrt{4 - r^2}$

In each of Exercises 21 to 24 describe the solid R in cylindrical coordinates and draw it.

21 R consists of the points above the disk bounded by $r = 2\sin\theta$ and below the paraboloid $z = x^2 + y^2$.

22 R consists of the points above the leaf of the eight-leaved rose $r = \cos 4\theta$ that lies on the polar axis and within the ball of radius 2 centered at the origin.

23 R is bounded by the planes $z = 0$, $x = 0$, $x = 1$, $y = 0$, $y = 1$, and the saddle $z = xy$.

24 R is the tetrahedron with vertices $(0, 0, 0)$, $(0, 0, 1)$, $(1, 0, 0)$, and $(1, 1, 0)$ in rectangular coordinates.

In each of Exercises 25 and 26 evaluate the repeated integral.

25 $$\int_0^{2\pi} \left[\int_0^1 \left(\int_r^1 zr^3 \cos^2\theta\, dz \right) dr \right] d\theta$$

26 $$\int_0^{2\pi} \left[\int_0^1 \left(\int_{-\sqrt{a^2-r^2}}^{\sqrt{a^2-r^2}} z^2 r\, dz \right) dr \right] d\theta$$

27 Let R be the solid region inside both the sphere $x^2 + y^2 + z^2 = 1$ and the cone $z = \sqrt{x^2 + y^2}$. Let the density at (x, y, z) be $f(x, y, z) = z$. Using cylindrical coordinates, find the mass of R.

28 Using cylindrical coordinates, find $\bar{z}$ for the center of mass of the region R in Exercise 27.

29 Using cylindrical coordinates, find $\bar{z}$ for the region below the paraboloid $z = x^2 + y^2$ and above the disk in the $r\theta$ plane bounded by the circle $r = 2$. (Include a drawing of the region.)

30 Using cylindrical coordinates, find the volume of the region below the plane $z = y + 1$ and above the circle in the xy plane whose center is $(0, 1, 0)$ and whose radius is 1. (Include a drawing of the region.) *Hint:* What is the equation of the circle in polar coordinates when the polar axis is along the positive x axis?

31 Using cylindrical coordinates, find the moment of inertia of a homogeneous cylinder of mass M, radius a, and height h around its axis.

32 Using cylindrical coordinates, find the moment of inertia of the cylinder in Exercise 31 around a line parallel to the axis and lying on the surface of the cylinder.

33 Using cylindrical coordinates, find the moment of inertia of the cylinder in Exercise 31 around a diameter of a base.

34 Using cylindrical coordinates, find the moment of inertia of a homogeneous cone of mass M, radius a, and height h around its axis.

35 Using cylindrical coordinates, find the moment of inertia of a homogeneous ball of mass M and radius a around a diameter.

36 A plane through the center of a homogeneous ball of radius a and mass M cuts it into two congruent pieces. Find the moment of inertia of one piece about any line through its center.

37 Find the moment of inertia of a homogeneous ball of mass M and radius a about a line tangent to the ball.

38 A plane through the center of a ball of radius a cuts it into two congruent pieces. Find the centroid of one piece.

39 A solid of varying density $\delta(P)$ occupies the region R in space. Let L_1 be a line through its center of mass and L_2 a line parallel to L_1 and at a distance r from it. Let I_1 be the moment of inertia of the solid around L_1 and I_2 the moment of inertia around L_2. Show that $I_2 = I_1 + r^2M$, where M is the mass of the solid.

40 Consider a plane $\mathcal{P}$ through the center of mass of a solid occupying the region R. The density at P is $\delta(P)$. Let $g(P)$ be the signed distance from P to the plane $\mathcal{P}$, positive if P is on one side of $\mathcal{P}$, negative if P is on the other side of $\mathcal{P}$. Show that $\int_R g(P)\,\delta(P)\,dV = 0$. ("The moment about $\mathcal{P}$ is 0.")

15.7 COMPUTING $\int_R f(P)\,dV$ USING SPHERICAL COORDINATES

The third standard coordinate system in space is **spherical coordinates**, which combines the θ of cylindrical coordinates with two other coordinates.

Spherical Coordinates

In spherical coordinates a point P is described by three numbers:

ρ	the distance from P to the origin O,
θ	the same angle as in cylindrical coordinates,
ϕ	the angle between the positive z axis and the ray from O to P.

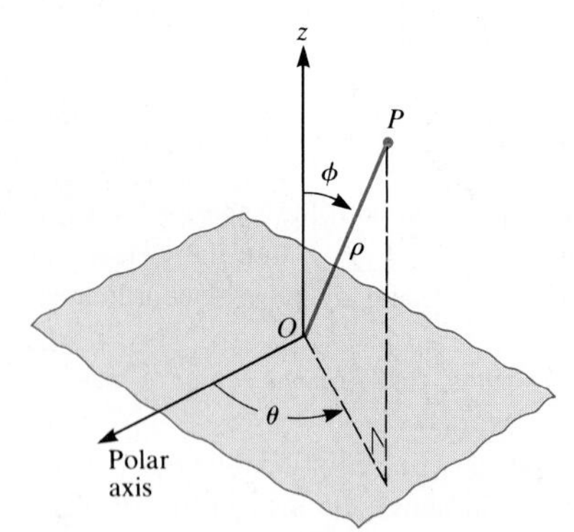

Figure 1

(ρ is pronounced "roe"; it is the Greek letter for r. The letter ϕ is pronounced "fee" or "fie.")

The point P is denoted $P = (\rho, \theta, \phi)$. Note the order: first ρ, then θ, then ϕ. See Fig. 1. Note that ϕ is the same as the direction angle of $\overrightarrow{OP}$ with $\mathbf{k}$, $0 \le \phi \le \pi$. The surfaces $\rho = k$ (a sphere), $\phi = k$ (a cone), and $\theta = k$ (a half plane) are shown in Fig. 2.

When ϕ and θ are fixed and ρ varies, we describe a ray, as shown in Fig. 3. The following table lists the equations of a sphere of radius a with center at the origin O and of a cone with vertex at O, half vertex angle α, and axis along the positive z axis, in the three coordinate systems described in this chapter.

Surface	*Rectangular*	*Cylindrical*	*Spherical*
Sphere	$x^2 + y^2 + z^2 = a^2$	$r^2 + z^2 = a^2$	$\rho = a$
Cone	$(\tan\alpha)z = \sqrt{x^2 + y^2}$	$(\tan\alpha)z = r$	$\phi = \alpha$

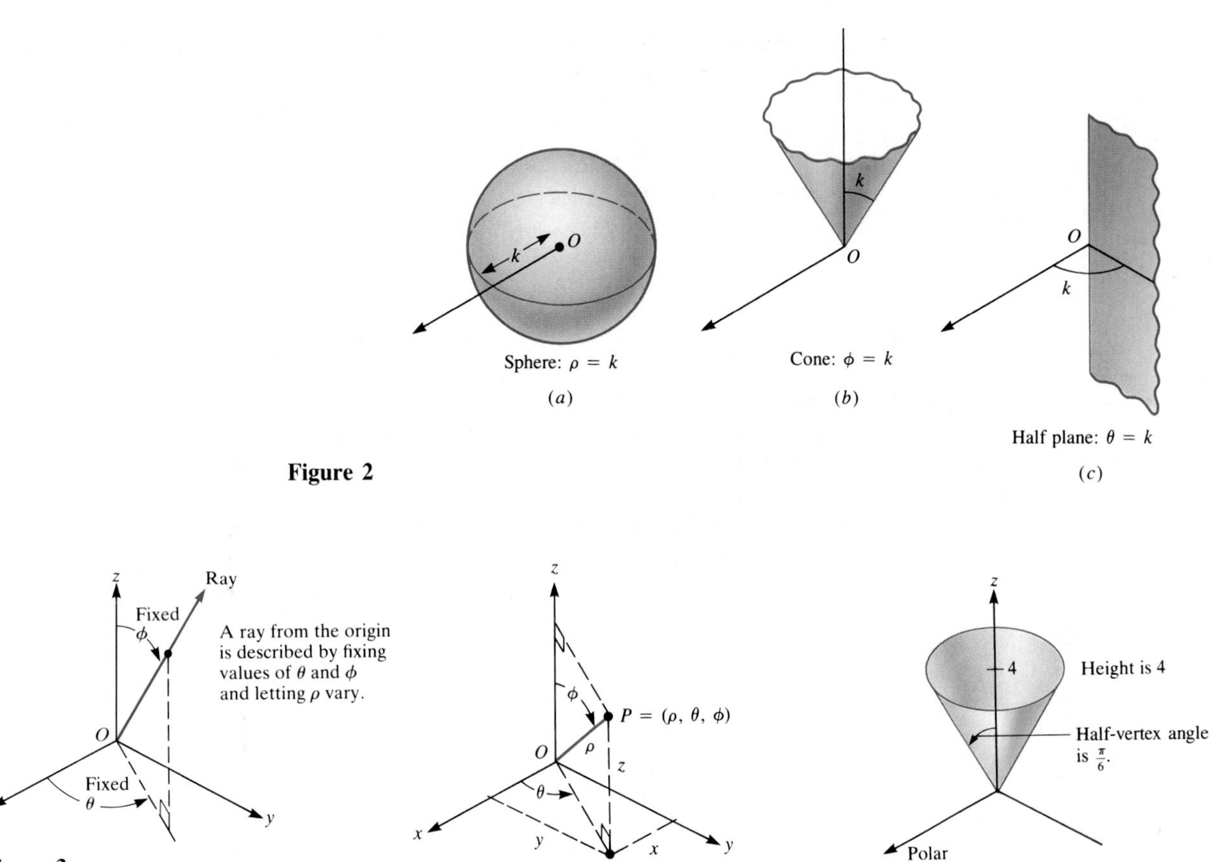

Figure 2

Figure 3

Figure 4

Figure 5

Clearly, spherical coordinates are preferable when working with spheres or cones.

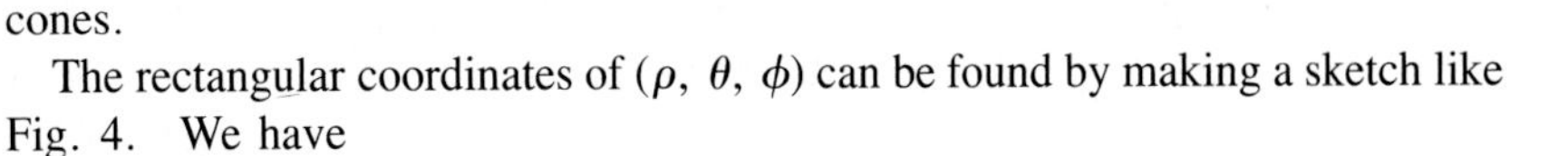

The rectangular coordinates of (ρ, θ, ϕ) can be found by making a sketch like Fig. 4. We have

$$x = \overline{OQ} \cos \theta = \rho \sin \phi \cos \theta$$

$$y = \overline{OQ} \sin \theta = \rho \sin \phi \sin \theta$$

$$z = \overline{OP} \cos \phi = \rho \cos \phi.$$

EXAMPLE 1 Find the equation of the plane $z = 4$ in spherical coordinates.

SOLUTION Since $z = \rho \cos \phi$ describes z in terms of spherical coordinates, the equation of the plane $z = 4$ is $\rho \cos \phi = 4$; hence $\rho = 4 \sec \phi$. ■

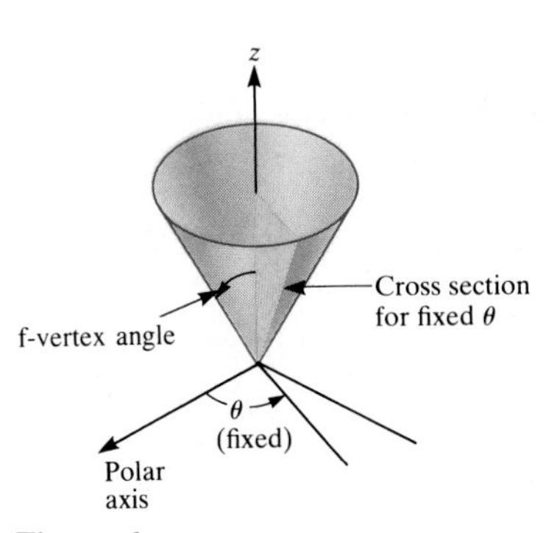

Figure 6

EXAMPLE 2 Describe in spherical coordinates the solid cone of height 4 and half-vertex angle $\pi/6$ shown in Fig. 5.

SOLUTION It is usually most convenient to examine first how θ varies. In this case, θ goes from 0 to 2π. For each fixed θ the cross section is a triangle, as shown in Fig. 6.

On this typical triangle ϕ varies from 0 to $\pi/6$. For each fixed θ and ϕ, the

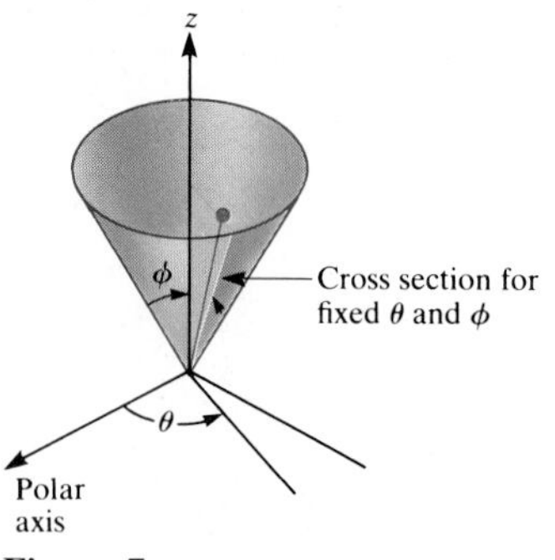

Figure 7

cross section is a segment on a ray emanating from the origin, as noted in Fig. 7. On this cross section ρ varies from 0 to its value where the ray meets the plane $z = 4$. By Example 1, this plane has the equation

$$\rho = 4 \sec \phi.$$

Hence for fixed θ and ϕ, ρ varies from 0 to $4 \sec \phi$.

This is the description of the cone:

$$0 \leq \theta \leq 2\pi, \qquad 0 \leq \phi \leq \frac{\pi}{6}, \qquad 0 \leq \rho \leq 4 \sec \phi. \quad \blacksquare$$

Notice the simplicity of the description of θ and ϕ for the cone in Example 2. Clearly, spherical coordinates are convenient for describing cones. They are also fine for a ball whose center is at the origin. To be specific,

$$0 \leq \theta \leq 2\pi, \qquad 0 \leq \phi \leq \pi, \qquad 0 \leq \rho \leq a$$

is the description of a ball of radius a whose center is at the origin.

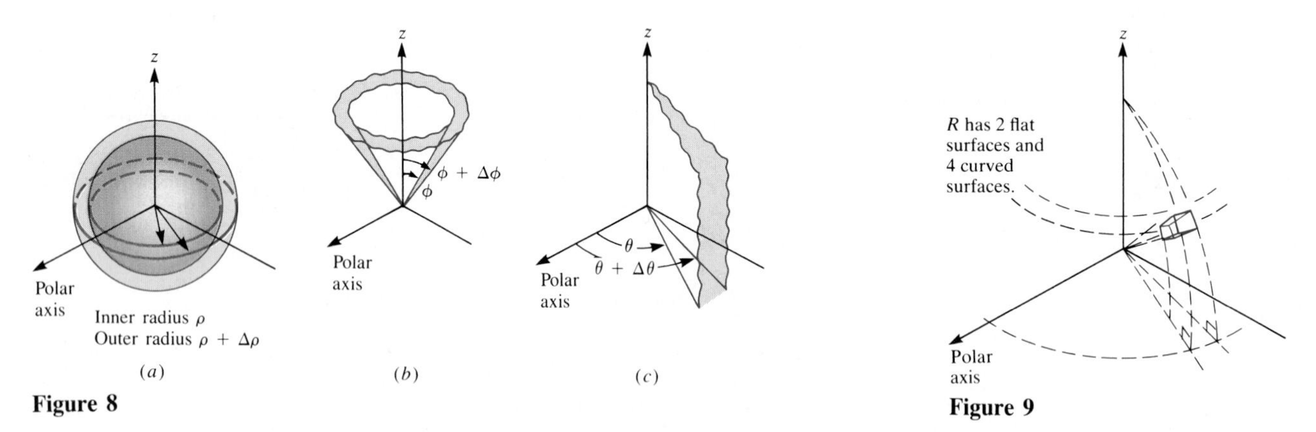

Figure 8

Figure 9

The set of all points whose ρ coordinates are between ρ and $\rho + \Delta\rho$, whose θ coordinates are between θ and $\theta + \Delta\theta$, and whose ϕ coordinates are between ϕ and $\phi + \Delta\phi$ is a solid R with two flat surfaces and four curved surfaces. We shall estimate its volume when $\Delta\rho$, $\Delta\theta$, and $\Delta\phi$ are small. This estimate will be needed when setting up repeated integrals in spherical coordinates.

R is bounded by the six surfaces shown in Fig. 8: by spheres of radii ρ and $\rho + \Delta\rho$; by cones of half-vertex angles ϕ and $\phi + \Delta\phi$; by half planes of polar angles θ and $\theta + \Delta\theta$.

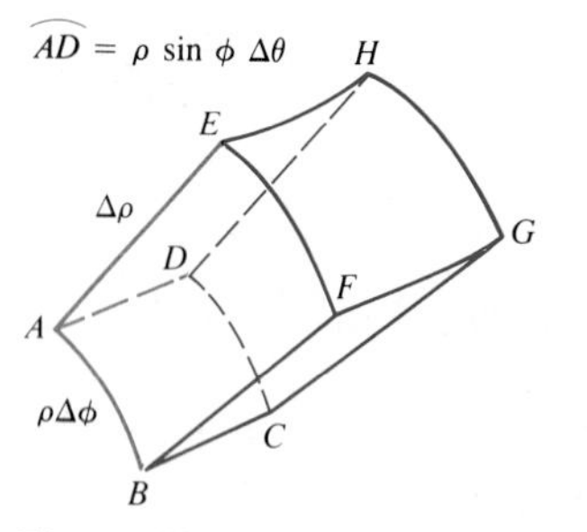

Figure 10

$\widehat{AB}$ denotes length of arc AB.

The little solid R appears as shown in Fig. 9. Label its eight corners as in Fig. 10. $ABCD$ and $EFGH$ are spherical. $BCGF$ and $ADHE$ are conical. $ABFE$ and $DCGH$ are flat. Since the small solid R resembles a rectangular box, its volume is approximated by the product

$$\overline{AE} \cdot \widehat{AB} \cdot \widehat{AD}.$$

First of all, $\overline{AE}$ is just the difference in the radii of the two spheres:

$$\overline{AE} = \Delta\rho.$$

Next, AB is an arc of a circle of radius ρ and subtends an angle $\Delta\phi$. Thus

$$\widehat{AB} = \rho \, \Delta\phi.$$

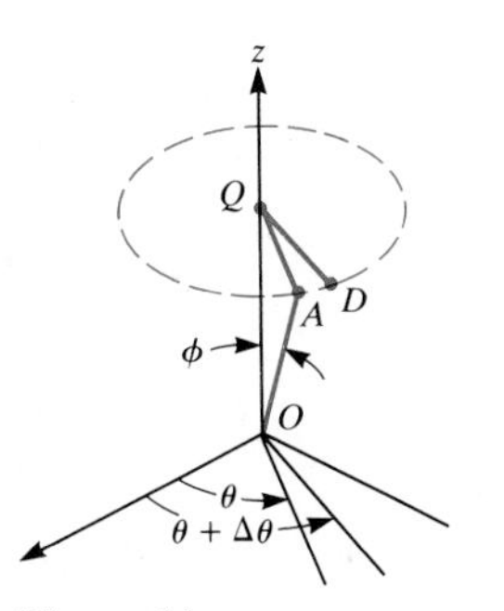

Figure 11

Finally, consider AD. It is an arc of a circle that is perpendicular to the z axis, as depicted in Fig. 11. AD subtends an angle $\Delta\theta$. The radius of the dashed circle, computed from right triangle AOQ in Fig. 12, is $\rho \sin \phi$. Hence

$$\widehat{AD} = \rho \sin \phi \; \Delta\theta.$$

The volume of R, ΔV, is therefore approximately

$$\overline{AE} \cdot \widehat{AB} \cdot \widehat{AD} = \Delta\rho(\rho \; \Delta\phi)(\rho \sin \phi \; \Delta\theta),$$

that is,

$$\boxed{\Delta V \approx \rho^2 \sin \phi \; \Delta\rho \; \Delta\phi \; \Delta\theta.} \tag{1}$$

The factor $\rho^2 \sin \phi$ will be needed in repeated integrals.

Notice the factor $\rho^2 \sin \phi$. It will be needed in forming repeated integrals in spherical coordinates. Note that in (1) ρ and $\Delta\rho$ have units of length, whereas $\Delta\phi$ and $\Delta\theta$ are dimensionless (radian measure is defined as the quotient of two lengths). Thus the units of (1) are *length cubed,* as they should be if (1) is to measure volume.

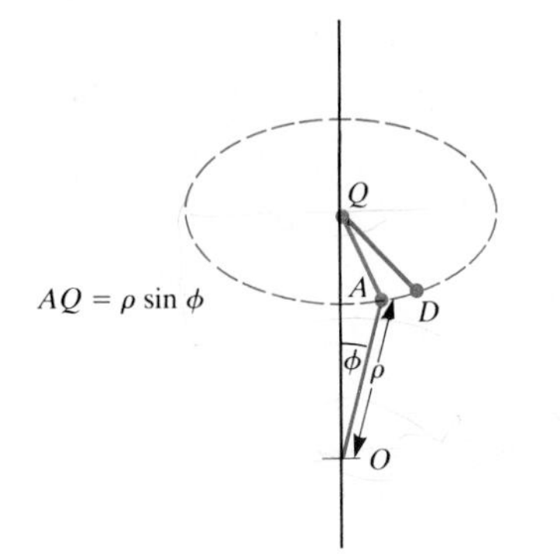

Figure 12

Warning to engineers and physicists: In your physics course *cylindrical* coordinates are labeled ρ, θ, and z. (ρ plays the role of our r.) Physicists prefer r for spherical coordinates (where we use ρ) since they already use r to denote the magnitude of the position vector $\mathbf{r} = \overrightarrow{OP}$. (They use ρ for density.) Finally, they use the letter θ to denote the angle we call ϕ, and they use θ for our ϕ. That means that their angle θ in cylindrical coordinates is different from their angle θ in spherical coordinates. (They assure us that this causes no problem, though it might complicate conversions from one coordinate system to another.)

Computing $\int_R f(P)\,dV$ in Spherical Coordinates

To evaluate $\int_R f(P)\,dV$ in spherical coordinates, first describe the region R in spherical coordinates. Usually this will be in the order:

$$\alpha \le \theta \le \beta, \qquad \phi_1(\theta) \le \phi \le \phi_2(\theta), \qquad \rho_1(\theta, \phi) \le \rho \le \rho_2(\theta, \phi).$$

Sometimes the order of ρ and ϕ is switched:

$$\alpha \le \theta \le \beta \qquad \rho_1(\theta) \le \rho \le \rho_2(\theta) \qquad \phi_1(\rho, \theta) \le \phi \le \phi_2(\rho, \theta).$$

Then set up an iterated integral, being sure to express dV as $\rho^2 \sin \phi \; d\rho \; d\phi \; d\theta$ (or $\rho^2 \sin \phi \; d\phi \; d\rho \; d\theta$).

EXAMPLE 3 Find the volume of a ball of radius a, using spherical coordinates.

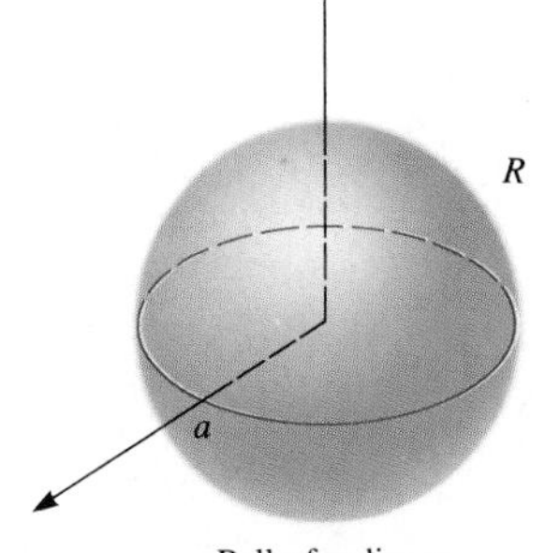

Figure 13

SOLUTION Place the origin of spherical coordinates at the center of the ball, as in Fig. 13. The ball is described by

$$0 \le \theta \le 2\pi, \qquad 0 \le \phi \le \pi, \qquad 0 \le \rho \le a.$$

Hence

$$\text{Volume of ball} = \int_R 1 \; dV$$

$$= \int_0^{2\pi} \int_0^{\pi} \int_0^{a} \rho^2 \sin \phi \; d\rho \; d\phi \; d\theta.$$

The inner integral is

$$\int_0^a \rho^2 \sin\phi\, d\rho = \sin\phi \int_0^a \rho^2\, d\rho = \frac{a^3 \sin\phi}{3}.$$

The next integral is

$$\int_0^\pi \frac{a^3 \sin\phi}{3}\, d\phi = \left.\frac{-a^3\cos\phi}{3}\right|_0^\pi = \frac{-a^3(-1)}{3} - \left[\frac{-a^3(1)}{3}\right] = \frac{2a^3}{3}.$$

The final integral is

$$\int_0^{2\pi} \frac{2a^3}{3}\, d\theta = \frac{2a^3}{3}\int_0^{2\pi} d\theta = \frac{2a^3}{3} 2\pi = \frac{4\pi a^3}{3}. \quad \blacksquare$$

EXAMPLE 4 Find the centroid of a homogeneous cone of height h and radius a.

SOLUTION By symmetry, the centroid lies on the axis of the cone. (If you spin the cone around the axis, the centroid must not move to another point; otherwise, the cone would have two centroids.)

Introduce a spherical coordinate system with the origin at the vertex of the cone and with the axis of the cone lying on the ray $\phi = 0$, as in Fig. 14.

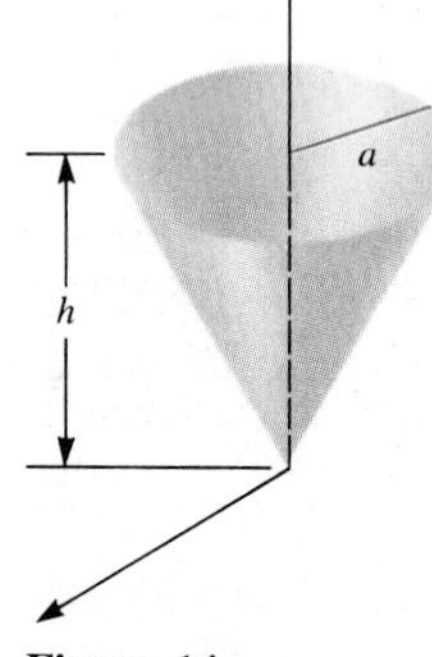

Figure 14

The description will be similar to that for the cone in Example 2. The half-vertex angle is $\tan^{-1}(a/h)$. The plane of the base of the cone is $z = h$ (in rectangular coordinates), hence

$$\rho\cos\phi = h.$$

In spherical coordinates, the cone's description is

$$0 \le \theta \le 2\pi, \qquad 0 \le \phi \le \tan^{-1}(a/h), \qquad 0 \le \rho \le h/\cos\phi.$$

To find the centroid of the cone we compute $\int_R z\, dV$ and divide the result by the volume of the cone, which is $\pi a^2 h/3$.

Now

$$\int_R z\, dV = \int_0^{2\pi}\int_0^{\tan^{-1} a/h}\int_0^{h/(\cos\phi)} \rho\cos\phi\,(\rho^2\sin\phi)\, d\rho\, d\phi\, d\theta.$$

In the first integration, ϕ and θ are constant; hence

$$\int_0^{h/(\cos\phi)} \rho\cos\phi\, \rho^2 \sin\phi\, d\rho = \cos\phi\sin\phi \int_0^{h/(\cos\phi)} \rho^3\, d\rho = \frac{h^4 \sin\phi}{4\cos^3\phi}.$$

See Exercise 25. (Fill in the details.)

The second integration is

$$\int_0^{\tan^{-1} a/h} \frac{h^4 \sin \phi}{4 \cos^3 \phi}\, d\phi = \frac{h^4}{4} \int_0^{\tan^{-1} a/h} \frac{\sin \phi}{\cos^3 \phi}\, d\phi$$

$$= \frac{h^4}{8} \frac{1}{\cos^2 \phi} \Bigg|_0^{\tan^{-1} a/h} = \frac{a^2h^2}{8}.$$

(Fill in the steps.)

The final integral is simply:

$$\int_0^{2\pi} \frac{a^2h^2}{8}\, d\theta = \frac{a^2h^2}{8} 2\pi = \frac{\pi a^2h^2}{4}.$$

Thus,
$$\bar{z} = \frac{\int_R z\, dV}{\text{Volume of } R} = \frac{\left(\dfrac{\pi a^2h^2}{4}\right)}{\left(\dfrac{\pi a^2h}{3}\right)} = \frac{3h}{4}.$$

The centroid of a cone is three-fourths of the way from the vertex to the base. ■

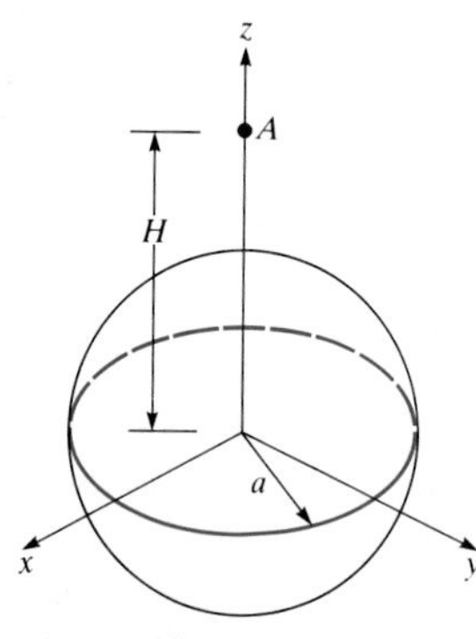

Figure 15

An Integral in Gravity

The next example is of importance in the theory of gravitational attraction. Students of the physical sciences will see later that it implies that a homogeneous ball attracts a particle (or satellite) as if all the mass of the ball were at its center.

EXAMPLE 5 Let R be a homogeneous ball of mass M and radius a. Let A be a point at a distance H from the center of the ball, $H > a$. Compute $\int_R (\delta/q)\, dV$, where δ is density and q is the distance from a point P in R to A. (See Figs. 15 and 16.)

SOLUTION First, express q in terms of spherical coordinates. To do so, choose a spherical coordinate system whose origin is at the center of the sphere and such that the ϕ coordinate of A is 0. (See Fig. 16.)

Let $P = (\rho, \theta, \phi)$ be a typical point in the ball. Applying the law of cosines to triangle AOP, we find that

$$q^2 = H^2 + \rho^2 - 2\rho H \cos \phi.$$

Hence
$$q = \sqrt{H^2 + \rho^2 - 2\rho H \cos \phi}.$$

Since the ball is homogeneous,

$$\delta = \frac{M}{\frac{4}{3}\pi a^3} = \frac{3M}{4\pi a^3}.$$

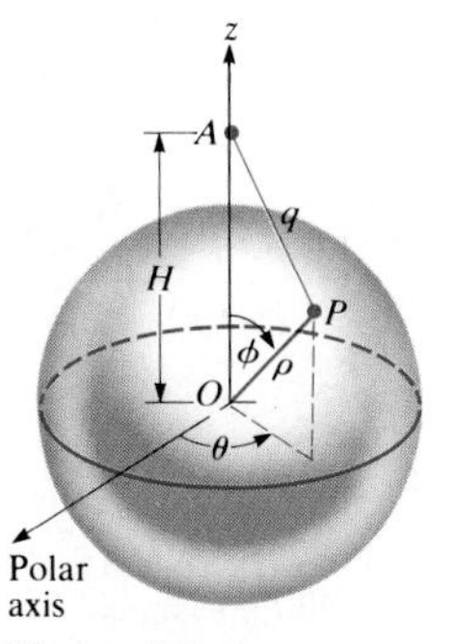

Figure 16

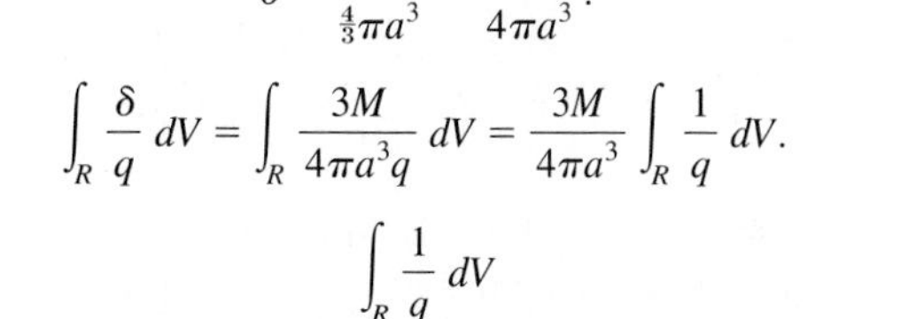

Hence
$$\int_R \frac{\delta}{q}\, dV = \int_R \frac{3M}{4\pi a^3 q}\, dV = \frac{3M}{4\pi a^3} \int_R \frac{1}{q}\, dV. \tag{2}$$

Now evaluate
$$\int_R \frac{1}{q}\, dV$$

by a repeated integral in spherical coordinates:

A case where integration with respect to ρ is not first

$$\int_R \frac{1}{q}\, dV = \int_0^{2\pi} \left[\int_0^a \left(\int_0^\pi \frac{\rho^2 \sin \phi}{\sqrt{H^2 + \rho^2 - 2\rho H \cos \phi}}\, d\phi \right) d\rho \right] d\theta.$$

(Integrate with respect to ϕ first, rather than ρ, because it is easier in this case.)

Evaluation of the first integral, where ρ and θ are constants, is accomplished with the aid of the fundamental theorem:

$$\int_0^\pi \frac{\rho^2 \sin \phi}{\sqrt{H^2 + \rho^2 - 2\rho H \cos \phi}}\, d\phi = \left. \frac{\rho\sqrt{H^2 + \rho^2 - 2\rho H \cos \phi}}{H} \right|_{\phi=0}^{\phi=\pi}$$

$$= \frac{\rho}{H}\left(\sqrt{H^2 + \rho^2 + 2\rho H} - \sqrt{H^2 + \rho^2 - 2\rho H}\right).$$

Now, $\sqrt{H^2 + \rho^2 + 2\rho H} = H + \rho$. Since $\rho \le a < H$, $H - \rho$ is positive and

$$\sqrt{H^2 + \rho^2 - 2\rho H} = H - \rho.$$

Thus the first integral equals

$$\frac{\rho}{H}[(H + \rho) - (H - \rho)] = \frac{2\rho^2}{H}.$$

Evaluation of the second integral yields

$$\int_0^a \frac{2\rho^2}{H}\, d\rho = \frac{2a^3}{3H}.$$

Evaluation of the third integral yields

$$\int_0^{2\pi} \frac{2a^3}{3H}\, d\theta = \frac{4\pi a^3}{3H}.$$

Hence

$$\int_R \frac{1}{q}\, dV = \frac{4\pi a^3}{3H}.$$

By Eq. (2)

$$\int_R \frac{\delta}{q}\, dV = \frac{3M}{4\pi a^3} \frac{4\pi a^3}{3H} = \frac{M}{H}.$$

Newton obtained this remarkable result in 1687.

This result, M/H, is exactly what we would get if all the mass were located at the center of the ball. ■

Section Summary

We introduced spherical coordinates ρ, θ, ϕ of point P. We may think of ρ as the magnitude of a position vector $\overrightarrow{OP}$, θ as the polar angle of the projection of $\overrightarrow{OP}$ on the "xy plane," and ϕ as the direction angle of $\overrightarrow{OP}$ with the vector **k**. When setting up an iterated integral in spherical coordinates we must replace dV by $\rho^2 \sin \phi\, d\rho\, d\phi\, d\theta$ (or $\rho^2 \sin \phi\, d\phi\, d\rho\, d\theta$).

EXERCISES FOR SEC. 15.7: COMPUTING $\int_R f(P)\, dV$ USING SPHERICAL COORDINATES

1 Fill in the blanks and explain with the aid of a sketch.
(*a*) Rectangular coordinates describe a point by specifying three ________ on which it lies.
(*b*) Spherical coordinates describe a point by specifying ________, ________, and ________ on which it lies.
(*c*) Cylindrical coordinates describe a point by specifying ________, ________, and ________ on which it lies.

2 What shapes of solid regions are best described in spherical coordinates?

In Exercises 3 to 8 sketch the set of points (ρ, θ, ϕ) that satisfy the given equations.

3 $\rho = 2$ **4** $\phi = \pi/6$

5 $\theta = \pi/2$ **6** $\phi = \pi/2$

7 $\phi = 0$ **8** $\phi = \pi$

9 What are the cylindrical coordinates of the point (ρ, θ, ϕ)?

10 Sketch the set of all points (ρ, θ, ϕ) such that $\phi = \pi/2$ and $\theta = \pi/2$.

In Exercises 11 to 16 describe in spherical coordinates the regions R.

11 R is the ball of radius a centered at the origin.

12 R is the top half of the ball of radius a centered at the origin.

13 R is the ice cream cone-shaped intersection of a solid cone and a ball shown in Fig. 17.

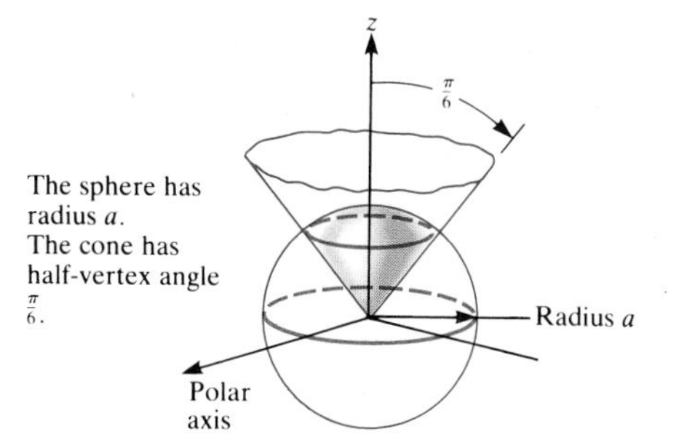

Figure 17

14 R is the region between two spheres, both with center at the origin, of radii a and b, $a < b$.

15 R is the region in the cone shown in Fig. 17 below the plane $z = 3a/5$.

16 R is the region bounded by the paraboloids $z = x^2 + y^2$ and $z = 1 - x^2 - y^2$.

In Exercises 17 to 20 sketch the regions described.

17 $0 \le \theta \le 2\pi$, $\pi/2 \le \phi \le \pi$, $1 \le \rho \le 2$

18 $0 \le \theta \le \pi/2$, $0 \le \phi \le \pi/2$, $0 \le \rho \le 1$

19 $0 \le \theta \le \pi/2$, $0 \le \phi \le \pi/4$, $0 \le \rho \le \sec \phi$

20 $0 \le \theta \le \pi/2$, $0 \le \phi \le \pi/4$, $1 \le \rho \le 2 \sec \phi$

21 Find the spherical coordinates of the point whose cylindrical coordinates are (r, θ, z), $r > 0$.

22 Find the spherical coordinates of the point whose rectangular coordinates are (x, y, z).

23 Sketch the solid whose description is $0 \le \theta \le \pi/2$, $\pi/4 \le \phi \le \pi/2$, $1 \le \rho \le 2$.

24 Find the equation in spherical coordinates of the plane (*a*) $x = 2$, (*b*) $2x + 3y + 4z = 1$.

25 Fill in the omitted steps in Example 4.

26 Compute the volume of a right circular cone of height h and radius r in spherical coordinates.

27 Find the volume of the region above the xy plane and below the paraboloid $z = 9 - r^2$ using cylindrical coordinates.

28 A right circular cone of radius a and height h has a density at point P equal to the distance from P to the base of the cone. Find its mass, using spherical coordinates.

In Exercises 29 and 30 draw the region R and give a formula for the integrand $f(P)$ such that $\int_R f(P)\,dV$ is described by the given iterated integrals.

29 $\int_0^{\pi/2} [\int_0^{\pi/4} (\int_0^{\cos \phi} \rho^3 \sin^2 \theta \sin \phi \, d\rho) \, d\phi] \, d\theta$

30 $\int_0^{\pi/4} [\int_{\pi/6}^{\pi/2} (\int_0^{\sec \theta} \rho^3 \sin \phi \cos \phi \, d\rho) \, d\phi] \, d\theta$

31 Let R be the solid region inside both the sphere $x^2 + y^2 + z^2 = 1$ and the cone $z = \sqrt{x^2 + y^2}$. Let the density at (x, y, z) be $f(x, y, z) = z$. Set up repeated integrals for the mass in R using (*a*) rectangular coordinates, (*b*) cylindrical coordinates, (*c*) spherical coordinates. (*d*) Evaluate the repeated integral in (*c*).

32 Find the average temperature in a ball of radius a if the temperature is the square of the distance from a fixed equatorial plane.

33 Find the average distance from the center of a ball of radius a to other points of the ball by setting up appropriate repeated integrals in the three types of coordinate systems and evaluating the easiest.

34 Show by using a repeated integral that the volume of the little solid in Fig. 9 is precisely

$$\frac{(\rho + \Delta\rho)^3 - \rho^3}{3} \Delta\theta \, [\cos \phi - \cos (\phi + \Delta\phi)].$$

35 A solid consists of that part of a ball of radius a that lies within a cone of half-vertex angle $\phi = \pi/6$, the vertex being at the center of the ball. Set up repeated integrals for $\int_R z\,dV$ in all three coordinate systems and evaluate the simplest.

36 Find the moment of inertia of a homogeneous ball of radius a and mass M about a diameter, using spherical coordinates.

37 Find the moment of inertia of a homogeneous solid hemisphere of radius a and mass M about a diameter in its circular base, (*a*) using cylindrical coordinates; (*b*) using spherical coordinates.

38 A homogeneous right circular cone has altitude h, radius a, and mass M. Using spherical coordinates, show that the moment of inertia about its axis is $3Ma^2/10$. *Suggestion:* Place the z axis on the axis of the cone and the origin at the vertex of the cone.

39 Using the method of Example 5 find the average value of q for all points P in the ball. Note that it is *not* the same as if the entire ball were placed at its center.

40 Show that the result of Example 5 holds if the density $\delta(P)$ depends only on ρ, the distance to the center. (This is approximately the case with the planet Earth, which is not homogeneous.) Let $g(\rho)$ denote $\delta(\rho, \theta, \phi)$.

41 Let R be a solid ball of radius a with center at the origin of the coordinate system.
(a) Explain why $\int_R x^2\,dV = \frac{1}{3}\int_R (x^2+y^2+z^2)\,dV$.
(b) Evaluate the second integral by spherical coordinates.
(c) Use (b) to find $\int_R x^2\,dV$.

42 Show that $\int_R (x^3+y^3+z^3)\,dV = 0$, where R is a ball whose center is the origin of a rectangular coordinate system. (Do not use a repeated integral.)

43 Show that the average of the reciprocal of the distance from a fixed point A outside a ball to points in the ball is equal to the reciprocal of the distance from A to the center of the ball.

✳ 44 (See Exercise 43.) Let A be a point in the plane of a disk but outside the disk. Is the average of the reciprocal of the distance from A to points in the disk equal to the reciprocal of the distance from A to the center of the disk?

45 A certain ball of radius a is *not* homogeneous. However, its density at P depends only on the distance from P to the center of the ball. That is, there is a function $g(\rho)$ such that the density at $P = (\rho, \theta, \phi)$ is $g(\rho)$. Using a repeated integral, show that the mass of the ball is

$$4\pi \int_0^a g(\rho)\,\rho^2\,d\rho.$$

46 Let R be the part of a ball of radius a removed by a cylindrical drill of diameter a whose edge passes through the center of the sphere.
(a) Sketch R.
(b) Notice that R consists of four congruent pieces. Find the volume of one of these pieces using cylindrical coordinates. Multiply by four to get the volume of R.

47 (See Exercise 46.) The following calculation obtains an incorrect value for the volume of R:

$$V = \int_{-\pi/2}^{\pi/2} \left[\int_0^{a\cos\theta} \left(\int_{-\sqrt{a^2-r^2}}^{\sqrt{a^2-r^2}} r\,dz \right) dr \right] d\theta$$

$$= \int_{-\pi/2}^{\pi/2} \left(\int_0^{a\cos\theta} 2r\sqrt{a^2-r^2}\,dr \right) d\theta$$

$$= \int_{-\pi/2}^{\pi/2} \left[-\tfrac{2}{3}(a^2-r^2)^{3/2} \right] \Bigg|_0^{a\cos\theta} d\theta$$

$$= \int_{-\pi/2}^{\pi/2} \frac{2a^3}{3}(1-\sin^3\theta)\,d\theta = \frac{2\pi a^3}{3}.$$

What is the error?

48 Let R be the ball of radius a. For any point P in the ball other than the center of the ball, define $f(P)$ to be the reciprocal of the distance from P to the origin. The average value of f over R involves an improper integral, since the function blows up near the origin. Does this improper integral converge or diverge? What is the average value of f over R? *Suggestion:* Examine the integral over the region between concentric spheres of radii a and t, and let $t \to 0^+$.

✳ 49 Let R be a region in a plane and P a point a distance $h > 0$ from the plane. P and R determine a cone with base R and vertex P, as shown in Fig. 18. Let the area of R be A. What can be said about the distance of the centroid of the cone from the plane of R?
(a) What is that distance in the case of a right circular cone?
(b) Experiment with another cone with any convenient base of your choice.
(c) Make a conjecture.
(d) Explain why the conjecture is true.

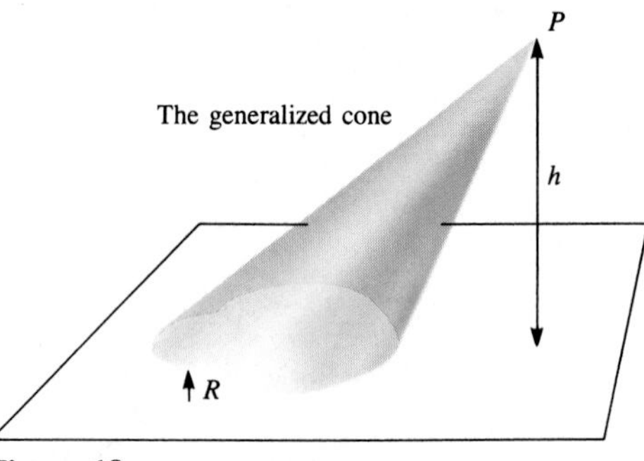

Figure 18

15.S SUMMARY

Let R be a region in the plane and f a function that assigns to each point P in R a number. Then the definite integral of f over R is defined with the aid of partitions of R and sampling points as

$$\lim_{\text{mesh}\to 0} \sum_{i=1}^{n} f(P_i)A_i.$$

This number is denoted $\int_R f(P)\,dA$.

The definite integral over a region R in space, $\int_R f(P)\,dV$, is defined similarly. Both definitions are analogous to the definition of $\int_a^b f(x)\,dx$, the integral over an interval.

These integrals are of use in physics for defining and/or computing mass, moments, the center of mass, the centroid, moment of inertia, and gravitational attraction. A few exercises illustrated some of the other interpretations and applications.

Most of the chapter was concerned with the computation of these integrals by repeated integrals over intervals.

When polar coordinates or cylindrical coordinates are used, an extra r must be put in the integrand. In the case of spherical coordinates, $\rho^2 \sin \phi$ must be inserted.

Vocabulary and Symbols

integrals over planar or spatial regions
$\int_R f(P)\,dA$ or $\int_R f(P)\,dV$ [Also denoted $\iint_R f(P)\,dA$ or $\iiint_R f(P)\,dV$.]
partition, diameter, mesh
average value
moment (first moment)
moment of inertia (second moment)
center of mass
center of gravity
centroid
repeated integral (rectangular, polar, cylindrical, spherical) (Also called iterated integral, double integral, and triple integral.)

Key Facts

Formula	*Significance*
$\int_R 1\,dA$	Area of R
$\int_R 1\,dV$	Volume of R
$\dfrac{\int_R f(P)\,dA}{\text{Area of } R}$ or $\dfrac{\int_R f(P)\,dV}{\text{Volume of } R}$	Average of f over R
$\int \sigma(P)\,dA$ or $\int \delta(P)\,dV$ where σ and δ denote density	M, total mass in R
$\int_R y\,\sigma(P)\,dA$, $\int_R x\,\sigma(P)\,dA$	Moments M_x and M_y about x and y axes respectively (A moment can be computed around any line in the plane.)
$\int_R f(P)\,\sigma(P)\,dA$, $\int_R f(P)\,\delta(P)\,dV$ where $f(P)$ = square of distance from P to some fixed line L	Moment of inertia around L for planar and solid regions respectively
$\left(\dfrac{M_y}{M}, \dfrac{M_x}{M}\right)$	Center of mass $(\bar{x}, \bar{y})$
$\int_R z\,\delta(P)\,dV$	Moment M_{xy}
$\int_R y\,\delta(P)\,dV$	Moment M_{xz}
$\int_R x\,\delta(P)\,dV$	Moment M_{yz}
$\left(\dfrac{M_{yz}}{M}, \dfrac{M_{xz}}{M}, \dfrac{M_{xy}}{M}\right)$	Center of mass of solid, $(\bar{x}, \bar{y}, \bar{z})$

If density is 1, the center of mass is called the centroid.

RELATIONS BETWEEN RECTANGULAR COORDINATES AND SPHERICAL OR CYLINDRICAL COORDINATES

$$x = \rho \sin \phi \cos \theta \qquad x = r \cos \theta$$

$$y = \rho \sin \phi \sin \theta \qquad y = r \sin \theta$$

$$z = \rho \cos \phi \qquad z = z$$

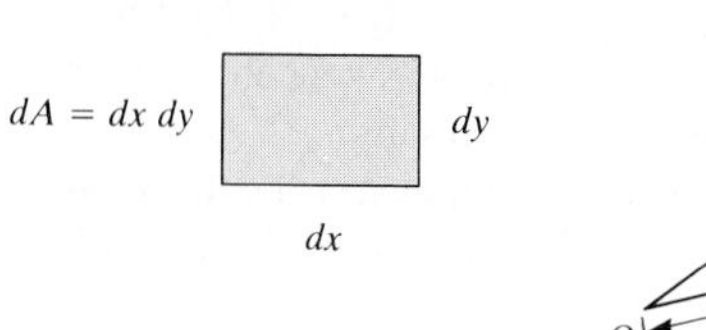

Figure 1

Repeated integrals can be used to evaluate $\int_R f(P)\,dA$. The repeated integrals are

$$\int_a^b \left[\int_{y_1(x)}^{y_2(x)} f(x, y)\,dy\right] dx \qquad \text{and} \qquad \int_c^d \left[\int_{x_1(y)}^{x_2(y)} f(x, y)\,dx\right] dy$$

in rectangular coordinates and

$$\int_\alpha^\beta \left[\int_{r_1(\theta)}^{r_2(\theta)} f(r, \theta)\,r\,dr\right] d\theta$$

in polar coordinates. (The other order is seldom convenient.) Remember the extra r in the integrand of the repeated integral in polar coordinates. It is present because $r\,dr\,d\theta$ (not $dr\,d\theta$) is the approximate area of the little region corresponding to changes of dr and $d\theta$ in the coordinates. Figure 1 serves as a reminder.

Similarly, $\int_R f(P)\,dV$ may be evaluated by a repeated integral in one of the three coordinate systems: rectangular, cylindrical, or spherical. Rectangular is usually best for boxes or polyhedra, cylindrical for right circular cylinders or regions whose projections are disks, and spherical for cones or spheres. However, the formula for the integrand may also influence the choice.

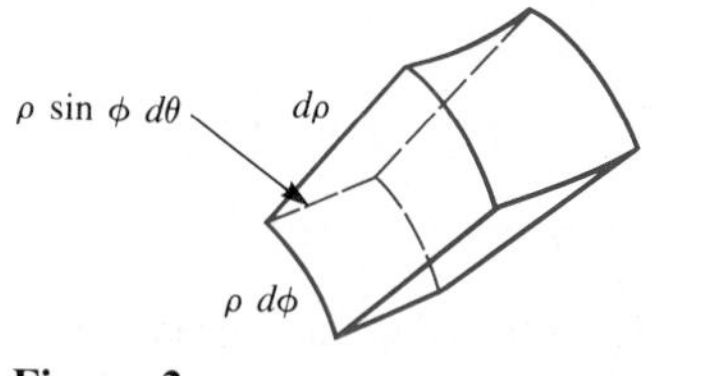

Figure 2

When using an iterated integral in cylindrical coordinates, an extra r must be put in the integrand, just as in the case of polar coordinates. In the case of spherical coordinates, $\rho^2 \sin\phi$ must be inserted, and the "almost box" in Fig. 2 serves as a reminder.

GUIDE QUIZ FOR CHAP. 15: DEFINITE INTEGRALS OVER PLANE AND SOLID REGIONS

1 (*a*) Describe in rectangular coordinates the region whose description in polar coordinates is

$$0 \le \theta \le \pi, \qquad 0 \le r \le a.$$

(*b*) Describe in polar coordinates the region whose description in rectangular coordinates is

$$1 \le x \le 2, \qquad \frac{x}{\sqrt{3}} \le y \le x.$$

2 Find the moment about the x axis of the triangle whose vertices are (0, 0), (2, 0), and (2, 2), using density $\sigma = 1$ and (*a*) a repeated integral in rectangular coordinates, (*b*) a repeated integral in polar coordinates.

3 (*a*) Find the average distance from points in a disk of radius a to the center.

(*b*) Why is the average larger than $a/2$? (Give an intuitive explanation.)

4 Transform this repeated integral to a repeated integral in polar coordinates, and evaluate the latter:

$$\int_0^a \left[\int_0^{\sqrt{a^2-x^2}} (x^2 + y^2)^{3/2}\,dy\right] dx.$$

5 (*a*) Find the moment of inertia of the region in one loop of the curve $r = \sin 2\theta$ about the z axis if the density is 1.

(*b*) Find the mass within the loop in (*a*) if the density at (r, θ) is r^2.

(*c*) Find the volume of a solid whose base is the loop in (*a*) and whose cross section above (r, θ) has length r^2.

(*d*) The temperature at the point (r, θ) inside the loop in (*a*) is r^2 degrees. What is the average temperature?

6 An agricultural sprinkler distributes water in a circle of radius 100 feet. By placing a few random cans in this circle, it is determined that the sprinkler supplies water at a depth of e^{-r} feet of water at a distance of r feet from the sprinkler in 1 hour. How much water does the sprinkler supply in 1 hour to the region within (*a*) 100 feet of the sprinkler? (*b*) 50 feet of the sprinkler?

7 A solid circular cylinder of radius a and height h is composed of a uniform material of mass M. Show that its moment of inertia about a line perpendicular to the axis and midway between the two ends of the cylinder is

$$\frac{Ma^2}{4} + \frac{Mh^2}{12}.$$

8 Two spheres, of radii a and b, $a < b$, have a common center. A plane passes through the center. Describe the centroid of the region that lies between the spheres and to one side of the plane.

9 (*a*) What is meant by the symbol $\int_R f(P)\, dV$?
(*b*) Show that, if $2 \le f(P) \le 3$ for all points P in R, then $2 \cdot$ Volume of $R \le \int_R f(P)\, dV \le 3 \cdot$ Volume of R.

10 (*a*) Find the cylindrical coordinates of the point whose rectangular coordinates are $(3, 4, -3)$.
(*b*) Find the rectangular coordinates of the point whose spherical coordinates are $(3, \pi/2, 2\pi/3)$.
(*c*) Find the spherical coordinates of the point whose cylindrical coordinates are $(2, \pi/4, 2)$.

11 Draw the set of points in a ball of radius 1, whose center is at the origin of the coordinate system, determined by (*a*) $x = \frac{1}{2}$, (*b*) $\phi = \pi/3$, (*c*) $\rho = \frac{1}{2}$, (*d*) $\theta = \pi/2$, (*e*) $z = -\frac{1}{2}$.

12 (*a*) What extra factor must be introduced when setting up a repeated integral in cylindrical or in spherical coordinates?
(*b*) Why?

13 (*a*) Draw the little solid region corresponding to changes $\Delta\rho$, $\Delta\theta$, and $\Delta\phi$ in the spherical coordinates.
(*b*) Show why its volume is approximately

$$\rho^2 \sin\phi \,\Delta\rho\, \Delta\theta\, \Delta\phi.$$

14 A solid right circular cylinder has radius a and height h. Find the average over R of the function f, where $f(P)$ is the square of the distance from the axis of the cylinder to P:
(*a*) Set up repeated integrals in at least two of the three coordinate systems.
(*b*) Evaluate the easier repeated integral in (*a*).

15 (*a*) Evaluate the repeated integral

$$\int_0^1 \left[\int_0^1 \left(\int_0^x ye^{x^2}\, dz\right) dy\right] dx.$$

(*b*) Draw the region R described by the ranges of integration in (*a*).

16 A solid homogeneous right circular cone of radius a and height h has mass M.
(*a*) What is meant by its "moment of inertia about a line through its vertex and parallel to its base"?
(*b*) Set up repeated integrals in all three coordinate systems for the moment of inertia in (*a*).
(*c*) Evaluate at least one of the repeated integrals in (*b*).

17 What is the equation of (*a*) the plane $z = 3$ in spherical coordinates, (*b*) the cylindrical surface $r = 2$ in rectangular coordinates, (*c*) the cylindrical surface $r = 2$ in spherical coordinates, (*d*) the spherical surface $\rho = 3$ in cylindrical coordinates?

18 Evaluate these integrals with as little work as possible:
(*a*) $\int_{-1}^{1} y^7 e^{-y^2}\, dy$
(*b*) $\int_R \sin\theta\, dA$, where R is the triangle in the xy plane with vertices $(1, 0)$, $(2, -1)$, and $(2, 1)$, and θ is a polar angle.
(*c*) $\int_R (2x + 3y + 4z)\, dV$ where R is a ball of radius a and center at the origin.

REVIEW EXERCISES FOR CHAP. 15: DEFINITE INTEGRALS OVER PLANE AND SOLID REGIONS

1 Describe the finite region between $y = x^2$ and $y = 4$ by (*a*) vertical cross sections, (*b*) horizontal cross sections.

2 Compute
(*a*) $\int_0^1 x^2 y\, dy$ (*b*) $\int_0^1 x^2 y\, dx$

3 Compute
(*a*) $\int_1^{x^2} (x + y)\, dy$ (*b*) $\int_y^{y^2} (x + y)\, dx$

4 Compute the easier of
(*a*) $\int_0^1 \sin(x^2 y)\, dy$ (*b*) $\int_0^1 \sin(x^2 y)\, dx$

5 Describe the region in Fig. 3 in terms of (*a*) rectangular coordinates and vertical cross sections, (*b*) rectangular coordinates and horizontal cross sections, (*c*) polar coordinates.

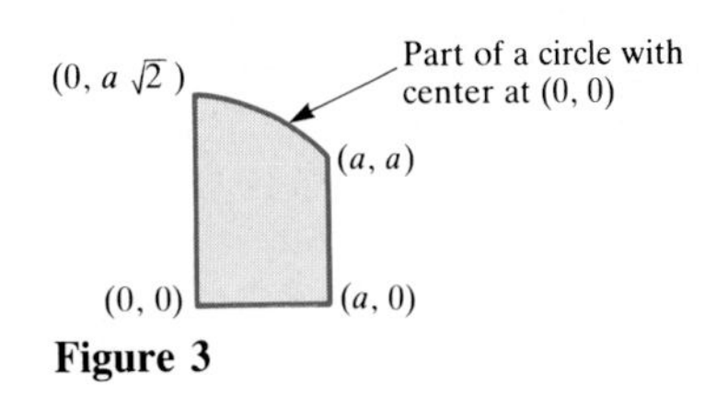

Figure 3

Translate the repeated integrals in Exercises 6 to 8 to repeated integrals in polar coordinates and in each case evaluate the latter.

6 $$\int_0^1 \left(\int_0^x x^2\, dy\right) dx$$

7 $\int_0^{1/\sqrt{2}} \left(\int_x^{\sqrt{1-x^2}} \sqrt{x^2+y^2}\, dy \right) dx$

8 $\int_0^2 \left(\int_0^{\sqrt{2x-x^2}} x\, dy \right) dx$

9 Consider the repeated integral $\int_{\pi/4}^{\pi/2} (\int_0^a r^2 \sin \theta\, dr)\, d\theta$.
(*a*) Draw R and describe f such that $\int_R f(P)\, dA$ is represented by the given repeated integral.
(*b*) Choose a convenient repeated integral in rectangular coordinates equal to $\int_R f(P)\, dA$.
(*c*) Evaluate $\int_R f(P)\, dA$ by the simplest method.

Translate the repeated integrals in Exercises 10 to 12 to repeated integrals in rectangular coordinates and evaluate. (Choose the more convenient direction.)

10 $\int_0^{\pi/4} \left(\int_0^a r^2 \cos \theta\, dr \right) d\theta$

11 $\int_0^{\pi/4} \left(\int_0^a r^3\, dr \right) d\theta$

12 $\int_{\pi/4}^{3\pi/4} \left(\int_0^a r^3\, dr \right) d\theta$

13 Find the moment of inertia of a homogeneous square of mass M and side a about (*a*) a side, (*b*) a diagonal.

14 Find the centroid of the region outside the circle $r = 1$ and inside the cardioid $r = 1 + \cos \theta$.

15 Find the moment of inertia of the finite region bounded by the curve $y = x^3$, $y = 8$, and the y axis about (*a*) the x axis, (*b*) the y axis, (*c*) the z axis. Assume that it is homogeneous and has mass M.

In Exercises 16 to 19 compute $\int_R f(P)\, dA$ if $f(x, y) = xy$ and R is described in coordinates as

16 $0 \le x \le 2$, $x^3 \le y \le 2x^3$

17 $0 \le x \le \pi/2$, $0 \le y \le \sin x$

18 $0 \le \theta \le \pi/4$, $0 \le r \le 2 \sin \theta$

19 $0 \le \theta \le \pi/4$, $0 \le r \le \cos 2\theta$

20 Find the centroid of the finite region bounded by $y = x^2$ and $y = \sqrt{x}$.

21 The depth of water provided by a water sprinkler is approximately 2^{-r} feet at a distance of r feet from the sprinkler. Find the total amount of water within a distance of a feet of the sprinkler.

22 Find the centroid of the region bounded by the curve $y = \cos x$, the x axis, and the lines $x = \pi/2$ and $x = -\pi/2$.

23 Let R be a triangle. Place an xy-coordinate system in such a way that its origin is at one vertex of the triangle and the x axis is parallel to the opposite side. Call the coordinates of the two other vertices (a, b) and (c, b). Show that $\bar{y} = 2b/3$.

24 Let R be a disk of radius a and center $(0, 0)$.
(*a*) Without evaluating them, explain why the integrals $\int_R x^2\, dA$ and $\int_R y^2\, dA$ are equal.
(*b*) Without evaluating any of these integrals, show that $\int_R x^2\, dA + \int_R y^2\, dA = \int_R r^2\, dA$.
(*c*) Evaluate $\int_R r^2\, dA$ by using polar coordinates.
(*d*) Combining (*a*), (*b*), and (*c*), compute $\int_R x^2\, dA$.

25 (*a*) Draw the region R whose description is

$$\frac{\sqrt{2}}{2} \le x \le 1, \qquad \sqrt{1-x^2} \le y \le x.$$

(*b*) Describe R in polar coordinates.
(*c*) Transform the repeated integral

$$\int_{\sqrt{2}/2}^{1} \left(\int_{\sqrt{1-x^2}}^{x} \frac{1}{\sqrt{x^2+y^2}}\, dy \right) dx$$

into polar coordinates.
(*d*) Evaluate the repeated integral in polar coordinates.

26 Evaluate $\int_R \ln (x^2 + y^2)\, dA$ over the region in Exercise 25.

27 A homogeneous right circular cylindrical shell has inner radius a, outer radius b, and height h. Its mass is M. Show that its moment of inertia (*a*) about its axis is $M(a^2 + b^2)/2$; (*b*) about a line through its center of mass and perpendicular to its axis is $M(a^2 + b^2 + h^2/3)/4$.

28 (*a*) A homogeneous solid of mass M occupies the space between two concentric spheres of radii a and b, $a < b$. Show that its moment of inertia around a diameter is $2M(b^5 - a^5)/[5(b^3 - a^3)]$.
(*b*) What is the limit of the moment of inertia in (*a*) as $a \to b$?

29 Let R be the region bounded by a circle of radius a. Let A be a point in the plane of the circle at a distance $H > a$ from the center of the circle. Define a function f by setting $f(P) = \overline{PA}^2$. Show that the average value of f over R is $H^2 + a^2/2$.

30 Find the moment of inertia of a homogeneous equilateral triangle of side a and mass M about a line through a vertex and perpendicular to the plane of the triangle.

31 A flat piece of homogeneous sheet metal of mass M forms a rectangle with sides of length a and b, as shown in Fig. 4. Find its moment of inertia about the line (*a*) AB, (*b*) AD, (*c*) AC, (*d*) through E parallel to AB, (*e*) through E perpendicular to the rectangle.

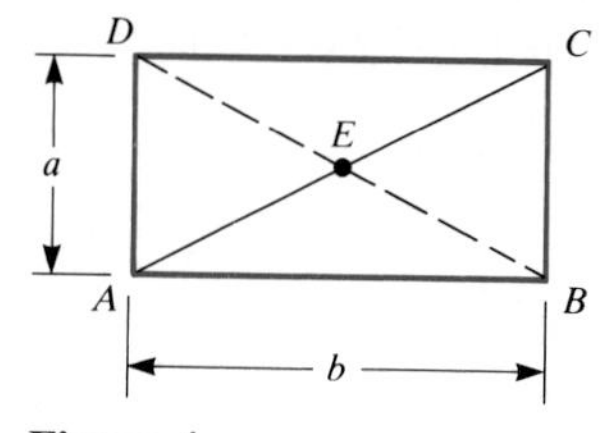

Figure 4

32 A flat piece of metal has mass M and forms a semidisk of radius a. (A semidisk is bounded by a semicircle and its diameter.) The density at a distance y from its diameter is $2y$.

(*a*) Find the total mass.
(*b*) Find the moment around its diameter.
(*c*) Describe its center of mass.
(*d*) Find the moment of inertia around its diameter.
(*e*) Find the moment of inertia around the line perpendicular to the metal and passing through the midpoint of its diameter.

33 A flat piece of homogeneous sheet metal has mass M and forms a disk of radius a.
(*a*) Find its moment of inertia around the line perpendicular to the metal and passing through its center.
(*b*) Find its moment of inertia around a line perpendicular to the metal and passing through its border.
(*c*) Find its moment of inertia around a diameter (that is, a line in the plane of the metal and passing through its center).
(*d*) Find its moment of inertia around a tangent line.

34 (A gravitational paradox) Consider a mass distributed uniformly throughout the entire plane and a point mass a distance a from the plane. The gravitational attraction of the planar mass on the point mass is directed toward the plane and has magnitude

$$\int_R \frac{a}{(\sqrt{a^2+r^2})^3}\,dA.$$

(The integral, which is taken over the entire plane, is improper in the sense that an integral over the x axis is improper. Treat it similarly by computing the integral over a circle of radius s centered at the origin, and letting s go to infinity.) In the integrand r refers to polar coordinates where the pole is the point in the plane closest to the point mass.
(*a*) Show that the integral has the value 2π.
(*b*) According to (*a*), the attractive force of the plane on the point mass is independent of the distance between the point mass and the plane. Does that make sense?

35 Define $f(t)$ to be $\int_t^1 e^{x^2}\,dx$. Find the average value of f over the interval $[0, 1]$.

36 Let $f(x, y) = e^{y^3}$.
(*a*) Devise a region R in the plane such that $\int_R f(P)\,dA$ *can* be evaluated with the aid of a repeated integral.
(*b*) Devise a region R in the plane such that $\int_R f(P)\,dA$ *cannot* be evaluated with the aid of a repeated integral, and describe the difficulty.

37 Let $f(x)$ and $g(x)$ be continuous functions on $[a, b]$. Let R be the square in the xy plane whose vertices are (a, a), (b, a), (b, b), and (a, b).
(*a*) Why is

$$\int_R [f(x)g(y) - f(y)g(x)]^2\,dA \geq 0?$$

(*b*) Deduce that

$$\left(\int_a^b f(x)g(x)\,dx\right)^2 \leq \int_a^b (f(x))^2\,dx \int_a^b (g(x))^2\,dx,$$

which is known as the Schwarz inequality.

In Exercises 38 and 39 find the volumes of the given solid regions. Sketch the regions.

38 The region between the saddle $z = xy$ and the cone $z^2 = x^2 + y^2$ and above the portion of the disk whose boundary is $x^2 + y^2 = 1$ that lies in the first quadrant.

39 The region between the planes $z = 3x + y$ and $z = 4x + 2y$ and above the triangle with vertices $(1, 1)$, $(2, 1)$, and $(2, 2)$.

40 Let $f(x)$ be a function with continuous derivatives $f^{(1)}(x)$ and $f^{(2)}(x)$. By the fundamental theorem of calculus,

$$f(b) = f(a) + \int_a^b f^{(1)}(x)\,dx.$$

Again, by the fundamental theorem of calculus,

$$f^{(1)}(x) = f^{(1)}(a) + \int_a^x f^{(2)}(t)\,dt.$$

Thus,

$$f(b) = f(a) + f^{(1)}(a)(b-a) + \int_a^b \left[\int_a^x f^{(2)}(t)\,dt\right] dx.$$

(*a*) By switching the order of integration, show that

$$f(b) = f(a) + f^{(1)}(a)(b-a) + \int_a^b f^{(2)}(x)(b-x)\,dx.$$

(*b*) By (*a*), $f(x) = f(a) + f'(a)(x-a) + \int_a^x f^{(2)}(t)(b-t)\,dt$. This implies that the error in using the Taylor polynomial of degree one to approximate $f(x)$ can be given as a definite integral. Repeat the process, replacing $f^{(2)}(x)$ in the above equation by $\int_a^x f^{(3)}(t)\,dt$. What formula for $f(b)$ results?
(*c*) Instead, apply integration by parts to the equation in (*b*). What formula for $f(b)$ results?

41 Let $z = g(y)$ be a decreasing function of y such that $g(1) = 0$. Let R be the solid of revolution formed by revolving about the z axis the region in the yz plane bounded by $y = 0$, $z = 0$, and $z = g(y)$. Using repeated integrals in cylindrical coordinates, show that $\int_R z\,dV = \int_0^1 \pi y[g(y)]^2\,dy$ and $\int_R z\,dV = \int_0^{g(0)} \pi[g^{-1}(z)]^2 z\,dz$.

42 (See Exercise 41.)
(*a*) Show that the z coordinate of the centroid of the solid described in Exercise 41 is

$$\frac{\int_0^1 x[g(x)]^2/2\,dx}{\int_0^1 xg(x)\,dx},$$

while the z coordinate of the centroid of the plane region that was revolved is

$$\frac{\int_0^1 [g(x)]^2/2\,dx}{\int_0^1 g(x)\,dx}.$$

(*b*) By considering

$$\int_0^1 \int_0^1 g(x)g(y)(x - y)[g(x) - g(y)]\, dx\, dy,$$

show that the centroid of the solid of revolution is below that of the plane region. *Hint:* Why is the repeated integral less than or equal to 0?

43 A right rectangular pyramid has a base of dimensions a by b and height h. Its mass is M. Show that the moment of inertia of the pyramid around the line that is perpendicular to the base and passes through its top vertex is $M(a^2 + b^2)/20$. (See Fig. 5.)

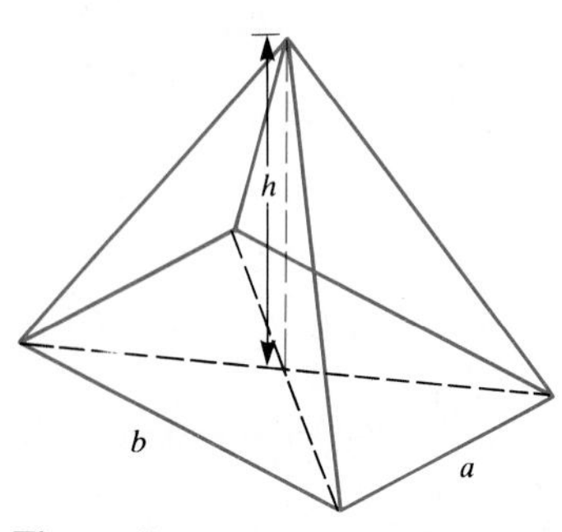

Figure 5

44 The gravitational attraction between a homogeneous ball of radius s and a point mass, as shown in Fig. 6, involves evaluation of the integral

$$\int_S \frac{\cos \alpha}{q^2}\, dV.$$

Show that its value is $4\pi s^3/3H^2$.

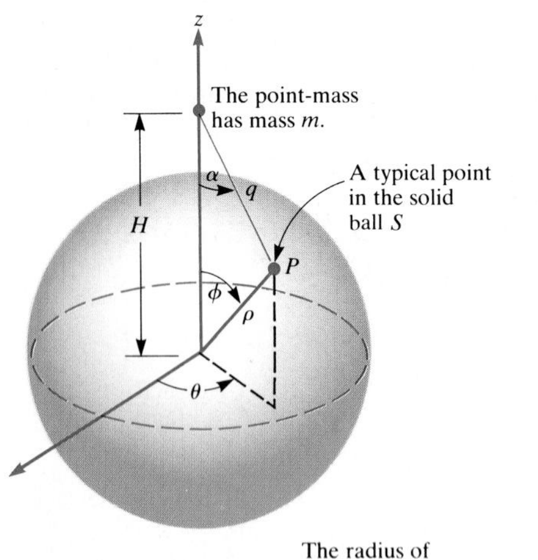

Figure 6

45 A doughnut (torus) is formed by revolving a circle of radius a in a plane around a line L in that plane that is a distance $b > a$ from the center of the circle. (See Fig. 7.) Its mass is M. Show that the moment of inertia of the doughnut around the line L is $M(b^2 + 3a^2/4)$. Assume that the density is constant.

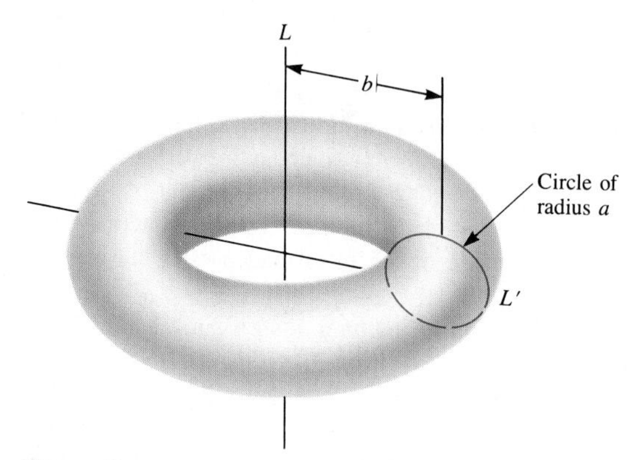

Figure 7

46 Show that the moment of inertia of the doughnut of Exercise 45 around line L' shown in Fig. 7 is $M(b^2/2 + 5a^2/8)$.

47 A homogeneous solid right circular cone of radius a and height h has mass M. Show that the moment of inertia of the cone around a line through its center of mass and parallel to its base is $3M(a^2 + h^2/4)/20$.

48 Find the moment of inertia around the z axis of a homogeneous flat piece of metal of mass M occupying the region R shown in Fig. 8.

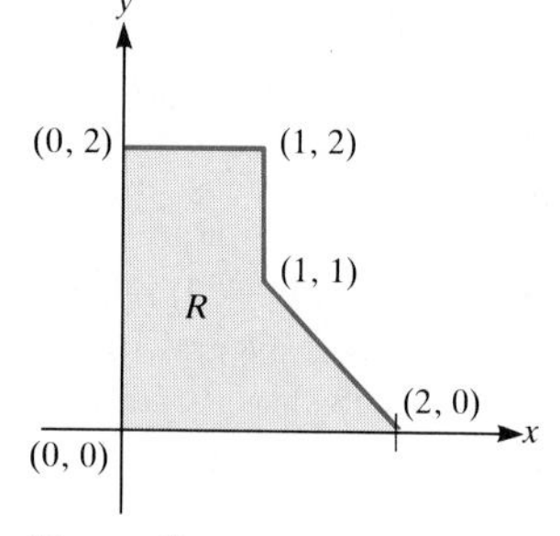

Figure 8

49 A hole of radius b is drilled along the axis of a cylinder of radius a. Assuming that the remaining solid is homogeneous and has mass M, find its moment of inertia around the axis.

50 Recall that $\int_R 1\, dA$ equals the area of R. Using this fact, develop the formula for area in polar coordinates given in Sec. 9.2, namely,

$$\text{Area} = \int_\alpha^\beta \frac{[f(\theta)]^2}{2}\,d\theta.$$

51 Using the fact (developed in Exercise 39 of Sec. 15.4) that $\int_0^\infty e^{-x^2}\,dx = \sqrt{\pi}/2$, evaluate

(*a*) $\displaystyle\int_0^\infty e^{-4x^2}\,dx$ (*b*) $\displaystyle\int_0^\infty \frac{e^{-x}}{\sqrt{x}}\,dx$

(*c*) $\displaystyle\int_0^\infty x^2e^{-x^2}\,dx$ (*d*) $\displaystyle\int_0^\infty \sqrt{x}e^{-x}\,dx$

(*e*) $\displaystyle\int_0^1 \frac{dx}{\sqrt{\ln(1/x)}}$ (*f*) $\displaystyle\int_0^1 \sqrt{\ln(1/x)}\,dx$

52 This exercise is based on "Sudden Expansion in a Pipeline," from *Introduction to Fluid Mechanics* by Stephen Whitaker, Krieger, 1981.

The velocity of a fluid at a distance r from the axis of a pipe of radius r_0 is given by a formula of the form $v(r) = a(1 - r/r_0)^{1/n}$, where a and n are constants and r_0 is the radius of the pipe. Let R be a cross section of the pipe perpendicular to its axis. Find the average over R of (*a*) the velocity of the fluid, (*b*) the square of the velocity of the fluid.

53 Consider a spherical coordinate system, ρ, θ, ϕ. At a given point let $\mathbf{u}_\rho$, $\mathbf{u}_\phi$, and $\mathbf{u}_\theta$ be unit vectors pointing in the directions of increasing ρ, ϕ, and θ, respectively. (See Fig. 9.) Let f be a function defined on space. Show that

$$\nabla f = \frac{\partial f}{\partial \rho}\mathbf{u}_\rho + \frac{1}{\rho}\frac{\partial f}{\partial \phi}\mathbf{u}_\phi + \frac{1}{\rho \sin\theta}\frac{\partial f}{\partial \theta}\mathbf{u}_\theta.$$

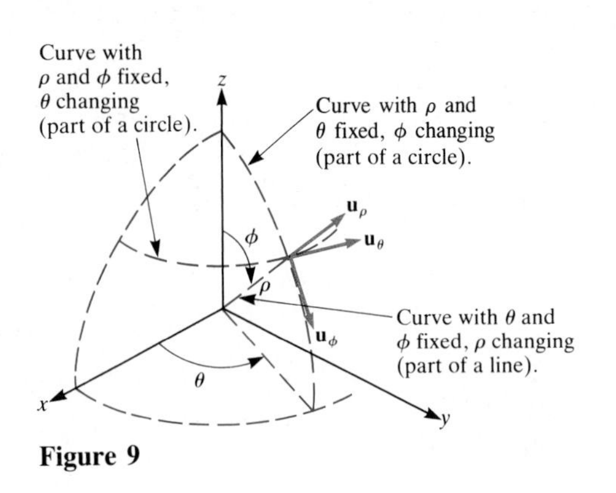

Figure 9

Hint: Think of directional derivatives.

54 In Example 5 of Sec. 15.7 it was assumed that $H > a$. Evaluate the same integral if $H < a$.

16

GREEN'S THEOREM

This chapter and the next develop the basic tools used in the study of gravity, electricity, magnetism, and fluid flow. In this chapter we develop the ideas in the plane. Chapter 17 extends the results to space.

In a sense both chapters generalize the fundamental theorem of calculus, which asserts that

$$\int_a^b \frac{df}{dx}\,dx = f(b) - f(a).$$

This theorem relates one function at the ends of the interval $[a, b]$, namely f, to another function defined throughout the interval $[a, b]$, namely df/dx. Green's theorem in this chapter and Stokes' theorem in Chap. 17 relate the integral of a function defined on a curve that bounds some surface $\mathcal{S}$ to an integral over the surface $\mathcal{S}$. The divergence theorem relates the integral of a function defined on a surface that bounds some spatial region $\mathcal{V}$ to the integral of another function over $\mathcal{V}$.

Sections 16.1 and 16.2 give most of the definitions needed in the rest of the chapter and in Chap. 17. The later sections will show their importance in applications. (It will *not* be assumed that the reader has studied fluid flow, gravitation, electricity, or magnetism. Any concepts borrowed from these areas will be quite intuitive.)

Section 16.3 presents some important applications of the mathematical concepts in Secs. 16.1 and 16.2. Section 16.4 develops Green's theorem, and Secs. 16.5 and 16.6 apply it.

The functions in this chapter will be assumed to have partial derivatives of all orders. The curves and surfaces will be assumed to be smooth (in the sense that the curves locally resemble straight lines and the surfaces locally resemble planes) or to be made up of a finite number of such curves or surfaces.

Engineering students, especially, should keep these words in mind, quoted from the preface of Stephen Whitaker's upper-division text, *Introduction to Fluid Mechanics,* "Vector notation is used freely throughout the text, not because it leads to elegance or rigor but simply because fundamental concepts are best expressed in a form which attempts to connect them with reality."

16.1 VECTOR AND SCALAR FIELDS

Consider an imaginary loop of wire C held firmly in place on the surface of a stream whose flow is horizontal. At some points of C water is entering the region bounded by C, and at some points of C water is leaving the region. In Fig. 1 the arrows indicate the velocity at various points on the stream, and the curve is the wire C. How can the net amount entering or leaving be computed? This chapter will develop a method for answering this question.

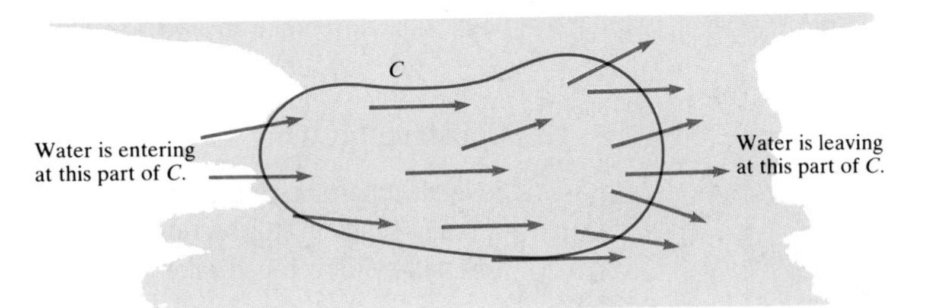

Figure 1

Definition *Vector field.* A function that assigns a vector to each point in some region in the plane (or space) is called a **vector field**. It will usually be denoted **F**.

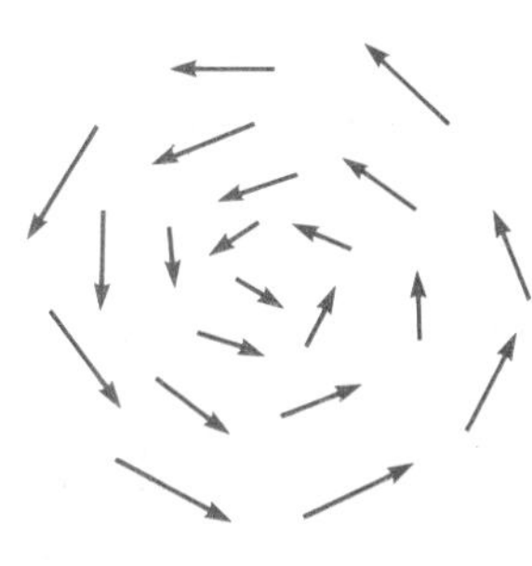

Figure 2

We use the term "field" instead of "function" in deference to physicists and engineers, who speak of "magnetic field" and "electric field," both of which are examples of "vector fields."

The daily weather map displays a few of the (vector) values of the vector field that assigns to each point on the surface of a portion of the earth (considered to be flat) the wind vector at that point. Figure 2 shows a few of the wind vectors of the vector field associated with a hurricane. Near the eye of the hurricane the wind vectors are shorter: The air is relatively calm.

"Scalar field" is just a name for $f(x, y)$ or $f(x, y, z)$.

Definition *Scalar field.* A function that assigns a number to each point in some region in the plane (or in space) is called a **scalar field**. It will usually be denoted f.

A scalar field is just a real-valued function. The function that assigns the temperature at a point is a scalar function; so is the function that describes the density at a point.

A vector field **F** in the plane is described by two scalar fields, the scalar components of **F**:

$$\mathbf{F}(x, y) = P(x, y)\mathbf{i} + Q(x, y)\mathbf{j}.$$

Both P and Q are scalar fields.

A vector field in space is described by three scalar fields:

$$\mathbf{F}(x, y, z) = P(x, y, z)\mathbf{i} + Q(x, y, z)\mathbf{j} + R(x, y, z)\mathbf{k}.$$

The single symbol $\mathbf{F}$ therefore should call to mind a picture of many arrows, such as in Figs. 1 and 2.

Note: Engineers and physicists use a different notation, namely

$$\mathbf{F} = F_x\mathbf{i} + F_y\mathbf{j} + F_z\mathbf{k}$$

or

$$\mathbf{F} = F_x\hat{x} + F_y\hat{y} + F_z\hat{z}$$

(where $\hat{x} = \mathbf{i}$, etc.). Since the theorems are stated in terms of $\mathbf{F}$, not its components, the difference in notation will cause no problem.

Constructing One Field from Another

Several important vector or scalar fields can be defined in terms of other vector or scalar fields. This section presents these definitions; their physical interpretation will be developed later in this and the next chapter.

In physics, f represents potential energy, and ∇f represents force.

Definition *The gradient field.* Let f be a scalar field. The vector field that assigns to each point (x, y) the gradient of f at (x, y), ∇f, is called the **gradient field** associated with f. A similar construction can be carried out on $f(x, y, z)$, again producing a vector field ∇f.

EXAMPLE 1 Compute and sketch the gradient field associated with the scalar field $f(x, y, z) = 1/\sqrt{x^2 + y^2 + z^2}$. (These two fields are of importance in gravitational and electromagnetic theory.)

SOLUTION

$$\nabla f = \frac{\partial f}{\partial x}\mathbf{i} + \frac{\partial f}{\partial y}\mathbf{j} + \frac{\partial f}{\partial z}\mathbf{k}$$

$$= \frac{-x}{(x^2 + y^2 + z^2)^{3/2}}\mathbf{i} + \frac{-y}{(x^2 + y^2 + z^2)^{3/2}}\mathbf{j} + \frac{-z}{(x^2 + y^2 + z^2)^{3/2}}\mathbf{k}.$$

Let $\mathbf{r}$ be the position vector of the point (x, y, z). Then

$$\|\mathbf{r}\| = \sqrt{x^2 + y^2 + z^2}.$$

The vector field $\mathbf{F} = -\mathbf{r}/\|\mathbf{r}\|^3$.

Thus

$$\nabla f = -\frac{\mathbf{r}}{\|\mathbf{r}\|^3}$$

and is therefore pointed toward the origin. Moreover,

$$\|\nabla f\| = \left\| -\frac{\mathbf{r}}{\|\mathbf{r}\|^3} \right\| = \frac{\|\mathbf{r}\|}{\|\mathbf{r}\|^3} = \frac{1}{\|\mathbf{r}\|^2}.$$

When $\mathbf{r}$ is short, ∇f is long, and when $\mathbf{r}$ is long, ∇f is short. In physics, the gradient field

$$-\frac{\mathbf{r}}{\|\mathbf{r}\|^3}$$

corresponds to a force of attraction that obeys an "inverse square" law. Figure 3 shows a few of the values of this vector field. ■

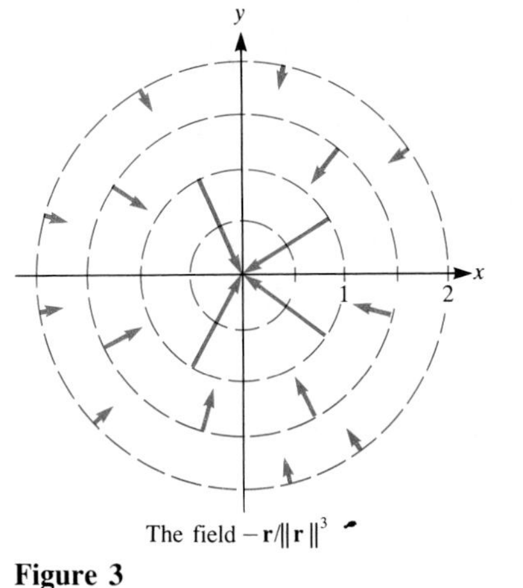

Figure 3

The Vector $\hat{\mathbf{r}}$

The length or norm of $-\mathbf{r}/\|\mathbf{r}\|^3$ is inversely proportional to $\|\mathbf{r}\|^2$. But just glancing at $-\mathbf{r}/\|\mathbf{r}\|^3$ does not give that impression. Engineers and physicists often use a slightly different notation that clearly displays the "inverse square" nature of the vector field. They introduce the vector

$$\hat{\mathbf{r}} = \frac{\mathbf{r}}{\|\mathbf{r}\|}.$$

This is a unit vector that records the direction of $\mathbf{r}$. (See Fig. 4.) Then we have

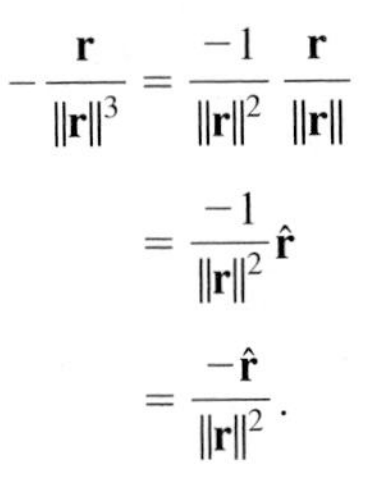

$$-\frac{\mathbf{r}}{\|\mathbf{r}\|^3} = \frac{-1}{\|\mathbf{r}\|^2}\frac{\mathbf{r}}{\|\mathbf{r}\|} = \frac{-1}{\|\mathbf{r}\|^2}\hat{\mathbf{r}} = \frac{-\hat{\mathbf{r}}}{\|\mathbf{r}\|^2}.$$

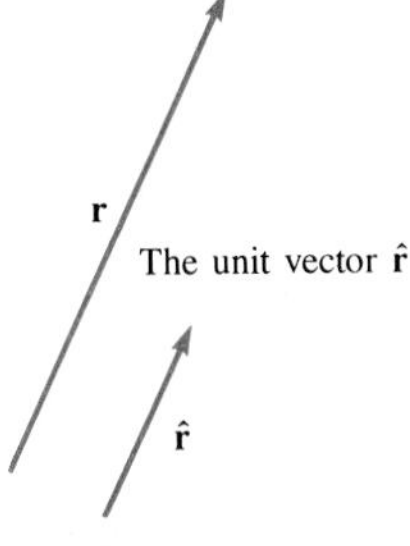

Figure 4

Since $\|\hat{\mathbf{r}}\| = 1$, the length of $-\hat{\mathbf{r}}/\|\mathbf{r}\|^2$ is clearly $1/\|\mathbf{r}\|^2$. Thus, if $r = \sqrt{x^2 + y^2 + z^2}$, we have

$$\nabla\left(\frac{1}{r}\right) = \frac{-\hat{\mathbf{r}}}{\|\mathbf{r}\|^2} = \frac{-\hat{\mathbf{r}}}{r^2}.$$

This notation for the vector field $-\mathbf{r}/\|\mathbf{r}\|^3$ is preferable because it displays the magnitude of the field, $1/\|\mathbf{r}\|^2$.

The Divergence of a Vector Field

In Example 1 a vector field was obtained from a scalar field by "taking the gradient." (However, not every vector field is the gradient of some scalar field, as Exercise 37 shows.) Also of importance in mathematics and physics is a procedure that constructs a scalar field from a vector field. It turns out that if a vector field $\mathbf{F}$ describes the motion of a fluid, then this associated scalar field, called the **divergence** of $\mathbf{F}$, describes the tendency of that fluid to accumulate or spread out at any given point.

The divergence of a vector field is a scalar field.

Definition *Divergence of a vector field.* Let

$$\mathbf{F}(x, y) = P(x, y)\mathbf{i} + Q(x, y)\mathbf{j}$$

be a vector field in the plane. The scalar field

$$\frac{\partial P}{\partial x} + \frac{\partial Q}{\partial y}$$

is called the **divergence** of $\mathbf{F}$. Similarly, if

$$\mathbf{F}(x, y, z) = P(x, y, z)\mathbf{i} + Q(x, y, z)\mathbf{j} + R(x, y, z)\mathbf{k},$$

a vector field in space, then the scalar field

$$\frac{\partial P}{\partial x} + \frac{\partial Q}{\partial y} + \frac{\partial R}{\partial z}$$

is called the **divergence** of **F**.

EXAMPLE 2 Compute the divergence of the vector field given by $5x^2y\mathbf{i} + xy\mathbf{j} + x^2z\mathbf{k}$.

SOLUTION By definition, the divergence of $5x^2y\mathbf{i} + xy\mathbf{j} + x^2z\mathbf{k}$ is

$$\frac{\partial}{\partial x}(5x^2y) + \frac{\partial}{\partial y}(xy) + \frac{\partial}{\partial z}(x^2z) = 10xy + x + x^2. \quad \blacksquare$$

Notation For convenience, introduce the formal "vector"

$$\nabla = \frac{\partial}{\partial x}\mathbf{i} + \frac{\partial}{\partial y}\mathbf{j}.$$

If $\mathbf{F} = P\mathbf{i} + Q\mathbf{j}$, compute $\nabla \cdot \mathbf{F}$ as if it were a dot product of two ordinary vectors:

$$\nabla \cdot \mathbf{F} = \frac{\partial}{\partial x} P + \frac{\partial}{\partial y} Q.$$

Interpret this to mean $\qquad \dfrac{\partial P}{\partial x} + \dfrac{\partial Q}{\partial y}.$

This explains the customary notation for the divergence of **F**,

$\nabla \cdot \mathbf{F}$ *is the usual notation for divergence.*

$$\nabla \cdot \mathbf{F}.$$

Similarly, to provide a short notation for the divergence of a vector field in space, introduce the formal "vector"

$$\nabla = \frac{\partial}{\partial x}\mathbf{i} + \frac{\partial}{\partial y}\mathbf{j} + \frac{\partial}{\partial z}\mathbf{k}.$$

Then, if $\mathbf{F} = P\mathbf{i} + Q\mathbf{j} + R\mathbf{k}$, $\nabla \cdot \mathbf{F}$ is a shorthand for the divergence

$$\frac{\partial P}{\partial x} + \frac{\partial Q}{\partial y} + \frac{\partial R}{\partial z}.$$

The divergence of **F** is also written "div **F**."

The next example, which is important in the study of gravitational and electrostatic forces, will be referred to later in the chapter.

EXAMPLE 3 Let $\mathbf{F}(P)$, for any point P in space other than the origin O, be defined as $\hat{\mathbf{r}}/\|\mathbf{r}\|^2$, where $\mathbf{r}$ is the position vector $\overrightarrow{OP}$, $\mathbf{r} = x\mathbf{i} + y\mathbf{j} + z\mathbf{k}$. Show that the divergence of **F** is 0, that is, $\nabla \cdot \mathbf{F} = 0$.

SOLUTION

$$\mathbf{F}(x, y, z) = \frac{\hat{\mathbf{r}}}{\|\mathbf{r}\|^2} = \frac{\mathbf{r}}{\|\mathbf{r}\|^3} = \frac{x\mathbf{i} + y\mathbf{j} + z\mathbf{k}}{r^3},$$

where $r = \|\mathbf{r}\| = \sqrt{x^2 + y^2 + z^2}$. By the definition of the divergence,

$$\nabla \cdot \mathbf{F} = \frac{\partial}{\partial x}\left(\frac{x}{r^3}\right) + \frac{\partial}{\partial y}\left(\frac{y}{r^3}\right) + \frac{\partial}{\partial z}\left(\frac{z}{r^3}\right).$$

It is necessary to compute three partial derivatives. The first one is

$$\frac{\partial}{\partial x}\left(\frac{x}{r^3}\right) = \frac{r^3 \dfrac{\partial x}{\partial x} - x \dfrac{\partial}{\partial x}(r^3)}{r^6} = \frac{r^3 - 3xr^2 \dfrac{\partial r}{\partial x}}{r^6} = \frac{r - 3x \dfrac{\partial r}{\partial x}}{r^4}.$$

Now,
$$\frac{\partial r}{\partial x} = \frac{\partial}{\partial x}(\sqrt{x^2 + y^2 + z^2}) = \frac{x}{\sqrt{x^2 + y^2 + z^2}} = \frac{x}{r}.$$

Thus
$$\frac{\partial}{\partial x}\left(\frac{x}{r^3}\right) = \frac{r - 3x \dfrac{x}{r}}{r^4} = \frac{1}{r^3} - \frac{3x^2}{r^5}.$$

Similarly,
$$\frac{\partial}{\partial y}\left(\frac{y}{r^3}\right) = \frac{1}{r^3} - \frac{3y^2}{r^5} \quad \text{and} \quad \frac{\partial}{\partial z}\left(\frac{z}{r^3}\right) = \frac{1}{r^3} - \frac{3z^2}{r^5}.$$

Consequently,

$$\nabla \cdot \mathbf{F} = \left(\frac{1}{r^3} - \frac{3x^2}{r^5}\right) + \left(\frac{1}{r^3} - \frac{3y^2}{r^5}\right) + \left(\frac{1}{r^3} - \frac{3z^2}{r^5}\right)$$
$$= \frac{3}{r^3} - \frac{3(x^2 + y^2 + z^2)}{r^5}$$
$$= \frac{3}{r^3} - \frac{3r^2}{r^5} = 0. \quad ■$$

How nicely everything canceled in Example 3 and reduced to 0. If a vector field in space has the form $\hat{\mathbf{r}}/\|\mathbf{r}\|^k$ for some constant k, the divergence of that field is generally not 0. As Exercise 31 asks you to check, only for $k = 2$ (the "inverse square law") is the divergence 0. The "zero divergence" of this field, which describes gravitational or electrical fields, has important consequences, as will be shown in Chap. 17.

Note how much information is packed into the compact notation $\nabla \cdot \mathbf{F}$, read as "del dot $\mathbf{F}$." When you read $\nabla \cdot \mathbf{F}$, think "$\mathbf{F}$ is a vector field in the plane or space. It is described by scalar fields P, Q, and, if in space, R:

$$\mathbf{F} = P\mathbf{i} + Q\mathbf{j} + R\mathbf{k}.$$

Then $\nabla \cdot \mathbf{F}$ is the sum of three partial derivatives,

$$\frac{\partial P}{\partial x} + \frac{\partial Q}{\partial y} + \frac{\partial R}{\partial z}."$$

The Curl of a Vector Field

We can derive from a vector field **F** another vector field, called the "curl" of **F**.

The curl of a vector field is again a vector field.

Definition *Curl of a vector field.* Let $\mathbf{F} = P\mathbf{i} + Q\mathbf{j} + R\mathbf{k}$ be a vector field. The function that assigns to each point the vector

$$\left(\frac{\partial R}{\partial y} - \frac{\partial Q}{\partial z}\right)\mathbf{i} + \left(\frac{\partial P}{\partial z} - \frac{\partial R}{\partial x}\right)\mathbf{j} + \left(\frac{\partial Q}{\partial x} - \frac{\partial P}{\partial y}\right)\mathbf{k}$$

is called the **curl** of **F** and is denoted **curl F**.

If you expand the formal determinant

$$\begin{vmatrix} \mathbf{i} & \mathbf{j} & \mathbf{k} \\ \dfrac{\partial}{\partial x} & \dfrac{\partial}{\partial y} & \dfrac{\partial}{\partial z} \\ P & Q & R \end{vmatrix},$$

you obtain **curl F**. The similarity to the cross product of vectors suggests the notation $\nabla \times \mathbf{F}$ for **curl F**. This second notation is read "del cross **F**."

The definition also applies to a vector field $\mathbf{F} = P(x, y)\mathbf{i} + Q(x, y)\mathbf{j}$ in the plane. Writing **F** as $P(x, y)\mathbf{i} + Q(x, y)\mathbf{j} + 0\mathbf{k}$, we find that

$$\nabla \times \mathbf{F} = \left(\frac{\partial Q}{\partial x} - \frac{\partial P}{\partial y}\right)\mathbf{k},$$

since $\partial Q/\partial z = 0$ and $\partial P/\partial z = 0$.

The physical meaning of **curl F** will be explored in Sec. 17.4. (It is related to rotational motion in a fluid. If **curl F** = **0**, then **F** is called **irrotational**.)

EXAMPLE 4 Compute the curl of $\mathbf{F} = xyz\mathbf{i} + x^2\mathbf{j} - xy\mathbf{k}$.

SOLUTION The curl of **F** is given by

$$\begin{vmatrix} \mathbf{i} & \mathbf{j} & \mathbf{k} \\ \dfrac{\partial}{\partial x} & \dfrac{\partial}{\partial y} & \dfrac{\partial}{\partial z} \\ xyz & x^2 & -xy \end{vmatrix},$$

which is short for

$$\left[\frac{\partial}{\partial y}(-xy) - \frac{\partial}{\partial z}(x^2)\right]\mathbf{i} - \left[\frac{\partial}{\partial x}(-xy) - \frac{\partial}{\partial z}(xyz)\right]\mathbf{j} + \left[\frac{\partial}{\partial x}(x^2) - \frac{\partial}{\partial y}(xyz)\right]\mathbf{k}$$

$$= (-x - 0)\mathbf{i} - (-y - xy)\mathbf{j} + (2x - xz)\mathbf{k}$$

$$= -x\mathbf{i} + (y + xy)\mathbf{j} + (2x - xz)\mathbf{k}. \quad \blacksquare$$

The Laplacian of a Scalar Field

There is another scalar field of general importance in engineering, physics, and mathematics. Say that you start with a scalar field f and form the gradient field

∇f. Then you may take the divergence of this vector field. All told, you form the divergence of the gradient of f.

Definition *Laplacian of a scalar field.* Let $f(x, y, z)$ be a scalar field. The scalar field formed by taking the divergence of ∇f is called the **Laplacian** of f.

The Laplacian of f is therefore

$$\nabla \cdot \nabla f = \nabla \cdot \left(\frac{\partial f}{\partial x}\mathbf{i} + \frac{\partial f}{\partial y}\mathbf{j} + \frac{\partial f}{\partial z}\mathbf{k} \right)$$
$$= \frac{\partial}{\partial x}\left(\frac{\partial f}{\partial x}\right) + \frac{\partial}{\partial y}\left(\frac{\partial f}{\partial y}\right) + \frac{\partial}{\partial z}\left(\frac{\partial f}{\partial z}\right)$$
$$= \frac{\partial^2 f}{\partial x^2} + \frac{\partial^2 f}{\partial y^2} + \frac{\partial^2 f}{\partial z^2}.$$

The Laplacian of f is usually denoted $\nabla \cdot \nabla f$ or, more briefly, $\nabla^2 f$. If f is a function of x and y, then its Laplacian is

$$\nabla^2 f = \frac{\partial^2 f}{\partial x^2} + \frac{\partial^2 f}{\partial y^2}.$$

If f describes temperature, the Laplacian gives information about the gain or loss of heat in a region. (See Sec. 16.5.)

EXAMPLE 5 Compute the Laplacian of $f(x, y) = x^3 - 3xy^2$.

SOLUTION First compute $\partial f/\partial x$ and $\partial f/\partial y$:

$$\frac{\partial f}{\partial y} = 3x^2 - 3y^2 \qquad \text{and} \qquad \frac{\partial f}{\partial y} = -6xy.$$

Then

$$\frac{\partial^2 f}{\partial x^2} = 6x \qquad \text{and} \qquad \frac{\partial^2 f}{\partial y^2} = -6x.$$

Consequently,

$$\nabla^2 f = \frac{\partial^2 f}{\partial x^2} + \frac{\partial^2 f}{\partial y^2} = 6x + (-6x) = 0.$$

The Laplacian in this special case has the constant value 0. ■

A function whose Laplacian is identically 0 is called **harmonic**. Harmonic functions are important in the study of electricity, temperature distributions, and functions of a complex variable.

Central Fields

The gravitational field $\hat{\mathbf{r}}/\|\mathbf{r}\|^2$ is an example of a "central field." A **central field** $\mathbf{F}$ is one whose magnitude at a point P depends only on the distance from P to some fixed point O and whose direction at P is parallel to $\mathbf{r} = \overrightarrow{OP}$. This means that there is a function f of one variable such that

$$\mathbf{F}(\mathbf{r}) = f(\|\mathbf{r}\|)\hat{\mathbf{r}}.$$

Such a field is called **radially symmetric**. If f is always positive, then $\mathbf{F}(\mathbf{r})$ points away from O. If f is always negative, then $\mathbf{F}(\mathbf{r})$ points toward O. (Figure 3 is a picture of a central field.) Often a central field is not defined at its center, O.

Section Summary

This table summarizes the key operations described in this section:

Type of Field	*Operation*	*Notation for New Field*	*Type*
Scalar, f	Gradient	∇f	Vector
Vector, $\mathbf{F}$	Divergence	$\nabla \cdot \mathbf{F} = \text{div } \mathbf{F}$	Scalar
Vector, $\mathbf{F}$	Curl	$\nabla \times \mathbf{F} = \mathbf{curl\ F}$	Vector
Scalar, f	Divergence of gradient	$\nabla \cdot \nabla f = \nabla^2 f$ (Laplacian of f)	Scalar

The magnitude of the vector field $\hat{\mathbf{r}}/\|\mathbf{r}\|^2$ varies inversely as the square of the length of $\mathbf{r}$. As a field defined in space ($\mathbf{r} = x\mathbf{i} + y\mathbf{j} + z\mathbf{k}$), its divergence is 0. (However, as a field in the xy plane its divergence is *not* 0. See Exercise 32.) We also defined central fields, which have a prominent role in applications.

EXERCISES FOR SEC. 16.1: VECTOR AND SCALAR FIELDS

In Exercises 1 to 8 compute the divergence of the given vector fields.

1 $x^2y\mathbf{i} + \sin xy\ \mathbf{j}$
2 $3y\mathbf{i} + 2x^{10}\mathbf{j}$
3 $\mathbf{r} = x\mathbf{i} + y\mathbf{j}$
4 $\mathbf{r} = x\mathbf{i} + y\mathbf{j} + z\mathbf{k}$
5 $e^{xy}\mathbf{i} + x \tan 2y\ \mathbf{j} + xz^2\mathbf{k}$
6 $z \sec 2xy\ \mathbf{i} + \dfrac{1}{\sqrt{x^2 + y^2}}\mathbf{j} + \ln(\tan^{-1} 2z)\mathbf{k}$
7 $\hat{\mathbf{r}}/\|\mathbf{r}\|$, $\mathbf{r} = x\mathbf{i} + y\mathbf{j}$
8 $\hat{\mathbf{r}}/\|\mathbf{r}\|$, $\mathbf{r} = x\mathbf{i} + y\mathbf{j} + z\mathbf{k}$

In Exercises 9 to 14 compute the curl of the given fields.

9 $(3x + 2y + 5z)\mathbf{i} + (2x - 3y + 4z)\mathbf{j} + (x + 6y + 7z)\mathbf{k}$.
10 $xy\mathbf{i} + \cos(x + 2y + z)\ \mathbf{j} + z^2\mathbf{k}$.
11 $\hat{\mathbf{r}}/\|\mathbf{r}\|$, where $\mathbf{r} = x\mathbf{i} + y\mathbf{j} + z\mathbf{k}$.
12 $\hat{\mathbf{r}}/\|\mathbf{r}\|^2$, where $\mathbf{r} = x\mathbf{i} + y\mathbf{j} + z\mathbf{k}$.
13 $\hat{\mathbf{r}}/\|\mathbf{r}\|^3$, where $\mathbf{r} = x\mathbf{i} + y\mathbf{j} + z\mathbf{k}$.
14 $\|\mathbf{r}\|^2\hat{\mathbf{r}}$, where $\mathbf{r} = x\mathbf{i} + y\mathbf{j} + z\mathbf{k}$.
15 If f is a scalar field and $\mathbf{F}$ is a vector field, which type of field is $f\mathbf{F}$?
16 If $\mathbf{F}$ and $\mathbf{G}$ are vector fields, which type of field is the function $\mathbf{F} \cdot \mathbf{G}$?
17 By a straightforward computation verify that the curl of the gradient of f is $\mathbf{0}$, that is,

$$\nabla \times (\nabla f) = \mathbf{0},$$

(*a*) In case $f(x, y)$ is defined on the plane;
(*b*) In case $f(x, y, z)$ is defined in space.

18 By a straightforward computation, verify that the divergence of the curl of $\mathbf{F}$ is 0, that is, $\nabla \cdot (\nabla \times \mathbf{F}) = 0$, where $\mathbf{F} = P\mathbf{i} + Q\mathbf{j} + R\mathbf{k}$. Thus div (**curl F**) = 0.
19 Which of the following are scalar fields? vector fields?
(*a*) **curl F** (*b*) $\|\mathbf{F}\|$
(*c*) $\mathbf{F} \cdot \mathbf{F}$ (*d*) div $\mathbf{F}$
(*e*) $\nabla \cdot \mathbf{F}$ (*f*) $\mathbf{F} \times \mathbf{i}$
(*g*) $\nabla \times \mathbf{F}$

20 Which of these expressions make sense? Which do not?
(*a*) The curl of the curl of **F**
(*b*) The curl of the gradient of f
(*c*) The divergence of the curl of **F**
(*d*) The gradient of the curl of **F**
(*e*) The divergence of the divergence of **F**

Exercises 21 to 24 produce an interesting contrast between the plane and space.

21 Show that $\ln(x^2 + y^2)$ is harmonic.

22 Show that $\ln(x^2 + y^2 + z^2)$ is not harmonic.

23 Show that $1/\|\mathbf{r}\|$, where $\mathbf{r} = x\mathbf{i} + y\mathbf{j} + z\mathbf{k}$, is harmonic.

24 Show that $1/\|\mathbf{r}\|$, where $\mathbf{r} = x\mathbf{i} + y\mathbf{j}$, is not harmonic.

25 (*a*) Prove that the divergence of a constant vector field is 0.
(*b*) Give an example of a nonconstant vector field **F** for which $\nabla \cdot \mathbf{F} = 0$.

26 (*a*) If f is a scalar field and **F** is a vector field, prove that

$$\nabla \cdot (f\mathbf{F}) = f\nabla \cdot \mathbf{F} + \nabla f \cdot \mathbf{F}.$$

(*b*) Express the equation in (*a*) in a sentence using such terms as "divergence," "gradient," and "dot product."

27 If f and g are scalar fields, so is fg, their product. Prove that the gradient of fg equals f times the gradient of g plus g times the gradient of f, that is,

$$\nabla(fg) = f\nabla g + g\nabla f.$$

28 Letting $\mathbf{F} = P\mathbf{i} + Q\mathbf{j} + R\mathbf{k}$, prove that

$$\nabla \times f\mathbf{F} = f\nabla \times \mathbf{F} + \nabla f \times \mathbf{F}.$$

29 Letting $\mathbf{F} = P\mathbf{i} + Q\mathbf{j} + R\mathbf{k}$ and $\mathbf{G} = L\mathbf{i} + M\mathbf{j} + N\mathbf{k}$, prove that

$$\text{div } (\mathbf{F} \times \mathbf{G}) = \mathbf{G} \cdot \textbf{curl } \mathbf{F} - \mathbf{F} \cdot \textbf{curl } \mathbf{G}.$$

30 For scalar fields f and g show that

$$\text{div } (\nabla f \times \nabla g) = 0.$$

Review Example 3, where it was shown that the divergence of
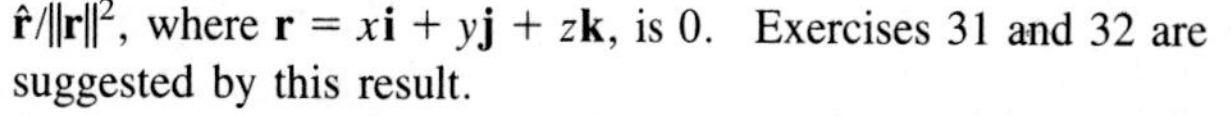
$\hat{\mathbf{r}}/\|\mathbf{r}\|^2$, where $\mathbf{r} = x\mathbf{i} + y\mathbf{j} + z\mathbf{k}$, is 0. Exercises 31 and 32 are suggested by this result.

31 Let k be a fixed real number other than 2. Show that the vector field $\hat{\mathbf{r}}/\|\mathbf{r}\|^k$, defined in *space* except at the origin, has a nonzero divergence.

32 Let **F** be the vector field $\hat{\mathbf{r}}/\|\mathbf{r}\|^k$, defined in the *plane* except at the origin. Show that (*a*) if $k = 1$, $\nabla \cdot \mathbf{F} = 0$, (*b*) if k is not 1, $\nabla \cdot \mathbf{F}$ is not 0. (This should be contrasted with Exercise 31 and Example 3.)

Exercises 33 to 36 concern central fields.

33 If **F** is the central force field with center at (0, 0) and $\mathbf{F}(5, 0) = 7\mathbf{i}$, find and draw (*a*) $\mathbf{F}(4, 3)$, (*b*) $\mathbf{F}(3, 4)$, (*c*) $\mathbf{F}(0, 5)$, (*d*) $\mathbf{F}(-5, 0)$, (*e*) $\mathbf{F}(0, -5)$.

34 Assume that $\mathbf{F}(x, y, z)$ is central, with the center at the origin. Hence $\mathbf{F}(\mathbf{r}) = f(r)\hat{\mathbf{r}}$, where $r = \|\mathbf{r}\|$.
(*a*) If $f(r) = kr^{-2}$, where k is a constant, what is div **F**?
(*b*) Show that, if $\nabla \cdot \mathbf{F} = 0$, then $f(r) = kr^{-2}$ for some constant k.

35 In Exercise 34(*b*) it was assumed that **F** is defined in space. What can you conclude if **F** is instead defined in the plane?

36 Show that the curl of the vector field $f(r)\hat{\mathbf{r}}$ is **0** if the field is defined in (*a*) the plane, (*b*) space.

The contrast of Exercises 34 and 36 shows that it is easy for a central force field to have curl **0**, but very hard to have divergence 0. Note that the field $\hat{\mathbf{r}}/\|\mathbf{r}\|^2$ in space can be described, up to a constant factor, as the central field with divergence equal to 0.

37 (*a*) Show that, if $\mathbf{F} = P\mathbf{i} + Q\mathbf{j}$ is a vector field and equals ∇f for some scalar field f, then

$$\frac{\partial P}{\partial y} = \frac{\partial Q}{\partial x}.$$

(*b*) Show that the vector field $x^2y\mathbf{i} + x^2y^3\mathbf{j}$ is not of the form ∇f for any scalar field f.

16.2 LINE INTEGRALS

This section introduces a type of definite integral in which the integrand is defined with the aid of a curve. The next section will illustrate some of its applications.

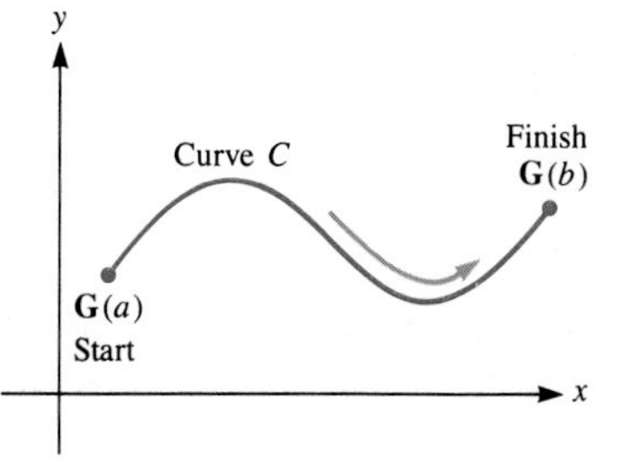

Figure 1

Curves

Let $\mathbf{r} = \mathbf{G}(t)$ for t in $[a, b]$ describe a curve C in the plane or space, as in Fig. 1. (The parameter t might be time, angle, arc length, x coordinate, etc.). $\mathbf{G}(a)$ is the **start**, or **initial point**, of the curve and $\mathbf{G}(b)$ is its **finish**, or **terminal point**. We assume that the curve is **smooth**; that is, $\mathbf{G}'$ is continuous and never zero.

Thus, the curve has a well-defined tangent at each point and the tangent changes direction continuously as we move along the curve.

If $\mathbf{G}(a) = \mathbf{G}(b)$, that is, the finish is the same as the start, the curve is called **closed**. If the curve does not intersect itself except, perhaps, at its endpoints, we call the curve **simple**. (See Fig. 2.)

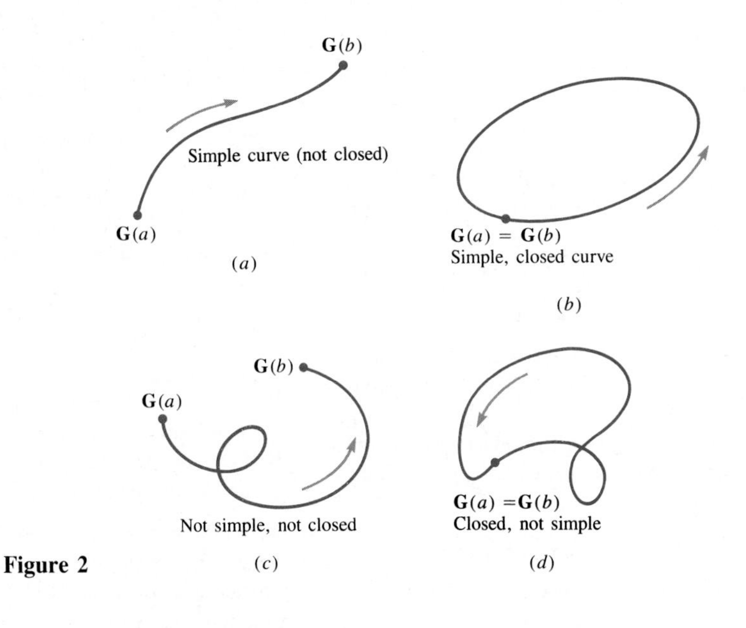

Figure 2

As t increases from a to b, the curve is swept out in a definite orientation, "from initial point to terminal point."

EXAMPLE 1 Draw the curves:

(*a*) $\mathbf{G}(t) = t\mathbf{i} + 2t\mathbf{j}$, t in $[0, 1]$

(*b*) $\mathbf{H}(t) = t^2\mathbf{i} + 2t^2\mathbf{j}$, t in $[0, 1]$

(*c*) $\mathbf{J}(t) = (1 - t/2)\mathbf{i} + (2 - t)\mathbf{j}$, t in $[0, 2]$.

SOLUTION

(*a*) In this case,

$$\begin{cases} x = t \\ y = 2t, \end{cases}$$

so $y = 2x$. The curve is part of the line $y = 2x$. The initial point is $\mathbf{G}(0) = 0\mathbf{i} + 0\mathbf{j}$, which is the origin. Its terminal point is $\mathbf{G}(1) = \mathbf{i} + 2\mathbf{j}$, or $(1, 2)$.

(*b*) In this case,

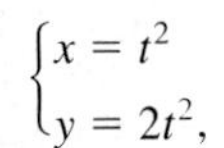

$$\begin{cases} x = t^2 \\ y = 2t^2, \end{cases}$$

so, again, $y = 2x$. The initial point is again the origin, and the terminal point is again $(1, 2)$. Therefore $\mathbf{G}$ and $\mathbf{H}$ are different parameterizations of the same curve, in the same orientation, *from* $(0, 0)$ *to* $(1, 2)$.

(*c*) In this case,

$$\begin{cases} x = 1 - \dfrac{t}{2} \\[2ex] y = 2 - t = 2\left(1 - \dfrac{t}{2}\right), \end{cases}$$

so once again $y = 2x$. Note that $\mathbf{J}(0) = \mathbf{i} + 2\mathbf{j}$, so this time the initial point is $(1, 2)$. The terminal point is $\mathbf{J}(2) = (1 - 2/2)\mathbf{i} + (2 - 2)\mathbf{j} = 0\mathbf{i} + 0\mathbf{j}$, which is the origin. Thus $\mathbf{J}$ sweeps out the same set of points as do $\mathbf{G}$ and $\mathbf{H}$, but with *opposite orientation, from* $(1, 2)$ *to* $(0, 0)$. (See Fig. 3.) ■

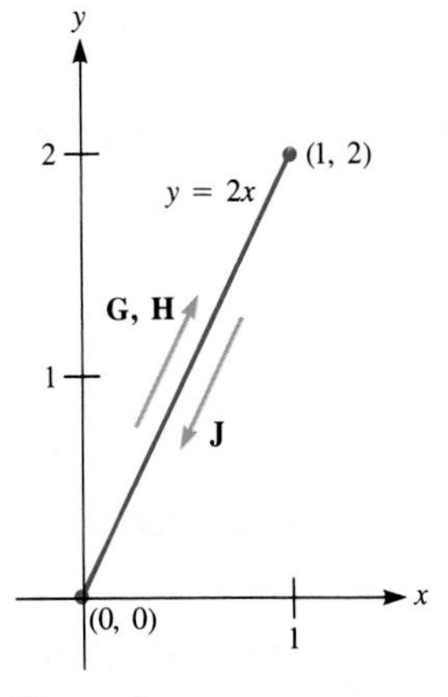

Figure 3

It may seem strange to call a straight line a "curve," but it is the custom to include a line as a special case of a curve.

The curves we will consider are simple (do not cross themselves), either closed or not closed. They will be smooth or, like a polygon, made up of smooth curves pieced together.

Given a parameterization $\mathbf{G}$, one can, as in Example 1, draw the curve it determines. (Even calculators can do this.) But given a curve, how does one construct a parameterization for it? Examples 2 and 3 show how.

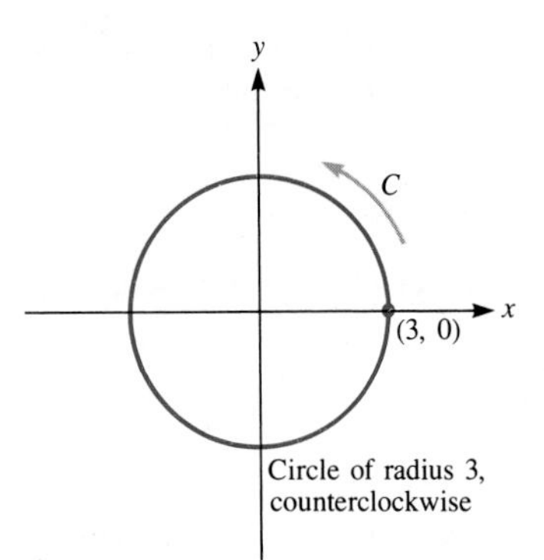

Figure 4

EXAMPLE 2 The curve C consists of all points in the xy plane at a distance 3 from the origin. Parameterize C so that the orientation is (a) counterclockwise, (b) clockwise.

SOLUTION (a) C is shown in Fig. 4, as is its orientation. We are free to choose the initial point (which will coincide with the terminal point). The natural parameter is the polar angle θ, which leads to a choice of $(3, 0)$ for our initial point and the parameterization

$$\begin{cases} x = 3\cos\theta \\ y = 3\sin\theta \end{cases} \qquad \text{for } \theta \text{ in } [0, 2\pi].$$

So $\mathbf{G}(\theta) = 3\cos\theta\,\mathbf{i} + 3\sin\theta\,\mathbf{j}$. As θ increases, the path goes counterclockwise from $(3, 0)$ back to $(3, 0)$.

(b) For each θ in $[0, 2\pi]$, define (x, y) to be

$$\begin{cases} x = 3\cos(-\theta) \\ y = 3\sin(-\theta), \end{cases}$$

as shown in Fig. 5.

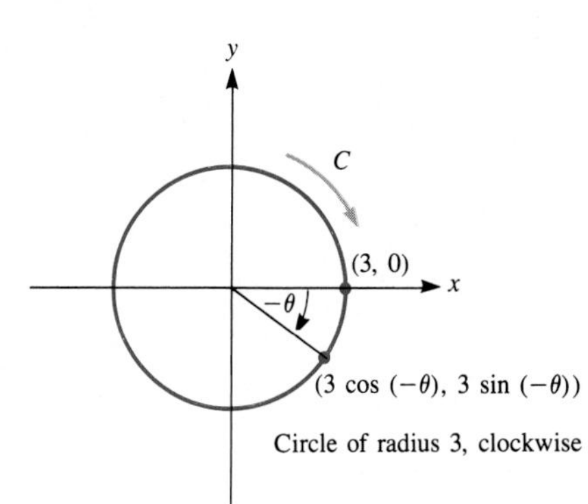

Figure 5

When θ is 0, the initial point is again $(3, 0)$. However, the motion is *clockwise*. For instance, as θ increases from 0 to $\pi/2$, (x, y) moves through the fourth quadrant from $(3, 0)$ to $(0, -3)$. ■

EXAMPLE 3 Parameterize the path shown in Fig. 6.

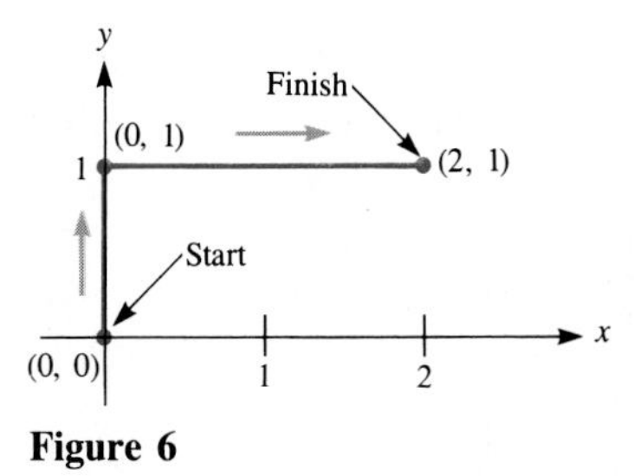

Figure 6

SOLUTION On the first part of the path, the vertical segment, x is 0 and y increases from 0 to 1. For this part of the path we can use the parameterization

$$\begin{cases} x = 0 \\ y = t \end{cases} \qquad t \text{ in } [0, 1].$$

Now we want to let t go beyond 1 to parameterize the horizontal part. From (0, 1) to (2, 1), y is 1 and x increases from 0 to 2. So when $t = 1$, we want $x = 0$, and, as t increases to some number b (which we are free to choose), we want x to increase to 2. A simple function that does this is

$$x = t - 1 \qquad t \text{ in } [1, 3].$$

The entire parameterization is

$$(x, y) = \begin{cases} (0, t) & \text{for } 0 \le t \le 1 \\ (t - 1, 1) & \text{for } 1 \le t \le 3. \end{cases}$$

The parameter t goes from 0 to 3 over the course of the entire curve. ■

With practice, finding a parameterization for arcs of circles or polygons should not be hard. Remember that the answer is not unique, as Example 1 shows.

Having parameterized a curve by some suitable choice of $\mathbf{G}(t)$, we will occasionally write $f(\mathbf{G}(t))$ to indicate that $f(x, y)$ is being evaluated at points (x, y) traced out by $\mathbf{G}(t)$. We use this notation in the following definition.

Line Integrals

We now define one of the most important concepts used in gravitational and electromagnetic theory.

Definition *Line integral.* Let $\mathbf{G}(t)$, $a \le t \le b$, describe a smooth curve C in the plane. Let the scalar function f be defined at least on every point of C. Then the definite integral

$$\int_a^b f(\mathbf{G}(t)) \frac{dx}{dt} \, dt$$

is called a **line integral** of f over C and is denoted

$$\int_C f \, dx \qquad \text{or} \qquad \int_C f(x, y) \, dx.$$

The line integral $\int_C f(x, y) \, dy$ is defined similarly:

$$\boxed{\int_C f(x, y) \, dy = \int_a^b f(\mathbf{G}(t)) \frac{dy}{dt} \, dt.}$$

If C is in space and $f(x, y, z)$ is defined at least on C, then the same definitions hold. In addition,

$$\boxed{\int_C f \, dz = \int_a^b f(\mathbf{G}(t)) \frac{dz}{dt} \, dt.}$$

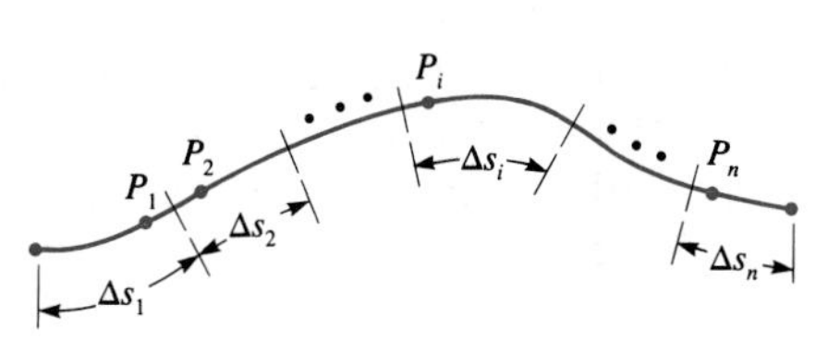

Figure 7

We also consider $\int_C f\,ds$, where s denotes arc length. It is defined just like a definite integral over an interval. First partition the curve C into n sections and choose a point in each section, as in Fig. 7. Let the length of the ith section be Δs_i and let P_i be the point chosen in the ith interval. Then

$$\int_C f\,ds = \lim_{\text{mesh}\to 0} \sum_{i=1}^{n} f(P_i)\,\Delta s_i.$$

For instance, $\int_C 1\,ds$ equals the length of C. If $f(P)$ is the linear density (in grams per centimeter) at P, then $\int_C f(P)\,ds$ is the mass of C. We may evaluate this line integral by parameterizing the curve C:

$$\int_C f\,ds = \int_a^b f\frac{ds}{dt}\,dt.$$

Hence there are four types of line integrals:

$$\int_C f\,dx, \qquad \int_C f\,dy, \qquad \int_C f\,dz, \qquad \text{and} \qquad \int_C f\,ds.$$

We will also consider sums of such line integrals, such as

$$\int_C (P\,dx + Q\,dy + R\,dz),$$

where P, Q, and R are scalar functions.

It may seem strange to use the term "line integral" when it is a curve that plays the key role. You might expect it to be called a "curve integral," but this term has never caught on.

The computation of the next example will bring the definition down to earth.

EXAMPLE 4 Compute $\int_C xy\,dx$, where C is each of the curves given in Example 1, namely,

(*a*) $\mathbf{G}(t) = t\mathbf{i} + 2t\mathbf{j}$, t in $[0, 1]$

(*b*) $\mathbf{H}(t) = t^2\mathbf{i} + 2t^2\mathbf{j}$, t in $[0, 1]$

(*c*) $\mathbf{J}(t) = (1 - t/2)\mathbf{i} + (2 - t)\mathbf{j}$, t in $[0, 2]$.

SOLUTION

(*a*)

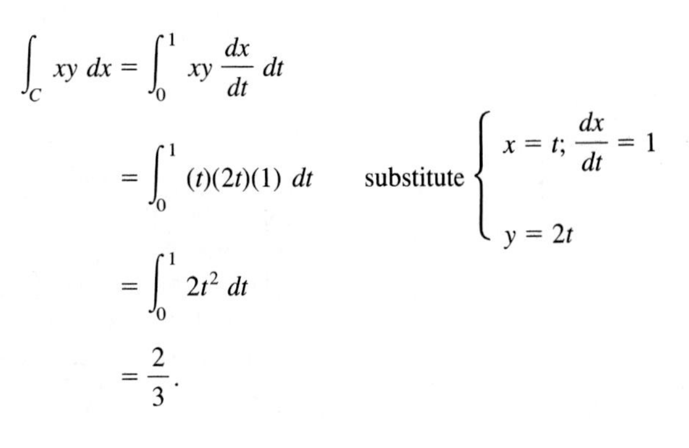

$$\int_C xy\,dx = \int_0^1 xy\frac{dx}{dt}\,dt$$

$$= \int_0^1 (t)(2t)(1)\,dt \qquad \text{substitute} \begin{cases} x = t; \ \dfrac{dx}{dt} = 1 \\ y = 2t \end{cases}$$

$$= \int_0^1 2t^2\,dt$$

$$= \frac{2}{3}.$$

(*b*)

$$\int_C xy\,dx = \int_0^1 xy\,\frac{dx}{dt}\,dt$$

$$= \int_0^1 (t^2)(2t^2)(2t)\,dt \qquad \text{substitute} \begin{cases} x = t^2; \ \dfrac{dx}{dt} = 2t \\ y = 2t^2 \end{cases}$$

$$= \int_0^1 4t^5\,dt$$

$$= \frac{2}{3}.$$

(*c*)

$$\int_C xy\,dx = \int_0^1 xy\,\frac{dx}{dt}\,dt$$

$$= \int_0^2 \left(1 - \frac{t}{2}\right)(2 - t)\left(-\frac{1}{2}\right)dt \qquad \text{substitute} \begin{cases} x = 1 - \dfrac{t}{2}; \ \dfrac{dx}{dt} = -\dfrac{1}{2} \\ y = 2 - t \end{cases}$$

$$= \int_0^2 \left(-\frac{t^2}{4} + t - 1\right)dt$$

$$= -\frac{2}{3}. \quad ■$$

In Example 4, (*a*) and (*b*) have the same answer, $\frac{2}{3}$, and the curve C is the line segment *from* (0, 0) *to* (1, 2), though with different parameterizations. In (*c*), the curve C is the *same* line segment but with the reverse orientation, *from* (1, 2) *to* (0, 0). (Start and finish are switched.) In this case the integral is $-\frac{2}{3}$, the negative of the integrals in (*a*) and (*b*).

Example 4 illustrates three basic principles for the line integrals $\int_C f\,dx$, $\int_C f\,dy$, and $\int_C f\,dz$.

Three basic principles

1 A line integral is a definite integral of the type defined in Chap. 5.
2 If two parameterizations sweep out the curve C with the *same orientation,* then the two line integrals are *equal*.
3 *Reversing the orientation* changes the sign of the line integral.

The first principle is immediate: That's how a line integral is defined. The second is proved in advanced calculus. The third is a consequence of the fact that dx/dt, for instance, changes sign when the orientation is switched. (If x increases in one orientation, it decreases in the opposite orientation.)

The magnitude of $\int_C f\,dx$ is *not* influenced by which parameterization you use, but the sign *is* influenced by the orientation you use to sweep out the curve C. If we use $-C$ to denote the curve C with the opposite parameterization, then

$$\int_{-C} f\,dx = -\int_{C} f\,dx.$$

Example 5 concerns a particularly simple line integral, which will be needed in Sec. 16.4.

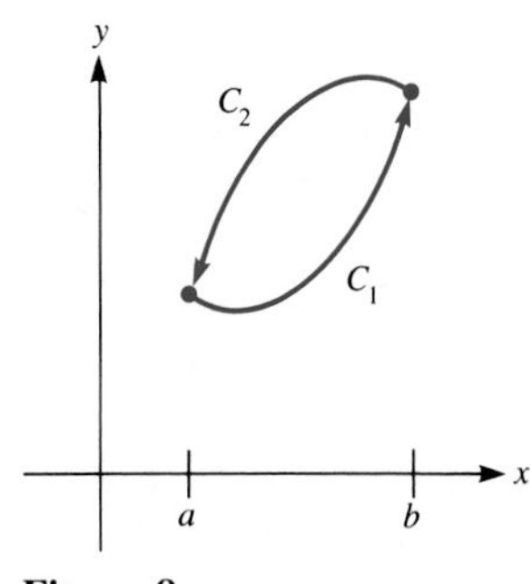

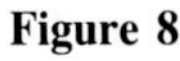
Figure 8

EXAMPLE 5 Compute (*a*) $\int_{C_1} y\,dx$ and (*b*) $\int_{C_2} y\,dx$ for the curves shown in Fig. 8, with the orientation as given.

SOLUTION

(*a*) We parameterize C_1 by x in $[a, b]$. For each x in $[a, b]$, let $\mathbf{G}(x)$ be the point on C_1 with x-coordinate x. Then

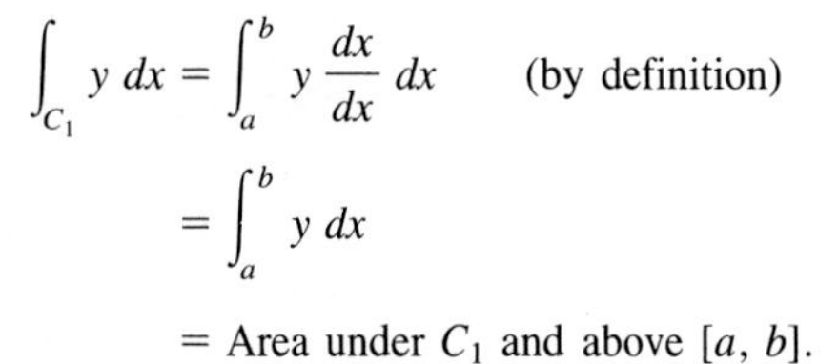

$$\int_{C_1} y\,dx = \int_a^b y\frac{dx}{dx}\,dx \qquad \text{(by definition)}$$

$$= \int_a^b y\,dx$$

$$= \text{Area under } C_1 \text{ and above } [a, b].$$

(*b*) We parameterize the curve $-C_2$, with x increasing from a to b. As with C_1,

$$\int_{-C_2} y\,dx = \text{Area under } C_2.$$

Thus,
$$\int_{C_2} y\,dx = -(\text{Area under } C_2 \text{ and above } [a, b]).$$

(See Fig. 9.)

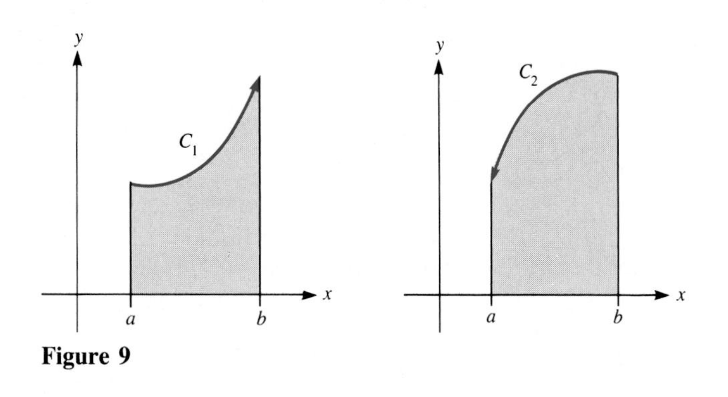

Figure 9 ■

Think of "y dx" as + or − the area of a thin rectangle.

We could think of C_2 being parameterized with x decreasing from b to a. Then dx/dt, hence dx, would be negative. The expression $y\,dx$ suggests a rectangle of height y and width $|dx|$, but with "negative area," $y\,dx$. We would therefore expect $\int_{C_2} y\,dx$ to be negative.

In part (*b*) it is helpful to think of "dx" as negative, since in the orientation of C_2, x is *decreasing* from b to a; no matter what parameterization is chosen, dx/dt is negative.

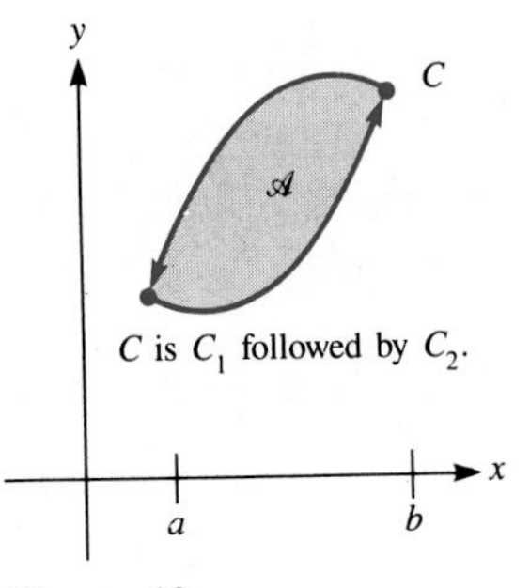

Figure 10

Now, let C be the closed curve consisting of C_1 followed by C_2, the curves defined in Example 5. It bounds a region $\mathscr{A}$ as shown in Fig. 10.

$$\int_C y\,dx = \int_{C_1} y\,dx + \int_{C_2} y\,dx$$

$$= \text{Area under } C_1 - \text{Area under } C_2$$

$$= -\text{Area of } \mathscr{A}.$$

This result is quite general: *The line integral of y dx around a simple closed curve in the xy plane swept out in a counterclockwise direction is the negative of the area enclosed by the curve.*

When C is a *simple closed curve*, $\int_C f\,dx$ is also written $\oint_C f\,dx$. The notations $\oint_C f\,dy$ and $\oint_C f\,dz$ are similarly defined. The absolute value of the line integral over a simple closed curve does not depend on the choice of the initial point (which equals the terminal point), nor on the parameterization. But the *sign* of the integral depends on the orientation of C. If C is a simple closed curve in the xy plane, we speak of the "clockwise" and "counterclockwise" orientations. The following theorem will be needed in Sec. 16.4.

Theorem If C is a simple closed curve in the xy plane, swept out *counterclockwise,* and A is the area of the region bounded by C, then

$$\oint_C y\,dx = -A \qquad \oint_C y\,dy = 0$$

$$\oint_C x\,dy = A \qquad \oint_C k\,dx = 0$$

$$\oint_C x\,dx = 0 \qquad \oint_C k\,dy = 0,$$

where k is any constant.

Proof We have already shown that $\oint_C y\,dx = -A$. That $\oint_C x\,dy = A$ is shown in a similar fashion. To evaluate $\oint_C x\,dx$, let $\mathbf{G}(t)$ be a parameterization of C, with t in $[a, b]$. Then

$$\oint_C x\,dx = \int_a^b x(t)\frac{dx}{dt}\,dt$$

$$= \int_a^b \frac{d}{dt}\left(\frac{1}{2}(x(t))^2\right)dt$$

$$= \frac{(x(t))^2}{2}\Bigg|_{t=a}^{t=b} \qquad \text{(fundamental theorem of calculus)}$$

$$= \frac{(x(b))^2}{2} - \frac{(x(a))^2}{2}$$

$$= 0. \qquad \text{(finish is same as start)}$$

That $\oint_C y\,dy$, $\oint_C k\,dx$, and $\oint_C k\,dy$ are all 0 can be shown the same way. ■

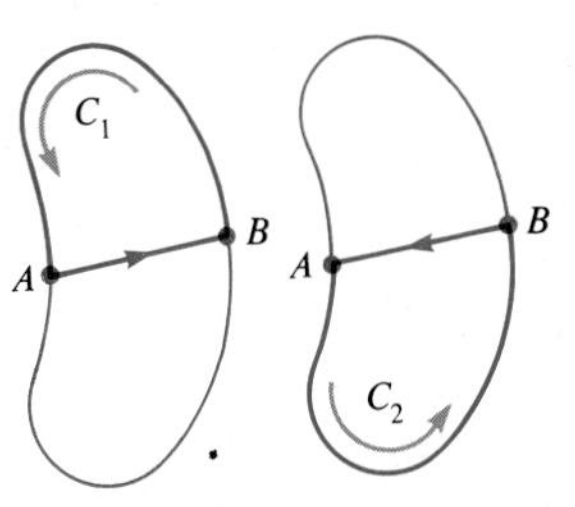

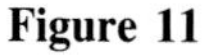

Figure 11

The Cancellation Principle

Consider two simple closed curves C_1 and C_2, both taken counterclockwise. Assume that they coincide in a section AB, as in Fig. 11. C_1 sweeps out the section between A and B from *left to right*, while C_2 sweeps it out in the opposite direction, from *right to left*. That means that the parts of

$$\oint_{C_1} f\,dx \quad \text{and} \quad \oint_{C_2} f\,dx$$

due to integration over the section between A and B have opposite signs. The parts of C_1 and C_2 other than AB together form C, the curve that bounds the entire region, shown in Fig. 12. This gives us the cancellation principle:

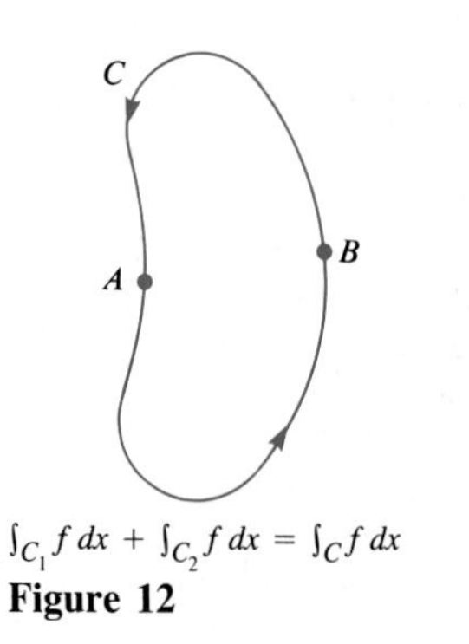

$\int_{C_1} f\,dx + \int_{C_2} f\,dx = \int_C f\,dx$

Figure 12

$$\boxed{\oint_{C_1} f\,dx + \oint_{C_2} f\,dx = \oint_C f\,dx.}$$

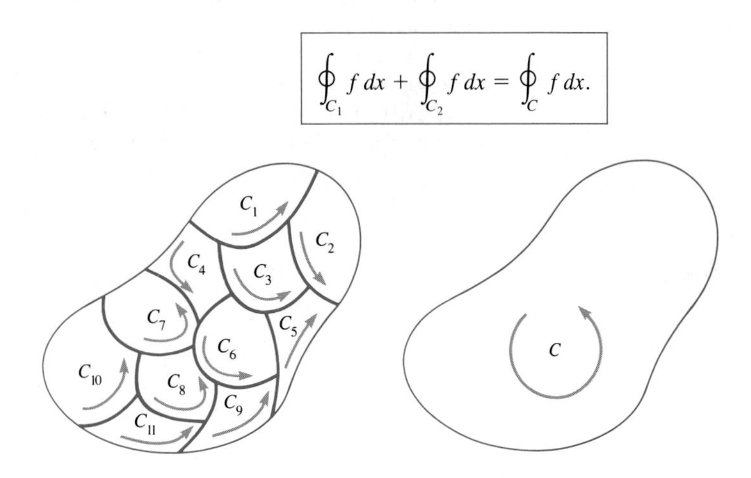

Figure 13

The cancellation principle extends to any finite number of curves, all with counterclockwise orientation (or all with clockwise orientation). Thus in Fig. 13, cancellations *inside* the region imply that

$$\sum_{i=1}^{11} \int_{C_i} f\,dx = \int_C f\,dx.$$

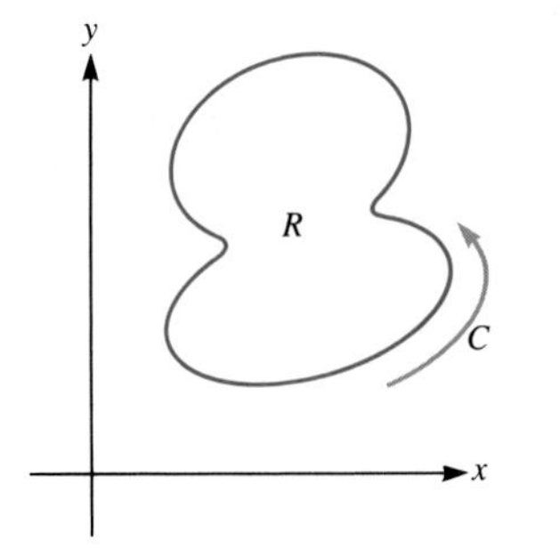

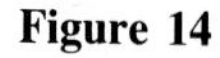

Figure 14

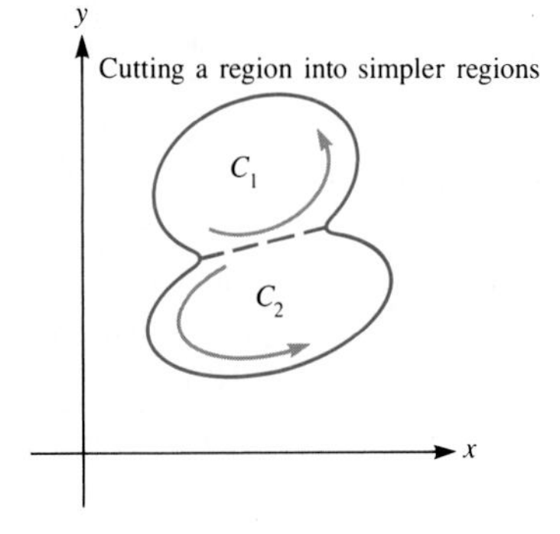

Figure 15

The cancellation principle applies to the other line integrals, $\oint_C f\,dy$ and $\oint_C f\,dz$, as well. However, it does not apply to $\oint_C f\,ds$, since this integral, defined in terms of Riemann sums, does not involve orientation. (See Exercises 32 to 35, where your choice of parameter will not affect the results.)

When we proved that $\oint_C y\,dx = -A$, we assumed that the curve C meets each vertical line at most twice. (That is why we could break C into C_1 and C_2.) The cancellation principle permits us to extend the proof to regions that can be broken into regions of that type. For instance, consider the region R in Fig. 14, bounded by the curve C. Some vertical lines meet the curve C as many as four times. However, the dashed line in Fig. 15 cuts R into two regions. The boundary curves C_1 and C_2 do meet each vertical line at most twice. Let the top region have area A_1 and the lower region have area A_2. Then,

$$\oint_{C_1} y\,dx + \oint_{C_2} y\,dx = -A_1 + (-A_2)$$

$$= -\text{Area of } R.$$

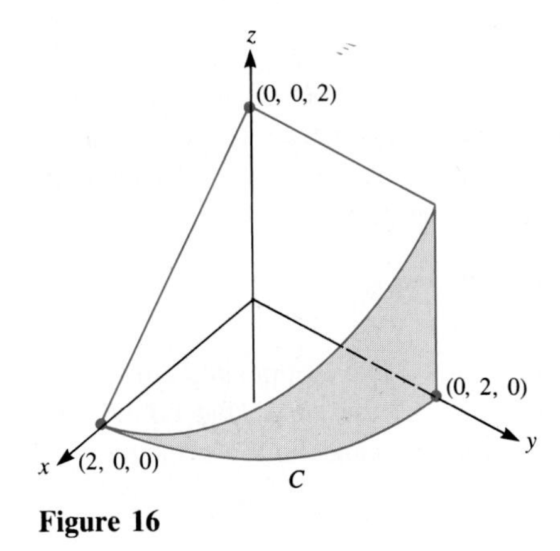

Figure 16

By the cancellation principle,

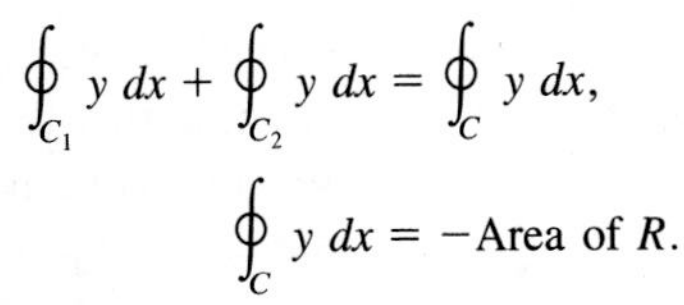

$$\oint_{C_1} y\,dx + \oint_{C_2} y\,dx = \oint_C y\,dx,$$

and we have
$$\oint_C y\,dx = -\text{Area of } R.$$

More generally, $\oint_C y\,dx = -\text{Area of } R$ for any region that can be cut into regions whose boundaries meet each vertical line at most twice.

EXAMPLE 6 Find the area of that portion of the surface $x^2 + y^2 = 4$ that lies above the first quadrant of the xy plane and below the plane $z = 2 - x$. (See Fig. 16.)

SOLUTION Figure 17 shows a local approximation of the surface area. Its area is approximately $z\,ds$. Then the area is

$$\int_C z\,ds,$$

where $z = 2 - x$. Using θ as a parameter, we see that $s = 2\theta$ and C in Fig. 16 has the description

$$\begin{cases} x = 2\cos\theta \\ y = 2\sin\theta \end{cases} \qquad 0 \le \theta \le \frac{\pi}{2}.$$

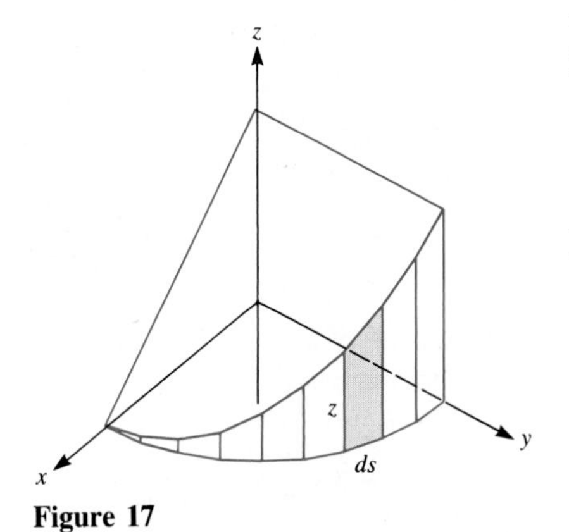

Figure 17

Hence
$$\begin{aligned} \int_C z\,ds &= \int_0^{\pi/2} z\,\frac{ds}{d\theta}\,d\theta \\ &= \int_0^{\pi/2} (2 - x)\,\frac{ds}{d\theta}\,d\theta \\ &= \int_0^{\pi/2} (2 - 2\cos\theta)\,2\,d\theta \\ &= 2\pi - 4. \quad \blacksquare \end{aligned}$$

Section Summary

We defined the line integral

$$\int_C f\,dx = \int_a^b f(\mathbf{G}(t))\,\frac{dx}{dt}\,dt,$$

and, similarly, $\int_C f\,dy$ and $\int_C f\,dz$. Here $\mathbf{G}$ parameterizes the curve C and f is defined at least for all points in C.

The line integral depends only on the orientation $\mathbf{G}$ gives the curve C. Switching the start and finish changes the sign of the line integral, which we record by writing

$$\int_{-C} f\,dx = -\int_C f\,dx.$$

We showed that

$$\oint_C y\,dx = -A,$$

where A is the area of the region bounded by C, if C is counterclockwise. Also,

$$\oint_C y\,dy = 0, \qquad \oint_C x\,dx = 0,$$

$$\oint_C k\,dx = 0, \qquad \oint_C k\,dy = 0. \qquad (k \text{ a constant})$$

We developed the cancellation principle for the line integrals $\int_C f\,dx$, $\int_C f\,dy$, and $\int_C f\,dz$. We also defined $\int_C f\,ds$.

EXERCISES FOR SEC. 16.2: LINE INTEGRALS

In Exercises 1 to 4, sketch the curve described by the given parameterization and label its start and finish.

1 $\mathbf{G}(t) = t\mathbf{i} + t^2\mathbf{j}$ $\quad t$ in $[0, 1]$.
2 $\mathbf{G}(t) = (1 - t)\mathbf{i} + (1 - t)^2\mathbf{j}$ $\quad t$ in $[0, 1]$.
3 $\mathbf{G}(t) = (2t + 1)\mathbf{i} + 3t\mathbf{j}$ $\quad t$ in $[0, 2]$.
4 $\mathbf{G}(t) = 4 \cos t\,\mathbf{i} + 5 \sin t\,\mathbf{j}$ $\quad t$ in $[0, \pi/2]$.

In Exercises 5 to 8, parameterize the given curve with the indicated orientation.

5 **6**

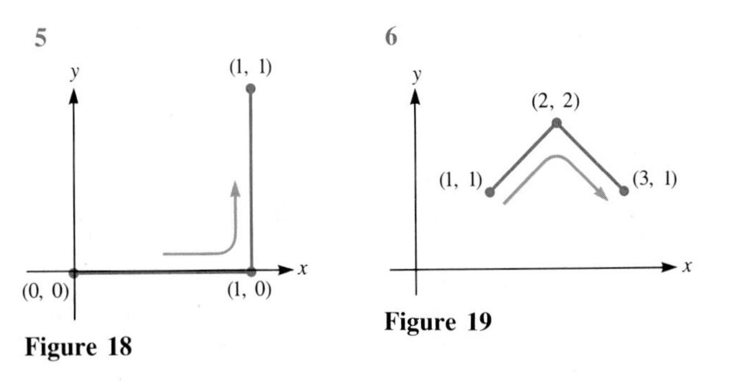

Figure 18 **Figure 19**

7 **8**

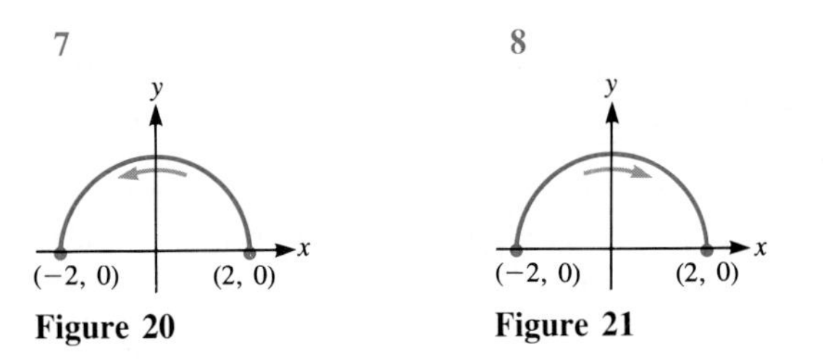

Figure 20 **Figure 21**

In Exercises 9 to 12 evaluate the given line integrals.

9 $\int_C xy\,dx$, where C is the straight line from $(1, 1)$ to $(3, 3)$.
10 $\int_C x^2\,dy$, where C is the straight line from $(2, 0)$ to $(2, 5)$.
11 $\int_C x^2\,dy$, where C is the straight line from $(3, 2)$ to $(7, 2)$.
12 $\int_C (xy\,dx + x^2\,dy)$, where C is the straight line from $(1, 0)$ to $(0, 1)$.

In Exercises 13 and 14, verify that $\int_C y\,dx$ (C counterclockwise) equals the negative of the area of the region R bounded by C. (Compute $\int_C y\,dx$ and also the area of R.)

13 C is the boundary of the rectangle with vertices $(0, 0)$, $(a, 0)$, (a, b), $(0, b)$, $a, b > 0$.
14 (*a*) C is the boundary of the triangle with vertices $(0, 0)$, $(a, 0)$, $(0, b)$, $a, b > 0$.
(*b*) C is the circle of radius a, center at the origin.
15 Show that $\oint_C x\,dy$, where the curve is counterclockwise, is equal to the area of the region bounded by the curve C.
16 Show that $\oint_C k\,dy = 0$.

In Exercises 17 to 20 evaluate with minimum effort. C is a counterclockwise curve bounding a region of area 5.

17 $\oint_C 3y\,dx$
18 $\oint_C (2y\,dx + 6x\,dy)$
19 $\oint_C [2x\,dx + (x + y)\,dy]$
20 $\oint_C [(x + 2y + 3)\,dx + (2x - 3y + 4)\,dy]$
21 Evaluate $\int_C x^3\,dx$ for all curves from $(1, 3)$ to $(2, 7)$.
22 Evaluate $\int_C (y + 2)\,dy$ for all curves from $(-1, 2)$ to $(3, 5)$.

In Exercises 21 and 22, the value of the line integral depends only on the endpoints, not on the particular path that joins them. Exercises 23 and 24 are examples where the path matters.

23 Evaluate $\int_C (xy\,dx + x\,dy)$ on (*a*) the straight path from $(1, 1)$ to $(2, 4)$; (*b*) the path from $(1, 1)$ to $(2, 4)$ that lies on the parabola $y = x^2$.
24 Evaluate $\int_C x\,dy$ on (*a*) the straight path from $(0, 0)$ to $(\pi/2, 1)$; (*b*) the path from $(0, 0)$ to $(\pi/2, 1)$ that lies on the curve $y = \sin x$.

In Exercises 25 and 26, the values of certain line integrals are given for curves oriented as shown. Use this information to find $\int_C f\,dy$. (Look at the orientations carefully.)

25

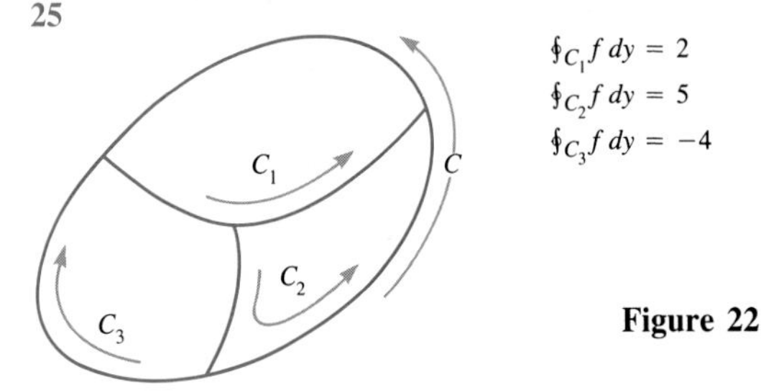

Figure 22

26

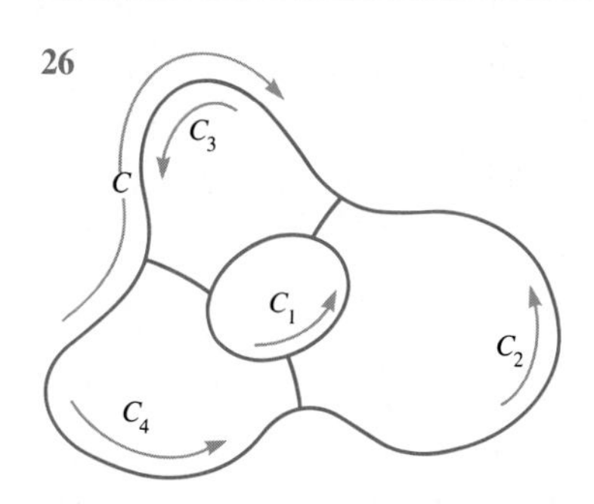

$\oint_{C_1} f\,dx = 1$
$\oint_{C_2} f\,dx = 2$
$\oint_{C_3} f\,dx = 4$
$\oint_{C_4} f\,dx = -8$

Figure 23

27 (*a*) Let $x = f(t)$, $y = g(t)$ be any parameterization of a curve C from $(0, 0)$ to $(1, 2)$. Show that $\oint_C (y\,dx + x\,dy)$ equals 2.
(*b*) Show that $\int_C (y\,dx + x\,dy)$ depends only on the endpoints of any curve C.

28 Compute $\oint_C (-y\,dx + x\,dy)$, where C is the ellipse

$$\begin{cases} x = a\cos t \\ y = b\sin t \end{cases} \qquad 0 \le t \le 2\pi.$$

29 Evaluate $\int_C (e^x y\,dx + x\sin \pi y\,dy + xy\tan^{-1} z\,dz)$, where C is composed of the line segment from $(0, 0, 0)$ to $(1, 1, 0)$ and the line segment from $(1, 1, 0)$ to $(1, 1, 1)$.

30 Evaluate

$$\int_C \left(\frac{x\,dx}{yz} + \frac{y\,dy}{xz} + \frac{8z\,dz}{xy}\right),$$

where C is parameterized by the vector function $\mathbf{G}(t) = \cos t\,\mathbf{i} + \sin t\,\mathbf{j} + \cos t\,\mathbf{k}$, for $\pi/6 \le t \le \pi/3$.

31 Compute

$$\int_C \left(\frac{-y\,dx}{x^2 + y^2} + \frac{x\,dy}{x^2 + y^2}\right),$$

where C goes from $(1, 0)$ to $(1, 1)$ along (*a*) the straight line $x = 1$, parameterized as $x = 1$, $y = t$; (*b*) the circular path parameterized as $x = \cos 2\pi t$, $y = \sin 2\pi t$, t in $[0, \frac{1}{4}]$, and then followed by the path $x = t$, $y = 1$, t in $[0, 1]$.

Exercises 32 to 35 involve $\int_C f\,ds = \oint_C f\,ds/dt\,dt$, where s is arc length. If s is not a convenient parameter, you may choose another, such as x, y, or θ, to serve as the parameter t. Recall that

$$\frac{ds}{dx} = \sqrt{1 + \left(\frac{dy}{dx}\right)^2}$$

and $$\frac{ds}{d\theta} = \sqrt{r^2 + \left(\frac{dr}{d\theta}\right)^2}. \qquad \text{(polar coordinates)}$$

In each of Exercises 32 and 33 find the mass of the wire.

32 Density at (x, y) is y; the wire occupies the portion of the cubic $y = x^3$ between $(1, 1)$ and $(2, 8)$.

33 Density at (x, y) is x; the wire occupies the portion of the parabola $y = x^2/2$ from $(0, 0)$ to $(1, \frac{1}{2})$.

34 A fence perpendicular to the xy plane has as its base the portion of the cardioid $r = 1 + \cos\theta$ in the first two quadrants. The height of the fence at the point (r, θ) is $\sin\theta$. Find the area of one side of the fence.

35 A wire occupies that part of the spiral $r = e^\theta$ corresponding to θ in $[0, 2\pi]$. At the point (r, θ) the temperature is r. Find the average temperature in the wire. [The average is $(\int_C f\,ds)/(\text{Length of } C)$.]

36 Let $\mathbf{r} = \mathbf{G}(t)$ describe a curve C in the plane or in space. What is the geometric interpretation of

$$\frac{1}{2}\int_C \|\mathbf{r} \times \mathbf{T}\|\,ds?$$

37 Compute $\int_C (xy\,dx + x^2\,dy)$ if C goes from $(0, 0)$ to $(1, 1)$ on (*a*) the line $y = x$, parameterized as $x = t$, $y = t$; (*b*) the line $y = x$, parameterized as $x = t^2$, $y = t^2$; (*c*) the parabola $y = x^2$, parameterized as $x = t$, $y = t^2$; (*d*) the polygonal path from $(0, 0)$ to $(0, 1)$ to $(1, 1)$, parameterized conveniently.

38 Verify that the integral $\oint_C (-y\,dx + x\,dy)$, where C is swept out counterclockwise, is twice the area of the region enclosed by C, when C is (*a*) the square path from $(a, 0)$ to $(0, a)$ to $(-a, 0)$ to $(0, -a)$ and back to $(a, 0)$; (*b*) the triangular path from $(0, 0)$ to $(a, 0)$ to $(0, b)$ and back to $(0, 0)$. Assume that a and b are positive.

39 (*Contributed by Daniel Drucker.*) The base of a fence is the curve $y = x^2/2$, for $-\sqrt{3} \le x \le \sqrt{3}$. The height of the fence above the point (x, y) is $1/(1 + x^2)$, where all distances are expressed in meters. Find the area of the fence. [*Hint:* First express the area as a line integral $\int_C f(P)\,ds$.]

40 (*Contributed by Daniel Drucker.*) A solid steel corkscrew-shaped sculpture has circular cross sections and tapers to a point. The centers of the cross sections lie on the helix $x = 2\cos 4t$, $y = 2\sin 4t$, $z = 6t$, for $0 \le t \le \frac{8}{3}$. The diameter of each cross section is $(16 - z)/32$, where all distances are given in feet. If the density of steel is 487 pounds per cubic foot, determine the weight of the sculpture to the nearest pound. (The cross sections are perpendicular to the helix. You may assume that volume dV is locally approximated by cross-sectional area times ds.)

41 Show that $\int f\,dx$ could be defined as the limit as all $\Delta t_i \to 0$ of sums of the form $\sum_{i=1}^n f(P_i)\,\Delta x_i$, where $\Delta x_i = x(t_i) - x(t_{i-1})$ and $a = t_0 < t_1 < \cdots < t_n = b$.

16.3 FOUR APPLICATIONS OF LINE INTEGRALS

In the previous section we defined line integrals and showed that $\oint_C y\,dx$ and $\oint_C x\,dy$ in the plane are related to the area of the region bounded by the curve C. In this section we show how line integrals occur in the study of work, fluid flow, and in the angle subtended by a curve.

Work Along a Curve

Consider a force **F** that remains constant (in direction and magnitude) and pushes a particle in a straight line from A to B. Let $\mathbf{r} = \overrightarrow{AB}$. The work accomplished by **F** is defined as $\mathbf{F} \cdot \mathbf{r}$:

$$\text{Work} = \mathbf{F} \cdot \mathbf{r}.$$

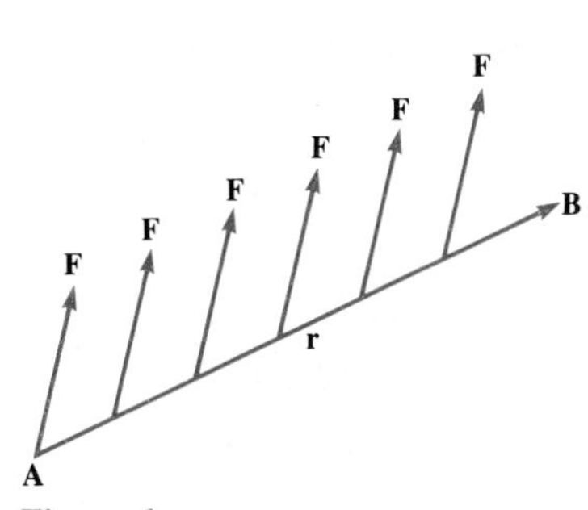

Figure 1

This is the product of the scalar component of **F** in the direction of **r** and the distance the particle moves. (See Fig. 1.) This definition generalizes the familiar formula "work equals force times distance" of Sec. 8.7.

But what if the force **F** varies and pushes the particle along a curve that is not straight? Then how do we represent the total work accomplished by the force? (See Fig. 2.)

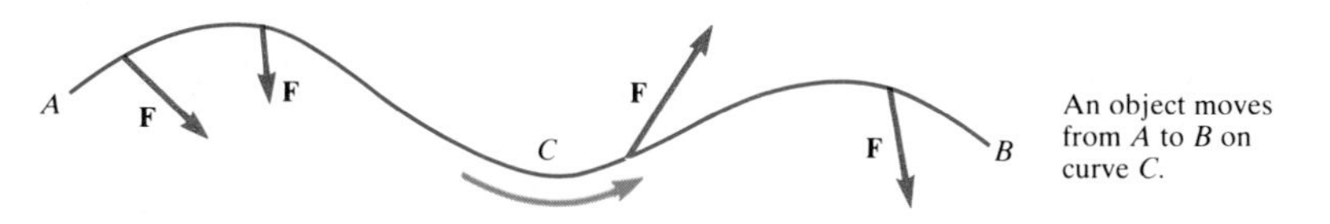

Figure 2

To find how much work is done, we look at the work done by **F** in pushing the particle over a very short piece of the curve. A short piece of the curve closely resembles a short line segment. Moreover, the force **F** is almost constant over a small segment. (See Fig. 3.)

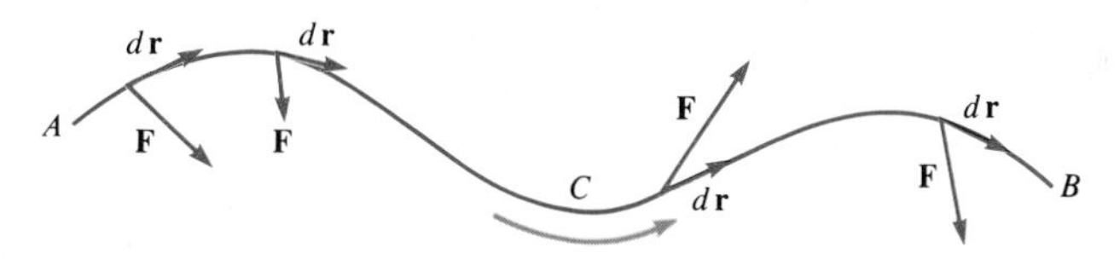

Figure 3

Then the work done by **F** in moving the particle from the tail to the head of $d\mathbf{r}$ is approximately

$$\mathbf{F} \cdot d\mathbf{r}.$$

If $\mathbf{F} = P\mathbf{i} + Q\mathbf{j}$, where P and Q are functions of x and y, and $d\mathbf{r} = dx\,\mathbf{i} + dy\,\mathbf{j}$, then

$$\mathbf{F} \cdot d\mathbf{r} = P\,dx + Q\,dy.$$

Since $P\,dx + Q\,dy$ is a *local estimate* of the work, the total work is represented by a line integral:

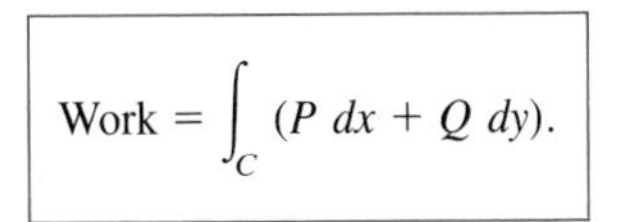

$$\text{Work} = \int_C (P\,dx + Q\,dy).$$

It is customary in physics and engineering to write this as

$$\boxed{\text{Work} = \int_C \mathbf{F} \cdot d\mathbf{r}.}$$

The vector notation $\mathbf{F} \cdot d\mathbf{r}$ is far more suggestive than the scalar notation $P\,dx + Q\,dy$. It reminds us that "work is the dot product of force and displacement."

For *any* vector field $\mathbf{F} = P\mathbf{i} + Q\mathbf{j}$ and curve C in the plane, we may consider the line integral

$$\int_C \mathbf{F} \cdot d\mathbf{r} = \int_C (P\,dx + Q\,dy).$$

It is called "the line integral of $\mathbf{F}$ along C." The notation $\int_C \mathbf{F} \cdot d\mathbf{s}$ is also used.

If $\mathbf{F}$ is a vector field in space, $\mathbf{F} = P\mathbf{i} + Q\mathbf{j} + R\mathbf{k}$, and C is a curve in space, then we have

$$\int_C \mathbf{F} \cdot d\mathbf{r} = \int_C (P\,dx + Q\,dy + R\,dz).$$

(This integral equals the work accomplished by force $\mathbf{F}$ along the path C.)

EXAMPLE 1 How much work is accomplished by the force $\mathbf{F}(x, y) = xy\mathbf{i} + y\mathbf{j}$ in pushing a particle from (0, 0) to (3, 9) along the parabola $y = x^2$?

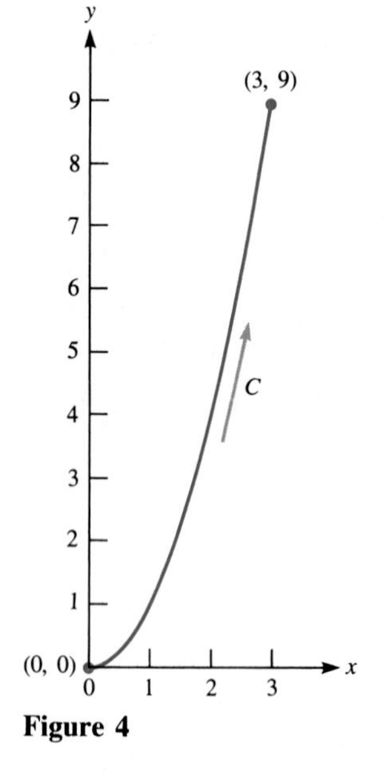

Figure 4

SOLUTION Figure 4 shows the path of the particle. Call this path C. Then

$$\begin{aligned}\text{Work} &= \int_C \mathbf{F} \cdot d\mathbf{r}\\ &= \int_C (xy\mathbf{i} + y\mathbf{j}) \cdot (dx\mathbf{i} + dy\mathbf{j})\\ &= \int_C (xy\,dx + y\,dy).\end{aligned}$$

To evaluate this line integral, let us use x as the parameter. Then $y = x^2$ and $dy = 2x\,dx$, so

$$\begin{aligned}\int_C (xy\,dx + y\,dy) &= \int_0^3 [x \cdot x^2\,dx + x^2(2x\,dx)]\\ &= \int_0^3 (x^3 + 2x^3)\,dx\\ &= \int_0^3 3x^3\,dx\\ &= \frac{243}{4}. \quad \blacksquare\end{aligned}$$

Circulation of a Fluid

Consider a fluid (liquid or gas) flowing on a portion of the xy plane. Let its density and velocity at the point P be given by $\sigma(P)$ and $\mathbf{v}(P)$, respectively. The product

$$\mathbf{F}(P) = \sigma(P)\mathbf{v}(P)$$

represents the rate and direction of the flow of the fluid at P. Now put an imaginary closed wire loop C on the fluid as in Fig. 5 or 6 and keep it fixed. In Fig. 5, C surrounds a whirlpool and there is a tendency for the fluid to flow along C rather than across it. This tendency arises because the component of $\mathbf{F}$ along the curve C in the whirlpool case is greater than the component of $\mathbf{F}$ perpendicular to C. The opposite case is shown in Fig. 6, where most of the fluid flow is *across* C rather than parallel to it. Now, $\mathbf{F} \cdot d\mathbf{r}$ represents flow in the *direction* of $d\mathbf{r}$, a small section of the curve C. Thus

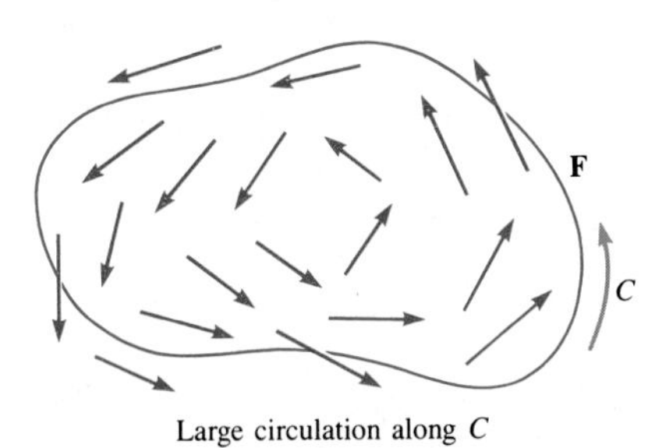

Large circulation along C

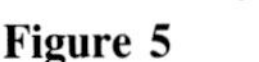
Figure 5

$$\oint_C \mathbf{F} \cdot d\mathbf{r}$$

represents the tendency of the fluid to flow along C. If C is counterclockwise and $\oint_C \mathbf{F} \cdot d\mathbf{r}$ is positive, the flow of $\mathbf{F}$ would be counterclockwise as well. If $\oint_C \mathbf{F} \cdot d\mathbf{r}$ is negative, the flow would tend to be clockwise. The line integral $\oint_C \mathbf{F} \cdot d\mathbf{r}$ is called the **circulation** of $\mathbf{F}$ along C.

Note that the very same integral, $\oint_C \mathbf{F} \cdot d\mathbf{r}$, occurs in the study of work and in the study of fluids.

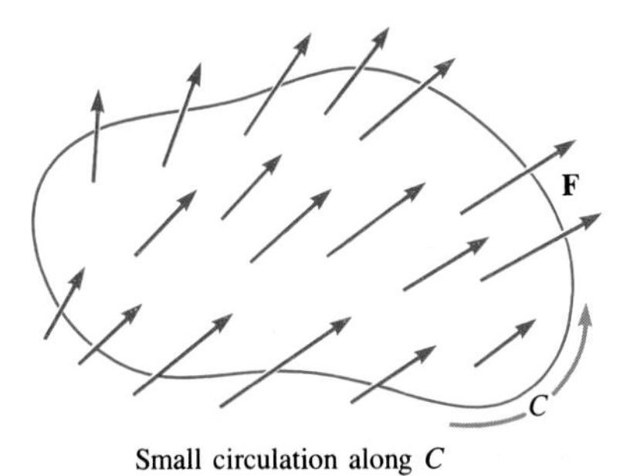

Small circulation along C

Figure 6

Loss or Gain of a Fluid (Flux)

Imagine again that we place an imaginary wire loop C on the surface of a stream.

We raise the question: At what rate is fluid escaping or entering the region R whose boundary is C?

If the fluid tends to escape, then it is thinning out in R, becoming less dense at some points. If the fluid tends to accumulate, it is becoming denser at some points. (Think of this ideal fluid as resembling a gas rather than a liquid; gases can vary widely in density while liquids tend to have constant density.)

Since the fluid is escaping or entering R only along its boundary, it suffices to consider the total loss or gain across C. Where $\mathbf{v}$, the fluid velocity, is tangent to C, fluid neither enters nor leaves. Where $\mathbf{v}$ is not tangent to C, fluid is either entering or leaving across C, as indicated in Fig. 7.

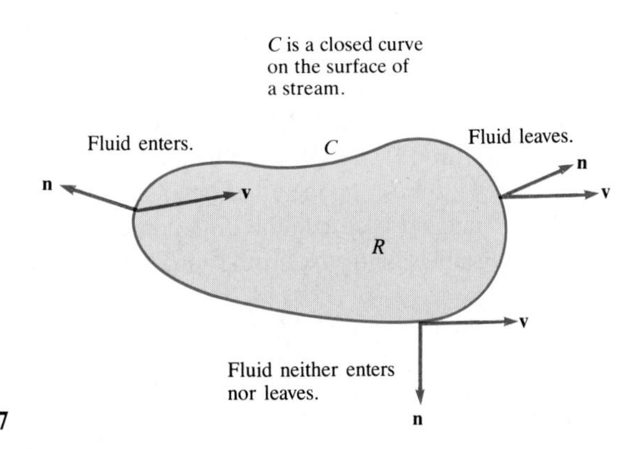

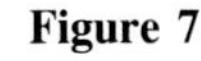
Figure 7

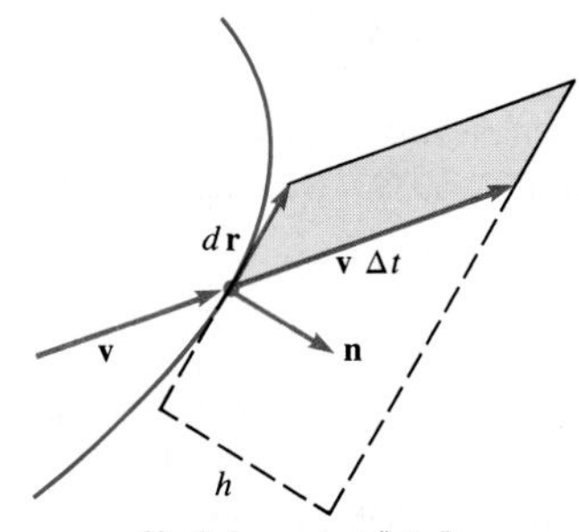

Shaded area is $h\|d\mathbf{r}\|$.

Figure 8

The vector $\mathbf{n}$ is a unit vector perpendicular to the curve C and pointing away from the region it bounds. It is called the **exterior normal**. Recall that $\mathbf{F} = \sigma\mathbf{v}$, the product of density and velocity, so $\mathbf{F}$ and $\mathbf{v}$ have the same direction.

To find the *total* loss or gain of fluid past C, let us look at a very short section of C, which we will view as a vector $d\mathbf{r}$. How much fluid crosses $d\mathbf{r}$ in a short interval of time, Δt?

During time Δt the fluid moves a distance $\|\mathbf{v}\|\ \Delta t$ across $d\mathbf{r}$. The fluid that crosses $d\mathbf{r}$ during the time Δt forms approximately the parallelogram shown in Fig. 8.

The area of the parallelogram is the product of its height h and its base $\|d\mathbf{r}\|$. That is,

$$\begin{aligned}\text{Area of parallelogram} &= \|\mathbf{proj_n}\ (\mathbf{v}\ \Delta t)\|\ \|d\mathbf{r}\| \\ &= (\mathbf{v}\ \Delta t)\cdot \mathbf{n}\ \|d\mathbf{r}\|.\end{aligned}$$

Since the density of the fluid is σ,

$$\begin{aligned}\text{Mass in parallelogram} &= \sigma(\mathbf{v}\ \Delta t)\cdot \mathbf{n}\ \|d\mathbf{r}\| \\ &= (\sigma\mathbf{v})\cdot \mathbf{n}\ \|d\mathbf{r}\|\ \Delta t \\ &= \mathbf{F}\cdot \mathbf{n}\ \|d\mathbf{r}\|\ \Delta t.\end{aligned}$$

Thus the rate at which fluid crosses $d\mathbf{r}$ per unit time is approximately

$$\frac{\mathbf{F}\cdot\mathbf{n}\ \|d\mathbf{r}\|\ \Delta t}{\Delta t} = \mathbf{F}\cdot\mathbf{n}\ \|d\mathbf{r}\|.$$

Since $d\mathbf{r}$ approximates a short piece of the curve, its length $\|d\mathbf{r}\|$ approximates the length of a short piece of the curve ds. Therefore, the rate at which the fluid crosses a short part of C, of length ds, is approximately

$$\mathbf{F}\cdot\mathbf{n}\ ds.$$

Hence the line integral

$$\oint_C \mathbf{F}\cdot\mathbf{n}\ ds$$

represents the rate of net loss or gain of fluid inside R. If it is positive, fluid tends to *leave* R, and the mass of fluid in R decreases. If it is negative, fluid tends to *enter* R and the mass of fluid in R increases. In short,

$$\boxed{\text{Net loss} = \oint_C \mathbf{F}\cdot\mathbf{n}\ ds.}$$

The quantity $\oint_C \mathbf{F}\cdot\mathbf{n}\ ds$ is called the **flux** of $\mathbf{F}$ across C (from the Latin for "flow"). So flux is "the integral of the normal component of $\mathbf{F}$." Circulation, $\oint_C \mathbf{F}\cdot d\mathbf{r}$, on the other hand, can be written as $\oint_C \mathbf{F}\cdot(\mathbf{T}\ ds)$, where $\mathbf{T}$ is the unit tangent vector in the direction of C. ($\mathbf{T}\ ds$ and $d\mathbf{r}$ have the same direction and same length ds, so they may be used interchangeably.) Hence

$$\boxed{\text{Circulation} = \oint_C \mathbf{F}\cdot\mathbf{T}\ ds.}$$

Circulation is "the integral of the tangential component of **F**."

EXAMPLE 2 Let $\mathbf{F} = (2 + x)\mathbf{i}$ describe the flow of a fluid in the xy plane. Does the amount of fluid within the circle C of radius 2 and center (0, 0) tend to increase or decrease?

Figure 9

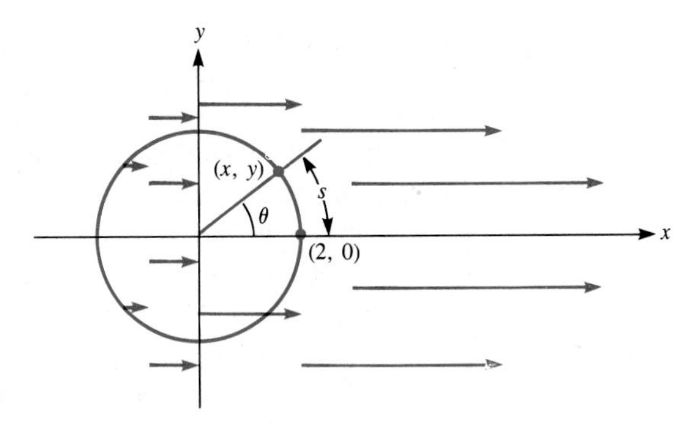

SOLUTION Figure 9 shows the circle and a few of the vectors of **F**, calculated by the formula $\mathbf{F}(x, y) = (2 + x)\mathbf{i}$. Since the flow increases as we move to the right, there appears to be more fluid leaving the disk than entering. We expect the flux $\oint_C \mathbf{F} \cdot \mathbf{n}\, ds$ to be positive. To compute $\oint_C \mathbf{F} \cdot \mathbf{n}\, ds$, introduce angle θ as the parameter. Then

$$x = 2\cos\theta, \qquad y = 2\sin\theta.$$

Since the circle has radius 2, $s = 2\theta$ and therefore

$$ds = 2\, d\theta.$$

The unit normal is parallel to the radius vector $x\mathbf{i} + y\mathbf{j}$. Therefore,

$$\mathbf{n} = \frac{x\mathbf{i} + y\mathbf{j}}{\|x\mathbf{i} + y\mathbf{j}\|} = \frac{2\cos\theta\,\mathbf{i} + 2\sin\theta\,\mathbf{j}}{2} = \cos\theta\,\mathbf{i} + \sin\theta\,\mathbf{j}.$$

Thus

$$\begin{aligned}
\text{Flux} &= \oint_C \mathbf{F} \cdot \mathbf{n}\, ds \\
&= \int_0^{2\pi} \underbrace{[(2 + x)\mathbf{i} \cdot \mathbf{n}]}_{\mathbf{F}\cdot\mathbf{n}}\ \underbrace{2\, d\theta}_{ds} \\
&= \int_0^{2\pi} (2 + 2\cos\theta)\mathbf{i} \cdot (\cos\theta\,\mathbf{i} + \sin\theta\,\mathbf{j})\, 2\, d\theta \\
&= \int_0^{2\pi} (4\cos\theta + 4\cos^2\theta)\, d\theta \\
&= \int_0^{2\pi} [4\cos\theta + 2 + 2\cos 2\theta]\, d\theta \qquad \text{(trig identity)} \\
&= [4\sin\theta + 2\theta + \sin 2\theta]\Big|_0^{2\pi} \\
&= 4\pi.
\end{aligned}$$

As expected, the flux is positive since there is a net flow *out* of the disk. ■

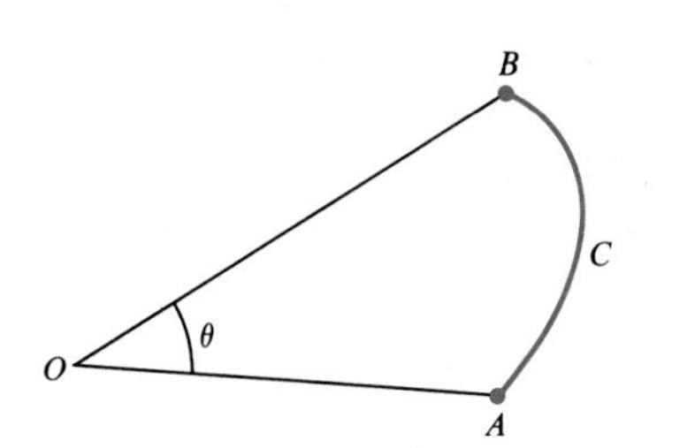

Figure 10

The Angle Subtended by a Curve

Our fourth illustration of a line integral concerns the angle subtended at a point O by a curve C in the plane. (We assume that each ray from O meets C in at most one point.) We include this example as background for "the solid angle subtended by a surface," an important concept in Chap. 17.

The curve C in Fig. 10 subtends an angle θ at the point O. We will show that θ can be expressed as a line integral of a suitable function. Of course, we do not need such an integral to find θ. Just knowing the points A, O, and B is enough. What is important is that θ can be expressed as a line integral. It is this idea that generalizes from a curve to a surface.

To develop the integral for θ we first estimate the angle $d\theta$ subtended by a very short section of the curve C. A short section is almost straight and we will approximate it by a line segment of length ds. (See Fig. 11.)

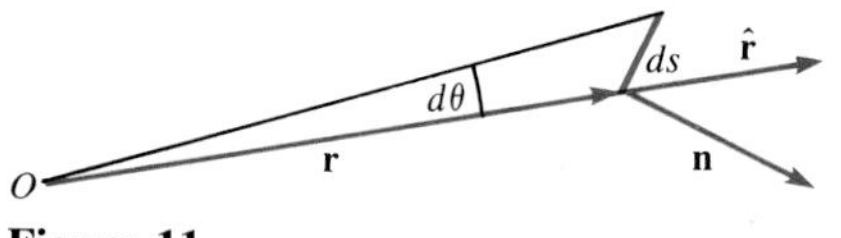

Figure 11

Let $\hat{\mathbf{r}}$ be the unit vector in the same direction as $\mathbf{r}$ and let $\mathbf{n}$ be the unit normal to C that makes an acute angle with $\hat{\mathbf{r}}$.

Let α be the angle between the line segment of length ds and a line L perpendicular to $\mathbf{r}$, as shown in Fig. 12. (Note that the angle between $\hat{\mathbf{r}}$ and $\mathbf{n}$ is also α.)

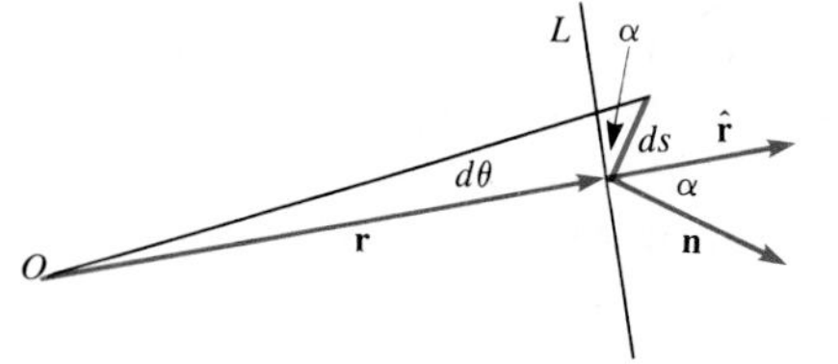

Figure 12

The projection of the line segment of length ds on L has length $\cos\alpha\, ds$. By the definition of radian measure,

$$d\theta = \frac{\cos\alpha\, ds}{\|\mathbf{r}\|}. \qquad \text{(Why?)}$$

But

$$\mathbf{n}\cdot\hat{\mathbf{r}} = \|\mathbf{n}\|\,\|\hat{\mathbf{r}}\|\cos\alpha = \cos\alpha.$$

Thus

$$d\theta = \frac{(\mathbf{n}\cdot\hat{\mathbf{r}})\, ds}{\|\mathbf{r}\|}. \tag{1}$$

In view of the local estimate (1) we conclude that θ can be expressed as an integral:

$$\boxed{\theta = \int_C \frac{\mathbf{n}\cdot\hat{\mathbf{r}}}{\|\mathbf{r}\|}\, ds.} \tag{2}$$

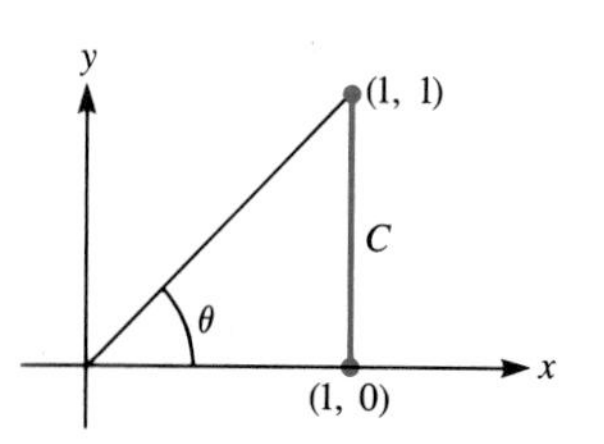

Figure 13

Therefore, *the angle subtended by C is the integral of the normal component of the vector function* $\hat{\mathbf{r}}/\|\mathbf{r}\|$.

EXAMPLE 3 Verify (2) for the angle subtended at the origin by the line segment that joins (1, 0) and (1, 1).

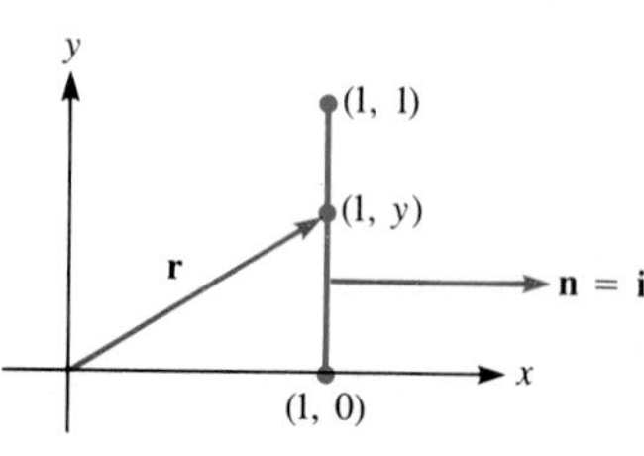

Figure 14

SOLUTION The subtended angle θ is shown in Fig. 13. Inspection of Fig. 13 shows that $\theta = \pi/4$.

Now let us evaluate the integral in (2) in this instance. Figure 14 shows that $\mathbf{n} = \mathbf{i}$ and $\mathbf{r} = \mathbf{i} + y\mathbf{j}$. Hence

$$\int_C \frac{\mathbf{n}\cdot\hat{\mathbf{r}}}{\|\mathbf{r}\|}\,ds = \int_C \frac{\mathbf{i}\cdot\left(\dfrac{\mathbf{i}+y\mathbf{j}}{\sqrt{1+y^2}}\right)}{\sqrt{1+y^2}}\,ds$$

$$= \int_C \frac{1}{1+y^2}\,ds$$

$$= \int_0^1 \frac{1}{1+y^2}\,dy \qquad \text{(since } s = y\text{)}$$

$$= \tan^{-1} y \Big|_0^1$$

$$= \frac{\pi}{4} - 0 = \frac{\pi}{4}.$$

This agrees with our observation, which was based on inspection of Fig. 13. ■

Section Summary

Line Integral	*Name*	*Application*	*Notation*
$\int_C \mathbf{F}\cdot d\mathbf{r}$	"Integral of **F** along C" "Integral of tangential component of **F**"	Work or circulation	$\int_C \mathbf{F}\cdot d\mathbf{r}$ $\int_C \mathbf{F}\cdot\mathbf{T}\,ds$ $\int_C (P\,dx + Q\,dy)$, if $\mathbf{F} = P\mathbf{i} + Q\mathbf{j}$ $\int_C (P\,dx + Q\,dy + R\,dz)$, if $\mathbf{F} = P\mathbf{i} + Q\mathbf{j} + R\mathbf{k}$.
$\oint_C \mathbf{F}\cdot\mathbf{n}\,ds$	"Integral of normal component of **F** along C" (usually C is a closed curve in the plane)	Flux	$\oint_C (-Q\,dx + P\,dy)$, if $\mathbf{F} = P\mathbf{i} + Q\mathbf{j}$ And C is oriented counterclockwise (this will be shown in Sec. 16.4).
$\int_C \frac{\mathbf{n}\cdot\hat{\mathbf{r}}}{\|\mathbf{r}\|}\,ds$	Integral of normal component of $\hat{\mathbf{r}}/\|\mathbf{r}\|$ along C	Angle subtended	

EXERCISES FOR SEC. 16.3: FOUR APPLICATIONS OF LINE INTEGRALS

In Exercises 1 to 6 evaluate $\int_C \mathbf{F}\cdot d\mathbf{r}$ for the given **F** and C.

1 $\mathbf{F}(x, y) = 2x\mathbf{i}$ and C is a semicircle, $\mathbf{G}(\theta) = 3\cos\theta\,\mathbf{i} + 3\sin\theta\,\mathbf{j}$, $0 \le \theta \le \pi$.

2 $\mathbf{F}(x, y) = x^2\mathbf{i} + 2xy\mathbf{j}$ and C is a line segment, $\mathbf{G}(t) = 2t^2\mathbf{i} + 3t^2\mathbf{j}$, $1 \le t \le 2$.

3 $\mathbf{F}(x, y, z) = x\mathbf{i} + y\mathbf{j} + z\mathbf{k}$ and C is a helix, $\mathbf{G}(t) = \cos t\,\mathbf{i} + \sin t\,\mathbf{j} + 3t\mathbf{k}$, $0 \le t \le 4\pi$.

4 $\mathbf{F}(x, y, z) = x^2\mathbf{i} + xy\mathbf{j} + 3\mathbf{k}$ and C is a line segment, $\mathbf{G}(t) = 2t\mathbf{i} + (3t + 1)\mathbf{j} + t\mathbf{k}$, $1 \le t \le 2$.

5 $\mathbf{F}(\mathbf{r}) = \hat{\mathbf{r}}/\|\mathbf{r}\|^2$ and C is a line, $\mathbf{r}(t) = 2t\mathbf{i} + 3t\mathbf{j} + 4t\mathbf{k}$, $0 \le t \le 2$.

6 $\mathbf{F}(\mathbf{r}) = \mathbf{r}$ and C is the circle, $\mathbf{r}(t) = \cos\theta\,\mathbf{i} + \sin\theta\,\mathbf{j} + 2\mathbf{k}$, $0 \le \theta \le 2\pi$.

In Exercises 7 to 10 compute the work accomplished by the force $\mathbf{F} = x^2y\mathbf{i} + y\mathbf{j}$ along the given curve.

7 From (0, 0) to (2, 4) along the parabola $y = x^2$.

8 From (0, 0) to (2, 4) along the line $y = 2x$.

9 From (0, 0) to (2, 4) along the path in Fig. 15.

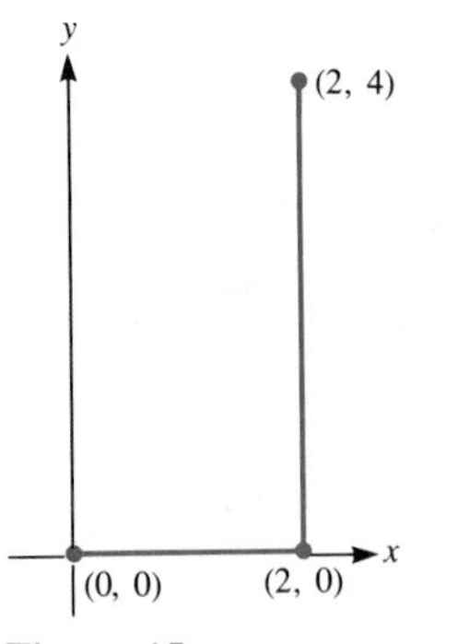

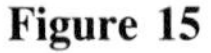
Figure 15

10 From (0, 0) to (2, 4) along the path in Fig. 16.

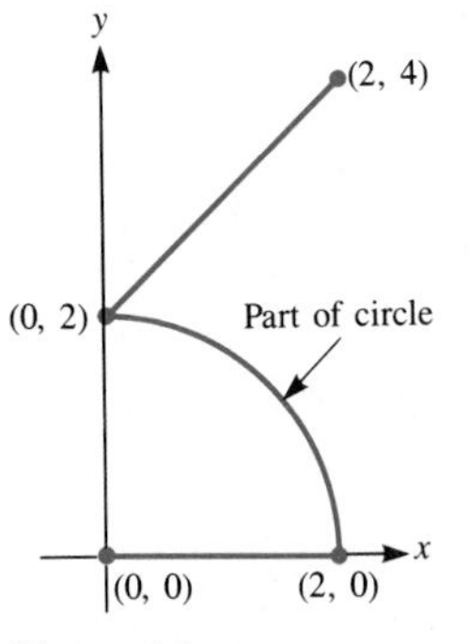

Figure 16

11 The gravitational force $\mathbf{F}$ of the earth, which is located at the origin (0, 0) of a rectangular coordinate system, on a particle at the point (x, y) is

$$\frac{-x\mathbf{i}}{(\sqrt{x^2+y^2})^3} + \frac{-y\mathbf{j}}{(\sqrt{x^2+y^2})^3} = \frac{-\mathbf{r}}{\|\mathbf{r}\|^3} = \frac{-\hat{\mathbf{r}}}{r^2}.$$

Compute the total work done by $\mathbf{F}$ if the particle goes from (2, 0) to (0, 1) along (*a*) the portion of the ellipse $x = 2\cos t$, $y = \sin t$ in the first quadrant; (*b*) the line parameterized as $x = 2 - 2t$, $y = t$.

12 (*a*) Let $W(b)$ be the work done by the force in Exercise 11 in moving a particle along the straight line from (1, 0) to $(b, 0)$.

(*b*) What is $\lim_{b\to\infty} W(b)$?

13 Let the vector field describing a fluid flow have at the point (x, y) the value $(x + 1)^2\,\mathbf{i} + y\mathbf{j}$. Let C be the unit circle described parametrically as $x = \cos t$, $y = \sin t$, for t in $[0, 2\pi]$.

(*a*) Draw $\mathbf{F}$ at eight convenient, equally spaced points on the circle.

(*b*) Is fluid tending to leave or enter the region bounded by C; that is, is the net outward flow positive or negative? [Answer on the basis of your diagram in (*a*).]

(*c*) Compute the net outward flow with the aid of a line integral.

14 Like Exercise 13 where $\mathbf{F}(x, y) = (2 - x)\mathbf{i} + y\mathbf{j}$ and C is the square with vertices (0, 0), (1, 0), (1, 1), and (0, 1).

15 Write up in your own words and diagrams why $\int_C \mathbf{F}\cdot d\mathbf{r}$ represents the work done by force $\mathbf{F}$ along the curve C.

16 Write up in your own words and diagrams why $\oint_C \mathbf{F}\cdot\mathbf{n}\,ds$ represents the net loss of fluid across C if $\mathbf{F}$ is the fluid flow and $\mathbf{n}$ is a unit external normal to C. Include the definition of $\mathbf{F}$.

17 Explain why $\oint_C \mathbf{F}\cdot d\mathbf{r}$ represents the tendency of a fluid to move along C, if $\mathbf{F}$ is the fluid flow.

18 Explain why $\int_C (\mathbf{n}\cdot\hat{\mathbf{r}})/\|\mathbf{r}\|\,ds$ represents the angle subtended by a curve C at the origin. (Assume that each ray from the origin meets C at most once.)

19 Let $\mathbf{F}(x, y) = \sigma\mathbf{v}$, the fluid flow, and C a closed curve in the xy plane. If $\oint_C \mathbf{F}\cdot d\mathbf{r}$ is positive and C is counterclockwise, does the motion along C tend to be clockwise or counterclockwise?

20 Let $\mathbf{F}(x, y) = \sigma\mathbf{v}$, the fluid flow, and C a closed curve in the xy plane. If $\oint_C \mathbf{F}\cdot\mathbf{n}\,ds$ is positive, is fluid tending to leave the region bounded by C or to enter it?

21 Verify (2) for the angle subtended at the origin by the line segment that joins (2, 0) to (2, 3).

22 Verify (2) for the angle subtended at the origin by the line segment that joins (1, 0) to (0, 1).

---

23 Find the work done by the force $-3\mathbf{j}$ in moving a particle from (0, 3) to (3, 0) along

(*a*) The circle of radius 3 with center at the origin.

(*b*) The straight path from (0, 3) to (3, 0).

(*c*) The answers to (*a*) and (*b*) are the same. Will they be the same for all curves from (0, 3) to (3, 0)?

24 Let C be a curve in space and C^* its projection on the xy plane. Assume that distinct points of C project onto distinct points of C^*. The line integral $\int_C 1\,ds$ equals the arc length of C. What integral over C equals the arc length of C^*?

25 Let C be a closed convex curve that encloses the point O. Let $\mathbf{r}$ be the position vector $\overrightarrow{OP}$ for points P on the curve. Determine the value of $\oint_C (\mathbf{n}\cdot\hat{\mathbf{r}})/\|\mathbf{r}\|\,ds$, where $\mathbf{n}$ is the external unit normal to C.

26 Let C be a closed convex curve. Let O be a point not on C and not in the region C bounds. Let $\mathbf{r}$ be the position vector $\overrightarrow{OP}$ for points P on the curve. Determine the value of $\oint_C (\mathbf{n}\cdot\hat{\mathbf{r}})/\|\mathbf{r}\|\,ds$, where $\mathbf{n}$ denotes the external unit normal to C. *Hint:* Draw a picture and pay attention to the angle between $\mathbf{n}$ and $\mathbf{r}$.

27 Let $\mathbf{F}(P) = \sigma(P)\mathbf{v}(P)$ represent the flow of a fluid as described in the discussion of circulation and flux. Let C be a closed curve that bounds the region R. Let $Q(t)$ be the total mass of the fluid in R at time t. Express dQ/dt in terms of a line integral.

28 Figure 17 shows some representative vectors for the vector field $\mathbf{F}$. Use this information to estimate (*a*) the circulation of $\mathbf{F}$ along the boundary curve C and (*b*) the flux of $\mathbf{F}$ across C. (Since you have no formula for $\mathbf{F}$, there is a range of "correct" answers.)

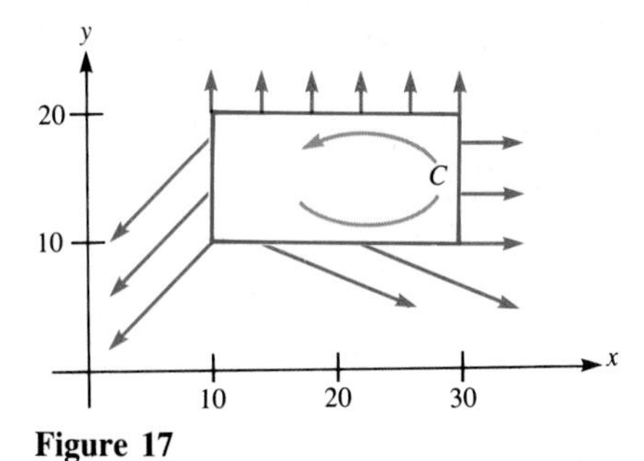

Figure 17

29 Repeat Exercise 28 for the vector field $\mathbf{F}$ and curve C represented in Fig. 18.

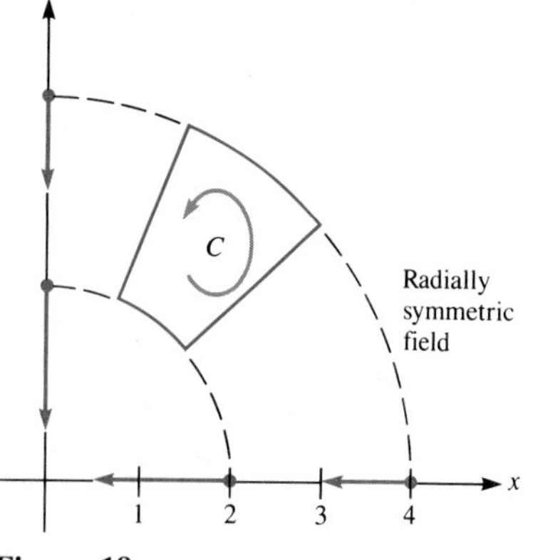

Figure 18

30 Differentiate for practice.

(*a*) $\dfrac{\sin^3 2x}{2x + 1}$

(*b*) $e^{2x} \tan 3x$

(*c*) $\ln(1 + \sec^2 4x)$

(*d*) $\tan^{-1}(2 \cos 5x)$

(*e*) $\sin^{-1} x^2$

16.4 GREEN'S THEOREM

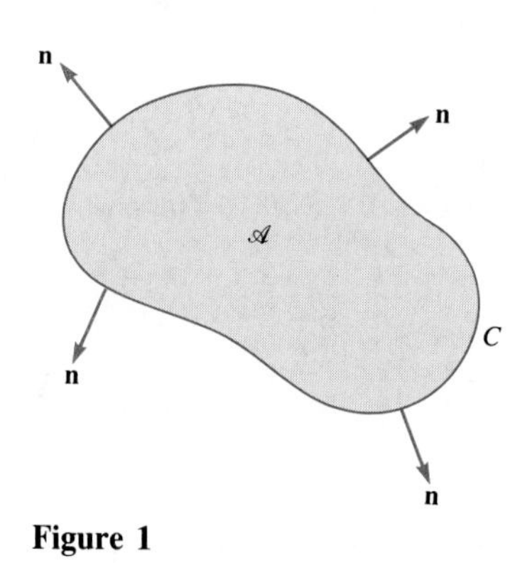

Figure 1

In the last section we met the line integral $\oint_C \mathbf{F} \cdot \mathbf{n}\, ds$, "the integral of the normal component of $\mathbf{F}$ around C." Green's theorem expresses this integral as the integral of the divergence of $\mathbf{F}$, $\nabla \cdot \mathbf{F}$, over the region $\mathcal{A}$ that C bounds. (See Fig. 1.)

Theorem 1 *Green's theorem.* Let C be a closed curve in the xy plane bounding a region $\mathcal{A}$ and let $\mathbf{n}$ be the unit exterior normal to C. Let $\mathbf{F}$ be a vector field defined on $\mathcal{A}$. Then

$$\oint_C \mathbf{F} \cdot \mathbf{n}\, ds = \int_{\mathcal{A}} \nabla \cdot \mathbf{F}\, dA.$$

Before showing why Green's theorem is true, we illustrate it by an example.

EXAMPLE 1 Let $\mathbf{F} = x\mathbf{i} + y\mathbf{j}$ and let C be the circle of radius a centered at the origin. Let $\mathcal{A}$ be the disk that C bounds. Verify Green's theorem in this case by computing $\int_{\mathcal{A}} \nabla \cdot \mathbf{F}\, dA$ and $\oint_C \mathbf{F} \cdot \mathbf{n}\, ds$, and checking that they are indeed equal.

SOLUTION First compute $\int_{\mathscr{A}} \nabla \cdot \mathbf{F}\, dA$. We have

$$\nabla \cdot \mathbf{F} = \nabla \cdot (x\mathbf{i} + y\mathbf{j}) = \frac{\partial x}{\partial x} + \frac{\partial y}{\partial y} = 2.$$

Thus
$$\int_{\mathscr{A}} \nabla \cdot \mathbf{F}\, dA = \int_{\mathscr{A}} 2\, dA$$
$$= 2(\text{Area of } \mathscr{A})$$
$$= 2\pi a^2. \tag{1}$$

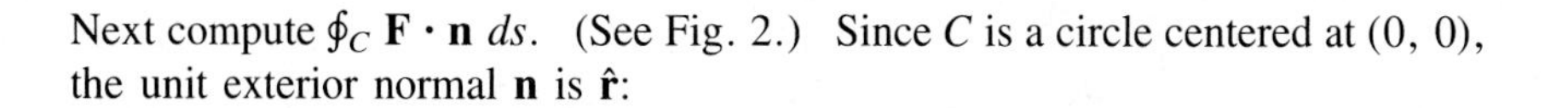

Next compute $\oint_C \mathbf{F} \cdot \mathbf{n}\, ds$. (See Fig. 2.) Since C is a circle centered at $(0, 0)$, the unit exterior normal $\mathbf{n}$ is $\hat{\mathbf{r}}$:

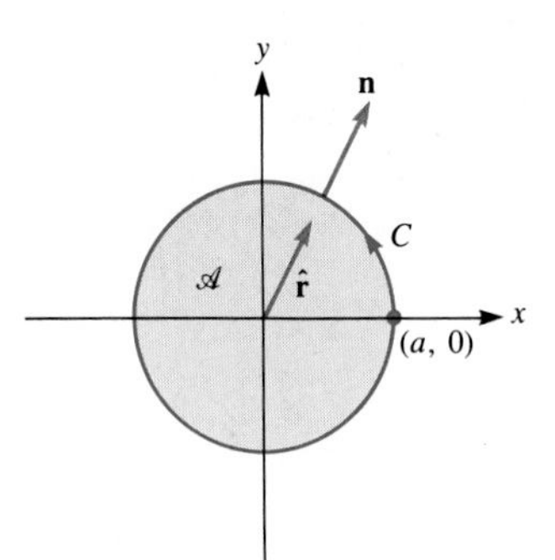

Figure 2

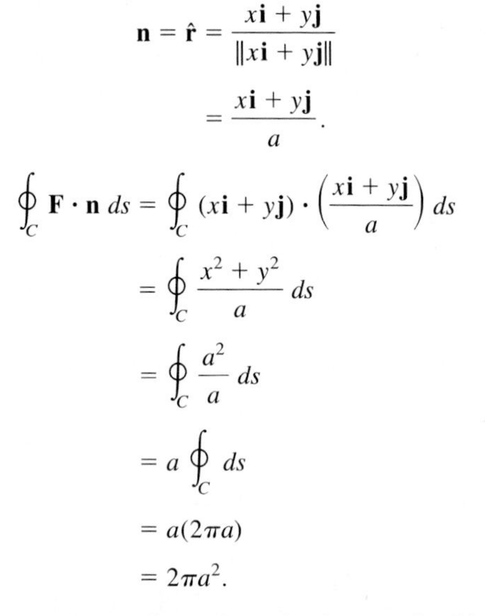

$$\mathbf{n} = \hat{\mathbf{r}} = \frac{x\mathbf{i} + y\mathbf{j}}{\|x\mathbf{i} + y\mathbf{j}\|}$$
$$= \frac{x\mathbf{i} + y\mathbf{j}}{a}.$$

Thus
$$\oint_C \mathbf{F} \cdot \mathbf{n}\, ds = \oint_C (x\mathbf{i} + y\mathbf{j}) \cdot \left(\frac{x\mathbf{i} + y\mathbf{j}}{a}\right) ds$$
$$= \oint_C \frac{x^2 + y^2}{a}\, ds$$
$$= \oint_C \frac{a^2}{a}\, ds$$
$$= a \oint_C ds$$
$$= a(2\pi a)$$
$$= 2\pi a^2. \tag{2}$$

Comparison of (1) and (2) confirms Green's theorem in this case. ■

Proof of Green's Theorem

We shall start with $\oint_C \mathbf{F} \cdot \mathbf{n}\, ds$ and transform it into $\int_{\mathscr{A}} \nabla \cdot \mathbf{F}\, dA$. The proof is instructive because it reviews important ideas:

The cancellation principle (Sec. 16.2).

The differential of a function of two variables (Sec. 14.8).

The definition of an integral over a plane region (Sec. 15.1).

As Steve Whitaker of the chemical engineering department at the University of California at Davis says, "The concepts that one must understand to *prove* a theorem are frequently the concepts one must understand to *apply* the theorem." So read the proof slowly at least twice. It is not here just to show that Green's theorem is true. After all, it has been around for over 150 years, and no one has said it is false. Studying a proof strengthens one's understanding of the fundamentals.

Proof of Green's Theorem We assume that C is swept out counterclockwise and express $\oint_C \mathbf{F} \cdot \mathbf{n}\, ds$ in terms of the components of $\mathbf{F} = P\mathbf{i} + Q\mathbf{j}$.

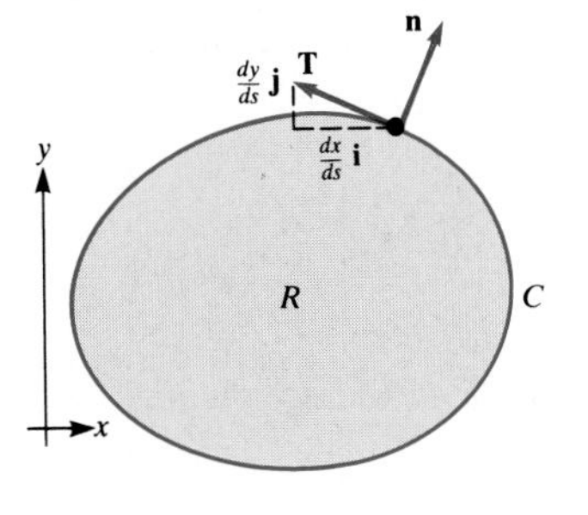

As shown in Sec. 13.4, the vector

$$\mathbf{T} = \frac{dx}{ds}\mathbf{i} + \frac{dy}{ds}\mathbf{j}$$

is tangent to the curve, has length 1, and points in the direction in which the curve is swept out. A typical **T** and **n** are shown in Fig. 3. As Fig. 3 shows, the exterior unit normal **n** has its x component equal to the y component of **T** and its y component equal to the negative of the x component of **T**. Thus

$$\mathbf{n} = \frac{dy}{ds}\mathbf{i} - \frac{dx}{ds}\mathbf{j}.$$

Consequently, if $\mathbf{F} = P\mathbf{i} + Q\mathbf{j}$, then

$$\oint_C \mathbf{F} \cdot \mathbf{n}\, ds = \oint_C (P\mathbf{i} + Q\mathbf{j}) \cdot \left(\frac{dy}{ds}\mathbf{i} - \frac{dx}{ds}\mathbf{j}\right) ds$$

$$= \oint_C \left(P\frac{dy}{ds} - Q\frac{dx}{ds}\right) ds$$

$$= \oint_C (P\, dy - Q\, dx).$$

Figure 3

Now express $\int_{\mathscr{A}} \nabla \cdot \mathbf{F}\, dA$ in terms of components:

$$\int_{\mathscr{A}} \nabla \cdot \mathbf{F}\, dA = \int_{\mathscr{A}} \left(\frac{\partial P}{\partial x} + \frac{\partial Q}{\partial y}\right) dA.$$

Thus, Green's theorem would follow from the two equations

$$\oint_C P\, dy = \int_{\mathscr{A}} \frac{\partial P}{\partial x}\, dA \tag{3}$$

and

$$\oint_C (-Q)\, dx = \int_{\mathscr{A}} \frac{\partial Q}{\partial y}\, dA. \tag{4}$$

Let us show why (3) holds.

Break $\mathscr{A}$ into n small regions $\mathscr{A}_1, \mathscr{A}_2, \ldots, \mathscr{A}_n$ with respective boundaries $C_1, C_2, \ldots, C_n$. Choose a point $P_i = (a_i, b_i)$ in $\mathscr{A}_i$, for $i = 1, 2, \ldots, n$. (See Fig. 4.)

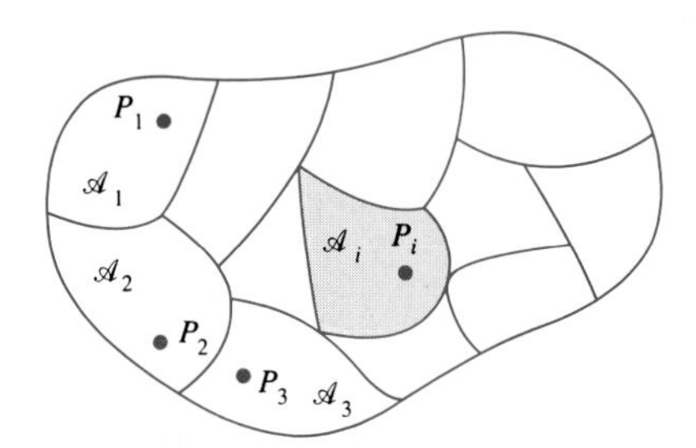

Figure 4

Orient each C_i counterclockwise. By the cancellation principle,

$$\oint_C P\, dy = \sum_{i=1}^{n} \oint_{C_i} P\, dy. \tag{5}$$

All that remains is estimating $\oint_{C_i} P\, dy$ where C_i is a very small curve. (See Fig. 5.)

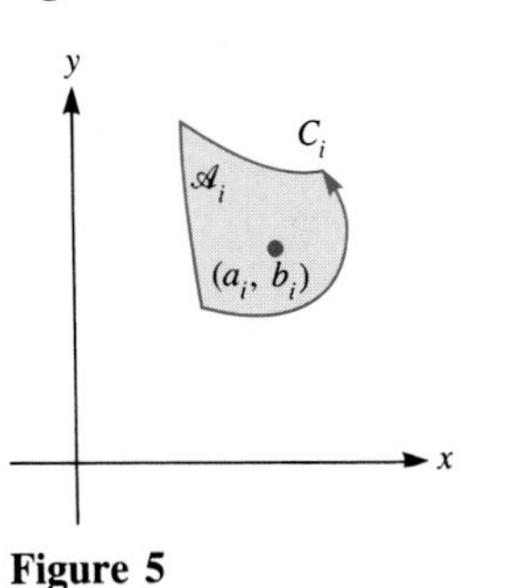

Figure 5

To estimate $P(x, y)$ for (x, y) near (a_i, b_i), let us use the differential of P. The differential of P at $(x, y) = (a_i, b_i)$ is

$$dP = \frac{\partial P}{\partial x}\,\Delta x + \frac{\partial P}{\partial y}\,\Delta y,$$

where the partial derivatives are evaluated at (a_i, b_i).

Thus

$$P(x, y) \approx P(a_i, b_i) + \frac{\partial P}{\partial x}(a_i, b_i)(x - a_i) + \frac{\partial P}{\partial y}(a_i, b_i)(y - b_i),$$

where $x - a_i$ plays the role of Δx and $y - b_i$ the role of Δy.

Consequently a good estimate of $\oint_{C_i} P\,dy$ is

$$\oint_{C_i} \left[P(a_i, b_i) + \frac{\partial P}{\partial x}(a_i, b_i)x - \frac{\partial P}{\partial x}(a_i, b_i)a_i + \frac{\partial P}{\partial y}(a_i, b_i)y - \frac{\partial P}{\partial y}(a_i, b_i)b_i \right] dy. \quad (6)$$

The line integral (6), in spite of its seeming complexity, is easy to evaluate.

In Sec. 16.2 it was shown that for a counterclockwise curve C bounding a region of area A,

$$\oint_C x\,dy = A, \qquad \oint_C y\,dy = 0, \qquad \text{and} \qquad \oint_C k\,dy = 0,$$

where k is constant.

The integral (6) breaks up into five integrals. Four of them are equal to 0. For instance,

$$\oint_{C_i} P(a_i, b_i)\,dy = 0,$$

since $P(a_i, b_i)$ is constant. The only integral that is not 0 is

$$\oint_{C_i} \frac{\partial P}{\partial x}(a_i, b_i)\,x\,dy = \frac{\partial P}{\partial x}(a_i, b_i) \oint_{C_i} x\,dy$$

$$= \frac{\partial P}{\partial x}(a_i, b_i)A_i,$$

where A_i is the area of $\mathcal{A}_i$.

By (5),

$$\oint_C P\,dy \approx \sum_{i=1}^{n} \frac{\partial P}{\partial x}(a_i, b_i)A_i. \quad (7)$$

But the sum on the right-hand side of (7) is an approximation of the definite integral

$$\int_{\mathcal{A}} \frac{\partial P}{\partial x}\,dA.$$

Letting the mesh of the partition approach 0, we conclude that

$$\oint_C P\,dy = \int_{\mathcal{A}} \frac{\partial P}{\partial x}\,dA.$$

A similar argument establishes (4), and therefore Green's theorem. ■

In the course of proving Green's theorem we saw it is equivalent to the assertion that

$$\oint_C P\,dy = \int_{\mathcal{A}} \frac{\partial P}{\partial x}\,dA$$

and

$$\oint_C (-Q)\,dx = \int_{\mathcal{A}} \frac{\partial Q}{\partial y}\,dA.$$

Hence Green's theorem can be stated without vector notation as follows.

Theorem 2 *Green's theorem (differential form).* Let C be a closed curve in the xy plane bounding a region $\mathcal{A}$. Let $P(x, y)$ and $Q(x, y)$ be scalar functions defined on $\mathcal{A}$. Then

$$\oint_C (-Q\,dx + P\,dy) = \int_{\mathcal{A}} \left(\frac{\partial P}{\partial x} + \frac{\partial Q}{\partial y}\right) dA,$$

where C is counterclockwise.

Cancellation Principle for F · n

In the preceding proof the cancellation principle for $\oint_C f\,dx$ and $\oint_C f\,dy$ was used to deal with $\oint_C \mathbf{F} \cdot \mathbf{n}\,ds$. But there is a cancellation principle that deals directly with the integrand $\mathbf{F} \cdot \mathbf{n}$, which we now describe.

CANCELLATION PRINCIPLE FOR F · n

Let C_1 and C_2 be simple closed curves that overlap in an arc AB. Let C be the simple closed curve formed by C_1 and C_2 together, without the arc AB. Then

$$\oint_{C_1} \mathbf{F} \cdot \mathbf{n}\,ds + \oint_{C_2} \mathbf{F} \cdot \mathbf{n}\,ds = \oint_C \mathbf{F} \cdot \mathbf{n}\,ds.$$

In each integrand, $\mathbf{n}$ describes the exterior unit normal to the corresponding curve.

Inspection of Fig. 6 shows why this principle holds.

Let C_1 and C_2 be two simple closed curves that overlap in the arc AB, as shown in Fig. 6. Let $\mathbf{n}_1$ denote the exterior unit normal to C_1 and $\mathbf{n}_2$ the exterior unit normal to C_2. Observe that on the arc AB, $\mathbf{n}_1 = -\mathbf{n}_2$.

Let $\mathbf{F}$ be a vector field defined on C_1 and C_2. Then $\mathbf{F} \cdot \mathbf{n}_1 = -\mathbf{F} \cdot \mathbf{n}_2$ for points on AB, and we conclude that

$$\int_{AB} \mathbf{F} \cdot \mathbf{n}_1\,ds + \int_{AB} \mathbf{F} \cdot \mathbf{n}_2\,ds = 0,$$

where integration is over the arc AB. This implies that

$$\int_{C_1} \mathbf{F} \cdot \mathbf{n}_1\,ds + \int_{C_2} \mathbf{F} \cdot \mathbf{n}_2\,ds = \int_C \mathbf{F} \cdot \mathbf{n}\,ds,$$

where C and $\mathbf{n}$ are shown in the right-hand part of Fig. 6. (The integrals over AB cancel, leaving the integral over C.)

The cancellation principle for $\oint_C \mathbf{F} \cdot \mathbf{n}\,ds$ concerns a special type of integrand. The cancellation principle for $\oint_C f\,dx$, $\oint_C f\,dy$, and $\oint_C f\,dz$ is far more general, because these integrals depend on the orientation of C.

Figure 6

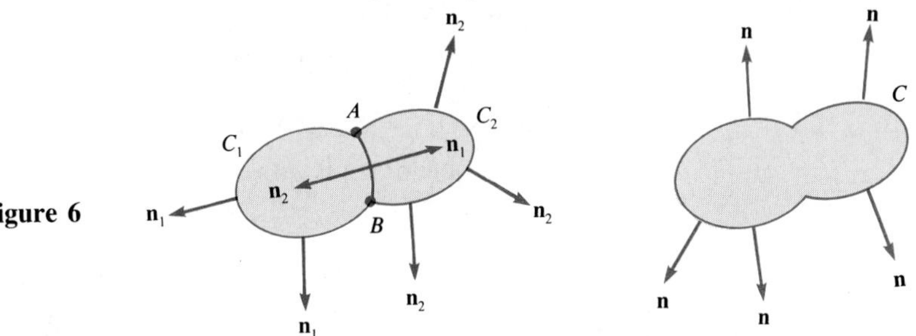

Why $\nabla \cdot \mathbf{F}$ Is Called Divergence

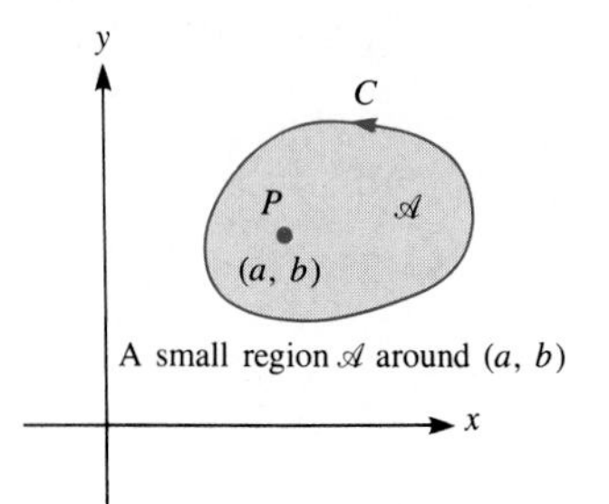

Figure 7

Let $P = (a, b)$ be a point in the plane and $\mathbf{F}$ a vector field describing fluid flow. Choose a very small region $\mathscr{A}$ around P of area A, and let C be its boundary, taken counterclockwise. (See Fig. 7.) Then the net flow out of $\mathscr{A}$ is

$$\oint_C \mathbf{F} \cdot \mathbf{n}\, ds.$$

By Green's theorem, the net flow is also

$$\int_{\mathscr{A}} \nabla \cdot \mathbf{F}\, dA.$$

Now, since $\nabla \cdot \mathbf{F}$ is continuous and $\mathscr{A}$ is small, $\nabla \cdot \mathbf{F}$ is almost constant throughout $\mathscr{A}$, staying close to the value at (a, b), namely, $\nabla \cdot \mathbf{F}(a, b)$, the divergence of $\mathbf{F}$ at (a, b). Thus

$$\int_{\mathscr{A}} \nabla \cdot \mathbf{F}\, dA \approx \nabla \cdot \mathbf{F}(a, b)A.$$

Thus

$$\frac{\text{Net flow out of } \mathscr{A}}{A} \approx \nabla \cdot \mathbf{F}(a, b). \tag{8}$$

This means that

$$\nabla \cdot \mathbf{F} \text{ at } P$$

is a measure of the rate at which fluid tends to leave a small region around P. Hence the name "divergence." If $\nabla \cdot \mathbf{F}$ is positive, fluid near P is tending to get less dense (diverge). If $\nabla \cdot \mathbf{F}$ is negative, fluid near P is tending to accumulate (converge).

Moreover, (8) suggests a new definition of the divergence $\nabla \cdot \mathbf{F}$ at (a, b), namely

$$\boxed{\nabla \cdot \mathbf{F}(a, b) = \lim_{\text{diameter}\,\mathscr{A} \to 0} \frac{\oint_C \mathbf{F} \cdot \mathbf{n}\, ds}{\text{Area of } \mathscr{A}},}$$

where $\mathscr{A}$ is any region enclosing (a, b) whose boundary C is a simple closed curve.

This definition appeals to our physical intuition. We began by defining $\nabla \cdot \mathbf{F}$ mathematically, as $\partial P/\partial x + \partial Q/\partial y$. We now see its physical meaning, which is independent of any coordinate system.

The next section presents applications of Green's theorem. We content ourselves at the moment with illustrating its use in computing $\oint_C \mathbf{F} \cdot \mathbf{n}\, ds$.

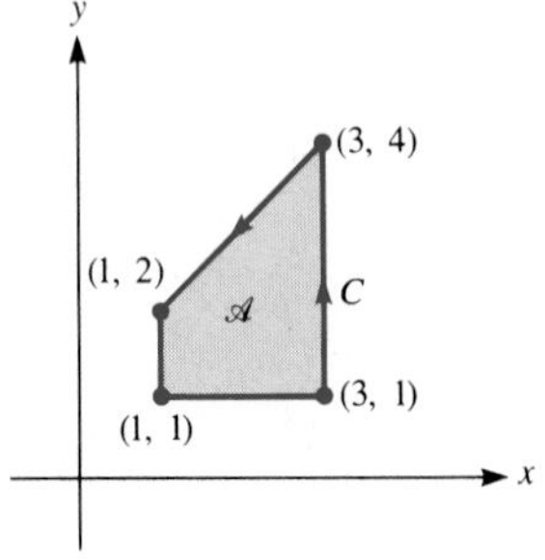

Figure 8

EXAMPLE 2 Let $\mathbf{F} = x^2\mathbf{i} + xy\mathbf{j}$. Evaluate $\oint_C \mathbf{F} \cdot \mathbf{n}\, ds$ over the curve that bounds the quadrilateral with vertices $(1, 1)$, $(3, 1)$, $(3, 4)$, and $(1, 2)$ shown in Fig. 8.

SOLUTION The line integral could be evaluated directly, but would require parameterizing the four edges of C. With Green's theorem we can instead evaluate an integral over a plane region.

Let $\mathcal{A}$ be the region that C bounds. By Green's theorem

$$\oint_C \mathbf{F} \cdot \mathbf{n}\, ds = \int_{\mathcal{A}} \nabla \cdot \mathbf{F}\, dA$$
$$= \int_{\mathcal{A}} \left(\frac{\partial(x^2)}{\partial x} + \frac{\partial(xy)}{\partial y} \right) dA$$
$$= \int_{\mathcal{A}} (2x + x)\, dA$$
$$= \int_{\mathcal{A}} 3x\, dA.$$

Then
$$\int_{\mathcal{A}} 3x\, dA = \int_1^3 \int_1^{y(x)} 3x\, dy\, dx,$$

where $y(x)$ is determined by the equation of the line that provides the top edge of $\mathcal{A}$. We easily find that the line through (1, 2) and (3, 4) has the equation $y = x + 1$. Therefore,

$$\int_{\mathcal{A}} 3x\, dA = \int_1^3 \int_1^{x+1} 3x\, dy\, dx.$$

The inner integration gives

$$\int_1^{x+1} 3x\, dy = 3xy \Big|_{y=1}^{y=x+1} = 3x^2.$$

The second integration gives

$$\int_1^3 3x^2\, dx = x^3 \Big|_1^3 = 26. \quad \blacksquare$$

A Different Proof of Green's Theorem

The opposite direction from the first proof

This second proof starts with $\int_{\mathcal{A}} \nabla \cdot \mathbf{F}\, dA$ and transforms it into $\oint_C \mathbf{F} \cdot \mathbf{n}\, ds$. It reviews:

The iterated integral in rectangular coordinates (Sec. 15.2).

The fundamental theorem of calculus (Sec. 5.7).

The line integral (Sec. 16.2).

Second Proof of Green's Theorem Let $\mathcal{A}$ be a region in the plane such that each vertical line and each horizontal line meets the boundary of $\mathcal{A}$ in at most two points.

Letting $\mathbf{F} = P\mathbf{i} + Q\mathbf{j}$, we wish to show that

$$\int_{\mathcal{A}} \left(\frac{\partial P}{\partial x} + \frac{\partial Q}{\partial y} \right) dA = \oint_C (P\, dy - Q\, dx).$$

We will prove that

$$\int_{\mathscr{A}} \frac{\partial Q}{\partial y}\, dA = \oint_C (-Q)\, dx. \tag{9}$$

A similar proof will show that

$$\int_{\mathscr{A}} \frac{\partial P}{\partial x}\, dA = \oint_C P\, dy.$$

Green's theorem follows immediately from these two equations.

Let the region $\mathscr{A}$ have the description:

$$a \le x \le b, \qquad y_1(x) \le y \le y_2(x),$$

as shown in Fig. 9. Then

$$\int_{\mathscr{A}} \frac{\partial Q}{\partial y}\, dA = \int_a^b \int_{y_1(x)}^{y_2(x)} \frac{\partial Q}{\partial y}\, dy\, dx.$$

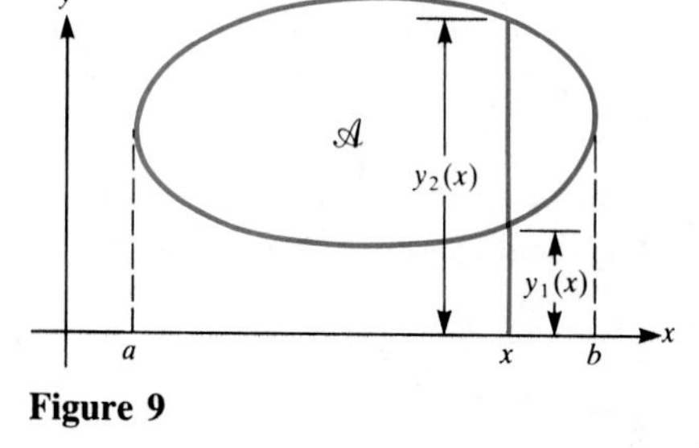

Figure 9

By the fundamental theorem of calculus,

$$\int_{y_1(x)}^{y_2(x)} \frac{\partial Q}{\partial y}\, dy = Q(x, y_2(x)) - Q(x, y_1(x)).$$

Hence

$$\int_{\mathscr{A}} \frac{\partial Q}{\partial y}\, dA = \int_a^b [Q(x, y_2(x)) - Q(x, y_1(x))]\, dx. \tag{10}$$

Now consider the right side of Eq. (9),

$$\oint_C (-Q)\, dx.$$

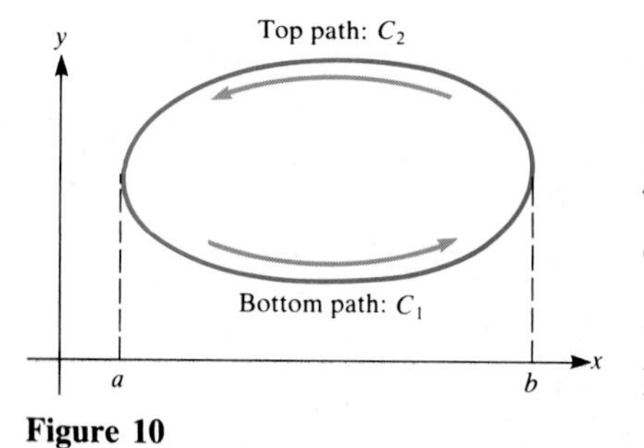

Figure 10

Break the closed path C into two successive paths, one along the bottom part of $\mathscr{A}$, described by $y = y_1(x)$, the other along the top part of $\mathscr{A}$, described by $y = y_2(x)$. Denote the bottom path C_1 and the top path C_2. (See Fig. 10.)

Then

$$\oint_C (-Q)\, dx = \int_{C_1} (-Q)\, dx + \int_{C_2} (-Q)\, dx. \tag{11}$$

But

$$\int_{C_1} (-Q)\, dx = \int_{C_1} [-Q(x, y_1(x))]\, dx = \int_a^b [-Q(x, y_1(x))]\, dx,$$

and

$$\int_{C_2} (-Q)\, dx = \int_{C_2} [-Q(x, y_2(x))]\, dx = \int_b^a [-Q(x, y_2(x))]\, dx,$$

$$= \int_a^b Q(x, y_2(x))\, dx.$$

Thus by Eq. (11),

$$\oint_C (-Q)\, dx = \int_a^b -Q(x, y_1(x))\, dx + \int_a^b Q(x, y_2(x))\, dx$$

$$= \int_a^b [Q(x, y_2(x)) - Q(x, y_1(x))]\, dx.$$

This is precisely the right side of Eq. (10) and concludes the proof. ■

Section Summary

In symbols, Green's theorem says that

$$\oint_C \mathbf{F} \cdot \mathbf{n}\, ds = \int_{\mathscr{A}} \nabla \cdot \mathbf{F}\, dA.$$

In words, Green's theorem says that "The integral of the normal component of **F** around a simple closed curve equals the integral of the divergence of **F** over the region that the curve bounds."

From this it follows that

$$\nabla \cdot \mathbf{F}(P) = \lim_{\text{diameter}\,\mathscr{A} \to 0} \frac{\oint_C \mathbf{F} \cdot \mathbf{n}\, ds}{\text{Area of } \mathscr{A}},$$

where C is the boundary of $\mathscr{A}$, which contains P.

Green's theorem also gives a way to compute $\oint_C \mathbf{F} \cdot \mathbf{n}\, ds$ (the flux across C) by integrating the divergence of **F** over $\mathscr{A}$ (and conversely).

EXERCISES FOR SEC. 16.4: GREEN'S THEOREM

In Exercises 1 to 4 compute $\int_{\mathscr{A}} \nabla \cdot \mathbf{F}\, dA$ and $\oint_C \mathbf{F} \cdot \mathbf{n}\, ds$ and verify Green's theorem. The differential form of Green's theorem (Theorem 2) will be helpful.

1 $\mathbf{F} = 3x\mathbf{i} + 2y\mathbf{j}$, and $\mathscr{A}$ is the disk of radius 1 with center (0, 0).

2 $\mathbf{F} = 5y^3\mathbf{i} - 6x^2\mathbf{j}$, and $\mathscr{A}$ is the disk of radius 2 with center (0, 0).

3 $\mathbf{F} = xy\mathbf{i} + x^2y\mathbf{j}$, and $\mathscr{A}$ is the square with vertices (0, 0), $(a, 0)$, (a, b), and $(0, b)$, $a, b > 0$.

4 $\mathbf{F} = \cos(x + y)\,\mathbf{i} + \sin(x + y)\,\mathbf{j}$, and $\mathscr{A}$ is the triangle with vertices (0, 0), $(a, 0)$, and (a, b), $a, b > 0$.

In Exercises 5 to 8 use Green's theorem to evaluate $\oint_C \mathbf{F} \cdot \mathbf{n}\, ds$ for the given **F**, where C is the boundary of the given region $\mathscr{A}$.

5 $\mathbf{F} = e^x \sin y\,\mathbf{i} + e^{2x} \cos y\,\mathbf{j}$, and $\mathscr{A}$ is the rectangle with vertices (0, 0), (1, 0), $(0, \pi/2)$, and $(1, \pi/2)$.

6 $\mathbf{F} = y \tan x\,\mathbf{i} + y^2\,\mathbf{j}$, and $\mathscr{A}$ is the square with vertices (0, 0), (1, 0), (1, 1), and (0, 1).

7 $\mathbf{F} = 2x^3y\mathbf{i} - 3x^2y^2\mathbf{j}$, and $\mathscr{A}$ is the triangle with vertices (0, 1), (3, 4), and (2,7).

8 $\mathbf{F} = \dfrac{-\mathbf{i}}{xy^2} + \dfrac{\mathbf{j}}{x^2y}$, and $\mathscr{A}$ is the triangle with vertices (1, 1), (2, 2), and (1, 2).

In Exercises 9 and 10, **F** is defined on the whole plane but indicated only at points on a curve C bounding a region $\mathscr{A}$. What can be said about $\int_{\mathscr{A}} \nabla \cdot \mathbf{F}\, dA$ in each case?

9 (*a*)

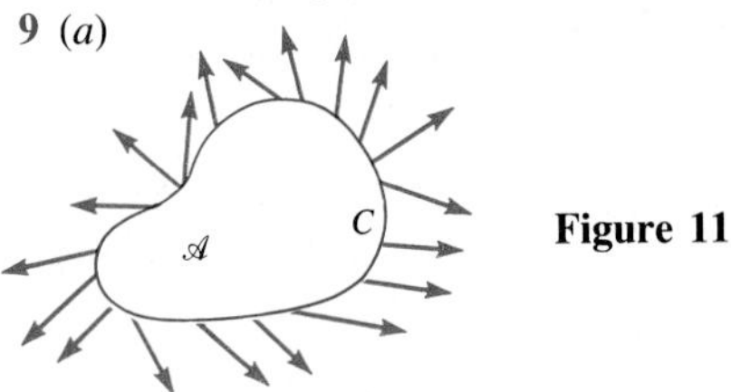

Figure 11

(*b*)

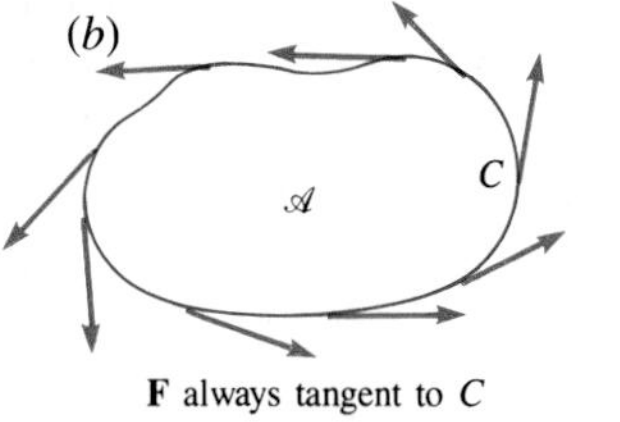

Figure 12

10

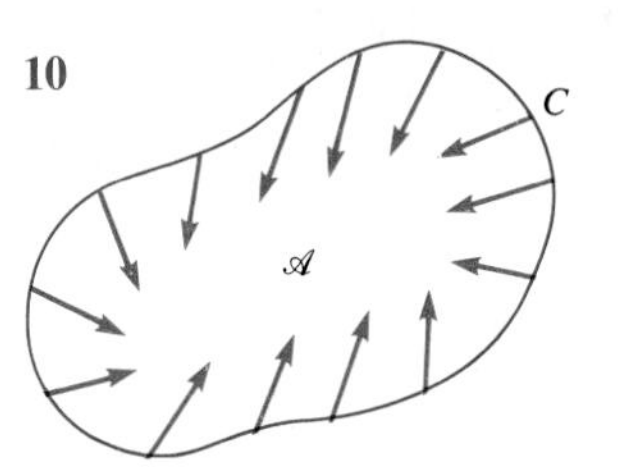

Figure 13

11 Use Green's theorem to evaluate $\oint_C (xy\, dx + e^x\, dy)$, where C is the curve that goes from (0, 0) to (2, 0) on the x axis and returns from (2, 0) to (0, 0) on the parabola $y = 2x - x^2$.

12 Let C be the circle of radius 1 with center (0, 0).

(*a*) What does Green's theorem say about the line integral

$$\oint_C [(x^2 - y^3)\, dx + (y^2 + x^3)\, dy]?$$

(*b*) Use Green's theorem to evaluate the integral in (*a*).

(*c*) Evaluate the integral in (*a*) directly.

13 If div $\mathbf{F} = 0$, **F** is said to be "divergence-free." If **F** describes the flow of a fluid, then the fluid is said to be "incompressible." For example, $\mathbf{F} = \hat{\mathbf{r}}/\|\mathbf{r}\|$ is divergence-free in the xy plane.

Let $\mathbf{F} = \hat{\mathbf{r}}/\|\mathbf{r}\|$ and let C be the square with vertices $(1, 0)$, $(2, 0)$, $(2, 1)$, and $(1, 1)$. Find $\oint_C \mathbf{F} \cdot \mathbf{n}\, ds$.

14 Let $\mathbf{F} = \hat{\mathbf{r}}/\|\mathbf{r}\|$ in the xy plane and let C be the circle of radius a and center $(0, 0)$.
(*a*) What does Green's theorem say about $\oint_C \mathbf{F} \cdot \mathbf{n}\, ds$?
(*b*) Evaluate $\oint_C \mathbf{F} \cdot \mathbf{n}\, ds$ without using Green's theorem.
(*c*) Let C now be the circle of radius 3 and center $(4, 0)$. Evaluate $\oint_C \mathbf{F} \cdot \mathbf{n}\, ds$, doing as little work as possible.

15 We proved that $\oint_C -Q\, dx = \int_{\mathcal{A}} \partial Q/\partial y\, dA$. Write up this proof in your own words. Your diagrams and exposition should differ enough from those in the text to show that you have thought through the proof.

16 Write up a complete proof that $\oint_C P\, dy = \int_{\mathcal{A}} \partial P/\partial x\, dA$, including diagrams. This is the part of our first proof of Green's theorem that is left to the reader.

17 The divergence of $\mathbf{F}$ at a certain point P is 4. If $\mathbf{F}$ describes fluid flow ($\mathbf{F} = \sigma\mathbf{v}$, where σ is density and $\mathbf{v}$ is velocity) at what rate is fluid leaving (or entering) the disk around P of radius 0.02? (This would be an estimate.)

18 The curve C encloses the point $(2, 1)$ and an extremely small disk of radius 0.1. Assume that $\oint_C \mathbf{F} \cdot \mathbf{n}\, ds = 0.06$, C counterclockwise. Use this information to estimate the divergence of $\mathbf{F}$ at $(2, 1)$.

19 The divergence of $\mathbf{F}$ at $(1, 1)$ is 5. Let $\mathcal{A}$ be the square around $(1, 1)$ with vertices $(0.99, 0.99)$, $(1.01, 0.99)$, $(1.01, 1.01)$, and $(0.99, 1.01)$. Let C be the boundary of $\mathcal{A}$, counterclockwise. Estimate the integral of the component of $\mathbf{F}$ along the external normal over the curve C.

20 Let f be a scalar function. Using Green's theorem, show that

$$\int_{\mathcal{A}} \left(\frac{\partial^2 f}{\partial x^2} + \frac{\partial^2 f}{\partial y^2} \right) dA = \oint_C \left(\frac{\partial f}{\partial x}\, dy - \frac{\partial f}{\partial y}\, dx \right),$$

where $\mathcal{A}$ is a convex region and C its boundary taken counterclockwise.

Exercises 21 to 27 are related.

21 Let $\mathcal{A}$ be a plane region with boundary C a simple closed curve swept out counterclockwise. Use Green's theorem to show that the area of $\mathcal{A}$ equals

$$\frac{1}{2} \oint_C (-y\, dx + x\, dy).$$

22 Use Exercise 21 to find the area of the region bounded by the line $y = x$ and the curve

$$\begin{cases} x = t^6 + t^4 \\ y = t^3 + t \end{cases}$$

for t in $[0, 1]$.

23 A curve is given parametrically by $x = t(1 - t^2)$, $y = t^2(1 - t^3)$, for t in $[0, 1]$.
(*a*) Sketch the points corresponding to $t = 0, 0.2, 0.4, 0.6, 0.8$, and 1.0, and use them to sketch the curve.
(*b*) Let $\mathcal{A}$ be the region enclosed by the curve. What difficulty arises when you try to compute the area of $\mathcal{A}$ by a definite integral involving vertical or horizontal cross sections?
(*c*) Use Exercise 21 to find the area of $\mathcal{A}$.

24 Repeat Exercise 23 for $x = \sin \pi t$ and $y = t - t^2$. In (*a*), let $t = 0, \frac{1}{4}, \frac{1}{2}, \frac{3}{4}, 1$.

25 Use Exercise 21 to obtain the formula for area in polar coordinates:

$$\text{Area} = \frac{1}{2} \int_\alpha^\beta r^2\, d\theta.$$

26 Assume that you know that Green's theorem is true when $\mathcal{A}$ is a triangle and C its boundary.
(*a*) Deduce that it therefore holds for quadrilaterals.
(*b*) Deduce that it holds for polygons.

27 Write up a proof that $\int_{\mathcal{A}} (\partial P/\partial x)\, dA = \oint_C P\, dy$ in the style of the second proof of Green's theorem.

28 Let $\mathcal{A}$ be a region in the xy plane bounded by the closed curve C. Let $f(x, y)$ be defined on the plane. Show that

$$\int_{\mathcal{A}} \left(\frac{\partial^2 f}{\partial x^2} + \frac{\partial^2 f}{\partial y^2} \right) dA = \oint_C D_{\mathbf{n}}(f)\, ds.$$

29 In engineering and physics the unit vectors $\hat{\mathbf{r}}$ and $\hat{\boldsymbol{\theta}}$ are often used. (See Fig. 14.)

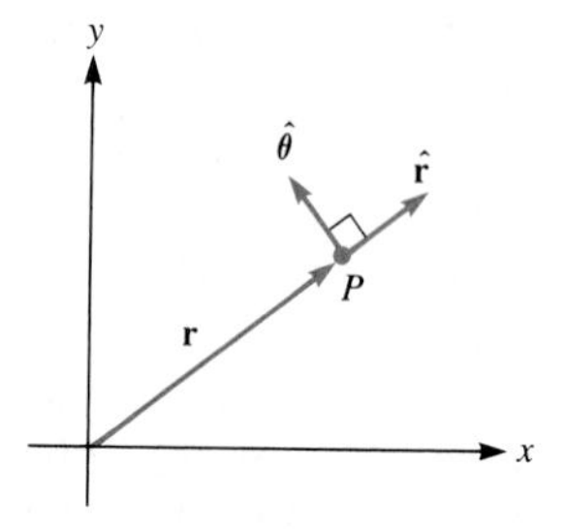

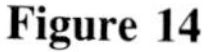
Figure 14

They depend on the point P in the plane: $\hat{\mathbf{r}} = \mathbf{r}/\|\mathbf{r}\|$ and $\hat{\boldsymbol{\theta}}$ is a unit vector obtained by rotating $\hat{\mathbf{r}}$ $\pi/2$ radians counterclockwise. If $\mathbf{F}$ is expressed relative to the vectors $\mathbf{i}$ and $\mathbf{j}$, $\mathbf{F} = P\mathbf{i} + Q\mathbf{j}$, then $\text{div}\, \mathbf{F} = (\partial P/\partial x) + (\partial Q/\partial y)$. If $\mathbf{F}$ is expressed relative to $\hat{\mathbf{r}}$ and $\hat{\boldsymbol{\theta}}$, $\mathbf{F} = A\hat{\mathbf{r}} + B\hat{\boldsymbol{\theta}}$, how would div $\mathbf{F}$ be computed?

We outline a way to find A and B, by exploiting the coordinate-free description of divergence as a limit:

$$\text{div}\, \mathbf{F} = \lim \frac{\oint_C \mathbf{F} \cdot \mathbf{n}\, ds}{\text{Area of } \mathcal{A}},$$

as the diameter of $\mathcal{A}$ approaches 0.

As the region $\mathcal{A}$ around the point P use the *small* patch shown in Fig. 15. The bounding curve C is broken into four

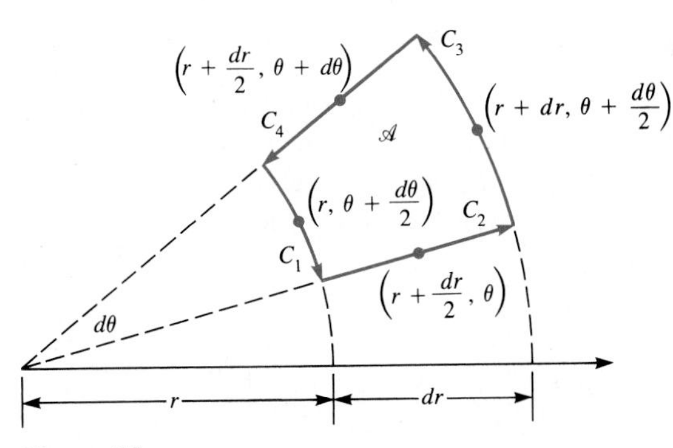

Figure 15

shorter curves, C_1, C_2, C_3, C_4. Estimate $\oint_C \mathbf{F} \cdot \mathbf{n}\, ds$ as follows.

(*a*) Why is a reasonable estimate of $\int_{C_1} \mathbf{F} \cdot \mathbf{n}\, ds$ the quantity $-A(r, \theta + d\theta/2)r\, d\theta$?

(*b*) Show that $\oint_C \mathbf{F} \cdot \mathbf{n}\, ds$ is approximately

$$\left[A\left(r + dr, \theta + \frac{d\theta}{2}\right) - A\left(r, \theta + \frac{d\theta}{2}\right)\right] r\, d\theta$$

$$+ A\left(r + dr, \theta + \frac{d\theta}{2}\right) dr\, d\theta$$

$$+ \left[B\left(r + \frac{dr}{2}, \theta + d\theta\right) - B\left(r + \frac{dr}{2}, \theta\right)\right] dr.$$

(*c*) Use the mean-value theorem to show that the sum in (*b*) is approximately

$$\frac{\partial A}{\partial r} r\, dr\, d\theta + A\, dr\, d\theta + \frac{\partial B}{\partial \theta}\, dr\, d\theta.$$

(*d*) The area of $\mathscr{A}$ is approximately $r\, dr\, d\theta$. Combine this with (*c*) to show that

$$\text{div}\, \mathbf{F} = \frac{\partial A}{\partial r} + \frac{A}{r} + \frac{1}{r}\frac{\partial B}{\partial \theta}.$$

30 Figure 16 shows four vector fields. Two are divergence-free and two are not. Decide which two are not, copy them onto a sheet of drawing paper, and sketch a closed curve C for which $\oint_C \mathbf{F} \cdot \mathbf{n}\, ds$ is not 0.

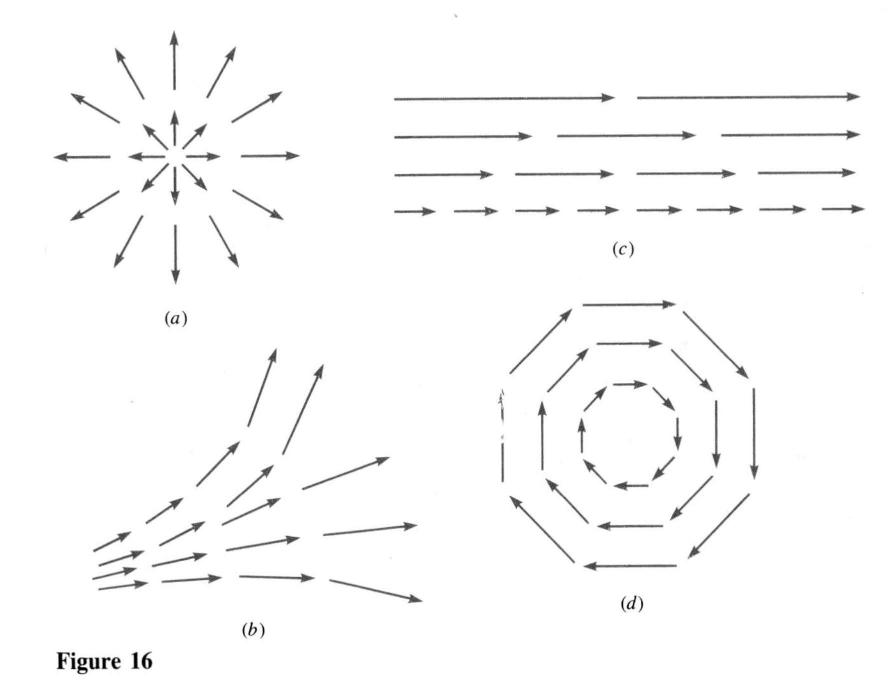

Figure 16

16.5 APPLICATIONS OF GREEN'S THEOREM

The previous section proved Green's theorem. This section presents a few of its applications. First, however, we generalize Green's theorem to a region bounded by two curves instead of one.

Green's Theorem for Two Curves

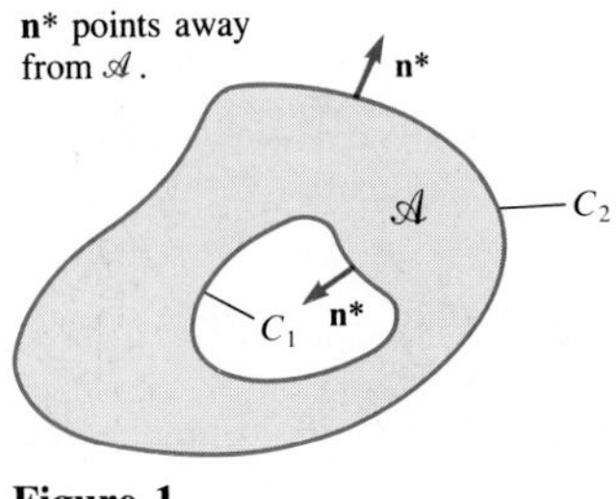

Figure 1

Green's theorem holds for regions in the plane whose boundaries consist of any finite number of curves. However, we state it in the special case in which the boundary consists of two nonoverlapping curves, since this case is especially useful. A region $\mathscr{A}$ is shown in Fig. 1. The region has a hole bounded by the curve C_1 and an outer boundary C_2.

> **Green's Theorem** *Two-curve case.* Let $\mathscr{A}$ be a region in the plane bounded by the curves C_1 and C_2. Let $\mathbf{n}^*$ denote the exterior unit normal along the boundary. Then
>
> $$\oint_{C_1} \mathbf{F} \cdot \mathbf{n}^* \, ds + \oint_{C_2} \mathbf{F} \cdot \mathbf{n}^* \, ds = \int_{\mathscr{A}} \nabla \cdot \mathbf{F} \, dA$$
>
> for any vector field $\mathbf{F}$ defined on $\mathscr{A}$.

Like Green's theorem in the case of a one-curve boundary, the two-curve version also records that the net loss of fluid in the region $\mathscr{A}$ is the same as the net flow past the boundary, which—in this case—consists of two curves. The proof is practically identical to our first proof of Green's theorem in Sec. 16.4, so we will not go through it here.

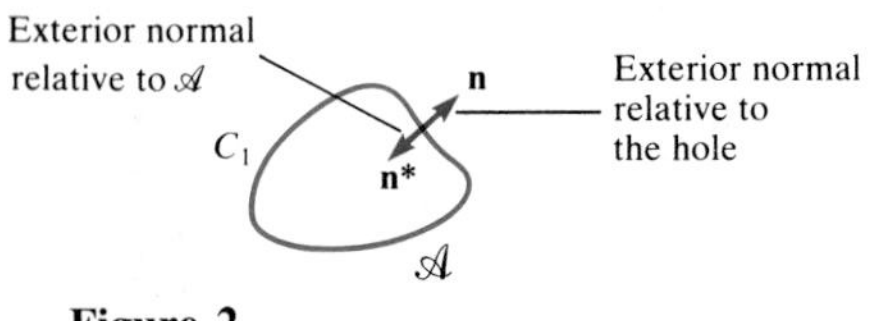

Figure 2

As shown in Fig. 1, the exterior normal $\mathbf{n}^*$ points into the hole for the inner curve. We call the normal $\mathbf{n}^*$ rather than $\mathbf{n}$ to simplify the proof of Corollary 1.

On the inner curve C_1, $\mathbf{n}^*$ is not the exterior normal $\mathbf{n}$ relative to the region that C_1 surrounds. Relative to the hole bounded by C_1, the exterior normal $\mathbf{n}$ points away from the hole, as shown in Fig. 2. Inspection of Fig. 2 shows that, on C_1, $\mathbf{n}^* = -\mathbf{n}$.

The two-curve case of Green's theorem is the basis of the following corollary.

> **Corollary 1** Let C_1 and C_2 be two closed curves that form the boundary of the region $\mathscr{A}$. Let $\mathbf{F}$ be a vector field defined on $\mathscr{A}$ such that the divergence of $\mathbf{F}$, $\nabla \cdot \mathbf{F}$, is 0 throughout $\mathscr{A}$. Then
>
> $$\oint_{C_1} \mathbf{F} \cdot \mathbf{n} \, ds = \oint_{C_2} \mathbf{F} \cdot \mathbf{n} \, ds.$$

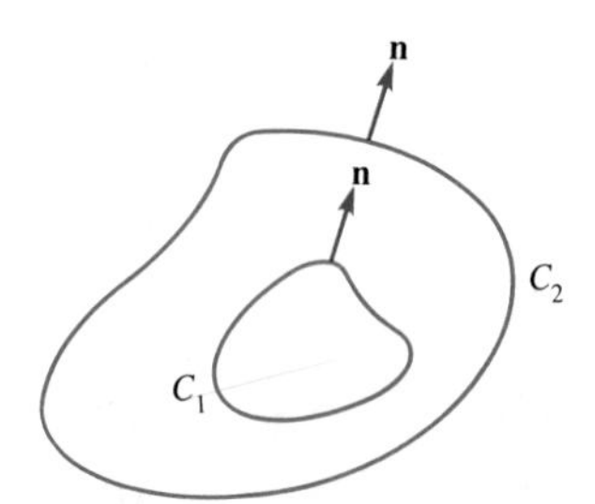

Figure 3

Proof Assume that C_1 is the inner curve and C_2 is the outer curve, as shown in Fig. 3. By Green's theorem in the two-curve case,

$$\oint_{C_1} \mathbf{F} \cdot \mathbf{n}^* \, ds + \oint_{C_2} \mathbf{F} \cdot \mathbf{n}^* \, ds = \int_{\mathscr{A}} \nabla \cdot \mathbf{F} \, dA$$

$$= \int_{\mathscr{A}} 0 \, dA = 0. \qquad (\text{since } \nabla \cdot \mathbf{F} = 0)$$

Thus

$$\oint_{C_1} \mathbf{F} \cdot (-\mathbf{n}) \, ds + \oint_{C_2} \mathbf{F} \cdot \mathbf{n} \, ds = 0,$$

or
$$-\oint_{C_1} \mathbf{F}\cdot\mathbf{n}\,ds + \oint_{C_2} \mathbf{F}\cdot\mathbf{n}\,ds = 0,$$
from which the corollary follows. ■

Some very important vector fields have divergence equal to 0. Hence Corollary 1 is important in applications. For example, the field $\hat{\mathbf{r}}/\|\mathbf{r}\|$, in the xy plane, has divergence equal to 0. (The field $\hat{\mathbf{r}}/\|\mathbf{r}\|^2$, in space, also has divergence equal to 0. This field, known as the **inverse-square law**, occurs in the study of electricity and gravity. In the next chapter we will obtain an analog of Corollary 1 for surfaces, which we will be able to apply to inverse-square fields.)

A field $\mathbf{F}$ for which div $\mathbf{F} = 0$ is called **divergence-free**. A fluid whose field is divergence-free is called **incompressible**.

The power of Corollary 1 is illustrated in the following example.

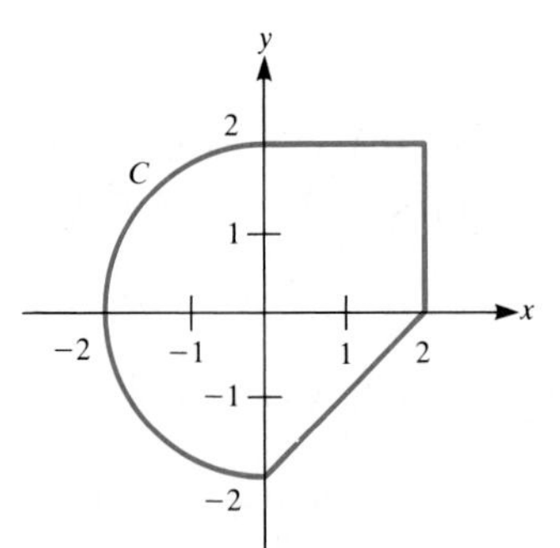

Figure 4

EXAMPLE 1 Let $\mathbf{F}(x, y) = \hat{\mathbf{r}}/\|\mathbf{r}\|$, where $\mathbf{r} = x\mathbf{i} + y\mathbf{j}$. Let C be the curve shown in Fig. 4. Evaluate $\oint_C \mathbf{F}\cdot\mathbf{n}\,ds$.

SOLUTION By Exercise 32 of Sec. 16.1, the divergence of $\mathbf{F}$ is 0 wherever $\mathbf{F}$ is defined. (It is not defined at the origin.) Motivated by Corollary 1, consider $\oint_{C_1} \mathbf{F}\cdot\mathbf{n}\,ds$, where C_1 is the unit circle with center at the origin. The curve C_1 is easier to deal with than the original curve C. (See Fig. 5.)

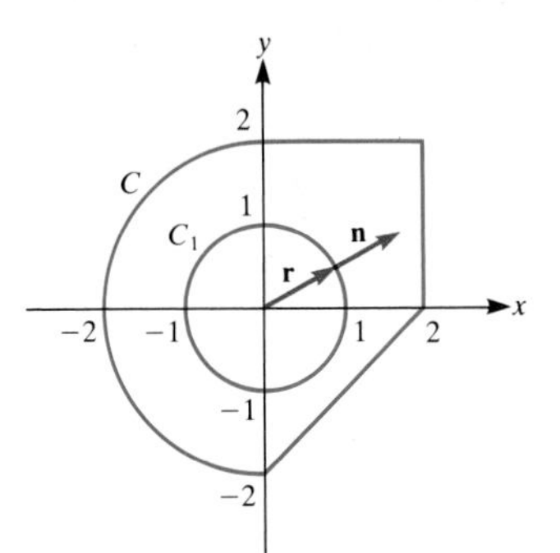

Figure 5

By Corollary 1,
$$\oint_C \mathbf{F}\cdot\mathbf{n}\,ds = \oint_{C_1} \mathbf{F}\cdot\mathbf{n}\,ds = \oint_{C_1} \frac{\hat{\mathbf{r}}\cdot\mathbf{n}}{\|\mathbf{r}\|}\,ds.$$

On the unit circle $\|\mathbf{r}\| = 1$. Also, $\mathbf{n}$ points in the same direction as $\mathbf{r}$. Hence, $\mathbf{n} = \mathbf{r}$, and $\mathbf{r}\cdot\mathbf{n} = 1$. Thus
$$\oint_{C_1} \frac{\hat{\mathbf{r}}\cdot\mathbf{n}}{\|\mathbf{r}\|}\,ds = \oint_{C_1} 1\,ds = 2\pi.$$

Consequently,
$$\oint_C \frac{\hat{\mathbf{r}}\cdot\mathbf{n}}{\|\mathbf{r}\|}\,ds = 2\pi. \quad ■$$

This integral required in Example 1 could have been computed directly, but only with much greater difficulty. [See also Eq. (2) on p. 964.]

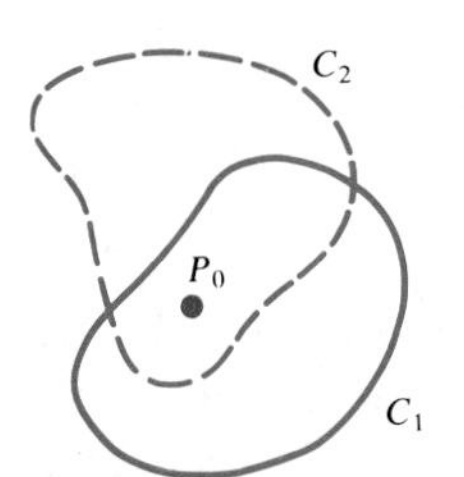

Figure 6

Corollary 2 Let $\mathbf{F}$ be a vector field defined everywhere in the plane except perhaps at the point P_0. Assume that the divergence of $\mathbf{F}$ is 0. Let C_1 and C_2 be two simple closed curves, each of which encloses the point P_0. Then
$$\oint_{C_1} \mathbf{F}\cdot\mathbf{n}\,ds = \oint_{C_2} \mathbf{F}\cdot\mathbf{n}\,ds.$$

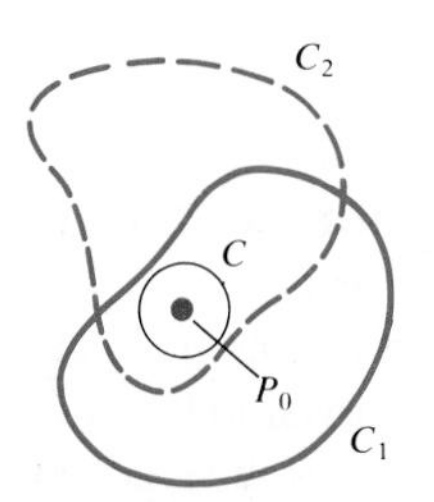

Figure 7

Proof The two typical curves, which may overlap, are shown in Fig. 6.

Choose a very small circle C around P_0, so small that it meets neither C_1 nor C_2. (See Fig. 7.) By Corollary 1,

$$\oint_{C_1} \mathbf{F} \cdot \mathbf{n}\, ds = \oint_{C} \mathbf{F} \cdot \mathbf{n}\, ds \quad \text{and} \quad \oint_{C_2} \mathbf{F} \cdot \mathbf{n}\, ds = \oint_{C} \mathbf{F} \cdot \mathbf{n}\, ds.$$

Thus
$$\oint_{C_1} \mathbf{F} \cdot \mathbf{n}\, ds = \oint_{C_2} \mathbf{F} \cdot \mathbf{n}\, ds. \quad \blacksquare$$

In practical terms, this is what Corollary 2 tells us:

> *If the divergence of a vector field is 0* in some planar region $\mathcal{A}$, then as you move a simple closed curve about in that region, even changing its shape, the integral $\oint_C \mathbf{F} \cdot \mathbf{n}\, ds$ stays constant.

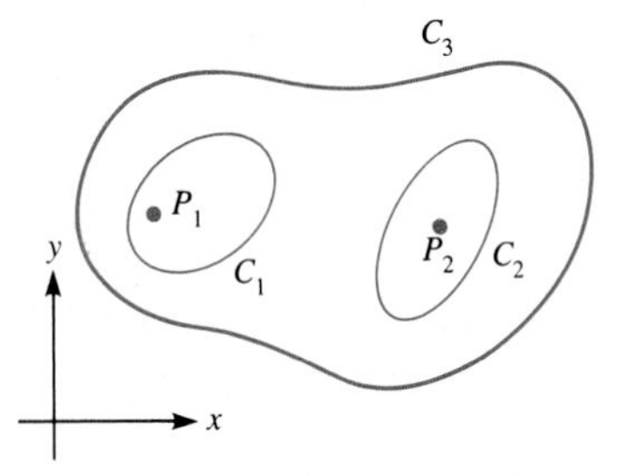

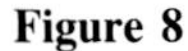
Figure 8

EXAMPLE 2 Assume that div $\mathbf{F} = 0$ at all points except at the points P_1 and P_2 where $\mathbf{F}$ is not defined. Assume that $\oint_{C_1} \mathbf{F} \cdot \mathbf{n}\, ds = 3$ and $\oint_{C_2} \mathbf{F} \cdot \mathbf{n}\, ds = 4$, where C_1 and C_2 are counterclockwise. Find $\oint_{C_3} \mathbf{F} \cdot \mathbf{n}\, ds$, C_3 counterclockwise. (See Fig. 8.)

SOLUTION Draw the dashed line shown in Fig. 9, cutting the region bounded by C_3 into two regions. Let C_4 be the curve bounding the right-hand region and C_5 the curve bounding the left-hand region.

By the cancellation principle,

$$\oint_{C_3} \mathbf{F} \cdot \mathbf{n}\, ds = \oint_{C_4} \mathbf{F} \cdot \mathbf{n}\, ds + \oint_{C_5} \mathbf{F} \cdot \mathbf{n}\, ds.$$

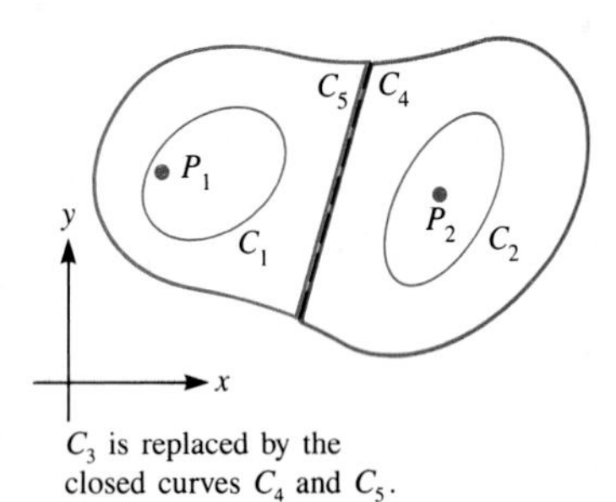

C_3 is replaced by the closed curves C_4 and C_5.

Figure 9

Corollary 2 tells us that

$$\oint_{C_4} \mathbf{F} \cdot \mathbf{n}\, ds = \oint_{C_2} \mathbf{F} \cdot \mathbf{n}\, ds = 4$$

and

$$\oint_{C_5} \mathbf{F} \cdot \mathbf{n}\, ds = \oint_{C_1} \mathbf{F} \cdot \mathbf{n}\, ds = 3.$$

Thus

$$\oint_{C_3} \mathbf{F} \cdot \mathbf{n}\, ds = 4 + 3 = 7. \quad \blacksquare$$

Green's Theorem Applied to Heat

A plane region $\mathcal{A}$ with boundary curve C is occupied by a sheet of metal. By various heating and cooling devices, the temperature along the border is held constant, independent of time. (See Fig. 10.) Assume that the temperature in $\mathcal{A}$ eventually stabilizes. This steady-state temperature at point P in $\mathcal{A}$ is denoted $T(P)$. We will show that the Laplacian of T is identically 0; that is,

$$\frac{\partial^2 T}{\partial x^2} + \frac{\partial^2 T}{\partial y^2} = 0.$$

This is an important step, since it reduces the study of the temperature distribution to solving a partial differential equation.

First of all, heat tends to flow "from high to low temperature," that is, in the direction of $-\nabla T$. According to Fourier's law, flow is proportional to the conductivity of the material k (a positive constant) and the magnitude of the gradient

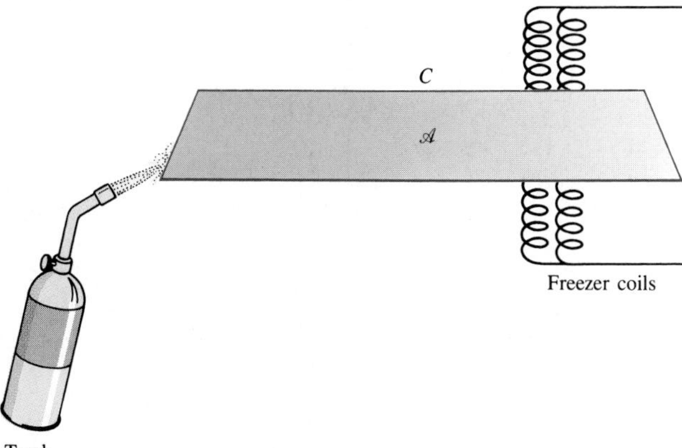

Figure 10

$\|\nabla T\|$. Thus

$$\oint_C (-k\,\nabla T)\cdot \mathbf{n}\,ds$$

measures the rate of heat loss across C.

Since the temperature in the metal is at a steady state, the heat in the region bounded by C remains constant. Thus

$$\oint_C (-k\nabla T)\cdot \mathbf{n}\,ds = 0.$$

Now, Green's theorem then tells us that

$$\int_{\mathscr{A}} \nabla\cdot(-k\nabla T)\,dA = 0$$

for any region $\mathscr{A}$ in the metal plate. Since $\nabla\cdot\nabla T$ is the Laplacian of T and k is not 0, we conclude that

$$\int_{\mathscr{A}} \left(\frac{\partial^2 T}{\partial x^2} + \frac{\partial^2 T}{\partial y^2}\right) dA = 0. \tag{1}$$

for all such $\mathscr{A}$. To show that the integrand itself must be 0, we call upon the "vanishing-integrals" principle used often in applications:

The Vanishing-Integrals Principle If $f(P)$ is a continuous function defined in some region $\mathscr{B}$, and its integral over every region $\mathscr{A}$ in its domain is 0, then $f(P) = 0$ for all points P in $\mathscr{B}$.

The idea behind the principle is simple: If $f(P)$ is not always 0, there would be at least one point, say P_0, where $f(P_0)$ is not 0. Assume that $f(P_0)$ is positive. Call its value k. Because f is continuous, there is a disk around P_0 such that $f(P)$ is close to $f(P_0)$ for all points P in the disk. In particular, there is a disk $\mathscr{D}$

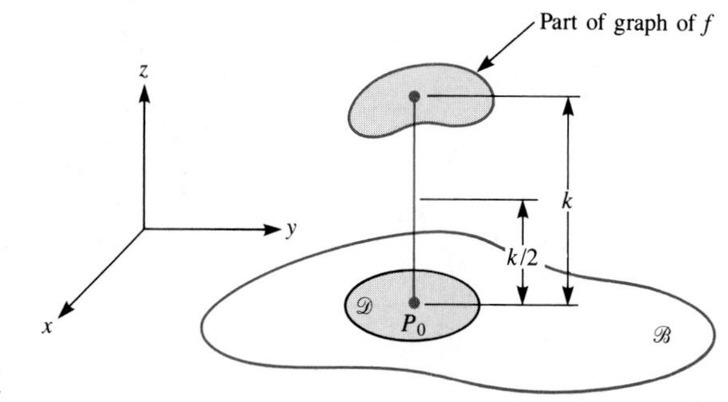

Figure 11

around P_0 such that $f(P) \geq k/2$ for all points P in $\mathcal{D}$. (See Fig. 11.) Then

$$\int_{\mathcal{D}} f(P)\, dA \geq \int_{\mathcal{D}} \frac{k}{2}\, dA = \frac{k}{2}\,(\text{Area of } \mathcal{D}) > 0.$$

Consequently $\int_{\mathcal{D}} f(P)\, dA$ is not 0, violating our assumption that the integral of f over every region in $\mathcal{B}$ is 0. We have a similar contradiction if we assume that $f(P_0)$ is ever negative. (Exercises 18 to 20 concern the "vanishing-integrals" principle.)

In view of the vanishing-integrals principle, we conclude from (1) that the Laplacian of T is identically 0.

A Different View of Green's Theorem

In Sec. 16.4 Green's theorem is expressed in terms of flux,

$$\oint_C \mathbf{F} \cdot \mathbf{n}\, ds = \int_{\mathcal{A}} \nabla \cdot \mathbf{F}\, dA,$$

and in differential form,

$$\oint_C (-Q\, dx + P\, dy) = \int_{\mathcal{A}} \left(\frac{\partial P}{\partial x} + \frac{\partial Q}{\partial y} \right) dA.$$

In fact,

$$\left.\begin{aligned} \oint_C Q\, dx &= \int_{\mathcal{A}} \left(-\frac{\partial Q}{\partial y} \right) dA \\ \text{and} \qquad \oint_C P\, dy &= \int_{\mathcal{A}} \frac{\partial P}{\partial x}\, dA. \end{aligned}\right\} \tag{2}$$

Green's theorem can also be expressed in terms of the circulation,

$$\oint_C \mathbf{F} \cdot d\mathbf{r}.$$

To formulate this expression, write

$$\mathbf{F} = A\mathbf{i} + B\mathbf{j}.$$

Then $$\oint_C \mathbf{F} \cdot d\mathbf{r} = \oint_C (A\mathbf{i} + B\mathbf{j}) \cdot (dx\mathbf{i} + dy\mathbf{j}),$$

or $$\oint_C \mathbf{F} \cdot d\mathbf{r} = \oint_C (A\,dx + B\,dy). \tag{3}$$

By (2) $$\oint_C A\,dx = \int_{\mathcal{A}} \left(-\frac{\partial A}{\partial y}\right) dA$$

and $$\oint_C B\,dy = \int_{\mathcal{A}} \frac{\partial B}{\partial x}\,dA.$$

Thus (3) becomes

$$\oint_C \mathbf{F} \cdot d\mathbf{r} = \int_{\mathcal{A}} \left(\frac{\partial B}{\partial x} - \frac{\partial A}{\partial y}\right) dA. \tag{4}$$

The integrand on the right side of (4) is reminiscent of the curl of **F**:

$$\nabla \times \mathbf{F} = \begin{vmatrix} \mathbf{i} & \mathbf{j} & \mathbf{k} \\ \frac{\partial}{\partial x} & \frac{\partial}{\partial y} & \frac{\partial}{\partial z} \\ A & B & 0 \end{vmatrix} = \left(\frac{\partial B}{\partial x} - \frac{\partial A}{\partial y}\right)\mathbf{k}.$$

Therefore the integral on the right side of (4) can be written as

$$\int_{\mathcal{A}} (\nabla \times \mathbf{F}) \cdot \mathbf{k}\,dA.$$

This shows that the circulation of **F** around a curve in the xy plane can be expressed in terms of an integral involving the curl of **F** over the region bounded by the curve. This result is known as **Stokes' theorem in the plane**. In the next chapter we extend Stokes' theorem to closed curves in space.

Theorem *Stokes' theorem in the plane*. Let C be a closed curve in the xy plane bounding a region $\mathcal{A}$. Let **F** be a vector field defined on $\mathcal{A}$. Then

$$\oint_C \mathbf{F} \cdot d\mathbf{r} = \int_{\mathcal{A}} (\nabla \times \mathbf{F}) \cdot \mathbf{k}\,dA, \qquad \text{where } C \text{ is counterclockwise.}$$

Green's theorem concerns the normal component of **F**, $\mathbf{F} \cdot \mathbf{n}$. Stokes' theorem concerns the tangential component $\mathbf{F} \cdot \mathbf{T}$. (Recall that $\mathbf{F} \cdot d\mathbf{r}$ denotes $\mathbf{F} \cdot \mathbf{T}\,ds$.)

Just as there is a two-curve version of Green's theorem, there is a similar one for Stokes' theorem. If it is assumed that $\mathbf{curl}\ \mathbf{F} = \mathbf{0}$, we have the following corollary:

Corollary 3 Let C_1 and C_2 be two closed curves taken counterclockwise and **F** a vector field. Assume that together they form the boundary of a region $\mathcal{A}$ in which $\mathbf{curl}\ \mathbf{F} = \mathbf{0}$. Then

$$\oint_{C_1} \mathbf{F} \cdot d\mathbf{r} = \oint_{C_2} \mathbf{F} \cdot d\mathbf{r}.$$

When **curl F** = **0**, Corollary 3 permits us to replace $\oint_{C_1} \mathbf{F} \cdot d\mathbf{r}$ by $\oint_{C_2} \mathbf{F} \cdot d\mathbf{r}$, where C_2 is more convenient. (Exercise 11 illustrates this.) Contrast this with Corollary 1, which permits us to replace $\oint_{C_1} \mathbf{F} \cdot \mathbf{n}\, ds$ by $\oint_{C_2} \mathbf{F} \cdot \mathbf{n}\, ds$ if div $\mathbf{F} = 0$.

Why Curl Is Called Curl

Let C be a small closed curve in the plane and let $\mathbf{F}$ describe fluid flow. Then $\oint_C \mathbf{F} \cdot d\mathbf{r}$, where C is counterclockwise, is a measure of the circulation along C. If this integral is positive, the fluid in the region $\mathcal{A}$ bounded by C tends to have a counterclockwise motion. If the integral is negative, there is a clockwise tendency. The larger the integral, the larger is the tendency of the fluid to have a rotational or spinning motion.

Stokes' theorem says that

$$\oint_C \mathbf{F} \cdot d\mathbf{r} = \int_{\mathcal{A}} (\nabla \times \mathbf{F}) \cdot \mathbf{k}\, dA.$$

Thus the curl, $\nabla \times \mathbf{F}$, provides information about fluid motion near P. The norm of $\nabla \times \mathbf{F}$, $\|\nabla \times \mathbf{F}\|$, is a measure of how rapidly the fluid near P tends to rotate. (In Chap. 17 we will see the significance of the direction of $\nabla \times \mathbf{F}$.) For this reason a field whose curl is **0** everywhere is called **irrotational**.

This is a good time to stop and think about vector notation. Symbols such as $\mathbf{B}$, $\mathbf{A} \times \mathbf{B}$, ∇f, $\mathbf{F}$, and $\nabla \cdot \mathbf{F}$ are easy to write and to read. We represent them in components only if we need to, such as when proving a theorem about them.

We may now be in a position to appreciate these remarks of the physicists James C. Maxwell and Oliver Heaviside early in the development of vector notation:

> The doctrine of vectors is a method of thinking, a method of saving thought. It calls upon us at every step to form a mental image of the geometrical figures represented by the symbols, so that in studying geometry by this method we have our minds engaged with geometrical ideas, and are not permitted to fancy ourselves geometers when we are only arithmeticians.
>
> Maxwell, *Nature* (1877)

> Ignorant men have long been in advance of the learned about vectors. Ignorant people, like Faraday, naturally think in vectors. They may know nothing of their formal manipulation, but if they think of vectors, they think of them *as* vectors, that is, directed magnitudes. No ignorant man could or would think about the three components of a vector separately, and disconnected from one another. For general purposes of reasoning the manipulation of the scalar components instead of the vector itself is entirely wrong.
>
> O. Heaviside, *Electromagnetic Theory* (1893)

In a letter to the mathematician Tait written on November 7, 1870, Maxwell offered some names for $\nabla \times \mathbf{F}$:

> Here are some rough-hewn names. Will you like a good Divinity shape their ends properly so as to make them stick? . . .
>
> The vector part $[\nabla \times \mathbf{F}]$ I would call the twist of the vector function. Here the

> word twist has nothing to do with a screw or helix. [T]he word *turn* . . . would be better than twist, for twist suggests a screw. Twirl is free from the screw motion and is sufficiently racy. Perhaps it is too dynamical for pure mathematicians, so for Cayley's sake I might say Curl (after the fashion of Scroll).
>
> His last suggestion, "curl," has stuck.

Section Summary

We applied a two-curve version of Green's theorem to divergence-free fields **F**. Then we showed how Green's theorem translated a physical situation, namely a temperature distribution, into a mathematical equation involving the Laplacian. Finally, we translated Green's theorem into a theorem concerning circulation, $\oint_C \mathbf{F} \cdot d\mathbf{r}$ (Stokes' theorem in the plane). In this form it involves **curl F**.

EXERCISES FOR SEC. 16.5: APPLICATIONS OF GREEN'S THEOREM

In Exercises 1 and 2 verify Green's theorem in its two-curve form.

1 $\mathbf{F} = x\mathbf{i} + y^2\mathbf{j}$ and $\mathcal{A}$ is the region between the circles of radii 1 and 2 with centers at the origin.

2 $\mathbf{F} = 3x^2\mathbf{i} + y\mathbf{j}$ and $\mathcal{A}$ is the region between the circle with equation $x^2 + y^2 = 1$ and the square with vertices (2, 2), (−2, 2), (−2, −2), and (2, −2).

In Exercises 3 and 4 evaluate $\oint_C \mathbf{F} \cdot \mathbf{n}\, ds$ for the given curves if **F** is *divergence-free,* but not defined at the indicated points.

3 Given that $\oint_{C_1} \mathbf{F} \cdot \mathbf{n}\, ds = 5$, find $\oint_{C_2} \mathbf{F} \cdot \mathbf{n}\, ds$ and $\oint_{C_3} \mathbf{F} \cdot \mathbf{n}\, ds$, where the curves are shown in Fig. 12.

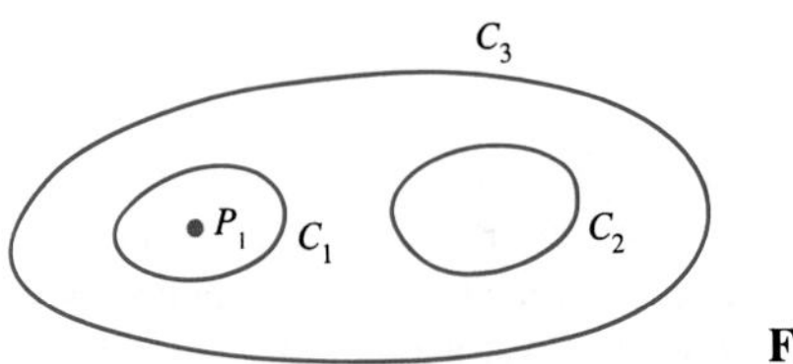

Figure 12

4 Given that $\oint_{C_1} \mathbf{F} \cdot \mathbf{n}\, ds = 3$ and $\oint_{C_2} \mathbf{F} \cdot \mathbf{n}\, ds = 4$, find $\oint_{C_3} \mathbf{F} \cdot \mathbf{n}\, ds$ and $\oint_{C_4} \mathbf{F} \cdot \mathbf{n}\, ds$, where the curves are shown in Fig. 13.

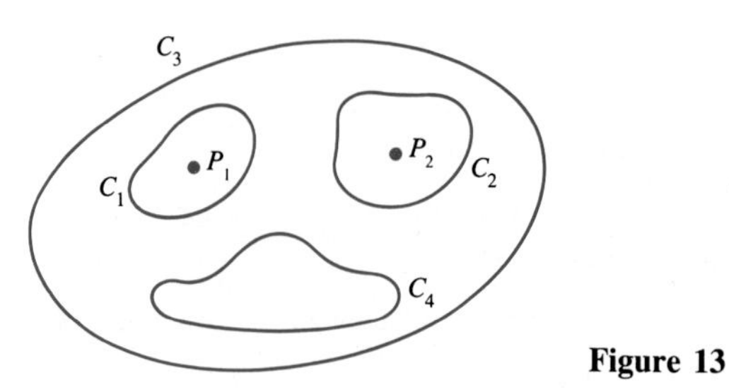

Figure 13

5 The vector field $\mathbf{F} = \hat{\mathbf{r}}/\|\mathbf{r}\|$, defined at every point in the plane except the origin, has divergence 0.

(*a*) Evaluate $\oint_C \mathbf{F} \cdot \mathbf{n}\, ds$ for a circle C of radius a and center (0, 0).

(*b*) The answer in (*a*) is not 0. Does that violate Green's theorem?

(*c*) The answer in (*a*) does not depend on the radius a. How could you have predicted this without any computations of integrals?

6 Let $\mathbf{F} = y\mathbf{i}/(x^2 + y^2)^3 - x\mathbf{j}/(x^2 + y^2)^3$.

(*a*) Compute $\nabla \cdot \mathbf{F}$.

(*b*) Evaluate $\oint_C \mathbf{F} \cdot \mathbf{n}\, ds$ over each of the curves C_1, C_2, and C_3 in Fig. 14 as easily as possible.

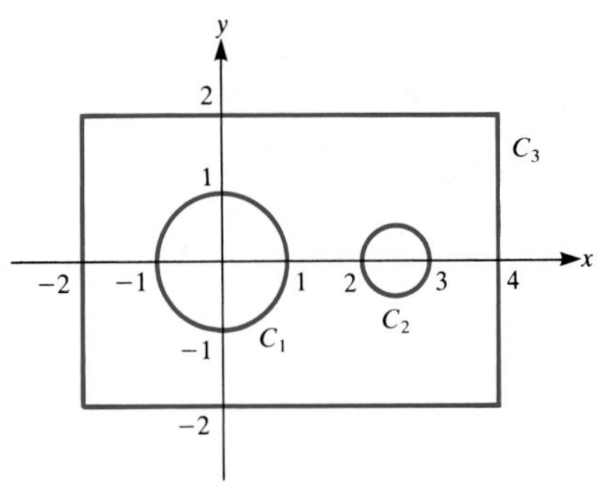

Figure 14

In Exercises 7 and 8 verify Stokes' theorem in the plane for the given **F**, where $\mathcal{A}$ is the region bounded by the curve C.

7 $\mathbf{F}(x, y) = 2x\mathbf{i}$ and C is the circle, $\mathbf{G}(\theta) = 3\cos\theta\,\mathbf{i} + 3\sin\theta\,\mathbf{j}$, θ in $[0, 2\pi]$.

8 $\mathbf{F}(x, y) = x\mathbf{i} + y\mathbf{j}$ and C is the ellipse, $\mathbf{G}(t) = a\cos t\,\mathbf{i} + b\sin t\,\mathbf{j}$, t in $[0, 2\pi]$.

9 Let **F** be a field defined everywhere except at (1, 2). As-

sume that the curl of **F** is **0**. What, if anything, can be said about the circulation of **F** around the three curves in Fig. 15?

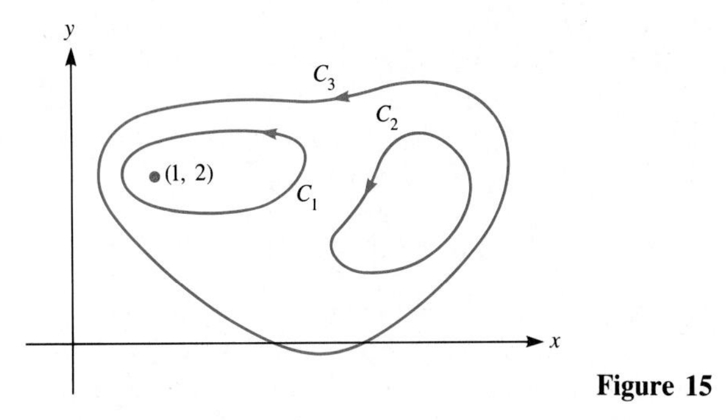

Figure 15

10 The curl of **F** is $(2x + 3xy)\mathbf{k}$. Evaluate $\oint_C \mathbf{F} \cdot d\mathbf{r}$ where C is the curve bounding the triangle with vertices (0, 0), (2, 0), and (1, 1).

11 Let $\mathbf{F}(x, y) = \hat{\mathbf{r}}/\|\mathbf{r}\|^3$, where $\mathbf{r} = x\mathbf{i} + y\mathbf{j}$.
 (*a*) What is **curl F**?
 (*b*) Evaluate $\oint_{C_1} \mathbf{F} \cdot d\mathbf{r}$, where C_1 is the circle of radius 1 and center (0, 0), counterclockwise.
 (*c*) Evaluate $\oint_{C_2} \mathbf{F} \cdot d\mathbf{r}$, where C_2 is the ellipse with equation $x^2/4 + y^2/9 = 1$, counterclockwise.

12 Let **F** be a vector field in the plane. Assume that the curl of **F** at (1, 2) is 5**k**. Use this fact to estimate the circulation of **F** around a very short closed curve C that has length L, encloses the point (1, 2), and is the boundary of a region $\mathscr{A}$ whose area is A.

13 Deduce the two-curve version of Green's theorem from the one-curve version, with the aid of the cancellation principle. *Hint:* Draw the extra lines shown in Fig. 16.

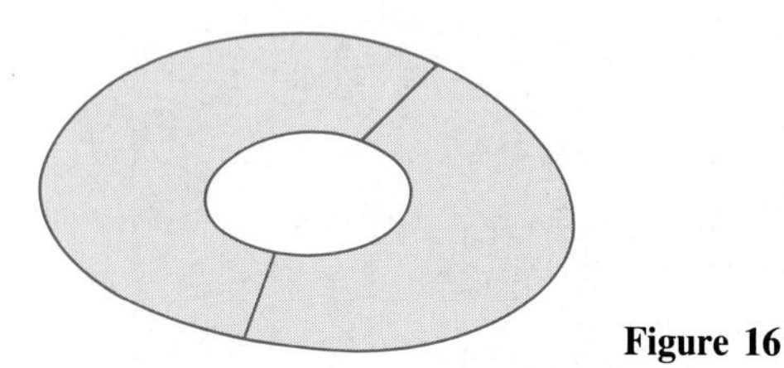

Figure 16

14 Use Green's theorem and the cancellation principle to extend Green's theorem to a region with two holes.

15 Express Stokes' theorem in the plane in words, using as few math symbols as you can.

16 Write up a complete proof of the two-curve version of Green's theorem using the method of our first proof of Green's theorem.

17 In Sec. 16.3 $\int_C (\hat{\mathbf{r}}/\|\mathbf{r}\|) \cdot \mathbf{n}\, ds$ was shown to represent the angle subtended by C at the origin. Using Green's theorem show that
 (*a*) If C encloses the origin, then the integral is 2π.
 (*b*) If C does not enclose the origin, the integral is 0.

18 Let $f(x)$ be a continuous function defined at least on some open interval (c, d). Assume that $\int_a^b f(x)\, dx = 0$ whenever $[a, b]$ is contained in (c, d). Use the first form of the fundamental theorem of calculus to show that $f(x) = 0$ for all x in (c, d).

19 Obtain the result in Exercise 18 *without* using the fundamental theorem of calculus, as follows: Assume that it is *not* the case that $f(x) = 0$ for all x in (c, d). Then there is a number x_0 in (c, d) such that $f(x_0) \neq 0$. Say that $f(x_0) > 0$. Since $f(x)$ is continuous, there is an interval $[e, f]$ around x_0 such that $f(x) > f(x_0)/2$. (Why?) But then $\int_e^f f(x)\, dx$ is not 0. (Why?) This is a contradiction. (Why?) A similar argument holds for the assumption that $f(x_0) < 0$. (Carry it out.) Thus, if $\int_a^b f(x)\, dx = 0$ for all intervals $[a, b]$ in (c, d), it follows that $f(x) = 0$ for all x in (c, d).

20 (See Exercise 19.) Let f be defined throughout some plane region $\mathscr{B}$. Assume that $\int_{\mathscr{A}} f\, dA = 0$ for all regions $\mathscr{A}$ in $\mathscr{B}$. Prove that $f(P) = 0$ for all P in $\mathscr{B}$. Complete the argument in the text by treating the case $f(P_0) < 0$.

21 Let **F** be a vector field with the property that for any closed curve C,

$$\oint_C \mathbf{F} \cdot \mathbf{n}\, ds = 0.$$

What, if anything, can we conclude about the divergence of **F**?

22 The moment of inertia about the z axis of a region $\mathscr{A}$ in the xy plane is defined as the integral $\int_{\mathscr{A}} (x^2 + y^2)\, dA$. Show that this equals

$$\frac{1}{3} \oint_C (-y^3\, dx + x^3\, dy),$$

where C bounds $\mathscr{A}$.

---

23 Let $\mathbf{F} = \sigma \mathbf{v}$ describe a fluid flow. Here σ is the density of the fluid and **v** the velocity. Both σ and **v** are functions of position (x, y) and time t. Let $\mathscr{A}$ be a fixed region with boundary C. Let $m(t)$ be the total mass of the fluid in $\mathscr{A}$ at time t.
 (*a*) Express $m(t)$ as an integral over $\mathscr{A}$ in terms of σ.
 (*b*) Express dm/dt as an integral over C.
 (*c*) Express dm/dt as an integral over $\mathscr{A}$ in terms of **F**.
 (*d*) Note that

$$\frac{m(t + \Delta t) - m(t)}{\Delta t} = \int_{\mathscr{A}} \frac{\sigma(x, y, t + \Delta t) - \sigma(x, y, t)}{\Delta t}\, dA.$$

Why does this suggest that

$$\frac{dm}{dt} = \int_{\mathscr{A}} \frac{\partial \sigma}{\partial t}\, dA?$$

(For a proof of this type of theorem see Appendix K.)
 (*e*) Deduce that

$$\frac{\partial \sigma}{\partial t} = -\nabla \cdot \mathbf{F}.$$

(This is called the "equation of continuity" in the study of motion of fluids.)

(*f*) Use physical intuition to argue that the equation in (*e*) is to be expected. (Think of a small patch $\mathscr{A}$ around the point where the functions are evaluated.)

24 (*a*) Show that the curve parameterized by

$$\mathbf{G}(t) = \frac{e^t + e^{-t}}{2}\mathbf{i} + \frac{e^t - e^{-t}}{2}\mathbf{j} = \cosh t\ \mathbf{i} + \sinh t\ \mathbf{j},$$

for t in $[0, a]$, $a > 0$, lies on the hyperbola $x^2 - y^2 = 1$ and joins the point $A = (1, 0)$ to the point $B = (\cosh a, \sinh a)$.

(*b*) Let $\mathscr{A}$ be the region bounded by the line segment OA, the line segment OB, and the curve in (*a*) joining A and B. (See Fig. 17.) Show that the area of $\mathscr{A}$ is $a/2$.

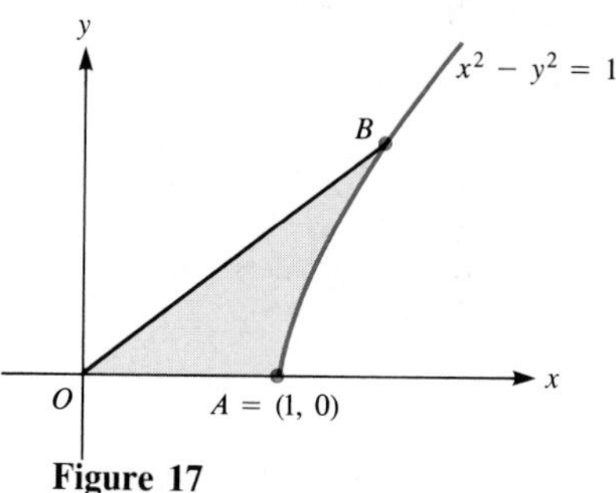

Figure 17

Part (*b*) brings out another analogy between the circular and the hyperbolic functions. An angle θ corresponds to a sector of area $\theta/2$ in the unit circle. So $\cos\theta$ and $\sin\theta$ could be defined as the coordinates of the point on $x^2 + y^2 = 1$ corresponding to the sector of area $\theta/2$ one arm of which lies on the positive x axis. Similarly $\cosh a$ and $\sinh a$ could be defined as the coordinates of the point on $x^2 - y^2 = 1$ corresponding to a "sector" of area $a/2$ as described in part (*b*).

25 (*a*) Find the curl of $\mathbf{F} = \hat{\mathbf{r}}/\|\mathbf{r}\|^3$.

(*b*) Find the circulation of $\mathbf{F}$ around the curves C_1 and C_2 in Fig. 18 as easily as possible.

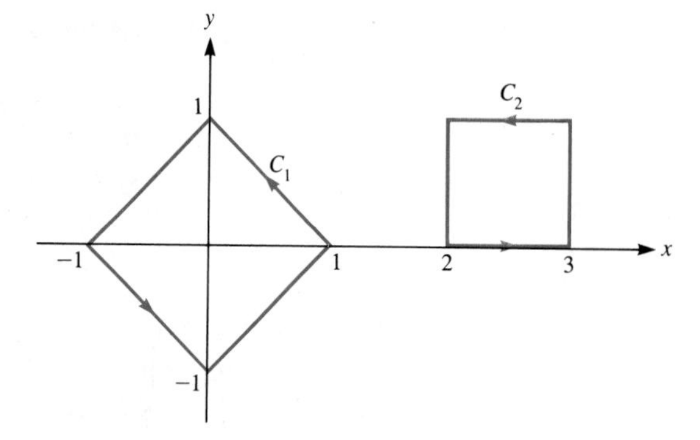

Figure 18

26 Prove Stokes' theorem in the plane without using Green's theorem (imitate the proof of Green's theorem, starting with $\oint_C \mathbf{F} \cdot d\mathbf{r}$ and using the cancellation principle).

27 What would the "two-curve" version of Stokes' theorem in the plane be?

28 Figure 19 shows three vector fields. One is irrotational, the other two are not. Decide which two are not, copy them onto drawing paper, and sketch a closed curve C for which $\oint_C \mathbf{F} \cdot d\mathbf{r}$ is not 0. (Explain why the integral is not 0.)

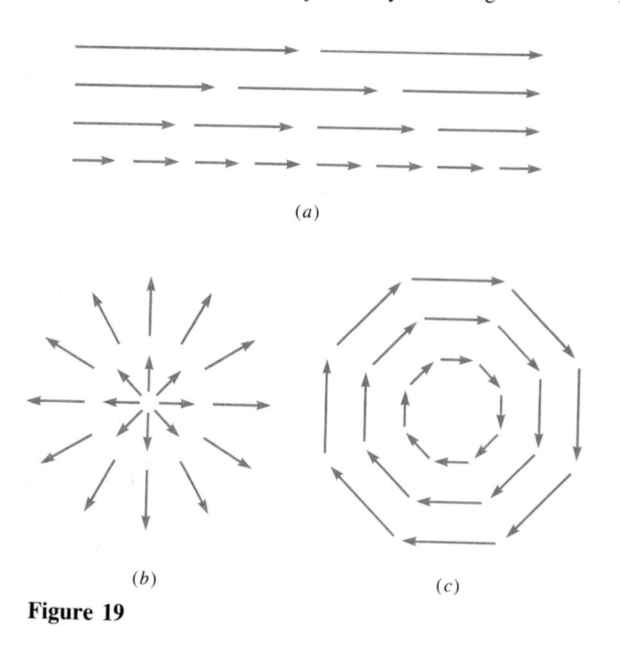

Figure 19

16.6 CONSERVATIVE VECTOR FIELDS

For some vector fields $\mathbf{F}$ the line integral $\int_C \mathbf{F} \cdot d\mathbf{r}$ depends only on the end-points of C. Such fields, which are important in physics and thermodynamics, are the subject of this section. Perhaps the best-known example in the physical world is gravity. (Keep in mind that $\int_C \mathbf{F} \cdot \mathbf{T}\, ds$ is often written as $\int_C \mathbf{F} \cdot d\mathbf{r}$; we will use both notations.)

Definition A vector field $\mathbf{F}$ defined in some planar or spatial region is called **conservative** if

$$\int_{C_1} \mathbf{F} \cdot \mathbf{T}\, ds = \int_{C_2} \mathbf{F} \cdot \mathbf{T}\, ds$$

whenever C_1 and C_2 are any two curves in the region with the same initial and terminal points.

Another way to view "conservative"

An equivalent definition of a conservative vector field $\mathbf{F}$ is that for any closed curve C in the region, $\oint_C \mathbf{F} \cdot \mathbf{T}\, ds = 0$. This alternative viewpoint is justified by the following theorem.

Theorem 1 A vector field $\mathbf{F}$ is conservative if and only if $\oint_C \mathbf{F} \cdot \mathbf{T}\, ds = 0$ for every closed curve in the region where $\mathbf{F}$ is defined.

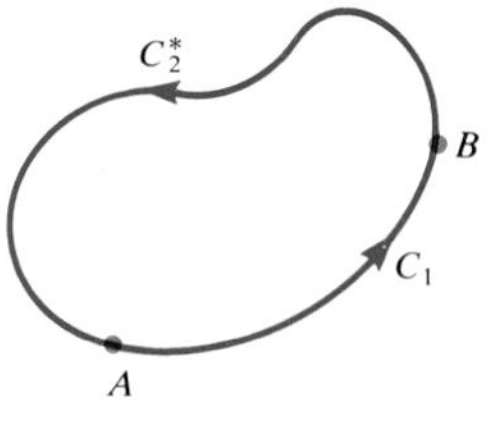

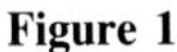
Figure 1

Proof Assume that $\mathbf{F}$ is conservative and let C be a closed curve that starts and ends at the point A. Pick a point B on the curve and break C into two curves: C_1 from A to B and C_2^* from B to A, as indicated in Fig. 1.

Let C_2 be the curve C_2^* traversed in the opposite direction, from A to B. Then

Note the sign change.

$$\oint_C \mathbf{F} \cdot \mathbf{T}\, ds = \int_{C_1} \mathbf{F} \cdot \mathbf{T}\, ds + \int_{C_2^*} \mathbf{F} \cdot \mathbf{T}\, ds$$

$$= \int_{C_1} \mathbf{F} \cdot \mathbf{T}\, ds - \int_{C_2} \mathbf{F} \cdot \mathbf{T}\, ds = 0,$$

since $\mathbf{F}$ is conservative.

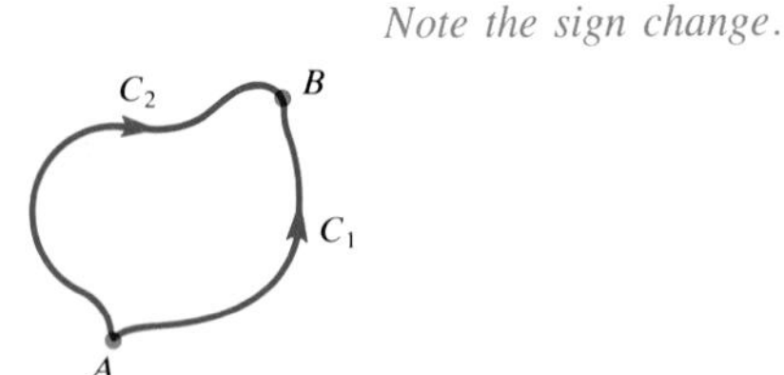

Figure 2

On the other hand, assume that $\mathbf{F}$ has the property that $\oint_C \mathbf{F} \cdot \mathbf{T}\, ds = 0$ for any closed curve C in the region. Let C_1 and C_2 be two curves in the region, starting at A and ending at B. Let C_2^* be C_2 taken in the reverse direction. (See Figs. 2 and 3.) Then C_1 followed by C_2^* is a closed curve C from A back to A. Thus

$$0 = \oint_C \mathbf{F} \cdot \mathbf{T}\, ds = \int_{C_1} \mathbf{F} \cdot \mathbf{T}\, ds + \int_{C_2^*} \mathbf{F} \cdot \mathbf{T}\, ds$$

$$= \int_{C_1} \mathbf{F} \cdot \mathbf{T}\, ds - \int_{C_2} \mathbf{F} \cdot \mathbf{T}\, ds.$$

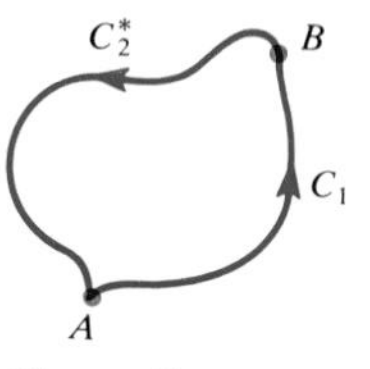

Figure 3

Consequently,
$$\int_{C_1} \mathbf{F} \cdot \mathbf{T}\, ds = \int_{C_2} \mathbf{F} \cdot \mathbf{T}\, ds.$$

This concludes both directions of the argument. ■

The criteria mentioned so far for determining that a vector field **F** is conservative are hardly practical. They require the evaluation of an infinite number of integrals. The next theorem provides a way of constructing conservative vector fields. Before discussing it, we mention two conditions that will be assumed about the domains of vector fields of interest.

An arcwise-connected region comes in a single piece.

First, let A and B be any two points in the region where the vector field **F** is defined. It will be assumed that there is a curve C lying completely in the region that joins A to B. Such a region is called **arcwise connected**. The region in Fig. 4 is arcwise connected, but the two-piece region shaded in Fig. 5 is not.

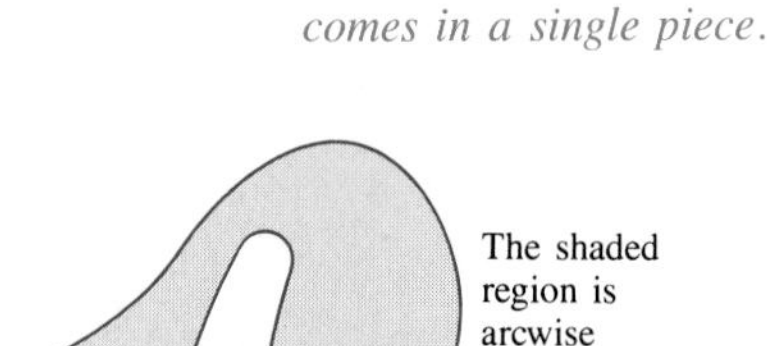

Figure 4

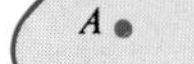

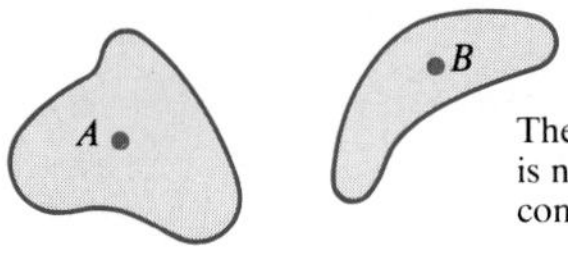

The shaded region is not arcwise connected.

Figure 5

Gradient Fields Are Conservative

The fundamental theorem of calculus asserts that $\int_a^b f'(x)\,dx = f(b) - f(a)$. The next theorem asserts that $\int_C \nabla f \cdot d\mathbf{r} = f(B) - f(A)$, where f is a function of two or three variables and C is a curve from A to B. Because of its resemblance to the fundamental theorem of calculus, Theorem 2 is sometimes called the fundamental theorem of vector fields.

Any vector field that is the gradient of a scalar field turns out to be conservative. That is the substance of Theorem 2.

Theorem 2 Let f be a scalar field defined in some region in the plane or in space. Then the gradient field $\mathbf{F} = \nabla f$ is conservative. In fact, for any points A and B in the region and any curve C from A to B in the region,

$$\int_C \nabla f \cdot d\mathbf{r} = f(B) - f(A).$$

Proof For simplicity take the planar case. Let C be given by the parameterization $\mathbf{r} = \mathbf{G}(t)$ for t in $[a, b]$. Let $\mathbf{G}(t) = x(t)\mathbf{i} + y(t)\mathbf{j}$. Then,

$$\int_C \nabla f \cdot d\mathbf{r} = \int_C \left(\frac{\partial f}{\partial x}\,dx + \frac{\partial f}{\partial y}\,dy \right)$$

$$= \int_a^b \left(\frac{\partial f}{\partial x}\frac{dx}{dt} + \frac{\partial f}{\partial y}\frac{dy}{dt} \right) dt.$$

The integrand $(\partial f/\partial x)(dx/dt) + (\partial f/\partial y)(dy/dt)$ is reminiscent of the chain rule in Sec. 14.6. To be specific, if we introduce the function H defined by the formula

$$H(t) = f(x(t), y(t)),$$

then the chain rule gives

$$\frac{dH}{dt} = \frac{\partial f}{\partial x}\frac{dx}{dt} + \frac{\partial f}{\partial y}\frac{dy}{dt}.$$

Thus
$$\int_a^b \left(\frac{\partial f}{\partial x}\frac{dx}{dt} + \frac{\partial f}{\partial y}\frac{dy}{dt} \right) dt = \int_a^b \frac{dH}{dt}\,dt = H(b) - H(a)$$

by the fundamental theorem of calculus. But

$$H(b) = f(x(b), y(b)) = f(B)$$

and $$H(a) = f(x(a), y(a)) = f(A).$$

Consequently, $$\int_C \nabla f \cdot d\mathbf{r} = f(B) - f(A),$$

So a gradient field is conservative.

and the theorem is proved. ■

In differential form Theorem 2 reads

$$\boxed{\int_C \left(\frac{\partial f}{\partial x}\,dx + \frac{\partial f}{\partial y}\,dy\right) = f(B) - f(A)} \tag{1}$$

or even $$\int_C df = f(B) - f(A),$$

where A is the start and B is the finish of C. To put it another way, if $P = \partial f/\partial x$ and $Q = \partial f/\partial y$, then

$$\int_C (P\,dx + Q\,dy) = f(B) - f(A).$$

For example, consider $\int_C x\,dx$ where C goes from $A = (a_1, a_2)$ to $B = (b_1, b_2)$. In this case the function $f(x, y) = x^2/2$ has the property that

$$df = \frac{\partial f}{\partial x}\,dx + \frac{\partial f}{\partial y}\,dy = x\,dx + 0\,dy = x\,dx.$$

Thus $$\int_C df = \int_C x\,dx = f(B) - f(A) = \frac{b_1^2}{2} - \frac{a_1^2}{2}.$$

In particular, if C is closed, $a_1 = b_1$, and $\oint_C x\,dx = 0$, a result obtained in Sec. 16.2. Theorem 2 also gives another reason why $\oint_C y\,dy = 0$ and $\oint_C k\,dx = 0$ for any constant k.

When the curve C is closed, Theorem 2 takes the following form.

Theorem 3 If f is a scalar function defined throughout some region, then the integral of the tangential component of the vector field ∇f around any closed curve in that region is 0, that is

$$\oint_C \nabla f \cdot d\mathbf{r} = 0.$$

Proof This is a consequence of Theorem 2. For in the case of a closed path, the initial point A coincides with the terminal point B. Hence $f(B) - f(A) = 0$. This proves the theorem. ■

EXAMPLE 1 Evaluate $\oint_C (y\,dx + x\,dy)$ around a closed curve C taken counterclockwise.

SOLUTION We will give two solutions.

In Sec. 16.2 it was shown that $\oint_C x\,dy = A$ and $\oint_C y\,dx = -A$, where A is the area of the region enclosed by C. Thus,

$$\oint_C (y\,dx + x\,dy) = -A + A = 0.$$

A second solution uses Theorem 3. Note that

$$\nabla(xy) = \frac{\partial(xy)}{\partial x}\mathbf{i} + \frac{\partial(xy)}{\partial y}\mathbf{j} = y\mathbf{i} + x\mathbf{j}.$$

Hence
$$\oint_C (y\,dx + x\,dy) = \oint_C \nabla(xy) \cdot d\mathbf{r}.$$

By Theorem 3, the integral is 0. ■

In the theory of gravitational attraction, the scalar function f defined by

$$f(x, y, z) = \frac{-1}{\sqrt{x^2 + y^2 + z^2}}$$

is of great importance. It is defined everywhere except at the origin and is called a **potential function**. The negative of its gradient ∇f is called the **force field**. A direct calculation shows that

$$-\nabla f = -\frac{x\mathbf{i} + y\mathbf{j} + z\mathbf{k}}{(\sqrt{x^2 + y^2 + z^2})^3} = -\frac{\mathbf{r}}{\|\mathbf{r}\|^3} = -\frac{\hat{\mathbf{r}}}{\|\mathbf{r}\|^2},$$

where $\mathbf{r}$ is the position vector $x\mathbf{i} + y\mathbf{j} + z\mathbf{k}$. Since

$$-\nabla f = -\frac{\hat{\mathbf{r}}}{\|\mathbf{r}\|^2},$$

$-\nabla f$ is pointed toward the origin, and its magnitude is inversely proportional to $\|\mathbf{r}\|^2$. (That is, the "force field" is nothing more than that described by the inverse-square law.)

Theorem 2 tells the physicist that the work accomplished by the gravitational field when moving a particle from one point to another is independent of the path. Theorem 3 says that the work accomplished by gravity on a satellite during one orbit is 0. (It is this fact that permits the satellite to remain in orbit indefinitely.)

The question may come to mind, "*If* $\mathbf{F}$ *is conservative, is it necessarily the gradient of some scalar function?*" The answer is "yes." This is the substance of the next theorem.

Theorem 4 Let $\mathbf{F}$ be a conservative vector field defined in some arcwise-connected region in the plane (or in space). Then there is a scalar function f defined in that region such that $\mathbf{F} = \nabla f$.

Proof Consider the case when $\mathbf{F}$ is planar, $\mathbf{F} = P(x, y)\mathbf{i} + Q(x, y)\mathbf{j}$. (The case where $\mathbf{F}$ is defined in space is similar.) For convenience, assume that the region contains the origin. Define a scalar function f as follows: Let (x, y) be a point in the region. Select a curve C in the region that starts at $(0, 0)$ and ends at (x, y).

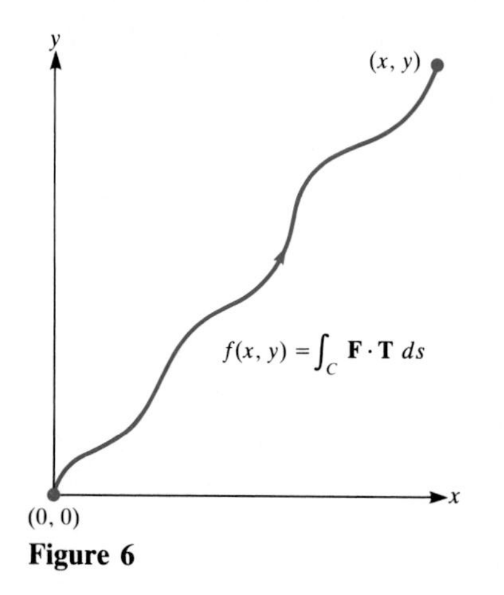

Figure 6

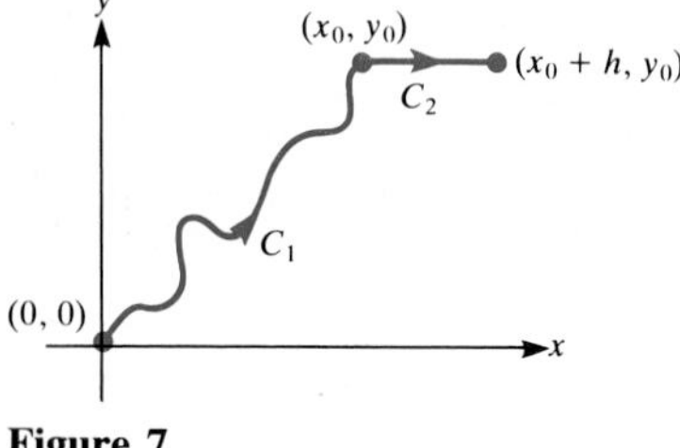

Figure 7

Define $f(x, y)$ to be $\int_C \mathbf{F} \cdot \mathbf{T}\, ds$. Since $\mathbf{F}$ is conservative, the number $f(x, y)$ depends only on the point (x, y) and not on the choice of C. (See Fig. 6.)

All that remains is to show that $\nabla f = \mathbf{F}$; that is, $\partial f/\partial x = P$ and $\partial f/\partial y = Q$. We will go through the details for the first case, $\partial f/\partial x = P$. The reasoning for the other is similar.

Let (x_0, y_0) be a fixed point in the region and consider the difference quotient whose limit is $\partial f/\partial x(x_0, y_0)$, namely,

$$\frac{f(x_0 + h, y_0) - f(x_0, y_0)}{h},$$

for h small enough so that $(x_0 + h, y_0)$ is also in the region.

Let C_1 be a curve from $(0, 0)$ to (x_0, y_0) and let C_2 be the straight path from (x_0, y_0) to $(x_0 + h, y_0)$. (See Fig. 7.) Let C be the curve from $(0, 0)$ to the point $(x_0 + h, y_0)$ formed by taking C_1 first and then continuing on C_2. Then

$$f(x_0, y_0) = \int_{C_1} \mathbf{F} \cdot \mathbf{T}\, ds,$$

and

$$f(x_0 + h, y_0) = \int_C \mathbf{F} \cdot \mathbf{T}\, ds = \int_{C_1} \mathbf{F} \cdot \mathbf{T}\, ds + \int_{C_2} \mathbf{F} \cdot \mathbf{T}\, ds.$$

Thus

$$\frac{f(x_0 + h, y_0) - f(x_0, y_0)}{h} = \frac{\int_{C_2} \mathbf{F} \cdot \mathbf{T}\, ds}{h} = \frac{\int_{C_2} (P(x, y)\, dx + Q(x, y)\, dy)}{h}.$$

But on C_2, y is constant, $y = y_0$; hence $dy = 0$. Thus $\int_{C_2} Q(x, y)\, dy = 0$. Also,

$$\int_{C_2} P(x, y)\, dx = \int_{x_0}^{x_0+h} P(x, y_0)\, dx.$$

Now,

$$\lim_{h \to 0} \frac{\int_{x_0}^{x_0+h} P(x, y_0)\, dx}{h}$$

is the derivative of $\int_{x_0}^{t} P(x, y_0)\, dx$ with respect to t at the value x_0. By the first fundamental theorem of calculus, this derivative is the value of the integrand when $x = x_0$. Consequently,

$$\frac{\partial f}{\partial x}(x_0, y_0) = P(x_0, y_0),$$

as was to be shown.

In a similar manner, we can show that

$$\frac{\partial f}{\partial y}(x_0, y_0) = Q(x_0, y_0). \quad \blacksquare$$

For a vector field $\mathbf{F}$ defined throughout some region in the plane (or space) the following three properties are therefore equivalent:

Three views of a conservative field

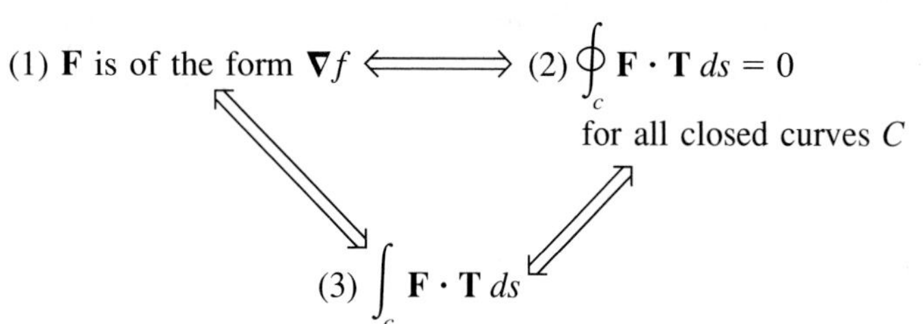

An arrow ⇒ means "implies." Any one of the three properties, (1), (2), or (3), describes a conservative field. In our approach, property (3) was used as the definition.

Conservative Fields and Curl

None of the three criteria offers a practical test to determine whether an arbitrary vector field is conservative. However, the next theorem does half the job.

Theorem 5 If **F** is a conservative vector field, then **curl F** = **0**.

The curl of a conservative field is **0**.

Proof If **F** is conservative, then it is of the form ∇f for some scalar function f. By Exercise 17 of Sec. 16.1, the curl of a gradient field is **0**. In other words, the curl of **F** is **0**. ■

According to Theorem 5, if $\mathbf{F}(x, y) = P(x, y)\mathbf{i} + Q(x, y)\mathbf{j}$ is a conservative vector field in the plane, then

$$\begin{vmatrix} \mathbf{i} & \mathbf{j} & \mathbf{k} \\ \dfrac{\partial}{\partial x} & \dfrac{\partial}{\partial y} & \dfrac{\partial}{\partial z} \\ P(x, y) & Q(x, y) & 0 \end{vmatrix} = \mathbf{0};$$

that is,

$$-\mathbf{i}\frac{\partial Q}{\partial z} + \mathbf{j}\frac{\partial P}{\partial z} + \mathbf{k}\left(\frac{\partial Q}{\partial x} - \frac{\partial P}{\partial y}\right) = \mathbf{0}.$$

Since Q and P are functions of x and y only, $\partial Q/\partial z$ and $\partial P/\partial z$ are both 0. Consequently, if $P(x, y)\mathbf{i} + Q(x, y)\mathbf{j}$ is conservative, then

$$\frac{\partial P}{\partial y} = \frac{\partial Q}{\partial x}. \tag{1}$$

Warning: The converse of Theorem 5 is false. A look at Appendix F would be timely.

All would be delightful if the converse of Theorem 5 were true. Unfortunately, it is not. There are vector fields **F** whose curls are **0** which are not conservative. Example 2 provides one such **F**.

EXAMPLE 2 Let $\mathbf{F} = \dfrac{-y\mathbf{i}}{x^2 + y^2} + \dfrac{x\mathbf{j}}{x^2 + y^2}$. Show that (*a*) $\nabla \times \mathbf{F} = \mathbf{0}$, but (*b*) **F** is not conservative.

SOLUTION

(*a*) To solve (*a*) it is necessary to show that **F** satisfies Eq. (1); that is,

$$\frac{\partial}{\partial y}\left(\frac{-y}{x^2+y^2}\right) = \frac{\partial}{\partial x}\left(\frac{x}{x^2+y^2}\right).$$

Straightforward computations show that both sides equal

$$\frac{y^2-x^2}{(x^2+y^2)^2}$$

and hence are equal.

(*b*) To show that **F** is *not* conservative, it suffices to exhibit a closed curve C such that $\oint_C \mathbf{F}\cdot d\mathbf{r}$ is not 0. As C we use the unit circle parameterized by

$$x = \cos\theta, \qquad y = \sin\theta, \qquad 0 \le \theta \le 2\pi.$$

Then
$$\oint_C \mathbf{F}\cdot d\mathbf{r} = \oint_C \left(\frac{-y\,dx}{x^2+y^2} + \frac{x\,dy}{x^2+y^2}\right)$$

$$= \int_0^{2\pi}\left[\frac{-\sin\theta\, d(\cos\theta)}{\cos^2\theta+\sin^2\theta} + \frac{\cos\theta\, d(\sin\theta)}{\cos^2\theta+\sin^2\theta}\right]$$

$$= \int_0^{2\pi}\frac{(\sin^2\theta+\cos^2\theta)\,d\theta}{\sin^2\theta+\cos^2\theta} = \int_0^{2\pi} d\theta = 2\pi.$$

This establishes (*b*). ■

Example 2 shows that even though **curl F** can be identically **0**, **F** need not be conservative. However, under suitable circumstances, if **curl F** = **0**, then **F** *is* conservative. The "suitable circumstances" concern the domain of **F**.

Definition A region $\mathscr{A}$ in the plane or in space is **simply connected** if all closed curves in $\mathscr{A}$ can be continuously shrunk down to a point in $\mathscr{A}$ while staying within $\mathscr{A}$.

A region in the plane bounded by a single curve is simply connected; a region bounded by at least two nonintersecting closed curves is not simply connected.

EXAMPLE 3 Which of these regions are simply connected?

(*a*) The plane without the origin
(*b*) The region between two concentric circles
(*c*) Space without the origin
(*d*) Space without the z axis
(*e*) The region between two concentric spheres

SOLUTION The regions are shown in Fig. 8. If a region is not simply connected, a curve is shown that cannot be shrunk to a point in the region.

(*a*) The curve C_1 *cannot* be shrunk to a point while still remaining in the region. The region is *not* simply connected.
(*b*) The curve C_2 *cannot* be shrunk to a point while still remaining in the region. The region is *not* simply connected.
(*c*) The deletion of the origin does not matter. Every closed curve can still be

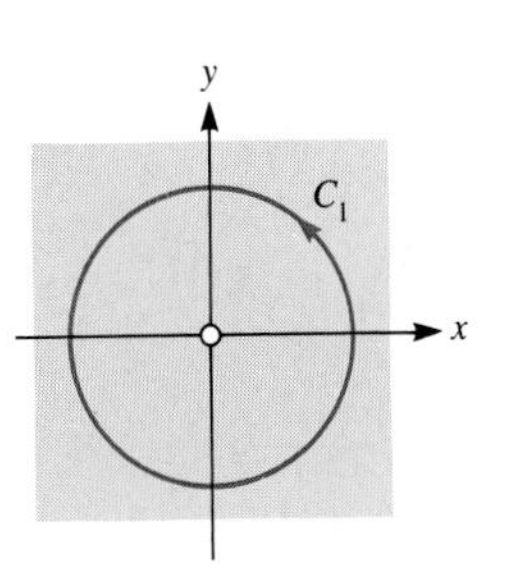

Plane without origin. Not simply connected.

(a)

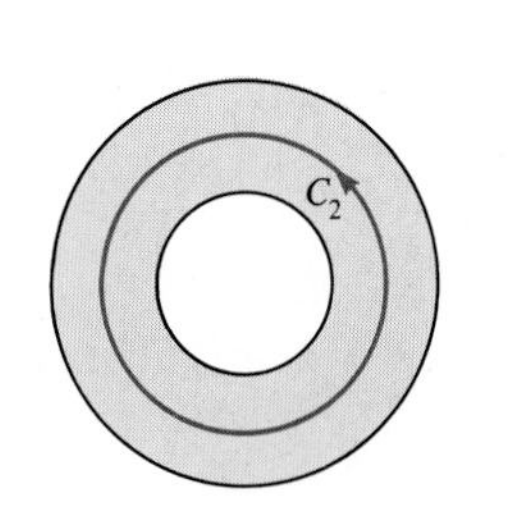

Band between two circles. Not simply connected.

(b)

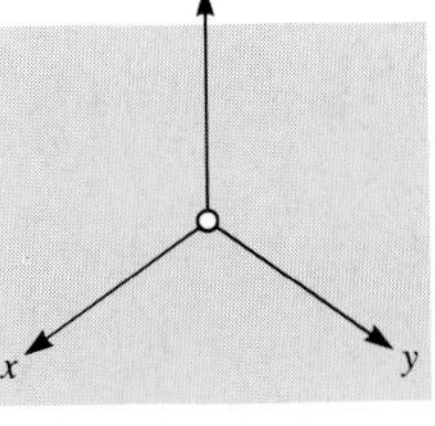

Space without origin. Simply connected.

(c)

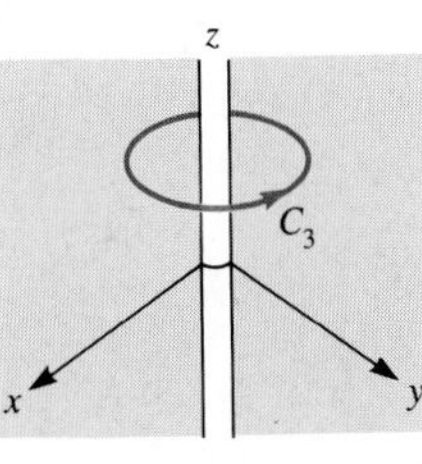

Space without z axis. Not simply connected.

(d)

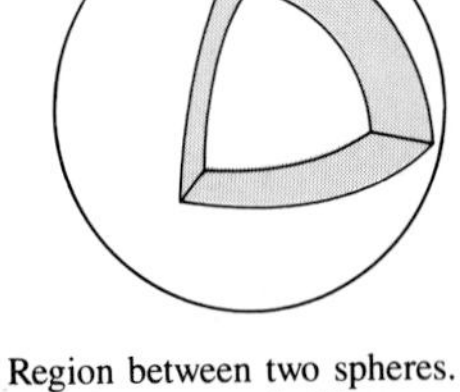
Region between two spheres. Simply connected. (Cutaway view)

(e)

Figure 8

shrunk to a point in the region while staying in the region. So "space minus a single point" is simply connected. [Contrast this with (*a*).]

(*d*) The curve C_3 cannot be shrunk to a single point without passing through the z axis. So "space minus the z axis" is *not* simply connected.

(*e*) The shell between the spheres is simply connected. [Contrast this with (*b*).] ■

In the plane "simply connected" is the same thing as "no holes"; but in space this is not true.

Now we are ready to state when **curl F** = **0** implies that **F** is conservative. We consider only **F** defined on the xy plane. (In Chap. 17 we consider **F** defined in space.)

Theorem 6 Let **F** be a vector field defined on a simply connected region in the plane. If **curl F** = **0**, then **F** is conservative.

Proof We wish to show that $\oint_C \mathbf{F} \cdot d\mathbf{r} = 0$ for every simple closed curve C in the region. (The result for more general closed curves is in advanced calculus.) Since C can be shrunk to a point while remaining in the region, the region $\mathcal{A}$ that C bounds has no holes. That is, C is the complete boundary of $\mathcal{A}$. By Stokes' theorem in the plane,

$$\oint_C \mathbf{F} \cdot d\mathbf{r} = \int_{\mathcal{A}} \mathbf{curl\ F} \cdot \mathbf{k}\, dA = \int_{\mathcal{A}} \mathbf{0} \cdot \mathbf{k}\, dA = 0.$$

This shows that **F** is conservative. ■

EXAMPLE 4 Let $\mathbf{F} = e^x y\mathbf{i} + (e^x + 2y)\mathbf{j}$.

(*a*) Show that **F** is conservative.

(*b*) Construct f such that $\mathbf{F} = \nabla f$.

SOLUTION

(*a*) The domain of **F** is the entire xy plane, which is simply connected. The curl of **F** is

$$\nabla \times \mathbf{F} = \begin{vmatrix} \mathbf{i} & \mathbf{j} & \mathbf{k} \\ \frac{\partial}{\partial x} & \frac{\partial}{\partial y} & \frac{\partial}{\partial z} \\ e^x y & e^x + 2y & 0 \end{vmatrix}$$

$$= \left(\frac{\partial(0)}{\partial y} - \frac{\partial}{\partial z}(e^x + 2y)\right)\mathbf{i} - \left(\frac{\partial(0)}{\partial x} - \frac{\partial}{\partial x}(e^x y)\right)\mathbf{j} + \left(\frac{\partial}{\partial x}(e^x + 2y) - \frac{\partial}{\partial y}(e^x y)\right)\mathbf{k}$$

$$= 0\mathbf{i} - 0\mathbf{j} + (e^x - e^x)\mathbf{k}$$

$$= \mathbf{0}.$$

Thus **F** is conservative.

(*b*) We demonstrate two different techniques. In the first, we use the construction that appeared in the proof of Theorem 4. Define $f(a, b)$ to equal $\int_C \mathbf{F} \cdot \mathbf{T}\, ds$, where C is any curve from $(0, 0)$ to (a, b). Any curve with the prescribed endpoints will do. For simplicity, choose C to be the curve that goes from $(0, 0)$ to (a, b) in a straight line. (See Fig. 9.) Then we can use x as a parameter, where $y = (b/a)x$ and $0 \le x \le a$. (If $a = 0$, we would use y as a parameter.) Then

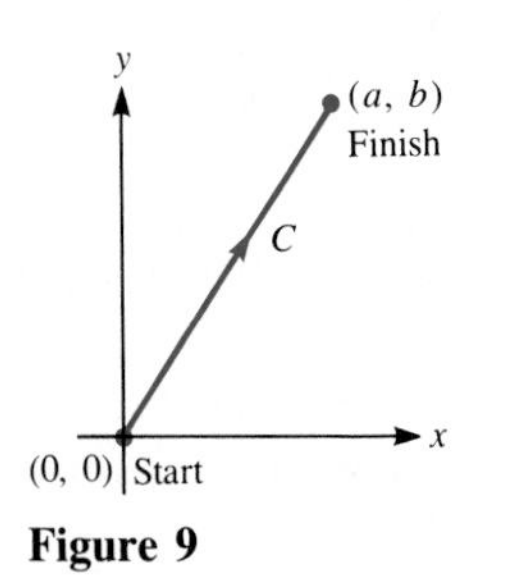

Figure 9

$$f(a, b) = \int_C [e^x y\, dx + (e^x + 2y)\, dy]$$

$$= \int_0^a \left[e^x \frac{b}{a} x\, dx + \left(e^x + 2\frac{b}{a} x\right)\frac{b}{a}\, dx\right]$$

$$= \frac{b}{a}\int_0^a \left(xe^x + e^x + 2\frac{b}{a} x\right) dx$$

$$= \frac{b}{a}\left[(x-1)e^x + e^x + \frac{b}{a}x^2\right]_0^a$$

$$= \frac{b}{a}\left[xe^x + \frac{b}{a}x^2\right]_0^a$$

$$= \frac{b}{a}[ae^a + ab]$$

$$= be^a + b^2.$$

$ye^x + y^2 + k$, for any constant k, also would be an answer.

Since $f(a, b) = be^a + b^2$, we see that $f(x, y) = ye^x + y^2$ is the desired function.

While the result is independent of the choice of C, the difficulty of the computations can vary significantly from path to path. In many cases it is convenient to break the integral into two straight-line segments: C_1 going from $(0, 0)$ to $(a, 0)$ and C_2 from $(a, 0)$ to (a, b). At the cost of doing two integrals, we get to have one variable held constant during each integration. (See Exercise 16.)

An alternate approach provides the same result without appealing directly to a line integral. Since we want $\nabla f = \mathbf{F}$, it follows that we must have

$$\frac{\partial f}{\partial x} = e^x y \qquad \text{and} \qquad \frac{\partial f}{\partial y} = e^x + 2y.$$

Integrating both sides of $\partial f/\partial x = e^x y$ with respect to x while holding y constant, we obtain

$$\int \frac{\partial f}{\partial x}\,dx = \int e^x y\,dx$$

$$f(x, y) = e^x y + C(y).$$

Notice that instead of writing the constant of integration as just C, we have to remember that it could possibly be a function of y. To find out what this function of y could be, we differentiate both sides of our result with respect to y (now holding x constant), to obtain

$$\frac{\partial f}{\partial y} = \frac{\partial}{\partial y}[e^x y + C(y)]$$

$$= e^x + C'(y).$$

But $\partial f/\partial y = e^x + 2y$, so we may write

$$e^x + C'(y) = e^x + 2y.$$

Hence $C'(y) = 2y$, $C(y) = y^2 + k$, for some constant k, and we have

$$f(x, y) = e^x y + y^2 + k,$$

as before. ■

A different view of a conservative vector field.

The vector field $\mathbf{F}(x, y) = P(x, y)\mathbf{i} + Q(x, y)\mathbf{j}$ is conservative if there is a scalar field $f(x, y)$ such that $\partial f/\partial x = P$ and $\partial f/\partial y = Q$. Thus the differential of f,

$$df = \frac{\partial f}{\partial x}\,dx + \frac{\partial f}{\partial y}\,dy,$$

coincides with the expression

$$P\,dx + Q\,dy.$$

The expression $P\,dx + Q\,dy$ is then said to be an **exact differential**. For instance, $y\,dx + x\,dy$ is exact since it is the differential of xy. Thus "exact differentials" and "conservative vector fields" are different views of the same concept.

For a random choice of P and Q, $P\,dx + Q\,dy$ is seldom exact.

If $P\,dx + Q\,dy$ is exact, then $\partial P/\partial y = \partial Q/\partial x$. (See Exercise 21.) This shows, for example, that $y^2\,dx + x\,dy$ is *not* exact. (Why?) The converse is not true, as Example 2 shows. The contrast between an exact differential and an inexact one is of fundamental importance in thermodynamics. (See Exercise 20.)

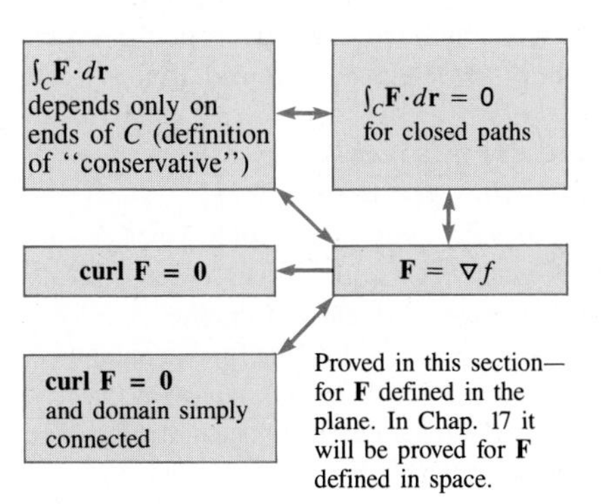

Section Summary

We examined vector fields $\mathbf{F}$ such that $\int_C \mathbf{F} \cdot d\mathbf{r}$ depends only on the endpoints of C. Such a field we call *conservative*. A conservative vector field can also be defined by the condition that $\oint_C \mathbf{F} \cdot d\mathbf{r} = 0$ for all closed curves C. We then showed that if $\mathbf{F}$ is conservative, then it is the gradient of some scalar function. The converse also holds: the gradient of a scalar function is always conservative. We showed that the curl of a conservative field is $\mathbf{0}$. However, the converse is not true unless the domain of $\mathbf{F}$ is simply connected, in which case **curl F** $= \mathbf{0}$ *does* imply that $\mathbf{F}$ is conservative.

EXERCISES FOR SEC. 16.6: CONSERVATIVE VECTOR FIELDS

In Exercises 1 to 4 let $\mathbf{F}$ be a vector field defined everywhere in the plane except at the point P shown in Fig. 10. Assume that $\mathbf{curl\ F} = \mathbf{0}$ and that $\int_{C_1} \mathbf{F} \cdot d\mathbf{r} = 5$.

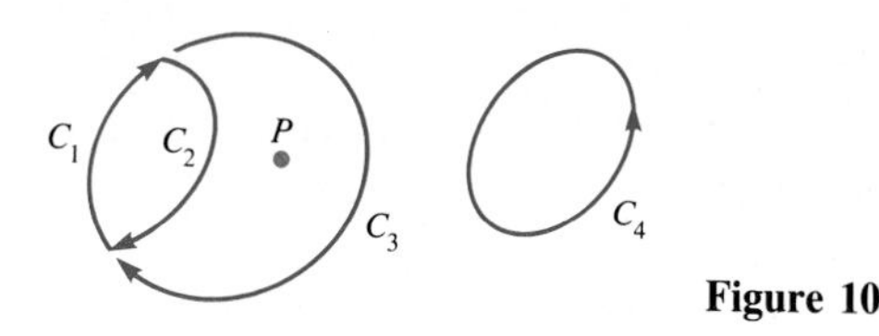

Figure 10

1 What, if anything, can be said about $\int_{C_2} \mathbf{F} \cdot d\mathbf{r}$?
2 What, if anything, can be said about $\int_{C_3} \mathbf{F} \cdot d\mathbf{r}$?
3 What, if anything, can be said about $\oint_{C_4} \mathbf{F} \cdot d\mathbf{r}$?
4 What, if anything, can be said about $\oint_C \mathbf{F} \cdot d\mathbf{r}$, where C is the curve formed by C_1 followed by C_3?

In Exercises 5 to 8 show that the vector field is conservative and then construct a scalar function of which it is the gradient. Use the method of Example 4.

5 $2xy\mathbf{i} + x^2\mathbf{j}$
6 $\sin y\ \mathbf{i} + (x \cos y + 3)\mathbf{j}$
7 $(y + 1)\mathbf{i} + (x + 1)\mathbf{j}$
8 $3y \sin^2 xy \cos xy\ \mathbf{i} + (1 + 3x \sin^2 xy \cos xy)\mathbf{j}$
9 Show that
(a) $3x^2y\, dx + x^3\, dy$ is exact.
(b) $3xy\, dx + x^2\, dy$ is not exact.
10 Show that $x\, dx/(x^2 + y^2) + y\, dy/(x^2 + y^2)$ is exact and exhibit a function f such that df equals the given expression.

In Exercises 11 to 14 answer "True" or "False" and explain.

11 "If $\mathbf{F}$ is conservative, then $\mathbf{curl\ F} = \mathbf{0}$."
12 "If $\mathbf{curl\ F} = \mathbf{0}$, then $\mathbf{F}$ is conservative."
13 "If $\mathbf{F}$ is a gradient field, then $\mathbf{curl\ F} = \mathbf{0}$."
14 "If $\mathbf{curl\ F} = \mathbf{0}$, then $\mathbf{F}$ is a gradient field."

---

15 Let $\mathbf{F}$ be the vector field in Example 2. For which curves C in the xy plane are you sure that $\oint_C \mathbf{F} \cdot d\mathbf{r} = 0$?
16 Solve Example 4(b) using a line integral along C, where C consists of two segments: C_1 going from (0, 0) to $(a, 0)$ and C_2 from $(a, 0)$ to (a, b).
17 State and prove a "two-curve" version of Stokes' theorem in the plane. *Hint:* Use the cancellation principle.
18 (See Exercise 17.) Let $\mathbf{F}$ be defined throughout the plane except at the origin. Assume that $\nabla \times \mathbf{F} = \mathbf{0}$. Show that if C_1 and C_2 are closed curves that enclose the origin, then the circulations of $\mathbf{F}$ around C_1 and around C_2 are equal.
19 Prove Theorem 5 with the aid of Stokes' theorem in the plane.
20 A gas at temperature T_0 and pressure P_0 is brought to the temperature $T_1 > T_0$ and pressure $P_1 > P_0$. The work done in this process is given by the line integral in the TP plane

$$\int_C \left(\frac{RT\, dP}{P} - R\, dT\right),$$

where R is a constant and C is the curve that records the various combinations of T and P during the process. Evaluate this integral over the following paths, shown in Fig. 11.

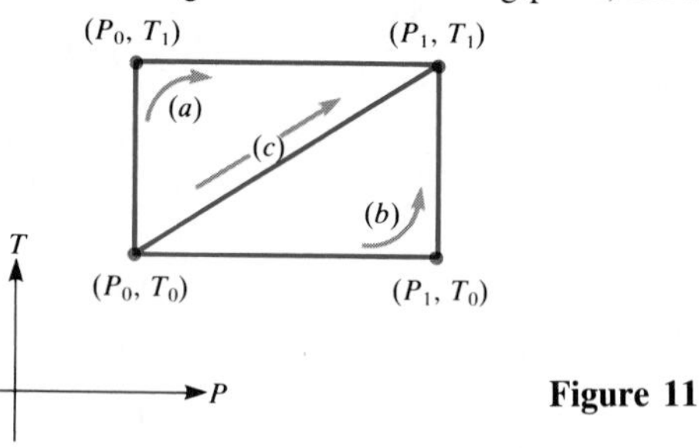

Figure 11

(a) The pressure is kept constant at P_0 while the temperature is raised from T_0 to T_1; then the temperature is kept constant at T_1 while the pressure is raised from P_0 to P_1.
(b) The temperature is kept constant at T_0 while the pressure is raised from P_0 to P_1; then the temperature is raised from T_0 to T_1 while the pressure is kept constant at P_1.
(c) Both pressure and temperature are raised simultaneously in such a way that the path from (P_0, T_0) to (P_1, T_1) is straight.

Because the integrals are path dependent, the differential expression $RT\, dP/P - R\, dT$ defines a thermodynamic quantity that depends on the process, not just on the state. Vectorially speaking, the vector field $(RT/P)\mathbf{i} - R\mathbf{j}$ is not conservative.

21 (a) Show that if $P\, dx + Q\, dy$ is exact, then $\partial P/\partial y = \partial Q/\partial x$.
(b) Use (a) to construct an example of a differential form $P\, dx + Q\, dy$ that is not exact.
22 A central field $\mathbf{F}(\mathbf{r})$ in the xy plane can be written in the form $\mathbf{F}(\mathbf{r}) = f(r)\hat{\mathbf{r}}$, where $r = \|\mathbf{r}\|$ and f is a scalar function which is defined for all positive inputs. Since $f(0)$ may not be defined, $\mathbf{F}$ may not be defined at the origin. (The field $\mathbf{r}/\|\mathbf{r}\|^3$ is an important example.) A straightforward computation shows that $\mathbf{curl\ F} = \mathbf{0}$. However, since the domain of $\mathbf{F}$ is not necessarily simply connected, we cannot apply Theorem 6 to assert that $\mathbf{F}$ is conservative. Instead, follow these steps:
(a) Show that $x\, dx + y\, dy = r\, dr$.
(b) Show that $\mathbf{F}(\mathbf{r}) \cdot d\mathbf{r} = f(r)\, dr$.
(c) Use (b) to explain why we would expect that $\oint_C \mathbf{F} \cdot d\mathbf{r} = 0$, where C is a closed curve that does not pass through the origin. (Include a diagram.)
23 (This continues Exercise 22.) By the first fundamental theorem of calculus, there is a function $g(r)$ such that $dg/dr = f(r)$. Use g to construct a function $h(x, y)$ such that $\mathbf{F} = \nabla h$.
24 If $\mathbf{F}$ and $\mathbf{G}$ are conservative, is $\mathbf{F} + \mathbf{G}$?

16.S SUMMARY

This chapter opened with a definition of various scalar and vector fields:

Given Field	*Derived Field*	*Name*	*Definition*	*Uses*
f	∇f	Gradient	$\frac{\partial f}{\partial x}\mathbf{i} + \frac{\partial f}{\partial y}\mathbf{j} + \frac{\partial f}{\partial z}\mathbf{k}$	$\nabla f \cdot \mathbf{u} = D_{\mathbf{u}} f$; ∇f indicates direction for maximal $D_{\mathbf{u}} f$, which equals $\lVert\nabla f\rVert$.
$\mathbf{F} = P\mathbf{i} + Q\mathbf{j} + R\mathbf{k}$	$\nabla \cdot \mathbf{F}$ or div $\mathbf{F}$	Divergence	$\frac{\partial P}{\partial x} + \frac{\partial Q}{\partial y} + \frac{\partial R}{\partial z}$	If $\mathbf{F}$ is fluid flow, $\nabla \cdot \mathbf{F}$ records rate of increase or decrease locally
$\mathbf{F} = P\mathbf{i} + Q\mathbf{j} + R\mathbf{k}$	$\nabla \times \mathbf{F}$ or **curl F**	Curl	$\begin{vmatrix} \mathbf{i} & \mathbf{j} & \mathbf{k} \\ \frac{\partial}{\partial x} & \frac{\partial}{\partial y} & \frac{\partial}{\partial z} \\ P & Q & R \end{vmatrix}$	Related to tendency of fluid to move along a closed curve
f	$\nabla \cdot \nabla f$ or $\nabla^2 f$	Laplacian	$\frac{\partial^2 f}{\partial x^2} + \frac{\partial^2 f}{\partial y^2} + \frac{\partial^2 f}{\partial z^2}$	See discussion of steady-state temperature in Sec. 16.5.

We defined line integrals. For instance, $\oint_C f(x, y, z)\,dx$ is defined as the definite integral $\int_a^b f(\mathbf{G}(t))\,(dx/dt)\,dt$, where $\mathbf{r} = \mathbf{G}(t)$ is a parameterization of C. The line integral depends only on the direction in which C is swept out. Switching the orientation of C changes the sign of the line integral (since dx/dt changes to its negative). When C is a closed curve, $\int_C f\,dx$ is denoted $\oint_C f\,dx$, and so on. If the closed curve lies in the xy plane, we may speak of its orientation being clockwise or counterclockwise; usually it is taken counterclockwise. $\int_C f\,ds$ is defined directly in terms of partitions of C.

Line Integral	*Application*
$\int_C \mathbf{F} \cdot d\mathbf{r}$	If $\mathbf{F}$ is force, integral is *work*. If $\mathbf{F}$ is fluid flow, integral is *circulation*.
$\oint_C \mathbf{F} \cdot \mathbf{n}\,ds$	If $\mathbf{F}$ is fluid flow, integral is *flux* across C (where C is usually closed).

Name	*Statement*
Green's theorem	$\oint_C \mathbf{F} \cdot \mathbf{n}\,ds = \int_{\mathcal{A}} \nabla \cdot \mathbf{F}\,dA$ (vector form) $\oint_C (-Q\,dx + P\,dy) = \int_{\mathcal{A}} \left(\frac{\partial P}{\partial x} + \frac{\partial Q}{\partial y}\right) dA$ (differential form) $\mathbf{F} = P\mathbf{i} + Q\mathbf{j}$; C counterclockwise

(continued)

Name	*Statement*
Stokes' theorem in the plane	$\oint_C \mathbf{F} \cdot \mathbf{T}\, ds = \int_{\mathscr{A}} (\nabla \times \mathbf{F}) \cdot \mathbf{k}\, dA$ (vector form)
	$\oint_C (P\,dx + Q\,dy) = \int_{\mathscr{A}} \left(\frac{\partial Q}{\partial x} - \frac{\partial P}{\partial y} \right) dA$ (differential form)
	$\mathbf{F} = P\mathbf{i} + Q\mathbf{j}$; C counterclockwise

These theorems also have versions when the region $\mathscr{A}$ is bounded by more than one curve. A two-curve version of Green's theorem shows that if C_1 and C_2 are two closed curves that together bound a region where $\nabla \cdot \mathbf{F} = 0$, then

$$\oint_{C_1} \mathbf{F} \cdot \mathbf{n}\, ds = \oint_{C_2} \mathbf{F} \cdot \mathbf{n}\, ds.$$

Computing the flux over a complicated curve can thus be replaced by computing the flux over a convenient curve.

If $\int_C \mathbf{F} \cdot \mathbf{T}\, ds$ depends only on the start and finish of C, $\mathbf{F}$ is called conservative. We obtained several descriptions of a conservative vector field, including one involving the curl of $\mathbf{F}$ and one involving a gradient.

Vocabulary and Symbols

scalar field f
vector field $\mathbf{F}$
gradient of scalar field ∇f
divergence of a vector field
$\nabla \cdot \mathbf{F}$, div $\mathbf{F}$
curl of a vector field
$\nabla \times \mathbf{F}$, **curl F**
line integral of a scalar field
$\int_C f(P)\, ds$, $\int_C f(P)\, dx$,
$\int_C f(P)\, dy$, $\int_C f(P)\, dz$
closed curve
simple closed curve
integral over a closed curve $\oint_C$
exterior normal $\mathbf{n}$
conservative vector field
exact differential
Green's theorem
divergence-free
arcwise-connected
simply connected
Stokes' theorem in plane
irrotational

GUIDE QUIZ ON CHAP. 16: GREEN'S THEOREM

1 (*a*) Give two physical interpretations of $\oint_C \mathbf{F} \cdot d\mathbf{r}$ and explain why they both lead to that integral.
(*b*) Why is $\oint_C \mathbf{F} \cdot d\mathbf{r}$ called the "circulation" of $\mathbf{F}$ along C?

2 (*a*) Give a physical interpretation of $\oint_C \mathbf{F} \cdot \mathbf{n}\, ds$ and explain why it leads to that integral.
(*b*) Why is $\oint_C \mathbf{F} \cdot \mathbf{n}\, ds$ called the "flux" of $\mathbf{F}$ across C?

3 Why is $\nabla \cdot \mathbf{F}$ called the divergence of $\mathbf{F}$?

4 Why is $\nabla \times \mathbf{F}$ called the curl of $\mathbf{F}$?

5 The field $\hat{\mathbf{r}}/\|\mathbf{r}\|^2$, $\mathbf{r} = x\mathbf{i} + y\mathbf{j} + z\mathbf{k}$, is the most important vector field in physics. Find its divergence and curl.

6 How could you produce lots of conservative vector fields?

7 How would you go about determining whether a given vector field is conservative?

8 How could div $\mathbf{F}$ be defined in a coordinate-free manner?

9 A disk $\mathscr{A}$ of radius 5 is bounded by the circle C. If $\oint_C \mathbf{F} \cdot \mathbf{n}\, ds = 20$, find the average value of the divergence of $\mathbf{F}$ on $\mathscr{A}$.

10 Let $\mathbf{F} = \hat{\mathbf{r}}/\|\mathbf{r}\|$, $\mathbf{r} = x\mathbf{i} + y\mathbf{j}$. Evaluate $\oint_C \mathbf{F} \cdot \mathbf{n}\, ds$ for the following curves:
(*a*) The circle of radius a, center at (0, 0), counterclockwise.
(*b*) The square with vertices $(a, 0)$, $(0, a)$, $(-a, 0)$, $(0, -a)$, counterclockwise, $a > 0$.
(*c*) The rectangle with vertices $(a, 0)$, $(a + b, 0)$, $(a + b, c)$, (a, c), counterclockwise, $a, b, c > 0$.

11 Evaluate $\int_C \mathbf{F} \cdot d\mathbf{r}$ where $\mathbf{F} = \sin x\, \mathbf{i} + \cos y\, \mathbf{j}$ and
(*a*) C is the straight path from (1, 1) to (2, 4).
(*b*) C is the path along the parabola $y = x^2$ from (1, 1) to (2, 4).

REVIEW EXERCISES FOR CHAP. 16: GREEN'S THEOREM

1 Check Green's theorem for $\mathbf{F} = (x - 5y)\mathbf{i} + xy\mathbf{j}$ and $\mathcal{A}$ the region bounded by $y = x^2$ and $y = \sqrt{x}$.

2 $\mathcal{A}$ is a convex set in the plane and C its boundary (taken counterclockwise). $\mathbf{F}$ is a vector field in the plane. Complete these equations.

(*a*) $\oint_C \mathbf{F} \cdot \mathbf{n}\, ds = \int_{\mathcal{A}}$ ________

(*b*) $\oint_C \mathbf{F} \cdot \mathbf{T}\, ds = \int_{\mathcal{A}}$ ________

3 (*a*) Show that if $\mathbf{F} = P\mathbf{i} + Q\mathbf{j}$ is a gradient field, then

$$\frac{\partial P}{\partial y} = \frac{\partial Q}{\partial x}.$$

(*b*) Show that $\mathbf{F}(x, y) = x^2y\mathbf{i} - xy^2\mathbf{j}$ is *not* a gradient field.

4 If a field $\mathbf{F}$ is divergence-free everywhere, what can be said about the flux of $\mathbf{F}$ across a closed curve?

5 (*a*) If a field $\mathbf{F}$ is irrotational, is it necessarily the gradient of some scalar field?

(*b*) If a field $\mathbf{F}$ is the gradient of a scalar field, is it necessarily irrotational?

6 The flux of $\mathbf{F}$ around a certain circle of radius 0.01 is -0.003. Estimate the divergence of $\mathbf{F}$ at points in the circle.

7 (*a*) Let $\mathbf{F}(x, y) = P(x, y)\mathbf{i} + Q(x, y)\mathbf{j}$. Show that the curl of $\mathbf{F}$ is parallel to $\mathbf{k}$.

(*b*) Assume that the circulation of $\mathbf{F}$ around a certain circle of radius 0.01, taken counterclockwise, is -0.002. Estimate the vector **curl** $\mathbf{F}$ at points within the circle.

8 Let $\mathbf{F} = P\mathbf{i} + Q\mathbf{j}$ be a vector field.

(*a*) Explain why $\int_C (P\, dx + Q\, dy)$ represents the work accomplished by the force $\mathbf{F}$ in moving a particle along C.

(*b*) Explain why $\int_C (-Q\, dx + P\, dy)$ represents the net loss of fluid past the closed curve C bounding a convex region if the flux is $\mathbf{F}$.

9 Using a diagram, show that if a closed planar curve C is swept out counterclockwise, then $(dy/ds)\mathbf{i} - (dx/ds)\mathbf{j}$ is a unit exterior normal.

10 Which of these vector fields in the plane are conservative? Explain.

(*a*) $\mathbf{F}(x, y) = 3\mathbf{i} + 4\mathbf{j}$

(*b*) $\mathbf{F}(x, y) = y \sin xy\, \mathbf{i} + x \cos xy\, \mathbf{j}$

(*c*) $\mathbf{F}(x, y) = x\mathbf{i} + y\mathbf{j}$

11 The electric field $\mathbf{E}$ at any point (x, y) due to a point charge q at $(0, 0)$ is equal to

$$\mathbf{E} = \frac{q\mathbf{u}}{4\pi\epsilon_0 r^2},$$

where r is the distance from the charge to the point (x, y) and $\mathbf{u}$ is the unit vector directed from the charge to the point (x, y). Evaluate the work done by the field when a unit charge is moved from $(1, 0)$ to $(2, 0)$ along (*a*) the x axis, (*b*) the rectangular path from $(1, 0)$ to $(1, \frac{1}{2})$ to $(2, \frac{1}{2})$ to $(2, 0)$. Is there a difference between the work done in (*a*) and that done in (*b*)?

12 Let $\mathbf{F}$ be the gradient of $f(x, y) = e^x y^2$. Determine the value of $\int_C \mathbf{F} \cdot d\mathbf{r}$ where C is the curve that starts at $(1, 1)$, goes to $(2, 3)$ on a straight line, and continues from $(2, 3)$ to $(3, 7)$ on a second straight line.

13 A field of the form $\mathbf{F}(x, y) = f(r)\hat{\mathbf{r}}$, where f is a scalar function, $\mathbf{r}$ is the position vector $\mathbf{r} = x\mathbf{i} + y\mathbf{j}$, $r = \|\mathbf{r}\|$, and $\hat{\mathbf{r}} = \mathbf{r}/r$ is called central (or radially symmetric). Consider the example $\mathbf{F}(x, y) = \hat{\mathbf{r}}/r^k$, where k is a constant.

(*a*) For which k is **curl** $\mathbf{F} = \mathbf{0}$?

(*b*) For which k is div $\mathbf{F} = 0$?

14 (See Exercise 13.) Let $\mathbf{F}(x, y, z) = \hat{\mathbf{r}}/r^k$, where $\mathbf{r} = x\mathbf{i} + y\mathbf{j} + z\mathbf{k}$, $r = \|\mathbf{r}\|$, $\hat{\mathbf{r}} = \mathbf{r}/r$, and k is a constant.

(*a*) For which k is **curl** $\mathbf{F} = \mathbf{0}$?

(*b*) For which k is div $\mathbf{F} = 0$?

15 Explain carefully, without diagrams, why $\oint_C \mathbf{F} \cdot \mathbf{n}\, ds$ can be thought of as the rate at which fluid crosses the curve C. How is $\mathbf{F}$ defined for a fluid flow?

16 Explain carefully, with diagrams, why $\int_C \mathbf{F} \cdot d\mathbf{r}$ represents work if $\mathbf{F}$ is interpreted as force.

17 Let C be the triangle with vertices $(0, 0)$, $(1, 0)$, and $(0, 1)$ oriented counterclockwise. Evaluate

$$\oint_C [(3x^2 + y)\, dx + 4y^2\, dy]$$

by (*a*) direct calculation, (*b*) Green's theorem.

18 Let $\mathbf{F}$ be a vector function defined at all points of a curve C and suppose $\|\mathbf{F}\| \le M$ on C, where M is some number. Show that $\left|\int_C \mathbf{F} \cdot d\mathbf{r}\right| \le Ml$, where l is the length of C.

19 A certain divergence-free field $\mathbf{F}$ is defined everywhere in the xy plane except at $(0, 0)$. At points on C_1, the unit circle with center $(0, 0)$, $\mathbf{F}$ is tangent to C_1. What, if anything, can be said about the flux across C_2 and C_3 in Fig. 1?

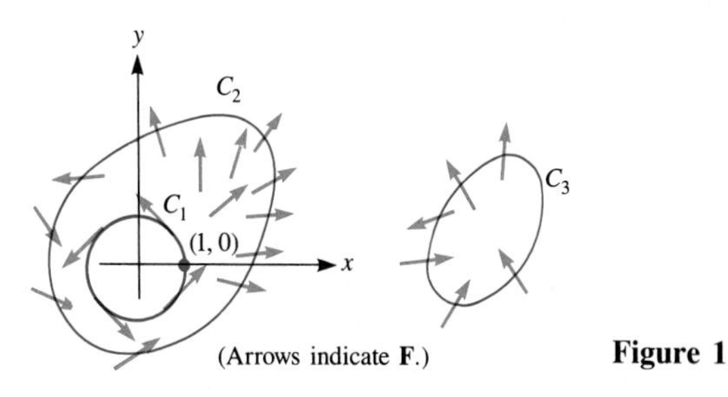

Figure 1

20 Let $f(x, y)$ be defined everywhere except at $(0, 0)$. Let C be a level curve of f that encloses the origin. What, if anything, can be said about

(*a*) The flux of $\mathbf{F} = \nabla f$ across C?

(*b*) The circulation of $\mathbf{F} = \nabla f$ around C?

21 Let C be the closed curve $x = 2 \cos t$, $y = 3 \sin t$, for t in $[0, 2\pi]$.

(*a*) Graph C.

(*b*) Compute $\int_C [x^2\, dx + (y + 1)\, dy]$.

(*c*) Devise a work problem whose answer is the integral in (*b*).

(*d*) Devise a fluid-flow problem whose answer is the integral in (*b*).

22 Let $\mathbf{F}(x, y)$ be a vector field defined everywhere except at the origin. Assume that the values of $\oint_C \mathbf{F} \cdot \mathbf{n}\, ds$ for all closed curves that enclose the origin are the same (not necessarily 0, however). Show that the divergence of $\mathbf{F}$ is 0.

23 (*a*) Show that the field $\mathbf{F} = 2xy\mathbf{i} + x^2\mathbf{j}$ is conservative by computing its curl.

(*b*) Construct a scalar function f such that $\mathbf{F} = \nabla f$, using line integrals.

24 At (x, y) let $\mathbf{r} = x\mathbf{i} + y\mathbf{j}$, $\hat{\mathbf{r}} = \mathbf{r}/\|\mathbf{r}\|$, and $\hat{\boldsymbol{\theta}}$ be the unit vector perpendicular to $\mathbf{r}$ as shown in Fig. 2.

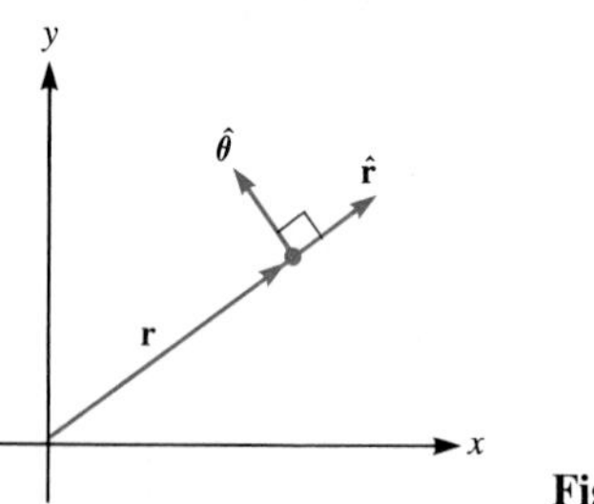

Figure 2

These vectors are of use in analyzing circular motion around the origin. Consider a pan full of water that is rotating around (0, 0) at a constant angular speed. Assume that the water rotates with the pan, as though it were a rigid body.

(*a*) Why is $\mathbf{F}(x, y) = \|\mathbf{r}\|\hat{\boldsymbol{\theta}}$ an appropriate representation of the flow of water at (x, y)?

(*b*) Compute the circulation of $\mathbf{F}$ around the closed curve C shown in Fig. 3.

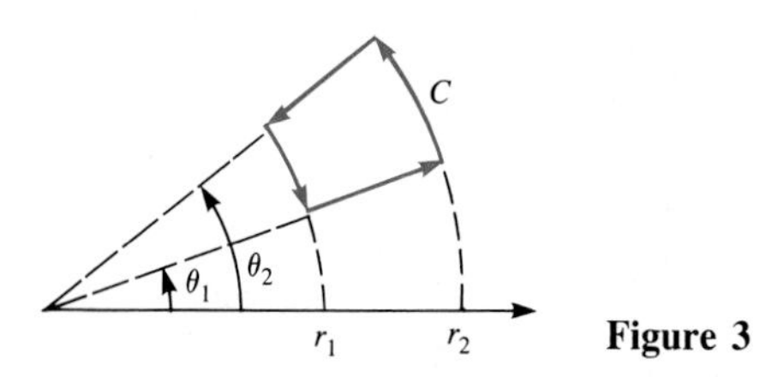

Figure 3

Suggestion: Break C into four parts, two of which are straight and two of which are curved.

(*c*) Imagine that you hold a little paddle wheel or propeller in the water, with its axis perpendicular to the surface of the water. Which way will the paddle wheel turn? counterclockwise? clockwise? not at all?

25 Like Exercise 24 with $\mathbf{F} = \|\mathbf{r}\|^2\hat{\boldsymbol{\theta}}$.

26 (See Exercise 24.) Let $\mathbf{F} = \|\mathbf{r}\|^k\hat{\boldsymbol{\theta}}$.

(*a*) For which k will the paddle wheel turn counterclockwise?

(*b*) For which k will it turn clockwise?

(*c*) For which k will it not turn at all?

27 This argument appears in Colin W. Clark, *Mathematical Bioeconomics*, Wiley, New York, 1976. Let $G(x, y)$ and $H(x, y)$ be two functions such that $\partial H/\partial x \geq \partial G/\partial y$. Let C_1 and C_2 be the two curves directed as shown in Fig. 4. (Each represents a way of harvesting a crop; x is time and y is the rate of harvest.) Show that

$$\int_{C_1} (G\, dx + H\, dy) \leq \int_{C_2} (G\, dx + H\, dy).$$

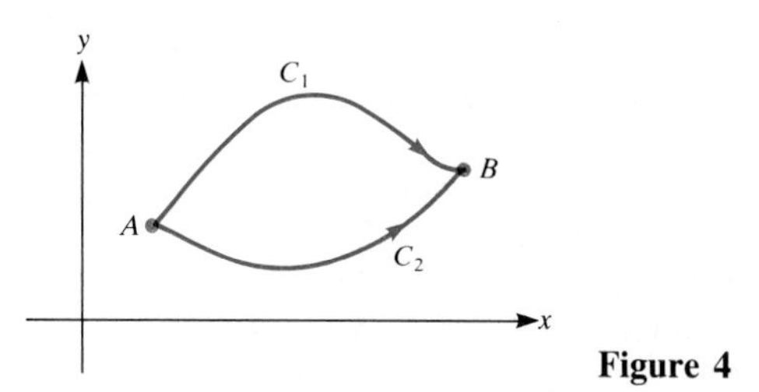

Figure 4

28 This exercise is an introduction to Exercise 29. A motorboat goes back and forth on a straight measured mile at a constant speed v relative to the water. Show that the time required for a round trip is always less when there is no current than when there is a constant current $\mathbf{W}$. (Assume that the component of $\mathbf{W}$ along the route is less than v and greater than 0. If the component along the route were larger than v, the boat could not make a round trip.)

29 (See Exercise 28.) An aircraft traveling at constant speed v relative to the air traverses a closed horizontal curve marked on the ground. Show that the time required for one complete trip is always less when there is no wind than when there is a constant wind $\mathbf{W}$. (Assume that $W = \|\mathbf{W}\|$ is less than v, so that the plane never meets an insuperable head wind.) *Hint:* Express the time as a line integral.

A harmonic function $f(x, y)$ is a function that satisfies the equation

$$\frac{\partial^2 f}{\partial x^2} + \frac{\partial^2 f}{\partial y^2} = 0.$$

Such a function has a remarkable property: its value at any point Q is the average of its values on any circle that has Q as a center and such that the disk it bounds lies in the domain of f. Exercises 30 to 33 outline why this is so.

30 Let r be a positive number and let C_r be the circle of radius r and center Q. For convenience, place the xy coordinate system so that Q is at the origin. Let

$$I(r) = \frac{1}{2\pi r} \oint_{C_r} f(P)\, ds.$$

We wish to show that $f(Q) = I(r)$. To begin, show that

$$I(r) = \frac{1}{2\pi} \int_0^{2\pi} f(r, \theta)\, d\theta,$$

where (r, θ) denotes polar coordinates.

31 We show $I(r)$ is constant, independent of r. For this we need the theorem in Appendix K that permits "differentiation under the integral sign,"

$$\frac{d}{dr}\int_0^{2\pi} f(r, \theta)\, d\theta = \int_0^{2\pi} \frac{\partial f}{\partial r}\, d\theta.$$

Show that

$$\frac{\partial f}{\partial r} = \nabla f \cdot \mathbf{n},$$

where $\mathbf{n}$ is the unit external normal to C_r.

32 Show that

$$\frac{d}{dr}(I(r)) = 0.$$

Thus, $I(r)$ is constant, say $I(r) = k$.

33 Why does $I(r)$ equal $f(Q)$? [*Hint*: What is the limit of $I(r)$ as r approaches 0?]

34 (See introduction to Exercise 30.)

(*a*) Show that $f(x, y) = \ln((x-2)^2 + (y-3)^2)$ is harmonic.

(*b*) What is the domain of f?

(*c*) Find the average value of f on the unit circle with center (0, 0).

35 Let $f(x, y)$ be defined on a region that includes the closed curve C and the region $\mathcal{A}$ that C bounds. Assume that $f(x, y)$ is harmonic and that $f(x, y) = 0$ for all points (x, y) on C. Show that $f(x, y) = 0$ for all points of $\mathcal{A}$, as follows:

(*a*) Assume $f(x, y)$ is not constant. Let M be its maximum and m its minimum for (x, y) in $\mathcal{A}$. We wish to show $M = 0 = m$. Assume $M > 0$ and let Q be a point where $f(x, y)$ reaches that maximum, $f(Q) = M$. Show that there is a disk (of nonzero radius) around Q such that $f(x, y) = M$ for all points in that disk.

(*b*) Continuing (*a*), why would $f(x, y) = M$ for all (x, y) in $\mathcal{A}$?

(*c*) Deduce that $f(x, y) = 0$ for all (x, y) in $\mathcal{A}$.

36 (See Exercise 35.)

(*a*) Let f and g be harmonic functions such that $f(x, y) = g(x, y)$ for points on the curve C in Exercise 35. Why must they be equal throughout $\mathcal{A}$? *Hint:* Consider $f - g$.

(*b*) Why is (*a*) important in the theory of steady-state temperature distributions? (See the discussion of heat in Sec. 16.5.)

37 Let $\mathbf{F}(x, y) = f(r)\hat{\mathbf{r}}$ where f is a scalar function defined for $r > 0$, $r = \|\mathbf{r}\|$, and $\hat{\mathbf{r}} = \mathbf{r}/r$. Assume that div $\mathbf{F} = 0$.

(*a*) Evaluate the flux of $\mathbf{F}$ around the curve C shown in Fig. 5.

(*b*) Deduce that $f(r)$ must be of the form $f(r) = k/r$, for some constant k.

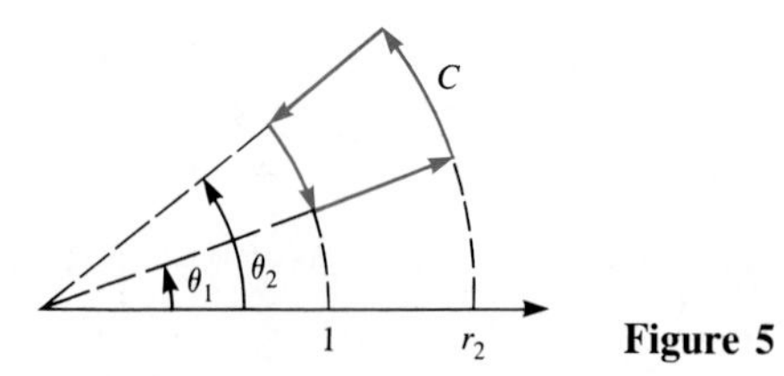

Figure 5

38 Figure 6 shows some representative vectors for the vector field $\mathbf{F}$. Use this information to estimate (*a*) the circulation of $\mathbf{F}$ along the boundary curve C and (*b*) the flux of $\mathbf{F}$ across C. (Since you have no formula for $\mathbf{F}$, there is a range of "correct" answers.)

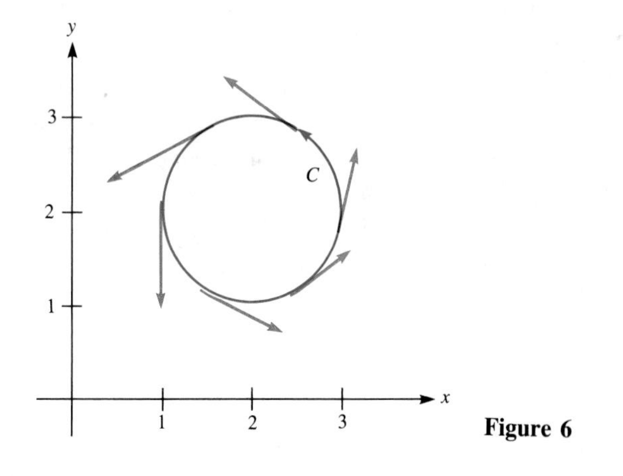

Figure 6

17

THE DIVERGENCE THEOREM AND STOKES' THEOREM

This chapter concerns two theorems that are standard tools in the study of electricity, magnetism, and gravity. The divergence theorem relates an integral over a solid region to an integral over its bounding surface. It is the analog in space of Green's theorem. Stokes' theorem relates an integral over a closed curve in space to an integral over any surface that the curve bounds. It is the analog in space of Stokes' theorem in the plane.

Section 17.1 discusses integrals over surfaces. These are needed in Sec. 17.2, which presents the divergence theorem. Stokes' theorem is the focus of Sec. 17.3. Section 17.4 uses Stokes' theorem to develop a geometric definition of curl.

17.1 SURFACE INTEGRALS

The main object of this section is to introduce the notion of an integral over a surface. This definition is very similar to that given in Sec. 15.1 for the integral of a function over a plane region, that is, over a flat surface. We shall assume that the surfaces we deal with are smooth, or composed of a finite number of smooth pieces, and that the integrals we define exist.

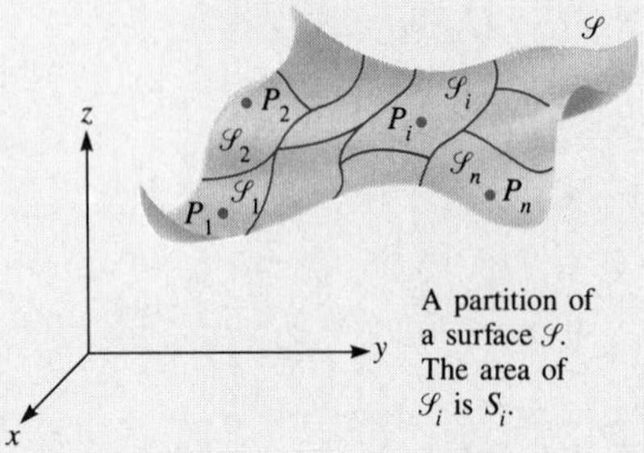

Definition *Definite integral of a function f over a surface $\mathcal{S}$.* Let f be a function that assigns to each point P in a surface $\mathcal{S}$ a number $f(P)$. Consider the typical sum

$$f(P_1)S_1 + f(P_2)S_2 + \cdots + f(P_n)S_n,$$

formed from a partition of $\mathcal{S}$, where S_i is the area of the ith region in the partition and P_i is a point in the ith region. (See Fig. 1.) If these sums approach a certain number as the mesh of the partitions shrinks to 0 (no matter how P_i is chosen in the ith region), the number is called the **definite integral** of f over $\mathcal{S}$ and is written

$$\int_{\mathcal{S}} f(P)\, dS.$$

A partition of a surface $\mathcal{S}$. The area of $\mathcal{S}_i$ is S_i.

Figure 1

The definitions of partitions and mesh used in the preceding definition are like those in Sec. 15.1. For instance, if $f(P) = 1$ for all points P in $\mathcal{S}$, then each sum $\sum_{i=1}^{n} f(P_i)\, S_i$ is equal to $\sum_{i=1}^{n} S_i$, which is the area of $\mathcal{S}$. So the limit of such sums as the mesh approaches 0 is the area of $\mathcal{S}$:

$$\int_{\mathcal{S}} 1\, dS = \text{Area of } \mathcal{S}.$$

For instance, if $\mathcal{S}$ is a sphere of radius a, then $\int_{\mathcal{S}} 1\, dS = 4\pi a^2$.

If $f(P)$ is the density of mass at P, then $\int_{\mathcal{S}} f(P)\, dS$ is the total mass. Similarly, if $f(P)$ is the density of electric charge, then $\int_{\mathcal{S}} f(P)\, dS$ is the total electric charge.

Integrals over surfaces, like other types of integrals we have encountered, are important for two reasons:

1 They provide a way to describe concepts in physics, engineering, mathematics, and other sciences.

2 They provide a tool for evaluating certain quantities. (Mass and charge are just two examples.)

First we show how to compute a surface integral if the surface is part of a sphere. Then we compute a special surface integral needed in the next section. Finally, we describe a general technique for evaluating surface integrals.

Integrating over a Sphere

If $\mathcal{S}$ is a sphere or part of a sphere, it is often convenient to evaluate an integral over it with the aid of spherical coordinates.

If the center of a spherical coordinate system (ρ, θ, ϕ) is at the center of a sphere of radius a, then ρ is constant on the sphere $\rho = a$. As Fig. 2 suggests, the area of the small region on the sphere corresponding to slight changes $d\theta$ and $d\phi$ is approximately

$$(a\, d\phi)(a \sin \phi\, d\theta) = a^2 \sin \phi\, d\theta\, d\phi.$$

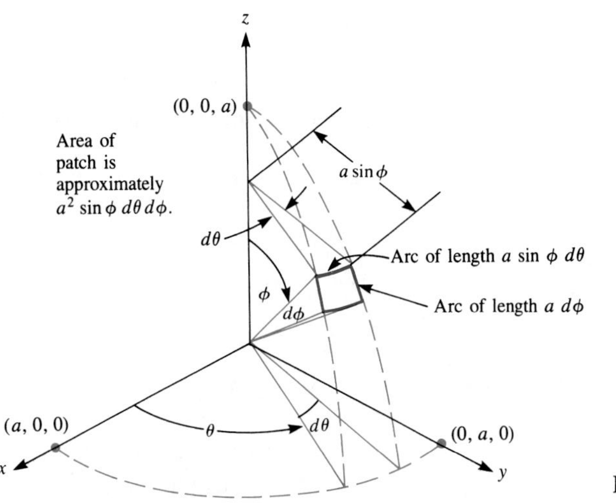

Figure 2

(See Sec. 15.7 for a similar argument, where ρ was not constant.) Thus we may write

$$dS = a^2 \sin\phi \, d\theta \, d\phi$$

and evaluate

$$\int_{\mathcal{S}} f(P)\, dS$$

in terms of a repeated integral in ϕ and θ. Example 1 illustrates this technique.

EXAMPLE 1 Let $\mathcal{S}$ be the top half of the sphere $x^2 + y^2 + z^2 = a^2$. Evaluate $\int_{\mathcal{S}} z\, dS$.

SOLUTION In spherical coordinates $z = \rho \cos\phi$. Since the sphere has radius a, $\rho = a$. Thus

$$\int_{\mathcal{S}} z\, dS = \int_{\mathcal{S}} (a\cos\phi)\, dS = \int_0^{2\pi} \left[\int_0^{\pi/2} (a\cos\phi)a^2 \sin\phi\, d\phi \right] d\theta.$$

Now,

$$\begin{aligned} \int_0^{\pi/2} (a\cos\phi)a^2\sin\phi\, d\phi &= a^3 \int_0^{\pi/2} \cos\phi \sin\phi\, d\phi \\ &= a^3 \frac{(-\cos^2\phi)}{2}\bigg|_0^{\pi/2} \\ &= \frac{a^3}{2}[-0-(-1)] \\ &= \frac{a^3}{2}, \end{aligned}$$

so that

$$\int_{\mathcal{S}} z\, dS = \int_0^{2\pi} \frac{a^3}{2}\, d\theta = \pi a^3. \quad \blacksquare$$

We can interpret the result in Example 1 in terms of average value or in terms of centroids, which we now define.

The **average value** of $f(P)$ over a surface $\mathcal{S}$ is defined as

$$\frac{\int_{\mathcal{S}} f(P)\, dS}{\text{Area of } \mathcal{S}}.$$

Example 1 shows that the average value of z over the given hemisphere is

$$\frac{\int_{\mathcal{S}} z\, dS}{\text{Area of } \mathcal{S}} = \frac{\pi a^3}{2\pi a^2} = \frac{a}{2}.$$

The z coordinate of the centroid of a surface $\mathcal{S}$ is defined as

$$\bar{z} = \frac{\int_{\mathcal{S}} z\, dS}{\text{Area of } \mathcal{S}}.$$

Example 1 shows that the centroid of the curved surface of a hemisphere of radius a is at a height $a/2$ above the center of the sphere.

Some Special Surface Integrals

In the next example we will express a certain surface integral as an integral over a plane region. The result will be essential in the next section.

Let $\mathcal{S}$ be a surface that is the boundary of a region in space. Such a surface is called a **closed surface**. Assume that any line meets $\mathcal{S}$ in at most two points. Let $\mathbf{n} = \mathbf{n}(P)$ be the exterior unit normal to $\mathcal{S}$ at the point P on $\mathcal{S}$. (See Fig. 3.) If the direction angles of $\mathbf{n}$ are α, β, and γ, then

$$\mathbf{n} = \cos\alpha\ \mathbf{i} + \cos\beta\ \mathbf{j} + \cos\gamma\ \mathbf{k}.$$

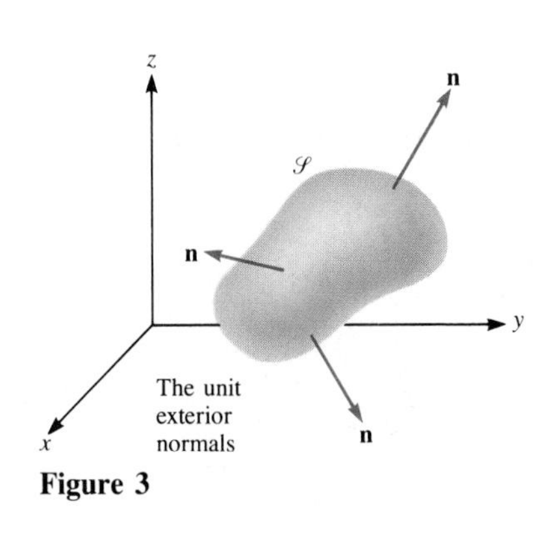

Figure 3

Now we are ready for Example 2, which will be used in the proof of the divergence theorem.

EXAMPLE 2 Let $R(x, y, z)$ be a scalar function defined over a closed surface $\mathcal{S}$. Express $\int_{\mathcal{S}} R(x, y, z) \cos\gamma\ dS$ as an integral over the projection of $\mathcal{S}$ on the xy plane.

SOLUTION The angle γ lies in the interval $[0, \pi]$. At the highest point of $\mathcal{S}$ it is 0 and at the lowest point it is π. Thus $\cos\gamma$ takes on both positive and negative values. Let $\mathcal{S}_2$ be the "top part" of $\mathcal{S}$, where $\gamma < \pi/2$. Let $\mathcal{S}_1$ be the "bottom part" of $\mathcal{S}$, where $\gamma > \pi/2$. The two parts are separated by the curve C, where $\gamma = \pi/2$, as shown in Fig. 4.

Now,

$$\int_{\mathcal{S}} R(x, y, z) \cos\gamma\ dS = \int_{\mathcal{S}_2} R(x, y, z) \cos\gamma\ dS + \int_{\mathcal{S}_1} R(x, y, z) \cos\gamma\ dS. \quad (1)$$

We will examine $\cos\gamma\ dS$ on $\mathcal{S}_2$ and $\mathcal{S}_1$ and express each of the integrals on the right side of (1) in terms of integrals over $\mathcal{A}$, where $\mathcal{A}$ is the projection of $\mathcal{S}$ on the xy plane, as shown in Fig. 4.

On $\mathcal{S}_2$, $\cos\gamma$ is positive; on $\mathcal{S}_1$, $\cos\gamma$ is negative. Consider a small patch $d\mathcal{S}$ on the top part, $\mathcal{S}_2$, where $\cos\gamma$ is positive, as shown in Fig. 5. Denote the area of $d\mathcal{S}$ by dS. Let the area of the projection of the small patch on the xy plane be

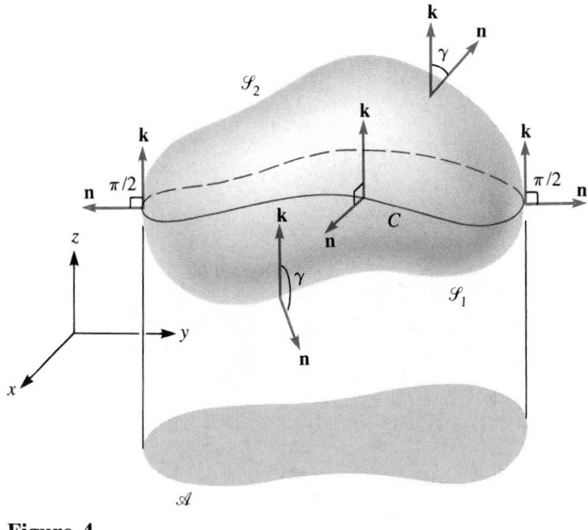

Figure 4

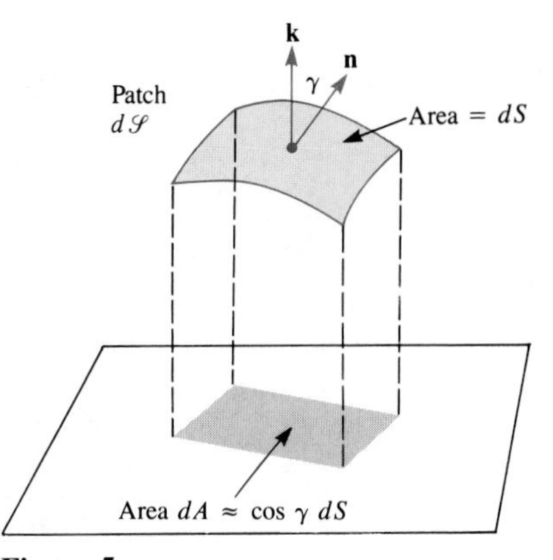

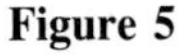

Figure 5

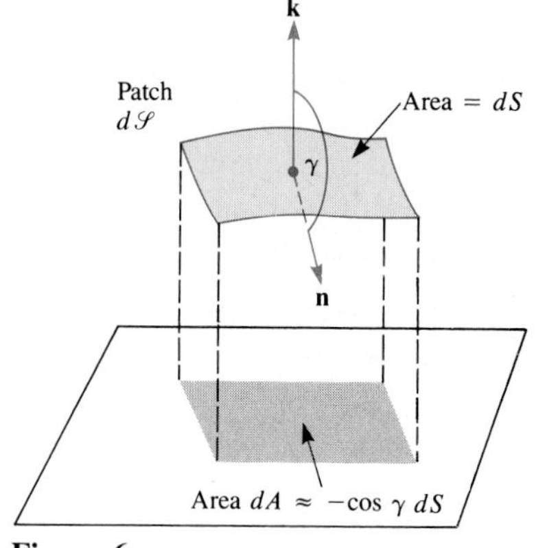

Figure 6

dA. Since the patch $d\mathcal{S}$ is nearly flat, the area of its projection on the xy plane is approximately $\cos\gamma\, dS$, as discussed in Sec. 12.2. That is,

$$dA \approx \cos\gamma\, dS. \tag{2}$$

Next consider a small patch $d\mathcal{S}$ from the bottom part, $\mathcal{S}_1$, where $\cos\gamma$ is negative, as shown in Fig. 6. The area of the projection of $d\mathcal{S}$ on the xy plane is now approximately

$$dA \approx -\cos\gamma\, dS. \tag{3}$$

The line through the point $(x, y, 0)$ in $\mathcal{A}$ and parallel to the z axis meets $\mathcal{S}_1$ in a point (x, y, z_1) and $\mathcal{S}_2$ in a point (x, y, z_2), as shown in Fig. 7. This means that (1) can be written as

$$\int_{\mathcal{S}} R(x, y, z)\cos\gamma\, dS = \int_{\mathcal{S}_2} R(x, y, z_2)\cos\gamma\, dS + \int_{\mathcal{S}_1} R(x, y, z_1)\cos\gamma\, dS. \tag{4}$$

On $\mathcal{S}_2$, $\cos\gamma\, dS = dA$ and on $\mathcal{S}_1$, $\cos\gamma\, dS = -dA$. Thus (4) becomes

$$\begin{aligned}\int_{\mathcal{S}} R(x, y, z)\cos\gamma\, dS &= \int_{\mathcal{A}} R(x, y, z_2)\, dA - \int_{\mathcal{A}} R(x, y, z_1)\, dA \\ &= \int_{\mathcal{A}} (R(x, y, z_2) - R(x, y, z_1))\, dA.\end{aligned} \tag{5}$$

Equation (5) expresses $\int_{\mathcal{S}} R(x, y, z)\cos\gamma\, dS$ as an integral over $\mathcal{A}$. ■

The analog of Example 2 holds for the other two direction angles of the unit exterior normal **n**. To be specific, let $P(x, y, z)$ and $Q(x, y, z)$ be defined on $\mathcal{S}$. Then

$$\int_{\mathcal{S}} P(x, y, z)\cos\alpha\, dS = \int_{\mathcal{A}} (P(x_2, y, z) - P(x_1, y, z))\, dA,$$

where $\mathcal{A}$ is the projection of $\mathcal{S}$ on the yz plane, and the line through $(0, y, z)$ parallel to the x axis meets $\mathcal{S}$ at (x_1, y, z) and (x_2, y, z), $x_1 \le x_2$. (See Fig. 8.)

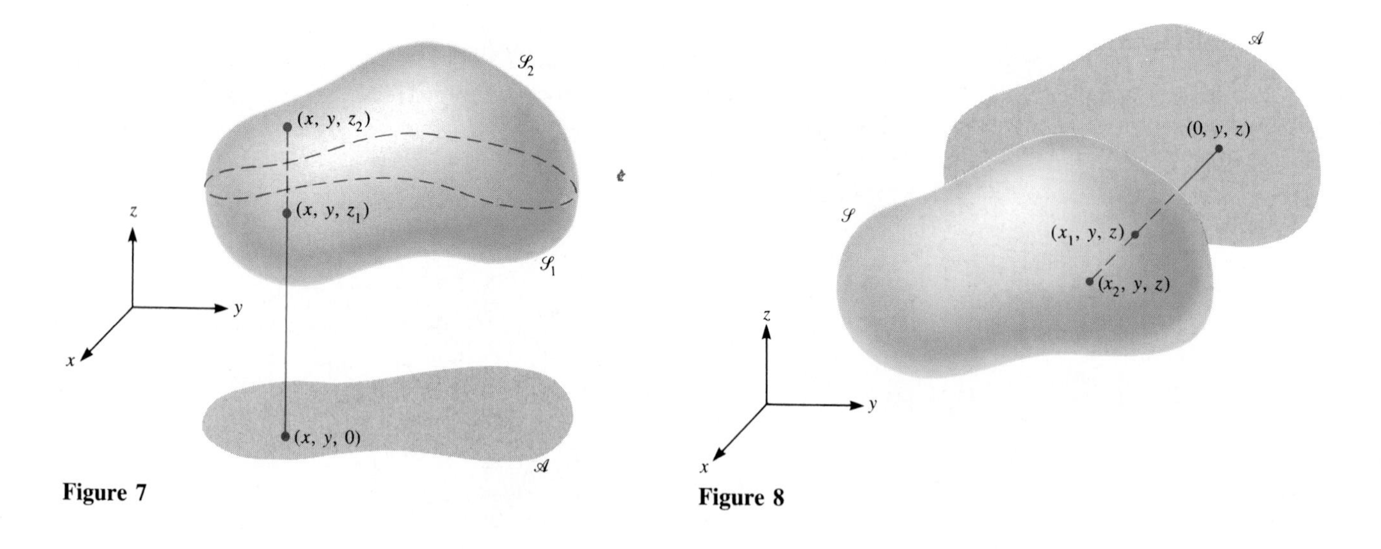

Figure 7

Figure 8

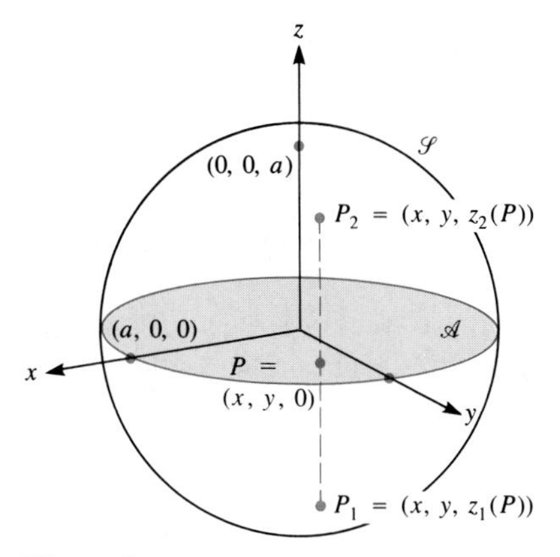

Figure 9

Similarly,

$$\int_{\mathcal{S}} Q(x, y, z) \cos \beta \, dS = \int_{\mathcal{A}} (Q(x, y_2, z) - Q(x, y_1, z)) \, dA,$$

where $\mathcal{A}$ is the projection of $\mathcal{S}$ on the xz plane, and the line through $(x, 0, z)$ meets $\mathcal{S}$ at (x, y_1, z) and (x, y_2, z), $y_1 \leq y_2$.

EXAMPLE 3 Let $\mathcal{S}$ be the sphere of radius a with center at the origin. Compute $\int_{\mathcal{S}} z \cos \gamma \, dS$.

SOLUTION Figure 9 shows the sphere $\mathcal{S}$ and its projection $\mathcal{A}$ on the xy plane. By Example 2,

$$\int_{\mathcal{S}} z \cos \gamma \, dS = \int_{\mathcal{A}} (z_2(P) - z_1(P)) \, dA,$$

where P, $z_2(P)$, and $z_1(P)$ are shown in Fig. 9. But $z_2(P) - z_1(P)$ is simply the length of the cross section of the ball corresponding to point P in $\mathcal{A}$. Consequently,

$$\int_{\mathcal{A}} (z_2(P) - z_1(P)) \, dA = \text{Volume of ball}$$

$$= \frac{4\pi a^3}{3}. \quad \blacksquare$$

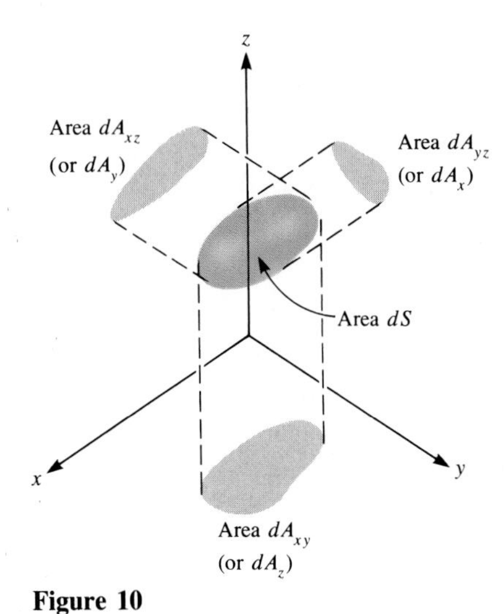

Figure 10

Remark: In applications in engineering and physics the small surface patch of area dS may be projected on any of the three coordinate planes. The area of the projection on the xy plane is denoted dA_z or dA_{xy}. Similar notations are used for the other two projections, as shown in Fig. 10.

A General Technique

In Example 2 we evaluated an integral over a surface $\mathcal{S}$ by projecting $\mathcal{S}$ on a coordinate plane. This technique is quite general, as shown in the following theorem.

Theorem 1 Let $\mathcal{S}$ be a surface and let $\mathcal{A}$ be its projection on the xy plane. Assume that for each point Q in $\mathcal{A}$ the line through Q parallel to the z axis meets $\mathcal{S}$ in exactly one point P. Let f be a function defined on $\mathcal{S}$. Define a function h on $\mathcal{A}$ by

$$h(Q) = f(P).$$

Then $$\int_{\mathcal{S}} f(P) \, dS = \int_{\mathcal{A}} \frac{h(Q)}{|\cos \gamma|} \, dA,$$

which replaces an integral over a surface by an integral over a planar region. In this equation γ denotes the angle between **k** and a vector normal to the surface of $\mathcal{S}$ at P. (See Fig. 11.)

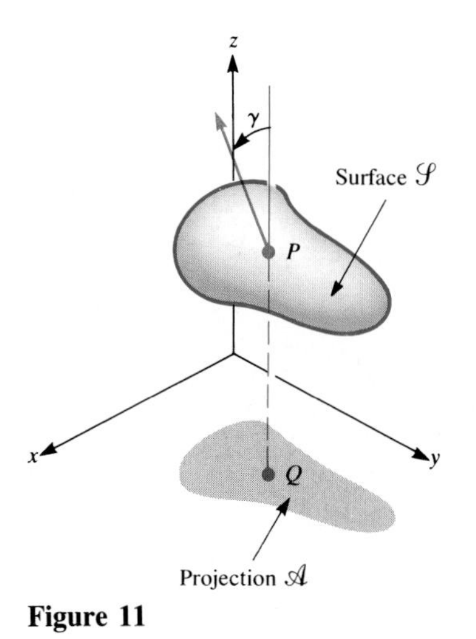

Figure 11

Our reasoning in Example 2 shows why Theorem 1 is true. Whether γ is acute or obtuse,

$$dA = |\cos \gamma|\, dS,$$

or equivalently,

$$dS = \frac{dA}{|\cos \gamma|}.$$

The integrand $f(P)$, where P is in $\mathcal{S}$, has to be rewritten as a function defined for points Q in $\mathcal{A}$.

To apply the technique it is necessary to find a vector perpendicular to the surface in order to compute $\cos \gamma$. If $\mathcal{S}$ is the level surface of $g(x, y, z)$,

$$g(x, y, z) = c,$$

then the gradient ∇g is such a vector.

If the surface $\mathcal{S}$ is given in the form $z = f(x, y)$, rewrite it as $z - f(x, y) = 0$. That means that $\mathcal{S}$ is a level surface of $g(x, y, z) = z - f(x, y)$. Theorem 2 shows what the formulas for $\cos \gamma$ look like. However, it is unnecessary, even distracting, to memorize them. Just remember that *a gradient provides a normal* to a level surface.

Theorem 2

(*a*) If the surface $\mathcal{S}$ is part of the level surface $g(x, y, z) = c$, then

$$|\cos \gamma| = \frac{\left|\dfrac{\partial g}{\partial z}\right|}{\sqrt{\left(\dfrac{\partial g}{\partial x}\right)^2 + \left(\dfrac{\partial g}{\partial y}\right)^2 + \left(\dfrac{\partial g}{\partial z}\right)^2}}.$$

(*b*) If the surface $\mathcal{S}$ is given in the form $z = f(x, y)$, then

$$|\cos \gamma| = \frac{1}{\sqrt{\left(\dfrac{\partial f}{\partial x}\right)^2 + \left(\dfrac{\partial f}{\partial y}\right)^2 + 1}}.$$

Proof

(*a*) A normal vector to $\mathcal{S}$ at a given point is provided by the gradient

$$\nabla g = \frac{\partial g}{\partial x}\mathbf{i} + \frac{\partial g}{\partial y}\mathbf{j} + \frac{\partial g}{\partial z}\mathbf{k}.$$

The cosine of the angle between $\mathbf{k}$ and ∇g is

$$\frac{\mathbf{k} \cdot \nabla g}{\|\mathbf{k}\|\, \|\nabla g\|} = \frac{\mathbf{k} \cdot \left(\dfrac{\partial g}{\partial x}\mathbf{i} + \dfrac{\partial g}{\partial y}\mathbf{j} + \dfrac{\partial g}{\partial z}\mathbf{k}\right)}{1 \cdot \sqrt{\left(\dfrac{\partial g}{\partial x}\right)^2 + \left(\dfrac{\partial g}{\partial y}\right)^2 + \left(\dfrac{\partial g}{\partial z}\right)^2}};$$

hence

$$|\cos \gamma| = \frac{\left|\dfrac{\partial g}{\partial z}\right|}{\sqrt{\left(\dfrac{\partial g}{\partial x}\right)^2 + \left(\dfrac{\partial g}{\partial y}\right)^2 + \left(\dfrac{\partial g}{\partial z}\right)^2}}.$$

(*b*) Rewrite $z = f(x, y)$ as $z - f(x, y) = 0$. The surface $z = f(x, y)$ is thus the level surface $g(x, y, z) = 0$ of the function $g(x, y, z) = z - f(x, y)$. Note that

$$\frac{\partial g}{\partial x} = -\frac{\partial f}{\partial x}, \qquad \frac{\partial g}{\partial y} = -\frac{\partial f}{\partial y} \qquad \text{and} \qquad \frac{\partial g}{\partial z} = 1.$$

By the formula in (*a*),

$$|\cos \gamma| = \frac{1}{\sqrt{\left(\frac{\partial f}{\partial x}\right)^2 + \left(\frac{\partial f}{\partial y}\right)^2 + 1}}. \quad ■$$

Theorem 2 is stated for projections on the xy plane. Similar theorems hold for projections on the xz or yz plane. The angle γ is then replaced by β or α, and the normal vector is dotted into **j** or **i**. Just draw a picture in each case; there is no point in trying to memorize formulas for each situation.

EXAMPLE 4 Find the area of the part of the saddle $z = xy$ inside the cylinder $x^2 + y^2 = a^2$.

SOLUTION Let $\mathcal{S}$ be the part of the surface $z = xy$ inside $x^2 + y^2 = a^2$. Then

$$\text{Area of } \mathcal{S} = \int_{\mathcal{S}} 1 \, dS.$$

The projection of $\mathcal{S}$ on the xy plane is a disk of radius a and center $(0, 0)$. Call it $\mathcal{A}$, as in Fig. 12. Then

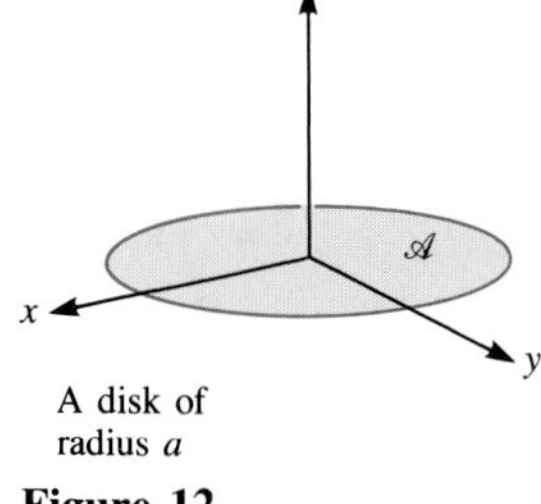

A disk of radius a

Figure 12

$$\text{Area of } \mathcal{S} = \int_{\mathcal{S}} 1 \, dS = \int_{\mathcal{A}} \frac{1}{|\cos \gamma|} \, dA. \tag{6}$$

To find a normal to $\mathcal{S}$ rewrite $z = xy$ as $z - xy = 0$. Thus $\mathcal{S}$ is a level surface of the function $g(x, y, z) = z - xy$. A normal to $\mathcal{S}$ is therefore

$$\begin{aligned} \nabla g &= \frac{\partial g}{\partial x}\mathbf{i} + \frac{\partial g}{\partial y}\mathbf{j} + \frac{\partial g}{\partial z}\mathbf{k} \\ &= -y\mathbf{i} - x\mathbf{j} + \mathbf{k}. \end{aligned}$$

Then

$$\begin{aligned} \cos \gamma &= \frac{\mathbf{k} \cdot \nabla g}{\|\mathbf{k}\| \, \|\nabla g\|} \\ &= \frac{\mathbf{k} \cdot (-y\mathbf{i} - x\mathbf{j} + \mathbf{k})}{\sqrt{y^2 + x^2 + 1}} \\ &= \frac{1}{\sqrt{y^2 + x^2 + 1}}. \end{aligned}$$

Note that the area of $\mathcal{S}$ *is* $\int_{\mathcal{A}} \sqrt{(\partial f/\partial x)^2 + (\partial f/\partial y)^2 + 1} \, dA.$

By (6),

$$\text{Area of } \mathcal{S} = \int_{\mathcal{A}} \sqrt{y^2 + x^2 + 1} \, dA. \tag{7}$$

Use polar coordinates to evaluate the integral in (7):

$$\int_{\mathcal{A}} \sqrt{y^2 + x^2 + 1} \, dA = \int_0^{2\pi} \int_0^a \sqrt{r^2 + 1} \, r \, dr \, d\theta.$$

The inner integration gives

$$\int_0^a \sqrt{r^2+1}\, r\, dr = \frac{(r^2+1)^{3/2}}{3}\bigg|_0^a = \frac{(1+a^2)^{3/2}-1}{3}.$$

The second integration gives

$$\int_0^{2\pi} \frac{(1+a^2)^{3/2}-1}{3}\, d\theta = \frac{2\pi}{3}[(1+a^2)^{3/2}-1]. \quad ■$$

Section Summary

After defining $\int_{\mathcal{S}} f(P)\, dS$, an integral over a surface, we showed how to compute it when the surface is part of a sphere. (Replace dS by $a^2 \sin\phi\, d\phi\, d\theta$, where a is the radius of the sphere.) In the case of a closed surface $\mathcal{S}$, whose projection on the xy plane is $\mathcal{A}$, we showed that

$$\int_{\mathcal{S}} R(x, y, z) \cos\gamma\, dS = \int_{\mathcal{A}} (R(x, y, z_2) - R(x, y, z_1))\, dA,$$

where z_2 and z_1 are the z coordinates of the top and bottom parts of $\mathcal{S}$. If each line parallel to the z axis meets the surface $\mathcal{S}$ in at most one point, an integral over $\mathcal{S}$ can be replaced by an integral over $\mathcal{A}$, the projection of $\mathcal{S}$ on the xy plane:

$$\int_{\mathcal{S}} f(P)\, dS = \int_{\mathcal{A}} \frac{h(Q)}{|\cos\gamma|}\, dA,$$

as shown in Theorem 1. If $\mathcal{S}$ is the graph of $z = f(x, y)$, then the area of $\mathcal{S}$ above $\mathcal{A}$ is

$$\int_{\mathcal{A}} \sqrt{(\partial f/\partial x)^2 + (\partial f/\partial y)^2 + 1}\, dA.$$

EXERCISES FOR SEC. 17.1: SURFACE INTEGRALS

1 A small patch of a surface makes an angle of $\pi/4$ with the xy plane. Its projection on that plane has area 0.05. Estimate the area of the patch.

2 A small patch of a surface makes an angle of 25° with the yz plane. Its projection on that plane has area 0.03. Estimate the area of the patch.

3 (*a*) Draw a diagram of the part of the plane $x + 2y + 3z = 12$ that lies inside the cylinder $x^2 + y^2 = 9$.
(*b*) Find as simply as possible the area of the part of the plane $x + 2y + 3z = 12$ that lies inside the cylinder $x^2 + y^2 = 9$.

4 (*a*) Draw a diagram of the part of the plane $z = x + 3y$ that lies inside the cylinder $r = 1 + \cos\theta$.
(*b*) Find as simply as possible the area of the part of the plane $z = x + 3y$ that lies inside the cylinder $r = 1 + \cos\theta$.

5 Let $f(P)$ be the square of the distance from P to a fixed diameter of a sphere of radius a. Find the average value of $f(P)$ for points on the sphere.

6 Find the area of that part of the sphere of radius a that lies within a cone of half-vertex angle $\pi/4$ and vertex at the center of the sphere, as in Fig. 13.

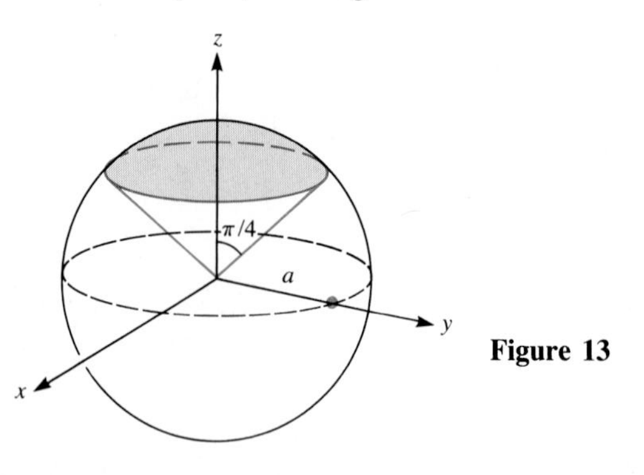

Figure 13

In Exercises 7 and 8 evaluate $\int_{\mathcal{S}} \mathbf{F} \cdot \mathbf{n}\, dS$ for the given spheres and vectors fields ($\mathbf{n}$ is the exterior unit normal).

7 The sphere $x^2 + y^2 + z^2 = 9$ and $\mathbf{F} = x^2\mathbf{i} + y^2\mathbf{j} + z^2\mathbf{k}$.

8 The sphere $x^2 + y^2 + z^2 = 1$ and $\mathbf{F} = x^3\mathbf{i} + y^2\mathbf{j}$.

9 Let $\mathcal{S}$ be a surface of the type in Example 2. Evaluate $\int_{\mathcal{S}} x \cos \gamma\, dS$.

10 Let $\mathcal{S}$ be a surface of the type in Example 2. Show that $\int_{\mathcal{S}} z \cos \gamma\, dS$ is equal to the volume of the solid bounded by $\mathcal{S}$.

11 Let $\mathcal{S}$ be a surface of the type in Example 2. Evaluate $\int_{\mathcal{S}} x \cos \alpha\, dS$, including clear diagrams and a complete, self-contained explanation.

12 Let $\mathcal{S}$ be a surface of the type in Example 2. Evaluate $\int_{\mathcal{S}} y \cos \beta\, dS$, including clear diagrams and a complete, self-contained explanation.

13 Find the z coordinate $\bar{z}$ of the centroid of the part of the saddle $z = xy$ that lies above the portion of the disk bounded by the circle $x^2 + y^2 = a^2$ in the first quadrant.

14 Find the area of the part of the spherical surface $x^2 + y^2 + z^2 = 1$ that lies within the vertical cylinder erected on the circle $r = \cos \theta$ and above the xy plane.

15 Find the area of that portion of the parabolic cylinder $z = \frac{1}{2}x^2$ between the three planes $y = 0$, $y = x$, and $x = 2$.

16 Evaluate $\int_{\mathcal{S}} x^2 y\, dS$, where $\mathcal{S}$ is the portion in the first octant of a sphere with radius a and center at the origin, in the following way:
(*a*) Set up an integral using x and y as parameters.
(*b*) Set up an integral using ϕ and θ as parameters.
(*c*) Evaluate the easier of (*a*) and (*b*).

17 A triangle in the plane $z = x + y$ is directly above the triangle in the xy plane whose vertices are (1, 2), (3, 4), and (2, 5). Find the area of
(*a*) the triangle in the xy plane,
(*b*) the triangle in the plane $z = x + y$.

18 Let $\mathcal{S}$ be the triangle with vertices (1, 1, 1), (2, 3, 4), and (3, 4, 5).
(*a*) Using vectors, find the area of $\mathcal{S}$.
(*b*) Using the formula

$$\text{Area of } \mathcal{S} = \int_{\mathcal{S}} 1\, dS,$$

find the area of $\mathcal{S}$.

19 Find the area of the portion of the cone $z^2 = x^2 + y^2$ that lies above one loop of the curve $r = \sqrt{\cos 2\theta}$.

20 Let $\mathcal{S}$ be the triangle whose vertices are (1, 0, 0), (0, 2, 0), and (0, 0, 3). Let $f(x, y, z) = 3x + 2y + 2z$. Evaluate $\int_{\mathcal{S}} f(P)\, dS$.

In Exercises 21 and 22 find $\bar{z}$ for the given surfaces.

21 The portion of the paraboloid $2z = x^2 + y^2$ below the plane $z = 9$.

22 The portion of the plane $x + 2y + 3z = 6$ above the triangle in the xy plane whose vertices are (0, 0), (4, 0), and (0, 1).

In Exercises 23 and 24 let $\mathcal{S}$ be a sphere of radius a with center at the origin of a rectangular coordinate system.

23 Evaluate each of these integrals with a minimum amount of labor.
(*a*) $\int_{\mathcal{S}} x\, dS$ (*b*) $\int_{\mathcal{S}} x^3\, dS$
(*c*) $\displaystyle\int_{\mathcal{S}} \frac{2x + 4y^5}{\sqrt{2 + x^2 + 3y^2}}\, dS$

24 (*a*) Why is $\int_{\mathcal{S}} x^2\, dS = \int_{\mathcal{S}} y^2\, dS$?
(*b*) Evaluate $\int_{\mathcal{S}} (x^2 + y^2 + z^2)\, dS$ with a minimum amount of labor.
(*c*) In view of (*a*) and (*b*), evaluate $\int_{\mathcal{S}} x^2\, dS$.
(*d*) Evaluate $\int_{\mathcal{S}} (2x^2 + 3y^2)\, dS$.

25 An electric field radiates power at the rate of $k(\sin^2\phi)/\rho^2$ units per square meter to the point $P = (\rho, \theta, \phi)$. Find the total power radiated to the sphere $\rho = a$.

26 A sphere of radius $2a$ has its center at the origin of a rectangular coordinate system. A circular cylinder of radius a has its axis parallel to the z axis and passes through the z axis. Find the area of that part of the sphere that lies within the cylinder and is above the xy plane.

Consider a distribution of mass on the surface $\mathcal{S}$. Let its density at P be $\sigma(P)$. The **moment of inertia** of the mass around the z axis is defined as $\int_{\mathcal{S}} (x^2 + y^2)\, \sigma(P)\, dS$. Exercises 27 and 28 concern this integral.

27 Find the moment of inertia of a homogeneous distribution of mass on the surface of a ball of radius a around a diameter. Let the total mass be M.

28 Find the moment of inertia about the z axis of a homogeneous distribution of mass on the triangle whose vertices are $(a, 0, 0)$, $(0, b, 0)$, and $(0, 0, c)$. Take a, b, and c to be positive. Let the total mass be M.

---

29 Let $\mathcal{S}$ be a sphere of radius a. Let A be a point at distance $b > a$ from the center of $\mathcal{S}$. For P in $\mathcal{S}$ let $\delta(P)$ be $1/q$, where q is the distance from P to A. Show that the average of $\delta(P)$ over $\mathcal{S}$ is $1/b$.

30 The data are the same as in Exercise 29 but $b < a$. Show that in this case the average of $1/q$ is $1/a$. (The average does *not* depend on b in this case.)

Exercises 31 to 33 concern integration over the curved surface of a cone. Spherical coordinates are also useful for integrating over a right circular cone. Place the origin at the vertex of the cone and the "$\phi = 0$" ray along the axis of the cone, as shown in Fig. 14 (next page). Let α be the half-vertex angle of the cone.

On the surface of the cone ϕ is constant, $\phi = \alpha$, but ρ and θ vary. A small "rectangular" patch on the surface of the cone corresponding to slight changes $d\theta$ and $d\rho$ has area approximately

$$(\rho \sin \alpha\, d\theta)\, d\rho = \rho \sin \alpha\, d\rho\, d\theta.$$

(See Fig. 14.) So we may write

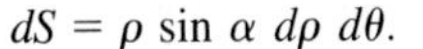
$$dS = \rho \sin \alpha \, d\rho \, d\theta.$$

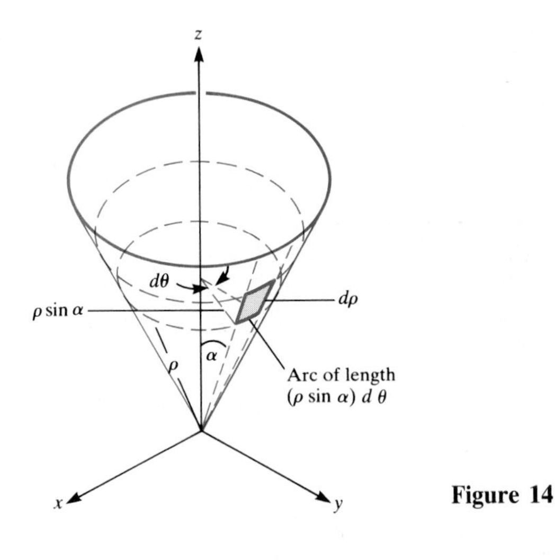

Figure 14

31 Find the average distance from points on the curved surface of a cone of radius a and height h to its axis.

32 Evaluate $\int_{\mathcal{S}} z^2 \, dS$, where $\mathcal{S}$ is the *entire* surface of the cone shown in Fig. 15, including its base.

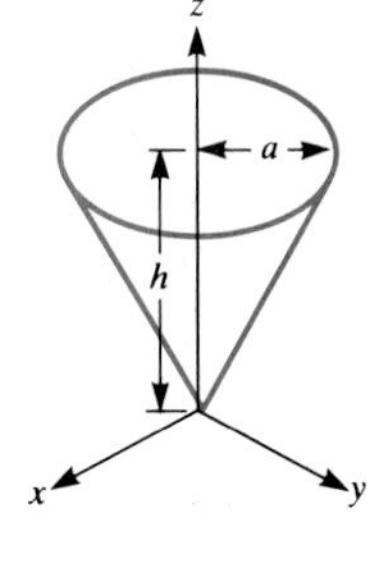

Figure 15

33 Evaluate $\int_{\mathcal{S}} x^2 \, dS$, where $\mathcal{S}$ is the curved surface of the right circular cone of radius 1 and height 1 with axis along the z axis.

Integration over the curved surface of a right circular cylinder is easiest in cylindrical coordinates. Consider such a cylinder of radius a and axis on the z axis. A small patch on the cylinder corresponding to dz and $d\theta$ has area approximately $dS = a \, dz \, d\theta$. (Why?) Exercises 34 and 35 illustrate the use of these coordinates.

34 Let $\mathcal{S}$ be the *entire* surface of a solid cylinder of radius a and height h. For P in $\mathcal{S}$ let $f(P)$ be the square of the distance from P to one base. Find $\int_{\mathcal{S}} f(P) \, dS$. Be sure to include the two bases in the integration.

35 Let $\mathcal{S}$ be the curved part of the cylinder in Exercise 34. Let $f(P)$ be the square of the distance from P to a fixed diameter in a base. Find the average value of $f(P)$ for points in $\mathcal{S}$.

36 The areas of the projections of a small flat surface patch on the three coordinate planes are 0.01, 0.02, and 0.03. Is that enough information to find the area of the patch? If so, find the area. If not, explain why not.

37 Let $\mathbf{F}$ describe the flow of a fluid in space. (See Sec. 16.3 for fluid flow in a planar region.) $\mathbf{F}(P) = \delta(P) \, \mathbf{v}(P)$, where $\delta(P)$ is the density of the fluid at P and $\mathbf{v}(P)$ is the velocity of the fluid at P. Making clear, large diagrams, explain why the rate at which the fluid is leaving the solid region enclosed by a surface $\mathcal{S}$ is $\int_{\mathcal{S}} \mathbf{F} \cdot \mathbf{n} \, dS$, where $\mathbf{n}$ denotes the unit exterior normal to $\mathcal{S}$.

38 We claimed in (3) that when $\gamma > \pi/2$, $dA = -\cos \gamma \, dS$. Drawing clear sketches, explain why this claim is correct.

17.2 THE DIVERGENCE THEOREM

The divergence theorem (also known as Gauss's theorem) is the analog in space of Green's theorem. Instead of a region in the plane and its bounding curve or curves, the divergence theorem concerns a region $\mathcal{V}$ in space and its bounding surface or surfaces. The region $\mathcal{V}$ may, for instance, be a ball, a cube, or a doughnut; in these cases the bounding surface consists of a single connected piece. Or perhaps the region $\mathcal{V}$ is bounded by two surfaces $\mathcal{S}_1$ and $\mathcal{S}_2$, one inside the other like two concentric spheres.

Just as Green's theorem was stated separately for a plane region $\mathcal{A}$ bounded by a single curve and for a plane region bounded by two curves, the divergence theorem will be stated in two forms. We first treat it for a solid region that has no holes, as in Fig. 1.

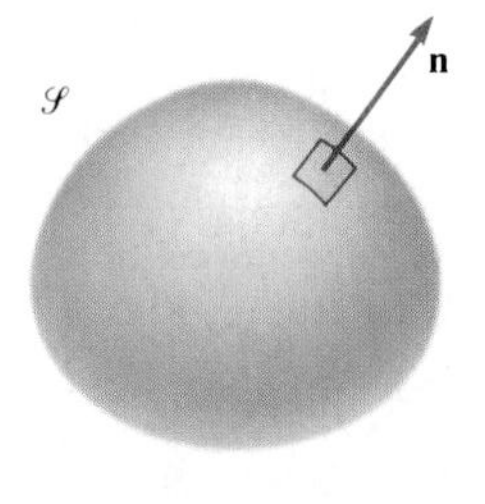

$\mathcal{S}$ is the surface of $\mathcal{V}$.

Figure 1

Divergence Theorem *One-surface case.* Let $\mathcal{V}$ be a region in space bounded by the single connected surface $\mathcal{S}$. Let $\mathbf{n}$ denote the exterior unit normal of $\mathcal{V}$ along the boundary $\mathcal{S}$. Then

$$\int_{\mathcal{S}} \mathbf{F} \cdot \mathbf{n}\, dS = \int_{\mathcal{V}} \nabla \cdot \mathbf{F}\, dV$$

for any vector field **F** defined on $\mathcal{V}$.

In words: "The integral of the normal component of **F** over a surface equals the integral of the divergence of **F** over the solid region that the surface bounds."

The integral $\int_{\mathcal{S}} \mathbf{F} \cdot \mathbf{n}\, dS$ is called the **flux** of the field **F** across the surface $\mathcal{S}$.

If $\mathbf{F} = P\mathbf{i} + Q\mathbf{j} + R\mathbf{k}$, then the divergence theorem reads

$$\int_{\mathcal{S}} (P\mathbf{i} + Q\mathbf{j} + R\mathbf{k}) \cdot (\cos\alpha\ \mathbf{i} + \cos\beta\ \mathbf{j} + \cos\gamma\ \mathbf{k})\, dS = \int_{\mathcal{V}} \left(\frac{\partial P}{\partial x} + \frac{\partial Q}{\partial y} + \frac{\partial R}{\partial z} \right) dV,$$

where $\cos\alpha$, $\cos\beta$, and $\cos\gamma$ are the direction cosines of the exterior normal. Evaluating the dot product puts the divergence theorem in the form

$$\int_{\mathcal{S}} (P\cos\alpha + Q\cos\beta + R\cos\gamma)\, dS = \int_{\mathcal{V}} \left(\frac{\partial P}{\partial x} + \frac{\partial Q}{\partial y} + \frac{\partial R}{\partial z} \right) dV.$$

When the divergence theorem is expressed in this form, we see that it would be the consequence of three scalar theorems:

$$\int_{\mathcal{S}} P\cos\alpha\, dS = \int_{\mathcal{V}} \frac{\partial P}{\partial x}\, dV, \qquad \int_{\mathcal{S}} Q\cos\beta\, dS = \int_{\mathcal{V}} \frac{\partial Q}{\partial y}\, dV, \tag{1}$$

and

$$\int_{\mathcal{S}} R\cos\gamma\, dS = \int_{\mathcal{V}} \frac{\partial R}{\partial z}\, dV.$$

As with Green's theorem, there are two ways to prove the divergence theorem. Either start with $\int_{\mathcal{S}} \mathbf{F} \cdot \mathbf{n}\, dS$ and show it equals $\int_{\mathcal{V}} \nabla \cdot \mathbf{F}\, dV$ or start with the integral $\int_{\mathcal{V}} \nabla \cdot \mathbf{F}\, dV$ and show it equals $\int_{\mathcal{S}} \mathbf{F} \cdot \mathbf{n}\, dS$. We will use the second method.

Proof of the divergence theorem We will prove the theorem in the special case that each line parallel to an axis meets the surface $\mathcal{S}$ in at most two points. We prove the third equation in (1). The other two are established the same way.

We wish to show that

$$\int_{\mathcal{S}} R\cos\gamma\, dS = \int_{\mathcal{V}} \frac{\partial R}{\partial z}\, dV. \tag{2}$$

Let $\mathcal{A}$ be the projection of $\mathcal{S}$ on the xy plane. Its description is

$$a \le x \le b, \qquad y_1(x) \le y \le y_2(x).$$

The description of $\mathcal{V}$ is then

$$a \le x \le b, \qquad y_1(x) \le y \le y_2(x), \qquad z_1(x, y) \le z \le z_2(x, y).$$

(See Fig. 2 on the next page.) Then

$$\int_{\mathcal{V}} \frac{\partial R}{\partial z}\, dV = \int_a^b \int_{y_1(x)}^{y_2(x)} \int_{z_1(x, y)}^{z_2(x, y)} \frac{\partial R}{\partial z}\, dz\, dy\, dx. \tag{3}$$

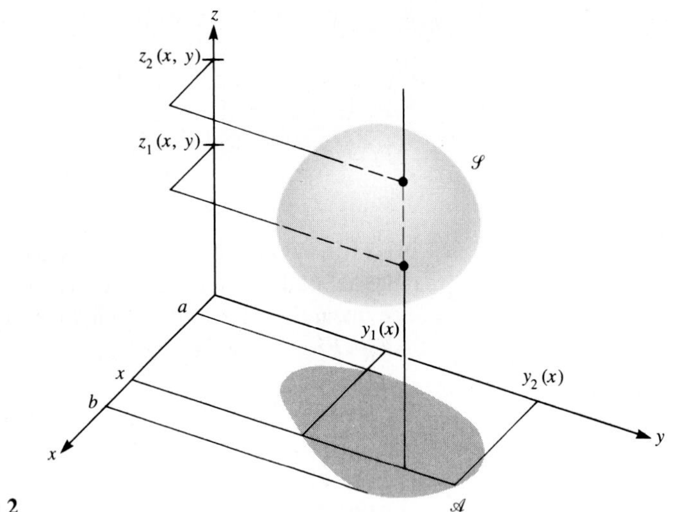

Figure 2

Now, the first integration gives

$$\int_{z_1(x,\,y)}^{z_2(x,\,y)} \frac{\partial R}{\partial z}\, dz = R(x, y, z_2) - R(x, y, z_1),$$

by the fundamental theorem of calculus. We have, therefore,

$$\int_{\mathcal{V}} \frac{\partial R}{\partial z}\, dV = \int_a^b \int_{y_1(x)}^{y_2(x)} [R(x, y, z_2) - R(x, y, z_1)]\, dy\, dx,$$

hence $$\int_{\mathcal{V}} \frac{\partial R}{\partial z}\, dV = \int_{\mathcal{A}} [R(x, y, z_2) - R(x, y, z_1)]\, dA.$$

By Example 2 in the preceding section, the last integral equals

$$\int_{\mathcal{S}} R(x, y, z) \cos \gamma\, dS.$$

Thus $$\int_{\mathcal{V}} \frac{\partial R}{\partial z}\, dV = \int_{\mathcal{S}} R \cos \gamma\, dS,$$

and (2) is established. As noted, similar arguments establish the other two equations in (1). ■

Using a cancellation principle similar to that of Sec. 16.4, we could show that the divergence theorem holds for solids that can be partitioned into solids of the type considered. You may therefore feel free to apply the divergence theorem in this more general case.

The next example is simply a check of the divergence theorem for a particular choice of $\mathcal{V}$, $\mathcal{S}$, and $\mathbf{F}$.

EXAMPLE 1 Let $\mathbf{F} = (z^2 + 2)\mathbf{k}$ and let $\mathcal{S}$ consist of $\mathcal{H}$, the top half of the sphere $x^2 + y^2 + z^2 = a^2$, together with its base $\mathcal{B}$, the disk of radius a centered at

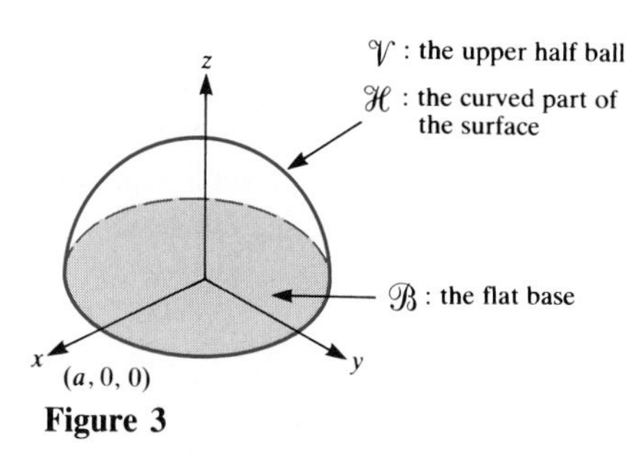

Figure 3

(0, 0) in the xy plane, as shown in Fig. 3. Verify the divergence theorem in this case.

SOLUTION Let $\mathcal{V}$ be the solid region below $\mathcal{H}$ and above $\mathcal{B}$. Then

$$\int_{\mathcal{V}} \nabla \cdot \mathbf{F}\, dV = \int_{\mathcal{V}} \left[\frac{\partial}{\partial x}(0) + \frac{\partial}{\partial y}(0) + \frac{\partial}{\partial z}(z^2+2)\right] dV = \int_{\mathcal{V}} 2z\, dV$$

$$= \int_0^{2\pi}\int_0^{\pi/2}\int_0^a 2(\rho\cos\phi)\rho^2 \sin\phi\, d\rho\, d\phi\, d\theta$$

$$= \int_0^{2\pi}\int_0^{\pi/2}\int_0^a 2\rho^3 \cos\phi \sin\phi\, d\rho\, d\phi\, d\theta.$$

Straightforward evaluation of the repeated integral gives the value $\pi a^4/2$.

First compute $\int_{\mathcal{B}} \mathbf{F}\cdot\mathbf{n}\, dS$.

Next, evaluate $\int_{\mathcal{S}} \mathbf{F}\cdot\mathbf{n}\, dS$. It is necessary to compute $\int_{\mathcal{B}} \mathbf{F}\cdot\mathbf{n}\, dS$ and $\int_{\mathcal{H}} \mathbf{F}\cdot\mathbf{n}\, dS$. On $\mathcal{B}$ the exterior normal is $-\mathbf{k}$. Thus

$$\int_{\mathcal{B}} \mathbf{F}\cdot\mathbf{n}\, dS = \int_{\mathcal{B}} (z^2+2)\mathbf{k}\cdot(-\mathbf{k})\, dS.$$

On $\mathcal{B}$, however, $z = 0$. So

$$\int_{\mathcal{B}} \mathbf{F}\cdot\mathbf{n}\, dS = \int_{\mathcal{B}} -2\, dS = -2\ (\text{Area of } \mathcal{B}) = -2\pi a^2.$$

Then compute $\int_{\mathcal{H}} \mathbf{F}\cdot\mathbf{n}\, dS$.

To evaluate $\int_{\mathcal{H}} \mathbf{F}\cdot\mathbf{n}\, dS$, recall that on a sphere of radius a centered at the origin,

$$\frac{x\mathbf{i}+y\mathbf{j}+z\mathbf{k}}{\|x\mathbf{i}+y\mathbf{j}+z\mathbf{k}\|} = \frac{x\mathbf{i}+y\mathbf{j}+z\mathbf{k}}{a}$$

is a unit exterior normal. Thus

$$\int_{\mathcal{H}} \mathbf{F}\cdot\mathbf{n}\, dS = \int_{\mathcal{H}} (z^2+2)\mathbf{k}\cdot\frac{(x\mathbf{i}+y\mathbf{j}+z\mathbf{k})}{a}\, dS$$

$$= \int_{\mathcal{H}} \frac{z^3+2z}{a}\, dS = \frac{1}{a}\int_{\mathcal{H}} (z^3+2z)\, dS.$$

Using spherical coordinates on the sphere of radius a, we have

$$\int_{\mathcal{H}} (z^3+2z)\, dS = \int_0^{2\pi}\int_0^{\pi/2} [(a\cos\phi)^3 + 2a\cos\phi]a^2 \sin\phi\, d\phi\, d\theta$$

$$= a^3 \int_0^{2\pi}\int_0^{\pi/2} (a^2\cos^3\phi \sin\phi + 2\cos\phi\sin\phi)\, d\phi\, d\theta$$

Fill in the details.

$$= a^3\left(\frac{a^2}{4}+1\right)(2\pi).$$

Thus

$$\int_{\mathcal{H}} \mathbf{F}\cdot\mathbf{n}\, dS = \frac{1}{a}\left[a^3\left(\frac{a^2}{4}+1\right)(2\pi)\right] = \frac{\pi a^4}{2} + 2\pi a^2.$$

All told,

$$\int_{\mathcal{S}} \mathbf{F}\cdot\mathbf{n}\, dS = \int_{\mathcal{B}} \mathbf{F}\cdot\mathbf{n}\, dS + \int_{\mathcal{H}} \mathbf{F}\cdot\mathbf{n}\, dS = -2\pi a^2 + \left(\frac{\pi a^4}{2} + 2\pi a^2\right)$$

$$= \frac{\pi a^4}{2},$$

which agrees with the result for $\int_{\mathcal{V}} \nabla\cdot\mathbf{F}\, dV$. ■

Two-Surface Version of the Divergence Theorem

The divergence theorem also holds if the solid region has several holes like a piece of Swiss cheese. In this case, the boundary consists of several separate connected surfaces. The most important case is when there is just one hole and hence an inner surface $\mathcal{S}_1$ and an outer surface $\mathcal{S}_2$, as shown in Fig. 4.

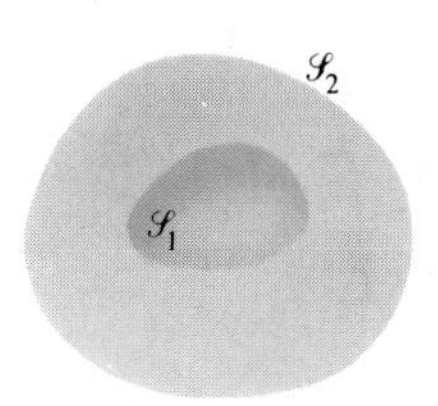

$\mathcal{V}$ is bounded by inner surface $\mathcal{S}_1$, and outer surface $\mathcal{S}_2$.

Figure 4

Divergence Theorem *Two-surface case.* Let $\mathcal{V}$ be a region in space bounded by the surfaces $\mathcal{S}_1$ and $\mathcal{S}_2$. Let $\mathbf{n}^*$ denote the exterior normal along the boundary. Then

$$\int_{\mathcal{S}_1} \mathbf{F} \cdot \mathbf{n}^* \, dS + \int_{\mathcal{S}_2} \mathbf{F} \cdot \mathbf{n}^* \, dS = \int_{\mathcal{V}} \nabla \cdot \mathbf{F} \, dV$$

for any vector field defined on $\mathcal{V}$.

The importance of this form of the divergence theorem is that it provides us with Corollary 1 in the same way that the two-curve form of Green's theorem provided Corollary 1 of Sec. 16.5.

Corollary 1 Let $\mathcal{S}_1$ and $\mathcal{S}_2$ be two connected surfaces that form the boundary of the region $\mathcal{V}$. Let $\mathbf{F}$ be a vector field defined on $\mathcal{V}$ such that the divergence of $\mathbf{F}$, $\nabla \cdot \mathbf{F}$, is 0 throughout $\mathcal{V}$. Then

$$\int_{\mathcal{S}_1} \mathbf{F} \cdot \mathbf{n} \, dS = \int_{\mathcal{S}_2} \mathbf{F} \cdot \mathbf{n} \, dS.$$

The proof, similar to that for Corollary 1 in Sec. 16.5, is omitted. The next corollary and its proof are similar to Corollary 2 of Sec. 16.5 and its proof.

Corollary 2 Let $\mathbf{F}$ be a vector field defined everywhere in space except perhaps at the point P_0. Assume that the divergence of $\mathbf{F}$ is 0. Let $\mathcal{S}_1$ and $\mathcal{S}_2$ be two surfaces each of which encloses the point P_0. Then

$$\int_{\mathcal{S}_1} \mathbf{F} \cdot \mathbf{n} \, dS = \int_{\mathcal{S}_2} \mathbf{F} \cdot \mathbf{n} \, dS.$$

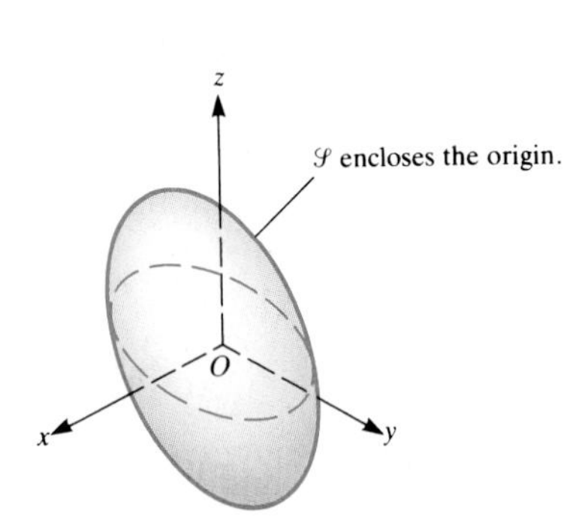

Figure 5

The next example, which plays an important role in electrostatics and gravitational theory, illustrates the use of Corollary 2.

EXAMPLE 2 Let $\mathbf{F} = \hat{\mathbf{r}}/\|\mathbf{r}\|^2$, where $\mathbf{r} = x\mathbf{i} + y\mathbf{j} + z\mathbf{k}$, and let $\mathcal{S}$ be a surface that surrounds the origin, as in Fig. 5. Show that $\int_{\mathcal{S}} \mathbf{F} \cdot \mathbf{n} \, dS = 4\pi$.

SOLUTION By Example 3 of Sec. 16.1, the divergence of $\mathbf{F}$ is 0. According to Corollary 2, then, we may learn the value of $\int_{\mathcal{S}} \mathbf{F} \cdot \mathbf{n} \, dS$ by computing $\int_{\mathcal{S}_1} \mathbf{F} \cdot \mathbf{n} \, dS$ for any surface $\mathcal{S}_1$ that encloses the origin. The simplest surface to use is a sphere $\mathcal{S}_1$ of radius 1 and center at the origin. On $\mathcal{S}_1$, $\|\mathbf{r}\| = 1$ and $\mathbf{n} = \hat{\mathbf{r}}$. Thus $\mathbf{F} \cdot \mathbf{n} = (\hat{\mathbf{r}} \cdot \hat{\mathbf{r}})/\|\mathbf{r}\|^2 = 1$ and

$$\int_{\mathcal{S}_1} \mathbf{F} \cdot \mathbf{n}\, dS = \int_{\mathcal{S}_1} 1\, dS = \text{Area of } \mathcal{S}_1 = 4\pi.$$

(The area of a sphere of radius a is $4\pi a^2$.) ■

The Divergence Theorem and Electrostatics

We will illustrate the importance of the divergence theorem by showing its use in electrostatics. (We do not assume that you already know electrostatics.) First, some basic information.

An electric charge can be negative (an electron, for example) or positive (a proton). Associated with each individual charge is an electric field $\mathbf{E}$. The field associated with a distribution of charges is the (vector) sum of the fields associated with the individual charges. In the case of a continuous distribution of charge, we must use an integral instead of a sum to find the electric field.

All that we need to know about any such field $\mathbf{E}$ is that

$$\nabla \cdot \mathbf{E} = \frac{q}{\epsilon_0}, \tag{4}$$

where ϵ_0 is a constant and q is the density of charge, which may vary from point to point. Equation (4) is called Coulomb's law. Gauss's law asserts that if $\mathcal{S}$ is a closed surface that encloses a total charge Q, then

$$\int_{\mathcal{S}} \mathbf{E} \cdot \mathbf{n}\, dS = \frac{Q}{\epsilon_0}.$$

EXAMPLE 3 Obtain Gauss's law from Coulomb's law.

SOLUTION Let $\mathcal{V}$ be the solid region whose boundary is $\mathcal{S}$. Then

$$\begin{aligned} \int_{\mathcal{S}} \mathbf{E} \cdot \mathbf{n}\, dS &= \int_{\mathcal{V}} \nabla \cdot \mathbf{E}\, dV && \text{divergence theorem} \\ &= \int_{\mathcal{V}} \frac{q}{\epsilon_0}\, dV && \text{Coulomb's law} \\ &= \frac{1}{\epsilon_0} \int_{\mathcal{V}} q\, dV \\ &= \frac{Q}{\epsilon_0}. \end{aligned}$$ ■

The next natural question is, "Does Gauss's law imply Coulomb's law?" Example 4 shows that the answer is yes.

EXAMPLE 4 Deduce Coulomb's law from Gauss's law.

SOLUTION Let $\mathcal{V}$ be any spatial region and let $\mathcal{S}$ be its surface. Let Q be the total charge in $\mathcal{V}$. Then

$$\begin{aligned} \frac{Q}{\epsilon_0} &= \int_{\mathcal{S}} \mathbf{E} \cdot \mathbf{n}\, dS && \text{Gauss's law} \\ &= \int_{\mathcal{V}} \nabla \cdot \mathbf{E}\, dV. && \text{divergence theorem} \end{aligned}$$

On the other hand,

$$Q = \int_{\mathcal{V}} q \, dV,$$

where q is the charge density. Thus

$$\int_{\mathcal{V}} \frac{q}{\epsilon_0} \, dV = \int_{\mathcal{V}} \nabla \cdot \mathbf{E} \, dV,$$

or

$$\int_{\mathcal{V}} \left(\frac{q}{\epsilon_0} - \nabla \cdot \mathbf{E} \right) dV = 0$$

for all spatial regions. Since the integrand is assumed to be continuous, the "vanishing-integral principle" tells us that it must be identically 0. That is,

$$\frac{q}{\epsilon_0} - \nabla \cdot \mathbf{E} = 0,$$

which gives us Coulomb's law. ■

EXAMPLE 5 Find the electric field $\mathbf{E}$ induced by a single point charge q.

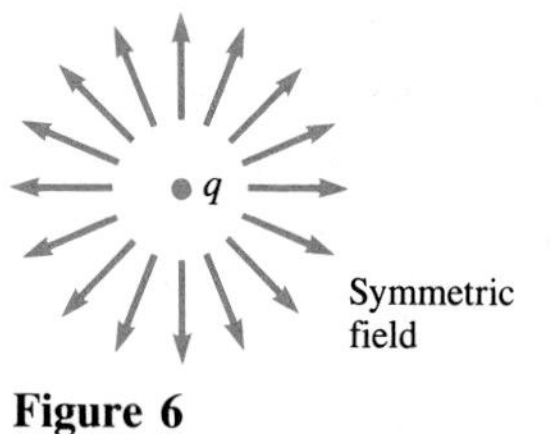

Figure 6

SOLUTION Because the charge is at a single point, its effect is the same in all directions. We conclude that $\mathbf{E}$ is a central field. (See Fig. 6.)

Place the origin of an xyz coordinate system at the charge q. Because $\mathbf{E}$ is a central field, there is a scalar function $f(r)$ such that

$$\mathbf{E}(\mathbf{r}) = f(r)\hat{\mathbf{r}},$$

where $r = \|\mathbf{r}\|$.

Now to find $f(r)$.

Let $\mathcal{S}_a$ be a sphere of radius a and center at the charge (the origin). By Gauss's law,

$$\int_{\mathcal{S}_a} \mathbf{E} \cdot \mathbf{n} \, dS = \frac{q}{\epsilon_0}.$$

$\mathbf{n} = \hat{\mathbf{r}}$, *so* $\hat{\mathbf{r}} \cdot \mathbf{n} = \hat{\mathbf{r}} \cdot \hat{\mathbf{r}} = 1$.

Thus

$$\frac{q}{\epsilon_0} = \int_{\mathcal{S}_a} \mathbf{E} \cdot \mathbf{n} \, dS = \int_{\mathcal{S}_a} f(r) \, \hat{\mathbf{r}} \cdot \mathbf{n} \, dS$$

$$= \int_{\mathcal{S}_a} f(a) \, dS$$

$$= f(a) \int_{\mathcal{S}_a} dS,$$

and we have

$$\frac{q}{\epsilon_0} = 4\pi a^2 f(a).$$

Thus

$$f(a) = \frac{q}{4\pi\epsilon_0} \frac{1}{a^2}.$$

This shows that $\mathbf{E}$ is an "inverse-square" field, and

$$\mathbf{E}(\mathbf{r}) = \frac{q}{4\pi\epsilon_0} \frac{1}{r^2} \hat{\mathbf{r}}. \quad ■$$

In Exercises 25 and 26 Gauss's law is used to determine the field associated with a uniform distribution of charge on an infinite line and also on an infinite plane.

The gravitational field associated with a mass distribution also satisfies Coulomb's law (with mass taking the place of charge). Hence gravitational fields also satisfy Gauss's law and must be inverse-square fields.

The Field $\hat{\mathbf{r}}/\|\mathbf{r}\|^2$ and Solid Angle

The total intensity of sunlight striking a surface $\mathcal{S}$ is proportional to the "solid angle" that the surface subtends as viewed from the sun. Let us define the notion of solid angle and see its relation to surface integrals. (You might review the relation between radians and line integrals described in Sec. 16.3.)

Let O be a point and $\mathcal{S}$ a surface such that each ray from O meets $\mathcal{S}$ in at most one point. Let $\mathcal{S}^*$ be the unit sphere with center at O. The rays from O that meet $\mathcal{S}$ intersect $\mathcal{S}^*$ in a set that we call $\mathcal{A}$, as shown in Fig. 7. Let the area of $\mathcal{A}$ be A. The solid angle subtended by $\mathcal{S}$ at O is said to have a measure of A steradians (from *stereo,* the Greek word for space, and *radians*). For instance, a closed surface $\mathcal{S}$ that encloses O subtends a solid angle of 4π steradians.

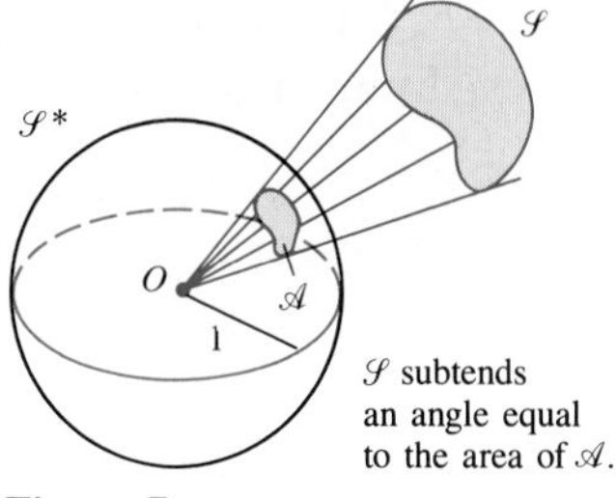

Figure 7

EXAMPLE 6 Let $\mathcal{S}$ be part of the surface of a sphere of radius a, $\mathcal{S}_a$, whose center is O. Find the angle subtended by $\mathcal{S}$ at O. (See Fig. 8.)

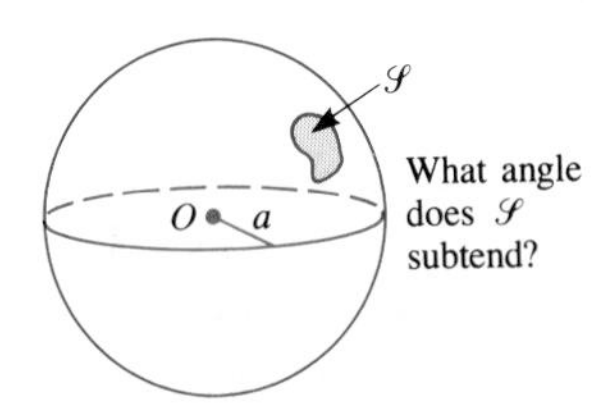

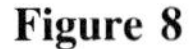

Figure 8

SOLUTION The entire sphere $\mathcal{S}$ subtends an angle of 4π steradians and has an area $4\pi a^2$. We therefore have the proportion

$$\frac{\text{Angle } \mathcal{S} \text{ subtends}}{\text{Angle } \mathcal{S}_a \text{ subtends}} = \frac{\text{Area of } \mathcal{S}}{\text{Area of } \mathcal{S}_a},$$

or

$$\frac{\text{Angle } \mathcal{S} \text{ subtends}}{4\pi} = \frac{\text{Area of } \mathcal{S}}{4\pi a^2}.$$

Hence

$$\text{Angle } \mathcal{S} \text{ subtends} = \frac{\text{Area of } \mathcal{S}}{a^2} \text{ steradians.} \quad \blacksquare$$

EXAMPLE 7 Let $\mathcal{S}$ be a surface such that each ray from the point O meets $\mathcal{S}$ in at most one point. Find an integral that represents in steradians the solid angle that $\mathcal{S}$ subtends at O.

SOLUTION Consider a very small patch of $\mathcal{S}$. Call it $d\mathcal{S}$ and let its area be dS. If we can estimate the angle that this patch subtends at O, then we will have the local approximation that will tell us what integral represents the total solid angle subtended by $\mathcal{S}$.

Let $\mathbf{n}$ be a unit normal at a point in the patch, which we regard as essentially flat, as in Fig. 9. Let $d\mathcal{A}$ be the projection of the patch $d\mathcal{S}$ on a plane perpendicular to $\mathbf{r}$, as shown in Fig. 9. The area of $d\mathcal{A}$ is approximately dA, where

$$dA = \mathbf{n} \cdot \hat{\mathbf{r}}\, dS.$$

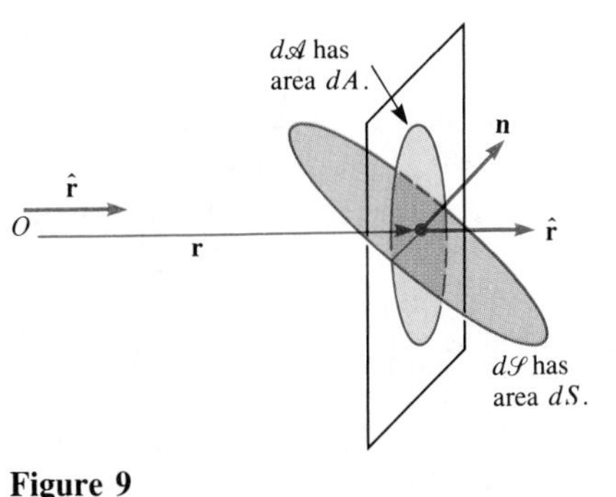

Figure 9

Now, $d\mathcal{S}$ and $d\mathcal{A}$ subtend approximately the same solid angle, which according to Example 6 is about

$$\frac{dA}{\|\mathbf{r}\|^2} = \frac{\mathbf{n} \cdot \hat{\mathbf{r}}}{\|\mathbf{r}\|^2}\, dS \text{ steradians.}$$

Consequently $\mathcal{S}$ subtends a solid angle of

$$\int_{\mathcal{S}} \frac{\mathbf{n} \cdot \hat{\mathbf{r}}}{\|\mathbf{r}\|^2}\, dS \text{ steradians.} \quad \blacksquare$$

In Example 2 it was shown with the aid of the divergence theorem that if $\mathcal{S}$ is a closed surface that encloses the origin, then

$$\int_{\mathcal{S}} \frac{\mathbf{n} \cdot \hat{\mathbf{r}}}{\|\mathbf{r}\|^2}\, dS = 4\pi.$$

Example 7 shows that this equation simply asserts that $\mathcal{S}$ subtends an angle of 4π steradians at the origin.

Remark: In applications $\mathbf{F} \cdot \mathbf{n}\, dS$ is often written $\mathbf{F} \cdot d\mathbf{S}$. The symbol $d\mathbf{S}$ is intended to suggest a vector in the direction of $\mathbf{n}$ with length dS. In this notation the divergence theorem reads

$$\int_{\mathcal{S}} \mathbf{F} \cdot d\mathbf{S} = \int_{\mathcal{V}} \nabla \cdot \mathbf{F}\, dV.$$

Section Summary

We generalized Green's theorem to the divergence theorem,

$$\int_{\mathcal{S}} \mathbf{F} \cdot \mathbf{n}\, dS = \int_{\mathcal{V}} \nabla \cdot \mathbf{F}\, dV,$$

where $\mathcal{S}$ bounds $\mathcal{V}$. We showed that if the divergence of $\mathbf{F}$ is 0 in the region between two closed surfaces $\mathcal{S}_1$ and $\mathcal{S}_2$, then $\int_{\mathcal{S}_1} \mathbf{F} \cdot \mathbf{n}\, dS = \int_{\mathcal{S}_2} \mathbf{F} \cdot \mathbf{n}\, dS$. This observation is important when working with the central field $\mathbf{F} = \hat{\mathbf{r}}/\|\mathbf{r}\|^2$, which appears in the study of gravity and electrostatics. We also showed that for this field, $\int_{\mathcal{S}} \mathbf{F} \cdot \mathbf{n}\, dS$ equals the steradians subtended by $\mathcal{S}$ at the origin.

EXERCISES FOR SEC. 17.2: THE DIVERGENCE THEOREM

1 State Green's theorem in words.

2 State the divergence theorem in words.

In Exercises 3 and 4 verify the divergence theorem for the given vector fields $\mathbf{F}$ and solid regions $\mathcal{V}$.

3 $\mathbf{F} = x\mathbf{i} + y\mathbf{j} + z\mathbf{k}$; $\mathcal{V}$ is the ball of radius a centered at the origin.

4 $\mathbf{F} = z\mathbf{k}$; $\mathcal{V}$ is the top half of a ball of radius a centered at the origin.

In Exercises 5 to 10 use the divergence theorem to evaluate $\int_{\mathcal{S}} \mathbf{F} \cdot \mathbf{n}\, dS$ for the given $\mathbf{F}$ where $\mathcal{S}$ is the boundary of the given region $\mathcal{V}$.

5 $\mathbf{F} = x^2\mathbf{i}$; $\mathcal{V}$ is the rectangular box bounded by the planes $x = 0$, $x = 2$, $y = 0$, $y = 3$, $z = 0$, and $z = 4$.

6 $\mathbf{F} = x^3\mathbf{i}$; $\mathcal{V}$ is the solid region between the spheres of radii a and b ($a < b$) with centers at the origin.

7 $\mathbf{F} = 3x\mathbf{i} + 2y\mathbf{j} + 6z\mathbf{k}$; $\mathcal{V}$ is the tetrahedron with the four vertices (0, 0, 0), (1, 0, 0), (0, 2, 0), and (0, 0, 3).

8 $\mathbf{F} = x\mathbf{i} + y\mathbf{j} + z\mathbf{k}$; $\mathcal{V}$ is bounded by the surface $z = 9 - x^2 - 2y^2$ and the xy plane.

9 $\mathbf{F} = x^3\mathbf{i} + y^3\mathbf{j} + 3z\mathbf{k}$; $\mathcal{V}$ is the region above the xy plane and below the surface $z = x^2$ which lies within the cylinder $r = \sin\theta$.

10 $\mathbf{F} = x^3\mathbf{i}$; $\mathcal{V}$ is bounded by the plane $z = 1$ and the cone $\phi = \pi/6$.

11 Show that, if $\mathbf{F}$ is a constant vector field and $\mathcal{S}$ is a surface bounding a region in space, then $\int_{\mathcal{S}} \mathbf{F} \cdot \mathbf{n}\, dS = 0$.

12 Let $\mathbf{F} = 2x\mathbf{i} + 3y\mathbf{j} + (5z + 6x)\mathbf{k}$, and let

$$\mathbf{G} = (3x + 4z^2)\mathbf{i} + (2y + 5x)\mathbf{j} + 5z\mathbf{k}.$$

Show that

$$\int_{\mathcal{S}} \mathbf{F} \cdot \mathbf{n}\, dS = \int_{\mathcal{S}} \mathbf{G} \cdot \mathbf{n}\, dS,$$

where $\mathcal{S}$ is any surface bounding a region in space.

In Exercises 13 to 20 use the divergence theorem.

13 Let $\mathcal{V}$ be the solid region bounded by the xy plane and the paraboloid $z = 9 - x^2 - y^2$. Evaluate $\int_{\mathcal{S}} \mathbf{F} \cdot \mathbf{n}\, dS$, where $\mathbf{F} = y^3\mathbf{i} + z^3\mathbf{j} + x^3\mathbf{k}$ and $\mathcal{S}$ is the boundary of $\mathcal{V}$.

14 Evaluate $\int_{\mathcal{V}} \text{div } \mathbf{F}\, dV$ for $\mathbf{F} = \sqrt{x^2 + y^2 + z^2}\,(x\mathbf{i} + y\mathbf{j} + z\mathbf{k})$ and $\mathcal{V}$ the ball of radius 2 and center at (0, 0, 0).

In Exercises 15 and 16 find $\int_{\mathcal{S}} \mathbf{F} \cdot \mathbf{n}\, dS$ for the given $\mathbf{F}$ and $\mathcal{S}$.

15 $\mathbf{F} = z\sqrt{x^2 + z^2}\,\mathbf{i} + (y + 3)\mathbf{j} - x\sqrt{x^2 + z^2}\,\mathbf{k}$ and $\mathcal{S}$ is the boundary of the solid region between $z = x^2 + y^2$ and the plane $z = 4x$.

16 $\mathbf{F} = x\mathbf{i} + (3y + z)\mathbf{j} + (4x + 2z)\mathbf{k}$ and $\mathcal{S}$ is the surface of the cube bounded by the planes $x = 1$, $x = 3$, $y = 2$, $y = 4$, $z = 3$, and $z = 5$.

17 Evaluate $\int_{\mathcal{S}} \mathbf{F} \cdot \mathbf{n}\, dS$, where $\mathbf{F} = 4xz\mathbf{i} - y^2\mathbf{j} + yz\mathbf{k}$ and $\mathcal{S}$ is the surface of the cube bounded by the planes $x = 0$, $x = 1$, $y = 0$, $y = 1$, $z = 0$, and $z = 1$, with the face corresponding to $x = 1$ removed.

18 Evaluate $\int_{\mathcal{S}} \mathbf{F} \cdot \mathbf{n}\, dS$, where $\mathbf{F} = x\mathbf{i} + y\mathbf{j} + 2z\mathbf{k}$ and $\mathcal{S}$ is the boundary of the tetrahedron with vertices (1, 2, 3), (1, 0, 1), (2, 1, 4), and (1, 3, 5).

19 Let $\mathcal{S}$ be a surface of area S that bounds a region $\mathcal{V}$ of volume V. Assume that $\|\mathbf{F}(P)\| \le 5$ for all points P on the surface $\mathcal{S}$. What can be said about $\int_{\mathcal{V}} \nabla \cdot \mathbf{F}\, dV$?

20 Evaluate $\int_{\mathcal{S}} \mathbf{F} \cdot \mathbf{n}\, dS$, where $\mathbf{F} = x^3\mathbf{i} + y^3\mathbf{j} + z^3\mathbf{k}$ and $\mathcal{S}$ is the sphere of radius a and center (0, 0, 0).

In Exercises 21 to 24 evaluate $\int_{\mathcal{S}} \mathbf{F} \cdot \mathbf{n}\, dS$ for $\mathbf{F} = \hat{\mathbf{r}}/\|\mathbf{r}\|^2$ and the given surfaces, doing as little calculation as possible.

21 $\mathcal{S}$ is the sphere of radius 2 and center (5, 3, 1).

22 $\mathcal{S}$ is the sphere of radius 3 and center (1, 0, 1).

23 $\mathcal{S}$ is the surface of the box bounded by the planes $x = -1$, $x = 2$, $y = 2$, $y = 3$, $z = -1$, and $z = 6$.

24 $\mathcal{S}$ is the surface of the box bounded by the planes $x = -1$, $x = 2$, $y = -1$, $y = 3$, $z = -1$, and $z = 4$.

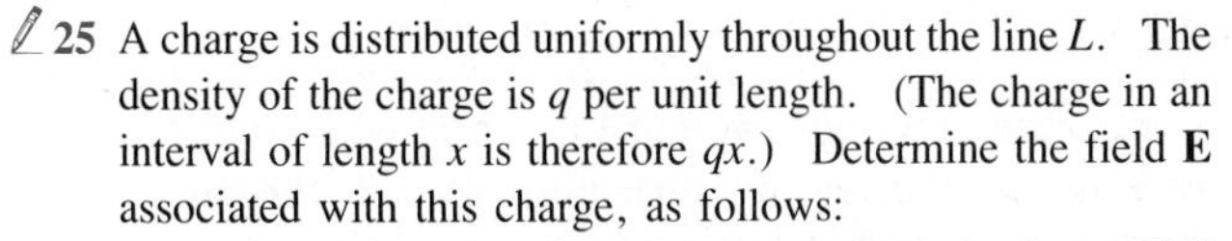

25 A charge is distributed uniformly throughout the line L. The density of the charge is q per unit length. (The charge in an interval of length x is therefore qx.) Determine the field $\mathbf{E}$ associated with this charge, as follows:

(*a*) Using symmetry, what can you say about the form of $\mathbf{E}$?

(*b*) To obtain more information about $\mathbf{E}$, apply Gauss's law to the closed cylindrical surface shown in Fig. 10.

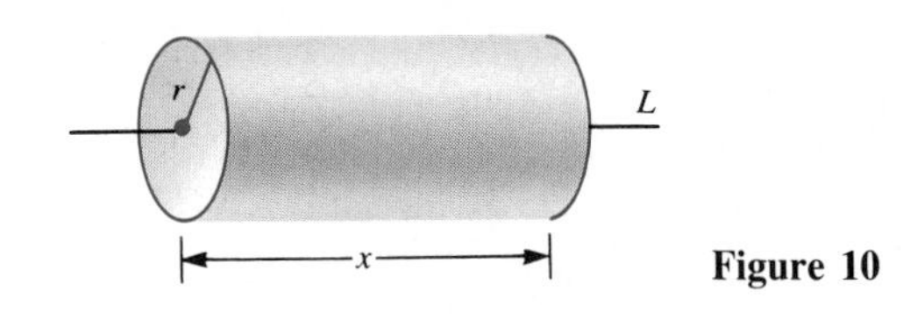

Figure 10

26 A charge is distributed uniformly throughout a plane with charge density q. (The charge in a region of area A is qA.) Determine the field $\mathbf{E}$ associated with this charge, as follows:

(*a*) Using symmetry, what can you say about the form of $\mathbf{E}$?

(*b*) To obtain more information about $\mathbf{E}$, apply Gauss's law to the closed cylindrical surface shown in Fig. 11. Note that the result also applies to the gravitational field associated with a uniform distribution of mass in a plane.

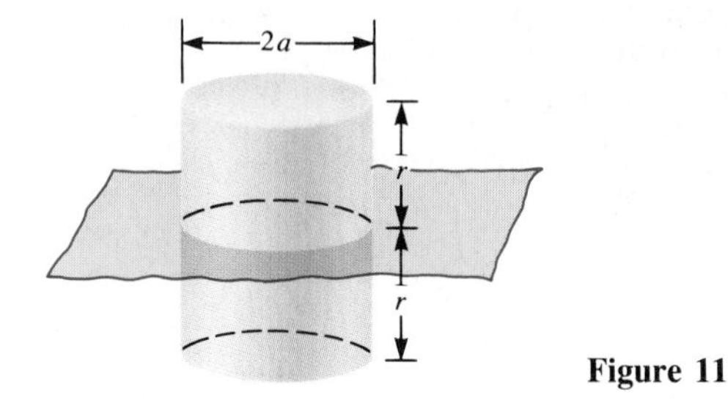

Figure 11

27 We established only one of the equations in (1). Establish that

$$\int_{\mathcal{S}} P \cos \alpha\, dS = \int_{\mathcal{V}} \frac{\partial P}{\partial x}\, dV.$$

Include large, clear diagrams.

28 See Exercise 27. Establish that

$$\int_{\mathcal{S}} Q \cos \beta\, dS = \int_{\mathcal{V}} \frac{\partial Q}{\partial y}\, dV.$$

Include large, clear diagrams.

29 Let $\mathcal{S}$ be a surface that meets each line parallel to a coordinate axis at most twice. It bounds a solid region $\mathcal{V}$. Let $\mathbf{F}(x, y, z) = z\mathbf{k}$.

(*a*) Use the divergence theorem to evaluate $\int_{\mathcal{S}} \mathbf{F} \cdot \mathbf{n}\, dS$.

(*b*) Evaluate the integral in (*a*) without using the divergence theorem.

30 Let $\mathbf{F}$ be a vector field in space such that for every closed surface $\mathcal{S}$ that does not surround the origin, $\int_{\mathcal{S}} \mathbf{F} \cdot \mathbf{n}\, dS = 0$.

(*a*) What can be said about div $\mathbf{F}$?

(*b*) If $\mathbf{F}$ is a central force field, $\mathbf{F}(\mathbf{r}) = f(r)\hat{\mathbf{r}}$, where $r = \|\mathbf{r}\|$, what can be said about $f(r)$?

31 Let $\mathbf{F}$ be defined everywhere in space except at the origin. Assume that

$$\lim_{\|\mathbf{r}\| \to \infty} \frac{\mathbf{F}(\mathbf{r})}{\|\mathbf{r}\|^2} = \mathbf{0}$$

and that div $\mathbf{F} = 0$. Evaluate $\int_{\mathcal{S}_1} \mathbf{F} \cdot \mathbf{n}\, dS$, where $\mathcal{S}_1$ is the unit sphere with center at the origin.

32 Let $\mathcal{S}$ be a closed surface that does not enclose the origin.
(*a*) Use the divergence theorem to evaluate

$$\int_{\mathcal{S}} \frac{\hat{\mathbf{r}} \cdot \mathbf{n}}{\|\mathbf{r}\|^2}\, dS.$$

(*b*) Use steradians to evaluate the integral in (*a*).

33 Let $\mathcal{S}$ be a sphere that passes through the origin. Discuss the value of

$$\int_{\mathcal{S}} \frac{\hat{\mathbf{r}} \cdot \mathbf{n}}{\|\mathbf{r}\|^2}\, dS.$$

34 A pyramid is made of four congruent equilateral triangles. Find the number of steradians subtended by one face at the centroid of the pyramid. (No integration is necessary.)

35 How many steradians does one face of a cube subtend at
(*a*) One of the four vertices not on the face?
(*b*) The center of the cube?
(No integration is necessary.)

36 Show that $\nabla \cdot \mathbf{F}$, evaluated at P_0, equals

$$\lim_{a \to 0} \frac{\int_{\mathcal{S}} \mathbf{F} \cdot \mathbf{n}\, dS}{\text{Volume of } \mathcal{V}},$$

where $\mathcal{V}$ is the ball of radius a and center P_0 and $\mathcal{S}$ is its surface. This gives a definition of the divergence of a vector field without referring to its components.

37 Use Exercise 36 to develop the analog of Exercise 29 of Sec. 16.4 for cylindrical coordinates.

38 Explain this reasoning, paraphrased from *Introduction to Fluid Mechanics,* by Stephen Whitaker, Krieger, Melbourne, Florida, 1981. Since

$$\int_{\mathcal{V}} \frac{\partial f}{\partial t}\, dV + \int_{\mathcal{S}} (f\mathbf{v} \cdot \mathbf{n})\, dS = 0,$$

for any solid region $\mathcal{V}$ with surface $\mathcal{S}$, it follows that

$$\frac{\partial f}{\partial t} + \nabla \cdot f\mathbf{v} = 0.$$

17.3 STOKES' THEOREM

Stokes' theorem in the xy plane asserts that

$$\oint_C \mathbf{F} \cdot d\mathbf{r} = \int_{\mathcal{A}} (\nabla \times \mathbf{F}) \cdot \mathbf{k}\, dA,$$

where C is counterclockwise and C bounds the region $\mathcal{A}$. The general Stokes' theorem extends this result to closed curves in space. (It is this version that is usually called Stokes' theorem.) It asserts that if the closed curve C bounds a surface $\mathcal{S}$, then

$$\oint_C \mathbf{F} \cdot d\mathbf{r} = \int_{\mathcal{S}} (\nabla \times \mathbf{F}) \cdot \mathbf{n}\, dS.$$

As usual, the vector $\mathbf{n}$ is a unit normal to the surface.

Stokes published his theorem in 1854 (without proof, for it appeared as a question on a Cambridge University examination). By 1870 it was in common use. It is the most recent of the three major theorems discussed in Chaps. 16 and 17, for Green published his theorem in 1828 and Gauss published the divergence theorem in 1839.

Choosing the Normal n

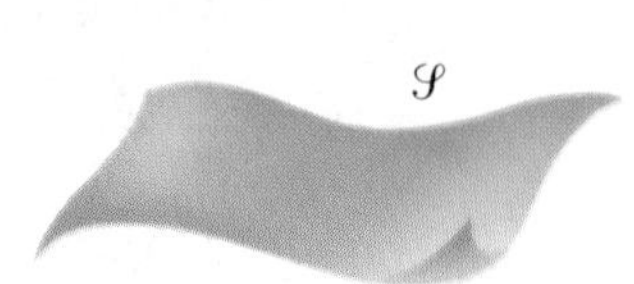

Figure 1

In order to state Stokes' theorem precisely, we must describe what kind of surface $\mathcal{S}$ is permitted and which of the two possible normals $\mathbf{n}$ is chosen.

In the case of a typical surface $\mathcal{S}$ that comes to mind, it is possible to assign at each point P a unit normal $\mathbf{n}$ in a continuous manner. On the surface shown in Fig. 1, there are two ways to do this. They are shown in Fig. 2. But, for the

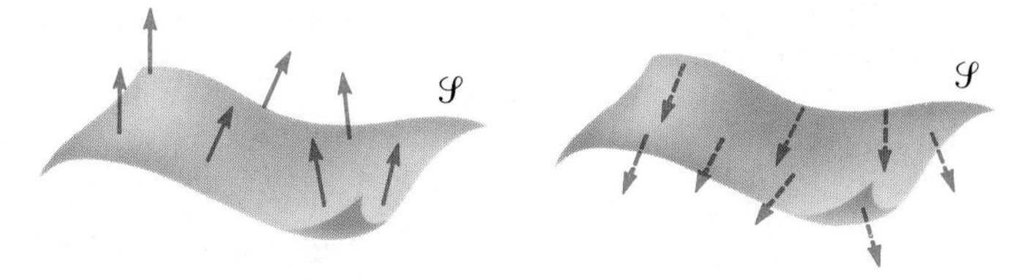

Figure 2

Follow the choices through nine stages—there's trouble.

surface shown in Fig. 3 (a Möbius band), it is impossible to make such a choice. If you start with choice (1) and move the normal continuously along the surface, by the time you return to the initial point on the surface at stage (9), you have the opposite normal. A surface for which a continuous choice *can* be made is called **orientable** or **two-sided**. Stokes' theorem holds for orientable surfaces, which include, for instance, any part of the surface of a convex body, such as a ball or cube.

Orientable surfaces

Consider an orientable surface $\mathcal{S}$, bounded by a parameterized curve C so that the curve is swept out in a definite direction. If the surface is flat or almost flat, we can simply use the right-hand rule to choose $\mathbf{n}$: The direction of $\mathbf{n}$ should match the thumb of the right hand if the fingers curl in the direction of C and the thumb and palm are perpendicular to the tangent plane to the surface. Figure 4 illustrates the choice of $\mathbf{n}$. For instance, if C is counterclockwise in the xy plane, this definition picks out the normal $\mathbf{k}$, not $-\mathbf{k}$.

In the case of an orientable surface that is not sufficiently "flat," the proper choice of the normal $\mathbf{n}$ may be made according to the following rule. Imagine walking along the curve C in the direction of its orientation but standing perpendicular to the surface and on the side of the surface such that *nearby points* on the surface are on your *left*. Then choose the normal $\mathbf{n}$ to be the one on the same side of the surface as you are. Figure 5 illustrates this choice for a surface that consists of five faces of a cube with a bounding curve oriented as shown. Now Stokes' theorem can be stated precisely.

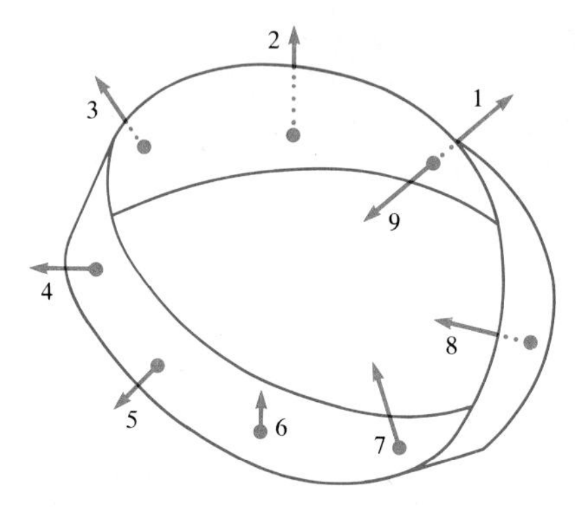

No way to choose a normal continuously

Figure 3

Theorem 1 *Stokes' theorem.* Let $\mathcal{S}$ be an orientable surface bounded by the parameterized curve C. At each point of $\mathcal{S}$ let $\mathbf{n}$ be the unit normal chosen by the right-hand rule. Let $\mathbf{F}$ be a vector field defined on some region in space including $\mathcal{S}$. Then

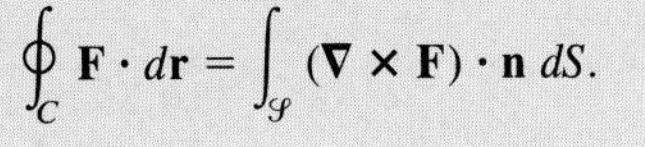

$$\oint_C \mathbf{F} \cdot d\mathbf{r} = \int_{\mathcal{S}} (\nabla \times \mathbf{F}) \cdot \mathbf{n}\, dS.$$

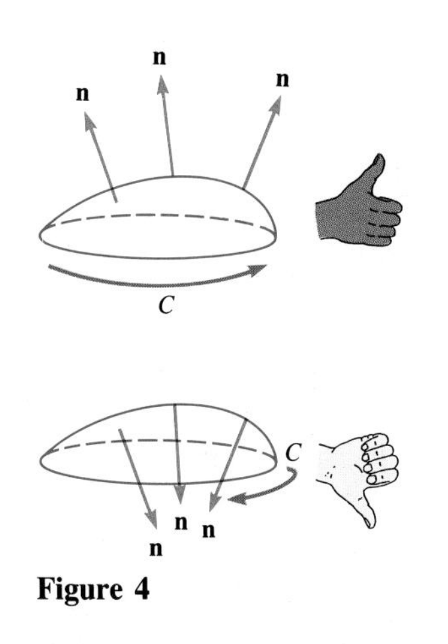

Figure 4

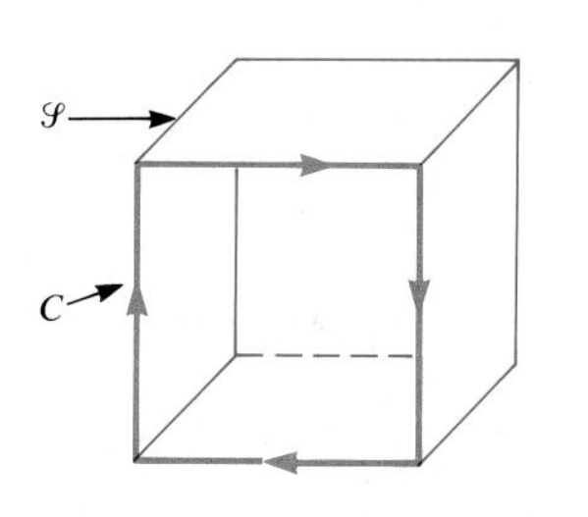

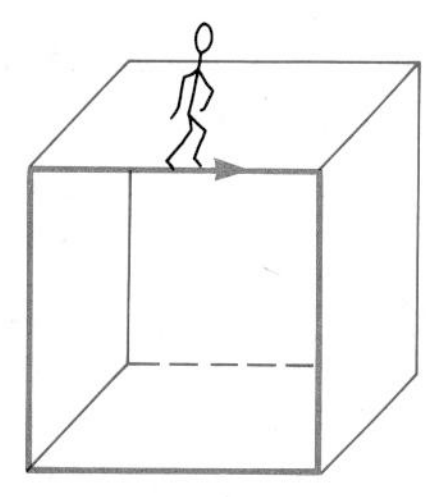

Walk in direction of C, with surface at your left.

Figure 5

The proof of Stokes' theorem is discussed at the end of the next section.

Some Applications of Stokes' Theorem

Choosing a simpler surface

First of all, Stokes' theorem enables us to replace $\int_{\mathcal{S}} (\nabla \times \mathbf{F}) \cdot \mathbf{n}\, dS$ by a similar integral over a surface that might be simpler than $\mathcal{S}$. That is the substance of the following corollary of Stokes' theorem.

> **Corollary 1** Let $\mathcal{S}_1$ and $\mathcal{S}_2$ be two surfaces bounded by the same curve C and oriented so that they yield the same orientation on C. Let $\mathbf{F}$ be a vector field defined on both $\mathcal{S}_1$ and $\mathcal{S}_2$. Then
>
> $$\int_{\mathcal{S}_1} (\nabla \times \mathbf{F}) \cdot \mathbf{n}\, dS = \int_{\mathcal{S}_2} (\nabla \times \mathbf{F}) \cdot \mathbf{n}\, dS.$$

(The two integrals are equal since both equal $\oint_C \mathbf{F} \cdot d\mathbf{r}$.)

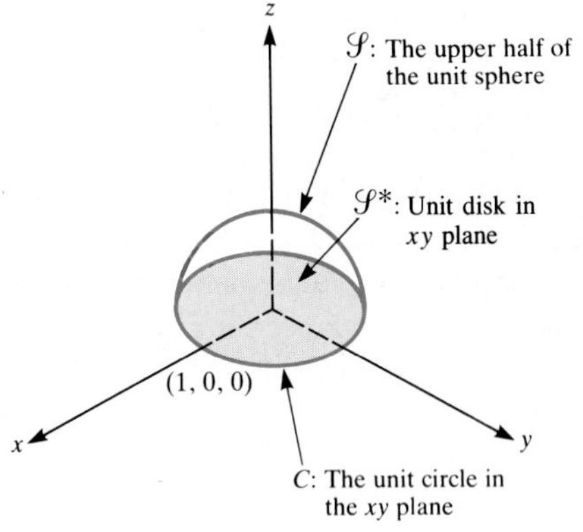

Figure 6

EXAMPLE 1 Let $\mathbf{F} = xe^z\mathbf{i} + (x + xz)\mathbf{j} + 3e^z\mathbf{k}$ and let $\mathcal{S}$ be the top half of the sphere $x^2 + y^2 + z^2 = 1$. Find $\int_{\mathcal{S}} (\nabla \times \mathbf{F}) \cdot \mathbf{n}\, dS$, where $\mathbf{n}$ is the outward normal. (See Fig. 6.)

SOLUTION By Corollary 1,

$$\int_{\mathcal{S}} (\nabla \times \mathbf{F}) \cdot \mathbf{n}\, dS = \int_{\mathcal{S}^*} (\nabla \times \mathbf{F}) \cdot \mathbf{k}\, dS,$$

where $\mathcal{S}^*$ is the flat base of the hemisphere. (On $\mathcal{S}^*$ note that $\mathbf{k} = \mathbf{n}$.)
A straightforward calculation shows that

$$\nabla \times \mathbf{F} = -x\mathbf{i} + xe^z\mathbf{j} + (z + 1)\mathbf{k},$$

hence $(\nabla \times \mathbf{F}) \cdot \mathbf{k} = z + 1$. On $\mathcal{S}^*$, $z = 0$, so

$$\int_{\mathcal{S}^*} (\nabla \times \mathbf{F}) \cdot \mathbf{k}\, dS = \int_{\mathcal{S}^*} dS = \pi.$$

Thus the original integral over $\mathcal{S}$ is π. ■

Just as there is a two-curve version of Green's theorem there is a two-curve version of Stokes' theorem.

> **Corollary 2** *Stokes' theorem (two-curve version).* Let $\mathcal{S}$ be an orientable surface whose boundary consists of the two closed curves C_1 and C_2. Give C_1 an orientation. Orient $\mathcal{S}$ consistent with the right-hand rule, as applied to C_1. Give C_2 the same orientation as C_1. (If C_2 is moved on $\mathcal{S}$ to C_1, the orientations agree.) Then
>
> $$\oint_{C_1} \mathbf{F} \cdot d\mathbf{r} - \oint_{C_2} \mathbf{F} \cdot d\mathbf{r} = \int_{\mathcal{S}} (\nabla \times \mathbf{F}) \cdot \mathbf{n}\, dS.$$

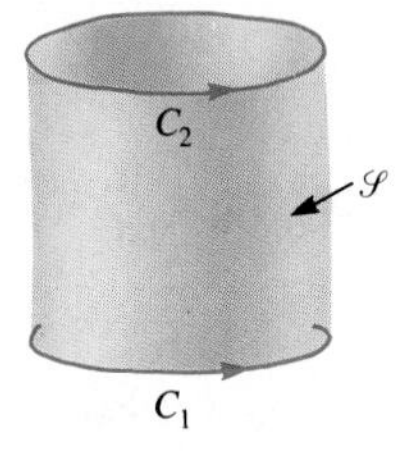

Figure 7

Proof Figure 7 shows the typical situation.

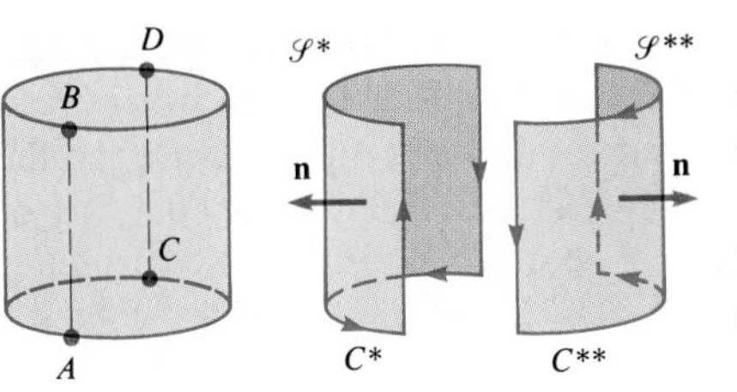

Figure 8

We will obtain the corollary from Stokes' theorem with the aid of the cancellation principle. Introduce lines AB and CD on $\mathcal{S}$, cutting $\mathcal{S}$ into two surfaces, $\mathcal{S}^*$ and $\mathcal{S}^{**}$. Now apply Stokes' theorem to $\mathcal{S}^*$ and $\mathcal{S}^{**}$. (See Fig. 8.)

Let C^* be the curve that bounds $\mathcal{S}^*$, oriented so that where it overlaps C_1 it has the same orientation as C_1. Let C^{**} be the curve that bounds $\mathcal{S}^{**}$, again oriented to match C_1. (See Fig. 8.)

By Stokes' theorem,

$$\oint_{C^*} \mathbf{F} \cdot d\mathbf{r} = \int_{\mathcal{S}^*} (\nabla \times \mathbf{F}) \cdot \mathbf{n}\, dS \tag{1}$$

and

$$\oint_{C^{**}} \mathbf{F} \cdot d\mathbf{r} = \int_{\mathcal{S}^{**}} (\nabla \times \mathbf{F}) \cdot \mathbf{n}\, dS. \tag{2}$$

Adding (1) and (2) and using the cancellation principle gives

$$\oint_{C_1} \mathbf{F} \cdot d\mathbf{r} - \oint_{C_2} \mathbf{F} \cdot d\mathbf{r} = \int_{\mathcal{S}} (\nabla \times \mathbf{F}) \cdot \mathbf{n}\, dS. \blacksquare$$

In practice, Corollary 2 is applied when $\mathbf{F}$ is irrotational, that is, when **curl** $\mathbf{F} = \mathbf{0}$.

Corollary 3 Let $\mathbf{F}$ be irrotational. Let C_1 and C_2 be two closed curves that together bound an orientable surface $\mathcal{S}$ on which $\mathbf{F}$ is defined. If C_1 and C_2 are similarly oriented, then

$$\oint_{C_1} \mathbf{F} \cdot d\mathbf{r} = \oint_{C_2} \mathbf{F} \cdot d\mathbf{r}.$$

Corollary 3 follows directly from Corollary 2 since $\int_{\mathcal{S}} (\nabla \times \mathbf{F}) \cdot \mathbf{n}\, dS = 0$.

EXAMPLE 2 Assume that $\mathbf{F}$ is irrotational and defined everywhere except on the z axis. Given that $\oint_{C_1} \mathbf{F} \cdot d\mathbf{r} = 3$, find ($a$) $\oint_{C_2} \mathbf{F} \cdot d\mathbf{r}$ and (b) $\oint_{C_3} \mathbf{F} \cdot d\mathbf{r}$. (See Fig. 9.)

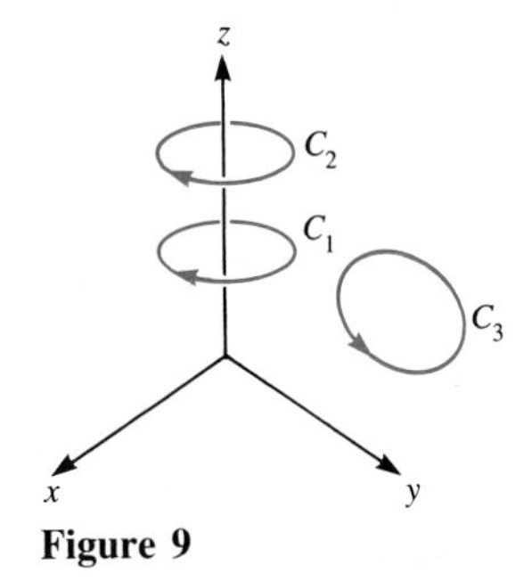

Figure 9

SOLUTION (a) By Corollary 3, $\oint_{C_2} \mathbf{F} \cdot d\mathbf{r} = \oint_{C_1} \mathbf{F} \cdot d\mathbf{r} = 3$. ($b$) By Stokes' theorem, $\oint_{C_3} \mathbf{F} \cdot d\mathbf{r} = 0$. ■

Curl and Conservative Fields

In Sec. 16.6 we learned that if $\mathbf{F} = P\mathbf{i} + Q\mathbf{j}$ is defined on a simply connected region in the xy plane and if **curl** $\mathbf{F} = \mathbf{0}$, then $\mathbf{F}$ is conservative. Now that we have Stokes' theorem, this result can be extended to a field $\mathbf{F} = P\mathbf{i} + Q\mathbf{j} + R\mathbf{k}$ defined on a simply connected region in space.

Theorem 2 Let $\mathbf{F}$ be defined on a simply connected region in space. If **curl** $\mathbf{F} = \mathbf{0}$, then $\mathbf{F}$ is conservative.

Sketch of proof Let C be a closed curve in the region. We wish to show that $\oint_C \mathbf{F} \cdot d\mathbf{r} = 0$. Since the region where $\mathbf{F}$ is defined is simply connected, C can be shrunk to a point in the region while staying in that region. Let $\mathcal{S}$ be the surface swept out as C shrinks to a point. (We assume that C is the boundary of $\mathcal{S}$ and that $\mathcal{S}$ does not intersect itself.)

By Stokes' theorem

$$\oint_C \mathbf{F} \cdot d\mathbf{r} = \int_{\mathcal{S}} (\nabla \times \mathbf{F}) \cdot \mathbf{n}\, dS = \int_{\mathcal{S}} \mathbf{0} \cdot \mathbf{n}\, dS = 0. \quad \blacksquare$$

EXAMPLE 3 Let $\mathbf{F} = \hat{\mathbf{r}}/\|\mathbf{r}\|^2$, where $\mathbf{r} = x\mathbf{i} + y\mathbf{j} + z\mathbf{k}$. Show that $\mathbf{F}$ is conservative.

SOLUTION The domain of $\mathbf{F}$ consists of all points in space except the origin. Thus the domain of $\mathbf{F}$ is simply connected. Moreover, $\mathbf{curl\ F} = \mathbf{0}$, as the reader may check. (See Exercise 36 in Sec. 16.1.) ■

An argument like that in Example 3 shows that every central field is conservative. (See also Exercise 31.)

Section Summary

We stated Stokes' theorem:

$$\oint_C \mathbf{F} \cdot d\mathbf{r} = \int_{\mathcal{S}} (\nabla \times \mathbf{F}) \cdot \mathbf{n}\, dS,$$

where C bounds the orientable surface $\mathcal{S}$ and $\mathbf{n}$ is chosen to be compatible with the orientation of C, according to the right-hand rule. Using a two-curve version, we showed that if C_1 and C_2 are similarly oriented and bound a surface where $\mathbf{curl\ F} = \mathbf{0}$, then

$$\oint_{C_1} \mathbf{F} \cdot d\mathbf{r} = \oint_{C_2} \mathbf{F} \cdot d\mathbf{r}.$$

Finally, we used Stokes' theorem to show that if $\mathbf{F}$ is defined on a simply connected region and its curl is $\mathbf{0}$, then it is conservative. Since central fields have curl $\mathbf{0}$, the field $\hat{\mathbf{r}}/r^2$, of major importance in physics, is conservative.

EXERCISES FOR SEC. 17.3: STOKES' THEOREM

1 State Stokes' theorem in words, not in mathematical symbols.

2 Assume that $\mathbf{F}$ is defined everywhere except on the z axis and is irrotational. What, if anything, can be said about

$$\oint_{C_1} \mathbf{F} \cdot d\mathbf{r},\ \oint_{C_2} \mathbf{F} \cdot d\mathbf{r},\ \oint_{C_3} \mathbf{F} \cdot d\mathbf{r}, \text{ and } \oint_{C_4} \mathbf{F} \cdot d\mathbf{r},$$

where the curves are shown in Fig. 10?

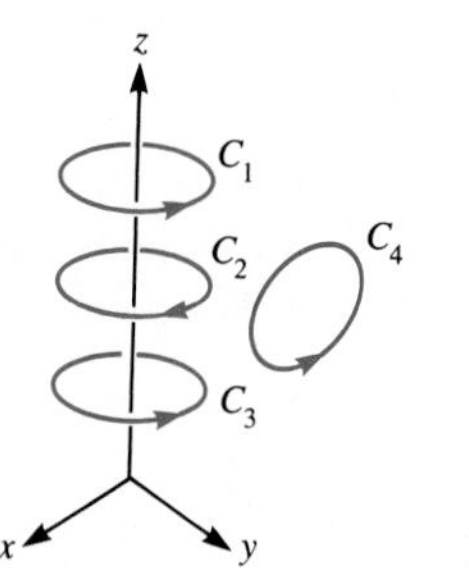

Figure 10

In Exercises 3 to 6 verify Stokes' theorem for the given $\mathbf{F}$ and surface $\mathcal{S}$.

3 $\mathbf{F} = xy^2\mathbf{i} + y^3\mathbf{j} + y^2z\mathbf{k}$; $\mathcal{S}$ is the top half of the sphere $x^2 + y^2 + z^2 = 1$.

4 $\mathbf{F} = y\mathbf{i} + xz\mathbf{j} + x^2\mathbf{k}$; $\mathcal{S}$ is the triangle with vertices (1, 0, 0), (0, 1, 0), and (0, 0, 1).

5 $\mathbf{F} = y^5\mathbf{i} + x^3\mathbf{j} + z^4\mathbf{k}$; $\mathcal{S}$ is the portion of $z = x^2 + y^2$ below the plane $z = 1$.

6 $\mathbf{F} = -y\mathbf{i} + x\mathbf{j} + z\mathbf{k}$, $\mathcal{S}$ is the portion of the cylinder $z = x^2$ inside the cylinder $x^2 + y^2 = 4$.

7 Evaluate as simply as possible $\int_{\mathcal{S}} \mathbf{F} \cdot \mathbf{n}\, dS$, where $\mathbf{F}(x, y, z) = x\mathbf{i} - y\mathbf{j}$ and $\mathcal{S}$ is the surface of the cube bounded by the three coordinate planes and the planes $x = 1$, $y = 1$, $z = 1$, exclusive of the surface in the plane $x = 1$. (Let $\mathbf{n}$ be outward from the cube.)

8 Using Stokes' theorem, evaluate $\int_{\mathcal{S}} (\nabla \times \mathbf{F}) \cdot \mathbf{n}\, dS$, where $\mathbf{F} = (x^2 + y - 4)\mathbf{i} + 3xy\mathbf{j} + (2xz + z^2)\mathbf{k}$, and $\mathcal{S}$ is the portion of the surface $z = 4 - (x^2 + y^2)$ above the xy plane. (Let $\mathbf{n}$ be the upward normal.)

In each of Exercises 9 to 12 use Stokes' theorem to evaluate $\oint_C \mathbf{F} \cdot d\mathbf{r}$ for the given $\mathbf{F}$ and C. In each case assume that C is oriented counterclockwise when viewed from above.

9 $\mathbf{F} = \sin xy\, \mathbf{i}$; C is the intersection of the plane $x + y + z = 1$ and the cylinder $x^2 + y^2 = 1$.

10 $\mathbf{F} = e^x\mathbf{j}$; C is the triangle with vertices (2, 0, 0), (0, 3, 0), and (0, 0, 4).

11 $\mathbf{F} = xy\mathbf{k}$; C is the intersection of the plane $z = y$ with the cylinder $x^2 - 2x + y^2 = 0$.

12 $\mathbf{F} = \cos(x + z)\, \mathbf{j}$; C is the boundary of the rectangle with vertices (1, 0, 0), (1, 1, 1), (0, 1, 1), and (0, 0, 0).

13 Let $\mathcal{S}_1$ be the top half and $\mathcal{S}_2$ the bottom half of a sphere of radius a in space. Let $\mathbf{F}$ be a vector field defined on the sphere and let $\mathbf{n}$ denote an exterior normal to the sphere. What relation, if any, is there between $\int_{\mathcal{S}_1} (\mathbf{curl\ F}) \cdot \mathbf{n}\, dS$ and $\int_{\mathcal{S}_2} (\mathbf{curl\ F}) \cdot \mathbf{n}\, dS$?

14 Let $\mathbf{F}$ be a vector field throughout space such that $\mathbf{F}(P)$ is perpendicular to the curve C at each point P on C, the boundary of a surface $\mathcal{S}$. What can one conclude about

$$\int_{\mathcal{S}} (\mathbf{curl\ F}) \cdot \mathbf{n}\, dS?$$

15 Let C_1 and C_2 be two closed curves in the xy plane that encircle the origin and are similarly oriented, as in Fig. 11.

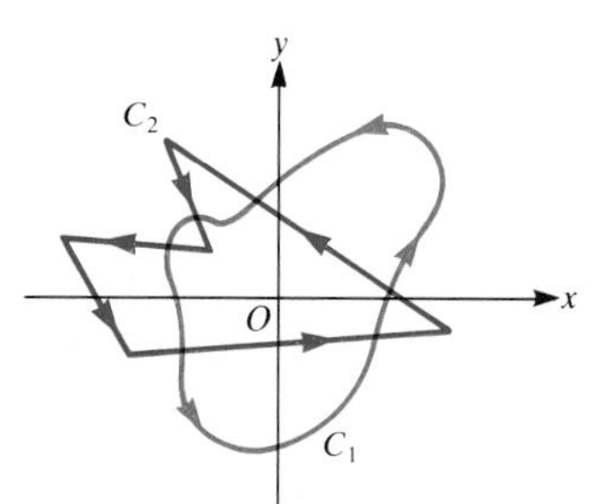

Figure 11

Let $\mathbf{F}$ be a vector field defined throughout the plane except at the origin. Assume that $\mathbf{curl\ F} = \mathbf{0}$.

(*a*) Must $\oint_{C_1} \mathbf{F} \cdot \mathbf{T}\, ds = 0$?

(*b*) What, if any, relation exists between $\oint_{C_1} \mathbf{F} \cdot \mathbf{T}\, ds$ and $\oint_{C_2} \mathbf{F} \cdot \mathbf{T}\, ds$?

16 Let $\mathbf{F}$ be defined everywhere in space except on the z axis. Assume also that $\mathbf{F}$ is irrotational, $\oint_{C_1} \mathbf{F} \cdot d\mathbf{r} = 3$, and $\oint_{C_2} \mathbf{F} \cdot d\mathbf{r} = 5$. (See Fig. 12.) What, if anything, can be said about (*a*) $\oint_{C_3} \mathbf{F} \cdot d\mathbf{r}$, (*b*) $\oint_{C_4} \mathbf{F} \cdot d\mathbf{r}$?

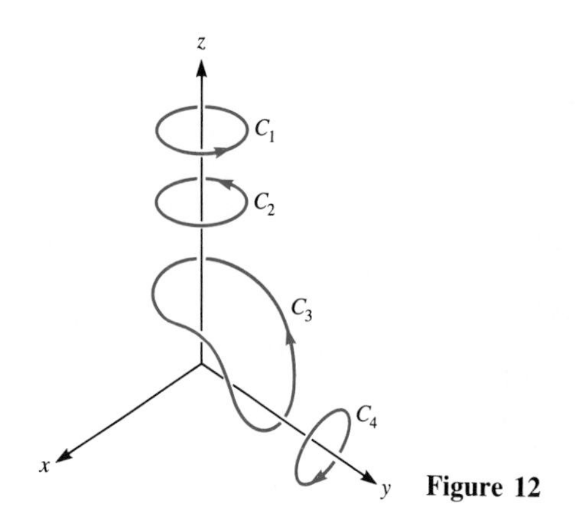

Figure 12

17 Which of the following sets are connected? simply connected?

(*a*) A circle ($x^2 + y^2 = 1$)

(*b*) A disk ($x^2 + y^2 \leq 1$)

(*c*) The xy plane from which a circle is removed

(*d*) The xy plane from which a disk is removed

(*e*) The xy plane from which one point is removed

(*f*) xyz space from which one point is removed

(*g*) xyz space from which a sphere is removed

(*h*) xyz space from which a ball is removed

(*i*) A solid torus (doughnut)

(*j*) xyz space from which a solid torus is removed

(*k*) A coffee cup with one handle

18 Which central fields have curl $\mathbf{0}$?

19 Let $\mathcal{V}$ be the solid bounded by $z = x + 2$, $x^2 + y^2 = 1$, and $z = 0$. Let $\mathcal{S}_1$ be the portion of the plane $z = x + 2$ that lies within the cylinder $x^2 + y^2 = 1$. Let C be the boundary of $\mathcal{S}_1$, with a counterclockwise orientation (as viewed from above). Let $\mathbf{F} = y\mathbf{i} + xz\mathbf{j} + (x + 2y)\mathbf{k}$. Use Stokes' theorem for $\mathcal{S}_1$ to evaluate $\oint_C \mathbf{F} \cdot d\mathbf{r}$.

20 (See Exercise 19.) Let $\mathcal{S}_2$ be the curved surface of $\mathcal{V}$ together with the base of $\mathcal{V}$. Use Stokes' theorem for $\mathcal{S}_2$ to evaluate $\oint_C \mathbf{F} \cdot d\mathbf{r}$.

21 Verify Stokes' theorem for the special case when $\mathbf{F}$ has the form ∇f, that is, is a gradient field.

22 Let $\mathbf{F}$ be a vector field defined on the surface $\mathcal{S}$ of a convex solid. Show that $\int_{\mathcal{S}} (\nabla \times \mathbf{F}) \cdot \mathbf{n}\, dS = 0$ (*a*) by the divergence theorem, (*b*) by drawing a closed curve C on $\mathcal{S}$ and using Stokes' theorem on the two parts into which C divides $\mathcal{S}$.

23 Evaluate $\oint_C \mathbf{F} \cdot \mathbf{T}\, ds$ as simply as possible if $\mathbf{F}(x, y, z) = -y\mathbf{i}/(x^2 + y^2) + x\mathbf{j}/(x^2 + y^2)$ and C is the intersection of the plane $z = 2x + 2y$ and the paraboloid $z = 2x^2 + 3y^2$ oriented counterclockwise as viewed from above.

24 Let $\mathbf{F}(x, y)$ be a vector field defined everywhere in the plane except at the origin. Assume that $\mathbf{curl\ F} = \mathbf{0}$. Let C_1 be the circle $x^2 + y^2 = 1$ counterclockwise; let C_2 be the circle $x^2 + y^2 = 4$ clockwise; let C_3 be the circle $(x - 2)^2 + y^2 = 1$ counterclockwise; let C_4 be the circle $(x - 1)^2 + y^2 = 9$ clockwise. Assuming that $\oint_{C_1} \mathbf{F} \cdot d\mathbf{r}$ is 5, evaluate (*a*) $\oint_{C_2} \mathbf{F} \cdot d\mathbf{r}$, (*b*) $\oint_{C_3} \mathbf{F} \cdot d\mathbf{r}$, (*c*) $\oint_{C_4} \mathbf{F} \cdot d\mathbf{r}$.

25 Let $\mathbf{F}(x, y, z) = \mathbf{r}/\|\mathbf{r}\|^a$, where $\mathbf{r} = x\mathbf{i} + y\mathbf{j} + z\mathbf{k}$ and a is a fixed real number.
(*a*) Show that $\mathbf{curl\ F} = \mathbf{0}$.
(*b*) Show that $\mathbf{F}$ is conservative.
(*c*) Exhibit a scalar function f such that $\mathbf{F} = \nabla f$.

26 Let $\mathbf{F}$ be defined throughout space and have continuous divergence and curl.
(*a*) For which $\mathbf{F}$ is $\int_{\mathcal{S}} \mathbf{F} \cdot \mathbf{n}\, dS = 0$ for all spheres $\mathcal{S}$?
(*b*) For which $\mathbf{F}$ is $\oint_C \mathbf{F} \cdot \mathbf{T}\, ds = 0$ for all circles C?
(*c*) If $\oint_C \mathbf{F} \cdot \mathbf{T}\, ds = 0$ for all circles C, must $\oint_C \mathbf{F} \cdot \mathbf{T}\, ds = 0$ for all closed curves?

27 Let C be the curve formed by the intersection of the plane $z = x$ and the paraboloid $z = x^2 + y^2$. Orient C to be counterclockwise when viewed from above. Evaluate $\oint_C (xyz\, dx + x^2\, dy + xz\, dz)$.

28 Assume that Stokes' theorem is true for triangles. Deduce that it then holds for the surface $\mathcal{S}$ in Fig. 13, consisting of the three triangles DAB, DBC, DCA, and the curve $ABCA$.

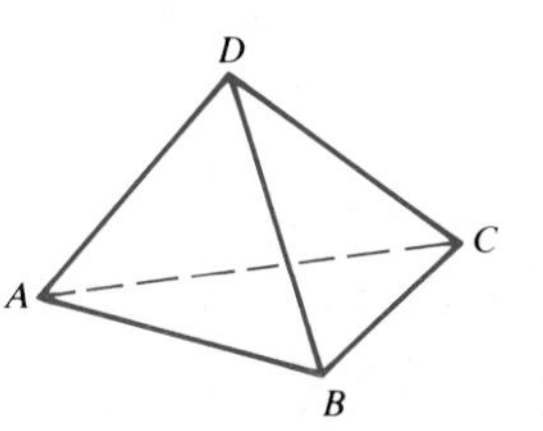

Figure 13

✱ 29 A Möbius band can be made by making a half-twist in a narrow rectangular strip, bringing the two ends together, and fastening them with glue or tape. See Fig. 14.

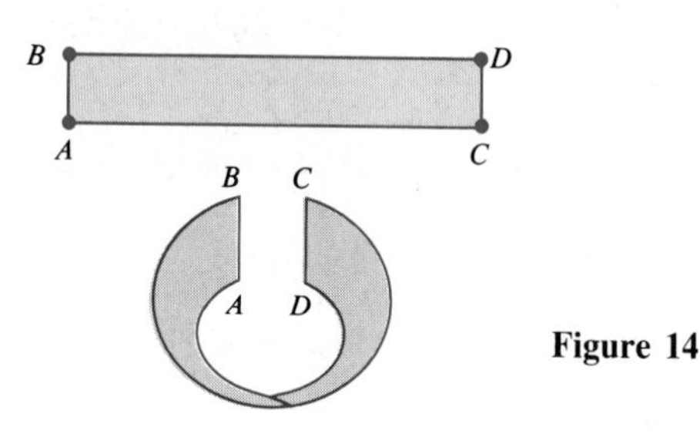

Figure 14

(*a*) Make a Möbius band.
(*b*) Letting a pencil represent a normal $\mathbf{n}$ to the band, check that the Möbius band is not orientable.
(*c*) If you form a band by first putting in a full twist (360°), is it orientable?
(*d*) What happens when you cut the bands in (*b*) and (*c*) down the middle? one third of the way from one edge to the other?

30 Let $\mathcal{S}$ be a surface bounded by the curve C. Assume that $\mathcal{S}$ can be cut into smaller pieces $\mathcal{S}_1, \mathcal{S}_2, \ldots, \mathcal{S}_n$ with boundary curves $C_1, C_2, \ldots, C_n$ in such a way that each curve C_i can be oriented so that on the arcs where two curves overlap their directions are opposite. (This is the basis of the cancellation principle.) (*a*) Show that $\mathcal{S}$ is orientable. (*b*) Is the converse true?

31 Let $\mathbf{F}$ be a central field with center at the origin and defined everywhere in space except at the origin. It can be written as $\mathbf{F}(\mathbf{r}) = f(\|\mathbf{r}\|)\hat{\mathbf{r}}$ for some scalar function f. A straightforward computation using components shows that $\mathbf{curl\ F} = \mathbf{0}$.
(*a*) Why is $\mathbf{F}$ conservative?
(*b*) According to (*a*) $\oint_C \mathbf{F} \cdot d\mathbf{r} = 0$ for every closed curve C that does not pass through the origin. Figure 15 suggests a direct argument for this result. It shows the parts of C contained between two concentric spheres of almost the same radii and with centers at the origin. Why would you expect the integral of $\mathbf{F} \cdot d\mathbf{r}$ over the parts of C between two such spheres to sum up to 0? (This suggests why $\oint_C \mathbf{F} \cdot d\mathbf{r} = 0$.)

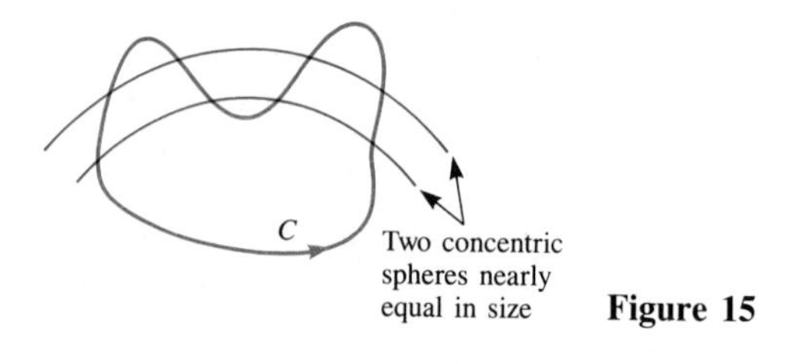

Figure 15

32 (Contributed by Sándor Szabó.) Figure 16 shows $\mathcal{S}$, a surface bounded by an oriented curve C. Let $\mathbf{F} = x\mathbf{i}$.
(*a*) Compute $\mathbf{curl\ F}$.
(*b*) Evaluate $\int_{\mathcal{S}} \mathbf{curl\ F} \cdot \mathbf{n}\, dS$.
(*c*) Evaluate $\oint_C \mathbf{F} \cdot d\mathbf{r}$.
(*d*) What does Stokes' theorem say about (*b*) and (*c*)?

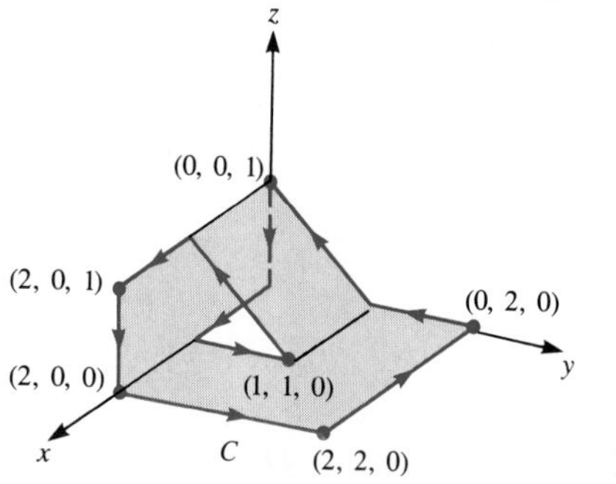

Figure 16

17.4 APPLICATIONS OF STOKES' THEOREM

In this section we analyze the physical meaning of **curl F** when **F** represents fluid flow. Then we present the laws of electricity and magnetism, for which the language of vector analysis was developed. Finally, we discuss the proof of Stokes' theorem.

Why Curl Is Called Curl

Let **F** be a vector field describing the flow of a fluid, as in Sec. 16.1. Stokes' theorem will give a physical interpretation of **curl F**.

Consider a fixed point P_0 in space. Imagine a *small* circular disk $\mathcal{S}$ with center P_0. Let C be the boundary of $\mathcal{S}$ oriented in such a way that C and **n** fit the right-hand rule. (See Fig. 1.)

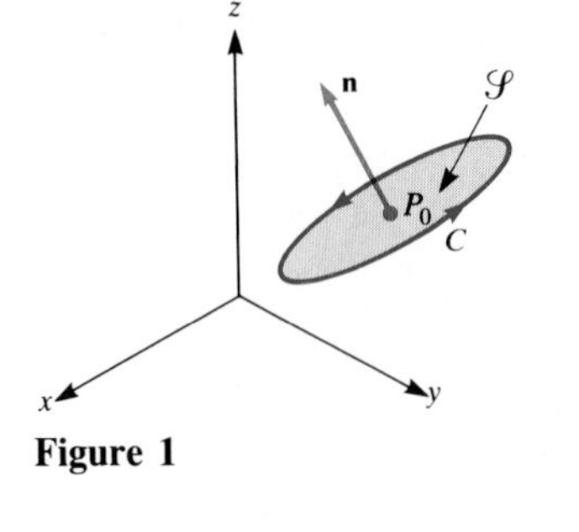

Figure 1

Now examine the two sides of the equation

$$\int_{\mathcal{S}} (\textbf{curl F}) \cdot \mathbf{n}\, dS = \oint_C \mathbf{F} \cdot \mathbf{T}\, ds. \tag{1}$$

The right side of Eq. (1) measures the tendency of the fluid to move along C (rather than, say, perpendicular to it). Thus $\oint_C \mathbf{F} \cdot \mathbf{T}\, ds$ might be thought of as the "circulation" or "whirling tendency" of the fluid along C. For each tilt of the small disk $\mathcal{S}$ at P_0—or, equivalently, each choice of unit normal vector **n**—$\oint_C \mathbf{F} \cdot \mathbf{T}\, ds$ measures a corresponding circulation. It records the tendency of a paddle wheel at P_0 with axis along **n** to rotate. (See Fig. 2.)

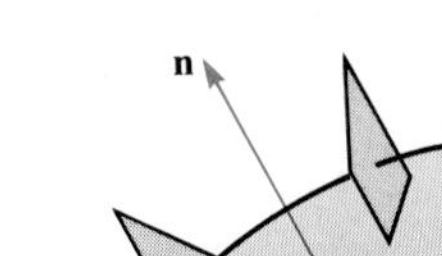
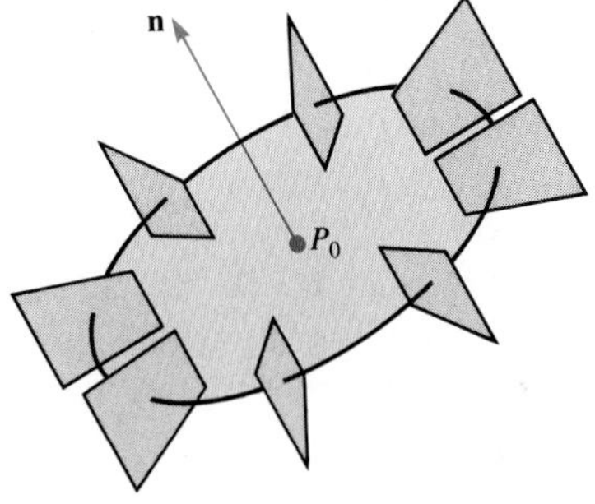

Figure 2

The physical interpretation of curl

Consider the left side of Eq. (1). If $\mathcal{S}$ is small, the integrand is almost constant and the integral is approximately

$$(\textbf{curl F})_{P_0} \cdot \mathbf{n} \cdot \text{Area of } \mathcal{S}, \tag{2}$$

where $(\textbf{curl F})_{P_0}$ denotes the curl at P_0.

Keeping the center of $\mathcal{S}$ at P_0, vary the vector **n** by tilting the disk $\mathcal{S}$. For which choice of **n** will (2) be largest? Answer: For that **n** which has the same direction as the fixed vector $(\textbf{curl F})_{P_0}$. With that choice of **n**, (2) becomes

$$\|(\textbf{curl F})_{P_0}\| \text{ Area of } \mathcal{S}.$$

Thus a paddle wheel placed in the fluid at P_0 *rotates most quickly when its axis is in the direction of* **curl F** at P_0. The *magnitude* of **curl F** is a measure of *how fast* the paddle wheel can rotate when placed at P_0. Thus **curl F** records the direction and magnitude of maximum circulation at a given point.

A Vector Definition of Curl

In Sec. 16.1 **curl F** was defined in terms of the partial derivatives of the components of **F**. By Stokes' theorem, **curl F** is related to the circulation, $\oint_C \mathbf{F} \cdot d\mathbf{r}$. We exploit this relation to obtain a new view of **curl F**, free of coordinates.

Let P_0 be a point in space and let **n** be a unit vector. Consider a small disk $\mathcal{S}_\mathbf{n}(a)$, perpendicular to **n**, whose center is P_0, and which has radius a. Let $C_\mathbf{n}(a)$ be the boundary of $\mathcal{S}_\mathbf{n}(a)$, oriented to be compatible with the right-hand rule. Then

$$\int_{\mathcal{S}_\mathbf{n}(a)} (\nabla \times \mathbf{F}) \cdot \mathbf{n}\, dS = \oint_{C_\mathbf{n}(a)} \mathbf{F} \cdot d\mathbf{r}.$$

As in our discussion of the physical meaning of curl, we see that

$$(\nabla \times \mathbf{F})_{P_0} \cdot \mathbf{n} \text{ (Area of } \mathcal{S}_{\mathbf{n}}(a)) \approx \oint_{C_{\mathbf{n}}(a)} \mathbf{F} \cdot d\mathbf{r},$$

or

$$(\nabla \times \mathbf{F})_{P_0} \cdot \mathbf{n} \approx \frac{\displaystyle\oint_{C_{\mathbf{n}}(a)} \mathbf{F} \cdot d\mathbf{r}}{\text{Area of } \mathcal{S}_{\mathbf{n}}(a)}.$$

Thus

$$(\nabla \times \mathbf{F})_{P_0} \cdot \mathbf{n} = \lim_{a \to 0} \frac{\displaystyle\oint_{C_{\mathbf{n}}(a)} \mathbf{F} \cdot d\mathbf{r}}{\text{Area of } \mathcal{S}_{\mathbf{n}}(a)}. \tag{3}$$

Equation (3) gives meaning to the component of $(\textbf{curl F})_{P_0}$ in any direction $\mathbf{n}$. So the norm and direction of **curl F** at P_0 can be described in terms of $\mathbf{F}$, without looking at the components of $\mathbf{F}$.

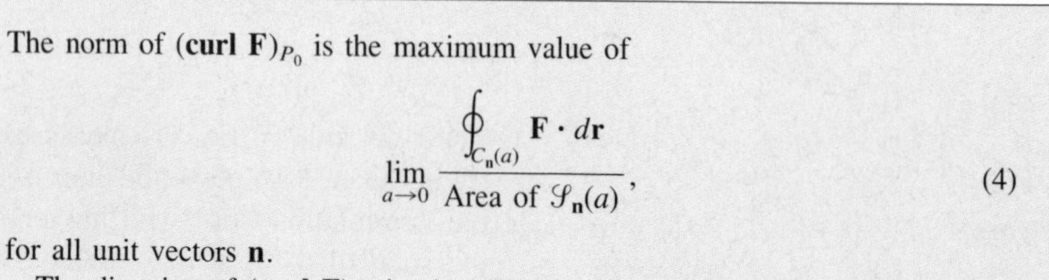
The norm of $(\textbf{curl F})_{P_0}$ is the maximum value of

$$\lim_{a \to 0} \frac{\displaystyle\oint_{C_{\mathbf{n}}(a)} \mathbf{F} \cdot d\mathbf{r}}{\text{Area of } \mathcal{S}_{\mathbf{n}}(a)}, \tag{4}$$

for all unit vectors $\mathbf{n}$.

The direction of $(\textbf{curl F})_{P_0}$ is given by the vector $\mathbf{n}$ that maximizes (4).

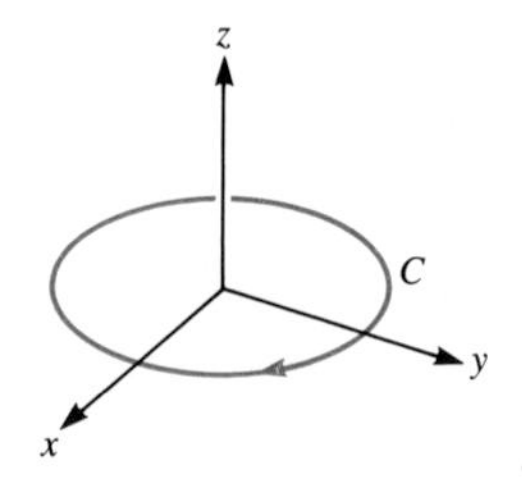

C has radius 0.01.

Figure 3

EXAMPLE 1 Let $\mathbf{F}$ be a vector field such that at (0, 0, 0) **curl F** $= 2\mathbf{i} + 3\mathbf{j} + 4\mathbf{k}$. Estimate $\oint_C \mathbf{F} \cdot d\mathbf{r}$ if C encloses a disk of radius 0.01 in the xy plane with center (0, 0, 0). C is swept out clockwise. (See Fig. 3.)

SOLUTION Let $\mathcal{S}$ be the disk whose border is C. Choose the normal to $\mathcal{S}$ that is consistent with the orientation of C and the right-hand rule. That choice is $-\mathbf{k}$. Thus

$$(\textbf{curl F}) \cdot (-\mathbf{k}) \approx \frac{\displaystyle\oint_C \mathbf{F} \cdot d\mathbf{r}}{\text{Area of } \mathcal{S}}.$$

The area of $\mathcal{S}$ is $\pi(0.01)^2$ and **curl F** $= 2\mathbf{i} + 3\mathbf{j} + 4\mathbf{k}$. Thus

$$(2\mathbf{i} + 3\mathbf{j} + 4\mathbf{k}) \cdot (-\mathbf{k}) \approx \frac{\displaystyle\oint_C \mathbf{F} \cdot d\mathbf{r}}{\pi(0.01)^2}.$$

From this it follows that

$$\oint_C \mathbf{F} \cdot d\mathbf{r} \approx -4\pi(0.01)^2. \quad \blacksquare$$

Maxwell's Equations

At any point in space there is an electric field $\mathbf{E}$ and a magnetic field $\mathbf{B}$. The electric field is due to charges (electrons, protons, or other charged particles) whether stationary or moving. The magnetic field is due to moving charges.

At the time that Newton published his *Principia* on the gravitational field (1687), electricity and magnetism were the subjects of little scientific study. But the experiments of Franklin, Oersted, Henry, Ampère, Faraday, and others in the eighteenth and early nineteenth centuries gradually built up a mass of information subject to mathematical analysis. All the phenomena could be summarized in four equations, which in their final form appeared in Maxwell's *Treatise on Electricity and Magnetism,* published in 1873. For a fuller treatment, see *The Feynman Lectures on Physics,* vol. 2, Addison-Wesley, Reading, Mass., 1964.

We first state Maxwell's equations in the static case, when neither $\mathbf{E}$ nor $\mathbf{B}$ varies with time. The equations had been developed before Maxwell's work.

In the following equations ϵ_0 and c are constants; c is the speed of light.

1 $\int_{\mathcal{S}} \mathbf{E} \cdot \mathbf{n}\, dS = Q/\epsilon_0$, where $\mathcal{S}$ is a surface bounding a spatial region and Q is the total charge in the region (Gauss's law).
(In particular, if there is no charge within a surface, then the "net flow" of $\mathbf{E}$ across the surface is 0.)

2 $\oint_C \mathbf{E} \cdot d\mathbf{r} = 0$, for any closed curve C.

3 $\int_{\mathcal{S}} \mathbf{B} \cdot \mathbf{n}\, dS = 0$, where $\mathcal{S}$ is any surface bounding a spatial region.

4 $c^2 \oint_C \mathbf{B} \cdot d\mathbf{r} = (1/\epsilon_0) \int_{\mathcal{S}} \mathbf{j} \cdot \mathbf{n}\, dS$, where $\mathcal{S}$ is a surface whose boundary is the closed curve C (Faraday's law). The vector $\mathbf{j}$ records the current density at points of $\mathcal{S}$ (and is not to be confused with the unit vector parallel to the y axis). (See Fig. 4.)

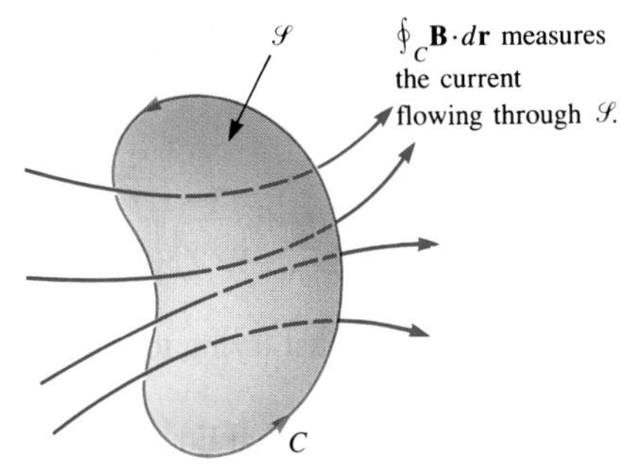

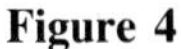
Figure 4

These equations are frequently expressed in terms of the behavior of $\mathbf{E}$ and $\mathbf{B}$ locally, at each point in space, rather than globally with integrals. The analogs of the four integral equations are the following four differential equations, which can be obtained from the integral equations with the aid of the divergence theorem and Stokes' theorem. (See Example 3 in Sec. 17.2, for instance.)

1′ $\nabla \cdot \mathbf{E} = q/\epsilon_0$, where q is the charge density per unit volume (Coulomb's law).
2′ $\nabla \times \mathbf{E} = \mathbf{0}$.
3′ $\nabla \cdot \mathbf{B} = 0$.
4′ $c^2\, \nabla \times \mathbf{B} = \mathbf{j}/\epsilon_0$, where $\mathbf{j}$ is the current density and c is the speed of light.

Maxwell's great contribution was to discover the laws joining $\mathbf{E}$ and $\mathbf{B}$ when they vary with time. It is in these laws that $\mathbf{E}$ and $\mathbf{B}$ become interrelated. [Each of the laws 1, 2, 3, and 4 refers to $\mathbf{E}$ or $\mathbf{B}$—but not to both.] We state Maxwell's laws in differential form:

When **E** *and* **B** *vary with time*

I $\nabla \cdot \mathbf{E} = q/\epsilon_0$.
II $\nabla \times \mathbf{E} = -\partial \mathbf{B}/\partial t$.
III $\nabla \cdot \mathbf{B} = 0$.
IV $c^2\, \nabla \times \mathbf{B} = \partial \mathbf{E}/\partial t + \mathbf{j}/\epsilon_0$.

All of electromagnetic theory is encompassed in these four equations.

Benjamin Franklin, in his book *Experiments and Observations Made at Phila-*

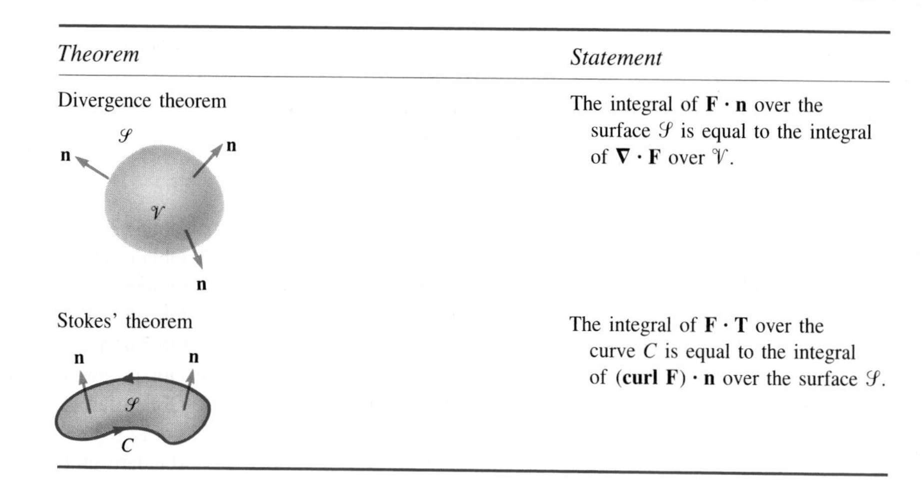

Theorem	*Statement*
Divergence theorem	The integral of $\mathbf{F} \cdot \mathbf{n}$ over the surface $\mathcal{S}$ is equal to the integral of $\nabla \cdot \mathbf{F}$ over $\mathcal{V}$.
Stokes' theorem	The integral of $\mathbf{F} \cdot \mathbf{T}$ over the curve C is equal to the integral of $(\mathbf{curl\ F}) \cdot \mathbf{n}$ over the surface $\mathcal{S}$.

The two-surface version of the divergence theorem shows that for any closed surface $\mathcal{S}$ that encloses the origin,

$$\int_{\mathcal{S}} \frac{\hat{\mathbf{r}} \cdot \mathbf{n}}{\|\mathbf{r}\|^2}\, dS = 4\pi. \tag{1}$$

This equation plays a key role in gravitational and electromagnetic theory, which involve the field $\hat{\mathbf{r}}/\|\mathbf{r}\|^2$.

The same field also appears in the measure of solid angles. Equation (1) is equivalent to the fact that a closed surface subtends a solid angle of 4π steradians at any point in the solid region that it bounds.

Stokes' theorem was used to show that if $\mathbf{F}$ is defined on a simply connected region and $\mathbf{curl\ F} = \mathbf{0}$, then $\mathbf{F}$ is conservative.

The chapter concluded with a paddle-wheel interpretation of the curl and the relation

$$(\nabla \times \mathbf{F})_{P_0} \cdot \mathbf{n} = \lim_{a \to 0} \frac{\oint_{C_\mathbf{n}} \mathbf{F} \cdot d\mathbf{r}}{\text{Area of } \mathcal{S}_\mathbf{n}},$$

where $\mathcal{S}_\mathbf{n}$ is a disk of radius a, perpendicular to $\mathbf{n}$, and $C_\mathbf{n}$ is its boundary, suitably oriented.

Vocabulary and Symbols

$\int_{\mathcal{S}} f(P)\, dS$

integral over a surface

divergence theorem

steradians

orientable

Stokes' theorem

Key Facts

These three conditions on $\mathbf{F}$ are equivalent: (1) $\int_C \mathbf{F} \cdot \mathbf{T}\, ds$ depends only on the endpoints of C, (2) $\oint_C \mathbf{F} \cdot \mathbf{T}\, ds = 0$ for closed curves C, and (3) $\mathbf{F}$ is of the form ∇f. If the domain of $\mathbf{F}$ is simply connected, then condition (4), $\mathbf{curl\ F} = \mathbf{0}$, is equivalent to any of the three conditions given.

To integrate over a sphere of radius a, replace dS by $a^2 \sin \phi \, d\phi \, d\theta$.

To integrate over a surface $\mathcal{S}$, replace dS by $dA/|\cos \gamma|$, and integrate over the projection of $\mathcal{S}$ on the xy plane. (We may also project on the other coordinate planes.)

If div $\mathbf{F} = 0$ throughout the region bounded by $\mathcal{S}$, then $\int_{\mathcal{S}} \mathbf{F} \cdot \mathbf{n} \, dS = 0$.

If div $\mathbf{F} = 0$ but $\mathbf{F}$ is not defined at a point P_0, then for two surfaces $\mathcal{S}_1$ and $\mathcal{S}_2$ that enclose P_0, $\int_{\mathcal{S}_1} \mathbf{F} \cdot \mathbf{n} \, dS = \int_{\mathcal{S}_2} \mathbf{F} \cdot \mathbf{n} \, dS$. This permits replacement of an inconvenient $\mathcal{S}_1$ with a convenient $\mathcal{S}_2$.

If **curl** $\mathbf{F} = \mathbf{0}$ throughout an orientable surface bounded by C, then $\oint_C \mathbf{F} \cdot d\mathbf{r} = 0$.

If **curl** $\mathbf{F} = \mathbf{0}$ on the tube bounded by the similarly oriented curves C_1 and C_2, then $\oint_{C_1} \mathbf{F} \cdot d\mathbf{r} = \oint_{C_2} \mathbf{F} \cdot d\mathbf{r}$. This permits the replacement of an inconvenient C_1 by a convenient C_2.

GUIDE QUIZ ON CHAP. 17: THE DIVERGENCE THEOREM AND STOKES' THEOREM

1 State in words (*a*) Stokes' theorem in the xy plane, (*b*) Stokes' theorem.

2 Let $\mathcal{S}$ be a closed surface that meets each line parallel to the coordinate axes at most twice. Drawing the appropriate diagrams, show that

(*a*) $\int_{\mathcal{S}} x \cos \alpha \, dS$ equals the volume of the region that $\mathcal{S}$ bounds,

(*b*) $\int_{\mathcal{S}} y \cos \alpha \, dS = 0$.

3 Verify that

$$\int_{\mathcal{V}} \nabla \cdot \mathbf{F} \, dV = \int_{\mathcal{S}} \mathbf{F} \cdot \mathbf{n} \, dS,$$

where $\mathcal{V}$ is the box bounded by the three coordinate planes and the three planes $x = a$, $y = b$, and $z = c$ (a, b, and c are positive), $\mathcal{S}$ is the surface of $\mathcal{V}$, and $\mathbf{F}(x, y, z) = x^2\mathbf{i} + y^2\mathbf{j} + z^2\mathbf{k}$.

4 Compute $\oint_C \mathbf{F} \cdot d\mathbf{r}$ for the vector field $\mathbf{F}(x, y, z) = xy\mathbf{i} + 3z\mathbf{j} + y\mathbf{k}$ and the polygonal curve C that starts at (0, 0, 0), goes to (1, 1, 1), then to (0, 1, 1), and then back to (0, 0, 0).

5 A ball of radius a is made of homogeneous material of total mass M. This mass creates a gravitational field $\mathbf{G}$. Assume that there is a constant k such that for any closed surface $\mathcal{S}$, $\int_{\mathcal{S}} \mathbf{G} \cdot \mathbf{n} \, dS = km$, where m is the mass in the region bounded by $\mathcal{S}$.

(*a*) Using symmetry arguments, discuss the form of $\mathbf{G}$.

(*b*) Using the integral of $\mathbf{G} \cdot \mathbf{n}$ over suitable surfaces, completely determine $\mathbf{G}$.

6 Let $\mathcal{S}$ be the top half of the ellipsoid whose equation is $x^2/4 + y^2/4 + z^2/25 = 1$. Let

$$\mathbf{F}(x, y, z) = (x^2 + y)\mathbf{i} + (y^2 + x)\mathbf{j} + (y^3 + x^3)\mathbf{k}.$$

Find the integral of the normal component of **curl F** over $\mathcal{S}$. (Use the exterior normal to the ellipsoid.)

7 (*a*) Which central fields in the plane have curl $\mathbf{0}$?

(*b*) Which central fields in space have curl $\mathbf{0}$?

(*c*) Which central fields in the plane have divergence 0?

(*d*) Which central fields in space have divergence 0?

8 (*a*) What integral represents the number of radians in an angle subtended by a curve in the plane?

(*b*) What integral represents the number of steradians in a solid angle subtended by a surface?

(*c*) Explain, with the aid of clear diagrams, why your answers to (*a*) and (*b*) make sense.

9 In what general circumstances can we conclude that

(*a*) $\oint_C \mathbf{F} \cdot \mathbf{T} \, ds = 0$?

(*b*) $\oint_C \mathbf{F} \cdot \mathbf{n} \, ds = 0$?

(*c*) $\int_{\mathcal{S}} \mathbf{F} \cdot \mathbf{n} \, dS = 0$? ($\mathcal{S}$ is the surface of a solid region.)

(*d*) $\int_{\mathcal{S}_1} \mathbf{F} \cdot \mathbf{n} \, dS = \int_{\mathcal{S}_2} \mathbf{F} \cdot \mathbf{n} \, dS$? ($\mathcal{S}_1$ and $\mathcal{S}_2$ are the inner and outer surfaces of a solid region $\mathcal{V}$.)

(*e*) $\int_{C_1} \mathbf{F} \cdot d\mathbf{r} = \int_{C_2} \mathbf{F} \cdot d\mathbf{r}$? ($C_1$ and C_2 are closed curves.)

10 Let $f(x, y, z, t)$ be the density of a fluid or gas at the point (x, y, z) at time t. Let $\mathcal{V}(t)$ be a solid region which depends on time, and let $\mathcal{S}(t)$ be the surface that bounds it. Let $\mathbf{W} = \mathbf{F}(x, y, z, t)$ be the velocity vector of the fluid at the point (x, y, z) at time t. The equation

$$\frac{d}{dt} \int_{\mathcal{V}(t)} f(P) \, dV = \int_{\mathcal{V}(t)} \frac{\partial f}{\partial t} \, dV - \int_{\mathcal{S}(t)} f(P)(\mathbf{W} \cdot \mathbf{n}) \, dS$$

is known as the **general transport theorem** in fluid mechanics.

(*a*) Interpret $\int_{\mathcal{V}(t)} f(P) \, dV$ physically.

(*b*) Interpret $\frac{d}{dt} \int_{\mathcal{V}(t)} f(P) \, dV$ physically.

(*c*) Interpret $\frac{\partial f}{\partial t}$ physically.

(*d*) Interpret $\int_{\mathcal{V}(t)} \frac{\partial f}{\partial t} \, dV$ physically.

(*e*) Interpret $\int_{\mathcal{S}(t)} f(P)(\mathbf{W} \cdot \mathbf{n})\, dS$ physically.

(*f*) Is the general transport theorem plausible?

11 Does the divergence theorem hold for regions in the shape of a doughnut?

12 Let F be defined throughout space and have the property that for every circle C of radius at most 0.01, $\int_C \mathbf{F} \cdot d\mathbf{r} = 0$. Does it follow that for every circle C, no matter how large, $\int_C \mathbf{F} \cdot d\mathbf{r} = 0$?

REVIEW EXERCISES FOR CHAP. 17: THE DIVERGENCE THEOREM AND STOKES' THEOREM

1 Outline the key steps in the proof of the divergence theorem.

2 Let

$$\mathbf{F}(x, y, z) = \frac{-y\mathbf{i}}{x^2 + y^2} + \frac{x\mathbf{j}}{x^2 + y^2}.$$

(*a*) Show that div $\mathbf{F} = 0$.
(*b*) Show that **curl F** = **0**.
(*c*) Let $f(x, y, z) = \tan^{-1}(y/x)$. Does $\mathbf{F}$ equal ∇f where both are defined?
(*d*) Evaluate $\oint_C \mathbf{F} \cdot d\mathbf{r}$, where C is the circle $x^2 + y^2 = 1$ in the xy plane.
(*e*) Show that $\mathbf{F}$ is not conservative.
(*f*) Does (*e*) contradict (*c*)? Explain.

3 Let $\mathcal{S}$ be the portion of the surface $z = x^2 + 3y^2$ below the surface $z = 1 - x^2 - y^2$.
(*a*) Set up a plane integral for the area of $\mathcal{S}$.
(*b*) Set up repeated integrals in rectangular coordinates and in polar coordinates for the integral in (*a*), but do not evaluate them.

4 Let C be the square with vertices (0, 0, 0), (0, 2, 2), (0, 0, 4), and (0, −2, 2) swept out in the given order. Evaluate $\oint_C \mathbf{F} \cdot d\mathbf{r}$, where

$$\mathbf{F} = x^2ye^z\mathbf{i} + (x + y + z)\mathbf{j} + x^2z\mathbf{k}.$$

5 Figure 1 shows two similarly oriented curves circling the line L.

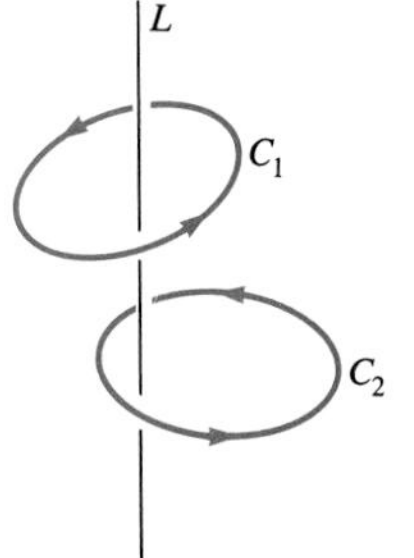

Figure 1

(*a*) If $\mathbf{F}$ is a vector field defined throughout space and has **curl F** = **0**, what can we conclude about $\oint_{C_1} \mathbf{F} \cdot \mathbf{T}\, ds$ and $\oint_{C_2} \mathbf{F} \cdot \mathbf{T}\, ds$?
(*b*) If $\mathbf{F}$ is a vector field defined everywhere in space except on the line L and **curl F** = **0**, what can we conclude about $\oint_{C_1} \mathbf{F} \cdot \mathbf{T}\, ds$ and $\oint_{C_2} \mathbf{F} \cdot \mathbf{T}\, ds$?

6 Using Stokes' theorem, evaluate $\int_{\mathcal{S}} (\mathbf{curl\ F}) \cdot \mathbf{n}\, dS$, where $\mathbf{F} = (x^2 + y)\mathbf{i} + 3xy\mathbf{j} + (2xz + z^2)\mathbf{k}$, $\mathcal{S}$ is the surface of the paraboloid $z = 4 - x^2 - y^2$ above the xy plane, and $\mathbf{n}$ has a positive z component.

7 Let $\mathbf{F}(x, y, z) = \mathbf{r}/\|\mathbf{r}\|^n$ for a fixed integer n and $\mathbf{r} = x\mathbf{i} + y\mathbf{j} + z\mathbf{k}$.
(*a*) For which n is **curl F** = **0**?
(*b*) For which n is div $\mathbf{F} = 0$?

In Exercises 8 to 11 evaluate the given line integrals $\int_C (P\, dx + Q\, dy + R\, dz)$, where C is described by the given parameterization $\mathbf{r}(t)$.

8 $P = x \sin y^2$, $Q = \dfrac{x}{y^2 + 1}$, $R = \tan^{-1} z$; $\mathbf{r}(t) = t\mathbf{i} + t\mathbf{j} + t\mathbf{k}$, $0 \le t \le 1$

9 $P = \ln z$, $Q = \dfrac{1}{xy}$, $R = \sqrt{xy + 1}$; $\mathbf{r}(t) = t\mathbf{i} + t^2\mathbf{j} + t^3\mathbf{k}$, $1 \le t \le 2$

10 $P = \dfrac{y}{x + 1}$, $Q = \dfrac{1}{y^2 + 1}$, $R = \dfrac{1}{z + 1}$; $\mathbf{r}(t) = t^2\mathbf{i} + t\mathbf{j} + t^4\mathbf{k}$, $0 \le t \le 1$

11 $P = -xy^2$, $Q = x^2y$, $R = z^2$; $\mathbf{r}(t) = \cos t\ \mathbf{i} + \sin t\ \mathbf{j} + \tan t\ \mathbf{k}$, $0 \le t \le \pi/4$

12 Let $\mathbf{F}$ be a vector field in space that is defined everywhere except at the origin and has div $\mathbf{F} = 0$. Assume that for all $r = \sqrt{x^2 + y^2 + z^2}$ sufficiently large, $\|\mathbf{F}(x, y, z)\| \le 1/r^3$. What can be said about $\int_{\mathcal{S}} \mathbf{F} \cdot \mathbf{n}\, dS$, where (*a*) $\mathcal{S}$ is a sphere that does not surround the origin? (*b*) $\mathcal{S}$ is a sphere that does surround the origin?

13 A closed surface $\mathcal{S}$ is cut into two pieces $\mathcal{S}_1$ and $\mathcal{S}_2$ by a closed curve C. Let $\mathbf{n}$ be the exterior normal to $\mathcal{S}$. What relation is there between

$$\int_{\mathcal{S}_1} (\mathbf{curl\ F}) \cdot \mathbf{n}\, dS \qquad \text{and} \qquad \int_{\mathcal{S}_2} (\mathbf{curl\ F}) \cdot \mathbf{n}\, dS?$$

14 Figure 2 shows a few typical samples for a vector field $\mathbf{F}$.

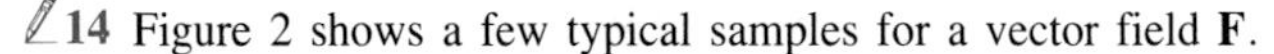

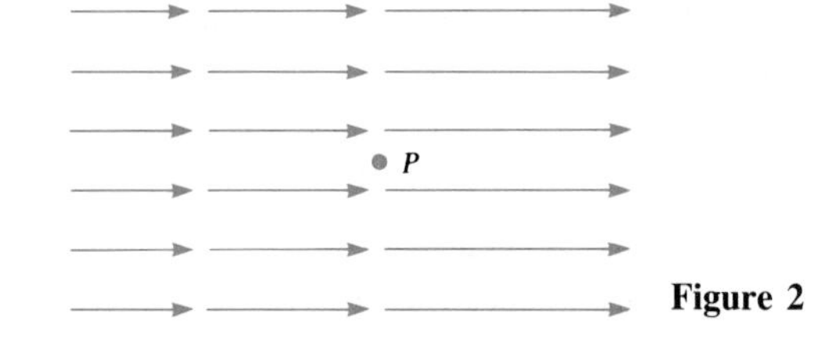

Figure 2

(*a*) What would you say about **curl F** at P?
(*b*) What would you say about div **F** at P?

15 Find the area of the portion of the cone $z^2 = x^2 + y^2$ that lies between the sphere $\rho = 1$ and $\rho = 2$, (*a*) using spherical coordinates, (*b*) by first projecting on the xy plane.

16 (*a*) Sketch the portion of the cylinder $x^2 + z^2 = a^2$ that lies above the xy plane and inside the cylinder $x^2 + y^2 = a^2$.
(*b*) Find its area.

17 (*a*) Sketch the solid bounded by the cylinders $z = y^2$, $z = x^2$, and the plane $z = 1$.
(*b*) Find the total surface area of the solid.

18 (*a*) Sketch the portion of the hyperboloid of one sheet $z^2 = x^2 + y^2 - 1$ that lies between the planes $z = 0$ and $z = \sqrt{3}$.
(*b*) Express the area of the surface in (*a*) as an integral over the xy plane.

19 Let $\mathbf{F} = xz\mathbf{i} + x^2\mathbf{j} + xy\mathbf{k}$. Let C be the path around the square whose vertices are (0, 0, 1), (1, 0, 1), (1, 1, 1), and (0, 1, 1). The path starts at (0, 0, 1), sweeps out the vertices in the indicated order, and returns to (0, 0, 1). Evaluate $\oint_C \mathbf{F} \cdot d\mathbf{r}$. (See Fig. 3.)

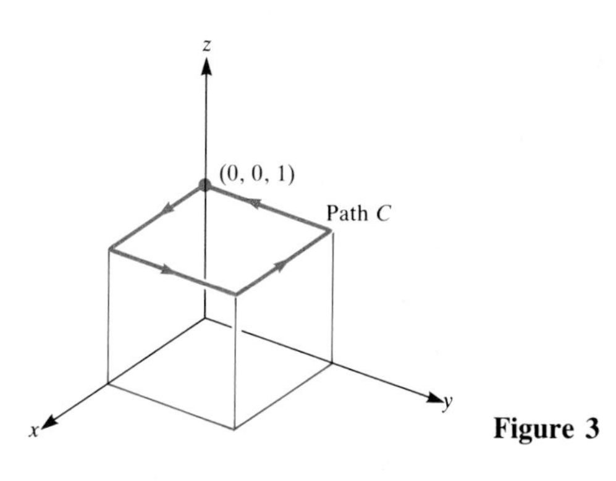

Figure 3

20 Let $f(x, y, z) = e^x yz$. Evaluate $\int_C \nabla f \cdot d\mathbf{r}$, where C is the curve shown in Fig. 4.

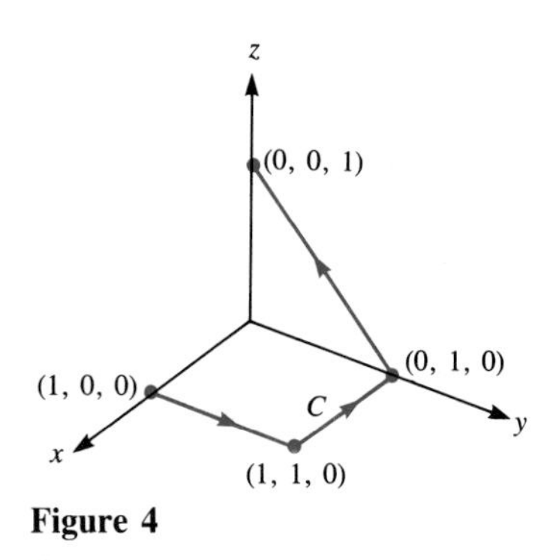

Figure 4

21 A charge Q is distributed uniformly on a sphere of radius a. It creates an electric field **E**.
(*a*) Explain why **E** is a central field.
(*b*) Find **E**(P) for points outside the sphere.
(*c*) Find **E**(P) for points inside the sphere. (These results also hold for the gravitational field created by a uniform distribution of mass on the sphere.)

22 Let $\mathcal{S}$ be the ellipsoid $x^2/a^2 + y^2/b^2 + z^2/c^2 = 1$ and **F** be $\hat{\mathbf{r}}/\|\mathbf{r}\|^2$.
(*a*) Draw $\mathcal{S}$.
(*b*) Find the flux $\int_{\mathcal{S}} \mathbf{F} \cdot \mathbf{n}\, dS$ by using the fact that **F** is divergence-free.
(*c*) Find $\int_{\mathcal{S}} \mathbf{F} \cdot \mathbf{n}\, dS$ by using steradians.

23 Evaluate $\int_C \mathbf{F} \cdot d\mathbf{r}$, where $\mathbf{F} = e^x\mathbf{i} + e^y\mathbf{j} + \sin z\,\mathbf{k}$ and C is the helix shown in Fig. 5.

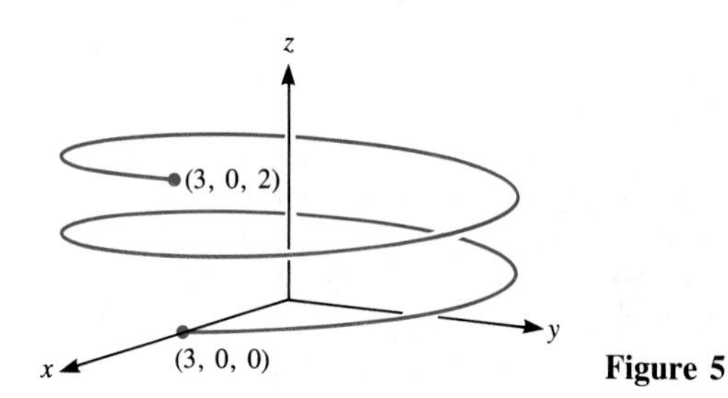

Figure 5

24 Let $P\,dx + Q\,dy + R\,dz$ be a differential form. Assume that there is a function f such that $df = P\,dx + Q\,dy + R\,dz$. What relations exist among the partial derivatives of P, Q, and R?
(*a*) Explain, using the equality of the mixed partial derivatives.
(*b*) Explain, using line integrals.

25 (See Gauss's law in Sec. 17.2.) A charge Q is distributed uniformly throughout a ball of radius a. It creates an electric field **E**. (*a*) Find **E**(P) for points P outside the ball. (*b*) Find **E**(P) for points inside the ball.

26 Find $\int_{\mathcal{S}} \mathbf{F} \cdot \mathbf{n}\, dS$, where $\mathcal{S}$ is the entire surface of the solid cylinder in Fig. 6 and $\mathbf{F} = z\mathbf{i}$, (*a*) without the aid of the divergence theorem, (*b*) with the divergence theorem.

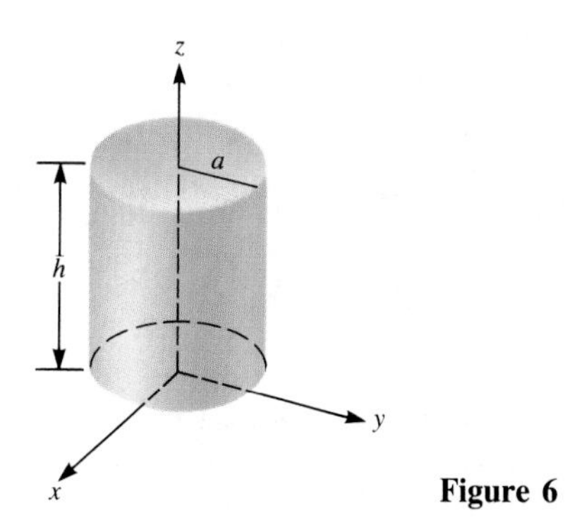

Figure 6

In Exercises 27 to 34 determine which vector fields are conservative and which are not.

27 (*a*) ∇f, where $f(x, y, z) = x^2ye^z$

(*b*) $\dfrac{-y\mathbf{i} + x\mathbf{j}}{(x^2 + y^2)^2}$

28 (*a*) $ye^z \cos xy\ \mathbf{i} + xe^z \cos xy\ \mathbf{j} + e^z \sin xy\ \mathbf{k}$

(*b*) $x^3\mathbf{i} + y^2\mathbf{j}$

29 (*a*) ∇f, where $f(x, y, z) = \dfrac{\sin 3xy}{1 + e^{x+z}}$

(*b*) $\dfrac{x\mathbf{i} + y\mathbf{j} - z\mathbf{k}}{x^2 + y^2 + z^2}$

30 (*a*) $\dfrac{x\mathbf{i} + y\mathbf{j} + z\mathbf{k}}{(x^2 + y^2 + z^2)^3}$

(*b*) $\sec^3 xz\ \mathbf{i} + \tan^3 yz\ \mathbf{j} + 3^x\mathbf{k}$

31 $(2xy^2 + xy^3)\mathbf{i} + (2x^2yz + \frac{3}{2}x^2y^2)\mathbf{j} + (x^2y + yz^2)\mathbf{k}$

32 $3\mathbf{i} - \dfrac{z\mathbf{j}}{y^2 + z^2} + \dfrac{y\mathbf{k}}{y^2 + z^2}$

33 $(\cos^2 r)\hat{\mathbf{r}}$, where $r = \|\mathbf{r}\|$.

34 (*a*) $x^3\mathbf{i} + y^3\mathbf{j} + z^3\mathbf{k}$, (*b*) $x^2\mathbf{i} + y^2\mathbf{j} + z^2\mathbf{k}$, (*c*) $x\mathbf{i} + y\mathbf{j} + z\mathbf{k}$.

35 Let $\mathbf{F}(\mathbf{r}) = \hat{\mathbf{r}}\|\mathbf{r}\|^2$, where $\mathbf{r} = x\mathbf{i} + y\mathbf{j} + z\mathbf{k}$.

(*a*) Why is $\mathbf{F}$ conservative?

(*b*) Construct a scalar function f so that $\mathbf{F} = \nabla f$.

36 The irrotational field $\mathbf{F}$ is defined everywhere except on the coordinate axes. Assume that $\oint_{C_1} \mathbf{F} \cdot d\mathbf{r} = 1$, $\oint_{C_2} \mathbf{F} \cdot d\mathbf{r} = 2$, and $\oint_{C_3} \mathbf{F} \cdot d\mathbf{r} = 3$. Find (*a*) $\oint_{C_4} \mathbf{F} \cdot d\mathbf{r}$, (*b*) $\oint_{C_5} \mathbf{F} \cdot d\mathbf{r}$, and (*c*) $\oint_{C_6} \mathbf{F} \cdot d\mathbf{r}$. (See Fig. 7.)

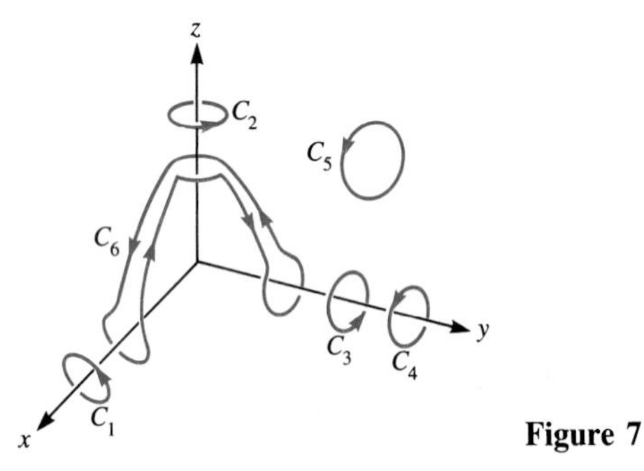

Figure 7

37 Let $\mathcal{S}$ be a surface that encloses a solid region $\mathcal{V}$. Assume that the origin O is in $\mathcal{V}$. Let $\mathbf{F}(P)$ be the position vector $\overrightarrow{OP}$. Using approximating sums and a sketch, show that $\int_{\mathcal{S}} \mathbf{F}(P) \cdot \mathbf{n}\, dS = 3 \cdot$ Volume of $\mathcal{V}$.

38 For the function $\mathbf{F}$ of Exercise 37, verify that

$$\int_{\mathcal{S}} \mathbf{F} \cdot \mathbf{n}\, dS = \int_{\mathcal{V}} \nabla \cdot \mathbf{F}\, dV.$$

39 State all the ways you can think of to show that the expressions $P\,dx + Q\,dy + R\,dz$ is (*a*) exact, (*b*) not exact.

40 Let $\mathcal{S}$ be the sphere of radius 3 centered at (4, 1, 2). Evaluate $\int_{\mathcal{S}} \mathbf{F} \cdot \mathbf{n}\, dS$, where $\mathbf{F} = 7x\mathbf{i} + x^2z\mathbf{j} + 5z\mathbf{k}$.

41 Let $\mathbf{F}(x, y, z) = ze^{xz}\mathbf{i} + xe^{xz}\mathbf{k}$.

(*a*) Exhibit a function f such that $\mathbf{F} = \nabla f$.

(*b*) Evaluate $\int_C \mathbf{F} \cdot \mathbf{T}\, ds$ for the curve given parametrically as

$$\mathbf{r}(t) = t^2\mathbf{i} + \cos t\ \mathbf{j} + e^t\mathbf{k} \qquad 0 \le t \le 2\pi.$$

42 Let $\mathcal{D}$ consist of those points (x, y, z) not on the z axis. Let

$$\mathbf{F}(x, y, z) = \frac{x\mathbf{i}}{x^2 + y^2} + \frac{y\mathbf{j}}{x^2 + y^2},$$

for (x, y, z) in $\mathcal{D}$. Let $f(x, y, z) = \ln(x^2 + y^2)$.

(*a*) Show that div $\mathbf{F} = 0$.

(*b*) Show that **curl** $\mathbf{F} = \mathbf{0}$.

(*c*) Show that $\mathbf{F} = \nabla f$.

(*d*) Is $\mathbf{F}$ conservative?

(*e*) Is the domain of $\mathbf{F}$ simply connected?

43 Let $\mathbf{F}(x, y, z) = 4xz\mathbf{i} - y^2\mathbf{j} + yz\mathbf{k}$. Let $\mathcal{V}$ be the cube bounded by the three coordinate planes and the planes $x = 1$, $y = 1$, $z = 1$. Let $\mathcal{S}$ be the surface of $\mathcal{V}$. Verify that

$$\int_{\mathcal{V}} \text{div}\, \mathbf{F}\, dV = \int_{\mathcal{S}} \mathbf{F} \cdot \mathbf{n}\, dS.$$

44 Which of these vector fields in space are conservative? Not conservative? Explain.

(*a*) $\mathbf{F}(x, y, z) = 3\mathbf{i} + 4\mathbf{j} + 5\mathbf{k}$

(*b*) $\mathbf{F}(x, y, z) = (2xy^2z + xy^3)\mathbf{i} + (2x^2yz + 3x^2y^2/2)\mathbf{j} + (x^2y^2 + yz^2)\mathbf{k}$

(*c*) $\mathbf{F}(x, y, z) = \sin^2 5x \cos^2 5x\mathbf{i} + y^2 \sin y\mathbf{j} + (z + 1)^2e^z\mathbf{k}$

45 A circle of radius a has its center at a distance $b > a$ from a line L. It is revolved about L to produce a surface of revolution (a **torus**, or surface of a doughnut). To integrate over this surface, it would be helpful to parameterize it, as we did the sphere and cone.

(*a*) Show how to parameterize the torus in terms of the angles u and v shown in Fig. 8.

(*b*) Find the average distance from points on the surface to the line L. (The answer is not b. Why not?)

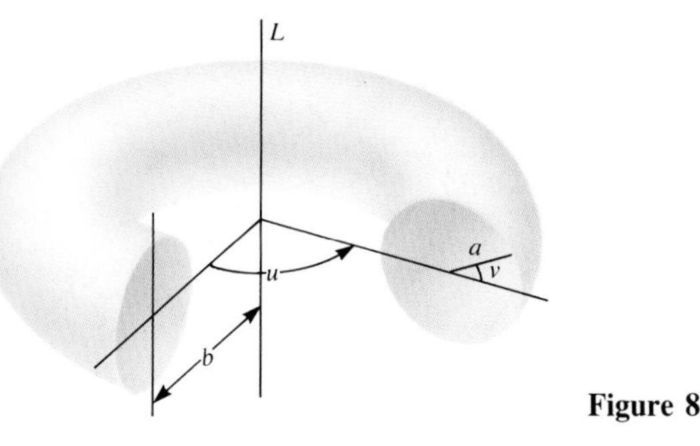

Figure 8

46 Let $\mathbf{F}(x, y, z) = 3xy\mathbf{i} + 4\mathbf{j} + z\mathbf{k}$. Let C be the closed polygonal curve that starts at (0, 0, 0), goes to (1, 0, 1), then to (1, 1, 1), then to (0, 1, 0), and back to (0, 0, 0).

(*a*) Evaluate $\int_C \mathbf{F} \cdot \mathbf{T}\, ds$.
(*b*) Is **F** conservative?

47 Consider an object submerged in water. It occupies a region $\mathcal{V}$ with surface $\mathcal{S}$. The force **F** of the water against the object at each point is perpendicular to $\mathcal{S}$. Assume that the surface of the water is the xy plane. Thus the z coordinate of a point in the water is negative. The force against a small patch of $\mathcal{S}$ of area ΔS and depth $|z|$ is approximately $cz\mathbf{n}\,\Delta S$, where **n** is the unit exterior normal where c is the density of water. The z component of this force is $cz\mathbf{k} \cdot \mathbf{n}\,\Delta S$.
(*a*) What integral over $\mathcal{S}$ represents the vertical component of the total force of the water against the submerged object?
(*b*) Show that the vertical component of the force against the object is equal to the weight of the water displaced by the object.
(*c*) Show that the x and y components of the total force against the object are 0.

48 Using Stokes' theorem, evaluate $\int_{\mathcal{S}} (\mathbf{curl\ F}) \cdot \mathbf{n}\, dS$, where $\mathbf{F} = (x^2 + y)\mathbf{i} + 3xy\mathbf{j} + (2xz + z^2)\mathbf{k}$, $\mathcal{S}$ is the surface of the paraboloid $z = 4 - x^2 - y^2$ above the xy plane, and **n** has a positive z component.

49 If $\int_{\mathcal{V}} (\nabla f \cdot \nabla f)\, dV = 0$, where $\mathcal{V}$ is a ball, what can we conclude about f? Explain.

Exercises 50 and 51 concern harmonic functions. A function f is harmonic if the Laplacian $\nabla^2 f = 0$.

50 Let $\mathcal{V}$ be a solid and let $\mathcal{S}$ be its surface. Let f and g be two scalar functions defined on $\mathcal{V}$.
(*a*) Show that

$$\int_{\mathcal{V}} \nabla \cdot (f\,\nabla g)\, dV = \int_{\mathcal{V}} (f\,\nabla^2 g + \nabla f \cdot \nabla g)\, dV = \int_{\mathcal{S}} (f\,\nabla g) \cdot \mathbf{n}\, dS.$$

(*b*) Show that

$$\int_{\mathcal{V}} \nabla \cdot (g\,\nabla f)\, dV = \int_{\mathcal{V}} (g\,\nabla^2 f + \nabla g \cdot \nabla f)\, dV = \int_{\mathcal{S}} (g\,\nabla f) \cdot \mathbf{n}\, dS.$$

(*c*) Deduce that

$$\int_{\mathcal{V}} (f\,\nabla^2 g - g\,\nabla^2 f)\, dV = \int_{\mathcal{S}} [(f\,\nabla g - g\,\nabla f) \cdot \mathbf{n}]\, dS.$$

This result is called **Green's theorem** too.

51 Assume that f and g are harmonic on $\mathcal{V}$ and equal on $\mathcal{S}$. This exercise will show that they must be equal on $\mathcal{V}$. Let $h = f - g$.
(*a*) Use Exercise 50(*a*) to show that

$$\int_{\mathcal{V}} h\,\nabla^2 h\, dV + \int_{\mathcal{V}} \nabla h \cdot \nabla h\, dV = \int_{\mathcal{S}} (h\,\nabla h) \cdot \mathbf{n}\, dS.$$

(*b*) Deduce that $\nabla h = \mathbf{0}$ throughout $\mathcal{V}$.
(*c*) Deduce that h is constant throughout $\mathcal{V}$.
(*d*) Show that $f = g$ throughout $\mathcal{V}$.

Throughout this chapter only integrals of scalar functions were considered. But integrals of vector functions are sometimes of use and are easily defined. If $\mathbf{F} = P\mathbf{i} + Q\mathbf{j} + R\mathbf{k}$ is defined throughout the solid $\mathcal{V}$, define $\int_{\mathcal{V}} \mathbf{F}(x, y, z)\, dV$ to be

$$\int_{\mathcal{V}} P\, dV\,\mathbf{i} + \int_{\mathcal{V}} Q\, dV\,\mathbf{j} + \int_{\mathcal{V}} R\, dV\,\mathbf{k}.$$

In short, to integrate a vector function, integrate each of its scalar components. The integral of a vector function over a surface is defined similarly. This concept is the subject of Exercises 52 to 57.

52 Let $\mathbf{F}(x, y, z) = f(x, y, z)(x\mathbf{i} + y\mathbf{j} + z\mathbf{k})$, where $f(x, y, z)$ denotes the density of matter at point (x, y, z) in the solid that occupies the region $\mathcal{V}$. What is the physical significance of the equation

$$\int_{\mathcal{V}} \mathbf{F}(x, y, z)\, dV = \mathbf{0}?$$

53 Let $\mathbf{F} = P\mathbf{i} + Q\mathbf{j} + R\mathbf{k}$ be a vector field and $\mathcal{V}$ a solid region. Let **c** be a fixed vector. Show that $\int_{\mathcal{V}} \mathbf{c} \cdot \mathbf{F}\, dV = \mathbf{c} \cdot \int_{\mathcal{V}} \mathbf{F}\, dV$. A similar theorem holds for surface integrals.

54 Let **A** and **B** be vectors such that for all vectors **c**, $\mathbf{c} \cdot \mathbf{A} = \mathbf{c} \cdot \mathbf{B}$. Show that $\mathbf{A} = \mathbf{B}$.

55 Let **c** be any fixed vector. Let $\mathcal{V}$ be a solid region and $\mathcal{S}$ the surface that bounds it. Let f be a scalar function on $\mathcal{V}$.
(*a*) Show that $\int_{\mathcal{V}} \nabla \cdot f\mathbf{c}\, dV = \int_{\mathcal{S}} f\mathbf{c} \cdot \mathbf{n}\, dS$.
(*b*) Show that $\int_{\mathcal{V}} \nabla \cdot f\mathbf{c}\, dV = \int_{\mathcal{V}} \mathbf{c} \cdot \nabla f\, dV$.
(*c*) Show that $\mathbf{c} \cdot \int_{\mathcal{V}} \nabla f\, dV = \mathbf{c} \cdot \int_{\mathcal{S}} f\mathbf{n}\, dS$.
(*d*) Deduce that $\int_{\mathcal{V}} \nabla f\, dV = \int_{\mathcal{S}} f\mathbf{n}\, dS$.

56 Let $\mathcal{S}$ be a surface bounding a solid region $\mathcal{V}$. Let **c** be a fixed vector.
(*a*) Show that $\int_{\mathcal{S}} \mathbf{c} \cdot \mathbf{n}\, dS = 0$.
(*b*) Deduce that $\int_{\mathcal{S}} \mathbf{n}\, dS = \mathbf{0}$.

57 (See Exercise 56.) On each of the four faces of a tetrahedron a vector is constructed that is perpendicular to the face, points outward, and has magnitude equal to the area of the face. Show that the sum of the four vectors is **0**.

58 A parallelepiped $\mathcal{V}$ is spanned by the three vectors $\mathbf{i} + \mathbf{j} + \mathbf{k}$, $2\mathbf{i} + \mathbf{j} + \mathbf{k}$, and $2\mathbf{i} + 3\mathbf{j} + 4\mathbf{k}$. Let $\mathbf{F} = 2x\mathbf{i} + 3y\mathbf{j} + 4z\mathbf{k}$. Evaluate $\int_{\mathcal{S}} \mathbf{F} \cdot \mathbf{n}\, dS$, where $\mathcal{S}$ is the surface of $\mathcal{V}$.

59 Evaluate $\int_{\mathcal{S}} \mathbf{curl\ F} \cdot \mathbf{n}\, dS$, where **F** is the field $xr^3\mathbf{i} - yr^3\mathbf{j} + x^2 \sin \pi y\, e^{\tan^{-1} z}\,\mathbf{k}$ and $\mathcal{S}$ is the top half of the ellipsoid $x^2/16 + y^2/9 + z^2/4 = 1$ and $r = \sqrt{x^2 + y^2 + z^2}$. (Use the simplest approach.)

APPENDIX **A**

REAL NUMBERS

This appendix discusses rational and irrational numbers, division by 0, inequalities, intervals, and absolute value.

The **positive integers** are the counting numbers, 1, 2, 3, . . . ; the **negative integers** are $-1, -2, -3, \ldots$. The set of positive and negative integers, together with 0, is called the set of **integers**. The integers appear on the number line as regularly spaced points, as shown in Fig. 1.

··· −4 −3 −2 −1 0 1 2 3 4 ···

Figure 1

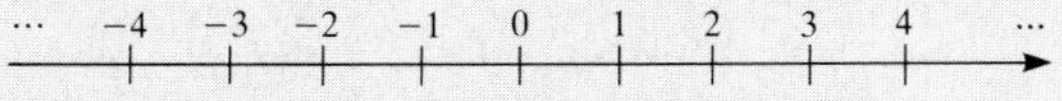

Every point on the number line corresponds to a **real number**. For instance, $\frac{11}{3}$, $-\sqrt{2}$, π, and 1.13 are real numbers and their positions are shown in Fig. 2.

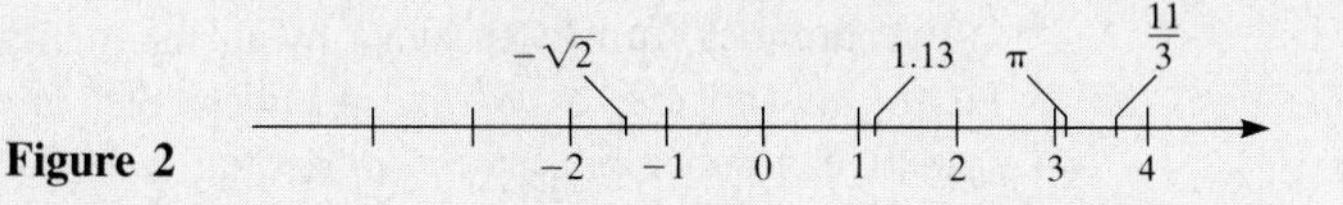

Figure 2

A real number that can be written as a fraction or ratio p/q, where p is an integer and q is a positive integer, is called a **rational number**. For instance, $\frac{11}{3}$ is rational, as are $8 = \frac{8}{1}$ and $7\sqrt{2}/(3\sqrt{2}) = \frac{7}{3}$. A real number that is not rational is called **irrational**. Greek mathematicians some 2,400 years ago showed that $\sqrt{2}$ is irrational; Johann Lambert in 1761 showed that π is irrational. This means that neither $\sqrt{2}$ nor π can be written as the quotient of two integers, p/q.

If a and b are real numbers, we may always form their sum $a + b$, their difference $a - b$, and their product ab. If b is not 0, then there is a unique number x such that $bx = a$. This number x, the **quotient** of a by b, is denoted a/b. For instance, since the equation $2x = 6$ has the unique solution 3, we write $6/2 = 3$.

Why division by zero is meaningless

However, when $b = 0$, solving the equation $bx = a$ runs into some problems. For instance, the equation

$$0x = 6$$

has no solution whatsoever, since the product of 0 and any real number is 0. Thus the symbol $\frac{6}{0}$ is totally meaningless. When both a and b are 0, the equation

$bx = a$ runs into a different trouble: there are too many solutions. The equation is now

$$0x = 0.$$

Every real number is a solution; for instance, $0 \cdot 5 = 0$, $0 \cdot \pi = 0$, and $0(-3) = 0$. The symbol $\frac{0}{0}$ is meaningless because it does not describe a single number.

WRONG WAY

$$\frac{a}{0}$$

In short, "division by zero" makes no sense. If you find yourself dividing by zero while working a problem, turn back, as though you had met a sign warning "WRONG WAY." In particular, resist the temptation to say that $\frac{0}{0}$ is equal to either 1 or 0. This temptation will be placed before you often in the study of limits and derivatives. The expression $\frac{0}{0}$ is utterly devoid of meaning.

Inequalities If the point that represents the number a lies to the left of the point that represents the number b on the number line, we write $a < b$ ("a is less than b") or $b > a$ ("b is greater than a"). For instance, $5 < 7$, $7 > 5$, $-8 < 2$, and $-8 < -3$.

The expression $a \le b$ means that a is either less than b or equal to b. Thus $3 \le 4$ and $4 \le 4$. Read "$a \le b$" as "a is less than or equal to b." If $a \le b$, we also write $b \ge a$.

Inequalities behave nicely with respect to addition and subtraction but demand great care in the case of multiplication and division. Inspection of the following list of properties of inequalities shows the difference.

The properties of inequalities

Multiplication or division by a positive number preserves an inequality but multiplication or division by a negative number reverses *an inequality.*

$2 < 3$ but $\frac{1}{2} > \frac{1}{3}$.

1 If $a < b$ and $c < d$, then $a + c < b + d$. (You can add two inequalities that are in the same direction.)

2 If $a < b$ and c is any number, then $a + c < b + c$. (You can add the same number to both sides of an inequality.)

3 If $a < b$ and c is any number, then $a - c < b - c$. (You can subtract the same number from both sides of an inequality.)

4 If $a < b$ and c is a *positive* number, then $ac < bc$. (Multiplication by a positive number preserves an inequality.)

5 If $a < b$ and c is a *negative* number, then $ac > bc$. (Multiplication by a negative number reverses an inequality.)

6 If $a < b$ and $c < d$ and a, b, c, and d are *positive,* then $ac < bd$. (You can multiply two inequalities, if they are in the same direction and involve only positive numbers.)

7 If $a < b$ and c is a *positive* number, then $a/c < b/c$. (Division by a positive number preserves an inequality.)

8 If $a < b$ and c is *negative,* then $a/c > b/c$. (Division by a negative number reverses the inequality.)

9 If a and b are positive numbers and $a < b$, then

$$\frac{1}{a} > \frac{1}{b}.$$

(Taking reciprocals of an inequality between positive numbers reverses the inequality.)

You cannot subtract one inequality from another; for instance, $5 < 8$ and $1 < 6$, but it is not true that $5 - 1$ is less than $8 - 6$.

The notation $a < x < b$ in the following definition is short for the two inequalities $a < x$ and $x < b$. Thus, $a < x < b$ is short for "x is larger than a and less than b."

Definition *Open interval.* Let a and b be real numbers, with $a < b$. The **open interval** (a, b) consists of all numbers x such that $a < x < b$.

Definition *Closed interval.* Let a and b be real numbers, with $a < b$. The **closed interval** $[a, b]$ consists of all numbers x such that $a \leq x \leq b$.

The meaning of ● and ○ in diagrams

The open interval (a, b) is obtained from the closed interval $[a, b]$ by the deletion of the endpoints a and b. Some intervals are shown in Fig. 3. The solid dot (●) indicates the presence of a point (included point); the hollow dot (○) indicates the absence of a point (excluded point). This notation is used in diagrams throughout the text.

a b — Closed interval $[a, b]$

a b — Open interval (a, b)

a b — Half-open interval $(a, b]$

Figure 3

Infinite intervals

Infinite intervals are also of use. For convenience they are listed in the following table:

Symbol	*Description*	*Picture*	*In Words*
$[a, \infty)$	$x \geq a$	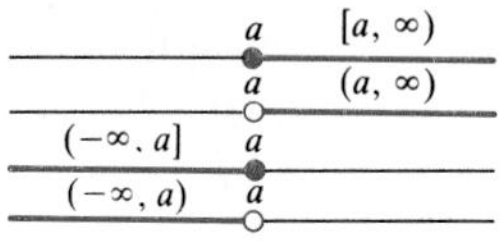 a $[a, \infty)$	Closed interval to the right of a
(a, ∞)	$x > a$	a (a, ∞)	Open interval to the right of a
$(-\infty, a]$	$x \leq a$	$(-\infty, a]$ a	Closed interval to the left of a
$(-\infty, a)$	$x < a$	$(-\infty, a)$ a	Open interval to the left of a
$(-\infty, \infty)$	all x		Entire x axis

The symbols ∞ and $-\infty$ do not refer to numbers but provide a convenient shorthand.

The following three examples apply the properties of inequalities and illustrate the interval notation.

EXAMPLE 1 Find all numbers x such that $3x + 1 < 5x + 2$.

SOLUTION Gradually transform the given inequality until one side of the inequality consists of just x and the other side does not involve x:

$$3x + 1 < 5x + 2 \qquad \text{given}$$

$$3x < 5x + 1 \qquad \text{(subtracting 1 from both sides, rule 3)}$$

$$-2x < 1 \qquad \text{(subtracting } 5x \text{ from both sides, rule 3)}$$

$$x > -\tfrac{1}{2}. \qquad \text{(dividing by the negative number } -2\text{, rule 8)}$$

Each of these steps can be reversed. So the given inequality is equivalent to the inequality $x > -\frac{1}{2}$ and so has as solutions all numbers in the interval $(-\frac{1}{2}, \infty)$. ■

EXAMPLE 2 Describe the set of numbers x such that

$$(x + 2)(x - 1)(x - 3) > 0.$$

SOLUTION First check where each of the three factors is positive or negative. For example, $x + 2$ is 0 if $x = -2$, positive if $x > -2$, and negative if $x < -2$. This information is recorded in Fig. 4. Similarly, $x - 1$ changes sign at $x = 1$ and $x - 3$ changes sign at $x = 3$. This information is collected in the following table:

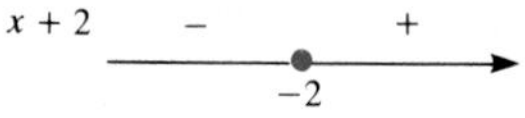

Figure 4

	-2	1	3	x
$x + 2$	−	+	+	+
$x - 1$	−	−	+	+
$x - 3$	−	−	−	+
$(x + 2)(x - 1)(x - 3)$	−	+	−	+

The product of the three factors is positive when x is in $(-2, 1)$ or $(3, \infty)$. The solutions of the inequality $(x + 2)(x - 1)(x - 3) > 0$ thus fill out the two intervals shown in Fig. 5.

Figure 5
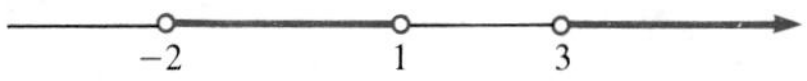

■

The absolute value of a real number tells "how far it is from 0." The following definition makes this precise.

> **Definition** *Absolute value.* The **absolute value** of a positive number x is x itself. The absolute value of a negative number x is its negative $-x$. The absolute value of 0 is 0.

The absolute value of x is denoted $|x|$. By definition, $|3| = 3$ and $|-3| = -(-3) = 3$. For all x, $|-x| = |x|$; x and $-x$ have the same absolute value. On some programmable calculators, the "absolute value" key is labeled "abs." On occasion, we use abs x to denote $|x|$.

The absolute value behaves better with respect to multiplication than with respect to addition. For any real numbers x and y,

$$|xy| = |x|\,|y|$$ (The absolute value of the product is the product of the absolute values.)

and $$|x + y| \le |x| + |y|.$$ (The absolute value of the sum is less than or equal to the sum of the absolute values.)

The second assertion is known as the **triangle inequality**. For instance, $|(-3) + 7| \le |-3| + |7|$, as a little arithmetic verifies.

EXAMPLE 3 Sketch on the number line the set of numbers x such that $|x| < 2$.

SOLUTION A positive number has an absolute value less than 2 if the number itself is less than 2. A negative number has an absolute value less than 2 if the number itself is greater than -2. Also, the absolute value of 0 is less than 2. All told, the set of numbers x such that $|x| < 2$ is the open interval $(-2, 2)$, sketched in Fig. 6. The interval $(-2, 2)$ consists of all numbers within a distance 2 of the origin. ■

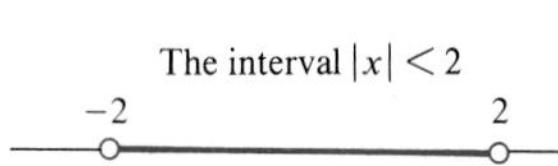

Figure 6

Observe that for any number $a > 0$, $|x| < a$ describes the open interval $(-a, a)$.

EXAMPLE 4 Sketch on the number line the set of numbers x such that $|5(x - 1)| < 3$.

SOLUTION

$\lvert 5(x-1)\rvert < 3$	given
$5\lvert x-1\rvert < 3$	absolute value of product
$\lvert x-1\rvert < \frac{3}{5}$	dividing by 5
$-\frac{3}{5} < x - 1 < \frac{3}{5}$	$x - 1$ has absolute value less than $\frac{3}{5}$
$1 - \frac{3}{5} < x < 1 + \frac{3}{5}$	adding 1 to an inequality
$\frac{2}{5} < x < \frac{8}{5}$	arithmetic

In short, x is in the open interval $(\frac{2}{5}, \frac{8}{5})$, sketched in Fig. 7. ■

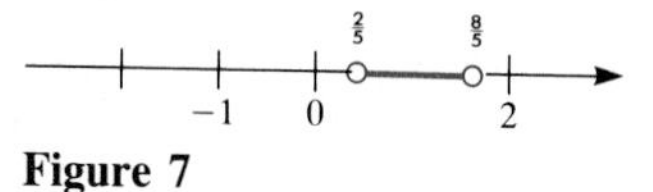

Figure 7

EXERCISES FOR APP. A: REAL NUMBERS

In Exercises 1 to 6 assume that a and b are positive numbers and $a < b$. Place the correct symbol, $<$ or $>$, in each blank.

1 $a + 3$ __ $b + 3$
2 $a - 2$ __ $b - 2$
3 $5a$ __ $5b$
4 $(-2)a$ __ $(-2)b$
5 $1/a$ __ $1/b$
6 $3 - a$ __ $3 - b$

In Exercises 7 to 20 describe where the inequalities hold. (Use interval notation.)

7 $2x + 7 < 4x + 9$
8 $3x - 5 < 7x + 11$
9 $-3x + 2 > 5x + 18$
10 $2x > 3x + 7$
11 $(x - 1)(x - 3) > 0$
12 $(x + 1)(x - 3) < 0$
13 $(x + 2)(x + 3) < 0$
14 $(2x + 6)(x + 3) < 0$
15 $x(x - 1)(x + 1) > 0$
16 $(x - 2)(x - 3)(x - 4) > 0$
17 $x(x + 3)(x + 5) > 0$
18 $(x - 1)^2(x - 2) > 0$
19 $(3x - 1)(2x - 1) > 0$
20 $x(2x + 1)(3x - 1) > 0$

In each of Exercises 21 to 24 sketch the intervals for which the inequality holds.

21 $|x - 3| < 2$
22 $|2x - 4| < 1$
23 $|3(x - 1)| < 6$
24 $|4(x + 2)| < 2$

25 For which x is $x^2(x - 3) > 0$?

26 For which x is $(x - 1)^2(x - 2)^2 < 0$?

27 For which positive numbers x is $x < x^2$?

28 For which positive numbers x is $x^2 < x^3$?

In Exercises 29 and 30 give an example of numbers a and b, $a < b$, neither 0, for which the stated inequality is *false*.

29 $a^2 < b^2$ **30** $1/a > 1/b$

31 Give an example of numbers a, b, c, and d such that $a < b$, $c < d$, and $ac > bd$.

32 Express 3.1416 in the form p/q, where p and q are integers. This shows that 3.1416 is a rational number. It is commonly used as a rational approximation of the irrational number π.

33 Let $a = 2.3474747 \ldots$, an endless decimal in which the block "47" continues to repeat without end. This exercise shows that a is rational.

(*a*) Compute $100a$.

(*b*) Compute $100a - a$.

(*c*) From (*b*) deduce that $a = 232.4/99$.

(*d*) From (*c*) deduce that a can be written as the quotient of two integers and is therefore rational. It can be shown that any number whose decimal representation from some point on consists of a block repeated endlessly is rational.

APPENDIX **B**

GRAPHS AND LINES

This appendix reviews coordinate systems, graphs, lines, and their slopes—concepts that play a key role in calculus.

B.1 COORDINATE SYSTEMS AND GRAPHS

Just as each point on a line can be described by a number, each point in the plane can be described by a pair of numbers. To do this, choose two perpendicular lines furnished with identical scales, as in Fig. 1. One is called the x **axis** and the other, the y **axis**. Usually the x axis is horizontal, as in Fig. 1. Any point P in the plane can then be described by a pair of numbers. The line through P parallel to the y axis meets the x axis at a number x, called the x **coordinate** or **abscissa** of P. The line through P parallel to the x axis meets the y axis at a number y, called the y **coordinate** or **ordinate** of P. P is then denoted (x, y), as in Fig. 2. The point $(0, 0)$, where the two axes cross, is called the **origin**.

The two axes cut the plane into four parts, called **quadrants**, numbered as in Fig. 3. In the first quadrant, both the x and y coordinates are positive; in the second, x is negative and y is positive; in the third, both x and y are negative; in the fourth, x is positive and y is negative.

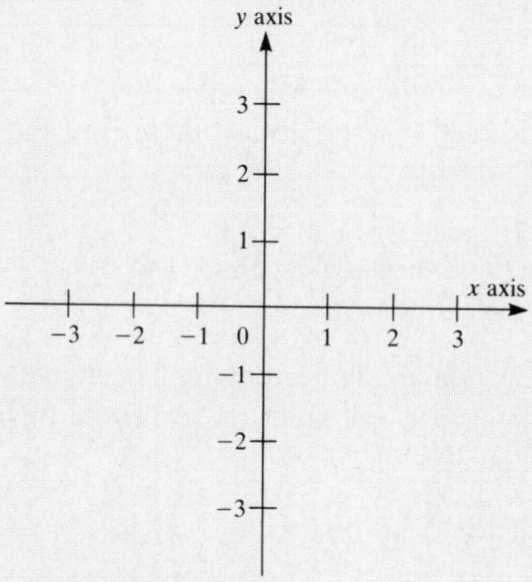

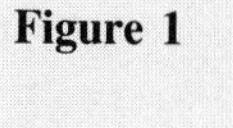
Figure 1

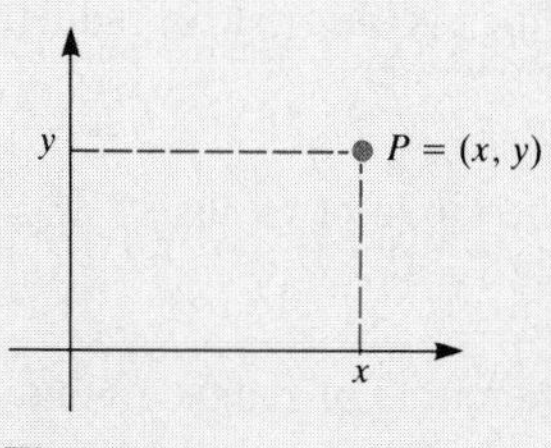

Figure 2

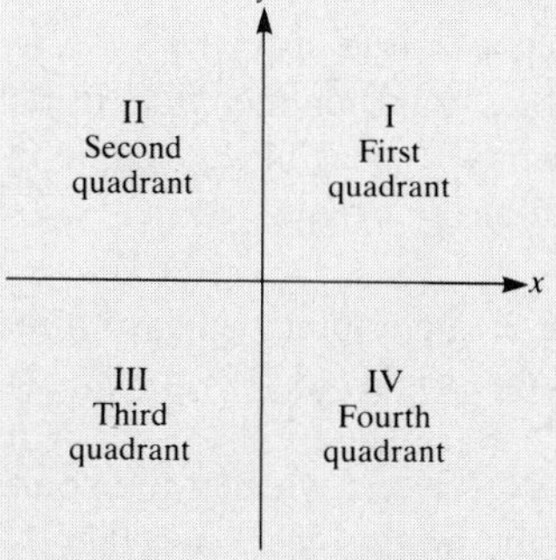

Figure 3

EXAMPLE 1 Plot the points $(1, 2)$, $(-3, 1)$, $(-\frac{1}{2}, -2)$, and $(3, -1)$ and specify the quadrant in which each lies.

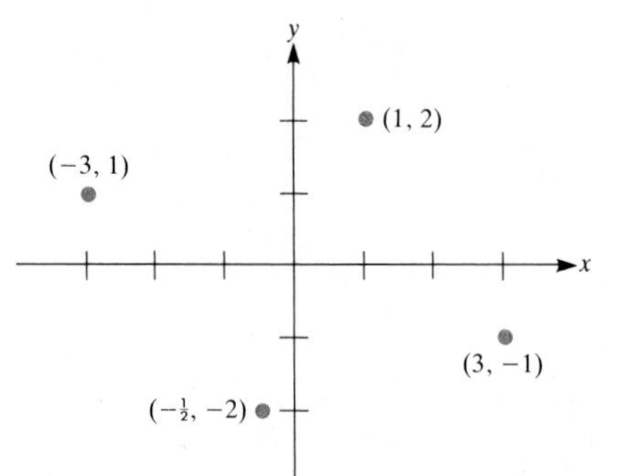

Figure 4

SOLUTION As shown in Fig. 4, the point (1, 2) is in the first quadrant, (−3, 1) is in the second, $(-\frac{1}{2}, -2)$ is in the third, and (3,−1) is in the fourth. ■

The distance d between two points $P_1 = (x_1, y_1)$ and $P_2 = (x_2, y_2)$ can be found with the aid of the pythagorean theorem. Form a right triangle whose hypotenuse is the line segment joining P_1 to P_2 and whose legs are parallel to the axes, as in Fig. 5.

The length of the horizontal leg is $x_2 - x_1$ if $x_2 \geq x_1$, as in Fig. 5. However, if $x_1 > x_2$, the length is $-(x_2 - x_1)$. Similarly, the length of the vertical leg is either $y_2 - y_1$ or its negative. In either case, since the negative of a number has the same square as the number, we have, by the pythagorean theorem,

$$d^2 = (x_2 - x_1)^2 + (y_2 - y_1)^2, \tag{1}$$

or

$$d = \sqrt{(x_2 - x_1)^2 + (y_2 - y_1)^2}. \tag{2}$$

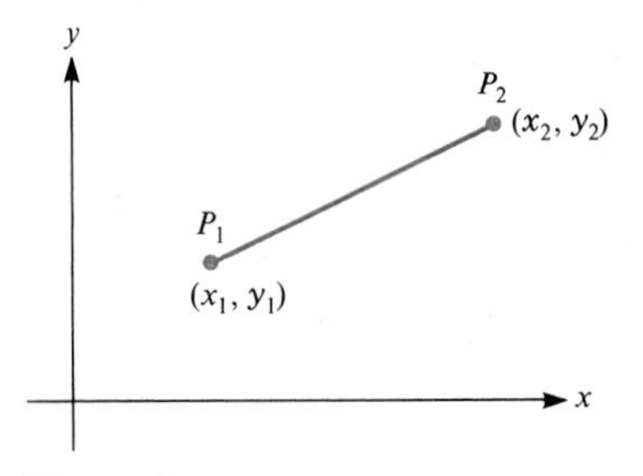

Figure 5

Keep both Eqs. (1) and (2) in mind. While Eq. (2) gives d explicitly, Eq. (1) is often preferable in computations since it does not involve square roots.

EXAMPLE 2 Find the distance between the points (−1, 3) and (2, −5).

SOLUTION Let $(x_1, y_1) = (-1, 3)$ and $(x_2, y_2) = (2, -5)$. Then

$$d^2 = [2 - (-1)]^2 + [(-5) - 3]^2 = 3^2 + (-8)^2 = 73.$$

Thus

$$d = \sqrt{73}.$$

We could just as well have labeled the points in the opposite order, $(x_1, y_1) = (2, -5)$ and $(x_2, y_2) = (-1, 3)$. The arithmetic is slightly different, but the result is the same:

$$d^2 = [(-1) - 2]^2 + [3 - (-5)]^2 = (-3)^2 + 8^2 = 73$$

and

$$d = \sqrt{73}. \quad ■$$

The distance formula gives us a way of dealing algebraically with the geometric notion of a circle of radius r and center (0, 0). A point (x, y) lies on this circle if its distance from (0, 0) is r, that is, if

$$\sqrt{(x - 0)^2 + (y - 0)^2} = r$$

or, more simply, if

$$x^2 + y^2 = r^2.$$

The converse is true also: If $x^2 + y^2 = r^2$, then $\sqrt{(x - 0)^2 + (y - 0)^2} = r$, so the point (x, y) lies on the circle of radius r and center (0, 0). For the sake of brevity, we may speak of "the circle $x^2 + y^2 = r^2$," which is the circle of radius r centered at the origin.

EXAMPLE 3 Determine which of these points lie on the circle of radius 13 and center at the origin: (5, 12), (−5, 12), (10, 7).

SOLUTION To test whether the point (x, y) lies on the circle of radius 13 and center at the origin, check whether

$$x^2 + y^2 = 13^2,$$

that is, whether $x^2 + y^2 = 169$.

Since
$$5^2 + 12^2 = 25 + 144 = 169,$$
the point (5, 12) lies on the circle. So does (−5, 12), since $(-5)^2 + 12^2 = 25 + 144 = 169$.

Does (10, 7) also lie on this circle? We find that $10^2 + 7^2 = 100 + 49 = 149$. Thus (10, 7) does not lie on the circle. (See Fig. 6.) ■

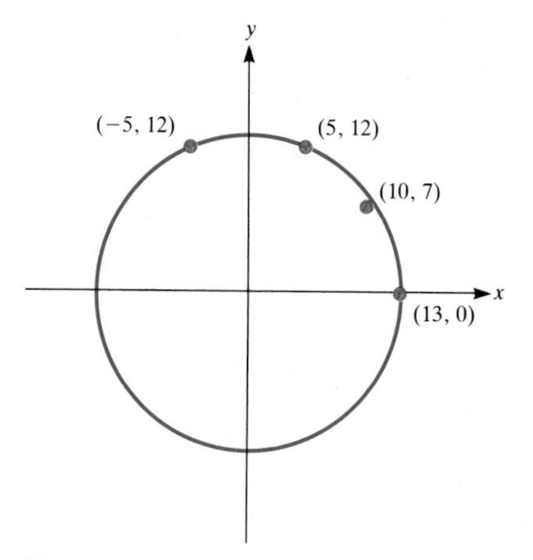

Figure 6

The circle shown in Fig. 6 is called the **graph** of the equation $x^2 + y^2 = 169$. The graph of any equation involving one or both of the letters x and y consists of those points (x, y) whose coordinates satisfy the equation. If the point $(a, 0)$ lies on the graph, a is called an **x intercept**. Similarly, b is called a **y intercept** if $(0, b)$ lies on the graph.

EXAMPLE 4 Graph the curve $y = x^2$.

SOLUTION To begin, find a few points on the graph by choosing some specific values of x and calculating the corresponding values of y. Let us use $x = 0, 1, 2, 3, -1, -2$, and -3 and fill in this table:

x	0	1	2	3	−1	−2	−3
$y = x^2$	0	1	4	9	1	4	9

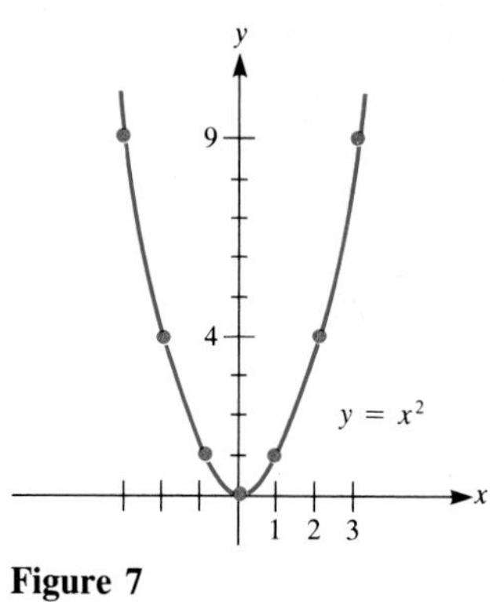

Figure 7

The table provides seven points on the graph of $y = x^2$, as shown in Fig. 7. Note that, as $|x|$ increases, so does $y = x^2$. The graph goes arbitrarily high. ■

The graph of $y = x^3$ can be sketched in a similar manner, by starting with a table of values. (Calculus provides additional techniques for graphing.) It is sketched in Fig. 8.

Near the point (0, 0) the graph of $y = x^3$ looks almost like the x axis. You can see why if you plot the points $(\frac{1}{2}, \frac{1}{8})$ and $(\frac{1}{4}, \frac{1}{64})$ on a large scale. Both points lie on $y = x^3$.

The graph of the equation
$$\frac{x^2}{a^2} + \frac{y^2}{b^2} = 1, \tag{3}$$
where a and b are positive, is called an **ellipse**.

Note that an ellipse is symmetric with respect to both axes. Moreover, its x intercepts are a and $-a$; its y intercepts are b and $-b$. For $a > b$, the ellipse is wider than high; for $b > a$, it is higher than wide.

EXAMPLE 5 Sketch the curve $\dfrac{x^2}{8} + \dfrac{y^2}{4} = 1$.

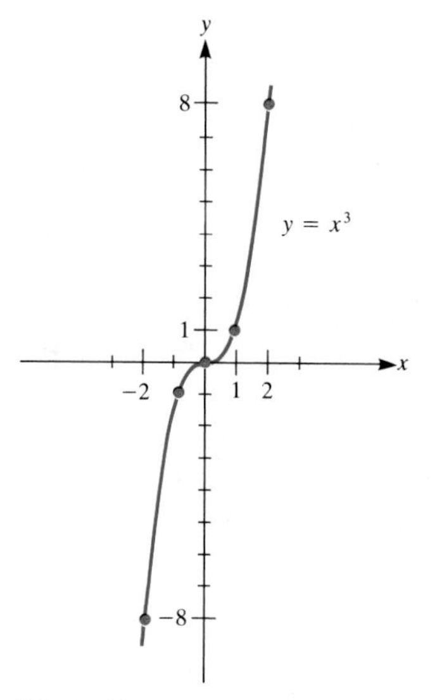

Figure 8

SOLUTION In this case, $a^2 = 8$ and $b^2 = 4$, so $a = \sqrt{8} \approx 2.8$ and $b = 2$.

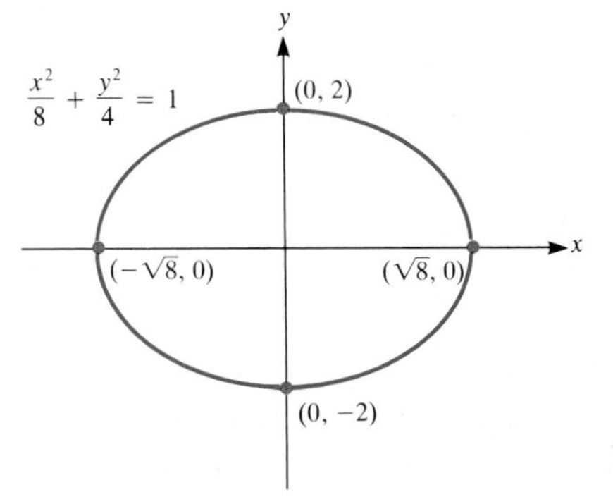

Figure 9

First plot the four points where the ellipse crosses the axes, then fill in the curve smoothly freehand, as shown in Fig. 9. ■

The graph of the equation

$$\frac{x^2}{a^2} - \frac{y^2}{b^2} = 1, \qquad (4)$$

where a and b are positive, is called a **hyperbola**.

Although Eq. (4) looks a lot like Eq. (3), its graph is quite different. First of all, the hyperbola has no y intercept. To see why, consider the equation

$$\frac{0^2}{a^2} - \frac{y^2}{b^2} = 1,$$

which reduces to

$$y^2 = -b^2.$$

Since b^2 is positive, $-b^2$ is negative. But there is no real number y whose square is negative. Thus the hyperbola (4) has no y intercept.

Second, the hyperbola extends arbitrarily far from the origin. To show this, solve for y:

$$\frac{y^2}{b^2} = \frac{x^2}{a^2} - 1$$

$$y^2 = b^2\left(\frac{x^2}{a^2} - 1\right)$$

$$y^2 = \frac{b^2}{a^2}(x^2 - a^2)$$

$$y = \pm\frac{b}{a}\sqrt{x^2 - a^2}.$$

For $|x| \geq a$, the expression in the radical is not negative. So, for $|x| \geq a$, hence for large x, there will be a corresponding value of y [namely, $y = \pm(b/a)\sqrt{x^2 - a^2}$].

When x is large, $\sqrt{x^2 - a^2}$ is very close to x. (Check this for $x = 20$ and $a = 3$.) So for large x, the graph is close to the line $y = (b/a)x$.

To graph the hyperbola $x^2/a^2 - y^2/b^2 = 1$, follow these steps:

1 Plot the points $(a, 0)$ and $(-a, 0)$, where the hyperbola meets the x axis.

2 Plot the point (a, b) and the line through it and the origin. Do the same for the point $(a, -b)$.

3 Sketch the hyperbola freehand, using the two lines in step 2 as guides. (As shown in Exercise 47, the hyperbola approaches these lines arbitrarily closely.) The two lines are called **asymptotes** of the hyperbola.

EXAMPLE 6 Graph the hyperbola $\frac{x^2}{9} - \frac{y^2}{4} = 1$.

SOLUTION In this case, $a^2 = 9$ and $b^2 = 4$, so $a = 3$ and $b = 2$. The x intercepts are 3 and -3. One asymptote passes through $(0, 0)$ and $(3, 2)$ and the

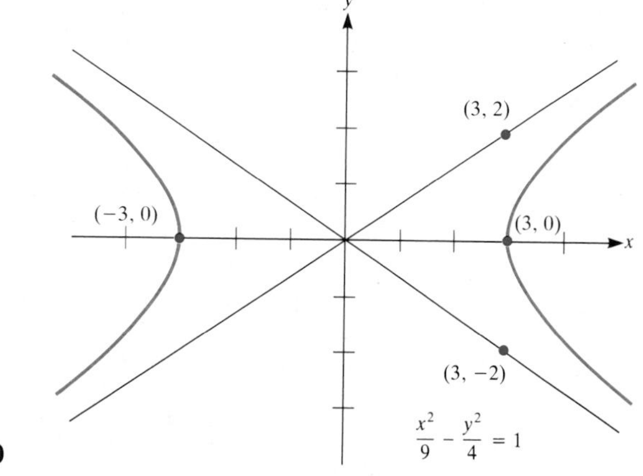

Figure 10

other through (0, 0) and (3, −2). With the aid of the intercepts and the asymptotes, the graph is easy to sketch freehand. (See Fig. 10.) Note that since x and y appear only to even powers, the graph is symmetric with respect to both axes. ■

Equations whose graphs are parabolas, ellipses, and hyperbolas are discussed in detail in Appendix G.

EXERCISES FOR SEC. B.1: COORDINATE SYSTEMS AND GRAPHS

In Exercises 1 and 2 plot the points and state in which quadrant each lies.

1 (*a*) (−3, 5) (*b*) (4, −3) (*c*) (2, 7) (*d*) (−2, −7)

2 (*a*) (1, 3) (*b*) $(-3, -\frac{1}{2})$ (*c*) (5, −3) (*d*) (−3, 4)

In Exercises 3 and 4 find the distances between the given points.

3 (*a*) (4, 1) and (−2,9) (*b*) (6, 3) and (1, 15) (*c*) (4, 0) and (7, 0)

4 (*a*) (3, 4) and (5, 6) (*b*) (−3, −2) and (4, −2) (*c*) (−3, −4) and (3, 4)

In Exercises 5 to 16 graph the equations.

5 $x^2 + y^2 = 49$ **6** $x^2 + y^2 = 1$

7 $y = 2x^2$ **8** $y = -2x^2$

9 $y = -x^2$ **10** $y = -x^2/2$

11 $y = -x^3$ **12** $y = 2x^3$

13 $y = x^4$ (Include the points for which $x = 0, \frac{1}{2}$, and 1.)

14 $y = x^5$ (Include the points for which $x = 0, \frac{1}{2}$, and 1.)

15 $y = x - 1$ **16** $y = -2x + 3$

In Exercises 17 to 22 find the x and y intercepts, if there are any.

17 $y = 2x + 6$ **18** $y = 3x - 6$

19 $xy = 6$ **20** $x^2 - y^2 = 1$

21 $y = 2x^2 + 5x - 3$ **22** $y = 4 - x^2$

In Exercises 23 to 32 graph the equations.

23 $\frac{x^2}{25} + \frac{y^2}{16} = 1$ **24** $\frac{x^2}{16} + \frac{y^2}{25} = 1$

25 $\frac{x^2}{16} - \frac{y^2}{25} = 1$ **26** $\frac{x^2}{25} - \frac{y^2}{16} = 1$

27 $xy = 8$ **28** $xy = -1$

29 $y = x^2 - 2x + 3$ **30** $y = -x^2 + 3x + 4$

31 $\frac{y^2}{4} - \frac{x^2}{9} = 1$ (Note that the negative sign is with x^2, not y^2.)

32 $y^2 - x^2 = 1$

In Exercises 33 to 36 graph the cubics.

33 $y = x^3 - 3x^2 + 3x$ **34** $y = 2x^3 - 3x^2$

35 $y = x^3 + x$ **36** $y = -x^3 + x$

In Exercises 37 to 44 graph the equations.

37 $y = x - x^2$ **38** $y = 1 + x^2$

39 $y = (x - 1)^2$ **40** $y = \sqrt{x}$

41 $y = |x|$ **42** $y = \sqrt[3]{x}$

43 $y = (x - 1)(x - 2)$ **44** $y = \sqrt[3]{x - 1}$

In Exercises 45 and 46 find an equation of the circle of given radius and center.

45 Radius 7, center (2, 1) **46** Radius $\frac{1}{2}$, center (−2, 3)

47 This exercise shows why the part of the hyperbola $x^2/a^2 - y^2/b^2 = 1$ in the first quadrant approaches the line $y = bx/a$ when x is large.

Consider the point $P = (x, y)$ on the hyperbola, $x \geq a$, $y \geq 0$.

(*a*) Show that $y = (b/a)\sqrt{x^2 - a^2}$.

(*b*) Let $Q = (x, bx/a)$ be the point on the line $y = bx/a$ with the same x coordinate as P. Show that Q is near P for large x, by showing that

$$\frac{bx}{a} - \frac{b}{a}\sqrt{x^2 - a^2}$$

approaches 0 when x is large. [*Hint:* Rationalize by multiplying by $(x + \sqrt{x^2 - a^2})/(x + \sqrt{x^2 - a^2})$.]

B.2 LINES AND THEIR SLOPES

Consider a line that is not parallel to the y axis. Select distinct points $P_1 = (x_1, y_1)$ and $P_2 = (x_2, y_2)$ on it as in Fig. 1. As we move from P_1 to P_2, the vertical change is $y_2 - y_1$ and the horizontal change is $x_2 - x_1$, as shown in Fig. 2. The quotient $(y_2 - y_1)/(x_2 - x_1)$ is a measure of the steepness of the line and is called its **slope**. The vertical change, $y_2 - y_1$, is the "*rise*"; the horizontal change, $x_2 - x_1$, is the "*run*"; slope is "rise over run."

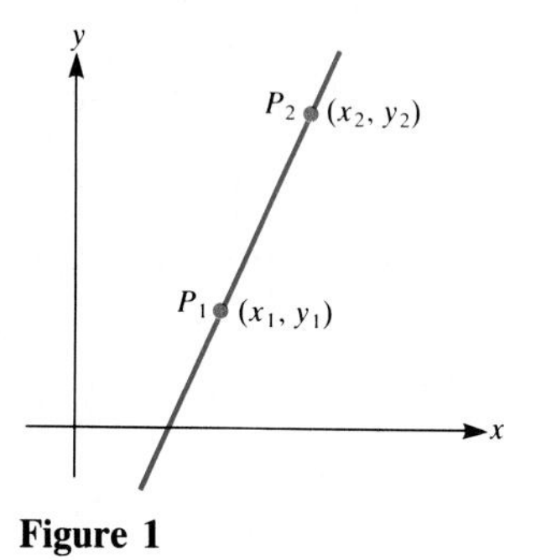

Figure 1

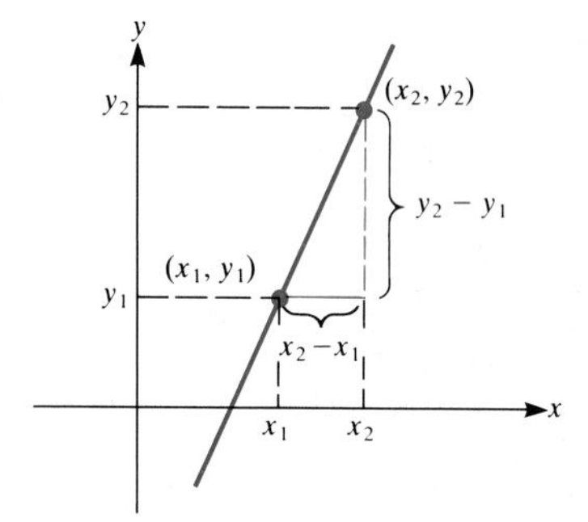

Figure 2

Definition *Slope of a line.* Consider a line that is not parallel to the y axis. Let $P_1 = (x_1, y_1)$ and $P_2 = (x_2, y_2)$ be two distinct points on the line. The **slope** of the line is

$$m = \frac{y_2 - y_1}{x_2 - x_1}.$$

The slope of a line parallel to the y axis is not defined. Accordingly, a vertical line is said to have "no slope."

EXAMPLE 1 Find the slope of the line through the points (1, 4) and (3, 5).

SOLUTION Let P_1 be (1, 4) and P_2 be (3, 5). Then

$$m = \frac{5 - 4}{3 - 1} = \frac{1}{2}. \quad ■$$

In finding the slope in Example 1 we could have chosen P_1 to be (3, 5) and P_2 to be (1, 4). The arithmetic would be slightly different, but the result would be the same, since in this case

$$m = \frac{4 - 5}{1 - 3} = \frac{-1}{-2} = \frac{1}{2}.$$

Order of points does not affect slope.

The order of the two points does not affect the slope, since

$$\frac{y_2 - y_1}{x_2 - x_1} = \frac{y_1 - y_2}{x_1 - x_2}.$$

Moreover, the slope of a line does not depend on the particular pair of points selected on it. If one pair is $P_1 = (x_1, y_1)$ and $P_2 = (x_2, y_2)$ and the other pair is $P_1' = (x_1', y_1')$ and $P_2' = (x_2', y_2')$, then

$$\frac{y_2 - y_1}{x_2 - x_1} = \frac{y_2' - y_1'}{x_2' - x_1'}.$$

This equation follows from the fact that the ratios between the lengths of corresponding sides of similar triangles are equal. (See Fig. 3, which shows the case where the slope is positive; a similar argument works for the case where the slope is negative.)

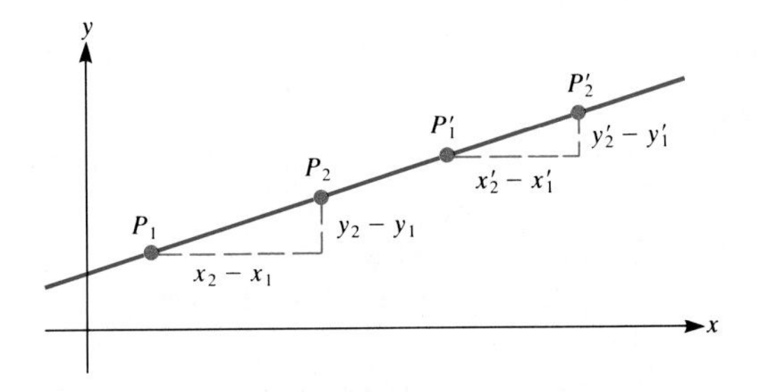

Figure 3

EXAMPLE 2 Find the slopes of the lines through the given points. Then draw the lines.

(*a*) (0, 2) and (1, 8) (*b*) (1, 1) and (5, 2)
(*c*) (2, 6) and (4, 3) (*d*) (5, 1) and (−2, 1)

SOLUTION

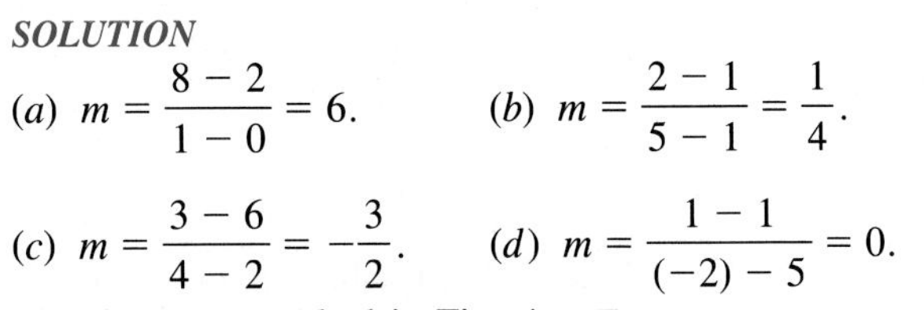

(*a*) $m = \dfrac{8 - 2}{1 - 0} = 6.$ (*b*) $m = \dfrac{2 - 1}{5 - 1} = \dfrac{1}{4}.$

(*c*) $m = \dfrac{3 - 6}{4 - 2} = -\dfrac{3}{2}.$ (*d*) $m = \dfrac{1 - 1}{(-2) - 5} = 0.$

The lines are graphed in Fig. 4. ■

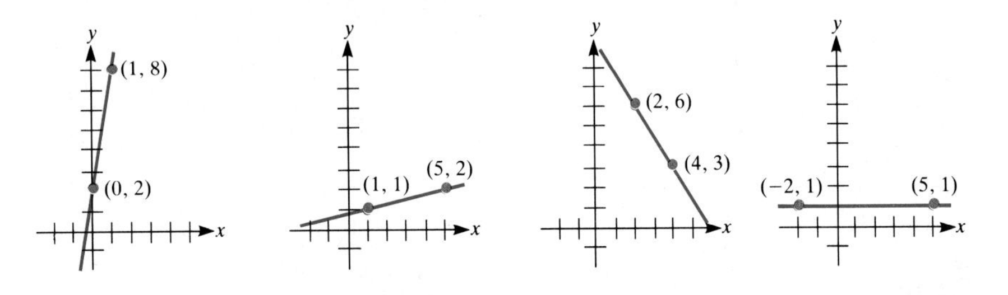

Figure 4

Meaning of positive and negative slopes

As suggested by the lines in Example 2, a *positive* slope tells us that as we move from left to right on the line, we go *up*. (As the x coordinate increases, so does the y coordinate.) *A negative* slope tells us that as we move from left to right on the line, we go *down*. (As the x coordinate increases, the y coordinate decreases.) A slope of 0 corresponds to a *horizontal* line, that is, a line parallel to the x axis.

Given the slopes of two lines, we can determine whether they are parallel or perpendicular.

Test for parallel lines

> A line of slope m_1 is *parallel* to a line of slope m_2 if and only if $m_1 = m_2$.

The test for perpendicularity is a little fancier.

Test for perpendicular lines

> A line of slope m_1 is *perpendicular* to a line of slope m_2 if and only if the *product of the two slopes* is -1: $m_1 m_2 = -1$.

(In other words,

$$m_2 = -\frac{1}{m_1};$$

the slope m_2 is the negative of the reciprocal of m_1.) It is assumed that neither line is parallel to the y axis.

Proofs of these two tests are outlined in Exercises 35 and 36.

EXAMPLE 3 Determine whether the line through $(-1, -2)$ and $(2, 2)$ is parallel to the line through $(6, 5)$ and $(1, 1)$.

SOLUTION The slope of the first line is

$$\frac{2-(-2)}{2-(-1)} = \frac{4}{3}.$$

The slope of the second line is

$$\frac{1-5}{1-6} = \frac{-4}{-5} = \frac{4}{5}.$$

Since $\frac{4}{3} \neq \frac{4}{5}$, the lines are not parallel. ■

EXAMPLE 4 Determine whether the line through $(1, 5)$ and $(4, 4)$ is perpendicular to the line through $(2, 1)$ and $(3, 4)$.

SOLUTION The first line has slope

$$\frac{4-5}{4-1} = \frac{-1}{3}.$$

The second line has slope

$$\frac{4-1}{3-2} = 3.$$

The product of the two slopes is $3(-\frac{1}{3}) = -1$, so the two lines are perpendicular. ■

Having discussed the geometry and slopes of lines, let us find equations whose graphs are lines.

A line parallel to the y axis has the equation $x = c$. If a line is not parallel to the y axis, it has a slope m and a y intercept b. The point $(0, b)$ is on the line. Now take any other point (x, y) on the line. Since both $(0, b)$ and (x, y) are on the line,

$$\frac{y - b}{x - 0} = m,$$

or simply,

$$y = mx + b. \tag{1}$$

The steps are reversible. That is, if (x, y) satisfies Eq. (1), then (x, y) lies on the line. Equation (1) is called the **slope-intercept equation** of a line. Each line not parallel to the y axis has a unique description of the form (1). If $m = 0$, the line is parallel to the x axis and has the equation $y = 0x + b$, which is just $y = b$. In this case, the symbol x does not appear.

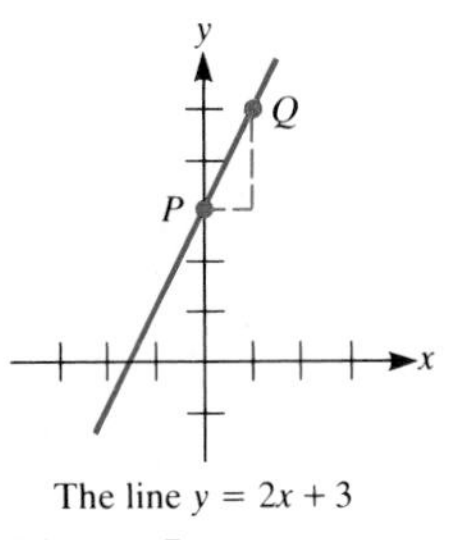

The line $y = 2x + 3$

Figure 5

EXAMPLE 5 Graph the lines (a) $y = 2x + 3$ and (b) $y = -4x/3 + 1$.

SOLUTION

(a) $y = 2x + 3$. Here the y intercept is 3, so the point $P = (0, 3)$ is on the line. Plot this point. To obtain a second point on the line, write the slope 2 in the form $\frac{2}{1}$. Plot a second point, Q, 1 unit to the right of P and 2 units above it, as shown in Fig. 5. The line through P and Q has slope 2 and y intercept 3.

(b) $y = -4x/3 + 1$. The y intercept is 1 and the slope is $-\frac{4}{3}$. The point $P = (0, 1)$ is on the line. Next draw the point Q that is 3 units to the right of P and 4 units lower, as shown in Fig. 6. The line through P and Q has slope $-\frac{4}{3}$ and y intercept 1. ■

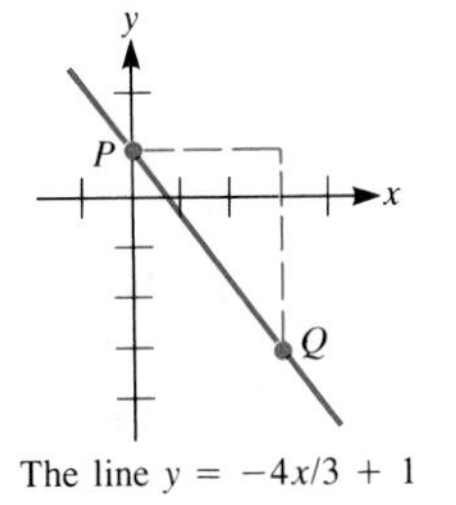

The line $y = -4x/3 + 1$

Figure 6

Sometimes we may wish to find an equation of the line that passes through a certain point $P_1 = (x_1, y_1)$ and has a certain slope m. To find such an equation, consider the typical point $P = (x, y)$ on the line, $x \neq x_1$. The slope determined by P and P_1 must be the same as the prescribed slope m, that is,

$$\frac{y - y_1}{x - x_1} = m.$$

Clearing the denominator provides an equation in x and y, which is called the **point-slope equation** of the line. It is useful if you know the slope and a point on the line. The line through (x_1, y_1) of slope m is described by the equation

$$y - y_1 = m(x - x_1).$$

EXAMPLE 6 Find the point-slope equation of the line through (1, 3) of slope -2.

SOLUTION The point-slope formula gives

$$y - 3 = -2(x - 1),$$

which could be rewritten in several ways. For example, $y - 3 = -2x + 2$ or $y = -2x + 5$ or $2x + y = 5$. ■

The point-slope formula gives a quick way to find an equation of the line through two given points $P_1 = (x_1, y_1)$ and $P_2 = (x_2, y_2)$. (Assume that $x_1 \neq x_2$. If $x_1 = x_2$, the line is parallel to the y axis and has the equation $x = x_1$.)

If $x_1 \neq x_2$, the slope of the line through P_1 and P_2 is

$$\frac{y_2 - y_1}{x_2 - x_1}.$$

How to find an equation of the line through two given points

Since the line passes through (x_1, y_1) and has slope $(y_2 - y_1)/(x_2 - x_1)$, its point-slope equation is

$$y - y_1 = \frac{y_2 - y_1}{x_2 - x_1}(x - x_1).$$

EXAMPLE 7 Find an equation of the line through $(-1, 4)$ and $(2, 3)$.

SOLUTION The slope of the line is

$$\frac{3 - 4}{2 - (-1)} = -\frac{1}{3}.$$

Since the point $(-1, 4)$ is on the line, the line has the point-slope equation

$$y - 4 = -\tfrac{1}{3}[x - (-1)].$$

Thus
$$y - 4 = -\frac{x}{3} - \frac{1}{3},$$

or
$$y = -\frac{x}{3} + \frac{11}{3},$$

which is the slope-intercept equation of the line. ■

Let A, B, and C be fixed numbers such that at least one of A and B is not zero. With the aid of the slope-intercept formula it will be shown that *the graph of the equation* $Ax + By + C = 0$ *is a line*. For this reason, the equation $Ax + By + C = 0$ is said to be "a linear equation in x and y."

To show that the graph of $Ax + By + C = 0$ is a line, consider two cases: $B = 0$ and $B \neq 0$.

Case 1: $B = 0$. In this instance, the equation is just $Ax + C = 0$, which is equivalent to $x = -C/A$. So the graph is a line parallel to the y axis.

Case 2: $B \neq 0$. In this case rewrite the equation $Ax + By + C = 0$ as follows:

$$By = -Ax - C,$$

or
$$y = -\frac{A}{B}x - \frac{C}{B}.$$

The graph is therefore the line of slope $-A/B$ and y intercept $-C/B$, by the slope-intercept formula.

EXAMPLE 8 Find the y intercept and slope of the line $2x + 3y - 6 = 0$.

SOLUTION To find the y intercept, set $x = 0$, obtaining

$$2 \cdot 0 + 3y - 6 = 0$$
$$3y = 6$$
$$y = 2.$$

The y intercept is 2.

To find the slope, rewrite the equation $2x + 3y - 6 = 0$ in the slope-intercept form by solving for y:

$$2x + 3y - 6 = 0$$
$$3y = -2x + 6$$
$$y = (-\tfrac{2}{3})x + 2.$$

The slope is $-\frac{2}{3}$. ■

EXERCISES FOR SEC. B.2: LINES AND THEIR SLOPES

In Exercises 1 to 6 plot the points and find the slopes of the lines through them.

1 $(-1, 1)$ and $(4, 2)$ **2** $(3, 1)$ and $(-2, -1)$
3 $(-1, 4)$ and $(3, -1)$ **4** $(1, 5)$ and $(5, 1)$
5 $(1, 7)$ and $(11, 7)$ **6** $(-2, -5)$ and $(0, -5)$

7 Describe the slope of each line in Fig. 7 as positive, negative, or 0.

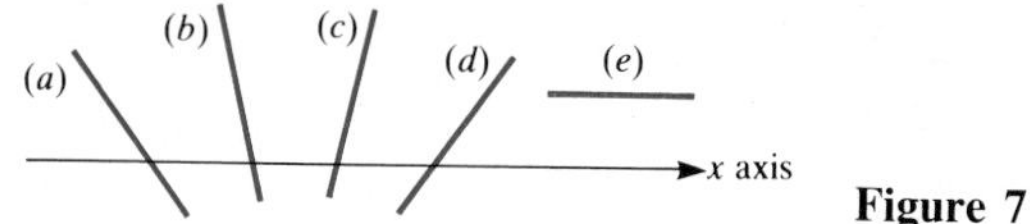

Figure 7

8 Describe the slope of each line in Fig. 8 as positive, negative, or 0.

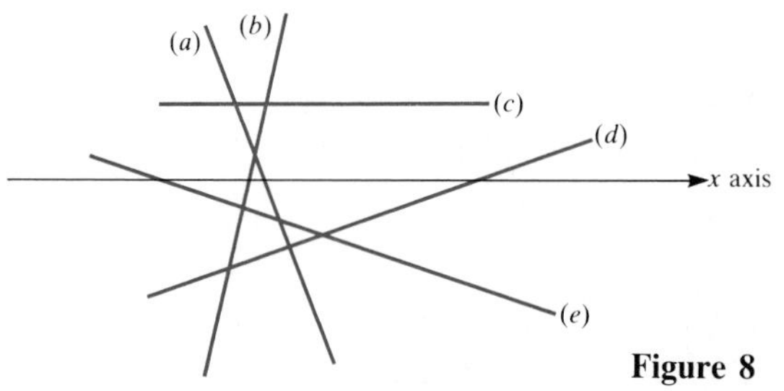

Figure 8

9 A line L has slope 4. (*a*) What is the slope of a line parallel to L? (*b*) What is the slope of a line perpendicular to L?

10 A line L has slope $-\frac{1}{2}$. (*a*) What is the slope of a line parallel to L? (*b*) What is the slope of a line perpendicular to L?

11 Is the line through $(2, 4)$ and $(3, 7)$ parallel to the line through $(1, 7)$ and $(-1, 1)$?

12 Is the line through $(1, 2)$ and $(-3, 5)$ parallel to the line through $(3, -1)$ and $(-1, 4)$?

13 Is the line through $(0, 0)$ and $(4, 6)$ perpendicular to the line through $(5, -1)$ and $(3, 2)$?

14 Is the line through $(-1, -2)$ and $(6, 2)$ perpendicular to the line through $(2, 3)$ and $(8, -7)$?

In Exercises 15 and 16 find the slope-intercept equations for the lines:

15 (*a*) with y intercept 2 and slope 3, (*b*) with y intercept -2 and slope $\frac{2}{3}$, (*c*) with y intercept 0 and slope -3.

16 (*a*) with y intercept $\frac{3}{4}$ and slope $-\frac{1}{2}$, (*b*) with y intercept -2 and slope 2, (*c*) with y intercept 3 and slope $\frac{3}{5}$.

In Exercises 17 and 18 find the slopes and y intercepts of the given lines and graph them.

17 (*a*) $y = 3x - 1$
(*b*) $y = -2x + 1$
(*c*) $y = 3x/5$

18 (*a*) $y = -x/2 + 3$
(*b*) $y = -3x + 4$
(*c*) $y = -2x/3 + 1$

In Exercises 19 and 20 find point-slope equations of the given lines:

19 (*a*) Slope 3, passing through $(1, 2)$
(*b*) Slope -2, passing through $(3, -1)$

20 (*a*) Slope $\frac{4}{3}$, passing through the origin
(*b*) Slope 0, passing through $(2, 5)$

In Exercises 21 and 22 find point-slope equations of the lines through the given points.

21 (*a*) $(1, 2)$ and $(5, 3)$
(*b*) $(-1, 2)$ and $(3, 1)$
(*c*) $(4, 5)$ and $(2, 3)$

22 (*a*) $(4, 4)$ and $(6, 6)$
(*b*) $(-2, -5)$ and $(3, 4)$
(*c*) $(0, 0)$ and $(3, 5)$

In Exercises 23 and 24 find the slope-intercept equations of the given lines and graph the lines.

23 (*a*) $x + y + 1 = 0$ (*b*) $-2x + y = 0$
(*c*) $2x + 3y - 12 = 0$

24 (*a*) $x - y = 0$ (*b*) $-2x - 5y + 10 = 0$
(*c*) $x - 2y = 4$

---

25 Does the line through $(4, 1)$ and $(7, 2)$ pass through the point (*a*) $(10, 3)$? (*b*) $(14, 5)$?

26 Find an equation of the line with y intercept 2 and perpendicular to the line $y = (-2x/3) + 7$.

27 Find an equation of the line through $(1, 3)$ and perpendicular to the line through $(4, 1)$ and $(2, 5)$.

28 Find an equation of the line parallel to the line $y = x + 6$ and passing through the origin.

29 Find an equation of the line parallel to $y = 3x + 2$ with y intercept 5.

Exercises 30 to 33 concern another form of an equation of a line, the "intercept form," which is convenient for working with the intercepts.

30 Let a and b be nonzero numbers. Show that the line $x/a + y/b = 1$ has x intercept a and y intercept b. (See Fig. 9.) The equation $x/a + y/b = 1$ is called the **intercept equation** of the line through $(a, 0)$ and $(0, b)$.

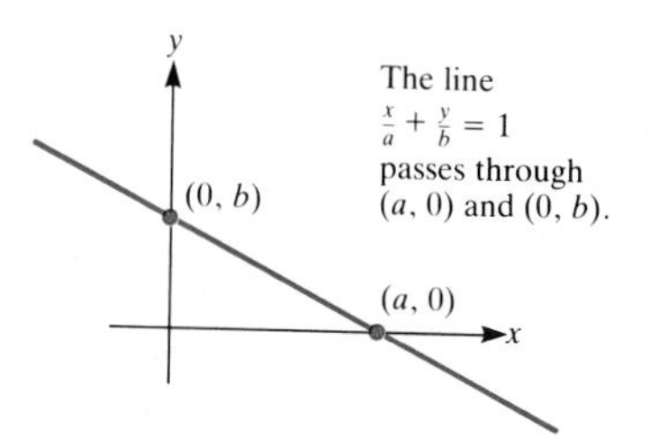

Figure 9

31 Find the slope of the line whose intercept equation is $x/a + y/b = 1$.

32 Find the intercept equation of the line with (a) x intercept 3 and y intercept 5, (b) x intercept -1 and y intercept 2, (c) x intercept $-\frac{1}{2}$ and y intercept -3.

33 Find the x intercept of the line (a) with slope 4 and y intercept -7, (b) through (1,3) with slope $-\frac{2}{3}$, (c) $x/3 - y/4 = 1$, (d) $-2x + 3y = 12$.

34 Find the intersection of the line through (0, 1) and (4, 7) with the line through (3, 3) and (5, 1).

35 This exercise outlines a proof that parallel lines have equal slopes. Let L and L' be parallel lines. Let the fixed distance between them, measured parallel to the y axis, be d. That is, if (x, y) is on L, then $(x, y + d)$ is on L'. Let $P_1 = (x_1, y_1)$ and $P_2 = (x_2, y_2)$ be points on L. Then $P_1' = (x_1, y_1 + d)$ and $P_2' = (x_2, y_2 + d)$ are points on L'.
(a) Find the slope of L using P_1 and P_2.
(b) Find the slope of L' using P_1' and P_2'.
Putting (a) and (b) together shows that the slopes of parallel lines are equal.

36 This exercise outlines a proof that the product of the slopes of perpendicular lines in the xy plane (neither parallel to the y axis) is -1. Assume that L_1 and L_2 are perpendicular lines. Let L_1 have slope m_1 and L_2 have slope m_2. For convenience, assume that both lines pass through the origin. A sketch shows that one line has positive slope and one line has negative slope. Say that m_1 is positive and m_2 is negative.
(a) Show that the point $(1, m_1)$ lies on L_1.
(b) Show that the point $(-m_1, 1)$ lies on L_2. (Recall that L_2 is perpendicular to L_1.)
(c) Deduce that $m_2 = -1/m_1$.
(d) If the two perpendicular lines do not pass through the origin, show that the product of their slopes is -1. [*Hint:* Use (c) and Exercise 35.]

37 (a) Sketch the lines $\dfrac{x}{3} + \dfrac{y}{4} = 1$ and $\dfrac{x}{4} - \dfrac{y}{3} = 1$.
(b) Find their slopes.
(c) Are they perpendicular?

38 Is the line through $(-2, 1)$ and $(3, 8)$ parallel to the line through $(3, -2)$ and $(10, 8)$?

39 A line has the equation $y = mx + b$. What can be said about m if the line (a) is nearly vertical (almost parallel to the y axis)? (b) is nearly horizontal (almost parallel to the x axis)? (c) slopes downward as you move to the right? (d) slopes upward as you move to the right?

40 Write in the form $y = mx + b$ the equation of the line through (1, 2) and (0, 5).

41 Find a point P on the x axis such that the line through P and (1, 1) is perpendicular to the line through P and $(-3, 4)$.

42 Estimate the slope of lines (a) and (b) in Fig. 10 as accurately as you can.

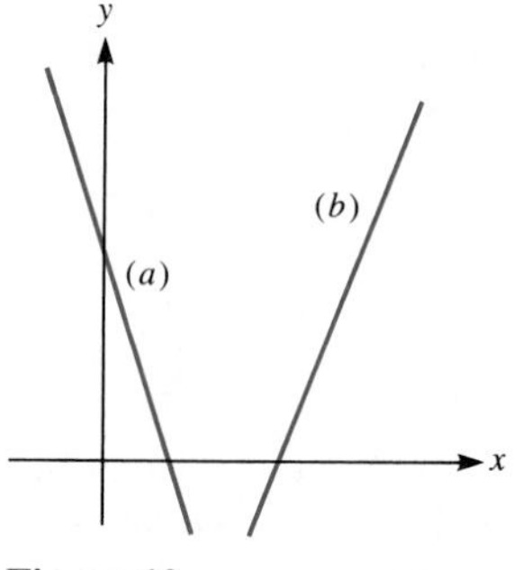

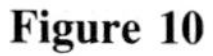
Figure 10

APPENDIX C

TOPICS IN ALGEBRA

This appendix presents several topics from algebra: rationalizing an expression, completing the square, the quadratic formula, the binomial theorem, geometric series, and roots of polynomials.

Rationalizing

The identities $\sqrt{a}\sqrt{a} = a$ and $(a + b)(a - b) = a^2 - b^2$ enable us *to remove a square root* from certain algebraic expressions. Three examples illustrate the technique.

EXAMPLE 1 Remove the square root from the denominator of the expression $4/\sqrt{2}$. ("Rationalize" the denominator.)

SOLUTION

$$\frac{4}{\sqrt{2}} = \frac{4}{\sqrt{2}} \cdot \frac{\sqrt{2}}{\sqrt{2}} = \frac{4\sqrt{2}}{2} = 2\sqrt{2} \quad \blacksquare$$

EXAMPLE 2 Remove the square root from the denominator of the fraction $2/(1 + \sqrt{5})$. (Rationalize the denominator.)

SOLUTION

$$\frac{2}{1+\sqrt{5}} = \frac{2}{1+\sqrt{5}} \cdot \frac{1-\sqrt{5}}{1-\sqrt{5}} = \frac{2-2\sqrt{5}}{1^2-(\sqrt{5})^2} = \frac{2-2\sqrt{5}}{1-5} = \frac{2-2\sqrt{5}}{-4}$$

$$= \frac{\sqrt{5}-1}{2}. \quad \blacksquare$$

In the next example the numerator is rationalized.

EXAMPLE 3 Remove the square roots from the numerator of the fraction $(\sqrt{x} - \sqrt{2})/(x - 2)$, $x \neq 2$. (Rationalize the numerator.)

SOLUTION

$$\frac{\sqrt{x}-\sqrt{2}}{x-2} = \frac{\sqrt{x}-\sqrt{2}}{x-2} \cdot \frac{\sqrt{x}+\sqrt{2}}{\sqrt{x}+\sqrt{2}} = \frac{(\sqrt{x})^2-(\sqrt{2})^2}{(x-2)(\sqrt{x}+\sqrt{2})}$$

$$= \frac{x-2}{(x-2)(\sqrt{x}+\sqrt{2})} = \frac{1}{\sqrt{x}+\sqrt{2}}. \quad \blacksquare$$

The Quadratic Equation

The expression $x^2 + bx + c$ can be written in the form $(x + k)^2 + d$ for a suitable choice of constants k and d. Finding the numbers k and d is called **completing the square**.

Completing the square

Recall that $$(x + k)^2 = x^2 + 2kx + k^2.$$

Therefore, if $$(x + k)^2 + d = x^2 + bx + c,$$

then we have $$x^2 + 2kx + k^2 + d = x^2 + bx + c.$$

Consequently, $2k$ must be b. In other words, k must be $b/2$. The next two examples show why this observation is the key to completing the square.

EXAMPLE 4 Complete the square in $x^2 + 6x + 11$.

SOLUTION In this case, $b = 6$ and $b/2 = 3$. Thus $k = 3$. To find d, proceed as follows:

Leaving space to complete the square

$$\begin{aligned} x^2 + 6x + 11 &= (x^2 + 6x \qquad) + 11 \\ &= (x^2 + 6x + 3^2) + 11 - 3^2 \\ &= (x + 3)^2 + 11 - 9 \\ &= (x + 3)^2 + 2. \quad \blacksquare \end{aligned}$$

EXAMPLE 5 Complete the square in $x^2 - 5x + 2$.

SOLUTION First of all, $k = -\frac{5}{2}$. Then

$$\begin{aligned} x^2 - 5x + 2 &= [x^2 - 5x + (-\tfrac{5}{2})^2] + 2 - (-\tfrac{5}{2})^2 \\ &= (x - \tfrac{5}{2})^2 + 2 - \tfrac{25}{4} \\ &= (x - \tfrac{5}{2})^2 - \tfrac{17}{4}. \quad \blacksquare \end{aligned}$$

A slight variation in the technique illustrated in Examples 4 and 5 can be used to complete the square in the expression $ax^2 + bx + c$, that is, to write it in the form $e(x + k)^2 + d$, for suitable constants k, d, and e. The next example shows how to do this.

EXAMPLE 6 Complete the square in $2x^2 + 5x + 3$.

SOLUTION First, factor the coefficient of x^2 out of the first two terms, obtaining

$$2x^2 + 5x + 3 = 2(x^2 + \tfrac{5}{2}x) + 3.$$

Then complete the square in the expression $x^2 + 5x/2$ as follows:

$$\begin{aligned} x^2 + \tfrac{5}{2}x &= x^2 + \tfrac{5}{2}x + (\tfrac{5}{4})^2 - (\tfrac{5}{4})^2 \\ &= (x + \tfrac{5}{4})^2 - (\tfrac{5}{4})^2. \end{aligned}$$

Thus

$$\begin{aligned} 2x^2 + 5x + 3 &= 2[(x + \tfrac{5}{4})^2 - (\tfrac{5}{4})^2] + 3 \\ &= 2[(x + \tfrac{5}{4})^2 - \tfrac{25}{16}] + 3 \\ &= 2(x + \tfrac{5}{4})^2 - \tfrac{25}{8} + 3 \\ &= 2(x + \tfrac{5}{4})^2 - \tfrac{1}{8}. \quad \blacksquare \end{aligned}$$

Completing the square permits us to analyze the real solutions, if there are any, of the equation

$$ax^2 + bx + c = 0.$$

The **quadratic formula**,

$$x = \frac{-b \pm \sqrt{b^2 - 4ac}}{2a}, \tag{1}$$

which describes the solutions, is obtained as follows:

$$ax^2 + bx + c = 0 \qquad \text{given}$$

$$a\left(x^2 + \frac{b}{a}x\right) + c = 0$$

$$a\left[\left(x + \frac{b}{2a}\right)^2 - \left(\frac{b}{2a}\right)^2\right] + c = 0 \qquad \text{completing the square}$$

$$a\left(x + \frac{b}{2a}\right)^2 - a\left(\frac{b}{2a}\right)^2 + c = 0$$

$$a\left(x + \frac{b}{2a}\right)^2 = a\left(\frac{b}{2a}\right)^2 - c$$

$$\left(x + \frac{b}{2a}\right)^2 = \left(\frac{b}{2a}\right)^2 - \frac{c}{a} = \frac{b^2}{4a^2} - \frac{c}{a} = \frac{b^2 - 4ac}{4a^2}$$

$$x + \frac{b}{2a} = \pm\sqrt{\frac{b^2 - 4ac}{4a^2}} \qquad \text{taking square roots}$$

$$= \pm\frac{\sqrt{b^2 - 4ac}}{2a}$$

$$x = \frac{-b}{2a} \pm \frac{\sqrt{b^2 - 4ac}}{2a}$$

and finally

$$x = \frac{-b \pm \sqrt{b^2 - 4ac}}{2a}.$$

The number $b^2 - 4ac$ which appears under the square root in Eq. (1) is called the **discriminant** of the quadratic expression $ax^2 + bx + c$. Note that if the discriminant is negative, the equation $ax^2 + bx + c = 0$ has no real solutions; complex solutions are treated in Sec. 11.6.

INFORMATION PROVIDED BY THE DISCRIMINANT $b^2 - 4ac$

If $b^2 - 4ac < 0$, $ax^2 + bx + c = 0$ has no real solutions.

If $b^2 - 4ac > 0$, $ax^2 + bx + c = 0$ has two distinct real solutions.

If $b^2 - 4ac = 0$, $ax^2 + bx + c = 0$ has one real solution.

EXAMPLE 7 Discuss the solutions of the equations
(*a*) $2x^2 + x + 3 = 0$,
(*b*) $2x^2 - 7x + 4 = 0$,
(*c*) $x^2 - 6x + 9 = 0$.

SOLUTION

(*a*) Here $a = 2$, $b = 1$, and $c = 3$. Therefore the discriminant $b^2 - 4ac$ equals $1^2 - 4 \cdot 2 \cdot 3 = -23$. Since the discriminant is negative there are no real solutions.

(*b*) Here $a = 2$, $b = -7$, and $c = 4$. The discriminant is thus $(-7)^2 - 4 \cdot 2 \cdot 4 = 17$, which is positive. In this case, there are two real solutions, $(7 + \sqrt{17})/4$ and $(7 - \sqrt{17})/4$.

(*c*) Here $a = 1$, $b = -6$, and $c = 9$. Thus

$$x = \frac{-(-6) \pm \sqrt{(-6)^2 - 4 \cdot 1 \cdot 9}}{2 \cdot 1} = \frac{6 \pm \sqrt{36 - 36}}{2} = \frac{6}{2}.$$

There is only one solution, namely, 3. This reflects the fact that when $x^2 - 6x + 9$ is factored, the factor $x - 3$ is repeated, that is, $x^2 - 6x + 9 = (x - 3)^2$. ■

The Binomial Theorem

Let n be a positive integer and let x be a real number. When $(1 + x)^n$ is multiplied out, it leads to a sum of terms, each of which is of the form "a coefficient times a power of x." For instance,

$$(1 + x)^2 = 1 + 2x + x^2,$$

$$(1 + x)^3 = 1 + 3x + 3x^2 + x^3,$$

and

$$(1 + x)^4 = 1 + 4x + 6x^2 + 4x^3 + x^4.$$

The **binomial theorem** provides a formula for the expansion of $(1 + x)^n$ for any positive integer n. It asserts that

$$(1 + x)^n = 1 + nx + \frac{n(n-1)}{1 \cdot 2}x^2 + \frac{n(n-1)(n-2)}{1 \cdot 2 \cdot 3}x^3 + \cdots + x^n. \tag{2}$$

The coefficient of x^k in the expansion of $(1 + x)^n$ is

$$\frac{n(n-1)(n-2)\cdots(n-k+1)}{1 \cdot 2 \cdot 3 \cdots k}. \tag{3}$$

Check whether your calculator has a key for C_k^n.

This coefficient is denoted $\binom{n}{k}$ or C_k^n and is called a **binomial coefficient**. To help remember formula (3) for $\binom{n}{k}$, note that the denominator is the product of the integers from 1 through k. The numerator has the same number of factors as the denominator, starting with n and going down to its smallest factor, which turns out to be $n - k + 1$. Simply write each factor in the numerator directly above a factor in the denominator, and you will stop at the right number. For instance,

$$\binom{7}{3} = C_3^7 = \frac{7 \cdot 6 \cdot 5}{1 \cdot 2 \cdot 3}.$$

In the $\binom{n}{k}$ notation the binomial theorem reads

$$(1 + x)^n = 1 + \binom{n}{1}x + \binom{n}{2}x^2 + \binom{n}{3}x^3 + \cdots + x^n,$$

where n is a positive integer.

EXAMPLE 8 Find the coefficient of x^4 in the expansion of $(1 + x)^9$.

SOLUTION In this case, $n = 9$ and $k = 4$. Thus the coefficient is

$$\frac{9 \cdot 8 \cdot 7 \cdot 6}{1 \cdot 2 \cdot 3 \cdot 4},$$

which equals 126. ■

The product of all the integers from 1 through k, $1 \cdot 2 \cdots k$, is called k **factorial** and is denoted "$k!$". Thus $5! = 1 \cdot 2 \cdot 3 \cdot 4 \cdot 5 = 120$. Formula (3) for the binomial coefficient can also be written

$$\frac{n(n-1)(n-2)\cdots(n-k+1)}{k!}. \tag{4}$$

In fact, formula (3) can be replaced by a formula that uses only factorials. To see this, note that

$$\begin{aligned} n(n-1)(n-2)\cdots(n-k+1) &= \frac{n(n-1)(n-2)\cdots(n-k+1)(n-k)\cdots 3 \cdot 2 \cdot 1}{(n-k)\cdots 3 \cdot 2 \cdot 1} \\ &= \frac{n!}{(n-k)!}. \end{aligned}$$

Combining this information with formula (3) shows that the binomial coefficient $\binom{n}{k}$ can be written completely in terms of factorials:

$$\binom{n}{k} = \frac{n!}{k!(n-k)!}. \tag{5}$$

When using a hand-held calculator you may find this the most useful of the three formulas, since many calculators have a factorial key that gives $k!$ for k up to around 69.

It is necessary to define $0!$ to be 1 in order that formula (5) remain valid even in the cases $k = 0$ and $k = n$. For instance, if $k = n$, formula (5) gives

$$\frac{n!}{n!0!} = 1,$$

which is what we want: the coefficient of x^n in the expansion of $(1 + x)^n$ is 1.

EXAMPLE 9 Find the coefficient of x^5 in the expansion of $(1 + x)^{11}$.

SOLUTION By formula (3),

$$\binom{11}{5} = \frac{11 \cdot 10 \cdot 9 \cdot 8 \cdot 7}{1 \cdot 2 \cdot 3 \cdot 4 \cdot 5} = 462.$$

If instead you use formula (5) with a calculator, the computations would be

$$\binom{11}{5} = \frac{11!}{5!6!} = \frac{39{,}916{,}800}{120 \cdot 720} = 462. \quad ■$$

The binomial theorem is frequently stated in terms of the expansion of $(a + b)^n$ instead of that of $(1 + x)^n$, which is the special case in which $a = 1$ and $b = x$. It asserts that

$$(a+b)^n = a^n + na^{n-1}b + \frac{n(n-1)}{1\cdot 2}a^{n-2}b^2 + \cdots + b^n. \tag{6}$$

The coefficient of $a^{n-k}b^k$ is the binomial coefficient

$$\binom{n}{k} = \frac{n!}{k!(n-k)!}.$$

For instance, $(a+b)^4 = a^4 + 4a^3b + 6a^2b^2 + 4ab^3 + b^4$.

Geometric Series

Another algebraic identity provides a short formula for the sum

$$a + ax + ax^2 + \cdots + ax^{n-1}, \tag{7}$$

So named by the Greeks because corresponding lengths in similar geometric figures are in the same ratio.

where a and x are real numbers and n is a positive integer. Note that there are n summands in (7). Such an expression is called a **finite geometric series** with **first term** a and **ratio** x. Each term in (7) is obtained from the one before by multiplying by the number x. We present a shortcut for evaluating the sum $a + ax + ax^2 + \cdots + ax^{n-1}$.

A short formula for the sum of a finite geometric series

> Let a and x be real numbers, with $x \neq 1$. Let n be a positive integer. Then
>
> $$a + ax + ax^2 + \cdots + ax^{n-1} = \frac{a(1-x^n)}{1-x}. \tag{8}$$

To show that the formula (8) is true, multiply both sides by $1-x$. The multiplication of $a + ax + ax^2 + \cdots + ax^{n-1}$ by $1-x$ leads, happily, to many cancellations:

$$\begin{aligned}(1-x)&(a + ax + ax^2 + \cdots + ax^{n-1})\\ &= (a + ax + ax^2 + \cdots + ax^{n-1}) - (ax + ax^2 + \cdots + ax^n)\\ &= a - ax^n\\ &= a(1-x^n).\end{aligned}$$

Division by $1-x$ then establishes formula (8).

EXAMPLE 10 Use formula (8) to evaluate $1 + \frac{1}{2} + (\frac{1}{2})^2 + (\frac{1}{2})^3 + (\frac{1}{2})^4$.

SOLUTION In this case, $a = 1$, $x = \frac{1}{2}$, and $n = 5$. Thus

$$\begin{aligned}1 + \tfrac{1}{2} + (\tfrac{1}{2})^2 + (\tfrac{1}{2})^3 + (\tfrac{1}{2})^4 &= \frac{1[1-(\frac{1}{2})^5]}{1-\frac{1}{2}}\\ &= \frac{1-(\frac{1}{2})^5}{\frac{1}{2}}\\ &= 2(1-\tfrac{1}{32}) = 2 - \tfrac{1}{16} = \tfrac{31}{16}. \quad ■\end{aligned}$$

The Factor Theorem

Next we examine the relation between the roots of a polynomial and its first-degree factors.

If $x - a$ is a factor, then a is a root.

Let a be a real number and $P(x)$ a polynomial that has $x - a$ as a factor. This means that there is a polynomial $Q(x)$ such that $P(x) = Q(x)(x - a)$. Replacing x by a gives us $P(a) = Q(a)(a - a)$. Since $a - a = 0$, it follows that $P(a) = 0$; that is, "a is a root of the polynomial $P(x)$."

How about the reverse direction? If a is a root of $P(x)$, that is, $P(a) = 0$, does it follow that $x - a$ must be a factor of $P(x)$?

In any case, whether $x - a$ is or is not a factor of $P(x)$, we may divide $x - a$ into $P(x)$ by long division to get a quotient $Q(x)$ and a remainder, which is 0 or a polynomial of lower degree than $x - a$. This means the remainder is just a constant k. So we may write

$$P(x) = Q(x)(x - a) + k. \tag{9}$$

Replacing x by a gives us

$$P(a) = Q(a)(a - a) + k.$$

If a is a root, then $x - a$ is a factor.

Since $P(a) = 0$ and $a - a = 0$, we see that $k = 0$. This means, according to Eq. (9), that $x - a$ divides $P(x)$.

Thus we have this neat test for deciding whether $P(x)$ has a first-degree factor: Look for a root of $P(x)$. If you can find one, say, a, then $x - a$ is a factor. If there are no roots, then $P(x)$ has no first-degree factors. This shortcut is known as the **factor theorem**.

Shortcut for finding first-degree factors

The factor theorem does *not* test for factors of degree higher than 1. For instance, $x^4 + 4$ has no real roots, yet it has a second-degree factor, since

$$x^4 + 4 = (x^2 + 2x + 2)(x^2 - 2x + 2),$$

as can be checked by multiplying out the right side.

There is also a shortcut for searching for rational roots of a polynomial that has integer coefficients. Example 11 illustrates the technique.

EXAMPLE 11 Does the equation $4x^3 + 3x + 2 = 0$ have any rational roots?

SOLUTION Let us see whether the rational number a/b is a root. (Here a and b are integers with no common divisor larger than 1. We may assume that b is positive.) If a/b is a root of the equation, then

Plug in $x = a/b$ and clear denominator.

$$4a^3 + 3ab^2 + 2b^3 = 0. \tag{10}$$

Now, a divides $4a^3 + 3ab^2$; hence a divides $2b^3 = -(4a^3 + 3ab^2)$. But a and b have no common divisors other than 1 and -1, so a divides 2. Similarly, b divides $4a^3$; hence b divides 4. Since $b > 0$, we have $b = 1, 2,$ or 4. All that remains is to check the various combinations of a and b, namely, the candidates

$$\frac{1}{1}, \frac{1}{2}, \frac{1}{4}, \frac{-1}{1}, \frac{-1}{2}, \frac{-1}{4}, \frac{2}{1}, \frac{2}{2}, \frac{2}{4}, \frac{-2}{1}, \frac{-2}{2}, \frac{-2}{4}.$$

(There are duplications, for instance, $\frac{1}{1} = \frac{2}{2}$.)

Test each of these possibilities by plugging them into Eq. (10). If none

$2^{x+y} = 2^x \cdot 2^y$ for *all real numbers* x and y will tell us how the exponential function 2^x must be defined. Note that $2^{x+y+z} = 2^{(x+y)+z} = 2^{x+y} \cdot 2^z = (2^x \cdot 2^y)2^z = 2^x \cdot 2^y \cdot 2^z$. Thus the basic law of exponents extends to the case when the exponent is the sum of three numbers: $2^{x+y+z} = 2^x \cdot 2^y \cdot 2^z$. In a similar manner, it can be extended to the case where the exponent is the sum of any finite number of numbers.

For instance, *what should the numerical value of* 2^0 *be?* If the basic law of exponents is to be true for all exponents, then, in particular, this equation must hold:

$$2^{0+1} = 2^0 \cdot 2^1;$$

hence

$$2^1 = 2^0 \cdot 2^1.$$

But $2^1 = 2$, so 2^0 must satisfy the equation

$$2 = 2^0 \cdot 2.$$

2^0 must be 1.

There is no choice; 2^0 must be 1.

What should $2^{-1}, 2^{-2}, 2^{-3}, \ldots$ *be?* To preserve the basic law of exponents, we must have, for instance,

$$2^3 \cdot 2^{-3} = 2^{3+(-3)}.$$

But $2^{3+(-3)} = 2^0 = 1$. Thus $2^3 \cdot 2^{-3} = 1$. This shows that if 2^{-3} is to have meaning, it must be the reciprocal of 2^3, that is,

$$2^{-3} = \frac{1}{2^3} = \frac{1}{8}.$$

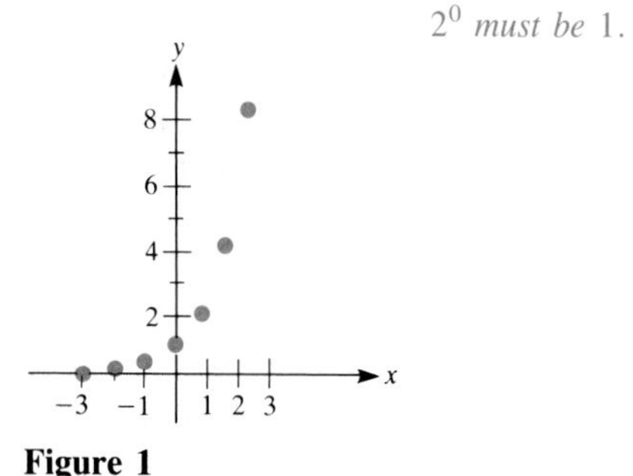

Figure 1

For this reason, *we define*

2^{-n} must be $1/(2^n)$.

$$2^{-n} \quad \textit{to be} \quad \frac{1}{2^n}$$

for any positive integer n.

The exponential 2^x has now been defined for any *integer x*. Let us graph $y = 2^x$ for integer values of x, as in Fig. 1.

It is tempting to draw a smooth curve through these points, but 2^x has not yet been defined if x is not an integer.

How should $2^{1/2}$ *be defined?* To preserve the basic law of exponents, it is necessary that

$$2^{1/2} \cdot 2^{1/2} = 2^{1/2+1/2};$$

thus

$$2^{1/2} \cdot 2^{1/2} = 2^1 = 2.$$

Hence, $2^{1/2}$ is a solution of the equation

$$x^2 = 2.$$

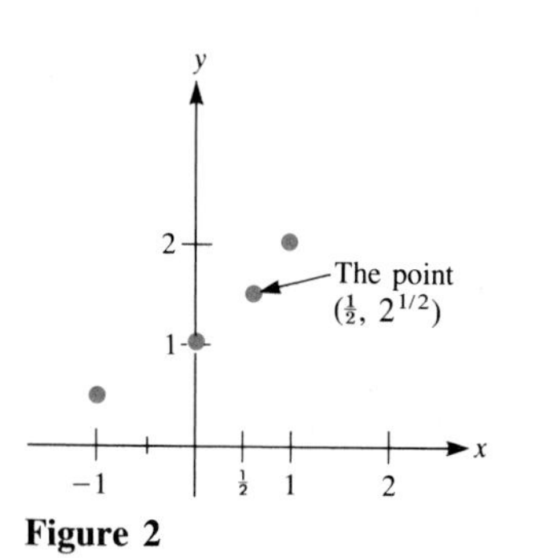

Figure 2

Why $2^{1/2}$ should be the square root of 2

Should it be the positive or the negative solution? The graph just sketched suggests that $2^{1/2}$ should be positive. Thus define $2^{1/2}$ to be $\sqrt{2}$, the square root of 2. Note that $2^{1/2} \approx 1.4$, which fits nicely into the preceding graph, as is shown in Fig. 2.

Similarly, $2^{1/3}$ can be determined by the basic law of exponents:

$$2^{1/3} \cdot 2^{1/3} \cdot 2^{1/3} = 2^{1/3+1/3+1/3} = 2^1 = 2.$$

Thus $2^{1/3}$ must be a solution of the equation

$$x^3 = 2.$$

Why $2^{1/3}$ should be the cube root of 2

There is only one solution, and it is called the **cube root** of 2, denoted also $\sqrt[3]{2}$. Thus define

$$2^{1/3} = \sqrt[3]{2},$$

which is about 1.26.

Similarly, *define $2^{1/n}$ for any positive integer n to be the positive solution of the equation $x^n = 2$*, that is,

$$2^{1/n} = \sqrt[n]{2}.$$

What should $2^{3/4}$ be?

How should $2^{3/4}$ be defined? To preserve the basic law of exponents, we must have

$$2^{1/4} \cdot 2^{1/4} \cdot 2^{1/4} = 2^{3/4}.$$

In short, we must have $$2^{3/4} = (2^{1/4})^3.$$

Defining $2^{m/n}$

This suggests that for *any integer m and positive integer n we should define*

$$2^{m/n} \quad \textit{to be} \quad (2^{1/n})^m.$$

With this step we have defined 2^x for every rational number x.

How should 2^x be defined when x is not rational? For instance, how should $2^{\sqrt{2}}$ be defined? To begin, consider the decimal representation of $\sqrt{2}$, namely,

$$\sqrt{2} = 1.41421356. \ . \ . \ .$$

The successive decimals 1.4, 1.41, 1.414, . . . are rational (for instance, $1.41 = \frac{141}{100}$). Thus $2^{1.4}$, $2^{1.41}$, $2^{1.414}$, . . . are already defined. This table indicates some of their values:

x	1.4	1.41	1.414	1.4142	1.41421
2^x	2.63901 . . .	2.65737 . . .	2.66474 . . .	2.66511 . . .	2.66513 . . .

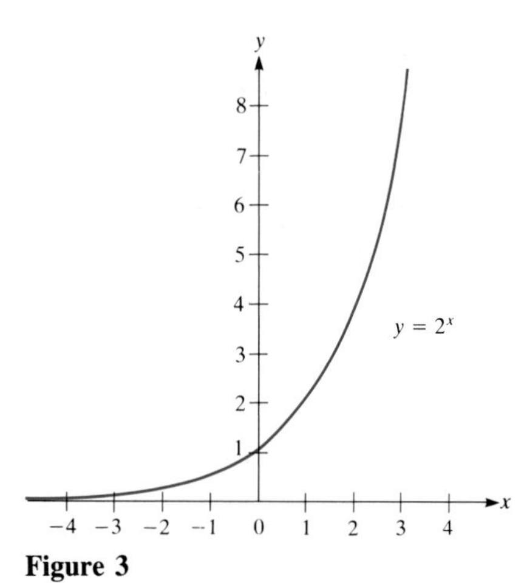

Figure 3

By going further in this table we can get better approximations of $2^{\sqrt{2}}$. This procedure, though cumbersome, would enable us to find $2^{\sqrt{2}}$ to any number of decimal places. To five decimal places,

$$2^{\sqrt{2}} \approx 2.66514.$$

The graph of $y = 2^x$ is shown in Fig. 3.

For any positive number b, the definition of b^x follows the same outline as the definition of 2^x. For $b > 1$, the graph of $f(x) = b^x$ resembles that of 2^x. Far to the left it gets near the x axis; far to the right it gets arbitrarily high. Moreover, as x increases, so does b^x.

For $0 < b < 1$, the graph of $f(x) = b^x$ resembles the graph of $(\frac{1}{2})^x$, which is shown in Fig. 4.

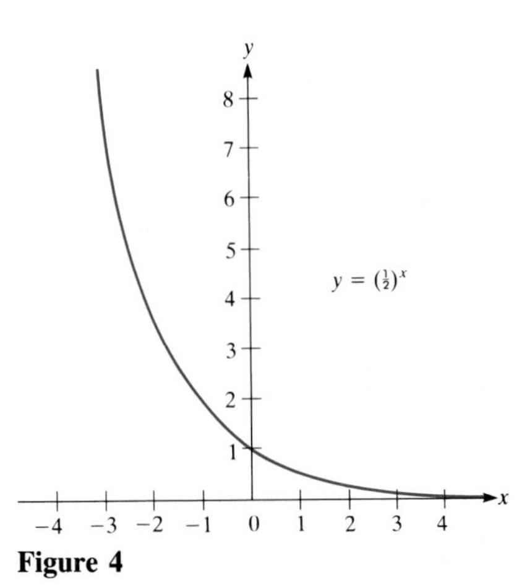

Figure 4

EXAMPLE 1 Evaluate $64^{1/3}$, $64^{2/3}$, $64^{1/2}$, and $64^{-2/3}$.

SOLUTION $64^{1/3}$ is the cube root of 64, $\sqrt[3]{64}$, which is 4.

$$64^{2/3} = (64^{1/3})^2 = 4^2 = 16;$$

$$64^{1/2} = \sqrt{64} = 8;$$

$$64^{-2/3} = (64^{1/3})^{-2} = 4^{-2} = \tfrac{1}{16} = 0.0625. \ \blacksquare$$

EXAMPLE 2 Find a decimal approximation of $2^{1/3}$, the cube root of 2.

SOLUTION The quickest way is to use a calculator. Either the y^x-key, with $y = 2$ and $x = \frac{1}{3}$, or the $\sqrt[x]{y}$-key, with $y = 2$ and $x = 3$, will do the trick and show that $2^{1/3} \approx 1.25992105$.

If a calculator is not available, brute-force arithmetic could be used. For instance, by straightforward multiplication, $1.2^3 = 1.728$ and $1.3^3 = 2.197$. Thus

$$1.2 < 2^{1/3} < 1.3.$$

To find the next decimal place, compute $1.21^3, 1.22^3, \ldots, 1.29^3$. Since $1.25^3 = 1.953125$ and $1.26^3 = 2.000376$,

$$1.25 < 2^{1/3} < 1.26.$$

With enough time and pencils, the curious could find as many decimal places of $2^{1/3}$ as needed. ■

In addition to the basic law of exponents, there are other laws that hold for the exponential functions for any positive base:

LAWS OF EXPONENTS

The bases are positive, the exponents any real numbers.

$b^{x+y} = b^x b^y$	basic law of exponents
$b^{x-y} = \dfrac{b^x}{b^y}$	difference of exponents
$(b^x)^y = b^{xy}$	power of a power
$(ab)^x = a^x b^x$	power of a product
$\left(\dfrac{a}{b}\right)^x = \dfrac{a^x}{b^x}$	power of a quotient
$b^0 = 1$	definition
$b^1 = b$	definition

The laws of exponents help in simplifying certain algebraic expressions, as illustrated by Example 3.

EXAMPLE 3 Let b be a positive number. Write each of the following in the form b^x for a suitable exponent x: (*a*) $(\sqrt{b})^3$, (*b*) b^7/b^3, (*c*) $\sqrt{b}\sqrt[3]{b}$, (*d*) $1/\sqrt{b}$.

Parts (a) and (d) use the "power of a power" law.

SOLUTION

(*a*) $(\sqrt{b})^3 = (b^{1/2})^3 = b^{(1/2)(3)} = b^{3/2}$.

(*b*) $\dfrac{b^7}{b^3} = b^{7-3} = b^4$.

(*c*) $\sqrt{b}\sqrt[3]{b} = b^{1/2}b^{1/3} = b^{1/2+1/3} = b^{5/6}$.

(*d*) $\dfrac{1}{\sqrt{b}} = (\sqrt{b})^{-1} = (b^{1/2})^{-1} = b^{-1/2}$. ■

APPENDIX F

THE CONVERSE OF A STATEMENT

Many mathematical assertions can be put into the form "if A, then C," where A is an assumption and C is a conclusion. For instance, the assertion "the product of a multiple of 2 and a multiple of 3 is a multiple of 6" can be restated as "if a is a multiple of 2 and b is a multiple of 3, then ab is a multiple of 6." (Here a and b refer to integers.) As another example, the assertion "the square of an even integer is always even" can be rephrased as "if a is even, then a^2 is even." (a refers to integers.)

Starting with a statement of the form "if A, then C," we can construct a new statement, "if C, then A," which is called the **converse** of "if A, then C." Even if the original statement, "if A, then C," is true, its converse may be false. Example 1 illustrates this case. In Example 2 the converse happens also to be true.

EXAMPLE 1 Is the converse of "if a is a multiple of 2 and b is a multiple of 3, then ab is a multiple of 6" true or false?

SOLUTION The converse is "if ab is a multiple of 6, then a is a multiple of 2 and b is a multiple of 3." (The symbols a and b refer to integers.) The counterexample $a = 6$ and $b = 1$ shows that it is not true for all possible pairs a and b under consideration. Hence the converse is false. ■

EXAMPLE 2 Is the converse of "if a is even, then a^2 is even" true or false?

SOLUTION The converse is "if a^2 is even, then a is even." (a refers to integers.) This assertion is true. If a were odd, a^2 would be odd, since the product of two odd numbers is always odd. ■

There are two important reasons to distinguish between a statement "if A, then C" and its converse "if C, then A." First of all, they make completely different claims. Second, although the first statement may be true, the converse may be true or it may be false. As it happens, calculus contains many statements whose converses are false. Therefore, failure to distinguish a statement from its converse can lead to confusion.

EXERCISES FOR APP. F: THE CONVERSE OF A STATEMENT

In each of Exercises 1 to 6 two true statements are given. However, the converse of one is true and the converse of the other is false. In each case write out the converse. If it is true, explain why it holds. If it is false, give a counterexample.

1 (*a*) If $a = b$, then $a^2 = b^2$. (a and b refer to real numbers.)
(*b*) If $a = b$, then $a^3 = b^3$. (a and b refer to real numbers.)

2 (*a*) If a is odd, then a^2 is odd. (a refers to integers.)
(*b*) If a is odd, then $a + a$ is even. (a refers to integers.)

3 (*a*) If $b = c$, then $ab = ac$. (a, b, and c refer to real numbers.)
(*b*) If $b = c$, then $a + b = a + c$. (a, b, and c refer to real numbers.)

4 (*a*) If a is rational, then a^2 is rational. (a refers to real numbers.)
(*b*) If a is rational, then $2a$ is rational. (a refers to real numbers.)

5 (*a*) If a and b are odd, then ab is odd. (a and b refer to integers.)
(*b*) If a and b are even, then ab is even. (a and b refer to integers.)

6 (*a*) If a is a multiple of 6, then a^2 is a multiple of 6. (a refers to integers.)
(*b*) If a is a multiple of 4, then a^2 is a multiple of 4. (a refers to integers.)

In each of Exercises 7 to 12 decide whether the converse of the given statement is true or false.

7 If the three sides of triangle T are equal, then the three angles of triangle T are equal.

8 If quadrilateral Q is a square, then all four sides of quadrilateral Q are equal.

9 If quadrilateral Q is a parallelogram, then opposite sides of quadrilateral Q are equal.

10 If hexagon H is regular, then opposite sides of hexagon H are equal.

11 $P(x)$ denotes a polynomial in x with real coefficients, and a denotes a real number. If $P(a) = 0$, then $x - a$ divides $P(x)$. (See "factor theorem" in Appendix C.)

12 If m is even, then the polynomial $x^2 - 1$ divides the polynomial $x^m - 1$. (m refers to nonnegative integers.)

Exercises 13 to 24 refer to concepts developed in the text. In each case, decide whether the converse of the given statement is true or false. The functions f and g are assumed to be defined on the entire x axis.

13 If f and g are one-to-one, then $f \circ g$ is one-to-one (Sec. 6.4).

14 If f is a one-to-one function, then $f \circ f$ is a one-to-one function (Sec. 6.4).

15 If $f'(x) = 0$ for all x, then $f(x)$ is constant (Sec. 4.1).

16 If $f'(x)$ is positive for all x, then f is increasing (Sec. 4.1).

17 If f is a polynomial of degree 2, then $f^{(3)}(x) = 0$ for all x (Sec. 4.3).

18 If f is differentiable at a, then f is continuous at a (Sec. 3.3).

19 If f has a relative maximum or minimum at a, then $f'(a) = 0$ (Sec. 4.1).

20 If f is elementary, then f' is elementary (Sec. 5.7).

21 If $\Sigma_{n=1}^{\infty} a_n$ converges, then $\lim_{n\to\infty} a_n = 0$ (Sec. 10.3).

22 If $\Sigma_{n=1}^{\infty} |a_n|$ converges, then $\Sigma_{n=1}^{\infty} a_n$ converges (Sec. 10.7).

23 If **F** is a conservative vector field, then **F** is a gradient field (Sec. 16.6).

24 If **F** is a conservative vector field, then **curl F** = **0** (Sec. 16.6).

APPENDIX G

CONIC SECTIONS

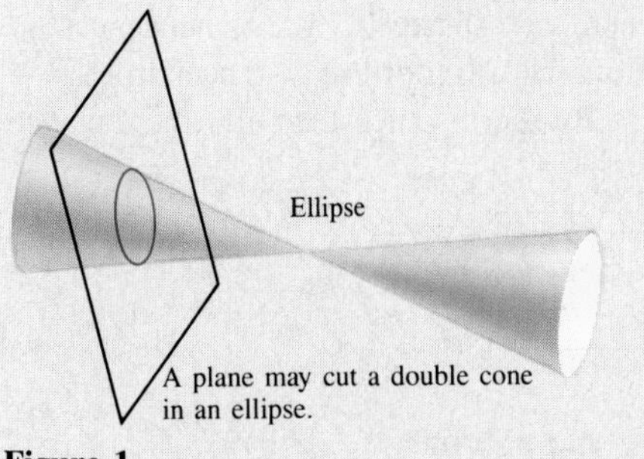

A plane may cut a double cone in an ellipse.

Figure 1

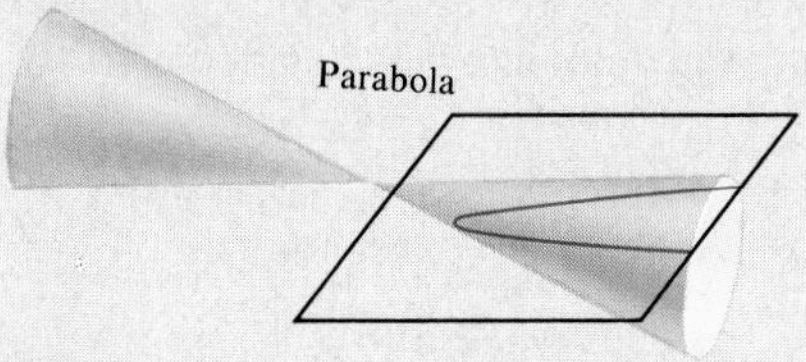

Parabola

Figure 2

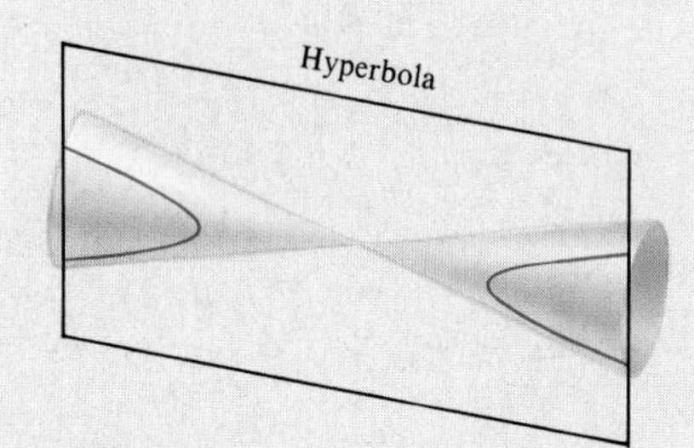

Figure 3

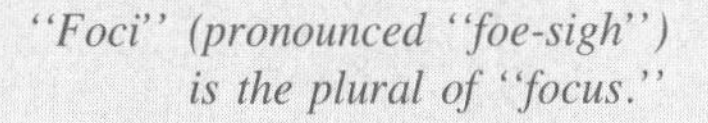
"Foci" (pronounced "foe-sigh") is the plural of "focus."

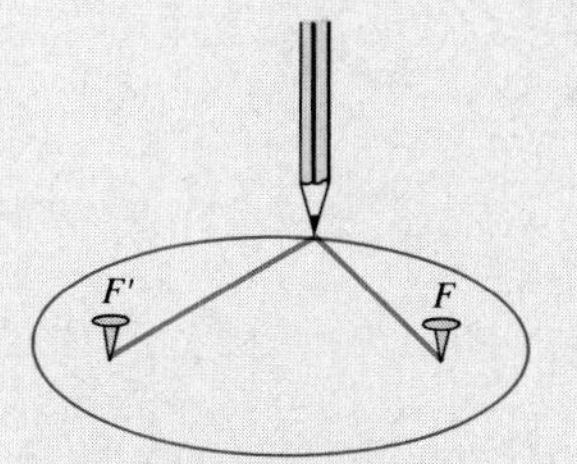

The string has length $2a$, greater than the distance between the tacks.

Figure 4

This appendix defines the ellipse, hyperbola, and parabola and develops their equations in rectangular coordinates (Secs. G.1, G.2, and G.3) and in polar coordinates (Sec. G.4).

G.1 CONIC SECTIONS

The intersection of a plane and the surface of a double cone is called a **conic section**. If the plane cuts off a bounded curve, that curve is called an **ellipse**. (In particular, a circle is an ellipse.) See Fig. 1.

If the plane is parallel to the edge of the double cone, as in Fig. 2, the intersection is called a **parabola**. In the cases of the ellipse and the parabola, the plane meets just one of the two cones.

If the plane meets both parts of the cone and is not parallel to an edge, the intersection is called a **hyperbola**. The hyperbola consists of two separate pieces. It can be proved that these two pieces are congruent and that they are *not* congruent to parabolas. (See Fig. 3.)

For the sake of simplicity, we shall use a definition of the conic sections that depends only on the geometry of the plane. It is shown in geometry courses that the two approaches yield the same curves.

Definition *Ellipse*. Let F and F' be points in the plane and let a be a fixed positive number such that $2a$ is greater than the distance between F and F'. A point P in the plane is on the **ellipse** determined by F, F', and $2a$ if and only if the sum of the distances from P to F and from P to F' equals $2a$. Points F and F' are the **foci** of the ellipse.

To construct an ellipse, place two tacks in a piece of paper, tie a string of length $2a$ to them, and trace out a curve with a pencil held against the string, keeping the string taut by means of the pencil point. (See Fig. 4.) The foci are at the tacks. (Note that when $F = F'$, the ellipse is a circle of radius a.) The four points on the ellipse that are farthest from or closest to the center are called **vertices**. A circle does not have vertices.

The equation of a circle whose center is at (0, 0) and whose radius is a is

$$x^2 + y^2 = a^2,$$

or

$$\frac{x^2}{a^2} + \frac{y^2}{a^2} = 1. \tag{1}$$

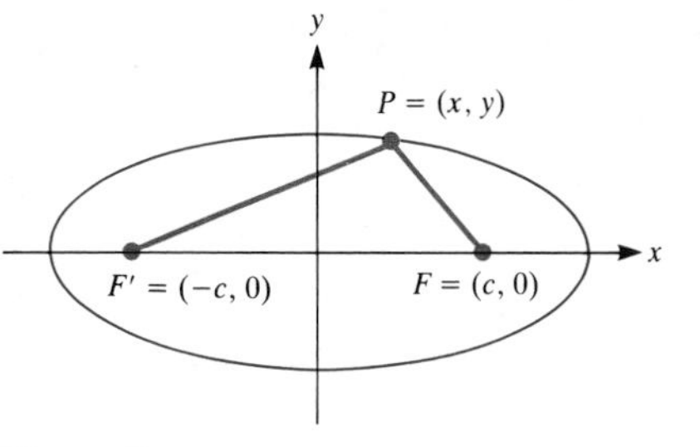

Figure 5

Let us generalize this result by determining the equation of an ellipse. To make the equation as simple as possible, introduce the x and y axes in such a way that the x axis contains the foci and the origin is midway between them, as in Fig. 5. Thus $F = (c, 0)$ and $F' = (-c, 0)$, where $c \geq 0$ and $2c < 2a$; hence $c < a$.

Now translate into algebra this assertion: The sum of the distances from $P = (x, y)$ to F and from P to F' equals $2a$. By the distance formula, the distance from P to F is

$$\sqrt{(x - c)^2 + (y - 0)^2}.$$

Similarly, the distance from P to F' is

$$\sqrt{(x + c)^2 + (y - 0)^2}.$$

Thus the point (x, y) is on the ellipse if and only if

$$\sqrt{(x - c)^2 + y^2} + \sqrt{(x + c)^2 + y^2} = 2a.$$

A few algebraic steps will transform this equation into an equation without square roots.

First, write the equation as

$$\sqrt{(x + c)^2 + y^2} = 2a - \sqrt{(x - c)^2 + y^2}.$$

Then square both sides, obtaining

$$(x + c)^2 + y^2 = 4a^2 - 4a\sqrt{(x - c)^2 + y^2} + (x - c)^2 + y^2.$$

Expanding yields

$$x^2 + 2cx + c^2 + y^2 = 4a^2 - 4a\sqrt{(x - c)^2 + y^2} + x^2 - 2cx + c^2 + y^2,$$

which a few cancellations reduce to

$$2cx = 4a^2 - 4a\sqrt{(x - c)^2 + y^2} - 2cx,$$

or

$$4cx - 4a^2 = -4a\sqrt{(x - c)^2 + y^2}.$$

Dividing by -4 yields

$$a^2 - cx = a\sqrt{(x - c)^2 + y^2}.$$

Squaring gets rid of the square root:

$$a^4 - 2a^2cx + c^2x^2 = a^2(x^2 - 2cx + c^2 + y^2),$$

$$= a^2x^2 - 2a^2cx + a^2c^2 + a^2y^2,$$

or

$$a^4 + c^2x^2 = a^2x^2 + a^2c^2 + a^2y^2.$$

This equation can be transformed to

$$(a^2 - c^2)x^2 + a^2y^2 = a^2(a^2 - c^2).$$

Dividing both sides by $a^2(a^2 - c^2)$ results in the equation

$$\frac{x^2}{a^2} + \frac{y^2}{a^2 - c^2} = 1. \tag{2}$$

Since $a^2 - c^2 > 0$, there is a number b such that

$$b^2 = a^2 - c^2 \qquad b > 0,$$

and thus Eq. (2) takes the shorter form

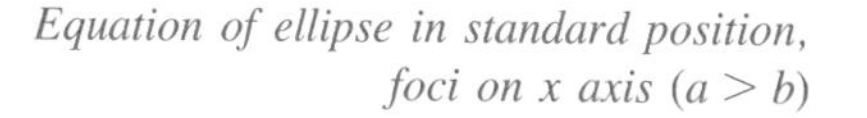

Equation of ellipse in standard position, foci on x axis $(a > b)$

$$\boxed{\frac{x^2}{a^2} + \frac{y^2}{b^2} = 1.} \tag{3}$$

[Note that Eq. (3) generalizes Eq. (1), the equation for a circle.]

We now find the four vertices of the ellipse by checking where the curve intersects the x and y axes. Setting $y = 0$ in Eq. (3), we obtain $x = a$ or $x = -a$; if we set $x = 0$ in Eq. (3), we obtain $y = b$ or $-b$. Thus the four ''extreme'' points of the ellipse have coordinates $(a, 0)$, $(-a, 0)$, $(0, b)$, and $(0, -b)$, as shown in Fig. 6. Observe that the distance from F or F' to $(0, b)$ is a, which is half the length of string. The right triangle in the diagram, with vertices F, $(0, b)$, and the origin, is a reminder of the fact that $b^2 = a^2 - c^2$. Keep in mind that in the above ellipse a is larger than b. The **semimajor axis** is said to have length a; the **semiminor axis** has length b. Observe that we could interchange the roles of x and y and produce an ellipse with foci on the y axis. In this case, y would have the larger denominator (the square of the semimajor axis) in the equation in standard form, which becomes $y^2/a^2 + x^2/b^2 = 1$, $a > b$.

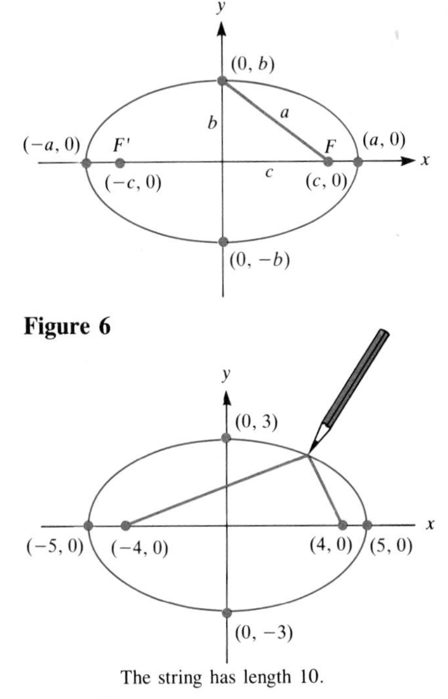

Figure 6

Figure 7

EXAMPLE 1 Discuss the foci and ''length of string'' of the ellipse whose equation is $\dfrac{x^2}{25} + \dfrac{y^2}{9} = 1$.

SOLUTION Since the larger denominator is with the x^2, the foci lie on the x axis. In this case, $a = 5$ and $b = 3$. The length of string is $2a = 10$. The foci are at a distance

$$c = \sqrt{a^2 - b^2} = \sqrt{25 - 9} = 4$$

from the origin, as shown in Fig. 7. ■

EXAMPLE 2 Discuss the foci and ''length of string'' of the ellipse whose equation is $\dfrac{x^2}{9} + \dfrac{y^2}{25} = 1$.

SOLUTION This is similar to Example 1. The only difference is that the roles of x and y are interchanged. The foci are at $(0, 4)$ and $(0, -4)$, and the ellipse is longer in the y direction than in the x direction. ■

The definition of the hyperbola is similar to that of the ellipse.

Definition *Hyperbola*. Let F and F' be points in the plane and let a be a fixed positive number such that $2a$ is less than the distance between F and

F'. A point P in the plane is on the **hyperbola** determined by F, F', and $2a$ if and only if the difference between the distances from P to F and from P to F' equals $2a$ (or $-2a$). Points F and F' are called the **foci** of the hyperbola. The points V_1 and V_2 are called the **vertices** of the hyperbola. (See Fig. 8.)

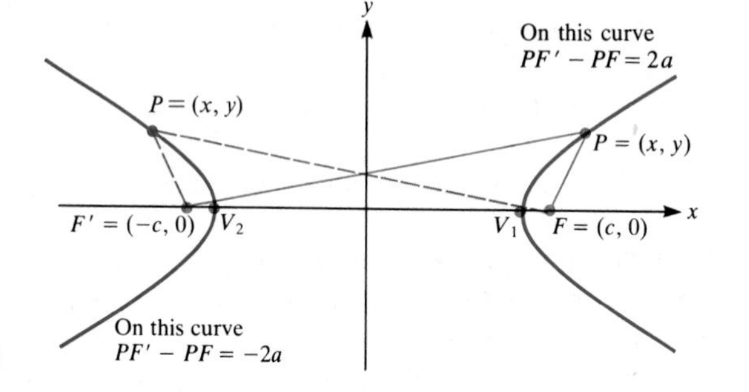

Figure 8

A hyperbola consists of two separate curves. On one of the curves we have $\overline{PF'} - \overline{PF} = 2a$; on the other, $\overline{PF'} - \overline{PF} = -2a$. If the distance $\overline{FF'}$ is $2c$, then $2a < 2c$; hence $a < c$. Again place the axes in such a way that $F = (c, 0)$ and $F' = (-c, 0)$. Let $P = (x, y)$ be a typical point on the hyperbola. Then x and y satisfy the equation

$$\sqrt{(x-c)^2 + y^2} - \sqrt{(x+c)^2 + y^2} = \pm 2a. \tag{4}$$

Some algebra similar to that used in simplifying the equation of the ellipse transforms Eq. (4) into

$$\frac{x^2}{a^2} + \frac{y^2}{a^2 - c^2} = 1. \tag{5}$$

But now $a^2 - c^2$ is *negative* and can be expressed as $-b^2$ for some number $b > 0$. Hence the hyperbola has the equation

Equation of hyperbola in standard position

$$\boxed{\frac{x^2}{a^2} - \frac{y^2}{b^2} = 1.} \tag{6}$$

If the foci are on the y axis, the equation is

$$\boxed{\frac{y^2}{a^2} - \frac{x^2}{b^2} = 1.}$$

In both cases, $c^2 = a^2 + b^2$.

EXAMPLE 3 Sketch the hyperbola $\frac{x^2}{9} - \frac{y^2}{16} = 1$.

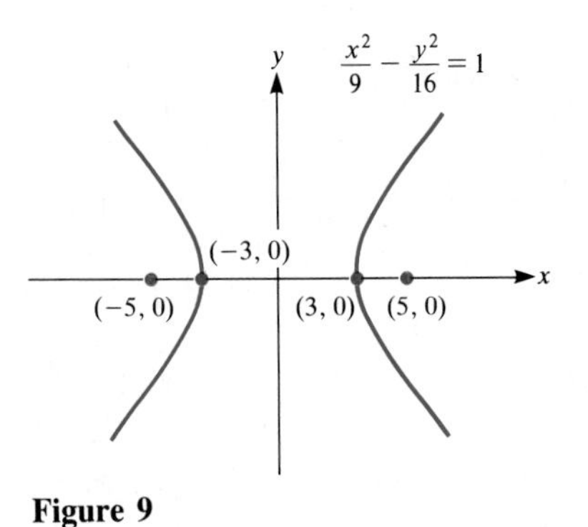

Figure 9

SOLUTION Since the minus sign is with the y^2, the foci are on the x axis. In this case, $a^2 = 9$ and $b^2 = 16$. Observe that the hyperbola meets the x axis at (3, 0) and (−3, 0). The hyperbola does not meet the y axis. The distance c from the origin to a focus is determined by the equation

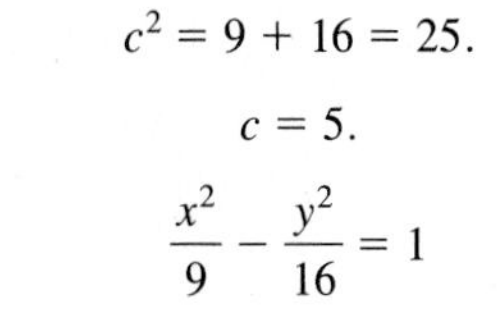

$$c^2 = 9 + 16 = 25.$$

Hence

$$c = 5.$$

The hyperbola

$$\frac{x^2}{9} - \frac{y^2}{16} = 1$$

is shown in Fig. 9. ■

The definition of a parabola involves the distance to a point and the distance to a line.

Definition *Parabola.* Let L be a line in the plane and let F be a point in the plane which is not on the line. A point P in the plane is on the **parabola** determined by F and L if and only if the distance from P to F equals the distance from P to the line L. Point F is the **focus** of the parabola; line L is its **directrix**. The point V in Fig. 10 is called the **vertex** of the parabola. It is the point on the parabola nearest the directrix.

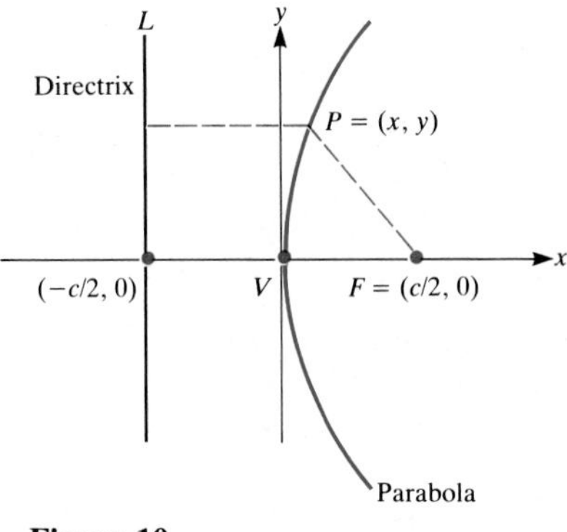

Figure 10

To obtain an algebraic equation for a parabola, denote the distance from point F to line L by c, and introduce axes in such a way that $F = (c/2, 0)$ and L has the equation $x = -c/2$ with $c > 0$. The distance from $P = (x, y)$ to F is $\sqrt{(x - c/2)^2 + (y - 0)^2}$. The distance from P to the line L is $x + c/2$. Thus the equation of the parabola is

$$\sqrt{\left(x - \frac{c}{2}\right)^2 + y^2} = x + \frac{c}{2}. \tag{7}$$

Squaring and simplifying reduces Eq. (7) to

$$\boxed{y^2 = 2cx \qquad c > 0,} \tag{8}$$

which is the equation of a parabola in "standard position."

If the focus is at $(0, c/2)$ and the directrix is the line $y = -c/2$, the parabola has the equation

$$\boxed{x^2 = 2cy.}$$

EXAMPLE 4 Sketch the parabola $y = x^2$, showing its focus and directrix.

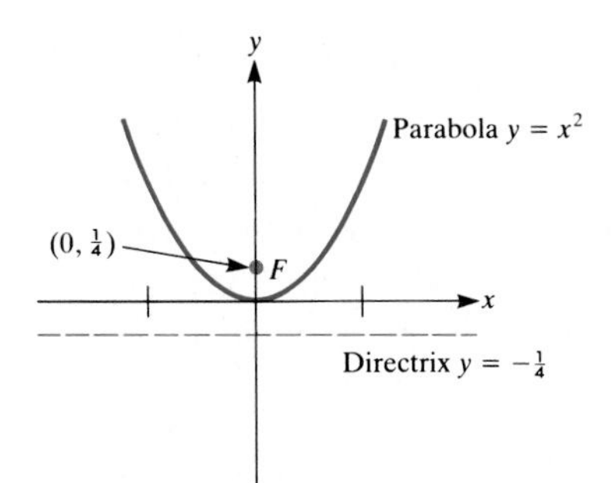

Figure 11

SOLUTION The equation $y = x^2$ is equivalent to $x^2 = 2cy$, where $c = \frac{1}{2}$. The focus is on the y axis at $(0, \frac{1}{4})$. The directrix is the line $y = -\frac{1}{4}$. The parabola is sketched in Fig. 11. ■

This table lists the equations of conics in standard position:

Conic Section	*Equation in Standard Position*	*Location of Foci*
Ellipse	$\frac{x^2}{a^2} + \frac{y^2}{b^2} = 1$	$a > b$, foci on x axis
	$\frac{y^2}{a^2} + \frac{x^2}{b^2} = 1$	$a > b$, foci on y axis
Circle	$\frac{x^2}{a^2} + \frac{y^2}{a^2} = 1$	Both foci at $(0, 0)$
Hyperbola	$\frac{x^2}{a^2} - \frac{y^2}{b^2} = 1$	Foci on x axis
	$\frac{y^2}{a^2} - \frac{x^2}{b^2} = 1$	Foci on y axis
Parabola	$y^2 = 2cx$ $x^2 = 2cy$	Focus at $(c/2, 0)$ Focus at $(0, c/2)$

EXERCISES FOR SEC. G.1: CONIC SECTIONS

1 Find the equation of the ellipse with foci at $(2, 0)$ and $(-2, 0)$ such that the sum of the distances from a point on the ellipse to the two foci is 10.

2 Find the equation of the ellipse with foci at $(0, 3)$ and $(0, -3)$ such that the sum of the distances from a point on the ellipse to the two foci is 14.

3 Sketch the hyperbola $x^2 - y^2 = 1$ and its foci.

4 Sketch the hyperbola $y^2 - x^2 = 1$ and its foci.

5 Sketch the parabola $y = 6x^2$, its focus, and its directrix.

6 Sketch the parabola $x = -6y^2$, its focus, and its directrix.

7 What is the equation of the parabola whose focus is at $(3, 0)$ and whose directrix is the line $x = -3$?

8 What is the equation of the parabola whose focus is at $(0, -5)$ and whose directrix is the line $y = 5$?

9 Obtain Eq. (5) from Eq. (4).

10 Obtain Eq. (8) from Eq. (7).

In Exercises 11 to 18 sketch the graphs of the given equations, the foci in the case of an ellipse or a hyperbola, and the focus and directrix in the case of a parabola.

11 $\frac{x^2}{49} + \frac{y^2}{25} = 1$ **12** $\frac{x^2}{4} + \frac{y^2}{36} = 1$

13 $\frac{x^2}{49} - \frac{y^2}{25} = 1$ **14** $\frac{y^2}{49} - \frac{x^2}{25} = 1$

15 $y^2 = 5x$ **16** $x^2 = 3y$

17 $y^2 = -5x$ **18** $x^2 = -3y$

---

19 (*a*) Using the definition of the hyperbola, show that the hyperbola that has its foci at $(\sqrt{2}, \sqrt{2})$ and $(-\sqrt{2}, -\sqrt{2})$ and for which $2a = 2\sqrt{2}$ has the equation $xy = 1$.
(*b*) Graph $xy = 1$ and show the foci.

20 How would you inscribe an elliptical garden in a rectangle whose dimensions are 8 by 10 feet? (The line through the foci is to be parallel to an edge of the rectangle.)

21 In the definition of the hyperbola it was assumed that $2a$ is less than the distance between the foci. Show that if $2a$ were greater than the distance between the foci, the hyperbola would have no points.

22 Find the equation of the parabola whose focus is $(2, 4)$ and whose directrix is the line $y = -3$.

Exercises 23 and 24 concern the determination of the position of an object with the aid of conic sections.

23 The location of a submarine can be found as follows. A small explosion is set off at point D. The sound from this explosion bounces off the submarine and is picked up by hydrophones at points A, B, and C. The time of the explosion and the times of reception of the sound at A, B, and C

are known. Show how to locate the submarine as the point where three ellipses meet.

24 The sound of the shooting of a cannon arrives 1 second later at point B than at point A.

(*a*) Show that the cannon is somewhere on a certain hyperbola whose foci are A and B.

(*b*) On which of the two pieces (branches) of the hyperbola is the cannon located?

With the aid of a third listening post the location of the cannon can be more precisely determined. LORAN, a system for long-range navigation, is based on a similar use of hyperbolas.

25 A plane intersects the surface of a right circular cylinder in a curve. Prove that this curve is an ellipse, as defined in terms of foci and sum of distances. *Hint:* Consider the two spheres inscribed in the cylinder and tangent to the plane, the spheres being on opposite sides of the plane. Let $2a$ denote the distance between the equators of the spheres perpendicular to the axis of the cylinder, and let F and F' be the points at which the spheres touch the plane.

G.2 TRANSLATION OF AXES AND THE GRAPH OF $Ax^2 + Cy^2 + Dx + Ey + F = 0$

L

Figure 1

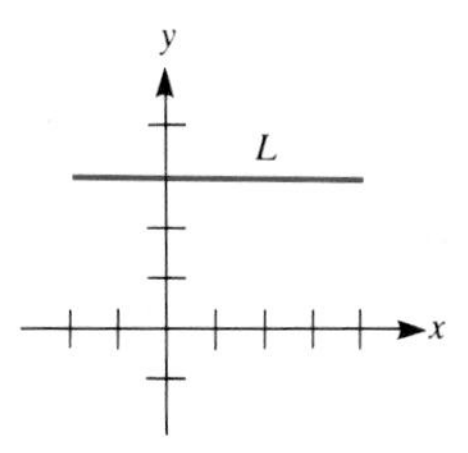

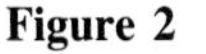

Figure 2

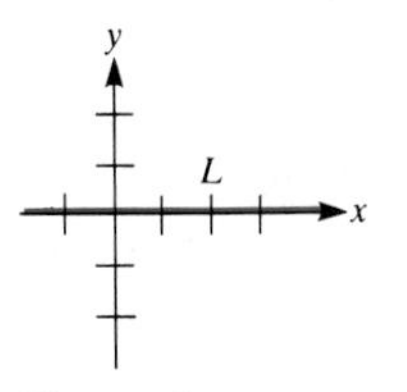

Figure 3

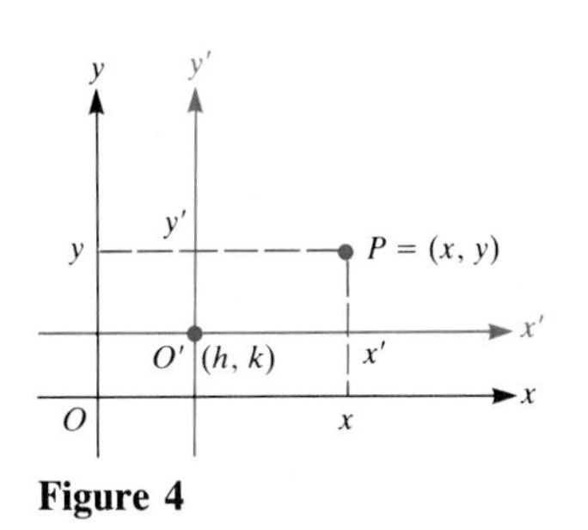

Figure 4

The equation of a particular geometric object such as a line, a circle, or a conic section depends on where we choose to place the axes. Consider, for instance, the line L in Fig. 1. Relative to the axes in Fig. 2 it has the equation $y = 3$. Relative to the axes in Fig. 3 it has the equation $y = 0$.

Clearly, a wise choice of axes may yield a simpler equation for a given line or curve. This section shows one way to choose convenient axes and uses the method to analyze equations of the form $Ax^2 + Cy^2 + Dx + Ey + F = 0$.

A point P has coordinates (x, y) relative to a given choice of axes. Another pair of axes is chosen parallel to the first pair with its origin at the point (h, k). Call the second pair of axes the $x'y'$ axes. (See Fig. 4.) Inspection of Fig. 4 shows that

$$x' = x - h \quad \text{and} \quad y' = y - k, \tag{1}$$

or, equivalently,

$$x = x' + h \quad \text{and} \quad y = y' + k. \tag{2}$$

The coordinates change by fixed amounts, h and k. This observation is applied in the following three examples.

EXAMPLE 1 Find an equation of the parabola whose focus is at $(3, 7)$ and whose directrix is the line $y = 1$, as shown in Fig. 5.

SOLUTION Introduce an $x'y'$-coordinate system whose origin is at the point $(3, 4)$, which is midway between the focus and the directrix, as shown in Fig. 6.

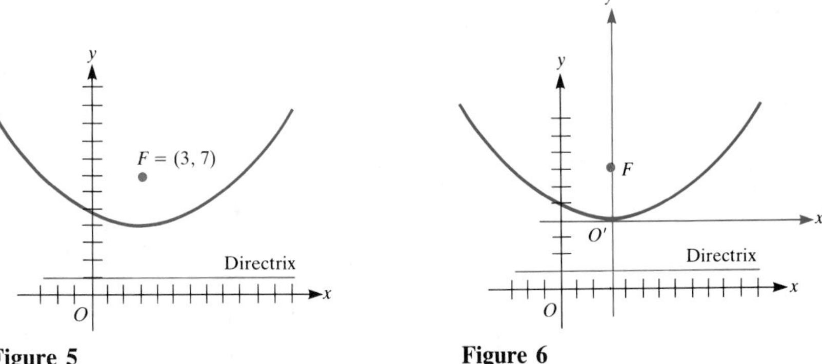

Figure 5 **Figure 6**

Relative to the $x'y'$ axes the parabola is in standard position. Since the distance from the focus to the directrix is $c = 6$, the parabola has an equation $(x')^2 = 2 \cdot 6y'$, that is, $(x')^2 = 12y'$ relative to the $x'y'$ axes. By Eq. (1),

$$(x - 3)^2 = 12(y - 4). \tag{3}$$

Equation (3) describes the parabola relative to the xy axes. It could be rewritten as

$$y = \frac{x^2 - 6x + 57}{12}. \quad \blacksquare$$

EXAMPLE 2 Graph the equation $x^2 - 2x + y^2 - 6y - 15 = 0$.

SOLUTION Complete the square in order to find axes relative to which the equation of the curve is simpler. We are looking for h and k, the coordinates of the "new" origin. The details run like this:

$$x^2 - 2x + y^2 - 6y - 15 = 0, \tag{4}$$

$$(x^2 - 2x \qquad) + (y^2 - 6y \qquad) = 15,$$

$$(x^2 - 2x + 1) + (y^2 - 6y + 9) = 15 + 1 + 9,$$

$$(x - 1)^2 + (y - 3)^2 = 25. \tag{5}$$

Equation (5) suggests that we introduce $x'y'$ axes with origin at the point $(1, 3)$. In this case, $h = 1$, $k = 3$, and $x' = x - 1$ and $y' = y - 3$. For (x', y') on the graph, x' and y' satisfy the equation

$$(x')^2 + (y')^2 = 25. \tag{6}$$

The graph of Eq. (6) is a circle of radius 5 centered at the origin of the $x'y'$-coordinate system. This is also the graph of Eq. (4). It is shown in Fig. 7. $\blacksquare$

The method of completing the square illustrated by Example 2 works equally well for any equation which has the general form $Ax^2 + Cy^2 + Dx + Ey + F = 0$. It can be shown that the graph is a conic section (perhaps empty or one or two intersecting lines) or a pair of parallel lines.

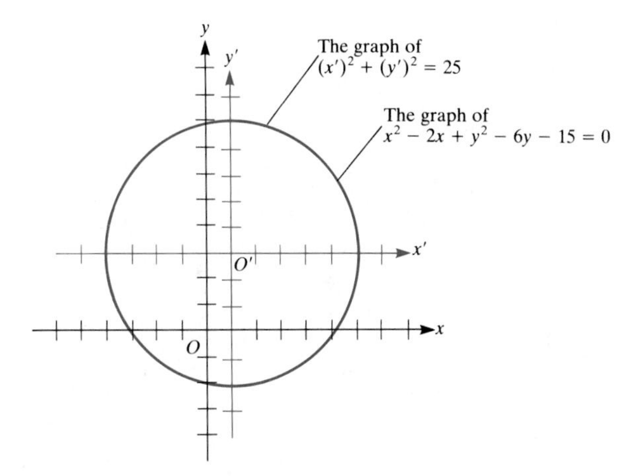

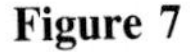
Figure 7

EXERCISES FOR SEC. G.2: TRANSLATION OF AXES AND THE GRAPH OF $Ax^2 + Cy^2 + Dx + Ey + F = 0$

Using a suitable translation of axes, graph the equations in Exercises 1 to 14 relative to the xy axes. Identify the type of conic and give the foci, vertices, and asymptotes, as appropriate.

1 $y = (x - 2)^2$
2 $y - 1 = 4(x - 1)^2$
3 $y + 1 = 2(x - 3)^2$
4 $y = x^2 + 6x + 9$
5 $y = 3x^2 + 12x + 13$
6 $y = 2x^2 - 12x + 20$
7 $x^2 + y^2 - 2x - 4y + 4 = 0$
8 $x^2 + y^2 + 6x - 7 = 0$
9 $x^2 - y^2 - 4x + 4y - 1 = 0$
10 $9x^2 - 4y^2 - 18x - 27 = 0$
11 $-4x^2 + 9y^2 + 24x + 36y - 36 = 0$
12 $4x^2 + y^2 - 16x + 12 = 0$
13 $25x^2 + 4y^2 + 100x + 24y + 36 = 0$
14 $x^2 + 4y^2 - 2x - 16y + 21 = 0$

In Exercises 15 to 22 a conic is described relative to xy axes. Choose $x'y'$ axes such that the equation of the conic is as simple as possible and give the equation in the x' and y' coordinates.

15 The ellipse that has vertices (1, 0), (4, 2), (1, 4), and (−2, 2)
16 The ellipse that has one focus at (1, 0) and vertices at (−1, 3) and (3, 3)
17 The hyperbola that has asymptotes $y = \frac{2}{3}x + \frac{1}{3}$ and $y = -\frac{2}{3}x + \frac{5}{3}$ and one focus at (3, 1)
18 The hyperbola that has vertices (1, 3) and (1, 1) and one focus at (1, 4)
19 The ellipse whose axes are parallel to the xy axes and which is inscribed in the rectangle defined by $x = 7$, $x = 1$, $y = -2$, and $y = 2$
20 The ellipse with foci (2, 1) and (8, 1) with constant sum of distances (string length) 10
21 The parabola with focus (7, 3) and directrix $x = 1$
22 The parabola with focus (5, 1) and directrix $x = 3$

---

23 This exercise concerns the graph of $Ax^2 + Cy^2 + F = 0$ for nonzero constants A, C, and F.
(*a*) Show that if A and C are positive and F is negative, then the graph is an ellipse.
(*b*) Show that if A, C, and F are positive, then the graph is empty.
(*c*) Show that if A and C have opposite signs, then the graph is a hyperbola.
(*d*) When is the graph a circle?

24 This exercise concerns the graph of the equation $Ax^2 + Cy^2 + Dx + Ey + F = 0$, where A, C, and F are nonzero constants and D and E are constants which may be 0.
(*a*) Show that if A and C have the same sign, then the graph is an ellipse or else is empty.
(*b*) Show that if A and C have opposite signs, then the graph is a hyperbola.
(*c*) When is the graph a circle?

25 Show that if A is 0 and C and D are not 0, then the graph of $Ax^2 + Cy^2 + Dx + Ey + F = 0$ is a parabola.

26 Show that the graph of $y = ax^2 + bx + c$, $a \neq 0$, is a parabola by finding the equation of the graph relative to $x'y'$ axes whose origin is at the point $(-b/2a, c - b^2/4a)$.

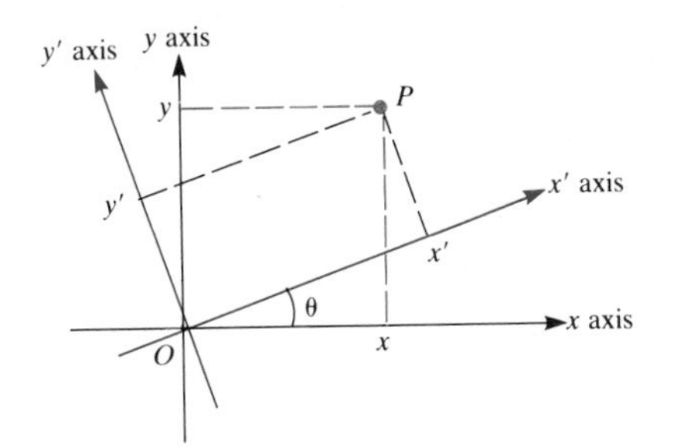

Figure 1

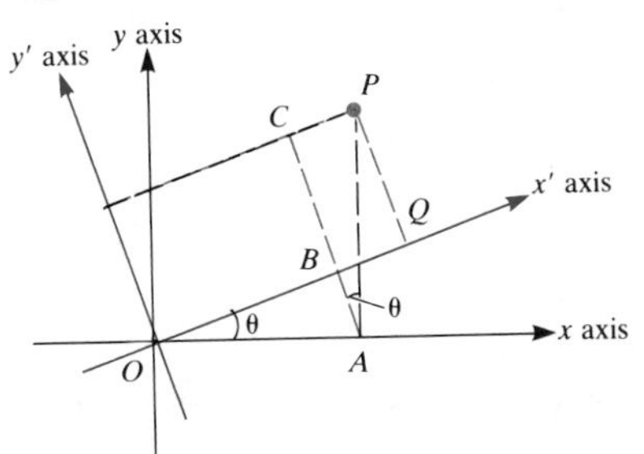

Figure 2

G.3 ROTATION OF AXES AND THE GRAPH OF $Ax^2 + Bxy + Cy^2 + Dx + Ey + F = 0$

This section examines the graph of $Ax^2 + Bxy + Cy^2 + Dx + Ey + F = 0$, which is a conic (including degenerate cases when it is a line, a pair of lines, a point, or empty). The key idea is to remove the term Bxy by choosing $x'y'$ axes that are tilted with respect to the xy axes.

A point P has coordinates (x, y) relative to a given choice of axes. Another pair of axes is chosen with the same origin but rotated by an angle θ. Call these tilted axes the $x'y'$ axes. How are x' and y', the coordinates of P in the $x'y'$ axes, related to x and y? (See Fig. 1.)

To determine the relation, introduce the line shown in Fig. 2. By Fig. 2,

$$x' = \overline{OB} + \overline{BQ} = \overline{OB} + \overline{CP} = \overline{OA}\cos\theta + \overline{AP}\sin\theta = x\cos\theta + y\sin\theta$$

and

$$y' = \overline{AC} - \overline{AB} = \overline{AP}\cos\theta - \overline{OA}\sin\theta = y\cos\theta - x\sin\theta.$$

Thus

$$\begin{cases} x' = x\cos\theta + y\sin\theta \\ y' = -x\sin\theta + y\cos\theta. \end{cases} \tag{1}$$

Just as the $x'y'$ axes are obtained from the xy axes through a rotation by the angle θ, the xy axes are obtained from the $x'y'$ axes by the rotation by the angle $-\theta$. In view of (1), then,

$$\begin{cases} x = x' \cos(-\theta) + y' \sin(-\theta) \\ y = -x' \sin(-\theta) + y' \cos(-\theta), \end{cases}$$

which reduces to

$$\begin{cases} x = x' \cos\theta - y' \sin\theta \\ y = x' \sin\theta + y' \cos\theta. \end{cases} \tag{2}$$

(These are the key equations of this section.)

EXAMPLE 1 Find an equation of the graph of $xy = 1$ relative to the $x'y'$ axes obtained by rotating the xy axes $\pi/4$ radians ($=45°$).

SOLUTION Use Eq. (2) with $\theta = \pi/4$ radians. We have

$$\begin{cases} x = x' \cos\dfrac{\pi}{4} - y' \sin\dfrac{\pi}{4} \\ y = x' \sin\dfrac{\pi}{4} + y' \cos\dfrac{\pi}{4}, \end{cases}$$

or

$$\begin{cases} x = \dfrac{\sqrt{2}}{2}(x' - y') \\ y = \dfrac{\sqrt{2}}{2}(x' + y'). \end{cases} \tag{3}$$

Substitution of Eqs. (3) into the equation $xy = 1$ yields

$$\frac{\sqrt{2}}{2}(x' - y')\frac{\sqrt{2}}{2}(x' + y') = 1,$$

$$\tfrac{1}{2}[(x')^2 - (y')^2] = 1, \qquad \text{multiplying out}$$

$$(x')^2 - (y')^2 = 2, \qquad \text{clearing denominator} \tag{4}$$

which is the equation of a hyperbola with foci on the x' axis.

To graph Eq. (4), rewrite it as

$$\frac{(x')^2}{(\sqrt{2})^2} - \frac{(y')^2}{(\sqrt{2})^2} = 1. \tag{5}$$

This is the hyperbola $(x')^2/a^2 - (y')^2/b^2 = 1$ with $a = \sqrt{2}$ and $b = \sqrt{2}$. Its foci are at $(x', y') = (c, 0)$ and $(-c, 0)$, where $c = \sqrt{a^2 + b^2} = \sqrt{2 + 2} = 2$. The hyperbola is sketched in Fig. 3. In the xy system, the foci are at $(\sqrt{2}, \sqrt{2})$ and $(-\sqrt{2}, -\sqrt{2})$. ■

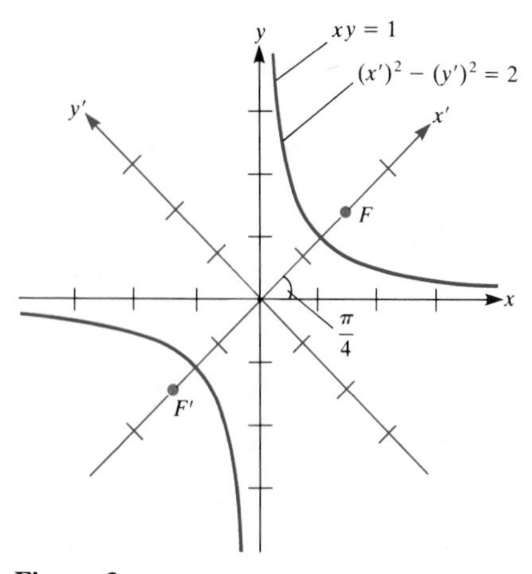

Figure 3

In Example 1 rotation of the axes by $\pi/4$ radians resulted in an equation with no $x'y'$ term. [In other words, the coefficient of the product $x'y'$ is 0 in (4).] Less precisely, we "got rid of the xy term." It turns out that for any equation of the form $Ax^2 + Bxy + Cy^2 + Dx + Ey + F = 0$, it is always possible to rotate the axes in such a way that the $x'y'$ term disappears.

We sketch the algebra, leaving the details to be filled in by the reader. Start with the equation

$$Ax^2 + Bxy + Cy^2 + Dx + Ey + F = 0, \tag{6}$$

and make the substitution (2). After some multiplications and collections, we obtain an equation in x' and y' of the form

$$A'(x')^2 + B'x'y' + C'(y')^2 + D'x' + E'y' + F' = 0 \tag{7}$$

for certain constants A', B', C', D', E', and F'.

At this point only the coefficient of $x'y'$, namely, B', draws our interest, since we want to find θ that makes B' equal to 0. The formula for B' is

$$B' = 2(C - A) \sin \theta \cos \theta + B(\cos^2 \theta - \sin^2 \theta), \tag{8}$$

which simplifies to

$$B' = (C - A) \sin 2\theta + B \cos 2\theta.$$

To make $B' = 0$, find θ such that

$$(C - A) \sin 2\theta + B \cos 2\theta = 0.$$

Thus

$$\frac{\sin 2\theta}{\cos 2\theta} = \frac{B}{A - C}$$

or

$$\tan 2\theta = \frac{B}{A - C}. \tag{9}$$

For θ in $(0, \pi/4)$, $\tan 2\theta$ sweeps through all positive numbers. For θ in $(\pi/4, \pi/2)$, $\tan 2\theta$ sweeps through all negative numbers. If $C = A$, use $\theta = \pi/4$; by Eq. (8), B' then is 0. Thus it is always possible to find a first-quadrant angle θ such that Eq. (9) holds. For that θ, B' is 0 and Eq. (7) takes the form

$$A'(x')^2 + C'(y')^2 + D'x' + E'y' + F' = 0. \tag{10}$$

This type of equation was discussed in the preceding section. Its graph is a conic section except in certain degenerate cases. Thus the same holds for the graph of Eq. (6).

Getting Rid of the *xy* Term We have seen that there is an angle θ such that, relative to $x'y'$ axes inclined at an angle θ to the xy axes, Eq. (7) takes the form of Eq. (10). First, find θ such that $\tan 2\theta = B/(A - C)$. Then, find $\cos \theta$ and $\sin \theta$ and use the substitution (2). The steps are

1 Find $\tan 2\theta$.
2 Find $\cos 2\theta$.
3 Find $\sin \theta = \sqrt{\frac{1 - \cos 2\theta}{2}}$ and $\cos \theta = \sqrt{\frac{1 + \cos 2\theta}{2}}$.

Since θ is in the first quadrant, the positive square roots are used.

Example 2 illustrates the technique.

EXAMPLE 2 Choose $x'y'$ axes so that the equation for the graph of

$$-7x^2 + 48xy + 7y^2 - 25 = 0 \tag{11}$$

has no $x'y'$ term. Then graph the curve.

SOLUTION In this case $A = -7$, $B = 48$, $C = 7$. Hence θ must be chosen so that

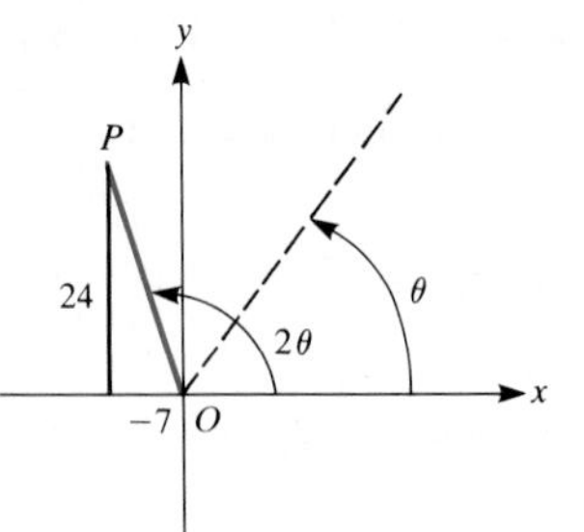

Figure 4

$$\tan 2\theta = \frac{48}{(-7)-7} = \frac{48}{-14} = \frac{24}{-7}. \tag{12}$$

Figure 4 shows 2θ and θ. By the pythagorean theorem, $\overline{OP}$ in Fig. 4 is $\sqrt{(24)^2+(-7)^2} = 25$. Thus $\cos 2\theta = (-7)/25$. Consequently,

$$\sin\theta = \sqrt{\frac{1-\left(\frac{-7}{25}\right)}{2}} = \frac{4}{5},$$

and

$$\cos\theta = \sqrt{\frac{1+\left(\frac{-7}{25}\right)}{2}} = \frac{3}{5}.$$

Thus (2) becomes

$$\begin{cases} x = \frac{3}{5}x' - \frac{4}{5}y' \\ y = \frac{4}{5}x' + \frac{3}{5}y'. \end{cases} \tag{13}$$

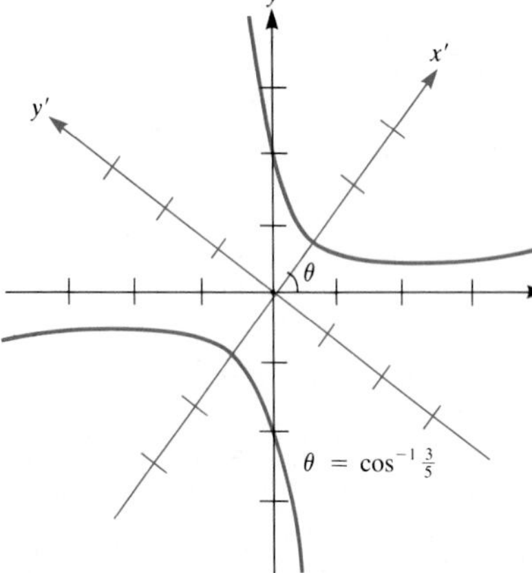

Figure 5

Substitution of (13) into Eq. (11) gives

$$-7(\tfrac{3}{5}x' - \tfrac{4}{5}y')^2 + 48(\tfrac{3}{5}x' - \tfrac{4}{5}y')(\tfrac{4}{5}x' + \tfrac{3}{5}y') + 7(\tfrac{4}{5}x' + \tfrac{3}{5}y')^2 - 25 = 0,$$

which simplifies to

$$625(x')^2 - 625(y')^2 = 625,$$

or

$$(x')^2 - (y')^2 = 1,$$

which describes a hyperbola in standard position relative to the $x'y'$ axes. Its foci are on the x' axis. It is sketched in Fig. 5. ■

Quick way to identify a curve: neph

The reasoning in this section and the one preceding shows that if the graph of (6) is not degenerate, it is a conic. *To determine what type of conic it is, compute the discriminant,* $\mathcal{D} = B^2 - 4AC$. If it is negative, the conic is an ellipse; if it is positive, the conic is a hyperbola; if it is zero, the conic is a parabola. Remember this with the nonsense word ''neph'' (*n*egative, *e*llipse, *p*ositive, *h*yperbola). Exercises 18 to 20 explain why this test works. For emphasis, we state this test formally: To determine what type of conic the equation

$$Ax^2 + Bxy + Cy^2 + Dx + Ey + F = 0$$

There are also degenerate cases.

describes, compute $\mathcal{D} = B^2 - 4AC$. When $\mathcal{D} < 0$, the conic is an ellipse; when $\mathcal{D} > 0$, the conic is a hyperbola. If $\mathcal{D} = 0$, the conic is a parabola.

EXERCISES FOR SEC. G.3: ROTATION OF AXES AND THE GRAPH OF $Ax^2 + Bxy + Cy^2 + Dx + Ey + F = 0$

Graph each of the curves in Exercises 1 to 8 using the technique described in this section. In the case of hyperbolas, draw the asymptotes too.

1 $x^2 - 4xy - 2y^2 - 6 = 0$
2 $41x^2 + 24xy + 34y^2 - 25 = 0$
3 $x^2 + xy + y^2 - 12 = 0$

4 $5x^2 + 6xy + 5y^2 - 8 = 0$
5 $23x^2 + 26\sqrt{3}\,xy - 3y^2 - 144 = 0$
6 $3x^2 + 2\sqrt{3}xy + y^2 + 2x + 2\sqrt{3}y = 0$
7 $6x^2 - 12xy + 6y^2 - \sqrt{2}x + \sqrt{2}y = 0$
8 $7x^2 - 48xy - 7y^2 - 25 = 0$

In Exercises 9 to 12 graph the equations.

9 $y = \dfrac{x^2 + 3}{x - 1}$ 10 $y = \dfrac{x + 1}{x + 3}$

11 $xy + x + y + 1 = 0$

12 $y = \dfrac{x^2 + 2x + 1}{x - 1}$

In each of Exercises 13 to 16 use the discriminant $\mathscr{D} = B^2 - 4AC$ to determine what type of conic the equation describes.

13 $-x^2 + 24xy + 6y^2 - 10x + 13y + 5 = 0$
14 $x^2 + xy + y^2 + 3x + 2y = 0$
15 $x^2 - 2xy + y^2 + x + 3y + 5 = 0$
16 $3x^2 - xy - y^2 = 1$
17 For each of the following equations the graph is degenerate; it is either one or two lines, a point, or empty. Graph each of them.
(*a*) $x^2 - y^2 = 0$ (*b*) $x^2 + 2xy + y^2 = 0$
(*c*) $3x^2 + 4y^2 = 0$
(*d*) $3x^2 + 2xy + 3y^2 + 1 = 0$

Exercises 18 to 20 are related.

18 Show that A', B', and C' in (7) are given by

$$A' = A\cos^2\theta + B\cos\theta\sin\theta + C\sin^2\theta,$$
$$B' = 2(C - A)\sin\theta\cos\theta + B(\cos^2\theta - \sin^2\theta),$$
$$C' = A\sin^2\theta - B\cos\theta\sin\theta + C\cos^2\theta.$$

19 Use Exercise 18 to show that $(B')^2 - 4A'C' = B^2 - 4AC$.
20 Assume that $x'y'$ axes have been chosen to make $B' = 0$. Thus the graph of (6) is given by the equation $A'(x')^2 + C'(y')^2 + D'x' + E'y' + F' = 0$. Assume this graph is a conic (that is, is not degenerate).
(*a*) Show that if $(B')^2 - 4A'C'$ is negative, the graph is an ellipse. (Keep in mind that $B' = 0$.)
(*b*) Show that if $(B')^2 - 4A'C'$ is positive, the graph is a hyperbola.
(*c*) Show that if $(B')^2 - 4A'C'$ is 0, the graph is a parabola.
(*d*) Combining these facts with Exercise 19, explain the "neph" memory device.
21 Show that if the graph of $Ax^2 + Bxy + Cy^2 + Dx + Ey + F = 0$ is a circle, then $B = 0$ and $A = C$.
22 Sketch the graphs of
(*a*) $(3x - y + 1)^2 = 0$
(*b*) $(x + y + 1)(x - y - 2) = 0$
(*c*) $x^2 - y^2 = 0$
[Note that (*a*) and (*b*), when multiplied out, produce equations of the form $Ax^2 + Bxy + Cy^2 + Dx + Ey + F = 0$. *Hint:* When is the product of two numbers equal to 0?]

In Exercises 23 to 26 give the coordinates of the foci relative to the $x'y'$ axes and their relation to the xy axes for the given conic.

23 The conic of Example 2
24 The conic of Exercise 1
25 The conic of Exercise 3
26 The conic of Exercise 6

G.4 CONIC SECTIONS IN POLAR COORDINATES

This section depends on Sec. 9.1 and is used in Sec. 13.5.

For the study of the conic sections in terms of polar coordinates, it is convenient to use definitions that depend on the ratios of distances rather than on their sums or differences. (Note that the definition of the parabola involves essentially the ratio of two distances being equal to 1.)

Consider the ellipse whose foci are at $F = (c, 0)$ and $F' = (-c, 0)$ and whose "length of string" is $2a$. In the algebraic treatment of the ellipse in Sec. G.1, the equation

$$a^2 - cx = a\sqrt{(x - c)^2 + y^2}$$

appears.

Some algebraic manipulations of this equation will show that this ellipse can be defined in terms of the focus F, a line, and a fixed ratio of the distances from P to F and from P to the line.

First, observe that this equation asserts that, for $P = (x, y)$,

$$a^2 - cx = a\overline{PF},$$

or, equivalently,

$$\overline{PF} = a - \frac{c}{a}x.$$

Denote the quotient c/a, which is less than 1, by e. The number e is called the **eccentricity** of the ellipse. (When $e = 0$, the ellipse is a circle.) Thus

$$\overline{PF} = a - ex = e\left(\frac{a}{e} - x\right),$$

an equation which is meaningful if the ellipse is not a circle. Now, $(a/e) - x$ is the distance from P to the vertical line through $(a/e, 0)$, a line which will be denoted L. Letting $Q = (a/e, y)$, the point on L and on the horizontal line through P, we have

$$\overline{PF} = e\overline{PQ}.$$

(See Fig. 1.)

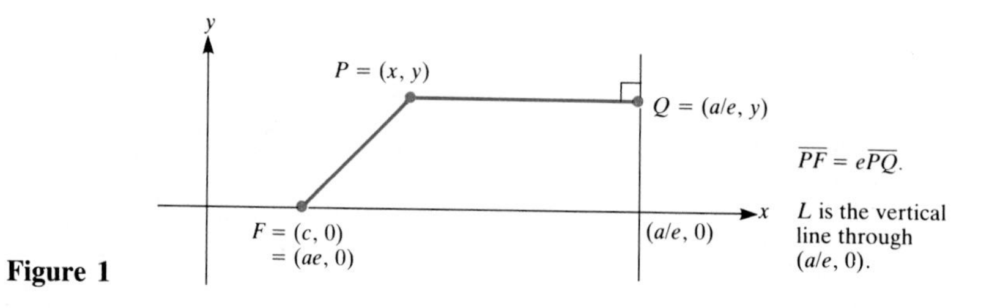

Figure 1

In other words, *the ratio $\overline{PF}/\overline{PQ}$ has a constant value*, e, less than 1. Thus the ellipse, like the parabola, can be defined in terms of a point F and a line L.

EXAMPLE 1 Find the eccentricity and draw the line L for the ellipse $\dfrac{x^2}{25} + \dfrac{y^2}{9} = 1$.

SOLUTION In this ellipse, $a = 5$ and $b = 3$. Thus $c = \sqrt{a^2 - b^2} = \sqrt{25 - 9} = 4$. Consequently, $e = c/a = \frac{4}{5}$. The line L has the equation $x = a/e$, or $x = 5/\frac{4}{5}$; hence

$$x = \tfrac{25}{4} = 6.25.$$

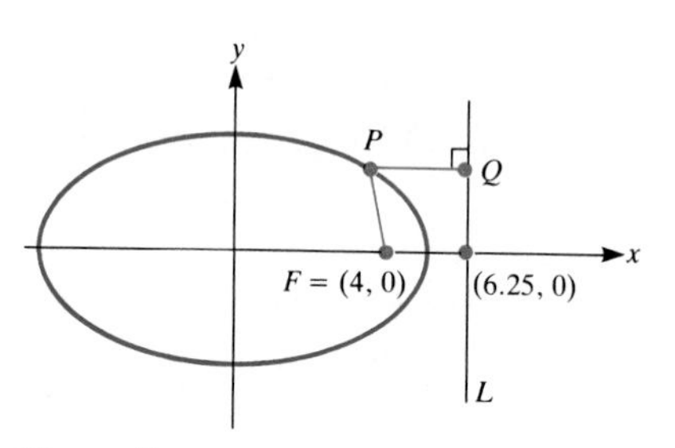

Figure 2

The ellipse is sketched in Fig. 2. Note that for each point P on the ellipse,

$$\frac{\overline{PF}}{\overline{PQ}} = \frac{4}{5}. \quad ■$$

The hyperbola can be treated in a similar manner. The main difference is that the eccentricity of a hyperbola, again defined as c/a, is greater than 1. With this background, we now describe the approach to the conic sections in terms of the ratios of certain distances.

Definition *Conic section.* Let L be a line in the plane and let F be a point in the plane but not on the line. Let e be a positive number. A point P in the plane is on the **conic section** determined by F, L, and e if and only if

$$\frac{\text{Distance from } P \text{ to } F}{\text{Distance from } P \text{ to } L} = e.$$

When $e = 1$, the conic section is a parabola (this is the definition of the parabola used in the preceding section). When $e < 1$, it is an ellipse. When $e > 1$, it is a hyperbola. The point F is called a **focus**; the line L is called the **directrix**.

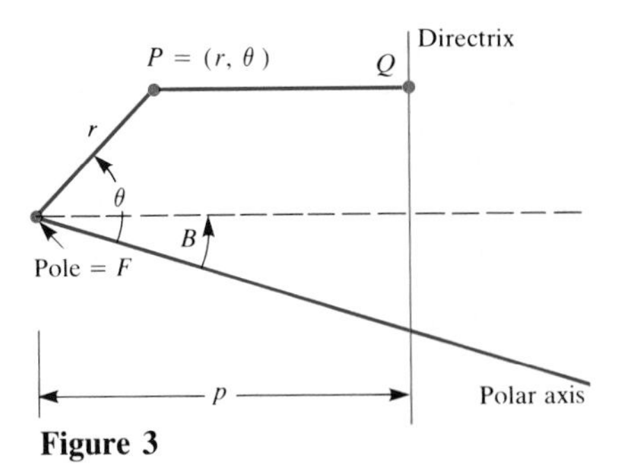

Figure 3

To obtain the simplest description of the conic sections in polar coordinates, place the pole at the focus F. Let the polar axis make an angle B with a line perpendicular to the directrix. Figure 3 shows a typical point $P = (r, \theta)$ on the conic section, as well as the point Q on the directrix nearest P. Let the distance from F to the directrix be p. Then $\overline{PF} = r$, $\overline{PQ} = p - r\cos(\theta - B)$, and $\overline{PF}/\overline{PQ} = e$. Thus

$$\frac{r}{p - r\cos(\theta - B)} = e. \tag{1}$$

Solving Eq. (1) for r yields *the equation of a conic section in polar coordinates,*

$$r = \frac{ep}{1 + e\cos(\theta - B)}. \tag{2}$$

Usually B is chosen to be 0 or π.

EXAMPLE 2 Show that the graph of the equation $r = \dfrac{8}{5 + 6\cos\theta}$ is a conic section.

SOLUTION This can be put in the form of Eq. (2) by dividing both numerator and denominator by 5:

$$r = \frac{\frac{8}{5}}{1 + \frac{6}{5}\cos\theta} = \frac{(\frac{6}{5})(\frac{8}{6})}{1 + \frac{6}{5}\cos\theta}.$$

Hence the graph is a conic section for which $p = \frac{8}{6}$ and $e = \frac{6}{5}$. It is a hyperbola, since $e > 1$. ■

EXERCISES FOR SEC. G.4: CONIC SECTIONS IN POLAR COORDINATES

1 (*a*) On the graph of $r = 10/(3 + 2\cos\theta)$ sketch the four points corresponding to $\theta = 0, \pi/2, \pi, 3\pi/2$.
(*b*) Using Eq. (2), show that the curve in (*a*) is an ellipse.

2 Obtain Eq. (2) from Eq. (1).

3 Find the eccentricities of these conics:
(*a*) $r = 5/(3 + 4\cos\theta)$ (*b*) $r = 5/(4 + 3\cos\theta)$
(*c*) $r = 5/(3 + 3\cos\theta)$ (*d*) $r = 5/(3 - 4\cos\theta)$

4 (*a*) Show that $r = 8/(1 - \frac{1}{2}\cos\theta)$ is the equation of an ellipse. *Hint:* Set $B = \pi$ in Eq. (2).
(*b*) Graph the ellipse and its foci.
(*c*) Find a, where $2a$ is the fixed sum of the distances from points on the ellipse to the foci.

5 In rectangular coordinates the focus of a certain parabola is at $(-1, 0)$ and its directrix is the line $x = 1$.
(*a*) Show that its equation is $y^2 = -4x$.
(*b*) Find the equation of the parabola relative to a polar coordinate system whose pole is at F and whose polar axis contains the positive x axis.
(*c*) Find the equation of the parabola relative to a polar coordinate system whose pole is at the origin of the rectangular coordinate system and whose polar axis coincides with the positive x axis.

APPENDIX H

LOGARITHMS AND EXPONENTIALS DEFINED THROUGH CALCULUS

This appendix provides a definition of logarithms and exponentials completely different from the definitions met in high school algebra. The main assumption is that $\int_1^x dt/t$ exists. In contrast to the algebraic approach, logarithms are defined first and exponentials after.

H.1 THE NATURAL LOGARITHM DEFINED AS A DEFINITE INTEGRAL

Recall how the logarithm and exponential functions were defined. First, the exponential function b^x $(b > 0)$ was built up in stages:

$$b^n = b \cdot b \cdot \cdots \cdot b \ (n \text{ times})$$

for $n = 1, 2, 3, \ldots$; $b^0 = 1$; $b^{-n} = 1/b^n$ for $n = 1, 2, 3, \ldots$; $b^{1/n} = \sqrt[n]{b}$, the positive nth root of b for $n = 1, 2, 3, \ldots$; $b^{m/n} = (b^{1/n})^m$ for m an integer and n a positive integer; and finally, for irrational x, b^x equals the limit of $b^{m/n}$ as $m/n \to x$.

A thorough treatment of the exponential functions based on this approach encounters many difficulties, such as: How do we know that b has an nth root? If $b > 1$, is b^x increasing? Does $\lim_{m/n \to x} b^{m/n}$ exist? After answering these questions, we would still be left with showing that b^x is continuous and that $b^x b^y = b^{x+y}$.

Assuming that b^x has these desired properties, we then defined $\log_b x$. Then, to obtain the derivative of $\log_b x$ we had to assume that

$$\lim_{h \to 0} (1 + h)^{1/h}$$

exists and that $\log_b x$ is a continuous function.

The present section takes a completely different approach. It first defines a function $L(x)$ with the aid of the definite integral. [It will turn out that $L(x) = \ln x$.] In Sec. H.2 the exponential function will be defined as the inverse of the function $L(x)$ and it will be shown that $\lim_{h \to 0} (1 + h)^{1/h}$ exists.

Basic to the following argument is the theorem (see Sec. 5.3) that a continuous function has a definite integral over each interval $[a, b]$ in its domain.

Definition *The function* $L(x)$. Define L by setting

$$L(x) = \int_1^x \frac{1}{t}\, dt, \qquad x > 0.$$

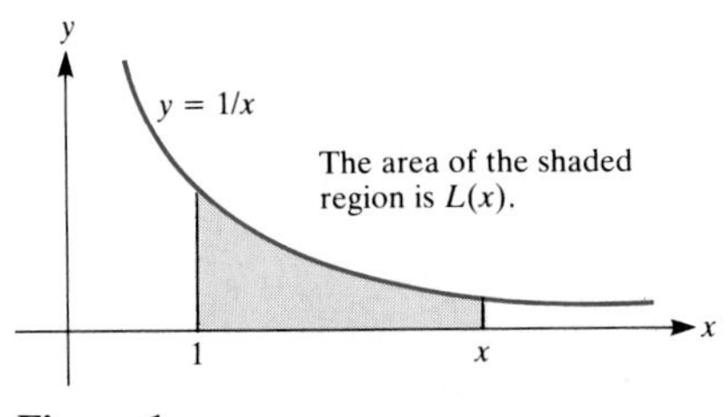

Figure 1

Observe that $L(1) = \int_1^1 (1/t)\, dt = 0$; if $x > 1$, $L(x) > 0$ and is the area of the shaded region in Fig. 1. If $0 < x < 1$, then $L(x) = \int_1^x (1/t)\, dt = -\int_x^1 (1/t)\, dt$, the negative of the area of the shaded region in Fig. 2. Thus, if $0 < x < 1$, we have $L(x) < 0$. $L(x)$ resembles a logarithm function also in that $L(x)$ is defined only for $x > 0$.

By the first fundamental theorem of calculus,

$$L'(x) = \frac{1}{x}. \tag{1}$$

The information that $L(1) = 0$ and that $L'(x) = 1/x$ already shows that $L(x) = \ln x$. After all, since $L(x)$ and $\ln x$ have the same derivative, they differ by a constant,

$$L(x) = \ln x + C. \tag{2}$$

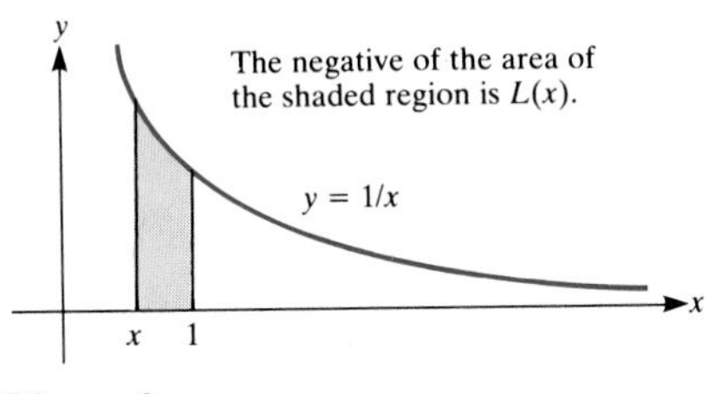

Figure 2

To find the constant C, set $x = 1$ in Eq. (2), obtaining

$$L(1) = \ln (1) + C.$$

Since both $L(1)$ and $\ln (1)$ equal 0, it follows that $C = 0$. Thus $L(x) = \ln x$.

It is reassuring to know that $L(x)$, though defined so differently, is $\ln x$, defined in Sec. 6.3. However, in this section, *no use will be made of this knowledge,* since the present goal is to build up the theory of logarithms from scratch, with minimal assumptions.

Incidentally, the fact that the area under the curve $y = 1/x$ behaves like a logarithm was first noted by Gregory St. Vincent and his friend A. A. de Sarasa in 1647, some twenty years before the work of Newton and Leibniz.

We now show that L behaves just like a logarithm.

Property 1 $L(xy) = L(x) + L(y)$, for $x, y > 0$.

Proof Hold y fixed at the value k. It will be shown that

$$L(xk) = L(x) + L(k), \qquad x > 0.$$

Introduce two functions of x,

$$f(x) = L(xk) \quad \text{and} \quad g(x) = L(x) + L(k).$$

We wish to show that $f(x) = g(x)$.

To begin, compute their derivatives:

$$\begin{aligned} f'(x) &= [L(xk)]' \\ &= \frac{1}{xk}(xk)' \qquad \text{by Eq. (1) and the chain rule} \end{aligned}$$

$$= \frac{k}{xk} = \frac{1}{x},$$

and $$g'(x) = [L(x) + L(k)]'$$

$$= L'(x) = \frac{1}{x}. \qquad k \text{ is a constant}$$

Since $f(x)$ and $g(x)$ have the same derivative, they differ by a constant, that is,

$$L(xk) = L(x) + L(k) + C. \qquad C \text{ constant} \tag{3}$$

To find C, set $x = 1$ in Eq. (3), obtaining

$$L(k) = L(1) + L(k) + C.$$

Since $L(1) = 0$, this equation shows that $C = 0$, and Property 1 holds. ■

We assume x^n is defined for n an integer.

Property 2 For any integer n, $L(x^n) = nL(x)$, $x > 0$.

Proof Differentiate both $L(x^n)$ and $nL(x)$:

$$[L(x^n)]' = \frac{1}{x^n} nx^{n-1} \qquad \text{by Eq. (1) and the chain rule}$$

$$= \frac{n}{x}$$

and $$[nL(x)]' = \frac{n}{x}. \qquad \text{by Eq. (1)}$$

Thus there is a constant C such that

$$L(x^n) = nL(x) + C. \tag{4}$$

Setting $x = 1$ in Eq. (4) shows that

$$L(1) = nL(1) + C,$$

or $$0 = 0 + C.$$

Hence $C = 0$ and Property 2 is established. ■

Property 2, with $n = -1$, implies that

$$L(x^{-1}) = (-1)L(x),$$

or simply, $$L\left(\frac{1}{x}\right) = -L(x),$$

an equation that should not come as a surprise.

Property 3 $\lim_{x\to\infty} L(x) = \infty$ and $\lim_{x\to 0^+} L(x) = -\infty$.

Proof Since $L'(x)$ is positive, $L(x)$ is an increasing function. Moreover, $L(2) > 0$ and $L(2^n) = nL(2)$. Thus, $L(2^n)$ gets arbitrarily large as $n \to \infty$, so

$$\lim_{x\to\infty} L(x) = \infty.$$

To show that

$$\lim_{x\to 0^+} L(x) = -\infty,$$

replace x by $1/t$, where $t \to \infty$. Then we have

$$\begin{aligned}\lim_{x\to 0^+} L(x) &= \lim_{t\to\infty} L\left(\frac{1}{t}\right) \\ &= \lim_{t\to\infty} [-L(t)] \qquad \text{by Property 2} \\ &= -\lim_{t\to\infty} L(t) = -\infty. \quad \blacksquare\end{aligned}$$

Thus the range of L consists of all real numbers.

Since $L(1) = 0$ and $L(x)$ is an increasing continuous function that takes on arbitrarily large values, there must exist a unique number x such that $L(x) = 1$. (The intermediate-value theorem guarantees that there is at least one x. Why is there only one?)

Definition e is the unique solution to the equation $L(x) = 1$, that is, $L(e) = 1$.

In Sec. H.2, after the exponential functions are defined, it will be shown that $e = \lim_{h\to 0} (1 + h)^{1/h}$.

EXERCISES FOR SEC. H.1: THE NATURAL LOGARITHM DEFINED AS A DEFINITE INTEGRAL

1 Show that L is continuous.

2 Show that the graph of $L(x)$ is concave downward.

3 (*a*) Compute the area of the eight rectangles of the same width in Fig. 3.

(*b*) From (*a*) deduce that $L(3) > 1$.

(*c*) From (*b*) deduce that $e < 3$.

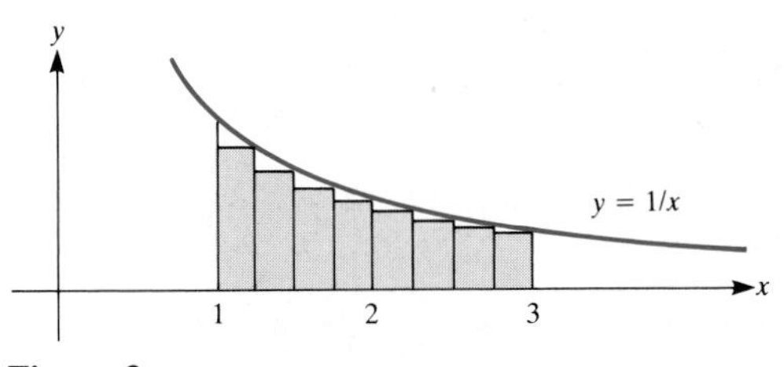

Figure 3

4 Using an argument like that in Exercise 3, show that $e > 2$.

5 Let n be an integer larger than 1. By considering appropriate rectangles of width 1, show that

$$L(n) > \frac{1}{2} + \frac{1}{3} + \frac{1}{4} + \cdots + \frac{1}{n}.$$

Exercises 6 and 7 outline a way to compute $L(2)$ with any degree of accuracy desired.

6 (*a*) By using a partition of $[1, 2]$ of n sections of equal length and left endpoints as sampling points, show that

$$L(2) < \frac{1}{n} + \frac{1}{n+1} + \cdots + \frac{1}{2n-1}.$$

(*b*) Using a "large" value of n, find a number greater than $L(2)$.

7 (*a*) Show that for any positive integer n,

$$L(2) > \frac{1}{n+1} + \frac{1}{n+2} + \cdots + \frac{1}{2n}.$$

Suggestion: Form a suitable partition of [1, 2] and use right endpoints as sampling points.

(*b*) Using a "large" value of n, find a number less than $L(2)$.

---

8 This exercise shows that $L(ab) = L(a) + L(b)$ without using the fundamental theorem of calculus.

(*a*) Prove that for $b > 1$, we have $\int_a^{ab} (1/x)\,dx = \int_1^b (1/x)\,dx$ by observing that if the sum

$$\Sigma_{i=1}^n (1/c_i)(x_i - x_{i-1})$$

is an approximation of the integral $\int_1^b (1/x)\,dx$, then the sum $\Sigma_{i=1}^n (1/(ac_i))(ax_i - ax_{i-1})$ is an approximation of $\int_a^{ab} (1/x)\,dx$.

(*b*) If a, $b > 1$, deduce that $\int_1^{ab} (1/x)\,dx - \int_1^a (1/x)\,dx = \int_1^b (1/x)\,dx$.

(*c*) From (*b*), show that if a, $b > 1$, then $L(ab) = L(a) + L(b)$.

H.2 EXPONENTIAL FUNCTIONS DEFINED IN TERMS OF LOGARITHMS

In Sec. H.1 the function $L(x) = \int_1^x dt/t$, $x > 0$, was introduced and shown to have the familiar properties of logarithms. It was pointed out that if all the gaps in the development of $\ln x$ were filled in, then $L(x)$ could be shown to coincide with $\ln x$. The number e was defined in Sec. H.1 by the condition $L(e) = 1$, that is, $\int_1^e dx/x = 1$.

The present section introduces a function E, which will turn out to be the exponential function to the base e. However, it is important in this section to make no use at all of b^x, as defined in Appendix D, since the object of this section is to construct these functions with the fewest assumptions.

Recall that the function L is increasing for $x > 0$. It thus has an inverse.

Definition *The function E.* The inverse of the function L will be denoted E.

Since L has domain $(0, \infty)$ and range $(-\infty, \infty)$, E has domain $(-\infty, \infty)$ and range $(0, \infty)$. Note that E is a one-to-one function.

Since $L(1) = 0$ and E is the inverse of L, $E(0) = 1$, just like an exponential function. Moreover, since $L(e) = 1$, it follows that $E(1) = e$. Keep in mind that $E(L(x)) = x$ for $x > 0$ and $L(E(x)) = x$ for all x on the x axis. Theorem 1 shows that the function E has the basic property of an exponential function.

Theorem 1 $E(x + y) = E(x)E(y)$.

Proof Write x as $L(u)$ and y as $L(v)$. This is possible because the range of L is $(-\infty, \infty)$. Then

$$\begin{aligned} E(x + y) &= E(L(u) + L(v)) \\ &= E(L(uv)) && L(uv) = L(u) + L(v) \\ &= uv && E \text{ and } L \text{ are inverses} \\ &= E(x)E(y). \quad \blacksquare \end{aligned}$$

Next, the exponential function b^x will be defined for $b > 0$. In the traditional approach, there is the identity

$$b^x = e^{x \ln b}.$$

This suggests how to define the function b^x in the present approach.

Definition *The exponential function b^x.* If $b > 0$, define b^x to be

$$E(xL(b)).$$

Theorem 2 If b is positive, then

$$b^0 = 1, \quad (1)$$

$$b^1 = b, \quad (2)$$

$$b^{x+y} = b^x b^y. \quad (3)$$

Proof

(1) $$b^0 = E(0 \cdot L(b)) = E(0) = 1.$$

(2) $$b^1 = E(1 \cdot L(b)) = E(L(b)) = b.$$

(3)

$b^{x+y} = E((x + y)L(b))$	definition of b^{x+y}
$= E(xL(b) + yL(b))$	algebra
$= E(xL(b))E(yL(b))$	Theorem 1
$= b^x b^y.$	definition of b^x and b^y ■

According to the definition of b^x just given, when the base b is chosen to be e, we have

$$e^x = E(xL(e)) = E(x \cdot 1) = E(x).$$

So e^x is another name for $E(x)$.

Theorem 3 $L(b^x) = xL(b)$.

Proof

$b^x = E(xL(b))$	definition of b^x
$L(b^x) = L(E(xL(b)))$	
$= xL(b).$	L and E are inverses of each other. ■

Theorem 4 $(b^x)^y = b^{xy}$.

Proof Since L is one-to-one, it suffices to show that $L(b^{xy}) = L((b^x)^y)$. This will be established by three applications of Theorem 3:

$$L(b^{xy}) = xyL(b), \qquad \text{Theorem 3}$$

while

$$\begin{aligned} L((b^x)^y) &= yL(b^x) && \text{Theorem 3} \\ &= y(xL(b)) && \text{Theorem 3} \\ &= xyL(b). && \text{algebra} \end{aligned}$$

Thus $L(b^{xy}) = L((b^x)^y)$, and the proof is complete. ■

If $b \neq 1$, the function b^x is one-to-one. To show this, assume that

$$b^{x_1} = b^{x_2}, \tag{4}$$

and show that $x_1 = x_2$. By the definition of b^x, Eq. (4) can be written

$$E(x_1L(b)) = E(x_2L(b)).$$

Because E is one-to-one, $x_1L(b) = x_2L(b)$. Since $b \neq 1$, $L(b) \neq 0$, and cancellation of $L(b)$ yields $x_1 = x_2$.

Definition *The function* $\log_b x$. If $b > 0$, define $\log_b x$ to be

$$\frac{L(x)}{L(b)}.$$

From the properties of $L(x)$, it can be shown that $\log_b x$ is the inverse of the function b^x. Theorem 5 shows that $\log_b x$ has the traditional properties of a logarithm.

Theorem 5 If $b > 0$, $b \neq 1$, then

$$\log_b 1 = 0, \tag{5}$$

$$\log_b b = 1, \tag{6}$$

$$\log_b xy = \log_b x + \log_b y \qquad x, y > 0, \tag{7}$$

$$\log_b x^y = y \log_b x. \qquad x > 0 \tag{8}$$

The proof is left to the reader.

Since E is the inverse of the increasing differentiable function L, it is also differentiable. To find its derivative, let

$$y = E(x). \tag{9}$$

Hence

$$x = L(y). \tag{10}$$

Implicit differentiation of Eq. (10) with respect to x yields

$$1 = \frac{1}{y}\frac{dy}{dx}.$$

Thus

$$\frac{dy}{dx} = y.$$

In short,
$$\frac{d}{dx}(E(x)) = E(x). \tag{11}$$

This proves the next theorem.

Theorem 6 The derivative of E is E:
$$\frac{d}{dx}(E(x)) = E(x).$$

With the help of Theorem 6 it is not hard to prove the following generalization.

Theorem 7 $\frac{d}{dx}(b^x) = b^x L(b).$

The proofs of Theorem 7 and of the following are left to the reader.

Theorem 8 $\frac{d}{dx}(\log_b x) = \frac{\log_b e}{x}.$

It is now possible to show that
$$\lim_{x \to 0} (1 + x)^{1/x} = e.$$

[Keep in mind that e is defined in Sec. H.1 by the demand that $L(e) = 1$, that is, $\int_1^e dt/t = 1$.]

Theorem 9 $\lim_{x \to 0} (1 + x)^{1/x} = e.$

Proof The derivative of $L(x)$ at $x = 1$ is $\frac{1}{1}$, or 1. On the other hand, by the definition of the derivative,

$$\begin{aligned} L'(1) &= \lim_{x \to 0} \frac{L(1 + x) - L(1)}{x} \\ &= \lim_{x \to 0} \frac{L(1 + x)}{x} \qquad \text{since } L(1) = 0 \\ &= \lim_{x \to 0} \frac{1}{x} L(1 + x) \\ &= \lim_{x \to 0} L[(1 + x)^{1/x}]. \qquad \text{Theorem 3} \end{aligned}$$

Hence
$$1 = \lim_{x\to 0} L[(1+x)^{1/x}],$$

so
$$E(1) = E\left\{\lim_{x\to 0} L[(1+x)^{1/x}]\right\}.$$

Since E is continuous, E and lim can be switched; thus

$$E(1) = \lim_{x\to 0} E[L(1+x)^{1/x}]. \tag{12}$$

Since $E(1) = e$, and E and L are inverse functions, Eq. (12) implies that

$$e = \lim_{x\to 0} (1+x)^{1/x}. \quad ■$$

Theorem 9 shows that the definition of e in Sec. 6.2 as a limit is consistent with the definition in Sec. H.1.

EXERCISES FOR SEC. H.2: EXPONENTIAL FUNCTIONS DEFINED IN TERMS OF LOGARITHMS

In the exercises prove the given statements using the definitions in this appendix.

1 Theorem 5

2 Theorem 7

3 Theorem 8

4 $L(x)/L(b)$ is the inverse of the function b^x.

APPENDIX I

THE TAYLOR SERIES FOR $f(x, y)$

The higher partial derivatives of $z = f(x, y)$ were introduced in Sec. 14.5. Recall that

$$\frac{\partial}{\partial y}\left(\frac{\partial z}{\partial x}\right) \quad \text{is denoted} \quad \frac{\partial^2 z}{\partial y\, \partial x}.$$

There are four partial derivatives of the second order:

$$\frac{\partial^2 z}{\partial y\, \partial x}, \quad \frac{\partial^2 z}{\partial x\, \partial y}, \quad \frac{\partial^2 z}{\partial x^2} = \frac{\partial}{\partial x}\left(\frac{\partial z}{\partial x}\right), \quad \text{and} \quad \frac{\partial^2 z}{\partial y^2} = \frac{\partial}{\partial y}\left(\frac{\partial z}{\partial y}\right).$$

If they are continuous, the order of differentiation does not matter:

$$\partial^2 z/\partial x\, \partial y = \partial^2 z/\partial y\, \partial x.$$

Each of these four partial derivatives may be differentiated with respect to x or with respect to y. Thus there are eight possible partial derivatives of order 3. For instance, two of them are

$$\frac{\partial}{\partial x}\left(\frac{\partial^2 z}{\partial x^2}\right) \quad \text{and} \quad \frac{\partial}{\partial y}\left(\frac{\partial^2 z}{\partial x^2}\right),$$

which will be denoted $\dfrac{\partial^3 z}{\partial x^3}$ and $\dfrac{\partial^3 z}{\partial y\, \partial x^2}$.

All told, there are eight partial derivatives of order 3. However, if they are continuous, the order of differentiation does not affect the result. For instance,

$$\frac{\partial^3 z}{\partial y^2\, \partial x} = \frac{\partial^3 z}{\partial y\, \partial x\, \partial y} = \frac{\partial^3 z}{\partial x\, \partial y^2}.$$

Similar statements and notations hold for partial derivatives of higher orders.

EXAMPLE 1 Compute the partial derivatives of $x^4 y^7$ up through order 3.

SOLUTION To begin

$$\frac{\partial z}{\partial x} = 4x^3 y^7 \quad \text{and} \quad \frac{\partial z}{\partial y} = 7x^4 y^6.$$

Then

$$\frac{\partial^2 z}{\partial y\,\partial x} = 28x^3y^6 = \frac{\partial^2 z}{\partial x\,\partial y}, \qquad \frac{\partial^2 z}{\partial x^2} = 12x^2y^7, \qquad \text{and} \qquad \frac{\partial^2 z}{\partial y^2} = 42x^4y^5.$$

The distinct third-order partial derivatives are

$$\frac{\partial^3 z}{\partial x^3} = 24xy^7, \qquad \frac{\partial^3 z}{\partial x^2\,\partial y} = 84x^2y^6,$$

$$\frac{\partial^3 z}{\partial x\,\partial y^2} = 168x^3y^5, \qquad \text{and} \qquad \frac{\partial^3 z}{\partial y^3} = 210x^4y^4.$$

Note that on account of duplication there are in practice only four partial derivatives of order 3. Similarly, there are in practice only five different partial derivatives of order 4, and $n + 1$ different partial derivatives of order n. ■

Just as many functions $f(x)$ of a single variable can be expressed as power series in $x - a$, so can many functions $f(x, y)$ of two variables be expressed as series in powers of $x - a$ and $y - b$. Such a series begins

$$c_1 + c_2(x-a) + c_3(y-b) + c_4(x-a)^2 + c_5(x-a)(y-b) + c_6(y-b)^2 + \cdots,$$

where the c's are constants. Frequently, in applications, only this much of the series is used to approximate the function near the point (a, b). The typical term has the form $c(x-a)^r(y-b)^s$. The **degree** of the term is $r + s$. The terms are written from left to right in increasing degree; within terms of a given degree, the terms are written in increasing degree of $y - b$.

We shall obtain a formula for this series in terms of the partial derivatives of f evaluated at (a, b). The argument begins by relating $f(x, y)$ to a function of one variable.

Let f be a function of x and y that possesses partial derivatives of all orders. Let a, b, h, and k be fixed numbers. Define a function g as follows:

$$g(t) = f(a + th, b + tk).$$

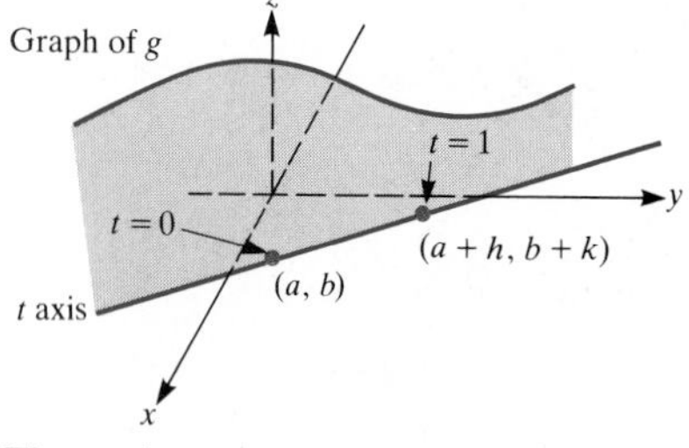

Figure 1

The Taylor series for f can be obtained from that for g by expressing the derivatives $g'(0)$, $g^{(2)}(0)$, $g^{(3)}(0)$, . . . in terms of partial derivatives of f.

To compute $g'(t)$, observe that g is a composite function:

$$g(t) = f(x, y), \qquad \text{where } x = a + th \text{ and } y = b + tk. \tag{1}$$

(See Fig. 1). By the chain rule (Sec. 14.6),

$$g'(t) = \frac{\partial f}{\partial x}\frac{dx}{dt} + \frac{\partial f}{\partial y}\frac{dy}{dt}.$$

Now by (1),

$$\frac{dx}{dt} = h \qquad \text{and} \qquad \frac{dy}{dt} = k.$$

Thus

$$g'(t) = \frac{\partial f}{\partial x}\cdot h + \frac{\partial f}{\partial y}\cdot k, \tag{2}$$

where $\partial f/\partial x$ and $\partial f/\partial y$ are evaluated at $(a + th, b + tk)$. In particular,

$$g'(0) = \frac{\partial f}{\partial x}(a, b)h + \frac{\partial f}{\partial y}(a, b)k. \tag{3}$$

Next, express $g^{(2)}(t)$ in terms of partial derivatives of the function f. To do so, differentiate Eq. (2) with respect to t:

$$g^{(2)}(t) = \frac{d}{dt}[g'(t)] = \frac{d}{dt}\left(\frac{\partial f}{\partial x}h + \frac{\partial f}{\partial y}k\right)$$

$$= \frac{\partial}{\partial x}\left(\frac{\partial f}{\partial x}h + \frac{\partial f}{\partial y}k\right)\frac{dx}{dt} + \frac{\partial}{\partial y}\left(\frac{\partial f}{\partial x}h + \frac{\partial f}{\partial y}k\right)\frac{dy}{dt}$$

$$= \frac{\partial}{\partial x}\left(\frac{\partial f}{\partial x}h + \frac{\partial f}{\partial y}k\right)h + \frac{\partial}{\partial y}\left(\frac{\partial f}{\partial x}h + \frac{\partial f}{\partial y}k\right)k$$

$$= \left(\frac{\partial^2 f}{\partial x^2}h + \frac{\partial^2 f}{\partial x\,\partial y}k\right)h + \left(\frac{\partial^2 f}{\partial y\,\partial x}h + \frac{\partial^2 f}{\partial y^2}k\right)k.$$

Hence

$$g^{(2)}(t) = \frac{\partial^2 f}{\partial x^2}h^2 + 2\frac{\partial^2 f}{\partial y\,\partial x}hk + \frac{\partial^2 f}{\partial y^2}k^2, \tag{4}$$

where all the partial derivatives are evaluated at $(a + th, b + tk)$. Thus

$$g^{(2)}(0) = \frac{\partial^2 f}{\partial x^2}(a, b)h^2 + 2\frac{\partial^2 f}{\partial y\,\partial x}(a, b)hk + \frac{\partial^2 f}{\partial y^2}(a, b)k^2. \tag{5}$$

Notice the similarity of the right side of Eq. (5) to the binomial expansion

$$(c + d)^2 = c^2 + 2cd + d^2,$$

in the coefficients, the powers of h and k, and the subscripts. To make use of this similarity, introduce the expression

$$(h\,\partial_x + k\,\partial_y)^2 f,$$

where $(h\,\partial_x + k\,\partial_y)^2$ is treated formally like an algebraic product: for instance, $(\partial_x\,\partial_x)f$ is interpreted as $\partial^2 f/\partial x^2$.

Thus Eq. (5) may be written in this shorthand as

$$g^{(2)}(0) = (h\,\partial_x + k\,\partial_y)^2 f\Big|_{(a,b)}. \tag{6}$$

[The symbol $|_{(a,b)}$ means "evaluated at (a, b)."]

Differentiating Eq. (4) with respect to t sufficiently often we can show similarly that

$$g^{(n)}(t) = (h\,\partial_x + k\,\partial_y)^n f\Big|_{(a+th,\,b+tk)} \tag{7}$$

for $n = 1, 2, 3, \ldots$.

Theorem *Taylor series for a function of two variables.* Let f have continuous partial derivatives of all orders up to and including $n + 1$ at and near the point (a, b). If $(x, y) = (a + h, b + k)$ is sufficiently near (a, b), then

$$f(x, y) = f(a, b) + (h\,\partial_x + k\,\partial_y)f\Big|_{(a,b)} + \frac{(h\,\partial_x + k\,\partial_y)^2 f}{2!}\Big|_{(a,b)} + \cdots + \frac{(h\,\partial_x + k\,\partial_y)^n f}{n!}\Big|_{(a,b)} + \frac{(h\,\partial_x + k\,\partial_y)^{n+1} f}{(n+1)!}\Big|_{(X,Y)},$$

where (X, Y) is some point on the line segment joining (a, b) and (x, y).
In sigma notation this reads

$$f(x, y) = \sum_{i=0}^{n} \frac{(h\,\partial_x + k\,\partial_y)^i f}{i!}\bigg|_{(a,b)} + \frac{(h\,\partial_x + k\,\partial_y)^{n+1} f}{(n+1)!}\bigg|_{(X,Y)}$$

Proof This theorem follows from Sec. 11.2. Once again we make use of the function g, defined as follows:

$$g(t) = f(a + th, b + tk).$$

(See Fig. 1.) Observe that

$$g(0) = f(a, b) \tag{8}$$

and
$$g(1) = f(a + h, b + k) = f(x, y). \tag{9}$$

By Lagrange's formula for the remainder $R_n(1; 0)$,

$$g(1) = g(0) + g'(0) \cdot 1 + \frac{g^{(2)}(0)}{2!}1^2 + \cdots + \frac{g^{(n)}(0)}{n!}1^n + \frac{g^{(n+1)}(T)}{(n+1)!}1^{n+1} \tag{10}$$

for a suitable number T, $0 \le T \le 1$. Combining Eqs. (7) and (10) completes the proof. ■

As a consequence of the theorem, the coefficient of $h^r k^s$ in the Taylor series for f is

$$\frac{1}{(r+s)!}\binom{r+s}{r}\frac{\partial^{r+s} f}{\partial x^r\,\partial y^s},$$

where the partial derivative is evaluated at (a, b), and $\binom{r+s}{r}$ denotes the binomial coefficient

$$\frac{(r+s)!}{r!\,s!}.$$

Observe that $(r + s)!$ can be canceled in the numerator and denominator. Thus the coefficient of $h^r k^s$ is

$$\frac{1}{r!\,s!}\frac{\partial^{r+s} f}{\partial x^r\,\partial y^s}.$$

EXAMPLE 2 Use the theorem to express $f(x, y) = x^2 y$ in powers of $x - 1$ and $y - 2$.

SOLUTION In this case, $a = 1$, $b = 2$, $h = x - 1$, and $k = y - 2$. To begin, compute the partial derivatives of f at $(1, 2)$. We have

$$\frac{\partial f}{\partial x} = 2xy, \quad \frac{\partial^2 f}{\partial x^2} = 2y, \quad \frac{\partial^2 f}{\partial x\,\partial y} = 2x, \quad \frac{\partial^3 f}{\partial x^2\,\partial y} = 2, \quad \text{and} \quad \frac{\partial f}{\partial y} = x^2.$$

All higher partial derivatives of f are identically 0. Thus $f(1, 2) = 2$, $\partial f/\partial x(1, 2) = 2 \cdot 1 \cdot 2 = 4$, and so on. Therefore,

$$
\begin{aligned}
f(x, y) &= f(1 + h, 2 + k) \\
&= f(1, 2) + \left[h\frac{\partial f}{\partial x}(1, 2) + k\frac{\partial f}{\partial y}(1, 2)\right] \\
&\quad + \left[\frac{h^2\frac{\partial^2 f}{\partial x^2}(1, 2) + 2hk\frac{\partial^2 f}{\partial x\,\partial y}(1, 2) + k^2\frac{\partial^2 f}{\partial y^2}(1, 2)}{2!}\right] \\
&\quad + \left[\frac{h^3\frac{\partial^3 f}{\partial x^3}(1, 2) + 3h^2k\frac{\partial^3 f}{\partial x^2\,\partial y}(1, 2) + 3hk^2\frac{\partial^3 f}{\partial x\,\partial y^2}(1, 2) + k^3\frac{\partial^3 f}{\partial y^3}(1, 2)}{3!}\right] \\
&= 2 + 4h + k + \frac{4h^2 + 4hk + 0k^2}{2!} + \frac{6h^2k}{3!};
\end{aligned}
$$

that is,

$$x^2y = 2 + 4(x - 1) + (y - 2) + 2(x - 1)^2 + 2(x - 1)(y - 2) + (x - 1)^2(y - 2). \quad (11)$$

This can be checked by expanding the right side of the equation. ■

According to the theorem, the Taylor series associated with $f(x, y)$ in powers of $x - a$ and $y - b$ begins

$$
\begin{aligned}
&f(a, b) + \frac{\partial f}{\partial x}(a, b)(x - a) + \frac{\partial f}{\partial y}(a, b)(y - b) \\
&\quad + \frac{\frac{\partial^2 f}{\partial x^2}(a,b)}{2!}(x - a)^2 + \frac{2\frac{\partial^2 f}{\partial x\,\partial y}(a, b)}{2!}(x - a)(y - b) + \frac{\frac{\partial^2 f}{\partial y^2}(a, b)}{2!}(y - b)^2 + \cdots.
\end{aligned}
$$

The test for extrema

With the aid of the Taylor series for $f(x, y)$ we now obtain the test for extrema of $f(x, y)$ stated in Sec. 14.9.

Assume that (a, b) is a critical point of $f(x, y)$ and that the first- and second-order partial derivatives of $f(x, y)$ are continuous at least in some disk around (a, b). Consider the case where

$$\frac{\partial^2 f}{\partial x^2}(a, b) > 0 \quad (12)$$

and

$$\frac{\partial^2 f}{\partial x^2}(a, b)\frac{\partial^2 f}{\partial y^2}(a, b) - \left[\frac{\partial^2 f}{\partial x\,\partial y}(a, b)\right]^2 > 0. \quad (13)$$

We will show that $f(x, y)$ has a relative minimum at (a, b). To do this, we begin with the equation

$$f(x, y) = f(a, b) + \frac{\partial f}{\partial x}(a, b)h + \frac{\partial f}{\partial y}(a, b)k + \left.\frac{(h\,\partial_x + k\,\partial_y)^2 f}{2}\right|_{(X, Y)}$$

Here $h = x - a$ and $k = y - b$. Since (a, b) is a critical point,

$$f(x, y) = f(a, b) + \left.\frac{(h\,\partial_x + k\,\partial_y)^2 f}{2}\right|_{(X, Y)}$$

All that remains is to show that

$$\left.\frac{(h\,\partial_x + k\,\partial_y)^2 f}{2}\right|_{(X, Y)} \geq 0 \quad (14)$$

when (x, y) is sufficiently close to (a, b).

Now, (14) equals

$$\frac{1}{2}\left[h^2\frac{\partial^2 f}{\partial x^2} + 2hk\frac{\partial^2 f}{\partial x\,\partial y} + k^2\frac{\partial^2 f}{\partial y^2}\right], \tag{15}$$

where the partial derivatives are evaluated at (X, Y), which is on the line segment joining (a, b) to (x, y).

Let

$$A = \frac{\partial^2 f}{\partial x^2}, \qquad B = \frac{\partial^2 f}{\partial x\,\partial y}, \qquad \text{and} \qquad C = \frac{\partial^2 f}{\partial y^2}$$

where the partial derivatives are evaluated at (X, Y). In view of (15), we must show that

$$Ah^2 + 2Bhk + Ck^2 \geq 0,$$

when (X, Y) is close to (a, b). By (12) and (13), and the continuity of the partial derivatives, we know that $A > 0$ and $AC - B^2 > 0$ for (X, Y) sufficiently close to (a, b). Since $A > 0$, it is enough to prove that

$$A(Ah^2 + 2Bhk + Ck^2) \geq 0.$$

We have

$$\begin{aligned} A(Ah^2 + 2Bhk + Ck^2) &= A^2h^2 + 2ABhk + ACk^2 \\ &= A^2h^2 + 2ABhk + B^2k^2 + ACk^2 - B^2k^2 \\ &= (Ah + Bk)^2 + (AC - B^2)k^2. \end{aligned} \tag{16}$$

Since the square of a real number is positive or 0 and $AC - B^2$ is positive, (16) is positive or 0. This completes the argument.

EXERCISES FOR APP. I: THE TAYLOR SERIES FOR $f(x, y)$

In each of Exercises 1 to 4 compute all eight partial derivatives of the third order of the given function.

1 x^5y^7

2 x/y

3 e^{2x+3y}

4 $\sin(x^2 + y^3)$

5 Verify Eq. (11) by expanding the right side of the equation.

6 Using Eq. (11), compute the difference in the volumes of these two boxes: One has a square base of side 1 foot and height 2 feet; the other has a square base of side 1.1 feet and height 2.1 feet.

7 (*a*) Using partial derivatives, obtain the first four nonzero terms in the Taylor series for e^{x+y^2} in powers of x and y.

(*b*) Noticing that $e^{x+y^2} = e^x e^{y^2}$ and using a few terms of the Maclaurin series for e^x and e^{y^2}, solve (*a*) again.

8 (*a*) Using partial derivatives, obtain the first four nonzero terms in the Taylor series for $\cos(x + y)$ in powers of x and y.

(*b*) Using the Maclaurin series for $\cos t$ and replacing t by $x + y$, obtain the terms described in (*a*).

9 Assume that f has continuous partial derivatives of all orders at $(0, 0)$, $f(0, 0) = 2$, $\partial f/\partial x(0, 0) = 3$, $\partial f/\partial y(0, 0) = -5$, $\partial^2 f/\partial x^2(0, 0) = 6$, $\partial^2 f/\partial x\,\partial y(0, 0) = 7$, and $\partial^2 f/\partial y^2(0, 0) = 1$. Write out the Taylor series in powers of x and y associated with f up through terms of degree 2.

10 The Taylor series for a certain function $f(x, y)$ begins $5 + 6x + 11y - 2x^2 - 3xy + 7y^2 + \cdots$. Use this information to determine $f(0, 0)$ and the first- and second-order partial derivatives of f at $(0, 0)$.

11 Verify that the expansion of $\sqrt{1 + x + y}$ begins with $1 + \frac{1}{2}x + \frac{1}{2}y - \frac{1}{8}x^2 - \frac{1}{4}xy - \frac{1}{8}y^2 + \cdots$ by using the theorem of this section.

APPENDIX J

THEORY OF LIMITS

In Sec. 2.10 a precise definition of $\lim_{x\to a} f(x)$ was given. This definition is the basis of this appendix, where the fundamental properties of limits will be obtained. In particular we will prove that any polynomial is continuous. With the aid of the ϵ, δ definition of limit in Sec. 2.10 we can establish that if

$$\lim_{x\to a} f(x) = A \quad \text{and} \quad \lim_{x\to a} g(x) = B,$$

then
$$\lim_{x\to a} [f(x) + g(x)] = A + B, \qquad \lim_{x\to a} [f(x)g(x)] = AB,$$

and
$$\lim_{x\to a} \frac{f(x)}{g(x)} = \frac{A}{B}. \qquad \text{(if } B \text{ is not 0)}$$

We will prove the first two assertions.

Theorem 1 If $\lim_{x\to a} f(x) = A$ and $\lim_{x\to a} g(x) = B$, then

$$\lim_{x\to a} [f(x) + g(x)] = A + B.$$

Proof It must be shown that given any $\epsilon > 0$, no matter how small, there exists a number $\delta > 0$ depending on ϵ such that

$$|[f(x) + g(x)] - (A + B)| < \epsilon \tag{1}$$

whenever $|x - a| < \delta$ and $x \neq a$.

Rewrite $[f(x) + g(x)] - (A + B)$ as $[f(x) - A] + [g(x) - B]$, which is a sum of two quantities known to be small when x is near a. Since the absolute value of the sum of two numbers is not larger than the sum of their absolute values,

$$|[f(x) - A] + [g(x) - B]| \leq |f(x) - A| + |g(x) - B|. \tag{2}$$

Since $\lim_{x\to a} f(x) = A$, there is a positive number δ_1 such that

$$|f(x) - A| < \frac{\epsilon}{2}$$

when $|x - a| < \delta_1$ and $x \neq a$. (Why we pick $\epsilon/2$ rather than ϵ will be clear in a moment.) Similarly, there is a positive number δ_2 such that

$$|g(x) - B| < \frac{\epsilon}{2}$$

when $|x - a| < \delta_2$ and $x \neq a$.

Now let δ be the smaller of δ_1 and δ_2. For any x (not equal to a) such that $|x - a| < \delta$, we have both

$$|x - a| < \delta_1 \qquad \text{and} \qquad |x - a| < \delta_2$$

and, therefore, simultaneously,

$$|f(x) - A| < \frac{\epsilon}{2} \qquad \text{and} \qquad |g(x) - B| < \frac{\epsilon}{2}. \tag{3}$$

Combining (2) and (3), we conclude that when $|x - a| < \delta$ but $x \neq a$,

$$|[f(x) + g(x)] - (A + B)| < \frac{\epsilon}{2} + \frac{\epsilon}{2} = \epsilon.$$

Thus for each $\epsilon > 0$ there exists a suitable $\delta > 0$, depending, of course, on $\epsilon, f,$ and g. This ends the proof. ■

Theorem 1 is the basis of the proof of the next theorem.

Theorem 2 The sum of two functions that are continuous at a is itself continuous at a.

Proof Let f and g be continuous at a. Let h be their sum; that is, let $h(x) = f(x) + g(x)$. We wish to show that h is continuous at a.

In view of the definition of continuity given in Sec. 2.8, it must be shown that $h(a)$ is defined and that $\lim_{x\to a} h(x) = h(a)$.

Since f and g are defined at a, so is h, and $h(a) = f(a) + g(a)$. All that remains is to show that $\lim_{x\to a} [f(x) + g(x)] = f(a) + g(a)$. By Theorem 1, $\lim_{x\to a} [f(x) + g(x)] = \lim_{x\to a} f(x) + \lim_{x\to a} g(x)$. Since f and g are continuous at a, $\lim_{x\to a} f(x) = f(a)$ and $\lim_{x\to a} g(x) = g(a)$. This concludes the proof. ■

Theorem 3 If $\lim_{x\to a} f(x) = A$ and $\lim_{x\to a} g(x) = B$, then

$$\lim_{x\to a} f(x)g(x) = AB.$$

Plan of proof: We know that $|f(x) - A|$ and $|g(x) - B|$ are small when x is near a. We wish to conclude that $|f(x)g(x) - AB|$ is small when x is near a. The algebraic identity

$$f(x)g(x) - AB = f(x)[g(x) - B] + B[f(x) - A] \tag{4}$$

will be of use. From Eq. (4) and properties of the absolute value, it follows that

$$|f(x)g(x) - AB| \le |f(x)||g(x) - B| + |B||f(x) - A|. \tag{5}$$

Now $|B|$ is fixed, and $|f(x) - A|$ and $|g(x) - B|$ are small when x is near a. The real problem is to control $|f(x)|$. Watch carefully the way in which $|f(x)|$ is treated in the proof.

Proof Consider the case $B \ne 0$. Let $\epsilon > 0$ be given. We wish to show that there is a number $\delta > 0$ such that $|f(x)g(x) - AB| < \epsilon$ when $|x - a| < \delta$ but $x \ne a$. Observe that

$$\begin{aligned} |f(x)g(x) - AB| &= |f(x)[g(x) - B] + B[f(x) - A]| \\ &\le |f(x)||g(x) - B| + |B||f(x) - A|. \end{aligned} \tag{6}$$

Since $\lim_{x\to a} f(x) = A$, there is a number $\delta_1 > 0$ such that

$$|f(x) - A| < \frac{\epsilon}{2|B|}$$

when $|x - a| < \delta_1$ but $x \ne a$. [It will be clear in a moment why we want $|f(x) - A|$ to be less than $\epsilon/(2|B|)$.] Thus the second summand on the right side of (6) is less than

$$\frac{|B|\epsilon}{2|B|} = \frac{\epsilon}{2}.$$

For x such that $|x - a| < \delta_1$, $|f(x)|$ does not become arbitrarily large, since $|f(x) - A| < \epsilon/(2|B|)$. Indeed, for such values of x

$$|f(x)| = |A + [f(x) - A]| \le |A| + |f(x) - A| < |A| + \frac{\epsilon}{2|B|}.$$

Letting $C = |A| + \epsilon/(2|B|)$, we have $|f(x)| < C$ when $|x - a| < \delta_1$ but $x \ne a$. [This controls the size of $|f(x)|$.]

Since $\lim_{x\to a} g(x) = B$, there is a $\delta_2 > 0$ such that $|g(x) - B| < \epsilon/(2C)$ when $|x - a| < \delta_2$ but $x \ne a$.

Now let δ be the smaller of δ_1 and δ_2. When $|x - a| < \delta$ and $x \ne a$, both $|x - a| < \delta_1$ and $|x - a| < \delta_2$ hold; hence $|f(x) - A| < \epsilon/(2|B|)$, $|f(x)| < C$, and $|g(x) - B| < \epsilon/(2C)$. Inspection of (6) then shows that for such x,

$$|f(x)g(x) - AB| < C\frac{\epsilon}{2C} + |B|\frac{\epsilon}{2|B|} = \frac{\epsilon}{2} + \frac{\epsilon}{2} = \epsilon.$$

The proof is completed. The case $B = 0$ is left to the reader as Exercise 7. ■

Theorem 4 The product of two functions that are continuous at a is itself continuous at a.

The proof is similar to that of Theorem 2, but depends on Theorem 3 instead of Theorem 1.

The fact that the sum and product of continuous functions are continuous is the basis of the proof that *any polynomial is continuous*. It is necessary to prove first that the function $f(x) = x$ is continuous and then that any constant function is continuous.

Theorem 5 The function f, such that $f(x) = x$, is continuous everywhere. So are the functions $x^2, x^3, x^4, \ldots$.

Proof Since $f(a) = a$, it must be shown that $|f(x) - a|$ is small whenever $|x - a|$ is sufficiently small. More precisely, for $\epsilon > 0$ we wish to exhibit $\delta > 0$ such that $|f(x) - a| < \epsilon$ whenever $|x - a| < \delta$. But $f(x) = x$; hence $|f(x) - a|$ is simply $|x - a|$. Let $\delta = \epsilon$. Thus whenever $|x - a| < \delta$, it follows that $|f(x) - a| < \epsilon$. This shows that the function x is continuous.

Since the function x^2 is the product of the function x, and the function x, Theorem 4 implies that x^2 is continuous. Similarly, x^3 is continuous, for $x^3 = x^2x$. Mathematical induction establishes the continuity of x^n for all positive integers n. ■

Theorem 6 Any constant function is continuous everywhere.

Proof Let $f(x) = c$ for all x. For any $\epsilon > 0$, choose $\delta = 722$, a perfectly fine positive number. Now $|f(x) - f(a)| = |c - c| = 0 < \epsilon$ for any x, and hence for x such that $|x - a| < 722$. (Any $\delta > 0$ will do.) Thus f is continuous at any number a. ■

Theorem 7 Any polynomial is continuous everywhere.

Proof We illustrate the idea of the proof by showing that $6x^2 - 5x + 1$ is continuous everywhere.

By Theorem 6, the constant functions $f(x) = 6$, $g(x) = -5$, and $h(x) = 1$ are continuous everywhere. By Theorem 5, the functions x and x^2 are continuous everywhere. By Theorem 4, the functions $6x^2$ and $(-5)x$ are continuous. By Theorem 2, the function $[6x^2 + (-5)x]$ is continuous. Again by Theorem 2, the function $[6x^2 + (-5)x] + 1$ is continuous. Thus $6x^2 - 5x + 1$ is continuous. The same argument applies to any polynomial. ■

EXERCISES FOR APP. J: THEORY OF LIMITS

1 Prove that $f(x) = 1/x$ is continuous at any number $a \neq 0$.

2 Prove that if g is continuous at a and f is continuous at $g(a)$, then the composite function $h = f \circ g$ is continuous at a.

3 Using Exercises 1 and 2, show that the function $1/(x^2 + 1)$ is continuous.

4 Using Exercises 1 and 2, show that if $g(x)$ is continuous at a and $g(a) \neq 0$, then $1/g(x)$ is continuous at a.

5 Let $f(x)$ and $g(x)$ be continuous at a and $g(a) \neq 0$. Show that $f(x)/g(x)$ is continuous at a. (See Exercise 4.)

6 Let $\lim_{x\to a} g(x) = B \neq 0$. Show that $\lim_{x\to a} [1/g(x)] = 1/B$.

7 Prove Theorem 3 in the case $B = 0$.

8 What theorems assure us that $(x^3 - 1)/(x^2 - 5)$ is continuous throughout its domain?

9 Assume that f is continuous at a and $f(a)$ is positive. Prove that there is an open interval (b, c) including a such that $f(x)$ is positive for all x in that interval.

In the remaining exercises use the definitions from Sec. 2.9.

10 Prove that if $\lim_{x\to\infty} f(x) = \infty$ and $\lim_{x\to\infty} g(x) = \infty$, then $\lim_{x\to\infty} [f(x) + g(x)] = \infty$.

11 Prove that if $\lim_{x\to\infty} f(x) = \infty$ and $\lim_{x\to\infty} g(x) = A$, then $\lim_{x\to\infty} [f(x) + g(x)] = \infty$.

12 Prove that if $\lim_{x\to\infty} f(x) = A$ and $\lim_{x\to\infty} g(x) = B$, then $\lim_{x\to\infty} [f(x) + g(x)] = A + B$.

13 Prove that if $\lim_{x\to\infty} f(x) = \infty$ and $\lim_{x\to\infty} g(x) = \infty$, then $\lim_{x\to\infty} f(x)g(x) = \infty$.

14 Prove that if $\lim_{x\to\infty} f(x) = \infty$ and $\lim_{x\to\infty} g(x) = A > 0$, then $\lim_{x\to\infty} f(x)g(x) = \infty$.

15 Prove that if $\lim_{x\to\infty} f(x) = A$ and $\lim_{x\to\infty} g(x) = B$, then $\lim_{x\to\infty} f(x)g(x) = AB$.

APPENDIX K

THE INTERCHANGE OF LIMITS

This appendix provides proofs of two theorems whose validity was assumed in the text. In Sec. K.1 the equality of the mixed partial derivatives is examined. It can be read after Sec. 14.5. Section K.2 concerns the differentiation of $\int_a^b f(x, y)\, dx$ with respect to y. Section K.3, besides presenting an example of a function $f(x, y)$ whose two mixed partials are not equal, gives a gentle introduction to advanced calculus.

K.1 THE EQUALITY OF $\frac{\partial^2 f}{\partial x\, \partial y}$ AND $\frac{\partial^2 f}{\partial y\, \partial x}$

For most common functions $f(x, y)$,

$$\frac{\partial}{\partial y}\left(\frac{\partial f}{\partial x}\right) \quad \text{equals} \quad \frac{\partial}{\partial x}\left(\frac{\partial f}{\partial y}\right);$$

that is, the order in which we compute partial derivatives does not affect the result. This assertion is justified in the following theorem, where for convenience we use subscript notation for the partial derivatives.

Theorem Let f be a function defined on the xy plane. If f_{xy} and f_{yx} exist and if at least one of them is continuous at all points, then they are equal.

Proof We assume f_{yx} is continuous. To keep the proof uncluttered, it will be shown that $f_{xy}(0, 0)$ equals $f_{yx}(0, 0)$. The identical argument holds for any point (a, b).

To begin, consider the definition of $f_{xy}(0, 0)$:

$$f_{xy}(0, 0) = \left.\frac{\partial(f_x)}{\partial y}\right|_{(0,\,0)} = \lim_{k \to 0} \frac{f_x(0, k) - f_x(0, 0)}{k}.$$

But, by definition of the partial derivative f_x,

$$f_x(0, k) = \lim_{h \to 0} \frac{f(h, k) - f(0, k)}{h} \qquad \text{and} \qquad f_x(0, 0) = \lim_{h \to 0} \frac{f(h, 0) - f(0, 0)}{h}.$$

Thus

$$f_{xy}(0, 0) = \lim_{k\to 0} \frac{f_x(0, k) - f_x(0, 0)}{k}$$

$$= \lim_{k\to 0} \frac{\displaystyle\lim_{h\to 0} \frac{f(h, k) - f(0, k)}{h} - \lim_{h\to 0} \frac{f(h, 0) - f(0, 0)}{h}}{k}$$

$$= \lim_{k\to 0} \left\{ \lim_{h\to 0} \frac{[f(h, k) - f(0, k)] - [f(h, 0) - f(0, 0)]}{hk} \right\}. \tag{1}$$

Let us focus our attention on the numerator in (1):

$$\text{Numerator} = [f(h, k) - f(0, k)] - [f(h, 0) - f(0, 0)]. \tag{2}$$

Note that the second bracketed expression is obtained from the first bracketed expression by replacing k by 0. Define, for fixed h, a function

$$u(y) = f(h, y) - f(0, y). \tag{3}$$

Then (2) takes the simple form

$$u(k) - u(0). \tag{4}$$

By the mean-value theorem (see Sec. 4.1),

$$u(k) - u(0) = u'(K)k \tag{5}$$

for some K between 0 and k. But by the definition of the function u, given in Eq. (3),

$$u'(K) = f_y(h, K) - f_y(0, K). \tag{6}$$

Thus, by the mean-value theorem applied to the function $f_y(x, K)$, for fixed K,

$$u'(K) = f_{yx}(H, K)h \tag{7}$$

for some H between 0 and h.

Thus Eq. (2) becomes

$$\text{Numerator} = f_{yx}(H, K)hk \tag{8}$$

for some point (H, K) in the rectangle with vertices $(0, 0)$, $(h, 0)$, (h, k), and $(0, k)$. (See Fig. 1.) Substituting Eq. (8) in Eq. (1), we obtain

$$f_{xy}(0, 0) = \lim_{k\to 0} [\lim_{h\to 0} f_{yx}(H, K)].$$

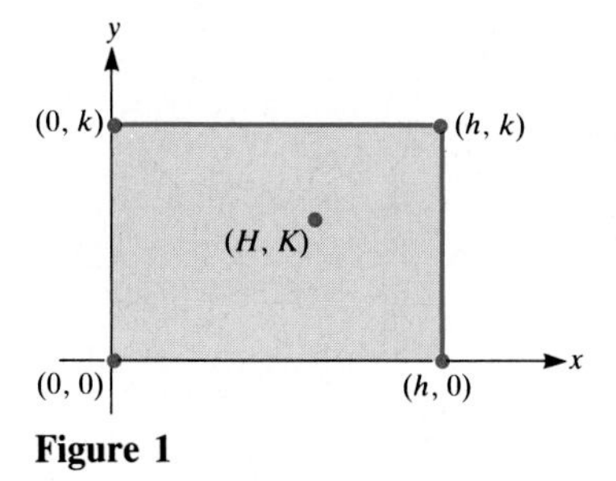

Figure 1

Since H is between 0 and h and K is between 0 and k, it follows that H and K both approach 0 as $h \to 0$ and $k \to 0$. By the continuity of f_{yx}, $f_{yx}(H, K)$ approaches $f_{yx}(0, 0)$. Hence,

$$f_{xy}(0, 0) = f_{yx}(0, 0),$$

as asserted. ■

Example 3 in Sec. K.3 presents a function whose two mixed partial derivatives f_{xy} and f_{yx} are *not* equal at (0, 0).

EXERCISES FOR SEC. K.1: THE EQUALITY OF $\frac{\partial^2 f}{\partial x\, \partial y}$ AND $\frac{\partial^2 f}{\partial y\, \partial x}$

1 Prove the theorem without looking at the text.
2 Prove the theorem at an arbitrary point (a, b).

K.2 THE DERIVATIVE OF $\int_a^b f(x, y)\, dx$ WITH RESPECT TO y

The integral

$$\int_a^b f(x, y)\, dx$$

depends on y. Let

$$F(y) = \int_a^b f(x, y)\, dx.$$

It makes sense to speak of the derivative of F with respect to y. It turns out that for most common functions f,

$$\frac{dF}{dy} = \int_a^b \frac{\partial f}{\partial y}\, dx.$$

That is,

$$\frac{d}{dy}\left[\int_a^b f(x, y)\, dx\right] = \int_a^b \frac{\partial f}{\partial y}\, dx.$$

Generally, it is safe to differentiate the integral by differentiating the integrand.

The reader may check this assertion for $f(x, y) = x^3 + xy^2$ before going through the proof of the theorem.

Theorem Let f be defined on the xy plane, and assume that f and f_y are continuous. Assume also that f_{yy} is defined on the xy plane and that it is bounded on each rectangle. [That is, if R is a rectangle, then there is a number M, depending on R, such that $|f_{yy}(x, y)| \leq M$ for all (x, y) in R.] Let F be defined by

$$F(y) = \int_a^b f(x, y)\, dx.$$

Then F is differentiable, and

$$\frac{dF}{dy} = \int_a^b \frac{\partial f}{\partial y}\, dx.$$

Proof To show that

$$\lim_{h \to 0} \frac{F(y + h) - F(y)}{h} = \int_a^b \frac{\partial f}{\partial y}\, dx,$$

consider, for a fixed y, the difference

$$\frac{F(y + h) - F(y)}{h} - \int_a^b \frac{\partial f}{\partial y}(x, y)\, dx, \qquad (1)$$

which, by the definition of F, equals

$$\frac{\int_a^b f(x, y+h)\,dx - \int_a^b f(x, y)\,dx}{h} - \int_a^b \frac{\partial f}{\partial y}(x, y)\,dx. \tag{2}$$

Now, (2) equals

$$\int_a^b \left[\frac{f(x, y+h) - f(x, y)}{h} - \frac{\partial f}{\partial y}(x, y)\right] dx. \tag{3}$$

To show that (1) approaches 0 as $h \to 0$, it suffices to show that the integrand in (3) is small when h is small. It may be assumed now that $|h| \le 1$.

First of all, the expression

$$\frac{f(x, y+h) - f(x, y)}{h}$$

in the integrand in (3) equals

$$\frac{h \dfrac{\partial f}{\partial y}(x, y+H)}{h} = \frac{\partial f}{\partial y}(x, y+H)$$

for some number H between 0 and h, by the mean-value theorem. (H depends on x, y, and h.)

Thus the integrand in (3) equals

$$\frac{\partial f}{\partial y}(x, y+H) - \frac{\partial f}{\partial y}(x, y). \tag{4}$$

By the mean-value theorem, (4) equals

$$H \frac{\partial^2 f}{\partial y^2}(x, y+H^*) \tag{5}$$

for some number H^* between 0 and H.

Since $|H^*| \le |H| \le |h| \le 1$, the point $(x, y + H^*)$ lies somewhere in the rectangle whose vertices are

$$(a, y-1), \qquad (a, y+1), \qquad (b, y-1), \qquad (b, y+1).$$

By assumption, $|\partial^2 f/\partial y^2| \le M$ on this rectangle. By (4) and (5), the integrand in (3) has absolute value at most

$$|H|\ M \le |h|\ M.$$

Thus the absolute value of (3) is at most

$$|h|\ M(b-a),$$

which approaches 0 as $h \to 0$, since M and $b - a$ are fixed numbers. This proves the theorem. ■

The assumption made in the theorem that $\partial^2 f/\partial y^2$ is bounded in each rectangle is satisfied if f_{yy} is continuous. So the theorem does cover the cases commonly encountered. In advanced calculus the theorem is proved without any assumption on f_{yy}. (See, for instance, R. C. Buck, *Advanced Calculus,* 3d ed., p. 118, McGraw-Hill, New York, 1978.)

EXERCISES FOR SEC. K.2: THE DERIVATIVE OF $\int_a^b f(x, y)\, dx$ WITH RESPECT TO y

1 Verify the theorem for $f(x, y) = x^3y^4$.
2 Verify the theorem for $f(x, y) = \cos xy$.
3 For what value of y does the function

$$F(y) = \int_0^{\pi/2} (y - \cos x)^2\, dx$$

have a minimum? (Use the theorem of this section.)
4 Let $G(u, v, w) = \int_u^v f(w, x)\, dx$. Find the partial derivatives (*a*) $\partial G/\partial v$, (*b*) $\partial G/\partial u$, (*c*) $\partial G/\partial w$.
5 Let $G(u) = \int_0^u f(u, x)\, dx$. Find dG/du.
6 Let $G(u, v) = \int_0^u e^{-vx^2}\, dx$. Find (*a*) $\partial G/\partial u$, (*b*) $\partial G/\partial v$.
7 Let $F(y) = \int_0^1 [(x^y - 1)/\ln x]\, dx$ for $y \geq 0$.
(*a*) Assuming that one may differentiate F by differentiating under the integral sign, show that $dF/dy = 1/(1 + y)$.
(*b*) From (*a*) deduce that $F(y) = \ln (1 + y) + C$.
(*c*) Show that the constant C in part (*b*) is 0 by examining the case $y = 0$.

K.3 THE INTERCHANGE OF LIMITS

Although the two theorems in Secs. K.1 and K.2 are independent, they both illustrate a certain type of problem which students who go on to advanced calculus will study, namely, the interchange of limits. To see what this means, let us take a new look at the theorem in Sec. K.1, which concerns the equality of the mixed partials. By (1) in Sec. K.1,

$$f_{xy}(0, 0) = \lim_{k\to 0} \left[\lim_{h\to 0} \frac{f(h, k) - f(0, k) - f(h, 0) + f(0, 0)}{hk} \right].$$

Similarly, from the definition of f_{yx}, it can be shown that

$$f_{yx}(0, 0) = \lim_{h\to 0} \left[\lim_{k\to 0} \frac{f(h, k) - f(0, k) - f(h, 0) + f(0, 0)}{hk} \right].$$

Note that the two quotients are identical, but that f_{xy} involves

$$\lim_{k\to 0} \left(\lim_{h\to 0} \right),$$

while f_{yx} involves

$$\lim_{h\to 0} \left(\lim_{k\to 0} \right).$$

It is tempting to claim that the order of taking limits should not matter. But it does. (In Example 3 it is shown that f_{xy} does not always equal f_{yx}.) This instance raises the general question, "When can one interchange limits?" Example 1 presents a simple case in which the order of taking the limits *does* matter.

EXAMPLE 1 Let $f(x, y) = x^y$ for $x > 0$ and $y > 0$. Evaluate the two "repeated limits"

$$\lim_{y\to 0^+} \left(\lim_{x\to 0^+} x^y \right) \quad \text{and} \quad \lim_{x\to 0^+} \left(\lim_{y\to 0^+} x^y \right).$$

SOLUTION

$$\lim_{y\to 0^+} \left(\lim_{x\to 0^+} x^y \right) = \lim_{y\to 0^+} 0 = 0.$$

On the other hand, $\lim_{x\to0^+}\left(\lim_{y\to0^+} x^y\right) = \lim_{x\to0^+} 1 = 1.$

This shows that *the order of taking limits may affect the result*. Moreover, it suggests why the symbol 0^0 is not given any meaning. ■

The next example also illustrates the effect of switching the order of taking limits. In Example 3 it becomes the basis of an illustration that shows the mixed partial derivatives f_{xy} and f_{yx} are not always equal.

EXAMPLE 2 Let

$$g(x, y) = \begin{cases} \dfrac{x^2 - y^2}{x^2 + y^2} & \text{if } (x, y) \neq (0, 0), \\ 0 & \text{if } (x, y) = (0, 0). \end{cases}$$

Show that
$$\lim_{x\to0}\left[\lim_{y\to0} g(x, y)\right] \neq \lim_{y\to0}\left[\lim_{x\to0} g(x, y)\right].$$

SOLUTION

$$\lim_{y\to0} g(x, y) = \lim_{y\to0} \frac{x^2 - y^2}{x^2 + y^2} = \frac{x^2}{x^2} = 1.$$

Thus
$$\lim_{x\to0}\left[\lim_{y\to0} g(x, y)\right] = \lim_{x\to0} 1 = 1.$$

On the other hand, $\lim_{x\to0} g(x, y) = \lim_{x\to0} \dfrac{x^2 - y^2}{x^2 + y^2} = \dfrac{-y^2}{y^2} = -1.$

Thus
$$\lim_{y\to0}\left[\lim_{x\to0} g(x, y)\right] = \lim_{y\to0} (-1) = -1. \quad ■$$

EXAMPLE 3 Let $f(x, y) = xyg(x, y)$, where g is given in Example 2. Show that $f_{xy}(0, 0) \neq f_{yx}(0, 0)$.

SOLUTION That f_x, f_y, f_{xy}, f_{yx} exist at all points is left for the reader to demonstrate. (See Exercise 8.) Note that $f(x, y) = 0$ whenever x or y is 0.

By (1) in Sec. K.1,

$$f_{xy}(0, 0) = \lim_{k\to0}\left\{\lim_{h\to0} \frac{[f(h, k) - f(0, k)] - [f(h, 0) - f(0, 0)]}{hk}\right\}$$

$$= \lim_{k\to0}\left[\lim_{h\to0} \frac{f(h, k)}{hk}\right] = \lim_{k\to0}\left[\lim_{h\to0} g(h, k)\right] = -1$$

from Example 2. Similarly,

$$f_{yx}(0, 0) = \lim_{h\to0}\left[\lim_{k\to0} g(h, k)\right] = 1.$$

Therefore, $f_{xy}(0, 0) \neq f_{yx}(0, 0)$. ■

The theorem in Sec. K.2, which asserts that in general

$$\frac{d}{dy}\left[\int_a^b f(x, y)\, dx\right] = \int_a^b \frac{\partial f}{\partial y}(x, y)\, dx,$$

also concerns the validity of switching limits. After all, both the derivative and the definite integral are defined as limits.

The exercises present more examples of the interchange of limits. The main point of the examples and the exercises is that the interchange of limits is a risky business. Fortunately, there are theorems that imply that sometimes the order of taking limits has no effect on the outcome.

EXERCISES FOR SEC. K.3: THE INTERCHANGE OF LIMITS

1 Let $f(x, y) = 1$ if $y \geq x$ and let $f(x, y) = 0$ if $y < x$.
(*a*) Shade in the part of the plane where $f(x, y) = 1$.
(*b*) Show that $\lim_{x\to\infty} [\lim_{y\to\infty} f(x, y)] = 1$.
(*c*) Show that $\lim_{y\to\infty} [\lim_{x\to\infty} f(x, y)] = 0$.

2 Show that $\lim_{x\to 0} [\lim_{n\to\infty} nx/(1 + nx)] = 1$, but that $\lim_{n\to\infty} [\lim_{x\to 0} nx/(1 + nx)] = 0$.

3 Let $f_n(x) = nx/(1 + n^2x^4)$. Show that

$$\int_0^\infty \lim_{n\to\infty} f_n(x)\, dx = 0,$$

but $$\lim_{n\to\infty} \int_0^\infty f_n(x)\, dx = \frac{\pi}{4}.$$

4 Show that $\lim_{x\to 0} [\lim_{y\to 0} x^2/(x^2 + y^2)]$ is not equal to $\lim_{y\to 0} [\lim_{x\to 0} x^2/(x^2 + y^2)]$.

5 Let $f_n(x) = n\pi \sin(n\pi x)$ if $0 \leq x \leq 1/n$ and 0 otherwise.
(*a*) Graph f_1, f_2, and f_3.
(*b*) Show that $\lim_{n\to\infty} \int_0^1 f_n(x)\, dx = 2$, but that $\int_0^1 \lim_{n\to\infty} f_n(x)\, dx = 0$.

6 Compare

$$\lim_{x\to\infty}\left(\lim_{y\to\infty} \frac{x^2}{x^2 + y^2 + 1}\right)$$

and $$\lim_{y\to\infty}\left(\lim_{x\to\infty} \frac{x^2}{x^2 + y^2 + 1}\right).$$

7 Let $f_n(x) = (1/n) \sin nx$ for all x and all positive integers n. Show that

$$\lim_{n\to\infty}\left[\lim_{h\to 0} \frac{f_n(h) - f_n(0)}{h}\right] = 1,$$

while

$$\lim_{h\to 0}\left[\lim_{n\to\infty} \frac{f_n(h) - f_n(0)}{h}\right] = 0.$$

8 Show that f_{xy} and f_{yx} exist at all points in the plane, where f is the pathological function in Example 3.

9 Show that

$$\int_0^\infty \left[\int_0^1 (2xy - x^2y^2)e^{-xy}\, dx\right] dy = 1,$$

but $$\int_0^1 \left[\int_0^\infty (2xy - x^2y^2)e^{-xy}\, dy\right] dx = 0.$$

10 Show that l'Hôpital's rule in the zero-over-zero case concerns the equality of these two limits:

$$\lim_{\Delta t\to 0}\left[\lim_{t\to a} \frac{f(t + \Delta t) - f(t)}{g(t + \Delta t) - g(t)}\right]$$

and $$\lim_{t\to a}\left[\lim_{\Delta t\to 0} \frac{f(t + \Delta t) - f(t)}{g(t + \Delta t) - g(t)}\right].$$

11 Let $f_n(x)$ be defined for each x in [0, 1] and each positive integer n. Assume that f_n is continuous for each n and that $\lim_{n\to\infty} f_n(x)$ exists for each x in [0, 1]. Call this limit $f(x)$,

$$\lim_{n\to\infty} f_n(x) = f(x).$$

(*a*) Show that the statement "f is continuous at a" is equivalent to the equation

$$\lim_{n\to\infty}\left[\lim_{x\to a} f_n(x)\right] = \lim_{x\to a}\left[\lim_{n\to\infty} f_n(x)\right].$$

(*b*) In particular, let $f_n(x) = x^n$. Show that in this case the two repeated limits in (*a*) are not necessarily equal.

APPENDIX L

THE JACOBIAN

In Sec. 3.2, we saw that magnification on a line is expressed in terms of the derivative. In Sec. 15.4, when evaluating an integral in polar coordinates, we had to insert an r in the integrand. When evaluating integrals in cylindrical or spherical coordinates (Secs. 15.6 and 15.7), we had to insert an r or a $\rho^2 \sin \phi$. It turns out that these extra factors are closely connected with magnification. This appendix first develops the notion of magnification in the plane or space, then applies it to the evaluation of integrals in various coordinate systems.

L.1 MAGNIFICATION AND THE JACOBIAN

Let $x = f(u)$ be a differentiable function. As we saw in Sec. 3.2, it can be viewed as a projection from the u axis to the x axis. Its derivative, dx/du, describes the local magnification of *lengths*. We now will consider functions that project from one plane to another. Section L.2 will compute local magnification of *areas*.

Mappings from a Plane to a Plane

A **mapping** is a function F that assigns to points in some plane region points in another plane. The inputs are points in the uv plane; the outputs are points in the xy plane. (See Fig. 1.)

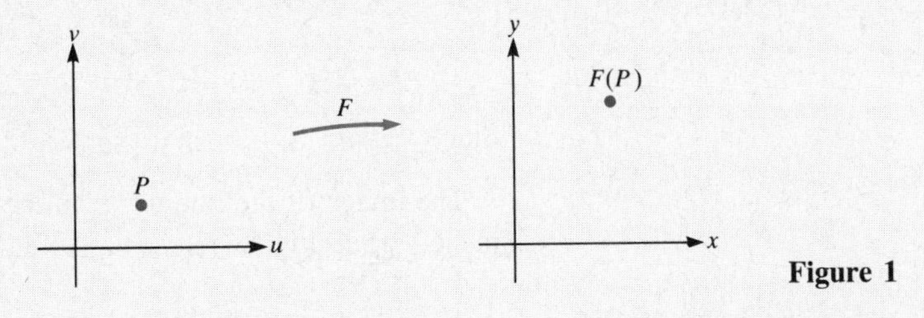

Figure 1

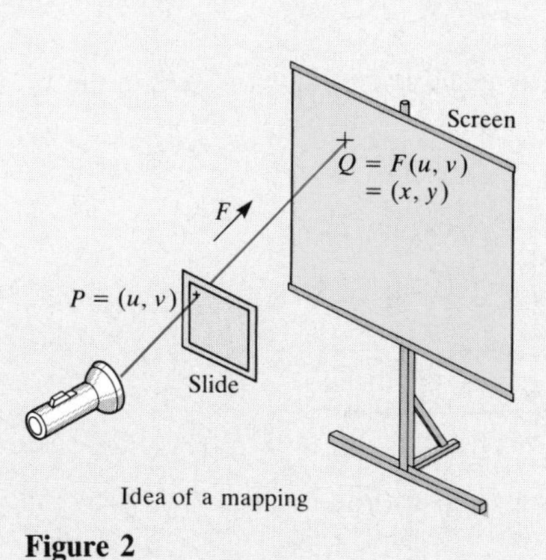

Idea of a mapping

Figure 2

We can think of the uv plane as a photographic slide, the function F as a projector, and the xy plane as a screen, as in Fig. 2. A point $P = (u, v)$ on the slide is projected by F onto the point $Q = (x, y) = F(u, v)$ on the screen.

We denote the input point (u, v) and the output point (x, y),

$$F(u, v) = (x, y).$$

Both x and y depend on u and v. So we may write

$$x = f(u, v) \qquad \text{and} \qquad y = g(u, v).$$

Notation for planar mapping F as a pair of real-valued functions

Thus
$$F(u, v) = (f(u, v), g(u, v)), \tag{1}$$
and F is shorthand for a pair, f and g, of real-valued functions of two variables. Any mapping from the uv plane to the xy plane is described by a pair of such functions, $x = f(u, v)$ and $y = g(u, v)$. It will be assumed that both f and g have continuous partial derivatives.

EXAMPLE 1 Let F be the mapping that assigns to the point (u, v) the point $(2u, 3v)$.

(*a*) Describe this mapping geometrically.
(*b*) Find the image of the line $v = u$.
(*c*) Find the image of the square in the uv plane whose vertices are $(0, 0)$, $(1, 0)$, $(1, 1)$, and $(0, 1)$.

SOLUTION

(*a*) In this case, $x = 2u$ and $y = 3v$. The table below records the effect of the mapping on the points listed in (*c*):

(u, v)	$(0, 0)$	$(1, 0)$	$(1, 1)$	$(0, 1)$
$(2u, 3v)$	$(0, 0)$	$(2, 0)$	$(2, 3)$	$(0, 3)$

In the notation $F(u, v) = (x, y)$, these data read

$$F(0, 0) = (2 \cdot 0, 3 \cdot 0) = (0, 0);$$
$$F(1, 0) = (2 \cdot 1, 3 \cdot 0) = (2, 0);$$
$$F(1, 1) = (2 \cdot 1, 3 \cdot 1) = (2, 3);$$
$$F(0, 1) = (2 \cdot 0, 3 \cdot 1) = (0, 3).$$

Note that the first coordinate of $(x, y) = F(u, v)$ is $x = 2u$, twice the first coordinate of (u, v). Thus the mapping magnifies horizontally by a factor of 2. Similarly, it stretches vertically by a factor of 3. (This causes a sixfold magnification of areas.)

(*b*) Let $P = (u, v)$ be on the line $v = u$. Then $F(P) = F(u, v) = (x, y)$, with $x = 2u$ and $y = 3v$. Thus

$$u = \frac{x}{2} \quad \text{and} \quad v = \frac{y}{3}.$$

Since $v = u$,
$$\frac{y}{3} = \frac{x}{2} \quad \text{or} \quad y = \tfrac{3}{2}x.$$

The image of the line $v = u$ in the uv plane is the line $y = 3x/2$ in the xy plane. (See Fig. 3.)

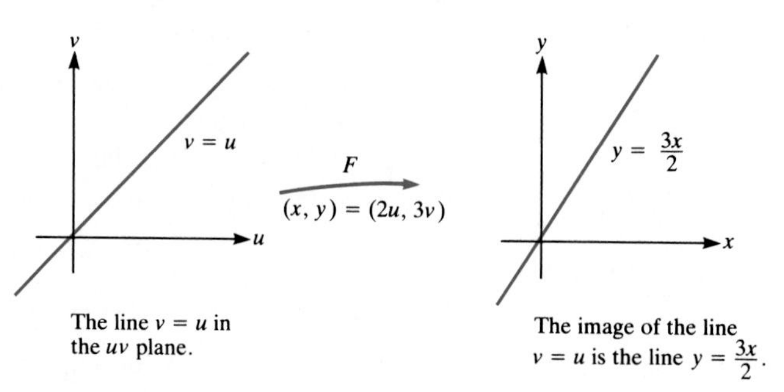

Figure 3 The line $v = u$ in the uv plane. The image of the line $v = u$ is the line $y = \frac{3x}{2}$.

This mapping takes any line into a line.

A similar argument shows that for this mapping the image of any line $Au + Bv + C = 0$ in the uv plane is a line in the xy plane, namely, the line $Ax/2 + By/3 + C = 0$.

(*c*) If P is a point in the square R whose vertices are

$$(0, 0), \quad (1, 0) \quad (1, 1) \quad (0, 1),$$

then the image of P is a point in the rectangle S whose vertices are

$$(0, 0), \quad (2, 0) \quad (2, 3) \quad (0, 3).$$

(See Fig. 4.)

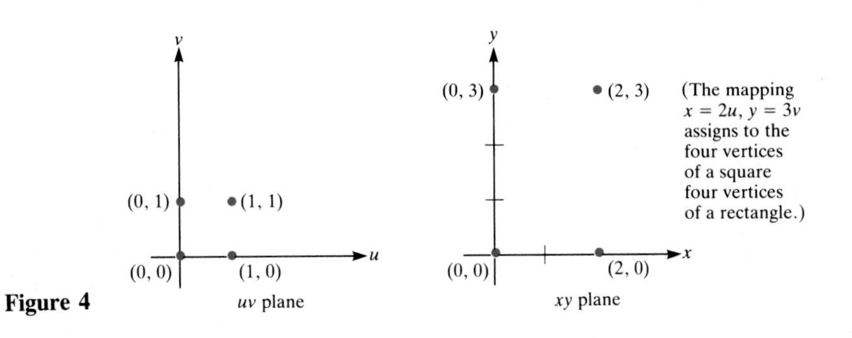

Figure 4

Think of (u, v) as a point on a slide and $(2u, 3v)$ as its image on the screen. Then the mapping F projects the square R on the slide onto a rectangle S on the screen. (See Fig. 5.)

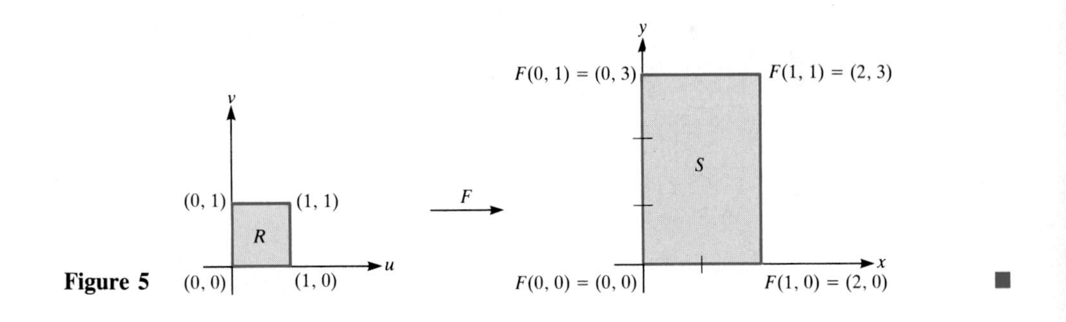

Figure 5

■

EXAMPLE 2 Let $F = (f, g)$ be the same as in Example 1. Let R be the set of points (u, v) in the uv plane (the slide) such that

$$u^2 + v^2 \leq 1.$$

In words, R is the disk of radius 1 centered at (0, 0). What is the image of R on the xy plane (the screen)?

Note once again how the image in the xy plane is calculated.

SOLUTION Since

$$u = \frac{x}{2} \quad \text{and} \quad v = \frac{y}{3},$$

the inequality

$$u^2 + v^2 \leq 1$$

implies that
$$\left(\frac{x}{2}\right)^2 + \left(\frac{y}{3}\right)^2 \le 1;$$

that is,
$$\frac{x^2}{4} + \frac{y^2}{9} \le 1.$$

Consequently, the image of the disk of radius 1 consists of all points inside the ellipse

$$\frac{x^2}{4} + \frac{y^2}{9} = 1.$$

For this mapping the image of a circle is an ellipse.

The image of the disk R is the ellipse S in Fig. 6.

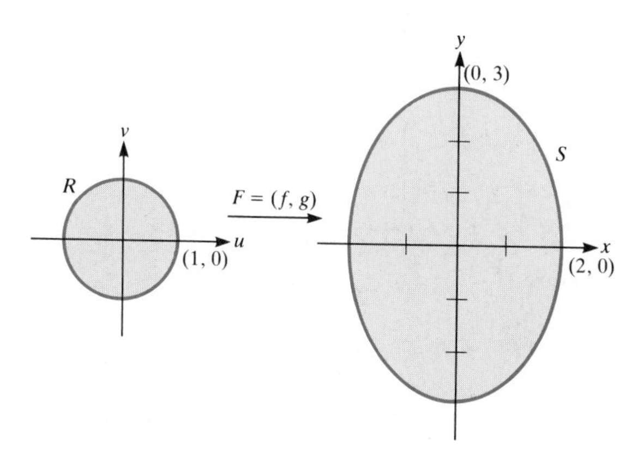

Figure 6

■

In the next mapping the image of a line is a parabola.

EXAMPLE 3 Let $(x, y) = (uv, v)$. If $P = (u, v)$ lies on the line $v = u$, where does $F(P) = (uv, v)$ lie?

SOLUTION In this case,

$$x = uv \qquad \text{and} \qquad y = v.$$

To find what the mapping does to the line $u = v$, find an equation linking x and y if (x, y) is on the image of the line $u = v$.

Observe that

$$x = uv = v^2 = y^2.$$

Thus as $P = (u, v)$ wanders about the line $u = v$, the image point (uv, v) wanders about the parabola $x = y^2$.

In order to make this fact more concrete, let us compute $F(P)$ for a few points on the line $v = u$. For instance, if $P = (2, 2)$,

$$F(P) = (2 \cdot 2, 2) = (4, 2).$$

P	*Image of P*
$P_1 = (-2, -2)$	$(4, -2)$
$P_2 = (-1, -1)$	$(1, -1)$
$P_3 = (0, 0)$	$(0, 0)$
$P_4 = (1, 1)$	$(1, 1)$
$P_5 = (2, 2)$	$(4, 2)$
$P_6 = (3, 3)$	$(9, 3)$

The table in the margin records the results of six such computations.

Figure 7 shows these data (the dashed line joins the images).

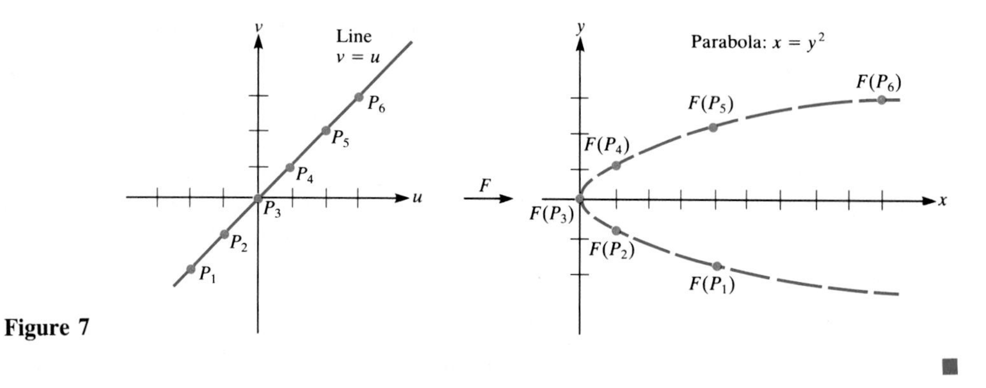

Figure 7

■

The next example introduces an important idea needed in the next section.

The restrictions on u and v make F one-to-one.

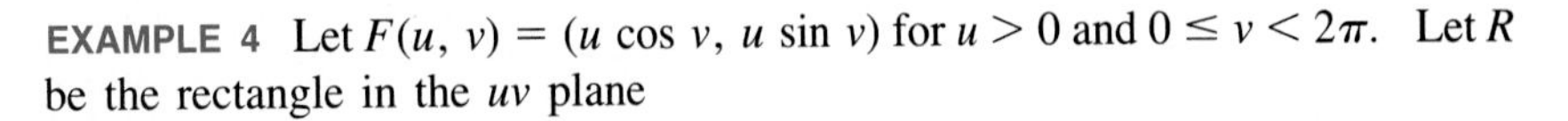

EXAMPLE 4 Let $F(u, v) = (u \cos v, u \sin v)$ for $u > 0$ and $0 \leq v < 2\pi$. Let R be the rectangle in the uv plane

$$1 \leq u \leq 2, \qquad \frac{\pi}{6} \leq v \leq \frac{\pi}{4}.$$

Sketch the image S of R under the effect of the mapping F.

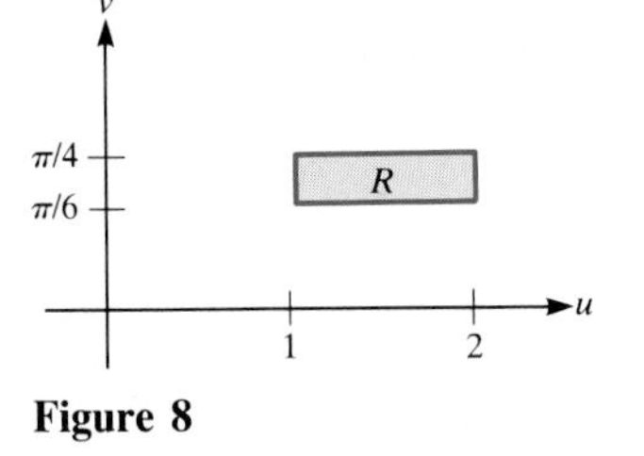

Figure 8

SOLUTION R is shown in Fig. 8. To find the image of R, examine the points $(x, y) = F(u, v)$ for (u, v) in the rectangle R.

By the definition of F,

$$x = u \cos v \qquad \text{and} \qquad y = u \sin v. \tag{2}$$

Note the resemblance between (2) and the equations that link polar and rectangular coordinates,

Assume $0 \leq \theta < 2\pi$.

$$x = r \cos \theta \qquad \text{and} \qquad y = r \sin \theta. \tag{3}$$

In fact, u must equal r and v must equal θ if we assume that $0 \leq \theta < 2\pi$. To see this, note from (2) that

$$\begin{aligned} x^2 + y^2 &= (u \cos v)^2 + (u \sin v)^2 \\ &= u^2(\cos^2 v + \sin^2 v) = u^2. \end{aligned}$$

So $u = \sqrt{x^2 + y^2}$. But $\sqrt{x^2 + y^2}$ equals the r of polar coordinates. So $u = r$. Thus $\cos v = \cos \theta$ and $\sin v = \sin \theta$; from this it follows that $v = \theta$, since both v and θ are between 0 and 2π.

So the image of R consists of those points in the xy plane whose polar coordinates satisfy the inequalities

$$1 \leq r \leq 2 \qquad \text{and} \qquad \frac{\pi}{6} \leq \theta \leq \frac{\pi}{4}.$$

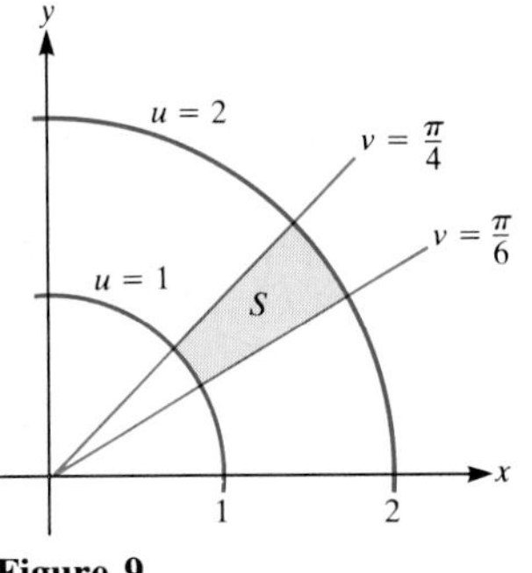

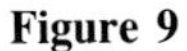

Figure 9

This region is shown in Fig. 9. ■

The main point of Example 4

The point of Example 4 is that while r and θ are polar coordinates in the xy plane, *they are also rectangular coordinates in their own right in an $r\theta$ plane.* (We may think of the r axis as horizontal and the θ axis as vertical.)

Magnification of a Linear Mapping

The mapping $F(u, v) = (2u, 3v)$ magnifies all areas by a factor of 6. In the next section we will need to know the magnification effect of a mapping of the form

$$F(u, v) = (au + bv, cu + dv) \tag{4}$$

for any constants a, b, c, and d, where $ad - bc \neq 0$. As is shown in Exercise 1, the image of a line in the uv plane is a line in the xy plane. Hence F is called a **linear mapping**.

> **Theorem** Let F be the mapping $F(u, v) = (au + bv, cu + dv)$. Let R be the rectangle
>
> $$u_1 \le u \le u_2, \qquad v_1 \le v \le v_2.$$
>
> Let S be the image of R. Then
>
> $$\frac{\text{Area of } S}{\text{Area of } R} = |ad - bc|.$$

Proof Let P_1, P_2, P_3, and P_4 be the vertices of the rectangle R and let Q_1, Q_2, Q_3, and Q_4 be the corresponding vertices of the parallelogram S, as shown in Fig. 10. Let $\Delta u = u_2 - u_1$ and $\Delta v = v_2 - v_1$.

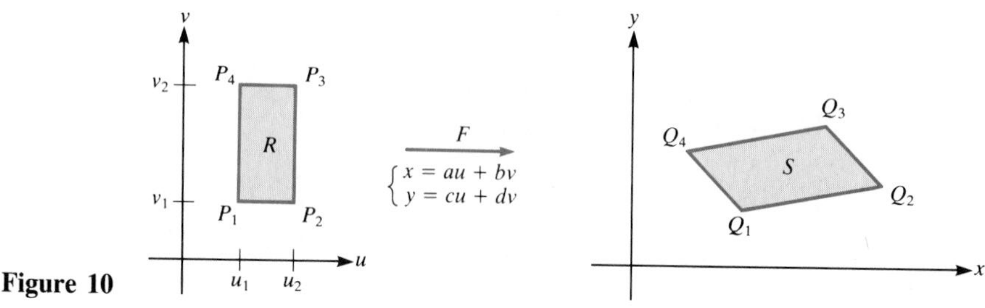

Figure 10

The area of R is $(u_2 - u_1)(v_2 - v_1) = \Delta u\, \Delta v$. All that remains is to compute the area of the parallelogram S.

The area of S is the magnitude of the cross product of the vectors $\overrightarrow{Q_1Q_2}$ and $\overrightarrow{Q_1Q_4}$. To find the cross product, first find the coordinates of Q_1, Q_2, and Q_4:

$$\begin{aligned} Q_1 &= F(P_1) = F(u_1, v_1) = (au_1 + bv_1, cu_1 + dv_1) \\ Q_2 &= F(P_2) = F(u_2, v_1) = (au_2 + bv_1, cu_2 + dv_1) \\ Q_4 &= F(P_4) = F(u_1, v_2) = (au_1 + bv_2, cu_1 + dv_2). \end{aligned} \tag{5}$$

From (5) it follows that

$$\overrightarrow{Q_1Q_2} = (au_2 - au_1)\mathbf{i} + (cu_2 - cu_1)\mathbf{j} = a\,\Delta u\,\mathbf{i} + c\,\Delta u\,\mathbf{j} = \Delta u\,(a\mathbf{i} + c\mathbf{j})$$

and $$\overrightarrow{Q_1Q_4} = (bv_2 - bv_1)\mathbf{i} + (dv_2 - dv_1)\mathbf{j} = b\,\Delta v\,\mathbf{i} + d\,\Delta v\,\mathbf{j} = \Delta v\,(b\mathbf{i} + d\mathbf{j}).$$

Thus the area of the parallelogram spanned by $\overrightarrow{Q_1Q_2}$ and $\overrightarrow{Q_1Q_4}$ is the absolute value of the determinant

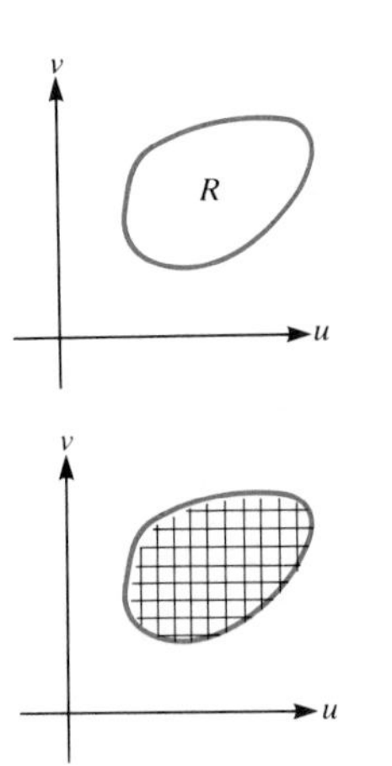

Figure 11

Figure 12

$$\begin{vmatrix} a\,\Delta u & c\,\Delta u \\ b\,\Delta v & d\,\Delta v \end{vmatrix} = \Delta u\,\Delta v(ad - bc).$$

The area of the parallelogram S is $|ad - bc|$ times the area of the rectangle R. This proves the theorem. ■

The mapping F in the theorem magnifies the area of *any* region, not just the rectangle described in the theorem, by the same factor, $|ad - bc|$. To see why, consider a typical region R in the uv plane, as in Fig. 11. This region may be approximated by many small rectangles of the type considered in the theorem, as shown in Fig. 12. Since the area of each of these small rectangles is magnified by the factor $|ad - bc|$, it is plausible that the area of the image of R is $|ad - bc|$ times the area of R.

EXERCISES FOR SEC. L.1: MAGNIFICATION AND THE JACOBIAN

1 Let a, b, c, and d be constants such that $ad - bc \neq 0$. Let F be the mapping.

$$x = au + bv, \qquad y = cu + dv.$$

(*a*) Solve for u and v in terms of x and y, obtaining

$$u = \frac{dx - by}{ad - bc} \quad \text{and} \quad v = \frac{ay - cx}{ad - bc}.$$

(*b*) Using the results in (*a*), show that the image of the line $Au + Bv + C = 0$ in the uv plane is a line in the xy plane.

2 The formula obtained in Exercise 1(*a*) defines a mapping G from the xy plane to the uv plane. (G is the inverse of the one-to-one mapping F.)
(*a*) Using the theorem in this section, compute the magnifying effect of G.
(*b*) The magnification of G is the reciprocal of the magnification of F. Why is this to be expected?

3 Let $F(u, v) = (3u + v, 4u + 2v)$.
(*a*) Sketch the image in the xy plane of the triangle in the uv plane whose vertices are (1, 1), (2, 1), (1, 2).
(*b*) Using the theorem, find the area of the image.

4 Let $F(u, v) = (3u + v, 4u + 2v)$.
(*a*) Plot the images of (1, 0), (0, 1), (−1, 0), (0, −1).
(*b*) Plot the image of the circle $u^2 + v^2 = 1$.
(*c*) Find an equation for the image of the circle in (*b*).
(*d*) Show that the image in (*c*) is an ellipse.
(*e*) What is its area?

Exercises 5 to 8 concern the mapping $F(u, v) = (2u, 3v)$.

5 (*a*) Plot the points $P_1 = (1, 0)$, $P_2 = (3, 0)$, and $P_3 = (6, 0)$ in the uv plane.
(*b*) Plot $Q_1 = F(P_1)$, $Q_2 = F(P_2)$, and $Q_3 = F(P_3)$ in the xy plane.
(*c*) What is the image in the xy plane of the line $v = 0$?

6 (*a*) Pick three points P_1, P_2, and P_3 on the line $u + v = 1$.
(*b*) Plot their images in the xy plane.
(*c*) Sketch the image in the xy plane of the line $u + v = 1$ and give its equation.

7 (*a*) Fill in this table:

(u, v)	$(x, y) = (2u, 3v)$
(0, 0)	
(2, 0)	
(1, 1)	

(*b*) Plot the three points in the left column of the table in (*a*) in the uv plane.
(*c*) Plot the three points entered in the right column of the table in (*a*) in the xy plane.
(*d*) Draw the triangular region R whose vertices are listed in (*a*) in the uv plane.
(*e*) Draw S, the image of R in the xy plane.
(*f*) Compute the areas of R and S, and show that S is six times as large as R.
(*g*) Show that R is a right triangle, but S is not.

8 Let R in the uv plane consist of all points (u, v) such that $u^2 + v^2 \leq 4$. (*a*) Draw R. (*b*) Draw the image in the xy plane of R.

In Exercises 9 and 10 let $F(u, v) = (u + v, u - v)$.

9 (*a*) Plot the points $P_1 = (1, 1)$, $P_2 = (3, 1)$, $P_3 = (1, 4)$, and $P_4 = (3, 4)$ in the uv plane.
(*b*) Plot in the xy plane the respective images of the points in (*a*) and label these images Q_1, Q_2, Q_3, and Q_4.
(*c*) Let R be the rectangle with vertices P_1, P_2, P_3, and P_4. Let S in the xy plane be the image of R. Sketch S. What kind of figure is S?

10 (*a*) Show that, if $F(u, v) = (x, y)$, then $u = (x + y)/2$ and $v = (x - y)/2$.
(*b*) Use (*a*) to find the image of the line $u + 2v = 1$.
(*c*) Use (*a*) to find the image of the circle $u^2 + v^2 = 9$.

11 Let $F(u, v) = (v, \sqrt{u})$ for positive u and v.
(*a*) The points (1, 2), (2, 4), (3, 6), and (4, 8) lie on the line $v = 2u$. Plot their images in the xy plane.
(*b*) Show that, if $(x, y) = (v, \sqrt{u})$, then $v = x$ and $u = y^2$.
(*c*) Use (*b*) to find the equation of the image of the half line $v = 2u$. (Recall that $u, v > 0$.)
(*d*) Sketch the image of the half line $v = 2u$.

12 Let $(f(u, v), g(u, v)) = (2u + 3v, u + v)$.
(*a*) Fill in this table:

(u, v)	$(x, y) = (2u + 3v, u + v)$
(1, 0)	
(0, 1)	
(−1, 0)	
(0, −1)	

(*b*) As (u, v) runs through the circle $u^2 + v^2 = 1$ counterclockwise, $(x, y) = (2u + 3v, u + v)$ sweeps out a curve C in the xy plane. Is C swept out clockwise or counterclockwise?

Exercises 13 to 16 concern the mapping given by $F(u, v) = (u \cos v, u \sin v)$ for $u > 0$ and $0 \le v < 2\pi$.

13 (*a*) The points $(5, 0)$, $(5, \pi/6)$, $(5, \pi/2)$, and $(5, \pi)$ lie on the line $u = 5$ in the uv plane. Plot their images in the xy plane.
(*b*) What is the image of the line segment $u = 5$?

14 (*a*) The points $(3, \pi/6)$, $(4, \pi/6)$, $(5, \pi/6)$, and $(6, \pi/6)$ lie on the line $v = \pi/6$. Plot their images in the xy plane.
(*b*) What is the image of the half line $v = \pi/6$, $u > 0$?

15 (*a*) Sketch the region R in the uv plane bounded by the lines $u = 2$, $u = 3$, $v = 1$, and $v = 4$.
(*b*) Sketch the image of R.

16 Find (u, v) if $F(u, v)$ is
(*a*) (3, 3) (*b*) $(4, 4\sqrt{3})$
(*c*) (6, 0) (*d*) (0, −7)

In Exercises 17 and 18 consider the mapping $x = uv$, $y = v$, for $u > 0$, $v > 0$. (See Example 3.)

17 Find the image of (*a*) the line segment $u + v = 1$, (*b*) the branch of the hyperbola $uv = 1$ in the first quadrant.

18 Sketch the image of the triangle with vertices (1, 1), (3, 3), (2, 4).

---

19 Let $F(u, v) = (u \cos v, u \sin v)$ for $u > 0$ and $0 \le v < 2\pi$. Let R be the rectangle $u_0 \le u \le u_0 + \Delta u$, $v_0 \le v \le v_0 + \Delta v$, where u_0, v_0, Δu, and Δv are constants.
(*a*) What is the area of R?
(*b*) What is the area of S, the image of R?
(*c*) Find the limit of the quotient

$$\frac{\text{Area of } S}{\text{Area of } R}$$

as Δu and Δv approach 0.
The limit in (*c*) depends on the point (u_0, v_0). It is called the **magnification** of the mapping at (u_0, v_0).

20 Let a_{ij}, $1 \le i \le 3$, $1 \le j \le 3$ be nine constants. (The double subscript is read in two parts; for example, a_{12} is read as "a sub one two" rather than as "a sub twelve.") Define a mapping F from uvw space to xyz space by

$$F(u, v, w) = (a_{11}u + a_{12}v + a_{13}w,\ a_{21}u + a_{22}v + a_{23}w,\ a_{31}u + a_{32}v + a_{33}w).$$

Find the factor by which it magnifies volumes.

L.2 THE JACOBIAN AND CHANGE OF COORDINATES

Review "magnification" in Secs. 3.1 and 3.2.

Let $x = f(u)$ be a differentiable function. The image of the interval $[u, u + \Delta u]$ is an interval $[x, x + \Delta x]$. The ratio of their lengths is $|\Delta x/\Delta u|$. As $\Delta u \to 0$, this ratio approaches $|f'(u)|$. So $f'(u)$ describes the local magnification of length. (See Secs. 3.1 and 3.2.) This magnification may vary from point to point.

In Sec. L.1 we found that the mapping

$$x = au + bv, \qquad y = cu + dv$$

magnifies all *areas* by the factor $|ad - bc|$. The local magnification is the same at all points.

We now define the local magnification of area for a mapping from the uv plane to the xy plane and obtain a formula for computing it. Then we will describe its applications.

Local Magnification of Area

Let $F(u, v) = (f(u, v), g(u, v))$ be a mapping such that f and g have continuous partial derivatives. Let $P_0 = (u_0, v_0)$ be a point in the domain of F. The magnification of F at P_0 is defined as follows.

The magnification defined as a limit. It tell how areas expand or shrink near a point.

Let R be the rectangle $u_0 \leq u \leq u_0 + \Delta u$, $v_0 \leq v \leq v_0 + \Delta v$. Let S be its image. Then the **magnification** of F at P_0 is

$$\lim_{\Delta u, \Delta v \to 0} \frac{\text{Area of } S}{\text{Area of } R}; \tag{1}$$

the limit, if it exists, is taken as both Δu and Δv approach 0.

To find the limit (1), we will approximate F near P_0 by a linear mapping. In the theorem, "abs" is short for "absolute value of."

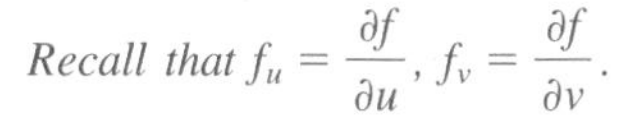

Recall that $f_u = \dfrac{\partial f}{\partial u}$, $f_v = \dfrac{\partial f}{\partial v}$.

Theorem 1 Let $F(u, v) = (f(u, v), g(u, v))$ be a mapping such that f and g have continuous partial derivatives. Let $P_0 = (u_0, v_0)$ be a point in the domain of F. Then the magnification of F at P_0 is

$$\text{abs} \begin{vmatrix} f_u(P_0) & f_v(P_0) \\ g_u(P_0) & g_v(P_0) \end{vmatrix},$$

that is, $|f_u g_v - f_v g_u|$ evaluated at P_0.

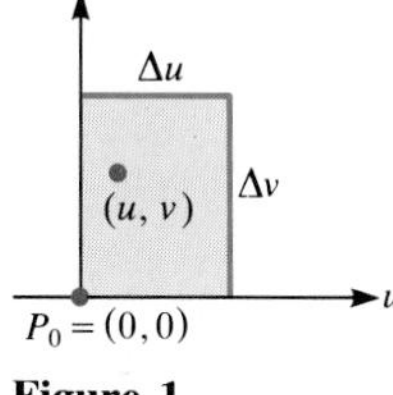

Figure 1

Proof (*Informal sketch*) In order to avoid a clutter of symbols, take P_0 to be (0, 0), the origin of the uv plane, and assume that $F(P_0)$ is the origin of the xy plane. Thus $f(P_0) = 0$ and $g(P_0) = 0$.

Let R be the rectangle $0 \leq u \leq \Delta u$, $0 \leq v \leq \Delta v$, where Δu and Δv are fixed numbers. Let S be its image in the xy plane.

For (u, v) in R approximate $f(u, v)$ and $g(u, v)$ with the aid of differentials. This is possible since we shall be interested in the case when Δu and Δv are small. (See Fig. 1.)

For (u, v) near P_0, $f(u, v)$ is approximated by $f(P_0)$ plus a differential:

Note that $f_u(P_0)$, $f_v(P_0)$, $g_u(P_0)$, $g_v(P_0)$ *are just constants.*

$$f(u, v) \approx f(P_0) + f_u(P_0)u + f_v(P_0)v.$$

Since $f(P_0) = 0$, $\quad f(u, v) \approx f_u(P_0)u + f_v(P_0)v;$

similarly, $\quad g(u, v) \approx g_u(P_0)u + g_v(P_0)v. \quad (2)$

Introduce the linear mapping F^* defined by

$$F^*(u, v) = (f_u(P_0)u + f_v(P_0)v,\ g_u(P_0)u + g_v(P_0)v). \tag{3}$$

By (2), F^* is a good approximation of F when (u, v) is near (0, 0). Since the image of R under the mapping F^* is a parallelogram, the image S of R under the mapping F will closely resemble a parallelogram, as shown in Fig. 2.

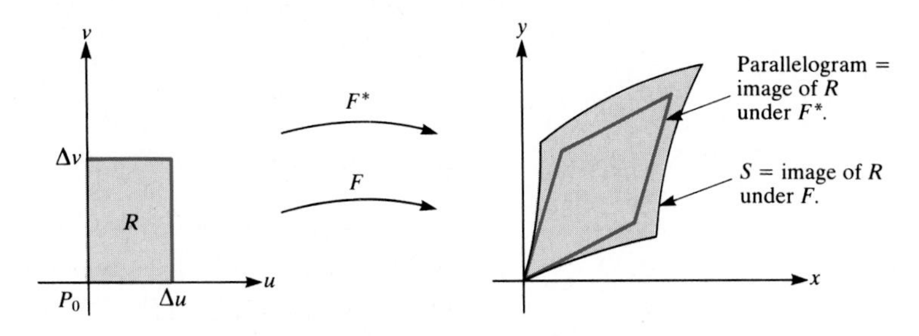

Figure 2

By the theorem in Sec. L.1, the area of the parallelogram is the absolute value of

$$\begin{vmatrix} f_u(P_0) & f_v(P_0) \\ g_u(P_0) & g_v(P_0) \end{vmatrix} \Delta u \, \Delta v. \tag{4}$$

Since F is closely approximated by F^* when Δu and Δv are small, it is reasonable to expect that the area of the image of R under F, though not usually a parallelogram, is close to (4). Consequently, we expect that

$$\lim_{\Delta u,\, \Delta v \to 0} \frac{\text{Area of } S}{\text{Area of } R} = \text{abs} \begin{vmatrix} f_u(P_0) & f_v(P_0) \\ g_u(P_0) & g_v(P_0) \end{vmatrix}.$$

This concludes the sketch of the argument. ■

The **Jacobian** of $F = (f, g)$ is defined as the determinant

$$\begin{vmatrix} f_u & f_v \\ g_u & g_v \end{vmatrix}.$$

It is denoted

Notations for the Jacobian

$$J, \qquad J(P), \qquad \frac{\partial(f, g)}{\partial(u, v)}, \qquad \text{or} \qquad \frac{\partial(x, y)}{\partial(u, v)}.$$

In the next example the Jacobian is computed for the mapping of Example 4 in the preceding section.

EXAMPLE 1 Let $F(u, v) = (u \cos v, u \sin v)$ for $u > 0$, $0 \le v < 2\pi$. Find the magnification of F at (u, v).

SOLUTION The magnification is the absolute value of the Jacobian,

$$\begin{aligned} \left| \frac{\partial(x, y)}{\partial(u, v)} \right| &= \text{abs} \begin{vmatrix} \dfrac{\partial}{\partial u}(u \cos v) & \dfrac{\partial}{\partial v}(u \cos v) \\ \dfrac{\partial}{\partial u}(u \sin v) & \dfrac{\partial}{\partial v}(u \sin v) \end{vmatrix} \\ &= \text{abs} \begin{vmatrix} \cos v & -u \sin v \\ \sin v & u \cos v \end{vmatrix} \\ &= |u \cos^2 v + u \sin^2 v| \\ &= |u|. \end{aligned}$$

Since $u > 0$, the local magnification at (u, v) is u. ■

Magnification and Integrals

Let F be a one-to-one mapping from the uv plane to the xy plane. Let R be a region in the uv plane and S its image in the xy plane. Let $h(x, y)$ be a real-valued function defined on S. (See Fig. 3.)

We use the letter h since f and g are used for F = (f, g).

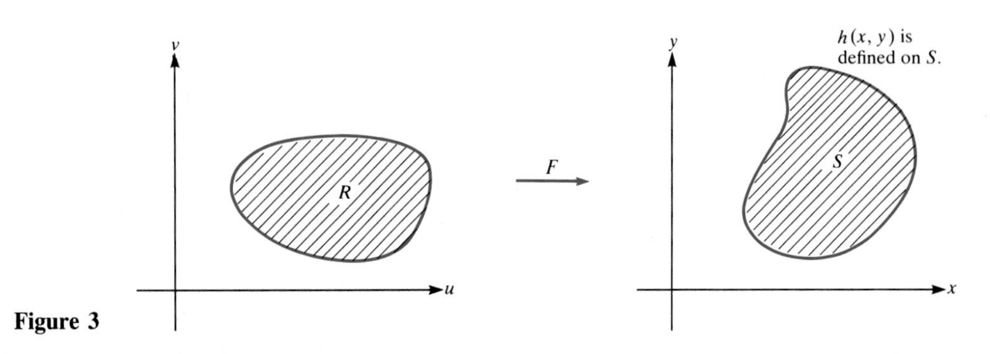

Figure 3

The integral $\int_S h(x, y)\,dA$ can be expressed as an integral of a different function over the region R. If R is a simpler region than S, then the new integral may be easier to compute. This is analogous to the substitution technique, which expresses a definite integral of a given function over one interval as the definite integral of another function over another interval. To see how to express an integral over S as an integral over R, go back to the approximating sums used in the definition of an integral.

Consider a typical partition $S_1, S_2, \ldots, S_n$ of S and corresponding sampling points $Q_1, Q_2, \ldots, Q_n$. Let $R_1, R_2, \ldots, R_n$ be the corresponding partition of R such that S_i is the image of R_i. Let P_i be the point in R_i whose image is Q_i, $1 \le i \le n$. (See Fig. 4.)

Use Q for points in S and P for points in R.

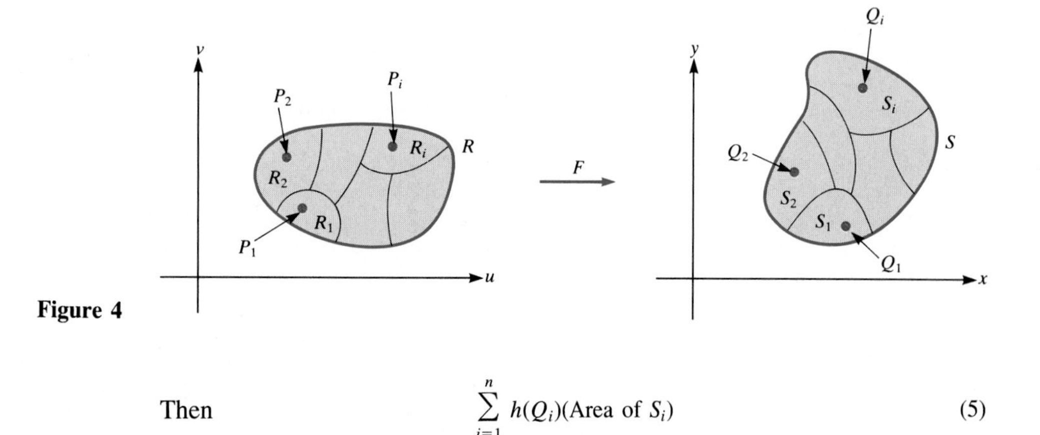

Figure 4

Then
$$\sum_{i=1}^{n} h(Q_i)(\text{Area of } S_i) \tag{5}$$

is an approximating sum for $\int_S h(Q)\,dA$. But the area of S_i is approximately $|J(P_i)|$ times the area of R_i. So (5) is approximately

$$\sum_{i=1}^{n} h(Q_i)|J(P_i)|\ (\text{Area of } R_i). \tag{6}$$

Since $Q_i = F(P_i)$, (6) equals

$$\sum_{i=1}^{n} h(F(P_i))|J(P_i)| \text{ (Area of } R_i). \tag{7}$$

The sum (7) is an approximating sum for the definite integral of the function $h(F(P))|J(P)|$ over the set R. Thus it is reasonable to expect that

$$\int_S h(Q)\, dA = \int_R h(F(P))|J(P)|\, dA^*. \tag{8}$$

(Here we use the symbol dA^* to indicate the element of area in the uv plane.)

This argument suggests the following theorem, the proof of which is part of an advanced calculus course.

Theorem 2 Let F be a one-to-one mapping from the uv plane to the xy plane, $x = f(u, v)$, $y = g(u, v)$. Let S be the image in the xy plane of the set R in the uv plane. Let h be a real-valued function defined on S. Then, if $J(P)$ is never 0 in R,

$$\int_S h(Q)\, dA = \int_R h(F(P))|J(P)|\, dA^*,$$

where dA^* is the element of area in the uv plane.

EXAMPLE 2 Let R be the triangle in the uv plane with vertices (0, 0), (1, 0), and (0, 1). Let F be the mapping

$$x = 2u - 3v, \qquad y = 5u + 7v.$$

Let S be the image of R under F. Evaluate the integral $\int_S x\, dA$ over the set S in the xy plane.

SOLUTION Since F is a linear mapping, S is a triangle. Its vertices are

$$F(0, 0) = (0, 0), \qquad F(1, 0) = (2, 5), \qquad F(0, 1) = (-3, 7).$$

(See Fig. 5.)

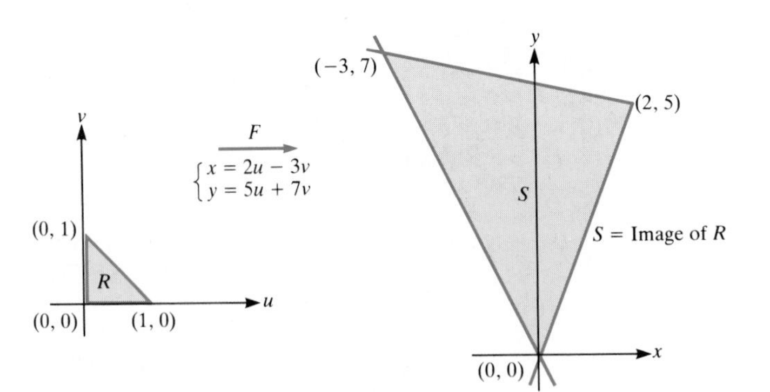

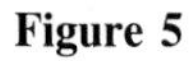
Figure 5

Since R is a simpler region over which to integrate, replace $\int_S x\, dA$ by an integral over R with the aid of Theorem 2:

$$\int_S x\, dA = \int_R (2u - 3v) \left| \frac{\partial(x, y)}{\partial(u, v)} \right| dA^*.$$

The Jacobian in this case is

$$\frac{\partial(x, y)}{\partial(u, v)} = \begin{vmatrix} \frac{\partial}{\partial u}(2u - 3v) & \frac{\partial}{\partial v}(2u - 3v) \\ \frac{\partial}{\partial u}(5u + 7v) & \frac{\partial}{\partial v}(5u + 7v) \end{vmatrix} = \begin{vmatrix} 2 & -3 \\ 5 & 7 \end{vmatrix} = 29.$$

Thus

$$\int_S x\, dA = \int_R (2u - 3v)(29)\, dA^*$$

$$= 29 \int_0^1 \left[\int_0^{1-u} (2u - 3v)\, dv \right] du.$$

The first integration gives

$$\int_0^{1-u} (2u - 3v)\, dv = \left(2uv - \frac{3v^2}{2} \right) \Bigg|_{v=0}^{v=1-u}$$

$$= 2u(1 - u) - \frac{3(1 - u)^2}{2}.$$

The second integration gives, as can be checked,

If in doubt about the usefulness of the Jacobian, compute $\int_S x\, dA$ directly.

$$\int_0^1 \left[2u(1 - u) - \frac{3(1 - u)^2}{2} \right] du = -\frac{1}{6}.$$

Thus

$$\int_S x\, dA = -\frac{29}{6}. \quad ■$$

Jacobian and Change of Variable

Why the extra r in polar coordinates

Theorem 2 is what lies behind the appearance of the extra r when integration is carried out with polar coordinates, as will now be shown.

Let F be the mapping $F(u, v) = (u \cos v, u \sin v)$. Let R be the rectangle in the uv plane, $a \le u \le b$, $\alpha \le v \le \beta$. Let S be the image of R in the xy plane. As shown in Example 4 of the preceding section, S is bounded by parts of two rays and two circles. Let $h(x, y)$ be a real-valued function defined on S. Then, by Theorem 2 and Example 1,

$$\int_S h(x, y)\, dA = \int_R h(u \cos v, u \sin v) \left| \frac{\partial(x, y)}{\partial(u, v)} \right| dA^*$$

$$= \int_R h(u \cos v, u \sin v)\, u\, dA^*.$$

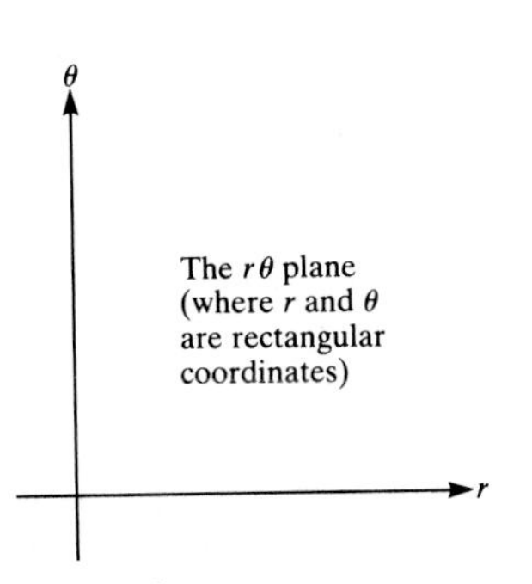

Figure 6

As was shown in Example 4 in the preceding section, u and v are the polar coordinates r and θ of the point (x, y). So let us relabel the uv plane the $r\theta$ plane, as in Fig. 6. Thus

$$\int_S h(x, y)\, dA = \int_R h(r \cos \theta, r \sin \theta)\, r\, dA^*. \tag{9}$$

(Here is where r enters the integrand, as a Jacobian.)

The second integral in (9) is over a set R in the $r\theta$ plane. But r and θ are the *rectangular* coordinates in the $r\theta$ plane. This second integral can be evaluated by repeated integration in rectangular coordinates:

$$\int_R h(r\cos\theta, r\sin\theta)\, r\, dA^* = \int_\alpha^\beta \left[\int_a^b h(r\cos\theta, r\sin\theta)\, r\, dr\right] d\theta. \tag{10}$$

The factor $\rho^2 \sin\phi$ introduced when integrating with spherical coordinates is also a Jacobian. (See Exercise 21.)

The next example shows how Theorem 2 can be used to simplify the computation of integrals.

EXAMPLE 3 Evaluate $\int_S x^2\, dA$, where S is the region bounded by the ellipse $x^2/4 + y^2/9 = 1$.

SOLUTION The given ellipse is the image of the disk of radius 1 and center $(0, 0)$ under the mapping $x = 2u$, $y = 3v$. The Jacobian of this mapping is 6 for all points (u, v). (Recall Example 2 in Sec. L.1.)

Let R be the disk mentioned. Then

$$\int_S x^2\, dA = \int_R (2u)^2\, 6\, dA^* = \int_R 24u^2\, dA^*.$$

Using polar coordinates in the uv plane, we evaluate the integral over R:

$$\begin{aligned}\int_R 24u^2\, dA^* &= \int_0^{2\pi}\left[\int_0^1 24(r\cos\theta)^2\, r\, dr\right] d\theta \\ &= \int_0^{2\pi} 6\cos^2\theta\, d\theta \\ &= 4\int_0^{\pi/2} 6\cos^2\theta\, d\theta = 4\cdot 6\cdot\frac{\pi}{4} = 6\pi. \quad \blacksquare\end{aligned}$$

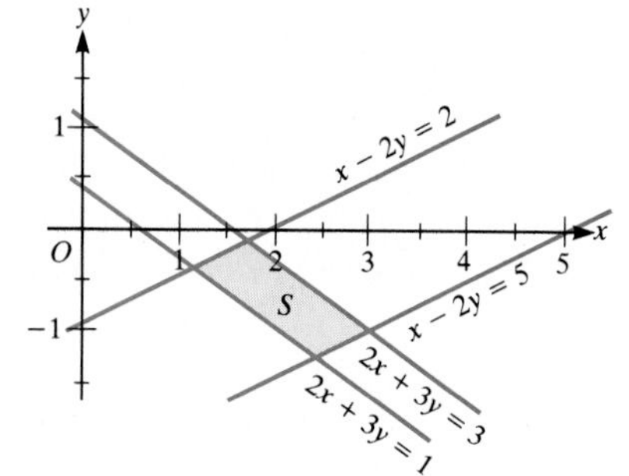

Figure 7

EXAMPLE 4 Find $\int_S y^2\, dA$, where S is the region bounded by the four lines $x - 2y = 2$, $x - 2y = 5$, $2x + 3y = 1$, $2x + 3y = 3$, shown in Fig. 7.

SOLUTION We wish to evaluate $\int_S y^2\, dA$. However, S is a somewhat messy region to describe by cross sections, so let us replace the integral by an integral over an easier region to work with.

Each point (x, y) in S is on a unique level curve of the function $x - 2y$ and on a unique level curve of the function $2x + 3y$. Let $u = x - 2y$ and $v = 2x + 3y$. Then each point (x, y) in S determines a pair of numbers u and v which describe which level curves (x, y) lies on.

As (x, y) varies over S, u varies from 2 to 5 and v varies independently from 1 to 3. That means that as (x, y) wanders over S, the point (u, v) wanders over the rectangle R in the uv plane described by

$$2 \le u \le 5, \qquad 1 \le v \le 3.$$

To replace $\int_S y^2\, dA$ by an integral over R, we need to express y^2 in terms of u and v and to find the Jacobian $\partial(x, y)/\partial(u, v)$. So we will find x and y in terms of u and v. To do this, solve the simultaneous equations

$$u = x - 2y \qquad \text{and} \qquad v = 2x + 3y$$

for x and y. By algebra, we obtain

$$x = \frac{3u + 2v}{7} \qquad \text{and} \qquad y = \frac{-2u + v}{7}$$

Hence
$$J = \frac{\partial(x, y)}{\partial(u, v)} = \begin{vmatrix} \frac{3}{7} & \frac{2}{7} \\ -\frac{2}{7} & \frac{1}{7} \end{vmatrix} = \frac{1}{7}.$$

Thus
$$\int_S y^2\, dA = \int_R \left(\frac{-2u + v}{7}\right)^2 \left(\frac{1}{7}\right) dA^*$$

$$= \int_2^5 \left(\int_1^3 \frac{4u^2 - 4uv + v^2}{7^3}\, dv\right) du$$

$$= \frac{1}{7^3} \cdot 170 = \frac{170}{343} \qquad \blacksquare$$

EXERCISES FOR SEC. L.2: THE JACOBIAN AND CHANGE OF COORDINATES

In Exercises 1 to 4 compute the local magnification of area by the given mappings at the given points in the uv plane.

1 $F(u, v) = (uv, v^2)$, $u, v > 0$, at
(a) $(1, 2)$ (b) $(3, 1)$

2 $F(u, v) = (1/u, 1/v)$, $u, v > 0$, at
(a) $(2, 3)$ (b) $(\frac{1}{2}, 4)$

3 $F(u, v) = (e^u \cos v, e^u \sin v)$, $0 \le v < 2\pi$, at
(a) $(1, \pi/4)$ (b) $(2, \pi/6)$

4 $F(u, v) = (u/(u^2 + v^2), v/(u^2 + v^2))$, $u^2 + v^2 \neq 0$, at
(a) $(3, 1)$ (b) $(1, 0)$

5 Let a, b, c, and d be constants such that $ad - bc \neq 0$. Let

$$x = au + bv, \qquad y = cu + dv.$$

Show that the Jacobian of the mapping is $ad - bc$ at all points.

6 Let the Jacobian of a mapping be 3 at $(2, 4)$. Let R be a small region around $(2, 4)$ of area 0.05. Approximately how large is the image of R under the mapping?

7 Let R be the disk $u^2 + v^2 \le 1$ in the uv plane. Let S be the region within the ellipse $x^2/a^2 + y^2/b^2 = 1$. Let $F(u, v) = (au, bv)$.
(a) Show that S is the image of R.
(b) Express $\int_S y^2\, dA$ as an integral over R.
(c) Evaluate the integral over R by polar coordinates.
(If you have some spare time, evaluate $\int_S y^2\, dA$ directly by a repeated integral in rectangular coordinates. The experience will deepen your appreciation of the Jacobian.)

8 (See Exercise 7.) Evaluate $\int_S y\, dA$ where S is the region within the ellipse $x^2/a^2 + y^2/b^2 = 1$ that lies in the first quadrant.

9 Let S be the image of the region R under the one-to-one mapping F. Show that the area of S equals $\int_R |J|\, dA^*$.

10 Let R be the square with vertices $(1, 1)$, $(2, 1)$, $(2, 2)$, and $(1, 2)$ in the uv plane. Let $F(u, v) = (u^3 + v, v^3/3)$. Find the area of the image of R.

11 Let R be the triangle bounded by $u = v$, $v = 0$, and $u = 1$. Let S be the image of R under the mapping $x = u^2$, $y = 2v$.
(a) Draw S, which has one curved side.
(b) Compute the area of S with the aid of the formula in Exercise 9.
(c) Compute the area of S directly.

12 Consider the mapping $x = u^2 - v^2$, $y = 2uv$. Let R be the square whose vertices are $(1, 0)$, $(2, 0)$, $(2, 1)$, and $(1, 1)$.
(a) Show that when $u = 1$, the image of (u, v) lies on the curve $x = 1 - (y/2)^2$.
(b) Show that when $u = 2$, the image of (u, v) lies on the curve $x = 4 - (y/4)^2$.
(c) Show that the image of the line $v = 0$ is the positive x axis.
(d) Show that the image of the line $v = 1$ is the curve $x = (y/2)^2 - 1$.
(e) Draw S, the image of R. (It has three curved sides and one straight side.)
(f) Find the area of S.

13 A mapping F is said to **preserve area** if the area of the image of R is equal to the area of R for all regions that have an area. Show that the mapping $F(u, v) = (4u + 3v, 3u + 2v)$ preserves area.

14 (See Exercise 13.)
(a) Show that the mapping defined by

$$x = u - v^2 - 2u^2v - u^4, \qquad y = v + u^2$$

preserves area.
(b) Sketch the image of the square whose vertices are $(0, 0)$, $(1, 0)$, $(1, 1)$, and $(0, 1)$.
(c) What is the area of the region sketched in (b)?

15 Consider only positive u, v, x, y. Define a mapping by setting $x = u^{1/3}v^{2/3}$, $y = u^{2/3}v^{1/3}$.
(*a*) Show that $x^2 = vy$ and $y^2 = ux$.
(*b*) Let R in the uv plane be the rectangle bounded by the lines $u = 1$, $u = 2$, $v = 3$, and $v = 4$. Let S be the image in the xy plane of R. Show that S is bounded by the four parabolas $y^2 = x$, $y^2 = 2x$, $x^2 = 3y$, and $x^2 = 4y$.
(*c*) Draw S.
(*d*) Compute the area of S by integrating the Jacobian over R.

16 Let $x = 2u + v$, $y = 3u + 2v$ be a mapping from the uv plane to the xy plane.
(*a*) Show that the mapping does not change areas [though, as (*b*) will show, it changes shapes].
(*b*) Sketch the image of the circle $u^2 + v^2 = 1$.
(*c*) Let R be the region in the uv plane bounded by the circle $u^2 + v^2 = 1$. Let S be its image in the xy plane. Evaluate $\int_S x^2 \, dA$.

17 Evaluate $\int_S x^2 \, dA$ of Example 3 directly, by vertical cross sections.

18 Evaluate the integral $\int_S y^2 \, dA$ in Example 4 directly by a repeated integral using vertical cross sections.

19 Evaluate the repeated integral in Example 4.

20 (*a*) For the mapping $x = u \cos v$, $y = u \sin v$, with $u > 0$ and $0 \le v < 2\pi$, compute u and v in terms of x and y and verify that

$$\frac{\partial(x, y)}{\partial(u, v)} \cdot \frac{\partial(u, v)}{\partial(x, y)} = 1.$$

(*b*) Why would you expect the equation in (*a*) to be true?

---

21 Let $x = f(u, v, w)$, $y = g(u, v, w)$, $z = h(u, v, w)$ be three functions with continuous partial derivatives. They determine a mapping F from uvw space to xyz space. The Jacobian of F is defined as the determinant

$$\begin{vmatrix} \dfrac{\partial f}{\partial u} & \dfrac{\partial f}{\partial v} & \dfrac{\partial f}{\partial w} \\ \dfrac{\partial g}{\partial u} & \dfrac{\partial g}{\partial v} & \dfrac{\partial g}{\partial w} \\ \dfrac{\partial h}{\partial u} & \dfrac{\partial h}{\partial v} & \dfrac{\partial h}{\partial w} \end{vmatrix}.$$

It can be shown that the absolute value of this Jacobian measures the local magnification of *volume* by F.
(*a*) Find the Jacobian of the mapping

$$x = \rho \sin \phi \cos \theta, \qquad y = \rho \sin \phi \sin \theta, \qquad z = \rho \cos \phi.$$

(*b*) Use (*a*) to explain the presence of $\rho^2 \sin \phi$ in integrals involving spherical coordinates.

22 (See Exercise 21.) Use a Jacobian to explain the presence of r in integrals using cylindrical coordinates.

23 (*a*) Show that

$$\begin{vmatrix} a & b \\ c & d \end{vmatrix} \cdot \begin{vmatrix} e & f \\ g & h \end{vmatrix} = \begin{vmatrix} ae + bg & af + bh \\ ce + dg & cf + dh \end{vmatrix}.$$

(*b*) Assume that the mapping F is one-to-one, so that u and v are functions of x and y. Use (*a*) to show that

$$\frac{\partial(x, y)}{\partial(u, v)} \frac{\partial(u, v)}{\partial(x, y)} = 1.$$

24 Find the moment of inertia of the region S in Example 4 about the y axis. Assume it has mass M and is homogeneous.

25 Let S be the region bounded by the lines $y = x$, $y = 2x$, $x + y = 3$, and $x + y = 4$. Find the moment of inertia of S around the x axis. Assume that S has mass M and is homogeneous. *Hint:* Model your argument on Example 4, first rewriting the first two lines as $y/x = 1$ and $y/x = 2$.

26 Let S be the region in the first quadrant bounded by the parabolas $y = x^2$, $y = 2x^2$, $x = y^2$, and $x = 2y^2$.
(*a*) Find the area of S. *Hint:* See the hint in Exercise 25.
(*b*) Find $(\bar{x}, \bar{y})$, the centroid of S.

27 Let S be the region in the first quadrant bounded by the hyperbolas $x^2 - y^2 = 1$ and $x^2 - y^2 = 4$ and the circles $x^2 + y^2 = 4$ and $x^2 + y^2 = 9$.
(*a*) Sketch S.
(*b*) Find $\int_S x \, dA$.

28 Let r and θ be rectangular coordinates in an $r\theta$ plane. (Do not think of them as polar coordinates.) Let $x = r \cos \theta$, $y = r \sin \theta$ be a mapping from the $r\theta$ plane to the xy plane. In each part of this exercise, sketch a region R in the $r\theta$ plane that is mapped onto S by the mapping.
(*a*) S is in the first quadrant and bounded by $y = x$, $y = \sqrt{3}x$, $x^2 + y^2 = 1$, $x^2 + y^2 = 9$.
(*b*) S is the disk of radius 3 centered at the origin.
(*c*) S is the disk of radius 2 centered at $(2, 0)$.

APPENDIX M

LINEAR DIFFERENTIAL EQUATIONS WITH CONSTANT COEFFICIENTS

"D.E." is short for "differential equation."

This section treats a type of differential equation that many engineering and physics students may meet even before they take a D.E. course. It is intended to serve as a reference. (In Theorem 4 it makes use of the complex numbers.)

The differential equation $dy/dx = ay$, where a is a constant, or equivalently,

$$\frac{dy}{dx} - ay = 0 \tag{1}$$

was solved in Sec. 6.7. Any solution has to be of the form $y = Ae^{ax}$ for some constant A. This section is concerned with generalizations of Eq. (1).

First, we consider differential equations of the form

$$\frac{dy}{dx} + ay = f(x), \tag{2}$$

where a is a real constant and $f(x)$ is some function of x. Equation (2) is called a **first-order linear differential equation with constant coefficients**

Second, we consider the second-order equation

$$\frac{d^2y}{dx^2} + b\frac{dy}{dx} + cy = f(x), \tag{3}$$

where b and c are real constants. For some b and c, solving Eq. (3) may use complex numbers even though the solution will be a real function.

An engineer or physicist will meet Eq. (3) in the form

$$L\frac{d^2q}{dt^2} + R\frac{dq}{dt} + \frac{q}{C} = V \sin \omega t$$

in the study of electric currents. Here q is a charge that varies with time, dq/dt is current, $V \sin \omega t$ describes an applied voltage, R is resistance, L is inductance, and C is a constant describing the capacitor. They also meet Eq. (3) in the study of motion in the form

$$m\frac{d^2x}{dt^2} + b\frac{dx}{dt} + kx = F_0 \sin \omega t.$$

Here x describes the location of a particle moving on a line, $F_0 \sin \omega t$ is an applied force, $b(dx/dt)$ describes a damping effect, kx describes the force of a spring, and m is the mass.

Solving $\frac{dy}{dx} + ay = f(x)$

Imagine for the moment that you have found a particular solution y_p of Eq. (2) and a solution y_1 of the associated so-called **homogeneous** equation obtained from Eq. (2) by replacing $f(x)$ by 0,

The homogeneous case

$$\frac{dy}{dx} + ay = 0. \tag{4}$$

A straightforward computation then shows that $y_p + y_1$ is a solution of Eq. (2), as follows:

If you know one solution of Eq. (2) and all solutions of Eq. (4), you know all solutions of Eq. (2).

$$\begin{aligned}\frac{d}{dx}(y_p + y_1) + a(y_p + y_1) &= \frac{dy_p}{dx} + \frac{dy_1}{dx} + ay_p + ay_1 \\ &= \left(\frac{dy_p}{dx} + ay_p\right) + \left(\frac{dy_1}{dx} + ay_1\right) \\ &= f(x) + 0 = f(x).\end{aligned}$$

Now, the function $y_1 = Ce^{-ax}$, for any constant C, is a solution of Eq. (4). Thus, if y_p is a solution of Eq. (2), then so is $y_p + Ce^{-ax}$. In fact, each solution of Eq. (2) must be of the form $y_p + Ce^{-ax}$. To see why, assume that y_p and y both satisfy Eq. (2). Then

$$\begin{aligned}\frac{d}{dx}(y - y_p) + a(y - y_p) &= \left(\frac{dy}{dx} + ay\right) - \left(\frac{dy_p}{dx} + ay_p\right) \\ &= f(x) - f(x) = 0.\end{aligned}$$

Thus $y - y_p$, being a solution of Eq. (4), must be of the form Ce^{-ax} for some constant C. Thus $y = y_p + Ce^{-ax}$.

These observations are summarized in the following theorem.

Theorem 1 Let y_p be a particular solution of the differential equation

$$\frac{dy}{dx} + ay = f(x).$$

Then the most general solution is

$$y_p + Ce^{-ax}.$$

EXAMPLE 1 Solve the differential equation

$$\frac{dy}{dx} + 3y = 12.$$

SOLUTION One solution is the constant function $y_p = 4$. The most general solution is, therefore, $4 + Ce^{-3x}$ for any constant C. ■

Once a particular solution y_p has been found, Theorem 1 provides the general solution. Example 2 illustrates one technique for finding y_p.

EXAMPLE 2 Find all solutions of the differential equation

$$\frac{dy}{dx} - y = \sin x. \tag{5}$$

Start by guessing what a solution might look like.

SOLUTION First find one solution. Since $f(x) = \sin x$, let us see if there is a solution of the form $y_p = A \cos x + B \sin x$, for some constants A and B. Substitution in Eq. (5) yields

$$\frac{d}{dx}(A \cos x + B \sin x) - (A \cos x + B \sin x) = \sin x.$$

So we want

$$-A \sin x + B \cos x - A \cos x - B \sin x = \sin x,$$

or simply,
$$(-A - B) \sin x + (B - A) \cos x = \sin x.$$

We are just equating coefficients of $\sin x$ *and* $\cos x$ *on both sides.*

Choose A and B such that $-A - B = 1$ and $B - A = 0$. It follows that $B = A$ and that $-A - A = 1$ or $A = -\frac{1}{2}$. Consequently,

$$y_p = -\tfrac{1}{2} \cos x - \tfrac{1}{2} \sin x$$

is a solution of Eq. (5), as may be checked by substitution in Eq. (5).

The general solution of the homogeneous equation $dy/dx - y = 0$ is Ce^x, so the general solution of Eq. (5) is

$$y = -\tfrac{1}{2} \cos x - \tfrac{1}{2} \sin x + Ce^x. \quad ■$$

Example 2 uses the method of undetermined coefficients: Guess a general form of the solution and see if the unknown constants can be chosen to yield a solution of the differential equation.

Solving $\frac{d^2y}{dx^2} + b\frac{dy}{dx} + cy = f(x)$

Before turning to solutions of Eq. (3), consider the special case when $f(x)$ is identically 0, the so-called **homogeneous** case.

Let us find all solutions of the homogeneous equation

Homogeneous linear differential equation of second order

$$\frac{d^2y}{dx^2} + b\frac{dy}{dx} + cy = 0. \tag{6}$$

If y_1 and y_2 are both solutions of Eq. (6), a straightforward computation shows that $C_1y_1 + C_2y_2$ is also a solution of Eq. (6) for any choice of constants C_1 and C_2. [Since Eq. (6) involves the second derivative of y, we expect the general solution for y to contain two arbitrary constants.]

EXAMPLE 3 Solve
$$\frac{d^2y}{dx^2} - 3\frac{dy}{dx} + 2y = 0. \tag{7}$$

SOLUTION Recalling our experience with Eq. (1), we are tempted to look for a solution of the form e^{kx} for some constant k.

Substitution of e^{kx} into Eq. (7) yields

$$\frac{d^2}{dx^2}(e^{kx}) - 3\frac{d}{dx}(e^{kx}) + 2e^{kx} = 0,$$

or

$$k^2e^{kx} - 3ke^{kx} + 2e^{kx} = 0,$$

which is equivalent to

$$k^2 - 3k + 2 = 0. \tag{8}$$

By the quadratic formula, $k = 1$ or $k = 2$. Thus $y_1 = e^x$ and $y_2 = e^{2x}$ are solutions of Eq. (7). Consequently,

$$C_1e^x + C_2e^{2x} \tag{9}$$

is a solution of Eq. (7) for any choice of constants C_1 and C_2. (It can be proved that there are no other solutions.) ■

The most general solution of the differential equation

$$\frac{d^2y}{dx^2} + 6\frac{dy}{dx} + 9y = 0 \tag{10}$$

is of a different form. If we try $y = e^{kx}$, we obtain

$$k^2e^{kx} + 6ke^{kx} + 9e^{kx} = 0$$

$$e^{kx}(k^2 + 6k + 9) = 0$$

$$(k + 3)^2 = 0$$

$$k = -3.$$

A repeated root produces only one solution of the form e^{kx}.

This gives only the solutions of the form $y = Ce^{-3x}$. However, a second-order equation should possess a solution containing *two* arbitrary constants. Let us seek all solutions of the form

$$y = v(x)e^{-3x},$$

hoping to find some not of the form Ce^{-3x}.

Straightforward computations give

$$\frac{dy}{dx} = v(x)(-3e^{-3x}) + v'(x)e^{-3x} = -3v(x)e^{-3x} + v'(x)e^{-3x}$$

and

$$\frac{d^2y}{dx^2} = 9v(x)e^{-3x} - 6v'(x)e^{-3x} + v''(x)e^{-3x}.$$

Substituting into Eq. (10) yields

$$9v(x)e^{-3x} - 6v'(x)e^{-3x} + v''(x)e^{-3x} - 18v(x)e^{-3x} + 6v'(x)e^{-3x} + 9v(x)e^{-3x} = 0,$$

Check the algebra

which simplifies to

$$v''(x)e^{-3x} = 0;$$

hence to

$$v''(x) = 0.$$

Therefore, $v(x) = C_1 + C_2x$, and our general solution is

$$y = C_1e^{-3x} + C_2xe^{-3x},$$

for arbitrary constants C_1 and C_2.

The key to the nature of the solutions of Eq. (6) lies in the **associated quadratic equation**

$$t^2 + bt + c = 0. \tag{11}$$

The type of solution to Eq. (6) depends on the nature of the roots of Eq. (11). There are three cases: two distinct real roots, a repeated root (necessarily real), and two distinct complex roots. Each case will be described by a corresponding theorem.

Distinct real roots

Theorem 2 If $b^2 - 4c$ is positive, Eq. (11) has two distinct real roots, r_1 and r_2. In this case, the general solution of Eq. (6) is

$$C_1e^{r_1x} + C_2e^{r_2x}.$$

The proof that $C_1e^{r_1x} + C_2e^{r_2x}$ is a solution is left to the reader. Theorem 2 covers the differential equation (7).

EXAMPLE 4 Solve $\dfrac{d^2y}{dx^2} - 5\dfrac{dy}{dx} + 6y = 0.$

SOLUTION In this case $b^2 - 4c = 1$, which is positive. Hence the roots of the associated quadratic equation, $t^2 - 5t + 6 = 0$, are real. We have

$$t^2 - 5t + 6 = (t - 3)(t - 2) = 0,$$

so $r_1 = 3$ and $r_2 = 2$. The general solution of the differential equation is

$$C_1e^{3x} + C_2e^{2x}. \quad \blacksquare$$

The next theorem concerns the special case when the associated quadratic equation $t^2 + bt + c = 0$ has a repeated root, r.

Repeated root

Theorem 3 If $b^2 - 4c = 0$, Eq. (11) has a repeated root r. In this case, the general solution of Eq. (6) is

$$C_1e^{rx} + C_2xe^{rx} = (C_1 + C_2x)e^{rx}.$$

That $(C_1 + C_2x)e^{rx}$ is a solution is left to the reader to check by substitution. Theorem 3 is illustrated by the solution of Eq. (10).

Distinct complex (nonreal) roots

Theorem 4 If $b^2 - 4c$ is negative, Eq. (11) has two distinct complex roots $r_1 = p + iq$ and $r_2 = p - iq$. In this case, the general solution of Eq. (6) is

$$(C_1 \cos qx + C_2 \sin qx)e^{px}.$$

Proof Just as in Theorem 2,

Here the complex numbers enter.

$$y = A_1e^{r_1x} + A_2e^{r_2x} \tag{12}$$

is a solution of Eq. (6) for any choice of constants A_1 and A_2, even complex. Unfortunately, (12) will usually be complex. In order to find a *real* function that satisfies Eq. (6), expand (12):

$$\begin{aligned} A_1e^{r_1x} + A_2e^{r_2x} &= A_1e^{(p+iq)x} + A_2e^{(p-iq)x} \\ &= A_1e^{px}e^{iqx} + A_2e^{px}e^{-iqx} \\ &= e^{px}[A_1(\cos qx + i \sin qx) + A_2(\cos(-qx) + i \sin(-qx))] \\ &= e^{px}[(A_1 + A_2)\cos qx + i(A_1 - A_2)\sin qx]. \end{aligned}$$

Appropriate choices of A_1 and A_2 will generate the desired solution.

Getting some real solutions

Choosing $A_1 = A_2 = \frac{1}{2}$ produces the real solution $e^{px}\cos qx$. Next, choose A_1 and A_2 so that $A_1 + A_2 = 0$ and $i(A_1 - A_2) = 1$. This will produce the real solution $e^{px}\sin qx$. [To find A_1 and A_2, solve the pair of simultaneous equations $A_1 + A_2 = 0$ and $i(A_1 - A_2) = 1$. The solutions are $A_1 = -i/2$ and $A_2 = i/2$, which may be found by algebra.]

Thus

$$C_1e^{px}\cos qx + C_2e^{px}\sin qx \tag{13}$$

is a real-valued solution of Eq. (6) for any choice of real constants C_1 and C_2. It can be proved that there are no other real solutions. ■

EXAMPLE 5 Find the general solution of the differential equation of **harmonic motion**,

$$\frac{d^2y}{dx^2} = -k^2y, \tag{14}$$

where k is a constant.

SOLUTION Rewrite Eq. (14) in the form

$$\frac{d^2y}{dx^2} + k^2y = 0,$$

which has the associated quadratic equation $t^2 + k^2 = 0$. The roots of this equation are $0 + ki$ and $0 - ki$. By Theorem 4, the general solution of Eq. (14) is

$$C_1e^{0x}\cos kx + C_2e^{0x}\sin kx,$$

or simply,

$$C_1\cos kx + C_2\sin kx. \quad ■$$

Equation (14) describes the motion of a mass bobbing at the end of a spring. The height of the mass at time x is y. Since the motion is oscillatory, it is plausible that it is described by a combination of $\cos kx$ and $\sin kx$.

The equation $\frac{d^2y}{dx^2} + b\frac{dy}{dx} + cy = f(x)$

If y_p is any particular solution of

$$\frac{d^2y}{dx^2} + b\frac{dy}{dx} + cy = f(x), \tag{15}$$

and y^* is a solution of the associated homogeneous equation (6), then $y_p + y^*$ is a solution of Eq. (15), as may be checked by a straightforward calculation. Since we know how to find the general solution of Eq. (6), all that remains is to find a particular solution of Eq. (15). This can often be accomplished by a shrewd guess and the use of undetermined coefficients, as illustrated by the following example.

EXAMPLE 6 Solve the differential equation

$$\frac{d^2y}{dx^2} + \frac{dy}{dx} + 2y = 2x^2 + 5. \tag{16}$$

If y has degree n why does the left side of (16) have degree n?

SOLUTION Since $2x^2 + 5$ is a polynomial, let us seek a polynomial solution. If there is such a solution, it cannot have degree greater than 2, since the right-hand side of Eq. (16) has degree 2. So try $y = Ax^2 + Bx + C$; hence $y' = 2Ax + B$ and $y'' = 2A$. Substitution in Eq. (16) gives

$$2A + (2Ax + B) + 2(Ax^2 + Bx + C) = 2x^2 + 5,$$

or

$$2Ax^2 + (2A + 2B)x + (2A + B + 2C) = 2x^2 + 5.$$

Comparing coefficients gives $2A = 2$, $2A + 2B = 0$, and $2A + B + 2C = 5$. Thus $A = 1$, $B = -1$, and $C = 2$.

Consequently, $y_p = x^2 - x + 2$ is a particular solution of Eq. (16).

Next, turn to solving the associated homogeneous equation

$$\frac{d^2y}{dx^2} + \frac{dy}{dx} + 2y = 0. \tag{17}$$

Here $b = 1$ and $c = 2$, so $b^2 - 4c = -7$. The roots of the associated quadratic equation $t^2 + t + 2 = 0$ are

$$\frac{-1 \pm \sqrt{-7}}{2} = \frac{-1}{2} \pm \frac{\sqrt{7}}{2}i.$$

By Theorem 4, the general solution of Eq. (17) is

$$y^* = C_1 e^{-x/2} \cos\frac{\sqrt{7}}{2}x + C_2 e^{-x/2} \sin\frac{\sqrt{7}}{2}x.$$

Putting everything together, we obtain the general solution of Eq. (16), namely,

$$y = x^2 - x + 2 + C_1 e^{-x/2} \cos\frac{\sqrt{7}}{2}x + C_2 e^{-x/2} \sin\frac{\sqrt{7}}{2}x. \quad ■$$

Guessing a particular solution of Eq. (15) depends on the form of $f(x)$. This table describes the most common cases:

Form of $f(x)$	*Guess for y_p*
A polynomial	Another polynomial
e^{kx} (k not a root of associated quadratic equation)	Ae^{kx}
xe^{kx} (k not a root of the associated quadratic equation)	$(A + Bx)e^{kx}$
$e^{kx} \sin qx$ or $e^{kx} \cos qx$ ($k + qi$ not a root of the associated quadratic equation)	$Ae^{kx} \cos qx + Be^{kx} \sin qx$

A complete handbook of mathematical tables includes several pages of specific solutions for a much wider variety of functions $f(x)$ that appear on the right side of Eq. (15).

EXERCISES FOR APP. M: LINEAR DIFFERENTIAL EQUATIONS WITH CONSTANT COEFFICIENTS

In Exercises 1 to 22 find all solutions of the given differential equations.

1 $y' + 2y = 0$
2 $y' + 2y = \cos x$
3 $3y' + 12y = x$
4 $y' - \frac{1}{3}y = e^{2x}$
5 $y' - y = x^2$
6 $y'/2 + y = xe^{2x}$
7 $y'' - 2y' - 3y = 0$
8 $y'' + 5y' + 6y = 0$
9 $2y'' - y' - 3y = 0$
10 $2y'' - y' + 3y = 0$
11 $4y'' - 12y' + 9y = 0$
12 $4y'' + 9y = 0$
13 $y'' - 3y' + y = 0$
14 $3y'' - 2y' + 3y = 0$
15 $y'' - 6y' + 9y = 0$
16 $y'' + y' + y = 0$
17 $y'' - 2\sqrt{11}y' + 11y = 0$
18 $y'' - 3y' + 4y = 0$
19 $y'' - 2y' - 3y = e^{2x}$
20 $y'' + y' + y = x^2$
21 $y'' - 4y' + y = \cos 3x$
22 $y'' + 3y' + 2y = e^{-2x} \sin x + \cos 3x$

---

23 (*a*) Show that $y = e^{-ax} \int e^{ax} f(x)\, dx$ is a solution of $y' + ay = f(x)$.
(*b*) Use (*a*) to find a solution of $y' + y = 1/(1 + e^x)$.
(*c*) Find all solutions of the equation in (*b*).

24 Check that $C_1e^{-3x} + C_2xe^{-3x}$ is a solution of Eq. (10).

25 (*a*) Show that if $b^2 - 4c = 0$, then $t^2 + bt + c = 0$ has only one root, r, and $(t - r)^2 = t^2 + bt + c$.
(*b*) Check that $C_1e^{rx} + C_2xe^{rx}$ is a solution of $y'' + by' + c = 0$.

26 Let k be a nonzero constant. Find all solutions of the equation $y'' = k^2y$.

In some tables of solutions to differential equations, y'' is written D^2y and y' is written Dy. The equation $y'' + by' + c = f(x)$ is then expressed as $(D^2 + bD + c)y = f(x)$. Similarly, $y'' - 2ay' + a^2y = (D^2 - 2aD + a^2)y = (D - a)^2y$.

27 Verify that $x^2e^{ax}/2$ is a solution of $(D - a)^2y = e^{ax}$.

28 Verify that $e^{rx}/(r - a)^2$ is a solution of $(D - a)^2y = e^{rx}$, if $r \neq a$.

29 Verify that $(-x \cos bx)/(2b)$ is a solution of $(D^2 + b^2)y = \sin bx$, $b \neq 0$.

30 Verify that $(\sin sx)/(b^2 - s^2)$ is a solution of $(D^2 + b^2)y = \sin sx$, $s^2 \neq b^2$.

ANSWERS TO SELECTED ODD-NUMBERED EXERCISES AND TO GUIDE QUIZZES

CHAPTER 1. AN OVERVIEW OF CALCULUS

Sec. 1.1. The Derivative

1 (*a*)

x	$\sqrt{x^2+2x}$	$\sqrt{x^2+2x}-x$
1	1.7320508	0.7320508
5	5.9160798	0.9160798
10	10.9544512	0.9544512
100	100.9950494	0.9950494
1000	1000.999500	0.9995000

(*b*) 1

3 (*a*)

x	x^3-1	$x-1$	$(x^3-1)/(x-1)$
0.5	−0.8750000	−0.500	1.7500000
0.9	−0.2710000	−0.100	2.7100000
0.99	−0.0297010	−0.010	2.9701000
0.999	−0.0029970	−0.001	2.9970010

(*b*) 3

5 (*a*) 0.0303 ft (*b*) 3.03 ft/min (*c*) 3.003 ft/min (*d*) 2.997 ft/min (*e*) 3 ft/min

7 (*a*)

x	2^x	2^x-1	$(2^x-1)/x$
1	2.0000000	1.0000000	1.0000000
0.5	1.4142136	0.4142136	0.8284272
0.1	1.0717735	0.0717735	0.7177346
0.01	1.0069556	0.0069556	0.6955550
0.001	1.0006934	0.0006934	0.6933870
−0.001	0.9993071	−0.0006929	0.6929070

(*b*) 0.693

9 (*a*) 0.84147, 0.99833, 0.99998 (*b*) It approaches 1.

11 (*a*) 10.451 m/sec, 9.942 m/sec

Sec. 1.2. The Integral

1 (*b*)

Rectangle	*Height*	*Width*	*Area*
First	$\frac{9}{16}$	$\frac{3}{4}$	$\frac{27}{64}$
Second	$\frac{9}{4}$	$\frac{3}{4}$	$\frac{27}{16}$
Third	$\frac{81}{16}$	$\frac{3}{4}$	$\frac{243}{64}$
Fourth	9	$\frac{3}{4}$	$\frac{27}{4}$

(*c*) $\frac{810}{64} = 12.62625$

3 (*c*) 11.88 **5** See Figs. 8 and 9. **7** (*a*) $\frac{25}{64} = 0.390625$

9 (*a*) 0.6688 (*b*) The areas are the same. **11** 2.9188

Sec. 1.3. Survey of the Text

1 (*a*) 0.83333333 (*b*) 0.84166667 (*c*) 0.84146285 (*d*) 0.84147101 (*e*) 0.84147098

CHAPTER 2. FUNCTIONS, LIMITS, AND CONTINUITY

Sec. 2.1. Functions

1 **3**

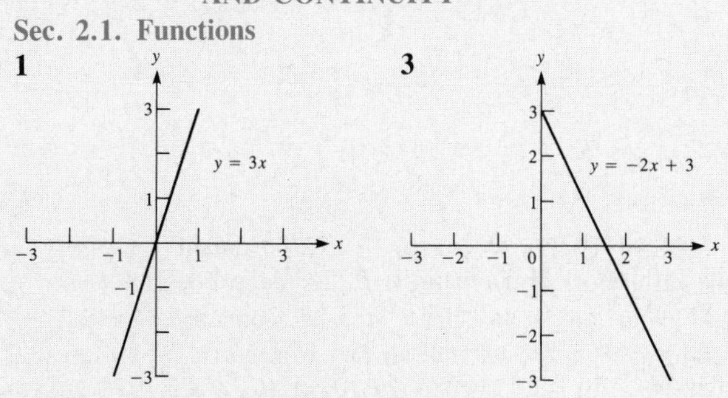

5 **7**

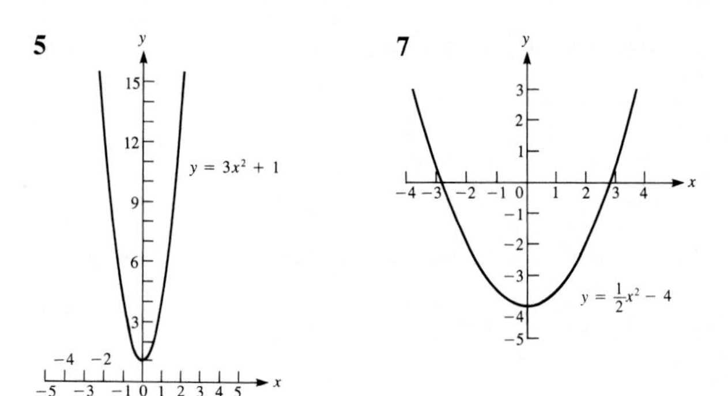

9

$y = x^2 - x$

11

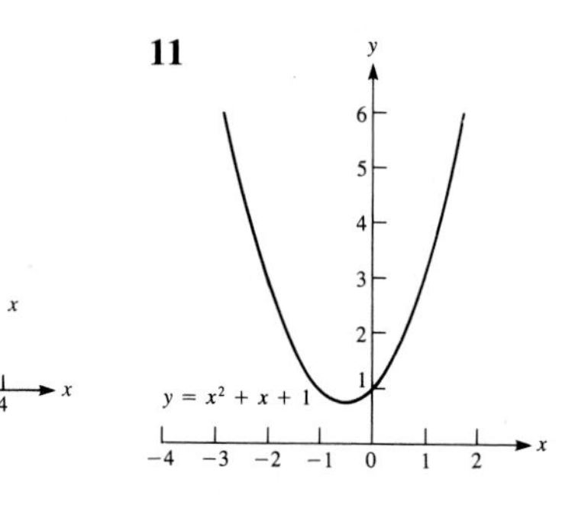

13

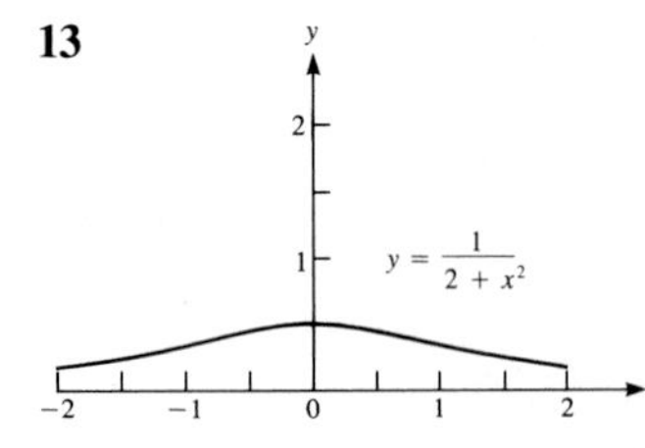

15

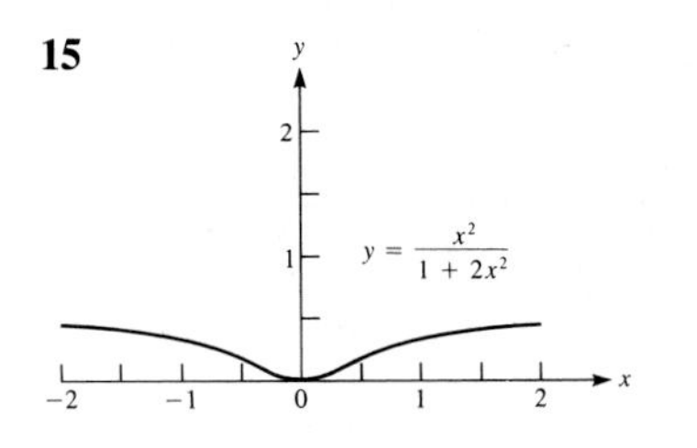

17 Domain: $[0, \infty)$; range: $[0, \infty)$ **19** Domain: $[-1, \infty)$; range: $[0, \infty)$ **21** Domain: $[-2, 2]$; range: $[0, 2]$ **23** Domain: $x \neq 0$; range: $y \neq 0$ **25** Domain: $x \neq -1$; range: $y \neq 0$ **27** Domain: all real x except $x = 1, -1$; range: $(-\infty, 0)$ and $[1, \infty)$ **29** (*a*) 0 (*b*) 4 (*c*) 2.25 (*d*) 1 **31** (*a*) 27 (*b*) 27 **33** (*a*) 7 (*b*) 6.01 (*c*) 5.99 (*d*) 6.0001 (*e*) It approaches 6. **35** $3a^2 + 3a + 1$

37 $(-c - d)/(c^2d^2)$
39 (*a*) No (*b*) Yes (*c*) Yes **41** $f(x) = 2\pi x$, $(0, \infty)$
43 $f(x) = 4x$, $(0, \infty)$ **45** $f(x) = 6x^2$, $(0, \infty)$
47 $f(x) = \sqrt{49 - x^2}$, $(0, 7)$
49 $f(x) = 2x + 4\sqrt{25 - \frac{x^2}{4}} = 2x + 2\sqrt{100 - x^2}$
51 $g(x) = \sqrt{4 + x^2} + \sqrt{25 + (6 - x)^2}$ **53** (*a*) $f(0) = 0$, $f(h) = \pi a^2$ (*b*) $f(x) = \pi a^2 x^2/h^2$ **55** (*a*) g/f (*b*) Most dangerous: 2 AM; Safest: 8 AM; Approximately 10
57 (*a*) $f(0) = 0$, $f(1) = 2$, $f(2) = 5$, $f(3) = 9$
(*b*)

x	0	1	2	3
$f(x)$	0	2	5	9

(*c*)

(*d*) $f(3) - f(2)$
59 (*b*), (*c*), and (*d*) **61** $f(x) = 0$, $f(x) = a2^x$, $f(x) = 2^{\lfloor x \rfloor}$, where $\lfloor x \rfloor$ is the largest integer less than or equal to x.
63 (*a*) $f(x) = x$, $f(x) = 2x$, $f(x) = x/2$ (*b*) Let f be a function that maps the positive integers to the real numbers. Let f satisfy the condition $f(x + y) = f(x) + f(y)$. Then $f(x) = cx$, where c is a real number. (*c*) By induction.

Sec. 2.2. Composite Functions

1 3, X^2, SIN **3** Algebraic: 4, Y^X, 3, =, +, 1, =, $\sqrt{X}$; Reverse Polish: 4, Enter, 3, Y^X, 1, +, $\sqrt{X}$ **5** Algebraic: 2, X^2, COS, Y^X, 3, 1/X, =; Reverse Polish: 2, X^2, COS, 3, 1/X, Y^X **7** $y = (1 + x)^2$ **9** $y = 1/x^3$
11 $y = x^6$ **13** $y = \sqrt{\cos x}$ **15** $y = (1 + \sqrt{x})^3$ **17** $y = \cos(1 + \tan^2 x)$ **19** $y = u^{50}$, $u = x^3 + x^2 - 2$ **21** $y = \sqrt{u}$, $u = x + 3$ **23** $y = \sin u$, $u = 2x$ **25** $y = u^3$, $u = \cos v$, $v = 2x$ **27** (*a*) 9 (*b*) 5 **31** $f(x) = x$
33 Infinitely many **35** Yes; one

Sec. 2.3. The Limit of a Function

1 12 **3** 4 **5** $\frac{4}{3}$ **7** $\frac{1}{5}$ **9** 25 **11** 0 **13** 1 **15** 3
17 $-\frac{1}{4}$ **19** 1 **21** (*a*) 2 (*b*) 1 (*c*) 1 (*d*) 2 **23** Yes, it is about 1.1.
25 (*a*)

x	1	0.1	0.01	0.001	−1	−0.01	−0.001
$f(x)$	2	1.618	1.588	1.585	1.333	1.582	1.585

(*b*) Yes, ≈1.585 **27** Yes; 1

29 (*a*) (*b*) For all a

31 (*a*)

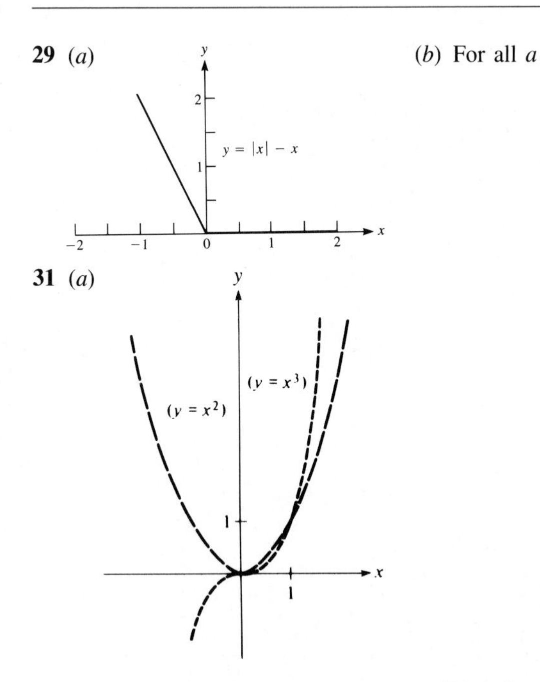

(*b*) No (*c*) $\lim_{x \to 1} f(x) = 1$ (*d*) $\lim_{x \to 0} f(x) = 0$ (*e*) $a = 0$ or $a = 1$ **33** (*a*) It is approximately 1.39. (*b*) 1.386 (*c*) It appears to be twice as large.

(*d*) $\lim_{x \to 0} \dfrac{4^x - 1}{x} = 2 \lim_{x \to 0} \dfrac{2^x - 1}{x}$

35 (*a*) 1.5, 1.0833, 0.95, 0.8845, 0.8456, 0.8199, 0.8016, 0.7879, 0.7773, 0.7688 (*b*) It decreases, but it is not approaching zero. (*c*) $\lim_{n \to \infty} f(n) = \ln 2$

Sec. 2.4. Computations of Limits

1 14 **3** $\frac{13}{5}$ **5** 20 **7** ∞ **9** ∞ **11** ∞ **13** $-\infty$ **15** ∞ **17** 0 **19** $\frac{1}{2}$ **21** 0 **23** ∞ **25** ∞ **27** (*a*) ∞ (*b*) $-\infty$ (*c*) Does not exist **29** $\frac{2}{3}$ **31** $\frac{2}{3}$ **33** (*a*) It is very near zero. (*b*) Not much **35** 0 **37** (*a*) $\frac{1}{2}$ (*b*) 0.5076 **39** All 3 are wrong. **41** (*a*) 1 (*b*) 0 (*c*) 0 (*d*) The limit does not exist. It may be ∞ or $-\infty$ or neither of these. (*e*) The limit is ∞; it does not exist. **43** (*a*) 0 (*b*) ∞ (*c*) Indeterminate

45 (*a*)

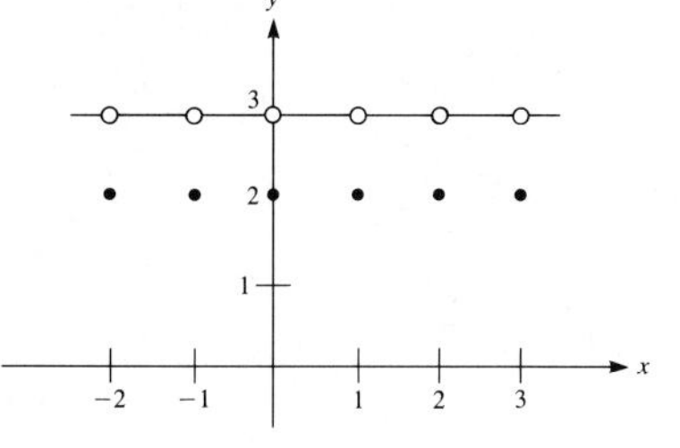

(*b*) Does not exist (*c*) 3

47 (*b*) $\lim_{x \to \infty} \sqrt{x^2 + 20x} - x = 10$

Sec. 2.5. Some Tools for Graphing

7 (*a*) Neither (*b*) Even (*c*) Even **9** $y = 3$, $x = -\frac{3}{2}$ **11** $y = 2$, $x = -2, -1$ **13** $y = 1$ **15** $x = 2$, $y = 1$ **17** $y = 0$ **19** $x = -\sqrt{3}$, $x = \sqrt{3}$, $y = 1$

21 **23** **25** **27**

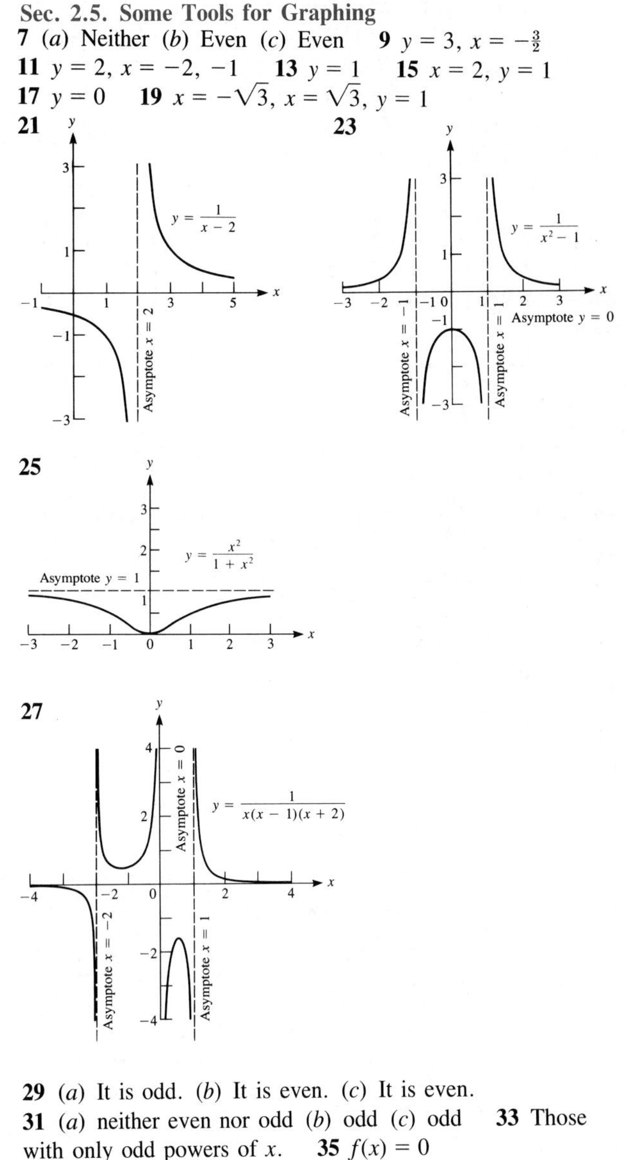

29 (*a*) It is odd. (*b*) It is even. (*c*) It is even. **31** (*a*) neither even nor odd (*b*) odd (*c*) odd **33** Those with only odd powers of x. **35** $f(x) = 0$

37 **39**

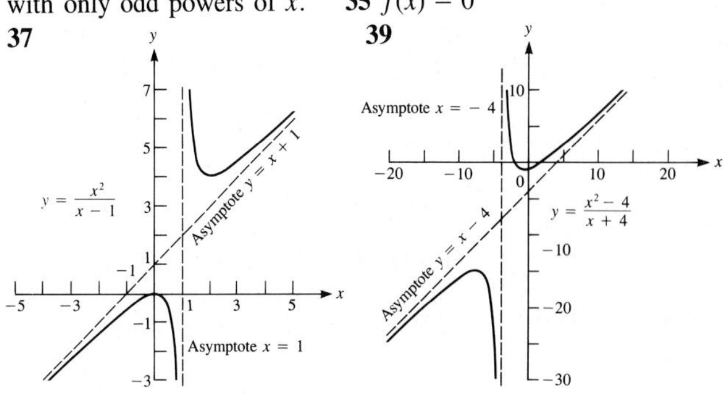

41 (*a*) Shift it up 2; that is, add 2 to every ordinate. (*b*) Shift it down 2. (*c*) Shift it to the right 2; that is, add 2 to the abscissa of each point. (*d*) Shift it to the left 2. (*e*) Double it; that is, multiply each ordinate by 2. (The doubling is vertical.) (*f*) Shift it to the right 2 and triple it (stretch vertically). **43** Yes, there is one: $f(x) = 0$.

Sec. 2.6. A Review of Trigonometry

1 (*a*) $\pi/2$ (*b*) $\pi/6$ (*c*) $2\pi/3$ (*d*) 2π **3** (*a*) 135 (*b*) 60 (*c*) 120 (*d*) 720 **5** (*a*) $\frac{5}{3}$ (*b*) $\approx 95.49°$ **7** (*a*) $5\pi/18$ (*b*) $\approx 114.59°$ **9** $\theta = \frac{2}{3}$

11 (*a*) (*b*)

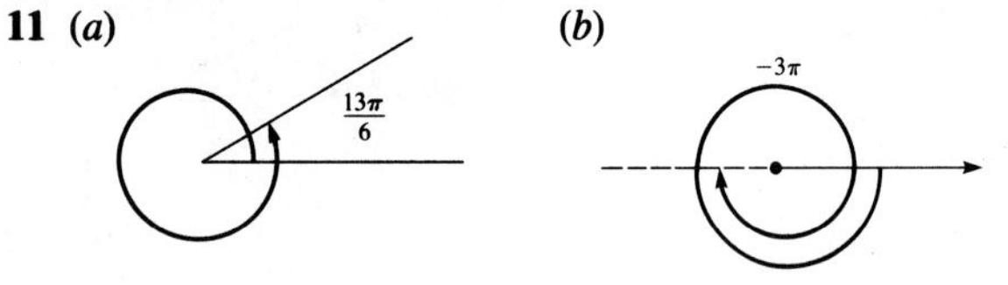

13 $\sin\theta = \cos\left(\frac{\pi}{2} - \theta\right)$; $\sin\frac{\pi}{6} = \cos\left(\frac{\pi}{2} - \frac{\pi}{6}\right) = \cos\frac{\pi}{3} = \frac{1}{2}$

15

θ	0	$\pi/6$	$\pi/4$	$\pi/3$	$\pi/2$	π	$3\pi/2$	2π
$\sin\theta$	0	$\frac{1}{2}$	$1/\sqrt{2}$	$\sqrt{3}/2$	1	0	-1	0

19 $\dfrac{1+\sqrt{3}}{2\sqrt{2}}$

27 (*a*) − (*b*) + (*c*) + (*d*) − **31** (*a*) ≈ 0.1763 (*b*) ≈ 2.7475 (*c*) ≈ -2.7475 (*d*) -1 (*e*) 0 **33** (*a*) $60° = \pi/3$ (*b*) $30° = \pi/6$ **37** (*a*) $\cos\alpha = b/c$ (*b*) $\sin\beta = b/c$ (*c*) $\tan\alpha = a/b$ **39** $\sqrt{3}/2$, $\frac{1}{2}$, $1/\sqrt{3}$

41 (*a*)

θ	0	$\pi/6$	$\pi/4$	$\pi/3$
$\sec\theta$	1	$2/\sqrt{3}$	$\sqrt{2}$	2

(*b*)

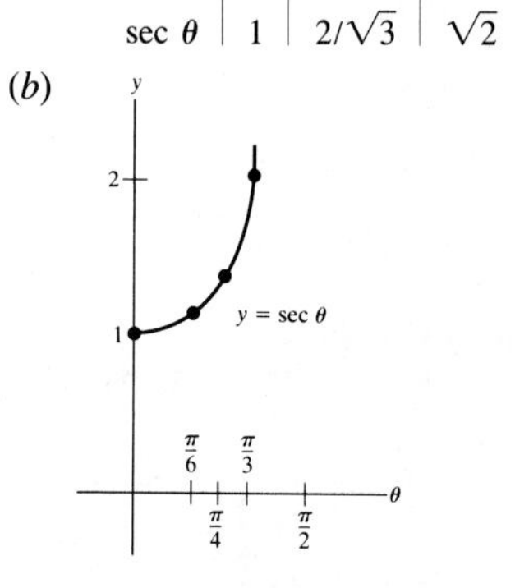

49 (*b*) $1/\sqrt{2}$ (*c*) $-1/\sqrt{2}$

55 (*a*)

θ	1	0.1	−0.1	0.01	−0.01	0.001
$\sin\theta$	0.841471	0.0998334	−0.0998334	0.0099998	−0.0099998	0.001
$(\sin\theta)/\theta$	0.841471	0.998334	0.998334	0.9998	0.9998	1

(*b*) Yes

Sec. 2.7. The Limit of $(\sin\theta)/\theta$ as θ Approaches 0

1 (*a*) $9\pi/4$ (*b*) $\theta/2$ (*c*) 2θ **3** $\frac{1}{2}$ **5** $\frac{3}{5}$ **7** 0 **9** 0 **19** Near 0.041667 or $\frac{1}{24}$

21 (*a*) All nonzero x (*c*) 0 (*d*) $x = n\pi$, n a nonzero integer (*e*)

x	0.1	$\pi/2$	$3\pi/2$	2π	$5\pi/2$	3π	$7\pi/2$
$\sin x$	0.10	1	−1	0	1	0	−1
$(\sin x)/x$	1.00	0.64	−0.21	0	0.13	0	−0.09

(*f*), (*g*) (*h*) 1

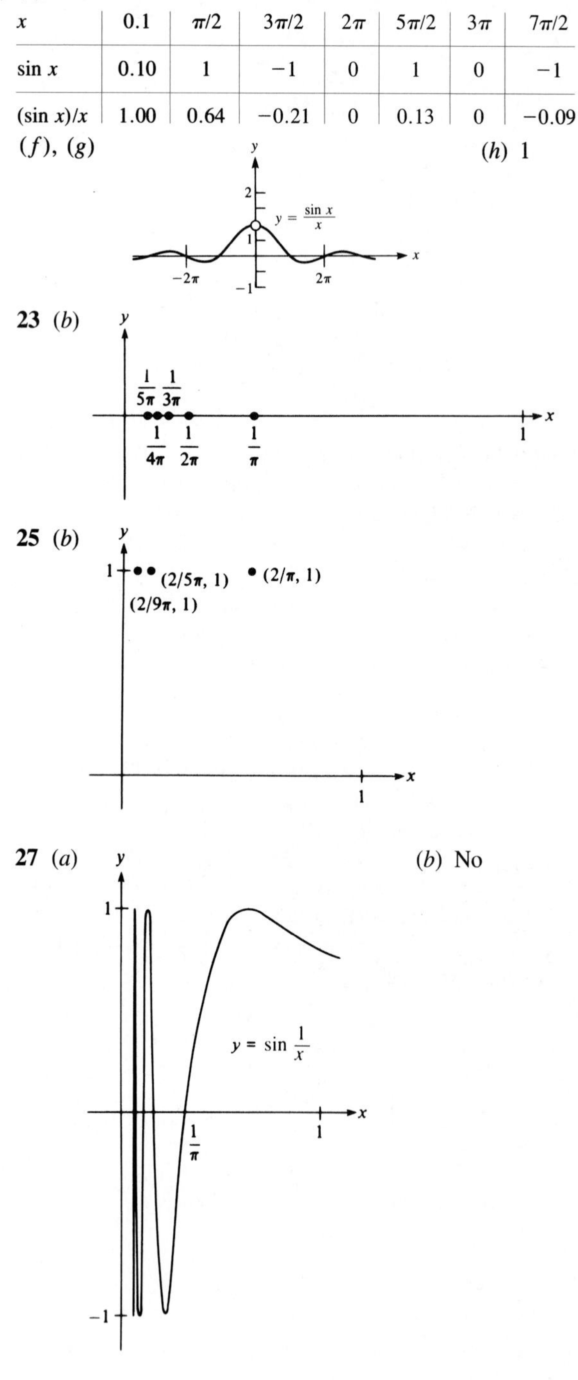

Sec. 2.8. Continuous Functions

1 1, 2, 3; continuous at $a = \frac{1}{2}$ **3** 1; not continuous **5** 1; not continuous **7** 1, 2; not continuous **9** not continous
11 (*a*)

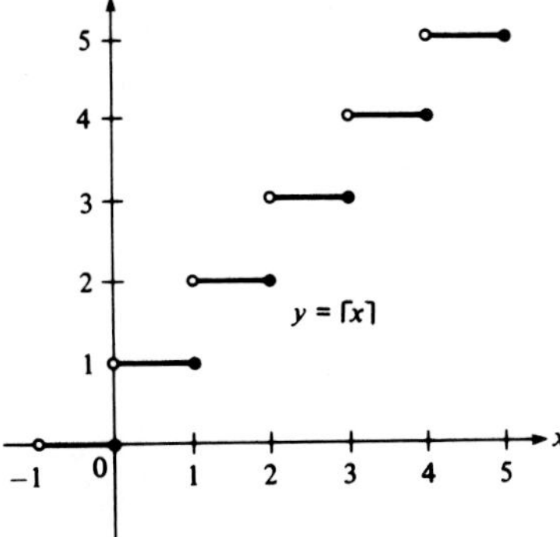

(*b*) Yes; 4 (*c*) Yes; 5 (*d*) No (*e*) No (*f*) All nonintegers (*g*) All integers **13** Yes **15** No **17** (*a*) At $x = \pi/4$, $\sin x = 1/\sqrt{2}$ (*b*) $\sin \pi/2 = 1$ **19** (*a*) Yes (*b*) Yes
21 (*a*) Yes; at $x = 4$ (*b*) Yes; at $x = 5$ (*c*) No
23 (*a*) Yes; at $x = 0$ (*b*) No **27** $c = \frac{5}{3}$
29 $c = 3\pi/2, 7\pi/2, 11\pi/2$
31 $c = -1, 0, 1$ **33** Yes
37 (*a*) (*b*) Yes

43 Yes **47** (*a*) Yes; infinitely many (*b*) Yes; infinitely many (*c*) Yes; one

Sec. 2.9. Precise Definitions of "$\lim_{x\to\infty} f(x) = \infty$" and "$\lim_{x\to\infty} f(x) = L$"

1 (*a*), (*b*) Any two numbers ≥ 200 (*c*) 200 **3** (*a*) 400 (*b*) 2,000 **5** Let $D = E/3$. **7** Let $D = E - 5$. **9** Let $D = (E - 4)/2$. **11** Let $D = (E + 100)/4$. **13** (*a*) 10 (*b*) $\sqrt{E}$ (*c*) Any value will work. (*d*) Let $D = \sqrt{|E|}$.
15 (*a*), (*b*) Any two numbers ≥ 10 (*c*) 10 (*d*) Let $D = 1/\epsilon$.
17 Let $D = 1/\epsilon$. **19** Let $D = 2/\sqrt{\epsilon}$. **21** Let $D = 100 + 1/\epsilon$. **23** Let $E = 1$. **25** Let $\epsilon = 3$. **27** For each number E there is a number D such that $f(x) < E$ for all $x > D$.

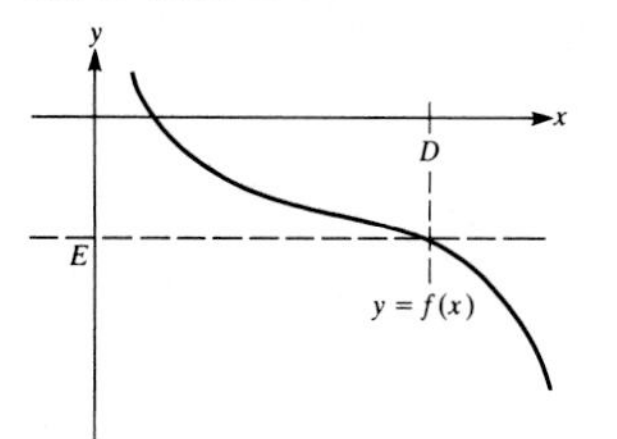

29 For each number E there is a number D such that $f(x) < E$ for all $x < D$.

31 (*a*), (*b*) Let D be any real number.

Sec. 2.10. Precise Definition of "$\lim_{x\to a} f(x) = L$"

1 $\delta = \epsilon/3$ **3** $\delta = \epsilon$ **5** $\delta = 0.02$ **7** $\delta = \sqrt{4\epsilon}$
9 $\delta = \epsilon/3$ **11** $\frac{1}{5}$ (or any positive $\delta \leq \sqrt{5} - 2$)

13 $\delta = 0.01$ **15** (*b*) Let δ be 1 or $\epsilon/7$, whichever is less.
17 (*b*) Let δ be 1 or $\epsilon/12$, whichever is less. **19** For each positive number ϵ there is a positive number δ such that $|f(x) - L| < \epsilon$ for all x between a and $a + \delta$. **21** For each number E there is a positive number δ such that $f(x) > E$ for all x satisfying $0 < |x - a| < \delta$. **23** For each number E there is a positive number δ such that $f(x) > E$ for all x satisfying $a < x < a + \delta$. **25** (*a*) $\frac{1}{30}$ (*b*) $\sqrt{\epsilon/3}$ **27** Let $\epsilon = \frac{1}{2}$.

Sec. 2.S. Summary: Guide Quiz

1 (*a*) $f(x) = x^2$, $g(x) = x$, $a = 0$ (*b*) $f(x) = g(x) = x$, $a = 0$ (*c*) $f(x) = x$, $f(x) = x^2$, $a = 0$

2 $f(x) = \dfrac{x^2 - 4}{x^2 - 1}$

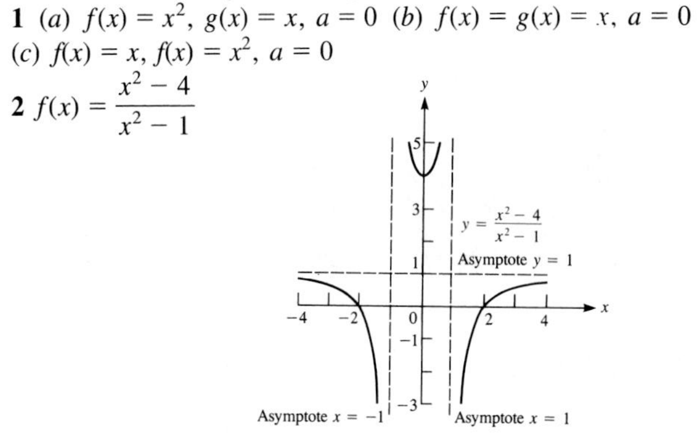

f is even, so it is symmetric with respect to the y axis. $\lim_{x\to\infty} f(x) = 1$, so $y = 1$ is a horizontal asymptote. Vertical asymptotes are at $x = \pm 1$ because of the denominator. $x = \pm 2$ are x intercepts; $y = 4$ is the y intercept.
4 (*a*) 1 (*b*) ≈ 0.0873 **5** (*a*) 0 (*b*) $\frac{1}{2}$ (See Exercise 16 of Sec. 2.7.) (*c*) 0.98 (*d*) 0.9391 (*e*) 0.9397 **7** (*a*) When $x = 1{,}000$, $\sqrt{x^2 + 4x} - x = 1.998003$ (*b*) 2 (*c*) $x \geq 0$ and $x \leq -4$ (*d*)

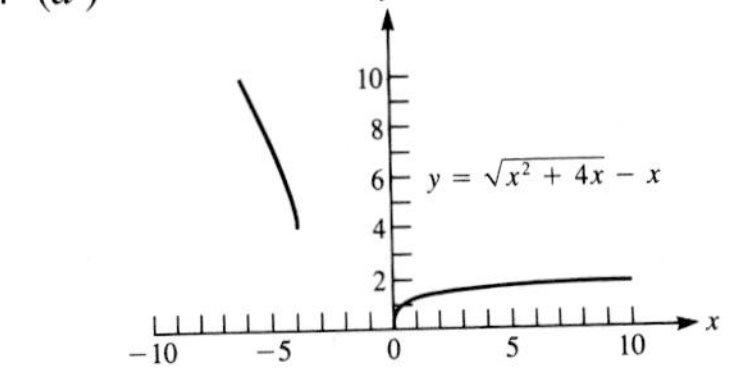

8 $3x^2 + 3xh + h^2$ (*b*) $-1/[x(x + h)]$
9 (*a*) $x \to \frac{1}{2}, f(x) \to f(\frac{1}{2})$ (*b*) $\lim_{x\to 0^+} f(x) = f(0)$ **10** (*a*) Yes (*b*) No (*c*) No (*d*) No (*e*) No (*f*) Yes **11** (*a*) 6 (*b*) $\frac{3}{5}$ (*c*) -3 (*d*) 0 (*e*) $\frac{1}{2}$ (*f*) ∞ (*g*) 0 (*h*) 4 (*i*) 12 (*j*) Does not exist (*k*) Does not exist (*l*) 0 (*m*) 1 (*n*) $\frac{1}{8}$ **12** (*a*) 12 (*b*) $\frac{3}{4}$ (*c*) 7 (*d*) Indeterminate (*e*) Indeterminate **13** (*a*) ∞ (*b*) 0 (*c*) Indeterminate (*d*) ∞ (*e*) Indeterminate **14** (*a*) f is continuous throughout some interval $[a, b]$. (*b*) For some number c in $[a, b]$, $f(c) \geq f(x)$ for all x in $[a, b]$. **15** (*a*) f is continuous throughout some interval $[a, b]$, and m is a number between $f(a)$ and $f(b)$. (*b*) There is at least one number c in $[a, b]$ such that $f(c) = m$. **16** (*a*) $V(x) = (10 - 2x)(15 - 2x)(x)$ (*b*) The domain of V is $[0, 5]$. **18** $P(\theta) = 3\theta + 6$

Sec. 2.S. Summary: Review Exercises

1 (*a*) $A = \theta r^2/2 = 25\pi/6$ (*b*) $s = r\theta = 5\pi/3$ **3** Domain: all x; range: all y **5** Domain: all x; range: $[-1, 1]$
7 Domain: $(-1, \infty)$; range: $(0, \infty)$ **9** 0.41
11 $-1/[(a + h + 1)(a + 1)]$ **13** $u^2 + uv + v^2 - 3$
15

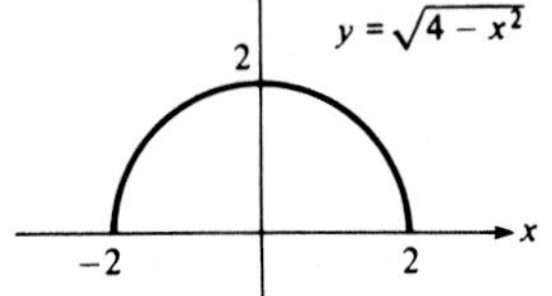

17
19
21

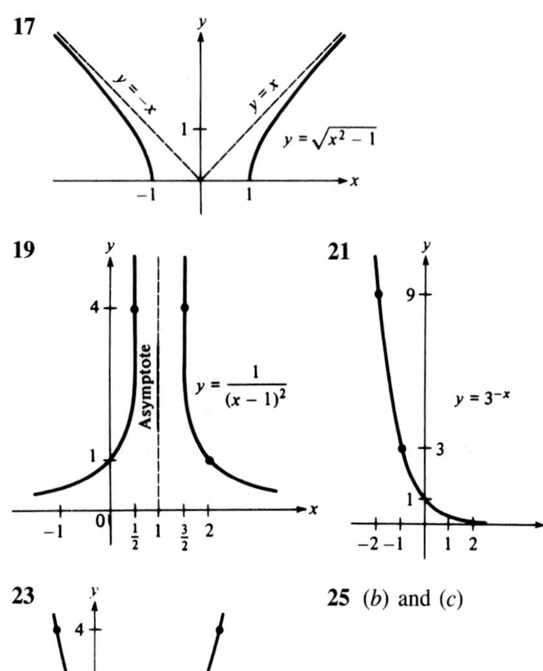

23

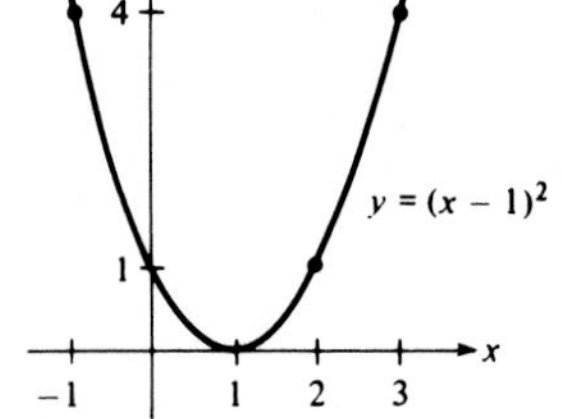

25 (*b*) and (*c*)

27 (*a*)

x	0.001	0.01	0.1	1	2	10	100
$f(x)$	2.717	2.705	2.594	2	1.732	1.271	1.047

(*b*)

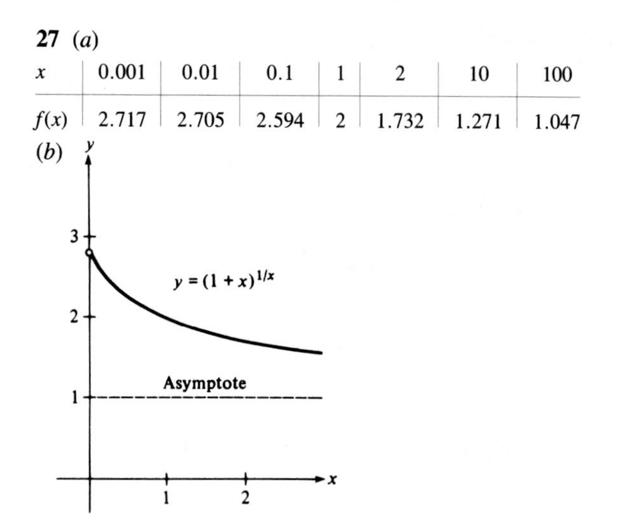

29 (*a*) $x(50 - x)$ (*b*) $(0, 50)$ **31** Yes **33** 1 **35** $\frac{8}{3}$ **37** $\frac{1}{2}$ **39** $-\infty$ **41** $\frac{1}{4}$ **43** 2 **45** ∞ **47** 5 **49** ∞ **51** 1 **53** 1 **55** 0 **57** $\frac{1}{3}$ **59** 0 **61** 0 **63** 0 **65** $\pi^2/(16\sqrt{2})$ **67** 1 **69** $f(x) = 5x$, $g(x) = x$ **71** $f(x) = 1/x$, $g(x) = x^2$ **73** $f(x) = x^2$, $g(x) = x$ **75** (*a*) Yes (*b*) No **77** (*a*) Yes (*b*) No **81** (*a*) Yes, 2 (*b*) Yes, 0 **83** $c = 2$ **85** $c = \pi/4$ **87** No **91** (*c*) No (*d*) Yes

CHAPTER 3. THE DERIVATIVE

Sec. 3.1. Four Problems with One Theme

1 (*a*) 45° (*b*) 63.43° **5** 6 **7** −4 **9** 12 **11** (*a*) 0 (*b*)

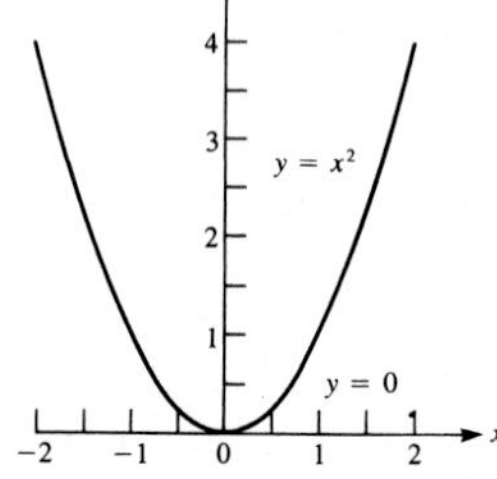

13 96 ft/sec **15** 32 ft/sec **17** (*a*) 1.261 ft (*b*) 12.61 ft/sec (*c*) $12 + 6h + h^2$ ft/sec (*d*) 12 ft/sec **19** (*a*) 3.99 (*b*) $4 + h$ **21** (*a*) 8 **23** (*a*) 2.1 (*b*) 2.01 (*c*) 2.001 (*d*) 2 **25** (*a*) 0.99 (*b*) 0.999 (*c*) 1 **27** (*a*) 0.0601 (*b*) 6.01 (*c*) 5.99 (*d*) 6 (*e*) 6 **29** (*a*) 3 **31** (*a*) 0.0401 (*b*) 4.01 (*c*) 4
33 (*a*)

(*d*) $5 + 2h$ (*e*) 5 **35** $2x$ **37** $2x$

39 (*a*)

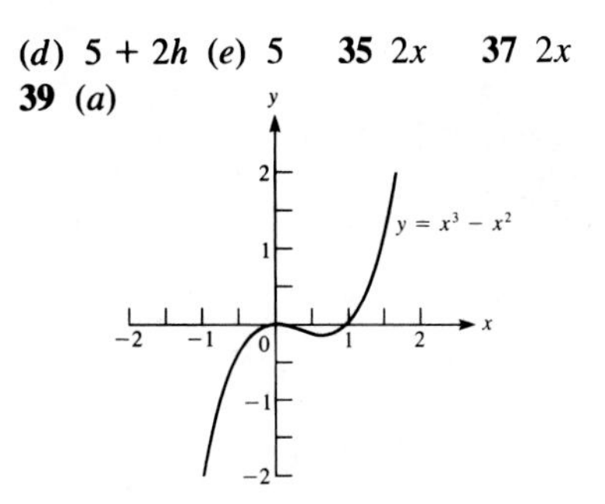

(*b*) $3x^2 - 2x$ (*c*) $(0, 0)$, $(\frac{2}{3}, -\frac{4}{27})$, (*d*) $(-\frac{1}{3}, -\frac{4}{27})$, $(1, 0)$
41 No **43** $(1, 0)$ and $(1 - \sqrt{3}, 9 - 5\sqrt{3})$
45 (*a*) 0.0609756 (*b*) 0.0623441 (*c*) 0.0624844 (*d*) 0.0625156

Sec. 3.2. The Derivative

1 2 **3** 4 **5** $10x$ **7** $2x + 2$ **9** $7/(2\sqrt{x})$
11 $2x + 3/(2\sqrt{x})$ **13** $3x^2 + 3$ **15** $2x - 1/x^2$
17 -4 **19** $5a^4$ **21** $\frac{1}{12}$ **23** $\pi 2^{\pi - 1}$ **25** $\frac{1}{3}$ **27** $\frac{5}{2}$
29 $-\frac{3}{16}$ **31** 8.772 **33** (*a*) 4.641 (*b*) 4
35 (*a*) 4.060401 (*b*) 4
37 (*b*)

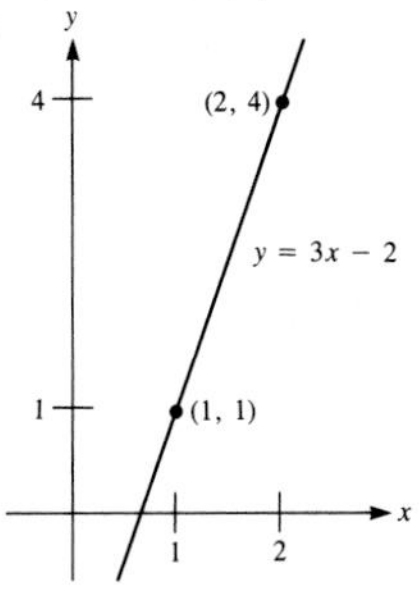

43 (*a*) ≈ -18.379 (*b*) ≈ 19.504

Sec. 3.3. The Derivative and Continuity

1 $3x^2$ **3** $1/(2\sqrt{x})$ **5** $10x$ **7** $-3/x^2$ **9** $-3/x^2 - 4$
11 (*a*) 0.21 (*b*) -0.59 **13** (*a*) 0.05 **15** (*a*) $-1, 2$ (*b*) 1, 3 **17** (*a*) 5 (*b*) 2, 3 **19** $6x^2 - 6x + 4$
21 (*a*) $x \geq 0$

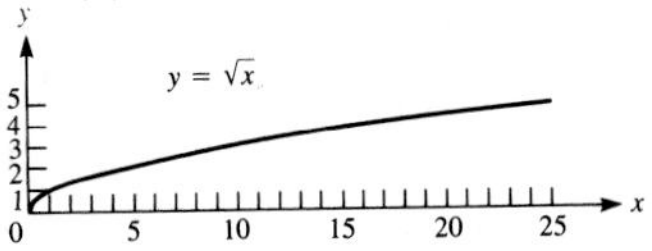

(*b*) $x > 0$

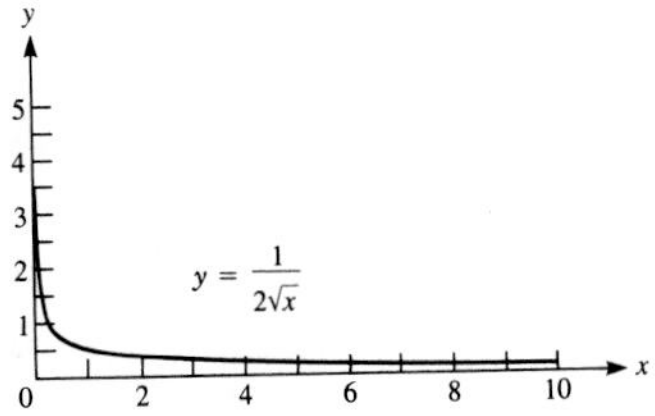

23 (*a*) x^5 (*b*) x^7
29 (*a*) x^4 (*b*) $1/x$ **31** (*a*) $x^3/3$ (*b*) $-1/(2x^2)$
33 2 **35** -0.4161

Sec. 3.4. The Derivatives of the Sum, Difference, Product, and Quotient

1 $2x + 5$ **3** $3x^2 + 10x$ **5** $5x^4 - 4x$
7 $12x^3 - 3/\sqrt{x}$ **9** $-3/x^2 + 1/(3x^{2/3})$ **11** $4x - 7$
13 $6x^2 + 6x + 1$ **15** $16x^7 - 24x^5 + 15x^2 - 10$
17 $-\frac{25}{2}x\sqrt{x} + \frac{45}{2}\sqrt{x} - 8x + 12$ **19** $\frac{1}{2} + \frac{1}{6\sqrt[3]{x}}$
21 $\frac{1}{(4 + x)^2}$ **23** $\frac{x^2 + 6x + 3}{(x + 3)^2}$ **25** $\frac{3t^2 - 3}{(t^2 + 1)^2}$
27 $\frac{-\sqrt{x} - 12 + 3/(2\sqrt{x})}{(2x + 3)^2}$ **29** $-\frac{1}{x^2} - \frac{2}{x^3}$
31 $\frac{-3x^2 - 2}{(x^3 + 2x + 1)^2}$
33 $\frac{30x^4 + 56x^3 - 25x^2 - 70x - 15}{(5x + 7)^2}$
35 $\frac{-2}{(x - 1)^2}$ **37** $y = 3x - 1$ **39** $y = 20.5x - 46$
41 13 **43** $\frac{1}{12}$ **45** $\frac{8}{3}$
47 (*a*) $\frac{1}{3}x^3 + 1, \frac{1}{3}x^3 + 2$ (*b*) $\frac{1}{4}x^4 + 1, \frac{1}{4}x^4 + 2$ (*c*) $\frac{-1}{x} + 1, 2 - \frac{1}{x}$ (*d*) $x^3 + \frac{1}{2}x^2 + 1, x^3 + \frac{1}{2}x^2 + 2$
51 $\frac{5}{6}x^{-1/6}$ **59** (*a*) $y = x - 1$ (*b*) $y = 1 - x$ (*c*) $(0, 1)$ and $(1, 0)$ **61** $1, -3, \frac{1}{9}(-13 \pm \sqrt{610})$

Sec. 3.5. The Derivatives of the Trigonometric Functions

1 $5 \cos x$ **3** $2 \sec^2 x$ **5** $3 \sec x \tan x$
7 $x^2 \cos x + 2x \sin x$ **9** $\frac{1 + \sin x}{\cos^2 x}$ **11** $\frac{-1 - 3 \cos x}{\sin^2 x}$
13 $\frac{-\csc x\,(3x \cot x + 1)}{3x^{4/3}}$ **15** $\sin x\,(\sec^2 x + 1)$
17 $\frac{-(1 + x^2) \csc^2 x - 2x \cot x}{(1 + x^2)^2}$ **19** $x \sin x$ **21** $x^2 \sin x$
23 $\tan^2 x$ **25** $x^2 \cos x$ **27** (*a*) $\sqrt{3}/2$ (*b*) 0.362 (*c*) $-1/\sqrt{2}$ **29** (*a*) 2 (*b*) $\frac{4}{3}$ (*c*) 1.020
31 $-\csc \theta \cot \theta$ **33** (*a*) 3 (*b*) -3 (*c*) 3, -3 (*d*) 3, 3 **35** $\pi/4$ **37** $3\pi/4$ **39** (*a*) $-3 \cos x$ (*b*) $4 \sin x$ **41** (*a*) $2 \sec x$ (*b*) $-7 \csc x$

Sec. 3.6. The Derivative of a Composite Function

1 $200(1+2x)^{99}$ **3** $240x^2(2x^3-1)^{39}$ **5** $\dfrac{3\cos\sqrt{x}}{2\sqrt{x}}$

7 $3\sin^2 x\cos x$ **9** $-12\cos^3 3x\sin 3x$

11 $6\tan 3x\sec^2 3x$ **13** $\dfrac{-\csc^2 x}{2\sqrt{\cot x}}$

15 $\dfrac{3x^2+1}{2\sqrt{x^3+x+2}}$ **17** $15(3x+2)^4\cos(3x+2)^5$

19 $\dfrac{(x^3+x^5)\sec^2 x+(3x^2+x^4)\tan x}{(1+x^2)^2}$

21 $(2x+1)^4(3x+1)^6(72x+31)$

23 $x\sin^4 3x\,(15x\cos 3x+2\sin 3x)$

25 $\dfrac{-10}{(2x+3)^6}$ **27** $\dfrac{2x(1-2x^2)}{(x^2+1)^4}$

29 $\dfrac{\sec^2 x}{(1+x)\sqrt{1-x^2}}[-1+2(1-x^2)\tan x]$

31 $\dfrac{5(2x^2-2x-13)(x^2+3x+5)^4}{(2x-1)^6}$

33 $2\cot^3 x^2\,(2x+1)^2[3\cot x^2-4x\csc^2 x^2\,(2x+1)]$

35 $\dfrac{1}{(1-x^2)^{3/2}}$ **37** $\dfrac{30(3x-2)^3}{\sqrt{5(3x-2)^4+1}}$

39 $4\cos 3x\cos 4x-3\sin 3x\sin 4x$

41 $x\sqrt{3x+1}$ **43** $\sin^4 5x$

45 $(1+x^2)^4$ **47** $\dfrac{(1+x^3)^5}{5}$ **49** 5, 28

51 (*a*) (*b*) $-\frac{4}{3}$

53 12°/min

Sec. 3.S. Summary: Guide Quiz

2 (*a*) $15x^2-2$ (*b*) $\dfrac{-15}{(3x+2)^2}+6$ (*c*) $6\cos 2x$ (*d*) $-2x^{-3}$

3 (*a*) $\dfrac{5}{2\sqrt{x}}$ (*b*) $\dfrac{6x(1-x^2)}{\sqrt{3-2x^2}}$ (*c*) $-5\sin 5x$

(*d*) $\dfrac{3x}{2}(1+x^2)^{-1/4}$ (*e*) $\dfrac{2\sec^2 6x}{(\tan 6x)^{2/3}}$

(*f*) $5x^3\cos 5x+3x^2\sin 5x$ (*g*) $-\dfrac{1}{(2x+1)^{3/2}}$

(*h*) $\dfrac{-4(10x^4-3x^2)}{(2x^5-x^3)^5}$ (*i*) $\dfrac{x^2}{(x^3-3)^{2/3}}$ (*j*) $\dfrac{6(2x^3+x^2-1)}{(3x+1)^2}$

(*k*) $\dfrac{-10x}{(5x^2+1)^2}$ (*l*) $\dfrac{-30}{(3x+2)^{11}}$

(*m*) $x^2(1+2x)^4\sec 3x\,[16x+3+3x(1+2x)\tan 3x]$

(*n*) $\dfrac{-\csc\sqrt{x}\cot\sqrt{x}}{2\sqrt{x}}$ (*o*) $\dfrac{24\csc^2 4x}{(1+3\cot 4x)^3}$

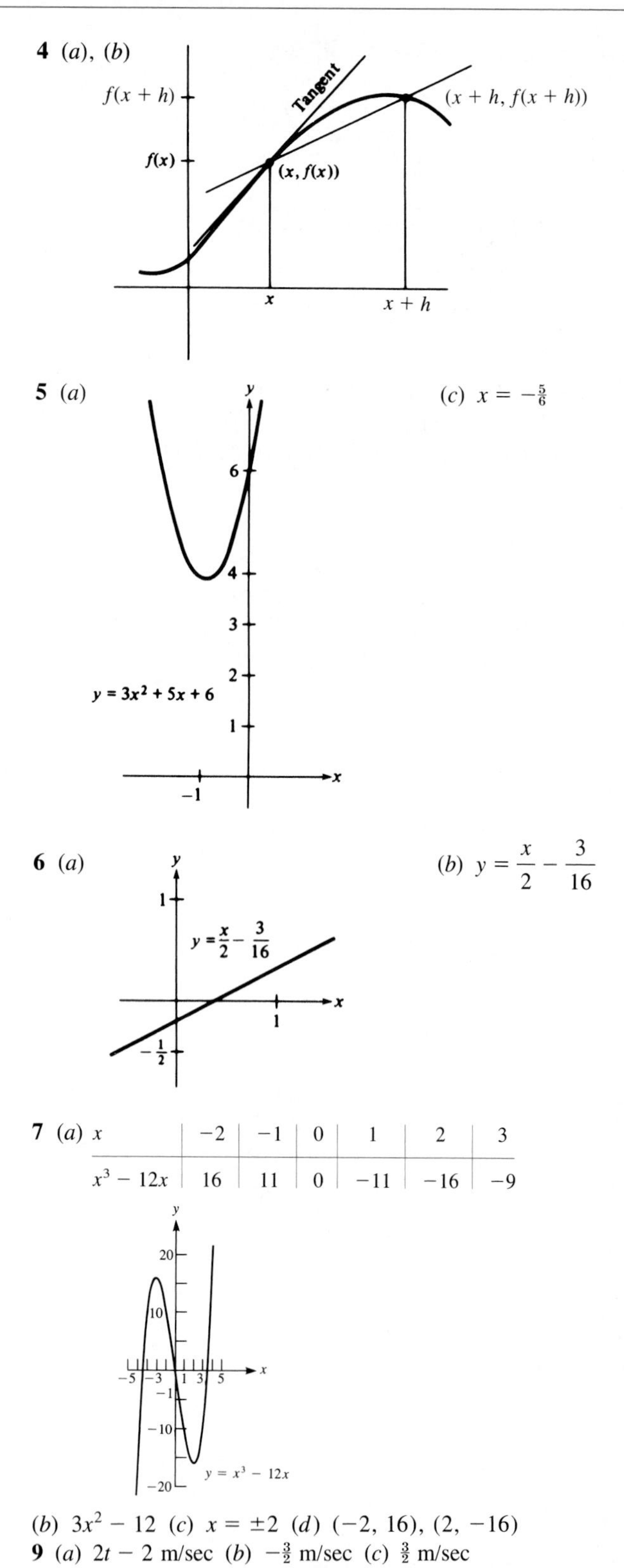

4 (*a*), (*b*)

5 (*a*) (*c*) $x=-\frac{5}{6}$

6 (*a*) (*b*) $y=\dfrac{x}{2}-\dfrac{3}{16}$

7 (*a*)

x	−2	−1	0	1	2	3
x^3-12x	16	11	0	−11	−16	−9

(*b*) $3x^2-12$ (*c*) $x=\pm 2$ (*d*) $(-2, 16)$, $(2, -16)$

9 (*a*) $2t-2$ m/sec (*b*) $-\frac{3}{2}$ m/sec (*c*) $\frac{3}{2}$ m/sec

(*d*) Left **10** (*a*) $x^6/6$ (*b*) $\dfrac{-\cos 2x}{2}$ (*c*) $\sin x^2$

Sec. 3.S. Summary: Review Exercises

1 $15x^2$ **3** $\dfrac{-1}{(x+3)^2}$ **5** $-3 \sin 3x$ **7** $10x^4 + 3x^2 - 1$

9 $\dfrac{4x^2 + 2x}{(4x+1)^2}$ **11** $\dfrac{3x+1}{\sqrt{3x^2 + 2x + 4}}$ **13** $\dfrac{4}{3\sqrt[3]{2t-1}}$

15 $10 \sin 5x \cos 5x$ **17** $\frac{20}{7}(5x+1)^3$

19 $3x \cos 3x + \sin 3x$ **21** $\dfrac{4 \tan \sqrt[3]{1+2x} \sec^2 (\sqrt[3]{1+2x})}{3(1+2x)^{2/3}}$

23 $\dfrac{(x^4 + 3x^2) \cos 2x - (2x^3 + 2x^5) \sin 2x}{(1+x^2)^2}$

25 $\dfrac{x}{3\sqrt{x^2+3}(1+\sqrt{x^2+3})^{2/3}}$ **27** $-\frac{35}{3} \csc^2 5x \, (\cot 5x)^{4/3}$

29 $\dfrac{4}{\sqrt{8x+3}}$ **31** $\dfrac{2x - x^4}{(x^3+1)^2}$

33 $\dfrac{-5[4(x^2+3x)^3(2x+3)+1]}{7[(x^2+3x)^4+x]^{12/7}}$

35 $\sqrt{2x+1} \cos 6x + \dfrac{x}{\sqrt{2x+1}} \cos 6x - 6x\sqrt{2x+1} \sin 6x$

37 $x \sin ax$ **39** $\sin^2 ax$ **41** (*b*) 64, 32, 0, −32 ft/sec (*c*) 64, 32, 0, 32 ft/sec (*d*) Rising: $0 < t < 2$; falling: $t > 2$

43 $y = -x$ **45** (*a*) 12 g/cm **47** (*a*) $\frac{3}{2}$ (*b*) $\frac{1}{2}$ (*c*) $\frac{1}{4}$ (*d*) $\frac{1}{6}$

49 (*a*)

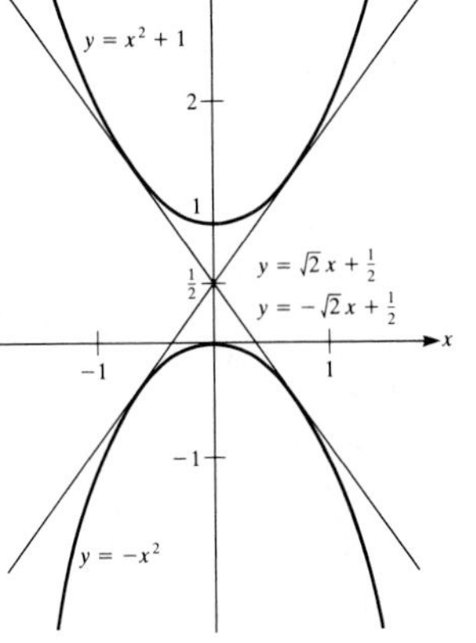

(*b*) $y = \sqrt{2}x + \frac{1}{2}$, $y = -\sqrt{2}x + \frac{1}{2}$

51 (*a*) 12.61 (*b*) 11.41 (*c*) 12 **53** (*a*) 1,000 (*b*) $5 + \dfrac{x}{100}$ (*c*) 5.1 (*d*) 5.105 **55** (*a*) 5 (*b*) 2.25 (*d*) 4

57 (*a*) $3y^2 \dfrac{dy}{dx}$ (*b*) $(-\sin y)\dfrac{dy}{dx}$ (*c*) $-\dfrac{1}{y^2}\dfrac{dy}{dx}$

59 In each part, pick any two values of c. (*a*) $x^4 + c$ (*b*) $\dfrac{x^4}{4} + c$ (*c*) $\dfrac{x^5}{5} + \dfrac{x^4}{4} + \sin x + c$ (*d*) $\dfrac{x^4}{4} - \cos x + c$ (*e*) $\dfrac{x^5}{5} + \dfrac{2x^3}{3} + x + c$

61 (*a*) It always exists. (*b*) $a = 1$ (*c*) $a = 0, 2, 3, 4$ **67** $\dfrac{\cos \sqrt{3}}{2\sqrt{3}}$

69 (*a*), (*b*)

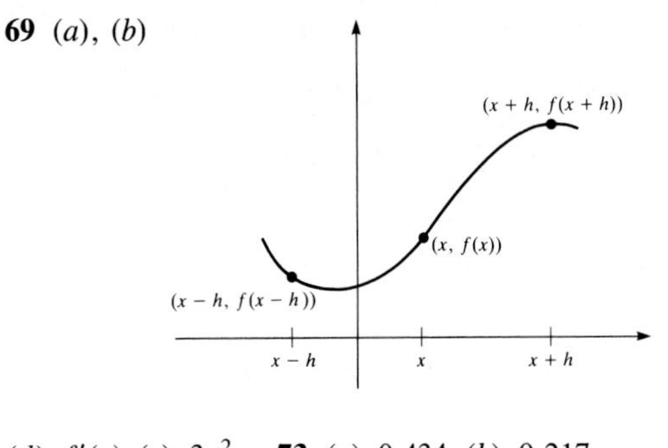

(*d*) $f'(x)$ (*e*) $3x^2$ **73** (*a*) 0.434 (*b*) 0.217

(*c*)

x	$\frac{1}{2}$	1	2	3	4
$f'(x)$	0.869	0.434	0.217	0.145	0.109

(*d*) $f'(x) \approx \dfrac{0.434}{x}$

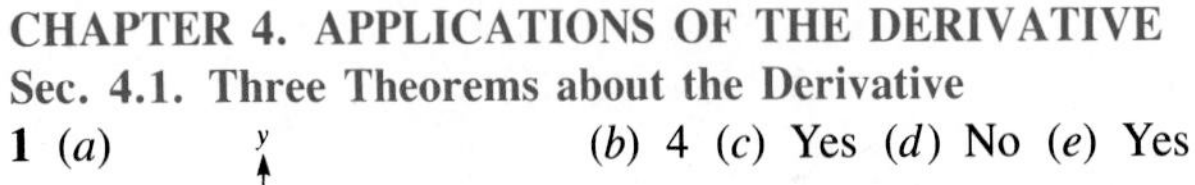

CHAPTER 4. APPLICATIONS OF THE DERIVATIVE

Sec. 4.1. Three Theorems about the Derivative

1 (*a*) (*b*) 4 (*c*) Yes (*d*) No (*e*) Yes

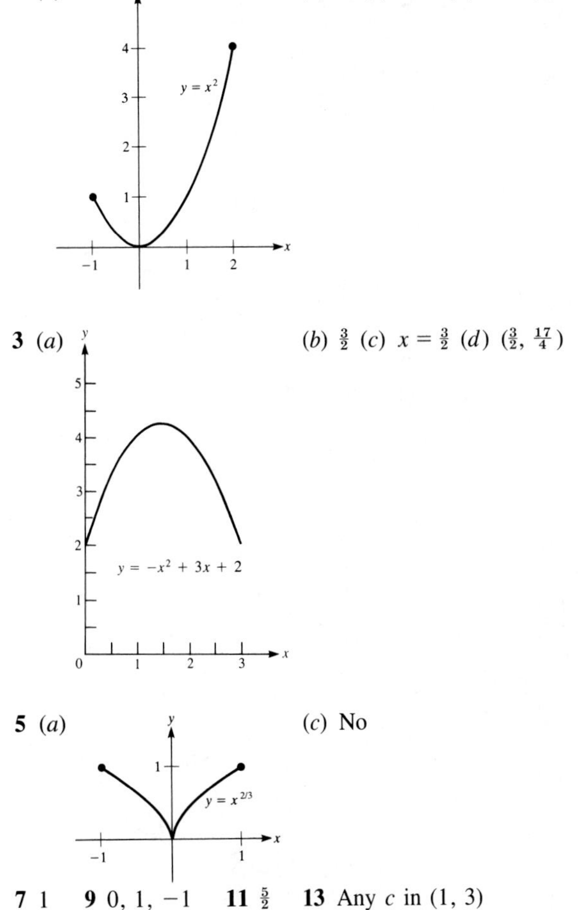

3 (*a*) (*b*) $\frac{3}{2}$ (*c*) $x = \frac{3}{2}$ (*d*) $(\frac{3}{2}, \frac{17}{4})$

5 (*a*) (*c*) No

7 1 **9** 0, 1, −1 **11** $\frac{5}{2}$ **13** Any c in (1, 3)

15 (*a*), (*b*), (*c*)

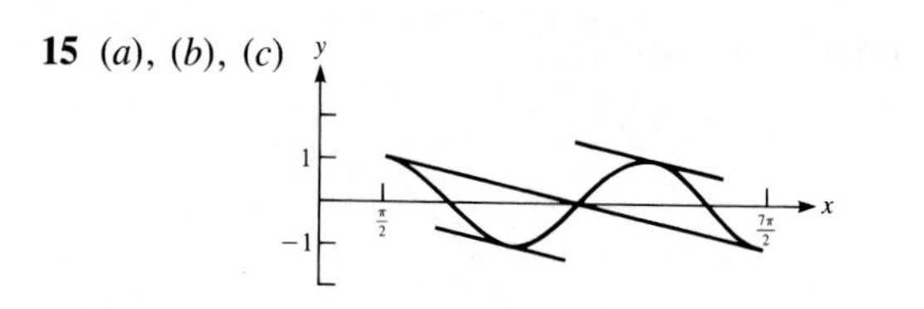

(*d*) 4 (*e*) 1.8, 4.5, 8.1, 10.8

17 (*a*) $2 \sec^2 x \tan x$ (*b*) 1

19

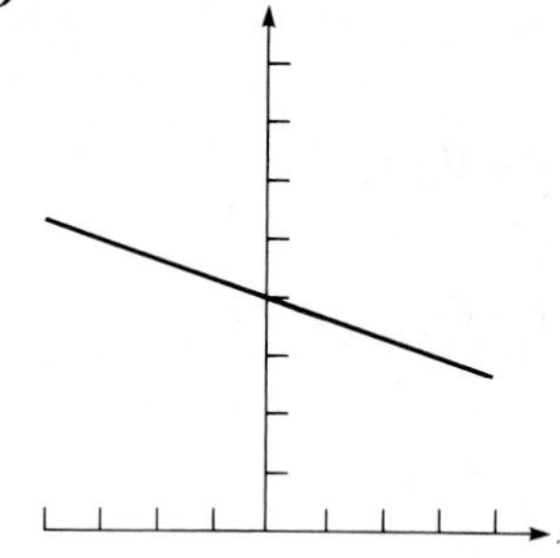

21 **23**

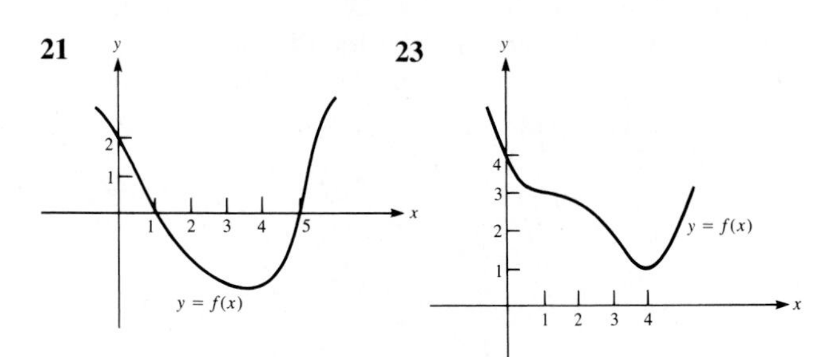

31 (*b*) [0, 1] **33**

35

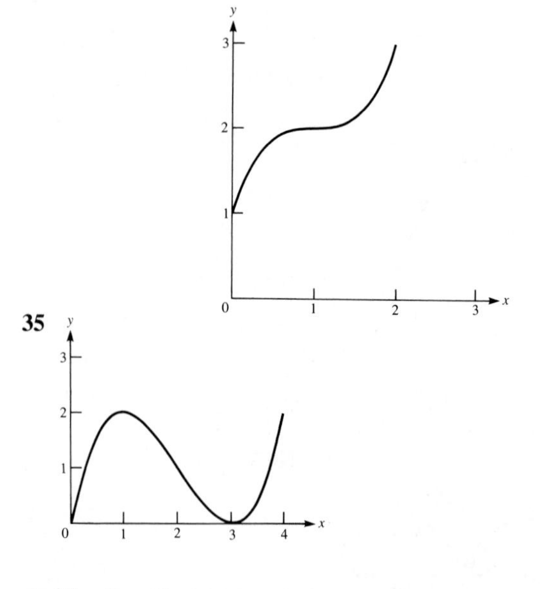

37 $|k| \leq \frac{7}{2}$ **39** (*a*) Yes (*b*) Not necessarily

43 (*b*) Velocity = 0 for some *t* in (0, 2). (*c*) 1

45 (*a*) $3\sqrt{1-x^2}\cos 3x - \dfrac{x}{\sqrt{1-x^2}}\sin 3x$

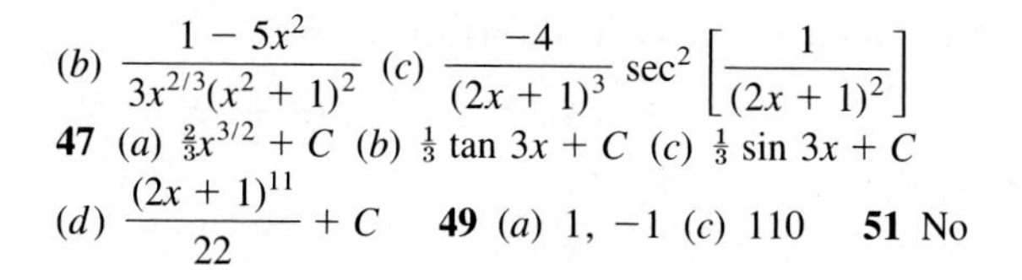

(*b*) $\dfrac{1-5x^2}{3x^{2/3}(x^2+1)^2}$ (*c*) $\dfrac{-4}{(2x+1)^3}\sec^2\left[\dfrac{1}{(2x+1)^2}\right]$

47 (*a*) $\frac{2}{3}x^{3/2} + C$ (*b*) $\frac{1}{3}\tan 3x + C$ (*c*) $\frac{1}{3}\sin 3x + C$

(*d*) $\dfrac{(2x+1)^{11}}{22} + C$ **49** (*a*) 1, −1 (*c*) 110 **51** No

Sec. 4.2. The First Derivative and Graphing

1 0, neither **3** 1, neither **5** $-\frac{1}{4}$, minimum; 0, neither

7 0, minimum; $\pm(4k+1)\pi/2$ is a maximum and $\pm(4k+3)\pi/2$ is a minimum, where *k* is a nonnegative integer.

9

13

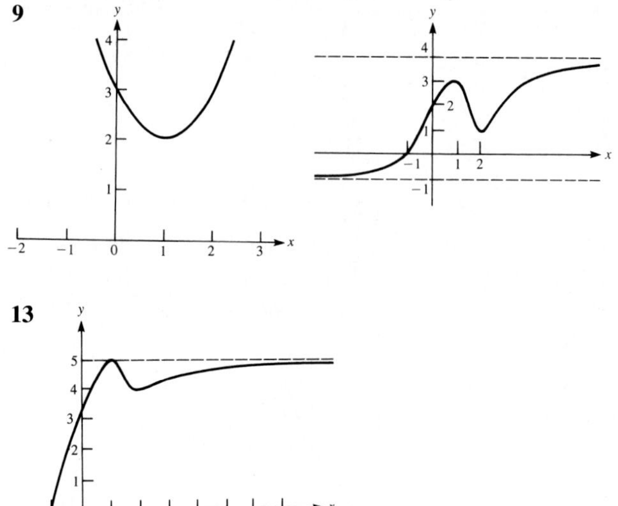

15

17 Critical point at (1, 1)

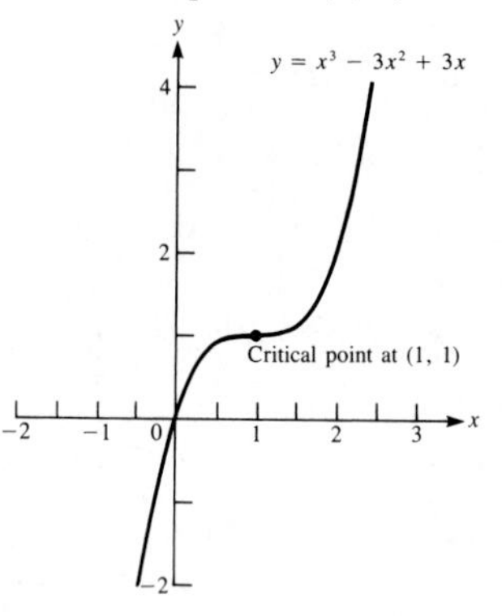

19 Critical point and global minimum at (1, 0)

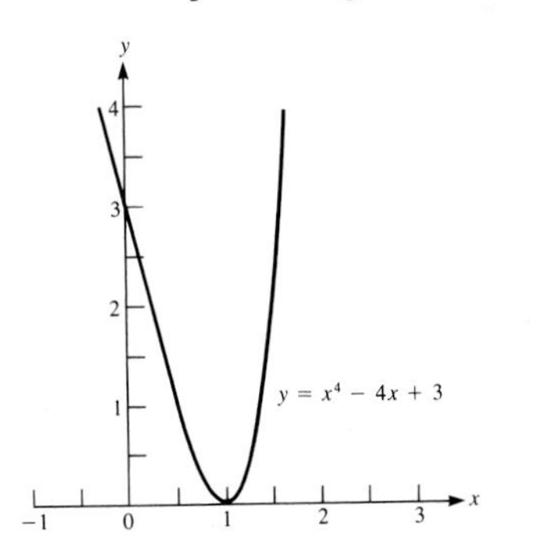

21 Critical point and global minimum at (3, −4)

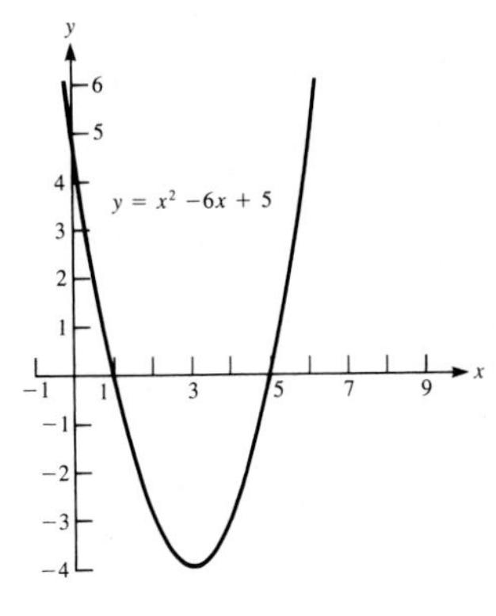

23 Critical point and local maximum at (0, 0); critical point and local minimum at $(b, f(b))$ for $b = (-3 + \sqrt{33})/4$; critical point and global minimum at $(a, f(a))$ for $a = (-3 - \sqrt{33})/4$.

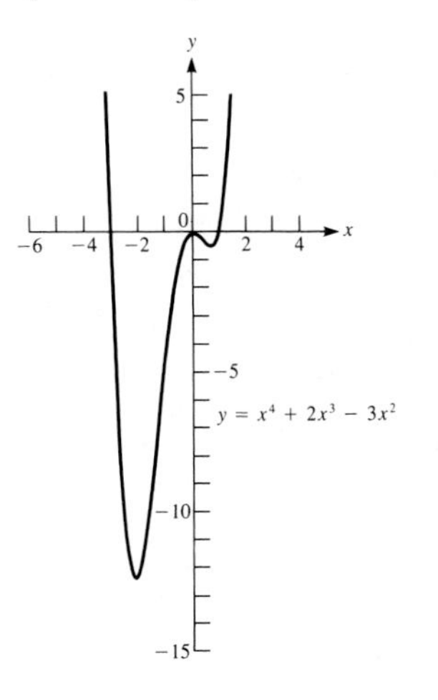

25 Asymptotes $x = \frac{1}{3}$ and $y = 1$

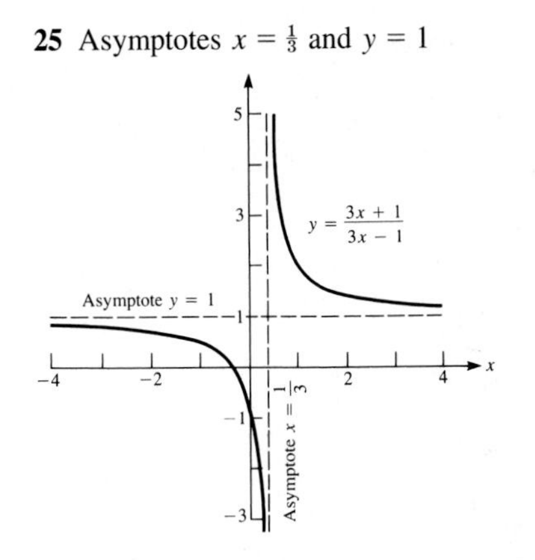

27 Critical point and global maximum at $(1, \frac{1}{2})$; critical point and global minimum at $(-1, -\frac{1}{2})$; asymptote $y = 0$.

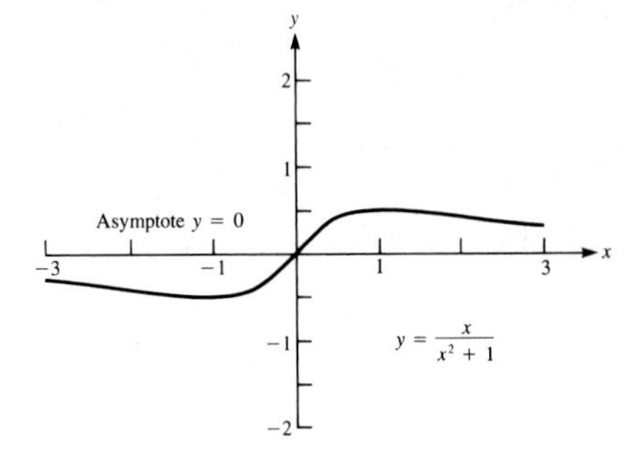

29 Critical point and local maximum at $(\frac{1}{4}, -8)$; asymptotes $x = 0$, $x = \frac{1}{2}$, and $y = 0$.

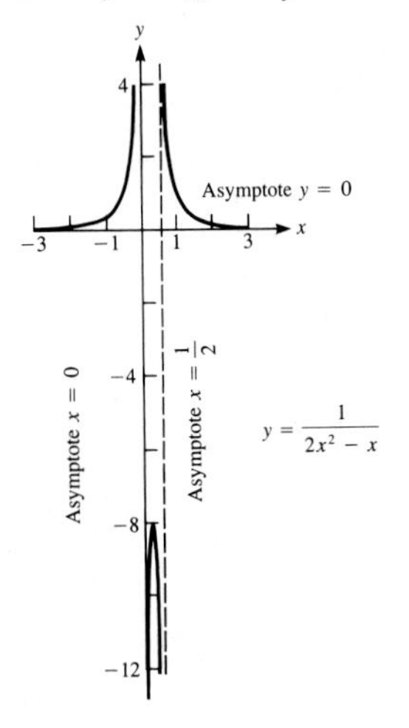

31 Critical point and local maximum at $(0, -\frac{3}{4})$; asymptotes $x = -2$, $x = 2$, $y = 1$.

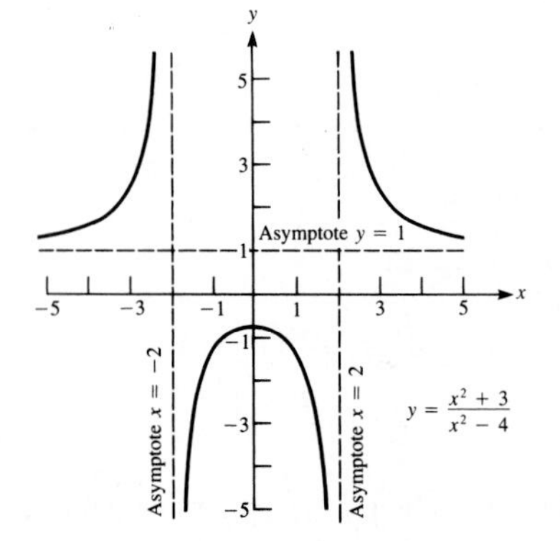

33 $\frac{1}{4}$, 0 **35** 3, 0 **37** 24, -8 **39** 2, 0 **41** $\sqrt{2}$, -1

43

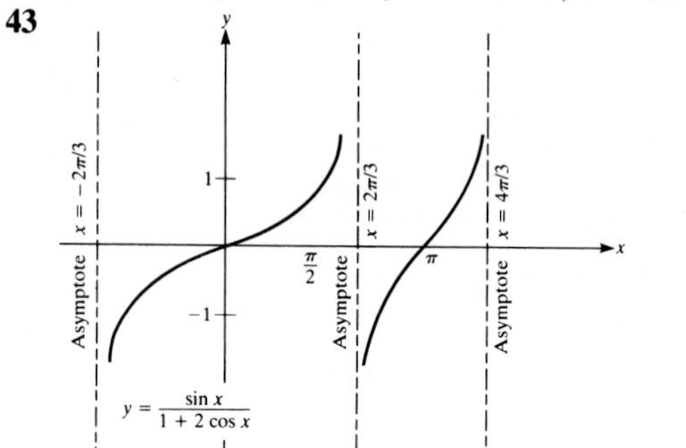

45

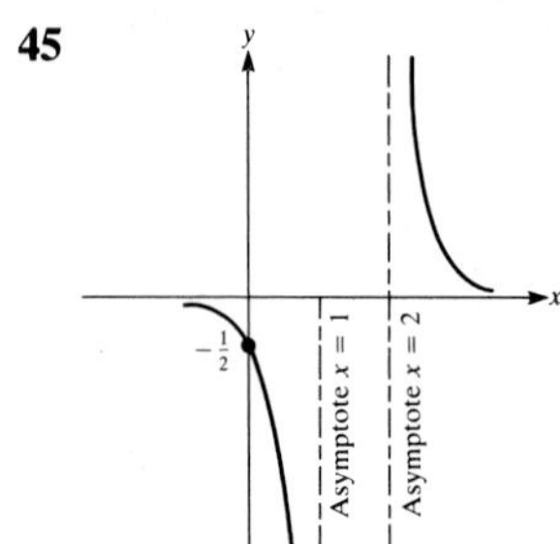
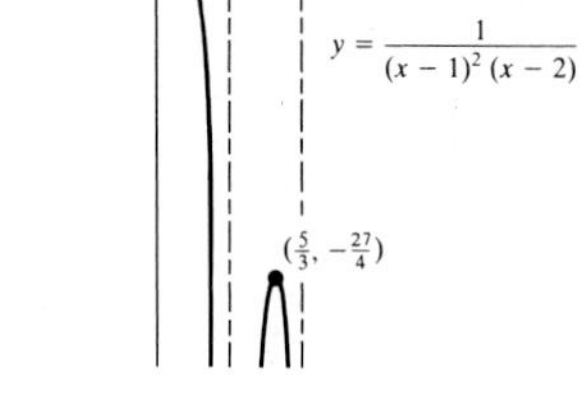

47

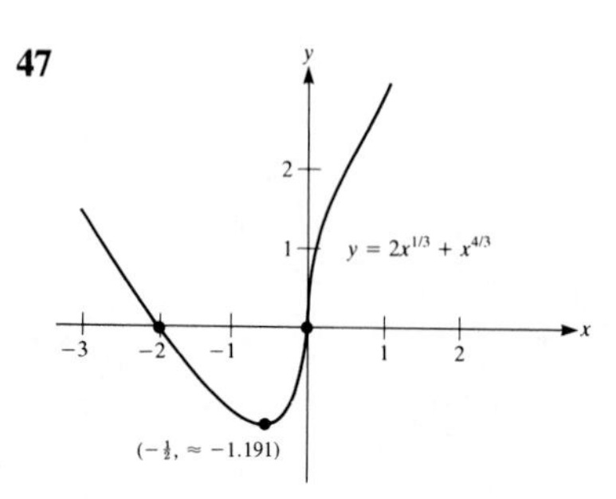

49

51 (*a*)

(*b*) Infinitely many

(*c*)

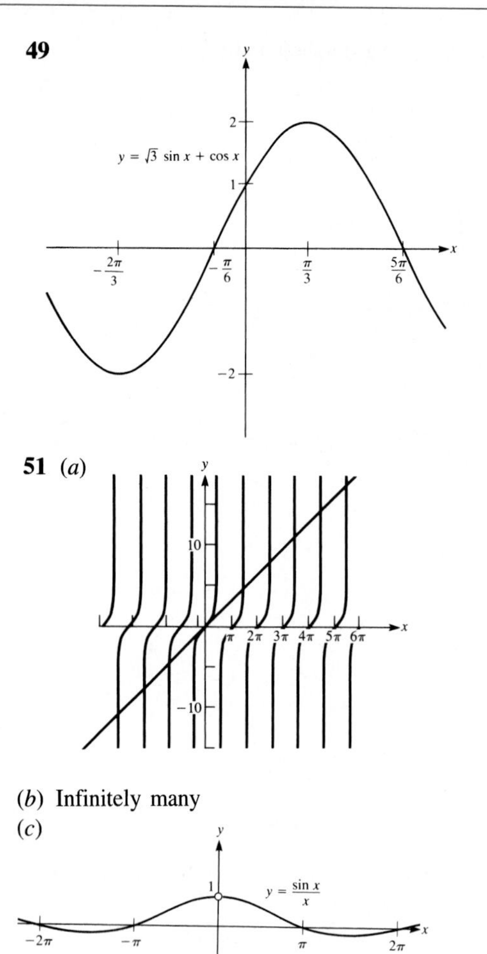

Sec. 4.3. Motion and the Second Derivative

1 $y' = 2$, $y'' = 0$ **3** $5x^4$, $20x^3$ **5** $6x^2 + 1$, $12x$

7 $\dfrac{1}{(x+1)^2}$, $\dfrac{-2}{(x+1)^3}$

9 $\cos x - x \sin x$, $-2 \sin x - x \cos x$

11 $\dfrac{x \cos x - \sin x}{x^2}$, $\dfrac{-x^2 \sin x - 2x \cos x + 2 \sin x}{x^3}$

13 $4(x-2)^3$, $12(x-2)^2$ **15** $3 \cos 3x$, $-9 \sin 3x$ **17** $f'' > 0$ **19** (*a*) 4 sec (*b*) Velocity = -64 ft/sec; speed = 64 ft/sec. **23** ≈ 21 min **25** $3t^2 - 3t + 3$ ft **27** $-16t^2$ **29** 660 ft

Sec. 4.4. Related Rates

1 $2\sqrt{901}$ ft/sec ≈ 60.03 ft/sec **3** (*d*) $\frac{3}{4}$ ft/sec, $\frac{4}{3}$ ft/sec, $9/\sqrt{19}$ ft/sec **5** (*b*) $x = 3 \tan \theta$ **7** (*a*) $1/(4\pi)$ ft/sec (*b*) $1/(16\pi)$ ft/sec **9** (*a*) $\dfrac{5}{2\pi}$ yd/hr (*b*) $\dfrac{1}{10\pi}$ yd/hr

11 $27/(2\sqrt{2})$ ft²/sec, increasing

13 (*a*) $-0.00014\ \text{ft/sec}^2$ (*b*) $-7.56\ \text{ft/sec}^2$
15 (*a*) $6x$ (*b*) 18 **17** $\frac{19}{4}$ ft/sec
19 (*a*) $-25/\sqrt{29}$ ft/sec (*b*) decreasing
21 (*a*) $\dfrac{5\pi}{2}$ ft/sec (*b*) $\dfrac{5\sqrt{3}\pi}{2}$ ft/sec
23 $\dfrac{211210}{\sqrt{42241}}$ ft/sec ≈ 1027.7 ft/sec

Sec. 4.5. The Second Derivative and Graphing

1 Concave up: $x > 1$, down: $x < 1$; inflection point: $x = 1$. **3** Concave up for all x **5** Concave up for all x **7** Concave up for $x < 0$ and $x > 2$; concave down for $0 < x < 2$; inflection points: $x = 0, 2$. **9** Concave up for $|x| > 1/\sqrt{3}$, down for $|x| < 1/\sqrt{3}$; inflection points at $x = \pm 1/\sqrt{3}$. **11** Concave up for $x > 2$, down for $x < 2$; inflection point at $x = 2$. **13** Inflection points at integral multiples of π; concave up for . . . , $0 < x < \pi/2$, $\pi < x < 3\pi/2$, . . . ; concave down for . . . , $-\pi/2 < x < 0$, $\pi/2 < x < \pi$, ... **15** Inflection points at $x = n\pi/2$, where n is an odd integer; concave up for . . . , $-3\pi/2 < x < -\pi/2$, $\pi/2 < x < 3\pi/2$, . . . ; concave down for . . . , $-\pi/2 < x < \pi/2$, $3\pi/2 < x < 5\pi/2$, ...

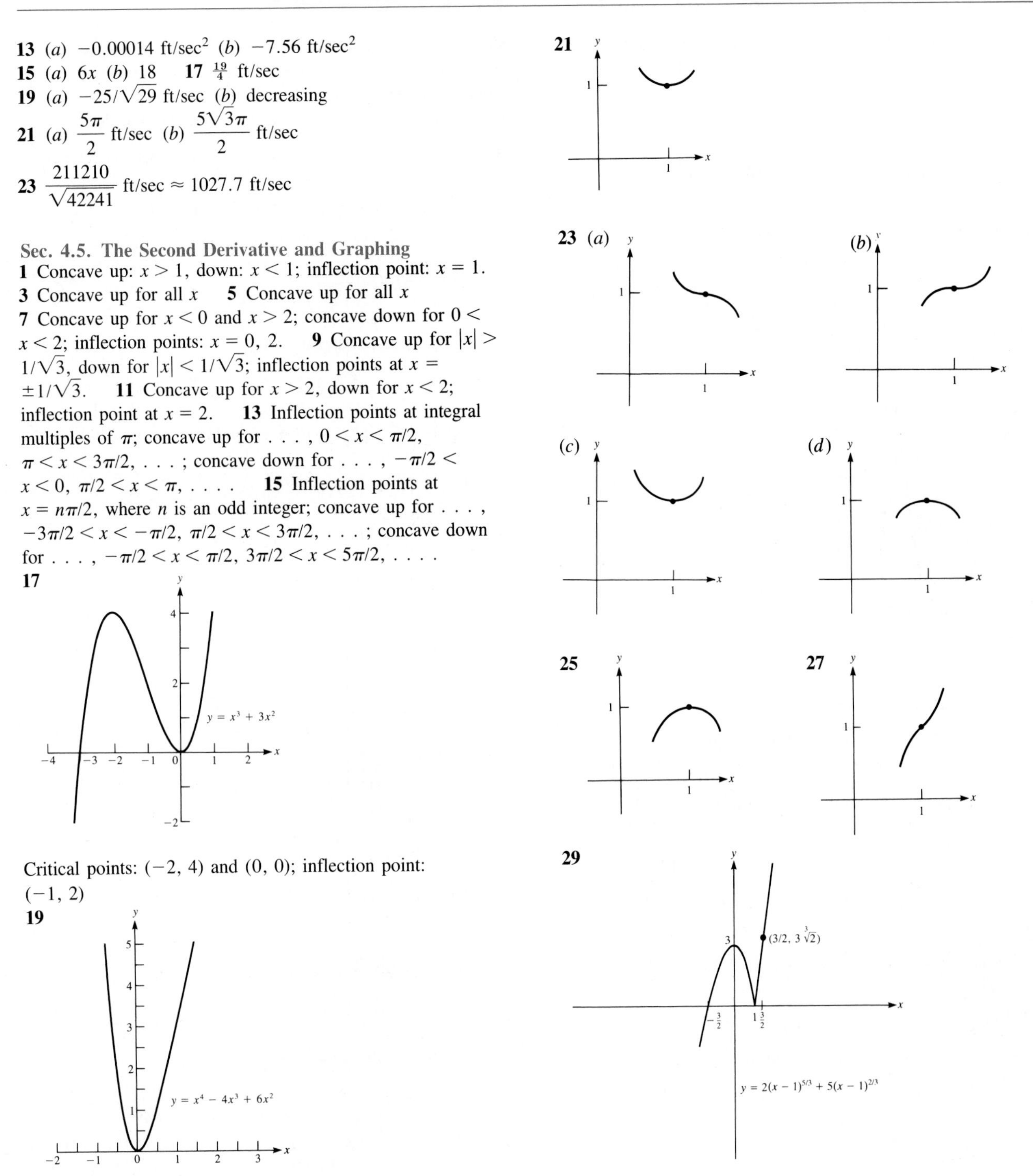

17 Critical points: $(-2, 4)$ and $(0, 0)$; inflection point: $(-1, 2)$

19 Critical point: $(0, 0)$

33 (*a*) Farm output is increasing. (*b*) Since 1957 the rate of increase has decreased. **35** Critical points at $x = 0$, 0.955, 2.186, π, 4.097, 5.328, 2π; inflection points at $x = 0.491$, $\pi/2$, 2.651, 3.632, $3\pi/2$, 5.792 **37** (*a*) No (*b*) Yes **39** (*a*) $x < 1$ and $x > 2$ (*b*) $1 < x < 2$ (*c*) 1, 2 (*d*) $\frac{x^4}{12} - \frac{x^3}{2} + x^2 + 3x + 1$ **41** (*a*) Yes (*b*) Yes

Sec. 4.6. Newton's Method for Solving an Equation

1 1.8 **5** $x_2 = 3.875$, $x_3 \approx 3.873$ **7** $x_2 \approx 1.917$, $x_3 \approx 1.913$ **9** (*a*) $x_4 \approx 2.236068$ (*b*) 2.236068 **11** (*b*) $x_2 \approx 0.857$ **13** $x_2 = 5.15$, $x_3 \approx 2.8663$, $x_4 \approx 1.9565$ **15** (*b*) ≈ 0.75; no (*c*) $x_3 \approx 0.7390851$ **17** (*b*) $x_2 = 1.5$, $x_3 \approx 1.316$ (*c*)

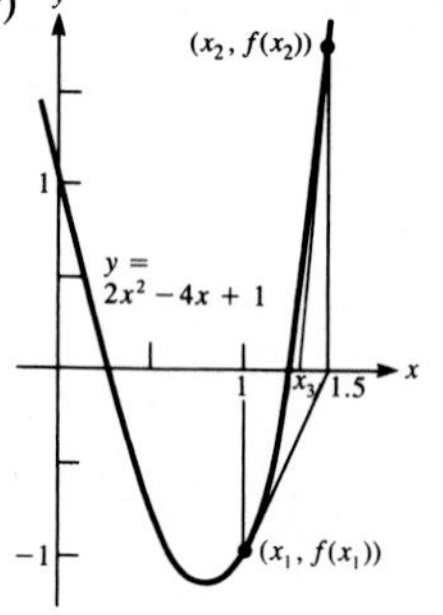

19 (*a*) $x_2 = \frac{-1}{\sqrt{5}}$, $x_3 = \frac{1}{\sqrt{5}}$ **21** 0.95 **23** (*a*) 1 (*b*) 0.86 **25** One; ≈ 0.68 **27** $x_{i+1} = \frac{4}{5}x_i + \frac{a}{5x_i^4}$

Sec. 4.7. Applied Maximum and Minimum Problems

1 $x = 50$ ft, $y = 25$ ft **3** $\frac{5}{6}$ in **5** $\frac{1}{3}(6 - 2\sqrt{3})$ in **7** $h = \sqrt[3]{\frac{400}{\pi}}$ in **9** 10 in × 10 in × 10 in **11** $\theta = \frac{\pi}{4}$; $w = h = \sqrt{2}a$ **15** 40 ft × 60 ft **17** (*a*) $x = 1$, $y = 0$ (*b*) $x = \frac{1}{2}$, $y = \frac{1}{2}$ **19** $x = 8$ ft **21** $r = 36/\pi$ in, $h = 36$ in **23** 18 in × 18 in × 36 in **25** (*c*) $\sqrt[3]{\frac{25}{\pi}}$ (*d*) $4\sqrt[3]{\frac{25}{\pi}}$ **27** Length $= 2\sqrt[3]{\frac{75}{7}}$ in, height $= \frac{7}{3}\sqrt[3]{\frac{75}{7}}$ in **29** 414 on the first road and 586 on the second **33** (*a*)

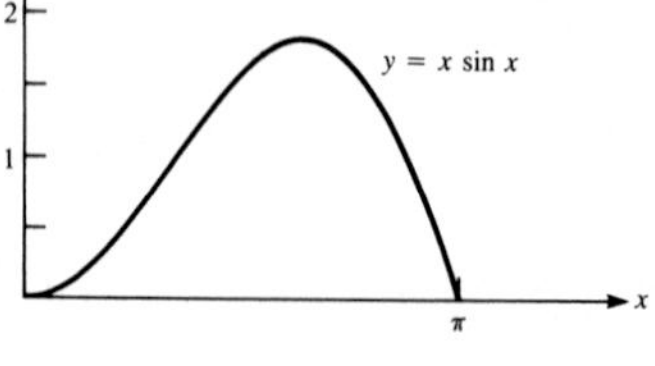

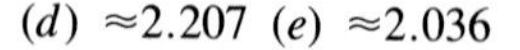
(*d*) ≈ 2.207 (*e*) ≈ 2.036

35 ≈ 0.8010 **37** (*a*) 10 ft × 20 ft (*b*) 25 feet square (*c*) 20 feet square **39** Walk from F to the midpoint of an edge containing neither F nor S, and then to S. **41** (*a*), (*b*) Walk across the grass to B. (*c*) Walk across the grass to the point on the sidewalk $\frac{1}{4}$ mile from B. **43** $r = \sqrt{\frac{2}{3}}\,a$, $h = \frac{2}{\sqrt{3}}\,a$ **45** Length = 20 ft, height = 30 ft **47** (*a*) *SD* (*b*) 63 **51** $13\sqrt{13}$ **53** $\sqrt{a(a+b)}$ ft **55** \$130 **57** (*b*) $\frac{ALc}{2v}x^3 + \frac{RLv}{2A}x - K$

Sec. 4.8. Implicit Differentiation

1 -4 **3** $-\frac{7}{15}$ **5** $-\pi/2$ **7** $-\frac{8}{9}$ **9** $r = \sqrt[3]{\frac{50}{\pi}}$ in, $h = 2\sqrt[3]{\frac{50}{\pi}}$ in **11** $x = 50$ ft, $y = 25$ ft **15** 18 in × 18 in × 36 in **17** $-\frac{y^3 + \sec^2(x+y)}{3xy^2 + \sec^2(x+y)}$ **19** $\frac{7x - 24y}{24x + 7y}$ **21** (*b*) $-\frac{3}{5}$ (*d*) $-\frac{168}{125}$ **23** -2, -6 **25** $(-2, 4)$, $(2, -4)$ **29** (*b*) 16 at $(\pm 4, 0)$ (*c*)

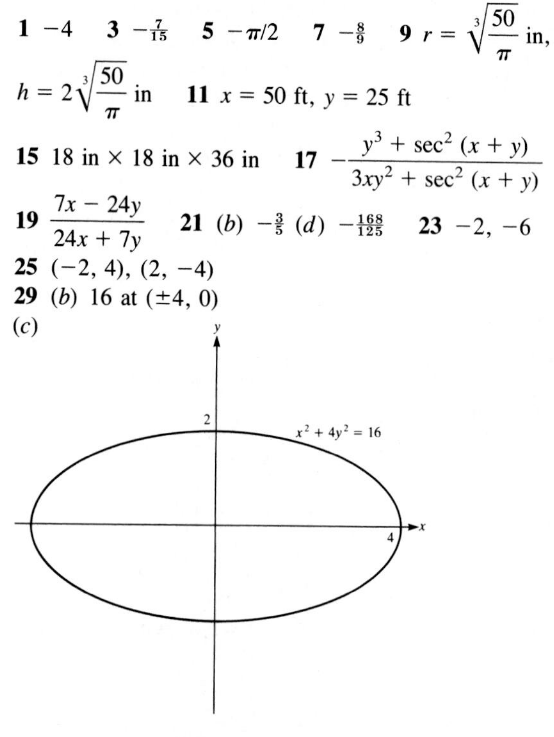

Sec. 4.9. The Differential and Linearization

1 $dy = 0.6$, $\Delta y = 0.69$ **3** $dy = -\frac{1}{3}$, $\Delta y = \sqrt{7} - 3 \approx -0.354$ **5** $dy = \frac{\pi}{9} \approx 0.349$, $\Delta y = 1 - \frac{1}{\sqrt{3}} \approx 0.423$ **7** (*a*) $y = \sqrt{x}$ (*b*) 100 (*c*) 9.9 **9** 10.91 **11** 2.9259 **13** 0.98 **15** 0.8560 **17** 0.13 **19** 0.248125 **21** 0.5302 **23** 0.8835 **29** $-(3/x^4)\,dx$ **31** $2\cos 2x\,dx$ **33** $-\csc x \cot x\,dx$ **35** $-(1/x^2)(5x\csc^2 5x + \cot 5x)\,dx$ **37** (*a*) $(x+1)/2$ (*b*) $(x+2)/(2\sqrt{2})$

39

x	$f(x) = \sqrt[3]{x}$	$p(x)$	$f(x) - p(x)$
1.5	1.1447142	1.1666667	−0.0219524
1.1	1.0322801	1.0333333	−0.0010532
1.01	1.0033223	1.0033333	−0.0000111
1.001	1.0003332	1.0003333	−0.0000001
1.0001	1.0000333	1.0000333	-1.1×10^{-9}

41 About $p/2$ percent

43 (*a*) $df = 2x\,\Delta x$, $\Delta f = 2x\,\Delta x + (\Delta x)^2$

(*b*)

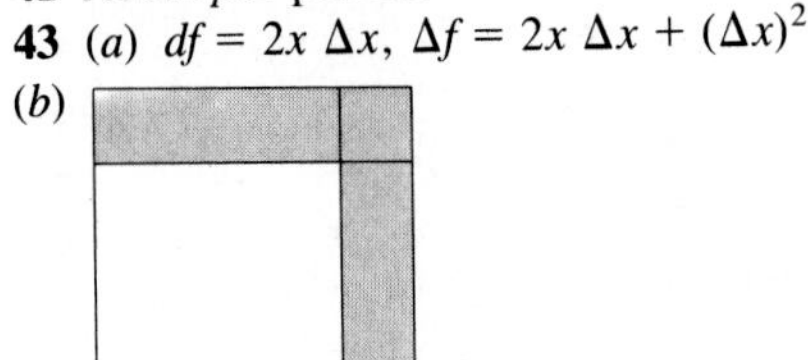

Δf

(*c*)

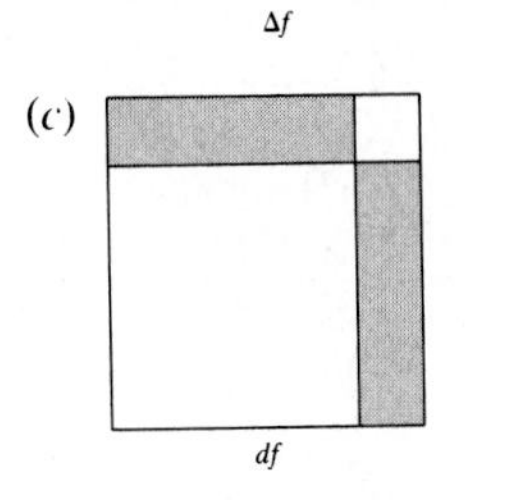

df

47 (*b*) It approaches 1.

Sec. 4.10. The Second Derivative and Growth of a Function

1 It is at most 1,800 ft. **3** It is between 40 and 96 miles.
5 ≤ 8 **7** $1.25 \le f(2) \le 1.3$
9 (*a*) $x/4 + 1$
(*b*)

x	$E(x)$	$(x-4)^2$	$\dfrac{E(x)}{(x-4)^2}$
5	−0.013932	1	−0.013932
4.1	−0.0001543	0.01	−0.0154327
4.01	−0.0000016	0.0001	−0.01561
3.99	−0.0000016	0.0001	−0.01564

(*c*) −0.015625
11 (*a*) $2(x-2)^2 \le E(x) \le 2.5(x-2)^2$
(*b*) $|x - 2| \le 1/\sqrt{250}$

Sec. 4.S. Summary: Guide Quiz

1 (*a*) f is continuous on $[a, b]$ and differentiable on (a, b). (*b*) There is at least one number c in (a, b) such that $f'(c) = \dfrac{f(b) - f(a)}{b - a}$. **2** (*a*) The graph crosses the x axis at $x = a$. (*b*) There is a maximum at $x = a$. (*c*) The concavity changes from upward to downward; $x = a$ is an inflection point. **3** $c = 3$ **5** (*a*) $x^3 + C$ (*b*) Corollary 2, Sec. 4.1 **6** $\frac{1}{3}(4 - \sqrt{7})$ is squared, $\frac{1}{3}(-1 + \sqrt{7})$ is cubed.

7 Maximum at $x = 0$; minimum at $x = L/2$ $(r = 0)$.
8 About 6 percent **9**

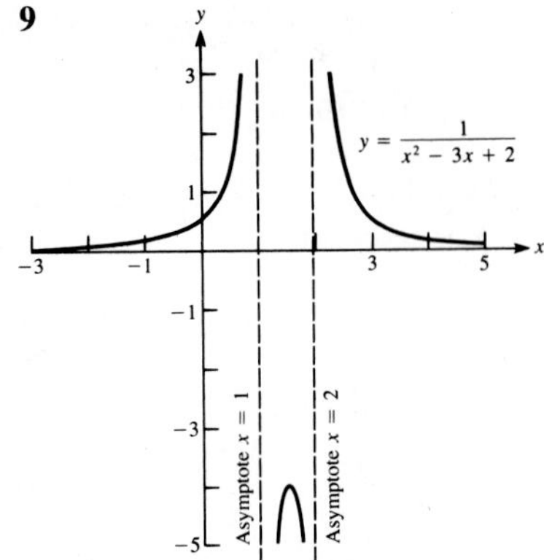

10

11

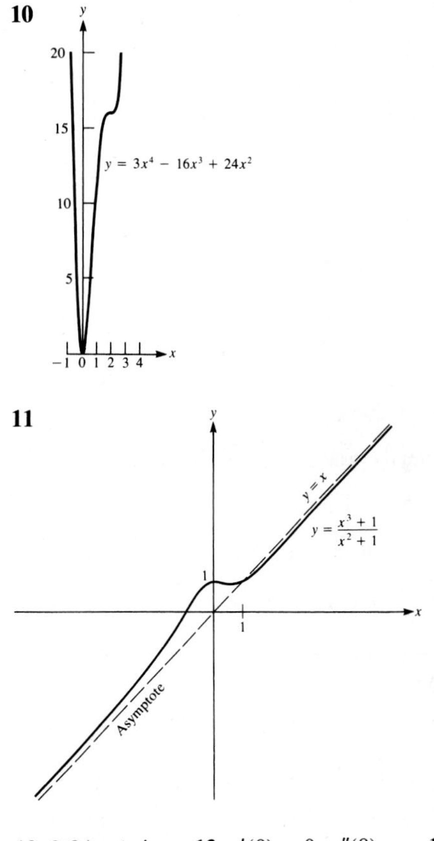

12 0.04 m/min **13** $y'(0) = 0$, $y''(0) = -1$

14 (*a*) $40x^3 - \dfrac{2}{x^3}$ (*b*) $-4\cos 2x$ (*c*) $102x$ (*d*) $-\dfrac{1}{4x^{3/2}}$

(*e*)

$$\frac{18(1 + 2x)^2(\sec^2 3x \tan 3x) - 12(1 + 2x)\sec^2 3x + 8\tan 3x}{(1 + 2x)^3}$$

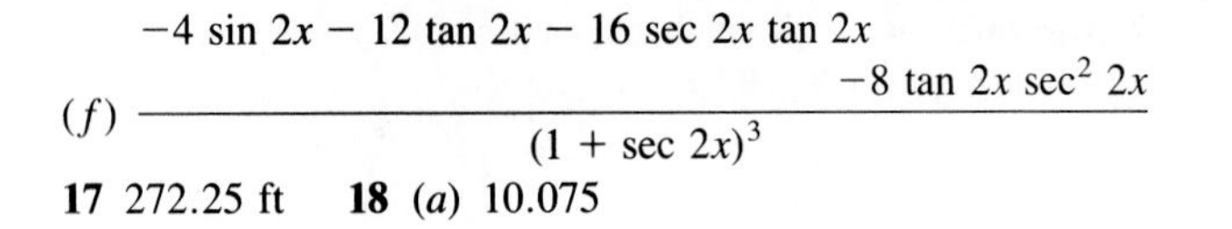

$(f)\ \dfrac{-4\sin 2x - 12\tan 2x - 16\sec 2x\tan 2x - 8\tan 2x\sec^2 2x}{(1+\sec 2x)^3}$

17 272.25 ft **18** (*a*) 10.075

Sec. 4.S. Summary: Review Exercises

3 (*b*) Not necessarily
5

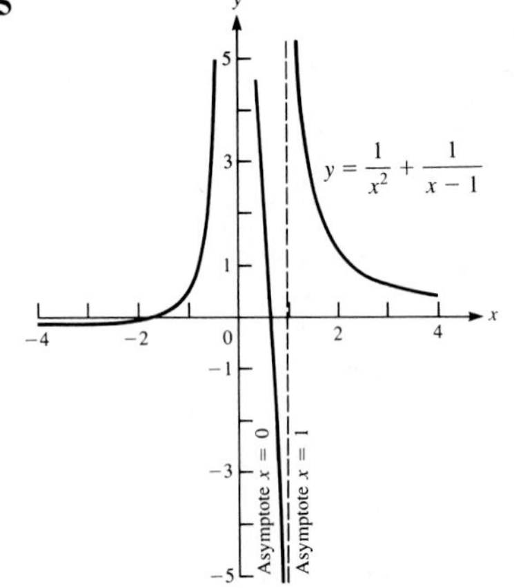

7 (*a*), (*d*)

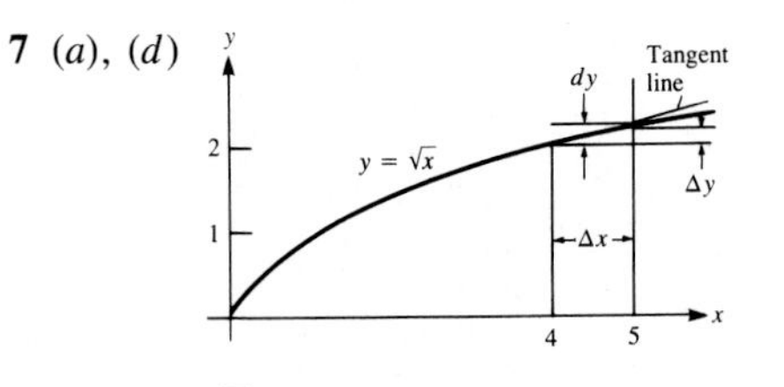

(*b*) $\frac{1}{4}$ (*c*) $\sqrt{5} - 2 \approx 0.236$

11 (*a*) $\dfrac{9L}{4\sqrt{3}+9}$ for the triangle, $\dfrac{4\sqrt{3}L}{4\sqrt{3}+9}$ for the square

(*b*) All for the square **13** $f(3) \ge 3$ **15** (*a*) $1 - x^2/2$

17 (*a*) $\sqrt{2} + \dfrac{7x}{2\sqrt{2}}$ **19** (*a*) 0.024184 (*b*) 0.0197461

(*c*) 0.0139626 **21** About 4% in S and 6% in V **23** −1

25

Δx	$f(a+\Delta x) - p(a+\Delta x)$	Δf	df	$\Delta f - df$
0.2	0.1084976	0.5084976	0.4	0.1084976
0.1	0.0230489	0.2230489	0.2	0.0230489
0.01	0.0002027	0.0202027	0.02	0.0002027
−0.01	0.0001974	−0.0198026	−0.02	0.0001974

27

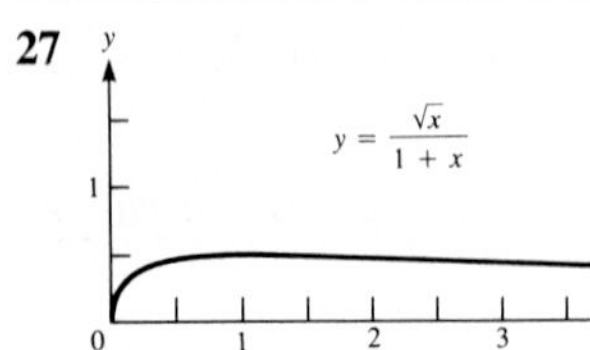

31 (*a*) $\dfrac{4x^3 + 12x^2 - 2}{(x+2)^2}$ (*b*) $\dfrac{3x^5}{2\sqrt{1+3x}} + 5x^4\sqrt{1+3x}$

(*c*) $\frac{10}{7}(2x-1)^4$ (*d*) $\dfrac{2}{\sqrt{x}}\sin^3\sqrt{x}\cos\sqrt{x}$ (*e*) $\dfrac{3}{x^4}\sin\dfrac{1}{x^3}$

(*f*) $\dfrac{-x}{\sqrt{1-x^2}}\sec^2\sqrt{1-x^2}$ **33** −12 **35** (*a*) 1.01

(*b*) 0.988

37 $y = x + 1$, $y = -x - 1$

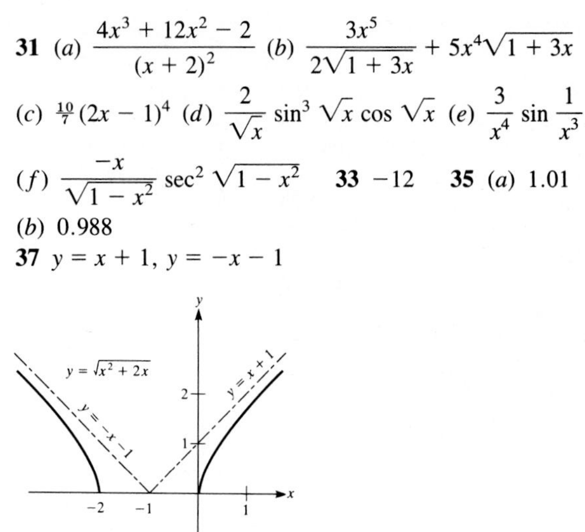

45 $\sqrt{2}a \times \sqrt{2}b$ **47** (1, 1) **49** (*b*) −1.671, 0.506, 1.402 **51** 2.104 **53** $-\dfrac{2}{e+1}$, $\dfrac{-(3e^2 + 2e - 5)}{(e+1)^3}$

55 $V = \frac{32}{81}\pi a^3$ **57** $h = \sqrt{8}r$ **59** (*a*) 5 (*b*) $\sqrt{A^2 + B^2}$

65 (*a*) Nothing (*b*) None **69** (*a*) $36x - 3x^2$ g/cm

(*b*) 6 cm from end

CHAPTER 5. THE DEFINITE INTEGRAL

Sec. 5.1. Estimates in Four Problems

1 (*c*) 12.65625 **3** (*a*) 11.88 (*b*) 6.48 (*c*) Less than 11.88 but larger than 6.48. **5** (*a*) $\frac{1}{5}$ (*c*) $\frac{1}{25}$, 9 (*d*) 9.92 **7** 2.64 **9** (*e*) 2.33 **11** 8.043 **13** (*c*) 0.7456 **15** (*d*) ≈6.75 **17** (*d*) ≈0.3403 **19** (*d*) ≈2.4777 **21** (*a*) 306 (*b*) 319.5 (*c*) 225 (*d*) 441 **23** (*a*) 9.2718 (*b*) 8.7318 **25** (*a*) 8.91 g (*b*) 11.88 g (*c*) 6.48 g **27** (*b*) 81 (*c*) 49 **29** $\frac{143}{16}$ = 8.9375 (millions of dollars) **31** ≈1.4 **33** ≈1.9 **37** $n \ge 2700$ **39** (*a*) $\frac{1}{4}(x+2)^4$

(*b*) $\frac{1}{5}x^5 + \frac{2}{3}x^3 + x$ (*c*) $-\frac{1}{2}\cos x^2$ (*d*) $\dfrac{1}{4}x^4 - \dfrac{1}{2x^2}$ (*e*) $2\sqrt{x}$

Sec. 5.2. Summation Notation and Approximating Sums

1 (*a*) 6 (*b*) 20 (*c*) 14 **3** (*a*) 4 (*b*) 1 (*c*) 450
5 (*a*) $\Sigma_{i=0}^{100} 2^i$ (*b*) $\Sigma_{j=3}^{7} x^j$ (*c*) $\Sigma_{k=3}^{102} 1/k$
7 (*a*) $\Sigma_{i=1}^{3} x_{i-1}^2(x_i - x_{i-1})$ (*b*) $\Sigma_{i=1}^{3} x_i^2(x_i - x_{i-1})$
9 (*a*) $2^{100} - 1$ (*b*) −99/100 (*c*) −100/101
13 13.50 **15** 10.00 **17** 1.07 **19** (*a*) 5050 (*b*) 1,501,500 **23** $11(b-a)$ **25** 6 **27** $\frac{117}{160}$ = 0.73125
29 About 32,000 years **31** (*a*) 9.13545 (*b*) 8.86545
33 (*a*) 2.283 (*b*) 0.542 (*c*) 1.545

Sec. 5.3. The Definite Integral

1 (*a*) 1 (*b*) 2 (*c*) 3 **3** (*a*) $\frac{125}{3}$ (*b*) $\frac{64}{3}$ (*c*) $\frac{61}{3}$ **5** (*a*) a

(*b*) $a + \dfrac{b-a}{n}$ (*c*) $a + 2\left(\dfrac{b-a}{n}\right)$, $a + i\left(\dfrac{b-a}{n}\right)$

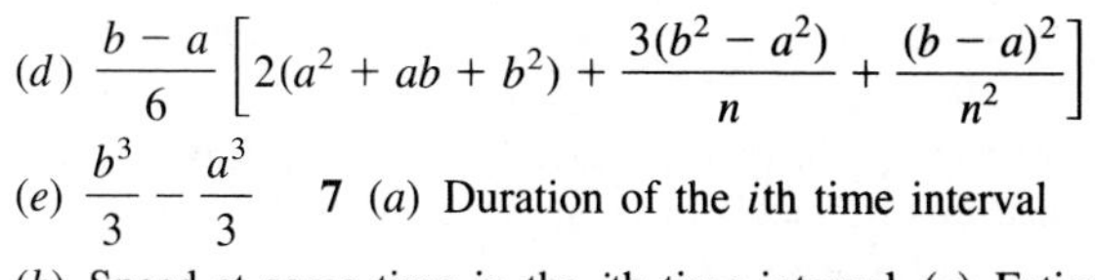

(*d*) $\dfrac{b-a}{6}\left[2(a^2+ab+b^2)+\dfrac{3(b^2-a^2)}{n}+\dfrac{(b-a)^2}{n^2}\right]$

(*e*) $\dfrac{b^3}{3}-\dfrac{a^3}{3}$ **7** (*a*) Duration of the *i*th time interval (*b*) Speed at some time in the *i*th time interval (*c*) Estimate of distance traveled during the *i*th time interval (*d*) Estimate of total distance traveled (*e*) Actual total distance traveled **9** It is a number (it happens to be $\frac{2}{3}$), the limit of certain sums. To be more specific it is $\lim_{\text{mesh}\to 0} \Sigma_{i=1}^{n} \Delta x_i/c_i^2$ on [1, 3]. You may think of it as the area under the curve $y=\dfrac{1}{x^2}$ and above the interval [1, 3]. **11** $\frac{26}{3}$ g **13** (*a*) $\frac{77}{60}\approx 1.2833$ (*b*) $\frac{19}{20}=0.95$ **15** (*a*) $\frac{4}{25}=0.16$ (*b*) $\frac{9}{25}=0.36$ **17** (*a*) 0.77 (*b*) 1.20 **19** (*a*) $\dfrac{b^4}{4}$ (*b*) $\frac{15}{4}$

21 $\frac{1669}{1800}\approx 0.9272$

27 (*a*)

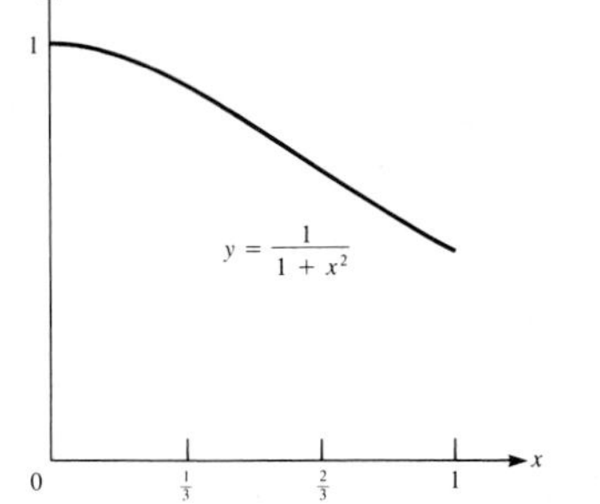

(*d*) $0.7337 < A < 0.8338$

29 (*c*) Logarithmic functions **33** $n \geq 200$

Sec. 5.4. Estimating the Definite Integral

1 1.1 **3** 1.335354 **5** 4.2653862 **7** −0.2338909 **9** 0.8425926 **11** 0.8671137 **13** (*a*) 1280 ft^2 (*b*) $1346\frac{2}{3}$ ft^2 **15** About 200 square miles **23** (*a*) $\frac{4}{3}\pi a^3$ (*b*) $\frac{1}{3}\pi a^2 h$ **27** (*d*) $\frac{3}{2}$ **31** $\approx$1740

Sec. 5.5. Properties of the Antiderivative and the Definite Integral

1 $\frac{5}{3}x^3 + C$ **3** $x^2 - \dfrac{x^4}{4} + \dfrac{x^6}{6} + C$ **5** (*a*) $\sin x + C$ (*b*) $\frac{1}{2}\sin 2x + C$ **7** (*a*) $-2\cos x + 3\sin x + C$ (*b*) $-\frac{1}{2}\cos 2x + \frac{1}{3}\sin 3x + C$ **9** $\tan x + C$ **13** (*a*) 39 (*b*) −39 (*c*) 0 **15** (*a*) $\dfrac{x^2}{2} + C$ (*b*) $\frac{7}{2}$ **17** It is between 10 and 15. **19** $f(c) = 6$, $c = 3$ **21** $f(c) = \frac{16}{3}$, $c = 4/\sqrt{3}$ **23** (*a*) −3 (*b*) 4

25 (*a*) $3(b-a) \leq \int_a^b f(x)\,dx \leq 7(b-a)$ (*b*) Between 3 and 7

27 1, 3, 2 **29** 4, 9, $\frac{19}{3}$ **31** $\frac{53}{18}\approx 2.944$

33 (*a*)

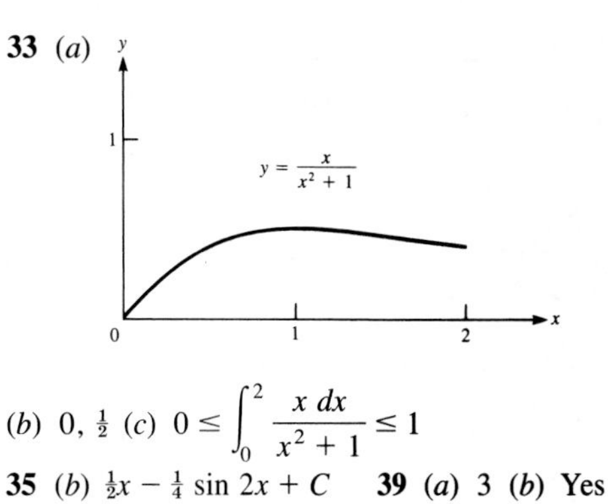

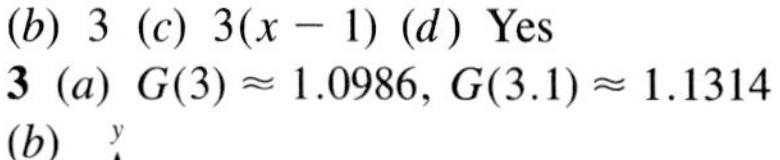

(*b*) 0, $\frac{1}{2}$ (*c*) $0 \leq \displaystyle\int_0^2 \frac{x\,dx}{x^2+1} \leq 1$

35 (*b*) $\frac{1}{2}x - \frac{1}{4}\sin 2x + C$ **39** (*a*) 3 (*b*) Yes

Sec. 5.6. Background for the Fundamental Theorem of Calculus

1 (*a*)

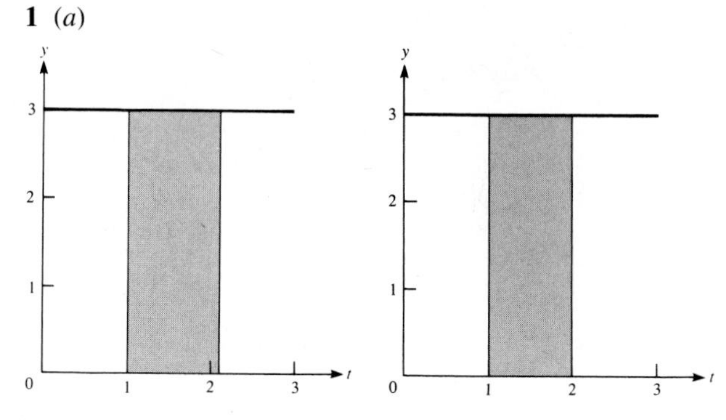

(*b*) 3 (*c*) $3(x-1)$ (*d*) Yes

3 (*a*) $G(3)\approx 1.0986$, $G(3.1)\approx 1.1314$

(*b*)

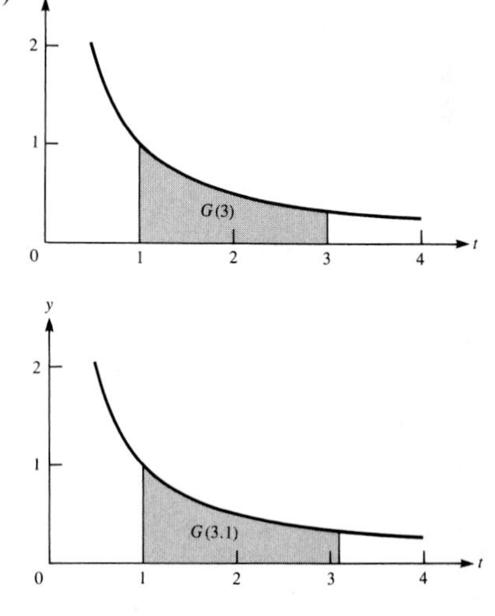

(*c*) 0.3280 (*d*) $\frac{1}{3}$

5 (*a*) 0.8357, 0.8820
(*b*)

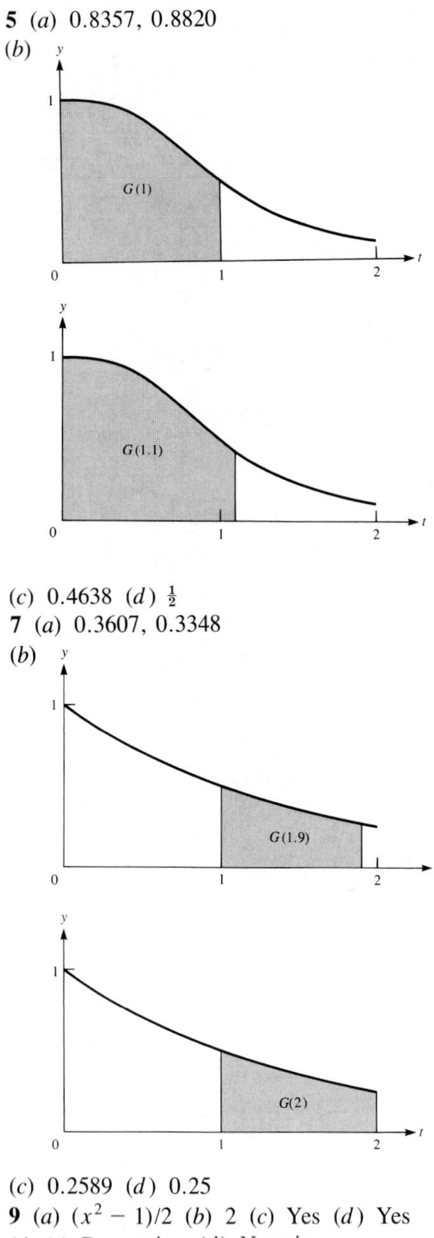

(*c*) 0.4638 (*d*) $\frac{1}{2}$
7 (*a*) 0.3607, 0.3348
(*b*)

(*c*) 0.2589 (*d*) 0.25
9 (*a*) $(x^2 - 1)/2$ (*b*) 2 (*c*) Yes (*d*) Yes
11 (*c*) Decreasing (*d*) Negative
13 (*a*)

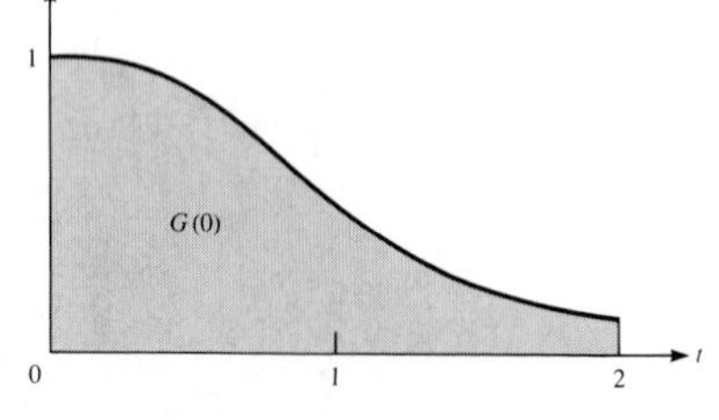

(*b*) $G(0)$ (*c*) Decreasing (*d*) Negative (*e*) -0.99999975
(*f*) Yes
15 (*d*) The area, which equals $G(2.01) - G(2)$, can be approximated by a rectangle of height $f(2)$ and width 0.01. Therefore, $\dfrac{G(2.01) - G(2)}{0.01} \approx f(2)$.

Sec. 5.7. The Fundamental Theorem of Calculus

1 (*a*) 7 (*b*) 9 (*c*) -2 **5** $\frac{75}{4}$ **7** $112\frac{1}{2}$ **9** $\frac{5}{2}(\sqrt{3} - 1)$
11 1 **13** $\frac{190}{3}$ **15** $\frac{93}{5}$ **17** $\frac{49}{3}$ **19** $2/\pi$
21 $\dfrac{12(\sqrt{3} - 1)}{\sqrt{3}\pi}$ **23** 63 **25** $\frac{12}{5}$ **27** 10.5 ft
29 18.75 g **31** 936 cm^3 **33** (*a*) Function (*b*) Number (*c*) Number **35** (*a*) True (*b*) False **37** (*a*) $\sin(x^2)$ (*b*) 3 (*c*) $\sin(x^2)$ **39** (*a*) $\sqrt[3]{1 + \sin x}$ (*b*) $2x\sqrt[3]{1 + \sin(x^2)}$ **41** $9x \tan 3x - 4x \tan 2x$
43

y
1
y = G(x)
0 1 2 3 x

45 The cross-sectional area is $V'(x)$, where $V(x)$ is the volume of water displaced when x inches are submerged.
47 $F(x) = \int_4^x \sqrt[3]{1 + t^2}\, dt$
49 (*b*)

h	*Simpson's Estimate*	*Error*
$\frac{1}{2}$	0.6380712	0.0285955
$\frac{1}{4}$	0.6565263	0.0101404
$\frac{1}{8}$	0.6630793	0.0035874
$\frac{1}{16}$	0.6653982	0.0012685
$\frac{1}{32}$	0.6662182	0.0004485
$\frac{1}{64}$	0.6665081	0.0001586

(*c*) They are all ≈ 0.354 (*d*) $A \approx 0.081$, $k \approx 1.5$ **51** 2

Section 5.S. Summary: Guide Quiz

2 (*a*) Divide the interval $[a, b]$ into n sections of equal width $h = (b - a)/n$, with $x_0 = a$, $x_i = a + ih$, $x_n = b$. Then the trapezoidal estimate of $\int_a^b f(x)\, dx$ equals $(h/2)[f(x_0) + 2f(x_1) + \cdots + 2f(x_{n-1}) + f(x_n)]$.
(*b*) First-degree polynomials: $f(x) = ax + b$. (*c*) You would expect it to decrease. (*d*) At least 339 **3** (*a*) Divide the interval $[a, b]$ into an even number of sections of width $h = (b - a)/n$, with $x_0 = a$, $x_i = a + ih$, $x_n = b$. Then Simpson's estimate of $\int_a^b f(x)\, dx$ is $(h/3)[f(x_0) + 4f(x_1) + 2f(x_2) + \cdots + 2f(x_{n-2}) + 4f(x_{n-1}) + f(x_n)]$.
(*b*) Polynomials of degree at most 3 (*c*) You would expect it to decrease. (*d*) At least 30 and even **4** (*a*) 1.3944

(*b*) 1.1533 (*c*) 1.1614 **5** (*a*) 0.1 (*b*) 5, 4 (*c*) $8 \leq G(3) \leq 10$ **7** (*a*) $\dfrac{-2}{(2x+3)^2}$ (*b*) $\frac{1}{15}$ **8** $\dfrac{4}{\pi}(3 - \sqrt{3})$
9 $6x \cos 3x^3 - 27x^4 \sin 3x^3 - 2 \cos 3x^2 + 12x^2 \sin 3x^2$
11 (*b*) $x \tan x$ (*c*) $\int_0^x u \tan u \, du$

Sec. 5.S. Summary: Review Exercises

1 $\frac{15}{2}$ **3** $\frac{1}{3}$ **5** $\frac{5}{72}$ **7** $\tan x + C$ **9** $\sec x + C$
11 $-4 \csc x + C$ **13** $\dfrac{x^7}{7} + \dfrac{x^4}{2} + x + C$ **15** $5x^{20} + C$
17 (*a*) $18x^2(x^3+1)^5$ (*b*) $\frac{1}{18}(x^3+1)^6 + C$
19 (*a*) $d + d^2 + d^3$ (*b*) $x + x^2 + x^3 + x^4$
(*c*) $0 + \frac{1}{2} + \frac{1}{2} + \frac{3}{8} = \frac{11}{8}$ (*d*) $\frac{3}{2} + \frac{4}{3} + \frac{5}{4} + \frac{6}{5} = \frac{317}{60}$
(*e*) $(\frac{1}{2} - \frac{1}{3}) + (\frac{1}{3} - \frac{1}{4}) + (\frac{1}{4} - \frac{1}{5}) = \frac{3}{10}$ (*f*) $\sin \dfrac{\pi}{4} + \sin \dfrac{\pi}{2} + \sin \dfrac{3\pi}{4} + \sin \pi = 1 + \sqrt{2}$ **21** $\frac{256}{3}$ **23** $\frac{214}{3}$ **25** $\frac{3}{2}$
27 $\sqrt{1 + x^2}$ **29** $-3 \tan 9x$ **31** $\cos(\tan x) \sec^2 x$
33 0 **35** -42
37 (*a*)

(*b*) 0,1 (*c*) -0.6 (*d*) -0.01
39 (*a*) 6.208333, 6.200521 (*b*) 0.008333, 0.000521
(*c*) 0.0625 (*d*) 6.20000083, 0.0001 times as large
41 (*a*) 3.03125, 4.4707 (*b*) -3.16875, -1.72930
(*c*) 0.54573 (*d*) 5.83083, 0.11650 times as large
43 (*a*) $\sqrt{1 + x^4}$ (*b*) $\sqrt{1 + x^4}$ (*c*) 0 **45** (*a*) $F(b) - F(a)$
(*b*) $\int_a^b f(x)\,dx$ **47** $\frac{3}{4}$ ft **49** $\frac{27}{32}$ **51** $\frac{8}{3}\sqrt{2}$ **53** (*a*) $\frac{2}{3}$
(*b*) 0.8140481 (*c*) $(\frac{3}{4})(2^{4/3} - 1)$ **55** (*b*) $1/n$ (*c*) $1/n$
(*d*) Right endpoint **57** (*a*) $\int_0^2 x^3\,dx$ (*b*) $\int_0^1 x^4\,dx$
(*c*) $\int_1^3 x^5\,dx$ **63** $\dfrac{2}{\pi}$ m/hr **67** (*a*) $\frac{1879}{2520}$ (*b*) $\frac{1879}{2520}$
(*c*) Yes (*g*) $\log_r x$ for any base r **69** (*b*) $f(x) = x$

CHAPTER 6. TOPICS IN DIFFERENTIAL CALCULUS

Sec. 6.1. Logarithms

1 (*a*) $5 = \log_2 32$ (*b*) $4 = \log_3 81$ (*c*) $-3 = \log_{10}(0.001)$
(*d*) $0 = \log_5 1$ (*e*) $\frac{1}{3} = \log_{1000}(10)$ (*f*) $\frac{1}{2} = \log_{49} 7$
3 (*a*)

x	$\frac{1}{9}$	$\frac{1}{3}$	1	3	9
$\log_3 x$	-2	-1	0	1	2

(*b*)

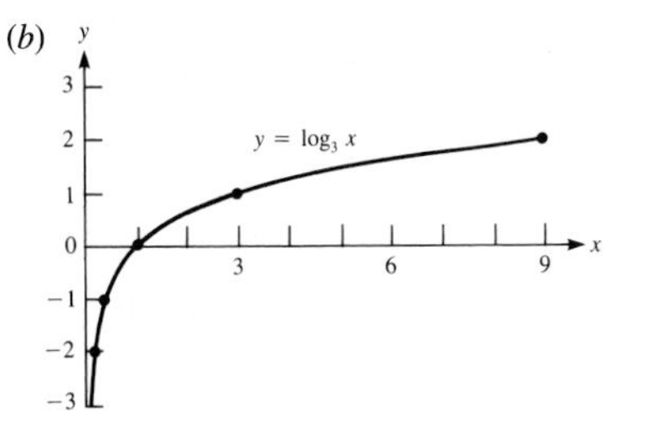

5 (*a*) $2^x = 7$ (*b*) $5^s = 2$ (*c*) $3^{-1} = \frac{1}{3}$ (*d*) $7^2 = 49$
7 (*a*) 16 (*b*) $\frac{1}{2}$ (*c*) 7 (*d*) g **9** (*a*) $\frac{1}{2}$ (*b*) 5 (*c*) -3
11 $x = \log_3 \frac{7}{2} \approx 1.1403$ **13** $x = 0$ **15** (*a*) 3, $\frac{1}{3}$, 1
17 (*a*) 0.60 (*b*) 0.70 (*c*) 0.78 (*d*) 0.90 (*e*) 0.96 (*f*) 0.18
(*g*) 0.08 (*h*) 0.12 (*i*) 1.3 (*j*) 2.3 (*k*) -2.22
19 (*a*) 1.46 (*b*) 1.58 **21** (*a*) $\log_{1/2} x = -\log_2 x$
(*b*)

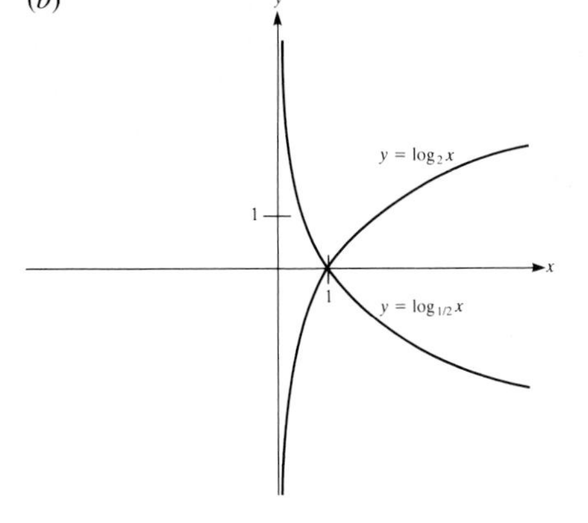

(*c*) Reflect across x axis.
25 $7 \log_{10}(\cos x) + \frac{3}{2} \log_{10}(x^2 + 5) - \log_{10}(4 + \tan^2 x)$
27 $x^2 [\log_{10} x + \frac{1}{2} \log_{10}(2 + \cos x)]$
29 $\log_2 10$ **33** (*b*) 501 (*c*) 63 (*d*) $E = 10^{1.5M + 11.4}$
(*e*) $10^{3/2}$ (*f*) 1.48 **35** (*a*) $-\infty$ (*b*) ∞

Sec. 6.2. The Number *e*

1

x	0.1	0.01	0.001	0.0005
$(1 + x)^{1/x}$	2.5937425	2.7048138	2.7169239	2.7176026

3 1 **5** $e^{3/4}$ **7** $e^{1/2}$ **9** Approaches ∞ **11** (*a*) \$1500
(*b*) \$1562.50 (*c*) \$1632.09 (*d*) \$1648.16 (*e*) \$1648.72
13

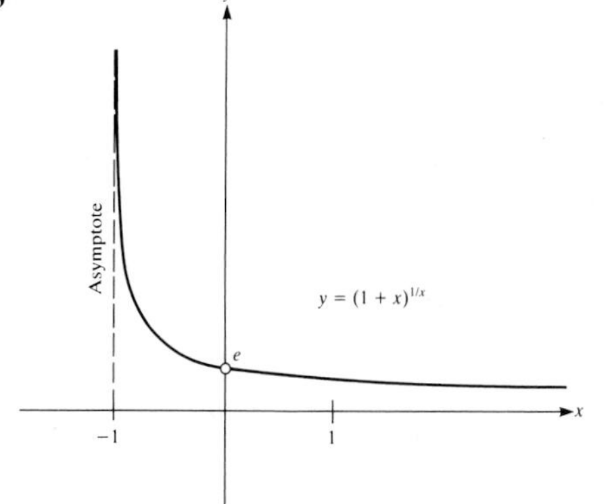

15 (*a*)

h	0.1	0.001	−0.001
$\dfrac{(2^h - 1)}{h}$	0.7177346	0.6933875	0.6929070

(*b*) 0.693 (*c*) 1.099 **17** Ae^{rt}

Sec. 6.3. The Derivative of a Logarithmic Function

1 $\dfrac{2x}{1+x^2}$ **3** $x + 2x \ln x$ **5** $\dfrac{1 - \ln x}{x^2}$

7 $2 \ln 5x \cos 2x + \dfrac{\sin 2x}{x}$ **9** $\cot x$ **11** $\dfrac{2}{2x+3}$

13 $\dfrac{x}{(5x+2)^2}$ **15** $\dfrac{1}{\sqrt{x^2-5}}$ **17** $\dfrac{1}{x(3x+5)}$

19 $\dfrac{1}{25 - x^2}$ **21** $\dfrac{6x}{x^2+1} + \dfrac{20x^4}{x^5+1}$ **23** $\dfrac{1}{(3 \ln 10)x}$

25

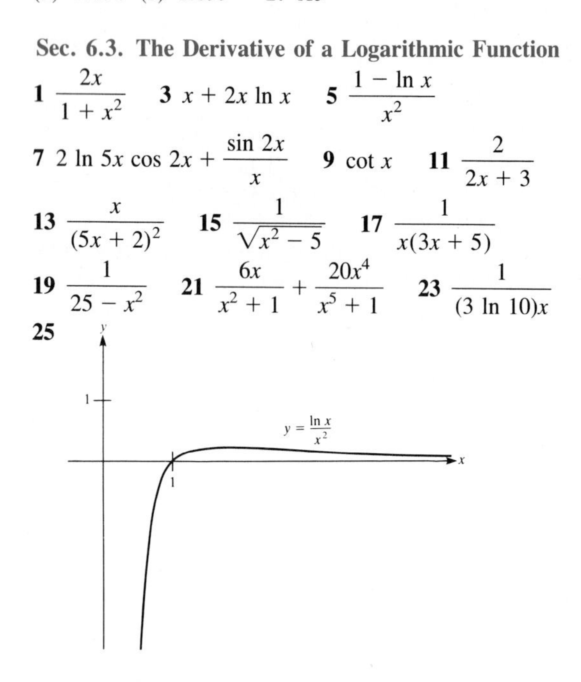

27 $(1 + 3x)^5(\sin 3x)^6\left(\dfrac{15}{1+3x} + 18 \cot 3x\right)$

29 $x^{-1/2}(\sec 4x)^{5/3} \sin^3 2x \left(\dfrac{20}{3} \tan 4x + 6 \cot 2x - \dfrac{1}{2x}\right)$

31 (*a*) 1.3876984 (*b*) $h < 0.188$ **33** (*c*) 0.5567209 (*d*) 0.5571456 **37** (*a*) $\ln |5x + 1| + C$ (*b*) $\frac{1}{2} \ln (x^2 + 5) + C$ (*c*) $\ln |\sin x| + C$ (*d*) $\ln |\ln x| + C$ **39** $\frac{3}{2}$

45 (*a*) $\ln b$ (*c*) $\pi\left(1 - \dfrac{1}{b}\right)$

Sec. 6.4. One-to-One Functions and Their Inverses

1 (*a*) No (*b*) Yes; $x = \sqrt[4]{y}$ **3** (*a*) Yes; $x = \sqrt[5]{y-1}$ (*b*) Yes; $x = \sqrt[5]{y-1}$ **5** (*a*) Yes; $x = \sqrt[3]{y^5 - 1}$ (*b*) Yes; $x = \sqrt[3]{y^5 - 1}$ **7** (*a*) Yes; $x = y^{3/5}$ (*b*) Yes; $x = y^{3/5}$

9 Yes

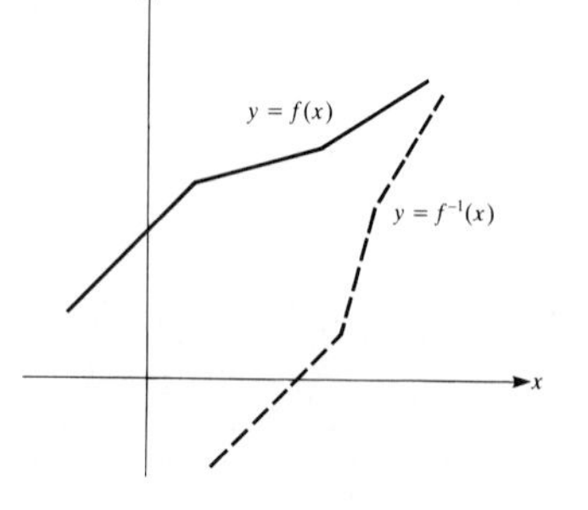

11 (*a*), (*b*)

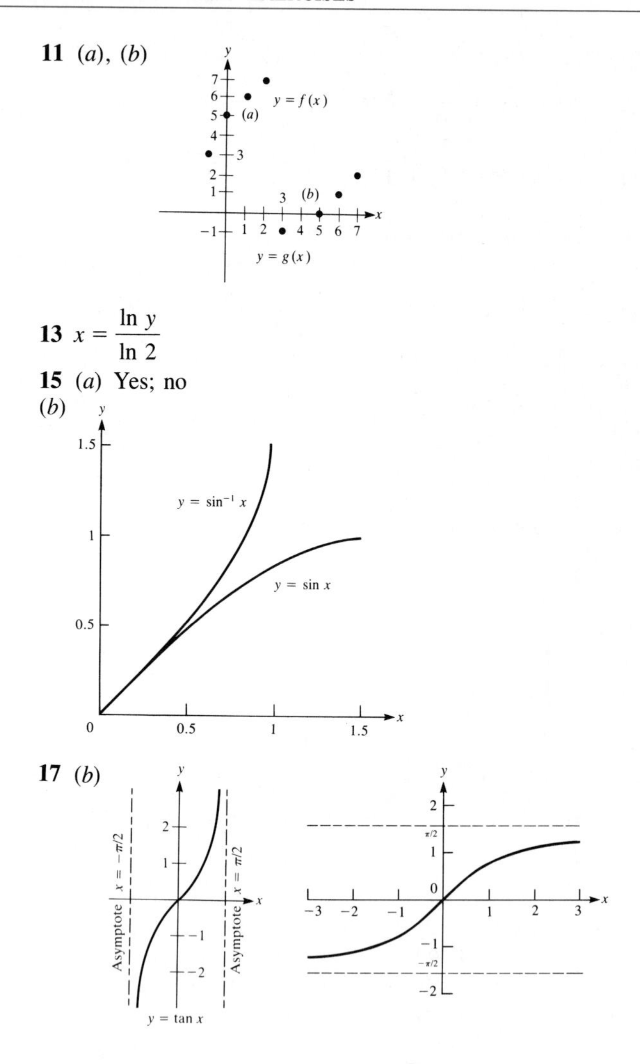

13 $x = \dfrac{\ln y}{\ln 2}$

15 (*a*) Yes; no (*b*)

17 (*b*)

19 (*a*) Yes (*b*) No **21** (*a*) $|k| \le \sqrt{3}$ (*b*) None **23** Increasing **25** (*a*) n odd (*b*) $n = 3$

Sec. 6.5. The Derivative of b^x

1 $2xe^{x^2}$ **3** $2(x^2 + x)e^{2x}$ **5** $-x(\ln 2)2^{-x^2+1}$

7 $x^{x^2}(x + 2x \ln x)$ **9** $x^{\tan 3x}\left(3 \sec^2 3x \ln x + \dfrac{1}{x} \tan 3x\right)$

11 $-\dfrac{4e^{-4x} + 5e^{-3x}}{(1 + e^x)^2}$

13 $x^{\sqrt{3}-1} e^{x^2}(2x^2 \sin 3x + 3x \cos 3x + \sqrt{3} \sin 3x)$

15 $\dfrac{1}{x + \sqrt{1 + e^{3x}}}\left(1 + \dfrac{3e^{3x}}{2\sqrt{1 + e^{3x}}}\right)$ **17** xe^{ax}

19 $e^{ax} \sin bx$ **21** $1 + x$ **23** $1 + 2.30x$ **25** x **27** $0.43x$ **29** (*a*) (0, 1), (−1, 0) (*b*) (0, 1) (*c*) (0, 1), global maximum (*d*) (1, 2/*e*) (*e*) $y = 0$

(*f*)

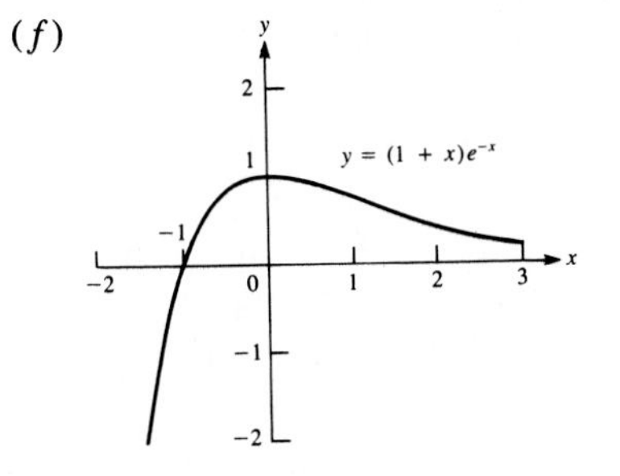

31 (*a*) (0, 0) (*b*) (0, 0), $(3, 27/e^3)$ (*c*) $(3, 27/e^3)$, global maximum (*d*) (0, 0), $(3 - \sqrt{3}, \approx 0.574)$, $(3 + \sqrt{3}, \approx 0.933)$ (*e*) $y = 0$

(*f*)

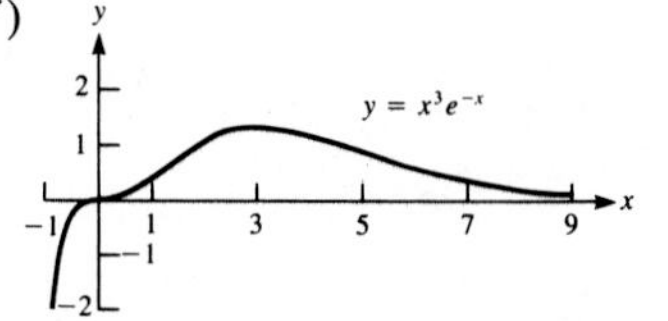

33 (*a*) (0, 0), (1, 0) (*b*) $\left(\dfrac{3 - \sqrt{5}}{2}, \approx 0.161\right)$, $\left(\dfrac{3 + \sqrt{5}}{2}, \approx -0.309\right)$ (*c*) The global maximum is approximately 0.161 and there is a local minimum of approximately −0.309. (*d*) (1, 0), $\left(4, -\dfrac{12}{e^4}\right)$ (*e*) $y = 0$

(*f*)

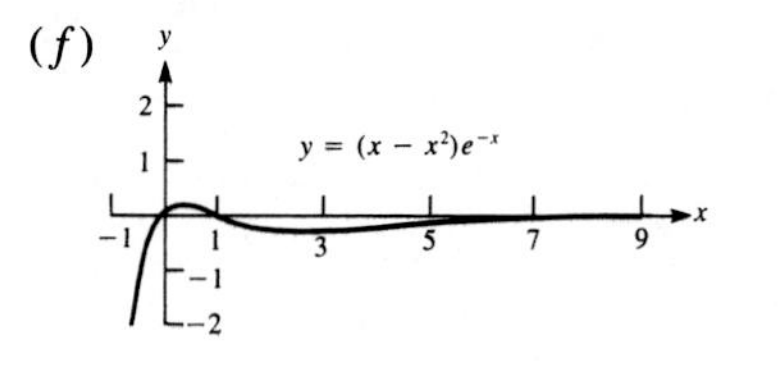

35 $\frac{1}{3}(e^{15} - e^3)$ **37** $\dfrac{999}{\ln 10}$ **39** (*a*) 1 (*b*) 2 ln 2 (*c*) ln 10

41 (*a*)

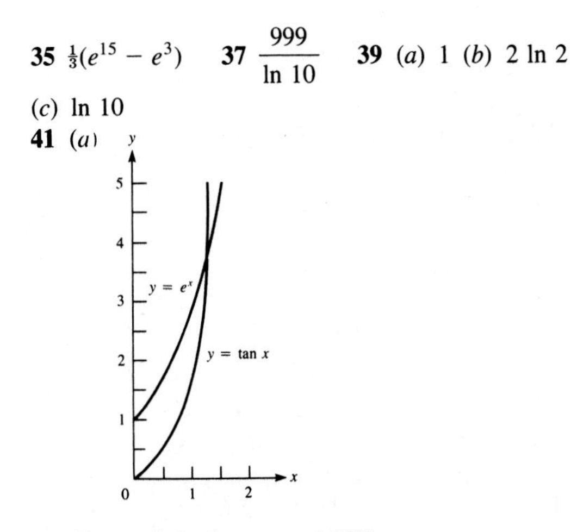

(*c*) If $x_1 = 1.3$, then $x_2 \approx 1.307$.

43 (*a*)

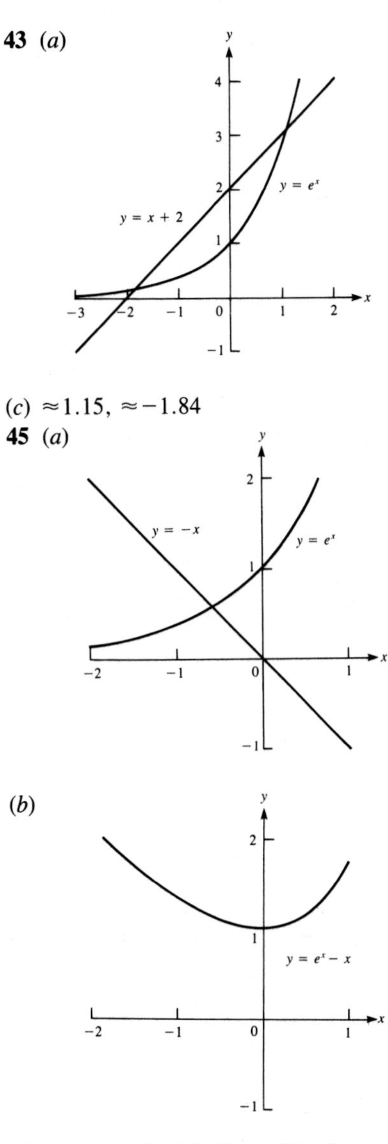

(*c*) ≈ 1.15, ≈ -1.84

45 (*a*)

(*b*)

(*c*) (0, 1) **47** (*a*) 60 m (*b*) 40 m

Sec. 6.6. The Derivatives of the Inverse Trigonometric Functions

1 (*a*)

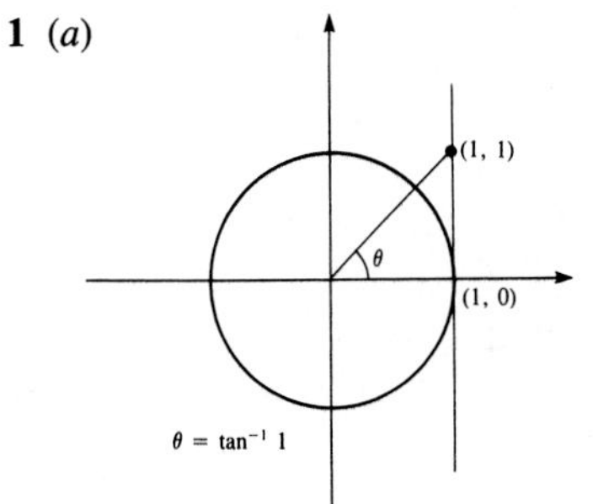

(*b*)

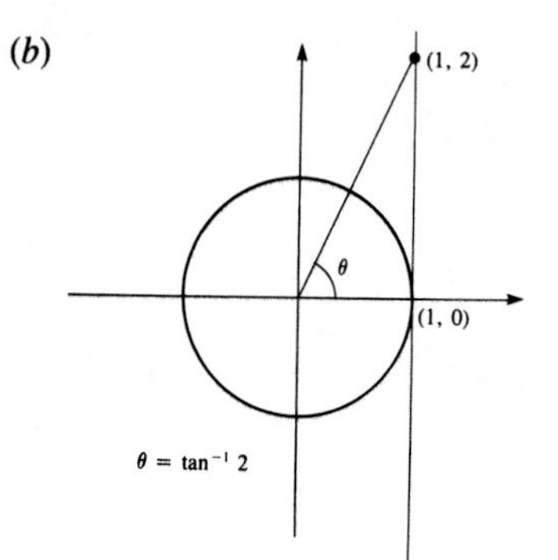

(*c*)

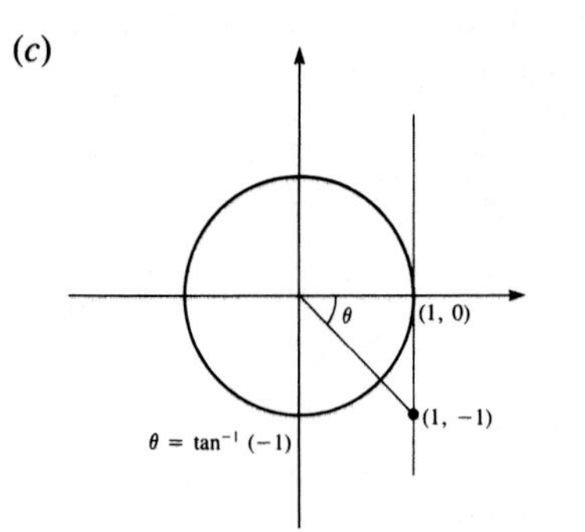

3

x	-4	-3	-2	-1	0	1	2	3	4
$\tan^{-1} x$	-1.326	-1.249	-1.107	-0.785	0	0.785	1.107	1.249	1.326

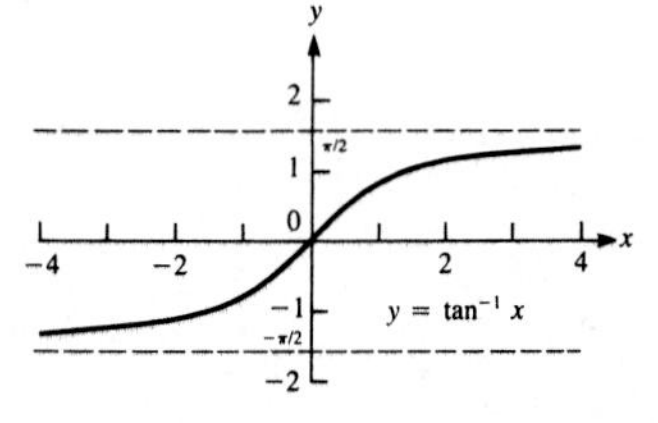

5 (*a*) $\alpha = \sin^{-1} \frac{1}{2}$ (*b*) $\beta = \sin^{-1} 1$ (*c*) $\gamma = \sin^{-1} (-1)$

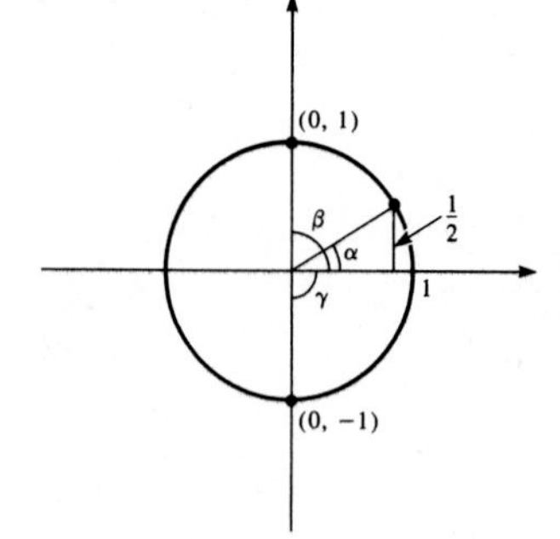

7

x	-1	-0.8	-0.6	-0.4	-0.2	0	0.2	0.4	0.6	0.8	1
$\sin^{-1} x$	$\frac{-\pi}{2}$	-0.927	-0.644	-0.412	-0.201	0	0.201	0.412	0.644	0.927	$\frac{\pi}{2}$

9

x	-4	-3	-2	-1	0	1	2	3	4
$\sec^{-1} x$	1.823	1.911	2.094	π	$-$	0	1.047	1.231	1.318

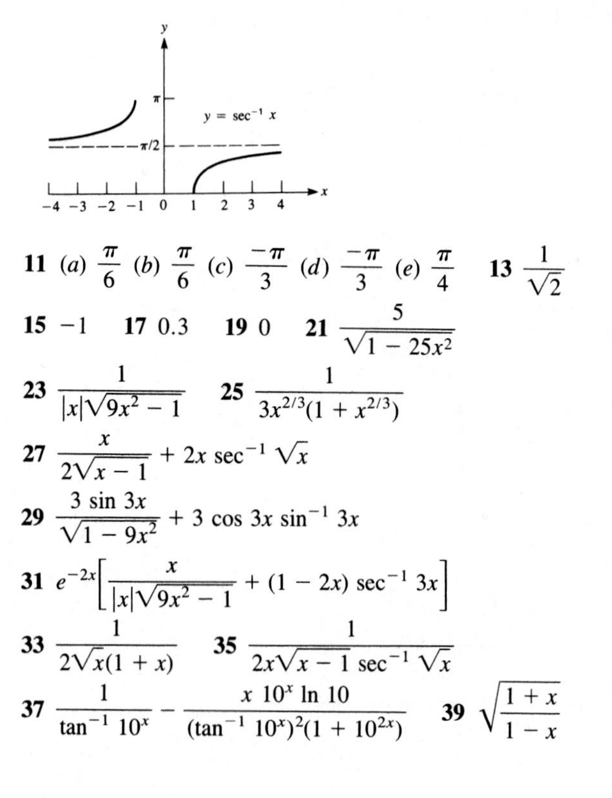

11 (*a*) $\frac{\pi}{6}$ (*b*) $\frac{\pi}{6}$ (*c*) $\frac{-\pi}{3}$ (*d*) $\frac{-\pi}{3}$ (*e*) $\frac{\pi}{4}$ **13** $\frac{1}{\sqrt{2}}$

15 -1 **17** 0.3 **19** 0 **21** $\frac{5}{\sqrt{1 - 25x^2}}$

23 $\frac{1}{|x|\sqrt{9x^2 - 1}}$ **25** $\frac{1}{3x^{2/3}(1 + x^{2/3})}$

27 $\frac{x}{2\sqrt{x - 1}} + 2x \sec^{-1} \sqrt{x}$

29 $\frac{3 \sin 3x}{\sqrt{1 - 9x^2}} + 3 \cos 3x \sin^{-1} 3x$

31 $e^{-2x}\left[\frac{x}{|x|\sqrt{9x^2 - 1}} + (1 - 2x) \sec^{-1} 3x\right]$

33 $\frac{1}{2\sqrt{x}(1 + x)}$ **35** $\frac{1}{2x\sqrt{x - 1} \sec^{-1} \sqrt{x}}$

37 $\frac{1}{\tan^{-1} 10^x} - \frac{x\, 10^x \ln 10}{(\tan^{-1} 10^x)^2(1 + 10^{2x})}$ **39** $\sqrt{\frac{1 + x}{1 - x}}$

41 $\dfrac{6(\tan^{-1} 2x)^2}{1 + 4x^2}$ **43** $\sqrt{2 - x^2}$ **45** $\dfrac{5}{3x\sqrt{3x^5 - 1}}$

47 $\sqrt{\dfrac{2 - x}{1 + x}}$ **49** $(\sin^{-1} 2x)^2$ **51** (*a*) $\dfrac{1}{\sqrt{x^2 - 9}}$

(*b*) $\dfrac{1}{\sqrt{9 - x^2}}$ **53** $\cos^{-1}(-1)$ **55** (*a*) $\dfrac{dx}{\sqrt{1 - x^2}}$

(*b*) ≈ 0.489 **59** (*b*) $\sin^{-1}\dfrac{x}{5} + C$ (*c*) $\sin^{-1}\dfrac{x}{\sqrt{5}} + C$

61 (*b*) 0.7853979 **63** $\pi/12$ **65** (*a*) $|x| \le 1$ (*b*) $|x| \le \pi/2$

Sec. 6.7. The Differential Equation of Natural Growth and Decay

1 $\frac{1}{3}y^3 = \frac{1}{4}x^4 + C$ **3** $\ln(y + 1) = \ln(x + 2) + C$
5 $\frac{1}{2}\sin 2y = -\frac{1}{3}\cos 3x + C$ **7** $-e^{-y} = \tan^{-1} x + C$
9 (*a*) 0.0953 (*b*) 7.2725 hr **11** (*a*) 10 g (*b*) 1.0986 (*c*) 200% **13** (*a*) 2028 (*b*) 2069 (*c*) 2163
15 (*a*) 6:18 P.M. (*b*) 0.2618 **17** (*a*) $Ae^{-0.005t}$ (*b*) 138.6 days **19** Sometime in the year 2083 **23** (*a*) 69

25 (*a*) $\dfrac{di}{dt} = \dfrac{E - Ri}{L}$ (*b*) $i = \dfrac{1}{R}[E - (E - Ri_0)e^{-Rt/L}]$

27 (*b*) $f(t) = f(0)e^{kt}$ **29** No, unless the relative growth rates are equal. **31** (*b*) $P(t) = \dfrac{1}{k}\{[kP(0) - h]e^{kt} + h\}$

33 $f(x) = 0$ **35** (*b*) The percentage increases during the five periods were 100%, 50%, 33.3%, 25%, and 40%.
37 $P(t) = P(0)2^{t/t_2}$

Sec. 6.8. l'Hôpital's Rule

1 3 **3** $\frac{3}{2}$ **5** 0 **7** $\frac{1}{2}$ **9** 3 **11** 0 **13** e^{-2} **15** 1

17 1 **19** 0 **21** $\dfrac{(\ln 3)}{(\ln 2)}$ **23** Does not exist **25** 1

27 -1 **29** Does not exist **31** 0 **33** 1 **35** Does not exist **37** 1 **39** 0 **41** $\ln \frac{5}{3}$ **43** $\frac{16}{9}$ **45** $\frac{1}{6}$
47 Approaches (3, 0) **49** (*b*) $\frac{3}{4}$ **53** $\sqrt{2}$
55

57

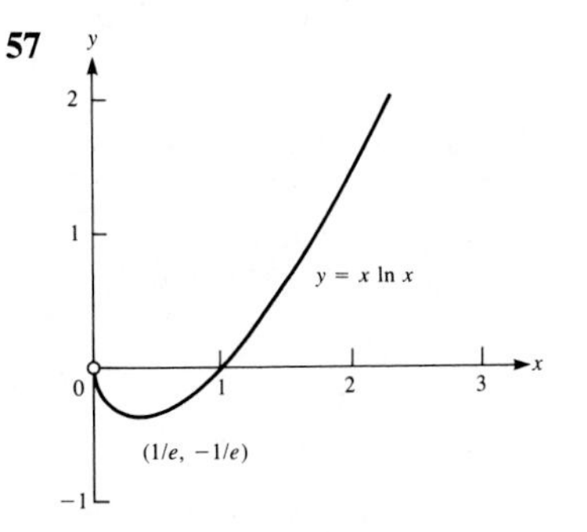

Sec. 6.9. The Hyperbolic Functions and Their Inverses

3 $\cosh t$ **5** $-\operatorname{sech} t \tanh t$ **7** $3 \sinh 3x$ **9** $\dfrac{\operatorname{sech}^2 \sqrt{x}}{2\sqrt{x}}$

11 $e^{3x}(\cosh x + 3 \sinh x)$
13 $(4 \sinh 4x)(\coth 5x)(\operatorname{csch} x^2) + (\cosh 4x)(-5 \operatorname{csch}^2 5x)(\operatorname{csch} x^2) + (\cosh 4x)(\coth 5x)(-2x \operatorname{csch} x^2 \coth x^2)$

29 $\pi/4$

31 (*a*)

t	0	1	2	3
$\cosh t$	1	1.543	3.762	10.068
$\sinh t$	0	1.175	3.627	10.018

33 (*a*)

x	0	1	2	3
$\tanh x$	0	0.762	0.964	0.995

(*b*)

35 (0, 0) **37** (*a*) $v(t) = V \tanh \dfrac{gt}{v}$

(*c*) $\dfrac{dv}{dt} = g \operatorname{sech}^2 \dfrac{gt}{V}$ (*e*) 0

Sec. 6.S. Summary: Guide Quiz

1 (*a*) $\dfrac{3x^2}{\sqrt{1 - x^6}}$ (*b*) $\dfrac{3e^{\sin^{-1} 3x}}{\sqrt{1 - 9x^2}}$

(*c*) $\dfrac{2 \cos[(\tan^{-1} x)^2] \tan^{-1} x}{1 + x^2}$

(*d*) $\left(\dfrac{2x}{x^2 + 5} + \dfrac{2 + \ln x}{2\sqrt{x}}\right)(x^2 + 5)x^{\sqrt{x}}$ (*e*) $\cot x$

(*f*) $(4 \ln 7)7^{4x-1}$ (*g*) $\dfrac{\sec^2 x}{\sec^{-1} 2x} - \dfrac{\tan x}{(\sec^{-1} 2x)^2|x|\sqrt{4x^2 - 1}}$

(*h*) $\dfrac{2x - 5}{x^2 - 5x + 4}$ (*i*) $\dfrac{-5(\cos 4x)^{2/3}}{4x^{9/4}} - \dfrac{8 \sin 4x}{3x^{5/4}(\cos 4x)^{1/3}}$

2 (*a*) $e^{ax} \cos bx$ (*b*) $\sin^{-1} ax$ (*c*) $\tan^{-1} ax$ (*d*) $\sec^{-1} ax$

3 (*a*) $\frac{e^{2y}}{2} = \frac{x^4}{4} + C$ (*b*) $\frac{\ln(4y^2+1)}{8} = x + C$
4 12.706 days **5** (*a*) $\frac{2}{3}$ (*b*) Does not exist (*c*) 0 (*d*) 1 (*e*) 1 **6** $\log_{10} x$, $\ln x$, x^3, $(1.001)^x$, 2^x **7** (*a*) $x > 0$ (*b*) $x = 1$ (*c*) $x = e$ (*d*) $(e, 1/e)$ is a maximum. (*e*) $0 < x < e$, $x > e$ (*f*) No; yes (*g*) $-\infty$ (*h*) 0
8 (*a*) $e = \lim_{x\to 0}(1+x)^{1/x}$ (*b*) $e \approx 2.718$
9 $\log_2 3 = \frac{(\log_{10} 3)}{(\log_{10} 2)}$ **10** (*a*) No (*b*) Yes; $y = 1 + \sqrt{x+1}$, $-1 \le x \le 0$ (*c*) Yes; $y = 1 - \sqrt{x+1}$, $0 \le x \le 3$ **11** (*a*) $2\sinh 2x$ (*b*) $\frac{\cosh\sqrt{x}}{2\sqrt{x}}$ (*c*) $12x^2\operatorname{sech}^2 4x^3$ (*d*) $\frac{e^x}{\sqrt{e^{2x}+1}}$

Sec. 6.S. Summary: Review Exercises

1 $\frac{3x^2}{2(1+x^3)^{1/2}}$ **3** $\frac{1}{2\sqrt{x}}$ **5** $-6\cos 3x\sin 3x$ **7** $\frac{3}{2}\sqrt{x}$
9 $\frac{\cos x}{2\sqrt{\sin x}}$ **11** $-2x\csc^2 x^2$
13 $\frac{x^{5/6}}{\sqrt{1-x^2}} + \frac{5}{6}x^{-1/6}\sin^{-1} x$ **15** $e^{3x}(3x^2+2x)$
17 $3\sec 3x$ **19** $\frac{-\sin\sqrt{x}}{2\sqrt{x}}$ **21** $\sec x$ **23** $\frac{-3x}{(6+3x^2)^{3/2}}$
25 $\sqrt{\frac{15}{2}(5x+7)}$ **27** $\frac{x}{3x+4}$
29 $(1+x^2)^4[(1+x^2)\,3\cos 3x + 10x\sin 3x]$
31 $(1+2x)^4[10\cos 3x - 3(1+2x)\sin 3x]$
33 $4\cos 3x\cos 4x - 3\sin 3x\sin 4x$
35 $-6x\csc 3x^2\cot 3x^2$ **37** $\left(\frac{x}{1+x}\right)^4(7x^2+4x^3)$
39 $(\sec 3x)(3x\tan 3x + 1)$ **41** $1/\sqrt{1+x^2}$
43 $e^{-x}\left(\frac{2x}{1+x^4} - \tan^{-1}x^2\right)$ **45** $e^{\sqrt{x}}$ **47** $\frac{1-2\ln x}{x^3}$
49 $(\sin x)(1+\sec^2 x)$ **53** $1 - \frac{2}{x-1} - \frac{1}{(x+1)^2}$
55 $\frac{-12x-3}{6x^2+3x+1}$ **57** $\frac{15}{5x+1} + \frac{12}{6x+1} - \frac{8}{2x+1}$
59 $\sqrt{9-x^2}$ **61** $3\tan^2 3x$ **63** $1/\sqrt{x^2+25}$
65 $\sin^{-1} x$ **67** $\sec 3x$ **69** $\sin^3 3x$ **71** $\sqrt{4x^2+3}$
73 $\frac{x^2}{(x+1)^3}(5x^5+7x^4-2x^2-x+9)$
75 $(1+3x)^{x^2}\left[\frac{3x^2}{1+3x} + 2x\ln(1+3x)\right]$
77 $\frac{1+\ln x + x^2 - 9x^2\ln x}{(1+x^2)^6}$ **79** (*c*) ∞ (*d*) $-\infty$
91 (*a*) $2/(1-x^2)$ (*b*) $\frac{1}{2}\ln 3$ **93** $\frac{ab}{\sqrt{ax+b}(ax+b-b^2)}$
95 (*a*) $\ln|x^3+x-6| + C$ (*b*) $\frac{1}{2}\ln|\sin 2x| + C$ (*c*) $\frac{1}{5}\ln|5x+3| + C$ (*d*) $\frac{-1}{5(5x+3)} + C$
99 (*a*) Does not exist (*b*) 0 **103** $\frac{1}{2}$ **105** (*a*) 0.1326 (*b*) e^{-2} **107** 2 **109** e^3 **111** $-\frac{1}{2}$ **113** e^2 **115** -1 **117** 0 **119** $\frac{2}{3}$ **121** 3 **123** 3 **125** 1 **127** $-\infty$ **129** 0 **131** $\ln\frac{5}{3}$ **133** 1 **135** 2 **137** 0 **141** (*a*) Cannot tell (*b*) 1 (*c*),(*d*),(*e*) Cannot tell (*f*) 0 **145** (*a*) e^x (*b*) $\ln x$ (*c*) $\sqrt[3]{x}$ (*d*) $x/3$ (*e*) x^3 (*f*) $\sin x$
147 (*b*) 0.0217 **149** -2 **151** 1 **153** (*b*) $-e/2$
157 (*a*) Does not exist. (*b*) 3 (*c*) 1 **161** It equals 1.

CHAPTER 7. COMPUTING ANTIDERIVATIVES

Sec. 7.1. Shortcuts, Integral Tables, and Machines

1 $\frac{5}{4}x^4 + C$ **3** $\frac{3}{4}x^{4/3} + C$ **5** $-\frac{6}{x} + C$ **7** $-\frac{5}{2}e^{-2x} + C$
9 $6\sec^{-1} x + C$ **11** $\ln(1+x^4) + C$
13 $-\ln(1+\cos x) + C$ **15** $\frac{-1}{x} + 2\ln|x| + C$
17 $\frac{1}{5}x^5 + 2x^3 + 9x + C$ **19** $\frac{1}{3}x^3 + \frac{3}{4}x^4 + C$ **21** 0
23 0 **25** $9\pi/2$ **27** (*a*) $\frac{-1}{9x+6} + C$ (*b*) $\frac{-1}{2}\ln\left(\frac{3x+2}{x}\right) + C$ **29** (*a*) $\tan^{-1}\frac{\sqrt{3x-4}}{2} + C$ (*b*) $\frac{\sqrt{3x-4}}{4x} + \frac{3}{8}\tan^{-1}\frac{\sqrt{3x-4}}{2} + C$
31 (*a*) $\frac{2}{\sqrt{11}}\tan^{-1}\frac{2x+3}{\sqrt{11}} + C$ (*b*) $\frac{1}{2}\tan^{-1}\frac{x+1}{2} + C$
35 (*a*) $\frac{1}{\sqrt{3}}\ln(6x+1+2\sqrt{3}\sqrt{3x^2+x+2}) + C$ (*b*) $\frac{1}{\sqrt{3}}\sin^{-1}\frac{6x-1}{5} + C$

Sec. 7.2. The Substitution Method

1 $\frac{1}{6}(1+3x)^6 + C$ **3** $\sqrt{2} - 1$ **5** $-\frac{1}{2}\cos 2x + C$
7 $\frac{1}{3}(e^6 - e^{-3})$ **9** $\frac{1}{3}\sin^{-1} 3x + C$ **11** $\frac{1}{3}$
13 $\frac{1}{5}(\ln x)^5 + C$ **15** $\frac{-1}{12}(1-x^2)^6 + C$
17 $\frac{3}{8}(1+x^2)^{4/3} + C$ **19** $2e^{\sqrt{t}} + C$ **21** $-\frac{1}{3}\cos 3\theta + C$
23 $\frac{2}{7}(x-3)^{7/2} + C$ **25** $\ln|x^2+3x+2| + C$ **27** $\frac{1}{2}e^{2x} + C$
29 $-\frac{1}{5}\cos x^5 + C$ **31** $\frac{1}{2}\tan^{-1} x^2 + C$ **33** $\frac{-2x-1}{2(x+1)^2} + C$
35 $\frac{1}{2}(\ln 3x)^2 + C$ **37** $\frac{1}{3}(e^8 - e)$ **39** $\frac{11}{12}$ **41** $\frac{1}{4}$
43 $\frac{1}{a^3}\left(\frac{1}{2}a^2x^2 - abx + b^2\ln|ax+b|\right) + C$
45 $\frac{1}{a^3}\left(ax - 2b\ln|ax+b| - \frac{b^2}{ax+b}\right) + C$
47 They are both right. **49** Jill is right.
53 (*a*) (*b*) ≈ 0.231

Sec. 7.3. Integration by Parts

1 $\frac{1}{4}e^{2x}(2x-1)+C$ **3** $-\frac{x}{2}\cos 2x+\frac{1}{4}\sin 2x+C$

5 $\frac{1}{2}x^2\ln 3x-\frac{1}{4}x^2+C$ **7** $\frac{5e-10}{e^2}$ **9** $\frac{\pi}{2}-1$

11 $\frac{1}{3}x^3\ln x-\frac{1}{9}x^3+C$

13 $3(\ln 3)^2-6\ln 3-2(\ln 2)^2+4\ln 2+2$ **15** $\frac{e-2}{e}$

17 $\frac{4}{13}(\frac{1}{2}\sin 2x+\frac{3}{4}\cos 2x)e^{3x}+C$

19 $-\frac{\ln(1+x^2)}{x}+2\tan^{-1}x+C$

21 $\frac{2}{9}(3x+7)^{3/2}\left(\frac{3x}{5}-\frac{14}{15}\right)+C$

23 $\frac{1}{20a^2}(ax+b)^4(4ax-b)+C$

25 (*a*) $\frac{1}{2}(x-\sin x\cos x)+C$
(*b*) $-\frac{1}{4}\sin^3 x\cos x+\frac{3}{8}(x-\sin x\cos x)+C$
(*c*) $-\frac{1}{6}\sin^5 x\cos x-\frac{5}{24}\sin^3 x\cos x+\frac{5}{16}(x-\sin x\cos x)+C$

29 (*a*)

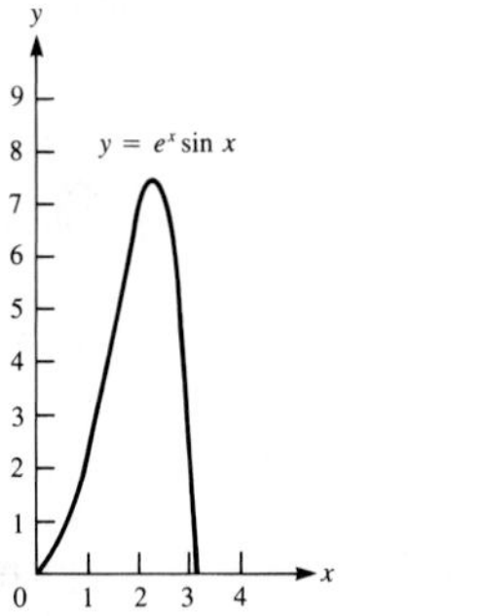

$y=e^x\sin x$

Maximum: $(3\pi/4, \approx 7.46)$
Inflection point: $(\pi/2, \approx 4.81)$

(*b*) $\frac{1}{2}(e^{\pi}+1)$ **31** $\frac{1}{4}(e^2-1)$

33 $2(\sin\sqrt{x}-\sqrt{x}\cos\sqrt{x})+C$

35 $2(\exp\sqrt{x})(\sqrt{x}-1)+C$ **39** (*a*) $I_0=\frac{\pi}{2}$, $I_1=1$

(*c*) $I_2=\frac{\pi}{4}$, $I_3=\frac{2}{3}$ (*d*) $I_4=\frac{3\pi}{16}$, $I_5=\frac{8}{15}$

41 $\frac{1}{a}x^ne^{ax}-\frac{n}{a}\int x^{n-1}e^{ax}\,dx=\int x^ne^{ax}\,dx$

43 $\int x^n\sin x\,dx=-x^n\cos x+nx^{n-1}\sin x-n(n-1)\int x^{n-2}\sin x\,dx$ **45** 0

Sec. 7.4. How to Integrate Certain Rational Functions

1 $\frac{1}{3}\ln|3x-4|+C$ **3** $\frac{-5}{2(2x+7)}+C$ **5** $\frac{1}{3}\tan^{-1}\frac{x}{3}+C$

7 $\frac{1}{2}\ln 2$ **9** $\ln(x^2+9)+\tan^{-1}\frac{x}{3}+C$

11 $\frac{1}{20}\tan^{-1}\frac{4x}{5}+C$ **13** $\frac{1}{32}\ln(16x^2+25)+C$

15 $\frac{1}{18}\ln(9x^2+4)+\frac{1}{3}\tan^{-1}\frac{3x}{2}+C$

17 $\frac{1}{\sqrt{6}}\tan^{-1}\frac{\sqrt{6}x}{3}+C$ **19** $\frac{1}{\sqrt{2}}\tan^{-1}\frac{x+1}{\sqrt{2}}+C$

21 $\frac{1}{\sqrt{2}}\tan^{-1}\frac{x-1}{\sqrt{2}}+C$ **23** $\frac{2}{\sqrt{23}}\tan^{-1}\frac{4x+1}{\sqrt{23}}+C$

25 $\frac{1}{\sqrt{3}}\tan^{-1}\frac{x+2}{\sqrt{3}}+C$ **27** $\frac{1}{\sqrt{10}}\tan^{-1}\frac{2x+2}{\sqrt{10}}+C$

29 $\ln(x^2+2x+3)-\sqrt{2}\tan^{-1}\frac{x+1}{\sqrt{2}}+C$

31 $\frac{3}{10}\ln(5x^2+3x+2)-\frac{9}{5\sqrt{31}}\tan^{-1}\frac{10x+3}{\sqrt{31}}+C$

33 $\frac{1}{2}\ln(x^2+x+1)+\frac{1}{\sqrt{3}}\tan^{-1}\frac{2x+1}{\sqrt{3}}+C$

35 $\frac{1}{2}\ln(3x^2+2x+1)+2\sqrt{2}\tan^{-1}\frac{3x+1}{\sqrt{2}}+C$

37 $\frac{\pi}{6\sqrt{3}}+\frac{\ln 3}{2}\approx 0.8516$ **43** (*a*) $(x-2)(x+2)$
(*b*) $(x-\sqrt{3})(x+\sqrt{3})$ (*c*) Irreducible (*d*) $(x+1)(2x+1)$
(*e*) Irreducible (*f*) $2[x+\frac{1}{4}(3-\sqrt{65})][x+\frac{1}{4}(3+\sqrt{65})]$
(*g*) Irreducible

45 $\frac{-(13+4x)}{72(4x^2+8x+13)}-\frac{1}{108}\tan^{-1}\left(\frac{2}{3}(x+1)\right)+C$

Sec. 7.5. Integration of Rational Functions by Partial Fractions

1 $\frac{k_1}{x+1}+\frac{k_2}{x+2}$ **3** $\frac{k_1}{x-1}+\frac{k_2}{(x-1)^2}+\frac{k_3}{x+2}$

5 $\frac{k_1}{x-1}+\frac{k_2}{x-2}+\frac{k_3}{2x-3}$

7 $\frac{k_1}{x+1}+\frac{k_2x+k_3}{x^2+x+1}+\frac{k_4x+k_5}{(x^2+x+1)^2}$

9 $\frac{k_1}{x+1}+\frac{k_2}{x-1}+\frac{k_3}{3x+5}+\frac{k_4}{(3x+5)^2}+\frac{k_5}{(3x+5)^3}$

11 $1-\frac{x+1}{x^2+x+1}$ **13** $x^2-x-\frac{x-1}{(x+1)(x^2+1)}$

15 $\frac{3}{x}+\frac{1}{x-1}$ **17** $\frac{2}{x}+\frac{1}{x^2}+\frac{3}{x-1}$

19 $\frac{-1}{x+1}+\frac{2}{x+2}$ **21** $\frac{1}{x-1}+\frac{1}{x+1}$

23 $\frac{3}{2x}-\frac{5}{x+1}+\frac{11}{2(x+2)}$

25 $\frac{-31}{x+1}+\frac{12}{(x+1)^2}+\frac{37}{x+2}$ **27** $\frac{2}{x+1}+\frac{3x+2}{x^2+2x+2}$

29 $x-1+\frac{1}{x}+\frac{5}{x+1}$ **31** $3+\frac{1}{x}+\frac{x}{x^2+1}$

33 $\frac{2}{3(x+1)}+\frac{1}{3(x-2)}$ **35** $\frac{2}{5(x-1)}-\frac{2}{5(x+4)}$

37 $c_1=\frac{11}{5}$, $c_2=\frac{9}{5}$ **39** $c_1=\frac{8}{13}$, $c_2=\frac{14}{13}$

41 $c_1=2$, $c_2=3$, $c_3=1$ **43** $c_1=3$, $c_2=-5$, $c_3=1$

45 $\frac{1}{2}\ln\frac{8}{5}$ **47** $(\frac{13}{6}+4\ln 3-8\ln 2)\pi$ **49** $\ln\frac{4}{3}$

51 $1 + \frac{1}{6} \ln \frac{4}{3} - \frac{\pi}{6\sqrt{3}}$

53 (*a*) $(x + 1)(x + 5)$ (*b*) $(x - \sqrt{5})(x + \sqrt{5})$
(*c*) $2\left(x - \frac{-3 - \sqrt{3}}{2}\right)\left(x - \frac{-3 + \sqrt{3}}{2}\right)$

55 (*a*) $-4, 3$ (*b*) $\frac{1}{2}, 2, 3$ (*c*) None (*d*) $-\frac{1}{3}$

57 $(x^2 + \sqrt{2}x + 1)(x^2 - \sqrt{2}x + 1)$

59 (*a*) $(x^2 + x + 1)(x^2 - x + 1)$
(*b*) $\frac{1}{4} \ln \left(\frac{x^2 + x + 1}{x^2 - x + 1}\right) + \frac{1}{2\sqrt{3}} \tan^{-1} \left(\frac{\sqrt{3}\,x}{1 - x^2}\right) + C$

Sec. 7.6. Special Techniques

1 $\frac{1}{4} \sin 2x - \frac{1}{16} \sin 8x + C$ **3** $-\frac{1}{10} \cos 5x + \frac{1}{2} \cos x + C$

5 $\frac{x}{2} - \frac{1}{12} \sin 6x + C$ **7** $-\frac{3}{2} \cos 2x + 2x - \frac{1}{5} \sin 10x + C$

9 $\frac{7x}{2} + \frac{1}{4\pi} \sin 2\pi x + C$ **11** $\frac{1}{2} \ln |\sec 2\theta| + C$

13 $\frac{1}{5} \tan 5x - x + C$ **17** $F(20°) \approx 0.3564$, $F(40°) \approx 0.7629$, $F(60°) = \ln (2 + \sqrt{3}) \approx 1.3170$, $F(80°) \approx 2.4362$

19 $\frac{1}{105}(2x + 1)^{3/2}(15x^2 - 6x + 2) + C$

21 $2[\sqrt{x} - 3 \ln (\sqrt{x} + 3)] + C$

23 $\frac{(3x + 2)^{7/3}}{21} - \frac{(3x + 2)^{4/3}}{6} + C$

25 $\frac{2}{3}x^{3/2} - 3x + 18\sqrt{x} - 54 \ln (\sqrt{x} + 3) + C$

27 $2x^{1/2} - 3x^{1/3} + 6x^{1/6} - 6 \ln (x^{1/6} + 1) + C$

29 (*b*) $\frac{1}{2} \tan^2 \theta + \ln |\cos \theta| + C$ (*c*) $\frac{1}{3} \tan^3 \theta - \tan \theta + \theta + C$

31 (*a*) $\ln |\csc \theta - \cot \theta| + C$ (*b*) $-\cot \theta + C$

33 (*a*) $\frac{1}{4} \sin^4 \theta - \frac{1}{6} \sin^6 \theta + C$ (*b*) $\frac{1}{5} \sin^5 \theta + C$ (*c*) $\frac{2}{35}$
(*d*) $\sin \theta - \frac{2}{3} \sin^3 \theta + \frac{1}{5} \sin^5 \theta + C$

35 (*a*) $\frac{1}{16}\theta - \frac{1}{64} \sin 4\theta - \frac{1}{48} \sin^3 2\theta + C$ (*b*) $\frac{\pi}{32}$

37 (*a*)

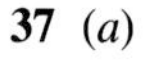

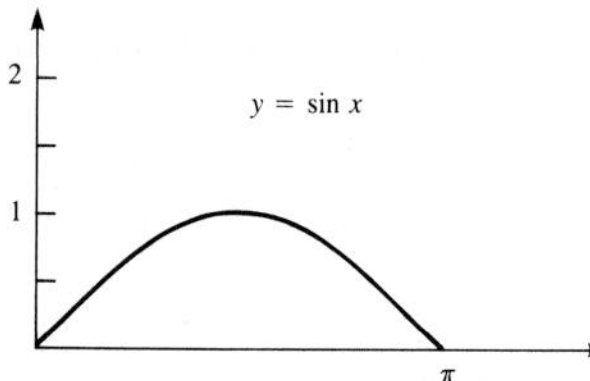

(*b*)

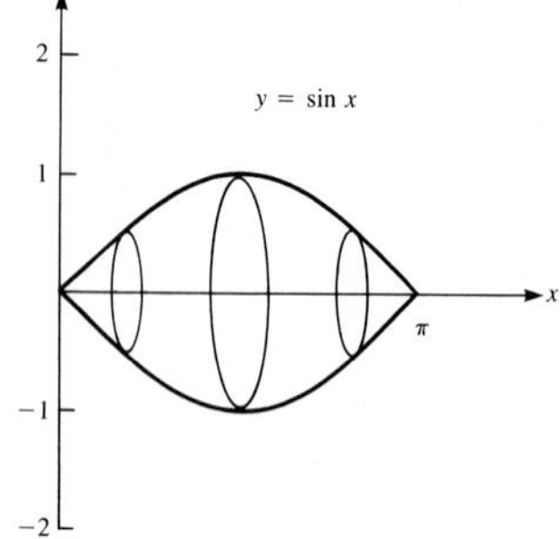

(*c*) $\int_0^\pi \sin x \, dx$ (*d*) $\pi \int_0^\pi \sin^2 x \, dx$ (*e*) $2, \frac{\pi^2}{2}$

41 $\ln \left(\frac{\sqrt{x^2 + 9} + x}{3}\right) + C$

43 $\frac{1}{5}(1 - x^2)^{5/2} - \frac{1}{3}(1 - x^2)^{3/2} + C$

45 $\frac{1}{2}\left(a^2 \sin^{-1} \frac{x}{a} + x\sqrt{a^2 - x^2}\right) + C$

47 $\frac{1}{2}x\sqrt{a^2 + x^2} + \frac{1}{2}a^2 \ln (\sqrt{a^2 + x^2} + x) + C$

49 $\frac{1}{5} \ln |5x + \sqrt{25x^2 - 16}| + C$

51 (*a*) $\int \frac{\sin \theta \cos \theta + \cos^2 \theta}{3 \sin^3 \theta} \, d\theta$
(*b*) $25\sqrt{5} \int \frac{\sin^3 \theta \cos^2 \theta}{1 + \sqrt{5} \cos \theta} \, d\theta$ **55** $\frac{1}{5} \ln 6$ **57** (*b*) No

Sec. 7.7. What to Do in the Face of an Integral

1 Division, power rule **3** Partial fractions **5** Integration by parts **7** Repeated integration by parts **9** Substitution **11** Substitution, power rule **13** Substitute $u = x^2 + x + 1$, power rule **15** Trigonometric identity **17** Substitution, integration by parts **19** Partial fractions **21** Substitution, power rule **23** Substitution, division, methods of Sec. 7.4 **25** Logarithm rules, power rule **27** Division, substitution, power rule **29** Substitution, power rule **31** Substitution, power rule **33** Substitution, partial fractions **35** Substitute $u = \sqrt[6]{x + 2}$, simplify, techniques of Sec. 7.4 **37** Substitution, power rule **39** Rules of logarithms and exponents **41** Break in two, substitution on both parts **43** Substitution, power rule **45** Substitution, power rule **47** Substitute $u = x^2 - 9$, power rule **49** Integration by parts, partial fractions **51** Substitute $u = \sin x$, integration by parts **53** Completing the square, substitution **55** Substitution, power rule **57** Substitution, power rule

59 Partial fractions **61** $n = 1$: $x + \frac{1}{3}x^3 + C$
$n = 2$: $\frac{x}{2}\sqrt{1 + x^2} + \frac{1}{2} \ln (\sqrt{1 + x^2} + x) + C$

63 $\frac{1}{6}(x + 2)^{3/2} + \frac{1}{6}(x - 2)^{3/2} + C$

65 $\frac{1}{3(9 - x^2)^{3/2}} + C$

67 $\frac{x}{27(9 - x^2)^{3/2}} + \frac{2x}{243\sqrt{9 - x^2}} + C$

69 $\sqrt{x^2 + 9} + C$ **71** $\frac{(x^2 - 5)^{5/2}}{5} + \frac{5(x^2 - 5)^{3/2}}{3} + \frac{x^3\sqrt{x^2 - 5}}{4} - \frac{5x\sqrt{x^2 - 5}}{8} - \frac{25}{8} \ln \left|\frac{x + \sqrt{x^2 - 5}}{\sqrt{5}}\right| + C$

Sec. 7.S. Summary: Guide Quiz

1 $\frac{1}{4} \ln (1 + x^4) + C$ **2** $\frac{1}{6}(4x + 3)^{3/2} + C$

3 $\frac{1}{10} \sec^5 2x + C$ **4** $\frac{1}{4} \tan^4 x + C$

5 $\frac{1}{4} \ln \left|\frac{x - 1}{x + 1}\right| - \frac{1}{2} \tan^{-1} x + C$

6 $x + \frac{1}{4}\ln\left|\frac{x-1}{x+1}\right| - \frac{1}{2}\tan^{-1}x + C$

7 $-\ln|\cos(\ln x)| + C$ **8** $\frac{1}{6}x^3 + \frac{1}{12}\sin 2x^3 + C$

9 $\frac{1}{3}\ln|\sec x^3 + \tan x^3| + C$ **10** $\frac{x^2}{2} + \frac{2}{3}x^{-3/2} + C$

11 $\frac{-2}{\sqrt{x+3}} + C$ **12** $1 - \frac{1}{3}\ln 2 - \frac{\pi}{3\sqrt{3}}$

13 $\frac{x}{3}\sin 3x + \frac{1}{9}\cos 3x + C$

14 $x\tan^{-1}5x - \frac{1}{10}\ln(1 + 25x^2) + C$

15 $\ln\frac{2}{\sqrt{3}}$ **16** $\sin^{-1}\frac{\sqrt{7}}{4}$ **17** $\frac{1}{4}\tan^{-1}x^4 + C$

18 $-e^{-2x}\left(\frac{x^3}{2} + \frac{3x^2}{4} + \frac{3x}{4} + \frac{3}{8}\right) + C$

19 $\frac{11\ln|x-3|}{14} + \frac{18\ln|x+4|}{35} - \frac{3\ln|x-1|}{10} + C$

20 $x + \frac{\ln|x-1|}{3} - \frac{\ln(x^2+x+1)}{6} - \frac{\tan^{-1}((2x+1)/\sqrt{3})}{\sqrt{3}} + C$

21 $\frac{2}{3}\sin^{-1}\frac{3x}{2} + \frac{x\sqrt{4-9x^2}}{2} + C$

22 $\frac{1}{3}\ln(\sqrt{9x^2+16} + 3x) + C$

23 $-\frac{1}{12}\csc^3 3x\cot 3x - \frac{1}{8}\csc 3x\cot 3x + \frac{1}{8}\ln|\csc 3x - \cot 3x| + C$ **24** $\frac{x\sqrt{x^2-9}}{2} + \frac{9}{2}\ln|x + \sqrt{x^2-9}| + C$

25 $\frac{1}{\sqrt{2}}\tan^{-1}\left[\frac{\tan(x/2)}{\sqrt{2}}\right] + C$

Sec. 7.S. Summary: Review Exercises

1 (*a*) $\int_1^2 u^{3/2}\,du$ (*b*) $\frac{2}{5}(4\sqrt{2} - 1)$ **3** (*a*) $\frac{373}{14}$ (*b*) $\frac{721}{9}$

5 (*a*) $-\frac{1}{2x^2} + C$ (*b*) $2\sqrt{x+1} + C$ (*c*) $\frac{1}{5}\ln(1 + 5e^x) + C$

7 (*a*) $\int_1^4 \frac{(u-1)^3}{\sqrt{u}}\,du = \frac{388}{35}$ (*b*) $2\int_1^2 (u^2-1)^3 du = \frac{388}{35}$

9 $\frac{1}{3}(x^3 + 1)\ln(1 + x) - \frac{1}{9}x^3 + \frac{1}{6}x^2 - \frac{1}{3}x + C$

11 $\frac{2}{x-1} + \frac{1}{x^2+x+1}$ **13** $\frac{-1}{x+1} + \frac{x}{x^2-x+1}$

15 $\frac{3}{x+1} - \frac{1}{x+2}$ **17** $5 + \frac{2x+1}{x^2+1} - \frac{1}{x+1}$

19 (*a*) $\frac{1}{2}x^2 + 2x + 3\ln|x-1| - \frac{1}{x-1} + C$

(*b*) $\frac{1}{2}(x-1)^2 + 3(x-1) + 3\ln|x-1| - \frac{1}{x-1} + C$

21 $\frac{3}{28}(x+1)^{4/3}(4x-3) + C$

23 $\frac{21}{8}$ **25** $2(\sqrt{2} - 1)$ **27** (*a*) Not elementary

(*b*) $2\sin\theta + C$, $-\pi/2 \le \theta \le \pi/2$ (*c*) $2\sqrt{2}\sin\frac{\theta}{2} + C$, $|\theta| \le \pi$ **29** $\frac{1}{5}\ln|\csc 5\theta - \cot 5\theta| + C$

31 $-\frac{1}{4}(1 + x^2)^{-2} + \frac{1}{6}(1 + x^2)^{-3} + C$

33 (*a*) $2\int (u^2 - 1)^2\,du$ (*b*) $\int \frac{(u-1)^2}{\sqrt{u}}\,du$

(*c*) $2\int \tan^5\theta\sec\theta\,d\theta$

(*d*) $\frac{2}{5}(x+1)^{5/2} - \frac{4}{3}(x+1)^{3/2} + 2(x+1)^{1/2} + C$

35 $-\frac{1}{7}(1 - x^2)^{7/2} + C$

37 $\frac{x}{8}(-2x^2 + 45)\sqrt{9 - x^2} + \frac{243}{8}\sin^{-1}\frac{x}{3} + C$

39 $x\sqrt{x^2 - 9/4} - \frac{9}{4}\ln|x + \sqrt{x^2 - 9/4}| + C$

41 (*a*) $\frac{1}{2}x - \frac{1}{2}\sin x\cos x + C$

(*b*) $2\sin\sqrt{x} - 2\sqrt{x}\cos\sqrt{x} + C$ (*c*) Not elementary

43 $\frac{-1}{8(x^4+1)^2} + C$ **45** $\frac{x^2}{2} - \frac{1}{2x^2} + \ln|x| + C$

47 $\frac{10^x}{\ln 10} + C$ **49** $-\ln(3 + \cos x) + C$

51 $\frac{2}{9}(x^3 - 1)^{3/2} + C$

53 $\frac{-\tan^{-1}x}{x} + \ln|x| - \frac{1}{2}\ln(x^2 + 1) + C$

55 $\frac{1}{10}e^x(\sin 3x - 3\cos 3x) + C$ **57** $\frac{1}{3}(x^2 + 4)^{3/2} + C$

59 $\frac{1}{3}\tan^{-1}x^3 + C$ **61** $-\frac{1}{3}\cos x^3 + C$

63 $\frac{x^5}{5}\ln x - \frac{1}{25}x^5 + C$ **65** $2e^{\sqrt{x}} + C$

67 $\frac{-1}{\sqrt{x^2+1}} + C$ **69** $\frac{-2}{\sqrt{x+1}} + C$

71 $\frac{1}{8}\ln\left|\frac{x^2-3}{x^2+1}\right| + C$

73 $\frac{3}{8}(x-1)^{8/3} + \frac{6}{5}(x-1)^{5/3} + \frac{3}{2}(x-1)^{2/3} + C$

75 $\sqrt{x^2+4} + 2\ln\left|\frac{\sqrt{x^2+4} - 2}{x}\right| + C$ **77** $\frac{1}{5}\sec^5\theta + C$

79 $-\frac{1}{3}\ln\left|\frac{3 + \sqrt{x^2+9}}{x}\right| + C$

81 $\frac{3}{2}x^{2/3} - \frac{6}{5}x^{5/3} + \frac{3}{8}x^{8/3} + C$ **83** $\frac{-1}{\sqrt{3}}\tan^{-1}(\sqrt{3}\cos x) + C$

85 $\frac{1}{2}e^{2x} - 2x - \frac{1}{2}e^{-2x} + C$

87 $\frac{1}{2}(x^2\sin^{-1}x^2 + \sqrt{1 - x^4}) + C$

89 $-\frac{x}{25} - \frac{1}{5}e^{-x} + \frac{1}{25}\ln(e^x + 5) + C$

91 $\frac{1}{135}(36x + 14)(3x + 2)^{3/2} + C$

93 $\ln|x-1| - \frac{2}{x-1} - \frac{1}{2(x-1)^2} + C$

95 $-x + 2\ln|e^x - 1| + C$ **97** $x + 2x^3 + \frac{9}{5}x^5 + C$

99 $x - \frac{1}{3}\ln|x+1| + \frac{1}{6}\ln(x^2 - x + 1) - \frac{1}{\sqrt{3}}\tan^{-1}\frac{2x-1}{\sqrt{3}} + C$

101 $\sqrt{2x+1} + C$ **103** $\frac{x}{8} - \frac{\sin 12x}{96} + C$

105 $\frac{1}{9}\tan^3 3\theta - \frac{1}{3}\tan 3\theta + \theta + C$ **107** $\frac{1}{4}(2x + \sin 2x) + C$

109 $\frac{x}{4\sqrt{4 - x^2}} + C$ **111** $\tan^{-1}e^x + C$

113 $\ln \left|\dfrac{x+2}{x+3}\right| + C$ **115** $2 \ln |x^2 + 5x + 6| + C$

117 $\dfrac{2}{\sqrt{23}} \tan^{-1} \dfrac{4x+5}{\sqrt{23}} + C$

119 $\dfrac{1}{\sqrt{73}} \ln \left|\dfrac{4x+5-\sqrt{73}}{4x+5+\sqrt{73}}\right| + C$ **121** $-\cot x + C$

123 $x \ln (x^2+5) - 2x + 2\sqrt{5} \tan^{-1} \dfrac{x}{\sqrt{5}} + C$

125 $\frac{1}{4}(\sin^{-1} 2x + 2x\sqrt{1-4x^2}) + C$ **127** $2\sqrt{x^2+1} + C$

129 $x + 3 \ln |x-1| + \ln |x+1| + 4 \tan^{-1} x + C$

131 $3 \ln |x| + \tan^{-1} 2x + C$

133 $\ln (x + \sqrt{1+x^2}) - \dfrac{\sqrt{1+x^2}}{x} + C$

135 $\frac{1}{3} \sin^{-1} 3x + C$ **137** $\dfrac{x}{2\sqrt{3x^2+2}} + C$

139 $\frac{1}{4} \ln |\sec 4x + \tan 4x| + C$

141 $\frac{1}{2}e^x - \frac{1}{10}e^x \cos 2x - \frac{1}{5}e^x \sin 2x + C$

143 $\frac{8}{315}(1 + \sqrt{1+\sqrt{x}})^{5/2}(61 + 35\sqrt{x} - 65\sqrt{1+\sqrt{x}}) + C$

145 $\frac{1}{3} \tan^3 x + \tan x + C$ **147** $\sin x - \frac{1}{3} \sin^3 x + C$

149 $\frac{1}{2}(\ln x)^2 + 2\sqrt{x} + C$ **151** $-e^{-x} + C$

153 $e^x - e^{-x} + C$

155 $\ln |x| + \dfrac{1}{2} \ln |x^2 + 2x + 3| + \dfrac{1}{\sqrt{2}} \tan^{-1} \dfrac{x+1}{\sqrt{2}} + C$

157 $\ln |\tan \theta| + C$ **159** $\frac{1}{8}(2x^2 - 2x \sin 2x - \cos 2x) + C$

161 $x \tan x - \frac{1}{2}x^2 + \ln |\cos x| + C$

163 (*a*) $\dfrac{1}{2} \ln \left|\dfrac{x+1}{x+3}\right| + C$ (*b*) $\dfrac{-1}{x+2} + C$

(*c*) $\tan^{-1}(x+2) + C$ (*d*) $\dfrac{1}{2\sqrt{6}} \ln \left|\dfrac{x+2-\sqrt{6}}{x+2+\sqrt{6}}\right| + C$

171 (*b*) $n = 1$: $\frac{2}{3}(1+x)^{3/2} - 2(1+x)^{1/2} + C$

$n = 2$: $\sqrt{1+x^2} + C$

$n = 4$: $\frac{1}{2} \ln (x^2 + \sqrt{1+x^4}) + C$

173 (*a*) n is odd. (*b*) $n = 3$: $\dfrac{1}{2}\sqrt{1+x^4} + C$;

$n = 5$: $\dfrac{x^2}{4}\sqrt{1+x^4} - \dfrac{1}{4} \ln (x^2 + \sqrt{1+x^4}) + C$

175 (*b*) 1.4629 **177** 0

179 (*a*) $\dfrac{1}{3}\left(\dfrac{1}{x-1} - \dfrac{x-1}{x^2+x+1}\right)$

(*b*) $\dfrac{1}{9}\left(-\dfrac{1}{x-1} + \dfrac{3}{(x-1)^2} + \dfrac{1}{x+2}\right)$

(*c*) $x^2 - \dfrac{1}{3}\left(\dfrac{1}{x+1} + \dfrac{2x-1}{x^2-x+1}\right)$

(*d*) $\dfrac{1}{12}\left(\dfrac{1}{x+2} - \dfrac{x-4}{x^2-2x+4}\right)$

181 No, there is no error. **187** $a = 0, -1$

189 $b = 0, 1, -1$

CHAPTER 8. APPLICATIONS OF THE DEFINITE INTEGRAL

Sec. 8.1. Computing Area by Parallel Cross Sections

1 (*a*)

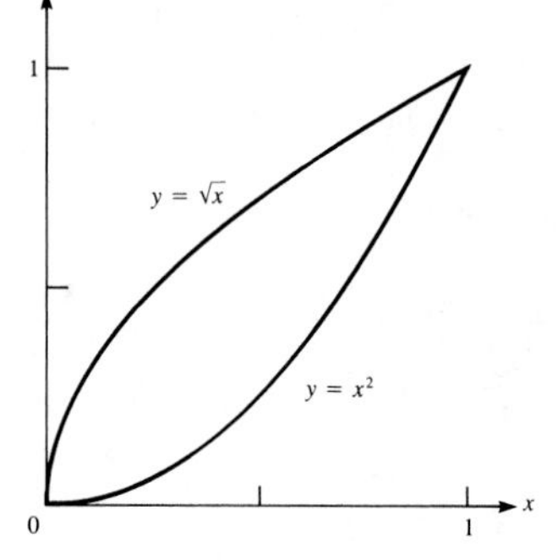

(*b*) $\sqrt{x} - x^2 \quad 0 \le x \le 1$ (*c*) $\sqrt{y} - y^2 \quad 0 \le y \le 1$

3 (*a*)

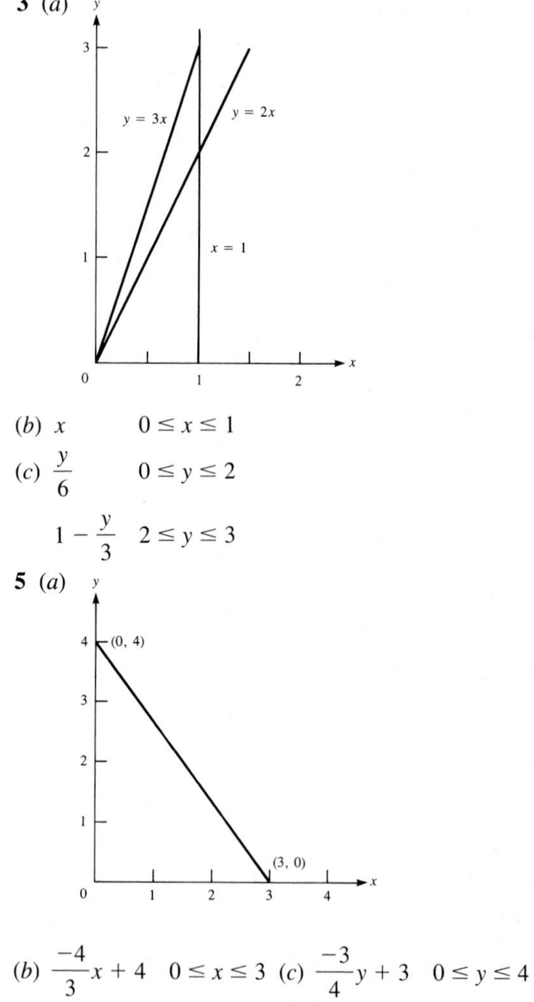

(*b*) $x \quad 0 \le x \le 1$

(*c*) $\dfrac{y}{6} \quad 0 \le y \le 2$

$1 - \dfrac{y}{3} \quad 2 \le y \le 3$

5 (*a*)

(*b*) $\dfrac{-4}{3}x + 4 \quad 0 \le x \le 3$ (*c*) $\dfrac{-3}{4}y + 3 \quad 0 \le y \le 4$

7 $\frac{1}{6}$ **9** $\frac{8}{3}$

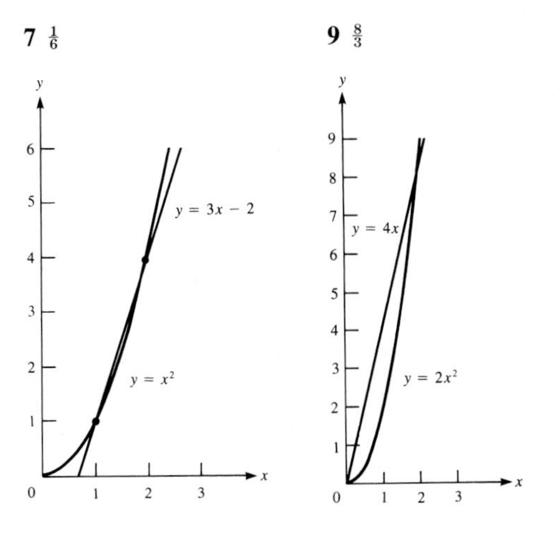

11 $\frac{2}{3}$

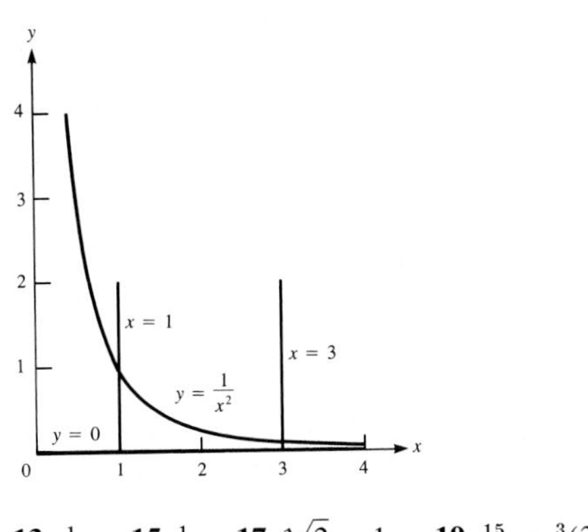

13 $\frac{1}{12}$ **15** $\frac{1}{3}$ **17** $\sqrt{2}-1$ **19** $\frac{15}{4}-\frac{3}{8}(3^{4/3}-1)$

21 $\dfrac{1}{\sqrt{3}}\tan^{-1}\left(\dfrac{2}{\sqrt{3}}\right)-\dfrac{\pi}{8}-\dfrac{1}{4}\ln\dfrac{7}{6}$

23 $\frac{9}{2}-\sqrt{2}\ln(3+\sqrt{8})$ **25** $\ln\frac{1125}{1024}$

27 $\ln\sqrt{2}$ **29** (a) $\dfrac{2-\sqrt{2}}{3}$ (b) $\dfrac{2-\sqrt{2}}{3}$

31 ≈ 2.86 **33** $\frac{1}{5}$ **39** $\dfrac{4\sqrt{2}-5}{3}a^2$

41 (b) $\dfrac{2b}{a}\sqrt{a^2-x^2}$ $-a\le x\le a$

(c) $\dfrac{2a}{b}\sqrt{b^2-y^2}$ $-b\le y\le b$ (d) πab

Sec. 8.2. Some Pointers on Drawing

1 (f) $A(x)=\left(\dfrac{a+2x}{2}\right)\left(\dfrac{ah-2hx}{a}\right)=\dfrac{a^2h-4hx^2}{2a}$

5 (b) $A(x)=h\sqrt{a^2-x^2}$ **7** The cross section is one-half of a hyperbola. **11** (d) $\pi(x^2-x^4)$

Sec. 8.3. Setting Up a Definite Integral

3 $g(r)(2\pi r\,dr)$ ft^3 (b) $\int_{1000}^{2000} 2\pi r g(r)\,dr$

7 (b) $\displaystyle\int_0^h \frac{\pi a^2x^2}{h^2}\,dx$ (c) $\dfrac{\pi a^2h}{3}$ **9** (b) $\displaystyle\int_0^h \frac{Ax^2}{h^2}\,dx$ (c) $\dfrac{Ah}{3}$

11 (a) 0 (b) 1 (d) Negative (e) $F(t)-F(t+dt)$ (f) $-\int_t^{t+dt}f(x)\,dx$ (g) -1 **13** (a) $c(t)\,e(t)\,dt$ (b) $\int_0^{24} c(t)e(t)\,dt$ **15** (a) $\int_a^b f(x)\,dx$ (b) $\int_a^b f(x)\,m(x)\,dx$

17 (a) $300\pi^3r^3\,dr$ joules (b) $\int_0^7 300\pi^3r^3\,dr$ joules (c) $180{,}075\pi^3$ joules **19** (b) $50\pi^2\dfrac{M}{a}x^2\,dx$

(c) $\displaystyle\int_0^a 50\pi^2\frac{M}{a}x^2\,dx$ (d) $\dfrac{50\pi^2Ma^2}{3}$ **21** (b) $\dfrac{\omega^2Mr^3\,dr}{a^2}$

(c) $\displaystyle\int_0^a \frac{\omega^2M}{a^2}r^3\,dr$ (d) $\dfrac{\omega^2Ma^2}{4}$ **23** $0.6471\,M\omega^2$

25 $\dfrac{M\omega^2}{6}\left(\dfrac{\pi+\ln 4-2}{\pi-\ln 4}\right)$

27 $\dfrac{18\sqrt{5}-\ln(2+\sqrt{5})}{16\sqrt{5}+8\ln(2+\sqrt{5})}M\omega^2$ **29** $\frac{1}{5}M\omega^2a^2$

31 (a) $2\pi r\,g(r)\,dr$ cm^3/sec (b) $\int_0^b 2\pi r\,g(r)\,dr$

Sec. 8.4. Computing Volumes

1 (d) $dV=\dfrac{\pi a^2x^2}{h^2}\,dx$ (e) $V=\displaystyle\int_0^h\frac{\pi a^2x^2}{h^2}\,dx$ (f) $V=\dfrac{\pi a^2h}{3}$

3 (a), (b) See Fig. 14. (d) $dV=4(9-x^2)\,dx$ (e) $V=\int_{-3}^{3}4(9-x^2)\,dx$ (f) 144 **5** (a), (b) See Figs. 7 and 8 in Sec. 8.2. (d) $\left(\dfrac{ax}{h}\right)^2 dx$ (e) $\dfrac{a^2}{h^2}\displaystyle\int_0^h x^2\,dx$ (f) $\dfrac{a^2h}{3}$

7 (a), (b) See Fig. 15. (d) $\dfrac{\sqrt{3}}{4}(2\sqrt{25-x^2})^2\,dx$

(e) $\sqrt{3}\displaystyle\int_{-5}^{5}(25-x^2)\,dx$ (f) $\dfrac{500\sqrt{3}}{3}$

9 $\dfrac{3\pi}{2}$ **11** $\pi(\ln 2-\frac{1}{2})$ **13** $(\pi/8)(10-3\pi)$

15 (d) $\dfrac{h(a+x)\sqrt{a^2-x^2}\,dx}{a}$

(e) $\displaystyle\int_{-a}^{a}\frac{h(a+x)\sqrt{a^2-x^2}\,dx}{a}$ (f) $\dfrac{\pi a^2h}{2}$

17 $\dfrac{\pi a^2h}{2}$ **19** $\frac{2}{3}a^2h$

21 (b) $A(x)=\begin{cases}9\pi & -4\le x\le 4\\ (25-x^2)\pi & -5\le x\le -4,\ 4\le x\le 5\end{cases}$

(c) $\dfrac{244\pi}{3}$ **23** $2\pi^2a^2b$

25 $V+2\pi kA$

Sec. 8.5. The Shell Method

1 $2\pi x^2\,dx$ **3** $4\pi x(1-x)\,dx$ **5** $\pi/10$ **7** $\pi/10$

9 $\frac{1}{3}\pi a^2h$ **11** (a) $\dfrac{256\pi}{15}$ (b) $\dfrac{433\pi}{30}$

13 (*a*) $2\pi(99\pi^2 + 2)$ (*b*) $\dfrac{81\pi^2}{2}$ **15** $\pi/2$
17 (*a*) $\pi(e^{10\pi} - 3e^{\pi})$ **19** $\pi \ln 6$
21 $29/(3\pi)$

Sec. 8.6. The Centroid of a Plane Region

1 $(0, \frac{12}{5})$ **3** $(2, \frac{8}{5})$ **5** $\left(\dfrac{e}{e-1}, \dfrac{e^2+e}{4}\right)$ **7** $(\frac{58}{35}, \frac{45}{56})$
9 (*b*) $90\pi^2$ in.^{3} **11** $\left(\dfrac{2a}{3}, \dfrac{b}{3}\right)$ **13** (*a*) $\frac{1}{20}$ (*b*) $\frac{1}{35}$ (*c*) $\frac{1}{12}$
(*d*) $\frac{3}{5}$ (*e*) $\frac{12}{35}$ **15** (*a*) 18.9741, (*b*) 73.1042, (*c*) 8.6950, (*d*) 2.1822, (*e*) 8.4076 **19** (*b*) $\bar{y}_R = \frac{29}{90}$, $\bar{y}_{R^*} = \frac{1}{3}$
(*c*) $\dfrac{2a^3+1}{3(a^2+1)}$ (*f*) 0.3222

Sec. 8.7. Work

1 (*a*) 0.5 joule (*b*) 1.125 joules **3** 0.512 joule **5** 2,000 mile·lb **7** 1,999.44 mile·lb **9** (*a*) $3.375 \times 10^{20}\pi$ ft·lb
(*b*) About 3.5×10^6 times as much
11 $\int_a^b 62.4(b - x)A(x)\,dx$

Sec. 8.8. Improper Integrals

1 Convergent, $\frac{1}{2}$ **3** Convergent, 1 **5** Divergent
7 Convergent, $\frac{1}{8}$ **9** Divergent **11** Divergent
13 Divergent **15** Convergent, $\pi/4$
17 Convergent, π **19** Convergent, $\frac{3}{2}$ **21** Convergent
23 Finite, ln 2 **31** Convergent for $p > 1$, divergent for $0 < p \le 1$. **37** $P(r) = 2/r^3$
39 $P(r) = \dfrac{1}{r^2+1}$

Sec. 8.S. Summary: Guide Quiz

1 (*a*)

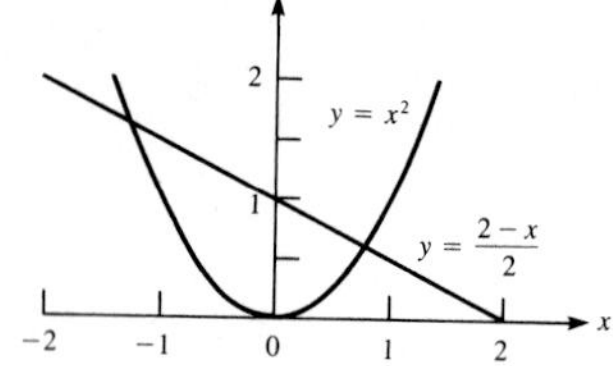

(*b*) $\displaystyle\int_{(-1-\sqrt{17})/4}^{(-1+\sqrt{17})/4} \left(1 - \frac{x}{2} - x^2\right) dx$
(*c*) $\displaystyle\int_0^{(9-\sqrt{17})/8} 2\sqrt{y}\,dy + \int_{(9-\sqrt{17})/8}^{(9+\sqrt{17})/8} (2 - 2y + \sqrt{y})\,dy$
(*d*) $\dfrac{17\sqrt{17}}{48}$
2 (*a*)

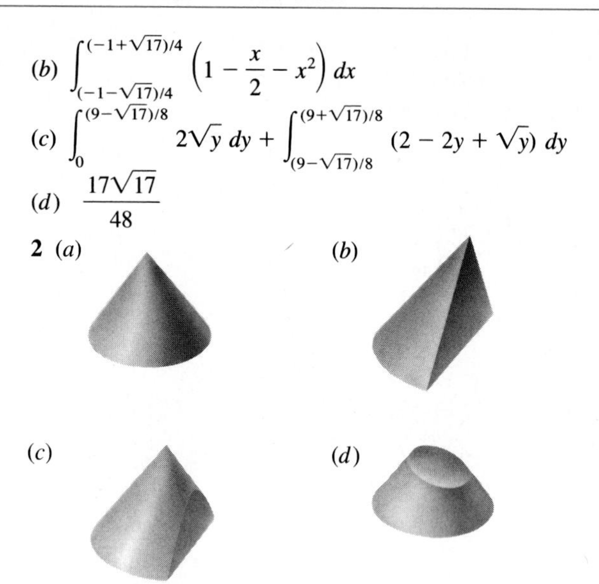

3 $\int_a^b \pi[(f(x) + 3)^2 - (g(x) + 3)^2]\,dx$ **5** $(\frac{19}{15}, \frac{7}{15})$ **6** $\frac{1}{4}W$
7 (*a*) $\displaystyle\int_0^1 \left(\frac{1}{\sqrt{y}} - 1\right) dy$ (*b*) 1 (*c*) $\displaystyle\int_1^{\infty} \frac{dx}{x^2}$ (*d*) 1
(*e*) $\displaystyle\pi \int_1^{\infty} \frac{dx}{x^4}$ (*f*) $\pi/3$ (*g*) $2\pi \int_0^1 (\sqrt{y} - y)\,dy$ (*h*) $\pi/3$

Sec. 8.S. Summary: Review Exercises

1 (*a*) $\int_0^2 (2x - x^2)\,dx$ (*b*) $\int_0^4 (\sqrt{y} - y/2)\,dy$
3 $\int_0^1 \pi y^2\,dy + \int_1^2 \pi y(2 - y)\,dy$ **5** (*a*) $1 + \ln\frac{2}{3}$
(*b*) $\pi(\frac{7}{6} + 2\ln\frac{2}{3})$ (*c*) $2\pi(\frac{1}{2} + \ln\frac{2}{3})$ (*d*) $\pi(\frac{19}{6} + 4\ln\frac{2}{3})$
7 (*a*) 1 (*b*) $\pi^2/4$ (*c*) $\pi^2/2$ (*d*) $\pi^2/4 + 2\pi$
9 (*a*) $\frac{1}{2}\ln 3$ (*b*) $\pi/3$ (*c*) $(\pi/2)(2 - \ln 3)$
(*d*) $(\pi/3)(1 + 3\ln 3)$
11 1 **13** $\sqrt{2} - 1$ **15** $\frac{1}{2}(2 + 2\ln 2 - \pi)$
17 $\frac{5}{6}\sqrt{2} - \frac{2}{3}$ **19** (*a*) $\frac{1}{2}\ln\frac{3}{2}$ (*b*) $\frac{1}{2}\ln 2$ **21** (*a*) $\dfrac{3\pi - 8}{12}$
(*b*) $\pi/64$ (*c*) $\dfrac{3\pi}{16(3\pi - 8)}$ **23** $4\pi/3$
25 (*a*) 1 (*b*) $\pi/2$ (*c*) 2π **35** $\frac{1}{5}(\sin 3 + 2\cos 3)$
41 $-\frac{1}{25}$ **45** $\frac{3}{2}, \frac{12}{5}$ **47** Convergent
49 (*a*) 1 (*c*) 1, 2, 6, 24 (*d*) $\Gamma(n + 1) = n!$
51 (*a*) $\pi r^2\sqrt{h}(2 - \sqrt{2})$ seconds (*b*) $2\pi r^2\sqrt{h}$ seconds
53 (*a*) $G(0) = G(1) = G(2) = \pi/4$ **55** $\dfrac{2\pi}{3} - \dfrac{\sqrt{3}}{2}$
57 $u = 1/x$ is not defined at $x = 0$. **59** $a = 0$ or 2π
61 (*b*) $\mu = 0$ (*c*) $\mu_2 = k^2$ (*d*) $\sigma = k$ **65** $\dfrac{n!}{a^{n+1}}$

CHAPTER 9. PLANE CURVES AND POLAR COORDINATES

Sec. 9.1. Polar Coordinates

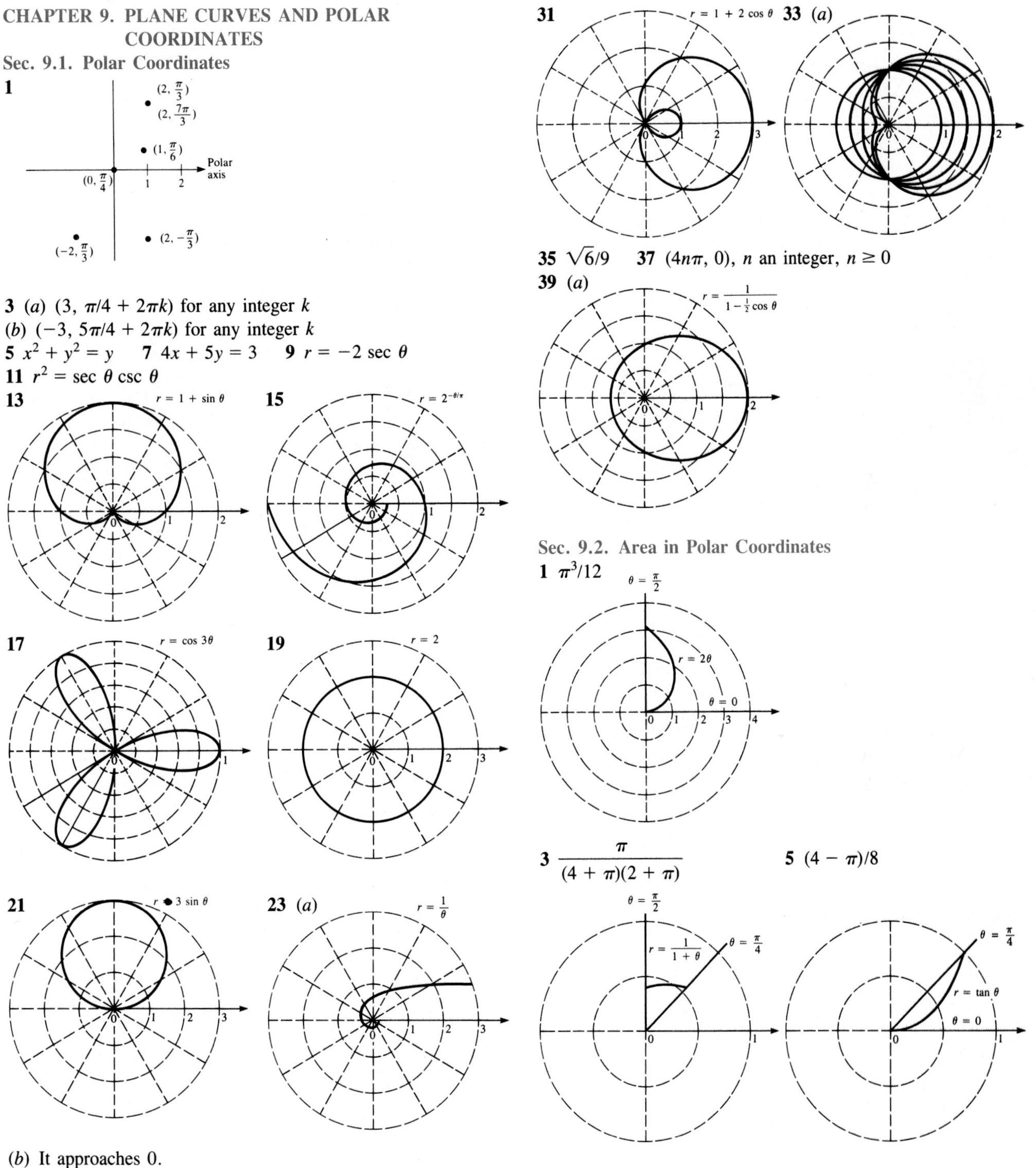

1 (graph)

3 (*a*) $(3, \pi/4 + 2\pi k)$ for any integer k
(*b*) $(-3, 5\pi/4 + 2\pi k)$ for any integer k

5 $x^2 + y^2 = y$ **7** $4x + 5y = 3$ **9** $r = -2 \sec \theta$

11 $r^2 = \sec \theta \csc \theta$

13 **15** **17** **19** **21** **23** (*a*)

(*b*) It approaches 0.

25 The curves are identical.

27 $(0, 0)$, $(1/\sqrt{2}, \pi/12)$, $(1/\sqrt{2}, 3\pi/4)$, $(1/\sqrt{2}, 17\pi/12)$

29 $(0, 0)$, $(1/2, \pi/6)$, $(1, \pi/2)$, $(1/2, 5\pi/6)$

31 **33** (*a*)

35 $\sqrt{6}/9$ **37** $(4n\pi, 0)$, n an integer, $n \geq 0$

39 (*a*)

Sec. 9.2. Area in Polar Coordinates

1 $\pi^3/12$

3 $\dfrac{\pi}{(4 + \pi)(2 + \pi)}$ **5** $(4 - \pi)/8$

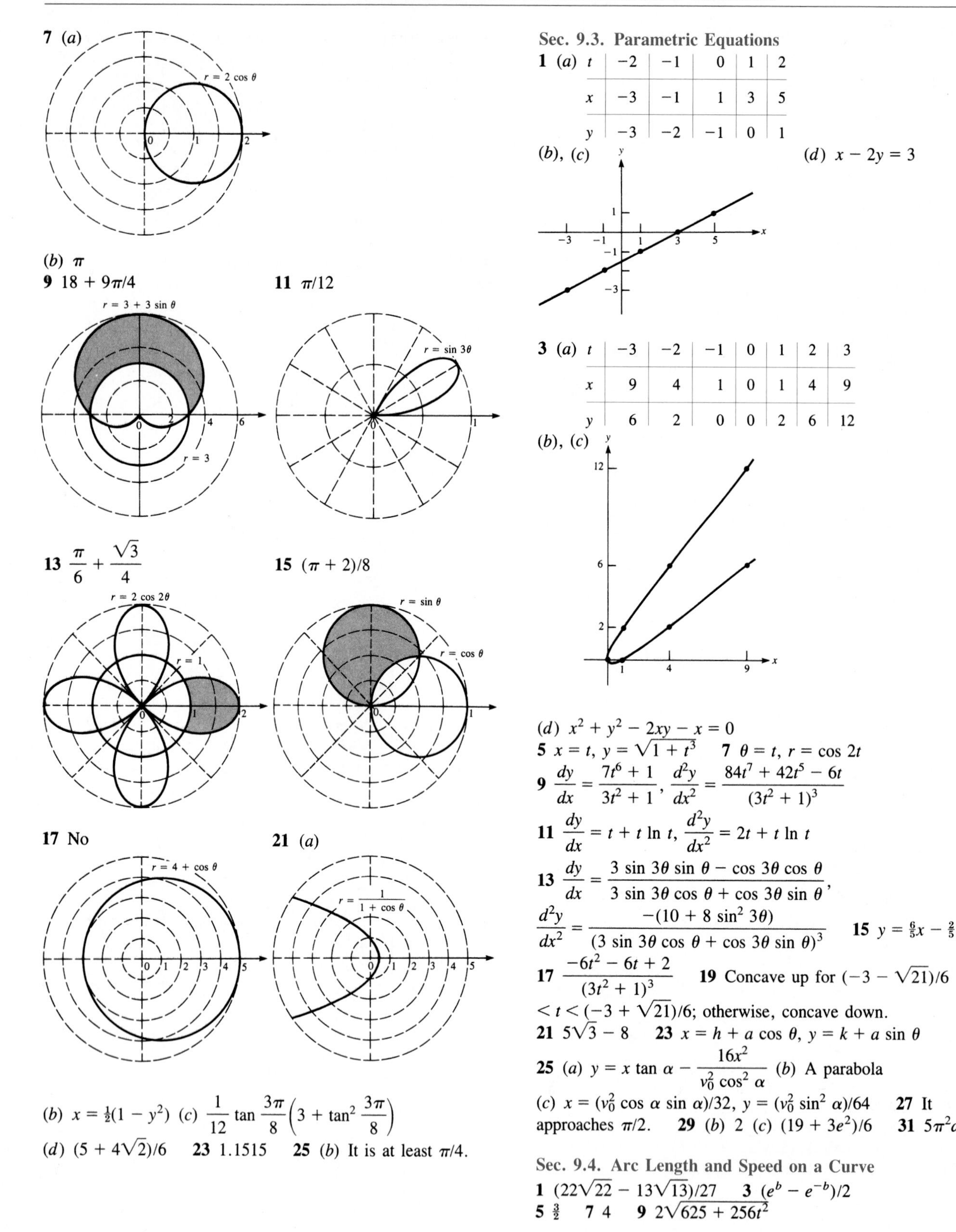

7 (*a*)

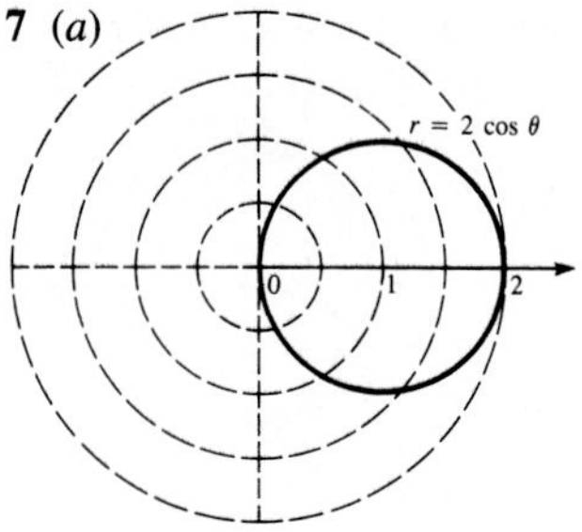

(*b*) π

9 $18 + 9\pi/4$

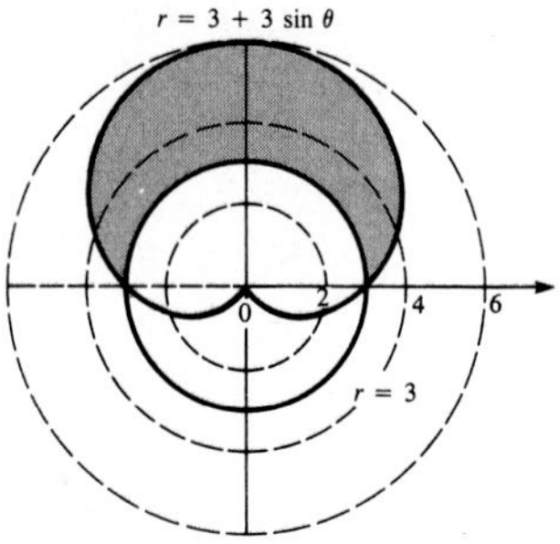

11 $\pi/12$

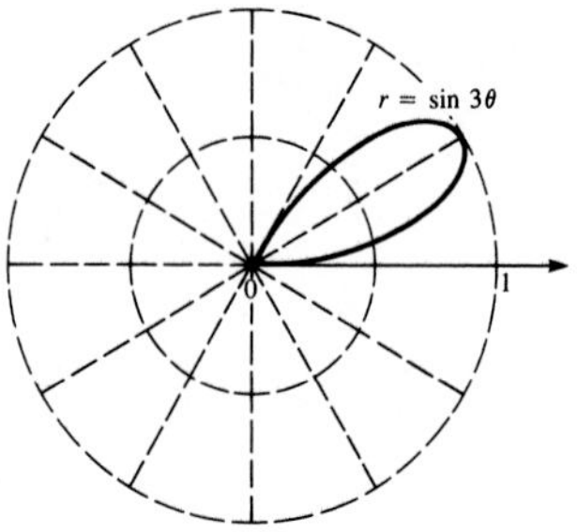

13 $\dfrac{\pi}{6} + \dfrac{\sqrt{3}}{4}$

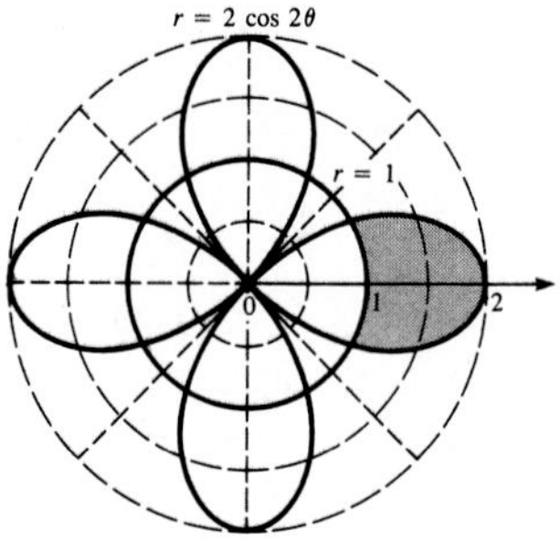

15 $(\pi + 2)/8$

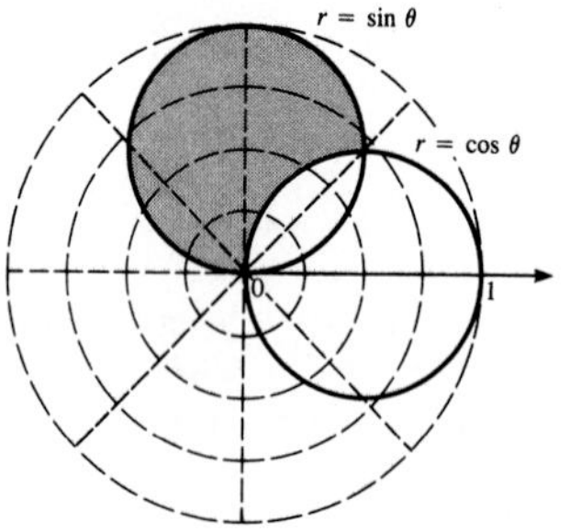

17 No

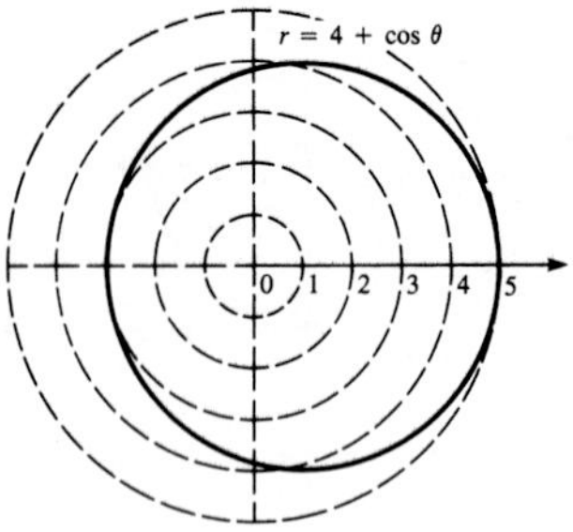

21 (*a*)

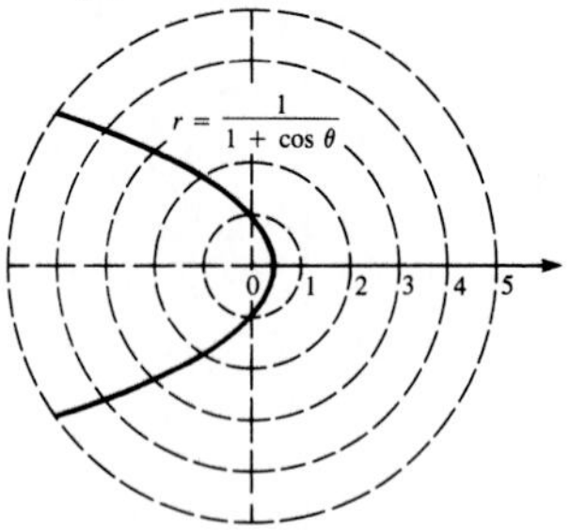

(*b*) $x = \frac{1}{2}(1 - y^2)$ (*c*) $\dfrac{1}{12}\tan\dfrac{3\pi}{8}\left(3 + \tan^2\dfrac{3\pi}{8}\right)$

(*d*) $(5 + 4\sqrt{2})/6$ **23** 1.1515 **25** (*b*) It is at least $\pi/4$.

Sec. 9.3. Parametric Equations

1 (*a*)

t	-2	-1	0	1	2
x	-3	-1	1	3	5
y	-3	-2	-1	0	1

(*b*), (*c*)

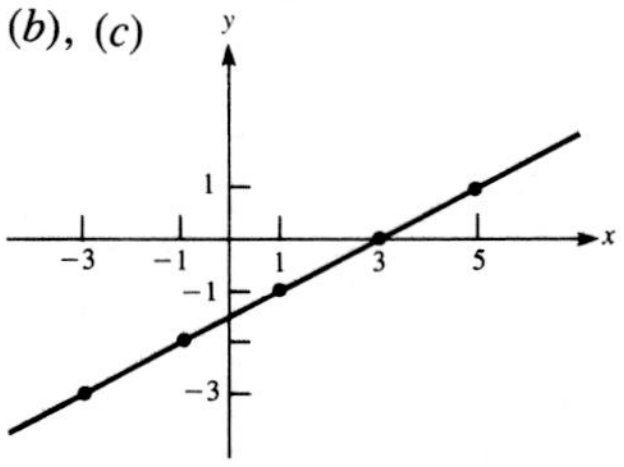

(*d*) $x - 2y = 3$

3 (*a*)

t	-3	-2	-1	0	1	2	3
x	9	4	1	0	1	4	9
y	6	2	0	0	2	6	12

(*b*), (*c*)

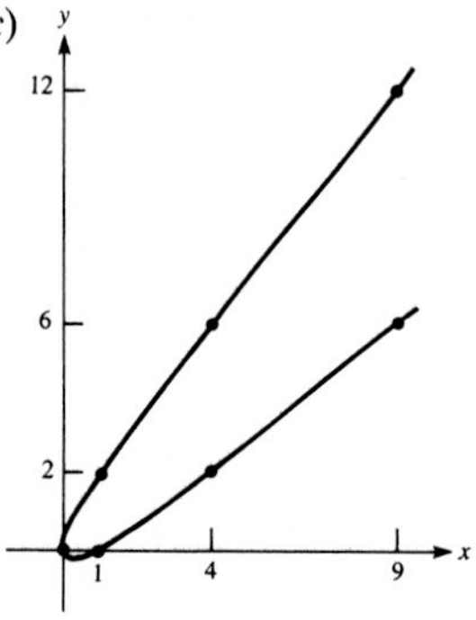

(*d*) $x^2 + y^2 - 2xy - x = 0$

5 $x = t,\ y = \sqrt{1 + t^3}$ **7** $\theta = t,\ r = \cos 2t$

9 $\dfrac{dy}{dx} = \dfrac{7t^6 + 1}{3t^2 + 1},\ \dfrac{d^2y}{dx^2} = \dfrac{84t^7 + 42t^5 - 6t}{(3t^2 + 1)^3}$

11 $\dfrac{dy}{dx} = t + t\ln t,\ \dfrac{d^2y}{dx^2} = 2t + t\ln t$

13 $\dfrac{dy}{dx} = \dfrac{3\sin 3\theta\sin\theta - \cos 3\theta\cos\theta}{3\sin 3\theta\cos\theta + \cos 3\theta\sin\theta}$,

$\dfrac{d^2y}{dx^2} = \dfrac{-(10 + 8\sin^2 3\theta)}{(3\sin 3\theta\cos\theta + \cos 3\theta\sin\theta)^3}$ **15** $y = \frac{6}{5}x - \frac{2}{5}$

17 $\dfrac{-6t^2 - 6t + 2}{(3t^2 + 1)^3}$ **19** Concave up for $(-3 - \sqrt{21})/6 < t < (-3 + \sqrt{21})/6$; otherwise, concave down.

21 $5\sqrt{3} - 8$ **23** $x = h + a\cos\theta,\ y = k + a\sin\theta$

25 (*a*) $y = x\tan\alpha - \dfrac{16x^2}{v_0^2\cos^2\alpha}$ (*b*) A parabola

(*c*) $x = (v_0^2\cos\alpha\sin\alpha)/32,\ y = (v_0^2\sin^2\alpha)/64$ **27** It approaches $\pi/2$. **29** (*b*) 2 (*c*) $(19 + 3e^2)/6$ **31** $5\pi^2a^3$

Sec. 9.4. Arc Length and Speed on a Curve

1 $(22\sqrt{22} - 13\sqrt{13})/27$ **3** $(e^b - e^{-b})/2$

5 $\frac{3}{2}$ **7** 4 **9** $2\sqrt{625 + 256t^2}$

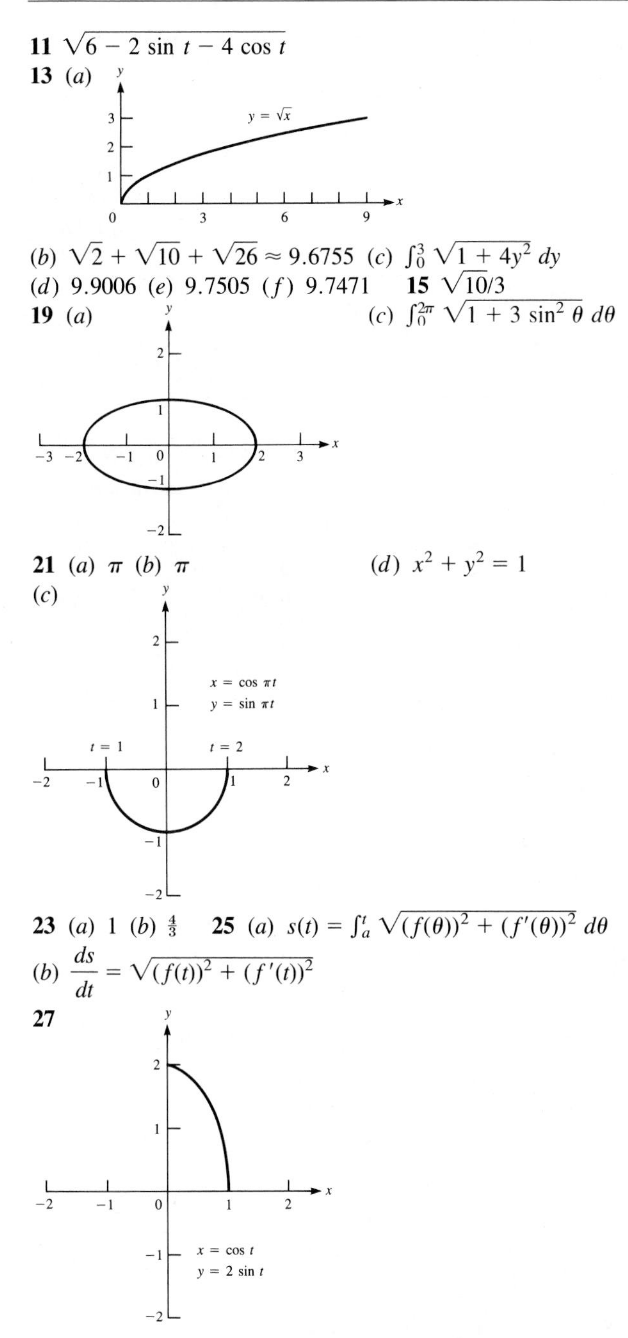

11 $\sqrt{6 - 2\sin t - 4\cos t}$
13 (*a*)

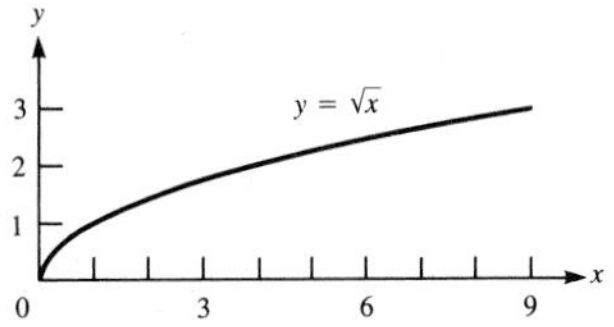

(*b*) $\sqrt{2} + \sqrt{10} + \sqrt{26} \approx 9.6755$ (*c*) $\int_0^3 \sqrt{1 + 4y^2}\, dy$
(*d*) 9.9006 (*e*) 9.7505 (*f*) 9.7471 **15** $\sqrt{10}/3$
19 (*a*)

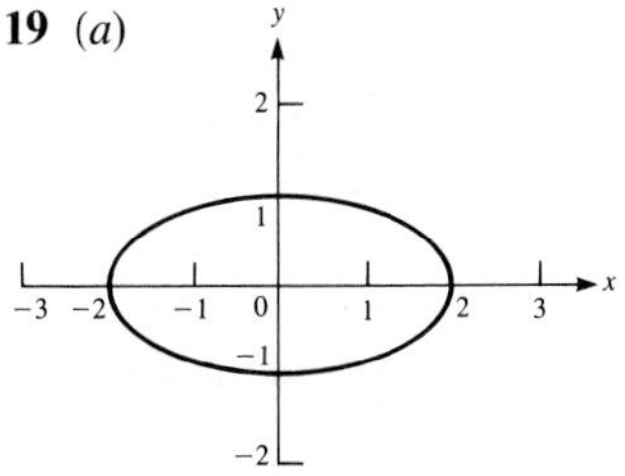

(*c*) $\int_0^{2\pi} \sqrt{1 + 3\sin^2\theta}\, d\theta$

21 (*a*) π (*b*) π (*d*) $x^2 + y^2 = 1$
(*c*)

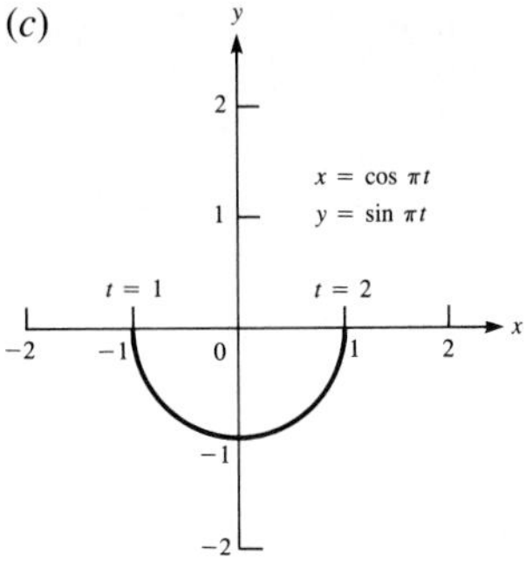

23 (*a*) 1 (*b*) $\frac{4}{3}$ **25** (*a*) $s(t) = \int_a^t \sqrt{(f(\theta))^2 + (f'(\theta))^2}\, d\theta$
(*b*) $\dfrac{ds}{dt} = \sqrt{(f(t))^2 + (f'(t))^2}$
27

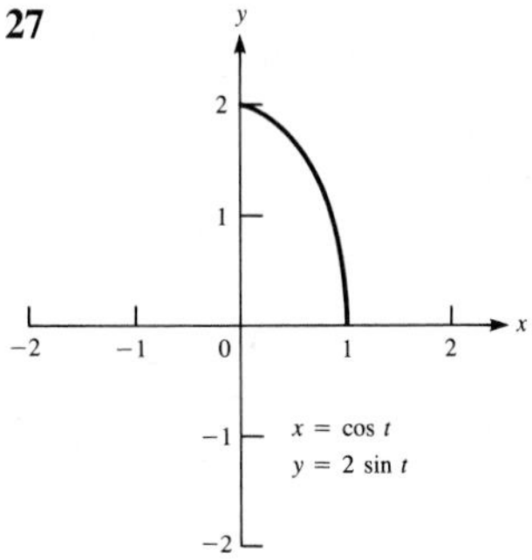

Sec. 9.5. The Area of a Surface of Revolution

1 $\int_1^2 2\pi x^3\sqrt{1 + 9x^4}\, dx$ **3** $\int_1^8 2\pi y^{1/3}\sqrt{1 + \frac{1}{9}y^{-4/3}}\, dy$
5 $\pi[e\sqrt{1 + e^2} + \ln(e + \sqrt{1 + e^2}) - \sqrt{2} - \ln(1 + \sqrt{2})]$
7 $64\pi/3$ **9** $\int_0^1 4\pi x^3\sqrt{1 + 36x^4}\, dx$; use $u = 1 + 36x^4$.
11 $\int_1^2 2\pi x^2\sqrt{1 + 4x^2}\, dx$; use $x = \frac{1}{2}\tan\theta$.
13 $\int_1^8 2\pi(x^{2/3} - 1)\sqrt{1 + \frac{4}{9}x^{-2/3}}\, dx$; use $x = u^{3/2}$ and $u = v^2 - \frac{4}{9}$. **15** $\displaystyle\int_1^2 2\pi\left(\frac{x^5}{3} + x^2 + \frac{x}{3} + \frac{1}{4x^2} + \frac{1}{16x^3}\right) dx$; use the power rule. **17** (*a*) The volume of the sphere is $\frac{2}{3}$ the volume of the cylinder. (*b*) Both surface areas are $4\pi a^2$.
19 15.0065 **23** $4\pi a^2$
25 $\int_0^{\pi/2} 2\pi \sin 2\theta \sin\theta \sqrt{1 + 3\cos^2 2\theta}\, d\theta$
27 $2\pi\left(b^2 + \dfrac{a^2 b}{\sqrt{a^2 - b^2}}\cos^{-1}\dfrac{b}{a}\right)$. Yes. **29** $\pi r L$
35 $\pi a\sqrt{a^2 + h^2}$

Sec. 9.6. Curvature

1 $2/(5\sqrt{5})$, $5\sqrt{5}/2$ **3** $e^2/(e^2 + 1)^{3/2}$, $(e^2 + 1)^{3/2}/e^2$
5 $4/(5\sqrt{5})$, $5\sqrt{5}/4$ **7** $\frac{1}{2}$ **9** $e^{\pi/6}/\sqrt{2}$
11 (*a*) $4/(e^x + e^{-x})^2$, $(e^x + e^{-x})^2/4$ **13** $-(\ln 2)/2$
17 At (1,000, 0), $A = \tan^{-1} 0.968 \approx 0.77$ radian ($\approx 44°$). At (0, 500), $A = \tan^{-1} 0.121 \approx 0.12$ radian ($\approx 7°$).
19 $(a^4y^2 + b^4x^2)^{3/2}/(a^4b^4)$
21

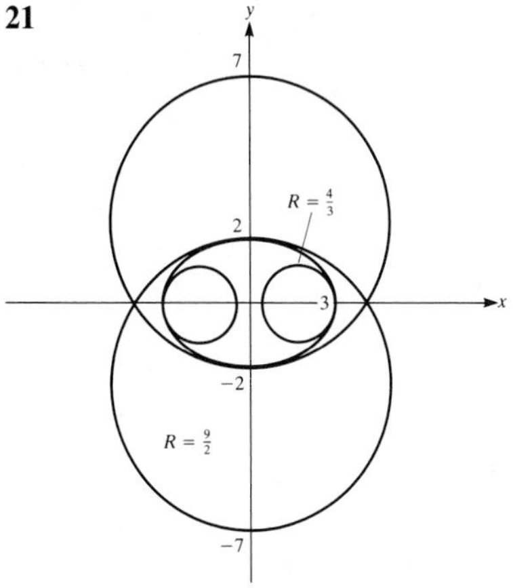

25 $\dfrac{3|y|^5}{(1 + y^6)^{3/2}}$

Sec. 9.7. The Reflection Properties of the Conic Sections

1 $\pi/2$ **3** 1 **5** $\frac{1}{7}$ **7** $-2\sqrt{2}$ **9** The tangent is undefined; the curves are perpendicular. **15** $\pi/4$
17 $f(\theta) = a\sin\theta$, for some constant a **19** $f(\theta) = Ae^{k\theta}$, for constants A and k

Sec. 9.S. Summary: Guide Quiz

1 $\int_{-\pi/10}^{\pi/10} \frac{1}{2}\cos^2 5\theta\, d\theta$ **2** $\int_1^2 \sqrt{1 + 16x^6}\, dx$
3 $\int_1^2 2\pi(20 - x^4)\sqrt{1 + 16x^6}\, dx$
4 $\int_1^2 2\pi(x - 1)\sqrt{1 + 16x^6}\, dx$ **5** $\int_0^{2\pi} \sqrt{5 + 4\sin\theta}\, d\theta$
6 $\int_0^{\pi} 2\pi(2 + \sin\theta)\sin\theta\sqrt{5 + 4\sin\theta}\, d\theta$
7 $\int_0^1 \sqrt{100t^2 + 1/(4t)}\, dt$ **8** $\sqrt{873}$ **9** $y^2 = x^2 + 1$
10 (*a*) $\pi/8 + \gamma \approx 3.07$ radians ($\approx 176°$) (*b*) See Fig. 12 in Sec. 9.1. **11** $\sqrt{2}$

Sec. 9.S. Summary: Review Exercises

7 (*a*) $\int_0^{\pi} \sqrt{1 + \cos^2 x}\, dx$, $\int_0^{\pi} 2\pi \sin x \sqrt{1 + \cos^2 x}\, dx$, $\int_0^{\pi} 2\pi x\sqrt{1 + \cos^2 x}\, dx$ (*b*) Surface area (about x axis) $= 2\pi[\sqrt{2} + \ln(1 + \sqrt{2})]$ **9** $3\sqrt{3}/4$ **11** Infinite

13 (*a*)

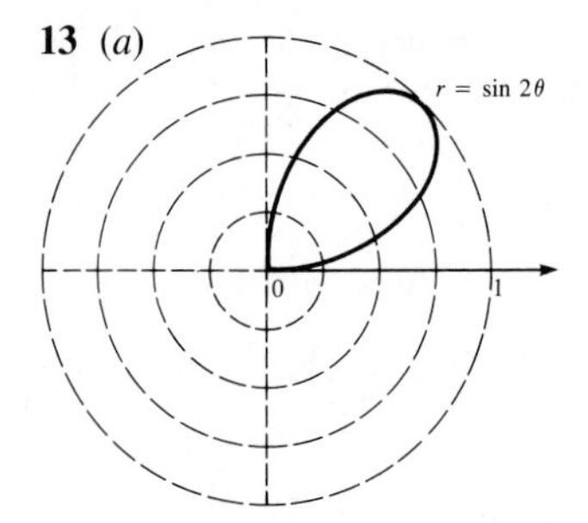

(*b*) $x = 4\sqrt{3}/9$ when $r = 2\sqrt{2}/3$ and $\theta = \sin^{-1}(1/\sqrt{3})$
(*c*) $y = 4\sqrt{3}/9$ when $r = 2\sqrt{2}/3$ and $\theta = \pi/2 - \sin^{-1}(1/\sqrt{3})$
(*d*) $r = 1,\ \theta = \pi/4$

15 $\dfrac{dy}{dx} = -\dfrac{3\cos 3t}{2\sin 2t}$, $\dfrac{d^2y}{dx^2} = \dfrac{18\sin 2t\sin 3t + 12\cos 2t\cos 3t}{(-2\sin 2t)^3}$

17 (*a*)

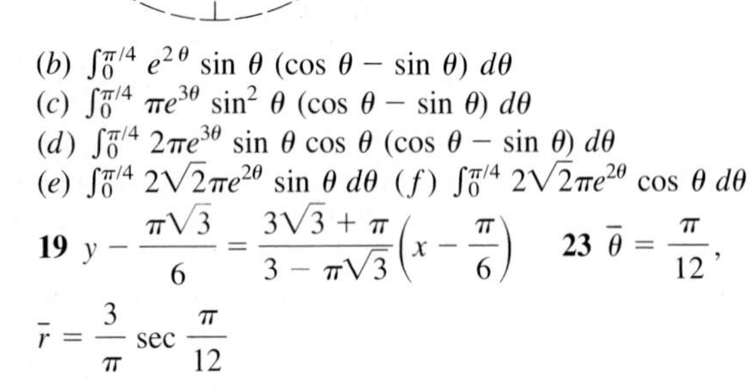

(*b*) $\int_0^{\pi/4} e^{2\theta}\sin\theta(\cos\theta - \sin\theta)\,d\theta$
(*c*) $\int_0^{\pi/4} \pi e^{3\theta}\sin^2\theta(\cos\theta - \sin\theta)\,d\theta$
(*d*) $\int_0^{\pi/4} 2\pi e^{3\theta}\sin\theta\cos\theta(\cos\theta - \sin\theta)\,d\theta$
(*e*) $\int_0^{\pi/4} 2\sqrt{2}\pi e^{2\theta}\sin\theta\,d\theta$ (*f*) $\int_0^{\pi/4} 2\sqrt{2}\pi e^{2\theta}\cos\theta\,d\theta$

19 $y - \dfrac{\pi\sqrt{3}}{6} = \dfrac{3\sqrt{3} + \pi}{3 - \pi\sqrt{3}}\left(x - \dfrac{\pi}{6}\right)$ **23** $\bar{\theta} = \dfrac{\pi}{12}$, $\bar{r} = \dfrac{3}{\pi}\sec\dfrac{\pi}{12}$

CHAPTER 10. SERIES

Sec. 10.1. An Informal Introduction to Series

1 (*a*) $\frac{113}{81} \approx 1.3950617$ (*b*) $\frac{2713}{1944} \approx 1.3955761$

3

x	$x - \dfrac{x^3}{6}$	$\sin x$	$\sin x - \left(x - \dfrac{x^3}{6}\right)$
0.1	0.0998333	0.0998334	0.0000001
0.2	0.1986667	0.1986693	0.0000027
0.3	0.2955000	0.2955202	0.0000202
0.4	0.3893333	0.3894183	0.0000850
0.5	0.4791667	0.4794255	0.0002589

5 (*a*) $\frac{11}{16} = 0.6875$ (*b*) $\frac{53}{72} \approx 0.7361111$

7 (*a*)

n	$\dfrac{1}{2} - \dfrac{(\frac{1}{2})^2}{2} + \dfrac{(\frac{1}{2})^3}{3} - \cdots + (-1)^n\dfrac{(\frac{1}{2})^n}{n}$	*Decimal*
1	$\frac{1}{2}$	0.5
2	$\frac{1}{2} - \frac{1}{8} = \frac{3}{8}$	0.375
3	$\frac{1}{2} - \frac{1}{8} + \frac{1}{24} = \frac{5}{12}$	0.4166666
4	$\frac{1}{2} - \frac{1}{8} + \frac{1}{24} - \frac{1}{64} = \frac{77}{192}$	0.4010416
5	$\frac{1}{2} - \frac{1}{8} + \frac{1}{24} - \frac{1}{64} + \frac{1}{160} = \frac{391}{960}$	0.4072916

(*b*) 0.0018264

9 (*a*) The second approach is better.

15 (*a*)

n	$\dfrac{1}{1\cdot 2} + \dfrac{1}{2\cdot 3} + \dfrac{1}{3\cdot 4} + \cdots + \dfrac{1}{n(n+1)}$	*Sum as Decimal*	*Sum as Fraction*
1	$\frac{1}{2}$	0.5	$\frac{1}{2}$
2	$\frac{1}{2} + \frac{1}{6}$	0.6666667	$\frac{2}{3}$
3	$\frac{1}{2} + \frac{1}{6} + \frac{1}{12}$	0.75	$\frac{3}{4}$
4	$\frac{1}{2} + \frac{1}{6} + \frac{1}{12} + \frac{1}{20}$	0.8	$\frac{4}{5}$
5	$\frac{1}{2} + \frac{1}{6} + \frac{1}{12} + \frac{1}{20} + \frac{1}{30}$	0.8333333	$\frac{5}{6}$

Sec. 10.2. Sequences

1 $0.999, 0.999^2, 0.999^3, \ldots; 0$ **3** $1, 1, 1, \ldots; 1$
5 $1, 2, 6, \ldots$; diverges **7** $4, \frac{11}{7}, \frac{7}{6}, \ldots; \frac{3}{5}$
9 $\cos 1, \frac{1}{2}\cos 2, \frac{1}{3}\cos 3, \ldots; 0$
11 $3, 4, \frac{125}{27}, \ldots; e^2$
13 (*a*)

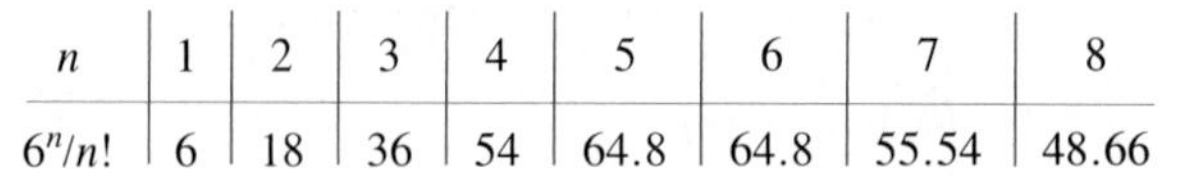

n	1	2	3	4	5	6	7	8
$6^n/n!$	6	18	36	54	64.8	64.8	55.54	48.66

(*b*)

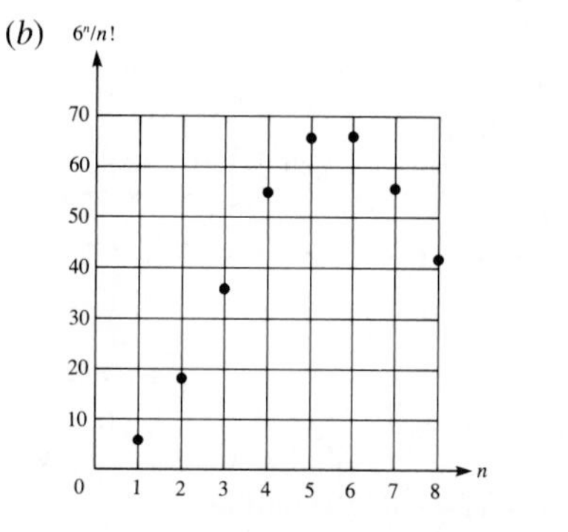

(*d*) 0

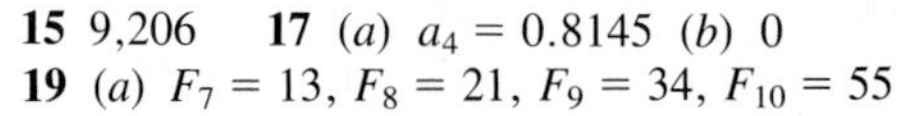

15 9,206 **17** (*a*) $a_4 = 0.8145$ (*b*) 0
19 (*a*) $F_7 = 13$, $F_8 = 21$, $F_9 = 34$, $F_{10} = 55$

(*b*)

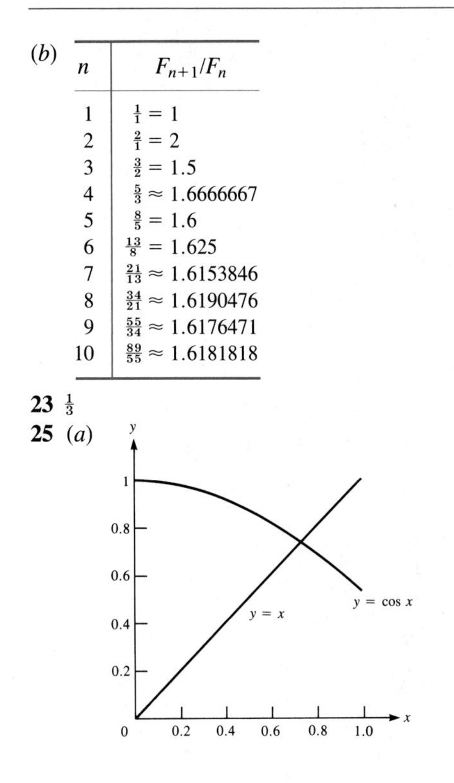

n	F_{n+1}/F_n
1	$\frac{1}{1} = 1$
2	$\frac{2}{1} = 2$
3	$\frac{3}{2} = 1.5$
4	$\frac{5}{3} \approx 1.6666667$
5	$\frac{8}{5} = 1.6$
6	$\frac{13}{8} = 1.625$
7	$\frac{21}{13} \approx 1.6153846$
8	$\frac{34}{21} \approx 1.6190476$
9	$\frac{55}{34} \approx 1.6176471$
10	$\frac{89}{55} \approx 1.6181818$

23 $\frac{1}{3}$
25 (*a*)

Sec. 10.3. Series
1 (*a*) False (*b*) False (*c*) False (*d*) True **3** (*a*) False (*b*) False (*c*) True (*d*) True (*e*) True (*f*) False **5** 2 **7** $\frac{1}{9}$ **9** 495 **11** 8 **13** Diverges **15** Diverges **17** 24 **19** $\frac{3}{2}$ **21** 4 ft **23** $3\frac{17}{99}$ **25** $4\frac{251}{1998}$ **29** \$5,000 **31** ≈23.25 sec **33** (*b*) 2 (*c*) $p/(1-p)^2$

Sec. 10.4. The Integral Test
1 Converges **3** Diverges **5** Diverges **7** Diverges **9** Diverges **11** Diverges **15** (*a*) 1.1777 (*b*) $\frac{1}{50} < R_4 < \frac{1}{32}$ (*c*) $1.1976 < \sum_{n=1}^{\infty} \frac{1}{n^3} < 1.2090$
17 (*a*) 0.8588 (*b*) $0.1974 < R_4 < 0.2449$ (*c*) $1.0562 < \sum_{n=1}^{\infty} \frac{1}{n^2+1} < 1.1037$
19 (*a*) $\frac{1}{101} < R_{100} < \frac{1}{100}$ (*b*) $n \geq 10{,}000$ **21** (*a*) $n \geq 15$ (*b*) $S_{15} \approx 1.082$ **29** It approaches 0.

Sec. 10.5. Comparison Tests
1 Converges **3** Converges **5** Converges **7** Diverges **9** Converges **11** Converges **13** Converges **15** Converges **17** Converges **19** Converges **21** Converges **23** Diverges **25** Diverges **27** Diverges **35** $0 < x < 2$; $x \geq 2$ **39** (*a*) Converges (*b*) $0.644 < \sum_{n=1}^{\infty} \frac{1}{1+2^n} < 0.769$

Sec. 10.6. Ratio Tests
1 Converges **3** Converges **5** Converges **7** Converges **9** (*a*) $0 < x < 1$ (*b*) $x \geq 1$ **11** (*a*) $0 < x < 2$ (*b*) $x \geq 2$ **13** Converges **15** The actual sum is 6, so any valid B must exceed 6. **17** The actual sum is $5e$, so any valid B must exceed 13.591409. **19** The actual sum is $\frac{3}{4}$, so any valid B must exceed $\frac{3}{4}$. **23** $m > \frac{\ln(11.01)}{\ln(1.01)} - 1$

Sec. 10.7. Test for Series with Both Positive and Negative Terms
1 Diverges **3** Converges **5** Diverges **7** Diverges **9** (*a*) $S_5 = \frac{47}{60} \approx 0.78333$, $S_6 = \frac{37}{60} \approx 0.61667$ (*b*) Larger (*c*) The sum lies between S_5 and S_6. **11** Converges conditionally **13** Converges conditionally **15** Converges absolutely **17** Converges absolutely **19** Converges absolutely **21** Converges conditionally **23** Converges absolutely **25** Converges absolutely **27** Converges for all x **29** (*a*) $n \geq 201$ (*b*) $S_{201} \approx 1.01$ **39** (*a*) Converges (*b*) Diverges

Sec. 10.S. Summary: Guide Quiz
2 No **3** (*a*) 44 (*b*) 2001 **5** (*a*) $S_8 = 0.918$ (*b*) $S_{10} = 0.920$ **6** (*a*) Converges (*b*) Diverges (*c*) Converges for $|x| < 1$, diverges for $|x| \geq 1$ (*d*) Converges **7** (*a*) False (*b*) False (*c*) False (*d*) True **8** (*a*) Yes; less than 1. (*b*) Yes; between -1 and 1. **9** (*a*) No; $R_9 < \frac{1}{9}$ (*b*) Yes

Sec. 10.S. Summary: Review Exercises
1 Converges **3** Converges **5** Diverges **7** Converges **9** Converges **11** Diverges **13** Converges **15** Converges **17** Diverges **19** Converges **21** Converges **23** Converges **25** Converges **27** Converges **29** Diverges **31** Converges **35** $S_4 = -0.1376$ **39** (*b*) $\sum_{n=1}^{\infty} \frac{1}{\sqrt{n}}$
45 (*a*) $I_0 = \pi/2$, $I_1 = 1$ **47** (*b*) 2.4329020×10^{18}, 0.9958423 **49** (*a*) $f(\frac{1}{2}) = 0$, $f(2) = 1$, $f(1) = \frac{1}{2}$
(*b*)

y
1
$\frac{1}{2}$
-1 $\frac{1}{2}$ 1 2 x

(*c*) f is continuous except at 1 and -1. **51** (*a*) Converges for $q > 1$ and all values of $p > 0$, diverges for $0 < q \leq 1$ and all values of $p > 0$. (*b*) Diverges for all $p > 0$ and $q > 0$. **53** Converges

CHAPTER 11. POWER SERIES AND COMPLEX NUMBERS

Sec. 11.1. Taylor Series

1 $P_1(x; 0) = 1 - x$, $P_2(x; 0) = 1 - x + x^2$

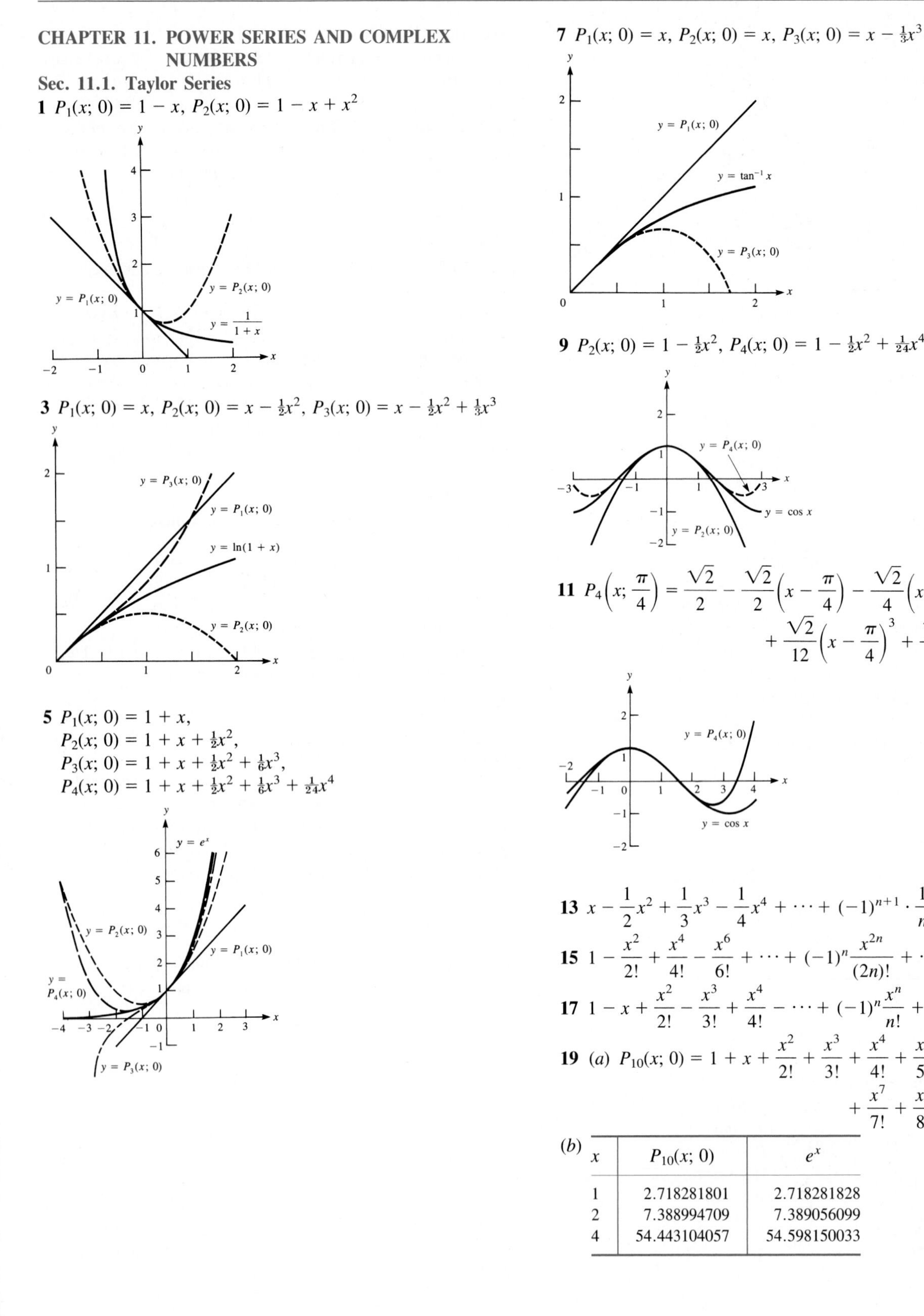

3 $P_1(x; 0) = x$, $P_2(x; 0) = x - \frac{1}{2}x^2$, $P_3(x; 0) = x - \frac{1}{2}x^2 + \frac{1}{3}x^3$

5 $P_1(x; 0) = 1 + x$,
$P_2(x; 0) = 1 + x + \frac{1}{2}x^2$,
$P_3(x; 0) = 1 + x + \frac{1}{2}x^2 + \frac{1}{6}x^3$,
$P_4(x; 0) = 1 + x + \frac{1}{2}x^2 + \frac{1}{6}x^3 + \frac{1}{24}x^4$

7 $P_1(x; 0) = x$, $P_2(x; 0) = x$, $P_3(x; 0) = x - \frac{1}{3}x^3$

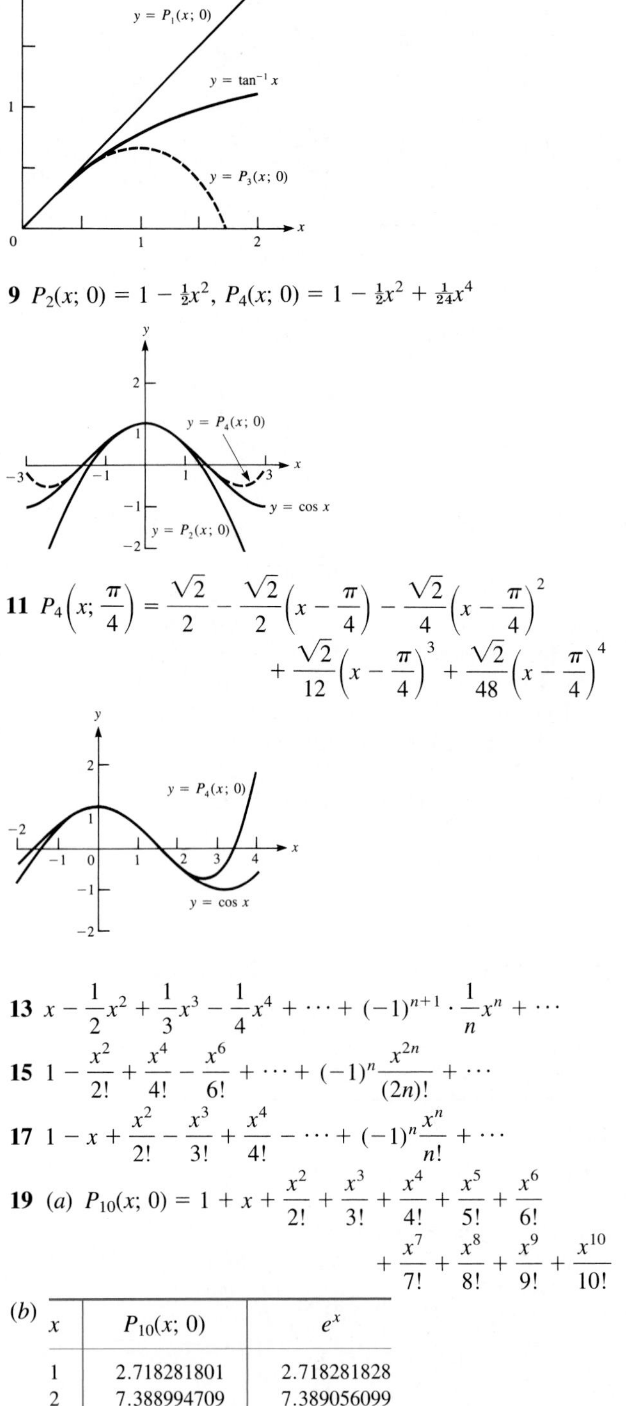

9 $P_2(x; 0) = 1 - \frac{1}{2}x^2$, $P_4(x; 0) = 1 - \frac{1}{2}x^2 + \frac{1}{24}x^4$

11 $$P_4\left(x; \frac{\pi}{4}\right) = \frac{\sqrt{2}}{2} - \frac{\sqrt{2}}{2}\left(x - \frac{\pi}{4}\right) - \frac{\sqrt{2}}{4}\left(x - \frac{\pi}{4}\right)^2 + \frac{\sqrt{2}}{12}\left(x - \frac{\pi}{4}\right)^3 + \frac{\sqrt{2}}{48}\left(x - \frac{\pi}{4}\right)^4$$

13 $x - \frac{1}{2}x^2 + \frac{1}{3}x^3 - \frac{1}{4}x^4 + \cdots + (-1)^{n+1} \cdot \frac{1}{n}x^n + \cdots$

15 $1 - \frac{x^2}{2!} + \frac{x^4}{4!} - \frac{x^6}{6!} + \cdots + (-1)^n \frac{x^{2n}}{(2n)!} + \cdots$

17 $1 - x + \frac{x^2}{2!} - \frac{x^3}{3!} + \frac{x^4}{4!} - \cdots + (-1)^n \frac{x^n}{n!} + \cdots$

19 (*a*) $$P_{10}(x; 0) = 1 + x + \frac{x^2}{2!} + \frac{x^3}{3!} + \frac{x^4}{4!} + \frac{x^5}{5!} + \frac{x^6}{6!} + \frac{x^7}{7!} + \frac{x^8}{8!} + \frac{x^9}{9!} + \frac{x^{10}}{10!}$$

(*b*)

x	$P_{10}(x; 0)$	e^x
1	2.718281801	2.718281828
2	7.388994709	7.389056099
4	54.443104057	54.598150033

21 (*a*) $P_1(x; 0) = 2 + 3x$, $P_2(x; 0) = 2 + 3x - 4x^2$, $P_3(x; 0) = 2 + 3x - 4x^2$ (*b*) $2 + 3x - 4x^2$
23 (*a*) $P_3(x; 0) = 1 + 3x + 3x^2 + x^3$
(*b*) $(1 + x)^3 = 1 + 3x + 3x^2 + x^3$

25 $P_6\left(x; \frac{\pi}{6}\right) =$

$$\frac{1}{2} + \frac{\sqrt{3}}{2}\left(x - \frac{\pi}{6}\right) - \frac{1}{4}\left(x - \frac{\pi}{6}\right)^2 - \frac{\sqrt{3}}{12}\left(x - \frac{\pi}{6}\right)^3 + \frac{1}{48}\left(x - \frac{\pi}{6}\right)^4 + \frac{\sqrt{3}}{240}\left(x - \frac{\pi}{6}\right)^5 - \frac{1}{1440}\left(x - \frac{\pi}{6}\right)^6$$

27 (*a*) $P_4(x; 0) = 1 + \frac{1}{2}x - \frac{1}{8}x^2 + \frac{1}{16}x^3 - \frac{5}{128}x^4$

(*b*)

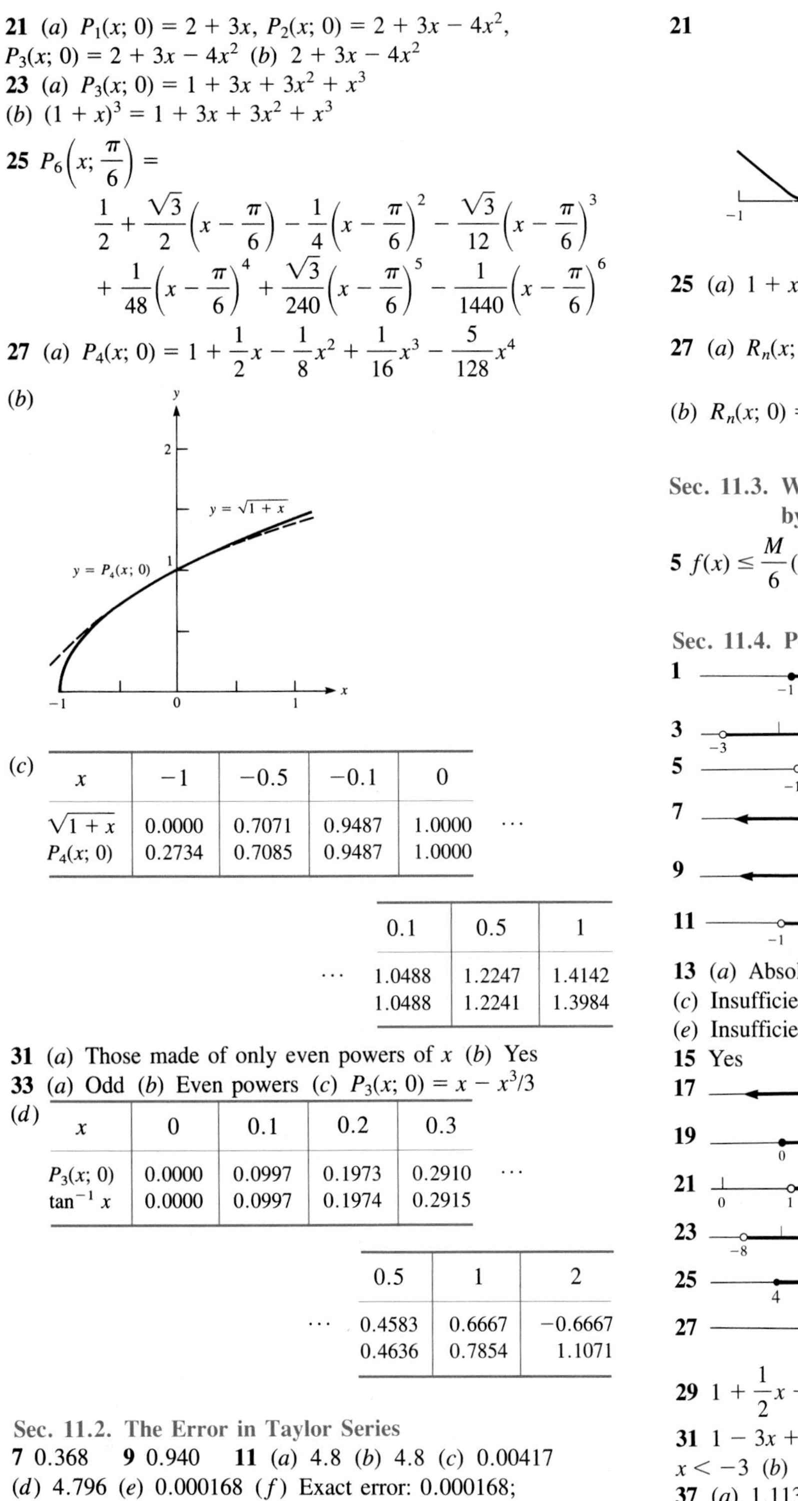

(*c*)

x	−1	−0.5	−0.1	0	0.1	0.5	1
$\sqrt{1+x}$	0.0000	0.7071	0.9487	1.0000	1.0488	1.2247	1.4142
$P_4(x; 0)$	0.2734	0.7085	0.9487	1.0000	1.0488	1.2241	1.3984

31 (*a*) Those made of only even powers of x (*b*) Yes
33 (*a*) Odd (*b*) Even powers (*c*) $P_3(x; 0) = x - x^3/3$
(*d*)

x	0	0.1	0.2	0.3	0.5	1	2
$P_3(x; 0)$	0.0000	0.0997	0.1973	0.2910	0.4583	0.6667	−0.6667
$\tan^{-1} x$	0.0000	0.0997	0.1974	0.2915	0.4636	0.7854	1.1071

Sec. 11.2. The Error in Taylor Series

7 0.368 **9** 0.940 **11** (*a*) 4.8 (*b*) 4.8 (*c*) 0.00417 (*d*) 4.796 (*e*) 0.000168 (*f*) Exact error: 0.000168; estimated error: 0.000197 **13** 3.684 **15** (*a*) $b = 1.2$ (*b*) 0.3963 (*c*) 0.40 **19** (*a*) $\pi/4$ (*b*) 0.7854 (*c*) 0.7842 (*d*) 0.7440

21

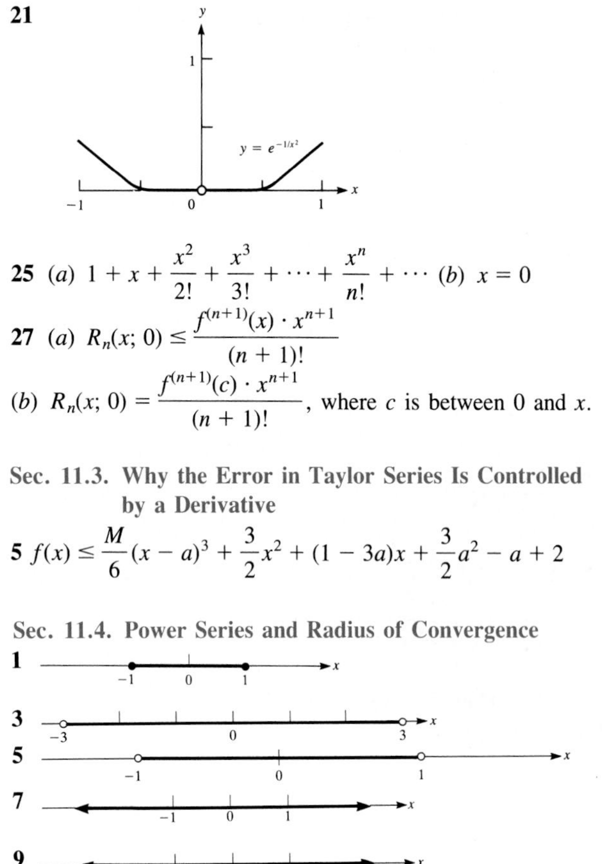

25 (*a*) $1 + x + \frac{x^2}{2!} + \frac{x^3}{3!} + \cdots + \frac{x^n}{n!} + \cdots$ (*b*) $x = 0$

27 (*a*) $R_n(x; 0) \leq \frac{f^{(n+1)}(x) \cdot x^{n+1}}{(n+1)!}$

(*b*) $R_n(x; 0) = \frac{f^{(n+1)}(c) \cdot x^{n+1}}{(n+1)!}$, where c is between 0 and x.

Sec. 11.3. Why the Error in Taylor Series Is Controlled by a Derivative

5 $f(x) \leq \frac{M}{6}(x - a)^3 + \frac{3}{2}x^2 + (1 - 3a)x + \frac{3}{2}a^2 - a + 2$

Sec. 11.4. Power Series and Radius of Convergence

1, **3**, **5**, **7**, **9**

11

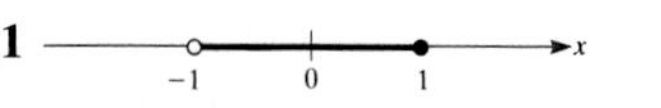

13 (*a*) Absolute convergence (*b*) Absolute convergence (*c*) Insufficient information (*d*) Insufficient information (*e*) Insufficient information (*f*) Diverges (*g*) Diverges
15 Yes

17

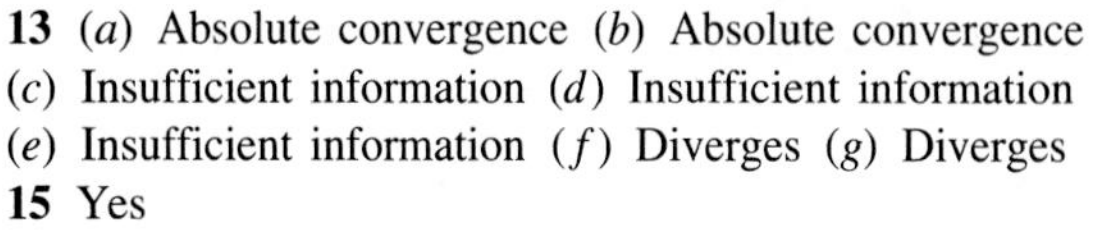

19, **21**, **23**, **25**, **27**

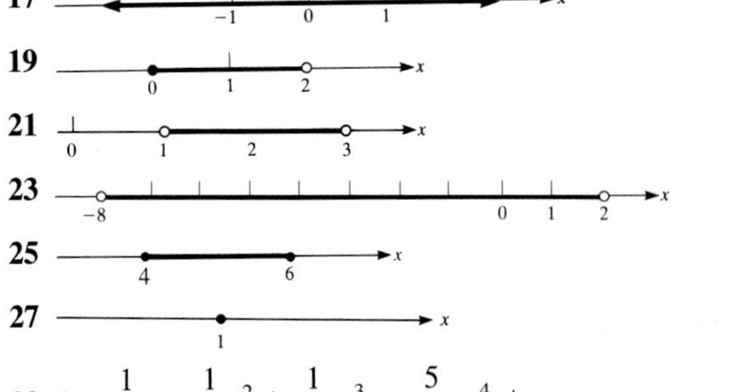

29 $1 + \frac{1}{2}x - \frac{1}{8}x^2 + \frac{1}{16}x^3 - \frac{5}{128}x^4 + \cdots$

31 $1 - 3x + 6x^2 - 10x^3 + 15x^4 - \cdots$ **33** (*a*) $x \geq 3$ and $x < -3$ (*b*) $x < -8$ and $x \geq -2$ **35** $R = \infty$
37 (*a*) 1.1134 (*b*) 1.111 **39** (*a*) $R = \infty$ (*b*) $R = \infty$ (*c*) $R = \infty$ (*d*) $R = 1$ (*e*) $R = 1$

Sec. 11.5. Manipulating Power Series

1 $\cos x = \sum_{n=0}^{\infty} \frac{(-1)^n x^{2n}}{(2n)!}$ **3** (*c*) $a_n(x) = \frac{(-1)^n x^{2n+1}}{2n+1}$ for $n = 0, 1, 2, 3, \ldots$ (*d*) 0.464 **5** $x + \frac{x^3}{3} + \frac{2x^5}{15} + \cdots$

7 $e^x \sin x = x + x^2 + \frac{x^3}{3} + \cdots$ **9** $\frac{1}{2}$ **11** ∞ **13** 1

15 (*a*) $\int_0^1 \sqrt{x} \sin x \, dx = \frac{2}{5} - \frac{1}{27} + \frac{1}{780} - \cdots$ (*b*) 0.3642

17 (*a*) $-1 < x < 1$ (*b*) $f^{(100)}(0) = 100^2 \cdot 100!$

19 (*a*) $\cos 2x = 1 - \frac{2^2 x^2}{2!} + \frac{2^4 x^4}{4!} - \frac{2^6 x^6}{6!} + \cdots + (-1)^n \frac{2^{2n} x^{2n}}{(2n)!} + \cdots$

(*b*) $\frac{\sin^2 x}{x^2} = \frac{2}{2!} - \frac{2^3 x^2}{4!} + \frac{2^5 x^4}{6!} - \frac{2^7 x^6}{8!} + \cdots + (-1)^n \frac{2^{2n+1} x^{2n}}{(2(n+1))!} + \cdots$ (*c*) $\frac{202}{225} \approx 0.898$ (*d*) $\frac{1}{2205} \approx 0.00045$

21 Yes **23** 1.61 **25** (*a*) $-1 < x < 1$

(*b*) $\sum_{n=0}^{\infty} n^2 x^n = \frac{x(1+x)}{(1-x)^3}$ (*c*) Yes

Sec. 11.6. Complex Numbers

1 (*a*) $7 + i$ (*b*) 13 (*c*) $\frac{2}{5} + \frac{1}{5}i$ (*d*) $\frac{10}{17} + \frac{11}{17}i$

3 (*a*)

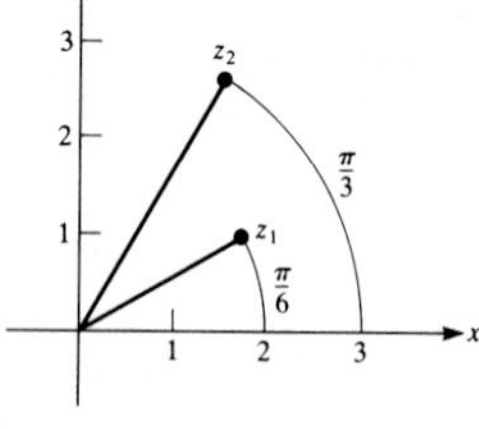

(*b*) $6i$ (*c*) $z_1 = \sqrt{3} + i$, $z_2 = \frac{3}{2} + \frac{3\sqrt{3}}{2}i$ (*d*) $6i$

5

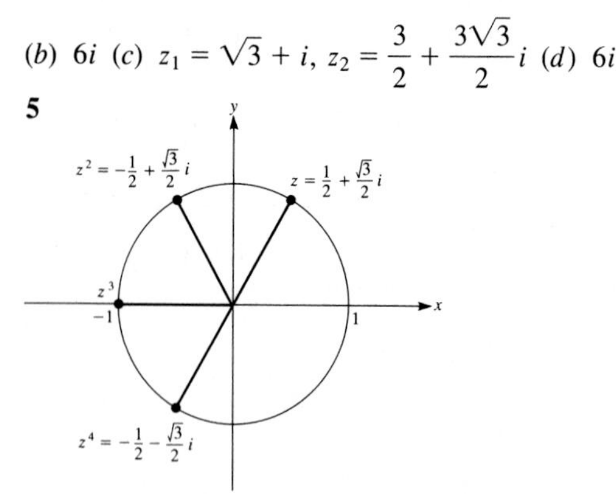

7 (*a*) $|z^2| = 4$, $\arg z^2 = \frac{\pi}{3}$ (*b*) $|z^3| = 8$, $\arg z^3 = \frac{\pi}{2}$

(*c*) $|z^4| = 16$, $\arg z^4 = \frac{2\pi}{3}$ (*d*) $|z^n| = 2^n$, $\arg z^n = \frac{n\pi}{6}$

(*e*)

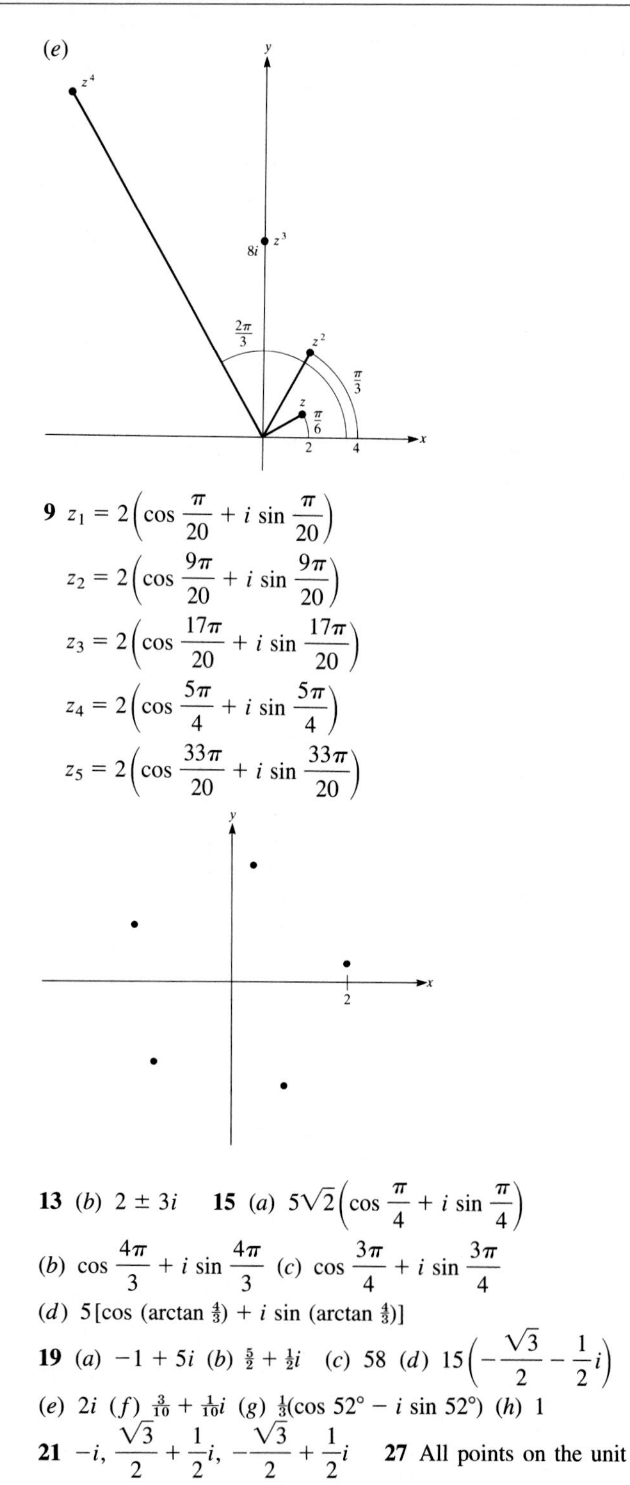

9 $z_1 = 2\left(\cos \frac{\pi}{20} + i \sin \frac{\pi}{20}\right)$

$z_2 = 2\left(\cos \frac{9\pi}{20} + i \sin \frac{9\pi}{20}\right)$

$z_3 = 2\left(\cos \frac{17\pi}{20} + i \sin \frac{17\pi}{20}\right)$

$z_4 = 2\left(\cos \frac{5\pi}{4} + i \sin \frac{5\pi}{4}\right)$

$z_5 = 2\left(\cos \frac{33\pi}{20} + i \sin \frac{33\pi}{20}\right)$

13 (*b*) $2 \pm 3i$ **15** (*a*) $5\sqrt{2}\left(\cos \frac{\pi}{4} + i \sin \frac{\pi}{4}\right)$

(*b*) $\cos \frac{4\pi}{3} + i \sin \frac{4\pi}{3}$ (*c*) $\cos \frac{3\pi}{4} + i \sin \frac{3\pi}{4}$

(*d*) $5[\cos(\arctan \frac{4}{3}) + i \sin(\arctan \frac{4}{3})]$

19 (*a*) $-1 + 5i$ (*b*) $\frac{5}{2} + \frac{1}{2}i$ (*c*) 58 (*d*) $15\left(-\frac{\sqrt{3}}{2} - \frac{1}{2}i\right)$

(*e*) $2i$ (*f*) $\frac{3}{10} + \frac{1}{10}i$ (*g*) $\frac{1}{3}(\cos 52° - i \sin 52°)$ (*h*) 1

21 $-i, \frac{\sqrt{3}}{2} + \frac{1}{2}i, -\frac{\sqrt{3}}{2} + \frac{1}{2}i$ **27** All points on the unit circle

29 (*a*)

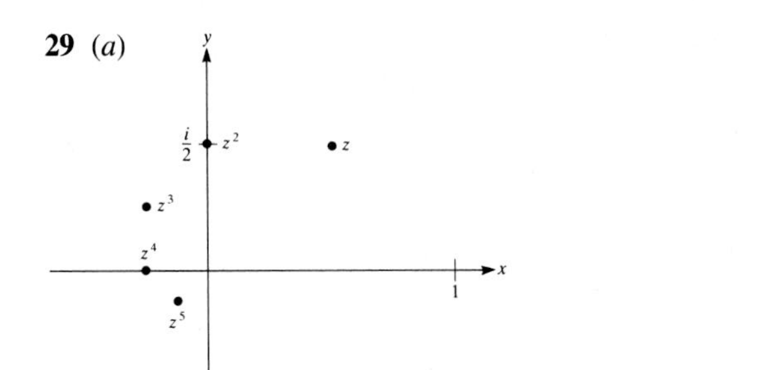

(*b*) $z^n \to 0$ as $n \to \infty$ **31** (*b*) $x = 0$, $y = 2t^2$
(*c*) Nonnegative y axis **33** (*b*) $x = t^2 - 1$, $y = 2t$
(*c*) Parabola, $x = y^2/4 - 1$

31 (*a*)

(*b*), (*c*)

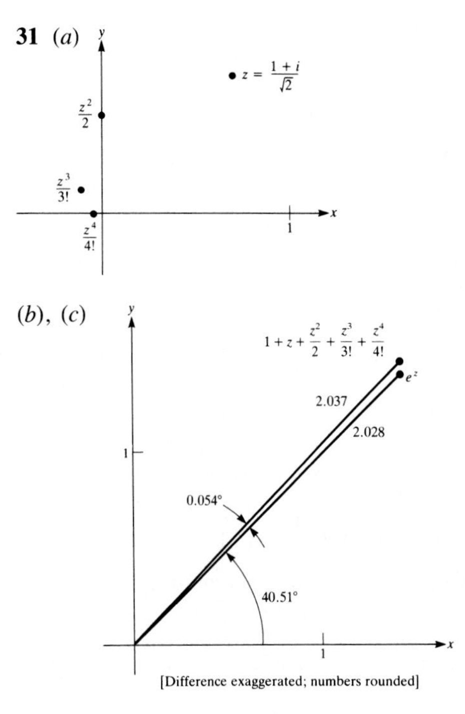

Sec. 11.7. The Relation between the Exponential and the Trigonometric Functions

1 **3** **5** **7** $e^2 \cdot e^{-\pi i/4}$ **9** $15e^{2\pi i/3}$ **11** **13**

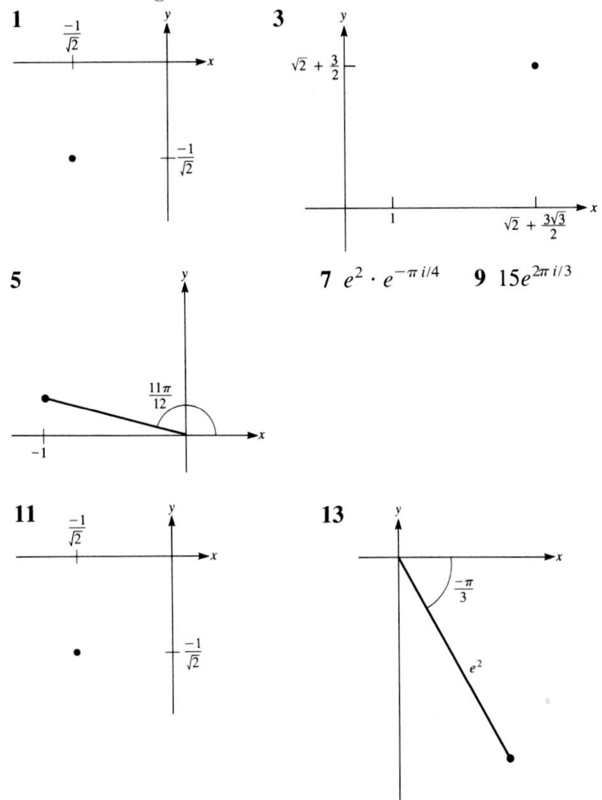

15 (*a*) e^a (*b*) e^{a-bi} (*c*) e^{-a-bi} (*d*) $e^a \cos b$ (*e*) $e^a \sin b$ (*f*) b **17** $z = 2\pi n i$ for $n = 0, \pm 1, \pm 2, \ldots$ **19** (*a*) e^3 **23** $z = \ln 5 + (2\pi n + \arctan \frac{4}{3})i$ for $n = 0, \pm 1, \pm 2, \ldots$

25 $\dfrac{4 - 2\cos\theta}{5 - 4\cos\theta}$

Sec. 11.S. Summary: Guide Quiz

1 Converges for $-2 < x < 2$, diverges for $|x| > 2$.
2 Diverges for $x > 5$ and $x < -1$.
3 $\dfrac{1}{1-x} = 1 + x + x^2 + x^3 + \cdots$ **4** 0.31
6 (*a*) $\dfrac{1}{1+x^2} = 1 - x^2 + x^4 - x^6 + \cdots$ for $|x| < 1$, $R = 1$.
(*b*) $e^{-x} = 1 - x + \dfrac{x^2}{2!} - \dfrac{x^3}{3!} + \cdots$ for all x, $R = \infty$.
(*c*) $\cos x = 1 - \dfrac{x^2}{2!} + \dfrac{x^4}{4!} - \dfrac{x^6}{6!} + \cdots$ for all x, $R = \infty$.
(*d*) $\ln(1 - x) = -x - \dfrac{x^2}{2} - \dfrac{x^3}{3} - \dfrac{x^4}{4} - \cdots$ for $-1 \le x < 1$, $R = 1$. (*e*) $\dfrac{1}{1-2x} = 1 + 2x + 4x^2 + 8x^3 + \cdots$ for $|x| < \frac{1}{2}$, $R = \frac{1}{2}$. (*f*) $1 + 3x + 5x^2 = 1 + 3x + 5x^2$ for all x, $R = \infty$.
(*g*) $\tan^{-1} 2x = 2x - \dfrac{2^3}{3}x^3 + \dfrac{2^5}{5}x^5 - \cdots$ for $|x| \le \frac{1}{2}$, $R = \frac{1}{2}$.
8 (*a*) Converges for $1 \le x < 3$, diverges for $x < 1$ and $x \ge 3$. (*b*) Converges for all x. **9** (*a*) All terms with n odd. (*b*) $1 + \frac{1}{2}x^2 + \frac{5}{24}x^4$

Sec. 11.S. Summary: Review Exercises

1 (*a*) $|R_n| < |a_{n+1}|$ (*b*) $R_n = ar^{n+1} \cdot \dfrac{1}{1-r}$
(*c*) $\int_{n+1}^{\infty} f(x)\,dx < R_n < \int_n^{\infty} f(x)\,dx$ (*d*) $|R_n| \le 27$
(*e*) $R_n = \dfrac{f^{(n+1)}(c)}{(n+1)!}x^{n+1}$, where c is between 0 and x

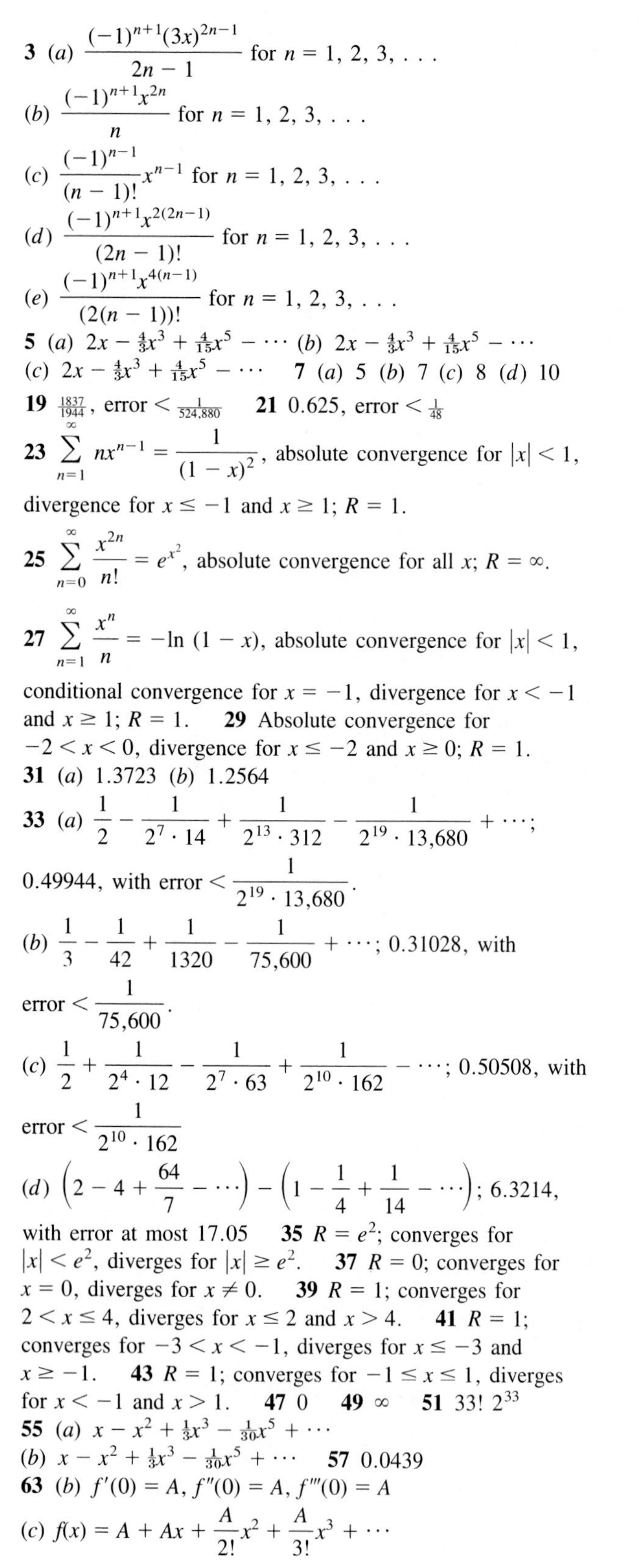

3 (*a*) $\dfrac{(-1)^{n+1}(3x)^{2n-1}}{2n-1}$ for $n = 1, 2, 3, \ldots$

(*b*) $\dfrac{(-1)^{n+1}x^{2n}}{n}$ for $n = 1, 2, 3, \ldots$

(*c*) $\dfrac{(-1)^{n-1}}{(n-1)!}x^{n-1}$ for $n = 1, 2, 3, \ldots$

(*d*) $\dfrac{(-1)^{n+1}x^{2(2n-1)}}{(2n-1)!}$ for $n = 1, 2, 3, \ldots$

(*e*) $\dfrac{(-1)^{n+1}x^{4(n-1)}}{(2(n-1))!}$ for $n = 1, 2, 3, \ldots$

5 (*a*) $2x - \frac{4}{3}x^3 + \frac{4}{15}x^5 - \cdots$ (*b*) $2x - \frac{4}{3}x^3 + \frac{4}{15}x^5 - \cdots$ (*c*) $2x - \frac{4}{3}x^3 + \frac{4}{15}x^5 - \cdots$ **7** (*a*) 5 (*b*) 7 (*c*) 8 (*d*) 10

19 $\frac{1837}{1944}$, error $< \frac{1}{524{,}880}$ **21** 0.625, error $< \frac{1}{48}$

23 $\displaystyle\sum_{n=1}^{\infty} nx^{n-1} = \frac{1}{(1-x)^2}$, absolute convergence for $|x| < 1$, divergence for $x \le -1$ and $x \ge 1$; $R = 1$.

25 $\displaystyle\sum_{n=0}^{\infty} \frac{x^{2n}}{n!} = e^{x^2}$, absolute convergence for all x; $R = \infty$.

27 $\displaystyle\sum_{n=1}^{\infty} \frac{x^n}{n} = -\ln(1-x)$, absolute convergence for $|x| < 1$, conditional convergence for $x = -1$, divergence for $x < -1$ and $x \ge 1$; $R = 1$. **29** Absolute convergence for $-2 < x < 0$, divergence for $x \le -2$ and $x \ge 0$; $R = 1$.

31 (*a*) 1.3723 (*b*) 1.2564

33 (*a*) $\dfrac{1}{2} - \dfrac{1}{2^7 \cdot 14} + \dfrac{1}{2^{13} \cdot 312} - \dfrac{1}{2^{19} \cdot 13{,}680} + \cdots$; 0.49944, with error $< \dfrac{1}{2^{19} \cdot 13{,}680}$.

(*b*) $\dfrac{1}{3} - \dfrac{1}{42} + \dfrac{1}{1320} - \dfrac{1}{75{,}600} + \cdots$; 0.31028, with error $< \dfrac{1}{75{,}600}$.

(*c*) $\dfrac{1}{2} + \dfrac{1}{2^4 \cdot 12} - \dfrac{1}{2^7 \cdot 63} + \dfrac{1}{2^{10} \cdot 162} - \cdots$; 0.50508, with error $< \dfrac{1}{2^{10} \cdot 162}$

(*d*) $\left(2 - 4 + \dfrac{64}{7} - \cdots\right) - \left(1 - \dfrac{1}{4} + \dfrac{1}{14} - \cdots\right)$; 6.3214, with error at most 17.05 **35** $R = e^2$; converges for $|x| < e^2$, diverges for $|x| \ge e^2$. **37** $R = 0$; converges for $x = 0$, diverges for $x \ne 0$. **39** $R = 1$; converges for $2 < x \le 4$, diverges for $x \le 2$ and $x > 4$. **41** $R = 1$; converges for $-3 < x < -1$, diverges for $x \le -3$ and $x \ge -1$. **43** $R = 1$; converges for $-1 \le x \le 1$, diverges for $x < -1$ and $x > 1$. **47** 0 **49** ∞ **51** $33!\,2^{33}$

55 (*a*) $x - x^2 + \frac{1}{3}x^3 - \frac{1}{30}x^5 + \cdots$

(*b*) $x - x^2 + \frac{1}{3}x^3 - \frac{1}{30}x^5 + \cdots$ **57** 0.0439

63 (*b*) $f'(0) = A$, $f''(0) = A$, $f'''(0) = A$

(*c*) $f(x) = A + Ax + \dfrac{A}{2!}x^2 + \dfrac{A}{3!}x^3 + \cdots$

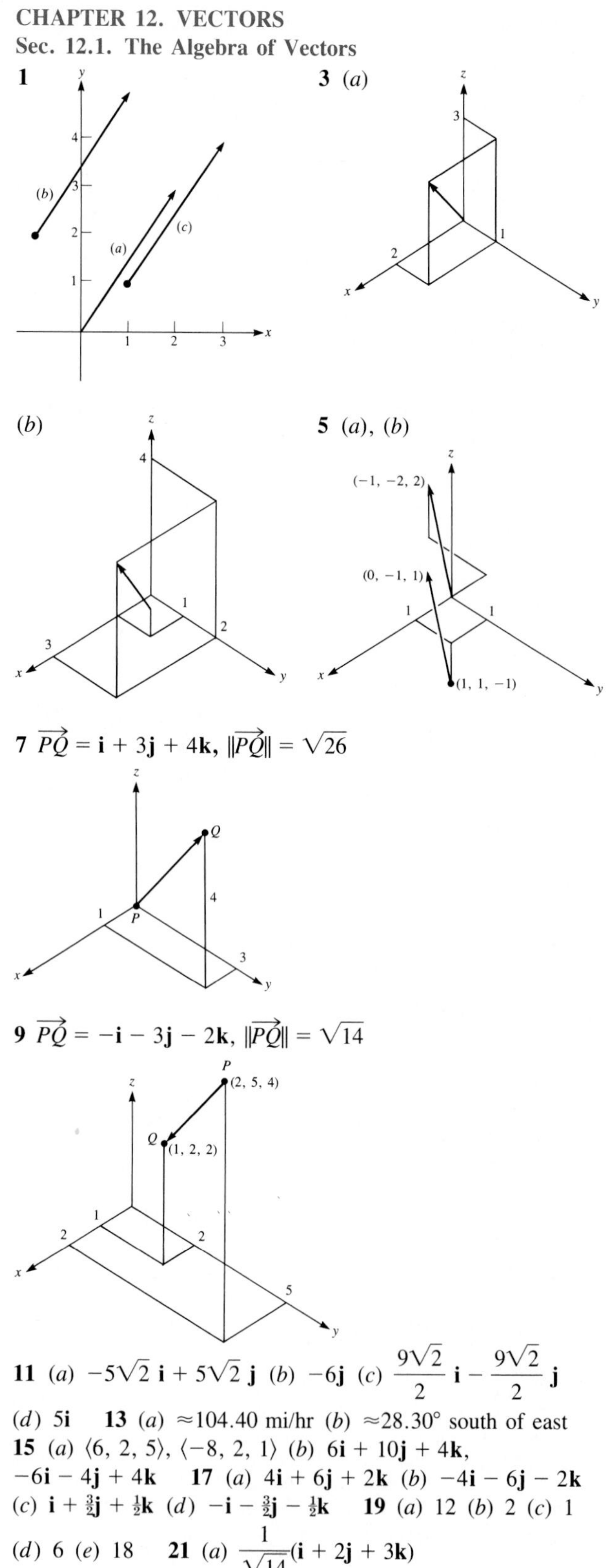

CHAPTER 12. VECTORS

Sec. 12.1. The Algebra of Vectors

1 **3** (*a*)

(*b*) **5** (*a*), (*b*)

7 $\overrightarrow{PQ} = \mathbf{i} + 3\mathbf{j} + 4\mathbf{k}$, $\|\overrightarrow{PQ}\| = \sqrt{26}$

9 $\overrightarrow{PQ} = -\mathbf{i} - 3\mathbf{j} - 2\mathbf{k}$, $\|\overrightarrow{PQ}\| = \sqrt{14}$

11 (*a*) $-5\sqrt{2}\,\mathbf{i} + 5\sqrt{2}\,\mathbf{j}$ (*b*) $-6\mathbf{j}$ (*c*) $\dfrac{9\sqrt{2}}{2}\mathbf{i} - \dfrac{9\sqrt{2}}{2}\mathbf{j}$ (*d*) $5\mathbf{i}$ **13** (*a*) $\approx$104.40 mi/hr (*b*) $\approx$28.30° south of east **15** (*a*) $\langle 6, 2, 5\rangle$, $\langle -8, 2, 1\rangle$ (*b*) $6\mathbf{i} + 10\mathbf{j} + 4\mathbf{k}$, $-6\mathbf{i} - 4\mathbf{j} + 4\mathbf{k}$ **17** (*a*) $4\mathbf{i} + 6\mathbf{j} + 2\mathbf{k}$ (*b*) $-4\mathbf{i} - 6\mathbf{j} - 2\mathbf{k}$ (*c*) $\mathbf{i} + \frac{3}{2}\mathbf{j} + \frac{1}{2}\mathbf{k}$ (*d*) $-\mathbf{i} - \frac{3}{2}\mathbf{j} - \frac{1}{2}\mathbf{k}$ **19** (*a*) 12 (*b*) 2 (*c*) 1 (*d*) 6 (*e*) 18 **21** (*a*) $\dfrac{1}{\sqrt{14}}(\mathbf{i} + 2\mathbf{j} + 3\mathbf{k})$

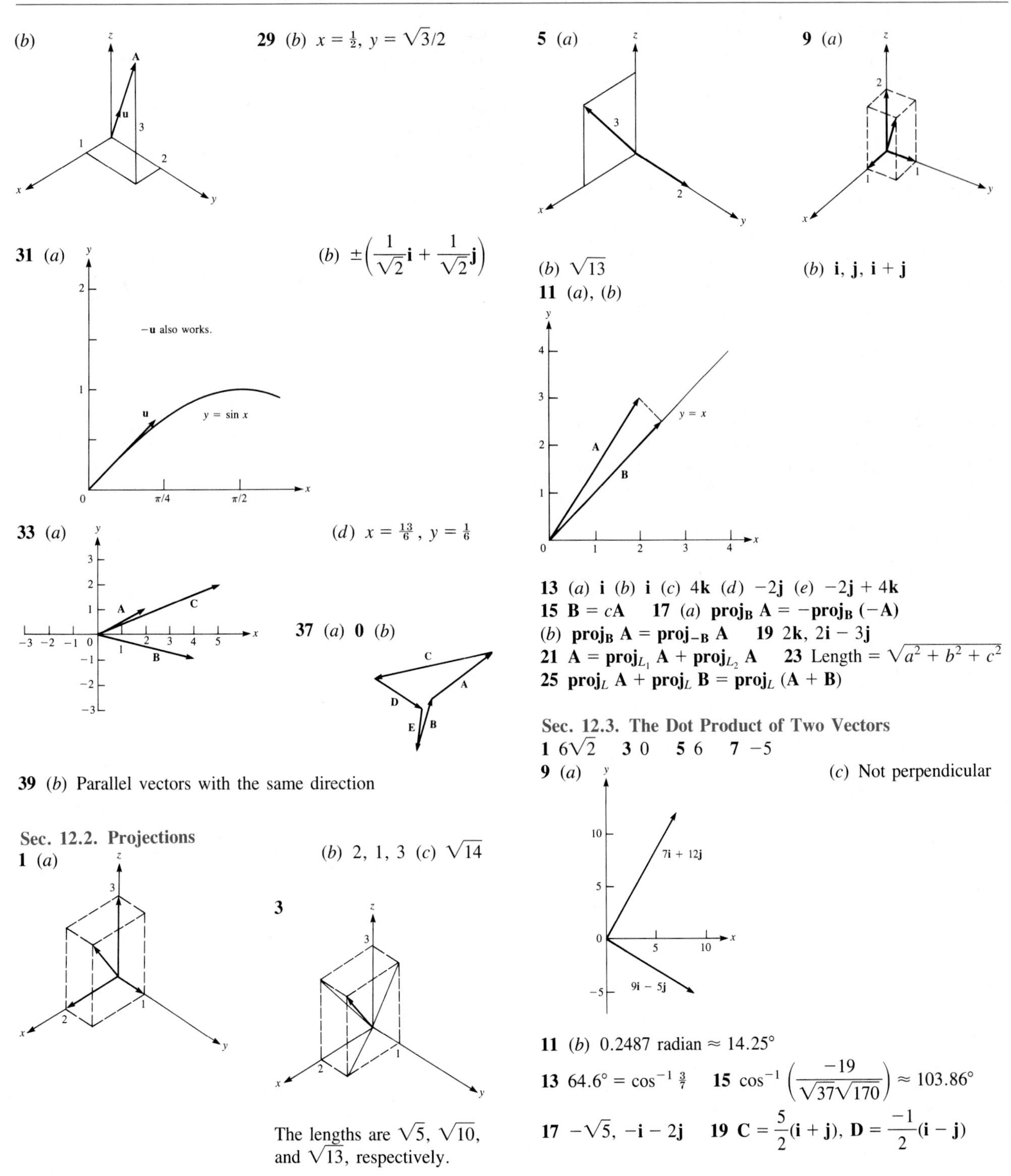

(*b*)

29 (*b*) $x = \frac{1}{2}$, $y = \sqrt{3}/2$

31 (*a*)

(*b*) $\pm\left(\frac{1}{\sqrt{2}}\mathbf{i} + \frac{1}{\sqrt{2}}\mathbf{j}\right)$

33 (*a*)

(*d*) $x = \frac{13}{6}$, $y = \frac{1}{6}$

37 (*a*) **0** (*b*)

39 (*b*) Parallel vectors with the same direction

Sec. 12.2. Projections

1 (*a*)

(*b*) 2, 1, 3 (*c*) $\sqrt{14}$

3

The lengths are $\sqrt{5}$, $\sqrt{10}$, and $\sqrt{13}$, respectively.

5 (*a*)

(*b*) $\sqrt{13}$

9 (*a*)

(*b*) **i**, **j**, **i** + **j**

11 (*a*), (*b*)

13 (*a*) **i** (*b*) **i** (*c*) 4**k** (*d*) −2**j** (*e*) −2**j** + 4**k**
15 **B** = *c***A** **17** (*a*) $\mathbf{proj_B\, A} = -\mathbf{proj_B}\,(-\mathbf{A})$
(*b*) $\mathbf{proj_B\, A} = \mathbf{proj_{-B}\, A}$ **19** 2**k**, 2**i** − 3**j**
21 $\mathbf{A} = \mathbf{proj}_{L_1}\,\mathbf{A} + \mathbf{proj}_{L_2}\,\mathbf{A}$ **23** Length = $\sqrt{a^2 + b^2 + c^2}$
25 $\mathbf{proj}_L\,\mathbf{A} + \mathbf{proj}_L\,\mathbf{B} = \mathbf{proj}_L\,(\mathbf{A} + \mathbf{B})$

Sec. 12.3. The Dot Product of Two Vectors

1 $6\sqrt{2}$ **3** 0 **5** 6 **7** −5

9 (*a*)

(*c*) Not perpendicular

11 (*b*) 0.2487 radian ≈ 14.25°

13 $64.6° = \cos^{-1}\frac{3}{7}$ **15** $\cos^{-1}\left(\frac{-19}{\sqrt{37}\sqrt{170}}\right) \approx 103.86°$

17 $-\sqrt{5}$, −**i** − 2**j** **19** $\mathbf{C} = \frac{5}{2}(\mathbf{i} + \mathbf{j})$, $\mathbf{D} = \frac{-1}{2}(\mathbf{i} - \mathbf{j})$

21 (*a*) $\mathbf{C} = \mathbf{i} + 2\mathbf{j}$, $\mathbf{D} = -4\mathbf{i} + 2\mathbf{j}$
(*b*)

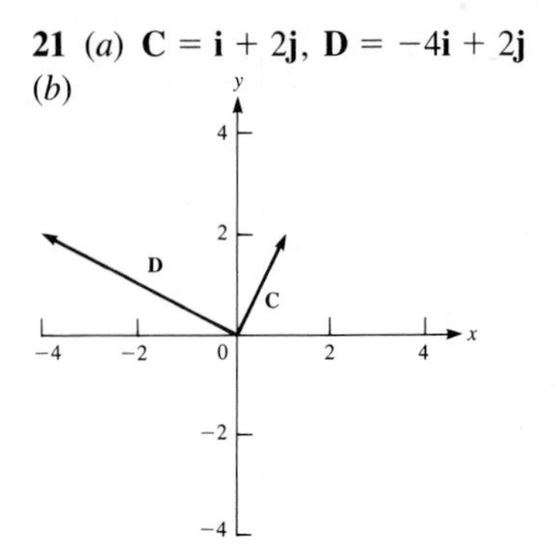

23 (*a*) 15 (*b*) −15 **25** $\mathbf{A} \cdot \mathbf{B} = 0$ or $\mathbf{A} = c\mathbf{B}$
27 (*a*) $1/\sqrt{13}$ (*b*) $-1/\sqrt{13}$
29 $\frac{24}{25}\mathbf{i} - \frac{18}{25}\mathbf{j}$, $\frac{51}{25}\mathbf{i} + \frac{68}{25}\mathbf{j}$ **31** $6\sqrt{26}/13$
33 $2\mathbf{i} + 3\mathbf{j}$ **35** −1/3 **37** $3/\sqrt{15}$ **39** 1/3
41 No **45** (*b*) $\mathbf{A} \cdot \mathbf{u}_1$ **47** (*a*) Income from chair sales (*b*) Income from sales of chairs and desks (*c*) Cost of production of chairs and desks (*d*) The firm makes a profit from the sales of chairs and desks. **49** The total profit is zero.

Sec. 12.4. Lines and Planes

1 $4(x-2) + 5(y-3) = 0$ **3** $2(x-4) + 3(y-5) = 0$
5 $2\mathbf{i} - 3\mathbf{j}$ **7** $3\mathbf{i} - \mathbf{j}$ **9** 2 **11** $2\mathbf{i} - 3\mathbf{j} + 4\mathbf{k}$,
$\frac{1}{\sqrt{29}}(2\mathbf{i} - 3\mathbf{j} + 4\mathbf{k})$ **13** $2/\sqrt{29}$ **15** $8/\sqrt{57}$
17 $2/\sqrt{29}$, $3/\sqrt{29}$, $4/\sqrt{29}$ **19** (*a*) Yes (*b*) No
21 (*a*) $x = 2t + \frac{1}{2}$, $y = -5t + \frac{1}{3}$, $z = 8t + \frac{1}{2}$
(*b*) $\mathbf{P} = \frac{1}{2}\mathbf{i} + \frac{1}{3}\mathbf{j} + \frac{1}{2}\mathbf{k} + t(2\mathbf{i} - 5\mathbf{j} + 8\mathbf{k})$
23 $\frac{x-1}{1} = \frac{y-0}{1} = \frac{z-3}{-2}$
25

z
γ_1
$\gamma_1 \approx 22.6°$
$\gamma_2 \approx 157.4°$
γ_2
y
x

27 $\theta \approx 1.898$ **29** $(\frac{11}{5}, \frac{19}{5})$ **31** No **33** $x = 1 + 2t$, $y = 3 - 3t$, $z = -5 + 4t$ **35** (*a*) $2a^2/\sqrt{17}$ (*b*) $2a^2/\sqrt{17}$ (*c*) $3a^2/\sqrt{17}$ **37** 2 **39** $\pi a^2/(2\sqrt{39})$
41 No **43** $\sqrt{3657}/69$ **45** $\cos^{-1} 1/\sqrt{3}$

Sec. 12.5. Determinants

1 −22 **3** 22 **5** 0 **7** 0 **9** −6 **11** 11 **13** 34
15 (*b*) $7x - y = 11$ **23** Yes

Sec. 12.6. The Cross Product of Two Vectors

1 $-\mathbf{i}$ **3** $-\mathbf{i} + \mathbf{j}$ **5** $-7\mathbf{i} - 3\mathbf{j} + 5\mathbf{k}$ **7** $\sqrt{507}$ **9** 2
11 5 **13** 18 **15** $-8\mathbf{i} - 2\mathbf{k}$ **17** $2\mathbf{i} - 2\mathbf{j} + 4\mathbf{k}$ **19** 0
25 (*a*) $\mathbf{0}$ (*b*) $-\mathbf{j}$ (*c*) $-\mathbf{k}$ **27** $\mathbf{A}$ and $\mathbf{B}$ are parallel.
29 (*a*) Yes (*b*) No **35** Four vectors are produced in the plane perpendicular to $\mathbf{u}$, each of length $\|\mathbf{u} \times \mathbf{B}\|$ and spaced 90° from each other. **37** (*b*) Yes (*c*) No, since we do not know that the cross product is distributive.

Sec. 12.7. More on Lines and Planes

1 One answer is $2\mathbf{i} + 8\mathbf{j} + 18\mathbf{k}$. **3** $\langle \frac{8}{7}, \frac{2}{7}, -\frac{4}{7} \rangle$
5 $x = 1 + 3t$, $y = 1 - t$, $z = 2 + t$ **7** $(\frac{143}{79}, -\frac{15}{79}, \frac{207}{79})$
9 (*b*) No **11** (*b*) Yes **13** 70.5° **15** (*b*) No
17 (*b*) $7x + y - 3z = 5$
19 (*c*) $\frac{1}{\sqrt{21}}(-4\mathbf{i} + 2\mathbf{j} + \mathbf{k})$ **21** (*b*) $(-1, 3, 1)$
23 $4/\sqrt{3}$ **25** $4\pi a^2/\sqrt{174}$ **27** (0, 0.645470, 0.763785)
29 93.405°

Sec. 12.S. Summary: Guide Quiz

1 (*a*) −3 (*b*) $\sqrt{6}$ (*c*) $\frac{1}{\sqrt{6}}(\mathbf{i} + 2\mathbf{j} - \mathbf{k})$ (*d*) $-3/\sqrt{6}$
(*e*) $-\frac{1}{2}\mathbf{i} - \mathbf{j} + \frac{1}{2}\mathbf{k}$ (*f*) $-3/\sqrt{14}$ (*g*) $\frac{5}{2}\mathbf{i} + \frac{5}{2}\mathbf{k}$ (*h*) $-3/(2\sqrt{21})$
(*i*) 109.1° (*j*) $5\mathbf{i} - 5\mathbf{j} - 5\mathbf{k}$ (*k*) $-5\mathbf{i} + 5\mathbf{j} + 5\mathbf{k}$
(*l*) $\frac{1}{\sqrt{3}}(\mathbf{i} - \mathbf{j} - \mathbf{k})$ (*m*) $5\sqrt{3}$ **3** $\frac{1}{3}, -\frac{2}{3}, \frac{2}{3}$ **5** (*a*) 8
6 (*b*) $\mathbf{A}_2 = \frac{\mathbf{A} \cdot (\overrightarrow{PQ} \times \overrightarrow{PR})}{\|\overrightarrow{PQ} \times \overrightarrow{PR}\|^2}(\overrightarrow{PQ} \times \overrightarrow{PR})$, $\mathbf{A}_1 = \mathbf{A} - \mathbf{A}_2$
7 $(\frac{11}{3}, \frac{2}{3}, 1)$ **11** $(1, \frac{4}{3}, \frac{1}{3})$ **12** (*a*) $4\mathbf{i} + 5\mathbf{j} - 3\mathbf{k}$
(*b*) (1, 1, 1) (*c*) $x = 1 + 4t$, $y = 1 + 5t$, $z = 1 - 3t$
13 (*a*) $17\mathbf{i} + 2\mathbf{j} - 7\mathbf{k}$ (*b*) (0, 0, 2) (*c*) $x = 17t$, $y = 2t$, $z = 2 - 7t$, (*d*) $\frac{x}{17} = \frac{y}{2} = \frac{z-2}{-7}$ **14** (*a*) $\mathbf{i} + 2\mathbf{j} + \mathbf{k}$, $3\mathbf{j} + 4\mathbf{k}$ (*b*) $5\mathbf{i} - 4\mathbf{j} + 3\mathbf{k}$ (*c*) $5x - 4y + 3z = 4$
15 (*a*) $11\mathbf{i} + 2\mathbf{j} - 7\mathbf{k}$ (*b*) 81.28° (*c*) $11x + 2y - 7z = -1$
18 (*c*) Parallelogram (*d*) $8\sqrt{2}$ (*e*) $8/\sqrt{7}$
19 (*a*) $\sqrt{2}\mathbf{i} + \sqrt{2}\mathbf{j}$ (*b*) $\sqrt{2}\mathbf{i}$, $\sqrt{2}\mathbf{j}$, $2\sqrt{3}\mathbf{k}$ (*c*) $\sqrt{2}/4$, $\sqrt{2}/4$, $\sqrt{3}/2$ (*d*) 69.30°, 69.30°, 30° (*e*) $\sqrt{2}\mathbf{j} + 2\sqrt{3}\mathbf{k}$

Sec. 12.S. Summary: Review Exercises

3 (*a*) $-2\mathbf{i}$ (*b*) $3\mathbf{j}$ (*c*) $0.72\mathbf{i} + 0.96\mathbf{j}$ (*d*) $-\frac{23}{41}(4\mathbf{i} + 5\mathbf{j})$
11 $-\frac{3}{2}, \frac{3}{2}, 3$ **15** (*a*) $\sqrt{\frac{2}{3}}$ (*b*) $1/\sqrt{2}$ (*c*) $\frac{1}{3}$
17 $\frac{|A(x_2 - x_1) + B(y_2 - y_1) + C(z_2 - z_1)|}{\sqrt{A^2 + B^2 + C^2}}$
19 $x = 1 + 9t$, $y = 1 + 3t$, $z = 2 - 5t$ **21** Yes
23 (*a*) $(\frac{72}{61}, \frac{48}{61}, \frac{36}{61})$ (*b*) $(-\frac{5}{61}, \frac{78}{61}, \frac{150}{61})$
27 $\overrightarrow{P_1P_2} \cdot (\overrightarrow{P_3P_4} \times \overrightarrow{P_3P_5}) = 0$
29 The line makes an angle of approximately 42.83° (0.7476 radian) with the normal to the plane.
35 $(\overrightarrow{PQ} \times \overrightarrow{PR}) \cdot (\overrightarrow{ST} \times \overrightarrow{SU}) = 0$

39 (*b*) $\mathbf{C}_2 = \dfrac{[\mathbf{C} \cdot (\mathbf{A} \times \mathbf{B})](\mathbf{A} \times \mathbf{B})}{\|\mathbf{A} \times \mathbf{B}\|^2}$

41 $\left(\dfrac{21 \pm \sqrt{15}}{9}, \dfrac{3 \pm \sqrt{15}}{9}, \dfrac{6 \pm \sqrt{15}}{9}\right)$

43 (*a*) $(-\frac{15}{17}, \frac{2}{17}, \frac{3}{17})$ (*b*) $(-\frac{111}{109}, \frac{250}{109}, \frac{186}{109})$ **47** (*b*) It is one-fourth. **49** (*a*) Yes

CHAPTER 13. THE DERIVATIVE OF A VECTOR FUNCTION

Sec. 13.1. The Derivative of a Vector Function

1 (*a*) $\mathbf{G}(1) = \mathbf{i} + \mathbf{j}$, $\mathbf{G}(2) = 2\mathbf{i} + 4\mathbf{j}$, $\mathbf{G}(3) = 3\mathbf{i} + 9\mathbf{j}$

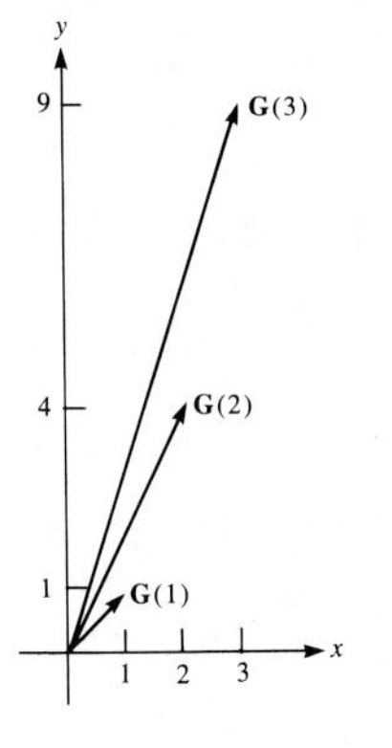

3 (*a*) $\mathbf{G}(1.1) = 2.2\mathbf{i} + 1.21\mathbf{j}$, $\mathbf{G}(1) = 2\mathbf{i} + \mathbf{j}$, $\Delta\mathbf{G} = 0.2\mathbf{i} + 0.21\mathbf{j}$

(*b*) $\Delta\mathbf{G}/0.1 = 2\mathbf{i} + 2.1\mathbf{j}$

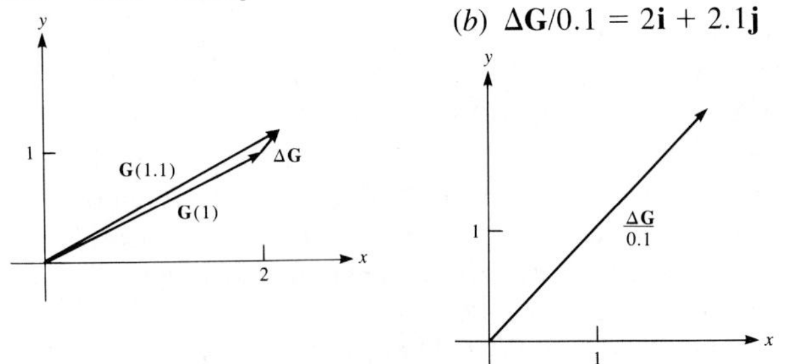

(*c*) $\mathbf{G}'(1) = 2\mathbf{i} + 2\mathbf{j}$

5 (*a*) $\mathbf{r}(1) = 32\mathbf{i} - 16\mathbf{j}$, $\mathbf{r}(2) = 64\mathbf{i} - 64\mathbf{j}$

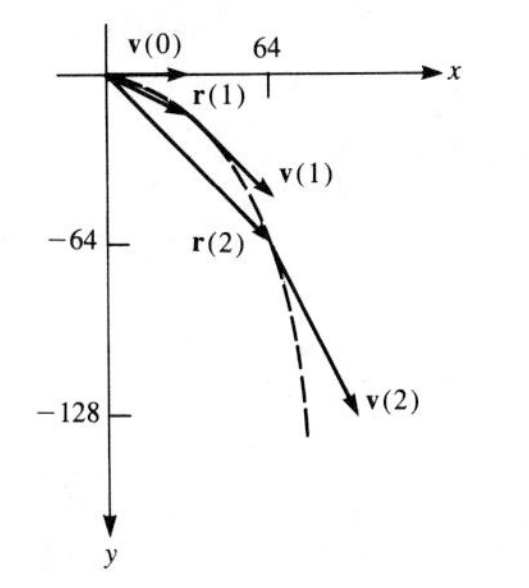

(*b*) See the dashed parabola in (*a*). (*c*) $\mathbf{v}(0) = 32\mathbf{i}$, $\mathbf{v}(1) = 32\mathbf{i} - 32\mathbf{j}$, $\mathbf{v}(2) = 32\mathbf{i} - 64\mathbf{j}$

7 (*a*) $\sqrt{26}/100$ (*b*) -5 (*c*) $0.05\mathbf{i} - 0.25\mathbf{j}$ (*d*) $\sqrt{26}/20$

9 $\mathbf{v}(t) = -3 \sin 3t\ \mathbf{i} + 3 \cos 3t\ \mathbf{j} + 6\mathbf{k}$, Speed $= 3\sqrt{5}$

11 $\mathbf{v}(t) = \dfrac{2t}{1+t^2}\mathbf{i} + 3e^{3t}\mathbf{j} + \dfrac{(1+2t)\sec^2 t - 2\tan t}{(1+2t)^2}\mathbf{k}$

$$\text{Speed} = \sqrt{\frac{4t^2}{(1+t^2)^2} + 9e^{6t} + \frac{[(1+2t)\sec^2 t - 2\tan t]^2}{(1+2t)^4}}$$

13 (*a*) $\Delta\mathbf{G} = 0.21\mathbf{i} + 0.331\mathbf{j}$ (*b*) $\Delta\mathbf{G}/\Delta t = 2.1\mathbf{i} + 3.31\mathbf{j}$ (*c*) $\mathbf{G}'(1) = 2\mathbf{i} + 3\mathbf{j}$

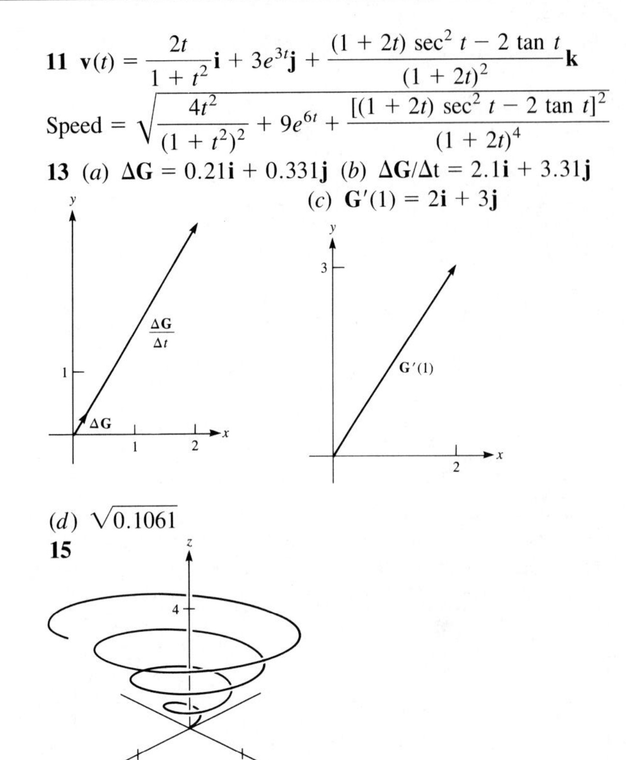

(*d*) $\sqrt{0.1061}$

15

17 (*a*)

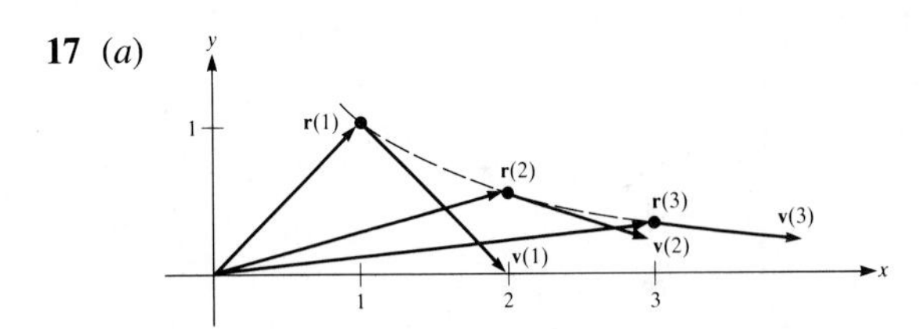

(*b*) $\mathbf{r}(1) = \mathbf{i} + \mathbf{j}$, $\mathbf{r}(2) = 2\mathbf{i} + (\frac{1}{2})\mathbf{j}$, $\mathbf{r}(3) = 3\mathbf{i} + (\frac{1}{3})\mathbf{j}$
(*c*) $\mathbf{v}(1) = \mathbf{i} - \mathbf{j}$, $\mathbf{v}(2) = \mathbf{i} - (\frac{1}{4})\mathbf{j}$, $\mathbf{v}(3) = \mathbf{i} - (\frac{1}{9})\mathbf{j}$
(*d*) $dx/dt \to 1$, $dy/dt \to 0$, $\mathbf{v}(t) \to \mathbf{i}$, $\|\mathbf{v}(t)\| \to 1$

19 (*a*) $\mathbf{r}(t) = 100 \cos (400\pi t)\ \mathbf{i} - 100 \sin (400\pi t)\ \mathbf{j}$, $\mathbf{v}(t) = -40{,}000\pi[\sin (400\pi t)\ \mathbf{i} + \cos (400\pi t)\ \mathbf{j}]$
(*b*) $\mathbf{r}(0) = 100\mathbf{i}$, $\mathbf{r}(\frac{1}{800}) = -100\mathbf{j}$, $\mathbf{v}(0) = -40{,}000\pi\mathbf{j}$, $\mathbf{v}(\frac{1}{800}) = -40{,}000\pi\mathbf{i}$

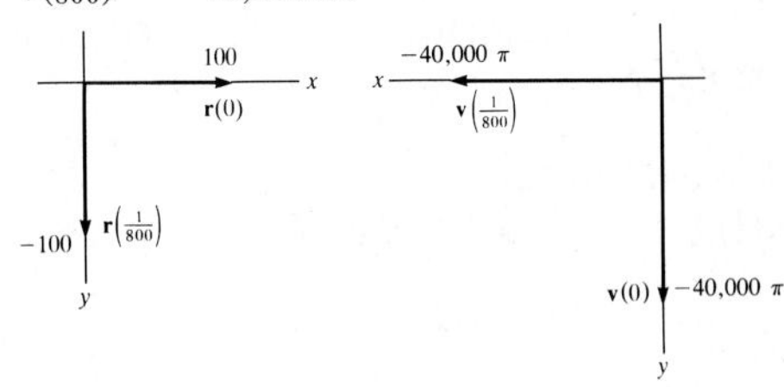

(*c*) They are constant.

21 (*b*) **23** (*a*) $\mathbf{r}(0) = a\mathbf{i}$, $\mathbf{r}(\frac{1}{4}) = a\mathbf{j}$

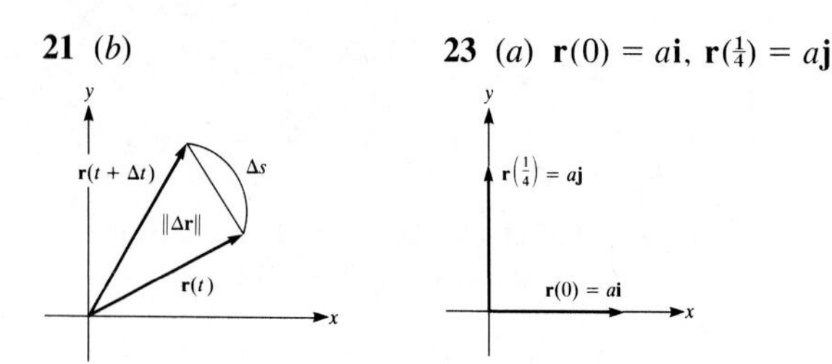

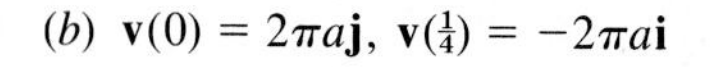

(*b*) $\mathbf{v}(0) = 2\pi a\mathbf{j}$, $\mathbf{v}(\frac{1}{4}) = -2\pi a\mathbf{i}$

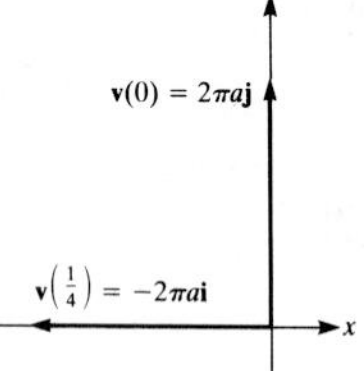

25 (*c*) $\theta = \pi/4$

27 $(\ln 2 + \frac{1}{2})\mathbf{i} + (\cos 6 - 3 \sin 6)\mathbf{j}$

31 $\left[\frac{1}{2}\ln(t^2 + t + 1) - \frac{1}{\sqrt{3}}\tan^{-1}\left(\frac{2t+1}{\sqrt{3}}\right) + 1 + \frac{\pi}{6\sqrt{3}}\right]\mathbf{i} + [t \tan^{-1} 3t - \frac{1}{6}\ln(1 + 9t^2) + 1]\,\mathbf{j}$

33 $\left[\frac{1}{13}e^{2t}(2 \sin 3t - 3 \cos 3t) + \frac{16}{13}\right]\mathbf{i} + [\frac{1}{9}t^3 - \frac{1}{9}t^2 + \frac{4}{27}t - \frac{8}{81}\ln|3t + 2| + \frac{8}{81}\ln 2 + 3]\,\mathbf{j}$

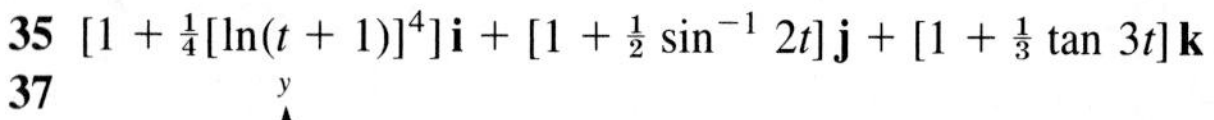

35 $[1 + \frac{1}{4}[\ln(t + 1)]^4]\mathbf{i} + [1 + \frac{1}{2}\sin^{-1} 2t]\,\mathbf{j} + [1 + \frac{1}{3}\tan 3t]\,\mathbf{k}$

37

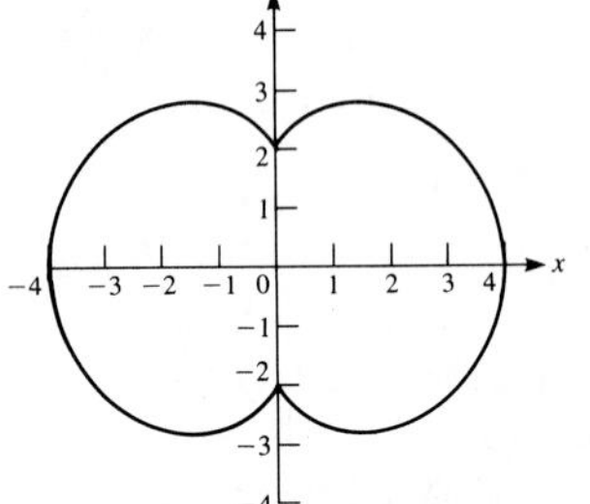

Sec. 13.2. Properties of the Derivative of a Vector Function

1 $9t^2 + 20t^4$ **3** $e^{2t}(2t + 1) + (3t^2 \cos 3t + 2t \sin 3t)$
5 $4\mathbf{i} - \mathbf{j}$ **9** (*b*) No. The cross product is not commutative. **13** (*b*) No

Sec. 13.3. The Acceleration Vector

1 $\mathbf{r}(1) = \mathbf{i} + \mathbf{j}$, $\mathbf{v}(1) = 2\mathbf{i} + 3\mathbf{j}$, $\mathbf{a}(1) = 2\mathbf{i} + 6\mathbf{j}$

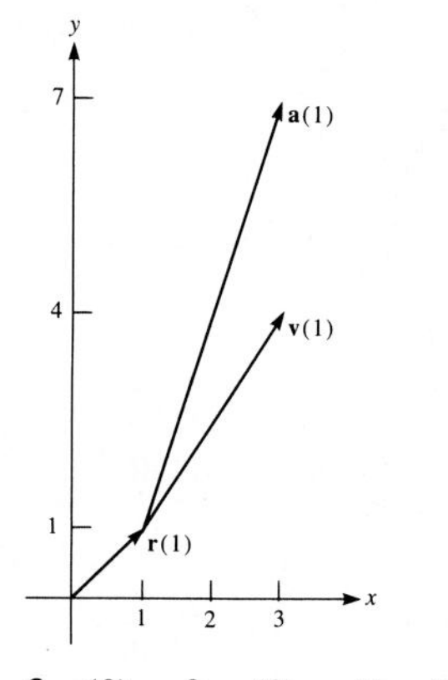

3 $\mathbf{r}(0) = \mathbf{0}$, $\mathbf{v}(0) = 4\mathbf{i}$, $\mathbf{a}(0) = -32\mathbf{j}$, $\mathbf{r}(1) = 4\mathbf{i} - 16\mathbf{j}$, $\mathbf{v}(1) = 4\mathbf{i} - 32\mathbf{j}$, $\mathbf{a}(1) = -32\mathbf{j}$, $\mathbf{r}(2) = 8\mathbf{i} - 64\mathbf{j}$, $\mathbf{v}(2) = 4\mathbf{i} - 64\mathbf{j}$, $\mathbf{a}(2) = -32\mathbf{j}$ **5** (*a*) A hyperbola

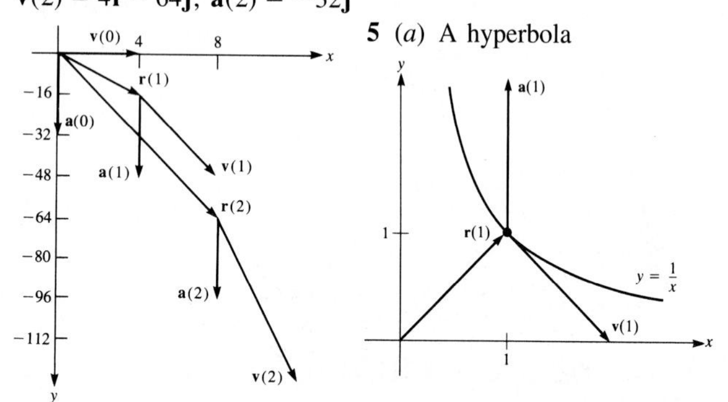

(*b*) $\mathbf{r}(1) = \mathbf{i} + \mathbf{j}$, $\mathbf{v}(1) = \mathbf{i} - \mathbf{j}$, $\mathbf{a}(1) = 2\mathbf{j}$ (*c*) $\mathbf{v}(t) \to \mathbf{i}$, $\mathbf{a}(t) \to \mathbf{0}$ (*d*) $ds/dt \to 1$
7 (*a*) $\mathbf{v}(t) = -15 \sin 3t\ \mathbf{i} + 15 \cos 3t\ \mathbf{j} + 6\mathbf{k}$, $\mathbf{a}(t) = -45 \cos 3t\ \mathbf{i} - 45 \sin 3t\ \mathbf{j}$ (*b*) The velocity in the z direction is constant. **9** (*a*) No, it could reverse direction. (*b*) No **15** Four times as great **17** (*b*) 4.4 mi/sec
19 22,300 miles **23** (*d*) A line through the origin.
27 (*a*) $\mathbf{v}(2) \approx 2\mathbf{i} - \mathbf{j} + 2\mathbf{k}$, $\mathbf{v}(2.01) \approx 2.03\mathbf{i} - 2.02\mathbf{j} + 2.04\mathbf{k}$, $\mathbf{a}(2) \approx 3\mathbf{i} - 102\mathbf{j} + 4\mathbf{k}$
29 (*a*) $\mathbf{v}(1.05) \approx 0.15\mathbf{i} + 3.20\mathbf{j}$, $\mathbf{v}(1.10) \approx 0.35\mathbf{i} + 3.45\mathbf{j}$, $\mathbf{v}(1.20) \approx 0.85\mathbf{j} + 4.15\mathbf{j}$ (*b*) $\mathbf{r}(1.05) \approx \mathbf{i} + 1.15\mathbf{j}$, $\mathbf{r}(1.10) \approx 1.0075\mathbf{i} + 1.31\mathbf{j}$, $\mathbf{r}(1.20) \approx 1.0425\mathbf{i} + 1.6550\mathbf{j}$

Sec. 13.4. The Components of Acceleration

1 $\mathbf{T}(1) = (\mathbf{i} - 2\mathbf{j})/\sqrt{5}$, $\mathbf{N}(1) = (-2\mathbf{i} - \mathbf{j})/\sqrt{5}$

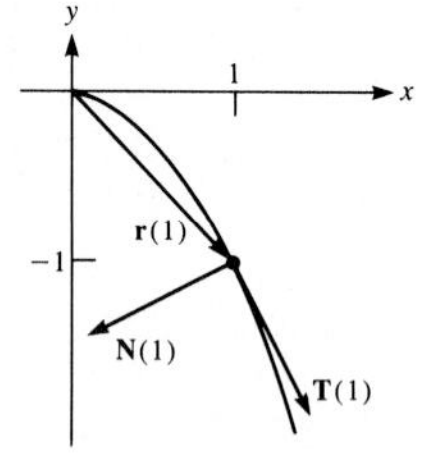

3 $\mathbf{T}(1) = -\sin 1\ \mathbf{i} + \cos 1\ \mathbf{j}$, $\mathbf{N}(1) = -\cos 1\ \mathbf{i} - \sin 1\ \mathbf{j}$

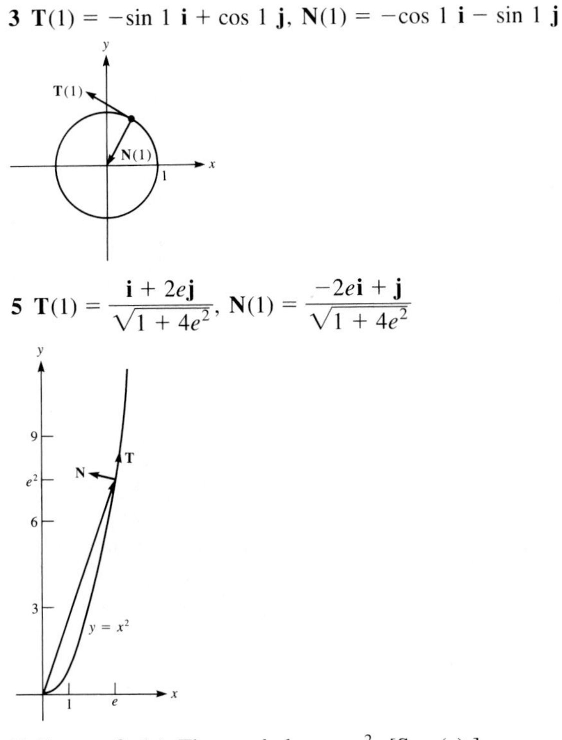

5 $\mathbf{T}(1) = \dfrac{\mathbf{i} + 2e\mathbf{j}}{\sqrt{1 + 4e^2}}$, $\mathbf{N}(1) = \dfrac{-2e\mathbf{i} + \mathbf{j}}{\sqrt{1 + 4e^2}}$

7 True **9** (*a*) The parabola $y = x^2$. [See (*c*).]
(*b*) $\mathbf{r}(1) = \mathbf{i} + \mathbf{j}$, $\mathbf{v}(1) = \mathbf{i} + 2\mathbf{j}$, $\mathbf{T}(1) = (\mathbf{i} + 2\mathbf{j})/\sqrt{5}$, $\mathbf{N}(1) = (-2\mathbf{i} + \mathbf{j})/\sqrt{5}$
(*c*)

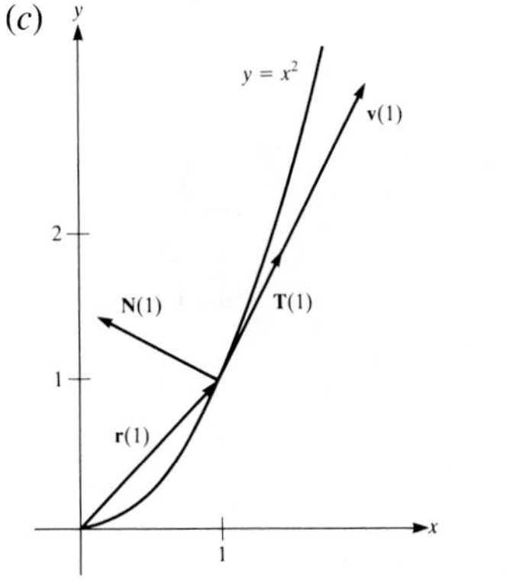

11 3 **13** $a_T = 3$, $a_N = \frac{16}{5}$ **15** $a_T = 0$, $a_N = \frac{1}{3}$,
17 $a_T = 0$, $a_N = 45\pi^2$, $\kappa = \frac{1}{5}$ **21** $a_T = \dfrac{2t(2 + 9t^2)}{\sqrt{4t^2 + 9t^4}}$,
$a_N = \dfrac{6t^2}{\sqrt{4t^2 + 9t^4}}$, $\kappa = \dfrac{6t^2}{(4t^2 - 9t^4)^{2/3}}$

23

Any **a** between **T** and **N**

a **N** **T**

25 (*a*)

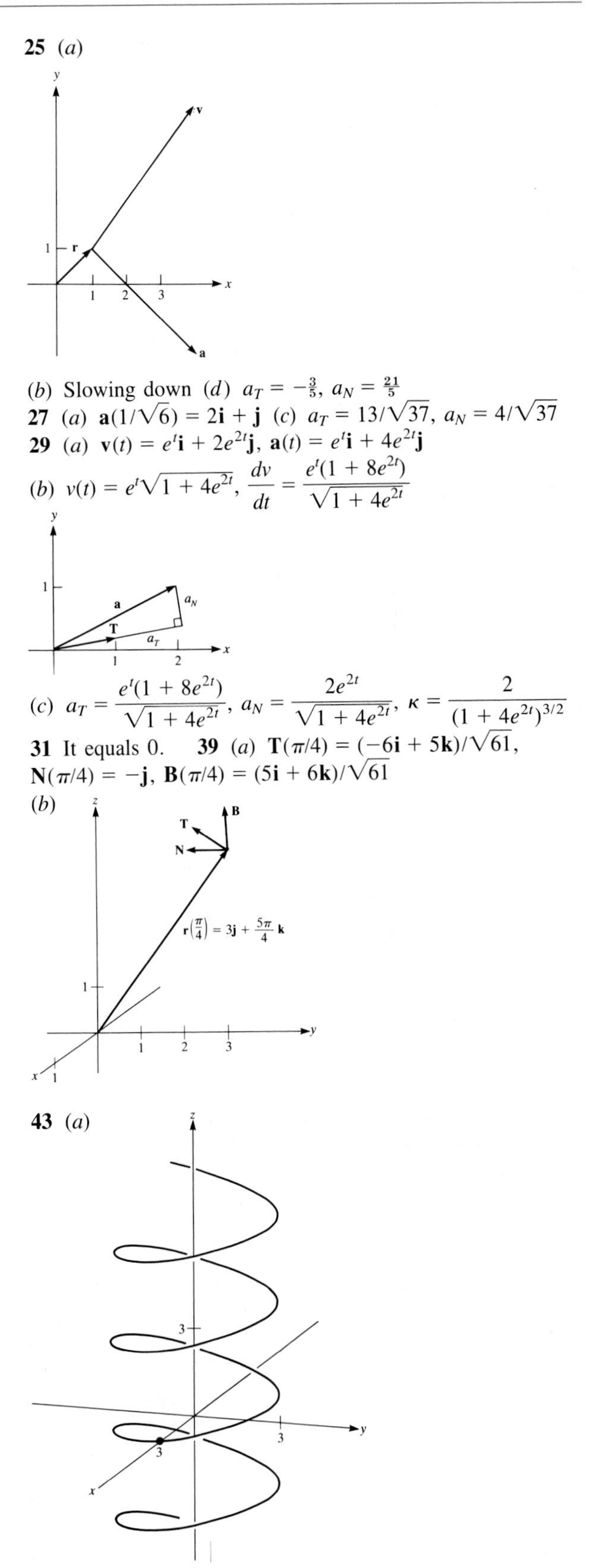

(*b*) Slowing down (*d*) $a_T = -\frac{3}{5}$, $a_N = \frac{21}{5}$
27 (*a*) $\mathbf{a}(1/\sqrt{6}) = 2\mathbf{i} + \mathbf{j}$ (*c*) $a_T = 13/\sqrt{37}$, $a_N = 4/\sqrt{37}$
29 (*a*) $\mathbf{v}(t) = e^t\mathbf{i} + 2e^{2t}\mathbf{j}$, $\mathbf{a}(t) = e^t\mathbf{i} + 4e^{2t}\mathbf{j}$
(*b*) $v(t) = e^t\sqrt{1 + 4e^{2t}}$, $\dfrac{dv}{dt} = \dfrac{e^t(1 + 8e^{2t})}{\sqrt{1 + 4e^{2t}}}$

(*c*) $a_T = \dfrac{e^t(1 + 8e^{2t})}{\sqrt{1 + 4e^{2t}}}$, $a_N = \dfrac{2e^{2t}}{\sqrt{1 + 4e^{2t}}}$, $\kappa = \dfrac{2}{(1 + 4e^{2t})^{3/2}}$
31 It equals 0. **39** (*a*) $\mathbf{T}(\pi/4) = (-6\mathbf{i} + 5\mathbf{k})/\sqrt{61}$, $\mathbf{N}(\pi/4) = -\mathbf{j}$, $\mathbf{B}(\pi/4) = (5\mathbf{i} + 6\mathbf{k})/\sqrt{61}$
(*b*)

$\mathbf{r}\left(\frac{\pi}{4}\right) = 3\mathbf{j} + \frac{5\pi}{4}\mathbf{k}$

43 (*a*)

(*b*) $\mathbf{v}(t) = -6\pi \sin 2\pi t\ \mathbf{i} + 6\pi \cos 2\pi t\ \mathbf{j} + 3\mathbf{k}$, $\mathbf{a}(t) = -12\pi^2 \cos 2\pi t\ \mathbf{i} - 12\pi^2 \sin 2\pi t\ \mathbf{j}$
(*c*) $4\pi^2/[3(4\pi^2 + 1)]$ (*d*) $a_T = 0$, $a_N = 12\pi^2$

Sec. 13.5. Newton's Law Implies Kepler's Three Laws
See the Student's Solutions Manual for a complete discussion.

Sec. 13.S. Summary: Guide Quiz
1 (*b*) $\mathbf{G}'(1) \approx 2\mathbf{i} + 3\mathbf{j}$ **6** (*a*) 1/25
(*b*) 25 (*c*) 2 (*d*) Speeding up (*e*) 1 **7** (*a*) $(2\mathbf{i} + \mathbf{j})/\sqrt{5}$
(*b*) $(\mathbf{i} - 2\mathbf{j})/\sqrt{5}$ (*c*) $2\mathbf{i} + \mathbf{j}$ (*d*) $4/\sqrt{5}$ (*e*) $2/(5\sqrt{5})$
(*f*) $5\sqrt{5}/2$

Sec. 13.S. Summary: Review Exercises
1 (*b*) $a_T = \dfrac{e^2 - e^{-2}}{\sqrt{2(e^2 + e^{-2})}}$, $a_N = \sqrt{\dfrac{2}{e^2 + e^{-2}}}$
3 (*b*)

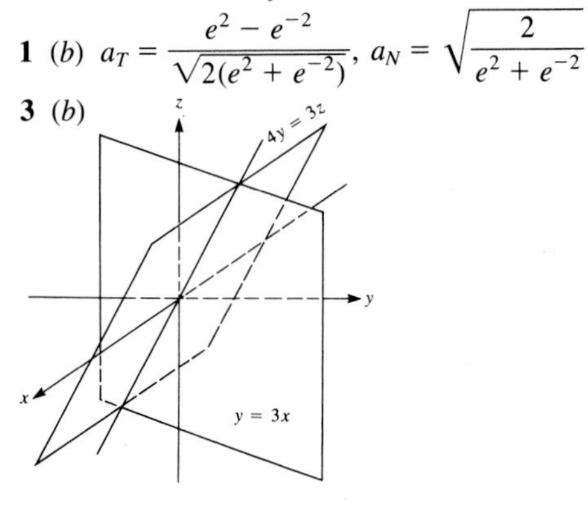

5 (*a*) $\mathbf{v}(1) = 3\mathbf{i} + e\mathbf{j}$, $\mathbf{a}(1) = 6\mathbf{i} + e\mathbf{j}$ (*b*) $v = \sqrt{9 + e^2}$, $a_T = \dfrac{18 + e^2}{\sqrt{9 + e^2}}$, $a_N = \dfrac{3e}{\sqrt{9 + e^2}}$, $r = \dfrac{(9 + e^2)^{3/2}}{3e}$
7 (*a*) $\sqrt{3}e^t$ (*b*) $\sqrt{3}(e - 1)$ **9** (*a*) $\frac{1}{2}\sqrt{26}$ (*b*) $\cos \alpha = 3/\sqrt{26}$, $\cos \beta = -1/\sqrt{26}$, $\cos \gamma = 4/\sqrt{26}$
(*c*) $(1 + \frac{3}{2}t)\mathbf{i} + (2 - \frac{1}{2}t)\mathbf{j} + (1 + 2t)\mathbf{k}$ **11** (−7.3953, 8.6390, 41.5578) **15** $a_N = \dfrac{\|\mathbf{v} \times \mathbf{a}\|}{\|\mathbf{v}\|}$

CHAPTER 14. PARTIAL DERIVATIVES
Sec. 14.1. Graphs

1 **3**

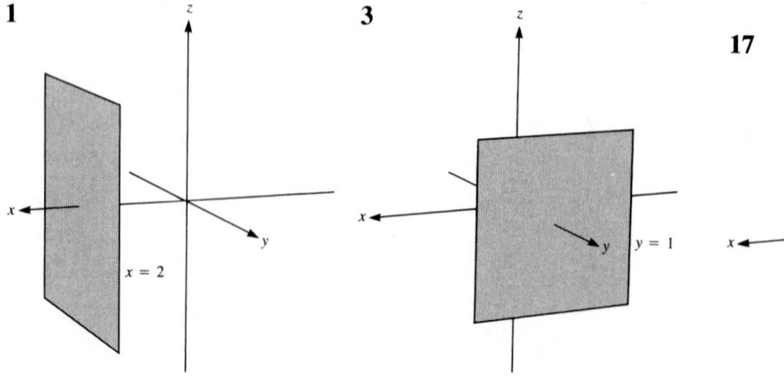

5 **7** **9** **11** **13** $x^2 + y^2 + z^2 = 9$ **15** $(x - 1)^2 + (y - 2)^2 + (z - 3)^2 = 1$

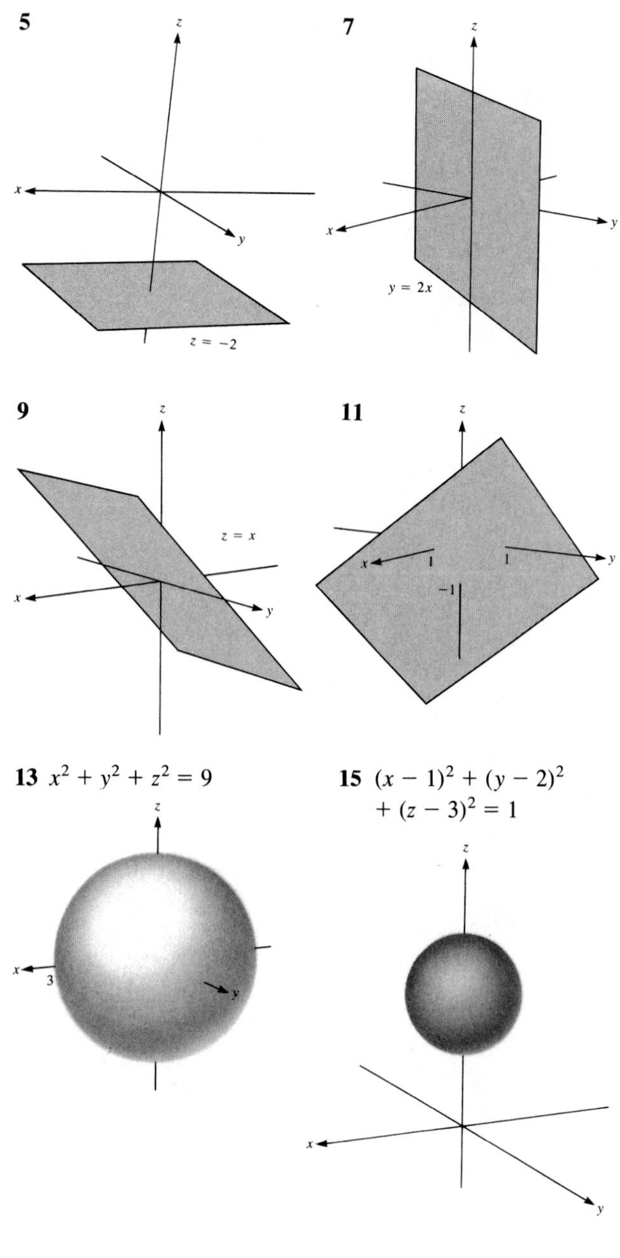

17
x² + y² = 5
√5

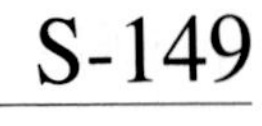

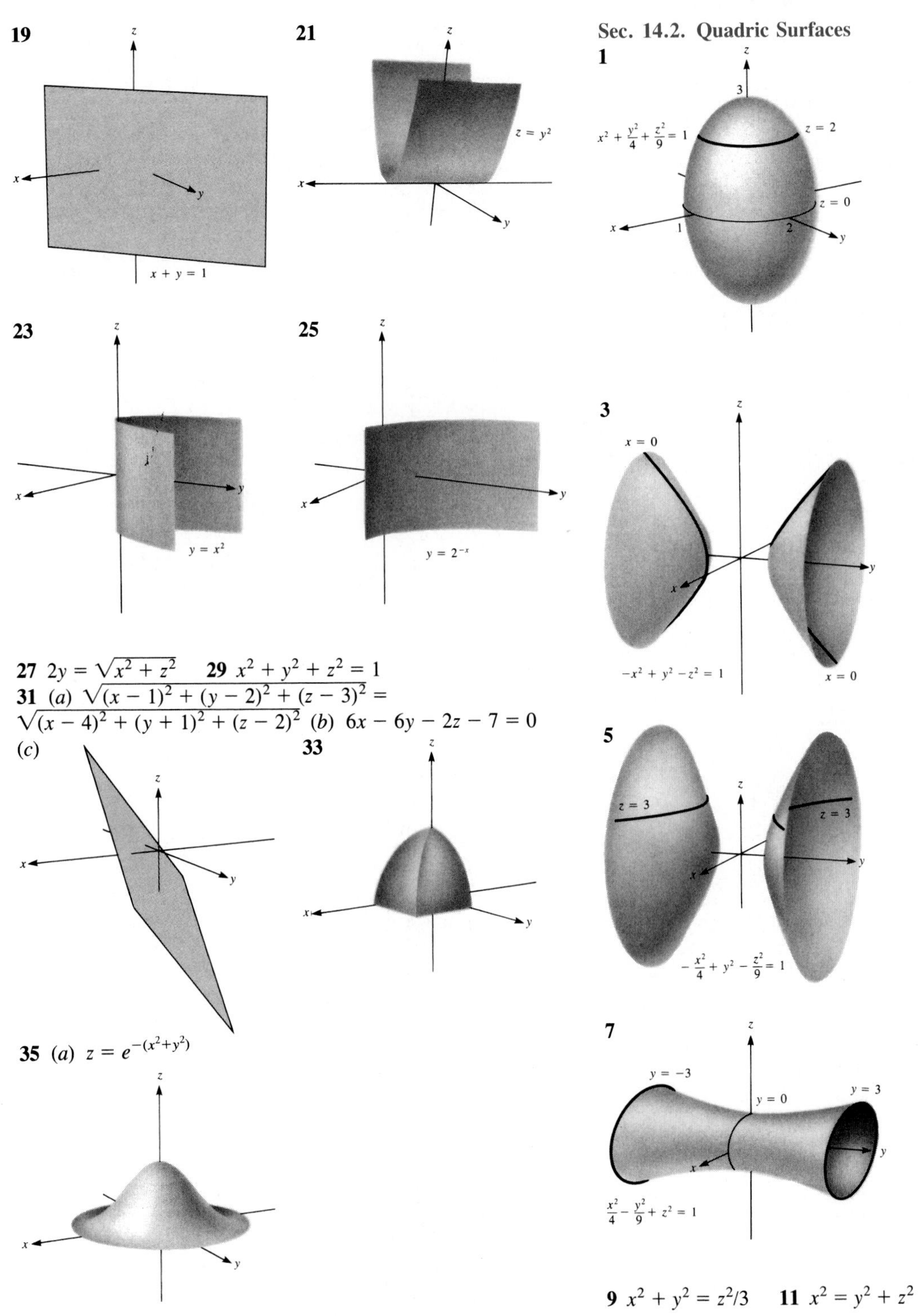

27 $2y = \sqrt{x^2 + z^2}$ **29** $x^2 + y^2 + z^2 = 1$
31 (*a*) $\sqrt{(x-1)^2 + (y-2)^2 + (z-3)^2} = \sqrt{(x-4)^2 + (y+1)^2 + (z-2)^2}$ (*b*) $6x - 6y - 2z - 7 = 0$
(*c*)

35 (*a*) $z = e^{-(x^2+y^2)}$

(*b*) $dV \approx 2\pi r e^{-r^2}\,dr$ (*c*) $V = \int_0^\infty 2\pi r e^{-r^2}\,dr$ (*d*) π

Sec. 14.2. Quadric Surfaces

9 $x^2 + y^2 = z^2/3$ **11** $x^2 = y^2 + z^2$

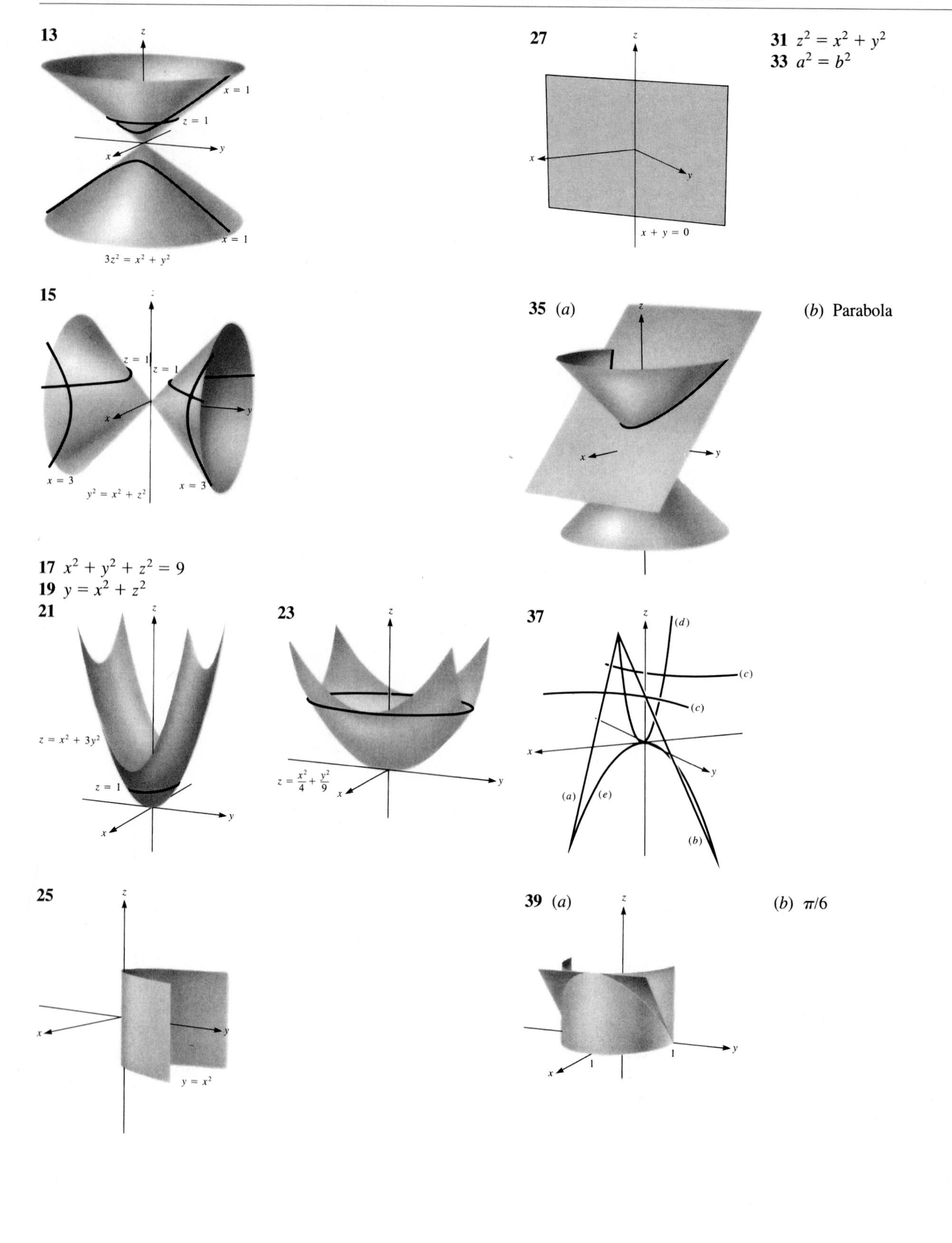

17 $x^2 + y^2 + z^2 = 9$

19 $y = x^2 + z^2$

31 $z^2 = x^2 + y^2$

33 $a^2 = b^2$

35 (*b*) Parabola

39 (*b*) $\pi/6$

41

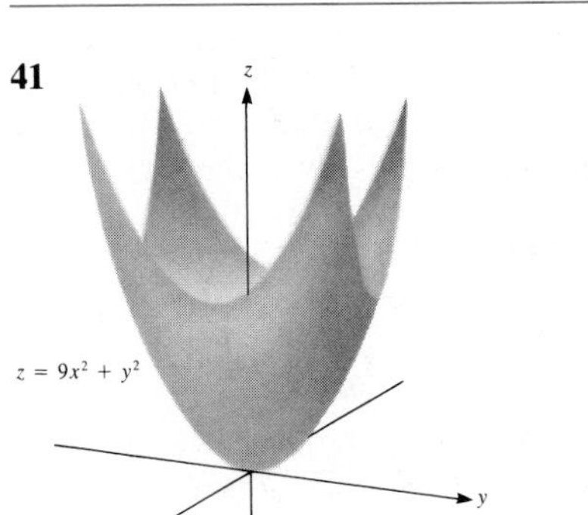

43

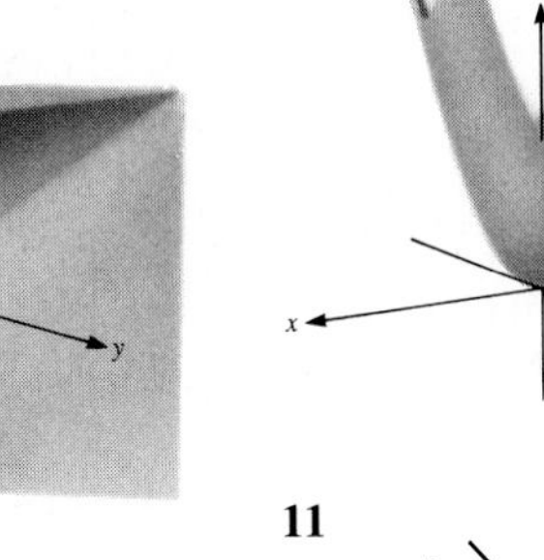

45

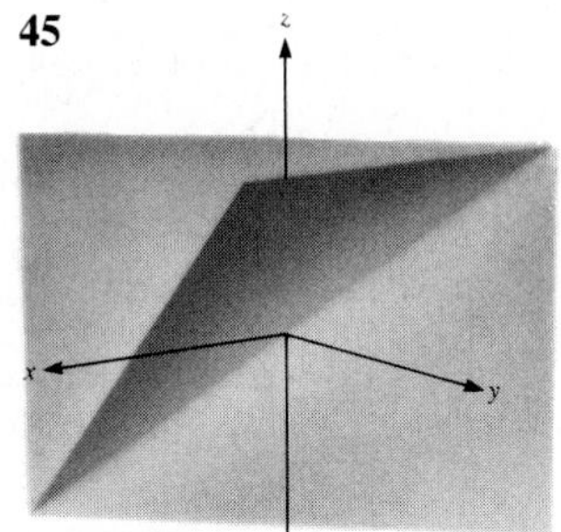

47 (*a*)

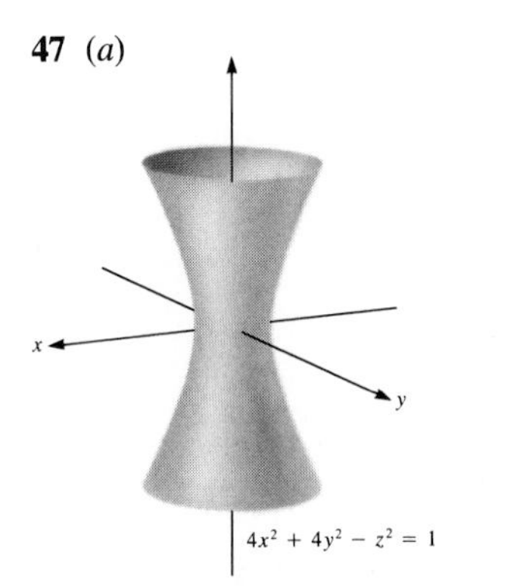

Sec. 14.3. Functions and Their Level Curves

1 **3**

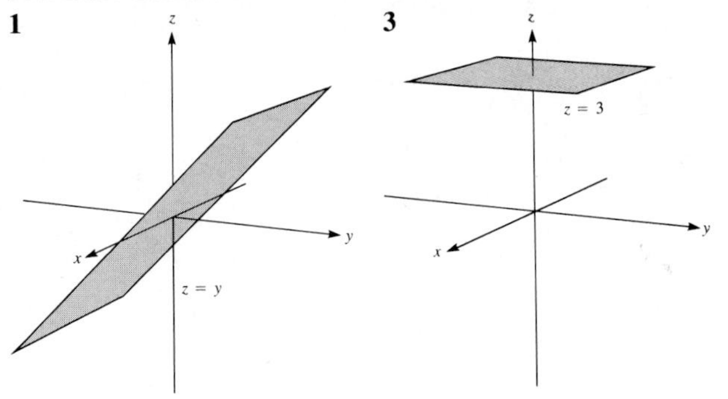

5 **7** **9** **11** **13** **15**

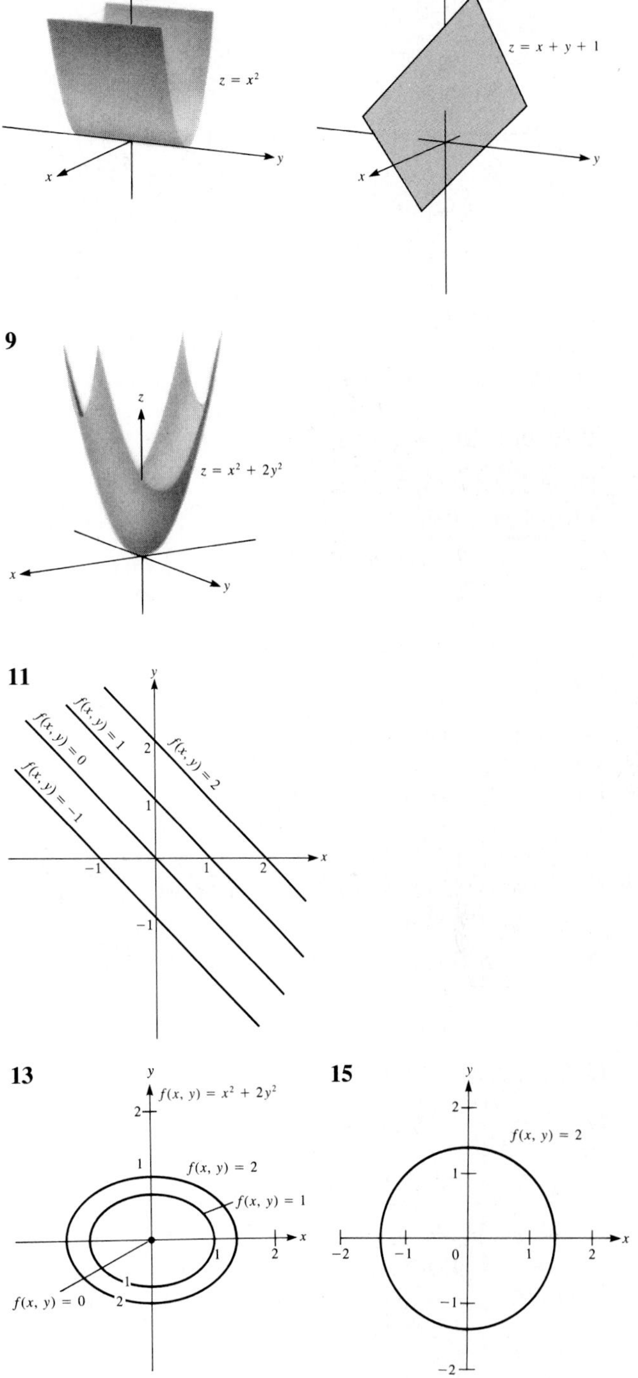

17 **19** (*a*)

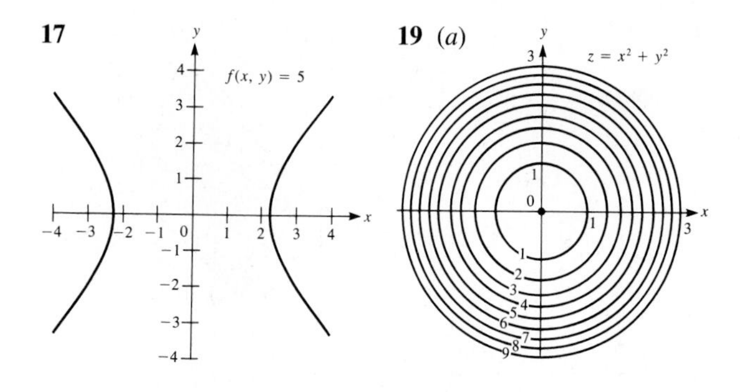

Sec. 14.4. Limits and Continuity

1 $\frac{5}{13}$ **3** Does not exist **5** 8 **7** e **9** (*a*) $x + y \neq 0$ (*b*) Continuous **11** (*a*) $x^2 + y^2 \neq 9$ (*b*) Continuous **13** (*a*) $x^2 + y^2 \leq 16$ (*b*) Continuous **15** $x^2 + y^2 = 1$ **17** (0, 0) **19** $y = x^2$ **21** (*e*) $\lim_{P \to (1,2)} f(P) = 3$

23 (*a*) $(x, y) \neq (0, 0)$

(*b*)

(x, y)	(0.01, 0.01)	(0.01, 0.02)	(0.001, 0.003)
$f(x, y)$	0.0050	0.0200	0.00017

(*d*)

(x, y)	(0.5, 0.25)	(0.1, 0.01)	(0.001, 0.000001)
$f(x, y)$	0.3333	0.3333	0.3333

(*f*) No

Sec. 14.5. Partial Derivatives

1 $f_x = 3, f_y = 2$ **3** $f_x = 3x^2y^4, f_y = 4x^3y^3$ **5** $f_x = \cos xy - xy \sin xy, f_y = -x^2 \sin xy$

7 $f_x = \dfrac{y}{1 + (xy)^2}, f_y = \dfrac{x}{1 + (xy)^2}$

9 $f_x = \sin^2 (x + y) + 2x \sin (x + y) \cos (x + y)$, $f_y = 2x \sin (x + y) \cos (x + y)$

11 $f_x = \dfrac{1}{2\sqrt{x}} \sec x^2y + 2xy\sqrt{x} \sec x^2y \tan x^2y$; $f_y = x^{5/2} \sec x^2y \tan x^2y$ **13** $f_x = \dfrac{e^x}{y^3}, f_y = -3e^xy^{-4}$

15 $f_{xx} = 10, f_{yy} = 12, f_{xy} = f_{yx} = -3$

17 $f_{xx} = \dfrac{2x^2 - y^2}{(x^2 + y^2)^{5/2}}, f_{yy} = \dfrac{2y^2 - x^2}{(x^2 + y^2)^{5/2}}$, $f_{xy} = f_{yx} = \dfrac{3xy}{(x^2 + y^2)^{5/2}}$ **19** $f_{xx} = 2 \sec^2 (x + 3y) \times \tan (x + 3y), f_{yy} = 18 \sec^2 (x + 3y) \tan (x + 3y), f_{xy} = f_{yx} = 6 \sec^2 (x + 3y) \tan (x + 3y)$ **21** 4 **23** 1 **25** $2e$

27 (*a*), (*c*)

(*b*) 2 **29** $f_x(1, 1) \approx 2.5; f_y(1, 1) \approx 20$ **31** $T_x(1, 2) \approx 2.5$; $T_y(1, 2) \approx 3$ **33** (*a*) $f(x, y) = 3$ (*b*) $f(x, y) = 3$ is the only one. **35** No **37** (*a*) $\sqrt{y + x}$ (*b*) $\sqrt{y + x} - \sqrt{y}$

Sec. 14.6. The Chain Rule

1 $13t^{12}$ **3** $-2e^{2t} \sec^2 3t \sin (e^{2t} \sec^2 3t)(1 + 3 \tan 3t)$
5 $135t^2 + 222ut + 56u^2$

7 (*c*) $\dfrac{\partial z}{\partial t_3} = \dfrac{\partial f}{\partial x_1}\dfrac{\partial x_1}{\partial t_3} + \dfrac{\partial f}{\partial x_2}\dfrac{\partial x_2}{\partial t_3} + \dfrac{\partial f}{\partial x_3}\dfrac{\partial x_3}{\partial t_3} + \dfrac{\partial f}{\partial x_4}\dfrac{\partial x_4}{\partial t_3} + \dfrac{\partial f}{\partial x_5}\dfrac{\partial x_5}{\partial t_3}$
(*d*) x_1, x_3, x_4, x_5 (*e*) t_1, t_2 **9** 19 **23** 17°/sec

Sec. 14.7. Directional Derivatives and the Gradient

1 (*a*) 4 (*b*) −4 (*c*) $9/\sqrt{2}$
3 (*a*) x^2z^3 (*b*) $3x^2yz^2$ (*c*) $-2xyz^3$ **5** (*a*) $4\mathbf{i} + 5\mathbf{j}$ (*b*) $\sqrt{41}$ (*c*) $\dfrac{4}{\sqrt{41}}\mathbf{i} + \dfrac{5}{\sqrt{41}}\mathbf{j}$ **7** (*a*) $20\mathbf{i} + 4\mathbf{j}$ (*b*) $6\mathbf{i} + 9\mathbf{j}$

9 −5 **11** (*a*) $\pm\dfrac{3\mathbf{i} - 2\mathbf{j}}{\sqrt{13}}$ (*b*) $\dfrac{2\mathbf{i} + 3\mathbf{j}}{\sqrt{13}}$ (*c*) $-\dfrac{2\mathbf{i} + 3\mathbf{j}}{\sqrt{13}}$

13 (*a*) $\mathbf{u} = \pm\left(-\dfrac{1}{\sqrt{2}}\mathbf{i} + \dfrac{1}{\sqrt{2}}\mathbf{j}\right)$ (*b*) $\mp\sqrt{2}$ **15** (*a*) 0.03 (*b*) ≈ 0.0023 (*c*) $\mathbf{u} = \pm\dfrac{3\mathbf{i} + 2\mathbf{j}}{\sqrt{13}}$ **17** 3.0990

19 $1/\sqrt{3}$, 1 **21** $\dfrac{5}{\sqrt{22}}e^{1/\sqrt{2}}, \sqrt{\dfrac{3}{2}}e^{1/\sqrt{2}}$ **23** $\dfrac{-1}{7\sqrt{2}}$, 1

25 (*a*) $2\mathbf{i} + 3\mathbf{j} + \mathbf{k}$
(*b*)

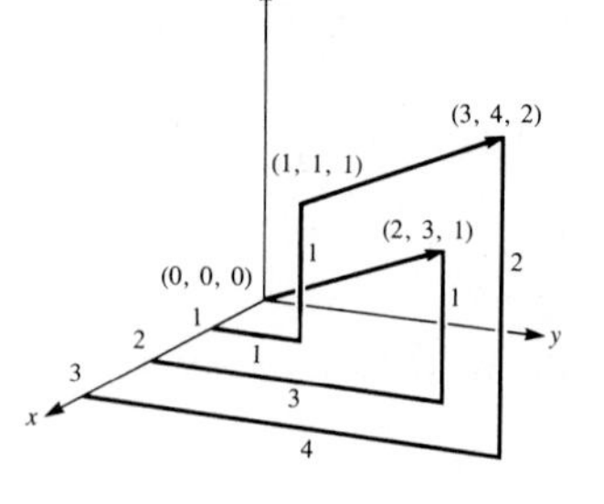

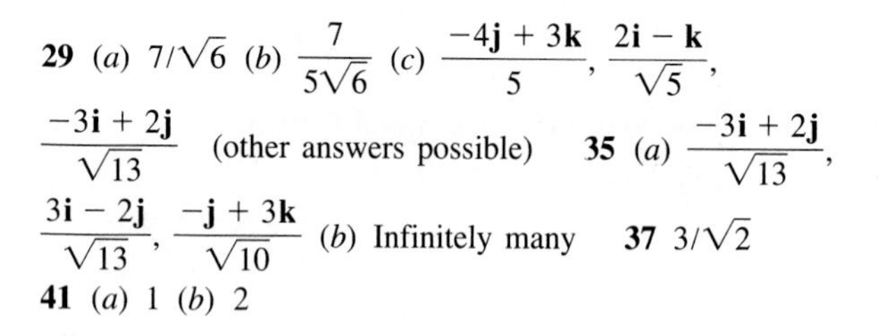

29 (*a*) $7/\sqrt{6}$ (*b*) $\dfrac{7}{5\sqrt{6}}$ (*c*) $\dfrac{-4\mathbf{j} + 3\mathbf{k}}{5}, \dfrac{2\mathbf{i} - \mathbf{k}}{\sqrt{5}}, \dfrac{-3\mathbf{i} + 2\mathbf{j}}{\sqrt{13}}$ (other answers possible) **35** (*a*) $\dfrac{-3\mathbf{i} + 2\mathbf{j}}{\sqrt{13}}, \dfrac{3\mathbf{i} - 2\mathbf{j}}{\sqrt{13}}, \dfrac{-\mathbf{j} + 3\mathbf{k}}{\sqrt{10}}$ (*b*) Infinitely many **37** $3/\sqrt{2}$
41 (*a*) 1 (*b*) 2

Sec. 14.8. Normals and the Tangent Plane

1 $4\mathbf{i} + 2\mathbf{j}$ **3** $\mathbf{i} + \frac{2}{3}\mathbf{j}$

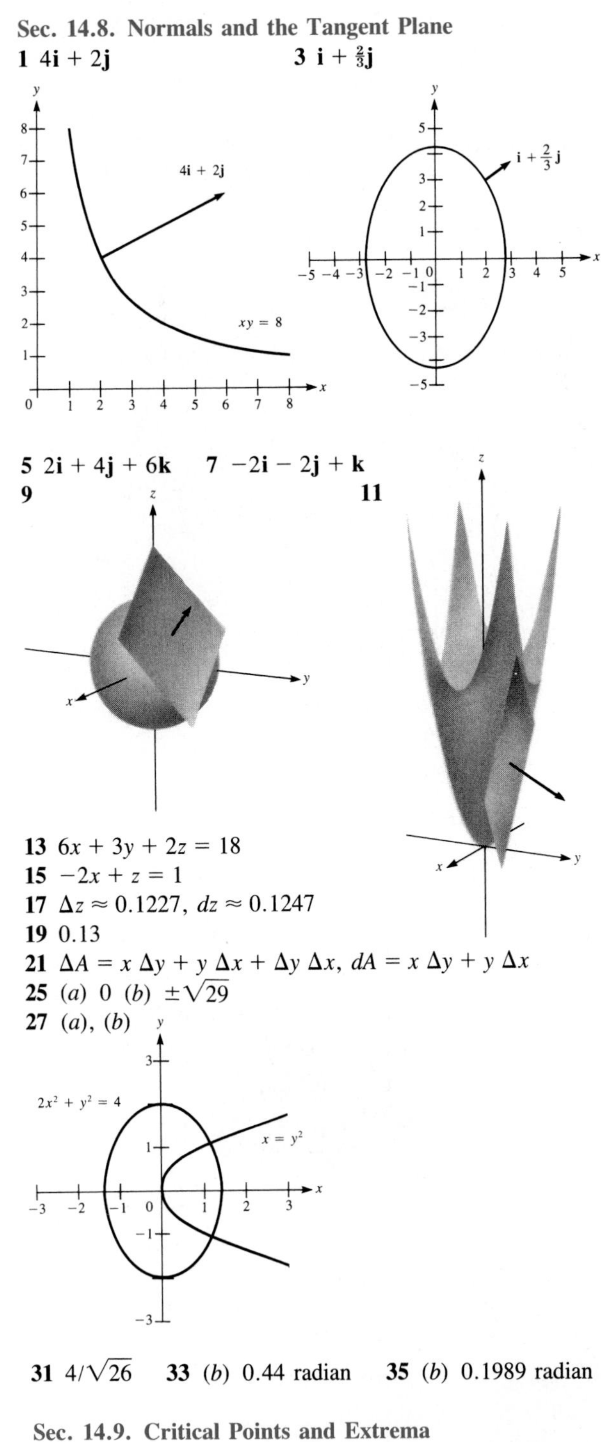

5 $2\mathbf{i} + 4\mathbf{j} + 6\mathbf{k}$ **7** $-2\mathbf{i} - 2\mathbf{j} + \mathbf{k}$

9

11

13 $6x + 3y + 2z = 18$

15 $-2x + z = 1$

17 $\Delta z \approx 0.1227$, $dz \approx 0.1247$

19 0.13

21 $\Delta A = x\,\Delta y + y\,\Delta x + \Delta y\,\Delta x$, $dA = x\,\Delta y + y\,\Delta x$

25 (*a*) 0 (*b*) $\pm\sqrt{29}$

27 (*a*), (*b*)

31 $4/\sqrt{26}$ **33** (*b*) 0.44 radian **35** (*b*) 0.1989 radian

Sec. 14.9. Critical Points and Extrema

1 None **3** Relative minimum at $(-4, -2)$ **5** Relative minimum at $(0, 0)$ **7** Relative minimum at $\left(\frac{-10}{9}, \frac{2}{9}\right)$ **9** Relative minimum at $(2, 1)$ **11** Cannot determine **13** Neither **15** Maximum **17** Relative maximum at $(-1, -1)$ **19** Saddle at (0,0), relative maximum at (4,4) **21** Saddle at (0,0) **23** Saddles at $(1,-1)$ and $(-2,1)$ **25** $l = w = 2^{1/3}$, $h = 2^{-2/3}$ **27** $((x_1 + x_2 + x_3 + x_4)/4, (y_1 + y_2 + y_3 + y_4)/4)$ **29** (*a*) $\frac{1}{3}$ (*b*) 1 **31** $x = y = z = \sqrt{\frac{1}{3}}$ **33** $k \geq \frac{25}{16}$ **35** (*a*) Maximum of $2/e$ at (1, 0) and $(-1, 0)$; minimum of $1/e$ at $(0, \pm 1)$ (*b*) Maximum of $8e^{-4}$ at $(\pm 2, 0)$, minimum of $4e^{-4}$ at $(0, \pm 2)$ **37** Maximum of $\frac{1}{4}$ at $(\frac{1}{2}, \frac{1}{2})$ **39** (*b*) $\pm e\sqrt{2}$ **41** $x = 15$, $y = 20$, $z = 2$ **43** (*b*) $f_m = -2\sum_{i=1}^{n} x_i[y_i - (mx_i + b)]$, $f_b = -2\sum_{i=1}^{n}[y_i - (mx_i + b)]$ (*d*) x_i's are not all equal. (*e*) $y = 2x - 1$ **49** (*b*) $\sqrt[4]{abcd} \leq \frac{1}{4}(a + b + c + d)$

Sec. 14.10. Lagrange Multipliers

1 2 at $(\pm\sqrt{2}, \pm\sqrt{2})$ **3** $2\sqrt{6}$ at $\left(\frac{\sqrt{6}}{2}, \frac{\sqrt{6}}{3}\right)$ **5** 9 for a 3×3 square **7** $(\frac{3}{7}, \frac{6}{7}, \frac{9}{7})$ **9** $\sqrt{\frac{2}{3}}$ at $(\frac{1}{3}, \frac{8}{3}, \frac{5}{3})$ **11** $\frac{1}{27}$ at $x = \pm 1/\sqrt{3}$, $y = \pm 1/\sqrt{3}$, $z = \pm 1/\sqrt{3}$ (eight different points) **13** $\frac{224}{75}$ at $(\frac{68}{75}, \frac{16}{15}, -\frac{76}{75})$ **15** $x = y = 2^{1/3}$, $z = 2^{-2/3}$ **17** $x = y = z = \sqrt{2}$ **19** (*a*) $t = u = v = \pi/6$ (*c*) $\frac{1}{8}$ **21** $\frac{224}{75}$ **23** (*b*) (0.5543, 0.6053, 2.3260) (*c*) 90°

25 Maximum is $\dfrac{27 + 14\sqrt{7}}{8}$ at $(\frac{3}{2}, 0, \frac{1}{2}\sqrt{7})$.

29 Maximum of n^{-n} at $x_i = \dfrac{1}{n}$ for all i.

31 $(\sum_{i=1}^{n} a_i^2)^{1/2}$ for $x_i = a_i\,(\sum_{j=1}^{n} a_j^2)^{-1/2}$

Sec. 14.11. The Chain Rule Revisited

1 $\left(\dfrac{\partial x}{\partial z}\right)_y = -(\partial u/\partial z)_{x,y}/(\partial u/\partial x)_{y,z}$ **9** $(\partial z/\partial r)_{s,t} = 2r$, $(\partial z/\partial r)_{s,u} = 2r + 2r(su)^2$

11 $\dfrac{\partial u}{\partial x} = \dfrac{(\partial F/\partial x)(\partial G/\partial u) - (\partial G/\partial x)(\partial F/\partial v)}{(\partial F/\partial v)(\partial G/\partial v) - (\partial F/\partial u)(\partial G/\partial u)}$

Sec. 14.S. Summary: Guide Quiz

4 (*a*) $\frac{2}{3}$ (*b*) 2 (*d*) 90° (*e*) $\frac{2}{3}\cos 37° + 2\sin 37°$

5 $f(x, y) = 2x + 3y$ is the only one.

6 (*a*) $2\sec^3(x + 2y)\left[3\tan(x + 2y)\ln(1 + 2xy) + \dfrac{x}{1 + 2xy}\right]$

(*b*) $\dfrac{3xy}{1 + 9x^2y^2} + \tan^{-1} 3xy$ (*c*) $3x^2y^2\,e^{x^3y}$

(*d*) $-6\cos(2x + 3y)$

8 (*a*)

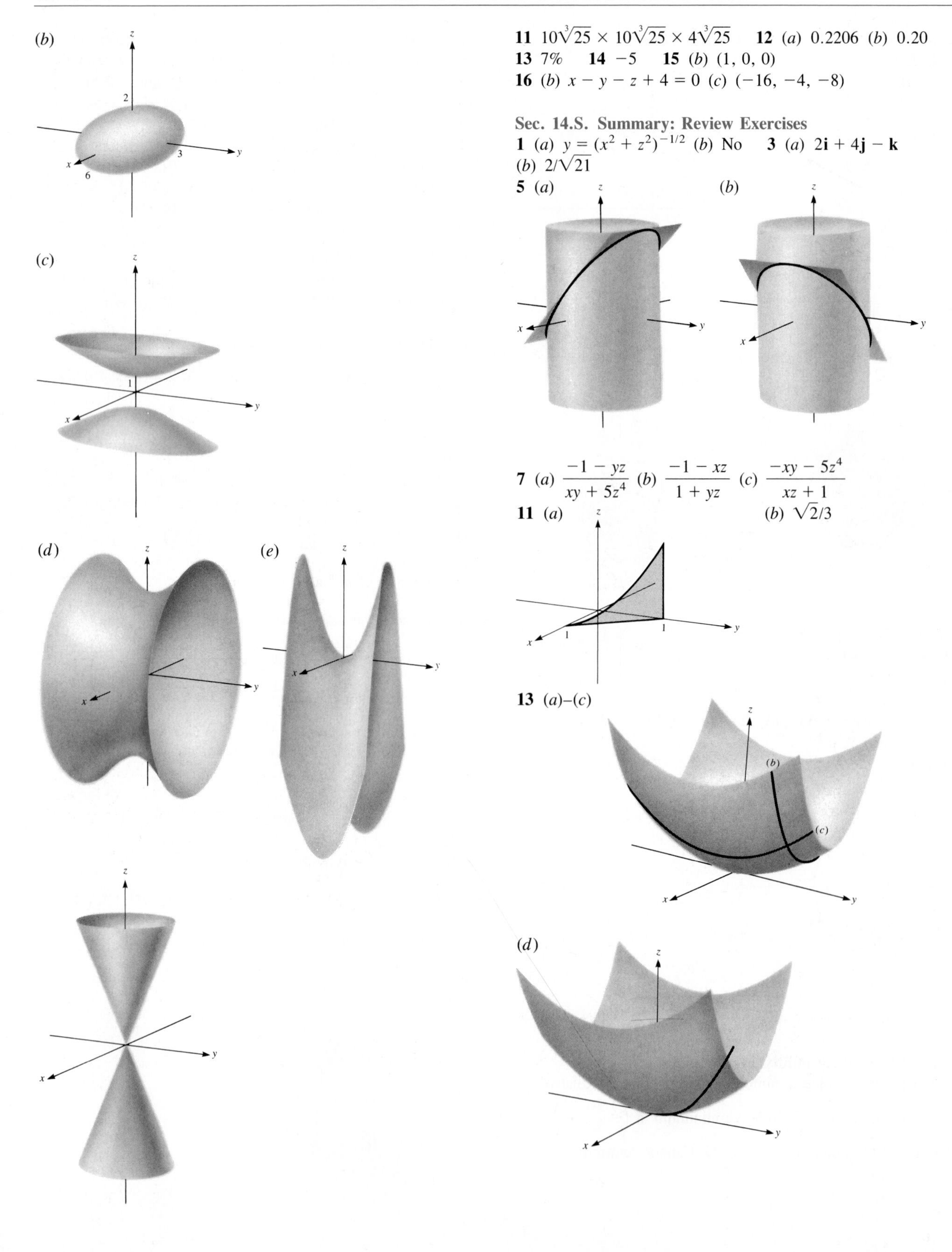

11 $10\sqrt[3]{25} \times 10\sqrt[3]{25} \times 4\sqrt[3]{25}$ **12** (*a*) 0.2206 (*b*) 0.20
13 7% **14** -5 **15** (*b*) (1, 0, 0)
16 (*b*) $x - y - z + 4 = 0$ (*c*) $(-16, -4, -8)$

Sec. 14.S. Summary: Review Exercises

1 (*a*) $y = (x^2 + z^2)^{-1/2}$ (*b*) No **3** (*a*) $2\mathbf{i} + 4\mathbf{j} - \mathbf{k}$ (*b*) $2/\sqrt{21}$

5 (*a*) (*b*)

7 (*a*) $\dfrac{-1 - yz}{xy + 5z^4}$ (*b*) $\dfrac{-1 - xz}{1 + yz}$ (*c*) $\dfrac{-xy - 5z^4}{xz + 1}$

11 (*a*) (*b*) $\sqrt{2}/3$

13 (*a*)–(*c*)

(*d*)

(*e*) $\mathbf{i} + \frac{2}{3}\mathbf{j} - \mathbf{k}$
(*f*) Empty
(*g*), (*h*)

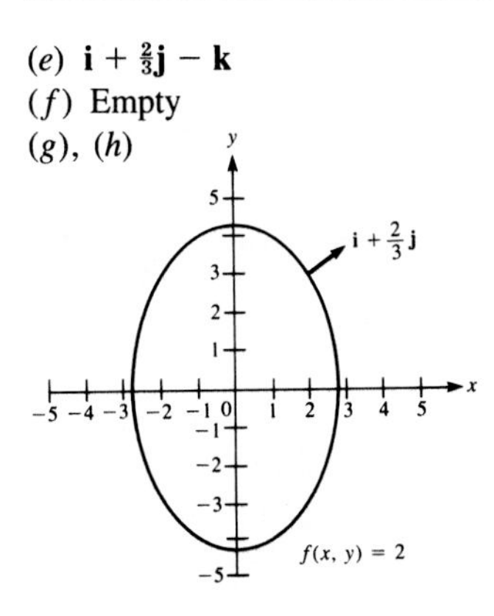

17 (*a*) 12 (*b*) $12/\sqrt{14}$ **19** (*b*) Closest: $(\pm 1, \pm 1)$, farthest: $(\pm\sqrt{3}, \mp\sqrt{3})$
29 (*a*) $\left(\frac{\partial P}{\partial V}\right)_T = -\frac{nRT}{V^2}$, $\left(\frac{\partial P}{\partial T}\right)_V = \frac{nR}{V}$
(*c*) $\left(\frac{\partial T}{\partial V}\right)_P = \frac{T}{V} = \frac{P}{nR}$ **31** $\Delta f = 0$, $df = 0.00064 \approx 0.001$
33 $\Delta f \approx 0.464$, $df \approx 0.436$ **35** $\Delta f \approx 0.036$, $df = 0.036$
37 $\Delta f = df = 1.8$ **39** 17% **41** $(3|m| + 4|n|)\%$
43 (*b*) 8.301594 **45** (*b*) 3.01251 **47** (*c*) $(\frac{1}{2}, \frac{1}{2}, \frac{1}{2})$
51 (*a*) $\frac{4}{3}\mathbf{i} + \frac{2}{3}\mathbf{j}$ (*b*) P (*c*) $\frac{\partial f}{\partial x} \approx \frac{4}{3}$ (*d*) $\frac{4}{3}\cos 20° + \frac{2}{3}\sin 20°$
53 $h = 0$, $\frac{4}{3}\pi r^3 = c$ **55** Maximum of 3 at $(\frac{1}{2}, 0)$; minimum of -2 at $(\frac{3}{2}, -1)$ **57** $g(y, x + 4) - g(y, x + 3)$
59 $f(x, y) = y$
65 (*a*), (*b*)

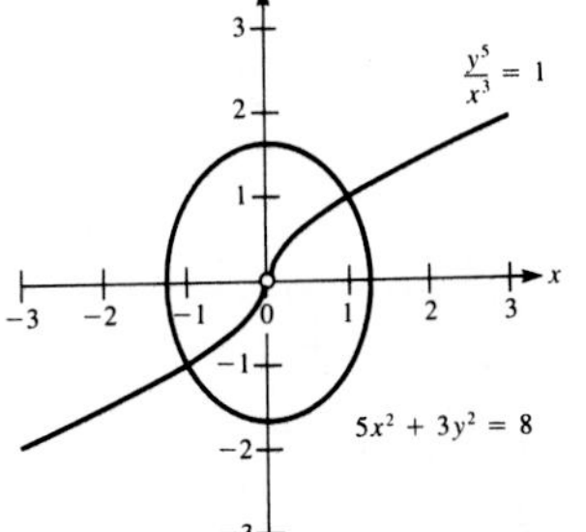

67 (*a*)

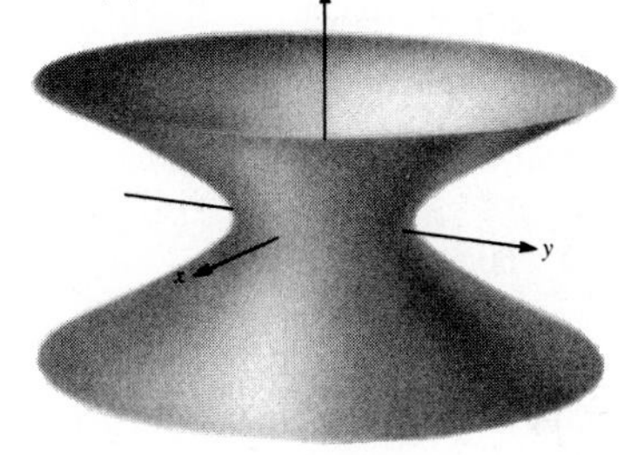

(*b*) (*c*)

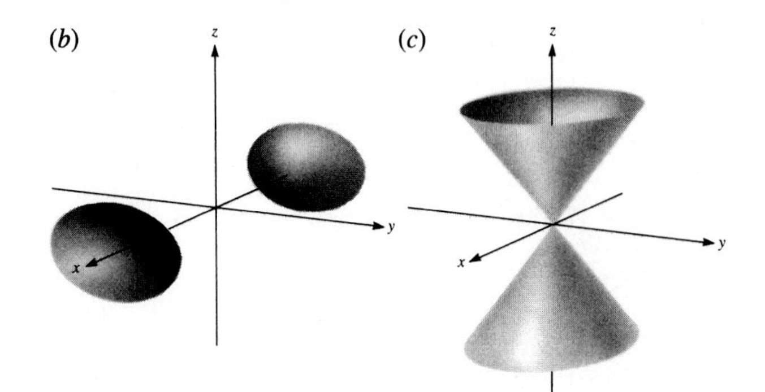

73 Maximum of 4 at (1, 0); minimum of -3 at (0, 1)
77 $\nabla f = f_r\mathbf{u}_r + (1/r)f_\theta\mathbf{u}_\theta$

CHAPTER 15. DEFINITE INTEGRALS OVER PLANE AND SOLID REGIONS

Sec. 15.1. The Definite Integral of a Function Over a Region in the Plane

1 (*a*) 20 in^3 (*b*) 100 in^3 (*c*) The volume is more than 20 but less than 100 in^3. **3** (*a*) It equals 5A (*b*) 5A
5 (*a*) 6.25 **7** (*a*) $3 + 2(\sqrt{2} + \sqrt{3}) \approx 9.29$ (*b*) ≈ 1.55
9 (*a*) 31 (*b*) $\frac{256}{27}$ (*c*) $\frac{2048}{27}$ **11** (*a*) Total rainfall on R (*b*) Average rainfall over R **13** (*a*) $2\pi/3$ (*b*) $\frac{2}{3}$
15 $\int_R f(P)\,dA \approx 3$ **19** $\sqrt{2}$

Sec. 15.2. Computing $\int_R f(P)\,dA$ Using Rectangular Coordinates

1 (*a*) $0 \le x \le 2$, $x/2 \le y \le 1$ (*b*) $0 \le y \le 1$, $0 \le x \le 2y$
3 (*a*) $0 \le x \le 1$, $0 \le y \le x$ or $1 \le x \le 2$, $x - 1 \le y \le 1$ (*b*) $0 \le y \le 1$, $y \le x \le y + 1$ **5** (*a*) $-5 \le x \le 5$, $-\sqrt{25 - x^2} \le y \le \sqrt{25 - x^2}$ (*b*) $-5 \le y \le 5$, $-\sqrt{25 - y^2} \le x \le \sqrt{25 - y^2}$ **7** (*a*) $-1 \le x \le 1$, $2 - x \le y \le (8 - x)/3$; $1 \le x \le 2$, $x \le y \le (8 - x)/3$ (*b*) $1 \le y \le 2$, $2 - y \le x \le y$; $2 \le y \le 3$, $2 - y \le x \le 8 - 3y$ **9** (*a*) $0 \le x \le 3$, $0 \le y \le (6 - 2x)/3$ (*b*) $0 \le y \le 2$, $0 \le x \le (6 - 3y)/2$ **11** (*a*) $1 \le x \le 2$, $1 \le y \le x$; $2 \le x \le 3$, $1 \le y \le 2$; $3 \le x \le 6$, $x/3 \le y \le 2$ (*b*) $1 \le y \le 2$, $y \le x \le 3y$
13 $0 \le y \le 4$, $y/3 \le x \le y/2$ or $4 \le y \le 6$, $y/3 \le x \le 2$

15 $0 \le y \le 1/\sqrt{2}$, $\sin^{-1} y \le x \le \cos^{-1} y$

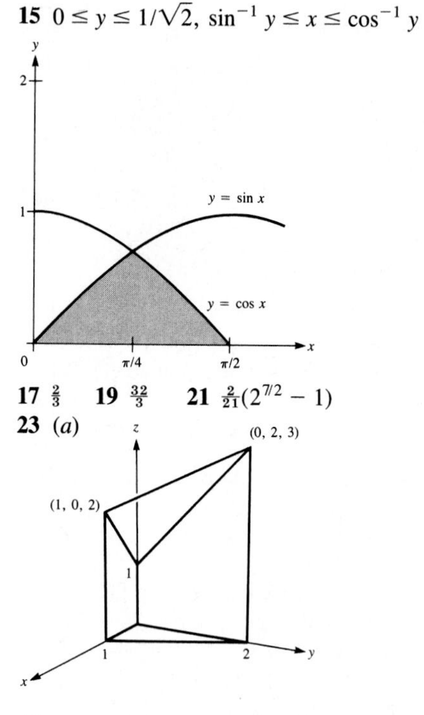

17 $\frac{2}{3}$ **19** $\frac{32}{3}$ **21** $\frac{2}{21}(2^{7/2} - 1)$

23 (*a*)

(1, 0, 2)

(0, 2, 3)

(*b*) $0 \le x \le 1$, $0 \le y \le 2 - 2x$

(*c*) $\int_0^1 \int_0^{2-2x} (x + y + 1)\, dy\, dx$ (*d*) 2 (*e*) $0 \le y \le 2$, $0 \le x \le 1 - y/2$, $V = \int_0^2 \int_0^{1-y/2} (x + y + 1)\, dx\, dy = 2$

25 (*a*)

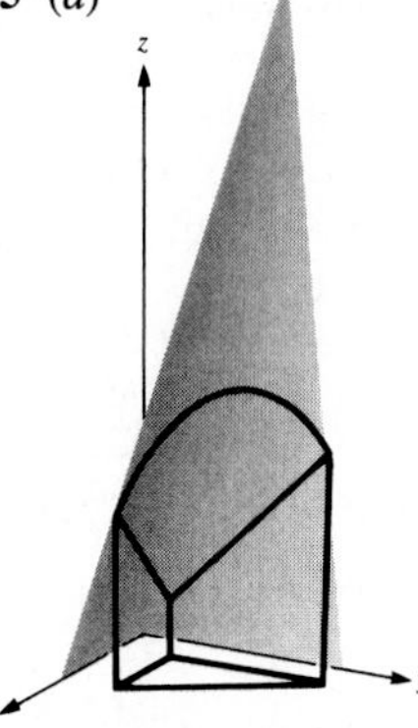

(*b*) $1 \le x \le 3$, $1 \le y \le -3x/2 + \frac{11}{2}$

(*c*) $\int_1^3 \int_1^{(-3x+11)/2} xy\, dy\, dx$ (*d*) $\frac{19}{2}$ (*e*) $1 \le y \le 4$, $1 \le x \le -\frac{2}{3}y + \frac{11}{3}$, $V = \int_1^4 \int_1^{(-2y+11)/3} xy\, dx\, dy = \frac{19}{2}$

27 $\frac{685}{384}$ **29** $\frac{2}{3}\cos 1 - \frac{1}{6}\cos 4 - \frac{1}{2} \approx -0.0309$

31 $\int_0^4 \int_0^{\sqrt{y}} x^3 y\, dx\, dy$ **33** $\int_0^1 \int_y^{2y} xy\, dx\, dy$

35 $(1 - \cos 1)/2$ **37** $(\sqrt{8} - 1)/6$

39 (*a*) $\int_0^a \int_{x/2}^{x} y^2 e^{y^2}\, dy\, dx + \int_a^{2a} \int_{x/2}^{a} y^2 e^{y^2}\, dy\, dx$, $\int_0^a \int_y^{2y} y^2 e^{y^2}\, dx\, dy$ (*b*) $\frac{1}{2}e^{a^2}(a^2 - 1) + \frac{1}{2}$ **43** (*b*) Symmetric about y axis **45** $\int_0^5 f(x, y)\, dy$

Sec. 15.3. Moments and Centers of Mass

1 $(\frac{3}{8}, \frac{3}{8})$ **3** $\left(\dfrac{\pi - 2}{\pi - 2\ln 2}, \dfrac{2}{3}\right)$ **5** $(\frac{4}{5}, \frac{152}{75})$ **7** $(\frac{57}{35}, \frac{75}{14})$

9 $5Ma^2/12$ **11** $2Ma^2/3$ **13** $\frac{8}{15}$ **15** (*b*) $I = I_x + I_y$

17 $(\bar{x}, \bar{y}) = \left(\dfrac{\bar{x}_1 M_1 + \bar{x}_2 M_2}{M_1 + M_2}, \dfrac{\bar{y}_1 M_1 + \bar{y}_2 M_2}{M_1 + M_2}\right)$ **19** (*a*) 0

Sec. 15.4. Computing $\int_R f(P)\, dA$ Using Polar Coordinates

1 $0 \le \theta \le 2\pi$, $0 \le r \le 3 + \cos\theta$ **3** $\pi/4 \le \theta \le \pi/3$, $0 \le r \le \sec\theta$ **5** $\pi/3 \le \theta \le 5\pi/3$, $0 \le r \le 5$

7 (*a*)

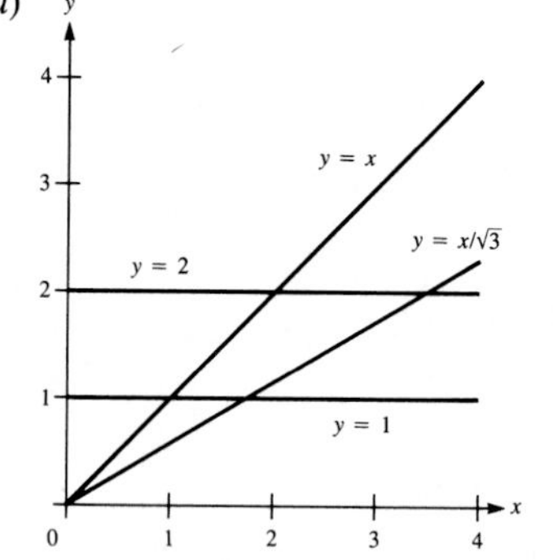

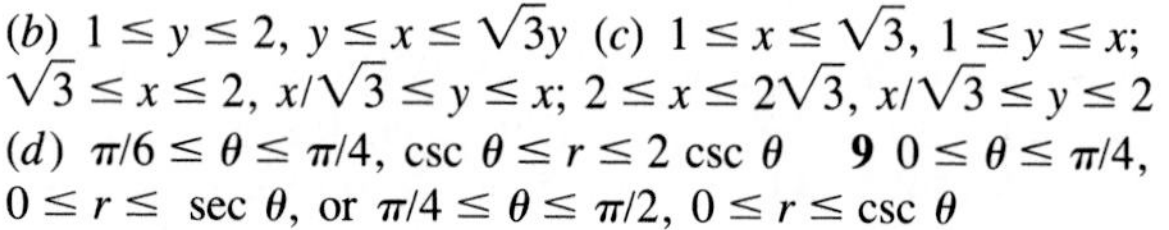

(*b*) $1 \le y \le 2$, $y \le x \le \sqrt{3}y$ (*c*) $1 \le x \le \sqrt{3}$, $1 \le y \le x$; $\sqrt{3} \le x \le 2$, $x/\sqrt{3} \le y \le x$; $2 \le x \le 2\sqrt{3}$, $x/\sqrt{3} \le y \le 2$

(*d*) $\pi/6 \le \theta \le \pi/4$, $\csc\theta \le r \le 2\csc\theta$ **9** $0 \le \theta \le \pi/4$, $0 \le r \le \sec\theta$, or $\pi/4 \le \theta \le \pi/2$, $0 \le r \le \csc\theta$

11 **13**

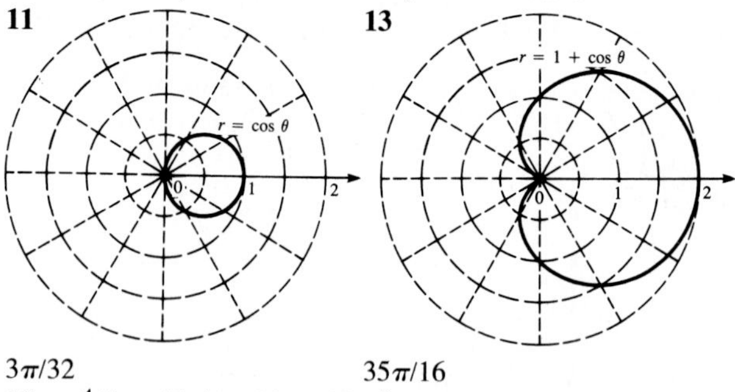

$3\pi/32$ $35\pi/16$

15 $\pi a^4/4$ **17** $49\pi/32$ **19** $(\frac{5}{6}, 0)$ **21** $16/(9\pi)$

23 $32a/(9\pi)$ **25** $\frac{1}{6}[\sqrt{2} + \ln(1 + \sqrt{2})]$ **27** $\frac{1}{16}$

29 (*a*) $\dfrac{\pi}{4}\sin a^2$ (*b*) $\frac{4}{3}[\sqrt{2} + \ln(1 + \sqrt{2})]$

31 $(0, 16a/15)$ **33** $Ma^2/2$ **35** $Ma^2/4$

37 (*a*) $\pi\Delta r(2r_0 + \Delta r)$ (*b*) $\Delta\theta/(2\pi)$ **41** On the line of symmetry, at distance $4a/(3\pi)$ from the center

43 (*b*) $4\pi/3$, $2\pi - \frac{32}{9}$ (*c*) The edge

Sec. 15.5. The Definite Integral of a Function Over a Region in Space

1 384, 1920 **3** $20\pi r^3/3$ **5** 960 **7** $2\sqrt{6}$

9

(3, 0, 3) (3, 2, 3) (3, 2, 0)

11 **13**

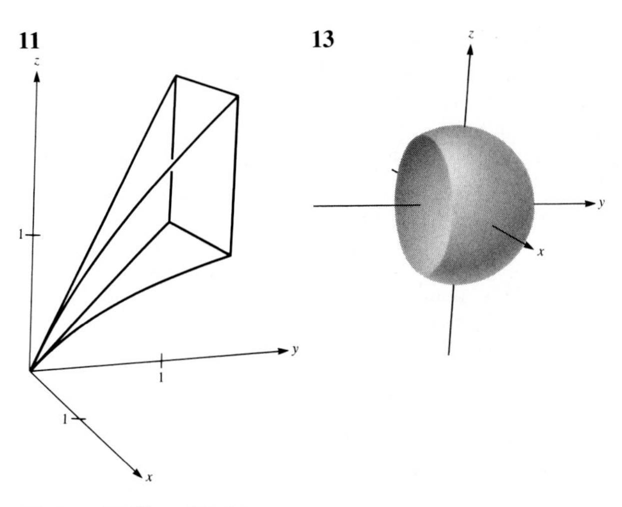

15 $\frac{1}{3}$ **17** $\frac{91}{12}$ **19** (*a*) $-a \le x \le a$, $-\sqrt{a^2 - x^2} \le y \le \sqrt{a^2 - x^2}$, $0 \le z \le h$ (*b*) $-a \le x \le a$, $0 \le z \le h$, $-\sqrt{a^2 - x^2} \le y \le \sqrt{a^2 - x^2}$
21 (*a*) $1 \le x \le 2$, $0 \le y \le 2 - x$, $2 - y \le z \le 2$
(*b*) $1 \le x \le 2$, $x \le z \le 2$, $2 - z \le y \le 2 - x$
23 (*a*)

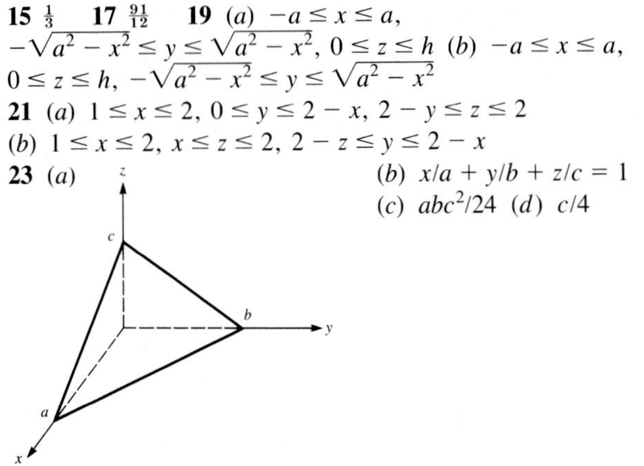

(*b*) $x/a + y/b + z/c = 1$
(*c*) $abc^2/24$ (*d*) $c/4$

25 1024 g **27** $a^2/4$ **29** $\frac{1}{120}$ **31** $\frac{3}{2}(1 - 4e^{-1} + 5e^{-2})$
33 $(M/3)(b^2 + c^2)$ **37** (*a*) $I_2 = I + k^2M$ (*b*) The x axis
39 $\int_M h(P)\, g(P)\, dV$, where M is the mountain

Sec. 15.6. Computing $\int_R f(P)\, dV$ Using Cylindrical Coordinates

1 **3**

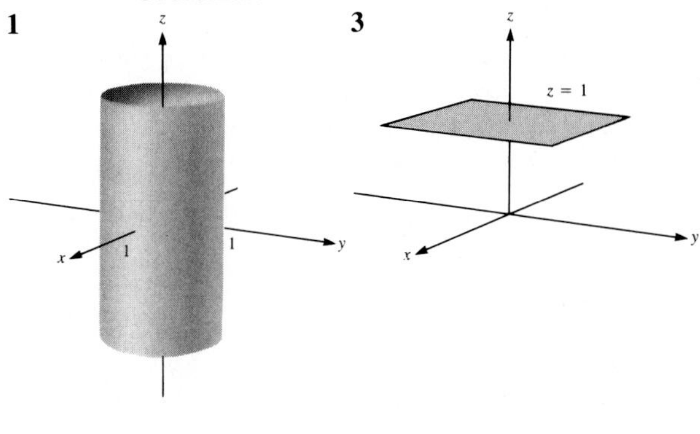

5 **7**

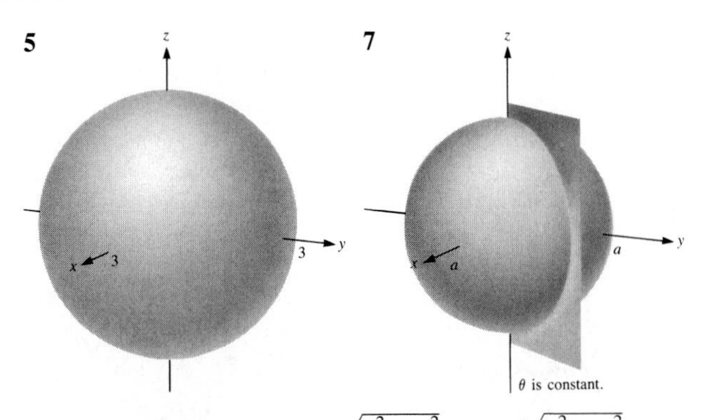

9 $0 \le \theta \le 2\pi$, $0 \le r \le a$, $-\sqrt{a^2 - r^2} \le z \le \sqrt{a^2 - r^2}$
11 (*a*) $r = \sqrt{x^2 + y^2}$, $\theta = \tan^{-1}(y/x)$, $z = z$ (*b*) $x = r\cos\theta$, $y = r\sin\theta$, $z = z$ **13** (*a*) $z = 0$ (*b*) $r = 0$ (and $\theta \le 0$)
(*c*) $r = 0$ **15** $0 \le r \le 1$, $0 \le \theta \le 2\pi$, $r \le z \le 1$
17 $0 \le r \le 1$, $0 \le \theta \le \pi/2$, $0 \le z \le r\cos\theta$
19 $1 \le r \le 2$, $\sin^{-1}(r - 1) \le \theta \le \pi/2$, $r \le z \le 2$
21 $0 \le \theta \le \pi$, $0 \le r \le 2\sin\theta$, $0 \le z \le r^2$
23 $0 \le \theta \le \pi/4$, $0 \le r \le \sec\theta$, $0 \le z \le r^2 \sin\theta\cos\theta$; $\pi/4 \le \theta \le \pi/2$, $0 \le r \le \csc\theta$, $0 \le z \le r^2\sin\theta\cos\theta$
25 $\pi/24$ **27** $\pi/8$ **29** $\frac{4}{3}$ **31** $Ma^2/2$ **33** $M\left(\frac{a^2}{4} + \frac{h^2}{3}\right)$
35 $2Ma^2/5$ **37** $7Ma^2/5$

Sec. 15.7. Computing $\int_R f(P)\, dV$ Using Spherical Coordinates

1 (*a*) Planes (*b*) A sphere, a half-plane, a cone
(*c*) Cylinder, a half-plane, a plane

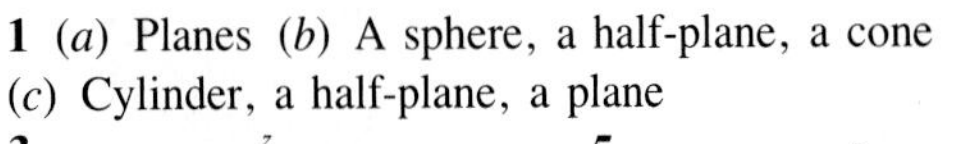

3 **5** **7**

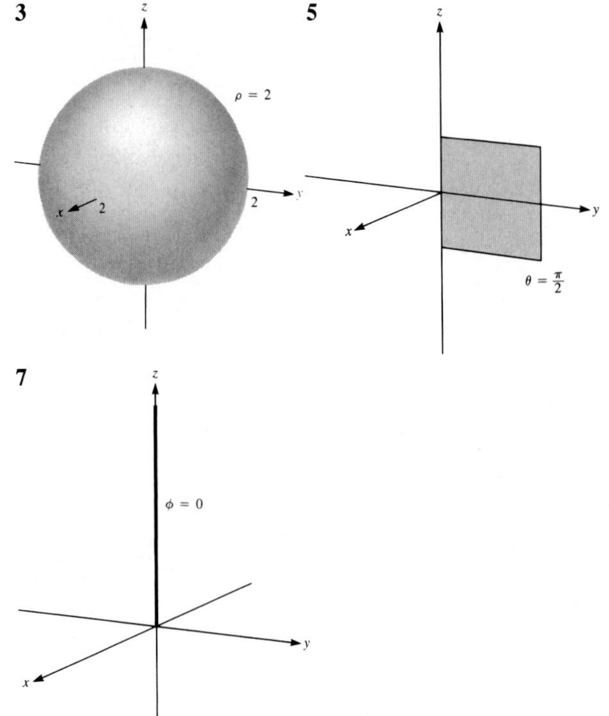

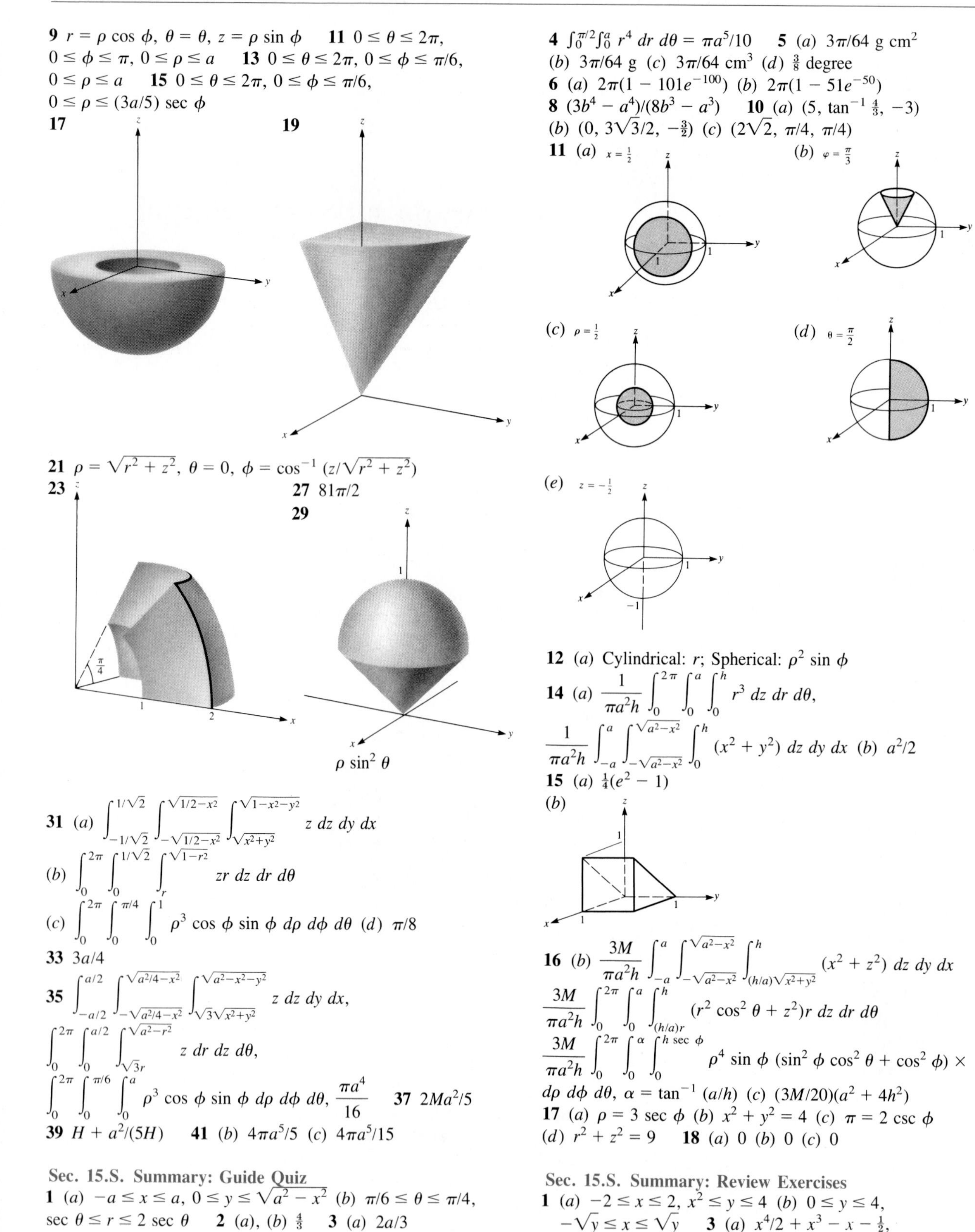

9 $r = \rho \cos \phi$, $\theta = \theta$, $z = \rho \sin \phi$ **11** $0 \le \theta \le 2\pi$, $0 \le \phi \le \pi$, $0 \le \rho \le a$ **13** $0 \le \theta \le 2\pi$, $0 \le \phi \le \pi/6$, $0 \le \rho \le a$ **15** $0 \le \theta \le 2\pi$, $0 \le \phi \le \pi/6$, $0 \le \rho \le (3a/5) \sec \phi$

17 **19**

21 $\rho = \sqrt{r^2 + z^2}$, $\theta = 0$, $\phi = \cos^{-1} (z/\sqrt{r^2 + z^2})$

23 **27** $81\pi/2$

29 $\rho \sin^2 \theta$

31 (*a*) $\int_{-1/\sqrt{2}}^{1/\sqrt{2}} \int_{-\sqrt{1/2-x^2}}^{\sqrt{1/2-x^2}} \int_{\sqrt{x^2+y^2}}^{\sqrt{1-x^2-y^2}} z\, dz\, dy\, dx$

(*b*) $\int_0^{2\pi} \int_0^{1/\sqrt{2}} \int_r^{\sqrt{1-r^2}} zr\, dz\, dr\, d\theta$

(*c*) $\int_0^{2\pi} \int_0^{\pi/4} \int_0^1 \rho^3 \cos \phi \sin \phi\, d\rho\, d\phi\, d\theta$ (*d*) $\pi/8$

33 $3a/4$

35 $\int_{-a/2}^{a/2} \int_{-\sqrt{a^2/4-x^2}}^{\sqrt{a^2/4-x^2}} \int_{\sqrt{3}\sqrt{x^2+y^2}}^{\sqrt{a^2-x^2-y^2}} z\, dz\, dy\, dx$,

$\int_0^{2\pi} \int_0^{a/2} \int_{\sqrt{3}r}^{\sqrt{a^2-r^2}} z\, dr\, dz\, d\theta$,

$\int_0^{2\pi} \int_0^{\pi/6} \int_0^a \rho^3 \cos \phi \sin \phi\, d\rho\, d\phi\, d\theta$, $\dfrac{\pi a^4}{16}$ **37** $2Ma^2/5$

39 $H + a^2/(5H)$ **41** (*b*) $4\pi a^5/5$ (*c*) $4\pi a^5/15$

Sec. 15.S. Summary: Guide Quiz

1 (*a*) $-a \le x \le a$, $0 \le y \le \sqrt{a^2 - x^2}$ (*b*) $\pi/6 \le \theta \le \pi/4$, $\sec \theta \le r \le 2 \sec \theta$ **2** (*a*), (*b*) $\frac{4}{3}$ **3** (*a*) $2a/3$

4 $\int_0^{\pi/2}\int_0^a r^4\, dr\, d\theta = \pi a^5/10$ **5** (*a*) $3\pi/64$ g cm^2 (*b*) $3\pi/64$ g (*c*) $3\pi/64$ cm^3 (*d*) $\frac{3}{8}$ degree

6 (*a*) $2\pi(1 - 101e^{-100})$ (*b*) $2\pi(1 - 51e^{-50})$

8 $(3b^4 - a^4)/(8b^3 - a^3)$ **10** (*a*) $(5, \tan^{-1} \frac{4}{3}, -3)$ (*b*) $(0, 3\sqrt{3}/2, -\frac{3}{2})$ (*c*) $(2\sqrt{2}, \pi/4, \pi/4)$

11 (*a*) $x = \frac{1}{2}$ (*b*) $\varphi = \frac{\pi}{3}$

(*c*) $\rho = \frac{1}{2}$ (*d*) $\theta = \frac{\pi}{2}$

(*e*) $z = -\frac{1}{2}$

12 (*a*) Cylindrical: r; Spherical: $\rho^2 \sin \phi$

14 (*a*) $\dfrac{1}{\pi a^2 h} \int_0^{2\pi} \int_0^a \int_0^h r^3\, dz\, dr\, d\theta$,

$\dfrac{1}{\pi a^2 h} \int_{-a}^{a} \int_{-\sqrt{a^2-x^2}}^{\sqrt{a^2-x^2}} \int_0^h (x^2 + y^2)\, dz\, dy\, dx$ (*b*) $a^2/2$

15 (*a*) $\frac{1}{4}(e^2 - 1)$

(*b*)

16 (*b*) $\dfrac{3M}{\pi a^2 h} \int_{-a}^{a} \int_{-\sqrt{a^2-x^2}}^{\sqrt{a^2-x^2}} \int_{(h/a)\sqrt{x^2+y^2}}^{h} (x^2 + z^2)\, dz\, dy\, dx$

$\dfrac{3M}{\pi a^2 h} \int_0^{2\pi} \int_0^a \int_{(h/a)r}^{h} (r^2 \cos^2 \theta + z^2) r\, dz\, dr\, d\theta$

$\dfrac{3M}{\pi a^2 h} \int_0^{2\pi} \int_0^{\alpha} \int_0^{h \sec \phi} \rho^4 \sin \phi\, (\sin^2 \phi \cos^2 \theta + \cos^2 \phi) \times$ $d\rho\, d\phi\, d\theta$, $\alpha = \tan^{-1} (a/h)$ (*c*) $(3M/20)(a^2 + 4h^2)$

17 (*a*) $\rho = 3 \sec \phi$ (*b*) $x^2 + y^2 = 4$ (*c*) $\pi = 2 \csc \phi$ (*d*) $r^2 + z^2 = 9$ **18** (*a*) 0 (*b*) 0 (*c*) 0

Sec. 15.S. Summary: Review Exercises

1 (*a*) $-2 \le x \le 2$, $x^2 \le y \le 4$ (*b*) $0 \le y \le 4$, $-\sqrt{y} \le x \le \sqrt{y}$ **3** (*a*) $x^4/2 + x^3 - x - \frac{1}{2}$, (*b*) $y^4/4 + y^3 - 3y^2/2$ **5** (*a*) $0 \le x \le a$, $0 \le y \le \sqrt{2a^2 - x^2}$ (*b*) $0 \le y \le a$, $0 \le x \le a$, $0 \le y \le a\sqrt{2}$, $0 \le x \le \sqrt{2a^2 - y^2}$ (*c*) $0 \le \theta \le \pi/4$, $0 \le r \le a \sec \theta$, $\pi/4 \le \theta \le \pi/2$, $0 \le r \le a\sqrt{2}$

7 $\int_{\pi/4}^{\pi/2} \int_0^1 r^2\, dr\, d\theta = \frac{\pi}{12}$

9 (*a*) $f(P) = y = r \sin \theta$ (*b*) $\int_0^{a/\sqrt{2}} \int_x^{\sqrt{a^2 - x^2}} y\, dy\, dx$ (*c*) $a^3/(3\sqrt{2})$

11 $\int_0^{a/\sqrt{2}} \int_y^{\sqrt{a^2-x^2}} (x^2 + y^2)\, dx\, dy = \frac{1}{16}\pi a^4$ **13** (*a*) $Ma^2/3$ (*b*) $Ma^3/12$ **15** (*a*) $128M/5$ (*b*) $8M/9$ (*c*) $1192M/45$

17 $(\pi^2 + 4)/32$ **19** $\frac{1}{80}$ **21** $\frac{2\pi}{(\ln 2)^2}[1 - 2^{-a}(1 + a \ln 2)]$

25 (*a*)

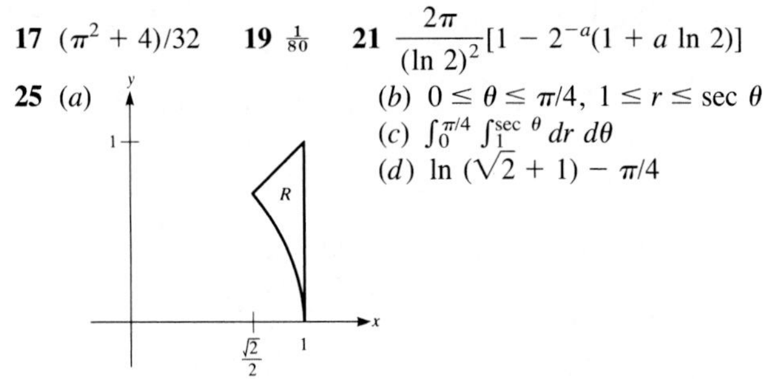

(*b*) $0 \le \theta \le \pi/4$, $1 \le r \le \sec \theta$ (*c*) $\int_0^{\pi/4} \int_1^{\sec \theta} dr\, d\theta$ (*d*) $\ln(\sqrt{2} + 1) - \pi/4$

31 (*a*) $Ma^2/3$ (*b*) $Mb^2/3$ (*c*) $\frac{Ma^2b^2}{6(a^2 + b^2)}$ (*d*) $Ma^2/12$ (*e*) $(M/12)(a^2 + b^2)$

33 (*a*) $Ma^2/2$ (*b*) $3Ma^2/2$ (*c*) $Ma^2/4$ (*d*) $5Ma^2/4$

35 $(e - 1)/2$ **37** (*a*) The integrand is nonnegative.

39 $\frac{3}{2}$

49 $M(a^2 + b^2)/2$ **51** (*a*) $\frac{\sqrt{\pi}}{4}$ (*b*) $\sqrt{\pi}$ (*c*) $\frac{\sqrt{\pi}}{4}$ (*d*) $\frac{\sqrt{\pi}}{2}$ (*e*) $\sqrt{\pi}$ (*f*) $\frac{\sqrt{\pi}}{2}$

CHAPTER 16. GREEN'S THEOREM

Sec. 16.1. Vector and Scalar Fields

1 $2xy + x \cos xy$ **3** 2 **5** $ye^{xy} + 2x \sec^2 2y + 2xz$ **7** 0 **9** $2\mathbf{i} + 4\mathbf{j}$ **11 0** **13 0** **15** Vector **19** (*a*) Vector (*b*) Scalar (*c*) Scalar (*d*) Scalar (*e*) Scalar (*f*) Vector (*g*) Vector **25** (*b*) $\mathbf{F} = x\mathbf{i} + y\mathbf{j} - 2z\mathbf{k}$ is one example.

33

35 $f(r) = kr^{-1}$

Sec. 16.2. Line Integrals

1, **3**

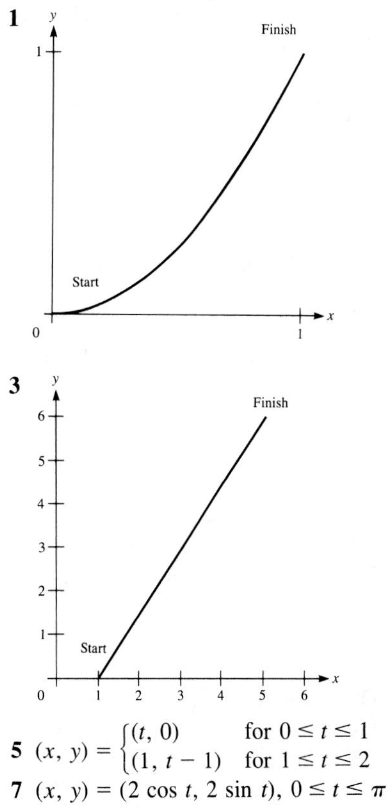

5 $(x, y) = \begin{cases} (t, 0) & \text{for } 0 \le t \le 1 \\ (1, t - 1) & \text{for } 1 \le t \le 2 \end{cases}$

7 $(x, y) = (2 \cos t, 2 \sin t)$, $0 \le t \le \pi$ **9** $\frac{26}{3}$ **11** 0

17 -15 **19** 5 **21** $\frac{15}{4}$ **23** (*a*) $\frac{17}{2}$ (*b*) $\frac{101}{12}$ **25** 11
29 $1/\pi + \pi/4 - 1/2 \ln 2 + 1$ **31** $\pi/4$ **33** $(2\sqrt{2} - 1)/3$
35 $(e^{2\pi} + 1)/2$ **37** (*a*) $\frac{2}{3}$ (*b*) $\frac{2}{3}$ (*c*) $\frac{3}{4}$ (*d*) $\frac{1}{2}$
39 $2 \ln (2 + \sqrt{3})$

Sec. 16.3. Four Applications of Line Integrals

1 0 **3** $72\pi^2$ **5** ∞ **7** $\frac{72}{5}$ **9** 8 **11** (*a*) $\frac{1}{2}$ (*b*) $\frac{1}{2}$
13 (*b*) Fluid is leaving; net outward flow is positive. (*c*) 3π
19 Counterclockwise **23** (*a*) 9 (*b*) 9 (*c*) Yes **25** 2π
27 $\dfrac{dQ}{dt} = -\oint_C \mathbf{F}(P) \cdot \mathbf{n}\, ds$

Sec. 16.4. Green's Theorem

1 5π **3** $\dfrac{ab^2}{2} + \dfrac{ba^3}{3}$ **5** $e - \dfrac{e^2}{2} - \dfrac{1}{2}$ **7** 0
9 (*a*) $\int_{\mathcal{A}} \nabla \cdot \mathbf{F}\, dA > 0$ (*b*) $\int_{\mathcal{A}} \nabla \cdot \mathbf{F}\, dA = 0$ **11** $\frac{8}{3}$ **13** 0
17 Rate of outflow $\approx 5.027 \times 10^{-3}$ **19** Rate of outflow $\approx 2.000 \times 10^{-3}$
23 (*a*)

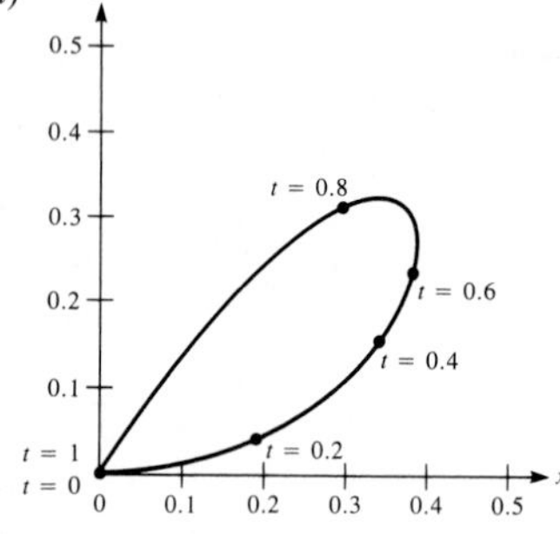

(*b*) Neither x nor y can be solved for explicitly. (*c*) $\frac{7}{120}$

Sec. 16.5. Applications of Green's Theorem

1 3π **3** $\oint_{C_2} \mathbf{F} \cdot \mathbf{n}\, ds = 0$; $\oint_{C_3} \mathbf{F} \cdot \mathbf{n}\, ds = 5$ **5** (*a*) 2π
(*b*) No **7** 0 **9** $\oint_{C_1} \mathbf{F} \cdot d\mathbf{r} = \oint_{C_3} \mathbf{F} \cdot d\mathbf{r}$; $\oint_{C_2} \mathbf{F} \cdot d\mathbf{r} = 0$
11 (*a*) **0** (*b*) 0 (*c*) 0 **21** div $\mathbf{F} = 0$ everywhere
23 (*a*) $m(t) = \int_{\mathcal{A}} \sigma\, dA$ (*b*) $\dfrac{dm}{dt} = -\oint_C \mathbf{F} \cdot \mathbf{n}\, ds$
(*c*) $\dfrac{dm}{dt} = -\oint_{\mathcal{A}} \nabla \cdot \mathbf{F}\, dA$ **25** (*a*) **0**
(*b*) $\oint_{C_1} \mathbf{F} \cdot d\mathbf{r} = \oint_{C_2} \mathbf{F} \cdot d\mathbf{r} = 0$

Sec. 16.6. Conservative Vector Fields

1 It equals -5. **3** It equals 0. **5** $x^2y + C$
7 $xy + x + y + C$ **11** True **13** True **15** Any curve C that does not enclose the origin **21** (*b*) $y^3 dx + x^3\, dy$ is one such example. **23** $h(x, y) = \int_0^{\sqrt{x^2+y^2}} f(u)\, du$ is one such function.

Sec. 16.S. Summary: Guide Quiz

5 $\nabla \cdot \mathbf{F} = 0$, $\nabla \times \mathbf{F} = 0$
8 div $\mathbf{F} = \lim\limits_{(\text{diameter of } \mathcal{A}) \to 0} \dfrac{\oint_C \mathbf{F} \cdot \mathbf{n}\, ds}{\text{Area of } \mathcal{A}}$ **9** $4/(5\pi)$
10 (*a*) 2π (*b*) 2π (*c*) 0
11 (*a*) $\sin 4 - \sin 1 + \cos 1 - \cos 2$
(*b*) $\sin 4 - \sin 1 + \cos 1 - \cos 2$

Sec. 16.S. Summary: Review Exercises

1 $\frac{29}{60}$ **5** (*a*) No (*b*) Yes **7** (*b*) $-20/\pi$
11 (*a*) $q/(8\pi\epsilon_0)$ (*b*) $q/(8\pi\epsilon_0)$ **13** (*a*) All integers k
(*b*) $k = 1$ **17** (*a*) $-\frac{1}{2}$ (*b*) $-\frac{1}{2}$
19 $\oint_{C_2} \mathbf{F} \cdot \mathbf{n}\, ds = \oint_{C_3} \mathbf{F} \cdot \mathbf{n}\, ds = 0$
21 (*a*) (*b*) 0 **23** (*b*) $f(x, y) = x^2y$

25 (*b*) $(\theta_2 - \theta_1)(r_2^3 - r_1^2)$ (*c*) Clockwise, as viewed from the origin **37** (*a*) 0

CHAPTER 17. THE DIVERGENCE THEOREM AND STOKES' THEOREM

Sec. 17.1. Surface Integrals

1 0.07 **3** (*b*) $3\pi\sqrt{14}$ **5** $2a^2/3$ **7** 0 **9** 0
11 Volume of $\mathcal{S}$ **13** $\dfrac{(a^2 - 2)(1 + a^2)^{3/2} + 2}{\pi[(1 + a^2)^{3/2} - 1]}$
15 $\frac{1}{3}(5^{3/2} - 1)$ **17** (*a*) 2 (*b*) $2\sqrt{3}$ **19** $1/\sqrt{2}$
21 $\dfrac{3 \cdot 19^{5/2} - 5 \cdot 19^{3/2} + 2}{10(19^{3/2} - 1)}$ **23** (*a*) 0 (*b*) 0 (*c*) 0
25 $8\pi k/3$ **27** $2Ma^2/3$ **31** $2a/3$ **33** $\pi/(2\sqrt{2})$
35 $\dfrac{a^2h}{2} + \dfrac{h^3}{3}$

Sec. 17.2. The Divergence Theorem

5 48 **7** 11 **9** $17\pi/256$ **13** 0 **15** 8π **17** $-\frac{1}{2}$
19 Its absolute value is at most 5 times the surface area of $\mathcal{S}$. **21** 0 **23** 0 **25** (*a*) Radially outward
(*b*) Magnitude of $\mathbf{E}$ is $q/(2\pi\epsilon_0 r)$, where r is the distance from L. **29** (*a*) Volume of $\mathcal{V}$ **31** 4π **33** 2π
35 (*a*) $\pi/6$ (*b*) $2\pi/3$

Sec. 17.3. Stokes' Theorem

3 Both integrals equal 0. **5** Both integrals equal $\pi/8$.
7 -1 **9** 0 **11** 0 **13** They are opposites. **15** (*a*) No
(*b*) Yes. They are equal. **17** (*a*), (*b*), (*d*), (*e*), (*f*), (*h*), (*i*), (*j*), and (*k*) are connected; (*b*), (*f*), and (*h*) are also simply connected; (*c*) and (*g*) are not connected. **19** $-\pi$
23 0 **25** (*c*) $f(x, y, z) = \dfrac{1}{2 - a}(x^2 + y^2 + z^2)^{(2-a)/2}$

27 $11\pi/64$ **29** (*c*) Yes (*d*) Middle: (*b*) gives 1 band with 4 half-twists; (*c*) gives 2 linked bands, each with a full twist. Thirds: (*b*) gives 2 linked bands: 1 short with a half-twist, 1 long with 4 half-twists; (*c*) gives 3 linked bands, each with a full twist.

Sec. 17.4. Applications of Stokes' Theorem

1 $\dfrac{0.0027\pi}{\sqrt{11}}$

13 (*a*)

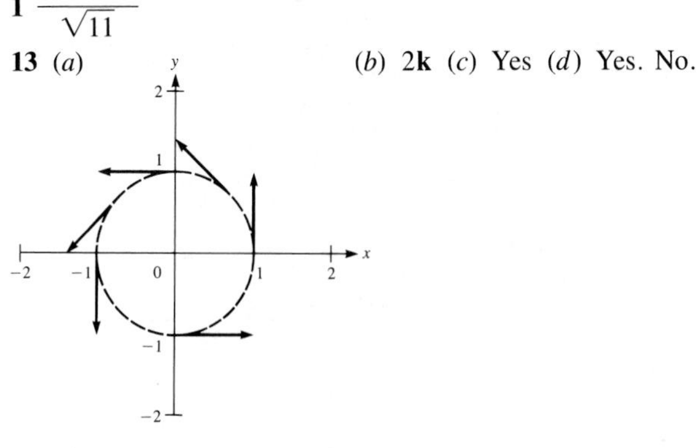

(*b*) $2\mathbf{k}$ (*c*) Yes (*d*) Yes. No.

Sec. 17.S. Summary: Guide Quiz

4 $-\frac{1}{6}$ **5** (*a*) Radially symmetric (*b*) $\mathbf{G} = \dfrac{km}{4\pi r^2}\hat{\mathbf{r}}$ **6** 0

7 (*a*) All (*b*) All central fields (*c*) $\hat{\mathbf{r}}/\|\mathbf{r}\|$ (*d*) $\hat{\mathbf{r}}/\|\mathbf{r}\|^2$

9 (*a*) $\mathbf{F}$ is conservative. (*b*) div $\mathbf{F} = 0$ (*c*) div $\mathbf{F} = 0$ (*d*) div $\mathbf{F} = 0$ on $\mathcal{V}$ **11** Yes **12** Yes

Sec. 17.S. Summary: Review Exercises

3 (*a*) $\int_{\mathcal{A}} \sqrt{4x^2 + 36y^2 + 1}\, dA$, $\mathcal{A}$ is $2x^2 + 4y^2 \le 1$
(*b*) $4\int_0^{1/\sqrt{2}} \int_0^{\sqrt{1-2x^2}/2} \sqrt{4x^2 + 36y^2 + 1}\, dy\, dx$,
$4\int_0^{\pi/2} \int_0^{1/\sqrt{2+2\sin^2\theta}} \sqrt{4r^2(1 + 8\sin^2\theta) + 1}\, r\, dr\, d\theta$

5 (*a*) $\mathbf{F}$ is conservative and $\oint_{C_1} \mathbf{F}\cdot\mathbf{T}\, ds = \oint_{C_2} \mathbf{F}\cdot\mathbf{T}\, ds = 0$. (*b*) $\oint_{C_1} \mathbf{F}\cdot\mathbf{T}\, ds = \oint_{C_2} \mathbf{F}\cdot\mathbf{T}\, ds$ **7** (*a*) All n (*b*) $n = 3$

9 $6\ln 2 + 16 - 2^{5/2}/3$ **11** $\frac{2}{3}$ **13** They are opposite in sign. **15** $3\pi/\sqrt{2}$

17 (*a*)

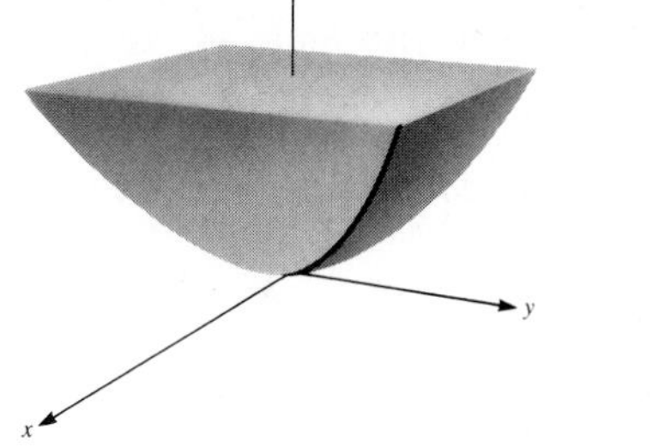

(*b*) $\dfrac{5\sqrt{5}}{6} + \dfrac{3}{2}$ **19** 1

21 (*a*) Symmetry (*b*) $\mathbf{E} = \dfrac{Q}{4\pi\epsilon_0 r^2}\hat{\mathbf{r}}$ (*c*) $\mathbf{0}$ **23** $-1 + \cos 2$

25 (*a*) $\dfrac{Q}{4\pi\epsilon_0 r^2}\hat{\mathbf{r}}$ (*b*) $\dfrac{Qr}{4\pi\epsilon_0 a^3}\hat{\mathbf{r}}$ **27** (*a*) Yes (*b*) No

29 (*a*) Yes (*b*) No **31** No **33** Yes

35 (*b*) $f(x, y, z) = \frac{1}{3}(x^2 + y^2 + z^2)^{3/2}$

41 (*a*) $f(x, y, z) = e^{xz}$ (*b*) $\exp(4\pi^2 e^{2\pi}) - 1$

45 (*a*) $\mathbf{r}(u, v) = (b + a\cos v)\cos u\,\mathbf{i} + (b + a\cos v)\sin u\,\mathbf{j} + a\sin v\,\mathbf{k}$ (*b*) $b + \dfrac{a^2}{2b}$ **47** (*a*) $\int_{\mathcal{S}} cz\mathbf{k}\cdot\mathbf{n}\, dS$

49 f is constant. **59** 0

APPENDIX A. REAL NUMBERS

1 $a + 3 < b + 3$ **3** $5a < 5b$ **5** $1/a > 1/b$ **7** $(-1, \infty)$

9 $(-\infty, -2)$ **11** $(-\infty, 1), (3, \infty)$ **13** $(-3, -2)$

15 $(-1, 0), (1, \infty)$ **17** $(-5, -3), (0, \infty)$ **19** $(-\infty, \frac{1}{3}), (\frac{1}{2}, \infty)$

21 0 1 2 3 4 5 x

23 −1 0 1 2 3 x

25 $x > 3$ **27** $x > 1$ **29** $a = -2, b = 1$ **31** $a = -1, b = 1, c = -2, d = 1$ **33** (*a*) $234.74747\cdots$ (*b*) 232.4 (*d*) $a = 2{,}324/990$

APPENDIX B. GRAPHS AND LINES

Sec. B.1. Coordinate Systems and Graphs

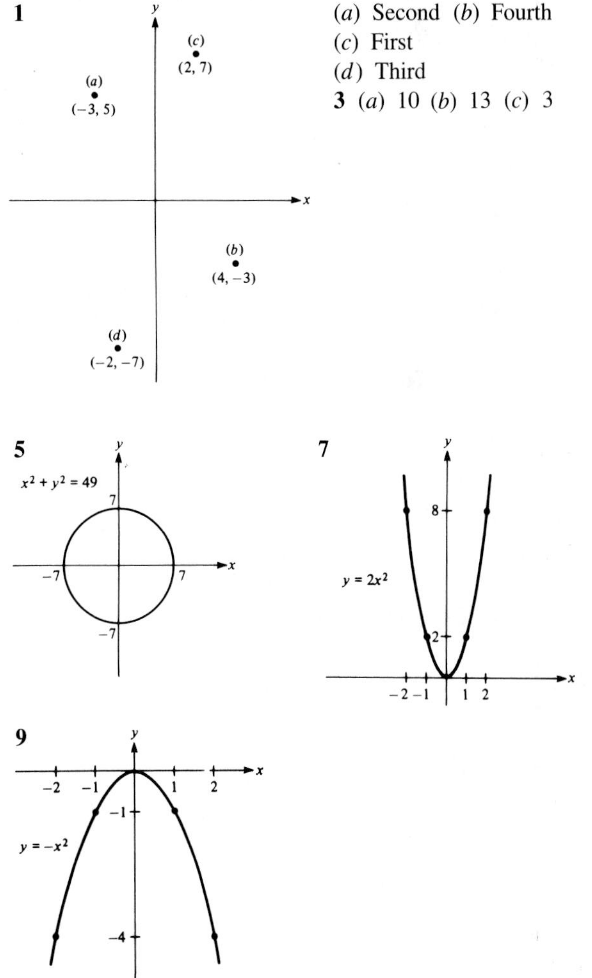

1 (*a*) Second (*b*) Fourth (*c*) First (*d*) Third

3 (*a*) 10 (*b*) 13 (*c*) 3

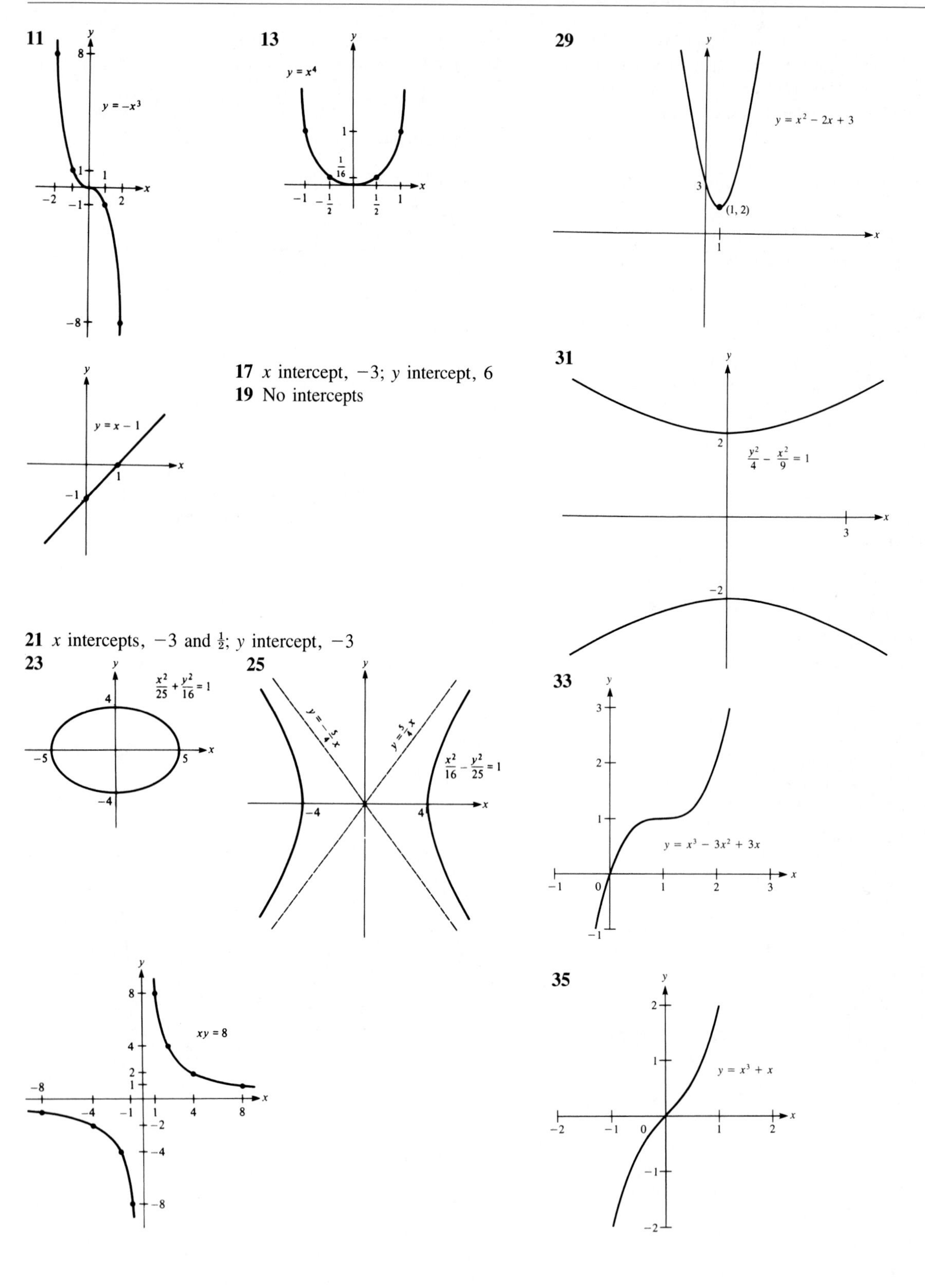

17 x intercept, -3; y intercept, 6

19 No intercepts

21 x intercepts, -3 and $\frac{1}{2}$; y intercept, -3

37

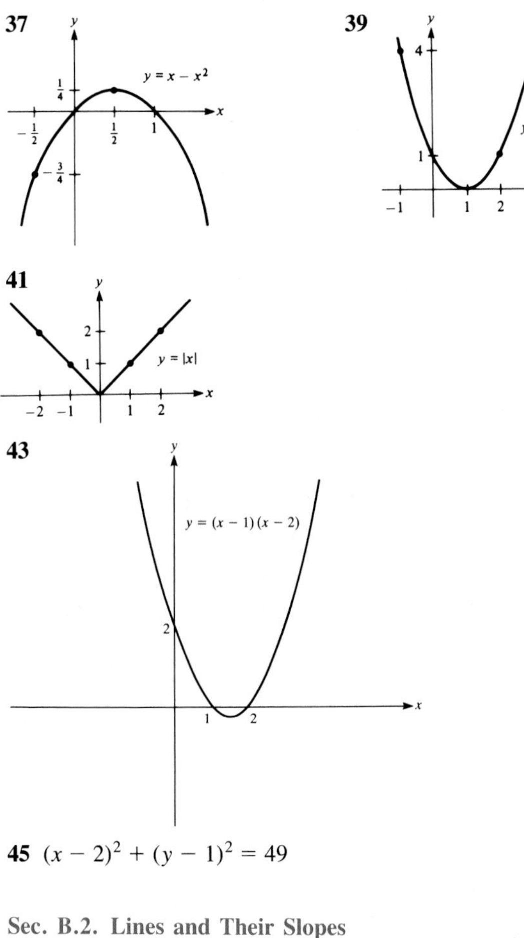

39

41

43

45 $(x-2)^2 + (y-1)^2 = 49$

Sec. B.2. Lines and Their Slopes

1 $m = \frac{1}{5}$ **3** $m = -\frac{5}{4}$

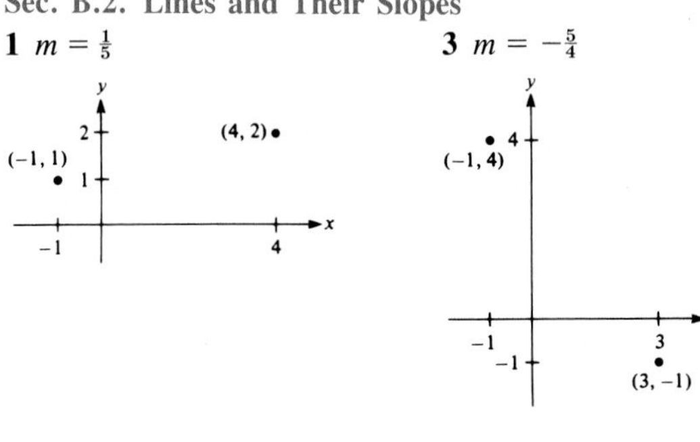

5 $m = 0$

7 (*a*) Negative (*b*) Negative (*c*) Positive (*d*) Positive (*e*) Zero

9 (*a*) 4 (*b*) $-\frac{1}{4}$ **11** Yes **13** No

15 (*a*) $y = 3x + 2$ (*b*) $y = \frac{2}{3}x - 2$ (*c*) $y = -3x$

17 (*a*) $m = 3$, $b = -1$ (*b*) $m = -2$, $b = 1$

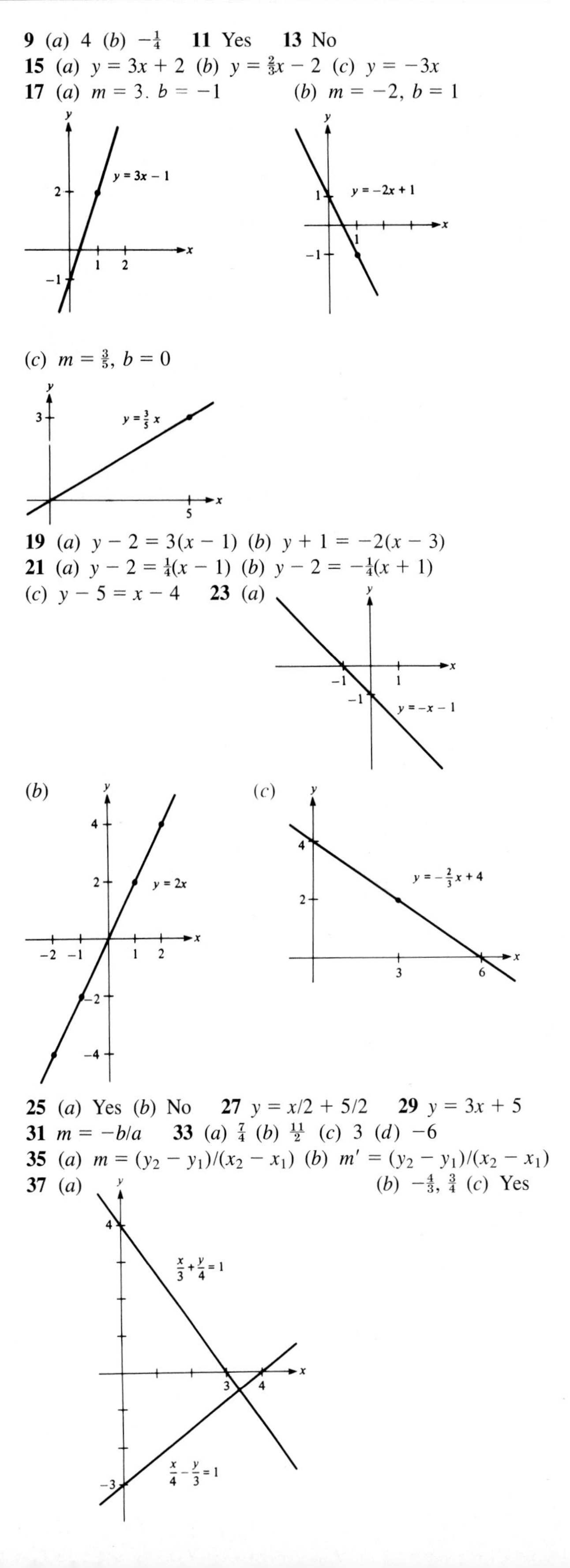

(*c*) $m = \frac{3}{5}$, $b = 0$

19 (*a*) $y - 2 = 3(x - 1)$ (*b*) $y + 1 = -2(x - 3)$

21 (*a*) $y - 2 = \frac{1}{4}(x - 1)$ (*b*) $y - 2 = -\frac{1}{4}(x + 1)$ (*c*) $y - 5 = x - 4$ **23** (*a*)

(*b*) (*c*)

25 (*a*) Yes (*b*) No **27** $y = x/2 + 5/2$ **29** $y = 3x + 5$

31 $m = -b/a$ **33** (*a*) $\frac{7}{4}$ (*b*) $\frac{11}{2}$ (*c*) 3 (*d*) −6

35 (*a*) $m = (y_2 - y_1)/(x_2 - x_1)$ (*b*) $m' = (y_2 - y_1)/(x_2 - x_1)$

37 (*a*) (*b*) $-\frac{4}{3}$, $\frac{3}{4}$ (*c*) Yes

39 (*a*) $|m|$ is large. (*b*) m is close to 0. (*c*) $m < 0$ (*d*) $m > 0$ **41** $P = (-1, 0)$

APPENDIX C. TOPICS IN ALGEBRA

1 $2\sqrt{5}$ **3** $\sqrt{2} + 4$ **5** $\frac{1}{3}(6 + 2\sqrt{3})$ **7** $\dfrac{x(1 + \sqrt{x})}{1 - x}$

9 $\dfrac{7}{5(3 - \sqrt{2})}$ **11** $\dfrac{1}{\sqrt{x} + \sqrt{5}}$ **13** (*a*) $(x + 4)^2 - 3$ (*b*) $(x - 4)^2 + 7$ (*c*) $(x - \frac{1}{2})^2 + \frac{7}{4}$ **15** (*a*) $(x + \frac{3}{2})^2 - \frac{17}{4}$ (*b*) $(x + \frac{3}{2})^2 + \frac{19}{4}$ (*c*) $(x + \frac{5}{4})^2 + \frac{39}{16}$ **17** (*a*) $2(x - \frac{5}{4})^2 - \frac{1}{8}$ (*b*) $2(x + \frac{3}{2})^2 + \frac{5}{2}$ (*c*) $3(x + \frac{5}{6})^2 - \frac{13}{12}$ **19** (*a*) No real solutions (*b*) $x = \frac{1}{2}(-1 \pm \sqrt{5})$ (*c*) $x = -1$ **21** (*a*) None (*b*) Two (*c*) One **23** (*a*) 10 (*b*) 15 (*c*) 45
25 $1 + 7x + 21x^2 + 35x^3 + 35x^4 + 21x^5 + 7x^6 + x^7$
27 45 **29** (*a*) 1,093/729 (*b*) 1,275 (*c*) 33/8
31 $x - (-2)$, $x - (-1)$, $x - 1$ **33** $x - 1$ **35** $x - (-1)$, $x - \frac{2}{3}$, $x - \frac{3}{2}$ **37** $\left(x + \dfrac{5 + \sqrt{17}}{2}\right)\left(x + \dfrac{5 - \sqrt{17}}{2}\right)$

39 $\dfrac{1}{3}\left(1 - \dfrac{1}{10^9}\right)$

APPENDIX D. EXPONENTS

1 (*a*) 1 (*b*) 32 (*c*) 2 (*d*) $\frac{1}{2}$ (*e*) 8
3 **5**

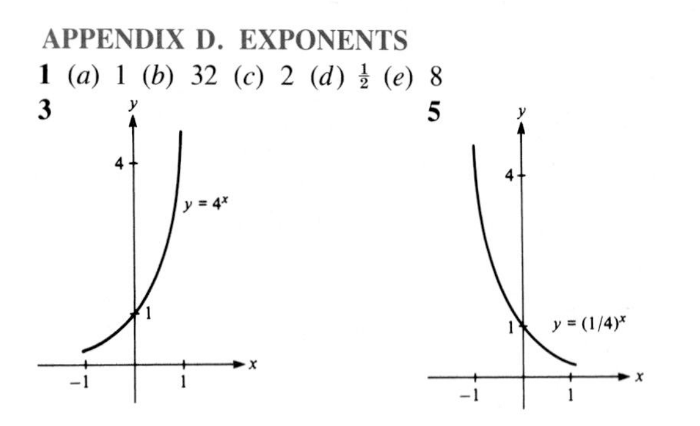

7 (*a*) 2^4 (*b*) 2^{-3} (*c*) $2^{1/2}$ (*d*) 2^0 (*e*) 2^{-2} **9** (*a*) $b^{3/2}$ (*b*) $b^{3/2}$ (*c*) $b^{-1/3}$ (*d*) $b^{2/3}$ (*e*) $b^{1/4}$ **11** (*a*) Domain: all x; range: $y > 0$ (*b*) Domain: all x; range: all y (*c*) Domain: $x \ge 0$; range: $y \ge 0$ (*d*) Domain: all x; range: $y \ge 0$
13 (*a*)

x	-32	-1	0	1	32
$x^{3/5}$	-8	-1	0	1	8

(*b*)

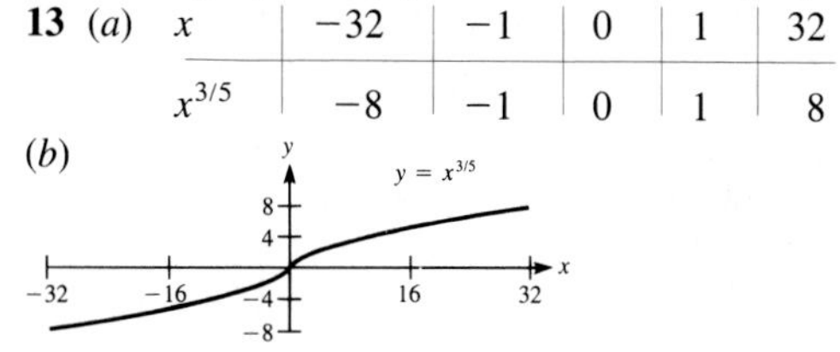

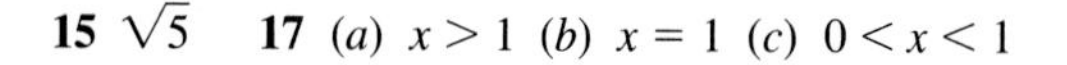

15 $\sqrt{5}$ **17** (*a*) $x > 1$ (*b*) $x = 1$ (*c*) $0 < x < 1$

19

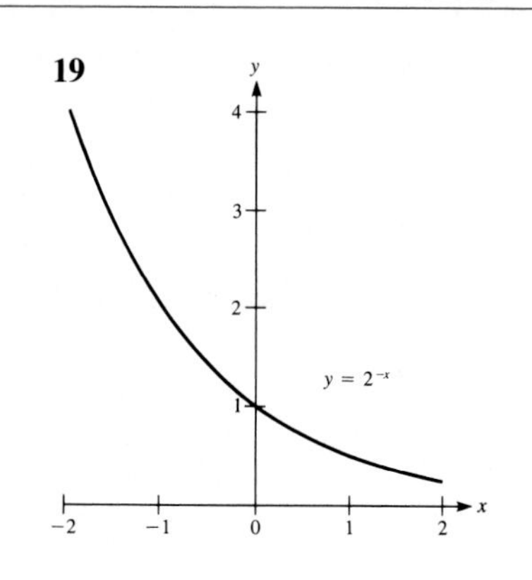

21 (*a*) 10^3 (*b*) 10^{-4} (*c*) 10^6 (*d*) 10^{-7} **23** The second way **25** (*b*) -1, 1

APPENDIX E. MATHEMATICAL INDUCTION

3 (*a*) 2π (*b*) $(n - 2)\pi$ **5** (*a*) k is even.

APPENDIX F. THE CONVERSE OF A STATEMENT

1 (*a*) "If $a^2 = b^2$, then $a = b$" is false. (*b*) "If $a^3 = b^3$, then $a = b$" is true. **3** (*a*) "If $ab = ac$, then $b = c$" is false. (*b*) "If $a + b = a + c$, then $b = c$" is true.
5 (*a*) "If ab is odd, then a and b are odd" is true. (*b*) "If ab is even, then a and b are even" is false. **7** True
9 True **11** True **13** False **15** True **17** False
19 False **21** False **23** True

APPENDIX G. CONIC SECTIONS

Sec. G.1. Conic Sections

1 $\dfrac{x^2}{25} + \dfrac{y^2}{21} = 1$

3

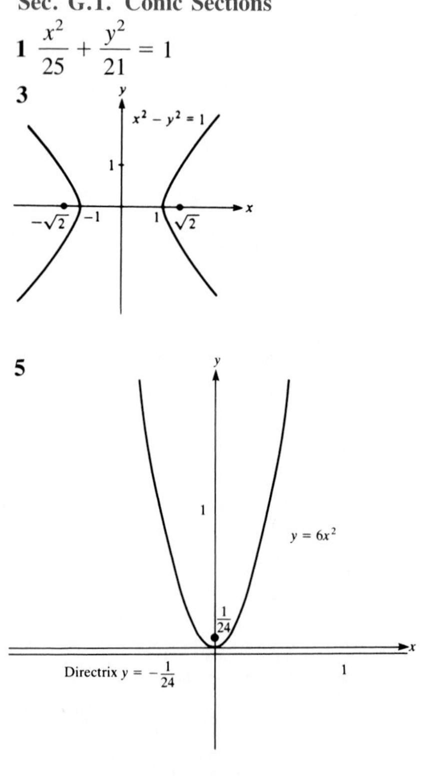

5

7 $y^2 = 12x$

11

13

15

17

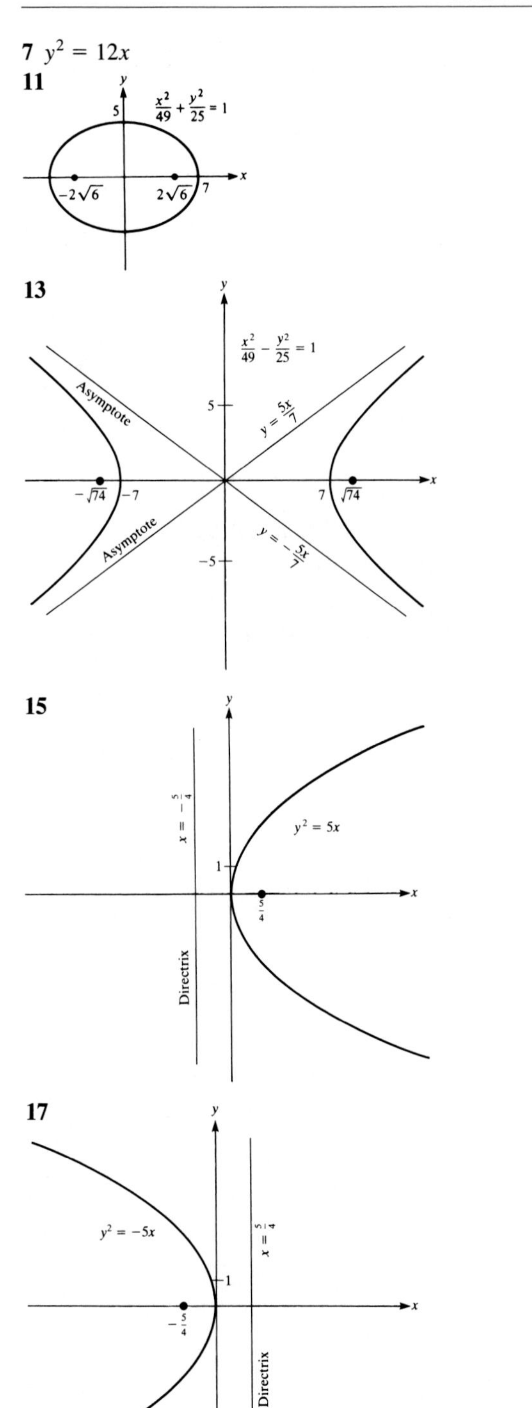

19 (*b*)

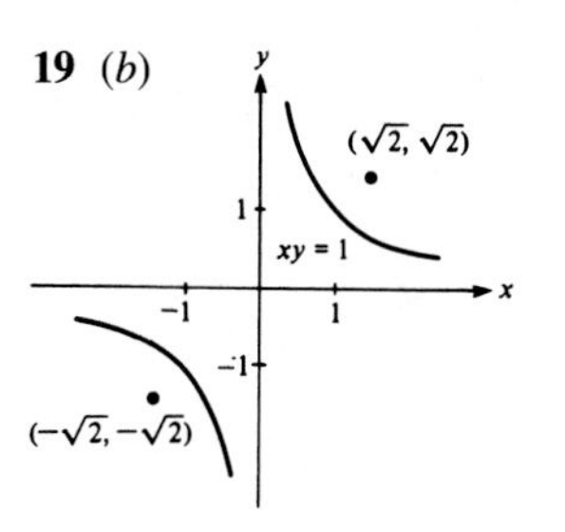

Sec. G.2. Translation of Axes and the Graph of $Ax^2 + Cy^2 + Dx + Ey + F = 0$

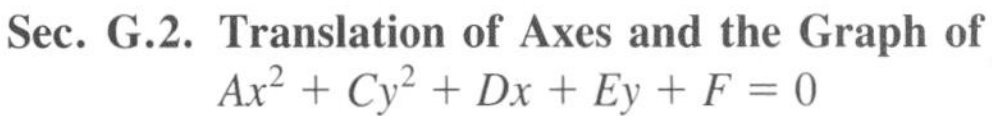

1

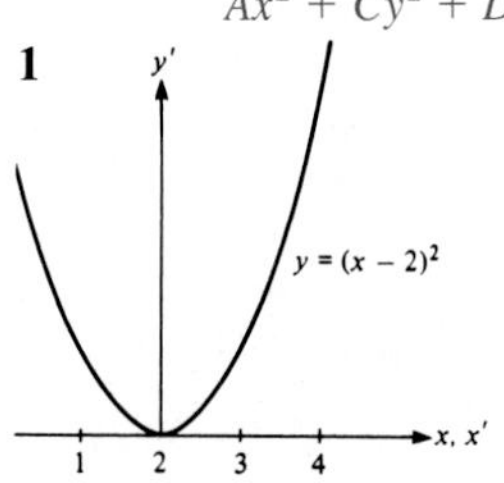

Parabola; focus = $(2, \frac{1}{4})$; vertex = $(2, 0)$

3

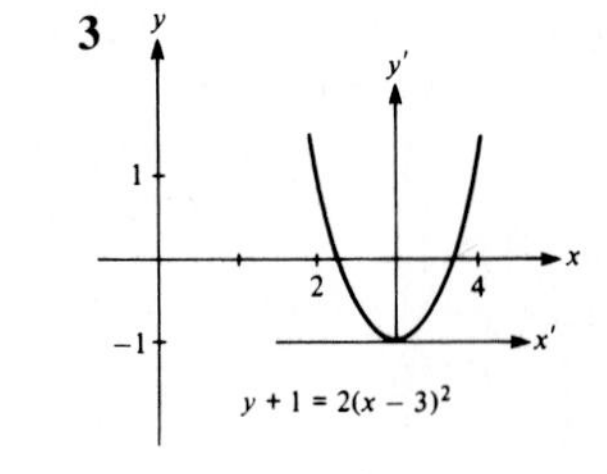

Parabola; focus = $(3, -\frac{7}{8})$; vertex = $(3, -1)$

5

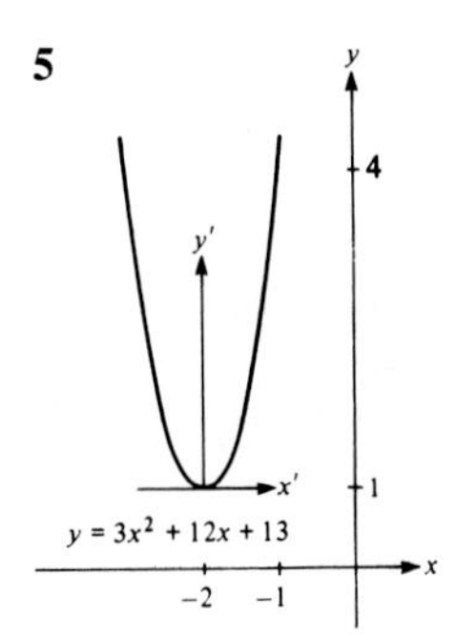

Parabola; focus = $(-2, \frac{13}{12})$; vertex = $(-2, 1)$

7

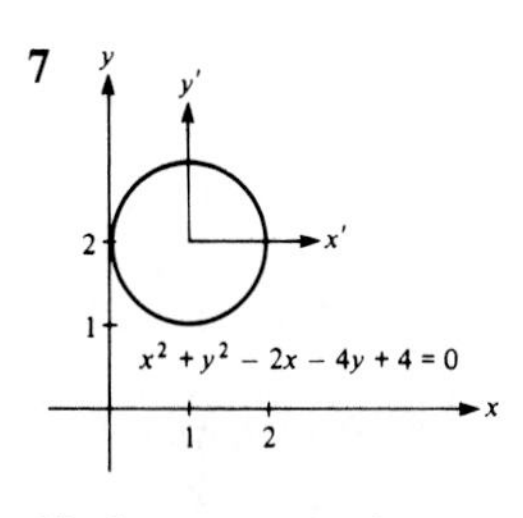

Circle; center (= focus) = $(1, 2)$

9

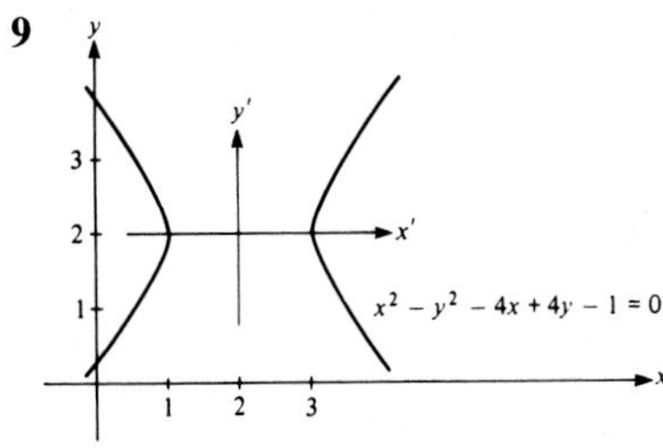

Hyperbola; foci = $(2 \pm \sqrt{2}, 0)$; vertices = $(1, 0)$, $(3, 0)$; asymptotes: $y = x$ and $x + y = 4$

11

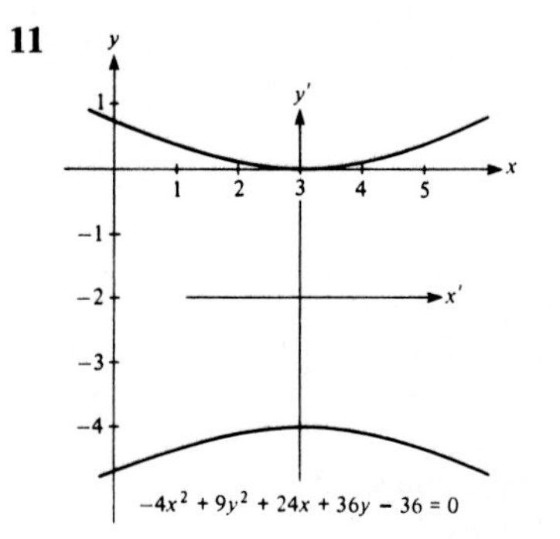

Hyperbola; foci = $(3, -2 \pm \sqrt{13})$; vertices = $(3, 0)$, $(3, -4)$; asymptotes: $y = \frac{2}{3}x - 4$ and $y = -\frac{2}{3}x$

13

$25x^2 + 4y^2 + 100x + 24y + 36 = 0$

Ellipse; foci = $(-2, -3 \pm \sqrt{21})$; vertices = $(-4, -3)$, $(0, -3)$, $(-2, -8)$, $(-2, 2)$

15 $(x')^2/9 + (y')^2/4 = 1$, where $x' = x - 1$, $y' = y - 2$.

17 $\frac{13}{36}(x')^2 - \frac{13}{16}(y')^2 = 1$, where $x' = x - 1$, $y' = y - 1$.

19 $(x')^2/9 + (y')^2/4 = 1$, where $x' = x - 4$, $y' = y$.

21 $(y')^2 = 12x'$, where $x' = x - 4$, $y' = y - 3$.

23 (d) When $A = C$ and their sign is opposite that of F

Sec. G.3. Rotation of Axes and the Graph of $Ax^2 + Bxy + Cy^2 + Dx + Ey + F = 0$

1

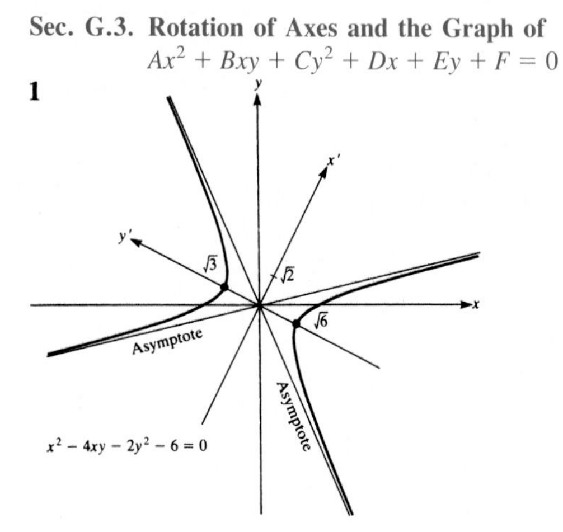

3

5

7

9

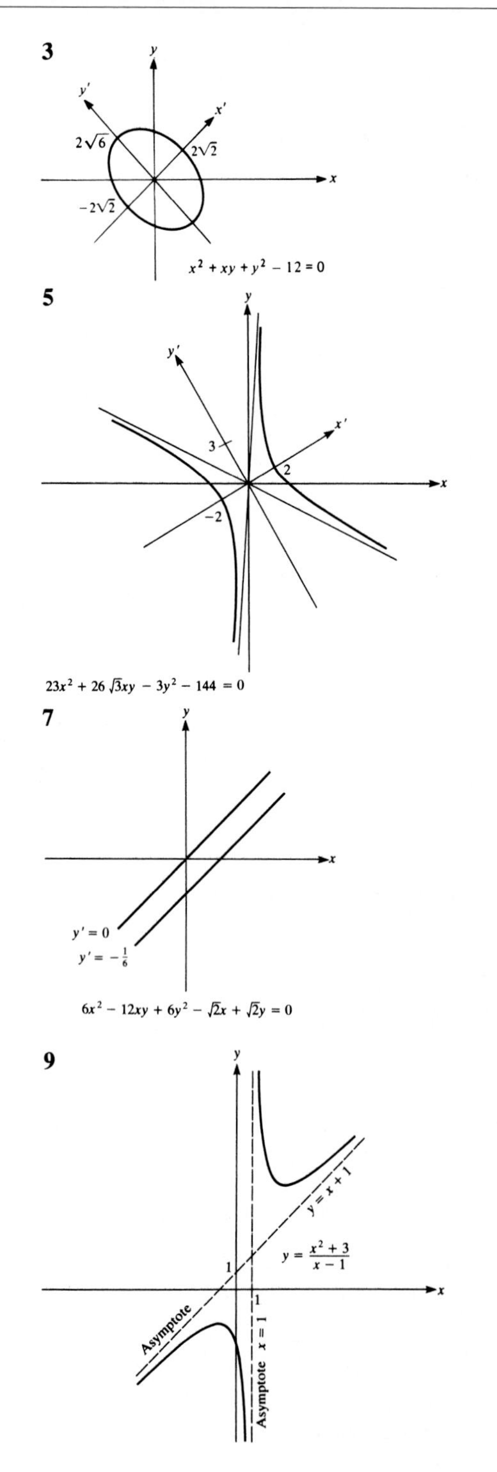

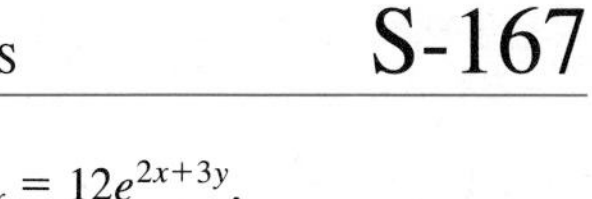

11

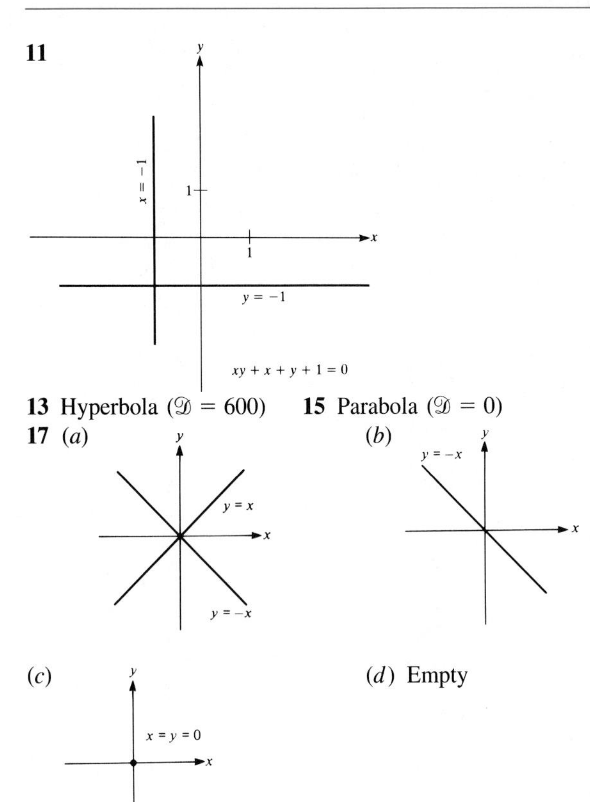

13 Hyperbola ($\mathscr{D} = 600$) **15** Parabola ($\mathscr{D} = 0$)

17 (*a*) (*b*)

(*c*) (*d*) Empty

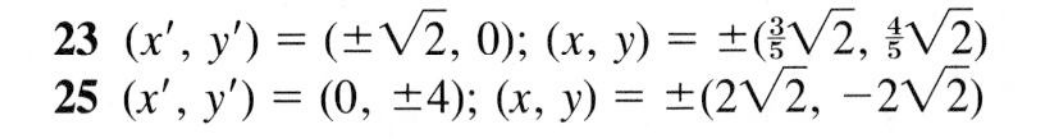

23 $(x', y') = (\pm\sqrt{2}, 0)$; $(x, y) = \pm(\frac{3}{5}\sqrt{2}, \frac{4}{5}\sqrt{2})$
25 $(x', y') = (0, \pm 4)$; $(x, y) = \pm(2\sqrt{2}, -2\sqrt{2})$

Sec. G.4. Conic Sections in Polar Coordinates

1 (*a*)

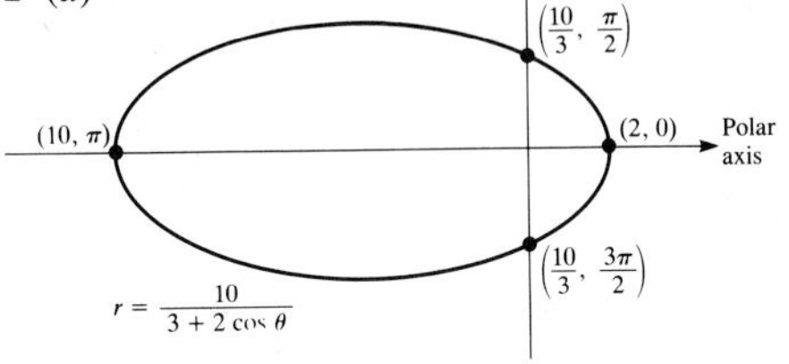

3 (*a*) $e = \frac{4}{3}$ (*b*) $e = \frac{3}{4}$ (*c*) $e = 1$ (*d*) $e = \frac{4}{3}$
5 (*b*) $r = 2/(1 + \cos\theta)$ (*c*) $r = -4 \csc\theta \cot\theta$

APPENDIX H. LOGARITHMS AND EXPONENTIALS DEFINED THROUGH CALCULUS

Sec. H.1. The Natural Logarithm Defined as a Definite Integral

3 (*a*) $28{,}271/27{,}720 \approx 1.0199$

APPENDIX I. THE TAYLOR SERIES FOR $f(x, y)$

1 $f_{xxx} = 60x^2y^7$, $f_{yyy} = 210x^5y^4$, $f_{xxy} = f_{xyx} = f_{yxx} = 140x^3y^6$, $f_{yyx} = f_{yxy} = f_{xyy} = 210x^4y^5$ **3** $f_{xxx} = 8e^{2x+3y}$, $f_{yyy} = 27e^{2x+3y}$, $f_{xxy} = f_{xyx} = f_{yxx} = 12e^{2x+3y}$, $f_{yyx} = f_{yxy} = f_{xyy} = 18e^{2x+3y}$
7 (*a*), (*b*) $e^{x+y^2} = 1 + x + (x^2/2) + y^2 + \cdots$
9 $2 + 3x - 5y + 3x^2 + 7xy + \frac{1}{2}y^2$

APPENDIX K. THE INTERCHANGE OF LIMITS

Sec. K.2. The Derivative of $\int_a^b f(x, y)\, dx$ with Respect to y

3 $y = 2/\pi$ **5** $f(u, u) + \int_0^u f_u(u, x)\, dx$

Sec. K.3. The Interchange of Limits

1 (*a*)

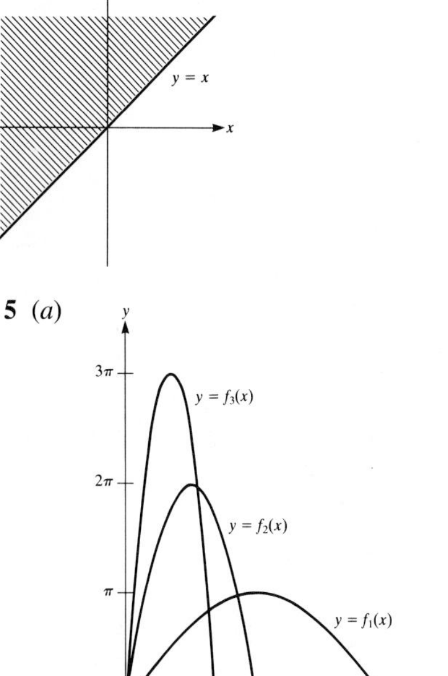

5 (*a*)

APPENDIX L. THE JACOBIAN

Sec. L.1. Magnification and the Jacobian

3 (*a*) (*b*) 1

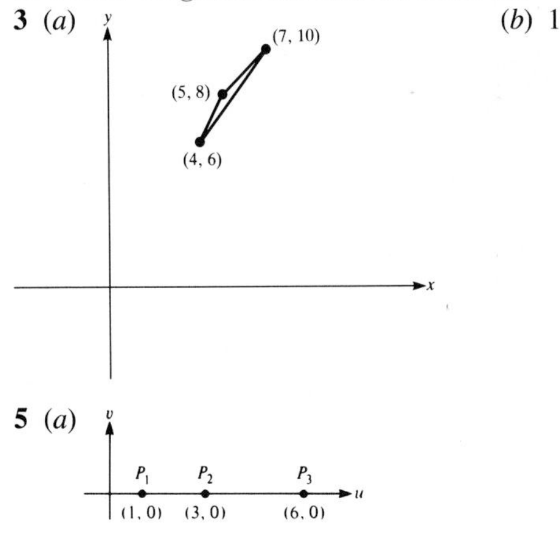

5 (*a*)

(*b*)

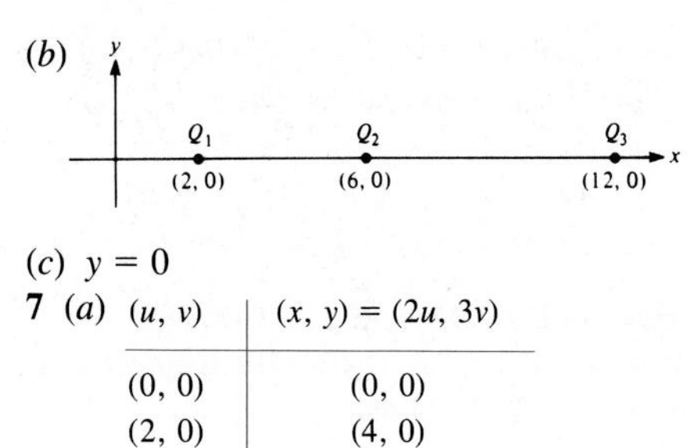

(*c*) $y = 0$

7 (*a*)

(u, v)	$(x, y) = (2u, 3v)$
(0, 0)	(0, 0)
(2, 0)	(4, 0)
(1, 1)	(2, 3)

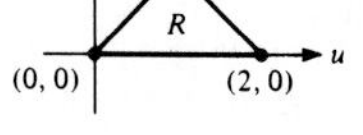

(*c*), (*e*)

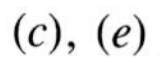

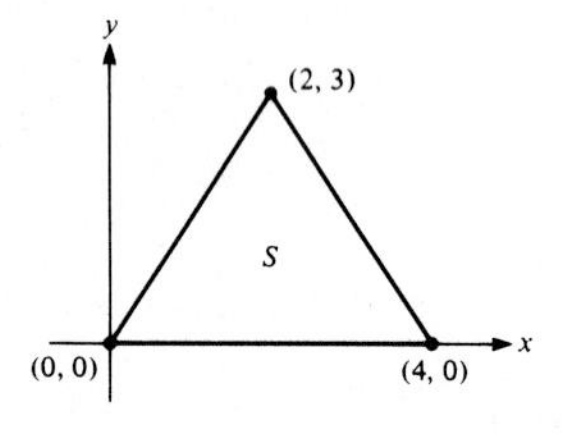

(*f*) Area of $R = 1$; Area of $S = 6$.

9 (*a*)

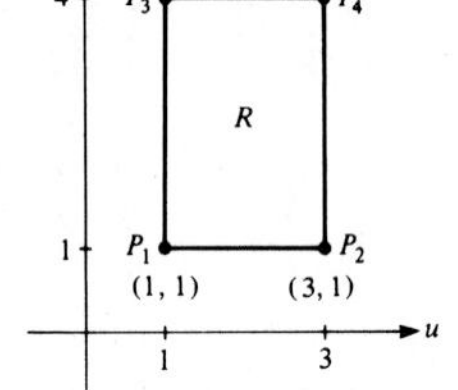

(*b*), (*c*)

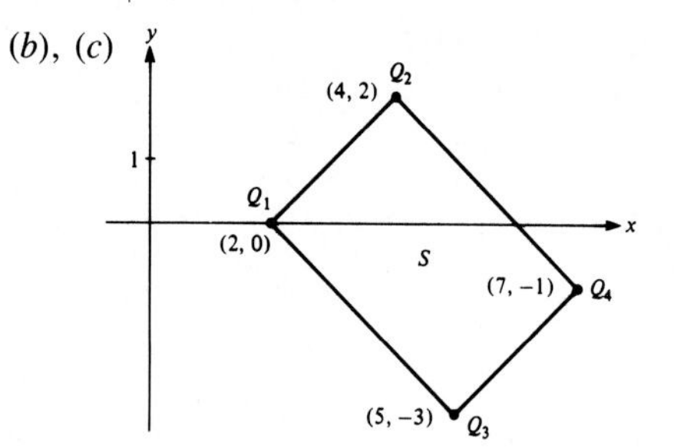

S is a rectangle.

11 (*a*), (*d*)

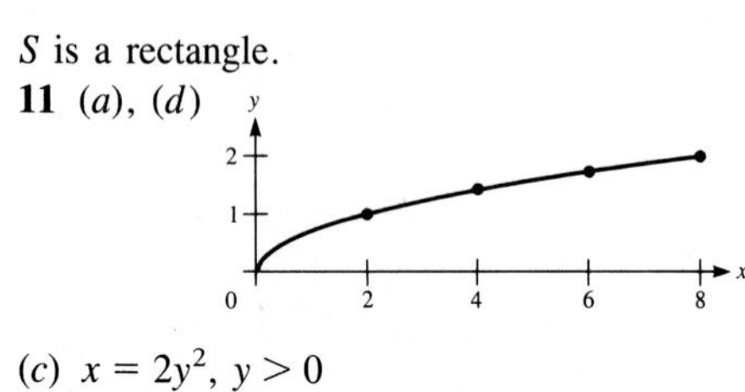

(*c*) $x = 2y^2$, $y > 0$

13 (*a*)

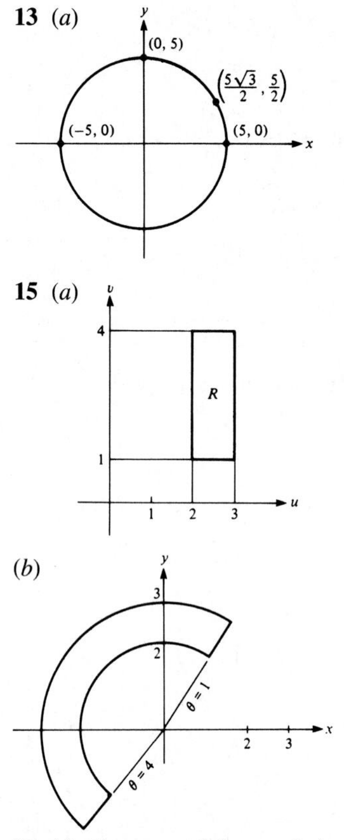

(*b*) $x^2 + y^2 = 25$

15 (*a*)

(*b*)

17 (*a*) That part of the parabola $x = y - y^2$ for which $0 < y < 1$. (*b*) The vertical ray $x = 1$, $y > 0$.

19 (*a*) $\Delta u \Delta v$ (*b*) $u_0 \Delta u \Delta v + \frac{1}{2}\Delta u^2 \Delta v$ (*c*) u_0

Sec. L.2. The Jacobian and Change of Coordinates

1 (*a*) 8 (*b*) 2 **3** (*a*) e^2 (*b*) e^4 **7** (*b*) $\int_R ab^3v^2 \, du \, dv$ (*c*) $\frac{1}{4}\pi ab^3$

11 (*a*)

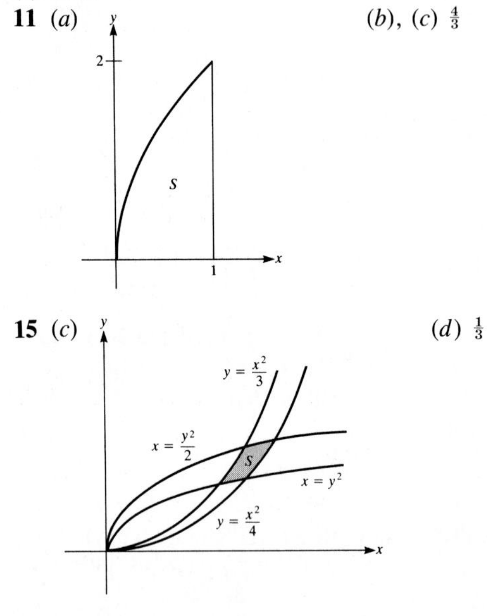

(*b*), (*c*) $\frac{4}{3}$

15 (*c*)

(*d*) $\frac{1}{3}$

17 6π **19** 170/343 **21** (*a*) $-\rho^2 \sin \phi$
25 $(925/216)M$
27 (*a*)

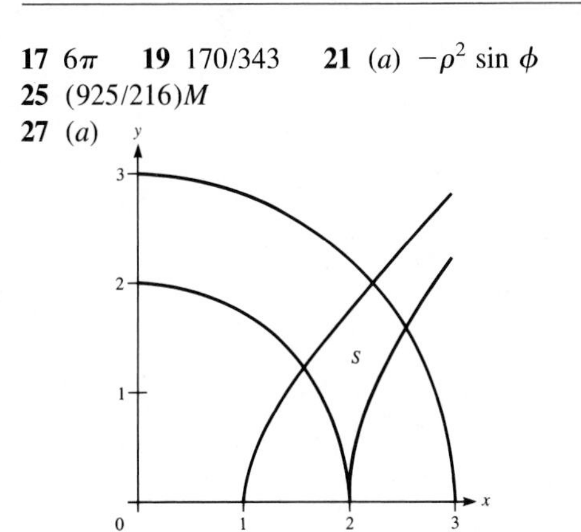

(*b*) $(32 - 5\sqrt{10} - 3\sqrt{6})/6$

APPENDIX M. LINEAR DIFFERENTIAL EQUATIONS WITH CONSTANT COEFFICIENTS

1 $y = Ce^{-2x}$ **3** $y = \dfrac{x}{12} - \dfrac{1}{48} + Ce^{-4x}$
5 $y = -x^2 - 2x - 2 + Ce^x$ **7** $y = C_1e^{-x} + C_2e^{3x}$
9 $y = C_1e^{3x/2} + C_2e^{-x}$ **11** $y = (C_1 + C_2x)e^{3x/2}$
13 $y = e^{3x/2}(C_1e^{\sqrt{5}x/2} + C_2e^{-\sqrt{5}x/2})$
15 $y = C_1e^{3x} + C_2xe^{3x}$ **17** $y = C_1e^{\sqrt{11}x} + C_2xe^{\sqrt{11}x}$
19 $y = -\frac{1}{3}e^{2x} + C_1e^{-x} + C_2e^{3x}$
21 $y = -\frac{1}{26}\cos 3x - \frac{3}{52}\sin 3x + e^{2x}(C_1e^{\sqrt{3}x} + C_2e^{-\sqrt{3}x})$
23 (*b*) $y_p = e^{-x}\ln(1 + e^x)$ (*c*) $y = e^{-x}\ln(1 + e^x) + Ce^{-x}$

LIST OF SYMBOLS

Note: Numbers in parentheses refer to exercises on the indicated pages. Numbers prefixed by "S" refer to the appendix pages.

Symbol	*Description*	*Page*
$\|x\|$, abs x	Absolute value of x	S-4
a_N	Normal component of acceleration	765
a_T	Tangential component of acceleration	765
$\overline{AB}$	Length of line segment AB	19
$\widehat{AB}$	Length of arc AB	924
$\to$	Approaches	28
$x \to a$	x approaches a	30
$x \to a^+$	x approaches a from the right	32
$x \to a^-$	x approaches a from the left	32
lim	Limit	30
D	Discriminant	848
$\bullet$	Included point	32, S-3
$\circ$	Excluded point	32, S-3
$\approx$	Approximate equality	192
$\stackrel{?}{=}$	Conjectured equality	822
ϵ	Epsilon (challenge)	91
δ	Delta (response)	94
$F(x)\Big\|_a^b$	$F(b) - F(a)$	303
$\Big\|_{x=a}^{x=b}$	Evaluation limits	892
$\Big\|_{(a,\, b)}$	Evaluation at (a, b)	809
FTC	Fundamental theorem of calculus	307
∞	Infinity	S-3
(a, b)	Open interval	S-3
$[a, b]$	Closed interval	S-3
$[a, b)$, $(a, b]$	"Half-open" intervals	S-3
J, $\dfrac{\partial(x, y)}{\partial(u, v)}$	Jacobian	S-88
$n!$	n factorial	S-23
$R(x, y)$	Rational function	443
$R(\cos\theta, \sin\theta)$	Rational function of $\cos\theta$ and $\sin\theta$	445
Σ	Summation notation	257, 586
$\{a_n\}$	Sequence	575

Symbol	*Description*	*Page*
S_n	nth partial sum	585
$P_n(x; a)$	nth Taylor polynomial at a	626
$R_n(x; a)$	Remainder or error	629
σ	Density in the plane	494
κ	Curvature	551
τ	Torsion	770(42)
ω	Angular speed	899
α, β, γ	Direction angles	711
$f(x)$	Function of x	13
$f(x, y)$	Function of x and y	795
$f(x, y, z)$	Function of x, y, and z	796
Δf	Change in f	123, 816
Δx	Change in x	122
df, dy, dz	Differentials	228, 843
f^{-1}	Inverse of f	348
$f \circ g$	Composition of f and g	23
$c(x), c(y)$	Cross-sectional length	462, 463
$A(x)$	Cross-sectional area	482
e	Base of natural logarithms	331
e^x	Exponential function	351
$\exp(u)$	Exponential function	420
$E(x)$	Exponential function	S-56
$\ln x$	Natural logarithm	337
$L(x)$	Natural logarithm	S-53
$\log_b x$	Logarithm base b	321
$\sin \theta, \cos \theta$	Sine and cosine functions	56
$\tan \theta$	Tangent function	60
$\sec \theta, \csc \theta, \cot \theta$	Secant, cosecant, and cotangent functions	63
$\tan^{-1} x, \arctan x$	Inverse tangent	357
$\sin^{-1} x, \arcsin x$	Inverse sine	359
$\sinh x, \cosh x, \tanh x$	Hyperbolic functions	387–388
$\sinh^{-1} x, \tanh^{-1} x$	Inverse hyperbolic sine and tangent	389
$\lfloor x \rfloor, [x]$	Greatest integer (floor) function	77
$\lceil x \rceil$	Ceiling function	84(11)
$f'(x)$	Derivative of f	114
$\dot{x}$	Newton's notation for derivative	124
$\frac{dy}{dx}, \frac{df}{dx}, D(f)$	First derivative	124
$\frac{d^2y}{dx^2}, \frac{d^2f}{dx^2}, D^2(f), f''$	Second derivative	185
$\frac{d^ny}{dx^n}, f^{(n)}, D^n(f)$	nth derivative	185
$\frac{\partial z}{\partial x}, f_x, f_1, \frac{\partial f}{\partial x}$	Partial derivative	809
$\left(\frac{\partial u}{\partial x}\right)_y$	Partial derivative	867
$\int f(x)\, dx$	Antiderivative	287
$\int_a^b f(x)\, dx$	Definite integral over interval	266
$\int_R f(P)\, dA$	Definite integral over plane region	883
$\int_R f(P)\, dV$	Definite integral over solid region	912
$\int_C f\, dx, \int_C f\, dy, \int_C f\, dz$	Line integrals	950

Symbol	*Description*	*Page*
$\int_C f\,ds$	Line integral with respect to arc length	951
$\int_C (P\,dx + Q\,dy + R\,dz)$	Line integral	951
$\oint_C f\,dx$	Line integral over closed path	954
$\int_C \mathbf{F}\cdot d\mathbf{r}$	Line integral	960
$\oint_C \mathbf{F}\cdot\mathbf{T}\,ds$	Line integral (circulation)	962
$\oint_C \mathbf{F}\cdot\mathbf{n}\,ds$	Line integral (flux)	962
$\int_{\mathcal{S}} f(P)\,dS$	Surface integral	1004
$\int_{\mathcal{S}} \mathbf{F}\cdot\mathbf{n}\,dS$	Surface integral (flux)	1015
i	$\sqrt{-1}$	657
z	Complex number	657
$\|z\|$	Magnitude of z	659
arg z	Argument of z	659
Im z	Imaginary part of z	658
Re z	Real part of z	658
z^{-1}	Reciprocal of z	664(11)
$\bar{z}$	Conjugate of z	658
$x + iy$	Complex number	657
$e^{i\theta}$	$\cos\theta + i\sin\theta$	667
cis θ	$\cos\theta + i\sin\theta$	659
$\mathbf{A}$	Vector	677
$\\|\mathbf{A}\\|$	Magnitude (norm) of **A**	677
$\overrightarrow{PQ}$	Directed line segment (vector)	677
$\langle x, y\rangle, \langle x, y, z\rangle$	Vector	682
$\mathbf{A}\cdot\mathbf{B}$	Dot product	693
$\mathbf{A}\times\mathbf{B}$	Cross product	721
i, j, k	Basic unit vectors	681
$\hat{\mathbf{r}}$	Unit radius vector	941
T	Unit tangent vector	761
N	Principal unit normal vector	763
B	Unit binormal	770(39)
$\mathbf{proj}_L\,\mathbf{A}$	Projection of **A** on L	689
$\mathbf{proj}_{\mathbf{B}}\,\mathbf{A}$	Projection of **A** on **B**	690
$\mathbf{proj}_{\mathcal{P}}\,\mathbf{A}$	Projection of **A** on $\mathcal{P}$	690
$\mathbf{orth}_{\mathbf{B}}\,\mathbf{A}$	Orthogonal component of **A** relative to **B**	691
$\mathrm{comp}_{\mathbf{B}}\,\mathbf{A}$	Scalar component of **A** relative to **B**	695
∇	Del	832, 942
∇f	Gradient of f	832
$\nabla\cdot\mathbf{F}$	Divergence of **F**	942
div **F**	Divergence of **F**	942
$\nabla\times\mathbf{F}$	Curl of **F**	944
curl F	Curl of **F**	944
$\nabla^2 f$	Laplacian of f	945
$D_{\mathbf{u}}f$	Directional derivative	830
(x, y)	Rectangular coordinates	S-7
(x, y, z)	Rectangular coordinates	678
(r, θ)	Polar coordinates	520
(r, θ, z)	Cylindrical coordinates	918

INDEX

INDEX

Note: Numbers in parentheses refer to exercises on the indicated pages. Page numbers prefixed with "S" refer to pages at the end of the book.

60. $\int x^n \cos ax\, dx = \frac{1}{a} x^n \sin ax - \frac{n}{a} \int x^{n-1} \sin ax\, dx \qquad n \text{ positive}$

61. $\int \sin ax \cos bx\, dx = -\frac{\cos (a-b)x}{2(a-b)} - \frac{\cos (a+b)x}{2(a+b)} \qquad a^2 \neq b^2$

Expressions Containing Exponential and Logarithmic Functions

62. $\int xe^{ax}\, dx = \frac{e^{ax}}{a^2}(ax - 1),$

$\int xb^{ax}\, dx = \frac{xb^{ax}}{a \ln b} - \frac{b^{ax}}{a^2(\ln b)^2} \qquad b > 0$

63. $\int x^n e^{ax}\, dx = \frac{1}{a} x^n e^{ax} - \frac{n}{a} \int x^{n-1} e^{ax}\, dx \qquad n \text{ positive}$

64. $\int e^{ax} \sin bx\, dx = \frac{e^{ax}}{a^2 + b^2}(a \sin bx - b \cos bx)$

65. $\int e^{ax} \cos bx\, dx = \frac{e^{ax}}{a^2 + b^2}(a \cos bx + b \sin bx)$

66. $\int x^n \ln ax\, dx = x^{n+1}\left[\frac{\ln ax}{n+1} - \frac{1}{(n+1)^2}\right] \qquad n \neq -1$

Expressions Containing Inverse Trigonometric Functions

67. $\int \sin^{-1} ax\, dx = x \sin^{-1} ax + \frac{1}{a}\sqrt{1 - a^2x^2}$

68. $\int \cos^{-1} ax\, dx = x \cos^{-1} ax - \frac{1}{a}\sqrt{1 - a^2x^2}$

69. $\int \csc^{-1} ax\, dx = x \csc^{-1} ax + \frac{1}{a} \ln |ax + \sqrt{a^2x^2 - 1}|$

70. $\int \sec^{-1} ax\, dx = x \sec^{-1} ax - \frac{1}{a} \ln |ax + \sqrt{a^2x^2 - 1}|$

71. $\int \tan^{-1} ax\, dx = x \tan^{-1} ax - \frac{1}{2a} \ln (1 + a^2x^2)$

72. $\int \cot^{-1} ax\, dx = x \cot^{-1} ax + \frac{1}{2a} \ln (1 + a^2x^2)$

Some special integrals

73. $$\int_0^{\pi/2} \sin^n x\, dx = \int_0^{\pi/2} \cos^n x\, dx = \begin{cases} \dfrac{1 \cdot 3 \cdot 5 \cdot 7 \cdots (n-1)}{2 \cdot 4 \cdot 6 \cdot 8 \cdots (n)} \dfrac{\pi}{2} & n \text{ even} \\ \dfrac{2 \cdot 4 \cdot 6 \cdot 8 \cdots (n-1)}{1 \cdot 3 \cdot 5 \cdot 7 \cdots (n)} & n \text{ odd} \end{cases}$$

74. $\int_{-\infty}^{\infty} e^{-x^2}\, dx = \sqrt{\pi}$

ALGEBRA

Roots of $ax^2 + bx + c = 0$; $x = \frac{-b \pm \sqrt{b^2 - 4ac}}{2a}$ (quadratic formula)

$a + ar + ar^2 + \cdots + ar^{n-1} = \frac{a(1 - r^n)}{1 - r}, r \neq 1$ (sum of finite geometric series)

$k! = 1 \cdot 2 \cdot 3 \cdots k$

$(1 + x)^n = 1 + nx + \binom{n}{2}x^2 + \cdots + \binom{n}{k}x^k + \cdots + x^n$ (binomial theorem)

$$\binom{n}{k} = \frac{n!}{k!(n-k)!} = \frac{n}{1}\,\frac{(n-1)}{2}\,\frac{(n-2)}{3} \cdots \frac{(n-k+1)}{k}$$

GEOMETRY

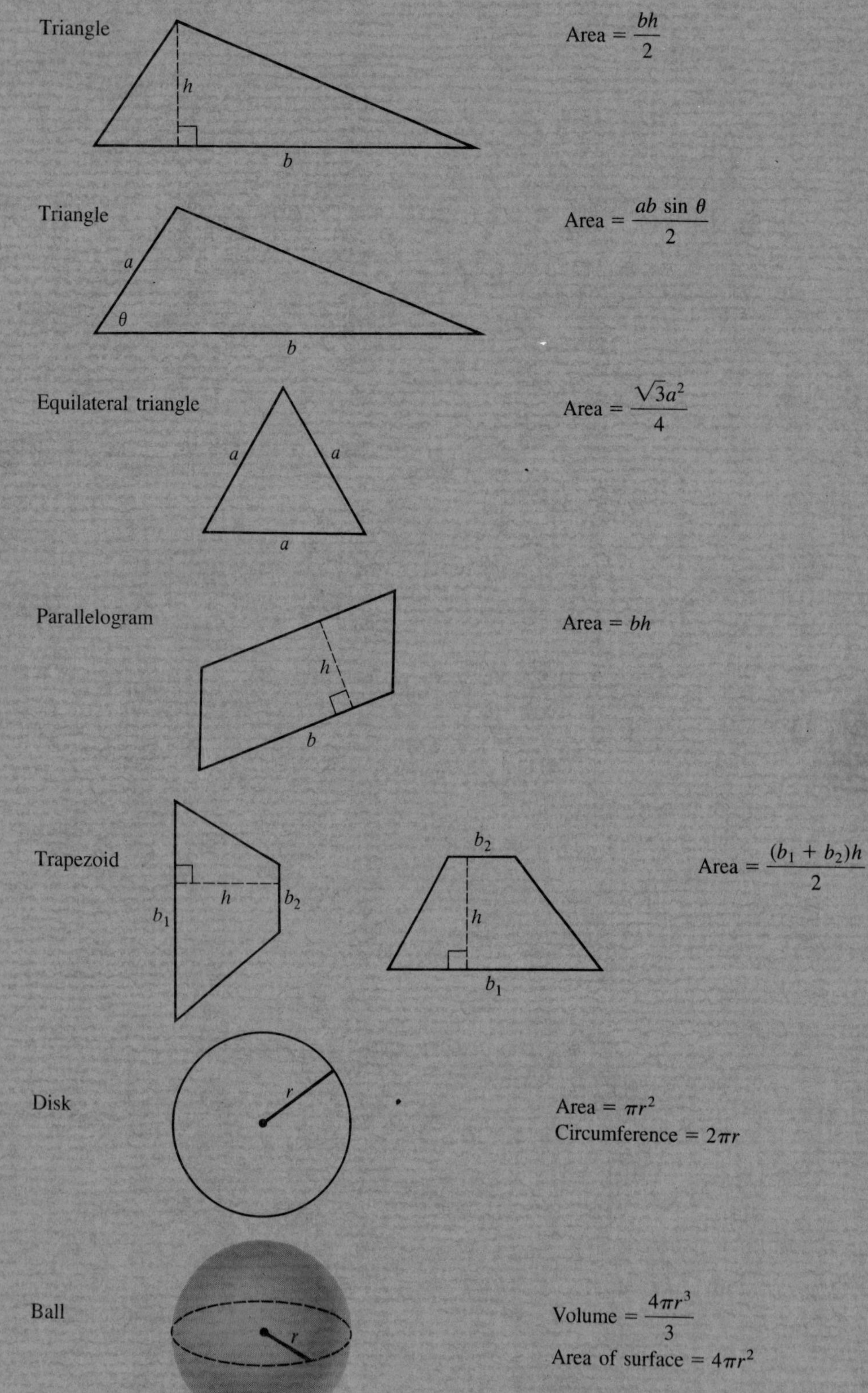

Triangle: $\text{Area} = \dfrac{bh}{2}$

Triangle: $\text{Area} = \dfrac{ab \sin \theta}{2}$

Equilateral triangle: $\text{Area} = \dfrac{\sqrt{3}a^2}{4}$

Parallelogram: $\text{Area} = bh$

Trapezoid: $\text{Area} = \dfrac{(b_1 + b_2)h}{2}$

Disk: $\text{Area} = \pi r^2$
$\text{Circumference} = 2\pi r$

Ball: $\text{Volume} = \dfrac{4\pi r^3}{3}$
$\text{Area of surface} = 4\pi r^2$